VICE COUNTIES OF THE BRITISH ISLES

100 km

Scotland

Great Britain

Ireland

England

Wales

Channel Isles

The Freshwater Algal Flora of the British Isles

An Identification Guide to Freshwater and Terrestrial Algae

The varied and often beautiful forms of freshwater algae have long held a fascination for naturalists and professional scientists alike, particularly those involved in monitoring and managing the freshwater environment. This Flora provides the first modern account and identification guide to more than 1700 out of a total of over 2200 species of freshwater algae (excluding diatoms) known in the British Isles, the majority of which also have a worldwide distribution. Non-technical descriptions are supported by clear line illustrations or photographs that emphasize the features significant for identification, most of which can be seen with a simple light microscope. In addition, user-friendly keys enable the accurate identification of specimens to the level of genus or species. Notes are included on ecology, world distribution and any taxonomic problems or identification difficulties. An accompanying CD-ROM photo catalogue of more than 500 spectacular colour images of freshwater algae and their habitats provides a valuable additional reference source and identification tool.

DAVID JOHN is a researcher in the Department of Botany at The Natural History Museum, London, UK.

BRIAN WHITTON is Emeritus Professor, School of Biological and Biomedical Sciences at the University of Durham, UK.

ALAN BROOK is Emeritus Professor of Biology at the University of Buckingham, UK.

The Flora is dedicated to

John W.G. Lund FRS

in thanks for his outstanding contribution to the knowledge of British freshwater algae and his continuing help and encouragement to phycologists throughout the world.

The Freshwater Algal Flora of the British Isles

An Identification Guide to Freshwater and Terrestrial Algae

Edited by **David M. John**
The Natural History Museum, London, UK

Brian A. Whitton
University of Durham, UK

Alan J. Brook
University of Buckingham, UK

with an accompanying CD
of images of algae and algal habitats
compiled by Peter V. York,
David M. John and Leslie R. Johnson

a collaborative project of the British Phycological Society and The Natural History Museum, London

PUBLISHED BY THE PRESS SYNDICATE OF THE UNIVERSITY OF CAMBRIDGE
The Pitt Building, Trumpington Street, Cambridge, United Kingdom

CAMBRIDGE UNIVERSITY PRESS
The Edinburgh Building, Cambridge CB2 2RU, UK
40 West 20th Street, New York, NY 1011–4211, USA
477 Williamstown Road, Port Melbourne, VIC 3207, Australia
Ruiz de Alarcón 13, 28014 Madrid, Spain
Dock House, The Waterfront, Cape Town 8001, South Africa

http://www.cambridge.org

First published 2002

Printed in the United Kingdom at the University Press, Cambridge

Typeface Trump Mediaeval 9.5/12pt *System* QuarkXPress™ [SE]

A catalogue record for this book is available from the British Library

Library of Congress Cataloguing in Publication data

The freshwater algal flora of the British Isles: an identification guide to freshwater and terrestrial algae / edited by David M. John, Brian A. Whitton, Alan J. Brook.
p. cm.
Includes bibliographical references (p.).
ISBN 0 521 77051 3
1. Freshwater algae – British Isles – Classification. 2. Freshwater algae – British Isles – Identification. 3. Algae – British Isles – Classification. 4. Algae – British Isles – Identification. I. John, David M. II. Whitton, Brian A. III. Brook, Alan J.
QK573 .F74 2002 579.8′176′0941–dc21 2001025917

ISBN 0 521 77051 3 hardback

CONTENTS

CONTRIBUTORS

A.J. BROOK
University of Buckingham, Buckingham MK18 1EG, UK

J.A. BRYANT
Department of Botany, The Natural History Museum, Cromwell Road, London SW7 5BD, UK

J.D. DODGE
University of London, Royal Holloway and Bedford New College, Egham Hill, Egham, Surrey TW20 0EX, UK

E.Y. HAWORTH
The Freshwater Biological Association, The Ferry House, Ambleside, Cumbria LA23 0LP, UK

R. HUXLEY
Department of Botany, The Natural History Museum, Cromwell Road, London SW7 5BD, UK

D.M. JOHN
Department of Botany, The Natural History Museum, Cromwell Road, London SW7 5BD, UK

J.F. JOHN
95 Hayes Road, Bromley, Kent BR2 9AE, UK

L.R. JOHNSON
Department of Botany, The Natural History Museum, Cromwell Road, London SW7 5BD, UK

M.G. KELLY
Bowburn Consultancy, 11 Monteigne Drive, Bowburn, Durham DH6 5QB, UK

J. KRISTIANSEN
University of Copenhagen, Botanical Institute, Department of Mycology and Phycology, Ø. Farimagsgade 2D, DK-1353, Copenhagen K, Denmark

G.W. LAWSON
Flat 3, 23 Sheffield Terrace, London W8 7NQ, UK

J.M. LEWIS
University of Westminster, School of Biological and Health Sciences, 115 New Cavendish Street, London W1M 8JS, UK

R. MERRITT
20 St John's Grove, Morecambe, Lancashire LA3 1ET, UK

Ø. MOESTRUP
University of Copenhagen, Botanical Institute, Department of Mycology and Phycology, Ø. Farimagsgade 2D, DK-1353, Copenhagen K, Denmark

G. NOVARINO
Department of Zoology, The Natural History Museum, Cromwell Road, London SW7 5BD, UK

A. PENTECOST
King's College London, Division of Life Sciences, Manresa Road, London SW3 6LX, UK

H.R. PREISIG
University of Zürich, Institut für Systematische Botanik und Botanischer Garten, Zollikerstrasse 107, CH-8008 Zürich, Switzerland

R.G. SHEATH
California State University San Marcos, San Marcos, California CA 92092-001, USA

A.R. SHERWOOD
University of Guelph, Department of Botany and Dean's Office, College of Biological Science, Guelph, Ontario N1G 2W1, Canada

N.F. STEWART
Chelwell Cottage, Posbury, Nr Crediton, Devon, EX17 3QE

I. TITTLEY
Department of Botany, The Natural History Museum, Cromwell Road, London SW7 5BD, UK

P.M. TSARENKO
M.G. Kholodny Institute of Botany, Department of Spore Plants, National Academy of Science of the Ukraine, Tereschenkivska Street 2, 252601 Kiev-MSP-1,Ukraine

J.D. WEHR
Louis Calder Center – Biological Field Station, Fordham University, 53 Whippoorwill Road, Box K, Armonk, NY 10504, USA

B.A. WHITTON
Department of Biological Sciences, University of Durham, Durham DH1 3LE, UK

K. WOŁOWSKI
Instytut Botaniki, Polska Akademia Nauk, Lubicz 46, 31-512 Kraków, Poland

P.V. YORK
Department of Botany, The Natural History Museum, Cromwell Road, London SW7 5BD, UK

Logos of societies and organizations which have sponsored and/or given significant support to the Flora Project.

FOREWORD

G.S.West's 1904 account of the British freshwater algal flora was the first to be published in the twentieth century and marked a very considerable advance on earlier ones. In Fritsch's revision (West & Fritsch, 1927) of this work there were approximately 250 genera dealt with. Now over 550 genera have been recorded from the British Isles, excluding those in the Euglenophyta and Pyrrophyta, which were not included in the 1927 Flora. These figures alone show how badly needed is the new Flora.

As a result of a proposal made by David John and Brian Whitton, the British Phycological Society set up a committee in 1991 to oversee the production of a new freshwater flora. The Natural History Museum, London, supported the undertaking financially and in practical terms, by allowing members of staff to work on the project. David John and Brian Whitton became its editors, to be joined later by Alan Brook.

The Flora Committee had a difficult decision to make. British marine phycologists started a comprehensive and now highly regarded series of volumes some 50 years ago, but it is still not yet quite complete. It was obvious that, no matter how urgent the need for a comprehensive replacement of the West and Fritsch volume, one on a similar scale to the marine Flora would take very many years to complete. Nevertheless, it was decided to include all genera containing photosynthetic taxa, apart from a few of doubtful validity. For some genera containing a relatively large number of species, a representative sample is described and figured and the others merely listed. In particular, only about 300 of the commonest of the 800 or so desmid species are described in full. This seems to me to be a sensible decision. Even so, it has taken some ten years to complete the present work. This may seem a long gestation time to some people, but not to those who, from experience, are able to appreciate the amount of work involved in and the complexities of producing multi-authored books. Moreover, the Flora is accompanied by a CD-ROM of photo-images.

Our knowledge of the distribution of the British freshwater algae has greatly increased since 1927 and there have also been some marked changes in taxonomy. Modern advances in taxonomy have been incorporated, even though the practical use of the Flora depends on light microscopy. Knowledge of the distribution and ecology of some species has also increased greatly, though there are still many about which is little is known. Hopefully, the Flora will help to correct this.

The representation of some groups has been enlarged as a consequence of studies made especially for the Flora, including the Euglenophyta by Konrad Wołowski and *Vaucheria* by Leslie Johnson and Roy Merritt. The much needed account of the Rhodophyta by Robert Sheath and Alison Sherwood benefited from a large number of collections at sites ranging widely over parts of the British Isles. Many illustrations have been drawn especially by Hilary Belcher and Erica Swale, who also added much floristic information. David Williamson has produced many original desmid figures. He and Alan Brook have continued the interest in and indeed love for desmids, which was pioneered so brilliantly by William and George West. It is noteworthy that over 10% of the descriptive part of the latter's 1904 Flora was devoted to desmids.

David John's account of the Ulotrichales and Chaetophorales is particularly useful for a different reason: namely, it gives us conclusions by someone with special experience and knowledge of groups in which the taxonomy is often in an unsatisfactory, or even very unsatisfactory state. A prime example is *Stigeoclonium*, a very common and abundant genus. Despite or as a result of a diversity of past studies, what he says about the difficulties of identifying species of this genus is only too true.

The diatoms are introduced by Martyn Kelly and Elizabeth Haworth, but otherwise are not included here. Their account contains useful references to introductory guides and Floras. There is also a valuable overview of the many taxonomic revisions and nomenclatural changes made in recent times at the level of genus. In view of the importance of diatoms in the British Isles, and their widespread use for a variety of purposes, I hope that it will not be long before there is also a diatom Flora of the British Isles. As in the case of the Chlorophyta referred to above, it is surely better to have even an imperfect account than no account at all.

It was decided to include only photosynthetic (pigmented) taxa, though there are references to one or two non-photosynthetic (colourless) genera. This lacuna is particularly large for the Euglenophyta where there are at least as many colourless as pigmented genera. Perhaps an account of such algae or protozoa can be produced by cooperation between our phycological and protozoological societies?

Is there amongst us a phycologist who has specialized knowledge of all groups of freshwater and terrestrial algae? Probably not, so that all phycologists can be grateful to those who have made this Flora possible. The book will also be invaluable to anyone who wants to know something about our freshwater algae or wishes to learn how to identify the algae they find. The Flora is much needed and will, I believe, be much used.

J.W.G. Lund FRS
March 2001

PREFACE

An increasing awareness of the importance of freshwater algae in environmental management and monitoring has led to an increased demand for their accurate naming. However, when two of us started to run an annual training course at the University of Durham for biologists in the water industry and water management, a number of problems about which we were already partially aware became very obvious. In order to make reliable identifications of all the species present in a range of mixed samples, well over 40 publications were needed including identification guides, Floras, monographs and research papers. Many of these are written in languages other than English and, even for the better linguists, are not always easy to use, since they are mostly written for professional taxonomists. Almost all focus on regions other than the British Isles, many are expensive and some are frequently difficult to obtain. The only comprehensive guide for the British Isles remains West and Fritsch's *A Treatise on the British Freshwater Algae*, first published in 1927.

It was evident that non-specialists require a user-friendly, well-illustrated identification guide written in English that should describe as many of the British freshwater algae as possible. Unfortunately, knowledge of the British freshwater algal flora remains far from complete. Many species recorded in the nineteenth century or early years of the twentieth century have not been recorded since. In a few cases this may be because of environmental changes, but more often than not it is simply because no one has resampled the localities and regions where these species grow. As a result of having no guide, we approached specialists in algal taxonomy and ecology with a proposal to produce a new and modern Flora dealing with all British freshwater algae. Everyone approached was very encouraging and offered strong support, despite a recognition that it was inevitable that the Flora would have many shortcomings.

In June 1991, the council of The British Phycological Society agreed to support the proposal submitted by two of us (DMJ, BAW) to prepare a modern synoptic account and identification guide to the freshwater and terrestrial algae of the British Isles. A committee was given the task of planning and organizing the 'Flora Project' and its first meeting took place at Durham on 25 September 1991. The Natural History Museum in London agreed to play a leading role by making staff time, workspace and other resources available. Because many of the experts required to produce such an authoritative work reside overseas, an early priority was to seek financial support to enable some overseas collaborators to visit the British Isles to collect material and to work with phycologists here. The first invitations were sent in 1993 and new specialists continued to join the project up until 1998.The successful outcome owes much to the contributors, who have given so freely of their time in preparing keys, descriptions and illustrative material. In 1996 it was decided to broaden the project to include a CD-ROM of mainly colour images. P.V. York agreed to be mainly responsible for compiling this photo catalogue.

Initially the committee hoped to cover all algal groups within a single volume, but it soon became clear that this was unrealistic. The decision about what to do was resolved in 1998 when the diatomists on the committee withdrew, believing it impossible to include diatoms until much further taxonomic research had been carried out. Fortunately, diatoms are one of the few algal groups for which there already exist fairly comprehensive identification guides. These guides and other useful literature on diatoms are introduced here by M.G. Kelly and E.Y. Haworth. Even with the omission of diatoms, there was still a problem over volume length, so the present work deals with only about one-third of the 1000 or so desmids recorded for the British Isles.

ACKNOWLEDGEMENTS

The Flora would not have been possible without the generous support of the British Phycological Society and The Natural History Museum in London. Financial support provided by the former covered the expenses of the Flora Committee, a 'Flora Workshop' held in July 1999 at the University of Durham, helped to employ research assistants for a few months, supported several visits by overseas collaborators and defrayed some of the production costs of the book. Special thanks go to all members of the 'British Freshwater Algal Flora Committee' who assisted in planning, organizing and running the project: B.S.C. Leadbeater, E.J. Cox, J.P.C. Harding, E.Y. Haworth,

D.G. Mann and R.G. Sheath. Many others helped the project: J.F. John, G.W. Lawson and D. Rose assisted in compiling information and edited key groups; B. Williamson, D. Page, M. Sadka, D. Sutton, J. Benfield and V. Noble gave advice, helped to develop the software or assisted in the production of the CD-ROM photo catalogue; M.D. Guiry and F. Rindi provided unpublished data and hosted a collecting trip to Clare Island; S.C. Hardiman redrew and scanned illustrations of blue-green algae, D.M. Balbi and P.J. Robinson also assisted in the preparation of these figures; P. Rye prepared the plates and labels for most drawings of eukaryotic algae, B. Steiner-Gafner did the same for the haptophytes; S. Blackmore and R.M. Bateman (Keeper of Botany, NHM) supported the Flora project by allowing one of us (DMJ) to devote considerable time to it; J.P.C. Harding maintained contacts with colleagues, initially in The National Rivers Authority and subsequently The Environment Agency; R.V. Smith and S.I. Heaney sought support for the project in Northern Ireland, R.A. Sweeting and J.T. Krokowski similarly obtained financial support from The Environment Agency (England and Wales). Contacts and help from algal specialists in addition to those listed on the contents page were very important. Among those from the British Isles who commented on text, keys and illustrations were A.E. Bailey-Watts, J.H. Belcher, J.A. Brodie, G. Eaton, B.S.C. Leadbeater, J.W.G. Lund, B. Moss, E.M.F. Swale and J.R. Young. Researchers from overseas who have rendered similar assistance were P.A. Broady (Department of Plant and Microbial Sciences, University of Canterbury, New Zealand), P.F.M. Coesel (Department of Aquatic Ecology, University of Amsterdam, The Netherlands), T.J. Entwisle (Royal Botanic Gardens, Sydney, Australia), O.M. Skulberg (Norwegian Institute for Water Research, Oslo, Norway) and C.G. Velez (Departamento de Ciencias Biologicas, Universidad de Buenos Aires, Argentina). Many others assisted with providing material, literature, and taxonomic advice, including D.M. Balbi, S.J. Brierley, J. Ellaway, J.E. Florey, M.D. Guiry, L.M. Irvine, J.P.C. Harding, J. Jamieson, S. Pritchard, W. Purvis, F. Rindi, N. Robson, F. Rumsey, L.E. Shubert, G.E. Stevens (fomerly Douglas) and R.Woods.

We owe special thanks to Hilary Belcher and Erica Swale, who contributed many new records, original illustrations, ecological data and showed great patience in revising and refereeing manuscripts. Another key person is David B. Williamson, who prepared about one-third of the plates, provided many new desmid illustrations, and redrew many pencilled illustrations from the Wests' archive of illustrations housed at The Natural History Museum, London. He has made many valuable contributions to the text of the desmid section and edited many of the illustrations. Throughout, John Lund has given us much valued advice and assistance especially during visits to the Freshwater Biological Association at Windermere to consult his card index of British freshwater algae. Special mention goes to Judith John, who has not only been a very patient wife during the long hours of preparing the Flora, but has also carried out a considerable amount of the literature research, edited text, produced the standardized list of author abbreviations and prepared the taxonomic index.

To Maria Murphy and Ward Cooper at Cambridge University Press we are indebted for the patient and efficient manner in which they have dealt with the many and varied problems encountered during production; special thanks go to Jane Bulleid for copy-editing the manuscript.

We are indebted to all those who have contributed to this book for their time, enthusiasm and patience. The name of each contributor is given under the heading of the taxonomic group(s) or section for which they are responsible. All introductory chapters are co-authored by the editors who also prepared the glossary.

Finally, we remember Dr T. Christensen ('Tyge'), who was to have prepared text and illustrations for the genus *Vaucheria* (p. 262), but sadly died before he could complete the task.

Financial Acknowledgements

In addition to the support from the British Phycological Society and The Natural History Museum, financial support from several outside sources was essential in several crucial ways. Unfortunately, the Flora Project never succeeded in obtaining finance to carry out the basic research needed to answer important taxonomic and ecological questions. We were both disappointed and surprised in the lack of support from most water companies and by some of the bizarre responses they provided to our correspondence. Grants from Northumbrian Water (then plc), Thames Water and the Department of Agriculture and Rural Development (Northern Ireland) received at an early stage were especially welcome and helped to get the project under way.

We are also most grateful to a number of other organizations and individuals. The Royal Society provided a research microscope for BAW and helped to finance visits by Dr K.Wołowski while The Linnean Society of London and The Natural Environment Research Council contributed towards the preparation of illustrations. Plantlife and the Freshwater Biological Association assisted by promoting the Flora while support towards the costs of publication has been provided by The Environment Agency, The British Phycological Society, The Natural Environment Research Council, The Systematics Association, The Countryside Council for Wales, English Nature, Scottish Natural Heritage, Accrofab Ltd, C. Wright (Director of Accrofab Ltd) and an anonymous donor.

David John, Brian Whitton, Alan Brook
London, Durham, Buckingham
June 2001

INTRODUCTION

The Editors

Freshwater algae have been a subject of interest for naturalists and professional scientists in the British Isles for more than two centuries. As can be seen from the long reference list, many made observations over long periods and their records range from notes in natural history periodicals to long ecological accounts in learned journals. Nowadays an increasing number of people are required to name algae accurately, particularly those involved in managing and monitoring the environment. All this suggests the need for readily available modern books, that are straightforward to use, on how to identify freshwater algae in the British Isles. However, such books do not exist and the present Flora sets out to fulfil this need for freshwater and terrestrial algae other than diatoms.

Before explaining the scope of the Flora, it is worth giving a little background. Following the 'Earth Summit' held in Rio de Janeiro in June 1992, various international and national strategies were prepared for the conservation and sustainable use of biological diversity. The oceanic planktonic algae received some attention because of their importance in global processes, but freshwater algae were mostly included simply as estimates of numbers. The UK Biodiversity Steering Group Report (Biodiversity, 1995a) estimated there to be 20,000 algal species in the UK, about 650 of which are seaweeds. However, a recent checklist of the freshwater and terrestrial algae of Great Britain and the Republic of Ireland (Whitton et al., 1998a) reported only about 5000 species. Since publication of this checklist, a few more records have come to light and there have been a number of new discoveries. However, it will require many detailed field surveys before it is possible to provide a reliable estimate of the total number of species and state whether there are really another 15,000 waiting to be found. Of those that are known, it is possible to comment reliably on the overall status of only a hundred or so species in the phyla covered here. In addition to the commonest species, a few others are sufficiently distinctive that enough is known to make a sound assessment of their general distribution. The red alga *Thorea hispida* and some charophytes are examples.

The situation is of course dynamic. Species may be lost or gained due to natural environmental changes or through human activity. This has been widely recognized for the charophytes, of which 12 species are included in the second volume of the UK Government's *Biodiversity: the UK Steering Group Report* (Biodiversity, 1995b) as 'priority species' in need of conservation action. They are the only algal group in the British Isles for which a Red Data book has been written (Stewart & Church, 1992). Other algae are known to have disappeared from specific sites and are vulnerable following the widespread loss or degradation of freshwater habitats. Unfortunately, these have mostly been neglected by those concerned with conservation in the British Isles, presumably because of their small size. This lack of interest contrasts with parts of continental Europe where freshwater algae have received more attention. 'Red Data Lists' of algae exist for, among others, the Czech Republic, Germany, Poland and Slovakia. The list for Germany is especially comprehensive, with over a thousand algal species mentioned as threatened (Ludwig & Schnittler, 1996).

New records for the British Isles of species previously entirely absent are sometimes suspected to be new invaders, but it is difficult to be sure about such cases. What is clear is that some potentially nuisance species have become much more abundant during the past century, including some bloom-forming blue-green algae and the macroalgae *Cladophora glomerata*, *Enteromorpha flexuosa*, *Hydrodictyon reticulatum* and *Vaucheria sessilis*. In at least some instances, these species show great morphological variation, suggesting that they may be undergoing rapid evolutionary changes. Such changes add to the need for a reliable flora to record what is now present.

The last fairly comprehensive account was West and Fritsch's *A Treatise on British Freshwater Algae*, published in 1927 by the same publisher as the present Flora. Much of this material had already been published by G.S. West in 1904 and was never revised. He died ten years before the appearance of the new book. More recent identification guides for the British Isles deal with the commoner species (Pentecost, 1984; Bellinger, 1992), specific habitats (Belcher & Swale, 1976, 1979) and specific groups (Lind & Brook, 1980; Barber & Haworth, 1981; Whitton et al., 2000). None is comprehensive enough to identify the majority of the species in most mixed field samples. Often those with little experience of algae fail to appreciate the limitations of such guides, and may ignore small but significant differences between their material and the text descriptions and illustrations. Sometimes these differences are highly significant and dismissing them inevitably leads to misidentification. Identification requires access to far more books, journals and microfiches than possessed by most institutes and universities, let alone individuals.

SCOPE OF FLORA

The Flora deals with the British Isles, comprising Great Britain (England, Scotland, Wales), Ireland (Northern Ireland, the Irish Republic), the Isle of Man and the Channel Islands. (The United Kingdom is Great Britain and Northern Ireland.) All major groups of freshwater and terrestrial algae are included (Table 1), apart from diatoms for which there is only a short introduction (p. 273). Species of slightly brackish environments are included, but the salinity limits differ slightly between groups. As in most Floras for other parts of the world, all marine blue-green algae are included, because some species occur in both freshwater and marine environments. Symbiotic algae, such as occur in lichens, are mentioned only briefly, and not listed as individual species. The same applies to those species which are colourless but clearly related to photosynthetic algae. The dinoflagellates are the only exception in which the few colourless species known from the British Isles are included.

Only 325 of the approximately 770 known desmid species are included. This is mainly for reasons of space, but also because the first volume of a monograph on this group has been published (Brook & Williamson, 2002). The desmids are in fact the only algal group where there is an older monograph for the British Isles. William and George West published a five-volume series between 1904 and 1923, with the last completed by Dr Nellie Carter: 690 species and 450 varieties of desmid were described and meticulously illustrated. As a consequence, more is known concerning the distribution and abundance of desmids than any other algal group, making it easier to select the commoner species. Those taxa recorded from the British Isles, but excluded from this Flora, are listed under each genus. A list of all species is given in *A Coded List of Freshwater Algae of the British Isles* (Whitton et al., 1998a).

Taxonomy

Because 22 contributors have prepared the entries for the different taxonomic groups, it is inevitable that there are differences in depth of cover. However, the editors have tried to ensure that the level of treatment is as similar as possible and that there is some consistency in the form of the entries.

The arrangement of the phyla follows the *Coded List of Freshwater Algae of the British Isles* (Whitton et al., 1998a). Orders are arranged alphabetically under a higher taxonomic category (usually phylum), genera are alphabetical under the order (or family in the Zygnematales), and species are alphabetical within the genus. Each species entry includes its current accepted name, authority, date of publication and basionym (if required). The synonyms listed are usually ones used widely in the literature on British freshwater algae or

Table 1. *Numbers of taxa of freshwater and terrestrial algae known from the British Isles; not included are 53 strictly marine Cyanophyta (although described in the Flora) and colourless species. Taxa in parentheses are those recorded from the British Isles, but not described here (mostly desmids). Doubtful includes species and subspecific taxa. Numbers of subspecific taxa are only very approximate since many are mentioned, although discounted as not worthy of recognition.*

PHYLUM	Taxa described in Flora		
	Species	Subspecific taxa	Doubtful
Cyanophyta (blue-green algae)	297	3	10
Rhodophyta (red algae)	22	1	1
Euglenophyta (euglenoids)	124	25	
Cryptophyta (cryptomonads)	15	1	
Pyrrophyta (dinoflagellates)	54	3 (13)	6
Rhaphidophyta	2		
Haptophyta	5		
Chrysophyta (golden(-brown) algae)	115 (36)	11 (5)	1
Xanthophyta (yellow-green algae)	73	1	6
Eustigmatophyta	3		
Phaeophyta (brown algae)	2		1
Prasinophyta	13		1
Chlorophyta (green algae)	992 (520)	117 (353)	68
Glaucophyta	2		
TOTAL IN FLORA	1719 (556)	162 (371)	94

Total number of British freshwater algae (excluding diatoms): 2275 species (2808 taxa).

in major taxonomic treatments. The author of each taxonomic name is given in full with initials or other conventions used following the recommendations given by Brummitt & Powell (1992). A list of standardized abbreviations and other conventions associated with the citation of authors of taxa is provided (p. 624).

The descriptions are mostly brief, diagnostic, and emphasize characters essential for identification. As few technical terms are included as possible, but some are essential. Specialist terms for a particular taxonomic group are sometimes discussed more fully in the introduction to that group (e.g. Charales, Oedogoniales), but otherwise are defined in the Glossary (p. 615). Little consideration is given to ultrastructural detail, since most users of this volume will not have access to scanning or transmission electron microscopy. Each taxonomic entry is accompanied by comments on world distribution and, where available, information on ecology. However, comments on world distribution should be treated with caution, since information on the freshwater algae of many regions is sparse, dubious or non-existent. Detailed distribution within the British Isles is mentioned where sufficient information is available, as is the case for many desmids. In some cases localities are mentioned including vice-counties. There are 111 vice-counties in Great Britain and 40 in Ireland, with the Isle of Man and the Channel Islands representing a further two (see end-papers). Clearly the amount of ecological information provided depends on the current state of knowledge. No doubt some species are much more widespread than is usually considered to be the case, because their characteristic habitat is seldom sampled or not sampled at the optimum time of year. Many records, such as those for some of the filamentous members of the Zygnematales, date back to the early 19th century and were unaccompanied by ecological information. Additional information is provided for environmentally important algae, such as those which can cause a nuisance or are important for monitoring water quality and long-term changes in the environment.

Inevitably material will be discovered that does not correspond to any of the taxa described and illustrated, possibly representing an entirely new record for the British Isles. Hence, there will be a need to consult literature from other parts of the world. Mention is made of culture-based taxonomic systems and in a few keys a character becomes evident only after material has been grown in the laboratory. Culturing can be quite simple (see Methods, p. 14), and is essential for organisms whose diagnostic features become apparent only under nutrient limitation or other stressful conditions. This applies to some blue-green algae, many filamentous chlorophytes and the xanthophyte *Vaucheria*. It is expected that most material will be identified from freshly collected field samples examined under bright-field illumination, although if available, differential interference microscopy can be helpful.

All species and infraspecific taxa recognized here are given a unique identifying 8-digit code number designed to assist researchers wanting to collect and store information on computer file. The full list of codes used was published as *A Coded List of Freshwater Algae of the British Isles* (Whitton et al., 1998a); this is also available at a website, though with less of the accompanying information. The code used is hierarchical with the first 2 digits representing the phylum or one or more major subgroups within a phylum, the next 2 digits the genus, the next 3 digits the species, and the last digit a variety or other infraspecific taxon. Doubtful taxa are only given a code number in the Flora if they were listed in the latest version of the *Coded List*. New codes have been generated for those taxa recognized here, but which do not appear in the 1998 version of the list.

Keys

The aim of the keys is to aid the accurate identification of any sample containing freshwater or terrestrial algae collected in the British Isles. The keys are based in large part on taxonomically important characters clearly visible with the light microscope; only where there is no alternative have we resorted to reproductive, ecological or more obscure characters. They are dichotomous and each character is presented as two descriptive choices or 'couplets'; the choice to follow is the one that most closely agrees with the specimen. Some characters are not absolute (e.g. free-living or attached, solitary or colonial), making it necessary on occasion to back-track and follow an alternative proposition. There are genera for which a character applies only to a subset of its species or a particular environmental condition. In such cases the genus is keyed out in more than one place and the generic name is followed by 'in part'. On occasion characters used in the key relate only to material sampled from the British Isles. Finally, the keys are not constructed in such a way that phylogenetically related taxa necessarily key out together. The keys are 'artificial' rather than 'natural' and are designed to make it as easy as possible to identify samples to the level of species.

Keys are at their least satisfactory when used to identify taxa where quantitative characters show continuous variation, with each character a graded series of overlapping expressions. In such cases, it is often more useful to use a table for comparing character combinations than a traditional key, and this has been adopted where appropriate. Blue-green algae present a particular problem, because so many of the characters are overlapping. Some of the keys in previous Floras are riddled with problems and have in practice seldom been used. Keys are given here only where they seem especially helpful. However, there is available separately an electronic, multi-access identification key for all the blue-green algae recorded in the British Isles (Whitton et al., 2002). This has been updated frequently and versions produced since March 2002 relate closely to the names and information included in the present flora.

Before using the keys it is necessary to be sure that the specimen is indeed an 'alga'. A number of bacteria can be a source of confusion, unless there is access to a fluorescence microscope to check whether or not chlorophyll *a* is present; if it is absent, then the organism is not a blue-green alga or eukaryotic alga. Organisms which may cause problems are green photosynthetic bacteria, purple photosynthetic bacteria and narrow colourless gliding filaments. Many of these live in environments with low oxygen and sometimes also hydrogen sulphide, the latter detected by its odour. However, a number of algae can also live in similar environments. Possible confusion between plates of the purple sulphur bacterium *Thiopedia* and the blue-green alga *Merismopedia* are discussed under the latter genus (pp. 50, 51). Superficially the two can look very similar, although they are far apart phylogenetically.

The green alga *Chlorella* occurs symbiotically with a number of common freshwater organisms, including amoebae, sponges and ciliated protozoans such as *Paramecium* and *Ophrydium*. The latter protozoan is rare in the British Isles (as *O. versatile* Müller) where it sometimes develops in huge quantities in ponds and lakes as irregular, spherical, ovoid or sausage-like, grey-green gelatinous masses that might easily be mistaken for an alga (see Eaton & Carr, 1980). Many protozoa feed on small-celled algae, so there may be doubt in a particular case as to whether intracellular algae are true endosymbionts or are merely ones which have been eaten recently. However, if the sample is still alive it should be quite easy to decide. All the cells of intracellular symbionts look bright green and healthy, whereas cells which have been eaten usually include some which are yellowish or starting to look disorganized. If doubt persists it is worth rechecking samples a day later when the difference between the two is likely to be very clear.

Lichen algae are mentioned only briefly in the Flora and in most cases it is easy to decide whether or not a lichen has been included in a sample. Simple lichens, known as protolichens, may show little structure macroscopically, but can be recognized under the microscope by the presence of many fungal hyphae. However, fragments of the more typical lichen *Verrucaria* often look like free-living green algae when sampled by scraping rocks in rivers. These are common in the middle stretches of many fast-flowing rivers, but may be partially smothered by other encrusting algae such as *Hildenbrandia* and so it may not be obvious that lichen is being removed in a sample. In this case the fungal component often looks more like parenchymatous tissue than obvious filaments.

Another potential source of confusion is the juvenile filaments of mosses and liverworts. These are sometimes abundant together with leafy shoots of the same species, such as on damp soils or the leading edge of rocks in rivers. In this case the bryophyte filaments (hyphae and protonema) are easy to distinguish, because the cross walls are oblique and the outer walls often have a dull brown colour. However, in highly acidic streams (below pH 3) or streams and mine spoils with very high levels of heavy metals such as zinc or copper, several mosses may exist only as juvenile filaments. In this case the cross walls are usually transverse like a typical green alga and sometimes the walls are not coloured brown. The most obvious distinguishing factor is that the chloroplasts are characteristic of mosses – numerous and disc-shaped, much like a higher plant. Published records from other countries of the green alga *Cladophora* in highly acidic streams have proved to be juvenile filaments of bryophytes.

Illustrations

Most species are illustrated by line drawings that emphasize important details needed for identification. However, the Rhodophyta (red algae) are illustrated by halftone photographs of their general habit and microscopic detail. The CD-ROM photo catalogue consists mostly of colour photographs of algal habitats, macroscopic growths and photomicrographs of living material taken under bright field illumination and/or differential interference contrast microscopy. Some scanning electron microscope photographs of dinoflagellates and euglenophytes are included to assist the interpretation of line drawings.

About 30% of the line drawings are new, with most of the remainder being taken from the literature and modified to a greater or lesser degree. If the original figure had thin lines or other unsatisfactory features, it has been redrawn or sometimes only a portion of it used. Inevitably, redrawing has sometimes resulted in changes in the relative thickness of lines or in the effects of shading. We have omitted from old figures any labelling, and in some cases have added or removed shading. If an original figure has been modified or redrawn then the person responsible is acknowledged (p. 630).

A problem in preparing the Flora has been to decide whether figures taken from the earlier literature are still under copyright. Furthermore, some authors state that their figure is 'after' an earlier author. In some such cases the figure has been little if at all changed from that of the earlier author, while in others it was changed considerably when redrawn. Sometimes earlier authors do not mention that their figure has been taken from another source, yet it is clear that it has. We have tried to acknowledge all the sources of illustrative material and other assistance, but apologize to anyone omitted inadvertently. Ideally every species should have been drawn or photographed afresh from British material, but to achieve this would have delayed publication by many years. Where an illustration is based on material from other countries, we have selected material which resembles what we believe to occur in the British Isles, though inevitably a few anomalies will have been introduced.

DISTRIBUTION AND ECOLOGY

The Editors

Introduction

Algae are almost ubiquitous in waters capable of supporting photosynthetic life and no waterbody in the British Isles has been reported to have conditions extreme enough to prevent algal growth. Providing some moisture is present, at least intermittently, then algae can also grow in a wide range of terrestrial habitats. Not surprisingly, such algae are much more conspicuous in the wetter parts of the region. A few algae, such as the algal components of lichens and several widespread microorganisms, are symbionts. Some algae are large enough to be treated as macrophytes (see Glossary). The charophyte *Chara hispida*, for instance, often exceeds 1 m in length. On the other hand, the great majority of species are microscopic and visible to the naked eye only when their numbers are large enough to discolour the water or form obvious surface growths. The type of habitat and size of the organisms determine what sort of method is best for collecting samples.

The algae living in the water column of lakes and larger rivers – the phytoplankton – have been studied for a long time. This is partly because of their ecological importance, but also because the larger species are easy to sample with a net and are also often attractive organisms to observe. However, the smaller species are more of a problem to sample (see Methods). Descriptive accounts of lake phytoplankton have divided the organisms into size classes, although the limits and terms for these classes have not always been the same. The *net plankton* are the largest planktonic algae to be retained in fine mesh nets (<60 μm across). The *picoplankton* are less than 2 μm in any dimension, while the *nannoplankton* range from above the size of picoplankton to 60 μm. Some species of phytoplankton are present in the water column throughout the year, though they may vary markedly in abundance according to the season. Other species exist for part of the year on bottom sediments or are associated with plant growths in shallower parts of the waterbody. Some planktonic algae are confined to the marginal shallows where they are normally associated with submerged higher plant vegetation (metaplankton). The flora associated with *Myriophyllum spicatum* is often especially rich.

When phytoplankton multiply to such an extent that they discolour the water they form what are known as waterblooms, especially when the cells rise to the surface, as with gas-vacuolate blue-green algae and lipid-rich organisms such as *Botryococcus*. The term waterbloom is now often used more widely to describe scums and mats of algae floating on or just beneath the water surface, though the two rather different uses of the term can be a source of confusion.

Benthic algae are those attached to all kinds of substratum, ranging from sheets of rock to fine silt and other living organisms. The charophytes or stoneworts are the largest benthic freshwater algae, growing over silt or fine sand and are anchored by rhizoids arising at intervals from creeping branches. In the past these algae have been collected and treated along with flowering plants, which is probably part of the reason why they have been given the most attention by conservationists. Other larger algae, such as the filamentous *Spirogyra* and *Oedogonium*, are often visible in the marginal shallows of lakes and ponds as free-floating masses or entangled around submerged aquatic plants or other underwater objects. Most fine-leaved aquatic plants provide suitable surfaces for filamentous algae.

The small algae attached to or associated with submerged surfaces are referred to as epilithic, epipelic, epipsammic, epiphytic or epizoic, according to whether they are growing on rocks, silt, sand, plants or animals, respectively. Many epilithic algae are capable of growing on a wide range of hard surfaces, as has been demonstrated when artificial substrates, such as glass and plastic, are examined after having been left submerged for some days or weeks. Epipelic organisms often form a distinctive community on silt and fine sand; many of these species show gliding motility, though some of them also produce mucilage and the whole community may form a mucilaginous layer. Especially in rivers, epipelic communities of diatoms and blue-green algae in late spring and early summer may rise from the bottom during the day and float at the surface. Epipsammic communities are less common in freshwater than in marine environments, but larger sand particles sometimes dominate the bottom of particular stretches of river and here may be found algae which are attached to individual particles rather than moving freely between smaller particles, as in epipelic communities.

The diverse assemblages of microscopic algae associated with surfaces are not necessarily all firmly attached, but may form a surrounding community which includes some attached organisms, others loosely associated with the surface, perhaps in a mucilaginous film, while others are small motile flagellates or phytoplankton floating in the water. Aufwuchs or periphyton are terms used frequently for this assemblage. At one time the term aufwuchs was applied only to plant surfaces, but now both terms are applied to any type of surface. Filamentous algae frequently associated with the periphyton include creeping or crustose forms (e.g. *Aphanochaete*, *Protoderma*) and those having an erect and/or a prostrate system of branches (e.g. *Stigeoclonium*, *Chaetophora*, *Draparnaldia*). On limestone rocks some are partly endolithic (e.g. *Gongrosira debaryana*, *Gomontia perforans*) or almost entirely so, such as some forms of narrow sheathed blue-green algae. Occasionally benthic forms are swept from the bottom by current or wave action to become temporary members of the plankton; in the case of lakes and ponds these are referred to as tychoplankton.

Algae occur in many terrestrial habitats and are broadly divided into those associated commonly with soil and algae on soil-free surfaces (subaerial algae). These subaerial algae not only grow on a very wide range of inanimate surfaces, but are associated also with other plants, especially the leaves (epiphyllic) or trunks (corticolous) of trees.

Distribution in the British Isles

The varied geology and geomorphology of the British Isles along with its Atlantic climate gives rise to a very diverse range of algal habitats. The inland waters of Great Britain cover about 1.04% (2404 km^2) of its total land area (Smith & Lyle, 1979). These comprise over 5500 lakes and reservoirs (<25 km^2), with by far the largest number in Scotland (3788). In the north and west of the British Isles, the climate is cooler, rainier and more oceanic and the geology dominated by hard Palaeozoic rocks. This contrasts to 'lowland' Britain where the climate is warmer and drier and the geology more varied.

Some early generalizations concerning algal distribution patterns in the British Isles remain true today (W. & G.S. West, 1909a; West & Fritsch, 1927). For example, desmids are more diverse in mountainous than lowland areas, and are more diverse in the relatively nutrient-poor waters on older Palaeozoic rocks or rocks of igneous origin than on calcareous rock formations. West & Fritsch (1927) concluded that filamentous zygnematalean algae (*Zygnema*, *Mougeotia*) are also well represented in mountainous areas where the water is reasonably soft and of an acid pH, whereas 'an abundance of the commoner unicellular forms are found in low-lying quiet waters'. The group of green algae belonging to the Chlorococcales are most abundant and diverse in the nutrient-enriched ponds and lakes characteristic of lowland areas of England, southern Scotland and parts of eastern Ireland.

During much of the nineteenth and twentieth centuries industrialization and the spread of urban areas brought about changes to the landscape which have had a major influence on water quality. The effect of intensive agriculture has been even greater than the influence of industry in many areas during the second half of the twentieth century. The input of sewage effluent, agricultural fertilizers and farm wastes has resulted in the artificial nutrient enrichment of waters, a process known as eutrophication. As a consequence of these inputs, waters naturally nutrient-poor, such as those draining mountainous areas, have become enriched with nutrients essential for plant growth. This has resulted in changes in the composition, diversity and abundance of algal species. The excessive growth of algae caused by nutrient enrichment results in the blockage of inland waterways, significantly affects the cost of treating drinking water, reduces the amenity and conservation value of water, and leads to the death of fish and other organisms. Changes in amounts and types of algae have been monitored over the long term in several of the larger lakes (e.g. Loch Leven in Scotland, Windermere in the English Lake District, Lough Neagh in Ireland), as well as the relatively small and shallow lakes of the Cheshire Meres and the Norfolk Broads. Many smaller water bodies, such as farm ponds, have disappeared or become degraded through in-filling or general lowering of the water table, drainage/land reclamation and nutrient enrichment and contamination by pollutants. To some extent these losses have been compensated for by the creation of new gravel pits, ornamental ponds, canals, fish ponds and reservoirs. Although the majority of algae are found in aquatic habitats, some of those most commonly encountered are on paving stones, walls, gutters, roofs, glasshouses, tree-trunks and wooden or concrete posts. Such subaerial algae are usually overlooked until they are present in sufficient quantity to form mats or crusts. In common with other cryptogamic groups these algae are more diverse and evident in the wetter western parts of the British Isles.

HISTORY OF FRESHWATER ALGAL STUDIES IN THE BRITISH ISLES

The Editors

Natural historians first started to look at algae carefully about two centuries ago, but no doubt many before then had noticed green slimes or algal blooms in their local pond. One of the earliest records for freshwater algae in the British Isles comes from Scotland. The twelfth century abbey of Soulseat, near Stranraer in Dumfries and Galloway, was described as 'monasterium viridis stagnii', or 'the Monastery of the Green Stank', as translated in a Victorian visitor's guide. Malodorous blooms, almost certainly caused by blue-green algae, frequently affected the Abbey, situated on a peninsula in a small lake; these blooms still persist to this day. As pointed out by Griffiths (1939b), the development of waterblooms could be a warning of impending tragedy. Two early chroniclers of a lake turning red at Finchampstead, Berkshire, forewarned of the untimely death of William Rufus in 1100.

There were a few scattered observations on freshwater algae during the eighteenth century, but mostly these were made on samples of common organisms used as demonstrations for the microscope. A 1777 record by Stephen Robson for *Chara hispida* in Hell Kettles, two well-known ponds in County Durham, is of interest, because this stonewort, which is about the largest alga in the British freshwater flora, still thrives in the same pond. Such old records are important, because they help us to assess the extent to which changes have taken place, but it was a century later before there were many records with sufficient detail to make such comparisons. This brief history of freshwater algal studies in the British Isles looks mostly at the older accounts, because these are the ones which readers are least likely to come across. A few of the more recent are mentioned, but for each one there are many more for which there is no space. Most research in recent decades by phycologists in the British Isles has been floristic or ecological and there are few critical taxonomic studies on freshwater algae other than diatoms. Hopefully, someone will one day write a full account of research on British freshwater algae during the past half century, especially the wide range of ecological studies.

The first attempt to bring together existing knowledge was the publication of Dillwyn's *British Confervae* between 1802 and 1809, which included freshwater as well as marine algae. A few records relating to freshwater algae in Scotland appeared in Greville's *Flora Edinensis* (Greville, 1824) and many more in a series of volumes on the *Scottish Cryptogamic Flora* published between 1823 and 1828. Greville also prepared a contribution on diatoms for the second volume of Sir William Hooker's *British Flora* (Hooker, 1833). Twelve years later Hassall (1845) published the first forerunner of the present volume, *A History of the British Freshwater Algae*.

Prior to the appearance of Hassall's book, John Ralfs, a surgeon turned mycologist and phycologist, had published seven papers in the *Transactions of the Botanical Society of Edinburgh* and the *Annals and Magazine of Natural History* on one particular group of freshwater algae, the desmids. His name was to be indelibly linked with this species-rich group of green algae with the publication in 1848 of the *British Desmidiae*. In the introduction to this volume, now accepted as the starting point for the naming of desmids, he writes, 'Until a recent period, the study of minute objects which form the subject of this work had been more neglected than almost any other branch of Natural History', and went on to express the hope that he 'might be able to present to the British Naturalist such a description of our species as seemed necessary towards making the knowledge of them at home keep pace with its advance on the Continent'. He further comments, 'I soon discovered not only that we possessed many species hitherto undescribed, but that various points in their economy, not devoid of interest, remained still unexplained or doubtful'.

Within the next few years another group of algae, the diatoms, attracted the attention of microscopists and these are still the favourites of many professional and amateur microscopists. Smith's *Synopsis of the British Diatomaceae* appeared between 1853 and 1856. The great improvements in design and availability of microscopes in the last two decades of the nineteenth century led to a considerable increase in interest in freshwater algae in many parts of Europe, including the British Isles. Within our own islands, the Irish botanist, Archer, was by far the most prolific, producing some 190 papers between 1858 and 1885. Most were published in the Proceedings of the Dublin Microscopical Club or the Annals and Magazine of Natural History and most were descriptions of desmids new to science. However, frequently there were also accounts of their sexual reproduction and often complex walls.

During the second half of the nineteenth century further papers on British desmids were written by the microscopist M.C. Cooke: there was even a description of a new *Cosmarium* from Trafalgar Square (Cooke, 1880). From 1882 to 1884 this prolific author issued his *British Fresh-Water Algae*, and published in 1886 as a supplement to the latter, *British Desmids*. Sadly, as West and Fritsch were to comment (West & Fritsch, 1927, p. 1), neither his work nor Wolle's *Freshwater Algae of the United States* (published in 1887), contributed much to further knowledge of the algae. Indeed they believed some of Cooke's illustrations were positively misleading and stated (*op. cit.*, p. 2) that 'It may be doubted if 25 per cent of the British freshwater algae could be identified with certainty from Cooke's book'!

William West and his son, George S. West, were the leading figures in freshwater botany in Britain over some 40 years, starting in the late 1870s. Together they made immense contributions to the taxonomy and distribution of the British freshwater algae. They travelled by rail, carriage, pony and steamer to many parts of the British Isles, making extensive collections not only in the more accessible parts of England, but also in the then remote regions of Scotland, Ireland and North Wales. As the Wests became increasingly acknowledged as leading world experts on freshwater algae, their opinions were sought on material from many countries – Spain, Portugal, Denmark, USA, the West Indies, Africa, the Far East, Australia and even Antarctica. Some 140 papers, monographs and books were published between 1888 and 1918. For the British Isles there were descriptions of more than 500 taxa, of which 210 were new to science. The great achievement of the Wests was to prepare a series of monographs in which all known British desmids were described and illustrated. This is the five-volume *Monograph of the British Desmidiaceae* (W. & G.S. West, 1904, 1905, 1908, 1911; W. West, G.S. West & Carter, 1923). Nellie Carter, one of G.S. West's students, completed the final volume following G.S. West's death in the 'flu epidemic of 1919 at just 43 years of age. Carter told one of the editors (DMJ) that, had he lived, G.S. West would almost certainly have written a British Freshwater Algal Flora in a format similar to that of the desmid monograph. He was proposing to describe all the genera and species of freshwater algae known to occur in the British Isles.

Researchers who made substantial contributions in the following decades included Carter herself, F.E. Fritsch, B.M. Griffiths, W.H. Pearsall, M.F. Rich and M. Rosenberg. Fritsch is best known internationally for his comprehensive two-volume *The Structure and Reproduction of the Algae*, but for the British Isles *A Treatise of the British Freshwater Algae* was as important. The first edition was published by G.S. West in 1904 and Fritsch became a co-author of a substantially revised second edition in 1927. Fritsch was Professor of Botany at Queen Mary College, London, and had a great influence not only through his own research, but also the training and encouragement he gave to many students. One of his most lasting contributions is the collection of published illustrations of freshwater algae, now housed at the Freshwater Biological Association at Windermere as 'The Fritsch Collection of Illustrations of Freshwater Algae'. This became of major assistance to those working on algal taxonomy worldwide (Lund, 1961e) and continues to grow as illustrations of newly described taxa are added. The Collection became available in 1964 in the form of a microfiche and regular supplements are prepared by J.W.G. Lund and colleagues; the microfiche edition is published by Interdocumentation Company AG, Switzerland (IDC). Fritsch also provided the first detailed accounts for the British Isles of algal communities in several types of environment, particularly fast-flowing streams and phytoplankton in rivers.

Early accounts of lake phytoplankton are especially valuable if they were detailed enough to permit comparisons between old and recent surveys. Some of those by the Wests, Griffiths and Pearsall are useful for this, although considerable care is needed in comparing different sets of records because of uncertainty about sampling methods, name changes and misidentification (Whitton, 1974). Some of the early phytoplankton studies included comparisons from a range of lakes and led to biological classifications of lake types. The survey by the Wests (W. & G.S. West, 1909a) was an early example, though the best known were those of Pearsall (1924, 1930, 1932) in the English Lake District, which increasingly combined environmental data with phytoplankton studies. A modern synthesis of this information has been given by Talling & Heaney (1988).

Fritsch and Pearsall succeeded in raising funds for the establishment in 1931 of a laboratory to study all aspects of the freshwaters of the British Isles. The birth of the Freshwater Biological Association on the shores of (Lake) Windermere in Cumbria, housed at first in the mock-tudor Wray Castle, later in the converted Ferry House Hotel, led to detailed, long-term surveys of the phytoplankton of the English Lake District. For instance, J.W.G. Lund and colleagues combined regular measurements starting in the late 1940s of algal cell densities in Windermere and other lakes with experimental studies on the diatoms *Asterionella formosa* and *Melosira italica*. By the 1970s these had not only provided considerable insight into the ecology of these widely distributed diatom species, but the regular monitoring of water chemistry and phytoplankton populations had led to a much improved understanding of the phytoplankton dynamics of temperate lakes in general (Lund, 1979). A more recent summary of the research on phytoplankton in Windermere has been provided by Reynolds & Irish (2000). Lund's own work has also included blue-green algae, palmelloid green algae, chrysophytes and desmids. Studies on the fungal parasites of phytoplankton and other algae carried out by H. Canter-Lund over the same period showed how important these can be in the rapid decline of algal populations (e.g. Canter, 1949a–c, 1954; Canter & Lund

1968, 1969). Protozoa can also have an important impact (Canter & Lund, 1968).

Similar studies to those on Windermere, though not quite so detailed, have since been applied to many other species of phytoplankton in the English Lake District and to lakes and reservoirs elsewhere in the British Isles. For instance, staff of the Freshwater Biological Association extended their interest to algae of the productive, eutrophic Cheshire Meres (Belcher & Storey 1968) and Shropshire Meres (Reynolds 1967, 1971, 1973a–c). These waters are very different from those of the English Lake District and support at certain times of the year considerable blooms of blue-green algae, especially *Microcystis aeruginosa*, whose vertical distribution and buoyancy was studied in Rosthene Mere (Reynolds & Rogers, 1976).

R.W. Butcher was the first to look closely at the attached algae in larger rivers. His studies started in the 1920s and continued to the 1950s, but his most detailed survey was made on the Tees in the mid-1930s. Several attached algae, such as *Ulvella frequens* (now *Protoderma frequens*), which appear to be widespread in the fast-flowing rivers of many countries, were first described from this river. Butcher worked together with a chemist and a fisheries specialist, so that his ecological information appears more relevant to a modern reader than most other research from the 1930s. The combined studies of this group (Butcher, J. Longwell and F.T.K. Pentelow: Butcher et al., 1937) led eventually to the development of biological methods for surveying British rivers. Strangely, however, Butcher became convinced that British rivers, or at least those in north-east England, were too small to allow a true algal plankton to develop. It is unclear how he got this idea, especially because the lower part of the non-tidal Tees has a quite well developed phytoplankton population (and almost certainly did so in the 1930s) and includes species scarcely found in the benthic community. Perhaps it was just a semantic matter of what is 'true', but it seems more likely that he never actually sampled the phytoplankton, or at least never did so in late spring when the population tends to be at its maximum abundance.

Butcher's particular interest was in the ecology of *Cladophora glomerata*, which formed huge nuisance growths in the River Tees. There was little further interest in this alga until the 1970s (Whitton, 1970; Bolas & Lund, 1974), although it must have reached nuisance proportions in many rivers during the century prior to this. It seems strange to a modern reader how little comment the older phycologists made on nuisance problems and that it was not until Butcher's studies that people became more aware of them. Butcher published further papers after the war, but as much on aquatic flowering plants as about algae. However, since the mid-1960s many other authors have studied the planktonic and benthic algal communities of rivers (e.g. Swale & Belcher, 1964; Swale, 1969; Lack, 1971; Moore, 1976; Moss, 1977; Holmes & Whitton, 1981; John & Moore, 1985a, b). Marker (1976a, b) and Marker & Casey (1982) described the benthic algae of chalk streams in southern England, but the communities of blue-green and green algae studied by Fritsch (1929a) in less hardwater streams have scarcely been looked at since then. Most other floristic studies of streams have dealt with diatoms or the effects of pollutants such as acid mine drainage (e.g. Hargreaves et al., 1975; Hargreaves & Whitton, 1976) and heavy metals (e.g. Benson-Evans et al., 1975; Say & Whitton, 1980). The establishment in 1949 of the Brown Trout Research Laboratory, later named the Freshwater Fisheries Laboratory, at Pitlochry in Perthshire was for a period a stimulus to research on the phytoplankton of Scottish freshwater lochs, especially the effects of fertilizing small hill lochs with the ultimate aim of increasing brown trout production. The weekly sampling led to increased understanding of phytoplankton periodicity in the largely nutrient-poor waterbodies, especially of the chrysophytes *Dinobryon* and *Stichogloea* (Brook & Holden, 1957); very small algae were found to be abundant during winter months under the ice. The studies also redirected attention to the rich and varied desmid flora of Scotland's freshwater lochs, virtually neglected since the earlier studies of the Wests (e.g. Brook 1957, 1958a–c, 1959b–d).

An important initiative in the study of British freshwater algae resulted when the large and comparatively shallow Loch Leven in Perthshire was chosen as one of the major study sites for the International Biological Programme (IBP) in the 1960s. Several short-term studies had been carried out on its phytoplankton from the early years of the twentieth century up to the end of the 1950s. As part of the IBP study, regular monitoring of the loch's phytoplankton was initiated and, although the Programme officially ended in 1971, studies have continued by A.E. Bailey-Watts and his colleagues. Approximately 300 species of blue-green and eukaryotic algae have been recorded from the loch, of which some 150 were listed by Bailey-Watts (1974, 1978).

The presence of a dense bloom of a small-celled species of the blue-green alga *Synechococcus* in Loch Leven (Bailey-Watts et al., 1968; subsequently named *S. capitatus*) drew attention to the importance of minute planktonic algae in the size range now known as nannoplankton. The ecological significance of these organisms had been realized as long ago as 1911 (Fogg, 1986), but no one was aware until the mid-1980s that in many large nutrient-poor lakes the most important algae were even more minute cells, the picoplankton. The dividing line between picoplankton and nannoplankton is now agreed to be 2 μm, so that picoplankton cells are less than 2 μm in every dimension. In most lakes, such as Wastwater in the English Lake District, blue-green algae dominate the picoplankton in summer (Hawley & Whitton, 1991), but elsewhere minute green algae or chrysophytes may also be important (Happey-Wood et al., 1988). The larger cells within this size range had been recognized as algae by earlier researchers, but appear to have been much underestimated in cell counts, while the even smaller cells were probably not recorded, if noticed at all; most phycologists probably assumed that they were bacteria. A good quality fluorescence microscope is needed to count picoplankton algae, and since many laboratories do not possess such a microscope, these minute organisms go unrecorded. Nevertheless they are almost certainly present in all lakes, so that old records which report large upland waterbodies as being dominated by desmids should be reinterpreted as referring only to the phytoplankton cells larger than 2 μm in at least one dimension.

The Norfolk Broads are a man-made, shallow wetland system that has been much influenced by human activities during the past century. The impact of these activities became apparent due to a variety of changes in their algal populations, both in species composition of particular Broads and also in more general effects, such as drainage channels being clogged by filamentous algae and the increased abundance of phytoplankton. B. Moss and his co-workers have tried to quantify the relative importance of factors, such as increased nutrients and the effects of boating on these changes, and also the ways in which the algal changes have influenced invertebrates, fish stocks and birds (e.g. Moss, 1983a; Moss et al., 1985, 1988).

Lough Neagh has also been the subject of many studies, because of its importance as the main water supply for Northern Ireland combined with many amenity uses. Many changes have been observed in the lake, including changes among the phytoplankton (e.g. Gibson et al., 1971; Gibson, 1993). Increases in two blue-green algae, *Oscillatoria agardhii* and *O. redekei*, are among the signs of a decrease in water quality due to increased nutrient levels (Foy & Smith, 1993). Studies have also been made on the benthic algae, though mainly the diatoms (Jewson & Briggs, 1993).

Early studies on the algae associated with pond sediments were undertaken by Lund (1942c), in which he found that some members of this community were neither recruited from the phytoplankton nor from epiphytes. These epipelic algae stabilize sediments and join sand grains together. F.E. Round later studied this community and the other benthic algae of lakes (e.g. Round, 1953, 1957b, c, d, f, h, 1959, 1960c, 1961a, 1966), ponds (Round, 1955) and springs (Round, 1957f, 1960b, d). These studies have focused largely on diatoms, but there are some accounts for other groups (e.g. Round, 1957c). Happey-Wood, who started research on benthic diatoms, subsequently looked at a range of other epipelic organisms, especially flagellates (Happey-Wood, 1980; Happey-Wood & Priddle, 1984).

Cores of sediment from lakes and ponds can provide a record of earlier algal populations in waterbodies. Most such studies in the British Isles have been based on diatoms, because of the persistence of their silica frustules. This has been applied, for instance, to lakes in the English Lake District (e.g. Haworth, 1969, 1984, 1985), Scotland (Pennington et al., 1972), the Norfolk Broads (Osborne & Moss, 1977; Phillips et al., 1978) and Lough Neagh (Battarbee & Carter, 1993). Other structures may also be preserved, such as the cysts of some chrysophytes, which have a siliceous wall, and some green algae in the order Chlorococcales (e.g. *Pediastrum*, *Coelastrum* and *Botryococcus*), which also have highly persistent walls, even though there is no silica present. In addition to non-living remains, the upper layers of sediment almost always include live algae. If there are several shifts in lake conditions, these may provide an inoculum for renewed growth of a population when conditions again become favourable. Considerable numbers of viable desmids of the genera *Closterium* and *Micrasterias* were found in sediments of a small lake in Leicestershire down to 5 cm (Brook & Williamson, 1988a). A few viable cells of a range of algae were found down to 35 cm in some lakes of the English Lake District, though they were scarce below 5 cm (Stockner & Lund, 1970).

Increasingly, many of the people looking at freshwater algae are involved in some way with water management, rather than researching on them or as enthusiastic amateurs. Unfortunately, those involved in management are mostly too busy to write about their observations, but a few have published useful contributions. For example, J.P.C. Harding (Harding & Hawley, 1991; Harding, 1996; Harding & Kelly, 1999) has provided accounts of the ways in which freshwater algae have been used in the UK for monitoring rivers, while Marsden et al. (1997) described the impact of nutrients on the algal vegetation of lowland rivers in Scotland. It is hoped that publication of the *Coded List of Freshwater Algae of the British Isles* (Whitton et al., 1998a), various interactive identification systems based on CD-ROMs, and the present Flora should together encourage many more people to make better use of the records collected in routine surveys. This means storing data on computer in a form which others can understand and assessing results in scientific journals rather than just in local reports.

FIELD METHODS

The Editors

Some of the larger freshwater algae can be identified in the field with the naked eye and rather more require the use of a hand lens (×10 or ×20), and sometimes fertile material can be recognized. However, most freshwater algae are microscopic and can only be identified using a reasonably good quality compound microscope. Small portable field microscopes are available, but are only suitable for identifying those algae for which critical microscopical examination is unnecessary.

Algae are generally most diverse and abundant during the summer and early autumn, but it is important to collect during different seasons as certain algal genera (e.g. *Tribonema*, *Microspora*) and groups (e.g. Chrysophyta) are more common at other times. Seasonality is therefore an important consideration when collecting in most aquatic habitats. Jane (1942) considered May to be best for collecting in some clear, soft waters near London, whereas June and July are good months in some Welsh lakes for 'cold-water' forms such as *Dinobryon*. In small ponds algal diversity is often greatest before the extensive development of aquatic macrophytes and the coming into leaf of shade-casting trees.

Aquatic algae

Macroalgae These can be collected by hand, wading in shallow water, and snorkelling or SCUBA-diving in deeper water. Forceps are easier than fingers to gather small delicate specimens. A grapnel or dredge is required for attached macroalgae (especially charophytes) and other aquatic plants growing in water too deep to be sampled by wading. These plants should be collected because small algae may grow attached to them or be associated with any mucilaginous covering. A suitable collecting implement for remote sampling of macroalgae, such as charophytes, is a three-pronged grapnel fashioned from wire coat-hangers suitably bent and a length of lead piping (Fig. 1). Other sampling devices include a right-angled digging fork, a long-handled garden rake or a scoop with a sharp edge and extending handle. These types of samplers may disrupt and cause damage to bottom-living communities. The conservation status of a site should therefore be considered before using them. In clear and shallow water it is possible to direct samplers to specific areas by viewing the bottom of the lake or pond using a glass-bottomed bucket or diving mask. Thick mats of algae can be collected by hand, as can surface-floating mats of such algae as *Oedogonium*, *Spirogyra*, *Zygnema*, *Mougeotia* and *Hydrodictyon*.

Large specimens should be shaken or gently squeezed on collection, to remove excess water before storing in polythene bags. Representative portions of macroscopic growths, or small individuals, should be placed in vials or jars along with a little water. Less conspicuous species are frequent amongst floating or attached masses of algae. Portions of such masses are best stored, without removing excess water, in a polythene bag or jars to be floated out and examined later for microscopic forms.

Algae associated with natural surfaces Various devices have been designed for brushing or scraping small algae from a surface into a collecting tube or bottle. An old toothbrush or teaspoon/tablespoon are very effective implements for removing material growing *in situ*. The spoon edge can be used to scrape the material carefully from a hard surface and by using

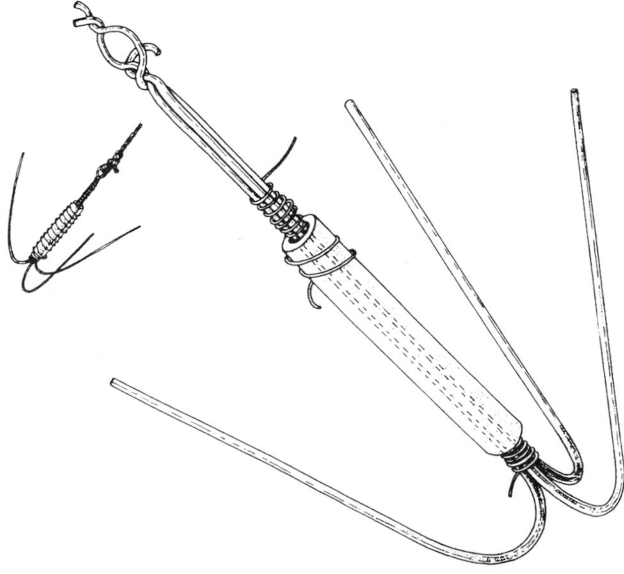

Figure 1. Design of grapnel ideal for sampling submerged aquatic vegetation including charophytes (after Moore, 1986).

a thumb the sample can be kept within the bowl of the spoon until transferring to a collecting vessel. By such means it is possible to collect successfully samples even from quite fast-flowing waters. The spoon method is especially used by diatomists, but is just as effective for other algae associated with surfaces. A knife is useful for scraping encrusting algae from stones and removing surface slivers of wood covered with algae. A geological hammer and stone chisel is necessary for sampling portions of rock when algae are firmly encrusting; safety glasses should be worn to protect the eyes.

Small epiphytes and other algae are commonly associated with filamentous forms, bryophytes and other aquatic macrophytes. These may be sampled by allowing excess water to drain away from collections of finely divided macrophytes (e.g. *Utricularia*, *Myriophyllum*), mosses (e.g. *Sphagnum*, *Fontinalis*, *Rhynchostegium*), or larger filamentous algae (e.g. *Cladophora*, *Vaucheria*), by shaking them a few times. These should then be gently squeezed and the water collected in a wide-mouthed container or filtered through a plankton net. Fine-leaved flowering plants, in particular, often have a diverse algal flora loosely associated with leaves rather than firmly attached to them. The flocculent material that sometimes tenuously coats such plants is best removed by very carefully inverting a jar or other wide-mouthed vessel over the macrophyte to be sampled, cutting the plant in the vessel and then capping it with the plant and associated floc contained inside. Sometimes microscopic forms are in very great abundance on brown, decaying *Sphagnum* plants with the result that the moss feels slippery or gelatinous to the touch. The squeezings from aquatic plants are often very concentrated algal samples, although not necessarily representative of all the algae that occur in the microenvironment of the macrophytes. After settlement of the algal-rich sediment to the floor of the container, the excess water can be poured off and the sample pipetted into small tubes or vials. Care should be taken to retain some of this water since it may contain smaller algae and flagellated forms. This squeezing technique is useful not only for sampling epiphytes, but free-living algae that occur among aquatic plants and reproduce in shallow water (tychoplankton) along with forms that are normally in open water (euplankton, true plankton).

Greenish or brownish slimy films of microalgae and other organisms frequently cover submerged parts of reeds and other emergent aquatics. These films can be removed with forceps or by running a thumb and forefinger up a stem and collecting the removed film. If the film is very loose, it can be removed by suction using a pipette with a large bulb, or by cutting a portion of film-bearing surface and carefully transferring it underwater to a container. Submerged twigs, dead culms, portions of the roots and stems of aquatic plants and non-living surfaces (e.g. stones, glass, plastic, wood) all provide surfaces for algae and therefore should be collected. These need not be stored in water, but just sufficient water should be added to the container to maintain a saturated atmosphere. In the laboratory microalgae can be removed by vigorously shaking with a little water, scraping with a blade, brushing using a toothbrush, washing with a jet of water, or by using fine forceps.

Several blue-green and other algae occur endophytically or are closely associated with floating bryophytes (e.g. *Riccia*, *Ricciocarpus*, *Sphagnum*), the fern *Azolla*, and the flowering plant *Lemna* (duckweed). Sometimes endophytes can be detected on decaying plants or on ageing plants that are discoloured. The endophyte often remains green and in the case of *Chlorochytrium lemnae* can be seen as minute, dark green areas on the leaves of *Lemna*.

Use of artificial surfaces to sample microalgae A variety of materials can be used to collect those algae that normally grow on opaque surfaces and which are almost impossible to remove undamaged. Glass or polyethylene surfaces are commonly used for this purpose. Slides have the disadvantage in requiring some kind of framework to hold them in place and are easily lost or broken. However, various devices have been designed to hold slides on riverbeds or to suspend them at a known depth beneath the water surface. Polyethylene strips or bags, or plastic petri dishes have the advantage that they can be easily fixed by a thumb tack or nail to wooden structures such as mooring posts or jetties. They need to be positioned vertically in order to reduce fouling with detrital material. On retrieval, molluscs and aquatic insect larvae should be removed to prevent significant loss by these grazing invertebrates. If the surfaces are left in position for too long then prolific algal growths often obscure the smaller forms. It is possible on collection to cut the artificial surface into small portions containing the selected algae and grow these on for examination and identification in culture. Other surfaces can be used for sampling, including wooden blocks, ceramic tiles and short lengths of cotton cloth.

Microalgae associated with sediments The free-living algae associated with mud or silt (epipelic algae) are predominantly diatoms, blue-green algae, green algae and euglenophytes. In shallow waters, the easiest collecting method is to draw a long glass tube (60–90 cm long and 2–4 mm diameter) slowly across the sediment surface. The tube should be stoppered with a finger at the upper end and the finger carefully released so that the tube slowly fills with a mixture of water, sediment and algae. Another method is to use a plastic tube with a large bulb over one end or, alternatively, a large plastic pipette. A useful sampling device is a giant pipette such as the type used for removing fat or oil from the surface of cooked meat and is available from most good kitchenware suppliers. Samples are transferred to a bottle and the supernatant drawn off after the solid particles have been allowed to settle for several hours in the dark. The remaining mixture is then thoroughly shaken to distribute evenly the sediment and algae.

Many motile algae are positively phototactic and can be sampled using the following method. Samples are concentrated by pouring into a petri dish illuminated from above. The algae migrate up to the light and become trapped amongst the fibres of small squares of lens tissue, or settle on the underside of cover glasses placed on the water-sediment surface. Many motile algae are known to show an endogenous rhythm, moving upwards through the sediment by the end of the night to reach maximum numbers at the surface before mid-day. To obtain the greatest diversity of epipelic algae, the entrapment surfaces need to be left in position overnight and removed before noon of the following day. Cover glasses can be examined directly, or mounted on slides in 40% glycerol to which Lugol's iodine solution has been added. Lens tissue squares can be similarly examined after macerating to separate the algal cells from the fibres. Another method is to place silt, mud and detritus in a small petri dish or watch glass and to provide unilateral illumination. If present in significant numbers, the algae will congregate against the side of the container next to the light source and can be removed by pipetting. In the case of euglenophytes, samples containing sediment or other debris (e.g. decaying leaves, organic matter) can be left undisturbed in their containers for a few days in a north-facing window. These motile algae have a tendency to aggregate in large numbers on surfaces and thus can be collected by pipetting. A 20 mL plastic pipette is ideal for this purpose and is a useful size for sampling neustonic (surface-living) algae.

Sediment samples containing sand should always be examined for epipsammic algae, that is forms growing attached to sand grains. Samples should be taken at depths down to 5 cm since such algae may occur well below the sediment surface. They can be detached from the grains by agitating the grains, vigorously shaking with a little water before placing the mixture in a second container and allowing the sediment to settle. The treatment should be repeated three or four times until the supernatant is clear and leave this undisturbed for 4–5 hours to allow all suspended matter to sink. Excess water must be carefully removed before the samples are examined live or stored in Lugol's solution or formalin. Ultrasonic vibration is another means of separating epipsammic algae from sand grains.

Phytoplankton Minute free-floating planktonic algae are best collected by towing a fine net of bolting silk through water. A 20-μm mesh is the smallest size of bolting silk usually available. The smaller the mesh, the more difficult it is to remove the algae and the net soon clogs if they are present in quantity. The advantage of this method over most others is that it enables a comparatively large volume of water to be sampled and few larger planktonic algae escape collection. Its principal disadvantage is that it allows many smaller forms to escape and also concentrates the grazing zooplankton. The zooplankton not only graze the algae but, if present in significant numbers, also cause the rapid deoxygenation of the sample through their respiration. Care is needed when using a plankton net in shallow water so as to avoid disturbing the bottom sediments; also, if drawn too rapidly through the water its fine mesh merely pushes the water ahead of the net rather than filtering it. Nets and other sampling equipment must be thoroughly cleaned after use to avoid contamination of the next sample and also to minimize the risk of transporting noxious algae or other pest organisms from one water body to another. It is recommended to disinfect the net and to take along large wash bottles of clean water for this purpose. When algae form a surface scum, or are present in such quantity that they form a 'waterbloom' (discolour the water), they can be collected by partly filling a container with the alga-rich water. It might be necessary to add some water since very dense samples often decay very rapidly and should be avoided.

The simplest method of obtaining a plankton sample from static or slow-moving water is to use a bucket, jug or a piece of flexible tubing. If tubing is used (about 25 mm in diameter), one end is placed just below the surface and the other is weighted and allowed to sink. The sunken end is attached to a line and the tube slowly raised until it forms the shape of a 'U'. It is removed from the water in this form and the column of water sampled poured into a suitable container. Similarly, electric rotary suction pumps can be used in most situations with a length of tubing lowered to the required sampling depth. Samples can be concentrated by filtering through a plankton net (if smaller plankton are not required) or by centrifuging (see below). Various types of collecting vessel can be used for quantitative sampling. These usually take the form of a weighted container that can be lowered to the required sampling depth before a messenger weight is sent down the line to operate a closing mechanism. Such samplers are made commercially and come in varying sizes although one litre is the smallest available. A marked cord or winch fitted with a depth indicator is necessary to record the sampling depth. Large plankton are very sparse in nutrient-poor lakes (oligotrophic) whereas the nannoplankton and picoplankton are too small to be sampled with a net. These latter plankton can only be collected using containers and then concentrating (see below).

Common methods used for concentrating phytoplankton samples are as follows:

Sedimentation This allows the algae simply to fall to the bottom of a vessel. Larger colonial and filamentous forms are sometimes very slow to sediment. It is usually possible to speed up sedimentation of most algae by adding Lugol's iodine solution. It may be necessary to use two sedimentation tubes for each sample, one with the normal addition of Lugol's iodine solution (see below) for the majority of algae and the other with a much higher concentration required for blue-green algae. It is possible to sediment many gas vacuole-containing algae by filling a small plastic tube completely with the sample then pushing in a cork. A sharp tap with a hammer or other hard object causes the collapse of the gas vacuoles and leads to rapid sedimentation. Sedimentation is usually performed in a separating funnel where 2–3 drops of concentrated Lugol's solution plus 10% acetic acid or sodium acetate are added until the sample turns the colour of strong tea, brandy or cognac. The iodine in the Lugol's solution kills, preserves and stains the algae and the acetic acid prevents loss of flagella (see under Preservation, p. 18); this treatment also makes the algae heavier, so that their rate of sinking is increased. After about 24 hours the algae accumulate at the bottom of the separating funnel and can be pipetted into tubes or vials. A dimple-bottomed plastic mineral water bottle can be used in place of a separating funnel and the sedimented algae removed by using a siphon made from a plastic drinking straw (Belcher, 1999). To the algal samples, 4% neutralized formalin or FAA (formalin-acetic-alcohol: see below) should be added before examination. Care should be used to prevent these preservatives contacting the skin and safety goggles should be used when diluting formaldehyde or pouring formalin in order to prevent them splashing into the eyes. One of the main problems with methods using separating funnels is that a significant number of algae fail to sediment, because they settle on the sides of the funnel.

Membrane concentration This is appropriate for the rarer and smaller cells. Different cell sizes and densities require membranes of different pore sizes and different ways of treating the material collected on the membrane. Sparse populations of intermediate-sized cells may require 500 mL or more of water sample to be gravity filtered onto 8-μm porosity, smooth membranes. The membrane is then brushed with a fine camel hair brush into a small volume of water for identification and counting. Samples for picoplankton counts are usually pre-filtered with a coarser filter before collecting them on a fine filter. In this case the cells are counted on the surface of the filter; the filter has to be treated before counts are made using epifluorescence microscopy.

Centrifugation Gentle centrifugation concentrates phytoplankton for identification and counting. It is, however, unsuitable for organisms whose density is less than that of water, a problem especially important in freshwaters when lipid-rich species or blue-green algae with gas vacuoles are present. A very effective miniature battery-operated centrifuge has been designed by Belcher (1993) for use in the field and laboratory. It is easy and cheap to construct, since it only requires a hand-held, battery operated cooling fan (Fig. 2). The blade of the fan should be replaced by a 25–30 mm length of 12-mm internal diameter polythene tube cut to accommodate at each end a push-in cap micro sample tube ('Eppendorf' tubes are ideal).

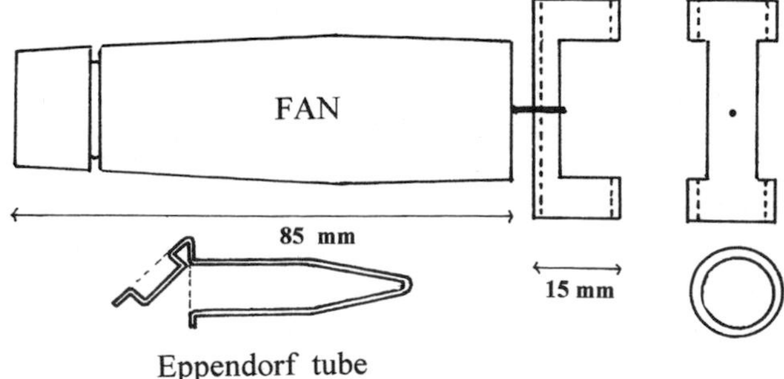

Figure 2. Hand-held electric centrifuge designed for concentrating small algal samples (after Belcher, 1993).

The polythene tube has a piece cut from the middle about three-quarters of the way round its circumference, leaving a ring at each end against which the flange of the centrifuge tube rests (Fig. 2). A small hole is made in the centre of the polythene tube to enable it to be pushed firmly onto the spindle which normally attached to the fan. It is estimated that these motors spin at about 6000 revolutions per minute; samples should be spun for at least a minute. The material concentrated at the bottom of each centrifuge tube should be agitated by tapping before removal using a pipette and then examined directly on a microscope slide.

Terrestrial algae

Some of these above techniques also apply to algae growing in terrestrial environments. Algae on the surface of peat in upland flushes or on damp mud can be skimmed into a receptacle, as can the greenish colouration on the drying beds of temporary pools or rainwater puddles. *Sphagnum* and other mosses sometimes contain algae that can be sampled by squeezing into a wide-mouthed bottle or jar. Algae can be scraped from the surface of the bark of trees or leaves, or whole portions can be removed. All such samples should be examined as soon after collection as possible, or preserved in 4% neutralized formalin or Lugol's solution.

Soil algae sometimes form visible growths in the form of gelatinous layers, crusts and mats that can be collected using a knife blade, scalpel, spatula or spoon. However, such samples are frequently mixed with fungi, moss protonema and lichens. Another method is to place microscope slides *in situ* on the soil surface and to collect them 2–4 weeks later. Certain soil algae grow on the underside of these slides and can be stored dried in paper packets before soaking in detergent solution or alcohol just prior to examination. Alternatively, the algae can be removed from the slides and grown on in laboratory culture. Samples of soil may also be mixed with a mineral agar medium to form an 'enrichment culture', or placed in a petri dish to which is added water that has been boiled and then cooled, then examined 2–3 weeks later. Algae can also be cultured from individual moss leaves after first macerating them before growing them on agar. Comprehensive information on culturing freshwater algae is given in the *Handbook of Phycological Methods* (Stein, 1973) and in a simple manual on methods for isolating, maintaining and handling smaller algae in schools and colleges (Belcher & Swale, 1982).

Some terrestrial and aquatic algae grow among crusts of calcium carbonate (travertine) and within limestone rocks. These can be examined by adding small pieces of rock to a jar containing 2.5% EDTA solution (ethylenediaminetetra-acetic acid) adjusted to pH 8 with sodium hydroxide. The carbonate is dissolved, leaving the algae free for direct examination a day or so later. It should be noted that the acid does cause some damage to the algae.

Snow algae are usually obtained simply by allowing the snow to melt in a jar. Once melted, the jar needs to be shaken and the algae allowed to sediment before pipetting off the excess water. The algae should be stored in vials or tubes using a suitable preservative.

LABORATORY METHODS

The Editors

Ideally, all samples should be examined while the algae are still alive and as soon after collection as possible. Field observations are important in the case of crustose algae. For example, the blue-green alga *Chamaesiphon* and the red alga *Hildenbrandia rivularis* are easily recognized in the field, but often overlooked among scraped fragments when examined back in the laboratory. Some algae are extremely fragile and survive for only a very short period, so it is advisable to preserve at least part of the sample immediately on collection. According to Jane (1942), many flagellates are so delicate that characteristic features may be lost even in carefully fixed material. If samples are not preserved immediately, then precautions must be taken to keep them in good condition and to minimize deterioration.

Many algae decompose rapidly if too much material is placed in a container, so this should not be filled to more than one-third capacity. Charophytes are an exception, since they require plenty of water for support. Algae from well-oxygenated habitats, such as fast-flowing streams, usually deteriorate more rapidly than those from habitats where oxygen levels are at least sometimes low. To be kept in good condition algae should be stored cool (below 15 °C) and never exposed to direct sunlight. If specimens cannot be examined on the day of collection, they should be kept overnight in a cool-box or refrigerator. Some material remains healthy for some days if stored in this way, whereas other material (e.g. lake samples collected in summer) may deteriorate quite rapidly. If in doubt as to how long material can be kept alive, it is best to assume that it will start to deteriorate after a day, even though some samples may prove to last much longer. Material stored overnight in a cool-box generally shows no obvious morphological changes, although there are a few algae (e.g. *Ulothrix zonata*) which undergo rapid changes when transferred to warmer conditions. There is always the risk in storing unpreserved samples that animals, such as *Daphnia* and other small Crustacea, will devour the algae, as mentioned for phytoplankton samples. It is advisable to open containers of living material as soon as possible on return to the laboratory, or to pour any large samples into petri dishes.

If material is to be stored for long, each bottle or plastic bag should be given a unique identifier using a waterproof marker pen with information concerning the collection (locality name, grid reference/coordinates, date, ecological data, collector) entered into a field notebook. A piece of card, good quality paper (e.g. herbarium mounting paper), or waterproof paper should be placed in each container with the unique identifier written in pencil or waterproof ink. Any aquatic macrophytes sampled should be identified and recorded.

Laboratory Study

Light microscopy It is often much easier to identify algae alive than preserved, although sometimes staining or special techniques are essential to observe key diagnostic characters. Live viewing is especially important in the case of motile specimens, since flagella and other delicate structures do not preserve well. Normally a small drop of a microalgal sample should be placed on a glass microscope slide and left for a minute to settle before carefully lowering a coverglass (coverslip) onto it. The slight delay is necessary otherwise many cells go to the sides of the cover glass and a false impression is given of overall abundance. Care must be taken not to put pressure on the coverglass since some very delicate flagellates are easily disrupted by even small pressure differences, and thin-walled filamentous algae distort thus resulting in errors in measuring filament diameter. Sometimes gentle pressure and careful tapping of the coverglass with a needle is useful for separating cell masses or in changing the orientation of algae whose appearance differs in each dimension. Less delicate algae can be examined on a microscope stage for an hour or more without showing abnormalities. One problem with temporary water mounts is the drifting of cells due to water evaporating from the edge of the coverglass. This can be overcome by ringing the coverglass with petroleum jelly, thus effectively checking evaporation. Such ringing should always be undertaken when an oil-immersion lens is to be used. Delicate and flagellate forms should be the first to be examined in a mixed sample, since many die within a short time due to the intense illumination and warming. To observe such algae over several days, the hanging-drop method is recommended. This requires a ring (usually glass or plastic) about 18 mm in diameter and 5 mm in height to be fixed firmly to a microscope slide with petroleum jelly ('vaseline'). A few drops of the sample is placed on a circular coverglass and this is quickly inverted and mounted over the ring whose margin has been smeared with petroleum jelly; the drop of liquid now hangs freely in the cavity. The petroleum-jellied ring effectively seals the cavity and prevents evaporation of the sample. An important advantage of the hanging-drop method is that motile algae tend to collect at one side of the drop and remain motionless without retracting their flagella. Its principal disadvantage is that the drop is too deep to permit the use of high power lenses.

General examination of macroscopic algae can be undertaken using a low power binocular dissecting microscope. For examination at a higher magnification a good quality compound microscope is essential. It is necessary to have ×10 and ×40 objectives, while a ×100 oil-immersion objective is essential for examination of smaller forms and to reveal cytological details such as nuclei, pyrenoids and mother cell wall remnants. Detailed and careful observations are necessary for accurate identification. A calibrated eyepiece is essential, since accurate measurements are crucial for the identification of most algae to species. It is useful to prepare accurate scale drawings of specimens using a 'camera lucida' or better a drawing attachment and to compare these with other illustrations. Some algae, such as many desmids, need to be viewed in all three planes. The re-orientation of such algae on the slide is possible with care and considerable patience by touching the cover glass with a sharp pencil or needle.

Every attempt should be made to examine living material under bright field illumination, since organisms have a characteristic appearance and colour. To assist in interpreting cytological features, it is recommended to 'optically stain' the material using differential interference contrast (Nomarski) that gives a 3-D effect. Phase-contrast and negative phase-contrast are useful for enhancing colourless structures such as mucilage and cell wall ornamentation, and in the case of flagellates for detecting and determining the number of their flagella. Dark-field illumination gives a black background and is also useful for examining flagella.

Photomicrography (BY P.V. YORK) Once the desired specimen is located, a permanent photographic record can be made using a photographic system fitted to the compound microscope. Optical staining techniques and methods for slowing down fast-moving cells may need to be used. Most photographic systems fitted to compound microscopes are of the miniature camera type (35 mm). This format is the one best suited for recording algae at the fastest exposure times in conjunction with low light levels. As most microscopes are fitted with a tungsten illumination system, an appropriate film must be used in order to equal the colour temperature of the light source. For example, a very suitable film is Kodak Ektachrome EPJ320T or, if plenty of light is available, Ektachrome EPY64T. It is essential that the microscope light is used at maximum power to record the correct colour rendition of the specimen; this should give an approximate colour temperature of 3200° Kelvin. Daylight films are unsuitable for this type of work unless a xenon or electronic flash light source is used. If an electronic flash is used, it is almost impossible to set up critically the microscope's illumination system and the specimen must not be observed during exposure as this will damage your eyesight.

Techniques used to aid identification Some diagnostic features are not evident without staining. Lugol's solution stains starch blue-black and is useful for determining pyrenoid number, itself a key diagnostic character in some algal groups. If the iodine stains cells too darkly, then adding a trace of sodium thiosulphate solution (photographic fixer) can decolourize them. Mounting specimens in Indian ink is a useful means of detecting mucilage sheaths or envelopes that otherwise are invisible; the ink particles do not penetrate the mucilage so it usually appears as a bright halo. A drop of 4% glutaraldehyde solution is especially useful for immobilizing euglenophytes and also considerably enhances cellular details for at least an hour (K. Wołowski, pers. comm.); considerable care is needed when handling glutaraldehyde since it is a toxic chemical. It is also possible to slow down swimming flagellates by mixing a sample containing them with an equal volume of aqueous 1% methylcellulose or glycerol. Another method for dealing with motile cells is to add a drop of chloroform to the sample on a slide. This works best if the chloroform is placed at the edge of the slide and allowed to diffuse in gradually, especially with some of the more delicate forms.

There are a number of simple techniques to aid identification. For example, most blue-green algae form more of the blue accessory pigment if kept under low light (but not in the dark) and small-celled forms (e.g. *Aphanocapsa*) may be easier to recognize when left for several days to become more blue. In many large genera, including those within the family Zygnemataceae (Order Zygnematales), identification to species requires the presence of reproductive structures. Simply leaving some filamentous green algae (e.g. *Spirogyra*) in an open container to allow evaporation sometimes induces sexual reproduction. The addition of a small amount of sugar has also been suggested. Keeping planktonic desmids at low light and low temperature can produce conjugation (J.W.G. Lund, pers. comm.). Identification of species of *Vaucheria* also requires the presence of reproductive organs. If living material is placed in a petri dish with sufficient water to cover the lower part of the sample, but not enough to flood it, and then left in a north-facing window for at least 5 days, reproductive structures commonly develop on filaments growing out of the liquid. Swarmers can sometimes be induced in green algae by shock treatment, such as the transfer from the field to the laboratory, from a dark refrigerator into a warm and lighted environment, or by adding a pinch of common salt. Terrestrial algae, in common with those grown on agar, often produce swarmers freely when flooded with water or culture solution.

Preservation

Drying Macroalgae should be washed gently in a tray of water, and a jet of water or forceps used to remove detritus and other extraneous matter. Washing provides an opportunity to search for small epiphytes and these can be stored separately. To make a herbarium mount of a large alga a typical specimen should be floated in a dish of shallow water and then manoeuvred over a sheet of good quality cartridge paper or herbarium paper supported by a piece of perforated zinc sheeting or aluminium. The support is raised out of the water and tilted gently to allow the water to drain slowly during which time the specimen can be carefully arranged on the mounting paper using needles, a paintbrush and jets of water from a small pipette or syringe. The mounted specimens should be dried between sheets of absorbent paper in a plant-press using moderate pressure. Most freshwater algae are mucilaginous, making it necessary to cover a specimen with muslin, nylon stocking, cloth, waxed paper or polythene to prevent it adhering to the drying paper. Every few days the drying paper has to be replaced until the specimen is dry. Care is needed with charophytes, because many are lime-encrusted and therefore brittle, and readily disintegrate if too much pressure is applied during drying. Large blue-green algae lacking gas vacuoles may be mounted and dried on microscope slides. Such dried algae can once more take on a life-like appearance when soaked in detergent or alcohol. The oospores of charophytes (<0.5 mm diameter) may be stored in a cardboard ring mounted on a glass slide and covered with a coverglass. Dried specimens are usually more difficult to identify than those preserved in liquid and for this reason it is advisable to preserve duplicates in spirit or as a permanent slide mount. Those blue-green algae which are subject to occasional drying in nature, usually remain alive for many years if stored in a dry condition; they are, however, likely to be killed if a desiccant like silica gel is added. Desmids, and some other green algae collected from dried algal crusts in puddles or birdbaths, remain viable for several months.

Some algae can only be identified with certainty based on features associated with surface ornamentation (e.g. walls of some desmids, zygospores of filamentous zygnematalean algae). Such surface features are most evident in cells that have lost their contents, often as a consequence of death, fungal infection or digestion. Careful treatment with 5% alcoholic sodium or potassium hydroxide enhances the details of cell wall ornamentation.

Formalin (Formaldehyde) Despite the fact that formalin poses a health hazard, as well as sometimes causing changes in cell dimensions, damage and distortion of chloroplasts, it remains the most commonly used liquid preservative. The concentrations recommended range between 2.5 and 4% formaldehyde. In general, the lower the concentration the less damaging it is as a preservative. However, higher concentrations are needed for long-term storage of organisms with cellulose walls. The ideal concentration is influenced by the amount of material in relation to the volume of preservative. Formaldehyde must be buffered and various buffers have been recommended. Borate and phosphate buffers have been widely used, though calcium carbonate can be used in an emergency; HEPES (*N*-2-hydroxymethylpiperazine-*N*-2′-sulphonic acid) is probably the best for freshwater use, though it has not yet been tested over very long periods. Glycerol is also commonly added to samples for long-term storage, because algae are readily damaged if a sample completely dries out.

Formalin mixtures The addition of copper nitrate (about 1 g L^{-1}) to formalin helps to maintain the green appearance of the chloroplasts of green algae. The addition of a 4% solution of potassium acetate, plus a trace of copper acetate, is useful for intensifying chloroplast details. These two treatments may aid routine cell counts of mixed populations. FAA is a mixture of formaldehyde, acetic acid and alcohol (10 mL 40% formaldehyde, 5 mL glacial acetic acid, 50 mL 95% ethanol, 35 mL water). It is particularly useful for flagellated algae, because it prevents the loss of the flagella.

Lugol's solution Several variants of this stain exist, differing in the ratio of potassium iodide to iodine. The Centre for Ecology and Hydrology/Freshwater Biological Association at Windermere use 96 g potassium iodide in 100 mL deionized water, to which is added 14 g of iodine to give a ratio of about 7:1, whereas other formulas give a ratio of 2:1 (2 g potassium iodide and 1 g iodine in 300 mL water to give a deep straw-colour); 10% acetic acid or glutaraldehyde (CARE) may be added to Lugol's solution to prevent the loss of flagella: however, this procedure is unsuitable for calcareous algae. Lugol's solution is often used for preserving planktonic algae by adding it to make a final concentration in the sample of 0.5%. However, blue-green algal gas vacuoles are largely destroyed by it, so taxonomic studies require a pair of samples, one preserved in formalin and the other in Lugol's solution. The iodine in Lugol's solution eventually sublimes from most types of container, so it can only be used for long-term storage with special care. Containers should be airtight and light tight (or stored in complete dark) and further Lugol's solution added if the samples decolourize. Only the smallest air space possible should be left in the containers and these must be stored in a cool place.

Glutaraldehyde (CARE) Normally 1 mL of 40% glutaraldehyde is added to each 100 mL of algal material to give a final concentration of 0.25–0.5%. This preservative offers minimal distortion, allows the use of epifluorescence on a sample and is recommended if the material is to be used for cytological study. Glutaraldehyde is highly toxic and great care should be taken to prevent contact with the skin and inhalation; eye protection and rubber gloves must be used when handling it.

Glass slide collection This is a useful way of storing microalgae, if it is important to have easily accessible reference samples. However, slide preparation can take several hours or days, and may involve cleaning, dehydration and staining before final mounting. An easy method for the preparation of permanent slides is to use corn syrup (commercially known as Karo), with formalin added as a preservative. This water-soluble syrup is added to the material and liquid placed on a microscope slide. A drop of formalin on the tip of a needle should be added in order to inhibit the growth of fungi. The mixture should be allowed to dry out under a petri dish to prevent dust and other airborne particles falling onto it. After about 12 hours a few drops of concentrated syrup are added and a cover glass is carefully positioned on the slide. In due course the syrup sets very hard, although further concentrated syrup often has to be added around the edge of the slide since it contracts on drying.

As diatom identification is based traditionally on cell wall morphology, preparations of acid-cleaned material are usually mounted on slides in a highly refractive medium (e.g. Naphrax, Styrax). A simple method, not involving the use of strong acids, but rather 20-volume hydrogen peroxide solution, is described by Belcher (1999). It can be used for cleaning not only diatoms, but also the siliceous cysts of chrysophytes and the loricas of the euglenophyte *Trachelomonas*. Such slides are permanent, while the high refractive index of the mountant allows resolution of fine detail on the silica valves. For the examination of features of living cells, diatoms should be treated just like other microalgae (Cox, 1996).

In the case of heavily encrusted charophytes, it is necessary to decalcify in undiluted vinegar or a weak solution of acid (e.g. 10% nitric acid) before examining microscopic features such as the cortex of *Chara*. After decalcification the material should be washed free of acid before mounting onto a glass slide.

Photographs These may be essential, because the characteristics of many small algae do not survive preservation. Photographic prints store well if kept in dry conditions and out of light. The long-term storage of mounted transparencies needs plastic wallets that are free from plasticizers (archival quality). Digital image technology makes it possible to store indefinitely archival images of algae from scanned 35-mm transparencies, monochrome photographs and line drawings. Images can also be stored directly by using a digital camera. It is important to label all colour slides and photographs with the name and origin of the alga/algae, together with details of the magnification used and lenses and/or techniques employed.

CLASSIFICATION

The Editors

The algae are oxygen-producing, mainly aquatic organisms possessing enormous morphological, cytological, molecular and reproductive diversity. Modern studies have led to the recognition that algae represent a number of evolutionary lines or lineages, almost all of which are represented in the British freshwater flora. Most of these lineages probably arose independently as a result of endosymbiosis. There has been a tendency by protozoologists to adopt what has been termed a 'protistan' view and to place these lineages into the Protista, a kingdom that itself is unnatural (Corliss, 1994). The majority of the lineages are eukaryotic, in which the cells have a double membrane surrounding the nucleus and most other organelles such as chloroplasts. Another evolutionary line, the Cyanophyta or Cyanobacteria, is often considered as algal (the blue-green algae), although these organisms are prokaryotic (membrane-bound organelles absent) and more closely related to the bacteria.

A few algae have photosynthetic structures known as cyanelles, which in many ways are intermediate between a chloroplast and a free-living blue-green alga. In most such cases the rest of the cell resembles quite closely other species of algae with normal chloroplasts. It is likely that the cyanelles have evolved relatively recently from free-living cyanobacteria. However, in the case of *Glaucocystis nostochinearum*, which has conspicuous cyanelles, there is still doubt about its relationships and it is here classified in its own phylum, the Glaucophyta. Some other organisms possess no photosynthetic structure, but are otherwise quite similar to those that have chloroplasts. These are often treated as protozoans, but clearly belong in the same phyla as the related photosynthetic organisms. Most examples are in the flagellated phyla (Euglenophyta, Cryptophyta, Pyrrophyta, Chrysophyta). Although it is quite logical to consider them as algae, for practical reasons they are usually treated separately and we have omitted them.

What constitutes an algal species has been the subject of much debate (see John & Maggs, 1997, for a review). Most species are recognized by discontinuities in sets of morphological characters observed with the light microscope, so that algal systematics is largely based on what has been termed 'the morphospecies concept'. Culture studies have frequently shown that species concepts based solely on characters observed in field-collected material are often too narrow and the taxonomic validity of many of the characters used is open to question (see Trainor, 1998, and his comments in a review of the genus *Scenedesmus*). As a result of culture studies, some taxonomic systems have developed in which species are defined using sets of characters evident only when clonal cultures are grown under carefully controlled laboratory conditions. Until considerably more is known of morphological variation under natural conditions, most algae will continue to be recognized explicitly or implicitly on morphological discontinuities in field samples. Some forms or morphospecies develop in response to particular environmental conditions and therefore have value as environmental indicators. Unfortunately, for the vast majority of algae we still have little detailed knowledge either of the environmental conditions responsible for the development of particular forms, or of the conditions under which they grow and reproduce. Soil algae are an exception since the majority of species are only known after isolation and subsequent growth in laboratory culture.

For the reasons outlined above, it can be difficult or even impossible to identify all species in a sample and forms will often be discovered that appear intermediate between two recognized species. It is always likely that detailed surveys will reveal genera or species that have not been recorded previously from the British Isles, making it necessary to refer to other taxonomic literature. The identification of species belonging to several important genera (e.g. *Vaucheria*, *Oedogonium*, *Mougeotia*, *Spirogyra*, *Zygnema*) depends on reproductive characters and yet fertile material is often difficult or almost impossible to find. Information is provided to increase the chance of successfully discovering fertile material in the field, also for inducing reproduction in the laboratory (see Methods, p. 16).

Much of the current understanding of algal systematics began in the 1950s. It is currently based on fine structural details, acceptance of the theory of endosymbiosis, information on life histories and morphological plasticity from culture-based studies, and data from biochemical and molecular biological research. As a result of the synthesis and evaluation of these data, the morphologically based traditional classification system underwent fundamental reorganization during the latter part of the twentieth century and has yet to achieve stability. For a review of progress made over much of the last decade of the century in understanding the systematics and evolution of algae (excluding cyanobacteria), see Preisig (1999). We decided to adopt a fairly conservative approach and follow the traditionally recognized phyla. Genera are arranged alphabetically within orders, although uncertainty surrounds the position of many chlorophyte genera due to the disassemblage of traditionally recognized orders based on molecular systematic analysis and the findings of ultrastructure studies. For this reason, several orders of filamentous green algae are combined (e.g. Chaetophorales, Ulotrichales, Klebsormidiales, Microsporales) and only the more distinctive orders are kept separate. The combined orders are no longer distinguished on morphological features evident only using the light microscope (LM).

The following is a synopsis of the characters used to distinguishing the major divisions. Only those characters are mentioned which can be observed readily with the light microscope. Care needs to be exercised because morphologically similar taxa may be related only very distantly.

Cyanophyta (Cyanobacteria) – blue-green algae
Variable in colour, especially when viewed macroscopically: blue-green, grey-green, violet, brown, purplish or red; colour depends on relative proportions of the photosynthetic pigments, chlorophyll (green), phycocyanin (blue) and phycoerythrin (red), and sheath pigments e.g. scytonemin (brown); unicellular, colonial or filamentous with filaments simple or branched; internal membrane-bound organelles lacking, nucleus, chloroplast.

Rhodophyta – red algae
Cells red due to predominance of phycocyanin and phycoerythrin in chloroplasts (one to several); unicellular to filamentous or pseudoparenchymatous (flagellated stages absent); storage material various, including floridean starch (stains yellow or brown with iodine); unique features associated with reproduction.

Euglenophyta – euglenoids
Cells usually green (occasionally colourless or with red pigment inside or outside the chloroplast) and chloroplasts (several) variously shaped; unicellular, motile, possessing 1 or 2 flagella (one species lacks a flagellum) arising in a flask-shaped invagination; eyespot at anterior, red and usually evident; walls often with longitudinal or spirally arranged striations, exhibiting characteristic squirming and undulating movements (euglenoid movement), sometimes surrounded by an envelope or lorica; food storage material usually paramylon (does not stain with iodine) and variously shaped (ovoid, star-, plate-, rod- or ring-shaped).

Cryptophyta – cryptomonads
Cells brown, blue, blue-green, red, red-brown, olive green, or yellow-brown due to accessory pigments (phycobiliproteins in the case of blue and red) in chloroplasts (often 1 or 2); unicellular and often bean-shaped, commonly dorsiventrally flattened and with an anterior invagination, with 2 or more unequal and subapical flagella, rarely colonial; storage material starch or starch-like.

Pyrrophyta (Dinophyta) – dinoflagellates
Cells usually brown since chlorophylls masked by accessory pigments (occasionally containing endosymbiotic algae); unicellular (rarely coccoid or filamentous) and biflagellate, one flagellum transverse and encircling the cell (usually in a groove-like girdle), the other directed posteriorly; cell walls firm or formed of regularly arranged polygonal plates; storage materials starch and oil.

Raphidophyta (Chloromonadophyta)
Cells yellow-green due to predominance of accessory pigment diatoxanthin in chloroplasts (2 or more); unicellular and dorsiventrally constructed, with 2 flagella inserted apically in a funnel-shaped invagination, one flagellum directed forwards, the other backwards; storage material oil; outer wall absent.

Haptophyta
Cells golden or yellow-brown, with chlorophyll masked by accessory pigments (principally fucoxanthin); unicellular flagellates with 2 flagella and a short coiled structure known as a haptonema; amoeboid, coccoid, palmelloid or filamentous stages; walls with scales, often calcified; storage material principally chrysolaminarin.

Chrysophyta – golden(-brown) algae
Cells golden to yellow-brown due to accessory pigments (principally fucoxanthin) masking the chlorophyll; single coccoidal cells and palmelloid, filamentous or parenchymatous forms, mostly uniflagellate or biflagellate with a long and a short flagellum; outer wall absent (naked) or cell(s) enclosed within an often urn-shaped envelope or lorica; silica scales often present; storage material oil or leucosin; chrysolaminarin gives cell a metallic lustre.

Xanthophyta (Tribophyta) – yellow-green algae
Cells usually yellow-green due to predominance of accessory pigment diatoxanthin in chloroplasts (2 or more chloroplasts per cell); unicellular, filamentous, colonial or coenocytic, motile forms with 2 subapical flagella; walls often with overlapping parts; storage material oil, fat or leucosin (never starch).

Eustigmatophyta
Cells yellow-green with principal accessory pigment violaxanthin masking chlorophyll in chloroplasts (one or more); unicellular and coccoid, motile forms with a single flagellum or 2 unequal flagella inserted near apex; eyespot unique and independent of chloroplast; pyrenoid unique; nature of food storage material unknown.

Bacillariophyta – diatoms (not included)
Cells usually yellowish-brown due to presence of various pigments (especially fucoxanthin) masking chlorophylls in chloroplasts (one to several); unicellular, essentially boat- or pillbox- shaped, solitary, colonial on mucilage stalks or forming pseudofilaments; walls of silica and in 2 overlapping halves; storage materials include fats and chryolaminarin.

Phaeophyta (Fucophyta) – brown algae
Cells brownish due to characteristic carotenoids (principally fucoxanthin) in chloroplasts (one to several per cell); only microscopic branched filamentous forms in freshwater, macroscopic parenchymatous forms in marine environments; motile stages pear-shaped with 2 laterally inserted flagella; walls often include alginic acid and fucinic acid; food storage materials laminarin and mannitol.

Prasinophyta
Cells with green (rarely yellow-green) chloroplasts; unicellular flagellates (rarely non-motile), with 1–8 laterally or apically placed flagella, usually (but not invariably) arising at base of a depression; walls and flagella mostly covered with organic scales; food storage materials starch or mannitol.

Chlorophyta – green algae
Cells with green chloroplasts (one to several per cell), rarely the colour of the chloroplast masked by orange or red carotenoid pigments; species in acidic environments sometimes have a violet or brown cytoplasm; unicellular, colonial, filamentous, coenocytic or macrophytes having robust axes bearing whorls of branches and branchlets; motile or non-motile, if motile then normally having (1–)2 or 4 usually apical flagella with the exception of some motile stages in the orders Charales and Oedogoniales; food storage material principally starch surrounding one or several pyrenoids (starch tests positive with iodine); sexual reproduction oogamous in some orders, with organs enclosed within a sheath of sterile cells in the order Charales.

Glaucophyta (not keyed out below)
Cells bright blue-green, containing phycocyanin in addition to other pigments, all confined to the cyanelles (not equivalent to chloroplasts); unicellular or small colonies; storage material starch, produced outside cyanelles. An artificial grouping of phylogenetically different organisms.

KEY TO PHYLA

1 Cells lacking chloroplasts; pigmentation (blue-green, grey-green, violet, olive-green, purplish or reddish) seemingly distributed throughout cell; sheaths, when present, often coloured and typically yellow-brown; nucleus absent (prokaryotic) .... **Cyanophyta** (p. 25)

1 Cells with chloroplasts to which pigments are normally confined; uni- or multinucleate (eukaryotic) 2

2 Plants green, macrophytes, often lime-encrusted; robust main branches bearing whorls of secondary branches and branchlets .... **Chlorophyta, Charales** (p. 593)

2 Plants of another form 3

3 Cell walls rigid due to silica, consisting of 2 overlapping halves, boat or pill-box shaped; flagellated stages absent (apart from male gametes of some genera) ..... **Bacillariophyta** (p. 273)

3 Cells of another form 4

4 Single-celled (rarely coccoid or filamentous) and having 2 flagella, one transverse and encircling cell (usually in groove-like equatorial girdle) and the other directed posteriorly; cell walls firm or formed of regularly arranged polygonal plates .... **Pyrrophyta** (p. 186)

4 Single- or multi-celled, with or without flagella and never having polygonal plates, never with a groove-like equatorial girdle.............. 5

5 Free-swimming, single-celled flagellates, typically with striations on wall and a single long flagellum emerging from an apical depression, sometimes cell surrounded by a firm envelope or lorica (often darkly coloured) through which the flagellum emerges via a pore (one species lacks a flagellum); chloroplasts generally green (very rarely masked by red or orange pigments), 2 to several, variously shaped; storage material paramylon (does not stain with iodine); eyespot often prominent; typically show a squirming or undulating movement............................. **Euglenophyta** (p. 144)

5 Motile or non-motile; if free-swimming then not as above... 6

6 Motile (rarely non-motile), with 1–8 apically or laterally inserted flagella; storage material mannitol or starch; chloroplast parietal and lobed; walls and flagella covered by organic scales (only visible using the electron microscope)............................ **Prasinophyta** (p. 281)

6 Motile or non-motile; if motile then having 1–4 usually apically inserted flagella (often swimming with a breast-stroke action).................. 7

7 Cells usually light to dark grass green, only few taxa red, orange or yellow-orange due to presence of masking pigments (e.g. *Haematococcus*, *Trentepohlia*, *Botryococcus*) .... **Chlorophyta** (p. 287)

7 Cells of another colour.. 8

8 Cells red or purplish-red unless bleached; storage material various including floridean starch (stains yellow or brown with iodine); unique features associated with reproduction (no flagellated stages); typically in flowing water .... **Rhodophyta** (p. 208)

8 Cells variously coloured (very rarely red); floridean starch absent; flagellated stages often present; still and flowing water habitats 9

9 Cells brown; motile stages pear-shaped and having 2 laterally inserted flagella; food storage materials laminarin and mannitol ....**Phaeophyta** (p. 278)

9 Cells not as above..10

10 Cells golden to yellow-brown, sometimes having a metallic lustre if storage material chrysolaminarin present in quantity 11

10 Cells of another colour...13

11 Single-celled flagellates, with 2 equal or unequal flagella and a short coiled flagellum, always free-swimming.............**Haptophyta** (p. 211)

11 Single-celled flagellated forms, if present then having a single or 2 unequal flagella, never accompanied by a short coiled flagellum 12

12 Solitary and coccoidal-celled (occasionally pallmelloid), filamentous or parenchymatous; flagellated cells with a small red eyespot within chloroplast; sometimes flagellated cell attached within an urn-shaped envelope; non-motile cells without a stalked pyrenoid.....**Chrysophyta** (p. 214)

12 Single-celled, non-motile or flagellated, if flagellated then having a conspicuous orange-red eyespot lying outside the chloroplast and never associated with an enclosing envelope; non-motile cells often having a stalked pyrenoid ....**Eustigmatophyta** (p. 271)

13 Cells brown, blue, blue-green, red, red-brown, olive green, or yellow-brown; single-celled and often bean-shaped, commonly dorsiventrally flattened, rarely colonial; anteriorly having an invagination and 2 or more unequal and subapical flagella; storage material starch or starch-like ....**Cryptophyta** (p. 180)

13 Cells usually yellow-green; single-celled, filamentous, colonial or coenocytic, motile forms having 2 or more flagella 14

14 Single-celled, filamentous, colonial or coenocytic, with motile forms having 2 subapical flagella; walls often in overlapping parts ....**Xanthophyta** (p. 245)

14 Single-celled and dorsiventrally organized, with 2 flagella inserted apically in a funnel-shaped invagination, one flagellum directed forwards and the other flagellum directed backwards.............................**Raphidophyta** (p. 208)

PHYLUM CYANOPHYTA (Cyanobacteria)

Brian A. Whitton

Introduction

The blue-green algae are unicellular or filamentous organisms that sometimes form structures recognizable with the naked eye, but usually require a microscope for identification. They differ from all other groups in this Flora in that they are prokaryotes: their cell contents are not differentiated into membrane-bound structures such as the nucleus, chloroplasts and mitochondria. The popular name for the group, blue-green algae, comes from the colour of the cells seen under the microscope. The photosynthetic pigments in the membranes inside the cells contain chlorophyll *a*, which gives a green colour, but almost all species can form the blue pigment, phycocyanin, under some conditions and some also form a red pigment, phycoerythrin. The cells therefore often appear blue-green, but sometimes shades of purple, when all three pigments are present. Visually obvious growths are also often blue-green, but sometimes they are brownish, purple or orange. This is because many species have a sheath around individual cells or the whole filament and this sheath is often golden or dark brown, though sometimes a shade of red.

Many blue-green algae are easy for someone without specialist knowledge to recognize under the microscope to the genus or even the species. However, the group has gained a reputation for being difficult to identify reliably. This is partly because some species are morphologically highly variable and in some cases groups of species which appear distinct in one place all tend to merge with each other elsewhere. However, there are also problems due to the different approaches which have been used to name the organisms. The most important difference is between the botanical and bacteriological approaches. The blue-greens are anomalous in that they are currently treated by some authors under the conventions of the International Code of Botanical Nomenclature, while others treat them under the International Code of Bacteriological Nomenclature.

The present account includes marine as well as freshwater and terrestrial species. Some species occur in both types of environment, many freshwater species extend into brackish environments and a few marine species also extend into brackish environments. Morphologically almost identical forms occurring in non-marine and marine environments have sometimes been given different names and in other cases the same name. At least with the current state of knowledge for the British Isles, there are relatively few strictly marine blue-green algae, so it seems best to provide an account of the whole Flora.

Morphology and Physiology

Single-celled blue-green algae range in size from about 0.6 μm to well over 30 μm in their largest dimension. Filamentous forms range in cell diameter from about 0.4 to 45 μm in the British Isles, but there are a few records of species well over 100 μm diameter from other countries. Although there are large subspherical unicells (*Chroococcus*), wide filaments always have very short cells. The sheath surrounding the cells or filaments of many species differs greatly in width, so it is important to record both the width of cells excluding the sheath and also the width with the sheath. In filamentous forms with a sheath e.g. *Lyngbya*, the cellular structure is known as a trichome, while the filament refers to the trichome plus the sheath. In genera which lack a sheath the terms trichome and filament are interchangeable.

Very small-celled blue-greens often occur as single cells in the plankton. Usually these are less than 2 μm in all dimensions and such cells are known as picoplankton. The slightly larger unicellular forms are almost always arranged in some form of colony – sometimes just a few cells e.g. *Chroococcus*, but more often large numbers of cells embedded in mucilage e.g. *Microcystis*. Filamentous forms are more likely to occur singly, but the majority of these also tend to occur in groups and, in many cases, form characteristic colonies.

A number of morphological features of cells and filaments are important in identification. In some cases they are also excellent indicators of past environmental conditions. Some filaments form characteristic cells known as heterocysts, which differentiate from an ordinary vegetative cell, and are the site of nitrogen fixation. The combined nitrogen formed can pass to the rest of the filament. In some genera e.g. *Anabaena* the heterocysts develop at regular intervals along the filament. However, there are other arrangements, such as in *Calothrix*, where there is often just a single heterocyst at one end of the filament. The two types of heterocyst are known as intercalary and terminal, respectively. Experimental studies have shown that heterocyst formation can almost always be prevented if there are high enough concentrations of nitrate and/or ammonium in the growth medium, so it is assumed that the presence of many heterocysts in a filament is an indication that the organism is growing in an environment relatively deficient in combined nitrogen in comparison to other nutrients.

The heterocyst is usually paler than adjacent vegetative cells because any phycocyanin is usually lost during its formation. However, sometimes in *Calothrix* this does not occur and the terminal heterocyst is bright blue. At the ends of heterocysts adjacent to vegetative cells there is often a glistening granule. This consists of a nitrogen storage material known as cyanophycin. The granule next to a heterocyst wall is especially prominent when the heterocyst is fixing nitrogen faster than the adjacent cell can use the nitrogen compounds formed.

Another feature of many filamentous forms is that part of the filament tapers, sometimes markedly so. In some cases the terminal cells of the tapered part become colourless, forming multicellular hairs. Well developed hairs apparently always develop in response to phosphorus limitation, though shorter hairs can be produced experimentally under iron deficiency. Some genera have both heterocysts and well developed hairs. These are characteristic of environments which are short of combined nitrogen for part of the time, and of phosphorus for the rest of the time.

Many authors have used laboratory cultures to aid taxonomic studies of blue-green algae. Unfortunately they have not always succeeded in reproducing the morphological features of a field sample in the laboratory. There is certainly much still to be found out about doing this, but two factors have often been overlooked. While most researchers are aware of the need to grow an organism to nitrogen limitation to check whether it can produce heterocysts, organisms have much less often been grown to phosphorus limitation and only very rarely so in taxonomic studies. This is in spite of the fact that phosphate insufficiency is recognized as a factor limiting algal growth in many freshwater habitats. Major taxonomic studies and routine subculturing of culture collections have often been (and still are) conducted in a medium with too much phosphate for it ever to becoming limiting during batch culture studies. This means that morphological features influenced by phosphorus limitation are never seen, and in the case of some strains in culture collections, may have been lost genetically. The other factor which needs to be considered more is the importance of shifting environments, where one or more factors are only important intermittently. This is widely appreciated for wet – dry habitats, but less often for nutrient shifts, such as the alternating nitrogen and phosphorus limitation mentioned above.

Most, if not all, filamentous blue-greens are capable of gliding movement. Some species are motile for much of the time, whereas in others motility is restricted to a particular growth stage. Motility always requires a surface (including soil or sediment particles), though in some species other filaments of the same species or mucilaginous colonies of other algae are adequate for this purpose. Specialized filaments formed under favourable growth conditions and capable of moving are known as hormogonia. These differ most from ordinary filaments in genera where the filaments show a marked polarity e.g. *Calothrix* and *Scytonema*. Addition of phosphate to a *Calothrix* filament with a well-developed hair leads to the hair falling off and the terminal part of the healthy filament developing into a hormogonium. After about a day the hormogonium develops a new terminal heterocyst at the end which breaks away from the mother filament and develops into a typical filament. Some unicellular species can also move. In the past it has been assumed that this is relatively uncommon, but it is now recognised that most unicellular species have motile stages. Multicellular plates of cells, such as *Microcrocis*, can move to the surface of sediments.

Many, but by no means all, nitrogen-fixing blue-green algae form specialized cells known as akinetes. They are largely restricted to filamentous forms with heterocysts, but occur in a few unicellular forms, being especially distinct in *Gloeocapsa magma*. The unicellular blue-greens forming akinetes are apparently also nitrogen-fixers, but further studies are needed to confirm whether this is always the case. In some filamentous forms almost all the filament differentiates into akinetes e.g. most *Nostoc* species, whereas in others few or only one akinete are formed per filament. The extreme case is in some species of *Cylindrospermum*, where a single very large akinete is formed adjacent to the terminal heterocyst. Akinetes differ considerably in morphology and some have highly characteristic wall structures.

As mentioned, cell colour is influenced by the presence of various pigments. The content of the blue pigment phycocyanin increases as the light intensity decreases. As the blue-green colour can help to confirm that an organism is a blue-green alga, it is recommended to include samples from shaded situations wherever possible. However, in some species or strains the phycocyanin is also important as a nitrogen store, so there are at least two factors which influence the extent of blue colour. In some species which form both phycocyanin and phycoerythrin, the relative proportions of the two pigments are influenced by the light spectrum. Cells living in microhabitats where the light has already passed through green filaments are enhanced in the red pigment. However, not all strains behave in this way and other factors may influence the relative proportions of the two pigments. Picoplanktonic blue-greens in large nutrient-poor lakes are often pink, as can be seen when they are concentrated on a filter.

Cells or filaments are sometimes yellowish and this

is usually a sign that they are somewhat unhealthy, often as a result of nitrogen deficiency or damage by high light intensity. Occasionally a single collapsed cell in a filament may appear bright blue. Sometimes this occurs because a cell has been damaged accidentally, but it can also indicate a necridium, a cell which appears to be programmed to degenerate, leading to the filament separating into two parts. This occurs routinely in the fragmentation of the trichome of some Oscillatoriales. The blue colour appears to be due to the water-soluble phycocyanin being released from the structures containing the pigment, but remaining inside the cell. However, necridia are not necessarily blue.

The brown colour in the sheaths of many species or strains is due to the pigment scytonemin. This is formed by the cells in the presence of ultra-violet light and also acts as a protection against ultra-violet light, so the sheaths of colonies or filaments capable of forming the pigment and exposed to high light are the most pigmented. Sheaths of the same strain in shaded conditions may appear colourless, so it is important to collect material from well illuminated situations where the presence or absence of scytonemin is a diagnostic feature for a particular species. The formation of scytonemin is most characteristic in slow-growing species. Most fast-growing forms lack scytonemin, even if they are exposed to high light; they have some other means of protecting the cells from damage by ultra-violet light. It should be pointed out that the colour of some terrestrial species viewed macroscopically often looks different according to whether the organism is wet or dry.

Although blue-green algae lack chloroplasts and other distinct organelles inside the cells, it is often possible to recognize one or more obvious structures. Two types of granule can be seen, though seldom both in the same cell. Granules of cyanophycin, the nitrogen store, can often be seen scattered in vegetative cells as well as at the ends of heterocysts. They form in almost all strains when the cells are nitrogen-rich, but are easiest to recognize in larger cells. Polyphosphate granules form when the cells are phosphorus-rich. These may be scattered through the cell like cyanophycin granules, but in some species of Oscillatoriaceae they form a row next to the cross wall. This is listed as a diagnostic feature in older accounts of some species, but presumably cannot be seen if the filaments are short of phosphorus. Another structure, the polyglucoside granule, is too small to be seen with the light microscope. However, their rapid formation during the daytime as a carbohydrate store can occupy sufficient cell volume to increase slightly cell dimensions.

A characteristic feature of many blue-green algae is the presence of gas vacuoles in the cells, which are often sufficient enough to make the cells positively buoyant. Similar structures occur in some planktonic bacteria, including photosynthetic bacteria, but they are completely absent in the eukaryotic algae. The gas vacuoles seen with the light microscope consist of many much smaller gas-filled vesicles. The gas vacuoles typically give regions of the cell a pinkish colour, though this colour is an optical artefact, as the walls of the vesicle lack any pigment. It is usually possible to demonstrate the presence of gas vacuoles by pressing carefully on the surface of a coverglass, which provides enough pressure to destroy many of the vesicles. Typically the gas vacuoles occur in most cells, though they normally disappear during differentiation of a heterocyst. A few species of *Oscillatoria* possess a single large gas vacuole either side of the cross wall, rather than a number of vacuoles scattered throughout the cell. Gas vacuoles are most characteristic of planktonic species in ponds and lakes, but they also occur in the hormogonia of some species living in quite different environments, such as streams, even though the normal growth form lacks gas vacuoles. Decaying cells of gas-vacuolate species are especially likely to release the water-soluble phycocyanin or phycoerythrin pigments into the surrounding water, sometimes sufficiently so to colour local areas of lakes blue or red.

Liquid-filled vacuoles sometimes occur. These are formed by separation of the two layers of the combined photosynthetic and respiratory membranes (thylakoids) inside the cell (intrathylakoidal vesicle). Sometimes the presence is a sign that the cell is becoming unhealthy, as in *Scytonema*, where the oldest cells in the filament often develop them. The colourless cells of hairs in *Calothrix*, *Homoeothrix* and other genera also consist largely of liquid vacuoles. Liquid vacuoles often occur in all cells of populations of *Oscillatoria bornetii*, *O. bourrellyi* and *Synechococcus aeruginosus*, though *O. bourrellyi* and perhaps also the others can also exist without the presence of obvious vacuoles. They also occur in the terminal regions of some Oscillatoriales, which taper towards the apex e.g. *Phormidium autumnale* (Plate 12B: reproduced from Geitler, 1932a). The vacuolated cells in the hairs of *Calothrix*, which form when the trichomes are phosphorus-limited, are the site of intense phosphatase activity and aid phosphate acquisition; studies are needed to establish how far this is a general phenomenon, but it seems likely that this will prove to be so.

Many – perhaps the majority of – blue-green algae occur for much of the time as distinct groups of cells or filaments. These are mostly termed 'colony' in the present account, though some other floras restrict the term colony to the more distinctive structures, such as occur in *Microcystis* and *Gloeotrichia*. The term 'thallus' is used here for groups of cells or filaments with ill-defined limits or where filaments are dispersed through another medium, such as endolithic growths. Some colonies develop from a single cell or filament e.g. *Nostoc commune* and probably most other *Nostoc*. However, other filamentous forms, such as *Rivularia* and *Gloeotrichia*, usually form their colonies as a result of a number of hormogonia aggregating together, and such hormogonia are not always genetically identical.

Taxonomy

A number of the Floras published during the latter part of the nineteenth century and the twentieth century are similar in layout to the botanical Floras published for other algal groups, with detailed lists and accounts of species based on morphology. The first very extensive accounts were those of Bornet & Flahault (1886–88) and Gomont (1892–93). However, the best known Flora is that of Geitler (1932a), who brought together much of the worldwide information known then, even though the work was nominally for Europe. In addition to morphological features, both macroscopic and microscopic, some authors have used environmental features as a justification for separating certain pairs of species, especially marine from freshwater, a practice which is less often followed nowadays.

However, in the 1970s Roger Stanier and his group at the Pasteur Institute in Paris became increasingly convinced not only that identification of these organisms should be done in similar ways as for bacteria, but also that they should be treated according to the rules of the International Code of Nomenclature of Bacteria (Stanier et al., 1978). This involved not just a change in the name of the group to Cyanobacteria, but also adopting the procedures used by bacteriologists to characterize and name organisms – isolation and culture of individual cells or filaments, influence of the growth medium on morphology and some biochemical features. Castenholz & Waterbury (1989) provide a thorough account along these lines for some of the more important genera. This approach has shown a number of cases where organisms, which had been considered in the past as closely related taxonomically, are not in fact so. However, the procedure is extremely time-consuming for general surveys and impossible for routine monitoring by water management bodies. Descriptive accounts of natural communities are still based on botanical species identified very largely from morphological features, whereas experimental research studies often use names based on the bacteriological approach. Frequently the names are the same whichever system is used, but this is not always the case. There is at present no easy way of avoiding two parallel systems of naming organisms, though the situation may eventually be resolved when molecular approaches to identification, such as the use of PCR (polymerase chain reaction), are streamlined for large-scale surveys.

A totally different approach to naming blue-green algae was put forward by Francis Drouet (1968) – essentially the exact opposite of Stanier's approach. Drouet considered that many of the species in floras were merely variations on a limited number of species. He took a lot of effort to check the type material of species, and found, not surprisingly, that many simple forms look much the same after drying or liquid preservation for long periods. In a series of monographs (e.g. Drouet, 1981) he gradually reduced the number of species recognized. This made it easy to provide lists of names to people who had conducted field surveys, but unfortunately the names did not mean much, as many habitats had almost identical lists. The system is now discarded, though there are still a few texts originating from the USA which include these names. The presence of *Anacystis* or *Agmenellum* in a list usually indicates that Drouet names have been used; although these genera were described long before Drouet, the names had largely fallen into disuse until Drouet's revisions. Apart from a marine study, the system of Drouet has seldom been used in the British Isles.

The other major difference in approach to naming is that introduced by K. Anagnostidis and J. Komárek in various monographs, especially for the Chroococcales (reviewed in Komárek & Anagnostidis, 1999) and Oscillatoriales (Anagnostidis & Komárek, 1988). The volume on the Chroococcales is essential for anyone wanting to look at this group critically, because it provides a comprehensive account of the literature, including descriptions of many species which probably occur in the British Isles, but have not yet been reported. The authors deal with the Chroococcales using their approach to naming diversity of these relatively simple, but often morphologically variable, forms. Most logical possibilities for pattern of cell division, polarity, cell release and arrangement of cells within a sheath or communal mucilage are given a different generic name. Although most of the generic names were already created by others, they were the first authors to take this approach for the whole group towards its extreme.

Some, but not all, of the generic definitions used by Komárek & Anagnostidis (1999) are adopted here. Among those not accepted is the genus *Cyanothece*, which was separated from *Synechococcus* by Komárek and Anagnostidis; however the main feature they used to distinguish the genus is the presence of large numbers of intrathylakoidal vacuoles, a feature which, as explained above, has developed in diverse blue-green algae. The break-up of *Dermocarpa* into four genera has not been adopted, because this would mean that many cells can be recognized to the genus only if they are undergoing division. The principle has been adopted throughout the present account that it should be possible to identify to the genus the great majority of blue-green algae in live field samples, even though further studies may be needed to identify the species.

In the case of the Oscillatoriales, it has long been recognized that the separation of species between the commonest genera, *Oscillatoria*, *Lyngbya*, *Phormidium* and *Plectonema*, is somewhat arbitrary and that, however it has been done in past floras, the genera include what are clearly very different organisms. Anagnostidis & Komárek (1988) solved this by splitting and rearranging the genera. The new generic limits have been adopted widely in eastern Europe and some other countries, and perhaps one-third of the accounts in recent literature use these names. The author considered whether these generic names should be adopted here, but decided that, on balance, it is better to retain the old names. There are a number of practical difficulties in using the new names (see introduction to

Oscillatoriales), but the main reason for caution is that culture and molecular studies are likely to provide much better understanding within the next few years of useful generic limits. It seems pointless to make changes now, when almost certainly there will soon be further major changes.

Similar reasoning has been applied here when considering whether or not to combine species. This has sometimes been done when two species are very similar, especially when one name has been used for a broadly defined species and the other name has been used infrequently. However, other sets of names are retained for the time being when it is likely that there will be substantial taxonomic revision in the future.

Some important original descriptions of blue-green algal taxa were made on material from the British Isles during the early and mid-nineteenth century. Further taxa have also been described subsequently, but many were based on very slight differences. Few of the species described in various papers by the West father and son or by Anand (1937) withstand critical scrutiny.

New taxa of blue-green algae are still being described from other countries, but these are published in diverse journals and often ones with very limited circulation. Sometimes the new taxon is very distinctive, although overlooked by previous researchers. However, in many cases the differences from previously described taxa are minor. Anyone making descriptive accounts of blue-green algal communities is likely to come across material which does not fit any of the species in this Flora and may not correspond exactly to any described species. It is the view of the author that ever more descriptions of new species or varieties hinder rather than advance understanding of the diversity of blue-green algae, because very few researchers use the full range of names available. This means that anyone assessing previous floristic accounts has to know the approach adopted by a particular author. It is suggested that the best way to report the range of form of blue-green algae in nature is to collate the information in a systematic manner on a web-site. This should be done in a way which would be easy for non-specialists to understand. Subsequently the description of new taxa could be restricted to forms which differ markedly from those described previously. For instance, if there is an internationally recognized list of characters to describe blue-green algae, it would be sensible for new species to differ from any previous description by at least three characters.

It is, however, helpful to have access to some of the earlier blue-green algal Floras. In addition to Komárek & Anagnostidis (1999) and the general algal texts mentioned elsewhere in this Flora, the following are the more widely available texts which are especially useful. Geitler's (1932a) Flora (in German) also exists as a 1985 reprint (with a few hand-written corrections). Starmach's (1966) Flora is essentially an update (in Polish) of Geitler's Flora. Frémy's (1929–32) account (in French) of species from the French coasts is still an excellent general Flora for marine species, while Coppejans (1995) has provided a more recent account for Belgium and the north coast of France. The Floras in English with the most useful general accounts are those of Desikachary (1959) and Prescott (1962).

The 1964 version of the check list of British marine algae (Parke & Dixon, 1964: Cyanophyceae by H.T. Powell) used classical ('Geitler') names, but unfortunately a later version (Parke & Dixon, 1976: Cyanophyceae by Parke & Dixon) shifted to Drouet names; the latter have not been incorporated here. These checklists include Ireland as well as the UK. Powell is quoted here wherever a species is in his checklist; it is quoted at the beginning of the ecological section for marine species or later if it also occurs in non-marine habitats.

Ideally, the present chapter should be used in conjunction with an interactive identification system for the same list of species, which is available on CD-ROM (Whitton et al., 2002, but frequently updated), especially versions produced from May 2002. An interactive approach using as many characters as possible is best for someone who is not a specialist to start to name blue-green algae, though a Flora is needed to check identifications. Past experience shows that predominantly dichotomous keys are of limited use for naming blue-green algae to the generic or species level, because of the frequently overlapping ranges of characters and hence the need to use information about a number of characters. Hence it was decided to include keys here in only a few cases where the key usually works well.

The genera are grouped here into four orders, the main features of which are:

Chroococcales never form true filaments
Oscillatoriales filaments, but not forming heterocysts
Nostocales filaments forming heterocysts, but never true branches, though false branches characteristic of some genera
Stigonematales filaments forming heterocysts and true branches.

These descriptions are explained more fully for the individual orders. The genera and species are arranged alphabetically within the individual orders. All the orders are probably phylogenetically diverse, although it is likely that the heterocyst evolved only once. It also seems likely that many genera and occasionally also species are phylogenetically diverse.

Some authors have also grouped genera into families, but these are difficult to define clearly and in practice not often used in general accounts. No attempt is made here to organize every genus into a family, but occasional mention is made of the Nostocaceae, Rivulariaceae and Scytonemaceae, which are quite easy to recognize. These are explained in the introduction to the Nostocales.

There are no confirmed records for the British Isles of photosynthetic prokaryotes with chlorophyll *b*, but no phycocyanin, the prochlorophytes. Such organisms are phylogenetically diverse, but are now usually grouped within a broadened definition of the cyanobacteria In view of the fact that *Prochlorothrix* was first recorded from The Netherlands (Burger-Wiesma et al., 1986) and looks somewhat like *Oscillatoria*, the lack of records for the British Isles is probably just a matter of the genus being overlooked.

The accounts for individual species sometimes include separate mention of a variety, where this is distinct and known to be recorded in the British Isles. However, in most cases varieties are distinguished from the type by little more than slight differences in the dimensions. Where there appears to be a continuum between the two and the whole range occurs in the British Isles, the general account is broadened to cover the variety (unless specifically stated otherwise). The description may therefore differ slightly from that in other Floras. Anyone wanting to make a critical taxonomic revision should consult the original type description. In addition several species are combined here, especially those which are more or less identical apart from the presence of calcification in one of them. In the case of several highly calcified forms with colourless sheaths, it is uncertain whether the absence of colour is simply because trichomes potentially capable of forming scytonemin are sufficiently shielded from the influence of ultra-violet light that they do not do so.

Ecology

Blue-green algae occur in a wide range of environments, but appear to be entirely absent at pH values below 4.0. In many habitats in the British Isles they tend to be more abundant in summer. For instance, picoplankton species are important in the summer phytoplankton of some large nutrient-poor lakes e.g. Wastwater in the English Lake District, while other species form dense blooms in mid- and late summer in nutrient-rich ponds and lakes throughout the British Isles. *Microcystis aeruginosa* is the most notorious bloom-former, because of the dense scums that can be formed and the fact that these are often highly toxic to humans and other mammals. However, all bloom-forming species appear to be capable of sometimes forming toxins.

Picoplanktonic blue-green algae, i.e. ones with all cell dimensions less than 2 μm, are widespread, and perhaps ubiquitous, in the plankton of ponds and lakes which are not highly acidic. A detailed account by Stockner et al. (2000) includes taxonomic, floristic and ecological information. However, although there are many records of picoplanktonic blue-green algae from the British Isles, there is usually too little information to fit these into the taxonomic categories recognized by Stockner et al. During use of the present account, specimens may sometimes fit well one of the listed species; where this is not so, it is recommended to record the genus and as much detail as possible on features such as dimensions, colour and any mucilaginous links between cells. The fact that picoplankton were largely overlooked until the mid-1980s, and in some studies still are so, means that generalizations in the older literature on the relative importance of different algal groups in some large lakes and reservoirs give a false impression of the true situation.

In general, blue-green algae tend to be more abundant and species-rich in calcareous environments, though some species, especially heterocystous ones, are characteristic of very soft waters, including pools with pH values as low as 4.1. Moist humid conditions favour the growth of species on soils and rocks. For instance, colonies of *Nostoc commune*, which typically grow on calcareous soils or depressions on limestone surfaces, are much more common in the west of Scotland and Ireland than in eastern Scotland and England. Many species of *Gloeocapsa* and sheathed Oscillatoriales depend on an intermittent water supply and their distribution is influenced by the frequency of water supply, and perhaps also of varying nutrient levels when water is flowing. The habitat note 'moist rock' for a particular species includes not only sites with a continuous trickle and ones like the vicinity of a waterfall subject to spray for much of the time, but also ones which are thoroughly wetted only intermittently. These are two very different types of environment for the organism, but it is often unclear in older accounts which was intended. More information on the influence of water regimes on species in calcareous environments is given by Pentecost & Whitton (2000).

Soils which show few or no obvious blue-green algal growths on the surface often reveal a diverse flora when cultured under a range of growth conditions in the laboratory, as shown for species of *Anabaena* and *Cylindrospermum* by Lund (1947). This suggests that some species may be much more widespread than commonly realized. In particular, the occurrence of visually obvious colonies of *Cylindrospermum* on the surface may represent uncommon situations where environmental factors for a particular species are especially favourable.

Some blue-green algae occur in symbiotic associations. Examples in the British Isles include about 10% of lichen species (with *Nostoc*, *Calothrix*/*Dichothrix*, *Scytonema*, *Stigonema*), some *Sphagnum* populations (with *Hapalosiphon*, though the association is loose), the water-fern *Azolla filiculoides* (with *Anabaena*) and the wetland flowering plant *Gunnera manicata* (with *Nostoc*). *Azolla filiculoides* is naturalized in the warmer parts of the British Isles and occasionally forms a dense cover on ponds and ditches. *Gunnera manicata* is largely restricted to gardens, but is naturalized on some stream banks in Cornwall and a few other places in the warmer parts of the British Isles. In all these associations the blue-green algal symbiont is a nitrogen-fixer. The widely researched phycomycete fungus *Geosiphon pyriforme*, the cells of which contain intracellular symbiotic *Nostoc*, is known from the soil surface of an arable field in Germany, so it is worth searching for it in the British Isles, especially where nitrogenous fertilizers are not used.

More detailed accounts of these and other aspects of the ecology of blue-green algae can be found in the book edited by Whitton & Potts (2000).

Floristics

The approximately 360 species listed here represent about 15% of the world total. The Flora includes all known species from the British Isles except the marine picoplankton. Several hundred other species have been recorded elsewhere in northern and central Europe and probably many of these will eventually be found in the British Isles. The groups most in need of further study are the smaller planktonic Chroococcales of ponds and lakes, especially those with humic-rich waters, and species in the upper intertidal and supralittoral zones. Some blue-green algal species are clearly restricted to the tropics or particular habitats such as hot springs, temperate rain forest or desert, which do not occur in the British Isles. However, many others appear to be cosmopolitan, but comments on world-wide distribution are omitted for most species, because too little is known for reliable generalization. The ecological notes refer to the British Isles, though sometimes depend heavily on comments in other Floras. Occasionally a species has also been recorded from quite different habitats in other countries; this is mentioned here, if it seems likely to have been overlooked in such habitats in the British Isles.

The listed species are based on not only published records, but also unpublished notes from several researchers and the author. The locations are mostly omitted, if the species are known to be widely distributed on mainland Europe, but are given for some of the less well-known species. Many of the general comments of West & Fritsch (1927) are also included, because the West father and son had earlier made extensive surveys of parts of the British Isles for which there are few, if any, subsequent studies. Several taxa are listed, which are known to be widespread elsewhere and which have apparently also been found in the British Isles, but without reliable records. In contrast, a few published records of species are omitted, where the organisms are rather obscure and it is doubtful whether the authors gave the material close study.

Practical matters

In addition to the general advice on sampling and storing algae, there are a number of matters which are especially relevant to blue-green algae. The following is only a brief guide; more detailed information is given by Whitton et al. (2002). Samples should be investigated live wherever possible, because of the importance of colour and observing motility. Soil and rock samples should always be dried rather than preserved. Gas vacuoles are often destroyed if cells are preserved with iodine. It is useful to keep live samples on a shelf in the light for a few days, because changes may take place which aid identification. Phycocyanin and gas vacuole formation may be enhanced, if cultures are kept under low light. Sheaths may develop around Oscillatoriales trichomes, especially if populations are allowed to dry slowly. Akinetes, which are essential for identification of *Anabaena* and *Cylindrospermum* to the species, may differentiate.

Care is needed, however, in interpreting information about live samples, if they cannot be viewed soon after collection. The trichomes of some sheathed Oscillatoriales tend to move out of their sheaths, with the risk of material being named *Oscillatoria* rather than *Lyngbya*. In such cases it is often possible to find empty sheaths elsewhere in the container. Selective grazing is a problem in studies of small-celled planktonic forms. A community rich in such species at the time of sampling may be almost entirely dominated by a few clumped colonies of *Aphanocapsa*, when viewed two days later. Selective filtration may help to slow down the effects of grazing, but it is difficult to remove protozoa and small rotifers without removing many of the blue-greens.

Blue-green algae in calcareous waters, especially streams, often have deposits of calcium carbonate in their sheaths. Other species penetrate into limestones and other calcareous materials, especially in the intertidal zone. In both cases samples should first be viewed without special treatment, but then it is essential to dissolve away the calcium carbonate to see cell or filament morphology, especially any branching. This can be done quickly with dilute hydrochloric acid, but it is preferable to use sodium ethylenediaminetetra-acetate solution, which takes longer, but has less drastic effects on the cells.

Epifluoresence microscopy is valuable for looking at very small cells and essential for making quantitative counts of picoplankton. Such counts require the small cells to be separated from larger cells and other particles. Counts can be made on preserved samples, but particular care is needed in storing the samples, especially keeping them away from the light.

Staining for polyphosphate granules is easy, if there is access to a laboratory with a fume cupboard. The method uses lead nitrate solution, followed by ammonium persulphide solution, which leads to deposition of lead sulphide in the granules. Details of this and other staining techniques for blue-greens are given by Fuhs (1973).

ORDER CHROOCOCCALES

Unicellular or colonial, sometimes forming a pseudofilamentous structure, but never forming true filaments. Individual cells are sometimes motile: examples include most, if not all, species associated with sediments, many spores, such as the terminal ones formed by *Chamaesiphon*, and the multicellular plates of *Merismopedia* and *Microcrocis*.

The Chroococcales are phylogenetically diverse,

but include almost all species incapable of forming hormogonia. The order was first described by Wettstein (1924), but has been described subsequently in slightly different ways in the various floristic accounts. The order sometimes includes all non-filamentous forms, as in the account by Komárek & Anagnostidis (1999; see also 1986), or one or more groups of the morphologically more complex forms are recognized as separate orders. The present account follows Komárek & Anagnostidis, since none of the other orders recognized can be separated unequivocally from other Chroococcales, at least based on morphological criteria. The most plausible subgroup to merit the status of order is Pleurocapsales Geitler 1925 emend. Waterbury *et* Stanier 1978, where some or all of the larger cells reproduce by multiple fission to form small spherical cells (often known as endospores, but better termed baeocytes). However, further molecular studies are needed to establish whether or not the Pleurocapsales consist of a single phylogenetically distinct group. In addition it is likely that the very small-celled forms will eventually be separated into more than one group based on molecular data.

The planktonic colonial forms with cells embedded in mucilage or linked together in various ways by mucilage strands are in urgent need of study in the British Isles. Records are lacking for a number of genera and species which appear to be quite common in other parts of Europe; some are characteristic of waters with humic materials. In some cases the cells are small enough to fall within the definition of picoplankton, yet the colony as a whole is well above the limit. Care is needed in the study of picoplankton to ensure that the colonial structure of those linked only weakly by mucilage strands is not overlooked. A set of colour figures by G. Cronberg (in Whitton & Potts, 2000) provides a useful introduction to the colonial genera with very small cells. A planktonic form which did not correspond exactly to any of these became frequent in the Kielder Reservoir, Northumberland, a few years after the reservoir was constructed and the outflow deposited a mucilaginous mass in the upper part of the downstream river, which caused nuisance problems for several years.

Komárek & Anagnostidis (1999) recognize a group of seven apparently closely related genera as the subfamily Gomphosphaerioideae. These are spherical or ovoid mucilaginous colonies with the cells located towards the periphery of the colony, and usually planktonic. Four have been recorded from the British Isles: *Coelosphaerium*, *Gomphosphaeria*, *Snowella*, *Woronichinia*. As at least some of these are common and the genera are defined in various ways in the commonly available Floras, their features (as recognized by Komárek & Anagnostidis, 1999) are summarized here.

- *Coelosphaerium* mature cells spherical; colony without central stalk system.
- *Gomphosphaeria* cells obovoid or slightly heart-shaped, at the ends of wide, branched stalks originating near centre of colony; cells clearly spaced from one another and often enveloped in the same gelatinous matrix as the stalks.
- *Snowella* cells spherical or slightly elongated in radial axis, at the ends of narrow, distinct, branched stalks originating near centre of colony; although the cells are towards the periphery of colony, they do not form a distinct single layer.
- *Woronichinia* cells spherical or obovoid, at the ends of rather indistinct stalks about as wide as the cell and which radiate from the centre of the colony; cells close to each other.

Aphanocapsa Nägeli 1849

01050000

Colony usually microscopic, but sometimes macroscopic in a few species, among which *A. grevillei* is the most likely in the British Isles; subspherical in some plankton forms, but terrestrial species are usually irregular in shape. Colony with many irregularly arranged cells embedded in mucilage, the outer margin of which is usually rather indistinct. Cells spherical or subspherical, with binary division, but successive divisions perpendicular to each. The cells tend to be packed quite closely in forms associated with terrestrial or submerged plants, but are more spaced apart in planktonic forms.

Many species have been recognized, with cell dimensions and habitat being key distinguishing characters. Komárek & Anagnostidis (1999) list 27 species, but it would be hard to distinguish all these in field samples based on the limited number of characters available using microscopy. The six species listed here for the British Isles have been recorded a number of times and cover a range of cell dimensions. However, the two planktonic species have often been considered as *Microcystis*. This account follows Komárek & Anagnostidis (1999) in restricting *Microcystis* to forms which are almost always gas-vacuolate when in the plankton.

Other record Kováčik (1988): *A. rivularis* (Carmichael) Rabenhorst 1865, based on sample isolated by Pringsheim (CCAP 1404–1).

Aphanocapsa delicatissima West *et* G.S.West 1912

Synonyms: *Microcystis pulverea* (Wood) Forti in De Toni 1907 fo. *delicatissima* West *et* G.S.West 1912; *M. delicatissima* (West *et* G.S.West) Starmach 1966

01050020

Colony spherical, ellipsoidal or irregular, but a fairly distinct outer sheath layer, up to 50 μm, with colourless or pale yellow mucilage. Cells spherical, (0.4–)0.5–0.8(–1.2 μm: Komárek & Anagnostidis, 1999), evenly spaced in the mucilage.

Plankton.

Aphanocapsa elachista West *et* G.S.West 1914

Synonyms: *Microcystis pulverea* (Wood) Forti in De Toni 1907 fo. *elachista* (West *et* G.S.West) Elenkin 1938; *M. elachista* (West *et* G.S.West) Starmach 1966; *Aph. conferta* (West *et* G.S.West) Komárková-Legnerová *et* Cronberg 1993; *Aph. planctonica* (G.M.Smith) Komárek *et* Anagnostidis 1995

01050030 **Pl. 1A** (p. 33)

Colony spherical or ellipsoidal, up to 100 μm in the largest dimension, though typically smaller, with

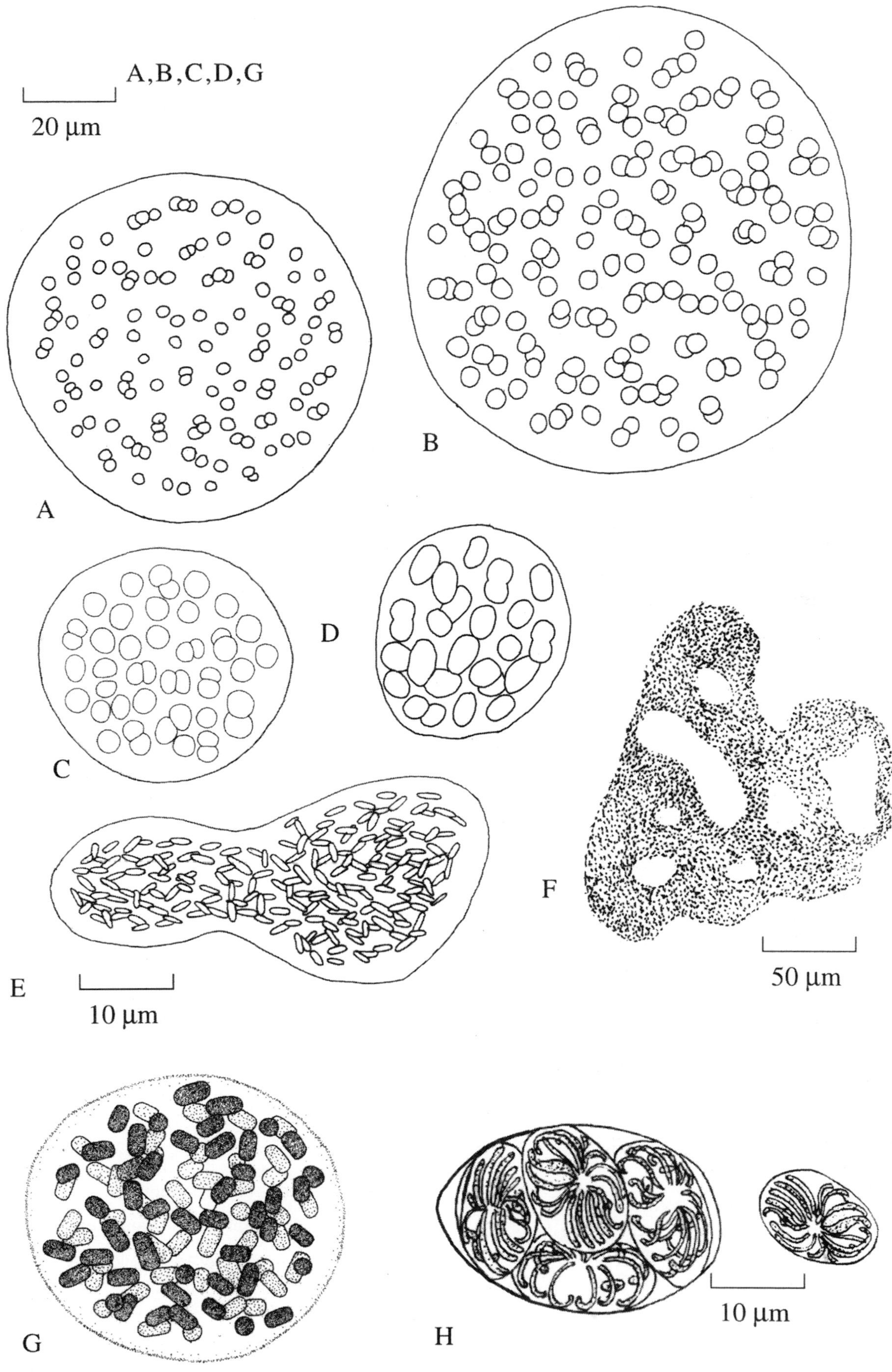

Plate 1 A–H
A. *Aphanocapsa elachista* (p. 32); **B**. *Aphanocapsa grevillei* (p. 34); **C**. *Aphanocapsa litoralis* (p. 34); **D**. *Aphanothece microscopica* (p. 35); **E,F**. *Aphanothece clathrata* (p. 35): E. Entire colony; F. Older, clathrate colony; **G**. *Aphanothece stagnina* (p. 35); **H**. *Glaucocystis nostochinearum* (p. 613): eukaryote containing cyanelles.

outer layer of the colourless mucilage weakly defined or even becoming diffluent. Cells 1.5–3 μm, single or in pairs, well spaced in the mucilage.

Plankton, especially in summer and moderately nutrient-rich waters.

Plate 1C shows *A. elachista* var. *conferta* West *et* G.S.West, where the cells are closer together than in the type. Komárek & Anagnostidis (1999) recognize this as a distinct species = *A. conferta* (West *et* G.S.West) Komárková-Legnerová *et* Cronberg 1993; this appears to be separated by a single character. Komárek & Anagnostidis (1999) also recognize *A. elachista* var. *planctonica* G.M.Smith 1932 as a distinct species = *A. planctonica* (G.M.Smith) Komárek *et* Anagnostidis 1995. In this case, the differences from the type are the slightly more distinct outer margin of the colony and the cell dimensions being in the upper part of the range. In the absence of experimental studies on representative populations and the fact it is often difficult to decide to which category a field sample belongs, it is suggested that the species should not be split in this way.

Aphanocapsa grevillei (Hassall) Rabenhorst 1865

Basionym: *Coccochloris grevellei* Hassall 1845
Synonym: *Microcystis grevillei* (Hassall) Elenkin 1938
01050040 **Pl. 1B** (p. 33)
Colony spherical or subspherical, with outer layer fairly well defined, though often with many colonies close together. Cells spherical, 3.2–5.6 μm, single or in pairs, quite closely packed.

On the bottom of shallow pools and on moist rocks and soil in a range of pH conditions, but probably often influenced by runoff from peaty areas.

Aphanocapsa litoralis Hansgirg 1892

Synonym: *Microcystis litoralis* (Hansgirg) Forti in De Toni 1907
01050010 **Pl. 1C** (p. 33) **CD**
Colony subspherical or irregular, often reaching 60 μm or more in largest dimension, typically with a number of subcolonies with a fairly distinct outer layer. Cells subspherical, 3–5 μm in diameter, usually packed quite closely.

Shallow saline waters near the sea or inland. Widely distributed and variable. Powell (1964) lists this based on Anand (1937); however, he also lists the synonym *Microcystis litoralis* as if it were another species.

Komárek & Anagnostidis (1999) list this as *A. litoralis* (Hansgirg) Komárek *et* Anagnostidis (1995) and provide more information than in the original description.

Aphanocapsa marina Hansgirg 1890

01050060 (new number)
Colony small, irregular, gelatinous, greyish-blue, sometimes united in a blue-green mass. Cells spherical, 0.4–0.5 μm wide, often in 2s following division, sometimes with thick, colourless gelatinous envelopes.

Powell (1964), without details. Brackish pools and wet rocks in supralittoral and upper intertidal.

The above largely follows the original description. The species differs from *A. delicatissima* only in that the latter was described as having a more distinct boundary to the colony and slightly larger cells. A range of small-celled forms occur in small brackish pools covering these features and the two species should probably be merged. However, the two names are retained here to avoid using *A. marina* (which has priority) for freshwater forms.

Aphanocapsa muscicola (Meneghini) Wille 1919

Basionym: *Coccochloris muscicola* Meneghini 1843
Synonym: *Microcystis muscicola* (Meneghini) Elenkin 1938
01050050
Colony irregular or united into an amorphous mass, with colourless mucilage. Cells spherical, 2–4(–4.5) μm, sometimes with their own individual sheath layer and often grouped in 2s or 4s.

On diverse terrestrial surfaces, including bryophytes, rocks and soil.

Aphanothece Nägeli 1849

01060000
Colony microscopic to macroscopic, reaching several centimetres in *A. stagnina*, with cells loosely or densely arranged in a common mucilage; colony multiplication by fragmentation of large colonies. Margins of colony may be diffluent or well delimited. Cells spherical or subspherical, with binary division & with successive divisions in the same plane, though sometimes one daughter cell shifts to another plane, giving the superficial appearance of cell division being more like *Aphanocapsa*; individual cells of some species have their own distinct envelopes. According to Komárek & Anagnostidis (1999) gas vacuoles occur in a few planktonic species, though the cells of most species are too small to be sure without electron microscopy.

Komárek & Anagnostidis (1999) give particular emphasis to the habitat within this genus, which means that the species limits are often more restricted than in other floras. Other species listed for the British Isles by West & Fritsch (1927) include *A. nidulans* P.G.Richter and *A. saxicola* Nägeli; they list the small cells often present in colonies of *Coelosphaerium* and some other planktonic blue-green algae as *A. nidulans* var. *endophytica* West *et* G.S.West = *A. endophytica* (West *et* G.S.West) Komárková-Legnerová *et* Cronberg 1994. Their status in the British Isles needs further study, as does a record of *A. saxicola* by W. & G.S.West (1894: forming floating gelatinous flocs) in the New Forest (Hampshire).

A. pallida (Kützing) Rabenhorst 1865 also poses problems. Powell (1964) includes this in the marine check-list, but gives *Coccochloris stagnina* Sprengel as a synonym, whereas in reality the latter is a synonym of *Aphanothece stagnina*. As there are apparently no other records, its status in the British Isles is uncertain. Frémy (1929–33) includes it in his flora, but has no records for coasts in France. However, the species has been reported as widespread on moist terrestrial surfaces in other countries, including brackish sites. It differs from *A. castagnei* mainly in its translucent pale green macrosopic appearance. Geitler (1932a) pointed out its similarities with *Gloeothece rupestris*.

Aphanothece castagnei (Brébisson) Rabenhorst 1865

Basionym: *Palmella castagnei* Brébisson in Kützing 1845
01060010
Colony subspherical or irregular, often visible macroscopically, reaching a few millimetres, margins of colony distinct; dirty blue-green, olive or brownish. Cells ovoid or cylindrical, 2.5–3.5(–5)×4–8 μm, surrounded by individual sheaths, which are usually lamellate and yellow-brown in those colonies occurring on rocks.

Komárek & Anagnostidis (1999) restrict this species to colonies on wet rocks and mosses, but other authors include colonies in standing fresh and brackish water.

Aphanothece clathrata West *et* G.S.West 1906

01060020 **Pl. 1E,F** (p. 33)
Colony microscopic, up to *c.* 400 μm, initially subspherical, later elongate, slightly flattened and sometimes clathrate; margins of colony distinct, with the mucilage sometimes very pale yellow, especially towards the margin. Cells rod-shaped or very slightly bent, 0.4–1(–2) × (1.5–)2.5–4.5 μm.

Plankton or benthos of moderately nutrient-rich shallow lakes, ponds or slow-flowing stretches of rivers.

Aphanothece elabens (Brébisson in Meneghini 1842) Elenkin 1938

0160050 (new number)
Colony initially microscopic, spherical or somewhat flattened, sometimes later becoming macroscopic with aggregates of colonies, usually pale bluish-green on mass. Cells ovoid to cylindrical, 1–2(–3) × 2.8–6.5 μm; often with gas vacuoles. Mucilage envelopes of individual cells usually distinct, though sometimes diffluent.

Typically loosely associated with submerged plants in oligotrophic or mesotrophic ponds and lakes, though sometimes on sediments or free-floating in the plankton. Komárek & Anagnostidis (1999) list the species as freshwater, but Powell (1964) includes it in the marine check-list, based on one of his own earlier records.

Aphanothece microscopica Nägeli 1849

01060030 **Pl. 1D** (p. 33)
Colony microscopic initially, subspherical, though later becoming amorphous and often reaching. about 2 mm; mucilage colourless and eventually diffluent when in water. Cells ovoid to cylindrical, (3–)3.5–5 × (4–)6–9 μm, sometimes with a poorly defined individual sheath in cells towards the outside of the colony.

Plankton, benthos or epiphytic in peaty non-calcareous waters, sometimes forming macroscopically visible flocs among submerged water-plants at the edge of nutrient-poor lakes; also on very moist peaty soils.

Aphanothece stagnina (Sprengel) A.Braun 1863

01060040 **Pl. 1G** (p. 33)
Colony initially small, but becoming large, often reaching 4 cm or more; mucilage with firm outer margin, colourless. Cells ovoid to cylindrical with rounded ends, 3–5 × 4–11 μm, sometimes with their own individual sheath in cells towards the periphery of the colony.

Floating, attached to submerged plants or benthic in slightly nutrient-rich fresh or brackish pools and lake margins, often forming conspicuous masses. Typically, colonies first develop loosely attached to submerged plants and then become free-floating, where they continue to grow and multiply. They are the most likely (but not the only) blue-green alga to be found as large gelatinous free-floating colonies, at least in the eastern parts of the British Isles.

This is probably a characteristic species of such water bodies throughout the world.

Chamaesiphon A.Braun *et* Grunow 1865

01100000
Mature cells single or arranged into a colony, the colonies of some species forming macroscopically visible structures which are sometimes sufficiently characteristic for recognition with the naked eye. The individual cell is differentiated into a basal end attached to a surface and the distal end either forms exospores or, in some colonies, acts as the surface for another mature cell, which may itself give rise to exospores. The mature cell, which develops from a small spherical exospore, is club-shaped, pear-shaped, ellipsoidal or almost cylindrical. A developing mature cell has a very thin sheath, which becomes more obvious and opens out (pseudovagina) when exospore formation commences. The sheath may be colourless, yellow-brown or rich brown, with the tendency for colour formation to be most obvious in species which form colonies. Some colonial forms produce obvious mucilage in addition to the sheaths. Colonies arise in one of two different ways or a combination of both. Exospores occur in various arrangements at the end of outer mother cells. Exospores may move to a position next to the base of a mother cell, eventually giving rise to a flat structure consisting of numerous cells adjacent to each other. In other cases some of the exospores persist in an empty sheath, giving rise to layered or shrub-like colonies.

Most of the species are variable and may sometimes show forms slightly outside the limits given here. However, unlike some other genera of Chroococcales, many species do appear to be well characterized and some of the larger colonial forms are highly characteristic of particular types of environment.

Geitler (1932a) separated the genus *Stichosiphon* and this is characterized by Komárek & Anagnostidis (1999) by the repeated division of exospores forming more or less long pseudofilamentous structures.

Other record Powell (1964), without detail: ? *Chamaesiphon* ? *marinus* Wille *et* Rosenvinge. Komárek & Anagnostidis (1999) comment that this is a problematic species.

Chamaesiphon britannicus (F.E.Fritsch) Komárek *et* Anagnostidis 1995

Basionym: *Oncobyrsa britannica* F.E.Fritsch 1927
Synonyms: *Chamaesiphonopsis regularis* F.E.Fritsch 1929; *Chamaesiphon regularis* (F.E.Fritsch) Geitler 1929; *Xenococcus britannica* West *et* F.E.Fritsch, 1927

01100100 **Pl. 2A** (p. 37)
Colony epilithic, forming brown spots up to 3 mm diameter, consisting of 1 or several layers of densely packed cells, the lower ones perpendicular to the surface and the upper ones arranged more irregularly. Because the cells are packed so tightly, they often appear polygonal

when viewed from above, especially at the younger stages of the colony when it is only one layer thick. Cell 3.5–4.5(–5) μm wide, 10.5–14 long; sheath very thin, diffluent and not giving rise to a persistent structure. Populations with olive or pinkish cells occasionally present, presumably indicating presence of phycoerythrin. The macroscopic colour of the spots suggests presence of scytonemin, but a brown colour is not obvious in the diffluent mucilage as seen by microscopy.

Fritsch (1927) described the species as occurring together with *Phormidium foveolarum*. West & Fritsch (1927) state that it is very abundant on pebbles in many Devon streams, forming small roughly circular patches or large irregular encrustations of a brownish-black or sometimes faintly purple colour; it is also widespread in streams elsewhere, though usually not as abundant as in the streams studied by Fritsch. Komárek & Anagnostidis (1999) list it as epiphytic on filamentous blue-green algae, but there appear to be no similar records from the British Isles.

The original description of the species by Fritsch has several times been modified slightly by subsequent authors. It was listed as *Chamaesiphon regularis* in the *Coded List* (Whitton et al., 1998a).

Chamaesiphon confervicolus A.Braun 1865

Synonym: *Chamaesiphon curvatus* Nordstedt 1878

01100010 (includes 01100020) **Pl. 2C** (p. 37)

Epiphytic, single, or, more usually, in loose clusters. Cells ranging from narrowly club-shaped to almost cylindrical apart from a marked narrowing towards base; often slightly curved and sometimes (var. *curvatus* Borzí) strongly curved. Mature cell: width 1–3 μm at base and 3–9(–12) μm at distal end; length 15–70 μm, with end part of sheath enclosing a short row of exospores or sometimes extending considerably more. Sheath thin and colourless, so it may be hard to distinguish dimensions of cell from that of the whole structure.

Freshwater, epiphytic on (especially) filamentous algae (e.g. *Cladophora*, *Oedogonium*), submerged mosses and vascular plants. Standing or flowing waters, especially moderately hard water; probably most characteristic of mesotrophic and absent from highly eutrophic waters. Quite common in the British Isles, though much less so than *C. incrustans*. Komárek & Anagnostidis (1999) note that it occurs rarely in slightly brackish water and also rarely as an epilith, but there are apparently no such records for the British Isles.

C. curvatus was listed as a separate species in the *Coded List* (Whitton et al., 1998a), but is treated as a synonym here. *C. confervicolus* was listed as *C. confervicola* in the *Coded List*.

Chamaesiphon curvatus Nordstedt 1878

01100020: see *Chamaesiphon confervicolus*

Chamaesiphon ferrugineus F.E.Fritsch 1929

01100030: see *Chamaesiphon fuscus*

Chamaesiphon fuscus (Rostafinski) Hansgirg 1888

Basionym: *Sphaerogonium fuscum* Rostafinski

Synonym: *Chamaesiphon ferrugineus* F.E.Fritsch 1929

01100040 (includes 01100030) **Pl. 2F,G** (p. 37)

Colony forming rusty-brown or brown black flecks on rocks; flecks typically a few mm in largest dimension, but occasionally exceeding 1 cm. Colony either 1–2(–3) layers of densely packed cells perpendicular to the surface or arranged in shrub-like (or *Dinobryon*-like) colonies. (Material of the latter was named *C. ferrugineus* by Fritsch, 1929.) Cells club-shaped to narrowly ovoid, cells 2.5–6(–8) μm wide, 5–20(–30) μm long. Sheath thin but very distinct, starting colourless, but soon becoming a rich brown; several texts state that the colour is due to iron, but it seems likely that much of the colour is due to scytonemin, though possibly iron is also sometimes deposited.

Epilithic in fast-flowing streams and small rivers; always soft water and attached to non-calcareous substrata; probably best developed in mesotrophic waters, though its absence in more eutrophic waters may be due simply to being outcompeted.

Chamaesiphon ferrugineus was described in detail from British streams by Fritsch (1929) and listed as a separate species in the Coded List; however, it is really just a well-developed shrub-like form of *C. fuscus*. Komárek & Anagnostidis (1999) treat it as a synonym as done here.

Chamaesiphon geitleri Luther 1954

01100050 **Pl. 2H** (p. 37)

Colony blackish-brown spots on rocks, irregular in shape but well delimited boundary, typically up to 2–3 mm in diameter, but when the organism is abundant, sometimes larger. Colony with densely packed cells perpendicular to the surface, usually combining several layers of cells and some cells organized into a bush-like arrangement. Cells club-shaped to almost cylindrical with narrowing towards the base. Cells 2.5–6.5(–9) μm wide, 5–20(–30) μm long. Sheath thick, lamellate, yellow-brown to very deep brown.

Epilithic in fast-flowing streams: records for the British Isles have been for highly calcareous, unpolluted waters, though with the organism typically attached to non-calcareous rocks; sites downstream of *Rivularia*-dominated stretches are a likely place to find it. Komárek & Anagnostidis (1999) make no mention of it being characteristic of hard waters, but they do state that it occurs rarely in submerged parts of the splash zone of the stony littoral of alpine tarns.

As least in the British populations, this is much like a form of *C. fuscus*, but occurring in calcareous water, forming particularly dark colonies and with a thicker and lamellate sheath.

The right-hand three figures of *C. geitleri* in Pl. 2H are reproduced from Geitler (1925: shown as *C. fuscus*). However, E. Rott (pers. comm.) comments that there are never small and large exospores together. The figures look as if there is another small *Chamaesiphon* in the open pseudovagina (cup).

Chamaesiphon incrustans Grunow in Rabenhorst 1865

Synonym: *Sphaerogonium incrustans* Rostafinski 1883

01100060 **Pl. 2E** (p. 37)

Common epiphyte in freshwater environments and occasional also in slightly brackish ones. Cells solitary or in small groups, cylindrical or slightly club-shaped. Cells 1.5–5 μm wide away from the base, 7–20(–30) μm long. Sheath thin, colourless, not lamellate. Exospores 1 or 2 subspherical cells at apex of mature cell. Populations inside dense growths of submerged plants sometimes have phycoerythrin as well as phycocyanin, and cells are then various shades of olive to pink.

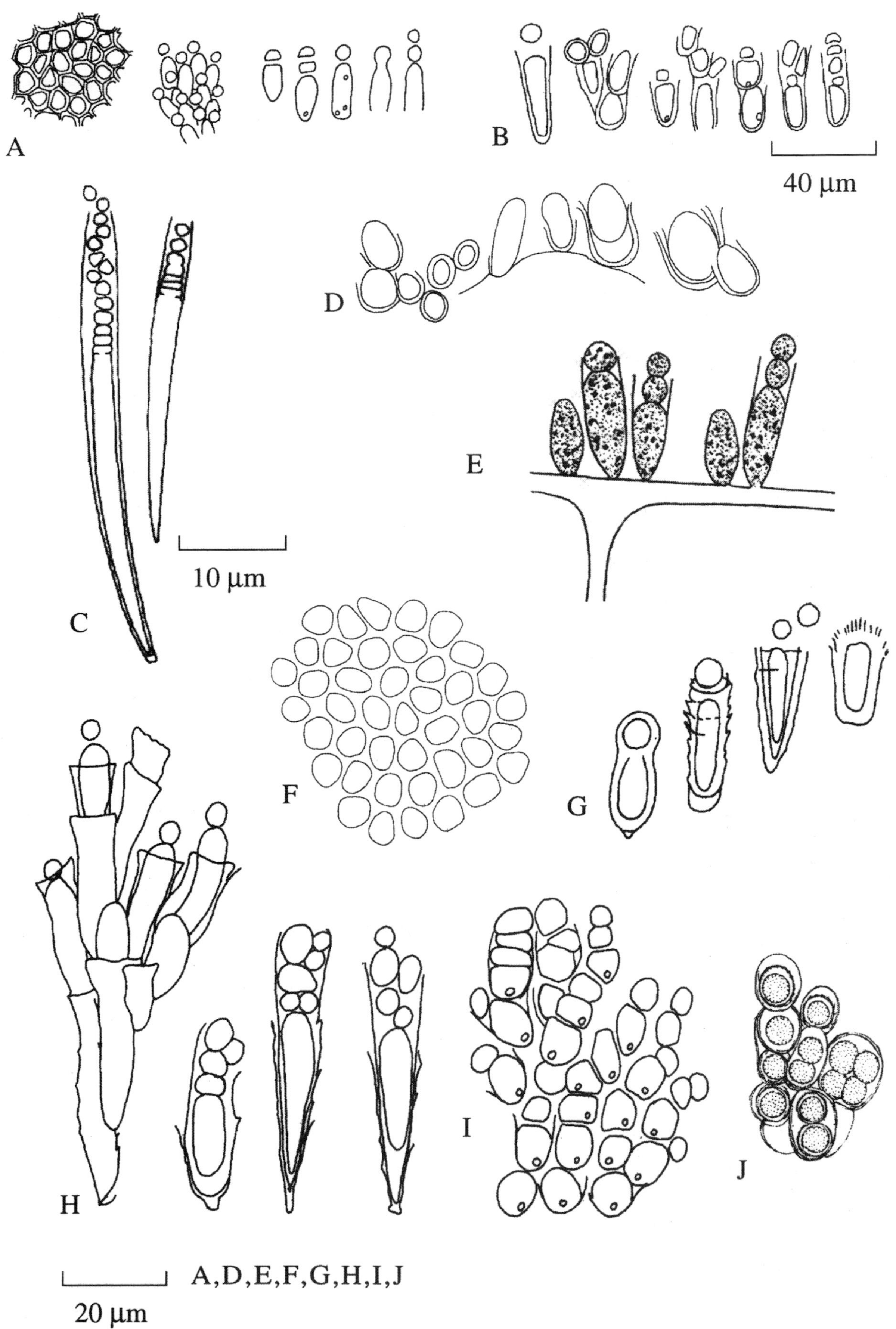

Plate 2 A–I
A. *Chamaesiphon britannicus* (p. 35): showing views from above and side; **B**. *Chamaesiphon pseudopolymorphus* (p. 38); **C**. *Chamaesiphon confervicolus* (p. 36); **D**. *Chamaesiphon polonicus* (p. 38); **E**. *Chamaesiphon incrustans* (p. 36); **F, G**. *Chamaesiphon fuscus* (p. 36): F. View from above; G. View from side; **H**. *Chamaesiphon geitleri*: left-hand figure shows typical form; for comments on the other three figures, see text (p. 36); **I**. *Chamaesiphon polymorphus* (p. 38); **J.** *Siphononema polonicum* (p. 56).

Epiphytic on many filamentous algae, submerged mosses and vascular plants in shallow standing and flowing water; typical of mesotrophic and slightly eutrophic conditions, but rare or absent if highly eutrophic.

Chamaesiphon minutus (Rostafinski) Lemmermann 1910

Basionym: *Sphaerogonium minutum* Rostafinski 1883

01100120 (new number) **CD**

Cells epiphytic, solitary or forming groups; ovoid, slightly pear-shaped or short cylindrical with rounded apex, 3.5–6(–7) × 1.5–3(–5) μm. Sheath very thin, colourless. Exospores found singly at apex.

Epiphyte on filamentous algae and mosses in low nutrient environments. This appears to form a continuum with *C. incrustans*, but is more restricted to standing and low nutrient waters, especially upland ponds.

Chamaesiphon polonicus (Rostafinski) Hansgirg 1892

Basionym: *Sphaerogonium polonicum* Rostafinski 1893

01100070 **Pl. 2D** (p. 37) **CD**

Colony visually obvious as irregular spot or patch on rocks, populations usually with examples reaching 1 cm diameter, occasionally much larger. Colony flat, yellow-orange to orange-brown, consisting of one or several layers of compact cells arranged irregularly or with tendency for vertical rows. Cells subspherical initially, but usually becoming ellipsoidal, 3–6 μm wide, (4–)6–10(–15) μm long. Sheath thick, firm, layered, pale orange-brown to deep brown, not gelatinizing. Exospores usually 1–2, occasionally a few more in a short vertical row.

Freshwater, mostly in large streams or upper stretches of fast-flowing rivers; absent from soft waters and sometimes present in hard waters, though apparently almost always on non-calcareous rocks; typical of mesotrophic and slightly eutrophic conditions, but absent if highly eutrophic. Occasionally also on submerged rocks near the margin of lakes with a water chemistry similar to that of river populations. This and *C. fuscus* are the two typical species of rocks in upper stretches of rivers, with *C. fuscus* only in soft water and *C. polonicus* in waters of a wide range of hardness. The two species only seldom occur at the same site.

Chamaesiphon polymorphus Geitler 1925

01100080 **Pl. 2I** (p. 37)

Colony epilithic, small, though usually large enough to be recognized macroscopically, typically brownish green. Colony consists of short irregular rows of cells enveloped by mucilage. Cells subspherical initially, but often becoming more pear-shaped or almost cylindrical, 3–6 μm wide, 3–9 μm long. Sheath thin, colourless initially, but, unless well shaded, soon becoming pale yellow or brownish-yellow. Exospores usually remain within the common mucilage and contribute to the growth of the colony.

Freshwater, typically on small rocks among other algae or mosses in fast-flowing streams or shallow water at edge of lakes. Water of relatively low hardness, being absent from hard water, [illegible]h not restricted to soft water. Unless sampling in the field is [illegible]out carefully, this species is easy to overlook: the colonies tend to be more short-lived than *C. fuscus*, and often occur on rocks with a range of other organisms; the colony structure is also less characteristic than other epilithic species and easily disrupted by scraping.

Chamaesiphon pseudopolymorphus F.E.Fritsch 1929

Synonym: *Stichosiphon pseudopolymorphus* (F.E.Fritsch) Komárek 1989

01100090 **Pl. 2B** (p. 37)

Colony macroscopic, forming brown or reddish-brown flecks on rocks; cells in various arrangements, sometimes with tendency to form layers with some cells perpendicular to surface and in others more irregular or spread out horizontally. Cells 6–8 μm wide, 9–10 μm long. Sheath distinct, sometimes thickened, cup-shaped, yellow-brown. Exospores often in rows of many cells. This is a variable species, but it is unclear whether due to genetic differences or to influence of environment on colony structure.

Epilithic in small, fast-flowing oligotrophic and mesotrophic streams.

Because the cells often form quite long rows, Komárek & Anagnostidis (1999) moved this species to *Stichosiphon*, although Geitler (1932), who originally separated the genus from *Chamaesiphon*, did not include this species. Komárek & Anagnostidis (1999) reproduce both Fritsch's diagrams and those of Kann (1972), which show a much greater tendency to form rows of cells. It is unclear whether all the cells of rows are formed as exospores at the end of the mother cell or whether the exospores sometimes divide further.

Chamaesiphon regularis

01100100: see *Chamaesiphon britannicus*

Chamaesiphon rostafinskii Hansgirg 1887

01100110

Cells epiphytic, solitary or forming groups; narrowly club-shaped; older ones typically slightly curved, 1–2.5(–3) μm wide, 10–60(–100) μm long. Sheath thin, colourless; cell often does not reach into the basal part of the sheath and is thus well separated from the surface of the host plant, but extending well into surrounding water.

Epiphytic on filamentous algae, including sheathed blue-green algae such as *Tolypothrix*, mostly in streams, but sometimes in shallow water at edge of lakes. Calcareous or at least moderately hard water, typically in mesotrophic, and absent from highly eutrophic conditions.

This is like an extremely long version of *C. confervicolus*, but sufficiently distinct to justify separation.

Chlorogloea Wille 1900

01110000

Colony mucilaginous, subspherical or irregular and with a granular surface when older due to the presence of many subcolonies; often developing into macroscopic structures. Cells in each subcolony arranged irregularly in common mucilage, though with a tendency for the outer cells to form radially orientated rows when the subcolony is large. Cells spherical or ovoid, sometimes with own relatively indistinct sheaths.

Other record
Powell (1964) lists *Chlorogloea conferta* (Kützing) Drouet *et* Daily 1948, = *Entophysalis conferta* (Kützing) Setchell *et* Gardner in Gardner 1918, without details. Komárek & Anagnostidis (1999) list it as *E. conferta*, but comment that the generic identification needs confirmation. The species was described as an epiphyte on marine algae. It consists of densely packed cells 0.8–1.2 μm diameter in dense common mucilage, which is colourless or pale yellow.

Chlorogloea microcystoides Geitler 1925

01110010
Colony mucilaginous, reaching about 1 mm diameter, dark green or brownish (at well illuminated sites), often including subcolonies. Cells arranged in usually densely packed irregular rows, spherical or ellipsoidal, 2–4 μm wide, with or without individual sheath, embedded in colourless or pale yellow-brown mucilage.

Freshwater, epilithic in shallow waters at the edges of lakes and streams or in very moist areas adjacent to them.

Chlorogloea tuberculosa (Hansgirg) Wille 1902

01110020
Colony a small epiphyte, up to 150 μm diameter (very rarely more). Cells initially arranged in short, irregular rows more or less perpendicular to the surface, but almost all organization into rows lost in older colonies. Cells spherical, ellipsoidal or ovoid, 1–1.5 × 1–2.5 μm, with or without individual sheath, embedded in colourless mucilage.

Powell (1964), without details. Epiphytic on submerged marine algae and bryozoans.

Geitler (1932a) lists *Palmella tuberculosa* Hansgirg as the basionym, but includes a query, presumably because of doubts about the original material.

Chroococcopsis Geitler 1925

01750000 (new number)
Cells single or in compact groups, usually with individual sheaths, but also firm common envelope; cells sometimes in short indistinct rows, which in larger colonies are radial, with distinctly larger cells at the outside. Cells subspherical, ovoid, irregular or pear-shaped with narrower end attached to substratum; sheaths firm, colourless, sometimes layered. Cells divide by binary division or form numerous baeocytes, especially in larger outer cells.
Other record Powell (1964), as *Pleurocapsa amethystea* Rosenvinge 1893, without details. Komárek & Anagnostidis (1999) list this as an unrevised species – *Chroococcopsis amethystea* (Rosinvinge) Geitler 1925.

Chroococcopsis chroococcoides (F.E.Fritsch) Komárek *et* Anagnostidis 1995

Basionym: *Xenococcus chroococcoides* F.E.Fritsch 1929
01750010 (new number; old number 01730010)
Pl. 6G–I (p. 53)
Cells initially, single, spherical, club- or pear-shaped, later forming groups of 2–4 cells or short rows, which eventually give rise to a 2-layered colony; cells 8.5–10 × 10–17 μm. Sheath thick, layered.

Epiphytic on filamentous green algae and higher plants in streams and small rivers.

This species was included under *Xenococcus* in the *Coded List*.

Chroococcopsis fluviatilis (Lagerheim) Komárek *et* Anagnostidis 1995

Basionym: *Pleurocapsa fluviatilis* Lagerheim 1888
Synonym: *Xenococcus fluviatilis* (Lagerheim) Geitler 1925
01750020 (new number; old number 01570020)
Pl. 7I–K (p. 55)
Cells single, small groups or macroscopically obvious colonies; the latter are hard, flattened hemispherical or developing into crusts; colony bluish or, more usually, dark brown. Unless care is taken to sample older colonies attached to the substrate, the species is likely to be seen as small irregular groups of cells of very differing sizes and shapes. Cells 4–10 μm in the main mass, but often a few considerably larger (–25 μm) ones, the largest of which give rise to endospores, which are only about 2 μm when first released. Sheath colourless. Pseudofilamentous structures often present, sometimes as short, radially oriented rows.

Freshwaters: streams, rivers, lake margins on rocks, wood and sometimes submerged mosses and macrophytes.

This species was included as *Pleurocapsa fluviatilis* in the *Coded List*.

Chroococcus Nägeli 1849

01130000
Small colonies of cells, each cell usually with its own distinct sheath, rarely single cells. Aggregations of usually 2–16 cells, occasionally many more; colonies sometimes with odd numbers of cells, when one cell of a pair of daughter cells divides before another. Cells subspherical, with divisions in three planes in successive generations, although this pattern of division may be less obvious in larger groups. Cells often form new sheath layer each time they divide, so that a 4-celled group has cells with 3 distinct sheath layers; further less distinct layers may also be present, especially in terrestrial forms; sheaths colourless or yellow. Some planktonic species have gas vacuoles.

A wide range of freshwater and brackish habitats, including lakes and shallow weedy ponds and on terrestrial rocks and among plants in moist situations.

A great range of forms exists in nature. Komárek & Anagnostidis (1999) list 35 species for Europe, some differing little from each other. However, it is also easy to find material which does not correspond exactly to any of these. It is therefore recommended to assign specimens to a limited number of widely used names, but to record the details of particular populations.

Chroococcus cohaerens (Brébisson) Nägeli 1849

Basionym: *Pleurococcus cohaerens* Brébisson in Meneghini 1846
Synonym: *Chroococcus calcicola* Anand 1937
01130010
Cells arranged into distinct groups of 2–8 cells, which are often grouped together into gelatinous, amorphous

colonies which can form obvious patches over extensive areas. Mucilage around the small groups of cells has distinct outer layer (like a sheath), but subsequent divisions lead to only very indistinct layers, so the cells are embedded in a common mucilage. Cells spherical or subspherical, 2–5(–6.5) μm; mucilage and any sheath colourless.

Moist terrestrial surfaces, such as rocks and greenhouses.

Anand (1937) described *C. calcicola*, which appears identical morphologically to *C. cohaerens*. He reported it as penetrating coastal chalk cliffs in Kent to a depth of about 1 mm and being common in the interior of caves, where it marks the limit of all visible growth. Komárek & Anagnostidis (1999) list the same reference, but quote the species as aerophytic from India.

Chroococcus dispersus (Keissler) Lemmermann 1904

Basionym: *Chroococcus minor* (Kützing) Nägeli var. *dispersus* Keissler 1902

01130020 **Pl. 3G** (p. 41)

Colony, small, floating, typically 4–32 cells. Cells subspherical or elliptical, 3–4.5(–5) μm, occasionally with a narrow indistinct sheath, but otherwise embedded in colourless, homogeneous mucilage, with an outer layer ranging from distinct to diffluent.

Plankton of less nutrient-rich lakes and ponds.

Reynolds & Irish (2000) record *C. planctonicus* Bethge in their summary of data from (Lake) Windermere. The description by Bethge (1935) differs from *C. dispersus* mainly in the presence of gas vacuoles. As Komárek & Anagnostidis (1999) describe the ecology of *C. planctonicus* as 'winter plankton of eutrophic ponds with muddy bottom and anaerobic layer', and as the author has found *C. dispersus* in Windermere, it needs to be confirmed that there are two distinct species in the lake.

Chroococcus giganteus West 1892

01130030 **Pl. 3E** (p. 41)

Colony 2–4 cells, not aggregated into larger groups. Cells 54–58(–65) μm, with distinct sheath that may reach almost 10 μm in width, though usually narrower. The relatively narrow range of cell dimensions is based on the description by West (1892).

This is essentially nothing more than a large form of *C. turgidus* and forms intermediate between the two size ranges occur. It is suggested that any cells above the size range of *C. turgidus* are included here.

In shallow freshwaters, usually among submerged macrophytes. Wide pH range (A. Pentecost, pers. comm.).

Chroococcus limneticus Lemmermann 1898

01130040 **Pl. 3D** (p. 41)

Colony, free-floating, ovoid or irregular, with up to about 40 cells, but usually fewer; these cells occur within the homogeneous mucilage in groups of 2s or 4s and with these mini-groups spaced from each other; outer margin of the colony usually distinct (though not layered), but sometimes very diffluent; mucilage colourless. Cells typically hemispherical, 6–12(–20) μm; gas vacuoles absent. This species shows considerable variation.

Planktonic in a range of lakes and ponds, though probably most characteristic of mesotrophic waters; present throughout the year, but usually forming only a minor component of the phytoplankton.

Chroococcus minor (Kützing) Nägeli 1849

Basionym: *Protococcus minor* Kützing 1843

01130050 **CD**

Colony small, gelatinous, mostly with cells in groups of 2–4; mucilage colourless; colony masses often form macroscopic layers. Cells spherical, hemispherical or elliptical, 3–4 μm wide, individual sheaths indistinct or not visible at all.

A wide range of freshwater environments, but especially shallow ponds and slow-flowing streams with submerged macrophytes.

Chroococcus minutus (Kützing) Nägeli 1849

Basionym: *Protococcus minutus* Kützing 1843

01130060 **Pl. 3F** (p. 41)

Colony small, usually 2–8 cells, though occasionally cells solitary; spherical or ovoid shape when (as is usual) the outer margin is firm, but shape may be more irregular if the outer margin is more diffluent. Cells spherical or hemispherical, 4–10(–12) μm, with a wide sheath, which usually appears homogeneous, though lamellations sometimes just detectable.

Freshwater or brackish, mostly shallower water bodies and often among submerged macrophytes, though sometimes free-floating in the plankton.

Chroococcus pallidus Nägeli 1849

01130070

Colony microscopic, an amorphous gelatinous structure made up of (usually) many groups of 2–4(–8) cells and often pairs of cells within their own sheath (rather than individual cells); sheath and surrounding mucilage colourless; the gelatinous masses can sometimes form extensive layers outside of the colonial mass which is indistinct, but the individual subcolonies have a fairly distinct outer margin. Cells subspherical, or hemispherical after division: 5–8 μm (original description); (5–)6–11(–13) μm (Komárek & Anagnostidis, 1999). These authors refer to the colour of the cells as pale blue-green, yellowish or yellow and it seems probable that some authors have used this name mainly because the material in question was yellowish. However, it is doubtful how useful this feature is for identification.

An aerophytic species of wet and moist rocks and sometimes forming extensive strata according to West & Fritsch (1927), though Komárek & Anagnostidis (1999) state that all records outside central Europe are unclear.

Chroococcus schizodermaticus West 1892

01130080

Colony microscopic, usually 2–4 cells, not embedded in common mucilage. Cells spherical or hemispherical, 5.8–11 μm, each cell with a highly lamellated yellow or yellow-brown sheath, which is usually much wider on the side away from adjacent cells and often exceeds the diameter of the cell. This is much like *C. turgidus*, apart from the width, lamellation and colour of the sheath.

Among water plants at the edge of shallow oligotrophic or mesotrophic water bodies.

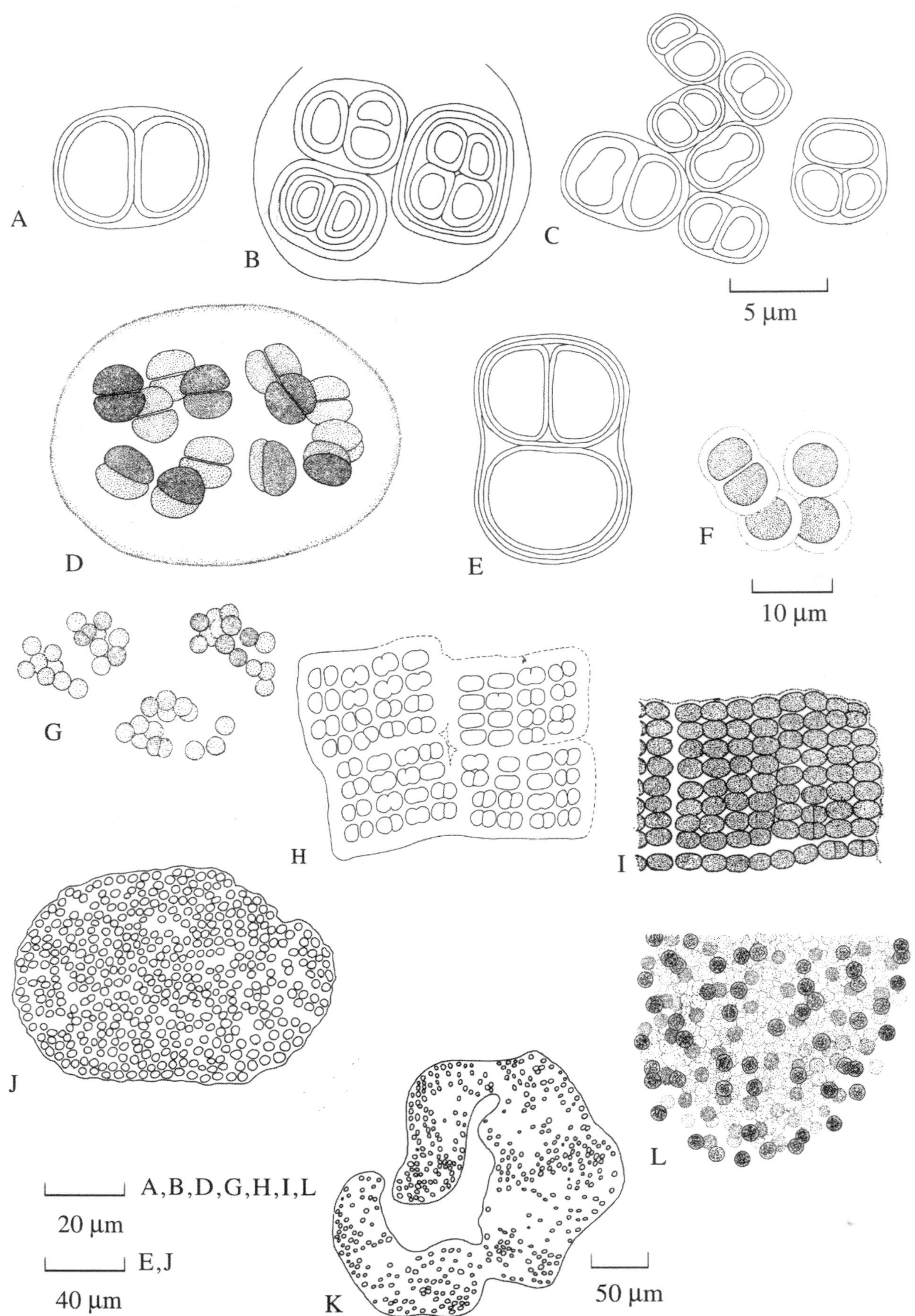

Plate 3 A–L

A, B. *Chroococcus turgidus* (p. 42); **C**. *Chroococcus varius* (p. 42); **D**. *Chroococcus limneticus* (p. 40); **E**. *Chroococcus giganteus* (p. 40); **F**. *Chrococcus minutus* (p. 40); **G**. *Chroococcus dispersus* (p. 40); **H**. *Merismopedia glauca* (p. 51); **I**. *Merismopedia elegans* (p. 51); **J**. *Microcystis flos-aquae* (p. 52); **K, L**. *Microcystis aeruginosa* (p. 52).

Chroococcus turgidus (Kützing) Nägeli 1849

Basionym: *Protococcus turgidus* Kützing 1843

01130090 **Pl. 3A,B** (p. 41) **CD**

Colony microscopic, usually 2–8 cells, though occasionally up to 32 and very rarely solitary cells, with the outermost sheath layer forming the margin of the colony, which is usually very distinct. Cells spherical, subspherical, ovoid or hemispherical (after division), (6–)8–32(–45) µm. Sheath layers fairly distinct, usually developed each time the cell divides, though occasionally with more lamellations. According to Geitler (1932) and Komárek & Anagnostidis (1999), the sheaths are colourless, but very occasionally cells from British populations, which otherwise fit this species well, have pale brown sheaths. This is a feature of *C. schizodermaticus* and there is probably a continuum of forms between the two species.

A widespread species in moist terrestrial habitats, sometimes frequent among wetter *Sphagnum* communities and among submerged macrophytes in shallow lakes and pools of a wide pH, hardness and nutrient range, though most frequent in oligotrophic or mesotrophic, slightly humic waters, and apparently absent from eutrophic waters. Typical forms also occur in brackish ponds and, according to West & Fritsch (1927) in saltmarshes. Powell (1964), without details.

Chroococcus varius A.Braun in Rabenhorst 1861

011300100 **Pl. 3C** (p. 41)

Colony initially microscopic, but often forming large gelatinous groups of subcolonies of 2–4 cells embedded in communal mucilage, the whole structure being olive or brown when viewed macroscopically; communal mucilage usually has faint lamellations and ranges from colourless to rich yellow. Cells spherical, subspherical or hemispherical, 2–4 µm, typically forming a sheath each time they divide, giving a rather indistinct lamellate appearance within the subcolony. According to Komárek & Anagnostidis (1999), cell division usually occurs in different oblique planes.

Common species on wet rocks and walls, including greenhouses.

Clastidium Kirchner 1880

01140000

Cells solitary or in groups, attached to a substratum by the basal end of the cell, elongate, ovoid or pear-shaped, narrowed towards the apex, from which a gelatinous strand extends. There is a distinct mucilaginous sheath, but it is thin and may be overlooked. Cell division by successive transverse division at the apical end and the resulting spore is released and attaches itself to a substrate; the spore shows slow gliding motility and may remain close to the parent cell, giving rise to groups of cells in one area.

Clastidium setigerum Kirchner 1880

01140010 **Pl. 5A** (p. 49)

Cells solitary or in groups, attached by the base to the submerged substratum, various shapes ranging from [illegible] cylindrical to slightly club-shaped; sheath very thin and sometimes difficult to detect. Cells 2–4×8–15(–40) µm, with the gelatinous strand extending a further 50+ µm.

Standing or flowing freshwater, on algae, submerged bryophytes and other macrophytes, also submerged rocks.

Komárek & Anagnostidis (1999) recognize *C. rivulare* (Hansgirg) Hansgirg (syn. *C. setigerum* var. *rivulare* Hansgirg) as a separate species. The only distinguishing morphological features appear to be the much greater tendency for *C. rivulare* to occur in gelatinous groups and have a shorter length: width ratio (<2); *C. rivulare* is listed as forming yellow or brown spots in fast-flowing water or benthic in mountain tarns. There appears to be a continuum between the two and it is suggested that *C. setigerum* should be used for them all; the name has priority (1880 vs. 1890).

Coelosphaerium Nägeli 1849

01150000

Colony free-floating in plankton, spherical, or subspherical, with numerous quite closely packed cells forming a more or less single layer in the outer part of the mucilage. Cells spherical or subspherical (after division), with or without their own individual sheaths, but not attached to distinct mucilage-stalks. Cell division perpendicular to outer surface of colony. Mucilage homogeneous.

Other record Benson-Evans & Antoine (1996), without detail.

Coelosphaerium kuetzingianum Nägeli 1849

01150010 **Pl. 5B** (p. 49) **CD**

Colony free-floating, spherical or subspherical, up to 100 µm diameter, with cells restricted to a zone near the surface, though a narrow layer of mucilage is always evident between the cells and the outside of the colony; cells in a layer mostly only one cell thick, though there are usually a few other cells just inside this layer. Cells spherical or subspherical, 2–4 µm; gas-vacuolate when in the plankton. It is important to view different planes within the colony, as a casual glance might suggest that these cells occur throughout.

Occasional in plankton of mesotrophic lakes and large ponds.

Cyanodermatium Geitler 1933

Colony gelatinous or crust, microscopic or macroscopic; composed of more of less parallel rows of cells perpendicular to the surface. Cells irregularly polygonal-rounded, sometimes slightly elongate and surrounded by a thin sheath or layer of mucilage. Reproduction by release of terminal cells (a form of exospore), which are sometimes enlarged in comparison with the underlying cell. The structure is close to *Hydrococcus*.

Cyanodermatium fluminense (F.E.Fritsch) Komárek *et* Anagnostidis 1995

Basionym: *Pseudoncobyrsa fluminensis* F.E.Fritsch 1929

01590010

Colony flat, microscopic, about 70 µm thick, epilithic, composed of numerous erect rows of cells (pseudofilaments), widenened slightly towards the apex, with narrow sheaths which are usually pale yellow. Cells elongate-rounded, 2–3.5 μm wide, (2–)3.5–5 μm long, each with its own individual thin layer of mucilage and a distinct sheath surrounding the whole pseudofilament. Apical cells (exospores) mostly subspherical, 2.8–4 μm, with distinct yellow-brown sheath.

Epilithic on rocks in fast-flowing streams and small rivers. First described by Fritsch (1929).

This species was included under *Pseudoncobyrsa* in the *Coded List* (Whitton et al., 1998a). The original code number is retained until the next overall upgrade to the list.

Cyanodictyon Pascher 1914

01170000

Colony free-floating, initially spherical but soon becoming irregular, reticulate and usually slightly elongate. Mucilage diffluent at the margin, colourless. Cell division in one plane in successive generations. Cells subspherical or (after division) hemispherical.

Cyanodictyon imperfectum Cronberg *et* Weibull 1981

01170010

Colony microscopic, initially with short rows of cells, but later developing loose, net-like arrangement. Cells 0.4–1 μm, with blackish precipitations in the forms of crusts and granules; the precipitate is reported to incorporate ferric iron.

Planktonic in mesotrophic and slightly eutrophic freshwater. The deposition of ferric iron suggests that the organism is associated with waters with low *E*h ('redox potential') values and/or chelated iron, with the soluble iron becoming oxidized and precipitated in the mucilage. According to Komárek & Anagnostidis (1999) the organism is sometimes dense enough to form a weak green colouration of the water, but there are no records of such dense populations for the British Isles.

Cyanodictyon reticulatum (Lemmermann) Geitler 1925

Basionym: *Polycystis reticulata* Lemmermann 1898

01170020 **Pl. 5E** (p. 49)

Colony subspherical or slightly elongate, with cells scattered in the gelatinous strands which make up the colony. Cells 1–1.5 μm; no gas vacuoles.

Planktonic in mesotrophic and slightly eutrophic freshwater.

UNCERTAIN TAXON

Cyanonephron Hickel 1985

01180000

Colony subspherical or ellipsoidal, sometimes composed of subcolonies, with a system of mucilaginous, pseudodichotomously branched stalks radiating from the centre of a colony. Cells more or less elongate, solitary and crosswise at the ends of the divaricate stalks, more or less on the periphery of the mucilage of the colony; with gas vacuoles. The species described, *C. styloides* Hickel, has cells 0.8–1.2×2–5.5 μm.

Although it has not been reported from the British Isles, it should be looked for in the habitat where it occurs elsewhere: shallow, hypertrophic water bodies near the sea.

Dermocarpa Crouan 1858

01220000

Cells solitary or in irregular groups forming a small colony, almost always attached to a substratum, usually with a narrow firm sheath. During reproduction the whole or most of the cell divides by multiple fission into numerous baeocytes. These are usually released separately, but sometimes as a group with the individual cells separating subsequently. Baeocytes are initially motile when in contact with a surface, but then attach and start to differentiate into the characteristic shape of the mature cell. The dimensions given for individual species refer to the mature cell. Usually epiphyte, though occasionally epizooic.

Komárek & Anagnostidis (1999) separate most of the forms reported as *Dermocarpa* in other accounts into four genera:

- *Stanieria* Komárek *et* Anagnostidis 1986 Cells more or less spherical, not polarized, usually in irregular groups, attached to the substratum by any side;
- *Cyanocystis* Borzí 1882 Cells various shapes, but most often subspherical or hemispherical, attached to the substratum by one end, with first division perpendicular to the substratum;
- *Dermocarpella* Lemmermann 1907 Cells various shapes, but most often subspherical or cylindrical, with first division horizontal to the substratum, though later in various planes; see also León-Tejera & Montejano (2000);
- *Chamaecalyx* Komárek *et* Anagnostidis 1986 Cells club-shaped, attached to the substratum with one end, with first division transverse and then the upper part develops into baeocytes.

Other records Powell (1964) lists *D. enteromorphae* Anand 1937. Anand reported an oval to spherical cell 30–35 μm wide epiphytic on *Enteromorpha*; he interpreted it as nearest to *D. sphaeroidea* Setchell *et* Gardner, but larger and with a thick stratified wall. His description is too brief to assess the organism further, but Komárek & Anagnostidis (1999) note ? *Xenococcus*.

Other record for English Channel Coppejans (1995) lists several records of *Dermocarpa minima* Geitler 1932 – as *Cyanocystis minima* (Geitler) Komárek *et* Anagnostidis – for Belgium/North France (including Ostend), although Frémy (1929–33) has no records; this has small (5–7 μm) subspherical cells.

Dermocarpa aquae-dulcis (Reinsch) Geitler 1925

Basionym: *Sphaenosiphon aquae-dulcis* Reinsch 1874

Synonym: *Cyanocystis aquae-dulcis* (Reinsch) Kann 1978

01220010 **Pl. 5F–K** (p. 49)

Colony small, flat or approaching hemispherical, occasionally single or few-celled group. Mature cells slightly pear-shaped or ellipsoid, arranged parallel to each other, 6–8.5 μm diameter at widest point, 11–17 μm high. First cell division is vertical, but eventually whole cell divides into baeocytes, which are released through opening at the top.

Unpolluted, fast-flowing streams.

Dermocarpa hemisphaerica Setchell *et* Gardner in Gardner 1918

Synonym: *Cyanocystis hemisphaerica* (Setchell *et* Gardner) Kaas 1985

Cells solitary, flattened hemispherical, attached to substrate by flat base; 18–21 μm diameter at base, 10–13 μm high. Sheath thin, firm, not lamellate, colourless. First cell division is vertical, but eventually whole cell divides into baeocytes, which are released through opening at the top.

Powell (1964), based on own record. Epiphytic on marine algae or epipsammic.

Dermocarpa leibleiniae (Reinsch) Bornet *et* Thuret 1880

Basionym: *Sphaenosiphon leibleiniae* Reinsch 1874

Synonyms: *Dermocarpella leibleiniae* (Reinsch) J. *et* G.Feldman 1953; *Chamaecalyx leibleiniae* (Reinsch) Komárek *et* Anagnostidis 1986

01220020

Cells solitary or united in small colony. Cells initially subspherical, but becoming pear-shaped or irregular, reaching 8–20 diameter at base, 13–35 μm high; often much narrower end attached to the substratum. First cell division is horizontal, typically leaving a larger cell near the substratum, while remainder of cell divides into baeocytes; baeocytes released in cluster and separate later.

Marine. Powell (1964), without details.

Komárek & Anagnostidis (1999) regard the cell division as more like *Chamaesiphon* (hence *Chamaecalyx*) than typical baeocyte formation, but it is unclear whether this is always so.

Dermocarpa olivacea (Reinsch) Tilden 1910

Basionym: *Sphaenosiphon olivacea* Reinsch 1874

Synonym: *Cyanocystis olivacea* (Reinsch) Komárek *et* Anagnostidis 1986

01220030

Cells usually forming a small flattened or flat-hemispherical colony, with individual cells initially subspherical but becoming pear-shaped, narrowing towards the base, sometimes almost enough to resemble a stalk. Mature cell 9.5–17 μm at base, 13–25 μm high. Sheath firm, thick or thin, but lamellate; colourless. Baeocytes spherical, released by rupture of the sheath at the apex.

Powell (1964), without details. Marine, epiphytic on various seaweeds, recorded from various locations, though perhaps more frequent on the west coast.

Komárek & Anagnostidis (1999) also list another species with the same name, but different authority: *Dermocarpa violacea* Crouan 1878 = *Cyanocystis violacea* (Crouan) Komárek *et* Anagnostidis 1986. This has larger cells (8–7 × 16–31 μm) and older cells obovoid. They list both species from the Atlantic coasts of England.

Dermocarpa prasina (Reinsch) Bornet *et* Thuret 1880

Basionym: *Sphaenosiphon prasinus* Reinsch 1874

01220040

Synonyms: *Cyanocystis prasina* (Reinsch) Komárek *et* Anagnostidis 1986; *Dermocarpella prasina* (Reinsch) Komárek *et* Anagnostidis 1995

Colony flat or hemispherical, with densely packed, parallel and radially arranged cells. Cells obovoid or slightly club-shaped, 4–24 μm at base, 15–30 μm high, blue-green, olive or purplish; sheath thin, colourless. Initial cell divisions in horizontal planes, but subsequently in various planes to form baeocytes; baeocytes released by sheath becoming gelatinous.

Marine: perhaps the commonest blue-green algal epiphyte on seaweeds round the coast of the British Isles. Powell (1964), with two forms. Coppejans (1995) lists several records for Belgium/North France.

Dermocarpa sphaerica Setchell *et* Gardner 1918

Synonym: *Stanieria sphaerica* (Setchell *et* Gardner) Anagnostidis *et* Pantazidou 1991

Cells solitary or loosely packed to form a colony. Cells spherical, 6–16(–20) μm diameter, typically pale blue-green. Sheath narrow, not lamellate, colourless. Multiple fission leads to formation of baeocytes, which are released from the gelatinized sheath.

Powell (1964), without details. Epiphytic on marine sheathed blue-green algae and various eukaryotic algae. Frémy (1929–33) states especially in salt marshes, but Komárek & Anagnostidis (1999) state probably only on (in vicinity of) calcareous rocks of the littoral zone.

Entophysalis Kützing 1843

01260000

Colony irregular, gelatinous, granular, microscopic or macroscopic (consisting of many subcolonies). Cells or groups of cells with a marked tendency to be arranged in rows. Each cell with a distinct sheath, often lamellate. Cells spherical or rounded-polygonal in outline. Cells in three different planes in successive generations (according to Komárek & Anagnostidis, 1999).

Entophysalis granulosa Kützing 1843

01260010

Colony often macroscopic, reaching 1 mm, usually firm gelatinous, brown-black, consisting of sheathed cells, which are mostly arranged irregularly, but usually with some in indistinct rows. Cells subspherical, 2–5 μm, each with individual lamellate sheath, but also communal sheaths around groups of cells.

Marine, widespread on rocks in the supralittoral zone, sometimes forming almost a continuous layer.

Eucapsis Clements *et* Schantz 1909

01270000

Colony usually free-floating among submerged plants or overlying sediments, forming a cube of mucilage, with cells arranged three-dimensionally in perpendicular rows. Large colonies are sometimes composed of sub-colonies. Colonies capable of slow movement through silt.

Only one species has been recorded from the British Isles. Komárek & Anagnostidis (1999) list six species and a number of unrevised taxa, separating the former largely on cell size and habitat.

Eucapsis alpina Clements *et* Schantz 1909

01270010 **Pl. 5L** (p. 49)

Colony a floating cube up to 80 μm in any one dimension, occasionally up to 128 cells, usually considerably less. Cells spherical, 2–7 μm.

Planktonic and on sediments of shallow lakes, often slightly peaty, possibly most frequent in winter (though this requires confirmation). Widely distributed in the English Lake District, where it is most frequent on the sediments of peaty boggy tarns, especially some distance from their margins (Lund, 1950b). Woodhead & Tweed (1954) reported it from North Wales as both planktonic and on *Sphagnum* at the edge of lakes.

Komárek & Anagnostidis (1999) treat var. *minor* Skuja as a separate species = *E. minor* (Skuja) Elenkin. The latter has smaller (1.5–3.2× 2–3.2(–4) μm) cells than the type (5–7.5 μm), but more cells; it is unclear whether both taxa have been recorded in the British Isles. Lund (1950b) reported that colonies in the English Lake District are usually composed of only 8–32 cells and their cubical arrangement may be rather irregular; aggregates of several small colonies are not uncommon. The combination of large size and extreme regularity of cubical arrangement in the original figure of Clements & Schantz (1909) was not seen.

Gloeocapsa Kützing 1843

01290000

Colony usually epilithic, sometimes epiphytic; initially microscopic, but often developing into an amorphous macroscopic structure; mucilaginous, multicellular. Cells and their immediate groups surrounded by sheaths, which are usually obviously lamellate, although scarcely or not at all so in several species (separated as *Gloeocapsopsis* by Komárek & Anagnostidis, 1999). Cells spherical, subspherical or (after division) hemispherical.

The binomials used in Geitler (1932) have been retained here, because it is sometimes difficult to equate the revisions made by Komárek & Anagnostidis (1999) with ones made by other authors. Komárek & Anagnostidis (1999) separate forms where division occurs in three perpendicular planes in successive generations (=*Gloeocapsa* Kützing 1843 sensu Komárek & Anagnostidis, 1999) and forms where cell division is irregular in successive generations (=*Gloeocapsopsi*s Geitler ex Komárek 1993). As several other characters tend to occur among the latter group, further study may justify separation on this or similar lines. However, it is unlikely that most users will be confident in separating the two, so all are retained here under the genus *Gloeocapsa*.

Gloeocapsa forms the phycobiont of the lichen *Pyrenocollema* (Pentecost & Whitton, 2000).

Gloeocapsa alpina (Nägeli) Brand 1900

Basionym: *Gloeocapsa ambigua* Nägeli in Kützing 1849

01290010 **Pl. 4A,B** (p. 46)

Colonies microscopic, 2–8-celled, but frequently aggregated into macroscopic, irregular blackish mass. Cells of (sub-)colony lamellate, with the mucilage blue to dark violet; (sub-)colony up to 40 μm. Cells spherical, 4–6(–8) μm. Much larger cells sometimes present, with narrow, very firm, dark sheaths.

Limestone cliffs and rocks, mainly in hill and mountain regions, in drier regions largely restricted to moist surfaces; parts of limestone cliffs with intermittent drainage often have obvious black stripes largely due to this species and visible at long distances.

Gloeocapsa crepidinum Thuret *sensu* Geitler 1932

01290020

Colony often becoming macroscopic, gelatinous, irregular. Cells subspherical, (2.5–)4–8 μm, with thin nonlamellate sheath, pale to rich yellow-brown in the outer part of the colony, almost colourless in the middle of large colonies.

Powell (1964), without details. On rocks and wooden surfaces close to the sea, reaching to the immediate supralittoral.

This species has been widely recorded following the detailed account in Geitler (1932) and Coppejans (1995) lists several records for Belgium/North France. However, Komárek & Anagnostidis (1999) equate *Gloeocapsopsis crepidinum* (Thuret) Geitler ex Komárek with *Gloeocapsa crepidinum* (Thuret) Thuret, not *sensu* Geitler (1932).

Gloeocapsa magma (Brébisson) Kützing 1846 emend. Hollerbach 1924

Basionym: *Protococcus magma* Brébisson in Brébisson *et* Goday 1836

Synonym: *Gloeocapsopsis magma* (Brébisson) Komárek *et* Anagnostidis 1986

01290030

Small colonies subspherical, larger ones irregular, consisting of many subspherical colonies, usually 30–60 μm. Cells subspherical or often polygonal rounded due to mutual pressure, 3–7(–18) μm; individual sheaths distinct, golden-brown and sometimes faintly lamellate in the outer part of a colony, but lacking individual sheaths and the communal mucilage scarcely coloured. Enlarged cells (12+ μm) with cyanophycin granules and a dark reddish-brown sheath, which are often present in old colonies, are sufficiently distinct to be termed akinetes.

Wet non-calcareous rocks and other surfaces; apparently frequent in upland areas. West & Fritsch (1927) reported that the species is very abundant in (mainland) West Scotland and the Hebrides in the form of lobed, brownish-purple patches among stones on wet ground and locally known as 'mountain dulse'. The species is probably a nitrogen-fixer (or includes nitrogen-fixing strains).

Gloeocapsa polydermatica Kützing *sensu* Geitler 1932

Synonym: *Gloeocapsa caldariorum* Rabenhorst 1865

01290040

Colony microscopic or macroscopic, consisting of aggregations of closely appressed subcolonies. Cells spherical, 3–4.5 μm, sheath broad, colourless, very distinct lamellations, usually several per cell.

Wet rocks and walls, common, probably most typical of surfaces which are neither highly calcareous nor entirely free of calcareous enrichment.

Komárek & Anagnostidis (1999) point out the type material deposited by Kützing is *Coccomyxa*, but the original name is retained here until various other confusions are resolved.

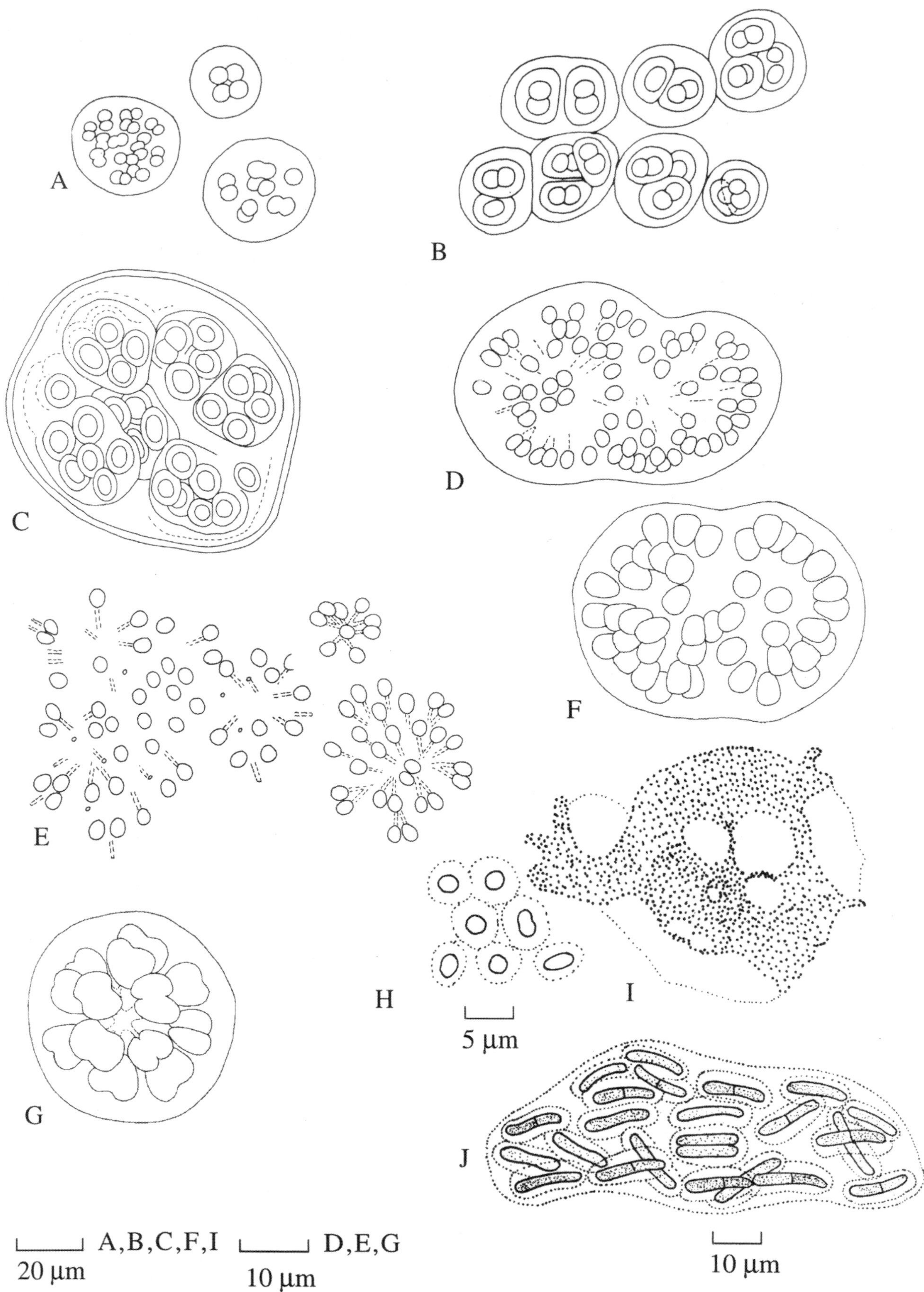

Plate 4 A–I
A, B. *Gloeocapsa alpina* (p. 45): A. Form lacking lamellation; B. Form showing sheath layers corresponding to all but most recent division; **C**. *Gloeocapsa sanguinea* (p. 47); **D, E**. *Snowella lacustris* (p. 56): D. General appearance of colony, E. Part of colony, showing stalks; **F, G**. *Gomphosphaeria aponina* (p. 47): F. Surface view of typical colony, G. Surface view, showing cordiform cells i.e. starting to divide; **H, I**. *Pannus spumosus* (p. 54); **J**. *Gloeothece linearis* (p. 47).

Gloeocapsa punctata Nägeli 1849

01290050

Colony microscopic or macroscopic, forming a dark green-grey mucilaginous mass, composed of numerous subcolonies. Cells 0.7–3 µm, with very wide colourless sheaths, not lamellate or with 1–2 faintly visible concentric layers; sheathed cell 7–17(–50) µm.

On limestone: wet rocks and walls, probably common.

The species is somewhat like a smaller-celled version of *G. polydermatica*.

Gloeocapsa sanguinea (C.Agardh) Kützing 1843

Basionym: *Haematococcus sanguineus* Agardh 1828–35

01290060 **Pl. 4C** (p. 46)

Colony microscopic or macroscopic, typically composed of many subcolonies, which each consist of many groups of cells. Often the innermost daughter cells have 2 or even 4 grouped within a single sheath system, but, outside this, each grouping of cells has its own lamellated sheath; sheaths vary markedly in how distinct is each lamellation and outer layer. The lamellated layers nearest the cells always have a reddish tinge and those of cells exposed to high light are an intense red colour. Larger (–10 µm), presumed resting cells, are sometimes present in old colonies.

Common, and the characteristic species of non-calcareous wet rocks and cliffs; probably more restricted to wet conditions than *G. alpina*, the species characteristic of calcareous rocks. West & Fritsch (1927) report both *G. sanguinea* and *G. ralfsii* (Harvey) Kützing = *G. ralfsii* (Harvey) Lemmermann. The main difference between the two is that the subcolonies of *G. sanguinea* occur in a variety of subspherical and ovoid shapes, often somewhat irregular, whereas the subcolonies of *G. ralfsii* are more or less spherical.

In view of the fact that *G. sanguinea* is a highly variable species and relatively little is known about the effects of environmental factors on its sheath and colony morphology, it is suggested that *G. ralfsii* should be included within *G. sanguinea*. Basionym taken from Kützing (1843, p. 175) and cited as '*Haematococcus sanguineus* Ag. (ex spec. orig.)'.

Gloeothece Nägeli 1849

01300000

Colony usually small, composed of sheathed cells or groups of cells. Cells broadly ellipsoidal or cylindrical, each with its own sheath, which is formed during or immediately after division; these sheathed cells are usually embedded in communal mucilage. Sheaths often lamellate; colourless, bluish or yellow-brown. According to Komárek & Anagnostidis (1999), cell division is transverse to the longitudinal axis, in one plane in successive generations.

Gloeothece confluens Nägeli 1849

01300030 (new number)

Colony an irregular mucilaginous mass, often macroscopic, consisting of single or paired cells inside a broad and quite well-defined sheath, but nevertheless confluent with sheaths of neighbouring groups. Cell cylindrical with rounded ends, 1.6–3×(4–)5.5–8 µm. Sheath not lamellate, colourless.

Aerophytic, widespread on wet rocks and walls and among mosses, though probably mostly pH values well below 7.

Gloeothece linearis Nägeli 1849

01300010 **Pl. 4J** (p. 46)

Cells usual single or in small groups with individual sheathed cells in a linear arrangement, but sometimes mass aggregations with partially gelatinized walls forming a macroscopic colony. Cell cylindrical, straight or slightly curved, 1.5–2.5×10–18 µm; sheath colourless, not lamellate.

Aerophytic, widespread on wet rocks and walls and terrestrial algae and mosses, especially old clumps which are starting to break down; apparently a wide pH range .

Gloeothece rupestris (Lyngbye) Bornet in Wittrock *et* Nordstedt 1880

Basionym: *Palmella rupestris* Lyngbye 1819

01300020

Colony a small group of typically 2–8(–32) cells, with a markedly lamellate sheath around individual cells and groups of cells within the colony; individual colonies form agglomerations, but not merging together to form a single large colony. Cell ellipsoidal or cylindrical, 3.6–4.6(–10)×5–9(–16) µm. Sheath colourless in shaded positions, but almost always yellow- or deep brown in well lit ones.

Aerophytic, widespread on wet rocks and walls and among mosses, though probably mostly at pH values well below 7.

Gomphosphaeria Kützing 1836

01320000

Colony spherical, subspherical or somewhat irregular in outline due to the presence of the subcolonies which form a stage in the formation of daughter colonies. Cells elongated in the radial axis and occurring at the ends of branching stalks which are much narrower than the width of the cell. Cells obovate, but remaining partially joined for part of the cell cycle, giving rise to a characteristic heart shape. There is always a distinct space beween adjacent single or paired cells.

Gomphosphaeria aponina Kützing 1836

01320010 **Pl. 4F,G** (p. 46) **CD**

Colony spherical, ellipsoidal or irregular when forming daughter colonies, up to 100 µm diameter, with moderately or densely arranged cells. Cells obovoid or slightly club-shaped, almost always with a number of heart-shaped cells present 4–6.5(–10 in heart-shaped state) µm wide, 8–12 µm long. Cells with or without gas vacuoles (see below).

Freshwaters, planktonic in mesotrophic or slightly eutrophic lakes and ponds or among algae and submerged vascular plants at the edge. The colony often includes other phototrophs (see Figs 391–394 in Canter-Lund & Lund, 1995).

Komárek & Hindak (1988) split this well-known species into two (*G. aponina* Kützing and *G. virieuxii* Komárek *et* Hindak 1988 = *G. aponina* var. *delicatula* Virieux 1916), but the descriptions in Komárek & Anagnostidis (1999) do not make really clear why they did this. They describe *G. aponina* (in the restricted sense) as a species without gas vacuoles occurring 'among algae and metaphytes (submerged macrophytes) in oligotrophic and mesotrophic swamps in littoral of freshwater lakes, springs, rarely tychoplanktonic or planktonic', while *G. virieuxii* occurs in 'plankton in clear or mesotrophic lakes, usually large ones'; no comment is made about gas vacuoles. However, they state that their account corresponds to the concepts of Huber-Pestalozzi (1938), whereas in the original description of the variety the cells were characterized by being pale and very narrow, with colonies 60–80 μm diameter; they suggest that it is probable that the taxon they describe is different from *G. virieuxii*. In view of the fact that there may be further nomenclatural changes and that there is no experimental evidence to show that the non-gas-vacuolate form widespread among submerged plants at the edges of many lakes cannot develop into the typical gas-vacuolate form of the plankton, it is recommended to retain, at least for the time being, the name *G. aponina* for the latter.

Hydrococcus Kützing 1833 (nomen cons.: see Komárek & Anagnostidis, 1999)

Synonym: *Oncobyrsa* Meneghini
01380000
Colony small, rarely exceeding 500 μm, more or less circular in outline, epiphytic or epilithic, consisting of densely packed cells which appear almost like a parenchyma. Colony typically flat initially, but becoming more or less hemispherical when old. Old margins of colony with short radially arranged rows of cells; cells in middle of colony are arranged in vertical rows, though this arrangement sometimes becomes less obvious in old colonies. Cell spherical, broadly ellipsoidal or more elongate and curved at the outer end of a radial row.

The code number was originally allocated to *Oncobyrsa*.

Other record Powell (1964): *Hydrococcus marinus* Grunow 1861. Komárek & Anagnostidis (1999) list this as one of two unrevised species in the genus; their brief description (based on Grunow's material from Greece) suggests the species is much like a slightly larger version of *H. rivularis*, but epiphytic on marine algae.

Hydrococcus cesatii Rabenhorst 1860

Synonym: *Oncobyrsa cesatiana* (Rabenhorst) Rabenhorst 1865
01380010
Colony epiphytic, starting as small group of cells growing radially, with some of the outer ones elongate, but irregular-arcuate shape; the cells then form parallel rows of about 8 cells vertical to the surface; eventually this organization partially disappears, giving a parenchymatous appearance in section; mature colony microscopic, largely flat, but flattened hemispherical at the edges. Cells in mature colony mostly 1.2–2.5 μm, but marginal cells of young colony reaching 4.5 μm; cells blue-green, olive or slightly purple; mucilage colourless, with sheaths of individual cells mostly too narrow to distinguish.

Epiphytic on larger filamentous algae, submerged mosses and some flowering plants in freshwaters, especially mesotrophic, fast-flowing calcareous rivers and larger streams.

Hydrococcus rivularis Kützing 1833

Synonym: *Oncobyrsa rivularis* (Kützing) Meneghini 1846
01380020 **Pl. 5M–Q** (p. 49)
This is essentially a larger version of *H. cesatii*. The colony develops in much the same way, but the vertical rows sometimes dichotomize, though soon become disorganized; in thicker parts of mature colonies there can be 10–15 layers of cells between the inner and outer surfaces; mature colonies spread out or hemispherical. Cells in mature colony (2.5–)3–6 × 3–6.5(–7.5) μm; cells blue-green, olive or slightly purple; mucilage colourless, with sheaths of some individual cells usually distinguishable in older colonies; occasionally slight calcite precipitation within the colony.

Epiphytic on larger filamentous algae, submerged mosses and flowering plants in freshwaters, especially in mesotrophic, fast-flowing calcareous rivers and larger streams; often also present on other substrata, especially where common as an epiphyte. Powell (1964), without details. Komárek & Anagnostidis (1999) state that records from marine biotopes (e.g. England) concern evidently other species, but provide no further explanation.

Hyella Bornet *et* Flahault 1888

01410000
Thallus ('colony') on calcareous rocks, mollusc shells or other calcareous substrata, usually clearly differentiated into a surface growth of irregular short branching pseudofilaments and endolithic growth reaching to various distances in the substrate. Pseudofilaments with one or more rows of cells, though the structure less obvious in older parts of the thallus. The endolithic pseudofilaments, which bore into the substrate, are usually morphologically quite different from the initial pseudofilaments. Mucilage sheaths are firm, usually layered and generally evident between adjacent cells.

In addition to the two marine species listed here, there are about a dozen other species described. Most records for these are for warmer climates, but studies on the possible diversity of populations in the British Isles are lacking.

Hyella balani Lehmann 1903

01410010 **Pl. 7L,M** (p. 55)
Thallus with a few cells at the surface, but largely endolithic. Irregularly arranged cells near surface, but further into the substrate forming rows vertical to the surface, multiseriate near the surface of the substratum, but a single row of cells (pseudofilament) further inside. Cells at surface 7–8 × 10–13 μm, cells in pseudofilament 4–8 μm wide, reaching 20 μm long in the terminal cell i.e. furthest into the substratum. Sheath firm, colourless, sometimes lamellate. Reproduction in larger cells at the surface typically by

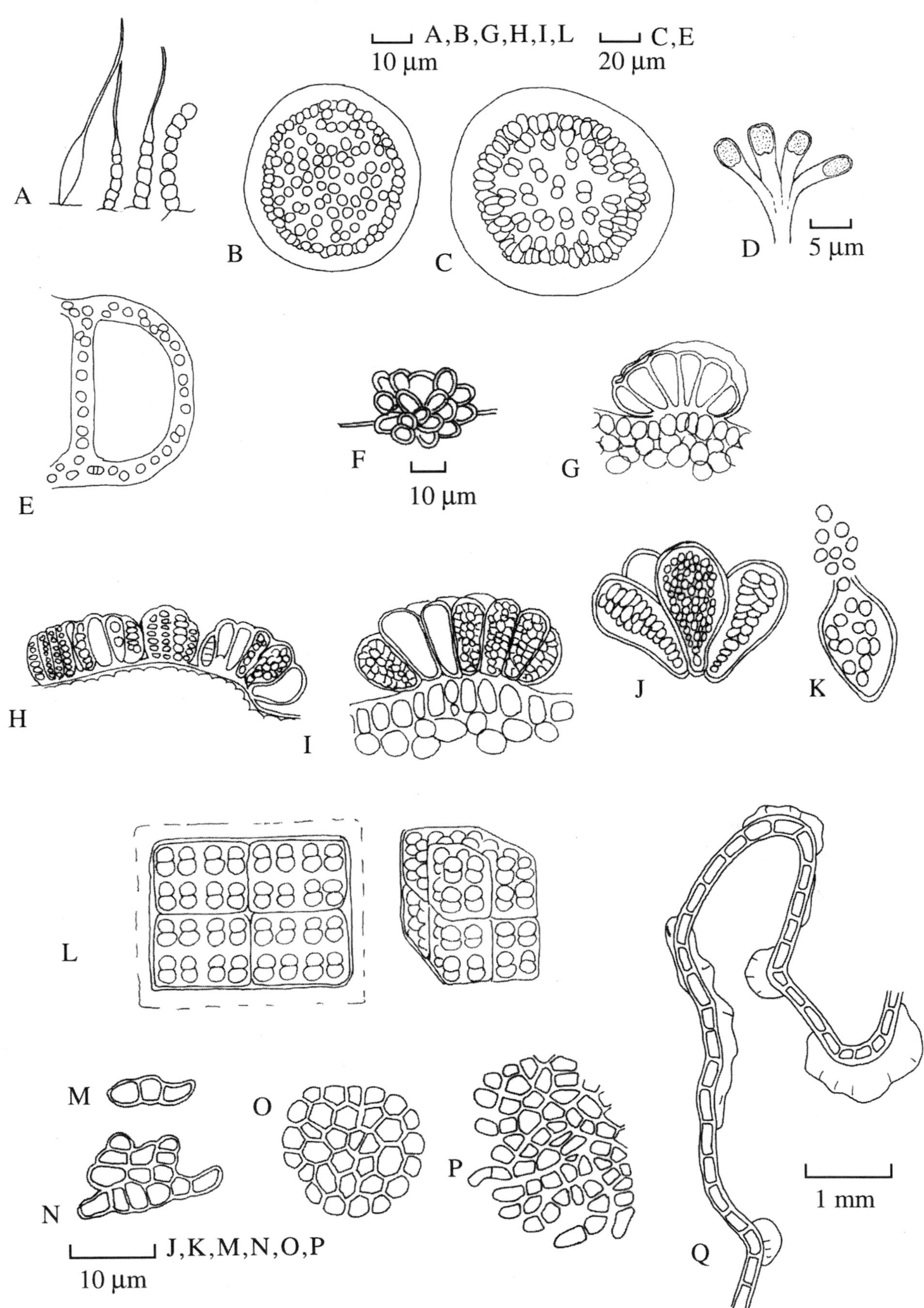

Plate 5 A–Q
A. *Clastidium setigerum* (p. 42); **B**. *Coelosphaerium kuetzingianum*; **C, D**. *Woronichinia naegeliana* (p. 58); **E**. *Cyanodictyon reticulatum* (p. 43); **F–K**. *Dermocarpa aquae-dulcis* (p. 43): F. Group of young cells, G. Groups of cells attached to host alga, H–J. Differentiation of baeocytes, K. Release of baeocytes; **L**. *Eucapsis alpina* (p. 45): as illustrated in original study: British material smaller and less regular; **M–Q**. *Hydrococcus rivularis* (p. 48): M–O. Young colonies viewed from above, P. Marginal part of characteristic colony, viewed from above, Q. Macro-view of colonies attached to filamentous alga.

multiple division to form spherical baeocytes 1.5–3.5 μm, but other patterns of cell division often occur.

Marine: intertidal, on limestone rocks and shells.

Hyella caespitosa Bornet *et* Flahault 1889

01410020 **Pl. 7N,O** (p. 55)

Thallus often with sufficient cells at the surface to form a 1–2 mm (macroscopic) yellow-brown spot, but largely endolithic; immediately below the surface the pseudofilaments often run mostly more or less parallel to the surface, though wavy and coiled, and eventually many of the cells dividing irregularly to form clusters. The horizontal filaments at surface also give rise to pseudofilaments which penetrate into the substratum; these are initially multiseriate, but further inside are unseriate and with the cells usually getting increasingly long, the further into the substratum. Cells in pseudofilaments 4–10 μm wide, reaching 40–60 μm long in the cell furthest from the surface; cells blue-green or (in deep sublittoral material) purple; sheath mostly firm, though sometimes gelatinizing at the margin, colourless, lamellate. Reproduction by release of individual cells from the surface without multiple division or by baeocyte formation.

Marine: lower intertidal and sublittoral on limestone rocks and shells. Listed by Powell (1964), but with note that this includes *H. balani*. Part of the chalk-boring community on cliffs at Westgate, Kent (Anand, 1937).

Hyella fontana Huber *et* Jardin 1892

01410030 **Pl. 7P,Q** (p. 55)

Thallus largely endolithic, though usually possible to distinguish macroscopically areas of rock with the species; the densely packed cells immediately below the surface superficially like parenchyma, but giving rise to pseudofilaments further into the substratum or at the side of the thallus; pseudofilaments with long branches nearer the surface and short ones further inside. Cells near the surface highly variable in shape and size, but mostly 5–10 × 7–18(–23) μm; pseudofilament cells 4–7 μm wide, 5–20 μm long, with the longer cells near the branches and furthest into the substratum; sheath firm, thin, colourless, not lamellate. Reproduction (at least usually) by baeocytes, which are often very small in comparison with mother cells, typically 1.2–1.5 μm. The species is highly variable, perhaps influenced by the nature of the substratum.

In limestones and other calcareous materials in oligotrophic fast-flowing streams, often at edge of waterfalls; widespread, but not always present where might be expected.

Johannesbaptistia De Toni 1934

01420000

Rows of cells embedded in mucilage and with mucilage separating adjacent cells (pseudofilament); pseudofilaments occasionally branched. Cells ovoid, with long axis perpendicular to the axis of the pseudofilament. Mucilage envelopes colourless, structureless, usually with the margin more or less distinct.

Johannesbaptistia pellucida (Dickie) Taylor *et* Drouet 1938

Basionym: *Hormospora pellucida* Dickie 1874

01420010

Colony a pseudofilament with a single row of cells embedded in mucilage, (6–)8–12(–20) × 300–800 μm, but occasionally much longer (to 2500 μm); pseudofilament sometimes dividing into two or even (rarely) anastomosing to form a more complex structure. Cells discoid, 4–5.5 × 2.3–3 μm as seen in side view. Mucilage usually with a distinct outer boundary, colourless; individual cells usually without their own envelope and, if present, indistinct.

Typically in brackish pools near the sea; occasionally also in small inland pools likely to dry at times and, usually, but not always, where evidence of enhanced salinity; probably more common in summer. Not listed by Frémy (1929–33), but apparently widespread in the British Isles.

Merismopedia Meyen 1839

01460000

Colony microscopic or rarely macroscopic, forming a square or rectangular plate of cells, often with several subcolonies remaining adjacent. Cells arranged in perpendicular rows, close or distant from each other. Cells spherical, elongate or ovoid, with the longer axis (where present) in the plane of the colony. Mucilage colourless and colony margin rather indistinct; cells sometimes with individual, though rather indistinct, sheath. Reproduction by break-up of colonies. Colonies capable of slow movement through silt.

There is a great range of forms and a large number of species have been recognized: 22 in Komárek & Anagnostidis (1999). However, few are well delimited and many appear to merge into one another. The names here have all been recorded a number of times in the British Isles and many, but not all, new records should fit quite well into one of these. Pink forms (due to dominance of phycoerythrin) are frequent in strongly shaded microhabitats, but these are very difficult to distinguish from the photosynthetic bacterium *Thiopedia* without the use of epifluorescence microscopy (presence of chlorophyll in the blue-green alga). If epifluorescence is unavailable, it may be possible to confirm *Merismopedia*, if a sample with the suspect colonies is transferred to a slightly higher light intensity for several days; *Merismopedia* (but not *Thiopedia*) colonies may become more olive.

The cells of some colonies are very pale (see Plate 1b in Whitton & Potts, 2000). The reasons for this are unclear. Possibly they are sometimes non-photosynthetic and hence not *Merismopedia*, but epifluorescence microscopy shows that (chlorophyll-containing) *Merismopedia* can often appear like this. Study is needed to establish whether this is simply a matter of low chlorophyll content or whether there is an additional factor, such as would occur if there were small liquid vacuoles.

Motility has been studied in *M. elegans* and *M. glauca* (Lund, 1950b). The maximum rate of movement appears to vary inversely with size of the colony, with movement of large colonies being hard to detect under the microscope. Movement is favoured by light, but under high light the colonies show a swinging movement such that the narrow edge of the plate is presented to the light source. Presumably movement results from the sum of contributions by individual cells. In that case it seems possible that individual cells might move and aggregate, leading to formation of new plates, in addition to the fragmentation mentioned above.

Of the species listed here, *M. elegans*, *M. glauca*, *M. punctata* and *M. tenuissima* are especially widespread and differ mainly in cell width; they more or less cover the cell size range found within the genus. *Merismopedia* has been reported to form substantial populations in the region of the thermocline of large lakes in some countries (though not British Isles), but at least some of these may have been *Thiopedia*.

Merismopedia convulata Brébisson in Kützing 1849

Synonym: *Merismopedia revoluta* Askenasy 1894
01460050 (new number)
Colony macroscopic, up to 4 mm (though usually 1–2 mm); the thin sheet usually with folds and convolute; green, bluish or bluish-yellow. Cells spherical, 4–5 μm diameter, or elongate 4–5 × 5–8 μm.

In shallow standing or slow-flowing, fresh or slightly brackish water; on sediments or mixed with other algae, usually at sites with submerged flowering plants. Powell (1964), without details. Powell also lists *M. revoluta*, which is treated as a synonym here, because the original descriptions of the two are almost identical, apart from a lower cell diameter limit of 3 μm in *M. revoluta*.

Merismopedia elegans A.Braun in Kützing 1849

01460010 **Pl. 3I** (p. 41)
Colony small to large (16–4000 cells), usually rectangular, with cells quite densely arranged in perpendicular rows; mucilage with a distinct margin about one cell length beyond outer cells. Cells subspherical, ellipsoidal or (after division) hemispherical, 4–7 × 5–9 μm.

Wide range of fresh- and brackish waters, usually on the surface of silt, but also often among submerged macrophyes and sometimes washed into the plankton.

Merismopedia glauca (Ehrenberg) Nägeli 1845

01460020 **Pl. 3H** (p. 41) **CD**
Colony up to 64 cells, regularly and quite densely packed. Margin of mucilage distinct, extending slightly beyond outer cells. Cells spherical, widely ellipsoidal or (after division) hemispherical, (2.8–)3–6 μm diameter; cells often rather pale and sometimes pinkish.

Wide range of fresh- and brackish waters, on the surface of silt, among submerged macrophytes or occasionally washed into the plankton; typically mesotrophic conditions.

This species is scarcely distinguishable from *M. punctata*, apart from a tendency for the cells to be larger and less spaced apart.

Merismopedia punctata Meyen 1839

01460030 **CD**
Colony up to 64 cells, regularly, but somewhat loosely arranged. Margin of mucilage distinct, extending slightly beyond the outer cells. Cells spherical, ellipsoidal or (after division) hemispherical, (2–)2.5–4 μm diameter; colour of cells often rather pale.

Wide range of fresh- and brackish waters, on the surface of silt, among submerged macrophytes or occasionally washed into the plankton; typically mesotrophic conditions.

This species is scarcely distinguishable from *M. glauca*, apart from a tendency for the cells to be smaller and more spaced apart.

Merismopedia tenuissima Lemmermann 1898

01460040
Colony flat, usually rectangular, often closely aggregated into groups of colonies, each typically with 16 cells, though sometimes more. Cells usually quite close to each other, spherical, ellipsoidal or (after division) hemispherical, 0.4–1.6 μm.

Wide range of eutrophic fresh- and brackish waters, near the bottom silt and in the water column, often above sediments with reducing conditions.

Merismopedia trolleri Bachmann 1920

01460060 (new number)
Colony usually 8–16 cells. Cells close to each other, spherical, ellipsoidal or (after division) hemispherical, 2–3 μm diameter, usually with narrow individual sheath; often with several gas vacuoles. This species is scarcely distinguishable from *M. glauca* apart from the presence of gas vacuoles.

Freshwater lakes, planktonic. Planktonic *Merismopedia* has been recorded in the British Isles, but it is unclear whether such records include forms with gas vacuoles.

Microcrocis P.G.Richter 1892

Synonym: *Holopedium* Lagerheim 1893
01350000
Colony microscopic or macroscopic, with cells forming a plate embedded in mucilage. Cells elongate, with their longer axis perpendicular to the plane of the colony, often arranged in perpendicular rows in young colonies, but later becoming irregular. Cells ellipsoidal, ovoid or rod-shaped, with rounded ends; cells divide lengthwise in two perpendicular planes in successive generations. Reproduction by break-up of colonies (but see comments on *M. geminata*). Colonies capable of slow movement through silt.

Cell division and colony reproduction appear to be more complex in *M. geminata*, but it is unclear whether the features of this species also apply to some or all of the other species.

Microcrocis was originally treated as a subgenus of *Merismopedia*, but differs in the elongated cells and the irregular arrangement of the densely packed cells in older colonies.

Microcrocis geminata (Lagerheim) Geitler 1942

Basionym: *Merismopedia geminata* Lagerheim 1883

01350010 **Pl. 6A** (p. 53)

Colony reaching 3 mm, flat, sometimes wavy or even rolled, with an irregular margin. Cells packed densely; in very young colonies in indistinct rows, but soon becoming irregular; the cell total may reach 3000 (Komárek & Anagnostidis, 1999). Cells rod-shaped, with rounded ends, 3.4–6(–7)×12–16 μm, with the long axis perpendicular to the plane of the colony. In addition to the longitudinal divisions, the rods sometimes divide transversely, but asymmetrically.

Freshwater, in ponds and shallow areas of lakes with submerged macrophytes, overlying the bottom silt or occasionally washed into the plankton. In a study of ponds in Richmond Park (near London), Lund (1942b) showed that the whole colony moved slowly back to the surface when it was buried in mud.

Detailed studies appear to be lacking on the fate of the smaller cells formed by transverse division, but it seems likely that these act as spores, which aid dispersal, and possibly also aggregate to form new colonies.

Microcrocis irregularis (Lagerheim) Geitler 1942

Basionym: *Merismopedium* (subgenus Holopedium) *irregulare* Lagerheim 1883

01350020 (new number)

Colony microscopic, flat or sometimes wavy or rolled up, with irregular outline. Cells in indistinct rows in very young colonies, but soon becoming irregular. Cells 1.5–3×2–4.5 μm.

Ponds and lakes, usually among submerged macrophytes, sometimes overlying sediments or planktonic. Benson-Evans & Antoine (1996): as *Holopedium irregulare*, without detail.

This is like a much smaller form of *M. geminata*.

Microcrocis subulicola (Lagerheim) Geitler 1942

Basionym: *Merismopedium* (subgenus Holopedium) *subulicolum* Lagerheim 1883

01350030 (new number)

Colony reaching a few hundred cells, at which stage it is flat or slightly wavy, with irregular outline. Smaller ones attached to sand particles, but larger ones detached, though still part of the general epipsammon community. Cells packed densely, initially in rows, but later irregular. Cells short rod-shaped, 3–4(–5)×(3–)5–6(–8) μm.

Powell (1964), without details. Sand and coarse silt (salt marsh) in sheltered brackish and marine environments.

It is unclear whether there is any real difference between this species and *M. geminata*.

Microcystis Kützing 1833 ex Lemmermann 1907 *nom. cons.*

01490000

Colony irregular or (less often) subspherical consisting of large numbers of cells embedded in mucilage; colony increase by fragmentation. Cells spherical or hemispherical after division; gas vacuoles present when the colonies are planktonic, though gas vacuoles apparently absent or few when colonies rest on bottom sediment. Cell division probably always in three planes perpendicular to each other, though hard to distinguish in field samples.

The descriptions adopted for the three species here reflect the accounts of Komárek & Anagnostidis (1999). The species differ in the extent to which the outer margin of the colony is firm, whether or not it contains an outer zone free of cells, how densely the cells are packed in the rest of the colony and, to a lesser extent, on the size of the cells. Forms are likely to be found which do not correspond particularly well to any of the three descriptions. Photographs and notes should be made of populations the taxonomic status of which is unsure. Komárek & Anagnostidis (1999) recognize ten species, but most are highly variable. Further, molecular studies have given little indication that the classical species correspond well with observed molecular differences. It seems likely that different species limits will eventually be adopted than those here, if in fact it is ever possible to recognize species adequately from the morphological features of field material.

Information about basionyms and synonyms for species in this genus is especially difficult to summarize and is best consulted in the accounts by Komárek and Anagnostidis (1999) and the references included there.

Microcystis aeruginosa (Kützing) Kützing 1846

01490010 **Pl. 3K,L** (p. 41) **CD**

Colonies in a well-developed plankton population typically 0.5–1 mm long, but may reach a much larger size. They may start with a subspherical shape, but eventually become more elongated and usually clathrate, sometimes markedly so. A large number of quite densely packed cells embedded in colourless mucilage, with outer margin of mucilage diffluent, though usually distinct enough to be recognized by light microscopy; cells usually scattered throughout the mucilage, but usually with a narrow outer zone free of cells. Cells spherical or subspherical, 4–6(–9) μm, gas-vacuolate while colonies are in the plankton.

Fresh to moderately brackish waters, often forming dense blooms in mid- to late summer, falling to the bottom sediments in autumn. Reynolds & Rogers (1976) describe its ecology in Rostherne Mere.

Microcystis flos-aquae (Wittrock) Kirchner 1898

01490020 **Pl. 3J** (p. 41)

Colonies usually only microscopic, spherical or irregular, occasionally with small holes, but without lobes or obvious clathrate structure. Cells very densely packed throughout the colony, with no outer cell-free layer. Cells spherical or subspherical, (2.5–)3.5–5 μm, gas-vacuolate while colonies are in the plankton.

Freshwaters, often forming a component of blooms, but typically of less nutrient-rich waters than *M. aeruginosa*.

Microcystis wesenbergii (Komárek) Komárek in Kondrateva 1968

Basionym: *Diplocystis wesenbergii* Komárek 1958

01490030

Colonies mostly macroscopic, spherical when young, but soon becoming elongate, much lobed, and clath-

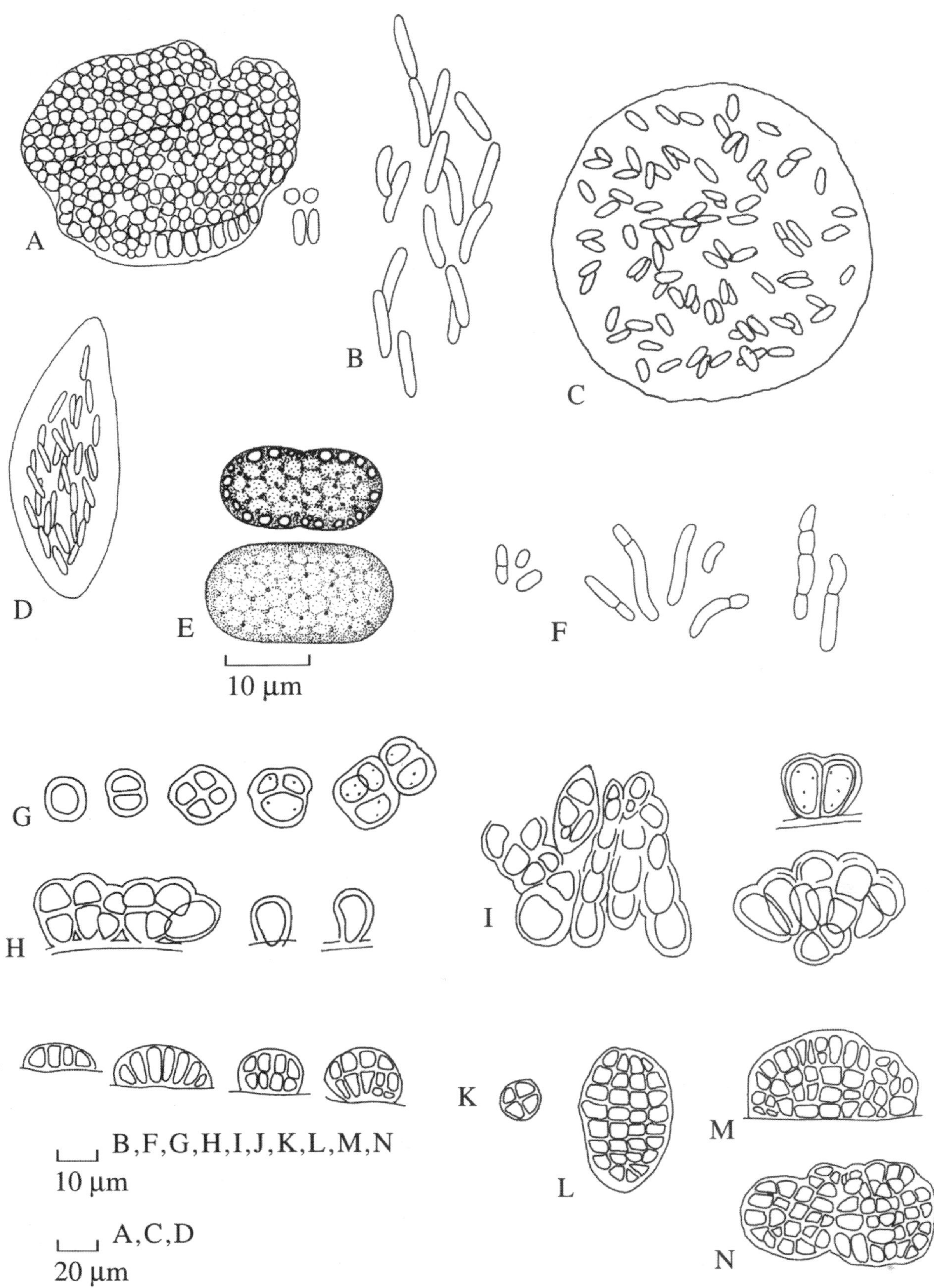

Plate 6 A–N

A. *Microcrocis geminata* (p. 52); **B**. *Rhabdoderma lineare* (p. 56); **C**. *Rhabdogloea planctonica* (p. 56); **D**. *Rhabdogloea smithii* (p. 56); **E**. *Synechococcus aeruginosus* (p. 57); **F**. *Synechococcus elongatus* (p. 57); **G–I**. *Chroococcopsis chroococcoides* (p. 39); **J–N**. *Xenococcus kerneri* (p. 58).

rate. Large colonies usually consist of a number of distinct subcolonies. Cells only moderately densely packed in colony; outer margin of colony clearly delimited. Cells spherical or subspherical, 4–7(–9) μm diameter, gas-vacuolate while colonies are in plankton.

Freshwaters, often forming a component of blooms, but seldom dominant.

Pannus Hickel 1991

01540000

Colony microscopic, free-floating, hollow, in young stages irregularly spherical with holes, later more irregular, in old stages usually elongate and lobate. Cells in 1–3 layers of often more or less radially, densely distributed cells, embedded in colourless mucilage; often a highly diffluent layer of mucilage external to the main part of the colony. Cells spherical, sometimes with gas vacuoles. Cell division in two planes in successive generations.

Pannus spumosus Hickel 1991

01540010 (new number) **Pl. 4H,I** (p. 46)

Colony more or less spherical when young, hollow, later irregular, with a number of holes, to 180 μm in largest dimension; cells forming a dense single layer at outside of colony. Cells spherical, 1–1.5 μm diameter, with own narrow indistinct mucilage layer; without gas vacuoles.

Planktonic in ponds and reservoirs. Hickel (1991) recorded it from brackish bays and reservoirs in North Germany, but Komárek & Anagnostidis (1999) mention two other (non-brackish) records, about which they comment 'probably refer to another species'. However, the author has seen material like Hickel's from two freshwater sites, only one of which might have been slightly brackish due to run-off from roads.

Pleurocapsa Thuret 1885

01570000

Colony attached to a surface, as an irregular individual, crust or a larger irregular group of more or less distinct individual colonies. Growth occurs in three different ways, two or all of which are usually present in the same colony:

(i) irregular groups of cells (chroococcalean stages);
(ii) radial rows of cells which form 'pseudofilaments' when well developed; pseudofilaments may be uni- or multiseriate;
(iii) baeocyte formation.

The generic limit is more restricted than that in Geitler (1932) and *P. fluviatilis* has been transferred to *Chroococcopsis*.

Other records Powell (1964) lists *P. entophysaloides* Setchell *et* Gardner in Gardner 1918, without details. Komárek & Anagnostidis (1999) list this as an unrevised species, but one which corresponds better to *Radaisia* Sauvageau, where the erect pseudofilaments are unified into a gelatinous mass.

Pleurocapsa crepidinum Collins 1901

not *Pleurocapsa crepidinum* (Thuret) Ercegović

01570010 **Pl. 7G,H** (p. 55)

Cells spherical or polygonal-rounded due to mutual pressure, 5–15(–25) μm; thin colourless sheaths. Cells often aggregated into groups. Short pseudofilaments usually present in older colonies; as these structures tend to grow into the substrate, they may be lost if samples are scraped from the surface of a rock or shell.

Powell (1964), without details. Marine: surfaces of rocks and shells, especially if calcareous; also sometimes on surface of algae.

Although *P. crepidinum* is the name most frequently recorded for marine populations in the British Isles, there are also records for the closely similar *P. fuliginosa* Hauck 1885. In addition there may well be confusion between this species and the organism treated as *Gloeocapsopsis crepidinum* (Thuret) Geitler ex Komárek.

Pleurocapsa fluviatilis Lagerheim 1888

11570020: see *Chroococcopsis fluviatilis*

Pleurocapsa fuliginosa Hauck 1885

01570040 (new number)

Colony a thin blackish crust, irregular in outline, up to 100 μm across and consisting of groups of cells and pseudofilaments. Cells irregular-spherical (3–)5–20 μm, often in groups, blue-green, brownish or violet. Sheath up to 100 μm thick, slightly lamellate, colourless.

Powell (1964), without details, but Frémy (1929–33) lists Dorset. Supralittoral and upper intertidal.

Pleurocapsa minor Hansgirg 1891

01570030 **Pl. 7A–F** (p. 55) **CD**

Colony consists of irregular groups of creeping or slightly erect cells, the latter pseudofilaments sometimes branching as they elongate. Cells in main mass 3–12 μm in the largest dimension, asymmetrical; cells of pseudofilaments 3–9 μm wide, some or all of which eventually form endospores; endospores 0.8–1.5 μm.

Freshwater, usually moderately calcareous. Records for the species come from two quite different types of environment: (i) streams, small rivers and edges of lakes, with the organism usually on or among calcareous substrata; (ii) on rocks among other algae in very nutrient-rich hardwater rivers, extending to the uppermost part of the tidal zone. Possibly these reflect two different species.

GENUS LISTED WITHOUT RECORDS

Pseudoncobyrsa Geitler 1925

01590000

Geitler (1932) included two species within this genus, but Komárek & Anagnostidis (1995) transfer the one recorded for the British Isles (Fritsch, 1929) to *Cyanodermatium fluminense* Geitler 1933 and this is followed here.

Pseudoncobyrsa fluminensis F.E.Fritsch 1929

01590010: see *Cyanodermatium fluminense*

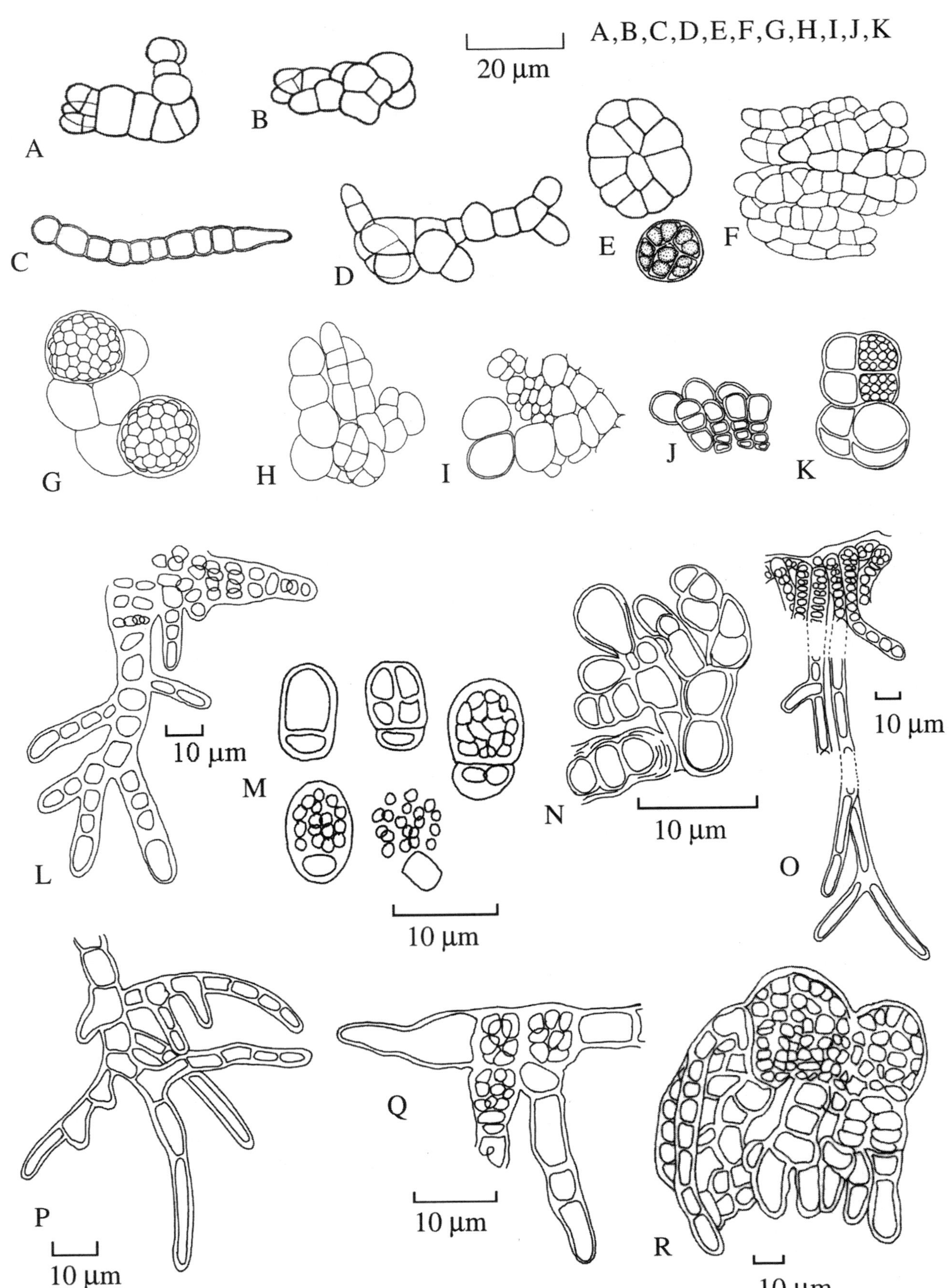

Plate 7 A–Q
A–F. *Pleurocapsa minor* (p. 54); **G, H.** *Pleurocapsa crepidinum* (p. 54); **I–K.** *Chroococcopsis fluviatilis* (p. 39); **L, M.** *Hyella balani* (p. 48); **N, O.** *Hyella caespitosa* (p. 50); **P–R.** *Hyella fontana* (p. 50).

Rhabdoderma Schmidle *et* Lauterborn 1900

01610000
Colony microscopic, more or less ovoid or elongate-irregular, mucilaginous. Cells usually well spaced in colony and often more or less oriented in the direction of the longer axis of the colony. Cells rod-shaped, straight, slightly curved or sigmoid, longer than broad, often several times so. The cells, which divide transversely to the long axis, sometimes remain in loose contact after division, forming a filament-like structure.

Rhabdoderma lineare Schmidle *et* Lauterborn 1900

01610010 **Pl. 6B** (p. 53)
Colony small, irregularly elongate, with mucilage boundary usually distinct, though sometimes diffluent; colonies typically with up to 16 cells, though occasionally considerably more. Cells cylindrical or slightly arcuate, (0.8–)1.5–3(–3.5) × 4–12(–22) μm; cells with marked or slight tendency to be orientated in the same direction; mucilage colourless, cells occasionally with thin individual sheath.

Plankton of shallow oligotrophic and mestrophic lakes and ponds, perhaps especially softer waters with submerged macrophytes. Benson-Evans & Antoine (1996), without detail.

Rhabdogloea Schröder 1917

01210000
Colony microscopic, mucilaginous, irregular in outline, though often slightly elongate, usually free-floating; cells spaced irregularly in the mucilage, with few to several hundred cells (depending on species). Cells elongate, narrow, spindle-shaped, ellipsoidal or more or less cylindrical with acute or acute-round poles, straight, slightly bow-shaped or sigmoidally coiled; gas vacuoles absent. Cells with tendency to be oriented in the same direction, especially in smaller colonies. Cell division binary, transverse to the long axis, giving rise to two more or less equal daughter cells. The only character separating the genus from *Rhabdoderma* is the shape of the cells.

Rhabdogloea planctonica (Teiling) Komárek 1983

Basionym: *Dactylococcopsis planctonica* Teiling 1942
01210010 **Pl. 6C** (p. 53)
Colony subspherical, usually with 50–200 cells spaced irregularly in the mucilage; mucilage colourless, boundary distinct; Cells elongate, usually slightly arcuate and asymmetrical, with one side more convex than the other, narrowed at the ends, 3–5 × 12–20 μm.

Planktonic in lakes.

Rhabdogloea smithii (R. Chodat *et* F.Chodat) Komárek 1983

Basionym: *Dactylococcopsis smithii* R. Chodat *et* F.Chodat 1925
01210020 **Pl. 6D** (p. 53)
Colony ellipsoidal-irregular, usually 4–30 cells widely spaced, but orientated in the direction of the long axis of the colony; mucilage colourless, usually diffluent at the boundary. Cells irregularly spindle-shaped, slightly curved, tapering to ends, 1.4–3 × 7–15(–25) μm.

Plankton of oligotrophic waterbodies, especially large lakes. Benson-Evans & Antoine (1996), without detail.

Other record (?) Benson-Evans & Antoine (1996) list *Dactylococcopsis montana* West *et* G.S.West, a species not in Komárek & Anagnostidis (1999).

Siphononema Geitler 1925

01650000
Colonies attached to substratum, microscopic to macroscopic. Initial cell cylindrical, subspherical or club-shaped, with colourless, yellow or orange sheath, subsequently dividing in upper part to form pseudofilaments or irregular masses of cells embedded in mucilage.

Siphononema polonicum (Raciborski) Geitler 1925

Basionym: *Pleurocapsa polonica* Raciborski 1910
01650010 **Pl. 2J** (p. 37)
Colony rust-orange to red-brown in colour, composed of clusters of cells arranged in a pseudofilament of one or many-celled rows, 15–30(–54) μm wide and up to 300 μm long. Cells subspherical, ovoid or long club-shaped, 2–7.5 μm wide, 2–90 μm long, surrounded by individual yellow-brown sheath, which is sometimes thick and lamellate. Akinete 7–8 μm wide, 10–11 μm long.

Epilithic in fast-flowing unpolluted calcareous streams.

Snowella Elenkin 1938

01750000 (new number)
Colony planktonic, spherical or irregularly ellipsoidal or ovoid mucilaginous mass, with cells forming an irregular layer some distance from the outside of the mucilage. Cells spherical or slightly elongated in radial axis (but not heart-shaped); at the ends of narrow, distinct, branched stalks originating near centre of colony. Cell division perpendicular to surface of colony, with successive divisions in two different planes. Cells are often spaced some distance from each other, and, even if quite close, never in direct contact.

Snowella lacustris (Chodat) Komárek *et* Hindák 1988

Basionym: *Gomphosphaeria lacustris* Chodat 1898
01750010 (new number; old number 01320020)
Pl. 4D,E (p. 46)
Colony spherical, ellipsoidal or irregular when forming daughter colonies; occasionally compound; up to 80 μm diameter. Cells ellipsoidal or obovoid, 1.5–3.5 μm × (1.5–)2–4 μm; usually, but not always, without gas

vacuoles. Cells well spaced from one another in young colonies, closer in old colonies. Stalks thin, usually easily visible without special techniques.

Widespread in plankton of mesotrophic or slightly eutrophic ponds and lakes, though usually present in low densities (Fig. 386–388 in Canter-Lund & Lund, 1995).

Synechococcus Nägeli 1849

01690000

Cells single or grouped into microscopic or macroscopic colony, which is largely free of obvious mucilage. Cells long ovoid or cylindrical. Cells divide transversely, always in one plane in successive generations, usually giving rise to two similar daughter cells, but occasionally asymmetrical.

The genus is retained here as a 'dumping ground' for diverse organisms. Detailed studies on some 50 isolates were reported by Rippka et al. (1979) and various subsequent studies at the Pasteur Institute (see Waterbury & Rippka, 1989), leading to recognition of distinct groupings; subsequently several new genera were recognized. However, most of the features require critical study in the laboratory. In addition it is unclear what relationship a particular isolate bore to the population from which it was isolated or how representative were the strains studied of the overall diversity in nature. Among these genera, Komárek & Anagnostidis (1999) adopted *Cyanobium* Rippka *et* Cohen-Bazire 1983, based on whether the cells are differentiated into an outer photosynthetic and inner non-photosynthetic region (chromatoplasm and centroplasm). Among other splits and rearrangements adopted by Komárek & Anagnostidis (1999), they separate three large-cell species into *Cyanothece* Komárek 1976, including the widespread *S. aeruginosus*. These species show a characteristic cytoplasmic structure, which Komarek & Anagnostidis (1999) describe as 'keritomized = irregular to radial arrangement of thylakoids, intrathylakoidal spaces'. As discussed in the Introduction, the presence of such liquid vacuoles is similar to that in blue-green algae in other genera and raises the question whether these species also exist at times without this appearance. However, there are no large-celled species of *Synechococcus* within the restricted definition used by Komárek & Anagnostidis (1999).

Synechococcus aeruginosus Nägeli 1849

Synonyms: *Coccochloris aeruginosa* (Nägeli) Drouet *et* Daily 1952; *Cyanothece aeruginosa* (Nägeli) Komárek 1976

01690010 **Pl. 6E** (p. 53)

Cells single or as a pair during division, short cylindrical with widely rounded ends. Cells 7–12(–15) μm wide, (9–)10–45(–50) μm long.

Freshwater, typically shallow waters influenced by humic waters, free-floating in vicinity of submerged plants; sometimes also in swamps and among other algae and bryophytes on wet rocks; occurs in waters over considerable hardness range, though apparently not in very soft water.

Synechococcus bacillaris Butcher 1952

Cells single or in short chains, spherical or widely ellipsoidal. Cells 1.2–1.5 × 1.5–2.8 μm. Mucilage colourless, wide, but indistinct layer.

Brackish reservoirs. Powell (1964), without details.

Synechococcus capitatus Bailey-Watts *et* Komárek 1991

Cells single or two together, more or less straight or slightly curved, rod-shaped, slightly widened at the ends, pale green or almost bluish; no mucilage. Cells 0.7–0.9 μm wide, 4–40 μm long.

Freshwater, planktonic. Original description by Bailey-Watts et al. (1968) from Loch Leven, Scotland, as *Synechococcus* sp. According to Komárek & Anagnostidis (1999), planktonic in large, cold lakes in several countries.

Synechococcus elongatus (Nägeli) Nägeli 1849

Basionym: *Synechococcus parvulus* Nägeli 1849

01690020 **Pl. 6F** (p. 53)

Cells cylindrical, usually in a cluster made up of single cells or pairs of recently divided cells, though occasionally with chains of 4 cells. Cells 1.2–3 μm wide, 2–9 μm long, occasionally with thin layer of mucilage. Cell division usually asymmetric in longer cells.

Wet soil or firm substratum such as rocks or a wall, often at the edge of springs or a lake.

Synechocystis Sauvageau 1892

01700000

Cells solitary spherical, broadly ovoid, or (after division) hemispherical, either lacking mucilage or with narrow indistinct mucilaginous layer associated with each cell. Cell division by binary division, with successive divisions in two different perpendicular planes.

This description is based on Komárek & Anagnostidis (1999), who recognize 16 species from Europe. The features separating species are largely size, presence or absence of mucilage, whether a tendency to aggregate in groups and habitat.

Synechocystis aquatilis Sauvageau 1892

Cells spherical or subspherical, solitary or (frequently) two together (1.8–)2.2–7(–7.5) μm. Freshwater forms sometimes have a thin gelatinous envelope.

Planktonic in fresh and brackish waters which are probably usually moderately eutrophic; the smaller forms typically occur in brackish waters.

Komárek & Anagnostidis (1999) recognize two species, *Synechocystis salina* Wislouch 1924 (synonym *Synechocystis aquatilis* var. *minor* Geitler 1935), with *S. salina* ranging in diameter up to about 4.5 μm and occurring in brackish, salty and alkaline environments, and *S. aquatilis* with the diameter usually 4.5 μm and over and occurring in relatively nutrient-rich freshwaters.

Synechocystis endobiotica (Elenkin *et* Hollerbach) Elenkin in Elenkin 1938

Basionym: *Synechococcus endobioticus* Elenkin *et* Hollerbach 1923

Cells spherical or elongated prior to division, solitary or in small groups inside mucilage of colonial planktonic blue-green algae, typically more towards outside of colony, but sometimes dispersed throughout colony; cells (1–)1.5–4 μm, usually with their own narrow gelatinous envelope confluent with mucilage of host colony.

Inside colonies of *Coelosphaerium*, *Microcystis* and other genera of mucilaginous, planktonic blue-green algae.

Morphologically there is little to justify the separation of this species from a broadly defined *S. aquatilis*, but as the habitat is well-defined and the organism common, it seems worth retaining a separate binomial until critical studies have been made.

Woronichinia Elenkin 1933

01760000 (new number)

Colony more or less spherical or irregularly ovoid, with up to several hundred cells in the outer region and each cell at the end of a stalk. The stalks radiate from the central region of the colony and mostly run parallel to each other, as the whole stalk typically divides when the cell divides; however, branches sometimes persist in the outer part of larger colonies; stalks may become somewhat diffluent in older colonies, making careful inspection necessary to distinguish from *Coelosphaerium*. Cells usually more or less ovoid or obovoid, though with the sides slightly flattened.

Woronichinia naegeliana (Unger) Elenkin 1933

Basionym: *Coelosphaerium naegelianum* Unger 1854

01760010 (new number; old number 01150020)

Pl. 5C,D (p. 49)

Colony free-floating, spherical, ellipsoidal or somewhat irregular in outline, up to 180 μm diameter, with cells in a single densely packed layer at some distance (typically 7–8 μm) from the surface; outside of colony distinct. Cells obovoid or ellipsoidal, 3.5–5 μm wide, 5–7 μm long, usually with an indistinct individual sheath, though this may be hard to recognize because of the close contact between adjacent cells; gas-vacuolate when in the plankton (condition in winter is uncertain, when the colonies are presumably in the benthos). Each cell occurs at the end of a branching gelatinous tubular, but rather indistinct stalk, which is the same width as the cell; these radially arranged stalks are often very difficult to distinguish in large colonies.

Common in plankton of moderately eutrophic lakes and ponds, often a minor component of blooms and occasionally the dominant.

The species is placed in *Woronichinia* following Komárek & Anagnostidis (1999), because of the gelatinous stalks, though the stalks of *W. naegelianum* are less distinct than those of several other species (not recorded for British Isles).

Xenococcus Thuret in Bornet *et* Thuret 1880

01730000

Cells single and then usually forming small groups which eventually grow together to form an extensive crust; usually single-layered, but sometimes several layered, almost always epiphytic. Cells usually polarized, sometimes slightly elongate along vertical axis; sheath usually evident around each cell. Mature cells all more or less same size. Cells divide irregularly in several planes, but usually perpendicular to substratum, or some or all with multiple divisions to form baeocytes. This broad definition includes *Xenotholos* Gold-Morgan et al., 1994, which forms several layers in a flattened hemispherical structure. However, there is considerable variation in populations from nature and it is not always easy to fit them into species descriptions. It seems best to retain a broad definition of *Xenococcus*.

Other records

* *Xenococcus rivularis* (Hansgirg) Geitler 1925. This was included in the *Coded List* (as 01730030), though its status is unclear. Komárek & Anagnostidis (1999) suggest that this probably belongs to *Chamaesiphon*.

In addition, Powell (1964) lists the following five species: *Xenococcus acervatus* Setchell *et* Gardner in Gardner 1918, the record from J. Komárek. However, Komárek & Anagnostidis (1999) list this as an unrevised species, whose generic classification is unclear (? *Xenolothos*).

* *Xenococcus crouanii* Feldmann. Komárek & Anagnostidis (1999) do not list this species, and it is unclear what organism Powell meant. He lists two forms, together with synonyms for each: *Dermocarpa violacea* (without authority) and *Sphaenosiphon roseus* Reinsch = *Dermocarpa rosea* (without authority) = ? *Xenococcus laminariae* Feldmann 1953. Komárek & Anagnostidis (1999) list two *Dermocarpa violacea* with different authorities and *D. rosea* (Reinsch) Batters 1889 as an unrevised poorly described species; the record for the latter was from Berwick (Frémy, 1929–33). Anand (1937) reported that the stratum of the green alga *Gomontia* in the chalk-boring community on cliffs at Westgate, Kent, was in places covered with a thick brownish-red growth of *D. violacea*.

* *Xenococcus gilleyae* Setchell *et* Gardner in Gardner 1918 (no detail). Komárek & Anagnostidis (1999) only list records for outside Europe and note *Dermocarpella gilleyae* (Setchell *et* Gardner) Komárek *et* Anagnostidis 1986 as a synonym. They report that the cells are solitary, spherical, 4–9 μm diameter, with binary division, *Chamaecalyx*-like baeocyte division.

* *Xenococcus violaceus* Anand 1937. Komárek & Anagnostidis (1999) mention this briefly, but as a species from India; however, the description was actually made from material collected on Beachy Head, Sussex, and epiphytic on green algae in the *Enteromorpha* belt. It is like a larger-celled form of *X. schousbei*.

Xenococcus chroococcoides F.E.Fritsch 1929

01730010: see *Chroococcopsis chroococcoides*

Xenococcus kerneri Hansgirg 1893

Synonyms: *Dermocarpa kerneri* (Hansgirg) Bourrelly 1970; *Xenotholos kerneri* (Hansgirg) Gold-Morgan et al. 1994

01730020 **Pl. 6J–N** (p. 53)

Cells initially single, but usually recognized as a colony, which is at first a 1-layered, but then a multilayered flattened hemispherical structure; cells usually irregular in arrangement, but sometimes a few irregular rows present; cells densely packed and inner ones often somewhat polygonal. Cells 3.5–6×3.5–10 μm; sheath thick, lamellate, colourless or pale brown-yellow. Reproduction by baeocytes, 2–3 μm.

Freshwater, epiphytic on larger algae and submerged mosses, occasionally epilithic, in fast-flowing mesotrophic streams and small rivers. Pentecost (1991b) describes its growth in Waterfall Beck, North Yorkshire. Powell (1964), without details.

Xenococcus pyriformis Setchell *et* Gardner in Gardner 1918

01730040 (new number)
Cells initially single, but usually recognized as a colony of cells with various types of radial arrangement. Cells initially subspherical, but often becoming obovoid or pear-shaped, 10–15 μm across at widest point, (10–)12–20 μm high; attached to substratum at narrower end. Sheath distinct, thin, colourless. Reproduction by baeocytes, 2.5–3.5 μm.

Epiphytic on various algae in the intertidal zone.

Xenococcus schousboei Thuret 1880

01730050 (new number)
Cells initially single, but usually recognized as a 1-layered colony of densely packed cells with a tendency to form rows. Cells subspherical, 4–9 μm, blue-green or violet; cells often appear somewhat polygonal when viewed from above. Reproduction by baeocytes, 1.5–2.5 μm.

Epiphytic on various algae in the upper part of the intertidal zone. Coppejans (1995) lists several records for Belgium/North France, mentioning *Lyngbya* as the host.

X. violaceus Anand (see above) appears to be very similar, apart from the larger cell size (8–15 μm). Probably the two should be combined.

ORDER OSCILLATORIALES

Filaments whose cells all undergo division in the same plane; heterocysts and akinetes absent. Gomont (1892) consolidated the earlier taxonomic accounts and his monograph is taken as the starting point for nomenclature for most genera in this order. Castenholz (1989) is an excellent source of general information on the order, while Anagnostidis & Komárek (1988) provide an exhaustive account of previous literature and illustrate the variation that occurs within the group.

Some species show a pattern of cell division in which further divisions commence before the previous one (or even two) has been completed. This is easiest to see with the electron microscope, but with larger-celled species it is often easy to see with the light microscope. Anagnostidis & Komárek (1988) use this feature to separate a distinct group within the order (family Oscillatoriaceae in their sense). This feature never occurs in the Nostocales, Stigonematales or eukaryotic algae. The complex pattern of division can make it difficult to decide what is the maximum cell length, a feature which is often quoted in species descriptions. It explains why the values for the ratio of cell length to width in the original accounts are often slightly less than would be expected for the ranges quoted for width and length. Care needs to be used throughout the keys, to differentiate between filament and trichome widths. Cell width refers to the diameter of a typical part of the trichome; it does not include the sheath.

Although the trichomes are mostly parallel-sided, they sometimes narrow towards one or both ends. The terms 'tapering' and 'attenuated' are often used interchangeably in identification manuals. The term 'taper' is reserved in this section to indicate a progressive and often marked narrowing towards the end. In some, but not all, cases it ends in a multicellular hair, such as in most species of *Homoeothrix*. 'Attenuated' is used more loosely, but includes instances where the narrowing is abrupt rather than progressive. The extent to which a trichome is attenuated is in at least some cases influenced by the extent to which the trichome is nutrient-limited. The liquid vacuoles in markedly tapered trichomes of *Phormidium autumnale* are reminiscent of those forming in hairs, suggesting that in this species attenuation may be related to phosphorus limitation. Experimental studies on Oscillatoriales should always include material grown to phosphorus limitation, just as Nostocales are grown to nitrogen limitation in order to observe heterocysts. It seems likely that not only the formation of liquid vacuoles and marked trichome attenuation may often be responses to phosphorus limitation, but probably also some (but not all) of the other features of the apical region used for species identification. If so, such features would only be expressed under phosphorus limitation. In addition several pairs of species listed here might merely be high and low phosphorus forms of the same species. An important question is whether the phosphorus-limited forms of species of Oscillatoriales are the usual ones in which species are found in nature, as is the case with phosphorus-limited forms of *Calothrix* and *Rivularia* in the Nostocales, or whether there are many species where both high and low phosphorus forms persist for some months.

Whether or not the trichome is attenuated towards the end, the terminal cell may differ markedly in shape from adjacent cells, being rounded, conical, forming a head, or various intermediate shapes; there may also be an additional modification at the outer end of the cell: cap, thickened membrane or calyptra. These features are important for identifying the species, especially in the genera *Oscillatoria* and *Phormidium*. However, as mentioned above, the development of some of these features may be influenced by the environment. In addition, a modified apical cell differentiates from an ordinary cell following trichome breakage, so not every end cell shows the special features, even if the environment is favourable. As wide a range of trichomes as possible should be checked before a feature is scored as negative, but it will probably always be difficult to be sure that a population is incapable of forming certain features without simple experimental or molecular tests. Nevertheless, it will be a lot easier to assess the taxonomic value of such characters when more is known about the functional reasons for their occurrence. Probably most cells retain the capacity to divide, though in some cases trichomes may need to return to a favourable environment before particular groups of cells do so. In long trichomes there are often regions where cell division is especially active. Other taxonomic characters are also likely to be influenced in individual strains by the environment. For instance, cell length to width ratio is influenced by light intensity in *Phormidium* cf *ectocarpi* (Wilmotte, 1988) and almost certainly some other species, with longer cells under higher light. As mentioned in the Introduction, many species of Oscillatoriales have conspicuous granules adjacent to the cross wall. This has often been regarded as a diagnostic feature, but even in those species forming them, their presence in any particular trichome is likely to be influenced by the recent environment. In cases where such granules have been checked by staining, they have proved to contain polyphosphate; the statement 'probably polyphosphate' in species descriptions here means that this has been confirmed by the author for at least one sample of the species. The original floristic descriptions often distinguish between trichomes with and without narrowing at the cross walls, and sometimes this is almost the only character separating two species. However, care is needed in using this character, as slight constrictions sometimes become apparent in dried or preserved material, when they were not evident in the living material. This seems to apply especially to *Lyngbya*.

All genera and probably all species are capable of motility under some conditions and at least some forms rotate on their axis as they move. Dense populations of motile forms often occur on the surface of sediments in ponds and rivers and sometimes float to the surface of the water during the day due to the presence of trapped oxygen bubbles; these may sink to the bottom again during the night. Even without such a marked shift in position within the waterbody, motile forms may shift their distribution on and within sediments depending on the prevailing light conditions.

The motility of the trichomes of some benthic species lacking sheaths is expressed not only as directional movement, but also a marked flexibility; the extent to which this occurs differs between species. It is often difficult to tell from the older descriptions whether 'flexuous' means the presence of persistent bends and undulations or whether it indicates the rapid changes of form which occur when trichomes are in contact with particles or other surfaces. Clearly it is impossible to make reliable comments if preserved material was used for the study. 'Flexible' is used here when it is known that such rapid changes in shape may occur in the species; the term is added in parentheses (with ?) if it seems likely that this is the situation, but it has not been confirmed for the species.

Many planktonic forms contain gas vacuoles. These may occur throughout the trichome (e.g. *Oscillatoria agardhii*) or localized near the cross wall (e.g. *O. redekei*). It seems likely that many of the truly planktonic species with gas vacuoles can also produce benthic trichomes with few or no gas vacuoles in response to particular environmental conditions, though studies are needed to check this for a range of populations in each species. *Lyngbya aestuarii* is one example which can occur with or without gas vacuoles. *Oscillatoria princeps* forms gas-vacuolate apical stretches of trichome, which are probably released and aid dispersal in the same way as the gas-vacuolate hormogonia produced by some species of *Calothrix* and Scytonemataceae.

The flattened trichomes characteristic of *Crinalium* are easy to overlook, so may be more widespread than reported. Study is needed to establish the extent to which this really is a distinctive genus or merely reflects populations which have evolved locally a flattened structure.

The separation of the genera *Oscillatoria*, *Lyngbya* and *Phormidium* has long worried taxonomists, especially as they came to realize that the characters used to distinguish these genera are often ones markedly influenced by environmental conditions. However, the problem is almost as great with separating other genera within the order, such as *Lyngbya*, *Plectonema* and *Schizothrix*. Anagnostidis & Komárek (1988) summarize the history of the problem and then present an extremely thorough account of features that differ within the order. They use this to make substantial changes to previously accepted generic limits in the order. However, the generic limits used in the present Flora are largely those of Geitler (1932a), as have also been those in some other recent Floras, such as for Israel (Nevo & Wasser, 2000). It seems sensible to adopt a conservative approach until there is a full revision based on combined molecular studies and critical experimental studies in culture, especially with respect to the nitrogen to phosphorus status (N : P ratio) of the organisms. The author finds some, but by no means all, of the recommendations of Anagnostidis & Komárek (1988) helpful. For instance, many authors now use the genus *Leptolyngbya* Anagnostidis *et* Komárek for a range of non-planktonic forms with trichomes up to 3 μm wide and usually with a firm narrow sheath. This brings together some very similar organisms otherwise classified into different genera. However, there is no evidence that the genus as a whole represents a genetically homogeneous group. The revisions introduce new problems, such as the separation of what are probably closely related species or even the same species into different genera; see, for instance, the accounts for *Oscillatoria bourrellyi*, *O. limnetica* and *O. redekei*. It is beyond the scope of this Flora to assess all the recommendations in detail.

As the absence of the heterocyst is the only distinct feature separating this order from the Nostocales, it seems likely that some non-heterocystous forms may have evolved from Nostocales. Several laboratory mutants of *Anabaena* have lost the ability to form a heterocyst, but these still look more like *Anabaena* than a member of the Oscillatoriales. Some *Homoeothrix* populations (e.g. *H. caespitosa*) look rather like *Calothrix*, but without heterocysts.

Ammatoidea West *et* G.S.West 1897

01330000

Filaments with trichomes which taper markedly at both ends when mature, often ending in long hairs. Filaments are usually bent in the middle, sometimes with a false branch.

The genus is somewhat like a form of *Calothrix* without heterocysts. However, populations appear to exist in this form for long periods, so it is probably not just a growth form of *Calothrix*. This is supported by an experimental study (Broady & Ingerfeld, 1999) of an Antarctic isolate, which did not form heterocysts in low nitrogen medium.

Ammatoidea normannii West *et* G.S.West 1897

01330010 **Pl. 8G** (p. 62)

Filaments 5.5–12.5 μm. Cells 3.5–5.5×2.5–5 μm in middle of trichome, but much longer at the tapered ends, especially where these are differentiated into a colourless hair; cross walls sometimes slightly narrowed in the middle of trichome. Sheath usually yellow-brown, though sometimes colourless. Geitler (1932a) states that hormogonia form towards the middle of the trichome, but the long stretch of empty sheath illustrated by West & Fritsch (1927) looks more like the situation in *Calothrix*, where the hormogonium is formed (in response to a phosphate pulse) in the region immediately beneath the terminal hair. Hormogonium development sometimes leads to a form of false branching rather than the hormogonium migrating from the mother filament. Filaments sometimes loop around the filament of a larger alga, with both ends running more or less parallel to each other.

Among other algae growing on rocks near the edges of ponds and in small streams with variable flow. Several sites, including Dartmoor (W. West & G.S.West, 1897).

UNCERTAIN STATUS

Amphithrix Kützing 1843 emend. Bornet *et* Flahault 1886

The status of this genus is unclear. It was described as crust-forming colonies, with upright, more or less straight, narrowly sheathed trichomes from a basal mass of cells. The latter separates the genus morphologically from *Homoeothrix* and most researchers now consider that the structure described was an artefact resulting from *Homoeothrix* forming a layer over another genus such as *Chamaesiphon*. Starmach (1966) retained one species, but included it within *Homoeothrix* and pointed out the doubts. *A. violacea* [Kützing] Bornet *et* Flahault was listed by Powell (1964). Molecular studies are needed to rule out unequivocally the possibility that some populations are highly modified forms of Stigonematales rather than a mixed population.

UNCERTAIN STATUS OF GENUS AND SPECIES IN BRITISH ISLES

Arthrospira [Stizenberger 1852] Gomont 1892

01070000

Trichome cylindrical, without heterocysts, mostly loosely and regularly coiled, usually of a relatively large diameter (>7 μm) and with wide spirals, with comparatively short cells; almost always gas-vacuolate. Apex of trichome sometimes (but not always) attenuated, terminal cell rounded, calyptra absent; cross walls distinct. Firm sheath absent, but sometimes a diffluent sheath or mucilage layer. Trichomes often show motility if in contact with a surface. Spirals of most (but not all) species are anti-clockwise, but mutants occasionally occur in cultures of some strains. Straight mutants of coiled forms also occur occasionally in culture; it is unknown whether these occur in nature, but, if they do, they would be difficult to distinguish from *Oscillatoria*.

This genus was included in *Spirulina* by Geitler (1932a) and some other authors, but there is now plenty of evidence to show that they are quite distinct. The most obvious difference seen with the light microscope is the distinct cell wall in *Arthrospira*, reflecting differences in wall layers shown by electron microscopy. Several species of *Arthrospira* provide the material sold as a health food under the name Spirulina. It is unclear how reliable are the records for *Arthrospira* in the British Isles: *A. jenneri* is the most likely. Probably almost all records are for *Spirulina*.

Arthrospira jenneri [Stizenberger 1852] Gomont 1892

Synonym: *Spirulina jenneri* (Stizenberger) Geitler 1925

01070010

Width of spiral 9–15 μm, distance between turns 21–31 μm. Cells 5–8×4–5.5 μm, granules at cross walls; trichome end cell rounded.

Griffiths (1936): Butterby Pond, Durham.

Crinalium Crow 1927

01160000

Trichomes flattened in cross-section; sheath thin or absent. In view of the fact that the flattened trichome is the only clear diagnostic feature, further studies are needed to establish whether the various records are closely related.

Crinalium endophyticum Crow 1927

01160010

Cells 3–4×3.5–6 μm.

Crow reported this growing inside *Aphanocapsa* colonies on moist rocks at Betws-y-Coed, North Wales. Woodhead & Tweed (1954) reported it from the same location and microhabitat.

Crinalium epipsammum de Winder, Stal *et* Mur 1992

01160020

Crust-forming colonies on sand. Trichomes non-motile, lacking a sheath, varying in length. Cells 4–6×1–1.5 μm (×c. 2.2 μm thick); end cells similar in morphology to intercalary cells. Reproduction by random trichome breakage.

Terrestrial on sand-dunes.

This species was described from sand-dunes in The Netherlands. Several samples from the east coast of England and Scotland appear to be similar, but detailed studies are needed.

Homoeothrix (Thuret) Kirchner 1898 *sensu* Komárek *et* Kann

01360000

Colony a slightly flattened hemisphere, crust or tuft. Trichome slightly tapered, unbranched or showing

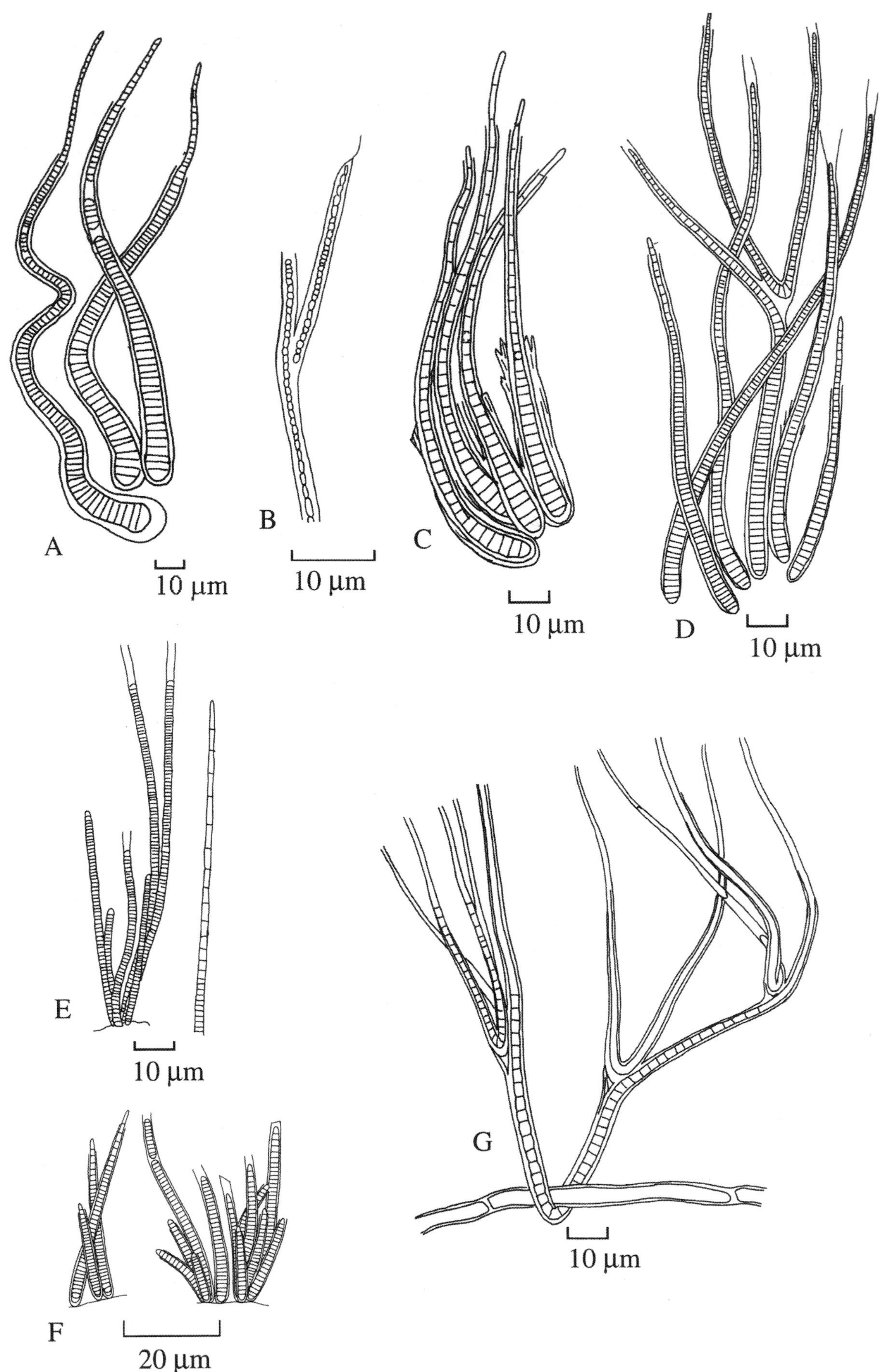

Plate 8 A–G
A. *Homoeothrix juliana* (p. 63); **B**. *Homoeothrix crustacea* (p. 63); **C**. *Homoeothrix fusca* (p. 63); **D**. *Homoeothrix rubra* (p. 63); **E, F**. *Homoeothrix varians* (p. 63); **G**. *Ammatoidea normannii* (p. 61).

false branching, usually with obvious sheath. Heterocysts and akinetes absent. The genus consists of a range of colony-forming structures with tapered trichomes, which sometimes look quite like a member of the Oscillatoriales and other times like a *Calothrix* lacking a heterocyst. Almost certainly the species included here are phylogenetically diverse.

Frémy (1929–33) listed *H. rubra* (Crouan) Frémy 1926, based on a description of material described by Crouan and material found by Frémy at Brest. This forms a short turf in the supralittoral, with filaments 3–5 μm wide, 1–3 mm long; and various degrees of branching, from little to much. The turf appeared coloured due to the reddish trichomes.

Homoeothrix balearica (Bornet *et* Flahault) Lemmermann 1910

Basionym: *Calothrix balearica* Bornet *et* Flahault 1886
01360010
Colony a spreading tuft of filaments, 3.5–12 μm wide, up to 1 mm long; prostrate towards base, then more or less upright, sparingly branched; olive-brown. Trichome 3–9 μm wide. Sheath thin, usually lamellate, yellow-brown when old.

On moist rocks and other surfaces.

The original description was for a population with filaments about 12 μm wide, but W. & G.S.West (1898) included material 3.5–5 μm wide (as var. *tenuis*).

Homoeothrix caespitosa (Rabenhorst) Kirchner 1900

Basionym: *Calothrix caespitosa* Rabenhorst 1873
Synonym: *Dichothrix nordstedtii* Bornet *et* Flahault 1886
01360020
Colony turf-bushy, up to 33 mm high, typically blue-green. Filaments 9–18 μm wide, prostrate at the base, then erect and then showing repeat branching. Trichome 8 μm wide towards base, usually a long tapering hair. Cells mostly 3.5–5 μm long. Sheath thin, not lamellate, colourless.

On rocks in fast-flowing streams and small rivers, probably not hard water. West & Fritsch (1927): (as *D. nordstedtii*) widespread.

Geitler (1932a) listed what was essentially the same form as both *H. caespitosa* and a non-heterocystous *Dichothrix*, *D. nordstedtii*, without making a firm recommendation. Apart from the absence of a heterocyst, *H. caespitosa* also looks quite like *Dichothrix orsiniana*.

Homoeothrix crustacea Woronichin 1923

01360030 **Pl. 8B** (p. 62) **CD**
Colony hemispherical or a flattened cushion, to 3(–4) mm diameter, olive-brown or milky brown, calcified. Filaments 2.7–3.2 μm wide, 150–400 μm long, upright, dense and richly branched. Trichomes showing only very slight tapering. Cells 1.6–2.4 μm wide, almost as long as wide or somewhat longer, cross walls slightly narrowed. Sheath thin, pale yellow-brown, though the colour is not always obvious because of the narrow sheath.

Widespread in highly calcareous streams and small rivers subject to slight nutrient enrichment, but phosphate not so high as to prevent calcification. Often occurs in flowing water systems downstream of the zone dominated by *Rivularia* and other Rivulariaceae, and frequently growing together with the green alga *Gongrosira incrustans*.

Homoeothrix fusca Starmach 1934

01360040 **Pl. 8C** (p. 62)
Filaments unbranched, straight or flexuous, 4.5–7.5 μm wide at base, 48–85 μm long, forming clumps up to 1 cm or single among other blue-green algae. Sheath thick, fimbriated at the apex, dark brown. Trichome 2–3.5 μm wide near the base, 1.5–2.5 μm in middle, usually ending in hair. Cells near the base shorter than wide, longer than wide towards the apex; grey blue-green.

In standing and flowing freshwaters. Margin of Windermere (Godward, 1937).

Homoeothrix juliana (Bornet *et* Flahault) Kirchner 1898

Basionym: *Calothrix juliana* Bornet *et* Flahault 1886
01360050 **Pl. 8A** (p. 62)
Filaments single or united into a small upright group, often olive-coloured, 10–15 μm wide, to 2 mm long, often thickened at the base. Trichome above any basal swelling 9–12.5 μm wide, tapering, usually ending in a long hair. Cells one-quarter to one-half as long as wide. Sheath thin, firm, not lamellate, colourless, covering basal cell and trichome as far as the hair.

On rocks and submerged plants in ponds and edge of lakes.

Homoeothrix rubra (Crouan 1867) Frémy 1926

Basionym: *Schizothrix rubra* Crouan 1867; not *S. rubra* [Meneghini] Gomont 1892
Synonym: *Calothrix rubra* Bornet *et* Flahault 1886
01360070: see account for genus **Pl. 8D** (p. 62)
Colony a turf, thin, spread out, red-brown. Filaments 3–5 μm wide, 1–3 mm long, packed densely, branching absent or sparse, with appearance of *Ammatoidea*. Cells 2–4 μm wide near the base, 1.2–2.5 μm long, typically half as long as wide. Sheath narrow, colourless.

Supralittoral, on submerged or very wet rocks and rotting algae.

None of the freshwater species of *Homoeothrix* is morphologically quite the same as this marine species.

Homoeothrix varians Geitler 1927

01360080 **Pl. 8E,F** (p. 62)
Colony ranging from a small bundle of filaments to an extended turf, somewhat mucilaginous; olive-green, dark violet or sometimes orange-yellow in old colonies Filaments 2.5–3 μm wide, young ones upright and straight, older ones flexuous and somewhat intermingled with one another. Older trichomes tapered

slightly towards the apex and often ending in a long colourless hair; if the hair has fallen off and the hormogonium been released at the apex, the sheath projects beyond the end of the trichome. Cells olive, purple or with an orange tinge. Sheath thin, colourless.

On rocks in slow and moderately fast-flowing small rivers, often in slightly shaded conditions.

Although obvious hairs can often be found, their presence is less the normal condition for this species than for most other hair-forming blue-green algae. It is therefore easy to mistake this for a narrow erect form of *Lyngbya*.

Hydrocoleum [Kützing 1843] Gomont 1892

01390000

Filament consists of a sheathed group of *Oscillatoria*-like trichomes, with several trichomes inside one sheath; filaments sometimes branching, though often with differing numbers of trichomes in the various parts; filaments themselves in some species grouped into bundles, forming a tuft or membranous thallus. End of trichome straight, end cell often with calyptra. Sheath usually colourless.

1 Marine or highly brackish.......................................2
1 Freshwater 3
2 Trichomes mostly ≤13 μm wide.....*H. lyngbyaceum*
2 Trichomes mostly >13 μm wide ....*H. glutinosum*
3 Trichomes mostly ≤10 μm wide.........*H. brebissonii*
3 Trichomes mostly >15 μm wide ...*H. heterotrichum*

Hydrocoleum brebissonii [Kützing 1849] Gomont 1892

Synonym: *Blennothrix brebissonii* (Gomont) Anagnostidis *et* Komárek

01390010 **Pl. 9A** (p. 65)

Filaments united in small, dark blue-green bundles, more or less straight, sparsely branched. Sheath in lower part of filament wide, lamellate, in upper part narrow and firm, sometimes indented, giving the appearance of rings or folds. Trichomes in filament 1–4, flexuous and loosely coiled round each other. Cells 8–10 μm wide, 2.4–5 μm long, cross walls not narrowed; end cell somewhat conical, often with calyptra.

In springs and streams.

Hydrocoleum glutinosum [C.Agardh 1824] Gomont 1892

01390020

Colony mucilaginous, not bushy, yellow-green to yellow-brown. Cells 14–21 μm wide, 2.5–3.5 μm long, not narrowed at the cross walls, often with granules at the cross walls; end cell narrowed, flattened end, often terminated by calyptra. Sheath uneven, almost amorphous.

Powell (1964). On rocks and old algal stipes in small pools in the supralittoral and uppermost part of the intertidal.

This merges with *H. lyngbyaceum*, differing only in the wider cells and perhaps greater tendency to produce mucilage. Both occur in similar marine habitats.

Hydrocoleum heterotrichum [(Kützing) Gomont 1890] Gomont 1892

Synonym: *Blennothrix heterotricha* (Gomont) Anagnostidis *et* Komárek

01390030 **Pl. 9B** (p. 65)

Colony forming bundles of filaments showing obvious and irregular branching, up to 5 mm, often with slight calcareous deposit on sheath. Trichomes in filament 1–4, flexuous and loosely coiled round each other. Cells 16–19 μm wide, 3–5 times shorter than wide, cross walls not narrowed; end cell truncate, with a slight cap. Sheath rather slimy, uneven, sometimes indented, giving the appearance of rings or folds.

In standing or flowing water, often among old clumps of moss.

Hydrocoleum lyngbyaceum [Kützing 1849] Gomont 1892

01390040 **Pl. 9C** (p. 65)

Colony a mat or turf, consisting of bundles of filaments, very dark green or almost black when viewed on mass. Filaments unbranched and with many trichomes near the base, branched further up and fewer or one trichome. Cells (8–)9–12(–16) μm wide, 3–6 times shorter than wide, often with granules at cross walls; end cell truncate, often with calyptra.

Powell (1964). In similar habitats to *H. glutinosum*, such as rocks and old algal stipes in small pools in the supralittoral and uppermost part of the intertidal, but also often an epiphyte and overall more common. Coppejans (1995) has two records for Belgium/North France.

See comments for *H. glutinosum*.

Lyngbya [C.Agardh 1824] Gomont 1892

01430000

Trichome typically inside a firm sheath, which ranges from thin to very thick according to species. Filaments free-floating or loosely or firmly attached to a substratum; if attached, this may be at the base, in the middle or the entire filament; sometimes coiled or spiral. Occasional false branching has been recorded in several species, thus resembling *Plectonema*. Sheath usually colourless, although occasionally pale yellow-brown or other colour. Hormogonia are formed as an ordered release of lengths of the trichome under conditions favourable for growth. The release of a hormogonium often leaves a length of sheath projecting beyond the remainder of the original trichome, though in some cases the 'extra' sheath may break down. In the case of species with very thin colourless sheaths, a small length of sheath projecting beyond the end of the trichome provides a useful diagnostic character to separate the genus from *Oscillatoria*. In some of the larger aquatic species the whole trichome sometimes moves

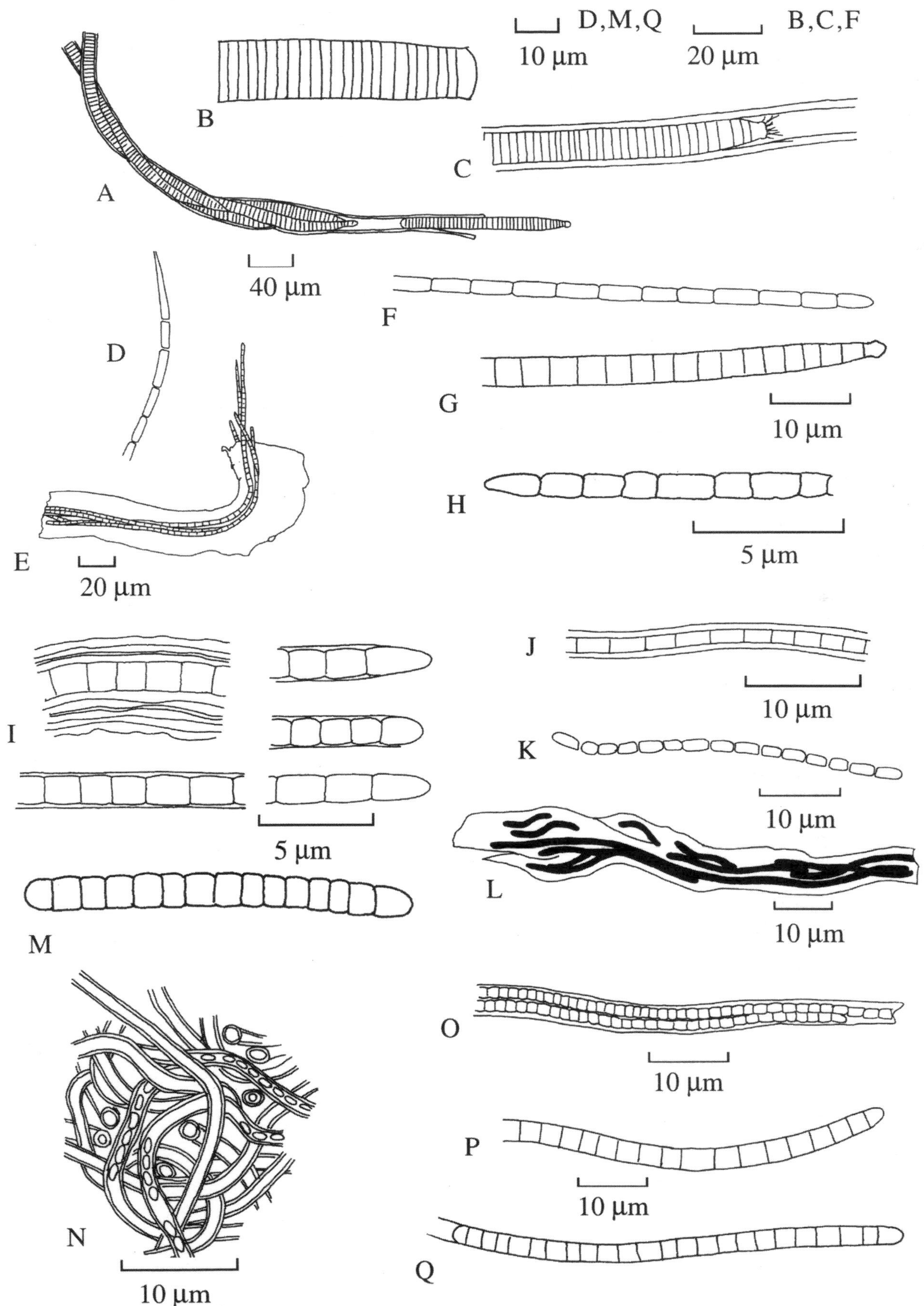

Plate 9 A–Q

A. *Hydrocoleum brebissonii* (p. 64); **B**. *Hydrocoleum heterotrichum* (p. 64); **C**. *Hydrocoleum lyngbyaceum*, (p. 64): showing epiphytic bacteria on apical cell, a common feature of the species (p. 64); **D, E**. *Microcoleus tenerrimus* (p. 71); **F**. *Microcoleus lacustris* (p. 71); **G**. *Microcoleus vaginatus* (p. 71); **H, J**. *Schizothrix coriacea* (p. 85); **I**. *Schizothrix friesii* (p. 85); **K, L**. *Schizothrix lateritia* (p. 86); **M**. *Schizothrix muelleri* (p. 86); **N**. *Schizothrix perforans* (p. 86); **O**. *Schizothrix tinctoria* (p. 86); **P**. *Symploca muralis* (p. 89); **Q**. *Symploca muscorum* (p. 89).

out of its sheath. This frequently occurs in response to the changed environment in a sample bottle, so it is likely that records for large *Oscillatoria* in pond samples may in fact sometimes be *Lyngbya*.

West & Fritsch (1927) stated that there were 12 freshwater British species, but mentioned only five by name.

Several changes have been made to the records in the *Coded List*. *Lyngbya birgei* Smith 1916 (Coded number 01430040) is removed, partly because of doubts about the British record, but also because the species described by Smith fits within *L. majuscula*, apart from the freshwater habitat.

Other records Benson-Evans & Antoine (1996) list, without detail:

- *Lyngbya nordgaardii* Wille 1918. Frémy (1929–33) states that this is synonymous with *L. epiphytica* Wille, not *L. epiphytica* Hieronymus 1898. However, the descriptions scarcely differ, apart from the fact *L. nordgaardii* has longer cells.
- *Lyngbya purpurea* [Hooker *et* Harvey] Gomont 1892. This was described originally only briefly (cells 1.4–1.8 µm wide, colourless sheath, violet cytoplasm); further study is needed before it can be accepted as a species.
- *Lyngbya taylorii* Drouet *et* Strickland.

The following key deals with the species listed here, but it is likely that forms of *Lyngbya* will be encountered which do not fit any of them, in addition to other genera such as *Phormidium* and *Plectonema*.

1 Marine, including supralittoral 2
1 Not marine: terrestrial, freshwater or slightly brackish 10
2 Sheaths of old, well-illuminated filaments yellow-brown .... *L. aestuarii*, in part
2 Sheaths of old filaments colourless, even if well-illuminated 3
3 Cells ≤8 µm wide 4
3 Cells >8 ≤16 µm wide 7
3 Cells >16 µm wide 9
4 Endophyte, very narrow ..... *L. rivulariarum*, in part
4 Not endophyte 5
5 Epiphyte more or less perpendicular to host surface .... *L. kuetzingii*, in part
5 Not as above 6
6 Sheath thin, not lamellate .... *L. gracilis*
6 Old sheath lamellate .... *L. semiplena*, in part
7 Old sheath lamellate; trichomes attenuated towards apex .... *L. semiplena*, in part
7 Not as above 8
8 Older sheath rough; supralittoral or upper intertidal .... *L. confervoides*
8 Older sheath smooth; lower intertidal or sublittoral .... *L. sordida*, in part
9 Cell length: width 0.2 .... *L. majuscula*
9 Cell length: width >0.2 .... *L. sordida*, in part
10 Endophyte, very narrow ..... *L. rivulariarum*, in part
10 Epiphyte more or less perpendicular to host surface .... *L. kuetzingii*, in part
10 Not as above 11
11 Cells ≤6 µm wide 12
11 Cells > 6 µm wide 24
12 Filament helical, planktonic .... *L. contorta*
12 Not as above 13
13 Cells ≤1 µm wide 4
13 Cells >1 µm wide 15
14 Many intertwined filaments, sheaths typically encrusted with iron oxide .... *L. ochracea*
14 Separate filaments, epiphytic on algae and aquatic plants .... *L. epiphytica*, in part
15 Well-illuminated old sheath yellow ..... *L. versicolor*
15 Well-illuminated old sheath colourless 16
16 Epiphytic, with most of filament firmly attached to live host 17
16 Not as above 18
17 Filament (not cells) width ≤2 µm wide .... *L. epiphytica*, in part
17 Filament (not cells) width >2 µm wide .... *L. lagerheimii*, in part
18 Calcite crystal deposits; in fast-flowing rivers .... *L. vandenberghenii*
18 Not as above 19
19 Cells ≤3.5 µm wide 20
19 Cells >3.5 µm wide 22
20 Colony a leathery mat .... *L. lutea*, in part
20 Attached tufts of fast-flowing streams, up to 1.5 cm long .... *L. lutescens*
20 Not as above 21
21 Forming a mat .... *L. diguetii*
21 Mostly single or small group of filaments, often planktonic .... *L. lagerheimii*, in part
22 Old sheath not lamellate .... *L. aerugineo-coerulea*
22 Old sheath lamellate .... *L. lutea*, in part
23 Well-illuminated old sheath usually lamellate, yellow-brown .... *L. aestuarii*, in part
23 Well-illuminated old sheath not lamellate or coloured 24
24 Filaments often forming a regular, stretched out helix .... *L. spirulinoides*
24 Filaments straight or irregularly flexuous 25
25 Cells ≤10 µm wide .... *L. martensiana*
25 Cells > 10 µm wide .... *L. maior*

Lyngbya aerugineo-coerulea [Kützing 1843] Gomont 1892

Synonym: *Phormidium aerugineo-coerulea* (Gomont) Anagnostidis *et* Komárek 1988

01430010

Filaments single or occasionally forming an expanded dull blue-green thallus, flexuous, fragile. Cells 4–6 µm wide, 2–4 µm long, half to as long as wide; cross walls not narrowed; end cell flattened, conical or rounded, with a slightly thickened outer membrane. Sheath thin, firm, not lamellate.

In standing and slow-flowing waters, sometimes on decaying plants. West & Fritsch (1927): one of the two most abundant British freshwater species.

Lyngbya aestuarii [Liebmann 1841] Gomont 1892

01430020 **Pl. 10J** (p. 69)

Filaments single or united in colonies, sometimes forming extensive floating masses. Cells 8–24 μm wide, typically 10–14 μm. Cells short, prior to division about one-quarter as long as wide; cross walls not narrowed; often with small (probably polyphosphate) granules adjacent to cross wall; cells often gas-vacuolate, especially in older colonies; end cell slightly narrowed, often with thickened membrane. Sheath at first thin and colourless, but often subsequently lamellate and yellow-brown; in old colonies the outer sheath layer may remain colourless, while the inner layers are brown. The trichomes of young filaments in samples returned to the laboratory often glide out of their sheaths.

Powell (1964). Freshwater or brackish waters, sometimes forming persistent masses throughout the summer, especially in slightly brackish ponds. Single filaments among submerged macrophytes or floating in the plankton; larger colonies may float at the surface. West & Fritsch (1927): the largest British freshwater species, although it is principally found in brackish or marine habitats.

This is a highly variable species and some forms recorded from other countries exceed the dimensions given here. It has been recorded in a range of marine environments in other countries, including hypersaline ones.

Lyngbya confervoides [C.Agardh 1824] Gomont 1892

01430230 (new number)

Colony tufts or turf, to 5 cm high, yellow-brown or dark green. Filaments at the base prostrate, upper upright, entangled, but otherwise straight. Trichomes not attenuated towards apex. Cells (8–)10–16(–25) μm wide, 2–4(–6.5) μm long, 2–8 times shorter than wide; often with small (presumably polyphosphate) granules on either side of cross walls; end cell rounded, calyptra absent. Sheath in older parts of filament becoming thick and rough on outside, colourless.

Powell (1964). Supralittoral and uppermost part of intertidal, typically in rock-pools.

Morphologically, but not usually ecologically, *L. martensiana* resembles *L. confervoides* apart from its narrower width (6–10 μm), though Frémy (1929–33) states that *L. martensiana* is more transparent and has very small granules.

Lyngbya contorta Lemmermann 1898

Synonym: *Planktolyngbya contorta* (Lemmermann) Anagnostidis *et* Komárek 1988

01430050 **Pl. 10B** (p. 69)

Filaments single, planktonic, forming a regular helix. Cells 1–2 μm wide, 3–5 μm long; cross walls not narrowed; end cell rounded. Helix 10–18 μm wide; coils of helix 6–12 μm apart.

Powell (1964). Planktonic in brackish water.

Lyngbya diguetii Gomont in Hariot 1895

Synonym: *Phormidium diguetii* (Gomont) Anagnostidis *et* Komárek 1988

01430060 **Pl. 10A** (p. 69)

Colony a mat of filaments, 2 mm thick. Filaments 2.5–3 μm wide, entangled towards base, but ends more or less straight towards apex. Cells 2–2.8 μm wide, 1–3.7 μm long, quadrate or shorter than wide; cross walls not narrowed; end cell rounded, without calyptra. Sheath thin, colourless.

Ponds, gravel pits and other standing waters.

Lyngbya epiphytica Hieronymus 1898

Synonym: *Leibleinia epiphytica* (Hieronymus) Anagnostidis *et* Komárek 1988

01430070 **Pl. 10H** (p. 69)

Filaments epiphytic on other algae, firmly attached and often coiled round the host. Filaments 1.5–2 μm wide Cells (0.8–)1–1.5 μm wide, 1–2 μm long; cells slightly shorter to slightly longer than wide; cross walls not narrowed; end cell rounded, not tapering. Sheath thin, colourless.

Epiphytic mainly on filamentous algae, such as *Oedogonium*, *Rhizoclonium* and *Lyngbya aestuarii*. Powell (1964).

Lyngbya gracilis [Rabenhorst 1865] Gomont 1892

Synonyms: *Lyngbya agardhii* [Crouan] Gomont 1892; *L. meneghiniana* [Kützing] Gomont 1892

01430080 **Pl. 10K** (p. 69)

Colony a slimy tuft, to 7 mm, often with a colour other than blue-green (purple or yellowish). Filaments flexuous. Trichomes not attenuated towards the end. Cells 5–8.5 μm wide, 2.8–4.6 μm long; end cell rounded, with (usually) or without outer membrane slightly thickened. Sheath thin, smooth, colourless.

Powell (1964). Intertidal and sublittoral, mostly epiphytic on other algae.

L. meneghiniana and *L. agardhii* are included here in *L. gracilis* and the above description covers all three original descriptions. The original description of *L. meneghiniana* differs from *L. gracilis* by little more than the tendency for its cells to be longer in relation to width. The original description of *L agardhii* differs from *L. gracilis* by the slightly wider trichomes (8–8.5 vs. 5–8.5 μm and the absence of slightly thickened membrane on the end cell. Frémy (1929–33) lists sites (for *L. meneghiniana*) near Edinburgh and Plymouth; Powell (1964) lists all three names.

Lyngbya kuetzingii Schmidle 1897

Synonyms: *Heteroleibleinii kuetzingii* (Schmidle) Compère; *Lyngbya infixa* Frémy 1932

01430090 **Pl. 10E** (p. 69)

Filaments epiphytic and typically perpendicular to the surface, either single or in groups. Filaments 2–3.5 μm wide, mostly up to 80 μm long, straight or slightly curved. Cells 1.5–2 μm wide, 0.5–1 μm long; cross walls not narrowed or with granules. Sheath thin, colourless and sometimes projecting slightly beyond the trichome, where a trichome has been released. The conditions favouring filament elongation versus release of an unsheathed trichome at the apex are unknown.

Epiphytic on *Cladophora*, other submerged filamentous algae with epiphytes and mosses in lakes and rivers, usually in moderately nutrient-rich water at neutral and higher pH values and especially in warmer months. Under such conditions, this is one of the most widespread organisms likely to be found in samples.

Lyngbya infixa is included here, because it scarcely differs from *L. kuetzingii* other than being marine. It was recorded as epiphytic on *Codium tomentosum* by Frémy (1929–33), but without British records; however, it is listed by Powell (1964).

Lyngbya lagerheimii [Mobius 1889 emend. Gomont 1890] Gomont 1892

Synonyms: *Lyngbya limnetica* Lemmermann 1898; *Leptolyngbya lagerheimii* (Gomont) Anagnostidis *et* Komárek 1988

01430100 **Pl. 10D,M** (p. 69) **CD**

Filaments single or occasionally intertwined, coiled irregularly or almost straight. Cells 1.8–2.5 μm wide, 1.2–4(–8) μm long; cross walls not narrowed, sometimes with a granule; end cell rounded, not tapering. Sheath thin, colourless.

Planktonic or epiphytic on submerged plants in fresh and brackish standing waters.

The original descriptions of this species and *L. limnetica* scarcely differ – slightly shorter cells in *L. lagerheimii* and a tendency to be slightly flexuous.

Lyngbya lutea [C.Agardh 1824 emend. Gomont 1890] Gomont 1892

01430130 **Pl. 10F** (p. 69)

Colony a leathery mat of densely packed filaments, yellow-brown to olive-green. Trichome not attenuated towards apex. Cells 2.5–6 μm wide, 1.5–5.5 μm long, cells one-third to as wide as long; cross walls not narrowed; end cell with rounded calyptra. Sheath up to 3 μm wide, lamellate, smooth.

In fresh and brackish waters and on rocks in well-illuminated humid situations, such as ones near the sea. Powell (1964).

Lyngbya lutescens (Meneghini) Hansgirg 1892

Basionym: *Calothrix lutescens* Meneghini in Kützing 1843
Synonym: *Phormidium lutescens* (Hansgirg) Anagnostidis *et* Komárek 1988

01430140

Colony consisting of bundles of filaments forming a floc or turf, up to 1.5 cm long, often yellowish. Filaments 2.5–3.3 μm wide. Sheath narrow, thin, colourless.

Fast-flowing streams, on rocks.

Lyngbya maior [Meneghini 1837] Gomont 1892

01430150 **Pl. 10P** (p. 69)

Colony a bundle of long, straight filaments. Cells 11–16 μm wide, 2–3.4 μm long; one-eighth to one-quarter as long as wide; cross walls sometimes slightly narrowed, often with irregularly spaced granules by the walls; end cell rounded, with slightly thickened membrane. Sheath colourless, lamellate.

In standing water and on wet mud. West & Fritsch (1927): Wimpole Park, Cambridgeshire.

There is little to separate this from *L. aestuarii* other than the sheath of the latter usually (? always) becoming yellow-brown in well-illuminated situations and more often occurring free-floating. *L. maior* has nomenclatural priority, but as *L. aestuarii* has been reported more frequently and covers a wider range of dimensions, the two names are retained.

Lyngbya majuscula [Harvey in Hooker 1833] Gomont 1892

01430160 **Pl. 10L** (p. 69)

Forming conspicuous benthic mats or floating flocs, which appear brownish blue-green or almost black when viewed on mass; mats up to 3 cm across, but floating material sometimes much more extensive. Filaments to 100 μm wide, very long, somewhat flexuous. Trichome not attenuated towards apex, usually blue-green, but sometimes grey or violet. Cells 16–50(–60) μm wide, 2–5 μm long; cross walls not narrowed; granules not aggregated near cross walls; end cell rounded, calyptra absent. Sheath to 10 μm wide, colourless, lamellate, often rough on outside.

WARNING! Material of this species from other countries has been shown to contain several biologically highly active molecules and should be treated as a possible carcinogenic risk. It is recommended to minimize contact with skin.

Powell (1964). Marine, upper intertidal and supralittoral pools, often floating. In the British Isles largely restricted to the Atlantic coast and most frequent in summer.

Lyngbya martensiana [Meneghini 1837] Gomont 1892

Synonym: *Porphyrosiphon martensianus* (Gomont) Anagnostidis *et* Komárek 1988

01430170 **Pl. 10G** (p. 69)

Filaments long, flexuous, usually forming blue-green tufts. Trichome not attenuated towards the end. Cells 6–10 μm wide, 1.7–3.3 μm long; one-quarter to half as long as wide; cross walls not narrowed, frequently with small (probably polyphosphate) granules on either side; end cell rounded, calyptra absent. Sheath thick, colourless, outer surface rough.

In standing fresh and slightly brackish water, flowing waters, very moist soil. West & Fritsch (1927): one of the two most abundant British freshwater species.

This is like a narrower version of *L. confervoides* (see above).

Lyngbya ochracea [Thuret 1875] Gomont 1892

Synonym: *Leptolyngbya ochracea* (Gomont) Anagnostidis *et* Komárek 1988

01430180 **Pl. 10N** (p. 69)

Colony a thickly intertwined mass of ochre- or brownish-yellow filaments. Cells 0.8–1 μm wide, 0.6–0.8 μm long, blue-green; cross walls narrowed; end cell rounded, calyptra absent. Sheath at first thin and colourless, later

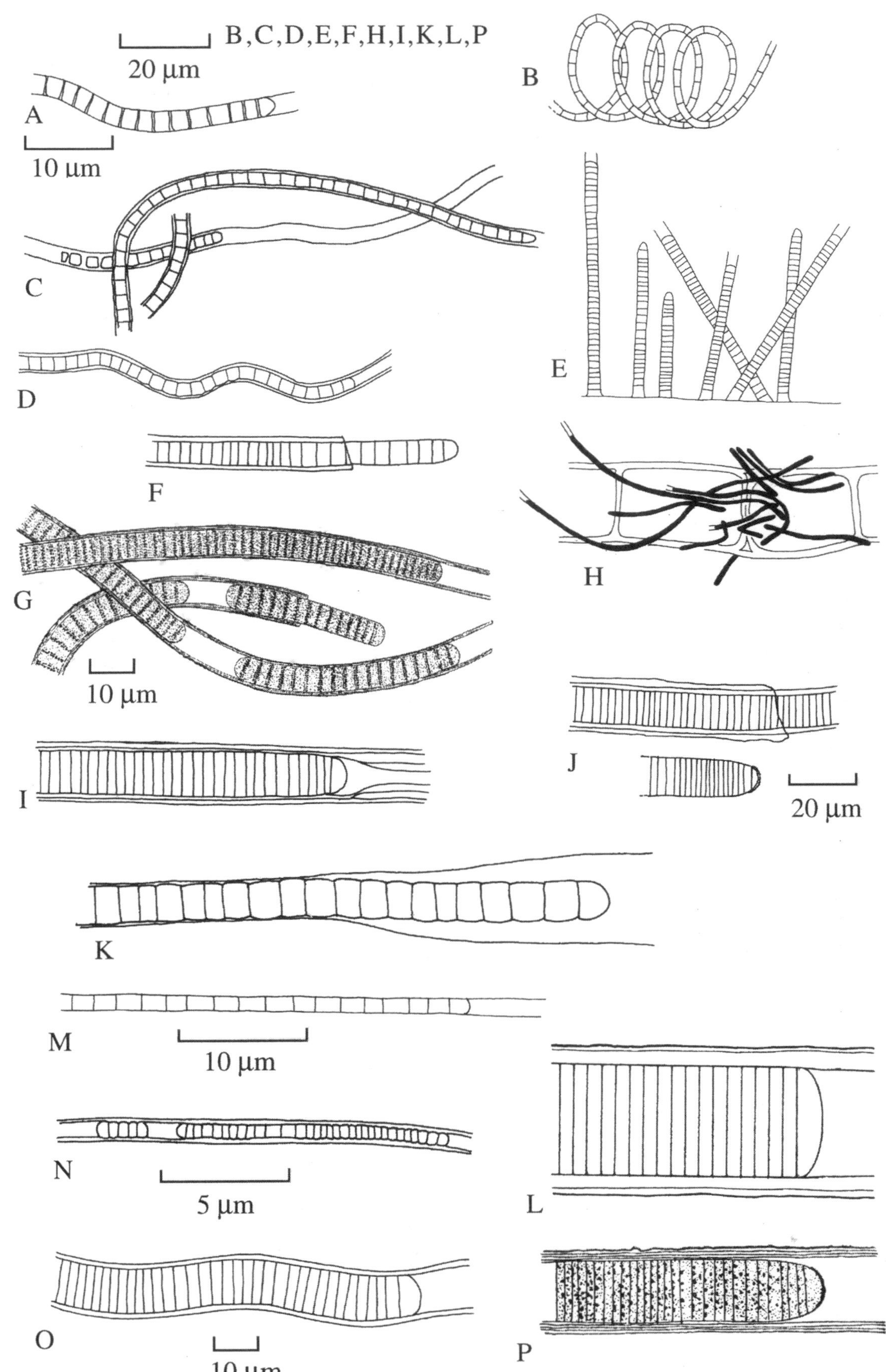

Plate 10 A–P
A. *Lyngbya diguetii* (p. 67); **B**. *Lyngbya contorta* (p. 67); **C**. *Lyngbya vandenberghenii* (p. 67); **D, M**. *Lyngbya lagerheimii* (p. 68); **E**. *Lyngbya kuetzingii* (p. 67); **F**. *Lyngbya lutea* (p. 68); **G**. *Lyngbya martensiana* (p. 68); **H**. *Lyngbya epiphytica* (p. 67); **I**. *Lyngbya semiplena* (p. 70); **J**. *Lyngbya aestuarii* (p. 67); **K**. *Lyngbya gracilis* (p. 67); **L**. *Lyngbya majuscula* (p. 68); **N**. *Lyngbya ochracea* (p. 68); **O**. *Lyngbya spirulinoides* (p. 70); **P**. *Lyngbya maior* (p. 68).

ochre-yellow due to deposition of iron oxide(s). (It is important to check that cells are photosynthetic, as sheathed colourless bacteria occur in similar habitats.)

In standing or flowing water, where reducing conditions permit iron-rich water to come into contact with the alga. West & Fritsch (1927): occurs in waters containing much iron.

Lyngbya rivularianum Gomont 1892

01430240 (new number)

Filament single, in mucilage of other algae, usually flexuous. Cells 0.7–0.9 μm wide, 2–3.5 μm long; cross walls not narrowed, or with adjacent granules; end cell rounded, not narrowed. Sheath narrow, colourless.

Very narrow sheathed trichomes, often in old colonies of *Rivularia* and, to a lesser extent, *Nostoc*. Batters reported it from Dorset inside sheaths of *Microcoleus chthonoplastes* (Frémy, 1929–33). Anand (1937): abundant in mucilage of Chrysophyceae–*Entocladia–Lyngbya* community in the Chysophyceae belt on chalk cliffs at Westgate, Kent. Powell (1964), without detail.

The above dimensions broaden the range given by Gomont (0.75–0.8 μm), but there is a considerable diversity of such filaments in nature. They perhaps merge with narrow forms of *Schizothrix*, though the commonest species (*S. coriacea*) can be distinguished by the conical shape of at least some end cells.

Lyngbya semiplena [(C.Agardh 1827) J.Agardh 1842] Gomont 1892

01430190 **Pl. 10I** (p. 69)

Colony a tuft or turf, to 3 cm high, typically dull yellow-green. Filaments prostrate near the base, then more or less upright, intermingled. Trichome yellow-green or blue-green, slightly attenuated towards the end. Cells (5–)7–10(–12) μm wide, 2–3 μm long; cross walls not narrowed, often with small (probably polyphosphate) granules on either side of cross walls; weakly conical or round calyptra. Sheath to 3 μm thick, colourless, lamellate in old parts of filament, slightly mucilaginous.

Powell (1964). Marine, especially on rocks in pools in the supralittoral, but also on rocks and larger algae in the upper intertidal.

L. semiplena is much like a narrower form of *L. confervoides*, though the size range of the former more or less comes within that of the latter. *L. semiplena* can form a calyptra, but many trichomes may not show this. Frémy (1929–33) states that the upright filaments of *L. semiplena* are more contorted. The two probably represent a continuum of forms, but both are retained for now as species.

Lyngbya sordida [Zanardini 1843] Gomont 1892

01430200

Colony a tuft, to 3 cm high, dark or dull yellow-green. Filaments more or less straight. Trichomes not attenuated towards apex, typically olive or pinkish green. Cells 14–31 μm wide, 4–10 μm long; cross walls narrowed; end cell rounded, calyptra absent. Sheath colourless, smooth.

Marine: on shaded rocks and algae in lower intertidal and sublittoral, extending to considerable depths.

This is quite like *L. majuscula*, but the cells are somewhat longer in relation to width and slightly narrowed at the cross walls. Where the size range overlaps, it is also quite similar to *L. confervoides*.

Lyngbya spirulinoides [Gomont 1890] Gomont 1892

01430210 **Pl. 10O** (p. 69)

Colony typically a free-floating dull green tuft of intertwined filaments. Filaments often show a regular pattern of well-spaced coils, which show as weak undulations on isolated filaments. Cells 14–16 μm wide, 4–6.8 μm long; cross walls not narrowed; end cell rounded, calyptra absent. Sheath thin, not lamellate, colourless.

In standing freshwaters.

Several species of *Lyngbya* show a tendency to form a weak helix when growing intertwined, so it is doubtful whether it is justified to recognize *L. spirulinoides* as a distinct species. Apart from the presence of the weak helix, *L. spirulinoides* is quite similar to *L. maior*.

Lyngbya vandenberghenii Symoens *et* van der Werff 1951

01430220 **Pl. 10C** (p. 69)

Colony a thin crust of lightly calcified filaments. Filaments 2.5–4 μm wide, more or less straight. Cells 2.2–3 μm wide, 2.5–4(–5) μm long; cross walls not narrowed; end cell round, calyptra absent. Sheath firm, colourless, but usually with attached calcite crystals.

At sites in the British Isles the species often gives a light-blue colouration to the surface of rocks in rivers, due to the combination of blue-green algal colour and moderate calcification. Typically it occurs on non-calcareous rocks or calcareous sandstone in fast-flowing rivers with calcareous water; it sometimes covers extensive stretches of small boulders and cobbles in summer under low flow conditions. Present in upper stretches of several rivers in north-east England, but especially River Tees; it is often the most downstream organism to show evidence of calcification.

The species was described from Belgium by Symoens & van der Werff (1951) as one of three *Lyngbya* species in travertine-forming springs. However, the description also fits well a common organism in some British rivers.

Lyngbya versicolor [Wartmann 1861] Gomont 1892

01430250 (new number)

Colony a mass of densely intertwined filaments, usually yellowish on outside, dirty green inside. Cells 2.8–3.2 μm wide, 2–6.4 μm long; cross walls neither narrowed nor with granules; end cell rounded, not narrowed, calyptra absent. Sheath to 3 μm, slightly lamellate, usually yellow on outside filaments.

Standing freshwaters, initially benthic, but often subsequently floating.

This differs from *L. diguetii* mainly in its sheath being thicker and yellowish when exposed to full light.

Microcoleus [Desmazières 1823] Gomont 1892

01480000

Filaments unbranched or sparsely branched, each sheath with a number of trichomes. Sheath colourless (or almost so), more or less cylindrical, not lamellate, but sometimes gelatinizing when old. Trichomes usually many in one sheath, densely aggregated and sometimes coiled and appearing like a rope; the trichomes are sometimes highly motile and may be seen moving backwards and forwards within the sheath. End cell usually conical, seldom capitate.

The genus is especially characteristic of saline environments, such as salt marshes and the soil surface in arid regions, but it sometimes occurs in non-saline environments. It is often associated with slightly reducing conditions and, in the case of non-saline sites, probably mostly so. A large number of species have been described from freshwater or non-saline terrestrial environments, but only three of the six British species have been reported from non-marine environments.

In addition to the species listed below, W. West & G.S.West (1896; figure reproduced in West & Fritsch, 1927) described *M. delicatulus*, with unbranched filaments and cells 1.5–2 μm wide, in standing water among other algae (Glen Tummel, Scotland). Further studies are needed to confirm whether it should be included in *Microcoleus* or one of the other genera with sheaths including a number of trichomes. Powell (1964) listed *M. vouki* Frémy 1932, but Frémy (1929–33) gives no location for the British Isles; as it differs from *M. tenerrimus* only in its pink colour and absence of narrowing at the cross walls, it should be treated as synonymous.

Other record Benson-Evans & Antoine (1996): *M. acutissimus* Gardner, without detail.

1 Marine or saline 2
1 Not obviously saline...3
2 Trichomes ≤2 μm wide.....................*M. tenerrimus*
2 Trichomes >2 ≤6 μm wide........*M. chthonoplastes*
2 Trichomes >8 μm wide....................*M. acutirostris*
3 End cells often capitate ....*M. vaginatus*
3 End cells not capitate...4
4 Trichome ≤6 μm wide..5
4 Trichome >6 μm wide...................*M. subtorulosus*
5 End cell rounded-conical; terrestrial habitats......................................*M. chthonoplastes*
5 End cell pointed-conical; freshwater ....*M. lacustris*

Microcoleus acutirostris Gomont 1892

01480050 (new number)
Colony a mat of flexuous filaments. Filament more or less straight, with many trichomes. Cells 9–11 μm wide, 3.5–5.5 μm long, 2–3 times shorter than wide; cross walls distinctly narrowed. End cell sharply conical, not capitate.

Powell (1964). Intertidal sediments in estuaries and brackish soils.

The species is quite like *M. subtorulosus*, apart from the considerably shorter cells.

Microcoleus chthonoplastes [Thuret 1875] Gomont 1892

01480010
Colony a mat of filaments, sometimes layered, dark green; filaments sometimes single. Filament a broad sheath with a tight bundle of trichomes towards the centre; trichomes sometimes almost fill the sheath, but other times there is considerable space between them and sheath. Cells 2.5–6 μm wide, 1–2 times longer than wide; cross walls slightly narrowed, without granules; end cell not capitate, pointed-conical. Sheath often diffluent.

In the upper intertidal and on sand and soil in the supralittoral and other saline habitats, including soils influenced by sea spray, where it sometimes covers a considerable area; occasionally also in inland saline areas. Powell (1964).

Microcoleus lacustris [(Rabenhorst) Farlow 1877] Gomont 1892

01480030 **Pl. 9F** (p. 65)
Colony a mat of filaments or single filament, dark blue-green. Filament a broad sheath with a tight bundle of trichomes towards the centre. Cells (3.6–)4–5 μm wide, 6–12 μm long; end cell not capitate, rounded-conical; cross walls narrowed. Sheath colourless.

In standing freshwater.

Apart from the environment and the tendency for the end cell to be less sharply conical, this description falls within the limits of *M. chthonoplastes*. However, the two species are left here, because both names have been used widely and freshwater forms in particular need more thorough observation.

Microcoleus subtorulosus [Brébisson in Kützing 1845–49 emend. Gomont 1890] Gomont 1892

01480060 (new number)
Colony a mat, highly mucilaginous. Filament more or less straight, with many trichomes. Cells 6–10 μm wide, 5–10 μm long, cross walls distinctly narrowed. End cell conical or cylindrical-conical, not capitate.

On fine sediments or submerged plants in slow-flowing or standing freshwater. West & Fritsch (1927): Scotland.

M. subtorulosus merges with wide forms of *M. chthonoplastes*.

Microcoleus tenerrimus [Crouan 1865] Gomont 1892

01480040 **Pl. 9D,E** (p. 65)
Colony a mat of interwoven filaments, dark green; filaments sometimes single. Filament a broad sheath with a tight bundle of trichomes towards the centre. Trichome apices with slight tapering. Cells 1.5–2 μm wide, 2–3 times longer than wide, often slightly barrel-shaped; cross walls slightly narrowed, without granules; end cell not capitate, sharply conical. Sheath often diffluent.

Powell (1964). Epiphytic on old algae in upper intertidal and in rock pools in the supralittoral.

This differs from *M. chthonoplastes* only in the narrower trichomes and tendency for cells to be slightly more elongated in relation to width.

Microcoleus vaginatus [Vaucher 1803] Gomont 1892

01480020 **Pl. 9G** (p. 65)
Colony a mat of interwoven filaments, dark green; filaments sometimes single. Filaments occasionally branched; trichomes sometimes almost fill the sheath; many trichomes in wider part of sheath, but often only one trichome in terminal branches. Cells 3.5–7 μm wide, 0.5–1.5 times longer than wide, cross walls not

narrowed, often with granules; end cell capitate, often with calyptra. Sheath diffluent.

On moist soil, often mixed with other algae or old moss clumps, probably often in slightly reducing environments.

Oscillatoria [Vaucher 1803] Gomont 1892

01530000

Trichome single or aggregated, free-floating or attached to a surface, sometimes forming macroscopic amorphous flocs or benthic mats. Sheath typically absent, though secreting a thin layer of mucilage when moving and sometimes a thin firmer sheath under conditions when growth is restricted. Trichomes usually capable of gliding motility when attached to a substratum and sometimes also when adjacent to each other, usually, if not always, with rotation of the longitudinal axis. In general, motility is most obvious when trichomes are nutrient-rich, though microscopy of populations apparently growing under favourable conditions may show considerable variation between individual trichomes in the rate of movement. The structure of the trichome, shape of its terminal region and the shape of the end cell are all important in identification of species. Increase occurs by trichome fragmentation or the formation of hormogonia. In some species these processes involve the formation of a necridial cell. Some species forming populations on bottom sediments may subsequently float to the surface, usually buoyed by trapped gas bubbles. Once at the surface some populations may continue to float for some days, whereas others may alternate between an attached and a free-floating state.

Various names have been given to very narrow forms of *Oscillatoria* (see account for *O. subtilissima*). However, such forms are probably more widespread than indicated by the literature and have probably often been overlooked as possible filamentous bacteria, such as flexibacteria. Unless the trichomes are clumped, the narrowest forms require an epifluorescence microscope to confirm their classification as a blue-green alga.

Among the genera separated by Anagnostidis *et* Komárek (1988) is *Tychonema*. They list a number of generic features, but the main ones are the presence of what are termed liquid vacuoles in the present flora and the absence of gas vacuoles in sometimes planktonic organisms. The two species in the British flora are *O. bornetii* and *O. bourrellyi*. The suggestion of the present author is that these are organisms adapted to live in environments which are phosphorus-rich for part of the year and phosphorus-limited for the rest. At least in the English Lake District, the former conditions tend to occur in winter and the latter in summer, so other features are perhaps related to other features the environment. Whether these and the other species included in *Tychonema* by Anagnostidis *et* Komárek really are a very distinct group or merely a collection of forms which have evolved morphological and physiological features to deal with phosphorus limitation remains to be established.

Most of the names recorded for the British Isles are listed, although there is considerable scope for treating some as synonyms. Two names in the *Coded List* are treated here as synonyms: *O. angustissima*, *O. subbrevis*. One is added: *O. bourrellyi*.

Other records

Benson-Evans & Antoine (1996) list *Oscillatoria acutissima* Kufferath 1914, stating that it occurs in flowing and standing water and in the mucilage of colonial diatoms; however, it seems unlikely that *Oscillatoria* would occur in this microhabitat, so the status is uncertain.

Powell (1964) lists *Oscillatoria rosea* [Crouan frat.] Gomont, but Gomont (1886) merely lists this among *Species inquirendae*. Frémy (1929–33) lists it as *O. rosea* (Crouan) Batters 1896.

Oscillatoria agardhii Gomont 1892

Synonym: *Planktothrix agardhii* (Gomont) Anagnostidis *et* Komárek 1988

01530010 **Pl. 11A** (p. 79) **CD**

Most populations in the British Isles can be distinguished as either the type or var. *isothrix*. The following description applies to the type; it does not incorporate the range shown by the variety, which is listed separately.

Trichomes usually planktonic, single or aggregated into a mass, straight or somewhat curved, slightly attenuated towards apex. Cells 4–6 μm wide, 2.5–4 μm long, mostly shorter than wide; cross walls not narrowed, sometimes with granules near cross walls; gas-vacuolate; end cell rounded, bluntly conical or more or less pointed, sometimes capitate, with rounded calyptra.

Planktonic in freshwater and brackish lakes, ponds and rivers, often forming dense populations, where it can be a serious nuisance, as in the brackish water of Preston Docks.

var. *isothrix* Skuja 1948

Synonym: *Plantothrix mougeotii* Anagnostidis *et* Komárek (see note below)

01530012

Trichomes planktonic, single, straight, neither attenuated nor curved towards apex. Cells 6–9.5 μm wide, gas-vacuolate; end-cell truncate-rounded, sometimes with thickened membrane; caplyptra absent. This differs from the type mainly in the greater cell dimensions (6–9.5 μm), one-third to three-quarters as long as wide (rarely quadrate), lack of attenuation at end of trichome and absence of calyptra.

Larger freshwater lakes. At least in the lakes where this occurs in the English Lake District, it is always green (rather than blue-green or other colour) (J.W.G. Lund, pers. comm.); Skuja (1948) did not comment on colour.

This organism is listed by Reynolds & Irish (2000) as synonymous with *Planktothrix mougeotii* (Bory ex Gomont) Anagnostidis *et* Komárek. However, there is no description of *Oscillatoria mougeotii* in Gomont (1892). *Oscillatoria mougeotii* Kützing 1849 (not in Gomont, but in Geitler, 1932a, based on description by Lemmermann) is slightly different: blue-green, with benthic and planktonic forms, cell width 5.5–7.5 μm.

Oscillatoria amoena [Kützing 1843] Gomont 1892

01530020

Trichomes growing on (usually) submerged surfaces. Trichome straight, attenuated towards apex. Cell 2.5–5 μm wide; cross walls usually slightly narrowed, often with granules at cross walls, tapering at ends, 2.5–5 μm wide, 2.5–4.2 μm long; end cell capitate, with cone-shaped calyptra.

On pond sediments, submerged plants and very wet mud.

Oscillatoria amphibia [C.Agardh 1827] Gomont

Synonym: *Phormidium acutissimum* (Gomont) Anagnostidis *et* Komárek

01530030

Trichome straight or curved, not attenuated towards apex, blue-green or pale green. Cells 2–3(–3.5) µm wide, 4–8.5 µm long, 2–3 times longer than wide, cross walls not narrowed; often with granules; end cell rounded, not capitate, calyptra absent.

On surfaces and planktonic in fresh and brackish standing waters, and on very wet soil. Powell (1964).

Oscillatoria anguina [Bory 1827] Gomont 1892

01530040

Forming dark green film on submerged objects, or intermingled with other algae. Trichome straight over most of the length, but bent and sometimes twisted in apical region, slightly attenuated towards apex. Cells 6–8 µm wide, 1.5–2.5 µm long, cross walls not narrowed, often with adjacent granules; end cell slightly narrowed, with calyptra.

In standing, shallow or muddy water, or on very moist soil.

Oscillatoria angusta Koppe 1923

Synonym: *Jaaginema metaphyticum* Anagnostidis *et* Komárek

01530050

Trichomes loosely entangled to form a thin film. Trichome straight. Cells 0.8–1.2 µm wide, 5–7 µm long, cross walls not narrowed, without adjacent granules; end cell rounded, calyptra absent.

Usually mixed with other blue-green algae, together forming a slimy layer in shallow water.

Oscillatoria angustissima W. *et* G.S.West 1897

01530060: see *Oscillatoria subtilissima*

Oscillatoria bonnemaisonii [Crouan in Desmazières 1858] Gomont 1892

01530310 (new number)

Colony dirty blue-green to violet-black. Trichomes slightly flexible. Cells 18–36 µm wide, 3–6 µm long; cross walls narrowed, without adjacent granules; end cell convex, without cap or calyptra.

Powell (1964). Sediments in highly brackish pools and on moist walls and rocks very close to the sea.

Frémy (1929–33) lists several sites on north coast of France, though none for British Isles.

Oscillatoria bornetii (Zukal) Forti 1907

Basionym: *Lyngbya bornetii* Zukal 1894

Synonym: *Tychonema bornetii* (Zukal) Anagnostidis *et* Komárek

01530070

Trichomes exist in two forms, which are at least partially influenced by their phosphorus status. The characteristic form by which the species has been recognized and which apparently reflects low phosphorus status is as follows. Trichomes usually straight, though occasionally slightly flexuous; single; trichome often slightly tapering towards one end. Cells (of type) 12–16 µm wide, 4–10 µm long, shorter than wide; very pale green or almost colourless, containing numerous liquid vacuoles, giving the cytoplasm a frothy appearance under the light microscope; cross walls not narrowed; end cell rounded, with a thickened membrane.

Under high phosphorus conditions in laboratory culture, the trichomes (of strains sent from Norway by O. Skulberg) appear quite different. Trichomes not tapered at the ends, pale red-brown, brownish-violet or green; frothy appearance largely or entirely absent, but sometimes with more intense contents running in lines between cells (? localized regions of photosynthetic material); sometimes, but not always, with granules (probably polyphosphate) adjacent to cross wall. No thickened membrane on end cell. Such trichomes resemble a more typical *Oscillatoria*, like those shown in the figures of Skuja (1964) reproduced by Anagnostidis & Komárek (1988).

Floristic descriptions report that the trichomes occasionally have a very thin sheath or that groups of trichomes can form slimy tufts; it is unclear whether these features apply to one or both stages.

Planktonic and benthic in lakes, ponds and ditches.

Geitler (1932a) questions whether *O. decolorata* G.S.West 1899 (in stagnant ditches: reference 1899b) is identical with *O. bornetii*. West & Fritsch (1927) stated that the trichomes are 'destitute of pigment and lead a saprophytic existence'. It seems more likely that Geitler's suggestion is correct, if West's organism was in fact a blue-green alga.

Two forms have been recognized, based almost entirely on trichome width: 8–10 µm wide = fo. *intermedia* (Woronichin) Elenkin 1938; 5.5–7(–8) µm wide = fo. *tenuis* Skuja 1929. The lower limit overlaps with the 3.8–6 µm range for *O. bourrellyi* (Skulberg & Skulberg, 1991). It seems probable that there is a continuum of forms with different widths, but cells of which contain numerous liquid vacuoles. It remains to be established whether these are a genetically homogeneous group or merely diverse *Oscillatoria* which share the feature of numerous liquid vacuoles, at least under some growth conditions.

Oscillatoria bourrellyi J.W.G.Lund 1955

Synonym: *Tychonema bourrellyi* (Lund) Anagnostidis *et* Komárek

01520310 **CD**

Trichomes individual in plankton. Cells 4–5(–6) µm wide, 2.5–6.6 µm long; cross walls not or very slightly narrowed; end cell rounded, with a slightly thickened membrane; gas vacuoles absent. The characteristic feature by which the species is recognized is the pale vacuolated appearance of the cytoplasm; this is probably similar to the condition in *O. bornetii* i.e. due to presence of many liquid vacuoles.

The organism exists in two forms in Windermere and Ullswater, where it has been studied in some detail (J.W.G. Lund, pers. comm.). The trichomes, when viewed on mass, are reddish-purple to yellow-brown in winter, lack the vacuolated appearance, possess obvious granules (polyphosphate) and are capable of movement; individual trichomes in summer, with their vacuolated appearance, are very pale (Fig. 423 in Canter-Lund & Lund, 1995), lack obvious granules and are incapable of movement; a mass of harvested material is pale yellow-green or greyish. Both conditions have a similar trichome width. (This information is largely based on studies prior to improvements in sewage treatment work effluents; the organism is now much less frequent.)

Planktonic.

The species is much like a narrower version of *O. bornetii*, at least when the cytoplasm shows the vacuolated appearance, and has sometimes been recorded as *O. bornetii* fo. *tenuis* (O. Skulberg, pers. comm.). It is unclear whether the winter form has ever been recorded as a distinct species. It seems likely that most of the general comments on *O. bornetii* also apply to *O. bourrellyi*.

Oscillatoria brevis [Kützing] Gomont 1892

01530080 **Pl. 11B** (p. 77)

Forming mats, sometimes quite extensive. Trichome straight, apart from slight attenuation and tendency to be bent in apical one or two cells. Cells 4–6.5 μm wide, 1.5–3 μm long, 0.3–0.5 times as long as wide; cross walls not narrowed, often with granules. End cell rounded-conical, calyptra absent.

Fresh and salt waters, on wet mud and moist walls. Powell (1964).

O. brevis is a variable species and trichomes in a population often fail to show attenuation; possibly the species merges with *O. subbrevis*. *O. brevis* is also quite like another variable species, *O. tenuis*, but the latter shows distinct narrowing at the cross walls.

Oscillatoria chalybea [Mertens in Jürgens 1822] Gomont 1892

Synonym: *Phormidium chalybeum* (Gomont) Anagnostidis *et* Komárek

01530090

Forming a dark green coherent mass. Trichome straight or slighly twisted, attenuated at the apex and somewhat bent. Cells 8–13 μm wide, 3.8–8 μm long, 0.3–0.5 times as long as wide; cross walls slightly narrowed, typically without adjacent granules, end cell conical or broadly rounded, calyptra absent.

Standing water, on mud, stones, brackish waters.

Oscillatoria corallinae [Kützing 1849] Gomont 1892

01530320 (new number)

Forming a thin film covering benthic organisms. Trichomes bright green, blue or pale brown; long, flexible, scarcely attenuated towards apices. Cells 6–10 μm wide, 2.7–4 μm long; cross walls without adjacent granules; end cell subcapitate, with slightly thickened, convex membrane.

Powell (1964). Marine, sublittoral, growing on larger algae and sessile colonial animals.

Oscillatoria cortiana [Meneghini 1837] Gomont 1892

Synonym: *Phormidium cortianum* (Gomont) Anagnostidis *et* Komárek

01530100 **Pl. 11C** (p. 77)

Forming a dark blue-green thallus. Trichome straight; ends gradually tapering, bent. Cells 5.5–8 μm wide, 5.4–8.2 μm long (to 14 μm at the ends); typically slightly shorter than wide in the middle of trichome, but considerably longer than wide in the apical few cells; cross walls narrowed, without adjacent granules, end cell blunt-conical, calyptra absent.

Freshwaters, on submerged plants.

This is one of a number of widespread species with rather similar dimensions: see comments under *O. formosa* and *O. laetevirens*.

Oscillatoria formosa [Bory 1827] Gomont 1892

Synonym: *Phormidium formosum* (Gomont) Anagnostidis *et* Komárek

01530110 **Pl. 11D** (p. 77)

Colony blue-green. Trichome straight, flexible, gradually tapering at ends, blue-green. Cells 4–6 μm wide, 2.5–5 μm long, half as long as wide or quadrate; cross walls slightly narrowed, sometimes with adjacent granules; end cell blunt-conical, calyptra absent.

Benthic or floating in standing fresh and brackish water; moist soil and rocks; springs with sulphide-rich water. Powell (1964).

This species scarcely differs from *O. cortiana*; the most distinctive feature is the greater tendency for the apex of the trichome to taper in *O. cortiana*, with the apical cells being considerably longer than those in the rest of the trichome.

Oscillatoria granulata Gardner 1927

Synonym: *Tychonema granulatum* (Gardner) Anagnostidis *et* Komárek

01530120

Trichomes single, long and flexible, or aggregated as small mucilaginous flakes. Cells 3.4–3.8 μm wide, cross walls not narrowed, but with adjacent granules; end cell not capitate; calyptra absent.

Typically, scattered among other algae, in standing freshwater.

Oscillatoria hamelii Frémy 1930

Synonym: *Phormidium hamelii* (Frémy) Anagnostidis *et* Komárek

01530130

Trichome single, somewhat flexuous, not attenuated towards apex. Cells 4.8–5 μm wide, 7.2–8 μm long, about 1.5 times longer than wide; cross walls scarcely narrowed, without adjacent granules; end cell rounded, not capitate, calyptra absent.

Standing water.

The species was originally recorded by Frémy from French Equatorial Africa, but trichomes fitting the description occasionally occur in samples from British Isles.

Oscillatoria irrigua [Kützing 1843] Gomont 1892

01530140

Forming dark blue-green colonies. Trichome straight. Cells 6–11 µm wide, 4–11 µm long, 1–1.5 times longer than wide; cross walls not narrowed, sometimes with adjacent granules; end cell convex with a dense apical layer.

Standing and fast-flowing water; moist rocks. West & Fritsch (1927): common in running water.

Oscillatoria laetevirens [Crouan] Gomont 1892

Synonym: *Phormidium laetevirens* (Gomont) Anagnostidis *et* Komárek

01530150

Attached growth thin, membranous, bright green. Trichome mostly straight, rigid, fragile, attenuated towards apex and bent near the very end; yellowish green. Cells 3–5 µm wide, 2.5–5 µm long, more or less quadrate; cross walls slightly narrowed, often with adjacent granules; end cell weakly rounded, not capitate, calyptra absent.

Powell (1964). Epilithic in marine and occasionally brackish.

Frémy (1929–33) pointed out the similarity with *O. brevis* and *O. formosa* (also *O. cortiana*: see above), but pointed out features which can at least sometimes separate them. The cross walls are slightly narrowed in *O. formosa* and *O. laetevirens*, but not *O. brevis*. The trichome colour helps to distinguish *O. laetevirens*.

Oscillatoria limnetica Lemmermann 1900

01530160 **Pl. 11E,F** (p. 77)

Trichomes straight or slightly flexuous, pale blue-green. Cells 1.3–2.2 µm wide, 4–12 µm long, 2.5–8 times as long as wide; cross walls distinctly narrowed; end cells rounded, not capitate, calyptra absent.

Planktonic and on sediments in eutrophic standing waters. Geitler (1932a) also reports the species in the mucilage of other algae.

This organism poses taxonomic and nomenclatural problems. At least some populations are almost certainly merely a growth form of *O. redekei*, as shown for material from St James's Park Lake, London (Whitton & Peat, 1969) and other data of the author. Trichomes recorded as *O. limnetica* and *O. redekei* in (Lake) Windermere are probably the same organism, although experimental studies are needed (J.W.G. Lund, pers. comm.). While it seems likely that *O. redekei* can always develop a characteristic *O. limnetica* growth form, it is uncertain whether the reverse is true. The two names should be retained until critical studies have been made on a wide range of single trichome isolates, in case *O. limnetica* proves to cover quite different organisms. *O. limnetica* has nomenclatural priority over *O. redekei*; the two organisms have also been considered as two different genera, *Pseudanabaena* and *Limnothrix*.

Oscillatoria limosa [C.Agardh 1812] Gomont 1892

01530170 **Pl. 11G,H** (p. 77)

Forming dark blue-green to brown mats. Trichome more or less straight, blue-green, brown or olive-green. Cells (11–)13–16(–22) µm wide, 2–5 µm long, one-sixth to one-third as long as wide; cross walls not or only slightly narrowed, frequently with adjacent granules; end cell rounded, with slightly thickened membrane.

Standing or slowly flowing fresh or slightly brackish nutrient-rich water, benthic or floating; also on moist nutrient-rich terrestrial habitats such as mud and base of trees. West & Fritsch (1927): (one of the two most) abundant species in British Isles. Powell (1964).

Some populations can form a sheath and then look rather like *Lyngbya aestuarii*. Possibly the two species are closely related.

Oscillatoria margaritifera [Kützing 1845] Gomont 1892

01530180

Forming very dark green mucilaginous masses. Trichome straight, though slightly attenuated and bent towards apex. Cells 17–29 µm wide, 3–6 µm long, one-sixth to one-third as long as wide; cross walls narrowed, often with adjacent granules; end cells slightly capitate, with a slightly convex calyptra.

Marine or highly brackish pools, shallow lakes and ditches; benthic or floating.

Oscillatoria nigra Vaucher 1803

Synonym: *Phormidium nigrum* (Vaucher) Anagnostidis *et* Komárek

01530190

Colony a somewhat leathery mass, olive, dark brown or blackish, glistening. Trichome straight or somewhat flexuous (? flexible). Cells (6–)8–9(–10) µm wide, one-third to as long as wide; cross walls sometimes slightly narrowed, often with adjacent granules; end cell rounded-flattened.

On sediments or free-floating in standing and slowly flowing freshwater.

Oscillatoria nigro-viridis [Thwaites in Harvey 1846–51] Gomont 1892

Synonym: *Phormidium nigro-viride* (Gomont) Anagnostidis *et* Komárek

01530200

Forming an extensive dark green mass. Trichome olive-green. Cells 7–11 µm wide, 3–5 µm long, 0.5–1.25 as long as wide; cross walls narrowed, sometimes with adjacent granules; end cell slightly capitate, with a slightly thickened convex outer membrane.

Marine or close to the sea: on moist or submerged rocks, walls and piers. Powell (1964).

Oscillatoria princeps [Vaucher 1803] Gomont 1892

01530210

Trichomes often forming extensive mats, blue-green, brownish, violet or reddish. Trichomes mostly straight, at the apex lightly and abruptly bent; gas vacuoles often present in apical region. Cells 16–60 µm wide, 3.5–7 μm long, cells 0.1–0.25 as long as wide; cross walls not narrowed, without adjacent granules, blue-green to dirty green, slightly tapered at apices and bent. Cells up to 0.5 times as long as wide. End cells flatly rounded, slightly capitate, with or without a slightly thickened membrane.

Benthic or floating in standing or slow-flowing moderately nutrient-rich freshwaters.

Oscillatoria prolifica [Greville 1828] Gomont 1892

Synonym: *Planktothrix prolifica* (Gomont) Anagnostidis *et* Komárek 1988

01530220

Trichomes single or forming bundles, planktonic. Trichomes straight or flexuous, usually slightly attenuated towards apex; often purple or reddish. Cells 2.2–5 μm wide, 4–6 μm long, quadrate or slightly longer than wide; cross walls not narrowed, often gas-vacuolate; end cell capitate, with calyptra.

Planktonic in standing water, sometimes forming dense populations, but apparently not as common in the British Isles as some other parts of Europe.

Oscillatoria redekei Goor 1918

Synonym: *Limnothrix redekei* (Goor) Meffert 1988

01530230 **Pl. 11I,J** (p. 77)

Trichome straight, not attenuated towards ends. Cells 1.2–2.3 μm wide, 8–14 μm long, 3–7 times as long as wide; cells narrowed at cross walls, appearing rectangular (rather than rounded at the ends). Gas vacuole localized either side of the cross wall, sometimes small, but appearing like a long stretched triangle when light is markedly limiting. Trichome division often (? always) involves loss of a cell, leaving a gas vacuole projecting from wall of cell at end of trichome (Fig. 424 in Canter-Lund & Lund, 1995). End cell of trichomes with very small or no gas vacuoles, blunt or somewhat rounded, not capitate, calyptra absent.

Planktonic in lakes, but especially abundant in shallow highly eutrophic ones. It is a characteristic co-dominant in summer of many city ponds where a large duck population is fed by the public, including St James's Park, London. The filaments show only a slight tendency to be positively buoyant, so dense scums on the surface of such ponds are usually due to other species.

At least some strains can appear without gas vacuoles when grown under high light intensity and then appear indistinguishable from *O. limnetica*, raising the question whether these two names should be treated as synonymous (see account for *O. limnetica*).

Oscillatoria rubescens [de Candolle 1825] Gomont 1892

Synonym: *Planktothrix rubescens* (Gomont) Anagnostidis *et* Komárek 1988

01530240

Trichomes planktonic, single or forming bundles; straight, gradually tapering at ends; crimson or violet. Cells (5–)6–8 μm wide, 2–4 μm long, one-third to one-half as long as wide; cross walls not narrowed, often with granules; gas-vacuolate; end cell slightly capitate, with convex calyptra.

Planktonic in standing water, occasionally forming blooms. West & Fritsch (1927): found in the British Isles, forming purplish floating masses. Only a few modern records.

This species differs morphologically from some forms of *O. agardhii* in nothing more than its reddish colour due to phycoerythrin.

Oscillatoria sancta [Kützing 1845–49] Gomont 1892

01530250 **Pl. 11L** (p. 77)

Forming a thin, dark blue, shining, gelatinous film. Trichomes straight or flexuous (? flexible); slight and abrupt tapering at the ends; dull blue-green or olive-green. Cells 10–20 μm wide, 2.5–6 μm long, up to one-third as long as wide; cross walls usually distinctly narrowed, often with adjacent granules; end cell weakly rounded, with thickened membrane.

Standing and flowing freshwater, attached or floating; moist terrestrial surfaces. Powell (1964).

The species looks much like *O. limosa*, but is distinguished by the narrowing of the terminal one or two cells and a tendency for the cross walls to be more clearly narrowed.

Oscillatoria splendida [Greville 1824] Gomont 1892

Synonym: *Phormidium splendens* (Gomont) Anagnostidis *et* Komárek 1988

01530260

Mass bright blue-green or olive-green. Trichome straight or flexuous (? flexible), gradually attenuated towards apex. Cells 2–3 μm wide, 3–9 μm long, 2–4 times longer than wide; cross walls not narrowed, sometimes with adjacent granules; end cell typically bent, slightly twisted or hook-shaped, capitate, more or less rounded.

On sediments of moderately nutrient-rich fresh and brackish waters.

W. West & G.S.West (1896) described var. *attenuata*, but this more or less falls within the description of the type; its key feature is the strongly narrowed and elongated end cell.

Oscillatoria subbrevis Schmidle 1901

01530270 **Pl. 11M,N** (p. 77)

Trichomes single, more or less straight, not attenuated towards the apex. Cells 5–6 μm wide, 1–2 μm long; cross walls without granules; end cell rounded.

Planktonic in ponds and ditches. Material from the British Isles occasionally key out to this species, though the earlier records were all for tropical countries.

Possibly the British records – or perhaps all records – are merely a form of *O. brevis*. G.S.West (1907) reported material from Lake Tanganyika as *O. subbrevis*, but this was well outside the trichome range for the type (10–10.5 vs. 5–6 μm wide) or material found in the British Isles.

Oscillatoria subtilissima Kützing 1845–49

Synonym: *Jaaginema subtilissimum* (Kützing) Anagnostidis *et* Komárek

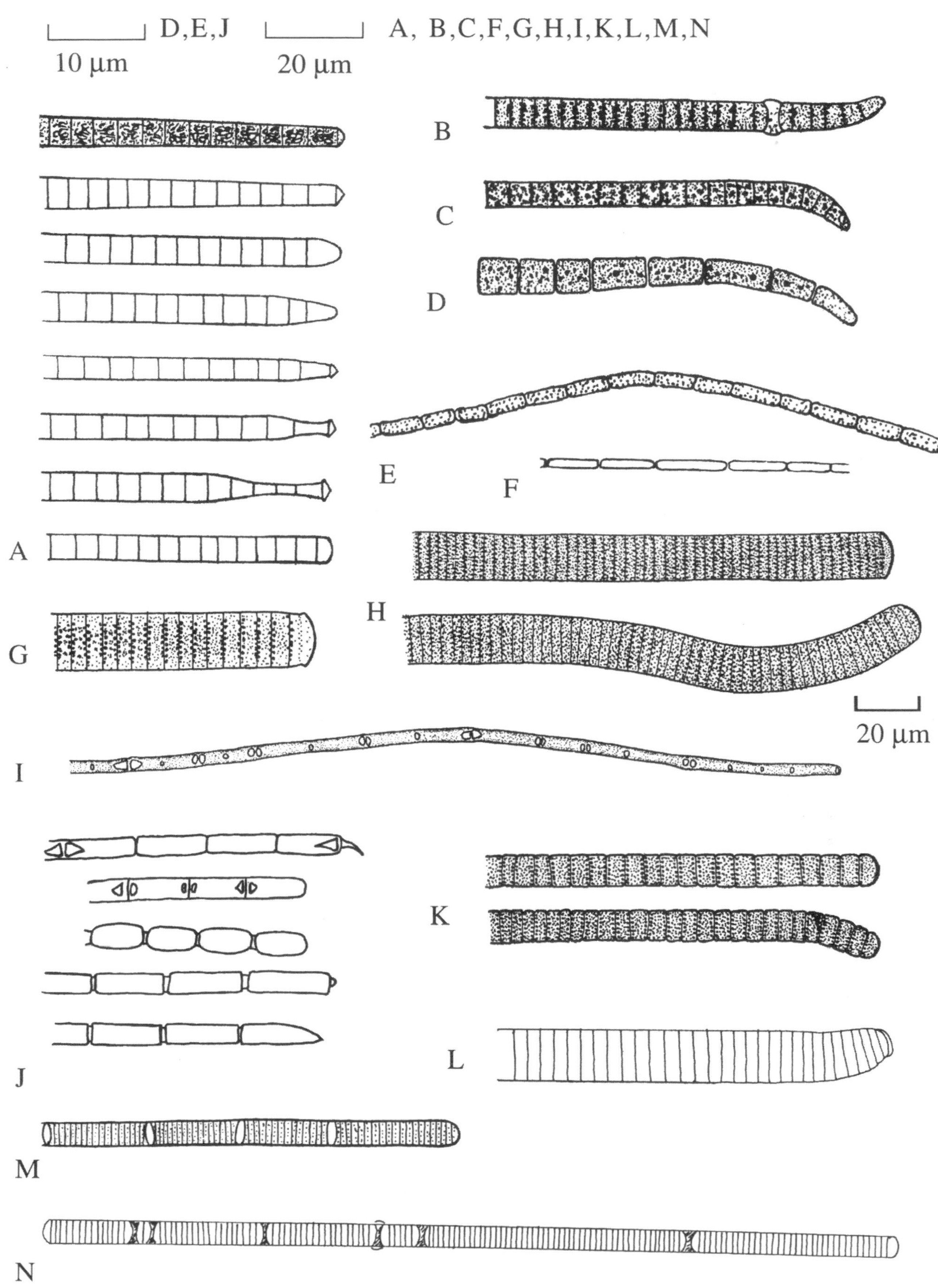

Plate 11 A–N
A. *Oscillatoria agardhii* (p. 72): upper five filaments common forms, others rarer; **B**. *Oscillatoria brevis* (p. 74); **C**. *Oscillatoria cortiana* (p. 74); **D**. *Oscillatoria formosa* (p. 74); **E, F**. *Oscillatoria limnetica* (p. 75); **G, H**. *Oscillatoria limosa* (p. 75); **I, J**. *Oscillatoria redekei* (p. 76); **K**. *Oscillatoria tenuis* (p. 78); **L**. *Oscillatoria sancta* (p. 76); **M, N**. *Oscillatoria subbrevis* (p. 76).

Synonym: *Oscillatoria angustissima* West *et* G.S.West 1897 (Coded List 01530060)
01530280
Trichome single or forming a blue-green or yellowish green layer; straight or flexuous (? flexible). Cells 0.5–1.5 μm wide; 1–2 times longer than wide.

A variety of reducing conditions, such as sulphide-rich springs and shallow water overlying organic-rich mud.

The type description was for trichomes 1–1.5 μm wide, while W. West & G.S.West (1897) described their species with trichomes 0.5–0.7 μm wide. Both descriptions give cells 1–2 times longer than wide, but little other detail. As very narrow trichomes of *Oscillatoria* are widespread in nature, it is suggested to include all those ≤1.5 μm wide and with cells only moderately longer than wide within *O. subtilissima*, while retaining *O. angusta* for cells of similar width, but much longer. Fluorescence microscopy may be needed to differentiate these from filamentous bacteria, particularly because some flexibacteria are yellow. Geitler (1932a) lists a number of other ill-defined species with rather similar dimensions, but these have apparently not been recorded for the British Isles.

Oscillatoria subuliformis [Kützing 1863] Gomont 1892

01530330 (new number)
Forming a green film. Trichomes flexible, undulate, long attenuated towards apex and especially flexible in this region. Cells in middle of trichome 4.7–6.5 μm wide, 4.7–6.3 μm long, but considerably narrower and longer towards apex (up 11 μm long); end cell obtuse, not capitate, calyptra absent.

Powell (1964). Details not known for British records, but Frémy (1929–33) lists a site near Brest, N. France at 20 m depth.

Oscillatoria tenuis [C.Agardh 1813] Gomont 1892

Synonym: *Phormidium tenue* (Gomont) Anagnostidis *et* Komárek 1988
01530290 **Pl. 11K** (p. 77)
Forming a thin, slimy, blue-green or olive-green film. Trichome straight, apart from sometimes showing a slight bend in the apical few cells; not attenuated towards apex. Cells 4–10 μm wide, 2.5–5 μm long, up to one-third as long as wide; cross walls narrowed, often with adjacent granules; end cell more or less hemispherical, with thickened outer membrane.

In fresh and slightly brackish waters; typically starting as benthic, but often subsequently floating. West & Fritsch (1927): (one of the two most) abundant species in British Isles. Powell (1964).

Oscillatoria terebriformis [C.Agardh 1827] Gomont 1892

Synonym: *Phormidium terebriforme* (Gomont) Anagnostidis *et* Komárek 1988
01530300
Typically a dull blue film. Trichome slightly attenuated and twisted at apex. Cells 4–6.5 μm wide, 2.5–6 μm long, one-half as long as wide to quadrate; end cell rounded or almost blunt, not capitate, calyptra absent.

Sulphide-rich springs.

Phormidium [Kützing 1843] Gomont 1892

01550000
Many trichomes aggregated to form a gelatinous or leathery mat, attached to a substratum or floating as a distinct colony. Sheath thin, more or less distinct, but often diffluent and merging with each other, especially when the trichomes are moving. Features of the end of the trichome and especially the end cell are important in distinguishing species. These include whether or not the trichome is attenuated towards the ends, the shape of the end cell and whether or not it is terminated by a calyptra.

In addition to visually obvious growths, trichomes of several species are frequent in enrichment cultures from soils, especially cultivated ones. These organisms are quantitatively much more important than routine sampling would suggest.

Other records
Benson-Evans & Antoine (1996) list:

- *Phormidium minnesotense* (Tilden) Drouet 1942

Powell (1964) lists:

- *P. minutum* A.Lindstedt 1943, as record from J. Komárek
- *P. monile* Setchell *et* Gardner 1930.

Phormidium ambiguum [Kützing 1843] Gomont 1892

01550010 **Pl. 12A** (p. 81)
Colony spread out, bright blue-green, dark or yellowish green. Trichomes flexuous (? flexible) and often intertwined, not or very slightly attenuated towards apex. Cells 4–6 μm wide, 1.5–2.7 μm long, one-quarter to half as long as wide; cross walls slightly narrowed, often with adjacent granules; end cell rounded, not capitate, calyptra absent; often with gas vacuoles. Sheath thin, firm or gelatinous, sometimes thick and more or less laminated.

In diverse environments, including fresh and brackish water, and also very moist soil. Powell (1964), without detail. Anand (1937): in summer among *Vaucheria* in the *Rhizoclonium–Vaucheria* community near and just above high-water level at Westgate, Kent.

Phormidium angustissimum West *et* G.S.West 1897

Synonym: *Leptolyngbya angustissimum* (West *et* G.S.West) Anagnostidis *et* Komárek 1988
01550020
Colony leathery, thin, pale blue-green. Trichomes bent and intertwined, not attenuated towards apex. Cells 0.6–0.8 μm wide, (2–)4–5(–8) times as long as wide; cross walls narrowed, without adjacent granules; end cell not capitate, calyptra absent.

On wet rocks by streams. Powell (1964) ? habitat.

This has similar cell dimensions to *Oscillatoria angusta*, but differs (in addition to sheath features) in the slight narrowing at cross walls.

Phormidium autumnale [C.Agardh 1812] Gomont 1892

Synonyms: *Phormidium uncinatum* [C.Agardh] Gomont 1892; *Phormidium subsalsum* Gomont 1899
01550030 **Pl. 12B,C** (p. 81)

Colony expanded, dark blue-green or olive-green, sometimes yellowish or violet. Trichome straight or slightly curved; attenuated to varying extents towards apex, sometimes rather abruptly so, other times more gradually; apical region very slightly or more strongly curved (corresponding to original descriptions of *P. autumnale* and *P. uncinatum*.). Cells 4–7(–9) μm wide, 2–5(–7) μm long, half as long as wide to slightly longer than wide; cross walls not narrowed, often with adjacent granules; end cell capitate, with rounded or weakly conical calyptra. Sheath firm or mucilaginous.

On wet soil, mud, base of walls; upper parts of rocks in rivers, especially microhabitats which are wetted intermittently. Very widespread, commoner in nutrient-rich environments, but not confined to them. West & Fritsch (1927): abundant on damp earth; also commonly encountered in cultivated soils (Bristol, 1920b). Frémy (1929–33) records all three from brackish habitats, his only record for *P. subsalsum* being brackish water in Norway. Powell (1964, as *P. autumnale, P. uncinatum* and *P. subsalsum*): British record (without location) identified by J. Komárek.

It seems likely that several morphological features are influenced by the extent to which material is phosphorus-limited, such as the degree of attenuation towards the apex, the extent of vacuolation (Plate 12B, p. 81) and possibly also the development of the calyptra. Geitler (1932a) followed Gomont (1892) in treating all three as distinct species, though he expressed doubt about *P. autumnale* versus *P. uncinatum* and strong doubt about *P. subsalsum* versus *P. uncinatum*. The differences between *P. autumnale* and *P. uncinatum* are the trichome width range, which overlaps (4–7 vs. 5.5–9 μm), and the more pronounced curve of *P. uncinatum* in the apical region. *P. subsalsum* resembles the original description of *P. autumnale* in trichome width (4–6 μm), of *P. uncinatum* in obvious curve of apical region and slightly longer cells than either. All three species are combined here.

Phormidium calcareum Kützing 1849

01550180 (new number)

Calcified, laminated crusts. Trichomes blue-green, slightly attenuated towards apex. Cells 3.7–4 μm wide, more or less quadrate; cross walls not narrowed; end cell rounded, with calyptra. Sheath thin, diffluent.

Calcareous streams. It occurs at similar sites to *P. incrustatum*, but is less common (A. Pentecost, pers. comm.).

This species differs from *P. incrustatum* only in the tendency to be slightly narrower and possessing a calyptra. Drouet (1968) gives the species as 'ex Gomont 1892', but it is apparently not listed by Gomont.

Phormidium cincinnatum Itzigsohn 1910

01550190 (new number) **CD**

Colony dark blue-green or green. Trichomes straight or weakly flexuous (? flexible), often with a number running parallel to each other, not attenuated towards the apex. Cells 15–18 μm wide, 2.5–4 μm long, much shorter than wide; cross walls slightly narrowed, often with adjacent granules; end cell rounded. Sheath firm, though gelatinous on the outside and fusing with adjacent ones.

On silt at shallow site in R. Medway, near Tonbridge, Kent (A. Pentecost, pers. comm).

Phormidium corium [C.Agardh 1812] Gomont 1892

Synonym: *Phormidium papyraceum* [C.Agardh 1824] Gomont 1892

01550040 **Pl. 12G** (p. 81)

Colony leathery-membranous, spread out, dark or blue-green. Trichomes thickly intertwined. Cells 3–4.5(–6.5) μm wide, 3.4–8 μm long; cross walls not narrowed, without adjacent granules; end cell rounded or shortly conical, without calyptra.

In standing and flowing waters; wet mud, base of walls. Powell (1964: as *P. corium* and *P. papyraceum*).

The above is based on the original description, apart from extending the width range to 6.5 μm. However, several varieties have been recognized, which together increase the limits of the species: width to 8 μm, cross wall slightly narrowed, cells considerably longer than broad, marked elongation of end cells. Combinations of some of these features overlap with those of other species.

Phormidium ectocarpi Gomont 1899

Synonym: *Leptolyngbya ectocarpi* (Gomont) Anagnostidis *et* Komárek 1988

01550200 (new number) **Pl. 13B** (p. 88)

Typically a thin film, often pinkish. Trichomes thickly intertwined or running parallel; not or only slightly attenuated towards apex. Cells 1.3–2 μm wide, 1.3–2.6 μm long, quadrate or slightly longer than wide; cross walls distinctly narrowed; end cell rounded or occasionally slightly elongated and narrowed, calyptra absent.

Powell (1964). Marine: epiphytic or on sediments. Frémy (1929–33): near Plymouth.

Phormidium favosum [Bory 1827] Gomont 1892

01550050 **Pl. 12L** (p. 81)

Colony dark blue-green (when wet). Trichome somewhat flexuous (? also flexible), towards the apex attenuated and sometimes also slightly twisted. Cells 4.5–9 μm wide, 3–7 μm long, half to as long as wide; cross walls not narrowed, often with adjacent granules; end cell capitate, with a weakly conical to almost hemispherical calyptra. Sheath mostly gelatinous.

Hard-water streams, occasionally penetrating limestone surfaces, often in microhabitats moistened intermittently; also on rocks at edge of calcareous lakes.

The illustration of the organism in Geitler (1932a) shows trichomes with both a calyptra and granules at the cross walls. The organism would therefore provide a particularly interesting study on the influence of environment on morphology.

Phormidium foveolarum [Montagne 1849] Gomont 1892

Synonyms: *Phormidium molle* [Kützing 1849] Gomont 1892; *Leptolyngbya foveolarum* (Montagne ex Gomont) Anagnostidis *et* Komárek 1988

01550060 **Pl. 12E** (p. 81)

Colony thin, dark blue-green. Trichome flexuous (? flexible), not attenuated towards apex. Cells 1.3–3.3 μm wide, 0.8–8 μm long, cells somewhat shorter to somewhat longer than wide to quadrate; cross walls narrowed, without adjacent granules; end cell rounded, calyptra absent. Sheath thin, mostly diffluent.

On moist soil and rocks, especially more calcareous ones; fresh and brackish standing water. Powell (1964: as *P. foveolarum* and *P. molle*).

The original descriptions for *P. foveolarum* and *P. molle* differ only in cell dimensions, with the former 0.8–2.2 µm wide (including fo. *maior* Elenkin) and the latter 2–3.3 µm wide (including fo. *tenuior* W. *et* G.S.West). In nature there appears to be a continuum of forms and sometimes trichomes of different widths occur in the same colony. The description therefore combines both species.

Phormidium fragile [Meneghini] Gomont 1892

Synonym: *Leptolyngbya fragilis* (Gomont) Anagnostidis *et* Komárek 1988

01550070 **Pl. 12D** (p. 81)

Colony gelatinous, layered, yellowish or brown blue-green. Trichomes somewhat flexuous (? flexible), intertwined or running parallel, attenuated towards apex. Cells 1.2–2.3 µm wide, 1.2–3 µm long; cross walls distinctly narrowed, without adjacent granules; end cell sharply conical, calyptra absent.

In fresh and brackish water. Powell (1964).

Phormidium incrustatum [Gomont 1889] Gomont 1892

01550080 **Pl. 12M** (p. 81)

Colony a calcified crust, brown-red, violet or blue-green; sometimes forming small oncoids. Trichomes intertwined or more or less parallel and upright. Cells 4–5 µm wide, 3.5–5.2 µm long; cross walls not narrowed, often with adjacent granules; end cell weakly conical, calyptra absent. Sheath thin, mucilaginous.

Widely distributed in highly calcareous streams and lake margins.

The species is very close to *Phormidium calcareum*.

Phormidium inundatum [Kützing 1849] Gomont 1892

01550090 **Pl. 12J** (p. 81)

Colony membranaceous, dark blue-green. Trichome more or less straight, slightly attenuated towards apex, but not curved. Cells 3–5 µm wide, 4–8 µm long, almost quadrate to longer than wide; cross walls not narrowed, often with adjacent granules; end cell weakly conical, calyptra absent. Sheath thin, gelatinous.

In flowing and standing water; wet rocks.

Phormidium laminosum [C.Agardh 1827] Gomont 1890 ex Gomont 1892

Synonym: *Leptolyngbya laminosa* (Gomont) Anagnostidis *et* Komárek 1988

01550100 **Pl. 12N** (p. 81)

Colony leathery, blue-green, yellow or almost brick-red. Trichome flexuous or coiled irregularly, attenuated towards apex. Cells 1–1.5(–2) µm wide, 2–4 µm long; cross walls not narrowed, often with an adjacent granule on either side; end cell sharply pointed, calyptra absent. Sheath narrow, soft or diffluent.

In standing water and on wet rocks.

P. laminosum shows similarities with *Schizothrix coriacea*, including dimensions and macroscopic colouration.

Phormidium lucidum [(C. Agardh 1827) Kützing 1843] Gomont 1892

01550110 **Pl. 12K** (p. 81)

Colony firm, thick, outer part dark green, inner very pale. Trichomes somewhat flexuous (? flexible), but often almost parallel, slightly attenuated towards apex, often ending in a hook. Cells 7–8 µm wide, 2–2.5 µm long; cross walls slightly narrowed, often with adjacent granules; end cell capitate, with round or almost conical calyptra. Sheath diffluent.

Fresh and brackish waters.

Phormidium luridum [Kützing 1849] Gomont 1892

Synonym: *Leptolyngbya lurida* (Gomont) Anagnostidis *et* Komárek 1988

01550120 **Pl. 12H** (p. 81)

Colony a mat, upper part deeply pigmented purple or deep blue-green, inner grey-blue-green. Trichome flexuous (? flexible), not attenuated or curved towards apex. Cells 1.7–2 µm wide, 1.8–4.7 µm long, quadrate or longer than wide; cross walls slightly narrowed, without adjacent granules. Sheath thin, soft, diffluent; end cell rounded, calyptra absent.

In standing freshwater.

Phormidium mucicola Huber-Pestalozzi *et* Naumann 1929

01550130

Trichomes short (mostly to 20 µm, rarely about twice this), in mucilage of other algae; straight, not attenuated towards apex. Cells (1.3–)1.5–2 µm wide, 2–4 µm long; cross walls distinctly narrowed, sometimes giving the trichome almost an articulated appearance; end cell rounded or slightly conical, calyptra absent.

In soft mucilage of colonial algae e.g. *Microcystis*, *Chaetophora*.

Not *Leptolyngbya mucicola* (Lemmermann) Anagnostidis *et* Komárek 1988.

Phormidium persicinum [Reinke] Gomont 1892

01550210 (new number)

Colony thin, rose coloured. Trichomes thickly intertwined, attenuated towards apex. Cells 1.7–2 µm wide, 2–7 µm long; cross walls distinctly narrowed, without adjacent granules; end cell sharply conical, calyptra absent. Sheath thin.

Powell (1964). Frémy (1929–33) records it as epiphytic on larger algae and epizooic on spirorbids and sponges.

This species is much like *P. fragile*; apart from the pink trichomes, it differs morphologically only in the longer cells. However, the two have not been combined here, because *P. persicinum*, which was described earlier, has been reported only from marine habitats.

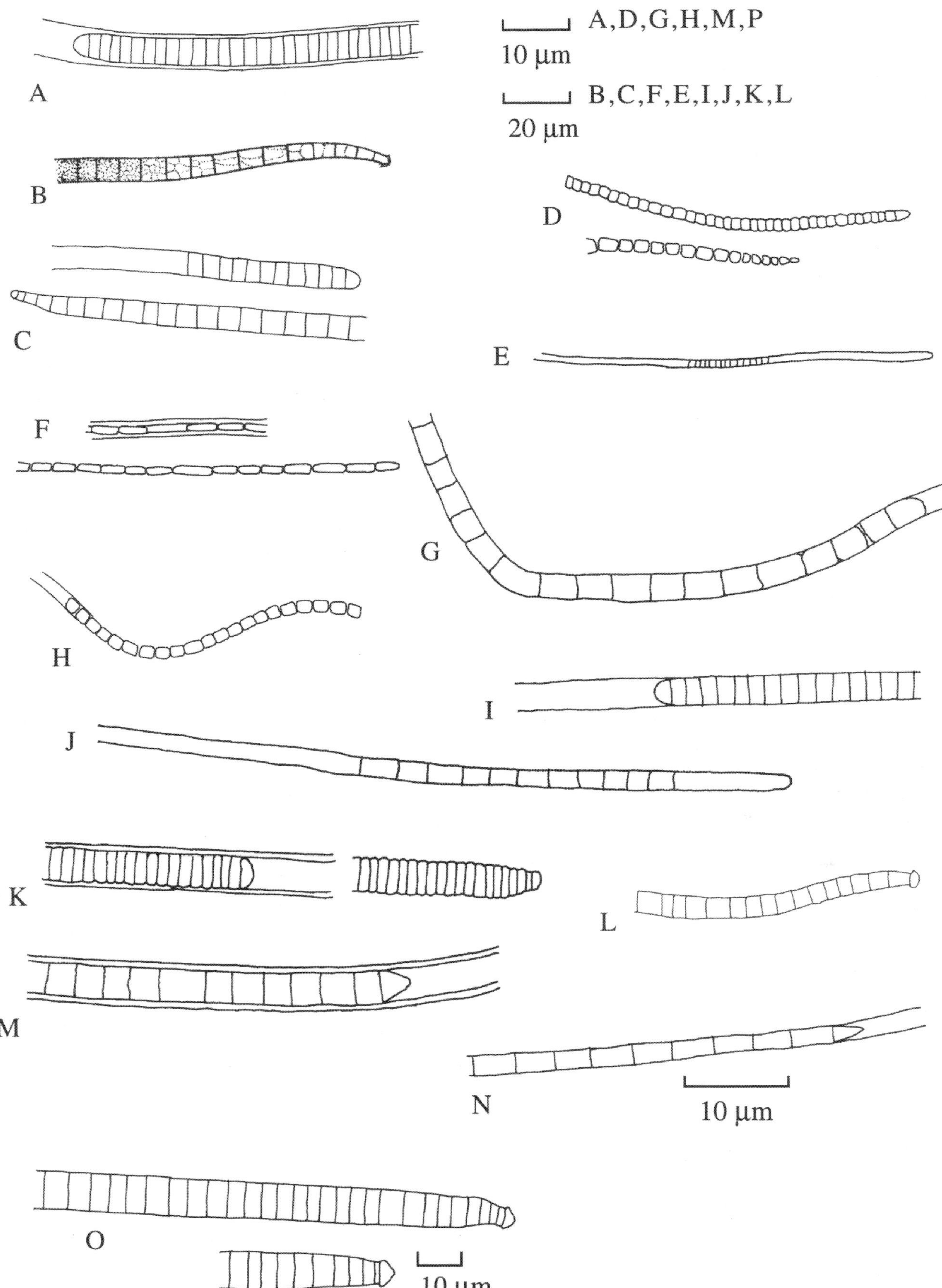

Plate 12 A–O

A. *Phormidium ambiguum* (p. 78); **B, C**. *Phormidium autumnale* (p. 78): B. Showing vacuolation and calyptra (probably phosphorus-limited state), C. Trichomes without calyptra; **D**. *Phormidium fragile* (p. 80); **E**. *Phormidium foveolarum* (p. 79); **F**. *Phormidium tenue* (p. 82); **G**. *Phormidium corium* (p. 79); **H**. *Phormidium luridum* (p. 80); **I**. *Phormidium retzii* (p. 82); **J**. *Phormidium inundatum* (p. 80); **K**. *Phormidium lucidum* (p. 80); **L**. *Phormidium favosum* (p. 79); **M**. *Phormidium incrustatum* (p. 80); **N**. *Phormidium laminosum* (p. 80); **O**. *Phormidium subfuscum*: as shown in Geitler (1932a), but see text (p. 82).

Phormidium retzii [(C.Agardh 1812) Gomont 1890] Gomont 1892

01550140 **Pl. 12I** (p. 81)

Colonies dark blue-green, steel blue or dark green sheets, occurring in a variety of forms, including thick mat or cushion, sometimes compact and well-defined, other times spread out; usually with the trichomes highly motile. Trichome not attenuated towards apex. Cells 4.5–12 μm wide, 4–9 μm long; cross walls not or only very slightly narrowed, without adjacent granules; end cell scarcely narrowed, blunt, with weakly thickened membrane, but typical calyptra absent.

Flowing or standing waters, on rocks, sediment or macrophytes, but especially characteristic of upland streams and rivers. West & Fritsch (1927): abundant in rapidly flowing water. Co-dominant in early spring with diatoms at site in River Caragh, Kerry (Heuff & Horkan, 1984).

Phormidium subfuscum [Kützing 1843] Gomont 1892

01550150 **Pl. 12O** (p. 81)

Colony a thin, weakly layered mat, dark to dirty green. Trichomes, often running parallel, attenuated, but straight, in the final few cells towards the apex. Cells 8–11.5 μm wide, 2–4 μm long; cross walls not narrowed, often with adjacent granules; end cell straight, sharply conical, often with a lightly rounded calyptra. Sheaths diffluent to form a layered gelatinous matrix.

Flowing or standing waters; rarely on wet rocks.

The figure shown in Pl. 12O corresponds to older interpretation of this species, but not Kann & Komárek (1970), where *P. subfuscum* has a conical apical cell and the figure on p. 81 would probably be interpreted as *P. autumnale*. However, all this group of organisms needs experimental study under conditions of nutrient limitation before the taxonomic value of the apical cell can be evaluated.

Phormidium submembranaceum [Ardissone *et* Strafforello 1877] Gomont 1892

01550220 (new number)

Colony leathery, dark green. Trichomes thickly intertwined, without distinct sheaths, straight for whole length, but attenuated towards the apex. Cells 4–5.5 μm wide, 4–10 μm long, almost quadrate to twice as long as wide; cross walls narrowed; end cell capitate, with somewhat conical calyptra.

Powell (1964), as fo. *minor*. Marine rocks.

Phormidium tenue [Meneghini 1837] Gomont 1892

Synonym: *Leptolyngbya tenuis* (Gomont) Anagnostidis *et* Komárek 1988

01550160 **Pl. 12F** (p. 81)

Colony a leathery mat, bright blue-green. Trichome straight or somewhat flexuous, attenuated towards apex. Cells 1–2 μm wide, 2.5–5 μm long; cross walls slightly narrowed, without adjacent granules; end cell long conical, calyptra absent. Sheath thin, gelatinous.

West & Fritsch (1927): frequent among other algae in ponds, ditches and rivers; common in cultivated soils (Bristol, 1920b). The name has sometimes been given to *Phormidium* populations causing nuisance odours in water supplies. Powell (1964): ? habitat.

Phormidium valderianum [Delponte 1857] Gomont 1892

01550170

Colony sometimes very slippery, spread out, to 1(–3) cm thick, laminated; outer part dirty green, inner almost colourless. Trichomes densely intertwined, not attenuated or curved towards apex. Cells 2–2.5 μm wide, 3–6.7 μm long; cross walls not narrowed, with 1 or 2 adjacent granules; end cell rounded, calyptra absent. Sheath narrow, but eventually becoming mucilaginous.

Fresh or brackish water, standing or flowing; occasionally on wet rocks. Powell (1964).

Plectonema [Thuret 1875] Gomont 1892

01560000

Trichome enclosed in a thin firm sheath, with branches single or geminate (appearing like a pair); heterocysts and akinetes absent. This genus is essentially like *Lyngbya*, but with false branching; in narrow forms it may be difficult to resolve from *Lyngbya*, but the pattern of branching in some wide species is easy to recognize and characteristic. The genus also merges with *Schizothrix*, where there are a number of trichomes within a single sheath.

Two of the species recorded for the British Isles are easy to name. *P. tomasinianum* has very wide filaments and occurs on wet rocks by streams and similar places, while *P. terebrans* has very narrow filaments which penetrate mollusc shells. Other forms of *Plectonema* likely to be found are all narrow. Seven species are listed here, but Geitler (1932a) and other Floras list many more morphologically rather similar species. Based on morphology alone there is scope for reducing the number of species, including those listed here, but it seems unwise to make nomenclatural changes without experimental study. Most researchers have treated names for marine forms separately, but Bristol (1920b) used *P. battersii* for cultivated soils because the dimensions were similar. *Plectonema* filaments 2–4 μm wide are an important component of the early stages of colonization of the moister parts of old spoil heaps in lead–zinc mining areas, occurring in varying proportions with moss protonema (usually *Dicranella*) and leafy shoots. The author has listed these in various accounts as *P. gracillimum*, but the choice is rather arbitrary. A range of other forms of *Plectonema* occur in different microhabitats in many metal mining areas, both narrower and wider than the 1.5–3 μm cell width of *P. gracillimum*.

Other record Benson-Evans & Antoine (1996) list *Plectonema carneum* (Kützing) Lemmermann; according to Geitler (1932a) this is a synonym of *Plectonema roseolum* [P.G.Richter] Gomont 1892.

Plectonema battersii Gomont 1899

Synonym: *Leptolyngbya battersii* (Gomont) Anagnostidis *et* Komárek 1988

01560010

Filaments long, intertwined, richly branched, with branches mostly arising in pairs and narrower than the main axis. Trichome slightly narrowed at point of branching. Cells 2–3.5 μm wide, 0.5–1 μm long; end cell rounded. Sheath in main axis quite thick, in branches much thinner, colourless.

Cliffs and rocks on the coast.

Plectonema boryanum Gomont 1899

Synonym: *Leptolyngbya boryana* (Gomont) Anagnostidis *et* Komárek 1988

01560020

Filaments flexuous, thickly intertwined, richly branched, with branches mostly arising in pairs and narrower than the main filament. Cells 1.3–2 μm wide, in the main axis slightly shorter or as long as wide, in branches somewhat longer than wide; cross walls strongly narrowed, without granules.

Freshwater ponds, loosely associated with submerged macrophytes.

Plectonema calothrichoides Gomont 1899

Synonym: *Leptolyngbya calothrichoides* (Gomont) Anagnostidis *et* Komárek 1988 (note changed spelling of species name)

01560030

Filaments quite short, strongly flexuous and intertwined, richly branched, with branches arising in pairs and often parallel. Cells weakly barrel-shaped, 2–2.5 μm wide, 0.5–0.8 μm long; end cell round. Sheath in middle of filament thick, yellow-brown, towards the ends narrowed and colourless.

Upper intertidal and immediate supralittoral, on rocks.

The narrowing sheath sometimes leads to resemblance with a form of *Homoeothrix*.

Plectonema gracillimum Zopf ex Hansgirg 1885

Basionym: *Glaucothrix gracillima* Zopf 1882

Synonym: *Leptolynbgya gracillima* (Hansgirg) Anagnostidis *et* Komárek 1988

01560040

Colony an often widely spread mat, mucilaginous, blue-, yellow-green or brownish. Filaments 2–4 μm wide, branches single or in pairs. Cells 1.5–3 μm wide. Sheath thin, colourless or pale yellow-brown.

On soil, base of moist walls in greenhouses. Material from old metal-rich spoil heaps has been referred to this species by the author (see genus entry). In areas where there is constant seepage of zinc-rich water, the alga may develop thick gelatinous mats.

Plectonema norvegicum Gomont 1899

Synonym: *Leptolyngbya norvegica* (Gomont) Anagnostidis *et* Komárek 1988

01560050

Colony a crust, brown or brownish green. Filaments quite short, intertwined, richly branched, with branches mainly in pairs and similar width to main axis. Cells torulose or submoniliform, 1.5–2 μm wide, up to half as long as wide. Sheath: main axis thick, rough yellow-brown; young branches smooth and colourless.

Maritime rocks, among gelatinous masses of other algae. Frémy (1929–33): east and south coasts of England. Anand (1937): component of Chrysophyceae–*Entocladia*–*Lyngbya* community in the Chysophyceae belt on chalk cliffs at Westgate, Kent.

Plectonema nostocorum [Bornet in Bornet *et* Thuret 1880] Gomont 1892

Synonym: *Leptolyngbya nostocorum* (Bornet) Anagnostidis *et* Komárek 1988

01560060

Filaments usually reported living inside mucilage of old colonies of other algae, straight or bent, branching sparse, usually single. Cells cylindrical, 0.7–1.5 μm wide, 2–3 μm long; end cell rounded. Sheath thin, colourless.

In mucilage of other algae; occasionally in streams. Anand (1937): component of Chrysophyceae–*Entocladia*–*Lyngbya* community in the Chysophyceae belt on chalk cliffs at Westgate, Kent.

Plectonema notatum Schmidle 1901

Synonym: *Leptolyngbya notata* (Schmidle) Anagnostidis *et* Komárek

01560070

Filaments richly branched, typically forming a thin mucous layer; branches occasional, single or a pair, 1.7–2 μm wide. Cells 0.8–1.3 μm wide, 2–3 times longer than wide; cross walls not narrowed, often with 1 or 2 granules at cross walls; end cell rounded. Sheath thin, colourless.

Old containers with water, such as troughs and sinks, often forming wefts on filamentous green algae.

Plectonema terebrans Bornet *et* Flahault 1889 ex Gomont 1899

Synonym: *Leptolyngbya terebrans* (Bornet *et* Flahault ex Gomont) Anagnostidis *et* Komárek 1988

01560080

Filaments long, flexuous, with sparse branching, branches single. Cells 0.9–1.5 μm wide; cross walls not narrowed. Sheath very thin, colourless.

On surface of and penetrating soft limestones and old calcareous shells of molluscs, very widespread in marine habitats, but occasionally also in freshwater. Part of the chalk-boring community on cliffs at Westgate, Kent (Anand, 1937).

Plectonema tomasinianum [Bornet 1889] Gomont 1892

01560090

Colony ranging from a small tuft to a felt-like or a bushy mass of flexuous filaments up to 2 cm high; blue-green, olive-green to brown. Filaments with rich or sparse branching, usually in pairs. Main axis cells 11–22 μm wide, 3–9(–12) μm long; cross walls narrowed, often with granules. Sheath up to 3 μm wide, lamellate; initially colourless, later brown.

Wet rocks and wood at edge of streams and ponds, widespread. West & Fritsch (1927): the largest and most frequent (*Plectonema*) occurring as greenish-brown felt-like masses on wet rocks. Thick-walled resting cells were noted by G.S.West (1904). Calcareous streams in Malham (Yorks) region (A. Pentecost, pers. comm.)

Pseudanabaena Lauterborn 1914–17

01580000

This is a genus of convenience comprising phylogenetically different organisms with rather similar morphological features. These consist of the following.

Short trichomes, which do not aggregate, though capable of motility. Cells cylindrical to barrel-shaped, deeply constricted at cross walls or appearing as if adjacent cells slightly separated. Apical cells usually with flattish end, though sometimes slightly rounded.

P. schmidlei Jaag 1938 (synonym: *Komvophoron schmidlei* (Jaag) Anagnostidis *et* Komárek) was included in the *Coded List* (01580020), but the 6–7 μm wide organism described and illustrated by Jaag is quite different, with cell walls rounded rather than rectangular (viewed from side) and end cell rounded-triangular rather than truncate. It looks rather like an *Anabaena* strain grown in culture and which has lost its heterocysts, or perhaps a hormogonium of some other heterocystous genus.

Other record Powell (1964) lists *P. brevis* N.Carter.

Pseudanabaena catenata Lauterborn 1914–17

01580010 **Pl. 13A** (p. 88)

Trichome single, not attenuated towards ends. Cells 1.8–2.2 μm wide, 2–5 μm long; cross walls narrowed, appearing almost rectangular when viewed from side; end cell truncate. Gas vacuoles absent.

Very widespread, especially among other algae and submerged plants in ponds, lake margins and slow-flowing rivers.

Schizothrix [Kützing 1843] Gomont 1892

01630000

Trichomes straight or slightly wavy, sometimes with false branching; usually many trichomes within one sheath, which may be thin or thick. The sheaths are often closely aggregated in various ways to form characteristic colony shapes, which used to be separated as subsections of the genus: upright bundles, flat leathery mat, cushion or hemispherical, bushy, often-floating flocs. Although these characters are useful for recognition of individual species, it is unlikely that these really do indicate distinct taxonomic groupings. Individual trichomes of wider species may also have their own sheath and the various sheath layers are often highly lamellate. The extent of branching by filaments ranges from rare to frequent, according to species and environment. Sheath colourless or with a range of colours, especially yellow-brown or red; the red colour may not always be obvious under the microscope in species with narrow trichomes, yet a mat as a whole may appear strongly coloured.

Species with upright bundles are associated with aquatic environments or ones which are humid for long periods, though, unless calcified, usually not submerged. Some mat-forming species grow permanently submerged, but many of those forming a leathery mat are subject to alternating periods of being very moist or even submerged with periods when they are almost or completely dry.

Two very narrow forms which W. & G.S.West described as species (*S. delicatissima, S. funalis*) are listed below. However, it is easy to find a range of very narrow forms in nature, which do not fit exactly either of these descriptions.

The key should separate characteristic forms of the species listed here, but examination of limestone and other areas where *Schizothrix* is frequent is likely to show other forms which do not fit exactly to any of them.

1 At least partially endolithic in limestone 2
1 Not endolithic 3
2 Cross walls narrowed .... *S. calcicola,* in part
2 Cross walls not narrowed .... *S. perforans*
3 Cells ≤1 μm 4
3 Cells >1 ≤6 μm wide (usually ≤4 μm) 5
3 Cells >6 μm wide .... *S. muelleri*
4 Sheath yellow-brown, lamellate .... *S. funalis*
4 Sheath colourless, not lamellate .... *S. delicatissima*
5 Filaments united in upright bundles 6
5 Filaments not united in upright bundles 7
6 Cells 2–3 μm wide, shorter than wide .... *S. vaginata,* in part
6 Cells 1.9–3.5 μm wide, longer than wide .... *S. cuspidata*
6 Cells 3–6 μm wide, mostly longer than wide... *S. friesii*
7 Marine .... *S. cresswellii*
7 Freshwater 8
8 End cells often distinctly conical 9
8 End cells rounded 11
9 Filaments ending in prominent fascicle of trichomes .... *S. fasciculata*
9 Not as above 10
10 Filaments with few branches, heavily calcified .... *S. coriacea*
10 Filaments with many branches, little or no calcification .... *S. lacustris,* in part
11 Calcified 12
11 Not calcified (or only weakly so) 16
12 Rose or red colour when view on mass 13
12 Not as above 14
13 Cells ≤2 μm wide .... *S. lateritia*
13 Cells >2 μm wide...................................... *S. rubella*
14 Cells ≤2 μm wide 15
14 Cells >2 μm wide .... *S. vaginata,* in part
15 Filaments often running parallel, with many trichomes .... *S. pulvinata*
15 Filaments with 2 to many trichomes in old part, but usually one in young part .... *S. calcicola,* in part
16 Cells quadrate or shorter than wide .... *S. tinctoria*
16 Cells 1.5 – 3 times as long as wide 17
17 Mat of moist terrestrial environments, becoming leathery when old ..... *S. calcicola,* in part
17 Submerged .... *S. lacustris,* in part

Schizothrix calcicola [C.Agardh 1812] Gomont 1892

01630010 **CD**

Colony a mat, initially thin and membranous-slimy, but later becoming more leathery-fleshy; initially bright blue-green, but later dark blue-green or yellowish-grey; not calcified. Filaments very richly branched, occasionally in pairs. Filament typically with 2 to many intertwined trichomes, though ultimate branches usually with only one. Cells 1–1.7 μm wide, 2–6 μm long; cross walls not narrowed, often with granules; end cell rounded. Sheath firm, colourless; usually narrowed towards the ends of branches, but sometimes fimbriated and spread out.

On moist rocks and walls; in greenhouses; widespread; sometimes penetrating the surface of very wet, soft calcareous rocks. The penetrating forms differ little from *S. perforans*.

Schizothrix coriacea [Kützing 1843] Gomont 1892

Synonym: *Schizothrix lardacea* [Cesati] Gomont 1892
01630020 (includes old number 01630070)
Pl. 9H,J (p. 65) **CD**

Colony a mat, leathery & folded, spread out & sometimes very thick (to 2 cm), sometimes incrusted with calcite; outer part of mat, where exposed to full daylight, brick or other reddish colour, inner colourless. Filaments densely intertwined, branches few, typically with up to about 6 trichomes, though often only one. Cells 1–2 μm wide, 3–6 μm long; cross walls slightly narrowed; end cell pointed conical. Sheath firm, not lamellate, usually colourless (under microscope, although probably source of colour seen macroscopically).

Often abundant on moist soil, cliffs and walls subject to seepage, and on rocks in small streams with varying flow; characteristic of intermittently wet surfaces in calcareous environments; sometimes forms a general cover of parts of the bed of streams with *Rivularia*, intermingling with *Rivularia* filaments at the edge of old colonies of the latter. *Schizothrix* usually does not calcify in streams where *Rivularia* is only weakly calcified. Powell (1964: as *S. lardacea*).

Gomont (1892) first suggested that *S. coriacea* (which has priority) and *S. lardacea* should be combined and most subsequent authors have made the same suggestion, without actually doing so. This account combines them, because the only clear-cut difference is the calcification in *S. coriacea* and lack of it in *S. lardacea*. *S. coriacea* also has many similarities with *Phormidium laminosum*.

Schizothrix cresswellii [Harvey 1846–51] Gomont 1892

Synonym: *Schizothrix fritschii* Anand 1937
01630120 (new number)

Colony a mat, soft, yellowish, up to 5 mm thick, calcified in the basal part. Filaments very long, intertwined, at the ends branched into distinct bundles. Filaments with many trichomes in basal region, but only one towards apex. Cells 1–1.6 μm wide, 2–3 μm long; cross walls usually slightly narrowed; end cell blunt-conical. Sheath yellow or colourless.

Rocks in the supralittoral. Marquand (1884): dripping cliffs below Mousehole Cave, Cornwall.

The original description of *S. cresswellii* is almost identical to *S. fasciculata*, but the two have presumably been kept separate because of the different habitats *S. cresswellii* was recorded originally from the marine 'floodzone' (supralittoral) near Sidmouth, Devon (Gomont, 1892), but in view of the calcification at the base, it seems probable that the site was influenced by a freshwater spring (which are frequent in the region). *S. fritschii* was described by Anand (1937) as a thin mat covering the algal mat of the Chrysophyceae belt in July and August in shaded situations on the chalk cliffs at Westgate, Kent. It differs from *S. cresswellii* in the sheaths (colourless versus yellow) and narrower width of tufts, filaments & trichomes (1–1.2 μm vs. 1.3–1.6 μm). These are all morphological shifts that might simply reflect the shaded situation, so the two species are combined here, with a slight extension of the original trichome width range of *S. cresswellii*. It was considered whether to combine these and *S. fasciculata* into a single species, or to combine *S. cresswellii* and *S. fasciculata*, but retain *S. fritschii*. The present solution retains different names for the marine and freshwater organisms.

Schizothrix cuspidata West *et* G.S.West 1896

Synonym: *Symploca cuspidata* West *et* G.S.West 1894
01630130 (new number)

Colony a mat of semi-erect bundles, 8–15 mm long, yellow-grey or rust colour. Filament with 1–3(–many) intertwined trichomes. Cells 1.9–3.5 μm wide, 2–4 times longer than wide; end cell rounded. Sheath 13.5–25(–40) μm wide, colourless, with or without lamellations

On rocks, tree bases, mosses and moist earth.

This species is little more than a narrower form of *S. friesii*.

Schizothrix delicatissima West *et* G.S.West 1897

01630140 (new number)

Colony soft, thin, brown. Filaments 5–6.5 μm wide, with 1–2 trichomes, flexuous, sparingly branched; forming ropes splayed out at the ends. Cells 0.6–0.8 μm wide, about 6–8 times as long as wide; cross walls narrowed. Sheath colourless, often long pointed at the ends.

On moist soil.

Schizothrix fasciculata [Nägeli in Kützing 1849] Gomont 1892

01630030

Colony usually a highly calcified mat, sufficiently so for innermost part of old ones to be hard; blue-green, reddish or brown; sometimes forming small oncoids; also hemispherical colonies (see below). Filaments bent, densely intertwined, richly branched and with the ends often consisting of a tuft of parallel filaments. Cells (1–)1.4–3 μm wide, 1.2–3.5 μm long; cross walls narrowed; end cell blunt-conical. Sheath thick, colourless or occasionally brownish, pointed at the end.

In calcareous lakes and fast-flowing streams. West & Fritsch (1927): responsible for formation of spherical calcareous pebbles on the bottom of Lough Belvedere, Mullingar, Ireland. However, A. Pentecost (pers. comm.) reports that this lake is now polluted, but oncoids are present in other lakes in the area. Geitler (1932a) describes a characteristic form (fo. *semiglobosa*) of hard-water lakes in central Europe, which has hemispherical, only slightly calcified colonies; this is present at Malham, North Yorkshire (A. Pentecost, pers. comm.).

A. Pentecost (pers. comm.) reports two other similar mat-forming species from the highly calcareous Waterfall Beck, Malham: *S. pulvinata* [Kützing] Gomont 1892 and *S. rubella* Gomont 1892. All three species are quite similar, but they are kept separate here.

Schizothrix friesii [C.Agardh 1817] Gomont 1892

01630040 **Pl. 9I** (p. 65)

Colony in rigid more or less upright bundles of filaments, to 1.5(–3) cm, dark green or greenish steel blue. Filaments in lower parts contorted, straighter above, with few trichomes. Cells 3–6 μm wide, 4–11 μm long, quadrate to twice as long as wide; end cell obtuse-conical. Sheath colourless, lamellate, pointed at the ends

On very moist soil, between mosses and rocks.

Schizothrix funalis West *et* G.S.West 1896

01630050

Colony soft, thin, brown. Filaments 8–12 µm wide, with 1–2 trichomes, flexuous, richly branched and intertwined to form rope-like bundles splayed out at the ends. Cells 0.5–0.7 µm wide, about twice as long as wide. Sheath firm, lamellate, dark brown in older parts, yellow in younger ones.

Ditches.

Schizothrix lacustris [A.Braun in Kützing 1849] Gomont 1892

01630060

Colony a cushion or warted mat, often soft spongy, not or only slightly calcified, dark green. Filaments united, often richly branched at ends; older parts often with many trichomes, young parts with single trichome. Cells 1–1.5 µm wide, to 4 µm long, cross walls slightly narrowed. Sheath thick, colourless or yellow.

On submerged rocks near the edge of hard-water lakes.

Schizothrix lardacea [Cesati] Gomont 1892

01630070: see *Schizothrix coriacea*

Schizothrix lateritia [Kützing 1849] Gomont 1892

01630080 **Pl. 9K,L** (p. 65)

Colony a membranaceous or leathery mat, often forming extensive cover, though sometimes as individual cushions up to 2 mm thick, calcified; outer surface grey to flesh red, inside pale. Filaments mostly densely packed together, branched to various extents, with many trichomes. Cells 1.3–1.6 µm wide, 2–5(–9) µm long; cross wall often with granule; end cell rounded. Sheath thin or thick, uneven on the outside, pointed at ends.

On moist rocks and tree bark, edge of lakes.

A number of varieties and forms have been reported, though it is uncertain how many from the British Isles. The only character which separates this range of forms from those grouped above as *S. coriacea* is the round end cell of *S. lateritia* and the more or less conical one of *S. coriacea*. It seems best to retain both taxa as species until there is a biological explanation for differences in the shape of the end cell.

Schizothrix muelleri [Nägeli in Kützing 1849] Gomont 1892

01630090 **Pl. 9M** (p. 65)

Colony a mat, brown to dark green. Filaments united in prostrate bundles of intertwined filaments or as free-floating tufts with little intertwining. Cells 7–13 µm wide, 4–9 µm long; cross walls slightly narrowed; end cell obtuse-conical. Sheath firm or somewhat diffluent, lamellate, golden yellow.

On moist soil or sediments in standing water.

Schizothrix perforans (Ercegović) Geitler 1927

Basionym: *Schizothrix coriacea* [Kützing] Gomont var. *endolithica* Ercegović 1925

01630100 **Pl. 9N** (p. 65)

Colony a thick meshwork in the upper part of calcareous rocks and shells. Filaments 1.5–1.8(–3) µm, branching sparse, mostly with only 1 trichome. Cells 0.8–1.4 µm wide, 1–2(–3) times longer than wide; cross walls narrowed. Sheath thin, firm, not lamellate, colourless.

In submerged or very wet soft or hard limestones.

As the characteristic branching pattern is absent (or hard to detect) the species can hardly be separated from *Lyngbya*.

Schizothrix pulvinata [Kützing] Gomont 1892

01630150 (new number)

Colony a mat or crust strongly calcified and with inner part very hard; blue-green. Filaments running parallel, more or less straight, densely intertwined, little branched; individual filaments with many trichomes. Cells 1–2 µm wide, 1.5–4 µm long; cross walls narrowed; end cell rounded. Sheath colourless, pointed at the end.

A. Pentecost: Waterfall Beck, Malham, North Yorkshire.

Schizothrix rubella Gomont 1892

01630160 (new number)

Colony a mat or crust, usually calcified; if calcified, appearing pink; if uncalcified, red. Filaments wound round each other, almost dichotomously branched. Cells 2.5–4.5 µm wide, 4–7 µm long, bright blue-green; cross walls not narrowed; end cell rounded. Sheath thick, lamellate, colourless or rose, pointed or frayed at the end.

A. Pentecost: Waterfall Beck, Malham, North Yorkshire.

Schizothrix tinctoria [Gomont 1890] Gomont 1892

01630110 **Pl. 9O** (p. 65)

Colony cushion-shaped tufts of brush-like filaments orientated in direction of flow, soft, slimy. Filaments branched near apex; trichomes many intertwined, or few. Cells 1.4–2.4 µm wide, almost as long as wide, blue-green or sometimes orange; cross walls distinctly narrowed. Sheath narrow, somewhat diffluent, not lamellate, colourless.

Fast-flowing streams and small rivers.

Schizothrix vaginata [Kützing 1849] Gomont 1892

01630150 (new number)

Colony a crusty mat, with or without calcification, grey-brown or dark green; also endolithic in Rivulariaceae colonies. Filaments straight, forming more or less upright bundles or intertwined, branched towards the upper ends. Cells 2–3 µm wide, mostly shorter than wide; cross walls not narrowed, often with adjacent granules; end cell rounded. Sheaths thick, usually laminated.

In standing and flowing waters, including brackish waters (Frémy, 1929–33).

Spirulina [Turpin 1829] Gomont 1892

01660000

Trichome cylindrical, loosely or tightly coiled into a regular helix; sheath absent; apex of trichome usually not attenuated, terminal cell rounded, calyptra absent; cross walls difficult to resolve with light microscopy. Trichomes often show motility if in contact with a surface. The species are mostly differentiated by little more than various combinations of the width of trichome, width of the helix and distance between consecutive coils of the helix. Where only a single value was listed in the original description, this has usually been entered here as a range extending about 10% either side of the listed value.

The genus is most widespread in brackish and saline environments, but also occurs in freshwater; often in microhabitats where reducing conditions are likely. Habitats are listed here only for species which are reported frequently.

Other record

- *Spirulina abbreviata* Lemmermann. Listed by Griffiths (1936) from Brasside Pond, Durham; no record from this site by author. The original diagrams by Lemmermann do not much resemble *Spirulina* – as much like free cells of *Rhabdoderma* as short lengths of *Spirulina*; it is hard to judge what the organism was.

Spirulina labyrinthiformis (Meneghini 1837) Gomont 1892

01660010

Trichome 0.8–1.2 μm wide, tightly coiled to form helix 2–2.7 μm wide.

Powell (1964, based on record by M. Parke). Brackish pools in supralittoral with reducing conditions (presence of decaying seaweed or overlying silt with black mud near surface). In other countries this species is common in thermal springs with sulphide-rich water.

Spirulina laxissima G.S.West 1907

01660020

Trichome 0.7–0.8 μm wide; helix 4.5–5.3 μm wide; coils of helix 17–22 μm apart.

G.S.West reported this from the plankton of two countries in Africa. Trichomes somewhat like this occasionally occur in samples from the British Isles, though intermingled with other algae rather than planktonic, and are perhaps just a form of *Lyngbya*.

Spirulina maior [Kützing 1843] Gomont 1892

01660030 **Pl. 13F–H** (p. 88)

Trichome 1–2 μm wide; helix 2.5–4 μm wide; coils of helix regular, 2.7–5 μm apart.

Freshwater and brackish pools, on bottom sediments or among other algae. Powell (1964).

Spirulina meneghiniana [Zanardini 1847] Gomont 1892

01660040

Trichome 1.2–1.8 μm wide; helix 3.2–5 μm wide; coils of helix somewhat irregular, 3–5 μm apart.

Marine and brackish habitats.

Spirulina nordstedtii [Nordstedt 1880] Gomont 1892

01660050

Trichome 1.8–2.2 μm wide; helix 5 μm wide; coils of helix 5 μm apart.

Marine and brackish habitats.

Spirulina subsalsa [Oersted 1842] Gomont 1892

01660060 **Pl. 13C–E** (p. 88)

Trichome 1–2(–3) μm wide; helix 3–5(–8) μm wide; coils of helix usually somewhat irregular and spaced apart, but occasionally quite close.

This is one of the species most likely to be found, mostly in brackish and marine habitats, such as rock pools and salt marshes, but occasionally elsewhere.

Spirulina subtilissima [Kützing 1843] Gomont 1892

01660070

Trichome 0.6–0.9 μm wide; helix 1.5–2.8 μm wide; coils of helix regular, 1.25–2 μm apart.

This is a frequent freshwater and marine species, usually as single trichomes among other algae in environments likely to have reducing conditions, such as among algal masses at the bottom of ponds, slightly organically polluted water, sulphide-rich seepages.

Powell (1964).

Symploca [Kützing 1843] Gomont 1892

01680000

Filaments initially prostrate on a surface; typically developing into a colony of upright bundles of filaments, though this does not occur in one of the two British species. Filaments occasionally showing false branching, but otherwise a single trichome within the narrow sheath; sheath firm or somewhat gelatinous. Trichomes more or less straight, though sometimes slightly narrowed towards the apex; heterocyst absent.

Some species in this genus are quite similar to some species of *Schizothrix*. The key differences are the tendency to form upright bundles in *Symploca* (although this occurs in some *Schizothrix*) and the fact that the trichome is always single inside the sheath. In relation to the number of species described, few have been recorded from the British Isles. Apart from those listed below, Fritsch (1906) reported *Symploca thermalis*: cells 1.2–2 μm wide, 2–3 times as long as wide; growing on chinks between the brickwork of the pits in the Royal Botanic Gardens, Kew.

Symploca atlantica Gomont 1892

01680030 (new number)

Colony a bushy tuft to 1 cm long, upright; dark green. Filaments densely intertwined, unbranched, bent irregularly. Cells 4–6 μm wide; cell length to width shorter or slightly longer than wide; cross walls thick, narrowed; end cell with bluntly rounded calyptra. Sheath thin.

Marine and highly brackish waters, on rocks, algae and other macrophytes. Frémy (1929–33) lists England, probably based on record by Batters. Powell (1964).

This differs from the better known *S. hydnoides* in narrower trichomes, the tendency for all cells to show slight narrowing at the cross walls and the presence of a weakly developed calyptra.

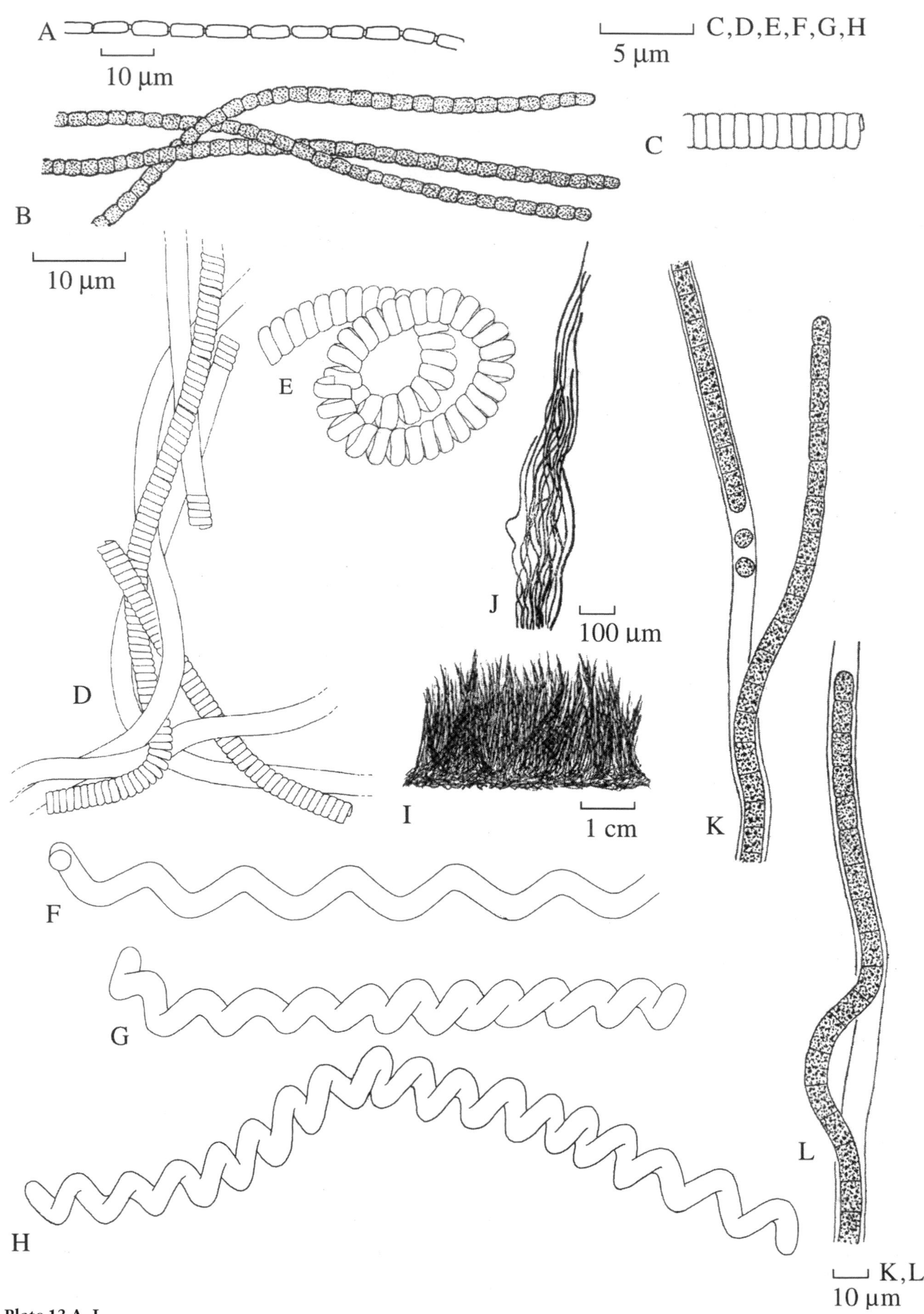

Plate 13 A–L

A. *Pseudanabaena catenata* (p. 84); **B**. *Phormidium ectocarpi* (p. 79); **C–E**. *Spirulina subsalsa* (p. 87); **F–H**. *Spirulina maior* (p. 87); **I–L**. *Symploca hydnoides* (p. 89).

Symploca dubia [Nägeli 1849] Gomont 1892

01680040 (new number)
Colony a mat, outer part yellowish, grey blue-green or red, inner colourless. Filaments more or less arranged in bundles, usually horizontal with tendency to anastomose, but sometimes more or less upright; individual filaments with many bends, especially when horizontal. Cells 1.5–2.5 μm wide, 3–8 μm long, up to 4 times as long as wide; cross walls not narrowed, often with an adjacent granule on each side.

Moist rocks by streams and waterfalls.

Powell (1964) without details of habitat.

Symploca hydnoides [Kützing 1849] Gomont 1892

01680050 (new number) **Pl. 13I–L** (p. 88)
Colony bushy tufts to 3 cm long, upright; bundles typically steel-blue to dark violet higher up, but often almost colourless near the base due to wide sheath in relationship to live cells. Filaments densely intertwined, occasionally branched, more or less stuck together, bent irregularly. Cells 6–14 μm wide; cell length to width as long as wide or longer in the narrower forms, but increasing tendency to be shorter than wide in the wider forms; cross walls near the end often narrowed or even these cells distinctly torulose; end cell slightly swollen, no calyptra. Sheath thin.

Powell (1964). Marine: on seaweeds and rocks in the intertidal.

The size range above includes both the type and varieties.

Symploca muralis [Kützing 1843] Gomont 1892

01680010 **Pl. 9P** (p. 65)
Colony a turf of upright bundles to 2 mm, each consisting of densely intertwined mass of flexuous filaments. Cells 3.4–4 μm wide, 1.6–4.3 μm long; cross walls usually not narrowed; end cell blunt-conical, with lightly thickened outer membrane. Sheath quite thick, firm.

Moist soil, walls and base of tree trunks.

Other forms which do not quite fit within this description have at times been included in the species.

Symploca muscorum [C.Agardh 1824 emend. Gomont 1890] Gomont 1892

01680020 **Pl. 9Q** (p. 65)
Colony a mass of intertwined filaments, usually prostrate, seldom becoming upright; usually in distinct bundles, but occasionally more like *Phormidium*; blue-green or yellow-brown. Cells 5–8 μm wide, 5–11 μm long; almost quadrate to twice as long as wide; cross wall not narrowed; end cell usually broadly rounded, occasionally rounded-conical; with lightly thickened outer membrane. Sheath to 2 μm wide, firm or slimy, colourless. The frequent yellow-brown colour of the colony is apparently always due to the colour of the trichomes.

Among mosses on moist soil and rocks; free-floating in standing or slowly flowing water.

Several varieties and many forms which do not quite fit within this description have at times been included in the species, though apparently none for the British Isles.

Trichodesmium [Ehrenberg 1830] Gomont 1892

01720000
Trichomes united in free-floating bundles. Trichome may or may not be gradually tapered towards the end; if not so, the end cell is often narrowed; gas-vacuoles usually obvious.

Mostly marine or highly brackish waters.

Trichodesmium erythraeum [Ehrenberg 1830] Gomont 1892

01720010
Trichomes united in free-floating bundle; bundle short, up to 1 mm long. Narrower trichomes are slightly tapered towards the end, while the wider ones are attenuated more abruptly. Cells 7–11(–21) μm wide, ranging from slightly shorter than wide to 3 times longer than wide, very slightly indented at cross wall; gas-vacuoles usually obvious. Trichomes usually a pinkish red, but sometimes pink-olive.

Marine plankton.

Trichodesmium lacustre Klebahn 1895

Synonym: *Oscillatoria lacustre* (Klebahn) Geitler 1925
01720020
Trichomes united in free-floating bundle. Trichome with end cell often longer than other cells, slightly narrowed. Other cells 5–7 μm wide, 3–7 μm long, with gas-vacuoles.

The status of this taxon is unclear. It is uncertain to what extent the different forms of bundle present in the marine plankton also occur in the freshwater plankton, or whether the freshwater form is in fact closely related to the two truly marine species. It is also unclear whether there are reliable records for the British Isles.

Trichodesmium thiebautii [Gomont 1890] Gomont 1892

Synonym: *Oscillatoria thiebautii* (Gomont) Geitler 1932
01720030
Trichomes united in free-floating bundle; bundle up to 6 mm long; trichomes often rolled into rope-like shapes in the middle of the bundle, often spread out at the ends. Trichomes not tapered towards the end; apical cell ranges from slightly attenuated to slightly inflated. Cells 7–16 μm wide, up to 2 times longer than wide, not indented at cross wall; gas-vacuoles usually obvious. Trichomes ranging from olive-pink to olive.

This species tends to occur further north in the Atlantic than *T. erythraeum* and is the species responsible for most records of the genus in seas off the British Isles.

ORDER NOSTOCALES

Filaments the cells of which divide in one plane and some or all of these cells with the potential to develop into heterocysts; the heterocysts may be terminal or intercalary. The heterocyst is usually easy to distinguish (see introductory notes on morphology). Species all possess the ability to form motile hormogonia, which lack heterocysts, but these are only a transitory stage. The trichomes of most species cease motility when or soon after heterocysts start to form. A number of genera show 'false' branching, which arises following trichome division, when one or both the new trichome ends grow(s) in a new direction.

The order broadly falls into three groups, recognized as families by Castenholz (1989). The Nostocaceae have trichomes with vegetative cells all more or less the same diameter and which do not show false branching. The Scytonemataceae have false branching prevalent in all genera; if obviously tapered trichomes occur, they are not the predominant form. The Rivulariaceae are characterized by tapered trichomes, apart from short phases of hormogonium formation. The mature trichome has a terminal heterocyst, though some species also have intercalary heterocysts; cell division is largely localized to a region near the heterocyst. However, some species of Rivulariaceae show Scytonemataceae-like growth forms under particular environmental conditions.

Other records Benson-Evans & Antoine (1996) list *Borzinema rupicola* De Toni (=*Diplonema rupicola* Borzí 1917), without detail. This is much like a form of *Tolypothrix*, but often forming chroococcal-type stages near the base. Borzí reported it from rocks and walls.

Anabaena [Bory 1822] Bornet *et* Flahault 1886

01020000

Large genus, usually easy to recognize under the microsope: characterized by unbranched trichomes, which almost always possess intercalary heterocysts. Trichomes are often aggregated into colonies, which may form tufts or flocs associated with aquatic plants or films on sediment. However, the aggregations do not form firm colonies with a distinct outer layer (characteristic of many species of *Nostoc*). Several gas-vacuolate species form water-blooms, usually with characteristic colonies. Field populations probably always form akinetes, though this feature is often lost in prolonged laboratory culture, almost certainly in response to genetic changes; presumably the strains are not cultured to the limiting conditions which lead to formation of akinetes. Many species can only be identified morphologically with certainty if the akinete is present. Features to look for include akinete shape and size, and whether or not the akinete forms next to a heterocyst and, if so, on both or only one side. Different forms of environmental stress lead to akinete formation and in some species there may be one specific factor (such as phosphorus limitation) bringing this about. In the case of phosphorus limitation, the first akinete almost always forms next to the heterocyst.

At least in the majority of strains, the heterocyst forms only under conditions of moderate nitrogen limitation, though trichomes are seldom seen in nature looking like *Anabaena*, but lacking heterocysts.

Bornet & Flahault (1886–8) recognized three sections (Trichormus, Dolichospermum, Sphaerozyga) within the genus, based on shape of the akinete and whether or not the akinete is adjacent to the heterocyst. Komárek & Anagnostidis (1989) incorporated this information into a break-up of the genus. This has several disadvantages. Although there are a few other features that can help, in practice most people can recognize material to the generic level only if akinetes are present, making it necessary to introduce an artificial taxon for records where akinetes cannot be viewed. It is doubtful whether the new taxa represent natural groupings. Species such as *Anabaena flos-aquae* are split between two genera.

Anabaena aequalis Borge 1906

01020010 **Pl. 14I** (p. 93)

Trichomes often aggregated to form a blue-green colony. Trichomes straight, without an obvious layer of mucilage. Cells short barrel-shaped, 4.5–5.5 μm wide, end cell rounded. Heterocyst 4.5–5.5 μm wide, 6.5–10.5 μm long; mature heterocyst almost always more elongated than vegetative cell. Akinete cylindrical, single or in pairs, distant from the heterocyst, 5–7 μm wide, 21–41 μm long, wall smooth, colourless.

Freshwaters, among submerged plants.

This species is very close to *A. inaequalis*, which name has priority.

Anabaena affinis Lemmermann 1897

01020020 (old number): see *Anabaena catenula* var. *affinis*

Anabaena augstumalis Schmidle 1899

01020030 **Pl. 14A,B** (p. 93)

Trichome single, more or less bent with indistinct sheath. Cells barrel-shaped or subspherical/ellipsoidal, 3.5–4.5 μm wide, 3–9 μm long, mostly longer than wide. Heterocyst cylindrical, 5.5–6.5 μm wide, 5.5–6.5 μm long. Akinete distant from heterocyst, 5.5–6.5 μm wide, 25–56 μm long.

Freshwaters, typically ponds and ditches.

A. augstumalis var. *incrassata* (Nygaardh) Geitler 1932 and var. *marchia* Lemmermann 1905 are listed in Plate 14, because these are the names associated with the original figures. Variety *incrassata* is overall slightly larger than the type and var. *marchica* slightly smaller; var. *incrassata* also differs in possessing gas vacuoles.

Anabaena catenula [Kützing 1849] Bornet *et* Flahault 1886

01020040 **Pl. 14C,15A** (p. 93, 96)

Trichomes straight, often aggregated to form a blue-green colony, typically in standing freshwaters, either free-floating or loosely attached to plants. Cells barrel-shaped, 5–8(–10) μm wide. Heterocyst 6–9 μm wide, 8–13 μm long. Akinete cylindrical or ellipsoidal, 7–10(–16) μm wide, 16–30(–35) μm long, wall smooth; akinetes often forming in rows away from the heterocyst, though occasionally recorded next to one.

Standing fresh or brackish water, plankton or loosely attached. The majority of samples of *Anabaena* loosely mixed with submerged plants in ponds or edges of lakes are likely to prove to be *A. catenula* if most akinetes are distant from the heterocyst, and *A. cylindrica* for most records with akinetes adjacent to heterocyst. However, other species are sufficiently widespread that all material sampled should be checked carefully.

Several infraspecific taxa have been recognized within this variable species, differing in the extent to which trichomes are single or aggregated, whether or not gas-vacuolate and the size of the cells and akinetes. All the forms appear to merge into each other, so the above description has been broadened to cover all three varieties recorded for the British Isles. One was treated as a variety and one as a species in the *Coded List*.

• var. *affinis* (Lemmermann) Geitler 1932 (new number 01020023; old number 01020020) (Pl. 15A, p. 95)

Cells 6–10 μm wide, mostly with gas vacuoles. Akinete ellipsoidal (? or cylindrical), 9.5–16 μm wide, 17–30 μm long, remote from the heterocyst; wall smooth.

Several rather different organisms have apparently been included under the specific name *A. affinis* and it is unclear which occur in the British Isles. A study of Japanese populations by Li et al. (2000) emphasized bundle formation and tapering apices as features distinguishing them from *A. viguieri*. However, populations sometimes occur in large weedy ponds in the British Isles which appear to provide a continuum with typical *A. catenula*. The material studied by Li et al. mostly had heterocysts which were single each side of the heterocyst, sometimes in pairs, but not in rows, the last being a frequent feature of typical *A. catenula*. Probably the type, the three varieties and *A. viguieri* should be treated as one large complex.

• var. *solitaria* (Klebahn) Geitler 1932 (Coded List 01020042; synonym *A. solitaria* Klebahn 1895) occurs in the plankton of Windermere and many other lakes as single filaments with gas vacuoles and with cell dimensions towards the upper limits in the above description. The original description stated akinetes either both sides of the heterocyst or distant from it, so the possibility must be considered that the material used may have been intermingled with *A. cylindrica*.

• var. *intermedia* Griffiths 1925 has akinetes which are more ellipsoidal than cylindrical; cells 8–10 μm wide; akinete 12–16 μm wide, 17–26 μm long; lowland lakes in England (Griffiths, 1925a); no *Coded List* number is allocated.

Anabaena circinalis [Rabenhorst 1852] Bornet *et* Flauhault 1886

01020050 **Pl. 14G** (p. 93), **15G** (p. 95) **CD**

Trichome coiled, seldom straight, usually with diffluent or no visible mucilage. Cells spherical, barrel-shaped or subspherical/ellipsoidal, (7–)8–14 μm wide; mostly shorter than wide, with gas vacuoles. Heterocyst almost spherical, 8–10 μm wide, 8–10 μm long. Akinete cylindrical, 14–18 μm wide, 25–34 μm long; wall smooth, colourless; forms with narrower and longer akinetes (9–10.5 × 28–42 μm) have been separated as var. *macrospora* (Wittrock) Forti 1907 (Pl. 14G).

Standing fresh or slightly brackish waters, often forming blooms.

This species differs from *A. flos-aquae* principally in the dimensions of vegetative cells and akinetes, with 8 μm trichome width being the key dimension separating the two in most Floras; the less distinct or absent mucilage in *A. circinalis* is another difference. Nevertheless the two species appear to form a continuum, though in view of the fact that lakes often contain populations with two distinct trichome widths, it is worth retaining both species.

Anabaena cylindrica Lemmermann 1896

Synonym: *Anabaena subcylindrica* Borge 1921 (Coded List 01020150)

01020060 **Pl. 14D,H** (p. 93) **CD**

Trichomes straight, often aggregated to form a blue-green colony. Mucilage thin and usually colourless, but sometimes with a pale yellow-brown tinge; mucilage more evident in aging material, especially around heterocysts and akinetes. Cells subspherical/ellipsoidal, 3–4(–4.5) μm wide, 5–7(–8.5) μm long. Heterocyst 4.5–5.5 μm wide, 5.5–8 μm long. Akinete cylindrical, next to the heterocyst, 5–8 μm wide, 21–30 μm long; wall smooth, colourless.

Standing fresh or slightly brackish water, either free-floating or loosely attached to plants. Akinete formation is an indication of phosphorus limitation.

Anabaena subcylindrica Borge differs from the above description only in longer akinetes (to 57 μm), which are often slightly narrowed in the middle, and sometimes abrupt at the ends. These features hardly merit separation as a species, especially since akinetes of clonal cultures of *A. cylindrica* are known to show considerable variation in culture, including occasionally a similar slight narrowing in the middle. It is suggested that formal recognition as a variety, let alone a species, should wait until further material is collected.

Anabaena delicatula Lemmermann 1898

01020070

Trichome single, straight or slightly bent. Cells, barrel-shaped, 3.5–4.5 μm wide, 5–7 μm long. Heterocyst 4–5 μm wide, 4–5 μm long. Akinete cylindrical, 7.5–8.5 μm wide, 17–19 μm long; wall smooth

Typically in standing freshwater associated with mucilaginous phytoplankton.

Anabaena elliptica Lemmermann 1904

01020080 **Pl. 15E** (p. 95)

Trichome single, straight or slightly bent with thick colourless mucilage. Cells ellipsoidal, 6.5–7.5 μm wide, 13.5–14.5 μm long, with gas vacuoles. Heterocyst subspherical/ellipsoidal, 6.5–7.5 μm wide, 7–8 μm long. Akinete 15–16 μm wide, 24–26 μm long, wall smooth.

Free-floating in freshwater.

Anabaena flos-aquae [(Lyngbye) Brébisson 1835] Bornet *et* Flauhault 1886

01020090 **Pl. 15H** (p. 95) **CD**

Trichomes coiled forming a solitary or, more usually, an entangled twisted mass in standing freshwaters; with colourless mucilage. Cells most often bent, ellipsoidal, rarely spherical, (2–)4–8 μm wide, 5.5–8 μm long with gas vacuoles. Heterocyst 4–9 μm wide, 6–10 μm long. Akinete cylindrical or slightly ellipsoidal and often very slightly bent, 6–13 μm wide, 20–50 μm long; wall smooth, colourless or yellowish; akinete usually adjacent to, but sometimes distant from heterocyst; akinetes adjacent to heterocyst are single or in rows.

One of the commonest bloom-forming species in the British Isles and in most other parts of the world.

The above broad description covers not just the original description, but most of the many varieties and forms. Although it has apparently never been recognized at the nomenclatural level, it seems likely that the most important difference within the species is whether or not akinetes develop next to the heterocyst; this may eventually justify separation at the species level. Some authors (e.g. Reynolds & Irish, 2000, but not Geitler, 1932a) recognize *A. lemmermanni* P.G.Richter 1903 as a separate species. This forms long rows of akinetes in the centre of the colony. Morphologically the difference from the type is slight, but the colonies have a characteristic appearance macroscopically. At least in Windermere, the colonies always have attached vorticellid ciliates (J.W.G. Lund, pers. comm.: Fig. 407 in Canter-Lund & Lund, 1995).

Anabaena inaequalis [Kützing] Bornet et Flahault 1886

01020100 **Pl. 15D** (p. 95)

Trichomes straight, with or without a diffluent mucilage layer, and usually aggregated to form a blue-green colony. Cells barrel-shaped, 4–6 μm wide; end cell rounded. Heterocyst 5.5–6.5 μm wide, 6–10 μm long. Akinetes single or many, cylindrical, 6–8 μm wide, 14–20 μm long; wall smooth, colourless or yellowish.

Standing freshwater, attached or free-floating.

Anabaena lemmermanni P.G.Richter 1903

see *Anabaena flos-aquae*

Anabaena macrospora Klebahn 1895

01020110 **Pl. 14F** (p. 93)

Trichome single, straight. Mucilage, thick and colourless. Cells subspherical or ellipsoidal, 5–6.5 μm wide, 5–9 μm long, with gas vacuoles. Heterocyst spherical or subspherical, 6–6.5 μm wide, 6–7 μm long. Akinetes single or in twos, at first rounded, later more elongated, with a rounded/6-sided outline, usually distant from the heterocyst, 16–18 μm wide, 17–26 μm long, wall smooth, colourless.

Free-floating in standing freshwaters.

Anabaena oscillarioides Bory 1822

01020120 **Pl. 15F** (p. 95)

Trichomes straight, aggregated to form a slimy dark blue-green coloured colony. Cells barrel-shaped, 4–6 μm wide, 4–7 μm long, mostly longer than wide; end cell rounded. Heterocyst spherical or ellipsoidal, 6–8 μm wide, 6–10 μm long. Akinete cylindrical, 8–10 μm wide, 20–40 μm long; wall smooth, brown.

Standing freshwaters.

Several varieties have been recognized (though apparently not for the British Isles), based on features such as larger akinetes and a tendency for the end cell to be conical.

Anabaena planctonica Brunnthaler 1903

01020130 **Pl. 15C** (p. 95)

Trichome straight, single, sometimes with a diffluent mucilage layer, though this tends to be most obvious adjacent to the akinete. Cells spherical or subspherical/ellipsoidal, 9–15 μm wide, 6–12 μm long, sometimes with gas vacuoles. Heterocyst spherical, 10–14 μm wide, 10–14 μm long. Akinete subspherical, distant from the heterocyst, 10–20 μm wide, 15–30 μm long; wall smooth, sometimes surrounded by conspicuous mucilage.

Free-floating in freshwaters. Although this is quite widespread in the plankton, it appears seldom to reach bloom-forming densities, at least in the British Isles.

Anabaena solitaria Klebahn 1895

see *Anabaena catenula*

Anabaena spiroides Klebahn 1895

01020140 **Pl. 15I** (p. 95)

Trichome single, helical, embedded in a wide, diffluent mucilage layer. Cells almost spherical, 6.5–8(–15) μm wide, 5–8.5(–12) μm long; cells usually shorter than wide. Heterocyst almost spherical, 6.5–8 μm wide, 6.5–8 μm long. Akinete at first rounded, later almost six-sided, next to or away from the heterocyst, 13–15(–25) μm wide, 13–18(–33) μm long; wall smooth.

Free-floating in standing freshwaters.

Several varieties have been recognized, differing in frequency and amplitude of the helix, cell and akinete dimensions. Some of the limits lie well outside the range for the type, especially var. *crassa* Lemmermann 1898, with cells 11–15 μm wide. It is unclear what is the status of these varieties in the British Isles.

Anabaena subcylindrica Borge 1921

01020150: see *Anabaena cylindrica*

Anabaena torulosa [Lagerheim] Bornet et Flahault 1886

01020160 **Pl. 14K** (p. 93)

Trichomes typically aggregated to form a thin, slimy, blue-green coloured colony. Cells barrel-shaped, 4.2–5 μm wide, 3–5 μm long; end cell conical. Heterocyst 5.5–6.5 μm wide. Akinete rounded-cylindrical, single or many on both sides of the heterocyst, 7–12 μm wide, 10–24 μm long; wall smooth, brown.

In standing fresh or brackish water, usually attached to submerged plants. Powell (1964).

Forms of this species probably merge into forms of *A. oscillarioides*.

Anabaena variabilis [Kützing 1843] Bornet *et* Flahault 1886

01020170 **Pl. 14J** (p. 93)

Trichomes single, straight, mostly without a sheath, forming a dark green colony. Cells barrel-shaped, 4–6 μm wide, 2.5–6 μm long, sometimes with gas vacuoles; end cell conical. Heterocyst spherical or ellipsoidal, 5.5–6.5 μm wide, 6–8 μm long. Akinetes forming rows distant from the heterocyst, 7–11 μm wide, 8–14 μm long; wall smooth or with fine warts, colourless or yellow-brown.

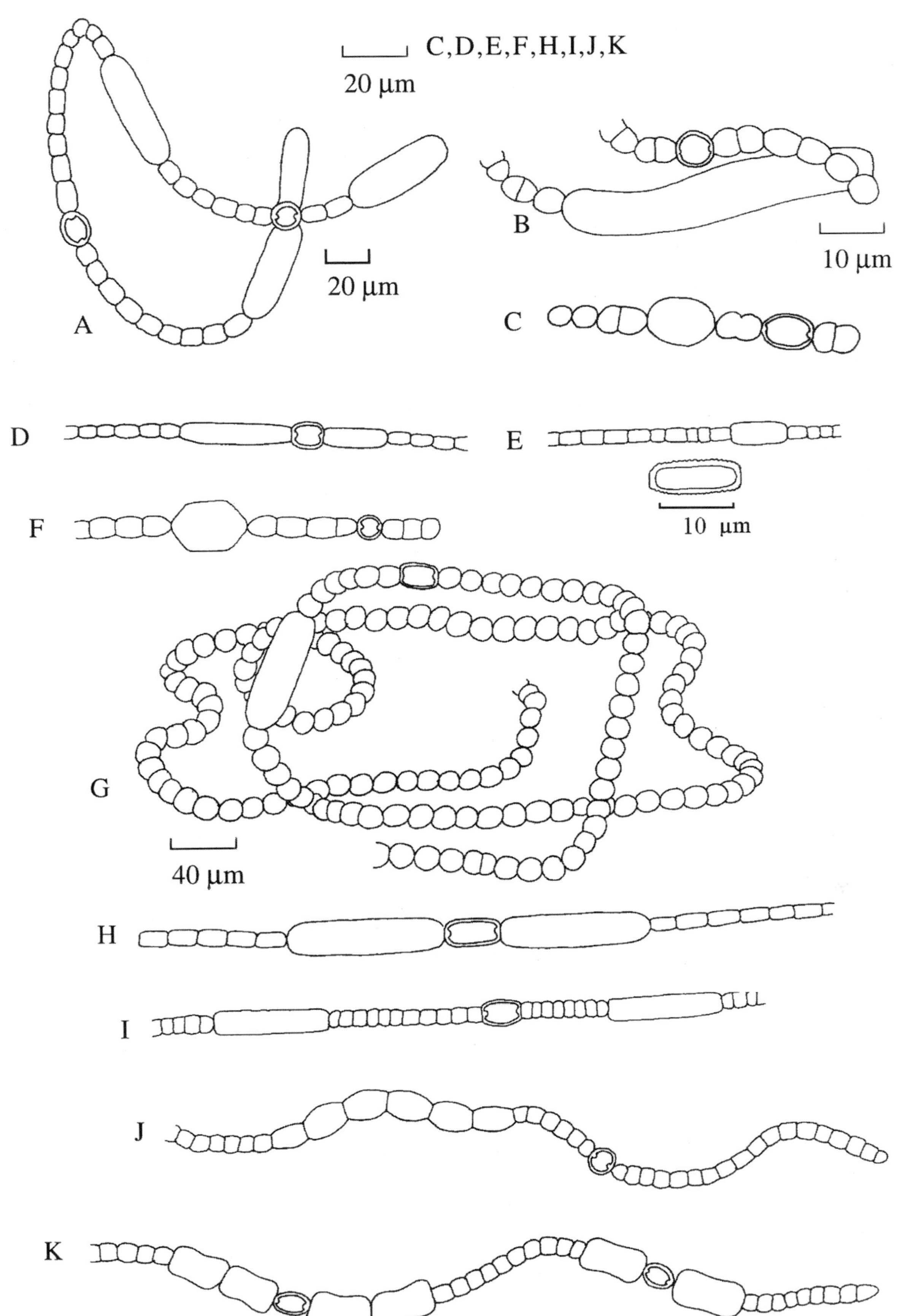

Plate 14 A–K
A. *Anabaena augstumalis* var. *incrassata* (p. 90); **B**. *Anabaena augstumalis* var. *marchica* (p. 90); **C**. *Anabaena catenula* (p. 90); **D**. *Anabaena cylindrica* (p. 91); **E**. *Anabaena verrucosa* (p. 94); **F**. *Anabaena macrospora* (p. 92); **G**. *Anabaena circinalis* var. *macrospora* (p. 91); **H**. *Anabaena cylindrica* (p. 91): large akinete form, originally described as *A. subcylindrica*; **I**. *Anabaena aequalis* (p. 90); **J**. *Anabaena variabilis* (p. 92); **K**. *Anabaena torulosa* (p. 92).

Attached or free-floating in standing bodies of fresh, brackish or sea water; also on very moist soil. Powell (1964).

This is a widespread and variable species. In addition, the name has probably often been selected when authors were unsure of the identification.

Anabaena verrucosa Boye-Petersen 1923

01020180 **Pl. 14E** (p. 93)
Trichome straight, often with a narrow, barely visible, layer of mucilage. Cells cylindrical, 3–4 μm wide, 4–8 μm long; end cell rounded. Heterocyst cylindrical, 3–4 μm wide, 5–8 μm long. Akinete cylindrical, single or in 2s, 6–7 μm wide, 12–15 μm long, rounded or slightly truncated at the ends; wall rough or slightly warted, yellow-brown.

The original record for Iceland was among *Sphagnum*, but this species has since been recorded in a variety of freshwater and moist terrestrial habitats, including ones in the British Isles.

Boye-Petersen's illustration suggests that akinetes form away from the heterocyst.

Anabaena viguieri Denis *et* Frémy 1923–24

01020190 **Pl. 15B** (p. 95)
Trichome straight, without obvious mucilage. Cells mostly barrel-shaped, 6–7 μm wide, 4–8.5 μm long, with gas vacuoles; end cell rounded. Heterocyst 5–7 μm wide, 5–7 μm long. Akinete ellipsoidal, single and away from the heterocyst, 12–13 μm wide, up to 17.5 μm long; wall smooth, colourless.

Planktonic in standing freshwaters, sometimes reaching high enough densities to form a major component of a water-bloom.

Anabaenopsis (Wołoszyńska) V.V.Miller 1923

01030000
Trichomes helical or forming loops, typically with a terminal heterocyst at each end, though occasionally at only one end or even absent. Akinetes single or in pairs, usually near the middle of the trichome and therefore distant from the heterocyst.

This genus is best known from the tropics and subtropics, but there are records for the British Isles.

Other record Galliford & Williams (1948), not identified to species: brackish pool, Cheshire.

Anabaenopsis elenkinii V.V.Miller 1923

01030010 **Pl. 15J** (p. 95)
Trichomes spiral, with 0.75 to 2.5 loops. Cells 4.6–6.7 μm wide, 5.5–10 μm long, ellipsoidal, gas-vacuolate; heterocysts terminal, rounded, 4.6–6.7 μm diameter, resulting from unequal cell division. Akinetes single or occasionally in pairs, spherical 8.3–10.7 μm diameter, or shortly ellipsoidal 8.3–10.5 μm wide, 9.3–12 μm long; wall smooth, colourless.

Planktonic in fresh and brackish waters. Several records from southern England including the one illustrated by Canter-Lund & Lund (1995, fig. 406).

Aphanizomenon [Morren 1838] Bornet *et* Flahault 1886

01040000
Trichomes straight or nearly so, single or in free-floating bundles; end cells tapered. Heterocyst intercalary. Akinete single.

Heterocysts are sometimes slightly difficult to distinguish from vegetative cells during casual inspection of a sample.

The best known species in the British Isles is *Aph. flos-aquae*, which forms characteristic bundles. Although they have seldom been reported in the literature for the British Isles, single trichome forms appear to be widespread in lakes and ponds and have probably often been overlooked. They show considerable differences in morphology between sites and do not always correspond closely to a described species; akinete morphology is a key feature to be noted. However, at least some populations correspond quite well to the three single trichome species listed (*Aph. aphanizomenoides*, *Aph. gracile*, *Aph. issatschenkoi*). It seems likely that some may at times occur without heterocysts and be confused with *Oscillatoria* (see entry for *Aph. issatschenkoi*). The taxonomic literature on the genus is of limited value, because of the lack of experimental studies controlling nutrient limitations; studies on the effects of alternating phosphorus and nitrogen limitation would probably help to resolve some of the uncertainties.

Separation of *Aphanizomenon* from *Anabaena* was for long based largely on the occurrence in the former of trichomes in bundles and ends of trichomes being attenuated and apical cells elongated (Komárek, 1958). There have since been a number of attempts to use other criteria, which are summarized by Hindák (2000), but the most convincing still appears to be the morphology at the end of the trichome. However, the taxonomic literature on the genus is of limited value, because of the lack of experimental studies on the influence of nutrients, including alternating nitrogen and phosphorus limitation.

The morphologically quite similar, *Cylindrospermopsis raciborskii* (Wolosz.) Seenayya *et* Subba Raju, which is best known from the tropics, appears to have become more widespread in temperate countries in recent years. If there is a succession of warm summers, it is likely that records will occur in the British Isles. *Cylindrospermopsis* differs by having a terminal heterocyst at one or both ends of the trichome and none in an intercalary position; trichomes occasionally lack any heterocysts and so might be confused with a form of *Raphidiopsis* or *Oscillatoria* with tapered ends of the trichome.

Aphanizomenon aphanizomenoides (Forti) Hortobágyi *et* Komárek 1979

Basionym: *Anabaena aphanizomenoides* Forti 1912
01040010
Trichomes straight or slightly curved, 1–2 mm long. Cells 4–5 μm wide in the main part of the trichome, 1–3 times longer than wide, with gas vacuoles. Trichome weakly narrowed at the cross walls. Heterocysts almost rounded or slightly ellipsoidal, 4–7.5 μm wide, 4.5–7 μm long. Akinete subspherical, single or several adjacent to the heterocyst, 8–14 μm; wall smooth colourless.

Plankton in freshwater ponds and lakes.

This species is compared with other single trichome species by Horecká & Komárek (1979) and Hindák (2000).

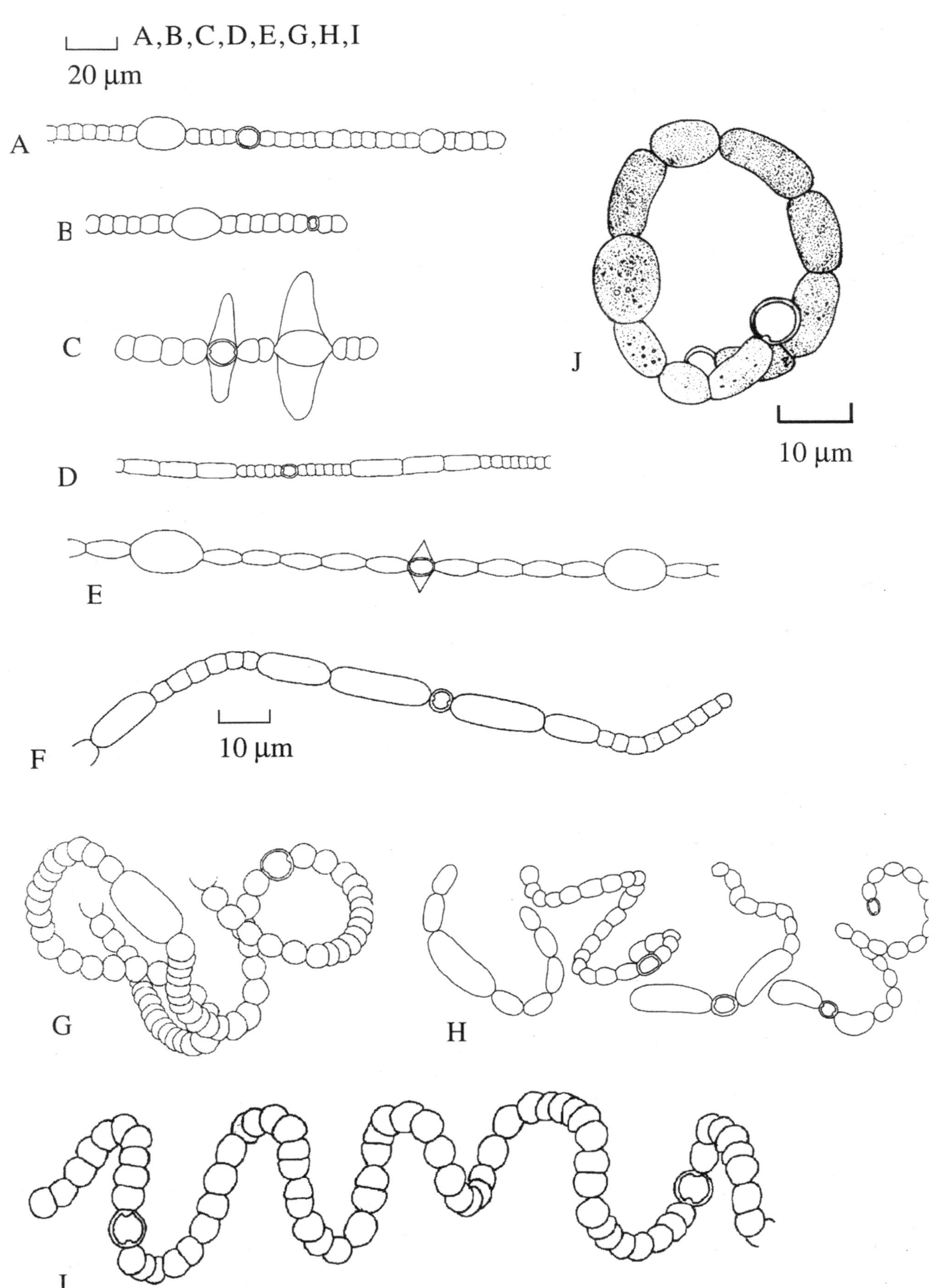

Plate 15 A–J
A. *Anabaena catenula* var. *affinis* (p. 91); **B**. *Anabaena viguieri* (p. 94); **C**. *Anabaena planctonica* (p. 92): showing mucilage around a heterocyst and an akinete (p. 92): **D**. *Anabaena inaequalis* (p. 92); **E**. *Anabaena elliptica* (p. 91); **F**. *Anabaena oscillarioides* (p. 92); **G**. *Anabaena circinalis* (p. 91); **H**. *Anabaena flos-aquae* (p. 91); **I**. *Anabaena spiroides* var. *crassa* (p. 92); **J**. *Anabaenopsis elenkinii* (p. 94)

Aphanizomenon flos-aquae [(Linnaeus 1753) Ralfs 1850] Bornet *et* Flahault 1886

01040020 **Pl. 18G–J** (p. 106) **CD**
Trichomes straight, aggregated in bundles or flakes to form macroscopic colonies of a few or hundreds of trichomes; single trichomes may also occur. Ends of trichomes usually showing slight tapering, with the cells towards the end slightly longer than those in the middle. Cells 5–6 μm wide, 5–15 μm long, with gas vacuoles; end cell conical. Heterocyst 5–7 μm wide, 7–20 μm long. Akinete cylindrical, 6–8 μm wide, 40–80 μm long; wall smooth, colourless.

Frequent component of freshwater blooms and sometimes highly abundant.

Aphanizomenon gracile Lemmermann 1910

Basionym: *Anabaena flos-aquae* (Lyngbye) Brébisson var. *gracile* Lemmermann 1898
Synonym: *Aphanizomenon flos-aquae* fo. *gracile* (Lemmermann) Elenkin 1938
01040030 **Pl. 18L** (p. 106)
Trichomes single, straight, seldom in bundles. Cells 2–3 μm wide, 2–6 μm long, with gas vacuoles; end cell conical. Heterocyst 2.5–3.5 μm wide, 5.5–7 μm long. Akinete cylindrical, with rounded end and often very slightly narrowed in the middle, 4.5–5.5 μm wide, 22–30 μm long; wall smooth, colourless.

Planktonic in fresh and brackish ponds and shallow lakes.

This species is quite like *Aph. flos-aquae*, but smaller in all dimensions and with a much greater tendency to occur as single trichomes. Detailed information is provided by Hindák (2000), who retained it as a separate species.

Aphanizomenon issatschenkoi (Usačev ex Proškina-Lavrenko 1968)

01040040 **Pl. 18K** (p. 106) **CD**
Trichomes straight or slightly bent. Cells, 2–6 μm wide with gas vacuoles; end cell conical. Heterocyst 3–7.5 μm wide, 9–18 μm long. Akinete cylindrical, 4–7 μm wide, 11–28.5 μm long; wall smooth, colourless.

Planktonic in lakes.

A population in flooded gravel workings (Denton Reservoir, nr Grantham, Lincolnshire) included trichomes which fit the species description well (records of D.M. Balbi and S.J. Brierley). However, the range of trichomes at times appeared to provide a continuum with ones indistinguishable morphologically from *Oscillatoria agardhii*, including single trichomes where one end resembled *Aph. issatschenkoi* and the other end *O. agardhii*.

Calothrix [C.Agardh 1824] Bornet *et* Flahault 1886

01090000
Filaments single or in small bundles, often united to form distinct colonies; such colonies may be cushions, mats or crusts. Trichomes tapered, running more or less parallel to each other and perpendicular to the surface in the more cushion-shaped colonies, but intermingled in flatter colonies. Trichomes single or with occasional false branches, surrounded by a firm sheath in the lower part of the trichome (though usually not the heterocyst, apart from false branching). Trichome often ending in a multicellular hair, though almost certainly not in all species; the hair develops in response to phosphorus limitation and (at least in the laboratory) sometimes also iron limitation. The trichome in most species tapers gradually almost to the heterocyst, but in a few species e.g. *C. contarenii* has a swollen shape towards the base; this is probably an important taxonomic character. The sheath is firm and in many species yellow-brown; sheaths are occasionally embedded in mucilage towards the base, but this appears to be atypical. Basal heterocyst and in some species also a few intercalary heterocysts. Collapsed former heterocysts sometimes persist under the active heterocyst; at least in some cases this is due to increased shortage of iron, and perhaps molybdenum. Akinetes have been recorded in one species, next to the heterocyst. Hormogonia form at the apical ends of trichomes under phosphorus-rich conditions; when a succession of hormogonia has been released, the sheath may extend beyond the remainder of the vegetative cells. Gas vacuoles occasionally present in hormogonia and such hormogonia can easily be mistaken for *Oscillatoria* when floating in the plankton. In laboratory cultures hormogonia show a marked tendency to aggregate, but it is unknown how important this is in nature.

Freshwater, brackish and marine environments, attached to rocks, other firm substrata or water plants.

Other records

Benson-Evans & Antoine (1996): *C. aeruginosa* Woronichin 1923
Powell (1964) lists (without details of locations): *C. cottonii* Parke; ? *C. fasciculatus* [C.Agardh] Bornet *et* Flahault 1886; *C. litoralis* Anand 1937; *C. vivipara* [Harvey 1858] Bornet *et* Flahault 1886. None of these is sufficiently distinctive to be accepted as a species without further study. Harvey's original material of *C. vivipara* was from New England, but Frémy (1929–33) lists Scotland.

Calothrix aeruginea [(Kützing 1836) Thuret 1878] Bornet *et* Flahault 1886

01090010
Filaments 9–10(–12) μm wide, up to 0.5 mm long, aggregated into films. Trichome tapered, often ending in a long colourless hair, embedded in a non-lamellated sheath, which is usually colourless, but sometimes pale yellow-brown. Lower cells 7–9 μm wide. Heterocyst basal and usually also a few intercalary ones.

Powell (1964). Marine, epiphytic on macroalgae in the sublittoral, often forming extensive thin films.

Calothrix braunii Bornet *et* Flahault 1886

01090020
Filaments straight, parallel, with a slight tendency to be swollen towards the base. Trichome tapered, often ending in colourless hair, sheath thin, colourless, non-lamellated, 9–10 μm wide. Cells cylindrical, in main part

of trichome 6–7 μm wide, mostly shorter than wide and slightly narrowed at cross-walls. Heterocyst basal only.

Freshwater lakes and ponds, on rocks and submerged plants. Widespread in the British Isles, including River Caragh, Kerry (Heuff & Horkan, 1984).

Calothrix confervicola [(Roth 1806) C.Agardh 1824] Bornet *et* Flahault 1886

01090030

A large species. Filaments 12–25 μm wide, up to 3 mm long, with colourless or yellow-brown non-lamellated sheath. Trichomes tapering in the upper part, often ending in long hairs, with hardly any swelling at the trichome base; cells in main part of trichome 10–18 μm wide, often with pink tinge. Heterocyst basal only.

Powell (1964). Marine, epiphytic on larger algae and old leaves of *Zostera*, frequent.

Calothrix contarenii [Zanardini 1839] Bornet *et* Flahault 1886

Synonym: *Calothrix parasitica* [Thuret] Bornet *et* Flahault 1886

010900110 (new number)

Colony usually a crust, firm, thin, dark green. Filaments densely packed, 9–15 μm wide, to 1 mm long; swollen at the base, often ending in long colourless hair. Cells (above the swollen base) 6–8 μm wide. Heterocyst basal only. Sheath quite thick, without or with lamellations, in the latter case splayed at apex, colourless to yellow-brown.

On rocks and wood in supralittoral. Powell (1964), with query.

Powell (1964) and Benson-Evans & Antoine (1996) both list *C. parasitica*. The morphology of individual filaments of this species is identical to that of *C. contarenii* (which has the nomenclatural priority). Frémy (1929–33) reports that this species lives inside fronds of *Nemalion*. The two species are treated as synonyms here. Frémy also points out how similar *C. contarenii* is to *C. scopulorum*, but they are kept separate here because of the swollen basal region in the former and its near absence in the latter.

Calothrix crustacea [Thuret 1878] Bornet *et* Flahault 1886

01090040

Colony often very dark, consisting of filaments aggregated to form a film or mat, with individual filaments up to 2 mm long; filaments 12–20(–40) wide, with colourless or (usually) yellow-brown lamellated sheath. Trichome tapered, often ending in long hair; cells in main part of trichome 8–15 μm wide. Heterocysts basal and intercalary. Akinetes have been reported several times in this species, but it is unclear whether all populations possess this feature, and there are apparently no records for akinete formation from the British Isles.

Powell (1964). Marine: rocks, especially in the upper intertidal zone, occasionally on submerged plants. Epilithic colonies are probably absent on parts of the coast with eutrophic water. This is a highly variable species and epiphytic forms probably merge with *C. confervicola*, which is differentiated by the non-lamellated sheaths and lack of intercalary heterocysts.

Calothrix epiphytica West *et* G.S.West 1897

01090050 **Pl. 16F** (p. 99)

Filaments single or in groups, 5–7.5 μm wide, 250(–350) μm long, with colourless sheath. Trichome tapered; no mention of hair in original description, but possibility of its occurrence cannot be ruled out; cells 3–4 μm wide, shorter than wide in the lower part of the trichome. Heterocyst basal.

Freshwater, epiphytic on filamentous algae, including *Tolypothrix*.

Calothrix fusca [Kützing 1843] Bornet *et* Flahault 1886

01090060 **Pl. 16E** (p. 99)

Filaments single or in small groups, unbranched or occasional false branching. Trichome straight or bent ending in a distinct hair; cells near base 7–8 μm wide, shorter than wide; colourless sheath, showing narrow lamellations. Heterocyst basal.

Freshwater, in mucilage of colonial algae, often species which also have multicellular hairs. It seems likely that at least some forms living in the mucilage of other algae are just isolated filaments of *C. parietina*. Powell (1964) ? habitat.

Calothrix parietina [Thuret 1875] Bornet *et* Flahault 1886

01090070 **Pl. 16A–C** (p. 99) **CD**

Colony a mat or crust on rocks and other firm substrates, consisting of a mass of intermingled filaments. Filaments 10–14(–18) μm wide, up to 1 mm long, often showing false branching. Cells 5–10 μm wide, shorter than wide towards the base of the trichome, but often as long or longer than wide in the middle of the trichome. Sheath yellow to brown, lamellated, often well splayed out towards the apical end. Heterocyst basal and occasionally intercalary, the latter feature probably differing between populations and not just an environmental effect. Gas vacuoles occur in hormogonium (at least in most populations).

Freshwater, typically occurring near the interface between water and dry land, favouring, but by no means confined to, harder waters and with colonies occasionally showing slight calcification. The species is frequent in garden ponds, which are not too nutrient rich, as a narrow brown zone just below the upper water level.

A *Calothrix* population forming oncoids in White Beck, Melmerby, North Yorkshire, in 1991 was referred to *C. viguieri* by A. Pentecost (pers. comm.). Apart from the intense calcification, this species differs from *C. parietina* mainly in the colourless sheaths; it seems possible that at least as this site, the lack of colour was merely due to reduced exposure to u-v light.

Calothrix pulvinata [C.Agardh 1824] Bornet *et* Flahault 1886

01090080 **Pl. 16G** (p. 99)

Colony consisting of a bushy mass of filaments 2–3 mm long. Filament 15–18 μm wide, with colourless to yellow-brown sheath. Trichome usually ending in a short hair; often slightly swollen at the base. Cells 8–12 μm wide, short, in the lower part of the trichome typically about one-third as long as wide. Heterocyst basal.

Powell (1964). Marine: epilithic and epiphytic, typically in pools of the mid-intertidal zone. A widespread species on parts of the coast where the water is not too eutrophic; easy to recognize with the naked eye.

Calothrix scopulorum [C.A.Agardh 1824] Bornet *et* Flahault 1886

01090090

Colony forming a (usually) dark olive-green turf of short (–1 mm) filaments. Filaments 10–18 μm wide, with colourless or, more usually, yellow sheath; sheath lamellate, this being most obvious in older filaments, with apical ends splayed out. Cells 8–15 μm wide. Heterocyst basal.

Powell (1964). Marine: on rocks and wood, occasionally also macroalgae, typically found near the upper limit of the intertidal zone; occasionally in brackish waters, near the water–air interface. It tends to occur in a rather similar position with respect to water level as *C. parietina* in freshwater environments, but is less prone to encountering long periods of desiccation. The wording of Geitler (1932a) makes it unclear whether he considered this also to be a freshwater species.

Calothrix stagnalis Gomont 1895

01090100 **Pl. 16D** (p. 99)

Filaments in small groups, sometimes with a star-shaped arrangement, up to 1 mm long. Filament 8–12 μm wide, with narrow colourless sheath. Trichome tapered, usually with a long hair. Cells 6–9 μm wide, sometimes shorter than wide towards base, in the middle part of the trichome slightly longer than wide, slightly narrowed at the cross walls. Heterocyst basal. This species is differentiated from other freshwater species by the ability to form akinetes: akinete more or less cylindrical, 10–11 μm wide, 26–40 μm long, yellowish; occasionally two akinetes present, but the inner one with its own basal heterocyst.

Freshwater, epiphytic on algae in ponds and at edge of shallow lakes, but it is unclear whether akinetes have been noted in British records. The filaments resemble isolated ones from a colony of a small species of *Gloeotrichia*. Studies are needed to confirm whether they are distinct.

Coleodesmium Borzí 1879 ex Geitler 1942

01230000

Colony a dichotomously branching group of filaments forming a bushy-cushion shape. Trichomes straight, several within one sheath, each with a basal heterocyst. The overall structure is like a form of *Tolypothrix* with multiple trichomes in the sheath.

The genus was listed by Geitler (1932a) and other authors of Floras (including the *Coded List*) as *Desmonema* Berkeley *et* Thwaites 1849. The reasons for the nomenclatural change, including the species below, are given by Komárek & Watanabe (1990).

Coleodesmium wrangelii [J.Agardh] Bornet *et* Flahault 1886 ex Geitler 1942

01230010 **Pl. 19A** (p. 110)

Colony cushion-shaped to bushy, to 6(–10) mm long, somewhat mucilaginous, dark green to brownish. Filaments more or less straight, with many branches. Cells short barrel-shaped, 9–10 μm wide, 2.5–3.5 μm long. Sheath quite thick, firm, colourless or, more usually pale yellow-brown. Heterocyst basal, mostly ellipsoidal.

Fast-flowing streams and lake margins, on rocks, pebbles and mosses. West & Fritsch (1927): rare. Woodhead & Tweed (1954): shingle at Bhychan, Merionethshire (oligotrophic lake).

Cylindrospermum [Kützing 1843] Bornet *et* Flahault 1886

01200000

Colony mucilaginous to varying degrees, consisting of many uniformly broad trichomes; trichomes usually quite narrow (often about 4 μm wide) and short; heterocyst terminal, at one or both ends. A large akinete develops adjacent to the heterocyst; usually single, but occasionally in pairs and, in *C. catenatum*, up to 8. Akinete shape highly characteristic for most species and it is usually possible to distinguish outer and inner wall layers. Akinetes frequently form at more or less the same time throughout a colony. There are often large rod-shaped bacteria attached to the heterocyst wall.

Many species are typical of moist soil and often appear as shining green mats with a very distinct margin; this usually makes it possible to distinguish *Cylindrospermum* from *Anabaena* macroscopically, when growing on soil. As mentioned in the Introduction, several species are probably much more widespread in soils than suggested by the occasional occurrence of colonies on the surface.

West & Fritsch (1927) list *C. catenatum*, *C. majus* and *C. stagnale*, but mention that there are three others; it is not known whether all these are included in the other four given here. Of these, *C. licheniforme* and *C. muscicola* are probably the most widespread, at least in lowland regions.

1 Vegetative cells ≤3 μm wide 2
1 Vegetative cells > 3 μm wide 5
2 Mature akinete with conspicuous wing ....*C. alatosporum*
2 Mature akinete without wing..................................3
3 Mature akinete colourless 4
3 Mature akinete yellow or brown ....*C. muscicola*
4 Akinetes never more than 1–2 in a row ....*C.minutissimum*
4 Akinetes often more than 2 in a row..*C. marchicum*
5 Akinetes never more than 1–2 in a row 6
5 Akinetes often more than 2 in a row..*C. catenatum*
6 Mature akinete with papillae ....*C. majus*
6 Mature akinete smooth 7
7 Akinete ovoid....................................*C. muscicola*
7 Akinete ellipsoidal or cylindrical 8
8 Akinete with truncated ends..........*C. licheniforme*
8 Akinete with rounded ends....................*C. stagnale*

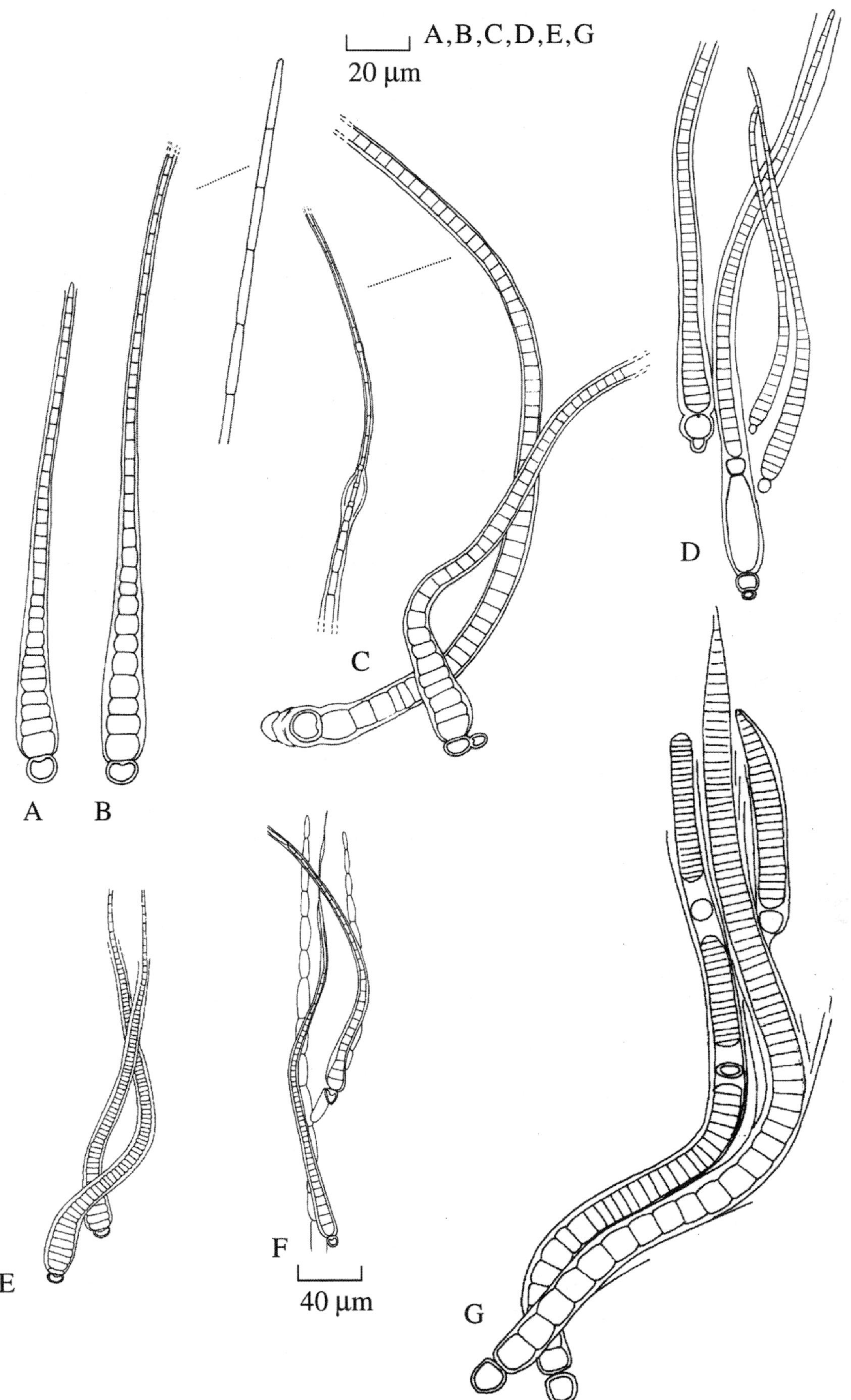

Plate 16 A–G
A–C. *Calothrix parietina* (p. 97): A, B. Laboratory culture showing moderately and markedly phosphorus-limited trichomes, C. Field sample, showing appearance when new basal heterocyst replaces earlier one(s); **D**. *Calothrix stagnalis* (p. 98); **E**. *Calothrix fusca* (p. 97); **F**. *Calothrix epiphytica* (p. 97); **G**. *Calothrix pulvinata* (p. 97).

Cylindrospermum alatosporum F.E.Fritsch

01200200 (new number)
Colony thin. Cells 3.5–4 µm wide, up to twice as long as wide, quadratic to cylindrical, narrowed at the cross walls. Heterocyst more or less cylindrical or slightly pointed, 4.5–5.5 µm wide, 7.5–10 µm long. Mature akinete with wall in two very distinct layers: inner thick, outer a colourless wing with conspicuous stripes perpendicular to surface; wall excluding wing, 9–11 µm wide, cylindrical-ellipsoidal, but truncated ends; with wing, 16–21 µm wide, 20–30 µm long.

Originally described by Fritsch from sediments in standing water in South Africa, but isolated by Lund (1947) from an English soil sample.

Cylindrospermum catenatum [Ralfs 1850] Bornet *et* Flahault 1886

01200010
Colony mucilaginous, dark green. Cells 4 µm wide, slightly longer than wide, quadratic to cylindrical, narrowed at the cross walls. Akinetes 2–8 in a row, 7–10 µm wide, 13–18 µm long; wall smooth, golden-brown.

On very moist mud or submerged sediments at the edge of standing waters and streams.

Cylindrospermum licheniforme [(Bory 1825) Kützing 1847] Bornet *et* Flahault 1886

01200020 **Pl. 17A** (p. 101) **CD**
Colony mucilaginous, dark green. Cells 3.5–4.2(–4.8) µm wide, slightly longer than wide, quadratic to cylindrical, narrowed at the cross walls. Heterocyst elongate, (4–)5–6 µm wide, 7–12 µm long. Akinete 10–14 µm wide, 20–30(–40) µm long, cylindrical or somewhat ellipsoidal but truncated ends, wall smooth, red-brown or dark golden-brown.

Damp firm soils, typically where obvious nutrient enrichment, such as contamination of the side of a moist track or stream bank by phosphate-rich fertilizer or dog excreta.

Cylindrospermum majus [Kützing 1843] Bornet *et* Flahault 1886

01200030
Colony sometimes covering several cm^2, mucilaginous, dark green. Cells 4–5 µm wide, narrowed at the cross walls. Heterocyst oblong in side view, 5–6 µm, up to 10 µm long. Akinete ellipsoidal, 10–15 µm wide, 20–30 µm long; outer part of wall brown, with distinct papillae.

On moist paths, mud or submerged sediments at the edge of standing waters and streams.

Cylindrospermum marchicum Lemmermann 1910

Basionym: *Cylindrospermum catenatum* Ralfs 1850 var. *marchicum* Lemmermann 1905
01200040
Colony mucilaginous, dark green. Cells 2.5–3 µm wide, 2.7–6 µm long, narrowed at the cross walls. Akinetes in a short row, 4.5–5.5 µm wide, 12–26 µm long; wall smooth, colourless.

On moist soil.

This differs from *C. catenatum* in little more than the distinctly narrow cells and akinetes; it is uncertain whether the colourless akinete wall is a reliable character.

Cylindrospermum minutissimum Collins 1896

01200070 **Pl. 17B** (p. 101)
Colony bright blue-green or olive green. Cells 2–2.7 µm, 4–7 µm long; end cell conical. Heterocyst elongated, 4 µm wide, 6–8 µm long. Akinete single or pair, 7–9 µm wide, 12–35 µm long; wall smooth, colourless.

On wet mud, especially in ditches. Lund (1947): isolated from soil.

Cylindrospermum muscicola [Kützing 1845] Bornet *et* Flahault 1886

01200080 **Pl. 17D,E** (p. 101) **CD**
Colony mucilaginous, dark green. Cells 3–4.7 µm wide, 4–5 µm long. Heterocyst 4–5 µm wide, 5–7 µm long. Akinete single, ovoid, 9–12 µm wide, 10–20 µm long, rounded at the ends; wall smooth, golden-brown.

On moist soil. Lund (1947): confined to soils not deficient in bases.

Cylindrospermum stagnale [Kützing 1843] Bornet *et* Flauhault 1886

01200090 **Pl. 17C** (p. 101)
Colony spread out, usually either forming suspended flocs or attached to submerged surfaces. Cells 3.8–4.5 µm wide, 4–6.5 µm long. Heterocysts various shapes from subspherical to markedly elongated, 6–7 µm wide, 7–16 µm long. Akinete cylindrical, 10–16 µm wide, 31–40 µm long, rounded at the ends; wall smooth, yellow-brown.

This is the most aquatic of the widespread species in the British Isles, sometimes on bottom sediments in ditches and ponds, but often mixed with submerged plants, either floating free or epiphytic. However, it has occasionally been reported from trees or other aerial habitats. It is less associated with eutrophic environments than some of the other species in this genus, but especially likely where the water contains humic materials. Fritsch & John (1942): oak wood soil.

Dichothrix [Zanardini 1858] Bornet *et* Flahault 1886

01250000
Colony a mat, cushion or bushy structure, consisting of many filaments with tapered trichomes, and abundant branching. False branching conspicuous and at the base of false branches there are often many trichomes in a common sheath. Trichomes with a basal heterocyst and sometimes also a few intercalary ones; usually ending in a long multicellular hair. Sheaths often yellow-brown. Hormogonia are released from the apical parts of trichomes as a result of phosphate enrichment.

This genus merges with *Calothrix* and is distinguished only by the greater tendency for false branching and the fact that there

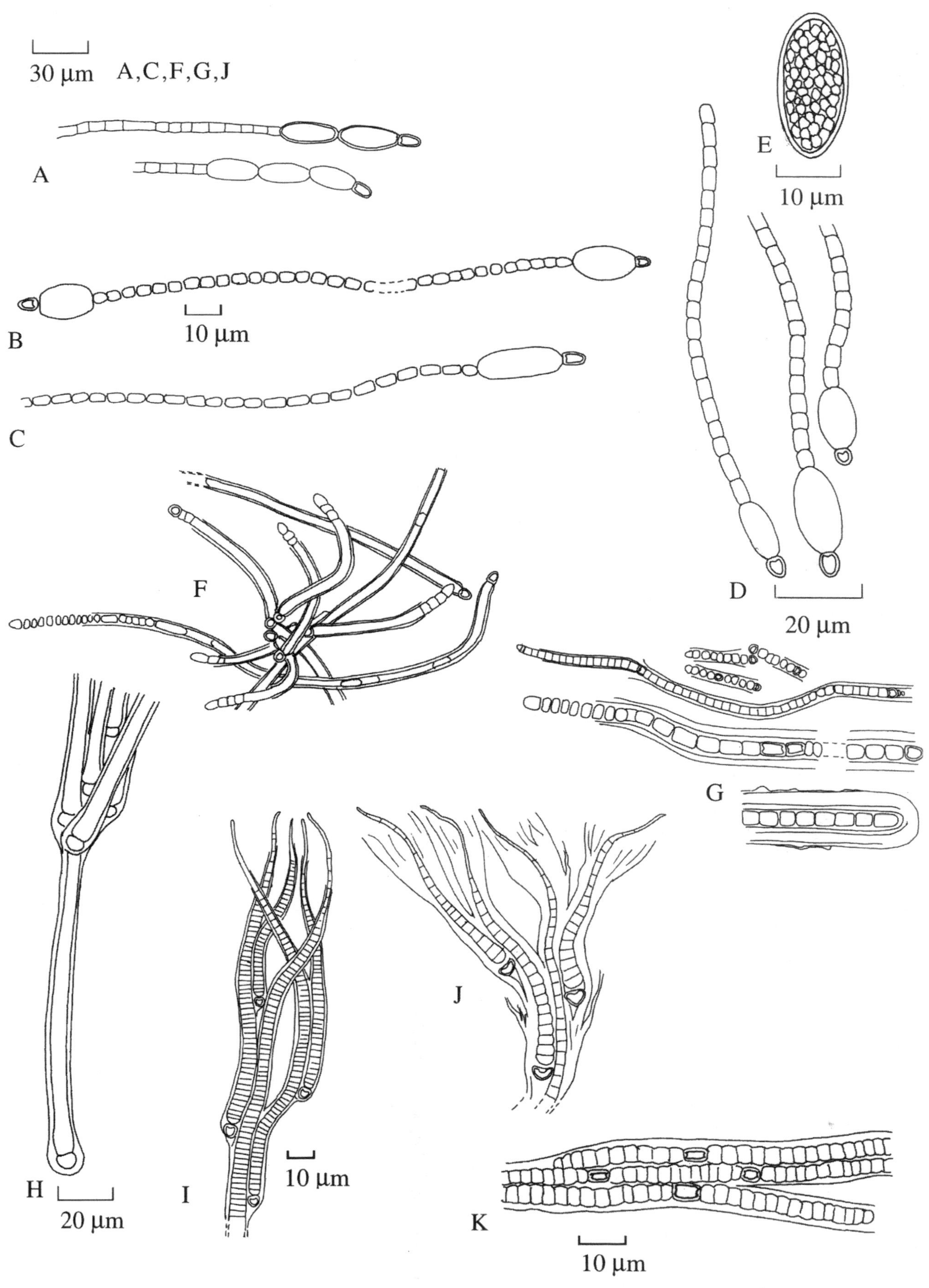

Plate 17 A–K
A. *Cylindrospermum licheniforme* (p. 100); **B**. *Cylindrospermum minutissimum* (p. 100); **C**. *Cylindrospermum stagnale* (p. 100); **D, E**. *Cylindrospermum muscicola* (p. 100): **F**. *Microchaete diplosiphon* (p. 104); **G**. *Microchaete tenera* (p. 104); **H**. *Dichothrix baueriana* (p. 102); **I**. *Dichothrix orsiniana* (p. 102); **J**. *Dichothrix gypsophila* (p. 102); **K**. *Hydrocoryne spongiosa* (p. 103).

may be many trichomes in a common sheath. The bushy (penicillate) habit in some species develops because the false branching is subdichotomous. The responses to phosphorus enrichment and limitation are broadly similar to those of *Calothrix*, with the presence of a hair indicating phosphorus limitation. However, the complex habit of the larger species makes it harder to relate the morphology in detail to its recent environment; a single colony may include many long hairs, but also some sheaths where hormogonia have apparently been released recently.

The phycobiont of the encrusting lichen *Placynthium nigrum* is possibly *Dichothrix* (Pentecost & Whitton, 2000).

Dichothrix baueriana [Grunow] Bornet *et* Flauhault 1886

01250010 **Pl. 17H** (p. 101) **CD**

Colony caespitose or penicillate, many filaments more or less upright, reaching 1 cm under the most favourable conditions. Filaments flexuous, ultimate branches 15–21 µm wide, sheath close to the trichome, uniform, colourless or yellowish. Trichome 5–7.5(–9) µm wide, cells half to as long as wide, slight constriction at the cross walls, usually gradually tapering to a hair. Heterocyst basal, subspherical or hemispherical.

On wet rocks at the edge of springs, lakes and rivers. River Caragh, Kerry (Heuff & Horkan, 1984).

Geitler (1932a) lists var. *minor* Hansgirg, which has much narrower filaments and is heavily calcified; there are apparently no records of this for the British Isles.

Dichothrix gypsophila [Kützing 1843] Bornet *et* Flauhault 1886

01250020 **Pl. 17J** (p. 101) **CD**

Colony caespitose or penicillate, dark brown, typically to 4–5 mm in diameter, but sometimes spread out further or even mixed as small clumps among other algae; many filaments more or less upright, usually (if not always) with calcite crystals and occasionally well calcified. Filaments with ultimate branches 15–18 µm wide, up to 2 mm long; false branches adpressed above and enclosed within a common sheath; sheath thick, lamellated, orange-brown, later broadened out and lacerated. Trichome 6–8 µm wide, usually tapered to a long hair; cells barrel-shaped, as long or slightly longer than wide. Heterocyst basal, typically subspherical, though usually a variety of shapes occur within a population.

On pebbles and rocks at the edge of, or just submerged in, shallow calcareous waters; moist calcareous cliffs, walls and soil, including near the sea. Benson-Evans & Antoine (1996), without detail.

Dichothrix interrupta West *et* G.S.West 1894

Synonym: *Calothrix westiana* (West *et* G.S.West) V.Poljanskij

01250030

Colony penicillate, up to 1.5 mm long, with many adjacent dichotomous branches. Filaments 19–23 µm wide where branching, 15–25 µm wide. Trichome blue-green, 2–2.7 µm wide near base, sometimes ending in a hair. Sheath broad, smooth, dark brown towards base, colourless near the apex. Heterocyst basal, spherical or subspherical, 2.4–2.7 µm diameter.

On moist rocks.

This was described originally by W. & G.S.West (1894) from Slieve Donard, Mourne Mts, Co. Down, but not mentioned by Geitler (1932a). The diagram reproduced by West & Fritsch (1927) and Starmach (1977), but not the above description, gives the impression of material which has recently been influenced by a pulse of phosphate. The validity of the species needs checking.

Dichothrix meneghiniana (Kützing 1847) Forti 1907

Basionym: *Schizosiphon meneghinianus* Kützing 1847
Synonym: *Calothrix meneghiniana* (Kützing) Kirchner 1878

01250040

Small colonies on submerged surfaces, with a structure approaching that of *Calothrix* or *Dichothrix*. Filaments up to 13 µm wide, short, initially unbranched, but later branching to form a bushy structure more like typical *Dichothrix*. Cells 6.5–7.5 µm wide, 4–8 µm long. Sheath yellow to brown at the base, more or less colourless further up, lacerated at the ends.

On submerged plants.

Field samples sometimes key out to this, but, as pointed out by Geitler (1932a), it is a questionable species.

Dichothrix orsiniana [Kützing 1847] Bornet *et* Flauhault 1886

01250050 **Pl. 17I** (p. 101)

Colony a bushy-cushion, mucilaginous, brown-green, 2–3 mm high. Filaments 10–12 µm wide, with the lower parts of the trichomes close together, but further up often flexuous or even bent. Trichome 6–7.5 µm wide, usually ending in long hair; cells short, mostly 2.5–3.5 µm in the lower part of the trichome. Sheath soft, sometimes thin and without obvious lamellations; occasionally thicker and then lamellated; not splayed out at ends; yellow to brown. Heterocyst subspherical or hemispherical.

On rocks in fast-flowing streams and moist rock faces with intermittent trickles of water. Geitler (1932a) stated that it also occasionally occurs in standing water, but it is unclear if there are records for this in the British Isles. In any case, West & Fritsch (1927) reported it to be widely distributed and this is still the situation. River Caragh, Kerry (Heuff & Horkan, 1984).

D. orsiniana is quite similar to *D. baueriana*, both of which are widespread in northern and western regions. The latter species has longer and wider filaments, though not necessarily wider trichomes; it also tends to have paler sheaths; they are not the rich yellow-brown of *D. orsiniana*. Both can be distinguished from *D. gypsophila*, because this has the ends of the sheaths splayed out. Some populations of *Homoeothrix caespitosa* are very similar in appearance, apart from the absence of a heterocyst. *Dichothrix calcarea* Tilden 1897 appears to differ from *D. orsiniana* only in being calcified.

Gloeotrichia [J.Agardh 1842] Bornet *et* Flahault 1886

01310000

Colony spherical or, if attached to a surface, hemispherical or part of a sphere; mucilage throughout the colony or in some species partially hollow when old. Trichome tapered, with basal heterocyst towards the inside of the colony, partially enclosed in a sheath. Trichome usually ends in a hair, which is sometimes very long. Sheath firm at the base, becoming soft or even diffluent further from the base. Filaments radial, more or less parallel, often with false branches. Akinete, large, often cylindrical or subcylindrical, forming immediately above the heterocyst. Hormogonia are released from the apical parts of trichomes as a result of phosphate enrichment. Colony formation occurs by aggregation of motile trichomes, either ones released by akinetes or the hormogonia released at the ends of trichomes under nutrient-rich conditions.

Fresh and slightly brackish waters.

Gloeotrichia echinulata (J.E.Smith) P.G.Richter 1894

01310010 **Pl. 19B** (p. 110) **CD**
Distinct spherical colony with colourless, firm sheath and mucilage. Trichomes straight, usually ending in a distinct hair. Cells 8–10 μm wide towards the base of the trichome, cylindrical. Heterocyst 7–10 μm wide, 7–12 μm long. Akinete cylindrical, smooth, colourless, 8–18 μm wide, 44–50 μm long. Apical cell usually conical. Gas vacuoles in mature trichome.

Freshwater lakes. Marquand (1884): Bosillow Moor, Cornwall.

Several studies in the literature together suggest that germinating akinetes on lake sediments release hormogonia, which eventually aggregate and form colonies, which in turn form gas vacuoles and become buoyant. It seems likely that *G. echinulata* is characteristic of lakes with moderately high levels of phosphate in the oxygenated surface sediments, but low levels in the water; however, further study is needed before this can be accepted as proven. Akinetes can persist in lake sediments for several decades. Canter-Lund & Lund (1995) illustrate a young colony which has not yet developed hairs (Fig. 419) and one with hairs (Fig. 418).

Gloeotrichia natans [(Hedwig) Rabenhorst 1847] Bornet *et* Flahault 1886

01310020 **Pl. 19C** (p. 110)
Distinct spherical colony with yellow to brown, firm sheath and mucilage. Trichome slightly bent/curved, usually ending in a distinct hair. Cells 7–9 μm wide, barrel-shaped. Heterocyst basal, 6–12 μm wide, 6–12 μm long. Akinete cylindrical, smooth, brown, 10–18 μm wide, 40–250 μm long.

Freshwaters.

Gloeotrichia pisum [(C.Agardh) Thuret 1875] Bornet *et* Flahault 1875

01310030 **Pl. 19D,E** (p. 110)
Distinct spherical colony with colourless, firm sheath and mucilage. Trichomes slightly bent/curved ending in a distinct hair. Cells 4–7 μm wide, 4–10(–14) μm long, cylindrical; apical cell conical. Heterocyst 8–12(–15) μm wide, 8–12(–14) μm long, usually basal. Akinete cylindrical, smooth, colourless, 9–15 μm wide, 60–400 μm long, next to the basal heterocyst.

Freshwaters, usually epiphytic on fine-leaved submerged plants such as *Myriophyllum spicatum*, occasionally free-floating; also in slightly brackish waters.

Gloeotrichia punctulata [Thuret 1875] Bornet *et* Flahault 1886

01310040 **Pl. 19F,G** (p. 110)
Distinct spherical colony with colourless, firm sheath and mucilage. Trichomes slightly bent/curved ending in a distinct hair. Cells 6–7 μm wide, cylindrical. Heterocyst usually basal. Akinete cylindrical, rough, warted, 15–18 μm wide, 100–110 μm long.

Epiphytic in brackish waters.

Hydrocoryne [Schwabe 1827] Bornet *et* Flahault 1886

01400000
Irregularly branched filaments, usually containing several heterocystous trichomes running parallel to each other. Heterocysts intercalary. Akinetes formed. The genus is somewhat like a heterocystous equivalent of the non-heterocystous genus *Schizothrix*.

Hydrocoryne spongiosa [Schwabe 1827] Bornet *et* Flahault 1886

01400010 **Pl. 17K** (p. 101)
Colony expanded, gelatinous, spongy, torn membranous structure, dirty green. Filaments with occasional false branches, 4–6.5(–7) μm wide; cells 3–4(–5) μm wide; sheath thin, colourless heterocyst typically barrel-shaped, 3.5–4 μm wide, 4–8 μm long. Akinete single, cylindrical or oblong (as seen in side view).

Fritsch & West (1927) refer to this as rare, but there are apparently no more recent reliable records. However, it is widespread in other countries, where it occurs among submerged plants in shallow water bodies, including brackish ones.

Isactis [Thuret 1875] Bornet *et* Flahault 1886

01760000 (new number)
Colony of sparingly branched upright filaments embedded in communal mucilage. Trichomes tapering markedly, usually ending in long multicellular hair; basal heterocyst. This is much like a form of *Rivularia* forming a stratum of upright, more or less parallel filaments and the genus should perhaps be recombined with *Rivularia*.

Isactis plana [Harvey 1833] Bornet *et* Flahault 1886

01760010 (new number) **Pl. 19M** (p. 110)
Colony flat, dark green or brownish, with the upper ends of the occasionally false-branched filaments terminating in the communal mucilage, which forms a layer up to about 0.5 mm from the underlying surface. Sheath thin, usually with several narrow layers in the lower part and many narrow layers in the upper part, these often spreading at the end in a fan-like arrangement; colourless or brown-yellow. Trichomes 7–9 μm towards the base, usually ending in a long hair, though this probably does not extend beyond the surface of the mucilage; cells blue-green or violet.

Intertidal region, epiphytic on larger algae (e.g. *Fucus*) and epilithic. Marquand (1884): wet rocks near Mousehole Cave, Cornwall.

Microchaete [Thuret 1875] Bornet *et* Flahault 1886

01470000

Filaments attached by one end to a substratum, single, or with colonies of many filaments, which are arranged irregularly or forming a turf. Filament consists of one trichome surrounded by a very distinct sheath; the sheath is mostly narrow, though very broad in one species. Trichome tapered towards the end; heterocyst basal and often also intercalary. Akinetes occur in some species, usually in short rows next to a heterocyst.

In contrast to many species of *Calothrix* (e.g. *C. parietina*), where cell division occurs mainly in the part of the trichome near the heterocyst, cell division in *Microchaete* apparently takes place towards the apex, and cell elongation takes place away from this region. *Microchaete* is in this respect more like *Tolypothrix*. Presumably the switch from growth leading to increase in trichome length to the formation of a hormogonium taking place is favoured by nutrient-rich conditions, but this needs investigation.

Microchaete aeruginea Batters 1892

01470040 (new number)

Typically in small tufts barely visible with the naked eye. Filament c. 12 μm wide, reaching 300 μm long, often curved and flexuous. Trichome 6–7 μm wide; cells away from the base about half as long as wide. Sheath quite thick, uniform, colourless. Heterocysts basal and sometimes also intercalary; basal heterocyst elongated or hemispherical.

Upper intertidal zone at Berwick (north-east England), epiphytic on two species of '*Rhodochorton*' (Batters, 1892).

Microchaete calothrichoides Hansgirg 1905

01470010

Filament 10–16(–20) μm wide, single or united in bundles which form a dirty grey-green colony. Cells at the base of the trichome 6–8 μm wide, 2.5–8 μm long, distinctly narrowed at the cross walls. Sheath thick, lamellate, colourless. Heterocysts mostly basal, 6 μm wide, ovoid to long ellipsoidal.

On submerged plants in shallow freshwaters.

Microchaete diplosiphon [Gomont 1885] Bornet *et* Flahault 1886

01470020 **Pl. 17F** (p. 101)

Filaments single or in small groups 20–40 μm wide towards the base. Trichome with moderate tapering towards the apex, sometimes extending beyond the sheath; cell walls at the base 4.4–6 μm wide, 5–8 μm long, towards the apex distinctly shorter, usually with slight indentation at the cross walls. Sheath with two distinct zones, colourless; outer sheath zone mucilaginous and irregular surface; inner sheath zone narrow and firm. Heterocysts intercalary or basal. Akinetes similar width to wider vegetative cells, in rows.

On submerged plants and other surfaces in shallow freshwaters.

The above description is based largely on Gomont. West (1892b, p. 739) described var. *cumbrica* (not *cambrica*, as listed by Geitler, 1932a; the quoted reference is also incorrect) with overall much larger dimensions, especially the outer sheath layer: the filaments sometimes reach 80 μm and the basal part of the trichome 8 μm; it was recorded from the English Lake District.

Microchaete tenera [Thuret 1875] Bornet *et* Flahault 1886

01470030 **Pl. 17G** (p. 101)

Filaments single, prostrate at the base, bent slightly, 6–7 μm wide. Trichome 4.5–5 μm wide at the base and only weakly tapered towards the apex; cells about twice as long as wide away from the base. Sheath narrow, firm, colourless, not lamellated. Heterocysts basal and intercalary. Akinete cylindrical, 6–7.5 μm wide, 13–17 μm long; wall brown.

On submerged plants and other surfaces in shallow freshwaters.

Nodularia [Mertens in Jürgens 1822] Bornet *et* Flahault 1886

01510000

Filaments unbranched, with single trichome and either a thin layer of diffluent mucilage or a narrow, but distinct sheath. Trichomes with intercalary heterocysts, occasionally two together; the mature trichomes usually fragment to one side of a heterocyst. Vegetative cells and heterocysts wider than long, sometimes markedly so (discoid). Akinetes in short rows, usually distant from, but sometimes adjacent to, the heterocyst.

This is a genus especially of brackish habitats, though there are records for truly marine and freshwater ones. West & Fritsch (1927) give the impression that it may be more often found in fresh than brackish water, but that does not fit the records of the present author.

Nodularia harveyana [Thuret 1875] Bornet *et* Flahault 1886

01510010

Filaments usually in small flocs; individual filament 4–5(–6) μm wide. Trichome only slightly narrower than the filament, because the sheath, though usually distinct, is very narrow. Terminal cells of young trichomes somewhat conical, though older trichomes usually have a heterocyst at each end. Cells one-third to two-thirds as long as wide. Sheath colourless. Akinetes discoid to subspherical, in rows; wall brownish, usually smooth, though occasionally papillate; mostly distant from the heterocyst, but occasionally adjacent.

Brackish standing waters, such as pools close to the coast; also in moist aerophytic communities near the sea, such as intermittently wettened soil, ditches and among mixed algal communities at the base of trees. Powell (1964).

N. harveyana var. *sphaerocarpa* (Thwaites) Thuret (Coded List 01510012; synonym *N. sphaerocarpa* Bornet *et* Flahault 1886) has slightly larger dimensions for trichome, sheath and akinete. West & Fritsch (1927) record it from Cambridgeshire (presumably a freshwater site).

Nodularia spumigena [Mertens 1822] Bornet *et* Flahault 1886

01510020 **CD**

Filaments in small flocs or larger mucilaginous masses; individual filament (including varieties) 8–16 μm wide. Cells discoid, mostly less than one-third as long as wide. Akinetes discoid, often forming quite long rows, wall brown; distant from the heterocyst.

Brackish pools and other water bodies, sometimes forming a bloom. Powell (1964).

N. spumigena differs from *N. harveyana* mainly in its wider trichome and in cells being much shorter than wide; *N. harveyana* is also more strictly aquatic. Frémy (1929–33) lists three varieties (described by Bornet & Flahault 1886), which differ only in trichome width range and slight differences in akinete shape.

Nostoc [Vaucher 1803] Bornet *et* Flahault 1886

01520000

Colony small and amorphous or, more usually, a macroscopic gelatinous or leathery structure; very small colonies subspherical or elongate, but usually becoming spherical during the early period of rapid growth; subsequently they may develop a variety of shapes which are often characteristic of the species. Colony consists of many trichomes embedded in communal mucilage; in older colonies the trichomes are mostly located towards the exterior. Individual colonies develop from single trichomes, though older colonies may include developing subcolonies. Many species form akinetes, each of which develops from one vegetative cell, which increases in size and develops an obvious wall; akinete morphology is a useful feature to aid identification.

Nostoc is sometimes closely associated with various other organisms and is the phycobiont of the majority of lichens containing blue-green algae.

The key features by which species are identified are colony shape, the extent to which there is differentiation inside the colony, width of trichome, whether akinetes occur in old colonies, shape of akinetes and habitat; the last broadly falls into aerophytic, standing water or flowing water. The colonies at any one site often have a highly characteristic appearance, but there is considerable variation within species between sites and most species tend to merge into one or more other species. In addition a considerable range of forms often develop in cultures inoculated with surface scrapes of habitats with only one macroscopically visible form of *Nostoc*. There is a need to reduce the number of recognized species, but it is not always easy to find clear-cut ways of doing this. A few changes are recommended here, but the genus would benefit from critical revision. In addition to the species listed below, there are records for several other not well distinguished species e.g. *N. minutum* (Bristol, 1920).

It is often difficult to equate field samples of large aquatic colonies with floristic accounts in the literature. *N. parmelioides* and *N. verrucosum* are usually easy to recognize, but *N. coeruleum*, *N. pruniforme* and *N. sphaericum* are more difficult. West & Fritsch (1932) give the impression that *N. pruniforme* is widespread (see account for species), but the only colonies found by the present author with a size matching the original description were those in Loch Bornish, South Uist, in July 2001. There is a great range of forms of *Nostoc* colony in this lake, including a continuum from ones which might be considered as *N. pruniforme* to ones typical of *N. coeruleum*. Similarly, *N. coeruleum* and *N. sphaericum* merge into each other, although at their extremes they are clearly different. (Typical *N. sphaericum* has not been found in or around Loch Bornish.) *N. sphaericum* has narrow trichomes, more often with pronounced brown pigment in the outer part of the colony, often forming akinetes and often in semi-terrestrial habitats.

Other records

Powell (1964) lists *N. entophytum* (Bornet in Thuret) Bornet *et* Flahault 1886. However, there is little to distinguish this from the widely recorded *N. punctiforme*, so the species is not listed separately.

Nostoc calcicola Brébisson 1843 emend. Hollerbach

01520010: see *Nostoc humifusum*

Nostoc carneum [C.Agardh 1824] Bornet *et* Flahault 1886

01520020 **CD**

Colony at first spherical, later irregular and eventually spread out, soft and slimy; old colonies a range of colours, including blue-green, but often red-brown or pink or violet. Filaments flexuous; sheath indistinct, colourless. Cells subspherical to barrel-shaped, 3–4 μm wide, longer than wide. Heterocyst 5.5–6.5 μm wide. Akinete ellipsoidal, 5.5–6.5 μm wide, 8–10 μm long; wall smooth, colourless.

Freshwaters, usually as flocs intermingled with submerged macrophytes.

Nostoc coeruleum [Lyngbye 1819] Bornet *et* Flahault 1886

01520210 **CD**

Colony initially spherical, to 1 cm diameter, solid, with firm outer layer; pale blue-green, seldom brownish; sometimes becoming considerably larger, but more irregular. Filaments flexuous, densely entangled, sheath mostly indistinct. Trichome 5–7 μm wide, with short barrel-shaped or subspherical cells. Akinetes not known.

In standing fresh or slightly brackish water, among submerged plants, free-floating or loose on bottom sediments. West & Fritsch (1927): *N. coeruleum* is widely distributed in ponds and ditches, generally occurring as free-floating masses.

Frémy (1929–33) and Geitler (1932a) both suggest that *Nostoc planctonicum* Poretsky *et* Tschernow 1929 may be identical, differing only in the presence of gas vacuoles, considerably smaller colony size and shorter cells. A sample of abundant colonies from Loch Bornish, South Uist (Outer Hebrides), in July 1983 had an identical appearance to that illustrated by Poretsky and Tschernow. It is suggested that this should now be recognized as a variety of *N. coeruleum*. The description below corresponds both to the original description and the material from Loch Bornish.

Loch Bornish is a highly calcareous machair loch (one associated with wind-blown sand), reaching 2.5 m depth and 1 km from the Atlantic. Not only do considerable quantities of *Nostoc* occur in this and other non-brackish machair lochs, but *N. commune* is also very common in the summer growing on the surface of machair grassland, though easily overlooked as it is under the sward (D.B. Jackson, pers. comm.).

No colonies resembling var. *planctonicum* were found in July 2001. However, some much larger colonies (about 1.5 cm diameter) of otherwise apparently typical *N. sphaericum* in this and two other machair lochs in

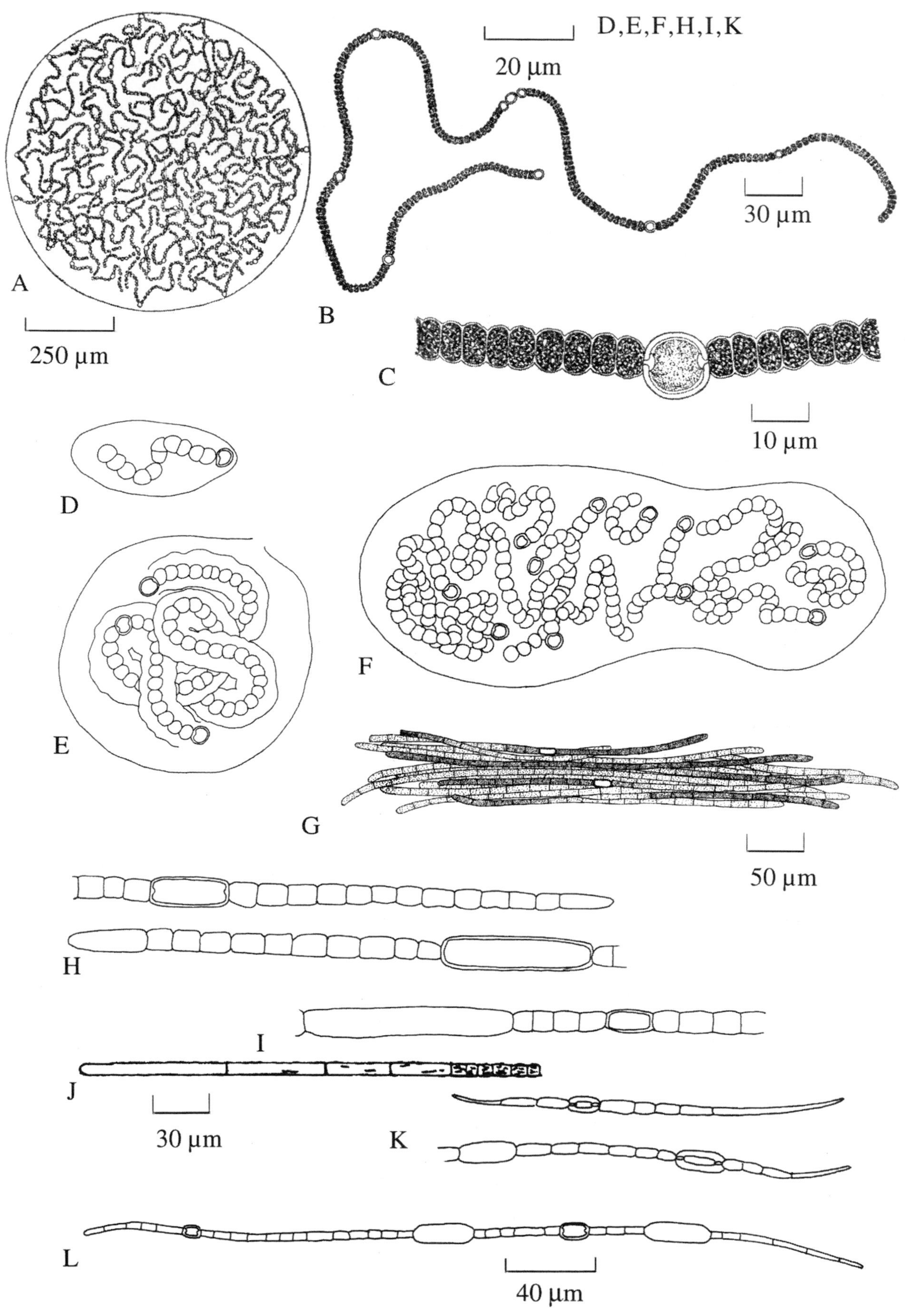

Plate 18 A–L
A–C. *Nostoc coeruleum* var. *planctonicum* (p. 107); **D–F**. *Nostoc linckia* (p. 108): young stages (many other species show quite similar stages); **G–J**. *Aphanizomenon flos-aquae* (p. 96): G. bundle of trichomes, in condition without much attenuation, H, I. regions of typical trichomes showing akinetes forming away from heterocyst, J. end of markedly attenuated trichome; **K**. *Aphanizomenon issatschenkoi* (p. 96); **L**. *Aphanizomenon gracile* (p. 96).

South Uist had gas-vacuolate trichomes in the centre of the colony, where the mucilage was much softer than usual. Slightly further from the centre other trichomes showed early stages of gas vacuole formation. Possibly gas vacuole formation is favoured by decreased light intensity or a change in the light spectrum as the colony increases in size. Initially the gas-vacuolate trichomes showed many bends, similar to those illustrated for all trichomes in the colony of var. *planctonicum* in Pl. 18A, but eventually fragmented into shorter straight lengths, largely free of heterocysts. On one occasion the surface of the water over a few square metres at the edge of the loch appeared like a thin milky film and this proved to consist of numerous such trichomes. Many of the floating and benthic *N. sphaericum* colonies showed a gap in their colony wall, with very soft mucilage by the gap. Presumably this gap in the wall indicates how the straight gas-vacuolate trichomes had been released, although wind action had probably concentrated trichomes from a larger area of the loch to this particular region.

var. *planctonicum* (V.S.Poretsky *et* Tschernow) B.A.Whitton, *nov. comb.*
V.S. Poretsky & V.K. Tschernow (1929) In: Bulletin Jardin Botanique 28(1–2): 549–560, including Figs. 1–10.
01520212 (new number; old number 01480130)
Pl. 18A–C (p. 106) **CD**

Colonies mostly free-floating, to 4.5 mm diameter, mostly spherical, occasionally other shapes, especially ones giving the appearance of budding from another sphere; solid. Colony bound by a distinct membrane, but no surface lamellation and little other differentiation towards the surface. Trichomes strongly flexuous, spread uniformly through colony, apart from sometimes being fewer towards the outer part. Cells short barrel-shaped, 5–7.5 μm wide, 3.2–6 μm long, gas-vacuolate. Heterocysts usually spherical, 6.8–10 μm diameter, though occasionally more ellipsoidal, quite often in rows of 2–3 cells. Akinete unknown.

Elenkin (1938) appears to regard *N. planctonicum* as synonymous with *Stratonostoc kilmani* (Lemmermann 1907) Elenkin *nov. comb.* However, his figures are very weak, in contrast to the excellent ones shown by Poretsky & Tschernow (1929).

Nostoc commune [Vaucher 1803] Bornet *et* Flahault 1886

01520030 **CD**

Colony starting spherical, subsequently flattened and spread-out, membranous and leathery, sometimes irregularly torn, in favourable conditions growing to many centimetres, olive-green to brown-green or yellowish to yellowish-brown, with firm outer layer. Sheaths of individual trichomes mostly distinct only at the periphery of the colony, thick yellowish-brown, often lamellated; trichomes often narrowed at the cross walls. Cells short barrel-shaped or nearly spherical, 4.5–6 μm wide, mostly shorter or slightly longer than wide, pale olive-green. Heterocysts nearly spherical, often growing up to 7 μm wide. Akinete only reported once in the literature (not for British Isles), as big as the vegetative cell, with smooth, colourless membrane.

Macroscopically visible colonies occur on soil, among mosses and grasses, and in depressions on limestone surfaces, especially where these occasionally hold water for short periods. Colonies sometimes form large masses on the floor of limestone quarries in the wetter parts of the British Isles; they are also a characteristic feature of depressions on limestone pavement in the wetter regions, especially in The Burren, Co. Clare, where it seems possible that the colonies themselves may aid the erosion of the limestone to form the depressions. Other characteristic habitats are among mosses at the edge of calcareous springs, small streams and pools, on the moister parts of calcareous sand-dunes and around the margins of salt marshes. Scattered colonies also occur in fields and on tracks, probably often favoured by phosphate fertilizer; in drier regions such occurrences seem to be largely restricted to calcareous soils, but in humid regions the colonies occur on a wide variety of soils. *N. commune* can frequently be isolated from garden and agricultural soils in regions where there are no macroscopically visible colonies. Powell (1964).

Forms among grasses on sand-dunes sometimes approach the morphology of *N. commune* var. *flagelliforme* (Berkeley *et* Curtis) Bornet *et* Flahault 1886 (treated as a variety in the Coded List = 01520032), which has thread-like colonies, with the trichomes mostly running more or less parallel to each other. However, the British material apparently never reaches the extreme forms found in some other countries, where the colonies may extend for many centimetres; in addition the occurrences are usually quite near areas with typical *N.commune* colonies. It seems probable that forms resembling var. *flagelliforme* occur in microhabitats which are very humid for part of the day, but always free from standing water.

Nostoc cuticulare (Brébisson) Bornet *et* Flahault 1886
01520040

Colony discoid, thin, at first with more or less distinct margin, but later irregular at the edges. Filaments flexuous, thickly aggregated; individual sheath indistinct, colourless. Cells barrel-shaped, 3.8–5 μm wide, heterocysts sometimes slightly larger. Akinete (in var. *ligericum* Bornet *et* Flahault 1886) spherical, 8–12 μm diameter; wall thick, smooth.

On submerged macrophytes.

Colony shape and the slightly wider cells are all that separate this from *N. humifusum*.

Nostoc ellipsosporum [Rabenhorst 1865] Bornet *et* Flahault 1886
01520050

Colony gelatinous, irregularly expanded, attached by the lower surface, reddish brown. Filaments flexuous, loosely entangled. Cells 3.8–4.2 μm wide, 6–14 μm long. Heterocyst subspherical or oblong, 6–7 μm wide, 6–14 μm long. Akinetes ellipsoidal to oblong-cylindrical, 6–8 μm wide, 14–19 μm long, developing in rows away from the heterocyst; wall smooth, colourless or (more usually) brownish.

Various moist aerial habitats, such as among mosses, calcareous sandy soils, rocks and greenhouses; often near the sea.

Several varieties have been reported, usually differing from the type by one or more of the features having smaller dimensions.

Nostoc humifusum [Carmichael 1833] Bornet *et* Flahault 1886

Synonym: *Nostoc calcicola* Brébisson 1843 emend. Hollerbach (Coded List 01520010)
01520060

The original descriptions of these two species were close and are combined here. Colony mucilaginous, slightly diffluent, spread out to form masses up to 5 cm across, but with no distinct shape; ranging in colour from blue-green to dark brown. Filaments flexuous, entangled. Individual sheaths often distinct towards the outside of the colony, colourless or yellow-brown, but indistinct or not visible at all towards the inside of the colony. Cells barrel-shaped or subspherical, 2.2–3 μm wide. Heterocyst subspherical, 3–5 μm diameter. Akinete subspherical or shortly ovoid, forming long rows, 4–5 μm wide; wall smooth, brown; eventually a high proportion of cells may differentiate into akinetes

Moist soils, base of damp old walls, conservatories. It is probably more widespread than suggested by the presence of macroscopic colonies, because scrapings from such habitats sometimes develop into typical colonies in laboratory culture.

The original descriptions of *N. humifus'um* and *N. calcicola* are not only close, but intermediate forms occur, so it is hardly worth even treating them as varieties. The original distinguishing features were the greater tendency for individual trichomes to have a distinct and pigmented sheath in *N. humifusum* and for the cells to be rather longer in relation to width. Both these features might be expected to occur in environments where growth was slower.

Nostoc linckia [(Roth) Bornet in Bornet *et* Thuret 1880] Bornet *et* Flahault 1886

Synonyms: *Nostoc piscinale* Kützing 1843 ex Bornet *et* Flahault 1886 (Coded List 01520130); *N. rivulare* Kützing 1850 ex Bornet *et* Flahault 1886 (Coded List 01520170)

01520070 **Pl. 17D–F** (p. 106)

Macroscopic colonies at first spherical, later becoming increasingly irregular and eventually torn; gelatinous; blue-green, brown or violet. Filaments flexuous or slightly coiled, highly entangled; individual sheaths distinct only towards the periphery and not always there. Cells usually barrel-shaped, 3.5–4 μm wide, though sometimes slightly more round. Heterocyst subspherical, 5–6 μm wide. Akinete subspherical or slightly elongate, forming long rows, 6–7 μm wide, 7–8(–10 μm) long; wall smooth and usually brown.

In fresh and brackish water, usually standing water, but sometimes also small slow-flowing rivers; initially attached, but later forming free-floating masses. Powell (1964).

The original description of *N. piscinale* differs from that of *N. linckia* by little more than its looser filaments. West & Fritsch (1927): widely distributed in ponds and ditches, generally occurring as free-floating masses. The original description of *N. rivulare* Kützing 1850 ex Bornet *et* Flahault 1886 also differs from that of *N. linckia* by relatively little: the vegetative cells are less barrel-shaped, and both these and the akinetes are slightly more elongate (or sometimes, almost barrel-shaped); its inclusion in *N. linckia* was first suggested by Elenkin (1938). The broadened description of *N. linckia* means that organisms with somewhat different shapes of vegetative cell and akinete are included, but it avoids the frequent need to decide the specific name of material which does not fit exactly any of the three descriptions.

Nostoc macrosporum [Meneghini 1843] Bornet *et* Fahault 1886

01520080: see *Nostoc microscopicum*

Nostoc microscopicum [Carmichael in Hooker 1833] Bornet *et* Flahault 1886

Synonym: *Nostoc macrosporum* Meneghini 1843 (Coded List 01520080)

01520090

Colony spherical or ellipsoidal, mostly only 1–2 mm, though much larger ones occasionally reported, soft, but with firm outer surface; initially blue-green, but later olive and usually eventually brownish. Filaments flexuous, initially entangled quite closely, later more spaced apart; sheath more or less distinct, at least in the outer part of the colony, yellow or brownish-yellow. Cells subspherical or barrel-shaped, 5–9 μm wide. Heterocysts subspherical, 7 μm diameter. Akinetes ellipsoidal, 6–7 × 9–15 μm; wall smooth

The species occurs in highly calcareous environments and Geitler (1932a) records *N. macrosporum* as often occurring together with *Scytonema myochrous*; the latter is highly characteristic of the vicinity of calcareous springs. West & Fritsch (1927): among mosses on wet rocks.

N. macrosporum was separated from *N. microscopicum* by little more than the slightly larger cells (8–9 versus 5–8 μm wide). *N. microscopicum* itself differs from *N. commune* by the typically smaller maximum colony size, slightly narrower trichome and the frequent presence of akinetes in old colonies.

Nostoc muscorum C.Agardh 1812

01520100

Colony initially spherical, but later irregularly expanded, sometimes extending for several centimetres (to 5 cm); olive or yellow-brown. Filaments thickly entangled. Sheath distinct only at the periphery of the colony, yellow-brown. Cells short barrel-shaped or cylindrical, 3.5–4(–5) μm wide, usually about as long as wide, but sometimes twice so. Heterocyst subspherical, 6–7 μm wide. Akinete oblong, 4–8 μm wide, 8–12 μm long; wall smooth, yellow.

West & Fritsch (1927): principally among mosses on wet rocks. Also on moist soil.

Nostoc paludosum [Kützing 1850] Bornet *et* Flahault 1886

01520110

Colony very small, hardly visible macroscopically, mucilaginous. Filaments flexuous, loosely entangled, with a broad, colourless or yellowish sheath. Cells barrel-shaped, 3–3.5 μm wide, 2.6–3.5 μm long. Heterocysts slightly larger than vegetative cells. Akinete ellipsoidal, 4–4.5 μm wide, 6–8 μm long; wall smooth, colourless.

Among submerged plants in standing water.

Nostoc parmelioides [Kützing 1843] Bornet *et* Flahault 1886

01520120

Colony sometimes initially subspherical, but soon becoming discoid, and then subsequently with irregular lobes, up to 3 cm. Filaments radiating from a centre, nearly parallel in the middle, densely entangled at the periphery; sheath of peripheral trichomes distinct, yellow; of inside ones diffluent, colourless. Cells short barrel-shaped or subspherical, 4–4.5 μm wide. Heterocyst spherical, 6 μm diameter. Akinete ellipsoidal, 4–5 μm wide, 7–8 μm long; wall smooth, yellowish. Mature colony usually (? always) associated with a midge larva.

Attached to rocks in moderately fast-flowing rivers, usually moderately hard water. Recorded from upper Tees, North Tyne, Tweed and three tributaries (Yarrow, Ettrick, Teviot) (Holmes & Whitton, 1975b).

This species has considerable resemblance to *N. verrucosum*. Apart from the association with a larva of the midge *Cricotopus*, it differs in the more complex external morphology and attachment to the substratum; perhaps related to the latter feature, *N. parmelioides* appears to be restricted to rivers, whereas *N. verrucosum* occurs in streams as well as rivers.

Nostoc planctonicum Poretsky *et* Tschernow 1929

01480130: see *Nostoc coeruleum* var. *planctonicum*

Nostoc pruniforme [C.Agardh 1812] Bornet *et* Flahault 1886

01520160

Colonies initially spherical, becoming subspherical or ellipsoidal, reaching the size of a hen's egg (size of largest British records uncertain), with a firm periphery and relatively firm mucilage, though sometimes larger colonies hollow in the centre. Filaments flexuous, loosely entangled and radiating from the centre. Periphery of colony showing striations, yellow-brown; trichome sheaths mostly distinct, colourless or pale yellow. Cells barrel-shaped, 4–6 μm wide. Heterocyst subspherical, 6–7 μm diameter. Akinete (when recorded) spherical, 10 μm diameter.

In standing water, free-floating. West & Fritsch (1927): widely distributed in ponds and ditches. Griffiths (1936): Durham (no records by present author from this locality). Geitler (1932a): colonies often impregnated with iron (presumably oxides).

Nostoc punctiforme (Kützing) Hariot 1891

Basionym: *Polycoccus punctiformis* Kützing 1843

01520140

Colonies very small, usually not visible macroscopically (though forms reported from other regions up to 2 mm), various shapes. Filaments so thickly entwined that it is usually difficult to recognize individual trichomes; in some cases groups appear to consist of individual cells, forming a stage resembling Chroococcales. Cells barrel-shaped to ellipsoidal, 3–4 μm wide. Heterocyst 4–6.5 μm wide. Akinete subspherical or ellipsoidal, 5–6 μm wide, 5–8 μm long; wall smooth, colourless.

On submerged macrophytes and among mucilaginous masses of other species of algae.

It is open to doubt how similar are all the records given this name, as some other species which form larger colonies produce stages resembling *N. punctiforme*. The presence of tiny isolated colonies producing akinetes is probably the most useful diagnostic character.

Nostoc sphaericum [Vaucher 1803] Bornet *et* Flahault 1886

01520180

Colonies remain spherical until quite large (typically 1 cm), but eventually become irregular, up to 2 cm across; olive-green or yellow-brown. Filaments thickly entangled. Trichomes mostly lacking individual sheath. Cells short barrel-shaped or rounded, 3.8–4.2(–5) μm wide. Heterocyst subspherical, 6 μm wide. Akinete ellipsoidal, 5 μm wide, 7 μm long; wall smooth, brown.

On moist soil, usually mixed with mosses, or in shallow standing water among macrophytes with floating leaves. West & Fritsch (1927) recorded it as commonly attached to rocks and stones in the beds of rapid streams and rivers, but this differs from other accounts or observations.

Nostoc spongiaeforme [C.Agardh 1824] Bornet *et* Flahault 1886

01520190

Colony initially spherical, later expanding and developing bumps (verrucose); blue-green or brownish. Filaments flexuous, loosely entangled. Trichomes with more or less distinct sheath near the periphery, but diffluent further towards the centre. Cells both cylindrical and barrel-shaped, 3.8–4.2 μm wide. Heterocyst subspherical or oblong, 7–8 μm wide, 8–10 μm long. Akinete oblong-ovoid, 6–7 μm wide, 10–12 μm long; wall smooth, yellowish.

In shallow standing freshwater: on bottom sediment, caught loosely between submerged free macrophytes or free-floating.

This is another form with considerable similarity to *N. linckia* and was also included in it by Elenkin (1938), but the development of a verrucose surface morphology is easy to recognize, so it seems worth retaining the species until critical studies have been made.

Nostoc verrucosum [Vaucher 1803] Bornet *et* Flahault 1886

01520200

Colony initially spherical or hemispherical, but subsequently developing a more irregular shape, with folds and bumps (verrucose) and eventually becoming hollow vesicles, soft and torn; a very old colony may therefore interface with the environment partly by the original firm periphery and partly by soft internal mucilage; colonies in British populations typically reach 2–3 cm, but much larger ones reported from other countries. Filaments flexuous, densely entangled towards the periphery, but much fewer towards the centre. Sheath thick, yellow-brown towards the periphery, colourless and diffluent towards the centre. Cells short barrel-shaped, 3–3.5 μm wide. Heterocyst subspherical, 6 μm wide. Akinete ellipsoidal, 5 μm wide, 7 μm long; wall smooth, yellow.

West &. Fritsch (1927): commonly attached to rocks and stones in the beds of rapid streams and rivers. Most records known to the author are for moderately hardwater streams on hills surrounded by sheep-grazing land in Derbyshire and Shropshire, where the colonies are not attached firmly to the surface and occur mostly in shallow stretches. Geitler (1932a) also records it from standing waters.

The species is mainly distinguished from *N. spongiaeforme* by the marked aggregation of filaments towards the periphery.

Rivularia [J.Agardh 1824] Bornet *et* Flahault 1886

01620000

Hemispherical or subspherical colonies containing a large number of filaments which are arranged radially or sometimes parallel to each other in part of the colony. Each filament contains a tapered trichome, which has a basal heterocyst and often ends in a long multicellular hair; occasional intercalary heterocysts present in some species. False branching sometimes evident. Sheath extends for much of the length of the trichome and is often frayed at the apical end; outer surface tends to merge with the communal mucilage of the colony. Colonies of some larger species can persist

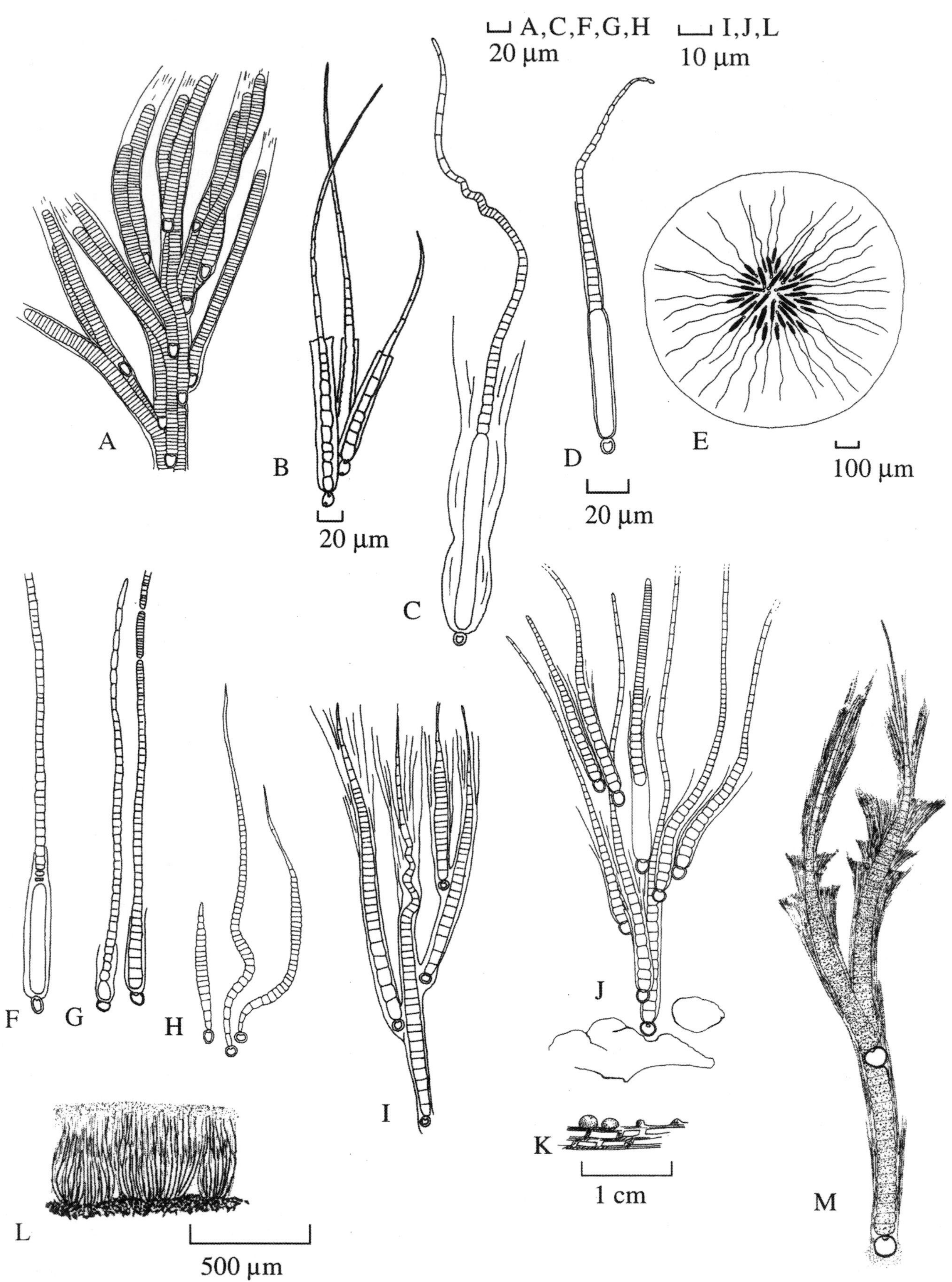

Plate 19 A–M

A. *Coleodesmium wrangelii* (p. 98); **B**. *Gloeotrichia echinulata* (p. 103); **C**. *Gloeotrichia natans* (p. 103); **D, E**. *Gloeotrichia, pisum* (p. 103); **F, G**. *Gloeotrichia punctulata* (p. 103); **H**. *Rivularia australis* (p. 111); **I–L**. *Rivularia atra* (p. 111): I, J. Typical trichomes with moderately well developed hairs (most shown only in part), K, L. macro-views. **M**. *Isactis plana* (p. 103): trichomes c. 8 µm wide towards base.

for several years and are often calcified. No akinetes. Hormogonia in those species studied are released from the apical parts of trichomes as a result of phosphate enrichment. Colony formation usually (though not always) occurs by aggregation of the released hormogonia, so a particular colony is not necessarily genetically homogeneous. Freshwater populations in the British Isles apparently all show at least slight calcification.

The occurrence of *Rivularia* is favoured by environments where there are intermittent short periods of higher phosphate levels in the water in an environment with otherwise low levels; at least at some stream sites this occurs in the British Isles especially in spring and, in the case of the marine *R. atra*, subsequent to deposits of rotting seaweed on the supralittoral. The layers that form inside the long-lived colonies of *R. haematites* reflect alternating favourable and unfavourable periods for growth, in some cases probably annual events. The fact that some forms of *Rivularia* in streams have colonies which persist for several years has usually been ascribed to an assumed effect of calcification on grazing, but several authors have recently suggested that toxic materials produced by the alga may play a role.

1 Freshwater 2
1 Marine or brackish 5
2 Section of colony shows marked zonation .... *R. haematites*
2 Section of colony shows little or no zonation 3
3 Mature colony ≤2 mm .... *R. dura*
3 Mature colony >2 mm 4
4 Trichomes towards base 5–9 μm, closely packed .... *R. biasolettiana*, in part
4 Trichomes towards base 9–12.5 μm, spaced apart by broad sheath .... *R. minutula*
5 Calcite crystals present; slightly brackish............................. *R. biasolettiana*, in part
5 Calcite crystals absent, marine or strongly brackish 6
6 Mature colony ≤4 mm .... *R. atra*
6 Mature colony >4 mm 7
7 Trichome at widest point ≤5 μm wide...... *R. nitida*
7 Trichome at widest point >5 μm wide 8
8 Colony firm .... *R. bullata*
8 Colony soft.. 9
9 Trichome at widest point 5–7.5 μm wide .... *R. australis*
9 Trichome at widest point 8–13 μm wide ...*R. polyotis*

Rivularia atra [Roth 1806] Bornet *et* Flahault 1886

01620010 **Pl. 19I–L** (p. 110) **CD**

Colonies hemispherical or (less usually) subspherical, single or confluent, to 3(–4) mm diameter; dark olive-green. Mucilage firm. Filaments densely packed in rows perpendicular to the upper surface. Trichome 2.5–5 μm towards the heterocyst at the base and usually a well-developed colourless hair at the apex; hairs not extending beyond surface of mucilage. Sheath close to the trichome, pale brown, usually reaching to the base of the hair region. A pulse of phosphate leads to the hair from many trichomes shrivelling to fragments, which remain embedded in the outer mucilage, together with the differentation of hormogonia which either migrate from the colony or form a new zone of filaments distal to the previous one; it is unclear what determines the two different types of response. The colonies apparently never persist long enough for more than several such zones to occur.

Powell (1964). Widely distributed round the coast of the British Isles, usually frequent where it does occur within a relatively narrow zone in the upper intertidal zone; probably always in regions where rotting seaweed or similar debris is deposited in the supralittoral. A well-developed population on calcareous sandstone at Tyne Sands, East Lothian, is known to have persisted for at least 20 years and been possible to find at all times of year, though most frequent in summer; the largest colonies probably take 2–3 months to form and seldom persist for much longer, but careful observations are needed. Coppejans (1995) has several records for Belgium/N. France.

Rivularia australis [Harvey] Bornet *et* Flahault 1886

01620020 **Pl. 19H** (p. 110)

Colony initially hemispherical, later cushion-shaped or more spread out, to 2 cm diameter. Mucilage soft. Trichomes with cells only about 3 μm wide towards the basal heterocyst, then widening to 5–7.5 μm; lower cells longer than broad, slightly narrowed at the cross walls, upper cells (excluding hair cells) one-third as long as broad. Sheath rather indistinct, colourless or yellow.

Powell (1964). Intertidal zone. Devon coast (Batters, quoted by Frémy, 1929–33).

Rivularia biasolettiana [Meneghini 1841] Bornet *et* Flahault 1886

01620030 **CD**

Colony initially hemispherical, later becoming more spread out and sometimes confluent, to 2.5 cm diameter, dark olive-green to golden-brown; subspherical or other forms may occur in standing water populations, and oncoids below small highly calcareous springs. Calcite crystals deposited inside mucilage, but extent varies markedly and this influences colony shape: the more calcified, the more the hemispherical shape is retained in older colonies, and perhaps also the larger the potential size; slight evidence of zonation sometimes present. Filaments up to 20 μm wide, orientated perpendicular to periphery. Trichome 5–9 μm wide towards the base, apex usually ending in long multicellular colourless hairs which do not extend beyond periphery of colony. Cells towards base of trichome slightly shorter than wide. Sheath wide, lamellated, splayed out at apical end, mostly yellow-brown, though usually very pale ones present. Heterocyst basal, quite often with one or two collapsed ones beneath (presumed) functional one adjacent to basal vegetative cell.

On rocks in calcareous springs and streams, especially in zone slightly downstream of the source, sometimes very abundant and sufficient to colour stretches of stream brown; on old wood and other substrata in shallow ponds; very moist terrestrial sites, especially near the sea, such as the base of walls and cliffs, probably usually where there is intermittent seepage. This is a characteristic species of unpolluted streams which combine highly calcareous spring water and variable contribution from peat drainage e.g. Malham region (Yorkshire), Upper Teesdale National Nature Reserve (Co. Durham) and The Burren (Co. Clare). Highly calcified forms appear to survive for periods with the colonies away from contact with water, but less calcified ones appear intolerant of drying. Young colonies are most obvious in spring, probably because this is the season when phosphate levels in waters draining peaty soils are highest.

Oncoids develop at several sites in the Pennines e.g. Sunbiggin (Cumbria) and near Bakethin Dam, Northumberland.

This species merges with *R. haematites* (see below).

Rivularia bullata [(Poir) Berkeley 1833] Bornet *et* Flahault 1886

01620040

Colony initially hemispherical, but becoming very irregular with surface bumps and folds and often becoming partially hollow inside, up to 6 cm across; firm, apart from very old colonies. Filaments densely packed together in outer part of colony, orientated perpendicular to periphery. Trichome 5–8(–10) µm wide towards the base, apex usually ending in a hair. Cells towards base of trichome typically 1–2 times longer than wide, but much narrower further towards the apex, though hair cells again elongating. Sheath narrow, rather indistinct, lamellations present, only visible where sheath opens out towards apex, colourless or pale yellow.

Powell (1964). Rock surfaces in uppermost part of intertidal region, probably higher than *R. atra*, but rarer and ecology not well understood.

Rivularia dura [Roth 1806] Bornet *et* Flahault 1886

01620050

Colony hemispherical, to 1(–2) mm diameter, highly calcified, dark green. Filaments densely compressed. Trichome 4–9 µm wide, apex usually ending in a long hair. Sheath thin, not lamellated. The combination of narrow sheath and calcification makes it difficult to separate filaments without dissolving calcite crystals.

Shallow, highly calcareous standing water, epiphytic on submerged macrophytes, including *Chara*. Records include West & Fritsch (1927).

Rivularia haematites [de Candolle 1806] Bornet *et* Flahault 1886

01620060

Colony initially hemispherical, later more spread out, to 4(–5) cm diameter, highly calcified; olive-green to brown. Section of colony showing obvious zonation, with majority of healthy trichomes in the outermost zone. Filaments up to 20 µm wide, orientated perpendicular to periphery. Trichome 4–7(–11) µm wide towards base, apex usually ending in long multicellular colourless hairs which do not extend beyond periphery of mucilage. Cells towards base or trichome slightly shorter than wide. Sheath wide, lamellated, splayed out at apical end, yellow-brown to dark brown, persistent in older parts of colony, even when trichome has disintegrated. Heterocyst basal.

Highly calcareous streams and ponds. West & Fritsch (1927) state that this is the most frequent British species, but the present author has found typical *R. biasolettiana* more often. It is unclear to what extent zonation inside the colony is simply a response to different photosynthetic rates at different times of year, or also reflects marked seasonal changes in the ambient water chemistry. Lund (1961a) and Pentecost (1987) discuss its ecology in Malham streams.

The species merges with *R. biasolettiana*, from which it differs only in the very high calcification and distinct zones. However, it is useful to retain both names, not least because *R. haematites*, which is widely known because of the conspicuous zonation, would be the name to take priority.

Rivularia minutula [Kützing 1843] Bornet *et* Flahault 1886

01620070

Colonies hemispherical, often confluent, to 5 mm diameter, usually (? always) scattered calcite crystals and sometimes more obvious calcification. Filaments radially arranged, but well spaced apart because of wide sheaths. Trichome 9–12.5 µm wide, usually ending in a hair, where the cells remaining relatively short. Sheath to 27 µm wide, lamellated, apex splayed out, colourless to (more usually) brown.

Moist calcareous surfaces besides or in shallow streams and ditches.

Rivularia nitida [C.Agardh 1817] Bornet *et* Flahault 1886

01620080

Colony initially hemispherical, but becoming spread out, with a somewhat folded surface, to 3 cm diameter; sometimes partially hollow in very old colonies. Filaments perpendicular to the surface, closely pressed together and not easy to separate. Trichomes 2–5 µm wide, apex usually ending in a very long hair. Sheath colourless or yellow-brown.

Powell (1964). Moist soil and rocks at brackish and sheltered marine sites. Benson-Evans & Antoine (1996): not uncommon.

Geitler (1932a) questions whether this is identical with *R. atra*. *R. nitida* resembles what *R. atra* might look like, if it continued to grow. However, *R. atra* populations known to the author never form colonies exceeding 3–4 mm diameter, so there is a big size difference. If they are genetically similar, perhaps the different sizes reflect different grazing regimes.

Rivularia polyotis [J.Agardh 1842] Bornet *et* Flahault 1886

Synonym: *Rivularia mesenterica* Thuret ex Bornet *et* Flahault 1886

01620090

Colony initially hemispherical to cushion-shaped, later irregular, with surface bumps, to 3 cm; hollow in older colonies. Filaments well spaced. Trichome near base 4–5 µm wide, considerably wider (8–13.5 µm) towards middle, apex usually ending in a hair. Cells twice as long as wide near base, much shorter in the middle. Sheath wide, lamellated, splayed at apex, colourless to yellow.

Rocks near upper limit of intertidal zone (? supra littoral); rare. Powell (1964): as *R. mesenterica*.

The only difference between the original descriptions of *R. polyotis* and *R. mesenterica* is the softer colonies of the latter; the two are therefore merged here.

Scytonema [C.Agardh 1824] Bornet *et* Flahault 1886

01640000

Filaments false branched, with branches single or geminate (appearing like a pair); occurrence of geminate branches is the diagnostic feature of the genus, but in

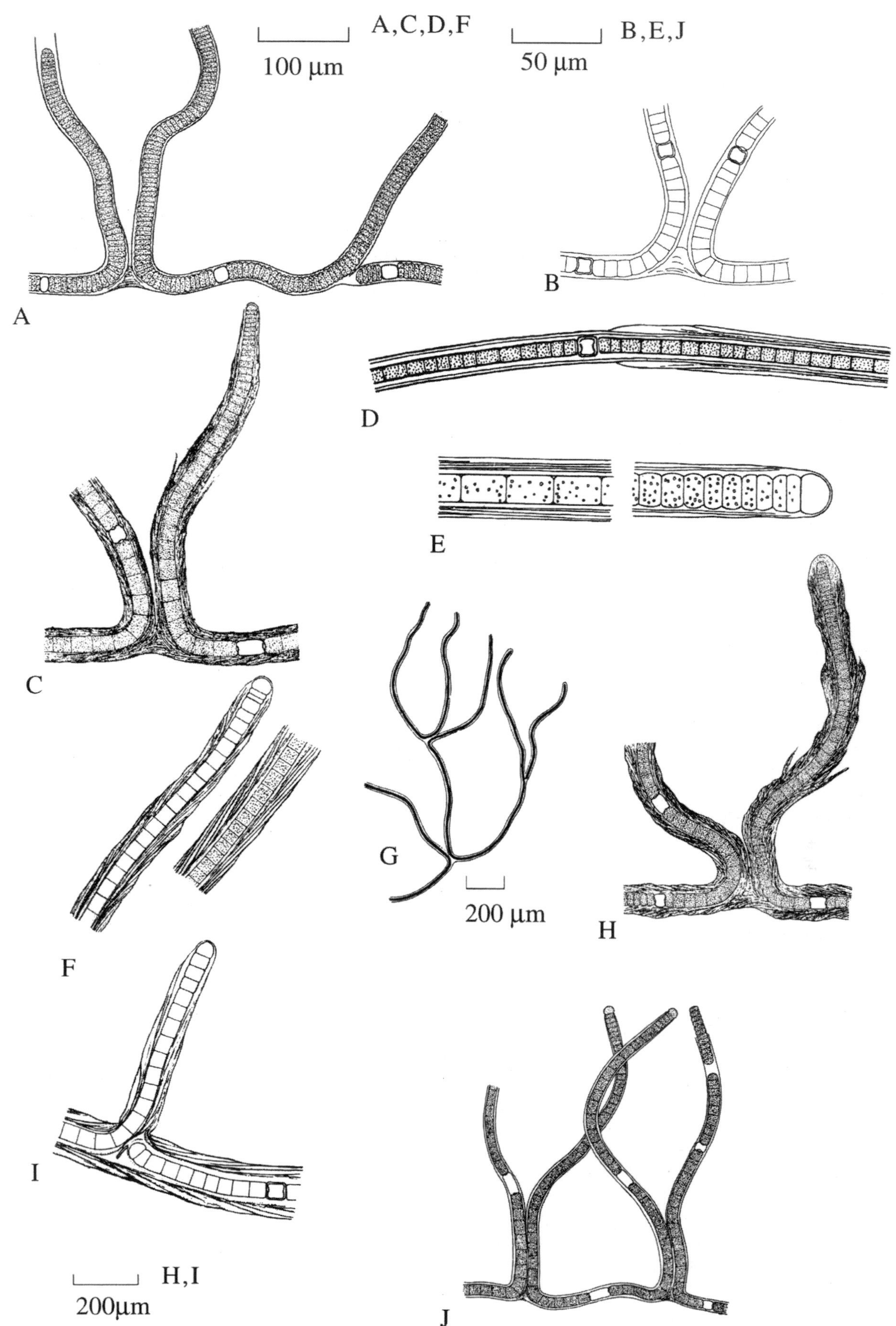

Plate 20 A–J
A. *Scytonema crispum* (p. 114); **B**. *Scytonema ocellatum* (p. 115); **C–E**. *Scytonema mirabile* (p. 114): C. Double branch, D. Part of trichome, showing lamellations, E. Older and apical parts of trichome, showing differing cell lengths and typical terminal cell; **F–I**. *Scytonema myochrous* (p. 114): F. General view of branching, G. Double branch, H. Apical part of branch, I. Atypical branch; **J**. *Scytonema hofmanni* (p. 114).

many species it occurs with only some (perhaps even a minority of) branches. Branches formed laterally generally in-between heterocysts. Trichome single in each sheath, straight. Hormogonia formed at the apex under certain favourable growth conditions. Akinetes known in a few species, spherical or ovoid, surface thin and smooth; it is unclear whether in such cases this ability is characteristic of the species as a whole.

Growth in *Scytonema* usually occurs towards the apex, though in turf-forming species the growth region is apparently sometimes slightly further back, permitting the organism to respond to grazing like grass in a meadow. The older parts of the trichome tend to break down, often leading to the cells becoming yellowish and the development of liquid (intrathylakoidal) vacuoles; however, the sheaths are often highly persistent, leading in some species to development of a miniature algal peat.

The forms sometimes recognized as a separate genus *Petalonema* Berkeley 1883 are included in *Scytonema*. The separation of *Scytonema* from *Tolypothrix* is largely artificial, but it is convenient to retain this for now. L. Hoffmann, who did (1986) combine *Scytonema* and *Tolypothrix* because of the influence of the environment on morphology, especially the levels of combined nitrogen, nevertheless proposed (1988) a more clear-cut taxonomic character, the development of 1- or 2-pored heterocysts. The 2-pored heterocysts develop in hormogonia, and this in turn leads to mainly geminate false branches.

Scytonema alatum [Borzí 1879] Bornet *et* Flahault 1886

Synonym: *Petalonema alatum* (Borzí) Berkeley 1883.

01640010

Colony forming dense tufts or single filaments intermingled with other algae (mostly species of *Scytonema*). Filaments curved or straight, 24–66 μm wide. False branching mostly in pairs, sometimes singly or in threes, mostly sparse. Cells 9–15 μm wide, at growing region short and barrel-shaped, in older parts long and cylindrical, up to twice as long as wide. Sheath very thick with distinct funnel-shaped pieces tucked into one another (divergently lamellate), at apices forming a very sharp angle with longitudinal axis of the filament, in older filaments and mostly on the inside parallel with the trichome; colourless when young, later becoming yellow to brown, often inside brown, outside colourless. Heterocyst to 17 μm wide, truncated spherical to long cylindrical.

On moist and submerged rocks.

Scytonema crispum (C.Agardh) Bornet 1889

Synonym: *Scytonema cincinnatum* [Thuret 1875] Bornet *et* Flahault 1886

01640020 **Pl. 20A** (p. 113)

Colony floccose, bushy, olive to brown or blue-green, often spread out. Filaments firm, up to 3 cm or more, with false branches mostly arranged in pairs. Sheath firm, colourless or brown. Cells 14–30 μm wide, mostly about one-third as long as wide, shorter in younger parts of trichome almost as long as wide in older parts, constriction at cross walls distinct or slight. Heterocyst rounded quadrate, short cylindrical or ellipsoidal, single or many.

In standing and flowing freshwaters and on very wet walls; also in hot-houses.

Scytonema crustaceum [C.Agardh 1824] Bornet *et* Flahault 1886

01640030

Colony crust-forming, black or dark brown. Filaments (15–)18–22 μm wide, in some stages swollen. False branches numerous, ascending, short, in pairs, joined at base and free above. Cells 6–8(–12) μm wide, distinctly shorter than wide or at most as long as wide and in older trichomes not as long as wide. Sheath to 30 μm wide, divergently lamellate, yellow brown to dark brown. Heterocyst cylindrical to disc-shaped.

On moist rocks.

Scytonema hofmanni [C.Agardh 1817] Bornet *et* Flahault 1886

01640040 **Pl. 20J** (p. 113)

Colony cushion-like, widely expanded, 1–2 mm high, blackish blue-green. Filaments often in bundles, each filament 7–8 μm wide, sometimes wider, false branches sparse. Cells 5–6 μm wide, in older parts of trichome mostly longer than wide, in meristem region shorter; typically blue-green. Sheath thin, firm, colourless or yellow to yellow brown. Heterocysts single or in pairs, rounded cylindrical.

Moist walls, soil and hot-houses.

Scytonema julianum (Kützing) Meneghini in Kützing 1849

Basionym: *Drilosiphon julianus* Kützing 1847

01640050

Filaments in more or less distinct erect bundles forming a cushion-shaped or almost tufted colony, impregnated with calcite. Filaments 7.5–12 μm wide, false branches sparse. Cells 7–9.5 μm wide, 2.5–4 μm long, when old as long as wide, blue-green. Sheath thin, firm, not lamellate, colourless or yellowish in older parts densely covered by calcite. Heterocyst spherical, quadrate or rounded cylindrical, up to 14 μm long.

On rocks, walls, plant pots, soil and abundant in hot-houses; cave entrances in Yorkshire (A. Pentecost, pers. comm.).

S. julianum only differs from *S. hofmanni* in its conspicuous calcification and perhaps slightly wider cells. It is suggested to retain the two names until the whole genus (and *Tolypothrix*) are subjected to a comprehensive revision.

Scytonema mirabile (Dillwyn 1809) Bornet 1889

Synonym: *Scytonema figuratum* [C.Agardh] Bornet *et* Flahault 1886

01640060 **Pl. 20C–E** (p. 113)

Colony bushy or tufted, blackish-brown or blackish-green or seldom more or less blue-green. Filaments entangled (13–)15–21 μm wide, 2–12 mm long, mostly false-branched. Cells 6–12 μm wide, cylindrical; at the end of trichome disc-shaped or more or less barrel-shaped, yellowish to blue- or olive-green. Sheath yellow-brown, with slightly divergent layers, sometimes outside colourless or slightly gelatinous. Heterocyst almost as long as wide or longer than wide.

On moist soil, in lakes sometimes forming large masses, with smaller forms in bogs.

Scytonema myochrous [(Dillwyn) C.Agardh 1812] Bornet *et* Flahault 1886

01640070 **Pl. 20F–I** (p. 113) **CD**

Colony cushion-like, membranous or crust-forming, to 3(–4) cm diameter, but often extending over larger areas as colonies become confluent; brownish-black or blackish-green. Filaments clumped together, entangled, 15–(18–36)–40 µm wide, 2–15 mm long. False branches richly or sometimes sparsely branched, mostly in pairs, long, thinner in comparison to the main filament. Cells 6–12 µm wide, quadrate, longer (up to three times as long) than wide, in parts of the trichome, disc- and barrel-shaped. Sheath yellowish-brown, with divergent striations. Heterocyst spherical or rounded-quadrate.

Highly calcareous sites, especially at the source of very small springs or the side of larger ones, occasionally free-floating in lakes. The species often grows where travertine is being deposited, though its own sheath is not calcified; its pattern of growth keeps it at the surface of the calcareous deposit, though remains of old sheaths may be evident further down. The forms by springs are easy to recognize macroscopically, the larger ones looking at a distance like small horse-droppings.

Scytonema ocellatum [Lyngbye 1819] Bornet *et* Flahault 1886

01640080 **Pl. 20B** (p. 113)

Colony cushion-like, blackish or grey-blue. Filaments 10–18(–19) µm wide up to 3 mm long, entangled. False branches short. Cells 6–14 µm wide, olive-green, as long as wide or shorter than wide. Sheath firm, lamellate, brown. Heterocyst almost quadrate to cylindrical.

On rocks, damp soil and walls. The sheath is sometimes calcite-encrusted, giving the colony a grey-green appearance like *S. julianum*.

Tolypothrix [Kützing 1843] Bornet *et* Flahault 1886

01710000

Filaments with a thin or thick firm sheath, colourless or frequently yellow- to deep brown. False branched, branches single, mostly subtended by a heterocyst. Hormogonia formed at the tips under especially favourable growth conditions. Growth mostly occurs apically; apices often broader, with shorter cells. Akinetes reported for a few species, but it is unclear whether this is a characteristic feature of the particular species as a whole.

A large number of species have been recognized – Geitler (1932a) listed 44 – but many are hard to distinguish from one another. Apart from habitat (aquatic versus terrestrial), the key features separating species are trichome width and aspects of branching and sheath. Species recorded for the British Isles are brought together here under three species names. *T. byssoidea* is terrestrial, while *T. distorta* and *T. tenuis* are aquatic. The descriptions for the latter two were originally fairly different, but as an increasing number of varieties and forms have been included within both species, there is now considerable overlap. However, some forms come clearly within the limits for one or the other. *T. distorta* var. *penicillata* and *T. tenuis* var. *calcarata* are particularly distinctive.

Tolypothrix byssoidea (Hassall) Kirchner 1900

Basionym: *Scytonema byssoideum* Berkeley 1832

01710010

Thallus like a felt when moist and well developed, but like a crust when dry; sometimes appears dark blue-green when moist, but black or brown when dry. Filaments branched irregularly, 10–15(–18) µm wide, to 1 mm long, with branches short, bent. Trichomes 9–11(–12) µm wide, 3.5–5.5 µm long. Sheath firm, usually quite narrow, but showing lamellations and sometimes slight folds corresponding to position of cross wall; yellow-brown to deep brown. Geitler (1932a) reports a single record of akinete formation: ellipsoidal, longer than vegetative cells, mostly in rows; wall ? yellow.

On walls of quarries, cliffs and rocks, especially more calcareous ones and sometimes also tree bases; appears to be characteristic of surfaces which are intermittently very wet, though not fully submerged, but then dry for much longer periods.

This is a highly variable species, especially with respect to trichome width and thickness of sheath Even material at one location may show considerable variation, often with a few filaments exceeding the above dimensions.

Tolypothrix calcarata Schmidle 1899

01710020 (old number): see *Tolypothrix tenuis* var. *calcarata*

Tolypothrix distorta [Kützing 1843] Bornet *et* Flahault 1886

01710030 **Pl. 21B,C** (p. 116) **CD**

Colony a mat or cushion; dark blue-green or brown, sometimes calcified. Filaments (8–)10–17 µm wide, up to 3 cm long, repeatedly branched, false branches often deeply or sharply erect. Cells 9–12 µm wide, 4–12 µm long, slightly narrowed at the cross walls. Sheath thin, close to the trichome, colourless initially, but later yellow-brown. Heterocysts usually single, but sometimes 2–3 in a row.

Attached or free-floating in ponds and shallow lakes, usually in area with submerged macrophytes; attached to rocks in rivers, including (var. *penicillata*) fast-flowing stretches.

var. *penicillata* (C.Agardh 1824) Lemmermann

Synonym: *Tolypothrix penicillata* [(C.Agardh 1824) Thuret 1875] Bornet *et* Flahault 1886

01710032 (new number: old number 01710050)

Pl. 21D–H (p. 116) **CD**

Shares the general features of the species, but colony forms a penicillate cushion or (when attached) tuft. The pattern of repeated branching, with branches running quite close to the original filaments, often leads to filamentous structures 3 cm long and sometimes 5 cm or more.

Free-floating in ponds and attached to rocks in fast-flowing, hard water, large streams and small rivers. In the case of forms in fast-flowing waters, this is probably the nearest the blue-green algae have come to evolving an attached form which occupies the same physical niche as filamentous eukaryotic algae like *Ulothrix* and *Lemanea*. Upstream river sites include Tees (Holmes & Whitton, 1981a) and Caragh (Heuff & Horkan, 1984).

Tolypothrix lanata [(Desvaux 1809) Wartmann in Rabenhorst 1858] Bornet *et* Flahault 1886

01710040 (old number): see *Tolypothrix tenuis*

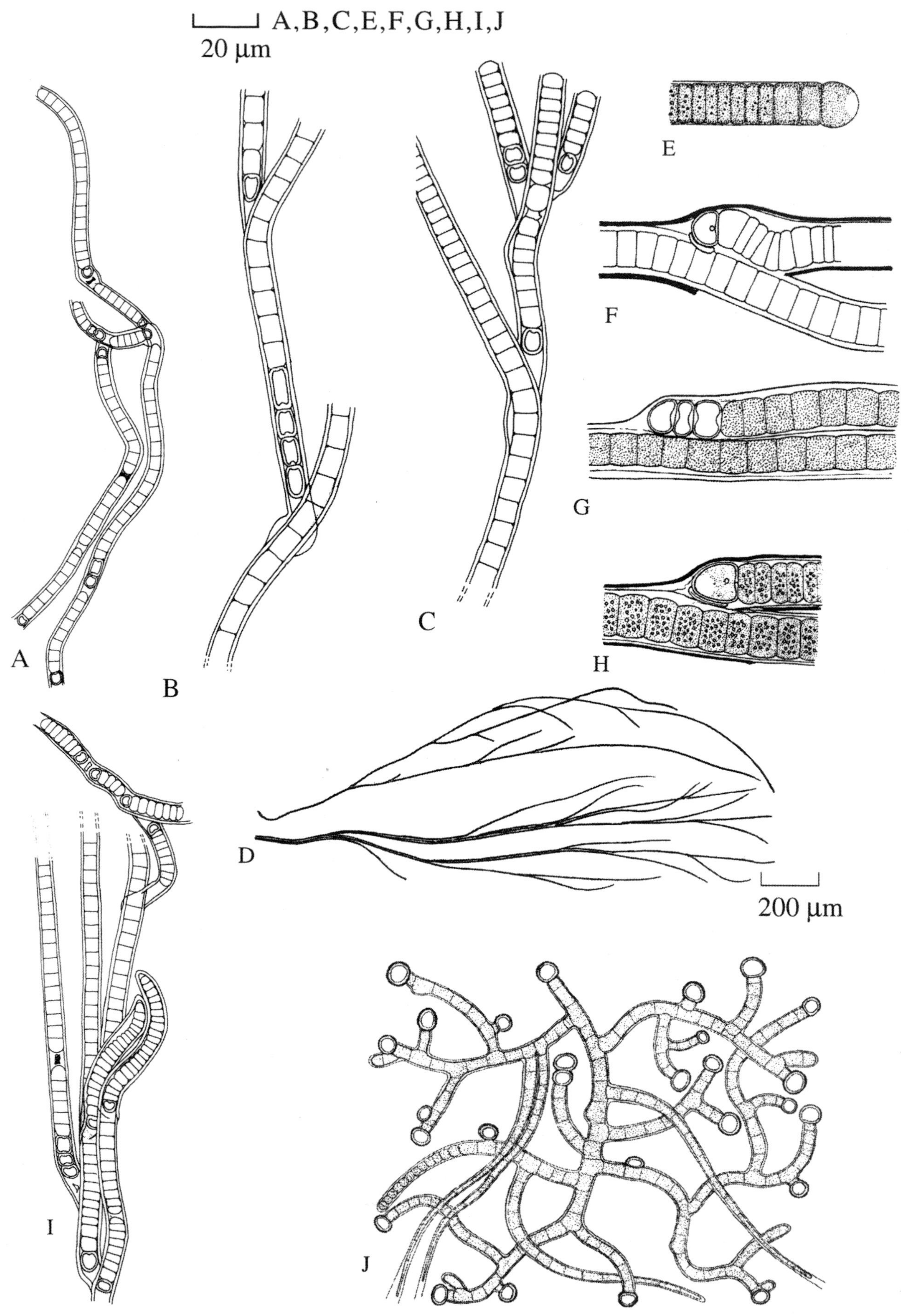

Plate 21 A–J
A. *Tolypothrix tenuis* (p. 117); **B–C**. *Tolypothrix distorta* (p. 115); **D–I**. *Tolypothrix distorta* var. *penicillata* (p. 115): D. General view, E–H. Views of filament branching and the repeat formation of heterocysts which sometimes occurs at branches. **J**. *Brachytrichia quoyi* (p. 117).

Tolypothrix penicillata [(C.Agardh 1824) Thuret 1875] Bornet *et* Flahault 1886

01710050 (old number): see *Tolypothrix distorta* var. *penicillata*

Tolypothrix tenuis Kützing 1843

Synonyms: *Tolypothrix lanata* Wartmann in Rabenhorst 1858; *Tolypothrix calcarata* Schmidle 1899

01710060 **Pl. 21A** (p. 116)

Colony a mat or cushion; dark blue-green or brown. Filaments (4–)6–13(–18) μm wide, up to 2 cm long, repeatedly branched. Cells (4–)5–12(–14) μm wide, ranging from much shorter to considerably longer than wide, slightly or not constricted at cross walls. Heterocysts 6–12 μm wide, 4–6 μm long, single or 2–5 in a row. Sheath thin, close to the trichome, often slight lamellations, colourless initially, but eventually yellow-brown.

On various substrata in ponds, rocks in streams and occasionally on very moist soils. West & Fritsch (1927), who kept *T. tenuis* and *T. lanata* separate, reported that they were the most frequent British species, occurring among various aquatic plants in ponds and lakes.

T. lanata was separated from *T. tenuis* largely because of slightly wider trichomes, but many authors have combined them, with various amendments to the description: Desikachary (1959) reviewed the literature. The above description covers both of the original species. *T. tenuis* var. *calcarata* is a strongly calcified form of moist highly calcareous surfaces and is treated here as a variety (see below). Wider forms of *T. tenuis* in terrestrial environments merge with *T. byssoidea*.

var. *calcarata* (Schmidle 1899) B.A.Whitton, *nov. comb.* W.Schmidle (1899) in Simmer's, III, Bericht über Kryptogamen-Flora der Kreuzeckgr. in Kärnt. Allgemeine Botanische Zeitschrift 12:4 (Figures 1, 2).
01710062 (new number; coded originally as 01710020).
Filaments forming a loosely interwoven mat, 6–8 μm wide, strongly calcified. Cells 5–6 μm wide, always longer than wide, cylindrical, slightly narrowed at cross walls. Sheath narrow, initially colourless, later yellow brown. Heterocyst long ellipsoidal or cylindrical.

Moist highly calcareous surfaces, such as edge of springs and cliffs.

ORDER STIGONEMATALES

These are the morphologically most complex blue-green algae. Longitudinal and sometimes oblique cross walls occur in addition to transverse walls, which lead to the formation of true branches in all genera; in some genera this also leads to filaments consisting of two or more rows of cells. The branches in some genera or a few species within other genera sometimes taper markedly and in *Brachytrichia* sometimes differentiate into colourless multicellular hairs. Heterocysts are usually intercalary, but in most (or perhaps all) genera may also be terminal. Most species can form hormogonia, but not enough have been studied thoroughly to be sure that all do so. Akinetes are formed in species of some genera. Many populations of Stigonematales show considerable morphological variation, even at one site. The two most widespread genera in the British Isles, *Hapalosiphon* and *Stigonema*, are characteristic of non-calcareous rocks and waters, often with low conductivity and low pH values, though several species (e.g. *S. mamillosum*) do occur in calcareous habitats.

Desmosiphon is included here in the Stigonematales, in spite of the apparent absence of heterocysts. It has otherwise considerable resemblance to *Stigonema*.

It seems likely that the ability for true branching has evolved a number of times in the blue-green algae and that the order is phylogenetically heterogeneous. *Brachytrichia*, which occurs on marine calcareous surfaces, is classified here with the Stigonematales, because of the occurrence of true branching, but otherwise has more similarities with the Rivulariaceae than with *Hapalosiphon* and *Stigonema*. It is unclear whether the closely related monospecific genus *Kyrtuthrix* Ercegovićhas been recorded from the British Isles. In countries where *K. dalmatica* Ercegović definitely occurs, such as around the Mediterranean, it grows almost entirely endolithically in soft limestone rocks in the intertidal region, where the surface of the limestone is subject to grazing by molluscs. The species is quite similar in overall morphology to *Brachytrichia quoyi* and should perhaps be incorporated within *Brachytrichia*, as done by Frémy (1929–33).

Attempts to find the limestone cave alga *Geitlerea* in the British Isles have become a challenge for a number of people, but so far without success. This was first described from Israel (Friedmann, 1955), but has since been reported from a number of other countries, including several in Europe. It is an aerial organism growing at very low light intensities, and is characterized by branched trichomes and a characteristic calcite sheath; the branches may be obviously lateral or appear to be dichotomous.

Brachytrichia [Zanardini 1872] Bornet *et* Flahault 1886

01080000

Colonies flat initially, but a more irregular surface when old, sometimes showing a hemispherical structure, but more usually developing a folded surface, with filaments embedded in mucilage. Filaments usually showing one or more bends in the inner part of the colony, with each end and their branches running more or less parallel to each other; true and false branching present, the latter with a reverse V-shaped structure. Trichomes with intercalary heterocysts and (when present) terminal multicellular colourless hairs; in contrast to colonial Rivulariaceae, the heterocysts are not restricted to the parts of the trichome further from the surface.

Epilithic on calcareous substrata in intertidal region (and rarely also epiphytic or partially endophytic on calcareous algae). The presence of heterocysts and hairs within a single filament suggests that the genus grows in environments with highly variable phosphate levels (see *Rivularia*).

Brachytrichia quoyi [C.Agardh] Bornet *et* Flahault 1886

Synonym: *Brachytrichia balani* [Lloyd] Bornet *et* Flahault 1886

01080010 **Pl. 21J** (p. 116)

Colony reaching 6 mm, but usually smaller, dark green, brownish or even almost black. Filaments in lower part of colony entangled, giving rise to more or less parallel arranged filaments or sometimes, where the surface is raised, to radially arranged filaments. Trichome 5–6 μm, sometimes narrower in the inner part of the colony, then broadening slightly before tapering towards the apex; cells of diverse shapes, ranging from subspherical to much longer than broad; heterocyst distinctly larger than vegetative cells; trichome often ending in short or long hairs, which usually remain inside the surface of the firm mucilage, though Bornet (reproduced by Frémy, 1929–33) shows a section of a colony with some emergent hairs; outer region of the mucilage usually pale brown; hormogonia formed in the terminal region of the photosynthetic part of the trichome below the hair, which falls off.

Powell (1964). On calcareous rocks and barnacles in the mid- and lower intertidal zone. Frémy (1929–33) states that it also occurs in fronds of *Codium*, but it is uncertain whether he had a record for this in the British Isles. Coppejans (1995) has two records for Belgium/North France.

This description largely follows Frémy (1929–33), who listed *B. balani* for 'Angleterre'. Desikachary (1959) regards *B. balani* as a synonym of *B. quoyi*, though the upper size limit for the thallus in his description (of *B. quoyi*) is much larger than given here.

Desmosiphon Borzí 1907

01240000

Colony crustose, forming flecks on rocks and (rarely) submerged plants. Upright filaments, sparingly branched, with branching sometimes appearing almost dichotomous. Cells inside the filaments usually in a single row, occasionally two rows in old parts of a colony, each cell appearing somewhat separated from the adjacent one. No akinete.

The genus *Stauromatonema* (not recorded for British Isles) possesses heterocysts, but is otherwise quite similar, so the possibility should be kept in mind that some non-heterocystous populations are merely environmentally induced forms.

Study of a soft-water stream in south-west France, led Bourrelly (1957) to describe a new species (*Desmosiphon vivieri*) and then to reorganize the genus into three species. More study would be needed to confirm that the morphologies described are not just environmental responses. As material from the British Isles has not been assessed critically, the features listed by Bourrelly are given here.

- *D. maculans*. Reproduction by planospores and terminal fragments of filaments; upright filaments freely branched;
- *D. vivieri* Bourrelly 1957. Reproduction by hormogonia; upright filaments little branched;
- *D. incrustans* (Geitler) Bourrelly 1957. Means of reproduction unknown; filaments little developed, remaining joined towards their apices.

Desmosiphon maculans Borzí 1907

01240010

Individual colonies often becoming confluent to form obvious flecks. Filament 4–5 μm wide. Cells 2–2.5 × 2–3 μm wide, spherical or barrel-shaped; individual cells rather distinct from each other (as in many Stigonematales). Sheath brown-yellow to brown. Reproduction by release of single cells at apex of branches, presumably motile.

On rocks in streams and lakes. River Caragh, Kerry (Heuff & Horkan, 1984).

Fischerella Gomont 1895

01280000

Colony usually forming mat on a moist terrestrial surface, but sometimes in shallow water or streams. Branched structure, with most branches derived from cell division in a plane perpendicular to the main axis i.e. true branching. Filament with one or several rows of cells, the main axis usually much broader than the branches; trichomes with intercalary heterocysts.

Hapalosiphon and *Fischerella* merge into each other and the genera are sometimes combined. However, it is suggested that both are retained for the time being. The only real difference is that the main axes of *Fischerella* tend to differ much more from the side branches than they do in *Hapalosiphon*.

Fischerella laminosus (Cohn) Castenholz

Basionym: *Mastigocladus laminosus* Cohn 1863
Synonym: *Hapalosiphon laminosus* [(Cohn 1863) Hansgirg 1885] Bornet *et* Flahault 1886

01280010 **Pl. 22A** (p. 119)

Colony firm, and sometimes quite hard, due to aragonite precipitation. Main filaments 5–6 μm wide, but with many branches which are often much narrower; thin sheath.

This species appears to be restricted to waters above about 35 °C and the only record for the UK is the warm spring at Bath (Pentecost, 1995). Geitler (1932a) mentions three records from cold waters in other parts of the world, but the possibility that these were associated with thermal pollution needs to be considered.

Fischerella muscicola (Thuret) Gomont 1895

Basionym: *Fishera muscicola* Thuret 1875

01280020

Colony a mass of interwoven filaments, mostly creeping, but often also many branches more upright; usually dark brown when viewed macroscopically. Cells in main axis 6–7.5 μm wide, 5–6 μm in branches, with thin sheath, which is usually distinct brown in older filaments; main axis often 2 cells wide, but branches only 1 cell wide; cells in the main axis often subspherical, but those in branches quadratic or shorter than wide; heterocysts mostly in the main axis, but rarely also in branches, usually smaller than adjacent vegetative cells. Hormogonia sometimes present, narrow (about 4 μm wide) and up to 100 μm long.

On moist soil.

Hapalosiphon [Nägeli 1849] Bornet *et* Flahault 1886

01340000

Colony forming mat on a terrestrial surface or a floating floc; branched structure, with most branches derived from cell division in a plane perpendicular to

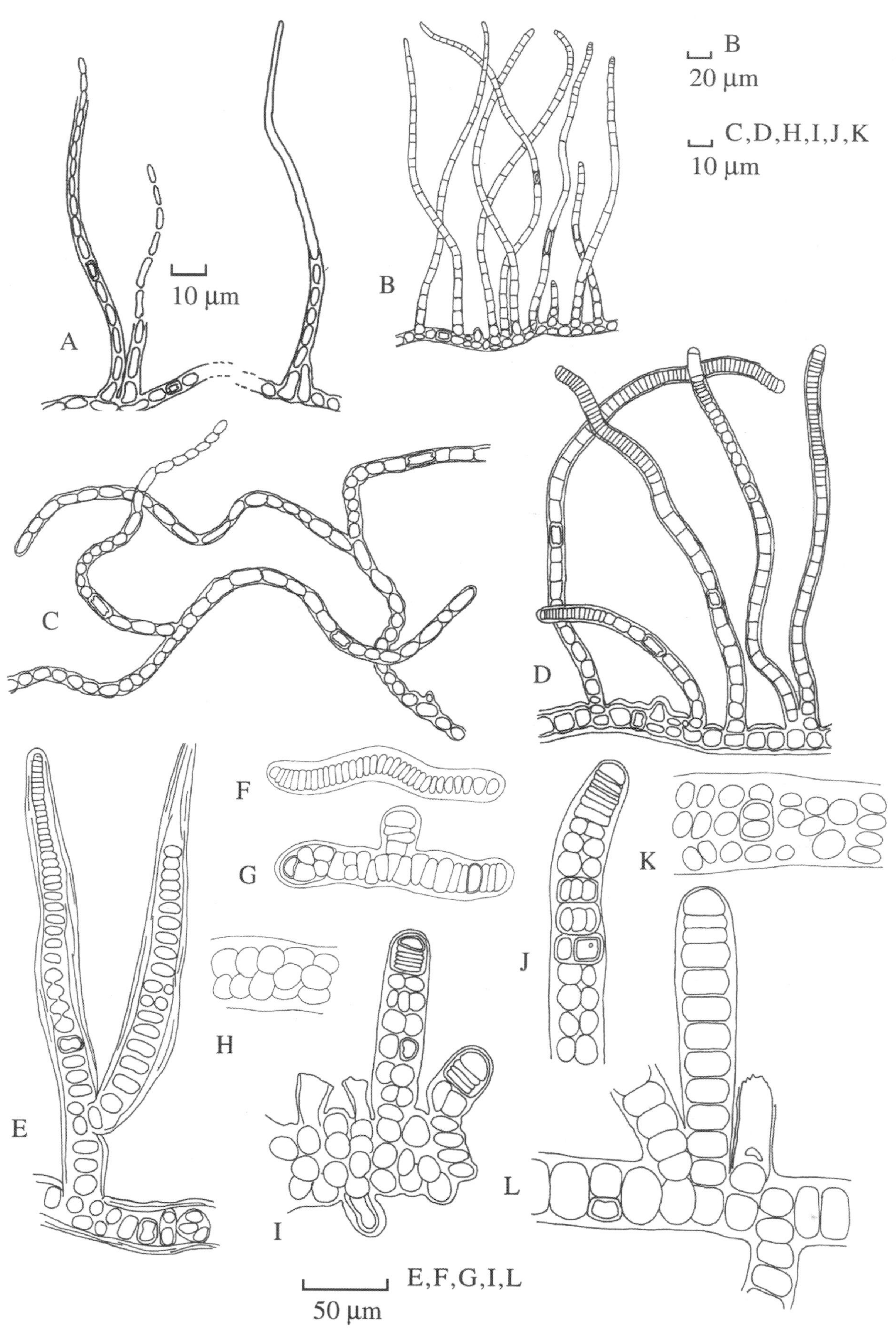

Plate 22 A–L

A. *Fischerella laminosus* (p. 118); **B**. *Hapalosiphon hibernicus* (p. 120); **C**. *Hapalosiphon intricatus* (p. 120); **D**. *Hapalosiphon fontinalis* (p. 120). **E**. *Stigonema panniforme* (p. 122); **F–H**. *Stigonema minutum* (p. 121): F. Young filament, G. Formation of first branch and first heterocyst, H. Section of mature filament; **I–K**. *Stigonema mamillosum* (p. 121); **L**. *Stigonema ocellatum* (p. 122).

the main axis i.e. true branching, but sometimes also false branches. Filament with one or two rows of cells; trichomes with heterocysts; heterocyst almost always intercalary, sometimes small and easy to overlook, but in the British species usually distinct and sometimes longer than the vegetative cells. Branches are usually narrower than the axis from which they arose, but this may be less obvious in floating forms, where the branching tends to occur repeatedly; the distinction into main axis and branches is most obvious in *H. fontinalis* and hardest to distinguish in *H. intricatus*.

Hapalosiphon and *Fischerella* merge into each other and the genera are sometimes combined. However, it is suggested that both are retained for the time being. For comments on the differences, see *Fischerella*.

Hapalosiphon fontinalis (C.Agardh) Bornet 1889

Basionym: *Conferva fontinalis* C.Agardh 1824

01340010 **Pl. 22D** (p. 119) **CD**

Colony a mass of interwoven filaments, consisting of creeping main axes and branches perpendicular to them, forming flocs. Main filament (8–)12–18(–24) μm wide, typically much of it one cell wide, but local regions 2 cells wide; branches narrower, mostly 8–12 μm wide; sheath obvious, especially in main axis, usually yellow-brown in older parts; heterocysts present in main axis and branches. Cells in branches mostly quadrate or slightly longer than broad, but sometimes much shorter in the apical region; it is uncertain whether this is merely a stage in the differentiation of a hormogonium; hormogonia 6–8 μm wide, often well over 100 μm long.

In shallow and often slightly peaty waters, free-floating, mixed with submerged plants or epiphytic.

The literature shows that this is a highly variable species, but it is unclear how much this is so within the British Isles. The species merges with *H. hibernicus* and *H. intricatus*.

Hapalosiphon hibernicus West *et* G.S.West 1896

01340020 **Pl. 22B** (p. 119)

Filaments mostly single among other algae; main axis with slight bends and numerous narrower branches perpendicular to it, which are themselves sometimes branched. Main filament 7.2–9.5 μm wide, branches 4.5–5.5 μm wide; cells of the branches usually much longer than wide, typically 3–4 times, occasionally up to 8 times. Akinete develops from several cells which form a common wall, which is thick and dark yellow-brown.

In shallow peaty waters, mixed with other algae or *Sphagnum*. The original record by W. & G.S.West (1896) was from Glen Caragh, Kerry, Ireland.

Although Geitler (1932a) comments that it resembles *H. intricatus*, it is perhaps even more like a narrow form of *H. fontinalis* which has not developed into a distinct floc.

Hapalosiphon intricatus West *et* G.S.West 1894

01340030 **Pl. 22C** (p. 119)

Filaments forming small bushy flocs; filaments 4–7 μm wide, with sparse branching; main axis and branches showing curves and bends; branches seldom showing the clear perpendicular development characteristic of *H. fontinalis* and sometimes only slightly distinguished from the main axis; sheath narrow, colourless, often hard to distinguish. Heterocysts in main axis and branches, sometimes quadrate, but more usually ellipsoidal, often up to about 2.5 times longer than wide.

In peaty pools and other shallow waters, usually mixed with other algae, *Sphagnum* or *Utricularia minor*; sometimes also on very moist moils, usually adjacent to pools where it is also present. Frequent in pools (pH 4.1–4.5) of the 'flow' country in North Scotland, together with *Tolypothrix* and *Sphagnum*.

Mastigocoleus [Lagerheim 1886] Bornet *et* Flahault 1886

01450000

Thallus growing predominantly endolithic on marine calcareous substrata. Filaments with true branching; branching irregular. Trichomes single within the sheath, with heterocysts terminal on short branches or lateral on a much reduced cell which looks like a stalk. In addition to the frequent short branches with heterocysts, there are actively growing branches and branches which are markedly tapered or forming multicellular hairs.

Mastigocoleus testarum [Lagerheim 1886] Bornet *et* Flahault 1886

01450010

Thallus initially a thin membranous structure on the surface of marine shells, but soon developing perforating filaments and eventually most of the thallus is endolithic, consisting of a complex pattern of curved and branched filaments. Endolithic filaments 5–10 μm wide; trichomes 3.5–6 μm wide; surface trichomes with similar dimensions, though sometimes with a different appearance towards the end, either a marked taper or differentiating into a hormogonium. Trichomes usually blue-green, but sometimes purplish or even rose-coloured; sheath colourless. Heterocysts almost entirely lateral or intercalary.

Powell (1964). On shells in the intertidal and sublittoral.

Stigonema [C.Agardh 1824] Bornet *et* Flahault 1886

01670000

Thallus a branched structure, with the branches derived from cell division in a plane perpendicular to the main axis i.e. true branching. The thallus is usually distinct enough to be justified in terming it a colony. Branches develop into a structure similar to that of the main axis, so that even in forms with a clear differentiation into prostrate and 'upright' filaments, they all have a broadly similar structure. Filaments with one or several rows of cells, the arrangement into distinct rows often difficult to recognize in the old filaments of the larger species. Individual cells within a row often appear slightly spaced apart from each other, though apparent protoplasmic connections are sometimes visible with the light microscope and these have been confirmed in several electron microscopy studies. However, it is not known what proportion of cells within a thallus are connected to each other or are completely independent.

Trichomes with heterocysts, which sometimes develop into shapes quite different from that of the original vegetative cell. Heterocysts mostly intercalary, but sometimes lateral. Sheath often yellow-brown and sometimes lamellated. Hormogonia, usually consisting of only a few cells (sometimes only 2) form at the end of short branches, and under favourable conditions this probably happens repeatedly.

The majority of species are characteristic of environments which are sometimes wet and sometimes completely dry, though usually still in a humid environment. The macroscopic appearance of colonies appears very different whether they are wet or dry and the descriptions below apply to wet material.

West & Fritsch (1927) list *S. hormoides*, *S. informe*, *S. mamillosum* and *S. ocellatum*, but state that there are three other British species. They also state that the species occur principally on damp or wet rocks, but are sometimes observed free-floating in ponds and lakes. It is not known whether their other three species are the same as the other three listed here (*S. minutum*, *S. panniforme*, *S. tomentosum*), nor which are the ones sometimes found free-floating. *Stigonema* occurs as a lichen symbiont.

1 Main filaments mostly with 1 row of cells.............2
1 Main filaments mostly with 2 or more rows of cells 5
2 Lateral branches prostrate or sub-erect, but not upright 3
2 Lateral branches upright, giving colony a turf-like appearance 4
3 Filaments 7–15(–18) µm wide ....*S. hormoides*
3 Filaments 30–50 µm wide....................*S. ocellatum*
4 Upright branches single.....................*S. panniforme*
4 Upright branches grouped into bundles...*S. tomentosum*
5 Main filaments 15–30(–40) µm wide ....*S. minutum*
5 Main filaments 40–70 µm wide.............................6
6 Hormogonia forming at apex of many branches ....*S. informe*
6 Hormogonia from numerous short branches on main branches giving rise to mamillate morphology; especially characteristic of streams ....*S. mamillosum*

Stigonema hormoides [Kützing 1843] Bornet *et* Flahault 1886

01670010

Filaments prostrate, densely intermingled, long, forming a thin tomentose, blackish-brown thallus. Filaments 7–15(–18) µm wide, irregularly and sparsely branched, with some lateral branches sub-erect, though often somewhat curved and often branching again. Cells subspherical, 3.5–6.5(–9) µm wide, largely or entirely in 1-cell rows, though occasional short sections 2 cells wide may occur; each cell in the row usually appears slightly distant from the adjacent cells, though joined by protoplasmic connections. Heterocysts both intercalary and lateral to a cell within the main row. Sheath thick, colourless initially and in shaded situations, but usually becoming pale to medium yellow-brown. Sometimes branches appear to become detached naturally; in other cases, Chroococcales-like stages develop due to irregular cell division in some branches. Many varieties and forms of this species have been reported in the literature, but it is not known how many occur in the British Isles. This is the smallest species in the genus.

On moist soil or wet non-calcareous rocks, often among other mucilaginous or sheath-forming algae in small depressions on the surface of a rock; sometimes on submerged rocks at the edge of lakes, especially where submergence is intermittent.

Stigonema informe [Kützing 1849] Bornet *et* Flahault 1886

01670020

Thallus usually a macroscopically visible dark brown crust, though occasionally found as single filaments. Filaments in the main axis 40–70 µm wide, 1–2 mm long, largely prostrate, but with sub-erect branches, which are straight or flexuous, 20–45 µm wide; the main axis is usually about 6 cells wide, but the number of cells at any position in the branches gradually decreases until there is usually only a single row in the apical part. The individual rows of cells are often hard to distinguish in the main axis, which sometimes appear more like chroococcoid masses, but the rows are usually distinct in the branches; cells (8–)12–18 µm wide. Secondary branches frequent, especially where the first branch is slightly bent, and these often (? always) give rise to a hormogonium.

On moist or submerged rocks.

Stigonema mamillosum [(Lyngbye 1819) C.Agardh 1824] Bornet *et* Flahault 1886

01670030 **Pl. 22I–K** (p. 119) **CD**

Colony forming a cushion up to 12 mm high, dull brown or dark blackish-green, with densely intermingled filaments. Main filaments usually up to about 70 µm wide, though occasionally wider, richly branched from near the base, with the branches slightly narrower than the main filaments; the branches in turn give rise to lateral branches or short 'mamilliform' branches typically 15–24 µm wide, which give rise to the hormogonia; these short branches are often numerous and give the species its characteristic appearance. The main filaments and the lower parts of the branches arising from them are densely packed with cells and it is often hard to distinguish individual rows. Sheath colourless initially, but usually becoming yellowish in submerged populations or yellow-brown in terrestrial populations.

On very moist or submerged rocks, especially in moderately fast-flowing streams, both soft and moderately hard water (Malham, Yorkshire). River Caragh, Kerry (Heuff & Horkan, 1984), where it covered rocks just below the average water level; during periods when dry it had the appearance of short, coarse dark grey hair.

The spelling *S. mammillosum* used by West & Fritsch (1927) is incorrect.

Stigonema minutum [(C.Agardh 1817) Hassall 1845] Bornet *et* Flahault 1886

01670040 **Pl. 22F–H** (p. 119)

Colony a crust or flattened cushion, brown to black, up to 1 mm high. Main filaments initially more or less prostrate, but flexuous and often becoming erect at the end; richly branched, with the branches mostly on one side as they become more or less erect; branches of markedly different lengths, with the longer branches often branching once or twice more, giving rise to a complex, irregular morphology; hormogonia arise from short branches on the longer lateral branches; main filaments 15–30(–40) μm wide. Cells usually densely packed in the filaments, often arranged in two distinct rows in the narrow parts of main branches, but appearance much more irregular in the main axis, which is 2–4 cells across; heterocysts intercalary or lateral. Outer sheath yellow-brown, but there is also a distinct sheath around individual cells in the older parts of filaments and this usually has particularly deep pigmentation.

On intermittently moistened rocks or, in more humid regions, rocks and other surfaces dependent only on rainfall; on bark of oak trees near waterfall, North Wales: A. Pentecost, pers. comm.). This species occurs in drier situations than some of the other *Stigonema* species and so presumably grows more slowly, which would explain why its sheath pigmentation is especially dark, though there is no information as to whether colonies in more shaded situations are as dark.

Stigonema ocellatum [(Dillwyn 1809) Thuret 1875] Bornet *et* Flahault 1886

01670050 **Pl. 22L** (p. 119)

Colony a crust or flattened cushion, though sometimes also found as single filaments among other algae and mosses; colonies usually brown, though tend to be slightly less dark than some of the other *Stigonema*. Filaments often very long, 30–50 μm wide, with irregular branching giving rise to complex structures; lateral branches often as wide as the main axis, though the ultimate branches are usually slightly narrower. Filaments mostly with only one row of cells, 18–30 μm wide, except where the heterocyst occurs; this is almost always lateral, differentiating from one of the vegetative cells in the main row; the trichome is therefore 2 cells at this point, with the heterocyst slightly smaller than the vegetative cell. Sheath yellow-brown, usually distinctly lamellated, and the individual cells also with their own distinct sheath, which becomes more obvious in the older parts due to the increasing yellow-brown colour; heterocyst apparently does not have its own sheath, in spite of the fact that it originates from the adjacent vegetative cell.

Recorded by West & Fritsch (1927), but without habitat details; general floras list it from a range of environments, including soil, wet rocks, among moss and submerged in standing water.

Stigonema panniforme [C.Agardh 1817] Bornet *et* Flahault 1886

01670060 **Pl. 22E** (p. 119)

Colony spread out, forming a turf; basal filament gives rise to erect filaments mostly reaching more or less the same height, typically about 1 mm; if the erect filaments branch, both parts are usually quite similar. Filaments 24–36 μm wide, with little difference between the prostrate and erect ones, except that the erect ones taper towards the end. Filaments mostly with one row of cells, though occasionally there are two adjacent cells; these arise by longitudinal division of the original one cell in the row. Cells 12–26 μm, sometimes filling all the space to the outer sheath, but usually with an obvious space between the cell and outer sheath; heterocyst intercalary. Main sheath thick, lamellated and yellow to brown; older cells with their own distinct sheath.

On moist soil and rocks.

This species is similar to *S. ocellatum* in general dimensions and the typical single row of cells. However, it differs in the upright bundles, greater tendency for the sheath to be lamellated and the fact that most heterocysts are intercalary.

Stigonema tomentosum (Kützing) Hieronymus 1895

01670070

Colony up to 2 mm thick, appearing like a turf, at least when moist; blackish-brown. Main filaments prostrate, 14–28(–35 μm) wide, mostly with 1 row of cells, though occasionally two cells across; erect lateral branches grouped in small bundles. Cells (8–)10–12 μm wide. Main sheath yellow-brown, older cells with their own narrow, but distinct sheath.

On moist rocks; according to Geitler (1932a) also on wood.

Acknowledgements

The author is most grateful to Dr Paul Broady, Dr John Lund, Dr Allan Pentecost, and Dr Olav Skulberg for helpful comments on various draft manuscripts. However, decisions about matters of nomenclature are entirely the author's. He also appreciates help from Dr Alan and Mrs Alison Donaldson with collecting field samples from northern Scotland, from Dr David Balbi, Isabel Douterelo and Neil Ellwood with culture studies on strains of *Oscillatoria bornetii* and *O. bourrellyi* (from Norway) supplied by Dr Skulberg, and from Dr Balbi and Sally Hardiman with assistance in preparing the figures.

PHYLUM RHODOPHYTA (Red Algae)

Robert G. Sheath and Alison R. Sherwood

Introduction

The red algae constitute a division (phylum) of organisms that share the following combination of attributes: eukaryotic cells, lack of flagella, floridean starch, phycobiliprotein pigments, unstacked thylakoids, and chloroplasts lacking an external endoplasmic reticulum (Woelkerling, 1990). They are primarily marine, with only approximately 3% of the over 5000 species in truly freshwater habitats (Sheath, 1984). Most of the inland red algae are restricted to streams and rivers (Sheath & Hambrook, 1990). A small number occur in habitats other than typical freshwaters, such as soils (Geitler, 1932b) and caves (Hoffmann, 1989). In spite of being an old account, Israelson (1942) still provides a very useful introduction to the ecology of freshwater red algae in mainland Europe and is especially relevant to the British Isles.

Vegetative Morphology

Most of the red algae in freshwater habitats tend to be macroscopic and benthic (Sheath & Hambrook, 1990). Nonetheless, these algae exhibit a smaller size range than do marine species, with 80% of freshwater red algae having a length range of 1–10 cm (the same percentage of marine species ranges up to 30 cm). There are 22 species recorded from the British Isles, of which nine are gelatinous filaments (*Batrachospermum*, Pl. 26D; *Sirodotia*, Pl. 30F; *Thorea*, Pl. 32F), three are free filaments without a prominent gelatinous matrix (*Compsopogon*, Pl. 23F; *Bangia*, Pl. 24B, C; *Balbiania*, Pl. 25E), three are pseudoparenchymatous forms (*Lemanea*, Pl. 31C–E, 32A, B; *Paralemanea*, Pl. 32C–E), two are tufts of short radiating filaments without a common matrix (*Audouinella*, Pl. 25C), two are unicells (*Porphyridium*, Pl. 23D, E), and there is one each of a pseudofilament consisting of loose chains of cells held together with a common gelatinous matrix (*Chroodactylon*, Pl. 23A, B), a colony with stalked unicells (*Chroothece*, Pl. 23C), and a crust of a flat thallus composed of compacted tiers of cells (*Hildenbrandia*, Pl. 33B, C). Among the filamentous forms, only *Thorea* has multiaxial growth (Pl. 33A); the rest are uniaxial, except *Bangia*, which has a uniaxial base and multiaxial apex at maturity (Pl. 24B, C). *Compsopogon* has uniaxial filaments that are covered by a layer of small cortical cells (Pl. 23F).

The various morphological forms encounter the stress caused by flow in riverine habitats in different ways, according to Sheath & Hambrook (1988, 1990). Crusts and short tufts occur within the boundary layer or at least in a region of reduced current velocity and hence avoid much of the flow-related stress. The remaining macroscopic species can be regarded as semi-erect, experiencing bending, tensile and compressive forces and perhaps torsional stresses in flowing waters (Vogel, 1984). This group includes mucilaginous and non-mucilaginous filaments, pseudoparenchymatous forms and tubes. It would be expected that the semi-erect forms possess adaptive mechanisms to tolerate flow, such as branch reconfiguration and extension of thalli in high water motion (Sheath & Hambrook, 1988, 1990).

Of the 23 infrageneric taxa of Rhodophyta in inland habitats in the British Isles, 17 are typically blue to olive-green in colour, while the rest are red. This trend contrasts with that of marine red algae, in which the great majority of species are red. There are a number of chloroplast morphologies among freshwater taxa: all members of the Florideophycidae (see taxonomic section) have multiple disc-shaped or ribbon-like chloroplasts without pyrenoids (Pl. 24E); the bangiophycidean taxa have this type as well as central star-shaped chloroplasts with a pyrenoid (Pl. 23D), a peripheral lamellate structure and a single peripheral disc (Sheath, 2002). However, some species which have what appear to be multiple disc-shaped or ribbon-like chloroplasts may actually have a complex, interconnected single chloroplast, such as *Compsopogon coeruleus* (Gantt et al., 1986).

Another characteristic that is useful in analysing the morphology of freshwater red algae is the external covering. Unicells and pseudofilaments typically have a gelatinous matrix surrounding the cells, which varies in thickness depending on the age and physiological state of the organisms (Sheath, 1984). The gelatinous filamentous members of the Batrachospermales, such as *Batrachospermum* and *Sirodotia*, have both distinct cell walls and an overall matrix surrounding the filament. The free filamentous and pseudoparenchymatous forms have only cell walls. Many freshwater red algae exhibit differential staining of external coverings with Alcian Blue, particularly of mucilaginous layers, rhizoids, sporangia and spermatangia (Sheath & Cole, 1990).

Reproduction

Freshwater red algal species exhibit a diversity of reproductive types, particularly in terms of dissemination. Cell division is the major mode of population increase among the unicellular forms. Cell division in the other forms is generally the mechanism by which the thallus is expanded (Sheath, 1984).

Monospore formation is the mode of asexual reproduction among the pseudofilamentous and filamentous taxa and typically involves the formation of single spores that germinate into the life history phase that produced them. In *Chroodactylon ornatum* and *Bangia atropurpurea*, spores are released by localized digestion of the common filamentous matrix and there is no obvious sporangium (Sheath, 1984) (Pl. 24D). The order Compsopogonales is characterized by its method of monospore production, which involves the cleavage of a relatively small cell from a larger vegetative cell by oblique cell division to form a monosporangium (Garbary et al., 1980) (Pl. 24A). Monosporangia of the Acrochaetiales, Balbianiales and Batrachospermales are specialized, enlarged and ovoid cells typically produced at the apices of vegetative branches (Sheath, 1984) (Pl. 24E, 25C, E). Monosporangia can be regenerated after spore release by protrusion and cleavage of cytoplasm from the subtending cell in *Audouinella hermannii* (Hymes & Cole, 1983). Monospores are also a mechanism by which certain life history phases perpetuate themselves in a complex life history alternation, such as the 'chantransia' phase (see below) of the Batrachospermales (Sheath, 1984).

Sexual reproduction and life history alternation is known for many species of freshwater red algae, although these phenomena have not been conclusively demonstrated for freshwater members of the Bangiophycidae or the Hildenbrandiales of the Florideophycidae. In the freshwater *Audouinella* species for which the life history has been analysed, the free-living gametophyte and tetrasporophyte are isomorphic, both having the same tuft-like morphology (Drew, 1935; Necchi et al., 1993a). The haploid gametophyte produces the gametangia. The female gametangium, the carpogonium, is a colourless cell with an inflated base and narrow tip, the trichogyne (Pl. 25A). The male gametangium, the spermatangium, is also colourless, ovoid in shape, and releases one spermatium at a time (Pl. 24F). Spermatia attach to the trichogyne and one eventually fertilizes the carpogonium. The zygote divides into a microscopic diploid phase, the carposporophyte (Pl. 25A), which remains attached to the gametophytic stage until its deterioration. Each carpospore germinates into the diploid tetrasporophyte, which at maturity forms tetrasporangia at the branch tips (Pl. 25B). Haploid tetraspores are formed by meiosis and germinate into the gametophytic stage, thereby completing the life history. The blue-coloured *Audouinella* species have not been observed to contain gametangia, carposporophytes or tetrasporangia (Necchi et al., 1993b). This finding may indicate that they are not in fact true *Audouinella* species, but rather one of the life history stages of the Batrachospermales, the 'chantransia' (see below; Necchi & Zucchi, 1997). *Balbiania investiens* has a life history similar to that of the red-coloured *Audouinella* species (Swale & Belcher, 1963).

In the British Isles, 59% of the freshwater Rhodophyta belong to the Order Batrachospermales, similar to the percentage in North America (Sheath, 2002). The haploid gametophyte is macroscopic and semi-erect, which alternates with a diploid microscopic to small (up to 6 cm) tufted 'chantransia' stage (Sheath, 1984). The latter stage is formed by germination of carpospores and resembles the blue-coloured forms of *Audouinella*, as noted above. Where it differs from the tetrasporophyte of the Order Acrochaetiales described above is in the process of meiosis. In the 'chantransia' stage, tetraspores are not produced but rather the haploid gametophyte is formed directly attached to this stage (Sheath, 1984). In the freshwater red algal species for which this process has been studied, it appears that meiosis takes place in an apical cell of the 'chantransia' filament; in each division a residual nucleus is extruded into a lateral protrusion which is then separated by wall formation (Sheath, 1984). The one remaining haploid nucleus forms the gametophyte. Sheath (1984) proposed that the 'chantransia' stage of the Batrachospermales may be an evolutionary adaptation for population maintenance in the upper portions of drainage basins. This stage is typically perennial, seasonally producing the attached gametophyte (Yoshida, 1959; Sheath, 1984; Necchi, 1993). Therefore, the population can continue to proliferate upstream while colonizing downstream with the release of carpospores (Sheath, 1984). If these algae possessed the typical red algal life history, as exhibited by the orders Acrochaetiales and Balbianiales, release of both tetraspores and carpospores would result in a gradual shift of populations downstream until they were solely in the larger trunk rivers which are too deep, turbid and sedimented to support the growth of most autotrophs. Raven (1993) demonstrated that the photosynthetic rates *in situ* of the 'chantransia' stage of *Lemanea* are one-twentieth of those of the gametophyte and the former phase has a negligible role in provisioning the growing gametophyte. He also concluded that the key role of the 'chantransia' stage is to occupy space throughout the year, including possible exposure during summer drawdown. This stage may also act in population dispersal through the production of monospores (Sheath, 1984; Raven, 1993).

Another key feature pertaining to the reproduction of the Batrachospermaceae (*Batrachospermum* and

Sirodotia) of the Batrachospermales is the formation of relatively enlarged and persistent trichogynes of the carpogonia (Pl. 26F, 27C, 28A, E, 29C, 30B, D), compared with those of red algae from other orders (Sheath, 1984). The larger surface area and longevity would enhance the probability of spermatia contact. Hambrook & Sheath (1991) demonstrated that the mean fertilization rate for various species of *Batrachospermum* was 45–72%, including dioecious taxa. This rate may be obtained because spermatia are released into turbulent eddies downstream of rocks, where they are carried through the female plants numerous times as the water moves back and forth (Sheath & Hambrook, 1990). Carpogonia are borne on carpogonial branches that may be little differentiated (Pl. 26F, 28A, E, 29C) to highly modified (Pl. 27C, 30B, D) from adjacent vegetative branches in the Batrachospermales. In the undifferentiated carpogonial branch of *Batrachospermum involutum*, the cells are uninucleate with abundant starch granules and several well-developed peripheral chloroplasts (Sheath & Müller, 1997). In contrast, the short carpogonial branch cells of *B. helminthosum* have no visible starch, the chloroplasts are highly reduced and cross walls break down among cells.

Freshwater populations of *Hildenbrandia* (Order Hildenbrandiales) are typically vegetative and reproduce by gemmae, dense aggregations of cells formed within the thallus (Starmach, 1969). The gemmae are eventually released from the thallus, leaving behind a cavity. They germinate into new crusts, presumably of the same ploidy level.

Studies of the British Isles

The study of freshwater red algae in the British Isles has a long history, dating back to the original description of *Batrachospermum atrum* by Hudson (1778) and early algal compilations, such as those of Lightfoot (1777), Sibthorp (1794) and Hull (1799). However, many of the studies have been part of larger ecological, physiological or survey investigations. Examples of specific taxonomic studies include those of Cooke (1882–84, 1890), Drew (1935, 1936, 1946), Belcher (1956, 1960), Belcher & Swale (1957, 1991), Swale (1962a, b, 1963b), and Swale & Belcher (1963). It is readily apparent that there is no modern, synthetic treatment of freshwater Rhodophyta in the British Isles. Hence, as part of the preparation of this section, we sampled 79 sites throughout much of the region. We have also employed the systematic treatment that we have used in North America (summarized in Sheath, 2002), as did Entwisle (1998) for France. From this activity, we have added several new taxa to the British flora.

1 Thallus pseudofilamentous, filamentous, pseudoparenchymatous or crustose 2
1 Thallus single-celled or of small masses of spherical cells in a common gelatinous matrix 3
2 Pseudofilaments with false branches .... *Chroodactylon* (p. 126)
2 True filaments, pseudoparenchymatous thalli or crusts 4
3 Cells each with a stratified, stalk-like matrix protruding to one side .... *Chroothece* (p. 126)
3 Cells with an ill-defined matrix without a 'stalk' .... *Porphyridium* (p. 126)
4 Thallus crustose .... *Hildenbrandia* (p. 140)
4 Thallus non-crustose, filamentous or pseudoparenchymatous 5
5 Thallus filamentous 6
5 Thallus pseudoparenchymatous 12
6 Multiaxial filaments with colourless medulla and determinate photosynthetic lateral branches................................ *Thorea* (p. 140)
6 Uniaxial filaments with no obvious medulla 7
7 Mature filament uniaxial at base and multiaxial in packets of cells at reproductively mature apices .... *Bangia* (p. 128)
7 Mature filament uniaxial throughout 8
8 Filaments with spherical or cuboidal cortical cells.......................... *Compsopogon* (p. 126)
8 Filaments without cortication or with rhizoid-like cortical filaments................................ 9
9 Simple branched filaments in small tufts or interwoven within other algae; isomorphic gametophytes and tetrasporophytes 10
9 Large, semi-erect gametophyte alternating with a small to microscopic 'chantransia' stage.... 11
10 Thallus composed of tight, radiating filaments in a small tuft; spermatangia uniform within clusters .... *Audouinella* (p. 128)
10 Thallus composed of loose filaments interwoven among fascicles of *Batrachospermum*; spermantangia in clusters including some with starch-containing cells .... *Balbiania* (p. 131)
11 Carposporophyte a distinct mass of determinate gonimoblast filaments .... *Batrachospermum* (p. 131)
11 Carposporophyte indistinct with indeterminate gonimoblast filaments .... *Sirodotia* (p. 137)
12 Main axis corticated; spermatangia in rings around the diameter of the thallus .... *Paralemanea* (p. 137)
12 Main axis without cortication, spermatangia in patches .... *Lemanea* (p. 135)

SUBCLASS BANGIOPHYCIDAE

ORDER PORPHYRIDIALES

Unicells, mucilaginous masses of cells or pseudofilaments. Pit connections and pit plugs absent. Sexual reproduction unknown and asexual reproduction by vegetative cell division.

Chroodactylon Hansgirg 1885

Synonym: *Asterocytis* (Hansgirg) Gobi *et* F.Schmitz

Pseudofilaments with 0–6 false branches; cells rectangular to ellipsoidal, loosely arranged in a linear fashion within a broad mucilaginous matrix; chloroplasts blue-coloured, axial and star-shaped; asexual reproduction by monospore release by localized digestion of matrix.

Scattered distribution in alkaline waters containing the chlorophytes *Cladophora* or *Rhizoclonium*.

Chroodactylon ornatum (C.Agardh) Basson 1979

Basionym: *Conferva ornata* C.Agardh

Synonyms: *Asterocytis smargadina* Reinsch, *Chroodactylon ramosum* (Thwaites) Hansgirg, *Asterocytis ramosa* (Thwaites) Gobi

03050010 **Pl. 23A,B** (p. 127) **CD**

Filaments 24–1240 μm long; cells 6–12 μm wide, 7–17 μm long.

Distribution like that of the genus. The species is occasional in hardwater ponds with submerged vegetation, where it occurs as an epiphyte and probably also free-floating among fine-leaved plants or larger filamentous algae. Although it is widespread, it never appears to be abundant at sites where it occurs. Published freshwater records include West (1912a) from Studley in Warwickshire and Belcher & Swale (1957) from the Lee Valley in Essex, where epiphytic on *Cladophora* growing in flowing water. Based on the dimensions, the latter record may be *C. ornatum*. *Chroodactylon halophila* is recorded by G.S.West (1912a), but also appears to be *C. ornatum* (see Vis & Sheath, 1993, for discussion of taxonomy).

Chroothece Hansgirg 1884

Synonym: *Petravenella* Kylin

Mucilaginous masses of cells, each with a stratified, stalk-like matrix protruding to one side; cells ellipsoidal and blue-coloured; chloroplast axial, star-shaped; reproduction by cell division and possibly colony fragmentation.

Scattered distribution, usually in flowing-water habitats.

Chroothece richteriana Hansgirg 1884

03060010 **Pl. 23C** (p. 127)

Cells 8–12 μm wide, 10–21 μm long, with a mucilaginous envelope about 8 μm thick.

Scattered distribution; only two records for the British Isles, West Yorkshire (W. & G.S.West, 1901) and Port Erin (Belcher & Swale, 1957).

Porphyridium Nägeli 1849

Solitary single cells to small clusters of cells in a mucilaginous matrix; cells spherical to ovoid; chloroplast axial, star-shaped; reproduction by cell division.

Probably cosmopolitan, scattered records on damp soils and walls.

Species distinguished by chloroplast colour.

1 Chloroplast blue-coloured ....*P. aerugineum*
1 Chloroplast red-coloured....................*P. purpureum*

Porphyridium aerugineum Geitler 1923

03110010 **Pl. 23E** (p. 127)

Cells 5–8.5(–12) μm in diameter; chloroplasts blue-coloured.

Scattered in moist terrestrial areas; only known in the British Isles from breckland soils in mixed wood–grass–heath habitats with pH range 6.2–7.8 (John, 1942).

Porphyridium purpureum (Bory) Drew *et* Ross 1965

Basionym: *Phytoconis purpurea* Bory

Synonym: *Porphyridium cruentum* (J.E.Gray) Nägeli

03110020 **Pl. 23D** (p. 127) **CD**

Cells 5–9(–15) μm in diameter; chloroplasts red-coloured.

Scattered in moist terrestrial areas, especially lower parts of walls with calcareous components, and permanent structures such as brickwork in shaded areas or intermittently submerged banks of nutrient-rich rivers. West & Fritsch (1927) reported it as common on damp ground and near the base of damp walls.

ORDER COMPSOPOGONALES

Filaments with an axial row of cells covered by a layer of cortication or saccate thalli. Pit plugs with no cap layers. Sexual reproduction unknown and asexual reproduction by monosporangia produced by an oblique cell division.

Compsopogon Montagne emend. Rintoul, Sheath *et* Vis 1999

Synonyms: *Pericystis* J.Agardh, *Compsopogonopsis* V.Krishnamurthy

Filaments branched, blue-coloured, uniaxial, corticated except for the growing tips of young branches; cortical cells typically small, spherical to cuboidal, 7–26(–55) μm long; chloroplasts in cortical cells disc-shaped or ribbon-like; axial cells colourless, sometimes deteriorate at maturity; reproduction by monosporangia, formed from cortical cells that undergo oblique division by a curved wall, outer cell becomes the sporangium with length 7.5–23 μm.

In warm, flowing or occasionally standing waters (mostly tropical and subtropical regions).

Probably monospecific (Rintoul et al., 1999).

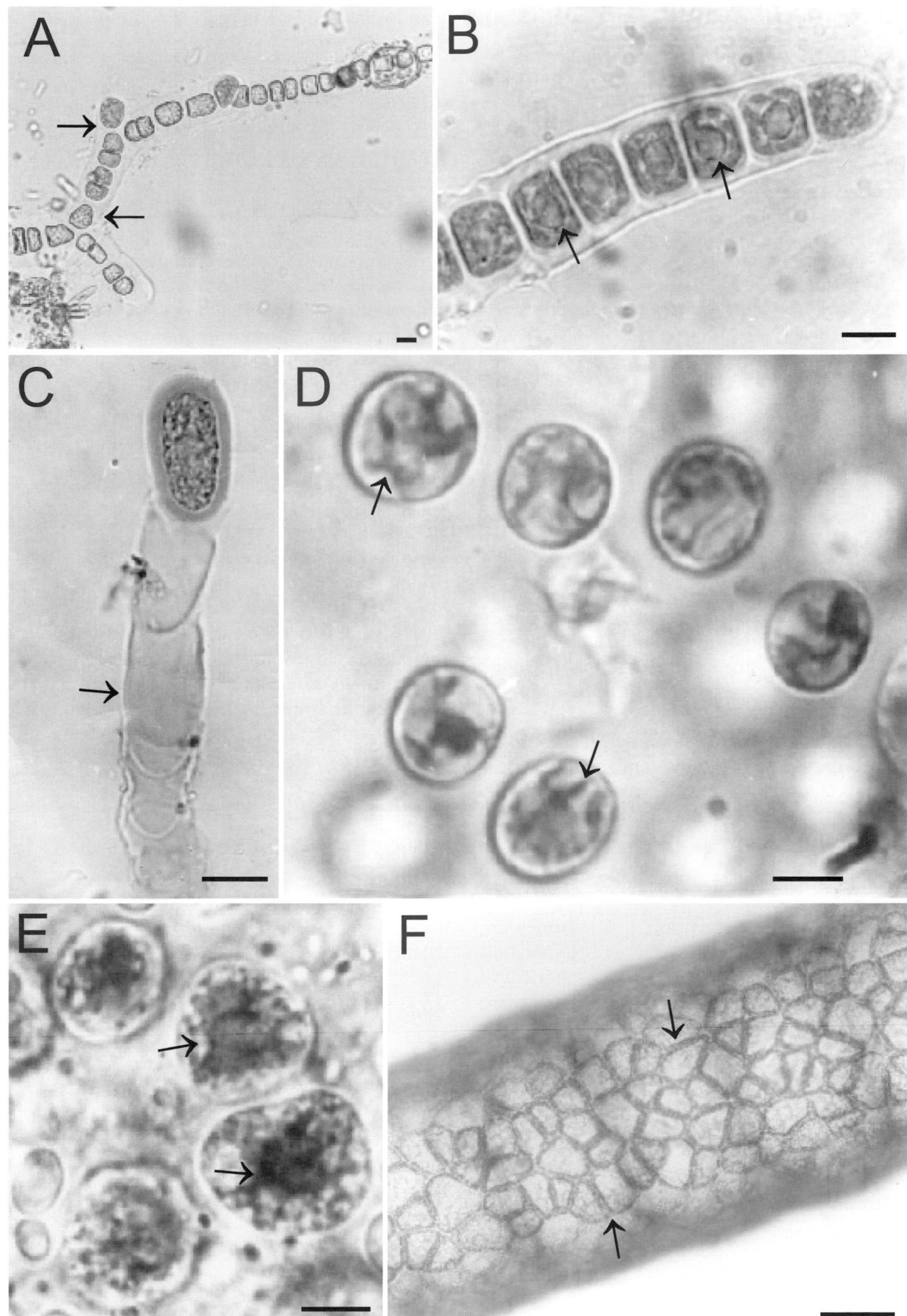

Plate 23 A–F

A, B. *Chroodactylon ornatum* (p. 126): A. Pseudofilamentous habit, showing false branches (arrows), B. Apex of a branch with a chain of cells in a broad, gelatinous matrix; cells with an axial star-shaped chloroplast with pyrenoids visible (arrows); **C**. *Chroothece richteriana* (p. 126): with a stratified, mucilaginous stalk (arrow); **D**. *Porphyridium purpureum* (p. 126): unicells in a broad, mucilaginous matrix, each with a reddish, star-shaped chloroplast (arrows); **E**. *Porphyridium aerugineum* (p. 126): unicells in a broad mucilaginous matrix, each with a bluish, star-shaped chloroplast (arrows); **F**. *Compsopogon coeruleus* (p. 128): uniaxial filament with single layer of small cortical cells (arrows). Scale bars: A–D, 10 µm; E, 4 µm; F, 100 µm.

Compsopogon coeruleus (Balbis) Montagne 1846

Basionym: *Conferva coerulea* Balbis in C. Agardh
Synonyms: *Compsopogon aeruginosus* (J. Agardh) Kützing, *C. chalybeus* Kützing, *C. corinaldii* (Meneghini) Kützing, *C. hookeri* Montagne, *C. lusitanicus* M.P.Reis, *C. oishii* Okamura, *Compsopogonopsis leptocladus* (Montagne) V.Krishnamurthy
03070010 **Pl. 23F, 24A** (pp. 127, 129)
Main axis 120–3000 μm wide, up to 40 cm long, rhizoids at base and occasionally at branching points.

Distribution as for genus; only reported in the British Isles from the Reddish Canal, Manchester (Weiss & Murray, 1909; West & Fritsch, 1927; Swale, 1962a) where it has been proposed to be an introduction. This site was investigated in our survey in 1999 and the canal is largely filled in. Hence, this alga may no longer exist as a permanent population in the British Isles. However, it formed extensive masses in a highly calcareous pond (Double Kettle) near Darlington, County Durham, in one summer, though was not refound in subsequent years (B.A.Whitton, pers. comm.). Almost certainly it was introduced to the pond from an aquarium tank by anglers.

See Vis et al. (1992) and Rintoul et al. (1999) for discussion of synonymies.

ORDER BANGIALES

Filaments with uniaxial bases and multiaxial apices or membranous sheets (blades). Gametophyte and a branched 'conchocelis' sporophytic phase in life history. Pit plugs with one cap layer present only in 'conchocelis' phase. Sexual reproduction only in the marine populations of *Bangia* and *Porphyra* and asexual reproduction by monosporangia.

Bangia Lyngbye 1819

Synonym: *Bangiella* Gaillon
Filaments unbranched, red-coloured, at maturity uniaxial in basal regions and multiaxial in upper portions, attached by basal rhizoid cells; cells spherical or cuboidal; chloroplast axial, star-shaped; asexual reproduction by monospores produced in apical packets and released by digestion of walls; sexual reproduction not observed in freshwater collections.

Cosmopolitan, freshwater, brackish-water and marine intertidal. All freshwater populations can be classified into a single species (Müller et al., 1998).

Bangia atropurpurea (Roth) C.Agardh 1824

Basionym: *Conferva atropurpurea* Roth
Synonyms: probably most of the varieties attributed to this species
03020010 **Pl. 24B–D** (p. 129) **CD**
Filaments 190–380 μm in diameter, up to 15 cm long; cells up to 15 μm in diameter, 13–60 μm long.

Scattered alkaline freshwater habitats; in the British Isles growing often on hard and somewhat shaded surfaces at scattered locations in ion-rich waters, such as the River Lea Navigation Canal, Hertfordshire (Belcher, 1956), the Leeds–Liverpool Canal at Wigan, Lancashire (Reed, 1980), navigable stretches of the River Thames in Surrey (John et al., 1990; our collection), and the River Shannon at Lough Derg in Ireland (Scannell, 1972; our collection). Measurements at the authors' two sites were pH 8–8.7 and 370–540 μS cm^{-1}.

SUBCLASS FLORIDEOPHYCIDAE

ORDER ACROCHAETIALES

Uniaxial, simple branched filaments. Pit plugs with two cap layers, outer layer sometimes dome-shaped and having a membrane. Sexual reproduction triphasic with free-living, isomorphic gametophyte and tetrasporophyte (see Introduction, p. 124). Asexual reproduction by monosporangia.

Audouinella Bory 1823

Synonyms: *Chantransia* (de Candolle) Fries, *Rhodochorton* Nägeli *pro parte*
Simple, branched filaments growing in tufts 2–6 mm in diameter, with erect filaments sometimes arising from a prostrate branching system; cylindrical cells with several parietal, ribbon-like chloroplasts; asexual reproduction by monospores formed as enlarged, ovoid cells at branch tips, singly or in clusters; red-coloured species sometimes have gametophytes with carpogonia, spermatangia or tetrasporophytes with tetrasporangia (see Introduction).

Probably cosmopolitan, widespread in streams and rivers.

See Garbary (1987) for discussion of other potential synonyms.

1 Filaments blue-coloured; reproduce only by monosporangia .... *A. pygmaea*
1 Filaments red-coloured; reproduce by monosporangia, gametangia and tetrasporangia .... *A. hermannii*

Audouinella hermannii (Roth) Duby in de Candolle 1830

Basionym: *Conferva hermannii* Roth
Synonyms: *Audouinella violacea* (Kützing) Hamel, *A. boweri* (G.Murray *et* E.S.Barton) Hamel
03010010 **Pl. 24E,F, 25A,B** (p. 128, 130)
Cells with red-coloured chloroplasts, (7.5–)10–15 μm wide, and 10–45(–60) μm long; monosporangia ovoid, 7.5–12 μm wide, 10–16 μm long; carpogonium basal diameter 5–7 μm, 18–25 μm long, with a thin trichogyne; tetrasporangia ovoid, 6–14.5 μm wide, 9.5–17 μm long.

Probably cosmopolitan; widely reported from streams and rivers in the British Isles (e.g. Cooke, 1882–84; Murray & Barton, 1891; West & Fritsch, 1927; Drew, 1935), but more common as an epiphyte on freshwater rhodophytes belonging to Family Lemaneaceae and bryophytes, especially *Rhynchostegium riparioides*. Conditions in our 21 sites were pH 6.8–8.8, 20–980 μS cm^{-1} and current velocity 11–119 cm s^{-1}.

See Necchi et al. (1993a) for a discussion of other potential synonyms. Measurements are from our survey and those of Drew (1936) and Necchi et al. (1993a).

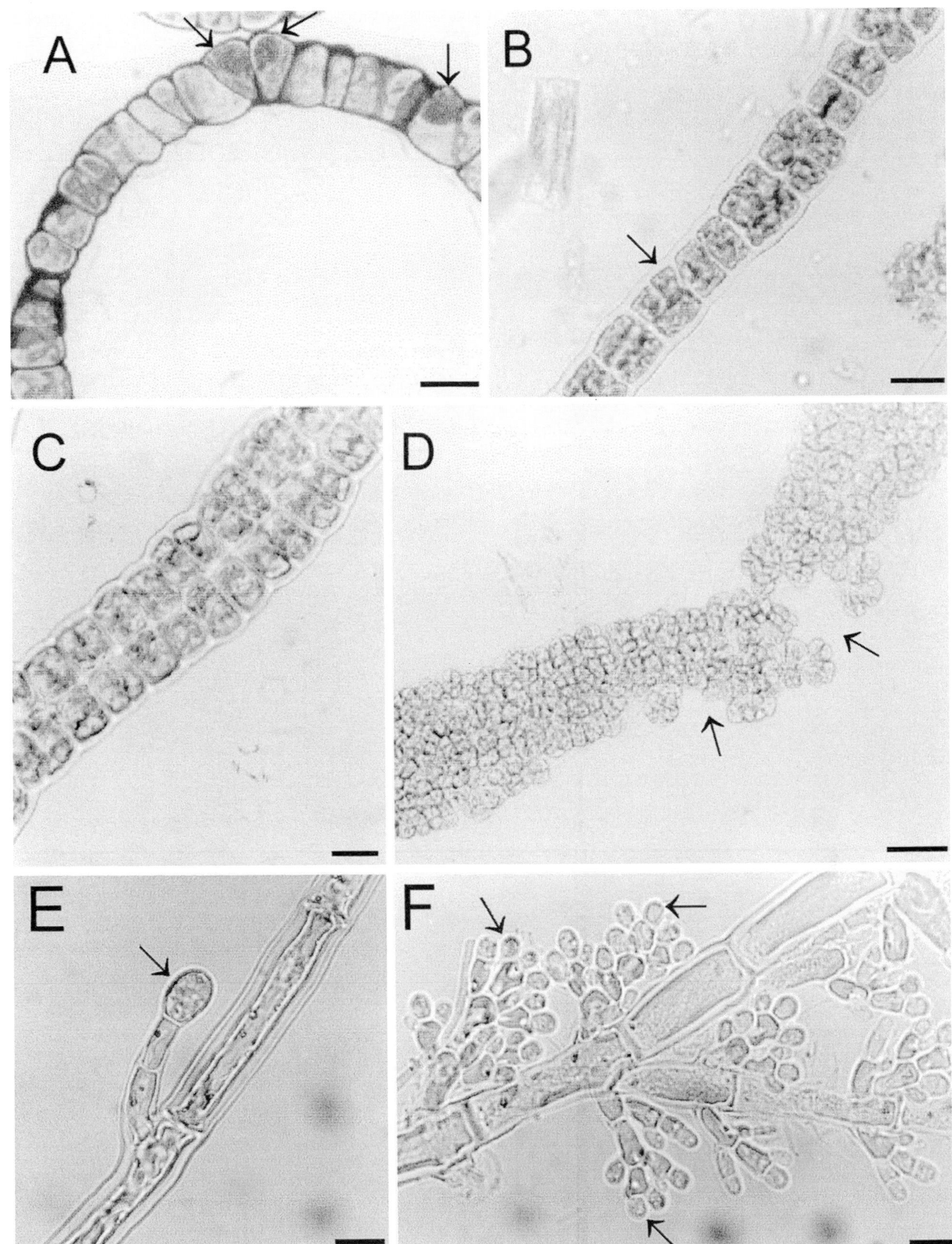

Plate 24 A–F

A. *Compsopogon coeruleus* (p. 128): toluidine blue O-stained transverse section showing layer of cortication and monosporangia (arrows); **B–D**. *Bangia atropurpurea* (p. 128): B. Young filament with uniaxial lower portion (arrow), C. Older, multiaxial portion of filament, D. Release of monospores by deterioration of gelatinous matrix (arrows); **E, F**. *Audouinella hermannii* (p. 128): E. Single monosporangium (arrow) on short lateral branch, F. Spermatangia produced in clusters (arrows) at the tips of short lateral branches.

Scale bars: A–C, E, F, 10 μm; D, 40 μm.

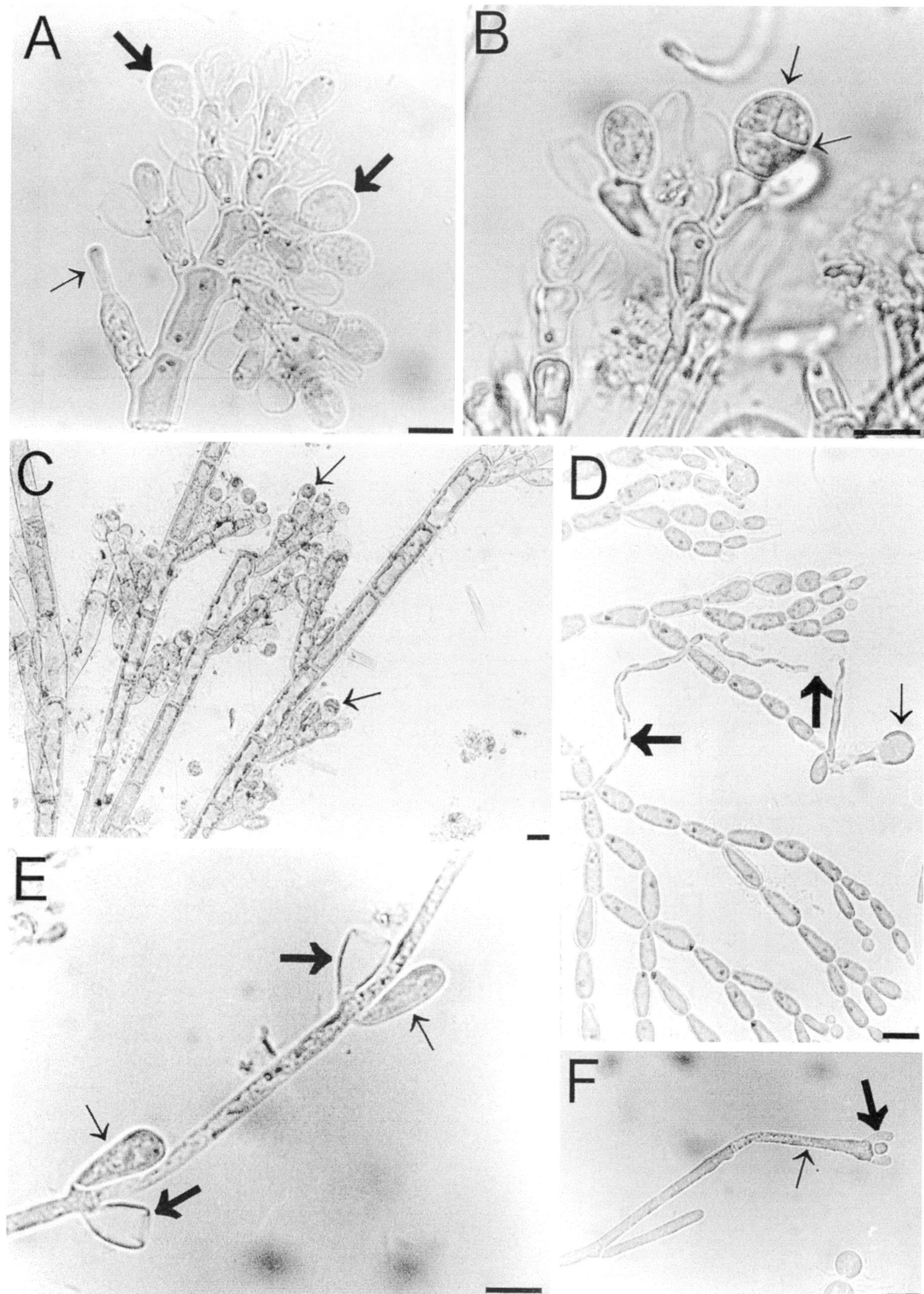

Plate 25 A–F

A, B. *Audouinella hermannii* (p. 128): A. Carpogonium with cylindrical base and narrow trichogyne (small arrow) and carposporophyte with carposporangia (large arrows), B. Tetrasporangium with obvious cleavage planes (arrows); **C**. *Audouinella pygmaea* (p. 131): showing tuft-like growth and apical monosporangia (arrows); **D–F**. *Balbiania investiens* (p. 131): D. Germinating monospore (small arrow) and filament (large arrows) among *Batrachospermum* fascicles, E. Monosporangia (small arrows) and empty sporangia walls (large arrows); F. Specialized cell (small arrow) bearing spermatangia (large arrow).

Scale bars: all 10 μm.

Audouinella pygmaea (Kützing) Weber-van Bosse 1921

Basionym: *Chantransia pygmaea* Kützing
Synonym: *Audouinella leibleinii* (Kützing) T.C.Palmer
03010030 **Pl. 25C** (p. 130)
Cells with blue-coloured chloroplasts, (6.5–)10–15 μm wide, 12.5–38 μm long; monosporangia ovoid, (6.5–)7.5–10 μm wide, (7.5–)10–13 μm long; gametangia or tetrasporangia unknown.

Probably cosmopolitan; only found at a single site in Scotland (River Deveron, west of Huntly, Strathbogie) during our surveys, conditions of which were pH 8.6, 110 μS cm^{-1} and current velocity 100 cm s^{-1}.

This species was included in Cooke's (1890) compilation. It is likely that the reports of *Audouinella chalybea* (Roth) Bory are in fact *A. pygmaea* (see Necchi et al., 1993b), such as those of Cooke (1882–84, 1890) and West (1899b). This and other blue-green coloured species possibly represent 'chantransia' stages of species in the order Batrachospermales (Necchi & Zucchi, 1997).

DOUBTFUL RECORD
Audouinella compacta (as *Chantransia compacta* Ralfs)
Mentioned in Cooke (1882–84, p. 286), but he noted that this taxon is uncertain and it appears to be synonymous with *A. pygmaea*).

ORDER BALBIANIALES

Uniaxial, simple branched filaments. Spermatangia borne on elongate differentiated cells and occurring together with similar starch-filled cells. Pit plugs with two cap layers, outer layer dome-shaped and lacking a membrane. Life history triphasic (see Introduction, p. 124), either with a tetrasporophyte or a 'chantransia' phase, alternating with a gametophyte. Asexual reproduction by monosporangia.

Balbiania (Lenormand) Sirodot 1876

Simple, branched, uniaxial, red-coloured filaments, exclusively epiphytic on species in Family Batrachospermaceae, basal stage sometimes observed; cells cylindrical, 3–5 μm wide, 30–90 μm long; chloroplasts parietal, several in each cell, ribbon-like; asexual reproduction by monosporangia, ovoid, 7–9 μm wide, 6–18 μm long; carpogonia on gametophytes with an ellipsoidal base, about 4 μm wide with a long, thin trichogyne; carpogonium 15–30 μm long; spermatangia in groups of 3–5 on the tips of specialized, elongate cells, some cells in cluster atypical of spermatangia, having plentiful floridean starch; carposporophytes compact, branched filaments, with terminal, ovoid carposporangia 8–9 μm wide, 14–17 μm long; tetrasporophytes infrequent and resemble gametophytes, with tetrasporangia produced on short branches.

Scattered distribution, usually in flowing water. Two currently recognized species, but only one in the northern hemisphere.

Descriptions from Swale & Belcher (1963) and Sheath & Müller (1999).

Balbiania investiens (Lenormand) Sirodot 1876

Basionym: *Chantransia investiens* Lenormand
03030010 **Pl. 25D–F, 26A** (p. 130, 132)
Description as for genus.

Distribution as for genus; known only in small streams from near Ambleside, Westmorland (Swale & Belcher, 1963) and Ireland (our survey at Glengarriff, Kerry: Sheath & Müller, 1999). Conditions at our sites were pH 7.2–7.4, 60–140 μS cm^{-1} and current velocity 37–43 cm s^{-1}.

ORDER BATRACHOSPERMALES

Macroscopic, uniaxial or multiaxial gametophytic filaments; uniaxial forms often producing cortical filaments around the main axis. Pit plugs with two cap layers, outer layer dome-shaped and having a membrane. Life history involving a macroscopic gametophyte phase and a small filamentous 'chantransia' phase that lacks tetrasporangia (see Introduction, p. 124).

Batrachospermum Roth 1797 **CD**

Synonyms: *Gelatinaria* Rousseau, *Charospermum* Link in Nees, *Batrachospermella* Gaillon
Mucilaginous gametophyte filaments, up to 40 cm long, with a beaded appearance, varying from blue-green, olive, violet, grey to brownish, uniaxial with large colourless axial cells, 4–6 pericentral cells produce repeatedly branched fascicles of determinate growth; rhizoid-like cortical filaments typically develop from the lower side of pericentral cells, grow downward and ensheath axial cells, often produce secondary fascicle branches; cells of each fascicle contain several ribbon-like, parietal chloroplasts with no pyrenoids; asexual reproduction by monosporangia (only a few species); sexual reproduction oogamous: spermatangia bud off from terminal primary and secondary fascicle cells or in some species from involucral filaments of the carpogonial branch, each spherical, colourless, 4–8 μm in diameter; carpogonial branch little modified or composed of short, colourless cells; carpogonia with a broad (5–18 μm) trichogyne and sometimes stalked on small base containing the nucleus; carposporophytes generally spherical, compact or of a loose mass of gonimoblast filaments, carposporangia formed at apices; carpospores germinate into a 'chantransia' stage of large basal cells and erect, sparsely branched filaments forming monosporangia or dividing meiotically to produce an attached gametophyte and two residual cells.

Cosmopolitan, in streams and rivers; occasionally in ponds and bogs.

Some species, such as *B. atrum* and *B. gelatinosum*, are broadly distributed in the British Isles (6 or 7 sites in our survey), while *B. boryanum*, *B. gelatinosum* fo. *spermatoinvolucrum* and *B. helminthosum* have only been found in one stream in our survey and the first two taxa are new records for the British Isles.

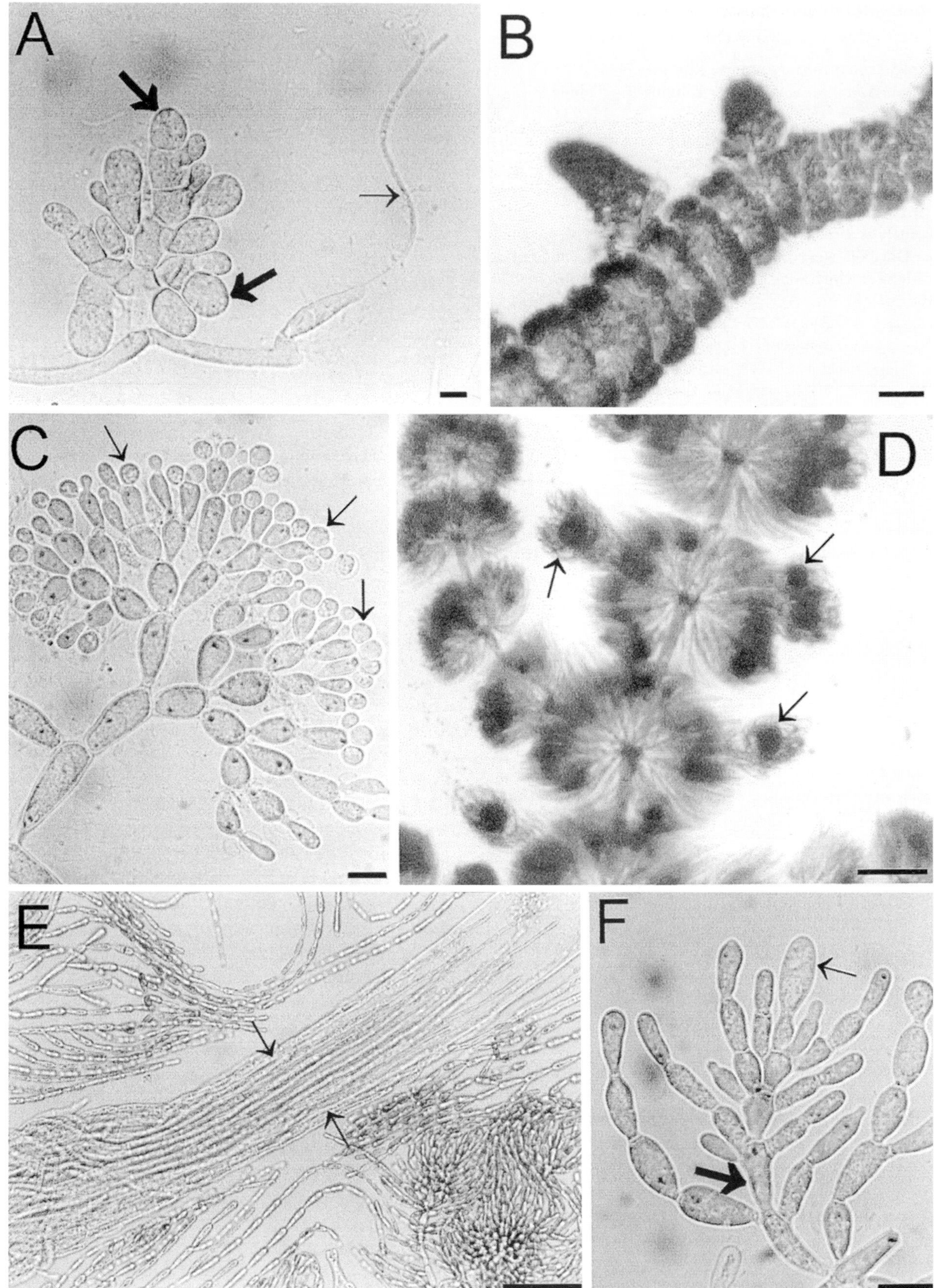

Plate 26 A–F

A. *Balbiania investiens* (p. 131): carpogonium with elongate, narrow trichogyne (small arrow) and carposporophyte with ovoid carposporangia (large arrows); **B–F**. *Batrachospermum arcuatum* (p. 131): B. Male filament with confluent whorls, C. Spermatangia (arrows) at tips of fascicle branches; spermatangia of other *Batrachospermum* species look similar so they are not shown, D. Female filament with carposporophytes evident in the whorls or on extended branches (arrows), E. Cortication consisting of cylindrical cells only (arrows), F. Carpogonium with a club-shaped trichogyne (small arrow) on an undifferentiated carpogonial branch (large arrow).

Scale bars: A, C, F, 10 µm; B, 100 µm; D, 250 µm; E, 75 µm.

1 Mature whorl <170 μm (mean about 120 μm) in diameter; carposporophyte >50% of whorl diameter .... *B. atrum*
1 Mature whorl >170 μm (mean about 500 μm) in diameter; carposporophyte <50% of whorl diameter 2
2 Carposporophytes axial, 1 or 2 per whorl 3
2 Carposporophytes scattered and multiple per whorl 4
3 Carpogonia with stalked trichogynes; cortication a single layer .... *B. helminthosum*
3 Carpogonia with sessile trichogynes; cortication in mature branches of several layers .... *B. turfosum*
4 Main axis with only cylindrical cortical cells 5
4 Main axis with some inflated, bulbous cortical cells 6
5 Filaments monoecious .... *B. gelatinosum*
5 Filaments dioecious .... *B. arcuatum*
6 Monoecious; spermatangia present on involucral filaments .... *B. confusum*
6 Dioecious; spermatangia absent from involucral filaments .... *B. boryanum*

Batrachospermum arcuatum Kylin 1912

03040050 **Pl. 26B–F, 27A** (pp. 132, 134)

Filaments dark blue-green, with confluent, barrel-shaped whorls 250–750 μm in diameter at reproductive maturity, 1–5.5(–15) cm long; main branches with cortication consisting of cylindrical cells only; dioecious; female plants with 1–3 peripheral, spherical carposporophytes, 100–150 μm in diameter, with gonimoblast filaments 2–4-celled; carpogonia 25–39 μm long with club-shaped trichogynes 6–8(–12) μm in diameter; carpogonial branch undifferentiated, 5–9 cells long, with carposporangia ovoid, 7–12 μm in diameter, 10–15 μm long.

Apparently widespread throughout the world, common in continental Europe (Vis et al., 1995); this is a first record for the British Isles where it was found at three sites in England (Blue Pool near Stanford Dingley, Berkshire), Scotland (east of Charlestown of Abelour, Moray), and Wales (Parkmill, Swansea) during our survey. Conditions in our sites were pH 7.2–8.2, 170–980 μS cm^{-1} and current velocity 24–85 cm s^{-1}.

Measurements are from our collections and those of Vis et al. (1995).

Batrachospermum atrum (Hudson) Harvey 1841

Basionym: *Conferva atra* Hudson
Synonyms: *Batrachospermum gallaei* Sirodot, *B. orthostichum* Skuja, *B. sertularia* (Bory) Bory

03040010 **Pl. 27B–D** (p. 134)

Filaments bluish to olive-green, with small whorls, 65–170 μm in diameter and 3–6 fascicle cells at reproductive maturity, 2–6(–10) cm long; main axis with cortication consisting of cylindrical cells only; monoecious; female plants of 1 or 2 axial carposporophytes per whorl, hemispherical, 70–170 μm in diameter, 6–125 μm in height, with carpogonia relatively small and having a sessile, club-shaped trichogyne, 6–10 μm in diameter, 25–30 μm long; carpogonial branch composed of short, colourless cells, carposporangia ovoid, 5.5–8 μm in diameter, 8–12 μm long.

Probably cosmopolitan; first described by Hudson (1778) from Gors Bach in Wales and since then reported by Dillwyn (1809), Harvey (1841), Hassall (1845), Jenner (1845), Landsborough (1851), Cooke (1882–84), Rich (1925) and Heuff & Korkan (1984). We have found it to be widespread in rivers and streams in the British Isles, including those in England, Ireland and Scotland. Conditions at our six collection sites pH 6.8–8.6, 60–460 μS cm^{-1} and current velocity 20–74 cm s^{-1}.

Measurements are from our collections and those of Sheath et al. (1993b).

Batrachospermum boryanum Sirodot 1874

Synonym: *Batrachospermum ectacarpoideum* Skuja *ex* Flint

03040060 **Pl. 27E,F, 28A,B** (pp. 134, 136)

Filaments dark blue to olive-green with confluent, barrel-shaped whorls, 340–1100 μm in diameter at reproductive maturity, 3–6(–15) cm long; main axis with inflated, irregular cortication; dioecious; female plants with 1–30 spherical carposporophytes scattered within the whorl, 70–140 μm in diameter with 2–4 gonimoblast cells, carpogonia 16–35 μm long with an inflated, club-shaped trichogyne, 6–10 μm in diameter; carpogonial branch undifferentiated, 4–8 cells in length, carposporangia ovoid, 7–14 μm in diameter, 8–17 μm long.

Probably widespread in the northern hemisphere; only known from a single locality in the British Isles (Hartsop, Cumbria), conditions at time of collection pH 7.8, 310 μS cm^{-1} and current velocity 8 cm s^{-1}.

This is a new record for the British Isles. Measurements from our sample and those of Vis et al. (1995).

Batrachospermum confusum (Bory) Hassall 1845

Basionym: *Batrachospermum ludibundum* var. *confusum* Bory
Synonyms: *Batrachospermum alpestre* Shuttleworth *et* Hassall, *B. crouanianum* Sirodot, *B. distensum* Kylin, *B. fruticulosum* Drew, *B. helminthosum* Sirodot *nom. illeg. non* Bory, *B. setigerum* Rabenhorst *ex* Sirodot

03040020 **Pl. 28C–F** (p. 136)

Filaments bluish to olive-green, with confluent, barrel-shaped whorls (456–)575–1125 μm in diameter at reproductive maturity, 2.5–6(–10) cm long; main axis with inflated, irregular cortication; monoecious; female plant with 1–19 spherical carposporophytes scattered within whorl, 60–125 μm in diameter with 2–4 gonimoblast cells, carpogonia 14–34 μm long with club-shaped trichogynes 5–11 μm in diameter; carpogonial branch undifferentiated, 4–9 cells long, with involucral filaments having apical spermatangia, carposporangia ovoid, 6–11 μm in diameter, 8–14 μm long.

Probably cosmopolitan; originally reported by Hassall (1845) from Devon, from near Swansea and as *Batrachospermum alpestre* from the River Lea, Hertfordshire. Drew (1946) described a new species, *B. fruticulosum* from Derbyshire but we have synonymized it with *B. confusum* (Vis et al., 1995). In our survey, we collected from three streams in Scotland with the conditions pH 8–8.9, 30–55 μS cm^{-1} and current velocity 20–74 cm s^{-1}.

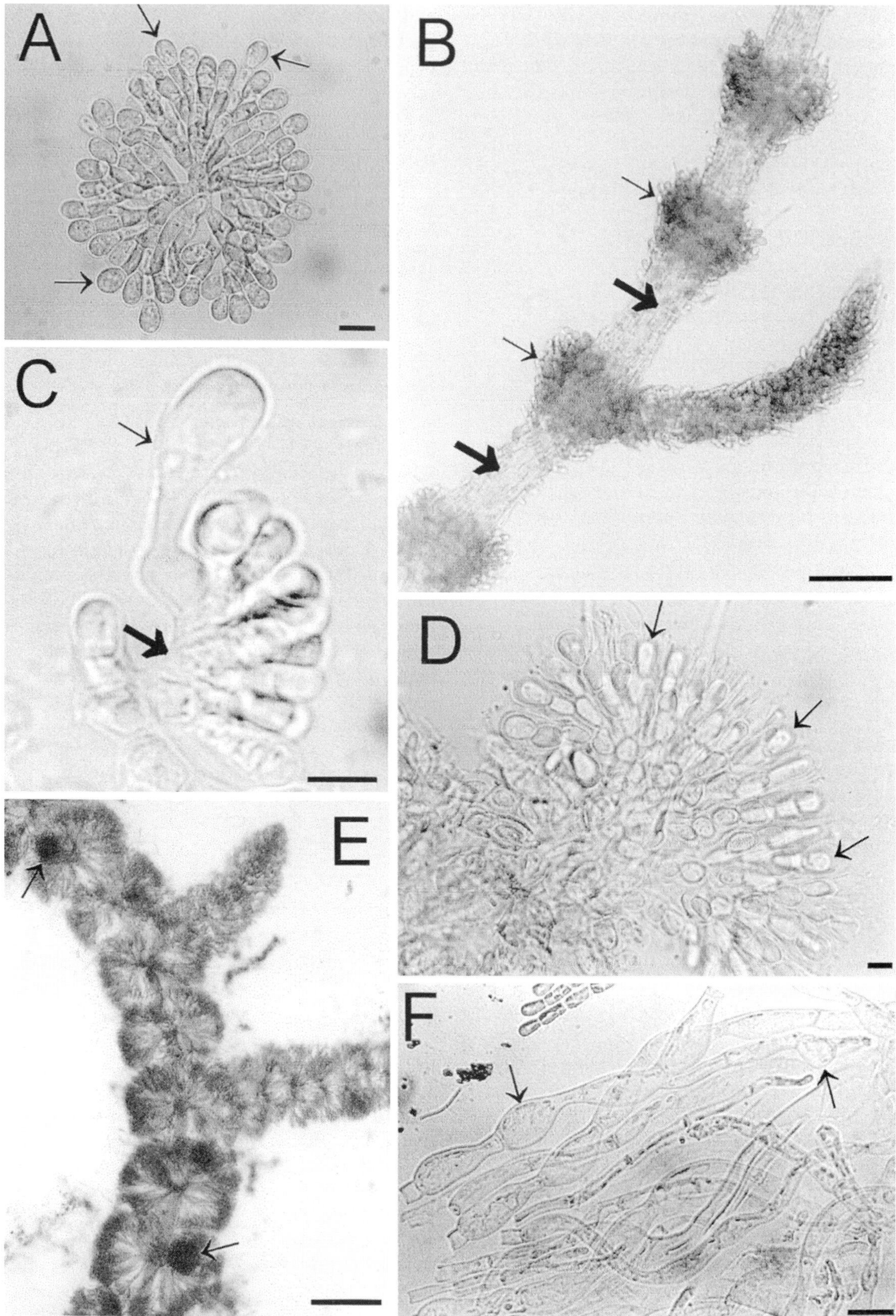

Plate 27 A–F

A. *Batrachospermum arcuatum* (p. 133): carposporophyte with radiating, determinate gonimoblast filaments and apical carposporangia (arrows); **B–D**. *Batrachospermum atrum* (p. 133): B. Filament with small, separated whorls (small arrows) and cortication with cylindrical cells only (large arrows), C. Carpogonium with a club-shaped trichogyne (small arrow) and a differentiated carpogonial branch composed of short, colourless cells (large arrow), D. Carposporophyte with radiating, determinate gonimoblast filaments and apical, ovoid carposporangia (arrows); **E, F**. *Batrachospermum boryanum* (p. 133): E. Female filament with confluent, barrel-shaped whorls and scattered carposporophytes (arrows), F. Cortication containing bulbous cells (arrows).

Scale bars: A, C, D, F, 10 μm; B, 100 μm; E, 500 μm.

See Vis et al. (1995) for discussion of synonymies. Measurements from our collections and those of Vis et al. (1995).

Batrachospermum gelatinosum (Linnaeus) de Candolle 1801

Basionym: *Conferva gelatinosa* Linnaeus
Synonyms: *Batrachospermum corbula* Sirodot, *B. decaisneanum* Sirodot, *B. densum* Sirodot, *B. ludibundum* var. *pulcherrimum* Bory, *B. ludibundum* var. *stagnale* Bory, *B. moniliforme* var. *chlorosum* Sirodot, *B. moniliforme* var. *rubescens* Sirodot, *B. moniliforme* var. *scopula* Sirodot, *B. moniliforme* var. *typicum* Sirodot *nom. inval.*, *B. pygmaeum* Sirodot, *B. pyramidale* Sirodot *nom. illeg.*, *B. radians* Sirodot, *B. reginense* Sirodot and their homotypic synonyms

03040030 **Pl. 29A–D** (p. 138)

Filaments blue, olive-green to brown with spherical or barrel-shaped whorls that are separated to confluent, 250–1000 μm in diameter at reproductive maturity, 1.5–10(–15) cm long; main axis with cortication consisting of cylindrical cells only; monoecious; female plants of 1–5(–11) spherical carposporophytes scattered within the whorl, (40–)70–140 μm in diameter with 2–5 gonimoblast cells per branch, carpogonia (20–)30–68 μm long with club-shaped or occasionally lance-shaped trichogynes 8–17 μm in diameter; carpogonial branch undifferentiated, 3–10 cells long; carposporangia ovoid, 6–12 μm in diameter, 8–16 μm long.

Probably widespread in the northern hemisphere; widely reported from streams and rivers throughout the British Isles (e.g. Cooke, 1882–84; Price, 1914; Goodwin, 1926). We found it in nine streams in England and Ireland with the conditions pH 6–8.4, 40–460 μS cm^{-1} and current velocity 0–74 cm s^{-1}.

See Vis et al. (1995) for a discussion of synonymy. Morphological measurements are from our samples and those of Vis et al. (1995).

fo. *spermatoinvolucrum* Vis *et* Sheath 1998

03040032 **Pl. 29E** (p. 138)

Differs from the type form in having spermatangia on the involucral filaments (Vis & Sheath, 1998).

Known previously from northern North America; in the British Isles, one sample was obtained from the Highlands of Scotland (Loch Cluanie outflow), a new record with the conditions pH 8.2, 20 μS cm^{-1} and current velocity 25 cm s^{-1}.

Batrachospermum helminthosum Bory 1808

Synonyms: *Batrachospermum helminthosum* Sirodot *nom. illeg.*, *B. bruziense* Sirodot, *B. graibussoniense* Sirodot, *B. sirodotii* Skuja *et* Flint, *B. testale* Sirodot, *B. virgatum* Sirodot *nom. illeg.* and *B. viride* Sirodot

03040070 **Pl. 29F, 30A,B** (pp. 138, 139)

Filaments olive-green with spherical or barrel-shaped confluent whorls, 300–800 μm in diameter at reproductive maturity; main axis with cortication consisting of cylindrical cells only; monoecious; female plant of one or two hemispherical, axial carposporophytes, 110–415 μm in diameter and 100–420 μm in height with 2–8 gonimoblast cells per branch, carpogonia 40–69 μm long with a stalk-like, cylindrical to slightly club-shaped trichogyne, 5–14 μm in diameter; carpogonial branch composed of 1–5 short, disc-shaped cells arising from pericentral cell or proximal cell of the fascicle branch; carposporangia ovoid, 5–17 μm in diameter, 10–28 μm long.

Probably widespread in the northern hemisphere; the one previous report of this species from the British Isles is that of Swale & Belcher (1963, as *B. virgatum*) from a small stream near Ambleside in the English Lake District; our sample is from the same site, with the conditions pH 7.4, 140 μS cm^{-1} and current velocity 25 cm s^{-1}.

See Sheath et al. (1994a) for a discussion of synonymies. Measurements are from our collection and those of Sheath et al. (1994a).

Batrachospermum turfosum Bory 1808

Synonyms: *Batrachospermum gulbenkianum* M.P.Reis, *B. keratophytum* Bory, *B. suevorum* Kützing *nom. illeg.*, *B. vagum* (Roth) C.Agardh and its varieties

03040080 (03040040) **Pl. 30C–E** (p. 139)

Filaments blue-green with confluent, indistinct whorls, 170–500(–1400) μm in diameter at reproductive maturity, 3–10(–20) cm long; main axis with cortication consisting of cylindrical cells only, but in multiple layers in mature branches; monoecious; plants often vegetative only, or periodically asexual with ovoid monosporangia, 7.5–11 μm in diameter, 9–12 μm long; monosporangia and gametangia on separate plants and originally thought to be different species (*B. keratophytum* and *B. turfosum*, respectively; see Sheath et al., 1994b); female plants 1 or 2 hemispherical, axial carposporophytes, 85–425 μm in diameter, carpogonia 32–64 μm in length with sessile, club-shaped trichogynes, 7–17 μm in diameter, with infrequent basal protuberances; carpogonial branches composed of 3–10 short, disc-shaped cells arising from pericentral cells or proximal cell of the fascicle branch, carposporangia ovoid, 6.5–12 μm in diameter, 7.5–17 μm long.

Probably widespread in the northern hemisphere; widely reported from streams and rivers in the British Isles; only one collection made by us is at Clifton, Stirling, Scotland, with the conditions pH 8, 30 μS cm^{-1} and current velocity 28 cm s^{-1}.

See Sheath et al. (1994b) and Müller et al. (1997) for a discussion of synonymies. Measurements from our sample and those of Sheath et al. (1994b).

DOUBTFUL RECORDS

Hassall (1845) described several species of *Batrachospermum* which we have included as synonyms of other species (e.g. *B. stagnale*, *B. moniliforme*, *B. proliferum*, *B. rubrum*, *B. pulcherrimum*). The other taxon that Hassall includes, *B. bombusinum* Bory, appears to correspond to *B. atrum* (Hudson) Harvey based on his description and figure (pl. XIII, fig. 3).

Lemanea Bory 1808 **CD**

Synonyms: *Apona* Adanson, *Polysperma* Vaucher, *Vertebaria* Roussel, *Trichogonium* P.Beauvois, *Nodularia* Link *ex* Lyngbye, *Gonycladon* Link, *Lemanella* Gaillon

Thallus (gametophyte) cartilaginous, sparingly to profusely branched, tubular, pseudoparenchymatous,

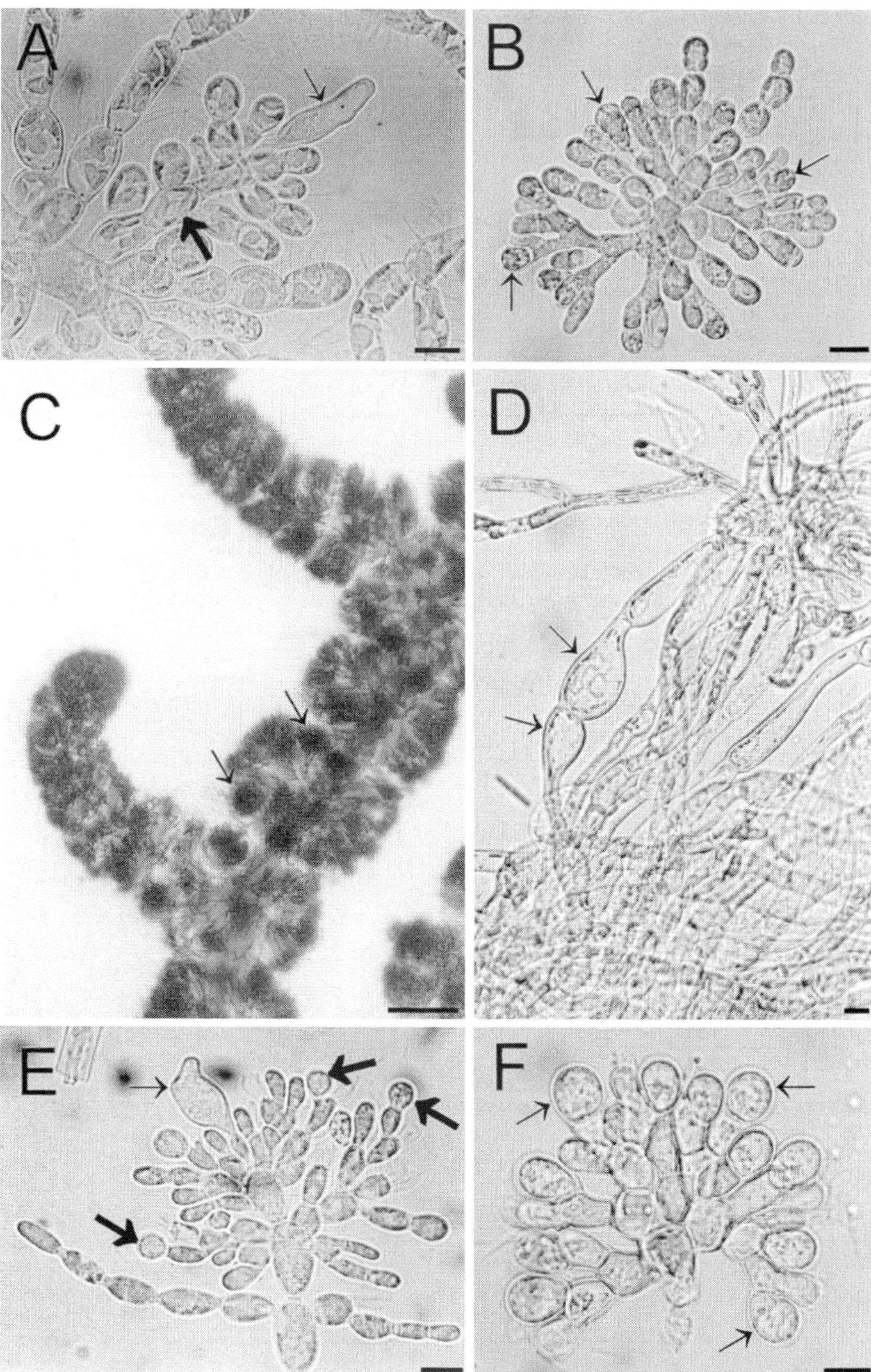

Plate 28 A–F

A, B. *Batrachospermum boryanum* (p. 133): A. Carpogonium with a club-shaped trichogyne (small arrow) on an undifferentiated carpogonial branch (large arrow), B. Carposporophyte with radiating, determinate gonimoblast filaments and apical, ovoid carposporangia (arrows); **C–F.** *Batrachospermum confusum* (p. 133): C. Filament with confluent, barrel-shaped whorls and scattered carposporophytes (arrows), D. Cortication with bulbous cells (arrows), E. Carpogonium with a club-shaped trichogyne (small arrow) on an undifferentiated carpogonial branch which has involucral branches with terminal spermatangia (large arrows), F. Carposporophyte with radiating determinate gonimoblast filaments and apical, ovoid carposporangia (arrows).

Scale bars: A, B, D–F, 10 μm; C, 125 μm.

lacking cortical filaments around central, uniseriate axis, up to 40 cm long and 0.2–2 mm in diameter, blue-green to olive when young, becoming rusty-brown to black at maturity; outer cortex having T- or L-shaped ray cells closely applied to it; chloroplasts parietal, several, disc-shaped, in outer cells only; sexual reproduction with spermatangia developing as yellowish circular patches and nearby small carpogonial branches entirely internal except for a thin trichogyne protruding beyond the outer cell layer; carposporophytes small, spherical masses of filaments forming large, ellipsoidal carpospores in a central cavity, released by thallus deterioration and germinating into the branched, uniseriate filaments of the 'chantransia' stage; 'chantransia' stage produces attached gametophyte seasonally after meiosis.

Probably cosmopolitan in the northern hemisphere; moderate to mostly fast-flowing streams.

Species characterized by quantity of branching, diameter and degree of basal constriction.

1 Relatively few plants per population branched (<50%); than 4 primary branches per plant ....*L. fluviatilis*
1 Most plants per population branched (>50%); more than 4 primary and secondary branches per plant ....*L. fucina*

Lemanea fluviatilis (Linnaeus) C.Agardh 1808

Basionym: *Conferva fluviatilis* Linnaeus
Synonym: *Lemanea corallina* Bory *nom. illeg.*
03100010 **Pl. 31C–F** (p. 141)
Thallus little-branched, 3–23 cm in length, with a maximum diameter (250–)350–800 μm at reproductive maturity, typically stalked with distinct, regular protuberances above stalk.

Probably widespread in the northern hemisphere; recorded throughout the British Isles (e.g. Hassall, 1845; Thwaites, 1851; Cooke, 1882–84; W. & G.S. West, 1900b; Thirb & Benson-Evans, 1982a,b, 1985), particularly in rivers or large streams in England and Wales growing under a wide range of conditions; it is possible that many earlier records should have been attributed to *L. fucina*. Conditions in our 14 collection sites were pH 7.1–8.8, 20–980 μS cm^{-1} and current velocity 31–311 cm s^{-1}.

Measurements are from our collections and those of Starmach (1977).

Lemanea fucina Bory 1808

Synonyms: *Lemanea mamillosa* Kützing, *L. rigida* (Sirodot) De Toni, *L. subtilis* C.Agardh
03100040 **Pl. 32A,B** (p. 142)
Thallus well-branched, tubular, 3–23 cm long, with a maximum diameter 280–700 μm at reproductive maturity, typically stalked with distinct, regular protuberances above stalk.

Probably widespread in the northern hemisphere; widely reported from streams and rivers in the British Isles (e.g. West & Fritsch, 1927), but tends to be more common in Wales and Scotland; our survey confirms this trend, with one sample from England, three from Scotland and seven from Wales with the site conditions pH 6.8–8.8, 30–160 μS cm^{-1} and current velocity 11–82 cm s^{-1}.

See Atkinson (1890) for analysis of synonymies. Measurements are from our collections and those of Starmach (1977).

Paralemanea Vis *et* Sheath 1992

Synonym: *Sacheria* Sirodot, *Entothrix* Wolle
Thallus (gametophyte) cartilaginous, branched, tubular, pseudoparenchymatous, with cortical filaments around a central uniseriate axis and without simple ray cells abutting outer cortical cells, up to 20 cm; outer cortical cells with several parietal, disc-shaped chloroplasts; sexual reproduction by spermatangia developing as yellowish nodal rings; carpogonial branches small and entirely internal except for a thin trichogyne protruding beyond outer layer; carposporophytes small, spherical masses of filaments forming large, ellipsoidal carpospores in a central cavity, released by thallus deterioration and germinating into a branched, uniseriate, filamentous 'chantransia' stage; 'chantransia' stage produces attached gametophyte seasonally after meiosis.

Probably widespread in the northern hemisphere, in flowing-water habitats.

See Vis & Sheath (1992) for a discussion on the basis for establishing this genus. Species characterized by degree of branching, length and diameter.

Paralemanea torulosa (Roth) Sheath *et* Sherwood, *nov. comb.*

Basionym: *Conferva torulosa* Roth 1797 Catalecta Botanica Vol. I: 200.
Synonym: *Lemanea torulosa* (Roth) C.Agardh; there may be several (e.g. some of those in Vis & Sheath, 1992) but no recent systematic study has included *P. torulosa*
03140010 **Pl. 32C–E** (p. 142)
Thallus (gametophyte) unbranched, pseudoparenchymatous, tubular, 3–10(–15) cm long, with a maximum diameter 500–650(–1000) μm at reproductive maturity, with no obvious stalk but having bands of spermatangia at regular intervals in regions of thallus swelling.

Probably widespread in the northern hemisphere; reported from scattered rivers and streams in the British Isles by Harvey (1841), Hassall (1845), Cooke (1882–84), W. & G.S. West (1900b), Rosenberg (1935), among others. Our collection is from Llangenny, Powys, Wales, with site conditions 200 μS cm^{-1} and current velocity 105 cm s^{-1}.

Sirodotia Kylin 1912

Mucilaginous gametophytic filaments, up to 17 cm long, with a beaded appearance, varying from blue-green to yellow-green in colour; central axis uniseriate with large, cylindrical cells; pericentral cells 4–6, producing repeatedly branched fascicles of limited growth, commonly rhizoid-like cortical filaments growing from lower side of pericentral cells of fascicles and each containing several parietal, ribbon-like chloroplasts with

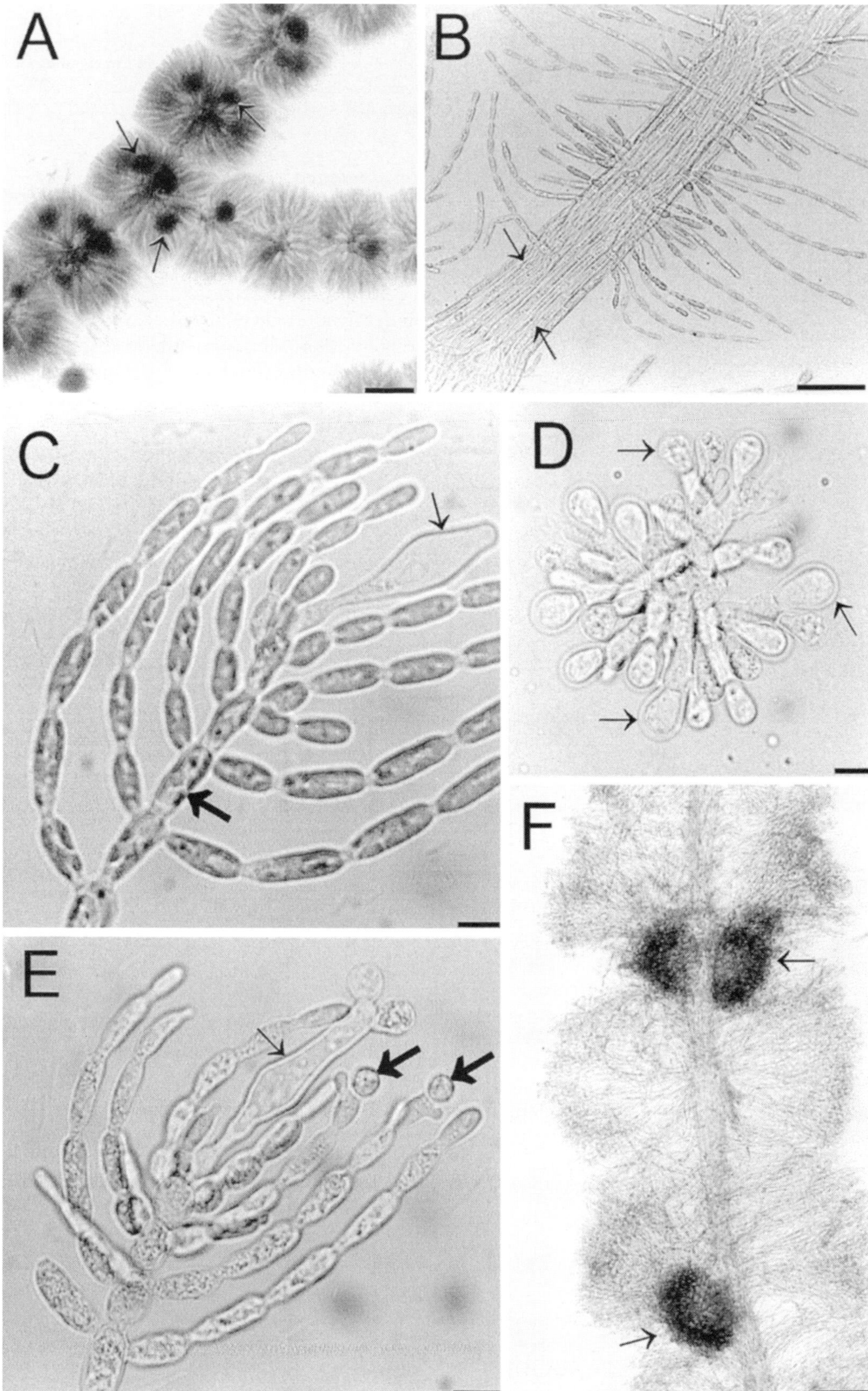

Plate 29 A–F

A–D. *Batrachospermum gelatinosum* (p. 135): A. Filament with barrel-shaped whorls and plentiful carposporophytes (arrows), B. Cortication with cylindrical cells only (arrows), C. Carpogonium with a club-shaped trichogyne (small arrow) and a small base on an undifferentiated carpogonial branch (large arrow), D. Carposporophyte with radiate, determinate gonimoblast filaments and apical, ovoid carposporangia (arrows); **E**. *Batrachospermum gelatinosum* fo. *spermatoinvolucrum* (p. 135): with a carpogonium having an elongate, club-shaped, trichogyne (small arrow) and an undifferentiated carpogonial branch with spermatangia at the tips of the involucral branches (large arrows); **F**. *Batrachospermum helminthosum* (p. 135): filament with globose whorls and large, axial carposporophytes (arrows). Scale bars: A, 400 μm; B, 75 μm; C–E, 10 μm; F, 100 μm.

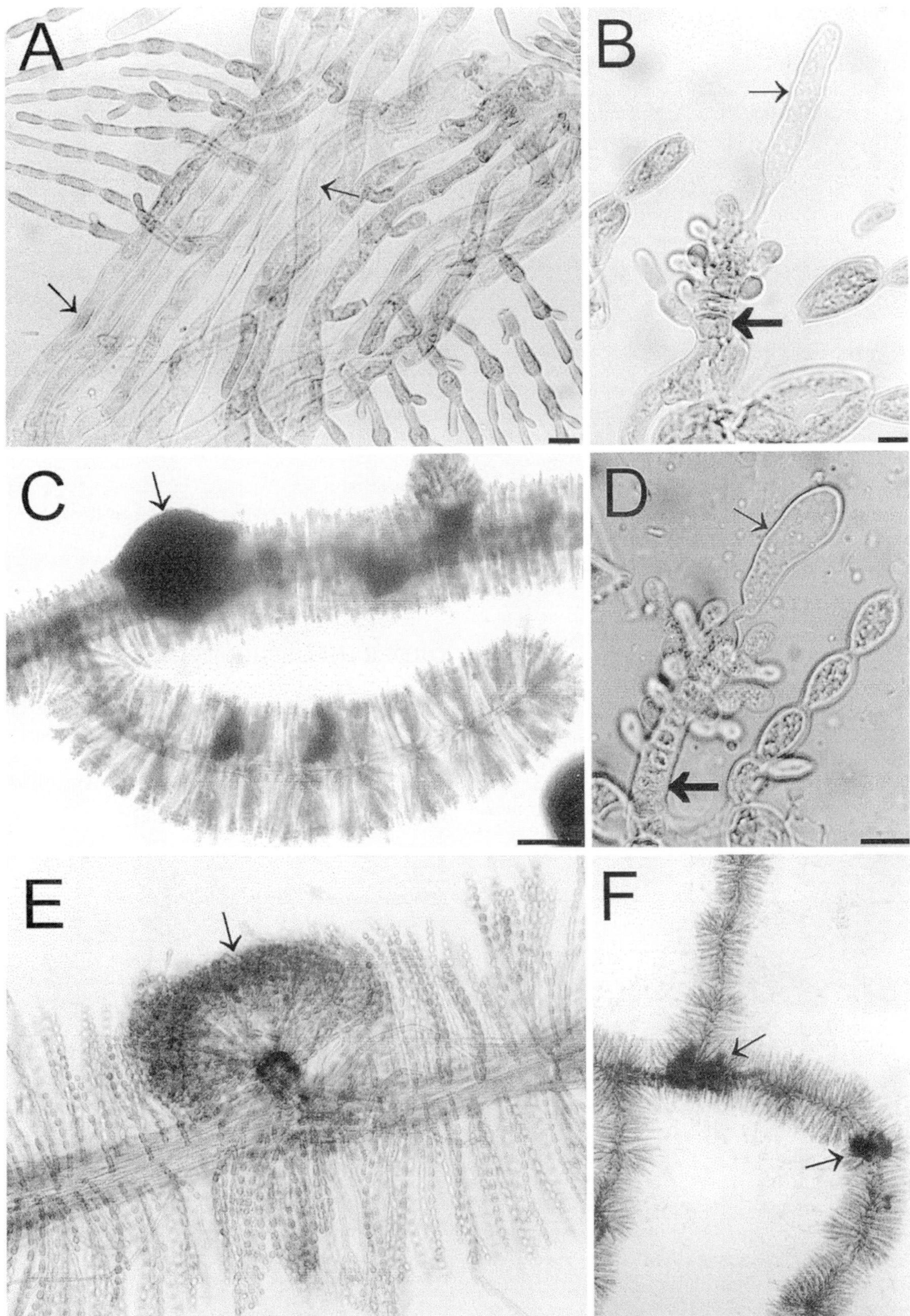

Plate 30 A–F

A, B. *Batrachospermum helminthosum* (p. 135): A. Cortication composed of cylindrical cells only (arrows), B. Carpogonium with an elongate, cylindrical, stalked trichogyne (small arrow) on a differentiated carpogonial branch composed of small, colourless cells (large arrow); **C–E**. *Batrachospermum turfosum* (p. 135): C. Filament with confluent, indistinct whorls with large, hemispherical carposporophyte (arrow), D. Carpogonium with club-shaped, sessile trichogyne (small arrow) on a differentiated carpogonial branch with small, colourless cells (large arrow), E. Axial carposporophyte with radiating, determinate gonimoblast filaments (arrow); **F**. *Sirodotia suecica* (p. 140): filament with truncate-pyramidal whorls and axial, indeterminate carposporophytes (arrows).

Scale bars: A, B, D, 10 μm; C, E, 100 μm; F, 300 μm.

no pyrenoids; sexual reproduction by spermatangia budding off from terminal fascicle cells, spherical, colourless, 4–7 μm in diameter; carpogonial branches somewhat differentiated with small cells, carpogonia each with a broad trichogyne attached off-centre to base, the latter structure having a definite protrusion; carposporophyte an indeterminate, branched filament creeping along main axis, carposporangia formed at branch apices singly or in clusters; carpospores germinate into a 'chantransia' stage composed of branched, uniserate filaments; meiosis and monosporangia not observed.

Probably cosmopolitan in flowing-waters.

Sirodotia suecica Kylin 1912

Synonyms: *Sirodotia acuminata* Skuja *et* Flint, *S. fennica* Skuja, *S. tenuissima* (Collins) Skuja *ex* Flint

03130010 **Pl. 30F, 31A,B** (pp. 139, 141)

Filaments having obconical to truncate-pyramidal whorls with plentiful secondary fascicles between them, bright blue-green in colour and whorl diameter at reproductive maturity (230–)260–450(–660) μm, 1.5–4(–10) cm long; monoecious; carpogonia 19–42 μm long, with a cylindrical to club-shaped, sessile trichogyne (4–)8–10 μm in diameter; carpogonial branch of 3–5 short cells; carposporophytes produce carposporangia on short lateral branches, carpospores ovoid, 7.5–10 μm in diameter, 8–14 μm long.

Probably cosmopolitan; in the British Isles known only from a single collection made by us in the Highlands of Scotland (Loch Cluanie outflow), with site conditions pH 8.2, 20 μS cm^{-1} and current velocity 25 cm s^{-1}.

This is a new record for the British Isles. See Necchi et al. (1993c) and Vis & Sheath (1999) for discussion of synonymy. Measurements are from our sample and those of Necchi et al. (1993c).

Thorea Bory 1808

Synonyms: *Polycoma* P.Beauvois, *Thorella* B.Gaillon *non* J.Briquet

Branched gametophytic filaments up to 20–200 cm long, 0.5–3 mm in diameter, olive-green, dark brown, reddish to black, composed of interwoven, colourless medullary filaments and dense, photosynthetic laterals of limited growth; chloroplasts in assimilatory filaments parietal, ribbon-like; asexual reproduction by monosporangia solitary or in clusters, formed at base of assimilatory filaments (spore-bearing branch-to-vegetative lateral length ratio <0.3); sexual reproduction rarely reported; spermatangia borne on specialized branches near base of assimilatory filaments, colourless, ellipsoidal or ovoid, 8–10×4–7 μm; carpogonia conical with elongate trichogyne 5–7 μm in diameter, carpogonial branch short and at base of assimilatory filaments; carposporophyte sparsely branched and compact, carposporangia terminal, 9–13 μm in diameter, 17–25 μm long; carpospores germinate into a branched, uniaxial filamentous 'chantransia' stage.

Scattered distribution in warm flowing-water habitats.

All British collections can be attributed to a single species.

Thorea hispida (Thore) Desvaux 1818

Basionym: *Conferva hispida* Thore

Synonyms: *Thorea andina* Lagerheim *et* K.Möbius, *T. lehmannii* Horneman, *T. ramosissima* Bory *nom. illeg.*

03120010 **Pl. 32F, 33A** (pp. 142, 143) **CD**

Branched, multiaxial filament with more than 10 primary and/or secondary branches per 30 mm, 500–400(–1900) μm in diameter and up to 1 m long; medullary region up to 600 μm broad; assimilatory, determinate lateral branches cylindrical with variable branching; monosporangia on short, basal branches, ovoid in shape, 9–30 μm long; gametangia not observed in specimens from the British Isles.

Widespread throughout the world; within British Isles known only from England and collected from several locations on the Rivers Thames, Lea and Great Ouse where of erratic appearance, with the large gametophytic plants absent for several years before reappearing again; always best developed where water movement is most rapid such as immediately below an open sluice gate (e.g. Cookham on River Thames). One record from a bog in the County Donegal Mountains in Ireland seems to have been in error (see John et al., 1989b).

See Sheath et al. (1993a) for a discussion of synonyms. Morphological measurements are those of Starmach (1977) and John et al. (1989a, b).

ORDER HILDENBRANDIALES

Pseudoparenchymatous crusts with or without an attached erect thallus. Pit plugs with one cap layer. Reproduction appears to be exclusively asexual by tetrasporangia, fragmentation, specialized propagules (gemmae) or stolon-like structures.

Hildenbrandia Nardo 1834 *orth. cons.* (=*Hildenbrandtia*)

Crustose thalli forming distinct, reddish patches closely adherent to rock substrata; cells cuboidal to polygonal in vertical files typically arising from a basal layer; reproduction in freshwater populations by gemmae, dense aggregations of cells formed in the thallus; marine populations in contrast form tetrasporangia.

Cosmopolitan in alkaline rivers and streams as well as in marine habitats.

Hildenbrandia rivularis (Liebmann) J.Agardh 1851

Basionym: *Erythroclathrus rivularis* Liebmann

03090010 **Pl. 33B,C** (p. 143) **CD**

Crusts in section composed of files of (5–)7–11(–16) cells, 40–70(–100) μm long, sometimes branched near apices; cells 5–8(–10) μm in diameter; basal layer 15–24 μm thick; gemmae occasionally observed in the British Isles.

Probably cosmopolitan in Europe; widely reported in streams and rivers in the British Isles (e.g. Cooke, 1882–84; West & Fritsch, 1927; Sherwood & Sheath, 2000) where it grows over stable rock surfaces provided they are not too overgrown by

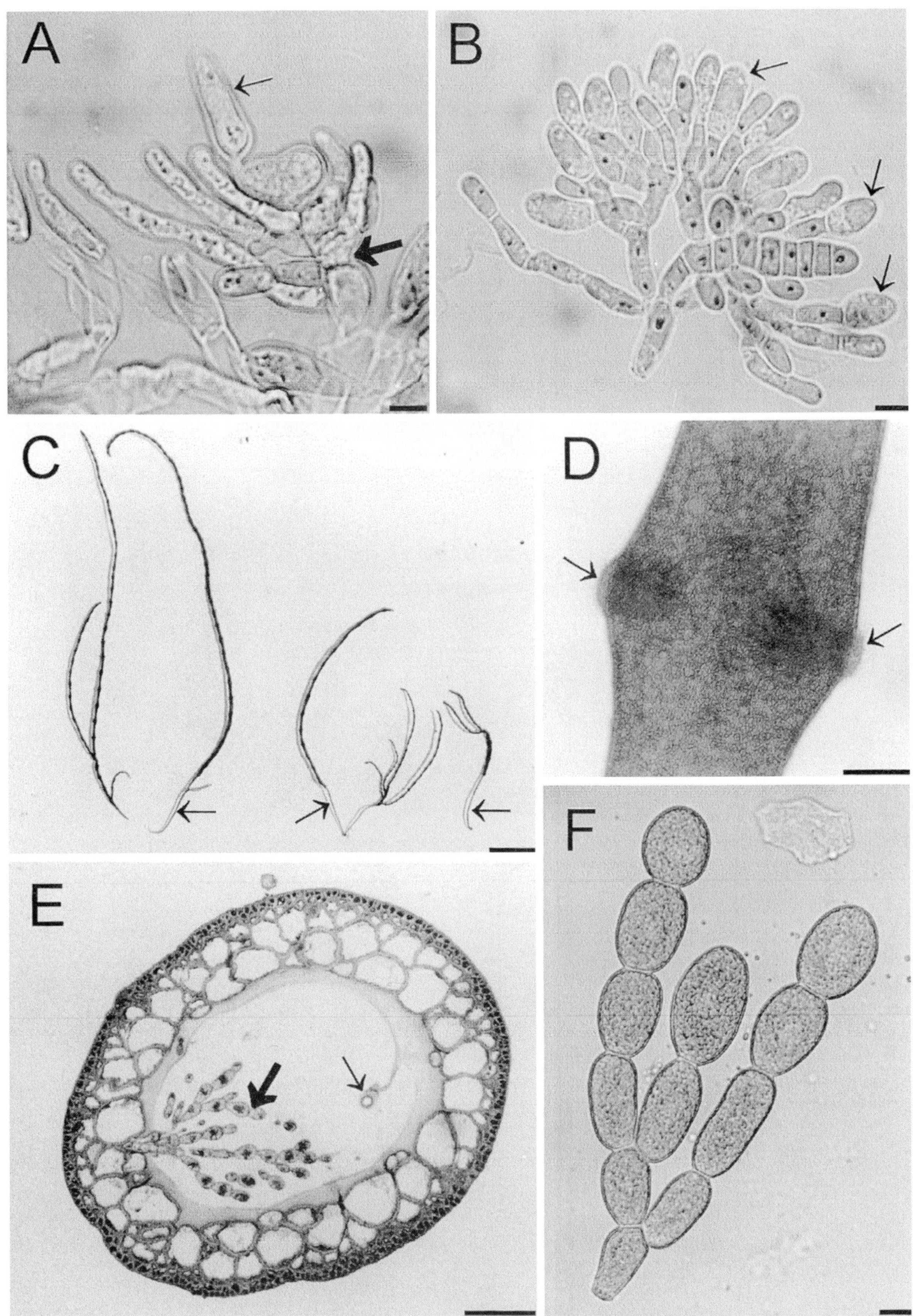

Plate 31 A–F

A, B. *Sirodotia suecica* (p. 140): A. Carpogonium with a cylindrical trichogyne (small arrow) attached off-centre to the base; carpogonial branch differentiated with colourless, short cells (large arrow), B. Carposporophyte lateral branch with a cluster of apical carposporangia (arrows); **C–F**. *Lemanea fluviatilis* (p. 137): C. Tubular, pseudoparenchymatous thalli which have primary branches, basal stalks (arrows) and regular protuberances, D. Close-up view of protuberance with spermatangial patches (arrows), E. Toluidine blue O-stained transverse section showing outer cortex layer, axial filament with no inner cortication (small arrow) and carposporophyte (large arrow), F. Chain of carpospores.

Scale bars: A, B, F, 10 μm; C, 1 cm; D, 200 μm; E, 100 μm.

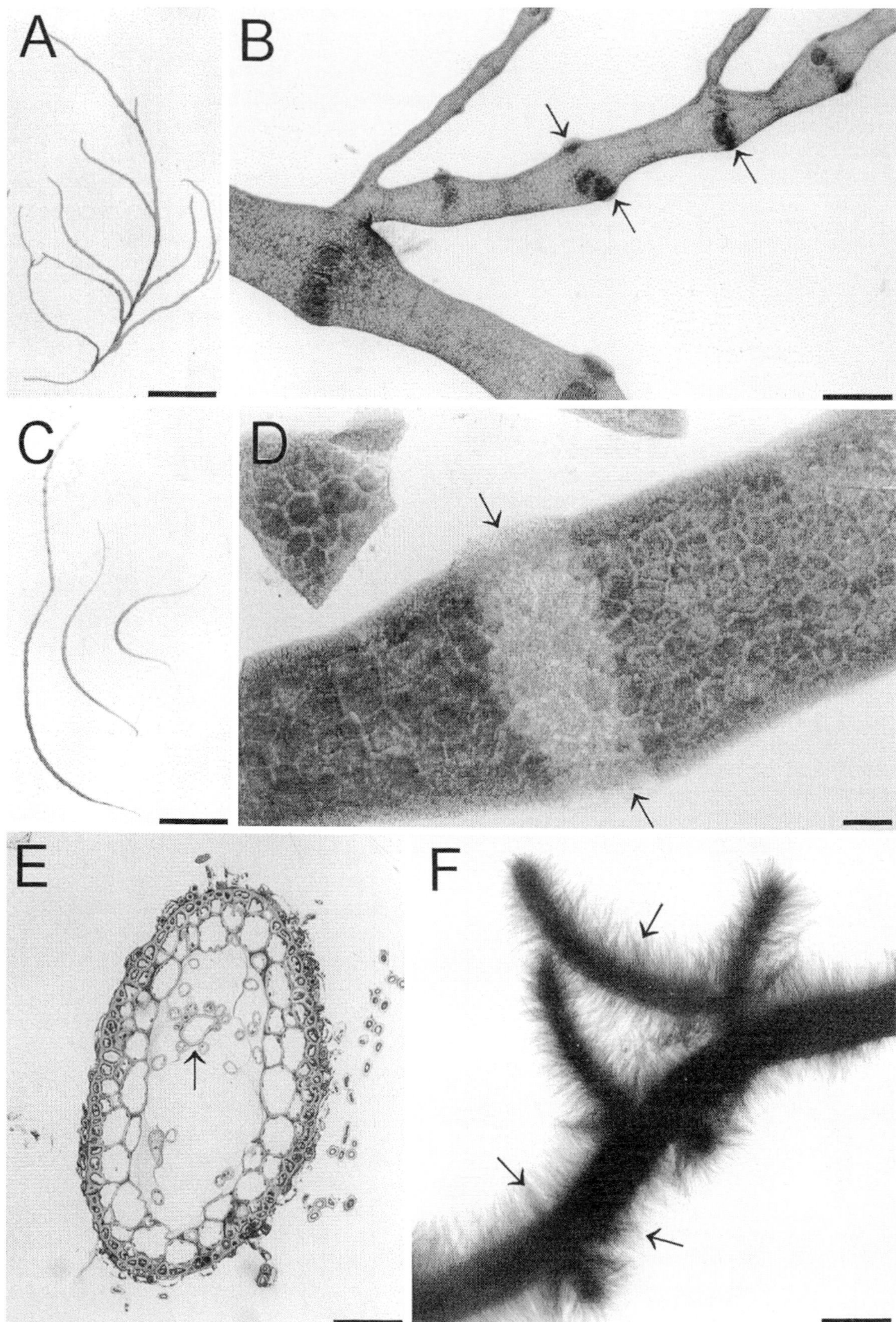

Plate 32 A–F

A, B. *Lemanea fucina* (p. 137): A. Tubular, pseudoparenchymatous thallus which has plentiful primary and secondary branches, B. Close-up view of branches and spermatangia patches on the protuberances (arrows); **C–E**. *Paralemanea torulosa* (p. 137): C. Tubular, pseudo-parenchymatous thalli that are unbranched and unstalked, D. Close-up of thallus showing distinctive ring of spermatangia (arrows), E. Toluidine blue O-stained transverse section, showing outer cortex layer and axial filament surrounded by inner cortical filaments (arrow); **F**. *Thorea hispida* (p. 140): branched thallus with determinate assimilatory filaments (arrows).

Scale bars: A, C, 1 cm; B, 500 μm; D, E, 100 μm; F, 500 μm.

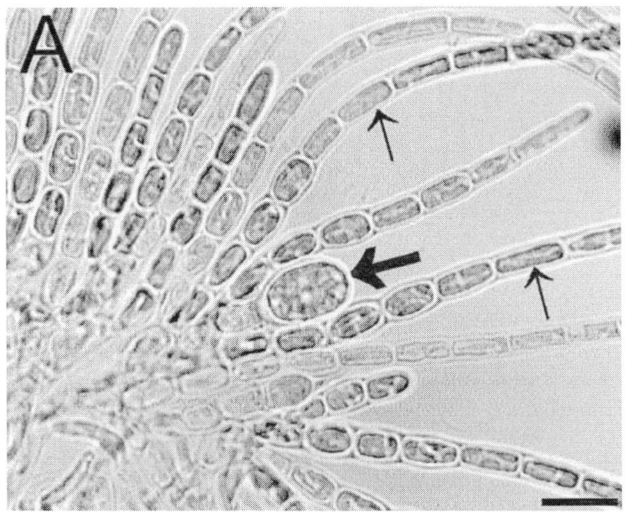

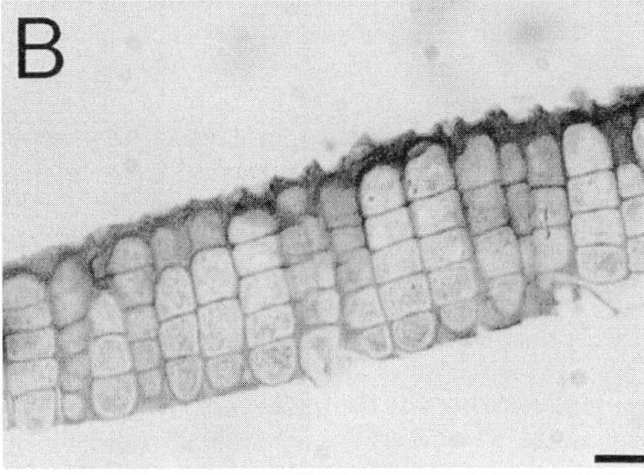

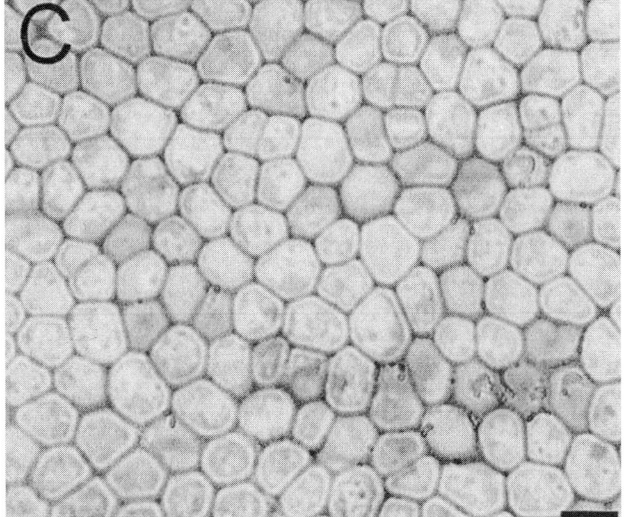

Plate 33 A–C
A. *Thorea hispida* (p. 140): part of multiaxial filament with central medulla and outer assimilatory filaments (small arrows); monosporangium (large arrow) on small basal branch; **B, C.** *Hildenbrandia rivularis* (p. 140): B. Transverse section showing vertical files of cells, C. Surface view showing polygonal-shaped apical cells of each file.
Scale bars: 10 μm.

mosses and other cryptogams or covered by mineral encrustations; collected by us in two streams in England, three in Ireland and three in Wales having the conditions pH 7.2–8.8, 70–460 μS cm^{-1} and current velocity 24–96 cm s^{-1}.

Acknowledgements
The authors thank the British Phycological Society for the invitation to write this chapter and for providing some of the travel funds necessary to do the survey. Other research funding was provided by NSERC grant 0183503 to RGS, NSERC scholarship A & B to ARS and the University of Guelph. Assistance in manuscript preparation from T. Pellizzari and collecting from D.M. John, L.R. Johnson, M. Koske, K. Müller and B.A. Whitton is greatly appreciated.

PHYLUM EUGLENOPHYTA

Konrad Wołowski

The euglenophytes or euglenoids contain about 44 free-living genera and more than 800 species. Most are colourless with phagotrophic or heterotrophic modes of nutrition, but about one-third are green and phototrophic. They are usually placed in the Phylum Euglenophyta by botanists who recognize a single class having two (Bourrelly, 1985), three (Asaul, 1975) or six (Leedale, 1967) orders. Zoologists place the euglenophytes in the Protozoa and divide the Order Euglenida into six suborders. Ultrastructural surveys of cell structure and subsequent molecular investigations of euglenophytes have revealed a close relationship to a group of flagellate protozoa.

The euglenophytes are a diverse group of largely solitary and motile flagellates, though a few species form non-motile, branched colonies when in the actively growing condition and, more rarely, they are epiphytes (e.g. *Colacium*). The cells are ovoid to spindle- or lance-shaped and most are naked with the plasma membrane surrounded by an often prominent pellicle with helical striations. Some forms are radially symmetrical (*Gyropaigne kosmos*), others are bilaterally symmetrical. The pellicle ranges from very flexible to semi- or completely rigid, and is usually striated. The striations are frequently twisted with respect to the longitudinal axis. There are muciferous bodies under the pellicle which eject mucilage that forms layers around the cell.

The genera *Trachelomonas*, *Strombomonas* and *Ascoglena* are surrounded by an extracellular matrix, termed a lorica, which often is impregnated by deposits of ferric and manganese compounds (Leedale, 1975; Dunlap et al., 1983). At the anterior end a narrow canal leads into a flask-shaped reservoir, the anterior invagination. Adjacent to the enlarged basal portion are one or more active contractile vacuoles. In certain genera, there may be a rod-shaped organelle parallel to the reservoir. The canal is a continuation of the cytoplasmic membrane. At higher magnification, microtubules are often seen to be in association with the canal. The contractile vacuole may be almost spherical before discharge into the reservoir, and is frequently reported as being surrounded by several 'subsidiary' vacuoles.

The flagella are inserted in the latero-posterior wall of the reservoir. There are usually two flagella, commonly one emergent and the second reduced and not emerging, although there may be one to several, each often associated with a swelling that is frequently referred to as the basal body. The red-orange eyespot (stigma) and photoreceptor together with the emergent flagellum form the photosensory transduction system, located on the surface of the reservoir (Walne & Kivic, 1989).

The chloroplasts are dark green, usually several in a cell. They vary in shape and may be disc-, plate-, star- or ribbon-shaped. Like the chlorophytes, euglenophytes contain both chlorophyll *a* and *b* (also carotene and several xanthophylls) in chloroplasts whose shape provides a character of taxonomic importance. Pyrenoids, when present, may be within the chloroplasts as a 'naked pyrenoid' or a 'double-sheathed pyrenoid', but are more often outside them. They vary in form and are sometimes, but not always, associated with cytoplasmic plates or caps of paramylon. Paramylon bodies consist of a β-1,3-linked glucan storage product (reserve photosynthate) and come in a large variety of shapes and sizes (disc-, rod- or ring-like, linked or granular). They are frequently near the pyrenoids, but neither are necessarily close to the chloroplasts. Sometimes the shape of paramylon bodies is used as a characteristic and diagnostic feature. The nucleus lies in the central or posterior region of the cell in all species but during euglenoid movement it may move. Usually the nucleus is spherical to ovoid in shape.

Euglenoid reproduction is asexual and by longitudinal division of motile or non-motile cells during the night. A sporadic meiosis that occurs in some euglenophytes is generally believed to be the result of non-sexual autogamy (Leedale, 1967).

Movement is by swimming brought about by helical undulations of the emergent flagella combined with a twisting of the cell (pellicle permitting; however, see *Euglena mutabilis*). This movement tends to be rotational and spiral, with, to a greater or lesser extent, a wobbling motion. Most euglenoid species are able to bend or contract the body to a greater or lesser degree; this type of movement is variously known as contractile body movement, euglenoid movement or 'metaboly'.

Most genera occur in water bodies, occupying a great diversity of ecological niches including freshwater, brackish water, acid and alkaline waters, aerobic and anaerobic (anoxic) conditions and water rich in decaying organic matter. Euglenoids are occasionally common enough to colour water bodies, causing green (*Euglena viridis*), red (*E. sanguinea*) or brown (*Trachelomonas*) blooms.

The text contains information about some of the green genera belonging to the orders Eutreptiales and Euglenales. All brackish-water, marine and colourless representatives are excluded. Information on the distribution and ecology of euglenophytes in the British Isles is scarce and therefore ecological remarks under individual entries refer to what is generally known from other parts of their distributional range unless stated otherwise.

The current classification of the Euglenophyta is based on variations within the cell architecture. Six orders are accepted here (after Leedale, 1967): Eutreptiales, Euglenales, Euglenamorphales, Rhabdomonadales, Sphenomonadales, Heteronematales. The first three orders contain both green and colourless species, the remainder have only heterotrophic species. This text covers the Eutreptiales with the genus *Eutreptia*, and the Euglenales with the genera *Euglena, Trachelomonas, Strombomonas, Colacium, Lepocinclis* and *Phacus*.

1 Cells with 2 emergent flagella of equal length .... *Eutreptia* (p. 145)
1 Cells with only one emergent flagellum 2
2 Cells attached to microfauna by gelatinous stalk, usually as colonies .... *Colacium* (p. 145)
2 Cells free-swimming, solitary and of another form 3
3 Cells naked, cylindrical to spindle-shaped, contractile body movement violent .... *Euglena* (p. 146)
3 Cells of different shape 4
4 Cells enclosed in a lorica (tests), movement euglenoid 5
4 Cells naked, flattened, rigid movement not euglenoid 6
5 Cells enclosed in a buff or brown-coloured lorica, regularly ornamented .... *Trachelomonas* (p. 169)
5 Cells enclosed in lorica, irregularly ornamented, wall usually rough .... *Strombomonas* (p. 168)
6 Cells flattened dorsiventrally, usually spirally twisted in at least part of the cell; paramylon bodies variable in number .... *Phacus* (p. 161)
6 Cells broadly ovoid, pear-shaped or ellipsoidal, usually with a short 'tail'; paramylon bodies 2 and large .... *Lepocinclis* (p. 158)

ORDER EUTREPTIALES

Cells having two emergent flagella of equal length and which undergo very active euglenoid movement. Contains the genera *Eutreptia, Eutreptiella, Distigma* and *Distigmopsis*, of which only *Eutreptia* occurs in the British Isles.

Eutreptia Perty 1852

Cells solitary, oblong, broadly spindle or egg-shaped, with very active euglenoid movement, free-swimming and rotating along axis of cell, sometimes creeping; pellicle slightly striated; flagella 2, emergent; eyespot present; chloroplasts disc- or ribbon-shaped with no pyrenoids; cysts observed.

Widespread but not common, in fresh and marine waters; usually in small and organically polluted water bodies, along the margins of lakes, ponds, puddles and peat bogs, also in sulphuric springs.

Eutreptia viridis Perty 1852

04030010 **Pl. 34C** (p. 149)

Cells 13–25 µm wide, 49–66 µm long, broadly spindle-shaped; anterior end rounded and posterior end narrowed, with a tail-piece; chloroplasts small, disc-like, numerous; paramylon bodies small, round or flat, cylindrical; eyespot large; flagella both emergent, as long as cell or a little shorter; pellicle thin, spirally striated; contractile body movement very rapid during swimming, sometimes creeping; palmelloid observed and cells undergo division in it.

Probably cosmopolitan; common, occurs in organically polluted freshwater and saltwater along the margins of lakes, puddles and swamps.

ORDER EUGLENALES

Cells having two flagella at the anterior, only one emerges from the reservoir and canal. Contains colourless genera: *Khawkinea, Astasia, Euglenopsis, Cyclidiopsis, Klebsiella, Ascoglena, Colacium* and *Hyalophacus*; only normally photosynthetic genera: *Euglena, Trachelomonas, Strombomonas, Colacium, Lepocinclis* and *Phacus* are described here.

Colacium Ehrenberg 1833

Cells sessile and periodically free-swimming, ovoid, obovoid, spindle-shaped, or cylindrical to ellipsoidal, surrounded by a mucilaginous wall, solitary or united into amorphous or dendritic colonies formed by the anterior end of each cell attaching to a dichotomously branched system of mucilaginous stalks; motile cells naked with a single emergent flagellum and eyespot, on settling the cells shed their flagellum, rarely divide and secrete a mucilaginous sheath; contractile vacuole present; eyespot and flagellar swelling present; pellicle slightly striated; paramylon bodies small and oval; chloroplasts numerous, disc-like, with or without pyrenoids; reproduction by division of uni- or multinucleate amoeboid stages; palmelloid colonies develop in culture, without stalks and each cell uninucleate or with 2–8 nuclei.

Cosmopolitan; generally found in ponds and puddles, epiphytic or epizoic on *Cyclops*, copepods, rotifers and other freshwater zooplankton; free-swimming for a brief period of time.

The dichotomously branched colony is a result of cell division, with each new daughter cell forming its own stalk on the anterior end and remaining attached to the stalk of the parent.

1 Free end of cell with a blunt or sharp papilla .... *C. epiphyticum*
1 Free end of cell rounded 2
2 Cells attached to substratum by a pillow-like foot .... *C. vesiculosum* fo. *cyclopicola*.
2 Cells attached to substratum by a mucilaginous stalk 3
3 Cells ovoid to spindle-shaped, mucilaginous stalks short and not branched .... *C. vesiculosum*
3 Cells ovoid, mucilaginous stalks long and branched .... *C. vesiculosum* fo. *arbuscula*

Colacium epiphyticum F.E.Fritsch 1933

04010010 **Pl. 41Z,AA** (p. 170)

Cells (8–)10.5–14 μm wide, 22–28 μm long, narrowly obovoid, rounded at anterior end and with blunt or sharp papilla at posterior end, attached by a single or branched mucilaginous stalk; pellicle very thin, faintly striated with transverse or longitudinal striations; chloroplasts occasionally irregular and not in series; eyespot not observed; nucleus large, from 5 μm in diameter, with a clear area around it, roughly central; free-swimming cells not observed.

Europe; occurs in the British Isles in puddles and ponds where epiphytic on filamentous green algae (e.g. *Oedogonium* spp.), rare.

Colacium vesiculosum Ehrenberg 1833

04010020 **Pl. 41T–W** (p. 170)

Cells 8–15(–19) μm wide, 18–25(–29) μm long, spindle-shaped to somewhat pear-shaped; anterior end attenuated when extended, solitary or 2, 4, or 8 cells together in a colony attached by a short, unbranched, mucilaginous stalk or foot-like structure; pellicle slightly striated, sometimes covered by a thin, mucilaginous membrane; motile cells resemble *Euglena*; flagellum slightly exceeding cell length.

Probably cosmopolitan; in the British Isles in ponds, puddles and ditches where usually growing as sessile colonies on microfauna such as Cladocera or Copepoda, common.

fo. *arbuscula* (F.Stein) Huber-Pestalozzi 1955

Basionym: *Colacium arbuscula* F.Stein

04010022 **Pl. 41X,Y** (p. 170)

Cells 7–13(–16) μm wide, 15.5–23(–35) μm long, ovoid, ellipsoidal or spindle-shaped, joined by a long, branched, gelatinous stalk to form 2–26-celled plume-like and bushy colonies; eyespot sometimes evident; flagellum absent except in single flagellated swarmers reported with an eyespot.

Europe and Asia, uncommon; reported in the British Isles by Lund (1937) where it occurs in puddles, ponds and lakes, often attached to Crustacea, copepods and rotifers, more rare than the type variety.

fo. *cyclopicola* (Gickelhorn) T.G.Popova in Woronichin & Popova 1939

Basionym: *Euglena cyclopicola* Gickelhorn

04010023 **Pl. 41P–S** (p. 170)

Cells 8–19 μm wide, (18–)20–27 μm long, obovoid or pear-shaped, sometimes almost ovoid, attached to substratum by a pillow-like foot, sometimes encrusted with brown iron compounds; paramylon bodies ring-like; solitary, swimming cells have a flagellum longer than cell.

Probably cosmopolitan since commonly reported in Europe, Asia and North America; rarely found in the British Isles (Wołowski, 1998b) where it occurs in lakes, ponds and puddles on Crustacea and rotifers.

Euglena Ehrenberg 1833

Cells solitary, elongate, oblong, lance- or spindle-shaped, spirally twisted, attenuate at posterior end, usually pointed, occasionally terminating in a short, fine point known as a cauda or tail-piece; flask-shaped reservoir at anterior end of cell, opens to exterior through a narrow canal; flagella 2, but only one responsible for locomotion and emerging from apical reservoir, second very short, reduced and not emerging, granular swelling at point of branching and each branch terminating within the reservoir in a basal granule; eyespot at anterior end; contractile vacuoles one or more, adjacent to and open into the reservoir; pellicle rigid or semi-rigid, limiting and controlling shape of cell, if pliable then allowing for changes in cell shape associated with euglenoid movement, usually marked by striations or rows of granules; haematochrome granules develop in cytoplasm of some species; chloroplasts numerous, disc- to band-shaped, sometimes contain pyrenoids that protrude slightly; paramylon bodies of variable number and shape, sometimes attached to chloroplasts or free in cytoplasm; nucleus usually relatively large and central; asexual reproduction by longitudinal division beginning at anterior end of motile cells, or by formation of thick- or thin-walled cysts, daughter cells at times forming palmelloid colonies, thick-walled resting cysts common.

Cosmopolitan; most commonly found in shallow and quiet waters, such as ponds and ditches; some species are confined to acid waters such as *Sphagnum* bogs, others usually in water rich in organic matter.

Actively motile due to beating of the flagellum, very occasionally creep and sometimes exhibiting a violent squirming known as euglenoid movement. Identification of species is not easy and relies on characters associated with cell shape, pellicle markings, form and numbers of chloroplasts and associated paramylon bodies, and presence or absence of pyrenoids. The genus was divided into six subgenera by Pringsheim (1956): Rigidae, Lentiferae, Catilliferae, Radiatae, Serpentes, Limpidae (colourless species of *Euglena*).

1 Cells rigid, almost completely inflexible usually with cylindrical colourless end piece (Rigidae group) 2
1 Cells showing varying degrees of flexibility 10
2 Cells obovoid ....*E. texta*
2 Cells long, cylindrical to ribbon-, spindle- or needle-shaped....................3
3 Cells long and thin, needle-shaped, elongate-truncate at apex of anterior end .....*E. acus*
3 Cells not as above....................4
4 Cells spindle-shaped or cylindrical to spindle-shaped, never twisted, slightly truncate at anterior end, tapering to a sharp straight tail-piece ....*E. limnophila*
4 Cells shortly spindle-shaped, ribbon-like, or if cylindrical then twisted....................5
5 Cells bent, anterior end slightly truncate at apex and tail-piece usually curved ....*E. limnophila* var. *swirenkoi*
5 Cells longitudinally spindle-shaped, cylindrical or flattened and twisted....................6
6 Cells with spirally arranged striations and without rows of granules 7
6 Cells with helices striations and having rows of shining granules or excrescences..............9
7 Cells longitudinally cylindrical and not triangular in cross section....................8
7 Cells ribbon-like, and in cross section triangular since having 3 ridges...............*E. tripteris*
8 Cells up to 250 µm long, with two paramylon bodies ....*E. oxyuris*
8 Cells up to 540 µm long, with several large paramylon bodies agglomerated in few groups ....*E. oxyuris* fo. *maior*
9 Cells with nearly parallel sides, pellicle of many rows of shining granules.............*E. spirogyra*
9 Cells sometimes curved or twisted, pellicle thickly covered by rows of excrescences....................*E. spirogyra* var. *fusca*
10 Chloroplast star-shaped, radiating from 1, 2 or 3 centres (Radiatae group) 32
10 Chloroplasts not star-shaped....................11
11 Cells elongated, usually not flattened, snake-like (Serpentes group) 36
11 Cells not snake-like 12
12 Cells moderately flexible; chloroplasts without pyrenoids (Lentiferae group) 13
12 Cells very flexible; chloroplasts without pyrenoids (*E. chlamydophora*, *E. elongata*, *E. repulsans*) and when present usually forming a 'double sheathed pyrenoid' (Catilliferae group) 16
13 Cells broadly spindle-shaped to cylindrical, almost cylindrical pellicle spirally striated, posterior end tapering to a long tail-piece; chloroplasts disc-shaped, located towards posterior end....................*E. hemichromata*
13 Cells spindle-shaped or cylindrical to ovoid....................14
14 Cells shortly spindle-shaped, pellicle distinctly spirally striated, short tail-piece at posterior end ....*E. proxima*
14 Cells shortly cylindrical to ovoid or elongated cylindrical to band-shaped..................15
15 Cells shortly cylindrical to ovoid with broadly rounded anterior end, tapering to a blunt short tail-piece at posterior end ....*E. variabilis*
15 Cells elongated cylindrical to band-shaped, long, flat, usually twisted in middle part of cell, rounded at ends; chloroplasts disc-shaped, tightly packed, sometimes with one long paramylon body ....*E. ehrenbergii*
16 Cells with a single chloroplast 17
16 Cells with more than one chloroplast 19
17 Cells with a single long and undissected chloroplast ....*E. elongata*
17 Cells with a plate-like and deeply dissected chloroplasts....................18
18 Cells with readily visible muciferous bodies; double pyrenoids not evident..............*E. splendens*
18 Cells without visible muciferous bodies; double pyrenoids clearly evident ....*E. magnifica*
19 Chloroplasts 2 (rarely 3), plate-shaped, curved; cells small, broadly oval to ovoid ...*E. agilis*
19 Cells with numerous chloroplasts 20
20 Chloroplasts concave discs or plate-shaped, deeply dissected into ribbon-shape, or irregularly lobed....................21
20 Chloroplasts not dissected 25
21 Cells with red carotenoid pigment, cylindrical to broadly spindle-shaped....................*E. sanguinea*
21 Cells lacking red carotenoid pigment....................22
22 Flagellum twice cell length ....*E. polymorpha*
22 Flagellum less than twice cell length 23
23 Chloroplasts more than 20 per cell; cells with a short narrow tail-piece bent to one side ....*E. velata*
23 Chloroplasts 15 or less; cells without a tail-piece....................24
24 Cells rounded at both ends ....*E. clara*
24 Cells with anterior rounded and passing into a small, obtuse, non-hyaline point..........*E. oblonga*
25 Chloroplasts plate-shaped, flat with double sheathed pyrenoid and folded edge, 4–14 per cell....................*E. anabaena*
25 Chloroplasts circular, shield-shaped, polygonal, oblong-ellipsoidal or disc-shaped 26
26 Cells cylindrical, chloroplasts polygonal, tightly packed, each with double sheathed pyrenoid ....*E. obtusa*
26 Chloroplasts not as above; cells with anterior end elongated 27
27 Cells elongate spindle-shaped to cylindrical; chloroplasts disc-shaped....................*E. gracilis*
27 Cells terminating at posterior end by a long tail-piece....................28
28 Chloroplasts saucer-shaped and lobed, numerous, filling whole cell; cells with

anterior end slightly extended and truncate ....*E. caudata*
28 Chloroplasts plate-shaped each with double sheathed pyrenoid..29
29 Flagellum one-sixth of cell length ......*E. communis*
29 Flagellum same length as cell or longer 30
30 Cells with spindle-shaped muciferous bodies arranged in regular spiral row; pyrenoids present; chloroplasts flat or ellipsoidal and lobed..*E. granulata*
30 Cells without visible muciferous bodies; pyrenoids absent; chloroplasts oblong or disc-shaped..31
31 Chloroplasts oblong; cell length >55 µm; pellicle thin and smooth ....*E. repulsans*
31 Chloroplasts disc-shaped; cell length <55 µm; pellicle distinctly striated ....*E. chlamydophora*
32 Cells with star-shaped chloroplasts 33
32 Cells with ribbon-like chloroplasts, radiating from a centre and densely packed 34
33 Cells with a single star-shaped chloroplast...*E. viridis*
33 Cells with 2 star-shaped chloroplasts, one anterior and one posterior to the nucleus ....*E. geniculata*
34 Cells with short ribbon-like chloroplasts placed in 2 star-like groups, one anterior and one posterior to nucleus...*E. geniculata* var. *terricola*
34 Cells with numerous chloroplasts, each ribbon-like and radiating from a centre................35
35 Cells spindle-shaped, twice attenuated towards posterior end.........................*E. contabrica*
35 Cells broadly cylindrical, rounded at anterior end, narrowing at posterior end...*E. cuneata*
36 Cells elongate cylindrical to ellipsoidal or elongate spindle-shaped, if spindle-shaped then with a large yellow-green and cylindrical chloroplast...37
36 Cells elongate spindle-shaped to longitudinally cylindrical, if longitudinally cylindrical then emergent flagellum never visible and with oval paramylon bodies 40
37 Chloroplasts with pyrenoids 38
37 Chloroplasts without pyrenoids..........................39
38 Chloroplasts lens-shaped or biconvex oval, with naked pyrenoids..................................*E. deses*
38 Chloroplasts large, appear like a split cylinder each part with pyrenoids, usually attached to substratum by posterior ....*E. mutabilis*
39 Cells 5–8.5(–15)×45–100(–130) µm; paramylon bodies small, short, rod-shaped ...*E. deses* fo. *klebsii*
39 Cells larger, usually 9–14(–18)×125(–140) µm; paramylon bodies large, oblong, ellipsoidal ....*E. deses* fo. *intermedia*
40 Chloroplasts plate-shaped and irregular at edges; cells sometimes bulge greatly due to much euglenoid movement; flagellum two-thirds cell length; pellicle smooth ....*E. elastica*
40 Chloroplasts disc-shaped; cells do not bulge; emergent flagellum not visible; pellicle slightly striated..................................*E. adhaerens*

Euglena acus Ehrenberg 1830

Synonyms: *Euglena acus* vars. *rigida* K.Hübner, *lata* Svirenko, *longissima* Deflandre, *E. acutissima* Lemmermann

04020010 **Pl. 34A** (p. 149) **CD**

Cells 7–28.3 µm wide, (52–)60–180(–311) µm long, needle-shaped, elongate spindle-shaped, sometimes bent and sometimes assuming an S-shape; anterior end narrowed and apically truncate; posterior end tapered to a long fine point; pellicle delicately striated; chloroplasts small, numerous, peripheral, disc-like without pyrenoids; paramylon bodies numerous, long, rod-shaped; flagellum short about one-sixth to one-third cell length; eyespot small, located towards end of canal; euglenoid movement absent; swims straight forwards rather rapidly rotating slowly (or gyrating rapidly); cysts not observed.

Probably cosmopolitan; widespread, in ditches, ponds, small flowing rivers, fish and village ponds and swamps, as well as in acid ponds and brackish-water; indicator of mild to moderately polluted water.

Euglena adhaerens Matvienko 1938

Synonym: *Euglena tatrica* Czosnowski

04020020 **Pl. 37P** (p. 157)

Cells 7.5–12 µm wide, 100–165 µm long, longitudinal cylindrical to spindle-shaped; anterior end narrowed and bluntly truncate; posterior end tapering; pellicle slightly striated; flagellum not visible; eyespot at front of reservoir; chloroplasts numerous, disc-shaped, sometimes without a pyrenoid; paramylon bodies oval; euglenoid movement violent, usually creeping, sometimes attaches to substratum by posterior end; cysts not observed.

Europe and Asia; in the British Isles reported by Wołowski (1998b), where uncommon and occurs in swamps, puddles and ponds.

Euglena agilis H.J.Carter 1856

Synonyms: *Euglena pisciformis* G.A.Klebs, *E. nana* L.P.Johnson, *E. pisciformis* vars. *fallax* E.G.Pringsheim, *lata* E.G.Pringsheim, *mucronata* E.G.Pringsheim, *obtusa* E.G.Pringsheim, *procera* E.G.Pringsheim, *striata* E.G.Pringsheim, *E. agilis* vars *pyrenoidea* J.Schiller, *circumsulcata* J.Schiller, *praexicisa* J.Schiller, *varians* J.Schiller

04020030 **Pl. 35 G,H** (p. 152) **CD**

Cells (4–)5–11.7(–15) µm wide, (13–)18–33(–60) µm long, when fully extended nearly oval, broadly ovoid; anterior end bluntly rounded, posterior end coming to a short, rather blunt point; pellicle very faintly striated; chloroplasts asymmetrical, consisting of 2(–3) elongated and widened plates, side by side and nearly lining cell, with a double sheathed pyrenoid; paramylon bodies ovoid, small, scattered, very variable in number; flagellum 1 to 2 times longer than cell; eyespot yellowish or

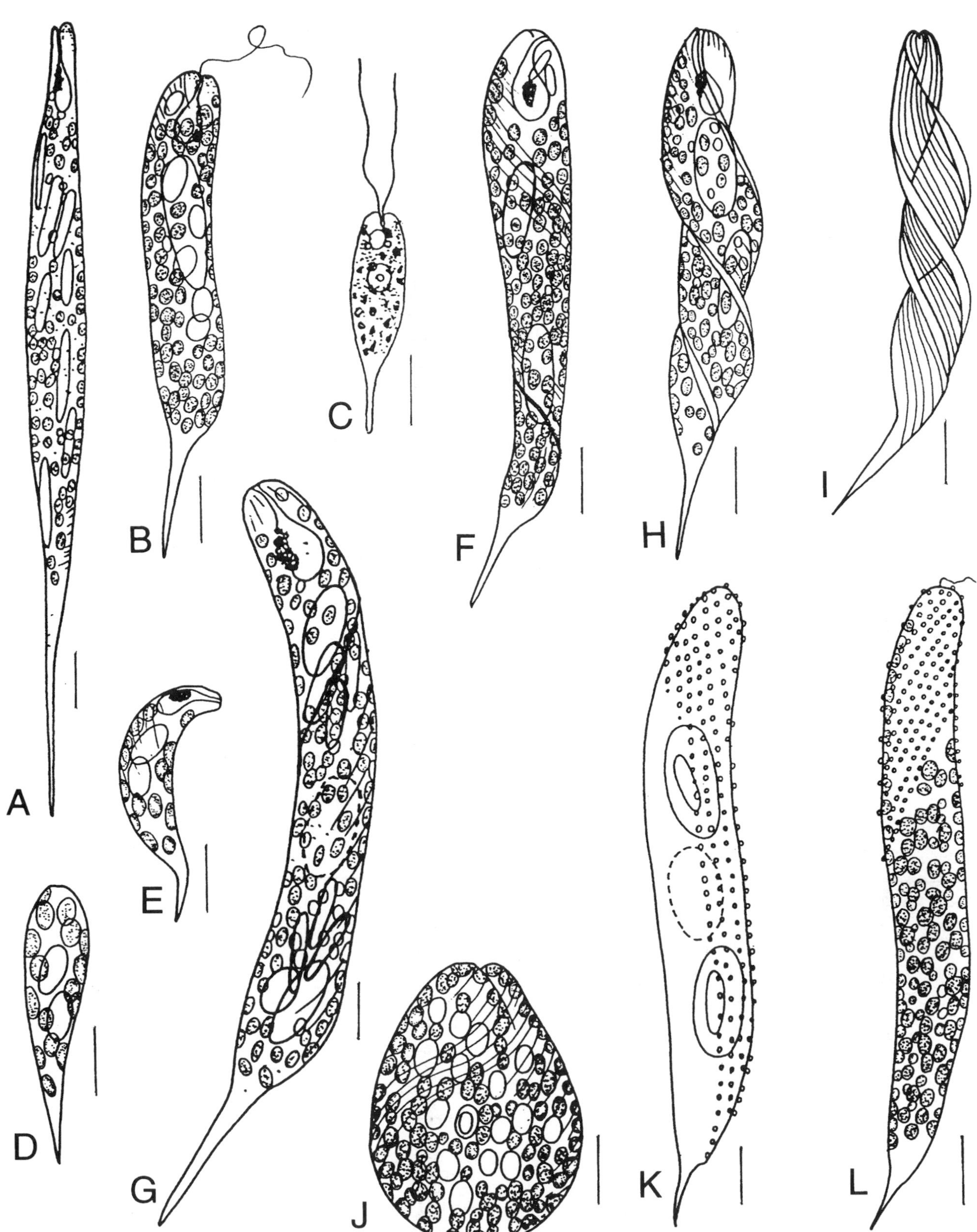

Plate 34 A–L
A. *Euglena acus* (p. 148); **B**. *Euglena limnophila* (p. 153); **C**. *Eutreptia viridis* (p. 145); **D, E**. *Euglena limnophila* var. *swirenkoi* (p. 153); **F**. *Euglena oxyuris* (p. 155); **G**. *Euglena oxyuris* fo. *maior* (p. 155); **H, I**. *Euglena tripteris* (p. 158); **J**. *Euglena texta* (p. 156); **K**. *Euglena spirogyra* var. *fusca* (p. 156); **L**. *Euglena spirogyra* (p. 156).
Scale bars: 10 µm.

red, rather faint, but relatively large; euglenoid movement present; swims very rapidly with wide gyrations and with a characteristic flip of posterior end; *Palmella*-stage observed.

Probably cosmopolitan, widespread; in permanent and temporary bodies of water including ditches, puddles, ponds, lakes and peat bogs, sometimes associated with sand; an indicator of moderately polluted water.

Cells of very variable shape.

Euglena anabaena Mainx 1926

Synonyms: *Euglena anabaena* Mainx, var. *minor* Mainx, *E. anabaena* var. *minima* Mainx

04020040 **Pl. 35O** (p. 152)

Cells 8–25 µm wide, 19.5–88(–94) µm long, spindle-shaped; anterior end rounded and posterior end narrowing to a tail-piece; pellicle slightly striated; chloroplasts 4–14, large, plate-shaped with folded edges, each with a double sheathed pyrenoid; paramylon bodies small and few; flagellum 1–1.5 times cell length; palmelloid stage observed; euglenoid movement violent.

Probably cosmopolitan; occurs in ponds, lakes and puddles where sometimes in sufficient quantity to discolour the water.

Mainx (1926) found three variants: the type, 88–94 µm long, 20–25 µm wide; var. *minor*, 36–43 µm long, 9–12 µm wide; var. *minima*, 26–30 µm long, 8–11 µm wide. According to Pringsheim (1956), they are very similar to one another and should be treated as conspecific.

Euglena caudata K. Hübner 1886

Synonym: *Euglena caudata* var. *minor* Deflandre

04020050 **Pl. 36B** (p. 154)

Cells (16–)28–38(–45) µm wide, 64–120 µm long, spindle-shaped or broadly spindle-shaped; anterior end slightly extended and apex truncate; posterior end tapering to a hyaline tail-piece; pellicle slightly diagonally striated; flagellum as long as cell or shorter; eyespot large, easily visible; chloroplasts numerous, filling whole cell, saucer-shaped, lobed, each with a double sheathed pyrenoid; paramylon bodies small, numerous, rod-shaped; euglenoid movement present; palmelloid stage observed.

Probably cosmopolitan; widespread, in lakes, ponds, puddles, ditches and slow-flowing rivers; known from unpolluted to heavily polluted water.

Euglena chlamydophora Mainx 1927

04020070 **Pl. 35I,J** (p. 152)

Cells 9–20 µm wide, 32.2–45(–54) µm long, spindle-shaped; anterior end rounded and posterior end narrowing to a tail-piece when swimming freely; euglenoid movement violent, shape sometimes changes to broadly cylindrical; pellicle distinctly striated; chloroplasts disc-shaped, numerous, without pyrenoids; paramylon bodies numerous, small, ellipsoidal or oblong; flagellum as long as cell; cysts unknown.

Europe, South America; occurs in ditches, lakes and ponds, rare; considered an indicator of moderately polluted water.

Euglena clara Skuja 1948

04020080 **Pl. 35N** (p. 152) **CD**

Cells (13–)16–20(–23) µm wide (33–)45–69.5 µm long, elongate oval or shortly spindle-shaped, sometimes with almost parallel sides or narrowly hexagonal, rounded at ends; pellicle gently spirally striated; chloroplasts 7–10, disc-shaped, irregularly lobed at margin, each with a double sheathed pyrenoid; paramylon bodies small not numerous, spherical or rod-like; euglenoid movement not very evident; eyespot conspicuous; flagellum equal to cell length; palmelloid stage observed.

Europe, Asia; occurs in ponds, ditches, swamps and peat bogs in slightly to moderately polluted water, uncommon.

Euglena communis Gojdics 1953

04020060 **Pl. 36E** (p. 154)

Cells 18–20.5 µm wide, 76–100 µm long, broadly spindle-shaped, rounded at anterior end, posterior end gradually tapering to a blunt point; pellicle weakly striated; chloroplasts 12, oval and plate-shaped, each with a double sheathed pyrenoid; paramylon bodies several, rectangular, small; flagellum about one-sixth cell length; eyespot large, yellowish, faint; moves very rapidly, euglenoid movement rather marked.

Europe, North and South America; in the British Isles occurs in ponds and streams, rare.

Shape of cell is very similar to that of *E. caudata* and *E. granulata*.

Euglena contabrica E.G.Pringsheim 1956

04020350 **Pl. 37E,F** (p. 157)

Cells 20–25 µm wide, 54–62 µm long, spindle-shaped and twice attenuated towards the posterior end, cylindrical in middle part of cell; euglenoid movement present; chloroplasts numerous, band-like and each radiating from a centre, often marked by a cluster of paramylon bodies; paramylon bodies near the cell surface, forming pseudopodium-like processes when viewed from above and appear surrounded by colourless, roundish sacks arranged in spirals parallel to striae (chloroplast system difficult to observe and describe); pellicle delicately striated; flagellum twice cell length; eyespot relatively large; palmelloid stage observed.

Europe; reported from the British Isles by Pringsheim (1956) in a ditch on clay soil, rare.

Euglena cuneata E.G.Pringsheim 1956

04020090 **Pl. 37G,H** (p. 157)

Similar to *E. contabrica* but cells smaller, broadly cylindrical, rounded at anterior and narrowed at posterior end; chloroplast bands radially arranged around a centre, densely packed, sometimes becoming wider and forked before their arms reach cell periphery, run parallel to striae at periphery; pellicle spirally striated; flagellum twice cell length; eyespot relatively large.

Europe; reported in the British Isles by Pringsheim (1956) from a polluted ditch.

Probably conspecific with *E. contabrica*, but requires to be checked by reference to material in culture.

Euglena deses Ehrenberg 1833

Synonym: *Enchelys deses* O.F.Müller

04020100 **Pl. 37I,J** (p. 157)

Cells 10–17(–24) µm wide, (83–)133(–163) µm long, elongate cylindrical to ellipsoidal, occasionally reported as slightly flattened; anterior end rounded or slightly truncate and posterior end narrowed to a blunt point; pellicle delicately striated; chloroplasts parietal and numerous, lens-shaped (biconvex and oval), each with a pyrenoid; paramylon bodies rod-shaped, generally scattered; flagellum relatively short, about one-sixth cell length, usually retracted; eyespot large; euglenoid movement violent, twisting and turning continuously, swimming rapid but sometimes weak, often creeping; division common when in the extended condition, with or without cysts.

Probably cosmopolitan; widespread, in puddles, ponds, swamps, lakes, in *Sphagnum* bogs on mud or organic detritus and in shallow freshwater or brackish water; indicator of moderate to heavy pollution and also known to be associated with radioactive waste.

Since cells tend to seek a substratum, they can lie more or less motionless for long periods.

fo. *intermedia* G.A.Klebs 1883

Synonym: *Euglena intermedia* (G.A.Klebs) F.Schmitz

04020102 **Pl. 37K** (p. 157)

Cells (7–)9–14(–18) µm wide, (84–)93–124(–145) µm long, cylindrical; anterior end blunt, posterior end tapering to a short point; pellicle very faintly striated; striae difficult to see; chloroplasts lens- to disc-shaped, without pyrenoids; paramylon bodies rod-shaped and numerous, or short and brick-shaped bodies with depressed centres; flagellum very short, about one-sixth cell length; eyespot conspicuous, purplish-red; euglenoid movement occurs, usually a slow, steady squirming, sometimes settles on a slide by its posterior tip and sways on it; cysts unknown.

Probably cosmopolitan; widespread, in pools, ponds, ditches, swamps and small rivers, tolerant of saline water; known from unpolluted to heavily polluted water.

fo. *klebsii* (Lemmermann) T.G.Popova 1966

Basionym: *Euglena intermedia* var. *klebsii* Lemmermann

Synonym: *E. klebsii* Lemmermann

04020103 **Pl. 37L,M** (p. 157)

Cells (5–)6–8.4(–15) µm wide, (45–)54–100(–130) µm long, cylindrical; anterior end slightly narrowed, rounded at apex; posterior end narrowing to a short, rounded tail-piece; flagellum about one-sixth cell length; pellicle finely striated; chloroplasts numerous, disc- to lens-shaped, without pyrenoids; paramylon bodies short, rod-shaped; euglenoid movement occurs, swims slowly; cysts not seen.

Probably cosmopolitan; widespread, in ponds, puddles and peat bogs; considered an indicator of moderately polluted water.

Differs from type form by the slight narrowing of the posterior end to a round tail-piece.

Euglena ehrenbergii G.A.Klebs 1883

Synonyms: *Euglena heimii* M.Lefèvre, *E. subehrenbergii* Skuja, *E. ehrenbergii* vars. *africana* Bourrelly, *minor* Hortobágyi

04020110 **Pl. 35F** (p. 152) **CD**

Cells 15–26(–35) µm wide, 140–286 µm long, elongate cylindrical to band-shaped, flattened, usually slightly twisted; anterior end slightly truncate, posterior end rounded; pellicle striated; chloroplasts numerous, small, disc-shaped, tightly packed, without pyrenoids; paramylon bodies generally numerous, oblong or ellipsoidal, sometimes with only a single, long, rod-shaped paramylon body; flagellum shorter than cell length; eyespot large; euglenoid movement violent, usually creeping; cysts known.

Probably cosmopolitan; widespread, occurs on the bottom of swamps, peat bogs, puddles and known associated with radioactive waste; ranges from unpolluted water to polluted freshwater or brackish water.

Euglena elastica Prescott 1944

04020120 **Pl. 37N** (p. 157)

Cells 9.5–11 µm wide, 76–100 µm long, usually spindle-shaped; anterior end narrowing abruptly, posterior end occasionally with a narrow, rounded apex; pellicle smooth; chloroplasts numerous, irregular, plate-shaped, with no pyrenoids; paramylon bodies short and rod-like, numerous, scattered throughout cell; flagellum about two-thirds cell length; eyespot at anterior end; euglenoid movement violent and sometimes greatly bulging.

North America, British Isles; known from small pools, rare.

Euglena elongata Schewiakoff 1891

04020130 **Pl. 35P** (p. 152)

Cells 5–8 µm wide, 55–60 µm long, elongate-cylindrical to spindle-shaped, tapering to a blunt point posteriorly; flagellum half to two-thirds cell length; chloroplast single, band-like, tending to be longitudinally positioned; pyrenoids probably not present; paramylon bodies numerous, small and rod-shaped; slight euglenoid movement present.

Europe, Asia, North America; in ponds and especially in *Sphagnum* bogs, sometimes a chance member of the plankton, rare.

Euglena geniculata (F.Schmitz) Dujardin 1841

Synonym: *Euglena schmitzii* Gojdics *et* Zakryś

04020140 **Pl. 37C** (p. 157)

Cells 9.5–12.5(–22) µm wide, 50–85 µm long, nearly cylindrical to bluntly spindle-shaped; anterior end rounded, posterior end narrowing to a sharp tail-piece; pellicle very finely and closely striated; chloroplasts 2, star-like with arms of different length, pyrenoids present; paramylon bodies numerous, short, rectangular, rod-shaped, located at pyrenoid centre; flagellum as long as cell, sometimes 1.5 times cell length; eyespot small, but visible; nucleus between 2 chloroplast

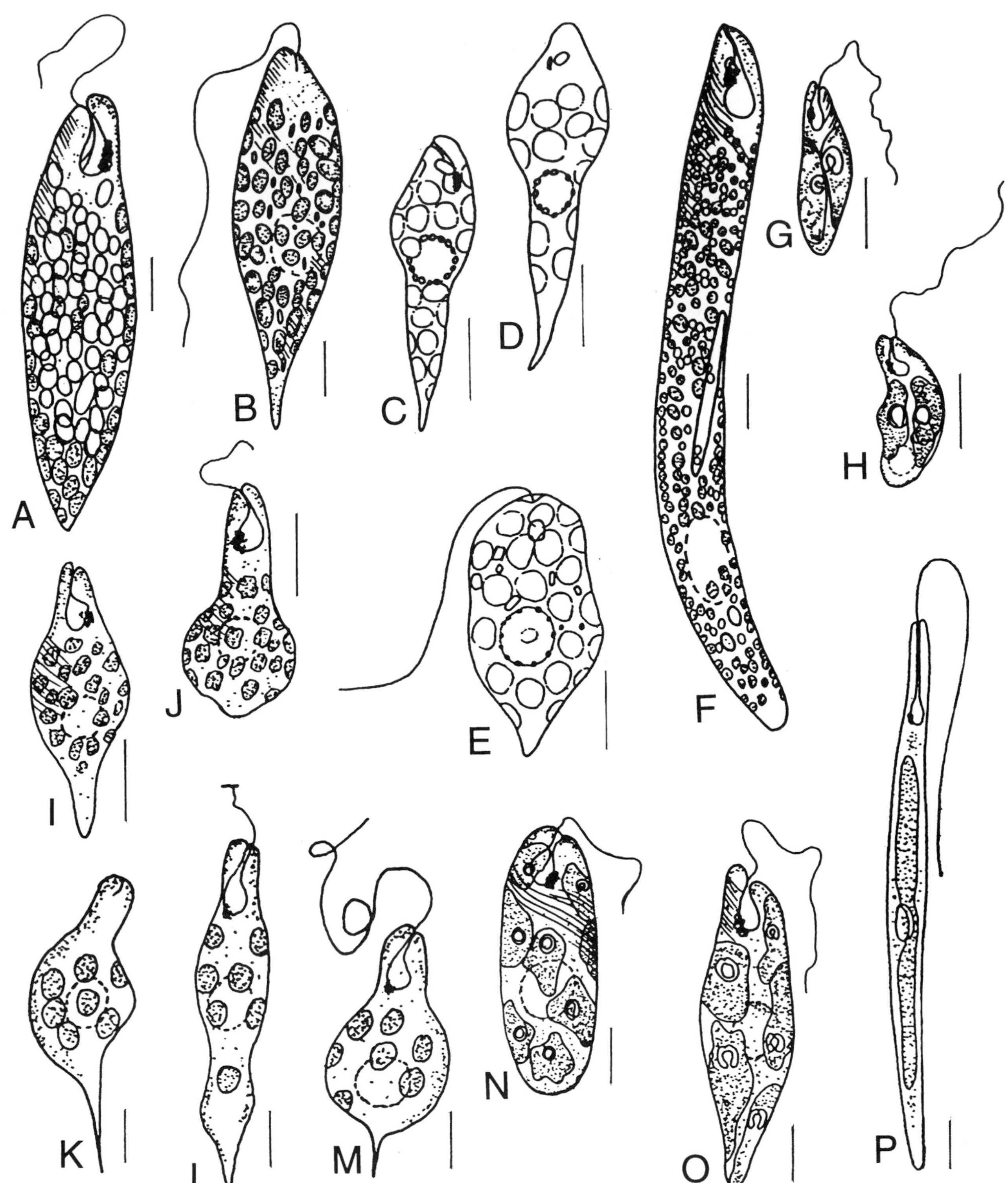

Plate 35 A–P

A. *Euglena hemichromata* (p. 153); **B**. *Euglena proxima* (p. 155); **C–E**. *Euglena variabilis* (p. 158); **F**. *Euglena ehrenbergii* (p. 151); **G, H**. *Euglena agilis* (p. 148); **I, J**. *Euglena chlamydophora* (p. 150); **K–M**. *Euglena repulsans* (p. 156); **N**. *Euglena clara* (p. 150); **O**. *Euglena anabaena* (p. 150); **P**. *Euglena elongata* (p. 151).

Scale bars: 10 μm.

groups; euglenoid movement occurs and cells sometimes twist; cysts observed, brownish.

Probably cosmopolitan; in ponds, ditches and puddles, on mud after rain and muddy banks; known from very pure to heavily polluted water, including water very rich in organic matter and sewage.

var. *terricola* P.A.Dangeard 1901

Synonym: *Euglena terricola* (P.A.Dangeard) Lemmermann

04020142 **Pl. 37D** (p. 157)

Cells 7–14 μm wide, 50–60 μm long, cylindrical; anterior end rounded and narrowing at posterior end; pellicle finely spirally striated; chloroplasts short, ribbon-like, arranged in two groups before and after the nucleus, with pyrenoids; paramylon bodies small and rod-like; posterior end having small mucilage-secreting apertures; euglenoid movement violent; cysts observed.

Probably cosmopolitan; common, in stagnant water including ponds and muddy puddles as well as in river plankton; considered an indicator of unpolluted water.

Chloroplasts are dispersed and are not star-shaped in arrangement as in the type variety.

Euglena gracilis G.A.Klebs 1883

04020150 **Pl. 36F** (p. 154)

Cells (6–)7–18(–22) μm wide, (31–)53–68(–70) μm long, nearly cylindrical to slightly spindle-shaped; anterior end rounded, slightly diagonally truncate at apex; posterior end tapering to a blunt point; pellicle very faintly spirally striated; chloroplasts large, 7–10(–14), disc- to lens-shaped, each with a double sheathed pyrenoid; paramylon bodies numerous, small, circular or as ring-like discs; flagellum usually same length as cell, occasionally half as long; eyespot large, prominent; euglenoid movement markedly violent; swims forward in a zigzag fashion, posterior end often gyrating in a wide circle; cysts known and undergo division in palmelloid stage.

Europe, Asia, North America; common, in *Sphagnum* bogs and ponds where there is much nitrogenous matter as well as in brackish-water and freshwater; known from unpolluted to heavily polluted water.

Several forms are known.

Euglena granulata (G.A.Klebs) F.Schmitz 1884

Synonym: *Euglena velata* var. *granulata* G.A.Klebs

04020160 **Pl. 36A** (p. 154)

Cells (11–)17–28 μm wide, 50–72.5(–115) μm long, spindle-shaped; anterior end conical, bluntly rounded at apex; posterior end varies with motion from a blunt to a gently tapering point; pellicle distinctly striated, mucilaginous bodies readily visible; chloroplasts 12–15, flat or ellipsoidal, lobed with a double sheathed pyrenoid; paramylon bodies ovoid, small, numerous; flagellum usually length of cell, sometimes twice as long; eyespot small, bright, yellow-red; euglenoid movement not very extensive; cysts mucilaginous, spherical or oval and irregular if compact.

Europe, Asia; common in the plankton and often forming surface blooms in ponds, slow-flowing water and peat bogs.

This species is similar to *Euglena caudata* but the posterior end of *E. granulata* is shorter.

Euglena hemichromata Skuja 1948

04020170 **Pl. 35A** (p. 152) **CD**

Cells 12–22 μm wide, 62–128 μm long, broadly spindle-shaped or cylindrical to spindle-shaped; anterior end slightly narrowed and rounded at apex, posterior end tapering to a long tail-piece; pellicle spirally striated; chloroplasts numerous, disc-shaped, usually located towards posterior end, without pyrenoids; paramylon bodies numerous, ellipsoidal, massed towards the anterior end; flagellum almost as long as cell; euglenoid movement occurs.

Probably cosmopolitan; widespread, in puddles, ditches, peat bogs as well as organically polluted waters, often forms water blooms; known in moderately to strongly polluted water.

Euglena limnophila Lemmermann 1898

04020180 **Pl. 34B** (p. 149) **CD**

Cells 7.5–12(–13.6) μm wide, 40–90 μm long, spindle-shaped, or cylindrical to spindle-shaped, slightly truncate at the anterior end and posterior end tapering to a sharp tail-piece; pellicle slightly striated; chloroplasts numerous, small, disc-shaped; paramylon bodies large, few, elongated rings or rod-shaped; flagellum shorter than cell length; eyespot small; cysts unknown.

Probably cosmopolitan; common but rare in the British Isles where it occurs in ditches, ponds and small slow-flowing rivers; known in unpolluted to moderately polluted water.

var. *swirenkoi* (Arnoldi) T.G.Popova 1955

Basionym: *Euglena swirenkoi* Arnoldi

Synonym: *E. limnophila* var. *minor* Drežepolski

04020182 **Pl. 34D,E** (p. 149)

Cells 7.5–12 μm wide, (24–)40–48 μm long, spindle-shaped or broadly spindle-shaped, sometimes bent; anterior end apically truncate; posterior end tapering to a sharp tail-piece, usually curved; pellicle slightly striated; chloroplasts numerous, small disc-shaped, without pyrenoids; paramylon bodies, ring-like or rod-shaped, several; flagellum shorter than cell length; eyespot small, located towards end of canal.

Europe, Asia; rare, only found on a few occasions in the British Isles where it occurs in moderately polluted water (e.g. ponds, rivers, sewage ditches).

The dimensions of the variety are smaller and the length to width ratio is different from the type variety.

Euglena magnifica E.G.Pringsheim 1956

04020190 **Pl. 36L,M** (p. 154)

Cells 25–35 μm wide, 90–120 μm long, broadly spindle-shaped to cylindrical; anterior end rounded, gradually narrowing to a short projection at the posterior end; chloroplasts numerous, deeply dissected, each with a double sheathed pyrenoid forming a centre

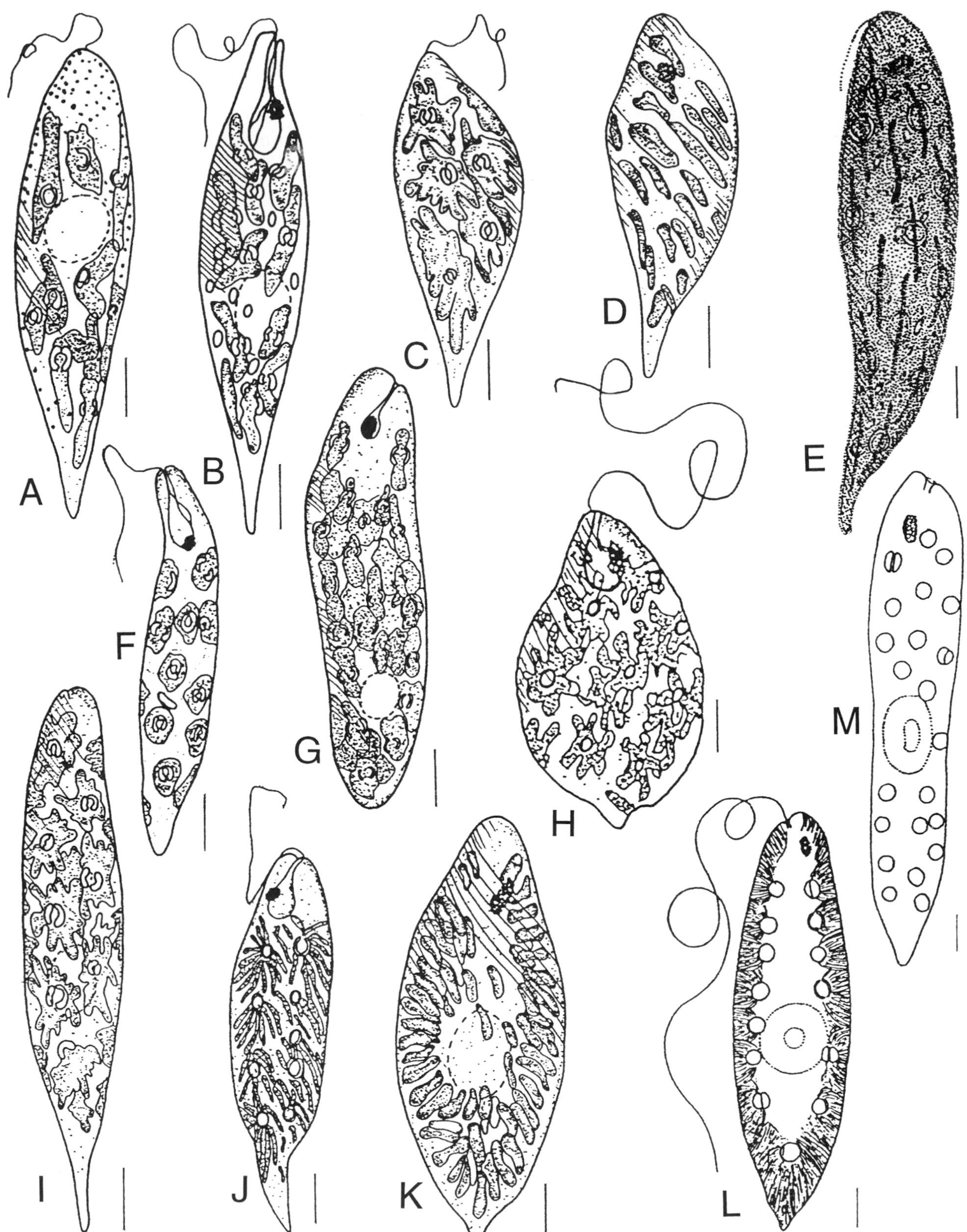

Plate 36 A–M

A. *Euglena granulata* (p. 153); **B**. *Euglena caudata* (p. 150); **C, D**. *Euglena polymorpha* (p. 155); **E**. *Euglena communis* (p. 150); **F**. *Euglena gracilis* (p. 153); **G**. *Euglena obtusa* (p. 155); **H**. *Euglena oblonga* (p. 155); **I**. *Euglena velata* (p. 158); **J**. *Euglena sanguinea* (p. 156); **K**. *Euglena splendens* (p. 156); **L, M**. *Euglena magnifica* (p. 153).

Scale bars: 10 μm.

from which curving chloroplasts radiate mainly towards the cell periphery; flagellum twice cell length; eyespot large; euglenoid body movement violent.

Europe; rare, in the British Isles only reported by Pringsheim (1956) in pools.

Similar in size and shape to *E. sanguinea* but never turns red, less likely to contract to form a ball and pyrenoids more evident and about 24 in each cell (Pringsheim, 1956).

Euglena mutabilis F.Schmitz 1884

Synonym: *Euglena acus* fo. *mutabilis* G.A.Klebs
04020200 **Pl. 37O** (p. 157)
Cells 4–9(–11) µm wide, 60–95(–122) µm long, narrowly cylindrical, slender or elongate spindle-shaped; anterior end tapering and bluntly truncate; posterior end extending to form an almost cylindrical and blunt tip; pellicle spirally striated; chloroplasts 2–8 (usually 4), delicate and thin, large, yellowish-green, closely pressed to wall, appear together like a split cylinder, each with a naked pyrenoid; paramylon bodies short, rectangular; flagellum not visible or very short; eyespot brilliant, crimson, shaped like a shallow cup and partly encircling canal, very close to pellicle; euglenoid movement moderate and reported to creep, coil, wriggle and loop; cysts unknown but palmelloid stage observed.

Probably cosmopolitan; common in small acidic water bodies including ditches, forest puddles, wet soil, village ponds, peat bogs, seepage areas and frequently abundant in heavy-metal contaminated waters, usually but not invariably where the pH is low; known in unpolluted to moderately polluted water. Populations growing among silt may move towards or away from the light according to its intensity, so they appear more dense in the early morning or early evening than at mid-day.

Euglena oblonga F.Schmitz 1884

04020210 **Pl. 36H** (p. 154)
Cells (23–)30–34(–44) µm wide, (48–)54–79 µm long, short cylindrical to broadly ellipsoidal; anterior end rounded and passing into a small, obtuse non-hyaline point; pellicle finely spirally striated; chloroplasts 8–15, irregularly lobed, each in the form of concave, dissected discs with 2 pyrenoids; paramylon bodies small, oval or ellipsoidal; flagellum about 1–1.5 times cell length; eyespot large; euglenoid movement violent; palmelloid stage observed.

Probably cosmopolitan; occurs in lakes, ponds and ditches.

Euglena obtusa F.Schmitz 1884

04020220 **Pl. 36G** (p. 154)
Cells 16–24 µm wide, 90–124 µm long, cylindrical and becoming pear-shaped towards both ends, with apices obtusely rounded; chloroplasts numerous, tightly packed, disc- or plate-shaped, polygonal, irregularly lobed at edges, each with a double sheathed pyrenoid; paramylon bodies small, rod-shaped; flagellum not emergent; eyespot round; euglenoid movement always by creeping, possibly assisted by mucilage production.

Probably cosmopolitan; rare in the British Isles, usually occurs on mud flats and pond margins, puddles and ditches; often in estuaries and brackish-water.

Euglena oxyuris Schmarda 1846

04020230 **Pl. 34F** (p. 149) **CD**
Cells (7.5–)16–46 µm wide, (36–)95–250 µm long, longitudinally cylindrical and slightly twisted, sometimes slightly flattened; anterior end rounded; posterior end tapering gently to a tail-piece; pellicle clearly spirally striated; chloroplasts numerous, ovoid or disc-shaped, without pyrenoids; paramylon bodies large, rectangular or ring-shaped, usually 2 with one anterior and the other posterior to the nucleus; flagellum short, up to one-third cell length; eyespot blood-red, lateral to cell reservoir; euglenoid movement slight, bends and rotates in a spiral fashion but does not squirm; cysts unknown.

Probably cosmopolitan; widespread, but not in large numbers; it occurs in swamps, ponds, ditches, rivers, streams and peat bogs; known from heavy metal contaminated waters and usually in moderately polluted water.

fo. ***maior*** (Woronichin) T.G.Popova 1966

Basionym: *Euglena oxyuris* var. *maior* Woronichin
Synonyms: *Euglena gigas* Drežepolski; *E. helicoidus* (C.Bernard) Lemmermann; *E. oxyuris* var. *helicoidea* Playfair
04020232 **Pl. 34G** (p. 149)
Cells 30–40 µm wide, 240–530 µm long, flattened with almost parallel sides, twisted at the posterior part of cell; anterior end rounded and posterior end narrowing to a sharp hyaline tail-piece; pellicle very finely striated, striae very close together and difficult to see; paramylon bodies large, 8–12, rod-shaped, agglomerated in a few groups; flagellum shorter than cell length; eyespot large; cysts unknown.

Probably cosmopolitan; in the British Isles occurs in ponds, ditches, lakes and small rivers, rare.

Euglena polymorpha P.A.Dangeard 1901

Synonym: *Euglena granulata* var. *polymorpha* T.G.Popova
04020240 **Pl. 36C,D** (p. 154)
Cells 20–26 µm wide, 80–90 µm long, almost spherical, ovoid to pear-shaped and narrowing gradually to a short, blunt, conical tail-piece; pellicle markedly spirally striated; chloroplasts 12–15, plate or disc-like with laciniate margins, each with a double sheathed pyrenoid; paramylon bodies small; flagellum twice cell length; euglenoid movement violent; division during palmelloid stage.

Europe, North America; usually common and occurs in permanent and temporary water bodies such as roadside ditches.

Euglena proxima P.A.Dangeard 1901

04020250 **Pl. 35B** (p. 152)
Cells 13–20(–25) µm wide, 52–80 µm long, spindle-shaped; anterior end slightly bluntly truncate, posterior end tapering to a short, hyaline tail-piece; pellicle

distinctly spirally striated; chloroplasts numerous, disc-shaped, without pyrenoids; paramylon bodies small, cylindrical or ellipsoidal, sometimes ring-like; flagellum about 1.5 times cell length; division in palmelloid stage; cysts observed.

Probably cosmopolitan; widespread, in puddles, edge of peat bogs, slow-flowing rivers, village ponds, and sometimes in surface water (neustonic); known in moderately polluted water.

Euglena repulsans J.Schiller 1952

04020260 **Pl. 35K–M** (p. 152)

Cells 20–22 μm wide, 60–63 μm long, spindle-shaped; anterior end slightly elongated at apex, posterior end tapering when swimming freely; pellicle thin and smooth; chloroplasts oblong, about 10 in each cell, without pyrenoids; paramylon bodies small, ellipsoidal; flagellum as long as cell; eyespot small and red; euglenoid movement very violent and rapid; cysts not observed.

Europe; in ponds and lakes, rare.

Euglena sanguinea Ehrenberg 1830

Synonyms: *Volvox calamus* A.Pritchard, *Euglena viridis* var. *sanguinea* F.Stein, *E. rubra* A.D.Hardy, *E. haematodes* (Ehrenberg) Lemmermann

04020270 **Pl. 36J** (p. 154) **CD**

Cells 22–42(–44) μm broad, 90–150 μm long, elongate cylindrical to broadly spindle-shaped; anterior end rounded; posterior end gradually tapering evenly to a blunt point or tip; pellicle conspicuously striated; chloroplasts numerous, dissected and radially arranged with arms extending to cell periphery, each with a double sheathed pyrenoid, green colour sometimes obscured by granules of the red pigment haematochrome; paramylon bodies spherical to ovoid or of other shapes, located in deeper parts of cell, often attached to chloroplasts; flagellum usually longer than cell, generally 1.5 times to twice as long; eyespot clearly evident and fairly large, towards base of canal; euglenoid movement markedly violent; swims very actively; cysts known.

Probably cosmopolitan; very widespread in ponds, puddles, ditches, lakes and edges of peat bogs, often accounts for red waterblooms; known in uncontaminated to moderately polluted water.

The chloroplasts and pyrenoids are very difficult to see in the living cell, but on staining with Lugol's solution the starch associated with the pyrenoids becomes visible at the ends or middle of the chloroplasts. Because the species has barely visible chloroplasts, it was described as a different taxon by several authors (Wołowski, 1998a).

Euglena spirogyra Ehrenberg 1838

04020280 **Pl. 34L** (p. 149) **CD**

Cells (6–)12–27(–35) μm broad, (45–)80–125(–250) μm long, longitudinally spindle-shaped and sometimes flattened, with sides nearly parallel; anterior end bluntly rounded, posterior end extended into a distinctly bent tail-piece; pellicle yellowish in colour and sometimes bearing rows of shining granules or beads; chloroplasts numerous, small, disc-shaped, lying close together, without pyrenoids; variable numbers of small, rectangular paramylon granules present, granules usually in 2(–3) clusters, one lying anterior to nucleus and another posterior to nucleus; flagellum about one-tenth to one-quarter cell length; eyespot bright, relatively large and prominent; euglenoid movement in form of squirming and markedly bending, or twists a little but does not shorten (usually seeks bottom of a dish); cysts unknown.

Probably cosmopolitan; widespread, in pools, ditches, fish ponds and field ponds including those in which the water is iron-rich, brackish and known to contain radioactive waste; considered an indicator of mildly to moderately polluted water.

var. *fusca* G.A.Klebs 1883

Synonym: *Euglena fusca* (G.A.Klebs) Lemmermann

04020282 **Pl. 34K** (p. 149)

Cells 17–25 μm wide, 153–198(–250) μm long, longitudinally flattened with nearly parallel sides, sometimes twisted or curved; anterior end rounded, posterior end with a sharp tail-piece; pellicle brownish and markedly patterned, with rows of variously sized and shaped excrescences lying almost parallel; chloroplasts numerous, small, ovoid, without pyrenoids; paramylon bodies usually forming two large clusters along with numerous smaller bodies; flagellum up to one-fifth cell length, sometimes longer or occasionally absent; eyespot dark red; euglenoid movement absent or very slow; cysts unknown.

Probably cosmopolitan; common, in swamps, puddles, ditches and in humic water, also known in radioactive waste; rare in the British Isles; indicates relatively uncontaminated to moderately polluted water.

Euglena splendens P.A.Dangeard 1901

04020290 **Pl. 36K** (p. 154)

Cells 21–29 μm wide, 84–104 μm long, spindle-shaped to broadly spindle-shaped; anterior end rounded, posterior end narrowing to a short tail-piece; pellicle striated, lying beneath are several small mucilage bodies; chloroplasts numerous, dissected and plate-like with arms lying parallel to striae, each with a double sheathed pyrenoid; paramylon bodies small, numerous, spherical or ellipsoidal; flagellum 1.5–2 times longer than cell; eyespot long, conspicuous; palmelloid stage observed, thick-walled cysts unknown.

Probably cosmopolitan; common in various water types, recorded from the British Isles by Pringsheim (1956).

Euglena texta (Dujardin) Hübner 1886

Basionym: *Crumenula texta* Dujardin

Synonym: *Lepocinclis texta* (Dujardin) Lemmermann

04020300 **Pl. 34J** (p. 149) **CD**

Cells (25–)36–43(–48.5) μm wide, 32–57 μm long, ovoid to spherical; anterior end slightly narrowed with a small depression at apex, posterior end broadly rounded; pellicle strongly spirally striated; chloroplasts

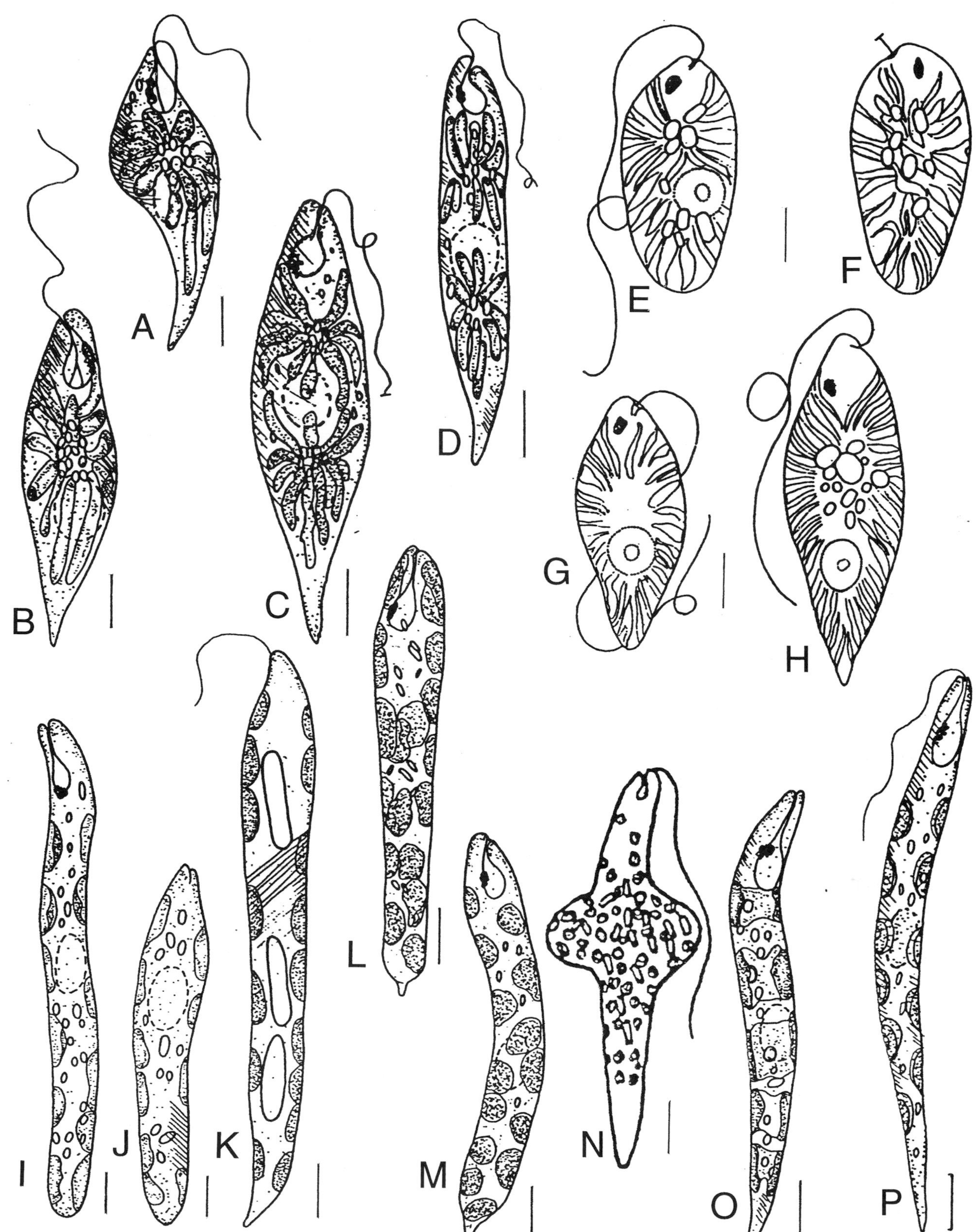

Plate 37 A–P

A, B. *Euglena viridis* (p. 158); **C**. *Euglena geniculata* (p. 151); **D**. *Euglena geniculata* var. *terricola* (p. 153); **E, F**. *Euglena contabrica* (p. 150); **G, H**. *Euglena cuneata* (p. 150); **I, J**. *Euglena deses* (p. 151); **K**. *Euglena deses* fo. *intermedia* (p. 151); **L, M**. *Euglena deses* fo. *klebsii* (p. 151); **N**. *Euglena elastica* (p. 151); **O**. *Euglena mutabilis* (p. 155); **P**. *Euglena adhaerens* (p. 148).
Scale bars: 10 μm.

small, numerous, without pyrenoids; paramylon bodies small, numerous, oval; flagellum 1–3 times longer than cell; eyespot red, close to reservoir; swims rapidly and not showing euglenoid movement.

Probably cosmopolitan; widespread, common in the British Isles in small rivers, fish ponds, village ponds and peat bogs, sometimes associated with surfaces; indicator of moderately polluted water.

Euglena tripteris (Dujardin) G.A.Klebs 1883

Synonyms: *Euglena torta* A.Stokes, *Phacus tripteris* Dujardin

04020310 **Pl. 34H,I** (p. 149) **CD**

Cells 8–16(–23) µm wide, (60–)70–190(–210) µm long, flattened, ribbon-shaped, spirally twisted and triangular in cross section, bowed inwards and giving the appearance of 3 ridges; anterior end slightly narrowed, sometimes slightly truncate; posterior end tapering to a long tail-piece, slightly spirally twisted; pellicle clearly striated, striae following twists; chloroplasts numerous, small, disc-shaped, without pyrenoids; paramylon bodies large, long and rod-like, 2 per cell with one usually anterior and the other posterior to the nucleus; flagellum about one-half to three-quarters length of cell; eyespot bright red and anterior to reservoir; swims by rotating rapidly, stopping frequently; longitudinal division known and not enclosed within a thickened membrane; cysts unknown.

Probably cosmopolitan; known from peat bogs, ditches, ponds, pools, ditches and lakes, occurs in plankton or associated with surfaces, usually scarce; known to range from unpolluted to moderately polluted water.

Euglena variabilis G.A.Klebs 1883

04020320 **Pl. 35C–E** (p. 152)

Cells (7–)9–15(–20) µm wide, (25–)31–46 µm long; shortly cylindrical to ovoid; anterior end broadly rounded; posterior end tapering to a blunt, short tail-piece, or sometimes slightly indented on side; pellicle very distinctly spirally striated; chloroplasts numerous, disc-shaped, without pyrenoids; paramylon bodies numerous, small, ellipsoidal or short and rod-shaped; flagellum 2 or 3 times as long as cell; eyespot large, dark red; swims very actively with flagellum, and shows considerable agility, during swimming cell shape remains unchanged; cell during division contracts somewhat but remains ovoid, unlike *E. viridis* where the cell becomes spherical.

Europe, Asia, North America; usually uncommon and in pools, puddles, ditches and fish ponds as well as the littoral zone of lakes; only reported in the British Isles by Lund (1942c) and Pringsheim (1956).

Euglena velata G.A.Klebs 1883

04020330 **Pl. 36I** (p. 154) **CD**

Cells 25–30 µm wide, 90–115 µm long, elongate, oval to broadly spindle-shaped, tapering to a short narrow tail bent to one side; pellicle spirally striated; chloroplasts over 20, flat, deeply lobed, often edges turned up towards cell periphery, parts of chloroplast sometimes visible at posterior end as ribbons, each having a pyrenoid surrounded by a double sheath (usually difficult to see because of paramylon); paramylon bodies numerous, short and rod-shaped, scattered; flagellum as long as cell; eyespot large; euglenoid movement very active; division in a spongy, slimy envelope.

Europe, Asia, Central America; known from small water bodies such as ponds, street drains, ditches and puddles.

Forms a slime sheath when stained with acid carmine.

Euglena viridis Ehrenberg 1830

Synonyms: *Enchelys viridis* E.Schrank, *Cercaria viridis* O.F.Müller, *Raphanella urbica* Bory

04020340 **Pl. 37A,B** (p. 157)

Cells (9–)14–18(–22) µm wide, (30–)38–62(–89) µm long, spindle-shaped to broadly spindle-shaped; anterior end rounded; posterior end usually tapering to a point of variable length; pellicle faintly spirally striated; chloroplast single, irregular and star-like, with pyrenoids; paramylon bodies ovoid, ring-like to brick-shaped, a mass of paramylon surrounds central area of chloroplast, some generally distributed throughout cell; flagellum slightly shorter or longer than cell, easily discarded; eyespot bright crimson, posterior to central mass of paramylon and surrounded by ribbons of chloroplast; euglenoid movement fairly frequent and also swims very rapidly; longitudinal division in a thick- or thin-walled cyst.

Probably cosmopolitan; widespread in freshwater streams, ponds, ditches as well as temporary water bodies and brackish-water, commonly forming green patches in farmyards and sewage outfalls; indicator of moderately to heavily polluted water.

It rounds up very readily to form a cyst when kept in culture, but the cysts have no clearly visible wall.

Lepocinclis Perty 1852

Cells spherical, ovoid, ellipsoidal, spindle-shaped, oval or nearly round in cross section and never compressed, radially symmetrical; pellicle rigid (without euglenoid movement), usually with numerous longitudinal or spiral striae, sometimes smooth; one flagellum emerging from apical reservoir and for locomotion, 1 to 2 times the length of cell, second very short and not emerging, granular swelling present; posterior end pointed and sometimes having an abruptly pointed caudus or tail-piece; chloroplasts parietal, numerous, disc-shaped, without pyrenoids; paramylum bodies large, usually 2, disc- or ring-shaped positioned laterally, sometimes almost encircling inner cell wall; eyespot large, anteriorly placed; reservoir and canal system similar to that of *Euglena*; reproduction by division in resting stage; cysts observed.

Cosmopolitan; usually occurs with other euglenoids in shallow water bodies often rich in organic acids and nitrogenous matter.

The description of pellicle decoration is essential for identification and may be partly obscured by the cell contents. Data for genus from Conrad (1934, 1935).

1 Cells with a smooth pellicle............*L. playfairiana*
1 Cells with a striated pellicle..................................2
2 Pellicle with left handed striations 3
2 Pellicle with longitudinal or slightly spiralled striations 6
3 Cells pear-shaped and conically narrowed at the posterior end, more than 2 paramylon bodies ....*L. teres*
3 Cells of a different shape and having 2 ring-shaped paramylum bodies 4
4 Anterior end rounded; paramylon bodies 2, ring-shaped ....*L. ovum*
4 Anterior end not rounded; both ends conically narrow 2 paramylon bodies.....................5
5 Cells lemon-shaped (cytriform), with a small concavity at apex and posterior end slightly conically produced..............................*L. fusiformis*
5 Cells club-shaped, with a small knob at the anterior end and a long tail at the posterior end ....*L. caudata*
6 Cells striated alternately with some thin and thick (rib-like) striae ....*L. steinii*
6 Cells striated only by thin striae ....*L. steinii* var. *suecica*

Lepocinclis caudata A.M.Cunha 1913

04040010 **Pl. 38J** (p. 160)

Cells 15–20 μm wide, 45–60 μm long, club-shaped to spindle-shaped; anterior end shortly elongated to a knob and posterior end narrowing to a long tail-piece; chloroplasts disc-shaped; paramylon bodies 2, large; pellicle with striae having a left-handed spiral.

Europe, Asia, North and South America; occurs in the British Isles in puddles and swamps, reported as rare by Wołowski (1998b).

Lepocinclis fusiformis (H.J.Carter) Lemmermann 1901

Basionym: *Euglena fusiformis* H.J.Carter
Synonym: *Lepocinclis sphagnophila* Lemmermann

04040020 **Pl. 38I** (p. 160)

Cells 15–32.5 μm wide, (15–)35–42.5 μm long, lemon-shaped, broadly oval; anterior end conically narrowing, with small concavity at apex; posterior end narrowing to a short, conical tail-piece; pellicle with left-handed striae, colourless to yellow; chloroplasts minute and numerous, disc-shaped; paramylon bodies large, usually 2, sometimes single.

Probably cosmopolitan; common, in swamps, ponds, ditches and small rivers, rare in the British Isles (Wołowski, 1998b).

A few varieties are known.

Lepocinclis ovum (Ehrenberg) Lemmermann 1901

Basionym: *Euglena ovum* Ehrenberg
Synonyms: *Chloropeltis ovum* F.Stein, *Phacus ovum* (Ehrenberg) G.A.Klebs

04040030 **Pl. 38B,C** (p. 160)

Cells 13–22(–25) μm wide, (20–)28–38(–43.3) μm long, broadly ovate; anterior end rounded, posterior end with a short blunt tail-piece, about 6–7 μm long; pellicle with striae having a left-handed spiral; paramylon as 2 rings, one on either side of cell; flagellum as long as cell and directed posteriorly when swimming; eyespot often very small.

Probably cosmopolitan; widespread, known from ditches, swamps and small ponds as well as in the shallow water of bays and lagoons, also in brackish waters, planktonic (rarely euplanktonic); indicator of moderately polluted water.

Several varieties are known.

Lepocinclis playfairiana Deflandre 1932

Synonyms: *Lepocinclis fusiformis* Playfair, *Crumenula playfairiana* Deflandre

04040040 **Pl. 38D,E** (p. 160)

Cells 19–26 μm wide, 32–49.5(–53) μm long, broadly spindle-shaped; anterior end slightly narrowed into a slender tip or point (rostrate); posterior end with a tail-piece 7–12 μm long; pellicle smooth; paramylon bodies 2, long, circular or oval rings; chloroplasts numerous, small, disc-shaped; flagellum (and reservoir) lateral to apex.

Probably cosmopolitan; in swamps, ponds and old river beds, uncommon.

Lepocinclis steinii Lemmermann 1901

04040060 **Pl. 38F** (p. 160)

Cells 8–15 μm wide, 21–33 μm long, shortly spindle-shaped; anterior end slightly narrowing and bluntly truncate with a small concavity at apex; posterior end with a sharp and distinct tail-piece; pellicle with longitudinal or slightly spiralled alternately thin and thickened (rib-like) striae, colourless to yellowish; chloroplasts minute; paramylon of 1–3 large bodies and a few small ones; eyespot large, 1–3 μm in size; flagellum twice cell length.

Probably cosmopolitan; widespread, in swamps, ditches, puddles, ponds and water tanks, occurs in plankton; indicator of moderately polluted water.

var. *suecica* Lemmermann 1901

04040062 **Pl. 38G,H** (p. 160)

Cells 8–13 μm wide, 23–26 μm long, shortly spindle-shaped; anterior end slightly concave; posterior end narrowing to a sharp and short tail-piece; pellicle longitudinally striated; chloroplasts small, disc-shaped; paramylon bodies 2, each ring-shaped.

Probably cosmopolitan; type variety in small water bodies, rare in the British Isles; indicator of moderately polluted water.

Lepocinclis teres (F.Schmitz) France 1897

Basionym: *Phacus teres* F.Schmitz

04040070 **Pl. 38A** (p. 160)

Cells 16–18 μm wide, 40–45 μm long, pear-shaped; anterior end broadly rounded and slightly concave at apex; posterior end narrowing to a conical tail-piece; pellicle having striae with a left-hand spiral; chloroplasts numerous, small, flattened at poles like an

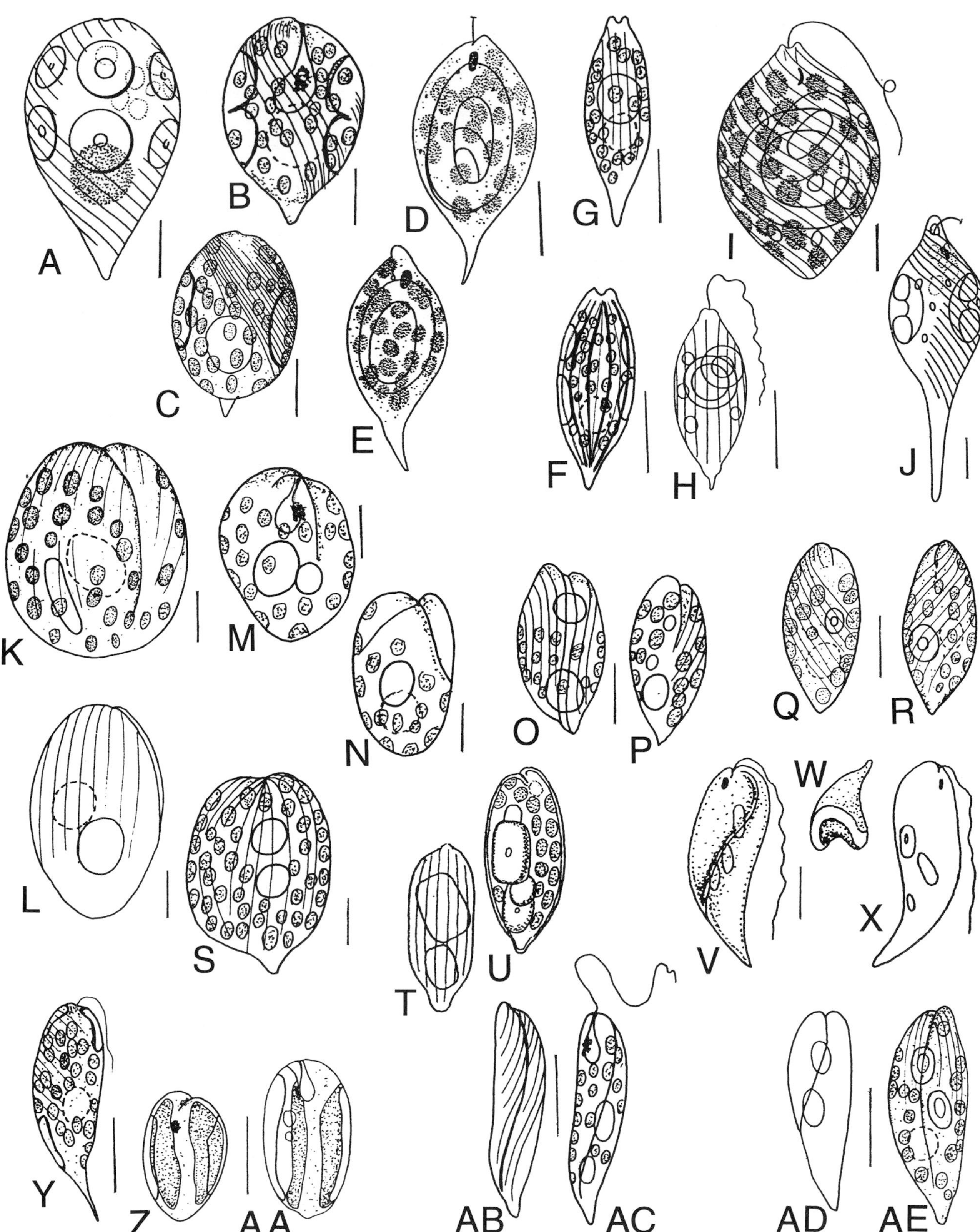

Plate 38 A–AE

A. *Lepocinclis teres* (p. 159); **B, C**. *Lepocinclis ovum* (p. 159); **D, E**. *Lepocinclis playfairiana* (p. 159); **F**. *Lepocinclis steinii* (p. 159); **G, H**. *Lepocinclis steinii* var. *suecica* (p. 159); **I**. *Lepocinclis fusiformis* (p. 159); **J**. *Lepocinclis caudata* (p. 159); **K, L**. *Phacus stokesii* (p. 168); **M, N**. *Phacus corculeum* (p. 164); **O, P**. *Phacus parvulus* (p. 167); **Q, R**. *Phacus pusillus* (p. 167); **S**. *Phacus acuminatus* (p. 162); **T, U**. *Phacus granum* (p. 165); **V–X**. *Phacus spiralis* (p. 168); **Y**. *Phacus striatus* (p. 168); **Z, AA**. *Phacus agilis* (p. 162); **AB, AC**. *Phacus oscillans* (p. 167); **AD, AE**. *Phacus skujae* (p. 167).

Scale bars: 10 μm.

orange (oblate); paramylon bodies ring-like and several; flagellum as long as cell; eyespot large and readily seen.

Europe, Asia, North America; commonly in puddles, village ponds, fish ponds, peat bogs and small rivers.

Phacus Dujardin 1841

Cells solitary, oval or ellipsoidal, pear- or spindle-shaped, considerably flattened, often twisted along longitudinal axis; one flagellum responsible for locomotion emerging from apical invagination, second very short and not emerging; canal opening slightly or obviously subapical; eyespot and flagellar swelling present; pellicle rigid, usually covered by distinct longitudinal or spirally arranged striations, granules, papillae or wart-like processes, transverse striae observed only after staining; posterior end sometimes bearing a straight, twisted or hooked (uncinate) tail-piece of variable length; some species have a longitudinal keel or an anterior furrow whose lateral edges spread and appear to almost divide the cell into two halves; anterior invagination (reservoir) with a canal similar to that in *Euglena*; chloroplasts numerous, without pyrenoids; paramylon bodies disc-shaped with raised margins or ring-shaped, variable in number; eyespot sometimes present at anterior end; reproduction by longitudinal division of motile or non-motile cells; cysts (resting cells) known in some species.

Cosmopolitan, most common in non-flowing waters especially in ponds, water bodies in thickets, swamps, peat bogs, and rivers; common in moderately polluted waters. To this genus belong about 150 taxa reported from freshwater and brackish water.

Often divided into the following sections: Dolichoplates, Phacus, Kampylopter, Chloropeltis, Monomorphina.

1 Pellicle spirally ribbed or ornamented by granules or spines 2
1 Pellicle not spirally ribbed or ornamented by granules or spines 8
2 Pellicle ornamented by granules or spines (Section Chloropeltis) 3
2 Pellicle spirally ribbed (Section Monomorphina) 4
3 Cells covered by lines of small or large granules .... *P. monilatus* var. *suecicus*
3 Cells covered by lines of fine spines ....*P. hispidulus*
4 Pellicle detached from protoplast 5
4 Pellicle firmly attached to protoplast 6
5 Cells ovate to broadly ovate .... *P. nordstedtii*
5 Cells spindle-shaped .... *P. cochleatus*
6 Anterior end slightly narrowing; pellicle spirally ribbed with ribs connected by transverse striae .... *P. costatus*
6 Anterior end broadly rounded; pellicle spirally ribbed but ribs not connected by transverse striae 7
7 Cells spindle-shaped to ovoid, slightly flattened; pellicle finely ribbed .... *P. pyrum*
7 Cells ovoid to broadly ellipsoidal and flattened dorsiventrally; pellicle roughly spirally ribbed .... *P. pseudonordstedtii*
8 Cells with a well-developed keel 9
8 Cells without a keel 14
9 Keel S-shaped 10
9 Keel of another shape 11
10 Cells slightly twisted, posterior end tapering to an oblique tail-piece .... *P. spiralis*
10 Cells not twisted but asymmetrically elongated, with dorsal keel undulate, tail-piece sharp and curved .... *P. hamelii*
11 Cells in transverse section symmetrically dumbbell-shaped .... *P. curvicauda*
11 Cells in transverse section ellipsoidal to slightly S-shaped 12
12 Anterior end of cell narrowly rounded and shallowly lobed (apical furrow up to half cell length), with tail-piece short (5–7 μm long), slender and turns obliquely to one side .... *P. pleuronectes*
12 Anterior end of cell rounded, furrow well-developed, triangular in cross section with tail-piece sharply pointed, straight or curved but not bent obliquely 13
13 Cells ovoid, pear-shaped to ovate in outline, often slightly twisted; posterior end tapering to a straight or sometimes curved, sharply pointed tail-piece (5–11 μm long) .... *P. caudatus*
13 Cells broadly ovoid in outline, not twisted, slightly asymmetrical; posterior end narrowing abruptly to a long, thin, curved tail-piece .... *P. triqueter*
14 Cells without a tail-piece or with a small, bluntly rounded extension or papilla 15
14 Cells with a tail-piece 22
15 Posterior of cells without a tail-piece 16
15 Posterior of cells with a small, bluntly rounded extension or papilla 17
16 Cells longitudinally heart-shaped, cut in at anterior end, <25 μm wide or long...... *P. corculeum*
16 Cells broadly oval to subspherical, flat, with an apical groove extending into posterior, >30 μm wide or long........ *P. stokesii*
17 Chloroplasts and paramylon bodies watchglass-shaped .... *P. agilis*
17 Chloroplasts and paramylon bodies of another shape 18
18 Cells almost cylindrical; paramylon bodies single, large and angular .... *P. granum*
18 Cells longitudinally oval to ovoid, obovoid, ellipsoidal, heart- or spindle-shaped in outline; paramylon bodies usually ring-shaped (occasionally rod-shaped) 19
19 Pellicle striations almost longitudinal; posterior end of cell narrowing very gradually .... *P. skujae*
19 Pellicle spirally striated; posterior end of cell narrowing quite abruptly 20

20 Cells flattened dorsiventrally, with a wide furrow running the full length of cell, slightly twisted .... *P. pusillus*
20 Cells not flattened and without a furrow 21
21 Anterior end rounded and obliquely truncate..*P. parvulus*
21 Anterior end forming 2 lobes, one markedly higher than the other, lateral edges fold downwards to form a broad shallow ventral groove..*P. oscillans*
22 Cells markedly spirally twisted through (1.5–)2–3 complete turns; flagellum 1.5 times cell length ....*P. helicoides*
22 Cells straight or only slightly twisted; flagellum cell length or less.................................23
23 Cells consisting of 2 unequal halves, each wing-like in appearance.......................................24
23 Cells of a different form 25
24 Cells ovate to pear-shaped in outline, <19 μm wide, <8 μm long, with two halves twisted in opposite directions ....*P. anomalus*
24 Cells broadly oval, usually >19 μm wide, >24 μm long, with two halves not twisted in opposite directions..........*P. alatus* var. *latviensis*
25 Posterior end with a short tail-piece or extension...26
25 Posterior end with a long tail-piece......................29
26 Cells with a short extension at posterior end ....*P. acuminatus*
26 Cells with a distinct short tail-piece 27
27 Cells spherical to broadly oval, >38 μm wide, >50 μm long.............................*P. orbicularis*
27 Cells irregularly rectangular, pear-shaped to obovoid or longitudinally oval, <20 μm wide, <36 μm long..28
28 Cells irregularly rectangular, pear-shaped to obovoid, flattened and slightly narrowed at middle, with tail-piece straight on; paramylon bodies 3–4 ....*P. striatus*
28 Cells longitudinally ovoid or oval, posterior end narrowing to a short, sharp tail-piece; paramylon body single, pseudo-ring-shaped ....*P. applanatus*
29 Cells with one rim folded ....*P. circumflexus*
29 Cells without a folded rim 30
30 Cells broadly obovoid to pear-shaped in outline; anterior end rounded, sometimes shallowly bilobed with lobes about equal in size ....*P. longicauda*
30 Cells obovoid or spindle-shaped, shallowly bilobed but one lobe usually slightly higher than the other 31
31 Cells usually >40 μm wide, <112 μm long, broadly spindle-shaped, spirally twisted in posterior part ....*P. tortus*
31. Cells >40 μm wide, >112 μm long, longitudinally obovoid, if twisted only very slightly 32
32 Cells not twisted; paramylon body single and ring-like ....*P. acutus*
32 Cells slightly twisted; paramylon bodies small, numerous, not ring-like ....*P. elegans*

Phacus acuminatus A.Stokes 1885

04070010 **Pl. 38S** (p. 160) **CD**

Cells (20–)23–30 μm wide, (23–)25–40 μm long, broadly ovoid to oval in outline (subspherical), sometimes triangular with straight lateral margins, very thin, greatest width below middle; shallow dorsal furrow extending half to three-quarters cell length, short extension at posterior end and incision at anterior end; pellicle longitudinally striated; chloroplasts parietal, disc-shaped, numerous; paramylon bodies 1–2, ring-like, sometimes second body near posterior of cell; flagellum approximately length of cell; eyespot present.

Probably cosmopolitan; occurs in swamps, ditches, puddles, ponds, fish ponds and lakes, often in the same habitats as *Euglena* or *Lepocinclis*; known in very clean water through to moderately polluted water.

Phacus acutus Pochmann 1941

Synonym: *Phacus longicauda* var. *ovata* Skvortsov

04070020 **Pl. 39M** (p. 163)

Cells 31–34 μm wide, 125–236 μm long, narrowly oval or obovoid, slightly asymmetrical; posterior end narrowing to a long thin tail-piece; chloroplasts disc-shaped, small, numerous; pellicle longitudinally striated; paramylon body single, ring-like; flagellum shorter than cell length.

Europe; in lakes, ponds and clay-pits, rare; only reported from the British Isles by Lund (1942c).

Phacus agilis Skuja 1926

04070030 **Pl. 38Z,AA** (p. 160)

Cells 8–11 μm wide, 11–17.5 μm long, about 5 μm thick, subovoid or coffee-bean-shaped; longitudinal groove full length of ventral surface; anterior end with a small depression, posterior end terminates in a short, blunt, wart-like extension; chloroplasts parietal, 2 and watchglass-shaped, one each side of groove; paramylon bodies usually 2; large, watchglass-shaped and lying opposed to lateral margins of cells, also smaller paramylon bodies present; eyespot prominent, to side of point of insertion of flagellum; flagellum about cell length.

Europe, Asia and Africa; in ponds, possibly with vegetation, puddles and water bodies in forests, occurs in plankton and benthic, common.

Information about the number of chloroplasts varies; according to Lund (1938, p. 275), Prowse (1958, p. 165) and Wołowski (1998a, p. 70) it has 2 chloroplasts, whereas Zakryś & Walne (1994) consider it to have several chloroplasts.

Phacus alatus G.A.Klebs 1883

var. *latviensis* Skvortsov 1928

Synonym: *Phacus macrostigma* Pochmann

04070052 **Pl. 39E,F** (p. 163)

Cells (10–)19–34 μm wide, 24–45 μm long, broadly oval, divided into 2 unequal halves, wing-like in appearance;

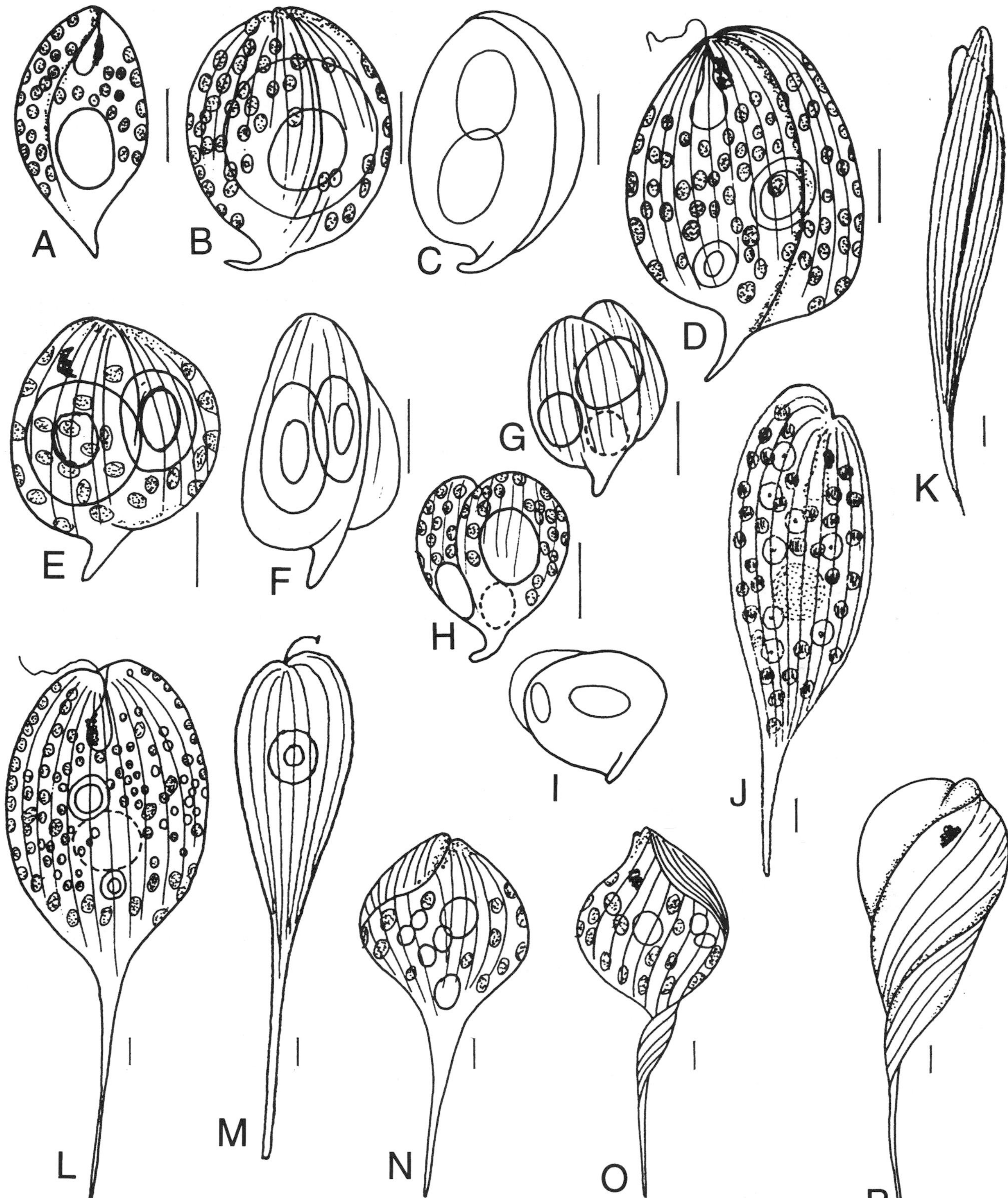

Plate 39 A–P
A. *Phacus hamelii* (p. 165); **B, C**. *Phacus curvicauda* (p. 164); **D**. *Phacus triqueter* (p. 168); **E, F**. *Phacus alatus* var. *latviensis* (p. 162); **G–I**. *Phacus anomalus* (p. 164); **J, K**. *Phacus elegans* (p. 164); **L**. *Phacus longicauda* (p. 165); **M**. *Phacus acutus* (p. 162); **N**. *Phacus circumflexus* (p. 164); **O**. *Phacus tortus* (p. 168); **P**. *Phacus helicoides* (p. 165).
Scale bars: 10 μm.

posterior end terminating in a strong, curved tail-piece; pellicle longitudinally striated; chloroplasts disc-shaped; paramylon bodies large, 2 in each cell; flagellum length of cell.

Europe; in small water bodies, puddles and ditches, rare.

Phacus anomalus F.E.Fritsch *et* M.F.Rich 1929

Synonym: *Phacus curvicauda* fo. *anomalus* (F.E.Fritsch *et* M.F.Rich) T.G.Popova *et* Safonova

04070060 **Pl. 39G–I** (p. 163)

Cells 16–18 μm wide, 19–22 μm thick, 23–27 μm long, ovate to pear-shaped, asymmetrical, a deep, broad ventral furrow or apical groove of varying length dividing cell into 2 unequal halves, wing-like, one half wider than the other, twisted in opposite directions; posterior end with a sharply bent tail-piece; chloroplasts parietal, disc-shaped, numerous; paramylon bodies 2; flagellum about cell length.

Europe, Africa, Asia; in lakes, ponds, fish ponds, ditches and puddles, uncommon.

This is a very variable species.

Phacus applanatus Pochmann 1941

04070040 **Pl. 40A** (p. 166)

Cells about 18 μm wide, 33–35 μm long, longitudinally ovoid or almost oval; apical groove short, only clearly visible in anterior part of cell; posterior end narrowing to a straight, short tail-piece; pellicle longitudinally and spirally striated; chloroplasts numerous, tightly packed; paramylon bodies one, pseudo-ring-shaped.

Europe; in ponds and puddles, rare.

Phacus caudatus K.Hübner 1886

04070070 **Pl. 40B** (p. 166)

Cells 16–27 μm wide, 31–50 μm long, pear-shaped to ovate, flattened, often slightly twisted, with a keel running full length of cell and triangular in cross section; anterior end broadly rounded; posterior end tapering to form a straight or sometimes curved, sharply pointed tail-piece 5–11 μm long; chloroplasts parietal, small, disc-shaped; pellicle longitudinally striated; paramylon 1 or 2 large ring-like bodies; flagellum about length of cell.

Probably cosmopolitan; in fish ponds, lakes, puddles and springs containing partly rotted leaves, a chance member of the plankton; known from clean water to moderately polluted water.

Several varieties are known. West & Fritsch (1927) reported a colourless variety (var. *hyalina*) in old cultures.

Phacus circumflexus Pochmann 1941

Synonyms: *Phacus longicauda* (Ehrenberg) Dujardin fo. *vixtorta* Kisselev, *P. longicauda* var. *torta* Skvortsov

04070080 **Pl. 39N** (p. 163)

Cells about 34–45 μm wide, 73–90 μm long, broadly ovoid, slightly twisted in posterior part with one rim folded, terminating in a long sharp tail-piece; chloroplasts numerous, disc-shaped, paramylon bodies ring-like and large; pellicle longitudinally striated; flagellum shorter than cell length.

Europe, Africa, Asia, South America; in ponds and puddles, rare.

According to Popova & Safonova (1976), it is a form of *P. longicauda*.

Phacus cochleatus Pochmann 1941

04070090 **Pl. 40F,G** (p. 166) **CD**

Cells 19–20 μm wide, 37–43 μm long, spindle-shaped, slightly asymmetrical, rounded at anterior end, with a sharp tail-piece at posterior end; pellicle spirally ribbed, detached from protoplast; paramylon bodies small and numerous.

Europe; known only from ponds, rare.

Phacus corculeum Pochmann 1941

Synonym: *Phacus brevicaudata* Lemmermann

04070100 **Pl. 38M,N** (p. 160)

Cells 12–13.6 μm wide, 17–20 μm long, heart-shaped, cut in at anterior end and slightly narrowed at posterior end; chloroplasts numerous; paramylon bodies 1–2, disc-shaped.

Europe, Asia, Africa; in ponds, ditches and puddles, rare.

Phacus costatus W.Conrad 1914

Synonyms: *Monomorphina pyrum* var. *costata* T.G.Popova, *Phacus pyrum* var. *costata* W.Conrad

04070330 **Pl. 40M,N** (p. 166)

Cells 6–16 μm wide, 20–36 μm long, longitudinally spindle-shaped; anterior end slightly narrowing, posterior end narrowing into a sharp, thin tail-piece; pellicle spirally ribbed and connected by transverse striae; chloroplasts numerous; paramylon of 2 large and several small bodies.

Europe, Asia; in lakes, ponds, swamps and small rivers, rare.

Phacus curvicauda Svirenko 1915

04070110 **Pl. 39B,C** (p. 163) **CD**

Cells (18.7–)20–25(–27) μm wide, 22.5–37.5 μm long, 10–12 μm thick, broadly oval, almost circular in outline, sometimes slightly wider than long, dorsal keel prominent and extending half to full length of cell, narrowing towards anterior end; posterior end terminating in a curved tail-piece; pellicle longitudinally striated; chloroplasts numerous, disc-shaped; paramylon bodies 2, large; flagellum shorter than cell length.

Probably cosmopolitan; known from small humic water bodies, puddles, ditches and fish ponds.

Several varieties are known.

Phacus elegans Pochmann 1941

04070120 **Pl. 39J,K** (p. 163)

Cells 38–40 μm broad, 127–147 μm long, longitudinally obovoid in outline, slightly twisted; posterior end narrowing to a long, sharp tail-piece; pellicle longitudinally striated; chloroplasts small, numerous, disc-shaped; paramylon bodies small; eyespot conspicuous; flagellum shorter than cell length.

Europe; in swamps and puddles, rare.

Phacus granum Dreżepolski 1925

04070130 **Pl. 38T,U** (p. 160)

Cells 7.5–12 μm wide, 18–23 μm long, cylindrical, narrowing at anterior end and with a short conical swelling (wart) at the posterior end; apical furrow short; pellicle longitudinally and sometimes slightly spirally striated; paramylon body large, single and angular; chloroplasts small, numerous; flagellum about same length as cell.

Probably cosmopolitan, known from Europe, Asia and North America; occurs in ponds and puddles, occurs in plankton, rare.

Phacus hamelii P.Allorge *et* M.Lefèvre 1930

Synonym: *Phacus pleuronectes* var. *hamelii* (P.Allorge *et* M.Lefèvre) T.G.Popova

04070340 **Pl. 39A** (p. 163)

Cells 12–20 μm wide, 25–37 μm long, asymmetrically elongated, ovoid to ellipsoidal, dorsal keel extending almost full length of cell, undulate; anterior end narrowly rounded, posterior end narrowing asymmetrically to a sharp tail-piece 4–6 μm long, slightly curved; pellicle longitudinally striated; chloroplasts minute, numerous; paramylon bodies 1–2, not large, circular and plate-like; flagellum shorter than cell length.

Europe, Asia, South America; in ponds, lakes and swamps, rare.

Phacus helicoides Pochmann 1941

Synonym: *Phacus longicauda* var. *torta* Lemmermann fo. *helicoides*

04070140 **Pl. 39P** (p. 163) **CD**

Cells 39–54 μm wide, 70–120 μm long, spindle-shaped, elongated or spindle-shaped to pear-shaped in outline, margins with 2 or 3 bulges, spirally twisted through 1.5 to 2 complete turns, usually 3; anterior end narrowing and bilobed; posterior end tapering into a twisted, long, straight tail-piece about half cell length (about 25 μm long); pellicle longitudinally striated and following turns; chloroplasts numerous and disc-shaped; paramylon one large body; flagellum about 1.5 times length of cell.

Probably cosmopolitan; known from ponds, lakes, swamps and ditches, freshwater and brackish water, commonly a chance member of the plankton in shallow water; indicator of mildly to moderately polluted water.

Compare with *P. tortus* (Lemmermann) Skvortsov, which is broader and twisted only in the posterior portion of the cell.

Phacus hispidulus (Eichwald) Lemmermann 1910

Synonyms: *Euglena hispidula* Eichwald, *Phacus hispidula* (Eichwald) Lemmermann, *P. hispidulus* var *steinii* Lemmermann, *Chloropeltis hispidula* F.Stein

04070150 **Pl. 40Q** (p. 166)

Cells 20–25(–36) μm wide, (34–)39–55(–63) μm long, ovoid or obovoid; anterior end with a slight furrow and a small tubular extension at apex; posterior end narrowing to a sharp tail-piece; pellicle longitudinally striated and covered by fine spines (setae); chloroplasts small, numerous; paramylon of 2 large and several small bodies; flagellum shorter than cell length; eyespot small.

Probably cosmopolitan; in lakes, ponds, ditches and swamps.

Phacus longicauda (Ehrenberg) Dujardin 1841

Basionym: *Euglena longicauda* Ehrenberg

Synonym: *Phacus longicauda* var. *maior* Svirenko

04070160 **Pl. 39L** (p. 163) **CD**

Cells (40–)45–70 μm wide, 85–170 μm long, broadly obovoid to pear-shaped in outline, slightly twisted; anterior end rounded, sometimes shallowly bilobed; posterior end gradually tapering to form a long, straight, sharply pointed tail-piece up to 70 μm long; pellicle longitudinally striated; chloroplasts disc-shaped; paramylon usually in form of a single large (or small) circular plate- or ring-like body, often accompanied by smaller ones; flagellum approximately cell length or shorter.

Probably cosmopolitan; in impoundments, lakes, ponds and swamps whether freshwater, saltwater, mineralized or with much humic material, planktonic or a chance member of the plankton; known from clean to moderately polluted water.

Several varieties are known.

Phacus monilatus A.Stokes

var. *suecicus* Lemmermann 1904

Synonyms: *Phacus suecica* Lemmermann, *P. hispidulus* fo. *suecicus* (Lemmermann) T.G.Popova

04070182 **Pl. 40O,P** (p. 166) **CD**

Cells 18–25 μm wide, (28–)30–38 μm long, ovoid to suborbicular or obovate in outline; anterior end slightly concave with a small prominent papilla at apex; posterior end narrowing to a long, sharp, straight or slightly curved tail-piece; pellicle longitudinally striated with rows of small granules, sometimes with alternate lines of small or large granules, number of rows variable; paramylon of 2 large, crescent-shaped or disc-like bodies; chloroplasts small, numerous, disc-shaped; flagellum shorter than cell; eyespot small.

Europe, Asia, Africa, North America; in ponds, lakes, swamps and small rivers, benthic, a chance member of the plankton, rare; indicator of unpolluted water.

Phacus nordstedtii Lemmermann 1904

04070350 **Pl. 40H,I** (p. 166)

Cells about 18.5 μm wide, 36 μm long, ovate to broadly ovate, slightly asymmetrical; anterior end broadly rounded, posterior end with a long, straight, sharply pointed tail-piece; chloroplasts numerous, ovoid and indistinctly disc-shaped; paramylon bodies small; pellicle slightly detached from protoplast, spirally ribbed.

Europe; known only from ponds, rare.

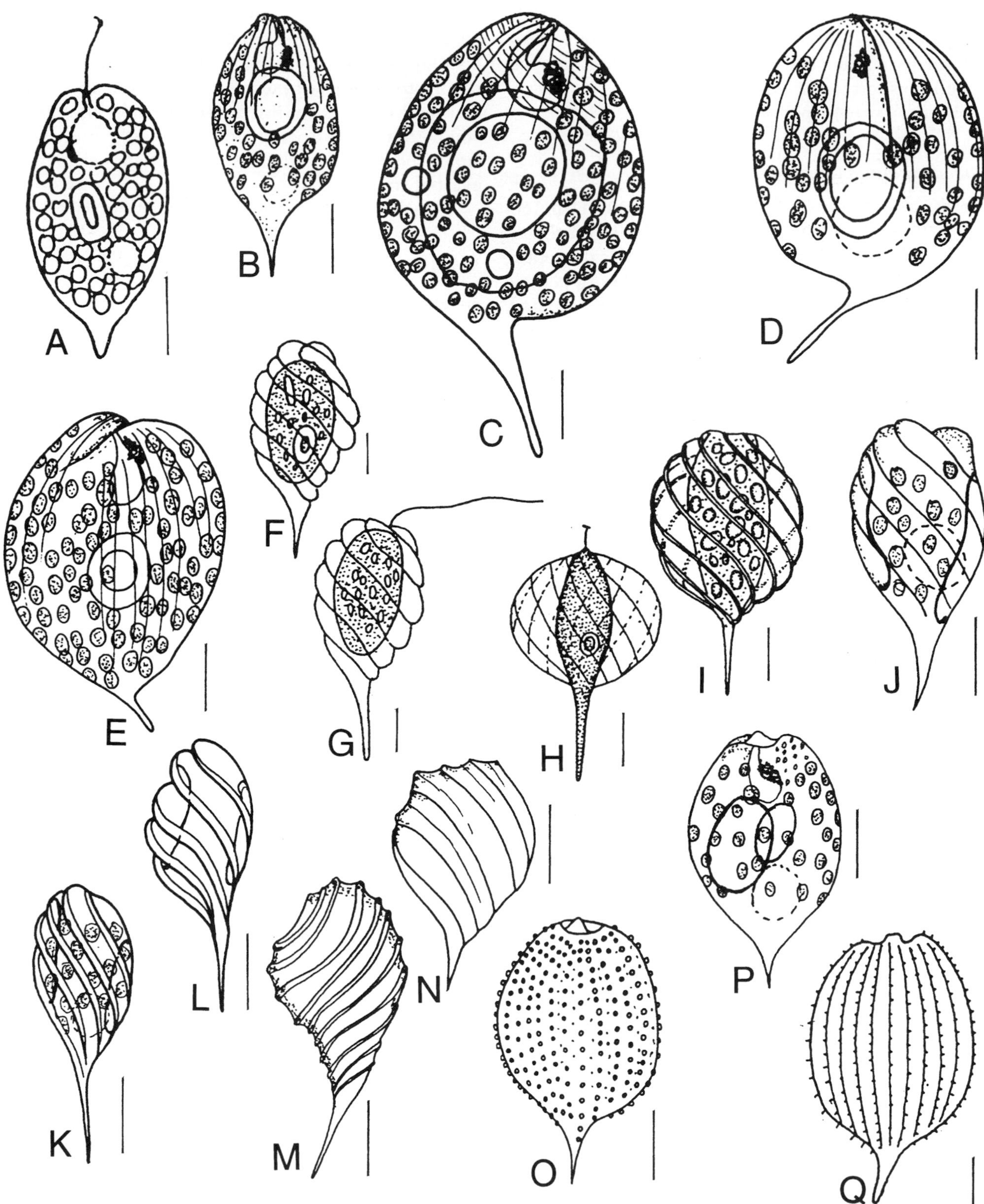

Plate 40 A–Q

A. *Phacus applanatus* (p. 164); **B**. *Phacus caudatus* (p. 164); **C, D**. *Phacus orbicularis* (p. 167); **E**. *Phacus pleuronectes* (p. 167); **F, G**. *Phacus cochleatus* (p. 164); **H, I**. *Phacus nordstedtii* (p. 165); **J**. *Phacus pseudonordstedtii* (p. 167); **K, L**. *Phacus pyrum* (p. 167); **M, N**. *Phacus costatus* (p. 164); **O, P**. *Phacus monilatus* var. *suecicus* (p. 165); **Q**. *Phacus hispidulus* (p. 165).
Scale bars: 10 μm

Phacus orbicularis K.Hübner 1886

04070190 **Pl. 40C,D** (p. 166) **CD**

Cells 39–46 μm wide, 50–80 μm long, orbicular to broad ovoid in outline; anterior end slightly narrowly rounded; posterior end broadly rounded with a short curved tail-piece; apical groove nearly full length of cell; pellicle longitudinally and roughly striated, short transverse striae present; paramylon of 1–2 disc- or circular plate-like bodies; flagellum same length as cell.

Probably cosmopolitan; in stagnant mineralized, mildly polluted and humic waters in ponds, swamps, ditches and lakes, also in brackish water; indicator of clean to mildly polluted water.

Phacus oscillans G.A.Klebs 1883

04070200 **Pl. 38AB,AC** (p. 160)

Cells (5–)7–10(–13) μm wide, (15–)20–26(–35) μm long, oval-cylindrical to ellipsoidal; anterior end slightly broader and with 2 lobes (one higher than other), posterior end narrowing slightly with a short (5 μm) blunt projection or rounded at end; lateral edges fold downwards to form a broad, shallow, ventral groove; pellicle spirally striated; chloroplasts parietal, small, numerous, crowded and disc-shaped; paramylon of 2 ring-shaped bodies, 3–6.5 μm in diameter, in about centre of cell; flagellum same length as cell.

Probably cosmopolitan; from ponds, lakes, forest water bodies and fish ponds, occurs in plankton or benthic, common; indicator of moderately polluted water.

Phacus parvulus G.A.Klebs 1883

Synonyms: *Phacus acuminatus* fo. *minor* Kisselev, *P. subsalsus* Proshkina-Lavrenko

04070210 **Pl. 38O,P** (p. 160)

Cells (7.6–)8–12 μm wide, 16–22(–30) μm long, obovoid to wedge- or heart-shaped; anterior end rounded and obliquely truncate; posterior end slightly tapering to a small, bluntly rounded, short extension; pellicle spirally striated; chloroplasts numerous, small; paramylon a single ring-shaped body located in centre of cell; flagellum equal to cell length; eyespot conspicuous.

Probably cosmopolitan; in ponds, ditches, pools, swamps and slow-flowing waters; considered an indicator of moderately polluted water.

Phacus pleuronectes (O.F.Müller) Dujardin 1841

Basionym: *Cerasteria pleuronectes* O.F.Müller

04070220 **Pl. 40E** (p. 166) **CD**

Cells (29–)37–46.8(–70) μm wide, (42–)50–80(–100) μm long, broadly ovoid to suborbicular, slightly asymmetrical; anterior end narrowly rounded and shallowly bilobed, apical furrow up to half cell length, slightly twisted, in transverse section ellipsoidal, slightly S-shaped; posterior end with a short (5–7 μm), slender tail-piece turning obliquely to one side; dorsal keel from anterior end; pellicle longitudinally striated; chloroplasts parietal, numerous, disc-shaped; paramylon usually one large or 2 ring- or disc-shaped bodies; flagellum equal to or longer than cell length; eyespot conspicuous.

Probably cosmopolitan; in marginal and quieter waters of moderately nutrient-enriched lakes, ponds and ditches, also in brackish water; considered an indicator of mildly to moderately polluted water.

Phacus pseudonordstedtii Pochmann 1941

04070230 **Pl. 40J** (p. 166)

Cells 10–16 μm wide, 31–42 μm long, ovoid to broadly ellipsoidal, flattened dorsiventrally; anterior end broadly rounded; posterior end narrowing into a long, straight, sharp tail-piece; pellicle roughly spirally ribbed; chloroplasts numerous, disc-shaped; paramylon 2 large bodies; flagellum 1.5 times cell length; eyespot large.

Europe, Asia, Africa, North America; in lakes, fish ponds and ditches, as well as marginal and quieter waters; indicator of relatively clean to moderately polluted water.

Popova & Safonova (1976) consider this to be a possible variety of *P. pyrum* (Ehrenberg) F.Stein.

Phacus pusillus Lemmermann 1910

Synonyms: *Phacus alatus* G.A.Klebs, *P. parvulus* var. *pusillus* (Lemmermann) T.G.Popova

04070240 **Pl. 38Q,R** (p. 160)

Cells 7–8.5 μm wide, (18–)20–33 μm long, ovoid to elongate oval, flattened dorsiventrally with a wide furrow running full length of cell, slightly twisted; anterior end rounded; posterior end slightly narrowed; pellicle spirally striated; chloroplasts parietal, small, numerous, disc-shaped; paramylon bodies 1–2, ring-shaped; flagellum same length as cell; eyespot conspicuous.

Probably cosmopolitan; known from the marginal and quieter waters of moderately nutrient-enriched lakes, ponds and ditches.

Phacus pyrum (Ehrenberg) F.Stein 1878

04070250 **Pl. 40K,L** (p. 166) **CD**

Cells (7–)15.6–21 μm wide, 27–30(–55) μm long, ovoid to spindle-shaped, possibly slightly flattened; anterior end broadly rounded and narrowing gradually at posterior end to a long, straight and finely pointed tail-piece; pellicle faintly spirally ribbed; paramylon 2 ring-like or disc-shaped bodies, laterally positioned, sometimes paramylon of several smaller disc-shaped bodies; chloroplasts small, numerous, disc-shaped; flagellum emerges between 2 papillae, about 1.5 times cell length; eyespot not prominent.

Probably cosmopolitan; from ditches, lakes and fish ponds, freshwater and brackish water, sometimes a chance member of the plankton; considered an indicator of clean to moderately polluted water.

Phacus skujae Skvortsov 1928

Synonym: *Phacus pusilla* Lemmermann *sensu* Skuja

04070280 **Pl. 38AD,AE** (p. 160)

Cells 5–17 μm wide, 13–30 μm long, longitudinally oval to spindle-shaped, slightly bent and flattened; anterior end slightly narrowing and obliquely truncate, shallowly bilobed with one side projecting; posterior end gradually narrowing to a small tail-piece and sides slightly swollen; pellicle almost longitudinally striated; chloroplasts numerous, disc-shaped; paramylon 1–3, ring-like or rod-like bodies; eyespot large.

Probably cosmopolitan; in ponds, lakes and slow-flowing rivers, occurs in plankton and benthic; considered an indicator of moderately polluted water.

Phacus spiralis Allegre *et* T.L.Jahn 1943

04070360 **Pl. 38V–X** (p. 160)

Cells about 10 μm wide, 24–26 μm long, longitudinally obovate, flattened, slightly twisted; anterior end rounded; posterior end tapering to an oblique tail-piece; concave in ventral view, with an S-shaped and rigid keel extending full length of cell; chloroplasts small, numerous, disc-shaped; paramylon 1 or 2 ring-shaped bodies; flagellum length of cell; eyespot small.

Europe, Asia, North America; known from ponds, lakes and stagnant bodies of water in forests, planktonic and benthic, rare.

Phacus stokesii Lemmermann 1901

04070290 **Pl. 38K,L** (p. 160)

Cells (30–)39–41.6(–48) μm wide, 37.2–46(–55) μm long, broadly oval, subspherical, flat; anterior end with a slight furrow at apex; posterior end rounded; apical groove extending into posterior end; pellicle longitudinally striated; paramylon a single, large, disc-shaped body, sometimes also many very small bodies; chloroplasts parietal, numerous, disc-shaped; eyespot distinctly cup-shaped; flagellum about cell length.

Europe, Asia, North and South America; in ponds, swamps, small lakes and ditches; indicator of unpolluted water.

Phacus striatus (Drezepolski) Francé 1893

Basionym: *Phacus aenigmatica* Drezepolski *non P. striatus* Lemmermann

04070300 **Pl. 38Y** (p. 160)

Cells 8.5–19 μm wide, 19–25 μm long, irregularly rectangular, pear-shaped to obovoid flattened, each cell slightly narrowed at middle; posterior end prolonged into straight or sometimes oblique tail-piece; pellicle firm and spirally striated; chloroplasts numerous, disc-shaped, polygonal when appressed; chloroplasts small, plate-shaped; paramylon 3–4 large bodies, peripherally arranged; flagellum about length of cell; eyespot near anterior end, often having a lens-shaped outer region; nucleus oval, usually located posteriorly.

Probably cosmopolitan; known from ditches, puddles, swamps and ponds, occurs in plankton and benthic; considered an indicator of clean to moderately polluted water.

Phacus tortus (Lemmermann) Skvortsov 1928

Basionym: *Phacus longicauda* var. *torta* Lemmermann

04070310 **Pl. 39O** (p. 163)

Cells (38–)40–50(–52) μm wide, (80–)85–95(–112) μm long, broadly spindle-shaped, broadest at anterior third of cell; anterior end conically rounded, tapering and spirally twisted in posterior region to form a long, straight or sometimes slightly curved tail-piece about length of cell; pellicle longitudinally or spirally striated; chloroplasts parietal, disc-shaped; paramylon 1 (or 2) large, circular plate-like body; flagellum two-thirds length of cell.

Probably cosmopolitan; in the shallow water of ditches, puddles, swamps and ponds.

Phacus triqueter (Ehrenberg) Dujardin 1841

Basionym: *Euglena triquetra* Ehrenberg

04070320 **Pl. 39D** (p. 163)

Cells 23–35 μm wide, 49–55 μm long, broadly ovoid, slightly asymmetrical, markedly concave to convex; anterior end broadly rounded; posterior end abruptly narrowing to a long, thin, sharp and curved tail-piece; pellicle longitudinally striated; dorsal keel prominently extending full length of cell, cell triangular in cross section, slightly hollowed on the ventral surface; chloroplasts disc-shaped; paramylon 2 ring-shaped bodies; flagellum same length as cell.

Probably cosmopolitan; occurs in lakes, dams, ponds, swamps and small rivers, freshwater and in brackish-water; indicator of clean to moderately polluted water.

Strombomonas Deflandre 1930

Cells similar to *Euglena*, enclosed in a lorica, free-swimming, with one emergent flagellum; chloroplasts numerous, lateral, plate-shaped or polygonal, usually without pyrenoids; paramylon bodies variously shaped; eyespot large, close to apical reservoir or invagination; protoplast tending to occupy entire lorica.

Cosmopolitan; in freshwater and brackish-water ponds, ditches and puddles, carried into the plankton by chance.

Separated from *Trachelomonas* by Deflandre (1930) by having a generally longer collar, tapering gradually into the lorica and not sharply demarcated from it, posterior end of lorica narrowing to a conical point, wall often variable in thickness but very rarely punctate, pitted, furrowed, perforated or ornamented by spines. The emergent flagellum is relatively shorter than in *Trachelomonas*, rarely exceeding length of lorica. The taxonomy is based generally on the structure of the lorica. Source data from Deflandre (1930).

1 Lorica triangular to trapezoid; walls smooth or irregularly warty, with or without an extension at the posterior end; flagellum 2–3 times cell length .... *S. acuminatus*
1 Lorica ellipsoidal to ovoid; flagellum less than twice cell length 2
2 Lorica with a short cylindrical collar, posterior end with a short conical point, broadly swollen, walls irregularly and minutely wrinkled .... *S. deflandrei*
2 Lorica with a short straight collar and the rim dentate, walls smooth, warty or punctate 3
3 Lorica <27 μm long, with posterior end broadly narrowed or rounded and apex

usually slightly conical; walls irregularly wrinkled and covered with warts........*S. eurystoma*
3 Lorica >38 μm long, posterior end narrowing to a long, straight or slightly curved point; walls smooth or punctate, with irregular transverse furrows......*S. tambowika*

Strombomonas acuminatus (Schmarda) Deflandre 1930

Basionym: *Trachelomonas acuminata* (Schmarda) F.Stein
Synonyms: *Strombomonas acuminata* var. *triangulata* Skvortzov, *S. verrucosa* formae *conspersa, dadai, ovalis* T.G.Popova
04090010 **Pl. 41A–H** (p. 170)
Lorica 18–27(–30) μm wide, (27–)35–48(–51) μm long, triangular or trapezoidal; anterior end distinctly narrowed; collar obliquely truncate, 3–4(–5) μm high, 5–7 μm wide; posterior end with a prominent straight or slightly curved extension 7–11 μm long, sometimes without an extension and posterior end rounded; walls smooth, pointed or irregularly warty, colourless or brown; chloroplasts numerous, without pyrenoids; paramylon bodies small, rod-like; flagellum 2–3 times cell length.

Probably cosmopolitan; in ponds, lakes, rivers and in brackish-water as well as freshwater.

Species is very variable.

Strombomonas deflandrei (Y.V.Roll) Deflandre 1930

Basionym: *Trachelomonas deflandrei* (Y.V.Roll) Deflandre
04090020 **Pl. 41M** (p. 170)
Lorica 23–26 μm broad, 39–45 μm long, broadly ellipsoidal to ovoid; anterior end slightly narrowed and having a short, cylindrical collar with a dentate margin, collar 6.8 μm wide, 3–7 μm high; posterior end rounded with a short conical point about 2.7 μm long; sides broadly swollen, sometimes nearly straight; walls irregularly and minutely wrinkled, brown; chloroplasts small and numerous, without pyrenoids; flagellum 1.5 times length of lorica.

Europe, Asia; in ponds, lakes and rivers, uncommon.

Strombomonas eurystoma (F.Stein) T.G.Popova 1966

Basionym: *Trachelomonas eurystoma* F.Stein
04090030 **Pl. 41I–L** (p. 170) **CD**
Lorica (14–)19–26 μm wide, (21–)23–26 μm long, broadly ellipsoidal or obovoid; anterior end gradually narrowing to a short cylindrical collar with a dentate rim, collar 5.5–8 μm wide and 2–3 μm high; posterior end broadly narrowed or rounded, usually with a slightly conical apex; walls covered with warts and irregularly wrinkled, colourless or brown.

Europe, Africa; known only from ponds, planktonic.

Strombomonas tambowika (Svirenko) Deflandre 1930

Basionym: *Trachelomonas tambowika* Svirenko
04090050 **Pl. 41N,O** (p. 170)
Lorica (20–)27–30 μm wide, (39–)47–55 μm long, ellipsoidal to obovoid; anterior end narrowing to a short, straight collar and with a dentate rim; posterior end narrowing to a long, straight or slightly curved point one-tenth to one-fifth length of lorica; walls thin, irregularly wrinkled, smooth or punctate with transverse irregular furrows, yellowish-brown; cells occupying entire lorica; chloroplasts plate-like; flagellum as long as lorica.

Europe; occurs in ponds and old river beds, rare.

Trachelomonas Ehrenberg 1833

Cells similar to *Euglena*, solitary, enclosed within a lorica, free-swimming; pellicle flexible, only loosely encased in a rigid lorica, capable of euglenoid movement within lorica, occasionally cell escapes through pore (especially during reproduction); one emergent locomotory flagellum (as in *Euglena*), emerges through a wide, circular, apical pore in lorica; eyespot and flagellar swelling at anterior end; chloroplasts 2 to many, rarely absent, disc-shaped, rarely parietal, with or without pyrenoids; paramylon bodies small, sometimes absent; lorica spherical, cylindrical or ellipsoidal, sometimes bell- or spindle-shaped, sometimes with a distinct collar (neck) arising at apical pore; walls smooth or net-like, punctate, scrobiculate, warty or ornamented with spines, becoming brown due to heavy impregnation with manganese and/or iron deposits; reproduction by division of immobile cells usually within lorica, one (or occasionally both) daughter cells emerge and secrete a new lorica; division of protoplast outside lorica forming a palmelloid stage; cysts known for several species

Cosmopolitan; shallow waters, especially where part of the environment has reducing conditions and low oxygen levels, in such environments the levels of soluble manganese and iron are often high, which presumably favours their deposition in the lorica; it seems likely that many species tend to move in and out of reducing conditions, so cells may also be found in well-oxygenated water.

The lorica (test, envelope) not only distinguishes this genus from related ones, but its ornamentation serves as the main taxonomic feature for identification. More recently it has been investigated by scanning and transmission electron microscopy.

1 Lorica always with a short, conical and acute projection .... *T. caudata*
1 Lorica without a posterior projection, or only rarely present and then very short..................2
2 Lorica spherical or slightly flattened along long axis, if a ring-like apical thickening present then not surrounded by warts...................3
2 Lorica not spherical, if broadly rounded then having a ring-like apical thickening surrounded by warts 20
3 Lorica usually smooth 4
3 Lorica ornamented, or if smooth then compressed and transversely oval (*T. curta*) or with a fold at the anterior end............................8

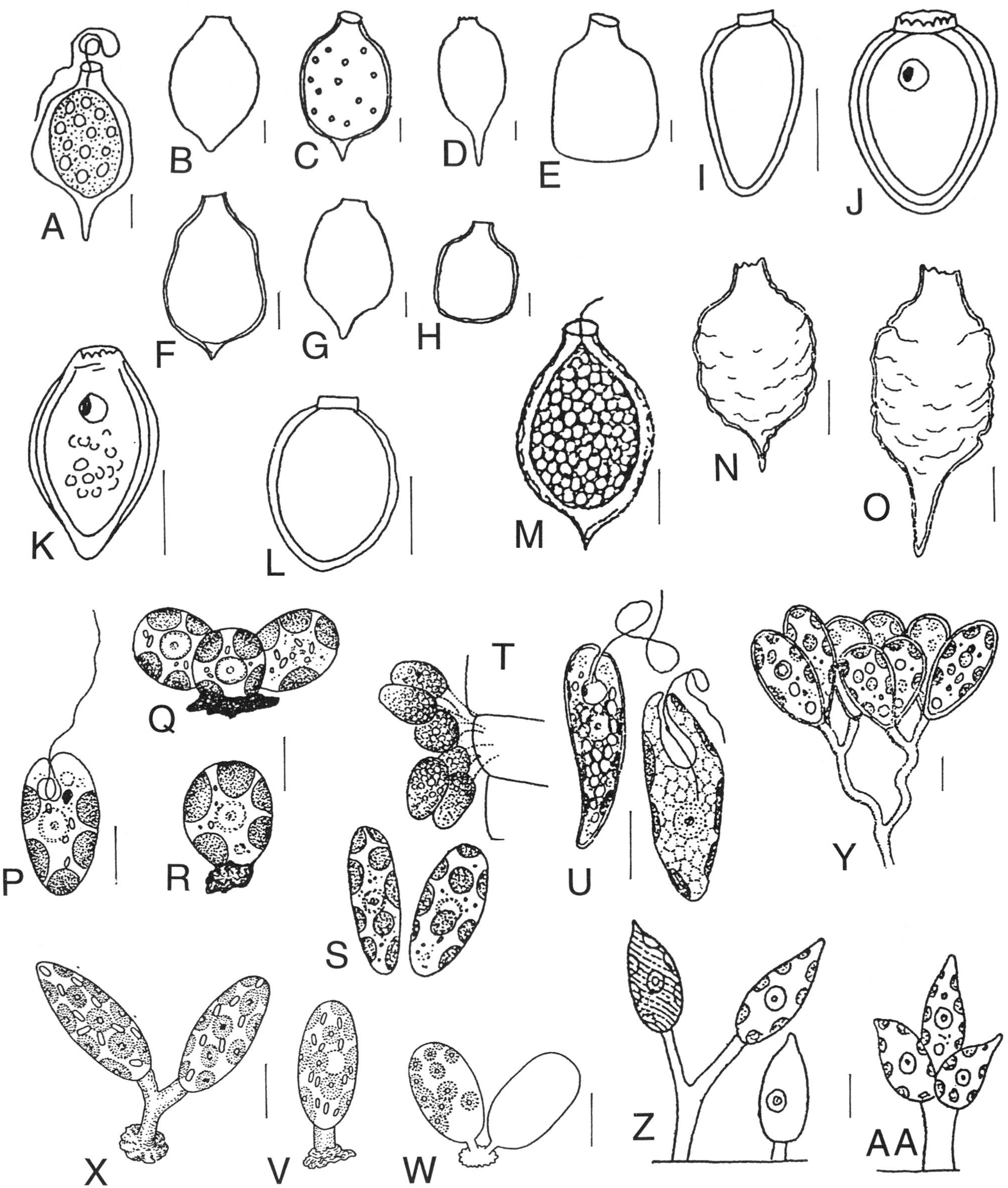

Plate 41 A–AA

A–H. *Strombomonas acuminatus* (p. 169); **I–L**. *Strombomonas eurystoma* (p. 169); **M**. *Strombomonas deflandrei* (p. 169); **N, O**. *Strombomonas tambowika* (p. 169); **P–S**. *Colacium vesiculosum* fo. *cyclopicola* (p. 146); **T–W**. *Colacium vesiculosum* (p. 146); **X–Y**. *Colacium vesiculosum* fo. *arbuscula* (p. 146); **Z, AA**. *Colacium epiphyticum* (p. 146).

Scale bars: 10 μm.

4 Lorica spherical or almost spherical; cells with 2 chloroplasts, each with a double sheathed pyrenoid 5
4 Lorica spherical; cells with several plate-shaped chloroplasts, without pyrenoids................7
5 Lorica with an apical pore surrounded by a thickening, rarely with a collar ....*T. volvocina*
5 Lorica with an apical pore surrounded by a collar...6
6 Collar low ....*T. volvocina* var. *derephora*
6 Collar depressed ....*T. volvocina* var. *subglobosa*
7 Lorica with an apical pore surrounded by a ring-like thickening or without it ...*T. volvocinopsis*
7 Lorica with an apical pore surrounded by a low flat thickening, with a cylindrical canal extending inwards into the lorica cavity ....*T. cervicula*
8 Lorica with spines, without a fold surrounding the anterior part of the lorica 9
8 Lorica mainly without spines, if with spines then having a fold surrounding anterior part, usually variously ornamented.............................11
9 Lorica covered by short, regularly distributed spines ....*T. globularis*
9 Lorica with spines confined to anterior and/or posterior part 10
10 Lorica with a few spines at both ends ....*T. rostafinskii*
10 Lorica with spines confined to anterior end ....*T. acanthostoma*
11 Lorica smooth or covered by granules, warts or short spines 12
11 Lorica covered by folds or ribs 14
12 Lorica covered with irregular warts or granules and without punctae; broadly oval and not compressed......*T. verrucosa* var. *irregularis*
12 Lorica smooth or covered with short spines, if covered with granules then irregularly punctate 13
13 Lorica compressed, transversely oval ....*T. curta*
13 Lorica not compressed, more or less spherical 18
14 Lorica covered with slightly spiralled and anastomosing ribs.................................*T. rugulosa*
14 Lorica with another type of ornamentation, or if having spiralled ribs then not anastomosing in a regular pattern 15
15 Lorica with slightly convex ribs in a regular pattern, not spiralled ....*T. rugulosa* var. *steinii*
15 Lorica covered with irregular ribs or folds, sometimes in a spiral pattern 16
16 Lorica with irregular, discontinuous and twisted folds, not spirally arranged ....*T. rugulosa* var. *meandriana*
16 Lorica covered with spiralled ribs or by folds forming a net-like pattern....................................17
17 Lorica with distinctly spiralled ribs, which generally do not anatomose ....*T. rugulosa* var. *obliqua*
17 Lorica covered with thick folds forming a net-like pattern.....................................*T. columba*
18 Lorica covered with granules and irregularly punctate, without a fold in the anterior part ....*T. subverrucosa*
18 Lorica smooth or covered with short spines, with a fold in the anterior part............................19
19 Lorica smooth, pore thickened at rim and surrounded by a thick crenate fold in the anterior part ....*T. zuberi*
19 Lorica covered with short spines, a few spines some distance from apical pore, pore surrounded by a ring-like thickening and with a fold surrounding anterior part of lorica ....*T. wislouchii*
20 Lorica smooth (minute spines sometimes present, scarcely visible) or variously ornamented...21
20 Lorica usually covered by variously developed spines 43
21 Walls smooth (sometimes with scarcely visible spines) or punctate, if warty then also punctate 22
21 Walls not as above but irregularly roughened, covered with regular pit-like depressions (scrobiculate), warty, granulate or with other small protruberances 36
22 Walls punctate and covered by groups of 2 or 4 warts ....*T. atrata* var. *pustulosa*
22 Walls smooth (sometimes with scarcely visible spines) or punctate, if punctate then without warts...23
23 Walls smooth or with minute and scarcely visible spines...24
23 Walls distinctly punctate....................................28
24 Lorica cylindrical, rectangular or ellipsoidal to cylindrical..25
24 Lorica broadly ellipsoidal 27
25 Lorica with a straight cylindrical collar (2.5–4 μm wide, 2.3–5 μm high) and reddish-brown..*T. dubia*
25 Lorica with apical pore collarless or with a small, thick or low collar, yellow-brown or light brown..26
26 Lorica oblong, almost rectangular (8–14 μm wide, 10–16 μm long).........*T. oblonga* var. *truncata*
26 Lorica broadly cylindrical with parallel sides (7.5–10 μm wide, 14.5–21 μm long).....*T. cylindrica*
27 Lorica usually >15 μm wide, apical pore surrounded by a cylindrical collar (2.3–3.3 μm high; chloroplasts without pyrenoids....*T. manginii*
27 Lorica usually ≤15 μm wide, apical pore with or without a ring-like thickening; chloroplasts each with a pyrenoid...........*T. oblonga*
28 Apical pore with a collar or ring-like thickening..32
28 Apical pore without a collar 29
29 Lorica obovoid, widened at anterior end ....*T. stokesii*

29 Lorica oval to ellipsoidal, or oblong......................30
30 Lorica oval with bluntly rounded apices, brown ....*T. intermedia*, in part
30 Lorica not as above..31
31 Lorica longitudinally ellipsoidal to oval, yellow or reddish-brown........*T. abrupta* var. *minor*
31 Lorica oblong, with small punctatae, brown ....*T. oblonga* var. *punctata*
32 Apical pore usually surrounded by a low collar ....*T. intermedia*, in part
32 Apical pore surrounded by a well-developed collar...33
33 Collar distinctly bent to one side or only very slightly oblique ....*T. similis*
33 Collar not bent to one side...................................34
34 Lorica dark brown, with collar very low and crenate at rim.......*T. intermedia* fo. *crenulatocollis*
34 Lorica reddish-brown, with collar straight or slightly oblique and rim irregularly toothed 35
35 Collar low and straight ....*T. lefevrei*
35 Collar high (2–5 μm), straight or oblique ....*T. planctonica*
36 Walls thick and with many small depressions ....*T. zorensis*
36 Walls rough and/or covered by warts and granules 37
37 Walls covered by worm-like (vermicular) granules...*T. gregussii*
37 Walls rough, covered by warts or regularly shaped granules...38
38 Lorica irregularly roughened, apical pore with a ring-like thickening surrounded by warts, ≤19 μm long ....*T. pseudofelix*
38 Lorica rough, granular or with generally distributed warts, with apical pore not surrounded by warts, usually >19 μm long.........39
39 Lorica cylindrical, with parallel sides and broadly rounded ends.............................*T. lacustris*
39 Lorica ellipsoidal..40
40 Walls with a few warts bearing small knobs ....*T. pulchella*, in part
40 Walls without warts bearing small knobs, if warts present then these minute and irregular, sometimes bearing larger warts............41
41 Walls evenly covered with small granules ....*T. granulosa*
41 Walls irregularly ornamented...............................42
42 Lorica with apical pore surrounded by a low, straight or slightly oblique collar, pale yellow to red-brown; walls rough and warty with irregularly arranged minute protuberances ....*T. scabra*
42 Lorica with apical pore surrounded by a low collar having a slightly thickened rim, yellow-brown to brown; walls with minute irregular warts sometimes bearing 2–7 larger warts...*T. granulata*
43 Lorica obovate, slightly narrowed at posterior end, wider at anterior end.......*T. amphora*
43 Lorica of another shape..44
44 Lorica an elongated bulb-shape or ovoid, anterior end slightly tapering; apical pore surrounded by a conical or almost cylindrical collar (4.6–7 μm wide, 6–7 μm high)...........*T. bulla*
44 Lorica of another shape; apical pore with a low or toothed collar, or lacking a collar 45
45 Walls with a few warts bearing short, blunt spines.......................................*T. pulchella*, in part
45 Walls without warts bearing spines 46
46 Lorica >26 μm wide 47
46 Lorica ≤26 μm wide 48
47 Collar low and toothed, scalloped or spiny; lorica at posterior end broadly rounded and bearing stout, long spines ....*T. armata*
47 Collar absent; lorica rounded at apices and evenly covered with sharp, pointed, dense conical spines...*T. superba*
48 Lorica cylindrical to ellipsoidal or oval, one or both ends flattened, truncate or conical...49
48 Lorica ellipsoidal or broadly oval, ends more or less rounded 51
49 Lorica with a low collar, without punctae ....*T. polonica*
49 Lorica without a collar, punctate 50
50 Apical pore surrounded by a ring-like thickening; lorica ≤15.5 μm wide, finely punctate ....*T. klebsii*
50 Apical pore without a ring-like thickening; lorica >15 μm wide, with spines coarsely punctate ....*T. abrupta*
51 Lorica covered with minute, sharp spines............52
51 Lorica covered by large spines 54
52 Apical pore without a collar or with short, slightly raised pore margins......................*T. hispida*
52 Apical pore surrounded by a low collar 53
53 Lorica with margin of collar bearing a circle of spines; walls covered by acute spines, finely punctate ....*T. hispida* var. *coronata*
53 Lorica with collar tooth-like at rim; walls covered with short sharp spines and with very small punctae ....*T. hispida* var. *crenulatocollis*
54 Lorica ellipsoidal to longitudinal; collar low with a scalloped or spiky margin......*T. sydneyensis*
54 Lorica ellipsoidal to subspherical or broadly oval; collar absent 55
55 Walls densely covered with blunt rod-like spines, dark brown ....*T. bacillifera*
55 Walls with a rare strong spines, yellowish-brown, apical pore surrounded by circle of spines ....*T. robusta*

Trachelomonas abrupta (Svirenko) Deflandre 1926

04100010 **Pl. 42P** (p. 174)

Lorica 15.5–20 μm wide, 25.3–31 μm long, ellipsoidal to subcylindrical, slightly truncate at anterior end; apical pore very wide; walls covered with small spines and coarsely punctate, light tan in colour; chloroplasts several, plate-shaped.

Probably cosmopolitan; occurs in swamps, puddles, ditches and ponds, uncommon.

Several varieties are known.

var. *minor* Deflandre 1926
04100012 **Pl. 42Q** (p. 174)
Lorica 9–11 μm wide, 16–20 μm long, longitudinal-ellipsoidal to oval; apical pore without a collar; walls densely punctate and without spines, yellow or reddish-brown; chloroplasts without pyrenoids.

Europe, Africa, North America; known from lakes, ponds, peat bogs, ditches and puddles, common.

Trachelomonas acanthostoma A.Stokes 1887

04100020 **Pl. 43H** (p. 178)
Lorica 15–36 μm wide, 18–42 μm long, ovoid, spherical or subspherical; walls minutely and densely punctate, rarely smooth, reddish-brown; apical pore surrounded by 1 or 2 rows of short blunt spines.

Europe, Asia, North and South America; in fish ponds, village ponds, shallow lakes and marshes; considered an indicator of mildly polluted water.

Several varieties are known.

Trachelomonas amphora Svirenko 1914

04100030 **Pl. 42N** (p. 174)
Lorica 21–25 μm wide, 30–34 μm long, obovate, slightly narrowing at posterior end; walls covered by small spines, reddish-brown; apical pore surrounded by a dentate collar, 5 μm wide, 2.5 μm high.

Europe; known only from ponds and puddles, rare.

Trachelomonas armata (Ehrenberg) F.Stein 1878

Basionym: *Chaetothyphla armata* Ehrenberg
04100040 **Pl. 42M** (p. 174) **CD**
Lorica 29–32 μm wide, (28–)30–45 μm long, broadly ellipsoidal to ovoid, thickened around apical pore (6–8 μm in diameter) or with a low and toothed, scalloped or spiny collar, posterior end broadly rounded; walls punctate, sometimes smooth, with stout long spines (3–10 μm long) at posterior end; flagellum twice length of lorica; chloroplasts numerous.

Probably cosmopolitan; in ponds, ditches and lakes.

Several varieties are known.

Trachelomonas atrata (Skvortsov) Deflandre 1926

var. *pustulosa* W.Conrad 1932
04100402 **Pl. 42AA** (p. 174) **CD**
Lorica 10.5–17.3 μm wide, 13–20 μm long, broadly ellipsoidal; walls covered with groups of 2 or 4 warts, punctate, brown; apical pore surrounded by a well-developed collar, 3.3–10 μm wide, 3.5 μm high, bent to one side.

Europe; known only from puddles, rare.

Trachelomonas bacillifera Playfair 1915

var. *minima* Playfair 1915
04100052 **Pl. 42I** (p. 174) **CD**
Lorica 18–26 μm wide, 22–30 μm long, broadly ellipsoidal to almost subspherical; walls covered with blunt rod-like spines, dark brown; apical pore without a collar.

Probably cosmopolitan; in lakes, puddles and peat bogs, common.

Trachelomonas bulla F.Stein emend. Deflandre 1926

Synonym: *T. pseudobulla* Svirenko
04100060 **Pl. 42X** (p. 174)
Lorica 21–25.5 μm wide, 39–46.4 μm long, elongated bulb-shaped or ovoid, with anterior end slightly tapering; walls covered by short, sharp spines and punctate, yellow to reddish-brown; apical pore surrounded by a conical or almost cylindrical collar 4.6–7 μm wide, 6–7 μm high, collar with a finely scalloped margin, diameter of apical pore same or slightly smaller than its base.

Europe, Asia; known from swamps, *Sphagnum* bogs, ditches and forest puddles, rare; a chance member of the plankton.

Trachelomonas caudata (Ehrenberg) F.Stein 1878

04100070 **Pl. 42AC** (p. 174)
Lorica 17–23 μm wide, 29–53 μm long, oval, narrowing slightly posteriorly and terminating in a short, straight or slightly bent, acute conical spine; walls covered by dense, short, straight spines (about 5 μm long); apical pore surrounded by a long cylindrical collar with a toothed rim (widening and with 5–6 teeth).

Probably cosmopolitan; widespread in fish ponds, ditches, puddles, small rivers and swamps; indicator of moderately polluted water.

Trachelomonas cervicula A.Stokes 1890

Synonyms: *Trachelomonas varians* (Lemmermann) Deflandre, *T. varians* fo. *globosa* Deflandre, *T. cervicula* var. *swirenkiana* Skvortsov
04100080 **Pl. 43E** (p. 178) **CD**
Lorica (16–)20–26 μm wide, (16–)22–29 μm long, spherical or subspherical; walls smooth or lightly punctate, golden or reddish-brown; apical pore surrounded by a low, flat ring with a cylindrical canal extending inwards to lorica cavity, 2.8–5 μm wide, (3.2–)8–9(–9.5) μm long; flagellum 1.5–2 times longer than lorica; chloroplasts numerous, plate-shaped, without pyrenoids.

Probably cosmopolitan; known from swamps, ditches, puddles and ponds, chance member of the plankton.

Trachelomonas columba T.C.Palmer 1925

04100410 **Pl. 43P** (p. 178) **CD**
Lorica (11.3–)18–24 μm in diameter, spherical, covered by thick folds giving surface an irregular net-like

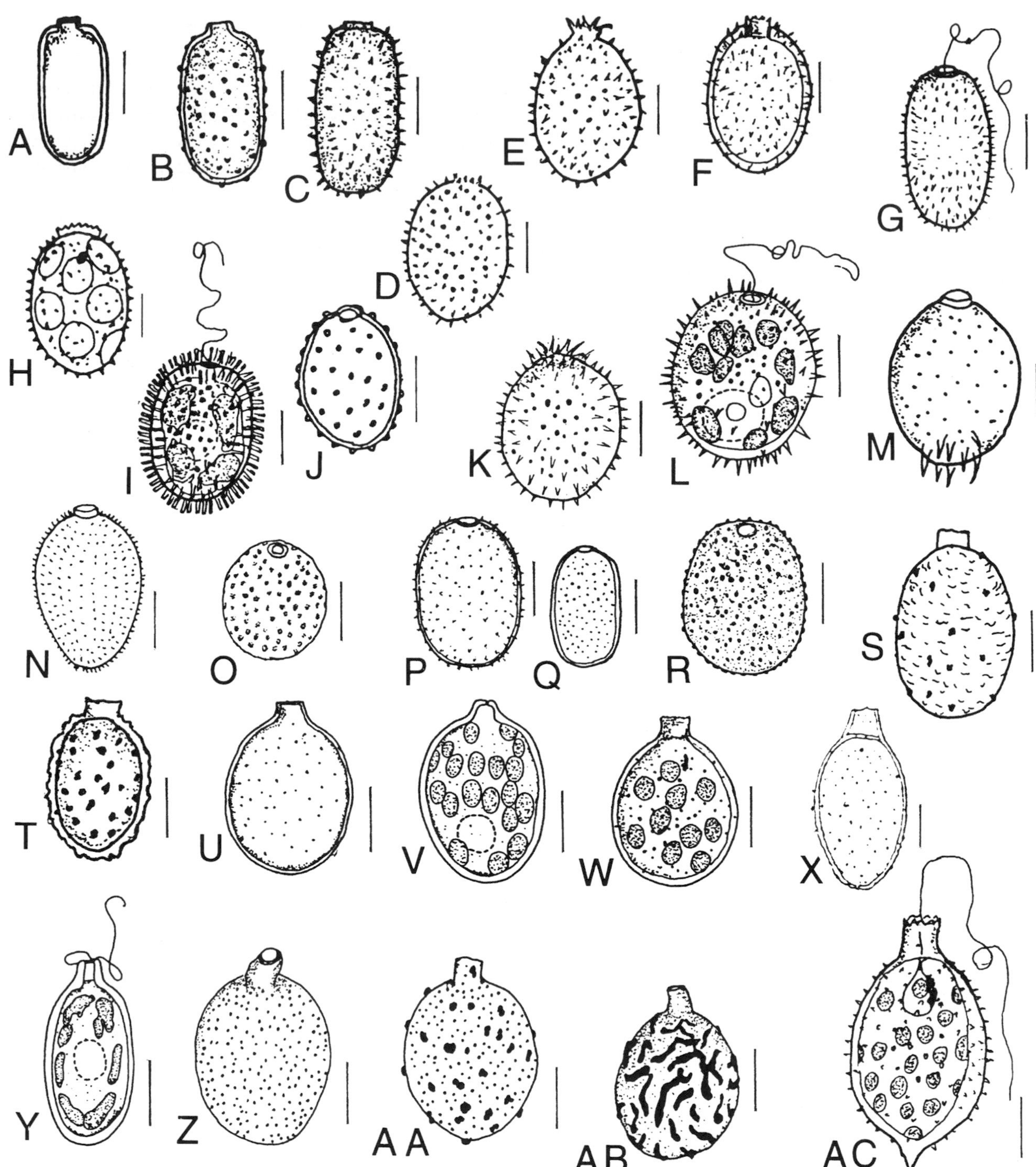

Plate 42 A–AC

A. *Trachelomonas cylindrica* (p. 175); **B**. *Trachelomonas lacustris* (p. 176); **C**. *Trachelomonas klebsii* (p. 176); **D**. *Trachelomonas hispida* (p. 175); **E**. *Trachelomonas hispida* var. *coronata* (p. 175); **F**. *Trachelomonas hispida* var. *crenulatocollis* (p. 176); **G**. *Trachelomonas polonica* (p. 177); **H**. *Trachelomonas sydneyensis* (p. 179); **I**. *Trachelomonas bacillifera* var. *minima* (p. 173); **J**. *Trachelomonas pulchella* (p. 177); **K**. *Trachelomonas robusta* (p. 177); **L**. *Trachelomonas superba* (p. 179); **M**. *Trachelomonas armata* (p. 173); **N**. *Trachelomonas amphora* (p. 173); **O**. *Trachelomonas zorenzis* (p. 179); **P**. *Trachelomonas abrupta* (p. 172); **Q**. *Trachelomonas abrupta* var. *minor* (p. 173); **R**. *Trachelomonas granulosa* (p. 175); **S**. *Trachelomonas scabra* (p. 177); **T**. *Trachelomonas granulata* (p. 175); **U**. *Trachelomonas lefevrei* (p. 176); **V**. *Trachelomonas manginii* (p. 176); **W**. *Trachelomonas planctonica* (p. 176); **X**. *Trachelomonas bulla* (p. 173); **Y**. *Trachelomonas dubia* (p. 175); **Z**. *Trachelomonas similis* (p. 177); **AA**. *Trachelomonas atrata* var. *pustulosa* (p. 173); **AB**. *Trachelomonas gregussii* (p. 175); **AC**. *Trachelomonas caudata* (p. 173).

Scale bars: 10 μm.

appearance; apical pore surrounded by a low, ring-like thickening.

Central Europe, North America; known only from *Sphagnum* bogs and ditches, rare.

Specimens from the British Isles were also smaller than described by Palmer (1925).

Trachelomonas curta A.M.Cunha 1913

04100090 **Pl.43Q** (p. 178)
Lorica 17–24 μm wide, 14–20 μm long, transversely oval, compressed in longitudinal direction and appearing wider than long, circular in cross-section, smooth, reddish-brown; apical pore surrounded by a ring-like thickening; flagellum 2–3 times longer than lorica; chloroplasts without pyrenoids.

Probably cosmopolitan; occurs in swamps, lakes and village ponds, uncommon; indicator of moderately polluted water.

Trachelomonas cylindrica (Ehrenberg) Playfair 1915

Basionym: *Trachelomonas cylindrica* Ehrenberg
Synonym: *Trachelomonas euchlora* var. *cylindrica* (Ehrenberg) Lemmermann
04100100 **Pl. 42A** (p. 174) **CD**
Lorica 7.5–10 μm wide, 14.5–21 μm long, broad, cylindrical with sides almost straight and parallel; walls smooth, light brown; apical pore without a collar or sometimes having a low collar (1 μm high, 1.4–3 μm wide), usually thickened at rim.

Probably cosmopolitan; in small, ephemeral water bodies, widespread.

Trachelomonas dubia (Svirenko) Deflandre 1926

04100110 **Pl. 42Y** (p. 174)
Lorica 11–15 μm wide, 21–28 μm long, ellipsoidal to cylindrical with blunt ends; walls smooth, reddish-brown; collar cylindrical, straight, 2.5–4 μm wide, 2.3–5 μm high, sometimes thickened at base; flagellum 2–2.5 times longer than lorica; chloroplasts 7–10 within each cell, pyrenoids absent.

Probably cosmopolitan; in ponds, swamps and forest puddles, commonly a chance member of the plankton.

This species is similar to *Trachelomonas cylindrica* and *T. abrupta* but *T. dubia* is larger than *T. cylindrica*, and *T. abrupta* lacks a collar or neck although these features are variable in other species. Several forms are known.

Trachelomonas globularis (Averintsev) Lemmerman 1910

04100120 **Pl. 43F** (p. 178)
Lorica 18–22 μm wide, 18–25 μm long, almost spherical, covered with short, thick and regularly distributed spines, yellow- or reddish-brown; apical pore surrounded by a ring-like thickening or thickening absent.

Probably cosmopolitan; known from swamps, lakes and field ponds; indicator of clean to mildly polluted water.

Trachelomonas granulata Svirenko 1914

04100130 **Pl. 42T** (p. 174)
Lorica 14–19 μm wide, 22–25 μm long, ellipsoidal; collar with a low and slightly thickened rim; walls very irregular in thickness, covered with minute and irregular warts, sometimes bearing 2–7 larger warts, yellow-brown to brown; flagellum 3–4 times longer than lorica.

Europe, Asia; known from ponds and small rivers occurs in plankton and also in mud, usually in overgrown water bodies.

Trachelomonas granulosa Playfair 1915

04100140 **Pl. 42R** (p. 174)
Lorica 17–22 μm wide, 22–26 μm long, ellipsoidal or almost spherical, covered with small granules; apical pore with or without a low collar.

Probably cosmopolitan; in ditches, ponds, puddles, lakes and swamps, widespread.

Varieties *subglobosa* and *oblonga* were described by Playfair (1915), but are considered along with the type variety since they only differ from it in size.

Trachelomonas gregussii Hortobágyi 1940

04100420 **Pl. 42AB** (p. 174)
Lorica 14–16.5 μm wide, 16.5–20 μm long, ellipsoidal; walls ornamented by vermicular granules, yellowish-brown; flagellar pore surrounded by a collar (2.5–5 μm high) bent to one side; chloroplasts small.

Europe; in rivers and ponds, rare.

Trachelomonas hispida (Perty) F.Stein emend. Deflandre 1926

04100150 **Pl. 42D** (p. 174) **CD**
Lorica 15–24(–26) μm wide, (20–)29–31(–42) μm long, ellipsoidal, ends more or less rounded or narrowed anteriorly; walls uniformly and densely covered with short, sharp, conical spines, sometimes finely punctate, yellowish to reddish-brown; apical pore without a collar or with a short and slightly raised pore 3–4 μm high; flagellum twice length of lorica; chloroplasts 8–10, plate-like, each with a pyrenoid.

Probably cosmopolitan; in swamps, ponds and ditches, widespread and in plankton hauls from stagnant water; indicator of clean to moderately polluted water.

var. *coronata* Lemmermann 1913
04100152 **Pl. 42E** (p. 174) **CD**
Lorica 19–24 μm wide, (29–)32–40 μm long, oblong-oval; apical pore surrounded by a short collar with margin bearing a circle of spines; walls uniformly beset with short (2–3 μm long) and uniformly distributed spines.

Probably cosmopolitan; in stagnant or slow-flowing waters including lakes, puddles, ditches and ponds, common.

var. *crenulatocollis* (Maskell) Lemmermann 1913
Basionym: *Trachelomonas crenulatocollis* Maskell
04100153 **Pl. 42F** (p. 174) **CD**
Lorica (15–)19–25 μm wide, (23–)26–35.5(–37) μm long, broadly oval; walls covered with minute, sharp spines and very small punctae, or only punctate, brown; apical pore surrounded by a toothed (dentate) collar.

Probably cosmopolitan; in puddles, peat bogs, swamps and ponds, widespread.

Trachelomonas intermedia P.A.Dangeard 1901

04100160 **Pl. 43X** (p. 178)
Lorica 10–16(–20) μm wide, 12–20(–25) μm long, oval with bluntly rounded apices; apical pore 2.5–3 μm in diameter, surrounded by a ring-like thickening (not always present); walls finely punctate, brown; flagellum 2–4 times longer than lorica; chloroplasts 4–5, each with a pyrenoid.

Probably cosmopolitan; occurs in lakes, ponds and ditches, widespread, a chance member of the plankton.

fo. *crenulatocollis* (M.Szabados) T.G.Popova 1966
Basionym: *Trachelomonas intermedia* var. *crenulatocollis* M.Szabados
04100162 **Pl. 43Y** (p. 178)
Lorica 17–18 μm wide, 21–24 μm long, broadly ellipsoidal, dark-brown; walls finely punctate as in type form but apical pore surrounded by a low collar with a crenate rim.

Europe; known from swamps, ditches and puddles, uncommon.

Trachelomonas klebsii Deflandre 1926

Synonym: *Trachelomonas hispida* fo. *cylindrica* G.A.Klebs
04100170 **Pl. 42C** (p. 174)
Lorica 12–15 μm wide, 27–35 μm long, distinctly cylindrical to oval, broadly rounded and slightly flattened at both ends, finely punctate and densely covered with short, sharp spines 1–1.5 μm high; apical pore surrounded by a ring-like thickening; flagellum 1.5 times lorica length.

Europe, Asia, Central America; known only from lakes and ponds, uncommon.

Trachelomonas lacustris Drezepolski 1925

Synonyms: *Trachelomonas cylindrica* Ehrenberg var. *hispidula* Skvortsov, *T. cylindrica* Ehrenberg var. *gordeievi* Skvortsov
04100180 **Pl. 42B** (p. 174)
Lorica 12–16.5 μm wide, 26–29.6(–33) μm long, cylindrical with sides almost parallel, broadly rounded both posteriorly and anteriorly; apical pore usually without a collar, but sometimes slightly raised at rim; walls covered with small warts, golden yellow-brown; flagellum 1.5 times lorica length; chloroplasts 7–8, each with a pyrenoid.

Probably cosmopolitan; occurs in puddles and forest ponds as well as ditches, uncommon, a chance member of the plankton.

Several varieties are known.

Trachelomonas lefevrei Deflandre 1926

04100190 **Pl. 42U** (p. 174)
Lorica 22–25.4 μm wide, 27–31 μm long, broadly ellipsoidal or obovoid; walls punctate, reddish-brown; flagellar pore with a low, straight and irregularly toothed collar.

Europe; in ponds, rare.

Trachelomonas manginii Deflandre 1926

04100200 **Pl. 42V** (p. 174)
Lorica (14–)15.5–20 μm wide, 20.5–26.5 μm long, broadly ellipsoidal; walls smooth or with minute spines (not clearly visible), yellowish-brown; apical pore surrounded by a cylindrical collar 2–2.6 μm wide, 2.3–3.3 μm high; chloroplasts without pyrenoids.

Probably cosmopolitan; in ponds, lakes, swamps and ditches.

Trachelomonas oblonga Lemmermann 1899

04100210 **Pl. 43U** (p. 178)
Cells 9–13(–15) μm wide, 11–19 μm long, ellipsoidal; walls smooth, yellowish- to reddish-brown, broadly rounded at apices; apical pore 2.5–4 μm wide, with or without a ring-like thickening (rare, with very low collar); flagellum about twice lorica length; chloroplasts 10 per cell, each with a pyrenoid.

Probably cosmopolitan; in field ponds, puddles and ditches, widespread, a chance member of the plankton or benthic; indicator of clean to moderately polluted water.

var. *punctata* Lemmermann 1898
04100212 **Pl. 43V** (p. 178)
Lorica 8–16.5 μm wide, 14–21 μm long, oblong; walls with numerous small punctae, brown; apical pore 2.5–3.5 μm wide, without a collar.

Probably cosmopolitan; in ponds, ditches and slow-flowing rivers, uncommon; indicator of mildly polluted waters.

var. *truncata* Lemmermann 1898
04100213 **Pl. 43W** (p. 178) **CD**
Lorica 8–14 μm wide, 10–16 μm long, oblong, almost rectangular; walls smooth, yellowish to brown; apical pore with a small thickened collar.

Probably cosmopolitan; in ponds, ditches and puddles in peat bogs, uncommon.

Trachelomonas planctonica Svirenko 1914

04100220 **Pl. 42W** (p. 174) **CD**
Lorica 16–22 μm wide, 18–32 μm long, almost oval, broadly ellipsoidal; walls densely and minutely punctate or with granules, reddish-brown; flagellar pore surrounded by a cylindrical collar, irregularly toothed at rim, collar 3.4–5.4 μm wide and 2–5 μm high, straight or oblique; chloroplasts 8–10 in each cell, without pyrenoids.

Probably cosmopolitan; in ponds, lakes, rivers and ditches, very common; indicator of mildly to moderately polluted water.

Several forms were reported by Popova (1966).

Trachelomonas polonica Drezepolski 1925

04100230 **Pl. 42G** (p. 174)
Lorica 14.5–22 μm wide, 26–35 μm long, cylindrical, ellipsoidal, sometimes posterior end with a conical point; walls densely covered with short, sharp spines, yellow; collar low, narrow and with a ring-thickening, 5 μm in diameter; flagellum about length of lorica.

Europe; known from fish ponds and village ponds, rare.

Trachelomonas pseudofelix Deflandre 1926

04100240 **Pl. 43Z** (p. 178)
Lorica 14.5–16 μm wide, 16–18 μm long, broadly oval; walls irregularly roughened, yellow-brown to brown; apical pore with a ring-like thickening surrounded by warts (verrucae); flagellum twice length of lorica.

Probably cosmopolitan; known from ditches, puddles and ponds, uncommon.

Trachelomonas pulchella Drezepolski 1925

04100430 **Pl. 42J** (p. 174)
Lorica 15–17 μm wide, 18–21 μm long, ellipsoidal; walls with a few warts bearing small knobs or short blunt spines; apical pore surrounded by a ring-like thickening or very low collar.

Europe; in puddles and ponds, rare.

Trachelomonas robusta Svirenko emend. Deflandre 1926

04100250 **Pl. 42K** (p. 174)
Lorica 15.5–25 μm wide, (20–)30–32 μm long, ellipsoidal to broadly ovoid; walls only rarely with strong spines, sometimes smooth or punctate, yellowish-brown; apical pore surrounded by a circle of spines; chloroplasts 8–10 in each cell.

Probably cosmopolitan; in ponds, puddles and peat bogs, rare.

Trachelomonas rostafinskii Drezepolski 1925

04100260 **Pl. 43G** (p. 178)
Lorica 13.5–15 μm wide, 16–17 μm long, almost spherical; walls smooth, with a few spines at posterior and anterior ends; collar absent.

Europe; known only from ponds, rare, occurs in plankton.

Trachelomonas rugulosa F.Stein emend. Deflandre 1926

Synonym: *Trachelomonas stokesiana* T.C.Palmer
04100270 **Pl. 43I,J** (p. 178) **CD**
Lorica 14–23 μm wide, 14–21 μm long, spherical or subspherical, covered with anastomosing striae or ribs slightly spirally arranged, yellow to reddish-brown; apical pore surrounded by a ring-like thickening, pore 1.3–2 μm in diameter; flagellum twice as long as lorica; chloroplasts 10 per cell, each with a pyrenoid.

Probably cosmopolitan; in marshes, ponds, puddles, ditches and peat bogs, common; indicator of moderately polluted water.

var. *meandriana* W.Conrad 1941
04100273 **Pl. 43L** (p. 178) **CD**
Lorica 16–18 μm in diameter, almost spherical, with irregular, discontinuous and twisted folds; apical pore without a collar.

Central Europe; known only from ponds, rare.

var. *obliqua* Bourrelly 1952
04100274 **Pl. 43K** (p. 178)
Lorica 20–22 μm in diameter, spherical, with distinctly spirally arranged ribs not anastomosing, dark yellow; apical pore with a short collar directly inside lorica, surrounded by a slight, ring-like thickening.

Europe, Asia; in puddles and ponds, rare.

var. *steinii* Deflandre 1927
04100272 **Pl. 43M** (p. 178) **CD**
Lorica 15–23 μm in diameter, almost spherical; walls with slightly convex ribs, brown; apical pore without a collar.

Europe; known from ponds and puddles, rare.

Trachelomonas scabra Playfair 1915

04100280 **Pl. 42S** (p. 174) **CD**
Lorica 14–19 μm wide, (8–)19–27 μm long, ellipsoidal; walls roughened (scabrous) with minute protuberances, irregularly arranged, pale yellow to reddish-brown; apical pore with a low collar, straight or slightly oblique, 2–3.5 μm high and 3–4 μm wide.

Probably cosmopolitan; in ponds, lakes and ditches, common, planktonic.

Trachelomonas similis A.Stokes 1890

04100290 **Pl. 42Z** (p. 174) **CD**
Lorica (12–)14–19(–27) μm wide, (19.2–)22.0–28.0(–40) μm long, broadly ellipsoidal or ovoid; walls with scattered punctae; apical pore surrounded by a well-developed collar, usually bent to one side, irregularly toothed, collar 3–5.4 μm wide and 3–3.5 μm high; chloroplasts numerous in each cell.

Probably cosmopolitan; known from humic waters, swamps, puddles, ponds, lakes and in brackish-water; indicator of clean to moderately polluted water.

Trachelomonas stokesii Drezepolski 1925

04100440 **Pl. 43AA** (p. 178)
Lorica 15–21 μm wide, 18–21 μm long, obovate, widened at anterior end; walls brown, punctate; apical pore 2.5–3 μm in diameter, without a collar.

Probably cosmopolitan; known from ponds, small rivers and swamps, uncommon.

According to Popova (1966), similar to *T. intermedia* Dangeard.

Trachelomonas subverrucosa Deflandre 1926

04100310 **Pl. 43AB** (p. 178)
Lorica 16–19.5 μm wide, 19.5–23 μm long, almost spherical; walls covered by granules and irregularly punctate; apical pore without a collar.

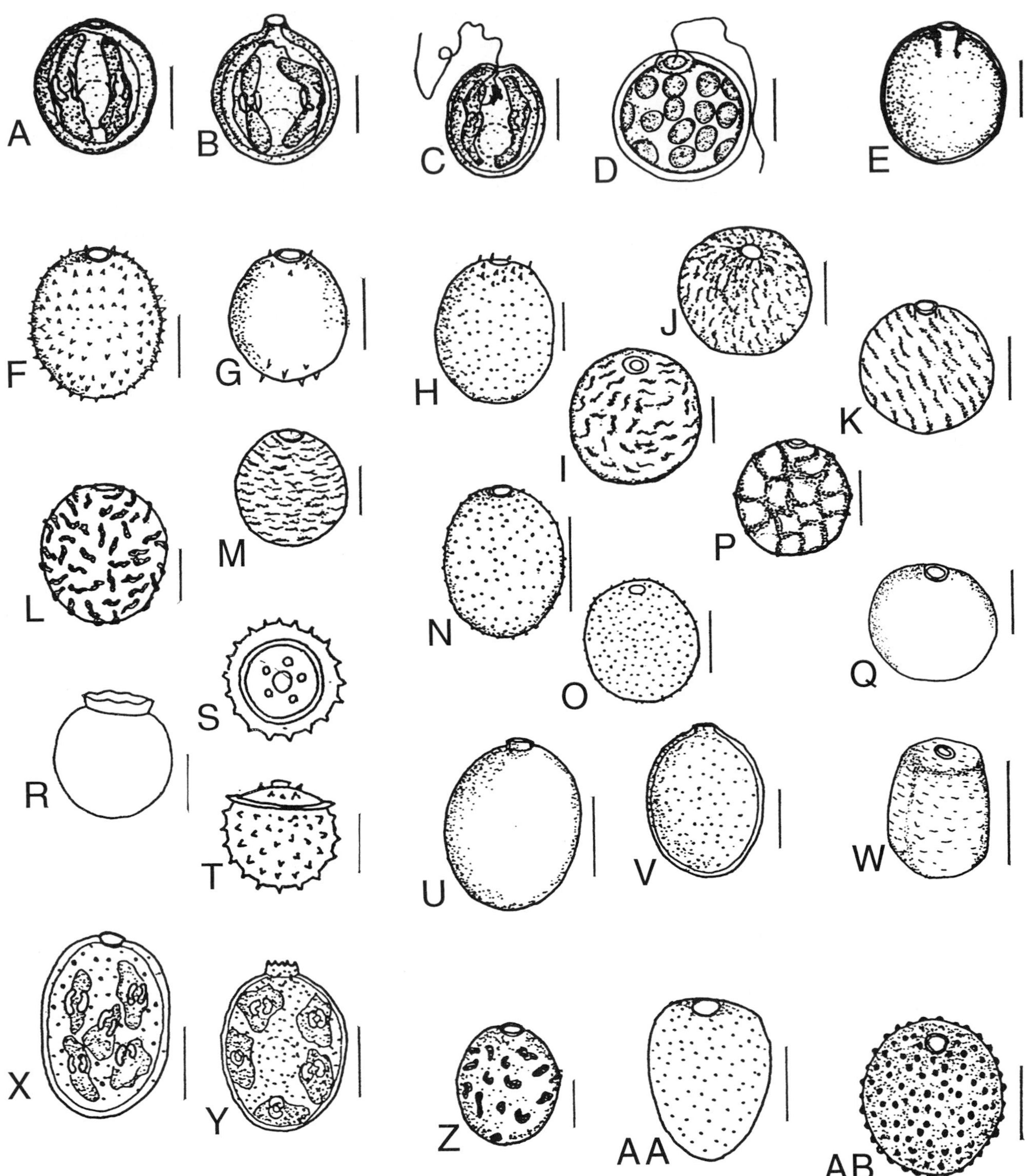

Plate 43 A–AB

A. *Trachelomonas volvocina* (p. 179); **B**. *Trachelomonas volvocina* var. *derephora* (p. 179); **C**. *Trachelomonas volvocina* var. *subglobosa* (p. 179); **D**. *Trachelomonas volvocinopsis* (p. 179); **E**. *Trachelomonas cervicula* (p. 173); **F**. *Trachelomonas globularis* (p. 175); **G**. *Trachelomonas rostafinskii* (p. 177); **H**. *Trachelomonas acanthostoma* (p. 173); **I, J**. *Trachelomonas rugulosa* (p. 177); **K**. *Trachelomonas rugulosa* var. *obliqua* (p. 177); **L**. *Trachelomonas rugulosa* var. *meandriana* (p. 177); **M**. *Trachelomonas rugulosa* var. *steinii* (p. 177); **N, O**. *Trachelomonas verrucosa* var. *irregularis* (p. 179); **P**. *Trachelomonas columba* (p. 173); **Q**. *Trachelomonas curta* (p. 175); **R**. *Trachelomonas zuberii* (p. 179); **S, T**. *Trachelomonas wislouchii* (p. 179); **U**. *Trachelomonas oblonga* (p. 176); **V**. *Trachelomonas oblonga* var. *punctata* (p. 176); **W**. *Trachelomonas oblonga* var. *truncata* (p. 176); **X**. *Trachelomonas intermedia* (p. 176); **Y**. *Trachelomonas intermedia* fo. *crenulatocollis* (p. 176); **Z**. *Trachelomonas pseudofelix* (p. 177); **AA**. *Trachelomonas stokesii* (p. 177); **AB**. *Trachelomonas subverrucosa* (p. 177).

Scale bars: 10 μm.

Europe, Asia; known from ponds, ditches and peat bogs, uncommon.

Trachelomonas superba Svirenko emend. Deflandre 1926

04100320 **Pl. 42L** (p. 174) **CD**

Lorica (27–)31–34 μm wide, 34–43 μm long, broadly ellipsoidal to ovoid, rounded at apices; walls finely punctate and evenly covered by sharply pointed, dense conical spines, yellowish-brown; flagellar pore without a collar; flagellum about lorica length; chloroplasts numerous.

Probably cosmopolitan; in lakes, ponds, puddles and swamps, common.

Several forms are known.

Trachelomonas sydneyensis Playfair 1915

04100330 **Pl. 42H** (p. 174)

Lorica 22–25 μm wide, 31–42.5 μm long, ellipsoidal to longitudinally oval; walls covered with sparsely distributed spines, densely punctate, yellow to brown; apical pore surrounded by a low collar, with a scalloped or spiky margin; flagellum 1.5–3(–4) times lorica length; chloroplasts few, each with a pyrenoid.

Probably cosmopolitan; in ponds, swamps and ditches, common.

According to Popova (1966), similar to *T. hispida* var. *crenulatocollis* but differs in having a thin lorica wall and rarely having spines.

Trachelomonas verrucosa A.Stokes

var. *irregularis* Deflandre 1926

04100343 **Pl. 43N,O** (p. 178) **CD**

Lorica 12.5–13.5 μm wide, 14–15.5 μm long, broadly oval, almost spherical, with irregular warts or granules, light brown.

Europe, North America; in ponds, puddles and ditches, rare.

Trachelomonas volvocina Ehrenberg 1833

04100350 **Pl. 43A** (p. 178) **CD**

Lorica 6–23(–32) μm wide, spherical; walls yellowish, sometimes colourless, smooth; apical pore without a collar, or collar only low and slightly thickened around margin; flagellum 2–3 times lorica length; chloroplasts 2 per cell, each with a double sheathed pyrenoid.

Probably cosmopolitan; in lakes, ponds, ditches, puddles, peat bogs and waste water, widespread; indicator of mildly to heavily polluted water.

var. *derephora* W.Conrad 1916

04100352 **Pl. 43B** (p. 178)

Lorica 11–21 μm in diameter, spherical, smooth; apical pore surrounded by a low collar, 2.6–3 μm wide and 3 μm high.

Europe, Asia; known from peat bogs and ponds, common.

var. *subglobosa* Lemmermann 1913

04100353 **Pl. 43C** (p. 178)

Lorica 9.5–22.8 μm wide, 20–22 μm long, almost spherical, smooth, reddish-brown; apical pore surrounded by a depressed collar.

Europe, Asia; known from fish ponds, village ponds, ditches and pools, common.

Trachelomonas volvocinopsis Svirenko 1914

Synonym: *Trachelomonas varians* (Deflandre) fo. *minor* J.W.G.Lund

04100360 **Pl. 43D** (p. 178) **CD**

Lorica 13.5–22(–33) μm wide, spherical, smooth or punctate; apical pore with a ring-like thickening; walls reddish-brown; flagellum 2–2.5 times longer than lorica; chloroplast plate-shaped, without pyrenoids.

Probably cosmopolitan; in lakes, ponds, swamps, ditches and in springs containing partly rotted leaves, widespread; indicator of mildly to moderately polluted water.

Several varieties are known.

Trachelomonas wislouchii Skvortsov 1917

04100370 **Pl. 43S,T** (p. 178)

Lorica 21–24 μm wide, 19–21 μm long, almost spherical; walls covered with short spines, yellow-brown; apical pore surrounded by a ring-like thickening and a few spines at some distance from apical pore, fold surrounding anterior part of lorica.

Europe, Asia; known from ponds, swamps and lakes, uncommon.

Trachelomonas zorensis M.Lefèvre 1933

04100380 **Pl. 42O** (p. 174)

Lorica 14–16 μm wide, 19–20 μm long, ellipsoidal; apical pore surrounded by a ring-like thickening; walls thick, covered with many small depressions (scrobiculate), yellow-brown; chloroplasts 2 per cell, each with a double pyrenoid.

Europe, North America; in ditches, puddles, ponds and lakes, uncommon.

Trachelomonas zuberi Koczwara 1915

04100390 **Pl. 43R** (p. 178) **CD**

Lorica 13–25 μm wide,10–25 μm long, almost spherical, smooth, brown or yellow-brown; apical pore thickened at rim and surrounded by a thick, scalloped fold at the anterior part.

Europe, Asia, Australia; in swamps, puddles and ponds, rare.

Acknowledgements

I am especially grateful to L.R. Johnson and D.M. John for providing me with the initial list of information about euglenophytes known from local literature and for assistance in collecting material. Also the following persons are thanked for their encouragement and assistance with some of the field collections: A.J. Brook, J.H. Belcher, J.F. John, B.A. Whitton, E.M.F. Swale. Special thanks go to Judith F. John to whom I am indebted for her help with the English.

PHYLUM CRYPTOPHYTA (Cryptomonads)

Gianofranco Novarino

In spite of being widespread in freshwater habitats, cryptomonads are a neglected group, with most reports in the British freshwater literature consisting of lists of taxa with no or few illustrations of the specimens observed.

Cryptomonads are notoriously difficult to identify with certainty from live or preserved natural samples using light microscopy alone. This is due to the scarcity of readily observable taxonomic characters, natural phenotypic variability, and the fact that in the last thirty years the traditional cryptomonad taxonomy and systematics have been revolutionized by a wealth of new ultrastructural information. There is also an increasing use of molecular sequence data for systematic purposes. Inevitably, all of this may overshadow the significance of existing light microscopy-based reports.

The following taxa are based mostly on the coded list of British freshwater algae by Whitton et al. (1998a). A 'nominalistic' approach is adopted as far as possible, with the reported taxa considered as being correctly identified *bona fide* although their ultrastructural identity may still be unknown. Each taxon is listed by the name under which it was reported. Listed features are based mostly on those given by the describing authors in their light microscopy-based diagnoses, and keys are based on characters visible with the light microscope. Although reports of all but one species are based on light microscopy only, reference is made to recent ultrastructural or systematic work where appropriate.

In the cryptomonads, cell shape is usually described based on dorsal, ventral, and lateral aspects, the dorsal and ventral faces being defined by the mutual insertion of the flagella. The terms length, thickness and width refer to the dimensions of the longitudinal, dorsiventral and lateral axes, respectively. When only the thickness or width values are reported, these are usually interchangeable, since the cells are circular in cross-section. Cells usually contain one or two chloroplasts and these can be blue-green, reddish, red-brown, olive-green, brown or yellow-brown due to the presence of various chlorophyll-masking pigments (mainly phycocyanin and phycoerythrin). Colour may vary between different cells of the same species, but the kind of the phycobilin pigment is now known to be taxonomically significant at the generic level. There are also colourless forms and these sometimes resemble coloured forms, e.g. the colourless *Chilomonas* and the chlorophyll-containing *Campylomonas* (=*Cryptomonas pro parte*) are similar. Trichocysts (Glossary, p. 622) lie immediately below the cytoplasmic membrane, larger ones in the gullet region and smaller ones elsewhere. They are easily visible with the light microscope in larger species of *Cryptomonas.*

Swimming behaviour is characteristic of the group as a whole and most cryptomonads swim with the flagella directed forwards along spiral-shaped paths, while the cell also revolves about its longitudinal axis; an exception is the colourless *Goniomonas* (=*Cyathomonas sensu auctorum*) which swims along irregular paths close to solid substrates, without rotating about the longitudinal axis.

About 200 cryptomonad species have been described and it is likely that many more species still await formal description. About 100 known species occur in freshwater, particularly in nutrient-poor lakes, although they also occur in smaller water bodies and rivers especially in the presence of nutrient enrichment or the vicinity of submerged macrophytes. Some are favoured by waters rich in organic substances and several are more common during the colder months the year.

1 Cells blue-green in colour .... *Chroomonas*
1 Cells variously coloured but never blue-green.......2
2 Cell length up to about 13–15 μm, with a single red chloroplast containing a large, centrally positioned pyrenoid .... *Rhodomonas*
2 Cell length usually 15–80 μm, with 1 or 2 red, brown, olive-green or yellowish-green chloroplasts containing 1–4 pyrenoids if present... *Cryptomonas*

Chroomonas Hansgirg 1885

Cells variously shaped; flagella mostly unequal, inserted apically or subapically; chloroplast 1 per cell, blue-green due to presence of phycocyanin as the principal accessory pigment, with a single, large and mostly central pyrenoid; eyespot usually absent.

Widely distributed; found in aquatic habitats especially in the plankton.

The features given above are visible with the light microscope. Although the eyespot is absent in *C. nordstedtii*, the type species, it has been reported for some others. In the scanning electron microscope the cell surface (periplast) is composed of discrete, hexagonal plates (Novarino & Oliva, 1998). This contrasts with the generally accepted concept that *Chroomonas* is characterized by rectangular periplast plates. There are other genera of blue-green cryptomonads which resemble *Chroomonas* using the light microscope, but whose ultrastructure differs substantially, e.g. *Komma* Hill and *Falcomonas* Hill (Hill, 1991b).

1 Cells with eyespot 2
1 Cells without an eyespot 3
2 Cells 6–7 μm wide; flagella equal, slightly shorter than cell .... *C. coerulea*
2 Cells about 4–6 μm wide; flagella subequal, longer flagellum about two-thirds cell length .... *C. rosenbergii*
3 Cells large (12–18 μm long), with a pair of highly refringent bodies close to pyrenoid .... *C. baltica*
3 Cells small (usually <12 μm long), without refringent bodies 4
4 Cells distinctly acute, with a ventrally bent posterior end or tail .... *C. acuta*
4 Cells not acute, without a posterior tail 5
5 Cells ellipsoidal to barrel-shaped; flagella subequal; pyrenoid slightly displaced towards cell posterior .... *C. nordstedtii*
5 Cells elongate-ellipsoidal; flagella distinctly unequal; pyrenoid central .... *Chroomonas* sp. *inedit.*

Chroomonas acuta Utermöhl 1925

05020010 **Pl. 44C,D** (p. 182)

Cells 4.5–5.5 μm wide and 7–10 μm long, anterior end rounded and acute posterior end forming a ventrally bent spike or tail; flagella slightly unequal, the longer (dorsal) one roughly as long as cell; pyrenoid in a dorsal position, roughly mid-way between centre and anterior end; nucleus posterior.

Widespread; in the plankton of lakes and small water bodies.

The ultrastructure of this species has never been examined in detail. It has been considered as a synonym of the newly recombined *Komma caudata* (Geitler) Hill (Hill, 1991b). Based on the presence of a posterior tail and available scanning electron micrographs, it has been hypothesized that Utermöhl's species may be an unusual, phycocyanin-containing representative of the genus *Plagioselmis* (Novarino et al., 1994).

Chroomonas baltica (J.Büttner) N.Carter 1937

Basionym: *Cyanomonas baltica* J.Büttner

05020020 **Pl. 44A** (p. 182)

Cells 4–5 μm wide, 12–18 μm long, 8–11 μm thick, widely ellipsoidal, laterally compressed, with an anterior vacuole; flagella nearly equal in length and considerably shorter than cell; pyrenoid central, one pair of refractive bodies close to pyrenoid, similar by light microscopy to the refractive bodies in other blue-green cryptomonads (Novarino & Lucas, 1993a).

Described from brackish-waters; rarely reported in freshwater habitats.

Chroomonas coerulea (Geitler) Skuja 1948

Basionym: *Cryptomonas coerulea* Geitler

05020030 **Pl. 44J** (p. 182)

Cells 6–7 μm wide, 8–10 μm long, ellipsoidal, with a contractile vacuole at anterior end; flagella equal in length, slightly shorter than cell; pyrenoid in dorsal position, slightly offset towards cell posterior; eyespot anterior to pyrenoid.

Widely distributed; occurs in the plankton of lakes and small water bodies.

Electron microscopy has shown that the cell surface of this species is composed of rectangular periplast plates (Hill, 1991b). Owing to this feature, *C. coerulea* ought to be transferred to a new genus, to be described (Novarino & Oliva, 1998).

Chroomonas nordstedtii Hansgirg 1885

05020040 **Pl. 44G,H** (p. 182)

Cells 6–8 μm wide, 9–12 μm long, ellipsoidal or slightly barrel-shaped; flagella slightly unequal in length, longer (dorsal) flagellum just longer than cell and ventral flagellum slightly shorter; pyrenoid up to 3 μm in diameter, slightly displaced towards posterior.

Widely distributed; frequently found in aquatic habitats especially in lake plankton.

The features listed above are taken from the original diagnosis and have been confirmed by an examination of Hansgirg's original type specimens (Novarino & Oliva, 1998). Later descriptions (Pascher & Lemmermann 1913; Skuja, 1939, 1948) report a considerable variation in cell size (up to 8–20×5–10 μm) and shape, ranging from narrow-elongate to oval-ellipsoidal. Scanning electron microscopy has shown that the cell surface of the type specimen(s) of *C. nordstedtii* is composed of hexagonal periplast plates (Novarino & Oliva, 1998). Since *C. nordstedtii* is the type species, the use of the generic name *Chroomonas* should be restricted to blue-green cryptomonads with this kind of periplast.

Chroomonas species, G.Novarino *inedit.*

05020060 **Pl. 44I** (p. 182)

Cells about 4 μm wide, 8 μm long, 5 μm thick, elongate-ellipsoidal; anterior end narrowly rounded and posterior end elongate, sometimes dorsally bent; pyrenoid prominent, usually in a central position; eyespot absent; vegetative cells perpetually motile, palmelloid stages absent, cysts occasionally formed.

British Isles; so far found only in Wales in lake plankton.

This species is distinguished from its closest ally, *C. nordstedtii*, by periplast features visible with the scanning electron microscope (Novarino & Oliva, 1998).

Chroomonas rosenbergii Huber-Pestalozzi 1950

05020050 **Pl. 44B** (p. 182)

Motile cells 4.2–6.3 μm wide, 7–10 μm long; non-motile cells 5–7 μm wide, 7–10 μm long; cells broadly ellipsoidal, with a broader posterior end; flagella subequal, the longer flagellum about two-thirds length of cell;

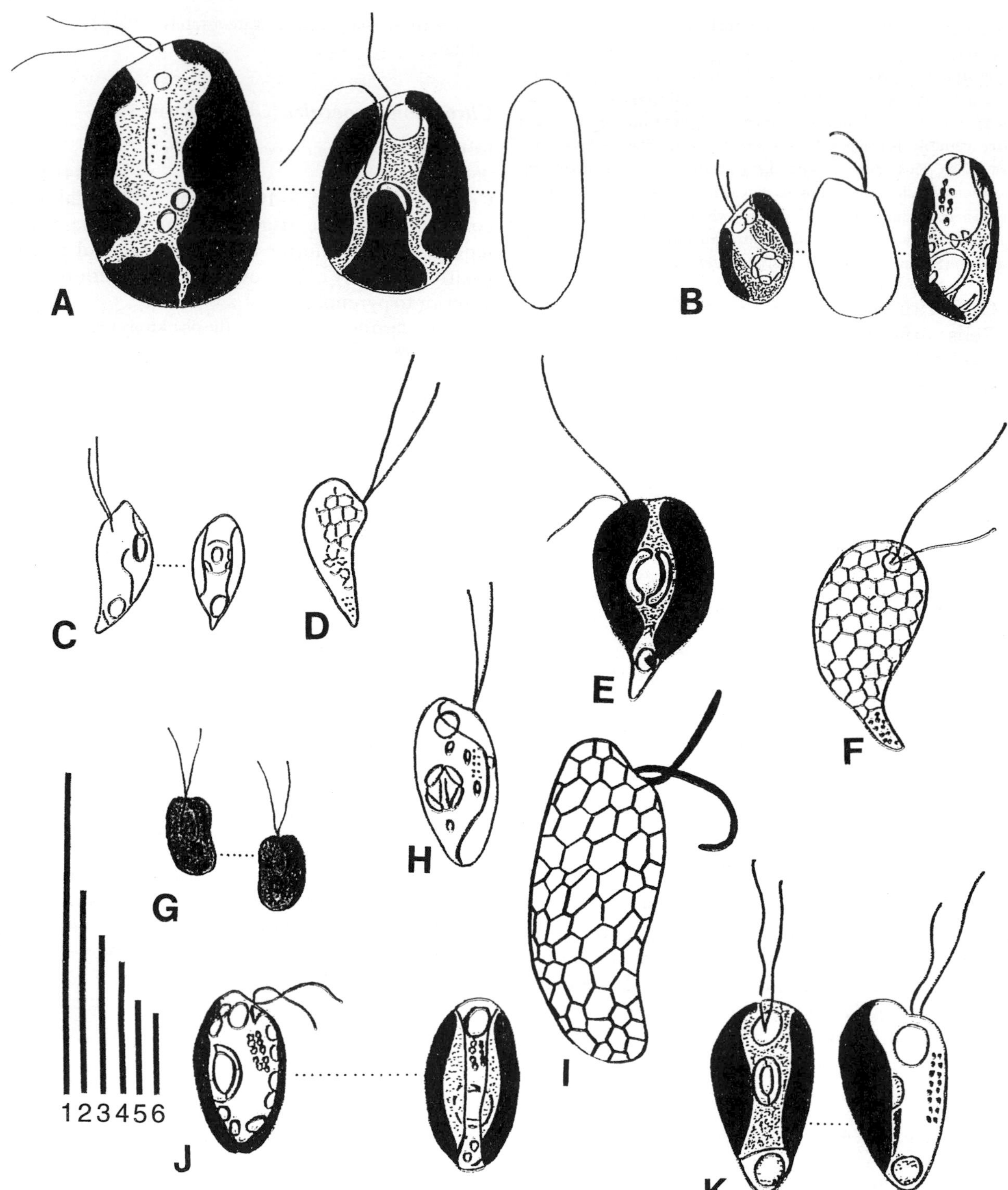

Plate 44 A–K

A. *Chroomonas baltica* (p. 181); **B**. *Chroomonas rosenbergii* (p. 181); **C, D**. *Chroomonas acuta* (p. 181); **E**. *Rhodomonas* (=*Plagioselmis*) *nannoplanctica* (p. 185); **F**. *Rhodomonas* (=*Plagioselmis*) *nannoplanctica* (p. 185): appearance under the scanning electron microscope; **G, H**. *Chroomonas nordstedtii* (p. 181); **I**. *Chroomonas* sp., inedit. (p. 181): appearance under the scanning electron microscope; **J**. *Chroomonas coerulea* (p. 181); **K**. *Rhodomonas lacustris* (p. 185).

Scale bars: 10 μm. 1 for I; 2 for E, F; 3 for A, D, J, K; 4 for B, C; 5 for H; 6 for G.

pyrenoid central or displaced towards cell posterior, with a starch sheath; eyespot present; nucleus displaced towards cell posterior.

British Isles (England); found in the plankton of Windermere.

This species is correctly referred to as *Chroomonas rosenbergii* (*'rosenbergae'* and *'rosenbergiae'* are incorrect spellings). It has been considered as a synonym of *Chroomonas coerulea* by Hill (1991b).

Cryptomonas Ehrenberg 1838

Cell shape diverse, often of considerable size (up to 80 μm in length); flagella usually equal or subequal, as long or shorter than cells; chloroplasts 1 or 2 per cell, containing phycoerythrin as the principal accessory pigment, and therefore red, brown, olive-green or yellowish-green in colour (never blue-green), with 1–4 pyrenoids (when present) often surrounded by a conspicuous starch sheath; eyespot reported in a few species based on light microsopy, but owing to a lack of ultrastructural information it is unclear if this actually corresponds to a true eyespot.

Cosmopolitan; widespread in aquatic habitats.

Ehrenberg's original generic descriptions (e.g. Ehrenberg, 1838) contain a limited amount of information and the characters listed above are based on the light microscope account of Huber-Pestalozzi (1950), as are the descriptions of Ehrenberg's species listed below. Such species were described by Ehrenberg on several occasions and for sake of simplicity the publication date given here is that of Ehrenberg (1838). Electron microscope information on *C. ovata*, the most widely reported species, has been summarized by Hill (1991a).

1 Pyrenoid(s) present....................2
1 Pyrenoid(s) absent....................4
2 Pyrenoids 2 per cell....................*C. stigmatica*
2 Pyrenoid 1 per cell....................3
3 Pyrenoid spherical....................*C. anomala*
3 Pyrenoid angular....................*C. richei*
4 Chloroplast usually single....................*C. platyuris*
4 Chloroplasts 2 per cell....................5
5 Paired refringent bodies present....................*C. ovata*
5 Paired refringent bodies absent....................6
6 Cells ovoid-ellipsoidal and posterior not bent....................*C. erosa*
6 Cells with a dorsally bent posterior and anterior forming a more or less pronounced protuberance....................7
7 Cells large (about 40 μm long); nucleus posterior....................*C. curvata*
7 Cells smaller (up to about 30 μm long); nucleus located between cell centre and posterior....................*C. marssonii*

Cryptomonas anomala F.E.Fritsch 1914

05040010 **Pl. 45D** (p. 184)

Cells 9–11 μm wide, 21–24 μm long, 11–12 μm thick, with a slight anterior dorsal protuberance; flagella equal, much shorter than cell; chloroplasts 2 and laterally placed with a circular pyrenoid, central or slightly displaced towards cell anterior; nucleus posterior; division occurs in a non-motile state with cells surrounded by abundant mucilage.

British Isles (England); known only from ponds near Leicester.

Cryptomonas curvata Ehrenberg 1838 emend. Penard 1921

05040020 **Pl. 45G** (p. 184)

Cells about 40 μm long, 20 μm thick, S-shaped in lateral view; anterior often with an obvious acute protuberance and containing a contractile vacuole; flagella subequal, longer one about two-thirds of cell length; chloroplasts 2 per cell, olive-green, without a pyrenoid or eyespot; nucleus in posterior half of cell.

Not widely distributed; found in aquatic habitats especially in the plankton.

Cryptomonas erosa Ehrenberg 1838

05040030 **Pl. 45A** (p. 184)

Cells 6–17 μm wide, 13–45 μm long, 6–26 μm thick, oval-ellipsoidal, often with a convex dorsal margin in lateral view and containing a contractile vacuole at cell anterior in a dorsal position; flagella equal or subequal, as long as or slightly shorter than cell; chloroplasts 2 per cell, very variable in colour but never blue-green, without a pyrenoid or eyespot.

Probably cosmopolitan; frequently found in aquatic habitats especially in the plankton.

Cryptomonas marssonii Skuja 1948

Synonym: *Cryptomonas erosa* Ehrenberg var. *reflexa* M.Marsson

05040040 **Pl. 45E** (p. 184)

Cells 8–14 μm wide, 16–33 μm long, anterior end forming a slight dorsal protuberance and containing a contractile vacuole, posterior end mostly acute and dorsally bent; flagella slightly unequal, longer one about two-thirds cell length; chloroplasts 2 per cell, without a pyrenoid; nucleus positioned between cell centre and posterior.

Not widely distributed; found in the plankton of lakes and small water bodies.

Cells assigned to this species have been examined briefly using electron microscopy (Novarino, 1991b).

Cryptomonas ovata Ehrenberg 1838

05040050 **Pl. 45B** (p. 184)

Cells 5–20 μm wide, 14–80 μm long, 6–26 μm thick, often with a moderate degree of lateral compression, anterior end often with a slight, acute, dorsal protuberance and 2 refringent bodies, posterior end rounded; contractile vacuole at cell anterior; flagella equal or subequal, as long as or shorter than cell; chloroplasts 2 per cell, olive-green, yellowish-green or yellowish-brown in colour, without a pyrenoid or eyespot; nucleus in posterior half of cell.

Probably cosmopolitan; frequently found in aquatic habitats especially in the plankton (e.g. Virginia Water lake system in Surrey; see Evans, 1988).

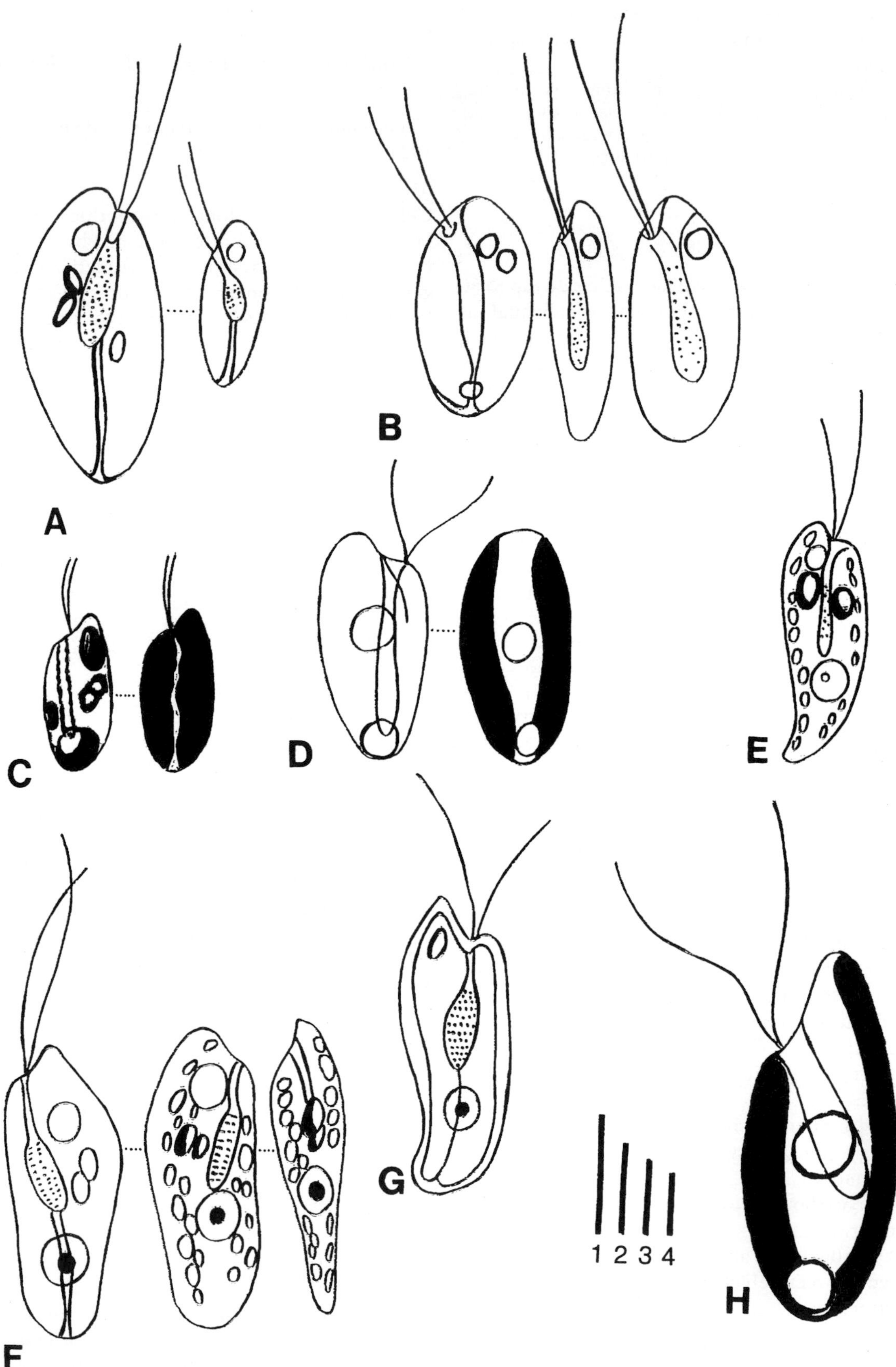

Plate 45 A–H

A. *Cryptomonas erosa* (p. 183); **B**. *Cryptomonas ovata* (p. 183); **C**. *Cryptomonas stigmatica* (p. 185); **D**. *Cryptomonas anomala* (p. 183); **E**. *Cryptomonas marssonii* (p. 183); **F**. *Cryptomonas platyuris* (p. 185); **G**. *Cryptomonas curvata* (p. 183); **H**. *Cryptomonas richei* (p. 185).

Scale bars: all 10 μm. 1 for H; 2 for D, E; 3 for A, F, G; 4 for B, C.

Cryptomonas platyuris Skuja 1948

05040060 **Pl. 45F** (p. 184)

Cells 9–14 μm wide, 31–50 μm long, 15–24 μm thick, laterally compressed, anterior end forming a dorsal protuberance and containing a large contractile vacuole, posterior end wide and rudder-like; flagella subequal, as long as cell or three-quarters cell length; chloroplasts 1(–2) per cell, occupying most of cell periphery, without pyrenoid; refringent bodies 2 in a dorsal position roughly mid-way along cell; nucleus posterior, with a distinct nucleolus.

Not widely distributed; found in the plankton of lakes and small water bodies.

Cryptomonas richei F.E.Fritsch 1914

05040080 **Pl. 45H** (p. 184)

Cells 9–10 μm wide, 18–22 μm long, with a marked anterior dorsal protuberance; flagella equal, slightly shorter than cell; chloroplasts 2 per cell, with angular pyrenoid in a central position; nucleus posterior.

British Isles (England); known only from ponds near Leicester.

Cryptomonas stigmatica Wislouch 1924

05040070 **Pl. 45C** (p. 184)

Cells 8.5–11 μm wide, 14–23 μm long, 6–7 μm thick, slightly compressed dorsiventrally, anterior end having a distinct rostrum, posteriorly rounded, anterior and posterior ends each containing a large lipid droplet; flagella equal, about one-half cell length; chloroplasts 2 per cell, pyrenoids 2 in a dorsal position mid-way along cell length, eyespot small, in a ventral position mid-way along cell length.

Described from hypersaline habitats in the Crimea, very rarely reported since.

Rhodomonas G.Karsten 1898

Cells with slightly unequal flagella, an anterior contractile vacuole and a single red (phycoerythrin-containing) chloroplast, with a central pyrenoid surrounded by a starch sheath.

Numerous reports exist of cryptomonads assigned to the genus *Rhodomonas* from all over the world, but owing to the uncertain application of this name their significance is very unclear.

The genus was described in the late 19th century based on a single species, *R. baltica* G.Karsten, a marine cryptomonad. The original generic description (see above) gives no other useful information and overall *Rhodomonas* is a very ill-defined genus. It has been argued that virtually any red cryptomonad can be assigned to *Rhodomonas* based on the original description (Novarino, 1991a), and closer examination of possible members of this genus has also shown that they can be ranked elsewhere. It has been recommended that the generic name *Rhodomonas* be abandoned (Novarino, 1991a; Novarino & Lucas, 1993b), but in view of its previous use it is retained here.

Rhodomonas lacustris Pascher *et* Ruttner in Pascher & Lemmermann 1913

05100010 **Pl. 44K** (p. 182)

Cells 5–8 μm wide, 10–13 μm long, oval, with broadly rounded anterior end and narrower posteriorly; flagella unequal, the longer (ventral) flagellum about as long as cell and dorsal one about two-thirds cell length; pyrenoid large and placed in centre; nucleus posterior.

Apparently widely distributed, although some reports probably pertain to var. *nannoplanctica* (see below); frequently reported from lake plankton.

var. *nannoplanctica* (Skuja) Javornický 1976

Basionym: *Rhodomonas minuta* var. *nannoplanctica* Skuja

05100012 **Pl. 44E,F** (p. 182)

Cells 3.5–5.5 μm wide, 8–12 μm long, anterior end rounded and posteriorly prolonged into an acute, conical, cylindrical or blade-like tail one-tenth to one-fifth cell length, ventrally bent and containing a refringent inclusion ('leucosin granule'); flagella unequal, about three-quarters and one-half cell length, respectively; chloroplast dorsal with ventral margins extending deeply into ventral region of cell, containing phycoerythrin, with a central, large pyrenoid surrounded by a thick starch sheath of 2 hemispherical halves; nucleus posterior.

Probably cosmopolitan; frequently found in the plankton of lakes and small water bodies. Lund (1962b) reported it from all around the British Isles: England, especially the English Lake District and the Shropshire Meres, as well as from Wales, Scotland, Northern Ireland and the Republic of Ireland.

Based on electron microscopy, this taxon can be assigned to the genus *Plagioselmis* Butcher emend. Novarino, I.A.N.Lucas *et* S.Morrall under which its correct name is *Plagioselmis nannoplanctica* (Skuja) Novarino, Lucas *et* Morrall (Novarino et al., 1994). Under the genus *Rhodomonas*, the correct name is *R. minuta* var. *nannoplanctica* Skuja. Alternatively, the name *R. lacustris* var. *nannoplanctica* (Skuja) Javornický is also nomenclaturally correct when the taxon is considered a variety of *R. lacustris* Pascher *et* Ruttner. However, taxonomically this seems hardly justifiable since there is no unequivocal evidence that *R. minuta* Skuja (1948) is identical to *R. lacustris* Pascher *et* Ruttner. The characters listed above are based on the original diagnosis of *Rhodomonas minuta* var. *nannoplanctica* (Skuja, 1948) and the description of *Plagioselmis nannoplanctica* (Novarino et al., 1994). The latter name is mentioned by Reynolds & Irish (2000) in a list of the phytoplankton of Windermere along with *Rhodomonas lens* Pascher *et* Ruttner. No further information is given concerning either of these records other than the latter is 'not distinguished in older records'. For further discussion on the taxonomic and nomenclatural problems associated with *Rhodomonas* and *Plagioselmis*, see Javornický (2001).

PHYLUM PYRROPHYTA (Dinoflagellates)

Jane M. Lewis and John D. Dodge

The study of freshwater dinoflagellates in the British Isles has been sporadic and partial. The majority of work describing species and their ecology took place between 1920 and 1960. Workers of note were the Wests (e.g. W. & G.S. West, 1906) and Harris (1940). Since that time there have been some detailed studies on individual genera, most notably *Ceratium* in the English Lake District (Chapman et al., 1981, 1985; Heaney et al., 1983, 1988), but little further descriptive work. As far as possible, all historical accounts of records for the British Isles have been brought together in the production of the present account, which aims to describe all the species recorded. However, it is likely that many more species remain to be discovered; indeed, in the course of writing this account a number of hitherto unrecorded species were found. So the keys provided here should be used with caution. Whilst they cover the species most likely to be found, studies on new water bodies may well lead to the discovery of species not described here. If an organism does not conform to any of the descriptions here, the most useful accounts for trying to identify it are those of Huber-Pestalozzi (1950) and Popovský & Pfeister (1990).

Dinoflagellates are characterized by their nucleus, which is distinctive by virtue of the permanently condensed chromosomes. Under the light microscope the nucleus frequently appears rather large in proportion to the cell and with obvious sausage-shaped chromosome,. Where the attribution of a cell is in doubt it is always worth checking this point as a number of unusual cell forms exist within the class. However, the most common form for dinoflagellates is a motile unicell, often with a defined groove (cingulum) encircling the cell body, and with two dissimilar flagella. One transverse flagellum is placed within the cingulum and the other, longitudinal, follows a second groove (sulcus) at right angles to the cingulum. These flagella produce a characteristic whirling swimming motion from which the name dinoflagellate is derived. The cell wall consists of vesicles, which in various genera contain differing amounts of cellulosic material. These vary from strongly armoured species, which have relatively thick plates, to those with little reinforcement, which are referred to as unarmoured. Under the light microscope these differences may be very evident, the armoured forms having strongly ornamented, angular outlines which are retained even after lengthy examination. This contrasts strongly with unarmoured forms, which change shape even as they are observed under the microscope. There are some genera where this distinction might not be so clear. For these it is necessary to squash or burst the cell under a cover slip and carefully check any wall remaining for plate structure. The plate pattern of armoured forms is used as a point of distinction between genera. Plate nomenclature recognizes the different series of plates around the cell, starting at the apex. Each series is denoted by the number of plates in that series. The apical pore plate (where present) is designated Po; the apical plates by ′; the intercalary plates by a; precingular plates by ″; the cingular plates by c; the sulcal plates by s; the postcingular plates by ‴; and the antapical plates by ⁗. For example, the plate formula of *Glenodinium* is Po, 4′, 4a, 7″, ?3c, ?6s, 6‴, 2⁗.

For the classification we follow Fensome et al. (1993), who divided the dinoflagellates into eleven orders of which four are used here: Gymnodiniales, Gonyaulacales, Peridiniales, Phytodiniales. The possession of plates forms a major distinction within the class and in this key the unarmoured genera (*Amphidinium, Gymnodinium, Katodinium*) are described first, followed by the armoured genera (*Ceratium, Glenodiniopsis, Glenodinium, Peridiniopsis, Peridinium*). Where plates are present, they are the major morphological features used for distinction of species. Their number, arrangement and ornamentation all provide important features for distinction. The overall shape of the cell and position of the cingulum also separates many genera. The cingulum divides the cell into the anterior epicone/epitheca and the posterior hypocone/hypotheca. Other features that may assist with identification are the presence, colour and arrangement of chloroplasts, the shape and position of the nucleus, and the presence of an eyespot. Figure 3 gives an overview of the genera in this account.

The chloroplasts of dinoflagellates are mostly yellow-brown, olive-brown or bistre-brown, while an orange-red eyespot occurs in some species. One quite common species, *Gymnodinium aeruginosum*, has bright blue chloroplasts. Some species accumulate carotenoids which colour particular regions of the cell orange or orange-red. Colourless dinoflagellates are also quite common in ponds and some of these have orange-red eyespots just like photosynthetic species.

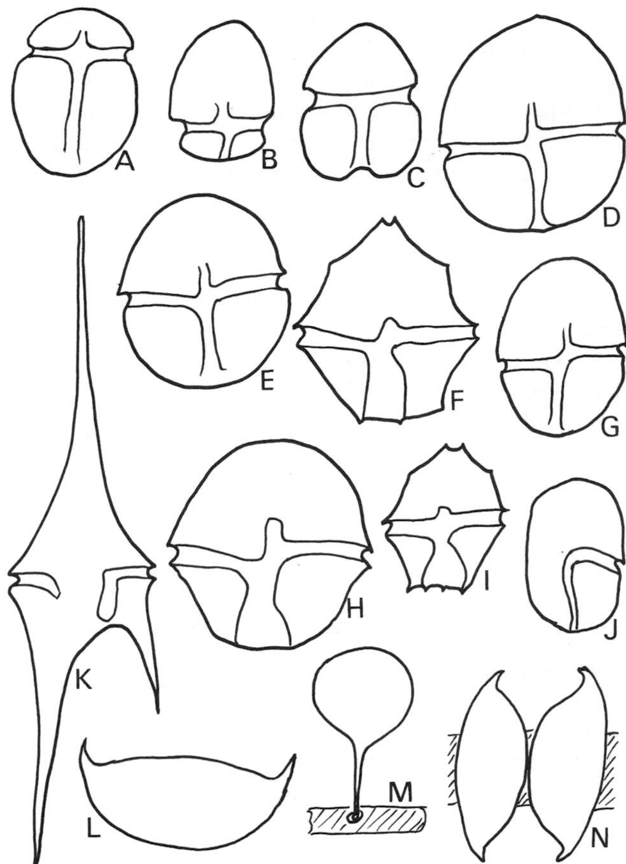

Figure 3. Outline diagrams to illustrate the typical shapes of the cells in the genera included in the Pyrrophyta. **A.** *Amphidinium* (p. 188); **B.** *Katodinium* (p. 193); **C.** *Gymnodinium* (p. 188); **D.** *Woloszynskia* (p. 195); **E.** *Glenodinium* (p. 198); **F.** *Peridiniopsis* (p. 199); **G.** *Glenodiniopsis* (p. 198); **H.** *Peridinium* (p. 201) (species with no apical pore); **I.** *Peridinium* (species with an apical pore); **J.** *Hemidinium* (p. 207); **K.** *Ceratium* (p. 196); **L.** *Cystodinium* (p. 205); **M.** *Stylodinium* (p. 207); **N.** *Dinococcus* (p. 205).

Dinoflagellates can be sampled in a similar manner to other phytoplankton. Water samples, net samples, squeezings of *Sphagnum*, other mosses or submerged flowering plants are all suitable (see Methods, p. 11). Because of their general paucity in water samples it can be advantageous to concentrate them in some fashion (gentle centrifugation for 5 minutes; filtration through a plankton net of 30–40 μm mesh size or a membrane of 10 μm pore size), or by allowing them to increase within a sample kept under suitable conditions. It is preferable to examine at least some of the material in the live condition, since cell shape often changes after fixation. The unarmoured dinoflagellates do not preserve well and once preserved identification is equivocal. Cells can be preserved in Lugol's iodine or formalin.

Relatively little is known of the ecology of freshwater dinoflagellates. Some species migrate vertically in the water column during summer, moving towards the light by day and to the bottom (where phosphate levels may be higher) at night. This applies to members of the genus *Ceratium*, which are normally found in large bodies of water such as lakes and reservoirs. Most dinoflagellates are found in smaller water bodies and even temporary pools, but even here the possibility should be borne in mind that cells may be most abundant at different positions within the waterbody at different times of day. *Ceratium* and some other motile dinoflagellates engulf bacteria under certain conditions as well as carrying out photosynthesis. Some species thrive in cold water (<10 °C) whilst others are only found in summer when the water is warm (>15 °C). Both types may be found to bloom in large numbers and colour the water brown; red-coloured blooms are also known from some mainland European lakes.

1 Vegetative phase motile and free-swimming (typical dinoflagellate) 2
1 Vegetative phase attached or non-motile (coccoid) 11
2 Cell naked and readily bursts open 3
2 Cell covered with firm or obviously armoured wall 4
3 Transverse groove (cingulum) near anterior end of cell; epitheca much smaller than hypotheca .... *Amphidinium* (p. 188)
3 Transverse groove median or below centre of cell; epitheca and hypotheca about equal in size .... *Gymnodinium* (p. 188)
4 Cell covered with obvious theca consisting of cellulosic plates 5
4 Cell covering not obvious 7
5 Cell drawn out into prominent apical and antapical horns; cell large, >100 μm long .... *Ceratium* (p. 196)
5 Cell shape almost compact or ovoid, sometimes with small projections at apex and antapex cell usually ≤100 μm long 6
6 Epithecal plates with 3 intercalary plates (see Pl. 51B) .... *Peridinium* (p. 201)
6 Epitheca with only one intercalary plate (see Pl. 52D) .... *Peridiniopsis* (p. 199)
7 Cell markedly asymmetric; epitheca and hypotheca markedly different in size 8
7 Cell almost symmetrical; epitheca and hypotheca similar in size 9
8 Hypotheca very much smaller than epitheca; cingulum complete .... *Katodinium* (p. 193)
8 Epitheca and hypotheca more or less similar in size; cingulum incomplete and sulcus situated to one side of cell .... *Hemidinium* (p. 207)
9 Cell covered with a large number (>100) of very delicate thecal plates; resting cyst stage angular .... *Woloszynskia* (p. 195)
9 Cell covered with a small number (about 20) of delicate thecal plates 10
10 Cell distinctly spherical; eyespot prominent .... *Glenodinium* (p. 198)
10 Cell almost ovoid; eyespot absent .... *Glenodiniopsis* (p. 198)
11 Cell almost spherical; attached to substrate by delicate stalk .... *Stylodinium* (p. 207)

11 Cell almost lunate; free-living or closely attached 12
12 Cell single; unattached .... *Cystodinium* (p. 205)
12 Cells often in pairs; attached to green algal filaments .... *Dinococcus* (p. 207)

ORDER GYMNODINIALES

Free-living, with well-developed transverse and sulcal grooves. Cells more or less compressed dorsiventrally and without thecal plates. Numerous, empty, polygonal, thecal vesicles around periphery of cell.

Amphidinium Claparède *et* J.Lachmann 1858–61

Vegetative cells free-swimming, generally ovoid in ventral view, sometimes dorsiventrally flattened; epicone shorter than hypocone and frequently narrower, separated by a deep cingulum with little or no offset; sulcus often extends into epicone and to bottom of hypocone; cell wall thin; chloroplasts absent or when present varying from yellow-brown to blue-green and scattered throughout cell; resting stage thick-walled, spherical, embedded within mucilage.

Cosmopolitan; inhabiting marine and freshwater, found in the water column or in interstitial waters.

Some brackish-water species are known to be toxic to fish.

1 Cell with numerous blue-green chloroplasts .... *A. amphidinioides*
1 Cell with several yellow-brown chloroplasts .... *A. elenkinii*

Amphidinium amphidinioides (Geitler) J.Schiller 1933

Basionym: *Gymnodinium amphidinioides* Geitler
Synonym: *Amphidinium geitleri* Huber-Pestalozzi
06010020 **Pl. 46A** (p. 189) **CD**

Cells ovoid, slight dorsiventral flattening, 11–20 µm wide and 18–30 µm long; epicone narrower than hypocone, one-quarter of cell length, shallowly conical in shape; hypocone straight sided with rounded antapex, three-quarters of cell long; cingulum wide, not offset; sulcus narrow, extends slightly into epicone, reaches antapex in hypocone sometimes; chloroplasts numerous, oval or irregularly ribbon-shaped and distinctively coloured blue-green; cytoplasm sometimes with red droplets resembling an eyespot; nucleus large, spherical, and in hypocone; resting stage spherical, surrounded by a thick membrane.

Worldwide; found in ponds, peat bogs, puddles and swamps; nutrition autotrophic to phagotrophic.

Amphidinium elenkinii Skvortsov 1925

Synonyms: *Amphidinium gyrinum* T.M.Harris, *A. larvale* Lindemann, *A. tenagodes* T.M.Harris
06010030 **Pl. 46B** (p. 189)

Cells round or oval, slightly dorsiventrally flattened, 5–15 µm wide, 6–16 µm long; epicone about one-quarter of cell length, slightly narrower than hypocone, variable in morphology from flattened to broadly conical; hypocone oval to rounded, with slight indentation at base; cingulum distinct, relatively deep, not offset; sulcus less visible than cingulum, fairly broad, reaching antapex in most cells; cytoplasm hyaline, colourless with small refractile bodies; chloroplasts (if present) 1 to several, oval, variable in size, yellow-brown; nucleus central in hypocone. Swims with a rapid rotating motion.

Europe; found largely in eutrophic habitats including reservoirs, rivers, pools, ponds and lakes.

In his original description Harris (1940) likens *A. gyrinum* to *A. elenkinii* but differentiates it by virtue of the lack of chromatophores and its possession of a shorter flagellum. Neither of these is considered reliable for distinguishing species in this group. Similarly Harris separates *A. tenagodes* and *A. elenkinii* on the basis of the 'shorter flagellum and shallower sulcus': again neither of these are good distinguishing features.

DOUBTFUL RECORD
Amphidinium aeschrum T.M.Harris 1940
06010010
Reported by Harris (1940) from amongst filamentous algae in a very nutrient-rich (eutrophic) lock at Mapledurham on the Thames, but unrecorded since. The misshapen cells might be more properly interpreted as an *Amphidinium* or *Gymnodinium* in the process of division (Popovský & Pfiester, 1990).

Gymnodinium F.Stein 1878

Vegetative cells free-swimming, single or in chains, rounded, squarish or ovoid in ventral view, sometimes dorsiventrally flattened; epicone and hypocone broadly similar, both long, separated by a well-defined cingulum offset by up to one-fifth of cell length; sulcus can extend into epicone and to bottom of hypocone; cell wall thin, generally smooth but sometimes ornamented with ridges, apical groove sometimes present; chloroplasts when present are numerous and variable; sexual reproduction sometimes results in formation of a spherical, thick-walled, hypnozygote.

Cosmopolitan; inhabiting marine and freshwater habitats.

1 Cell without chloroplasts 2
1 Cell containing chloroplasts 3
2 Cell large, rather wedge-shaped, with vertical striations in wall .... *G. helveticum*
2 Cell small, broadly spherical, without such striations .... *G. colymbeticum*
3 Cell large, rather flattened with rounded epicone and pointed hypocone .... *G. fuscum*
3 Cell not as above 4
4 Chloroplasts radially arranged 5
4 Chloroplasts parietal or scattered........................... 7
5 Cell with an eyespot and oval in profile .... *G. paradoxum*
5 Cell without an eyespot and with conical epicone or indented hypocone 6

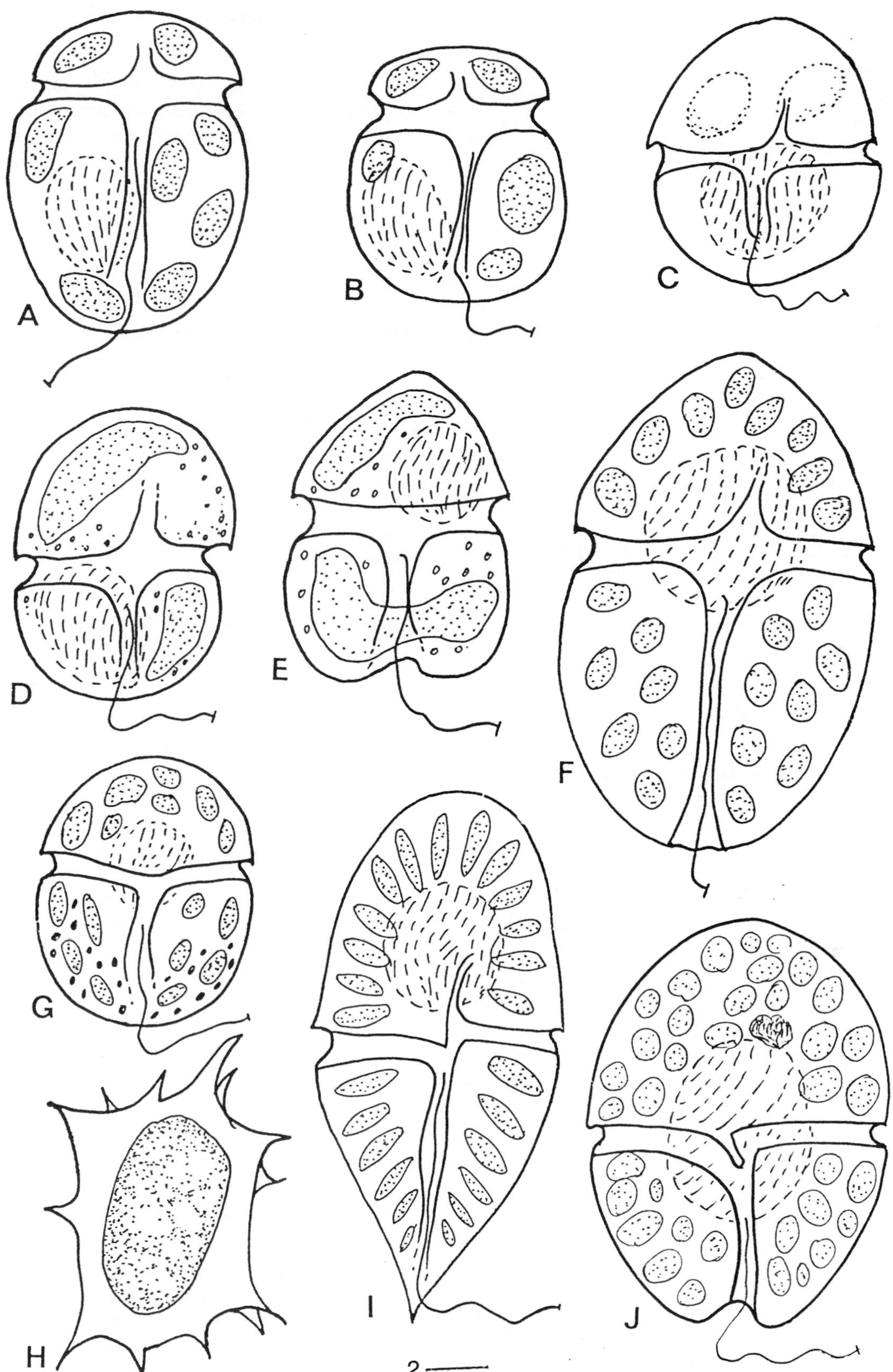

Plate 46 A–J

A. *Amphidinium amphidinioides* (p. 188); **B**. *Amphidinium elenkinii* (p. 188); **C**. *Gymnodinium colymbeticum* (p. 190); **D**. *Gymnodinium discoidale* (p. 190); **E**. *Gymnodinium cnecoides* (p. 190); **F**. *Gymnodinium aeruginosum* (p. 190); **G, H**. *Gymnodinium chiatosporum* (p. 190): motile cell and cyst; **I**. *Gymnodinium fuscum* (p. 190); **J**. *Gymnodinium pseudopalustre* (p. 192).

Scale bars: all 10 μm. 1 for A–H, J; 2 for I.

6 Epicone rather conical; cingulum deeply incised .... *G. inversum*
6 Epicone rounded; hypotheca incised by sulcus .... *G. uberrimum*
7 Hypocone with 1–3 projections .... *G. triceratium*
7 Hypocone without projections and rounded or incised by sulcus 8
8 Hypocone distinctly incised at base of sulcus .... *G. pseudopalustre*
8 Hypocone rounded 9
9 Epicone conical 10
9 Epicone rounded 12
10 Chloroplasts blue-green .... *G. aeruginosum*
10 Chloroplasts yellow-brown 11
11 Chloroplasts ribbon-like .... *G. cnecoides*
11 Chloroplasts rounded .... *G. saginatum*
12 Epicone much larger than hypotheca (about two-thirds cell length) .... *G. palustre*
12 Epicone shorter than or equal to hypotheca 13
13 Epicone not more than one-third cell length .... *G. hippocastanum*
13 Epicone more or less equal to length of hypotheca 14
14 Chloroplasts numerous, disc-shaped .... *G. chiastosporum*
14 Chloroplasts few, parietal, band-shaped .... *G. discoidale*

Gymnodinium aeruginosum F.Stein 1883

06070010 **Pl. 46F** (p. 189)

Cells ovoid, tear-shaped or squarish, strongly dorsiventrally flattened, with a postmedian cingulum, 9–35 µm wide, 13–44 µm long; epicone slightly longer than hypocone; cingulum distinct, sulcus extends slightly into epicone and continues to antapex of hypocone; chloroplasts small and numerous, bright blue-green, arranged around cell periphery; eyespot absent; nucleus large and in epicone.

Worldwide; in ponds, lakes, bogs and rivers, from nutrient-poor to slightly nutrient-rich water, present throughout the year, especially in the vicinity of decaying aquatic vegetation.

Rounded cells embedded in mucilage are possibly a stage associated with nutrient limitation.

Gymnodinium chiastosporum (T.M.Harris) Cridland 1958

Basionym: *Tetradinium chiastosporum* T.M.Harris

06070020 **Pl. 46G,H** (p. 189)

Cells ovoid, slightly dorsiventrally flattened, 10–12 µm wide, 12–25 µm long; epicone slightly shorter than hypocone and separated by a wide cingulum; sulcus variable, sometimes extends into epicone and almost reaches base of hypocone; chloroplasts numerous; eyespot absent; starch grains abundant in hypocone; resting stage thick-walled, squarish, flattened, 13–40 µm × 13–25 µm in size, bearing 2 short spines at each corner.

Europe; in pools and ponds; originally described from a tub of the duckweed *Lemna*.

Gymnodinium cnecoides T.M.Harris 1940

06070030 **Pl. 46E** (p. 189)

Cells rounded, very slightly dorsiventrally flattened, 8–11 µm wide, 9–14 µm long; epicone and hypocone of equal size, separated by a wide cingulum; sulcus wide, extends to base of hypotheca and not into epicone; chloroplasts 1 or 2, very pale, yellow, ribbon-like, arranged around periphery of cell leaving a hyaline region in cell centre; eyespot absent; nucleus in epicone.

Europe; in woodland pools, peat bogs, ponds and lakes.

Gymnodinium colymbeticum T.M.Harris 1940

06070040 **Pl. 46C** (p. 189)

Cells broadly spherical, about 16 µm in diameter; epicone slightly smaller than hypocone, separated by a well-defined cingulum and offset by one cingulum width; sulcus wide, extends into epicone and not reaching antapex of hypocone; chloroplasts and eyespot absent.

British Isles; in small pools, a rapid-swimming species.

Gymnodinium discoidale T.M.Harris 1940

Synonym: *Glenodinium eurystomum* T.M.Harris

06070050 **Pl. 46D** (p. 189)

Cells rounded to ovoid in ventral and dorsal view, dorsiventrally flattened, 7–16 µm wide, 8–18 µm long; epicone and hypocone equal in size, separated by cingulum with no offset; sulcus extends into epicone by about one-third of its height and not to base of hypocone; cell wall firm and not separable into plates; chloroplasts 1 or 2; eyespot; refractive droplets within epi- and hypocones; nucleus in hypocone.

Europe; in temporary pools, lakes, peat bogs and roadside ditches; nutrition autotrophic, holozoic.

Gymnodinium fuscum (Ehrenberg) F.Stein 1878

Basionym: *Peridinium fuscum* Ehrenberg

06070060 **Pl. 46I** (p. 189)

Cells large, 29–80 µm wide, 44–118 µm long, tear-shaped, slight dorsiventral flattening; epicone hemispherical, hypocone conical with a slight horn at base, separated by a narrow, deep-set cingulum with overhanging upper margin; cingulum offset by one cingulum width, left-hand spiral and deeply grooved; sulcus narrow, protruding slightly into epicone to form a characteristic notch, extends halfway down hypocone; chloroplasts numerous, yellow to yellow-brown, rod-shaped to oval, arranged radially throughout cell; trichocysts comma-shaped in hypocone; pusule system coiled; nucleus just above cell centre in epicone; numerous pigmented granules in central cytoplasm; two morphologically dissimilar, non-motile stages (produce mucilage in unfavourable conditions): spherical resting stage with thick, irregular mucilage, and '*Cystodinium*' stage with a large nucleus in epicone, sometimes 4 non-motile zoospores produced in epicone.

Worldwide; in peat bogs and acid nutrient-poor lakes.

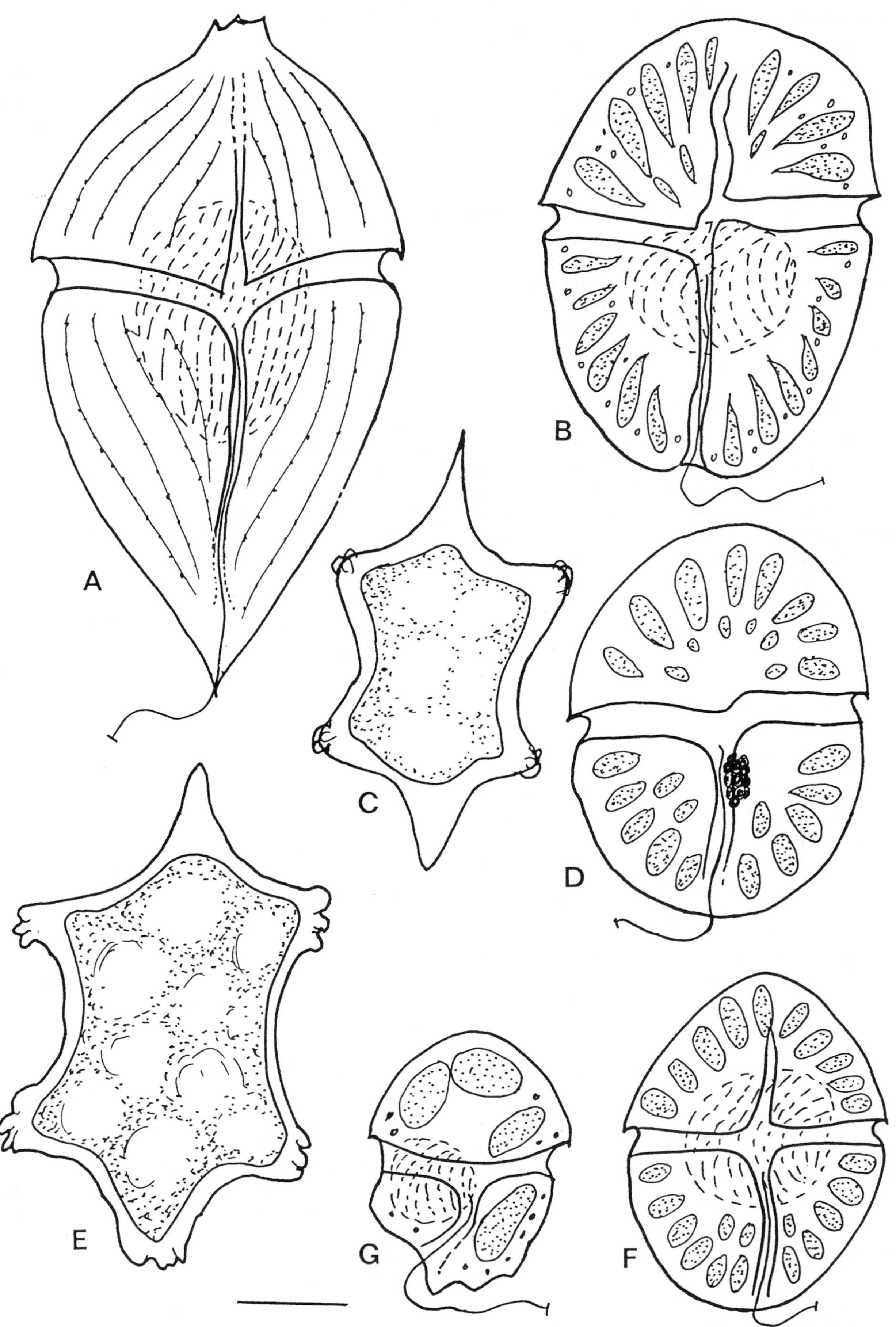

Plate 47 A–G
A. *Gymnodinium helveticum* (p. 192); **B**. *Gymnodinium uberrimum* (p. 193); **C, D**. *Gymnodinium paradoxum* (p. 192): cyst and motile cell; **E, F**. *Gymnodinium inversum* (p. 192): cyst and motile cell; **G**. *Gymnodinium triceratium* (p. 192).
Scale bar: 10 μm.

Gymnodinium helveticum Penard 1891

06070110 **Pl. 47A** (p. 191)
Cells wedge-shaped, 23–30 μm wide, 42–57 μm long; epicone broadly conical with a slight apical protuberance sometimes with 3 small projections at apex; hypocone, longer than epicone, elongated conical shape with pointed antapex; cingulum in upper half of cell, broad, deeply incised and not offset; sulcus narrow and extends three-quarters of length of both epicone and hypocone; cell covering lightly ridged and ornamented with rows of fine papillae; chloroplasts and eyespot absent; cytoplasm pale pink with rod-shaped refractile granules; nucleus ovoid, above cell centre.

Europe; plankton of nutrient-poor lakes; nutrition phagotrophic.

Gymnodinium hippocastanum Cridland 1958

06070120 **Pl. 50C,D** (p. 200)
Cells small, about 15 μm wide and 25 μm long, ovoid, rapidly swimming; epicone smaller than hypocone, sometimes hypocone compressed towards anterior end; cingulum and sulcus at right angles; sulcus extends well into epicone and almost to base of hypocone; chloroplasts numerous, yellow and eyespot absent. Resting stage (cyst) thick-walled, squarish, bearing about 10 simple or bifurcate spines around its edge and across the faces, about 25 × 40 μm with spines 13 μm long.

British Isles; found in winter in a shallow pond with rotting oak and beech leaves.

Gymnodinium inversum Nygaard 1949

06070070 **Pl. 47E,F** (p. 191) **CD**
Cells ellipsoidal, slightly dorsiventrally compressed, 16–26 μm wide, 20–30 μm long, both apex and antapex broadly rounded; hypocone slightly larger than epicone; sulcus extending length of hypocone and a short distance onto epicone; cingulum slightly offset on right side; nucleus broadly oval, above cingulum; chloroplasts numerous, rounded and brown; resting stages thick-walled, dumb-bell-shaped with knob-like appendages at corners, apex and antapex, contents granular with red globules 26 μm across and 42 μm long.

Europe; in small ponds, a cold-water species appearing as motile cells in late winter and then forming cysts as the water temperature warms (mid-March to April in the UK).

Gymnodinium palustre A.J.Schilling 1891

06070130 **Pl. 50A** (p. 200)
Cells small, 22–37 μm wide, 35–60 μm long, elongate-elliptical; epicone two-thirds of cell length, divided from the hypocone by a deep cingulum; cingulum not offset; sulcus narrow, extends into the epicone and reaches base of hypocone; chloroplasts numerous, rod-shaped, yellow or dark brown, arranged radially; eyespot absent.

Europe; in the summer plankton of pools, ponds and lakes.

Gymnodinium paradoxum A.J.Schilling 1891

06070080 **Pl. 47C,D** (p. 191)
Cells ovoid and slightly dorsiventrally flattened, 61–67 μm wide, 60–75 μm long; epicone slightly larger than hypocone, separated by a poorly defined cingulum; sulcus indistinct, extending to base of hypocone; chloroplasts numerous, yellow-green and radially arranged; zoospores with eyespot present in hypocone, smaller than vegetative cells, 34–35 μm wide and 36–40 μm long; hypnozygotes bispindle-shaped with equatorial constriction, slightly flattened with 3 rounded structures on each corner, 25 μm wide, 36–40 μm long.

Europe; in vegetation and plankton of swamps, ponds, lakes and reservoirs; nutrition autotrophic and holozoic.

Several forms are described within this species. *G. paradoxum* var. *major* has been recorded from Ireland (W. & G.S.West, 1906). The cyst of this species resembles some species of *Tetradinium*.

Gymnodinium pseudopalustre (Wołoszyńska) A.J.Schiller 1933

Basionym: *Gymnodinium palustre* Wołoszyńska
Synonym: *Gymnodinium excavatum* Nygaard
06070140 **Pl. 46J** (p. 189)
Cells ovoid with rounded apex and distinctly indented (excavated) antapex, 21–34 μm wide, 26–42 μm long; epicone hemispherical, hypocone conical with a deep groove at base of sulcus; cingulum more or less median, very slightly offset on right side, small finger-like process at intersection of cingulum and sulcus; cell wall fairly tough; nucleus large and central; chloroplasts numerous, rounded, yellow-brown, scattered throughout cell periphery; orange body often present adjacent to anterior side of nucleus.

Europe; in ponds, a warm-water (summer) species.

Cell division is by transverse division, daughter cells swim together, one in front of the other, for at least 12 hours. Under adverse conditions cells rapidly lose their motility and round off. The life history has been described in detail by von Stosch (1973): small cells (gametes) fuse in pairs to form planozygotes which can be recognized by their larger size, 2 longitudinal flagella, presence of red body beneath sulcus and extended epicone (the form originally described as *G. palustre* by Wołoszyńska); these eventually form spheroidal cysts with fine spines over the wall. Nygaard described the vegetative stage as *G. excavatum*.

Gymnodinium saginatum T.M.Harris 1940

06070150 **Pl. 50B** (p. 200)
Cells very rounded, 18–20 μm wide, 18–23 μm long; epicone and hypocone of equal size, separated by a narrow, deep cingulum; sulcus rather narrow, extends to base of hypocone; cell wall quite firm; chloroplasts pale; eyespot absent; cytoplasm granular and without colour.

British Isles; in plankton of nutrient-poor and peaty pools in autumn and winter.

This species is especially distinct because of its poor swimming ability.

Gymnodinium triceratium Skuja 1939

Synonym: *Gymnodinium impar* T.M.Harris
06070090 **Pl. 47G** (p. 191)

Cells slightly dorsiventrally flattened, 8–20 μm wide, 10–20 μm long; epicone hemispherical; hypocone irregularly conical, with 1–3 large projections at antapex; epicone wider than hypocone and separated by a wide, deeply incised cingulum; sulcus short, poorly defined, visible only in hypocone, descending to right side of antapex between antapex and right-sided projection when present; chloroplasts 1–5, oval or irregularly plate-shaped, bright yellow or yellow-brown and mainly in epicone; eyespot absent; red oil droplets sometimes in cytoplasm; nucleus spherical, in centre of cell.

Europe; in peat bogs, pools, lakes and humic-rich (dystrophic) waters in winter or spring; nutrition autotrophic, holozoic and holophytic.

Gymnodinium uberrimum (G.J.Allman) Kofoid *et* Swezy 1921

Basionym: *Peridinium uberrima* G.J.Allman
Synonyms: *Melodinium uberrimum* Kent, *Gymnodinium rufescens* Lemmermann, *G. rotundatum* G.A.Klebs
06070100 **Pl. 47B** (p. 191)
Cells broadly spherical, 19–75 μm in diameter, 24–90 μm long, although antapex sometimes rather square, flattened dorsiventrally, unusually forms chains; cingulum wide, deep and divides cell into equal sized epicone and hypocone; sulcus deep, narrow, extends into epicone where it sometimes finishes in point deflected to left of cell and descends almost to antapex of hypocone where it sometimes widens and deflects right of cell; chloroplasts numerous, yellow-brown, radially arranged; eyespot absent; red droplets numerous in cytoplasm; nucleus large and in epicone or centre of cell; trichocysts when present (rare) just below cell membrane; pusule small, spherical, present at proximal junction of furrows.

Europe; in lakes and nutrient-poor waters with low alkalinity and pH, sometimes forming blooms discolouring the water brown.

The original description appears to have been overlooked by Lemmermann when describing *G. rufescens*. The description of *G. rotundatum* indicates it has a rounded cell and is less symmetrical in the hypocone; these slight differences (which might be caused by observation under the microscope) are not thought sufficient to retain their separation.

Katodinium Fott 1957

Vegetative cells free-living, ellipsoidal, inverted top- or mushroom-shaped; epitheca at least two-thirds cell length and broader than hypotheca, the cingulum spanning this difference in width; sulcus sometimes poorly defined but occasionally extends into epitheca; cell wall thin, probably of delicate and sometimes striated plates, triangular scales on outer membrane of one species; chloroplasts absent or yellow, brown or pale green when present; eyespot in some species; sexual reproduction isogamous.

Cosmopolitan; known from marine and freshwater environments.

Genus may not be valid, improved fixation and electron microscopy has led to transfer of some species to other thecate genera.

1 Cell with a very much reduced hypotheca 2
1 Cell with the hypotheca forming about one-third of cell 5
2 Eyespot present 3
2 Eyespot absent 4
3 Epitheca rounded, cingulum horizontal .... *K. campylops*
3 Epitheca conical, cingulum offset .... *K. tetragonops*
4 Hypotheca small and flattened .... *K. fungiforme*
4 Hypotheca rounded .... *K. molopicum*
5 Chloroplasts small, green .... *K. vorticella*
5 Chloroplasts yellow-brown .... *K. hyperxanthum*

Katodinium campylops (T.M.Harris) A.R.Loeblich III 1965

Basionym: *Massartia campylops* T.M.Harris
06090060 **Pl. 48G** (p. 194)
Cells ovoid, dorsiventrally flattened, 10–21 μm wide, 12–25 μm long; epicone larger than rounded hypocone, separated by a deep cingulum offset by one cingulum width; sulcus protrudes into epicone and descends almost to base of hypocone; chloroplasts absent and eyespot occasionally present.

Europe; in moderately nutrient-rich ponds and lakes.

Katodinium fungiforme (Anisimova) A.R.Loeblich III 1965

Basionym: *Gymnodinium fungiforme* Anisimova
Synonym: *Gymnodinium blax* T.M.Harris
06090010 **Pl. 48A** (p. 194) **CD**
Cells small, subovoid, 5–12 μm wide, 4–15 μm long, flattened dorsiventrally; epicone larger than hypocone, both broadly rounded and divided by a well-marked cingulum, overhung by epicone; sulcus descends to antapex of hypocone in an oblique manner; chloroplasts and eyespot absent; nucleus in hypocone.

Europe, USA; in plankton of ponds, lakes and also brackish-water; nutrition holozoic.

Katodinium hyperxanthum (T.M.Harris) A.R.Loeblich III 1965

Basionym: *Massartia hyperxantha* T.M.Harris
Synonyms: *Katodinium hyperxanthoides* (T.M.Harris) A.R.Loeblich III, *Massartia hyperxanthoides* T.M.Harris
06090020 **Pl. 48C** (p. 194)
Cells ovoid, dorsiventrally flattened, about 10 μm wide, 12–14 μm long; epicone about twice length of hypocone, separated by a well-defined cingulum, offset by nearly one-third cell length; sulcus not well-defined, narrow and extends most of length of epicone and hypocone; chloroplasts yellow-brown and eyespot absent.

Worldwide; in plankton of nutrient-enriched ponds in winter and spring; nutrition autotrophic, holozoic and holophytic.

Harris described two species, *M. hyperxantha* and *M. hyperxanthoides*. He comments that they are very similar but separated them on the basis of the firmer cell wall, the curved nature of the sulcus, and the presence of an eyespot in *M. hyperxanthoides*. However, all of these are variable characters and it seems justified to synonymize these two species. Popovský & Pfiester (1990) state that these are dividing monads of *Amphidinium* or *Gymnodinium* but there is no direct evidence of this.

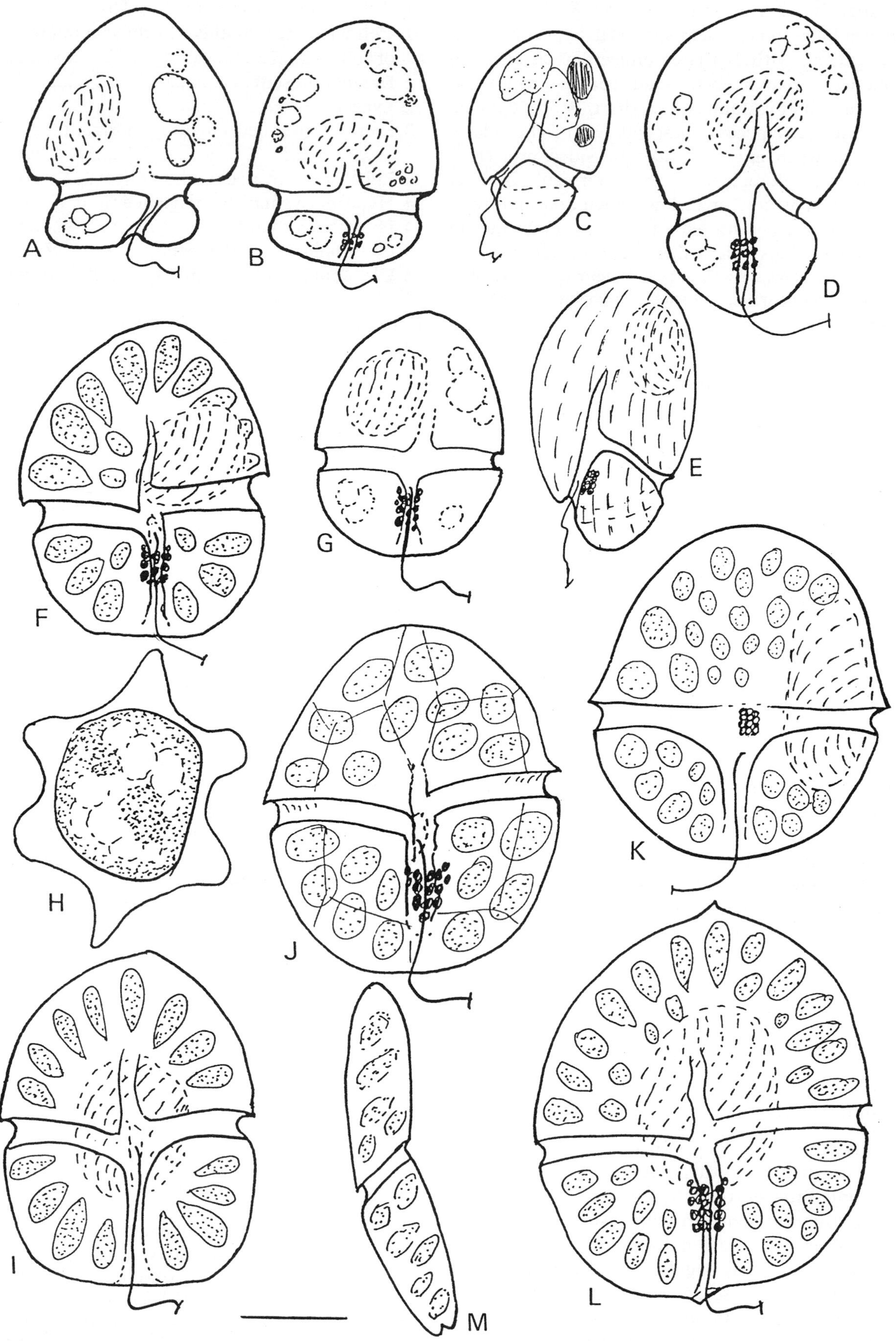

Plate 48 A–M

A. *Katodinium fungiforme* (p. 193); **B**. *Katodinium vorticella* (p. 195); **C**. *Katodinium hyperxanthum* (p. 193); **D**. *Katodinium tetragonops* (p. 195); **E**. *Katodinium molopicum* (p. 195); **F**. *Woloszynskia coronata* (p. 195); **G**. *Katodinium campylops* (p. 193); **H, I**. *Woloszynskia tylota* (p. 196): cyst and motile cell; **J**. *Glenodinium cinctum* (p. 199); **K**. *Woloszynskia neglecta* (p. 195); **L, M**. *Woloszynskia tenuissima* (p. 196): ventral and side views.

Scale bar: 10 μm.

Katodinium molopicum (T.M.Harris) A.R.Loeblich III 1965

Basionym: *Massartia molopica* T.M.Harris

06090070 **Pl. 48E** (p. 194)

Cells ovoid, about 15 μm wide, 21 μm long; epicone about twice length of hypocone, separated by a spiralling cingulum offset by one-third cell length; sulcus deep, extends obliquely most of length of cell; cell wall fragile but striated; chloroplasts absent, red elongate eyespot in hypocone.

British Isles; in the plankton of nutrient-enriched ponds.

Katodinium tetragonops (T.M.Harris) A.R.Loeblich III 1965

Basionym: *Massartia tetragonops* T.M.Harris

Synonyms: *Massartia ptyrtica* T.M.Harris, *Katodinium ptyrticum* (T.M.Harris) A.R.Loeblich III

06090080 **Pl. 48D** (p. 194)

Cells small, ovoid, 8–16 μm wide, 8–17 μm long; epicone conical, longer than rounded hypocone, separated by a wide cingulum and displaced by one cingulum width; sulcus deep, penetrates epicone and descends nearly to base of antapex; chloroplasts absent, eyespot occasionally large but sometimes small or absent.

British Isles, USA; in nutrient-poor and very nutrient-enriched ponds and lakes.

Katodinium vorticella (F.Stein) A.R.Loeblich III 1965

Basionym: *Gymnodinium vorticella* F.Stein

Synonyms: *Peridinium vorticella* F.Stein, *Massartia vorticella* (F.Stein) J.Schiller

06090050 **Pl. 48B** (p. 194)

Cells ovoid to spherical, slightly dorsiventrally flattened, 11–30 μm wide, 14–33 μm long; epicone variable in shape, rounded to conical, larger than hypocone; epicone overhangs a wide cingulum offset by up to 1.5 cingulum widths; sulcus penetrates epicone and descends, narrowing almost to base of hypocone; chloroplasts, if present, small and green; eyespot if present bright red, square and located in hypocone; nucleus in epicone; refractive bodies numerous and reddish droplets often present in cytoplasm.

Europe, USA; in peat bogs, ponds, pools and melted snow; nutrition includes photosynthetic (if with chloroplasts) and phagotrophic.

Woloszynskia R.H.Thompson 1950

Vegetative cells free-swimming, ovoid to spherical; epicone and hypocone generally equal in length, separated by a shallow and slightly offset cingulum; sulcus sometimes extends into epicone and descends to base of hypocone; cell wall thin and composed of a network of delicate hexagonal fields sometimes forming into a distinct line across apex; chloroplasts yellow to brown and eyespot sometimes present; sexual reproduction results in either thick-walled, spherical to quadrangular resting stages with extended poles, or spherical, spinose resting stages.

World-wide; occurs in plankton of pools and lakes of varied nutrient status.

1 Cell markedly dorsiventrally flattened with small apical point .... *W. tenuissima*
1 Cell more or less ovoid with rounded apices and not flattened 2
2 Eyespot absent .... *W. tylota*
2 Eyespot present 3
3 Eyespot behind sulcus; nucleus ovoid .... *W. coronata*
3 Eyespot at junction of cingulum and sulcus; nucleus elongate ovoid .... *W. neglecta*

Woloszynskia coronata (Wołoszyńska) R.H.Thompson 1950

Basionym: *Gymnodinium coronatum* Wołoszyńska

06140040 **Pl. 48F** (p. 194) **CD**

Cells spherical to ovoid in ventral view, very slight dorsiventrally flattening, 14–32 μm wide, 19–35 μm long; epicone rounded to conical, slightly longer than rounded hypocone; cingulum shallow, offset by one cingulum width, lower margin not clearly defined; sulcus inconspicuous forms groove onto epicone and for a short distance into hypocone; cell wall composed of irregularly hexagonal plates (5–6 μm across), rectangular in cingulum, visible with light microscope when covering removed from cell; short carina on epicone; chloroplasts numerous, radially arranged, light to dark brown and a bright red eyespot behind sulcus; nucleus central, hypnozygotes dumbbell-shaped, pointed, knobbly, containing orange globules.

Europe, USA; in plankton of ponds and lakes, a cold-water species appearing as motile cells in late winter and then forming cysts as water temperature warms (mid-March to April in the British Isles).

Woloszynskia neglecta (A.J.Schilling) R.H.Thompson 1950

Basionym: *Glenodinium neglectum* A.J.Schilling

Synonym: *Gymnodinium neglectum* (A.J.Schilling) Lindemann

06140010 **Pl. 48K** (p. 194)

Cell broadly rounded in dorsal and ventral view, dorsiventrally flattened, 28–30 μm wide, 32–36 μm long, 20 μm deep; epitheca slightly longer than hypotheca, separated by a deep and slightly offset cingulum; sulcus shallow, extending for about half the hypotheca; theca composed of small hexagonal fields visible when cells empty of contents; chloroplasts numerous, small, yellow-brown, with a red eyespot towards ventral surface of cell just behind junction of cingulum and sulcus; nucleus sausage-shaped, to one side of cell, mostly in epicone, characteristic for species.

Europe, USA; in ponds, pools and ditches.

Woloszynskia tenuissima (Lauterborn) R.H.Thompson 1950

Basionym: *Gymnodinium tenuissimum* Lauterborn

060140020 **Pl. 48 L,M** (p. 194) **CD**

Cells squarish in ventral view and strongly dorsiventrally flattened, curved in lateral view and ventral face concave; apex with short point and antapex pointed or bearing a deep furrow, 33 μm wide, 32–40 μm long (some records of cells 60–72 μm wide × 66–80 μm long, which possibly refer to planozygotes); epicone slightly longer than hypocone, divided by a fairly wide cingulum; sulcus narrow and not reaching antapex; cell wall delicate, composed of many hexagonal plates (4–5 μm), within cingulum plates become rectangular (seen under light microscope by separating from cell); chloroplasts oval or irregular, yellow or yellow-brown, radially arranged mostly in cell periphery; eyespot orange and behind sulcus; nucleus quite large, irregularly shaped and lies in hypocone; tubular pusules (2) seen near cell centre; hypnozygotes dumb-bell-shaped, usually with long processes at each end and shorter, thicker spine-like processes over cell surface.

Europe; in plankton of ponds and lakes; this is a low temperature species (optimum 2–4.5 °C) and tolerant of low light intensities. It has a high phosphate requirement and has been recorded in a lake in high numbers post phosphate-reduction treatment (Steinburg, 1981).

Woloszynskia tylota (Mapletoft, J.Waters *et* P.Wells) B.T.Bibby *et* J.D.Dodge 1972

Basionym: *Gymnodinium tylotum* Mapletoft, J.Waters *et* P.Wells

06140030 **Pl. 48H,I** (p. 194) **CD**

Cells ovoid with a blunt conical epicone slightly smaller than rounded hypocone, 25–30 μm wide, 30–35 μm long; cingulum slightly offset; sulcus extends slightly into epicone and on hypocone reaches antapex; cell wall delicate, composed of tiny hexagonal plates, on ecdysis wall composition visible under light microscope; chloroplasts numerous, disc-like yellow-brown; eyespot absent; nucleus ovoid and in epicone; hypnozygotes variable in shape, often square with variable knob-like projections at corners and apex, contain dark bodies and large vesicles with vibrating granules, 40–50 μm long.

British Isles; in nutrient-poor pools, a cold-water species found before May.

ORDER GONYAULACALES

Armoured cells with asymmetrical first apical plate and lacking a canal plate.

Ceratium Schrank 1793

Vegetative cells free-swimming; epitheca forms a long horn and hypotheca forms 1–3 horns, dorsiventrally flattened; cingulum equatorial; sulcus narrow and to one side of indented ventral area, confined to hypotheca; theca composed of plates with pores, smooth to heavily ornamented, plate formula: Po, 4′, 5–6″, 5–6c, ?5s, 6‴, 2⁗; chloroplasts ovoid, numerous, pale yellow-brown, eyespot absent; vegetative reproduction by oblique fission of cell with theca shared between daughter cells; sexual reproduction anisogamous with some freshwater species producing thick-walled, 3–4-horned hypnozygotes.

Cosmopolitan; inhabiting marine and freshwater habitats, sometimes forms blooms.

1 Cell narrow, spindle-shaped, with 2 posterior horns more or less parallel .... *C. furcoides*
1 Cell not as above 2
2 Posterior horns normally 3, widely spreading .... *C. hirundinella*
2 Posterior horns 2 only 3
3 Anterior horn considerably curved to right; cell rather delicate .... *C. carolinianum*
3 Anterior horn slightly curved to right; cell thick and stocky .... *C. cornutum*

Ceratium carolinianum (Bailey) Jørgensen 1911

Basionym: *Peridinium carolinianum* Bailey
Synonym: *Ceratium curvirostre* Huitfeldt-Kaas

06020010 **Pl. 49D** (p. 197)

Cells broadly spindle-shaped in outline; epitheca very wide just above cingulum, narrowing abruptly especially on right-hand side to form a stout curved apical horn, 73–105 μm wide, 125–213 μm long; cingulum deep and narrow, with solid flanges; hypotheca broad below cingulum with short divergent horn on right and central horn longer, wider and obliquely directed; thecal plates robust, distinct, deep and with net-like ornamentation; plate formula: 4′, 5″, 5c, ?s, 5‴, 2⁗; hypnozygotes irregular tetrangular in shape, within thick multilayered wall.

Europe, North America; common in nutrient-poor lakes though rare in ones rich in humic materials (dystrophic lakes).

Ceratium cornutum (Ehrenberg) Claparède *et* J.Lachmann 1858

Basionym: *Peridinium cornutum* Ehrenberg

06020020 **Pl. 49A** (p. 197)

Cells stout and broadly spindle-shaped, 48–75 μm wide, 97–150 μm long; epitheca broad above cingulum, narrows rapidly but not as abruptly as *C. carolinianum* and forms a short, stout, curved horn appearing cut-off at apex; hypotheca broad below cingulum, extending into 2 horns: one short lateral horn and one longer median horn, all horns dorsiventrally flattened; cingulum deep, wider than in *C. carolinianum*; thecal plates robust, with a distinct, deep net-like surface ornamentation; plate formula: 4′, 5″, 5c, ?s, 5‴, 2⁗; hypnozygotes triangular or tetrangular with rounded ends, within a thick multilayered wall.

Europe, North America; in plankton of ponds, shallow lakes, reservoirs, littoral zone and peat bogs; abundant in spring, dependent on low temperatures and adequate oxygen supply; nutrition both photosynthetic and phagotrophic.

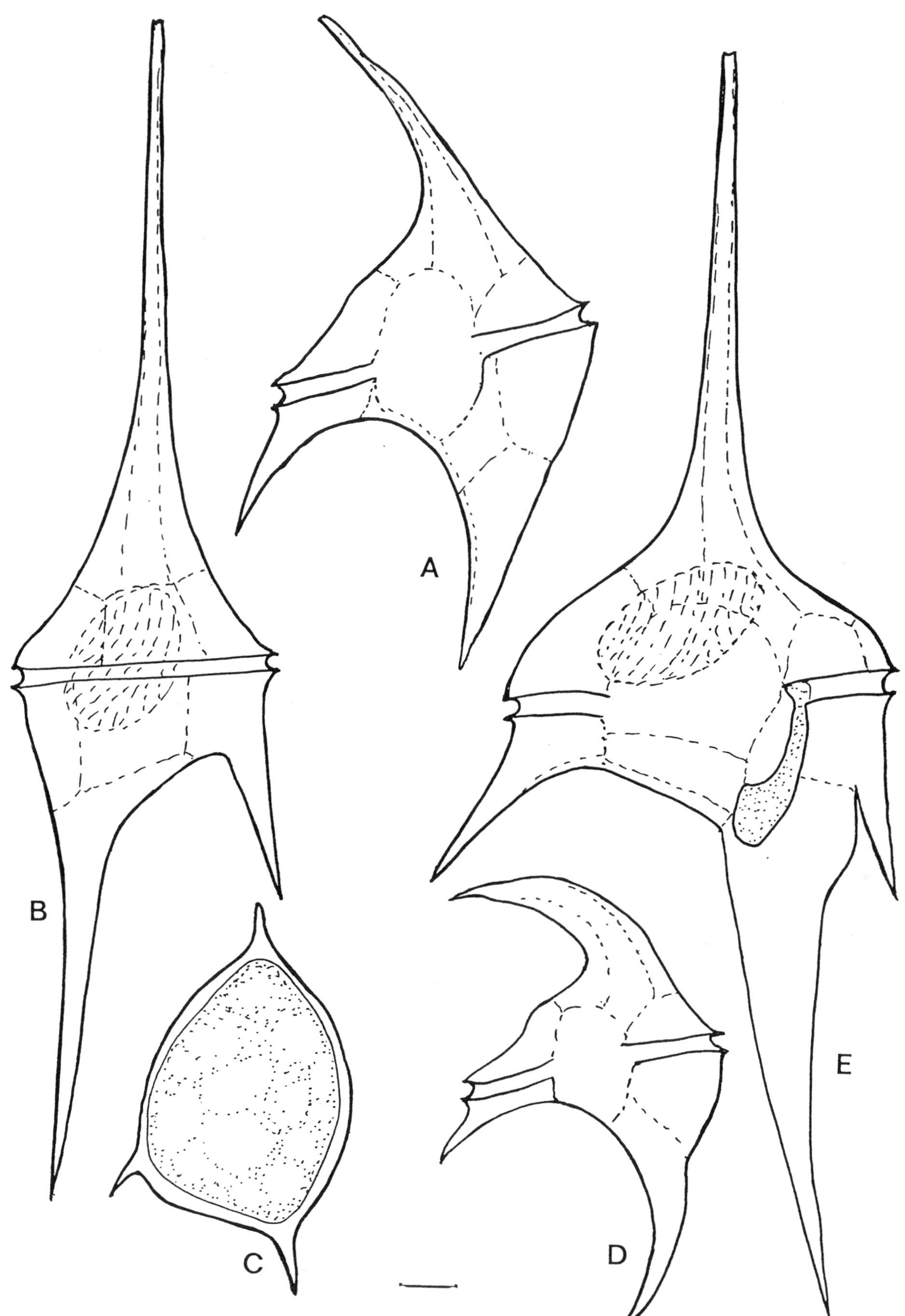

Plate 49 A–E

A. *Ceratium cornutum* (p. 196); **B, C**. *Ceratium furcoides* (p. 198): vegetative cell and cyst; **D**. *Ceratium carolinianum* (p. 196); **E**. *Ceratium hirundinella* (p. 198).

Scale bar: 10 μm.

Cell division in motile stage, after division of dinokaryon and cytoplasm, theca divides obliquely between daughter cells. Each daughter cell then forms the remaining theca.

Ceratium furcoides (Levander) Langhans 1925

Basionym: *Ceratium hirundinella* var. *furcoides* Levander
06020030 **Pl. 49B,C** (p. 197) **CD**
Cells narrowly spindle-shaped, strongly dorsiventrally flattened, 28–42(–56) μm wide, 123–222 μm long; epitheca formed into a narrow horn without shoulders; hypotheca broad and short, drawn out into 2 (occasionally 1 or 3) posterior horns of different lengths; plates smooth and with shallow net-like ornamentation; plate formula: Po, 4′, 6″, ?c, ?s, 6‴, 2⁗ with fourth apical plate not reaching apex of epitheca; chloroplasts numerous and oval; nucleus prominent in epicone; hypnozygotes smooth-walled, triangular, with one narrow horn extending from each corner.

Europe; from nutrient-poor and moderately nutrient-rich lakes and reservoirs, a summer species.

Confusion about the distribution of this species has resulted in it often being attributed to *C. hirundinella* (e.g. Dodge & Crawford, 1970). It is likely that at least some of the British records for *C. hirundinella* should be attributed to *C. furcoides*.

Ceratium hirundinella (O.F.Müller) Dujardin 1841

Basionym: *Bursaria hirundinella* O.F.Müller
Synonym: *Ceratium tetraceros* Schrank
06020040 **Pl. 49E** (p. 197) **CD**
Cells broadly or narrowly spindle-shaped depending on degree of horn divergence, strongly dorsiventrally flattened, 28–55 μm wide, 40–450 μm long, one apical and 2 or 3 antapical horns; epitheca with distinct shoulders just above cingulum and tapering to a long horn; hypotheca broad and short, divided into 2–3 posterior horns; cingulum narrow; plates smooth and with net-like ornamentation on cell body, coarsening on horns; plate formula: Po, 4′, 5″, 5c, ?s, 6‴, 2⁗, fourth apical plate not reaching apex of cell (in contrast to *Ceratium furcoides*); chloroplasts numerous, parietal, oval and yellow-brown; red corpuscular bodies often seen; hypnozygotes, smooth, triangular and with one horn at each corner.

World-wide; common in deeper ponds and lakes that are nutrient-poor to nutrient-enriched, a summer species tolerant of low oxygen levels; nutrition sometimes phagotrophic in addition to photosynthetic.

A very variable species in terms of spread of antapical horns and ornamentation of thecal plates. Several forms and varieties can be distinguished; *C. hirundinella* typus *scotticum* and *C. hirundinella* var. *gracile* have been recorded from British Isles.

ORDER PERIDINIALES

Armoured cells often with well-developed cingulum and sulcus. Typically apical plate symmetrical, connected to pore plate with a canal plate.

Glenodiniopsis Wołoszyńska 1916

Vegetative cells free-swimming, generally ovoid in ventral view, with epitheca slightly larger than hypotheca, separated by a well-defined cingulum with slight offset; sulcus confined to hypotheca and not extending to pole, apical pore absent; cell wall consisting of smooth plates; plate formula: 3–4′, 3–4a, 8″, ?c,?s, 7‴, 2⁗; chloroplasts numerous, yellow to brown; eyespot absent; resting stages spherical, reported embedded in mucilage.

Cosmopolitan; widespread and reported from peat bogs, vegetated pools and lakes rich in humic material (dystrophic lakes).

Glenodiniopsis uliginosa (A.J.Schilling) Wołoszyńska 1916

Basionym: *Glenodinium uliginosum* A.J.Schilling
Synonym: *Glenodiniopsis steinii* (Lemmermann) Wołoszyńska
06060020 **Pl. 52A** (p. 206)
Cells ellipsoidal, dorsiventrally flattened, without apical pore, 21–44 μm wide, 26–50 μm long; epitheca larger and wider than hypotheca; cingulum offset by one cingulum width; sulcus transgresses slightly onto epitheca and not reaching antapex on hypotheca; thecal plates smooth; plate formula: 4′, 4a, 8″, ?c, ?s, 7‴, 2⁗; nucleus spherical, occasionally ellipsoidal and in anterior portion of cell; chloroplasts yellow to dark brown and parietal; resting stages 33–40 μm wide.

Europe; found in vegetated pools and peat bogs often in great abundance; maximum numbers in summer months, preferring temperatures of 12–17 °C, overwinters as a non-motile resting stage.

Glenodinium Ehrenberg 1837

Vegetative cells free-swimming, generally round in ventral view, sometimes dorsiventrally flattened; epitheca and hypotheca equal in size, separated by a shallow cingulum and occasionally offset by up to one cingulum width; sulcus protrudes slightly into epitheca and extends to pole of hypotheca; cell wall thin, well-defined, consisting of delicate, smooth or lightly ornamented plates (not always seen with the light microscope); plate formula: Po, 4′, 4a, 7″, ?3c, ?6s, 6‴, 2⁗; chloroplasts numerous, yellow-brown to brown, some species possess a large eyespot; sexual reproduction sometimes results in a thick-walled hypnozygote.

Cosmopolitan, known from marine and freshwater habitats, found in plankton of pools and peat bogs.

This genus comprises a variety of small round species with poorly defined plates. Fensome et al. (1993) consider the genus well-defined by its type *Glenodinium cinctum*. Wołoszyńska (1916) erected the genus *Sphaerodinium* and later (Wołoszyńska, 1917) transferred *Glenodinium cinctum* into the genus. However, as *G. cinctum* is the type species of *Glenodinium*, *Sphaerodinium* is a later homonym of *Glenodinium*. As improved fixation techniques and electron microscopy studies are brought to bear on species currently grouped in this genus, many will probably be assigned to other genera.

1 Cell more or less spherical; eyespot large present .... *G. cinctum*
1 Cell ovoid, eyespot absent .... *G. pulvisculus*

Glenodinium cinctum Ehrenberg 1838

Synonym: *Sphaerodinium cinctum* (Ehrenberg) Wołoszyńska

06050070 **Pl. 48J** (p. 194)

Cells spherical, slightly dorsiventrally flattened, 25–48 μm wide, 25–50 μm long; epitheca and hypotheca of equal length divided by cingulum offset by one cingulum width; sulcus extends into epitheca by about one third of its length and descends the hypotheca, narrowing as it approaches antapex; cell wall composed of delicate, smooth plates, occasionally bearing soft papillae; plate formula: Po, 4′, 4a, 7″, 5c, ?6s, 6‴, 2⁗; apical pore present; epithecal plate arrangement fairly symmetrical, third apical plate 6 sided and surrounded by other apical plates and 4 intercalary plates; antapical plates 2 and uneven in size; chloroplasts numerous, small, brown, oval; eyespot large, in upper centre of hypotheca; hypnozygotes spherical with a thick membrane.

Worldwide; found in plankton of pools, ponds, swamps and peat bogs.

Glenodinium pulvisculus (Ehrenberg) F.Stein 1883

Basionym: *Peridinium pulvisculus* Ehrenberg

06050050 **Pl. 52J** (p. 206)

Cells rounded with equal epitheca and hypotheca, 13–29 μm wide, 23–35 μm long; cingulum median, somewhat offset by up to one cingulum width; sulcus runs to antapex of hypocone, narrowing towards base; nucleus spherical, situated centrally towards base of epitheca; chloroplasts rounded and parietal.

Europe, Africa, North America; in plankton of pools and lakes; this species has been recorded as blooming in numbers sufficient to discolour water in a Canadian river towards the end of summer.

DOUBTFUL RECORDS

Glenodinium helicozoster T.M.Harris 1940

06050010

Cells ovoid, slightly dorsiventrally flattened, 9–10 μm long; cingulum shallow, offset by more than one cingulum width; sulcus protrudes into epicone and penetrates about halfway down into hypocone; epicone rounded and hypocone slightly pointed. It was found in water with rotting leaves between October and May in ditches and ponds. Species changes slightly as the season progresses, autumn cells small and pale, mid-winter cells larger and browner, spring cells smaller and rarer. Reported by Harris (1940) in Reading University grounds, but he was doubtful about this species, stating that it was 'intermediate between *Glenodinium* and *Gymnodinium* being impossible to demonstrate the presence of plates'. It seems possible this is a *Gymnodinium* species.

Glenodinium inaequale Chodat 1921

06050020

Cells small and ovoid, 14–16 μm wide, 20 μm long; epitheca considerably reduced in size (narrower and shorter) than hypotheca. This is very reminiscent of a *Katodinium* species. Considered doubtful as it has not been found since Woodhead & Tweed (1954) recorded it in North Wales.

Glenodinium leptodermum T.M.Harris 1940

06050030

Cells ovoid, equal epicone and hypocone, dorsiventrally compressed, 12–16 μm wide, 14–24 μm long; cingulum deep; sulcus penetrates high into epicone and about halfway down hypocone; chloroplasts absent and eyespot present. From the description it seems possible this is a *Gymnodinium*. This species has not been recorded since Harris's description (Harris, 1940) of it from a very nutrient-enriched water butt in Reading University grounds.

Glenodinium limos T.M.Harris 1940

06050040

Cells rounded, flattened bottom to hypocone; heavily dorsiventrally flattened and banana-shaped; epicone and hypocone equal in length, cells 40 μm wide, 46 μm long; cingulum clear on dorsal side but less obvious on ventral side. Individual chloroplasts not obvious, cytoplasm appears evenly brown; cell wall rigid but not possible to separate a theca from the cell. Found by Harris (1940) in a very nutrient-rich pond as a single specimen. It has not been recorded again from the British Isles. He states it was close to *Glenodinium foliaceum* but could be distinguished by the lack of an eyespot.

Glenodinium punctulatum T.M.Harris 1940

06050060

Cell oval, not dorsiventrally flattened, 12–18 μm wide, 15–24 μm long; epitheca and hypotheca even in length, shallow cingulum, poorly defined sulcus; cell wall thin but well-defined with faint striations; chloroplasts and eyespot absent. Found by Harris (1940) in a large very nutrient-rich pond, it has not been recorded since. Nutrition includes phagotrophy.

Peridiniopsis Lemmermann 1904

Vegetative cells free-swimming; cells ovoid to rounded in ventral view, rounded to bean-shaped in apical view; epitheca shorter or equal to hypotheca in length, separated by a well-defined cingulum with little or no offset; sulcus sometimes protrudes marginally into epitheca and extends to base of hypotheca; cell wall composed of well-defined, porate plates, often bearing various types of ornamentation; plate formula rather varied across genus: Po, 3–5′, 0–1a, 6–8″, 6c, 6s, 5‴, 2⁗; epitheca conical or rounded, hypotheca sometimes bears spines or wings and occasionally shaped into horns; chloroplasts sometimes absent, when present numerous; eyespot sometimes present; sexual reproduction results in formation of thick-walled hypnozygotes.

Widespread; in the plankton of freshwater pools and lakes.

This genus is one to which species of *Glenodinium* are likely to be transferred.

1 Cell ovoid with rounded outlines; eyespot present .... *P. oculatum*
1 Cell top-shaped and angular; eyespot absent 2
2 Cell length and breadth more or less equal, hypotheca angular .... *P. borgei*
2 Cell much longer than broad, small spines on antapex .... *P. cunningtonii*

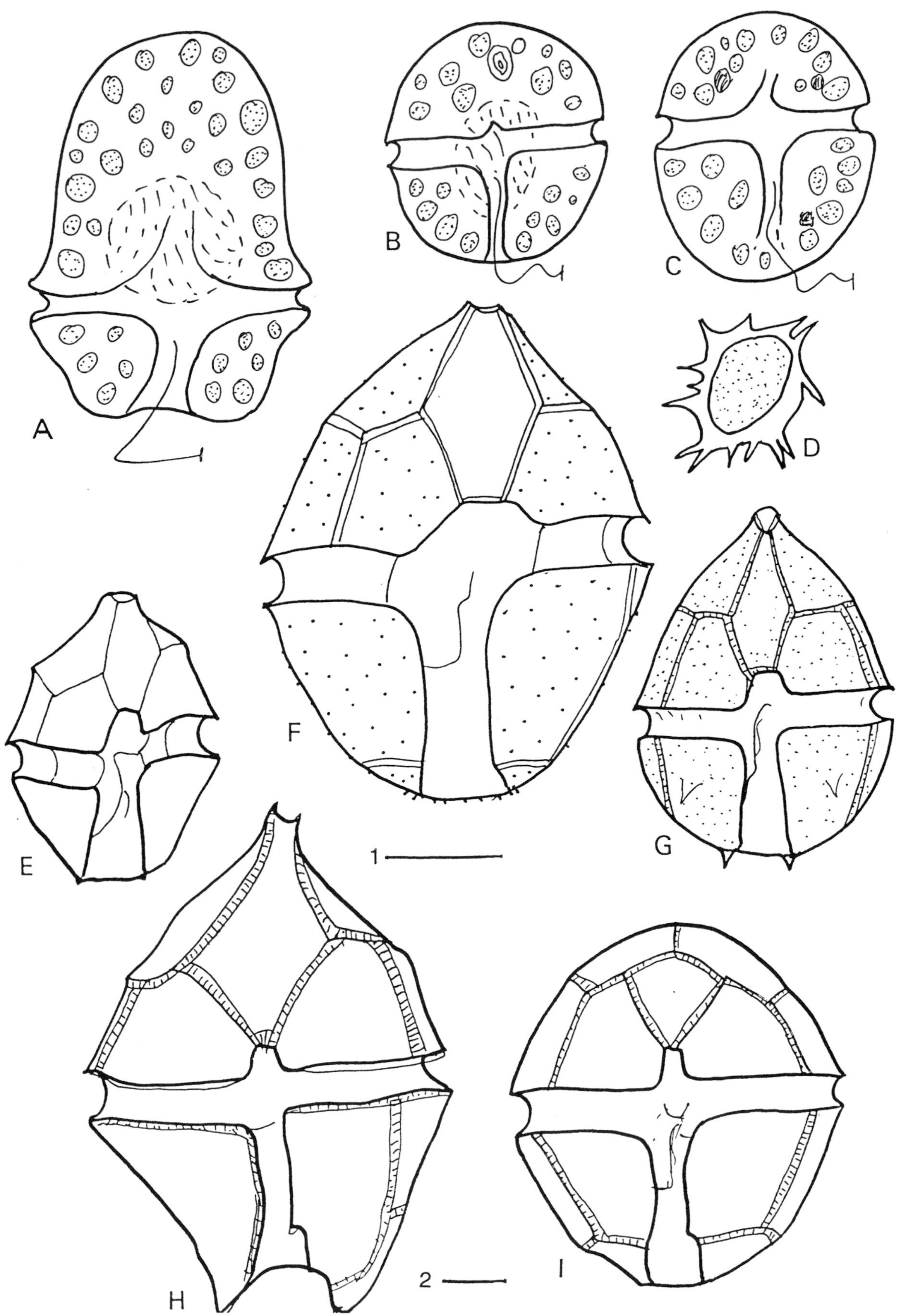

Plate 50 A–I

A. *Gymnodinium palustre* (p. 192); **B**. *Gymnodinium saginatum* (p. 192); **C, D**. *Gymnodinium hippocastanum* (p. 192): motile cell and cyst (not to scale); **E**. *Peridinium inconspicuum* (p. 203); **F**. *Peridinium anglicum* (p. 203); **G**. *Peridiniopsis cunningtonii* (p. 201); **H**. *Peridinium limbatum* (p. 204); **I**. *Peridinium volzii* (p. 205).

Scale bars: 10 μm. 1 for A–G; 2 for H, I.

Peridiniopsis borgei (Lemmermann) Lemmermann 1910

Basionym: *Peridinium borgei* Lemmermann
Synonym: *Glenodinium borgei* (Lemmermann) J.Schiller
06100030 **Pl. 52C,D** (p. 206) **CD**
Cells rather kite-shaped, sometimes partly rounded; slightly dorsiventrally flattened, 35–41 μm wide, 40–55 μm long; epitheca broadly conical with a distinct apex, hypotheca slightly smaller and broadly rounded; cingulum wide and displaced to right; sulcus wider on hypotheca and with a flange on its left side; thecal plates covered with net-like ornamentation, often arranged in vertical lines, pimpled ridges at margins of plates; plate formula: Po, 3′, 1a, 6″, ?c, ?s, 5‴, 2⁗; chloroplasts numerous and elongated.
Worldwide; occurs in lakes and ponds.

Peridiniopsis cunningtonii (Lemmermann) Popovský *et* Pfiester 1990

Basionym: *Peridinium cunningtonii* Lemmermann
Synonym: *Peridinium suttoni* B.M.Griffiths
06100010 **Pl. 50G** (p. 200)
Cells ovoid, dorsiventrally flattened, 20–30 μm wide, 20–40 μm long; epitheca and hypotheca equal in length; epitheca with a slight apical horn and shoulders, hypotheca hemispherical and bearing 2–6 spines on antapex; cingulum wide and well-defined; sulcus extends two-thirds of length of hypotheca with slight penetration into epitheca; cell wall composed of delicately net-like ornamented plates; plate formula: Po, 4′, 1a, 6″, 6c, ?5s, 5‴, 2⁗; chloroplasts present; resting stages oval and thick-walled.
Worldwide; occcurs in the plankton of lakes and ponds.

Peridiniopsis oculatum (F.Stein) Bourrelly 1968

Basionym: *Glenodinium oculatum* F.Stein
06100020 **Pl. 52B** (p. 206)
Cells spherical, slightly dorsiventrally flattened, 21–36 μm wide, 23–36 μm long; epitheca and hypotheca equal in length, both rounded and separated by a well-defined cingulum offset by one cingulum width; sulcus wide and not reaching antapex of hypotheca; cell wall formed of thecal plates bearing minute punctae; plate formula: Po, 3′, 1a, 7″, ?s, ?c, 5‴, 2⁗, third apical plate characteristically 5-sided and spineless; chloroplasts parietal, numerous, yellow-brown, disc- or club-shaped; eyespot present.
Europe, USA; in ponds and pools.
Vegetative cells often embedded in mucilage.

Peridinium Ehrenberg 1832

Vegetative cells free-swimming, generally ovoid or rounded in ventral view, sometimes circular to bean-shaped in apical view; epitheca and hypotheca usually fairly even in size, separated by a well-defined cingulum with little or no offset; sulcus extends from cingulum to bottom of hypotheca; cell wall of porate plates with a wide range of ornamentation; plate formula: (Po, x), 4′, 2–3a, 7″, 5c, 5s, 5‴, 2⁗; apical pore complex present or absent, antapex sometimes bearing spines or wings and shaped into horns; chloroplasts numerous, yellow-brown to dark brown; resting cysts thick-walled, formed following sexual reproduction.
Cosmopolitan freshwater genus, few marine representatives; free-living in pools, ponds, ditches and lakes.
Some species can be prolific and discolour water.

1 Cell rounded to ovoid...2
1 Cell angular or with apical/antapical projections...6
2 Apical pore present ....*P. lomnickii*
2 Apical pore absent...3
3 Cell ovoid; epitheca and hypotheca of equal length ....*P. pseudolaeve*
3 Cell flattened or rounded; epitheca at least slightly longer than hypotheca 4
4 Epithecal intercalary plates not symmetrical ....*P. cinctum*
4 Epithecal intercalary plates symmetrical 5
5 First apical plate small and pointed towards apex ....*P. volzii*
5 First apical plate large and almost flat towards apex ....*P. willei*
6 Cell angular with ridges along plate margins; apical pore present or absent.................................7
6 Cell rounded and without ridges; apical pore present...9
7 Apical horn and pore absent..................*P. anglicum*
7 Apical horn and pore present.................................8
8 Antapex formed into 2 horns ....*P. limbatum*
8 Antapex rounded without horns ....*P. palatinum*
9 Cell top-shaped with 3 short antapical spines..*P. aciculiferum*
9 Cell rhomboidal or angular and without such spines..10
10 Thecal plates with vertical surface striations ....*P. africanum*
10 Thecal plates smooth or net-like surface ornamentation 11
11 Cell large (>40 μm), more or less polyhedral; posterior end of sulcus with 2 short projections ....*P. bipes*
11 Cell smaller (≤40 μm); epitheca rounded, hypotheca angular with no projections at posterior 12
12 Apical pore structure prominent .....*P. umbonatum*
12 Apical pore inconspicuous ....*P. inconspicuum*

Peridinium aciculiferum Lemmermann 1900

Synonym: *Glenodinium aciculiferum* (Lemmermann) Lindemann
06110010 **Pl. 51A–D** (p. 202)
Cells ovoid, dorsiventrally flattened, 29–42 μm wide, 35–51 μm long; epitheca conical with apical horn equal in length to hemispherical hypotheca; cingulum wide, offset by one cingulum width; sulcus wide and

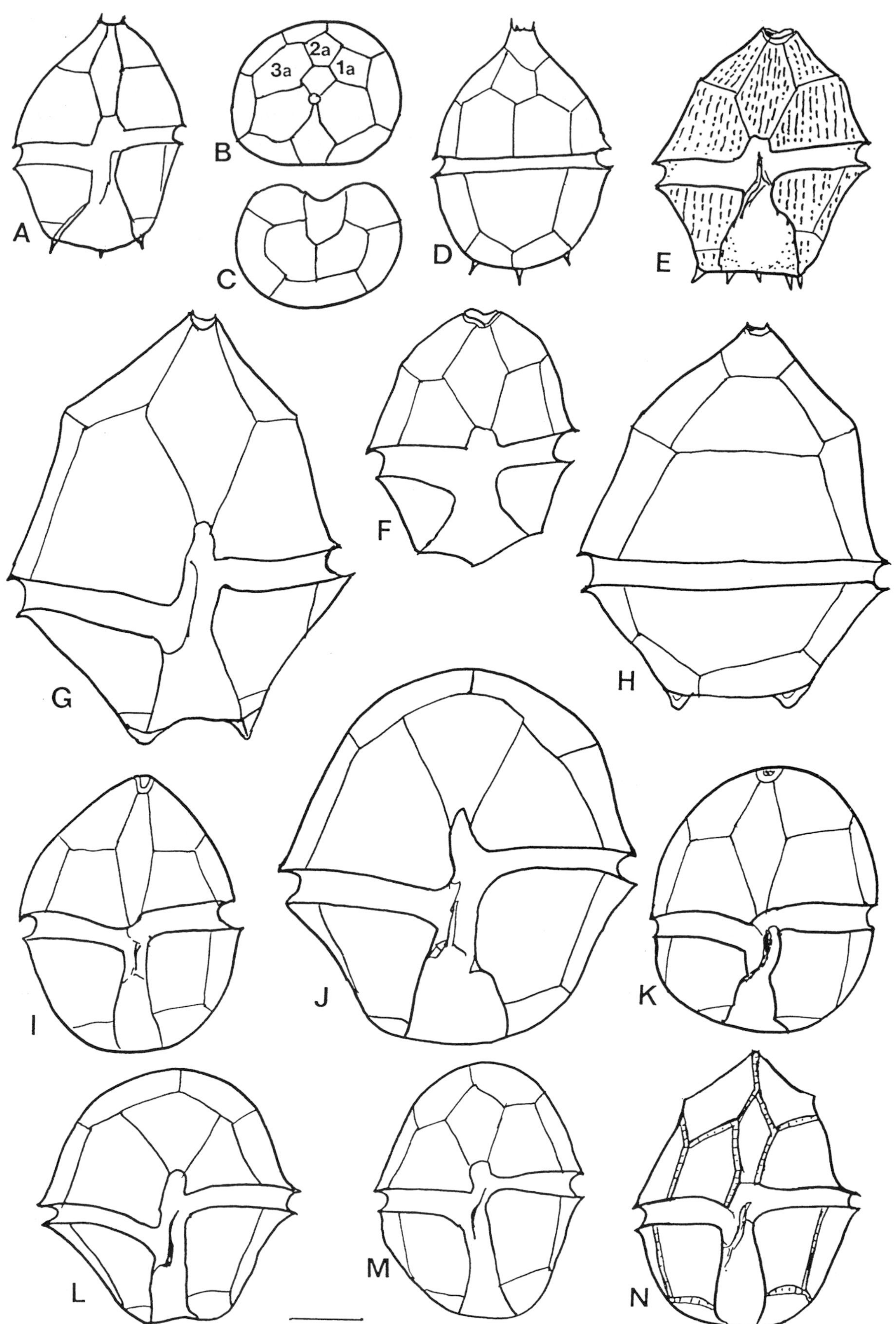

Plate 51 A–N

A–D. *Peridinium aciculiferum* (p. 201): A. Ventral view, B. Apical view with intercalary plates 1–3a indicated, C. Antapical view of theca, D. Dorsal view of theca; **E**. *Peridinium africanum* (p. 204); **F**. *Peridinium umbonatum* (p. 204); **G, H**. *Peridinium bipes* (p. 203): ventral and dorsal views; **I**. *Peridinium lomnickii* (p. 204); **J**. *Peridinium cinctum* (p. 203); **K**. *Peridinium lomnickii* var. *weirzejskii* (p. 204); **L**. *Peridinium willei* (p. 205); **M**. *Peridinium pseudolaeve* (p. 204); **N**. *Peridinium palatinum* (p. 204).

Scale bar: 10 μm.

broadens as it descends to base of hypotheca, 1–4 flattened spines on base of hypotheca; cell wall composed of thecal plates with a net-like ornamentation; plate formula: Po, 4′, 3a, 7″, 6c, ?6s, 5‴, 2⁗; resting stages spherical, thick-walled, sometimes bearing spines.

Europe; in lakes, ponds and reservoirs; winter species present in plankton for a short time (February–March), although it can be a dominant species in March.

Peridinium africanum Lemmermann 1907

06110020 **Pl. 51E** (p. 202)

Cells pentangular in ventral view, slightly dorsiventrally flattened, 20–35 μm wide, 20–40 μm long; epicone angular with slight shoulders and a small apical horn, epicone marginally longer than hypocone; hypocone flattened at base and bearing numerous spines, (some prominent); cingulum wide, horizontal and well-defined with no offset; sulcus wide, protrudes slightly into epicone, widens further as it descends to base of hypocone; cell wall divided into thecal plates bearing rows of fine papillae on their surface; plate formula: Po, 4′, 2a, 7″, 6c, ?s, 5‴, 2⁗, sometimes plates separated by wide intercalary bands; chloroplasts dark brown; resting stages heart-shaped with thick, smooth walls.

Worldwide; frequent in both very nutrient-rich and nutrient-poor ponds and lakes in the British Isles where found in late winter/early spring.

Peridinium anglicum G.S.West 1909

06110030 **Pl. 50F** (p. 200)

Cells ovoid, very slightly dorsiventrally flattened, 42–48 μm wide, 50–58 μm long; epitheca and hypotheca of equal length, epitheca rounded to conical, short apical horn, hypotheca rounded, base ornamented with small spines; cingulum well-defined, offset by one cingulum width; sulcus protrudes into epitheca and descends to base of hypotheca; cell wall composed of thecal plates, punctate and ornamented with minute spines; plate formula: 3′, 1a, 7″, ?c, ?s, 5‴, 2⁗, apical pore absent, single intercalary plate on right-hand side of cell, 2 antapical plates even in size and surrounded by numerous small spines; resting stage thin-walled and spherical, 35–41 μm wide.

British Isles; found in pools late in winter but reaches maximum in April.

Division takes place in a thecate, non-motile stage. Popovský & Pfiester (1990) synonomize this species with *P. palatinum*; however, their plate tabulations are different and this does not seem justified.

Peridinium bipes F.Stein 1883

06110040 **Pl. 51G,H** (p. 202) **CD**

Cells rounded to ovoid, slightly dorsiventrally flattened, 38–90 μm wide, 40–95 μm long; epitheca broadly conical with slight shoulders towards apex, apical horn not pronounced, epitheca longer than hypotheca; hypotheca with flattened antapex bearing 2 stout, winged spines; cingulum well-defined with narrow cingular lists and striate ornamentation, offset by one cingular width; sulcus penetrates epitheca and descends, widening, to antapex of hypotheca; cell wall composed of thecal plates, covered by a net-like ornamentation, surrounding pores (junctions of nets sometimes drawn out into small spines in older specimens); plates sometimes slightly concave; plate formula: Po, x, 4′, 3a, 7″, 5c, ?4s, 5‴, 2⁗, apical pore distinct, surrounded by a low wing, sometimes extending across apex; intercalary bands can be wide; chloroplasts numerous, radially arranged; resting stage thick-walled, spherical.

Europe; common in ponds, lakes and reservoirs; species can become sufficiently abundant to discolour water.

A large number of forms and varieties have been recorded. In the British Isles *P. bipes* var. *excisum* Lemmermann, *P. bipes* fo. *travectum* Lefèvre and *P. bipes* fo. *tabulatum* (Ehrenberg) Lefèvre have been found.

Peridinium cinctum (O.F.Müller) Ehrenberg 1838

Basionym: *Vorticella cincta* O.F.Müller

Synonyms: *Peridinium tabulatum* Penard, *P. westii* Lemmermann

06110050 **Pl. 51J** (p. 202) **CD**

Cells rounded to ovoid, sometimes dorsiventrally flattened, 35–73 μm wide, 40–78 μm long; hypotheca shorter than epitheca, separated by narrow, winged, cingulum offset by one cingular width; sulcus with narrow wings extending about one-third length of epitheca, descending to antapex of hypotheca; cell wall consists of thick thecal plates with coarsely net-like ornamentation, each reticulation surrounding 1–3 pores and in older cells ornamentation extending to small spines; plate formula: 4′, 3a, 7″, 5c, ?5s, 5‴, 2⁗, narrow striated intercalary bands, antapical plates frequently uneven in size; apical pore absent; chloroplasts numerous, dark brown and arranged around periphery of cell; eyespot absent; resting stage having a thickened 'warty' wall and enlarged, striated intercalary bands.

World-wide; found in plankton of pools, lakes, ponds and reservoirs; predominant in winter and spring.

Extensive number of forms and varieties of which *P. cinctum* var. *lemmermannii* West, *P. cinctum* fo. *irregulatum*, *P. cinctum* fo. *regulatum* and *P. cinctum* fo. *westii* have been recorded in the British Isles. *Peridinium tabulatum contactum* var. *excavatum* (Playfair) Lefèvre has also been recorded from British Isles

Peridinium inconspicuum Lemmermann 1899

Synonym: *Peridinium minimum* A.J.Schilling

06110110 **Pl. 50E** (p. 200)

Cells broadly spherical, flattened at anterior end, 15–19 μm wide, 12–16 μm long; cingulum wide and straight, slightly postmedian hence epitheca slightly longer than hypotheca; cell wall composed of delicate, smooth plates; plate formula: Po, 4′, 2a, 7″, ?c, ?s, 5‴, 2⁗, first apical plate wide, 2 intercalary plates separated by 3′ and 4″ being joined, edge of posterior sulcal plate sometimes bearing small spines.

Worldwide; found in plankton of lakes.

Peridinium limbatum (A.Stokes) Lemmermann 1900

Basionym: *Protoperidinium limbatum* A.Stokes

06110120 **Pl. 50H** (p. 200)

Cells distinctively five-sided in ventral view, flattened in cross section, 60–82 μm wide, 74–93 μm long; epitheca with a prominent apical horn sometimes bent to left, divided from hypotheca by deep, striated cingulum offset by one cingulum width; hypotheca smaller than epitheca, with 2 antapical horns occasionally somewhat splayed; sulcus extends to epitheca and descends, widening to base of cell between antapical horns; cell wall composed of concave plates with a net-like ornamentation; plate formula: Po, 4′, 3a, 7″, 5c, ?s, 5‴, 2⁗, intercalary bands striated and in older cells horns sometimes winged; chloroplasts numerous and parietal; resting cyst thick-walled with enlarged striated intercalary bands and a large red accumulation body, 62–64 μm wide, 85–103 μm long.

Europe, North America; found in plankton of peaty pools with *Sphagnum* and other mosses, a summer species.

Peridinium lomnickii Wołoszyńska 1916

Synonym: *Glenodinium lomnickii* (Wołoszyńska) Lindemann

06110060 **Pl. 51I** (p. 202)

Cells ovoid to spherical, sometimes slightly dorsiventrally flattened, 22–50 μm wide, 20–50 μm long; epitheca rounded to conical, slightly longer than rounded hypotheca, separated by a wide, shallow, smooth cingulum offset by up to one-half cingulum width; sulcus hardly penetrates epitheca and not reaching antapex of hypotheca; cell wall composed of smooth plates ornamented with tiny papillae and often more pronounced on hypotheca, narrow intercalary bands very weakly striated; plate formula: Po, x, 4′, 3a, 7″, 5c, ?6s, 5‴, 2⁗, apical pore not proud of cell, pore plate ringed; narrow canal plate; chloroplasts numerous, tiny and brown; hypnospores thick-walled and obovoid.

Worldwide; found in plankton of pools and lakes in winter.

var. *wierzejskii* (Wołoszyńska) Lindemann 1928

Basionym: *Peridinium wierzejskii* Wołoszyńska

06110062 **Pl. 51K** (p. 202) **CD**

Epithecal plate arrangement more symmetrical and papillae less evident on hypotheca compared with type form.

Worldwide; found in neutral–acid ponds, early spring form.

Peridinium palatinum Lauterborn 1896

Synonym: *Peridinium marssonii* Lemmermann

06110070 **Pl. 51N** (p. 202) **CD**

Cells ovoid, dorsiventrally flattened, 25–48 μm wide, 30–55 μm long; epitheca angularly conical, longer than rounded hypotheca, separated by distinct cingulum offset by half a cingulum width; sulcus wide and widens as it descends hypotheca, narrows in its short entry to epitheca; cell wall composed of plates, sometimes slightly concave, ornamented with ridges along margins and small spines, more pronounced on hypotheca; plate formula: 4′, 2a, 7″, 6c, 4s, 5‴, 2⁗, first apical plate narrow, short not reaching apex, sutures pronounced with delicate striations and occasionally flanged on either side, those reaching apex give overall impression of an apical horn; chloroplasts numerous, radially arranged; eyespot sometimes present; cysts smooth-walled and with red body.

Worldwide; found in ponds and pools mainly in winter and spring, sometimes forms unialgal blooms.

Vegetative division, takes place following ecdysis, produces 2 naked zoospores which form new thecae; plates of new thecae deeply concave with sutures raised as prominent sharp flanges.

Peridinium pseudolaeve Lefèvre 1925

06110080 **Pl. 51M** (p. 202)

Cells slightly angular, tending to spherical, slight dorsiventral flattening, 31–35 μm wide, 35–48 μm long; epitheca and hypotheca equal in length and divided by a wide, vertically reticulate (net-like) cingulum; sulcus broad and widens slightly down hypotheca, sometimes bearing a flange on right side; cell wall composed of thecal plates with a vertically arranged net-like ornamentation (often pronounced in older specimens); plate formula: Po, x, 4′, 2a, 7″, ?c, ?s, 5‴, 2⁗, epithecal plate arrangement symmetrical; plate boundaries marked with a ridge, intercalary bands often striated.

Europe; found in pools, ponds, lakes and gravel pits in nutrient-poor, but not peaty water.

Peridinium umbonatum F.Stein 1883

Synonym: *Peridinium pusillum* (Penard) Lemmermann

This species and its forms and varieties have become very widely circumscribed and from the literature it seems certain that present descriptions cover more than one species. For simplicity only included are those forms described from the British Isles.

var. *umbonatum* F.Stein *sensu* Popovský *et* Pfiester 1990

06110090 **Pl. 51F** (p. 202)

Cells broadly ovoid in ventral view, slightly dorsiventrally flattened, 18–22 μm long; epitheca longer than hypotheca, rounded, hypotheca flattened; cingulum wide and deep; sulcus protrudes very slightly into epitheca and widens down hypotheca where it reaches antapex of cell; cell wall composed of thecal plates with rugose ornamentation and arranged in striae running from top to bottom of cell, sutures emphasized by formation of ridges around plate boundaries, anterior plates and posterior sulcal plate sometimes bearing spines around edges; plate formula: Po, x, 4′, 2a, 7″, 6c, ?s, 5‴, 2⁗; apical pore surrounded by pronounced ridge hence pore plate appears sunken, 2 intercalary plates contact each other; chloroplasts numerous, pale brown, radially arranged, eyespot absent.

Europe; found in small and large standing waters from nutrient-poor to very nutrient-rich water.

Vegetative reproduction of armoured vegetative cells occurs: fusion of 2 thecate isogametes occurs with zygote formation completed outside fusing thecae, ecdysis of growing spherical zygote takes place several times until cell becomes peanut-shaped. Vegetative division takes place following ecdysis, produces 2 naked zoospores which form new thecae; plates of new thecae deeply concave with sutures raised as prominent sharp flanges.

var. *inaequale* Lemmermann 1910

06110092 **CD**

Cells smoothly oval with rounded epitheca and hypotheca, 15–20 µm long; epitheca longer than hypotheca, separated by a wide excavated cingulum; sulcus protrudes very slightly into epitheca and widens down hypotheca where it reaches antapex of cell; cell wall composed of delicate thecal plates with a faint ornamentation, large pores towards plate edges, pores most marked below cingulum on hypotheca, plate boundaries marked by smooth edges; plate formula: Po, x, 4′, 2a, 7″, 6c, ?s, 5‴, 2⁗, apical pore surrounded by a pronounced ridge hence pore plate appears sunken, 2 intercalary plates contact each other; chloroplasts pale brown; eyespot absent.

Worldwide; from nutrient-poor pools.

As commented by other authors, it seems likely that this form should have specific status.

Peridinium volzii Lemmermann 1905

06110130 **Pl. 50I** (p. 200)

Cells spherical, 39–59 µm wide, 40–56 µm long; epitheca marginally larger than hypotheca, both rounded; cingulum median and wide, excavated and offset by one cingulum width; sulcus extends into epitheca and descends to base of hypotheca; sulcus and cingulum bear narrow wings; cell wall composed of plates with fine net-like ornamentation; plate formula: 4′, 3a, 7″, 5c, ?s, 5‴, 2⁗, apical pore absent, first apical plate smaller and more symmetrical than that of *P. willei*, anterior sulcal plate extends well into epitheca, 2 antapical plates even in size; resting stage with wide striated intercalary bands.

Worldwide; found in plankton of lakes.

A number of forms have been described of which *P. volzii* var. *australe* Huitfeld-Kaas has been recorded in the British Isles.

Peridinium willei Huitfeldt-Kaas 1900

06110100 **Pl. 51L** (p. 202)

Cells rounded, often wider than long and dorsiventrally flattened, 36–80 µm wide, 38–78 µm long; epitheca longer than hypotheca, separated by a deep, narrow cingulum sometimes bearing a narrow wing, offset by one cingulum width; sulcus indented, protrudes into epitheca and widens as it descends to base of hypotheca; sulcus bears narrow wings which appear with the light microscope as tooth-like structures at base of cell; cell wall composed of fine net-like ornamentation on thecal plates; plate formula: 4′, 3a, 7″, 5c, 6s, 5‴, 2⁗; apical pore absent; intercalary and third apical plates narrow and laterally elongate, first apical plate relatively large and asymmetric, anterior sulcal plate shorter than that of *P. volzii*, 2 antapical plates different in size, intercalary bands sometimes broad and striated, apical and antapical plates sometimes bearing very pronounced wings; chloroplasts numerous, dark brown, parietal; eyespot inside hypotheca to right side of cell; hypnozygote non-motile, resembling thickened form of vegetative stage.

Worldwide; found commonly in plankton of pools, lakes, reservoirs and peat bogs in all seasons.

A number of forms described of which *P. willei* fo. *stagnale* has been described from the British Isles.

ORDER PHYTODINIALES

Coccoid organization. Reproduce by non-motile spores or by zoids resembling in structure *Gymnodinium* or *Gonyaulax*.

Cystodinium G.A.Klebs 1912

Vegetative cells non-motile, free-floating, frequently crescent-shaped, apices sometimes pointed or bearing spines; cell wall thick and smooth; chloroplasts sometimes absent and when present variable in colour; eyespot absent in non-motile stage but often present in zoospores or gametes; vegetative reproduction by formation of zoospores, autospores or amoebae within parental cell; sexual reproduction isogamous with thick-walled resting stage sometimes formed.

Worldwide; peat and clay pools, vegetated pools and marshy areas.

Complex life histories (including *Gymnodinium*-like stages) within this genus have led to some taxonomic confusion in the literature.

Cystodinium cornifax (A.J.Schilling) G.A.Klebs 1912

Basionym: *Glenodinium cornifax* A.J.Schilling

Synonyms: *Cystodinium iners* Geitler, *C. steinii* G.A.Klebs

06030010 **Pl. 52H,I** (p. 206) **CD**

Cells free-floating or attached, halfmoon-shaped (lunate), elongate or shortly ovate, 17–60 µm wide, 33–130 µm long, one pole or occasionally both terminated with a large spine, sometimes a cluster of very short bristles on one spine; cells often having large vacuoles and refractile bodies; chloroplasts parietal, small, numerous, disc- or rod-shaped, dark red-brown; reproduction by formation of 2 or 4 motile zoospores, 20–25 µm in size with eyespot (similar in appearance to *Gymnodinium fuscum*), sulcus transgresses slightly into epicone and reaches antapex on hypocone; epicone bell-shaped and higher than conical hypocone.

Europe; widespread in moderately nutrient-rich (mesotrophic) ponds with submerged plants.

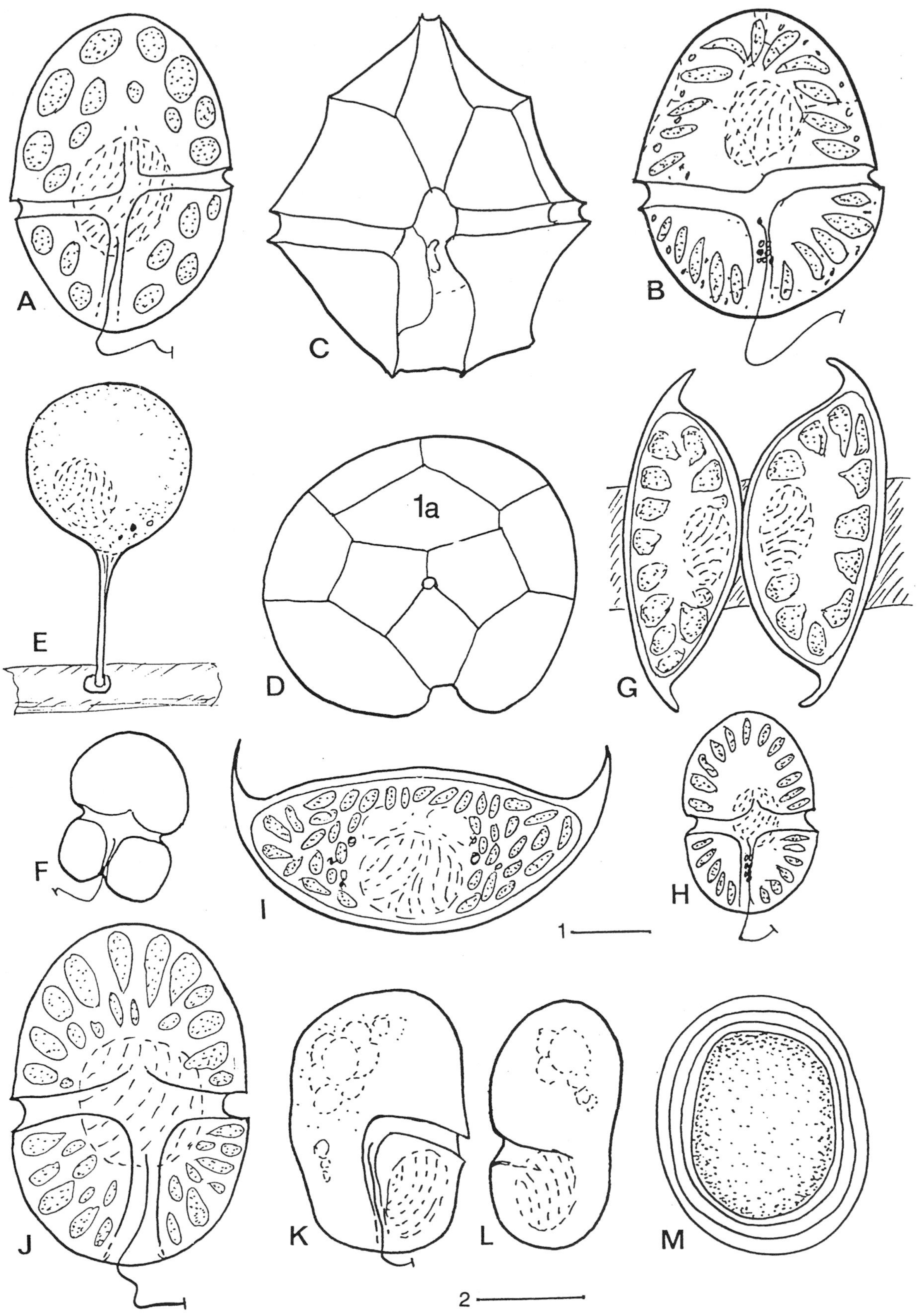

Plate 52 A–M

A. *Glenodiniopsis uliginosa* (p. 198); **B**. *Peridiniopsis oculatum* (p. 201); **C, D**. *Peridiniopsis borgei* (p. 201): ventral view and apical view with a single intercalary plate 1a indicated; **E, F**. *Stylodinium grande* (p. 207): vegetative stage attached to algal filament and a motile zoospore; **G**. *Dinococcus oedogonii* (p. 207): pair of vegetative cells attached to an algal filament; **H, I**. *Cystodinium cornifax* (p. 205): a motile zoospore and vegetative cell; **J**. *Glenodinium pulvisculus* (p. 199); **K, L**. *Hemidinium nasutum* (p. 207): ventral and side views; **M**. *Hemidinium nasutum* (=*Gloeodinium montanum*) (p. 207): cyst stage.

Scale bars: 10 µm. 1 for A–I, 2 for J–M.

With the variability and complex life history of this and the succeeding genera, much more taxonomic study remains to be done.

Dinococcus (=*Woloszynska*) Fott 1960

Cells attached, in lateral view attachment disc visible (extends to become a stout stalk), roundly rectangular body often with 2 extended points, uppermost margin sometimes concave and lower margin convex, viewed from above broadly ellipsoidal; cell wall thin; chloroplasts oval or star-shaped; eyespot sometimes present; asexual reproduction by zoospores or autospores.

Europe; among filamentous algae and aquatic mosses, possibly epiphytic.

Dinococcus oedogonii (P.G.Richter) Fott 1960

Basionym: *Rhizophydium oedogonii* P.G.Richter
Synonyms: *Raciborskia oedogonii* (P.G.Richter) Pascher, *Cystodinium brevipes* Geitler *pro parte*

06040010 **Pl. 52G** (p. 206)

Cells variable in shape, rounded to ellipsoidal, poles sometimes ending in spines, 13–20 μm wide, 20–35 μm long, attached by broad disc, up to 7 μm high; chloroplast star-shaped, dark brown with a central pyrenoid and numerous starch grains in older cells; orange-red body in some cells; nucleus central; zoospores produced singly or in pairs and gymnodinioid in form, pale golden-yellow, 10–18 μm wide and 20–33 μm long; autospores non-motile, in aggregations of 2–16 cells.

Europe; nutrient-rich ponds on filamentous algae (possibly epiphytic); nutrition not fully known.

Detailed observations of this species and a discussion of its taxonomic status are to be found in Williams & Lund (1972).

Hemidinium F.Stein 1879

Cells free-swimming or non-motile, motile cells ovoid in ventral view, dorsiventrally flattened, cingulum not fully circling cell and descends to left; sulcus narrow and descends to bottom of hypotheca; cell wall thin, consisting of smooth plates; chloroplasts yellow-brown to brown; reproduction not fully known, spherical non-motile stage embedded in mucilage has been described.

Worldwide; found in peat bogs, lakes and ponds rich in humic material.

The non-motile stage has been described under the genus *Gloeodinium*.

Hemidinium nasutum F.Stein 1878

Synonym: *Gloeodinium montanum* G.A.Klebs *pro parte*

06080010 **Pl. 52K–M** (p. 206)

Cells kidney-shaped with rounded poles, dorsiventrally flattened, 14–30 μm wide, 22–36 μm long, cingulum only on left half of cell, descends cell in left-hand spiral; sulcus reaches antapex; cell wall composed of thin plates covered with delicate ornamentation; plate formula: 6′, 6″, ?c, ?s, 5‴, 1p, 1⁗; chloroplasts numerous, small, yellow-brown to brown, arranged radially; eyespot absent; nucleus lies centrally in hypocone; resting stage thick-walled, embedded in gelatinous material.

Worldwide; in ponds, pools, peat bogs and vegetated swamps occurs year-round, most abundant in late summer.

Division takes place in non-motile ('*Gloeodinium*') phase.

Stylodinium G.A.Klebs 1912

Cells live attached to substratum, generally spherical to bean-shaped, with stalk of variable length terminating in an attachment disc; cells thick-walled, mostly wall appears smooth but some life history stages become more dinoflagellate-like in appearance; chloroplasts yellow-brown; eyespot sometimes present; life histories within genus are very complex: division of *Stylodinium*-phase described above sometimes results in one of three motile phases (2 amoeboid cells, 2 helizoid cells, 2 *Gyrodinium*-like and 2 *Gymnodinium*-like cells or 2–4 *Gymnodinium*-like cells) which interchange with other phases; resting stage thick-walled, spherical and stalked.

Worldwide; mainly reported attached to filamentous algae.

Complex life history within this genus has led to taxonomic confusion. Some taxa attributed to this genus may be a stage in the life history of another genus.

Stylodinium grande Jane ex J.D.Dodge *et* Lewis, *inedit.*

Basionym: *Stylodinium grande* Jane 1945

06130010 **Pl. 52E,F** (p. 206)

Cells usually epiphytic, spherical, 50 μm wide, attached by a stalk between 4–8 μm long and of variable thickness, terminates in a brownish-coloured disc; chloroplasts numerous, brown and disc-shaped.

British Isles; recorded as an epiphyte on *Zygogonium ericetorum*, detached cells may be also be found.

PHYLUM RAPHIDOPHYTA

Allan Pentecost

Small group of single-celled flagellates with 2 apical flagella differing in function and structure. One flagellum directed forwards when cells are in motion, often longer than cell and easily visible, the second thinner, rising close to first but trailing, possibly acting as rudder, initially in apical groove. The large cells have no distinct wall and are often metabolic. Trichocysts or muciferous bodies are normally present in abundance just below the protoplast surface. Pigmented genera have numerous yellow-green, disc-like chloroplasts containing chlorophylls *a* and *c* as well as several xanthophylls. The cells have oil as a storage product, an eyespot is usually absent, contractile vacuoles are numerous (sometimes in a prominent angular structure near apex), and the nucleus is large with sometimes an endomembrane cap present.

Named the Chloromonadida by Klebs (1893) but changed to Rhaphidophyta by Bourrelly (1970) who believed the Chloromonadida could be confused with the genus *Chloromonas* (Chlorophyta, Volvocales). There are some similarities with the Cryptophyta, Pyrrophyta and Euglenophyta but insufficient to warrant inclusion within any of them. Potter et al. (1997), on the basis of nucleotide sequencing, concluded that the Raphidophyceae are monophyletic, while 28S ribosomal sequencing had previously suggested affinities with the Chrysophyceae (Perasso et al., 1989). There are two families: Vacuolariaceae containing photosynthetic genera, and the colourless Thaumatostigaceae whose members have pseudopodia. Only the former family is considered here.

Nine genera are classified in the Raphidophyta and these occur in freshwater and marine habitats. Freshwater species are usually to be found in rather acid waters of some ponds and pools.

Some of the most important references on the phylum are Mignot (1967), Huber-Pestalozzi (1968), Bourrelly (1970), Spencer (1971), Heywood (1990) and Canter-Lund & Lund (1995).

The cells are extremely delicate and can be disrupted even by gentle pressure on a cover slip. Observation using 'optical staining', phase- or anoptral contrast is recommended.

1 Cells with small, needle-shaped (3–5 μm long) muciferous bodies (trichocysts) scattered among the chloroplasts .... *Gonyostomum*

1 Cells with minute, spherical (1–1.5 μm in diameter) muciferous bodies among the chloroplasts.. *Vacuolaria*

Gonyostomum Diesing 1866

Cells rounded apically, usually attenuated basally, flattened laterally with 2 apical flagella, one flagellum directed forwards and the other backwards; chloroplasts numerous, densely crowded at periphery of cytoplasm, disc-shaped, 2–3 μm in diameter, yellow-green, usually in a single layer; trichocysts refractive, needle-shaped, associated with chloroplasts, frequently in small groups, about 0.4 μm in diameter, up to 5 μm in length, explosively discharging contents on stimulation, function unknown.

Probably cosmopolitan; usually in small acid water bodies. Five species have been described but only one is recorded from the British Isles.

Gonyostomum semen (Ehrenburg) Diesing 1866

Basionym: *Monas semen* Ehrenberg

07010010 **Pl. 53F** (p. 209)

Motile cells obovoid, 23–70 μm wide, 32–92 μm long, often metabolic, eyespot absent, non-motile resting stages common, with cells becoming spherical and losing flagella; trichocysts scattered over cell surface or concentrated near base and apex.

Europe, North America; numbers have recently increased significantly in parts of northern Europe (possibly as a result of acid rain); in the British Isles most records from stagnant and humus-rich acidic waters; sporadic in appearance and occasionally abundant.

Malin Smith (1933) noted that it often swims steadily forwards, occasionally reversing and that the chloroplasts are best observed in dilute iodine solution which discharges the trichocysts. After a few minutes' observation under the microscope, some cells disintegrate, releasing their characteristic trichocysts (about 70 per cell). *Gonyostomum* has been likened to a flat green bottle when motile.

Vacuolaria Cienkowski 1870

Cells rounded and elongate, flagella arising from rounded apex in a small depression, one flagellum directed forwards and the other backwards; chloro-

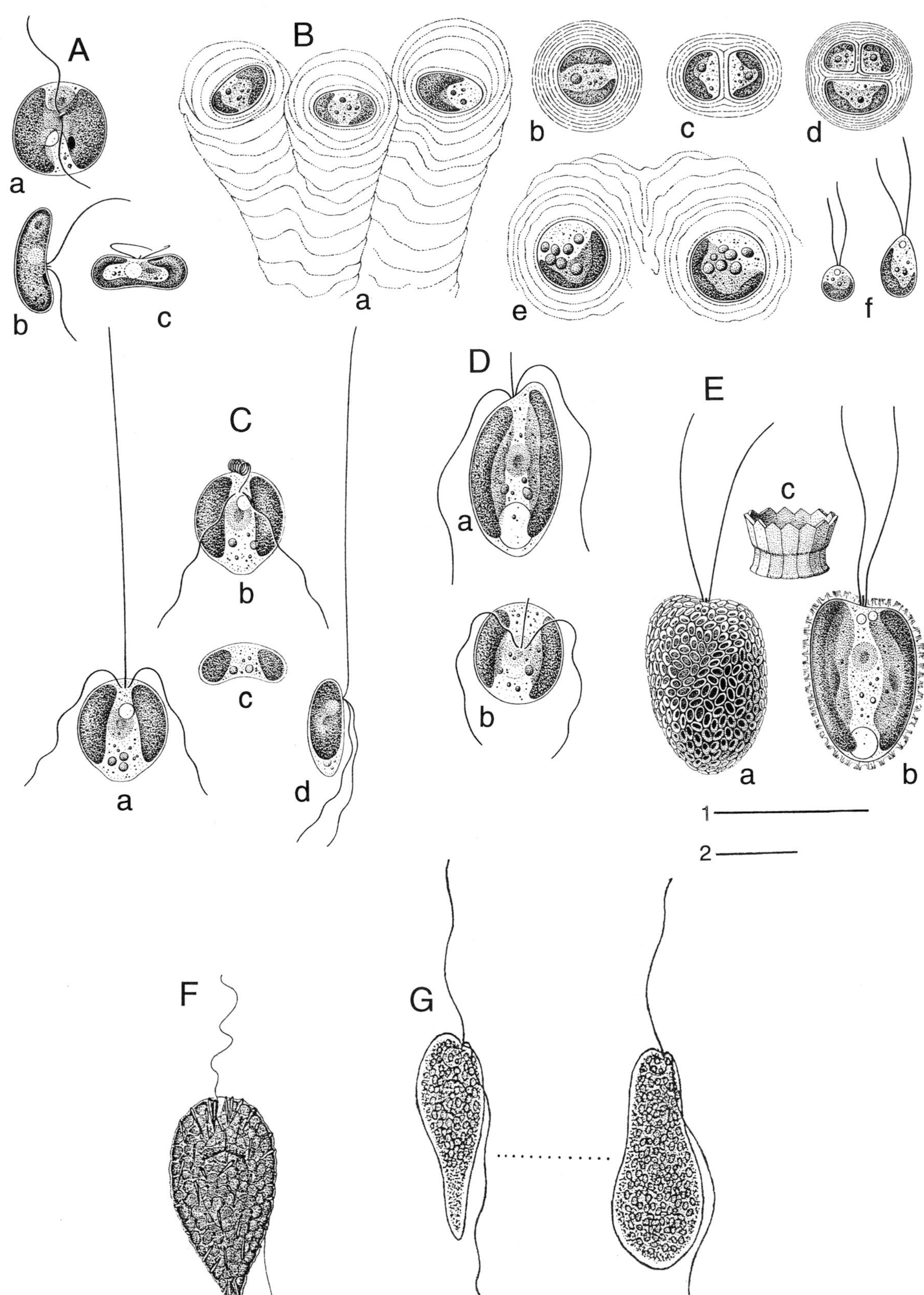

Plate 53 A–G

A. *Diacronema vlkianum* (p. 212): a. cell from broad side, b. cell in lateral view, c. cell from posterior end; **B**. *Chrysotila lamellosa* (p. 212): a–e. different stages of vegetative cells with stratified mucilaginous sheaths, the latter sometimes developed asymmetrically and forming stalks, f. zoospores; **C**. *Chrysochromulina parva* (p. 212): a. cell from broad side with an extended haptera, b. cell from broad side with a coiled haptonema, c. cells in cross section, d. cells from a lateral view; **D**. *Prymnesium parvum* (p. 213): a. cell from broad side, b. cell from anterior end; **E**. *Hymenomonas roseola* (p. 213): a. cell from broad side showing arrangement of coccoliths on surface, b. organelles inside the cell, c. a single coccolith at higher magnification (drawn from an electron micrograph); **F**. *Gonyostomum semen* (p. 208); **G**. *Vacuolaria virescens* (p. 210): showing two cells.

Scale bars: 1, 20 μm for B, Ea,b; 10 μm for A, C, D; 2 μm for Ec; 2, 10 μm for F, G.

plasts numerous, disc-shaped, bright green in 2–3 layers near cell periphery; refractive muciferous bodies present, small and spherical, scattered among chloroplasts, 1–1.5 μm in diameter (more evident on staining with ruthenium red); contractile vacuoles 1–6, usually anterior and coalescing to form large spherical or triangular vacuoles prior to discharge; asexual palmelloid stage formed by binary fission after cells round-off and produce a thick mucilaginous coat from discharge of muciferous bodies; spherical resting stages known.

Probably cosmopolitan; tends to occur most commonly in small and usually rather acidic waters. Four species of which only one is known from the British Isles.

The muciferous bodies may represent rudimentary trichocysts although with a different form from those in euglenoids. In other respects, the cells of *Vacuolaria* closely resemble those of *Gonyostomum*.

Vacuolaria virescens Cienkowski 1870

07020010 **Pl. 53G** (p. 209)

Cells elongate-ellipsoidal, occasionally narrowing basally to a short tail, sometimes laterally flattened, 20–30(–40) μm wide and 50–160 μm long; eyespot absent.

Probably cosmopolitan; in the British Isles occasional in small muddy ponds or pools, often among water plants or planktonic, possibly more frequent during colder months. According to Woodhead & Tweed (1954), mainly in lowland sites with pH range 4.4–8.4.

Appears to be less common than *Gonyostomum*, but occurs in similar sites, often more common in acidic water. The apparent lack of association with *Gonyostomum* is interesting considering their similar ecological requirements.

PHYLUM HAPTOPHYTA (Prymnesiophyta)

Hans R. Preisig

Most species of Haptophyta (also known as Prymnesiophyta) are single-celled flagellates, but some may have amoeboid, coccoid, palmelloid or filamentous stages. The flagellate cells have usually two flagella of equal or subequal length (unequal in the order Pavlovales), both of which are smooth. In most members of the Pavlovales the forward one is covered with small knob-like scales and fine non-tubular hairs that are invisible in the light microscope; tubular flagellar hairs as in the Chrysophyta or other heterokont flagellates do not exist in the Haptophyta. The two flagella may have a similar action (homodynamic) or move differently (heterodynamic). Between them there is a characteristic additional appendage, the haptonema. Depending on the species, this organelle is long, thread-like, exceeding the cell in length, or shorter, bulbous, or vestigial. It consists of 6 or 7 microtubules enclosed in two or three sheathing membranes (not visible with light microscopy). Haptonemata are capable of bending and flickering and, if long, can coil rapidly and uncoil slowly in response to shock. Haptonemata do not play a role in locomotion, but may be used in prey capture or anchoring the cell. The 1–2(–4) chloroplasts are usually golden or yellow-brown, because accessory photosynthetic pigments mask the green of the chlorophyll (chlorophylls a, c_1/c_2 and sometimes c_3). The surface of the cell is covered with minute scales or granules of organic composition (not visible in light microscope) and, in addition, there may be calcified scales (coccoliths) visible in the light microscope; species with such scales are called coccolithophorids.

The Haptophyta comprises at present some 80 genera and 300 species; a large number of additional genera and species of fossil coccolithophorids have been described. Especially common in marine waters are species of *Chrysochromulina*, *Phaeocystis* and coccolithophorids (e.g. *Emiliania huxleyi*). Only about a dozen haptophyte species are known from freshwater or terrestrial habitats. Five have been recorded from the British Isles.

1 Cells usually in groups, sessile, surrounded by stratified mucilaginous sheaths; flagella present only in zoospores........................*Chrysotila*
1 Cells solitary, free-swimming, normally flagellate..2
2 Cell surface covered with distinct calcareous structures (coccoliths)...*Hymenomonas*
2 Cell surface appearing naked with the light microscope..3
3 Haptonema longer than cell when extended (usually coiled during rapid swimming) ....*Chrysochromulina*
3 Haptonema always short, sometimes hardly visible in the light microscope 4
4 Cells distinctly flattened, 3.5–8 μm long; haptonema short, but hardly visible in the light microscope..................................*Diacronema*
4 Cells not or only slightly flattened, 8–15 μm long; haptonema short but always clearly visible in the light microscope...........*Prymnesium*

ORDER PAVLOVALES (=Pavlovophyceae *sensu* Cavalier-Smith)

Cells with two unequal flagella, the longer one often covered by knob-like scales and fine hairs, with a reduced haptonema.

Pavlova granifera (Mack) J.C.Green and *Pavlova noctivaga* (Kalina) Veer *et* Leewis (=*Corcontochrysis noctivaga* Kalina) have been observed in some European freshwaters, but there are no records from the British Isles.

Diacronema Prauser 1958

Cells solitary, free-swimming, compressed, asymmetrical; flagella 2, unequal, heterodynamic, flagellar insertion mid-ventral; haptonema present, short, hardly visible with the light microscope; chloroplast single, with two parietal lobes, yellow-green to olive-green, with an eyespot underlying posterior flagellum; contractile vacuole single, sometimes near flagellar insertion; asexual reproduction by cell fission.

Europe, reported only a few times from fresh and marine waters in Europe including the British Isles, but certainly not rare since found in many lakes, ponds and temporary puddles (own unpublished records from Cambridge area and Central Europe). Monospecific.

Diacronema vlkianum Prauser 1958

08020010 **Pl. 53A** (p. 209)
Cells orbiculate to heart-shaped in dorsiventral view, 3.5–8×3–6 μm, 1.5–3 μm thick; long (anterior) flagellum 7–10 μm, short (posterior) flagellum 3–9 μm; haptonema about 1 μm long.

Distribution and ecology as for genus.

ORDER PRYMNESIALES (Haptophyceae/ Prymnesiophyceae *sensu stricto*)

Cells with two smooth, subequal or equal flagella, with haptonema evident, reduced or absent.

Chrysotila P.L.Anand 1937 emend. J.C.Green *et* Parke 1975

Cells non-motile, young cells forming regular cuboidal masses, older cells with a thick, stratified, mucilaginous sheath, latter frequently developing asymmetrically to form lamellated, colourless, unbranched stalks with one or more cells situated distally; extracellular calcareous bodies often occur associated with mucilage; chloroplasts parietal, 1(–2) per cell, with a single pyrenoid, eyespot lacking, but frequently with a small orange-red pigment body (pseudoeyespot); asexual reproduction by vegetative cell division, aplanospore or zoospore formation; zoospores with 2 subequal, subapically inserted, homo- or heterodynamic flagella; haptonema between flagella, but not visible with the light microscope; unmineralized body scales detectable only on electron microscopic examination; sexual reproduction unknown.

Europe; two species and mainly marine, growing on calcareous cliff faces (e.g. reports from Devon and Kent); one species also known from inland terrestrial localities.

Chrysotila lamellosa P.L.Anand emend. J.C.Green *et* Parke 1975

Synonyms: *Ruttnera spectabilis* Geitler, *Ruttnera chadefaudii* Bourrelly *et* F.Magne, *Isochrysis maritima* Billard *et* Gayral
08050010 **Pl. 53B** (p. 209) **CD**
Colonies at first spherical, later becoming laminar, forming crusts several mm thick and up to some metres long; young vegetative cells spherical or hemispherical, 4.5–11 μm wide, surrounded by mucilaginous sheaths or on short stalks (up to 30 μm long); extracellular calcareous bodies at first rod- or cross-shaped, later with striated, conical protuberances (up to 50 μm across); zoospores 4.5–8×3–6 μm, compressed and somewhat metabolic; flagella subequal, 6–10 μm and 4–8 μm long.

Europe; known in the British Isles from marine habitats, usually splash zone of calcareous cliffs, also terrestrial habitats probably where high local concentrations of salts (damp walls, rocks, etc.); scattered distribution.

Flagellate cells known as *Isochrysis maritima*, before life cycle was resolved.

Chrysochromulina Lackey 1939

Cells solitary, free-swimming, somewhat metabolic; cell shape variable (spherical, subspherical or saddle-shaped); biflagellate (rarely quadriflagellate), flagella equal to subequal, homo- or heterodynamic, 1.5 to 3.5 times as long as cell; haptonema usually long, sometimes attaching to surfaces with its tip or at any point along its length; flagellar position and haptonema size vary: during rapid swimming, with rotation and gyration of the cell, flagella held posteriorly with haptonema fully coiled, during slower swimming one flagellum directed forwards and other directed backwards trailing behind cell, during gliding movements, without rotation, flagella more widely separated or laterally disposed with haptonema fully or partly extended in direction of motion; cells usually covered by small unmineralized scales (not visible with the light micoscope); chloroplasts parietal, 1–4 per cell (rarely none), each with a pyrenoid (sometimes difficult to see); eyespot absent; contractile vacuole in freshwater species; commonly phagotrophic; asexual reproduction by longitudinal fission into 2 or 4.

Cosmopolitan and planktonic; mostly brackish and marine, sometimes bloom-forming. More than 50 species described but only four freshwater species known: *C. breviturrita* K.H.Nicholls, *C. inornamenta* Wujek *et* W.E.Gardiner, *C. laurentiana* H.J.Kling, and *C. parva* Lackey, with only the last found in the British Isles.

Life history not well-known. Old cultures sometimes produce an amoeboid phase giving rise to walled resting cells, some with a thick sculptured wall. In one marine species (*C. polylepis*) the life history has an alternation of a haploid and diploid motile generation. 'Pseudofilaments' of 4 cells also observed.

Chrysochromulina parva Lackey 1939

08010010 **Pl. 53C** (p. 209)
Cells bilaterally symmetrical, spheroidal to ovoid and often flattened, 2–7 μm wide and 2.5–5 μm thick; flagella 2 per cell, 6–17 μm long; haptonema up to 20 times cell-length when extended; chloroplasts 2 (4–6 in fission stages), pyrenoids often difficult to detect with the light microscope.

Probably cosmopolitan, in Europe widely distributed in lakes, ponds and puddles; few reports from the British Isles (e.g. lakes and tarns in the English Lake District), but certainly often escapes detection because of its extreme fragility.

Erkenia subaequiciliata Skuja is likely to be identical to *C. parva*. Some marine species of *Chrysochromulina* (especially *C. polylepis* Manton *et* Parke) are notorious toxin producers and *C. parva* is suspected of causing extensive fish mortality in Danish freshwater lakes. In a small freshwater lake in Lincolnshire (Normanby Hall Lake) very high densities of *Chrysochromulina* (not identified to species) also coincided with a large fish kill in October/November 1997, but a link between these events is so far only tentative.

Hymenomonas F.Stein 1878 emend. Gayral *et* Fresnel 1979

Life history with alternation of haploid and diploid generations, both motile; cell of diploid generation covered by a single layer of open tube-shaped calcareous plates (coccoliths) and an underlayer of unmineralized scales (the latter not visible with the light microscope); haploid generation of same general organization but lacking coccoliths and possessing unmineralized scales morphologically distinct from those of coccolith-bearing stage; flagella 2 per cell, subequal, heterodynamic in both motile generations; haptonema short or vestigial; chloroplasts parietal, 1 or 2 per cell, each with a bulging pyrenoid on inner face (sometimes difficult to see with the light microscope), eyespot absent; contractile vacuoles 1 or 2, anterior (in freshwater species); non-motile stages also occur, often forming clumps; cell division predominantly in non-motile stage.

Probably cosmopolitan; several species occurring in marine and brackish waters, one freshwater species currently recognized.

Hymenomonas roseola F.Stein 1878

08030010 **Pl. 53E** (p. 209) **CD**
Motile cells metabolic, subspherical, ellipsoidal to elongate, 13–50×10–24 μm; slightly emarginate at anterior end; flagella 1.3 to 2 times cell length; haptonema short, bulbous (just visible with the light microscope); coccoliths circular to elliptical with a jagged top, 1–2 μm long, 1–1.2 μm wide and 0.7–0.9 μm high.

Apparently cosmopolitan; scattered distribution in Europe (including British Isles), not rare but often only in small cell numbers among other flagellates in freshwater lakes, ponds and ditches.

Haploid motile stages without coccoliths have been described as '*H. roseola* fo. *glabra* G.A.Klebs' in the literature. Other species of *Hymenomonas* described from freshwater are now generally considered to be conspecific with *H. roseola*: *H. danubiensis* Kamptner, *H. flava* A.Stokes, *H. scherffelii* W.Conrad, *H. stagnicola* (Chodat *et* R.Rosillo) Kamptner (=*Pontosphaera stagnicola* Chodat *et* R.Rosillo), whereas the species described as *H. prenantii* Lecal-Schlauder is in fact *Gyromitus disomatus* Skuja (Thaumatomastigidae, Protista *incertae sedis*). Other coccolithophorids described from European freshwaters, such as *Acanthoica schilleri* or *Anacanthoica ornata*, have not been observed in the British Isles.

Prymnesium J. Massart *ex* W.Conrad 1926

Cells elongate, pear-shaped to almost spherical, not compressed; flagella equal or subequal, heterodynamic, arising subapically from a slight depression in truncate apex, one at right angles to body during swimming and the other curved backwards close to body surface, or both flagella directed posteriorly; haptonema relatively short, non-coiling, directed forwards and capable of attaching to other objects; cells covered by small unmineralized scales (not visible with the light microscope); chloroplasts 2–4, parietal, dissected or lobed, with pyrenoids often difficult to detect with the light microscope and eyespot absent; contractile vacuole at flagellar pole (if present); cysts ovoid to ellipsoidal, with a thin, delicately rugose wall and a simple sub-anterior pore.

Cosmopolitan, mainly in brackish and marine waters; few freshwater records.

Prymnesium parvum N.Carter emend. J.C.Green, D.J.Hibberd *et* Pienaar 1982

08040020 **Pl. 53D** (p. 209) **CD**
Cells subspherical to elongate, 8–15×4–10 μm, not or only slightly compressed; flagella 12–20 μm long; haptonema 3–5 μm long.

Scattered distribution in Europe (including British Isles) and elsewhere; mainly in brackish-waters, but able to grow in waters of very low salinity.

This species cannot be distinguished by the light microscope from several other species of *Prymnesium* occurring in brackish and marine waters. *Prymnesium parvum* and *P. patelliferum* J.C.Green, D.J.Hibberd *et* Pienaar are now known to be part of the same life history. The type species, *P. saltans* J. Massart *ex* W.Conrad, has not yet been characterized ultrastructurally, but is likely to be identical to *P. parvum*. Species of *Prymnesium* (especially *P. parvum*) may form toxins causing massive fish mortalities during bloom developments. There are reports of it from lakes at Barton-on-Humber (North Lincolnshire), North Fleet in Kent, the Norfolk Broads, Lincolnshire Fenland (old clay pits) and in drains and rivers flowing into the Wash and North Sea (S.J.Brierley, pers. comm.).

PHYLUM CHRYSOPHYTA (Golden Algae)

Jørgen Kristiansen

The chrysophytes ('Golden Algae') constitute a group consisting mainly of the classes Chrysophyceae and Synurophyceae. They have in common their unequal flagella (a long hairy and a short smooth flagellum) and pigment composition with the brown fucoxanthin giving the chloroplasts a characteristic golden-brown colour. They form specific resting stages, stomatocysts, which have a silicified wall with a pore closed by a non-silicified stopper. The two classes diverge in the position of the flagella, chlorophyll composition, and in the way the silica scales are formed. A new class, the Phaeothamniophyceae, includes most coccoid and filamentous forms, but is not considered here. The small Class Dictyochophyceae is represented by the radially symmetric pedinellids.

The chrysophytes number about 1000 species. Most are unicellular or colonial flagellates, predominantly occurring in freshwater plankton. This treatment includes about 120 species found in the British Isles. Most of the silica-scaled forms require electron microscopy for identification and are not included; only treated are those species which can be identified by means of the light microscope.

Some genera include representatives that are not only phototrophic but phagotrophic. Sometimes remnants of engulfed organisms are visible.

1 Cells with radial symmetry; single flagellum surrounded by tentacles ..Class Dictyochophyceae, Order Pedinellales (p. 240)
1 Cells not radially symmetrical 2
2 Silica scales always present, not differentiated into spine scales and plate scales ....Class Synurophyceae, Order Synurales (p. 240)
2 Silica scales mostly absent, if present then having spine scales and/or plate scales..................3
3 Mostly flagellated forms with a single or pair of visible flagella, but also amoeboid forms, forms with gelatinous envelopes or with walls, or filamentous forms ....Class Chrysophyceae, Orders Chromulinales and Hibberdiales (p. 236)
3 Cells arranged in crustose or bushy thalli .....Class Chrysophyceae, Order Hydrurales (p. 239)

DOUBTFUL TAXON

Microglena butcheri J.H.Belcher 1966

The position of the genus is problematical and on the basis of electron microscope investigations (Couté & Preisig, 1981) it might belong perhaps to the Raphidophyta rather than to the Chrysophyta. This species (Pl. 54I, p. 217) is characterized by its thick periplast with small lenticular bodies. It has a single yellowish-green chloroplast with an eyespot and a funnel-shaped, non-contractile reservoir behind which the large nucleus is located. Recorded from the English Lake District (Belcher, 1966c) and known also from France.

CLASS CHRYSOPHYCEAE

ORDERS CHROMULINALES AND HIBBERDIALES

Mostly flagellated forms with 1 or 2 visible flagella; also amoeboid forms, forms with gelatinous envelopes or walls, and filamentous forms.

1 Flagellate 2
1 Not flagellated 21
2 Cells naked..............................3
2 Cells with lorica or scales 8
3 Solitary 4
3 Colonial..............................6
4 Chloroplasts 4..............................*Amphichrysis*
4 Chloroplasts 1 or 2 5
5 Flagella, 2 visible..............................*Ochromonas*
5 Flagella, 1 visible ....*Chromulina* and Order Hibberdiales (p. 236)
6 Colonies flat, ring-shaped ....*Cyclonexis*
6 Colonies spherical 7
7 Cells elongated, joined by their tails in the centre of colony ....*Synochromonas*
7 Cells round to elongated, peripherally arranged, on more or less visible radiating stalks .....*Uroglena*
8 Cells with lorica..............................9
8 Cells with scales 18
9 Lorica spherical, with a narrow pore for the flagellum ....*Chrysococcus*
9 Lorica with wide opening 10
10 Colonial..............................11
10 Solitary 13
11 Spherical colonies..............................*Sphaerobryon*
11 Branched or dendroid colonies..............................12
12 Colonies free-swimming; lorica in one piece attached to inner side of parent lorica ....*Dinobryon*, in part

12 Colonies attached to a substrate; lorica scaly, mostly attached to outer side of parent lorica .... *Epipyxis*, in part
13 Lorica attached to a substrate 14
13 Cells free-living 15
14 Lorica scaly, attached directly .... *Epipyxis*, in part
14 Lorica not scaly, with stalk .... *Poteriochromonas*
15 Lorica curved, with projection on convex side .... *Chrysolykos*
15 Lorica straight 16
16 One visible flagellum .... *Kephyrion*
16 Two unequal flagella visible 17
17 Cell attached to lorica base by stalk .... *Dinobryon*, in part
17 Cell not attached to lorica base ... *Pseudokephyrion*
18 Solitary 19
18 Colonial .... *Chrysosphaerella*
19 Cells and flagella covered with flower-pot like, organic scales .... *Chrysolepidomonas*
19 Only cells covered with siliceous scales 20
20 Cells covered with spine scales and plate scales .... *Spiniferomonas*
20 Cells covered only with plate scales .... *Polylepidomonas*
21 Cells amoeboid, naked 22
21 Cells immotile, with wall, mucilage or lorica 24
22 Cells solitary or in small aggregates ... *Chrysamoeba*
22 Cells in colonies 23
23 Cells in linear colonies .... *Chrysidiastrum*
23 Cells regularly arranged in flattened circular colonies .... *Chrysostephanosphaera*
24 Cells with lorica or wall 25
24 Cells with mucilage 26
25 Cells with lorica ...... see Order Hibberdiales (p. 236)
25 Cells with wall 32
26 Cells in flat, often monostromatic colonies 27
26 Cells in round, ellipsoidal or irregular colonies 29
27 Cells bearing gelatinous hairs projecting from the colony .... *Naegeliella*
27 Cells without such hairs 28
28 Cell division in two directions ...... *Dermatochrysis*
28 Cell division in three directions .... *Gloeochrysis*
29 Cells in very irregular, branched, cylindrical colonies .... *Phaeosphaera*
29 Cells in spherical or ellipsoid colonies 30
30 Chloroplast single, star-shaped .... *Phaeaster*
30 Chloroplasts 2–4, plate-like 31
31 Swarmers uniflagellate .... *Chrysocapsa*
31 Swarmers biflagellate .... *Chrysocapsella*
32 Rounded cells 33
32 Filamentous 34
33 Epiphytic .... *Chrysosphaera*
33 Free-floating .... *Stichogloea*
34 Filaments creeping .... *Apistonema*
34 Filaments upright .... *Phaeothamnion*

ORDER CHROMULINALES

Amphichrysis Korshikov 1929

Cells solitary, naked, with a single flagellum; chloroplasts 4 per cell and 2 eyespots.

Only found in a few places world-wide; planktonic. Monospecific.

It has been characterized as an incomplete double-organism.

Amphichrysis compressa Korshikov 1929

09010010 **Pl. 54A** (p. 217)

Cells 15–18 µm wide, 27–30 µm long.

Distribution as for genus; in the British Isles reported only from a small pool near Elterwater, English Lake District (Lund, 1962a).

Apistonema Pascher 1925

Filaments creeping, branched, consisting of barrel-shaped cells, each cell containing 1 or 2 chloroplasts.

Four species are mentioned in freshwater by Starmach (1986), but their status is unclear, and thus statement concerning world distribution is not possible. One species has been recorded from the British Isles.

Several species have been described. However, they are very similar to benthic phases of some prymnesiophytes, which have swarmers of the *Hymenomonas*-type.

Apistonema pyrenigerum Pascher 1931

09020010 **Pl. 62E** (p. 238)

Thallus mostly single-celled or as very short filaments; chloroplasts 2 per cell; swarmers and cysts not observed.

Central Europe, on wet rocks; in the British Isles known only on wet limestone walls along a stream at Roche Abbey, near Rotherham, Yorkshire (Lund, 1942a).

Chromulina Cienkowski 1870

Cells solitary, surface smooth or warty, with a single visible flagellum; chloroplasts 1 or 2 per cell, eyespot usually present.

About 140 species, most very difficult to identify, thus distribution and ecology are uncertain.

Those species that are neustonic might better be classified with *Chromophyton* (such as *Chromophyton rosanoffii*), which is placed in the Order Hibberdiales.

1 Neustonic (associated with surface film) 2
1 Not neustonic 3
2 Chloroplast with a pyrenoid .... *C. aerophila*
2 Chloroplast without a pyrenoid .... *C. ferrea*
3 Sessile stage forms a sporangium with numerous daughter cells .... *C. sporangifera*
3 Sessile stages absent 4
4 Cell surface warty .... *C. wislouchiana*
4 Cell surface smooth 5

5 Cell with an eyespot 6
5 Cell without an eyespot ....*C. nebulosa*
6 Eyespot small, apical in a chloroplast lobe ....*C. ovalis*
6 Eyespot large, central in a chloroplast ....*C. placentula*

Neustonic species

Chromulina aerophila J.W.G.Lund 1942

09050010 **Pl. 54B** (p. 217)

Cells free-swimming or neustonic as a palmelloid aggregate; motile cells ellipsoidal, 4–5 µm wide, 5–9 µm long, flagellum about length of cell; non-motile cells more spherical, 6–8 × 8.5 µm, with a short stalk and division stages up to 4 cells on same stalk, fusing into larger aggregates; chloroplast with a single pyrenoid and without an eyespot.

British Isles (England); so far seen only in cultures from Richmond Park, Surrey (Lund, 1942a).

Chromulina ferrea J.W.G.Lund 1942

09050020 **Pl. 54C** (p. 217)

Cells free-swimming or neustonic; motile cells ellipsoidal, 4.3–5 µm wide, 6.8–11.9 µm long; non-motile cells with a short brown stalk, positioned on top of water film where division takes place; chloroplast single and without a pyrenoid or eyespot.

British Isles (England); known only from ponds in Richmond Park, Surrey (Lund, 1942a).

Planktonic species

Chromulina nebulosa Cienkowski 1870

09050030 **Pl. 54E** (p. 217)

Cells free-swimming, ovoid to spindle-shaped, 4–6 µm wide, 12–16 µm long; chloroplast single, without a pyrenoid or eyespot; cysts spherical with fine longitudinal striations.

Europe; widely distributed in swamps (Pentecost, 1984).

Chromulina ovalis G.A.Klebs 1893

09050040 **Pl. 54F** (p. 217)

Cells free-swimming, almost ellipsoidal, 6–7 µm wide, 9–14 µm long; chloroplast single, with an eyespot and without a pyrenoid.

Widely distributed in Northern Europe; in the British Isles known only from ponds in Richmond Park, Surrey (Lund, 1937).

Chromulina placentula J.H.Belcher *et* Swale 1967

09050050 **Pl. 54G** (p. 217)

Cells free-swimming, kidney-shaped, 4.5–8.5 µm long; chloroplast single, brown, positioned towards convex side of cell, without pyrenoid but with a very conspicuous and centrally located eyespot.

British Isles (England); known only in the English Lake District from the small lake, Priest Pot (Belcher & Swale, 1967a).

Chromulina sporangifera J.W.G.Lund 1942

9050060 **Pl. 54D** (p. 217)

Motile cells ellipsoidal, 6.8–10.2 µm long; chloroplast single and spiralled, without a pyrenoid or eyespot; non-motile stages on water plants, forming sporangia containing 40 swarmers; stomatocysts spherical and smooth.

British Isles (England); only identified from a pond in Richmond Park, Surrey (Lund, 1942a, c).

Chromulina wislouchiana (Wislouch) Bourrelly 1957

Basionym: *Chrysoglena verrucosa* Wislouch

09050070 **Pl. 54H** (p. 217)

Cells irregularly ellipsoidal, 20–43 × 13–27 µm, with walls densely covered with rounded warts; chloroplast single and large, with an eyespot and without pyrenoid.

Russia; reported in the British Isles from peaty pools in Cumbria (Lund, 1949a) and in small bog pools in Sussex during the spring (Pentecost, pers. comm.).

Chrysamoeba G.A.Klebs 1893

Cells solitary or aggregated, naked, forming lobed or branched rhizopodia or pseudopodia (rhizopodial stage); stomatocysts known; motile stage known.

Europe; planktonic in ponds. About 16 species, only one reported from the British Isles.

Chrysamoeba radians G.A.Klebs 1893

Synonym: *Chrysamoeba scherffelii* Pascher

09060010 **Pl. 61Q** (p. 233)

Cells with one curved chloroplast, a pyrenoid but no eyespot; rhizopodial stage predominant, 4.5–8.5 × 5–12 µm, with delicate, radiating and scarcely branching rhizopodia, sometimes bead-like in appearance, very short flagellum present; motile stage *Chromulina*-like, obovoid, with one long, visible flagellum, highly metabolic.

Widely distributed; known in the British Isles from acid pools and ponds in England (Lund, 1937; Hibberd, 1971) and Wales (Jane, 1945).

Chrysidiastrum Lauterborn 1913

Cells arranged in linear, chain-like colonies, naked, with a tendency to form lobed or branched rhizopodia; stomatocysts known.

Widely distributed; planktonic. Only 3 species, one recorded from the British Isles.

Chrysidiastrum catenatum Lauterborn 1913

09080010 **Pl. 61R** (p. 233)

Cells in chain-like colonies, up to 10 cells long, each spherical, 15–21 × 12–14 µm, with about 10 radiating rhizopodia; chloroplast single and without an eyespot;

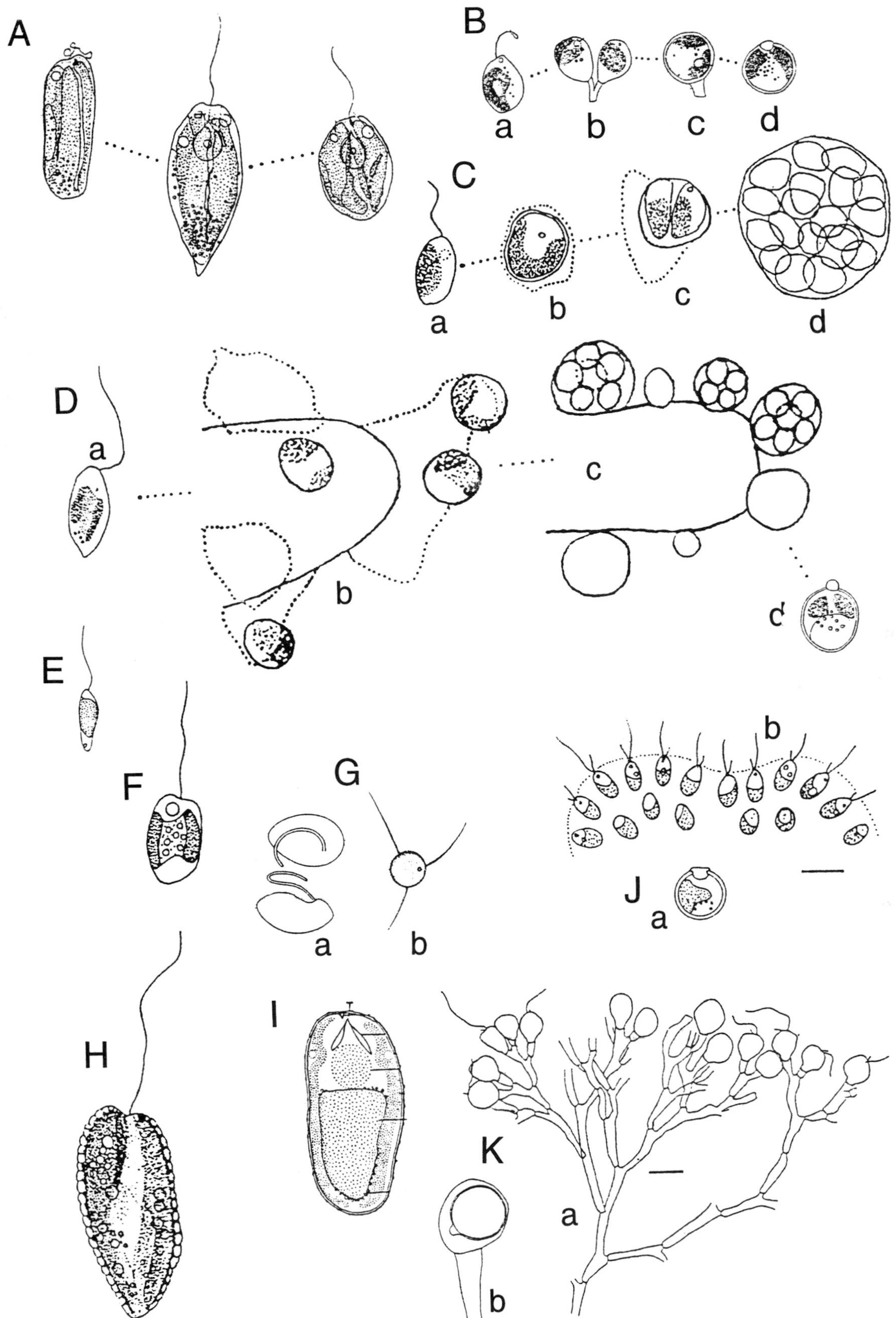

Plate 54 A–K

A. *Amphichrysis compressa* (p. 215); **B**. *Chromulina aërophila* (p. 216): a. motile stage, b, c. stalked non-motile stage, d. stomatocyst; **C**. *Chromulina ferrea* (p. 216): a. motile stage, b. non-motile stage, c. sporangium, d. stomatocyst; **D**. *Chromulina sporangifera* (p. 216): a, b. vegetative stages, c. sporangia and stomatocyst; **E**. *Chromulina nebulosa* (p. 216); **F**. *Chromulina ovalis* (p. 216); **G**. *Chromulina placentula* (p. 216): a. motile vegetative cells, b. stomatocyst; **H**. *Chromulina wislouchiana* (p. 216); **I**. *Microglena butcheri* (p. 214); **J**. *Uroglena americana* (p. 236): a. stomatocyst, b. part of a colony; **K**. *Uroglena eustylis* (p. 236): a. part of colony, b. stomatocyst.
Scale bars: 10 μm.

stomatocysts surrounded by a mucilaginous envelope; motile stage unknown.

Northern Europe, North America; reported in the British Isles from ponds in Richmond, Surrey (Lund, 1942b, c) and in the English Lake District (J.W.G.Lund, pers. comm.).

Chrysocapsa **Pascher 1913**

Cells naked, embedded within mucilage to form spherical or ellipsoidal colonies; chloroplasts 1 or 2(–4), with or without an eyespot; contractile vacuoles often present; reproduction by uniflagellate swarmers or aplanospores; cysts unknown.

Europe; planktonic or periphytic. Six species, only two reported from the British Isles.

1 Cells ovoid; epiphytic .... *C. epiphytica*
1 Cells spherical or subspherical; planktonic .... *C. planctonica*

Chrysocapsa epiphytica **J.W.G.Lund 1949**

09090010 **Pl. 56B** (p. 221)

Cells ovoid, embedded in surface of spherical or irregularly shaped mucilaginous colonies; chloroplasts 2 per cell, without an eyespot; reproduction by aplanospores.

British Isles (England); reported growing on water plants at the type locality of Blelham Beck, near Windermere, English Lake District (Lund 1949a). Also frequent in nutrient-poor ponds in Ashdown Forest, Sussex during the spring (Pentecost, pers. comm.).

Chrysocapsa planctonica **Pascher 1913**

09180020

Cells spherical or subspherical, usually in multiples of 2 (up to 64-celled), enclosed within a mucilaginous envelope with radiating fibrils or homogeneous; chloroplast 1 or 2; plate-shaped.

Probably cosmopolitan; in the British Isles reported from lake plankton in the English Lake District (Reynolds & Irish, 2000).

Confusion surrounds the identity of *Phaeococcus planctonicus* (Pl. 56D) described by W. & G.S.West's (1905b) in Scottish and Irish plankton. Fritsch (in West & Fritsch, 1927) combined it in the genus *Chrysocapsa* and stated (p. 331) that Pascher's *C. planctonica* 'is not the same as West's species' and Pascher's species 'should be renamed *C. Pascheri*'. Bourrelly (1957) mentions that Fritsch's '*C. Pascheri*' is the same as *C. planktonica* Pascher and that *C. planktonica sensu* Fritsch is distinct and does not belong to his new genus *Chrysocapsella*. However, Starmach (1985) introduces the unwarranted combination *Chrysocapsella planctonica* (West *et* G.S.West) Bourrelly 1957.

Chrysocapsella **Bourrelly 1957**

Cells ovoid or ellipsoidal and within ellipsoidal or spherical colonies; chloroplasts parietal, 2 per cell, one with an eyespot; swarmers with 2 flagella; cysts unknown.

Europe, North America; planktonic. Three species of which two are reported from the British Isles.

1 Colonies of groups of 2, 4, or 8 cells ...... *C. paludosa*
1 Colonies of 16–64 cells .... *C. planctonica*

Chrysocapsella paludosa **(West *et* G.S.West) Bourrelly 1957**

Basionym: *Phaeococcus paludosus* West *et* G.S.West

09100010 **Pl. 56C** (p. 221)

Cells ovoid, groups of 2–4–8 united to form spherical to ellipsoidal colonies up to 100 μm in size.

Europe, North America; reported in the British Isles from West Yorkshire (W. & G.S.West, 1903b; West, 1904; West & Fritsch, 1927).

Chrysocapsella planctonica **(West *et* G.S.West) Bourrelly 1957**

Basionym: *Phaeococcus planctonicus* West *et* G.S.West

09100020 **Pl. 56D** (p. 221)

Colonies 50–250 μm in diameter, containing 16–64 globular cells; chloroplasts 2 per cell, with irregular margins.

Europe, North America; In the British Isles reported in the plankton of lakes in Scotland and Ireland (W. & G.S.West, 1905b; West & Fritsch, 1927)

Chrysococcus **G.A.Klebs 1892**

Cells surrounded by a spherical or subspherical and sometimes compressed lorica, stained golden-brownish by iron compounds; flagella 2 per cell, one flagellum emerges through pore in lorica, second flagellum reduced and remains within lorica, sometimes 1 or 2 other pores present; chloroplasts 1 or 2 per cell, eyespot usually present, division within lorica with one daughter cell escaping through pore and forming its own lorica.

World-wide distribution; planktonic. At least 25 species of which several are dubious. Six species reported from the British Isles.

1 Lorica 2–3 μm wide........................ *C. punctiformis*
1 Lorica 6–16 μm wide... 2
2 Pore single 3
2 Pores 2–3 4
3 Lorica smooth.......................... *C. rufescens*, in part
3 Lorica tessellate in front part .... *C. tessellatus*
4 Lorica with 2 pores.. 5
4 Lorica with 3 pores.. 7
5 Lorica with attached empty lorica .... *C. cystophorus*
5 Lorica always single.. 6
6 Lorica 8–12 μm wide; chloroplasts longitudinally orientated .... *C. rufescens*, in part
6 Lorica 14–16 μm wide; chloroplasts transversely orientated .... *C. diaphanus*
7 Lorica flattened, heart-shaped .... *C. cordiformis*
7 Lorica spherical.................. *C. rufescens* fo. *tripora*

Chrysococcus cordiformis **Naumann 1921**

09130010 **Pl. 58Ba,b** (p. 225)

Lorica heart-shaped and flattened, about 5–10 μm wide and 1.5–5 μm thick, with 2 apical pores, through one of

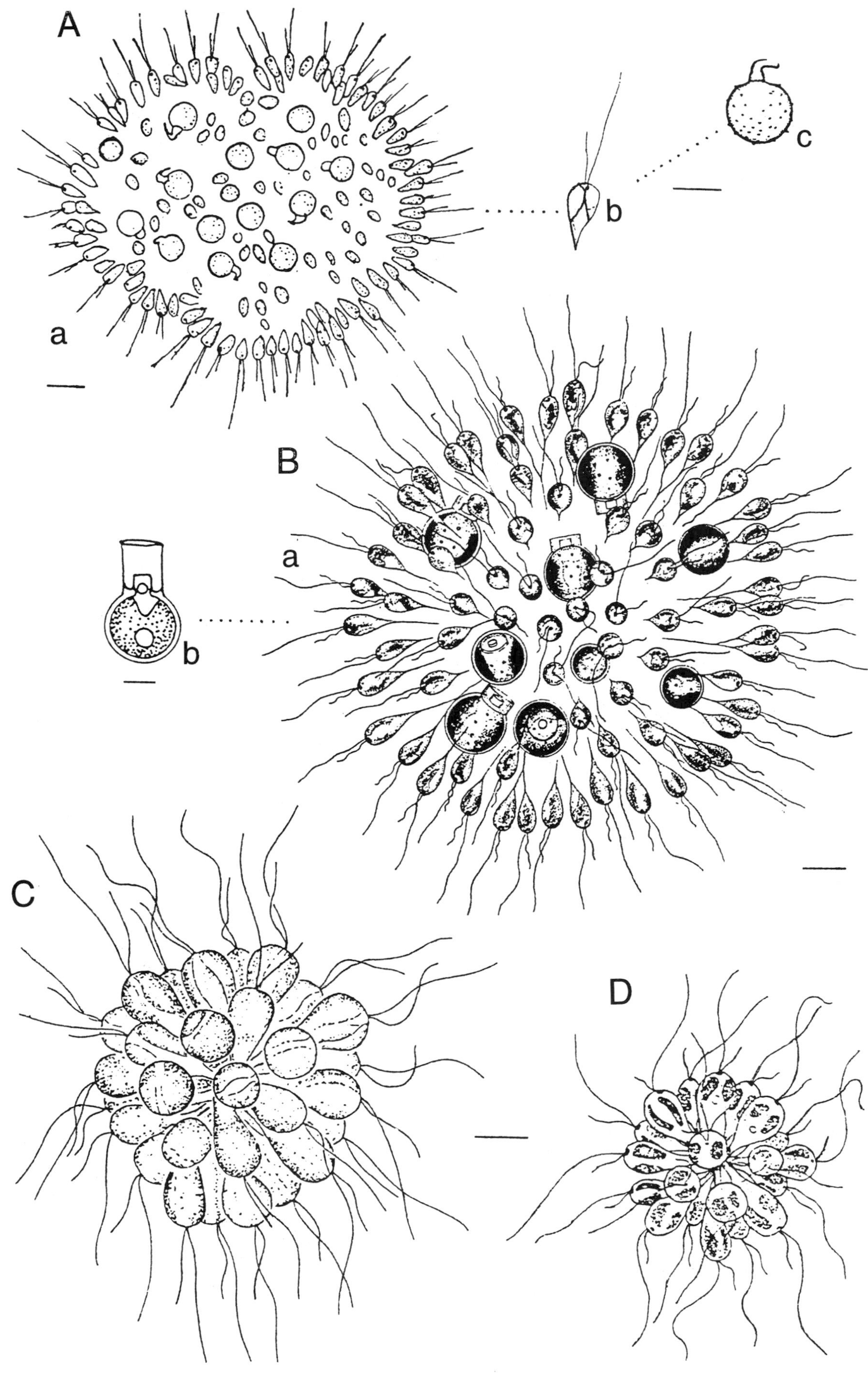

Plate 55 A–D
A. *Uroglena lindii* (p. 236): a. colony, b. single cell, c. stomatocyst; **B**. *Uroglena volvox* (p. 236): a. colony, b. stomatocyst; **C**. *Synochromonas elaeochrus* (p. 235); **D**. *Synochromonas janei* (p. 236).
Scale bars: 10 µm. 1 for B; 2 for A, C, D.

which the long flagellum protrudes, a third pore antapical; chloroplast single and without an eyespot.

Widely distributed in Europe; known only in the British Isles from Priest Pot, a small lake in the English Lake District (Belcher & Swale, 1972).

Chrysococcus cystophorus Skuja 1956

fo. *astigmatus* J.W.G.Lund 1960

09130022 **Pl. 58C** (p. 225)

Lorica subspherical, 10–13×10–12 μm, smooth-walled, 2 opposite pores and anterior pore occupied by a flagellum, with a second, slightly smaller lorica (9–10 ×8–10 μm in size) attached by posterior pore; second lorica with 2 pores, a granular surface and empty; chloroplasts 1–2 per cell and no eyespot in the British specimens (fo. *astigmatus*).

Sweden; reported in the British Isles in the plankton of Blelham Tarn, Cumbria (Lund,1960b).

The role of the second lorica in the life history of this species is hardly understood.

Chrysococcus diaphanus Skuja 1950

09130030 **Pl. 58D** (p. 225)

Lorica subspherical, 14–16 μm wide, colourless or slightly brownish with 2 opposite pores, flagellum projecting through apical pore; chloroplasts 2 per cell, transversely positioned, one possessing an eyespot; contractile vacuoles 2, equatorial.

Northern and Central Europe; reported in the British Isles from Abbot's Pond, Somerset (Happey & Moss, 1967), where present in large numbers.

Chrysococcus punctiformis Pascher 1911

09130040 **Pl. 58E** (p. 225)

Lorica spherical, 2–3 μm in diameter, reddish, with a single pore and flagellum seven times longer than cell.

Northern and Central Europe; reported from the British Isles in the nannoplankton of ponds (Scourfield, 1930).

Chrysococcus rufescens G.A.Klebs 1893

09130050 **Pl. 58F** (p. 225)

Lorica spherical, 8–12 μm in diameter, brown, with 1 or 2 pores; chloroplasts 2 per cell and a single eyespot.

Widely reported from the northern hemisphere, but records may sometimes represent different species; widespread in England and Ireland where common in ponds.

A form with 3 pores, fo. *tripora* J.W. Lund 1942 (09130052) (Pl. 58G), has been described from the Richmond area, Surrey (Lund, 1942b). It might be identical with *C. triporus* B.Mack described from Austria.

Chrysococcus tessellatus F.E.Fritsch 1914

09130060 **Pl. 58H** (p. 225)

Lorica spherical, about 14 μm in diameter, with a single pore, anterior half of wall with net-like ornamentation.

British Isles (England); so far known only from a pond at Keston, Kent (Fritsch, 1914).

Chrysolepidomonas N.Peters *et* R.A.Andersen 1993

Cells and flagella covered with an armour of organic scales of two sorts: canistrate (flowerpot-like) scales and dendritic (branched) scales.

Three known species of which only one is found in the British Isles.

Chrysolepidomonas anglica N.Peters *et* R.A.Andersen 1993

Synonym: *Sphaleromantis tetragona* Skuja *sensu* K.Harris

09140010 **Pl. 61P** (p. 233)

Cells spherical, flattened, 6–9 μm wide, 6–6.5 μm long and 34 μm thick; flagella 2 per cell but only one apparent, the other short (bears photoreceptor) and closely adpressed to eyespot region of chloroplast and hence not visible.

British Isles (England); known only during early spring in ponds near Reading, Berkshire (Harris, 1963; Manton & Harris, 1966).

Chrysolykos B.Mack 1951

Cells free-living, solitary, surrounded by a curved, hyaline lorica with a projection on the convex side, biflagellate and with a single chloroplast and eyespot; stomatocyst at apex of lorica; sexual reproduction by hologamic copulation (vegetative cells act as gametes).

Widely distributed in the northern temperate and subarctic regions; planktonic. Five species known of which two occur in the British Isles.

1 Lorica sickle-shaped, with straight projection ....*C. planctonicus*

1 Lorica with both basal part and projection straight and spine-like...............................*C. skujae*

Chrysolykos planctonicus B.Mack 1951

09150010 **Pl. 58I** (p. 225)

Lorica sickle-shaped, 11–20 μm long, flattened, somewhat twisted, with a straight central projection from convex side; flagellum of same length as cell and second much shorter; stomatocyst 6.5–7.5 μm, zygote a little larger, with 2 empty loricae remaining attached.

Northern Europe, North America; known in the British Isles from ponds in England.

Chrysolykos skujae (Nauwerck) Bourrelly 1957

Basionym: *Diceras skujai* Nauwerck

Synonym: *Chrysolykos gracilis* J.W.G.Lund

09150020 **Pl. 58J** (p. 225)

Lorica slightly curved, 20–28 μm long, basally with 2 opposite, straight spine-like projections (one perhaps homologous with basal part of lorica in *C. planctonicus*); stomatocyst oblong, within a special lorica; zygote 7–8.5 μm wide, placed at apex of lorica.

Temperate and subarctic regions of Europe and America where it occurs in clear-water lakes and ponds; known in the British Isles from Wise E'en Tarn (Cumbria) (Lund, 1960b).

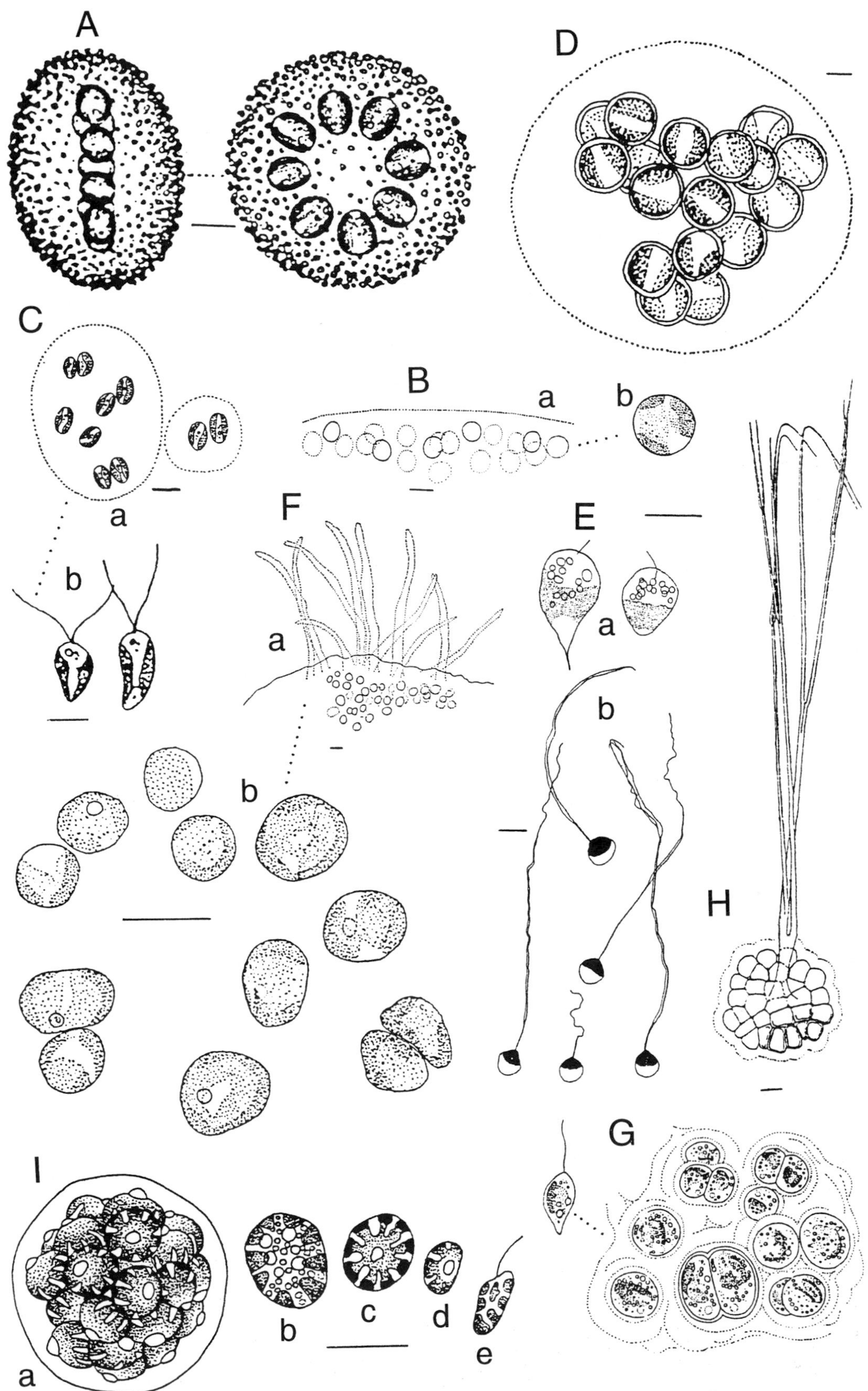

Plate 56 A–I

A. *Chrysostephanosphaera globulifera* (p. 222); **B**. *Chrysocapsa epiphytica* (p. 218): a. part of colony, b. details of a single cell; **C**. *Chrysocapsella paludosa* (p. 218): a. colonies, b. two swarmers; **D**. *Chrysocapsella planctonica* (p. 218); **E**. *Chrysonebula holmesii* (p. 239): a. motile cells, b. cells within a gelatinous matrix; **F**. *Dermatochrysis pseudofenestrata* (p. 222): a. edge of colony, b. details of single cells; **G**. *Gloeochrysis turfosa* (p. 231): colony and a swarmer; **H**. *Naegeliella flagellifera* (p. 232); **I**. *Phaeaster pascheri* (p. 234): a. colony, b–d. single cells each with a star-shaped chloroplast, e. a swarmer.

Scale bars: 10 μm.

Chrysosphaera Pascher emend. Bourrelly 1957

Cells solitary or aggregated, epiphytic, with a distinct wall; chloroplasts parietal, 1–8 per cell; contractile vacuoles 1–2; asexual reproduction by autospores or uniflagellate swarmers.

Europe, North America; epiphytic on other algae. Several species, only one reported from the British Isles.

Chrysosphaera paludosa (Korshikov) Bourrelly 1957

Basionym: *Phaeocapsa paludosa* Korshikov

09180010 **Pl. 62B** (p. 238)

Cells solitary or in 16-celled groups, each about 10 µm wide; chloroplast single, plate-shaped; reproduction by swarmers.

Central Europe; in the British Isles, so far known only from Llandegfan Pools in Wales where epiphytic on *Tribonema* filaments (Jane, 1945).

Chrysosphaerella Lauterborn 1896

Cells united basally to each other to form spherical colonies, covered with plate scales and spine scales of silica; flagella 2 per cell, one flagellum smooth and very short; chloroplasts 2, one with an eyespot; stomatocyst spherical.

Widely distributed mainly in temperate regions; planktonic. Six species, most of which are identified by characters associated with the scales and spines that are only observed with the electron microscope. Only one species known from the British Isles.

Chrysosphaerella brevispina Korshikov 1942

09190010 **Pl. 61N** (p. 233)

Colonies about 35 µm in diameter; spines long, with a double basal disc, up to 18 µm long; scales oval, about 4 × 3 µm, with a scalloped (crenulated) border structure just visible with the light microscope.

Widely distributed; in the British Isles reported from ponds in England and Scotland during the spring.

Chrysostephanosphaera Scherffel 1911

Cells united in flattened and disc-like colonies (30–100 µm in diameter), naked and somewhat amoeboid, with numerous fine rhizopodia extending from colony; contractile vacuoles 2 per cell; colony surrounded by numerous symbiotic, spherical bacteria; zoospores with one visible flagellum.

Probably cosmopolitan; planktonic, mainly in turf pits. Two species, one found in the British Isles.

Chrysostephanosphaera globulifera Scherffel 1911

09200010 **Pl. 56A** (p. 221)

Cells sometimes without rhizopodia, otherwise as above; chloroplasts 1–2 per cell; stomatocysts smooth.

North and Central Europe; reported in the British Isles from *Sphagnum* pools in Wales (Jane, 1945).

Cyclonexis A.Stokes 1886

Cells in ring-shaped colonies, consisting of 10–20 laterally joined cells, each with 2 unequal flagella; chloroplasts 1 or 2, without an eyespot; stomatocysts unknown.

Europe, North America; planktonic. Three species, all of which are recorded from the British Isles but their taxonomy is still unclear. A beautiful illustration is given in Canter-Lund & Lund (1995).

The colonies swim rapidly rotating. They are very fragile and difficult to examine.

1 Cells with 1 chloroplast; eyespot present or absent 2
1 Cells with 2 chloroplasts; eyespot present .... *C. annularis*
2 Dorsal side of cell undulate; eyespot present .... *C. uraliensis*
2 Dorsal side of cell straight; eyespot absent .... *C. erinus*

Cyclonexis annularis A.Stokes 1886

09210010 **Pl. 57A** (p. 223)

Cells have 2 chloroplasts and an eyespot.

Europe and North America; only a few dubious records from the British Isles (e.g. from Warwickshire; see Grove et al., 1920).

Cyclonexis erinus Jane 1940

09210020 **Pl. 57B** (p. 223)

Cells with a single chloroplast.

Europe and North America, in acid waters; in the British Isles known from various localities in England including a *Sphagnum* swamp in Hertfordshire (Jane, 1940), Wise E'enTarn in Cumbria (Lund, 1949a) and from the Reading area (Jane, 1945), as well as from the Bangor area of North Wales (Jane, 1945).

Cyclonexis uraliensis Pochmann 1957

09210030 **Pl. 57C** (p. 223)

Cells with single chloroplast and an eyespot, with dorsal side of cell undulated.

Russia and Germany; in the British Isles recorded only from a pond near North Moss Tarn in Lancashire (Lund, 1960b), but certainly more widely distributed.

Dermatochrysis Entwisle *et* R.A.Andersen 1990

Colonies single-layered, formed by cell division in 2 planes, upper surface bearing numerous gelatinous projections; contractile vacuoles present but no eyespot; swarmers and cysts not observed.

Northern hemisphere; gelatinous colonies in the periphyton. Three species known, one in the British Isles.

Dermatochrysis pseudofenestrata (J.W.G.Lund 1960) Entwisle *et* R.A.Andersen 1990

Basionym: *Tetrasporopsis pseudofenestrata* J.W.G.Lund

09220010 **Pl. 56F** (p. 221)

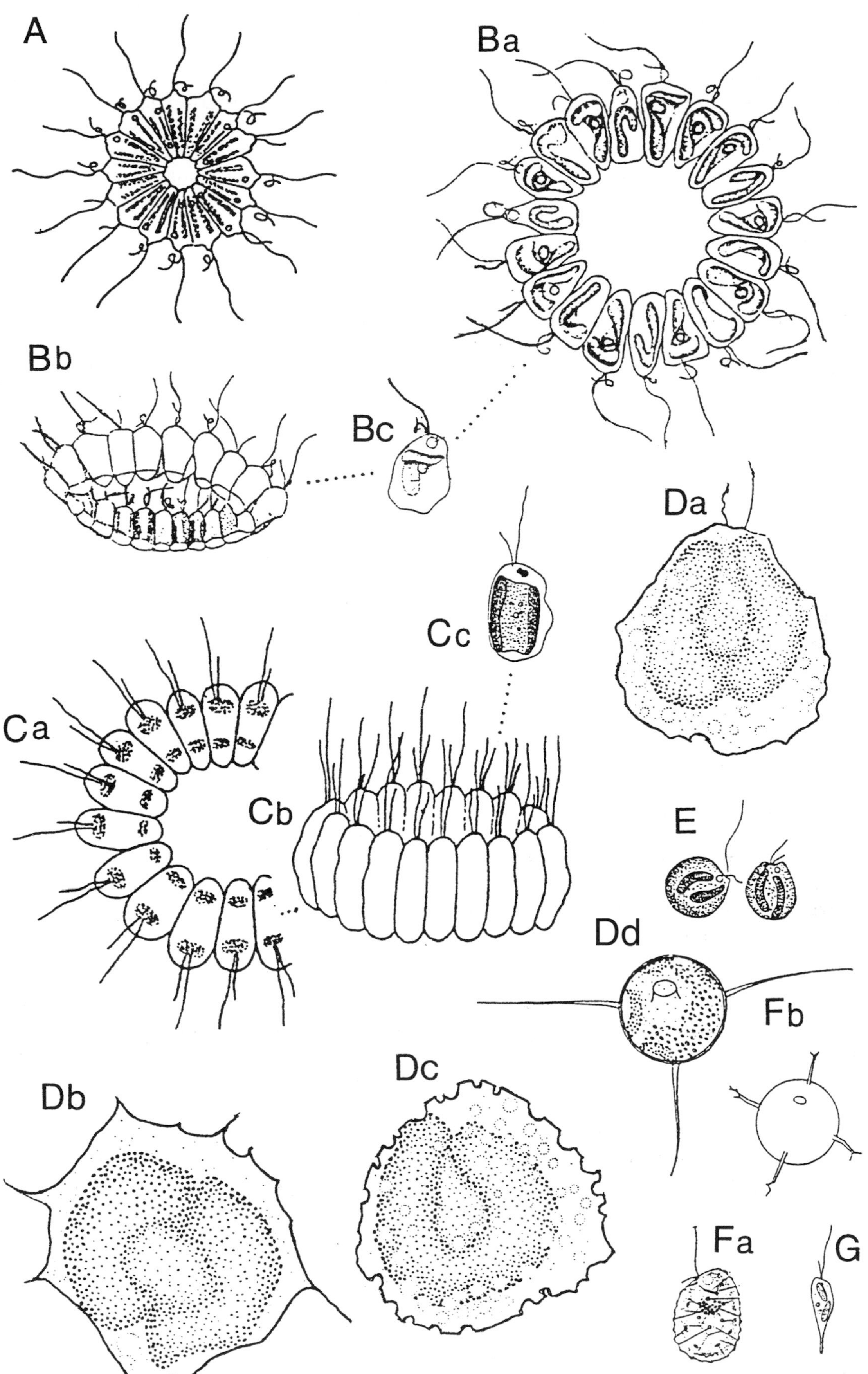

Plate 57 A–G
A. *Cyclonexis annularis* (p. 222); **B**. *Cyclonexis erinus* (p. 222): a. colony from edge, b. colony from side, c. single cell; **C**. *Cyclonexis uraliensis* (p. 222): a. colony from edge, b. colony from side, c. single cell; **D**. *Ochromonas ostreaeformis* (p. 232): a, b, c. different vegetative aspects, d. stomatocyst; **E**. *Ochromonas variabilis* (p. 234); **F**. *Ochromonas tuberculata* (p. 234): a. flagellate stage, b. stomatocyst; **G**. *Ochromonas pyriformis* (p. 232).
Scale bar: 10 μm.

Colonies up to 1.5 mm long; cells randomly arranged in a meshwork; gelatinous projections only visible with Indian Ink.

British Isles (England); known only from a small pool near Green Tarn, Claife Heights, Cumbria (Lund, 1960b).

Dinobryon Ehrenberg 1834

Cell surrounded by a lorica, attached with a stalk to its base, solitary or in colonies; colonies branched, after division a new cell settles on the edge of the mother cell lorica and constructs its own lorica; biflagellate; chloroplasts usually 2 per cell, with a large basal chrysolaminaran granule; eyespot present; stomatocysts on apex of lorica, surrounded by a special envelope; sexual reproduction in solitary forms by hologamy and in colonial form by male gametes from other colonies.

Worldwide; planktonic. Forty-one species of which 9 (not including varieties) are known from the British Isles.

1 Solitary 2
1 Colonial..4
2 Epiphytic on planktonic algae; lorica cone-shaped ....*D. calyciforme*
2 Planktonic; lorica of a different shape (except *D. sociale*) 3
3 Lorica hyaline, scalloped (crenulated) ....*D. crenulatum*
3 Lorica brown, with spiral thickening ...*D. suecicum*
4 Lorica cone-shaped ....*D. sociale*
4 Lorica of another shape 5
5 Lorica vase-shaped (convex lower part and concave upper part, or cylindrical, smooth-walled) 6
5 Lorica not vase-shaped, with a more or less undulate wall 7
6 Lorica vase-shaped, short and broad; colony dense; cyst with a short neck ....*D. sertularia*
6 Lorica vase-shaped to cylindrical, slender; colony loose; cyst with a long, curved neck ....*D. cylindricum*
7 Lorica cylindrical, with undulate walls and an irregular, oblique basal part ....*D. pediforme*
7 Lorica with a cone-shaped base and cylindrical upper part, more or less undulated walls................8
8 Lorica clearly separated into a basal conical part and an upper cylindrical part; colonies spreading ....*D. divergens*
8 Lorica with a long, stalk-like basal part and a long undulated upper part; colonies most often with parallel loricae..................*D. bavaricum*

Solitary species

Dinobryon calyciforme H.Bachmann 1907

09230020 **Pl. 58K** (p. 225)
Lorica cone-shaped, up to 30 µm long, almost fully occupied by cell.

Europe; reported in the British Isles from Scotland in the gelatinous envelope of the cyanophyte *Gomphosphaeria* (Bachmann, 1907).

Dinobryon crenulatum West *et* G.S.West 1909

09230030 **Pl. 58L** (p. 225)
Lorica almost cylindrical, with a basal spine, 8–10 µm wide, 30 µm long, walls scalloped and with a basal spine, hyaline; chloroplasts parietal, 2 per cell; stomatocysts in a special funnel-shaped envelope at mouth of old lorica. Few-celled colonies may occur.

Europe; Denmark, known in the British Isles from the English Lake District (W. & G.S.West, 1909) and Scotland (Brook, 1955a).

Dinobryon suecicum Lemmermann 1904

09230090 **Pl. 58M** (p. 225)
Lorica vase-shaped, 4–5 µm wide, 32–43 µm long, brownish, with a conspicuous spiral thickening, basally terminating in a spine, 5–15 µm long; stomatocyst at lorica mouth; zygote with 2 empty loricae remaining attached.

Northern Europe, North America; in the British Isles reported only from a tarn in Lancashire (Lund, 1952).

Colonial species

Dinobryon bavaricum O.E.Imhof 1890

09230010 **Pl. 58N** (p. 225)
Loricae very prolonged, 6–10 µm wide, 50–120 µm long, lying parallel in often few-celled colonies, with a long stalk-like base and an upper cylindrical to slightly funnel-shaped part with undulate walls; stomatocysts about 11 µm in diameter, with a short and backwardly directed neck.

Temperate and subarctic parts of the northern hemisphere, known from nutrient-poor to moderately nutrient-rich ponds and lakes all around the British Isles (West, 1904; West & G.S.West, 1905b, Teiling, 1916, Griffiths, 1926; Round & Brook, 1959; Brook, 1964; Williams, 1965; Kristiansen, 1998).

var. *vanhoeffenii* (H.Bachmann) Willi Krieger 1930
Basionym: *Dinobryon vanhoeffenii* H.Bachmann
09230012 **Pl. 58O** (p. 225)
Differs from type by loricae having an almost 90° bend so colony has a very spreading appearance.

Temperate, subarctic and Arctic parts of the northern hemisphere (not common); known in the British Isles only from Irish lochs (Round & Brook, 1959).

Dinobryon cylindricum O.E.Imhof 1887

09230040 **Pl. 58P** (p. 225) **CD**
Lorica almost cylindrical, with a somewhat trumpet-shaped mouth, 40–115 µm long, arranged in rather open, few-celled colonies; stomatocyst spherical, with a forwardly directed hooked neck.

Widely distributed in Northern Europe; known in the British Isles from nutrient-poor and moderately nutrient-rich lakes and ponds, especially in Scotland.

var. *alpinum* (O.E.Imhof) H.Bachmann 1911
Basionym: *Dinobryon sertularia* Ehrenberg var. *alpinum* O.E.Imhof
09230042 **Pl. 58Q** (p. 225)

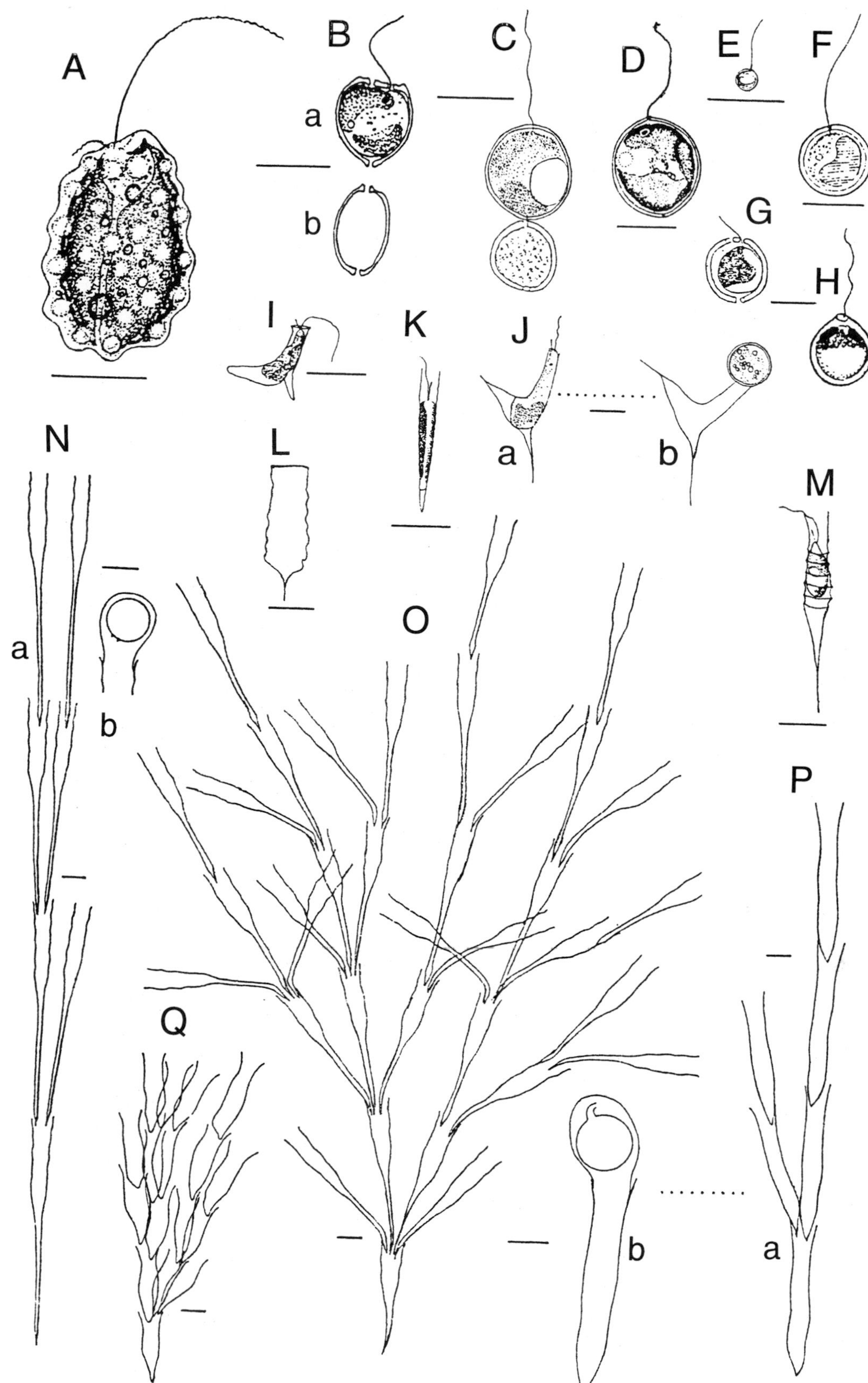

Plate 58 A–Q

A. *Ochromonas verrucosa* (p. 234); **B**. *Chrysococcus cordiformis* (p. 218): a. side view of cell, b. edge view; **C**. *Chrysococcus cystophorous* fo. *astigmatus* (p. 220); **D**. *Chrysococcus diaphanus* (p. 220); **E**. *Chrysococcus punctiformis* (p. 220); **F**. *Chrysococcus rufescens* (p. 220); **G**. *Chrysococcus rufescens* fo. *tripora* (p. 220); **H**. *Chrysococcus tessellatus* (p. 220); **I**. *Chrysolykos planktonicus* (p. 220); **J**. *Chrysolykos skujae* (p. 220): a. vegetative cell, b. zygote; **K**. *Dinobryon calyciforme* (p. 224); **L**. *Dinobryon crenulatum* (p. 224); **M**. *Dinobryon suecicum* (p. 224); **N**. *Dinobryon bavaricum* (p. 224): a. colony, b. stomatocyst; **O**. *Dinobryon bavaricum* var. *vanhoeffenii* (p. 224); **P**. *Dinobryon cylindricum* (p. 224): a. colony, b. stomatocyst; **Q**. *Dinobryon cylindricum* var. *alpinum* (p. 224). Scale bars: 10 μm.

Lorica shorter than type, 40–64 μm long, not cylindrical but more vase-shaped; colonies many-celled and rather dense.

Northern Europe in nutrient-poor to moderately nutrient-rich lakes; so far reported in the British Isles from ponds and lakes in Scotland (Brook, 1959c) and Ireland (Round & Brook, 1959; Kristiansen, 1998).

var. *palustre* Lemmermann 1900
09230043 **Pl. 59A** (p. 227)
Lorica very long, cylindrical (as in var. *cylindricum*), 80–100 μm long, but because of oblique base the large colonies are conspicuously divergent.

Northern Europe and Northern Asia where characteristic of nutrient-poor lakes; known in the British Isles from lakes in Ireland (Round & Brook, 1959; Kristiansen, pers. obs.) and Scotland (Brook, 1964).

Dinobryon divergens O.E.Imhof 1887

09230050 **Pl. 59B** (p. 227) **CD**
Lorica consists of a lower conical and an upper cylindrical part, transition between the two parts broad and marked by 1–2 undulations; colonies spreading in appearance; stomatocysts with a short, straight, backwardly directed neck.

Widely distributed in the northern hemisphere; reported in moderately to very nutrient-rich ponds and lakes throughout the British Isles.

var. *schauinslandii* (Lemmermann) Brunnthaler 1901
Basionym: *Dinobryon schauinslandii* Lemmermann
09230052 **Pl. 59C** (p. 227)
Differs from type variety in having undulated lorica walls, both in the transition region and the upper part.

Distributed as var. *divergens* but less common, reported from lochs in Scotland and Ireland (Brook 1959c; Round & Brook, 1959).

Dinobryon pediforme (Lemmermann) Steinecke 1916

Basionym: *Dinobryon protuberans* Lemmermann var. *pediforme* Lemmermann
09230060 **Pl. 59D** (p. 227)
Lorica cylindrical, 8–9 μm wide and 35–45 μm long, with an asymmetrical base and 1–2 lateral processes; stomatocyst spherical, without neck and within an envelope not opening towards lorica mouth.

Temperate to subarctic regions in humic acid water; reported in the British Isles only from Scotland (Teiling, 1916).

Dinobryon sertularia Ehrenberg 1838

Synonym: *Dinobryon sertularia* var. *protuberans* (Lemmermann) Willi Krieger
09230070 **Pl. 59E** (p. 227)
Lorica rather short and broad, 10–14 μm wide and 30–40 μm long, often with an oblique basal outgrowth; colonies very dense and bushy; stomatocysts rare, spherical, with a short neck.

Probably cosmopolitan since recorded from most of the world, mainly from very nutrient-rich ponds (not all reports are reliable); widely distributed in the British Isles with many reports from England, Scotland and Ireland.

Dinobryon sociale Ehrenberg 1838

09230080
Lorica more or less conical, often more so towards the apex of colony where longest, 8–10 μm wide, 30–70 μm long.

var. *americanum* (Brunnthaler) H.Bachmann 1911
Basionym: *Dinobryon stipitatum* F.Stein var. *americanum* Brunnthaler
09230082 **Pl. 59F** (p. 227)
Differs from type variety in having the lorica divided into a lower stalk-like and an upper cylindrical part.

var. *stipitatum* (F.Stein) Lemmermann 1903
Basionym: *Dinobryon stipitatum* F.Stein
09230083 **Pl. 59G** (p. 227)
Differs from type in its very marked conical shape.

All varieties in northern temperate regions in moderately to very nutrient-rich ponds and lakes; reported in the British Isles from ponds and lakes in England, Scotland and Ireland (West, 1904; Grove et al., 1920; Williams, 1965; Brook, 1964; Round & Brook, 1959; Kristiansen, pers. obs.).

Epipyxis Ehrenberg 1838

Loricae solitary, clustered or forming colonies with loricae mostly attached to outer side of parent lorica, constructed of overlapping scales (often only visible using phase contrast or after staining with methylene blue); stomatocysts surrounded by a mucilaginous layer sometimes containing scales, located inside lorica, within lorica mouth or externally on lorica.

Widely distributed in the northern hemisphere but distribution not well-known; epiphytic on other algae, solitary, clustered or forming colonies. Thirty-nine species, 16 of these reported from the British Isles. Frequently encountered in the British Isles but seldom identified to species.

1 Lorica smooth (as viewed with the light microscope) 2
1 Lorica not smooth, with flaring scales 9
2 Lorica base acute and spindle-shaped 3
2 Lorica base not acute and of another shape 5
3 Lorica base with disc-like holdfast; cyst with a complicated plug .... *E. utriculus*
3 Lorica without a disc-like holdfast; cyst with a simple plug 4
4 Scales few, regularly arranged .... *E. utriculus* var. *acuta*
4 Scales numerous, in irregular rows .... *E. tabellariae*
5 Lorica cylindrical with a rounded base 6
5 Lorica not cylindrical and of another form basally 8
6 Lorica short (13–24 μm); cyst external on basal part of lorica .... *E. condensata*
6 Lorica long (24–82 μm); cyst otherwise 7
7 Lorica mouth divergent .... *E. tubulosa*
7 Lorica mouth not divergent .... *E. turgida*

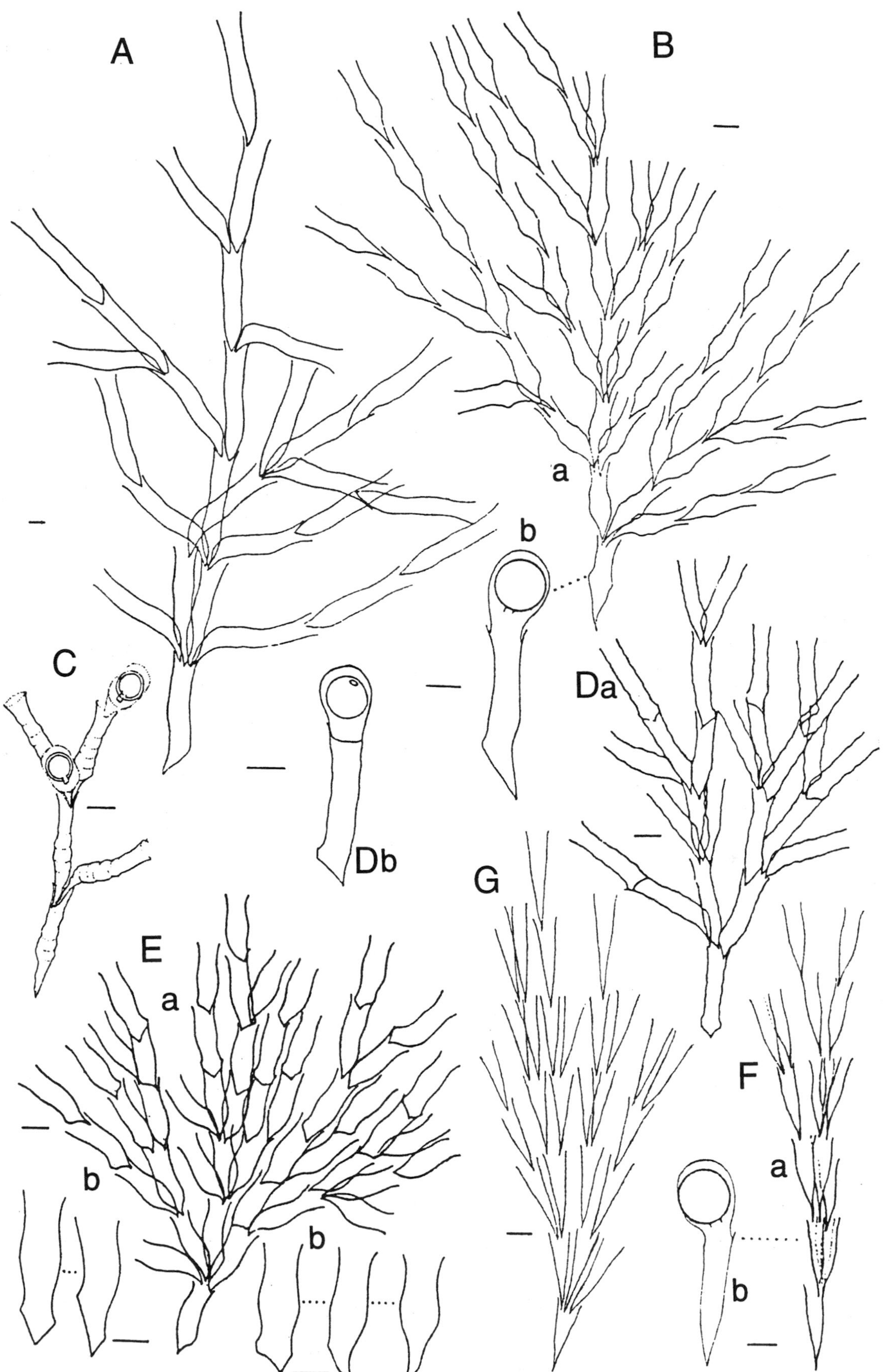

Plate 59 A–G

A. *Dinobryon cylindricum* var. *palustre* (p. 226); **B**. *Dinobryon divergens* (p. 226): a. colony, b. stomatocyst; **C**. *Dinobryon divergens* var. *schauinslandii* (p. 226); **D**. *Dinobryon pediforme* (p. 226): a. colony, b. stomatocyst; **E**. *Dinobryon sertularia* (p. 226): a. colony, b. stomatocyst; **F**. *Dinobryon sociale* var. *americanum* (p. 226): a. colony, b. stomatocyst; **G**. *Dinobryon sociale* var. *stipitatum* (p. 226).

Scale bars: 10 μm.

8 Lorica conical ....*E. proteus*
8 Lorica vase-shaped...........................*E. kodiakensis*
9 Lorica with flaring scales apically, mouth constricted by inwardly flaring scales ....*E. diplostomum*
9 Lorica with flaring scales at base or all over, mouth not constricted 10
10 Scales flaring at base, or mostly at base................11
10 Scales flaring at base and mouth, or all over.........15
11 Lorica spindle-shaped, with a long stalk ....*E. lauterbornii* var. *mucicola*
11 Lorica not spindle-shaped, without a long stalk 12
12 Lorica regularly cylindrical; cyst basal in lorica 13
12 Lorica not regularly cylindrical; cyst otherwise 14
13 Lorica with acute base, fastened to substratum by cytoplasmic thread through lorica base; scales in oblique rows of about 2, or alternate.......................................*E. polymorpha*
13 Lorica base subacute, with no cytoplasmic threads; scales in oblique rows of 3–4.......*E. gracilis*
14 Lorica short (21–28 μm), cylindrical to bell-shaped, with subacute base; scales in rows of 2; basal scales sigmoid in optical section ....*E. singoalikensis*
14 Lorica long (36–45 μm), cylindrical to attenuated; scales pair-wise, overlapping and in rows of 2–3; cyst in lorica mouth ....*E. sinuosa*
15 Scales flaring at lorica base and mouth, those at mouth flaring inwards; cyst within lorica..*E. kenaiensis*
15 Scales flaring all over lorica 16
16 Lorica short (16–17 μm)...........................*E. anglica*
16 Lorica long (25–72 μm) 17
17 Lorica 25–50 μm, broadest at mouth ....*E. lauterbornii*
17 Lorica 42–72 μm, broadest in middle.......*E. ramosa*

Epipyxis anglica D.K.Hilliard 1966

09250010 **Pl. 60A** (p. 229)

Lorica shortly cylindrical with attenuated acute base, 7–8 μm wide, 16–17.4 μm long, consisting of overlapping (imbricate) scales arranged in a spiral pattern, slightly flaring; stomatocyst unknown.

British Isles (England); solitary as epiphytes on diatoms in swampy pools in Lancashire (Hilliard, 1966).

Epipyxis condensata (B.Mack) D.K.Hilliard *et* Asmund 1963

Basionym: *Dinobryon condensatum* B.Mack
09250020 **Pl. 60B** (p. 229)

Lorica short and cylindrical, 4.8–6.5 μm wide,13–24.4 μm long, with rounded base and scales overlapping, hardly flaring and if so then at mouth; stomatocysts spherical, 7.2–8 μm wide, with no collar around the pore, surrounded by loosely clustered scales.

Northern Europe, North America; in the British Isles known only from swampy pools in Lancashire (Hilliard, 1966).

Epipyxis diplostoma (Jane) D.K.Hilliard *et* Asmund 1963

Basionym: *Hyalobryon diplostoma* Jane
09250030 **Pl. 60C** (p. 229)

Lorica elongate and slightly curved, scales slightly flaring and some directed inwards at mouth, brown-coloured by iron salts; stomatocyst ovoid, with a pore at broad end, located in the swollen lorica mouth.

British Isles; known only growing solitary or in clusters on filamentous algae in temporary pools in Wales (Jane, 1945).

Epipyxis gracilis D.K.Hilliard *et* Asmund 1963

09250040 **Pl. 60D** (p. 229)

Lorica cylindrical with subacute base, 5.6–6.1 μm wide, 18.3–21.8 μm long, basal part with slightly flaring scales; scales all in oblique rows of 3–4; stomatocysts located in lorica base.

North America, northern Europe, northern Asia; in the British Isles so far reported only from swampy pools in Lancashire (Hilliard, 1966).

Epipyxis kenaiensis D.K.Hilliard *et* Asmund 1963

09250050 **Pl. 60E** (p. 229)

Lorica cylindrical to bell-shaped with subacute base, 4.4–5.7 μm wide, 26.2–29.6 μm long, 7–9 rows of scales, each row of about 4 scales, flaring at base and again recurved, apical scales moderately flaring, attached small scales curving inwards and constricting lorica mouth; stomatocyst lodged inside lorica.

Alaska; in the British Isles reported only as an epiphyte on *Oedogonium* and *Dinobryon* growing in pools in Lancashire (Hilliard, 1966).

Epipyxis kodiakensis D.K.Hilliard 1959

09250060 **Pl. 60F** (p. 229)

Lorica vase-shaped, 6.5–8.7 μm wide, 26.2–34.8 μm long, with attenuated oblique base; scales circular, not flaring; stomatocysts unknown.

Alaska; in the British Isles reported only from swampy pools in Lancashire (Hilliard, 1966).

Epipyxis lauterbornii (Lemmermann) D.K.Hilliard *et* Asmund 1963

Basionym: *Hyalobryon lauterbornii* Lemmermann
09250070 **Pl. 60G** (p. 229)

Lorica with broadened mouth and attenuated base, 7–8 μm wide, 25–50 μm long, with scales flaring especially at mouth; stomatocysts unknown.

Northern Europe; known in the British Isles as an epiphyte on plankton algae in Wales and Scotland (West & G.S.West, 1905b; Bachmann, 1907; Jane, 1945; Woodhead & Tweed, 1955).

var. *mucicola* (Lemmermann) D.K.Hilliard *et* Asmund 1963
Basionym: *Hyalobryon mucicola* Lemmermann
09250072 **Pl. 60H** (p. 229)

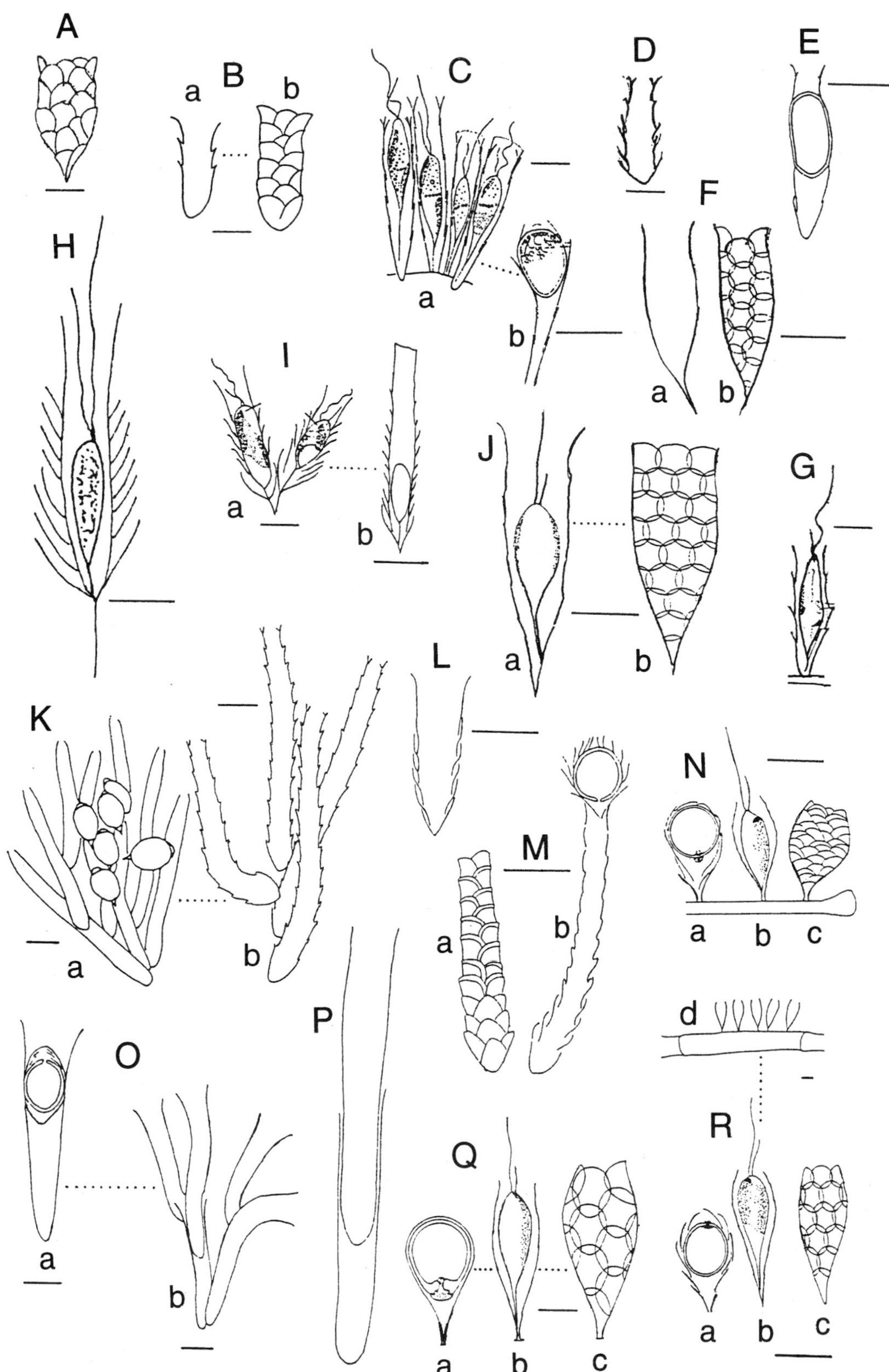

Plate 60 A–R

A. *Epipyxis anglica* (p. 228); **B**. *Epipyxis condensata* (p. 228): a. lorica, b. scale arrangement; **C**. *Epipyxis diplostoma* (p. 228): a. vegetative cells, b. stomatocyst; **D**. *Epipyxis gracilis* (p. 228); **E**. *Epipyxis kenaiensis* (p. 228): a stomatocyst; **F**. *Epipyxis kodiakensis* (p. 228): a. lorica shape, b. scale arrangement; **G**. *Epipyxis lauterbornii* (p. 228); **H**. *Epipyxis lauterbornii* var. *mucicola* (p. 228); **I**. *Epipyxis polymorpha* (p. 230): a. vegetative cells, b. stomatocyst; **J**. *Epipyxis proteus* (p. 230): a. lorica shape, b. stomatocyst; **K**. *Epipyxis ramosa* (p. 230): a. colony with stomatocysts, b. colony without stomatocysts; **L**. *Epipyxis singoalikensis* (p. 230); **M**. *Epipyxis sinuosa* (p. 230): a. scale arrangement, b. lorica with stomatocyst; **N**. *Epipyxis tabellariae* (p. 230): a. stomatocyst, b. vegetative cell, c. scale arrangement; **O**. *Epipyxis tubulosa* (p. 230): a. single cell, b. colony; **P**. *Epipyxis turgida* (p. 230); **Q**. *Epipyxis utriculus* (p. 230): a. stomatocyst, b. vegetative cell, c. scale arrangement; **R**. *Epipyxis utriculus* var. *acuta* (p. 231): a. stomatocyst, b. vegetative cell, c. scale arrangement, d. envelopes on an algal filament.

Scale bars: 10 μm.

Lorica spindle-shaped, 4–8 μm wide, 23–50 μm long, basally with a long stalk (4–16 μm long); scales flaring except in upper third; stomatocysts unknown.

Northern Europe; in the British Isles known as an epiphyte on the blue-green alga *Gomphosphaeria* and other plankton algae in Loch Earn, Scotland (Bachmann, 1907; West & Fritsch, 1927).

Epipyxis polymorpha (J.W.G.Lund) D.K.Hilliard *et* Asmund 1963

Basionym: *Hyalobryon polymorphum* J.W.G.Lund
09250080 **Pl. 60I** (p. 229)
Lorica almost cylindrical, broadest at its base, 3.5–5.5 μm wide, 20–38 μm long, attached to substratum by a thin cytoplasmic thread passing through the acute base of the lorica, with scales flaring and basal ones most flared; stomatocysts spherical to oblong, 7–12 × 5–8.5 μm, sometimes with a short collar, located basally within lorica.

North America and Europe; reported in the British Isles growing solitary or as colonies on plankton algae (e.g. *Gomphosphaeria*, *Dinobryon*, inside *Uroglena* colonies) in tarns in Lancashire (Lund, 1953).

Epipyxis proteus (Wislouch) D.K.Hilliard *et* Asmund 1963

Basionym: *Dinobryon proteus* Wislouch
09250090 **Pl. 60J** (p. 229)
Lorica almost conical, attenuated towards base, 9.6–14 μm wide, 32–40 μm long, with scales circular, overlapping, in regular rows and not flaring; stomatocysts unknown.

Russia and Alaska; in the British Isles reported only growing solitary and epiphytic on *Oedogonium* and diatoms in lakes in Lancashire (Hilliard, 1966).

Epipyxis ramosa (Lauterborn) D.K.Hilliard *et* Asmund 1963

Basionym: *Hyalobryon ramosum* Lauterborn
09250100 **Pl. 60K** (p. 229)
Lorica very long, broadest at middle, somewhat curved, with a rounded base, 3.5–6.5 μm wide, 42–73 μm long, with scales oblong, numerous and overlapping; stomatocyst ovoid, attached externally on the lorica wall.

Northern parts of North America and Europe; in the British Isles reported from swampy areas in Scotland and England (Lancashire) in bushy colonies sometimes consisting of hundreds of cells (Hilliard, 1966).

Epipyxis singoalikensis D.K.Hilliard *et* Asmund 1963

09250110 **Pl. 60L** (p. 229)
Lorica cylindrical to bell-shaped with an attenuated base, 7–8 μm wide and 21–28 μm long, scales overlapping and in 4–5 transverse rows, each row of 2 scales, with scales in posterior half flaring and S-shaped in optical section; stomatocysts unknown.

America, Europe; in the British Isles epiphytic, solitary, on coccoid cyanophytes in swampy areas in Lancashire (Hilliard 1966).

Epipyxis sinuosa D.K.Hilliard 1966

09250120 **Pl. 60 M** (p. 229)
Lorica very long, often curved, 3.5–6.5 μm wide, 32.6–45.7 μm long, with scales overlapping and in rows of 2–3, flaring in posterior half and S-shaped in optical section, almost covering each other in pairs; stomatocysts located in lorica mouth, covered with scales.

British Isles (England); solitary or in pairs on diatoms in swamps in Lancashire (Hilliard, 1966).

Epipyxis tabellariae (Lemmermann) G.M.Smith 1950

Basionym: *Dinobryon utriculus* Ehrenberg var. *tabellariae* Lemmermann
09250160 **Pl. 60N** (p. 229)
Lorica short and broad, 6.5–9 μm wide, 15.5–22.6 μm long, base attenuated into a stalk and mouth a little narrowed, almost smooth, with scales overlapping, oblong, not flaring and in several irregular transverse rows; stomatocyst at mouth of lorica, surrounded by somewhat flaring scales.

Northern parts of Europe and America; in the British Isles only known solitary or in clusters on plankton algae (e.g. *Asterionella formosa*) in a loch in Perthshire (Brook, 1957).

Epipyxis tubulosa (B.Mack) D.K.Hilliard *et* Asmund 1963

Basionym: *Dinobryon tubulosum* B.Mack
09250130 **Pl. 60O** (p. 229)
Lorica cylindrical with broader mouth and rounded base, 4–7.6 μm wide, 26–76 μm long, smooth and covered by transverse rows of overlapping, oval scales; stomatocyst located at lorica mouth, surrounded by scales; large colonies sometimes with loricae fastened to mother lorica externally or internally, often in long series.

Alaska, Northern Europe; in the British Isles known from swampy areas in Lancashire (Hilliard, 1966).

Epipyxis turgida D.K.Hilliard 1966

09250140 **Pl. 60P** (p. 229)
Lorica very long, 6.5–8.7 μm wide, 23–34 μm long, straight, cylindrical, with a rounded base, smooth, composed of 12–23 rows of overlapping, pointed scales; stomatocysts unknown; solitary or in colonies of up to 50 cells with loricae attached to either inner or outer side of mother lorica.

British Isles (England); known only from swamps in Lancashire (Hilliard 1966).

Epipyxis utriculus Ehrenberg 1838

09250150 **Pl. 60Q** (p. 229) **CD**
Lorica spindle-shaped, apically somewhat constricted and basally attenuated, 5–6 μm wide, 24.8–28.2 μm long, attached to substratum by a small disc-shaped holdfast, smooth, regular pattern of oblong scales only visible by staining, with scales overlapping each other

on all sides; stomatocyst located in upper part of lorica, with pore directed downwards and provided with a plug of complicated shape.

Widely distributed in Europe and North America where solitary or clustered on filamentous algae and widespread in ponds and swamps; in the British Isles known only from Lancashire (Hilliard, 1966).

var. *acuta* (J.Schiller) D.K.Hilliard *et* Asmund 1963

Basionym: *Dinobryon utriculus* Ehrenberg var. *acutum* J.Schiller

09250152 **Pl. 60R** (p. 229)

Similar to type variety but somewhat smaller (7.8 μm wide, 19.6–21.8 μm long), attenuated base without holdfast; stomatocyst with a simple plug.

Widespread in Europe and North America; in the British Isles known on filamentous algae in ponds and swamps in Scotland (Brook, 1959c), Wales (Woodhead & Tweed, 1955) and England (Hilliard, 1966).

Gloeochrysis Pascher 1925

Cells forming flat colonies embedded in a layered mucilage, with cell division in 2 planes; chloroplasts 1–2 per cell; swarmers uniflagellate.

Europe; on stones in streams or on peaty soil. Five species, only one recorded from the British Isles; an unidentified species has been mentioned by Lund (1942a).

Gloeochrysis turfosa (Pascher) Bourrelly 1957

Basionym: *Geochrysis turfosa* Pascher

09260010 **Pl. 56G** (p. 221)

Forms paper-like sheets on drying turfy soil; on rotting leaves and other surfaces in highly acidic streams.

Central Europe; in the British Isles so far known only from highly acidic streams in Northern England (Hargreaves et al., 1975); forms from acid streams require critical study.

Kephyrion Pascher 1911 and *Pseudokephyrion* Pascher emend. G.Schmid 1934

Synonym: *Stenocalyx* (of *Kephyrion*)

Small solitary flagellates, with a transparent or brownish lorica; chloroplasts 1–2 per cell, sometimes with an eyespot. Difference between the genera is only in the number of visible flagella (1 or 2). There are 2 flagella in both, but in *Kephyrion* the short flagellum is so reduced that it cannot be seen. As a result, the genera cannot be satisfactorily delimited from one another and so are treated together. Sexual reproduction has been observed in many species; the 2 loricae remain attached to the zygote.

Scattered records mainly in Europe, but their distribution is largely unknown; usually in the plankton of ponds, especially in spring. Sixty-two species of *Kephyrion* and 52 species of *Pseudokephyrion* have been described of which only 8 are reported so far from the British Isles.

The generic concept is unclear, and several of the species have also been included in the genera *Stenocalyx* and *Kephyriopsis*.

1 Lorica smooth-walled 2
1 Lorica with annular or spiral ridges or constrictions 7
2 Lorica oblong, with equatorial bulge ....*K. rubri-claustri*
2 Lorica ovoid to ellipsoidal, without a bulge 3
3 Lorica with apical wall thickenings ....*K. littorale*
3 Lorica without wall thickenings 4
4 Lorica cylindrical, with an apical constriction ....*P. ruttneri*
4 Lorica ovoid, ellipsoidal or barrel-shaped..............5
5 Lorica ellipsoidal ....*P. ellipsoideum*
5 Lorica ovoid or barrel-shaped 6
6 Lorica ovoid, with walls not undulate ....*P. ovum*
6 Lorica barrel-shaped, with undulate walls ....*P. undulatum*
7 Lorica with a ring-like collar...........*K. moniliferum*
7 Lorica with a helical thickening............................8
8 Lorica almost cylindrical, broadest basally ....*P. klarnetii*
8 Lorica broadest in middle..............*P. pseudospirale*

Kephyrion littorale J.W.G.Lund 1942

09290010 **Pl. 61A** (p. 233)

Lorica ovoid to bowl-shaped, with a thickening around mouth, 4.8–5.1 μm wide and 6.0–6.8 μm long, brownish; single flagellum, visible and inserted laterally; chloroplasts 2 per cell, one with an eyespot.

var. *constricta* J.W.G.Lund 1942

09290012 **Pl. 61B** (p. 233)

Similar to type but with a more rounded lorica with a flattened base, equatorially slightly constricted.

British Isles, reported only from England (Lund,1942a), but may be identical with *P. ovum* (see below).

Kephyrion moniliferum (G.Schmid) Bourrelly 1957

Basionym: *Stenocalyx monilifera* G.Schmid

09290020 **Pl. 61C** (p. 233)

Lorica with a characteristic undulated collar around mouth; flagellum one visible; chloroplast one per cell; sexual reproduction known.

Central and Northern Europe; reported in the British Isles from a pond in Richmond Park, Surrey (Lund, 1942a).

Kephyrion rubri-claustri W.Conrad 1939

09290030 **Pl. 61D** (p. 233)

Lorica barrel-shaped with an equatorial bulge, 5–7× 5 μm; single flagellum; chloroplast parietal, one per cell.

Central and Northern Europe; reported only from ponds in the English Lake District (Scourfield, 1930).

Pseudokephyrion ellipsoideum (Pascher) G.Schmid 1934

Basionym: *Kephyrion ellipsoideum* Pascher

09430020 **Pl. 61E** (p. 233)

Lorica ellipsoidal, 4–8 × 4–8 μm; flagella 2, visible.

Central and Northern Europe; in the British Isles known from the English Lake District (Lund, 1960b).

Pseudokephyrion klarnetii Bourrelly 1957

09430030 **Pl. 61F** (p. 233)
Lorica almost cylindrical, broadest basally, 3.5–4 μm wide and 7–8 μm long, with a delicate helical thickening, hyaline; flagella 2; chloroplast 1 per cell, with an eyespot; zygotes have been observed.

Central and Northern Europe; known in the British Isles from ponds in the English Lake District (Scourfield, 1930; Lund, 1960b).

Pseudokephyrion ovum (Pascher *et* Ruttner) G.Schmid 1934

Basionym: *Kephyrion ovum* Pascher *et* Ruttner
09430040 **Pl. 61G** (p. 233)
Lorica rather similar to that of *Kephyrion littorale*, only thickening around lorica mouth mostly absent.

Central Europe; reported in the British Isles only from the English Lake District (Lund, 1960b).

It is possible that *Kephrion littorale* should be included in this species; the second flagellum may have been overlooked.

Pseudokephyrion pseudospirale Bourrelly 1957

09430050 **Pl. 61H** (p. 233)
Lorica cylindrical to somewhat ovoid, broadest at middle or near base, about 5.5 μm wide, 6–8 μm long, with a helical thickening making 3 turns, brown in colour; chloroplast single; flagella 2, unequal, the short one very difficult to detect (compare the uniflagellate *Kephyrion spirale*); sexual reproduction known.

Northern Europe; known in the British Isles from ponds in the English Lake District and a lake ('The Serpentine') in Eaton Park, Chester (Williams, 1965).

Pseudokephyrion ruttneri (J.Schiller) W.Conrad 1939

Basionym: *Kephyrion ruttneri* J.Schiller
09430060 **Pl. 61I** (p. 233)
Lorica cylindrical with an apical constriction, 5–6 μm wide, 7–8 μm long, hyaline; flagella 2 per cell.

Central and Northern Europe; reported in the British Isles from a lake ('The Serpentine') in Eaton Park, Chester (Williams, 1965).

Pseudokephyrion undulatum (G.A.Klebs) Pascher 1913

Basionym: *Dinobryon undulatum* G.A.Klebs
09430080 **Pl. 61J** (p. 233)
Lorica ovoid to barrel-shaped, undulate because of 2–4 transverse constrictions, 18–25 μm long; brownish; chloroplasts parietal, 2 per cell, one with eyespot.

Central Europe; reported in the British Isles from Scotland (Brook, 1955a).

Naegeliella Correns 1883

Cells forming flat colonies, each cell bearing a long and sometimes branched gelatinous bristle or hair; reproduction by biflagellate swarmers.

Central Europe, British Isles; epiphytic on water plants. Monospecific.

Naegeliella flagellifera Correns 1883

Synonyms: *Naegeliella britannica* Godward, *Chrysochaete britannica* Rosenberg
09340010 **Pl. 56H** (p. 221)
Description as for genus.

Distribution as for genus; so far reported from the British Isles as an epiphyte along shores of ponds and lakes in Epping Forest (Godward, 1933) and the English Lake District (Rosenberg, 1941).

Ochromonas Vysotskij 1887

Cells solitary, naked, with 2 unequal flagella; chloroplasts 1–3 per cell and eyespot present in some species; stomatocysts with spines.

Probably cosmopolitan; planktonic, some species may be sessile for short periods. About 100 species, most of them very difficult to identify, only 5 species recorded from the British Isles.

Ochromonas malhamensis has been transferred to the genus *Poteriochromonas*.

1 Cells large and flattened................*O. ostreaeformis*
1 Cells small, ovoid to pear-shaped..........................2
2 Cell surface with conspicuous warts 3
2 Cell surface smooth 4
3 Cells 12–18 μm long.........................*O. tuberculata*
3 Cells 19–30 μm long ....*O. verrucosa*
4 Cells pear-shaped...............................*O. pyriformis*
4 Cells ovoid-globular............................*O. variabilis*

Ochromonas ostreaeformis Swale *et* J.H.Belcher 1966

09350010 **Pl. 57D** (p. 223)
Cells disc-like, 25–50 × 25–50 μm, with angular apex; muciferous bodies within cell membrane; chloroplast single, large and flattened, without a pyrenoid and eyespot; contractile vacuoles 2, apical in position; rhizopodial stages known; stomatocysts with very long unbranched spines.

Denmark (Kristiansen, unpublished); in the British Isles only recorded during spring from the plankton of Priest Pot in the English Lake District (Swale & Belcher, 1966).

Ochromonas pyriformis Matvienko 1938

09350020 **Pl. 57G** (p. 223)
Cells pear-shaped, attached to substratum by tail, 4.5–7.5 μm wide,18 μm long; chloroplasts 2, one with an eyespot; contractile vacuoles 2 per cell.

Ukraine; recorded from the British Isles by Pentecost (1984).

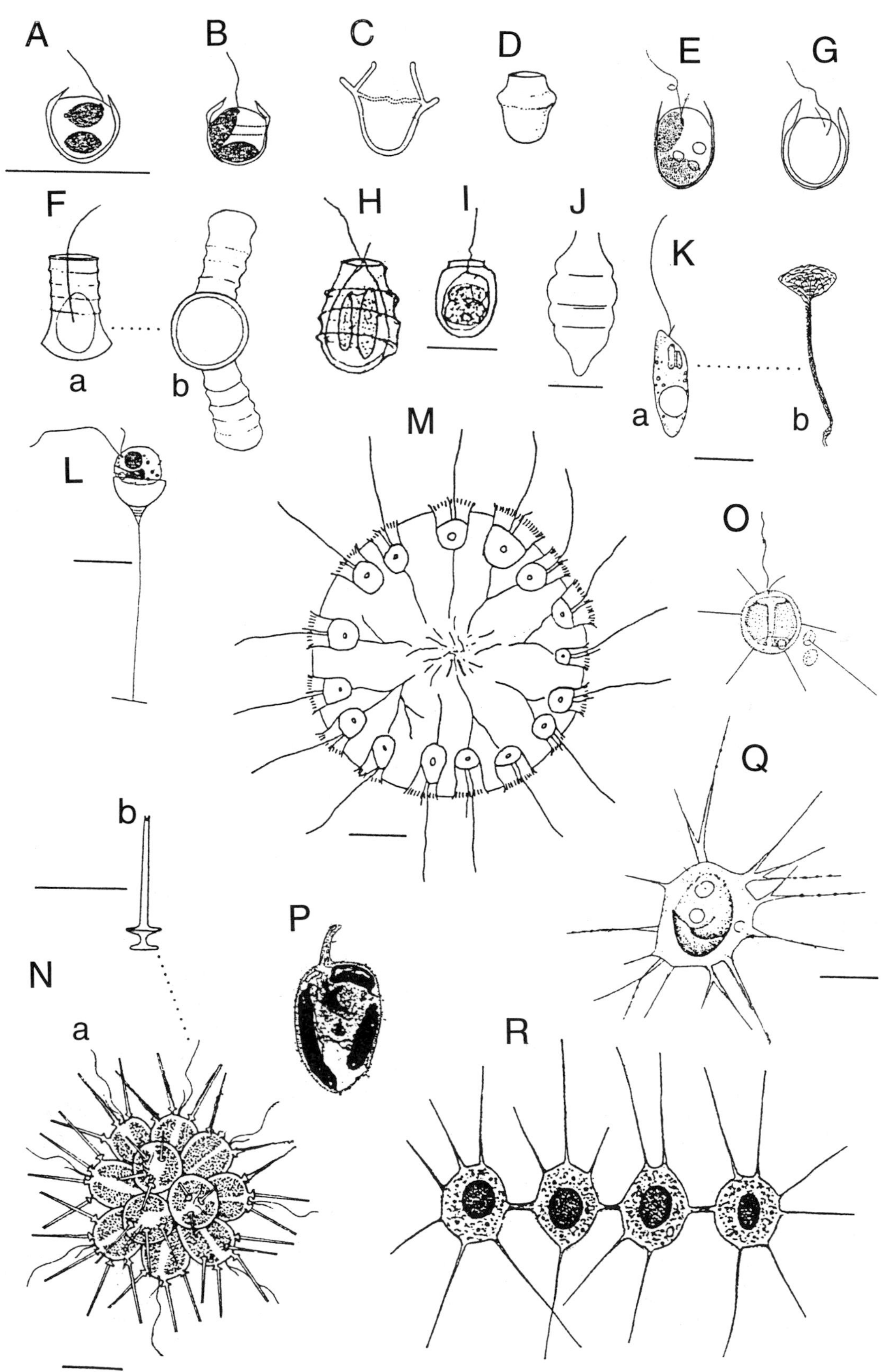

Plate 61 A–R

A. *Kephyrion littorale* (p. 231); **B**. *Kephyrion littorale* var. *constricta* (p. 231); **C**. *Kephyrion moniliferum* (p. 231); **D**. *Kephyrion rubri-claustri* (p. 231); **E**. *Pseudokephyrion ellipsoideum* (p. 231); **F**. *Pseudokephyrion klarnetii* (p. 232): a. vegetative cell, b. zygote; **G**. *Pseudokephyrion ovum* (p. 232); **H**. *Pseudokephyrion pseudospirale* (p. 232); **I**. *Pseudokeyrion ruttneri* (p. 232); **J**. *Pseudokephyrion undulatum* (p. 232); **K**. *Poteriochromonas malhamensis* (p. 235): a. flagellated vegetative cell, b. lorica; **L**. *Poteriochromonas nutans* (p. 235); **M**. *Sphaerobryon fimbriatum* (p. 235); **N**. *Chrysosphaerella brevispina* (p. 222): a. colony, b. scale and spine; **O**. *Spiniferomonas bourrellyi* (p. 235); **P**. *Chrysolepidomonas anglica* (p. 220); **Q**. *Chrysamoeba radians* (p. 216); **R**. *Chrysidiastrum catenatum* (p. 216).

Scale bars: 10 μm (A–H, O and P are same scale).

Ochromonas tuberculata D.J.Hibberd 1970

09350030 **Pl. 57F** (p. 223)

Cells ovoid, 12–18×10–14 μm; flagella 2, unequal, close to narrow end; surface warty due to numerous muciferous bodies; chloroplast single, with a large, plate-shaped eyespot anteriorly; contractile vacuoles 2, anterior; cysts spherical, 14–17 μm in diameter, with 3–7 dendroid spines.

British Isles (England); planktonic from Yorkshire (Hibberd, 1970).

Very similar to *Ochromonas verrucosa*.

Ochromonas variabilis H.Meyer 1897

09350040 **Pl. 57E** (p. 223)

Cells spherical, 6–9×5–8 μm; chloroplasts 2 per cell, without an eyespot; contractile vacuole single and apical.

Switzerland; in the British Isles known only in the plankton of a garden pond in Cambridge (Pringsheim,1952); probably considerably more widely distributed than records indicate.

Ochromonas verrucosa Skuja 1939

09350050 **Pl. 58A** (p. 225)

Cells obovate, 19–27×12–18 μm, walls with conspicuous warts (muciferous bodies); chloroplast single, without a pyrenoid but with an oval eyespot; stomatocysts not known.

Northern and Eastern Europe; so far only recorded in the British Isles from Scotland (Brook, 1959c).

Very similar to *O. tuberculata*, but a little larger.

Phaeaster Scherffel 1927

Cells spherical to angular, naked and embedded in mucilage to form small colonies; chloroplast single, star-shaped, with an eyespot; swarmers with one visible flagellum, smooth flagellum very short and hidden in a pocket, with flagellar pit surrounded by a system of 18–24 rhizopodia; stomatocysts with an irregular ornamentation.

Europe, North America; planktonic. Two species, one recorded from the British Isles.

Phaeaster pascheri Scherffel 1927

09370010 **Pl. 56I** (p. 221)

Cells spherical, 10 μm in diameter.

Europe (Czech Republic); recorded in England as common in the plankton of lakes in the English Lake District (Belcher, 1969b).

Phaeosphaera West *et* G.S.West 1903

Cells spherical, division in 3 planes; arranged in small groups within a mucilaginous cylindrical, branched or net-like colony; chloroplasts parietal, 2–4 per cell; swarmers not known.

British Isles (England); periphytic. Monospecific.

Phaeosphaera gelatinosa West *et* G.S.West 1903

09390010 **Pl. 62A** (p. 238)

Description as for genus.

Distribution as for genus; so far known only from Tremethick Moor and other *Sphagnum* bogs in Cornwall (W. & G.S.West, 1903b; West, 1904; West & Fritsch, 1927).

Phaeothamnion Lagerheim 1884

Branched upright filaments attached by a basal disc to other algae, following cell division the 2 new protoplasts secrete their own walls within mother cell and upper cell is then extended; asexual reproduction by biflagellate swarmers, flagella lateral, swarmers liberated through a pore in cell wall.

Epiphytic in freshwaters. Three species, one species and variety recorded from England.

For ultrastructural reasons, supported by DNA sequence analysis, this genus has now been removed from the Chrysophyceae. It has some affinities to both the Xanthophyta and Phaeophyta, but it has now been placed in a separate class, the Phaeothamniophyceae (see Andersen et al., 1998; Bailey et al., 1998).

Phaeothamnion confervicolum Lagerheim 1884

Pl. 62F (p. 238)

var. ***britannica*** Godward 1933

09400012 **Pl. 62G** (p. 238)

Thallus consisting of upright and much-branched filaments; cells 3–6 μm wide, 10–20 μm long, with walls distinctly thickened at ends, especially at lower end where thickening is in layers (distinguishes variety from type); chloroplast lobed.

British Isles (England); so far found only in ponds in Epping Forest by Godward (1933).

Phaeothamnion confervicolum has also been reported from upper Elf Loch, Midlothian by M'Keever (1912).

Polylepidomonas Preisig *et* D.J.Hibberd 1983

Cells *Ochromonas*-like, but covered with plate-like scales. It can only be identified by electron microscopic examination of the scales and so is outside the scope of this volume. The only known species, *P. vacuolata* (Thompsen) Preisig *et* D.J.Hibberd (09410010), has been reported from the British Isles (Preisig & Hibberd, 1983) as well as elsewhere in northern Europe.

Poteriochromonas Scherffel 1901

Cells of the *Ochromonas*-type, with 2 flagella and within a delicate and goblet-shaped lorica, but also occurring in a free-swimming state.

Europe; epiphytic in ponds. Three species described of which 2 are known from the British Isles.

1 Lorica cup-shaped with a transverse septum ....*P. nutans*
1 Lorica goblet-shaped, without a septum and very delicate ....*P. malhamensis*

Poteriochromonas malhamensis (E.G.Pringsheim) L.S.Péterfi 1969

Basionym: *Ochromonas malhamensis* E.G.Pringsheim
09420010 **Pl. 61K** (p. 233)
Lorica cup-shaped, 4–8 μm wide and 4–10 μm deep, terminal on a 10–40 μm long stalk, attached to a substrate.

British Isles; known only from Malham Tarn, Yorkshire (Pringsheim, 1952; Lund, 1961a).

This species was considered previously to be a species of *Ochromonas*; careful studies have revealed the very delicate lorica which is best seen after staining, or with anoptral contrast; ultrastructural details require examination by an electron microscope.

Poteriochromonas nutans Jane 1944

09420020 **Pl. 61L** (p. 233)
Lorica shaped like a shallow cup, septum present across base of cone-shaped transitional region, 7–7.6 μm wide, terminal on a long stalk, attached to algal filaments.

British Isles; only known as an epiphyte on *Oedogonium* filaments collected at Bangor, Wales (Jane, 1944b).

Lorica is very delicate, only obvious after staining.

Sphaerobryon F.J.R.Taylor 1954

Loricate, with loricae located on dichotomously branched, mucilaginous stalks within spherical colonies united by a common mucilage; loricae bell-shaped (campanulate), with fringed edge, 4–6 μm wide, 6–8 μm long; flagella 2 per cell; chloroplast single, with an eyespot.

British Isles (England); planktonic in stagnant water. Monospecific.

Sphaerobryon fimbriatum F.J.R.Taylor 1954

09440010 **Pl. 61M** (p. 233)
Colonies 50–75 μm in diameter.

Distribution as for genus; so far only reported on one occasion as a few colonies in Saddington Reservoir, Leicestershire (Taylor, 1954).

Spiniferomonas E.Takahashi 1973

Cells covered with an armour of plate scales and spine scales. The species are very small and can only be identified by electron microscopic examination of their scales.

Planktonic in freshwater. About 16 species known of which 3 have been recorded from the British Isles (Kristiansen, 1979; Preisig & Hibberd, 1982a, 1983). *Spiniferomonas bourrellyi* E.Takahashi 1973 (09450010) is shown (Pl. 61O, p. 233).

Stichogloea Chodat 1897

Cells surrounded with a wall, often in 4s and united by mucilaginous strands to form free-floating colonies; chloroplasts 1–2 per cell; asexual reproduction by autospores and biflagellate swarmers; swarmer cells each with a single chloroplast and eyespot; stomatocyst spherical, smooth, with a short neck.

Europe; planktonic.Three species of which 2 are known from the British Isles.

Retained here in the Chrysophyceae although the genus has been moved along with *Phaeothamnion* to the Phaeothamniophyceae by Bailey et al. (1998).

1 Cells only partly embedded within the colonial mucilage ....*S. olivacea*
1 Cells totally embedded within the colonial mucilage, united by mucilaginous strands ....*S. doederleinii*

Stichogloea doederleinii (Schmidle) Wille 1911

Basionym: *Oodesmus doederleinii* Schmidle
09480010 **Pl. 62C** (p. 238)
Cells ellipsoidal, 8×6 μm, united by mucilaginous strands to form irregular chains of about 4 cells embedded within a colonial mucilage.

Widely distributed in Central and Northern Europe; reported from the plankton of nutrient-poor lakes in Ireland (Round & Brook, 1959), Loch Doon, Ayrshire (West & Fritsch, 1927) and in the English Lake District (J.W.G.Lund, unpublished).

Stichogloea olivacea Chodat 1897

09480020 **Pl. 62D** (p. 238)
Cells ellipsoidal, 6–10×4–6 μm, partly projecting from colonial mucilage.

Central and Northern Europe; reported in the British Isles from the plankton of Scottish lakes (West, 1904; West & Fritsch, 1927).

Synochromonas Korshikov 1929

Cells *Ochromonas*-like, with smooth surface, united by their tails to form spherical colonies; chloroplasts parietal and 2 per cell; chrysolaminaran vacuole present basally.

Northern Europe; planktonic.

Three species, but unclear delimitation against several similar genera, with which it sometimes was united under the name of *Syncrypta* which has been replaced by *Synuropsis*.

1 Cells pear- or club-shaped.............................*S. janei*
1 Cells pear-shaped with prolonged tails ....*S. elaeochrus*

Synochromonas elaeochrus Jane 1940

09520010 **Pl. 55C** (p. 219)
Cells pear-shaped with long tails, about 9.5 μm wide and 24 μm long, with colonies of 20–50 cells; chloroplasts parietal and 2 per cell, mostly without an eyespot; stomatocysts not known.

British Isles (England); known only from Middlesex (Jane, 1940) where among aquatic macrophytes in stagnant water rich in organic matter.

Synochromonas janei Bourrelly 1950

09520020 **Pl. 55D** (p. 219)

Cells pear- to club-shaped, about 9 μm in width and to 21 μm long, in spherical colonies of about 60 cells; chloroplasts parietal and 2 per cell, often with an eyespot.

France; only found in the British Isles at Reading (Jane, 1940; as *S. pallida*) but probably considerably more widely distributed.

Uroglena Ehrenberg 1834

Cells *Ochromonas*-like, rounded to pear-shaped, located at periphery of spherical colonies (often sufficiently large as to be visible to naked eye), attached to a radiating system of more or less visible, branched, gelatinous stalks; flagella 2 per cell, one flagellum several times longer than the other; chloroplasts 1–2 with a conspicuous eyespot; contractile vacuoles 1–3 per cell, apical; stomatocysts spherical, often very distinctive. In most cases mature stomatocysts are indispensable for species identification.

Widely distributed; occurs in the plankton of lakes and ponds. Altogether 20 known species of which 4 are reported from the British Isles.

If present in large numbers *Uroglena* may give the water a disagreeable fishy odour.

1 Cells attached to very distinct, branched, articulated and radially arranged filaments .... *U. eustylis*
1 Cells attached to not visible or scarcely visible, radially arranged filaments 2
2 Cells pear-shaped; stomatocysts with a double collar .... *U. volvox*
2 Cells rounded; stomatocysts without a double collar 3
3 Stomatocysts spiny, with a hooked neck ... *U. lindii*
3 Stomatocysts smooth, without a neck .... *U. americana*

Uroglena americana G.N.Calkins 1892

09540010 **Pl. 54J** (p. 217)

Colonies 150–300(–700) μm in diameter; cells ellipsoidal, gelatinous stalks not visible; chloroplast single with an eyespot; contractile vacuoles 1–2 per cell; stomatocysts spherical, smooth, without a collar.

Widely recorded in North America and Europe (identifications not always reliable); planktonic in ponds and in the British Isles known from the English Lake District (Lund, 1949a; Pentecost, 1984a), Malham Tarn in Yorkshire (Lund, 1961a) and from Scotland (Brook, 1955a, 1964).

Uroglena eustylis Skuja 1948

09540050 **Pl. 54K** (p. 217)

Colonies up to 300 μm in diameter, more or less regular in shape; cells pear-shaped, attached to thick, branched, articulated stalks; chloroplast single and with an eyespot; stomatocysts spherical, smooth, without a collar.

Europe: Sweden and Denmark, planktonic; known in the British Isles only from the English Lake District (Lund, 1954a).

Perhaps synonymous with *Uroglena articulata* Korshikov.

Uroglena lindii Bourrelly 1957

Synonym: *Uroglena soniaca* H.Conrad *sensu* Lind

09540030 **Pl. 55A** (p. 219)

Colonies up to 100 μm in diameter, with an uneven outline; cells pear-shaped, 8 μm wide, 12–15 μm long, attached to scarcely visible gelatinous stalks; chloroplast single, spiralled and with an eyespot; cysts spherical, warty to spiny, with collar prolonged into a hooked neck.

France and Denmark; in the British Isles only reported as planktonic in a pond near Sheffield (Lind, 1939a, c, as *Uroglena soniaca*).

According to J.W.G.Lund (pers. comm.), he has discovered true *U. soniaca* H.Conrad in Priest Pot, English Lake District.

Uroglena volvox Ehrenberg 1834

09540040 **Pl. 55B** (p. 219)

Colonies up to 400 μm in diameter; cells attenuated towards base, 8–13 μm wide, 12–20 μm long, attached to just visible, dichotomously branched stalks; chloroplast single, with an eyespot; cysts spherical, smooth, pore surrounded by a double collar.

Probably cosmopolitan; several British records but only one with reliable documentation (Pentecost, 1984a).

ORDER HIBBERDIALES

1 With colonial immotile stages 2
1 Without colonial immotile stages 3
1 Immotile stage in neuston .... *Chromophyton*
2 Immotile stage in mucilaginous colonies .... *Hibberdia*
3 Free-living .... *Bitrichia*
3 Epiphytic 4
4 Attached to mucilaginous envelopes of other algae .... *Stylochrysalis*
4 Epiphytic on algal filaments 5
5 Lorica riding on algal filament, secured with a string around .... *Chrysopyxis*
5 Lorica flask-shaped, often laterally fastened to algal filament .... *Lagynion*

Bitrichia Wołoszyńska 1914

Cell surrounded by a hyaline lorica, consisting of 2 layers, outer layer drawn out into terminal processes, with a ramified rhizopodium projecting from the protoplast through a lateral pore; chloroplast single and without an eyespot; after cell division one daughter cell leaves lorica; cyst formation just outside pore.

Widely distributed; planktonic. Eight species, only 2 known from the British Isles.

1 Lorica with its processes straight........*B. longispina*
1 Lorica with its processes curved.............*B. chodatii*

Bitrichia chodatii (Reverdin) Chodat 1926

Basionym: *Diceras chodatii* Reverdin

09030010 **Pl. 62J** (p. 238)

Lorica with 2 curved processes, one often more curved than the other, 40 μm and 20 μm long respectively.

Central and Northern Europe; in the British Isles so far known only from Scottish lochs (Brook 1955a, 1964).

Bitrichia longispina (J.W.G.Lund 1949) Bourrelly 1957

Basionym: *Diceras longispina* J.W.G.Lund

09030020 **Pl. 62K** (p. 238)

Lorica 4.5–5 μm wide and 10–12 μm long, with very long and straight terminal processes of equal length (53–65 μm).

Europe; known in the British Isles from Wise E'en Tarn (Cumbria) (Lund, 1949a).

Chromophyton Woronin 1880

Flagellate stage in life history consists of solitary, naked cells, with a long hairy flagellum and a very much reduced smooth flagellum; non-motile stage of cells within a mucilaginous envelope; flagellated stage breaks through water surface film and forms spherical resting stages lying on top of a 'life-belt' structure; stomatocysts germinate to form small colonies.

Central and Northern Europe; neustonic (on water surface), widely distributed in ephemeral waters. Monospecific.

The genus has been previously also referred to *Chromulina* or *Ochromonas*.

Chromophyton rosanoffii Woronin 1880

09040010 **Pl. 62H** (p. 238)

Resting cells 4–7 μm in diameter.

Distribution as for genus; neustonic, single layer of resting cells may give water surface a golden reflection when seen from a certain angle because all the chloroplasts are arranged at same angle to sun or light rays, mainly reported in the British Isles from aquaria (see West & Fritsch, 1927).

Chrysopyxis F.Stein 1878

Epiphytic, lorica with basal part double, outer part ring-shaped and surrounding algal filament, branched rhizopodium projecting through apical pore of lorica; chloroplasts 1–2 per cell; after cell division one daughter cell leaves the lorica and constructs its own lorica while moving around an algal filament; stomatocysts spherical.

Widely distributed; epiphytic on filamentous algae. Fifteen species, only 2 known from the British Isles.

1 Lorica spherical ....*C. globosa*
1 Lorica ovoid, narrowed at apex.........*C. stenostoma*

Chrysopyxis globosa Pascher

09170010 **Pl. 62L** (p. 238)

Lorica spherical, with a very small pore and without a surrounding collar.

Europe; known in the British Isles only from England where reported from ponds in Richmond Park, Surrey (Lund, 1942c) and Miles Rough Bog, Bradford, Yorkshire (Malin Smith, 1942).

Chrysopyxis stenostoma Lauterborn 1911

09170020 **Pl. 62M** (p. 238) **CD**

Lorica ovoid, 10–16 μm wide and 12–22 μm long, apically pointed and without a collar around terminal pore; chloroplasts and contractile vacuoles 2 in each cell.

Europe; known in the British Isles from pools at Llandegfan Common, Anglesey in Wales where it grew on the filaments of *Mougeotia* and *Zygnema* (Jane, 1945).

Hibberdia R.A.Andersen 1985

Flagellate stage in life history consisting of solitary, naked cells, with a long hairy flagellum and a very much reduced smooth flagellum, placed at a wide angle (160°) to each other, short flagellum hidden in a pocket close to eyespot; non-motile stage of cells within a mucilaginous envelope, flagella also present and beats within mucilage; chloroplast curved.

British Isles (England); planktonic in ponds.

One species described, transferred from the genus *Chrysosphaera*.

Hibberdia magna (J.H.Belcher) R.A.Andersen 1989

Basionym: *Chrysosphaera magna* J.H.Belcher

09270010 **Pl. 62I** (p. 238)

Colonies up to 60 μm in diameter, consisting of up to 256 cells within a thick (15 μm) mucilaginous envelope; swarmers about 10 μm long and having 1 or 2 chloroplasts.

Distribution as for genus; so far known only from a pond at Wimpole, Cambridgeshire (Belcher, 1974).

Lagynion Pascher 1912

Epiphytic, lorica brownish, bottle-shaped with a distinct neck, attached basally or by one side to a surface (an algal filament), a branched rhizopodium emerges through neck; chloroplasts 1–2 per cell; after division one daughter cell leaves lorica and constructs its own lorica.

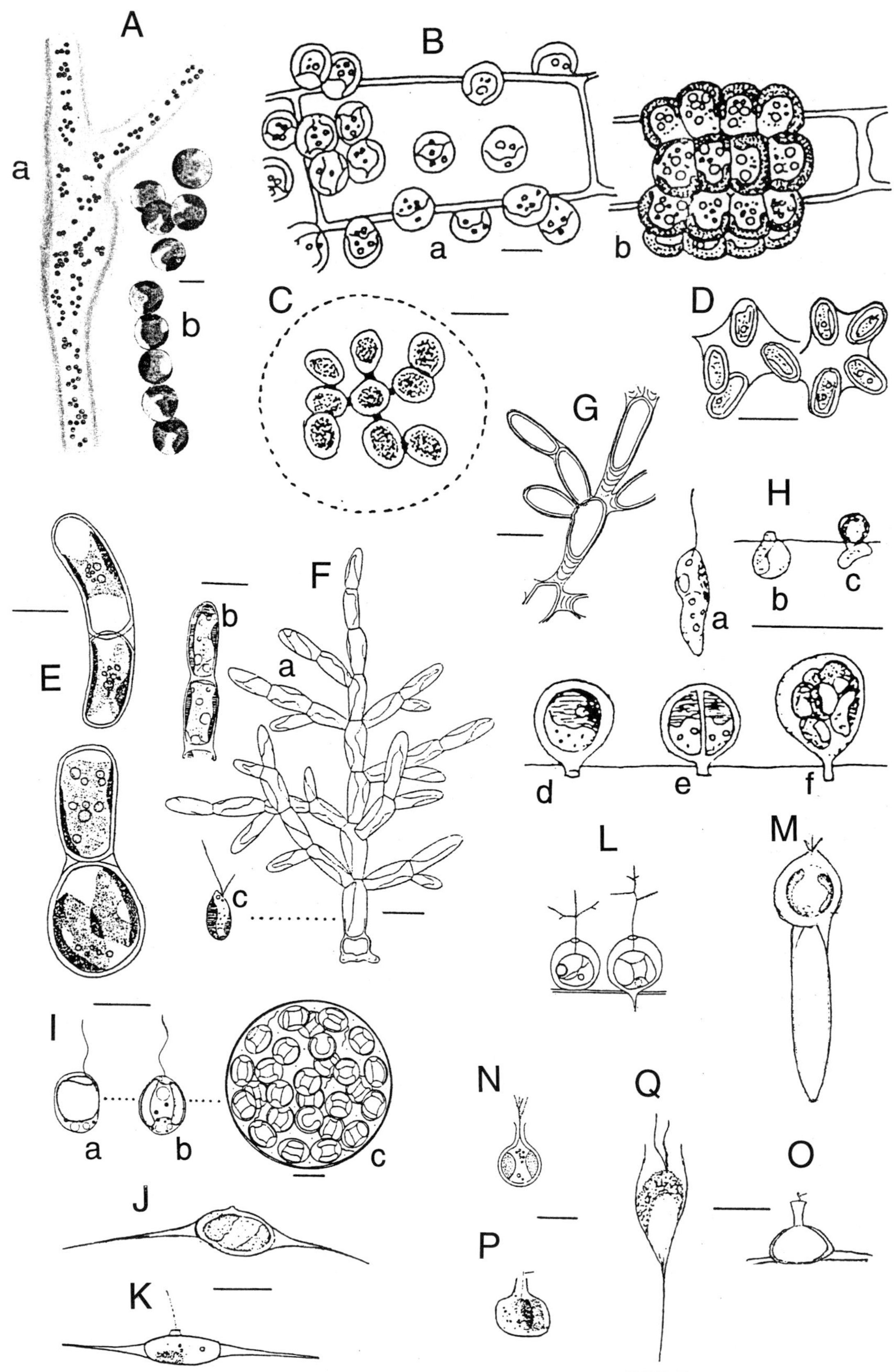

Plate 62 A–Q

A. *Phaeosphaera gelatinosa* (p. 234): a. colony, b. single cells; **B**. *Chrysosphaera paludosa* (p. 222): a-b. cells epiphytic on algal filaments; **C**. *Stichogloea doederleinii* (p. 235); **D**. *Stichogloea olivacea* (p. 235); **E**. *Apistonema pyrenigerum* (p. 215); **F**. *Phaeothamnion confervicolum* (p. 234): a. habit, b. two cells from tip of branch, c. swarmer; **G**. *Phaeothamnion confervicolum* var. *britannica* (p. 234); **H**. *Chromophyton rosanoffii* (p. 237) a. motile cell, b, c. stages in cyst formation, d-f. mature cysts and germination; **I**. *Hibberdia magna* (p. 237): a, b. motile vegetative cells, c. sporangium; **J**. *Bitrichia chodatii* (p. 237); **K**. *Bitrichia longispina* (p. 237); **L**. *Chrysopyxis globosa* (p. 237); **M**. *Chrysopyxis stenostoma* (p. 237); **N**. *Lagynion ampullaceum* (p. 239); **O**. *Lagynion macrotrachelum* (p. 239); **P**. *Lagynion scherffelii* (p. 239); **Q**. *Stylochrysalis aurea* (p. 239). Scale bars: 10 μm.

Widely distributed in Europe and North America; on filamentous algae. Twenty-four species, only 3 species recorded from the British Isles.

The closely related but colourless genus *Heterolagynion* is reported from the British Isles with Butcher et al. (1937) mentioning *H. oedogonii* Pascher as abundant at all sites studied by them in the River Tees.

1 Lorica base rounded .... *L. ampullaceum*
1 Lorica base more or less flat 2
2 Lorica broadly conical .... *L. macrotrachelum*
2 Lorica subspherical .... *L. scherffelii*

Lagynion ampullaceum Pascher 1912

09300010 **Pl. 62N** (p. 238)
Lorica bottle-shaped, 6–12 μm wide, 12–22 μm long, with rounded base and long neck (up to 12 μm).

Widely distributed in Europe, epiphytic on filamentous algae; in the British Isles only listed by Pentecost (1984a).

Lagynion macrotrachelum Pascher 1912

09300020 **Pl. 62O** (p. 238)
Lorica bottle-shaped, broadly conical with a flattened base, 14–17 μm wide, about 8.4 μm long, with a long neck (up to 11 μm).

Europe, North America; known in the British Isles only on *Oedogonium* filaments in Wales where discovered in a pool and tubs at Bangor and at Llandegfan Common (Jane, 1945).

Lagynion scherffelii Pascher 1912

09300030 **Pl. 62P** (p. 238)
Lorica bottle-shaped, with rounded or flattened base, 8–12 μm broad, 10–15 μm long, with neck 4–4.5 μm long.

Central Europe; reported in the British Isles with some doubt only from ponds in Richmond Park, Surrey (Lund, 1942a).

Stylochrysalis F.Stein 1878

Lorica delicate, with a long stalk attaching it to the colonial mucilaginous envelopes of various algae, 2 rhizopodia project from mouth of lorica; chloroplasts 2 per cell; cysts formed inside lorica.

Distribution imperfectly known. Six species, only one reported from the British Isles.

Stylochrysalis aurea (Chodat) H.Bachmann 1907

Basionym: *Stylotheca aurea* Chodat
09500010 **Pl. 62Q** (p. 238)
Lorica broadly spindle-shaped, 7–10 μm wide, 13–22 μm long, with stalk up to 20 μm long.

Central and Northern Europe; recorded from Scotland by Bachmann (1907) but its identity was questioned by West & Fritsch (1927).

ORDER HYDRURALES

Cells arranged in prostrate or bushy thalli.

1 Thallus bushy, darkish brown .... *Hydrurus*
1 Thallus prostrate 2
2 Cells square, regularly arranged .... *Phaeodermatium*
2 Cells rounded, irregularly arranged, some with a long posterior filament .... *Chrysonebula*

Chrysonebula J.W.G.Lund 1953

Macroscopic masses, prostrate colonies covering stones, containing crystals of calcium carbonate; cells naked, variable in shape (often spherical), 8–25×8–14 μm, irregularly arranged, some having a long gelatinous filament from posterior end; chloroplasts 1–2 per cell, with a single pyrenoid; eyespot absent; asexual reproduction by tetrahedral uniflagellate swarmers.

British Isles (England); attached to rocks in fast-flowing streams in the Pennines with relatively hard water, where it sometimes forms a mass cover in late winter (typically February) under low flow conditions and can give long stretches of stream a rich brown colour. It is especially obvious in streams enriched with zinc from old mines.

Preisig (1994) included the genus in *Celloniella*.

Chrysonebula holmesii J.W.G.Lund 1953

09160010 **Pl. 56E** (p. 221)
Description as for genus.

Reported only from the British Isles; originally recorded from surface of stones in Gordale Beck, near Malham Tarn, Yorkshire (Lund, 1953); also Waterfall Beck and Howgill Beck in the same area where it forms a dense mucilaginous coating on stones in spring (A. Pentecost, pers. comm.) and from north-east England (Hibberd, 1977a). In the last case it is sometimes very abundant in late winter in streams combining calcareous and peaty waters and also in some cases zinc contamination from old mines (B.A. Whitton, pers. comm.).

Hydrurus C.Agardh 1824

Macroscopic, consisting of mucilaginous, branched and bushy to feathery, dark brown thalli up to 30 cm long; growth apical, cells arranged peripherally in mucilage; swarmers tetrahedral, with one long flagellum visible and no eyespot; stomatocysts spherical, with an equatorial ring.

Widely distributed in mountainous and northern regions, growing attached to stones in mountain streams and rivers. Monospecific.

The alga has a foetid smell, especially when rubbed between the fingers.

Hydrurus foetidus (Villars) Trevisan 1848

Basionym: *Conferva foetida* Villars
09280010 **Pl. 63A** (p. 241)
Description as for genus.

Known in the British Isles from upland areas of northern England, southern Scotland and Devon especially during the colder part of the year and often in flushes combining humic and calcareous inputs; widespread and occasionally abundant in the English Lake District (J.W.G.Lund, pers. comm.).

Phaeodermatium Hansgirg 1889

Macroscopic, crustose on stones; cells almost square, radially arranged in one to few layers to form a thick crust, with a single chloroplast in each cell; swarmers tetrahedral, uniflagellate and without an eyespot.

Europe, North America; on stones in cold mountain ponds and streams. Monospecific.

Phaeodermatium rivulare Hansgirg 1889

09380010 **Pl. 63B** (p. 241)

Description as for genus.

Distribution as above; reported in the British Isles from a pool on Llandegfan Common, Wales (Jane, 1945), streams in the English Lake District (Canter-Lund & Lund, 1995), rare in the upper stretches of the River Tees (Butcher et al., 1937); occasional on rocks in fast-flowing rivers in north-east England, but almost all records made in winter (B.A. Whitton, pers. comm.).

CLASS DICTYOCHOPHYCEAE

ORDER PEDINELLALES

Unicellular flagellates with a peculiar radial symmetry.

Pedinella Vysotskij 1887

Cells free-swimming or attached to substratum by a thin stalk, apple-shaped, 6–11×8–11 μm, with 6 radially arranged chloroplasts; flagella 2 per cell, long flagellum provided with a lateral wing, short flagellum reduced, about 12 tentacles arranged around flagella.

Europe; usually found in stagnant freshwater and brackish-water habitats. Monospecific.

Pedinella hexacostata Vysotskij 1887

09360010 **Pl. 63C** (p. 241)

Description as for genus.

Distribution as above; in the British Isles reported in nearly all lakes in the English Lake District, certainly common but never very numerous (Swale, 1969).

CLASS SYNUROPHYCEAE

ORDER SYNURALES

Silica scales, not differentiated into spine scales and plate scales.

1 Cells with 2 flagella of almost equal length, in colonies; bristles not present ....*Synura*
1 Cells with one visible flagellum, solitary; scales with attached bristles ....*Mallomonas*

Mallomonas Perty 1851

Cells solitary with one visible flagellum (in few cases 2 flagella), covered with an armour of silica-scales, in most species with some scales bearing silica bristles hinged to a special area in their distal part; chloroplast 2-lobed, brownish or greenish; cysts initially surrounded by scales.

Widely distributed; planktonic. In the British Isles 45 species have been recorded based upon electron microscopic examination mainly by Bradley, Harris, Harris & Bradley and Kristiansen.

Taxonomy is based on the fine structure of the scales and bristles, hence electron microscopy is necessary for the identification of most species. For several described species scale structure is not known, they are considered dubious and not accepted in modern taxonomic treatments. More than 120 species have been recognized. Only a few of the species can, however, be reliably identified by light microscopy. A key to the species recognized by light microscopy is given below, but it should be used with caution. All species recorded from the British Isles, which can only be identified with the electron microscope, are listed below the descriptive text.

1 Cells without bristles, elongated, with apical and caudal spines..........................*M. insignis*
1 Cells with bristles..2
2 Bristles covering cells 3
2 Bristles at only one or at both ends.........................6
3 Bristles curved and serrated; cells often pear-shaped..*M. caudata*
3 Bristles not serrated 4
4 Bristles straight, smooth, with apical scales forming a crown; cells cylindrical-oblong ....*M. punctifera*
4 Bristles with hooked apex, straight needle-shaped bristles sometimes also present; cells ellipsoidal or almost spherical 5
5 Bristles all hooked.............................*M. multiunca*
5 Bristles both hooked and needle-like ....*M. heterospina*
6 Bristles only at front of cell....................................7
6 Bristles at both ends of cell 8
7 Cells spindle-shaped, very slender ...*M. akrokomos*
7 Cells ovoid. ....*M. tonsurata*
8 Bristles all delicate and needle-shaped ....*M. teilingii*
8 Bristles at apex long and straight, caudal bristles short and hooked........................*M. hamata*

Mallomonas akrokomos Ruttner in Pascher 1913

09310030 **Pl. 63D** (p. 241)

Cells spindle-shaped, slender and long, 3–15 μm wide, 8–78 μm long, broadest near blunt apex, tail attenuated and pointed; bristles present as an apical tuft, short bristles 4–20 μm in length and long bristles 19–39 μm; cysts oblong, 6–17×6–14 μm.

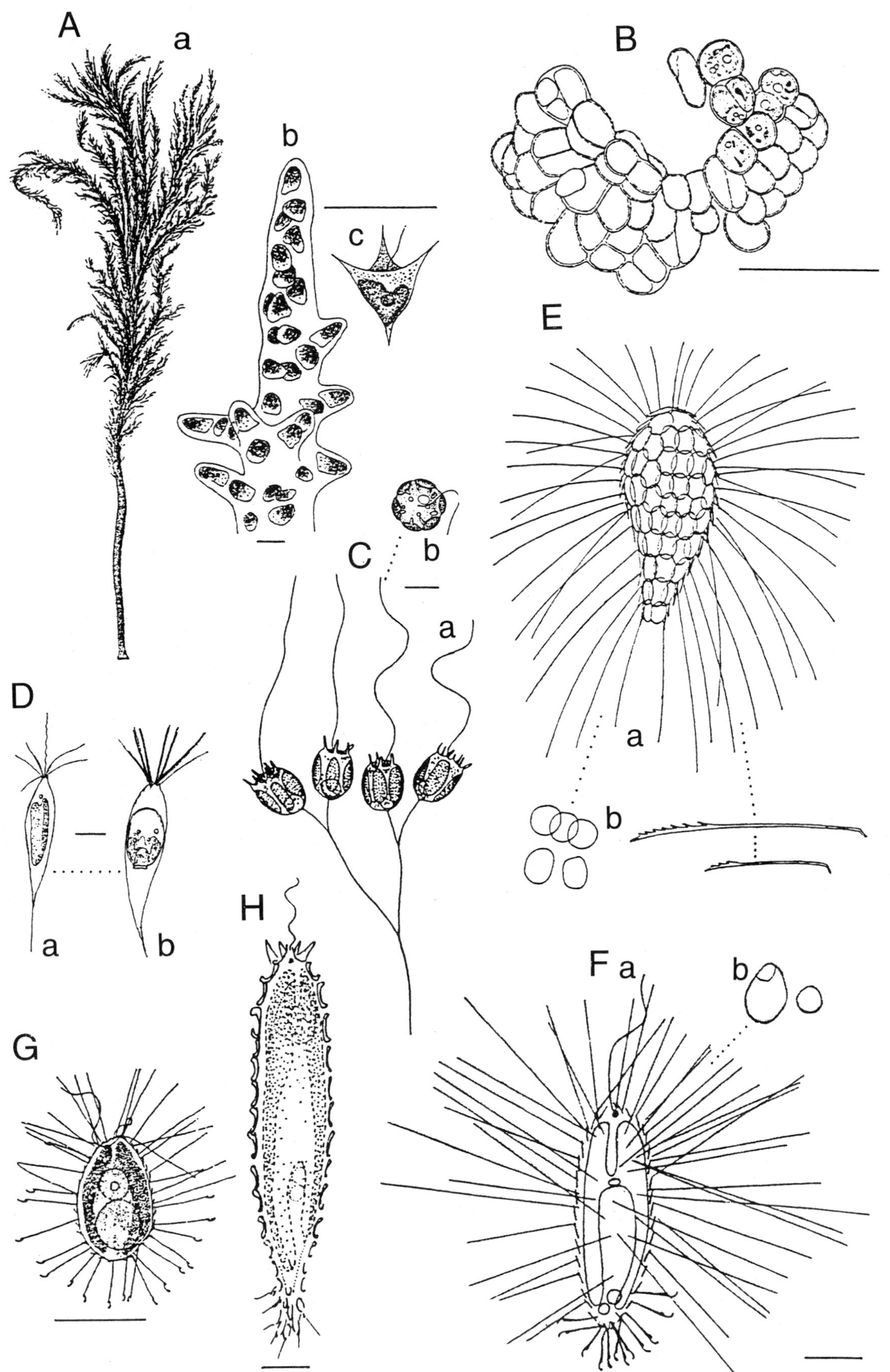

Plate 63 A–H

A. *Hydrurus foetidus* (p. 239): a. thallus in natural size, b. apical part of thallus, c. swarmer; **B.** *Phaeodermatium rivulare* (p. 240); **C.** *Pedinella hexacostata* (p. 240): a. colony of cells attached by a branched stalk, b. flagellated cell; **D.** *Mallomonas akrokomos* (p. 240): a. vegetative cell, b. stomatocyst; **E.** *Mallomonas caudata* (p. 242): a. whole cell, b. scales and bristles; **F.** *Mallomonas hamata* (p. 242): a. whole cell, b. scales; **G.** *Mallomonas heterospina* (p. 242); **H.** *Mallomonas insignis* (p. 242).
Scale bars: 10 μm.

Probably cosmopolitan; mainly in winter and spring plankton of standing bodies of water, commonly reported in southern England (Lund, 1937, 1942c; Harris, 1958), the English Lake District (Lund, 1949a) and Scotland (Bradley, 1966b).

Mallomonas caudata Ivanov 1899 emend. Willi Krieger 1930

09310080 **Pl. 63E** (p. 241)
Cells ovoid to pear-shaped, 10–30 μm wide and up to 100 μm long, sometimes with a 'tail'; bristles densely covering cells, sometimes as long as cell or longer, curved and coarsely serrated; scales large, elliptical to almost circular, without any visible structure; cyst smooth, 20–35 μm in width.

Widely distributed, mainly in the northern hemisphere; in the British Isles reported from Scotland (Teiling, 1916; Brook, 1964), the English Lake District (Canter-Lund & Lund, 1995), southern England (Harris & Bradley, 1957), Wales (Happey-Wood, 1978) and Ireland (Round & Brook, 1959; Kristiansen, pers. obs.).

Mallomonas hamata Asmund 1959

09310140 **Pl. 63F** (p. 241)
Cells elongate, 8–12 μm wide and 20–35 μm long, with bristles at both ends, forming an apical tuft of long, straight and needle-shaped bristles (up to 30 μm), caudally tufted and short (3–7 μm long), curved and hooked; cysts oblong, 17–23 × 12–19 μm.

Temperate and Arctic regions of Europe and America; in the British Isles known from ponds in England (Harris, in Asmund, 1959), Wales and Ireland (Kristiansen, 1979, pers. obs.).

Mallomonas heterospina J.W.G.Lund 1942

09310150 **Pl. 63G** (p. 241)
Cells small and ellipsoidal, 12–15 × 7–8 μm; bristles of two types: needle-shaped and hooked bristles, the latter mainly on posterior half of cell; cyst oblong with a pointed apex, 17–21 × 13–16 μm.

Widely distributed in temperate and subarctic–subantarctic regions; in the British Isles known so far from Scotland (Bradley, 1966b) and southern England (Lund, 1942a, c; Harris, 1967).

Mallomonas insignis Penard 1919

09310160 **Pl. 63H** (p. 241)
Cells elongate to ellipsoidal, very long and slender, up to 100 μm in length, the blunt apex with forwardly directed spines, posteriorly cell drawn out into a spiny tail; bristles absent; cysts 10–20 × 8–30 μm.

Widely distributed but records very scattered, mainly in the northern hemisphere; known in the British Isles from ditches, ponds and lakes during colder part of the year in south-east England (Harris, 1958).

Mallomonas multiunca Asmund 1956

09310240 **Pl. 64A** (p. 243)
Cells ovoid to nearly spherical, 12–22 × 8–14 μm, covered all over with short hooked bristles; cyst about 15 μm in diameter.

Widely distributed in the northern hemisphere; known in the British Isles only during autumn from small acid pools in Scotland (Bradley, 1966b) and southern England (Harris & Bradley, 1960).

Mallomonas punctifera Korshikov 1941

09310350 **Pl. 64B** (p. 243)
Cells ovoid to elongate-oblong, almost cylindrical, 10–13 μm wide, 25–60 μm long, apically with a crown of pointed scales; bristles very long, straight, up to 50 μm long, most often directed backwards.

Widely distributed; in the British Isles common in the English Lake District and also recorded from ponds in south-east England (Harris & Bradley, 1957).

Mallomonas teilingii H.Conrad 1927

09310380 **Pl. 64C** (p. 242)
Cells almost cylindrical, 19–21 μm wide, 30–56 μm long, each end with a tuft of bristles; scales oblong with a net-like pattern on surface, spirally arranged; stomatocysts spherical or ellipsoidal, with spines 20–37 μm long.

Northern Europe, North America; in the British Isles recorded from ponds near Reading in southern England (Harris, 1953), often containing a conspicuous parasite (*Phagodinium tridentatum* (Nygaard) Kristiansen).

Mallomonas tonsurata Teiling 1912

09310400 **Pl. 64D** (p. 243)
Cells spherical to ovoid, 11–30 × 6–14 μm; bristles forming an apical tuft, a group of short curved bristles surrounded by some longer straight ones, the latter often directed backwards; cysts oblong, 12–15 μm wide and up to 25 μm long.

Probably cosmopolitan; in the British Isles known from ponds near Reading in south-east England (Harris & Bradley, 1960), Wales (Kristiansen, 1979) and Ireland (Kristiansen, pers. obs.).

TAXA RECORDED FROM THE BRITISH ISLES BUT NOT DESCRIBED IN TEXT

All taxa that can only be identified using the electron microscope. Species in parentheses cannot be identified because scale structure unknown.

Mallomonas acaroides Perty emend. Iwanoff 09310010; *M. adamas* K.Harris *et* D.E.Bradley 09310020; *M. allorgei* (Deflandre) W.Conrad 09310040; *M. alpina* Ruttner 09310050; *M. annulata* K.Harris 09310060; *M. calceolus* D.E.Bradley 09310070; *M. clavus* D.E.Bradley 09310090; *M. cratis* K.Harris *et* D.E.Bradley 09310100; *M. doignonii* Bourrelly emend. Asmund *et* Cronberg 09310110; *M. eoa* E.Takah. 09310120; *M. flora* K.Harris *et* D.E.Bradley 09310130; *M. intermedia* Kisselev 09310170; *M. leboimii* Bourrelly 09310180; *M. lelymene* K.Harris *et* D.E.Bradley 09310190; (*M. limnicola* J.W.G.Lund 09310200; *M. longiseta* Lemmermann 09310420); *M. lychenensis* W.Conrad 09310210; *M. mangofera* K.Harris *et* D.E.Bradley 09310220; *M. mangofera* var. *foveata* Dürrschmidt 09310222; *M. matvienkoae* (Matv.) Asmund *et* Kristiansen 09310230; *M. ouradion* K.Harris *et* D.E.Bradley 09310250; *M. oviformis* Nygaard 09310260; *M. papillosa* K.Harris *et* D.E.Bradley

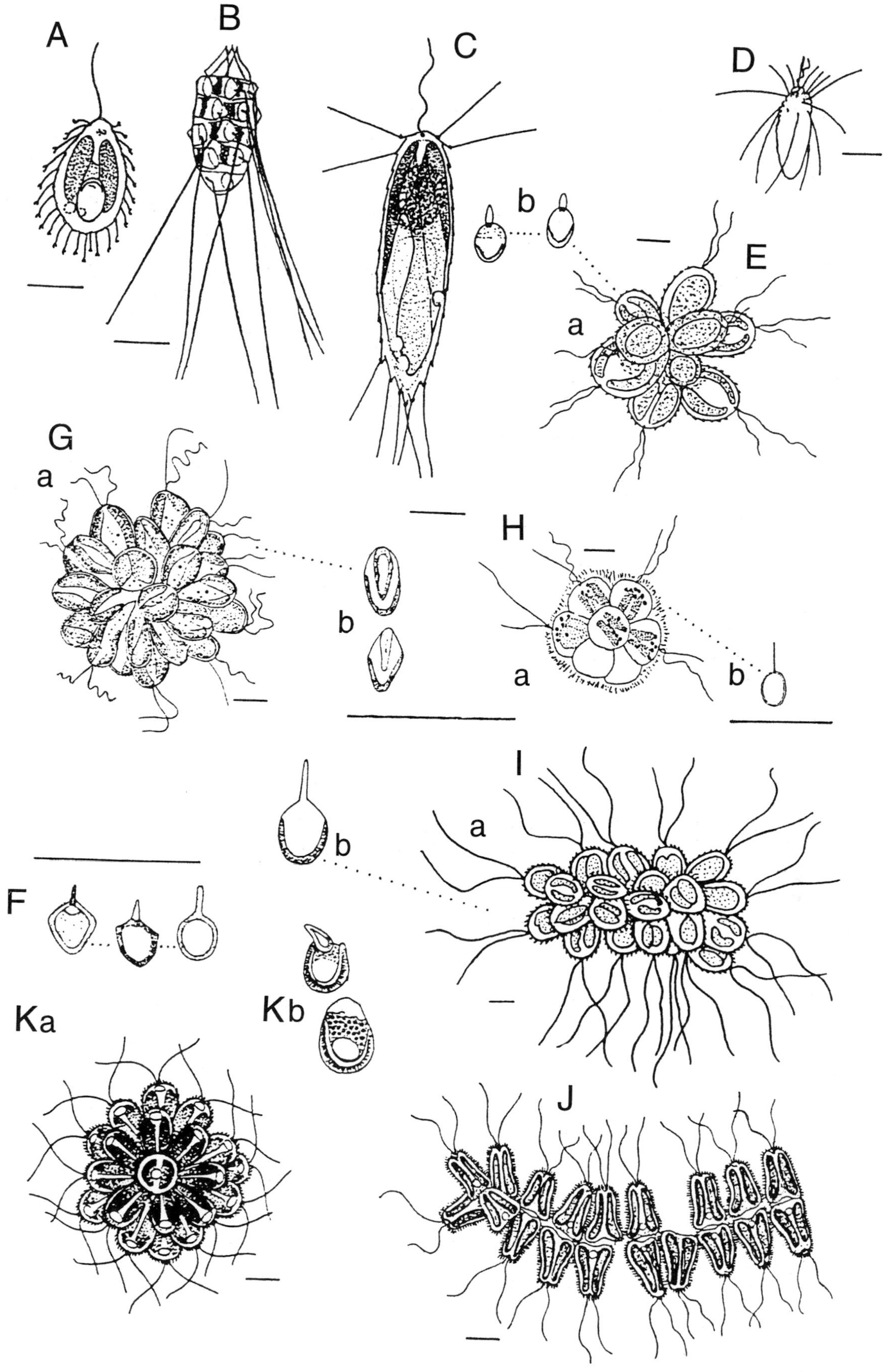

Plate 64 A–K

A. *Mallomonas multiunca* (p. 242); **B**. *Mallomonas punctifera* (p. 242); **C**. *Mallomonas teilingii* (p. 242); **D**. *Mallomonas tonsurata* (p. 242); **E**. *Synura curtispina* (p. 244): a. colony, b. single scales; **F**. *Synura echinulata* (p. 244): single scales; **G**. *Synura petersenii* (p. 244): a. colony, b. single scales; **H**. *Synura sphagnicola* (p. 244): a. colony, b. single scale; **I**. *Synura spinosa* (p. 244): a. colony, b. single scale; **J**. *Synura spinosa* (p. 244); **K**. *Synura uvella* (p. 244): a. colony, b. single scales.
Scale bars: 10 μm.

09310270; *M. papillosa* var. *ellipsoida* K.Harris 09310272; *M. papillosa* var. *monilifera* K.Harris (09310273); *M. parvula* Dürrschmidt 09310280; *M. paxillata* (D.E.Bradley) L.S.Péterfi *et* Momeu 09310290; *M. peronoides* (K.Harris) Momeu *et* L.S.Péterfi 09310300; *M. phasma* K.Harris *et* D.E.Bradley 09310310; *M. pillula* K.Harris 09310320; *M. pugio* D.E.Bradley 09310330; *M. pumilio* K.Harris *et* D.E.Bradley emend. Asmund et al. 09310340; *M. pumilio* var. *silvicola* K.Harris *et* D.E.Bradley 09310342; *M. reginae* Korshikov 09310410; *M. splendens* (G.S.West) Playfair 09310360; *M. striata* Asmund 09310370; *M. striata* var. *serrata* K.Harris *et* D.E.Bradley 09310372; *M. tolerans* Asmund *et* Kristiansen 09310390.

Synura Ehrenberg 1835

Cells in spherical or oblong colonies, each cell surrounded by an armour of minute silica scales (only detected by phase contrast or in dry mounts); flagella 2 per cell and almost of equal length; chloroplasts 2, brownish, eyespot absent; cysts and sexual reproduction known.

Widely distributed; planktonic. Seventeen species of which 6 are known from the British Isles.

Some species can be identified based upon scale characters observed with the light microscope, but electron microscopic examination of the scales is most often necessary and also required for checking light microscope identifications. The key to the species is based on light microscope characters.

1 Cells appearing almost smooth; colonies spherical; scales oval, without spines but with a median ridge .... *S. petersenii*
1 Cells spiny; colonies spherical or oblong; scales with apical spines 2
2 Cells with axial chloroplast lobes..... *S. sphagnicola*
2 Cell with parietal chloroplast lobes 3
3 Scales each with a very stout, oblique and conical spine .... *S. uvella*
3 Scales each with a delicate apical spine .... *S. curtispina, S. echinulata, S. spinosa*

Synura curtispina (J.B.Petersen *et* J.B.Hansen) Asmund 1968

Basionym: *Synura spinosa* Korshikov var. *curtispina* J.B.Petersen *et* J.B.Hansen
Synonym: *Synura favus* D.E.Bradley
09530010 **Pl. 64E** (p. 243)
Colonies spherical or oblong; cells with a spiny appearance; scales with a short and delicate apical spine, basal scales smaller and slipper-shaped; chloroplasts parietal.

Widely distributed in many parts of the world; in the British Isles known from aquatic habitats in Scotland (Bradley, 1966b).

Synura echinulata Korshikov 1929

09530020 **Pl. 64F** (p. 243)
Colonies spherical or oblong, rather few-celled; cells often with apical red oil droplets; scales with delicate apical spine; chloroplasts parietal.

World-wide; in the British Isles widely distributed in nutrient-poor localities in Scotland (Bradley, 1966b), Wales (Jane, 1945; Kristiansen 1979), England (Harris & Bradley, 1958; Hibberd, 1973) and Ireland (Kristiansen, pers. obs.).

Synura petersenii Korshikov 1929

09530030 **Pl. 64G** (p. 243) **CD**
Colonies spherical; cells with apparently an almost smooth surface; scales oval, without spines but having a dorsal ridge, chloroplasts parietal.

Probably cosmopolitan; one of the most widely distributed chrysophytes, known from water bodies throughout the British Isles (Bradley, 1966b; Jane, 1945; Kristiansen, 1979; Harris & Bradley, 1956; Hibberd, 1973).

If large populations are present, they impart a fishy, cucumber-like smell to water.

Synura sphagnicola (Korshikov) Korshikov 1929

Basionym: *Skadowskiella sphagnicola* Korshikov
09530040 **Pl. 64H** (p. 243)
Colonies spherical; cells often with red apical oil droplets, and decidedly spiny in appearance; scales tennis racket-shaped, apical scales having a long spine and basal scales spineless; chloroplasts axial.

Widely distributed; throughout much of the British Isles where reported from *Sphagnum*-dominated pools and other acid localities (Bradley, 1966b; Jane, 1945; Kristiansen 1979; Hibberd, 1978).

Synura spinosa Korshikov 1929

Synonym: *Chlorodesmus hispidus* F.W.Phillips
09530050 **Pl. 64I,J** (p. 243)
Colonies often elongated, sometimes almost thread-like and then previously considered a separate taxon (namely *Chlorodesmus hispidus* F.W.Phillips); cells decidedly spiny, scales with a delicate apical spine; chloroplasts parietal.

Widely distributed; recorded in the British Isles from Scotland (Bradley 1966b), Wales (Kristiansen, 1979), the English Lake District (Canter-Lund & Lund, 1995), Hertfordshire (Phillips, 1884, as *Chlorodesmus hispidus*), southern England (Harris & Bradley, 1958) and Ireland (Kristiansen, pers. obs.).

Synura uvella Ehrenberg 1835 emend. F.Stein 1878

09530060 **Pl. 64K** (p. 243)
Colonies spherical or slightly elongated, up to 400 μm diameter; cells elongate, with a distinct spiny covering of thick scales; scales with a very stout, conical, oblique spine; chloroplasts parietal.

Widely reported from many parts of the world (most records are undocumented); numerous reports from the British Isles, but the only reliable records with illustrations of scale structure are from Scotland (Bradley, 1996b), Wales (Kristiansen, 1979) and England (Pentecost, 1984a); listed as frequent by Reynolds & Irish (2000) in an account of the phytoplankton of the English Lake District.

PHYLUM XANTHOPHYTA (Yellow-Green Algae)

Leslie R. Johnson

The Xanthophyta are much less species-diverse than the Chlorophyta, with about 600 species and many of the 100 known genera containing only a few species. However, they show a wide range of form and include biflagellate and non-motile unicells, simple or branched uniseriate filaments, and others are coenocytic or siphonous (large multinucleate cells). Colonial forms may or may not have a well-defined shape. Some species are epiphytes and sessile or attached by a stalk.

Most are non-motile, single-celled or colonial, although there a few more advanced filamentous forms and coenocytic forms such as *Vaucheria*. If motile, they are biflagellate, and often possess associated photoreceptors. Asexual reproduction is mainly by fragmentation into portions of one or more cells in multicellular species, aplanospores or zoospores that each have two unequal flagella and sometimes an eyespot. Sexual reproduction is apparently comparatively rare, though is well-known in *Vaucheria*, where it is distinctive and provides important taxonomic characters. Resting structures or cysts are known and often have walls impregnated with silica. Heterogamy is uncommon and isogametes known only for a few genera.

The distinction between the Xanthophyta and the Chlorophyta rests largely on chloroplast pigments and food storage products rather than on morphological characters. Traditionally, there are three features that distinguish the xanthophytes: (i) yellow or yellow-green colour of the chloroplasts; (ii) carbohydrate storage as oil droplets or chrysolaminarin (usually termed leucosin) granules, with starch and pyrenoids rare; (iii) walls of pectin or pectic acid (sometimes in association with cellulose or siliceous substances) and consisting of two spliced and overlapping sections (most conspicuous in *Tribonema*), which on dissociation of the filaments tend to break into H-shaped sections or pieces. Although a useful taxonomic character, these sections are not present in all genera, and certainly not always readily visible even in genera where they occur.

In practice, the yellow-green colour of these plants is not always easy to distinguish, and in the case of *Vaucheria*, the colour is indistinguishable from that of green algae (*Vaucheria* was initially described as a green alga). The xanthophytes differ in containing no chlorophyll *b*, but have in addition to chlorophyll *a*, chlorophylls c_1 and c_2; other pigments are carotenoids (especially β-carotene) and at least three xanthophylls. These xanthophylls can give these algae a blue-green colour, when treated with hot hydrochloric acid in the laboratory.

Although more often found in aquatic habitats, many occur in subaerial habitats or on damp soil. They are especially frequent in spring in waters which are both hard and enriched with humic materials. A few genera (e.g. *Vaucheria*) are not wholly freshwater, as some include brackish and marine species.

1 Thallus filamentous 2
1 Thallus not filamentous 8
2 Microscopic growths 3
2 Macroscopic growths..4
3 Filaments branched and cells contiguous ....*Heterococcus* (p. 254)
3 Filaments branched and cells often slightly separated......................................*Neonema* (p. 256)
4 Thallus unbranched, not coenocytic; aquatic........5
4 Thallus branched or a swollen vesicle, coenocytic (large-celled and multinucleate); aquatic or terrestrial 7
5 Filaments without thinner areas of overlapping wall in mid-region of cell ....*Xanthonema* (=*Heterothrix*) (p. 261)
5 Filaments with thinner areas of wall in mid-region of cell and H-shaped sections clearly visible...6
6 Cells often quadrate, with H-shaped sections every 2 or 3 cells ....*Bumilleria* (p. 260)
6 Cells usually much longer than broad, with one H-shaped section in each cell ....*Tribonema* (p. 260)
7 Thallus a spherical vesicle attached by rhizoids.....................................*Botrydium* (p. 246)
7 Thallus spongy and of ramifying tubular branches.....................................*Vaucheria* (p. 262)
8 Thallus amoeboid with many fine pseudopodia and within the hollow cells of *Sphagnum* ....*Chlamydomyxa* (p. 260)
8 Thallus not as above and attached or free-floating...9
9 Thallus usually attached 10
9 Thallus mostly free-living 15
10 Thallus attached by a short stalk..........................11

10 Thallus not attached by a stalk............................13
11 Cells in a thick envelope (lorica) and attached with rhizoid-like processes (not always evident)......................*Stipitococcus* (p. 260)
11 Cells not in a lorica 12
12 Cells cylindrical, often curved, usually with 1 short spine ....*Ophiocytium*, in part (p. 256)
12 Cells variously shaped, without a spine ....*Characiopsis*, in part (p. 248)
13 Cells not enclosed by mucilage ....*Characiopsis*, in part (p. 248)
13 Cells enclosed by mucilage..................................14
14 Cells embedded at intervals usually in dichotomously branched mucilaginous tubes ....*Mischococcus* (p. 254)
14 Cells single or in 2s or 4s as a single layer within a mucilaginous sheath ....*Chlorosaccus* (p. 252)
15 Cells cylindrical or ellipsoidal, with a single spine at one or both rounded ends 16
15 Cells cylindrical, ellipsoidal or commonly polygonal and angular, commonly with projecting arms at corners but stouter than spines..17
16 Cells with 1 short spine, often strongly curved ....*Ophiocytium*, in part (p. 256)
16 Cells with 2 long spines, one at each pole, straight or slightly bent ....*Centritractus* (p. 248)
17 Cells cylindrical, ovoid, ellipsoidal to spindle-shaped (never tetrahedral), usually rounded and without protuberances; walls smooth (except *Trachydiscus*).............................18
17 Cells polygonal, frequently tetrahedrally angular, usually with pronounced potruberances or arms; walls usually ornamented..24
18 Chloroplasts band- or plate-shaped; cells of 2 halves although not always visible in living material..19
18 Chloroplasts disc-shaped; cells not of 2 halves......20
19 Cells spherical or compressed spherical; chloroplasts 1–3 or more ....*Botrydiopsis* (p. 248)
19 Cells subcylindrical to spindle-shaped; chloroplasts 2 ....*Bumilleriopsis* (p. 248)
20 Cells colonial or commonly incidentally forming colonies ....*Gloeobotrys* (p. 252)
20 Cells not in colonies 21
21 Cells essentially cylindrical ....*Monallanthus* (p. 256)
21 Cells non-cylindrical 22
22 Cells circular, always flattened and lens-shaped; walls ornamented with warts;chloroplasts 2–6 (cells in one species 4–6-sided but without arms) ....*Trachydiscus* (p. 259)
22 Cells not usually flattened, often ovate; walls not ornamented; chloroplasts 1–2 23
23 Cells ellipsoidal to croissant- or kidney-shaped, in one section; chloroplasts 1–2 per cell ....*Nephrodiella* (p. 256)
23 Cells ellipsoidal to spindle-shaped, number of sections uncertain; chloroplasts 2 per cell..*Pleurogaster* (p. 258)
24 Cells more or less perfectly tetrahedral, distinct angles between planes and arms which are only slightly concave or convex, sides slightly thickened, arms or corner extensions much reduced, almost spines ....*Tetraedriella* (p. 259)
24 Cells tetrahedral or polygonal, but not perfectly regular, and not as above.....................................25
25 Cells arms bifurcating...*Pseudostaurastrum* (p. 259)
25 Cells arms simple 26
26 Cells often irregular, polygonal with 3–10 sides or arms, variable from subordinate to central body becoming the major component of cell, occasionally with spines ....*Goniochloris* (p. 252)
26 Cells somewhat irregular, almost circular to polygonal with 4–6 sides or corners, with small rounded protuberances subordinate to central body ....*Polygoniochloris* (p. 258)

ORDER BOTRYDIALES

Multicellular vesicles. Sexual reproduction isogamous or anisogamous.

Botrydium Wallroth 1815

Macroscopic, coenocytic, with a conspicuous aerial thallus and pear-shaped or spherical, with a much branched, subterranean, colourless rhizoidal system; chloroplasts parietal, numerous and disc-shaped; oil or leucosin present; walls sometimes encrusted with calcium carbonate; asexual reproduction by uninucleate biflagellate zoospores, released when thallus flooded, or by aplanospores; sexual reproduction by iso- or anisogametes.

Cosmopolitan; clustered together on moist soil or mud at margins of recently flooded ponds, or on such areas when drying out; occasionally on damp soil or the shaded lowest parts of tree trunks, often associated with a superficial crust of calcium carbonate.

Botrydium granulatum (Linnaeus) Greville 1830

Basionym: *Ulva granulata* Linnaeus

10020010 **Pl. 65A** (p. 247)

Aerial portion spherical, ovoid or pear-shaped, up to 2.5(–3) mm wide; walls thin, often lime encrusted.

Probably cosmopolitan; as dull dark green beads on drying mud, especially at edges of lakes, ponds, rivers or ditches, usually in spring and autumn; sometimes on damp soil in greenhouses and on the shaded lowest parts of tree trunks, often associated with a superficial crust of calcium carbonate.

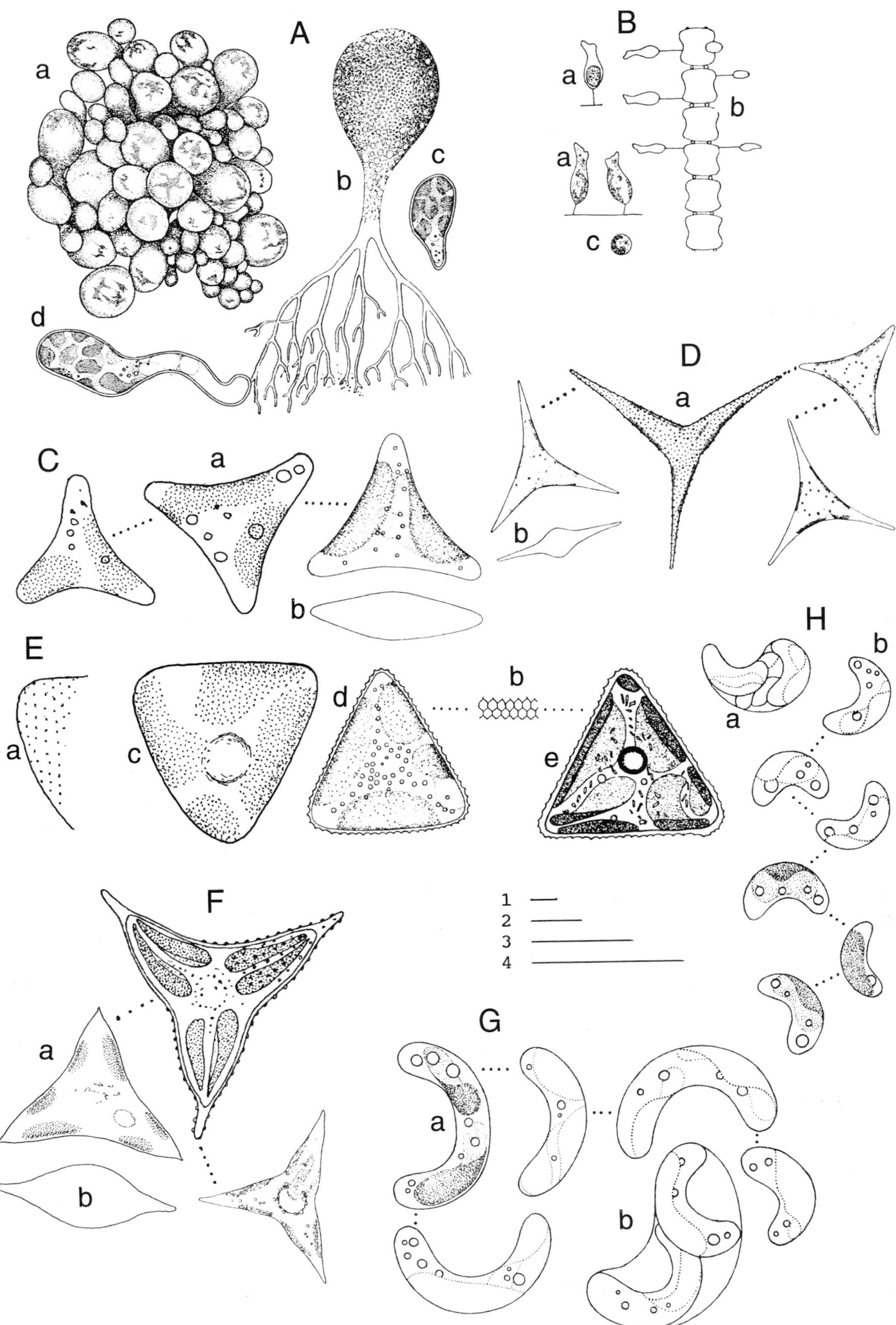

Plate 65 A–H
A. *Botrydium granulatum* (p. 246): a. colony, b. single plant, c, d. germlings; **B**. *Stipitocccus urceolatus* (p. 260): a. single cells, b. cells attached to a filamentous desmid, c. spore; **C**. *Goniochloris mutica* (p. 254): a. single cells, b. cell in profile; **D**. *Goniochloris fallax* (p. 254): a. single cells, b. cell in profile; **E**. *Goniochloris sculpta* (p. 254): a, b. details of cell ornamentation, c–e. single cells; **F**. *Goniochloris smithii* (p. 254): a–c. details of single cells, d. cell in profile; **G**. *Nephrodiella lunaris* (p. 256): a. single cells, b. cell dividing; **H**. *Nephrodiella nana* (p. 256): a. dividing cell, b. single cells.
Scale bars: 10 μm. 1 for Bc; 2 for Ba,b, D, F; 3 for C, E, G; 4 for Ac, d, H; Aa, largest vesicles about 2.5 mm wide; Ab, vesicle about 2.5 mm wide.

ORDER MISCHOCOCCALES

Coccoidal forms, solitary or colonial, not palmelloid. Zoospores uni- or biflagellate.

Botrydiopsis Borzí 1889

Cells non-motile, solitary or clustered, spherical, often comparatively large; walls smooth and in 2 overlapping halves (junction often not visible); chloroplasts parietal, 1–2(–3) when young and numerous in older plants; oil globules present; reproduction by numerous biflagellate zoospores and aplanospores.

Cosmopolitan; on soil and in aquatic habitats especially acidic humic waters. Four known species.

Botryidiopsis arrhiza Borzí 1895

10010010 **Pl. 67B** (p. 251)

Cells spherical, young cells 8–10 μm in diameter, older ones up to 70 μm; chloroplasts 2 to many, plate-like, without a pyrenoid.

Probably cosmopolitan; occurs on moist soil and aquatic at the edge of ponds, especially *Sphagnum* bogs, occasionally forms yellowish water blooms due to the presence of oil reserves which cause cells to rise to the surface as a film in calm weather or in sheltered waters.

DOUBTFUL RECORD

Botrydiopsis anglica F.E.Fritsch *et* R.P.John 1942

According to Ettl (1978), this is probably a green alga close to the genus *Dictyococcus*.

Bumilleriopsis Printz 1914

Cells solitary, occasionally in small clumps or chains, cylindrical or spindle-shaped, sometimes curved, apices rounded or acute, sometimes asymmetrical; walls thin to thick, without ornamentation, divided into 2 parts with junction not usually visible except during reproduction; chloroplasts parietal, 2 to several; asexual reproduction by biflagellate zoospores.

North America and Europe; mostly isolated from soils.

1 Apices somewhat tapered, one more than the other, often pointed .... *B. brevis*
1 Apices both bluntly rounded, not tapered .... *B. petersiana*

Bumilleriopsis brevis (Gerneck) Printz 1914

Basionym: *Ophiocytium breve* Gerneck

10040020 **Pl. 66I** (p. 250)

Cells cylindrical, straight or curved, mostly with ends asymmetrical, one end sometimes pointed, 4–10 μm wide, 10–30(–45) μm long; walls thin to fairly thick, of 2 unequal pieces; chloroplasts parietal, numerous; oil reserves present; reproduction by biflagellate zoospores.

Widespread; on damp mud and soil.

Bumilleriopsis petersiana Vischer *et* Pascher 1936

10040010 **Pl. 66J** (p. 250)

Cells cylindrical, straight or slightly curved, with both ends smoothly rounded, 7–10 μm wide, 20–40 μm long; chloroplasts parietal, numerous; oil droplets present; reproduction by biflagellate zoospores and aplanospores.

Widespread; occurs in bogs, still waters and soils.

Centritractus Lemmermann 1900

Cells solitary, ellipsoidal to cylindrical, straight or slightly curved, with a spine at each apex; walls often visible in 2 parts, jointed in mid-region or near one end, or in more elongated cells at both ends so forming a 'cap' at each apex; chloroplasts parietal, 1 to several; oil droplets present; reproduction by biflagellate zoospores, thin-walled autospores or thick-walled aplanospores.

North America and Europe, planktonic or amongst other algae.

Centritractus belenophorus Lemmermann 1900

10050010 **Pl. 67A** (p. 251)

Cells ellipsoidal when young with apical spines longer than cell, older cells up to 8 times longer than broad; wall visible in 2 parts, separated by a thin-walled area as the cell lengthens; reproduction by biflagellate zoospores or aplanospores.

Probably cosmopolitan; often known from acid bogs.

Characiopsis Borzí 1895

Cells solitary, ovoid, cylindrical, ellipsoidal, spindle-shaped, reverse pear-shaped or sickle-shaped, attached by a long or short stalk, often with basal disc; walls firm but usually thin and in one piece; chloroplasts parietal, disc- or band-shaped, 1 to several, without pyrenoids; oil globules present; usually uninucleate; asexual reproduction by zoospores, usually 2–4 per cell, with 2 unequal flagella, one usually very short.

Cosmopolitan; with many species, usually epiphytic or epizoic, but little recorded in the British Isles.

Characiopsis may be easily confused with the chlorophycean *Characium*; whose species are similar in size, shape and habit. However, the latter usually has one denser green chloroplast (due to its greater thickness and denser microstructure), with starch grains and also one or more pyrenoid(s) having a starch sheath.

1 Cells essentially sessile 2
1 Cells with stalk, sometimes very short.................4
2 Cells with a rounded apex.......................*C. naegelii*
2 Cells with an acute apex..3
3 Cells ellipsoidal, with a prominent spine .....*C. avis*
3 Cells spindle-shaped, sometimes almost cylindrical, without a spine..................*C. subulata*
4 Cells with a rounded apex......................................5

4 Cells with an acute apex 6
5 Cells halfmoon-shaped, 2–5 μm wide, frequently curved stem at angle to surface ....*C. curvata*
5 Cells ovoid, ellipsoidal to lance-shaped, 5–12(–24) μm wide, tapering at base to the disc ....*C. pyriformis*
6 Cells usually with 1 or 2 chloroplasts 7
6 Cells usually with 2–4 or more chloroplasts 9
7 Cells >30 μm wide ....*C. longipes*
7 Cells <30 μm wide 8
8 Stalk slender, with an attachment disc ......*C. acuta*
8 Stalk stout, often without an attachment disc ....*C. minuta*
9 Cells usually spindle-shaped to apiculate ovoid, 6–8 μm wide, 10–15 μm long ....*C. minor*
9 Cells spindle-shaped, cylindrical or broadly lance-shaped, >8 μm wide, >15 μm long 10
10 Cells spindle-shaped or cylindrical, about 10 μm wide ....*C. saccata*
10 Cells cylindrical to broadly lance-shaped, 10–18 μm wide ....*C. turgida*

Characiopsis acuta (A.Braun) Borzí 1895

Basionym: *Characium acutum* A.Braun
Synonym: *Hydrianum acutum* Rabenhorst
10060080 **Pl. 69G** (p. 255)
Cells ellipsoidal to spindle-shaped, tapering to an acute apex and to the base which narrows into a short stalk arising from a small basal disc, 6–10 μm wide and 15–28 μm long with stalk; walls thin; chloroplasts 1 or 2, parietal and thin; reproduction by 2–4 unequally biflagellate zoospores.

Europe and North America; epiphytic and epizoic, often in soft waters.

Characiopsis avis Pascher 1939

10060010 **Pl. 69B** (p. 255)
Cells obovoid, without stalk, but attached by a broad basal disc with a laterally directed spine arising near the apex, 6–9 μm wide, 12–23 μm long, 2 to 2.5 times longer than broad; walls thin; chloroplasts several, parietal and ellipsoidal; reproduction unknown.

Europe; epiphytic on filamentous algae, mainly in acid waters, uncommon.

Characiopsis curvata (G.M.Smith) Skuja 1948

Basionym: *Characium curvatum* G.M.Smith
10060020 **Pl. 69D** (p. 255)
Cells broadly lunate, attached by a short thick stalk arising laterally from the lower end; about 2.5 μm wide, 4–13 μm long and 1.5–3 μm thick, not including the 1.3 μm stalk; walls thin; chloroplasts 1–4, parietal; reproduction unknown.

Probably cosmopolitan; epiphytic on planktonic algae.

Characiopsis longipes (Rabenhorst) Borzí 1895

Basionym: *Characium longipes* Rabenhorst
10060090 **Pl. 69F** (p. 255)
Cells spindle-shaped, straight or curved on a slender stalk about as long as the cell, usually without a basal attachment disc, 4–7.5 μm wide, 40–50 μm long; walls thin; chloroplasts 1–3; reproduction by unequally biflagellate zoospores.

Probably cosmopolitan; epiphytic on filamentous algae and aquatic macrophytes, also epizoic.

Characiopsis minor Pascher 1925

10060030 **Pl. 69I** (p. 255)
Cells variable in shape, usually spindle-shaped to apiculate, oviform, at the base tapering into a short stalk without a basal disc, 6–8 μm wide, 10–15 μm long (without stalk); walls thin; chloroplasts 2–4, parietal, often pale.

Probably cosmopolitan; epiphytic on filamentous algae, mostly *Zygnema*, in fairly acidic waters.

Characiopsis minuta (A.Braun) Lemmermann 1914

Basionym: *Characium minutum* A.Braun
Synonyms: *Characium acutum* Schröder, *C. ambiguum* Harman, *C. tenue* Harman, *C. subulatum* G.S.West *non Characiopsis minuta sensu* Borzí
10060040 **Pl. 69K** (p. 255)
Cells cylindrical to spindle-shaped, tapering at apices, with a short stalk having a small basal attachment disc, apex sometimes somewhat hooked, up to 7 μm wide and 12–20 μm long, 2.5 to 4.5 times as long as broad; chloroplasts 1 or 2, parietal, sometimes elongated or band-shaped, pale; reproduction by biflagellate zoospores, usually 4 per cell.

Probably cosmopolitan; widespread as an epiphyte on filamentous algae, diatoms and aquatic macrophytes.

Characiopsis naegelii (A.Braun) Lemmermann 1914

Basionym: *Characium naegelii* A.Braun
10060050 **Pl. 69A** (p. 255)
Cells ellipsoidal, smoothly rounded at apex, with a very short stalk and a small basal disc, 20–35(–60) μm long; chloroplasts numerous, disc-shaped; reproduction by biflagellate zoospores.

Probably cosmopolitan; widespread, epiphytic or attached to various substrata.

Characiopsis pyriformis (A.Braun) Borzí 1895

Basionym: *Characium pyriforme* A.Braun
Synonym: *Hydrianum pyriforme* Rabenhorst
10060100 **Pl. 69E** (p. 255)
Cells obovoid to pear-shaped, rounded at apex, with a long slender stalk and a small basal disc, 10–14 μm wide, 12–24 μm long, stalk up to 15 μm long; chloroplasts 2–5, parietal, thin; reproduction usually by 4 flagellated and amoeboid zoospores produced in each cell.

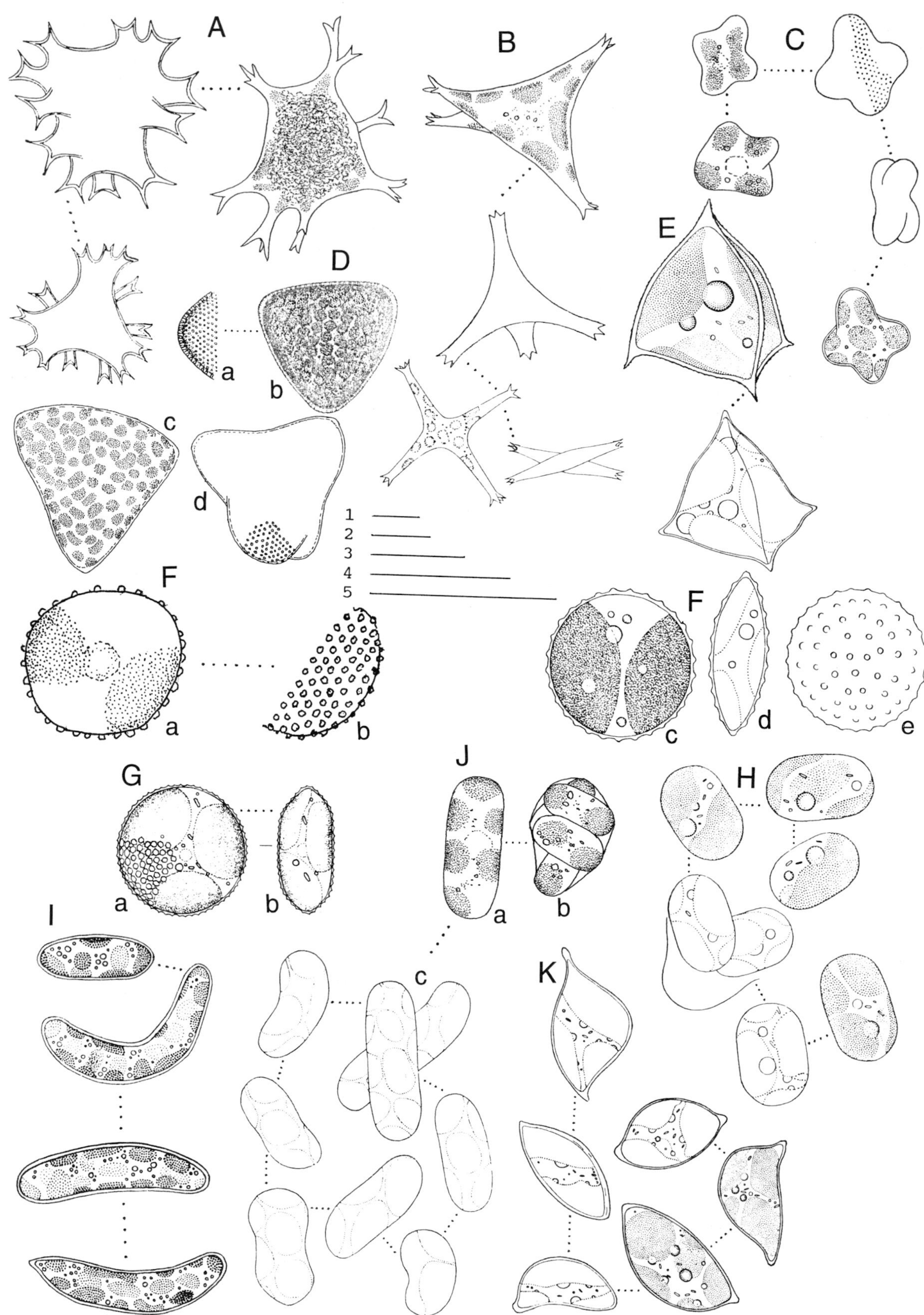

Plate 66 A–K

A. *Pseudostaurastrum enorme* (p. 259); **B**. *Pseudostaurastrum hastatum* (p. 259); **C**. *Tetraedriella jovetti* (p. 259); range of form of single cells containing chloroplasts and one cell (top right) showing wall ornamentation; **D**. *Tetraedriella polychloris* (p. 259): a, c. detail of ornamentation, b, d. single cells; **E**. *Tetraedriella limbata* (p. 259); **F**. *Trachydiscus lenticularis* (p. 260): a, c, e, single cells, b, e. details of ornamentation, d. cell in profile; **G**. *Polygoniochloris circularis* (p. 258): a. single cell, b. cell in profile; **H**. *Monallanthus brevicylindrus* (p. 256); **I**. *Bumilleriopsis brevis* (p. 248): showing variation on cell shape; **J**. *Bumilleriopsis petersiana* (p. 248): a, c. single cells, b. dividing cell; **K**. *Pleurogaster lunaris* (p. 248): showing variation in cell shape.

Scale bars: 10 μm. 1 for D, J; 2 for A, B, C, K; 3 for G, H, I; 4 for E; 5 for F.

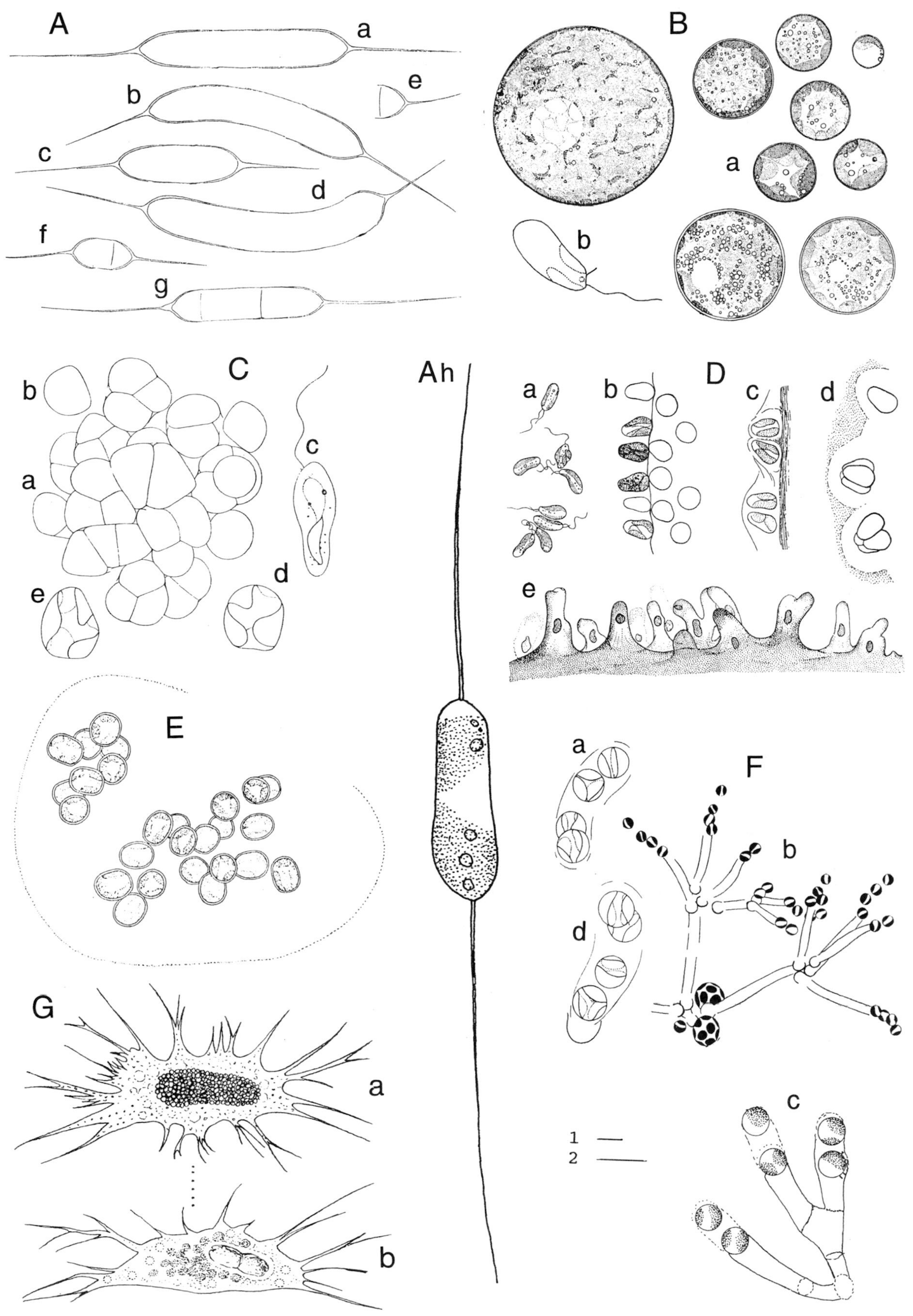

Plate 67 A–G

A. *Centritractus belenophorus* (p. 248): a-d, h. single cells, e. detached cap, f, g. cells forming caps; **B.** *Botrydiopsis arrhiza* (p. 248): a. individual cells, b. zoospore; **C.** *Chlorellidiopsis separabilis* (p. 252): a. colony, b. single cell, c. zoospore; d, e. details of separate cells; **D.** *Chlorosaccus fluidus* (p. 252): a. zoospores, b. details of individual cells, c, d. detail of mucilaginous envelope, e. profile of colony; **E.** *Gloeobotrys limneticus* (p. 252); **F.** *Mischococcus confervicola* (p. 254): a, d. details of individual cells, b, c. typical colonial structure; **G.** *Chlamydomyxa labrynthuloides* (p. 260): a.,b. vegetative stages.

Scale bars: 10 µm. 1 for D, Fb, G; 2 for A, B, C, E, Fa,c,d.

Probably cosmopolitan; attached to volvocalean algae, diatoms, filamentous algae and aquatic macrophytes.
Eight varieties have been reported for this species.

Characiopsis saccata N.Carter 1919

10060110 **Pl. 69J** (p. 255)
Cells spindle-shaped or cylindrical, tapering at ends, acute apex, with very short stalk and a small basal disc, about 10 μm wide, 20–50 μm long; chloroplasts numerous, parietal; reproduction unknown.

Europe; epiphytic, including on dead leaves in pools.

Characiopsis subulata (A.Braun) Borzí 1895

Basionym: *Characium subulatum* A.Braun
10060060 **Pl. 69C** (p. 255)
Cells spindle-shaped with acute apex, base rounded and arising from a well-developed basal disc, 4–6 μm wide, up to 20 μm long; chloroplasts 2–4(–6), parietal; reproduction by 2–4 unequal biflagellate zoospores.

Europe and America; epiphytic on various filamentous algae in water with a wide pH range.

Characiopsis turgida West *et* G.S.West 1903

10060070 **Pl. 69H** (p. 255)
Cells cylindrical to broadly lance-shaped, with acute apex and tapering at base to a very short stalk and a small basal disc, 10–18 μm wide, 30–50 μm long; chloroplasts numerous, parietal; reproduction unknown.

British Isles and Europe; epiphytic on various algae, particularly in more acid waters.

DOUBTFUL RECORDS

Characiopsis cylindrica (Lambert) Lemmermann 1914
A very doubtful species that is not recognized in the most recent taxonomic treatments, possibly an *Ophiocytium*. It has been renamed *Rhopalosolen* by Fott (see Komárek & Fott, 1983) with the figure showing little resemblance to the taxon as originally described. The only record for the British Isles is in a publication by Benson-Evans & Antoine (1996) on the freshwater, brackish-water and marine algae of South Wales. Considerable doubt attaches to the record since it is not accompanied by any mention of a locality and the illustration is inadequate. According to Benson-Evans & Antoine, it grows on *Daphnia* and other planktonic animals in small bodies of standing water.

Chlorellidiopsis separabilis Pascher 1939
Only isolated from soil in the Prague botanical garden (Pl. 67C, p. 251) and so its single record from 'static water' in South Wales (Benson-Evans & Antoine, 1996) is doubtful in the absence of a locality or adequate illustration.

Chlorosaccus Luther 1899

Colonial, spherical to irregular, mucilaginous, with cells embedded at the periphery mostly in 4s in conical or finger-like projections; cells ellipsoidal to pear-shaped, outermost ones tending to point away from centre of colony; walls very thin; chloroplasts 2 or more, large and parietal; oil droplets present; reproduction by unequally biflagellate zoospores.

Europe; epiphytic and aquatic. Monospecific.

Chlorosaccus fluidus Luther 1899

Synonyms: *Chlorosaccus ulvaceus* Messkommer *et* Vischer, *C. concarnensis* Bourrelly
10080010 **Pl. 67D** (p. 251)
Colony soft, light green, up to 15 μm or more across; cells in 4s, 5–6 × 18–12 μm, frequently with short papillae.

Europe; often on aquatic macrophytes, possibly favours the cooler months and organically rich waters.

Gloeobotrys Pascher 1930

Colonies free-floating or attached, 2- to many hundred-celled, within an unstratified or weakly stratified soft mucilaginous envelope; cells spherical to ellipsoidal; walls thin to thick; chloroplasts parietal, 1 to many, plate- or disc-shaped, without a pyrenoid; reproduction by zoospores or aplanospores.

North America and Europe; occurs usually in acidic or soft water.

Gloeobotrys limneticus (G.M.Smith) Pascher 1939

Basionym: *Chlorobotrys limneticus* G.M.Smith
10220010 **Pl. 67E** (p. 251)
Mucilaginous colonies 40–200 μm in diameter, with 10–30 cells in small groups or scattered; cells ellipsoidal, 5–6 × 6–8 μm; chloroplasts 3–4 per cell.

Possibly cosmopolitan; occurs in lake plankton.

Goniochloris Geitler 1928 emend. Bourrelly 1951

Cells solitary, regularly or irregularly triangular, more or less lens-shaped in side view, sometimes twisted; walls thick and firm, smooth or sometimes regularly sculptured with pits or warty; apices acute or rounded, sometimes with small spines or processes; chloroplasts 2–6, parietal, plate- or ribbon-shaped; oil bodies and leucosin present, 1–2 red pigment-spots usually present; reproduction by autospores.

Probably cosmopolitan; a wide range of aquatic habitats. Some 20 known species.

Bourrelly (1981) notes that Ettl (1978) reserves *Goniochloris* for cells with a triangular outline, and *Polygoniochloris* for polygonal or circular cells.

1 Corners of the triangular cells more or less rounded; walls regularly pitted 2
1 Corners of cells tapering into acute spices; walls not pitted 3
2 Cells 10–12 μm across, with concave sides; pits <1 μm in diameter .... *G. mutica*
2 Cells 12–16 μm across, straight sided; pits about 1.4 μm in diameter .... *G. sculpta*

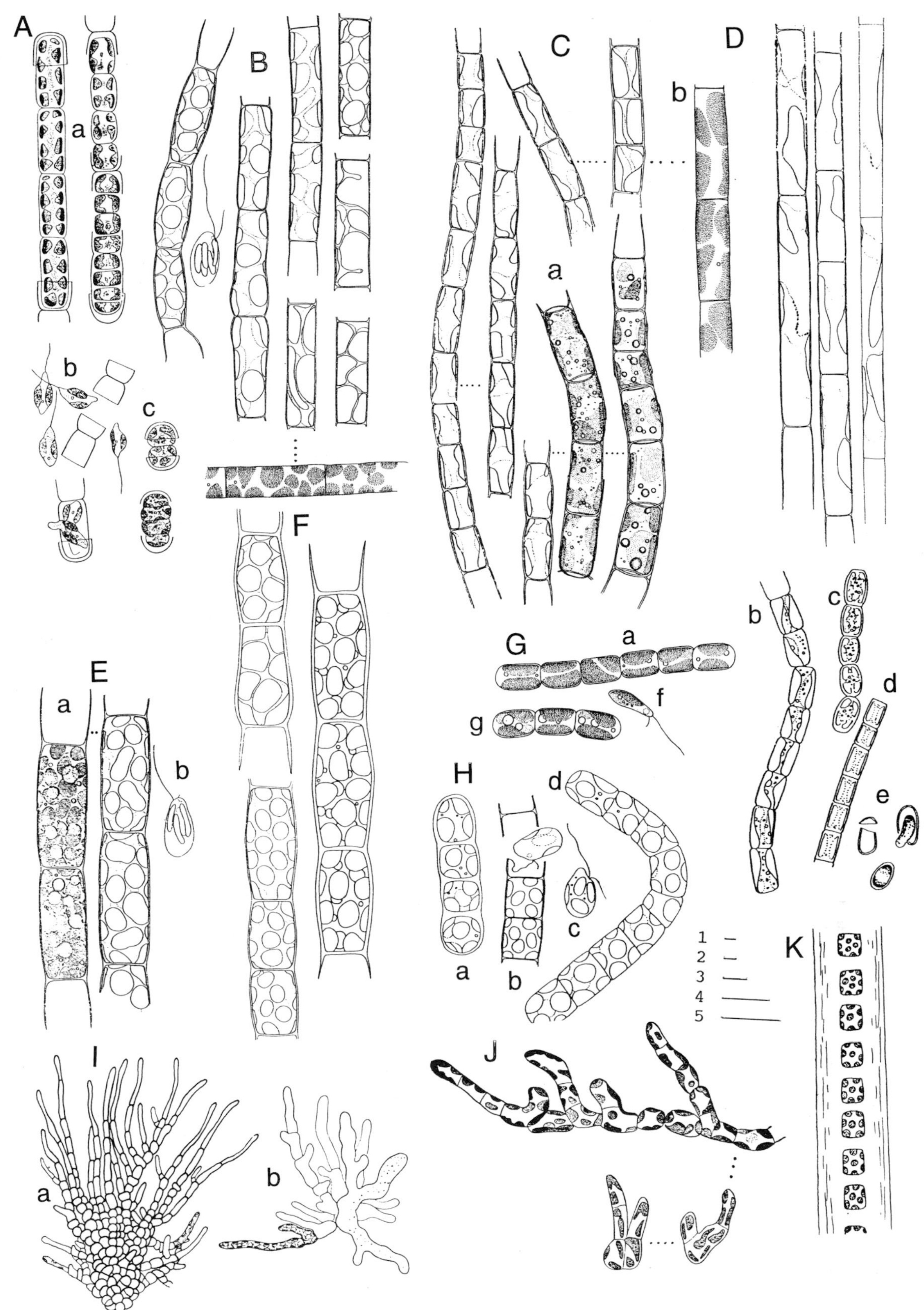

Plate 68 A–K

A. *Bumilleria sicula* (p. 260): a. filaments, b. swarmer and swarmer release, c. germination of a swarmer; **B**. *Tribonema vulgare* (p. 261); **C**. *Tribonema minus* (p. 261): a. filaments, b. details of cell contents; **D**. *Tribonema affine* (p. 261); **E**. *Tribonema viride* (p. 261): a. filaments, b. swarmer; **F**. *Tribonema utriculosum* (p. 261); **G**. *Xanthonema exile* (p. 261): a–d, g. filaments, e. details of germination, f. swarmer; **H**. *Xanthonema quadrata* (p. 261): a, d. filaments, b. filament with release of swarmer, c. swarmer; **I**. *Heterococcus chodatii* (p. 254): a. radiating filaments with cells coccoid towards centre, b. small individual with details of 2 cells; **J**. *Heterococcus viridis* (p. 254): mature and young filaments; **K**. *Neonema pumilum* (p. 256).

Scale bars: 10 μm. 1 for Ia; 2 for A; 3 for Ca, J; 4 for Cb, D, E–H, Ib, K; 5 for B.

3 Corners of cells forming stout arms with rather convex sides .... *G. smithii*
3 Corners of cells forming slender, almost spine-shaped arms with sides smoothly concave .... *G. fallax*

Goniochloris fallax Fott 1960

Synonym: *Tetraedron trigonum* auct.
10090010 **Pl. 65D** (p. 247)
Cells mostly 20–28 μm across, more or less regularly triangular, with strongly concave sides and acute, almost spine-shaped arms; walls delicate, colourless or pale brown, sometimes finely punctate (although this is not always visible); chloroplasts parietal, 4–6, disc-shaped; reproduction unknown.

Europe; occasional, planktonic especially in still or flowing nutrient-enriched waters.

Goniochloris mutica (A.Braun) Fott 1960

Basionym: *Polyedrium muticum* A.Braun
Synonyms: *Pseudostaurastrum muticum* (A.Braun) Bourrelly; *Tetraedron muticum* (A. Braun) Hansgirg
10090020 **Pl. 65C** (p. 247)
Cells mostly 10–12 μm in size, in the shape of an equilateral triangle, with slightly concave sides and rounded angles; walls thin, ornamention in the form of a honeycomb-like pattern of small pits less than 1 μm across and not readily visible; chloroplasts 2–3, disc-shaped; oil globules often one or more, sometimes orange-tinted; reproduction unknown.

Possibly cosmopolitan; planktonic in pools, lakes and slow-flowing rivers.

Goniochloris sculpta Geitler 1928

Synonym: *Pseudostaurastrum sculpta* (Geitler) Bourrelly
10090030 **Pl. 65E** (p. 247)
Cells mostly 12–16 μm across, in the shape of an equilateral triangle, with more or less straight sides and rounded angles; walls firm, silicified, ornamented with pits about 1.4 μm across and arranged in a honeycomb-like pattern; chloroplasts 4 or more, disc-shaped or slightly elongated; reproduction by 4 autospores.

Probably cosmopolitan; in bogs, ponds and lakes, less frequent than *G. mutica*.

Goniochloris smithii (Bourrelly) Fott 1960

Basionym: *Pseudostaurastrum smithii* Bourrelly
10090040 **Pl. 65F** (p. 247)
Cells 25–30(–50) μm across, triangular with stout and often slightly convex-sided arms tapering to an acute point having an almost nipple-shaped extension; walls thin to fairly thick, with an almost invisible ornamentation; chloroplasts typically 6, disc-shaped; oil droplets concentrated around the nucleus; reproduction unknown.

Europe; sporadic in the plankton of small ponds and lakes.

Heterococcus Chodat 1908

Multicellular and attached, consisting of a central and more or less irregular aggregation of cells from which arise radiating branched filaments, with filaments prostrate or erect; cells of central mass spherical to irregular and those of filaments shortly cylindrical to elongated; chloroplasts parietal, disc-shaped, usually several per cell; asexual reproduction by zoospores and aplanospores.

Probably cosmopolitan; subaerial alga of damp soil, bases of tree trunks or on the underside of stones.

A difficult genus to identify to species level, especially since some species are known only from culture. Lokhorst (1992) monographed this genus based on laboratory culture studies and recognized some 20 species and varieties. In culture the central area of the plant may consist solely of rounded empty cells after zoospore production.

1 Cells up to 10 times longer than broad, especially at tips of filaments ....*H. chodatii*
1 Cells usually less than 4 times longer than broad ....*H. viridis*

Heterococcus chodatii Vischer 1937

10230010 **Pl. 68I** (p. 253)
Cells about 5 μm wide, up to 10 times longer than broad, with central cells becoming more rounded with age, eventually coccoidal and possibly 20 μm across.

Probably cosmopolitan; usually subaerial in damp or sheltered situations in the British Isles where known from alkaline soils.

Heterococcus viridis Chodat 1908

10230020 **Pl. 68J** (p. 253)
Cells 6–14(–16) μm wide, 12–20 μm long, with central cells becoming more rounded in maturity, eventually coccoidal.

Probably cosmopolitan; usually subaerial in damp or sheltered situations, found in alkaline soils in the British Isles.

A poorly defined species. Pascher (1939) reduces *H. chodatii* to a synonym of it.

Mischococcus Nägeli 1849

Plants attached, consisting of a dichotomously branched mucilaginous tube enclosing or bearing spherical cells at the dichotomies or ends, with cells single or in 2s or 4s; chloroplasts parietal, 1–4, disc-shaped, pale yellow-green; asexual reproduction by aplanospores or flagellate zoospores; sexual reproduction by isogametes, only known in one species.

North America and Europe; frequently attached to filamentous algae as well as on inert surfaces, common in calcareous waters.

Mischococcus confervicola Nägeli 1849

10120010 **Pl. 67F** (p. 251)
Cells spherical, 5–8(–10) μm in diameter; walls firm; chloroplasts 1 or 2; reproduction by unequally biflagellate zoospores.

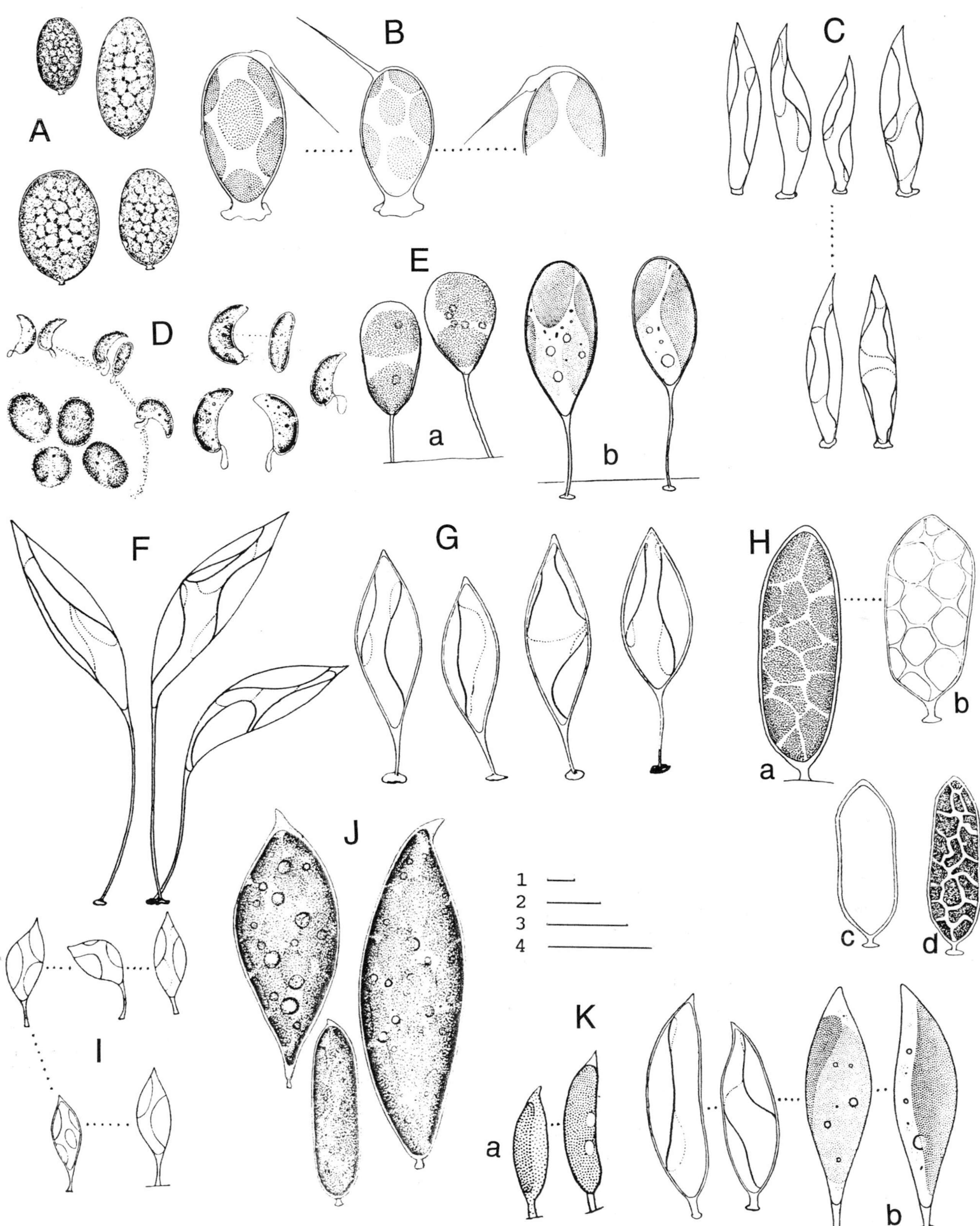

Plate 69 A–K

A. *Characiopsis naegelii* (p. 249); **B**. *Characiopsis avis* (p. 249); **C**. *Characiopsis subulata* (p. 252); **D**. *Characiopsis curvata* (p. 249); **E**. *Characiopsis pyriformis* (p. 249); **F**. *Characiopsis longipes* (p. 249); **G**. *Characiopsis acuta* (p. 249); **H**. *Characiopsis turgida* (p. 252); **I**. *Characiopsis minor* (p. 249); **J**. *Characiopsis saccata* (p. 252); **K**. *Characiopsis minuta* (p. 249).

Scale bars: 10 μm. 1 for A, Ka; 2 for D–F, H; 3 for B, G, I, Kb; 4 for C, J.

Probably cosmopolitan; widespread, mainly as an epiphyte, rare on various filamentous algae or on submerged macrophytes in small ponds and shallow lake margins, usually in alkaline or neutral water.

Monallanthus Pascher 1939

Cells solitary, or after division briefly united in 2s or 4s in a group or filament, regularly cylindrical with rounded or flattened ends, occasionally slightly curved; walls thin to thick, sometimes reddish in one piece, without ornamentation; chloroplasts parietal, mostly 2, sometimes 1 or 4, disc-shaped or elongated, with pyrenoids in one species; oil droplets often present; asexual reproduction mostly by 2 unequally biflagellate zoospores, sometimes zoospores amoeboid, or by 2 autospores.

North America and Europe; known from shallow waters.

Similar to *Bumilleriopsis* but cells shorter and straight, and with numerous chloroplasts.

Monallanthus brevicylindrus Pascher 1939

10240010 **Pl. 66H** (p. 250)

Cells shortly cylindrical with hemispherical ends, 6–8 μm wide and 9–12 μm in length; walls very thin, sometimes with a thin mucilaginous layer; chloroplasts (1–)2–4, large, usually lining most of cell wall.

Europe and North America; from shallow acid or organically enriched waters.

Neonema Pascher 1925

Filamentous, forming slimy yellow-green masses; filaments containing a single row of short separate cells surrounded by a stratified mucilaginous sheath; cells spherical to shortly cylindrical; walls thin, H-shaped sections not apparent; chloroplasts parietal, disc-shaped, several present; asexual reproduction by unequally biflagellate zoospores.

Europe, rare. Only 2 known species.

Neonema pumilum (West *et* G.S.West) Pascher 1932

Basionym: *Bumilleria pumila* West *et* G.S.West

10250010 **Pl. 68K** (p. 253)

Filaments mucilaginous with a single row of unconnected cells; cells 4.8–5.7 μm wide and 5–6 μm long.

Central Europe; rare in the British Isles (only known from Cornwall).

Nephrodiella Pascher 1939

Cells normally solitary, occasionally temporarily surrounded by mucilage and in 2s or 4s; cells kidney-shaped to lunate, sometimes S-shaped, spherical or ellipsoidal in cross section; walls thin, without ornamentation, probably in one piece; chloroplasts parietal, disc- or cup-shaped, 1–3 but mostly single; asexual reproduction by autospores or zoospores with a very short lateral flagellum.

Europe, in boggy water of low pH or on soil.

A poorly defined genus of about 6 species.

1 Cells about 1.5 μm wide, up to 6 μm long; chloroplast always single............................*N. nana*
1 Cells 3–6 μm wide, up to 18 μm long; chloroplasts 1 or 2 per cell........................*N. lunaris*

Nephrodiella lunaris Pascher 1939

10130020 **Pl. 65G** (p. 247)

Cells solitary except soon after reproduction; cells 3–6 μm wide, 7–18 μm long, kidney-shaped to lunate, rounded or somewhat tapered at ends; chloroplasts 1 or 2; asexual reproduction by 2 or 4 autospores, attached for a short time after release.

Europe; in acidic to mildly alkaline water, especially in moorlands and bogs; common.

Nephrodiella nana H.Ettl 1978

10130010 **Pl. 65H** (p. 247)

Cells 1.5 μm wide, up to 6 μm long, kidney-shaped, rounded or slightly tapered at ends; chloroplast single.

Europe; from soil samples.

Ophiocytium Nägeli 1849

Cells attached or free-floating, solitary or forming branched colonies, cylindrical or club-shaped, usually elongated, straight, curved or even spirally coiled, sometimes with a capitate apex, sometimes with one apex or both bearing a spine; walls thin to fairly thick; chloroplasts parietal, disc-shaped and elongated or irregularly star-shaped; oil globules present; asexual reproduction by aplanospores or unequally biflagellate zoospores, released by the end part of the cell lifting away as a cap, with daughter cells sometimes becoming attached to the mother cell wall near point of exit.

Cosmopolitan; epiphytic on other algae or free-floating.

1 Cells epiphytic with a short stalk, long, often forming branched colonies where daughter cells become attached to the mother cell wall ....*O. arbusculum*
1 Cells free-floating, not as above.............................2
2 Cells without a spine...........................*O. parvulum*
2 Cells with 1 or 2 spines..3
3 Cells with a spine at one end.................................4
3 Cells with a spine at both ends..............................5
4 Cells 8–17 μm in diameter, with a small knob at tip of spine ....*O. majus*
4 Cells 6–9.5 μm in diameter, no knob at tip of spine ....*O. cochleare*
5 Cells 5–10 μm wide ....*O. capitatum*
5 Cells 14–21 μm wide ....*O. bicuspidatum*

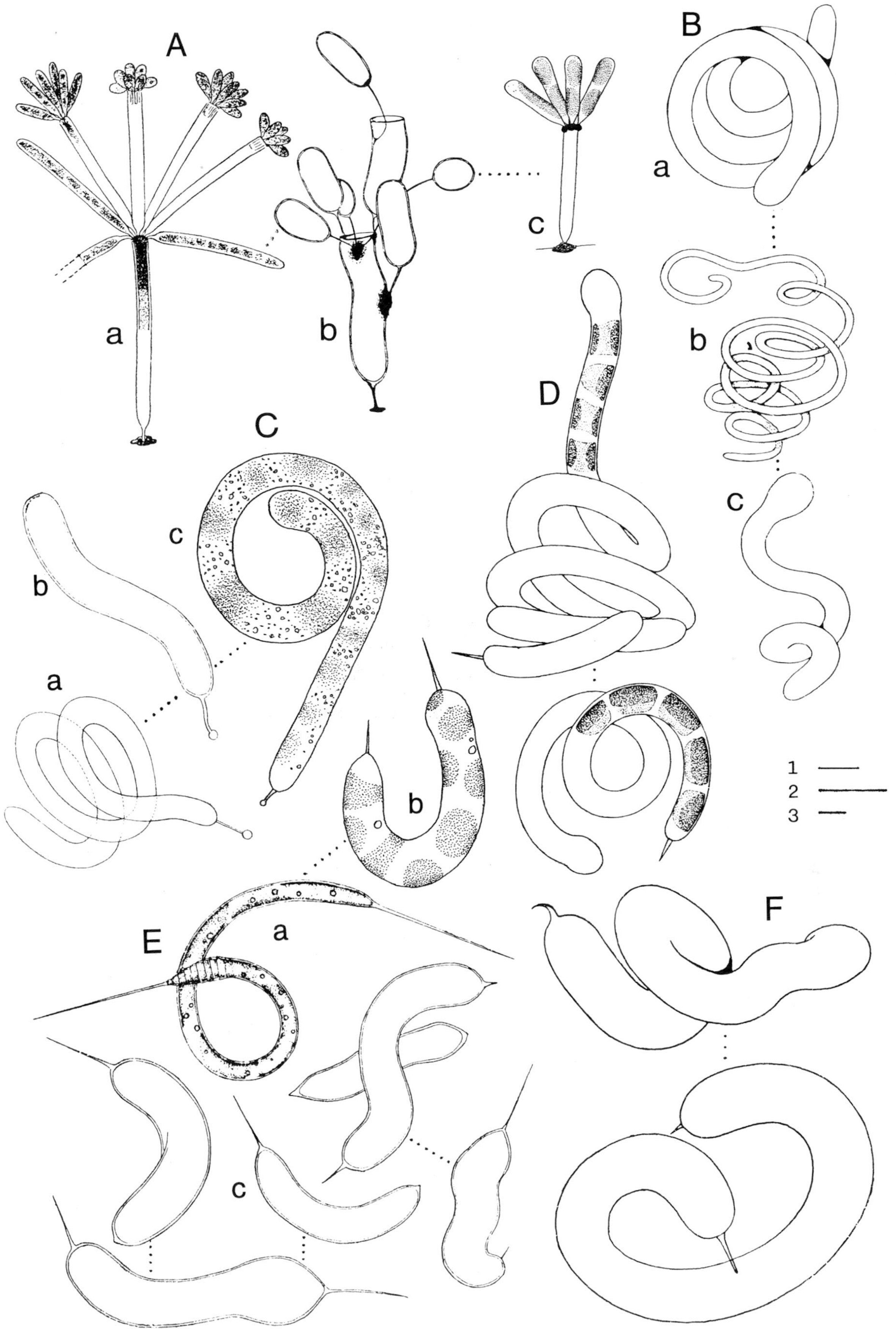

Plate 70 A–F

A. *Ophiocytium arbusculum* (p. 258): a. multiple colony, b. sporelings each with a stalk, c. young colony; **B**. *Ophiocytium parvulum* (p. 258): a–c. variation in cell shape; **C**. *Ophiocytium majus* (p. 258): a, b. variation in cell shape, c. details of cell content; **D**. *Ophiocytium cochleare* (p. 258); **E**. *Ophiocytium capitatum* (p. 258): a–c. variation in cell shape; **F**. *Ophiocytium bicuspidatum* (p. 258).

Scale bars: 10 μm. 1 for Aa,c, C, D, Eb,c, F; 2 for Ab, Ba,c; 3 for Ea.

Ophiocytium arbusculum (A.Braun) Rabenhorst 1868

Basionym: *Sciadium arbuscula* A.Braun

10140010 **Pl. 70A** (p. 257) **CD**

Cells epiphytic, forming colonies consisting of an empty cylindrical mother cell wall, straight or curved, 3–5(–8) μm wide, up to 100 μm long, attached at base by a short stalk and about whose apex similar daughter cells are attached as a result of the settlement and growth of released zoospores, these daughter cells are sometimes branched again; walls firm and persistent, with stalk and basal disc often iron-encrusted; chloroplasts irregularly elliptical and at intervals along cell; oil globules often present.

Probably cosmopolitan; usually on filamentous algae, especially in calcareous water.

Ophiocytium bicuspidatum (Borge) Lemmermann 1899

Basionym: *Ophiocytium majus* Nägeli var. *bicuspidatum* Borge

10140040 **Pl. 70F** (p. 257)

Robust free-floating cells, cylindrical, arcuate or twisted, 15–20 μm wide, up to 55 μm long, with a spine up to 10 μm long at each end; chloroplasts disc-shaped and numerous; colony formation not observed.

North America, Europe; uncommon, in acidic waters especially those of lakes and pools.

West & Fritsch (1927, p. 309, fig. 120 H, I) describe and illustrate forma *longispina* Lemmermann, with longer spines than the type; probably no useful purpose is served by recognizing such forms.

Ophiocytium capitatum Wolle 1887

10140020 **Pl. 70E** (p. 257)

Free-floating cells, cylindrical, straight, arcuate or twisted, with a short spine at each end; cells 5–7(–10) μm wide, up to 150 μm long; chloroplasts disc-shaped and numerous; oil droplets present.

Probably cosmopolitan; common in lakes and acidic waters.

Differs from *O. bicuspidatum* particularly in its much narrower cells.

Ophiocytium cochleare (Eichwald) A.Braun 1855

Basionym: *Spirodiscus cochlearis* Eichwald

0140030 **Pl. 70D** (p. 257)

Cells free-floating, strongly curved or spirally twisted, 5–9 μm wide, 10–500 μm long, with an expanded bulb-like portion at one end and a stout sharp spine up to 12 μm long at the other; chloroplasts disc-like at intervals along cell.

Probably cosmopolitan; common in the plankton of pools, lakes and swamps, especially on moorlands.

Ophiocytium majus Nägeli 1849

10140050 **Pl. 70C** (p. 257)

Cells free-floating, cylindrical, often with a wavy margin or spiralled, 5–10 μm wide, up to 600 μm long or even longer, rounded at each end, with a spine on one end terminating in a spherical process; chloroplasts disc-shaped or net-like.

Probably cosmopolitan; common among other algae and water plants in small ponds and slow-flowing ditches; it is most characteristic of hard waters and especially common on the magnesian limestone in north-east England.

Ophiocytium parvulum (Perty) A.Braun

Basionym: *Brochidium parvulum* Perty

10140060 **Pl. 70B** (p. 257)

Cells free-floating, cylindrical, straight, curved or contorted, (3–)4–6(–9) μm wide, up to 1000 μm long, without spines but with both apices rounded, one often rather swollen.

Probably cosmopolitan; in the plankton of lakes, pools and puddles, often in rather acid waters.

Pleurogaster Pascher 1939

Cells solitary or sometimes in 2s and 4s for a short time after division, asymmetrically ellipsoidal or broadly halfmoon-shaped (lunate); walls smooth, fairly thick, thickened at each end and prolonged to give cell more or less the shape of an orange pip; chloroplasts parietal, (1–)2–4, pyrenoids absent; oil droplets often present, usually reddish in colour; asexual reproduction by 2 or 4 aplanospores, or occasionally by unequally biflagellate zoospores.

Europe and America; usually in standing water and on soil, often in slightly acidic water.

A poorly known genus.

Pleurogaster lunaris Pascher 1939

10260010 **Pl. 66K** (p. 250)

Cells as above, 8–15 μm wide, 16–25 μm long; chloroplasts several and oil droplets present.

North America, Europe; planktonic in muddy pools and ponds as well as other bodies of standing water and on soil.

Polygoniochloris H.Ettl 1965

Cells solitary, flattened and cushion-like, square, polygonal, almost round or sometimes somewhat irregular in form; wall firm, often somewhat pigmented yellowish or reddish, with surface ornamented with regularly arranged pits; chloroplasts parietal, 2–5 in each cell; oil droplets present; asexual reproduction only by autospores.

Most reports from Europe, possibly cosmopolitan; planktonic and usually in standing water of bogs and moors.

Polygoniochloris circularis (Bourrelly *et* Georges) H.Ettl 1965

Basionym: *Pseudostaurastrum circulare* Bourrelly *et* Georges

10150010 **Pl. 66G** (p. 250)

Cells flattened, cushion-like, circular or nearly circular, 2.5–5 μm thick, 7–15 μm wide, in cross-section biconvex, almost ellipsoidal or lens-shaped; walls firm, ornamented with small depressions arranged in a regular honeycomb-like pattern; reproduction not observed.

Probably cosmopolitan; planktonic in various types of water body including very nutrient-rich ponds.

Considered common and widespread, but must be distinguished from the very similar *Trachydiscus lenticularis* H.Ettl, whose wall is ornamented with regularly arranged wart-like tubercles rather than pits.

Pseudostaurastrum Chodat 1921

Cells solitary, basically tetragonal with a central portion and 4 arms, either arms long and thin or short and much subdivided, with walls concave between arms; chloroplasts parietal, disc- or band-shaped, numerous, yellowish-green, without pyrenoids; reproduction not observed.

Probably cosmopolitan; planktonic in acid water lakes and ponds. Of the 3 or 4 species, only 2 recorded from the British Isles.

1 Cells with a stout central body and relatively short stout arms, much branched ....*P. enorme*
1 Cells with a relatively small central body and slender arms terminating in 3–4 claw-like spines...*P. hastatum*

Pseudostaurastrum enorme (Ralfs) Chodat 1921

Basionym: *Staurastrum enorme* Ralfs
Synonym: *Tetraedron enorme* (Ralfs) Hansgirg
10160010 **Pl. 66A** (p. 250)
Cells (20–)25–45 µm across, irregularly tetrahedral or polyhedral, with a stout central body and short thick bifurcating arms ending in 5–10 µm long spines; cytoplasm sometimes contains many granules of an unknown nature; chloroplasts occasionally extend into arms.

Probably cosmopolitan; planktonic or amongst other algae in acid to alkaline waters including Scottish lochs and moorland ponds.

A relatively variable species, of which several varieties and forms have been described, many of which remain little known.

Pseudostaurastrum hastatum (Reinsch) Chodat 1921

Basionym: *Polyedrium tetraedricum hastatum* Reinsch
Synonym: *Tetraedron hastatum* (Reinsch) Hansgirg
10160020 **Pl. 66B** (p. 250)
Cells 28–38 µm across, tetrahedral, with a relatively small central body and tapering arms dividing at the ends into small claw-like spines; oil globules present; chloroplasts tending not to extend into arms.

Europe; planktonic in standing water such as small lakes (e.g. Chartwell Lake, Kent), possibly in organically polluted waters.

Tetraedriella Pascher 1930

Cells solitary, free-living, basically tetrahedral, sometimes with 4 rounded corners and concave or convex sides, corners occasionally extended into slender spines or more robust arms; walls thin to robust, sometimes coarsely ornamented, punctae regular or more random in some species; chloroplasts parietal, disc- or cup-shaped, usually several; oil droplets and chrysolaminarin sometimes present; asexual reproduction by autospores and zoospores.

Possibly cosmopolitan; usually in standing or slow-flowing water, fresh or brackish.

A poorly known genus, reported only from Europe.

1 Cells irregularly tetrahedral, with broadly rounded corners and sides more or less concave..*T. jovetii*
1 Cells more regularly tetrahedral, with less obtuse corners and sides not concave.....................2
2 Cells tetrahedral, sometimes somewhat flattened, with rounded corners, without ribs, along edges................................*T. polychloris*
2 Cells tetrahedral with slightly acute corners and ribs along edges ....*T. limbata*

Tetraedriella jovetti (Bourrelly) Bourrelly 1968

Basionym: *Pseudostaurastrum jovetti* Bourrelly
10180010 **Pl. 66C** (p. 250)
Cells 10–12 µm across, tetrahedral, sometimes rather flattened, with broadly rounded corners and often somewhat concave sides, occasionally with a short spine; wall thin, finely ornamented with sparse punctae; reproduction unknown.

Europe; planktonic in pools and ponds.

Tetraedriella limbata Pascher 1939

10180020 **Pl. 66E** (p. 250)
Cells 10–12 µm across, irregularly tetrahedral, with distinct ridges along the often slightly convex edges and coming together at each corner into a short spine; wall fairly thick, with a fine ornamentation; reproduction by aplanospores.

Europe; sometimes in acidic or brackish-water ponds and ditches.

Tetraedriella polychloris Skuja 1964

10180030 **Pl. 66D** (p. 250)
Cells 20–42 µm across, basically tetrahedral but flattened, so shape approaches that of *Goniochloris*, with rounded corners and slightly convex sides, rounded corners without spines; wall fairly thick, with ornamentation of coarse punctae in intersecting diagonal rows; oil droplets present; reproduction not observed.

Europe; among filamentous and other algae in ditches and pools.

Trachydiscus H.Ettl 1964

Single-celled, usually distinctly flattened, with a broad and a narrow side, broader side almost round to ellipsoidal or polygonal (4–6-sided), sometimes irregular, narrower side elliptical or spindle-shaped and without spines; walls usually delicate, not obviously in 2 parts, with warts, dimples and large depressions in small numbers, smaller warts or granules more numerous, probably slightly silicified; chloroplasts parietal, 2–6, disc- to slightly cup-shaped, often polygonal on compaction, pyrenoid absent; oil sometimes present; asexual reproduction by 2–4 autospores, released by fracture of mother cell wall.

Europe, planktonic; little known genus.

Trachydiscus lenticularis H.Ettl 1964

10200010 **Pl. 66F** (p. 250)

Cells 6.5–10 μm across, lens-shaped in cross-section, almost circular in outline, ornamented by warts, relatively firm walled; chloroplasts (2–)3, usually at wider side of cell.

Europe; planktonic and possibly in very nutrient-rich waters.

ORDER RHIZOCHLORIDALES

Amoeboid stage dominant, unicellular or multicellular. Zoospores biflagellate.

Stipitococcus West *et* G.S.West 1898

Single-celled, with cells enclosed within an ellipsoidal, ovoid or flask-shaped lorica, attached to objects by a slender stalk sometimes arising from a basal disc; chloroplasts parietal, 1 or 2, faintly yellow-green; asexual reproduction by zoospores and these sometimes become amoeboid (only known in some species). North America and Europe; attached to various submerged objects including filamentous algae.

Stipitococcus urceolatus West *et* G.S.West 1878

10170010 **Pl. 65B** (p. 247)

Cell enclosed within an ovoid flask-shaped or jug-shaped lorica attached by a thin stalk, without a basal disc; lorica 3–4 μm wide and 6–11 μm long, stalk 4–6 μm long; chloroplast single, with a red eyespot.

North America and Europe; epiphytic mainly on *Mougeotia* and filamentous desmids, often in acid waters such as *Sphagnum* bogs.

ORDER TRIBONEMATALES

Simple or branched filaments. Zoospores uniflagellate or biflagellate.

Bumilleria Borzí 1888

Filamentous, with cells cylindrical, thin-walled, often constricted at cross wall where filaments easily break into short lengths, with H-shaped pieces often conspicuous between each group of 2–4 cells; chloroplasts parietal, 1 to several, disc-shaped, with pyrenoids only visible on staining; oil globules often present; asexual reproduction by biflagellate zoospores, released by disruption of cell wall.

Probably cosmopolitan; terrestrial or aquatic. Three species known, only one reported from the British Isles.

Bumilleria sicula Borzí 1888

10030010 **Pl. 68A** (p. 253)

Cells 15–20 μm wide and up to 35 μm long, 1.5 to 2 times longer than broad; chloroplasts 2 to several per cell depending on size, sometimes with a metallic lustre.

Probably widespread; little recorded in the British Isles where it occurs in swamps and ponds, often entangled with other algae, or on soil.

Chlamydomyxa W.Archer 1875

Cells amoeboid, with many fine pseudopodia; walls absent; chloroplasts numerous, disc-like, clustered together at centre; contractile vacuoles many and in outer hyaline area; nuclei many and scattered; oil globules present; nutrition phototrophic and holozoic, with food vacuoles containing desmids, diatoms and other algae; reproduction by division during which pseudopodia withdraw and a wall is formed to give an aplanospore.

Europe; often in the hollow cells of *Sphagnum*. Monospecific.

Chlamydomyxa labrynthuloides W.Archer 1875

10070010 **Pl. 67G** (p. 251)

Same characters as the genus.

Distribution and ecology as for genus.

Tribonema Derbès *et* Solier 1856

Filaments uniseriate and cells cylindrical or sometimes slightly barrel-shaped, with the cells divided into two equal sections, slightly overlapping in mid-region and forming H-shaped pieces, not always clearly evident but more evident on disruption of filaments, in empty cells or during reproduction; chloroplasts parietal, 2 to several, disc-shaped, light yellow-green, pyrenoids absent; asexual reproduction by 1 or 2 zoospores per cell, liberated on separation of cells into H-pieces; sexual repoduction isogamous; asexual reproduction by aplanospores and statospores.

Probably cosmopolitan; free-floating and only occasionally attached when young, common in standing or temporary waters particularly during colder period of the year.

It may be mistaken for *Microspora*, which also has H-pieces, but the latter has bright green chloroplasts which contain starch. Settled zoospores secrete a cell wall and attach to a surface and may resemble *Characiopsis*. *Tribonema obsoletum* G.S.West is referred to *Microspora* by Ettl (1978).

1 Cells usually <6 µm wide 2
1 Cells usually 6 µm wide 4
2 Chloroplasts usually >4 per cell ....*T. vulgare*
2 Chloroplasts 2–4 per cell...3
3 Cells >30 µm long.....................................*T. minus*
3 Cells ≤30 µm long......................................*T. affine*
4 Cells 6–11(–20) µm wide, 15–30(–100) µm long, thin-walled ....*T. viride*
4 Cells 10–17 µm wide, 15–63 µm long, thick-walled.....................................*T. utriculosum*

Tribonema affine (G.S.West) G.S.West 1904

Basionym: *Conferva affinis* G.S.West
10190010 **Pl. 68D** (p. 253)
Cells elongate, cylindrical, 5–6 µm wide, 35–40 µm long, thin-walled; chloroplasts 2 or 3, plate-shaped, partly encircling cell, pale yellow-green.

Possibly cosmopolitan; in ponds, small lakes and ditches, especially in moorlands.

Tribonema minus (G.A.Klebs) Hazen 1902

Basionym: *Conferva minor* G.A.Klebs
Synonym: *Conferva bombycina* fo. *mina* Wille
10190020 **Pl. 68C** (p. 253)
Cells cylindrical, 5–6 µm wide, 23–27 µm long; chloroplasts 2–4, relatively large, disc-like, arranged almost symmetrically around cell.

Possibly cosmopolitan; in shallow lakes and ponds, often associated with humic acids.

Tribonema utriculosum (Kützing) Hazen 1902

Basionym: *Tribonema bombycinum* var. *utriculosum* Hazen
Synonyms: *Conferva utriculosa* Kützing, *C. bombycina* var. *utriculosum* Wille
10190040 **Pl. 68F** (p. 253)
Cells 10–17 µm wide, 15–53 µm long, thick-walled; chloroplasts numerous, small.

North America and Europe; loose masses at the margins of standing waters and free-floating in marshes.

Tribonema viride Pascher 1925

Synonyms: *Tribonema bombycinum* (C.Agardh) Derbès *et* Solier, *Conferva bombycina* C.Agardh
10190050 **Pl. 68E** (p. 253) **CD**
Cells cylindrical or slightly constricted at cross walls, 6–11(–20) µm wide, 15–38(–100) µm long, thin-walled; chloroplasts numerous, pale yellow-green, occasionally in contact with each other.

Probably cosmopolitan; common in relatively cold water, forms a greyish-yellow cloudy mass, frequently where slightly shaded and in still or gently flowing water such as that of ditches and springs, especially in humus-rich waters (usually entangled in quieter waters), occasionally on soil; most abundant in colder months of the year.

West (1904, p. 258, figs. 121 H, I, p. 257) mentions a small form of this species (fo. *minor* (Wille) West) with cells 5–6.5 µm wide and considered to be very common. This narrow form is probably not worthy of recognition.

Tribonema vulgare Pascher 1923

10190030 **Pl. 68B** (p. 253)
Cells cylindrical although occasionally slightly swollen at ends, 5–6 µm wide and 18–48 µm long; chloroplasts numerous, disc-like, becoming elongated and even distorted in shape in mature cells.

Europe and America; in shallow waters of lakes and ponds.

Xanthonema P.C.Silva 1979

Synonyms: *Heterothrix* Pascher, *Bumilleria pro parte sensu* G.A.Klebs, *Tribonema pro parte sensu* Pascher
Filaments unbranched, occasionally forming pseudobranches where filaments break; cells oval with truncate apices, clearly constricted at cross walls but walls in one piece; chloroplasts parietal, 1 or 2 per cell, disc-shaped, folded.

Cosmopolitan; in a wide variety of aquatic habitats, also subaerial.

1 Cells 3.5–4.5 µm wide, 6–8 µm long; chloroplasts 2 per cell; commonly found in soil...*X. exile*
1 Cells 10–14 µm wide, almost quadrate; chloroplasts several per cell; commonly aquatic ....*X. quadrata*

Xanthonema exile (G.A.Klebs) P.C.Silva 1979

Basionym: *Bumilleria exilis* G.A.Klebs
Synonym: *Heterothrix exilis* (G.A.Klebs) Pascher
10270010 **Pl. 68G** (p. 253)
Filaments short and frequently fragmenting, but without H-shaped pieces; cells 3.5–4.5 µm wide, 6–8 µm long; chloroplasts 2 (variable in younger cells), not fully encircling cell; zoospores with 2 chloroplasts and a single flagellum.

Europe; essentially a soil alga, sometimes common.

Xanthonema quadrata (Pascher) P.C.Silva 1979

Basionym: *Heterothrix quadrata* Pascher
10270020 **Pl. 68H** (p. 253)
Filaments relatively long; cells 10–14 µm wide, almost quadrate; chloroplasts several; zoospores with a single chloroplast and 2 unequal flagella.

Europe; in various aquatic habitats including tanks, ponds and ditches.

ORDER VAUCHERIALES

BY LESLIE R. JOHNSON AND ROY MERRITT

Multinucleate vesicles and filaments, lacking cross walls. Simple or compound zoospores. Sexual reproduction oogamous

This section was to have been written by Tyge Christensen, formerly of the Sporeplanter Intitut at

Copenhagen, Denmark. Christensen had a lifetime's interest in *Vaucheria*, and studied not only Danish material, but British and other foreign material. He sadly died in Spring 1996 before his work on the Flora could be completed. He had also intended to prepare a monograph on the *Vaucheria* of western Europe, and some of this section is based on his ideas for that work. The section is dedicated to Tyge Christensen.

Vaucheria de Candolle 1805 **CD**

Macroscopic, forms green felty patches consisting of a microscopic network of branched, cylindrical, multinucleate filaments (coenocytic) attached by colourless rhizoids; chloroplasts parietal, numerous, disc-shaped, oval or ellipsoidal, with or without pyrenoids, starch rare, usual storage product oil or fat; asexual reproduction by zoosporangia cut-off by formation of a septum, each producing a single, large and multinucleate (nuclei associated with pairs of flagella) zoospore liberated through an apical pore; akinetes and aplanospores known; sexual reproduction oogamous, antheridia and oogonia often develop on short stalks or occasionally at end of a filament; oogonia ovoid and at maturity have a distinctive beak with a pore, antheridia cylindrical or saccate, straight or curved (circinate) and twisted to one side, sessile or stalked, with zoospores liberated through a terminal or lateral pores or longitudinal slits; androphore, a special spherical structure at base of an antheridium and attached to filament by means of an empty cell, rare; monoecious or dioecious.

Probably cosmopolitan; most species below are widespread in the British Isles, very common as green felts, almost akin to moss when growing on damp surfaces such as along river banks and on moist undisturbed soil (particularly noticeable in the cooler and damper seasons of late autumn to spring), or attached to submerged surfaces in rivers, streams and waterfalls, sometimes forming free-floating mats. Some species grow throughout the year, for example around stream banks or lock areas, and in greenhouses where they may form green coatings on seedbeds or trays if left undisturbed. The terminal parts of aerial filaments of at least the majority of species grow towards the light (positively phototrophic), especially when growing under shaded, humid conditions. This gives the mats of many species on damp surfaces a highly characteristic appearance. Many species are common in maritime areas, especially salt marshes.

Those species which are only known from marine habitats are excluded from this freshwater Flora. Most of the species described here have been personally collected by one of us (RM).

Most of the characters of taxonomic importance are associated with the organs of sexual reproduction; these may be formed in any season and are generally distinctive and well differentiated. It is quite easy to produce sex organs in some species by incubating pieces of mat in a petri dish or similar glass container on a window ledge for a few days. This genus was long regarded as a green alga, first described by Vaucher in 1803 as *Ectosperma*, subsequently renamed *Vaucheria* by de Candolle in 1805. Hassall (1845) describes 13 species of British *Vaucheria* in his volume on British freshwater algae.

1 Antheridia with supporting cell (androphore) ..... *V. synandra*
1 Antheridia without a supporting cell 2
2 Antheridia sessile or on a short stalk 3
2 Antheridia distinctly stalked 5
3 Antheridia club- or lance-shaped, with an apical pore ..... *V. dichotoma*
3 Antheridia cylindrical, opening by a slit 4
4 Oospore filling oogonium; oogonia 80–150 μm wide ..... *V. fontinalis*
4 Oospore not filling oogonium; oogonia 125–250 μm wide ..... *V. aversa*
5 Oogonia sessile or on a short stalk, with an antheridium beside one oogonium or between 2 oogonia 6
5 Oogonia distinctly stalked, 1 to many on a lateral branch and terminating in an antheridium 9
6 Oogonial beak oblique or perpendicular to filament; filaments 50–110 μm wide; oogonia 70–85 × 75–100 μm ..... *V. sessilis*
6 Oogonial beak tending to align with length of filament or directed slightly towards it 7
7 Filaments generally ≤40 μm wide, although up to 47 μm; oogonia generally ≤70 μm ..... *V. prolifera*
7 Filaments generally ≥40 μm wide; oogonia generally ≥70 μm 8
8 Filaments 40–123 μm wide; oogonia 69–160 × 69–220 μm; oogonia truncate to turnip-shaped, with wall lamellate and rugose ..... *V. dillwynii*
8 Filaments 60–141 μm wide; oogonia 112–140 × 150–165 μm; oogonia ovoid to kidney-shaped, with wall evenly textured and smooth ... *V. borealis*
9 Antheridia opening by both terminal and lateral pores ..... *V. compacta* var. *dulcis*
9 Antheridia opening by terminal or lateral pores 10
10 Antheridia opening by lateral pores, much broader at distal end than base 11
10 Antheridia opening by terminal pores, tubular and hooked 12
11 Filaments 22–55 μm wide; antheridia with 1–4 pores ..... *V. cruciata*
11 Filaments 55–99 μm wide; antheridia with 2 pores ..... *V. canalicularis*
12 Oospore filling oogonium except for beak 13
12 Oospore filling the oogonium completely 15
13 Oogonial beak a short tube (8–14 μm long) ..... *V. alaskana*
13 Oogonial beak not a short tube 14
14 Oogonial beak conical, about 23–31 μm wide at base, tapering to about 8–16 μm at pore; oogonia 50–67 × 67–93 μm ..... *V. pseudogeminata*
14 Oogonial beak twisted, about 9–14 × 16–27 μm; oogonia 35–57 × 43–60 μm ..... *V. lii*
15 Oogonium more or less erect on filament, more or less longitudinal, usually pointing upwards 16
15 Oogonium more or less pendant on filament, frequently pointing downwards 18
16 Branch bearing a single oogonium ..... *V. longata*

16 Branch bearing several oogonia 17
17 Branch usually bearing a pair of oogonia ....*V. geminata*
17 Branch bearing (2–)3–8 oogonia in a whorl ....*V. taylorii*
18 Antheridia usually >25 μm in diameter; oogonia >80×100 μm ....*V. frigida*
18 Antheridia usually <25 μm in diameter; oogonia <80×100 μm 19
19 Antheridia 14–26×30–65 μm; oogonia 42–78×43–94 μm.......................................*V. prona*
19 Antheridia 17–21×48–88 μm; oogonia 55–78×63–93 μm ....*V. racemosa*

Vaucheria alaskana Blum 1953

10210080 **Pl. 72E** (p. 267)

Filaments 42–62(–71) μm wide; monoecious; antheridium and oogonia on special fruiting branch; antheridial stalk curves downwards towards oogonium making 0.5 to 0.8 turns, or lying between 2 oogonia; antheridium 16–24×38–53 μm, conspicuously narrowing towards pore; oogonia 1 or 2 per branch, ovoid, (62–)71–81(–92)×(83–)85–102(–120) μm, with very small tubular beak (8–14 μm wide and 5–16 μm in length), reflexing back towards base of oogonium; oospore filling oogonium except for beak, with a conspicuous 2-layered wall, outer wall sometimes thickened, patterned or with net-like ridges.

Probably cosmopolitan; on soil and in freshwater habitats, in the British Isles known only from Inverness, Scotland.

Vaucheria aversa Hassall 1843

10210010 **Pl. 71D** (p. 265)

Filaments coarse, 42–98(–131) μm wide; monoecious; antheridia 21–35×56–112 μm, cylindrical, sessile or on a short stalk, parallel to or at an acute angle with filament and usually directed towards the neighbouring oogonium; oogonia usually in pairs (sometimes solitary) or in short series (2–4) in vicinity of antheridium, ovoid to subspherical, 86–176×100–190(–250) μm; one oogonium of a pair, or end one of series, often oriented with pore close to but directed away from the adjacent antheridium; oospores ovoid, 68–105×68–112 μm, not filling the oogonium, membrane 3-layered.

Probably cosmopolitan; submerged in ponds and ditches, and also in moist trickles and stream banks, known from south and north-west England as well as Scotland.

According to Blum (1972), in this species the oogonia are commonly of two types which are sometimes mixed in a single collection or limited to one type per collection: type 1 is short and spindle-shaped, erect or somewhat oblique, with an apical pore; type 2 is subspherical with a bent, downward-directed or reflexed beak, which is sometimes parallel to the filament and directed towards the base of the oogonium.

Vaucheria borealis Hirn 1900

10210090 **Pl. 72A** (p. 267)

Filaments 60–141 μm wide; oogonia essentially sessile or scarcely stalked, single or rarely paired, ovoid to oblique, with a horizontal beak and longest diameter parallel to length of filament; oospores 111–138 μm in diameter and 148–163 μm long, ovoid to kidney-shaped, with wall evenly textured and smooth.

Probably cosmopolitan; rare in Britain, grows on damp soils and in peat bogs or outflows.

A slightly problematic species similar to *V. dillwynii* and *V. pachyderma*. In the past many specimens have been mistakenly identified as *V. dillwynii*. According to Entwistle (1987), this species and *V. dillwynii* are separated on shape and wall characters associated with the oogonia.

Vaucheria canalicularis (Linnaeus) T.A.Christensen 1968

Basionym: *Conferva canalicularis* Linnaeus
Synonym: *Vaucheria woroniniana* Heering

10210110 **Pl. 72D** (p. 267)

Filaments 35–85 μm wide; monoecious; oogonia and antheridia often terminal on vegetative filaments, or borne on a special fruiting branch; antheridia 31–64×56–94 μm, coiled with a saccate-deltoid (equilateral triangular in shape) tip, formed terminally on a fruiting branch and stalked, bent laterally and downward between 2 oogonia, with an oval to circular pore at each opposing corner; oogonia 50–112×71–140 μm, 1 or 2, occasionally 3, on a fruiting branch, elongate-ovoid, sometimes with a small distal prominence, erect on a stalk or somewhat inclined towards antheridial stalk; oospore relatively thin-walled, filling oogonium.

Probably cosmopolitan; widespread in ponds and ditches, known in north-west England, the English Midlands and Scotland.

We have followed Christensen (1968, 1969) in accepting *V. canalicularis* as the earliest legitimate name, although in some subsequent works (e.g. Rieth, 1980) *V. woroniniana* Heering continues to be recognized despite being a later name.

Vaucheria compacta (Collins) Collins in Taylor 1937

var. *dulcis* Simons 1974

10210120 **Pl. 72B** (p. 267)

Filaments 25–70 μm wide; dioecious, with occasional monoecious filaments; antheridia 72–224 μm long, often in groups of 2 or 3 on a stalk, usually solitary or occasionally in small groups on one side of filament; oospore subspherical, (210–)260–340(–460)× 280–380 μm, nearly filling distal end of oogonium, with a relatively thin wall.

Europe; not common in the British Isles (R. Merritt, unpublished records), essentially freshwater but sometimes in brackish-water.

Separated as a variety by Simons (1974) after studying material from the Netherlands.

Vaucheria cruciata (Vaucher) de Candolle in Lamarck *et* de Candolle 1805

Basionym: *Ectosperma cruciata* Vaucher
Synonym: *Vaucheria debaryana* Woronin

10210130 **Pl. 72C** (p. 267)

Filaments 17–22(–55) μm wide; monoecious; oogonia and antheridia borne on special and relatively short fruiting branches; antheridial stalk at end of fruiting branch; antheridia 13–36 × 13–24 μm, inclined or erect, cushion-shaped, opening by 2 pores at opposite sides of distal end or by 1–4 lateral pores; oogonia 39–50(–60) × 43–60(–79) μm, ovoid or spherical, 1 or 2 (occasionally 3) per fruiting branch, borne on short stalks lateral to the antheridial stalk, erect or slightly inclined to one side; oospore 43–56 × 49–77 μm, filling oogonium, wall thin and 3-layered.

Probably cosmopolitan; grows in freshwater habitats and terrestrial on damp soil, known from north-west England to East Anglia, as well as in Scotland.

We have followed Christensen (1969) and Blum (1972) in considering *V. cruciata* and *V. debaryana* to be conspecific, with the former the earliest legitimately published name and hence having priority. Despite this, Rieth (1980) recognizes *V. debaryana* and places *V. cruciata* into its synonymy.

Vaucheria dichotoma (Linnaeus) Martius 1817

Basionym: *Conferva dichotoma* Linnaeus
Synonym: *Vaucheria dichotoma* (Linnaeus) C.Agardh
10210020 **Pl. 71B** (p. 265)
Filaments (80–)100–135(–225) μm wide; dioecious; antheridia (75–)100–144 × (110–)137–225 μm, commonly in groups of 3 or more, often in irregular series on one side of filament, sessile, erect or a little oblique, oval to ovoid with a slightly pointed apex; oogonia 260–360 × 280–380 μm, subspherical to spherical, solitary or in small groups on one side of filament, sessile and usually erect; oospore subspherical, (210–)260–340(–460) × 280–380 μm, nearly filling oogonium, with a relatively thin wall.

Probably cosmopolitan; not common in the British Isles, growing in freshwater and brackish-water habitats, ponds and ditches; records from Tyge Christensen (*pers. comm.*) largely from south-east England.

Vaucheria dichotoma is described as dioecious although Christensen (1952) has recorded oogonia and antheridia on the same filament. It comes very close to *V. sescuplicaria* T.A.Christensen and Rieth (1954) questioned whether they should be regarded as separate species. Rieth (1980) continues to recognize these as two separate species and this is accepted here. The species *V. sescuplicaria* is recorded in the British Isles, but is excluded from the present text as it is essentially a brackish-water and marine species. According to Stafleu (1967), as reported in Christensen (1968), *V. dichotoma* (L.) Martius was published on 20 April 1817 and *V. dichotoma* (L.) C.Agardh on 1 May 1817, hence the former takes precedence.

Vaucheria dillwynii (Weber *et* Mohr) C.Agardh 1812

Basionym: *Conferva dillwynii* Weber *et* Mohr
Synonym: *Vaucheria pachyderma* Walz
10210140 **Pl. 71G** (p. 265)
Filaments 40–128 μm wide; monoecious; antheridia 24–31 × 50–80 μm, borne at upper end of a short erect stalk perpendicular to filament, or stalk somewhat curved in continuation with the antheridium and making 0.2 to 0.6 turns, saccate or shortly cylindrical, beside oogonium or more rarely between 2 oogonia; oogonia usually single, adjacent to an antheridial stalk, each 69–160 × 69–220 μm, truncate to turnip-shaped, virtually sessile, erect with its long axis horizontal, commonly with a reflexed beak and pore opening directed towards filament; oospore 145 × 180 μm, filling oogonium except for oogonial beak, with a thick 7-layered wall and frequently ornamented (punctate or net-like).

Probably cosmopolitan; in freshwater habitats and on soil, occasionally reported from brackish-water habitats; known from north-east to southern England, Wales and Scotland.

See note under *V. borealis*. We have followed Christensen (1969) and Blum (1972) in placing *V. pachyderma* under the synonymy of *V. dillwynii* since the latter is the earlier legitimately published name. However, some other authors such as Heering (1921), Venkataraman (1961) and Rieth (1980) recognize *V. pachyderma* and place *V. dillwynii* in its synonymy. Reith (1980) recognizes that *V. dillwynii* is the earliest published name but points out that the author of *V. pachyderma* Walz 'was the first to give an unequivocal description and picture of the characteristic antheridium'. He considers Weber and Mohr's description to be unacceptable.

Vaucheria fontinalis (Linnaeus) T.A.Christensen 1968

Basionym: *Conferva fontinalis* Linnaeus
Synonyms: *Vaucheria ornithocephala* C.Agardh, *V. polysperma* Hassall
10210030 **Pl. 71C** (p. 265)
Filaments (16–)22–40(–75) μm wide; monoecious; antheridia 20–32 × 50–92(–100) μm, tubular or slightly club-shaped and essentially straight, sometimes on a short stalk, bent horizontally parallel to or forming an acute angle with filament, nearly always directed towards adjacent oogonia; oogonia 37–60(–74) × (56–)70–106 μm, commonly in regular unilateral series of 2–5(–6) on filament, shortly stalked or almost sessile, with 1 or 2 antheridia at one end or both ends of a series of oogonia, beak commonly lateral and turned to same end of filament as the series of oogonia, thus pore tending to point towards one antheridium or antheridial group, sometimes essentially erect or else depressed towards filament; oospore spherical, 41–59 × 47–75 μm, filling oogonium at its widest point but not at ends, ovoid with relatively thin wall, reddish in colour, 3-layered.

Probably cosmopolitan; in cold streams in spring, shaded rivers, and free-floating in non-flowing waters, sometimes on damp soil, known from south and south-west England as well as Scotland.

Christensen (1968) recognizes that *Vaucheria ornithocephala* C.Agardh and *V. polysperma* Hassall are conspecific when placing them under the synonymy of *V. fontinalis* (Linnaeus) T.A.Christensen. Some later authors (e.g. Rieth, 1980)) continue to recognize *V. ornithocephala* and rather place *V. fontinalis* in its synonymy despite the latter having priority.

Vaucheria frigida (Roth) C.Agardh *sensu* T.A.Christensen 1969

Basionym: *Conferva frigida* Roth
Synonyms: *Vaucheria terrestris sensu* Götz *non V. terrestris sensu* de Candolle, *V. hamata* (Vaucher) de Candolle *non V. hamata sensu* Götz
10210150 **Pl. 73D** (p. 269)
Filaments 42–100(–1200) μm wide; monoecious; antheridia terminal on bisexual branches, overgrown at maturity by oogonial stalk so appearing borne on side of

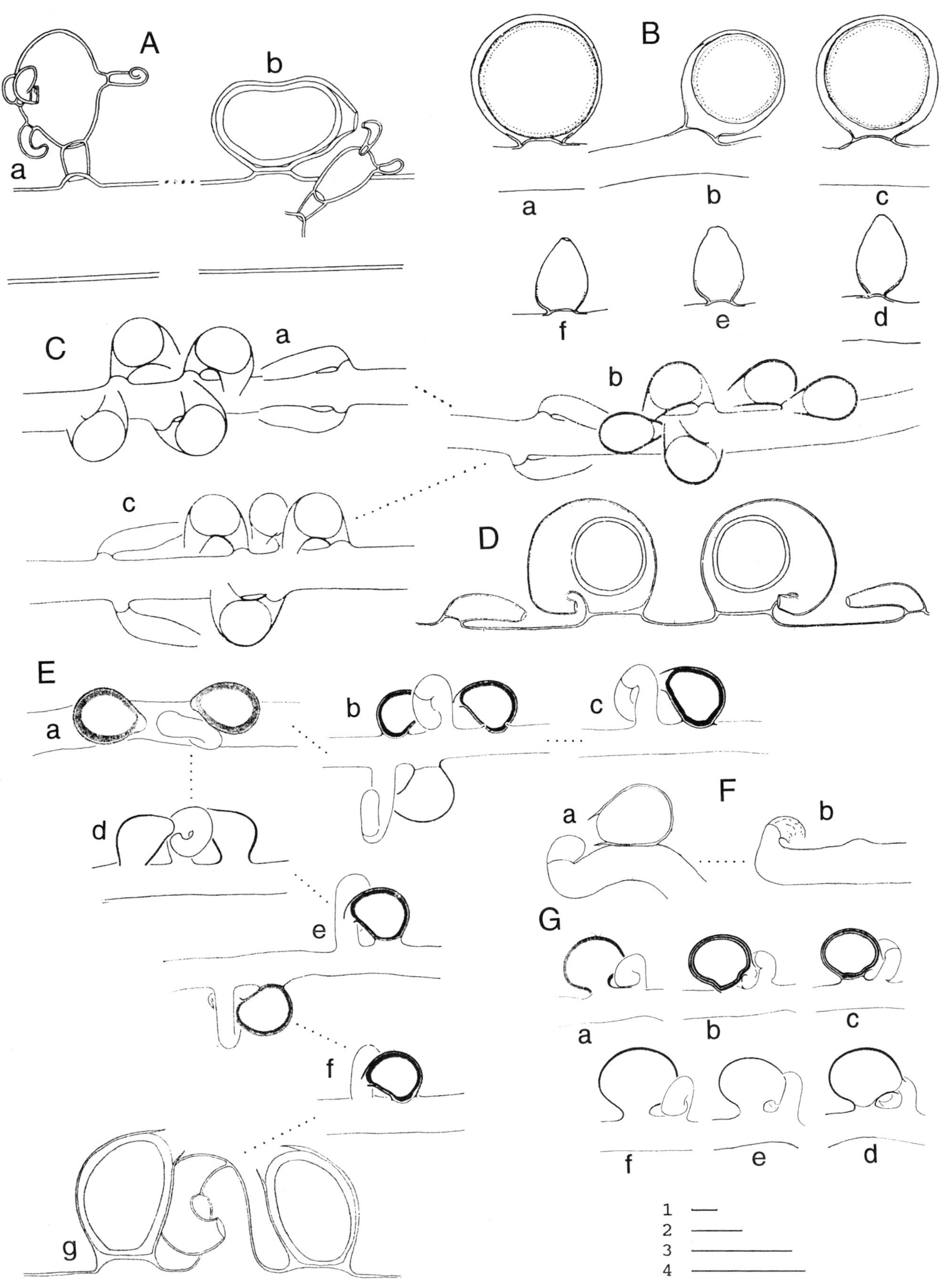

Plate 71 A–G

A. *Vaucheria synandra* (p. 268): a. hooked antheridia on an androspore, b. sessile oogonium and 2 hooked antheridia on an androspore; **B**. *Vaucheria dichotoma* (p. 264): a–c. oospores, d–f. antheridia after dehiscence; **C**. *Vaucheria fontinalis* (p. 264): a–c. reproductive stages; **D**. *Vaucheria aversa* (p. 263); **E**. *Vaucheria sessilis* (=*V. bursata*) (p. 268): a–g: reproductive stages; **F**. *Vaucheria prolifera* (p. 266): a. oospore, b. antheridium; **G**. *Vaucheria dillwynii* (=*V. pachyderma*) (p. 264): a–f. reproductive stages.

Scale bars: 75 µm. 1 for B, D, Ea–f, G; 2 for Eg; 3 for C, F; 4 for A.

fruiting branch below oogonium, stalk curves laterally and downwards before terminating in a coiled antheridium; antheridia (10–)24–38 × 64–93 μm, making 0.5 to 0.75 turns; oogonia 80–135 × 108–135(–165) μm, spherical to ovoid or kidney-shaped with a distal prominence, solitary, rarely 2 together, sessile or with a short stalk, long axes of oospore horizontal to vegetative filament or depressed on side towards it; oospore filling oogonium, wall relatively thick, sometimes black.

Probably cosmopolitan; widespread in freshwater habitats where it grows as tangled mats in quiet waters or on damp soil; known from England, Wales and Scotland.

Christensen (1969) and Blum (1972) recognize *V. frigida*, and include in its synonymy *V. terrestris sensu* Götz. Rieth (1980) takes an opposing view, placing *V. frigida* (Roth) C.Agardh under *V. terrestris sensu* Götz var. *terrestris*. We have followed Christensen (1968, 1969) in recognizing *V. frigida*.

Vaucheria geminata (Vaucher) de Candolle in Lamarck *et* de Candolle 1805

Basionym: *Ectosperma geminata* Vaucher *non Vaucheria racemosa sensu* Götz

10210040 **Pl. 73B** (p. 269)

Filaments (35–)41–78(–110) μm wide; monoecious; antheridium central on bisexual branch, stalk rarely much longer than oogonia and curved to one side, stalk terminating in a coiled antheridium making 0.3 to 1 turns; oogonia 64–85(–110) × 71–120 μm, ovoid to kidney-shaped, with a small distal prominence, in pairs (very rarely 1 or 3) and on a short stalk, long axes erect to nearly horizontal with filament and pores therefore often directed vertically; oospores 52–225 × 64–190 μm, filling the oogonium, wall thick and 3-layered.

Probably cosmopolitan; fairly common in freshwater habitats, often as trapped floating mats or on very moist soil; known from England, Wales and Scotland.

Vaucheria lii Reith 1959

10210160 **Pl. 72G** (p. 267)

Filaments 18–25 μm wide; monoecious; antheridium terminal on bisexual branch, antheridial stalk curved or coiled laterally and downwards, lateral to an oogonium or between 2 oogonia; antheridia somewhat coiled and making 0.3 to 0.7 turns; oogonia 35–57 × 43–60 μm, 1–4 but usually in pairs, clearly stalked, with a coiled beak (9–14 μm in diameter and 16–27 μm long); oospore filling oogonium with the exception of the beak.

Probably cosmopolitan; very rare since based upon a single specimen collected in freshwater habitats in Perth, Scotland (R. Merritt, unpublished record).

Vaucheria longata Blum 1953

10210170 **Pl. 73A** (p. 269)

Filaments 43–75 μm wide; monoecious; usually 3 antheridia and 1 oogonium borne on stalks of special fruiting branches; antheridia coiled with a single terminal pore, narrowing towards tip, 16–20 μm in diameter, antheridial stalk terminating on fruiting branch, upper portion of stalk bent towards oogonial aperture; oogonia 56–74 × 83–102 μm, subovoid, borne on end of a short stalk arising at base of antheridial stalk, vertical or parallel to antheridial stalk; oospores 55–73 × 82–100 μm, thin-walled.

Europe; rare in the British Isles since only known from freshwater and brackish-water habitats in Cheshire.

Described by Blum (1953) from herbarium material labelled as *V. geminata*. Christensen (1956) reported some proliferating fruiting stages of this species from Denmark.

Vaucheria prolifera P.A.Dangeard 1939

10210180 **Pl. 71F** (p. 265)

Filaments (10.5–)27(–47) μm wide; monoecious; sexual organs formed as bisexual pairs, oogonium subterminal on the vegetative filament; antheridia terminal, (15.5)–21(–31) μm in diameter making about 0.25 of a turn, saccate, directed towards the oogonium; oogonia (34.0–)45.5(–62.5) × (54.5–)64.0(–78.0) μm, ovoid, with beak directed toward the antheridium, sessile.

Probably cosmopolitan; apparently normally from brackish-water or marine influenced habitats, found only twice in the British Isles (R. Merritt, unpublished records) and then well above high water level.

Vaucheria prona T.A.Christensen 1970

Synonym: *Vaucheria hamata* (Vaucher) de Candolle *sensu* Götz *pro parte*

10210190 **Pl. 73E** (p. 269)

Filaments 28–60(–110) μm wide; monoecious; antheridia 14–16 × 30–65 μm, stalk central on bisexual fruiting branch, curved to one side and downwards between paired oogonia, coiled antheridium terminal on a stalk typically making 0.25 to 0.6 turns; oogonia 42–78 × 43–94 μm, ovoid to kidney-shaped with a small distal prominence, usually 2(–3) on each fruiting branch, long axes horizontal or depressed on side towards vegetative filament, filling oogonium.

Europe; in freshwater habitats and on moist soil in north-west England and Scotland.

Christensen (1969) accepted *V. hamata* (Vaucher) de Candolle *sensu* Götz as conspecific with a species he later (Christensen, 1970) described as *V. prona*. He established this new species to avoid further confusion with the often misapplied *V. hamata* (Vaucher) de Candolle and *V. hamata sensu* Götz. Rieth (1980) appears not to accept Christensen's opinion and places *V. prona* under the synonymy of *V. hamata sensu* Götz.

Vaucheria pseudogeminata P.A.Dangeard 1939

Synonym: *Vaucheria hamata sensu* Walz, *non V. hamata* (Vaucher) de Candolle

10210200 **Pl. 72F** (p. 267)

Filaments (28–)33–50 μm wide; monoecious; antheridium terminal on bisexual fruiting branch, stalk curved laterally and downwards, lateral to oogonium or between 2 oogonia; antheridia 16–21 × 38–58 μm and making 0.2 to 0.8 turns; oogonium 50–67 × 67–93 μm, 1–2 per fruiting branch, sessile or borne on a short stalk, kidney-shaped with a conspicuous beak directed

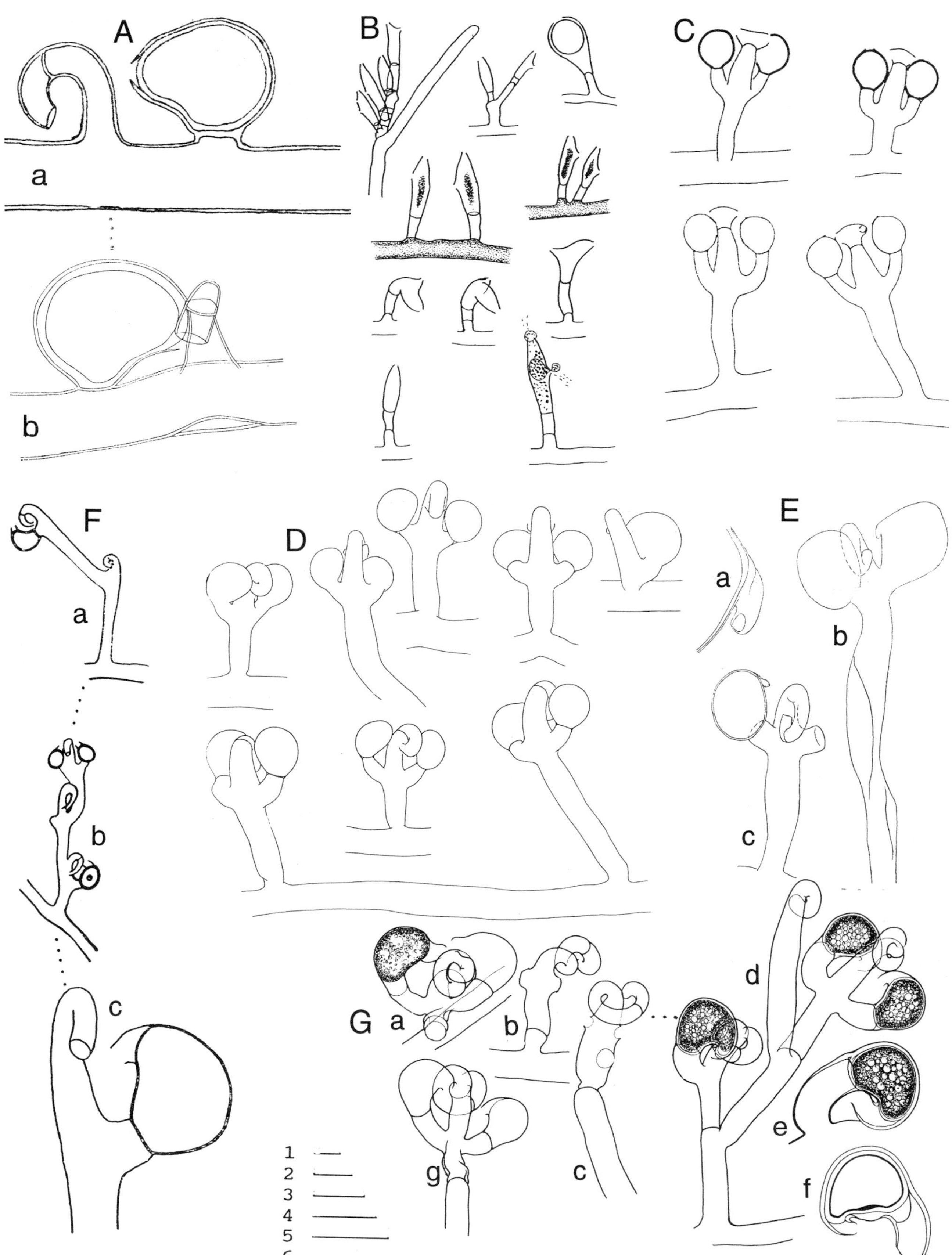

Plate 72 A–G

A. *Vaucheria borealis* (p. 263): a, b. oogonia and hooked antheridia; **B**. *Vaucheria compacta* var. *dulcis* (p. 263): stages in reproduction; **C**. *Vaucheria cruciata* (p. 263); **D**. *Vaucheria canalicularis* (p. 263); **E**. *Vaucheria alaskana* (p. 263): a. open oogonial beak, b, c. reproductive structures; **F**. *Vaucheria pseudogeminata* (p. 266): a–c. stages in reproduction; **G**. *Vaucheria lii* (p. 266): a–g. stages in reproduction.

Scale bars: 50 µm. 1 for A, B, D, Fa,b; 2 for C; 3 for Eb; 4 for G; 5 for Fc; 6 for Ea,c.

towards base of fruiting branch, oogonia 23–31 μm wide at base and tapering to 8–16 μm at pore, 6–18 μm long; oospore filling oogonium except for beak, subovoid to kidney-shaped, wall relatively thin.

Probably cosmopolitan; not common, grows on damp soil in woods and terrestrial habitats as well as on the sides of ditches; also known from brackish-water habitats.

Vaucheria racemosa (Vaucher) de Candolle in Lamarck *et* de Candolle 1805

Basionym: *Ectosperma racemosa* Vaucher
Synonym: *Vaucheria walzi* Rothert
10210210 **Pl. 73F** (p. 269)
Filaments 42–126 μm wide; monoecious; antheridia 17–31 × 48–88 μm, central on bisexual branch, stalk curved laterally and downwards between paired oogonia, or between ranks (usually 2) of oogonia, terminal, coiled and making 0.3 to 1 turns; oogonia 55–78 × 63–93 μm, 2– 4 (–7) on each fruiting branch, if 2 then both at same level and opposite or if more then oogonia borne in 2 or more whorls (usually in 2 vertical ranks), thus fruiting branches having a racemose or corymbose appearance; oogonium ovoid with a small distal prominence, oogonial axis usually horizontal or making an acute angle with vegetative filament; oospore filling oogonium, wall conspicuous.

Probably cosmopolitan; not common in the British Isles where known from freshwater habitats in England and Scotland.

Christensen (1969) and Blum (1972) accept *V. walzi* Rothert as conspecific with *V. racemosa* (Vaucher) de Candolle, with the latter the earlier published name. Despite this fact, Rieth (1980) only recognized *V. walzi* and considers *V. racemosa* to be a synonym.

Vaucheria sessilis (Vaucher) de Candolle in Lamarck *et* de Candolle 1805

Basionym: *Conferva bursata* O.F.Müller
Synonyms: *Conferva bursata* (O.F.Müller) C.Agardh, *Ectocarpus sessilis* Vaucher, *Vaucheria bursata* (O.F.Müller) C. Agardh, *V., ovoidea* Vaucher, *V. repens* Hassall (see under 'Doubtful Records')
10210060 **Pl. 71E** (p. 265) **CD**
Filaments 42–126 μm wide; monoecious; antheridia 21–29(–32) × 50–78 μm, located between oogonial pairs, coiled, usually curved in a vertical plane making 0.3 to 0.7 turns, borne at upper end of a short erect stalk, perpendicular to filament and straight or curved in continuation of curve of antheridium; oogonia usually in pairs, ovoid to subspherical, 50–78 × 71–104 μm, usually sessile, upper end not usually radially symmetrical due to eccentric position of pore; axis of pore making an acute angle with filament and generally turned in direction of adjacent antheridium, ovoid, with a conical beak 3–10 μm long; oospore 51–71 × 73–96 μm, filling the oogonium except for beak, walls relatively thick, dark, spotted, and 3-layered.

North America and Europe; in freshwater habitats (e.g. ditches and pools) and subaerial under trickling water or in damp areas; records from more or less everywhere if soil or water is at least moderately calcareous. Common in fast-flowing rivers with water which is at least moderately hard and nutrient-rich, where it sometimes reaches lengths of 50 cm flowing in the water. It is also very frequent in streams which are nutrient-rich, but where iron oxide or other types of sediment are being deposited. Often mixed with *Cladophora glomerata* and probably often misrecorded as *Cladophora*.

Rieth (1963) and Christensen (1969) found difficulty in separating the *bursata–sessilis* complex. Christensen (1969) considers '*V. bursata* as a single entity' with this the earliest published. Venkataraman (1961) and Blum (1972) recognize *V. sessilis* and make no mention of *V. bursata*. Rieth (1980) places *V. bursata* under the synonymy of *V. sessilis*, noting that O.F. Müller had a 'bryophyte' in 1779 and named it *Conferva bursata* in 1788. The figure and description are of historical interest, but we cannot be persuaded to change the name '*sessilis*' to '*bursata*'. We accept the view that the two species are conspecific with *V. sessilis*, the correct name to apply, although Entwisle (1987) preferred to use the epithet *V. bursata*.

Vaucheria synandra Woronin 1869

10210220 **Pl. 71A** (p. 265)
Filaments 35–100 μm wide; monoecious; antheridia sessile, hooked, (2–)4–7 in a group on an androspore attached by a small empty cell to filament, 175 μm long and 45 μm in diameter; oogonia sessile, spherical, separated from filament by a cross wall, side facing antheridia slightly drawn out into a downwardly hooked beak; oospore oval, filling oogonium, 100–120 μm wide, wall thick, 2–3-layered.

Probably cosmopolitan; often considered a marine species but found at the limits of the freshwater to brackish-water junction in river estuaries; known from coastal areas in the north and south of England, Wales and Scotland.

Vaucheria taylorii Blum 1972

Synonym: *Vaucheria verticillata* Meneghini
10210230 **Pl. 73C** (p. 269)
Filaments 56–90(–140) μm wide; monoecious; antheridia on bisexual branch, 21–33 μm in diameter by 62–96 μm long, conspicuously surpassing oogonia, curved to one side, making 0.5 to 0.9 turns; oogonia 64–85 × 71–106 μm, ovoid to oblong-ovoid, with a small distal prominence, (2–)3–8 borne in whorl on short stalks attached near frequently inflated mid-point of fruiting branch, long axes erect or forming an acute angle with filament, oospore filling oogonium.

North America, Europe; very rare in freshwater in the British Isles.

DOUBTFUL SPECIES

Vaucheria clavata Vaucher 1803

De Toni (1889) and Heering (1921) place *Ectosperma clavata* Vaucher and *V. clavata* de Candolle in the synonymy of *V. sessilis* (Vaucher) de Candolle. The report by Hassall (1845, p. 59) is based on records by Berkeley (1833, as *V. ungeri* Thuret) and by Harvey (1841, as *V. clavata* de Candolle) who mentions it as occurring 'in ponds and ditches, Yarwell, Northamptonshire'.

Vaucheria ovoidea (Vaucher) Hassall 1845

Reported in Hassall (1845, pp. 57–58) based on earlier records. Placed by De Toni (1889) under the synonymy of *V. sessilis*. Mentioned in Heering (1921) under doubtful species as '*V.*

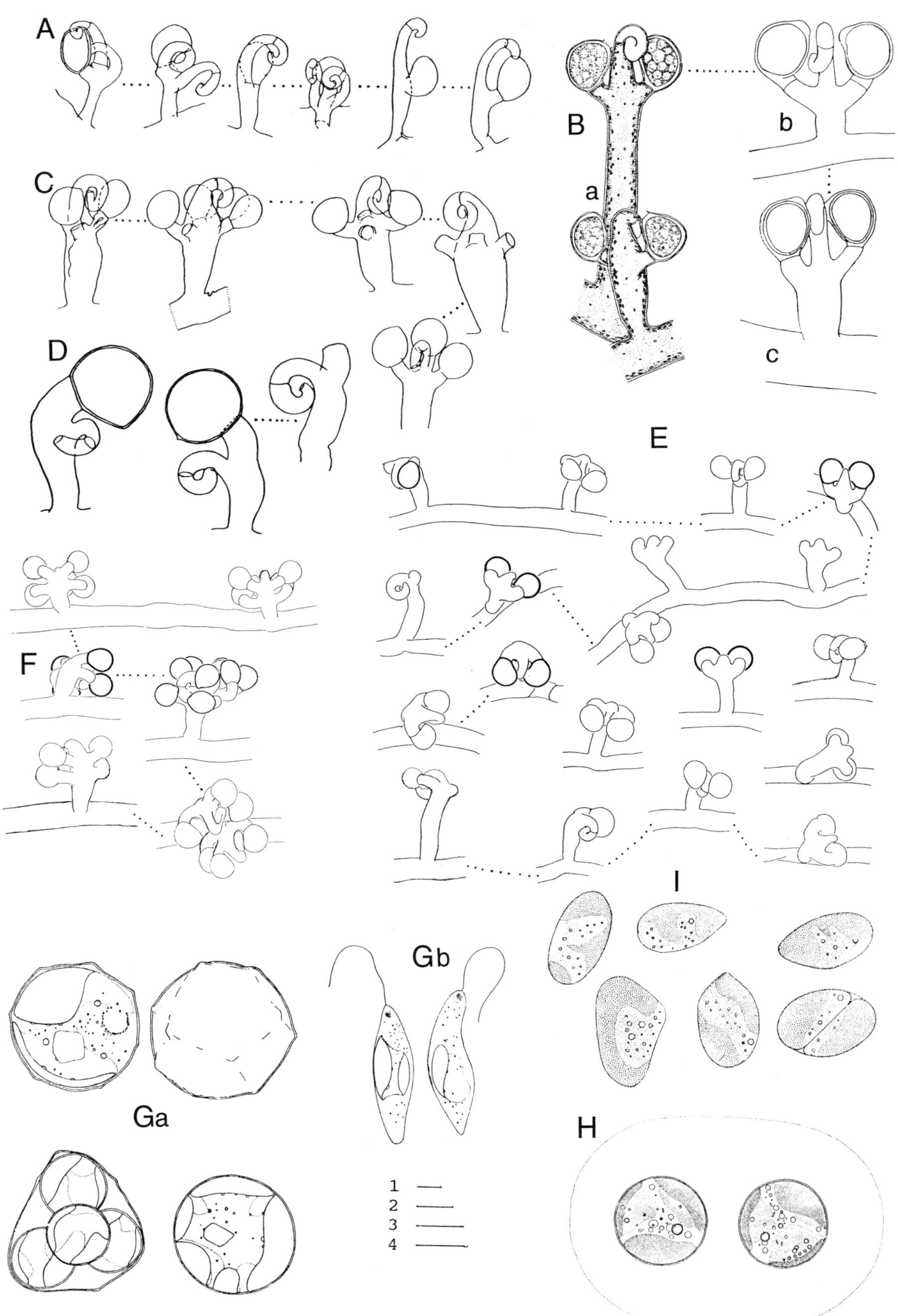

Plate 73 A–I

A. *Vaucheria longata* (p. 266); **B**. *Vaucheria geminata* (p. 266): a–c. pairs of oogonia with a hooked antheridium between them; **C**. *Vaucheria taylorii* (p. 268); **D**. *Vaucheria frigida* (p. 264); **E**. *Vaucheria prona* (p. 266): mostly paired oogonia and an antheridium beteween each pair; **F**. *Vaucheria racemosa* (p. 268): multiple pairs of oogonia associated with an antheridium; **G**. *Polyedriella helvetica* (p. 272): a. non-motile cells, b. two zoospores; **H**. *Chlorobotrys regularis* (p. 271); **I**. *Monodus subterraneus* (p. 272): showing cells with one cell producing autospores.

Scale bars: 1–4, 50 μm for A–F. 1 for Bb, E; 2 for C, D; 3 for A, F; 4 for Ba, c. 4, 5 μm for Ga, b, I; 4, 10 μm for H.

ovoidea Hassall: Art der *Corniculatae Racemosae'*, and not mentioned in more recent monographs. For this reason we considered it to be doubtful.

Vaucheria repens Hassall 1843
Described by Hassall (1843) as growing on a footpath near Roydon, Essex. Recognized by Blum (1972) although considered by many authors to be a form of *V. sessilis*, *V. sessilis* fo. *repens* (Hassall) Hansgirg (see Heering, 1921; Venkataraman, 1961). Christensen (1969) places it in the synonymy of *V. bursata* and Rieth (1980) makes no mention of it. We regard this little-known species to be doubtful.

Acknowledgements

Special thanks go to Dr H.J. Belcher for critically reading and correcting the Xanthophyta text.

PHYLUM EUSTIGMATOPHYTA

David M. John

The Eustigmatophytes are small single-celled coccoidal algae, the term 'eustigma' referring to a conspicuous orange-red eyespot that is possessed only by flagellated forms. In such forms the flagella are inserted near the cell apex, and on the anteriorly directed flagellum are two rows of tripartite hairs (not visible with the light microscopy). A second flagellum, if present, is usually very short, posteriorly directed and smooth. The eyespot has a unique structure and is at the anterior end of the cell where it is not associated with the chloroplast. It consists of a number of carotenoid-containing globules above which lies the basal expansion of the long flagellum. The one or more yellow-green chloroplasts contain chlorophyll *a* and violaxanthin is the principal accessory pigment. The pyrenoid characteristically projects on a short stalk from the inner side of the chloroplast and is surrounded by flat plates of an unknown food storage material; pyrenoids are not present in all eustigmatophycean algae and are never present in zoospores. Another unusual feature is the presence of a red-pigmented body whose role is unknown. Reproduction is by the formation of 2 or 4 autospores or zoospores.

The phylum was created to include a number of algae previously classified in the Xanthophyta (Hibberd, 1980; Santos, 1996). It currently consists of a single order (Eustigmatales) divided into four families containing about seven genera and 15 species; most are freshwater or soil-dwelling, although there are some marine representatives. One unusual genus (*Corvomyenia*) is a symbiont within a freshwater sponge. Other eustigmatophytes reported from the British Isles are only known from brackish waters (e.g. *Nannochloris ocelata* Droop).

The eustigmatophytes are commonly mistaken for green algae when examined with the light microscope. Features useful for distinguishing flagellated forms are the separation of the prominent eyespot from the chloroplast and, in the case of coccoidal cells, the presence of a stalked pyrenoid although this is not always very evident. The carbohydrate storage material, which does not stain with Lugol's solution or iodine, is another useful distinguishing feature. Otherwise the identification of some genera and subgeneric taxa requires examination by transmission electron microscopy.

1 Cells usually in regular groups of 2 or 3, rarely more, surrounded by a thick mucilaginous envelope....................... *Chlorobotrys*
1 Cells solitary or in irregular groups, not surrounded by mucilage 2
2 Cells spherical or polyhedral, (7–)8–11(–20) μm across.. *Polyedriella*
2 Cells variously shaped, one end often narrowing to a point, other broadly rounded, 3–5 μm wide, 7–9 μm long .... *Monodus*

Chlorobotrys Bohlin 1901

Cells spherical, subspherical or broadly oval, solitary or more commonly in groups of 2, 4, 8 or 16, rarely more, surrounded by a colourless, homogeneous mucilaginous envelope; cell walls firm, smooth, thick and often prominently silicified; chloroplasts parietal, disc-shaped, single in young cells, several in older cells; prominent red pigment spot commonly present; reproduction by autospores; cysts cylindrical but flattened, germinating to produce aplanospores.

Europe, North America; usually planktonic or associated with other algae in softwater (acidic) bogs.

Chlorobotrys regularis (West) Bohlin 1901

Basionym: *Chlorococcum regulare* West

11010010 **Pl. 73H** (p. 269)

Cells usually 2 or 3, rarely 4 or more, spherical, (12–)15–20 μm in diameter, surrounded by a wide (about 34–90 μm across) mucilaginous envelope about 34–90 μm broad, the outline of envelope distinct and faintly striated around groups of cells.

Europe; usually associated with aquatic plants and larger algae in acidic waters; according to West & Fritsch (1927, p. 307), it is 'often abundant in *Sphagnum*-bogs of the British Isles'.

Monodus Chodat 1913

Cells solitary, free-living, thin-walled, frequently very irregularly shaped with one pole often rounded, the other blunt, sometimes somewhat pear- or spindle-shaped; chloroplasts parietal, 1 or several, pyrenoids present or absent; reproduction by autospores.

Europe, Antarctic; most frequently in soil cultures.

Monodus subterraneus J.B.Petersen 1932

11020010 **Pl. 73I** (p. 269)

Cells obovoid to ellipsoidal, also irregularly oblong and occasionally almost spindle-shaped, one end often narrowing to a point, other end broad and commonly irregularly rounded, 3–5 μm wide, 7–9 μm long; chloroplast always single, parietal, large and often with a deeply lobed margin, without a pyrenoid; autospores 2 per cell.

Probably cosmopolitan; in the British Isles known in cultures of soil or associated with subaerial surfaces; rare.

Polyedriella Pascher 1930

Cells usually solitary, spherical, angular and polyhedral in outline (7–20 μm across); chloroplasts parietal, one or several, without pyrenoids; reproduction by autospores or zoospores.

Europe; rare and usually in soil cultures.

Polyedriella helvetica Vischer *et* Pascher 1937–38

11040010 **Pl. 73G** (p. 269)

Cells single or aggregated, spherical or polyhedral with a variable number of thickened angles, (7–)8–11(–20) μm across, often with a large central crystal present; chloroplast usually single and deeply lobed; zoospores 1 or 2 per cell, spherical or ovoid, single flagellum longer than cell, eyespot linear and at anterior end; autospores 2–4 per cell.

Distribution as for genus; a soil alga only recorded in cultures of different soils (see Fritsch & John, 1942).

PHYLUM BACILLARIOPHYTA (Diatoms)

Martyn G. Kelly and Elizabeth Y. Haworth

Introduction

The Bacillariophyta are a highly distinctive phylum which exist as single cells, colonies or filaments. Their characteristics, as seen under the light microscope, include a unique cell wall of silica (termed a 'frustule') and golden-brown plastids due to the presence of the carotenoid fucoxanthin alongside chlorophylls *a* and c_2. The frustule is essentially of two parts, each having a valve adjoined by one or more girdle bands (termed 'cingula') that allow for expansion and cell division. When seen in a medium of high refractive index, these valves are extremely beautiful and this has attracted microscopists to their study almost since microscopes were invented. The vegetative cells lack flagella, but some diatoms are capable of gliding across surfaces. This depends upon a longitudinal slit in the valve, termed a 'raphe' although the precise mechanism for this motility remains unclear. The Bacillariophyta are also characterized by several features visible only with electron microscopy: these are described at greater length in Round et al. (1990) and Hoek et al. (1995).

Two main groups of diatoms are recognized: centric diatoms, which are radially symmetrical, and pennate diatoms which show longitudinal symmetry in at least one of three planes. The arrangement of the frustule around these planes is an important diagnostic character for pennate genera. Arrangement of structures on the valve surface also plays an important role, for identifying both genera and species. These structures include the raphe, as well as many smaller perforations through which the diatom is connected to its environment.

Microscopists have studied diatoms since the late 1700s; however, the small size of diatom cells (most are <30 μm), coupled with the reliance on distinguishing surface ornamentation on these cells, has meant that developments in diatom taxonomy have been limited by the technology available. Invention of apochromatic lenses and phase contrast techniques, availability of better light sources and the development of mountant with high refractive indices all contributed to the ease with which features could be observed and species distinguished from one another. This evolving taxonomic interpretation has caused many name changes – with lumping, splitting and recombination of both species and genera.

The last comprehensive account of freshwater diatoms from the British Isles was Smith's *A Synopsis of the British Diatomaceae*, published between 1853 and 1856. This included 43 genera and 284 species. Since this time, there has been a substantial number of studies on freshwater diatoms from the British Isles. Notable workers include A.S. Donkin (1869), R.F. Bastow (1949, 1954, 1957), J.R. Carter (see Hartley, 1994) among many others (see below). By the end of the twentieth century, the number of freshwater taxa recorded from the British Isles had risen to 108 genera, 1652 species and 453 infraspecific taxa (Whitton et al., 1998a). This figure is likely to be a considerable underestimate of the true situation (Anderson, 1992; Mann & Droop, 1996).

The most significant development from the 1970s onwards has been the application of the scanning electron microscope to issues of diatom taxonomy. The recognition that morphological characteristics not visible with light microscopy could help to refine taxonomic boundaries was accompanied by a re-evaluation of the work of nineteenth century diatomists, such as that of Mereschkowsky who used other morphological characteristics (e.g. plastid arrangement) rather than simply relying upon features of the frustule. A brief flavour of this debate is necessary to appreciate why diatoms cannot be included in this Flora at the present time.

In 1930, Hustedt published a single-volume diatom Flora of central Europe which included three closely related genera: *Fragilaria*, *Ceratoneis* and *Synedra*. *Ceratoneis* was readily differentiated from *Fragilaria* and *Synedra* by its curved valves whilst the classical separation between *Fragilaria* and *Synedra* was based on the types of colonies that each formed. Krammer & Lange-Bertalot (1986–91), in their revision of Hustedt (1930), argued that these distinctions were tenuous and 'lumped' the three genera into a single genus, *Fragilaria*. However, at about the same time Williams & Round (1986, 1987) re-evaluated *Fragilaria* and *Synedra* using characters visible only with the scanning electron microscope and split these genera into ten new or resurrected genera. Meanwhile, *Ceratoneis* had undergone a name change, purely on nomenclatural grounds (Patrick & Reimer, 1966), and was now *Hannaea*. So the modern diatomist is faced with three separate sets of conventions, all of which reflect the same broad taxonomic reality, but which differ in their understanding of where generic, specific and subspecific divisions should lie.

Similar situations exist for other genera – notably *Navicula* and *Achnanthes*.

The genus *Navicula* was one of the earliest to be described (Bory, 1822) and was therefore the initial home for many of the bilaterally symmetrical raphid diatoms. Although many of these were moved to new genera as distinguishing features were recognized, by the 1960s *Navicula* had become 'a dump for all bilaterally symmetrical raphid diatoms lacking particularly distinctive features' (see Round et al., 1990). Within this *mélange*, there were many that deviated quite markedly from the type in both frustule and protoplast features and efforts have been made to establish (or, in many cases, re-establish) more natural generic boundaries. Whilst Hustedt (1930) subsumes the genus *Placoneis* (erected by Mereschkowsky in 1903) into *Navicula*, Cox (1987) argues for its reinstatement on the basis of several characteristics but particularly because of its single large plastid (in contrast to the two plastids of *Navicula sensu stricto*). Krammer & Lange-Bertalot (1986–91) adopt a more conservative approach to *Navicula*, but many of their subdivisions broadly coincide with these new or resurrected genera.

Alongside this re-evaluation of genera, there has also been a proliferation of new species, although the criteria on which these are based are often purely morphological and on studies of only one or a few populations (see Round, 1997) rather than on demonstrations of a link between morphological characters and reproductive behaviour. This, in turn, has generated a vigorous debate about how best different taxonomic ranks should be applied to diatoms (Round, 1996; Cox, 1997).

Although a new diatom Flora of the British Isles is long overdue, the issues mentioned above are not mere distractions. Rather, without their resolution, such a Flora will not be based upon secure foundations. However, this is of little comfort for ecologists and others who need to identify diatoms. For this reason, we have provided a brief guide to the most useful literature for identifying diatoms from the British Isles and provide a table to aid reconciliation of the various taxonomic systems (Table 2).

Checklists and monographs

Many papers have been written which include lists of diatoms from different freshwater habitats. However, the number of such papers has declined markedly in recent years and most of the records are included in Hartley's checklist of all marine and freshwater diatoms recorded from the British Isles (Hartley, 1986). Many of these taxa are illustrated in the *Atlas of British Diatoms* (Hartley et al., 1996). Hartley's checklist was the basis for the diatom section of the coded list of British freshwater algae (Whitton et al., 1998a), although the nomenclature was updated and some new records were included. Diatoms are also included in the algal checklist for Wales (Benson-Evans & Antoine, 1996).

Papers which include good lists of taxa from different freshwater habitats, or which contain useful information on particular taxa, include:

England and Wales: Bastow (1949, 1954, 1957); Belcher & Swale (1977, 1978, 1981); Bennion & Appleby (1999); Evans (1970); Harris (1930); Haworth et al. (1988); Juggins (1992); Knudson (1952–3, 1954); Rich (1925); Round (1957d, f, 1961b); Speller (1990); Woodhead & Tweed (1947).

Ireland: Adams (1908); Battarbee (1978); Foged (1977); Round (1959); W.West (1912).

Scotland: Carter & Bailey-Watts (1981); Haworth (1972; 1976); McCall (1933); Pennington et al. (1972); W. & G.S.West (1905a).

In addition, Carter wrote many other useful papers (often including keys). These are listed in his obituary (Hartley, 1994).

Introductory guides

Barber & Haworth (1981) and Kelly (2000) provide keys to freshwater genera and illustrations of common species, based on the morphology of the diatom frustule. Cox (1996), however, designed a guide based on living cells, specifically for use on samples that have not been acid-cleaned (although for some genera cleaned material is still necessary to confirm identifications). In this she includes the most common species, approximately 25% of the total number recorded. One recent European introductory guide (which includes a CD-ROM) is that by Prygiel & Coste (2000).

Floras

British diatomists have relied on various European Floras. These include those by Frederich Hustedt (1930, 1927–66), Cleve-Euler (1951–55) and Germain (1981), although not all are still in print. Patrick & Reimer (1966, 1975) is a useful English language flora but is incomplete (notably missing *Nitzschia* and centric diatoms). The four volumes of the *Süßwasserflora von Mitteleuropa* (Krammer & Lange-Bertalot, 1986–91) are the most recent and very useful, largely because of the large number of photographs. Although written in German, it is quite easy to pick out the most important diagnostic characteristics from the text. However, users should check the names used in these volumes against the names in the checklists above. Finally, Hartley et al. (1996), though not strictly a Flora since it lacks keys and descriptions, contains illustrations of most diatoms recorded from the British Isles.

Table 2. *Relationships between generic names used in recent taxonomic works on freshwater diatoms. All genera likely to be found in freshwater or brackish-water habitats in the British Isles are included along with the names under which they are classified. Fourtanier & Kociolek (1999) provide a catalogue of diatom genera from which information on genera not included in recent floras can be obtained. This table should be treated as a broad indication, rather than definitive, as precise boundaries between genera occasionally vary between authors.*

Hustedt (1927–66)	Krammer & Lange-Bertalot (1986–91)	Round *et al.* (1990)	Changes post-Round *et al.* (1990)
Centric diatoms			
Actinocyclus	*Actinocyclus*	*Actinocyclus*	
Attheya	*Acanthoceras*	*Acanthoceras*	
Chaetoceros	*Chaetoceros*	*Chaetoceros*	
Cyclotella	*Cyclotella*	*Cyclotella*	
Cyclotella or *Stephanodiscus*	*Cyclostephanos*	*Cyclostephanos*	
Hyalodiscus	—	*Hyalodiscus*	
—	—	*Hydrosera*	
Melosira	*Melosira*	*Melosira*	
Melosira	*Aulacoseira*	*Aulacoseira*	
Melosira	*Ellerbeckia*	*Ellerbeckia*	
Melosira	*Orthoseira*	*Orthoseira*	
Melosira	—	*Paralia*	
Rhizosolenia	*Rhizosolenia*	*Urosolenia*	
Stephanodiscus	*Stephanodiscus*	*Stephanodiscus*	
Stephanodiscus	*Stephanodiscus*	*Stephanodiscus*	
—	*Stephanocostis*	—	
Thallassiosira	*Thallassiosira*	*Thallassiosira*	
Araphid pennate diatoms			
Asterionella	*Asterionella*	*Asterionella*	
Centronella	*Fragilaria*	*Centronella*	
Ceratoneis	*Fragilaria*	*Hannaea*	
Diatoma	*Diatoma*	*Diatoma*	
Fragilaria	*Fragilaria*	*Fragilaria*	
Fragilaria	*Fragilaria*	*Fragilariforma*	
Fragilaria	*Fragilaria*	*Pseudostaurosira*	
Fragilaria	*Fragilaria*	*Punctastriata*	
Fragilaria	*Fragilaria*	*Staurosira*	
Fragilaria	*Fragilaria*	*Staurosirella*	
Meridion	*Meridion*	*Meridion*	
Opephora	*Fragilaria*	*Martyana*	
Opephora	*Opephora*	*Opephora*	
Synedra	*Fragilaria*	*Ctenophora*	
Synedra	*Fragilaria*	*Synedra*	
Synedra	*Synedra*	*Tabularia*	
Tabellaria	*Tabellaria*	*Oxyneis*	
Tabellaria	*Tabellaria*	*Tabellaria*	
Tetracyclus	*Tetracyclus*	*Tetracyclus*	
—	—	—	*Distrionella*
Raphid pennate diatoms			
Achnanthes	*Achnanthes*	*Achnanthes*	
Achnanthes	*Achnanthes*	*Achnanthes*	*Lemnicola*
Achnanthes	*Achnanthes*	—	*Karayevia*
Achnanthes	*Achnanthes*	—	*Psammothidium*
Achnanthes	*Achnanthes*	—	*Rossithidium*
Achnanthes	*Achnanthes*	—	*Kolbesia*
Achnanthes	*Achnanthes*	*Achnanthidium*	
Achnanthes	*Achnanthes*	*Achnanthidium*	*Planothidium*
—	*Actinella*	*Actinella*	

Table 2 (*cont.*)

Hustedt (1927–66)	**Krammer & Lange-Bertalot (1986–91)**	**Round *et al.* (1990)**	**Changes post-Round *et al.* (1990)**
Amphicampa	*Eunotia*	*Semiorbis*	
Amphipleura	*Amphipleura*	*Amphipleura*	
Amphiprora	*Entomoneis*	*Entomoneis*	
Amphiprora	—	—	
Amphora	*Amphora*	*Amphora*	
Anomoensis	*Anomoensis*	*Brachysira*	
Anomoeoneis	*Anomoeoneis*	*Anomoeoneis*	
Bacillaria	*Bacillaria*	*Bacillaria*	
Brebissonia	—	*Brebissonia*	
Caloneis	*Caloneis*	—	
Campylodiscus	*Campylodiscus*	*Campylodiscus*	
Cocconeis	*Cocconeis*	*Cocconeis*	
Cylindrotheca	*Cylindrotheca*	*Cylindrotheca*	
Cymatopleura	*Cymatopleura*	*Cymatopleura*	
Cymbella	*Cymbella*	*Cymbella*	
Cymbella	*Cymbella*	*Encyonema*	
Cymbella	*Cymbella*	*Reimeria*	
Cymbella	*Cymbella*	*Encyonema*	*Encyonopsis*
—	*Cymbellonitzschia*	*Cymbellonitzschia*	
Denticula	*Denticula*	*Denticula*	
Diatomella	*Diatomella*	*Diatomella*	
Didymosphenia	*Didymosphenia*	*Didymosphenia*	
Diploneis	*Diploneis*	*Diploneis*	
Entomoneis	*Entomoneis*	*Entomoneis*	
Epithemia	*Epithemia*	*Epithemia*	
Eucocconeis	*Achnanthes*	*Eucocconeis*	
Eunotia	*Eunotia*	*Eunotia*	
Frustulia	*Frustulia*	*Frustulia*	
Gomphocymbella	*Cymbella*	*Gomphocymbella*	
Gomphonema	*Gomphoneis*	*Gomphoneis*	
Gomphonema	*Gomphonema*	*Gomphonema*	
Gyrosigma	*Gyrosigma*	*Gyrosigma*	
Hantzschia	*Hantzschia*	*Hantzschia*	
Mastogloia	*Mastogloia*	*Mastogloia*	
Navicula	*Navicula*	*Aneumastus*	
Navicula	*Navicula*	*Cavinula*	
Navicula	*Navicula*	*Cosmioneis*	
Navicula	*Navicula*	*Craticula*	
Navicula	*Navicula*	*Diadesmis*	
Navicula	*Navicula*	*Fallacia*	
Navicula	*Navicula*	*Luticola*	
Navicula	*Navicula*	*Navicula*	
Navicula	*Navicula*	*Petroneis*	
Navicula	*Navicula*	*Placoneis*	
Navicula	*Navicula*	*Scolioneis*	
Navicula	*Navicula*	*Sellaphora*	
Navicula	*Navicula*	*Navicula*	*Eolimna*
Navicula	*Navicula*	*Navicula*	*Fistulifera*
Navicula	*Navicula*	*Navicula*	*Geissleria*
Navicula	*Navicula*	*Navicula*	*Hippodonta*
Navicula	*Navicula*	*Navicula*	*Mayamaea*
Neidium	*Neidium*	*Neidium*	
Nitzschia	*Nitzschia*	*Nitzschia*	
Nitzschia	*Nitzschia*	*Psammodictyon*	
Nitzschia	*Simonsenia*	*Simonsenia*	

Table 2 (*cont.*)

Hustedt (1927–66)	**Krammer & Lange-Bertalot (1986–91)**	**Round *et al.* (1990)**	**Changes post-Round *et al.* (1990)**
Nitzschia	*Nitzschia*	*Tryblionella*	
Peronia	*Peronia*	*Peronia*	
Pinnularia	*Pinnularia*	*Pinnularia*	
—	*Pinnularia*	–	*Kraskella*
Pleurosigma	*Pleurosigma*	*Pleurosigma*	
Rhoicosphenia	*Rhoicosphenia*	*Rhoicosphenia*	
Rhopalodia	*Rhopalodia*	*Rhopalodia*	
Stauroneis	*Stauroneis*	*Stauroneis*	
Stauroneis	*Stauroneis*	*Staurophora*	
Stenopterobia	*Stenopterobia*	*Stenopterobia*	
Surirella	*Surirella*	*Surirella*	

PHYLUM PHAEOPHYTA (Brown Algae)

John D. Wehr

Members of the Phaeophyta or brown algae are mostly macroscopic seaweeds, some of which attain lengths of more than 50 metres. Freshwater species are more modest in size, occurring mainly as microscopic filaments or macroscopic tufts or crusts, several mm to a few cm in size (Starmach, 1977). Members are recognized as a class, the Fucophyceae (Christensen, 1978) or a division, the Phaeophyta (Papenfuss, 1951; Bold & Wynne, 1985), with more than 265 genera and perhaps 2000 species worldwide. Their classification is being re-evaluated in light of emerging ultrastructural and molecular data, as well as affinities with the Tribophyceae (=Xanthophyta) and other golden algae (De Reviers & Rousseau, 1999). Members of the Phaeophyta are distinguished by a brown or yellow-brown color, which results from the carotenoid pigment, fucoxanthin. Major photosynthetic pigments are chlorophylls a, c_1, and c_2, plus varying amounts of β-carotene, violaxanthin, and diatoxanthin (Goodwin, 1974; Hoek et al., 1995). Their colour is also influenced by phaeophycean tannins, which occur in physodes (Svedelius, 1930; Chadefaud, 1950). The major photosynthetic storage product in not starch, but laminarin, a water-soluble $\beta_{1\text{–}3}$ glucan. Cell walls are composed of cellulose, alginic acid and sulphated polysaccharides; relative proportions vary among species and environmental conditions (Mackie & Preston, 1974). Ultrastructural features of cells and reproductive structures for the group have been described by Hoek et al. (1995) and Graham & Wilcox (2000).

Less than 1% of all species are known from freshwater habitats. There are also several species which colonize estuarine and saltmarsh environments (Waern, 1952; Wilce, 1966; Dop, 1979; Wehr & Stein, 1985). Despite some species having been described nearly 150 years ago (e.g. *Pleurocladia* described by A. Braun in 1855), brown algae remain poorly understood and very likely under-reported. All known forms are benthic, although flagellated cells are produced (sexual and asexual zoospores; Kumano & Hirose, 1959; Dop, 1979). In addition to pigments and morphology, nearly all freshwater species may be recognized by possessing large unilocular sporangia or clusters of plurilocular sporangia (Israelsson, 1938; Starmach, 1977). Different authors recognize between 3 and 7 freshwater brown algal genera, with as many as 12 species (West & Fritsch, 1927; Smith, 1950; Starmach, 1977; Dop, 1979; Bourrelly, 1981). Nearly all published reports have been based on European and North American collections (see Wehr & Stein, 1985).

The present author recognizes six freshwater genera and seven species of Phaeophyta worldwide, with two species confirmed from habitats in the British Isles: *Heribaudiella fluviatilis* and *Pleurocladia lacustris.* Both species colonize rocks, wood or other solid substrata, and occur in streams, larger rivers, and lakes. A third possible species in the British flora is *Porterinema fluviatile*, a species recognized elsewhere in Europe (Waern, 1952; Dop, 1979), and described by Belcher (1959) from a pond in England, but as *Apistonema pyrenigerum* (putative member of the Chrysophyta).

1 Thallus forming crusts, as dark brown patches mainly on stones; microscopic appearance in two forms: branched basal filaments and densely packed upright filaments .... *Heribaudiella*
1 Thallus branched, on rocks or plants, with a spreading filamentous morphology 2
2 Basal filaments curved or arching and erect filaments spreading with a single (rarely 2) parietal chloroplast; plurilocular sporangia elongate-linear, rare .... *Pleurocladia*
2 Basal and erect filaments arranged irregularly; cells with 2 parietal chloroplasts; plurilocular sporangia in crown-shaped clusters of 4 or more .... *Porterinema*

ORDER ECTOCARPALES

Uniseriate filaments, branched; prostrate filaments grow by apical cell division and erect ones by intercalary division. Asexual reproduction by motile spores produced in uni- or plurilocular sporangia. Sexual reproduction by anisogamy or oogamy.

Heribaudiella Gomont 1896

Thallus forming olive-brown to dark brown crusts, with an irregular-rounded outline but with distinct margins; microscopic morphology in two components: a horizontal (basal) system of frequently branched filaments and

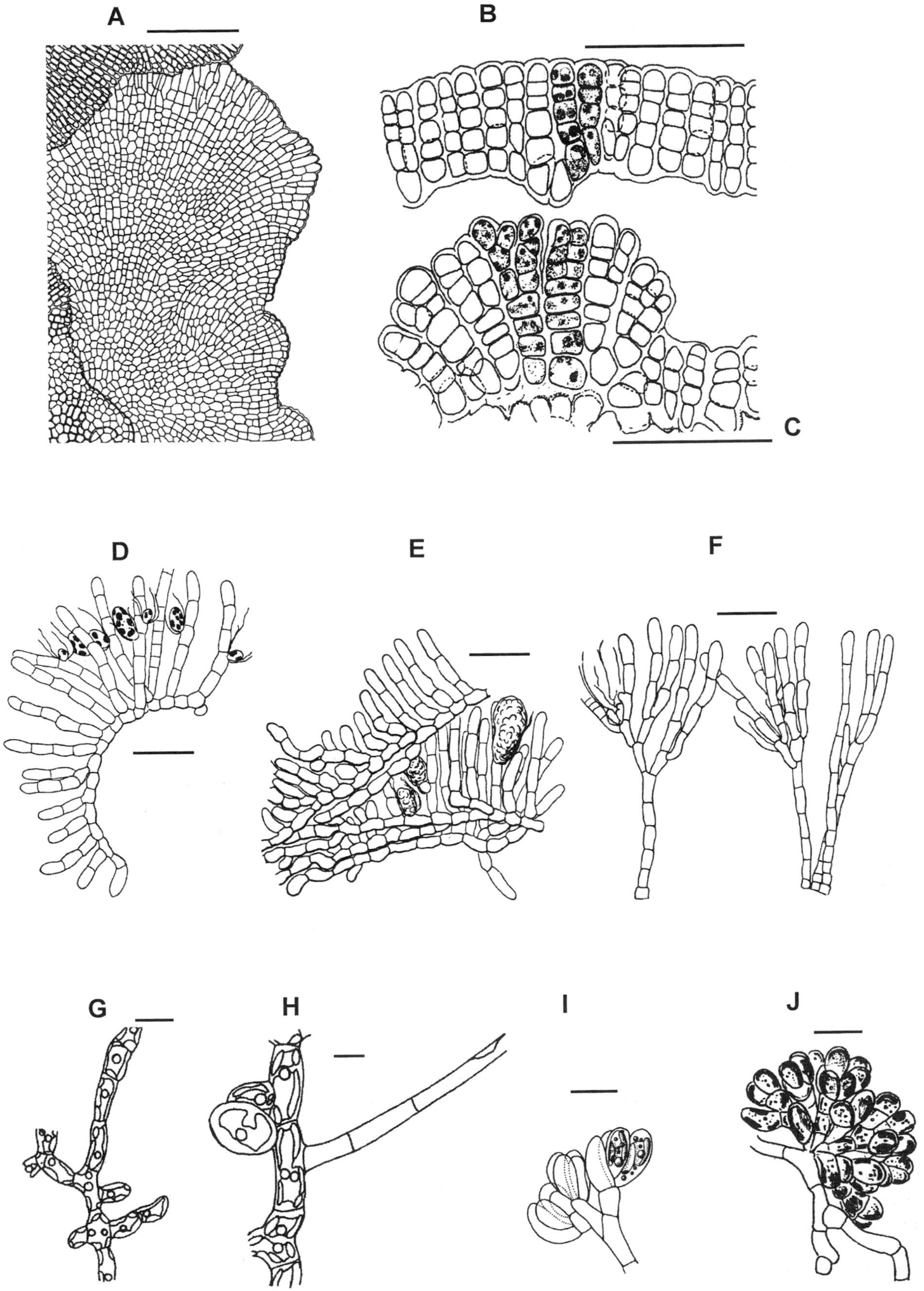

Plate 74 A–J
A–C. *Heribaudiella fluviatilis* (p. 280): A. Microscopic appearance of branched, horizontal filaments or crusts on a rock surface, B. Vertical filaments and details of contents of cells, C. Branching filaments showing chloroplasts and unilocular sporangia forming at apex; **D–F**. *Pleurocladia lacustris* (p. 280): D. Typical arched or 'comb-like' growth pattern and lateral unilocular sporangia, E. Network of branched filaments with terminal unilocular sporangia, F. Apical branching pattern; **G–J**. *Porterinema fluviatile* (p. 280): G, H. Close-up of portion of a filament, I. Detail showing cluster or crown of intercalary plurilocular sporangia, J. Multiple clusters of plurilocular sporangia.
Scale bars: A, D–F, 50 μm; B, C, 100 μm; G–J, 10 μm.

an erect system of tightly packed filaments (8–15 cells long); upright system revealed after applying pressure to the cover slip; filaments occasionally dichotomously branched; chloroplasts 4–10 per cell, oval or discoid; physodes common; details described for the species; unilocular sporangia terminal, ovoid or club-shaped, swarmers pear-shaped, biflagellate; plurilocular sporangia rare.

Northern hemisphere, widely distributed; individual colonies in patches or coalesce to cover entire rocks or boulders in streams and lakes. Monospecific.

Macroscopic appearance on rocks similar to and may be confused with species of *Chamaesiphon*, but margins of the latter appear as indistinct smudges (see Holmes & Whitton, 1975b). Swarmers are produced if kept in an incubator (Kumano & Hirose, 1959).

Heribaudiella fluviatilis (Areschoug) Svedelius 1930

Basionym: *Ralfsia fluviatilis* Areschoug
Synonyms: *Lithoderma fluviatile* Areschoug, *L. fontanum* Flahault, *L. zonatum* Jão

14010010 **Pl. 74A–C** (p. 279)

Crusts 1–5 (up to 20) cm across; cells typically thick-walled, 8–15 μm wide; unilocular sporangia, 10–25 μm wide, 15–35 μm long, swarmers 6–8 μm in size.

Distribution as for genus, most widely recorded freshwater phaeophyte (Israelson, 1938; Waern, 1938; Kann, 1945, 1993) occurring in streams, rivers (usually fast-flowing) and in rocky littoral areas of lakes (e.g. Windermere; N.T.H. Holmes, pers. comm.). In the British Isles Fritsch (1929a) and Holmes & Whitton (1975b, 1977a, b) describe populations in Devon and north-east England to south-east Scotland, respectively. N.T.H. Holmes (pers. comm.) reports that at least 143 populations have been recorded from the British Isles. Often occurs with *Audouinella*, *Chamaesiphon*, *Cladophora*, *Lemanea*, *Nostoc* and *Hildenbrandia rivularis*; at least in the latter case, even colonizing the same rock (Fritsch, 1929a; Geitler, 1932b; Kusel-Fetzmann, 1996).

Often overlooked (mistaken for diatoms, lichens or encrusting cyanobacteria), but can be abundant; occasionally dominant in streams (Kann, 1978).

Pleurocladia A.Braun 1855

Thalli brown or pale brown, forming hemispherical tufts or cushions 1–3 mm in diameter, sometimes almost whitish depending on calcification; consisting of two growth forms, usually occurring together: a microscopic basal system of infrequently branched filaments composed of inflated (occasionally elongate) cells having an arched growth pattern, giving rise to upright, irregularly branched, narrower filaments; hairs long, terminal, multicellular and common in field populations; unilocular sporangia lateral or terminal and single, club-shaped or spherical; plurilocular sporangia uncommon, linear-elongate (Waern 1952).

Probably worldwide; an epiphyte on submerged or emergent aquatic plants (e.g. *Phragmites*, *Typha*, *Sparganium*), occasionally endophytic, or on rocks in lakes or rivers. Colonies encrusted with $CaCO_3$ in limestone districts, or may be gelatinous in less calcareous regions (Waern, 1952). Monospecific.

Pleurocladia lacustris A.Braun 1855

Synonyms: *Pleurocladia ralfsioides* Bornet, *Streblonema longiseta* Arnoldi

14020010 **Pl. 74D–F** (p. 279)

Cells of basal system 8–16 μm wide, cells of upright system 6–12 μm wide, more elongate (12–35 μm long) and more nearly isodiametric; multicellular hairs 100–300 μm long and 5–7 μm in diameter; unilocular sporangia, 15–30 μm wide, 25–60 μm long.

Distribution as for genus; known from a few locations in the British Isles, but possibly widespread in suitable habitats; an apparent calciphile. The first population in England recorded by Kirkby et al. (1972) from the littoral zone of a nutrient-rich (eutrophic) pond near Brasside, County Durham. The authors mention that *Pleurocladia* was observed on rotting (but not recently dead) *Typha* leaves. It has also been collected from a hard-water pond, Hell Kettles, near Darlington (A.J. Marker and B.A. Whitton, pers. comm.). No populations have been reported from flowing water in the British Isles. It occurs in calcareous benthic habitats where *Chaetophora incrassata*, *Gloeotrichia pisum*, *Oedogonium* sp., *Tolypothrix tenuis* or *Rivularia* also occur (Israelson, 1938). Other European collectors have reported *Pleurocladia* together with *Heribaudiella* on rocks in lakes (Kann, 1993) and in streams (Kusel-Fetzmann, 1996); *Pleurocladia* also has been observed in brackish-water or intermittently marine habitats (Waern, 1952).

Porterinema Waern 1952

Thalli monostromatic, forming brown disc-shaped colonies of irregularly branched and loosely arranged filaments; basal filaments of inflated or proximally enlarged cells, occasionally bearing short, erect filaments of more elongate cells; chloroplasts parietal and lobed, 1–3 per cell; hairs terminal, multicellular (sometimes sheathed at base), common in some populations; plurilocular sporangia intercalary, on stalks or sessile, often forming crown-like structures (Waern, 1952); unilocular sporangia rare.

World distribution incompletely known, usually epiphytic on macroalgae and higher plants, endophytic in other algae (e.g. *Rhizoclonium*), occasionally epilithic. Mainly a brackish-water species in Europe and North America, but has been recorded from freshwater ponds and lakes in Europe (Waern, 1952; Dop, 1979).

Porterinema fluviatile (Porter) Waern 1952

Basionym: *Streblonema fluviatile* Porter

02030010 **Pl. 74G–J** (p. 279)

Cells of basal filaments 6–12 μm wide, 6–12 μm long; cells of erect filaments 20–40(–50) μm long; hairs 3–8 μm wide, up to 200 μm long; plurilocular sporangia in 4-celled clusters (up to 32).

World distribution incompletely known; in the British Isles known from a single record from a small pond in a gravel pit in the Lee Valley, England where it was epiphytic on *Cladophora* and initially recorded as *Apistonema pyrenigerum* Pascher (Chrysophyta) by Belcher (1959); this specimen was later referred to *P. fluviatile* by Dop (1979) and Bourrelly (1981). Resampling is needed to verify this record.

PHYLUM PRASINOPHYTA

Øvind Moestrup

The primitive green algae may be classified into two classes, the Prasinophyceae and the Pedinophyceae. The class name Prasinophyceae goes back to Christensen (1962), and is based on the Greek word πρασου, meaning a leek. It refers to the yellowish-green colour of the cell in many prasinophyceans, reflecting the presence of pigments in the cell which are slightly different from those of most other green algae. Prasinophycean green algae are mainly flagellates. The cell surface and the flagella are covered with minute organic scales but these are usually too small to be visible under the light microscope (exception: the largest scales in *Mesostigma*). For critical species determination, scale structure must therefore often be ascertained by electron microscopy. In some genera all traces of scales have been lost and recently an increasing number of minute flagella-lacking species of the marine plankton have proved by gene sequencing to be prasinophyceans. Prasinophyceans are generally accepted to be the oldest group of green algae from whose ancestors all other groups of green algae have arisen. They are thought to be a very ancient group, a theory based also on the presence of prasinophycean-like cells in fossil material dating perhaps as far back as the PreCambrian.

Ultrastructurally the flagellated prasinophyceans usually have parallel and notably long flagellar bases, which are inserted in a groove. The genera are very different, probably reflecting the ancient origin of the group, thus the number of flagella varies from one in a few genera, to 2 and 4 in several genera, to 8 and even 16 flagella in one species, the only 16-flagellated algal flagellate known. The total number of extant species of prasinophyceans is probably around 100.

The ultrastructural features separating the prasinophytes from the chlorophytes, eustigmatophytes and raphidophytes are not distinguished using the light microscope. For this reason the prasinophyte genera are keyed out with other flagellates in Section IV of the keys to chlorophytes (principally the Volvocales) and the phyla mentioned above (see p. 290).

CLASS PEDINOPHYCEAE

The Pedinophyceae, named by Moestrup (1991), comprises three genera of small, naked (scale- or wall-less) flagellates, with around 10 species. The cells show a number of unusual traits, e.g. during mitosis, where the nuclear envelope remains intact throughout, surrounding the entirely internal spindle apparatus.

ORDER PEDINOMONADALES

Cells less than 10 μm long, with one laterally inserted flagellum but without body scales.

Pedinomonas Korshikov 1923

Single-celled asymmetrical flagellates, elongate, more or less compressed; flagellum single, emergent, inserted laterally at the posterior end, directed posteriorly behind the cell during swimming; chloroplast single with a pyrenoid in anterior of cell; eyespot located near pyrenoid, on opposite side of the flagellar insertion.

Cosmopolitan; in freshwater, brackish-water and seawater. Fewer than 10 species, only two species known from freshwater habitats in the British Isles.

1 Pyrenoid surrounded by 2 starch grains ....*P. major*
1 Pyrenoid surrounded by more than 2 starch grains ....*P. minor*

Pedinomonas major Korshikov 1923

150600100 **Pl. 75B** (p. 283)
Cells 3–5 μm wide, 3.5–6.5 μm long and 1.5–2 μm thick; pyrenoid surrounded by 2 large starch grains.

Europe; in small, particularly in nutrient-rich water bodies; found on a single occasion in England in a flooded cart rut near Cartmel, north Lancashire.

Pedinomonas minor Korshikov 1923

15060020 **Pl. 75A** (p. 283)
Cells 2–4.5 μm wide, 3.5–7 μm long and 1–2.5 μm thick; pyrenoid surrounded by 3–5 (usually 4) starch grains.

Widely distributed; in the British Isles particularly in nutrient-rich water bodies, if present in large numbers imparting a green colour to the water.

CLASS PRASINOPHYCEAE

ORDER CHLORODENDRALES

Cells compressed, with 4 flagella arising from bottom of an anterior pit. Scales coalescing to form a wall.

Myochloris J.H.Belcher *et* Swale 1961

Single-celled naked flagellates with 2 unequal (anisokont) flagella; cells ovoid, one side convex (dorsal side) and the other flat or slightly concave (ventral side), the anterior end pointed or narrowing to a slender tip (rostrate); chloroplast parietal, dorsal, containing a dorsal eyespot and 1–3 pyrenoids; contractile vacuole anterior; flagella inserted ventrally about one-third the cell length from the anterior end, the long flagellum trailing and the short flagellum held laterally; movement described as typically darting, often revolving on longitudinal axis when swimming, frequently attaching temporarily to a substratum by front end, cells then moving vigorously around point of attachment for some time.

British Isles (England); puddles and a canal. Monospecific.

The phylogenetic affinities of *Myochloris* remain obscure since no ultrastructural examinations have been made. It resembles *Nephroselmis* and *Scourfieldia* in having anisokont flagella, but differs in the orientation of the flagella during swimming and, in particular, by its mode of temporary attachment. *Nephroselmis* and *Scourfieldia* do not attach to surfaces.

Myochloris collorhynchus J.H.Belcher *et* Swale 1961

15040010 **Pl. 75F** (p. 283)

Cells 4–7 μm wide and 7–9 μm long, with one flagellum about two-thirds the length of the other.

Distribution as for genus; found first in a temporary puddle of water collected in a cow's hoof-print at Malham, Yorkshire, in 1960 and subsequently among weeds from the Kendal Canal at Tewitfield, Lancashire.

This unusual flagellate is known only from England. It should be sought for and established in culture for studies on ultrastructure and phylogenetic relationships.

Nephroselmis F.Stein 1878

Synonyms: *Heteromastix* Korshikov, *Bipedinomonas* H.J.Carter

Single-celled flagellates, flattened, more or less bean-shaped, with 2 unequal (anisokont) flagella extending from invagination near the middle of one side, anterior flagellum short and held forwards during swimming, posterior flagellum longer and trailing behind cell; chloroplast parietal, single, eyespot usually near the short flagellum; pyrenoid present on side opposite the flagellar insertion; contractile vacuole near the flagellar insertion; surfaces all covered with layers of submicroscopical organic scales; sexual reproduction by 2 cells similar to normal vegetative cells functioning as gametes, and after fusion forming a thick-walled zygote, germination results in formation of 4 cells.

Cosmopolitan; sometimes in freshwaters but the majority of species occur in brackish-water and marine environments. Less than 10 species described of which two freshwater species are recorded from the British Isles.

Nephroselmis is one of the few prasinophytes in which sexual reproduction has been demonstrated.

Nephroselmis olivacea F.Stein 1878

Synonyms: *Sennia commutata* Pascher, *Heteromastix angulata* Korshikov, *Nephroselmis angulata* (Korshikov) Skuja

15050010 **Pl. 75E** (p. 283)

Cells 7–15 μm wide, 6–10 μm long and 3–6 μm thick; from side semicircular or bean-shaped, occasionally rounded, quadrangular or almost hexagonal; flagella in resting cells bent backwards in a characteristic way along the 2 narrow sides of cell; pyrenoid very characteristic, surrounded by 2 starch grains formed like watch-glasses with concave sides facing each other, lowermost grain (near plasmalemma) larger, as they age grains grow considerably in size and eventually cause cell to become angular in shape; sexual reproduction as described above.

Widely distributed in many types of freshwater in temperate and subtropical regions; in the British Isles recorded from England where it occurs in rivers (e.g. Great Ouse, St Ives, Cambridgeshire, and River Cam, Queen's College, Cambridge) and ponds in Richmond Park, Surrey.

An unidentified freshwater species of *Nephroselmis* was illustrated by Manton et al. (1965), isolated by J. Belcher and maintained as culture no. 148 at the Windermere laboratory. It has apparently not been studied in detail.

Scherffelia Pascher 1911

Single-celled flagellates with 4 flagella emerging from an anterior pit; cells markedly compressed and more or less twisted; chloroplasts 2 and lateral, often incised into 2 lobes anteriorly, without pyrenoids; eyespot single, located anteriorly near middle or posterior of cell; contractile vacuoles 2(3?), near flagellar pit; cells surrounded by a periplast of fused organic scales resembling a cell wall, flagella covered with submicroscopical organic scales; asexual reproduction by longitudinal division of cell, the two new cells often inverted with respect to each other within the parental periplast.

British Isles, Europe, Russia, North America; in smaller bodies of stagnant water. About 5 species.

Scherffelia dubia (Perty) Pascher 1911

Basionym: *Cryptomonas dubia* Perty

Synonyms: *Carteria dubia* (Perty) Scherffel, *Scherffelia ovata* Pascher, *S. opisthostigma* Skuja, *S. pelagica* Skuja, *Platymonas*

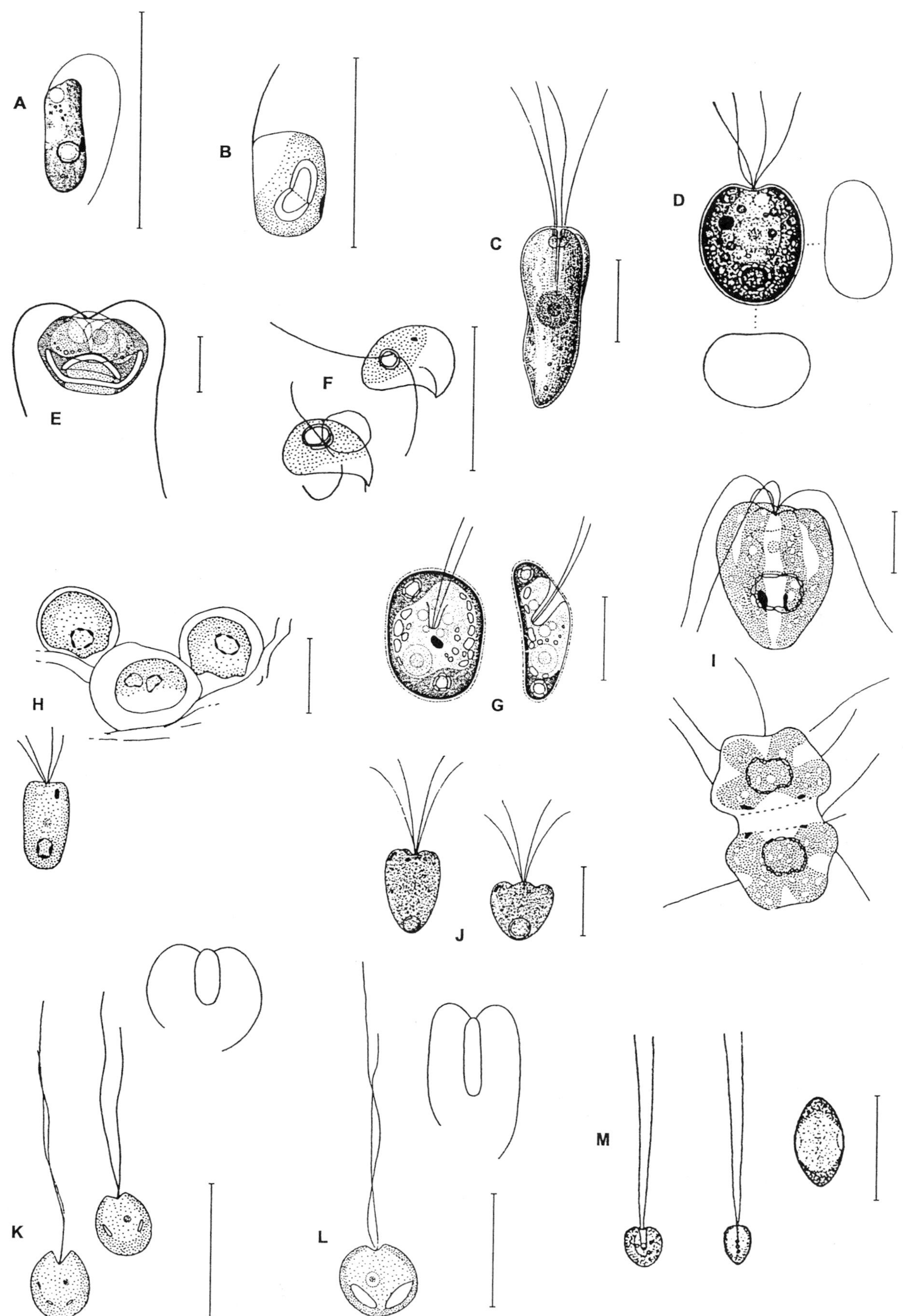

Plate 75 A–M

A. *Pedinomonas minor* (p. 281); **B**. *Pedinomonas major* (p. 281); **C**. *Scherffelia dubia* (p. 282); **D**. *Tetraselmis cordiformis* (p. 284); **E**. *Nephroselmis olivacea* (p. 282); **F**. *Myochloris collorhynchus* (p. 282); **G**. *Mesostigma viride* (p. 284); **H**. *Prasinochloris sessilis* (p. 284); **I**. *Pyramimonas tetrarhynchus* (p. 285); **J**. *Pyramimonas inconstans* (p. 285); **K**. *Scourfieldia complanata* (p. 286); **L**. *Scourfieldia caeca* (p. 286); **M**. *Scourfieldia cordiformis* (p. 286).

Scale bars: 10 μm.

bichlora H. *et* O.Ettl, *Tetraselmis bichlora* (H. *et* O.Ettl) R.E.Norris et al.
16640020 **Pl. 75C** (p. 283)
Cells typically 7–12 μm wide, 10–18 μm long and 4–6 μm thick, often with a short posterior protuberance, lacking wing-like edges or barb-like anterior protrusions.

Widely distributed in Europe and Russia; in the British Isles known only from England where recorded from the Kendal Canal, Lancashire, Priest Pot in the English Lake District, and puddles in the north of England.

Tetraselmis F.Stein 1878

Synonyms: *Platymonas* G.S.West, *Aulacochlamys* Margalef
Single-celled flagellates with 4 flagella emerging from an anterior pit; cells heart- or somewhat cherry-shaped, ellipsoidal or almost spherical, more or less compressed, often somewhat curved when seen from the side but never twisted; flagella beating as 2 pairs, one pair along each of the narrow sides of cell during forward swimming; chloroplast single (rarely 2), cup-shaped, usually with a basal pyrenoid; eyespot located on one of the flat sides of cell; contractile vacuoles 2, anterior; cells with a close-fitting periplast of fused scales resembling a cell wall and flagella covered with submicroscopical organic scales; asexual reproduction in non-motile stage, the two new cells often inverted within the parent periplast.

Probably cosmopolitan; widely distributed in freshwaters but more especially in marine waters.

Several species recorded from brackish-water and marine habitats in the British Isles (e.g. *Tetraselmis contracta* (H.J.Carter) Butcher, *T. apiculata* (Butcher) Butcher, *T. levis* Butcher).

Tetraselmis cordiformis (H.J.Carter) F.Stein 1878

Basionym: *Cryptoglena cordiformis* H.J.Carter
Synonyms: *Carteria cordiformis* (H.J.Carter) Dill, *Platymonas cordiformis* (H.J.Carter) Korshikov, *Pyramichlamys cordiformis* (H.J.Carter) H. *et* O.Ettl
15140010 **Pl. 75D** (p. 283) **CD**
Cells 16–23 μm wide, 14–20 μm long and 9–13 μm thick, cherry-shaped, a large basal pyrenoid with a starch sheath; contractile vacuoles 2 in number and near flagellar pit; eyespot red.

Probably cosmopolitan; widely distributed in the British Isles where it occcurs in nutrient-enriched puddles, ponds, lakes, canals (e.g. Lancaster–Kendal Canal at Tewitfield) and in slow-flowing reaches of rivers including the Cam, Ouse, Blyth, Waveney, Witham, Trent and Severn (Belcher & Swale, 2000).

Tetraselmis cordiformis was described from water tanks in Bombay by Carter in 1859. It was subsequently found to be very widely distributed throughout the world in many types of freshwater.

ORDER PYRAMIMONADALES

Cells having 4 (rarely 8, very rarely 16) flagella arising from bottom of a pit at the anterior end (rarely lateral). Body scales 3-layered, with outer scale layer complex.

Mesostigma Lauterborn 1894

Single-celled, flat to saddle-shaped flagellates with 2 flagella emerging from an invagination in the convex dorsal side, ventral side more or less concave; flagellar pit near front end and flagella pointing more or less anteriorly in swimming cells, cells rotate during swimming; chloroplast parietal, single, with or without pyrenoids; eyespot on ventral side near mid-axis of cell; contractile vacuoles several and near flagellar insertion; surfaces covered with organic scales, outermost layer of scales large enough to give the cell a finely punctate appearance under the light microscope.

Europe, USA, Russia, in various types of water body. Two species, only one recorded from the British Isles.

Mesostigma viride Lauterborn 1894

15010010 **Pl. 75G** (p. 283)
Cells 11–13 μm wide, 15–18 μm long and 5–7 μm thick; pyrenoids 2, almost opposite one another.

Europe, USA, Russia; in the British Isles seems to prefer muddy parts of ponds (e.g. Richmond Park and Royal Botanic Gardens, Kew, both in Surrey, England) rich in bottom-living diatoms and with good growth of aquatic plants (Lund, 1937).

A second species, *M. grandis*, was described from the Ukraine (Korshikov, 1939) and differs in being slightly larger (27 μm long and 18–20 μm wide, about 8 μm thick) and lacking pyrenoids. It occurred together with *M. viridis*. *Mesostigma grandis* has been found subsequently in the USA together with *M. viridis* but may have been overlooked in Europe.

Prasinochloris J.H.Belcher 1966

Single-celled sessile stage on dead leaves and zoospores the only motile stage; sessile cells spherical or ovoid, surrounded by a loose transparent membrane, with a single cup-shaped chloroplast, a pyrenoid and 2 contractile vacuoles; zoospores 1 or 2 in each cell, obpear-shaped or heart-shaped, 4 flagella inserted in an anterior invagination, sometimes flagella held in a *Pyramimonas* fashion to form a cross in anterior view, wall absent but apparently covered with minute scales, with a basal pyrenoid and lateral eyespot in a cup-shaped chloroplast; 2 contractile vacuoles present.

British Isles (England), known only from springs. Monospecific.

Prasinochloris appears to be closely related to *Pyramimonas*, the difference being the thin-walled sedentary phase in *Prasinochloris*, which contrasts with the thick-walled spherical cysts known in several species of *Pyramimonas*.

Prasinochloris sessilis J.H.Belcher 1966

15090010 **Pl. 75H** (p. 283)
Cells 7–10 μm wide; zoospores 5–8 μm wide, 9–12 μm long.

Distribution as for genus; described from a spring in a field near Hawkshead, Cumbria.

Attempts to establish *Prasinochloris* in culture by Belcher failed and the species remains very poorly known. A *Prasinochloris*-like organism was subsequently reported from Iceland soil by Broady (1982), differing mainly in the production of copious mucilage around the cells and in the formation of 2–8 zoospores in the cells.

Pyramimonas Schmarda 1850

Single-celled flagellates, usually more or less inversely pyramidal, in end view quadrangular or rounded; flagella 4, emerge from an apical depression, extend crosswise in four different directions (see *Tetraselmis* whose 4 flagella form 2 pairs); chloroplast cup-shaped, divided anteriorly into 4 or 8 lobes extending almost to anterior end of cell, with a basal pyrenoid; eyespot at posterior or anterior end; surfaces entirely covered with submicroscopical organic scales, whose detailed structure (seen in the electron microscope) provides important species characteristics.

Cosmopolitan; freshwater, brackish-water and marine. Twenty-three species recorded by Ettl (1983), but only 2 described here and both from freshwater habitats in the British Isles.

1 Cells 12–18 μm wide, 18–27 μm long; eyespots located in 2 adjacent chloroplast lobes in posterior half of cell..........*P. tetrarhynchus*
1 Cells 7.5–10 μm wide, 10–15 μm long; eyespot elongate, located in one of the anterior lobes.......................................*P. inconstans*

Pyramimonas inconstans Hodgetts 1920

15110020 **Pl. 75J** (p. 283)

Cells 7.5–10 μm wide and (8.8–)10–15 μm long, highly variable in shape, most commonly sub-pyramidal, also heart-shaped or even subspherical, in anterior view sub-quadrate with rounded corners and in transverse optical section circular, with 4 anterior lobes; eyespot elongate, located in one of the anterior lobes; contractile vacuoles 2(–3) in anterior part of the cell; *Palmella*-stage forms readily under unfavourable conditions, flagella then disappear and the cells assume irregular shapes, generally subspherical; many cells often present within the same mucilage.

British Isles (England); only reported among the filaments of *Spirogyra* and *Tribonema* in the overflow ditch of a small pond at Quinton, near Birmingham.

Differs from *P. tetrarhynchus* particularly in cell size and location of the eyespot in the cell.

Pyramimonas tetrarhynchus Schmarda 1850

15110030 **Pl. 75I** (p. 283)

Cells 12–18 μm wide, 18–27 μm long, obovoid to heart-shaped in side view, bluntly pointed or smoothly rounded at base, truncate at apex, widest part usually about one-third to half the cell length from the posterior end; flagellar pit up to 5 μm deep; chloroplast divided into 8 lobes anteriorly; eyespots located in 2 adjacent chloroplast lobes in the posterior half of cell; contractile vacuoles 2, near base of the flagella; cysts spherical, 20–30 μm in diameter, with a smooth transparent wall.

Probably cosmopolitan; in the British Isles found almost throughout the year, especially in muddy and peaty pools and ponds where often attached temporarily by the anterior end to water plants or other surfaces, sometimes when in great numbers forming a green coating on decaying leaves that if disturbed result in the cells immediately becoming detached and swimming away.

Bacteria commonly adhere to the cyst surface, giving it a spiny appearance. Cells are positively phototactic.

DOUBTFUL RECORD
Pyramimonas delicatula Griffiths 1909
15110010
This species was first described from England by Griffiths in Stanklin Pool, near Kidderminster, Worcestershire, and was rediscovered by Lund (1937) in ponds at Richmond Park, Surrey, and at Pickmere, Cheshire. It was described as 20–26 μm long and 11–16 μm wide with no mention made of an eyespot nor of any contractile vacuoles. Griffiths described the presence of 4 chloroplast lobes but noticed that the lobes split into 2 at the anterior end. *P. delicatula* is almost certainly conspecific with *P. tetrarhynchus*. It is probably significant that all sightings of *Pyramimonas* in freshwater anywhere in the world have been proved to be *P. tetrarhynchus* when the structure of the scaly covering has been examined.

ORDER SCOURFIELDIALES

Cells with 2 unequal and laterally inserted flagella. Scales absent.

Scourfieldia G.S.West 1912

Synonym: *Cardiomonas* Korshikov

Single-celled flagellates lacking walls; cells flattened, usually ovoid, heart-shaped or conical in lateral view, rarely round or quadrangular, with an anterior indentation from which emerge 2 very long flagella (2.5–6 times body length); flagella noticeably of different lengths (anisokont), during swimming they point backwards and lie parallel behind cell, while in resting cells the flagella are directed forwards and only distal ends curved backwards; chloroplast parietal, single, without a pyrenoid and stigma; contractile vacuole single and anterior; cell lacking organic scales on both flagella and body (contrasts to many other prasinophyceans).

British Isles, northern and Central Europe, Ukraine; small bodies of freshwater.

The taxonomy of *Scourfieldia* is unsatisfactory and the species concept presently rather uncertain. Ettl (1960) merged the 3 described species (*S. complanata*, *S. caeca*, *S. cordiformis*) but his illustrations differ from all of them in showing cells which have a distinct pyrenoid with 2 starch grains. Its identity is uncertain. Subsequently Ettl (1980) illustrated, in material from Denmark, a cell which in top view was curved (convex on one side, concave on the other) and identified it as *S. cordiformis* although the original description of *S. cordiformis* states the cells to be obovate. There is a need for a taxonomic revision of the genus, including the use of molecular methods.

1 Cells obovate in side view ....*S. cordiformis*
1 Cells flat in side view..2
2 Flagella in resting cells bend back in a regular curve......................................*S. complanata*
2 Flagella in resting cells bend back over the cell, the middle part parallel to the cell sides which are parallel or almost so....................*S. caeca*

Scourfieldia caeca (Korshikov) J.H.Belcher *et* Swale 1963

Synonyms: *Cardiomonas caeca* Korshikov, *Scourfieldia magnopyrenoidea* Huber-Pestalozzi

15120030 **Pl. 75L** (p. 283)

Cells round to heart-shaped, 5–7 μm wide, in lateral view 1 μm thick, long flagellum 20–25 μm and short flagellum 15–20 μm long; flagellum in resting cells bent back over body, the middle part parallel with cell body whose sides are parallel or almost so; pyrenoid absent but 2 flattened starch grains in basal part of cell, located on either side of cell midline, rarely joined to form a single crescent-shaped body, grains not clearly visible in living cells but only after light staining with iodine or after disintegration of cell.

Northern and Central Europe, Ukraine; in the British Isles known from ponds and pools in England.

Scourfieldia complanata G.S.West 1912

15120010 **Pl. 75K** (p. 283)

Cells round to heart-shaped, 4–5 μm wide in lateral view, 1–1.5 μm thick, in lateral view cells have almost parallel sides; flagella 12–17 μm long, one always 2–3 μm longer than the other, in resting cells the flagella bend back over the body in a smooth curve; pyrenoids absent, but 2 or 3 small starch grains present in basal part of chloroplast.

Central Europe in small water bodies; in the British Isles known from ponds in England, e.g. Leg-of-Mutton Pond, Richmond Park, Surrey (Lund, 1937, 1942c).

Scourfieldia cordiformis Takeda 1916

15120020 **Pl. 75M** (p. 283)

Cells heart-shaped, 3.5–4 μm wide, 4–4.5 μm long and 2.3–2.6 μm thick, in side view obovate (in contrast to flat cells of other 2 species); flagella unequal, up to 20 μm long, emerging from anterior notch in cell; pyrenoid absent.

British Isles (England); described from a *Sphagnum* marsh at Keston, Kent, which remains its only known locality.

PHYLUM CHLOROPHYTA (Green Algae)

The green algae or chlorophytes are the most species-rich and morphologically diverse group to be covered here. Division of the generic key into sections is made in order to facilitate identification of the more than 250 chlorophyte genera known from the British Isles. The first step in identifying an unknown sample is to decide to which of the sections (broad morphological groupings) it belongs. The next step is to use the appropriate sectional key to identify the genus. Finally, the sample should be compared with the description and illustrations of the genus before proceeding to the separate species identification keys.

Keys prepared by J.A. Bryant and N. Stewart (Section I), A.J. Brook (Section II), A. Pentecost and D.M. John (Section IV), and D.M. John (Sectional Key, Sections III, V–X).

KEY TO SECTIONS

1 Macroscopic, with whorls of branches and branchlets divided into alternating 'nodes' and 'internodes' .... **Section I**

1 Macroscopic or microscopic, if macroscopic then not as above...2

2 Cells constricted into two 'semi-cells' by a shallow or deep median constriction, if not so constricted then cellular contents bisymmetrically organized or with a star-shaped, ribbon-like or spiralled chloroplast ....**Section II**

2 Cells not as above...3

3 Unicellular, solitary and/or loosely clustered........4

3 Multicellular...6

4 Terrestrial or aquatic, if aquatic then always growing attached to surfaces ....**Section III**

4 Aquatic and commonly planktonic.......................5

5 Motile, single-celled or colonial.............**Section IV**

5 Non-motile in vegetative state, colonial ....**Section V**

6 Tubular or membrane-like, parenchymatous in structure, always macroscopic and easily visible to the unaided eye.......................**Section VI**

6 Not as above, micro- or macroscopic 7

7 Filamentous, usually a single row of cells.............8

7 Colonial...9

8 Filaments unbranched, occasionally with short rhizoidal branches ....**Section VII**

8 Filaments branched (rarely pseudoparenchymatous)......................**Section VIII**

9 Motile ....**Section IV**

9 Non-motile in vegetative state............................10

10 Colonies normally of cells in multiples (2, 4, 8, 16, etc.), numbers remain fixed, often reproducing asexually by means of autospores..**Section IX***

10 Colonies not of fixed cell numbers, frequently enclosed within a mucilaginous envelope or attached by mucilage strands ....**Section X**

SECTION I Macroscopic, whorls of branches bearing simple or divided whorls of branchlets and having alternate 'nodes' and 'internodes' (Charales)

1 Plants corticate; spine-cells and stipulodes present, although these may be rudimentary ....*Chara* (p. 596)

1 Plants ecorticate; spine-cells absent; with or without stipulodes..2

2 Branchlets undivided; bract cells at their nodes; oogonium with a single apical ring of 5 cells forming the coronula 3

2 Branchlets divided (sometimes only minutely so); bract-cells and stipulodes absent; oogonium with a coronula of two rings, each of 5 cells...........4

3 Stipulodes forming a single ring, each acuminate and downward-pointing; upper branchlet whorls forming dense foxtail-like heads; bulbils absent; monoecious.......*Lamprothamnium* (p. 605)

3 Stipulodes absent; upper branchlet whorls lax and diffuse; bulbils star-shaped and developing from lower nodes of branches; dioecious ....*Nitellopsis* (p. 609)

4 Branchlets forked one or more times into similar single-celled segments, with end segment (dactyl) divided into one or more cells; branchlets without a dominant axis (sympodial); oospore laterally compressed ....*Nitella* (p. 605)

* Colonies of fixed cell numbers ('coenobia') characterize the order Chlorococcales; often it is difficult to determine in chlorococcalean algae whether the number is fixed, hence some are included in Section X.

4 Branchlets divided into unequal, multicellular segments; branches with a dominant main axis (monopodial); oospore circular in cross section .... *Tolypella* (p. 609)

SECTION II Unicellular, colonial or filamentous, non-motile; most genera with cells constricted in middle to form two semicells; if not constricted then chloroplast usually divided into two symmetrical portions

1 Cells solitary, or readily separating into short filaments 2
1 Cells not solitary but united in firm filaments or colonies 27
2 Cells cylindrical, thus circular in apical view 3
2 Cells not cylindrical but ovoidal or angular in apical view 20
3 Cells with median constriction 4
3 Cells without median constriction 10
4 Median constriction extremely slight 5
4 Median constriction conspicuous 6
5 Walls with fine pores in oblique rows or irregularly disposed .... *Actinotaenium* (p. 530)
5 Walls with pores, sometimes warty, often in regular rows .... *Penium*, in part (p. 529)
6 Cells <2.5 times longer than wide, ovoidal in apical view .... *Cosmarium*, in part (p. 532)
6 Cells >5 times longer than wide 7
7 Semicell apex with a deep, narrow notch .... *Tetmemorus* (p. 582)
7 Semicell apex entire 8
8 Isthmus encircled by rings of tooth-like thickenings .... *Docidium* (p. 551)
8 Isthmus without thickenings or with only a ring of wall material 9
9 Chloroplast single, ridged, axile or ribbon-like; terminal vacuoles absent .... *Haplotaenium* (p. 556)
9 Chloroplasts several longitudinal, parietal ribbons; terminal vacuoles distinct .... *Pleurotaenium* (p. 562)
10 Cells mostly 10–20 times longer than wide; commonly geniculate and often joined to form short filaments 11
10 Cells not as above 12
11 Chloroplasts 2–3 tightly spiralling, narrow longitudinal bands .... *Genicularia* (p. 527)
11 Chloroplast 2 loosely spiralling, narrow longitudinal bands .... *Gonatozygon* (p. 527)
12 Cells sometimes in 2 parts; walls with pores and/or granulations and striations 13
12 Cells never in 2 parts; walls always smooth, without obvious pores 14
13 Cells short, cylindrical .... *Penium*, in part (p. 529)
13 Cells crescent-shaped (some nearly straight), always tapering towards apices where terminal vacuoles occur; walls sometimes with striae and pores .... *Closterium* (p. 516)
14 Chloroplast single, except prior to division 15
14 Chloroplasts 2 per cell 18
15 Chloroplast a parietal, twisted ribbon .... *Spirotaenia*, in part (p. 515)
15 Chloroplasts of another form 16
16 Chloroplast with a central axis from which ridges spiral longitudinally along length of cell; ends of chloroplasts in some species with red tips .... *Tortitaenia* (p. 515)
16 Chloroplasts not as above but simple plates or ribbons 17
17 Chloroplast with a median notch, mostly with 1, sometimes 2 pyrenoids; cells usually with broad, rounded apices (in one species conical) .... *Mesotaenium* (p. 511)
17 Chloroplast with a median notch, mostly with a row of 4 or more pyrenoids; cells comparatively narrow, often slightly curved .... *Roya* (p. 514)
18 Chloroplasts 2, parietal, intertwined, narrow spiralling ribbons .... *Spirotaenia*, in part (p. 515)
18 Chloroplasts otherwise 19
19 Chloroplasts with elongated dissected ridges, or with star-like processes or ridges radiating from a conspicuous, central pyrenoid .... *Cylindrocystis* (p. 510)
19 Chloroplasts with longitudinal ridges radiating from a spherical or elongate pyrenoid; margins of ridges often elaborately dissected .... *Netrium* (p. 513)
20 Cells in apical view compressed, so appearing ellipsoidal 21
20 Cells in apical view radially symmetrical, appearing 3–4-angular 25
21 Cell apices with a cleft of varying depth .... *Euastrum* (p. 551)
21 Cells without an apical cleft 22
22 Lateral margins of cells deeply incised .... *Micrasterias* (p. 557)
22 Lateral margins of cells not incised 23
23 Cells with marginal spines 24
23 Cells without marginal spines 26
24 Cell face with a central protuberance .... *Xanthidium* (p. 583)
24 Cell face without a central protuberance, angles with 1 spine or mucro .... *Staurodesmus*, in part (p. 577)
25 Angles of cell with a single spine or mucro .... *Staurodesmus*, in part (p. 577)
25 Angles of cell extending into hollow processes .... *Staurastrum*, in part (p. 563)
26 Angles of cell not continued into processes, mostly rounded .... *Cosmarium*, in part (p. 532)
26 Angles of cell not continued into processes, body often ornamented with granules or spines .... *Staurastrum*, in part (p. 563)
27 Colonies not filamentous 28
27 Colonies as filaments 29

28 Colonies of *Cosmarium*-like cells joined by gelatinous strands..............*Cosmocladium* (p. 551)
28 Colonies of *Cosmarium*-like cells embedded in a branched, radiating system of gelatinous strands and encrusted in lime...*Oocardium* (p. 561)
29 Adjoining cells united into filaments by apical granules, tubercles or rods..........................30
29 Adjoining cells not united by apical granules, tubercles or rods 32
30 Cells united by 4 apical granules or tubercles ....*Teilingia* (p. 593)
30 Cells united by 2 apical tubercles or rods............31
31 Apical tubercles in middle of apex; short and not overlapping adjoining cell ....*Sphaerozosma* (p. 592)
31 Apical tubercles remote on apex, long and capitate and overlapping adjoining cells ....*Onychonema* (p. 592)
32 Cells with girdle-like thickenings at time of division...33
32 Cells without girdle-like thickenings at time of division 34
33 Cell walls not striated; semicells not markedly swollen above isthmus ....*Desmidium* (p. 587)
33 Cells walls at least partially striated; semicells slightly swollen above isthmus ....*Bambusina* (p. 587)
34 Sinus very distinct; cells *Cosmarium*-like; 2- or 3-radiate.......................*Spondylosium* (p. 592)
34 Sinus indistinct or absent 35
35 Chloroplasts star-shaped; cells about as long as wide..............................*Hyalotheca* (p. 589)
35 Chloroplasts ribbon-like; cells longer than wide ....*Groenbladia* (p. 589)

SECTION III Unicellular, loosely or firmly attached to surfaces (only secondarily free-living), if grouped then individual cells each with its own attachment; sometimes within tissues of plants or animals

1 Associated with soil and soil-free surfaces (subaerial), or if aquatic then attached 2
1 Lichen symbiont, endophytically within tissues of aquatic plants and animals, or associated with mucilage of cyanobacteria..........28
2 Cells attached by a very slender stalk 3
2 Cells with a short and thick stalk or without a stalk 6
3 Cells growing on planktonic crustacea, often attached by a short triangular anchor ....*Korshikoviella* (p. 362)
3 Cells not on crustacea, without an anchor-like attachment structure.....................................4
4 Cells attached to planktonic algae and with 2 contractile vacuoles, with or without an eyespot ....*Stylosphaeridium* (p. 301)
4 Cells usually attached to filamentous algae or aquatic angiosperms, without contractile vacuoles or an eyespot 5
5 Cells narrowly spindle-shaped, with stalk an extension of apex ....*Podohedra* (p. 380)
5 Cells of another form ....*Characium*, in part (p. 333)
6 Shortly-stalked...7
6 Stalk absent...11
7 Apical wall with a ring-like thickening 8
7 Apical wall without a ring-like thickening 9
8 Zoospore persisting within mother cell wall to produce next generation (sometimes remnants of walls of successive generations evident); zoospores biflagellate ....*Hydrianum* (p. 358)
8 Zoospores all released and no successive generations develop in original mother cell wall; zoospores quadriflagellate ....*Pseudocharacium*, in part (p. 380)
9 Cell usually bearing apically 2 short vestigial flagella; chloroplast longitudinally divided ....*Characiochloris* (p. 333)
9 Cells without vestigial flagella; chloroplast not longitudinally divided 10
10 Zoospores biflagellate...*Characium*, in part (p. 333)
10 Zoospores quadriflagellate ....*Pseudocharacium*, in part (p. 380)
11 Cells associated with simple or divided hairs or bristles..12
11 Cells without hairs or bristles 17
12 Cells narrowly spindle-shaped and apices prolonged to form a long hair or bristle ....*Keratococcus* (p. 360)
12 Cells not as above..13
13 Cells loaf-shaped, bearing 1 or 2 very fine and forked hairs...................*Dicranochaete* (p. 472)
13 Cells of another shape, each bearing one or more undivided hairs or bristles 14
14 Cells bearing sheathed bristles 15
14 Cells bearing unsheathed bristles.........................16
15 Cells subspherical or depressed and embedded within mucilage; bearing several radiating bristles......................*Conochaete* (p. 472)
15 Cells spherical, sometimes connected by mucilage tubes, each cell bearing a single, sheathed bristle ....*Chaetosphaeridium* (p. 470)
16 Cells each bearing 8–12 bristles, walls thick and striated......................*Polychaetophora* (p. 472)
16 Cell each bearing 2–4 long and flexuose bristles, walls thin..........*Oligochaetophora* (p. 472)
17 Cells obovoid, ellipsoidal, broadly oval, spindle- or pod-shaped 18
17 Cells of another shape (often spherical)................19
18 Cells bearing spirally twisted ribs and wall thickened apically ....*Scotiellopsis* (p. 399)
18 Cell not as above 21
19 Cells obovoid to almost spherical, or if another shape then chloroplast inverted, cup-shaped and confined to apical portion of cell ....*Fernandinella* (p. 353)

19 Cells of another shape, if obovoid then inversely obovoid, with chloroplast not as above 20
20 Cells spindle-shaped, straight to crescent-shaped or ellipsoidal, apices acute or broadly rounded and without a mucilage cap ....*Chlorolobion*, in part (p. 338)
20 Cells inversely obovoid to ellipsoidal, apices rounded and mucilage caps present at one or both apices (often only visible with Indian ink) ....*Pseudococcomyxa*, in part (p. 381)
21 Chloroplasts not star-shaped and, if parietal, then not cup-shaped........22
21 Chloroplasts star- or cup-shaped 25
22 Cell protoplast divided into separate portions, each associated with a polygonal chloroplast........*Dictyococcus*, in part (p. 348)
22 Cell protoplast undivided 23
23 Chloroplast initially a dissected hollow sphere, later a network of anastomosing and interlacing strands ....*Dictyochloris* (p. 348)
23 Chloroplast 2 or more and disc-shaped, trough-shaped, angular or polygonal 24
24 Chloroplasts in pairs and plate-like; reproduction by zoospores and aplanospores ....*Bracteacoccus* (p. 333)
24 Chloroplasts usually several,separate or densely packed, variously shaped; reproduction by aplanospores and akinetes ....*Muriella* (p. 368)
25 Cells often gregarious, forming a greenish layer on soil-free surfaces, commonly producing cuboidal packets of 2 or 4 cells; chloroplast parietal, often lobed........26
25 Cells usually solitary, commonly associated with terrestrial surfaces, sometimes aquatic, never producing packets of cells; chloroplast parietal and cup-shaped 27
26 Cells with a small pyrenoid ..*Desmococcus* (p. 438)
26 Cells without a pyrenoid ....*Apatococcus* (p. 433)
27 Cells large (>10 μm) in diameter, often thick-walled, with a prominent pyrenoid; reproduction by zoospores ....*Chlorococcum*, in part (p. 337)
27 Cells small (≤10 μm) in diameter, generally thin-walled, often without a pyrenoid; reproduction by autospores ....*Chlorella*, in part (p. 335)
28 Cells confined in vase- or cup-shaped loricae of colonies of the chrysophyte *Dinobryon* ....*Apodochloris* (p. 332)
28 Cells not associated with *Dinobryon* colonies 29
29 Chloroplasts cup-shaped and living within cells of lower animals (e.g. *Hydra*) ....*Chlorella*, in part (p. 335)
29 Chloroplasts of another form and not associated with animals 30
30 Associated with tissues of aquatic macrophytes and filamentous cyanobacteria;.chloroplast radiating from centre with arms expanding to form broad lobes or a complete parietal layer ....*Chlorochytrium* (p. 336)
30 Associated with a fungus to form a symbiotic association known as a lichen; chloroplasts parietal and if axile then of another form 31
31 Cells spherical or ellipsoidal to broadly pear-shaped; chloroplast axile, marginally lobed and with a pyrenoid ....*Trebouxia*, in part (p. 407)
31 Cells elongate-ovoid or ellipsoidal; chloroplast parietal and disc-like, without a pyrenoid........*Coccomyxa*, in part (p. 339)

SECTION IV Unicellular or colonial, motile, possessing one or more flagella; without distinct longitudinal or spirally arranged striations in wall and not exhibiting squirming movement

1 Multicellular and colonial........2
1 Unicellular 10
2 Cells with long axes parallel to one another, arranged in 2 or 4 alternating tiers, without a mucilage envelope........3
2 Cells arranged otherwise and surrounded by a mucilage envelope........4
3 Colonies of 2–4 cells, each cell containing a pyrenoid ....*Pascherina* (p. 320)
3 Colonies usually of 8–16 cells, rarely 4, without pyrenoids........*Pyrobotrys* (p. 323)
4 Cells elongate, with sharp processes or extensions, arranged in a circle within the envelope ....*Stephanosphaera* (p. 326)
4 Cells usually rounded, ovoid or pear-shaped, without extension, radially arranged or as a flat plate 5
5 Colony flat 6
5 Colony usually spherical or oval 7
6 Cells separated, not basally united, surrounded by a mucilaginous envelope ....*Gonium* (p. 318)
6 Cells united only basally by mucilage, normally only two rows of cells ....*Basichlamys* (p. 304)
7 Colony large, forming hollow spheres of more than 200 cells, often containing small daughter colonies........*Volvox* (p. 326)
7 Colonies small, composed of fewer than 200 cells, without daughter colonies........8
8 Cells closely packed, somewhat pear-shaped, with the broader end directed toward surface ....*Pandorina* (p. 320)
8 Cells arranged towards periphery of a mucilage envelope and of another shape........9
9 Cells tend to be arranged in several rows, each with 1 to several pyrenoids ....*Eudorina* (p. 317)

9 Cells irregularly arranged, without pyrenoids ..... *Volvulina* (p. 327)

10 Cells with a wide sheath-like wall, protoplast connected to wall by radiating extensions, often orange to red in colour (especially in non-motile encysted stage); common in rock pools and limestone basins ..... *Haematococcus* (p. 318)

10 Cells of another form, with green chlorophyll not masked by a reddish pigment 11

11 Cells enclosed in a wall-like envelope (lorica), separate from the chloroplast, often impregnated with iron or calcareous granules 12

11 Cells not enclosed by such a wall 16

12 Lorica spherical, irregularly spherical or ellipsoidal 13

12 Lorica compressed 15

13 Lorica with a transverse equatorial groove ..... *Hemitoma* (p. 319)

13 Lorica without such a groove 14

14 Flagella emerge from a common opening ..... *Coccomonas* (p. 314)

14 Flagella emerge from separate openings ..... *Dysmorphococcus* (p. 315)

15 Lorica smooth, colourless, wing-like, with corners prolonged into horns ... *Pteromonas* (p. 322)

15 Lorica irregularly granulate, often brown composed of two portions, lens-like ..... *Phacotus* (p. 320)

16 Cells with processes or lobed, longitudinally folded or with extensions, sometimes wing-like 17

16 Cells without distinct processes 22

17 Cells with distinct, blunt or pointed processes 18

17 Cells of another form 20

18 Cell with 2 backwardly directed processes ..... *Furcilla* (p. 317)

18 Cell with processes otherwise 19

19 Cells with a distal point, with 4 backwardly directed processes ..... *Brachiomonas* (p. 304)

19 Cells almost square, each distal and apical corner prolonged to form a recurved spine-like process ..... *Diplostauron* (p. 314)

20 Cells with a wavy outline, usually 4–8-lobed, a protoplast often withdrawn from wall; biflagellate ..... *Lobomonas* (p. 319)

20 Cells of another form 21

21 Cells oval or strawberry-shaped, with longitudinal folds and 4-lobed in end view ..... *Pyramimonas* (p. 285)

21 Cells strongly flattened, with 2 flared, wing-like lateral processes ..... *Scherffelia* (p. 282)

22 Cells spindle-shaped, with apices bluntly pointed ..... *Chlorogonium* (p. 312)

22 Cells not spindle-shaped 23

23 Cells with a broad envelope, with protoplast of a different shape to envelope and without extensions ..... *Sphaerellopsis* (p. 324)

23 Cells normally without such an envelope, if present then the protoplast of similar shape to it 24

24 Cells with 4 flagella 25

24 Cells with 1 or 2 flagella 29

25 Cells (zoospores) obpear- or heart-shaped; flagella together in an anterior invagination, often cross-shaped in anterior view; single-celled sessile stage in life history ..... *Prasinochloris* (p. 284)

25 Cells of another form and with no sessile life-history stage, if heart-shaped then apical flagella in 2 separate pairs 26

26 Cells spherical, shortly cylindrical or heart-shaped; flagella arising apically in 2 separate pairs ..... *Carteria* (p. 304)

26 Cells of another form 27

27 Cells banana-shaped, swimming in spirals ..... *Spermatozopsis* (p. 323)

27 Cell shape and swimming motion otherwise 28

28 Cells narrow and cylindrical, square in end view ..... *Hafniomonas* (p. 319)

28 Cells broadly oval to slightly pear-shaped, flattened in end view ..... *Tetraselmis* (=*Platymonas*) (p. 284)

29 Cells with a single flagellum, arising laterally at posterior end of cell ..... *Pedinomonas* (p. 281)

29 Cells with 2 flagella, usually arising apically or subapically 30

30 Cells not compressed, usually spherical in end or polar view 31

30 Cells compressed in end or side view 33

31 Cells with a large apical depression from which the flagella arise ... *Chlamydonephris* (p. 311)

31 Cells without any end depression 32

32 Cells variously shaped, with one to several pyrenoids ..... *Chlamydomonas* (p. 306)

32 Cells spherical, elongate-ellipsoidal, with or without a papilla, lacking a pyrenoid ..... *Chloromonas* (p. 306)

33 Cells bean-shaped, flagella inserted laterally; swimming sideways ..... *Nephroselmis* (=*Heteromastix*) (p. 282)

33 Cells of another shape 34

34 Cells compressed apically to give a rubber dinghy-like appearance, with flagella in a subapical depression and about the same length as cell ..... *Mesostigma* (p. 284)

34 Cells not as above 35

35 Cells ovoid, compressed with one side convex and the other flat or slightly concave, anterior end pointed or narrowing to a slender tip; flagella of unequal length ..... *Myochloris* (p. 282)

35 Cells of another shape, equatorially compressed and lens-shaped in end view 36

36 Cells ellipsoidal, heart-shaped or rectangular, a depression at apex; flagella more than twice as long as cell ..... *Scourfieldia* (p. 285)

36 Cells somewhat pear-shaped, bluntly pointed at apex and distal end; flagella no more than twice length of cell ..... *Phyllocardium* (p. 322)

SECTION V Unicellular (rarely united after division), non-motile, free-living and frequently planktonic

1 Cells narrowly or broadly cone-shaped or cylindrical to elongate spindle-shaped; outer wall sheath with a distinct equatorial ridge and longitidinal ridges ....*Desmatractum* (p. 345)
1 Cells of another form and without an outer wall sheath 2
2 Cells spindle-shaped, pointed or more rarely blunt-tipped or needle-like, straight, curved or spirally twisted 3
2 Cells rounded, ovoid or angular, crescent-shaped, pyramidal, trapeziform or angular, if spindle-shaped then curved 7
3 Cells with one apex prolonged into a long hair-like point and other bifid and anchor-like ....*Ankyra* (p. 331)
3 Cells with apices rounded, finely acute or extended into a long hair-like point or bristle 4
4 Cells straight, very narrowly spindle- or needle-shaped, with a single row of more than 2 pyrenoids ....*Closteriopsis* (p. 338)
4 Cells of another form and pyrenoids, if present, no more than 1 or 2 5
5 Pyrenoids absent ....*Monoraphidium* (p. 365)
5 Pyrenoids present..6
6 Cells pod- or spindle-shaped, with apices acute or bluntly rounded; reproduction by autospores, released by transverse fracture of mother cell wall into 2 cone-shaped fragments ....*Chlorolobion*, in part (p. 338)
6 Cells narrowly spindle-shaped, tapering equally towards each apex and terminating in a long spine or bristle; reproduction by zoospores released by lateral fracture of mother cell wall.......................*Schroederia* (p. 398)
7 Spines, spine-like appendages or papilla-like wall thickenings at cell corners.............................8
7 Not as above..20
8 Cells angular, pyramidal, trapezoidal or polygonal...9
8 Cells spherical, ovoid, lemon-shaped or ellipsoidal...12
9 Cells with several tapering spines arising from each rounded corner......*Polyedriopsis* (p. 380)
9 Cells with a spine, spine-like appendage or papilla-like wall thickening at each corner..........10
10 Spine or spine-like appendage present.................11
10 Spines absent and pointed papilla-like wall thickenings present ....*Tetraedron*, in part (p. 404)
11 Cells triangular or pyramidal, bearing a stout spine-like appendage at each corner ....*Treubaria* (p. 407)
11 Cells irregularly lobed, each lobe bearing apically a short tapering spine ....*Tetraedron*, in part (p. 404)
12 Cells ovoid, ellipsoidal or lemon-shaped 13
12 Cells spherical...17
13 Spines confined to apices or at apices and cell equator...14
13 Spines often randomly distributed over cell wall...15
14 Spines confined to apices, one very long spine at each opposite apex.......*Diacanthos* (p. 347)
14 Spines at apices or apices and equator, often several in each position ....*Lagerheimia* (p. 362)
15 Walls often brownish and warty; spines fine, hyaline and each emerging from a granule-like base ....*Siderocystopsis* (p. 401)
15 Walls colourless and smooth; spines evident and not emerging from a granule.........................16
16 Coenobia of 2 closely united cells; mucilaginous envelope absent .....*Dicellula* (p. 347)
16 Coenobia of 2 or 4 loosely united cells, more commonly single-celled, mucilaginous envelope often present ....*Franceia* (p. 355)
17 Spines shorter than cell diameter; walls often decorated with ridges.................................18
17 Spines usually longer than cell diameter; walls unornamented..19
18 Cells enclosed within a mucilaginous envelope; chloroplast net-like or deeply incised ....*Acanthococcus* (p. 327)
18 Cells without a mucilaginous envelope; chloroplast single or several and disc-like ....*Troschiscia*, in part (p. 408)
19 Walls often thin; pyrenoid spherical to ellipsoidal..........................*Golenkiniopsis* (p. 357)
19 Walls usually thick; pyrenoid cup- or kidney-shaped ....*Golenkinia* (p. 356)
20 Cells sometimes with a thickened cap at one apex; living at air–water interface (neustonic) ....*Nautococcus* (p. 370)
20 Cells without an apical cap; planktonic...............21
21 Walls warty or with net-like ridges 22
21 Walls smooth or with a different type of ornamentation 23
22 Walls brownish and warty, mucilaginous sheath usually present ....*Siderocelis* (p. 401)
22 Walls colourless and warty or with net-like ornamentation; mucilaginous sheath absent ....*Troschiscia*, in part (p. 408)
23 Cells with fused or unfused greyish-brown granules at apices or equator ....*Amphikrikos* (p. 328)
23 Cells without such wall ornamentation 24
24 Cells each containing several chloroplasts 25
24 Cells each with a single chloroplast 26
25 Cells having a segmented protoplast and no pyrenoids; associated with aquatic macrophytes (or soil) and usually in acidic environments ....*Dictyococcus*, in part (p. 348)
25 Cells without a segmented protoplast and each chloroplast with 1–3 minute pyrenoids; associated with acid waters ....*Eremosphaera* (p. 353)

26 Adult cells spherical 27
26 Adult cells elongate and oval 29
27 Chloroplast axile and bearing radiating lobes (star-shaped) .....*Actinochloris*, in part (p. 328)
27 Chloroplast parietal, covering entire wall or cup-shaped 28
28 Cells usually small (<10 μm wide); reproducing by autospores ....*Chlorella*, in part (p. 335)
28 Cells larger (≥10 μm wide); reproducing by autospores and zoospores ....*Chlorococcum* (p. 337)
29 Cells with 1 or more pyrenoids and without a mucilage cap ....*Chlorella*, in part (p. 335)
29 Cells without pyrenoids and having a mucilage cap at each apex (often only visible using Indian ink) ....*Pseudococcomyxa* (p. 381)

SECTION VI Parenchymatous, forming tubular or flat plates of cells, often membrane- or blade-like

1 Thallus consisting of tubular branches ....*Enteromorpha* (p. 479)
1 Thallus flat and forming a single-layered blade 2
2 Single-layered initially and becoming polystromatic with onset of reproduction; cells in 2s or more and separated by mucilage; chloroplast star-shaped, with a single pyrenoid; terrestrial ....*Prasiola* (p. 473)
2 Single-layered throughout life history; cells not in any orderly arrangements and always closely packed; chloroplasts cup-shaped and each with 1 to several pyrenoids; aquatic ....*Monostroma* (p. 455)

SECTION VII Filamentous, single cell series throughout (rarely partly multiseriate), simple or occasionally with simple rhizoid-like laterals

1 Filaments partly of several rows of cells (multiseriate) above 2
1 Filaments a single row of cells (uniseriate) throughout 3
2 Chloroplasts parietal, encircling about two-thirds of cell; aquatic................*Schizomeris* (p. 461)
2 Chloroplasts star-shaped; terrestrial ....*Rosenvingiella* (p. 473)
3 Cells occasionally having 1 or several ring-like growths at distal end ....*Oedogonium* (p. 413)
3 Cells without ring-like growths 4
4 Cells very long with few cross walls (sometimes up 90 times longer than wide), with each cell numerous ring-like (annular) or net-like chloroplasts ....*Sphaeroplea* (p. 475)
4 Cells and chloroplasts of another form 5
5 Chloroplasts parietal, plate or ribbon-like, usually but not always spirally twisted..6
5 Chloroplast otherwise 7
6 Conjugating tubes present, no special gametangial cells formed; chloroplast(s) spiral in most species.............................*Spirogyra* (p. 491)
6 Conjugation directly between filaments, conjugation tubes absent, special gametangial cells formed; chloroplasts almost straight or only slightly spiralled ....*Sirogonium* (p. 490)
7 Chloroplasts in pairs, centrally positioned (axile) and spherical, ovoid or star-shaped.............8
7 Chloroplasts of another form 10
8 Chloroplasts in pairs and ovoid, sometimes bearing irregular radiating processes; cell sap usually purplish in only known British species (*Z. ericetorum*) ....*Zygogonium* (p. 510)
8 Chloroplasts a pair of spherical to quadrangular or star-shaped bodies; cell sap not purplish in colour 9
9 Chloroplasts star-shaped; zygospores ovoid or compressed, develop within a conjugation tube or in one gametangium; cell wall not thickening during conjugation.....*Zygnema* (p. 505)
9 Chloroplasts spherical to quadrangular; zygospores frequently quadrate, ovoid or pillow-like, developing in conjugation tube and often extending into gametangia; cell wall thickens during conjugation ....*Zygnemopsis* (p. 509)
10 Chloroplasts 1 or 2, plate- or ribbon-like, axile or lying on opposite sides of cell 11
10 Chloroplasts in pairs and not as above 14
11 Pyrenoids absent ....*Mougeotiopsis* (p. 490)
11 Pyrenoids 2 to several per cell..............................12
12 Prior to conjugation vegetative cells divide, with smaller portion serving as a gametangium ....*Temnogametum* (p. 505)
12 Prior to conjugation no special gametangium formed by cell division 13
13 Cytoplasmic residue remains in the gametangia subsequent to conjugation ....*Mougeotia* (p. 481)
13 Cytoplasmic residue not left in the gametangia subsequent to conjugation ....*Debarya* (p. 481)
14 Cell walls composed of two overlapping halves; pyrenoids absent...........*Microspora* (p. 452)
14 Cell walls never so constructed; pyrenoids present or absent 15
15 Filaments often divided into units by thick and transversely striated walls, often each containing 2 subspherical protoplasts ....*Binuclearia* (p. 434)
15 Filaments of a different structure 16
16 Chloroplast usually a dense network of interconnecting discs, each containing a pyrenoid 17

16 Chloroplasts of another form 19
17 Filaments <40 μm in diameter, 1–3-celled rhizoidal branches sometimes arising at right angles to main axis.......*Rhizoclonium* (p. 470)
17 Cells >40 μm in diameter, without rhizoidal branches..18
18 Cells less than twice as long as broad, cylindrical to barrel-shaped, slightly to markedly constricted at cross walls ....*Chaetomorpha* (p. 468)
18 Cells more than twice as long as broad, cylindrical, without constrictions at cross walls ....*Cladophora*, in part (p. 468)
19 Filaments without a mucilaginous sheath...........20
19 Filaments enclosed within a distinct mucilaginous sheath..26
20 Filaments readily fragmenting into short lengths (<10 cells) or solitary cells......................21
20 Filaments not readily fragmenting......................23
21 Short filaments forming zigzag growths on surfaces; cells spherical, oval or short and blunt-ended cylinders, normally without pyrenoids...................*Stichococcus*, in part (p. 461)
21 Short filaments not forming such a pattern; cells of another shape, with or without pyrenoids..22
22 Cells spindle-shaped, thread-like or very elongated; aquatic and terrestrial ...*Koliella* (p. 451)
22 Cells elongate cylindrical and apical cells acuminate or gradually tapering to a point; snow alga..............................*Raphidonema* (p. 460)
23 Chloroplast filling cells, structure not discernible due to density of food storage material; cells ovoid and thick-walled; reproductive cells swollen and red or orange ....*Cylindrocapsa* (p. 437)
23 Chloroplast and cells of another form; reproductive cells not swollen............................24
24 Filaments terminating in a slightly curved and pointed apical cell.................*Uronema* (p. 467)
24 Filaments without a tapering apical cell 25
25 Chloroplast girdle-shaped, usually somewhat lobed, partly (often >80%) or fully encircling the cell.................*Ulothrix* (p. 466)
25 Chloroplast ellipsoidal, disc-like or girdle-shaped, not usually marginally lobed, normally covering 80% or less of cell ....*Klebsormidium* (p. 447)
26 Cells spherical to subspherical, bead-like, shorter than broad, with wall divided into two equal halves and a ring-like transverse rim at join, always contiguous...*Radiofilum* (p. 460)
26 Cells not as above, contiguous or separated and equidistant 27
27 Cells having swollen apices often encircled by unequal rings or caps, or with minute stick-like appendages or granules, always contiguous......................................*Catena* (p. 436)
27 Cells not as above, contiguous or separated and equidistant 28
28 Filaments of contiguous cells, tending to fragment easily; pyrenoids absent...*Gloeotila* (p. 442)
28 Filaments of contiguous or almost equidistant cells, not readily fragmenting; pyrenoids normally present 29
29 Filaments of contiguous cells or cells almost equidistant; apices not containing vacuoles ....*Geminella* (p. 440)
29 Filaments of cell almost equidistant spaced; apices often containing vacuoles with conspicuous oil droplets ....*Planctonema* (p. 457)

SECTION VIII Filamentous, single series of cells throughout, usually branched, prostrate and erect filaments or only an erect/prostrate system

1 Filaments perforating limestone rock, wood, or mollusc shells; emergent filaments numerous and branched.............*Gomontia* (p. 442)
1 Filaments not perforating surfaces........................2
2 Thallus macroscopic, almost spherical and cushion-like or having irregularly divided, nodulose branches; microscopic filaments embedded in a firm or soft mucilage ....*Chaetophora* (p. 436)
2 Thallus of another form...3
3 Filaments very closely aggregated to form a pseudoparenchyma (at least in part).......................4
3 Filaments not so closely aggregated 12
4 Prostrate filaments only 5
4 Prostrate and erect filaments................................9
5 Filaments without bristles or hairs 6
5 Filaments having bristle- or hair-like appendages 7
6 Subaerial, forming lobed crusts on surface of the leaves of angiosperms and the fronds of ferns...*Phycopeltis* (p. 476)
6 Aquatic, forming lobed, disc- or star-shaped crusts on a wide variety of surfaces ....*Protoderma* (p. 458)
7 Crust-like with fine hairs scattered over surface ....*Chaetopeltis* (p. 300)
7 Crust-like bearing bristle-like appendages 8
8 Bristles having basal sheaths, often sheath persists after the loss of a bristle ....*Coleochaete*, in part (p. 471)
8 Bristles not possessing a basal sheath ....*Aphanochaete*, in part (p. 433)
9 Subaerial on various plants; several-layered subepidermal expansion bearing numerous erect, unbranched filaments ...*Cephaleuros* (p. 475)
9 Aquatic and of another form...............................10
10 Prostrate system having a concentric appearance in surface view due to alignment of cross walls and chloroplasts of adjacent cells; erect system very reduced ....*Stigeoclonium*, in part (p. 462)
10 Prostrate and erect systems not as above 11
11 Thallus cushion-like, with a several-layered

prostrate system bearing sparingly divided erect filaments; zoospores biflagellate ....*Gongrosira*, in part (p. 444)
11 Thallus crust-like, with a single-layered prostrate system bearing little to profusely developed erect filaments; zoospores quadriflagellate (where known) ....*Pseudendoclonium*, in part (p. 458)
12 Erect filaments only..13
12 Erect and prostrate filaments, or only a system of prostrate filaments 19
13 Filaments bearing bristles 14
13 Filaments not possessing bristles 15
14 Filaments less than 6 cells long, with branches rudimentary or absent; minute and within mucilage of *Gloeotrichia* (blue-green alga) ....*Thamniochaete* (p. 466)
14 Filaments more than 6-celled, distinctly branched and these lying in one plane; aquatic and free-living.............*Bulbochaete* (p. 409)
15 Branches of main filaments irregularly contorted and bearing short laterals; embedded in the mucilaginous envelope of macroalgae ....*Chaetonella* (p. 468)
15 Branches of another form and free-living 16
16 Microscopic (<100 μm in height), often freely branched with cross wall distal to point of branch origin; pyrenoids absent ....*Microthamnion* (p. 455)
16 Macroscopic, coarse or delicate and gelatinous; pyrenoids present...............................17
17 Filaments enclosed within a delicate mucilaginous envelope; main filaments distinct and bearing laterally soft tufts of narrower-celled branchlets....*Draparnaldia* (p. 438)
17 Filaments not enclosed in mucilage and branching of another form 18
18 Filaments usually coarse, dark green; chloroplasts net-like; main filaments commonly >50 μm in diameter ....*Cladophora* (p. 468)
18 Filaments usually soft, bright green; chloroplasts plate-like; main branches <50 μm in diameter ....*Stigeoclonium*, in part (p. 462)
19 Soil algae or associated with soil-free surfaces...20
19 Aquatic algae...21
20 Soil alga (usually only detected in culture); filaments relatively compact, only very short lateral branches and not arising from middle of cell (sometimes forming colonies of single, paired or tetrads of cells); always green in colour.............*Pleurastrum*, in part (p. 457)
20 Subaerial alga; prostrate and erect filaments having side branches arising from middle of cell commonly orange or orange-red in colour ....*Trentepohlia*, in part (p. 477)
21 Prostrate filaments only 22
21 Prostrate and erect filaments................................24
22 Filaments without bristles, often richly branched, creeping on surface of *Cladophora* and *Rhizoclonium* ....*Entocladia*, in part (p. 440)
22 Filaments with bristles, loosely branched or forming subfilamentous masses on filamentous algae or aquatic angiosperms 23
23 Filaments forming a subfilamentous mass of thick-walled cells, each cell bearing 8–12 flexuous bristles ....*Polychaetophora* (p. 472)
23 Filaments loosely branched, with cells thin-walled, each usually bearing less than 6 straight bristles ....*Aphanochaete*, in part (p. 433)
24 Growing within the mucilage of other algae; cells bearing basally swollen hairs or these arising laterally on short prolongations ....*Chaetonema* (p. 436)
24 Free-living, if hairs present then not as above 25
25 Commonly cushion-like and the prostrate filaments several layers thick; zoospores biflagellate 26
25 Crust-like and prostrate filaments forming a single layer; zoospores quadriflagellate when known... 27
26 Cells without pyrenoids; only known in the British Isles amongst plants in bog pools ....*Leptosira* (p. 451)
26 Cells with 1 to several pyrenoids; commonly found on hard surfaces in flowing water and the shallow margins of standing bodies of water.............................*Gongrosira*, in part (p. 444)
27 Reproduces in nature by akinetes (often form dark green masses) and aplanospores ....*Dilabifilum* (p. 438)
27 Reproduces commonly by zoospores ....*Pseudendoclonium*, in part (p. 458)

SECTION IX Colonies of fixed cells numbers (multiples of 2, 4, 8, 16, etc.), known as 'coenobia', non-motile, sometimes of distinctive form (e.g. net-like, plate-like)

1 Coenobia macroscopic, hollow, consist of cylindrical cells repeatedly joined at end walls to form a network ....*Hydrodictyon* (p. 358)
1 Coenobia of another form......................................2
2 Coenobia flat and typically single-layered. 3
2 Coenobia not as above 13
3 Coenobia 2-celled; cells deeply notched at margin ....*Euastropsis* (p. 353)
3 Coenobia not as above 4
4 Coenobia usually 2- or 4-celled; cells separated from each other by a thick dark brown or black band of calcite ....*Gloeotaenium* (p. 356)
4 Coenobia 4- or more celled, without a distinctive calcite band separating cells 5
5 Cells in 4s, outer cells kidney- or heart-

shaped, inner ones ellipsoidal and straight ....*Dimorphococcus*, in part (p. 351)
5 Cells usually all of similar shape 6
6 Cells each having a remnant of mother cell wall attached laterally or apically in the form of an appendage 7
6 Cells without such an appendage 8
7 Cells often connected by short mucilaginous strands, each bearing at one apex a cone- or horn-like appendage..............*Coronastrum* (p. 344)
7 Cells connected directly by walls, each bearing laterally an appendage (often connected to form square-shaped syncoenobia) ....*Komarekia* (p. 361)
8 Coenobia usually 4-celled, inner cells almost parallel to one another and alternately arranged (apices in contact), outer cells at angle to central axis and parallel to one another (contact inner cell by their sides) ....*Tetrachlorella* (p. 404)
8 Coenobia not arranged as above 9
9 Coenobia plate-like and marginal cells with 1, 2 or more projections, rarely notched (cf. *Pediastrum tetras*) ....*Pediastrum* (p. 376)
9 Coenobia of another form 10
10 Coenobia ovoid, triangular or square, usually 4-celled, with outer cell wall spiny, granular or warty 11
10 Coenobia square to elongate and trapezoid, quadrangular or lozenge-shaped, usually more than 4-celled, with walls smooth or warty 12
11 Coenobia obovoid or triangular and cell walls typically spiny (rarely warty or smooth) ....*Tetrastrum* (p. 406)
11 Coenobia square and cell walls warty ....*Pseudotetrastrum* (p. 381)
12 Daughter coenobia orientated at an angle of 45° with respect to mother cell ....*Crucigenia* (p. 344)
12 Daughter coenobia orientated in same direction as mother cell ....*Crucigeniella* (p. 345)
13 Cell groups consisting of 2 pairs of cells, broadly rounded base of each cell pair adposed, sometimes narrowing apically and somewhat curved ....*Dichotomococcus* (p. 347)
13 Cells of a different arrangement and shape 14
14 Colony of cells usually grouped in 4s and terminal on di- or tetrachotomously divided mucilage strands radiating from its centre...........15
14 Colony of a different structure 17
15 Pyrenoids absent.....*Pseudodictyosphaerium* (p. 381)
15 Pyrenoids present..16
16 Cells usually in pairs and with a refractive granule in colourless area towards apices ....*Lobocystis* (p. 364)
16 Cells usually arranged in tetrads and without a granule-containing colourless area ....*Dictyosphaerium* (p. 348)
17 Spines present 18
17 Spines absent...27
18 Coenobia spherical or many-sided; cells closely joined or cells slightly separated and connecting by evident mucilaginous extensions from wall 19
18 Coenobia never in the form of a hollow sphere 20
19 Cell walls very thick with thickened, parallel, coarse, serrated ribs usually present along margin and these connected by minute transverse ribs ....*Coelastropsis* (p. 339)
19 Cell walls relatively thin and not ornamented as above ....*Coelastrum* (p. 339)
20 Cells spindle- or club-shaped, often 4–8 radiating from a common centre ....*Actinastrum* (p. 328)
20 Cells of a different shape and arrangement 21
21 Cells usually elongate, 2 or more joined laterally to form a single flat row..........................22
21 Cells of another form 23
22 Coenobia of 2, 4, 8 or more cells (rarely single), one daughter colony formed in a cell; pyrenoids present ......*Scenedesmus*, in part (p. 384)
22 Coenobia always 2-celled, with 1 or 2 daughter coenobia formed in a cell; pyrenoids absent ....*Didymocystis* (p. 349)
23 Coenobia 4-celled, form inside often 2-celled daughter coenobia and sometimes released; cells elongate and straight or curved, crossing over one another or bearing long thick spines ....*Didymogenes* (p. 351)
23 Coenobia of another form 24
24 Coenobia of 4 spindle-shaped or cylindrical cells, arranged as a quadrate packet ....*Tetradesmus* (p. 404)
24 Coenobia of another form 25
25 Coenobium 4-celled, outermost pairs of cells kidney- or heart-shaped, inner cell-pairs straight ovoid or slightly cylindrical, often united in a zigzag manner with bridge-like connections between cells ....*Dimorphococcus*, in part (p. 351)
25 Coenobium otherwise 26
26 Coenobia of groups of spherical cells, each bearing on outer wall 1–8 long, slender bristles ....*Micractinium* (p. 364)
26 Coenobia of laterally attached ovoid, ellipsoidal or spindle-shaped cells, possessing variously positioned spines or horns ....*Scenedesmus*, in part (p. 384)
27 Coenobia of cells joined together by remnants of mother cell walls and within a common mucilaginous envelope 28
27 Coenobia of cells joined directly to one another 29
28 Coenobia of 4 or 8 cells joined by common mucilage and loop-like remnants of mother cell wall ....*Westella* (p. 408)

28 Coenobia of 4 cells attached in pairs to ends of mother cell wall, sometimes forming syncoenobia of up to 64 cells ....*Quadricoccus* (p. 381)
29 Coenobia of 1 or 2 rows of 2, 4, 8, 16 or more laterally attached cells ....*Scenedesmus*, in part (p. 384)
29 Coenobia of another form 30
30 Cells terminal on short, radiating strands of mucilage, halfmoon-, horseshoe-, kidney- or wedge-shaped to pyramidal, with a stalk and 1–4 apical spines.........................*Sorastrum* (p. 401)
30 Coenobia of a different form 31
31 Cells straight, curved or spirally twisted, joined near the middle (sometimes loosely entangled) ....*Ankistrodesmus*, in part (p. 330)
31 Cells strongly curved and sometimes almost forming complete circle, each cell connected equatorially by convex side.....*Selenastrum* (p. 399)

SECTION X Colonies not of fixed cell number, although cells sometimes grouped (often in 4s), growing by further cell division or addition of further generations (represent syncoenobia in case of chlorococcalean algae); cells enclosed within a common mucilaginous envelope and sometimes connected by mucilaginous threads

1 Colonies attached or usually closely associated with surfaces 2
1 Colonies free-floating, only incidentally associated with surfaces 14
2 Colonies macroscopic, forming irregular, sac-like or intestine-like masses, usually mucilaginous 3
2 Colonies microscopic and of another form 5
3 Colony of an irregular linear series of cells connected by wall material to form a dendroidal structure enclosed within a stratified mucilaginous envelope ....*Echallocystis* (p. 353)
3 Colony of another form 4
4 Cells in 2- or 4-celled groups towards periphery of a homogeneous mucilaginous envelope, pseudocilia sometimes present; aquatic ....*Tetraspora* (p. 301)
4 Cells not arranged in groups, surrounded by concentrically striated mucilage, without pseudocilia; terrestrial ....palmelloid stage of *Chlamydomonas*, in part (p. 306)
5 Colonies hollow and pear-shaped or spherical; cells marginally distributed, thick-walled and having pseudocilia ....*Apiocystis* (p. 299)
5 Colonies of a different form 6
6 Colonies usually of ellipsoidal cells aligned with long axes parallel and scattered within a mucilaginous envelope, often concentric striations surround cell groups; commonly terrestrial.....................*Coccomyxa*, in part (p. 339)
6 Cells rarely elongate and ellipsoidal, not arranged as above; terrestrial or aquatic 7
7 Cells positioned terminally or at irregular intervals in branching cylinders of homogeneous or transversely striated mucilage; terrestrial ....*Hormotila* (p. 358)
7 Colonies of another form 8
8 Cells ovoid to spherical, surrounded by concentric striations and embedded within a mucilaginous envelope of indefinite extent .....*Gloeocystis* (=*Chlamydocapsa*), in part (p. 355)
8 Cells surrounded by a thick or thin and completely homogeneous mucilaginous envelope 9
9 Colonies of 2–4(–8) cells, each bearing 1 or 2 long gelatinous hairs (pseudocilia) ....*Gloeochaete* (p. 300)
9 Colonies usually having more than 8 cells and without pseudocilia 10
10 Cells scattered within a mucilaginous mass, sometimes each with an eyespot; usually terrestrial or along drying margins of water bodies..*Palmella*-stage of *Chlamydomonas*, in part (p. 306)
10 Cells densely packed and irregularly arranged, grouped or in rows within a low or hemispherical mucilaginous envelope; aquatic on hard surfaces in flowing water............11
11 Colonies usually crimson or red; cells densely packed and irregularly arranged ....*Rhodoplax* (p. 384)
11 Colonies green; cells usually having a definite arrangement 12
12 Cells ellipsoidal or ovoid, usually in tetrads or pairs, within a thick mucilaginous envelope to form hemispherical colonies ....*Sporotetras* (p. 402)
12 Cells ellipsoidal to inversely pear-shaped, often in short vertical rows, within a thin mucilaginous envelope to form low expansions ....*Sphaerobotrys* (p. 402)
13 Colonies initially ellipsoidal, spindle-shaped or ellipsoidal and outline well-defined ....*Elakatothrix* (p. 439)
13 Colony otherwise...14
14 Colonies of branched anastomosing mucilaginous threads, or forming irregularly shaped mucilaginous masses of sometimes ill-defined outline 15
14 Colonies spherical or ovoid, on ageing sometimes becoming more irregular in shape 19
15 Colonies of branched and anastomosing mucilaginous threads containing linear rows of cells; common in pools within *Sphagnum* bogs ....*Palmodictyon* (p. 375)
15 Colonies of irregularly shaped mucilaginous

masses containing scattered cells, or if in 2s or 4s then associated with pseudocilia 16

16 Cells spherical or slightly flattened or depressed at one end, each partly enclosed within circular or semicircular remnants of mother cell wall or having pseudocilia 17

16 Cells spherical, ovoid to broadly ellipsoidal, if mother cell wall remnants present then not partly enclosing cell and pseudocilia absent 18

17 Pseudocilia arising in tufts from flattened pole of each cell ....*Schizochlamys* (p. 301)

17 Pseudocilia absent........*Schizochlamydella* (p. 398)

18 Cells partly or almost completely embedded within a tough oily mucilage (often brownish) and without persistent remnant of mother cell wall ....*Botryococcus* (p. 332)

18 Cells embedded in a soft and non-oily mucilage (usually colourless), sometimes connected by strands of delicate mucilage and cells associated with remnants of mother cell wall ....*Botryosphaerella* (p. 332)

19 Cells each bearing 2 long pseudocilia, sometimes extending beyond the mucilaginous envelope...........*Paulschulzia* (p. 300)

19 Cells without pseudocilia....................................20

20 Cells spherical to ovoid.......................................21

20 Cells usually of another shape.............................31

21 Cells surrounded by a distinct and concentrically striated mucilaginous sheath.......22

21 Cells without a sheath and mucilage homogeneous or with ray-like striations 24

22 Chloroplast star-shaped; common in *Sphagnum*-dominated bog pools ....*Asterococcus* (p. 299)

22 Chloroplast parietal and cup-shaped; plankton of lakes, ponds and rivers 23

23 Cells associated with remnants of mother cell wall................*Planktosphaeria*, in part (p. 379)

23 Cells not associated with cell wall remnants ....*Gloeocystis*, in part (p. 355)

24 Mucilage envelope with ray-like striations surrounding cells ....*Radiococcus*, in part (p. 382)

24 Mucilage homogeneous throughout 25

25 Cells never associated with mother cell wall remnants 26

25 Cells associated with mother cell wall remnants although only evident for a short period immediately following daughter cell release..29

26 Cells usually in groups of 2, 4, 8, or 16 within a mucilaginous envelope 27

26 Cells usually randomly distributed and only grouped immediately after release from mother cell 28

27 Cells in pairs, often united into subgroups of 4 or 8 cells, each with contractile vacuoles ....*Pseudosphaerocystis* (p. 301)

27 Cells not grouped in pairs and without contractile vacuoles ....*Coenococcus* (p. 342)

28 Asexual reproduction by autospores and zoospores ....*Sphaerocystis* (p. 402)

28 Asexual reproduction only by autospores ....*Coenochloris*, in part (p. 342)

29 Cell only associated with remnants of mother cell wall for a short time immediately following release of daughter cells (only evident in actively growing and reproducing populations) ....*Coenochloris*, in part (p. 342)

29 Cells often associated with the persistent remnants of the mother cell wall 30

30 Remnants of mother cell wall irregularly shaped ....*Planktosphaeria*, in part (p. 379)

30 Remnants of mother cell wall often curved and cap-like................*Radiococcus*, in part (p. 382)

31 Cells ellipsoidal, ovoid to spindle-shaped, sometimes cylindrical 32

31 Cells of another shape..35

32 Chloroplast of several band-like ribbons radiating from cell centre ....*Neglectella* (p. 370)

32 Chloroplasts of another form 33

33 Cells in 4– or 8-celled groups and associated with remnants of mother cell wall ....*Coenocystis* (p. 343)

33 Cells not grouped, or if occasionally grouped, then not associated with mother cell wall remnants 34

34 Cell walls smooth.........................*Oocystis* (p. 371)

34 Cell walls granulate, often granules confined to rounded apices and sometimes fused to form a ring ....*Granulocystopsis* (p. 357)

35 Cells notched, slightly curved and sausage-shaped, asymmetrically ovoid to kidney-, horseshoe- or crescent-shaped, with apices sometimes almost touching................................36

35 Cells of another shape..42

36 Cell walls granulate 37

36 Cell walls without granules.................................38

37 Cells usually solitary; pyrenoid present ....*Juranyiella* (p. 360)

37 Cells usually forming 4– or more-celled colonies; pyrenoid absent ....*Raphidocelis*, in part (p. 383)

38 Pyrenoids present..39

38 Pyrenoids absent 41

39 Colony shape not similar to that of cells enclosed within its mucilaginous envelope ....*Kirchneriella* (p. 360)

39 Colony of similar shape to its cells since surrounded by the partly gelatinized wall of the original mother cell 40

40 Chloroplast parietal and normally filling cell, not sponge-like ....*Nephrocytium*, in part (p. 371)

40 Chloroplast star-shaped, arms anastomosing and sponge-like in appearance....*Oonephris* (p. 375)

41 Cells very small (<7 μm long), kidney- or crescent-shaped, within an expanded and wrinkled mother cell (no mucilage) ....*Nephrochlamys* (p. 370)

41 Cells generally larger, variously curved and twisted, within mucilage produced by gelatinization of mother cell wall ....*Raphidocelis*, in part (p. 383)
42 Cells usually curved or spirally twisted, often united equatorially to form bundles and these usually surrounded by an inconspicuous mucilaginous envelope ....*Ankistrodesmus*, in part (p. 330)
42 Cells straight and lying parallel to form clusters of cells within an often conspicuous mucilaginous envelope 43
43 Clusters of cells having long axes parallel but not lying in-line ....*Pseudoquadrigula* (p. 381)
43 Clusters of cells always parallel and lying in-line ....*Quadrigula* (p. 382)

ORDER TETRASPORALES

BY ALLAN PENTECOST

Members of this artificial order are united by the presence of contractile vacuoles in the cells and the absence of flagellated vegetative stages. In many respects they are similar to members of the Order Volvocales, to which they are closely related. It should be noted that non-motile resting stages of the Volvocales sometimes possess contractile vacuoles. Some genera belonging to the Tetrasporales possess pseudocilia, which resemble flagella but are not used for swimming since they are not observed to beat. They lack the two axial microtubules found in most flagella and Pickett-Heaps (1975) suggests that some pseudocilia are disorganized flagella. Under the light microscope they appear as extremely fine bristles radiating from the cells. The order consists mainly of colonial algae invested in copious mucilage with some attaining a large size and easily visible to the naked eye. Others are microscopic and often found as epiphytes. A few are planktonic and sometimes form a significant component of lake plankton.

Apiocystis Nägeli in Kützing 1849

Colonial, forming mucilaginous and usually club-shaped colonies containing about 50 to several hundred cells regularly distributed about the periphery, up to 1 mm or more in length, attached by a narrow basal area of mucilage, cells occasionally in pairs and sometimes in circular zones, spherical, each bearing 2 long pseudocilia extending beyond the mucilage envelope; chloroplast parietal, usually with a single pyrenoid; contractile vacuoles not reported; asexual reproduction by 4 non-motile daughter cells or by direct metamorphosis of cells into zoospores; sexual reproduction isogamous.

Probably cosmopolitan; epiphytic on aquatic plants including algae. Two species, of which one is recorded from the British Isles.

Apiocystis brauniana Nägeli 1849

16010010 **Pl. 76G** (p. 302) **CD**

Cells 6–8 μm in diameter.

Europe, USA; in the British Isles attached to the bladderwort *Utricularia* and other aquatic macrophytes including larger filamentous algae (e.g. *Cladophora, Vaucheria*), usually in shallow water with relatively low nutrient levels, locally frequent in the spring.

Early stages of *Tetraspora gelatinosa* closely resemble *Apiocystis*.

Asterococcus Scherffel 1908

Cells spherical to subspherical, enclosed within a hyaline unlaminated or laminated mucilage considerably exceeding protoplast diameter, single or 2, 4, 8 or 16 in a common mucilage and forming a small colony; chloroplast dense, star-shaped with radiating arms sometimes expanded at cell periphery, with pyrenoid central; contractile vacuoles 2 and peripheral when present; eyespot present or absent; asexual reproduction by formation of 2–8 daughter cells or by zoospores.

Probably cosmopolitan; usually in ponds and lakes. Two species, both recorded in the British Isles.

1 Envelopes not lamellate; cells 10–25(–30) μm in diameter...*A. limneticus*
1 Envelopes lamellate; cells 30–43 μm in diameter ....*A. superbus*

Asterococcus limneticus G.M.Smith 1918

16020010 **Pl. 76A** (p. 302)

Small colonies of 4–16 cells, clearly separated by mucilage, 50–125 μm in diameter; cells 10–25(–35) μm in diameter; eyespot normally absent.

Europe, USA; common in the English Lake District and some Irish loughs.

Asterococcus superbus (Cienkowski) Scherffel 1908

Basionym: *Pleurococcus superbus* Cienkowski
Synonyms: *Gloeocystis infusionum* West *et* G.S.West, *Asterococcus korschikoffii* Ettl, *Chlamydomonas scherfelii* Korshikov

16020020 **Pl. 76B** (p. 302)

Cells solitary or in small 2–4-celled colonies, with a broad lamellate and mucilaginous envelope, often several times diameter of protoplast; cells 30–35(–43) μm in diameter; eyespot present.

Europe, USA; frequently associated with marginal vegetation and in the shallows (tychoplankton) of nutrient-poor lakes such as many Irish loughs and in bog pools, more rarely in slow-flowing rivers.

The cells sometimes contain large numbers of starch grains obscuring the chloroplast.

Chaetopeltis Berthold 1878

Circular and mucilaginous, a single or occasionally a multilayered disc of cells radiating from a common centre; cells rounded or angular, with some cells bearing 1 or 2 long pseudocilia orientated approximately perpendicular to disc; chloroplast parietal, often perforated and having a pyrenoid; contractile vacuoles 2, sometimes in cytoplasm close to substratum; asexual reproduction by 2–8 quadriflagellate zoospores; sexual reproduction isogamous.

Europe, USA; epiphyte on aquatic plants. Monospecific.

The genus should be compared with *Coleochaete* which is morphologically similar, but lacks contractile vacuoles and possesses basally sheathed setae.

Chaetopeltis orbicularis Berthold 1878

Synonym: *Chaetopeltis megalocystis* Schmidle

16080010 **Pl. 76L** (p. 302)

Disc up to 1 mm in diameter; cells 15–20 μm wide, 15–30 μm long, with well-defined and often mucilaginous walls.

Distribution as for genus; in the British Isles occurs as an epiphyte on aquatic plants including larger filamentous algae; frequent, at least in faster-flowing rivers, reported as abundant at sites in the upper and middle reaches of the River Tees (Butcher et al., 1937).

DOUBTFUL GENUS

Chlamydocapsa Fott 1972

Synonym: *Gloeocystis* Nägeli *pro parte*

Cells spherical or subspherical in small (1–)2–8-celled mucilaginous colonies, often in a tetrahedral arrangement, with mucilage forming a series of layers around cells; chloroplast usually cup-shaped, sometimes dense, with a single pyrenoid; contractile vacuoles 2 or absent and sometimes obscured by starch grains or lipid droplets; asexual reproduction by 4–8 zoospores. There has been considerable debate as to the validity of this genus as there is little to separate its species from palmelloid stages of the Chlamydomonadaceae. Three species have been recorded from the British Isles: *Chlamydocapsa bacillus* (Teiling) Fott (16170030), with broadly ellipsoidal cells irregularly arranged within a mucilage envelope, each cell having a cup-shaped chloroplast and a basal pyrenoid; *C. planctonica* (W. *et* G.S.West) Fott (16170020) with spherical to subspherical, tetrahedrally arranged cells in small planktonic colonies of 4–32 cells, each cell having a cup-shaped chloroplast with a basal pyrenoid; *C. ampla* (Kützing) Fott (16170010) (=*Gloeocystis gigas* (Kützing) Lagerheim), with obovoid cells in groups of 8 forming small and usually attached colonies, each cell having a lobed and lateral pyrenoid (Pl. 76C). See also *Gloeocystis* in the Order Chlorococcales (pp. 355, 356).

Gloeochaete Lagerheim 1883

Single-cells or colonies of 2–4(–8) spherical cells, embedded within a broad hyaline mucilage, each cell having 1 or 2 long and erect gelatinous hairs (pseudocilia); cells usually containing ovoid or sausage-shaped cyanelles, lying close together in a cup-shaped region; contractile vacuoles absent or 2; small starch grains and lipid droplets present; asexual reproduction by biflagellate zoospores.

Probably cosmopolitan; subspherical mucilaginous colonies attached to walls of larger algae and other submerged aquatics including aquatic mosses.

Often placed along with *Cyanophora* (chloromonad) in the phylum Glaucophyta (Kies, 1976).

Gloeochaete wittrockiana Lagerheim 1883

16280010 **Pl. 76M** (p. 302)

Cells (4–)6–10(–21) μm in diameter, pseudocilia up to 250 μm long; cyanelles 1 or more often difficult to distinguish due to food reserves.

Distribution as for genus; widely distributed in the British Isles as an occasional epiphyte on aquatic plants and larger filamentous algae such as *Cladophora* and *Vaucheria*, occasionally on *Sphagnum* in acidic boggy sites.

Paulschulzia Skuja 1948

Synonyms: *Tetraspora* Link *pro parte*, *Schulziella* Teiling

Cells spherical or ellipsoidal, enclosed within a hyaline, homogeneous or faintly lamellate mucilage, forming small free-floating colonies of 2, 4, 8, 16, 32, . . . to 256 cells, the cells each having 2 long pseudocilia sometimes extending beyond mucilage investment; cells in small tetrahedrally arranged groups, each group with its own mucilage coat; chloroplast cup-shaped, sometimes perforated with a basal pyrenoid; eyespot present or absent; contractile vacuoles 2; asexual reproduction by biflagellate zoospores.

Probably cosmopolitan, usually in lake plankton.

Differs from *Tetraspora* in the tetrad groups being surrounded by their own mucilaginous investment. However, the two genera must be closely related. The occurrence of the two species in the British Isles and their taxonomy is described by Lund (1956a, 1961b).

1 Pseudocilia extending beyond the outermost mucilaginous investment ....*P. pseudovolvox*
1 Pseudocilia contained within the outermost mucilaginous investment.........................*P. tenera*

Paulschulzia pseudovolvox (Schulz) Skuja 1948

Basionym: *Tetraspora pseudovolvox* Schulz

Synonyms: *Schulziella pseudovolvox* (Schulz) Teiling, *Tetraspora nygaardii* Teiling

16490010 **Pl. 76D** (p. 302)

Colonies spherical or broadly ellipsoidal of 2, 4, 8, 16, 32, 64 . . . to 256 cells, 40–350 μm in diameter, when in larger colonies cells often in sub-groups of 2–6 surrounded by an inner mucilage investment; cells spherical, 6–13 μm in diameter, thin-walled, each with 2 pseudocilia up to 200 μm long and often extending beyond the mucilage envelope; chloroplast cup-shaped, occasionally lobed; eyespot sometimes present, small.

Germany; occasionally reported in the British Isles from nutrient-poor and moderately nutrient-rich lakes and ponds, truly planktonic (euplankton) or in the shallows and then often associated with marginal aquatic vegetation (tychoplankton).

Paulschulzia tenera (Korshikov) J.W.G.Lund 1960

Basionym: *Tetraspora tenera* Korshikov
Synonyms: *Paulschulzia elegans* (Woronichin) Fott, *Tetraspora elegans* Woronichin
16490020 **Pl. 76E** (p. 302)

Colonies spherical, up to 200 μm in diameter, of 4, 8, 16, 32 or 64 cells, cells spherical, 6–13 μm in diameter, each with 2 pseudocilia much longer than cells,but frequently twisted and contained within an envelope; chloroplast cup-shaped with a pyrenoid; eyespot small.

Czech Republic, Ukraine; North America; fairly frequent in the plankton of nutrient-poor and moderately nutrient-rich lakes.

Possibly a form of *P. pseudovolvox*. The plant should be compared with *Radiococcus* in which pseudocilia are absent but the mucilage is radially striated and should be compared carefully with *Tetraspora lemmermanii*.

Pseudosphaerocystis Woronichin 1931

Synonyms: *Gemellicystis* Teiling, *Planctogloea* Skuja

Cells spherical or ellipsoidal, in groups of 2 to 32 within a common mucilaginous envelope forming small free-floating colonies, infrequently furnished with 2 long pseudocilia; chloroplast cup-shaped, with a basal pyrenoid; eyespot present or absent; contractile vacuoles 2 in cytoplasm; cells usually closely paired with anterior ends containing contractile vacuoles facing each other; asexual reproduction by autospores or biflagellate zoospores.

Europe, India; planktonic and usually in lakes.

Pseudosphaerocystis lacustris (Lemmermann) Novákova 1965

Basionym: *Tetraspora lacustris* Lemmermann *non sensu* Lemmermann
Synonyms: *Planctogloea lundii* (Bourrelly) Bourrelly, *P. planctonica* Woronichin, *Gemellicystis lundii* Bourrelly, *Pseudosphaerocystis planktonica* Woronichin
16590010 **Pl. 76F** (p. 302)

Colonies of 2, 4, 8, 16 or 32 cells in a rounded, often flattened homogeneous mucilaginous investment; cells spherical to subspherical, without a papilla, 8–12 μm wide, 7–10 μm long, colonies 20–100 μm in diameter; eyespot anterior; pseudocilia absent.

Europe; frequent in the plankton of moderately nutrient-rich lakes, less common in nutrient-poor waters.

The paired cells are often united into subgroups of 4 or 8. The opposing cell pairs, presence of contractile vacuoles, and unstratified mucilage are characters which distinguish the species from *Paulschulzia*, *Radiococcus* and *Sphaerocystis*. The confused taxonomy of this genus is discussed by Lund (1961b). *Pseudosphaerocystis neglecta* (Teiling emend. Skuja) Bourrelly, distinguished by its longitudinally striated chloroplasts, is probably a synonym since the two forms intergrade (J.W.G. Lund, pers. comm.).

Schizochlamys A. Braun 1849

Cells irregularly arranged with a structureless mucilage to form micro- or macroscopic colonies; cell spherical or slightly ellipsoidal, often slightly flattened or with a depression at one end, usually in 2s or 4s, original cell wall persists as a single, 2 or 4 distinct fragments, each cell with 4 extremely fine pseudocilia arising close together from the flattened pole; chloroplast parietal, cup-shaped, sometimes with a pyrenoid; eyespot absent; contractile vacuoles usually 2 and near the cell pole; reproduction by divisions of a cell to form 2 or 4 daughter cells or by quadriflagellate (rarely biflagellate) zoospores.

Cosmopolitan, attached to various submerged surfaces and sometimes becoming detached and free-floating.

Schizochlamys gelatinosa A. Braun

16650010 **CD**

Colonies forming mucilaginous masses sometimes reaching several cm in diameter; cells 9.5–15 μm, somewhat compressed longitudinally and anteriorly flattened. associated with cell wall remnants.

Distribution as for genus; known from ditches, ponds and rivers (e.g. River Caragh, Kerry, Ireland), most common in *Sphagnum* bogs.

Readily distinguished from *Schizochlamydella delicatula* (Order Chlorococcales) by the presence of pseudocilia.

DOUBTFUL RECORD
Sphaerellocystis stellata Ettl 1960
Synonym: *Asterococcus siderogloeus* (Pascher *et* Jahoda) Novákova
Recorded from the British Isles by Schlösser in 1982 and held in the Göttingen Culture Collection, but the location of the original sample is unknown.

Stylosphaeridium Geitler *et* Gimesi 1925

Cells ellipsoidal, attached by a short, fine stalk; chloroplast parietal, cup-shaped, with a single pyrenoid; contractile vacuoles 2 in cytoplasm; eyespot present or absent; asexual reproduction by biflagellate zoospores.

Probably cosmopolitan; epiphytic on planktonic algae.

Several species distinguished on the cell form and eyespot.

Stylosphaeridium stipitatum (Bachmann) Geitler *et* Gimesi 1925

Basionym: *Chlamydomonas stipitata* Bachmann
16790010 **Pl. 76K** (p. 302)

Cells 4–8 μm wide, 5–10 μm long, stalk of similar length; chloroplast cup-shaped, with a basal pyrenoid; eyespot present or absent.

Europe, USA; epiphytic on planktonic algae, particularly the blue-green alga *Coelosphaerium*, where large numbers of cells are sometimes attached to the surface mucilage of a single colony, also on *Anabaena*.

Tetraspora Link 1809

Cells united by mucilage to form large, often macroscopic colonies, pale green in colour, jelly-like in consistency and spherical, sac-like or irregularly lobed,

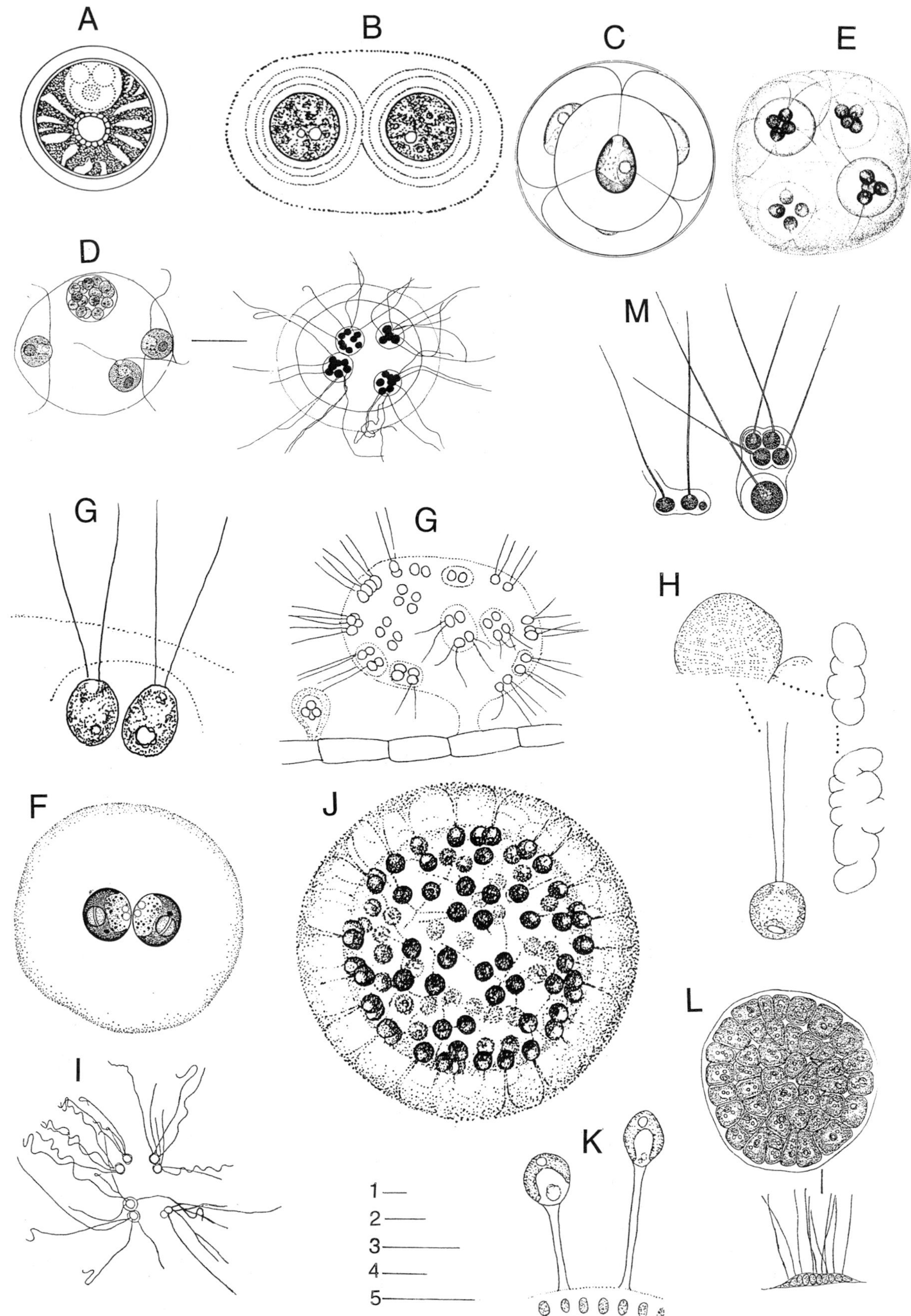

Plate 76 A–M

A. *Asterococcus limneticus* (p. 299); **B**. *Asterococcus superbus* (p. 299); **C**. *Chlamydocapsa ampla* (p. 300); **D**. *Paulschulzia pseudovolvox* (p. 300): right stained with gentian violet; **E**. *Paulschulzia tenera* (p. 301); **F**. *Pseudosphaerocystis lacustris* (p. 301); **G**. *Apiocystis brauniana* (p. 299); **H**. *Tetraspora gelatinosa* (p. 303): colonies on right of natural size; **I**. *Tetraspora hexanematoidea* (p. 303); **J**. *Tetraspora lemmermannii* (p. 303); **K**. *Stylosphaeridium stipitatum* (p. 301): on colony of *Coelosphaerium*; **L**. *Chaetopeltis orbicularis* (p. 300): surface and lateral views; **M**. *Gloeochaete wittrockiana* (p. 300).
Scale bars: 1–3, 10 µm; 4, 5, 100 µm; 1 for B, F, L (surface view); 2 for C, D (left); 3 for A, G (left), H (cell), K; 4 for D (right), E; 5 for G (right), I, J, L (lateral view), M.

membrane-like, or convoluted and several cm across (exceptionally up to 1 m); cells spherical, often arranged in 2s or 4s, close to surface of colonies so that 2 long pseudocilia usually extend free of mucilage; chloroplast parietal, cup-shaped and covering most of cell, with a single pyrenoid but sometimes obscured by starch; eyespot absent; contractile vacuoles usually present in the cytoplasm; asexual reproduction by formation of 2 or 4 daughter cells or by biflagellate zoospores; akinetes occasionally with sculptured walls; sexual reproduction isogamous.

Cosmopolitan; attached to leaves, twigs or water plants when young, later becoming free-floating as large gelatinous masses, with one species planktonic. Often in shallow ditches, pools and in the littoral zone of lakes in the spring, usually disappearing in the summer.

The several species are distinguished on the form of the colony; they should be compared carefully with *Apiocystis* and *Paulschulzia*.

1 Thalli microscopic, <1 mm in diameter and planktonic .... *T. lemmermannii*
1 Thalli macroscopic, ≥1 cm in diameter, often attached 2
2 Cells each with 2–6 pseudocilia; thalli to 2 cm across .... *T. hexanematoidea*
2 Cells with no more than 2 pseudocilia; thalli often large and convoluted................. *T. gelatinosa*

Tetraspora gelatinosa (Vaucher) Desvaux 1818

Basionym: *Ulva gelatinosa* Vaucher
Synonyms: *Tetraspora explanata* C.Agardh, *T. ulvacea* Kützing

16740010 **Pl. 76H** (p. 302) **CD**

Colonies vesicular and sac-like, containing many hundreds of cells at periphery, long pseudocilia extending beyond mucilage envelope; pseudocilia usually evident and far exceed diameter of cells, occasionally absent from part or whole of colony; cells 6–12 μm in diameter.

Probably cosmopolitan; most frequently reported *Tetraspora* species in the British Isles where found in the shallow water of ditches, ponds, lakes and rivers, often under trees; sometimes associated with *Draparnaldia* during late winter and spring, often disappearing during the summer.

Tetraspora lubrica (Roth) C.Agardh, which has a more tubular thallus, is probably a growth form of *T. gelatinosa*.

Tetraspora hexanematoidea J.W.G.Lund 1957

16740040 **Pl. 76I** (p. 302)

Colonies subspherical, up to 2 cm in diameter, with cells in 2s or 4s, sometimes enclosed within a series of mucilage layers and possessing 2–6 pseudocilia up to 220 μm in length, with pseudocilia in places split a short way along length; cells spherical or subspherical, 10–12 μm in diameter; contractile vacuoles absent.

British Isles; described only from a garden pond in the English Lake District (Lund, 1957).

Possibly a form of *T. gelatinosa* since the splitting of pseudocilia has occasionally been observed in other members of the Tetrasporales. The presence of multiple envelopes recalls *Paulschulzia* and the two genera must be closely related.

Tetraspora lemmermannii Fott 1972

Synonym: *Tetraspora lacustris* Lemmermann *non* Lemmermann 1898

16740020 **Pl. 76J** (p. 302)

Colonies small and planktonic, spherical, 50–500 μm in diameter, with cells scattered near periphery of colony, often in open groups of 4 with pseudocilia directed radially and usually confined within outer homogeneous mucilaginous layer; cells spherical or subspherical, 6–12 μm in diameter; contractile vacuoles 2 and apical.

Europe, USA; occasional in lake plankton in the British Isles.

The colonies should be carefully compared with those of *Paulschulzia tenera*, which possess inner mucilage envelopes and frequently have twisted pseudocilia. *Sphaerocystis schroeteri* is also similar, but lacks pseudocilia. *Tetraspora limnetica* West *et* G.S.West described from Ennerdale Water, Cumbria, with cells just 4 μm in diameter, may be attributed to this species.

ORDER VOLVOCALES

BY ALLAN PENTECOST

Vegetative cells motile with 2 or 4 anterior, smooth, equal flagella and 2 or more contractile vacuoles present near their base. Cells usually with thin cell wall which is often contiguous with the protoplast, but may be separated by a well-defined space to form an envelope. In a few genera the protoplast is distended by protoplasmic extensions. The cell wall is normally smooth, thin and colourless but may be angled, ridged or nodular, and tinted brown or yellow. The thin and perhaps 'leaky' walls may explain why contractile vacuoles occur in practically all members of this order. Cells are usually ellipsoidal or spherical but they may be pear-shaped, flattened and twisted, furrowed or spindle-shaped. The cells are single or united into colonies of various shapes. Point of entry of the flagella into cell sometimes marked by a small protuberant papilla. Chloroplast normally single, usually cup-shaped but may be lateral or perforated and parietal, eyespot usually visible. Pyrenoid(s) present or absent, if present then usually single. Asexual reproduction by division into 2, 4 or rarely more daughter cells, or by zoospores which may act as autospores (2–8 per cell). Cells normally uninucleate. Sexual reproduction mainly isogamous, occasionally anisogamous or oogamous. Non-motile palmelloid states are often encountered. Monographs by Pascher (1927), Korshikov (1938), Huber-Pestalozzi (1961), Ettl (1976, 1983) and Dillard (1989).

The Volvocales include several well-known and common freshwater algae. Among the unicellular forms the genus *Chlamydomonas* is the most abundant and widely distributed. They are more often encountered in small water bodies than in the plankton of large lakes. Colonial forms such as *Eudorina* and *Volvox* are fascinating microscopic objects (Starr, 1980). The number of cells per colony is fixed at an early stage of development and such colonies are referred to as *coenobia*. The examination of Volvocales, in common with other flagellates, requires much patience as the careful preparation of collections and their observation is usually required for determination of species. The student should note that even given the extensive literature available, frequently specimens are found which cannot be readily assigned to species. Where possible numerous cells or coenobia should be examined since individuals often display considerable variability. The species included here are those for which reports appear to be reliable, but details are provided only for those most frequently reported or especially distinctive. Members of the Prasinophyta may easily be mistaken for unicellular Volvocales since their distinguishing characters (naked cells except for the presence of minute scales) cannot be seen with the light microscope.

Basichlamys Skuja 1956

Coenobium a flat plate of 4 cells with flagella directed all in one direction or radially arranged, surrounded by a mucilaginous envelope extending basally into a cup-shape, in lateral view the mucilage extending around the cell anteriorly (difficult to observe without staining); cells obovoid, about 1.5 times as long as wide, with a prominent hemispherical papilla and flagella up to three times as long as cells; chloroplast cup-shaped, with a basal pyrenoid and an anterior eyespot; contractile vacuoles 2 in each cell and anterior; asexual reproduction by formation of daughter coenobia within a single mother cell.

Europe; aquatic habitats. Monospecific.

Basichlamys sacculifera (Scherffel) Skuja 1956

Basionym: *Gonium sacculifera* Scherffel

16040010 **Pl. 80P** (p. 321)

Coenobia up to 50 µm in diameter; cells 4–13 µm wide, 9–20 µm long.

Central Europe; in a muddy ditch in Cambridge but probably widespread (J.H.Belcher & E.M.F.Swale, pers. comm.).

This species should be compared carefully with *Gonium* which it closely resembles.

Brachiomonas Bohlin 1897

Cells with 4–5 large, backwardly directed cellular protrusions (horns), usually with a large posterior horn surrounded by 4 equally spaced and slightly smaller lateral horns; cells anchor-shaped in lateral view with rounded anterior, bearing 2 flagella as long as or longer than cell; papilla present or absent; contractile vacuoles not reported in British specimens; chloroplast parietal, poorly defined, sometimes with an eyespot and pyrenoids present.

Cosmopolitan; in waters of elevated salinity, usually near the coast. Two of the four species known in the British Isles are distinguished by the number of pyrenoids and form of the horns.

1 Cells in lateral view with small horns, not clearly drawn out from anterior end of cell........*B. simplex*
1 Cells in lateral view with pronounced lateral horns, clearly drawn out on both sides of cell ....*B. submarina*

Brachiomonas simplex Hazen 1922

16050020 **Pl. 79O** (p. 316)

Cells anteriorly rounded and smooth, abruptly narrowing toward posterior end to produce a pronounced horn a little less than half total length of cell, with 4 much smaller lateral protrusions present; quadrate in polar view; cells 18–24 µm wide, 30–48 µm long; papilla absent; chloroplast parietal with a single pyrenoid and an eyespot in anterior end of cell.

Northern and Central Europe; largest species, rarely observed in coastal brackish-water pools and freshwater lake plankton in the British Isles.

Brachiomonas submarina Bohlin 1897

Synonym: *Brachiomonas westiana* Pascher

16050010 **Pl. 79P** (p. 316)

Cells anteriorly drawn out into a small papilla and in lateral view the anterior wall wedge-shaped back to tips of 4 lateral horns which protrude clearly from cell; horns together with a single acute posterior horn containing much of cell volume; lateral horns give a cruciform outline in polar view; cells 13–36 µm wide, 15–30(–40) µm long; chloroplast sometimes entering part of the horns, with 1 pyrenoid and eyespot.

Europe, North America; most frequently reported species from coastal and inland sites, sometimes in nutrient-rich waters such as Regent's Park Lake, London, and in rainwater-filled troughs; not confined to brackish water or those prone to dry out, although sometimes found associated with *Haematococcus pluvialis*.

Jane (1938) found it variable with the cell contents often obscure. In some collections cells answering the description of *B. simplex* are found, suggesting that it is merely a form of *B. submarina*.

Carteria Diesing 1866 emend. Francé 1893

Cells spherical, ellipsoidal or pear-shaped, with 4 equally sized apical flagella, usually longer than cells; papilla present or absent; contractile vacuoles 2 or 4 per cell and near flagellum base; chloroplast parietal, usually cup-shaped, sometimes lateral and incompletely surrounding cell or H-shaped in optical section; pyrenoids absent or 1, rarely more; eyespot usually present; asexual reproduction by zoospores or simple cleavage; sexual reproduction isogamous; palmelloid stages known.

Cosmopolitan; less frequently encountered than *Chlamydomonas*, but occurs in similar habitats. Most species recorded from soil cultures, rainwater pools and among *Sphagnum* with some in nutrient-rich ponds and ditches, occasionally in large numbers and also abundant in ice-covered pools.

Carteria closely resembles *Chlamydomonas* morphologically, but differs in having 4 rather than 2 flagella. Over 100 species of *Carteria* have been described, but many are poorly defined and of uncertain status.

The species have been divided into several subgenera based on the form of the chloroplast as in *Chlamydomonas* (see below). The chloroplast types parallel those of *Chlamydomonas* suggesting that the two genera are closely related. Most species possess a simple cup-shaped chloroplast (Eucarteria) while a considerable number have a lateral chloroplast and pyrenoid (Corbiera). A few have H-shaped chloroplasts (Agloe, Pseudagloe) or possess more than one pyrenoid (*Carteriopsis*); none of the last group of species has been recorded from the British Isles. Cells lacking pyrenoids are assigned by some authors to the genus *Provasoliella* A.R.Loeblich.

The colourless genus *Polytomella* Aragão is closely related to *Carteria*. Several species have been recorded in the British Isles from organically enriched sites but none is common (Pringsheim, 1955). They give a positive starch test with iodine. *Polytomella agilis* has ellipsoidal cells and an eyespot, *P. magna* has oviform cells and an eyespot and both have a papilla. *Polytomella papillata* has egg-shaped cells, a large papilla but no eyespot; *P. parva* has narrowly egg-shaped cells, a small papilla and no eyespot.

1 Pyrenoid absent and cells obovoid with a minute papilla .... *C. ovata*
1 Pyrenoids present and cells not as above 2
2 Chloroplast cup-shaped, with a single basal pyrenoid (Subgenus Eucarteria) 3
2 Chloroplast lateral or H-shaped, with a pyrenoid 6
3 Chloroplast divided into 5–7 lobes, basally positioned and approximately radial when viewed laterally .... *C. radiosa*
3 Chloroplast continous around cell and not divided into lobes 4
4 Papilla absent; cells spherical (or weakly ellipsoidal) .... *C. globosa*
4 Papilla present; cells spherical to pear-shaped 5
5 Cells ellipsoidal, 6–17 µm wide .... *C. klebsii*
5 Cells spherical, about 20 µm in diameter .... *C. multifilis*
6 Chloroplast H-shaped or central 7
6 Chloroplast cup-shaped and pyrenoid lateral (Subgenus Corbiera) 9
7 Chloroplast H-shaped with a pyrenoid in its centre (Subgenus Pseudagloe) 8
7 Chloroplast central and massive with a central pyrenoid .... *C. peterhofiensis*
8 Nucleus of the cell anterior .... *C. arenicola*
8 Nucleus of the cell posterior .... *C. micronucleolata*
9 Cells 12–28 µm long, elongate-ellipsoidal, 4 or more times as long as wide .... *C. acidicola*
9 Cells 25–34 µm long, ellipsoidal, about twice as long as wide .... *C. obtusa*

Carteria acidicola F.E.Fritsch *et* R.P.John 1942

16060040 **Pl. 80E** (p. 321)

Cells markedly elongate, 4.5–16 µm wide, 12–28 µm long, with a pronounced dome-shaped papilla; chloroplast lateral, extending completely around cell, with a lateral pyrenoid lacking a starch sheath in a median position; eyespot anterior.

British Isles (England); isolated from soil with pH 3–4.

Carteria arenicola F.E.Fritsch *et* R.P.John 1942

16060050 **Pl. 80F** (p. 321)

Cells ellipsoidal, 10–14 µm wide, 18–20 µm long, with a dome-shaped papilla; chloroplast H-shaped, with pyrenoid in centre of broad bar; eyespot small; nucleus anterior, above pyrenoid.

Central Europe; isolated from dune soils in North Wales.

Carteria globosa Korshikov 1927

16060060 **Pl. 80G** (p. 321)

Cells spherical or weakly ellipsoidal, 14–18(–28) µm in diameter, lacking a papilla; chloroplast cup-shaped, thickened basally with a large pyrenoid; eyespot small and anterior.

Europe, Asia, North America; especially in cold water, widely distributed in lowland pools, occasionally forming blooms; reported from nitrate-rich waters.

Carteria klebsii (P.A.Dangeard) Francé emend. Troitzkaja 1921

Basionym: *Pithiscus klebsii* P.A.Dangeard

16060070 **Pl. 80H** (p. 321)

Cells ellipsoidal, 6–17 µm wide, 8–35 µm long, about 1.5 times as long as wide, with a small hemispherical papilla; chloroplast cup-shaped with a basal pyrenoid; eyespot absent or anterior.

Central Europe in organically enriched lakes and bogs; rare in the British Isles where found as a chance member of the plankton.

Smooth zygotes have been observed.

Carteria micronucleolata Korshikov in Pascher 1927

16060080 **Pl. 80I** (p. 321)

Cells ellipsoidal to shortly cylindrical, 9–16 µm wide, 16–23 µm long, about 3 times as long as wide, with a pronounced papilla; chloroplast H-shaped with a pyrenoid in bar; eyespot small, apical; nucleus posterior.

India, Russia; recorded from a shallow nutrient-poor pool on heathland in south-east England.

The species should be compared with *C. arenicola* as it is similar, apart from the position of the nucleus.

Carteria multifilis (Fresenius) Dill 1895

Basionym: *Chlamydomonas multifilis* Fresenius

16060090 **Pl. 80J** (p. 321)

Cells spherical to weakly ellipsoidal, 9–22 µm in diameter with a hemispherical papilla; chloroplast cup-shaped surrounding most of cell, with a basal pyrenoid and an anterior eyespot.

Probably cosmopolitan; occasional in boggy and temporary rainwater pools

Differs from *C. klebsii* only by being more spherical and having slightly smaller cells. Zygotes reported to have smooth walls.

Carteria obtusa Dill 1895

16060100 **Pl. 80K** (p. 321)

Cells elongate-ellipsoidal, 15 μm wide, 25–32 μm long, with a broad, low papilla; chloroplast cup-shaped, surrounding most of cell with pyrenoid in median position; eyespot anterior.

Europe; rare in the British Isles in moorland pools.

Zygote spherical with a smooth wall. *Carteria oliveri* G.S.West 1915 is very similar and probably belongs here.

Carteria ovata H.C.Jacobsen 1910

Synonym: *Provasoliella ovata* (H.C.Jacobsen) A.R.Loeblich

16060020 **Pl. 80L** (p. 321)

Cells obovoid to pear-shaped, basally narrowed, 8–15 μm wide, 15–25 μm long, 1.5 to 2 times as long as wide, with a minute papilla; chloroplast cup-shaped, without a pyrenoid, eyespot median or anterior.

Northern and Central Europe; a variable species associated with bird-manured pools and other organically enriched waters.

Carteria peterhofiensis Kisselev 1931

16060110 **Pl. 80M** (p. 321)

Cells broadly ellipsoidal, 15–38 μm wide, 24–55 μm long, about 1.5 times as long as wide, with a low, broad papilla; chloroplast dense and massive, enclosing a large central pyrenoid; eyespot median or apical; contractile vacuoles numerous, scattered.

Widely distributed in organically enriched waters of Central Europe; recorded on several occasions with aquatic bacteria in Priest Pot in the English Lake District.

Some authors include this species in *Pseudocarteria* H.Ettl because of the chloroplast structure.

Carteria radiosa Korshikov 1927

Synonyms: *Carteria regularis* Korschikoff, *C. stellifera* Nygaard

16060030 **Pl. 80N** (p. 321)

Cells spherical or weakly ellipsoidal, 13.5–25 μm in diameter, with a broad and very low papilla; chloroplast axile with nucleus lying in a small cup-like anterior depression, lobes in lateral view extending to cell periphery without attenuating, with a central pyrenoid; eyespot anterior.

Europe; a distinctive species reported by Lund (1954a) from a garden pond in the English Lake District, where it was found in great numbers associated with decaying leaves; also known from Cambridge.

Lund (1954) found the contractile vacuoles difficult to observe.

Chlamydomonas Ehrenberg 1833

Synonyms: *Chlamydococcus* A. Braun, *Chloromonas* Gobi emend. Wille, *Gloeomonas* G.A.Klebs emend. H. *et* O.Ettl, *Isococcus* F.E.Fritsch, *Platychloris* Pascher, *Chlorogonium pro parte sensu* Pascher

Cells spherical, ellipsoidal, shortly cylindrical or pear-shaped in lateral view, from above circular or slightly flattened in outline, with papilla present or absent; cell wall usually thin and smooth except *C. reinhardtii* with 7 layers and cellulose seems absent); flagella as long, longer or rarely, shorter than cells, inserted apically; chloroplasts green or rarely obscured by red carotenoid (haematochrome) pigment, usually cup-shaped, entire or with lateral or radiating lobes or broken into small 'islands' (in some species band-shaped, lateral or more rarely H-shaped), with pyrenoids absent or 1 to many; eyespot usually present, median or anterior; contractile vacuoles usually 2 in each cell and apical; asexual reproduction by division of protoplast into 2–16 daughter protoplasts released from parent as zoospores after parent wall gelatinization; sexual reproduction isogamous, occasionally anisogamous or more rarely oogamous; zygote spherical, with a thickened and sometimes ornamented wall; palmelloid stage common, often intermingled with motile cells.

Cosmopolitan; little is known of the ecology of the majority of the species but many are abundant in small, very or extremely nutrient-rich waters, such as sewage oxidation ponds, particularly in the spring and early summer. Lund (1942b) describes the occurrence of several species in just one English pond. Detailed studies by Happey-Wood (1980), Happey-Wood & Priddle (1984) have shown that many species are important members of the epipelic flora of ponds and lakes, responding to elevated levels of bicarbonate alkalinity. A large number of species are known from soil cultures and are important members of the soil flora. They are also found on the bark of trees. A small number inhabit snowfields and are red due to the presence of a high concentration of carotenoid pigments in the cells.

More than 100 species have been recorded from the British Isles, but only the commoner or more interesting ones are described here. Unless rich material is available and careful observations made, accurate identification to species can be difficult. Several British species are described by West (1915, 1916a) and Fritsch & John (1942). Lund (1947b) and Ettl (1965) provide a summary of the variability of major taxonomic characters, including cell morphology.

Members of this cosmopolitan genus are the most frequently observed Volvocales and may be found in practically all water types. Several hundred species have been described and are usually placed into subgenera according to chloroplast morphology and the number of pyrenoids (see Fig. 4A–G). The most common species belong to the subgenus Euchlamydomonas which have a cup-shaped chloroplast and a single basal pyrenoid. Species without a pyrenoid are placed into a separate genus *Chloromonas* Gobi emend Wille by some specialists. Species identification is sometimes difficult due to the variability of the taxonomic characters within individual collections. *Chlamydomonas* is distinguished from *Chlorogonium* by its zoosporogenesis, and from *Sphaerellopsis* by its thin, firm cell wall. A few species of *Chlamydomonas* possess a short acuminate 'tail' and resemble *Phyllocardium*, but in that genus the cells are strongly flattened. The zygotes of some species of *Chlamydomonas* are ornamented (e.g. *C. ehrenbergii*, *C. platystigma*), and may provide another taxonomically important character, but are not considered here as information is lacking for the majority of species.

Members of the colourless genus *Polytoma* Ehrenberg are closely related to *Chlamydomonas*. They are frequently encountered and often abundant in very nutrient-enriched sites. The cells contain starch grains (positive iodine test) and pyrenoids may be present. The species are distinguished by their size and shape, presence/absence of papilla, eyespot and cell wall characters. Cells containing only oil and having a negative starch reaction are referrable to *Tussettia*, a genus which does not appear to have been recorded from the British Isles. Much variability is evident among collections and *Polytoma* requires thorough revision. Probably ubiquitous in organically enriched waters. *Polytoma uvella* Ehrenberg with a small

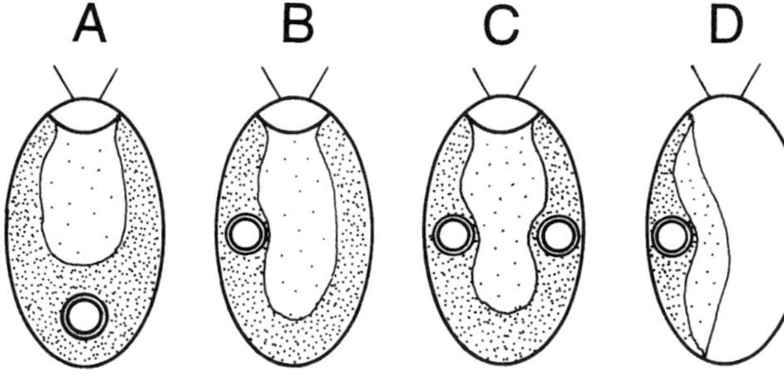

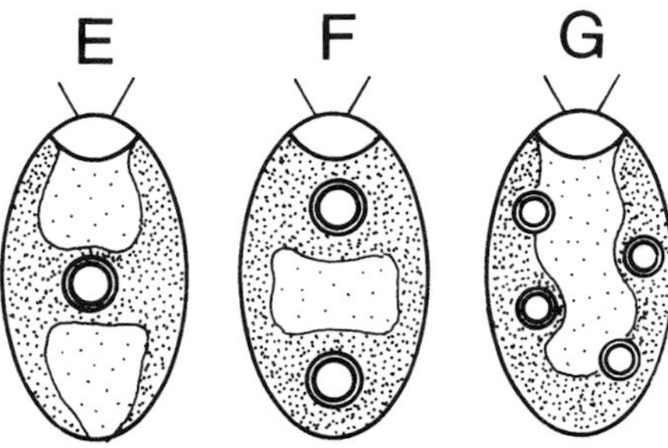

Figure 4. Illustrations of the pyrenoid-containing subgenera of *Chlamydomonas*. **A**. Euchlamydomonas; **B**. Chlamydella; **C**. Bicocca; **D**. Chlorogoniella; **E**. Agloe: with nucleus posterior; **F**. Pseudagloe: with nucleus anterior; **F**. Amphichloris; **G**. Pleiochloris.

hemispherical papilla is supposedly the most frequently recorded species, found in organically enriched ponds. Two other species have been recorded: *Polytoma* cf. *obtusum* Pascher is similar to the above species but lacks an eyespot and papilla, and *P. granuliferum* Lackey is unusual in having cells 14–20 μm in diameter, covered in particles of silt, and not having an eyespot or papilla. Colourless forms are outside the scope of this treatment and the genus is not considered further.

1 Snow algae coloured red with carotenoid pigments (on persistent snow patches in Scotland) .... *C. nivalis*
1 Algae living in other habitats 2
2 Pyrenoids 1 to several 5
2 Pyrenoids absent 3
3 Chloroplast perforated with small net-like openings .... *C. mirabilis*
3 Chloroplast entire, not perforated or divided 4
4 Papilla absent (chloroplast cup-shaped); cells very small .... *C. grovei*
4 Papilla present; broadly wedge-shaped ...... *C. crassa*
5 Pyrenoids 1 6
5 Pyrenoids 2 or more 19
6 Chloroplast cup-shaped 7
6 Chloroplast otherwise 15
7 Pyrenoid lateral (Subgenus Chlamydella) 8
7 Pyrenoid basal (Subgenus Euchlamydomonas) 9
8 Cells spherical; pyrenoid half-ring-shaped .... *C. monadina*
8 Cells ellipsoidal; pyrenoid spherical .... *C. gloeopara*
9 Papilla rounded or subconical 10
9 Papilla absent 13
10 Pyrenoid half-ring-shaped .... *C. braunii*
10 Pyrenoid spherical 11
11 Chloroplast smooth and not striated or perforated; cells almost spherical .... *C. debaryana*
11 Chloroplast minutely longitudinally striated or perforated; cells spherical or shortly cylindrical 12
12 Chloroplast minutely longitudinally striated; cells shortly cylindrical to obovoid .... *C. snowiae*
12 Chloroplast with minute scattered perforations; cells almost spherical .... *C. anticontata*
13 Cells obovoid or pear-shaped and attenuated apically .... *C. ehrenbergii*
13 Cells spherical to weakly ellipsoidal 14
14 Cells 5–8 μm in diameter .... *C. globosa*
14 Cells 8–22 μm in diameter .... *C. reinhardtii*
15 Chloroplast and pyrenoid lateral (Subgenus Chlorogoniella) 16
15 Chloroplast H-shaped or divided into lobes around cell, with a pyrenoid median 18
16 Cells minute (up to 5 μm), inhabiting the empty loricas of *Dinobryon* .... C. *dinobryonis*
16 Cells larger, not growing within other algae 17
17 Chloroplast not divided into lobes; cells without a papilla; species usually an extreme acidophile .... *C. acidophila*
17 Chloroplast divided into lobes; cells with a small papilla; species not an extreme acidophile .... *C. macrostellata*
18 Nucleus anterior (Subgenus Pseudagloe); papilla small .... *C. mutabilis*
18 Nucleus posterior (Subgenus Agloe); papilla absent .... *C. lundii*
19 Pyrenoids more than 2; cells usually >30 μm long, spherical with a small biconical papilla (subgenus Pleiochloris) .... *C. sphagnicola*
19 Pyrenoids 2 per cell; cells usually ≤30 μm long 20
20 Chloroplast cup-shaped, with 2 lateral pyrenoids lying on opposite sides of cell (Subgenus Bicocca) .... *C. platyrhyncha*
20 Chloroplast parietal and of 2 transverse bands in lateral view, each band with a pyrenoid .... *C. pertusa*

Chlamydomonas acidophila Negoro 1944

Synonym: *Chlamydomonas applanata* var. *acidophila* Fott

16180160 **Pl. 77K** (p. 310)

Cells ellipsoidal to obovoid, broadly rounded, 3–5 μm wide and 6–10 μm long, about twice as long as wide, without a papilla; chloroplast lateral, surrounding at least half of cell, with a median and lateral pyrenoid; eyespot anterior.

Japan, Central Europe; in the British Isles recorded from highly acidic mine waters associated with pyrite oxidation (Hargreaves et al., 1975).

Chlamydomonas anticontata J.Schiller 1952

16180170 **Pl. 77A** (p. 310)
Cells broadly ellipsoidal, almost spherical, 13–16 μm wide with a low, broad papilla; chloroplast cup-shaped with minute, irregular perforations, with a basal pyrenoid; eyespot median to apical.

Austria, Czech Republic; a sediment-associated species of the North Wales lakes where over 9×10 cells cm^{-2} were found in nutrient-rich Llyn Maelog (Happey-Wood & Priddle, 1984); also recorded from pools in southern England.

The var. *perforata* Ettl differs in having obovoid cells; recorded from polluted water in the British Isles.

Chlamydomonas braunii Gorozhankin 1890

16180180 **Pl. 77B** (p. 310)
Cells broadly ellipsoidal to nearly spherical, 14–26 μm wide, with a small but prominent papilla; chloroplast cup-shaped, dense, with a large basal pyrenoid in form of a semicircular ring; eyespot apical.

Europe; an occasional species of ponds and lakes, often planktonic, reported from ponds and rivers in south-east England.

Cell shape combined with the semicircular pyrenoid distinguishes this species. The cells should compared with those of *C. monadina*.

Chlamydomonas crassa Christen 1958

Synonym: *Chloromonas crassa* (Christen) Gerloff *et* H.Ettl
16180140 **Pl. 78A** (p. 313)
Cells subspherical, slightly longer than wide, 19–21 μm wide, 24–26 μm long, with a low, broad papilla appearing rounded in vertical view; chloroplast cup-shaped, appearing dense and smooth; eyespot anterior.

Europe (Switzerland); in pools and ponds in south-east England.

Similar to *C. platystigma* but cells are more spherical and the papilla more regular in shape.

Chlamydomonas debaryana Gorozhankin 1891

16180190 **Pl. 77C** (p. 310)
Cells subspherical to ellipsoidal, 8–15 μm wide, 12–22 μm long, 1.2 to 1.5 times as long as wide, with a large subconical papilla; chloroplast cup-shaped, basally thickened with a single large basal pyrenoid; eyespot median or anterior.

Probably cosmopolitan; a frequent species of nutrient-enriched pools and ponds.

Collections often display some variation in the form of the papilla and position of the eyespot. Two varieties have been recorded: var. *atactogama* (Korshikov) Gerloff with ellipsoidal cells and thicker cell wall and var. *micropapillata* Gerloff with smaller cells and papilla.

Chlamydomonas dinobryonis G.M.Smith 1920

16180200 **Pl. 77L** (p. 310)
Cells ellipsoidal or slightly pear-shaped, about 3 μm wide, 2–5 μm long, about 1.5 times as long as wide, papilla absent; chloroplast lateral with a minute pyrenoid; eyespot apical.

Cosmopolitan; probably common but overlooked, recorded from ponds in southern England and from the English Lake District, Cumbria and Scotland; associated with empty thecae of *Dinobryon* spp., often present in large numbers. It is one of the smallest species of *Chlamydomonas*.

Chlamydomonas ehrenbergii Gorozhankin 1891

Synonym: *Chlamydomonas pulvisculus* Ehrenberg *pro parte sensu auct.*
16180050 **Pl. 77D** (p. 310)
Cells pear-shaped, basally rounded and apically attenuated, 10–22 μm wide, 14–26 μm long, without a papilla; protoplast sometimes detached from cell wall; chloroplast cup-shaped, often irregularly thickened with a basal pyrenoid; eyespot median or anterior.

Probably cosmopolitan; found in a wide range of water types, but mainly in nutrient-rich ponds where sometimes common.

The zygote of this species is notable for its granular outer wall.

Chlamydomonas globosa J.Snow 1902

Synonym: *Chlamydomonas reinhardtii* var. *minor* Nygaard, *C. microsphaerella* Pascher *et* Jahoda
16180210 **Pl. 77E** (p. 310)
Cells spherical, 5–8 μm in diameter, without a papilla; chloroplast cup-shaped, basally slightly thickened, with a small basal pyrenoid; eyespot median or anterior.

Probably cosmopolitan with a broad range of habitats; widely distributed in lakes, small ponds and puddles, especially near the surface of sediment; reported from Priddy Pool in Somerset where Happey-Wood (1980) found up to 2.8×10^5 cells cm^{-2} sediment surface and is common in North Wales lakes (Happey-Wood & Priddle, 1984).

Chlamydomonas gloeopara Rodhe *et* Skuja 1948

16180220 **Pl. 77H** (p. 310)
Cells ellipsoidal, 3–9 μm wide, 5–14 μm long, with a small conical papilla; chloroplast cup-shaped, even or slightly irregular, with a small lateral pyrenoid; eyespot apical.

Sweden, Czech Republic; recorded frequently from nutrient-poor lakes and ponds in England and Wales (see Happey-Wood, 1980).

Chlamydomonas grovei G.S.West 1916

Synonym: *Chloromonas grovei* (G.S.West) Gerloff *et* H.Ettl
16180230 **Pl. 78B** (p. 313)
Cells spherical to subspherical, 2–4 μm in diameter, 2.5–5 μm long, slightly pointed apically but without a papilla; chloroplast cup-shaped, smooth; eyespot absent.

Europe; an uncommon species in the British Isles where recorded from the plankton of ponds and lakes.

It is notable for its flagella which are up to 3 times as long as the cell.

Chlamydomonas lundii H. *et* O.Ettl 1959

Synonym: *Chlamydomonas astigmata* J.W.G.Lund *non C. astigmata* Spargo
16180240 **Pl. 77N** (p. 310)

Cells ellipsoidal, often somewhat asymmetrical but basally and apically rounded, 5–9 μm wide, 11–15 μm long, without a papilla; chloroplast parietal, presenting either an H-form or substellate form in lateral view, with a central pyrenoid; eyespot absent.

British Isles; found in soil cultures.

This species belongs to the subgenus Agloe (see Fig. 4E) where the nucleus is situated below the pyrenoid.

Chlamydomonas macrostellata J.W.G.Lund 1947

16180070 **Pl. 77M** (p. 310)

Cells subspherical to ellipsoidal, basally rounded but often slightly asymmetric with one side more convex than the other in lateral view, 8–12(–22) μm wide, 14–19(–22) μm long, with a small hemispherical papilla; chloroplast parietal, divided into 4–6 radial lobes and a single lateral pyrenoid in an approximately median position; eyespot anterior.

Central Europe; an uncommon but widespread species first described from a British soil culture.

Chlamydomonas mirabilis (Korshikov) Pascher 1927

Basionym: *Chloromonas mirabilis* Korshikov

16180250 **Pl. 78C** (p. 313)

Cells ellipsoidal, 8–23 μm wide, 11–27 μm long, about 1.5 times as long as wide, with a truncated papilla; contractile vacuoles present or absent; chloroplast parietal covering most of protoplast, with a net-like pattern of tears or thinnings on surface; eyespot anterior.

Europe (Austria, Russia, Sweden); the var. *minor* (J.H.Belcher) Gerloff *et* H.Ettl, with cells in the lower size range was found by Belcher in large numbers under ice in Priest Pot in the English Lake District.

Net-like aplanospores have been observed.

Chlamydomonas monadina F.Stein 1878

16180080 **Pl. 77I** (p. 310) **CD**

Cells spherical to subspherical, 18–35 μm wide, with a small, apically truncated papilla; chloroplast cup-shaped, often with an annular thickening extending part-way around cell, with a median pyrenoid and usually half-ring shaped; eyespot median or anterior.

Probably cosmopolitan; widely distributed in ponds, lakes and lowland rivers. In Abbot's Pond, Somerset, a small nutrient-rich water body, the species was recorded throughout the year with maximum numbers of 4000 per mL but often underwent rapid decline (Hickman, 1974), also frequent in a nutrient-poor pond known as Priddy Pool in Somerset (Happey-Wood, 1976).

Sexual reproduction anisogamous. Characterized by the form of the papilla and pyrenoid. The cells should be compared with those of *C. braunii*. Swale & Belcher (1964) give a description of the pyrenoid.

Chlamydomonas mutabilis Gerloff 1940

16180260 **Pl. 77O** (p. 310)

Cells ellipsoidal to almost cylindrical, 8–15 μm wide and 17–25 μm long, 2 to 3 times as long as wide, with a small hemispherical papilla; chloroplast H-shaped, often irregular, pyrenoid in centre of bar of H; eyespot and nucleus apical.

Germany; reported from south-west England with nothing known of its ecology, the original plant appearing in enrichment cultures.

The only species belonging to the subgenus Pseudagloe so far reported from the British Isles.

Chlamydomonas nivalis (Bauer) Wille 1903

Basionym: *Uredo nivalis* Bauer

Synonym: *Sphaerella nivalis* (Bauer) Sommerfelt

16180270 **Pl. 77R** (p. 310)

Motile cells pear-shaped or obovoid, basally rounded, 8–20 μm wide, 12–26 μm long, papilla absent; chloroplast cup-shaped, sometimes lobate, with a basal pyrenoid but often obscured by dense red pigmentation (carotenoid pigments); eyespot sometimes present but obscured by pigment; aplanospores (?) common, spherical, blood-red and often with oil reserves, 10–40(–50) μm in diameter.

Cosmopolitan on snowfields; in the British Isles known only from persistent snow patches in the Cairngorm Mountains, Scotland, and usually found as brick-red aplanospores; the motile stage sometimes developing as the snow surface melts in the sun. One of the major components of 'red snow' elsewhere and the base of an unusual ecosystem whose top predators are raptors.

Chlamydomonas pertusa Chodat 1896

16180090 **Pl. 77P** (p. 310) **CD**

Cells ellipsoidal or weakly pear-shaped, basally rounded, 10–14 μm wide, 16–24 μm long, about twice as long as wide, with a small rounded papilla; chloroplast parietal, dense, surrounding most of protoplast with a central thinner region between the 2 pyrenoids, one anterior and the other posterior; eyespot median.

Probably cosmopolitan; an occasional member of the river plankton (e.g. Great Ouse) but no doubt more widespread.

Zygote smooth-walled and 10–14 μm in diameter.

Chlamydomonas platyrhyncha Korshikov in Pascher 1927

16180280 **Pl. 77J** (p. 310)

Cells ellipsoidal, 10–18 μm wide, 17–25 μm long, 1.5 to 2 times as long as wide, with a broad, low and truncated papilla; chloroplast parietal, with an approximately median thickening containing 2 pyrenoids; eyespot anterior; nucleus in lower half of cell.

Germany, Sweden, Ukraine; recorded from ditches in Kent, England, probably more widely distributed.

This species could be confused with the green flagellate *Monomastix opisthostigma* Scherffel (Pl. 78D, p. 313) which also has 2 median and opposite pyrenoids and a parietal chloroplast. The latter is distinguished by having a single flagellum which enters the cell via a weakly depressed and slightly offset apical groove. *Monomastix opisthostigma* has been reported by Belcher & Swale (1951) from temporary pools on limestone pavement at Malham Tarn, Yorkshire, similar pools on Scout Scar, Lancashire, several times in the Kendal Canal at Tewitfield, and in a nitrogenous puddle near Sawrey, Lancashire. The British material is generally smaller than the type and refractive bodies have been rarely observed. Its taxonomic position is uncertain, although it is usually referred to the Prasinophyta.

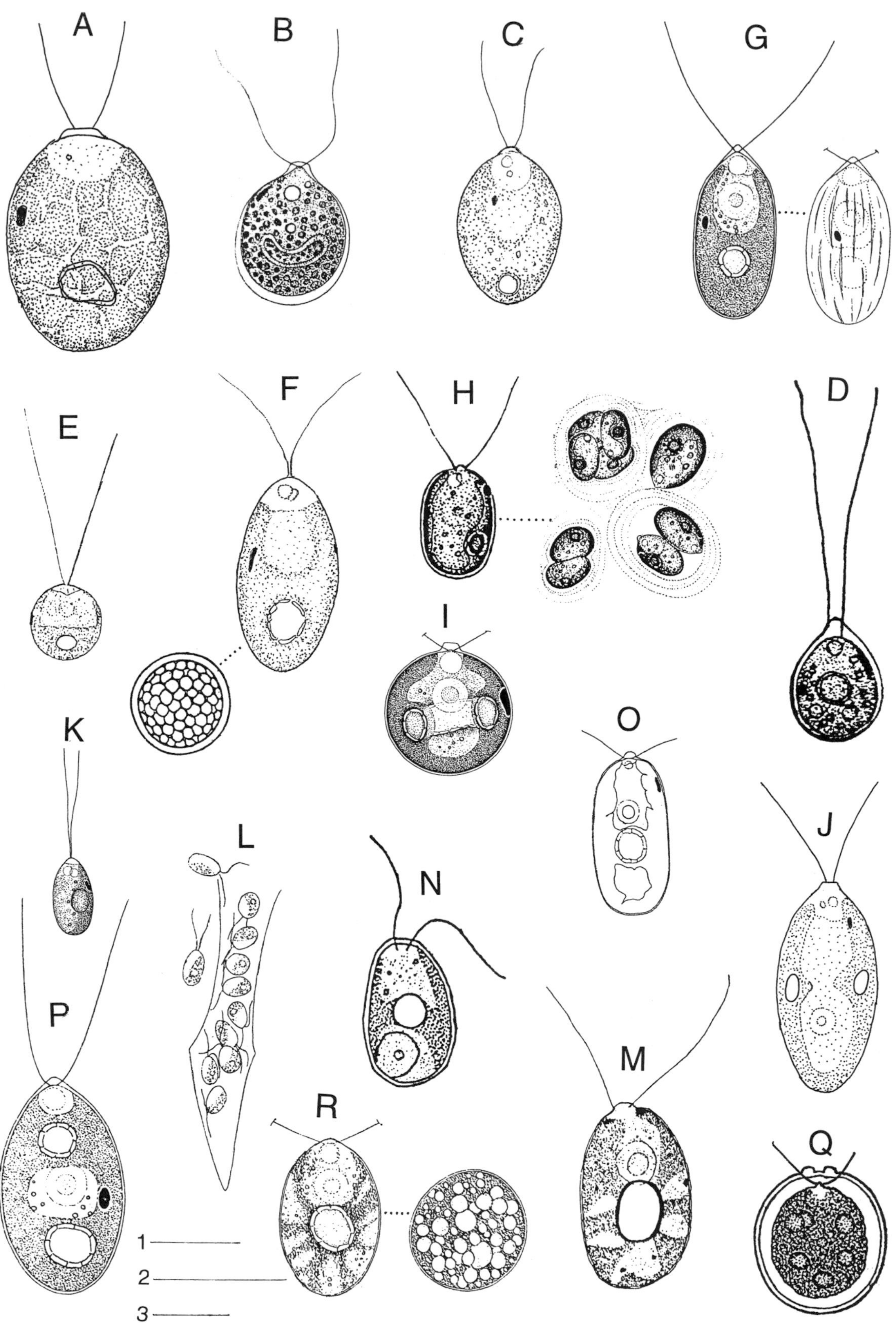

Plate 77 A–R

A. *Chlamydomonas anticontata* (p. 308); **B**. *Chlamydomonas braunii* (p. 308); **C**. *Chlamydomonas debaryana* (p. 308); **D**. *Chlamydomonas ehrenbergii* (p. 308); **E**. *Chlamydomonas globosa* (p. 308); **F**. *Chlamydomonas reinhardtii* (p. 311): zygote; **G**. *Chlamydomonas snowiae* (p. 311): two cells; **H**. *Chlamydomonas gloeopara* (p. 308): with palmelloid stage; **I**. *Chlamydomonas monadina* (p. 309); **J**. *Chlamydomonas platyrhyncha* (p. 309); **K** . *Chlamydomonas acidophila* (p. 307); **L**. *Chlamydomonas dinobryonis* (p. 308): in *Dinobryon* theca; **M**. *Chlamydomonas macrostellata* (p. 309); **N**. *Chlamydomonas lundii* (p. 309); **O**. *Chlamydomonas mutabilis* (p. 309); **P**. *Chlamydomonas pertusa* (p. 309); **Q**. *Chlamydomonas sphagnicola* (p. 311); **R**. *Chlamydomonas nivalis* (p. 309): with aplanospores containing oil bodies.
Scale bars: 10 μm. 1 for B–K; 2 for A, M–R; 3 for L.

Chlamydomonas reinhardtii P.A.Dangeard 1888

16180120 **Pl. 77F** (p. 310)

Cells spherical or subspherical, basally rounded, 8–22 µm wide and 10–22 µm long, papilla absent; chloroplast cup-shaped, thickened basally and containing a large basal pyrenoid; eyespot median or anterior; asexual reproduction by 4 zoospores; sexual reproduction isogamous.

Cosmopolitan; one of the most frequently encountered species found in a wide range of water types but usually in small water bodies subject to nutrient enrichment.

Zygote about 12 µm in diameter with a smooth wall, often reddish.

Chlamydomonas snowiae Printz 1914

Synonyms: *Chlamydomonas communis* J.Snow *non C. communis* Perty, *C. pluristigma* Bristol

16180290 **Pl. 77G** (p. 310)

Cells pear-shaped or slightly cylindrical, rounded basally, 4–18 µm wide, 10–24 µm long, about twice as long as wide, with a small conical papilla; chloroplast cup-shaped, with a series of fine lateral striations and basally thickened with a large basal pyrenoid; eyespot 1 or several, median or anterior.

Europe, Australia, USA; scarce in puddles and pools in the British Isles, usually in soil cultures.

Two varieties have been recorded, var. *pluristigma* (Bristol) Gerloff and var. *palmelloides* J.W.G. Lund. The former has several eyespots and the latter possesses the smaller size range and occurs mostly in the palmelloid stage.

Chlamydomonas sphagnicola (F.E.Fritsch) F.E.Fritsch *et* Takeda 1916

Basionym: *Isococcus sphagnicola* F.E.Fritsch

16180300 **Pl. 77Q** (p. 310)

Cells spherical or weakly ellipsoidal, 15–20 µm wide and 21–35 µm long, with a small biconical papilla and thick-walled; chloroplast parietal, with 2–6 scattered pyrenoids forming small bulges into cytoplasm; eyespot apical.

Europe, Russian Federation, USA; originally described from marshes and nutrient-rich (eutrophic) ponds in south-east England where it was found to be variable with respect to cell wall thickness and the form of the papilla.

Perhaps referable to *Sphaerellopsis* on account of the wall structure.

TAXA RECORDED FROM THE BRITISH ISLES, BUT NOT DESCRIBED IN TEXT

These have not been considered individually due to either their scarcity or lack of detailed information. Many are previously unpublished records.

Chlamydomonas acuta Korshikov 16180310; *C. acutata* Korshikov 16180320; *C. altera* Skuja 16180330; *C. angulosa* Dill 16180010; *C. annulata* Nygaard 16181180; *C. asymmetrica* Korshikov 16180020; *C. botryopara* Rodhe *et* Skuja 16180340; *C. bourrelleyi* H.Ettl 16180350; *C. britannica* J.W.G.Lund 16180360; *C. calcicola* F.E.Fritsch *et* R.P.John 16180370; *C. callosa* Gerloff 16180380; *C. carpatica* Ettl 16181160; *C. cienkowskii* Schmidle 16181170; *C. clathrata* (Korshikov) Pascher 16180390; *C. cingulata* var. *seligeriensis* Korshikov16180033; *C. coccifera* Gorozhankin 16180400; *C. cribrum* H.Ettl 16180410; *C. depauperata* (Pascher) Gerloff *et* H.Ettl 16180420; *C. elegans* G.S.West 16180430; *C. elliptica* var. *britannica* F.E.Fritsch *et* R.P.John 16180442; *C. flos-aquae* var. *minor* H. *et* O.Ettl 16180452; *C. foveolarum* Skuja 16180460; *C. gallica* H.Ettl 16180470; *C gigantea* Dill 16180480; *C. gloeogama* Korshikov 16180490; *C. gordalensis* J.W.G.Lund 16180510; *C. hebes* Pascher 16180530; *C. humiphilos* Gerloff 16180540; *C. incerta* Pascher 16180550; *C. inflexa* Pringsheim 16180560; *C. intermedia* Chodat 16180570; *C. inversa* Pascher 16180580; *C. isogama* Korshikov 16180590; *C. kleinii* Schmidle 16180600; *C. klinobasis* Skuja 16180610; *C. kniepii* F.Moewus 16180620; *C. komarekei* H.Ettl 16180630; *C. korschikoffii* Pascher 16180640; *C. kuwadae* Gerloff 16180650; *C. longistigma* Dill 16180660; *C. latifrons* Nygaard 16180670; *C. macroplastida* J.W.G.Lund 16180680; *C. media* G.A.Klebs 16180690; *C. mexicana* Lewin 16180700; *C. micropapillata* F.Moewus 16180710; *C. microscopica* G.S.West 16180720; *C. microsphaera* Pascher *et* Jahoda 16180730; *C. minuta* Pringsheim 16180750; *C. minutissima* Korshikov 16180760; *C. moewusii* Gerloff 16180770; *C. multitaeniata* Korshikov 16180780; *C. muriella* J.W.G.Lund 16180790; *C. oblongella* J.W.G.Lund 16180800; *C. palatina* Schmidle 16180810; *C. parietaria* Dill 16180820; *C. parvula* Gerloff 16180830; *C. perpusilla* Gerloff 16180840; *C. pertyi* Gorozhankin 16180100; *C. peterfi* Gerloff 16180850; *C. pisum* Pascher 16180860; *C. planktogloea* Skuja 16180870; *C. planoconvexa* J.W.G.Lund 16180880; *C. polypyrenoides* F.Moewus 16180890; *C. primoveris* Skuja 16180900; *C. proboscigera* var. *conferta* (Korshikov) H.Ettl 16180912; *C. proboscigera* var. *minor* H.Ettl 16180913; *C. protracta* Pascher *et* Jahoda 16180920; *C. pseudoelegans* F.E.Fritsch *et* R.P.John 16180930; *C. pseudoelegans* var. *minor* J.W.G.Lund 16180932; *C. pseudogloeogama* Gerloff 16180940; *C. pseudomacrostigma* L.S.Péterfi 16180110; *C. pseudomutabilis* H.Ettl 16180950; *C. pseudotarda* Bourrelly 16180960; *C. pyramidalis* J.W.G.Lund 16180970; *C. quiescens* Skuja 16180980; *C. rapida* Brabez 16180990; *C. reisiglii* H.Ettl 16181000; *C. reticulata* Gorozhankin 16181010; *C. sagittula* Skuja 16181020; *C. similis* Korshikov 16180130; *C. sphagnophila* Pascher 16181030; *C. spinifera* H.Ettl 16181040; *C. subangulosa* F.E.Fritsch *et* R.P.John 16181050; *C. subreticulata* Pascher 16181060; *C. subtilis* Pringsheim 16181070; *C. terricola* Gerloff 16181080; *C. tetras* J.W.G.Lund 16171090; *C. ulvaensis* Lewin 16181100; *C. umbonata* Pascher 16181110; *C. variabilis* P.A.Dangeard 16181120; *C. varians* J.W.G.Lund 16181130; *C. vulgaris* Arachin 16181140; *C. westiana* Pascher 16181150.

Chlamydonephris H. *et* O.Ettl 1959

Cells broadly ellipsoidal to spherical but with a large apical depression from which the 2 flagella arise, papilla absent; chloroplast cup- or band-shaped, with or without a pyrenoid; contractile vacuoles 2 and apical in position.

Probably cosmopolitan; rare.

Similar in general form to *Chlamydomonas*, but with a clear apical depression reminiscent of *Pyramimonas*.

Chlamydonephris pomiformis (Pascher) H. *et* O.Ettl 1959

Basionym: *Chlamydomonas pomiformis* Pascher

16120010 **Pl. 78E** (p. 313)

Cells ellipsoidal to almost spherical apart from apical depression, 10–19 μm wide and 8–14 μm long; chloroplast cup-shaped, surrounding most of cell, with a basal pyrenoid; eyespot small, apical.

Czech Republic, Germany, USA; Woodhead & Tweed (1948) describe the form *submontana* in several upland sites in North Wales. These authors mention that Jane discovered it it in Buckinghamshire, so the species is not restricted to mountainous areas.

Chlorogonium Ehrenberg 1830

Cells spindle-shaped, apically and basally attenuated, (1.5–)4–12(–17) μm wide, (14–)20–80(–170) μm long, 5 to 15(–30) times as long as wide, walls thin, papilla indistinct; flagella 2, normally shorter than cell, apically inserted; contractile vacuoles 2 or many, apical or scattered; chloroplast parietal, single, covering most of cell wall, entire or perforated, outline sometimes indistinct, with or without pyrenoids; eyespot anterior; asexual reproduction by 2–8 zoospores formed by transverse division of parent protoplast; sexual reproduction isogamous or oogamous, zygote spherical; aplanospores and akinetes known.

Probably cosmopolitan; most frequently recorded from small nutrient-poor pools, those rich in humic material (dystrophic) or those containing decaying leaves, often found in the winter but rarely forming blooms.

About 20 species of *Chlorogonium* have been described, most of which lack a pyrenoid. The cells of some species are among the largest of the unicellular Volvocales and when motile they often swim in a graceful, slightly undulating fashion. Slight metaboly of the cells may be apparent. Distinguished from *Chlamydomonas* by the zoosporogenesis, which results solely from transverse divisions of the parent protoplast. The cells are also spindle-shaped and normally more than 4 times as long as wide. Jane (1938) comments on the variability of some species which can lead to identification problems. In the latest UKNCC (2001) list *C. kasakii* Nozaki is listed as isolated from Priest Pot (Cumbria) by Jaworski in 1985. The colourless genus *Hyalogonium* is structurally similar with the cells often containing starch and an eyespot. They are rarely recorded from organically enriched ponds. *H. fusiforme* (Korshikov) H.Ettl has been reported from stagnant ponds.

1 Chloroplasts with 1 to several pyrenoids 3
1 Chloroplasts without pyrenoids.............................2
2 Cells 36–55 μm long ....*C. elegans*
2 Cells 20–45 μm long...........................*C. minimum*
3 Cells with a single pyrenoid 4
3 Cells with more than one pyrenoid 5
4 Cells 12–37 μm long, posteriorly rounded ....*C. fusiforme*
4 Cells up to 33 μm long, posteriorly pointed ....*C. tetragamum*
5 Cells with 2 pyrenoids and 36–55 μm long ....*C. elongatum*
5 Cells with 4 or more pyrenoids and 25–70(–100) μm long..........................*C. euchlorum*

Chlorogonium elegans Playfair 1918

16190030 **Pl. 78F** (p. 313)

Cells spindle-shaped, more attenuated basally, 6–10 μm wide and 36–55 μm long, about 7 times as long as wide; chloroplast covering most of the cell, thin; pyrenoid absent; eyespot anterior; contractile vacuoles 2(–3?), anterior.

Australia, Austria; reported in the British Isles from the plankton of the River Dee, Cheshire, England, probably widespread.

Chlorogonium elongatum (P.A.Dangeard) P.A.Dangeard 1899

16190010 **Pl. 78G** (p. 313)

Cells spindle-shaped, more attenuated basally, 4–10(–17) μm wide, 20–80(–120) μm long, about 10 times as long as wide; chloroplast parietal, covering most of cell wall, sometimes with a median constriction when viewed laterally and corresponding with position of nucleus, with 2 pyrenoids, one anterior and the other basal; eyespot anterior; contractile vacuoles many and scattered.

Frequent in mainland Europe and reported from highly acidic waters in the USA; widely distributed but usually in small numbers in the plankton of ditches, ponds, lakes and nutrient-rich rivers. Jane (1938) frequently found this species in temporary heathland puddles, particularly in spring and autumn. The species tolerates acidic, nutrient-poor and nutrient-rich waters.

Several varieties have been described, but none except the type form reported from the British Isles.

Chlorogonium euchlorum Ehrenberg 1833

16190020 **Pl. 78H** (p. 313)

Cells spindle-shaped, variable in shape, occasionally centrally swollen, 4–15(–20) μm wide, 25–70(–100) μm long, (3–)5 to 15 times as long as wide; chloroplast covering most of cell, thin, sometimes irregular in thickness with 4 or more scattered pyrenoids; eyespot present; contractile vacuoles numerous, scattered.

Probably cosmopolitan; occasionally reported from nutrient-rich rivers and stagnant water among *Sphagnum*. Evans (1958) reported a pH range of 4.4–6.6.

The morphological variability of the species is noteworthy.

Chlorogonium fusiforme Matvienko 1938

16190040 **Pl. 78I** (p. 313)

Cells spindle-shaped, 1.5–4 μm wide, 12–37 μm long, 4 to 8 times as long as wide; chloroplast parietal, enclosing only lateral part of cell with a single, approximately median pyrenoid; eyespot anterior; contractile vacuoles 2 or 3, 2 near flagellum base and other usually basal.

Europe, Russian Federation; recorded from the plankton of Loch Leven, Scotland; on mainland Europe in bogs and associated with *Sphagnum*.

Identical to *Chlorogonium tetragamum* apart from minor morphological differences and in the number of contractile vacuoles.

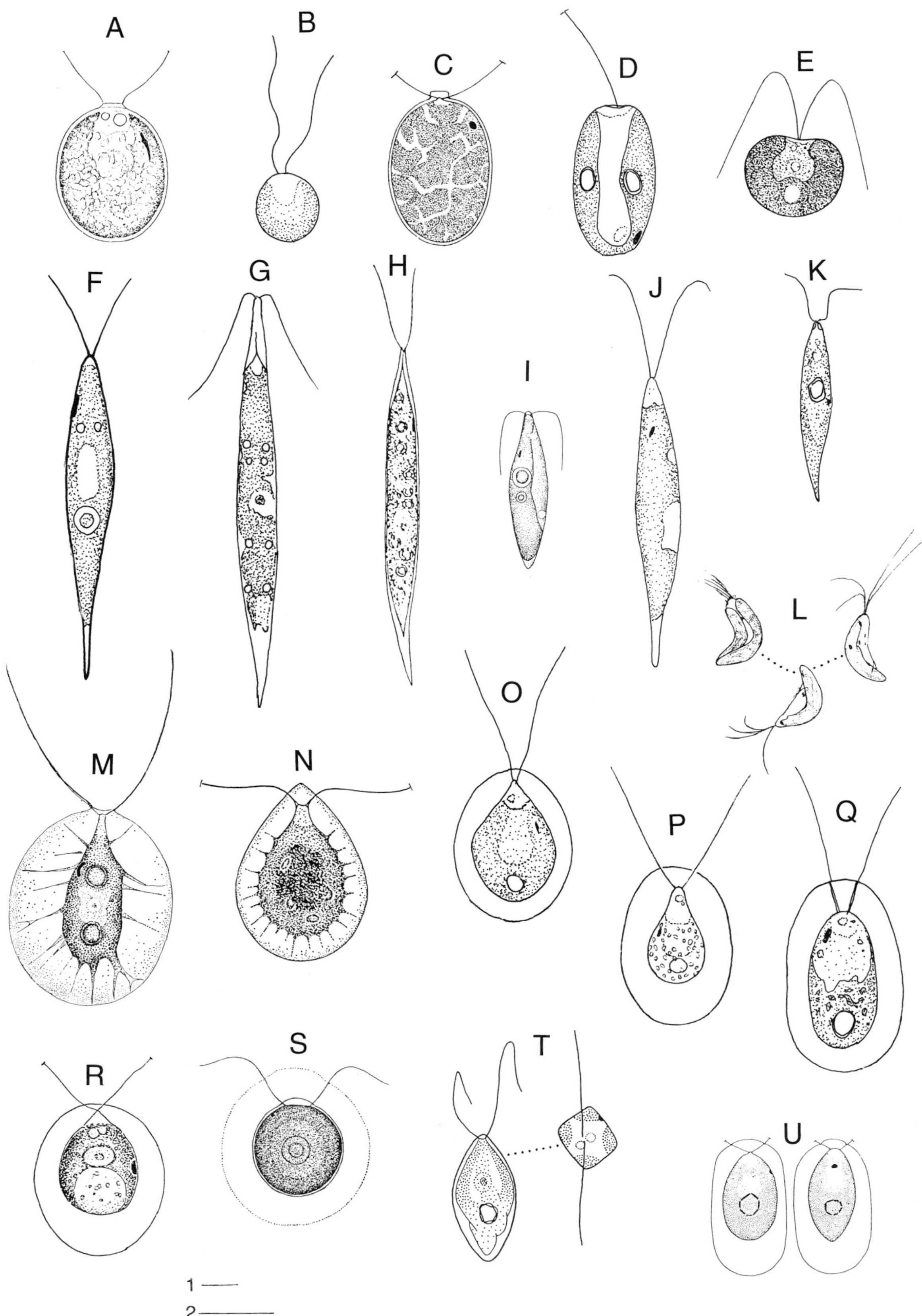

Plate 78 A–U

A. *Chlamydomonas crassa* (p. 308); **B**. *Chlamydomonas grovei* (p. 308); **C**. *Chlamydomonas mirabilis* (p. 309); **D**. *Monomastix opisthostigma* (p. 309); **E**. *Chlamydonephris pomiformis* (p. 312); **F**. *Chlorogonium elegans* (p. 312); **G**. *Chlorogonium elongatum* (p. 312); **H**. *Chlorogonium euchlorum* (p. 312); **I**. *Chlorogonium fusiforme* (p. 312); **J**. *Chlorogonium minimum* (p. 314); **K**. *Chlorogonium tetragamum* (p. 314); **L**. *Spermatozopsis exsultans* (p. 324); **M**. *Haematococcus droebakensis* (p. 318); **N**. *Haematococcus pluvialis* (p. 319); **O**. *Sphaerellopsis fluviatilis* (p. 324); **P**. *Sphaerellopsis gloeocystiformis* (p. 324); **Q**. *Sphaerellopsis lefevrei* (p. 324); **R**. *Sphaerellopsis montana* (p. 324); **S**. *Sphaerellopsis mucosa* (p. 324); **T**. *Sphaerellopsis spiralis* (p. 326); **U**. *Sphaerellopsis velata* (p. 326): two cells. Scale bars: 10 μm. 1 for M; 2 for A–L, N–U.

Chlorogonium minimum Playfair 1918

16190050 **Pl. 78J** (p. 313)
Cells needle-shaped, 1.5–4 μm wide, 20–30(–45) μm long, 10 or more times as long as wide; chloroplast parietal, sometimes enclosing only lateral part of cell, with no pyrenoid; eyespot anterior; contractile vacuoles 2, apical.

Australia, northern Europe; sometimes common in heathland pools in the south of England, also from a nutrient-rich lake in Cheshire.

Chlorogonium tetragamum Bohlin 1897

Synonym: *Chlamydomonas tetragama* (Bohlin) H.Ettl
16190060 **Pl. 78K** (p. 313)
Cells short, spindle-shaped, up to 33 μm long, 4 to 6 times as long as wide, basally attenuated with a minute flattened papilla; chloroplast band-shaped, surrounding most of the protoplast, with a median pyrenoid and eyespot; contractile vacuoles 2 per cell.

Europe, USA; frequent in English river plankton.

J.H. Belcher (pers. comm.) has observed up to 3 pyrenoids in some cells.

DOUBTFUL SPECIES
Chloromonas cryophila Hoham *et* Mullet 1977
This species is reported from snow by Light & Belcher (1968) as *Scotiella nivalis/cryophila* (see p. 399). It has an elongate-ellipsoidal cell with a small tail and papilla, but no pyrenoid. The species should probably be transferred to *Chlamydomonas*.

Coccomonas F.Stein 1878

Cell enclosed within a broad, pale to yellow-brown envelope encrusted with calcium carbonate, with envelope up to about twice the cell diameter and touching protoplast only where flagella pass through a circular opening in envelope; envelope ellipsoidal, slightly angled or box-shaped, often flattened laterally and sometimes indented at the flagellar insertion, about 10–20 μm wide and 15–30(–50) μm long; protoplast pear-shaped with 2 apical contractile vacuoles; flagella as long as or longer than envelope; chloroplast cup-shaped, with a basal pyrenoid, with or without an eyespot; asexual reproduction by zoospores or simple cleavage.

Probably cosmopolitan; mostly in small lowland calcareous ponds. Seven species are known, two of which have been found in the British Isles.

On the Continent, some lakes contain substantial deposits of marl full of *Coccomonas* envelopes (Müller & Oti, 1981). The pattern of calcification and the envelope shape is considered important in determining the species but this appears to be variable. Cross-polarized light is useful in detecting cells of *Coccomonas*, and also *Phacotus*. Because of their brownish colouration, *Coccomonas* could be mistaken for *Trachelomonas*, but the envelopes of the latter are uncalcified and the cells possess a single flagellum.

1 Envelope slightly flattened laterally, less than twice as long as wide; with or without a slight apical indentation................. *C. orbicularis*
1 Envelope strongly laterally flattened, twice or more times longer than wide, sometimes with an apical indentation................ *C. platyformis*

Coccomonas orbicularis F.Stein 1878

Synonym: *Coccomonas lunzensis* Huber-Pestalozzi
16210010 **Pl. 79R** (p. 316)
Envelope ellipsoidal, sometimes slightly irregular, often flattened laterally, rounded or slightly indented apically, surface finely granular, encrusted with calcite and possibly other minerals, pale to yellow-brown, 16–18 μm wide, 18–25 μm long, 14–15 μm thick; protoplast weakly ellipsoidal to pear-shaped, dimensions about half that of envelope; chloroplast cup-shaped, with a basal pyrenoid and anterior eyespot.

Probably cosmopolitan; the best known species, widely distributed in calcareous waters of Continental Europe; only reported in the British Isles from a few ponds in southern England where usually associated with sediments.

Coccomonas platyformis Jane 1944

16210020 **Pl. 79S** (p. 316)
Envelope rounded, 12–20 μm wide, 16–20 μm long, 6–8 μm thick, at least twice as long as wide in lateral view, surface finely granular and encrusted with calcite, pale to yellow-brown, strongly indented at apex; protoplast dimensions about half that of envelope; chloroplast cup-shaped, with a basal pyrenoid; eyespot lateral.

Europe (Czech Republic); in the British Isles known only from calcareous dune pools at Newborough Warren in Gywnedd, Wales, the type locality.

The cells are slightly smaller than those of *C. orbicularis*, but there is no mention of calcification in the original description. Perhaps a growth form of *C. orbicularis*.

Diplostauron Korshikov 1925

Synonym: *Lobomonas* P.A.Dangeard *pro parte sensu* Hazen
Cells angular with 7–8 protrusions in 2 sets of 4 (one set anterior, the other posterior); cell apex rounded, conical or depressed, with or without a papilla, approximately equidimensional or a little longer than wide; protoplast not always extending into angles of protrusions; flagella as long as or longer than cell, inserted apically; contractile vacuoles 2, apical; chloroplast parietal, cup-shaped, with pyrenoid usually present but position variable; eyespot anterior; asexual reproduction by 2–8 zoospores per cell; sexual reproduction isogamous with spherical zygospores.

Cosmopolitan; uncommon and in small puddles or pools and occasionally lakes. Five of the nine species described have been recorded from the British Isles.

Similar in some respects to *Brachiomonas* in having regularly arranged lobes, but in *Diplostauron* there are 8 rather than 4 lobes. Linked to *Lobomonas* through *D. panduriforme* where the regularity of the ornamentation appears to be breaking down. Some of the British species are described in detail by Belcher & Swale (1961) and Belcher (1965a).

1 Cells regularly shaped, quadrate or octagonal -star-shaped in vertical view 3
1 Cells irregularly shaped, cross-like or polygonal in vertical view 2
2 Cells in vertical view with large rounded lobes, deep intervening depressions about one-third of cell width, irregular to cross-shaped in outline ....*D. elephas*
2 Cells in vertical view with small protrusions, depressions between protrusions slight and much less than one-third cell width ....*D. panduriforme*
3 Cell protuberances not vertically aligned and appearing as an 8-pointed star in vertical view 4
3 Cell protuberances vertically aligned and appearing quadrate in vertical view ....*D. platyrhynchum*
4 Cells laterally with concave lateral margins, protuberances acute................................*D. elegans*
4 Cell laterally with almost linear lateral margins, protuberances quadrate or obtuse ....*D. pentagonium*

Diplostauron elegans Skuja 1927

16230010 **Pl. 79E** (p. 316)

Cells 10–14 μm wide, 12–16 μm long, with 8 moderately acuminate protrusions, approximately quadrate laterally, sides and posterior concave between spines; apex broadly conical, with a small papilla; spines in vertical view in two sets, an anterior and a posterior set, twisted out of plane by about 45° and appearing as an 8-pointed star; protoplast not filling entire spine, lobate; chloroplast cup-shaped, with pyrenoid basal; eyespot anterior.

Russia, USA; recorded in the British Isles from Crosemere, a mildly nutrient-rich lake in Shropshire, and in marshy puddles.

Diplostauron elephas J.H.Belcher 1965

16230020 **Pl. 79F** (p. 316)

Cells about 15–25 μm long, with 7 or 8 rounded but elongate lobes in two sets (one anterior and other posterior), with a deep concave depression between each lobe; papilla absent; lobes in vertical view are twisted out of plane and irregular both in length and position; protoplast approximately spherical or angular, not entering lobe extremities, 6–15 μm in diameter; chloroplast cup-shaped, with pyrenoid basal; eyespot anterior.

British Isles (England); described from limestone solution-hollows in the north of England (Belcher, 1965a), sometimes associated with *Stephanosphaera pluvialis.*

Diplostauron panduriforme J.H.Belcher 1965

16230030 **Pl. 79G** (p. 316)

Cells 5–8 μm wide, 7–12 μm long, about 1.5 times as long as wide laterally, lateral view showing 2 sets of weak, broadly rounded protrusions and weak concave depressions in between them, cell posterior with small rounded lobe; apex broadly rounded with a wedge-shaped papilla depressed centrally; protoplast filling most of cell; chloroplast parietal, with pyrenoid lateral; eyespot anterior.

British Isles (England); described from a puddle near Priest Pot, English Lake District (Belcher, 1965a).

Similar to *D. haccardii* Bourrelly and some species of *Lobomonas*, but the protuberances are more regularly arranged.

Diplostauron pentagonium (Hazen) Pascher 1927

Basionym: *Lobomonas pentagonia* Hazen

16230040 **Pl. 79H** (p. 316)

Cells 9–10 μm wide, 10–13 μm long, laterally pentagonal, about 1.5 times as long as wide, with parallel or slightly diverging sides at anterior end, apex broadly conical and base flat or slightly recurved, with 8 'corners' forming a regular octagonal pattern in vertical view, the 4 anterior corners alternating with the 4 posterior ones; papilla absent; protoplast ellipsoidal or angular and entering the 8 protrusions; chloroplast cup-shaped or indistinct and parietal, with pyrenoid basal or lateral; eyespot anterior.

Europe, North America; known in the British Isles from a bog pool in northern England and from Ham Common, Surrey, but probably more widely distributed. In a range of water types on the continent of Europe but uncommon.

Similar to *D. elegans*, but lacking the pronounced acute protuberances of that species.

Diplostauron platyrhynchum J.H.Belcher 1965

16230050 **Pl. 79I** (p. 316)

Cells approximately quadrate laterally, 4–11 μm wide, 6–11 μm long, with pronounced and slightly attenuated quadrate posterior corners and more rounded anterior corners, sides parallel, straight or slightly incurved with apex broadly conical to almost flat with a wide, truncated papilla, quadrate in vertical view with sides weakly concave and corners rounded; protoplast filling most of cell; chloroplast cup-shaped, with an approximately basal pyrenoid; eyespot anterior.

British Isles (England); described from Crosemere, a calcareous and mildly nutrient-rich lake in Shropshire (Belcher, 1965a).

The species has a distinctive box-like form so that in vertical view the basal corners are hidden by the anterior corners.

Dysmorphococcus Takeda 1916

Synonyms: *Chlamydomonas pro parte sensu* Pascher, *Thorakomonas pro parte sensu* Bourrelley, *Chlorothorakus* H.Ettl

Cells with envelope spherical or ellipsoidal, often slightly flattened with an apical papilla-like protrusion, brownish but uncalcified, smooth or granular, flagella entering via 2 apical openings and as long as or longer than envelope; protoplast spherical, ellipsoidal or pear-shaped; chloroplast cup-shaped, with pyrenoids 1 to

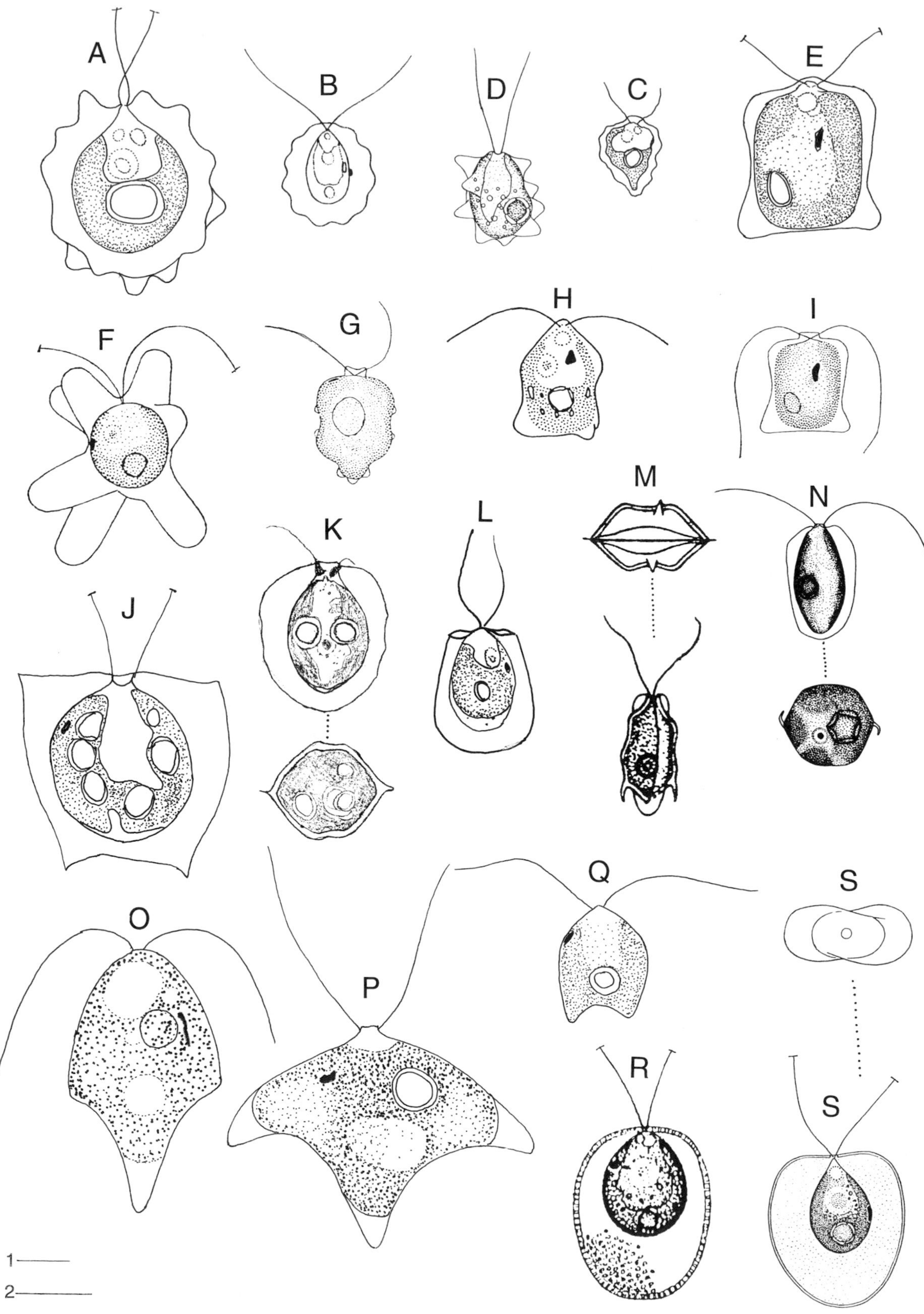

Plate 79 A–S

A. *Lobomonas ampla* var. *okensis* (p. 320); **B**. *Lobomnas ampla* (p. 320); **C**. *Lobomonas francei* (p. 320); **D**. *Lobomonas rostrata* (p. 320); **E**. *Diplostauron elegans* (p. 315); **F**. *Diplostauron elephas* (p. 315); **G**. *Diplostauron panduriforme* (p. 315); **H**. *Diplostauron pentagonium* (p. 315); **I**. *Diplostauron platyrhynchum* (p. 315); **J**. *Pteromonas aculeata* (p. 322); **K**. *Pteromonas aequiciliata* (p. 322): lateral and vertical views; **L**. *Pteromonas angulosa* (p. 323); **M**. *Pteromonas spinosa* (p. 323): lateral and vertical views; **N**. *Pteromonas tenuis* (p. 323): lateral and vertical views; **O**. *Brachiomonas simplex* (p. 304); **P**. *Brachiomonas submarina* (p. 304); **Q**. *Furcilla stigmatophora* (p. 318); **R**. *Coccomonas orbicularis* (p. 314); **S**. *Coccomonas platyformis* (p. 314): lateral and vertical views.
Scale bars: 10 μm. 1 for B; 2 for A, C–S.

several; eyespot lateral or apical; contractile vacuoles apical, 2 to several; asexual reproduction by 2–4 zoospores.

Asia, Europe, USA; rare and found mainly in rain pools, bogs or organically enriched waters.

Species distinguished mainly on form and ornamentation of the envelope and number of pyrenoids. They should be compared with *Trachelomonas* which has a single strong flagellum. The genus differs from *Coccomonas* by lack of calcification and the double pore at anterior end of the envelope, although this can be difficult to see.

1 Chloroplast with numerous small, scattered pyrenoids and many contractile vacuoles ....*D. coccifer*

1 Chloroplast with a single basal pyrenoid and 2 apical contractile vacuoles................*D. variabilis*

Dysmorphococcus coccifer Korshikov 1925

16250020 **Pl. 80A** (p. 321)

Cell envelope flattened and ovoid in lateral view, often finely punctate, deep brown to black, about 22 μm in diameter; protoplast spherical; chloroplast cup-shaped, dense, with pyrenoids small and numerous; eyespot lateral; contractile vacuoles numerous, scattered.

Czech Republic, Ukraine; an occasional species of small water bodies in the north of England; it has usually been found in organically enriched sites.

Dysmorphococcus variabilis Takeda 1916

Synonym: *Dysmorphococcus fritschii* Takeda *sensu* Pascher

16250010 **Pl. 80B** (p. 321)

Cell envelope 10–16 μm wide and 14–20 μm long, spherical to subspherical, slightly flattened laterally, sometimes with a slight apical protrusion, yellow-brown, smooth or finely granular; flagella longer than envelope, inserted about 2 μm apart; protoplast slightly more than half the length of envelope, pear-shaped; chloroplast cup-shaped, with pyrenoid basal; eyespot apical; contractile vacuoles 2, apical.

Russia, USA; the best known species but recorded infrequently in the British Isles from stagnant ponds and in the plankton of the River Dee, Cheshire.

Eudorina Ehrenberg 1830

Synonym: *Pleodorina* Shaw

Coenobia spherical or ellipsoidal, 8-, 16-, 32–64-celled and cells more or less regularly arranged on periphery of a hollow mucilaginous envelope, often with an anterior and posterior group of 4 cells, (50–)60–180(–200) μm in diameter; cells (5–)8–20(–24) μm in diameter, size often varying within coenobium, subspherical to weakly pear-shaped, walls thin; papilla absent, fine protoplasmic strands sometimes interconnecting cells; flagella longer than cells, visible passing through canals in mucilage; contractile vacuoles 2 and apical; chloroplast cup-shaped or irregular, sometimes dense, with 1 to several pyrenoids; eyespot present; asexual reproduction by daughter cells which form within parent coenobium; sexual reproduction oogamous or anisogamous, coenobia homo- or heterothallic, thus similar to *Volvox*.

Cosmopolitan; variously sized standing bodies of water and slow-flowing rivers. Only one species is currently recognized from the British Isles.

Several species have been distinguished by cell size, cell arrangement and pyrenoid number. There is considerable variability within individual populations. *Eudorina* is distinguished from *Volvulina* by the ellipsoidal to spherical coenobium and regular arrangement of cells. The genus *Pleodorina*, recognized by some algologists, is distinguished by having the cells differentiated into vegetative and reproductive types. *Pleodorina illinoisensis* Kofoid has been recorded from the British Isles (Grove, 1915), but is nowadays regarded as a form of *Eudorina elegans*. In *Pyrobotrys* the cells are contiguous and not clearly separated by mucilage and the coenobia are smaller than those of *Eudorina*. Young coenobia of *Eudorina* may appear similar to *Pandorina*.

Eudorina elegans Ehrenberg 1831

Synonyms: *Eudorina stagnale* Wolle, *E. pluricocca* G.M.Smith

16260010 **Pl. 81G** (p. 325)

Coenobia spherical or ellipsoidal, 60–200 μm in size, 4-, 8-, 16-, 32–64-celled; cells 12–24 μm in diameter, spherical or slightly pear-shaped, regularly arranged at periphery, often in tier-like rows; chloroplast cup-shaped, with up to 5 pyrenoids per cell; eyespot anterior.

Cosmopolitan; widespread in the British Isles in lakes, ponds, puddles, ditches and slow-flowing rivers, particularly frequent on clay or calcareous soils associated with nutrient enrichment; normally in spring or autumn when blooms with 3000 cells per mL have been reported.

The most commonly observed member of the colonial Volvocales.

Furcilla A.Stokes 1890

Cells in lateral view lunate and winged, with a broad posterior depression, often shorter than wide, anterior broadly rounded to weakly acuminate and terminating in a small papilla, strongly flattened in side view and usually broader in mid-region, somewhat asymmetrical and about twice as long as wide; flagella longer than cells; chloroplast parietal, often pale, thin and more or less cup-shaped or absent, with pyrenoids and eyespot present or absent; contractile vacuoles 2, apical; asexual reproduction by 2–8 zoospores per cell; sexual reproduction isogamous.

Europe, Russia; found in small puddles, typically humic-rich (dystrophic) water.

The genus is morphologically close to *Brachiomonas*, but the two 'wings' lie in the same plane and the cells are distinctly flattened. Distinguished from *Pteromonas* by the absence of a continuous hyaline wing surrounding the cell and the broad basal depression. In some publications the name is misspelt *Furcilia*. The colourless flagellate *Aulacomonas* Skuja is similar in some respects, being flattened, but in polar view it is lunate or winged due to the presence of a vertical furrow (see Belcher & Swale, 2000). Unlike *Furcilla*, the cells of *Aulacomonas* can extend into pseudopodia and it occurs rarely in organically enriched sites.

Furcilla stigmatophora Korshikov 1939

Synonyms: *Selenochloris stigmatophora* (Korshikov) H.Ettl, *Pseudofurcilla lobosa* (Skuja) Jane

16270010 **Pl. 79Q** (p. 316)

Cells in lateral view flattened with posterior depression ranging from deep to shallow so cells often winged, 8–14 μm wide and 10–18 μm long, weakly conical at apex with flagella closely inserted into a minute papilla; chloroplast thin, occasionally absent so cells either colourless or pale green, with an anterior eyespot and a pyrenoid sometimes present.

Brazil, Ukraine; found in small, organically enriched pools and other waters.

Described from a vase containing *Aesculus* stems as *Pseudofurcilla lobosa* by Jane (1944a) and more fully investigated by Belcher (1968b) from rain-filled rock hollows polluted by bird excreta where it was associated with *Polytoma* and *Stephanosphaera*.

Gonium O.F.Müller 1773

Synonym: *Tetragonium* West *et* G.S.West

Coenobia 4-, 8-, 16–32-celled, with cells arranged in a regular pattern and held together by a hyaline plate-like area of mucilage, flagella directed outwards in same or different directions from each other; cells spherical, ellipsoidal or pear-shaped, with or without a papilla; flagella at least twice as long as cells, passing though mucilage by a narrow canal; chloroplast cup-shaped, with 1 to several pyrenoids; contractile vacuoles 2, apical; asexual reproduction by daughter coenobia; sexual reproduction iso- or anisogamous.

Cosmopolitan; in variously sized standing bodies of water and slow-flowing rivers. Two species are currently recognized in the British Isles.

The several species are distinguished by the number of cells per coenobium, cell arrangement and shape, presence or absence of papilla and pyrenoids.

1 Coenobium of 4 cells, each cell pointing in same direction; cells with a prominent papilla ....*G. sociale*

1 Coenobium 8-, 16- or 32-celled, flagella directed radially; cells without a papilla ....*G. pectorale*

Gonium pectorale O.F.Müller 1773

16330010 **Pl. 80Q** (p. 321) **CD**

Coenobium a slightly curved plate, 70–100 μm wide; usually of 16 cells, 12 at periphery with radially directed flagella and remainder in centre, cells usually weakly pear-shaped or spherical, up to 18 μm wide, 15–20 μm long; chloroplast cup-shaped, with a basal pyrenoid and an anterior eyespot.

Cosmopolitan; widely distributed and sometimes common in nutrient-rich lakes, ponds, ditches and slow-flowing rivers; pH range 4.6–8.4; usually found in spring.

Gonium sociale (Dujardin) Warming 1876

Basionym: *Cryptomonas socialis* Dujardin

Synonyms: *Gonium tetras* A.Braun, *Tetragonium lacustre* West *et* G.S.West

16330020 **Pl. 80R** (p. 321) **CD**

Coenobium of 4 cells and 20–50 μm in diameter, with cells cruciately arranged and flagella pointing in same direction; cells pear-shaped, basally rounded, 6–16 μm wide, 10–22 μm long, with a prominent papilla sometimes displaying a median cleft; chloroplast cup-shaped, with a basal pyrenoid and median or anterior eyespot.

Cosmopolitan; frequent in moderately rich and very nutrient-rich lakes, occasionally forming blooms with up to 3000 coenobia per mL, also in ditches, small ponds and slow-flowing rivers; occurring throughout the year.

The colonies should be compared with those of *Basichlamys*.

Haematococcus C.Agardh emend. Flotow 1844

Synonyms: *Sphaerella* Sommerfelt *pro parte*, *Chlamydococcus* A.Braun *pro parte*, *Hysginum* Perty *sensu auct.*

Cells ellipsoidal to subspherical, with a broad mucilaginous hyaline wall, with or without a papilla; flagella closely inserted into cell; protoplast spherical to pear-shaped, sometimes irregular owing to often numerous protoplasmic extensions permeating the broad cell wall, extensions radial, fine and unbranched to coarse and dichotomously branched; chloroplast parietal, sometimes cup-shaped, with 1 to several pyrenoids and an anterior eyespot; contractile vacuoles often numerous, thoughout cytoplasm; cytoplasm usually containing an abundance of a red carotenoid (astaxanthin) obscuring contents; asexual reproduction by 4–8 zoospores per cell; aplanospores blood-red, common; sexual reproduction isogamous, zygote smooth and thick-walled.

Cosmopolitan; common in many types of water body and slow-flowing rivers. Two of the five known species have been recorded in the British Isles.

They are noteworthy for their blood-red pigmentation and cell structure.

1 Cells with 2 pyrenoids and protoplasmic extensions about 2 μm wide at their base, usually branched distally...............*H. droebakensis*

1 Cells with several pyrenoids and extremely fine protoplasmic extensions ....*H. pluvialis*

Haematococcus droebakensis Wollenweber 1908

16350010 **Pl. 78M** (p. 313) **CD**

Cells 34 μm wide, up to 70 μm long, subspherical to slightly longer than wide, sometimes with a pronounced bidentate papilla, flagella at least as long as cell, closely inserted; protoplast subspherical, with 6 to many radiating protoplasmic extensions in lateral view, usually branching distally; chloroplast parietal, sometimes partly entering protoplasmic extensions, hollow central region containing nucleus, with pyrenoids usually 2, one anterior and one posterior; eyespot anterior; contractile vacuoles throughout cytoplasm.

Europe (Scandinavia, Hungary); an uncommon species in small pools and slow-flowing, nutrient-rich rivers in southern England.

The var. *fastigiatus* Wollenweber (16350012) with its basally attenuated pear-shaped cells has been recorded by Belcher & Swale (1963).

Haematococcus pluvialis Flotow emend. Wille 1903

Synonyms: *Sphaerella pluvialis* (Flotow) Wittrock, *S. lacustris* (Girod) Wittrock, *Haematococcus lacustris* (Girod) Rostafinski

16350020 **Pl. 78N** (p. 313) **CD**

Cells spherical to ellipsoidal, (8–)10–30(–51) µm wide, (10–)15–50(–63) µm long, up to 1.5 times as long as broad, papilla absent; protoplasmic extensions extemely fine so that cytoplasmic detail is not visible within them, usually unbranched; chloroplast parietal, sometimes cup-shaped with several scattered pyrenoids; eyespot anterior; contractile vacuoles numerous, scattered; cytoplasm becomes crimson due to carotenoids developing under conditions of moderate nutrient limitation; aplanospores 30–50 µm in diameter, forming brick red crusts on rocks when partially or completely dry.

Cosmopolitan; a common and widely distributed species in small rock pools, cattle troughs, ponds and ornamental birdbaths, especially in harder waters, including pools enriched with zinc where the red colour is especially pronounced; densities up to 230,000 cells per mL have been reported from small shallow water bodies.

The organism is grown commercially in other countries for astaxanthin and other carotenoids.

Hafniomonas H.Ettl *et* Moestrup 1980

Synonym: *Pyramimonas* Schmarda 1850 *pro parte sensu auct.*

Cells in lateral view cylindrical, (6–)14–18 µm wide, (8–)12–25(–35) µm long; about twice as long as wide, basally rounded but with an anterior depression from which 4 flagella arise; polar view quadrate with rounded corners; walls thin sometimes with slight metaboly; chloroplast parietal, cup-shaped, entire or perforated, pyrenoid usually present; eyespot median or anterior; contractile vacuoles 2, apical.

Probably cosmopolitan; usually standing bodies of water.

The 10 species show a strong resemblance to *Pyramimonas* (Prasinophyta) which has naked cells clothed in minute scales. The two genera cannot normally be distinguished with certainty using the light microscope.

Hafniomonas reticulata (Korshikov) H.Ettl *et* Moestrup 1980

Basionym: *Pyramimonas reticulata* Korshikov

16360010 **Pl. 80O** (p. 321)

Cells cylindrical or broadening apically, with linear margins in lateral view, 6–15 µm wide, 10–25 µm long; chloroplast net-like, covering most of cell, with a large basal pyrenoid; eyespot median or anterior.

Central Europe, Ukraine; found in flooded cart tracks in Cumbria; widely distributed in small pools on the Continent.

A further species, *H. botryodes* (Jane) H.Ettl, has been described by Jane (1943) from Hertfordshire but its position is uncertain since it has not been investigated with the electron microscope. It possessed a nodular and cylindrical cell wall, with cells 9–11 µm wide and 12–23 µm long.

Hemitoma Skuja 1939

Cells spherical to weakly ellipsoidal, with a thick, pale yellow or brown and minutely net-like envelope (about 1 µm wide) in 2 hemispherical sections joining midway along cell when seen in lateral view; flagella about twice as long as cell, entering envelope by two minute pores; chloroplast cup-shaped, with an anterior eyespot and lacking pyrenoid; contractile vacuoles 2, apical.

Europe; usually in standing water. Monospecific.

Hemitoma meandrocystis Skuja 1939

16370010 **Pl. 80D** (p. 321)

Cells 13–15 µm in diameter.

Germany, Sweden; recorded from Priest Pot in the English Lake District and also from Cheshire; thought to be a cold-water species.

Resembles *Trachelomonas*, but has 2 flagella and a divided envelope. Autospores may be formed within the envelope.

Lobomonas P.A.Dangeard 1898

Synonym: *Tylomonas* Korshikov

Cells subspherical or ellipsoidal, more or less equidimensional in polar view, lobed or angular; papilla present or absent; wall usually broad and up to half diameter of protoplast; flagella as long as or longer than cell; protoplast subspherical to pear-shaped; chloroplast cup-shaped, with pyrenoid and eyespot present or absent; contractile vacuoles 2 to many; asexual reproduction by 2–8 zoospores per cell.

Probably cosmopolitan; in variously sized bodies of standing water and slow-flowing rivers.

The species are morphologically similar to *Diplostauron*, but the wall processes of *Lobomonas* are more numerous and not arranged in 2 tiers of 4 (but cf. *D. panduriforme*). Considerable variation may be encountered within collections. A dozen species known from Europe and distinguished on presence/absence of a papilla and pyrenoid, number of contractile vacuoles and wall morphology. The three British species are widely distributed on the Continent. Two have been described from Britian by Belcher & Swale (1961) and Belcher (1965a).

1 Cells ≤15 µm long, papilla present or absent, in lateral view with 4–20 shallow or deep indentations 2

1 Cells >15 µm long, papilla absent, in lateral view with 7–12 shallow indentations and broadly rounded protuberances.................*L. ampla*

2 Papilla small and broad, in lateral view with 10–20 shallow indentations or small cusp-like protuberances in side view .... *L. rostrata*
2 Papilla absent, with 4–10 strong indentations in side view .... *L. francei*

Lobomonas ampla Pascher 1927

16420010 **Pl. 79B** (p. 316)
Cells subspherical, 15–22(–39) µm in diameter, slightly longer than wide, with a thick hyaline wall about half as wide as protoplast; protoplast usually pear-shaped; chloroplast cup-shaped, with pyrenoid basal; eyespot median or anterior.

Europe (Austria, Czech Republic); a widely distributed but uncommon and sporadic species of puddles, pools and slow-flowing rivers.

Var. *okensis* Korshikov differs from the type in having more pronounced cell wall nodules and an apically attenuated protoplast; occasionally associated with the type (Pl. 79A, p. 316).

Lobomonas francei P.A.Dangeard 1898

16420040 **Pl. 79C** (p. 316)
Cells obovoid to ellipsoidal, 8–10 µm wide and 10–12 µm long, with 4–10 strong, rounded or conical protrusions in side view, papilla absent; protoplast obovoid or ellipsoidal; chloroplast cup-shaped, with pyrenoid basal; eyespot small, anterior.

France, Romania, Russian Federation; an uncommon species reported from Sussex in nutrient-rich puddles and in northern England.

Lobomonas rostrata Hazen 1922

Synonyms: *Lobomonas denticulata* Korshikov, *L. piriformis* Pringsheim, *Tylomonas irregularis* Korshikov
16420030 **Pl. 79D** (p. 316)
Cells subspherical to pear-shaped in lateral view, 4–8 µm wide and 5–12 µm long, papilla low and broad with minute cusp-like indentations fairly regularly disposed around wall; protoplast subspherical or pear-shaped; chloroplast cup-shaped, with pyrenoid basal; eyespot apical.

Scandinavia, Ukraine, USA; a minute species occasionally found in slow-flowing rivers, boggy pools and lakes in the British Isles; easily overlooked.

Pandorina Bory 1824

Coenobia spherical or subspherical, consisting of 4, 8, 16 or 32 closely appressed cells surrounded by a layer of hyaline mucilage; cells spherical, sometimes basally conical, without a papilla; flagella 2, longer than cells, radially arranged around coenobium; chloroplast cup-shaped, sometimes striated with an anterior eyespot and 1 to several pyrenoids; contractile vacuoles 2, apical; asexual reproduction by daughter coenobia; sexual reproduction iso- or anisogamous.

Cosmopolitan; common in standing bodies of water and slow-flowing rivers. Only one species recorded from the British Isles.

The several species are distinguished by chloroplast morphology, number of pyrenoids and number of cells in the coenobium. Coenobia often show polarity when motile.

Pandorina morum (O.F.Müller) Bory 1824

Basionym: *Volvox morum* O.F.Müller
16470010 **Pl. 81E** (p. 325) **CD**
Coenobia subspherical, consisting of 8–16 cells compressed into a dense spherical aggregate, 20–60 µm in diameter; cells 8–17 µm long, with a longitudinally striated chloroplast containing a basal pyrenoid and anterior eyespot.

Cosmopolitan; a common plant of lowland rivers, ponds, lakes and ditches with circumneutral pH, occasionally forming blooms with up to 1000 cells per mL; often found in stratified, shallow lakes rich in nitrate.

Pascherina P.C.Silva 1959

Coenobia consisting of 4 cells in close proximity and not separated by mucilage, flagella all pointing in same direction; cells spheroidal, about 1.5 times as long as wide, with a minute hemispherical papilla; flagella about twice as long as cells; chloroplast cup-shaped, thickened basally with a pyrenoid and anterior eyespot; contractile vacuoles 2 and apical; asexual reproduction by formation of daughter colonies within mother cell; sexual reproduction isogamous.

Ukraine, Europe, USA; standing bodies of water and slow-flowing rivers. Monospecific.

Similar to *Pyrobotrys* but with fewer cells in the colony. The coenobia display a jiggling motion when swimming.

Pascherina tetras (Korshikov) P.C.Silva 1959

Basionym: *Pascheriella tetras* Korshikov
16480010 **Pl. 81A** (p. 325)
Cells 10 µm wide, up to 15 µm long.

Distribution as for genus; an uncommon plant known from the plankton of lowland nutrient-rich rivers.

Phacotus Perty 1852

Cells with a calcified envelope fitting together in two equal halves and united in a circular rim, surface of envelope smooth, granular or furrowed, hyaline or brownish, with 2 flagella each passing through a separate pore; cells in lateral view circular with rim forming a distinct outer disc, in side view cells flattened and lens-shaped with rim protruding basally; protoplast spherical to pear-shaped, usually flattened and occupying about half or more of volume of envelope; chloroplast cup-shaped, with pyrenoids present or absent, eyespot median or anterior; contractile vacuoles 2, apical; asexual reproduction by 2–16 zoospores per cell.

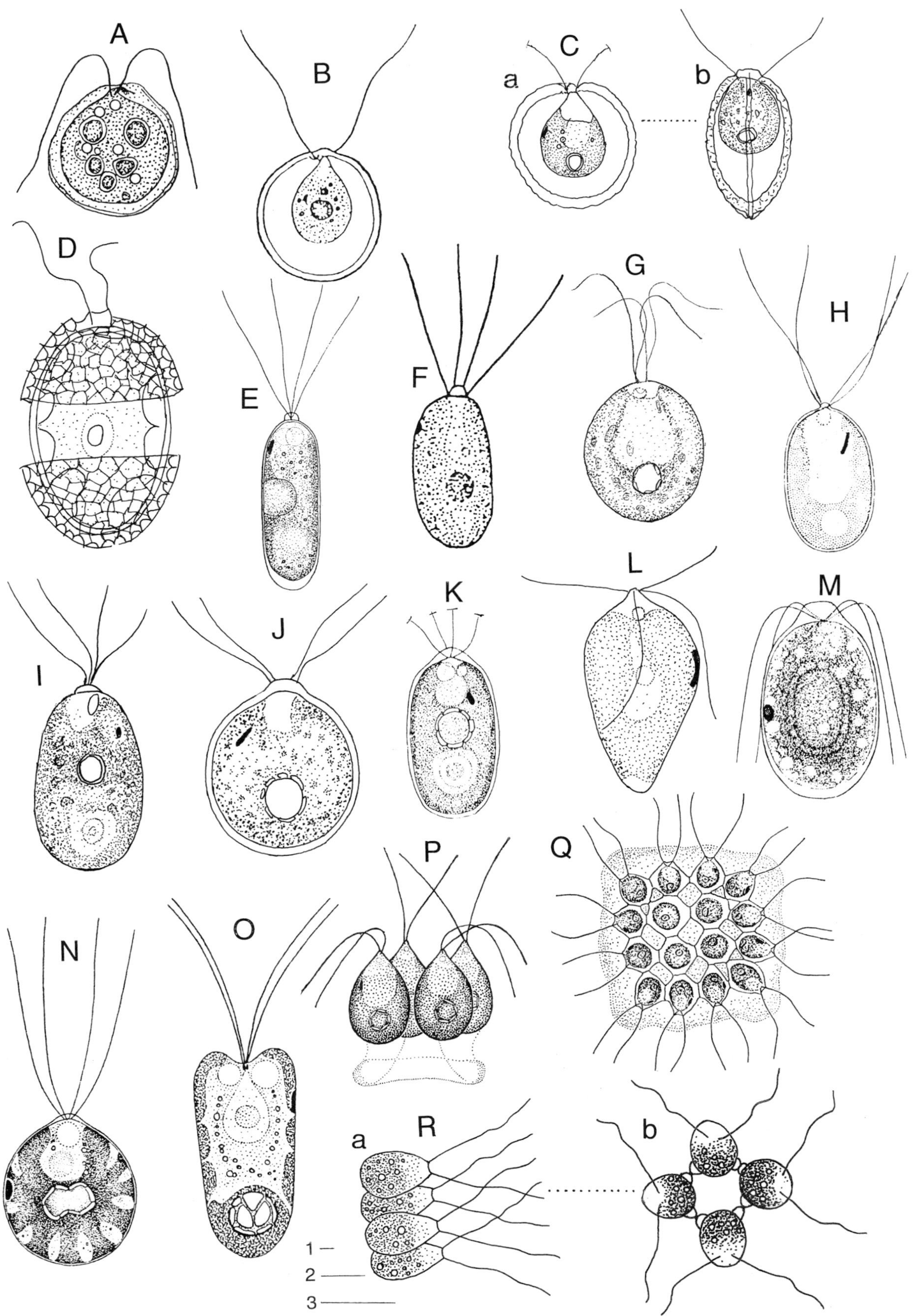

Plate 80 A–R

A. *Dysmorphococcus coccifer* (p. 317); **B**. *Dysmorphococcus variabilis* (p. 317); **C**. *Phacotus lenticularis* (p. 322): a. front view of cell, b. side view; **D**. *Hemitoma meandrocystis* (p. 319); **E**. *Carteria acidicola* (p. 305); **F**. *Carteria arenicola* (p. 305); **G**. *Carteria globosa* (p. 305); **H**. *Carteria klebsii* (p. 305); **I**. *Carteria micronucleolata* (p. 305); **J**. *Carteria multifilis* (p. 305); **K**. *Carteria obtusa* (p. 306); **L**. *Carteria ovata* (p. 306); **M**. *Carteria peterhofiensis* (p. 306); **N**. *Carteria radiosa* (p. 306); **O**. *Hafniomonas reticulata* (p. 319); **P**. *Basichlamys sacculifera* (p. 304); **Q**. *Gonium pectorale* (p. 318); **R**. *Gonium sociale* (p. 318): a. side view of colony, b. polar view.

Scale bars: 10 μm. 1 for Q; 2 for P, R; 3 for A–O.

Cosmopolitan; occurs in variously sized bodies of standing water.

The several species are distinguished by envelope and chloroplast characters. The calcified envelope is well seen in crossed-polarized light. Distinguished from *Coccomonas* and *Hemitoma* by the raised lateral margin.

Phacotus lenticularis (Ehrenberg) F.Stein 1878

Basionym: *Cryptomonas lenticularis* Ehrenberg
Synonym: *Phacotus lenderi* Chodat
16510020 **Pl. 80C** (p. 321)

Envelope circular in lateral view with an outer rim, 13–20 μm wide and in side view biconvex to lemon-shaped, about twice as long as wide, minutely granular and several μm in thickness; protoplast pear-shaped, flattened, usually occupying only anterior end of cell, with flagella arising from 1 (or 2?) canals in envelope; chloroplast usually with a pyrenoid; eyespot median.

India, Europe, USA; widely distributed in hard-water ditches, ponds, lakes and rivers, but usually in small numbers; often a chance member of the plankton (tychoplankton).

Phacotus lendneri Chodat was recorded from Malham Tarn by Belcher & Swale (1961), but more recent work by them (unpublished) indicates that the species is identical to *P. lenticularis*.

DOUBTFUL SPECIES

Phyllocardium complanatum Korshikov 1927
16530010

Unicellular flagellate distinguished by short, flattened unfurrowed cells and pronounced 'tail' but otherwise similar to *Chlorogonium*. A plant resembling this species was observed in an old algal culture by T.M. Harris in 1935, but the original location was unknown. An unidentified species of *Phyllocardium* was recorded in large numbers from Priddy Pool, Somerset, by Happey-Wood (1978).

Pteromonas Seligo 1887

Cells consisting of an envelope consisting of 2 halves joining in a wing-like extension of cell wall only interrupted by insertion point of 2 flagella, in lateral view circular, often apically truncated or occasionally quadrangular in outline, sometimes broadening apically, with wing-like wall extension as an outer hyaline rim up to half as wide as protoplast, basally rounded or occasionally weakly pointed, distinctly flattened in side view with edge of 'wing' appearing as an anterio-posterior line often curved at the extremity, in polar view cell usually angular with 4–6 sides, corners rounded and wings bent in opposite directions on either side of cell; protoplast obovoid, spherical or pear-shaped, usually with a papilla and 2 flagella arising from either side; chloroplast parietal, cup-shaped or band-shaped, with 1 to several pyrenoids and usually an eyespot; contractile vacuoles 2, apical; asexual reproduction by 2–4 zoospores per cell; sexual reproduction isogamous.

Probably cosmopolitan; mainly in very nutrient-rich waters.

Members of this genus contain morphologically complex cells. The wings in side view are curved and not linear and calcified as in *Phacotus*. Seen in lateral view the species could be mistaken for *Sphaerellopsis* but when motile the wings are clearly seen. There is overlap between some species and also considerable variability within collections. About 20 species are known, distinguished mainly on envelope morphology; most of the British species have been described by Belcher & Swale (1967b).

1 Cells in lateral view quadrate with wings drawn out into 4 processes, 2 anterior and 2 posterior .... *P. aculeata*
1 Cells in lateral view broadly rounded basally and not as above 2
2 Cells ellipsoidal and without truncated wings in lateral view, in side view with 4 small spines .... *P. spinosa*
2 Cells with wings truncated apically in lateral view, in side view spines are absent 3
3 Cell walls in side view curved, not parallel .... *P. angulosa*
3 Cell walls in side view straight and more or less parallel 4
4 Chloroplast band-like with one pyrenoid .... *P. tenuis*
4 Chloroplast cup-shaped with 2–3 pyrenoids .... *P. aequiciliata*

Pteromonas aculeata Lemmermann 1900

Synonym: *Pteromonas danubialis* Hortobágyi
16600010 **Pl. 79J** (p. 316)

Cells in lateral view rectangular, 4 corners drawn out into short, acute spines, 19–37 μm wide, 24–35 μm long in side view about 4 times as long as wide with a slight median waist; in polar view strongly flattened with wings extending about half width of protoplast; protoplast flattened and in lateral view ellipsoidal or irregular, occupying half or less total cell width; chloroplast cup-shaped, with 1–6 scattered pyrenoids; eyespot anterior.

Europe, Russian Federation, USA; in the British Isles an occasional species of lowland nutrient-rich lakes and slow-flowing rivers.

In lateral view the cells look like a *Brachiomonas*.

Pteromonas aequiciliata (Gricklhorn) Bourrelly 1947

Basionym: *Amphitropis aequiciliata* Gricklhorn
Synonyms: *Pteromonas varians* Jane, *P. vexilliformis* (Playfair) L.S.Péterfi
16600020 **Pl. 79K** (p. 316)

Cells in lateral view variable in shape, ranging from ellipsoidal to slightly asymmetrical, apically rounded and broadened, 11–18 μm wide, 15–26 μm long, in side view 2 to 3 times as long as wide with an irregular somewhat undulate margin and 6–12 μm thick, in polar view hexagonal to pentagonal with sides slightly indented and wings weakly hooked and short; protoplast spheroidal or pear-shaped, laterally flattened; chloroplast cup-shaped, with 1 to several lateral pyrenoids; eyespot anterior; contractile vacuoles 1 or 2.

France, Scandinavia; in the British Isles in nutrient-rich ponds and slow-flowing rivers, uncommon.

Jane (1944a) describes in detail a collection from a Hertfordshire pond, noting that morphological variability is dependent upon the reproductive state.

Pteromonas angulosa (H.J.Carter) Lemmermann 1900

Basionym: *Cryptoglena angulosa* H.J.Carter
Synonyms: *Pteromonas alata* (Cohn) Seligo, *Phacotus angulosus* (H.J.Carter) F.Stein, *Pteromonas takedana* G.S.West
16600030 **Pl. 79L** (p. 316)
Cells in lateral view ellipsoidal, anterior end truncated, 9–20 µm wide, 12–18 µm long, in side view walls more or less parallel or slightly indented with angular corners, about 7 µm thick, in polar view rhomboidal to hexagonal and angular; protoplast obovoid, laterally flattened; chloroplast cup-shaped, with basal pyrenoid; eyespot anterior.

Europe, USA; in the British Isles frequent but sporadic in nutrient-rich lakes, ponds and slow-flowing rivers.

West (1916b) and Swale (1963a) discuss the taxonomy of this species.

Pteromonas spinosa Nygaard 1943

16600040 **Pl. 79M** (p. 316)
Cells in lateral view broadly ellipsoidal with an apical indentation, 14–17 µm wide, 18–22 µm long, in side view approximately quadrangular but narrowing apically with 2 narrow apical spines and 2 larger posterior spines equidistant from wings, laterally flattened and about 3 times as long as wide (9–11 µm thick), in polar view more or less hexagonal with wings 2–3 µm long and protruding spines; protoplast in lateral view pear-shaped, flattened laterally; chloroplast cup-shaped, with basal pyrenoid; eyespot anterior.

Europe (Denmark); in the British Isles an uncommon species of nutrient-rich ponds and lakes.

Pteromonas tenuis J.H.Belcher *et* Swale 1967

16600050 **Pl. 79N** (p. 316)
Cells in lateral view broadly ellipsoidal but truncated apically, 5–10 µm wide, 11–15 µm long, about twice as long as wide, with protoplast occupying most of cell, in side view side walls more or less parallel and rounded basally except for wings (about 1 µm long), in polar view somewhat hexagonal with rounded angles; protoplast in lateral view obovoid with small papilla; chloroplast band-shaped, with pyrenoid lateral and approximately median; eyespot median.

British Isles (England); described from small puddles in flooded fields near Hawkshead, Cumbria (Belcher & Swale, 1967b), probably more widely distributed.

DOUBTFUL SPECIES
Pteromonas chodatii Lemmermann 1900
Recorded by G.S.West (1912) from a lake in Great Barr Park (Staffordshire) but the status of this species is doubtful.

Pyrobotrys Arnoldi 1916

Synonyms: *Uva* Playfair *non Uva* Burman ex Kuntze, *Chlamydobotrys* Korshikov
Coenobium 8–16-celled, forming mulberry- or grape-like clusters (15–)20–50(–60) µm in diameter with 2 flagella directed outwards; cells contiguous, ellipsoidal or pear-shaped, with or without a papilla; chloroplast cup-shaped, with pyrenoid absent; eyespot median or anterior; contractile vacuoles 2, apical; asexual reproduction by daughter coenobia; sexual reproduction isogamous.

Cosmopolitan; usually recorded from organically enriched water.

About 7 species are recognized; distinguished by form of the coenobium and cells. Some species are heterotrophic and utilize acetate.

1 Cells pear-shaped and not basally attenuated, all facing same direction ....*P. casinoensis*
1 Cells pear-shaped and slightly attenuated basally, facing different directions.........................*P. stellata*

Pyrobotrys casinoensis (Playfair) P.C.Silva 1972

Basionym: *Uva casinoensis* Playfair
Synonym: *Pyrobotrys gracilis* (Korshikov) Korshikov
16800010 **Pl. 81B** (p. 325) **CD**
Coenobia 45–60 µm in diameter, with cells and flagella all pointing in same direction; cells pear-shaped and narrowed basally, 10–25 µm long, papilla absent; chloroplast cup-shaped, with pyrenoid absent, with a median or anterior eyespot.

Ukraine, Australia, USA; in puddles polluted by cattle dung, occasional although often dominant where it occurs.

Pyrobotrys stellata (Korshikov) Korshikov 1938

Basionym: *Chlamydobotrys stellata* Korshikov
16800020 **Pl. 81C** (p. 325)
Coenobium 8-celled in a grape-like cluster, about 40 µm in diameter; cells more or less radially arranged with flagella pointing outwards, pear-shaped and basally attenuated, about 14 µm long and with a small papilla; chloroplast cup-shaped, with pyrenoid present or absent and having an anterior eyespot.

Europe (Czech Republic, Ukraine), USA; reported by Jane (unpublished) from south-east England.

Spermatozopsis Korshikov 1913

Cells asymmetric, banana-shaped and 5 or more times as long as wide in lateral view, often S-shaped, basally attenuated and apically rounded; flagella 2 or 4, longer than cell, closely inserted apically, papilla absent; chloroplast parietal, band-like, occupying at least half of cell in lateral view, with pyrenoids absent; eyespot anterior; contractile vacuoles 2 and apical; asexual reproduction by longitudinal cleavage.

Probably cosmopolitan; various bodies of standing water and slow-flowing rivers. Monospecific.

Spermatozopsis exsultans Korshikov 1913

16680010 **Pl. 78L** (p. 313)

Cells 2–4 µm wide, 7–10(–12) µm long.

Widely distributed though infrequently reported and usually present in small numbers; in the British Isles; known from nutrient-rich pools, ponds, lakes and slow-flowing rivers.

When swimming the cells describe a distinctive helical path.

Sphaerellopsis Korshikov 1925

Synonyms: *Chlamydococcus* F.Stein *pro parte*, *Chlamydomonas* Ehrenberg *pro parte sensu auct.*

Cells spherical, ellipsoidal, obovoid, pear-shaped or cylindrical, with a broad hyaline envelope up to a quarter of width of protoplast in lateral view, occasionally with protoplasmic extensions (cf. *Haematococcus*); flagella as long as or longer than cell, passing through envelope in 2 small canals; protoplast usually pear-shaped, occasionally ellipsoidal, papilla rarely present; chloroplast parietal, cup-shaped, band- or more rarely H-shaped, pyrenoids present or absent; eyespot usually present; contractile vacuoles 2, apical; asexual reproduction by 4 zoospores per cell; sexual reproduction isogamous.

Cosmopolitan; usually from nutrient-poor and humic-rich water (dystrophic water).

Similar to *Chlamydomonas* except for a broad smooth envelope. For most of the species the protoplast diameter ranges from 50–70% of cell diameter. In *Chlamydomonas* the protoplast diameter is usually 85–95% of cell diameter. A few intermediate forms occur such as *Chlamydomonas sphagnicola*. The species of *Sphaerellopsis* are distinguished by cell morphology and chloroplast characters.

1 Chloroplast parietal and twisted .... *S. spiralis*
1 Chloroplast cup-shaped, basal or lateral, not twisted 2
2 Chloroplast lateral or cup-shaped, without a pyrenoid 3
2 Chloroplast basal, cup-shaped, with a pyrenoid 4
3 Chloroplast cup-shaped; flagella widely inserted .... *S. mucosa*
3 Chloroplast lateral; flagella narrowly inserted .... *S. montana*
4 Cells cylindrical to ellipsoidal, 2–3 times as long as wide 5
4 Cells subspherical to ellipsoidal, less than twice as long as wide 6
5 Cells cylindrical to ellipsoidal, twice as long as wide, with a low broad papilla .... *S. velata*
5 Cells cylindrical, 2–3 times as long as wide, papilla absent .... *S. lefevrei*
6 Cell wall lamellate .... *S. gloeocystiformis*
6 Cell wall hyaline .... *S. fluviatilis*

Sphaerellopsis fluviatilis (F.Stein) Pascher 1927

Basionym: *Chlamydococcus fluviatilis* F.Stein

16700030 **Pl. 78O** (p. 313)

Cells ellipsoidal, 10–20 µm wide, 14–30 µm long, about 1.5 times as long as wide, envelope broad, about half width of protoplast, hyaline; protoplast usually pear-shaped; chloroplast cup-shaped, with basal pyrenoid and apical eyespot.

Scandanavia, Ukraine, USA; in the British Isles only reported from small pools.

Sphaerellopsis gloeocystiformis (Dill) Gerloff 1940

Basionym: *Chlamydomonas gloeocystiformis* Dill

16700040 **Pl. 78P** (p. 313)

Cells ellipsoidal, 10–17 µm wide, 13–22 µm long, area between flagella sometimes enlarged to form a broadly conical papilla, envelope sometimes lamellate and about half width of pear-shaped protoplast; chloroplast cup-shaped, with pyrenoid basal or absent; eyespot median or apical.

Europe; recorded in England by Rich (1925) from some Leicestershire ponds.

The cell wall is not consistently lamellate, otherwise the species is similar to *S. fluviatilis*.

Sphaerellopsis lefevrei Bourrelly 1951

16700050 **Pl. 78Q** (p. 313)

Cells cylindrical with rounded ends, 10–18 µm wide, 21–34 µm long, 2.5–3 times as long as wide, with envelope about half width of protoplast, hyaline; protoplast pear-shaped to cylindrical; chloroplast cup-shaped, with pyrenoid basal; eyespot apical, occasionally absent.

Austria, France, Switzerland; in the British Isles common in a small, nutrient-poor pond in Sussex during spring.

Sphaerellopsis montana Christen 1962

16700060 **Pl. 78R** (p. 313)

Cells spherical to subspherical, 13–18 µm in diameter, envelope hyaline, about half protoplast width, occasionally thickened basally; protoplast broadly pear-shaped with chloroplast lateral; without pyrenoid; eyespot median.

Europe (Switzerland); in the British Isles known only from a nutrient-rich village pond in Kent, likely to be more widely distributed.

Sphaerellopsis mucosa (Korshikov) Pentecost, *nov.comb.*

Basionym: *Chloromonas mucosa* Korshikov in Pascher 1917, p. 296, Fig. 263, p. 297.

Synonym: *Chlamydomonas mucosa* (Korshikov) Pascher

16700070 **Pl. 78S** (p. 313)

Cells spherical with thin wall but a broad outer mucilaginous envelope, 17–23 µm in diameter; flagella widely inserted into cell, separated by a broad, low papilla; chloroplast cup-shaped, without pyrenoid; eyespot median.

Ukraine, Czech Republic, Denmark; usually in nutrient-poor and often peaty water, only reported once in the British Isles from Cheshire.

Species with broadly inserted flagella and lacking pyrenoids are placed in the genus *Gloeomonas* G.A.Klebs emend. H. *et* O.Ettl by some authors,

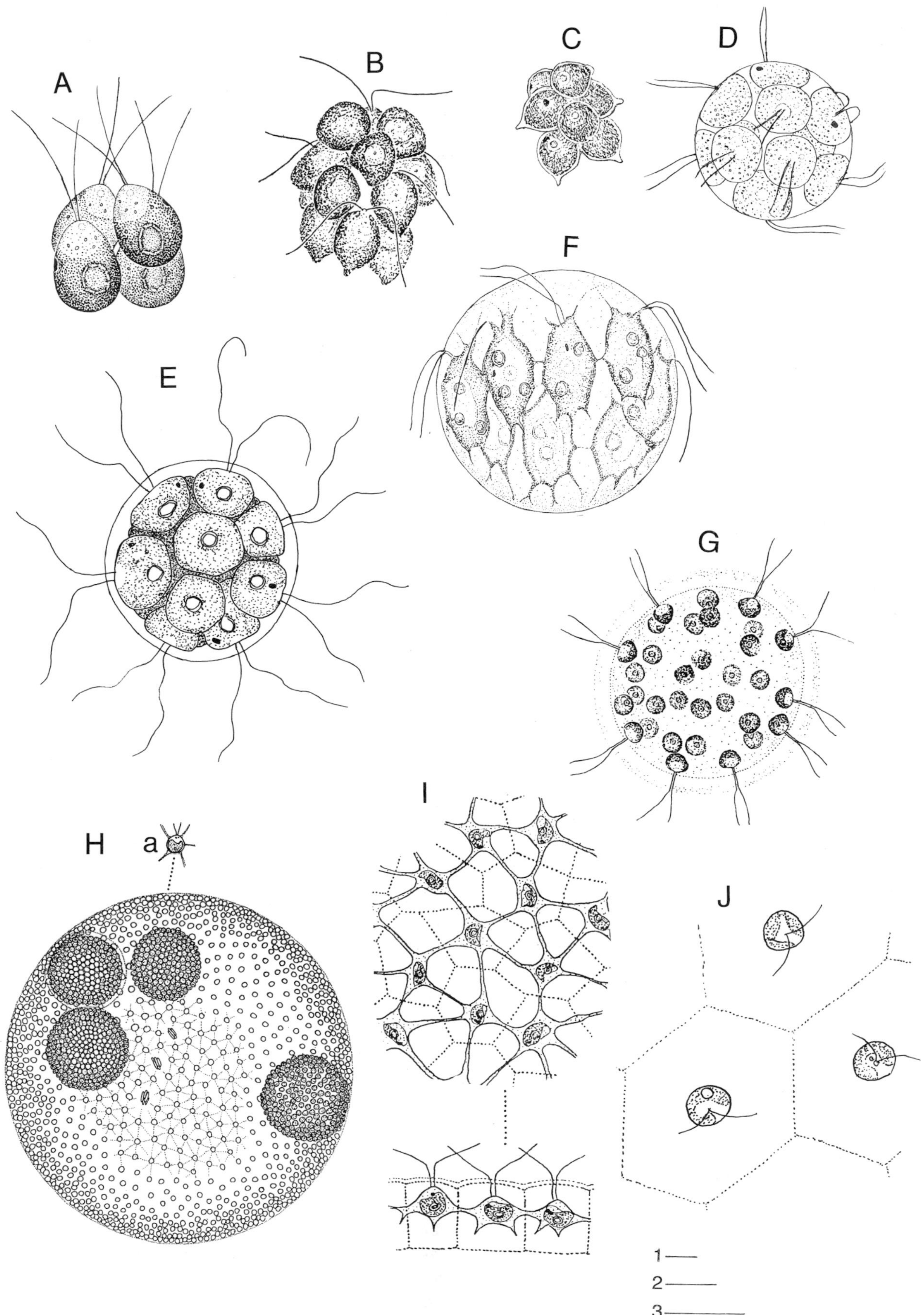

Plate 81 A–J

A. *Pascherina tetras* (p. 320); **B**. *Pyrobotrys casinoensis* (p. 323); **C**. *Pyrobotrys stellata* (flagella not shown) (p. 323); **D**. *Volvulina steinii* (p. 327); **E**. *Pandorina morum* (p. 320); **F**. *Stephanosphaera pluvialis* (p. 326); **G**. *Eudorina elegans* (p. 317); **H**. *Volvox aureus* (p. 326): colony (1.5 mm in diameter) showing 4 gonidia, sperm packets and an individual cell at a; **I**. *Volvox globator* (p. 327): part of a coenobium in vertical and lateral view; **J**. *Volvox tertius* (p. 327): part of a coenobium in vertical view.
Scale bars: 10 μm. 1 for B-D, G, H (cell); 2 for A, F; 3 for E, I, J.

but flagellar insertion often shows a considerable range of variation in both *Chlamydomonas* and *Sphaerellopsis*. This species is best placed in *Sphaerellopsis* on account of its broad envelope which is well within the range described for other species of the genus.

Sphaerellopsis spiralis J.H.Belcher *et* Swale 1963

16700010 **Pl. 78T** (p. 313)

Cells obovoid with a locally thin atypical envelope, 6–8.5 μm wide, 11–15 μm long, in polar view box-shaped; protoplast irregular, with a twisted parietal chloroplast, and basal pyrenoid; eyespot small, median.

British Isles (England); an unusual plant described from small puddles and ponds in Lancashire (Belcher & Swale, 1963).

The species requires further investigation as the envelope is not typical of the genus.

Sphaerellopsis velata (Korshikov) Gerloff 1940

Basionym: *Chlamydomonas velata* Korshikov

16700020 **Pl. 78U** (p. 313)

Cells cylindrical in lateral view, 8–20 μm wide, 15–38 μm long, about twice as long as wide, envelope thickened basally and rounded apically, with 2 flagella widely inserted; protoplast spheroidal; chloroplast cup-shaped, with basal pyrenoid and anterior eyespot.

Russian Federation; in the British Isles known only from Lancashire (Belcher & Swale, 1963).

Stephanosphaera Cohn 1852

Colonies consisting of 1, 2, 4, 8 or 16 protoplasts within a common mucilaginous envelope, protoplasts arranged in tiers of 4, 8 or 16 separated by mucilage and all pointing in same direction, flagella bending outwards and often radially arranged; protoplasts 2 to 3 times longer than wide, irregular in shape due to formation of many fine, unbranched protoplasmic strands permeating mucilage; chloroplast parietal, with one anterior and one posterior pyrenoid and a median to (?basal) eyespot; cytoplasm sometimes tinged pink by presence of a carotenoid pigment; contactile vacuoles several and scattered; asexual reproduction by longitudinal division into 4–8 individuals; sexual reproduction isogamous; zygotes red-pigmented.

Cosmopolitan; usually to be found in variously sized bodies of standing water. Monospecific.

Stephanosphaera pluvialis Cohn 1852

16720010 **Pl. 81F** (p. 325)

Colonies 30–70 μm in diameter; protoplasts 9–13 μm wide, 9–35 μm long.

Distribution as for genus; known from small rock pools and depressions and well-known in the British Isles from the Malham Tarn area of North Yorkshire where found on limestone and in the marginal shallows of the tarn (Lund, 1961a); also found in roof gutters in towns.

The protoplasts are similar to *Haematococcus* and the two genera can be found in association, though *Stephanosphaera* is less common.

Volvox (Linnaeus) Ehrenberg 1830

Synonyms: *Janetosphaera* Shaw, *Merrillosphaera* Shaw

Coenobia large and hollow (150–)500–1000(–3000) μm in diameter (larger ones visible to naked eye), containing several hundred to many thousand cells; cells approximately equidistant, separated by hyaline mucilage from nearest neighbours, sometimes interconnected by fine protoplasmic strands (they can be stained with methylene blue); flagella 2 per cell, directed away from coenobium, papilla absent; cells spherical, ellipsoidal, pear-shaped or angular, (3–)4–8(–9) μm in diameter; chloroplast parietal, cup-shaped or irregular, usually with 1 or more pyrenoids; eyespot anterior, size varying according to position of cell in coenobium; contractile vacuoles 2 to many, scattered; asexual reproduction by formation of daughter colonies within hollow centre of coenobium by specialized gonidial cells; sexual reproduction oogamous: male gametes formed from enlarged vegetative cells (sperm packets), minute and spindle-shaped with 2 flagella; female cells large and spherical with 10 or more differentiating within a coenobium; zygote orange on maturity, sometimes developing a thickened and ornamented wall; monoecious or dioecious; in latter there may be sexual dimorphism.

Cosmopolitan; most often found in lowland lakes and ponds during the late summer, overwintering as zygotes.

Several species have been described and differentiated on coenobium size and cell and zygote morphology. The number of vegetative cells in the parent coenobium is a useful character and can be estimated by the formula $15\,r^2/a^2$, where *r* is colony radius and *a* is average distance between adjacent cells using the same unit of measurement.

1 Coenobium of (1500–)5000–17,000 cells, each angular or star-shaped in polar view and having fine interconnecting protoplasmic strands; monoecious; zygote rough and spiny .... *V. globator*
1 Coenobium of (500–)2000–3500 cells, each rounded in polar view, protoplasmic strands present or absent; dioecious; zygote smooth.......... 2
2 Coenobia always with well-defined protoplasmic strands between adjacent cells; individual mucilage zones absent around cells .... *V. aureus*
2 Coenobia without protoplasmic strands between cells or only apparent in young coenobia, individual mucilage zones around cells sometimes present............ *V. tertius*

Volvox aureus Ehrenberg 1832

Synonyms: *Volvox dioica* Cohn, *V. lismorensis* Playfair, *Janetosphaera aurea* (Ehrenberg) Shaw

16770010 **Pl. 81H** (p. 325) **CD**

Coenobia spherical or ellipsoidal, (170–)400–600(–850) μm in diameter containing (250–)500–1500(–3500) cells each interconnected with protoplasmic strands, mucilage hyaline and undivided; cells circular in polar view, 5–9 μm in diameter, pear-shaped in lateral view; chloroplast parietal; pyrenoid present or absent; eyespot anterior; contractile vacuoles 2 and apical; dioecious; sperm packets numerous in male coenobia, oospheres up to 15 in each female coenobium; zygote orange, with a thick smooth wall, about 60 µm in diameter.

Probably cosmopolitan; the most frequently recorded species in the British Isles, commonly in lowland lakes, ponds, ditches and puddles especially in late summer with up to 10 coenobia per mL; sometimes in nitrate-enriched water. Some British collections are described by Pentecost (1983).

Volvox globator Linnaeus 1758

16770020 **Pl. 81I** (p. 325) **CD**

Coenobium spherical, 500–1000 µm in diameter and consisting of 1500–17,000 cells, each cell with fine protoplasmic strands; cells spherical or pear-shaped, 3–5(–8) µm in diameter with mucilage usually distinctly zoned around each cell, forming a polygonal network superimposed on the interconnecting strands; chloroplast parietal, with pyrenoids several and minute; eyespot small; contractile vacuoles 2–6; coenobia usually contain 4 or more daughter colonies; monoecious; oospheres 20 or more and numerous sperm packets; zygote reddish-orange, with rough, spinose wall, 40–60 µm in diameter.

Probably cosmopolitan; an uncommon alga in the British Isles where sometimes found with *V. aureus* in lowland lakes and ponds.

Volvox tertius Meyer 1896

Synonyms: *Volvox mononae* G.M.Smith, *Merillosphaera tertia* (Meyer) Shaw

16770030 **Pl. 81J** (p. 325) **CD**

Coenobia spherical, 250–600 µm in diameter, containing about 500–2000 spherical or weakly ellipsoidal cells 5–8 µm in diameter, protoplasmic strands present only in younger, smaller coenobia, with each cell surrounded by a polygonal pattern of zoned mucilage; chloroplast parietal, band-shaped with a basal pyrenoid and small anterior eyespot; contractile vacuoles 2, anterior; dioecious; oospheres 3–8 or with up to 60 sperm packets; zygote smooth-walled, 60–70 µm in diameter.

Europe, Australia, USA; in the British Isles rarely reported from lowland lakes and ponds.

Distinguished from *V. aureus* by the presence of zoned mucilage, lack of protoplasmic strands in the mature coenobia and the slightly smaller coenobia.

Volvulina Playfair 1915

Coenobia spherical or ellipsoidal, containing 4, 8, 16 or 32 flattened cells regularly arranged at periphery of a hollow colony, usually cells occur in tiers of 4 separated locally by hyaline mucilage, with flagella narrowly inserted and directed away from coenobium; cells with a broad and apical depression, papilla absent; chloroplast cup-shaped, with a pyrenoid and eyespot sometimes present; contractile vacuoles 2 and apical; asexual reproduction by formation of daughter coenobia; sexual reproduction isogamous.

Cosmopolitan; but rarely reported.

One species known to be mixotrophic.

Volvulina steinii Playfair 1915

16780010 **Pl. 81D** (p. 325)

Coenobia 50–100 µm in diameter, usually with 16 cells 10–25 µm in diameter; position of flagella variable, those at anterior of colony displaced backwards and those at posterior displaced forwards; chloroplast sometimes pale, with pyrenoid absent but small grains of starch occur in cytoplasm; eyespot anterior, often confined to anterior of coenobium.

Russian Federation, Switzerland, Australia, Africa; rare in the British Isles, where in organically enriched temporary puddles, often associated with other members of the Volvocales (Belcher, 1965b).

Distinguished from *Eudorina* and *Pyrobotrys* by its flattened, almost hemispherical cells; often pale in colour due to the scarcity of chlorophyll.

ORDER CHLOROCOCCALES

BY DAVID M. JOHN AND PETRO M. TSARENKO

Traditionally coccoid unicells and non-motile colonies, often the colonies of fixed numbers of cells (cells 2, 4, 8, 16 etc.) and known as coenobia. Asexual reproduction is frequently by the formation of spores (autospores) morphologically similar to the mother cell in which they develop. The concept of the order was broadened by Hoek et al. (1995) to include all green algae in which the daughter cells remain within the parental wall, thus multicellular sarcinoid and filamentous forms were transferred to it along with siphonous forms. This remains one of the largest chlorophyte orders and is beginning to be disassembled as the traditional classification, based on morphological features observed by the light microscope, gradually becomes replaced by one more accurately reflecting phylogenetic relationships (see Graham & Wilcox, 2000). For convenience, a traditional approach has been adopted, since a consensus has still to be reached on newly proposed classifications.

Acanthococcus Lagerheim 1883

Cells solitary, spherical or slightly ovoid, thick-walled, covered by large and regularly arranged spines, sometimes enclosed in a mucilaginous envelope; chloroplast parietal, massive, net-like or deeply incised, with a single (rarely 2), large pyrenoid; asexual reproduction by 4, 8 or 16 hemizoospores (with eyespot and contractile vacuoles) and autospores (?), released by gelatinization of the mother cell wall.

Probably cosmopolitan; planktonic or amongst the periphyton of lakes, bogs or rivers, also known from soil, moist rocks and other terrestrial surfaces. Only 3 known species.

Often *Acanthococcus* is considered a synonym of the ill-defined genus *Trochiscia*. Several species in this latter genus are probably incorrectly placed and others might be dinoflagellate cysts or volvocalean zygotes. In our opinion, there is convincing evidence that *Acanthococcus* is a separate genus from *Trochiscia* as currently understood. *Acanthococcus* has a mucilaginous envelope, a parietal and net- or bladder-like chloroplast and reproduces by hemizoospores. In contrast, species of *Trochiscia* do not have a mucilaginous envelope or pyrenoids, the cell wall is warty, the chloroplasts are star-shaped and reproduction is rather by autospores, so *Acanthococcus* is more closely related to other genera (e.g. *Cymatococcus* Hansgirg).

Acanthococcus aciculiferus Lagerheim 1883

Synonyms: *Trochiscia aciculifera* (Lagerheim) Hansgirg var. *aciculifera*, *T. aciculifera* var. *pulchra* Hansgirg

17010010 **Pl. 82A** (p. 329) **CD**

Cells spherical to ovoid, (9–)20–30(–33) μm wide; spines straight, (2.5–)4–10 μm long, not extending beyond outermost edge of mucilaginous envelope; asexual reproduction by 4–8 hemizoospores.

Probably cosmopolitan; usually associated with submerged objects and amongst filamentous algae in small pools and lakes in boggy areas, so far reported only from upland areas, including Westmorland (Cumbria) in England and in the north and west of Ireland.

A further unidentified species is mentioned by West (1892a) from The Burren, County Clare, Ireland, in which the cells are 34–42 μm wide.

Actinastrum Lagerheim 1882

Coenobia free-floating, usually of 4, 8 or 16 elongated cells radiating in all directions from a common centre, with or without a mucilage envelope; chloroplast parietal, plate-like, with or without a pyrenoid; asexual reproduction usually by 4, 8 or 16 autospores.

Cosmopolitan; planktonic, widely distributed in aquatic habitats.

Doubt attaches to the validity of some of the 6 known species of which one is reported from the British Isles. Single cells are often found in iodine-preserved samples and when grown in culture.

Actinastrum hantzschii Lagerheim 1882

17020010 **Pl. 97H** (p. 396) **CD**

Coenobia 4- or 8-celled; cells (1.5–)2–6 μm wide, (7–)10–25 μm long, spindle-shaped, club-shaped or cylindrical towards base and tapering to a rounded or bluntly pointed apex; chloroplast with a single pyrenoid, often not reaching extremities of cell.

Probably cosmopolitan; usually occurs throughout the year in the plankton of rivers, ditches, ponds and in some nutrient-enriched lakes in the British Isles, but rarely abundant.

Little useful purpose is served by recognizing varieties characterized by minor differences in cell size and shape. Two varieties are reported in the British Isles, var. *fluviatile* Schröder (17020012) has more spindle-shaped and more pointed cells than the type, and var. *subtile* Wołoszyńska (17020013) is a very narrow form (<2.5 μm wide).

Actinochloris Korshikov 1953

Synonym: *Radiosphaera* J.Snow *sensu* R.C.Starr

Cells solitary, initially broadly ellipsoidal or ovoid and later spherical, wall smooth and thickening with age, multinucleate and multivacuolate; chloroplast star-shaped, consisting of densely packed and branched lobes radiating from cell centre and adpressed against cell wall, with a single central pyrenoid; asexual reproduction by biflagellate zoospores and non-motile hemizoospores.

Probably cosmopolitan; occurs in soil cultures and small puddles.

1 Chloroplast with narrow lobes, richly and finely divided towards ends, granular or punctate in surface view; pyrenoids surrounded by tangentially arranged starch grains .... *A. terrestris*

1 Chloroplast with broad lobes, little divided towards ends and polygonal in surface view; pyrenoid surrounded by radially arranged starch grains .... *A. sphaerica*

Actinochloris sphaerica Korshikov 1953

Synonyms: *Radiosphaera sphaerica* (Korshikov) Fott, *R. dissecta* (Korshikov) R.C.Starr

17030010 **Pl. 82B** (p. 329) **CD**

Cells 30–150 μm wide, spherical though initially ellipsoidal; chloroplast 1 per cell,very broadly lobed, lobes little divided towards ends and polygonal in surface view, with pyrenoid surrounded by radially arranged starch grains; vacuoles present; zoospores 4–7 μm wide, 10–15 μm long, nucleus confined to apical portion.

Probably cosmopolitan; in the British Isles known only from soil cultures.

Actinochloris terrestris (Vischer) H.Ettl *et* G.Gärtner 1995

Basionym: *Asterococcus terrestris* Vischer

Synonyms: *Macrochloris dissecta* Korshikov *sensu* F.E.Fritsch *et* R.P.John, *Radiococcus dissecta* (Korshikov) R.C.Starr

17030020 **Pl. 82D** (p. 329)

Cells 30–50(–70) μm wide, spherical; chloroplast lobes narrow, many times finely divided towards ends and appearing granular or finely punctate in surface view, with chloroplasts and pyrenoids increasing in number as cells age; pyrenoids with tangentially arranged starch grains; vacuoles absent; zoospores 3–4 μm wide, 9–10 μm long, nucleus confined to distal end.

Europe; usually isolated in the British Isles from more alkaline soils (Fritsch & John, 1942).

Amphikrikos Korshikov 1953

Cells solitary, barrel-shaped, ellipsoidal to broadly ovoid, or short and cylindrical with rounded apices, with or without a thick mucilage envelope; cells thin-walled, with greyish-brown granules at the apices or

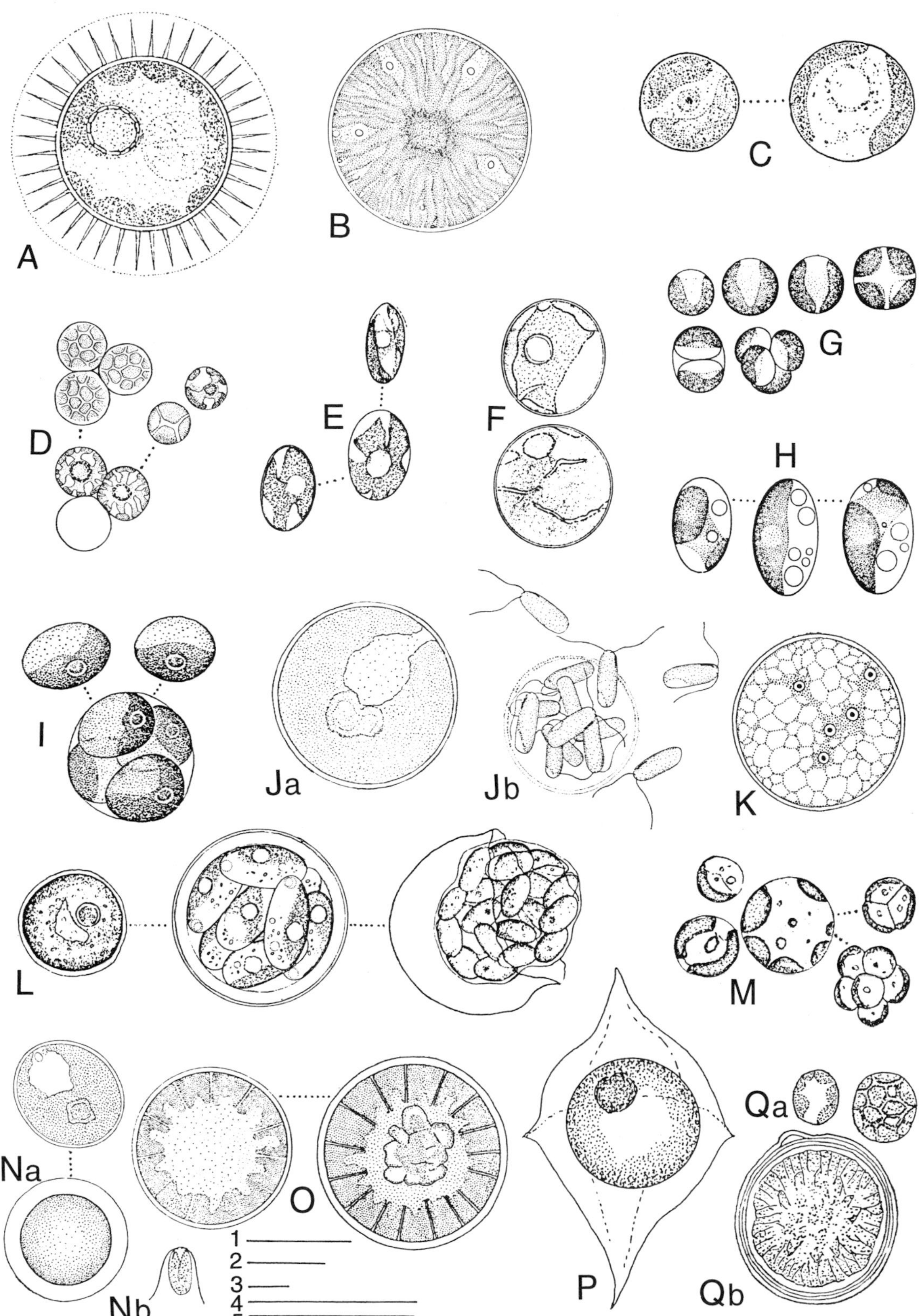

Plate 82 A–Q

A. *Acanthococcus aciculiferus* (p. 328); **B**. *Actinochloris sphaerica* (p. 328); **C**. *Bracteacoccus anomalus* (p. 333); **D**. *Actinochloris terrestris* (p. 328); **E**. *Chlorella ellipsoidea* (p. 335); **F**. *Chlorella emersonii* (p. 335); **G**. *Chlorella minutissima* (p. 335): cells at different stages and producing autospores; **H**. *Chlorella saccharophila* (p. 336); **I**. *Chlorella vulgaris* (p. 336): vegetative cells and enlarged cell containing autospores; **J**. *Chlorococcum hypnosporum* (p. 337): a. details of vegetative cell, b. release of biflagellate zoospores; **K**. *Dictyochloris fragrans* (p. 348); **L**. *Chlorococcum infusionum* (p. 337): vegetative cells (left), swollen cells containing zoospores and early release of zoospores (right); **M**. *Muriella terrestris* (p. 370): vegetative cells and stages in autospore development; **N**. *Chlorococcum minutum* (p. 338): a. two vegetative cell, b. biflagellate zoospore; **O**. *Dictyococcus varians* (p. 348); **P**. *Desmatractum bipyramidatum* (p. 347); **Q**. *Muriella magna* (p. 370): a. vegetative cells at different stages of development; b. thick-walled akinete.

Scale bars: 1–4, 10 μm; 5, 5 μm. 1 for C, D, F, H–K, M, P; 2 for A, E, Nb; 3 for B, L, O, Q; 4 for Na; 5 for G.

equator, sometimes fusing to form a complete ring; chloroplast parietal, often disc-like, with or without a pyrenoid; asexual reproduction by 2, 4 or 8 autospores, released by fracture of wall at one or both apices.

Probably cosmopolitan; planktonic and often associated with the periphyton, most common in non-flowing water.

Amphikrikos minutissimus Korshikov 1953

Synonym: *Siderocelis minutissima* (Korshikov) Bourrelly
17040010 **Pl. 87L** (p. 354)
Cells 2–5 μm wide, (4–)5–7 μm long, cylindrical, sometimes slightly concave in centre, with apices blunt to broadly rounded, usually with mucilage; granules raised and confined to apices, sometimes fused to form a ring.

Probably cosmopolitan; reported in the British Isles from nutrient-rich lakes, occasionally in rivers.

DOUBTFUL RECORD
Amphikrikos nanus (Fott *et* Heynig) Hindák 1977
17040020 **Pl. 87J** (p. 354)
Considered doubtful since the mention of 'England' in Komárek & Fott (1983) is the only record we have traced for the British Isles. Differs from *A. minutissimus* in cell shape (ellipsoidal to shortly oval) and having irregularly arranged equatorial granules that are sometimes fused to form a double ring.

Ankistrodesmus Corda 1838

Colonies of bundles or fan-shaped groups of (1–)2, 4, 8 or 16 cells often only united equatorially, with or without a mucilaginous envelope; cells usually very elongate, narrowly spindle-shaped to cylindrical, curved to spirally twisted or more rarely straight, gradually or abruptly narrowing to acute apices; chloroplast parietal, single and covering most of cell, with or without a pyrenoid; asexual reproduction by 2, 4 or 8(–16) autospores lying parallel to one another, sometimes autospores in two tiers with each tier 4-celled, released by fracture of wall in middle of mother cell.

Probably cosmopolitan; planktonic or associated with detritus and sediment in various types of water body.

See remarks under *Monoraphidium*.

1 Cells arranged often at right angles to one another (cross-wise) .... *A. fusiformis*
1 Cells not so arranged 2
2 Cells spirally twisted around one another, equatorially closely united and spreading apart at apices 3
2 Cells almost parallel to one another and straight or slightly curved 4
3 Cells gradually narrowing to finely acute apices .... *A. spiralis*
3 Cells cylindrical and suddenly narrowing to acute apices .... *A. densus*
4 Cells needle-like to narrowly spindle-shaped .... *A. falcatus*
4 Cells cylindrical and abruptly narrowing to apices .... *A. fasciculatus*

Ankistrodesmus densus Korshikov 1953

Synonym: *Ankistrodesmus spiralis* (W.B.Turner) Lemmermann var. *fasciculatus* G.M.Smith
17050020 **Pl. 97K** (p. 396)
Colonies usually of 4 or 8(–32) cells, equatorially united or overlapping, curved and twisted around one another, with cells in middle often straight and spread apart towards apices, enclosed within a thick mucilaginous envelope; cells 1.5–5 μm wide, (35–)40–60(–100) μm long, almost straight except for a slight bend in centre, cylindrical and suddenly narrowing to acute apices.

Probably cosmopolitan; often associated with *Sphagnum* and other mosses in boggy areas or lakes in Scotland and Ireland where the water is neutral or acid; probably more widely distributed in the British Isles than records suggest.

Ankistrodesmus falcatus (Corda) Ralfs 1848

Basionym: *Micrasterias falcatus* Corda
17050030 **Pl. 97J** (p. 396) **CD**
Colonies of 2, 4, 8 or more cells, usually united by convex sides if curved, sometimes in bundles with cells more or less parallel to one another, enclosed within a mucilaginous envelope; cells (1–)1.5–7 μm wide, 20–165 μm long, needle-like to narrowly spindle-shaped, usually slightly curved or sigmoid, tapering gradually to acute apices.

Probably cosmopolitan; widely distributed in many flowing and still water habitats in the British Isles where planktonic, or more usually associated with detritus and submerged marginal vegetation; known to be pollution tolerant, often very abundant in sewage effluent and in fish-farm ponds where nutrient levels are high.

It has been hypothesized by Grochowski & Trainor (1987) that high bicarbonate concentrations and pH lead to *A. falcatus* becoming dominant, providing nutrient levels are also elevated. They report it to be a very efficient remover of orthophosphate and therefore possibly a useful organism to incorporate into tertiary filtration systems in sewage plants for this purpose. All largely single-celled varieties of *A. falcatus* recorded from the British Isles are now referred to the genus *Monoraphidium* (e.g. *Ankistrodesmus falcatus* var. *acicularis* (A.Braun) G.S.West, now *Monoraphidium griffithii* (Berkeley) Komárková-Legnerová). See remarks under *A. fusiformis*.

Ankistrodesmus fasciculatus (Lundberg) Komárková-Legnerová 1969

Basionym: *Quadrigula fasciculata* Lundberg
Synonym: *Ankistrodesmus biplexus* (Reinsch) Hortobágyi
17050040 **Pl. 97L** (p. 396)
Colonies 2- or 4-celled with cells connected equatorially by convex side, surrounded by a spherical mucilaginous envelope; cells 1.4–5 μm wide, 15–32 μm long, cylindrical and abruptly narrowing to acute apices, slightly curved.

Europe; associated with the periphyton in small acid pools, only rarely recorded in the British Isles.

Ankistrodesmus fusiformis Corda ex Korshikov 1953

17050050 **Pl. 97N** (p. 396) **CD**

Colonies of (1–)2, 4, 8 or 16(–32) cells, with cells crosswise (often at right angles to one another) and connected by mucilage or enclosed within a mucilaginous envelope; cells (1–)1.4–4.3 μm wide, (19–)22–48.6(–57) μm long, almost cylindrical in centre and gradually narrowing to acute apices.

Probably cosmopolitan; planktonic or associated with detritus and sediment in a wide variety of aquatic habitats in the British Isles.

We have chosen to recognize this species as distinct from *A. falcatus* although Hindák (1988) considers them to be conspecific. He believes that the principal diagnostic feature (cross-like orientation of cells) is present to a lesser or greater extent in almost all species including *A. falcatus*, the species it most closely resembles.

Ankistrodesmus spiralis (W.B.Turner) Lemmermann 1908

Basionym: *Raphidium spirale* W.B.Turner

17050060 **Pl. 97M** (p. 396)

Colonies of (2–)4, 8 or 16 cells, united equatorially, strongly helically twisted at the centre and free towards apices, enclosed in a mucilage sheath; cells 1–3.5(–4.3) μm wide, 20–45(–68) μm long, needle-like and gradually narrowing to acute points.

Probably cosmopolitan; in the British Isles most common in more acid waters in Scotland and Wales.

Similar to *Ankistrodesmus densus* whose cells are of more even width, cylindrical, and narrow rapidly only close to apices.

DOUBTFUL RECORDS

Ankistrodesmus cucumiformis J.H.Belcher *et* Swale 1962

17050010

Belcher & Swale (1962) described this species in a sample from a farm pond at Preston Montford in Shropshire. Its cells are solitary, curved (about a quarter-circle) and slightly spiralled, 2–3 μm wide, 4–10 μm long, with one apex obtusely rounded and the other acute. Starch was not detected and Belcher and Swale believe that it might attributable to the genus *Nephrodiella*, a xanthophyte. According to Hindák (pers. comm., in Komárková-Legnerová, 1969), the description and drawings suggests that it closely resembles *Koliella sempervirens* (Chodat) Hindák. Further information is needed to determine whether Hindák is correct.

Ankistrodesmus stipitatus (Chodat) Komárková-Legnerová 1969

17050070

Considered doubtful since it is easily mistaken for *Ankistrodemus falcatus*. Only mentioned in a list of algae from the River Wear (see Holmes & Whitton, 1981a) and not accompanied by an illustration or description.

Ankyra Fott 1957

Cells solitary, subcylindrical or spindle-shaped, curved or bent, gradually narrowing apically to a long spine and basally to a narrow, stout, forked or bifid extension; chloroplast parietal with a single pyrenoid; asexual reproduction by zoospores released by splitting of the mother cell wall into two halves; aplanospores known.

Probably cosmopolitan; planktonic and in many aquatic habitats where the water is slow or non-flowing.

Closely related to *Characium*, *Schroederia* and *Korshikoviella* from which it is distinguished by its cell wall dividing into two almost equal parts to release the zoospores (see Fott, 1957, 1974). Of the 7 recognized species only two are known from the British Isles, both of which have a basal bifid (anchor-like) extension. Sometimes the bifid extension may be difficult to detect in small specimens.

1 Chloroplast H-shaped and thickened in centre..*A. ancora*
1 Chloroplast plate- or trough-shaped ....*A. judayi*

Ankyra ancora (G.M.Smith) Fott 1957

Basionym: *Schroederia setigera* var. *ancora* G.M.Smith

Synonyms: *Schroederia ancora* G.M.Smith, *Characium ancora* (G.M.Smith) Fott

17060010 **Pl. 90B** (p. 367)

Cells 5–14 μm wide, 35–150 μm long, straight or slightly curved, with one side straight and the other convex, narrowing apically to a curved spine and basally to a straight or curved bifid extension, the extension sometimes poorly developed or very reduced; chloroplast H-shaped with pyrenoid in thickened central portion.

Probably cosmopolitan; planktonic and reported in the British Isles from ponds, lakes and reservoirs, rare.

Various forms have been recognized based on differences in cell size and the dimensions of the bifid basal extension. Little purpose is served by recognizing these forms since the characters used to distinguish them are known to be very variable and probably dependent upon environmental conditions (see Komárek & Fott, 1983).

Ankyra judayi (G.M.Smith) Fott 1957

Basionym: *Schroederia judayi* G.M.Smith

Synonyms: *Lambertia judayi* (G.M.Smith) Korshikov, *Korshikoviella judayi* (Korshikov) P.C.Silva

17060020 **Pl. 90A** (p. 367)

Cells (2–)3–9 μm wide, (30–)50–75(–100) μm long (including spine), straight, slightly or markedly curved depending on age, linear to spindle-shaped, gradually narrowing apically to a long spine and basally to a bifid extension whose curvature may be slight to marked; chloroplast plate- or trough-like.

Probably cosmopolitan; recorded occasionally in the British Isles during the summer months in the plankton of ponds, lakes and rivers (e.g. Thames, Lea, Great Ouse, Cam); recorded in Oak Mere, Cheshire where it usually occurs from midsummer to autumn with the maximum number of cells recorded in 1966 as 2250 per mL.

According to Hindák (1988), the dimensions and shape of the cells and basal extension vary with age. He considers it to be distinguished from *A. ancora* (the type species) principally by the different form of its chloroplast and smaller cell size.

Apodochloris Komárek 1959

Cells solitary, pear-shaped, ovoid, spindle-shaped, almost cylindrical, ellipsoidal or irregularly rounded (rarely spherical); chloroplast parietal, usually filling at least two-thirds of each cell, margin slightly wavy to deeply incised, with or without perforations, with a single pyrenoid; asexual reproduction by 4, 8 or 16 biflagellate zoospores and autospores (more rare), released through median, subapical or apical pore; sexual reproduction unknown.

Probably cosmopolitan; associated with other algae (including blue-green algae) in lakes and on soil.

Apodochloris dinobryonis (J.W.G.Lund) H.Ettl *et* G.Gärtner 1995

Basionym: *Chlorochytrium dinobryonis* J.W.G.Lund

18040010 **Pl. 91J** (p. 369)

Cells up to 8 μm wide, 48 μm long, very variable in shape, ellipsoidal, obovoid, pear-shaped or very irregular, occupying the cup-shaped lorica of the chrysophyte *Dinobryon*; chloroplast almost filling cell, pyrenoid often confined to distal end of cell; zoospores up to 32 per cell, 3 μm wide, 7–9 μm long.

Europe; only known in the British Isles growing within the loricas of *Dinobryon divergens* at Blelham Tarn in the English Lake District.

Botryococcus Kützing 1849

Colonies green to yellow or orange, even reddish-orange, consisting of irregularly arranged cells protruding or embedded within the periphery of a tough, folded, wrinkled mucilage (sometimes end drawn out into lobes) containing numerous oil globules; cells spherical, ovoid, ellipsoidal, obovate or heart-shaped; chloroplast parietal, net-like, with a single pyrenoid and starch; numerous oil globules within cells as well as mucilage; asexual reproduction by 4 or 8 autospores, produced by radial division.

Probably cosmopolitan; most common and abundant in ponds and lakes, occasionally in brackish-water.

The colonies under a cover slip release oil and individual cells. The oil contains hydrocarbons whose derivatives are present in commercial oils and are either linear olefins or yellow-coloured branched olefins (known as botryococcenes) depending on the physiological form of *Botryococcus* (see Tenaud et al., 1989). It occurs in certain geological formations and fossilized *Botryococcus* is believed to have contributed significantly to deposition of high-grade coals, bitumen and oil shales around the world. Oil from *Botryococcus* is the principal component of boghead coal and torbanite, the latter named after the Scottish town of Torbanhill. In the latter half of the 19th century much of the mineral oil industry in Scotland was based upon torbanite.

1 Cells projecting prominently from mucilage ....*B. protuberans*
1 Cells embedded in mucilage, only slightly protruding 2
2 Cells ovoid, without an indentation of the outer wall...*B. braunii*
2 Cells ovoid to wedge-shaped, outer wall usually indented ....*B. calcareus*

Botryococcus braunii Kützing 1849

Synonym: *Ineffigiata neglecta* West *et* G.S.West

17080010 **Pl. 85H** (p. 350) **CD**

Cells oval or obovoid, (2.5–)3–6(–9) μm wide, 5.7–10.5 (–12) μm long, usually embedded in the periphery of mucilaginous masses.

Probably cosmopolitan; widely distributed in many types of aquatic habitat in the British Isles including ditches, bog pools, open water tanks, ponds and lakes (e.g. Loch Leven, where always at low population densities). Sometimes appearing as masses of suspended brown granules each up to 1 mm across; if present in vast numbers (often in autumn or winter) then forming an almost brick-red, brown or orange surface bloom.

Often develops during a period of low algal population density. This observation has led Swale (1968) to suggest that such a slow-growing alga might only be capable of becoming dominant under conditions of minimal competition although animal grazing may also have an effect. Long after the bloom has subsided it is still possible to detect conspicuously coloured colonies (often with most of the cells missing) suspended in the water column plankton or forming a bottom deposit since bacterial decay of the oily mucilage is extremely slow.

Botryococcus calcareus West 1892

17080020 **Pl. 85I** (p. 350)

Cells ovoid to wedge-shaped, with indented apical margin, 7.5–10 μm wide, 10–12.5 μm long, not very densely packed within mucilaginous mass.

British Isles (England, Ireland); only known from two areas in England (Malham area, Yorkshire; the English Lake District) and in Ireland where 'frequent amongst *Spirogyra* in small limestone pools on the Burren Hills, Co. Clare' (West, 1892a, p. 92).

We have included it since cell shape readily distinguishes *B. calcareus* from all other species. Some uncertainty inevitably attaches to it since recorded in widely separated localities (see above). According to West (1892a), it is distinguished from *B. braunii* by the shape of its cells which are also less densely packed.

Botryococcus protuberans West *et* G.S.West 1905

17080030 **Pl. 85J** (p. 350)

Cells elongate, ellipsoidal or ovoid, 7.2–14.4 μm wide, 12–20 μm long, prominently projecting from periphery of a stratified mucilaginous mass.

Probably cosmopolitan; known in the British Isles from the west of Ireland and numerous lochs in north-west Scotland and the Outer Hebrides.

Botryosphaerella P.C.Silva 1970

Synonym: *Botryosphaera* Lemmermann

Colonies spherical or grape-like, consisting of cells embedded along the periphery of mucilaginous masses or united by irregularly fused mucilaginous strands; cells spherical or broadly ovoid, containing starch and oil droplets; chloroplast parietal, cup-shaped, without pyrenoids; asexual reproduction by 4 autospores produced in each cell.

Probably cosmopolitan; common in nutrient-poor (oligotrophic) water or where discoloured with brown humic materials (dystrophic).

Botryosphaerella sudetica (Lemmermann) P.C.Silva 1970

Basionym: *Botryococcus sudeticus* Lemmermann
Synonym: *Botryosphaera sudetica* (Lemmermann) Chodat
17090010 **Pl. 86A** (p. 352)
Colonies about 80 μm wide; cells ovoid, up to 10 μm wide, with thin remnants of mother cell wall material persisting within mucilage following autospore release.

Probably cosmopolitan; reported on only few occasions in the British Isles from *Sphagnum* bogs and other bodies of water in England (including a moat at Groombridge, Kent) and Ireland.

Bracteacoccus Tereg 1923

Cells solitary or forming irregular masses, spherical, with walls thick or thin; chloroplasts parietal, 1 to several, plate-like to angular or polygonal, without pyrenoids; starch or oil present; multinucleate; asexual reproduction by aplanospores and naked biflagellate zoospores, sometimes flagella of slightly unequal length; akinetes unknown.

Probably cosmopolitan; associated with soil and subaerial surfaces.

A little-known genus which is in need of revision. Attributed to the Order Sphaeropleales as defined by Deason et al. (1991).

Bracteacoccus anomalus (E.J.James) R.C.Starr 1955

Basionym: *Pleurochloris anomalus* E.J.James
17100010 **Pl. 82C** (p. 329)
Cells 7.5–16 μm wide, thin-walled; chloroplasts in pairs, curved and plate-shaped, yellow-green or pale green; oil droplets present; zoospores 2–3 μm wide, 6–8 μm long, elongate, chloroplast single and pale green, with a lateral and streak-like eyespot; aplanospores 3 or more per cell.

British Isles (England); known only from cultures of sandy soil from woodland in Surrey.

Characiochloris Pascher 1927

Cells solitary, more or less cylindrical, narrowing to a rounded or conical apex and attached basally by a mucilage pad; chloroplast divided longitudinally and possessing 1 to several pyrenoids; contractile vacuoles distributed throughout cell and 2 short, vestigial flagella often projecting from distal end; asexual reproduction by biflagellate zoospores and aplanospores, often sporangia swollen and ovoid.

Probably cosmopolitan; grows on filamentous algae, dead material and crustaceans in various aquatic habitats, occasionally isolated from soil.

Often the mucilage attachment pad is brown due to the presence of iron oxides. Nozaki (1993) has provided an identification key and reviewed the genus when describing a new species from Japan. Lee & Bold (1974) have studied *Characium* and *Characium*-like algae (including *Characiochloris*) in culture and use zoospore characters to separate some of the species. Often placed in the Order Tetrasporales.

1 Cells shortly stalked, with a nipple-shaped apical wall thickening .... *C. apiculata*
1 Cells without a distinct stalk, without an apical wall thickening 2
2 Cells spindle-shaped, narrowing equally towards base and apex .... *C. characioides*
2 Cells shaped like a tear-drop, rounded basally and conical above..................................... *C. sessilis*

Characiochloris apiculata Korshikov 1953

17110010 **Pl. 99C** (p. 403)
Cells about 11 μm wide, up to 40 μm long, varying from elongate (maximum width in lower two-thirds) through to elongate-oval or somewhat cylindrical, with apex bluntly rounded or cone-like and having a nipple-shaped terminal wall thickening, narrowing basally to a short stalk.

Probably cosmopolitan; known in the British Isles only on roots of the duckweed *Lemna minor*; probably more widely distributed than records indicate.

Characiochloris characioides (Korshikov) Pascher 1927

Basionym: *Chlamydomonas characioides* Korshikov
17110020 **Pl. 99D** (p. 403)
Cells about 15 μm wide, up to 70(–120) μm long, regularly spindle-shaped, with apex acute and thin-walled, also narrowing basally but lacking a distinct stalk.

Probably cosmopolitan; growing in the British Isles on other algae and less commonly on other submerged surfaces.

Characiochloris sessilis (Korshikov) Pascher 1927

Basionym: *Chlamydomonas sessilis* Korshikov
17110030 **Pl. 99E** (p. 403)
Cells up to 24 μm long, initially spindle-shaped, later becoming shortly conical or tear-drop in shape, broadly rounded basally, sessile and tapering above, with apex blunt and thin-walled.

Probably cosmopolitan; in the British Isles often epiphytic on filamentous algae in various aquatic habitats.

Characium A.Braun in Kützing 1949

Cells solitary, sessile or more frequently with a short stalk, usually attached by an expanded, mucilaginous disc (often brown due to presence of iron oxides); cells variously shaped, straight or curved, usually asymmetrical, apex acute or obtuse, with distal wall thin or thickened and sometimes with a nipple-like protrusion or point; chloroplast parietal and containing 1 or more pyrenoids; asexual reproduction by naked biflagellate zoospores, usually released through lateral splitting or gelatinization of wall.

Probably cosmopolitan; most commonly epiphytic on algae, aquatic macrophytes and animals although sometimes on other surfaces, widely distributed in freshwater habitats; more rarely associated with soil.

Many *Characium* species closely resemble members of the xanthophyte genus *Characiopsis*, but are distinguished by staining positively for starch except for young and actively growing cells which do not always stain. A few of the species recorded from the British Isles have been transferred to *Characiopsis* on closer examination (e.g. *Characium subulatum* to *Characiopsis subulata* (A.Braun) Borzí). The principal characters used to separate species are differences in cell shape and size although little is known of the variability of these in response to age and ecological conditions.

1 Soil alga; stalk short and basally pointed; attachment disc absent ....*C. pseudopyriforme*
1 Aquatic algae; stalk absent, if present then not pointed; mucilaginous attachment disc present 2
2 Cells symmetrical, initially elongate-oval or linear-ellipsoidal, later pear-shaped or obovoid, frequently >20 μm wide..........................3
2 Cells asymmetrical, more or less curved, usually of another shape, always <20 μm wide 4
3 Cells narrowly lance-shaped; apex acute and thick-walled..*C. angustum*
3 Cells elongate-oval; apex more or less obtuse or acute, usually thin-walled.................*C. sieboldii*
4 Cells very elongate and stalk exceedingly long (>50 μm long) ....*C. westianum*
4 Cells crescent-shaped to almost semicircular, arising obliquely on a relatively short (<50 μm long) stalk.................*C. ornithocephalum*

Characium angustum A.Braun 1855

Synonym: *Characium apiculatum* Rabenhorst

17120010 **Pl. 99F** (p. 403)

Cells 25–100 μm long, up to 5.5 times longer than broad, symmetrical, narrowly lance-shaped, arising on a short, thick stalk; apex short, acute and thick-walled.

Probably cosmopolitan; in the British Isles reported growing on filamentous green algae.

Characium ornithocephalum A.Braun 1855

17120030 **CD**

Cells 6–16 μm wide, (11.5–)25–33(–46) μm long (excluding stalk), asymmetrical, crescent-shaped or almost semicircular, arising obliquely on a narrow, short stalk and attached by a distinct disc; apex acute with a short colourless point.

Probably cosmopolitan; in the British Isles recorded on filamentous green algae in England including at Sheep's Green in Cambridge and Keighley Moor in Yorkshire (West, 1899b).

var. *harpochytriforme* Printz 1914

Synonym: *Characium falcatum* Schröder

17120032 **Pl. 99H** (p. 403)

Cells 3.8–6.5(–7.8) μm wide, 36–50 μm long, spindle- to crescent-shaped, markedly asymmetrical; apex with a short or long point.

Probably cosmopolitan; similar ecology to other varieties including the type variety.

var. *pringsheimii* (A.Braun) Komárek 1979

Basionym: *Characium pringsheimii* A.Braun

Synonym: *Characium ornithocephalum* A.Braun *sensu* Korshikov

17120033 **Pl. 99G** (p. 403)

Cells 5–12 μm wide, (13–)18–35 μm long (excluding stalk), asymmetrical, generally ovoid, arising from a relatively thick and short stalk; apex acute and bent slightly to one side.

Probably cosmopolitan; frequently occurs in the British Isles on other algae, aquatic plants and inanimate surfaces in a variety of aquatic habitats.

Characium pseudopyriforme (J.W.G.Lund) Philipose 1967

Basionym: *Characium pyriforme* J.W.G.Lund *non* A.Braun

Synonym: *Characium naegeli* A.Braun *sensu* Bristol

17120040 **Pl. 99I** (p. 403)

Cells up to 13 μm wide, 34 μm long, broadly pear-shaped to ovoid and arising abruptly from a short and pointed stalk (no basal disc), occasionally elongate and irregularly ellipsoidal; apex usually broadly rounded and occasionally asymmetrical.

British Isles (England); isolated from soil (see Lund, 1947a).

Hindák (1988) attributes it to the genus *Apodochloris*, considering it a possible synonym of *A. polymorpha* although the latter is ellipsoidal or sac-like and without a stalk.

Characium sieboldii A.Braun 1855

17120050 **Pl. 99J** (p. 403)

Cells (17–)20–33 μm wide, 40–70 μm long, up to about 3 times longer than broad, symmetrical, initially elongate-oval or linear-ellipsoidal and later pear-shaped or obovoid, with a short, thick and not sharply delimited stalk; apex obtuse or slightly acute and thin-walled.

Probably cosmopolitan; grows in the British Isles as an epiphyte on other algae, mosses and aquatic macrophytes.

Characium westianum Printz 1914

Synonym: *Characium ensiforme sensu* G.S.West

17120060 **Pl. 99K** (p. 403)

Cells (2.5–)3.8–10(–12.5) μm wide, (16–)50–65(–86) μm long, exceedingly elongated, slightly asymmetrical in apical part, narrowing to a blunt point and with a very long, slender stalk.

Probably cosmopolitan; in the British Isles a rarely recorded epiphyte on filamentous algae and other aquatic plants (e.g. from Pilmoor, Yorkshire).

A little-known and possibly doubtful taxon.

Chlorella Beijerinck 1890

Cells solitary, more rarely grouped, spherical or ellipsoidal, generally thin- and smooth-walled; chloroplast parietal, usually single, cup-, band- or trough-shaped, rarely with narrow incisions or net-like, usually with a single pyrenoid and sometimes surrounded by a starch envelope; asexual reproduction by 2, 4 or 8 (rarely 16, 32 or 64) autospores, released by equal or unequal fracture of cell wall.

Cosmopolitan; widely distributed and free-living in freshwater and marine water, soil and subaerial habitats; endosymbionts of freshwater invertebrate animals including sponges, *Hydra viridis*, and several protozoa (e.g. *Ophrydium*, *Paramecium bursaria*); also forms a symbiotic relationship with ascomycete fungi in lichens; isolated from plant sap.

The genus is highly diverse genetically and many of the more than 100 traditionally defined species are not closely related. Molecular phylogenetic analyses have shown that the genus can be separated into at least three clades, one of which has given rise to the colourless genus *Prototheca* (see Friedl, 1995). The absence of morphological characters (visible with the light microscope), combined with autospores as the only means of reproduction, has led to considerable difficulty in defining and identifying species. Various approaches have been adopted, e.g. Fott & Nováková (1969) combined a traditional approach using morphological and physiological characters. Other approaches involve serological cross-reactions or species and strains are defined based on ultrastructural characters and chemical composition of the cell wall. Huss et al. (1999) took account of biochemical, physiological and ultrastructural characters along with molecular data. They demonstrated a complete lack of congruence between the molecular and the classical morphology-based systematics with traditionally defined species distributed over at least two classes of green algae. Only 4 species were regarded as belonging to the genus *Chlorella* and these all possessed glucosamine as the dominant cell wall component and a double thylakoid 'bisecting' the pyrenoid. It is outside the scope of this volume to adopt the approach of Huss and his colleagues to identify algae traditionally placed in the genus *Chlorella*. The morphologically defined species considered are very artificial and this applies equally to many other green algal taxa. We have not considered *C. protothecoides*, since it is only mentioned in the catalogue of the *Culture Collection of Algae and Protozoa* (Tompkins et al., 1995) as an isolate from a 'Plant; Cambridge, England'.

1 Cells with a pyrenoid 2
1 Cells without a pyrenoid 5
2 Cells spherical or subspherical; chloroplast usually cup-shaped or broadly band-shaped when mature 3
2 Cells ellipsoidal, cylindrical-ellipsoidal or ovoid-ellipsoidal; chloroplast saucer-like, trough-shaped, band-shaped or disc-like and concave 4
3 Chloroplast with a few longitudinal and narrow fissures or splits .... *C. emersonii*
3 Chloroplast without any splits .... *C. vulgaris*
4 Chloroplast small, trough-like or band-shaped, with a smooth margin; pyrenoid without starch sheath; autospores equal in size .... *C. saccharophila*
4 Chloroplast large, saucer-shaped, band-shaped or disc-like and concave, margin irregularly undulate and lobed; pyrenoid with starch sheath; autospores unequal in size .... *C. ellipsoidea*
5 Cells 1–2.5(–3.5) µm wide; autospores no more than 8 in each sporangium .... *C. minutissima*
5 Cells 3–15 µm wide; autospores more than 16 in each sporangium .... *C. miniata*

Chlorella ellipsoidea Gerneck 1907

Synonym: *Chlorella saccharophila* (W.Krüger) Migula var. *ellipsoidea* (Gerneck) Fott *et* Nováková

17130010 **Pl. 82E** (p. 329)

Cells 1.5–13 µm wide, 2–15 µm long, symmetrical; chloroplast initially saucer-, trough- to band-shaped, becoming more lobed with age, margins irregularly undulate and occasionally incised; pyrenoid associated with numerous starch grains; autospores ellipsoidal, 2, 4, 8, 16 or 32 in each sporangium, unequal in size with smaller ones more numerous.

Probably cosmopolitan; in the plankton of pools, small lakes and other aquatic habitats as well as associated with soil and subaerial surfaces (e.g. wall of building at the University of Cork, Ireland; see Schlichting, 1975) in the British Isles.

Chlorella emersonii Shihira *et* R.W.Krauss 1965

17130020 **Pl. 82F** (p. 329)

Cells 3–17 µm wide, spherical to subspherical, solitary or grouped; chloroplast initially trough-shaped, later band-shaped and covering most of wall, with fissures or a split present; pyrenoid large, surrounded by 2–4 starch grains; autospores ellipsoidal, 2, 4 or 8(–16) in each sporangium.

Probably cosmopolitan; associated with subaerial surfaces in the British Isles.

Chlorella miniata (Nägeli) Oltmanns 1904

Basionym: *Pleurococcus miniatus* Nägeli

17130030

Cells 3–15 µm wide, spherical, single or in 2- or 4-celled groups; chloroplast cup-shaped, spherical with pore in its wall relatively small; pyrenoid absent; autospores 16, 32 or 64 in each sporangium, released by fracture of wall.

Europe; known in the British Isles from various subaerial surfaces, including walls and flower pots in greenhouses, rare.

Chlorella minutissima Fott *et* Nováková 1969

17130040 **Pl. 82G** (p. 329)

Cells 1–2.5(–3) µm wide, spherical; chloroplast cup- or saucer-shaped; pyrenoid not visible; autospores broadly ellipsoidal, 2, 4 or 8 per sporangium, released by gelatinization of wall.

Probably cosmopolitan; often associated in the British Isles with submerged objects in a diverse range of aquatic habitats, sometimes on subaerial surfaces.

Chlorella saccharophila (W.Krüger) Migula 1907

Basionym: *Chlorothecium saccharophilum* W.Krüger
Synonyms: *Chlorella ellipsoidea* Gerneck *sensu* Shikira *et* Krauss, *Palmellococcus saccharophilus* (W.Krüger) Chodat
17130050 **Pl. 82H** (p. 329)

Cells 2–7 μm wide, 3.5–10.2 μm long, cylindrical-ellipsoidal, ellipsoidal or ovoid-ellipsoidal, with broadly rounded apices; chloroplast trough-like or band-shaped, filling one half of cell and open at apex, with an indistinct pyrenoid and lacking a starch sheath; autospores ovoid to ellipsoidal, 4 or 8 (rarely 2 or 16) per sporangium.

Probably cosmopolitan; in the British Isles associated with subaerial surfaces, soil and isolated from tree sap.

Chlorella vulgaris Beijerinck 1890

Synonym: *Chlorella pyrenoidosa* var. *duplex* (Kützing) West
17130060 **Pl. 82I** (p. 329)

Cells 1.5–10(–13.3) μm wide, spherical or almost spherical; chloroplast broadly band-shaped or cup-shaped, pyrenoid spherical to broadly ellipsoidal and usually surrounded by 2–4 starch grains; autospores spherical, 2, 4 or 8(–16) in each sporangium, released by wall breaking into 2–4 parts.

Probably cosmopolitan; in the British Isles very widely distributed in the plankton, associated with surfaces in a wide range of aquatic habitats as well as in soil and subaerial habitats; most frequent in slightly to moderately polluted water.

A very variable species represented by a large number of local physiological and ecological races growing under very different conditions. Most literature records need treating with caution since it has been common practice to refer to this species almost any small 'coccoid' green alga.

DOUBTFUL RECORD

Chlorella botryoides J.B.Petersen

Doubt attaches to the report by James (1935) who isolated it from clayey and sandy soils at Redlands Woods, Shoreham in Kent. Bold (1958) suggests that it might be identical with *Neochloris alveolaris* H.C.Bold (now *Ettlia alveolaris* (H.C.Bold) H.Ettl *et* G.Gärtner).

Chlorochytrium Cohn 1872

Synonym: *Kentrosphaera* Borzí

Cells solitary, spherical, ellipsoidal or irregularly curved or lobed, with walls usually thick and stratified; chloroplast radiating from cell centre, towards periphery the arms expand into broad parietal lobes or form an almost complete parietal layer, with 1 or several scattered pyrenoids. Reproduction by biflagellate zoospores and gametes; aplanospores and akinetes known.

Probably cosmopolitan; often in intercellular spaces of aquatic macrophytes and animals, sometimes associated with soil or other algae (including blue-greens); rarely free-living.

The taxonomic status of *Chlorochytrium* requires re-evaluation with some species considered to be stages in the life history of filamentous algae. According to Smith (1933), *Kentrosphaera* is separated from *Chlorochytrium* by its free-living habit, lack of gametes and irregular thickening of the wall. We have followed Bristol (1920a) in not considering these differences sufficient justification for separating the two genera.

1 Cells with one or more striated external wall projections and internal projections 2
1 Cells without wall projections (only associated with zoosporangia in *C. facciolae*) or having button-like excresences 3
2 Cells <65 μm in size; associated with mosses or an endophyte ....*C. paradoxum*
2 Cells usually >65 μm in size; associated with mats of cyanobacteria ....*C. bristolae*
3 Cells usually <40 μm (26–42 μm) in size, each with a pyrenoid; free-living and associated with filaments of cyanobacteria ....*C. facciolae*
3 Cells usually >40 μm in size, each with several pyrenoids; endophytic in aquatic macrophytes...*C. lemnae*

Chlorochytrium bristolae (G.M.Smith) D.M.John *et* Tsarenko, *nov. comb.*

Basionym: *Kentrosphaera bristolae* G.M.Smith, 1933. *The Freshwater Algae of the United States* p. 476, fig. 318.
Synonym: *Chlorochytrium paradoxum* (G.A.Klebs) G.S.West *sensu* Bristol
18050010 **Pl. 113D** (p. 450)

Cells spherical, ellipsoidal, triangular or very irregular, 35–63 μm wide and 50–165 μm long; walls irregularly thickened, frequently with a single or several striated external and internal projections; chloroplast possessing several pyrenoids; zoosporangia often containing bright orange granules.

Probably cosmopolitan; reported in the British Isles associated with mats of the blue-green alga *Phormidium* growing on chalk soils.

Smith (1933) considers Bristol's *Chlorochytrium paradoxum* to be a new species and places it in the synonymy of his new species *Kentrosphaera bristolae* G.M.Smith (see above).

Chlorochytrium facciolae (Borzí) Bristol 1920

Synonym: *Kentrosphaera facciolae* Borzí
18050020 (24080010) **Pl. 113B** (p. 450)

Cells spherical, ellipsoidal or irregularly shaped, 26–42 μm in size; walls thick and stratified; chloroplast with a single pyrenoid; red haematochrome pigment sometimes present; zoosporangia having thick and stratified walls, bearing 1–3 internal projections and a single curved external projection.

Europe and North America; reported in the British Isles associated with filaments of blue-green algae belonging to the Oscillatoriales.

Culture-based studies are required to establish whether *C. facciolae* and *C. paradoxum* should continue to be considered as separate species.

Chlorochytrium lemnae Cohn 1872

Synonyms: *Chlorochytrium knyanum* Cohn *et* Szymański, *Chlorosphaeropsis lemnae* F.Moewus
18950030 (24090020) **Pl. 113C** (p. 450)

Cells spherical, oblong, ellipsoidal or irregularly shaped, 40–100 μm in size, sometimes with a tubular neck; walls thin or occasionally thickened, often bearing button-like excrescences; choroplast possessing several pyrenoids; zoosporangia often remaining thin-walled and without projections.

Europe where known to be endophytic in several aquatic plants (e.g. *Ceratophyllum, Elodea*); reported in the British Isles to be common in the duckweed *Lemna* (*L. minor, L. trisulca*) where visible as green specks on dead and bleached leaves.

A very comprehensive culture-based investigation was carried out by Lewin (1984) on material isolated principally from the River Cherwell in Oxford.

Chlorochytrium paradoxum (G.A.Klebs) G.S.West 1916

Synonym: *Scotinosphaera paradoxa* G.A.Klebs
18050040 (24080030)
Cells very variable in shape, 35–65 μm in size; walls irregularly thickened and frequently one or several striated external and internal projections; chloroplast with 1–4 pyrenoids; zoosporangia often containing red granules.

Europe; reported in the British Isles growing as an endophyte within the duckweed *Lemna trisulca* and the moss *Hypnum*.

Doubt attaches to material isolated from soil by James (1935) who attributed it to this species. Komárek & Fott (1983) believe James's material might belong to the genus *Spongiochloris*.

DOUBTFUL SPECIES

Chlorochytrium grande Bristol 1917 (=*Kentrosphaera grandis* (Bristol) G.M.Smith)
Doubtful since known only from the original description of an isolate from soil collected in West Yorkshire by Bristol (1917). Similar and possibly conspecific with *C. facciolae* from which it differs only in cell size.

Chlorococcum Meneghini 1842

Cells solitary or initially in groups of indefinite form, without a mucilaginous envelope; cells spherical to ellipsoidal; walls smooth, usually thin but sometimes thickening with age; uni- or multinucleate; chloroplast parietal, cup-shaped or spherical and having a lateral pore, with 1 to several pyrenoids; starch and oil present; contractile vacuoles present; asexual reproduction by biflagellate zoospores, flagella of equal length, eyespots and contactile vacuoles present (resemble *Chlamydomonas*); aplanospores also produced; sexual reproduction by biflagellate isogametes; zygospores thick-walled, walls smooth or ornamented; hypnospores thick-walled, smooth or spiny.

Cosmopolitan; recorded from a wide range of aquatic habitats, frequently associated with soil and subaerial surfaces.

There is considerable confusion and disagreement regarding the limits of the 100 or more described species. Many earlier descriptions are known to have been of mixed populations collected from the field. It is now recognized that to reliably identify species of *Chlorococcum* and morphologically similar genera (e.g. *Chlorella*) requires isolation into unialgal culture. Starr (1955) was one of the first to adopt such a culture-based approach to the study of this genus and considered that 18 species were not sufficiently distinct to be positively identified. Often old cells (including ageing cultures) are no longer green or yellow-green but rather orange or red due to the abundance of carotenoid pigment, principally haematochrome. Doubt attaches to all species recorded from the British Isles. Analyses of sequence data indicates the genus to be polyphyletic (Nakayama et al., 1996).

1 Cells 4–10(–20) μm wide; zoospores ellipsoidal-ovoid .... *C. minutum*
1 Cells usually 10–40 μm wide; zoospores cylindrical or ellipsoidal 2
2 Hypnospores present (spiny-walled); pyrenoid surrounded by 2–5 starch grains .... *C. hypnosporum*
2 Hypnospores unknown; pyrenoids with many small starch grains .... *C. infusionum*

Chlorococcum hypnosporum R.C.Starr 1955

17140010 **Pl. 82J** (p. 329)
Cells 10–20(–30) μm wide, spherical to ovoid, always thin-walled; chloroplast a hollow sphere, sometimes having a lateral pore in which are often 2 contractile vacuoles; pyrenoid eccentric in position and associated with 2–5 discontinuous starch grains, with or without oil droplets; zoospores and aplanospores formed by successive bipartition of sporangial mother cell contents; zoospores ellipsoidal, 2.5–5.5 μm wide, 7.5–13 μm long, with 2 contractile vacuoles and an eyespot; aplanospores smooth-walled or occasionally spiny-walled and similar to hypnospore; sexual reproduction oogamous and the zygote smooth-walled; hypnospores spherical, spiny- and thick-walled, filled with oil droplets.

Europe and North America; in the British Isles known only from southern Ireland where identified by Schlichting (1975) in cultures of algae isolated from subaerial surfaces.

Chlorococcum infusionum (Schrank) Meneghini 1842

Basionym: *Lepraria infusionum* Schrank
Synonyms: *Chlorococcum humicolum* (Nägeli) Rabenhorst, *Cystococcus humicola* Nägeli *pro parte*, *Gloeocystis infusionum* Schrank
17140020 **Pl. 82L** (p. 329)
Cells (5–)10–40(–60) μm wide, spherical, rarely ovoid, walls thickening with age; chloroplast a hollow sphere with a lateral pore; pyrenoid eccentrically positioned,usually associated with several small starch grains; zoospores cylindrical or oval-cylindrical, 2–4 μm wide, (3–)6–12 μm long, with papilla and possessing contractile vacuoles and an eyespot; hypnospores unknown.

Probably cosmopolitan; recorded neutral to acid soils, porous acidic rocks and aquatic habitats in the British Isles including lakes (e.g. from Lough Guitane, Ireland).

Considerable doubt attaches to the records (as *C. humicolum*) from British soils by West & Fritsch (1927), James (1935), Bristol (1920a), Lund (1947a) and Fritsch & John (1942); the latter went so far as to describe a new variety (var. *incrassata*) distinguished from the type by its thick, striated walls. The dimensions given by James (1935) for the zoospores (4.5–6 μm wide, 9–12 μm long) are larger than those normally given thus bringing into question his identification. Petersen (1932) points out that 'It is questionable whether it will ever be fully cleared up what Nägeli actually had before him when he described this species. Both the description and the figure are quite insufficient as a means of identifying the alga, and later authors have tried to complete the diagnosis each in his own way'. We have followed most authors and consider *C. infusionum* and *C. humicolum* to be conspecific.

Chlorococcum minutum R.C.Starr 1955

17140030 **Pl. 82N** (p. 329)
Cells 4–10(–20) μm wide, spherical to ovoid; walls thicker in older cells; chloroplast a hollow sphere with a lateral pore, often 2 contractile vacuoles present in opening, pyrenoid eccentric and covered by a continuous starch sheath; zoospores, aplanospores and gametes produced by successive bipartitions; zoospores ellipsoidal-ovoid, 3–5 μm wide and 6–7 μm long, usually with 2 parietal chloroplasts, 2 anterior contractile vacuoles and an eyespot; sexual reproduction by isogametes of similar form and size to zoospores, zygotes smooth-walled; hypnospore spherical and thick-walled.

Probably cosmopolitan; in the British Isles known from soil and in small nutrient-enriched lakes where it occurs in quantity during the summer months, especially when circulation ceases and the water stratifies.

Chlorolobion Korshikov 1953

Cells solitary or in groups, attached or free-living, pod- or spindle-shaped to crescent-shaped, or ellipsoidal, asymmetrical, more convex on one side, with apices bluntly rounded, pointed or with a small nipple-like wall thickening; chloroplast parietal, plate-shaped and having 1 or 2 pyrenoids; oil droplets sometimes present; asexual reproduction by (2–)4, 8 or 16 autospores in each sporangium, arranged successively, released by fracture or gelatinization of mother cell wall.

Probably cosmopolitan; in various aquatic habitats where attached initially but often becoming planktonic.

The relationship of *Chlorolobion* to *Podohedra* is discussed by Komárek & Fott (1983) together with the relationship of *Dactylococcus caudatus* var. *bicaudatus* (A.Braun) Hansgirg to *Podohedra bicaudata* Geitler. According to them, it seems that Geitler (Geitler, 1965) was in error when considering this variety of *Dactylococcus* to be a synonym of *Podohedra bicaudata*; the latter has not been reported in the British Isles and was included in error in the *Coded List of Freshwater Algae of the British Isles* (Whitton et al., 1998a). *Chlorolobion* is a very ill-defined genus which closely resembles *Keratococcus* (see p. 360).

Chlorolobion braunii (Nägeli) Komárek 1979

Basionym: *Raphidium braunii* Nägeli
Synonyms: *Ankistrodesmus braunii* (Nägeli) Collins, *Monoraphidium braunii* (Nägeli) Komárková-Legnerová, *Keratococcus braunii* (Nägeli) Hindák
17150010 **Pl. 90C** (p. 367)
Cells solitary or temporarily united, usually free-living, (1–)2–4.5(–7.8) μm wide and (10–)13–52(–56) μm long, cylindrical or spindle-shaped, straight or rarely slightly asymmetrical, gradually narrowing to acute or somewhat obtuse apices; chloroplast covering most of inner surface and centrally notched, with a pyrenoid and oil droplets; autospores 2, 4 or 8 in each sporangium, released by fracture of mother cell wall into two cone-shaped halves.

Probably cosmopolitan; planktonic or associated with submerged surfaces in various aquatic habitats, recorded on only a few occasions in the British Isles.

Surprisingly, Korshikov (1953) did not transfer this species himself when creating the genus *Chlorolobion*. Doubt attaches to most records from the British Isles since the presence of pyrenoids is seldom mentioned.

DOUBTFUL RECORD
Chlorolobion tatrae Hindák
Doubt attaches to this species since its only mention is from the River Wye in an unpublished PhD thesis (Furet, 1979).

Closteriopsis Lemmermann 1899

Cells solitary, slender and needle-like or narrowly spindle-shaped, straight or slightly curved, apices needle- or spine-like or acute, without a mucilaginous envelope; chloroplast parietal, elongate and band-like, sometimes spirally twisted, usually not reaching extremities of each apex, with 4–42 pyrenoids (number increasing with age) in a single series; asexual reproduction by 2, 4 or 8 autospores in each sporangium, arranged successively and released by longitudinal fracture of mother cell wall.

Probably cosmopolitan; planktonic in various aquatic habitats.

Often young autospores contain a single pyrenoid and the number increases rapidly during cell growth. The increase in pyrenoid number with age is a characteristic feature of the genus.

1 Cells usually 2–3 μm wide, 12–150 μm long; pyrenoids always fewer than 9 per cell ....*C. acicularis*
1 Cells 3–7.5 μm wide, 190–240(–570) μm long; pyrenoids up to 16(–42) per cell..........*C. longissima*

Closteriopsis acicularis (G.M.Smith) J.H.Belcher *et* Swale 1962

Basionym: *Closteriopsis longissima* var. *acicularis* G.M.Smith
Synonym: *Ankistrodesmus falcatus* var. *tumidus* G.S.West
17170010 **Pl. 90D** (p. 367)
Cells 2–3(–6.5) μm wide, (10–)20–150(–210) μm long, narrowly spindle-shaped, straight or slightly curved, tapering to sharply or bluntly pointed apices; chloroplast banded to grooved, straight or spirally twisted, with 2–8 pyrenoids; autospores 2, 4 or 8 per sporangium, each containing a single pyrenoid.

Probably cosmopolitan; widely distributed in the British Isles where associated with water plants and plankton in small water bodies (e.g. animal drinking troughs), as well as in ponds, lakes, reservoirs and flowing water.

Morphologically very similar to *Schroederia setigera*, although more slender and not reproducing by zoospores.

Closteriopsis longissima (Lemmermann) Lemmermann 1899

Basionym: *Closterium pronum* var. *longisssima* Lemmermann
Synonyms: *Ankistrodesmus longissimus* (Lemmermann) Wille, *Rhaphidium pyrenogerum* Chodat
17170020 **Pl. 90E** (p. 367)
Cells 3–7.5(–17) μm wide, 190–240(–570) μm long, very narrowly spindle-shaped, straight or slightly curved, tapering to needle- or spine-like apices; chloroplast banded or spirally twisted, 2–16(–42) pyrenoids; reproduction unknown.

Probably cosmopolitan; widely distributed in the British Isles in the plankton of Irish rivers and loughs (e.g. Upper River Bann and Lough Larne, Kerry), as well as in similar habitats in Scotland (e.g. Loch Lomond, Loch Leven, lochs in the Orkney and Shetland Islands) and England.

var. *tropica* West *et* G.S.West 1905
17170022
Similar to the type variety, although apices acute rather than needle- or spine-like; chloroplast band-like and spirally twisted.

Probably cosmopolitan; in the British Isles known only from the Orkneys (West & G.S.West, 1905a) and from Ireland (Round & Brook, 1959); planktonic.

Coccomyxa Schmidle 1901

Cells cylindrical, elongate-ovoid or ellipsoidal, sometimes asymmetrical, thin-walled, solitary or grouped within a mucilaginous envelope, with or without concentric striations surrounding cell groups; chloroplast parietal, plate-like and lateral in position, pyrenoids absent; asexual reproduction by 2–4(–8) autospores produced in each sporangium.

Probably cosmopolitan; microscopic or in macroscopic mucilaginous masses in terrestrial and aquatic habitats, rarely free-floating, also a lichen phycobiont.

Coccomyxa confluens (Kützing) Fott 1975

Basionym: *Gloeocapsa confluens* Kützing
Synonym: *Coccomyxa dispar* Schmidle
17180010 **Pl. 86B** (p. 352) **CD**
Cells 1.2–8 μm wide, 2.5–16 μm long, ellipsoidal, sometimes slightly asymmetrical, often with long axes lying parallel to each other, concentric striations surrounding cell groups most conspicuous immediately following autospore formation.

Probably cosmopolitan; very common in the British Isles on wood, stones, bark (especially near base of woodland trees), soil and mosses where it often forms extensive green to olive-green mucilaginous masses in wet summers.

REJECTED RECORDS

Coccomyxa subellipsoidea E.Acton 1909

West & Fritsch (1927) provide measurements (4–6 μm wide, 6–10 μm long) and illustrations of material attributed to this species that was associated with stones collected in the Botanical Gardens at Birmingham. It is clear from West & Fritsch's (*op. cit.*) and Acton's (1909) description, that it closely resembles *C. confluens*. According to Acton (*op. cit.*), the new species differs only by 'the greater regularity in the form of the cells, and in the presence of pyrenoids', and is widely distributed in the British Isles (including Clare Island, Co. Mayo) where it occurs 'in subaerial situations as a dark-green stratum on damp rocks and stones'. Its exact taxonomic placement remains uncertain since all other species of *Coccomyxa* lack pyrenoids.

Coccomyxa solorinae Chodat 1913

Only reported in the British Isles growing in soil cultures (Bristol, 1920b). Current status uncertain since not mentioned in recent reviews of chlorococcalean algae (e.g. Komárek & Fott, 1983); usually considered to be a phycobiont of the lichen genus *Solorina*.

Coelastropsis Fott *et* Kalina 1979

Coenobia spherical or broadly ovoid, sometimes producing two rows of cells similar to *Scenedesmus* (usually in culture), consisting of (2–)4, 8 or 16 spherical or broadly ellipsoidal cells, 5–6 adjacent cells connected by very short protuberances; walls very thick, usually with thickened, parallel, coarse, serrated ribs along margin and connected by minute, transverse ribs; chloroplast parietal with internal projections and a single large pyrenoid; asexual reproduction by (2–)4, 8 or 16 autospores, with daughter colonies remaining connected on release from the sporangia.

Europe and New Zealand; usually in bogs amongst mosses and filamentous algae in upland areas. Monospecific.

Coelastropsis costata (Korshikov) Fott *et* Kalina 1979

Basionym: *Coelastrum costatum* Korshikov
18060010
Coenobia up to 40(–58) μm in diameter and cells 10–15 μm in diameter.

Europe; in the British Isles known only from Loch Lubnaig from where it has been reported by Hindák (1984).

Coelastrum Nägeli 1849

Coenobia hollow, spherical or less commonly pyramidal, quadrate or many-sided, consisting of (2–)4, 8, 16 or 32(–64) spherical, ovoid, polygonal or angular cells, with or without surface ornamentation, closely interconnected by blunt processes or protuberances arising from the mucilaginous sheath surrounding each cell; chloroplast parietal, cup-shaped or diffuse, with a single pyrenoid; asexual reproduction by autospores and akinetes.

Probably cosmopolitan; planktonic and widely distributed in various aquatic habitats.

1 Cells usually spherical, neighbouring cells attached directly to one another, without connecting wall-protuberances or surface ornamentation .... *C. microporum*
1 Cells spherical, ovoid or angular, with or without connecting wall-outgrowths, sometimes with surface ornamentation 2

2 Cells ovoid, outer end of each narrowed and obtuse, with connecting protuberances not evident 3
2 Cells of another shape, with evident connecting protuberances 6
3 Cells not hexagonal in surface view, very small intercellular spaces .... *C. pseudomicroporum*
3 Cells hexagonal in surface view, large intercellular spaces (to about half cell width or more) 4
4 Cells with a distinct apical wall thickening and surface smooth .... *C. sphaericum*
4 Cells without an apical wall thickening and with or without surface projections, if wall thickened then having warty surface projections 5
5 Walls having warty surface projections .... *C. proboscideum*
5 Walls always smooth .... *C. astroideum*
6 Cells somewhat angular, connected to neighbouring cells by 5–6 truncate protuberances .... *C. pulchrum*
6 Cells usually spherical, connecting protuberances of another form 7
7 Cells connected laterally by 1(–2) wall outgrowth, without free wall outgrowths .... *C. reticulatum*
7 Cells connected laterally by 2–3 wall outgrowths, with many warty wall outgrowths .... *C. verrucosum*

Coelastrum astroideum De Notaris 1867

Synonyms: *Coelastrum microporum* fo. *astroidea* (De Notaris) Nygaard, *C. microporum* Nägeli *sensu pro parte auct. post.*

17200010 **Pl. 83H** (p. 341)

Coenobia spherical, 4–32-celled, up to about 100 µm in diameter; cells (3.5–)5–20×(3.5–)5–20 µm, obovoid with outer wall strongly protruding, connected basally to neighbouring cells but without obvious wall projections, often hexagonal in surface view due to mutual compression; intercellular spaces often more than half the width of the cell.

Probably cosmopolitan; planktonic in many types of water body in the British Isles but most common in ponds and lakes.

More common and widely distributed than records suggest since frequently misidentified as *Coelastrum microporum*, a species from which it can often be separated on cell shape alone.

Coelastrum microporum Nägeli in A.Braun 1855

17200020 **Pl. 83I** (341) **CD**

Coenobia spherical, (4–)8-, 16- or 32(–64)-celled, up to about 100 µm in diameter; cells (3.5–)6–10.7(–17) µm wide, spherical to ovoid, joined without connecting projections; intercellular spaces small, triangular or rectangular.

Probably cosmopolitan; widely distributed throughout the British Isles where common during the summer months in ponds, lakes and rivers. Often most abundant in April and May when *Pediastrum boryanum* is also present in quantity.

Coelastrum proboscideum Bohlin in Wittrock & Nordstedt 1896

17200030 **Pl. 83J** (p. 341)

Coenobia more or less pyramidal, usually 4-, 8- or 16-celled, up to about 110 µm wide; cells 4.5–10.5(–14)× 4.5–10(–12.5) µm, conical or truncate and hexagonal in surface view, poles distinctly thickened, each cell connected to 3 neighbouring ones by short, blunt projections; intercellular spaces half width of cell or greater, often polygonal; surface warty in appearance.

Probably cosmopolitan; in the British Isles known from ponds, especially those associated with bogs, rare.

Young colonies might be easily mistaken for *C. sphaericum* when the apical thickenings are weakly developed. Various authorities have misidentified the two species, so care is needed accepting earlier records. We consider all records from the British Isles to be based upon Bohlin's concept of the species.

Coelastrum pseudomicroporum Korshikov 1953

17200040 **Pl. 83K** (p. 341) **CD**

Coenobia spherical, 8–16(–32)-celled, up to about 100 µm in diameter; cells 5.2–10(–14.2)×5.2–7.1 µm, ovoid, strongly outwardly protruding, each cell with a narrowing and sometimes obtuse apex, connected at broad end to neighbouring cells by 4–6 short protuberances (often not very evident), sometimes with surface ornamentation; intercellular spaces very small and usually triangular.

Europe; reported on a few occasions from rivers in South-East England, probably more widely distributed in the British Isles than records would suggest.

See remarks under *Coelastrum verrucosum*.

Coelastrum pulchrum Schmidle 1892

Synonym: *Coelastrum cambricum* W.Archer

17200090 **Pl. 83L** (p. 341)

Coenobia spherical, tetrahedral or cuboidal, (4–)8–16 (–32)-celled, with intercellular spaces up to half cell width and triangular or rounded; cells more or less angled, connected to neighbouring cells by 5–6 lateral and truncated projections, each cell possessing a blunt polar projection.

Probably cosmopolitan; recorded only rarely in ponds and lakes in widely separated localities (e.g. Lough Shannacloontippen and Lough Aunierin in Ireland, Cambridge area in England).

Coelastrum reticulatum (P.A.Dangeard) Senn 1899

Basionym: *Hariotina reticulata* P.A.Dangeard

17200100 **Pl. 83M** (p. 341)

Coenobia spherical to ovoid, 8-, 16- or 32-celled, up to about 100 µm in diameter; cells (3.3–)5–15 µm, spherical to slightly ellipsoidal, neighbouring cells connected by a single (rarely 2) wall outgrowth and by 5–9(–12) hollow protuberances arising from near outer pole of each cell; daughter coenobia sometimes remaining attached to colony by remnants of mother cell wall and mucilaginous strands.

Probably cosmopolitan; widely distributed in ponds throughout the British Isles.

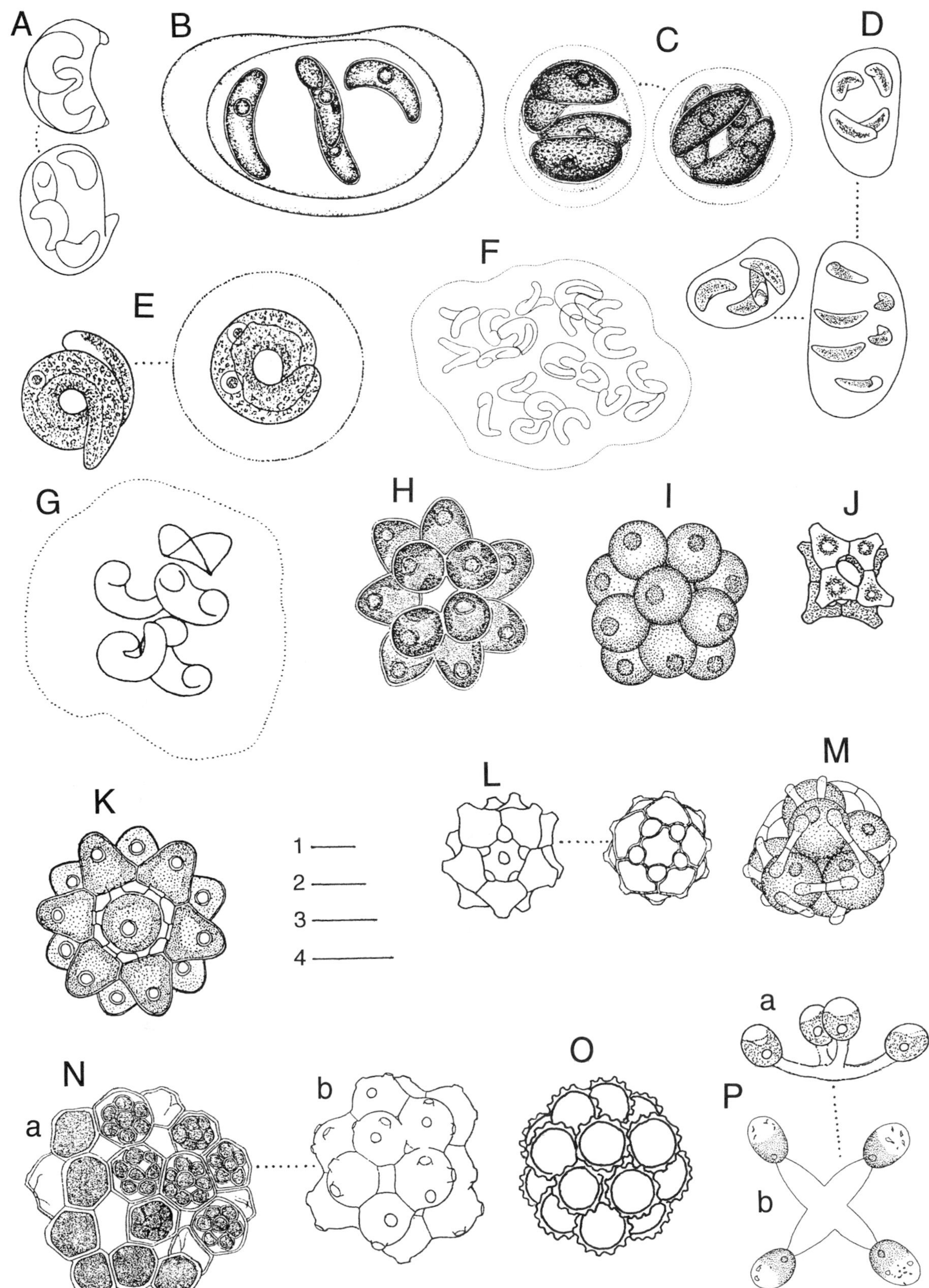

Plate 83 A–P

A. *Nephrochlamys rostrata* (p. 371): autospores in mother cells; **B**. *Nephrocytium agardhianum* (p. 371); **C**. *Nephrocytium limneticum* (p. 371); **D**. *Nephrocytium lunatum* (p. 371); **E**. *Raphidocelis rotunda* (p. 384); **F**. *Raphidocelis subcapitata* (p. 384); **G**. *Raphidocelis danubiana* (p. 383); **H**. *Coelastrum astroideum* (p. 340); **I**. *Coelastrum microporum* (p. 340); **J**. *Coelastrum proboscideum* (p. 340); **K**. *Coelastrum pseudomicroporum* (p. 340); **L**. *Coelastrum pulchrum* (p. 340); **M**. *Coelastrum reticulatum* (p. 340); **N**. *Coelastrum sphaericum* (p. 342): a. some cells of coenobium containing autospores, b. protuberances evident on surface of cells; **O**. *Coelastrum verrucosum* (p. 342); **P**. *Dictyosphaerium chlorelloides* (p. 349): details of a group of 4 cells in side (a) and surface (b) view connected by remnants of mother cell wall.

Scale bars: 1, 2, 3, 10 μm; 4, 20 μm. 1 for O; 2 for A, B, F, H, J, L–Na,b; 3 for D; 4 for C, E, G, I, K, Pa,b.

Coelastrum sphaericum Nägeli 1849

Synonymy: *Coelastrum cubicum* Nägeli

17200070 **Pl. 83N** (p. 341)

Coenobia spherical, ellipsoidal or cuboid, 4-, 8-, 16- or 32-celled, up to about 100 μm wide; cells 5–18(–30) μm wide, ovoid with outer wall protruding, 6-sided in surface view due to mutual pressure, each cell bearing 3, 6 or more tuberculate protuberances (not evident); intercellular spaces about half cell width; surface smooth.

Probably cosmopolitan; in the British Isles known from small ponds and lakes in Ireland (e.g. Ballynahinch, near Westport in the west of Ireland) and the English Midlands; probably more widely distributed than records indicate.

Coelastrum verrucosum (Reinsch) Reinsch 1878

Basionym: *Sphaerastrum verrucosum* Reinsch

Synonym: *Coelastrum morus* West *et* G.S.West

17200080 **Pl. 83O** (p. 341)

Coenobia spherical to irregularly shaped, 4-, 8- or 16-celled (rarely single-celled), up to 36(–64) μm; cells 10–15 (–24) μm, spherical or somewhat flattened, each cell bearing 6–11(–12) low and obtuse or bluntly pointed protuberances, connected laterally to neighbouring cells by (1–)2–3 wall ougrowths, polar end bearing free wall outgrowths; intercellular spaces small and irregularly shaped.

Probably cosmopolitan; planktonic and reported in the British Isles from a few Scottish localities (e.g. Loch Lubnaig) and in the west of Ireland (Adrigole).

Hajdu et al. (1976) considered *C. verrucosum* and several other species (including *C. morus* West *et* G.S.West) to be synonyms of *C. sphaericum* Nägeli on the grounds that they fell within the form range of a strain investigated in culture. We have followed Hindák (1992) in continuing to recognize *C. verrucosum* as a separate species. Comas (1989) considers *C. morus* to be a rare tropical form of *C. verrucosum*, but the report of its discovery in Scotland and elsewhere in the northern hemisphere (as *C. morus*) suggests that it is widely distributed.

UNCERTAIN RECORD

Coelastrum schizodermaticum Rich 1921

Doubt attaches to this *Coelastrum* described from a sample collected from a ditch near Leicester (Rich, 1921). The coenobia (usually 8-celled) have faintly hexagonal cells, small intercellular spaces, a well developed mucilage envelope, and little 4-cornered cap-like pads of wall material detached from the outer surface of each cell. This unusually structured *Coelastrum* is incompletely known and has not been reported again since its original description.

Coenochloris Korshikov 1953

Coenobia microscopic, of (2–)4, 8 or more cells in closely packed groups, embedded within a spherical, ellipsoidal or irregularly shaped envelope of mucilage; cells spherical to broadly ellipsoidal, thin-walled, smooth surfaced or sometimes with warts; chloroplast single, parietal, with or without a pyrenoid; asexual reproduction by (2–)4–8(–16) tetrahedrally or octahedrally arranged autospores, released by splitting of mother cell wall into two irregularly shaped fragments, wall remnants persisting for a short time.

Probably cosmopolitan; planktonic in a wide range of aquatic habitats.

Only two species recognized by Korshikov (1953) when creating the genus and 10 species subsequently described. Considerable confusion exists between this and morphologically similar genera such as *Coenococcus, Coenocystis* and *Sphaerocystis*. Various attempts have been made to clarify its identity by sometimes redefining the characters used to distinguish between them (e.g. Hindák, 1988). Little or no information exists on the influence of environment on many of the characters (e.g. persistence of mother cell wall remnants) considered to be of taxonomic importance. We have attempted to follow the currently accepted view concerning these difficult genera, but recognize the situation to be far from satisfactory since many characters are obscure and often evident only in actively growing and reproducing colonies.

Coenochloris fottii (Hindák) Tsarenko 1990

Basionym: *Coenococcus fottii* Hindák

Synonyms: *Sphaerocystis schroeteri* Chodat *sensu auct. post. pro parte, Eutetramorus fottii* (Hindák) Komárek

17210010 **Pl. 86N** (p. 352)

Coenobia of (4–)8, 16 or 32(–64) cells in 4- or 8-celled groups (sometimes wreath-shaped) within a spherical, ellipsoidal or irregularly shaped mucilaginous envelope; cells (2–)4–10(–16.5) μm in diameter, spherical, smooth surfaced or occasionally granular; chloroplast parietal, cup-shaped, with a single pyrenoid; asexual reproduction by (4–)8(–16) autospores arranged in geometrical groups, remnants of mother cell wall gelatinize soon after release of daughter cells.

Probably cosmopolitan; planktonic in water bodies of widely different nutrient status in the British Isles including ponds, lakes, gravel pits and slow-flowing reaches of rivers.

Probably considerably more common in the British Isles than published records suggest and frequently mistaken for *Sphaerocystis schroeteri* Chodat, a species considered to differ from *Coenochloris fottii* only by the dubious distinction of reproducing by both zoospores and autospores (see Komárek & Fott, 1983) rather than by autospores alone. Doubt therefore attaches to all records of *S. schroeteri* from the British Isles and elsewhere unless zoospores are recorded (see remarks under *Planktosphaeria gelatinosa*). The persistence of mother cell wall remnants (albeit for a very short time) immediately following release of daughter cells into the mucilaginous envelope is a character that has been used to distinguish *Coenochloris* (see Hindák, 1984) from *Sphaerocystis*. Tsarenko (1990) considers that *Coenococcus fottii* Hindák has persistent mother cell wall remnants and for this reason transferred it to the genus *Coenochloris*.

Coenococcus Korshikov 1953

Colonies microscopic, cells embedded within a structureless mucilage of clearly delimited outline, spherical, elongate-ellipsoidal, with age of indefinite form; cells spherical to broadly ovoid, usually grouped into tetrads or sometimes wreath-shaped; chloroplast parietal, single, with 1 to several pyrenoids or sometimes absent; asexual reproduction by 4 or 8 autospores arranged in tetrads, released by gelatinization of mother cell wall or disintegration of colony.

Probably cosmopolitan; planktonic in various types of water body.

The genus has been considered synonymous with *Eutetramorus* Walton (see Bourrelly, 1966; Komárek, 1979; Komárek & Fott, 1983), the latter having priority. Fott (1974), Hindák (1977, 1984) and others have pointed out that the original description of *Eutetramorus* was based on a single specimen and included no mention of most of the important characters (e.g. method of autospore release, form of residual spore, mother cell wall) of the family to which it belongs (Radiococcaceae). We follow Hindák (1984) who, in accepting Korshikov's *Coenococcus* (see Korshikov, 1953), states that the 'name of genus *Eutetramorus* must be rejected according to the International Code of Botanical Nomenclature, because it is the source of errors because of incomplete diagnosis'. He was of the opinion that the spore mother cell wall in Korshikov's *Coenococcus polycoccus* remains following autospore release and therefore transferred it to the genus *Coenochloris*. Most others (including ourselves) consider that the mother cell wall disappears immediately upon release of the daughter cells and hence continue to recognize *Coenococcus polycoccus* Korshikov (see remarks under *Coenococcus polycoccus*). Tsarenko (1990) has noted that mother cell wall remnants persist in the mucilage of Hindák's *Coenococcus fottii* and thus confirms its transfer to the genus *Coenochloris*. Two species of the genus *Coenococcus*, *C. planctonicus* Korshikov and *C. polycoccus* (Korshikov) Hindák, are widespread in the British Isles.

1 Cells 5.5–10 µm wide, usually in tetrads; chloroplast with a single pyrenoid ....*C. planctonicus*
1 Cells 10–20 µm wide, often in wreath-shaped groups; chloroplast with 2–8 pyrenoids ....*C. polycoccus*

Coenococcus planctonicus Korshikov 1953

Synonym: *Eutetramorus planctonicus* (Korshikov) Bourrelly
17220010 **Pl. 86C** (p. 352)
Coenobia of 4, 8, 16 or 32(–64) cells and cells forming dense tetrads lying within a spherical, ellipsoidal or irregularly-shaped mucilaginous envelope; cells 5.5–8 (–10) µm wide, spherical, broadly oval in young colonies, smooth-walled; chloroplast cup-shaped to goblet-shaped, with a single large pyrenoid; autospores (2–)4(–8) in each sporangium.

Europe; planktonic or associated with marginal aquatic plants in the British Isles where reported from ponds, lakes, rivers and canals.

Coenococcus polycoccus (Korshikov) Hindák 1977

Basionym: *Sphaerocystis polycocca* Korshikov
Synonym: *Eutetramorus polycoccus* (Korshikov) Komárek
17220020 **Pl. 86D** (p. 352)
Coenobia 4-celled to multicellular, sometimes with cells in wreath-shaped groups (8, 16 or 32 cells) embedded within a spherical to irregularly spherical mucilaginous envelope; cells (8–)12–20 µm in diameter, spherical and smooth-walled; chloroplast parietal, dense, with 2–8 pyrenoids; autospores spherical or slightly oval, 4 or 8(–16) in each mother cell.

Probably cosmopolitan; planktonic in a wide range of aquatic habitats in the British Isles although most common in ponds and lakes.

Hindák (1984) considered *Coenococcus polycoccus* to be more correctly attributed to the genus *Coenochloris* believing that its mother cell wall persists on release of the daughter cells. However, it is known that the maternal cell wall disappears rapidly in colonies that have several pyrenoids in each cell (corresponding to *Coenococcus polycoccus*) and only persists in those having a single pyrenoid. Colonies in which the cells have a single pyrenoid are regarded as belonging to *Coenochloris fottii* as currently recognized. See remarks under the genus *Coenococcus*.

Coenocystis Korshikov 1953

Colonies microscopic, consisting of 4- or 8-celled groups within an ellipsoidal to almost spherical or irregularly-shaped mucilaginous envelope; cells ellipsoidal, ovoid to spindle-shaped, straight or curved, with rounded apices; chloroplast parietal, massive, with a single pyrenoid; asexual reproduction by (2–)4, 8 or 16 autospores, remnants of mother cell wall lying along side of cell or forming a tetrad, persisting for some time but eventually disappearing; large colonies sometimes fragment into smaller ones.

Probably cosmopolitan; planktonic in various aquatic habitats.

See remarks under *Coenochloris*.

1 Cells symmetrical and ovoid; pyrenoid towards one pole ....*C. planctonica*
1 Cells slightly asymmetrical and elongate-ovoid; pyrenoid near cell centre..........................*C. obtusa*

Coenocystis obtusa Korshikov 1953

Synonym: *Kirchneriella obtusa* (Korshikov) Komárek
17230010 **Pl. 86E** (p. 352)
Coenobia of 8 or 16 cells, randomly distributed or in 3- or 4-celled groups surrounded by a common mucilage sheath and embedded within an elongate-ovoid mucilaginous envelope; cells 3.5–4.6 µm wide, 9–16 µm long, elongate-ellipsoidal, asymmetrical, convex on one side and the other side straight or slightly concave, sometimes with a bulge in the centre; chloroplast parietal, grooved and lying against convex side of cell, with pyrenoid near centre; autospores (2–)4 in each mother cell, long axes parallel to each other.

Europe; known in the British Isles only from the River Thames, probably much more widely distributed.

Coenocystis planctonica Korshikov 1953

Synonym: *Coenochloris korshikovii* Hindák
17230020 **Pl. 86F** (p. 352)
Coenobia of 4 or 8 cells located towards centre of spherical to broadly ellipsoidal mucilaginous envelope with an indistinct margin; cells 8–10 µm wide, spherical or ovoid, symmetrical, walls smooth and thin; chloroplast parietal, thickened at apex where there is a single, indistinct pyrenoid; autospores 4 or 8 in each mother cell, released by splitting of wall.

Europe; in the British Isles planktonic in various aquatic habitats including ponds, lakes and rivers.

Coronastrum R.H.Thompson 1938

Coenobia of 4 cells linked equatorially by hyaline processes, cells frequently orientated with long axes at right angles to plane of attachment and remnants of mother cell wall (horn- or cone-shaped) attached to end of each cell, often coenobia linked to form up to 16-celled syncoenobia with cells in groups of 3; cells ovoid, ellipsoidal or crescent-shaped; chloroplast parietal and with a single pyrenoid; asexual reproduction by 4 autospores, released by mother cell splitting into 4 parts, a wall remnant remaining attached to end of each daughter cell.

Probably cosmopolitan; planktonic in various types of water body.

Coronastrum ellipsoideum Fott 1946

Synonym: *Coronastrum anglicum* E.A.Flint

17240010 **Pl. 84A** (p. 346)

Coenobia 4-celled, more rarely forming 16-celled syncoenobia then 'crown-like' in appearance; cells 2.5–6 μm wide, (3.8–)5–8 μm long, ovoid to ellipsoidal, often horn-like mother cell wall fragments present.

Europe; in the British Isles reported during the summer months in the plankton of moderately nutrient-rich ponds, lakes and reservoirs, rare.

Crucigenia Morren 1830

Synonym: *Staurogenia* Kützing

Coenobia 4-celled, flat and quadrate, with or without an internal space, sometimes joined by remnants of mother cell wall to form syncoenobia, often surrounded by a wide mucilaginous envelope; cells ellipsoidal, spindle-shaped, cylindrical or crescent-shaped, frequently symmetrical, with apices acute and sometimes prolonged into a spine; chloroplast parietal, single or fragmented, with a single pyrenoid (sometimes indistinct); asexual reproduction by 2 or 4 autospores, daughter coenobia orientated at 45° to mother cell, released by breakage of mother cell wall into 2 or 4 parts, with wall remnants often persisting.

Probably cosmopolitan; in aquatic habitats or on moist terrestrial surfaces.

The principal feature for separating *Crucigenia* from the genus *Crucigeniella* is the position of cells within the daughter coenobium (see Komárek, 1974). Otherwise the two genera are indistinguishable, although the coenobia of *Crucigenia* are typically quadrate rather than rhomboidal or rectangular. Komárek (1974) placed it in the genus *Hofmania* Chodat (now *Komarekia* Fott), a genus whose daughter coenobia adopt a similar position in the mother cell to those of *Crucigenia* but which are liberated by the splitting of the mother cell wall into four rather than two parts. The remnants of the mother cell wall are usually only visible in large and actively growing populations.

1 Coenobia usually having remnants of mother cell wall associated with outermost wall of each cell .... *C. lauterbornii*

1 Coenobia without remnants of spore mother cell wall, if present then as two fragments associated with each coenobium 2

2 Cells trapezoid, with a large quadrate internal space .... *C. fenestrata*

2 Cells triangular, internal space minute if present .... *C. tetrapedia*

Crucigenia fenestrata (Schmidle) Schmidle 1900

Basionym: *Staurogenia fenestrata* Schmidle

Synonym: *Crucigenia apiculata* (Lemmermann) Schmidle *sensu* Williams

17250010 **Pl. 84B** (p. 346)

Coenobia rhomboidal, about (6–)8–14(–16) μm across, with a relatively large internal space, sometimes forming irregularly shaped multiple colonies (syncoenobia); cells 1.8–6 × 5–12(–16) μm, trapezoid, straight or slightly curved on outermost side.

Probably cosmopolitan; in the British Isles recorded in lake and river plankton in England and some Scottish lochs.

This species shows considerable morphological variation. Williams (1965) reported that samples from a lake in Chester (The Serpentine, Eaton Park) ranged from the typical form through to cells with a much larger internal space. According to Hindák (1984), the principal feature separating this species and *C. tetrapedia* is the size and shape of the internal space. He mentions that there is no internal space in *C. tetrapedia* or it is minute if present, whereas in *C. fenestrata* the space is relatively large (up to 10 μm across), square, and has concave sides.

Crucigenia lauterbornii (Schmidle) Schmidle 1900

Basionym: *Staurogenia lauterbornii* Schmidle

Synonyms: *Komarekia lauterbornii* (Schmidle) Fott, *Hofmania lauterbornii* (Schmidle) Wille

17250020 **Pl. 84C** (p. 346) **CD**

Coenobia quadrate, 15–30.5 μm across, with a large, quadrate, internal space; sometimes forming multiple colonies of 16 or more cells and attached by remnants of mother cell wall which often persist on outermost side of each cell; cells 4–9 × 6–12(–15) μm, almost semicircular, outer wall very convex and the inner wall almost straight or slightly convex.

Probably cosmopolitan; recorded from just a few lakes in the British Isles.

Crucigenia tetrapedia (Kirchner) West *et* G.S.West 1902

Basionym: *Staurogenia tetrapedia* Kirchner

17250030 **Pl. 84D** (p. 346) **CD**

Coenobia regularly quadrate, about (8–)10.5–15.5 μm across, with internal space minute, often forming colonies (syncoenobia) of 16 or more cells; cells (2.5–)4–9.5(–12) μm, tightly packed, triangular, with straight or slightly concave sides and rounded ends.

Probably cosmopolitan; widely distributed during the summer months in the plankton of pools, ponds, large lakes (e.g. Lough Neagh, Loch Leven, lakes in the English Lake District) and slow-flowing rivers in the British Isles.

Williams (1965) mentioned finding in a Cheshire lake a peculiar form asymmetrical due to 'a shortening of the line of division at the outer corners of the coenobium'. He referred to these irregularly shaped cells as 'forma *asymmetrica*' although provided no formal diagnosis. See also remarks under *C. fenestrata*.

Crucigeniella Lemmermann 1900

Coenobia 4-celled, flat, rectangular, with or without a rectangular or rhombic internal space, sometimes joined by remnants of mother cell wall to form syncoenobia, enclosed within a mucilaginous envelope; cells elongate or ovoid, with wall occasionally thickened at apices; chloroplast parietal, solitary and usually with a pyrenoid; asexual reproduction by 4 autospores, daughter coenobia lying in same position as original mother cell, released by rupture of the mother cell wall into two parts, with the wall remnants often persisting.

Probably cosmopolitan; mostly planktonic in a wide range of freshwater habitats.

See remarks under *Crucigenia*.

1 Cells inclined to one side, often 3-cornered with a short apical projection ....*C. apiculata*
1 Cells of another form 2
2 Cells rhomboidal, outermost wall slightly concave ....*C. crucifera*
2 Cells ovoid, elongate-ovoid or almost cylindrical, the outermost wall straight or convex 3
3 Cells elongate-oval or cylindrical, adjacent ones lying at a slight angle to one another, with a large rhombic space in centre ....*C. rectangularis*
3 Cells more or less oval, adjacent ones irregularly arranged, with a small central space ....*C. irregularis*

Crucigeniella apiculata (Lemmermann) Komárek 1974

Basionym: *Staurogenia apiculata* Lemmermann
Synonyms: *Crucigenia apiculata* (Lemmermann) Schmidle, *C. pulchra* West *et* G.S.West, *Crucigeniella pulchra* (West *et* G.S.West) Komárek, *Tetrastrum apiculatum* (Lemmermann) Brunnthaler, *T. pulloideum* Teiling

17260010 **Pl. 84E** (p. 346) **CD**

Coenobia rectangular, 6–12.5 μm wide, 9–18 μm long, often with a rectangular internal space, sometimes in multiples (syncoenobia), with remnants of mother cell wall visible (4 wall remnants per coenobium); cells 2–7 ×4–10 μm, oval to elongate-ovoid, inclined to one side, often 3-cornered with a short apical projection.

Probably cosmopolitan; a distinctive species common during the summer months in a wide range of aquatic habitats (e.g. ditches, small pools, moats, lakes and slow-flowing rivers) and widespread in the British Isles.

Crucigeniella crucifera (Wolle) Komárek 1974

Basionym: *Staurogenia crucifera* Wolle
Synonyms: *Crucigenia cruciata* Schmidle, *C. crucifera* (Wolle) Collins

17260040 **Pl. 84M** (p. 346)

Coenobia rhomboidal, with sides slightly concave, 9–11(–22) μm wide, 14–16(–24) μm long, internal space small and rectangular, sometimes coenobia joined to form syncoenobia of 16 or more cells; cells 3–5× 5–9(–10) μm, elongate, rhomboidal, outer side concave, inner side straight or slightly convex, with blunt apices.

Probably cosmopolitan; doubt attaches to the report of its occurrence over a 30-year period in a natural freshwater lake in Eaton Park, Chester (see below), rare in the British Isles.

According to Williams (1965), the material from Eaton Park is tentatively attributed to this species. Doubt inevitably attaches to his records since Williams's illustrations appear to fall within the form range of *Crucigenia fenestrata*.

Crucigeniella irregularis (Wille) Tsarenko *et* D.M.John, *nov. comb.*

Basionym: *Crucigenia irregularis* Wille 1898. *Bot. Zbl. Leipzig* 18: 302.
Synonym: *Willea irregularis* (Wille) Schmidle

17260020 **Pl. 84L** (p. 346)

Coenobia rectangular or rhomboidal, 4-celled to many hundreds of cells (syncoenobia), usually irregularly arranged due to slippage; cells 4–9×6–14 μm, more or less oval, often more convex on one side than the other, wall slightly thickened at apical end, lying parallel or more usually adjacent cells irregularly arranged.

Probably cosmopolitan; planktonic and/or associated with submerged surfaces in the British Isles where widespread in various aquatic habitats although most common in nutrient-poor lakes.

Crucigeniella rectangularis (Nägeli) Komárek 1974

Basionym: *Crucigenia rectangularis* (Nägeli) F.Gay

17260030 **Pl. 84F** (p. 346) **CD**

Coenobia rhomboidal, 7–14 μm wide, 8–20 μm long, often forming multiple colonies of 16 or more cells (syncoenobia); cells 3–7×4–10 μm, elongate ovoid or somewhat cylindrical, lying at a slight angle to one another and having a large rhombic internal space, adjacent cells attached apically and laterally.

Probably cosmopolitan; the most frequently encountered and widely distributed *Crucigeniella* species in the British Isles where known from ponds, lakes and slow-flowing rivers.

REJECTED RECORD

Dactylococcus lacustris Chodat 1897

Doubt attaches to the only record (as *Coccomyxa lacustris* (Chodat) Pascher) from North Wales by Woodhead & Tweed (1954) since this lacks an accompanying description and illustration.

Desmatractum West *et* G.S.West 1902

Synonyms: *Bernardinella* Chodat, *Calyptrobactron* Geitler

Cells solitary, the protoplast spherical to ellipsoidal, initially uninucleate and later multinucleate, surrounded by a shortly cylindrical to elongate and spindle-shaped outer wall, sometimes enclosed within a mucilaginous envelope; outer cell wall smooth or ridged, more or less transparent (occasionally brownish), consisting of two halves united in the equatorial region (bipyramidal), sometimes narrowing to pointed apices and bearing spiny ridges; chloroplast parietal, trough-shaped to cup-shaped and with a pyrenoid; asexual reproduction by biflagellate zoospores, hemizoospores and sometimes

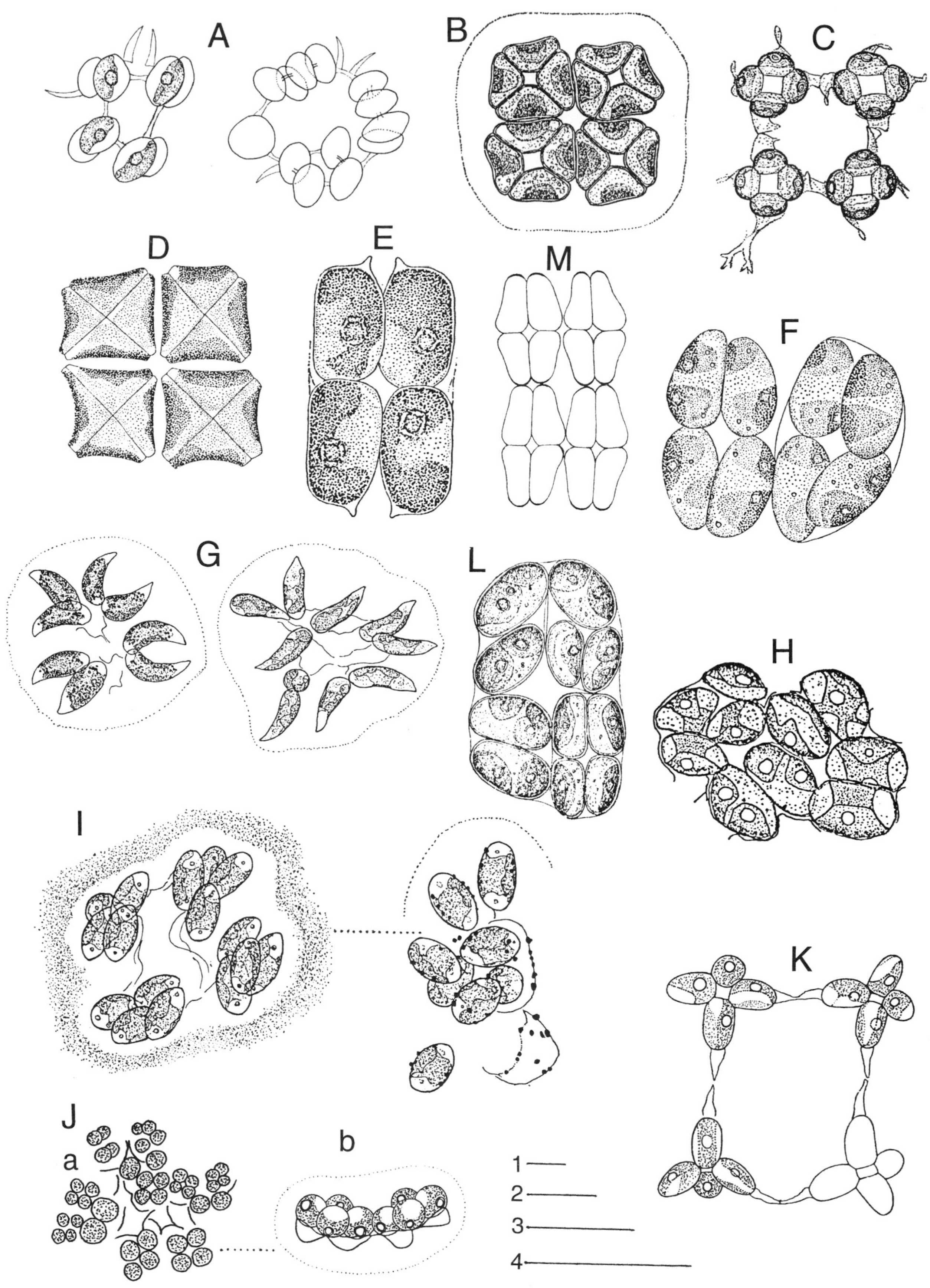

Plate 84 A–M

A. *Coronastrum ellipsoideum* (p. 344); **B**. *Crucigenia fenestrata* (p. 344); **C**. *Crucigenia lauterbornii* (p. 344); **D**. *Crucigenia tetrapedia* (p. 344); **E**. *Crucigeniella apiculata* (p. 345); **F**. *Crucigeniella rectangularis* (p. 345); **G**. *Dichotomococcus curvatus* (p. 347); **H**. *Tetrachlorella alternans* (p. 404); **I**. *Quadricoccus verrucosus* (p. 382); **J**. *Westella botryoides* (p. 408): a. large colony showing fragments of mother cell wall, b. small colony showing details of the chloroplast and pyrenoid in each cell; **K**. *Komarekia appendiculata* (p. 362); **L**. *Crucigeniella irregularis* (p. 345); **M**. *Crucigeniella crucifera* (p. 345).

Scale bars: 10 μm. 1 for C, Ja, M; 2 for A, B, G, I, Jb, K; 3 for D, H, L; 4 for E, F.

autospores, all developing within a widened mother cell and released by its wall separating into two halves; akinetes produced.

Probably cosmopolitan; free-floating in still and slow-flowing water. Six species, of which only one has been found in the British Isles to date.

Desmatractum bipyramidatum (Chodat) Pascher 1930

Basionym: *Bernardinella bipyramidatum* Chodat
17270010 **Pl. 82P** (p. 329)
Cells 8.5–19 μm wide, 17–44 μm long, ellipsoidal to spindle-shaped or narrowly or broadly cone-shaped, sometimes within a mucilaginous envelope; outer wall distinct, thick, divided into two halves by an equatorial ridge and bearing 5–12 longitudinal ridges, each half cap-shaped to bell-shaped and narrowing abruptly towards apex to a spine-like point; protoplast 8–12 μm wide, spherical, with a cup-shaped chloroplast.

Probably cosmopolitan; in the British Isles reported throughout the year associated with bottom deposits and on aquatic plants in slow-flowing water and ponds, especially in the acid waters of bogs and in a pond in Richmond Park (Leg-of-Mutton pond) where the pH rarely rose above 6.5 (Lund 1942b).

Diacanthos Korshikov 1953

Cells solitary, ovoid to ellipsoidal, thin-walled, bearing a long spine from each apex and these pointing in opposite directions, each spine gradually narrowing to a fine point, without a mucilaginous envelope; chloroplast parietal, cup-shaped, slightly asymmetrical and possessing a single pyrenoid; asexual reproduction by 4 or 8 autospores, released by splitting of mother cell wall.

Europe, planktonic and usually in standing water. Monospecific.

Diacanthos belenophorus Korshikov 1953

17280010 **Pl. 88C** (p. 359)
Cells (3.3–)4.5–5.5 μm wide and (5–)8–10 μm long, each tapering spine reaching 30(–55) μm long.

Europe; in a few small ponds in England (e.g. Priest Pot, Cumbria) where usually found in August and September, probably much more widely distributed than records indicate.

Dicellula Svirenko 1926

Coenobia of 2 broadly ovoid to ellipsoidal cells joined by long axes, enclosed within a mucilaginous envelope; cells bearing many very fine and sometimes basally thickened spines; chloroplasts parietal, plate-shaped, initially a pair but increasing in number with age, each containing a single pyrenoid; asexual reproduction by 4 or 8 autospores, released by gelatinization of wall of enlarged mother cell.

Probably cosmopolitan; planktonic in lakes and slow-flowing rivers.

Closely related to the genus *Franceia* from which it differs in having 2-celled coenobia of firmly united cells rather than cells loosely united and embedded in soft mucilage. According to Korshikov (1953), *Dicellula* may become single-celled under unfavourable conditions when it is impossible to separate from the genus *Franceia*.

Dicellula geminata (Printz) Korshikov 1953

Basionym: *Franceia geminata* Printz
Synonyms: *Dicellula planktonica* Svirenko, *Franceia tuberculata* G.M. Smith
17290010 **Pl. 91I** (p. 369) **CD**
Cells 5–9(–12) μm wide, 8–13(–22) μm long, ovoid to ellipsoidal, slightly flattened at point of contact; spines 9–20(–25) μm long, fine and often inconspicuous, each spine having an extended thickened base which often persists after loss of upper portion of spine.

Probably cosmopolitan; in the British Isles probably widely distributed although so far recorded from a few ponds, lakes and rivers (e.g. Thames, Great Ouse).

The thickened basal part of each spine is sometimes brownish and the cell wall similarly coloured. We are in agreement with Hindák (1984) who considers *Dicellula geminata* and *D. planktonica* to be conspecific. He believes the extent of thickening at the base of the spines to be insufficient justification for separating the species.

Dichotomococcus Korshikov 1928

Colonies of somewhat radially arranged cells attached to pseudodichotomously divided remnants of mother cell walls; daughter cells often in pairs, enclosed within an indistinctly delimited mucilaginous envelope; cells elongate ovoid, cylindrical or irregularly spindle-shaped, straight or curved; chloroplast parietal, without pyrenoids although starch and oil droplets present; asexual reproduction by 2(–4) autospores, released by longitudinal splitting of the mother cell wall, often with the basal portion of each cell remaining attached to wall remnants so that repeated autospore production gives rise to a pseudodichotomous branching system.

Probably cosmopolitan; planktonic in ponds, lakes and flowing water, rare.

Dichotomococcus curvatus Korshikov 1939

Synonym: *Dichotomococcus elongatus* Fott
17300010 **Pl. 84G** (p. 346)
Colonies of 4, 8, 16 or 32 cells attached to radially arranged remnants of mother cell walls, with cells generally in pairs towards periphery of a spherical or ovoid mucilaginous envelope; cells 2–4 μm wide and (5.5–)7–11.6 μm long, broadly rounded bases often adpressed, narrowing apically and somewhat curved.

Probably cosmopolitan; so far reported in the British Isles on a few occasions from ponds, lakes (e.g. Loch Leven, Perthshire) and flowing water.

Often placed in the Xanthophyta since starch not detected in cells although known to contain chlorophyll *b*.

Dictyochloris Vischer 1945

Cells solitary or in groups, spherical, smooth-walled; chloroplast initially a dissected hollow sphere, with age becomes a network of anastomosing and interlacing strands penetrating to cell centre, lacking pyrenoids; cells multinucleate when mature; asexual reproduction by zoospores and aplanospores produced by progressive cleavage of mother cell contents; zoospores naked, biflagellate, with flagella of slightly different lengths, eyespot present.

Europe; grows associated with soil. Only two known species.

Dictyochloris fragrans Vischer 1945

17310010 **Pl. 82K** (p. 329)

Cells 5–38(–75) μm wide, spherical, sometimes protruding on one side; walls not appreciably thickening with age; chloroplast net-like in older cells, with starch present; zoospores ovoid, 2.5 μm wide, 6 μm long, with an eyespot at opposite pole to flagella; aplanospores released by gelatinization of mother cell wall.

Europe; isolated in the British Isles from a large number of soil types.

Dictyococcus Gerneck 1907

Cells solitary, spherical, walls smooth and sometimes thickening with age, the protoplast divided into separate portions (number increases with age) and each associated with a polygonal chloroplast; chloroplasts initially parietal, later developing a variable number of centripetally projecting processes, lacking pyrenoids but with starch or sometimes oil; asexual reproduction by zoospores and aplanospores; zoospores naked, biflagellate, flagella of equal length, eyespot present.

Europe; isolated from soil and water plants. Only three species.

Dictyococcus varians Gerneck 1907

17320010 **Pl. 82O** (p. 329)

Cells reaching up to 60 μm wide, walls thickening with age (3–4 μm thick) and outermost layers sometimes becoming shed; zoospores 3 μm wide, 7.5 μm long, ovoid, with an anterior eyespot; aplanospores usually more than 25 μm wide.

Europe; reported in the British Isles from a small pond near Glasgow, Scotland (see Starr, 1955) where it was associated with water plants; elsewhere isolated from soil.

Dictyosphaerium Nägeli 1849

Coenobia consisting of 4, 8, 16, 32 or 64 cells in groups towards periphery of a spherical or ovoid mucilaginous envelope; cells spherical, oval to cylindrical, obovoid, spindle-shaped, or kidney-shaped, walls smooth or occasionally granulate, forming (2–)4-celled groups arising terminally on dichotomously or tetrachotomously branched, ribbon-like remnants of mother cell walls which radiate from centre of colony; chloroplasts parietal, cup-shaped (trough-shaped in longish cells) and possessing a single pyrenoid; asexual reproduction by 4 autospores, more rarely 2 or 8, released by splitting or occasionally by gelatinization of mother cell wall, the spores sometimes remaining attached to wall remnants. Sexual reproduction known. Coenobia sometimes disintegrate under unfavourable conditions to release individual cells.

Probably cosmopolitan; widely distributed in the plankton and a few species associated with soil.

The numbers of autospores, size of coenobium and cellular dimensions are known to vary depending upon environmental conditions (see Hindák, 1977; Komárek & Perman, 1978; Komárek & Fott, 1983). Cell form is also age-dependent, with young cells elongate or narrowly oval and adult cells usually broadly rounded, obovoid or spherical. It is therefore necessary to examine colonies of different ages to follow changes in cell shape when using it as the principal character for distinguishing between species. Doubt inevitably attaches to the validity of many of the characters traditionally used to separate species, e.g. cell shape, cell arrangement and number of cells forming a colony (see *D. chlorelloides* (Naumann) Komárek *et* Perman and *D. pulchellum* H.C.Wood). The release of autospores by fracture of the mother cell wall is considered to be an important generic character, although in *D. subsolitaria* van Goor and *D. anomalum* Korshikov release is by gelatinization. Hindák (1978) established the genus *Pseudodictyosphaerium* to accommodate those *Dictyosphaerium* species not possessing pyrenoids. Considerable doubt attaches to this new genus since we consider the presence or absence of pyrenoids to be a questionable character for distinguishing genera. The genus is recognized here (see p. 381) although on further investigation the new species described by Hindák (1978) from Scotland (*Pseudictyosphaerium scoticum*) might prove to be no more than a form of *D. chlorelloides*.

1 Older cells obovoid to spherical; young cells (also autospores) spherical, obovoid, irregularly oval or spindle-shaped 2
1 Older cells ovoid, broadly ovoid, subcylindrical, ellipsoidal and subspherical; young cells inversely ovoid (in respect to point of attachment), asymmetrically ovoid, triangular or irregularly spindle-shaped 4
2 Older cells spherical and covered with granules ....*D. granulatum*
2 Older cells not spherical or if spherical then not granulate 3
3 Colonies 2–4-celled (rarely 16-celled); older cells oval to spherical, attached laterally to inverted cup-shaped remnants of mother cell wall ....*D. chlorelloides*
3 Colonies (4–)8-celled or up to 64-celled; older cells spherical, attached to narrow and tetrachotomously branched thread-like remnants of mother cell wall ....*D. pulchellum*
4 Young cells asymmetrically ovoid, inversely obovoid to almost triangular, or irregularly spindle-shaped, attached laterally to remnants of mother cell wall ....*D. tetrachotomum*
4 Young and older cells ovoid to ellipsoidal, or subcylindrical and attached on longitudinal side to remnants of mother cell wall ....*D. ehrenbergianum*

Dictyosphaerium chlorelloides (Naumann) Komárek *et* Perman 1978

Basionym: *Brachionococcus chlorelloides* Naumann
Synonym: *Dictyosphaerium minutum* J.B. Petersen, ?*D. simplex* Korshikov

17330010 **Pl. 83P** (p. 341)

Colonies (10–)25–34(–46) µm wide, tetrahedral to spherical, 2- or 4-celled, rarely 16-celled; cells (3.5–)4–6.9(–9) µm, young cells spherical, older cells spherical to obovoid, walls thin and smooth, attached laterally to cup-like remnants of mother cell wall; chloroplast cup-shaped and basal; autospores 2 or 4 in each sporangium, 3–6×2–6 µm, asymmetrically ovoid or subspherical, released by fracture of mother cell wall.

Probably cosmopolitan; planktonic or associated with submerged surfaces in the British Isles where known so far from a few aquatic habitats and in low pH soils.

According to Hindák (1988), some of the earlier confusion surrounding this and other species of *Dictyosphaerium* was due to failure to recognize that cell shape changes as the colony matures. He discovered that the asymmetrically ovoid or subspherical autospores of *D. chlorelloides* all become subspherical when mature. A characteristic of this very variable species is its readiness to disintegrate under stress into individual cells surrounded by mucilage which then closely resemble *Coenochloris* (see Hindák, 1984). Very similar to *Pseudodictyosphaerium scoticum* except for the presence of a pyrenoid and attachment of cells to remnants of mother cell wall, possibly conspecific with the latter but maintained separately here pending further investigation.

Dictyosphaerium ehrenbergianum Nägeli 1849

17330020 **Pl. 85A** (p. 350) **CD**

Colonies 48–80 µm wide, 4-, 8-, 16- or 32-celled, spherical, broadly ovoid; cells (3.5–)4–5.5×2–3 µm, oval to ellipsoidal or subcylindrical, attached laterally to dichotomously branched threads; chloroplast a curved plate; autospores (2–)4 in each sporangium, 2–5.5 µm and similar in shape to vegetative cells, released by gelatinization of mother cell wall.

Probably cosmopolitan; widely distributed in the British Isles where planktonic or on submerged objects in a diverse range of aquatic habitats; considered in Europe to be an indicator of mildly to moderately nutrient-rich water.

Very similar to *D. tetrachotomum* and it is possible that most records of *D. ehrenbergianum* should be attributed to this other species (J.W.D.Lund, pers. comm.).

Dictyosphaerium granulatum Hindák 1977

17330030 **Pl. 85B** (p. 350) **CD**

Colonies 4-, 8- or 16-celled and spherical; cells 4–6(–8) µm wide, young cells ovoid or broadly ovoid and older cells spherical, covered with many small granules, initially cells attached obliquely and later terminally on a branched thread; chloroplast cup-shaped and basal; autospores 4 in each sporangium, 3–6×3–4 µm and similar in form to young vegetative cells.

Europe; planktonic in the British Isles where reported but rarely from small lakes in southern England (e.g. Keston Ponds, Bromley, Kent).

Dictyosphaerium pulchellum H.C.Wood 1872

17330040 **Pl. 85C** (p. 350)

Colonies 80–100 µm wide, 4-, 8-, 16-, 32- or 64-celled or more, spherical, ovoid or of indefinite shape; cells (4–)5–8(–10) µm wide, young cells obovoid, ellipsoidal to spindle-shaped, older cells spherical, walls thin and smooth, attached to tetrachotomously branched threads which are often in contact with one another (depends on colony age); chloroplast basal and cup-shaped, fills half to two-thirds of cell, lateral in young cells; autospores 2 or 4 in each sporangium, (2.8–)4–6(–7)×(2.5–)4–5.5(–6.5) µm and similar in shape to vegetative cells, released by gelatinization of mother cell wall.

Probably cosmopolitan; common and widely distributed in the plankton or associated with submerged macrophytes in the British Isles where in slow-moving water as well as in ponds and lakes (abundant in Loch Leven only when phosphate levels are high; A. Bailey-Watts, pers. comm.).

The relatively small-celled (cells 4–5 µm) colonies found on occasion in ponds and reservoirs (e.g. Wistlandpound Reservoir, Devon) closely resemble the variety *minutum* Deflandre.

Dictyosphaerium tetrachotomum Printz 1914

Synonym: *Dictyosphaerium pulchellum* var. *ovatum* Korshikov

17330050 **Pl. 85D** (p. 350) **CD**

Colonies 4-, 8-, 16- or 32-celled (rarely 64-celled), 16-celled colonies reaching about 50 µm in size, spherical to irregular in form; cells (3–)4–9(–11)×3–8(–8.5) µm, young cells asymmetrically ovoid, ovoid to almost triangular or irregularly spindle-shaped, older cells oval, broadly oval to almost spherical, often in tetrahedral groups with each cell attached laterally to a narrow terminal branch of the tetrachotomously branched system of threads; chloroplast cup-shaped, lateral or almost lateral in young cells, basally positioned; autospores 4 (rarely 2) in each sporangium, each 3–6(–7)×2–5.5 µm and similar in form to young vegetative cells, released by fracture of mother cell wall.

Probably cosmopolitan; planktonic or associated with submerged objects in the British Isles where it occurs in various types of aquatic habitat with widely differing nutrient status.

DOUBTFUL RECORD

Dictyosphaerium terrestre F.E.Fritsch *et* R.P.John 1942

Cells up to 12 µm wide, spherical, ovoid when young, without surface ornamentation and connected basally to threads. Reproduction by biflagellate zoospores, usually 4 in each cell. Autospores unknown. Isolated in the Brecklands area of Norfolk from grass heath soils with pH range 3.7–7.8. Its position is doubtful since there is no information on autospore formation. Further investigation is needed of this little-known soil alga.

Didymocystis Korshikov 1953

Coenobia of 2 laterally joined cells enclosed within a mucilaginous envelope; cells ovoid to cylindrical, somewhat asymmetrical, slightly flattened where

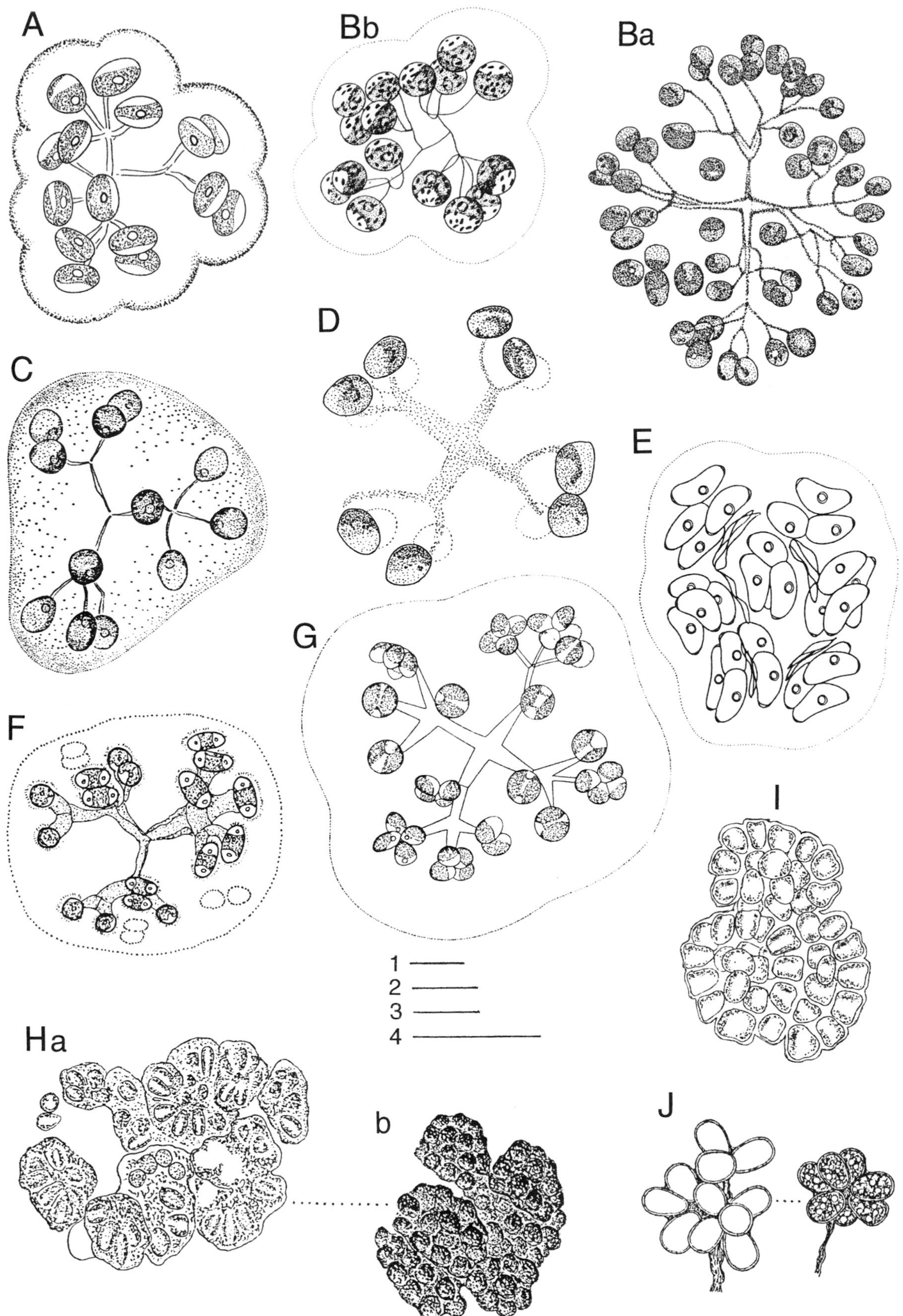

Plate 85 A–J

A. *Dictyosphaerium ehrenbergianum* (p. 349); **B**. *Dictyosphaerium granulatum* (p. 349): a. general view of colony and connecting mucilaginous threads, b. details of cells showing granular nature of wall; **C**. *Dictyosphaerium pulchellum* (p. 349); **D**. *Dictyosphaerium tetrachotomum* (p. 349); **E**. *Dimorphococcus lunatus* (p. 351); **F**. *Lobocystis planctonica* (p. 364); **G**. *Pseudodictyosphaerium scoticum* (p. 381); **H**. *Botryococcus braunii* (p. 332): a. a squashed colony showing details of cells, b. unsquashed colony; **I**. *Botryococcus calcareus* (p. 332); **J**. *Botryococcus protuberans* (p. 332).

Scale bars: 1, 2, 4, 10 μm; 3, 20 μm. 1 for Bb, C, E, F; 2 for Ba, D, G; 3 for H–J; 4 for A.

joined, with apices broadly rounded; cell wall randomly or regularly covered by relatively large granules of varying size, sometimes granules occurring in short rows; chloroplast parietal, laterally positioned in young cells, lacking a pyrenoid; asexual reproduction by 4 autospores giving rise to two 2-celled coenobia, released by lateral or subapical fracture of mother cell wall.

Probably cosmopolitan; planktonic in various types of water body. Monospecific.

The involved taxonomic history of the genus is summarized by Hindák (1990). Only one of the four species originally recognized by Korshikov retained since the rest have been transferred to *Pseudodidymocystis* E.H.Hegewald *et* Deason (see Hegewald & Deason, 1996).

Didymocystis inermis (Fott) Fott 1973

Basionym: *Dicellula inermis* Fott
Synonyms: *Didymocystis tuberculata* Korshikov, *Scenedesmus inermis* (Fott) E.H.Hegewald

17340010 **Pl. 91K** (p. 369)

Cells 3–6 μm wide, 6–10(–13) μm long.

Probably cosmopolitan; only found in the British Isles on a few occasions in lakes and rivers, probably more widely distributed than records suggest.

Didymogenes Schmidle 1905

Coenobia flat and of two rows of 4, 8 or 16 cells, occasionally forming multicellular syncoenobia, sometimes disintegrating into cell pairs or solitary cells; cells cylindrical to spindle-shaped, straight or curved and crossing over one another, adjacent cells in contact by convex side and having apices slightly curved, walls smooth or granulate, with or without 1–3 long spines arising from each apex; chloroplast parietal with a single pyrenoid; asexual reproduction by 2 or 4 autospores in each sporangium, 2 daughter coenobia produced in mother cell containing 4 autospores, released by breakage of wall.

Probably cosmopolitan; planktonic. Only three known species.

1 Apex of outer cell usually bearing 1–3 spines .... *D. anomala*
1 Apex of outer cell without spines .... *D. palatina*

Didymogenes anomala (G.M.Smith) Hindák 1974

Basionym: *Tetrastrum anomalum* G.M.Smith
Synonym: *Scenedesmus anomalus* (G.M.Smith) Ahlstrom *et* Tiffany

17350010 **Pl. 91N** (p. 369)

Cells (1.2–)2.5–4(–5) μm wide, (6–)7.4–12(–17.8) μm long, cylindrical to spindle-shaped, slightly curved to arc-shaped, smooth-walled, usually bearing 1–2(–3) spines on each outermost apex (rarely absent), spines frequently not present on inner cells; coenobia released by mother cell wall splitting into 4 parts, wall remnants often persist.

Probably cosmopolitan; known in the British Isles only from a few lakes but probably more common than records indicate; common in continental Europe where frequently planktonic or associated with surfaces in ponds, lakes, rivers and bogs.

Didymogenes palatina Schmidle 1905

17350020 **Pl. 91M** (p. 369)

Cells 1.3–3.7 μm wide, (4–)6–11.7 μm long, cylindrical and somewhat narrowed towards rounded apices, curved towards one another by convex sides, walls smooth and not bearing spines.

Probably cosmopolitan; mainly planktonic and recorded on a few occasions in lakes and rivers in the British Isles.

Dimorphococcus A.Braun 1855

Coenobia 4-celled and often united by remnants of mother cell walls to form a syncoenobium, consisting of an outer pair of curved cells and an inner pair of straight cells enclosed within a mucilaginous envelope; cells alternate and adjacent pairs having a bridge-like connection, outermost cells kidney- or heart-shaped, inner cells straight and more or less ovoid, ellipsoidal or cylindrical with rounded apices; chloroplast parietal, almost covering entire wall, with a single pyrenoid; asexual reproduction by autospores, released by fracture of mother cell wall and remaining attached to thread-like wall remnants.

Probably cosmopolitan; planktonic or associated with surfaces such as submerged aquatic macrophytes. Only three known species.

Dimorphococcus lunatus A.Braun 1855

Synonym: *Dictyosphaerium reniforme* Bulnheim

17360010 **Pl. 85E** (p. 350)

Cells in groups of 4, usually in syncoenobia of up to 64 cells, 32–100(–150) μm wide; cells (4–)4.5–8.5(–25) μm wide and 9–17(–25) μm long, each group consisting of 2 kidney- or heart-shaped outer cells and 2 elongate-ellipsoidal (straight) inner cells, bridge-like outgrowths connect adjacent cells, apices broadly rounded with bluntly truncated and disc-like wall thickenings.

Probably cosmopolitan; a relatively rare planktonic alga most frequently encountered in small tarns in mountainous districts, e.g. Westmorland area of northern England, the Highlands of Scotland, Lewis in the Outer Hebrides; probably an indicator of acid waters.

DOUBTFUL RECORD

Diplochloris lunata (Fott) Fott 1979
17370010

A single unpublished record (J.H. Belcher & E.M.F. Swale, pers. comm.) from Walsham Broad in Norfolk. Characterized by having cells in pairs, these attached to middle part of the dorsal side of the remnants of the mother cell wall. Further observations needed to establish it as a new member of the British algal flora.

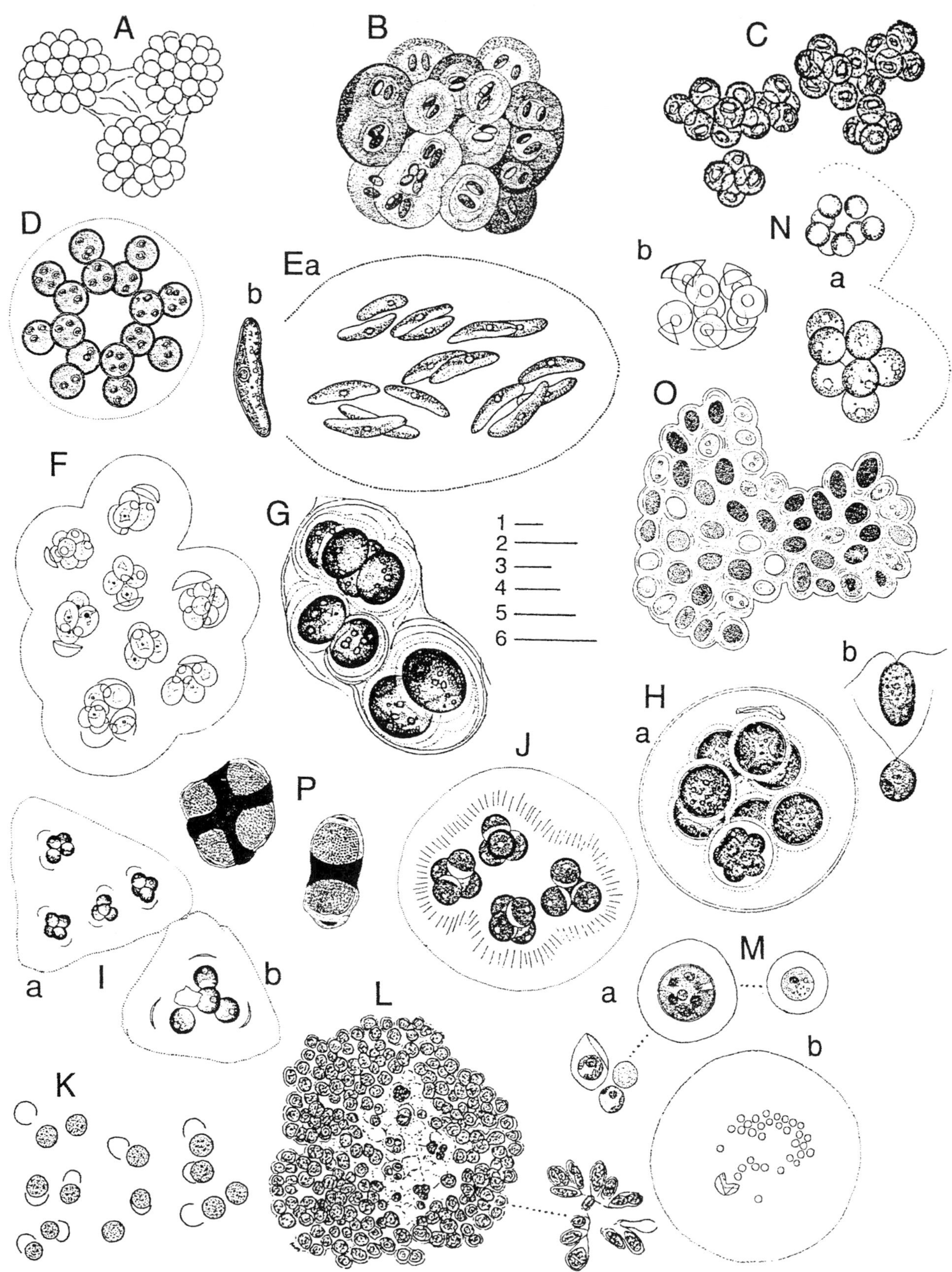

Plate 86 A–P

A. *Botryosphaerella sudetica* (p. 333); **B**. *Coccomyxa confluens* (p. 339); **C**. *Coenococcus planctonicus* (p. 343): mucilaginous envelope not shown; **D**. *Coenococcus polycoccus* (p. 343); **E**. *Coenocystis obtusa* (p. 343): a. colony, b. details of cell contents; **F**. *Coenocystis planctonica* (p. 343); **G**. *Gloeocystis vesiculosa* (p. 356); **H**. *Sphaerocystis schroeteri* (p. 402): a. colonies with one cell producing spores, b. biflagellate zoospores; **I**. *Radiococcus planktonicus* (p. 383): a. coenobium of 4 groups of cells, b. coenobium of just 4 cells; **J**. *Radiococcus nimbatus* (p. 383); **K**. *Schizochlamydella delicatula* (p. 398): mucilaginous envelope not shown; **L**. *Sphaerobotrys fluviatilis* (p. 402); **M**. *Planktosphaeria gelatinosa* (p. 379): a. cells with thick walls, b. colony with cells within a mucilaginous envelope; **N**. *Coenochloris fottii* (p. 342): a. colony, b. remnants of mother cell walls associated with daughter cells; **O**. *Rhodoplax schinzii* (p. 384); **P**. *Gloeotaenium loitelsbergerianum* (p. 356).

Scale bars: 1, 50 μm; 2, 30 μm; 3–6, 10 μm. 1 for Mb; 2 for A, B, D, Ia, K, Ma, O, P; 3 for F, J; 4 for Ea, Ha, Ib, N; 5 for C, L; 6 for Eb, G, Hb.

Ecballocystis Bohlin 1897

Colonies micro- or macroscopic, consisting of cells often partly enclosed by remnants of mother cell wall and embedded within a mucilaginous envelope; cells ovoid, ellipsoidal or cylindrical and broadly rounded apically, sometimes slightly irregular in shape; chloroplast parietal, with or without a pyrenoid; asexual reproduction by 2 or 4 autospores, each ellipsoidal, oval or cylindrical and lying obliquely within the mother cell wall, released by apical splitting of wall, daughter cells remain attached to wall and partly enclosed by it.

Probably cosmopolitan; forming mucilaginous masses on rocks in shallow water or seepage areas. Five known species, all tropical except *E. fluitans*.

Ecballocystis fluitans F.E.Fritsch 1933

17380010 **Pl. 99P** (p. 403)

Cells in 1 or 2(–4) irregular linear series, connected by wall material to form a dendroid colony surrounded by an unevenly stratified mucilaginous sheath; cells 3–3.5 μm wide, 9–18 μm long, 3–4 times longer than broad, apically rounded and sometimes wider at one apex, confining sheath 14–18 μm wide; chloroplasts plate- to band-shaped, usually 4 in each cell and lying in pairs, without pyrenoids or starch; autospores 2 in each sporangium, released by apical fracture of mother cell wall and remaining within the mucilage tube formed by gradual gelatinization of wall remnants.

British Isles; so far known only from two moorland locations, one in North Wales (Tal y Bont, Merionethshire), where *Ecballocystis* formed a mucilaginous layer over stones in a very shallow stream, the other in the English Lake District (Helm Crag, Grasmere) where it grew along with *Oedogonium* in a seepage area.

Requires further investigation.

Eremosphaera de Bary 1858

Cells solitary or forming a 2–4-celled coenobium enclosed within the mother cell wall, sometimes enclosed within a thin layer of mucilage; cells broadly ellipsoidal, ovoid, spherical to polygonal, walls smooth, thick, lamellated and sometimes thickened apically; chloroplasts numerous, disc-like or angular, mostly within a thin peripheral layer of protoplasm, each having 1–3 minute pyrenoids; nucleus large, suspended by protoplasmic strands within a large central vacuole; asexual reproduction by 2 or 4 autospores released by breakage of mother cell wall; hypnospores forming under unfavourable growing conditions, usually thick-walled and red in colour due to high concentration of haematochrome; oogamy known.

Probably cosmopolitan; often associated with surfaces and aquatic macrophytes in waters with an acid pH.

1 Cells broadly ellipsoidal with rounded apices .... *E. gigas*
1 Cells normally spherical .... *E. viridis*

Eremosphaera gigas (W.Archer) Fott *et* Kalina 1962

Basionym: *Oocystis gigas* W.Archer

17390010 **Pl. 87A** (p. 354)

Cells (32.5–)40–80(–120) μm wide, (41–)50–87(–130) μm long, about 1.25–1.5 times longer than broad, broadly ellipsoidal with rounded apices.

Probably cosmopolitan; usually associated with submerged surfaces in acid waters in the British Isles including loughs in Connemara, Ireland and elsewhere in boggy moorland pools.

Little purpose seems to be served by recognizing forms and varieties. Two have been described by the Wests (W. & G.S. West, 1894): fo. *minor* with cells just '26–28.5 μm wide' and 'up to 36.3 μm long' (from Connemara), and fo. *incrassata* whose cells are thick-walled especially at the apex (from Devon).

Eremosphaera viridis de Bary 1858

17390020 **Pl. 87B** (p. 354)

Cells 30–150(–200) μm in diameter, spherical, only very rarely oval.

Probably cosmopolitan; widely distributed in acid waters throughout the British Isles, normally associated with *Sphagnum*-lined boggy pools.

DOUBTFUL RECORD

Eremosphaera eremosphaeria (G.M.Smith) R.L.Smith *et* H.C.Bold 1966

Considerable doubt attaches to Flint's record (Flint, 1950, as *Oocystis eremosphaeria*) from a reservoir at Hammersmith, London of a species only known previously from the southern parts of the USA and from Cuba. Flint's material also differs from the usual description of this taxon in its smaller size and lack of apical thickening of the wall.

Euastropsis Lagerheim 1894

Synonym: *Euastrum* Schmidle *pro parte*

Coenobia 2-celled and flat; cells trapezoidal, deeply incised apically and joined by their broad bases, walls thin and smooth; chloroplast parietal, with a single pyrenoid; asexual reproduction by 2, 4, 8, 16 or 32 zoospores, released within a mucilage vesicle and then uniting in pairs to form daughter coenobia.

Probably cosmopolitan; planktonic or associated with periphyton, usually in non-flowing water. Monospecific.

Euastropsis richteri (Schmidle) Lagerheim 1894

Synonym: *Euastrum richteri* Schmidle

1740010 **Pl. 91L** (p. 369)

Coenobia 6–25 μm wide, (10–)12–40 μm long.

Probably cosmopolitan; planktonic or associated with periphyton in the British Isles where found in neutral water or the more acidic waters of pools in marshy areas or peat bogs, rare.

Little-known genus with uncertainty surrounding its taxonomic status.

Fernandinella Chodat 1922

Cells attached, usually in crowded groups, almost spherical, obovoid, inversely obovoid or shortly pear-shaped,

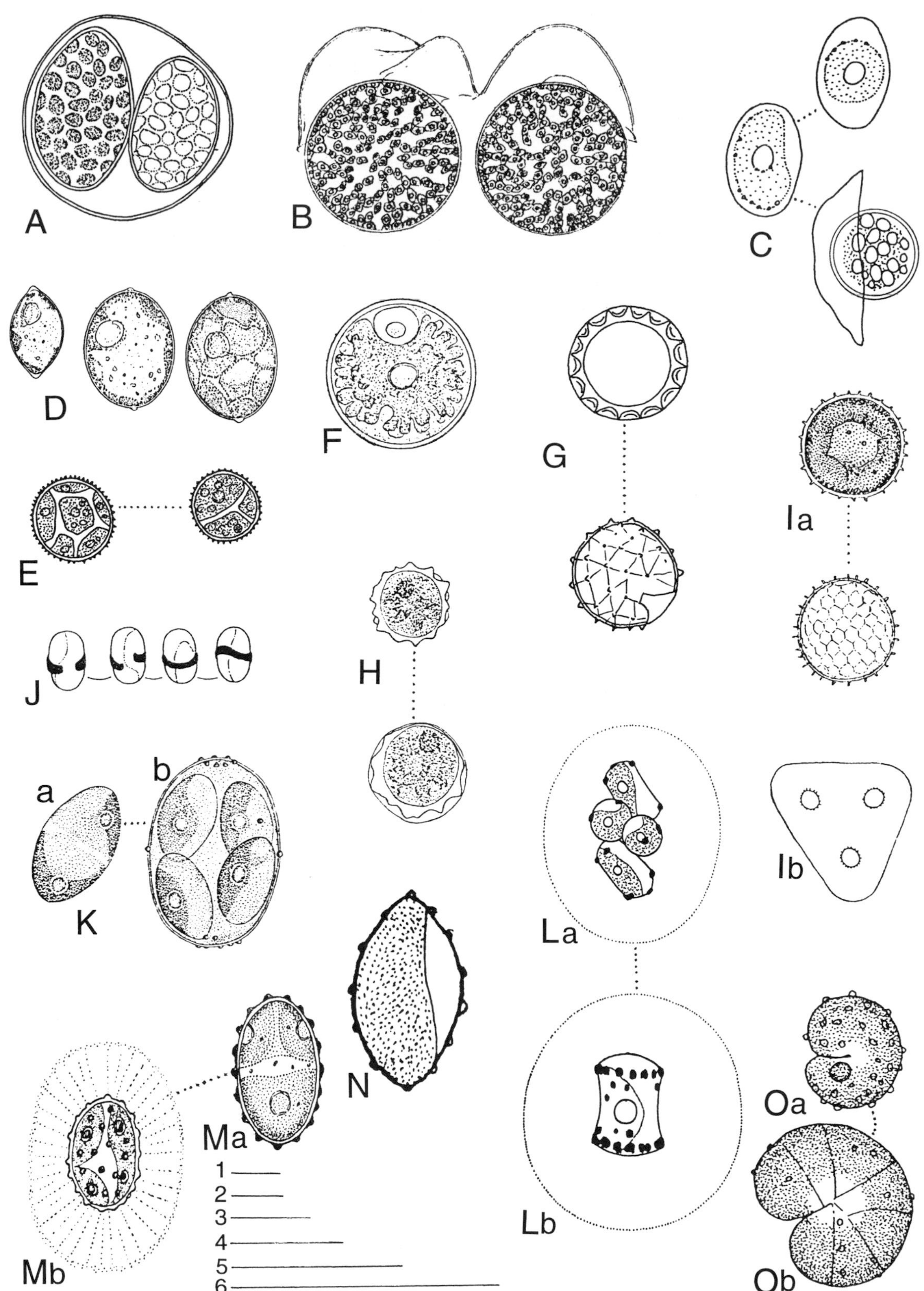

Plate 87 A–O

A. *Eremosphaera gigas* (p. 353): two daughter cells within mother cell wall; **B**. *Eremosphaera viridis* (p. 353): daughter cells undergoing release by breakage of mother cell wall; **C**. *Nautococcus pyriformis* (p. 370): two young vegetative cells and older cell showing cap-like extracellular layer, vacuoles and thick-walls; **D**. *Scotiellopsis oocystiformis* (p. 399); **E**. *Trochiscia hirta* (p. 408); **F**. *Trebouxia arboricola* (p. 407); **G**. *Trochiscia aspera* (p. 408); **H**. *Trochiscia reticularis* (p. 408); **I**. *Trochiscia planctonica* (p. 408): a. close-up of vegetative cells, b. general view of cells enclosed within a pyramid-shaped, mucilaginous envelope; **J**. *Amphikrikos nanus* (p. 330); **K**. *Granulocystopsis decorata* (p. 357): a. details of a vegetative cell, b. autospores or daughter cells within mother cell wall; **L**. *Amphikrikos minutissimus* (p. 330): a. coenobium of 4 cells, b. details of a single cell surrounded by a mucilaginous envelope; **M**. *Siderocelis ornata* (p. 401): a. details of a cell without a sheath, b. single cell surrounded by a mucilaginous sheath; **N**. *Siderocelis minor* (p. 401); **O**. *Juranyiella javorkae* (p. 360): a. vegetative cell, b. dividing cell.

Scale bars: 1, 30 μm; 2–6, 10 μm. 1 for A, B, E, Ib; 2 for C, G; 3 for D, H, Ia, La, Oa; 4 for F, J, K, M, Ob; 5 for Lb; 6 for N.

apex rounded or narrowing to a point, basal disc frequently associated with an incrustation; chloroplast inverted, cup-shaped, confined to distal end of cell, with a distinct pyrenoid; asexual reproduction by 2–4(–8) quadriflagellate zoospores; non-motile spores with contractile vacuoles (hemizoospores) have been reported.

Probably cosmopolitan; grows on the surface of soil.

Fernandinella alpina Chodat 1922

var. *semiglobosa* F.E.Fritsch *et* R.P.John 1942
18060010 **Pl. 99B** (p. 403)
Cells single or densely aggregated, each 3.5–10 μm wide, 3.5–12 μm long, initially obovoid and becoming broadly obovoid to almost spherical; chloroplast almost covering entire cell.

British Isles (England); mainly isolated from calcareous soils, only rarely observed on base-deficient soils.

Differs from the type variety (known from Europe) by not having the chloroplast confined to the distal end of the cell and never pear-shaped but often with a cone-shaped apex.

Franceia Lemmermann 1898

Synonym: *Bohlinia* Lemmerman

Single-celled or forming 2- or 4-celled coenobia of autospores enclosed within the expanded mother cell wall, sometimes enclosed within a mucilaginous envelope; cells oval, broadly ovoid to almost spherical, bearing regularly distributed and thick or very thin (scarcely visible) spines immersed in mucilage from one-fifth to one-third of their length; chloroplasts parietal, plate-like, single or 4–8 in number depending on age, each with a pyrenoid; asexual reproduction by 2, 4 or 8 autospores, released by gelatinization of enlarged mother cell wall.

Probably cosmopolitan; common in plankton of ponds and lakes.

Doubt attaches to the validity of many of the characters traditionally used for separating species in this ill-defined genus. The principal features are cell size and length and thickness of the spines. Similar to the genus *Siderocystopsis* from which *Franceia* can be separated by the marked expansion of its mother cell wall just prior to autospore release.

1 Cells bearing 6 regularly curved spines .... *F. breviseta*
1 Cells bearing 10 or more straight, wavy or curved spines 2
2 Spines thin and very indistinct .... *F. ovalis*
2 Spines thick and distinct .... *F. amphitricha*

Franceia amphitricha (Lagerheim) E.Hegewald 1980

Basionym: *Oocystis ciliata* var. *amphitricha* Lagerheim

Synonyms: *Chodatella ciliata* var. *amphitricha* (Lagerheim) Chodat, *Lagerheimia amphitricha* (Lagerheim) Wille, *Franceia droescheri* (Lemmerman) G.M.Smith, *F. radians* (West *et* G.S.West) Fott, *Oocystis ciliata* West *et* G.S.West, *O. ciliata* var. *radians* West *et* G.S.West

17410010 **Pl. 88D** (p. 359) **CD**

Cells ovoid, 4–6(–13.5) μm wide, 8–12(–18) μm long; spines 10 or more, (10–)12–20 μm long, irregularly distributed or in 3–5 rows, thick, straight or wavy.

Probably cosmopolitan; widely distributed in the British Isles where known from small bodies of water such as ponds and lakes.

We have followed Hegewald et al. (1980) who reduce *Franceia radians* (West *et* G.S.West) Fott (=*Oocystis ciliata* West *et* G.S.West) to synonymy under this species.

Franceia breviseta (West *et* G.S.West) Fott 1981

Basionym: *Chodatella breviseta* West *et* G.S.West
17410020 **Pl. 88E** (p. 359)
Cells ovoid or broadly ellipsoidal, 8–9.5 μm wide, 12–12.5 μm long; spines 6 in number, 11.5–17.5 μm long, curved.

British Isles (Ireland); known only from the type locality of Lough Garten in County Donegal, Ireland.

Considered to be very fragile and probably unable to survive prolonged preservation in spirit.

Franceia ovalis (Francé) Lemmermann 1898

Basionym: *Phythelios ovalis* Francé
Synonym: *Franceia elongata* Korshikov
17410030 **Pl. 88F** (p. 359)
Cells oblong to broadly spindle-shaped, 7–12.5 μm wide, 13–17 μm long, apices blunt or slightly acute; spines 10–20 in number, 15–23(–30) μm long, very thin and delicate, straight or curved.

Probably cosmopolitan; so far reported in the British Isles only from rivers.

We have followed Hegewald et al. (1980) in placing *Franceia elongata* within the synonymy of this species.

Gloeocystis Nägeli 1849

Colonies micro- or macroscopic, consisting of dispersed cells or cell groups surrounded by concentric striations and embedded within a mucilaginous mass of indefinite form; cells ellipsoidal, ovoid or almost spherical, sometimes slightly asymmetrical; chloroplast parietal, cup-shaped and with a single pyrenoid; asexual reproduction by 2, 4 or 8(–16) autospores in each sporangium, often autospores becoming re-orientated and remaining enclosed within the stratified mother cell wall.

Probably cosmopolitan; free-floating, associated with submerged surfaces in stagnant water or growing on damp soil.

1 Cells ellipsoidal or oval; chloroplast trough-shaped .... *G. polydermatica*
1 Cells broadly oval or almost spherical; chloroplast cup-shaped .... *G. vesiculosa*

Gloeocystis polydermatica (Kützing) Hindák 1978

Basionym: *Gloeocapsa polydermatica* Kützing
17420020
Colony of single cells or groups of 2, 4 or 8 cells surrounded by concentrically striated mucilage; cells

(3.5–)6–11 × (2.5–)3.7–6.2(–8) μm, ovoid to ellipsoidal, sometimes slightly asymmetrical; chloroplast trough-shaped.

Probably cosmopolitan; very rarely reported in the British Isles where it forms macroscopic gelatinous masses on soil and lawns.

Gloeocystis vesiculosa Nägeli 1849

17420010 **Pl. 86G** (p. 352)
Colony of ovoid or almost spherical groups of 1–8(–16) cells surrounded by concentrically striated mucilage; cells (4–)6–8(–12) μm, broadly oval to almost spherical, sometimes slightly asymmetrical; chloroplast cup-shaped.

Probably cosmopolitan; widely distributed in the British Isles where it occurs in non-flowing or stagnant water; an early colonizer of burnt heathland.

DOUBTFUL RECORDS

Gloeocystis rupestris Rabenhorst and ***Gloeocystis gigas*** (Kützing) Lagerheim

Considerable confusion surrounds the identity of species attributed to this genus including those mentioned in earlier studies of British freshwater algae. Only *G. gigas* (Kützing) Lagerheim and *G. vesiculosa* Nägeli are mentioned in West & Fritsch (1927) although the Wests had recorded other species but without providing detailed descriptions and illustrations; *Gloeocystis infusionum* was recorded by W. & G.S.West (1897) from Barnes Common, Surrey and was reduced by them to synonymy under *Asterococcus superbus*. Doubt also attaches to the identity of '*G. rupestris* Lyngbye' that has been recorded by Rich (1925), West (1880) and W. & G.S. West (1901). In referring to *G. gigas* and *G. vesiculosa*, West & Fritsch (1927, p. 91) make the following comment: 'These commonly encountered forms are not altogether free from the suspicion (stronger in the case of other sp. of the genus) that they belong, as *palmella*-stages, to the life-cycles of other green Algae' (possibly an indirect reference to *G. infusionum* and *G. rupestris*). Considerable doubt attaches to the identity of *G. gigas*, a species not considered in Komárek & Fott's (1983) review of the Chlorococcales. West & Fritsch (1927, p. 90) make the statement 'The most abundant of the two Brit. sp. is *G. gigas* (Kütz.) Lagerh. (*Gl. ampla* (Kütz.) Rabenh., *Chlorococcum gigas* Grun.), found in stagnant waters among other Algae; it has globose cells (10–17 br.) which are often grouped in tetrads'. According to Starr (1955), information is so incomplete on species such as *Chlorococcum gigas* Grunow (=*Gloeocystis gigas*) that they should be considered as uncertain.

Gloeotaenium Hansgirg 1890

Single-celled or spherical 2–4(–8)-celled colonies of ovoid or ellipsoidal cells, within a wide, stratified sheath containing calcite crystals; cells arranged in the form of a cross within the persistent mother cell wall, usually separated from each other by a thick dark brown or black band of calcite, deposit in 'I'-shapes in 2-celled colonies and a Maltese cross in 4-celled colonies; chloroplast parietal, massive; reproduction by autospores and akinetes.

Probably cosmopolitan; in moorland waters and calcareous waters. Monospecific.

Gloeotaenium loitlesbergerianum Hansgirg 1909

18090010 **Pl. 86P** (p. 352)
Single cells 25 μm in diameter, 2-celled colonies 70–90 μm in diameter, or in 4-celled colonies.

Distribution as genus; in the British Isles known only from mud of a pond near Sheffield that had a rich bottom flora (including *Hippuris* and *Chara*) and lay in an area of limestone.

Lind (1939b) discovered this alga on the sides of a jar in which mud from the pond had been kept for several weeks.

Golenkinia Chodat 1894

Cells solitary, spherical or slightly ovoid, thick-walled, covered by regularly distributed spines and surrounded by a thin or thick mucilaginous envelope; spines of more or less equal width throughout or broader at the base, extending beyond the mucilage or occasionally mucilage confined to base of spines and scarcely visible; chloroplast parietal, bell- or cup-shaped or indistinctly net-like, pyrenoid single and often cup- or kidney-shaped (rarely spherical), a starch sheath present or absent; asexual reproduction by 2, 4 or 8(–16) autospores, non-motile spores with contractile vacuoles (hemizoospores) or naked biflagellate zoospores released into a mucilage vesicle following rupture of mother cell wall; akinetes known; oogamous reproduction, 8 or 16 biflagellate male gametes per cell and a vegetative-like cell functioning as a female gamete, zygotes with spiny walls.

Probably cosmopolitan; usually planktonic in ponds, lakes and slow-flowing reaches of rivers.

1 Bristles, <20 μm long, more than 20 per cell ....*G. paucispina*
1 Bristles >20 μm long, fewer than 20 per cell ....*G. radiata*

Golenkinia paucispina West *et* G.S.West 1902

Synonyms: *Golenkenia brevispina* Korshikov, *G. radiata* Chodat var. *brevispina* Tiffany *et* Ahlstrom, *Micractinium paucispinum* (West *et* G.S.West) Wille

17430010 **Pl. 88G** (p. 359)
Cells (5.5–)8–19 μm wide, spherical, smooth-walled, without mucilage; spines 8–16 μm long, more than 20 per cell; pyrenoid spherical and surrounded by starch.

Probably cosmopolitan; recorded in the British Isles only from the plankton of Irish rivers.

Golenkinia radiata Chodat 1894 emend. Korshikov 1953

Synonyms: *Golenkinia radiata* var. *longispina* G.M.Smith, *Micractinium radiatum* (Chodat) Wille

17430020 **Pl. 88H** (p. 359) **CD**
Cells (6.4–)9–21 μm wide, spherical, smooth-walled or rough, with mucilage hardly visible or thick and well-developed; spines (21–)24–45(–65) μm long, usually fewer than 15 per cell; pyrenoid cup-shaped to spherical, with starch confined to convex side only if

cup-shaped (pyrenoid difficult to discern if starch absent)

Probably cosmopolitan; widely distributed in the British Isles in the plankton of still and flowing water habitats.

Golenkiniopsis Korshikov 1953

Cells solitary, spherical to almost ovoid, thin-walled, regularly covered by thin and sometimes slightly basally-thickened spines, with or without a mucilaginous envelope; chloroplast parietal, often cup-shaped; pyrenoid spherical to ellipsoidal, with a distinct starch sheath; asexual reproduction by 2, 4 or 8 autospores, released by fracture of mother cell wall; sexual reproduction oogamous, male gametes biflagellate and zygote with a warty outer wall or covered by short spines.

Probably cosmopolitan; planktonic in flowing and non-flowing freshwater habitats.

It might be easily mistaken for a single-celled *Micractinium* Fresenius or *Golenkinia* Chodat, but differs from these two genera in that *Golenkiniopsis* very rarely produces new coenobia except in laboratory culture. *Golenkinia* usually has thick walls whereas *Golenkiniopsis* is thin-walled and both genera typically have a distinct starch sheath surrounding the pyrenoid. Such a starch sheath is absent in *Micractinium* which is usually colonial.

1 Cells broadly ovoid, usually with more than 20 spines .... *G. chlorelloides*
1 Cells spherical, usually with fewer than 20 spines 2
2 Cells always >7 μm in diameter; spines 20–65 μm long, sometimes slightly curved .... *G. longispina*
2 Cells usually <7 μm in diameter; spines typically <20 μm long and always straight .... *G. parvula*

Golenkiniopsis chlorelloides (J.W.G.Lund) Fott 1981

Basionym: *Golenkinia chlorelloides* J.W.G.Lund

17440010 **Pl. 88I** (p. 359)

Cells 4–10 μm in diameter, broadly ovoid, sometimes irregularly shaped, thin-walled, bearing more than 20 very delicate spines (5–10 μm long), each with a swollen warty (tuberculate) base.

British Isles (England); recorded only from a garden pond at Ashford in Kent where collected in summer.

According to Lund (1954a), the irregularly disposed spines are so delicate that they are difficult to see without staining.

Golenkiniopsis longispina (Korshikov) Korshikov 1953

Basionym: *Golenkinia longispina* Korshikov

17440020 **Pl. 88K** (p. 359)

Cells 7.2–11 μm in diameter, spherical; spines (20–)40–65 μm long, curved or sometimes straight, less than 20 per cell; zygote 15–16 μm wide, covered by warty wall outgrowths.

Probably cosmopolitan; in the British Isles planktonic in various types of water body.

Golenkiniopsis parvula (Woronichin) Korshikov 1953

Basionym: *Golenkinia parvula* Woronichin

Synonyms: *Golenkiniopsis minutissima* (M.O.P.Iyengar *et* M.S.Balakrishnan) Philipose, *Golenkinia minutissima* M.O.P. Iyengar *et* M.S.Balakrishnan

17440030 **Pl. 88J** (p. 359)

Cells 4–7 μm in diameter, spherical; spines (4–)10.5–21 μm long, narrow, 6–20(–25) per cell; zygotes thick-walled, covered by short spines (1–1.5 μm long).

Probably cosmopolitan; in the British Isles found in rain puddles and the plankton of ponds or slow-flowing reaches of rivers (e.g. Great Ouse).

Granulocystopsis Hindák 1977

Cells solitary or forming 2-, 4- or 8-celled coenobia, surrounded by a mucilaginous envelope; cells broadly ellipsoidal or ovoids, with a crown of brown granules at the rounded apices and sometimes fused to form an apical ring; chloroplasts 1–4, parietal, trough-shaped, each containing a single pyrenoid; asexual reproduction by 2 or 4(–8) autospores, released by gelatinization of expanded mother cell wall.

Probably cosmopolitan; planktonic in various aquatic habitats.

1 Cells elongate-ellipsoidal to cylindrical-oval .... *G. coronata*
1 Cells broadly oval .... *G. decorata*

Granulocystopsis coronata (Lemmermann) Hindák 1977

Basionym: *Oocystis coronata* Lemmermann

Synonym: *Oocystopsis coronata* Lemmermann

17450010 **CD**

Cells (3.5–)4.5–9 μm wide, 4.5–19.2 μm long, elongate-ellipsoidal to cylindrical-ovoid, apices rounded, each with a subapical ring of isolated or almost continuous granules, sometimes granules confined to very tip.

Probably cosmopolitan; usually planktonic in small bodies of water in the British Isles.

According to Hindák (1988), *G. coronata* and *G. decorata* (as *G. pseudocoronata*) are sufficiently different to be regarded as separate species although sometimes they have been considered conspecific. We have followed Hindák (*op. cit.*) and have proposed a new combination after discovering an earlier and legitimately described species that is identical to *G. pseudocoronata* (see below).

Granulocystopsis decorata (Svirenko) Tsarenko *et* D.M.John, *nov. comb.*

Basionym: *Oocystis decorata* Svirenko 1931. *Trudy Gos. Ihtiol. Opitn. Stant.* 6(2): 130, figs 16–17.

Synonyms: *Oocystis pseudocoronata* Korshikov, *Granulocystopsis pseudocoronata* (Korshikov) Hindák

17450030 (17450020) **Pl. 87K** (p. 354) **CD**

Cells (3.5–)5–9 μm wide, (7–)9–14.5 μm long, broadly ovoid, with rounded apices each having a ring of often fused granules.

Europe; in small ponds and lakes, very rare in the British Isles.

Hormotila Borzi 1883

Colony attached, consisting of cells positioned terminally or at irregular intervals within branched cylinders of structureless or transversely layered mucilage; cells spherical, oval or broadly ellipsoidal, rarely oblong, walls thin and mucilaginous; chloroplast parietal, indistinctly net-like, with or without a pyrenoid; asexual reproduction by 8–64 naked, biflagellate zoospores; akinetes known.

Probably cosmopolitan; on damp surfaces and submerged surfaces in standing water.

Hormotila mucigena Borzí 1883

17460010 **Pl. 99O** (p. 403)

Cells 4–12 µm in size, spherical or oval, solitary or in pairs along mucilaginous branches and at branch apices; chloroplast with a single pyrenoid; zoosporangia reaching 30 µm wide, zoospores 1–2.5(–4) µm wide, 3–5(–8) µm long, ovoid, with an eyespot.

Probably cosmopolitan; recorded on a few occasions in the British Isles as a dull green layer on damp calcareous rocks and within glass houses; probably more common and widely distributed than records indicate.

Hydrianum Rabenhorst 1868

Synonym: *Chlororhabdion* Jane

Cells attached and solitary or in groups, straight or curved, narrowly spindle-shaped, ellipsoidal, pear-shaped, or flask-shaped to cylindrical, rounded apically and sometimes having a ring-like wall thickening, either gradually narrowing distally to a short, thick stalk or attached directly by a mucilaginous pad; chloroplast parietal and thin, with or without (majority) a pyrenoid, often with minute starch grains when pyrenoids absent; asexual reproduction by biflagellate zoospores, released by apical or subapical splitting of mother cell wall, often a single zoospore remains in the sporangium and gives rise to a new generation (successive generations arise by this means).

Probably cosmopolitan; often in nutrient-poor or brownish-coloured humic waters (dystrophic), commonly epiphytic on aquatic vegetation and filamentous algae.

Hydrianum coronatum Fott 1957

17470010 **Pl. 99L** (p. 403)

Cells 6–8 µm wide, up to 12 µm long, obovoid, often broader basally and having a ring-like wall thickening at apex, attached by a flattened base and a brownish mucilaginous pad; chloroplast thin, light green, with a single pyrenoid.

Europe; in the British Isles usually on algae, mosses (especially *Sphagnum*) and aquatic macrophytes growing in pools within fens and boggy areas, very rare.

Hydrodictyon Roth 1800

Coenobium macroscopic, free-floating or secondarily attached, forming sac-like or flat plate-like networks of several thousand cells; cells cylindrical to oblong-oval or spherical (depending on age), (2–)3(–4) adjacent cells connected by their edges to form 5–6-angled meshes, thin-walled, multinucleate; chloroplast parietal, perforate or forming an irregular net, with many pyrenoids within protoplasmic layer surrounding a central vacuole; asexual reproduction by biflagellate, uninucleate zoospores that form a daughter coenobium within the mother cell wall, released by gelatinization of cell wall; sexual reproduction isogamous, gametes similar to zoospores only smaller, released through lateral opening in mother cell wall and fuse to form a thick-walled zygote (hypnospore) which produces 2–5 zoospores, each zoospore develops into a star-shaped cell ('polyhedral-stage') once settled and from which a new miniature net-like coenobium arises.

Probably cosmopolitan; usually free-floating in ponds, small lakes and rice paddies, or secondarily attached in slow-moving streams, rivers and irrigation ditches.

Of the 5 species only *Hydrodictyon reticulatum* (Linnaeus) Lagerheim and *H. patenaeforme* Pocock are known in the northern and southern hemispheres; the latter species is sometimes considered no more than a developmental stage of the former (Pickett-Heaps, 1975). The remaining species are known from temporary basins or lakes in South Africa (*H. africanum* Yamanouchi), South America (*H. major* Kühnemann) and India (*H. indicum* M.O.P.Iyengar). For details of its complicated life history and variations (e.g. hypnospores developing into a new coenobium), see Pickett-Heaps (1975). Considered by Deason et al. (1991) to belong to the broadly defined Order Sphaeropleales which includes some other Chlorococcalean algae such as *Pediastrum*, *Scenedesmus* and *Coelastrum*.

Hydrodictyon reticulatum (Linnaeus) Lagerheim 1883

Basionym: *Conferva reticulatum* Linnaeus

Synonyms: *Hydrodictyon utriculatum* Roth, *H. pentagonum* Vaucher

17480010 **Pl. 99S** (p. 403) **CD**

Coenobia closed, a sac-like network of cells reaching to over 50 cm long, meshes up to 10(–15) mm across and clearly visible in large colonies; cells vary considerably in size, sometimes reaching to 10(–15) mm long in large colonies, cylindrical, 3 commonly connected by their edges to form 5-sided meshes.

Probably cosmopolitan; over the last decade of the twentieth century the 'water net' became a widely distributed nuisance alga in the British Isles where it formed massive surface mats from about July to September in ponds and often hard-water lakes, slow-flowing rivers and trout-farm lakes (see John et al.,1998). Commonly free-floating, but it sometimes undergoes daily changes in buoyancy resulting in its sudden disappearance from the water surface only to quickly reappear. In slow-flowing rivers, it accumulates in the marginal shallows where it sometimes secondarily attaches. If present in large quantities *Hydrodictyon* blocks waterways and irrigation ditches, clogs boat engine intakes, causes economic losses to trout fisheries, taints drinking water, and adversely affects the amenity value of water especially if rotting masses accumulate. The water net is difficult to eradicate since physical removal offers but a short-term solution.

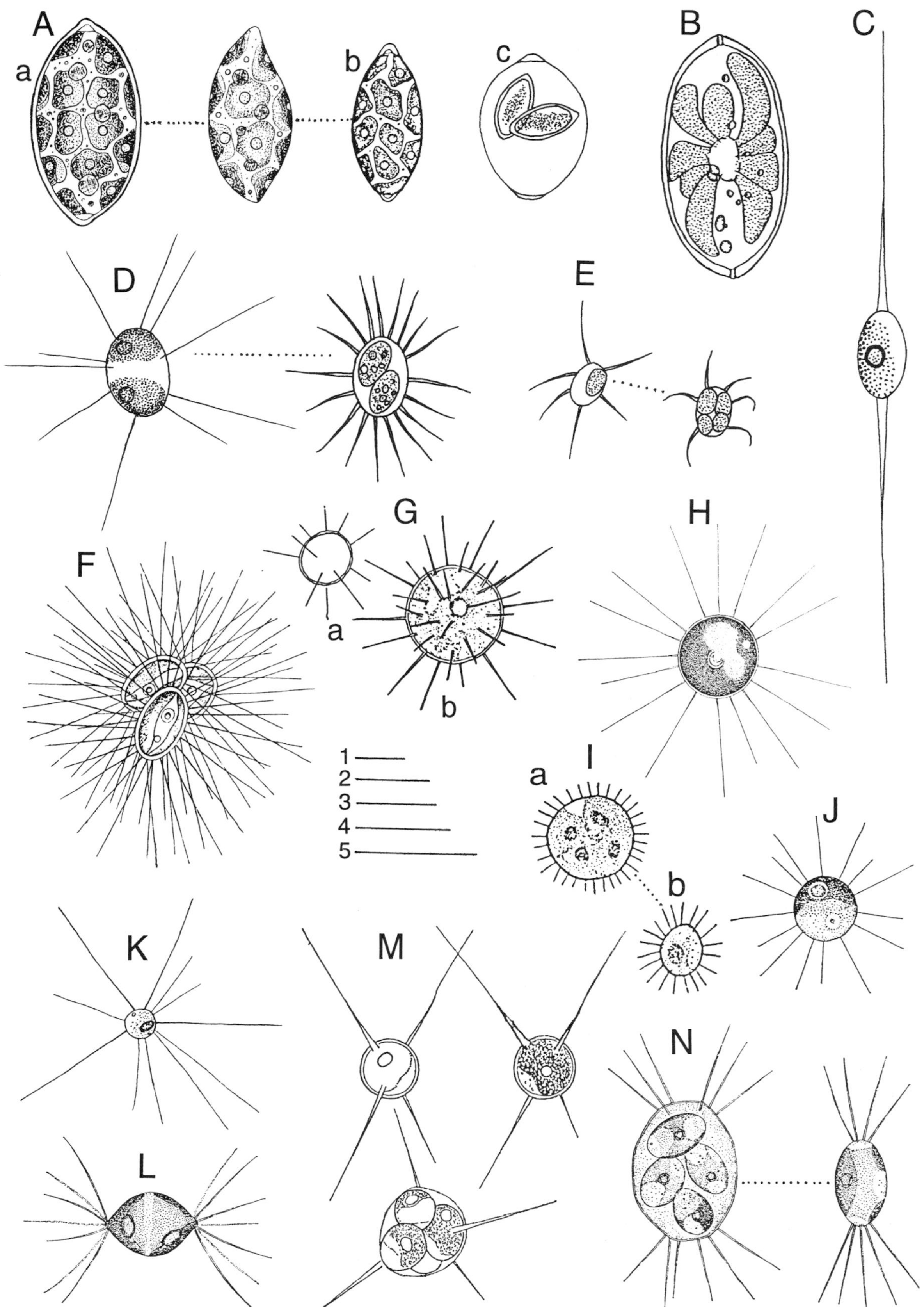

Plate 88 A–N

A. *Oocystis solitaria* (p. 375): a,b. details of individual cells, c. two daughter cells within mother cell wall; **B**. *Neglectella asterifera* (p. 370); **C**. *Diacanthos belenophorus* (p. 347); **D**. *Franceia amphitricha* (p. 355): vegetative cell (left) and cell containing autospores (right); **E**. *Franceia breviseta* (p. 355): right cell containing autospores; **F**. *Franceia ovalis* (p. 355); **G**. *Golenkinia paucispina* (p. 356); **H**. *Golenkinia radiata* (p. 356); **I**. *Golenkiniopsis chlorelloides* (p. 357); **J**. *Golenkiniopsis parvula* (p. 357); **K**. *Golenkiniopsis longispina* (p. 357); **L**. *Lagerheimia citriformis* (p. 362); **M**. *Lagerheimia chodatii* (p. 362): two vegetative cells (above) with one below containing autospores; **N**. *Lagerheimia ciliata* (p. 362): vegetative cell (right) and cell containing autospores (left).
Scale bars: all 10 μm. 1 for Ab, K; 2 for Aa, E, F, Ga, H, L, N; 3 for Ac, Gb, Ib, M; 4 for B–D; 5 for Ia, J.

Juranyiella Hortobágyi 1962

Cells solitary, more rarely forming a 4-celled coenobium; cells crescent- or kidney-shaped, circular with a broad to very narrow, lateral notch, thick-walled, covered by numerous, irregularly arranged warts; chloroplast parietal, almost completely covering wall, with a single pyrenoid; asexual reproduction by (2–)4(–8) autospores, released by splitting of convex side of mother cell wall.

Europe; planktonic and usually in nutrient-rich lakes. Monospecific.

Juranyiella javorkae (Hortobágyi) Hortobágyi 1962

Basionym: *Kirchneriella javorkae* Hortobágyi
17490010 **Pl. 87O** (p. 354)
Cells (4–)7–14 μm wide, covered by variable numbers of often brownish warts.

Distribution as for genus; so far recorded in the British Isles only from a pond near Cambridge, almost certainly more widely distributed.

Keratococcus Pascher 1915

Synonym: *Dactylococcus* Nägeli *sensu* Hansgirg
Cells solitary or loosely connected at extremities, ellipsoidal, spindle-shaped, cylindrical or crescent-shaped, often asymmetrical, apices acute and sometimes prolonged into a hair or spine; chloroplast parietal, single or fragmented, with pyrenoid present but sometimes indistinct; asexual reproduction by 2 or 4 autospores, produced by oblique division of mother cell protoplasm.

Probably cosmopolitan; aquatic or forming a green layer on damp subaerial surfaces.

An ill-defined genus distinguished from *Monoraphidium* mainly due to the presence of pyrenoids. We do not consider the presence or absence of pyrenoids to be a satisfactory criterion for the separation of genera unless linked to other characters. For convenience, it has been decided to follow Hindák (1988) in recognizing *Keratococcus* although future studies might well lead to it becoming reduced to synonymy under *Monoraphidium*. Recognized by Komárek (1979) and Komárek & Fott (1983) is the unsatisfactorily defined genus *Chlorolobion* (p. 388) which morphologically resembles *Keratococcus* and also possesses pyrenoids.

Keratococcus bicaudatus (A.Braun) J.B.Petersen 1928

Basionym: *Dactylococcus bicaudatus* A.Braun
Synonyms: *Dactylococcus caudatus* Hansgirg *sensu* G.S.West *et* F.E.Fritsch, *D. bicaudatus* var. *subramosus* West *et* G.S.West, *D. bicaudatus* var. *exilis* West *et* G.S.West
17500010 **Pl. 91G** (p. 369)
Cells 2.5–5.8 μm wide, 13–19(–36) μm long, spindle-shaped, slightly asymmetrical, with apices prolonged into a long hair or spines.

Probably cosmopolitan; recorded on several occasions in the British Isles where mainly on wet stones, puddles, damp soil and on wood.

Considerable doubt attaches to the species (*Dactylococcus dispar*) and varieties (*Dactylococcus bicaudatus* vars. *subramosus* and *exilis*) described by the Wests towards the end of the nineteenth century (see W. & G.S.West, 1897, 1898). We have reduced the two varieties to synonymy under *Keratococcus bicaudatus* in the belief that small differences in form (var. *subramosus*: cells remaining attached to form colonies) and overall length (var. *exilis*: reaching 36 μm long) are insufficient justification for recognizing new varieties. Doubt inevitably attaches to the placement of var. *exilis* since its original description made no mention of pyrenoids.

DOUBTFUL RECORDS

Keratococcus dispar (West *et* G.S.West) Tsarenko *et* D.M.John, *nov. comb.*
Basionym: *Dactylococcus dispar* West *et* G.S.West, 1897, p. 500.
17500020 **Pl. 91H** (p. 369)
We are recombining *Dactylococcus dispar* into the genus *Keratococcus* despite having strong reservations as to whether it should continue to be regarded as a species distinct from *K. bicaudatus*. The Wests' material from 'old wood' at Dorking in Surrey was described by them as having cells 2–5 μm wide and 8.5–21 μm long, spindle-shaped, oblique and almost halfmoon-shaped, with the apices subequally developed; they regarded *Keratococcus raphidioides* (as *Dactylococcus raphidioides*) as the 'nearest species' to their *Dactylococcus dispar*. Examination of further material is required to establish whether it should be regarded as a separate species.

Keratococcus raphidioides (Hansgirg) Pascher 1915
17500030
W. & G.S.West (1900b) attributed material collected from Bradford, West Yorkshire, to this species (as *Dactylococcus raphidioides* Hansgirg). The Wests state (*op. cit.*, p. 118) that its cells were 'mostly fusiform, sigmoid, or sublunate, and were 2–2.5 μm wide' but do not provide an illustration. An unpublished line drawing in the Wests' archive at The Natural History Museum in London shows cells in which the apices are prolonged into a spine, a feature of the species. Doubt must inevitably attach to this record since no mention is made of a pyrenoid.

Kirchneriella Schmidle 1893

Synonyms: *Kirchneria* Hindák *non* A.Braun, *Pseudokirchneriella* Hindák
Coenobia spherical, ovoid or irregularly shaped, consisting of 2, 4, 8, 16 or 32 irregularly arranged cells, usually enclosed within a mucilaginous envelope; cells broadly or narrowly arc- or crescent-shaped, sigmoid or spirally twisted, cylindrical or narrowing from centre, with apices acute, rounded or bluntly pointed; chloroplast parietal, single, with or without a pyrenoid; asexual reproduction by autospores, released by rupture of mother cell wall.

Probably cosmopolitan; a common component of the plankton in many freshwaters.

Hindák (1988) established a genus he called *Kirchneria* to accommodate those *Kirchneriella* species not known to have pyrenoids. He was unaware at the time that the generic name *Kirchneria* had been used for both a fossil plant and a living member of the family Leguminosae. After having been informed of these earlier names by P.C. Silva, Hindák (1990) proposed replacing *Kirchneria* with *Pseudokirchneriella*. Considerable doubt attaches to the justification for separating genera on the presence or absence of a pyrenoid, especially if based on light microscope observations alone. We have decided not to follow Hindák but to recognize *Kirchneriella* and to consider the presence or absence of pyrenoids as a character for separating species not genera.

1 Cells relatively narrow, without a pyrenoid ....*K. irregularis*
1 Cells short and wide, with a pyrenoid 2
2 Cells with sides of 'concavity' or notch so strongly curved that apices almost touching or overlapping...3
2 Cells with sides of concavity only slightly curved; apices often well separated 4
3 Cells with sharply pointed apices, never turning inwards ....*K. lunaris*
3 Cells with bluntly rounded apices turning inwards ....*K. incurvata*
4 Cells with sides of concavity almost parallel ....*K. obesa*
4 Cells concavity usually V-shaped, sides rarely almost parallel ....*K. aperta*

Kirchneriella aperta Teiling 1912

Synonyms: *Kirchneriella obesa* (West) Schmidle var. *aperta* (Teiling) Brunnthaler, *K. obesa* (West) Schmidle var. *pygmaea* West *et* G.S.West

17510010 **Pl. 98F** (p. 400)

Cells 6–12 μm wide, almost circular in outline, concavity usually broadly V-shaped, only occasionally with almost parallel sides, apices rounded or bluntly pointed; pyrenoids present.

Probably cosmopolitan; usually in small water bodies, only very rarely recorded in the British Isles.

Kirchneriella incurvata J.H.Belcher *et* Swale 1962

Synonym: *Kirchneriella irregularis* (G.M.Smith) Korshikov *sensu* Korshikov *sine typo*

17510020 **Pl. 98G** (p. 400)

Cells up to about 8 μm wide and 20 μm long, half-moon-shaped to almost circular, slightly spiralled, weakly curved centrally and strongly curved towards apices where they turn inwards and bend slightly to one side, regularly narrowing to bluntly rounded apices, one apex sometimes more acute than the other; pyrenoids present.

Europe; so far only reported in the plankton or associated with aquatic macrophytes in several pools in England, including the Kendal Canal in Lancaster.

Korshikov (1953) made the new combination *Kirchneriella irregularis* (G.M.Smith) Korshikov (not valid according to Ross, pers. comm., in Belcher & Swale, 1962), with his description and illustration not corresponding to G.M.Smith's *Kirchneriella lunaris* (Kirchner) Möbius var. *irregularis* G.M.Smith. Belcher & Swale (1962) discovered an alga corresponding almost exactly to Korshikov's description and established *Kirchneriella incurvata*. They studied unialgal cultures of this new species and discovered that arcuate apices were a constant feature during all phases of growth.

Kirchneriella irregularis (G.M.Smith) Korshikov 1953

Basionym: *Kirchneriella lunaris* (Kirchner) K.Möbius var. *irregularis* G.M.Smith
Synonym: *Kirchneria irregularis* (G.M.Smith) Hindák

17510030 **Pl. 98H** (p. 400) **CD**

Cells 3–5(–6) μm wide, (6–)21(–26) μm long, almost circular, markedly curved or sigmoid, narrowing to acute or bluntly pointed apices, with apices inwardly curving and slightly bent and overlapping; pyrenoid absent.

Probably cosmopolitan; often in the plankton and associated with aquatic vegetation in rivers, ponds and lakes (e.g. Loch Leven and other Perthshire lochs) as well as a moat at Groombridge in Kent; probably more widely distributed in the British Isles than records indicate.

There is no good evidence that a pyrenoid exists in this species (see Hindák, 1988, p. 226).

Kirchneriella lunaris (Kirchner) K.Möbius 1894

Basionym: *Raphidium convolutum* (Corda) Rabenhorst var. *lunare* Kirchner

17510040 **Pl. 98I** (p. 400)

Cells (1.2–)3–8 μm wide, (4–)6–15 μm long, flattened, crescent- or sickle-shaped, concavity with its sides curved, apices sharply pointed to slightly rounded; pyrenoid present.

Probably cosmopolitan; known from small ponds and frequent in lake plankton in the British Isles (including Loch Leven, Perthshire).

Kirchneriella obesa (West) Schmidle 1893

Basionym: *Selenastrum obesum* West
Synonym: *Kirchneriella intermedia* Korshikov

17510050 **Pl. 98J** (p. 400) **CD**

Cells 2–8(–16) μm wide, 6–16 μm long, almost circular in outline, strongly crescent-shaped, concavity usually with almost parallel sides, apices rounded or bluntly pointed; pyrenoid present.

Probably cosmopolitan; one of the most widely distributed species in the British Isles and sometimes frequent in the plankton of ponds and lakes; considered by the Wests (W. & G.S.West, 1903a) as 'sparingly distributed in the summer plankton of the lakes of Western Scotland'.

Komarekia Fott 1981

Synonym: *Hofmania* Chodat

Coenobia flat, cross- to wreath-shaped, consisting of 4 cells surrounding a rounded or square central space, sometimes forming 16- or 32-celled syncoenobia, enclosed within a thin mucilaginous envelope; cells spherical, ovoid to ellipsoidal, thin-walled, laterally attached and rarely free; chloroplast parietal, with a single pyrenoid (sometimes scarcely visible); asexual reproduction by 4(–8) autospores, released by mother cell wall splitting into 4 parts, remnants of wall remain attached laterally to each daughter cell as a colourless appendage.

Probably cosmopolitan; planktonic or associated with submerged surfaces.

Komarekia appendiculata (Chodat) Fott 1981

Basionym: *Hofmania appendiculata* Chodat
Synonym: *Crucigenia appendiculata* (Chodat) Schmidle

17520010 **Pl. 84K** (p. 346)

Coenobia 12–14 μm wide, cross-shaped, with remnants of mother cell wall persisting as colourless appendages attached to outer cell wall, the remnants connecting successive generations to form square syncoenobia; cells 2.5–6 μm wide, 4–9 μm long, ovoid, wall remnants 4–8 μm long.

Probably cosmopolitan; very rare in the British Isles since recorded on a few occasions in slow-flowing water (e.g. Ulverston Canal) and in ponds or lakes (e.g. Loch Leven, Perthshire).

Korshikoviella P.C.Silva 1959

Cells elongate spindle-shaped to cylindrical, thin-walled, narrowing apically to a needle-like point or broadly rounded apex, often narrowing basally to a short, triangular, anchor-shaped or divided end; chloroplast parietal, continuous or subdivided into several segments lying in one or many rows, each segment with a single cup-shaped chloroplast, pyrenoid and nucleus; asexual reproduction by biflagellate zoospores, released by lateral splitting of the mother cell wall.

Probably cosmopolitan; planktonic or more usually on various planktonic crustaceans in ponds and lakes.

Korshikoviella michailovskoensis (Elenkin) P.C.Silva 1959

Basionym: *Characium michailovskoense* Elenkin
Synonyms: *Lambertia michailovskoensis* (Elenkin) Korshikov, *Characium gracilipes* F.D.Lambert
17530010 **Pl. 90F** (p. 367)
Cells 5–23 μm wide, 30–140 μm long, cylindrical to spindle-shaped, slightly or strongly curved, distal end with a short or more frequently a long, straight or somewhat curved spine, basally terminating in two short branches or branches very reduced or absent.

Probably cosmopolitan; in the British Isles epizoic on crustaceans (often *Daphnia*), or planktonic in ponds, lakes and reservoirs, rare.

Lagerheimia Chodat 1895

Synonym: *Chodatella* Lemmermann
Cells solitary or sometimes forming 2-, 4- or 8-celled coenobia, often enclosed within a thin mucilaginous envelope; cells ovoid to almost spherical, ellipsoidal or slightly cylindrical with rounded apices, smooth-walled; spines apical, subapical or equatorial and straight or curved, often basally thickened or convex, equal to cell length or more usually longer; chloroplasts parietal, plate-like, increasing up to 8 with age and just prior to autospore formation, each with a pyrenoid; asexual reproduction by 2, 4 or 8 autospores, released by splitting or gelatinization of mother cell wall.

Probably cosmopolitan; planktonic or associated with beds of aquatic macrophytes in non-flowing water.

The presence or absence of a basal thickening of the spines is no longer considered a character of any taxonomic significance. It was formerly the principal character used for separating *Lagerheimia* from *Chodatella*, the latter now reduced to synonymy under *Lagerheimia* (see Hindák, 1983).

1 Cells with spines opposite one another in the form of a cross, or at equator and apex 2
1 Cells with 2–12 spines at each apex 3
2 Four spines arising opposite one another and in form of a cross .... *L. chodatii*
2 Single spines at each apex and 2(–3) opposite each other at the cell equator ...... *L. wratislaviensis*
3 Cells lemon-shaped (4–8 spines at each apex) .... *L. citriformis*
3 Cells ovoid or ellipsoidal to spindle-shaped........... 4
4 Spines slightly subapical, a pair at each apex, lying in one plane .... *L. genevensis*
4 Spines usually apical, 3–10 at each apex, lying in different planes 5
5 Spines (2–)3–4(–5), as long as or shorter than cell .... *L. subsalsa*
5 Spines 4–8(–10), longer than cell............................ 6
6 Spines up to 40 μm long .... *L. ciliata*
6 Spines 40–55(–80) μm long .... *L. longiseta*

Lagerheimia chodatii C.Bernard 1908

Synonym: *Chodatella chodatii* (C.Bernard) S.H.Li
17540010 **Pl. 88M** (p. 359)
Cells 5–10.8 μm wide, spherical to broadly ovoid, with 4(–6) perpendicular spines lying in one plane and arranged in the form of a cross; spines 8–20(–30) μm long, with pronounced concavity at base, comparatively thick and narrowing terminally to become needle-like, sometimes brown.

Probably cosmopolitan; usually in the plankton of large lakes (e.g. Loch Leven, Scotland) and widespread in smaller water bodies in many parts of the British Isles.

Lagerheimia ciliata (Lagerheim) Chodat 1895

Basionym: *Oocystis ciliata* Lagerheim
Synonym: *Chodatella ciliata* (Lagerheim) Lemmermann
17540020 **Pl. 88N** (p. 359) **CD**
Cells (5.8–)9–18 μm wide, (10–)12–21 μm long, oval, with 3–18 spines forming a subapical ring; spines 12–20(–40) μm long, usually longer than cell.

Probably cosmopolitan; common in the plankton of slow-flowing rivers (e.g. Thames, Great Ouse), ponds and lakes (e.g. Loch Leven) in many parts of the British Isles.

Very similar to *Lagerheimia longiseta*, but differs by having somewhat shorter and fewer spines, characters that show considerable variation and bring into question the separation of these two species (see Hindák, 1984).

Lagerheimia citriformis (J.Snow) Collins 1909

Basionym: *Chodatella citriformis* J.Snow
17540030 **Pl. 88L** (p. 359) **CD**
Cells usually lemon-shaped, 8–20 μm wide, (10–)13–26 μm long, with (3–)4–8 spines forming a ring at each apex; spines 22–60 μm long, longer than cell.

Probably cosmopolitan; recorded from rivers and lake plankton in the British Isles.

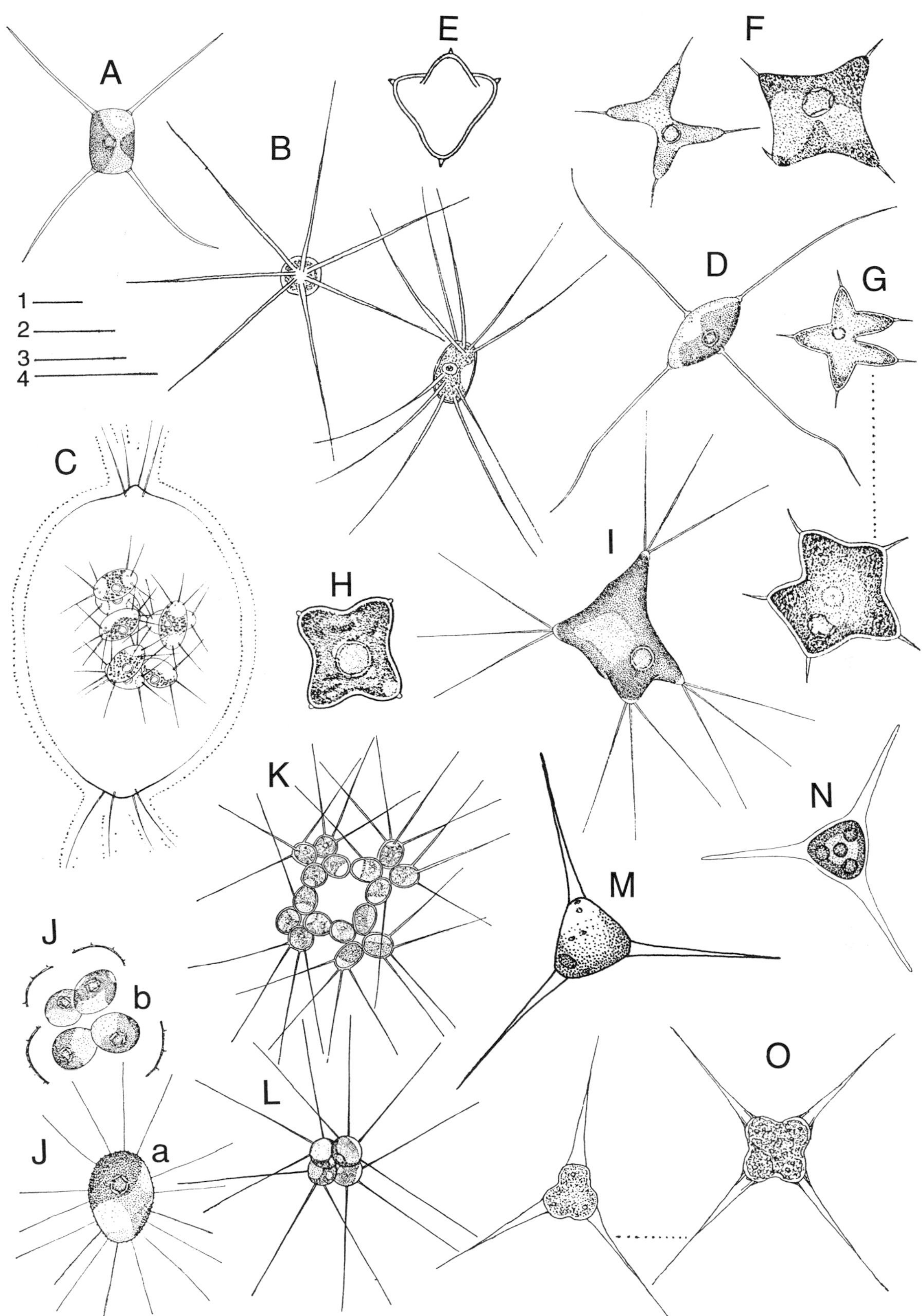

Plate 89 A–O

A. *Lagerheimia genevensis* (p. 364); **B**. *Lagerheimia longiseta* (p. 364): cell in polar view (left) and side view (right); **C**. *Lagerheimia subsalsa* (p. 364): daughter colonies in mother cell wall; **D**. *Lagerheimia wratislaviensis* (p. 364); **E**. *Tetraedron regulare* (p. 405); **F**. *Tetraedron incus* (p. 405); **G**. *Tetraedron caudatum* (p. 405): showing range of form variation; **H**. *Tetraedron minimum* (p. 405); **I**. *Polyedriopsis spinulosa* (p. 380); **J**. *Siderocystopsis punctifera* (p. 401): a. single vegetative cell, b. mother cell wall fracturing to release daughter colonies; **K**. *Micractinium quadrisetum* (p. 365); **L**. *Micractinium pusillum* (p. 365); **M**. *Treubaria setigera* (p. 407); **N**. *Treubaria schmidlei* (p. 407); **O**. *Treubaria triappendiculata* (p. 407): vegetative cell with one cell (right) at early stage of dividing to produce autospores.

Scale bars: 10 μm. 1 for A, L; 2 for B, C, G, Jb, K, G, N; 3 for D, I, F, H, Ja, M, O; 4 for E.

Lagerheimia genevensis (Chodat) Chodat 1895

Basionym: *Tetraceras genevensis* Chodat
Synonyms: *Chodatella genevensis* (Chodat) S.H.Li, *Lagerheimia quadriseta* (Lemmermann) G.M.Smith, *Chodatella quadriseta* Lemmermann
17540040 **Pl. 89A** (p. 363)
Cells 2–6.5(–8) µm wide, 3.5–12(–15) µm long, broadly ovoid, ellipsoidal to subcylindrical or cylindrical, with 2(–3) slightly subapical spines lying more or less in a single plane; spines (4–)6–18(–23) µm long, with or without basal thickening and needle-like, often more than twice cell length.

Probably cosmopolitan; widely distributed and common in the British Isles where in the plankton of pools, lakes, rivers and small water bodies (e.g. ponds, ditches), rarely abundant although Swale (1965) records a very dense population (230,000 cells per mL) reached in March 1965 in Oak Mere, Cheshire, declining rapidly to complete extinction in May due at least in part to a protozoan parasite; considered an indicator in Europe of slightly to moderately nutrient-rich water.

Lagerheimia longiseta (Lemmerman) Wille 1909

Basionym: *Chodatella longiseta* Lemmermann
17540050 **Pl. 89B** (p. 363)
Cells 5–8(–15) µm wide, 9–13(–21) µm long, ovoid, with 4–10 spines forming a ring at each apex; spines 40–55(–80) µm long and thus much longer than cell.

Probably cosmopolitan; recorded on only few occasions in the plankton of British lakes (e.g. Loch Leven).

Lagerheimia subsalsa Lemmermann 1898

Synonym: *Chodatella subsalsa* Lemmermann
17540060 **Pl. 89C** (p. 363) **CD**
Cells 2.5–9 µm wide, 4.5–16 µm long, ovoid or ellipsoidal with rounded ends, rarely almost lemon-shaped, with (2–)3–4(–5) spines at each apex almost forming a ring; spines 2–20(–26) µm long, gradually narrowing apically to become needle-like, straight or slightly curved, as long as or slightly shorter than cell, colourless or brown.

Probably cosmopolitan; recorded only occasionally in the plankton of British rivers and lakes.

Number and length of spines are very variable characters and probably influenced by environmental factors. Only with considerably more information on variation in spine length and number will it be possible to determine whether *L. subsalsa* and related species (e.g. *L. ciliata*, *L. citriformis*, *L. longiseta*) should continue to be recognized.

Lagerheimia wratislaviensis Schröder 1897

17540070 **Pl. 89D** (p. 363)
Cells (2.5–)4–9(–11) µm wide, (5–)7–14 µm long, ovoid, somewhat narrowed towards apices, with a single spine at each apex and 2(–3) opposite each other at equator, arranged crosswise in one plane; spines 8–31 µm long, with or without a basal thickening.

Probably cosmopolitan; frequently recorded from small water bodies (e.g. ponds) in the British Isles, occasionally in rivers.

Little purpose is served by recognizing varieties such as var. *trisetigera* G.M.Smith. Some material studied in the British Isles corresponds to this variety in having relatively small cells (2.5–5.5 µm wide, 5–9 µm long) and 3 regularly distributed equatorial spines.

Lobocystis R.H.Thompson 1952

Coenobia initially 2-celled, later multicellular with cells in groups or pairs connected by elongate, band-like, dichotomously divided strands of persistent mother cell wall, enclosed within a mucilaginous envelope; chloroplasts parietal, 1 or 2 per cell with a single pyrenoid; asexual reproduction by 2 autospores, released by splitting of mother cell wall, daughter cells remaining attached to wall remnants.

Probably cosmopolitan; frequently planktonic in lakes and ponds.

Lobocystis planctonica (Tiffany *et* Ahlstrom) Fott 1975

Basionym: *Dictyosphaerium planctonica* Tiffany *et* Ahlstrom
Synonyms: *Lobocystis planctonica* var. *mucosa* Bourrelly, *L. dichotoma* R.H.Thompson
17560010 **Pl. 85F** (p. 350)
Coenobia 34–45 µm wide, usually up to 24-celled; cells 5–7 µm wide, 6.5–8 µm long, ovoid to broadly ellipsoidal; chloroplast central, colourless area towards each apex containing a refractive granule.

Probably cosmopolitan; recorded in the British Isles only associated with aquatic macrophytes in a shallow and nutrient-enriched pond in pastureland.

Recorded in the British Isles as a variety, var. *mucosa*, since smaller than the type (especially in cell length), with a wider mucilaginous envelope. We consider that no useful purpose is served by recognizing the variety, which is probably no more than a growth form.

Micractinium Fresenius 1858

Synonyms: *Richteriella* Lemmermann, *Golenkinia* Chodat *pro parte*, *Errerella* W.Conrad
Coenobia cuboidal, tetrahedral, polyhedral, or some other geometric form, 4-, 8-, 16-, 32- or 64-celled; cells spherical, ovoid, walls thin and smooth, bearing 1–8(–18) spines, rarely enclosed within a mucilaginous envelope; spines evident, thin and needle-like; chloroplast parietal, cup-shaped, with a single and sometimes indistinct pyrenoid; asexual reproduction by 2, 4, 8 or 16 autospores, usually released by splitting of mother cell wall; sexual reproduction oogamous, only known in *M. pusillum*.

Probably cosmopolitan; most common in the plankton of non-flowing water.

1 Coenobia regularly tetrahedral and united into pyramidal syncoenobia .... *M. bornhemiense*
1 Coenobia of a different form 2
2 Coenobia cuboidal or polyhedral; cells spherical .... *M. pusillum*
2 Coenobia flat and wreath-shaped; cells obovoid to polygonal .... *M. quadrisetum*

Micractinium bornhemiense (W.Conrad) Korshikov 1953

Basionym: *Errerella bornhemiensis* W.Conrad

17570030

Coenobia in regular tetrads, usually producing regular pyramidal (tetrahedral) syncoenobia, often (4–)8-, 16-, 32-, 64-, 128- or 256-celled; cells 6–8 μm in diameter, spherical, bearing a single spine (rarely absent, or 2–3) from 30–60(–90) μm in length; autospores 4 in each sporangium.

Probably cosmopolitan; rare in the British Isles since known only from the plankton of lakes in the Norfolk Broads.

Micractinium pusillum Fresenius 1858

Synonyms: *Golenkinia botryoides* Schmidle, *Staurogenia multiseta* Schmidle, *Crucigenia multiseta* (Schmidle) Schmidle, *Tetrastrum multisetum* (Schmidle) Chodat, *Richteriella botryoides* (Schmidle) Lemmermann, *Micractinium pusillum* var. *longisetum* Tiffany *et* Ahlstrom, *M. pusillum* var. *mucosa* Korshikov, *M. eriense* Tiffany *et* Ahlstrom

17570010 **Pl. 89L** (p. 363) **CD**

Coenobia cuboid or polyhedral, 4-, 8- or 16(–32)-celled, sometimes disintegrating into single cells, with or without a mucilaginous envelope; cells (3–)7–13 μm wide, spherical and thin-walled, bearing 2, 4 or 8 spines (occasionally 30–40 in culture); setae 0.5–2.5 μm wide, sometimes wider at base, (12–)24–51(–100) μm long; autospores 2, 8 or 16 in each sporangium; zygospores 14–17 μm wide and covered in warty projections.

Probably cosmopolitan; widely distributed in the British Isles where planktonic or associated with submerged surfaces in many types of aquatic habitat including slow-flowing rivers, ditches, moats, ponds, reservoirs, and small water bodies (e.g. ponds, dune-slack pools, water barrels); considered an indicator of slight to moderately nutrient-enriched water.

Similar to *Golenkiniopsis* when single-celled, but distinguished from it by not having a prominent pyrenoid with a surrounding starch sheath (see remarks under *Golenkiniopsis*).

Micractinium quadrisetum (Lemmermann) G.M.Smith 1916

Basionym: *Richteriella quadriseta* Lemmermann

Synonyms: *Richteriella botryoides* (Schmidle) Lemmermann var. *quadriseta* (Lemmermann) Chodat, *R. botryoides* var. *quadriseta* (Lemmermann) West

17570020 **Pl. 89K** (p. 363) **CD**

Coenobia flat, wreath-shaped, sometimes tetrahedral, consisting of a 4-angled central space and radially arranged cells in 4-, 8- or 16-celled groups, without a mucilage envelope; cells 3.5–7 μm wide, 5–10 μm long, obovoid to polygonal, bearing (1–)2–4 setae; setae 0.5–1.5 μm wide (slightly wider at base) and 23–50 μm long; autospores usually 4 in each sporangium; sexual reproduction unknown.

Probably cosmopolitan; rare in the British Isles where usually in the plankton of rivers, reservoirs, canals and temporarily pools or ponds.

Monoraphidium Komárková-Legnerová 1969

Cells free-living, without a specialized organ; if attached (very rare), more or less spindle-shaped, straight, curved or spirally twisted, gradually or abruptly narrowing to acute apices, thin-walled and without a mucilaginous envelope; chloroplasts parietal, single and without a visible pyrenoid when examined with the light microscope; asexual reproduction by 2, 4, 8 or 16 autospores, produced by oblique division and serially arranged, released by longitudinal split of mother cell wall or its disintegration at apices.

Probably cosmopolitan; planktonic or associated with aquatic macrophytes or other submerged surfaces in still or slow-flowing water.

Komárková-Legnerová (1969) separated this genus from *Ankistrodesmus* and transferred to *Monoraphidium* those single-celled species of *Ankistrodesmus* having serially arranged autospores that slightly overlap immediately prior to release. A more restrictive view of the genus has been taken subsequently. For example, those species whose pyrenoids are visible with the light microscope have been transferred by Hindák (1970) to the genus *Keratococcus* (see comments under that genus). Species are largely recognized by cell shape, size and characters associated with cell curvature. Not surprisingly, intermediate forms will be encountered which are impossible to attribute to one or other species. Curvature is a frequently used character and yet many of those that are strongly curved or sigmoid are occasionally almost straight or only slightly curved (e.g. *Monoraphidium mirabile*, *M. convolutum*). Such almost straight forms are rare and difficult, if not impossible, to separate from other species and are not considered in the key. When further investigation makes clearer species limits then many of those currently recognized may well prove to be conspecific.

1 Cells straight or only slightly curved 2
1 Cells markedly curved to sigmoid or spirally twisted....................7
2 Cells abruptly narrowing to obtusely rounded apices....................*M. obtusum*
2 Cells narrowing to acute or sometimes long, spine-like or finger-like apices....................3
3 Cells narrowly spindle-shaped and cylindrical in centre, usually 60–182 μm long, gradually narrowing to long needle-like apices ....*M. komarkovae*
3 Cells narrowly or broadly spindle-shaped and usually not cylindrical in centre, generally less than 70 μm long, narrowing to acute or finger-like apices....................4
4 Cells usually 50–72 μm long ....*M. griffithii*
4 Cells always <40 μm long....................5
5 Cells frequently attached to sufaces, each apex tapering to an acute point....................*M. litorale*
5 Cells always free-floating, each apex with an acute, bluntly rounded or needle-like point....................6
6 Cells always gradually tapering to needle-like apices ....*M. tortile*
6 Cells tapering to bluntly rounded and long finger-like apices ....*M. pusillum*
7 Cells arched, horseshoe-shaped, crescent-shaped or slightly contorted 8
7 Cells sigmoid, markedly screw-shaped or spirally contorted....................10

8 Cells with rounded apices, horseshoe-shaped to crescent-shaped .... *M. minutum*
8 Cells with pointed ends, arched to crescent-shaped or slightly sigmoid and screw-shaped 9
9 Cells usually < 60 µm long.................*M. arcuatum*
9 Cells always 60 µm long or longer ....*M. mirabile*
10 Cells broadly spindle-shaped, 4–7(–17) µm long, usually less than 5 times longer than broad ....*M. convolutum* (including var. *pseudosabulosum*)
10 Cells narrowly spindle-shaped, 7–72 µm long, always more than 5 times longer than broad........11
11 Cells 7–40 µm long, usually sigmoid to spirally twisted 0.5–1.5 times.........................*M. contortum*
11 Cells usually 40–72 µm long, sigmoid to spirally twisted 1–2 times ....*M. irregulare*

Monoraphidium arcuatum (Korshikov) Hindák 1970

Basionym: *Ankistrodesmus arcuatus* Korshikov
Synonyms: *Ankistrodesmus pseudomirabilis* Korshikov, *A. sabrinensis* J.H.Belcher *et* Swale
17580010 **Pl. 90I** (p. 367)
Cells (0.8–)1–3(–4.5) µm wide, 18–60(–90) µm long, narrowly spindle-shaped, arched to slightly sigmoid and screw-shaped, narrowing equally to each finely pointed apex, sometimes slightly curved.

Probably cosmopolitan; frequent in nutrient-rich ponds and rivers in many parts of the British Isles.

Monoraphidium contortum (Thuret) Komárková-Legnerová 1969

Basionym: *Ankistrodesmus contortus* Thuret
Synonyms: *Ankistrodesmus angustus* C.Bernard, *A. falcatus* (Corda) Ralfs *var. spirilliformis* G.S.West and var. *duplex* (Kützing) G.S.West, *A. pseudomirabilis* var. *spiralis* Korshikov, *Raphidium polymorphum* var. *spirale* West *et* G.S.West
17580020 **Pl. 90J** (p. 367)
Cells 1–2.2(–5.2) µm wide, 7–40(–45) µm long, narrowly spindle-shaped, straight or more usually slightly bent to spirally twisted (up to 1.5 spirals), gradually narrowing to pointed apices; autospores 2, 4 or 8 per cell.

Probably cosmopolitan; common throughout the year in the plankton or associated with submerged surfaces in nutrient-poor to moderately to heavily polluted waters. In many types of aquatic habitat in the British Isles including rivers, ponds, lakes, reservoirs, moats, ditches and animal drinking troughs.

Monoraphidium convolutum (Corda) Komárková-Legnerová 1969

Basionym: *Ankistrodesmus convolutus* Corda *pro parte*
Synonym: *Ankistrodesmus curvulus* J.H.Belcher *et* Swale
17580030 **Pl. 90M** (p. 367)
Cells 1.4–5(–6) µm wide, 3–7(–11) µm long, 2–5 times longer than broad, irregular, elongated and broadly spindle-shaped, slightly asymmetrical, usually strongly curved or spirally twisted with apices lying in different planes and gradually tapering to acute points, becoming rounded with age and sometimes forming short finger-like prolongations; autospores 2 or 4 per cell, occasionally attached for short periods by apices.

var. *pseudosabulosum* Hindák 1970

Synonym: *Dactylococcus infusionum* Nägeli *sensu auct. post.*
17580032
Solitary or sometimes attached by extremities to form chains; cells (1.6–)2.8–6 µm wide and (5.5–)7.5–19 µm long, short to broadly spindle-shaped, mostly asymmetrical, slightly pointed or rounded at apices; autospores 4 or 8 per cell.

Probably cosmopolitan; usually planktonic in moderately to highly nutrient-rich ponds, lakes, reservoirs, and rivers (e.g. River Lee, Hertfordshire). In the British Isles the variety *pseudosabulosum* is only known in quantity in aquarium tanks and standing water within flower pots.

The variety has been considered a stage in the life history of *Scenedesmus obliquus* (see West & Fritsch, 1927; Trainor, 1963). According to Komárková-Legnerová (1969), specimens attributed to this species, in which the cells are 18–20 µm long, in reality belong to the species *M. pusillum*.

Monoraphidium griffithii (Berkeley) Komárková-Legnerová 1969

Basionym: *Closterium griffithii* Berkeley
Synonyms: *Ankistrodesmus falcatus* (Corda) Ralfs var. *acicularis* (A.Braun) G.S.West, *A. falcatus* (Corda) Ralfs var. *spirale* (Kützing) G.S.West, *A. acicularis* (A.Braun) Korshikov
17580040 **Pl. 90K** (p. 367)
Cells (1–)2–4.5 µm wide, (28–)50–72(–110) µm long, usually about 12 times longer than wide, narrowly spindle-shaped, straight, each apex gradually narrowing and terminating in an acute point or short spine.

Probably cosmopolitan; in the British Isles planktonic or associated with submerged surfaces where in various types of water body including ditches, ponds, lakes and reservoirs.

Cell length and form of the apices are very variable. Differs from the closely related *M. komarkovae* in having slightly wider cells and without seta-like apices, and usually in size from *M. tortile* whose cells sometimes terminate in a short spine.

Monoraphidium irregulare (G.M.Smith) Komárková-Legnerová 1969

Basionym: *Dactylococcopsis irregularis* G.M.Smith
Synonym: *Ankistrodesmus angustus* C.Bernard *sensu* Korshikov
17580050 **Pl. 90L** (p. 367)
Cells 1.3–2.2(–5) µm wide, 40–72(–150) µm long, usually more than 10 times longer than broad, narrowly spindle-shaped but sometimes cylindrical at centre, sigmoid to spirally twisted (0.5 to 2.5 spirals), gradually narrowing to pointed apices.

Probably cosmopolitan; in the British Isles planktonic and associated with submerged surfaces in many aquatic habitats.

One of the most widely distributed species especially in temperate regions, but not so common as *M. contortum*.

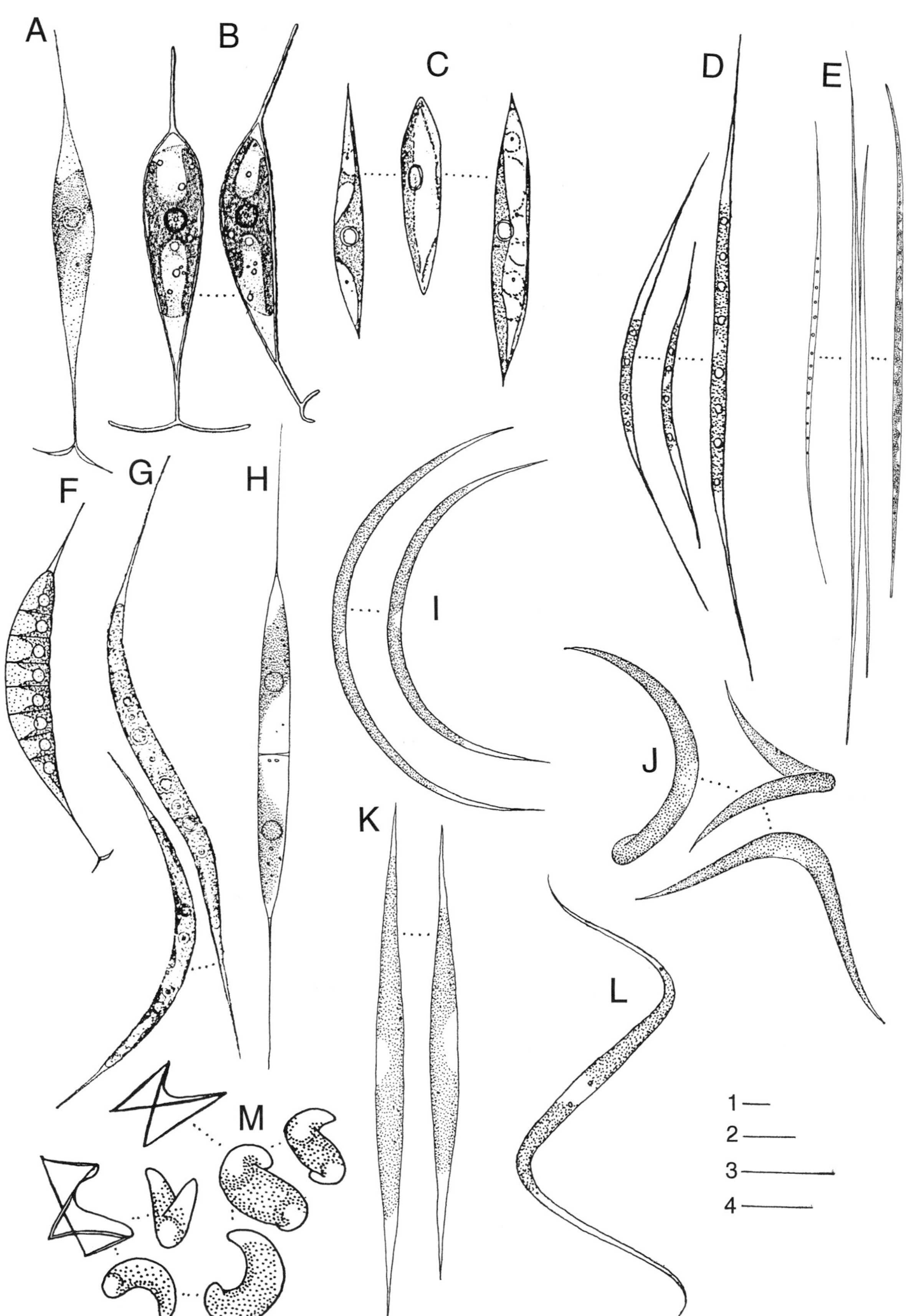

Plate 90 A–M

A. *Ankyra judayi* (p. 331); **B**. *Ankyra ancora* (p. 331); **C**. *Chlorolobion braunii* (p. 338); **D**. *Closteriopsis acicularis* (p. 338); **E**. *Closteriopsis longissima* (p. 339); **F**. *Korshikoviella michailovskoensis* (p. 362); **G**. *Schroederia robusta* (p. 398); **H**. *Schroederia setigera* (p. 399); **I**. *Monoraphidium arcuatum* (p. 366); **J**. *Monoraphidium contortum* (p. 366); **K**. *Monoraphidium griffithii* (p. 366); **L**. *Monoraphidium irregulare* (p. 366); **M**. *Monoraphidium convolutum* (p. 366): vegetative cells and remnants of mother cell wall.

Scale bars: 1-4, 10 μm. 1 for B, E, F; 2 for A, D; 3 for C, H–M; 4 for G.

Monoraphidium komarkovae Nygaard 1979

Synonym: *Monoraphidium setiforme* (Nygaard) Komárková-Legnerová

17580070 **Pl. 91A** (p. 369)

Cells 1.4–3.5 μm wide, (25–)70–182 μm long, more than 20 times longer than broad, straight or slightly curved, narrowly spindle-shaped, almost cylindrical in centre, gradually narrowing to long spine-like apices.

Probably cosmopolitan; known in the British Isles from dune-slack pools (e.g. Braunton Burrows, Devon), moats, slow-flowing rivers (e.g. River Lee, Hertfordshire) and lakes in the Windermere drainage basin; probably more widely distributed in the British Isles than the few records suggest.

Easily confused with *M. griffithii* but usually much longer. Similar to some slightly curved forms of *M. mirabile*.

Monoraphidium litorale Hindák 1977

17580060 **Pl. 91B** (p. 369)

Free-living or frequently attached directly or by a small mucilaginous pad; cells 2–3(–4.5) μm wide, (16–)18–30 μm long, spindle-shaped, mostly straight but slightly asymmetrical, gradually narrowing to pointed apices.

Europe; known in the British Isles only from the marginal shallows of Loch Lomond in Scotland.

Very similar to *Monoraphidium affixum* Hindák and to *M. pusillum* (Printz) Komárková-Legnerová (1969, p. 108).

Monoraphidium minutum (Nägeli) Komárková-Legnerová 1969

Basionym: *Raphidium minutum* Nägeli

Synonyms: *Ankistrodesmus minutissimus* Korshikov, *A. lunulatus* J.H. Belcher *et* Swale, *Raphidium convolutum* (Corda) Rabenhorst var. *minutum* (Nägeli) Rabenhorst, *Selenastrum minutum* (Nägeli) Collins

17580080 **Pl. 91C** (p. 369)

Cells 1–7.2 μm wide, (3.5–)5–17(–20) μm long, horse-shoe-shaped to crescent-shaped, occasionally with extremities touching, sometimes slightly spirally twisted, slightly narrowing to rounded apices.

Probably cosmopolitan; in the British Isles planktonic in rivers, canals and other aquatic habitats.

Hindák (1988) is of the opinion that it should be transferred to the closely related genus *Choricystis* (as *C. minuta* (Nägeli) Hindák). Easily mistaken for small-celled species of *Nephrochlamys* or *Kirchneriella*.

Monoraphidium mirabile (West *et* G.S.West) Pankow 1976

Basionym: *Ankistrodesmus falcatus* (Corda) Ralfs var. *mirabilis* West *et* G.S.West

Synonyms: *Ankistrodesmus mirabilis* (West *et* G.S.West) Lemmermann, *Raphidium polymorphum* var. *mirabile* West *et* G.S.West

17580090

Cells 2–3.5(–6) μm wide, 60–130(–155) μm long, more than 20 times longer than broad, narrowly spindle-shaped, crescent-shaped or sigmoid, gradually narrowing to acute or long spine-like apices.

Probably cosmopolitan; in the British Isles planktonic and amongst detrital material in ponds (e.g. Wimbledon Common, Surrey), rare.

Monoraphidium obtusum (Korshikov) Komárková-Legnerová 1969

Basionym: *Ankistrodesmus obtusus* Korshikov

17580100 **Pl. 91D** (p. 369)

Cells 4.5–5 μm wide, 32–62 μm long, straight or slightly curved, cylindrical to broadly spindle-shaped, abruptly narrowing to obtusely rounded apices.

Europe; in the British Isles planktonic and associated with submerged surfaces or the periphyton in lakes, rivers, marshes and other acidic water bodies.

Hindák (1988) has transferred this species to the genus *Choricystis*, as *C. obtusus* (Korshikov) Hindák.

Monoraphidium pusillum (Printz) Komárková-Legnerová 1969

Basionym: *Ankistrodesmus braunii* (Nägeli) Collins var. *pusilla* Printz

17580110 **Pl. 91E** (p. 369)

Cells 1.4–7.6 μm wide, 12–25 μm long, narrowly or broadly spindle-shaped, initially narrowing to an acute point but with age each apex becoming bluntly rounded and finger-like, straight or nearly straight, slightly curved or sigmoid.

Europe; in the British Isles known from the plankton of various types of water body, rare.

Monoraphidium tortile (West *et* G.S.West) Komárková-Legnerová 1969

Basionym: *Ankistrodesmus tortilis* West *et* G.S.West

Synonym: *Ankistrodesmus pseudobraunii* J.H. Belcher *et* Swale

17580120 **Pl. 91F** (p. 369)

Cells 0.5–3(–4) μm wide, 11–33(–40) μm long, more than 60 times longer than broad, narrowly spindle-shaped, sometimes asymmetrical, straight, rarely slightly curved or sigmoid, gradually narrowing to needle-like apices.

Europe; in the plankton and associated with aquatic macrophytes in rivers, canals, and lakes including the Norfolk Broads; rare in the British Isles.

Komárek & Fott (1983) consider *Monoraphidium pseudobraunii* (J.H.Belcher) Heynig (=*Ankistrodesmus pseudobraunii* J.H.Belcher *et* Swale) to be a synonym of the above species.

Muriella J.B.Petersen 1932

Cells solitary or grouped, almost spherical or broadly ellipsoidal, thin-walled but thickening with age; chloroplast parietal, initially single but soon dividing into 2 or more disc-, lens-, cup- or trough-shaped portions, without pyrenoids; asexual reproduction by autospores, released by splitting of mother cell wall; akinetes known.

Probably cosmopolitan; subaerial or associated with soil.

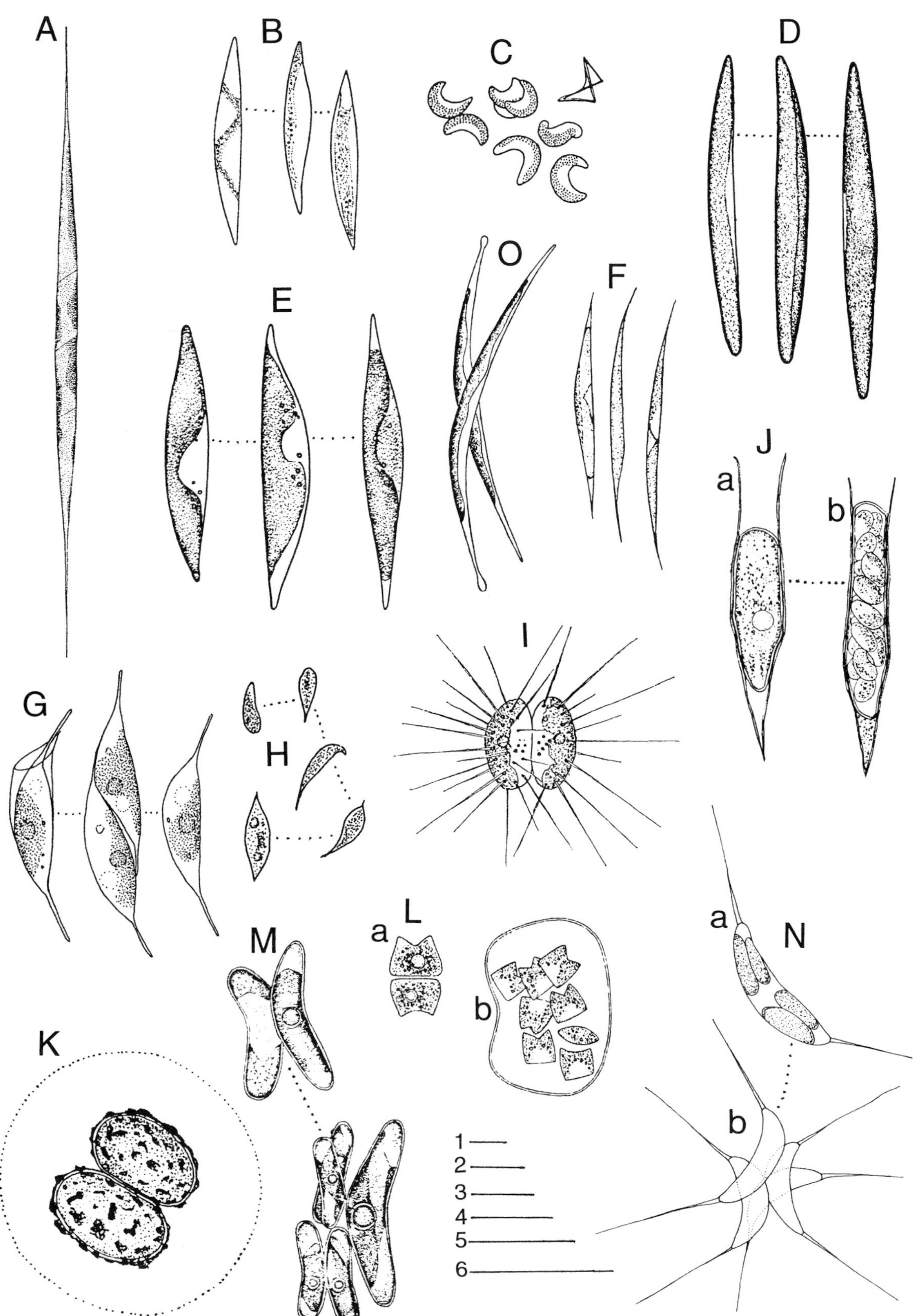

Plate 91 A–O

A. *Monoraphidium komarkovae* (p. 368); **B**. *Monoraphidium litorale* (p. 368); **C**. *Monoraphidium minutum* (p. 368); **D**. *Monoraphidium obtusum* (p. 368); **E**. *Monoraphidium pusillum* (p. 368); **F**. *Monoraphidium tortile* (p. 368); **G**. *Keratococcus bicaudatus* (p. 360); **H**. *Keratococcus dispar* (p. 360); **I**. *Dicellula geminata* (p. 347); **J**. *Apodochloris dinobryonis* (p. 332): a. single cell within the lorica of *Dinobryon*, b. cell containing autospores; **K**. *Didymocystis inermis* (p. 351); **L**. *Euastropsis richteri* (p. 353): a. two-celled coenobium, b. autospores within mother cell wall; **M**. *Didymogenes palatina* (p. 351); **N**. *Didymogenes anomala* (p. 351): a. cell containing developing autospores, b. coenobium of 4 cells; **O**. *Podohedra falcata* (p. 380).

Scale bars: 1, 3, 30 μm; 2, 4–6, 10 μm. 1 for Lb; 2 for A, C, D, G, H, J, O; 3 for La; 4 for B, I; 5 for E, F, Nb, M; 6 for K, Na.

1 Chloroplasts densely packed; cells up to 20 μm wide ....*M. magna*
1 Chloroplasts usually distinctly separate; cells usually up to 8 μm wide ....*M. terrestris*

Muriella magna F.E.Fritsch *et* R.P.John 1942

18070010 **Pl. 82Q** (p. 329)
Cells up to 20 μm wide, spherical or broadly ellipsoidal; chloroplasts initally in pairs, later dividing into several trough-shaped portions, with age disc-like and polygonal, always densely packed; autospores unknown; akinetes up to 90 μm wide, with thick, stratified walls.

British Isles (England); isolated in the British Isles from a wide range of soil types.

On agar cultures it forms rough, dark green patches which become orange on drying due to accumulation of coloured oil globules.

Muriella terrestris J.B.Petersen 1932

18070020 **Pl. 82M** (p. 329)
Cells 3–8(–13) μm wide, almost spherical; chloroplasts 2 to several, plate-shaped, distinctly separate even in young cells; autospores (2–)4 or 8 in each sporangium; akinetes unknown.

Europe; in the British Isles known only from soil cultures.

Like *Muriella magna*, it often becomes orange-red with age due to accumulation of carotenoid pigments. The two species are very similar and further studies are required to establish whether they should be regarded as conspecific.

Nautococcus Korshikov 1926

Cells solitary and semi-emergent at water surface, spherical or pear-shaped, with walls irregularly thickened, often with a cap-like extracellular layer and sometimes inwardly curved at margins; chloroplast central, massive, with a single pyrenoid; contractile vacuoles common; sexual reproduction by biflagellate gametes, sometimes hemizoospores (without flagella) develop within mother cell to form a multicellular complex; akinetes thick-walled, reddish due to presence of haematochrome.

Probably cosmopolitan; usually on the surface film (neuston) of nutrient-rich puddles, pools and small ponds.

Nautococcus pyriformis Korshikov 1926

17600020 **Pl. 87C** (p. 354)
Cells up to 23 μm in size, spherical to irregularly pear-shaped, frequently with an extracellular layer (often brownish) covering part or entire cell, occasionally the layer becoming cap-shaped; zoospores (8–)9–12(–13) μm long, somewhat flattened, irregularly oval, with nucleus anterior to the centrally positioned chloroplast.

Probably cosmopolitan; very rarely recorded in the British Isles where it forms dense, yellow-green layers during the summer months on the surface of small nutrient-rich ponds and pools.

Very detailed observations were made by Pentecost (1984b) on a bloom that developed in a pond (Holden Pond) near Tunbridge Wells, Kent. The extracellular layer he examined was often faintly brown in colour, sometimes it formed a sheet connecting several cells, and cap-shaped thickenings were absent or, if present, then more frequently found towards the decline of the bloom (October).

DOUBTFUL RECORD
Nautococcus caudatus Korshikov
17600010
Belcher & Swale (1996) mention the resemblance of an alga forming an iridescent surface film in November and December in buckets of rainwater at Girton (near Cambridge) to this species despite it possessing several chloroplasts and lacking a pyrenoid. *Nautococcus caudatus* is distinguished from *N. pyriformis* by its bulb-shaped cells and the presence of a well-developed cap on the thick-walled apex. We believe the Cambridge material was probably a xanthophyte.

Neglectella Vodeničarov *et* Benderliev 1971

Cells solitary or forming 2-, 4- or 8-celled coenobia of autospores remaining enclosed within a swollen mother cell; cells almost ellipsoidal, ovoid to nearly spherical, apically rounded or shortly pointed, with or without an apical pore, enclosed within mucilage; walls becoming thicker with age, lamellated, developing small pores; chloroplast of radiating, ribbon-like lobes and with a central pyrenoid; asexual reproduction by 2, 4 or 8 autospores, often persisting for long periods within the mother cell.

Europe; common in standing bodies of water.

We have decided to continue to recognize *Neglectella* although Vodeničarov (1989) has created the new genus *Skujaster* based upon *Oocystis asterifera*.

Neglectella asterifera (Skuja) Fott 1976

Basionym: *Oocystis asterifera* Skuja
17610010 **Pl. 88B** (p. 359)
Cells 10–22 μm wide, 12–32 μm long, ellipsoidal, thick-walled, each apex perforated by a pore.

Europe; known in the British Isles only from a small pond at Madingley Hall, near Cambridge.

Nephrochlamys Korshikov 1953

Synonyms: *Kirchneriella* Schmidle *pro parte*, *Nephrocytium* Nägeli *pro parte*
Coenobia of 2, 4 or 8 randomly arranged cells retained within a kidney-shaped or irregularly wrinkled and expanded mother cell; cells kidney-shaped or strongly curved, with rounded apices, thin-walled and lacking a mucilaginous envelope; chloroplast parietal, plate or trough-shaped and on convex side of cell, pyrenoids absent; asexual reproduction by 2, 4 or 8 autospores, remaining until fully mature within the enlarged and sac-like mother cell.

Probably cosmopolitan; planktonic in ponds, lakes and rivers.

Cells similar in appearance to *Kirchneriella*, but differ in that the daughter cells remain within a persistent mother cell wall which does not become mucilaginous.

Nephrochlamys rostrata Nygaard et al. 1986

Synonyms: *Kirchneriella subsolitaria* G.S.West, *Nephrochlamys subsolitaria* (G.S.West) Korshikov

17620010 **Pl. 83A** (p. 341) **CD**

Solitary or forming 2- or 4-celled coenobia surrounded by persistent mother cell wall, similar in shape to daughter cells; cells 4–6.5(–12) µm wide, 5–14 µm long, initially crescent-shaped but becoming strongly curved with age until apices almost touch, one apex broad and obtuse, the other shortly narrowed and acute.

Probably cosmopolitan; reported from ponds and lakes in the British Isles.

Nephrocytium Nägeli 1849

Coenobia of 4, 8 or 16 irregularly or spirally arranged cells and often retained prior to release within a thickened and well-defined gelatinizing mother cell wall, usually surrounded by a mucilaginous envelope; cells ovoid, elongate ovoid, ellipsoidal and with obtuse apices, or spindle-shaped with almost acute apices, asymmetrical or distinctly curved; chloroplast parietal, with 1 to several pyrenoids; asexual reproduction by 4, 8 or 16 autospores, released by gelatinization of expanded mother cell wall.

Probably cosmopolitan; common in the plankton of ponds and lakes, also associated with filamentous algae and mosses in marginal shallows.

1 Cells crescent- or sickle-shaped, with acute apices ....*N. lunatum*
1 Cells of another shape, with rounded apices..........2
2 Cells broad, elongate and somewhat kidney-shaped ....*N. agardhianum*
2 Cells sausage-shaped and slightly curved ....*N. limneticum*

Nephrocytium agardhianum Nägeli 1849

Synonym: *Nephrocytium naegelii* Grunow

17630010 **Pl. 83B** (p. 341)

Coenobia usually of 4 or 8 spirally or irregularly arranged cells embedded within mucilage, often asymmetrically ovoid, 40–90 µm in size; cells initially (2–)4–7(–12) µm wide, 3 to 6 times longer than broad, older cells 8–22 µm wide, twice as long as broad, elongate, somewhat kidney-shaped and with rounded apices.

Probably cosmopolitan; widely distributed in the British Isles especially in ponds, lakes and moats where usually associated with aquatic macrophytes rather than planktonic.

Nephrocytium limneticum (G.M.Smith) G.M.Smith 1933

Basionym: *Gloeocystopsis limneticum* G.M.Smith

17630020 **Pl. 83C** (p. 341)

Coenobia usually 4- or 8-celled, daughter colonies tending to remain united, 20–50 µm in diameter, spherical to broadly ovoid; cells 4–7 µm wide, 10–15 µm long, sausage-shaped, slightly curved, with rounded apices.

Probably cosmopolitan; in the British Isles recorded from several Scottish lochs (e.g. Lomond, Leven) and lakes in north-west Ireland (Lind & Pearsall, 1945).

Nephrocytium lunatum West 1892

17630030 **Pl. 83D** (p. 341)

Coenobia of 4, 8 or 16 cells arranged in an indistinct spiral within an ellipsoidal to asymmetrically ovoid mucilaginous envelope (25–70 µm in size); cells (3–)4–7 µm wide, (12–)14–18(–21) µm long, crescent- to sickle-shaped, with acute apices.

Probably cosmopolitan; in the British Isles locally abundant in ponds and pools, especially in north-west England and lochs in the Scottish Highlands; frequently associated with the moss *Sphagnum* and submerged aquatic macrophytes.

DOUBTFUL RECORD

Nephrocytium allantoideum Bohlin 1897

Known only from tropical America (see Komárek & Fott, 1983), hence the record by W. & G.S. West (1901) from Yorkshire must be considered doubtful.

Oocystis A.Braun 1855

Synonym: *Oocystella* Lemmermann

Cells solitary or forming 2-, 4-, 8-, 16-, 32- or 64-celled coenobia, one or several generations contained within the expanded mother cell wall or a mucilaginous envelope; cells ellipsoidal, ovoid to almost spherical or rhomboidal, spindle- or fiddle-shaped, smooth-walled; apices rounded or acute, sometimes with a nipple-like thickening of wall; chloroplast parietal, 1 to several, plate- or trough-shaped with margin entire or lobed and with or without a pyrenoid (sometimes difficult to distinguish); asexual reproduction by 2, 4 or 8(–16) autospores, often retained for a period within the expanding mother cell, released by fracture or gelatinization of wall, sometimes several generations present; akinetes known.

Probably cosmopolitan; very common in the plankton of ponds and lakes.

Several studies have demonstrated that the genus is very unnatural or heterogeneous with the result that new genera (e.g. *Oonephris* Fott, *Neglectella* Vodeničarov *et* Benderliev, *Granulocystopsis* Hindák, *Granulocystis* Hindák) have been created to accommodate taxa originally placed within it. Hindák (1988) transferred to the genus *Oocystella* those species of *Oocystis* not known to possess pyrenoids. We believe that presence or absence of pyrenoids is by itself insufficient justification to warrant such a separation.

As a result of variation in cell size and form, degree of apical wall thickening, and number of pyrenoids (if present) doubt attaches to many of the more than 90 currently recognized species of *Oocystis*. Often reliable species identification requires examination of large populations to determine the full range of potential morphological variation. Considerable doubt attaches to some of the 22 species reported from the British Isles. Several were originally described by the Wests (e.g. West, 1892b, 1893; W. & G.S.West, 1894, 1898), who frequently failed to mention key characters such as chloroplast structure and presence or absence of pyrenoids.

Subsequent authors have expanded the Wests' descriptions by providing additional information, although usually based upon morphologically similar material from other areas. It is difficult to know whether all such reinterpretations are correct. For this and other reasons, several records from Britain and Ireland remain doubtful and future investigations will undoubtedly lead to many species becoming reduced to synonymy. The identification key is not satisfactory, because many quantitative characters overlap and comparisons are impossible if character states are known for some but not all species. All identifications arrived at using the key need to be regarded as provisional and checked against the illustrations.

1 Cells fiddle-shaped, >50 μm long .... *O. panduriformis*
1 Cells of another shape, <50 μm long 2
2 Apical wall thickening often in form of a nodule 3
2 Apical wall thickening absent 8
3 Chloroplasts usually more than 5 per cell (typically 10–20), not perforate .... *O. solitaria*
3 Chloroplasts fewer than 5 per cell, sometimes perforated 4
4 Cells 16–40 μm long; pyrenoids absent .... *O. naegelii*
4 Cells usually <16 μm long; pyrenoids present 5
5 Opposite apices of different shape, one broadly rounded and the other bluntly pointed with thickened wall .... *O. apiculata*
5 Opposite apices similar, both rounded or bluntly pointed and thickened 6
6 Chloroplasts usually 1 per cell .... *O. parva*, in part
6 Chloroplasts usually 1–4 per cell 7
7 Chloroplasts grooved, 2–4 (only rarely 8) per cell; autospores released by fracture of mother cell wall .... *O. marssonii*
7 Chloroplasts trough-shaped and not grooved, 1–4 per cell; autospores often released by gelatinization of mother cell wall .... *O. lacustris*
8 Cells broadly oval or ellipsoidal, up to 1.5 times longer than broad .... *O. borgei*
8 Cells oblong, ellipsoidal or oval to subcylindrical, always more than 1.5 times longer than broad 9
9 Chloroplasts 4–20 per cell, if 4 then lobed and star-shaped 10
9 Chloroplasts 1–4 per cell, never star-shaped 11
10 Chloroplasts (2–)4–8 per cell, star-shaped, each with a pyrenoid .... *O. natans*
10 Chloroplasts usually 10–20 per cell, often polygonal or irregularly shaped, without pyrenoids .... *O. elliptica*
11 Cells with obtusely rounded apices and loosely arranged within mother cell; pyrenoids absent .... *O. pusilla*
11 Cells with bluntly pointed apices and often densely packed in mother cell; pyrenoid present .... *O. parva*, in part

Oocystis apiculata West 1893

17640010 **Pl. 92A** (p. 373)
Coenobia 2-, 4-celled or more, sometimes two generations within mother cell wall; cells 5–6(–15) μm wide and 11–15(–25) μm long, about 1.5 to 2 times longer than broad, oblong, spindle-shaped or ovoid-ellipsoidal, one apex broadly rounded, the other apex bluntly pointed with a nodular wall thickening; chloroplasts 1 or 2, each with a pyrenoid.

Probably cosmopolitan; recorded on a few occasions in the British Isles from small water bodies widely separated, from Kent to the Orkney Islands.

The original diagnosis has been amended based on observations of material from other geographical regions because West (1893) failed to mention chloroplast structure, presence or absence of pyrenoids, or to provide data on maximum cell size. See remarks under another even more doubtful species, *O. asymmetrica* (p. 375).

Oocystis borgei J.Snow 1903

17640030 **Pl. 92C** (p. 373)
Cells solitary or consisting of coenobia of 2, 4 or 8 densely packed and tetrahedrally arranged cells; cells (6–)8.5–17(–20) μm wide, (9–)10.5–16(–20) μm long, usually less than 1.5 times longer than broad, broadly ovoid or ellipsoidal, apices broadly rounded and usually without polar wall thickenings; chloroplasts 1–4 depending on age, grooved, each with a pyrenoid. Reproduction by 2–4(–8) autospores, each autospore with a chloroplast, released by fracture of slightly inflated mother cell wall.

Probably cosmopolitan; reported on a few occasions in the British Isles in the plankton or associated with submerged surfaces in ponds, lakes, rivers and flooded basins.

Reported as having small cell wall thickenings at apices (see Hindák 1980).

Oocystis elliptica West 1892

Synonym: *Oocystis elliptica* fo. *minor* West
17640040 **Pl. 92D** (p. 373)
Coenobia of (2–)4 or 8 almost touching cells, mother cell slightly swollen and almost spherical; cells (6.8–)8–16.5 μm wide, (11–)15–20(–25) μm long, 2–2.25 times longer than broad, oval to subcylindrical, apices broadly rounded and without polar wall thickenings; chloroplasts large, (6–)10–20 per cell, without pyrenoids; autospores (2–)4 or 8 per sporangium, each with several chloroplasts.

Probably cosmopolitan; planktonic in the British Isles where sometimes on submerged objects in bogs or in slow-flowing rivers and lakes.

A form, fo. *minor*, described by W. & G.S.West (1894) from Ireland (Derryclare Lough, Connemara), has cells 8.2–10 μm wide and 18–20.8 μm long; it probably does not warrant recognition.

Oocystis lacustris Chodat 1897

Synonym: *Oocystella lacustris* (Chodat) Hindák
17640050 **Pl. 92E** (p. 373) **CD**
Coenobia of 2, 4 or 8 cells, occasionally cells in tiers, sometimes containing up to 3 generations, with age mother cell wall slightly or markedly expanding; cells (1.5–)3.2–9.2(–10) μm wide, (4–)6.4–15 μm long, about twice as long as broad, narrow to broadly ellipsoidal or spindle-shaped, slightly asymmetrical, apices rounded

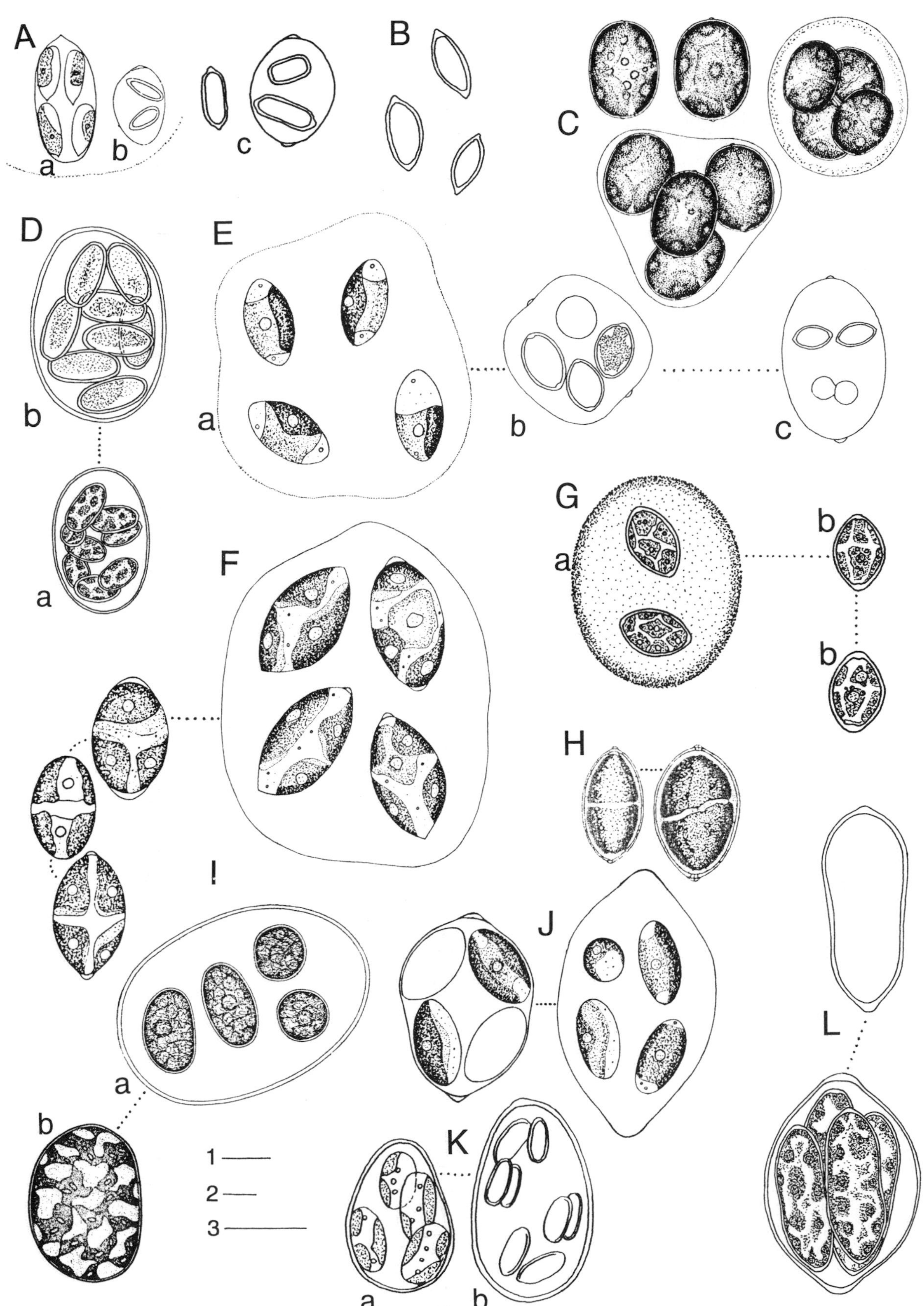

Plate 92 A–L

A. *Oocystis apiculata* (p. 372); **B**. *Oocystis asymmetrica* (p. 375); **C**. *Oocystis borgei* (p. 372); **D**. *Oocystis elliptica* (p. 372): a, b. autospores within mother cell wall; **E**. *Oocystis lacustris* (p. 372): a. coenobium of 4 cells, b, c. autospores within mother cell wall; **F**. *Oocystis marssonii* (p. 374); **G**. *Oocystis natans* (p. 374): a. coenobium of 2 cells, b. details of individual cells; **H**. *Oocystis naegeli* (p. 374); **I**. *Oonephris obesa* (p. 375): a. details of single cell, b. four autospores within mother cell wall; **J**. *Oocystis parva* (p. 374); **K**. *Oocystis pusilla* (p. 374): autospores within mother cell wall showing details in one figure (a); **L**. *Oocystis panduriformis* (p. 374).

Scale bars: 10 µm. 1 for A, B, Da, Eb,c, G, Ib, Kb; 2 for Db, H, Ia, L; 3 for C, Ea, F, J, Ka.

to obtuse and wall distinctly thickened; chloroplasts single (increasing to 4 with age), trough-shaped, with a single pyrenoid; autospores 2, 4 or 8 per sporangium, each possessing a chloroplast, released by gelatinization of mother cell wall and only rarely by its breakage.

Probably cosmopolitan; so far reported in the British Isles only from moderately nutrient-rich (mesotrophic) lakes where it occurs in the plankton or amongst moss and associated with other surfaces.

According to Hindák (1988), the species can be reliably separated from *O. parva* only by differences in cell size. Despite the considerable overlap in size between *O. lacustris* and *O. marssonii*, it is considered by Hindák (1984) and others to be the only reliable character for separating the two species.

Oocystis marssonii Lemmermann 1898

Synonym: *Oocystis crassa* var. *marssonii* (Lemmermann) Printz

17640060 **Pl. 92F** (p. 373) **CD**

Cells solitary or, more rarely, forming 2-, 4- or 8-celled coenobia which sometimes contain up to 3 generations, with cells initially lying against mother cell wall but separating from it as wall expands; cells (4–)6.3–12(–14.4) μm wide, (6.4–)8.5–18(–20.8) μm long, broadly spindle-shaped to almost ovoid, with each apex narrow, obtuse or slightly rounded and its wall thickened; chloroplasts 2–4(–8), grooved, each with a pyrenoid; autospores 2, 4 or 8 per sporangium, each with a chloroplast, released by fracture of mother cell wall.

Probably cosmopolitan; planktonic and associated with surfaces in different water bodies, only recorded from the the British Isles on a few occasions in moats and lakes including the shallow lakes of the Norfolk Broads; probably considerably more widely distributed than records indicate, but easily mistaken for the more common and morphologically very similar *O. lacustris*.

Several authors have questioned the validity of this species which closely resembles *O. lacustris* Chodat *sensu auct. post.* (see remarks under this entry) since cell size is regarded to be the only reliable distinguishing feature (see Hindák, 1984). According to Reháková (1969), second or third cell generations of *O. marssonii* develop in colonies grown in culture and the cells have up to 12 chloroplasts and often a thickened apical cell wall.

Oocystis naegelii A. Braun 1855

17640070 **Pl. 92H** (p. 373)

Coenobia of 2, 4 or 8 cells, tightly packed and tetrahedrally arranged, more rarely solitary; cells 10–22(–26) μm wide, 16–40 μm long, 1.5 to 2 times longer than broad, ovoid to ellipsoidal or spindle-shaped, with apices rounded and having thickened walls; chloroplasts 1–2(–8), without pyrenoids; autospores 2 or 4(–8) per sporangium, each with a chloroplast, released by fracture of mother cell wall.

Probably cosmopolitan; planktonic or associated with submerged objects, only known from large lochs in Scotland; probably more widely distributed in the British Isles than the few literature records suggest.

Oocystis natans (Lemmermann) Lemmermann 1908

Basionym: *Oocystella natans* Lemmermann

17640080 **Pl. 92G** (p. 373)

Coenobia of 2, 4 or 8(–16) cells, each in contact with mother cell wall and only separating from the wall as it expands; cells (10–)12–16(–25) μm wide, (12–)23–38 μm long, 1.5–2 times longer than broad, ovoid to broadly spindle-shaped, without apical wall thickening; chloroplasts 4–8 per cell, lobed and star-shaped, each with a pyrenoid; autospores 2, 4 or 8 per sporangium.

Probably cosmopolitan; common in the British Isles where recorded from the plankton of large productive lakes, including Loch Leven in Scotland.

Oocystis panduriformis West *et* G.S.West 1894

Synonyms: *Oocystis panduriformis* fo. *major* West *et* G.S.West, *O. panduriformis* var. *pachyderma* West *et* G.S.West

17640110 **Pl. 92L** (p. 373)

Cells solitary and usually 23–27(–32.5) μm wide, 50–61.5(–77) μm long, 2–2.5 times longer than broad, fiddle-shaped, with apices thickened and subacute; chloroplasts 15–25 per cell, plate-like and each usually associated with a pyrenoid; reproduction unknown.

Probably cosmopolitan; recorded in the British Isles from a few widely separated lakes in Ireland and England (Yorkshire).

The original description of this distinctively shaped species failed to mention either its chloroplasts or pyrenoids. We consider that no useful purpose is served by recognizing infraspecific taxa based on trivial and highly variable differences in cellular morphology. Two such taxa were described by the Wests (W. & G.S.West, 1894), namely fo. *major* (cells 29–32.5 μm wide, 77 μm long) and var. *pachyderma* (wall 2.5–2.8 μm thick).

Oocystis parva West *et* G.S.West 1898

Synonym: *Oocystella parva* (West *et* G.S.West) Hindák

17640120 **Pl. 92J** (p. 373)

Cell solitary or forming coenobia of 2, 4 or 8 densely packed cells in contact by their long axes and only separating from mother cell wall on its expansion; cells (1.5–)3–7(–10) μm wide, (3.2–)6–12(–17) μm long, 1.5 to almost twice as long as broad, ellipsoidal to broadly spindle-shaped, with apices rounded or bluntly pointed and each with a nipple-shaped thickening which often develops during autospore formation; chloroplast single (2–4 with age), 1 pyrenoid present; autospores 2–4(–8) per cell, each with a chloroplast, released by fracture of mother cell wall.

Probably cosmopolitan; in the British Isles planktonic or associated with submerged surfaces in rivers, lakes and bogs, often most abundant in pools.

Oocystis pusilla Hansgirg 1890

17640130 **Pl. 92K** (p. 373)

Cells solitary or forming coenobia of 2, 4 or 8 loosely arranged cells within a mother cell wall; cells 3–7.5 μm wide, 6–12 μm long, almost twice as long as broad, ellipsoidal to cylindrical, with obtusely rounded apices and unthickened walls; chloroplasts 2–3 in each cell, without a pyrenoid; autospores 2, 4 or 8 per sporangium, each autospore having a chloroplast and released by fracture of mother cell wall.

Probably cosmopolitan; planktonic in small and often temporary bodies of water; known in the British Isles only from one small pond in High Elms Nature Centre, near Downe village in Kent, but undoubtedly widely distributed.

Oocystis solitaria Wittrock in Wittrock & Nordstedt 1879

Synonyms: *Oocystis crassa* Wittrock, *O. solitaria* var. *notabile* West *et* G.S.West

17640140 **Pl. 88A** (p. 359)

Cells solitary or forming coenobia of 2, 4 or 8 cells, sometimes as multiple colonies of 2 or 3 generations, with cells initially densely packed but separating on expansion of stratified mother cell wall; cells (3–)9.5–25(–29.7) µm wide, (7–)15–48(–52.8) µm long, 1.5 to 2 times longer than broad, ovoid to ellipsoidal, apices broadly rounded to slightly acute, each with a nodular wall thickening; chloroplasts large, (4–)8–20 per cell, with margins often lobed, occasionally polygonal when closely packed, each possessing a pyrenoid; autospores 2, 4 or 8 per sporangium, each with 4 or more chloroplasts, released by fracture of mother cell wall.

Probably cosmopolitan; reported on few occasions in the British Isles from widely separated localities (e.g. Lanlivery Moor, Cornwall; Orkney Islands) where planktonic or associated with aquatic macrophytes, often in water bodies with an acid pH (bogs, peat lakes, puddles).

Two forms are reported from the British Isles (W. & G.S.West, 1894): fo. *solitaria* with cells (3–)4.9–18.2 µm wide and (7–)12.8–28(–31) µm long, and fo. *major* Wille with cells 8.2–25(–29.7) µm wide and 16.5–48(–52.8) µm long (both from Scarf Gap Pass, English Lake District). Probably little useful purpose is served by recognizing forms based upon such trivial differences.

DOUBTFUL SPECIES

Oocystis asymmetrica West *et* G.S.West 1894

17640020 **Pl. 92B** (p. 373)

Described by W. & G.S.West (1894) based on material collected from English and Irish lakes. Cells are 7–8.6 µm wide, 15–18 µm long, 2.25 times as long as wide, oblong to ellipsoidal, asymmetrical (one side more convex than the other) and with an apically thickened wall. In the original description there is no mention of chloroplast structure and of the presence or absence of pyrenoids. Printz (1913) considers it to have several chloroplasts and to be a form of *O. solitaria*, and Playfair (1916) reduces it to a variety under *O. apiculata*.

Oocystis geminata Nägeli in Kützing 1849

Only recorded by W. & G.S.West (1902) from Northen Ireland and by Woodhead & Tweed (1947b). These two records are considered doubtful for uncertainty surrounds the identity of the species. Korshikov (1953) considers that it resembles his new species *Oocytidium ovale*, but its original description was too incomplete for him to be certain that they are identical.

Oocystis nodulosa West *et* G.S.West 1894

17640090

Described by W. & G.S.West (1894) based on material collected in Ireland. Its cells are 16–17 µm wide, 25–26 µm long and about 1.5 times longer than wide, oblong to ellipsoidal, with rounded apices of which each has a nodular wall thickening. Chloroplasts and pyrenoids were not mentioned although material attributed to the species from other regions is reported to have several chloroplasts and pyrenoids (see Printz, 1916). Until further information is available, we regard the species as doubtful.

Oocystis novae-semliae Wille 1879

17640100

W. & G.S.West (1894) described under this species a new form (fo. *major*) and a new variety (var. *maxima*) based on material collected in the New Forest in Hampshire. The new form had cells 7–7.9 µm wide, 11.1–12.5 µm long whereas the variety had cells almost twice these dimensions. Doubt attaches to the species since there is no information on either its chloroplast structure or pyrenoids.

Oonephris Fott 1964

Cells solitary or forming coenobia of usually irregularly arranged cells enclosed within the partially gelatinized mother cell wall; cells ellipsoid, ovoid or almost kidney-shaped, with walls moderately thick and sometimes lamellate, with or without nodular wall thickenings at apices; chloroplast star-shaped with arms anastomosing so giving it a spongy appearance, with a single pyrenoid; asexual reproduction by 2, 4, 8 or 16 autospores, often autospores retained for a period before their release from mother cell wall.

Probably cosmopolitan; common in the plankton of ponds and lakes.

Oonephris obesa (West) Fott 1964

Basionym: *Nephrocytium obesum* West

Synonyms: *Oocystis gigas* W.Archer fo. *minor* West *et* G.S.West, *Nephrocytium ecdysiscepanum* West *et* G.S.West

17650010 **Pl. 92I** (p. 373)

Cells 24–28 µm wide, 34–42 µm long, asymmetrically ovoid, with one side straight or slightly concave and the other strongly curved, surrounded by a compact mucilaginous envelope (about 10 µm wide).

Probably cosmopolitan; in the British Isles planktonic, occurs in small and slightly acid water bodies (e.g. boggy pools in East Anglian fenlands and upland moorland areas); probably more widely distributed than the few records indicate.

Palmodictyon Kützing 1845

Synonym: *Palmodactylon* Kützing

Colonies attached or free-floating, consisting of simple or branched and anastomosing cylinders of structureless or slightly stratified mucilage (sometimes red or brown), containing rows of disorganized cells or cells in pairs or groups of 4, each cell surrounded by a mucilaginous envelope; cells spherical to ovoid, with walls smooth and thin; chloroplasts parietal, 1 or several per cell, with or without pyrenoids; asexual reproduction by 2–4 autospores, released by breakage of mother cell wall into 2 or 4 parts or by gelatinization, the persistent cell wall remnants often remaining connected and in close proximity to the cells; akinetes thick-walled.

Probably cosmopolitan; usually relatively rare and often in stagnant pools.

Some uncertainty surrounds its systematic position. In a revision of the genus, Hoek (1963b) concluded that *P. viride* Kützing and *P. varium* (Nägeli) Lemmermann are distinct species and observed that all the diagnostic characters can be observed in dried specimens. These species have often been considered conspecific under *P. viride*. Uncertainty therefore attaches to earlier records unless accompanied by an illustration showing unequivocally the characters used to separate them.

1 Chloroplast 1 per cell, cup-shaped, with a pyrenoid .... *P. viride*
1 Chloroplasts 4–6 per cell, disc-shaped, without a pyrenoid .... *P. varium*

Palmodictyon varium (Nägeli) Lemmermann 1915

Basionym: *Palmodactylon varium* Nägeli
Synonyms: *Palmodactylon subramosum* Nägeli, *Palmodictyon viride* Kützing *sensu auct. post. pro parte*
17670010 **Pl. 99R** (p. 403) **CD**

Cells (4–)6.5–9.5(–12) μm wide, spherical, arranged in a single row or grouped, surrounded by a usually structureless mucilaginous envelope (11.5–33 μm wide); chloroplasts 4–6 per cell, disc-shaped, without pyrenoids.

Probably cosmopolitan; more common than *P. viride* in the British Isles where it usually occurs in peaty ditches and bog pools.

Palmodictyon viride Kützing 1845

Synonym: *Palmodictyon varium* (Nägeli) Lemmermann *sensu auct. post.*
17670020 **Pl. 99Q** (p. 403)

Cells 5.5–9(–21.5) μm wide, spherical to spherical-ovoid, solitary or in groups of 2 or 4, surrounded by an often stratified mucilaginous envelope (28–52 μm wide); chloroplast 1 per cell, cup-shaped, containing a pyrenoid.

Europe and North America; very rare in the British Isles where planktonic or associated with surfaces in ponds, lakes and bogs.

Pediastrum Meyen 1829

Coenobia disc- or star-shaped, flat and single-layered, usually 4-, 8-, 16-, 32- or 64-celled; cells closely adpressed or with intercellular spaces, marginal cells having 1, 2 or 4 projections or lobes, sometimes emarginate, walls smooth or variously ornamented, occasionally bearing tufts of mucilaginous spines or with mucilaginous caps at tips of lobes; chloroplast parietal, massive and with a single pyrenoid; asexual reproduction by zoospores; sexual reproduction isogamous.

Probably cosmopolitan; planktonic or associated with aquatic macrophytes and *Sphagnum* moss, often common in moderately nutrient-enriched water bodies.

Only mentioned are the more common and distinctive of the large number of described forms and varieties. Many described have little significance and probably simply reflect colony age (e.g. *Pediastrum boryanum* var. *brevicorne*). Some details of the cell surface (e.g. granule density) are very variable and are of limited value as taxonomic criteria for separating infraspecific taxa (see Nielsen, 2000). Its cell wall is very resistant to decay resulting in recognizable *Pediastrum* colonies present in pond and lake sediments; fossil species are known. For a review of those species preserved in wetland sediments and an identification key, see Komárek & Janlovská (2001). It is regarded as belonging to the more broadly defined Order Sphaeropleales (see Deason et al., 1991) based largely on ultrastructural features of the body associated with each flagellum of its motile cells. Until further studied it has been decided to keep the genus in the Chlorococcales, the order in which it has traditionally been placed.

1 Inner cells irregularly arranged, with these often similar in shape to marginal cells; marginal cells rounded or somewhat angled, without lobes or with 1 or 2 very short lobes .... *P. integrum*
1 Inner cells concentrically arranged and not similar in shape to marginal cells; marginal cells of another form 2
2 Outermost wall of each marginal cell in the form of a single tapering lobe .... *P. simplex*
2 Outermost wall of each marginal cell with 2 or 4 distinct lobes 3
3 Coenobia with very evident spaces between cells 4
3 Coenobia without evident spaces.......................... 6
4 Margin cells bilobed, each lobe divided into two projections .... *P. biradiatum*
4 Marginal cells bilobed, each lobe truncate............. 5
5 Intercellular spaces inconspicuous and somewhat lens-shaped when present .... *P. boryanum* (only var. *cornutum*)
5 Intercellular spaces large and very conspicuous .... *P. duplex*
6 Marginal cells with a narrow central incision dividing each cell into into two lobes, each lobe truncate, slightly emarginate .... *P. tetras*
6 Marginal cells of a different form 7
7 Cell walls with net-like ridges; projections broad and with or without short lobes .... *P. angulosum*
7 Cell walls smooth, granular or warty; projections narrow, long and lobed........................ 8
8 Marginal lobes lying in one plane........ *P. boryanum*
8 Marginal lobes in different planes, bent in opposite directions .... *P. kawraiskyi*

Pediastrum angulosum Ehrenberg ex Meneghini 1840

Synonym: *Pediastrum araneosum* (Raciborski) Raciborski
17680010 **Pl. 93Aa–c** (p. 378)

Coenobia 8-, 16-, 32-, 64- or 128-celled, not usually with intercellular spaces, (28–)60–320(–420) μm across; marginal cells (7–)10–30(–36) × 8–27.6(–38) μm broad, slightly to deeply indented, having wide projections, with or without short lobes, the lobes if present short and rounded or pointed at apices; inner cells (7–)11–23 × (8–)11–22.5(–27) μm, with a depression in outer wall corresponding in position to indentation on outer wall of colony; cell walls with net-like ridges.

Probably cosmopolitan; in the British Isles in the plankton and associated with surfaces in a wide range of aquatic habitats (e.g. Capel Curig in North Wales and various Irish loughs), rare.

Very polymorphic species in which some trivial growth forms have been given species status in the past.

Pediastrum biradiatum Meyen 1829

Synonym: *Pediastrum rotula* (Ehrenberg) Kützing *pro parte*, *P. biradiatum* var. *longecornutum* Gutwinski

17680020 **Pl. 93B** (p. 378) **CD**

Coenobia (4–)8-, 16- or 32-celled, 36–124 μm across; marginal cells (8–)12.4–27(–30) μm wide, adjoined near base and divided into 2 broad lobes, having a V-shaped sinus separating lobes and almost reaching to middle of cell, with each lobe dichotomously divided into two conical to cylindrical projections; inner cells 6–10.3(–21) ×(6–)8.2–20.7(–26), with intercellular spaces and each usually bilobed; cell walls smooth, dotted or granulate.

Probably cosmopolitan; occurs commonly in various aquatic habitats including spring-fed ponds, lakes, flooded brick pits and rivers especially in lowland areas of the British Isles. Often considered in Europe to be an indicator of mildly to moderately nutrient-enriched water.

Pediastrum boryanum (Turpin) Meneghini 1840

Basionym: *Hierella boryana* Turpin

Synonyms: *Pediastrum bidentulum* A.Braun, *P. constrictum* Hassell

17680030 **Pl. 93C** (p. 378) **CD**

Coenobia (16–)25–180(–208) μm wide; marginal cells (5–)8–30(–40)×(5–)9–21(–31) μm, slightly to deeply notched and with two projections; projections lying almost parallel to each other, often projections equal to or shorter than cells bearing them; inner cells (5–)6–20(–26)×(4–)5.7–22.5(–27) μm, polygonal, without intercellular spaces; walls smooth or granulate.

Probably cosmopolitan; most common and widely distributed species in the British Isles where especially abundant in ponds, lakes, reservoirs, ditches and slow-flowing rivers in spring and early summer although present throughout the year; most abundant in nutrient-rich water (meso-eutrophic).

var. *cornutum* (Raciborski) Sulek 1969

Basionym: *Pediastrum duplex* var. *cornutum* Raciborski

17680032 **Pl. 93D,F** (p. 378) **CD**

Coenobia 20–111(–118) μm wide; marginal cells 4.5–27×4–24 μm, with irregular, small and often triangular or lens-shaped intercellular spaces.

Probably cosmopolitan; sometimes in the same types of water body as the typical variety.

Very closely related to var. *perforatus* Raciborski from which it differs by the greater depth of the indentation of the marginal cells, shape of projections and arrangement of cell wall granules. We agree with Nielsen (2000) in believing the distinction between var. *perforatus* and var. *cornutum* to be questionable with the two possibly synonymous. Nielsen goes so far as to believe that both are merely morphotypes of type variety.

var. *longicorne* Reinsch 1867

Synonym: *Pediastrum glanduliferum* A.W.Bennett

17680033 **Pl. 93E** (p. 378)

Coenobia 30–150(–312) μm wide; marginal cells 8.1–30(–41)×(4–)5.7–21(–35) μm, bearing cylindrical projections that exceed cell length and sometimes terminate in a globular cap; inner cells 5–27×4–22.5 μm, only very rarely with small intercellular spaces.

Probably cosmopolitan; in the British Isles associated with marginal aquatic macrophytes in small ponds and lakes, rare.

Most of subspecific taxa of *P. boryanum* do not warrant recognition since characters associated with the marginal cells and nature of walls are known to exhibit considerable variation depending on the environment and age of the coenobium. For this reason the following varieties recorded for the British Isles are not described here, namely var. *productum* West, var. *granulatum* Kützing and var. *brevicorne* A.Braun. We have followed Nielsen (2000) in recognizing var. *longicorne*.

Pediastrum duplex Meyen 1829

Synonyms: *Pediastrum napoleonis* Ralfs, *P. pertusum* Kützing

17680050 **Pl. 93I** (p. 378) **CD**

Coenobia (28–)35–135(–212) μm across, with cells adjoining intercellular spaces usually having concave side walls; marginal cells 7–24(–28)×6–24(–28) μm, bearing short, cylindrical or slightly tapering truncate projections lying almost parallel to each other; inner cells (4–)6–21(–26)×(4–)5–21(–30) μm, varying from H-shaped to angular, often with centre of wall of adjacent cells not in contact resulting in very evident intercellular spaces; cell walls smooth, without granulations.

Probably cosmopolitan; very common throughout the British Isles in a wide range of aquatic habitats including ditches, small ponds, lakes, reservoirs and rivers, especially in the English Midlands and southern England.

var. *gracillimum* West *et* G.S.West 1895

Synonym: *Pediastrum gracile* A.Braun

17680052 **Pl. 93L** (p. 378) **CD**

Coenobia up to 125 μm across; cells with concave, straight or convex side walls; marginal cells 9–17(–30) ×4–6.6(–22.5) μm, often very long projections sometimes bent in different directions or at an angle; inner cells 24×22.5 μm; cell walls smooth, without granulations.

Probably cosmopolitan; much less common in the British Isles than the type variety, usually in ponds and other small water bodies.

var. *rugulosum* Raciborski 1890

17680053 **Pl. 93G** (p. 378)

Coenobia 76–198 μm across; marginal cells 11–27.5 × (10–)11.5–25.5 μm, inner cells 13.8–25.5× 10–21(–24) μm, similar in form to var. *subgranulatum* but walls covered by randomly arranged granules and also finely wrinkled.

Probably cosmopolitan; very rare in the British Isles where planktonic or associated with submerged surfaces in a range of aquatic habitats.

var. *subgranulatum* Raciborski 1890

17680054 **Pl. 93H** (p. 378)

Coenobia (35–)81–112.5(–161) μm across; inner cells broad and H-shaped, with indentation along sides of cells; marginal cells (5.5–)15–25(–28.5)×(5–)12–25 μm,

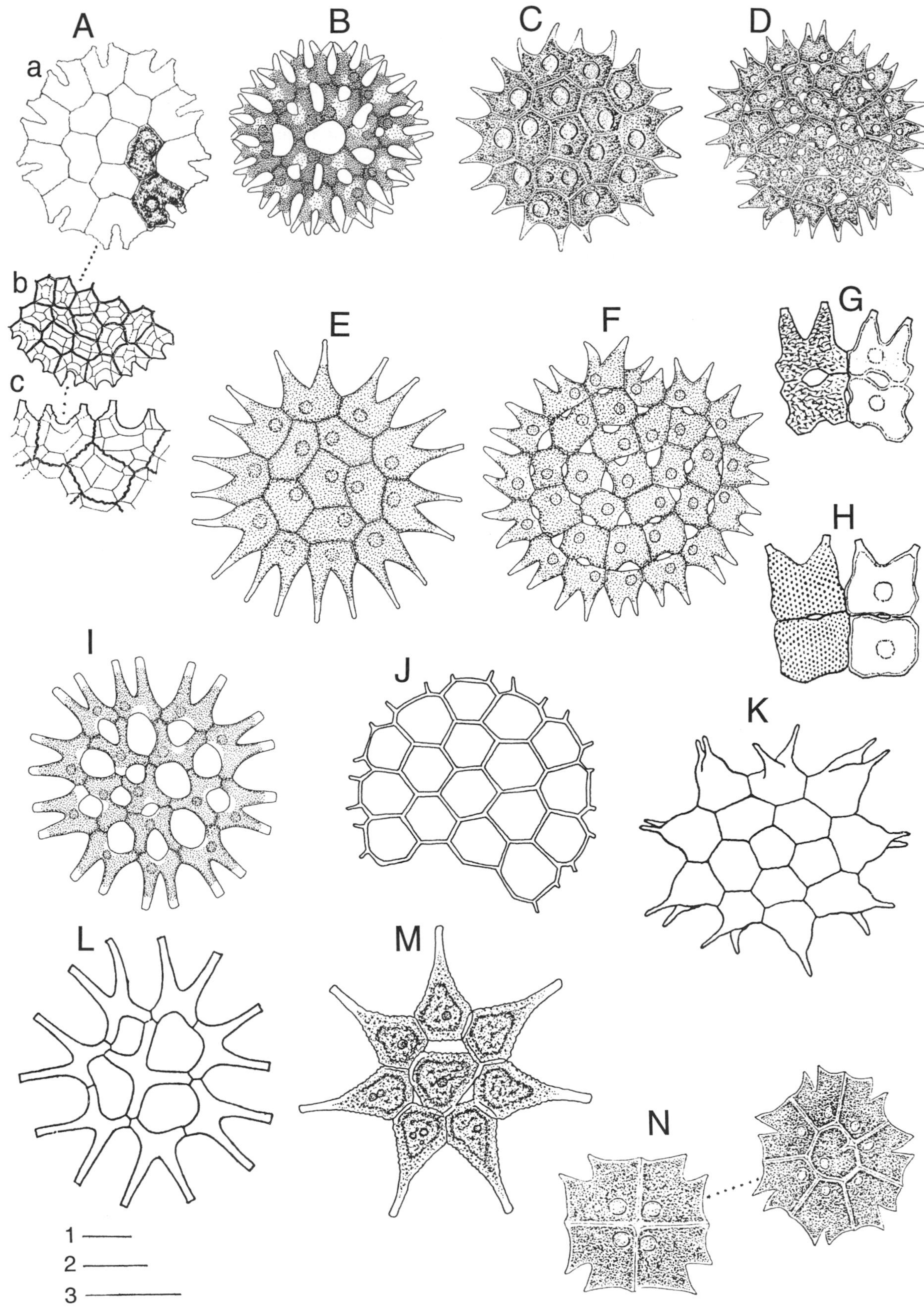

Plate 93 A–N

A. *Pediastrum angulosum* (p. 376): a. colony showing contents of two cells, b, c. details of cell wall; **B**. *Pediastrum biradiatum* (p. 377); **C**. *Pediastrum boryanum* (p. 377); **D, F**. *Pediastrum boryanum* var. *cornutum* (p. 377); **E**. *Pediastrum boryanum* var. *longicorne* (p. 377); **G**. *Pediastrum duplex* var. *rugulosum* (p. 377); **H**. *Pediastrum duplex* var. *subgranulatum* (p. 377); **I**. *Pediastrum duplex* (p. 377); **J**. *Pediastrum integrum* (p. 379); **K**. *Pediastrum kawraiskyi* (p. 379); **L**. *Pediastrum duplex* var. *gracillimum* (p. 377); **M**. *Pediastrum simplex* (p. 379); **N**. *Pediastrum tetras* (p. 379).

Scale bars: 1, 3, 30 μm; 2, 10 μm. 1 for Ab, B; 2 for Ac, G, H, K, J, M, N; 3 for Aa, C–F, I, L.

having a moderately deep and wide incision, with walls parallel where joined and covered by small, pointed granules; inner cells 12–15(–20) × (5–)10.5–13 (–25) μm.

Probably cosmopolitan; very rare in the British Isles where planktonic or associated with submerged surfaces in various aquatic habitats.

Pediastrum integrum Nägeli 1849

17680060 **Pl. 93J** (p. 378) **CD**

Coenobia 4-, 8-, 16- or 32(–64)-celled; marginal cells rounded or somewhat angled, without projections or with 1 or 2 very short projections; inner cells irregularly arranged, similar in shape to marginal ones, without intercellular spaces; walls smooth or granular.

Probably cosmopolitan; widely distributed in the British Isles but more frequently in quiet fen and bog pools in the east of England.

Pediastrum kawraiskyi Schmidle 1897

17680070 **Pl. 93K** (p. 378)

Coenobia (4–)8-, 16- or 32(–64)-celled, (30–)75–189 μm across; marginal cells (12–)22.5–31 × (9–)15–30 μm, extending into two cone-shaped or finger-like projections, one usually lying above the other and bent in opposite directions; inner cells 7.5–16.5 × 7.5–16.5 μm, polygonal, without intercellular spaces; cell walls usually smooth, sometimes spiny, warty, wrinkled or rough.

Europe, North America; rare in the British Isles where reported from small ponds and lakes (e.g. the Norfolk Broads; Loch Shurrey, Caithness, north of Scotland).

Pediastrum simplex Meyen 1829

Synonyms: *Pediastrum clathratum* (Schröder) Lemmermann, *P. enoplon* West *et* G.S.West, *P. sturmii* Reinsch, *P. simplex* var. *sturmii* (Reinsch) Wolle

17680080 **Pl. 93M** (p. 378) **CD**

Coenobia usually 4-, 8-, 16- or 32(–128)-celled, 65–250 μm across; marginal cells (6–)10.5–40 × (6–)7–36 μm, extending to form a single and gradually tapering projection; inner cells (8–)11.7–46.5(–57) × 6–30(–38) μm, 3–5-sided to polygonal, obovoid or spherical, with or without intercellular spaces or occasionally with a single central space; cell walls smooth, granular or spiny.

Probably cosmopolitan; widespread in the British Isles, especially in southern England and the English Midlands where occasionally in small water bodies, river backwaters and nutrient-rich lakes during the summer months.

Pediastrum tetras (Ehrenberg) Ralfs 1844

Basionym: *Micrasterias tetras* Ehrenberg

Synonym: *Pediastrum ehrenbergii* (Corda) A.Braun

17680090 **Pl. 93N** (p. 378) **CD**

Coenobia 4-, 8-, 16- or 32-celled, 15–35(–74) μm across; cells mostly inversely trapezoid with 4–7 angles; marginal cells (4–)8–14(–18) × (5–)8–14(–18) μm, laterally wall united to apex, divided by a narrow central incision into two lobes, each lobe truncate and slightly emarginate, further divided into two obtuse-rounded lobes; inner cells (4–)8–11(–14) × (5–)7–11(–16) μm, straight-sided, without intercellular spaces; cell walls smooth.

Probably cosmopolitan; widely distributed but rarely abundant in the British Isles where planktonic or associated with surfaces in many types of aquatic habitat; considered in Europe to be an indicator of mildly to moderately nutrient-enriched water.

Another very polymorphic species exhibiting considerable variation in length of marginal processes, shape and depth of incision, and other features.

DOUBTFUL RECORD

Pediastrum braunii Wartmann 1862

17680040

A species having 4 short and small projections on outer side of each marginal cell, 2 projections close together near centre of cell and the other 2 towards its edge, neighbouring cells connected by short lateral processes. The illustration in West & Fritsch (1927, p. 114, as *Pediastrum tricornutum* Borge) of a specimen from Glen Tummel, Perthshire (see also West, 1893), is inadequate to confirm the identification of this rare and little-known species.

Planktosphaeria G.M.Smith 1918

Cells solitary or in colonies of spherical cells associated with remnants of mother cell wall, enclosed within a firm, hyaline or concentrically layered mucilaginous envelope; chloroplast parietal, initially cup- or trough-shaped and with a large basal pyrenoid, older cells having several polygonal chloroplasts and each with a pyrenoid. Asexual reproduction by autospores and zoospores; autospores (4, 8) 16 or 32 in each mother cell, released by fracture of wall following its expansion and thickening, the remnants of mother cell wall persisting in vicinity of daughter cells; zoospores naked, with 2 flagella of equal length.

Europe and North America; usually planktonic in moderately nutrient-rich lakes.

Similar to many other chlorococcalean algae in producing several chloroplasts (often polygonal) and pyrenoids just prior to autospore formation. For further information on the genus and its separation from several closely related genera, see Starr (1954, 1955). Komárek & Fott (1983) considered Starr's interpretation as doubtful, since no mention was made of zoospores in the original diagnosis, although they have been discovered subsequently in laboratory culture (see Hindák, 1984).

Planktosphaeria gelatinosa G.M.Smith 1918

17690010 **Pl. 86M** (p. 352)

Cells solitary, or more commonly as colonies with cells surrounded by a firm and homogeneous mucilaginous envelope (up to 250 μm across); cells 4–25(–46) μm in diameter, spherical, with walls moderately thick; zoospores 5–7 μm wide, spherical to broadly oval and having flagella about 3 times longer than cell, with dash- or dot-like eyespot.

Europe, North America; recorded in many parts of the British Isles where widespread in lake and reservoir plankton, especially common in the English Lake District and Scottish lochs; sometimes present in great abundance.

Zoospores known to be produced only in laboratory culture. See Hindák (1984) for a detailed account of the species and his remarks concerning morphologically similar algae with which it might be easily mistaken (e.g. *Coenococcus, Coenochloris, Sphaerocystis*). Earlier Lund (1952) had pointed out that the original description of *Sphaerocystis schroeteri* by Chodat (1897) might have contained several currently recognized taxa with one of his illustrations clearly attributable to *Planktosphaeria gelatinosa*. Lund believes this alga to have been frequently misidentified as *Sphaerocystis schroeteri* which in his view is an aggregate species (see comments under *Coenochloris fottii*, p. 342).

Podohedra Düringer 1958

Cells attached or secondarily free, solitary or in 2-, 4- or 8-celled colonies; cells narrowly spindle-shaped, straight, slightly bent or sigmoid, thin and smooth-walled, narrowing to sharply or bluntly pointed apices, with one apex sometimes slightly extended and the other of a different shape (heteropolar); chloroplast parietal, usually trough-shaped, with or without a pyrenoid; asexual reproduction by 2, 4 or 8 autospores, released by fracture of mother cell wall, often remaining attached to form a colony.

Probably cosmopolitan; mostly attached to submerged surfaces and rarely planktonic or secondarily free.

A conservative approach has been adopted here since species not possessing a pyrenoid are retained in this genus. Hindák (1988) considers the presence or absence of a pyrenoid (visible with light microscopy) to be a taxonomically important character and thus established the genus *Podohedriella* to accommodate those *Podohedra* species in which it is absent. For a discussion of the problems concerning the closely related *Keratococcus* and *Chlorolobion*, see under the latter two genera. If secondarily free then difficult to separate from the genus *Monoraphidium* although still heteropolar. We consider the presence or absence of pyrenoids to be an unreliable character to be used for distinguishing genera.

Podohedra falcata Düringer 1958

Synonym: *Podohedriella falcata* (Düringer) Hindák

17700020 **Pl. 91O** (p. 369)

Cells 1.5–2(–5) µm wide, 30–35(–68) µm long, narrowly spindle-shaped, sometimes asymmetrical and slightly bent, tapering to each apex with the attached one sometimes slightly longer and less pointed.

Europe; in the British Isles known only as an epiphyte on *Cladophora glomerata* growing in a pond at Cambridge, undoubtedly more widely distributed; elsewhere reported attached to detritus and associated with the mucus surrounding periphytic algae in wet subaerial habitats.

Some doubt attaches to the only record from the British Isles since the cells illustrated by Belcher & Swale (1996) were only slightly heteropolar and the chloroplast was not the usual trough-shape. For further information on this species, see Hindák (1988).

Polyedriopsis Schmidle 1899

Cells solitary, pyramidal, tetrahedral or polyhedral, with groups of hollow and somewhat tapering spines usually arising from rounded corners, walls thin and smooth, lacking a mucilaginous envelope; chloroplast parietal, with a single lateral pyrenoid; asexual reproduction by 2, 4 or 8 autospores or biflagellate zoospores; records of akinetes doubtful.

Probably cosmopolitan; planktonic in a wide range of freshwater habitats.

The genus has been reviewed by Kováčik (1975a).

Polyedriopsis spinulosa (Schmidle) Schmidle 1899

Basionym: *Tetraedron spinulosum* Schmidle

Synonyms: *Polyedrium spinulosum* Schmidle, *P. spinulosum* var. *exavatum* (Playfair) G.M.Smith, *Tetraedron spinulosum* var. *exavatum* Playfair

17710010 **Pl. 89I** (p. 363)

Cells (10–)16.4–25 µm across, with 4 or 5 curved sides and rounded corners, each corner bearing a cluster of (1–)3–5(–10) spines with length range 21–40 µm.

Probably cosmopolitan; infrequently recorded but widely distributed in the British Isles in small ponds, moats, rivers and lakes.

The number of spines is very variable and sometimes they are absent or scarcely visible. Young cells have fairly acutely angled corners but with age these become more rounded resulting in almost spherical cells.

Pseudocharacium Korshikov 1953

Cells solitary, oval or obovoid to broadly ellipsoidal, elongate to spindle- or club-shaped, each with a spine at the apex or having a ring-like apical wall thickening, attached by a sole-like extension to the stalk; chloroplast parietal, minutely and indistinctly net-like, usually with a single pyrenoid but several present in ageing cells; cells uninucleate with older ones becoming multinucleate; asexual reproduction by naked quadriflagellate zoospores, released by lateral fracture of mother cell wall or through an apical opening.

Probably cosmopolitan; usually epiphytic on submerged water plants in various water bodies.

Pseudocharacium obtusum (A.Braun) Petry-Hesse 1968

Basionym: *Characium obtusum* A.Braun

Synonym: *Characium ovum* Matvienko

17720010 **Pl. 99M** (p. 403)

Cells (6–)7–22(–25) µm wide, (12–)20–40 µm long, elongate-ovoid to club-shaped or broadly oval, thick-walled, sessile or with a short stalk, basally with a well-developed, brownish-coloured sole-like attachment portion and apically rounded with a ring-like wall thickening; chloroplast in ageing cells containing several pyrenoids.

Probably cosmopolitan; usually on submerged surfaces in ponds, lakes and slow-flowing water, very rarely reported from the British Isles.

Pseudococcomyxa Korshikov 1953

Cells solitary, elongate and inversely ovoid to ellipsoidal, slightly asymmetrical, thin-walled, without a mucilaginous envelope but having a mucilaginous cap at one or both apices (often only visible with Indian ink); chloroplast parietal, trough-shaped and with a small depression in centre, without pyrenoids; asexual reproduction by 2 or 4(–8) autospores, released by apical splitting of the mother cell wall.

Europe; in many types of aquatic habitat as well as associated with soil and subaerial surfaces. Monospecific.

Pseudococcomyxa simplex (Mainx) Fott 1981

Basionym: *Coccomyxa simplex* Mainx
Synonym: *Pseudococcomyxa adhaerens* Korshikov

17730010 **Pl. 99N** (p. 403) **CD**

Cells 2–4.5(–6.5) μm wide, (4.6–)6–11(–12.6) μm long.

Europe; in the British Isles only in small bodies of water and soil; thrives commonly in neglected containers of distilled or deionized water such as wash bottles.

Pseudodictyosphaerium Hindák 1978

Synonym: *Dictyosphaeriopsis* Hindák

Coenobia of (2–)4, 8, 16, 32 or 64 cells in groups towards periphery of a mucilaginous envelope; cells spherical, ellipsoidal, cylindical-ovoid or ovoid, with walls thin and smooth, usually in groups of 4 arising terminally on irregularly or regularly branched remnants of mother cell walls radiating from centre of colony; chloroplast parietal, broadly girdle-, cup- or trough-shaped, without a pyrenoid; asexual reproduction by (2–)4 autospores, on release remaining attached to persistent remnants of mother cell wall.

Europe; planktonic in various aquatic habitats.

Pseudodictyosphaerium scoticum Hindák 1988

17740010 **Pl. 85G** (p. 350)

Coenobia 20–50 μm wide, spherical, tetrahedral or oval; cells 4–6 μm in diameter, spherical, with a 5–8 μm thick hyaline wall around cells; chloroplast broadly girdle-shaped; autospores (2–)4 in each sporangium, asymmetrically oval to ovoid, 2–2.5 μm wide, 3–4 μm long.

British Isles (Scotland); so far known only from the type locality of Loch Lomond.

See remarks under *Dictyosphaerium chlorelloides.*

Pseudoquadrigula E.N.Lacoste 1973

Coenobia 4-, 8-, 16- or 32-celled, with cells grouped in tetrads and lying parallel to each other, the cells gradually move apart along their longitudinal axes once released from the mother cell, enclosed within a structureless mucilaginous envelope; cells narrowly cylindrical to spindle-shaped, straight to slightly curved, narrowing abruptly often to bluntly pointed apices; chloroplast parietal and divided in centre, with or without a pyrenoid; asexual reproduction by 2, 4 or 8 autospores, if the latter number then occurring in two tiers each of 4 spores, released by gelatinization of mother cell wall or its transverse fracture into two parts.

Probably cosmopolitan; planktonic in various water bodies.

Pseudoquadrigula britannica Hindák 1988

17750010 **Pl. 98A** (p. 400)

Coenobia often ovoid and 4-celled, 24–30 μm wide, 45–55 μm long, rarely up to 32-celled; cells 4–5 μm wide, 24–28 μm long, spindle-shaped, apices acute or bluntly pointed; autospores released by fracture of mother cell wall into two parts, the wall remnants soon disappearing.

British Isles (Scotland); so far known only from a plankton sample collected in Loch Tay in September 1984.

Pseudotetrastrum Hindák 1977

Coenobia flat, quadrate and 4-celled, sometimes forming 16-celled syncoenobia; cells often irregularly arranged and either in contact or separate, enclosed within a mucilaginous envelope; chloroplast parietal,with a single pyrenoid; asexual reproduction by 4 autospores, released by disintegration of mother cell wall.

Europe; planktonic or associated with submerged surfaces in various types of water body.

Pseudotetrastrum punctatum Hindák 1977

Synonyms: *Tetrastrum punctatum* Ahlstrom *et* Tiffany, *T. multisetum* (Schmidle) Ahlstrom *et* Tiffany var. *punctatum* Brunnthaler

17760010 **Pl. 94H** (p. 389)

Coenobia 7–18(–20) μm wide; cells 2.5–7 μm in diameter, spherical to subspherical, with walls covered by brown warts that are denser along outermost margin; aplanospores 5 × 3 μm.

Europe; reported in the British Isles from the plankton of ponds and lakes including The Serpentine in Eaton Park, Chester (see below), rare.

The illustration of '*Siderocystopsis ornata*' (with autospores) from The Serpentine in Chester by Williams (1965, fig. 6b) is almost certainly the above species.

Quadricoccus Fott 1948

Synonym: *Tetratomococcus* Korshikov

Coenobia of groups of (2–)4 cells attached in pairs to tips of mother cell wall remnants, the repetition of autosporulation and release of daughter cells gives rise to 64-celled colonies, with older cell groups often separating, enclosed within a structureless mucilaginous envelope; cells ellipsoidal or broadly ovoid to cylindrical, with rounded apices, walls thin and smooth or covered by warts; chloroplast parietal, plate- or trough-shaped, with a single pyrenoid; asexual reproduction by (2–)4(–8) autospores, released by a longitudinal split of mother cell wall and remaining attached to persistent remnants of wall.

Probably cosmopolitan; planktonic and associated with submerged aquatic vegetation in still and slow-flowing water.

Quadricoccus verrucosus Fott 1948

Synonyms: *Tetratomococcus ornatus* Korshikov, *Quadricoccus laevis* Fott *sensu* Williams

17770010 **Pl. 84I** (p. 346) **CD**

Cells about 4 μm wide and 8–9 μm long, oval or cylindrical, with apices rounded and walls smooth or occasionally covered with dark brown granules.

Probably cosmopolitan; so far recorded in the British Isles only from the Serpentine Lake in Eaton Park, Chester, England.

The Chester material, described and illustrated by Williams (1968, as *Q. laevis*), is unusual since lacking pyrenoids, possessing highly refractive granules in a colourless area towards each apex, and exhibiting considerable variation in its wall ornamentation. The presence of brown warts was considered by Fott (1948) to be a character of sufficient importance for him to recognize a separate species, *Q. verrucosus*. It is possible that Williams's record is of two species (*Q. laevis, Q. verrucosus*) if the presence of a warty ornamentation (possibly due to iron oxide staining) is considered a character of sufficient importance to separate the species.

Quadrigula Printz 1915

Coenobia 4-, 8-, 16- or 32-celled, cells in groups of 4 with their long axes lying parallel to one another and all orientated in one direction, the daughter cells gradually moving apart laterally following gelatinization of mother cell wall, enclosed within a structureless mucilage; cells narrowly cylindrical to spindle-shaped, usually more than 7 times longer than broad, straight to slightly curved, abruptly narrowing to acute or sometimes rounded apices; chloroplast parietal, notched in centre, with or without a pyrenoid; asexual reproduction by (2–)4 or 8 autospores arranged parallel to long axis of cell, sometimes in two tiers each with 4 autospores, released by gelatinization of mother cell wall.

Probably cosmopolitan; planktonic and often in ponds and lakes, especially in marshy or boggy areas.

The pyrenoid is sometimes very difficult to distinguish, especially if not associated with starch.

1 Cells usually with broadly rounded apices .... *Q. pfitzeri*
1 Cells with acute or bluntly pointed apices 2
2 Cells always straight, narrowing gradually to acute apices .... *Q. closterioides*
2 Cells straight or slightly curved, cylindrical and narrowing abruptly to blunt apices .... *Q. korshikovii*

Quadrigula closterioides (Bohlin) Printz 1915

Basionym: *Nephrocytium closterioides* Bohlin
Synonym: *Ankistrodesmus closterioides* (Bohlin) Printz

17780010 **Pl. 98B** (p. 400)

Cells 2–6 μm wide, 20–30 μm long, spindle-shaped, straight, narrowing gradually to acute apices; chloroplast with an indistinct pyrenoid.

Probably cosmopolitan; not uncommon in the plankton of Scottish lochs, elsewhere in the British Isles planktonic and associated with submerged aquatic macrophytes especially in ponds, lakes and moats.

Quadrigula korshikovii Komárek 1979

Synonym: *Ankistrodesmus closterioides* (Printz) Korshikov *sensu* Korshikov

17780030

Cells 2–4 μm wide, 20–32 μm long, cylindrically spindle-shaped, straight or slightly curved, narrowing abruptly to short and blunt apices; chloroplast sometimes without a pyrenoid.

Europe; material collected by Williams (1965) from a small lake in Cheshire, described as '?*Quadrigula pfitzeri* (Schröder) G.M.Smith; ?*Ankistrodesmus closteroides* (Printz) Korshikov', is morphologically identical to this species.

Quadrigula pfitzeri (Schröder) G.M.Smith 1920

Basionym: *Raphidium pfitzeri* Schröder
Synonym: *Ankistrodesmus pfitzeri* (Schröder) G.S.West

17780020 **Pl. 98C** (p. 400)

Cells 1.7–3.5(–6) μm wide, 10–22.5(–45) μm long, spindle-shaped to almost cylindrical, straight or very slightly curved, gradually narrowing to broadly rounded apices; chloroplast without a pyrenoid.

Probably cosmopolitan; planktonic or associated with submerged surfaces in the British Isles where found in ponds and lakes including lochs in the Outer Hebrides.

DOUBTFUL SPECIES

Quadrigula chodatii (Tanner-Füllemann) G.M.Smith 1920 and ***Quadrigula lacustris*** (Chodat) G.M.Smith 1920

These species have each been recorded on only a single occasion in the British Isles (see Williams, 1950; Pearsall & Lind, 1942). Komárek & Fott (1983) consider them to be doubtful species and probably attributable to the genus *Pseudoquadrigula*. Neither species is known to produce autospores and Korshikov (1953) believes that *Quadrigula lacustris* may be identical with *Elakatothrix lacustris* Korshikov (Order Ulotrichales).

Radiococcus Schmidle 1902

Colonies micro- or macroscopic, consisting of randomly arranged groups of 2, 4 or 8(–16) cells, sometimes remnants of mother cell wall are present or ray-like striations surround each cell group, enclosed within a spherical, ellipsoidal, tetrahedral or irregularly-shaped mucilaginous envelope; cells spherical to broadly ellipsoidal; chloroplast parietal, cup-shaped and with a single basal pyrenoid; asexual reproduction by 4 or 8(–16) autospores, released by fracture of the mother cell wall whose remnants persist in some species.

Probably cosmopolitan; planktonic or epiphytic on aquatic macrophytes, most frequent in bodies of standing water.

Considered a rather ill-defined genus in need of taxonomic revision (see Hindák, 1988).

1 Cells 6–9(–15) µm in diameter, in groups surrounded by ray-like striations in the mucilage ....*R. nimbatus*
1 Cells 4–7 µm in diameter, in groups often associated with remnants of mother cell wall (sometimes remnants 'cap-like'), without striations in mucilage....................*R. planktonicus*

Radiococcus nimbatus (De Wildeman) Schmidle 1902

Basionym: *Pleurococcus nimbatus* De Wildeman

17790010 **Pl. 86J** (p. 352)

Colonies of randomly arranged 4- or 8-celled groups surrounded by distinctive ray-like striations, mucilaginous envelope usually spherical; cells 6–9(–15) µm in diameter, spherical.

Probably cosmopolitan; widely distributed in the British Isles where sometimes planktonic but most commonly attached to surfaces, frequently on the undersurface of water lily leaves in still water.

Radiococcus planktonicus J.W.G.Lund 1956

17790020 **Pl. 86I** (p. 352)

Colonies most commonly of cells grouped in tetrads, more rarely single-celled or as 2- or 8-celled groups, sometimes persistent remnants of the mother cell wall associated with cell groups (sometimes remnants cap-like on cells), with a structureless mucilaginous envelope initially almost spherical but later becoming irregular and of considerable size (up to 250 µm across); cells 4–7 µm in diameter, spherical.

Probably cosmopolitan; sometimes abundant in the plankton of Windermere and other lakes in the English Lake District, probably widely distributed in the British Isles but easily confused with other species (see comments below).

The alga is easily confused with other planktonic algae whose cells are grouped within what on superficial examination appears to be a structureless mucilaginous envelope (e.g. *Sphaerocystis, Coenococcus, Planktosphaeria*). It is separated from such algae by having its cells grouped in tetrads, mother cell wall remnants associated with young daughter cells, and the colonies eventually losing their spherical shape and reaching a considerable size.

Raphidocelis Hindák emend. Marvan et al. 1984

Cells solitary or more usually forming coenobia of 4, 8 or 16(–64) randomly arranged cells, enclosed within a homogeneous and often indistinct mucilaginous envelope; cells spindle-shaped, halfmoon- to horseshoe-shaped, elongate-cylindrical to variously curved, spiralled and contorted, with apices acute, rounded or somewhat swollen and capitate, walls relatively thick, smooth, colourless or brown and sometimes granular; chloroplast parietal, often filling most of cell, without a pyrenoid; asexual reproduction by 2, 4 or 8 autospores, released by fracture of mother cell wall.

Probably cosmopolitan; planktonic or sometimes loosely associated with aquatic macrophytes growing in the marginal shallows of a wide range of different water bodies.

Development of granules is considered a secondary character and dependent on environmental conditions and age (e.g. walls of young autospores always smooth). Cell morphology is the principal feature used to distinguish species (see Komárek, 1979; Marvan et al., 1984).

1 Cells each forming a flat spiral of 1 to 1.5 turns and therefore somewhat ring-shaped ....*R. rotunda*
1 Cells of another form 2
2 Cell apices usually somewhat inflated and broadly rounded (capitate) ....*R. subcapitata*
2 Cell apices usually broadly rounded but never inflated 3
3 Cells 1.8–5 µm wide and 3–6 times longer than broad, not usually contorted; remnants of mother cell wall in mucilage ....*R. danubiana*
3 Cells 0.6–2 µm wide and 6–10 times longer than broad, spirally curved to variously contorted; without remnants of mother cell wall in mucilage ....*R. contorta*

Raphidocelis contorta (Schmidle) Marvan et al. 1984

Basionym: *Kirchneriella obesa* var. *contorta* Schmidle

Synonyms: *Kirchneriella contorta* (Schmidle) Bohlin, *Kirchneria contorta* (Schmidle) Hindák, *Kirchneriella gracillima* Bohlin

17800010 **Pl. 98K** (p. 400)

Cells 0.6–2(–3) µm wide, (7–)8–14(–16) µm long, usually 6–10 times longer than broad, narrowly cylindrical, halfmoon-shaped or semicircularly curved and spirally or irregularly twisted and much contorted, with broadly rounded apices; autospores (2–)4 or 8 in each sporangium, the mother cell wall not persisting after release.

Probably cosmopolitan; in the British Isles found only on rare occasions in river plankton (e.g. River Thames), but probably more widely distributed than records suggest.

Raphidocelis danubiana (Hindák) Marvan et al. 1984

Synonym: *Kirchneriella danubiana* Hindák

17800020 **Pl. 83G** (p. 341)

Cells (1.4–)1.8–5 µm wide, 6–18(–21) µm long, 3–6 times longer than broad, halfmoon- to horseshoe-shaped or somewhat sausage-shaped (circumscribing about two-thirds of a circle), of almost equal thickness throughout, slightly narrowing to broadly rounded apices; autospores 2 or 4 in each mother cell, remnants of mother cell wall often persisting after release.

Probably cosmopolitan; rarely recorded in the British Isles where planktonic in ponds, lakes and slow-flowing rivers.

Raphidocelis rotunda (Korshikov) Marvan et al. 1984

Basionym: *Ankistrodesmus rotundus* Korschikov
Synonyms: *Kirchneria rotunda* (Korshikov) Hindák, *Kirchneriella rotunda* (Korshikov) Hindák

17800030 **Pl. 83E** (p. 341)

Cells 2.5–4.5(–5.5) µm wide, (15–)20–43(–50) µm long, less than 10 times longer than broad, usually narrowing regularly to broadly rounded apices, sometimes with one end larger than the other, forming a flat spiral of 1 to 1.5 turns and therefore ring-like with apices overlapping; autospores 2 or 4 in each mother cell, with remnants of mother cell wall not persisting after release.

Europe; associated with aquatic macrophytes in rivers and still waters in the British Isles, rare but probably more widely distributed than the few records suggest.

Raphidocelis subcapitata (Korshikov) Nygaard et al. 1986

Basionym: *Ankistrodesmus subcapitatus* Korshikov
Synonyms: *Kirchneria subcapitata* (Korshikov) Hindák, *Kirchneriella subcapitata* Korshikov, *Selenastrum capricornutum* Printz *sensu* Skulberg

17800040 **Pl. 83F** (p. 341)

Cells 1.2–3.2(–5) µm wide, (7–)10–20(–23) µm long, narrowly cylindrical, horseshoe-shaped or slightly spirally contorted, usually less than 10 times longer than broad, equally thick throughout, with apices widely rounded and often slightly inflated and obtuse (capitate), sometimes having the wall thickened at one or both apices; autospores (2–)4 or 8 in each mother cell and its walls not persisting after release.

Probably cosmopolitan; frequent in the British Isles where planktonic or associated with aquatic macrophytes in a range of moderately to very nutrient-riched habitats, including water tanks, ditches, ponds and lakes.

'Selenastrum capricornutum' is one of the most widely used bioassay test algae found in various culture collections. According to Nygaard et al. (1986), this alga is not a *Selenastrum* but rather *R. subcapitata* (see comments under *Kirchneriella*).

Rhodoplax Schmidle *et* Wellheim 1901

Thallus initially green, becoming red or orange in colour with age, consisting of a layer of irregularly arranged cells, closely packed and forming a parenchyma-like membrane enclosed within a thick mucilage; cells ellipsoidal, 10–18 m in diameter; chloroplast cup-shaped, with a basal pyrenoid; reproduction by autospores and biflagellate zoospores.

Europe; on rocks in cataracts and fast-flowing stretches of rivers. Monospecific.

For a detailed account of the genus, see Jaag (1932).

Rhodoplax schinzii Schmidle *et* Wellheim 1901

18080010 **Pl. 86O** (p. 352)

Description as for genus.

Europe; in the British Isles only known from tributaries of the River Tweed in north-east England (Holmes & Whitton, 1975b) and the River Caragh in Ireland (Heuff & Horken, 1984) where it forms crimson patches on boulders, often blood red when on smaller boulders in stretches with torrential currents, particularly associated with waterfalls.

Scenedesmus Meyen 1829

Coenobia flat, straight or slightly curved, with cells in 1 or 2 rows (rarely 3), usually 2-, 4- or 8-celled, more rarely 16- or 32-celled, joined laterally and lying parallel to each other (linearly arranged), or alternately arranged and joined towards ends; cells elongate or cylindrical, ovoid, ellipsoidal to ovoid, with apices rounded, truncate, capitate, or narrowing to an obtuse point, often bearing long spines or teeth; cell wall smooth, granular, spiny or toothed, with nipple-like projections or ribs present and often in various combinations; chloroplast parietal, with a single pyrenoid; asexual reproduction by autospores, released by fracture of lateral cell wall.

Cosmopolitan; planktonic or associated with sediments or aquatic vegetation, usually in standing and slow-flowing water.

One of the commonest and most taxonomically diverse of the chlorococcalean genera in which over 200 species and almost 1200 infraspecific taxa have been recognized (see review by Hegewald & Silva,1988). Trainor and co-workers (see Trainor, 1992; Trainor & Egan, 1991) consider that the great majority of these species are environmentally-induced forms and the true number probably ranges from 15–30 species. It is now recognized that characters associated with cell wall ultrastructure can assist in separating species that are traditionally defined on morphological characters observed with the light microscope (e.g. *'S. quadricauda'*, *S. abundans*). Such cell wall features have been used to prove the independent existence of other morphologically defined species (e.g. *S. verrucosus, S. subspicatus*). Species continue to be described (e.g. Hegewald et al., 1988, 1990, 1994, 1998; Hindák, 1995), many having been studied in culture and the fine structural details of their cell walls determined (see Komárek & Fott, 1983). It is of our opinion that to construct a more acceptable taxonomic framework for the genus requires data from a variety of sources, especially important are molecular data (see comments by Kessler et al., 1997; Hegewald & Hanagata, 2000).

Molecular data have lent support to the division of the genus into two subgenera or genera and to the merging of these with other genera. An et al. (1999) concluded, based on ITS-2 rDNA sequence analysis, that *Scenedesmus* and *Scenedesmus*-like taxa (e.g. *Arthrodesmus*) should be divided into two separate genera, *Scenedesmus sensu stricto* and *Desmodesmus*. In a later paper Hegewald (2000) transferred a further 32 species and 22 varieties to the genus *Desmodesmus* from *Scenedesmus* subgenus *Desmodesmus*. Of the 42 species recorded from the British Isles the following 21 have been transferred to the genus *Desmodesmus* (including 3 varieties): *S. abundans, S. armatus, S. armatus* var. *bicaudatus, S. asymmetricus, S. brasiliensis, S. communis, S. costato-granulatus, S. dispar, S. flavescens, S. grahneisii, S. hystrix, S. insignis, S. intermedius, S. kissii, S. maximus, S. opoliensis, S. opoliensis* vars. *carinatus* and *mononensis, S. pannonicus, S. protuberans* (Hegewald, 2000); *S. subspicatus* (Hegewald and A.Schmidt in Hegewald, 2000); *S. arthrodesmus, S. denticulatus, S. serratus* (An et al., 1999). Five British species of *Scenedesmus* have transferred to the new genus *Acutodesmus*: *S. acuminatus, S. dimorphus, S. incrassatulus, S. obliquus, S. pectinatus* (Tsarenko, 2000; Tsarenko & Petlevanyi, 2001). We have chosen to continue to recognize *Scenedesmus* as a single genus in the belief that further molecular analysis and ultrastructural observations might well lead to further taxonomic or nomenclatural changes, possibly the recognition of further genera and subgenera.

This is one of the least satisfactory genera treated here and yet many of the problems associated with it are to be found in most morphologically simple taxa. One problem concerns that of interpretation with the concept of a species often changing over time. We have followed Hegewald (1987) and accept some of the changes proposed by Komárek & Fott (1983), but not those concerning infrageneric taxa. It is considered premature to evaluate fully and include here taxonomic conclusions based upon the new molecular data (see above).

We recognize 42 species in the British Isles. To identify these 'species' using the light microscope, it is necessary to give special attention to the arrangement of cells forming a coenobium, cell morphology (particularly of marginal cells), wall ornamentation, presence or absence of a mucilaginous envelope, whether mucilage forms a connective 'covering' between cells, and the presence of small spines or teeth on apices and sides of cells (more important than number of projections). It is possible that in nature unicells believed to belong to *Lagerheimia* ('*Chodatella*-stage') might be more correctly attributed to a *Scenedesmus* which has failed to form colonies, a condition well-known when clones of *Scenedesmus* are grown in laboratory culture. We are grateful to Dr E. Hegewald for critically commenting on the artificial key to the species. All identifications arrived at using the key need to be regarded as provisional and should be checked against the illustrations.

1 Cells spindle-shaped, ovoid to ovoid-cylindrical, without spines or horns; wall smooth, without ornamentation or sometimes with thin longitudinal ridges formed by wrinkling of hemicellulose layer....2
1 Cells ovoid to cylindrical, with or without long spines or horns; wall smooth (under light microscope) to granular, sometimes with small spines, teeth or ribs in various combinations 14
2 Cells spindle-shaped to ellipsoidal, straight or curved, sometimes tapering to acute and curved apices; wall smooth or with thin longitudinal ridges....3
2 Cells ovoid, cylindrical or obovoid, straight or curved, with broadly rounded and sometimes slightly thickened apices; walls smooth, without ridges; coenobium surrounded by a mucilaginous envelope....10
3 Cell wall completely smooth 4
3 Cell wall bearing 1 or several longitudinal ridges....9
4 Coenobia formed of a single row of cells arranged in a linear or alternating series; cells straight and spindle-shaped....*S. obliquus*
4 Coenobia formed of 1 or 2 rows of cells in linear or zigzag series; all cells or only marginal cells arc-shaped 5
5 Coenobia zigzag in shape; cells united by their apices to the median or apical portion of adjacent cell....*S. bernardii*
5 Coenobia of a single row of linear or alternately arranged cells, or cells in 2 rows; cells all connected laterally 6
6 Cells asymmetrically spindle-shaped, most apices nipple-shaped and marginal cells inwardly curved (concave)....*S. raciborskii*
6 Cells regularly spindle-shaped, arc-shaped or curved, without nipple-shaped apices....7
7 Cells broadly spindle-shaped, only subapical part of marginal cells curved, inner cells straight, linear to alternately arranged ....*S. dimorphus*
7 Cells narrowly spindle-shaped, sickle-shaped, arc-shaped to crescent-shaped, outermost side of marginal cells convex, only internal cells with subapical part curved (sometimes straight) 8
8 Coenobia flat, linear or 8-celled coenobia of regularly alternating cells in 2 rows; marginal cells crescent- or arc-shaped, internal cells straight or only slightly curved, without a mucilaginous envelope....*S. falcatus*
8 Coenobia curved, a single twisted row of cells; cells all sickle-shaped, enclosed in a thin mucilaginous envelope ....*S. acuminatus*
9 Coenobia linear, or cells only slightly alternating; cells broadly spindle-shaped, with 1–3(–4) longitudinal ridges and without nipple-like prominences ....*S. acutiformis*
9 Coenobia of markedly alternating cells arranged in two rows; cells oval, with 6–10 longitudinal ridges and nipple-like prominences at apices....*S. costatus*
10 Cells ovoid, with widely rounded apices and each terminating in a nipple-shaped polar thickening....*S. apiculatus*
10 Cells ellipsoidal-cylindrical to broadly ovoid, apices broadly rounded, without any thickening or ornamentation 11
11 Coenobia of a single and usually linear row of cells; cells ellipsoidal to cylindrical 2
11 Coenobia of 2 linear rows of cells, or of a single row of alternating cells; cells oval to almost cylindrical 13
12 Coenobia of 2 cells, outermost wall markedly convex ....*S. planktonicus*
12 Coenobia of 2, 4 or 8(–32) cells, outermost wall not markedly convex....*S. ellipticus*
13 Coenobia of markedly alternating cells, almost united subapically, without perforations....*S. obtusus*
13 Coenobia of almost linear or subalternating cells in 4-celled coenobia, usually in 2 rows, plate-like and curved, with or without perforations along sides of cells or forming only proximally 2 rows of perforations at apices ....*S. arcuatus*
14 Cells with or without small spines/teeth on rounded apices; cell walls granular, rarely smooth, with small teeth or granules variously dispersed or in groups....15
14 Cells with short or long spines or a single massive individual spine; cell walls smooth or with grouped or variously dispersed granules/small spines and wing-shaped ribs or other surface structures....24
15 Cell wall without spines or teeth (except occasionally *S. costato-granulatus*) 16
15 Cell wall with small spines or teeth regularly

or irregularly distributed all over surface (except very occasionally *S. serratus*) 19

16 Cells ellipsoidal to almost cylindrical; granules on marginal cells arranged in shape of the letter 'C' and as a wavy line on any internal cells; coenobia often only 2-celled ....*S. grahneisii*

16 Cells ovoid to obovoid or slightly cylindrical; granules uniformly or randomly distributed, or in discontinuous or interrupted rows; coenobia 2-, 4-, or 8-celled 17

17 Cell wall not brown-tinged, sometimes with small teeth at apices; coenobia usually 2-celled and in a single row ....*S. costato-granulatus*

17 Cell wall brownish, without teeth; coenobia 4- or 8-celled and cells in 1 or 2 rows 18

18 Coenobium a single linear row of cells, granules not infrequently in longitudinal rows ....*S. granulatus*

18 Coenobium in 2 rows of alternating cells, granules not regularly distributed ......*S. verrucosus*

19 Cell wall uniformly covered by small teeth, spines always absent ....*S. hystrix*

19 Cell wall smooth (viewed with light microscope) or granular, granules arranged in longitudinal rows and sometimes fused to form rib-like structures, or present as a row of small teeth, with small spines at apices 20

20 Coenobia of alternately or linearly arranged cells; cells broadly ovoid to cylindrical, with 1–4 teeth on rounded or slightly truncate apices, sometimes with longitudinal rows of granules....21

20 Coenobia of linear or feebly alternately arranged cells; cells elongate-ovoid to cylindrical, with short spines/teeth arranged in longitudinal rows on rounded apices, or teeth or granules in longitudinal rows or forming ribs along sides of cells 22

21 Coenobia of two cell rows or of alternating cells; cells broadly ovoid, with teeth confined to outermost edge of each apex ....*S. denticulatus*

21 Coenobia a single linear row of cells; cells cylindrical, sometimes gradually tapering to a rounded or slightly truncate apex bearing 1–4 teeth, with or without granules, if granules present then sometimes in a longitudinal row ....*S. aculeolatus*

22 Cells elongate-ovoid to cylindrical or asymmetrically oval (marginal cells), with 1–4 teeth on apex, sometimes short spines confined to apices of marginal cells and longitudinal ribs on sides of each cell ....*S. brasiliensis*

22 Cells elongate-ovoid/ellipsoidal, with 1–4 teeth or small spines at apices and longitudinal rows of small teeth/granules on sides 23

23 Single row of granules on each cell ....*S. arthrodesmiformis*

23 Single row of teeth on inner cells but confined only to outer convex wall of marginal cells ....*S. serratus*

24 Coenobia of alternately or linearly arranged cells; cells oval, cylindrical to spindle-shaped, marginal cells having straight or convex sides; spines diagonally or linearly-symmetrically arranged; wall smooth or with granules, teeth, or ribs...25

24 Coenobia of linear rows of cells; cells elongate-ellipsoidal to cylindrically-ovoid, marginal cells having diagonally or linearly-symmetrically arranged spines on convex side, with additional spines of same length or shorter on some/all cells or only on sides of marginal cells; cell wall smooth or punctate, rarely with thin rib-shaped lines (visible in light microscope), punctae, granules or 'rosettes' 38

25 Cells with longitudinal rows of granules, sometimes small obtuse teeth on marginal cells, usually smaller than those on internal cells, with spines no longer than width of cells (except var. *bicaudatus*), sometimes only teeth present ....*S. circumfusus*

25 Cells with longitudinal ribs or rows of granules, with a smooth cell wall without teeth, spines shorter or longer than cell width 26

26 Spines tooth-like and of similar length; cell wall smooth 27

26 Spines of different lengths (main spines very long, supplementary ones shorter); cell wall smooth, granular or ribbed..................................29

27 Cells elongate-ovoid to cylindrical, marginal cell with an almost straight outermost side; spines short, arising on rounded or polygonal apices...*S. dispar*

27 Cells ellipsoidal to broadly ovoid, all apices rounded; spines tooth-shaped and usually on apices of each cell, with or without 1 or 2 spines on outermost wall of marginal cells..........28

28 Cells ellipsoidal to broadly ovoid and linearly arranged, marginal cells having 1–3 spines on outermost side of marginal cells ....*S. subspicatus*, in part

28 Cells broadly ovoid and linearly or alternately arranged, spines marginal and confined to apices ....*S. kissii*

29 Coenobia of alternating to almost linearly arranged cells; spines on apices of marginal cells, sometimes apices of inner cells with spines..30

29 Coenobia of linearly arranged cells (slightly alternating in *S. armatus*); spines variously arranged, linear or diagonal and symmetrical 31

30 Coenobia straight; cells ovoid, with delicate, curved or straight spines with main spines on outermost edge of apices of marginal cells ....*S. intermedius*

30 Coenobia curved; cells cylindrical-ovoid or ellipsoidal, with stout and almost straight spines on one apex of each inner cell ....*S. pannonicus*

31 Cells with diagonally-symmetrically arranged

spines, except for *S. insignis* which may have spines on each apex 32

31 Cells with linear-symmetrically arranged spines..35

32 Cells elongate-ovoid to cylindrical; cell wall smooth, without decoration ....*S. bicaudatus*

32 Cells elongate-ellipsoidal, ovoid to cylindrical; cell wall smooth or with ridges, rows or evenly dispersed granules, sometimes with rows of large granules 33

33 Cell wall evenly granular, sometimes with rows of larger granules; cells elongate-ovoid to nearly cylindrical, sometimes ovoid, with small spines or teeth on all the apices, or only on 2 apices ....*S. insignis*

33 Cell wall smooth, with longitudinal ridges or only rows of granules 34

34 Cells elongate-ellipsoidal to cylindrical, apices rounded, sometimes with a single tooth at apex; cell wall bearing rows of granules or 1–2 longitudinal ridges of granules ....*S. semipulcher*

34 Cells elongate-cylindrical, apices broadly-rounded, with 1–2 short teeth on apices of internal cells; cell wall bearing longitudinal lines or crested ridges ....*S. armatus*, in part

35 Cells elongate-ovoid to cylindrical, with small spines or teeth on rounded apices.........................36

35 Cells ellipsoidal to spindle-shaped, with obtusely pointed or stretched out capitate apices..37

36 Cells ovoid to ovoid-cylindrical and wall smooth; apices with 1–3 teeth or 1–2 short spines..................................*S. caudato-aculeolatus*

36 Cells elongate-cylindrical and wall with longitudinal, continuous or interrupted lines or crest-shaped ridges; apices of internal cells bearing 1–2 short teeth ....*S. armatus*, in part

37 Marginal cells with obtusely pointed or often truncate apices; cell wall sometimes bearing thin ridges, small teeth, small nipples or spines ....*S. opoliensis*

37 Marginal cells slightly narrowing towards capitate apices, with main spines on straight or curved apices................................*S. protuberans*

38 Cells ellipsoidal to cylindrical, with rounded apices, outer sides of marginal cells with 1–3 additional spines (as well as main apical spines); cell wall smooth...39

38 Cells cylindrically-ovoid, with rounded or truncated apices, outer side of marginal cells without additional spines; cell wall smooth or with punctae, sometimes with granules in the form of 'rosettes' 42

39 Cells elongate-ellipsoidal to cylindrical, with a single spine in the middle (only on one side of coenobium) of each marginal cell; inner cells without spines...............................*S. asymmetricus*

39 Cells ellipsoidal to cylindrical or broadly ovoid, with 1–3 spines in middle part of outermost margin; inner cells with or without apical spines..40

40 Marginal cells with 1–3 spines in middle part and a single, short, obliquely positioned spine on apex; inner cells frequently having a single apical tooth or spine ....*S. abundans*

40 Marginal cells usually with 2–3 spines in middle part, sometimes in different planes; inner cells with 1–2 apical spines...41

41 Marginal cells with 2 spines towards middle and shorter than apical spines; inner cells often with 1–2 apical spines..................................*S. flavescens*

41 Marginal cells with 1–3 spines (sometimes more) arranged in middle and similar in length to apical spines; inner cells often with a single apical spine...........................*S. subspicatus*, in part

42 Cell wall with granules in form of 'rosettes', rarely with longitudinal ridges or smooth, with additional spines sometimes of similar length to main spines and spines arise on apices of inner cells, thin layer of mucilage around cells; cells >17 μm long..................................*S. magnus*

42 Cell wall smooth, without additional long spines on lateral sides of cells, with or without small spines or teeth at cell apices; mucilage layer absent 43

43 Cells elongate-cylindrical, usually with mucilage confined to area between apices of inner cells and base of long spines (spines usually longer than cell width), with or without teeth on apices of inner cells ....*S. communis*

43 Cells oval-cylindrical, without mucilage between apices of inner cells and base of short spines (spines equal to or less than half cell width), without teeth at apices of inner cells ....*S. microspina*

Scenedesmus abundans (Kirchner) Chodat 1913

Basionym: *Scenedesmus caudatus* fo. *abundans* Kirchner
Synonyms: *Scenedesmus quadricauda* var. *abundans* (Kirchner) Hansgirg, *S. sempervirens* Chodat, *Chlorella fusca* Shihara *et* Krauss

17810010 **Pl. 95D** (p. 391)

Coenobia of 2 or 4(–8) linearly or alternately arranged cells; cells (2–)4–6(–8) μm wide and (5–)11–15(–20) μm long, ellipsoidal to cylindrical, attached along two-thirds of length, slightly tapering and rounded at apices, not infrequently with a single spine or tooth on apex of each inner cell, a single main spine slightly obliquely positioned on apex of each marginal cell and 1–3 spines in middle part of outer convex wall.

Probably cosmopolitan; in the British Isles reasonably common in the plankton of ponds and lakes, more rarely recorded in rivers.

Similar to *Scenedesmus subspicatus* Chodat *pro parte* from which it is separated by the absence of teeth on the marginal cells, the presence of a spine in the middle of outer side of each marginal cell, and by a very different cell wall ultrastructure.

Scenedesmus aculeolatus Reinsch 1877

Synonym: *Scenedesmus denticulatus* var. *linearis* Hansgirg
17810020 **Pl. 95J** (p. 391)
Coenobia a single row of 2, 4 or 8 linearly arranged cells; cells 2.5–6 μm wide, 9–14(–21) μm long, ovoid-cylindrical to cylindrical, tightly packed, with outermost side of each marginal cell almost straight or slightly convex, apices rounded or more rarely slightly truncate and having 1–3 teeth on each apex; cell walls smooth.

Probably cosmopolitan; in the British Isles known only from a moat and lakes in south-east England.

Scenedesmus acuminatus (Lagerheim) Chodat 1902

Basionym: *Selenastrum acuminatum* Lagerheim
Synonym: *Scenedesmus acuminatus* var. *elongatus* G.M.Smith
17810030 **Pl. 97A** (p. 396) **CD**
Coenobia a single row of 2 or 4(–8) linearly to slightly alternately arranged cells, curved to semi-circular or contorted, often surrounded by mucilage; cells 2–5.5 μm wide, 9.2–34.5(–40) μm long, spindle-shaped, arc-like to sigmoid, with the outermost side concave, gradually tapering to extended apices.

Probably cosmopolitan; one of the most common planktonic species in the British Isles, where present in very different aquatic habitats including slow-flowing rivers, animal drinking troughs, ponds, lakes, reservoirs, moats and canals.

Very similar to *Scenedesmus falcatus* Chodat with which it might easily be confused, except that *S. acuminatus* has a distinctly contorted coenobium. Form of the coenobium (straight or curved) is a good diagnostic feature for distinguishing species since it remains relatively constant and unchanging during growth and development (ontogenesis). Unfortunately, the form of the coenobium is often not taken into account and is seldom shown in illustrations.

Scenedesmus acutiformis Schröder 1897

17810040 **Pl. 96C** (p. 394)
Coenobia of 2, 4 or 8 linearly or slightly alternately arranged cells; cells (3–)3.5–8 μm wide and (9–)11–21 (–22.4) μm long, broadly spindle-shaped to narrowly ellipsoidal, bearing on convex side 1–3(–4) longitudinal ridges of thickened wall material, tapering to obtuse or rounded apices.

Probably cosmopolitan; recorded on only a few occasions from the British Isles in ponds, lakes (e.g. Loch Astra, Shetlands) and slow-flowing rivers.

var. *costatus* (Huber-Pestalozzi) Pankow 1986
Basionym: *Scenedesmus bijugatus* var. *costatus* Huber-Pestalozzi
Synonym: *Scenedesmus acutiformis* var. *tricostatus* Schröder
17810042
Coenobia of 4 narrowly ellipsoidal cells; cells 3–4 μm wide and 9–11 μm long, slightly narrowing to rounded apices and with 3 longitudinal ridges.

Asia, Europe; in the plankton or associated with submerged surfaces in lakes and bogs in the British Isles.

A very variable species with data absent on variability of cell/coenobium shape and other morphological characters. It is possible that *S. acutiformis* might represent a 'species complex' and may need to be divided into several infraspecific taxa. It was transferred to the genus *Enallax* by Hindák (1990) and this has found support from the phylogenetic analysis of molecular data (18S-RNA) carried out by Hegewald & Hanagata (2000).

Scenedesmus apiculatus (West *et* G.S.West) Chodat 1926

Basionym: *Scenedesmus alternans* var. *apiculatus* West *et* G.S.West
17810060 **Pl. 96J** (p. 394)
Coenobia a single row of 4 or 8 linearly or alternately arranged cells; cells 4–9.4 μm wide, (7.5–)9.5–17(–18.8) μm long, oval, ovoid to cylindrical, connected together for a short distance (to one-third length) or sometimes to more than half length, apices broadly rounded and having nipple-like thickenings only on protruding apices of alternately arranged cells, on both apices of linearly arranged cells.

Probably cosmopolitan; in the British Isles in the plankton of rivers, ponds, lakes and temporary water bodies, rare.

Scenedesmus arcuatus (Lemmermann) Lemmermann 1899

Basionym: *Scenedesmus bijugatus* Kützing var. *arcuatus* Lemmermann
Synonym: *Scenedesmus bijugatus* fo. *arcuatus* (Lemmermann) West *et* G.S.West
17810070 **Pl. 97C** (p. 396) **CD**
Coenobia of 4, 8 or 16 cells in 2 rows, sometimes in 3 rows or only inner cells lying above each other, when 2- or 3-rowed then evident spaces or perforations between sides or only between apices of cells, enclosed within a mucilaginous envelope; cells (2–)3–9.5 μm wide, 7–18 μm long, ovoid to slightly cylindrical, slightly curved, with broadly rounded apices, outermost side of marginal cells usually convex.

Probably cosmopolitan; in the British Isles frequent in the plankton of slow-flowing reaches of rivers, such as the Thames and Great Ouse, also often observed in ponds and lakes in the English Midlands.

Scenedesmus armatus (Chodat) Chodat 1913

Basionym: *Scenedesmus hystrix* Chodat var. *armatus* Chodat
Synonyms: *Scenedesmus helveticus* Chodat, *S. columnatus* Hortobágyi
17810080 **Pl. 94I** (p. 389) **CD**
Coenobia of 2 or 4(–8) linearly to slightly alternately arranged cells; cells 3–8(–9.5) μm wide, 7–17(–26.5) μm long, long-cylindrical or ovoid-cylindrical, narrowing to rounded apices, only inner cells frequently bearing small teeth, longitudinal ridges continuous or interrupted and surrounding cell, longitudinal ridges confined to middle portion, sometimes as continuous wing-shaped structures on outermost convex side of marginal cells, main spine on outermost side of apices of marginal cells, or only one on each marginal cell and diagonally symmetrical, with outermost wall of marginal cells straight or slightly convex.

Probably cosmopolitan; in the British Isles planktonic in various water bodies (e.g. the River Wye); from mildly to moderately nutrient-enriched water.

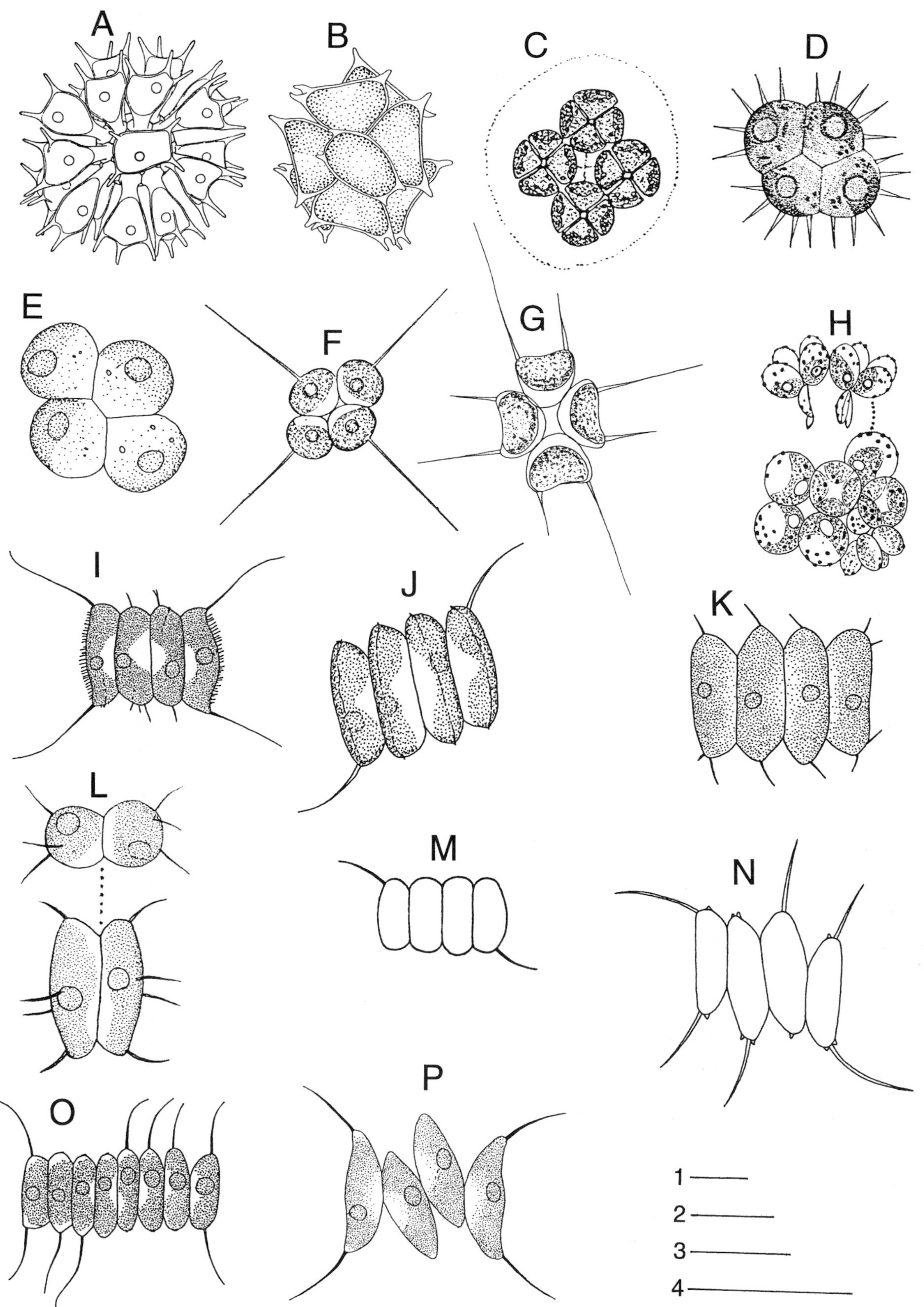

Plate 94 A–P

A. *Sorastrum americanum* (p. 401); **B**. *Sorastrum spinulosum* (p. 402); **C**. *Tetrastrum komarekii* (p. 406); **D**. *Tetrastrum staurogeniaeforme* (p. 406); **E**. *Tetrastrum triangulare* (p. 406); **F**. *Tetrastrum elegans* (p. 406); **G**. *Tetrastrum heteracanthum* (p. 406); **H**. *Pseudotetrastrum punctatum* (p. 381); **I**. *Scenedesmus armatus* (p. 388); **J**. *Scenedesmus armatus* var. *bicaudatus* (p. 390); **K**. *Scenedesmus dispar* (p. 393); **L**. *Scenedesmus flavescens* (p. 393): two cells in polar view (top); **M**. *Scenedesmus bicaudatus* (p. 390); **N**. *Scenedesmus caudato-aculeolatus* (p. 390); **O**. *Scenedesmus magnus* (p. 395); **P**. *Scenedesmus opoliensis* (p. 397).

Scale bars: 1, 3, 4, 10 μm; 2, 20 μm. 1 for B, P; 2 for H, O; 3 for A, C, F, G, I–N; 4 for D, E.

var. *bicaudatus* (Guglielmetti) Chodat 1926

Basionym: *Scenedesmus acutiformis* var. *bicaudatus* Guglielmetti

Synonyms: *Scenedesmus semicristatus* Uherkovich, *S. columnatus* var. *bicaudatus* Hortobágyi

17810082 **Pl. 94J** (p. 389)

Cells 3–6 μm wide, 8–15 μm long, with one main spine on apex of each marginal cell, diagonally symmetrical.

Probably cosmopolitan; only rarely recorded in the British Isles from ponds and lakes (e.g. Chartwell Lake, Kent).

Scenedesmus arthrodesmiformis Schröder 1920

17810090 **CD**

Coenobia a row of 2(–4) cells; cells 10 μm wide, 12 μm long, elongate-ellipsoidal, apices rounded and each bearing an inwardly directed tooth or short spine, with a single row of granules on side of each cell.

Europe; so far recorded in the British Isles only from a single pond in Gillingham, Kent, containing koi carp.

The material from England is unusual in consisting of 4-celled coenobia.

Scenedesmus asymmetricus (Schröder) Chodat 1902

Basionym: *Scenedesmus quadricauda* var. *asymmetricus* Schröder

17810100 **Pl. 95F** (p. 391)

Coenobia a row of 4 linearly arranged cells; cells 2.5–4.5 μm wide, 12–15 μm long, elongate-ellipsoidal to cylindrical, with rounded apices, marginal cells having a slightly convex outer wall and bearing a long curved spine on each apex, with an additional spine in middle of each cell only on one side of coenobia.

Europe; recorded in the British Isles on very few occasions in ponds.

Scenedesmus bernardii G.M.Smith 1916

Synonym: *Scenedesmus acuminatus* var. *bernardii* (G.M.Smith) Dedusenko

17810110 **Pl. 96P** (p. 394)

Coenobia of (2–)4 or 8(–16) cells in a zigzag, alternate arrangement, each cell united by its apex to the subapical or median portion of adjacent cell; cells (1.8–)3–7(–7.3) μm wide, 7–40(–48) μm long, spindle-shaped to arc-like, tapering to acute or sometimes slightly obtuse apices.

Probably cosmopolitan; rare in the British Isles where found in the plankton of ponds, lakes and rivers.

Scenedesmus bicaudatus Dedusenko 1925

Synonyms: *Scenedesmus quadricauda* var. *bicaudata* Hansgirg, *S. bicaudatus* (Hansgirg) Chodat

17810120 **Pl. 94M** (p. 389)

Coenobia of 2 or 4(–8) linearly arranged, or sometimes slightly alternately arranged, tightly packed cells; cells (2–)3–5.4 μm wide, 7.8–12.3(–14.5) μm long, long-ellipsoidal to ovoid-cylindrical, with rounded apices and a diagonally symmetrical spine on one apex of each marginal cell, walls smooth and without ornamentation.

Probably cosmopolitan; recorded in the British Isles in the plankton of lakes and moats.

Sometimes a small tooth may be present at cell apex.

Scenedesmus brasiliensis Bohlin 1897

Synonym: *Scenedesmus brasiliensis* var. *cinnamomeus* Y.V.Roll

17810130 **Pl. 96E** (p. 394) **CD**

Coenobia of 2, 4 or 8 linearly arranged cells; cells 2–8(–10) μm wide, (8–)9–28 μm long, narrowly ovoid to cylindrical, with 1–4 longitudinal and often continuous ridges on sides of each cell, with apices slightly tapering, rounded and bearing 1–4 teeth of equal length (some forms, not recorded in the British Isles, have a single spine on one apex of each marginal cell, usually diagonally symmetrical), walls colourless to dark brown.

Probably cosmopolitan; in the British Isles in the plankton of ponds, lakes and gravel pits; considered to range from mildly to moderately nutrient-enriched waters.

Scenedesmus caudato-aculeolatus Chodat 1926

Synonym: *Scenedemus quadricauda* var. *dentatus* Dedusenko

17810140 **Pl. 94N** (p. 389)

Coenobia of (2–)4 or 8 linearly arranged cells; cells 2.4–7.2(–9.6) μm wide, 7.2–15(–22.5) μm long, ovoid to ovoid-cylindrical, with apices rounded and each of the inner cells with 1–3 teeth or 1–2 bent spines, marginal cells bearing 2 apical spines which are often curved, oblique to long axis of coenobia and linearly symmetrical.

Probably cosmopolitan; in the British Isles in the plankton of many different types of waters, mostly recorded from southern England and the English Midlands.

Scenedesmus circumfusus Hortobágyi 1960

17810150 **Pl. 96F** (p. 394)

Coenobia a row of 2 or 4 linearly arranged cells; cells 2–4.5(–5.5) μm wide, 8–14(–16.8) μm long, elongate-ovoid, with rounded apices bearing 1–2 straight or slightly curved, short teeth (long spines absent), a single longitudinal row of warty teeth (sometimes merging to form a continuous ridge-like structure) on side of each cell; marginal cells with outermost wall often less convex in centre and having 1 or more spines (1–1.6 μm wide and 1.3–4.3 μm long) arising on outermost side of each marginal cell apex, diagonally opposite each other, the spines oblique to long axis of coenobia.

Probably cosmopolitan; in the British Isles in the plankton of ponds, lakes and slow-flowing reaches of rivers, mostly recorded in the south and east of England.

Very probably a form of *S. armatus*.

var. *bicaudatus* Hortobágyi 1960

Synonym: *Scenedesmus pseudogranulatus* Masjuk 1962 *pro parte*

17810152

Cells bearing 1 or 2(–3) straight or slightly curved, short teeth, with a single long spine arising from diagonally

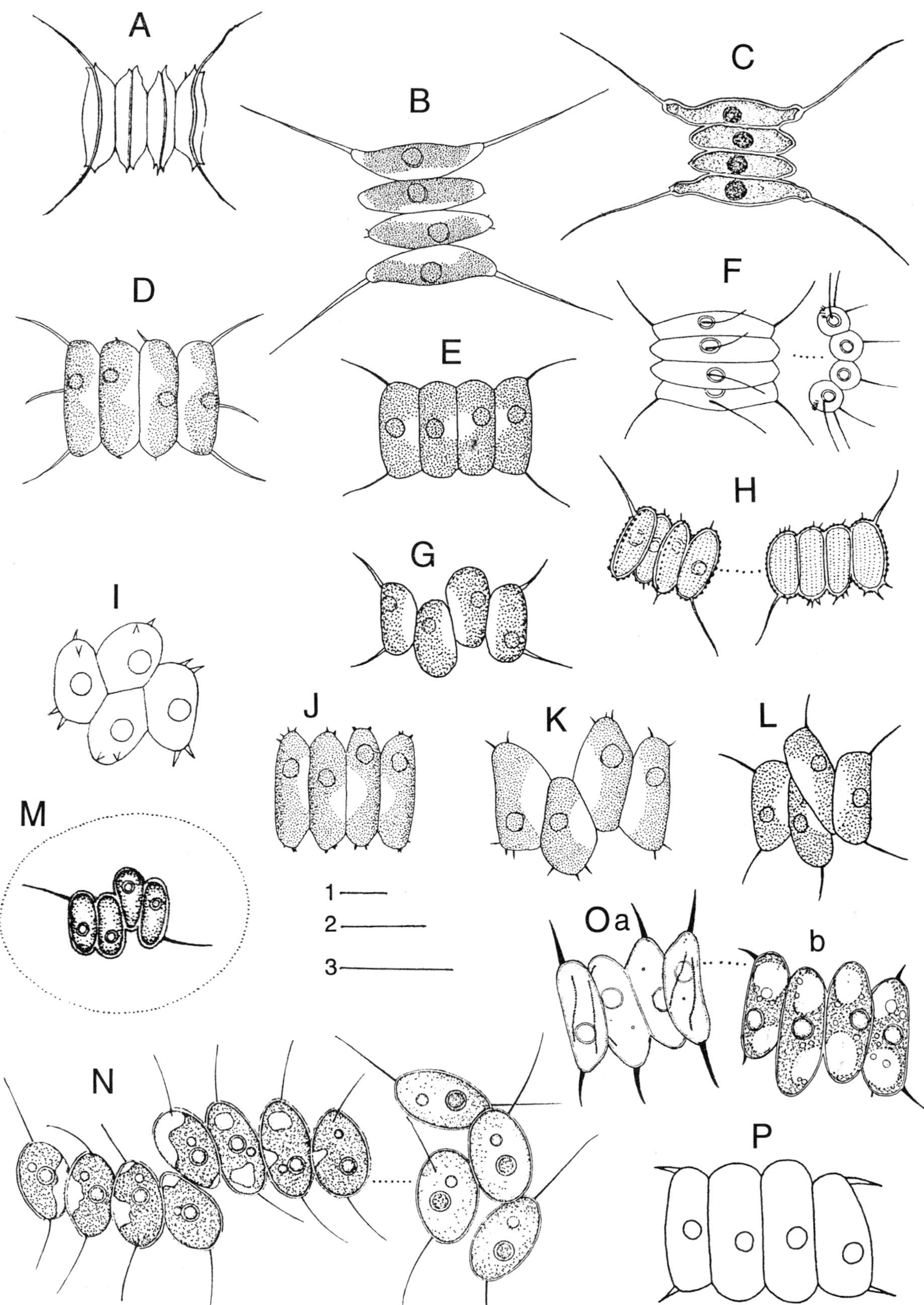

Plate 95 A–P

A. *Scenedesmus opoliensis* var. *carinatus* (p. 397); **B**. *Scenedesmus opoliensis* var. *mononensis* (p. 397); **C**. *Scenedesmus protuberans* (p. 397); **D**. *Scenedesmus abundans* (p. 387); **E**. *Scenedesmus communis* (p. 392); **F**. *Scenedesmus asymmetricus* (p. 390): coenobium in side (left) and polar view (right); **G**. *Scenedesmus intermedius* (p. 395); **H**. *Scenedesmus insignis* (p. 395); **I**. *Scenedesmus denticulatus* var. *disciformis* (p. 392); **J**. *Scenedesmus aculeolatus* (p. 388); **K**. *Scenedesmus denticulatus* (p. 392); **L**. *Scenedesmus intermedius* var. *balatonicus* (p. 395); **M**. *Scenedesmus intermedius* var. *acutispinus* (p. 395); **N**. *Scenedesmus kissii* (p. 395); **O**. *Scenedesmus pannonicus* (p. 397): details of cell contents shown in one figure (b); **P**. *Scenedesmus microspina* (p. 395).

Scale bars: 10 μm. 1 for C, H, I; 2 for B, D–F, K, N; 3 for A, G, J, L, M, O, P.

opposite apices of each marginal cell and usually oblique to long axis of coenobia; cells each with a single longitudinal and frequently interrupted row of warts, obtuse teeth sometimes present and only rarely joining to form a continuous ridge on side of cell.

Europe, Asia; in the British Isles recorded in the plankton of a wide range of aquatic habitats, mostly in the English Midlands and southern counties.

Scenedesmus communis E.H.Hegewald 1977

Synonyms: *Scenedesmus quadricauda* Chodat, *S. quadricauda* (Turpin) Brébisson *sensu auct. post. pro parte*

17810160 **Pl. 95E** (p. 391) **CD**

Coenobia consisting of up to 4(–8) linearly arranged cells; cells (2.3–)3.5–6.6 μm wide, (7.8–)9–17(–20.4) μm long, elongate-cylindrical, with the inner cells having rounded apices; main spine on each of the slightly outwardly curved apices of marginal cells whose outermost wall is almost straight; outer mucilaginous envelope present or confined to area between cell apices; cell walls smooth.

Probably cosmopolitan; most commonly recorded species of *Scenedesmus* (usually as '*S. quadricauda*'), planktonic or occasionally associated with submerged surfaces in a wide variety of different types of waters throughout the British Isles; considered to range (as *S. quadricauda*) from the purest, to mildly to moderately nutrient-enriched.

Often the name '*S. quadricauda*' has been associated with various species differing from one another in cell shape, spine-associated characters, cell wall structure and other characters. These species only resemble one another in having 4 symmetrically disposed main spines that arise apically on the marginal cells. As a result, most records in the literature of '*S. quadricauda*' must remain suspect. In an attempt to prevent further confusion Hegewald (1977) has described *S. communis*, a species that he believes corresponds closely to Chodat's original concept of *S. quadricauda*. It is likely that *S. communis* and *S. microspina* represent no more than part of a morphological series or cline of variation relating to age and environmental conditions.

Scenedesmus costato-granulatus Skuja 1948

Synonyms: *Scenedesmus transilvanicus* Kirjakov, *S. elegans* L.S.Péterfi, *S. elegans* L.S.Péterfi fo. *regularis* L.S.Péterfi

17810170 **Pl. 96L** (p. 394) **CD**

Coenobia of 2(–4) linearly or slightly alternating cell rows; cells 2–4 μm wide, (4–)6–10.2 μm long, elongate-ovoid to ellipsoidal or ovoid, slightly narrowing and with rounded apices, connected for almost all of length; outermost wall of marginal cells convex; cell wall bearing 1–6 (most commonly 3) longitudinal and interrupted rows of granules forming a ridge-like structure, sometimes with other granules dispersed between ridges and small teeth at apices.

Probably cosmopolitan; widespread and reasonably common in the British Isles where planktonic and/or associated with surfaces in many types of aquatic habitat including slower flowing reaches of rivers, ponds, lakes and small bodies of water such as open tanks, animal drinking troughs and fish ponds.

Scenedesmus costatus Schmidle 1895

17810180 **Pl. 96K** (p. 394)

Coenobia flat, consisting of 2, 4 or 8(–16) densely packed, distinctly alternating cells in 2 rows; cells (4–)7–10(–12) μm wide, (7.5–)13–19.8(–22) μm long, broadly spindle-shaped to ovoid or slightly ovoid, with 4–10(–12) longitudinal ridges and bluntly pointed apices each with a nipple-like polar thickening.

Probably cosmopolitan; only recorded in the British Isles from a bog pool near Inverness, Scotland and a similar habitat on Pilmoor, North Yorkshire.

Due to its well-documented distribution in Northern and Central European mountains (principally the Alps), it is considered by Komárek & Fott (1983) to be a boreal-alpine species.

Scenedesmus denticulatus Lagerheim 1882

17810190 **Pl. 95K** (p. 391)

Coenobia of 2 or 4 distinctly alternating cells in one row, with 2-rowed cells tightly packed or having a space between apices, straight or slightly curved, without a mucilaginous envelope; cells 4–11 μm wide, (4–)6–15(–20) μm long, broadly ovoid, with (1–)2–4 short teeth, often outwardly directed and confined to the outer apices.

Probably cosmopolitan; reported from the plankton of various aquatic habitats in England and Ireland; considered an indicator of mildly to moderately nutrient-enriched water.

var. *disciformis* Hortobágyi 1973

17810192 **Pl. 95I** (p. 391)

Coenobia of 4–8 cells arranged in form of cross, or in 2 rows; cells (5–)6–8.2 μm wide, (8–)10–16.5 μm long, broadly ovoid to ovoid, tightly packed, with 2 or 3 obliquely arranged teeth near cell apex and confined to outermost side of marginal cell apices.

Europe; recorded in the British Isles only from a lake at Chartwell House, Kent.

Scenedesmus dimorphus (Turpin) Kützing 1833

Basionym: *Achnanthes dimorpha* Turpin

Synonyms: *Scenedesmus acutus* Meyen, *S. antennatus* Brébisson in Ralfs, *S. costulatus* Chodat, *S. acutus* vars. *dimorphus* (Turpin) Rabenhorst, *obliquus* Rabenhorst, *S. obliquus* var. *dimorphus* (Turpin) Hansgirg

17810440 (17810050) **Pl. 96M** (p. 394)

Coenobia of 2, 4 or 8 linearly or distinctly alternately arranged cells in 1 or 2 rows; cells 2–9.4(–14) μm wide, (5–)6–25(–27) μm long, broadly spindle-shaped, tapering to slightly extended apices, with inner cells straight, marginal cells slightly outwardly curved but only in subapical part.

Cosmopolitan; a commonly encountered species in the plankton and associated with submerged surfaces in aquatic habitats ranging from open water tanks in farmyards to large shallow lakes such as those characteristic of the Norfolk Broads; considered to range from pristine water to mildly or moderately nutrient-enriched water.

Many intraspecific taxa have been recognized which are distinguished by features of little or no taxonomic significance (e.g. alternate or costulate coenobia), but which probably reflect local ecological conditions or stages in development. For no more than convenience, it is sometimes useful to recognize linear or alternately arranged cells, or 2-rowed coenobia but little useful purpose is served by giving them formal taxonomic ranking (see Tsarenko, 1990). It grows well in liquid culture and on agar, often becoming unicellular and the cells gradually becoming coccoidal and thick-walled.

Scenedesmus dispar Brébisson 1868

17810200 **Pl. 94K** (p. 389)

Coenobia of (2–)4 linearly to slightly alternately arranged and tightly packed cells; cells (2.3–)4–9 μm wide, (8–)8.6–16.5 μm long, elongate-ovoid to cylindrical, tapering to rounded or polygonal apices bearing 1 or 2 short spines, with spines arising laterally on apices and diagonally opposite on adjacent cells and lying almost perpendicular to long axis of coenobia, spines on adjacent cells often facing in opposite direction.

Probably cosmopolitan; in the British Isles recorded from the plankton of ponds and lakes in England and Ireland.

Scenedesmus ellipticus Corda 1835

Synonyms: *Scenedesmus ecornis* (Ehrenberg ex Ralfs) Chodat var. *flexuosus* Lemmermann, *S. bijugatus* (Turpin) Kützing *sensu auct. post.*, *S. flexuosus* (Lemmermann) Ahlstrom, *S. linearis* Komárek

17810210 **Pl. 97B** (p. 396)

Coenobia of 2, 4 or 8 (more rarely 16 or 32) cells in a single linear or sometimes slightly alternate row, straight or only rarely curved; cells (2–)3.5–9 μm wide, (3.5–)7–21(–23) μm long, ovoid, ellipsoidal or cylindrical, the outer wall slightly convex, with apices broadly rounded, without ornamentation.

Probably cosmopolitan; fairly common in the British Isles in the plankton of various types of aquatic habitat including the Cheshire Meres and Norfolk Broads.

Scenedesmus falcatus Chodat 1894

Synonyms: *Scenedesmus acuminatus* var. *biseriatus* Reinhard, *S. acuminatus auct. plur. non S. acuminatus* (Lagerheim) Chodat

17810220 **Pl. 96N** (p. 394) **CD**

Coenobia of (2–)4 or 8 cells, usually alternating in 2 rows or sometimes linear, ridged, flat, without mucilage; cells 3.4–9 μm wide, 18–35.5 μm long, the inner cells narrowly spindle-shaped and straight or only slightly outwardly curved, the marginal cells extremely curved and crescent-shaped or falcate, gradually tapering from centre to acute apices.

Probably cosmopolitan; fairly common in the British Isles in the plankton of a various ponds, lakes and rivers.

A very polymorphic species with the arrangement of cells known to be dependent on external conditions (including water chemistry) and the physiological condition of the population. There seems little justification for affording formal taxonomic status to the linear, alternate or costulate condition, although such forms might be useful for indicating ecological conditions (see remarks under *S. dimorphus*).

Scenedesmus flavescens Chodat 1913

Synonym: *Scenedesmus tenuispina* Chodat

17810230 **Pl. 94L** (p. 389)

Coenobia of 2 or 4(–8) linearly arranged cells; cells (1.7–)2.5–4.5(–7) μm wide, (5.5–)7.5–12(–15) μm long, cylindrical to ovoid, the marginal cells sometimes with parallel sides and narrowing to slightly curved apices bearing main spines which are straight or slightly curved and almost equal to cell length, 2 shorter equatorial spines on outer side of each marginal cell; inner cells straight, tapering to rounded apices and bearing 1or 2 spines.

Probably cosmopolitan; recorded only occasionally in ponds and lakes in England.

Scenedesmus grahneisii (Heynig) Fott 1973

Basionym: *Didymocystis grahneisii* Heynig

17810240 **Pl. 96H** (p. 394)

Coenobia a linear row of 2 or 4(–8) linearly arranged cells; cells 2–4(–7) μm wide, 5–10(–13) μm long, ellipsoidal to cylindrical, apices slightly rounded or sometimes asymmetrical and narrowing, the outermost wall of marginal cells slightly convex; cell wall often with groups of granules or granules form continuous or interrupted longitudinal rows, on inner cells form a curved line and on marginal cells in the shape of the letter 'C'.

Probably cosmopolitan; reasonably common in the plankton and/or associated with surfaces in ponds, lakes, canals and recorded from slow-flowing reaches of various rivers in the southern and eastern counties of England.

Scenedesmus granulatus West *et* G.S.West 1897

17810250 **Pl. 96I** (p. 394)

Coenobia straight, consisting of a single row of 4 cells with the ends in line or of alternating cells; cells (2–)3.5–12 μm wide, (5–)7–12(–24) μm long, elongate-ovoid to cylindrical-ovoid, with apices broadly rounded and outermost wall of marginal cells convex; cell walls yellow, granular, with granules randomly arranged or forming longitudinal rows (rarely ridge-like) that are sometimes more clearly visible at apices.

Europe; mostly common in the plankton of ponds and small lakes in southern England (e.g. Abbots Pond, Somerset; ponds in Richmond Park, Surrey).

Differs from the closely related species *S. verrucosus* Y.V.Roll both in the form of the coenobium and in cell wall ultrastructure (Hegewald & Schnepf, 1974); see description of *S. verrucosus*.

Scenedesmus hystrix Lagerheim 1882

17810260 **Pl. 96G** (p. 394)

Coenobia of 2 or 4 linearly arranged cells attached for almost whole of length, lacking a mucilaginous envelope; cells 2–6 μm wide, 10–18(–20) μm long, oval-cylindrical, with acute apices having a single tooth, outermost wall of marginal cell straight, walls covered with small teeth.

Probably cosmopolitan; only recorded on a few occasions in the British Isles where it occurs in the plankton of pools and ponds.

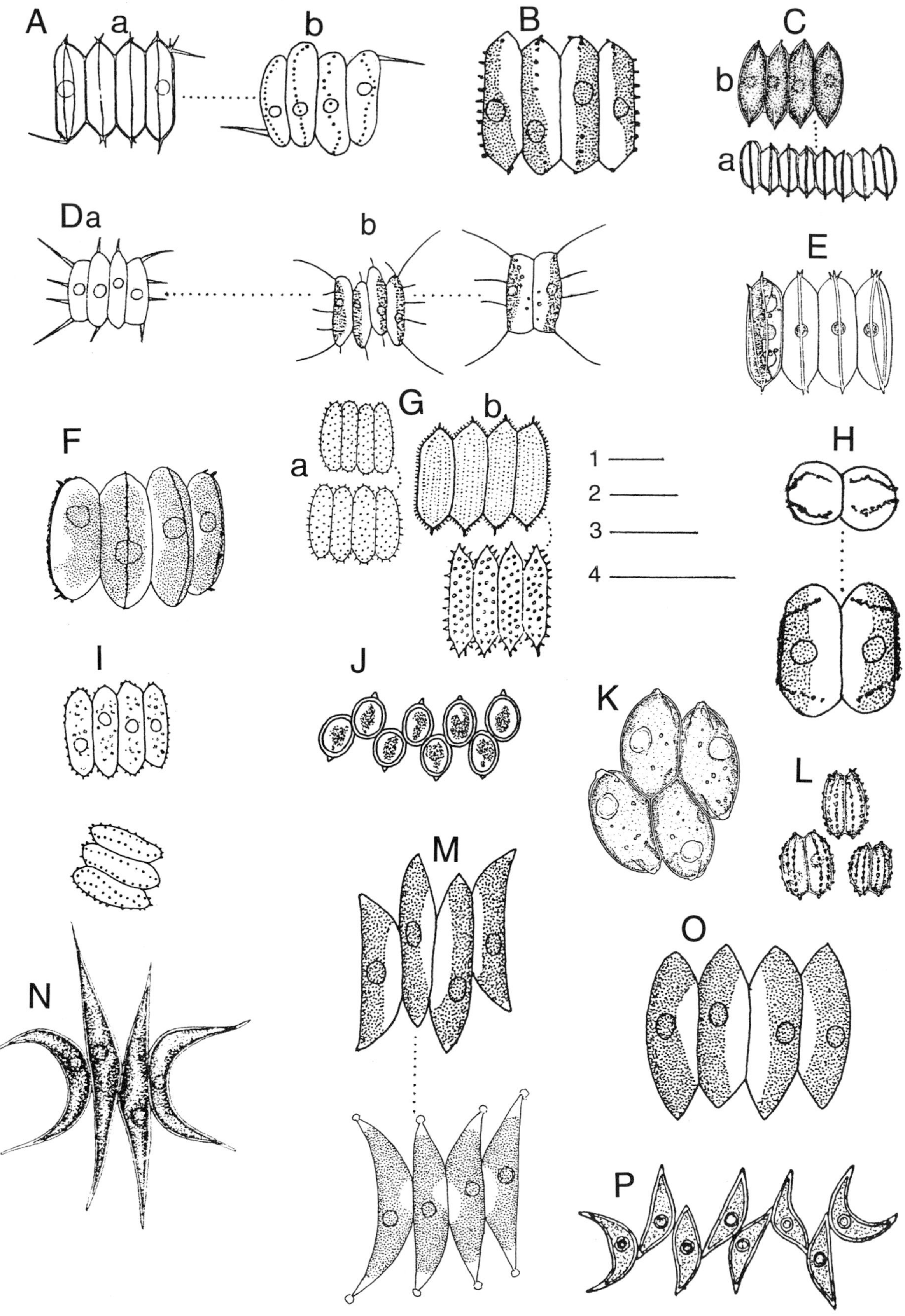

Plate 96 A–P

A. *Scenedesmus semipulcher* (p. 397); **B**. *Scenedesmus serratus* (p. 398); **C**. *Scenedesmus acutiformis* (p. 388): a, b. coenobia showing cells having a longitudinal thickening; **D**. *Scenedesmus subspicatus* (p. 398); **E**. *Scenedesmus brasiliensis* (p. 390); **F**. *Scenedesmus circumfusus* (p. 390); **G**. *Scenedesmus hystrix* (p. 393): a, b. coenobia showing surface ornamentation; **H**. *Scenedesmus grahneisii* (p. 393): coenobium in side view (below) and polar view (top); **I**. *Scenedesmus granulatus* (p. 393); **J**. *Scenedesmus apiculatus* (p. 388); **K**. *Scenedesmus costatus* (p. 392); **L**. *Scenedesmus costato-granulatus* (p. 392); **M**. *Scenedesmus dimorphus* (p. 392); **N**. *Scenedesmus falcatus* (p. 393); **O**. *Scenedesmus obliquus* (p. 395); **P**. *Scenedesmus bernardii* (p. 390).

Scale bars: 1, 3, 4, 10 µm; 2, 20 µm. 1 for Ca, E, Ga; 2 for Cb, J, P; 3 for A, B, Da,b, Gb, I, K–N; 4 for F, O, H.

Scenedesmus insignis (West *et* G.S.West) Chodat 1913

Basionym: *Scenedesmus quadricauda* var. *insignis* West *et* G.S.West

17810270 **Pl. 95H** (p. 391)

Coenobia a linear to slightly alternating row of (2–)4(–8) cells; cells 4–5(–7.5) μm wide, (10–)12–19.5 μm long, elongate-ovoid to ovoid-cylindrical, apices slightly tapering and rounded, with 1–3 warts and short straight or oblique teeth sometimes on one apex of each cell, or small spines on each apex, with main spines on only one apex of each marginal cell and diagonally symmetrical; cell walls with evenly arranged granules or rows of granules, sometimes with longitudinal ridges or large warts.

Probably cosmopolitan; in the plankton or associated with submerged surfaces in ponds, lakes and rivers.

Scenedesmus intermedius Chodat 1926

17810280 **Pl. 95G** (p. 391)

Coenobia a row of (2–)4(–8) alternately arranged cells; cells 2–4(–5.5) μm wide, (3.5–)5–7.5(–10) μm long, ovoid, apices rounded and the main spines (curved or straight) projecting from outermost edge of marginal cell apices, or on only one apex and then diagonally symmetrical; cell walls smooth, sometimes with discontinuous rows of warts, often irregularly arranged at apices.

Probably cosmopolitan; in the British Isles planktonic and associated with submerged surfaces in ponds, lakes and rivers.

var. *acutispinus* (Y.V.Roll) E.Hegewald *et* An in Hegewald, An & Tsarenko 1998

Basionym: *Scenedesmus quadricauda* var. *acutispinus* Y.V. Roll

Synonym: *Scenedesmus intermedius* var. *bicaudatus* Hortobágyi

17810284 (17810283) **Pl. 95M** (p. 391)

Spines arising at apex of marginal cells and diagonally opposite.

Probably cosmopolitan; recorded in the British Isles only in the plankton of lakes and other water bodies in south-east England.

Taxonomic position of this variety and var. *balatonicus* is not clear (see Hegewald et al. 1988).

var. *balatonicus* Hortobágyi 1943

17810282 **Pl. 95L** (p. 391)

Mainly straight and symmetrically arranged spines arising on the marginal cells, additional spines of the same or shorter length on apices of inner cells.

Probably cosmopolitan; recorded in the British Isles only from moats and ponds in south-east England.

Scenedesmus kissii Hortobágyi 1975

17810290 **Pl. 95N** (p. 391) **CD**

Coenobia of 4 or 8 linearly or alternately arranged cells, sometimes 4-celled coenobia having cells very irregularly arranged; cells 3–8 μm wide, 7.2–12 μm long, broadly ovoid, with a single straight or curved seta-like spine at each apex, sometimes with inner cells having a spine only at one apex.

Probably only in Europe; known from 2 localities in the British Isles, an open rainwater collecting tank in Jersey, Channel Islands and an ornamental pond at Histon, near Cambridge.

Scenedesmus magnus Meyen 1829

Synonyms: *Scenedesmus longus* Meyen, *S. naegelii* Brébisson, *S. quadricauda* var. *maximus* West *et* G.S.West, *S. maximus* (West *et* G.S.West) Chodat, *S. oahuensis* (Lemmermann) G.M.Smith, *S. quadricauda* var. *oahuensis* Lemmermann, *S. quadricauda* var. *setosus* Kirchner *pro parte*, *S. magnus* var. *naegelii* (Brébisson) Tsarenko

17810300 **Pl. 94O** (p. 389) **CD**

Coenobia of (2–)4–8 linearly arranged cells; cells (4–)6–11(–14) μm wide, (18–)27–33(–43) μm long, elongate-ovoid to cylindrical with rounded apices, main spines subapical on marginal cells (straight, or slightly curved and oblique) and almost equal to cell length, not infrequently with an additional diagonally symmetrical spine on adjacent apices of inner cells along only half of coenobia; mucilaginous layers around cells; cell walls with granules on sides, longitudinal thin ridges or ridge-like rows of paired dots, small spines and separate 'rosette-like' structures at apices, sometimes with rounded and continuous borders formed by surface layer of cell wall.

Probably cosmopolitan; widespread in the British Isles in the plankton of ponds, lakes and slow-flowing reaches of rivers.

It is sometimes difficult to distinguish this species from *Scenedesmus communis* with the main characters used for separation including cell dimensions, peculiarities of cell structure and features of the cell wall. The cell wall characters of *S. magnus* are similar to those of *S. perforatus* Lemmermann, but differ markedly from those of *S. communis*.

Scenedesmus microspina Chodat 1926

17810310 **Pl. 95P** (p. 391)

Coenobia of (2–)4 or 8 linearly arranged cells; cells (2.3–)4–6 μm wide, (5–)9–14 μm long, cylindrical to ovoid, with apices broadly rounded; inner cells bearing at apex straight or slightly curved spines which are long but less than width of cell, marginal cells whose outermost wall is usually convex, bear short and oblique spines; cell walls smooth, sometimes with warts at apices of inner cells, lacking a mucilaginous envelope.

Probably cosmopolitan; recorded in the British Isles only from the River Wye.

Scenedesmus obliquus (Turpin) Kützing 1833

Basionym: *Achnanthes obliquus* Turpin

Synonyms: *Achnanthes bijuga* Turpin, *Scenedesmus bijugatus* Kützing, *S. chlorelloides* Chodat, *S. dactylococcoides* Chodat

17810320 **Pl. 96O** (p. 394)

Coenobia of 2 or 4(–8) linearly or alternately arranged cells, more often unicellular when in culture; cells 2.2–9.6(–11) μm wide, (4–)6–15(–25) μm long, all spindle-shaped, straight, with acute apices and sometimes slightly asymmetrical.

Probably cosmopolitan; reasonably common in plankton, recorded from a variety of aquatic habitats including animal drinking troughs, ditches, puddles, ponds, reservoirs and rivers in many parts of the British Isles.

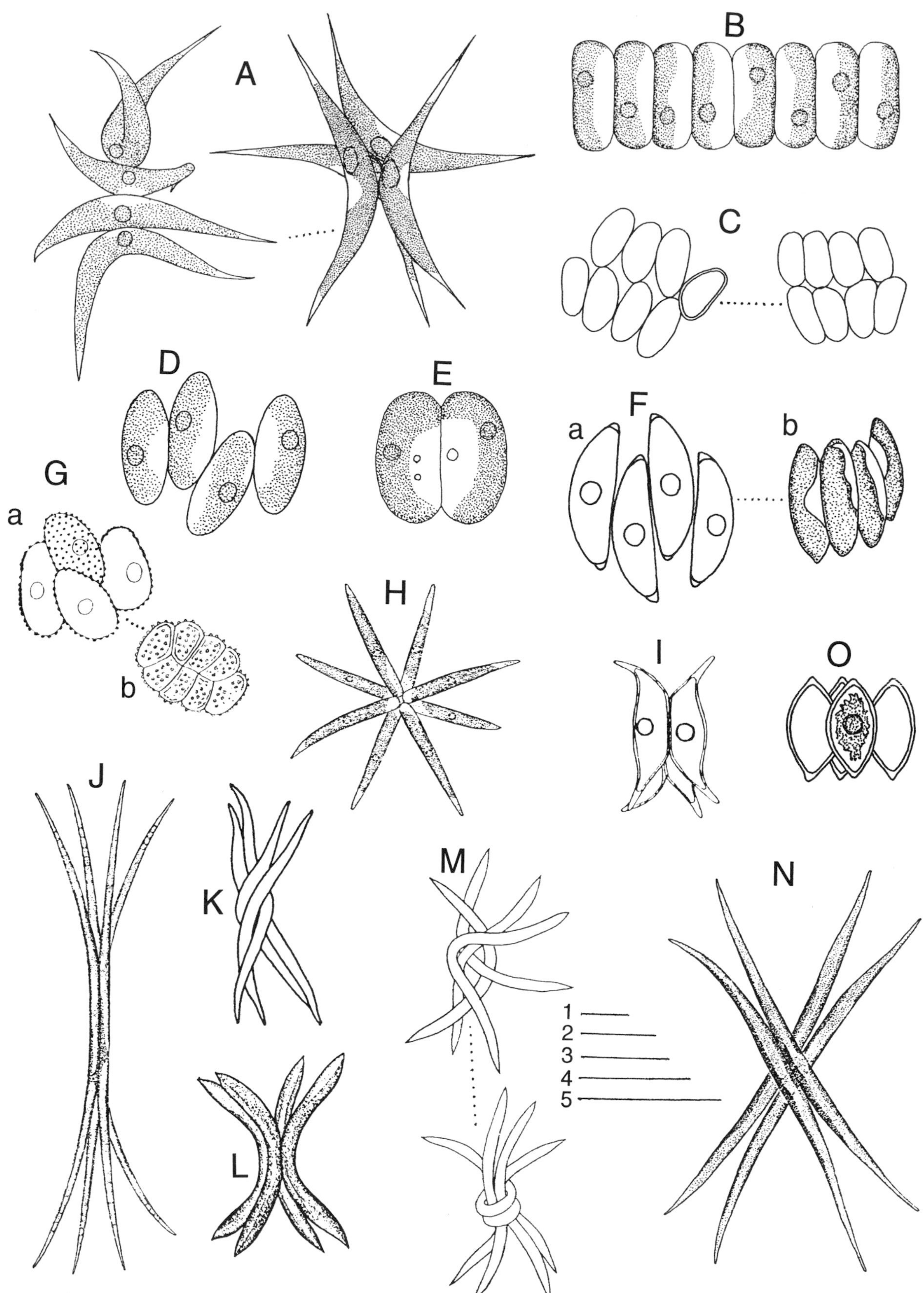

Plate 97 A–O

A. *Scenedesmus acuminatus* (p. 388); **B**. *Scenedesmus ellipticus* (p. 393); **C**. *Scenedesmus arcuatus* (p. 388); **D**. *Scenedesmus obtusus* (p. 397); **E**. *Scenedesmus planctonicus* (p. 397); **F**. *Scenedesmus raciborskii* (p. 397); **G**. *Scenedesmus verrucosus* (p. 398); **H**. *Actinastrum hantzschii* (p. 328); **I**. *Tetradesmus wisconsinensis* (p. 404); **J**. *Ankistrodesmus falcatus* (p. 330); **K**. *Ankistrodesmus densus* (p. 330); **L**. *Ankistrodesmus fasciculatus* (p. 330); **M**. *Ankistrodesmus spiralis* (p. 331); **N**. *Ankistrodesmus fusiformis* (p. 331); **O**. *Tetradesmus cumbricus* (p. 404).

Scale bars: all 10 μm. 1 for C, I–M, O; 2 for B, D, F, H; 3 for A, G; 4 for N; 5 for E.

Scenedesmus obtusus Meyen 1829

Synonyms: *Scenedesmus obtusus* fo. *alternans* (Reinsch) Compère, *S. alternans* Reinsch, *S. bijugatus* var. *alternans* (Reinsch) Hansgirg, *S. ovalternus* Brébisson

17810330 **Pl. 97D** (p. 396)

Coenobia of 2 rows of 4 or 8 distinctly alternating cells which are in contact over a short distance (up to one-third of length) in subapical part where flat to slightly curved, usually surrounded by a mucilaginous envelope; cells 3–13 µm wide, (4–)7–21 µm long, oval to ellipsoidal, rarely ovoid, with rounded apices and usually convex on outermost side.

Probably cosmopolitan; fairly common in the British Isles in the plankton of a wide range of aquatic habitats including ponds, lakes and rivers.

Scenedesmus opoliensis P.G.Richter 1897

Synonym: *Scenedesmus opoliensis* var. *setosus* Dedusenko

17810340 **Pl. 94P** (p. 389) **CD**

Coenobia a single row of 2, 4 or 8 linearly arranged cells, not infrequently with obliquely arranged inner cells; cells (3.4–)4.6–6.5(–9) µm wide, (8–)10–29.5 µm long, spindle-shaped; inner cells oblique to straight, with straight walls and narrowing to slightly rounded apices bearing 1–2 teeth; marginal cells narrowing to truncate apices.

Probably cosmopolitan; widely distributed in the British Isles where common in the plankton and/or associated with submerged surfaces in many types of aquatic habitat ranging from dune-slack pools, ponds, lakes, reservoirs to rivers.

var. *carinatus* Lemmermann 1899

Synonym: *Scenedesmus carinatus* (Lemmermann) Chodat

17810342 **Pl. 95A** (p. 391)

Cells (2.5–)5.2–8 µm wide, (8–)12–28.3 µm long, parallel to each other or inner cells oblique, narrowing to truncate apices with each usually 2-toothed and having longitudinal and continuous smooth ridges.

Probably cosmopolitan; only rarely reported in the British Isles.

var. *mononensis* Chodat 1926

17810343 **Pl. 95B** (p. 391) **CD**

Cells (2–)3.5–8 µm wide, (8–)10–19(–25) µm long, straight not spindle-shaped, narrowing to somewhat truncate apices with each bearing 1 or 2 teeth.

Probably cosmopolitan; rarer in the British Isles than the type and most frequent in the plankton of ponds and lakes.

Scenedesmus pannonicus Hortobágyi 1944

17810350 **Pl. 95O** (p. 391)

Coenobia curved, a row of 4 or 8 cells, sometimes cells slightly alternately arranged; cells 3–4.5 µm wide, 10.5–12 µm long, cylindrical-ovoid or ellipsoidal, apices conically rounded, outer cells having a short spine and sometimes other minute spines at each apex, inner cells having an almost straight and stout spine at one apex.

Probably cosmopolitan; so far known only from a small pool in sand dunes at Braunton Burrows, near Barnstaple, Devon (DMJ unpublished record).

Scenedesmus planctonicus (Korshikov) Fott 1973

Basionyn: *Didymocystis planctonica* Korshikov

17810360 **Pl. 97E** (p. 396)

Coenobia of 2 cells connected by a straight wall; cells 2.4–6.5 µm wide, 6–12 µm long, broadly ovoid, with inner wall straight and outer wall convex, apices broadly rounded.

Probably cosmopolitan; recorded only from a few ponds, lakes and a moat in Kent.

Possibly this taxon is a member of the Oocystaceae, as *Pseudodidymocystis planctonica* (Korshikov) E.Hegewald *et* Deason.

Scenedesmus protuberans F.E.Fritsch *et* M.F.Rich 1929

Synonym: *Scenedesmus protuberans* fo. *minor* S.H.Li

17810370 **Pl. 95C** (p. 391)

Coenobia a row of 4(–8) linearly arranged cells; cells (3.5–)4–7 µm wide, (11.2–)12–34 µm long, inner cells spindle-shaped with rounded apices and sometimes bearing small spines; marginal cells rounded in middle and narrowing to pin-headed (capitate) apices; main spines on apices of marginal cells, straight or curved and almost equal to cell length.

Probably cosmopolitan; fairly common in the British Isles in the plankton of ponds and lakes.

Scenedesmus raciborskii Wołoszyńska

Synonyms: *Scenedesmus incrassatulus* var. *mononae* G.M.Smith, *S. incrassatulus* var. *alternans* (Bohlin) Dedusenko

17810380 **Pl. 97F** (p. 396)

Coenobia of 2 or 4 linearly or alternately arranged cells, straight or slightly curved, often enclosed within a mucilaginous envelope; cells (2.5–)3.5–8(–11.3) µm wide, 9–25(–28) µm long, asymmetrically spindle-shaped, with inner side straight to slightly inwardly curved and outer side convex; marginal cells slightly inwardly curved, narrowing to obtuse apices, each bearing a small nipple-shaped projection.

Probably cosmopolitan; so far reported in the British Isles on only a few occasions in the plankton of ponds (e.g. Wilden Pool, Worcestershire) and lakes.

Scenedesmus semipulcher Hortobágyi 1960

17810390 **Pl. 96A** (p. 394)

Coenobia of (2–)4 linearly arranged closely connected cells; cells 2–4 µm wide, (8.4–)10–12.8 µm long, elongate-ellipsoidal to cylindrical, with a longitudinal ridge; marginal cells not infrequently shorter than inner ones, with apices rounded and sometimes each with a short tooth, bearing a main spine on one apex of each marginal cell and diagonally symmetrical, with spines lying perpendicular to long axis of coenobium; cell walls smooth or warty to granular and these continuous or interrupted, sometimes with a longitudinal row of large warts.

Probably cosmopolitan; in the British Isles planktonic in various types of aquatic habitat including ponds, rivers and flooded basins.

Easily confused with *Scenedesmus armatus* var. *bicaudatus* (Guglielmetti) Chodat from which it differs in having a longitudinal ridge, or an interrupted row of rough warts confined to only one side of cell. Possibly only an ecological form of *S. armatus*.

Scenedesmus serratus (Corda) Bohlin 1902

Basionym: *Arthrodesmus serratus* Corda
Synonyms: *Scenedesmus serratus* fo. *minor* Chodat, *S. hystrix* var. *regularis* H.Alten, *S. denticulatus* var. *linearis* fo. *granulatus* Hortobágyi
17810400 **Pl. 96B** (p. 394)
Coenobia of 2, 4 or 8 linearly arranged cells; cells 2.5–7 μm wide, (7.8–)9–20 μm long, elongate-ovoid, with 1–4 teeth on rounded or sometimes almost truncate apices, bearing longitudinal and continuous rows of teeth on inner cells, teeth confined to outer convex wall of marginal cells (sometimes teeth small and indistinct, or as longitudinal rows of granules); cell wall smooth except for teeth in longitudinal rows or occasionally scattered.

Probably cosmopolitan; known in the British Isles in the plankton of various water bodies.

Scenedesmus subspicatus Chodat 1926

Synonyms: *Scenedesmus spicatus* West et G.S.West, *S. quadricauda* var. *abundans* Kirchner *pro parte*, *S. abundans* (Kirchner) Chodat *sensu auct. post sine typo*, *S. spinosus* Chodat *sensu auct. post sine typo*, *S. gutwinskii* Chodat, *S. gutwinskii* var. *heterospina* Bodrogsközy
17810420 **Pl. 96D** (p. 394) **CD**
Coenobia of 2, 4 or 8, linearly to almost alternately arranged, tightly compact cells; cells 2.4–5(–7) μm wide, (5–)7–10(–12) μm long, ellipsoidal to broadly ovoid, with apices broadly rounded; inner cells often bearing a single apical spine; marginal cells bearing the a single large spine on each apex and these diagonal to long axis of coenobium, with 1–3 additional spines often present on outermost side of marginal cells; all spines almost equal in length or sometimes main spines longer.

Probably cosmopolitan; reported in the British Isles only in the plankton of ponds and small lakes (e.g. Priest Pot, English Lake District).

Differs from *Scenedesmus abundans* (Kirchner) Chodat by absence of short spines at the cell apices and in features of cell wall ultrastructure.

Scenedesmus verrucosus Y.V.Roll 1925

Synonyms: *Scenedesmus bijugatus* var. *granulatus* Schmidle, *S. granulatus* var. *verrucosus* Dedusenko, *S. disciformis* (Chodat) Fott *et* Komárek
17810430 **Pl. 97G** (p. 396)
Coenobia usually of 2 rows of 4 or 8 linearly arranged and tightly or loosely packed cells, with a single marginal cell for the 2 rows, very rarely a row of distinctly alternately arranged cells; cells (2–)2.5–6(–7) μm wide, 5–12(–14.4) μm long, broadly ovoid; cell walls greyish-brown, with warty surface, granules irregularly arranged, or in rows that are sometimes barely visible.

Probably cosmopolitan; in the British Isles known only from ponds and moats in southern England.

Schizochlamydella Korshikov 1953

Colonies microscopic, consisting of randomly scattered cell pairs or 4-celled groups associated with mother cell wall remnants, enclosed within a structureless and irregularly shaped mucilaginous envelope; cells spherical, smooth-walled, closely associated with circular or semi-circular remnants of wall; chloroplast parietal, cup-shaped with a single pyrenoid; asexual reproduction by 2 or 4 autospores, usually tetrahedrally arranged within mother cell and released by fracture of its wall, remnants of which persist.

Probably cosmopolitan; planktonic in various aquatic habitats, also isolated from soil.

Schizochlamydella delicatula (G.S.West) Korshikov 1953

Basionym: *Schizochlamys delicatula* G.S.West
17820010 **Pl. 86K** (p. 352)
Cells (4.5–)5–9.4 μm in diameter, spherical.

Probably cosmopolitan; often planktonic, especially in the soft, acid waters of moorland pools dominated by *Sphagnum* in western and north-western regions of the British Isles.

Schroederia Lemmermann 1898

Cells solitary, straight or curved, narrowly spindle-shaped, tapering equally to apices and each terminating in a long spine; chloroplast parietal, plate- or band-shaped, with 1 to several pyrenoids; uni- or multinucleate; asexual reproduction by biflagellate zoospores, released by lateral fracture of mother cell wall; akinetes thick-walled.

Probably cosmopolitan; planktonic and widely distributed in slow-flowing and standing water.

The genus has been re-interpreted by Hegewald & Schnepf (1986), but further experimental work is required to establish the justification for their recognition of *Pseudoschroederia* as a new and independent genus.

1 Cells and setae usually straight .... *S. setigera*
1 Cells slightly to strongly arched or in form of a long indistinct spiral .... *S. robusta*

Schroederia robusta Korshikov 1953

17830020 **Pl. 90G** (p. 367)
Cells 3–8 μm wide, 30–140 μm long, slightly to strongly arched or forming a long indistinct spiral (S-shaped twist), sometimes almost straight, spindle-shaped, cylindrical in centre, narrowing apically with each apex terminating in a delicate spine (20–33 μm long); chloroplast with 1–5 distinct pyrenoids.

Probably cosmopolitan; planktonic and associated with submerged surfaces in the British Isles in slow-flowing water as well as in ponds and lakes.

Schroederia setigera (Schröder) Lemmermann 1898

Basionym: *Reinschiella setigera* Schröder
Synonyms: *Ankistrodesmus setigerus* (Schröder) G.S.West, *Characium setigerum* (Schröder) Bourrelly
17830030 **Pl. 90H** (p. 367)
Cells 2.5–7(–10) μm wide, (41–)85–200 μm long, straight or rarely slightly curved, narrowly spindle-shaped, each apex terminating in a delicate, straight (rarely curved) spine (13–27(–50) μm long); chloroplast with 1(–2) pyrenoids.

Probably cosmopolitan; widespread in the British Isles especially in the plankton of ponds, lakes and reservoirs.

DOUBTFUL RECORD
Scotiella fritschii B.M.Griffiths in West & Fritsch 1927
17840010
Mentioned by West & Fritsch (1927, p. 124) as 'a new sp., *S. Fritschii* B.M.Griff., found by Dr Griffiths in a pool occupying a rocky depression in Fallowfield Fell, near Hexham (Northumberland) (cells 34–37 l.)'. The West & Fritsch illustration (fig. 41, E–F) is of an elongate-ovoid cell with a longitudinal ridge or wing on each side. We have been unable to trace a description or any of the material collected by Griffiths. Until new collections are made of specimens corresponding to this little-known alga it is not possible to comment further on it. Most other species attributed to the genus are snow algae and many have been discovered subsequently to be the resting stages of flagellated algae. For example, Light & Belcher (1968) record *Scotiella nivalis* (Shuttleworth) F.E.Fritsch and *S. cryophila* Chodat from a snowfield in the Cairngorm Mountains. These two species are now regarded as life-history stages of the volvocalean alga *Chloromonas cryophila* Hoham *et* Mullet (see Komárek & Fott, 1983).

Scotiellopsis Vinatzer 1975

Synonyms: *Scotiellopsis* Fott, *Scotiellocystis* Fott
Cells solitary or more rarely grouped, lemon-shaped, ellipsoidal or broadly ovoid, thin-walled, bearing 6–12(–18) often spirally-twisted ribs which sometimes meet apically where the wall is frequently thickened; chloroplast parietal, initially single but later dividing into crowded fragments each associated with a pyrenoid; asexual reproduction by 2, 4 or 8(–16) spindle-shaped or broadly oval autospores, arranged 3-dimensionally in mother cell and released by fracture of mother cell wall.

Probably cosmopolitan; mostly on damp or wet terrestrial surfaces including rocks, sand and soil.

Scotiellopsis oocystiformis (J.W.G.Lund) Punčochárová *et* Kalina 1981

Basionym: *Scotiella oocystiformis* J.W.G.Lund
17850010 **Pl. 87D** (p. 354)
Cells (4–)12–15 μm wide, (7–)16–20 μm long, with 8–12 very fine ribs and a wart-like wall thickening at each apex.

Europe; on wet rocks and associated with soil, very rarely recorded in the British Isles (e.g. Windermere, UKNCC 2001, strain CCAP 277/1).

Note added in proof: Hegewald & Hanagata (2000) have transferred the species to *Coelastrella* Chodat based on a phylogenetic analysis of 18S-RNA.

Selenastrum Reinsch 1867

Coenobia of 2, 4, 8 or 16 cells connected equatorially by convex sides, rarely single-celled, enclosed within a mucilaginous envelope; cells spindle-, sickle- or half-moon-shaped or so curved as to form an almost complete circle, gradually narrowing to acute apices; chloroplast parietal, single and with a pyrenoid (not visible with light microscope); asexual reproduction by 2, 4 or 8 (–16) autospores, if 8 then in two tetrads one above the other, released by breakage of mother cell in middle to leave 2 horn-like wall remnants connected to daughter cells.

Probably cosmopolitan; planktonic usually in nutrient-enriched water bodies. Six species of which 2 are recorded from the British Isles.

1 Cells usually lie across one another centrally in various directions .... *S. gracile*
1 Cells connected by inner convex sides and usually in tetrads .... *S. bibraianum*

Selenastrum bibraianum Reinsch 1867

Synonym: *Ankistrodesmus bibraianus* (Reinsch) Korshikov
17860010 **Pl. 98D** (p. 400) **CD**
Coenobia of 4, 8 or 16 cells (rarely more), usually cells in tetrads and connected by inner convex sides, enclosed within a broad mucilaginous envelope (40–70 μm wide); cells (1.5–)3.5–8 μm wide, 16–40(–44) μm long, spindle-shaped, gradually narrowing to acute or slightly obtuse apices, broadly halfmoon-shaped to curved and almost circular.

Probably cosmopolitan; distributed throughout the British Isles in the plankton of ponds, lakes and rivers as well as associated with aquatic vegetation.

Colonies in cross section are cross-shaped.

Selenastrum gracile Reinsch 1867

Synonym: *Ankistrodesmus gracilis* (Reinsch) Korshikov, *Selenastrum westii* G.M.Smith
17860020 **Pl. 98E** (p. 400) **CD**
Coenobia of (1–)2, 4 or 8(–16) cells which cross one another centrally in various directions, connected by mucilage, enclosed within a mucilaginous envelope; cells 1–5(–6) μm wide, (13–)15–50(–55) μm long, narrowly spindle-shaped and gradually narrowing to acute apices, markedly curved (arcuate), semicircular to subcircular.

Probably cosmopolitan; planktonic and associated with submerged surfaces in the British Isles found in rivers (e.g. Wye, Stour), lakes and reservoirs.

Solitary cells are morphologically similar to *Monoraphidium arcuatum*, but differ in having a mucilaginous envelope and in the arrangement of autospores within the mother cell.

DOUBTFUL RECORD
Selenastrum bifidum Bennett 1887
17860030
Doubt attaches to this species that has not been reported since first described by Bennet (1887) from a plankton sample collected

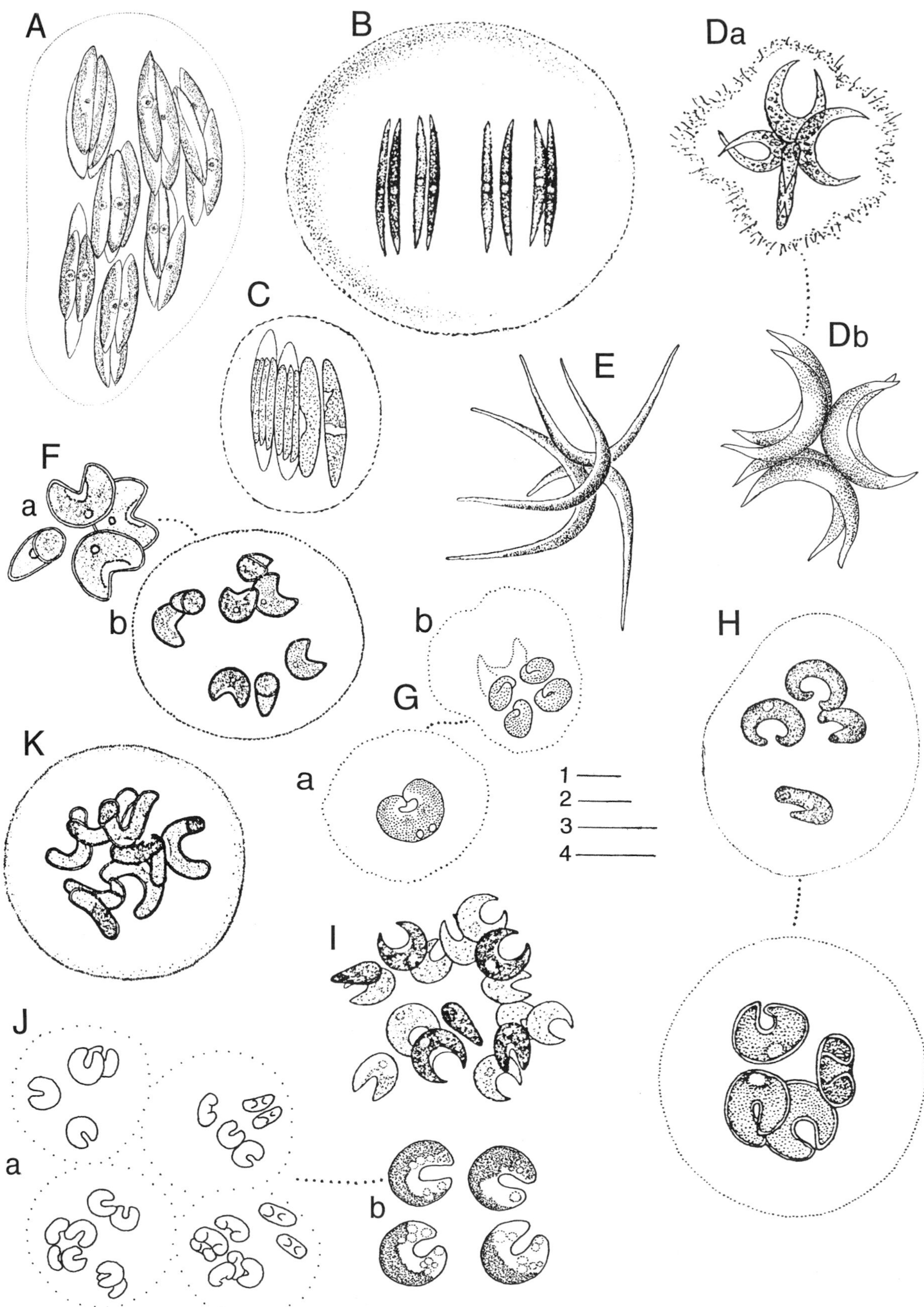

Plate 98 A–K

A. *Pseudoquadrigula britannica* (p. 381); **B**. *Quadrigula closterioides* (p. 382); **C**. *Quadrigula pfitzeri* (p. 382); **D**. *Selenastrum bibraianum* (p. 399): a. coenobium of 4 cells connected by the convex surface and within a mucilaginous envelope, b. coenobium of 6 cells; **E**. *Selenastrum gracile* (p. 399); **F**. *Kirchneriella aperta* (p. 361): a. details of a group of 4 cells, b. coenobium of 8 cells within mucilaginous envelope; **G**. *Kirchneriella incurvata* (p. 361): a. details of a single cell, b. four daughter cells and remnant of mother cell wall, all surrounded by a mucilaginous envelope; **H**. *Kirchneriella irregularis* (p. 361); **I**. *Kirchneriella lunaris* (p. 361); **J**. *Kirchneriella obesa* (p. 361): a. coenobia surrounded by mucilaginous envelopes, b. details of a group of 4 cells; **K**. *Raphidiocelis contorta* (p. 383).

Scale bars: 1, 2, 4, 10 μm; 3, 20 m. 1 for A, E, Fb, Gb; 2 for B, C, Db, H, Jb, K; 3 for Da, I, Ja; 4 for Fa, Ga.

in England. Its most distinctive feature is the presence of 2 hyaline points at each apex of the curved cells which form 4-, 8- or 16-celled coenobia. Other authors consider that it should be attributed to the genus *Sorastrum* (see Komárek & Fott, 1983).

Siderocelis (Naumann) Fott 1934

Cells solitary, spherical, ellipsoidal, ovoid to cylindrical, with walls colourless to dark-brown, thin and covered by granules or warts, with or without a structureless mucilaginous envelope; chloroplasts parietal, 1–4, cup-shaped, lacking pyrenoids or these difficult to observe; asexual reproduction by 2, 4 or 8 autospores, released by fracture of mother cell wall into 2–4 or more parts.

Probably cosmopolitan; planktonic and associated with submerged surfaces, usually in standing water.

1 Cells ellipsoidal, <5 μm long; chloroplast single, without a pyrenoid .... *S. minor*
1 Cells ellipsoidal to ovoid, >5 μm long; chloroplasts usually 4 or more, each with a pyrenoid .... *S. ornata*

Siderocelis minor (Naumann) Fott 1934

Basionym: *Chlorella minor* Naumann
17870010 **Pl. 87N** (p. 354)
Cells less than 5 μm long, ellipsoidal with acute apices, walls covered by granules; chloroplast single, without a pyrenoid.

Europe; in the British Isles planktonic and usually in pools.

Siderocelis ornata (Fott) Fott 1934

Basionym: *Oocystis ornata* Fott
Synonym: *Siderocelis balatonica* Hortobágyi
17870020 **Pl. 87M** (p. 354) **CD**
Cells 4–8.5 μm wide, 6–14 μm long, broadly ovoid to ellipsoidal, apices rounded or gradually becoming acute, walls covered by large and irregularly shaped granules which are sometimes in lines; chloroplast occasionally single, more often 4(–8), each with a pyrenoid (usually scarcely visible).

Probably cosmopolitan; in the British Isles planktonic and associated with submerged surfaces usually in ponds and lakes.

Siderocystopsis Swale 1964

Synonym: *Siderocystis* Korshikov *non Siderocystis* Naumann
Cells solitary, ovoid to spherical-oval or ellipsoidal, usually within a mucilaginous envelope; walls thin, dark brown, warty, covered by long and delicate spines each with a dark granule at its base; chloroplast parietal, cup-shaped, with a single pyrenoid; asexual reproduction by 2, 4, or 8 smooth-walled autospores, released by fracture of mother cell wall into 2 or 4 almost equal parts.

Europe; planktonic usually in nutrient-rich water bodies.

Siderocystopsis punctifera (Bolochonzew) E.H.Hegewald *et* Schnepf 1984

Basionym: *Golenkinia punctifera* Bolochonzew
Synonyms: *Siderocystis fusca* Korshikov, *Siderocystopsis fusca* (Korshikov) Swale
17880020 (17880010) **Pl. 89J** (p. 363)
Cells (4–)5.5–10(–18) μm wide, (6–)7–12(–18) μm long, ovoid or broadly ellipsoidal to slightly asymmetrical, young daughter cells sometimes roughly triangular, with walls becoming dark brown due to encrustation by iron salts and roughly warty on ageing; spines 10–15(–18) μm long, each arising from a dark conical to hemispherical granule.

Probably cosmopolitan; from small water bodies such as ponds and animal drinking troughs as well as in lakes and rivers (e.g. Thames and Severn); probably very widely distributed in the British Isles.

This species closely resembles *Siderocelis ornata* in cell shape, presence of granules, chloroplast structure, method of autospore release and ecology (widespread in nutrient-rich water), but is distinguished by having spines which arise from a tubercle or wart-like granule. Young cells of *Siderocystopsis punctifera* might be confused with *Franceia tenuispina* Korshikov, but differ from it by releasing autospores by fracture of the mother cell rather than gelatinization of the wall. Additionally, the cell wall of *Siderocystopsis punctifera* is moderately thick, often dark brown and bears spines arising from granules, unlike *Franceia tenuispina* in which the wall is thin, colourless, and there is no granule associated with each spine.

Sorastrum Kützing 1845

Coenobia spherical, consisting of 4, 8, 16, 32 or 64(–128) cells on short, thick, mucilaginous stalks arising from a common centre or united by their disclike bases to form a faceted and hollow sphere; cells halfmoon-, horseshoe-, kidney- or wedge-shaped to pyramidal, with 1–4 apical spines and stalks arising from centre of convex side, walls thin, smooth and transparent; chloroplast parietal, with a single pyrenoid; multinucleate; asexual reproduction by biflagellate, pear-shaped zoospores, released in a mucilaginous vesicle; akinetes formed from old and enlarged cells.

Probably cosmopolitan; planktonic and sometimes associated with bottom sediments or marginal vegetation in non- or slowflowing reaches of nutrient-rich rivers, usually in acid or neutral waters.

1 Cells broadly halfmoon-shaped to roundly triangular, with a distinct stalk .... *S. spinulosum*
1 Cells wedge-shaped, heart-shaped or pyramidal, with a stalk gradually merging with cell .... *S. americanum*

Sorastrum americanum (Bohlin) Schmidle 1900

Basionym: *Selenosphaerium americanum* Bohlin
Synonym: *Selenastrum americanum* (Bohlin) Schmidle
17890010 **Pl. 94A** (p. 389)
Cells 4.8–20 μm wide, about equal in length and width, wedge- to heart-shaped or pyramidal, the outer side

concave with a pair of blunt apical spines on each side of cell, gradual transition between stalk and cell, away from stalk the cell widens to become polygonal in shape, sides of cells join to form a hollow and sometimes perforated central sphere.

Probably cosmopolitan; so far reported in the British Isles associated with surfaces and in the plankton of lochs in the Outer Hebrides and a few lochs in the Scottish Highlands.

Sorastrum spinulosum Nägeli 1849

Synonyms: *Sorastrum spinulosum* Nägeli var. *crassispinosum* Hansgirg, *S. crassispinosus* (Hansgirg) Bohlin, *S. cornutum* Reinsch

17890020 **Pl. 94B** (p. 389)

Cells rather tightly grouped, 8–20 μm wide and 6–22(–26) μm long, broadly halfmoon-shaped or roundly triangular with the outermost margin concave, a pair of spines arising on outer margins of each cell (spines 3.5–8 μm long), stalks clearly differentiated and slightly wider near base, adjacent basal cells often fused to form a small and often indistinct central sphere.

Probably cosmopolitan; usually found in bog pools or associated with aquatic vegetation growing in the quiet, shallow margins of ponds and lakes in north-western parts of the British Isles; generally most common in slightly acid water.

The most widely distributed species within the genus *Sorastrum*.

Sphaerobotrys Butcher 1932

Colonies forming low expansions over rocks, consisting of short cells in vertical and oblique rows, often with cells attached to remnants of mother cell wall, enclosed within a thin mucilaginous envelope; cells ellipsoidal or inversely pear-shaped; chloroplast parietal, with a single pyrenoid; asexual reproduction by 2(–4) autospores, which remain on release at end of empty mother cell wall or within the mucilaginous envelope.

Europe; attached to rocks, boulders and other hard surfaces in lakes, streams and rivers. Monospecific.

Colonies often thick and brown towards centre where older cells have lost their contents.

Sphaerobotrys fluviatilis Butcher 1932

17900010 **Pl. 86L** (p. 352)

Cells about 5 μm long, initially ellipsoidal, later more pear-shaped, arranged in oblique rows.

Distribution as for genus; grows in the British Isles on rocks, glass and other artificial surfaces in lakes, streams and shallow stretches of rivers, probably widely distributed, but easily overlooked.

Sphaerocystis Chodat 1897

Colony microscopic, consisting of 4, 8 or 16(–32) regularly arranged cells within a spherical to ellipsoidal and homogeneous mucilaginous envelope; cells spherical, with walls thin and smooth; chloroplast parietal, cup-shaped, with a single pyrenoid; asexual reproduction by biflagellate zoospores and autospores; autospores 4 or 8(–16) per cell and zoospores 2, 4, 8 or 16 per cell, both often with an eyespot, autospores and zoospores released by gelatinization of mother cell wall which persists.

Probably cosmopolitan; planktonic in water bodies having widely different nutrient status.

Considerable taxonomic confusion continues to surround this genus and its separation from others with a similar morphology (see notes under *Coenochloris* and *Planktosphaeria*). Korshikov (1953) emended the description of the genus but discounted Chodat's (Chodat, 1897) mention of zoospores and considered it to reproduce only by autospores. It is now recognized that Korshikov was in error and that *Sphaerocystis* reproduces by both autospores and zoospores. The principal character separating *Coenochloris fottii* from *Sphaerocystis schroeteri* is that the former reproduces only by autospores and remnants of the mother cell wall persist only for a very short time following release of the daughter cells. Most British records of *Sphaerocystis schroeteri* should probably be attributed to *Coenochloris fottii*.

1 Cells 4–9 μm wide; sporangia up to 12.5 μm wide .... *S. planctonica*
1 Cells 6–15 μm wide; sporangia up to 22 μm wide .... *S. schroeteri*

Sphaerocystis planctonica (Korshikov) Bourrelly 1966

Basionym: *Palmellocystis planctonica* Korshikov

17910010 **Pl. 99A** (p. 403)

Coenobia of 4, 8 or 16(–32) cells within a spherical mucilaginous envelope ((35–)50–124(–150) μm wide) which sometimes becomes indistinct with age; cells 4–9 μm wide and sporangia up to 12.5 μm wide.

Probably cosmopolitan; in the British Isles planktonic in lakes and the slow-flowing reaches of rivers (e.g. Thames, Great Ouse).

It seems very probable that this species is only a form of *Sphaerocystis schroeteri*.

Sphaerocystis schroeteri Chodat 1897

17910020 **Pl. 86H** (p. 352)

Colony of 4, 8, 16 or 32(–64) cells within a spherical mucilaginous envelope (up to 200 μm wide); cells 6–15 μm wide and sporangia up to 22 μm wide.

Europe; in the British Isles considered common and widespread in lake plankton, but difficult to know whether this is correct since many records are probably based on misidentifications (see comment under *Coenochloris fottii*, p. 352).

Sporotetras Butcher 1932

Colonies low and cushion-like, consisting of cells arranged in 4s or occasionally in pairs within a mucilaginous envelope; cells oval, ellipsoidal or obovoid, sometimes slightly asymmetrical or laterally flattened; chloroplast parietal and having a lobed margin, with a single pyrenoid; asexual reproduction probably by 2 or 4 autospores.

Europe; attached to hard surfaces in rivers and streams.

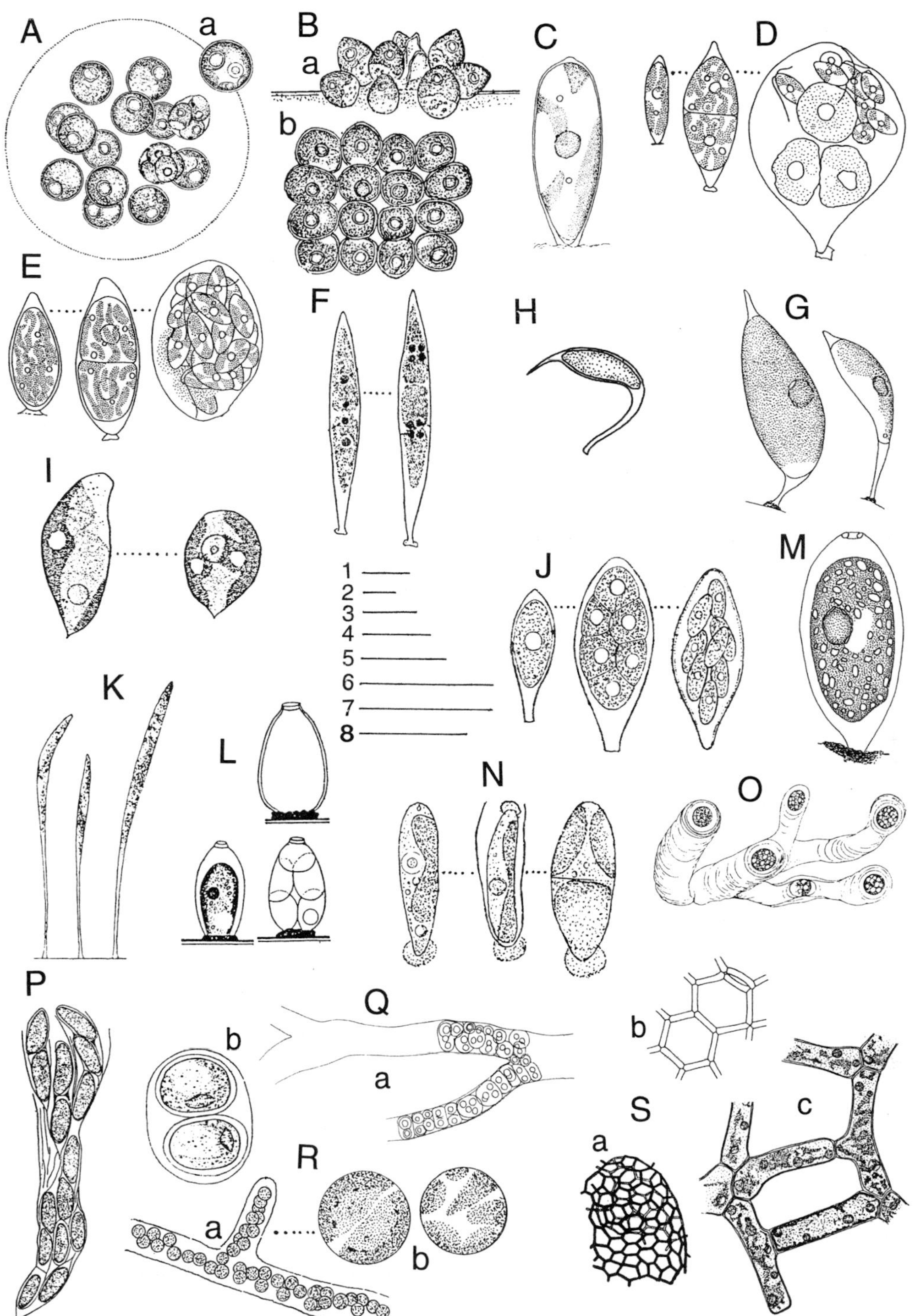

Plate 99 A–S

A. *Sphaerocystis planctonica* (p. 402): vegetative cells and mother cells containing autospores, with eyespots associated with these spores scarcely visible, a. detail of a single cell; **B**. *Fernandinella alpina* var. *semiglobosa* (p. 355): a. side view, b. surface view; **C**. *Characiochloris apiculata* (p. 333); **D**. *Characiochloris characioides* (p. 333): single vegetative cell, division of contents into 2 portions and a swollen cell containing many biflagellate zoospores; **E**. *Characiochloris sessilis* (p. 333): single vegetative cell, division of contents into 2 and a swollen cell containing biflagellate zoospores; **F**. *Characium angustum* (p. 334); **G**. *Characium ornithocephalum* var. *pringsheimii* (p. 334); **H**. *Characium ornithocephalum* var. *harpochytriforme* (p. 334); **I**. *Characium pseudopyriforme* (p. 334); **J**. *Characium sieboldii* (p. 334): single vegetative cell, division of contents and a swollen cell containing many zoospores; **K**. *Characium westianum* (p. 334); **L**. *Hydrianum coronatum* (p. 358): characteristic colar around cell apex of an empty cell (top), vegetative cell (lft) and a cells containing zoospores; **M**. *Pseudocharacium obtusum* (p. 380); **N**. *Pseudococcomyxa simplex* (p. 381); **O**. *Hormotila mucigena* (p. 358); **P**. *Ecballocystis fluitans* (p. 353); **Q**. *Palmodictyon viride* (p. 376): a. cells within a mucilaginous sheath, b. cells each containing a cup-shaped chloroplast and pyrenoid; **R**. *Palmodictyon varium* (p. 376): a. cells within a mucilaginous sheath, b. individual cells containing more than one chloroplast and without a pyrenoid; **S**. *Hydrodictyon reticulatum* (p. 358): a–c. views at different magnifications of a portion of the net-like coenobium.

Scale bars: 1, 30 μm; 2–6, 10 μm; 7, 1 mm; 8, 5 mm. 1 for D, Qa, Ra, Sc; 2 for B, F, H, J, K, O, P; 3 for Aa, C, I; 4 for A, G, M, Qb; 5 for E, L; 6 for Rb, N; 7 for Sb; 8 for Sa.

Sporotetras pyriformis Butcher 1932

17920010 **Pl. 108I** (p. 435)

Cells 4–7 μm wide, elongate in profile but ovoid or pear-shaped when viewed from above.

Europe; in the British Isles on rocks and other hard surfaces (including glass), often in swiftly flowing streams and shallower stretches of rivers; probably widely distributed but easily overlooked unless sufaces such as glass are examined.

Tetrachlorella Korshikov 1939

Coenobia flat, rhomboidal, usually consisting of (2–)4 cells, occasionally forming 16-celled syncoenobia, inner cells almost parallel to one another and alternately arranged (apices in contact), outer cells at an angle to central axis and parallel to one another, outer cells contacting the inner ones by their sides, enclosed within a thick mucilaginous envelope; cells elongate, ovoid to ellipsoidal with rounded apices, smooth-walled or a brown ring or pair of cylinders at apices; chloroplast parietal, initially single and later dividing into 2 or 4, each with a pyrenoid (sometimes difficult to distinguish); asexual reproduction by 2 or 4 autospores, released by fracture of mother cell wall which soon gelatinizes.

Probably cosmopolitan; planktonic in a wide range of water bodies. Four species of which only one is recorded from the British Isles.

Tetrachlorella alternans (G.M.Smith) Korshikov 1939

Basionym: *Crucigenia alternans* G.M.Smith
Synonym: *Scenedesmus arcuatus* (Lemmermann) Lemmermann var. *irregularis* E.A.Flint

17940010 **Pl. 84H** (p. 346)

Cells (4–)4.5–8(–10) μm wide, 6.5–12(–18) μm long, ellipsoidal, with broadly rounded or bluntly pointed apices, grouped in pairs or 4s, within a mucilaginous envelope up to 220 μm across.

Probably cosmopolitan; planktonic and associated with submerged surfaces, only reported in the British Isles from a reservoir at Barnes Elms, Hammersmith, London (Flint, 1950); undoubtedly more widely distributed.

The illustration by Flint (1950, fig. 4E) is very characteristic and the record is accepted here although she considers the identification as 'uncertain' (as *Scenedesmus arcuatus* var. *irregularis*).

Tetradesmus G.M.Smith 1913

Synonym: *Victoriella* Wołoszyńska

Coenobia (2–)4-celled, sometimes with cells in contact by long axis for much of their length, each cell spherical in dorsal or ventral view, forming a square and without a mucilaginous envelope; cells spindle-shaped or cylindrical, the inner wall straight where in contact and the outermost wall straight or concave, with apices directed away from centre; chloroplast parietal, with a single pyrenoid; asexual reproduction by (2–)4 autospores, released by fracture of mother cell wall.

Probably cosmopolitan; planktonic in various types of aquatic habitat.

1 Cells straight or sometimes slightly asymmetrical; apices not curved, conically acute .... *T. cumbricus*
1 Cells slightly outwardly curved towards apices; apices often horn-like .... *T. wisconsinensis*

Tetradesmus cumbricus G.S.West 1915

Synonym: *Scenedesmus cumbricus* (G.S.West) Chodat

17950010 **Pl. 97O** (p. 396)

Cells 4.2–13.5 μm wide, 14–30 μm long, broadly spindle-shaped, straight or sometimes slightly asymmetrical, with the apices conically acute and cell wall thickened or sometimes forming a short papilla.

British Isles (England); known from the plankton of several lakes in the English Lake District.

Tetradesmus wisconsinensis G.M.Smith 1913

Synonym: *Scenedesmus wisconsinensis* (G.M.Smith) Chodat

17950020 **Pl. 97I** (p. 396)

Cells (2–)4–6(–8) μm wide, (8–)12–18(–25) μm long, with sides almost parallel in central region and apices conical with a short, slightly outwardly bent horn.

Probably cosmopolitan; in the British Isles recorded from the plankton of nutrient-poor lakes and ponds as well as in slow-flowing water.

Tansferred to the genus *Acutodesmus* by Tsarenko (2001).

Tetraedron Kützing 1845

Synonym: *Astericium* Corda

Cells solitary, flat or slightly twisted, 3-, 4- or 5-sided (triangular, pyramidal, polyhedral), with corners blunt or rounded and sometimes terminating in a papilla-like wall thickening or spine; walls 3-layered, thick or thin, smooth to wrinkled or distinctly granular; chloroplast parietal, 1 to several, often massive, with several pyrenoids; asexual reproduction by autospores, released in a gelatinous envelope following splitting of the mother cell wall into two parts.

Probably cosmopolitan; planktonic or associated with bottom sediments or aquatic macrophytes in ponds, lakes and rivers.

Easily confused with certain xanthophyte genera which have a similar cell morphology, but differ in a number of important characters (e.g. xanthophytes normally lack starch although pyrenoids sometimes present). Several species of *Tetraedron* reported from the British Isles by the Wests have been transferred to various xanthophyte genera (see Bourrelly, 1951). For example, *Tetraedron trigonum* (Nägeli) Hansgirg is now *Goniochloris fallax* Fott, *T. enorme* is *Pseudostaurastrum enorme* (Ralfs) Chodat, *T. hastatum* (Reinsch) Hansgirg is *Pseudostaurastrum hastatum* (Reinsch) Chodat, and *T. lobulatum* Nägeli has been placed in the synonymy of *Isthmochloron lobulatum* (Nägeli) Skuja. According to West & Fritsch (1927, p. 126), only those species known to produce autospores should be referred to *Tetraedron* since morphologically similar stages are known to occur in *Pediastrum*, *Hydrodictyon* and *Oocystis*. Considerable doubt attaches to the current status of many of the early records of *Tetraedron* species (including forms and varieties). Many have been reported on just a single occasion and not always accompanied by an illustration (e.g. *Tetraedron horridum* West *et* G.S.West (W. & G.S.West, 1897), *T. limneticum* Borge, *T. octaedricum* (Reinsch) Hansgirg, *T. muticum* (A.Braun) Hansgirg, *T. punctulatum* (Reinsch) Hansgirg, *T. decussatum* (Rabenhorst) Hansgirg, *T. dodecaedricum* (Reinsch) Hansgirg, *T. tetragonum* (Nägeli) Hansgirg). Some are keyed out, described and illustrated by Philipose (1967) who recognizes that many are parallel morphological forms belonging to different algal classes.

1 Cells pyramidal or cushion-shaped, each rounded corner terminating in a papilla-like wall thickening 2
1 Cells 4- or 5-sided and with a spine arising at each corner 3
2 Cells flat and cushion-shaped ....*T. minimum*
2 Cells pyramidal ....*T. regulare*
3 Cells 4-sided, flat or slightly twisted, with each corner drawn out into a lobe bearing a spine ....*T. incus*
3 Cells 5-sided, flat and not twisted, with a distinct spine at each corner ....*T. caudatum*

Tetraedron caudatum (Corda) Hansgirg 1888

Basionym: *Astericium caudatum* Corda
Synonyms: *Polyedrium pentagonum* Reinsch, *P. caudatum* Corda, *Tetraedron caudatum* vars. *incisum* (Lagerheim) Brunnthaler, *punctatum* Lagerheim
17960010 **Pl. 89G** (p. 363)
Cells (6–)9–23 μm wide, 5-sided, flat and lobed, frequently almost star-shaped, often with one indentation deeper than the other 4 (most evident in young cells) giving the cells bilateral rather than radial symmetry, with sides more or less concave and corners rounded, each corner often bearing a spine; spines of different lengths (up to 6 μm long) and not all lying in same plane; walls relatively thick and smooth, rough, slightly warty or with small spots (punctae).

Probably cosmopolitan; fairly common in the British Isles where planktonic and associated with submerged surfaces in a wide range of aquatic habitats including lakes (e.g. Loch Leven, Scotland), canals and rivers.

The taxonomic significance of the orientation of the spines is unclear although is a character that has been used to distinguish var. *longispinum* (2 spines bent to one side, 3 spines bent to the other side). Differences in wall structure are no longer considered sufficient justification for recognizing infraspecific taxa; a warty cell wall may be net-like when viewed with the electron microscope (see Kováčik, 1975b).

Tetraedron incus (Teiling) G.M.Smith 1926

Basionym: *Tetraedron regulare* Kützing var. *incus* Teiling
Synonyms: *Tetraedron incus* G.M.Smith var. *irregulare* G.M.Smith, *T. minimum* Hansgirg var. *longispinum* Deflandre *sensu* Korshikov, *Chlorotetraedron incus* (Teiling) MacEntee et al.
17960020 **Pl. 89F** (p. 363) **CD**
Cells (3–)5–23(–33) μm wide, 4-sided, sometimes slightly twisted to almost flat, sides concave (convex in older cells), with one side longer than other 3 and corners drawn out into long lobes, each corner terminating in a spine up to 7 μm long; walls smooth.

Probably cosmopolitan; in the British Isles planktonic and also associated with the bottom or submerged objects in ponds and ditches, most common in the quiet backwaters of slow-flowing rivers.

It differs from other *Tetraedron* species in having a smooth cell wall (slightly wrinkled under the electron microscope, see Kováčik, 1975b) and possessing zoospores and multinucleate cells. Komárek & Kováčik (1985) have transferred it to the genus *Chlorotetraedron* MacEntee et al. on the basis that it reproduces both by zoospores and autospores. We consider there is insufficient justification for its transfer and prefer to retain it in the genus *Tetraedron*.

Tetraedron minimum (A.Braun) Hansgirg 1888

Basionym: *Polyedrium minimum* A.Braun
Synonyms: *Tetraedron platyisthmum* (W.Archer) G.S.West, *T. quadratum* (Reinsch) Hansgirg
17960030 **Pl. 89H** (p. 363) **CD**
Cells 5–25 μm wide, 4-sided, flat and cushion-shaped, with opposite sides of young cells often deeply concave and older cells with flat or slightly convex sides, sometimes with each rounded corner bearing 1–5 teeth and terminating in a papilla-like wall thickening; walls thick and smooth, wrinkled or warty.

Probably cosmopolitan; widely distributed and very common in the British Isles in the plankton and associated with submerged surfaces in many types of aquatic habitat, most common in ponds and small lakes.

Smooth or granular-walled forms are known from the British Isles and are sometimes recognized as distinct varieties (see comments in Kováčik, 1975b), for example var. *scrobiculata* (granular-walled) and the type (smooth-walled). Another variety recorded from the British Isles, var. *apiculato-scrobiculatum* (Reinsch) Skuja fo. *elegans* Hortobágyi, is characterized by having 3–7 fine teeth arising at each cell corner. These infraspecific taxa are a few of many characterized by minor morphological differences that are believed to have little or no taxonomic significance.

Tetraedron regulare Kützing 1845

Synonyms: *Polyedrium tetraedricum* Nägeli *sensu* West *et* G.S.West *pro parte*, *Tetraedron quadrilobum* G.M.Smith
17960040 **Pl. 89E** (p. 363)
Cells 5–12(–15) μm wide, pyramidal with sides concave, straight or slightly convex, each corner slightly rounded and with a papilla-like wall thickening.

Probably cosmopolitan; planktonic or associated with submerged objects in a wide range of habitat types in the British Isles. Difficult to determine just how widespread and common in the British Isles because in the past *Tetraedron regulare* has been confused with the xanthophyte *Tetraedriella regularis* (see below).

Considerable confusion surrounds this species since many literature records are undoubtedly misidentfications of the xanthophyte *Tetraedriella regularis* (Kützing) Fott, which is very similar in form but distinguished by its greater cell size (30–40 μm wide) along with an absence of pyrenoids and starch (see Komárek & Fott, 1983). It is clear from the size range given by West & Fritsch (1927) for *T. regulare* (13.5–40 μm wide) that many earlier records of the Wests' *Polyedrium tetraedricum* were misidentifications of *Tetraedriella regularis*. *Polyedrium tetraedricum* was regarded as a synonym of *Tetraedron regulare* when West & Fritsch's *Treatise* was published in 1927. Only later was it correctly placed in the synonymy of *Tetraedriella regularis* (see Ettl, 1978). In many of the Wests' publications, the taxa mentioned do not have accompanying illustrations or sufficient information to decide to which of the 2 genera they correctly belong. Only very occasionally is the information sufficiently detailed, for example, the record of *Polyedrium tetraedricum* from the west of Ireland (West, 1892a) is definitely attributable to *Tetraedriella regularis* (cells range from 30–38 μm wide), whereas the record from Cambridge by G.S. West (1899b) of *Tetraedron regulare* is correct since 'a very minute form; diam. sine spin. 8 μ, cum spin. 14 μ'. Various forms described by the Wests (fo. *pachyderma* West (West, 1892b) and fo. *minima* West *et* G.S.West (W. & G.S. West, 1899) have little or no significance, although the small form *minima* is almost certainly attributable to *Tetraedron regulare*. Unfortunately, the Wests seldom mention starch or pyrenoids in their descriptions, characters essential if records are to be correctly assigned. The new combination *Tetraedriella regularis* by Fott (1967) was based upon *Tetraedron regulare*, since he considered Kützing's original description (Kützing, 1845) to correspond to the xanthophyte genus *Tetraedriella* rather than to *Tetraedron*. If his interpretation proves correct then a new name is required for *Tetraedron regulare* Kützing *sensu* West & G.S.West *pro parte*.

Tetrastrum Chodat 1895

Coenobia 4-celled, quadrate or rhomboidal, with cells arranged in a flat plate, sometimes having an internal space, usually with a thin mucilaginous envelope; cells ovoid, semi-circular or broadly triangular, with rounded apices and bearing from outer wall one or more spines or papillae; chloroplast single, plate-shaped, with or without a pyrenoid; asexual reproduction by 4 autospores, released by rupture of mother cell wall.

Probably cosmopolitan; usually planktonic in a wide range of aquatic habitats.

Distinguished from *Crucigenia* by the presence of spines (more rarely papillae) on the outer cell margin (except *T. triangulare* = *T. glabrum*) and an inability to produce a complex of several closely associated coenobia (syncoenobia). Like many other chlorococcalean genera, it is artificial and 'its maintenance is purely a matter of convenience' (see West & Fritsch, 1927, p. 138).

1 Outermost convex cell wall not bearing a spine(s) 2
1 Outermost convex or concave cell wall bearing one or more spines 3
2 Pyrenoids present .... *T. triangulare*
2 Pyrenoids absent .... *T. komarekii*
3 Cells ovoid, with a single long spine on the outer cell wall .... *T. elegans*
3 Cells triangular and/or trapezoid, with more than one spine on the outer cell wall 4
4 Cells with flat or slightly concave outer walls, each bearing one short and one long spine .... *T. heteracanthum*
4 Cells with convex outer walls, each usually bearing 4–6 spines .... *T. staurogeniaeforme*

Tetrastrum elegans Playfair 1917

17970010 **Pl. 94F** (p. 389)

Coenobia 4–16.5 μm across, quadrate to rectangular in outline; cells 2–9.2 μm wide, ovoid, with an indistinct space between cells and a long median spine (10–20 μm long) arising from convex outermost surface; pyrenoid present.

Probably cosmopolitan; recorded in the British Isles on very few occasions in small lakes and ponds.

Williams's (1965) illustration of his doubtful record (p. 442, fig. 6Q) in a sample taken from a natural freshwater pool, The Serpentine in Eaton Park, Chester, is typical of this species as currently understood.

Tetrastrum heteracanthum (Nordstedt) Chodat 1895

Basionym: *Staurogenia heteracantha* Nordstedt

17970030 **Pl. 94G** (p. 389)

Coenobia up to 23 μm across, quadrate in outline; cells 2–11.5 μm wide, almost triangular, with the outermost wall often straight and only more rarely concave, bearing 2 curved or straight spines of unequal length, the short spine 1–24 μm long and the long spine 8–46 μm long; pyrenoid present.

Probably cosmopolitan; recorded in the British Isles in the plankton of ponds and rivers (e.g. River Gowy in Cheshire; Keston Ponds in Kent), rare.

Tetrastrum komarekii Hindák 1977

17970040 **Pl. 94C** (p. 389)

Coenobia 5–15 μm across, quadrate to rhomboidal in outline; cells 3–6 μm wide, triangular and trapezoid, outermost wall convex, with or without a small square internal space; pyrenoid absent.

Probably cosmopolitan; certainly more widely distributed in the plankton of ponds, small lakes and rivers in the British Isles than records indicate (see below).

Williams's (1965) figure (p. 436, fig. 3H) of '*Crucigenia quadrata*' from the Serpentine, Eaton Park in Chester is attributable to this species as are probably earlier records (including fo. *octagona* (Schmidle) Schmidle) of the Wests (W. & G.S.West, 1900b, 1906) from Yorkshire and southern Ireland. *Crucigenia quadrata* Morren is generally regarded as a tropical species. It is possible that some early records under this name should be more correctly attributable to *Tetrastrum triangulare* (see Komárek & Fott, 1983).

Tetrastrum staurogeniaeforme (Schröder) Lemmermann 1900

Basionym: *Cohniella staurogeniaeforme* Schröder

17970050 **Pl. 94D** (p. 389) **CD**

Coenobia 7–15 μm across, rhomboidal in outline; cells 3–6(–10) μm wide, triangular to trapezoid, outermost wall straight or convex, bearing 4–6 spines, each spine sometimes thickened at base and short or very long (occasionally 2–3 times length of cell).

Probably cosmopolitan; most widely distributed and commonly encountered species in the British Isles, where in the plankton of ponds, lakes, moats and slow-flowing rivers.

Tetrastrum triangulare (Chodat) Komárek 1974

Basionym: *Staurogenia triangularis* Chodat

Synonyms: *Tetrastrum glabrum* (Y.V.Roll) Ahlstrom *et* Tiffany, *T. staurogeniaeforme* var. *glabrum* Y.V.Roll, *Crucigenia minima* (Fitschen) Brunnthaler

17970060 (17970020) **Pl. 94E** (p. 389)

Coenobia about 7–15 μm across, rhomboidal or more rarely quadrate in outline; cells 2–8 μm in breadth, more or less triangular, outermost wall convex and spineless; pyrenoid present.

Probably cosmopolitan; widely distributed in the British Isles in the plankton of slow-flowing rivers, more rare in ponds and lakes (e.g. Loch Leven, Scotland).

DOUBTFUL RECORD

Tetrastrum rocklandiensis B.M.Griffiths 1927

Komárek & Fott (1983) consider this species, described from the Norfolk Broads by Griffiths (1927), to be attributable probably to *Micractinium quadrisetum*.

Trebouxia Puymaly 1924

Cells solitary or in groups of 2, 4, 8 or more cells, spherical, sometimes ellipsoidal to broadly pear-shaped; chloroplast axile (suspended in centre), massive, star-shaped, with numerous short outgrowths and a single central pyrenoid; asexual reproduction by 8, 16 or 32 naked and biflagellate zoospores or autospores, released by fracture of lateral wall of mother cell; sexual reproduction by isogamous gametes similar to zoospores.

Cosmopolitan; majority are lichen phycobionts, rarely free-living on subaerial surfaces such as tree bark.

Trebouxia arboricola Puymuly 1924

Basionym: *Cystococcus humicola* Treboux *non* Nägeli
Synonym: *Trebouxia humicola* (Treboux) G.S.West *et* F.E.Fritsch
17990020 (17990010) **Pl. 87F** (p. 354)

Cells 2–25 μm wide, spherical, walls relatively thin; chloroplast hollow with a lateral notch and one or more pyrenoids.

Probably cosmopolitan; a common terrestrial alga on tree trunks, woodwork, masonry and other subaerial surfaces (often with *Desmococcus* and *Apatococcus*), as well as on soil where it sometimes forms an 'almost pure strata' (West & Fritsch, 1927) although Ahmadjian (1967) considers such 'strata' to be most probably leprose lichen.

Considerable doubt attaches to this and other species in this ill-defined genus. Most information on the genus concerns algae associated with lichens and is based upon laboratory culture studies.

Treubaria C.Bernard emend. O.Reymond 1979

Synonym: *Borgea* G.M.Smith

Cells solitary, elongate-triangular, tetrahedral to spherical, sometimes flat, walls thin and colourless or brownish, with age outer wall swells and separates from protoplast to form 3–8 long, acute appendages lying in one or more planes, sometimes enclosed within a mucilaginous envelope; chloroplast parietal, 1 to several, not infrequently spreading over cell surface, with 1–4 pyrenoids (depending on age); uni- or multinucleate; asexual reproduction by 2 to 4(–8) naked hemizoospores, released by disintegration and gelatinization of mother cell wall.

Probably cosmopolitan; planktonic and more rarely associated with submerged surfaces and periphyton in standing water or the slow-flowing reaches of very nutrient-rich rivers.

1 Cells triangular and sides straight or convex .... *T. schmidlei*
1 Cells triangular and sides concave ....2.
2 Cells roundly triangular to almost spherical, without a mucilaginous envelope ....*T. setigera*
2 Cells distinctly triangular, with a mucilaginous envelope ....*T. triappendiculata*

Treubaria schmidlei (Schröder) Fott *et* Kováčik 1975

Basionym: *Polyedrium schmidlei* Schröder
Synonyms: *Tetraedron schmidlei* (Schröder) Lemmermann, *Echinosphaerella limnetica* G.M.Smith, *Treubaria limnetica* (G.M.Smith) Fott *et* Kováčik
18010030 **Pl. 89N** (p. 363)

Cells (6–)8–19 μm wide, roundly tetrahedral to spherical, with sides straight or slightly convex and having broadly rounded apices, with (3–)4 appendages in one or more planes; appendages (5–)12–40(–60) μm long, noticeably wider at base, sometimes with minute (10–20(–50)) and uniformly spaced spines, without mucilage except during reproduction.

Probably cosmopolitan; only a few records in the British Isles from the plankton of ponds and small lakes.

Treubaria setigera (W.Archer) G.M.Smith 1933

Basionym: *Tetrapedia setigera* Archer
Synonyms: *Polyedrium trigonum* Nägeli var. *setigerum* (W.Archer) Schröder, *Tetraedron trigonum* (Nägeli) Hansgirg var. *setigerum* (W.Archer) Lemmermann
18010010 **Pl. 89M** (p. 363)

Cells 5–13 μm wide, always distinctly triangular, with sides slightly concave and with broadly rounded apices bearing 3 relatively narrow, spine-like appendages lying in a single plane; appendages 6–20 μm long, without a mucilaginous envelope.

Probably cosmopolitan; widely distributed throughout the British Isles especially in lake plankton, never abundant.

Treubaria triappendiculata C.Bernard 1908

Synonyms: *Polyedrium schmidlei* Schröder var. *euryacanthum* Schmidle, *Tetraedron schmidlei* (Schröder) Lemmermann var. *euryacanthum* (Schmidle) Lemmermann, *T. triappendiculatum* (C.Bernard) Wille, *Treubaria euryacantha* (Schmidle) Korshikov
18010020 **Pl. 89O** (p. 363)

Cells (4–)6–13(–14) μm wide, roundly triangular to spherical, with sides slightly concave and having broadly rounded apices bearing 3 appendages in one plane, or more rarely 4 appendages lying in two planes; appendages (0.9–)2–4(–5) μm wide at base, (6–)8–21 (–40) μm long, surrounded by a mucilaginous layer (5 μm wide).

Probably cosmopolitan; fairly common and widely distributed in the British Isles in a variety of aquatic habitats including ponds, small lakes, drainage ditches, moats, and slow-flowing rivers; cell densities of 2700 per mL were recorded in Oak Mere, Cheshire when it appeared in May and June 1994 (see Swale, 1968).

Trochiscia Kützing 1845

Synonym: *Glochiococcus* De Toni

Cells solitary or clustered, spherical; walls often somewhat thickened, usually covered by spines, warts, or net-like thickenings; chloroplast parietal, 1 to several, disc-like, each with a pyrenoid; uni- or multinucleate; asexual reproduction by autospores; palmelloid stages known in some species.

Probably cosmopolitan; commonly associated with damp terrestrial surfaces and in undisturbed water, rarely planktonic.

Many of the large number of described 'species' are very doubtful and some are now known to be the resting stages of other algae (e.g. zygospores of volvocalean algae, dinoflagellate cysts, a stage in the life history of *Chlorococcum*). According to West & Fritsch (1927, p. 120), 'Some eight or nine Brit. sp., distinguished by their external ornamentation have been recorded, but all of these are not likely to be independent forms'. The relationship of some of these species to other algae remains uncertain with many not mentioned in Komárek & Fott's (1983) monograph of chlorococcalean algae. Only the most common and widely distributed species in the British Isles (according to West & Fritsch, 1927), and *Trochiscia planktonica* E.M.Lind *et* Pearsall, are described below. The following are very rarely reported, often considered doubtful, and a few are in the 'Fritsch Collection' of illustrations: *Trochiscia insignis* fo. *minor* West (W. & G.S.West, 1899, 1904), *T. paucispinosa* West (West, 1893), *T. uncinata* West (West, 1892b), *T. stagnalis* Hansgirg (W. & G.S.West, 1897), *T. granulata* (Reinsch) Hansgirg, *T. pachyderma* (Reinsch) Hansgirg (Woodhead & Tweed, 1947) and *T. prescottii* Siemińska (Belcher & Swale, 1999).

1 Cells with a net-like surface ornamentation, without spines or papillae....................*T. reticularis*
1 Cells with or without a net-like surface ornamentation and covered by spines or papillae....................2
2 Cells with net-like ornamentation and minute spines or papillae, often within a mucilaginous envelope....................*T. planktonica*
2 Cells without net-like ornamentation and covered by fine papillae or minute spines, without mucilage....................3
3 Walls covered by fine papillae....................*T. aspera*
3 Walls covered by minute spines....................*T. hirta*

Trochiscia aspera (Reinsch) Hansgirg 1888

Basionym: *Acanthococcus asper* Reinsch

18020010 **Pl. 87G** (p. 354)

Cells 13–29 μm in diameter, spherical, walls thick and covered by numerous fine papillae.

Probably cosmopolitan; fairly common in the British Isles on or within soil and in sheltered non-flowing aquatic habitats.

Trochiscia hirta (Reinsch) Hansgirg 1888

Basionym: *Palmella hirta* Reinsch

18020020 **Pl. 87E** (p. 354)

Cells 17–27 μm in diameter, spherical, covered with very minute spines.

Probably cosmopolitan; in the British Isles reported on soil especially from near base of trees.

Trochiscia planctonica E.M.Lind *et* Pearsall f. 1945

18020040 **Pl. 87I** (p. 354)

Cells solitary or in groups of 4–32 cells, within an almost spherical to pyramidal mucilaginous envelope; cells 10–14(–20) μm in diameter, spherical, walls brownish and covered by net-like ridges and short warts or spines (1–2 μm long).

British Isles (Ireland); recorded only from the plankton of lakes in north-west Ireland.

Komárek & Fott (1983) consider that this species might be a xanthophyte or spore of another chlorococcalean alga.

Trochiscia reticularis (Reinsch) Hansgirg 1888

Basionym: *Acanthococcus reticularis* Reinsch

18020050 **Pl. 87H** (p. 354)

Cells (11.5–)18–32(–39) μm in diameter, spherical, walls moderately thick and covered by a network of irregular ridges forming polygons.

Probably cosmopolitan; reported in the British Isles only from Kent, where it occurs in ponds on Keston Common, and cultures from chalk, sand and clay soils; undoubtedly considerably more frequent than records indicate.

Similar, or possibly identical, to *T. prescottii* Siemińska. The latter has been reported from Vision Park Pool, Histon, near Cambridge (Belcher & Swale, 1999).

Westella De Wildemann 1897

Synonym: *Tetracoccus* W.West *non Tetracoccus* Engelmann ex Parry

Coenobia quadrate to longitudinally elongate, almost flat or concave, consisting of 4, 8 or 16 cells or forming multicellular syncoenobia, cells sometimes arranged in the form of a cross (2 cells in one plane and 2 in other plane) and connected by remnants of mother cell wall, enclosed within a mucilaginous envelope; cells spherical to broadly ovoid, smooth-walled; chloroplast parietal, single and cup-shaped, with a pyrenoid enclosed by starch and with grains of volutin and oil droplets also present; asexual reproduction by (1–)2 or 4 autospores, released by fracture of outermost wall of mother cell, the daughter cells remain attached to wall remnants and consist of 2(–4) approximately equal parts.

Probably cosmopolitan; usually planktonic in various types of water.

Westella botryoides (W.West) De Wildeman 1897

Basionym: *Tetracoccus botryoides* W.West

18030010 **Pl. 84J** (p. 346)

Coenobia of 4(–8) quadrately arranged cells lying in one plane, spherical to ovoid, up to 90 μm across; cells (3–)5–7(–13) μm wide, spherical.

Probably cosmopolitan; widely distributed in the British Isles where planktonic and associated with submerged surfaces, especially in nutrient-poor ponds and lakes.

ORDER OEDOGONIALES

BY ROBERT HUXLEY AND ALLAN PENTECOST

A very distinctive order containing three genera of branched or unbranched filamentous species having a specialized and unique form of sexual reproduction. Two genera (*Oedogonium* and *Bulbochaete*) are known from the British Isles. Species identification depends on examining fertile material showing different stages in reproduction. To identify species it is necessary to know whether the sexual reproductive organs develop on the same (monoecious) or different plants (dioecious), whether male gametes develop in cells on normal filaments (macrandrous forms) or in one to few-celled filaments known as dwarf males (nannandrous forms), or if the spores giving rise to dwarf males (androspores) are either formed in cells on the same filaments bearing the oogonia (gynandrosporous species) or in cells on different filaments (idioandrosporous species). Another important character relates to the ornamentation of the thick-walled and often dark-coloured zygospores resulting from fertilization. In *Bulbochaete* there are additional characters associated with a cell that is cut-off below the oogonium during development, the so-called suffultory cell (see Fig. 5).

Bulbochaete C.Agardh 1817 CD

BY ALLAN PENTECOST

Filaments richly branched, each cell bears a long hair with a bulbous base and two such hairs on a terminal cell; chloroplasts as in *Oedogonium* (see p. 413); sexual reproduction oogamous, similar to *Oedogonium* (see generic description) but differing in the following details: suffultory cell normally divides into 2 cells immediately below it, each cell normally contains no chloroplasts and is shorter than vegetative cells though similar in width.

Probably cosmopolitan, most commonly attached to aquatic plants and occasionally on stones at the edges of ponds, lakes, streams and marshes, although often becoming detached and free-floating on exposed shores; occurs over a wide pH range in nutrient-poor to very nutrient-rich waters, although little else is known about their ecology. Sexual reproduction occurs from spring to summer.

The genus is characterized by its branched habit and long hairs with bulbous bases. Although with fewer species than *Oedogonium*, identification can be difficult since some key characters can vary within collections. Important characters are the size, shape of the oogonia and oospore and ornamentation of the oospore, whether the species is macrandrous or nannandrous, and the division of the suffultory cell (Fig. 5). A number of species are distinguished on the position of the transverse septum dividing these cells which may be median (Fig. 5A), basal (Fig. 5B), superior (Fig. 5B) in position or lacking. Length to breadth ratio of the vegetative cells is sometimes used to distinguish species. Oospores develop from almost any vegetative cell and may be terminal, intercalary or patent (see Fig. 5). The position of the oospores is usually of little taxonomic value but in patent oospores the lower suffultory cell may be pentagonal in shape. Unlike most *Oedogonium* species, the oospores of *Bulbochaete* are similar in shape to the oogonia. All oospores possess pore-like (porous) openings and are reddish or orange. The vegetative cells form mainly from a basal meristem and the cells rarely possess more than one cap. A few species have minutely granular walls, but none can be identified with certainty in the sterile, vegetative condition.

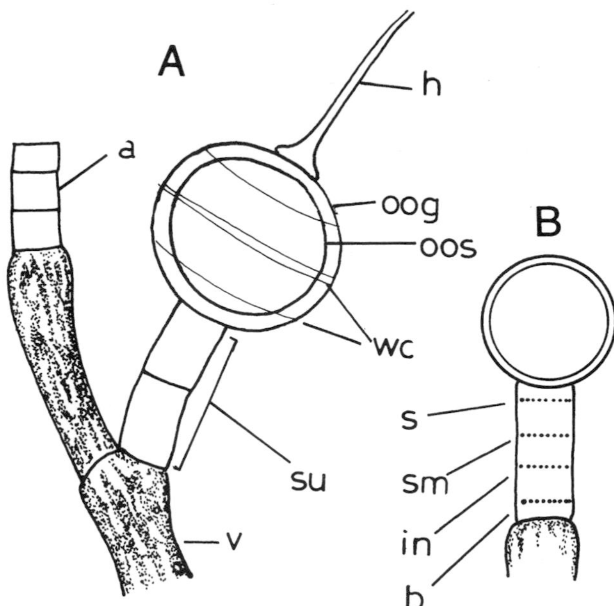

Figure 5. A. Illustrating the specialized reproductive structures of the genus *Bulbochaete*: a, antheridium; h, hair; oog, oogonium; oos, oospore; su, suffultory cell showing a median division into upper and lower cells; v, vegetative cell; wc, wall caps of oogonium. The illustration shows a patent oospore. **B.** Divisions of the suffultory cell in *Bulbochaete*: b, basal; in, inframedian; s, superior; sm, supramedian.

1 Macrandrous species 2
1 Nannandrous species 4
2 Oogonia <40 μm long ....*B. nana*
2 Oogonia >40 μm long..3
3 Oospore 25–31 μm wide.......................*B. mirabilis*
3 Oospore 18–26 μm wide....*B. mirabilis* var. *gracilis*
4 Oospores smooth, without ornamentation 5
4 Oospores ornamented with striations or pits or both 7
5 Oogonia angular, often biconical but approximately equidimensional ....*B. angulosa*
5 Oogonia rounded, not angular 6
6 Division of suffultory cell basal; cell walls smooth ....*B. elatior*
6 Division of suffultory cell approximately median; cell walls granular.....................*B. obliqua*
7 Oospores with longitudinal ribs 8
7 Oospores without longitudinal ribs but with pits or net-like ornamentation 13
8 Vegetative cells approximately rectangular in cross section...................................*B. rectangularis*
8 Vegetative cells circular in cross section................9
9 Vegetative cells with wavy edges...........*B. repanda*

9 Vegetative cells without wavy edges....................10
10 Oogonia >40 μm in diameter...............................11
10 Oogonia <40 μm in diameter...............................12
11 Oogonia 46–56(–60) μm in diameter; oospore ornamented with broadly denticulate ribs ....*B. insignis*
11 Oogonia 68–70 μm in diameter; oospore ornamented with irregularly anastomosing ribs ....*B. imperialis* var. *regalis*
12 Oogonia 22–26 μm in diameter.............*B. pygmaea*
12 Oogonia 26–42 μm in diameter.................*B. minor*
13 Oospores with a net-like wall ornamentation ....*B. gigantea*
13 Oospores with a pitted cell wall 14
14 Division of suffultory cell basal.........*B. brebissonii*
14 Division of suffultory cell median or superior.....15
15 Division of suffultory cell median........................16
15 Division of suffultory cell superior 18
16 Oospores weakly angular......................*B. quadrata*
16 Oospores subspherical...17
17 Oogonia 60–80 μm wide, 50–65 μm long ....*B. setigera*
17 Oogonia 39–51 μm wide, 31–44 μm long ....*B. intermedia*
18 Oogonia 36–48 μm wide, 29–38 μm long ....*B. borealis*
18 Oogonia 51–56 μm wide, 44–50 μm long ....*B. sessilis*

Bulbochaete angulosa Wittrock *et* Lundell 1874

19010080 **Pl. 100A** (p. 411)

Cells 3–18 μm wide, 1.5 to 2.5 times as long as wide; oogonium angular, 36–42 μm wide, 33–39 μm long, usually below a terminal hair; oospore angular, 34–40 μm wide, 30–36 μm long, wall smooth; suffultory cell division supramedian or superior; gynandrosporous.

Europe, USA, Central Africa; in the British Isles known only from the Birmingham and Warwickshire areas.

Clearly distinguished by the angular oogonium and unornamented oospore.

Bulbochaete borealis Wittrock 1870

19010090 **Pl. 100B** (p. 411)

Cells 6–21 μm wide, 1.3 to 2 times as long as wide; oogonium subspherical, 40–48 μm wide, 35–40 μm long, below a vegetative cell; oospore 38–46 μm wide, 33–38 μm long, wall pitted; suffultory cell division superior; gynandrosporous.

USA, Scandinavia; only recorded in the British Isles by Harris (1933) from Chudleigh Knighton, Devon who considered the short vegetative cells and superior suffultory division to be distinctive.

Bulbochaete nordstedtii Wittrock, recorded from several sites in south-west England, Wales and Ireland (see W & G.S.West, 1902; Harris, 1933), probably belongs here. It differs only in the slightly smaller oogonia and longer vegetative cells.

Bulbochaete brebissonii Kützing 1854

19010100 **Pl. 100C** (p. 411)

Cells 17–20 μm wide, 3 to 4.5 times as long as wide; oogonium subspherical, 42–50 μm wide, 37–45 μm long; oospore 40–48 μm wide, 35–43 μm long, wall pitted; suffultory cell division basal; gynandrosporous.

USA, China, Russian Federation; known in the British Isles only from Devon and from Clare Island, Co. Mayo, Ireland.

The only species known from the British Isles with pitted oogonia and the basal division of the suffultory cell.

Bulbochaete elatior Pringsheim 1858

19010110 **Pl. 100D** (p. 411)

Cells 3–18 μm wide, 2 to 3.5 times as long as wide; oogonium subspherical, 34–44 μm wide, 31–38 μm long; oospore 32–42 μm wide, 29–36 μm long, wall smooth; suffultory cell division basal; gynandrosporous.

Probably cosmopolitan; only recorded in the British Isles from Hartlebury Common, Worcestershire, England (G.T. Harris, unpublished record).

Close to *Bulbochaete obliqua* P.Lundell ex Hirn and differing only in that the antheridia are internal and do not have a cap in *B. obliqua*. In *B. elatior*, the antheridia are external and possess a cap (see Leonardi et al., 1998).

Bulbochaete gigantea Pringsheim 1858

19010120 **Pl. 100E** (p. 411)

Cells 24–32 μm wide, 2 to 3.5 times as long as wide; oogonium subspherical, 60–70 μm wide, 50–58 μm long, usually situated below a terminal hair; oospore 58–68 μm wide, 48–56 μm long, wall surface net-like and pitted; suffultory cell division submedian; idioandrosporous.

Probably cosmopolitan; known in the British Isles from the west of Ireland and North Wales.

Distinguished by the size of the oospores, vegetative cells and oospore ornamentation.

Bulbochaete imperialis (Wittrock) Hirn 1874

var. ***regalis*** Wittrock 1874

Synonym: *Bulbochaete regalis* (Wittrock) Tiffany

19010132 **Pl. 100F** (p. 411)

Cells 24–26 μm wide; oogonium broadly ellipsoidal, 68–70 μm wide, 88–90 μm long, terminated by a hair; oospore with longitudinal ribs irregularly connected by lateral ribs; suffultory cell division superior; gynandrosporous.

USA, Brazil; only known in the British Isles from South Wales.

Differs from the type in the slightly narrower cells, though this hardly justifies varietal status.

Bulbochaete insignis Pringsheim 1858

19010140 **Pl. 100G** (p. 411)

Cells 19–25 μm wide, 2.5 to 3.5(–4.5) times as long as wide; oogonium ellipsoidal, 46–69 μm wide, 70–100 μm long; oospore ellipsoidal, 44–54 μm wide, 68–88 μm

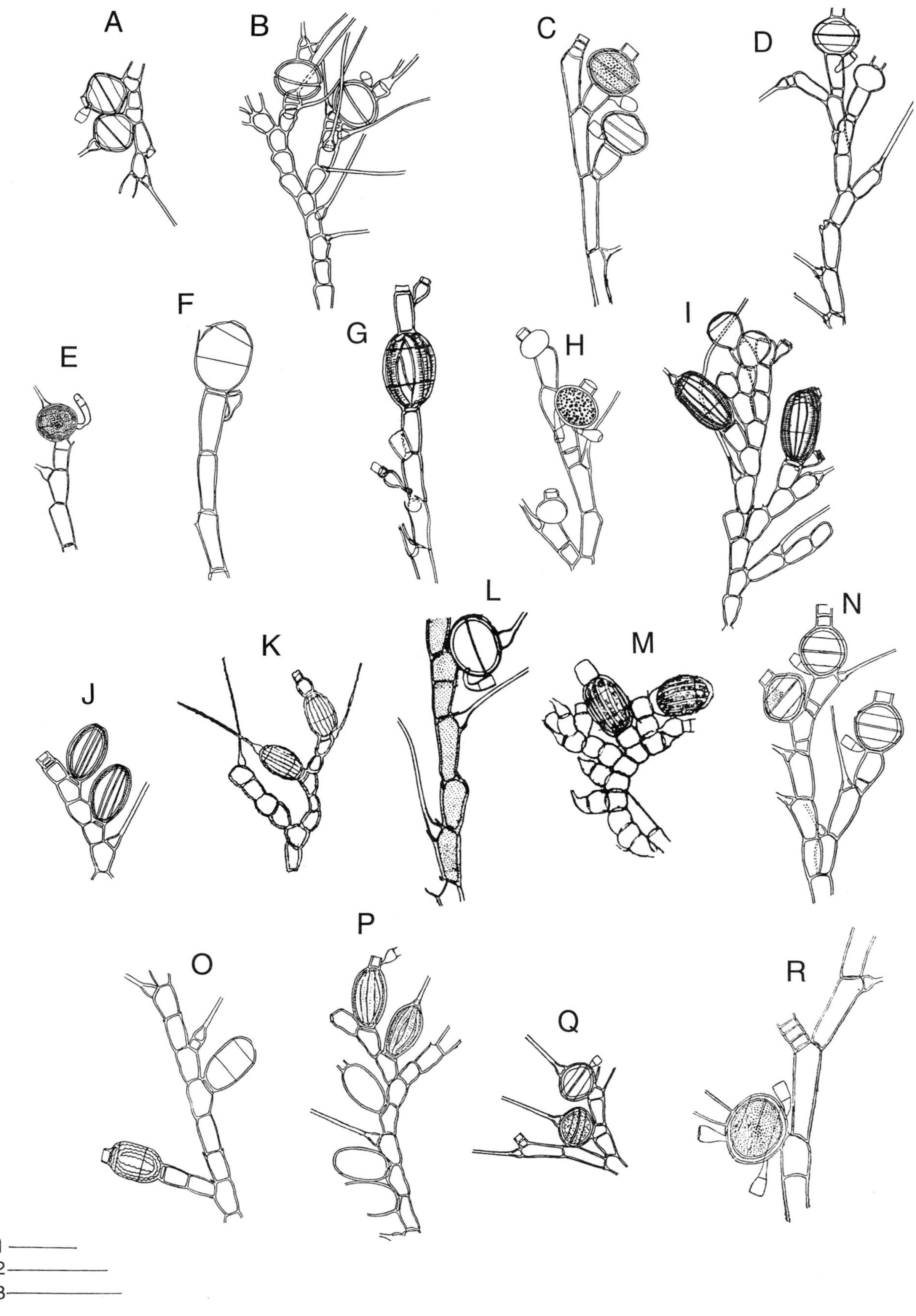

Plate 100 A–R

A. *Bulbochaete angulosa* (p. 410); **B**. *Bulbochaete borealis* (p. 410); **C**. *Bulbochaete brebissonii* (p. 410); **D**. *Bulbochaete elatior* (p. 410); **E**. *Bulbochaete gigantea* (p. 410); **F**. *Bulbochaete imperialis* (p. 410); **G**. *Bulbochaete insignis* (p. 410); **H**. *Bulbochaete intermedia* (p. 412); **I**. *Bulbochaete minor* (p. 412); **J**. *Bulbochaete mirabilis* (p. 412); **K**. *Bulbochaete nana* (p. 412); **L**. *Bulbochaete obliqua* (p. 412); **M**. *Bulbochaete pygmaea* (p. 412); **N**. *Bulbochaete quadrata* (p. 412); **O**. *Bulbochaete rectangularis* (p. 413); **P**. *Bulbochaete repanda* (p. 413); **Q**. *Bulbochaete sessilis* (p. 413); **R**. *Bulbochaete setigera* (p. 413). Scale bars: all 100 μm. 1 for E–G, L, Q; 2 for H–K, N–P, R; 3 for A–D, M.

long, ornamented with broadly toothed longitudinal ribs; suffultory cell division superior; gynandrosporous.

Probably cosmopolitan; recorded in the British Isles from a few habitats in Devon, and from Ireland; Mrozinska (1985) notes it to be typical of weedy lake margins mainly in the pH range 5.8–7.6.

Bulbochaete intermedia de Bary 1854

19010020 **Pl. 100H** (p. 411)

Cells 17–20 μm wide, 1.5 to 3.5 times as long as wide; oogonium subspherical, 40–48 μm wide, 31–40 μm long; oospore subspherical, 38–46 μm wide, 30–38 μm long, wall pitted; suffultory cell division median; gynandrosporous.

Probably cosmopolitan; recorded in the British Isles along the shores of Windermere in Cumbria (Godward, 1937), Warwickshire and Devon, also from Ireland (West & West, 1902; Lund, 1942c); in Windermere it occurred on water plants and stones, with a maximum abundance from June to September.

The var. *depressa* Wittrock has been recorded from King's Norton, Leicestershire, and differs from the type in having slightly flattened oogonia. *Bulbochaete subintermedia* Elfving has been recorded by Lund (1942c) and only differs from *B. intermedia* by having a slightly wider oogonium (44–51 μm wide) and a submedian rather than a median division of the suffultory cell. *Bulbochaete polyandria* Cleve, recorded from Slapton in Devon, has similarly sized oogonia (32–42 μm long, 39–46 μm wide) to *B. intermedia* and the division of the suffultory cell ranges from submedian to superior. The principal difference is that it is idioandrosporous unlike *B. intermedia* which is gynandrosporous; however, doubt has been raised over the taxonomic validity of this character. Since the differences between these species are so minor they are grouped here under *B. intermedia* pending further investigation. Harris (1933) also noted the close similarity between *B. polyandria*, *B. borealis* and *B. quadrata*.

Bulbochaete minor A.Braun 1849

19010150 **Pl. 100I** (p. 411)

Cells 15–25 μm wide, 1.5 to 2 times as long as wide; oogonium obovoid, 32–42 μm wide, 59–69 μm long; oospore 30–40 μm wide, 57–67 μm long, wall with longitudinal ribs; gynandrosporous; suffultory cell division superior (see note under *B. rectangularis*).

Probably cosmopolitan; widely distributed in the British Isles according to Harris (1933).

Bulbochaete varians Wittrock, whose oospores are 28–34 μm wide, 42–52 μm long and have a wall of serrated longitudinal ribs, is probably no more than a form of *B. minor*, as might also be *B. varians* var. *subsimplex* (Wittrock) Hirn; these taxa are widely distributed in the British Isles.

Bulbochaete mirabilis Wittrock 1870

19010030 **Pl. 100J** (p. 411)

Cells 15–20 μm wide, 1.25 to 2 times as long as wide; oogonia ellipsoidal, 26–35 μm wide, 46–58 μm long; oogonia ellipsoidal, 25–31 μm wide, 44–56 μm long, with longitudinal ribs; suffultory cell division superior; macrandrous; monoecious.

Probably cosmopolitan, in freshwater and brackish water; recorded in the British Isles from Wales, Devon and Leicestershire as well as in the west of Ireland (see Rich, 1925; Harris, 1933; Woodhead & Tweed, 1954).

The var. *gracilis* (Pringsheim) Hirn differs from the type in its narrower oospores (18–26 μm in diameter) and has been recorded from Ireland by Cooke (1884). One of two macrandrous species so far recorded from the British Isles.

Bulbochaete nana Wittrock 1872

19010040 **Pl. 100K** (p. 411)

Cells 10–17 μm wide, up to 1.5 times as long as wide; oogonia ellipsoidal, 20–25 μm wide, 33–40 μm long; oospore 18–23 μm wide, 30–38 μm long, with longitudinal ribs; suffultory cell division superior; macrandrous; monoecious.

Probably cosmopolitan; recorded from Marlborough Deep Pools, Hampshire, from Oxfordshire and from Wales; noted by Harris (1933) as widespread in Devon.

Differs from *B. mirabilis* only in its shorter oogonia.

Bulbochaete obliqua P.Lundell ex Hirn 1900

19010160 **Pl. 100L** (p. 411)

Cells 21–27 μm wide, 2 to 4 times as long as wide, with walls minutely granular; oogonia subspherical, 55–64 μm wide, 43–51 μm long; oospore 53–62 μm wide, 40–49 μm long, wall smooth; suffultory cell division median; gynandrosporous or idioandrosporous.

USA, Europe; in the British Isles recorded only by Harris (1933) from Rippon Tor, Devon where it was rare.

Notable for its minutely granular wall.

Bulbochaete pygmaea Pringsheim 1858

19010170 **Pl. 100M** (p. 411)

Cells 11–15 μm wide, 0.6 to 1 times as long as wide; oogonia ellipsoidal, 22–25 μm wide, 32–40 μm long; oospore 20–23 μm wide, 30–38 μm long, walls with longitudinal ribs surmounted by small granules; suffultory cell not divided; gynandrosporous.

Probably cosmopolitan; recorded in the British Isles from Devon and the west of Ireland (Cooke, 1884; Harris, 1933).

Noteworthy for the lack of division of the suffultory cell and by the ornamentation of the oospore. From Windermere there is an unconfirmed record of *Bulbochaete tenuis* (Wittrock) Hirn which is similar but has longer oospores.

Bulbochaete quadrata Wittrock 1872

19010180 **Pl. 100N** (p. 411)

Cells 19–25 μm wide, 1.5 to 2 times as long as wide; oogonia subspherical and slightly angled, 40–50 μm wide, 40–45 μm long; oospore 38–48 μm wide, 38–43 μm long, wall pitted; suffultory cell division median or inframedian; gynandrosporous.

Africa, USA, Russia, Europe; recorded in the British Isles only from Kentismoor, Devon by Harris (1933) who considered it close to *B. polyandria*.

The oospores are very weakly angular according to some descriptions.

Bulbochaete rectangularis Wittrock 1870

19010050 **Pl. 100O** (p. 411)

Filaments sparingly branched, cells 16–23 μm wide, 1.25 to 3 times as long as wide, rectangular to subrectangular in cross section; oogonia ellipsoidal, 31–39 μm wide, 45–63 μm long; oospore 29–37 μm wide, 43–61 μm long, wall longitudinally ribbed, ribs slightly wavy; suffultory cell division superior; gynandrosporous.

Probably cosmopolitan; widespread in the British Isles on aquatic plants including *Nitella* in England, Wales and Ireland but no records from Scotland.

The shape of the cell in cross section is a distinctive feature in this species. Almost all species with ellipsoidal oogonia have superior division of the suffultory cell; the exceptions are small species such as *B. pygmaea*

Bulbochaete repanda Wittrock 1874

19010190 **Pl. 100P** (p. 411)

Cells 24–26 μm wide, 2 to 3.5 times as long as wide, lateral wall often slightly wavy (repand); oogonia ellipsoidal, 26–36 μm wide, 43–58 μm long; oospore 21–33 μm wide, 40–50 μm long, wall with longitudinally and slightly flattened ribs bearing delicate spines; division of suffultory cell superior; gynandrosporous.

Probably cosmopolitan; recorded in the British Isles from two sites in Devon by Harris (1933).

The species is distinguished partly by the wavy cell outline (repand) but this is said to be variable even within individual plants. A more reliable character may be the ornamentation of the oospore wall. The division of the suffultory cell is not given in descriptions but in illustrations it appears to be superior.

Bulbochaete sessilis Wittrock 1872

19010060 **Pl. 100Q** (p. 411)

Cells 19–22 μm wide, 2 to 3.25 times as long as wide; oogonia subspherical, slightly angular, 51–56 μm wide, 44–50 μm long; oospore 48–54 μm wide, 42–48 μm long, wall pitted; suffultory cell division superior; gynandrosporous.

Europe, USA; recorded in the British Isles from Pil Tor and Slapton in Devon, and from Cambridgeshire (West, 1899b; Harris, 1933).

Bulbochaete setigera (Roth) C.Agardh 1817

Basionym: *Conferva setigera* Roth

19010200 **Pl. 100R** (p. 411)

Cells 25–28 μm wide, 2.5 to 5 times as long as wide; oogonia subspherical, 70–80 μm wide, 56–65 μm long; oospore 67–77 μm wide, 53–62 μm long, wall pitted; suffultory cell division median; gynandrosporous.

Probably cosmopolitan; widely distributed in Great Britain and Ireland.

Bulbochaete crassiuscula Nordstedt differs from it only in being idioandrosporous and has been collected in western England. Harris (1933) considered it to be more widespread than *B. setigera* in Devon. Whether the former taxon deserves specific rank is debatable. *Bulbochaete ellipsospora* West is a synonym of *B. crassiuscula* (see Mrozinska, 1985).

Oedogonium Link 1820 **CD**

BY ROBERT HUXLEY

Filaments unbranched, composed of cylindrical, capitellate or occasionally almost hexagonal cells, terminal cell rounded but sometimes acute or drawn out into a long hair; chloroplast parietal and net-like, usually with several pyrenoids; sexual reproduction oogamous; oogonia arise by division of a supporting, sometimes inflated, suffultory cell, occasionally bearing projections such as ribs or warts, spermatozoids enter the oogonium through either a simple pore or cleft-like opening (rimiform), or a split in the wall that forms a lid-like operculum, opening positioned in middle of oogonium (median), above middle (supramedian), below middle (inframedian) or high up on oogonium (superior), occasionally opening at very top and immediately below a following cell (supreme); spermatozoids in macrandrous species, 1 or 2 in each antheridium, formed by a horizontal or vertical division; dwarf males in nannandrous species vary in size, shape of the stipe and number of cells; oospores often different in shape from the oogonia and not always completely filling them, smooth-walled or having ribs, pores or another type of wall ornamentation.

Cosmopolitan; generally in freshwater habitats, attached to aquatic plants and on stones and rocks at the edge of ponds, lakes, streams and marshes sometimes free-floating; occasionally on animals and in snow to which it imparts a red or brown colour. Grows over a wide pH range in nutrient-poor to nutrient-rich waters, but little is known of its ecology. Sexual reproduction commonly occurs from spring into summer. Occasionally forming massive growths in slow-flowing streams and small ponds where it can be mistaken for the more common 'blanketweed' *Cladophora glomerata*.

The genus is distinguished from *Bulbochaete* and *Oedocladium* by its unbranched habit. It lacks the long hairs with bulbose bases and secondary division of the suffultory cell that characterizes *Bulbochaete*. Characters defining the species include ornamentation and shape of the vegetative cells, shape of the oogonium, position and nature of the opening in the oogonium wall that allows entry of spermatozoids, and shape and ornamentation of oospores.

Article 13 of the International Code of Botanical Nomenclature states that the start date for valid publication of names for the Oedogoniaceae is Hirn's monograph published in 1900. For the purpose of this account it has been decided to follow Mrozińska (1985) and other recent taxonomic treatments in which the authors who described taxa in the nineteenth century are recognized.

1 Cells distinctive in shape and/or ornamented, i.e. undulate, swollen at distal end (capitellate), with rows of small dots (punctae), hexagonal-ellipsoidal 2
1 Cells not distinctive, mainly cylindrical, usually with smooth and unornamented walls ...11
2 Cells normally cylindrical; walls undulate............3
2 Cells cylindrical or swollen at distal end (capitellate); walls with rows of small dots or punctae 4
2 Cells distinctly swollen at distal end (capitellate); walls without rows of punctae 6

2 Cells often hexagonal to ellipsoidal; walls without rows of punctae; dwarf males absent .... *O. reinschii*
3 Cells 15–22 μm wide and wall undulations 4 per cell; oogonium opening by an inferior operculum; dwarf males present ...... *O. undulatum*
3 Cells 20–27 μm wide and wall undulations 2 per cell; oogonium opening by a superior operculum; dwarf males absent........ *O. nodulosum*
4 Cells slightly swollen at distal end (capitellate), 9–13 μm wide .... *O. minus*
4 Cells cylindrical, 12–22 μm wide 5
5 Punctae spirally arranged; basal cell hemispherical; 1 spermatozooid per antheridium .... *O. punctatostriatum*
5 Punctae not spirally arranged; basal cell not hemispherical; 2 spermatozooids per antheridium .... *O. curtum*, in part
6 Oogonial wall with median projections .... *O. platygynum*
6 Oogonium without projections 7
7 Cells <13 μm wide 8
7 Cells >13 μm wide 10
8 Oogonium a little wider than vegetative cells .... *O. microgonium*
8 Oogonium at least twice as wide as vegetative cells 9
9 Basal cell hemispherical; oogonium 20–25 μm wide; dwarf males absent .... *O. capitellatum*
9 Basal cell not hemispherical; oogonium 34–39 μm wide; dwarf males present .... *O. areschougi*
10 Oogonium ellipsoidal or spherical-ellipsoidal, opening by a supreme operculum .... *O. obtruncatum*
10 Oogonium obovoid or quadrangular-obovoid, opening by a superior sometimes median pore .... *O. borisianum*
11 Oogonia and antheridia on same filament (monoecious) 12
11 Oogonia and antheridia on different filaments which may differ slightly in size (dioecious, macrandrous) 29
11 Ooogonia and antheridia on different filaments; males considerably smaller than female (dwarf males) and growing attached to female filaments (dioecious nannandrous) 43
12 Oogonium opening by a pore 13
12 Oogonium opening by an operculum 21
13 Pore supra-median on oogonium .... *O. obsoletum*
13 Pore median on oogonium 14
13 Pore superior on oogonium 16
14 Oogonium 21–28 μm wide; oospore wall smooth 15
14 Oogonium 30–40 μm wide; oospore wall covered with small pores .... *O. cymatosporum*
15 Filament sometimes curved; oogonium depressed-spherical, 18–24 μm wide .... *O. curvum*
15 Filament not curved; oogonium spherical, 26–31 μm in diameter .... *O. cryptoporum*
16 Filaments up to 8 cells long .... *O. zigzag*
16 Filaments many more than 8 cells long 17
17 Cells 8–13 μm wide; oogonium 32–37 μm wide, 32–39 μm long .... *O. hirnii*
17 Cells 15–30 μm wide; oogonium 40–58 μm wide, 45–63 μm long 18
18 Oospore clearly does not fill oogonium; oospore spherical 19
18 Oospore fills or nearly fills oogonium; oospore obovoid-spherical 20
19 Cells 12–16 μm wide .... *O. urbicum*
19 Cells 20–30 μm wide .... *O. vaucherii*
20 Oogonium single; cells 4–7 times as long as wide and walls without rows of small dots (punctae); suffultory cell not swollen .... *O. fragile*
20 Oogonium single to 4 in a series; cells 2–4 times as long as wide, walls sometimes with spirally arranged punctae; suffultory cell sometimes swollen .... *O. curtum*, in part
21 Oospore wall with 30–35 longitudinal ribs .... *O. nobile*
21 Oospore smooth-walled 22
22 Oogonium with projections or wall longitudinally folded (plicate) 23
22 Oogonium without projections, wall not plicate 24
23 Oospore constricted in middle, 9–12 μm in diameter; oogonium without projections and wall plicate .... *O. excisum*
23 Oospore not constricted, 20–23 μm in diameter; oogonium with conically obtuse projections and wall not plicate .... *O. itzigsohnii*
24 Oogonium 15–29 μm long, operculum median or slightly supramedian 25
24 Oogonium 41–69 μm long, operculum superior 27
25 Oospore noticeably constricted in middle .... *O. pusillum*
25 Oospore not constricted in middle 26
26 Cells 3–5 μm wide; oogonium 15–19 μm wide, operculum median; basal cell almost hemispherical .... *O. tapeinosporum*
26 Cells 6–7 μm wide; oogonium 21–23 μm wide, operculum slightly supramedian; basal cell not hemispherical .... *O. petri*
27 Oogonium oblong-ellipsoidal, 20–26 μm wide; oospore not filling oogonium .... *O. oblongum*
27 Oogonium ellipsoidal spherical or almost spherical, 35–45 μm wide; oospore filling oogonium 28
28 Oogonium spherical or almost spherical, 41–53 μm long .... *O. crispum*
28 Oogonium ellipsoidal, 57–69 μm long .... *O. ahlstrandii*
29 Oogonium opening via a pore 30
29 Oogonium opening via an operculum 41
30 Oospore wall bearing spines .... *O. suecicum*
30 Oospore wall covered in small pores .... *O. giganteum*
30 Oospore wall bearing ribs 31
30 Oospore wall smooth 32

31 Oospore ribs toothed; division of spermatozoids horizontal .... *O. crenulato-costatum*
31 Oospore ribs undulating; division of spermatozoids vertical.............................. *O. boscii*
32 Oogonium <20 μm wide, pore median 33
32 Oogonium >20 μm wide, pore inframedian-superior 34
33 Cells 8–10 μm wide, not encrusted with lime .... *O. rufescens*
33 Cells 11–14 μm wide, sometimes encrusted with lime .... *O. calcareum*
34 Oogonium little wider than vegetative cells .... *O. capillare*
34 Oogonium markedly wider than vegetative cells 35
35 Oogonium almost ovoid-spherical to heart-shaped, with pore supramedian 36
35 Oogonium ovoid-obovoid or ellipsoidal, with pore superior..37
36 Oospore spherical, not filling oogonium; oogonium almost spherical and sometimes heart-shaped, 48–70 μm in diameter........... *O. cardiacum*
36 Oospore spherical-quadrangular, filling oogonium; oogonium obovoid to ovoid, 40–49 μm in diameter .... *O. lautumniarum*
37 Spermatozooids divide horizontally 38
37 Spermatozoids divide vertically 39
38 Oogonium <50 μm wide................. *O. capilliforme*
38 Oogonium >50 μm wide........................ *O. rivulare*
39 Terminal cell with a short point; oogonium single .... *O. princeps*
39 Terminal cell without a short point; oogonium up to 3-seriate 40
40 Oospores ovoid to ellipsoidal ..... *O. landsboroughii*
40 Oospores ellipsoidal to spherical-ellipsoidal .... *O. crassum*
41 Terminal cell without a short point or hair; oogonium 37–63 μm wide; oospore walls smooth or with granulate ribs .... *O. tumidulum*
41 Terminal cell with a short point or bearing a hair; oogonium 23–43 μm wide; oospore walls smooth 42
42 Terminal cell bearing a hair; oogonium ellipsoidal, 23–29 μm wide .... *O. pisanum*
42 Terminal cell obtuse or with a short point; oogonium spherical-ovoid, 35–43 μm wide .... *O. pringsheimii*
43 Oogonium opening by a pore 44
43 Oogonium opening by an operculum 54
44 Oospore wall covered in small pores, with teeth, ribbed or with spines 45
44 Oospore wall smooth...48
45 Pore inferior; oospore wall with spiral rows of teeth .... *O. cleveanum*
45 Pore superior 46
45 Pore median; oospore wall spiny 47
46 Oogonium 63–83 μm wide; oospore wall covered in small pores .... *O. concatenatum*
46 Oogonium 57–66 μm wide, oospore wall with curved ribs...................................... *O. cyathigerum*
47 Oospore ellipsoidal, pore not noticeably wide .... *O. hystrix*
47 Oospore spherical, pore noticeably wide .... *O. echinospermum*
48 Pore median 49
48 Pore superior 52
49 Oogonium 28–37 μm wide 50
49 Oogonium 40–52 μm wide 51
50 Oogonium hexagonal to ellipsoidal... *O. sexangulare*
50 Oogonium ovoid-spherical .... *O. braunii*
51 Terminal cell with a short point or bearing a hair; cells sometimes swollen at distal end (capitellate); oogonium sometimes ovoid-quadrangular, 55–90 μm long; suffultory cell inflated... *O. borisianum*
51 Terminal cell without a short point or bearing a hair; cells sometimes swollen at distal end (capitellate); oogonium spherical-ellipsoidal, inflated, 51–60 μm long; suffultory cell not inflated .... *O. flavescens*
52 Oogonium 24–35 μm wide .... *O.multisporum*
52 Oogonium 48–60 μm wide 53
53 Gynandrosporous; oogonium spherical-ellipsoidal.................................... *O. crassiusculum*
53 Idioandrosporous; oogonium ovoid-ellipsoidal .... *O. idioandrosporum*
54 Operculum supreme; oogonium and oospore ribbed; oogonium always terminal .... *O. acrosporum*
54 Operculum median or superior, sometimes almost supreme; oogonium and oospore smooth; oogonium rarely terminal....................................55
55 Operculum superior..56
55 Operculum median...58
56 Terminal cell with a seta or hair; oogonium ovoid-ellipsoidal, 55–70 μm long.......... *O. ciliatum*
56 Terminal cell not as above; oogonium ovoid-spherical, 43–50 μm long.....................................57
57 Cells 10–16 μm wide; dwarf male stipe curved, 24–33 μm long .... *O. macrandrium*
57 Cells 22–29 μm wide; dwarf male stipe obovoid, 14–15 μm long....................................... *O. pluviale*
58 Filaments noticeably contorted; oogonium spherical to pear-shaped, at least 4.5 times wider than vegetative cells .... *O. contortum*
58 Filaments not contorted; oogonium spherical, 2 to 3.5 times wider than vegetative cells............59
59 Cells 15–20 μm wide .... *O. macrospermum*
59 Cells 6–12 μm wide 60
60 Cells 6–10 μm wide; oogonium 20–27 μm wide .... *O. rothii*
60 Cells 9–12 μm wide; oogonium 25–38 μm wide .... *O. decipiens*

Oedogonium acrosporum de Bary 1854

19020240 **Pl. 107F** (p. 431)

Cells cylindrical, 12–21 μm wide, 3 to 6 times as long as wide; oogonium single, terminal, ellipsoidal, 35–48 μm wide, 50–63 μm long, with longitudinal ridges on inner

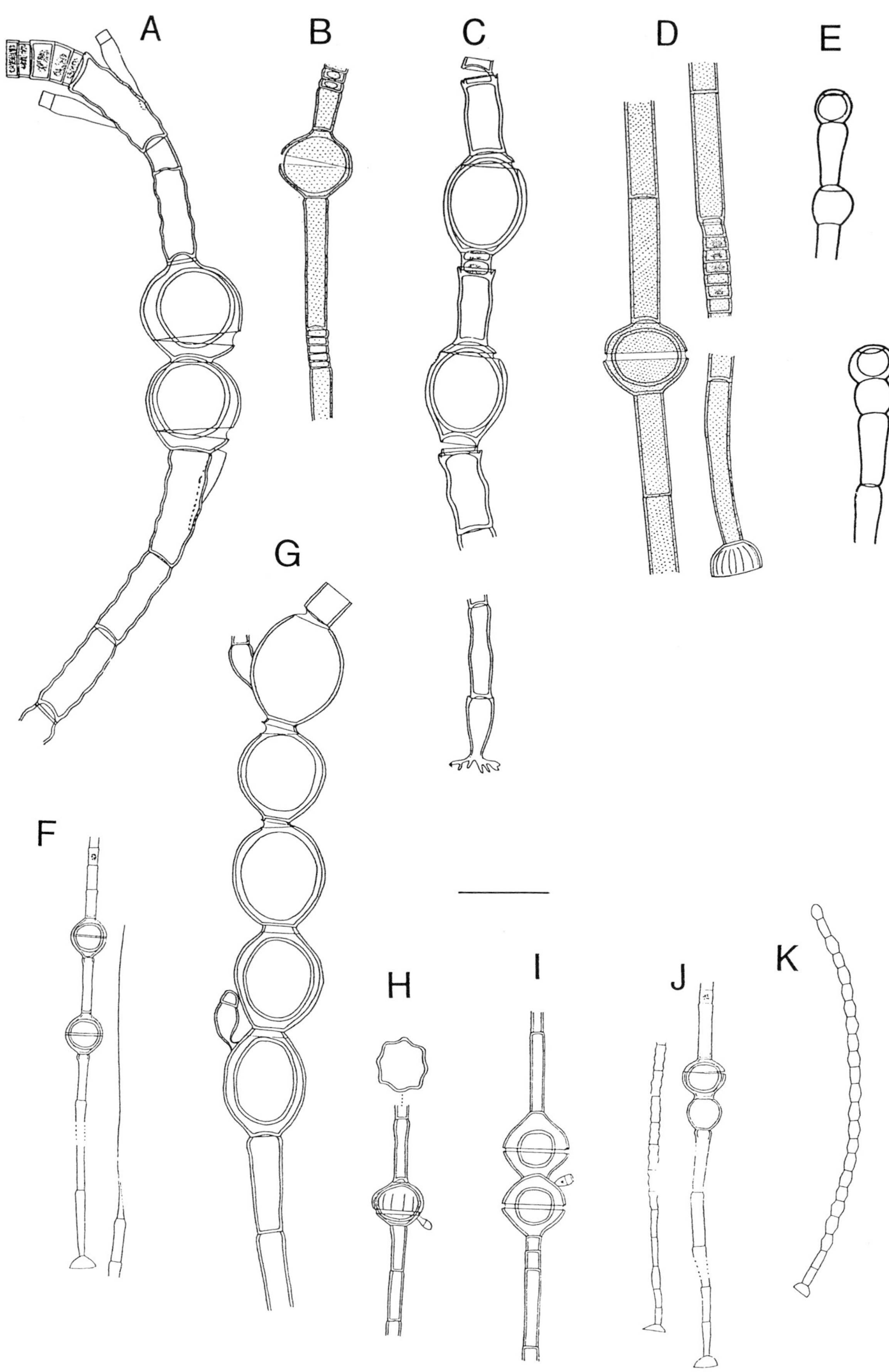

Plate 101 A–J
A. *Oedogonium undulatum* (p. 432); **B**. *Oedogonium minus* (p. 426); **C**. *Oedogonium nodulosum* (p. 426); **D**. *Oedogonium puncatostriatum* (p. 428); **E**. *Oedogonium microgonium* (p. 424); **F**. *Oedogonium capitellatum* (p. 418); **G**. *Oedogonium obtruncatum* (p. 426); **H**. *Oedogonium platygynum* (p. 428): a. cell in cross section, b. filament with oogonium and dwarf male; **I**. *Oedogonium areschougii* (p. 417); **J**., **K** *Oedogonium reinschii* (p. 430).
Scale bar: 50 μm.

surface, opening by a superior, wide operculum; oospore ellipsoidal, 35–48 μm wide, 50–63 μm long, filling oogonium entirely, walls with 23–30 ribs fitting into ridges on oogonial wall; suffultory cell inflated; dwarf male stipe with upper cells 6–8 μm wide, 55–71 μm long, lower cells 9–12 μm wide, 30–38 μm long; gynandrosporous or idioandrosporous.

Probably cosmopolitan; recorded in the British Isles by W. & G.S.West (1900b) from Yorkshire, by Harris (1933) from Devon, and by Cooke (1882–84) from Ireland.

A very distinctive species with large terminal, ribbed oogonium and long, dwarf males with a 2–3-celled stipe.

Oedogonium ahlstrandii Wittrock in Wittrock & Nordstedt 1882

19020230 **Pl. 103F** (p. 422)

Cells cylindrical, 10–18 μm wide, 3 to10 times as long as wide; oogonium single, elliposidal, 35–42 μm wide and 57–69 μm long, opening by an operculum at the very top of the cell (supreme); oospore ellipsoidal, 34–41 μm wide, 53–62 μm long, filling oogonium, walls smooth; suffultory cell similar to adjacent vegetative cells; spermatozoids 2 per antheridium, division horizontal; monoecious, macrandrous.

Europe, North Africa, USA, India; recorded by W. & G.S.West (1900b) from Yorkshire who noted it as a first record for the British Isles.

Oedogonium areschougii Wittrock 1870

19020540 **Pl. 101I** (p. 416)

Cells swollen at distal end (capitellate), 8–13 μm wide, 4 to 6 times as long as wide; oogonium single-6-seriate, mostly terminal, almost depressed or depressed to pear-shaped, 34–39 μm wide, 36–40 μm long opening by a wide, median operculum; oospore spherical, rarely almost depressed-spherical, 22–26 μm wide, 22–25 μm long, not filling oogonium, walls smooth; suffultory cell similar to other vegetative cells; dwarf males obovoid, unicellular; gynandrosporous.

Europe, Africa, Brazil, USA, Greenland, Siberia; recorded in the British Isles by Cook (1882–84) from Ireland, and from North Wales (Woodhead & Tweed, 1954).

Oedogonium borisianum (Le Clerc) Wittrock 1870

Basionym: *Prolifera borisiana* Le Clerc

19020250 **Pl. 106G** (p. 429)

Cells cylindrical, sometimes swollen at the distal end (capitellate), 15–23 μm wide, 3 to 6 times as long as wide; terminal cell ending abruptly in a short broad point, sometimes bearing a hair, sometimes bearing an oogonium; oogonium single to 5-seriate, sometimes terminal, obovoid or quadrangular-obovoid, (34–)40–50 μm wide, (45–)55–90 μm long, opening by a superior or sometimes median pore; oospore ovoid to obovoid (often quadrangular-ovoid), (28–)35–46 μm wide, (38–)48–60 μm long, not filling oogonium, walls smooth; suffultory cell inflated; dwarf males 35–47 μm long; gynandrosporous or idioandrosporous.

Cosmopolitan; in the British Isles recorded in Cornwall by W. & G.S.West (1903b), Devon by Harris (1933), with Cooke (1882–84) mentioning its distribution as 'Britain' and 'Ireland'.

Oedogonium boscii (Le Clerc) Wittrock 1870

Basionym: *Prolifera boscii* Le Clerc
Synonym: *Oedogonium paludosum* (Hassall) Wittrock

19020010 **Pl. 103O** (p. 422)

Cells cylindrical, 14–23 μm wide (female), 13–18 μm wide (male), 3 to 6 times longer than wide (female), 4 to 6 times longer than wide (male); oogonium single, rarely in 2s, oblong-ellipsoidal, 39–51 μm wide, 75–110 μm long, opening by a superior pore; oospore ellipsoidal, 36–43 μm wide, 56–70 μm long, walls with 27–35 longitudinal ribs, undulating in cross section; suffultory cell similar to adjacent vegetative cells; spermatozoids 2 per antheridium, division vertical; dioecious (sometimes monoecious), macrandrous.

Cosmopolitan; recorded in the British Isles from Yorkshire (W. & G.S.West, 1900b), Cambridgeshire (West, 1899b), Henfield in West Sussex (Hassall, 1845), Hertfordshire (Fritsch & Rich, 1913), Devon (Harris, 1933) and North Wales (Woodhead & Tweed, 1954).

Oedogonium braunii Kützing 1849

19020020 **Pl. 106D** (p. 429)

Cells cylindrical, 13–15 μm wide, 2 to 4 times as long as wide; oogonium single, ovoid to almost spherical, 30–37 μm wide, 33–43 μm long, opening by a median pore; oospore spherical, not filling oogonium, 27–33 μm wide, 27–33 μm long, walls smooth; suffultory cell similar to adjacent vegetative cells; dwarf male, stipe 7–12 μm wide, 20–28 μm long; gynandrosporous.

Africa, North America, most of Europe; recorded in the British Isles from Yorkshire (W. & G.S.West, 1900b), Middlesex (W. & G.S.West, 1897), Hertfordshire (Howland, 1931), and from Ireland (West, 1892a; Cooke, 1882–84).

Oedogonium calcareum Cleve *et* Wittrock 1870

Synonym: *Vesiculifera compressa* Hassall

19020260 **Pl. 104D** (p. 425)

Cells cylindrical, 11–14 μm wide, 2 to 4 times as long as wide; cells sometimes encrusted with lime; oogonium single, rarely 2-seriate, depressed-spherical, 27–30 μm wide and 21–23 μm long, opening by a median pore; oospore depressed-spherical, 26–28 μm wide and 20–21 μm long, filling oogonium, walls smooth; suffultory cell similar to adjacent vegetative cells; spermatozoids 1 per antheridium; dioecious, macrandrous

Africa, Asia, New Zealand, various parts of Europe; recorded in the British Isles from Yorkshire (W. & G.S.West, 1900b), the English Lake District (West, 1892b), Hertfordshire (Hassall, 1845) and from Ireland (West, 1892a).

Oedogonium capillare (Linnaeus) Kützing 1843

Basionym: *Conferva capillare* Linnaeus

19020270 **Pl. 104A** (p. 425)

Cells cylindrical, 35–56 μm wide (female), 35–50 μm (male), 1 to 2 times as long as wide; oogonium single, cylindrical to almost cylindrical, 40–60 μm wide, 45–75 μm long, little wider than vegetative cells, opening by a superior pore; oospore spherical to cylindrical-spherical, 30–52 μm wide, 35–65 μm long, either completely or not filling oogonium, walls smooth; suffultory cell similar to adjacent vegetative cells; spermatozoids 2 per antheridium, division horizontal; dioecious, macrandrous.

Cosmopolitan; only reference is to 'Britain' by Cooke (1882–84), but widespread throughout Europe and included here as a likely species.

It is distinctive in that the oogonia appear similar to adjacent vegetative cells.

Oedogonium capilliforme Kützing 1849

19020030 **Pl. 104C** (p. 425)

Cells cylindrical, 28–38 μm wide (female), 25–30 μm (male), 1.5 to 3 times as long as wide; terminal cell sometimes ending abruptly in a short point; oogonium single, obovoid to almost ovoid, slightly swollen, 42–50 μm wide, 51–62 μm long, opening by a superior pore; oospores variable in shape: ovoid-spherical, cylindrical-spherical, almost spherical or spherical, 37–45 μm wide, 40–50 μm long, not filling oogonium, walls smooth; suffultory cell similar to adjacent vegetative cells; spermatozoids 2 per antheridium, division horizontal; dioecious, macrandrous.

Cosmopolitan; known in the British Isles from Hampshire (Woodhead & Tweed, 1938), Devon (Harris, 1933) and North Wales (Woodhead & Tweed, 1954).

Oedogonium capitellatum Wittrock 1874

19020680 **Pl. 101F** (p. 416)

Cells swollen at the distal end (capitellate), 6–9 μm wide, 3 to 7 times as long as wide; terminal cell hair-like, basal cell almost hemispherical; oogonium single, depressed or almost depressed-spherical, 20–25 μm wide, 17–23 μm long, opening by a median operculum; oospore depressed-spherical, 18–23 μm wide, 15–19 μm long, filling or nearly filling oogonium, walls smooth; suffultory cell similar to adjacent vegetative cells; spermatozoids 1 per antheridium; monoecious, macrandrous.

Probably cosmopolitan; recorded in the British Isles from Hampshire (Woodhead & Tweed, 1938) and North Wales (Woodhead & Tweed, 1954).

Although only two records for the British Isles, its capitellate cells, hair-like terminal cell and hemispherical basal cell make it especially distinctive and thus readily identified.

Oedogonium cardiacum (Hassall) Wittrock 1870

Basionym: *Vesiculifera cardiaca* Hassal

19020040 **Pl. 104F** (p. 425)

Cells cylindrical, 18–30 μm wide (female), 15–25 μm wide (male), 3 to 7 times as long as wide; oogonium single, rarely 2, almost spherical to almost heart-shaped or inverted heart-shaped, 48–70 μm wide, 58–78 μm long, opening by a supramedian pore; oospore spherical, 42–60 μm wide, 42–60 μm long, not filling oogonium, walls smooth; suffultory cell similar to adjacent vegetative cells; spermatozoids 2 per antheridium, division horizontal; dioecious, macrandrous.

Probably cosmopolitan; recorded in the British Isles from Middlesex (W. & G.S.West, 1897; Hassall, 1845), 'Britain' (Cooke, 1882–84), Devon (Harris, 1933) and Surrey (W. & G.S.West, 1897, as var. *carbonicum*).

var. *carbonicum* Wittrock in Wittrock & Nordstedt 1882

19020042

Mainly differs from the type variety in having obovoid-spherical or ovoid oogonia, and oogonia more commonly in pairs. There are also slight differences in cell and oogonium dimensions.

Oedogonium ciliatum (Hassall) Pringsheim 1856

Basionym: *Vesiculifera ciliata* Hassall

19020050 **Pl. 107C** (p. 431)

Cells cylindrical, 14–24 μm wide, 2.5 to 4 times as long as wide; terminal cells each with a long hair-like seta or bristle; oogonium single to 7-seriate, ovoid to ovoid-ellipsoidal, 43–50 μm wide, 55–72 μm long, opening by a superior, almost supreme operculum; oospore ovoid to almost ellipsoidal, rarely spherical, 40–47 μm wide, 44–57 μm long, nearly filling oogonium, walls smooth; suffultory cell similar to adjacent vegetative cells, dwarf male stipe 10–15 μm wide, 24–31 μm long; gynandrosporous.

Probably cosmopolitan; recorded in Great Britain and Ireland from Yorkshire (W. & G.S.West, 1900b), Cornwall (West & Fritsch, 1927), 'Britain' and 'Ireland' (Cooke, 1882–84), Cheshunt, Herts (Hassall, 1845) and Devon (Harris, 1933). One of the Devon collections was from near 'peat works' on Dartmoor and is illustrated with a terminal hair.

Oedogonium cleveanum Wittrock 1870

19020280 **Pl. 105D** (p. 427)

Cells cylindrical, sometimes swollen at the distal end (capitellate), (14–)18–26 μm wide, 3 to 7 times as long as wide; oogonium single, sometimes terminal, almost ovoid-spherical or almost spherical, 45–60 μm wide, 48–63 μm long, opening by an inferior pore; oospore spherical, 44–57 μm wide, 45–59 μm long, filling oogonium, walls with 4–6 rows of teeth; suffultory cell sometimes inflated; dwarf males single-celled, 10–11 μm wide, 29–32 μm long; dioecious, nannandrous, gynandrosporous.

Africa, South America, Europe; recorded in 'Britain' by (Cooke, 1882–84) and from Devon (Harris, 1933) and Ireland (Cooke, 1882–84).

Oedogonium concatenatum (Hassall) Wittrock 1874

Basionym: *Vesiculifera concatenata* Hassall

Synonym: *Oedogonium rectangulare* (F.Rich) Tiffany

19020290 **Pl. 105G** (p. 427)

Cells cylindrical, 25–40 μm wide, 3 to 10 times as long as wide; oogonium single to 6-seriate, almost ovoid or quadrangular-ellipsoidal, 63–83 μm wide, 76–108 μm long, opening by a superior pore; oospore almost ovoid or quadrangular-ellipsoidal, 60–76 μm wide, 67–95 μm long, nearly filling oogonium, walls with 30–35 longitudinal rows of pits; suffultory cell inflated; dwarf male stipe 17–25 μm wide, 50–75 μm long; gynandrosporous.

Africa, North America, Asia, Europe; recorded in 'Britain' by (Cooke, 1882–84) and more specifically from Middlesex (Hassall, 1845), Leicester (Rich, 1933, as var. *rectangulare*) and Ireland (Cooke, 1882–84, as *O. hutchinsiae*).

var. *hutchinsiae* (Wittrock) Hirn 1900

Basionym: *Oedogonium hutchinsiae* Wittrock

19020292 **Pl. 105B** (p. 427)

Differs from the type variety in having oospores more spherical and pits within the oospore wall not in longitudinal rows (a distinguishing feature of the type variety).

This probably does not warrant varietal status.

var. *rectangulare* Rich 1925

Synonym: *Oedogonium rectangulare* (F.Rich) Tiffany

19020293 **Pl. 105H** (p. 427)

Although differing only slightly from the type in cell dimensions, this variety is idioandrosporic in habit and the small pits in the oospore wall are in transverse as well as longitudinal rows.

This was considered a separate species by Tiffany (1937a) and Gonzalves (1981), but not by Mrozinska (1985).

Oedogonium contortum Hallas 1905

19020300 **Pl. 107B** (p. 431)

Cells cylindrical, filaments contorted, 5–7 μm wide, 4 to 9 times as long as wide; oogonium single or rarely 2-seriate, depressed-spherical, rarely pear-shaped, 23–35 μm wide, 16–35 μm long, opening by an inframedian operculum; oospore depressed-spherical, rarely spherical, 16–28 μm wide, 12–21 μm long, filling oogonium, walls smooth; suffultory cell similar to other vegetative cells; dwarf males single-celled, obovoid, stipe about 4 μm wide and 13 μm long; gynandrosporous.

North America (Florida), Denmark; a record by Harris (1933) from Devon is its only mention from the British Isles.

Oedogonium crassiusculum Wittrock 1870

19020310 **Pl. 106F** (p. 429)

Cells cylindrical, 27–30 μm wide, 3.5 to 5 times as long as wide; oogonium single or 2-seriate, spherical-ovoid or almost spherical, 54–60 μm wide, 60–75 μm long opening by a superior pore; ellipsoidal-spherical or spherical, 51–57 μm wide, 52–63 μm long, nearly filling oogonium, walls smooth and distinctly thickened; suffultory cell similar to other vegetative cells; dwarf male stipe 10–11 μm wide 60 μm long; gynandrosporous.

North America, Asia, Australia, various parts of Europe (Finland, Germany, Sweden); recorded in the British Isles from Epping Forest (Cooke, 1882–84) and North Wales (Woodhead & Tweed, 1954).

Oedogonium crassum (Hassall) Wittrock 1872

Basionym: *Vesiculifera crassa* Hassall

19020320 **Pl. 104E** (p. 425)

Cells cylindrical, 30–50 μm wide (female), 24–36 μm wide (male), 2 to 7 times longer than wide (female), 3 to 8 times longer than wide (male); oogonium single or 2-seriate, obovoid-ellipsoidal to ovoid, 40–60 μm wide, 45–75 μm long, opening by a superior pore; oospore ellipsoidal to spherical-ellipsoidal, 58–76 μm wide, 65–96 μm long, either completely or not filling oogonium, walls smooth; suffultory cell similar to adjacent vegetative cells; spermatozoids 2 per antheridium, produced by a vertical division; dioecious, macrandrous.

Africa, North America, Asia; recorded as in 'Britain' by Cooke (1882–84) and from Hertfordshire (Hassall, 1845).

This species is very similar to *O. landsboroughii* (*q.v.*).

Oedogonium crenulato-costatum Wittrock 1878

19020330 **Pl. 103P** (p. 422)

Cells cylindrical,10–18 μm wide (female), 9–13 μm wide (male), 2.5 to 7 times longer than wide (female), 3.5 to 6 times longer than wide (male); oogonium single to 6-seriate, obovoid to almost ellipsoidal, 30–36 μm wide, 40–65 μm long, opening by a superior pore; oospore obovoid to almost ellipsoidal, 28–34 μm wide, 37–55 μm long, either completely or nearly filling oogonium, walls with 14–20 distinctly toothed ribs; suffultory cell similar to adjacent vegetative cells; spermatozoids 2 per antheridium, division horizontal; dioecious, macrandrous.

North America, South America (Brazil), Asia (China), parts of Europe; recorded in the British Isles from Devon (Harris 1933) and the Ashdown Forest, Sussex (Pentecost, pers. comm.).

Oedogonium crispum (Hassall) Wittrock 1874

Basionym: *Vesiculifera crispa* Hassall

Synonym: *Oedogonium vernale* (Hassall) Wittrock *non O. vernale* Wittrock

19020060 **Pl. 103H** (p. 422)

Cells cylindrical, 12–16 μm wide (female), 3 to 5 times as long as wide; oogonium single, obovoid-spherical, 37–45 μm wide, 41–53 μm long, opening by a superior operculum; oospore spherical or almost spherical, 35–43 μm wide, 37–43 μm long, completely filling oogonium, walls smooth; suffultory cell similar to adjacent vegetative cells; spermatozoids 2 per antheridium, division horizontal; monoecious, macrandrous.

British Isles; recorded as from 'Britain and Ireland' (Cooke, 1882–84); various localities in England, including Yorkshire (W. & G.S.West, 1900b) and Cambridgeshire (West, 1899b), also Scotland (Hassall, 1845) and Clare Island in Ireland (W.West, 1912).

fo. *obesum* (Wittrock) Mrozinska 1985

Basionym: *Oedogonium pyrulum* var. *obesum* Wittrock

Synonym: *Oedogonium obesum* (Wittrock) Hirn

19020162 **Pl. 103B** (p. 422)

Differs from the type in having slightly smaller cells and in the oospore not filling the oogonium.

North and South America, Asia, Europe; recorded in the British Isles as *O. pyrulum* var. *obseum* from Cambridgeshire (West, 1899b).

Oedogonium cryptoporum Wittrock 1870

Synonyms: *Oedogonium cryptoporum* var. *vulgare* Wittrock, *O. vulgare* (Wittrock) Tiffany

19020070 **Pl. 102I** (p. 421)

Cells cylindrical, 7–10 μm wide, 4 to 6 times as long as wide; oogonium single, almost depressed-obovoid to spherical, 23–28 μm wide, 26–31 μm long, opening by a median pore; oospore spherical, 19–27 μm in diameter, almost or sometimes completely filling oogonium, walls smooth; suffultory cell similar to adjacent vegetative cells; spermatozoids 1 per antheridium; monoecious, macrandrous.

Cosmopolitan; recorded in Great Britain and Ireland from Yorkshire (W. & G.S.West, 1900b), North Wales (West, 1890), the English Lake District (West, 1892b), Devon (Harris, 1933), Hampshire (Woodhead & Tweed, 1938, as var. *vulgare*), Middlesex (W. & G.S.West, 1897), Ireland (West, 1892a), and Clare Island (W.West, 1912).

The variety *vulgare* differs only in having slightly smaller cell and oogonium dimensions and is not considered sufficiently distinct to warrant varietal status.

Oedogonium curtum Wittrock *et* Lundell in Wittrock 1870

19020340 **Pl. 102G** (p. 421)

Cells cylindrical, sometimes ornamented with spirally arranged dots, 12–22 μm wide, 2 to 5 times as long as wide; oogonium single, obovoid-spherical or almost spherical, 38–55 μm wide, 37–54 μm long, opening by a superior pore; oospore obovoid-spherical or almost spherical, 38–45 μm wide, 48–60 μm long, nearly filling oogonium, walls smooth; suffultory cell occasionally inflated; spermatozoids 2 per antheridium, divison horizontal; monoecious, macrandrous.

Africa, North America, Europe; recorded in the British Isles only from South Wales (Benson-Evans & Antoine, 1996).

Similar to *O. zigzag* apart from the lack of a zigzag cell arrangement, the greater number of cells in a filament (10–20), and the occasional presence of spirally arranged dots on the cell wall.

Oedogonium curvum Pringsheim 1858

19020080 **Pl. 102C** (p. 421)

Cells cylindrical, 5–10 μm wide, 1.5 to 4 times as long as wide; filaments regularly curved; oogonium single to 6-seriate, depressed-spherical, 21–25 μm wide, 18–24 μm long, opening by a wide median pore; oospore depressed-spherical, 19–23 μm wide, 14–19 μm long, filling or not filling the oogonium, walls smooth; suffultory cell similar to adjacent vegetative cells; spermatozoids 1 per antheridium; monoecious, macrandrous.

Africa, North America, Asia, much of Europe; recorded in the British Isles from Cambridgeshire (West, 1899b) and Ireland (Cooke, 1882–84).

Oedogonium cyathigerum Wittrock 1870

19020090 **Pl. 105C** (p. 427)

Cells cylindrical, 21–30 μm wide, 2 to 10 times as long as wide; oogonium single to 3-seriate, often terminal, almost ovoid or quadrangular-ellipsoidal, 57–66 μm wide, 77–100 μm long, opening by a superior pore; oospore almost ovoid or quadrangular-ellipsoidal, 52–62 μm wide, 60–75 μm long, filling or not filling the oogonium, walls with 16–25 longitudinal and often curved ribs; suffultory cell inflated; dwarf males goblet-shaped, 12–15 μm wide, 50–58 μm long; idioandrosporous.

Africa, North America, Asia (India), most of Europe; recorded in the British Isles from Yorkshire (W. & G.S.West 1900b; West & Fritsch, 1927) and Middlesex (W. & G.S. West,1897).

Oedogonium cymatosporum Wittrock *et* Nordstedt in Wittrock 1870

19020350 **Pl. 102B** (p. 421)

Cells cylindrical, 8–10 μm wide, 4 to 7 times as long as wide; oogonium single, rarely 2, almost depressed-spherical, 30–40 μm wide, 27–40 μm long, opening by a median cleft-like pore; oospore depressed-spherical, 27–35 μm wide, 22–33 μm long, filling or not filling oogonium, walls covered in small pits; suffultory cell similar to adjacent vegetative cells; spermatozoids 1 per antheridium; monoecious, macrandrous

North Africa, North America, Asia, much of Europe; only recorded in the British Isles by Cooke (1882–84) and then simply as 'Britain', although mentioned from Wales and England by Gonzalves (1981) but with no indication as to source.

Oedogonium decipiens Wittrock 1870

19020360 **Pl. 107E** (p. 431)

Cells cylindrical or slightly swollen at the distal end (capitellate), 9–12 μm wide, 3 to 5 times as long as wide; oogonium single or 3-seriate, almost depressed-spherical, 30–38 μm wide, 27–40 μm long, opening by a median operculum; oospore almost depressed or depressed-spherical, 25–34 μm wide, 23–28 μm long, nearly filling oogonium, walls smooth; suffultory cell similar to other vegetative cells; dwarf males single-celled, 6–7 μm wide, 13–15 μm long; gynandrosporous.

Africa, North and Central America, Puerto Rico, Asia, much of Europe; recorded in the British Isles from Cornwall (W. & G.S.West, 1903b) and North Wales (Woodhead & Tweed, 1954).

Oedogonium echinospermum A.Braun in Kützing 1849

19020370 **Pl. 105E** (p. 427)

Cells cylindrical, 18–30 μm wide, 2.5 to 4 times as long as wide; oogonium single, ellipsoidal-spherical, 39–50 μm wide, 41–57 μm long, opening by a large median pore; oospore spherical, 38–49 μm in diameter, completely filling oogonium, walls with spines 1–5 μm in length; suffultory cell sometimes slightly swollen; dwarf male stipe 10–15 μm wide, 26–35 μm long; gynandrosporous or idioandrosporous.

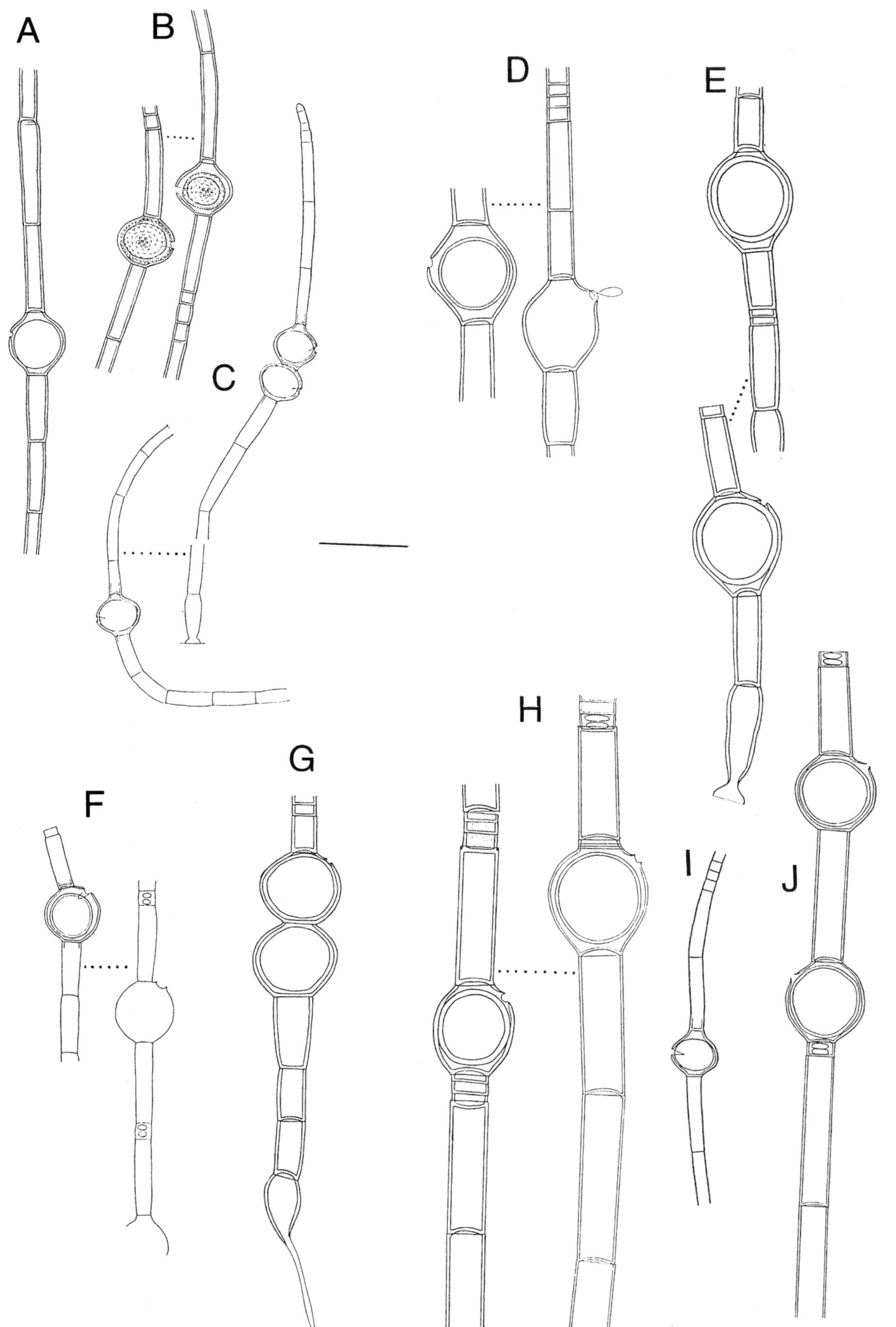

Plate 102 A–J
A. *Oedogonium obsoletum* (p. 426); **B**. *Oedogonium cymatosporum* (p. 420); **C**. *Oedogonium curvum* (p. 420); **D**. *Oedogonium urbicum* (p. 432); **E**. *Oedogonium zigzag* (p. 433); **F**. *Oedogonium hirnii* (p. 423); **G**. *Oedogonium curtum* (p. 420); **H**. *Oedogonium vaucherii* (p. 432); **I**. *Oedogonium cryptoporum* (p. 420); **J**. *Oedogonium fragile* (p. 423).
Scale bar: 50 μm.

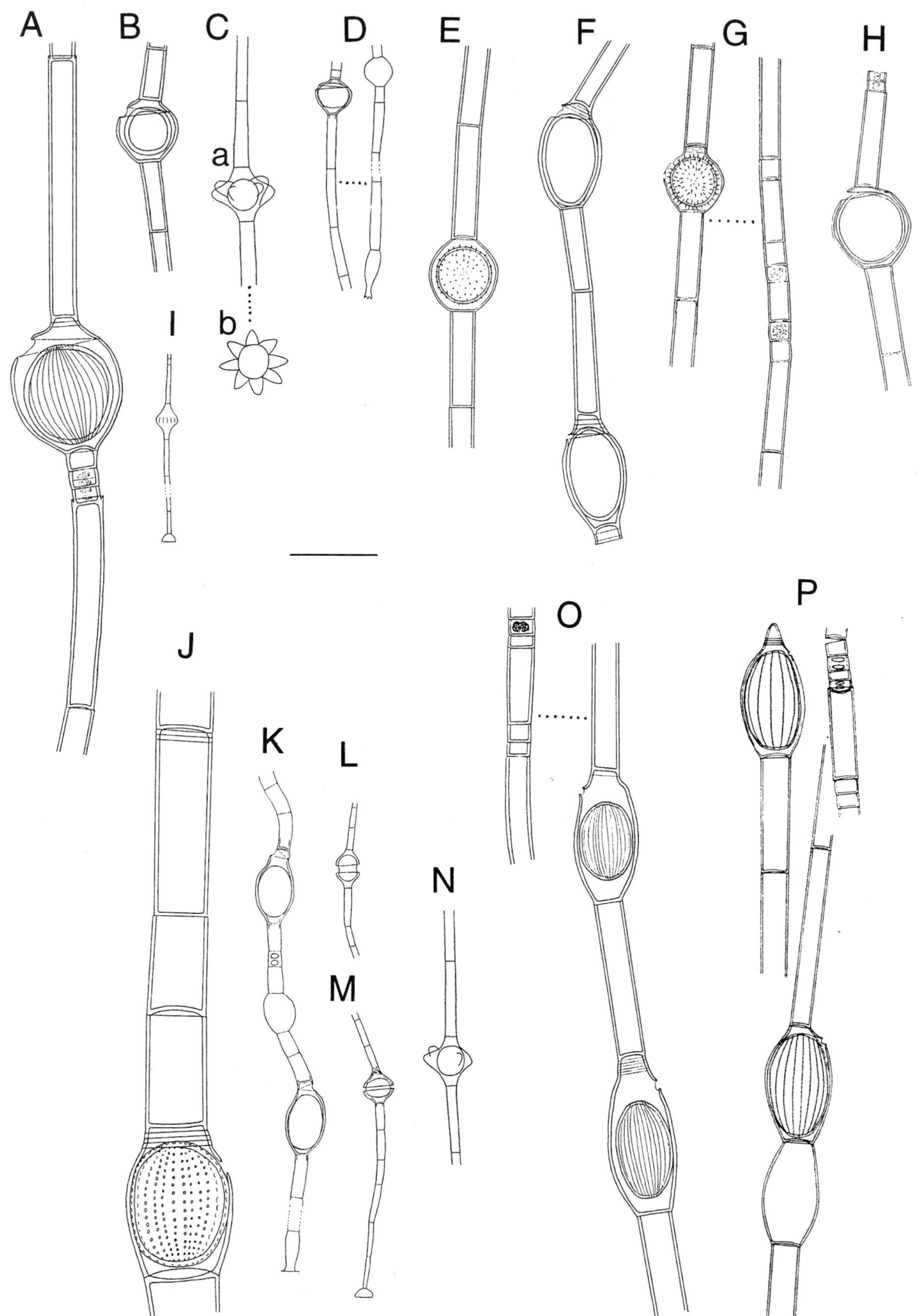

Plate 103 A–P

A. *Oedogonium nobile* (p. 426); **B**. *Oedogonium crispum* fo. *obesum* (p. 419); **C**. *Oedogonium itsigsohnii* (p. 424): a. filament with oogonium, b. oogonium in vertical section; **D**. *Oedogonium petri* (p. 428); **E**. *Oedogonium suecicum* fo. *australe* (p. 432); **F**. *Oedogonium ahlstrandii* (p. 417); **G**. *Oedogonium suecicum* (p. 432); **H**. *Oedogonium crispum* (p. 419); **I**. *Oedogonium excisum* (p. 423); **J**. *Oedogonium giganteum* (p. 423); **K**. *Oedogonium oblongum* (p. 426); **L**. *Oedogonium pusillum* (p. 430); **M**. *Oedogonium tapeinosporum* (p. 432); **N**. *Oedogonium itzigsohnii var. minus* (p. 424); **O**. *Oedogonium boscii* (p. 417); **P**. *Oedogonium crenulato-costatum* (p. 419).

Scale bar: 50 μm.

Widely distributed in North America and Europe; recorded from Ireland and Scotland (Cooke, 1882–84) as well as from England (Harris, 1933; Devon).

Cooke's description and drawing (Cooke, 1882–84) suggests it to be very similar to *O. cleveanum*. It is distinguished from that species by clearly having spines rather than teeth on the oospore.

Oedogonium excisum Wittrock *et* Lundell in Wittrock 1872

19020380 **Pl. 103I** (p. 422)
Cells cylindrical, 3–6 μm wide, 4 to 8 times as long as wide; oogonium single, almost conical-oblong, 3–15 μm wide, 18–26 μm long, opening by a median operculum, walls with 9 longitudinal undulations around middle; oospore ellipsoidal, constricted in the middle, 9–12 μm wide, 15–18 μm long, nearly filling oogonium, walls smooth; suffultory cell similar to other vegetative cells; spermatozoids 1 per antheridium; monoecious, macrandrous.

Europe (Belgium, Finland, Sweden, Latvia); recorded in the British Isles from Ireland where rare (Cooke, 1882–84).

Although only recorded once in the British Isles, the constriction in the oospore makes this species very distinctive.

Oedogonium flavescens (Hassall) Wittrock 1870

Basionym: *Vesiculifera flavescens* Hassall
19020390 **Pl. 106E** (p. 429)
Cells cylindrical, 18–23 μm wide, 4 to 6 times as long as wide; oogonium single, ellipsoidal-spherical to almost spherical, 49–52 μm wide, 52–60 μm long, opening by a median pore; oospore spherical, 45–49 μm in diameter, almost filling oogonium, walls smooth; suffultory cell similar to other vegetative cells; dwarf male stipe 11–12 μm wide, 36–45 μm long; gynandrosporous or idioandrosporous.

Probably cosmopolitan; recorded in the British Isles only from England where known from Yorkshire (W. & G.S. West, 1900b), Cornwall (Hassall, 1845) and Devon (Harris, 1933).

Oedogonium fragile Wittrock 1870

19020100 **Pl. 102J** (p. 421)
Cells cylindrical, 12–17 μm wide, 4 to 7 times as long as wide; oogonium single, spherical or almost ovoid-spherical, 42–50 μm wide and 44–55 μm long, opening by a superior pore; oospore spherical, 39–46 μm wide and 39–46 μm long, filling oogonium, walls smooth; suffultory cell similar to adjacent vegetative cells; spermatozoids 2 per antheridium, division horizontal; monoecious, macrandrous.

Cosmopolitan; recorded in the British Isles from Cambridgeshire (West, 1899b), Hertfordshire (Fritsch & Rich, 1913; Howland, 1931) and North Wales (Woodhead & Tweed, 1954).

var. *robustum* (West *et* G.S.West) Tiffany 1930

Basionym: *Oedogonium zigzag* Cleve var *robustum* W. *et* G.S.West 1903b
Synonym: *Oedogonium robustum* Tiffany
19020102
Differs from the type variety in having wider vegetative cells (19–20 μm wide) and larger oogonium (53–58 μm wide, 50–60 μm long).

Only known from Middlesex (W. & G.S.West, 1903b).

Oedogonium giganteum Kützing 1845

19020400 **Pl. 103J** (p. 422)
Cells cylindrical, 30–50 μm wide, 2 to 4.5 times as long as wide; oogonium single, cylindrical-obovoid, 53–69 μm wide, 67–106 μm long, opening by a superior pore; oospore cylindrical-ellipsoidal or ellipsoidal, 51–65 μm wide, 65–103 μm long, nearly filling oogonium, walls covered in 25 to 30 rows of small pits; suffultory cell 10 μm wider than adjacent vegetative cells; dioecious and macrandrous, both uncertain.

Africa, North America, Europe; recorded in Great Britain and Ireland from Yorkshire (W. & G.S.West, 1900b; Cooke, 1882–84) and Clare Island (W. West, 1912).

An incompletely described species. Its large cells and ornamented oospore make it distinctive, however.

Oedogonium hirnii Gutwinski 1896

19020410 **Pl. 102F** (p. 421)
Cells cylindrical, sometimes slightly swollen at the distal end (capitellate), 8–13 μm wide, 4 to 5 times as long as wide; oogonium single, almost spherical or almost ovoid, 32–37 μm wide, 33–37 μm long, opening by a superior pore; oospore spherical, 28–31 μm in diameter, not filling oogonium, walls smooth; suffultory cell similar to adjacent vegetative cells; spermatozoids 2 per antheridium, division horizontal; monoecious, macrandrous.

Probably cosmopolitan; recorded only from Donegal in Ireland (W. & G.S.West, 1902).

Oedogonium hystrix Wittrock 1870

19020420 **Pl. 105F** (p. 427)
Cells cylindrical, 17–28 μm wide, 1.5 to 4.5 times as long as wide; oogonium single, ellipsoidal, 38–48 μm wide, 45–65 μm long, opening by a median pore; oospore almost hexagonally spherical, 37–46 μm wide, 43–55 μm long, nearly filling oogonium, walls with spines; suffultory cell similar to other vegetative cells; dwarf male stipe 10–11 μm wide, 22–25 μm long; gynandrosporous.

North Africa, North America, Europe; only recorded in the British Isles from Devon (Harris, 1933).

Oedogonium idioandrosporum (Nordstedt *et* Wittrock) Tiffany 1934

Basionym: *Oedogonium crassiusculum* Wittrock var. *idioandrosporum* Nordstedt *et* Wittrock
Synonym: *Oedogonium crassipellitum* G.S.West
19020110 **Pl. 106B** (p. 429)
Cells cylindrical, 25–36 μm wide, 2.5 to 5.5 times as long as wide; oogonium single to 3-seriate, spherical-ovoid to spherical, 48–59 μm wide, 57–90 μm long, opening by a superior pore; ellipsoidal-spherical, ovoid,

angular-spherical, rarely spherical, 42–57 µm wide, 50–66 µm long, nearly filling oogonium, walls smooth; suffultory cell similar to other vegetative cells; dwarf male stipe 14–16 µm wide, 60–70 µm long; idioandrosporous.

North America, Asia, Europe; only recorded in the British Isles from Cambridgeshire (West, 1899b, as *O. crassipellitum*).

Oedogonium itzigsohnii de Bary 1854

19020430 **Pl. 103C** (p. 422)
Cells cylindrical, 8–10 µm wide, 3 to 6 times as long as wide; oogonium single, ellipsoidal, with median and conically obtuse projections, appearing star-shaped with 7–10 points in vertical section, 34–40 µm wide, 32–40 µm long, opening by an inframedian operculum; oospore spherical, 20–23 µm in diameter, not filling oogonium, walls smooth; suffultory cell similar to adjacent vegetative cells; monoecious, macrandrous.

Africa, North America, Asia, widely distributed in Europe; recorded in the British Isles from Yorkshire (W. & G.S. West, 1900b), the English Lake District (West, 1892b), North Wales (Woodhead & Tweed, 1954) and 'Ireland and Scotland' (Cooke, 1882–84).

var. *minus* West 1893
19020432 **Pl. 103N** (p. 422)
Distinguished from the type variety by having longer, thinner cells and rounded projections on each oogonium.

Africa, Asia, parts of Europe; recorded in the British Isles from the Orkneys (West, 1893) and Clare Island, Ireland (W.West, 1912).

Oedogonium landsboroughii (Hassall) Wittrock 1874

Basionym: *Vesiculifera landsboroughii* Hassall
Synonym: *Oedogonium grande* Kützing
19020120 **Pl. 104J** (p. 425)
Cells cylindrical, 31–40 µm wide (female), 30–37 µm wide (male), 3 to 6 times as long as wide (female), 4–6 times as long as wide (male); oogonium single to 2, rarely 3-seriate, obovoid-ovoid, 63–78 µm wide, 85–115(–120) µm long, opening by a superior pore; oospore ovoid to ellipsoidal, (55–)59–70 µm wide, 73–120 µm long, filling or not filling oogonium, walls smooth; suffultory cell similar to adjacent vegetative cells; spermatozoids 2 per antheridium, division vertical; dioecious, macrandrous.

Africa, North and South America, Asia, much of Europe; recorded in the British Isles from Yorkshire (W. & G.S. West, 1900b), Cambridgeshire (West, 1899b), Hertfordshire (Hassall, 1845) and mentioned as from 'Britain and Ireland' by Cooke (1882–84).

This species is very similar to *O. crassum* in most respects, differing only in having ellipsoidal to spherical-ellipsoidal oospores compared to ovoid-ellipsoidal in *O. landsboroughii*. Its specific status is doubtful. A fo. *gemelliparum* is described by Cooke (1882–84) with almost hyaline terminal cells. This is considered to be a synonym of *O. grande* by Mrozinska (1985). *Oedogonium grande* differs only slightly in its cell dimensions and number of antheridia from *O. landsboroughii* and is probably a synonym of it.

Oedogonium lautumniarum Wittrock in Wittrock & Nordstedt 1877

19020440 **Pl. 104G** (p. 425)
Cells cylindrical, 16–22 µm wide (female), 15–20 µm wide (male), (2–)3 to 5 times as long as wide; oogonium single to rarely 2-seriate, almost obovoid-ovoid, 40–49 µm wide, 45–51 µm long, opening by a supramedian (rarely superior) pore; oospore almost spherical or sometimes almost angular, 36–46 µm wide, 35–47 µm long, filling oogonium, walls smooth; suffultory cell similar to adjacent vegetative cells; spermatozoids 2 per antheridium, division horizontal; dioecious, macrandrous.

North America, Asia, Europe; recorded in the British Isles from Middlesex (West & Fritsch, 1927; W. & G.S.West, 1897) and Devon (Harris, 1933).

Oedogonium macrandrium Wittrock 1870

19020450 **Pl. 107G** (p. 431)
Cells cylindrical, 15–20 µm wide, (2–)3 to 5 times as long as wide; oogonium single to 4-seriate, spherical-ovoid, 36–42 µm wide, 43–54 µm long, opening by a superior operculum; oospore spherical, rarely ovoid-spherical, 31–37 µm wide, 33–39 µm long, not completely filling oogonium, walls smooth; suffultory cell similar to other vegetative cells; dwarf male stipe 12–13 µm wide, 24–33 µm long; not known if gynandrosporous or idioandrosporous.

North Africa, North America, West Indies, Asia, widely distributed in Europe; recorded in the British Isles from Yorkshire (W. & G.S. West, 1900b), Essex (W. & G.S. West, 1897), Devon (Harris, 1933), North Wales (Woodhead & Tweed, 1954), Clare Island (W.West, 1912) and from 'Ireland' and 'Britain' by Cooke (1882–84).

Oedogonium macrospermum West *et* G.S.West 1897

19020460 **Pl. 107D** (p. 431)
Cells cylindrical, 10–16 µm wide, 4 to 5 times as long as wide; oogonium single, rarely 2-seriate, almost depressed to depressed-spherical, 39–46 µm wide, 34–44 µm long, opening by a narrow median operculum; oospore almost depressed-spherical, 38–47 µm wide, 38–49 µm long, filling oogonium, walls smooth; suffultory cell similar to other vegetative cells; dwarf male stipe 9–14 µm wide, 6–30 µm long, scattered and at right angles to female filament; gynandrosporous.

North and South America, Asia, parts of Europe; recorded in the British Isles from Middlesex (W. & G.S.West, 1897) and North Wales (Woodhead & Tweed, 1954).

Oedogonium microgonium Prescott 1944

19020470 **Pl. 101E** (p. 416)
Cells distinctly swollen at the distal end (capitellate), 8–9.5 µm wide, 2 to 3.5 times as long as wide; oogonium single, rarely 2-seriate, possibly terminal, spherical or depressed-spherical, little wider than vegetative cells, 11–13 µm wide, 10–11 µm long, opening by a

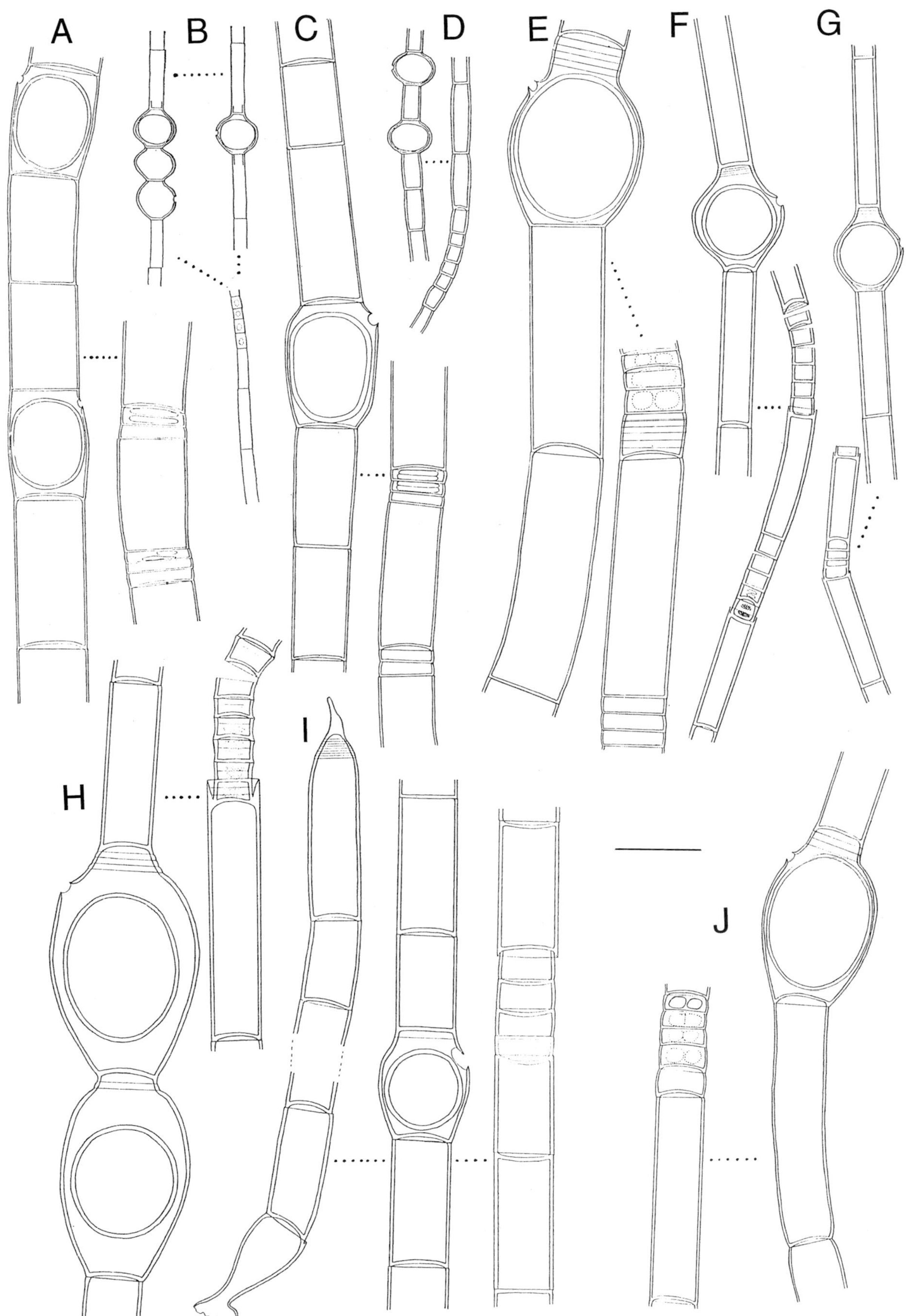

Plate 104 A–J

A. *Oedogonium capillare* (p. 417); **B**. *Oedogonium rufescens* (p. 430); **C**. *Oedogonium capilliforme* (p. 418); **D**. *Oedogonium calcareum* (p. 417); **E**. *Oedogonium crassum* (p. 419); **F**. *Oedogonium cardiacum* (p. 418); **G**. *Oedogonium lautumniarum* (p. 424); **H**. *Oedogonium rivulare* (p. 430); **I**. *Oedogonium princeps* (p. 428); **J**. *Oedogonium landsboroughii* (p. 424).

Scale bar: 50 μm.

superior operculum; oospore spherical, 11–12 μm in diameter, filling oogonium, walls smooth; suffultory cell similar to adjacent vegetative cells; antheridia not observed but as none seen in same filaments as oogonium plant may be dioecious, macrandrous.

Only described from Wisconsin in the USA (Prescott, 1944); in Europe only recorded from South Wales by Benson-Evans & Antoine (1996).

Oedogonium minus Wittrock 1874

19020480 **Pl. 101B** (p. 416)
Cells slightly swollen at the distal end (capitellate), walls ornamented with spirally arranged dots (punctate), 9–13 μm wide, 3 to 6 times as long as wide; basal cell depressed-spherical or almost hemispherical, with vertically folded walls; oogonium single, pear-shaped to spherical or depressed-spherical, 34–36 μm wide, 28–42 μm long, opening by a wide median operculum; oospore depressed-spherical, 30–42 μm wide, 26–36 μm long, not filling oogonium, walls smooth; suffultory cell similar to adjacent vegetative cells; spermatozoids 1 per antheridium; monoecious, macrandrous.

Africa, North America, Asia, widely distributed in Europe; only recorded in the British Isles from Ireland (Cooke, 1882–84).

Similar to *Oedogonium punctatostriatum* in most characters including basal cell shape and size. The latter is heterothallic, however. It is possible that these two species have been confused in the past.

Oedogonium multisporum H.C.Wood 1869

19020490 **Pl. 106C** (p. 429)
Cells cylindrical, 10–15 μm wide, 1.25 to 1.75 times as long as wide; oogonium single, rarely 3-seriate, almost ovoid or almost spherical, 24–35 μm wide, 27–33 μm long, opening by a superior pore; oospore spherical, 24–30 μm in diameter, nearly filling oogonium, walls smooth; suffultory cell similar to other vegetative cells; dwarf male stipe 10–11 μm wide, 26–30 μm long; gynandrosporous?

Africa, North America, Asia; only European record is from Hertfordshire (Fritsch & Rich, 1913) in England.

Oedogonium nobile Wittrock 1874

19020500 **Pl. 103A** (p. 422)
Cells cylindrical, 16–20 μm wide, 5 to 9 times as long as wide; oogonium single, very rarely 2, ellipsoidal or almost ovoid, 57–65 μm wide, 67–90 μm long, opening by a superior operculum; oospore ellipsoidal-spherical or spherical, 48–55 μm wide, 50–58 μm long, filling or not filling oogonium, walls with 30–35 longitudinal ribs; suffultory cell similar to adjacent vegetative cells; spermatozoids 2 per antheridium, division horizontal; monoecious, macrandrous.

North America, Asia, Europe (Finland, Poland); recorded in the British Isles from Middlesex (W. & G.S. West, 1897).

Oedogonium nodulosum Wittrock 1872

19020510 **Pl. 101C** (p. 416)
Cells undulate with two constrictions, 20–29 μm wide, 1.5 to 4.5 times as long as wide; oogonium single to 2-seriate, obovoid-spherical or rarely obovoid-ellipsoidal, 48–57 μm wide, 56–73 μm long, opening by a superior operculum; oospore spherical or almost spherical, 46–53 μm wide, 49–56 μm long, nearly filling oogonium, walls smooth; suffultory cell similar to adjacent vegetative cells; spermatozoids 2 per antheridium, division horizontal; monoecious, macrandrous.

North and South America, Asia, Australia, parts of Europe; recorded in the British Isles from Yorkshire (W. & G.S. West, 1900b).

Oedogonium oblongum Wittrock 1872

19020130 **Pl. 103K** (p. 422)
Cells cylindrical, 6–11 μm wide, 3 to 8 times as long as wide; oogonium single, oblong-ellipsoidal, 20–26 μm wide, 41–60 μm long, opening by a superior operculum; oospore ellipsoidal, 19–23 μm wide, 30–36 μm long, not filling oogonium, walls smooth; suffultory cell similar to adjacent vegetative cells; spermatozoids 2 per antheridium, division horizontal; monoecious, macrandrous.

Africa, Asia, Australia, New Zealand, widely distributed in Europe; recorded in the British Isles from Middlesex (W. & G.S. West, 1897) and North Wales (Woodhead & Tweed, 1954).

Oedogonium obsoletum Wittrock 1874

Synonym: *Oedogonium vernale* Wittrock
19020520 **Pl. 102A** (p. 421)
Cells cylindrical, 9–15 μm wide, 3 to 5 times as long as wide; oogonium single, almost spherical or almost depressed-spherical, 34–39 μm wide, 34–43 μm long, opening by a supramedian pore; oospore almost depressed-spherical, 30–34 μm wide, 28–32 μm long, not filling oogonium, walls smooth; suffultory cell similar to adjacent vegetative cells; spermatozoids 1 per antheridium; monoecious, macrandrous.

Africa, North America, parts of Europe; recorded in the British Isles from Oxfordshire (W. & G.S. West, 1897) and Devon (Harris, 1933).

Oedogonium obtruncatum Wittrock 1874

19020530 **Pl. 101G** (p. 416)
Cells swollen at the distal end (capitellate), 18–22 μm wide, 3 to 5 times as long as wide; oogonium single to 6-seriate, ellipsoidal or spherical-ellipsoidal, 45–55 μm wide, 56–68 μm long, opening by a supreme operculum; oospore ellipsoidal or spherical-ellipsoidal, 15–17 μm wide, 31–36 μm long, nearly filling oogonium, walls smooth; suffultory cell similar to other vegetative cells; dwarf male stipe 17–20 μm wide, 36–40 μm long; gynandrosporous or idioandrosporous.

Cosmopolitan; recorded in the British Isles from Yorkshire (W. & G.S. West, 1900b) and Middlesex (W. & G.S. West, 1897).

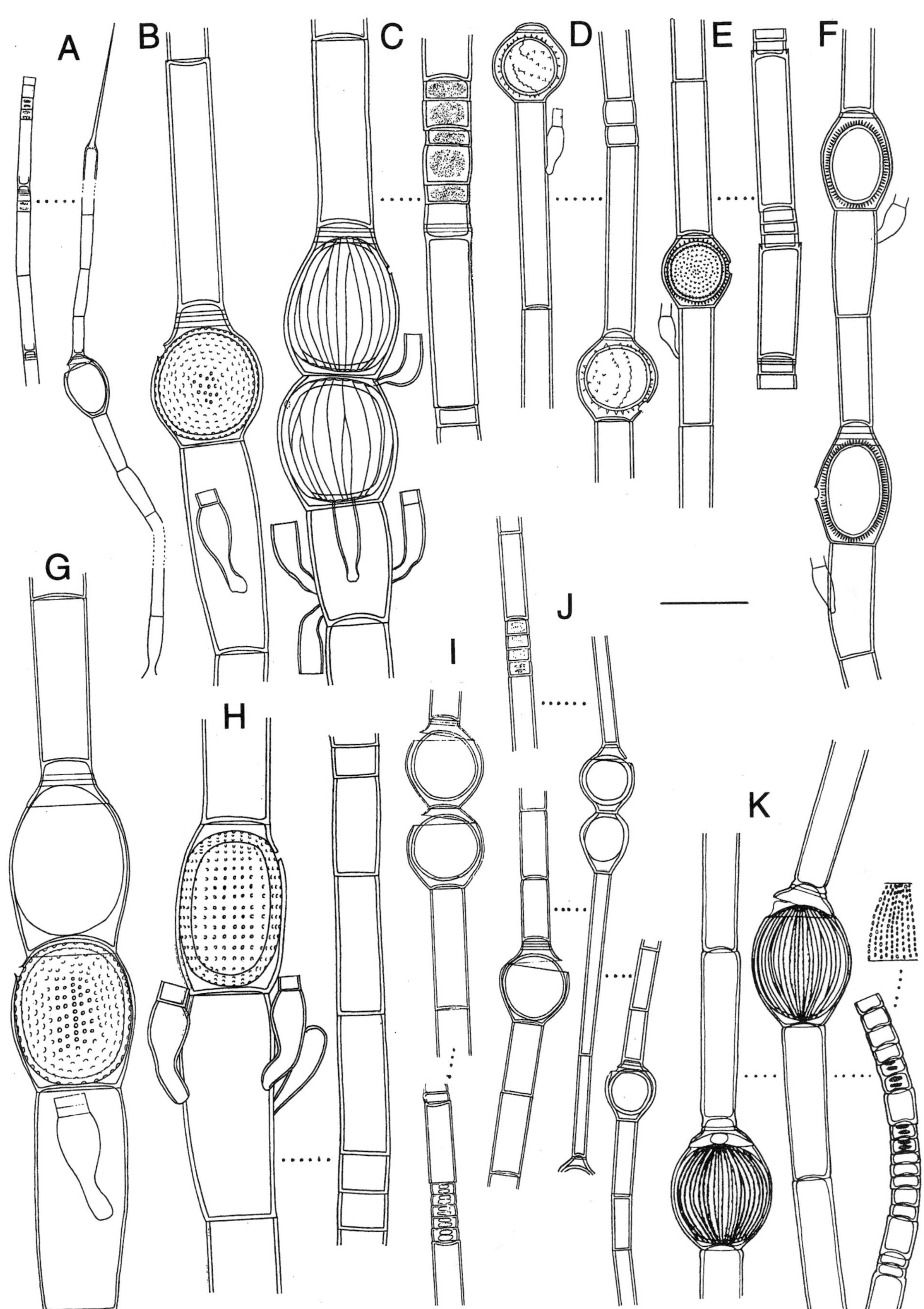

Plate 105 A–K

A. *Oedogonium pisanum* (p. 428); **B**. *Oedogonium concatenatum* var. *hutchinsiae* (p. 419); **C**. *Oedogonium cyathigerum* (p. 420); **D**. *Oedogonium cleveanum* (p. 418); **E**. *Oedogonium echinospermum* (p. 420); **F**. *Oedogonium hystrix* (p. 423); **G**. *Oedogonium concatenatum* (p. 418); **H**. *Oedogonium concatenatum* var. *rectangulare* (p. 419); **I**. *Oedogonium pringsheimii* (p. 428); **J**. *Oedogonium pringsheimii* var. *norstedtii* (p. 428); **K**. *Oedogonium tumidulum* (p. 432).

Scale bar: 50 μm.

Oedogonium petri Wittrock 1874

19020550 **Pl. 103D** (p. 422)

Cells cylindrical, 6–7 μm wide, 5 to 7 times as long as wide; oogonium single, rarely 2-seriate, pear-shaped, or pear-shaped to spherical, 21–24 μm wide, 22–29 μm long, opening by a supramedian operculum; oospore almost depressed or depressed-spherical, 20–23 μm wide, 16–19 μm long, filling oogonium, walls smooth; suffultory cell similar to adjacent vegetative cells; spermatozoids one per antheridium; monoecious, macrandrous.

Africa, Asia, Europe; recorded in the British Isles from Surrey (W. & G.S. West, 1897) and Ireland (Cooke, 1882–84).

Oedogonium pisanum Wittrock 1876

19020560 **Pl. 105A** (p. 427)

Cells cylindrical, 5–12 μm wide, 2 to 6 times as long as wide, terminal cell with a hair; oogonium single, rarely to 3-seriate, ellipsoidal, 23–29 μm wide, 34–43 μm long, opening by a superior operculum; oospore ovoid to ellipsoidal, 28–34 μm wide, 37–55 μm long, nearly filling oogonium, walls smooth; suffultory cell similar to adjacent vegetative cells; spermatozoids 2 per antheridium, produced by a horizontal division; dioecious, macrandrous.

Africa, North America, Asia, Europe; recorded in the British Isles only from Devon (Harris, 1933).

Oedogonium platygynum Wittrock 1872

19020140 **Pl. 101H** (p. 416)

Cells swollen at the distal end (capitellate), 5–11 μm wide, 2 to 5 times as long as wide; oogonium single to 2 together, depressed-ovoid, with 7–12 (usually 8) rounded median projections arranged in a whorl, 21–30 μm wide, 16–24 μm long, opening by an inframedian operculum; oospore depressed or almost depressed-spherical, 17–24 μm wide, 13–20 μm long, nearly filling oogonium, walls smooth; suffultory cell sometimes inflated; dwarf males single-celled, obovoid, 4–5 μm wide, 8–10 μm long; gynandrosporous or idioandrosporous.

Cosmopolitan; recorded in the British Isles from Yorkshire (W. & G.S. West, 1900b), Cornwall (W. & G.S. West, 1897), Scotland (West, 1893), and Ireland (West, 1892a; Cooke, 1882–84) including Donegal/Antrim (W. & G.S. West, 1902) and Clare Island (W. West, 1912).

Oedogonium pluviale Nordstedt 1872

19020570 **Pl. 107A** (p. 431)

Cells cylindrical, 22–29 μm wide (female), 18–27 wide (male), 0.75 to 3 times longer than wide (female), 1 to 2 times longer than wide (male); oogonium single, rarely 2- to 3-seriate, obovoid-spherical or almost spherical, 34–45 μm wide, 34–50 μm long, opening by a superior operculum; oospore almost spherical or almost ellipsoidal-spherical, 32–40 μm wide, 24–30 μm long, nearly filling oogonium, walls smooth; suffultory cell similar to other vegetative cells; dwarf males obovoid, 2-celled, a very reduced lens-shaped stipe cell and a larger (14–15 μm long) antheridial cell; idioandrosporous.

Africa, North and South America, Asia, widely distributed in Europe; recorded in the British Isles from North Wales (West, 1890; Woodhead & Tweed, 1954) and Devon (Harris, 1933).

Cooke (1882–84) notes that the androsporangia are slightly narrower than other cells.

Oedogonium princeps (Hassall) Wittrock 1874

Basionym: *Vesiculifera princeps* Hassall
Synonym: *Vesiculifera capillaris* Hassall

19020580 **Pl. 104I** (p. 425)

Cells cylindrical, 33–42(–45) μm wide, 1.25 to 3 times as long as wide; oogonium single, slightly inflated to almost ovoid, 51–63 μm wide and 54–80 μm long, opening by a superior pore; oospore spherical or almost spherical, 47–58 μm wide and 47–65 μm long, not filling oogonium, walls smooth; suffultory cell similar to adjacent vegetative cells; spermatozoids 2 per antheridium, formed by a vertical division; dioecious, macrandrous.

North and South America, Asia, New Zealand, Europe (England); recorded as from 'England' (Cooke, 1882–84) where known from Yorkshire (W. & G.S. West, 1900b) and Hertfordshire (Hassall, 1845).

Oedogonium pringsheimii C.E.Cramer 1859

19020590 **Pl. 105I** (p. 427)

Cells cylindrical, 14–20 μm wide (female), 12–16 μm wide (male), 2 to 5 times as long as wide (female), 2 to 4 times as long as wide (male); oogonium single to 6-seriate, almost ovoid-spherical, 35–43 μm wide, 36–46 μm long, opening by a superior operculum; oospore spherical, 30–37 μm wide, 30–37 μm long, nearly filling oogonium, walls smooth; suffultory cell similar to adjacent vegetative cells; spermatozoids 2 per antheridium, division horizontal; dioecious, macrandrous.

Cosmopolitan; recorded from Ireland (West, 1892a) and 'Britain' (Cooke, 1882–84).

var. *nordstedtii* Wittrock in Wittrock & Nordstedt 1877

19020592 **Pl. 105J** (p. 427)

Cells slightly smaller than the type variety with the oospore not filling the oogonium.

Cosmopolitan; recorded in the British Isles from Yorkshire (W. & G.S. West, 1900b) and Hertfordshire (Fritsch & Rich, 1913).

Oedogonium punctatostriatum de Bary 1854

19020150 **Pl. 101D** (p. 416)

Cells cylindrical, walls ornamented with spirally arranged dots (punctate), 18–22 μm wide (female), 16–19 μm wide (male), 2 to 6 times as long as wide; basal cell depressed-spherical or almost hemispherical, walls vertically folded (plicate); oogonium single, depressed-spherical, 48–55 μm wide, 38–48 μm long,

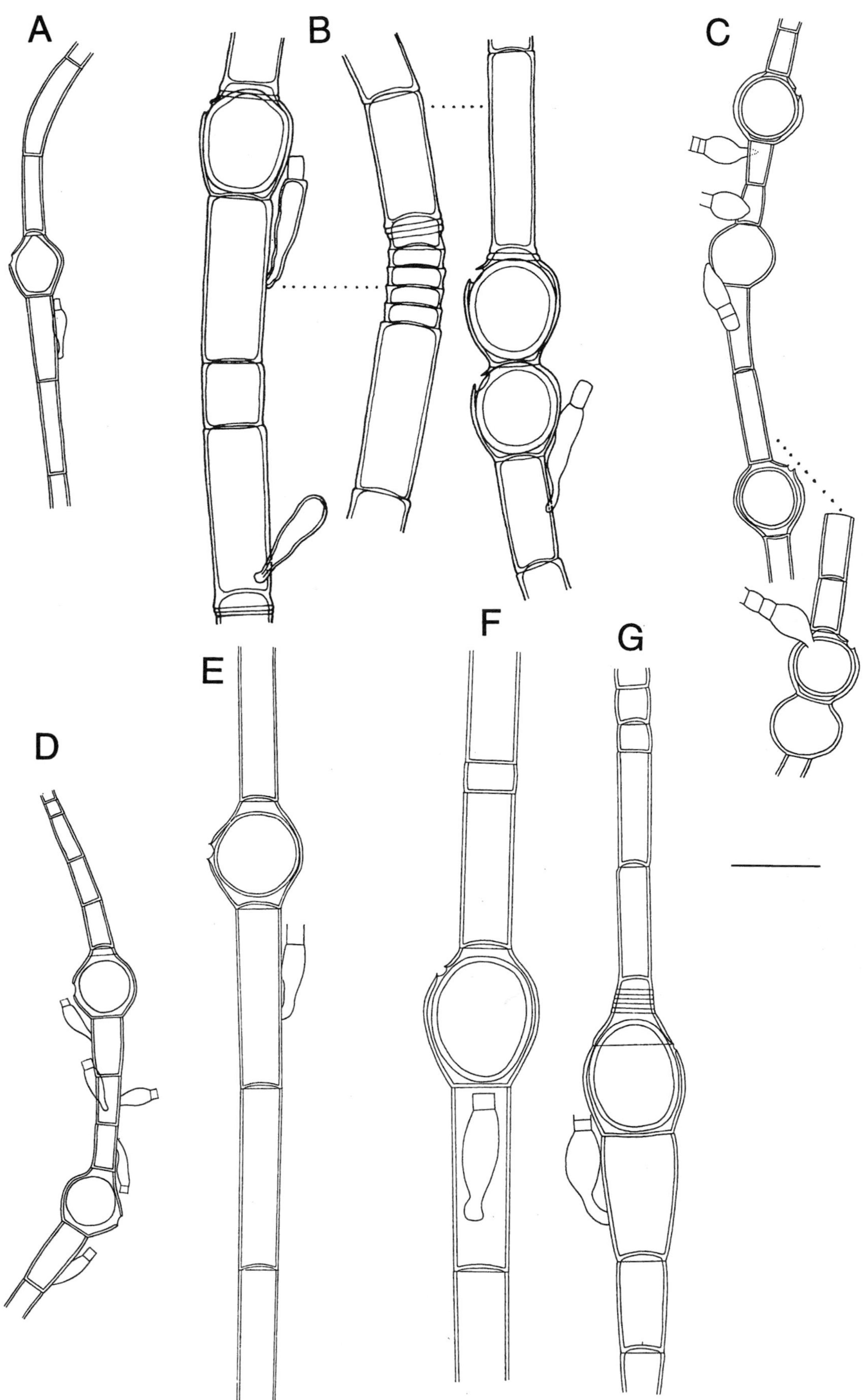

Plate 106 A–G
A. *Oedogonium sexangulare* (p. 430); **B**. *Oedogonium idioandrosporum* (p. 423); **C**. *Oedogonium multisporum* (p. 426); **D**. *Oedogonium braunii* (p. 417); **E**. *Oedogonium flavescens* (p. 423); **F**. *Oedogonium crassiusculum* (p. 419); **G**. *Oedogonium borisianum* (p. 417).
Scale bar: 50 µm.

opening by a wide, distinct median operculum, walls ornamented with spirally arranged dots (punctate); oospore spherical, 40–51 μm wide, 35–43 μm long, not filling oogonium, walls smooth; suffultory cell similar to adjacent vegetative cells; spermatozoids 1 per antheridium; dioecious or monoecious, macrandrous.

Cosmopolitan; recorded in the British Isles from Yorkshire (W. & G.S. West, 1900b), and Ireland (West, 1892a; Cooke, 1882–84) including Donegal (W. & G.S. West, 1902) and Clare Island (W. West, 1912).

Oedogonium pusillum Kirchner 1878

19020160 **Pl. 103L** (p. 422)

Cells cylindrical, 3–6 μm wide, 3 to 10 times as long as wide; basal cell almost hemispherical; oogonium single, rarely 2-seriate, almost biconical-spherical or almost biconical-ellipsoidal, 14–16 μm wide, 15–25 μm long, opening by a very wide median operculum; oospore ellipsoidal or spherical, 11–13 μm wide, 13–15 μm long, not filling oogonium, constricted in middle, walls smooth; suffultory cell similar to adjacent vegetative cells; spermatozoids 1 per antheridium; monoecious?, macrandrous.

Cosmopolitan; recorded in the British Isles from Devon (Harris, 1933) and from Clare Island (W.West, 1912).

This species is described by Tiffany (1937a) as monoecious and by Mrozinska (1985a) as dioecious.

Oedogonium reinschii J.Roy ex Cooke 1882–84

19020600 **Pl. 101J** (p. 416)

Cells cylindrical, nearly cylindrical, nearly hexagonal or almost ellipsoidal, 5–11 μm wide, 2 to 2.5 times as long as wide; basal cell almost hemispherical; oogonium single to 2-seriate, spherical, depressed-spherical or ovoid, 17–20 μm wide, 15–21 μm long, opening by a wide median operculum, walls smooth; oospores ellipsoidal, 13–18 μm wide, 14–17 μm long, filling or not filling oogonium, walls smooth; suffultory cell similar to other vegetative cells; spermatozoids 2 per antheridium, produced by a vertical division; monoecious, macrandrous.

Africa, North, Central and South America, Asia, widely distributed in Europe; in the British Isles recorded only from Scotland (Cooke, 1882–84).

A very distinctive species since its cells are almost bead-like and hexagonal.

Oedogonium rivulare (Le Clerc) A.Braun 1855

Basionym: ?*Prolifera rivularis* Le Clerc

19020610 **Pl. 104H** (p. 425)

Cells cylindrical, 35–45 μm wide (female), 30–36 μm wide (male), 3 to 8 times longer than wide (female), 4 to 8 times longer than wide (male); oogonium single to 7-seriate, obovoid, walls smooth, 70–85 μm wide, 130–160 μm long, opening by a superior pore; oospore obovoid, ellipsoidal or almost spherical, 55–70 μm wide, 65–100 μm long, not filling oogonium longitudinally, walls smooth; suffultory cell similar to adjacent vegetative cells; spermatozoids 2 per antheridium, division horizontal; dioecious, macrandrous.

Africa, North America, Asia, Europe; recorded in the British Isles from Yorkshire (W. & G.S. West, 1900b) and Scotland (Cooke, 1882–84).

Clearly distinguishable from other related species by its very large cells.

Oedogonium rothii (Le Clerc) Pringsheim 1858

Basionym: ?*Prolifera rothii* Le Clerc

19020190 **Pl. 107H** (p. 431)

Cells cylindrical, 6–10 μm wide, 3 to 8 times as long as wide; oogonium single to 3-seriate, almost depressed-spherical, 20–27 μm wide, 16–27 μm long, opening by a median operculum; oospore depressed-spherical, 17–25 μm wide, 14–20 μm long, nearly filling oogonium, walls smooth; suffultory cell similar to other vegetative cells; dwarf male stipe 4 μm wide and 11–12 μm long; gynandrosporous.

Africa, North America, Asia, much of Europe; recorded in the British Isles from Yorkshire (W. & G.S. West, 1900b), Cambridge (West, 1899b), Devon (Harris, 1933), Clare Island (W. West, 1912), North Wales (Woodhead & Tweed, 1954); from 'Scotland, Ireland, England' by Cooke (1882–84).

Oedogonium rufescens Wittrock 1870

19020610 **Pl. 104B** (p. 425)

Cells cylindrical, 8–10 μm wide (female), 7–9 μm wide (male), 4 to 7 times as long as wide (female), 4 to 6 times as long as wide (male); oogonium single to 3-seriate, obovoid or depressed obovoid-spherical, 22–44 μm wide, 22–30 μm long, opening by a median, rimiform (cleft-like) pore; oospore spherical or depressed-spherical, 21–23 μm wide, 17–22 μm long, filling or nearly filling oogonium, walls smooth; suffultory cell similar to adjacent vegetative cells; spermatozoids 1 per antheridium; dioecious, macrandrous.

Africa, North America, Asia, Atlantic islands, widely distributed in Europe; recorded in the British Isles from the Scilly Isles (W. & G.S West, 1903b; West & Fritsch, 1927), Devon (Harris, 1933) and North Wales (Woodhead & Tweed, 1954).

var. *lundellii* (Wittrock) Tiffany 1929

Basionym: *Oedogonium lundellii* Wittrock

19020612

As type with cells slightly wider in proportion to oogonium (Tiffany, 1929).

Oedogonium sexangulare Cleve ex Wittrock 1870

Synonym: *Oedogonium hexagonum* Kützing

19020620 **Pl. 106A** (p. 429)

Cells cylindrical, 9–16 μm wide, 3 to 7 times as long as wide; oogonium single, rarely 2-seriate, sexangular-ellipsoidal, 29–33 μm wide, 33–39 μm long, opening by a median-supramedian pore; oospore sexangular-ellipsoidal, 27–31 μm wide, 31–36 μm long, nearly filling oogonium, walls smooth; suffultory cell similar to other vegetative cells; dwarf males having stipes 7–9 μm wide and 21–30 μm long; gynandrosporous.

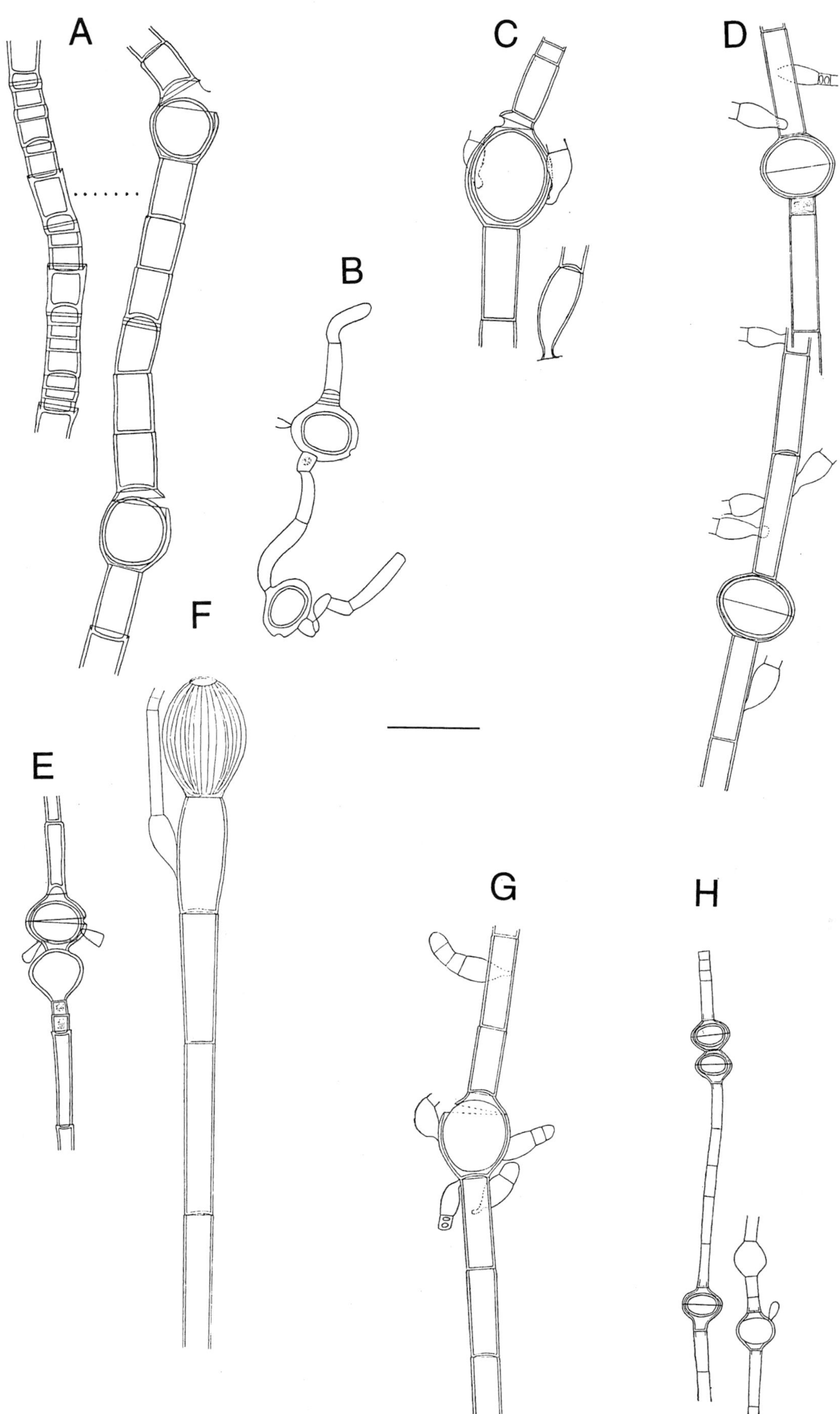

Plate 107 A–H
A. *Oedogonium pluviale* (p. 428); **B**. *Oedogonium contortum* (p. 419); **C**. *Oedogonium ciliatum* (p. 418); **D**. *Oedogonium macrospermum* (p. 424); **E**. *Oedogonium decipiens* (p. 420); **F**. *Oedogonium acrosporum* (p. 415); **G**. *Oedogonium macrandrium* (p. 424); **H**. *Oedogonium rothii* (p. 430). Scale bar: 50 μm.

North Africa, North America, Asia, Europe; recorded in the British Isles from Devon (Harris, 1933), Hertfordshire (Hassall, 1845), and Deeside in Scotland (Cooke, 1882–84).

Oedogonium suecicum Wittrock 1872

19020630 **Pl. 103G** (p. 422)

Cells cylindrical, 9–14 μm wide, 3 to 7 times as long as wide; oogonium single, almost spherical, 32–38 μm wide, 34–41 μm long, opening by median pore; oospore spherical, 30–37 μm wide, 30–37 μm long, nearly filling oogonium, outer wall with spines; suffultory cell similar to adjacent vegetative cells; spermatozoids 1 per antheridium; dioecious, macrandrous.

Cosmopolitan; recorded in Ireland (West, 1892a) and from Devon in England (Harris, 1933) and in North Wales (Woodhead & Tweed, 1954).

fo. *australe* G.S.West 1909

19020632 **Pl. 103E** (p. 422)

Differs from the type form in having crooked and more scattered spines on the outer oospore wall with cells slightly broader and oogonium larger.

Africa, North America, Australia, Asia; only recorded in the British Isles from North Yorkshire (Pentecost, pers. comm.).

Oedogonium tapeinosporum Wittrock 1874

19020640 **Pl. 103M** (p. 422)

Cells cylindrical, 3–5 μm wide, 2.5 to 5 times as long as wide; basal cell almost hemispherical; oogonium single to 2-seriate, depressed-spherical, 14–19 μm wide, 17–23 μm long, opening by a median operculum; oospore depressed-spherical, 13–16 μm wide, 8–14 μm long, rarely filling oogonium, walls smooth; suffultory cell similar to adjacent vegetative cells; antheridia unknown; dioecious?, macrandrous.

Africa, North America, Asia, Europe; recorded in the British Isles from Yorkshire (W. & G.S.West, 1900b) and Hampshire (Woodhead & Tweed, 1938).

Oedogonium tumidulum (Kützing) Wittrock 1874

Basionym: *Conferva tumidula* Eng. Bot. No. 1670 in Kützing

19020650 **Pl. 105K** (p. 427)

Cells cylindrical, 18–25 μm wide (female), 15–20 μm wide (male), 3.5 to 5 times as long as wide (female), 4 to 6 times as long as wide (male); oogonium single, almost ellipsoidal, 52–63 μm wide, 78–90 μm long, opening by superior operculum; oospore ellipsoidal or almost spherical-ellipsoidal, 49–58 μm wide, 61–75 μm long, filling oogonium, walls with 40–50 longitudinal rows of granulate ribs; suffultory cell similar to adjacent vegetative cells; spermatozoids 2 per antheridium, division horizontal; dioecious, macrandrous.

Europe (Germany, Poland, Russia, Ukraine); recorded in the British Isles from Yorkshire (W. & G.S.West, 1900b) and from Ireland (Cooke, 1882–84).

Oedogonium undulatum (Brébisson) A.Braun in de Bary 1854

Basionym: *Conferva undulata* Brébisson

19020200 **Pl. 101A** (p. 416)

Cells with 4 undulate constrictions, 15–20 μm wide, 3 to 5 times as long as wide; oogonium single to 2-seriate, sometimes terminal, almost spherical or ellipsoidal-spherical, 48–56 μm wide, 50–75 μm long, opening by an inferior operculum; oospore spherical or almost spherical, 42–50 μm wide, 42–52 μm long, filling oogonium, walls smooth; suffultory cell similar to other vegetative cells; dwarf males 8–10 μm wide, 48–65 μm long; gynandrosporous or idioandrosporous.

Cosmopolitan; recorded from North Yorkshire (West & Fritsch, 1927), the English Lake District (West, 1892b), Ireland (West, 1892a; Heuff & Horkan, 1984), Cambridgeshire (West, 1899b), Cornwall (W. & G.S. West, 1897), Scotland (Cooke, 1882–84), Clare Island (W.West, 1912), North Wales (Woodhead & Tweed, 1954) and Yorkshire (W. & G.S. West, 1900b).

The variety *moebiusii* Schmidle has been recorded in the British Isles (W. & G.S.West, 1900b). This is not recognized by more recent authors as warranting varietal status (Tiffany, 1930; Mrozinska, 1985). One of the most commonly recorded species, probably because it can be readily identified in the vegetative condition on account of its distinctive cell wall undulations.

Oedogonium urbicum Wittrock 1874

19020660 **Pl. 102D** (p. 421)

Cells cylindrical, 12–16 μm wide, 2.5 to 6 times as long as wide; oogonium single, rarely more, obovoid depressed or almost depressed pear-shaped-spherical, 34–50 μm wide, 34–55 μm long, opening by a superior pore, walls smooth; oospore spherical, 31–41 μm wide, 30–41 μm long, not filling oogonium, walls smooth; suffultory cell similar to other vegetative cells; spermatozoids 2 per antheridium, division horizontal; monoecious, macrandrous.

Asia (India), Europe (Germany, Poland, Ukraine); recorded in Ireland (Cooke, 1882–84) including Clare Island (W. West, 1912), and from Devon (Harris, 1933).

Oedogonium vaucherii (Le Clerc) A.Braun 1855

Basionym: ?*Prolifera vaucherii* Le Clerc

19020210 **Pl. 102H** (p. 421)

Cells cylindrical, 20–30 μm wide, 1.5 to 4 times as long as wide; oogonium single, obovoid to almost obovoid-spherical, 40–58 μm wide and 45–65 μm long, opening by a superior pore; oospore spherical to almost spherical, 35–54 μm wide and 35–55 μm long, not filling oogonium, walls smooth; suffultory cell similar to adjacent vegetative cells; spermatozoids 2 per antheridium, division horizontal; monoecious, macrandrous.

Africa, Asia, North America, widely distributed in Europe; recorded in the British Isles from Yorkshire (W. & G.S. West, 1900b), Cambridgeshire (West, 1899b) and Hertfordshire (Hassall, 1845).

Oedogonium zigzag Cleve ex Wittrock 1870

19020670 **Pl. 102E** (p. 421)

Cells cylindrical, 15–20 μm wide, 2 to 4 times as long as wide; oogonium single, spherical or obovoid-spherical, 45–63 μm wide and 42–63 μm long, opening by a superior pore; oospore spherical or obovoid-spherical, 43–60 μm wide and 48–58 μm long, filling oogonium, walls smooth; suffultory cell similar to adjacent vegetative cells; spermatozoids 2 per antheridium, produced by a vertical divison; monoecious, macrandrous.

Europe (Sweden, Ukraine); recorded in the British Isles from Middlesex (W. & G.S. West, 1897).

This species has close affinities with *Oedogonium curtum* and *O. fragile*. It differs mainly in its zigzag pattern of cells and limited growth with filaments no more than 6–8 cells long.

DOUBTFUL TAXA

Included here are unsubstantiated and single early records from the British Isles, or species not recognized in recent monographs. Some doubtful taxa have been described and keyed out above, because they are either particularly distinctive or widely distributed in mainland Europe.

Oedogonium alternans Wittrock *et* Lundell, ***Oedogonium depressum*** Pringsheim, ***Oedogonium flexuosum*** Hirn, ***Oedogonium gallicum*** Hirn, ***Oedogonium grande*** Kützing, ***Oedogonium inerme*** Hirn, ***Oedogonium irregulare*** Wittrock, ***Oedogonium londinense*** Wittrock, ***Oedogonium longatum*** Kützing, ***Oedogonium longicolle*** var. ***senegalense*** Nordstedt, ***Oedogonium magnusii*** Wittrock, ***Oedogonium moniliforme*** Wittrock, ***Oedogonium oviforme*** (Lewin) Hirn, ***Oedogonium paludosum*** (Hassall) Wittrock, ***Oedogonium pilosporum*** W.West, ***Oedogonium pseudoboscii*** Hirn, ***Oedogonium richterianum*** Lemmermann, ***Oedogonium rugulosum*** Nordstedt, ***Oedogonium rupestre*** Hirn, ***Oedogonium sociale*** Wittrock, ***Oedogonium subcetaceum*** Kützing, ***Oedogonium tenellum*** Kützing, ***Oedogonium vesicatum*** (Lyngbye) Wittrock.

Acknowledgements

Special thanks go to Dr C.G. Velez (Universidad de Buenos Aires, Argentina) for critically reading and correcting the Oedogoniales.

ORDERS CHAETOPHORALES, KLEBSORMIDIALES, MICROSPORALES, ULOTRICHALES

BY DAVID M. JOHN

Mostly branched or unbranched filamentous, prostrate, erect or with variously developed prostrate and erect systems, only rarely solitary or membrane-like. Molecular and ultrastructural evidence has resulted in the disaggregation of what were traditionally the two large chlorophyte orders, the Chaetophorales and Ulotrichales. These orders are still recognized, the position of many genera currently placed within them remaining uncertain. Molecular systematic analyses and molecular architectural evidence indicate a close relationship of members of the Klebsormidiales to the ancestry of land plants (see Graham & Wilcox, 2000).

Apatococcus V.F.Brand 1925

Cells solitary, sometimes producing short uniseriate filaments, or dividing to form cube-like packets of cells; chloroplast often lobed, pyrenoids absent; asexual reproduction by zoospores and autospores; zoospores biflagellate and slightly flattened, 8–32(–64) produced in each zoosporangium; sexual reproduction unknown.

Probably cosmopolitan; a very common and widely distributed subaerial alga, usually growing on tree trunks where it is a component of the ubiquitous '*Protococcus–Pleurococcus* assemblage'; sometimes confused with other members of the assemblage (e.g. *Chlorococcum*, *Desmococcus*, *Trebouxia*).

Currently considered to be represented by a single very polymorphic species though traditionally several species have been recognized. Retained in the Order Chaetophorales to which it is has been traditionally assigned, but ultrastructural features suggest that the genus should be more correctly placed in the Chlorococcales *sensu lato* (see Hoek et al., 1995).

Apatococcus lobatus (Chodat) J.B.Petersen 1928

Basionym: *Pleurococcus lobatus* Chodat
Synonyms: *Apatococcus vulgaris* Brand, *Pleurococcus lobatus* Chodat

24010010 **Pl. 113A** (p. 450)

Cells single or forming 2-, 3- or 4-celled packets, often clustered together; cells spherical to somewhat compressed, 6–12(–15) μm across, walls frequently thickening with age.

Probably cosmopolitan; one of the commonest terrestrial algae and sometimes the principal component of the widespread '*Pleurococcus–Protococcus* assemblage' that forms a pale green, powdery coating over walls, tree bark, wooden palings and other surfaces. Very evident in urban environments where atmospheric pollution results in the absence or reduced abundance of many pollution-sensitive lichens that often compete with algae on soil-free surfaces. It is well developed in agricultural areas where fertilizers applied to the land have drifted, resulting in nutrient-enrichment leading to rapid growth of subaerial algae at the expense of lichens. Often most conspicuous in the spring and autumn when humid, damp and reasonable light conditions result in the optimal growth of many subaerial algae.

Aphanochaete A.Braun 1849

Synonyms: *Herposteiron* Nägeli, *Gonatoblaste* Huber

Epiphytic, consisting of creeping, uniseriate filaments, irregularly branched or more rarely unbranched; cells inflated or cylindrical, sometimes bearing 1 or several long, unicellular, basally-inflated bristles on upper surface; chloroplasts parietal, disc-shaped, with 1 to several pyrenoids; asexual reproduction by quadriflagellate zoospores or aplanospores; sexual reproduction oogamous and rarely observed: oogonium large and stalked, male gametes quadriflagellate; zygotes thick-walled, containing red or yellow oil droplets.

Probably cosmopolitan; widespread in hard and often nutrient-enriched flowing and still-water habitats (sometimes brackish-water), commonly epiphytic on submerged aquatic plants (e.g. *Myriophyllum*, *Sagittaria*, *Najas*, *Chara*) and filamentous green algae.

Culture investigations by Tupa (1974) have demonstrated that the extent of bristle and branch development is dependent upon environmental conditions. She considered 4 species to be valid and another 2 doubtful. Phosphorus limitation appears to be the main factor leading to bristle formation in the field (B.A.Whitton, pers. comm.). One of the species recognized in the British Isles, *A. polychaete*, has sometimes been considered a mere growth form of *Stigeoclonium*. The genus is in need of further revision.

1 Plants closely branched and forming disc-like crusts or expansions................................*A. magna*
1 Plants not as above...2
2 Plants having a distinct mucilaginous envelope; bristles basally sheathed ....*A. pilosissima*
2 Plants without a mucilaginous envelope; bristles unsheathed...3
3 Cells generally shorter than wide, each typically bearing more than 2 bristles..............*A. polychaete*
3 Cells usually longer than wide, each bearing a single bristle (rarely 2) ....*A. repens*

Aphanochaete magna Godward 1934

24020010 **Pl. 108A** (p. 435)

Filaments closely branched and forming irregularly-shaped, pseudoparenchymatous, disc-like crusts; cells 6–10 µm wide, cylindrical to slightly inflated, longer than wide, each bearing 1 or several slightly basally swollen bristles.

Europe, North America; frequently reported in the British Isles on submerged portions of aquatic macrophytes (e.g. *Callitriche aquatica*, *Oenanthe fluviatilis*, *Mentha aquatica*) growing in the marginal shallows of ponds and larger waterbodies, often most common during the winter.

The only *Aphanochaete* species in which the zoospores are reported to grow initially into a 4-celled rather than a 2-celled germling (Tupa, 1974).

Aphanochaete pilosissima Schmidle 1897

Synonym: *Herposteiron pilosissima*

24020020 **Pl. 108C** (p. 435)

Filaments loosely arranged within a mucilaginous envelope; cells about 4 µm wide, spherical or polygonal, shorter or longer than wide, each bearing 2–6 basally sheathed bristles, these bristles having slender processes along margin (fimbriate).

Probably cosmopolitan; in the British Isles in a wide range of aquatic habitats on various filamentous green algae including *Mougeotia*, *Debarya*, *Zygnema* and *Oedogonium*.

Aphanochaete polychaete (Hansgirg) F.E.Fritsch 1902

Basionym: *Herposteiron polychaete* Hansgirg

24020030 **Pl. 108B** (p. 435)

Filaments loosely branched; cells 9–15 µm wide, spherical, polygonal or oblong to somewhat rectangular, rarely longer than wide, each bearing (1–)2–6 unsheathed bristles.

Probably cosmopolitan; in the British Isles in ponds and lakes on aquatic angiosperms and filamentous green algae, rare.

The species is recognized here although West (1904, p. 72) thought that it 'is most probably identical with *Herposteiron pilosissima* Schmidle [=*Aphanochaete pilosissima*]' despite the latter having bristles with a distinctive basal sheath. West's doubt was not shared by Fritsch (1902) when making the combination *Aphanochaete polychaete*.

Aphanochaete repens A.Braun 1849

Synonyms: *Aphanochaete confervicola* (Nägeli ex Kützing) Rabenhorst, *Gonatoblaste rostrata* Huber, *Herposteiron repens* (A.Braun) Wittrock

24020040 **Pl. 108D** (p. 435) **CD**

Filaments loosely branched; cells 8–10 µm wide, irregularly swollen to subcylindrical, usually longer than wide, each bearing a single (rarely 2) unsheathed bristle.

Probably cosmopolitan; most common and widely distributed *Aphanochaete* in the British Isles, frequently growing along the shallow margins of ponds, lakes and rivers on various filamentous green algae; most common during the summer months.

A variety (var. *gracilis*) was described from southern England by the Wests (W. & G.S.West, 1897, as *Herposteiron repens*), its cells were 4–7 µm wide, 2 to 3 times longer than wide and the bristles were distinctly articulate. The variety has not been recorded again and it is difficult to see what purpose is served by recognizing varieties based upon relatively minor morphological differences when little or nothing is known of character plasticity.

Binuclearia Wittrock 1886

Filaments initially attached, later free-floating, uniseriate, unbranched, enclosed within a thick mucilaginous sheath; cells cylindrical to ellipsoidal with rounded apices and often in pairs immediately following division, later widely separating, the older transverse septa becoming greatly thickened and prominently striated; independently walled daughter cells give rise to 'H-shaped' sections; chloroplast parietal, usually band-shaped and occupying middle of cell, pyrenoid inconspicuous; asexual reproduction by quadriflagellate zoospores; akinetes and aplanospores known.

Probably cosmopolitan; widespread especially in the soft and acidic waters mostly of marshes, moorland pools and mountain lakes.

Taxonomic position of the genus remains uncertain. The parental wall enclosing the independently walled daughter cells, and the 'H-shaped' remains of earlier generations, have been interpreted as a linear row of autospores. For this reason, Hoek et al. (1995) consider the genus to belong to the Order Chlorococcales. Until ultrastructural and molecular data become available the genus is retained here in the Ulotrichales, the order to which it has traditionally been assigned.

Binuclearia tectorum (Kützing) Berger ex Wichmann 1937

Basionym: *Gloeotila tectorum* Kützing

Synonym: *Binuclearia tatrana* Wittrock

24030010 **Pl. 115B** (p. 456)

Cells (4.5–)6–9(–13) µm wide, 0.5 to 8 times longer than wide, oval or ellipsoidal, often in pairs.

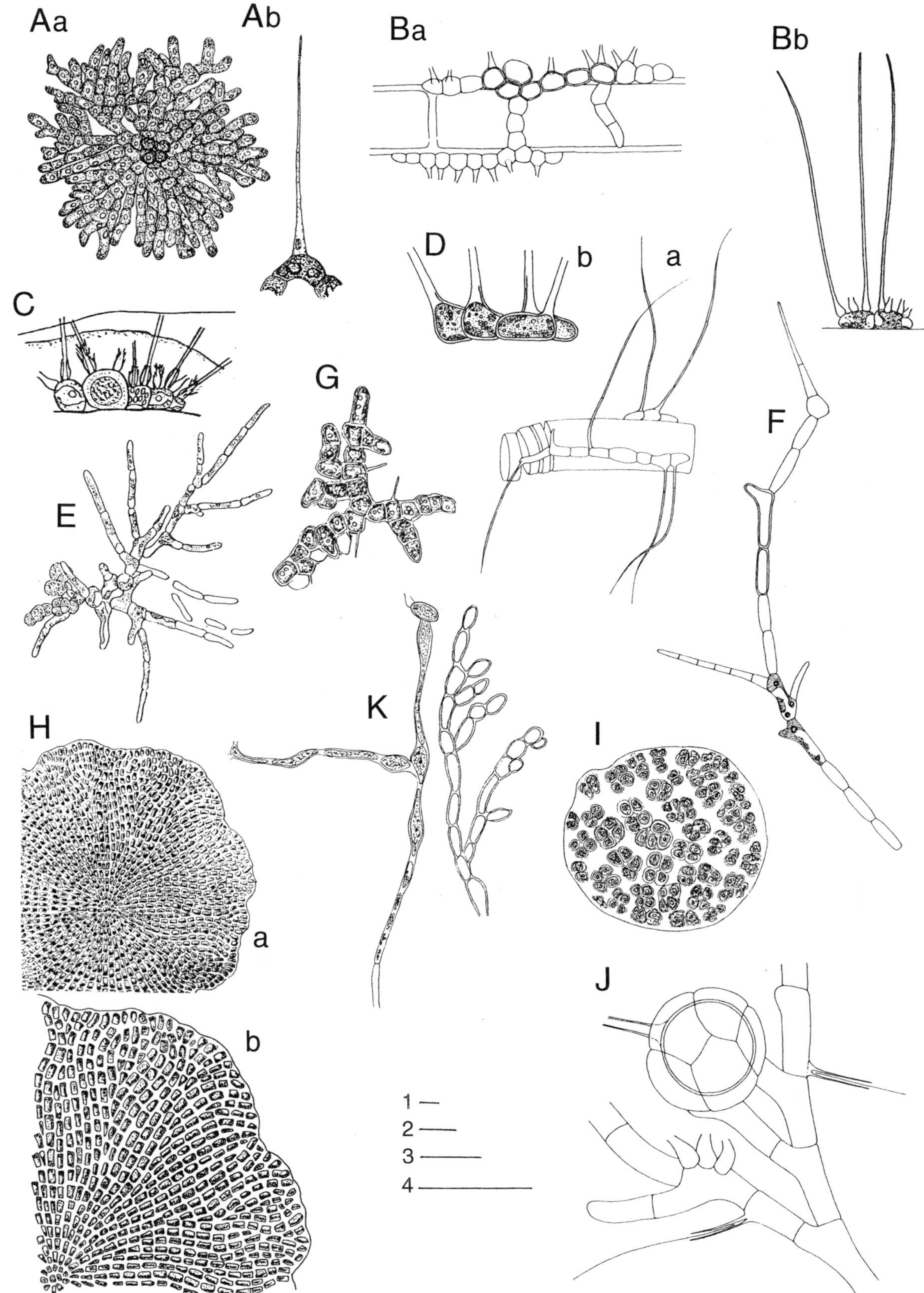

Plate 108 A–K

A. *Aphanochaete magna* (p. 434): a. prostrate system of filaments, b. details of a single cell with a bristle; **B**. *Aphanochaete polychaete* (p. 434): a. filaments creeping over a filamentous alga, b. details of 2 cells showing 3 intact bristles; **C**. *Aphanochaete pilosissima* (p. 434); **D**. *Aphanochaete repens* (p. 434): a. filaments creeping over a filamentous alga, b. details of cells bearing bristles; **E**. *Dilabifilum printzii* (p. 438); **F**. *Chaetonema irregulare* (p. 436); **G**. *Coleochaete irregularis* (p. 471); **H**. *Coleochaete orbicularis* (p. 471): a, b. prostrate, pseudoparenchymatous expansion; **I**. *Sporotetras pyriformis* (p. 404); **J**. *Coleochaete pulvinata* (p. 471): branch bearing an oogonium and sheathed bristles; **K**. *Gomontia aegagropila* (p. 444).

Scale bars: 1–3, 20 μm; 4, 10 μm. 1 for Ab, Db, E, F, Ha, J; 2 for Bb, Hb, G, K; 3 for Aa, Ba, Da, I; 4 for C.

Probably cosmopolitan; not uncommon throughout the British Isles where there are boggy margins to tarns and lakes, also in bog pools and often associated with *Sphagnum*.

Catena Chodat 1900

Filaments uniseriate, less than 16-celled, straight or curved, with cells connected by conical outgrowths from cross wall (not always very evident), enclosed within a mucilaginous sheath; cells cylindrical to oval, apices swollen and often encircled by unequal rings or caps, or with minute stick-like appendages or granules; chloroplast parietal, single, band-like and without pyrenoids; vegetative reproduction by fragmentation; asexual and sexual reproduction unknown.

Europe, North America; planktonic in various freshwater habitats.

Catena viridis Chodat 1900

24050010 **Pl. 115C** (p. 456)

Filaments 4–16-celled, enclosed within a thin mucilaginous sheath; cells 3–4 μm wide, 1 to 2 times longer than wide, often reddish-brown; chloroplast usually central.

Distribution as for genus; reported only from the River Dee (near Chester), but undoubtedly more widely distributed in the British Isles.

According to Williams (1941), the ring-like thickenings do not develop immediately, with the result that they may be absent from the apices of newly formed cells. He noted that the ring-like thickening is often reddish-brown in colour and sometimes broken into granules.

Chaetonema Nowakowski 1876

Epiphytic, consisting of irregularly branched and uniseriate filaments, short lateral branches arise from middle of cells and usually lie perpendicular to main axes; branches narrowing to a terminal hair; basally swollen hairs sometimes borne laterally on short projections; chloroplasts parietal, plate-like, encircling more than half cell circumference, with 1 or 2 pyrenoids; asexual reproduction by quadriflagellate zoospores formed in pairs in each cell, with flagella of equal length; sexual reproduction oogamous: 8 biflagellate male gametes produced in a barrel-shaped gametangium, female gametes single and developing within a swollen cell; palmelloid stage known.

Probably cosmopolitan; grows within the mucilaginous envelope of various algae. Monospecific.

Chaetonema irregulare Nowakowski 1876

24060010 **Pl. 108F** (p. 435)

Cells 9–15 μm wide, 2 to 4 times longer than wide, walls occasionally rust-red, often bearing curved hairs.

Probably cosmopolitan; only reported on a few occasions in the British Isles as an epiphyte within the mucilaginous envelope surrounding macroalgae (e.g. *Batrachospermum*, *Chaetophora*, *Tetraspora*, palmelloid stage of *Chlamydomonas*) and on grass found in springs (e.g. Cherry Hinton, Cambridge), canals and some Scottish lochs.

Chaetophora F.Schrank 1783

Macroscopic, branched and lobed/nodulated, spherical, hemispherical or tubercular, enclosed within a soft or firm mucilaginous envelope; microscopic filaments forming a feebly developed prostrate system of rounded cells and an erect system of intertwined, uniseriate filaments, usually more densely branched towards their apices and terminating in a bluntly pointed cell or a long, multicellular hair; chloroplast parietal, sometimes band-like, with 1 to several pyrenoids; reproduction by quadriflagellate zoospores and biflagellate isogametes; akinetes frequently produced in upper cells of branches, brown in colour.

Probably cosmopolitan; on aquatic plants and other submerged surfaces in standing or flowing water.

Requires re-investigation, since little is known about the plasticity of the morphological characters used to distinguish such species as *C. pisiformis*, *C. elegans* and *C. tuberculosa*. Phosphorus limitation leads to increased hair formation in all the species listed here, while addition of phosphate to phosphorus-limited material usually leads to zoospore formation. Although other factors have some influence on hair formation, phosphate limitation is probably the key factor at most field sites (see Gibson & Whitton, 1987a, b; Whitton, 1988).

1 Thallus subdichotomously branched or lobed .... *C. incrassata*
1 Thallus not as above 2
2 Mucilage soft; thalli very light green; microcopic branches lax and densely clustered at summit .... *C. elegans*
2 Mucilage firm; thalli often dark green; microscopic branches in upper parts very densely clustered or not clustered 3
3 Thallus warty or nodular; basal system not rhizoidal and branches densely clustered towards summit .... *C. tuberculosa*
3 Thallus spherical or hemispherical; basal system rhizoidal and branches not clustered ... *C. pisiformis*

Chaetophora elegans (Roth) C.Agardh 1812

Basionym: *Rivularia elegans* Roth

24070010 **Pl. 110I** (p. 443) **CD**

Thallus spherical or hemispherical, sometimes forming pale green, irregular masses enclosed within a soft mucilaginous envelope; microscopic filaments di- or trichotomously divided, forming irregular, spreading, entangled masses radiating from colony centre, prostrate and erect branches well developed, attached by rhizoids; uppermost branches densely clustered, tapering towards apices or terminating abruptly in an acute apex.

Probably cosmopolitan; in the British Isles common along the marginal shallows of rivers, lakes and pools where it grows on submerged rocks, roots, and stems as well as on the leaves of grasses and sedges.

Chaetophora incrassata (Hudson) Hazen 1902

Basionym: *Ulva incrassata* Hudson
Synonyms: *Chaetophora cornu-danae* (Roth) C.Agardh, *C. endiviaefolia* C.Agardh
24070020 **Pl. 110K** (p. 443) **CD**
Thallus bushy, consisting of subdichotomously divided and somewhat flattened branches, often lobed or nodulated; microscopic filaments 4–5 μm wide, embedded in a very firm mucilage, loosely arranged in centre, towards periphery dense, outwardly directed and forming curved clusters, apices acute or forming a hair.

Probably cosmopolitan; in the British Isles grows on rocks and vascular plants in flowing water and common in the shallow margins of lakes, ponds, gravel pits and bogs where individuals reach up to 15 cm long; free-floating individuals frequently accumulate along the downwind shores of ponds and lakes. Tolerant of different types of water and when growing in hard water sometimes lime-encrusted.

Chaetophora pisiformis (Roth) C.Agardh 1812

Basionym: *Rivularia pisiformis* Roth
24070030 **Pl. 110J** (p. 443) **CD**
Thallus almost spherical or hemispherical, dark green; microscopic filaments embedded in a relatively firm mucilaginous envelope, dichotomously divided and radiating from a common centre,with the basal system exclusively rhizoidal; erect branches 5.5–8 μm wide, with uppermost branches clustered and with acute apices.

Probably cosmopolitan; in the British Isles on rocks, shells of aquatic gastropods, roots, wood and culms of rushes in slow-flowing and stagnant water, including lakes, springs and marshy areas.

West (1904) considers the 'slightly torulose character of the cells of the branches' to be characteristic of this species.

Chaetophora tuberculosa (Roth) C.Agardh 1824

Basionym: *Rivularia tuberculosa* Roth
24070040 **Pl. 111A** (p. 446) **CD**
Thallus warty (tuberculate) or nodular, usually dark green; microscopic filaments about 8–9 μm wide, embedded in a firm mucilage, alternately or dichotomously divided, radiating from a common centre, uppermost branches densely clustered and all tapering slightly to an acute apex or, more rarely, terminating in a multicellular hair.

Probably cosmopolitan; recorded in a few widely separated localities in England, colonies about 1 cm across on submerged macrophytes growing in marshy or boggy areas.

A feature of New Zealand plants is the presence of shiny droplets in the cells (Sarma, 1986); these have yet to be recorded in British material.

Cylindrocapsa Reinsch 1867

Filaments initially attached and later free-floating, uniseriate but often becoming biseriate, multiseriate, or as a result of oblique and longitudinal divisions producing irregular, pseudoparenchymatous cell masses; cells cylindrical, ellipsoidal, rounded or triangular, sometimes arranged in pairs, enclosed within a thick, rough, stratified mucilaginous sheath; chloroplast axile or star-shaped in young filaments, sometimes parietal and filling cell, occasionally obscured by starch masses, with a single pyrenoid; asexual reproduction by biflagellate zoospores and aplanospores; sexual reproduction oogamous and monoecious: oogonia spherical or oval, dark green, with a thick stratified wall and opening by a lateral pore; antheridia in series, 2–4 produced in enlarged vegetative cells, 2 spindle-shaped biflagellate gametes formed in each antheridial cell; zygote spherical, thick-walled and bright reddish when mature.

Probably cosmopolitan; known from a wide range of aquatic habitats.

The ultrastructural diversity of species attributed to this genus led Lokhorst (1991) to suggest that it is probably polyphyletic as currently circumscribed. For this reason uncertainty surrounds its placement in classification schemes designed to reflect our current understanding of phylogenetic relationships. Often assigned to its own order or to the Order Ulotrichales. Hoek et al. (1995) regard it as a filamentous representative of the Order Chlorococcales *sensu lato*, based on ultrastructual details of cell division and if the vegetative cells are correctly interpreted as being equivalent to autospores (Sluiman et al., 1989). It is considered here as belonging to the Ulotrichales, an order to which it has been assigned traditionally.

1 Filaments usually less than 35 cells in length; cells generally much shorter than wide .... *C. conferta*
1 Filaments usually more than 100 cells in length; cells generally longer than wide 2
2 Cells 22–30 μm wide, 0.5–1.5 times longer than broad .... *C. involuta*
2 Cells usually 9–25 μm wide, 1–2 times longer than broad .... *C. geminella*

Cylindrocapsa conferta West 1892

24100010 **Pl. 116E** (p. 464)
Filaments rarely more than 35 cells long, sometimes irregular cell division gives rise to irregular cell masses; cells (16–)21–26(–29) μm wide and usually shorter than wide, quadrate, subquadrate or ellipsoidal, often densely packed, enclosed within a mucilaginous sheath (24–30 μm thick).

Probably cosmopolitan; reported in the British Isles from widely separated localities (e.g. Kent, the English Lake District, the Isle of Mull in Scotland, Clare Island in Ireland) where it grows amongst other algae in softwater lakes, ponds, ditches and swiftly-flowing streams.

Cylindrocapsa geminella Wolle 1887

24100020
Filaments over 150 cells long, sometimes multiseriate and pseudoparenchymatous; cells 9–25 μm wide, 1–2 times longer than wide, cylindrical, spherical, ovate or oblong, mucilaginous sheath thickening and becoming stratified with age.

var. *minor* Hansgirg 1888

24100022 **Pl. 116F** (p. 464)

Filaments generally straight but sometimes twisted or contorted; cells (9–)12–15(–25) µm wide (including sheath), reaching twice as long as wide, ovate or ellipsoidal.

Europe, North America; reported on only few occasions in various parts of the English Midlands and mentioned by Benson-Evans & Antoine (1996: 149) from South Wales where it is 'usually in soft water habitats where desmids occur' although they do not distinguish between var. *minor* and the type variety.

Cylindrocapsa involuta Reinsch 1867

24100030 **Pl. 116G** (p. 464)

Filaments usually more than 100 cells long; cells 22–30 µm wide, 0.5 to 1.5 times longer than wide, cylindrical, spherical or ellipsoidal, slightly or deeply constricted at cross walls, with sheath thick and stratified.

Probably cosmopolitan; reported from Ireland and England where usually associated with other filamentous algae in non-flowing water.

Desmococcus F.Brand 1925

Synonyms: *Pleurococcus* Meneghini *pro parte sensu auct.*, *Protococcus* C. Agardh *pro parte sensu auct.*

Cells solitary or in packets formed by division in 3 planes, occasionally as short uniseriate filaments; cells spherical to widely ellipsoidal, sometimes angular through mutual compression, usually fairly thick-walled; chloroplast parietal, single, massive, more or less lobed at margin, with a single pyrenoid; asexual reproduction by 1 or several aplanospores formed in large, spherical cells with a warty or punctate wall.

Cosmopolitan; the most widely distributed and common subaerial alga world-wide.

In the past there has been considerable taxonomic confusion surrounding the genus. Many species formerly attributed to it have been transferred to *Apatococcus*, *Chlorococcum*, *Trebouxia* and *Chlorella*. Until more information becomes available on its relationship to other genera it is regarded as belonging to the Chaetophorales, the order in which it has been traditionally placed.

Desmococcus olivaceum (Persoon ex Acherson) Laundon

Basionym: *Lepraria olivacea* Persoon ex Acherson

Synonyms: *Desmococcus vulgaris* Brand, *Pleurococcus naegelii* Chodat, *P. angulosus* (Corda) Meneghini, *P. vulgaris* Nägeli, *Protococcus viridis* C.Agardh

24110010 **Pl. 113E** (p. 450) **CD**

Cells solitary or in indefinite clusters, often forming 2-, 3- or 4-celled cuboidal packets; cells spherical or angular, 9–20 µm across, sometimes giving rise to unbranched filaments.

Cosmopolitan; sometimes growing in great profusion on damp, soil-free surfaces, often as a thin green powdery coating particularly on the windward side of stone walls, wooden palings and fencing, tree trunks and wooden telegraph poles; often common in shaded and polluted habitats where lichens are absent or sparse.

Dilabifilum Tschermak-Woess 1971

Synonym: *Pseudopleurococcus* Snow *pro parte*

Filamentous, uniseriate, consisting of a prostrate primary filament bearing erect or creeping secondary filaments; cells cylindrical in young filaments, frequently swelling to form akinetes, in cultures grown on agar often forming dark green masses of coccoidal cells; chloroplast parietal, single, with a pyrenoid; asexual reproduction by quadriflagellate zoospores, developing within swollen intercalary sporangia, 16, 32 or 64 zoospores or aplanospores in each sporangium.

Europe, North America; isolated from many types of surface (e.g. lichens, mollusc shells, calcareous crusts, stones), only rare in marine environments.

It can be separated with certainty from closely related genera only if zoospores are present, since flagellar number is a key character (see remarks under *D. printzii*). It is usually necessary to isolate and grow material in culture (see Johnson & John, 1990) to identify to species. Yet to be confirmed is a report of sporangia and quadriflagellate zoospores observed in cultures of *D. arthropyreniae* (Vischer) Tschermak-Woess, the type species.

Dilabifilum printzii (Vischer) Tschermak-Woess 1971

Basionym: *Pseudopleurococcus printzii* Vischer

Synonym: *Pseudendoclonium printzii* (Vischer) Bourrelly

24120010 **Pl. 108E** (p. 435) **CD**

Filaments sparingly branched and loosely arranged; cells cylindrical, 4–8 µm wide, (2–)4–6(–8) times longer than wide in the prostrate filaments and often longer (up to 12 times) in the erect filaments, terminal cell usually straight; akinetes frequently with thickened walls (>1 µm); zoospores unknown.

Europe, North America; in the British Isles known only from a small stream flowing through a coastal valley near Branscombe, South Devon, where it formed a greenish-coloured layer on calcareous encrustations covering submerged rocks and twigs.

Uncertainty inevitably surrounds the placement of this *Dilabifilum* since motile stages are unknown.

Draparnaldia Bory 1808

Filamentous, erect, uniseriate filaments attached basally by rhizoids and enclosed within a soft mucilaginous envelope; primary branches bearing oppositely, alternately or in whorls, tufts of richly divided and smaller-celled secondary branches whose apices terminate in a blunt cell or multicellular hair; cells of primary branches barrel-shaped or cylindrical, each with a parietal, band-shaped, entire or net-like chloroplast with a smooth or narrowly lobed (laciniate) margin and several pyrenoids; cells of secondary branches each possess a single, layered (laminate) chloroplast filling most of wall and 1–3 pyrenoids; asexual reproduction by biflagellate zoospores and thick-walled aplanospores; sexual reproduction by quadriflagellate isogametes.

Probably cosmopolitan; in a wide range of habitats including streams, ditches, springs, canals and shallow peaty lakes.

The genus exhibits considerable morphological plasticity and thus there is doubt about the validity of characters traditionally used for taxonomic determination (e.g. size and shape of cells, degree of development and form of side branches). Hair formation is an environmental effect, occurring under conditions when growth of the alga is phosphorus-limited (Whitton, 1988), though other factors can also influence the extent of hair formation. As the alga is normally found in nature with well developed hairs, this suggests that its typical habitat is one where phosphorus is limiting for considerable periods. A culture-based study by Beem & Simons (1988) suggests that *Stigeoclonium*-like forms of *Draparnaldia* may develop in nitrogen-rich habitats.

1 Secondary branches narrowly lance-like in outline, each with a distinct main axis running beyond its distal end ....*D. mutabilis*
1 Secondary branches orbicular or elongated in outline, each without a distinct main axis ....*D. glomerata*

Draparnaldia glomerata (Vaucher) C.Agardh 1824

Basionym: *Conferva glomerata* Vaucher
24130010 **Pl. 111C** (p. 446)
Cells of primary branches 50–100 µm wide, typically barrel-shaped; secondary branches alternately divided, sometimes opposite or whorled, spreading and tufted, orbicular or ellipsoidal in outline and without a distinct main axis.

Probably cosmopolitan; widely distributed in the British Isles where it forms bright green gelatinous tufts (commonly 5–20 cm long) of somewhat feathery appearance growing on various surfaces (e.g. stones, aquatic macrophytes) in a wide range of aquatic habitats, including the clear, soft water of shallow peaty pools in the Scottish Highlands, quiet pools on the drying beds of seasonally-flowing 'winterbourne' streams (e.g. near source of River Thames, Gloucestershire), and slow-flowing and relatively pollution-free streams.

Draparnaldia mutabilis (Roth) Cedergren 1920

Basionym: *Batrachospermum glomerata* Roth
Synonym: *Draparnaldia plumosa* (Vaucher) C.Agardh
24130020 **Pl. 111B** (p. 446) **CD**
Cells of primary branches 40–70 µm wide, cylindrical or slightly constricted at cross walls; secondary branches arising alternately, oppositely or in whorls, each narrowly lance-shaped in outline and with a distinct main axis extending beyond the distal end of each branch.

Probably cosmopolitan; a beautiful, delicate alga with a similar ecology to *D. glomerata*.

Doubt attaches to the separation of these two *Draparnaldia* species due to the presence of intermediate forms. In the opinion of Beem & Simons (1988) 'the various fascicle morphologies of *D. acuta*, *D. plumosa* and *D. glomerata* are merely developmental expressions influenced by unknown growth conditions'. They point out that the principal character separating *D. mutabilis* from *D. glomerata* (distinct axis in secondary branches) is not evident in culture and for this reason consider them conspecific. It has been decided to recognize both species until more is known of the environmental conditions responsible for the phenotypic responses of British material. No useful purpose is served by recognizing trivial growth forms such as the *pulchella*, a variety described by West (1892a) from near 'Leename' in Ireland.

Elakatothrix Wille 1898

Cells solitary or in colonies of cells scattered within a mucilaginous envelope; cells spindle-shaped, cylindrical, oval or ellipsoidal, apices blunt or acute, walls thin and hyaline, forming pseudofilaments and contiguous or separate; chloroplast parietal, layered (laminate), cup- or girdle-shaped, straight or spiralled, 1 or 2 per cell, each with a single or pair of pyrenoids; asexual reproduction by aplanospores; sexual reproduction unknown.

Probably cosmopolitan; widely distributed in the plankton although juveniles are sometimes epiphytic (see Lokhorst, 1991).

Traditionally placed in the Order Ulotrichales.

1 Cells spindle-shaped ....*E. gelatinosa*
1 Cells shortly cylindrical, tapering with pencil-like apices ....*E. acuta*

Elakatothrix acuta Pascher 1915

25010020 **Pl. 114M** (p. 454)
Colonies usually of 2–10 cells arranged in an irregular row within a mucilaginous envelope, shortly cylindrical to spindle-shaped, up to 120 µm long; cells 3–4 µm wide and 8–21 µm long, shortly cylindrical, tapering to acute points ('pencil-like'), with daughter cells initially attached by oblique end wall.

Europe; so far only reported in the British Isles from St James's Park, London (Whitton, 1969).

A well-defined species recognized on account of the shape of its cells (Hindák, 1987).

Elakatothrix gelatinosa Wille 1898

25010010 **Pl. 114K** (p. 454) **CD**
Colonies of cells (up to 80) randomly distributed or arranged in irregular rows within a mucilaginous envelope, initially envelopes small and ellipsoidal, spindle-shaped or cylindrical, becoming larger, very wide and irregular; cells 3–4 µm wide and 12–26 µm long, spindle-shaped, with long axis of each cell tending to lie parallel to envelope.

Probably cosmopolitan; reported on just a few occasions in the British Isles in the plankton of lakes (e.g. Wastwater in the English Lake District and several Scottish lochs) and other aquatic habitats including flooded clay pits.

var. *aplanospora* J.W.G.Lund 1956
25010012 **Pl. 114L** (p. 454)
Differs from the type species in having spherical aplanospores, about 11 µm in diameter, one per cell, released by gelatinization of middle part of mother cell wall.

British Isles (England); according to Lund (1956a), to be found in most years as a member of lake plankton in the English Lake District.

Entocladia Reinke 1897

Synonyms: *Endoderma* Wille, *Ectochaete* (Huber) Wille, *Epicladia* Reinke

Endo- or epiphytic, consisting of creeping, irregularly divided, uniseriate filaments which sometimes coalesce to form compact pseudoparenchymatous expansions; branches frequently short, usually develop from a bluntly pointed terminal cell; hairs or bristles when present long, often basally swollen and tapering, sometimes arising on the dorsal side of cells; chloroplasts parietal, plate-like, with 1 or several pyrenoids; asexual reproduction by quadriflagellate zoospores; sexual reproduction by biflagellate isogametes.

Probably cosmopolitan; mostly in marine rather than freshwater environments.

The taxonomic placement of freshwater species belonging to the genus remains uncertain as does their relationship to those marine species whose ultrastructure has been investigated. This genus is in need of further investigation.

1 Filaments often anastomose to form compact masses of cells; bristles absent .... *E. cladophorae*
1 Filaments spreading and often unbranched; bristles present .... *E. endophytica*

Entocladia cladophorae (Hornby) G.S.West *et* F.E.Fritsch 1927

Basionym: *Endoderma cladophorae* Hornby

24140010 **Pl. 110G** (p. 443)

Compact masses consisting of 2–3 layers of filaments; cells (3.5–)6–17.5 μm wide, initially ellipsoidal, spherical or ovoid, later becoming angular as filaments anastomose, thick-walled; bristles absent.

Europe; epiphytic or endophytic on filamentous green algae (e.g. *Cladophora glomerata, Rhizoclonium hieroglyphicum*) growing in a wide range of aquatic habitats.

Entocladia endophytica (K.Möbius) D.M.John, *nov. comb.*

Basionym: *Bulbocoleon endophytica* K.Möbius 1892. Biol. Centrabl. 12: 71.
Synonyms: *Endoderma jadinianum* Huber, *Ectochaete endophyticum* (K.Möbius) Wille

24140020 **Pl. 110H** (p. 443)

Loosely arranged and consisting of simple or, more rarely, dichotomously divided filaments; cells 8–20(–30) μm wide and up to 4 times longer than wide, mostly cylindrical; bristles long and tapering.

Europe; in the British Isles only known on filamentous algae including *Cladophora glomerata.*

DOUBTFUL RECORD

Entocladia polymorpha (G.S.West) G.M.Smith 1933

Doubt attaches to the record by Benson-Evans & Antoine (1996) from 'freshwater or brackish' water in South Wales where it grows as 'thin, cushion-like masses on *Cladophora* and *Rhizoclonium*'. There is no indication given as to the source of the record and the description given along with its dimensions ('cells 9–20 μm wide') suggests that it might be more correctly attributed to *Entocladia cladophorae.*

Geminella Turpin 1823

Synonym: *Hormospora* Nägeli *non* Brébisson

Filaments attached or free-floating, uniseriate and unbranched; cells in loose rows and equidistant, in pairs or united end to end, enclosed within a thick mucilaginous sheath; cells generally longer than wide, cylindrical with rounded ends or inflated and ellipsoidal, oval or barrel-shaped; chloroplast parietal, girdle-like or layered (laminate), often in cell centre, with a single pyrenoid; reproduction by fragmentation and the formation of thick, brown-walled akinetes.

Probably cosmopolitan; planktonic and in terrestrial habitats, often in soft and acid waters where desmids are usually abundant.

The presence of a thick mucilaginous sheath is a character often used to separate *Geminella* from closely related genera. Most authors consider it distinct from *Gloeotila,* the latter established to accommodate species which have a strict contiguous cell arrangement, a tendency to fragment and lacking pyrenoids (see Lokhorst, 1991). Several species of *Geminella* and/or *Gloeotila* that lack pyrenoids and sheaths have been transferred to *Stichococcus* by Hindák (1996). Extensive comparative study is needed to clarify species limits and those between *Geminella* and closely related genera. Its taxonomic position is uncertain with Hoek et al. (1995) assigning it to the Chlorococcales *sensu lato.*

1 Cells mostly contiguous 2
1 Cells separated or in pairs 3
2 Mucilage sheath usually <15 μm wide; cells 2–10 μm wide .... *G. minor*
2 Mucilage sheath >15 μm wide; cells usually 16–19 μm wide .... *G. mutabilis*
3 Cells in pairs, usually separated by large gaps, 2 to 3 times longer than wide .... *G. interrupta*
3 Cells separate and usually equidistant, often 1.5 times longer than wide .... *G. ordinata*

Geminella interrupta (Turpin) Lagerheim1883

Basionym: *Hormospora interrupta* Turpin

24150010 **Pl. 114A** (p. 454)

Cells in pairs, usually adjacent pairs separated by large gaps, enclosed within a 15–33 μm wide mucilaginous sheath; cells 6.5–7(–15) μm wide, 2 to 3 times longer than wide (9–25 μm long), cylindrical with broadly rounded apices; chloroplast filling at least two-thirds of cell, with 1 or a pair of pyrenoids.

Probably cosmopolitan; in boggy areas especially in the acid waters of the western regions of the British Isles, rare in pools.

Geminella minor (Nägeli) Heering 1914

Basionym: *Hormospora minor* Nägeli

24150020 **Pl. 114C** (p. 454)

Cells in a contiguous row, within a 8–18 μm wide mucilaginous sheath; cells 2–10 μm wide, 2 to 4 times longer than wide, cylindrical; chloroplast filling part of cell, with 1 or more pyrenoids.

Probably cosmopolitan; usually in the British Isles reported free-floating in small, soft water pools, sometimes in *Sphagnum* bogs in the western regions of England, Scotland and Wales.

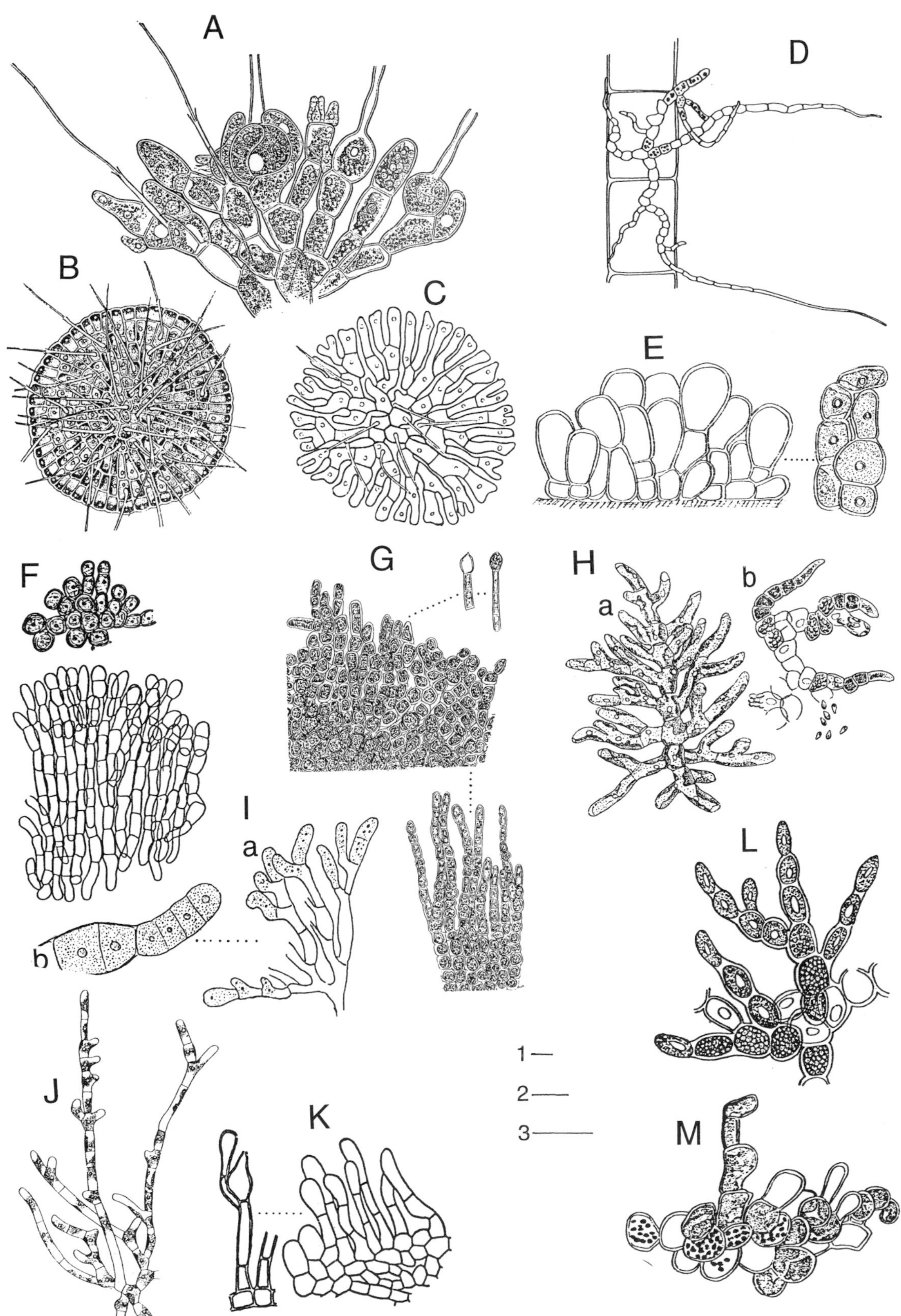

Plate 109 A–M

A. *Coleochaete pulvinata* (p. 471); **B**. *Coleochaete scutata* (p. 471); **C**. *Coleochaete soluta* (p. 471); **D**. *Chaetonella goetzii* (p. 468); **E**. *Gongrosira debaryana* (p. 444): erect filaments (left) with cells (right) each having a single chloroplast and pyrenoid; **F**. *Gongrosira fluminensis* (p. 444): prostrate filaments (above) and erect filaments (below); **G**. *Gongrosira incrustans* (p. 445): erect filaments (below), compact prostrate filaments (above) and terminal akinetes; **H**. *Gongrosira papuasica* (p. 445): a. vegetative filaments, b. zoospore release; **I**. *Gongrosira schmidlei* (p. 445): a. erect filaments, b. terminal cells dividing, possibly to form akinetes; **J**. *Gongrosira scourfieldii* (p. 445); **K**. *Gongrosira stagnalis* (p. 447): erect filaments terminating in inflated and irregularly shaped cells; **L**. *Leptosira mediciana* (p. 452); **M**. *Pleurastrum terricola* (p. 457).

Scale bars: all 20 µm. 1 for A–C, Ia, J, K; 2 for D–H, L, M; 3 for Ib.

Geminella mutabilis (Brébisson) Wille 1911

Basionym: *Hormospora mutabilis* Brébisson

24150030 **Pl. 114B** (p. 454)

Cells generally contiguous within a mucilaginous sheath (about 39 μm wide); cells (9–)16–19(–20) μm wide, (1–)1.5 to 2 times longer than wide, spherical to cylindrical with broadly rounded apices; chloroplast girdle-shaped, completely encircling cell, with a single pyrenoid.

Probably cosmopolitan; most commonly in boggy areas and often amongst *Sphagnum*.

Geminella ordinata (West *et* G.S.West) Heering 1914

Basionym: *Hormospora ordinata* West *et* G.S.West

24150040 **Pl. 114D** (p. 454)

Cells separate and equidistant, within a mucilaginous sheath (15–20 μm wide); cells 5–8 μm wide, about 1.5 times longer than wide, oval with broadly rounded apices; chloroplast filling more than half of cell, with a single pyrenoid.

Probably cosmopolitan; very rare in the British Isles where confined to boggy areas in western regions.

DOUBTFUL RECORD

Reynold & Irish (2000) mention '*Ulothrix* aff. *amphigranulata*' as frequent in a list of Windermere phyoplankton. The species *Ulothrix amphigranulata* was described by Skuja (1948) from lake plankton in Sweden and Ramanathan (1964) later transferred it to the genus *Geminella*. He gave as its principal distinguishing features the presence of vacuoles at the end of each cell 'containing moving granules or crystals and the broad mucilaginous sheath'.

Gloeotila Kützing 1843

Filaments attached or free-floating, uniseriate, unbranched, often breaking into short lengths enclosed within a structureless mucilaginous sheath; cells generally longer than wide, oblong or ellipsoidal, with blunt apices; chloroplast parietal, single, laminate or girdle-shaped, restricted to cell centre or almost filling cell, without pyrenoids; asexual reproduction by biflagellate zoospores; sexual reproduction unknown.

Probably cosmopolitan; usually planktonic or occasionally associated with soil.

Sometimes considered to be congeneric with *Geminella*, but usually separated from it because *Gloeotila* lacks pyrenoids and its filaments readily fragment.

1 Filaments rarely more than 4-celled, with a very wide sheath (>100 μm wide) .... *G. monospora*
1 Filaments always more than 4-celled and with a narrower sheath 2
2 Cells >5 μm wide, 3 to 5 times longer than wide .... *G. subconstricta*
2 Cells <5 μm wide, 1 to 3 times longer than wide .... *G. protogenita*

Gloeotila monospora (J.W.G.Lund) Hindák 1996

Basionym: *Geminella monospora* J.W.G.Lund

24170010 **Pl. 114F** (p. 454)

Filaments with small gaps between cells, rarely more than 4 cells long, embedded within a mucilaginous sheath (100–130 μm wide); cells 4.5–5 μm wide, 14–40 μm long, 3 to 10 times longer than wide, cylindrical with rounded apices; chloroplast single or rarely in pairs, often in the form of a curved plate or confined to one side of cell; aplanospores 7–10 μm across, spherical, 1 per cell.

British Isles (England); only reported in the plankton of a few lakes in the English Lake District.

Gloeotila protogenita Kützing 1849

Synonyms: *Geminella protogenita* (Kützing) G.S.West, *Gloeotila mucosa* Borzí *non* Kützing

24170020 **Pl. 114H** (p. 454)

Filament a contiguous row of cells, shorter and longer cells together, or just long or just short cells, within a mucilaginous sheath (about 15 μm wide); cells 3–4.5 μm wide, 1 to 2(–3) times longer than wide, cylindrical, sometimes constricted at cross walls; chloroplast girdle-shaped, extending along much of cell length.

Probably cosmopolitan; widely distributed in England (e.g. Devon, Yorkshire) where reported from boggy pools and damp soil, rare.

Gloeotila subconstricta (G.S.West) Printz 1964

Basionym: *Ulothrix subconstricta* G.S.West

24170040 **Pl. 114G** (p. 454)

Filament a contiguous row of cells, sometimes enclosed within a mucilaginous sheath; cells 5.7–9 μm wide, 10–36 μm long, 3 to 5 times longer than wide, slightly inflated at apices, thin-walled, moderately constricted at cross walls; chloroplast in middle of cell, extending about two-thirds around cell circumference, only occasionally with a pyrenoid visible.

Europe, North America; reported in the British Isles from pond plankton.

Gomontia Bornet *et* Flahault 1888

Filamentous, consisting of uniseriate, irregularly branched masses of filaments immediately below the surface of shells, wood, limestone or turtle carapace, sometimes densely crowded and pseudoparenchymatous, with downwardly growing rhizoidal filaments; cells cylindrical, ovoid or polygonal, multinucleate, with walls thick and stratified; chloroplast single, dense, layered (laminate), sometimes net-like or lobed, with several pyrenoids; asexual reproduction by zoospores and aplanospores formed within enlarged cells of short erect branches, zoospores bi- or quadriflagellate; akinetes known.

Probably cosmopolitan; perforating shells of molluscs, turtles, limestone, submerged wood and '*Cladophora* balls', mostly marine.

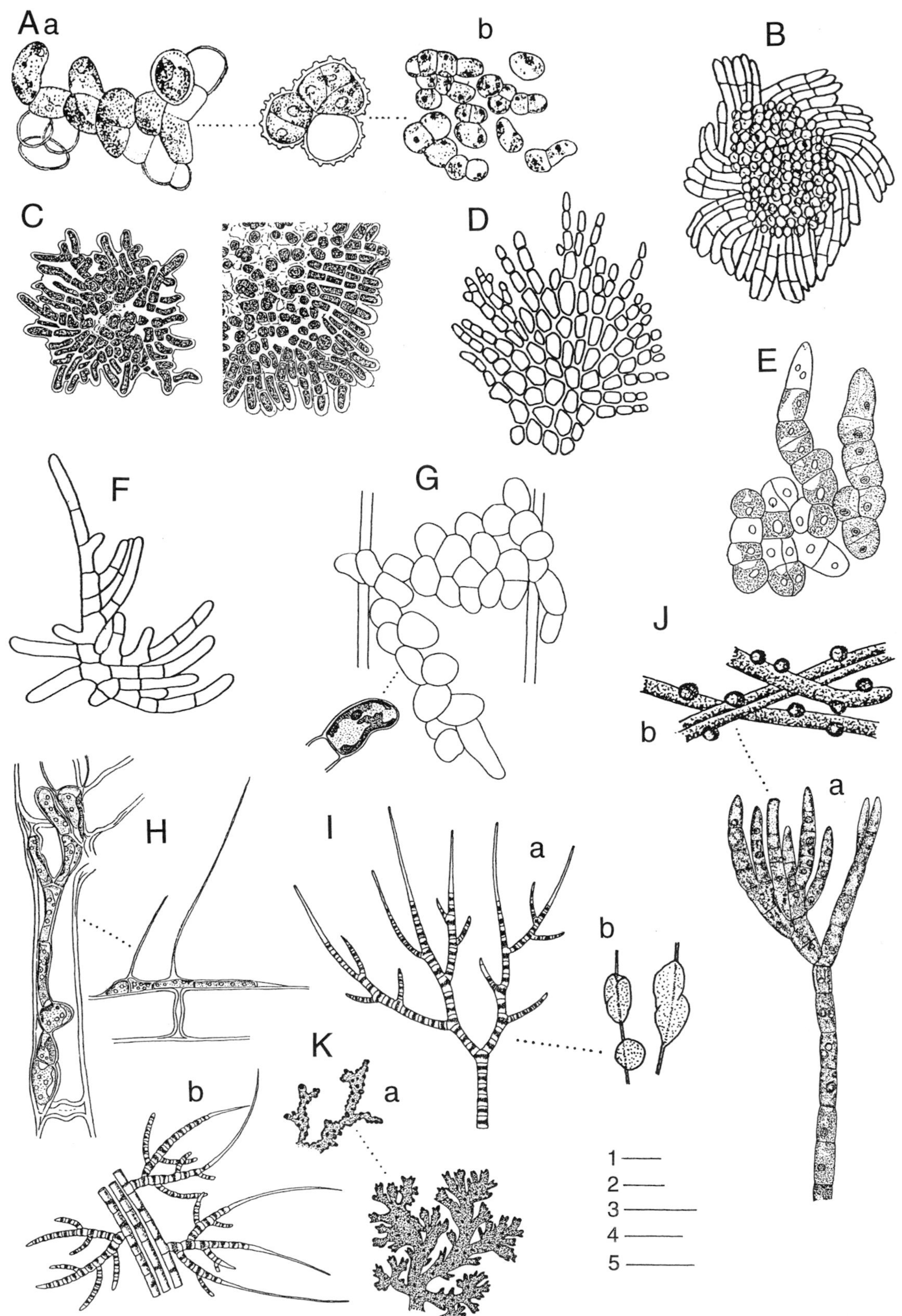

Plate 110 A–K
A. *Pleurastrum insigne* (p. 457): a. vegetative filaments (left) and hypnospores with ornamented walls (right), b. filaments dissociating into individual cells; **B**. *Protoderma beesleyi* (p. 458); **C**. *Protoderma frequens* (p. 458); **D**. *Protoderma viride* (p. 458); **E**. *Pseudendoclonium basilense* var. *brandii* (p. 459); **F**. *Pseudendoclonium prostratum* var. *radiatum* (p. 459); **G**. *Entocladia cladophorae* (p. 440): vegetative filaments and details of a terminal cell (lower left); **H**. *Entocladia endophytica* (p. 440); **I**. *Chaetophora elegans* (p. 436): a. habit of 3 soft hemispherical thalli, b. microscopic filaments terminating in hairs; **J**. *Chaetophora pisiformis* (p. 437): a. microscopic filaments, b. habit of several hemispherical thalli; **K**. *Chaetophora incrassata* (p. 437): a. bushy habit of macroscopic thallus, b. microscopic view of filaments.
Scale bars: 1, 50 µm; 2, 3, 20 µm; 4, 10 µm; 5, 1 cm. 1 for H; 2 for B–D, Ia, Kb; 3 for Ab, G, F, Ja; 4 for Aa, E; 5 for Ib, Jb, Ka.

Gomontia aegagropila Acton 1916

24180010 **Pl. 108K** (p. 435)

Filaments radially arranged, densely crowded, and dichotomously branched; cells at or near the surface barrel-shaped, 8–10 μm wide, usually 10–16 μm long, with the colourless penetrating cells about 2 μm wide and up to 110 μm long; fragmentation the only known form of reproduction.

British Isles (Scotland); only known on and within the wall of '*Cladophora* balls' (*Cladophora aegagropila*) collected from Loch Kildona, South Uist in the Hebrides.

Some doubt inevitably attaches to this little-known species for which there is little information on reproduction. It has yet to be discovered again since Acton (1916b) reported it growing under the glaze of an earthenware pie-dish containing a culture of *Cladophora aegagropila*. She found it on examination of preserved balls of *Cladophora aegagropila* (as *C. holsatica*).

DOUBTFUL RECORD

Gomontia codiolifera (Chodat) Wille 1909

24180020

Rich's (1925) record of this limestone-penetrating alga from Leicester in England is considered doubtful since neither accompanied by an illustration nor a detailed description.

Gongrosira Kützing 1843

Commonly cushion-like or crustose, occasionally lime-encrusted, consisting of an erect and prostrate system of uniseriate filaments; prostrate system single or multilayered, loose or pseudoparenchymatous; erect system often of short branches with blunt apices, hairs absent; cells cylindrical or somewhat inflated, usually having thick, lamellate walls; chloroplasts parictal, with 1 to several pyrenoids; asexual reproduction by aplanospores or biflagellate zoospores produced within terminal or intercalary sporangia; akinetes reddish in colour.

Probably cosmopolitan; often appears as a greenish layer or crust on stones, snails, water plants and other hard surfaces along the marginal shallows of ponds, lakes and rivers (rarely in soil).

Many species must be regarded as doubtful since there is often considerable variation in many vegetative features regarded as taxonomically important. Only a few species have been examined in laboratory culture (e.g. *G. debaryana*, *G. scourfieldia*) and one species (*G. pseudoprostrata*) has been defined on culture-based characters. The genus is in need of re-examination and revision. Uncertainty surrounds its current placement in newly proposed schemes of classification. It has been decided here to retain the genus in the Order Chaetophorales.

1 Thallus impregnated with lime 2
1 Thallus not lime-impregnated 5
2 Erect filaments developing unilaterally; grows on larger algae .... *G. schmidlei*
2 Erect system not branching as above; grows on various surfaces but not on other algae 3
3 Erect and prostrate systems usually conspicuous and well-defined; erect filaments parallel and sparsely branched .... *G. incrustans*
3 Prostrate system indistinct or erect and prostrate systems of filaments not readily delimited; erect filaments sometimes growing obliquely upwards and branching from almost every cell 4
4 Thallus crustose or nodular, 4–6 mm thick; cells of erect filaments 1–9 times longer than wide .... *G. scourfieldii*
4 Thallus crustose, about 1 mm thick; cells of erect filaments almost equal in length and breadth .... *G. sclerococcus*
5 Erect branches >15 μm wide..................................6
5 Erect branches <15 μm wide..................................7
6 Grows on the water snail *Limnaea*; cells of erect filaments subcylindrical, 2–6 times longer than wide .... *G. stagnalis*
6 Grows on hard, inanimate surfaces; cells of erect filaments often somewhat rounded, 1–2.5 times longer than wide .... *G. debaryana*
7 Erect filaments terminate in a single cell or pair of cells separated by an oblique wall; pyrenoids 1–3 per cell.. *G. fluminensis*
7 Erect filaments terminate in a single, often swollen cell; pyrenoid single, prominent 8
8 Swollen cells >12 μm wide, not dissociating; erect system variously developed (at least in laboratory culture) .... *G. papuasica*
8 Swollen cells 12 μm wide or less, often dissociate in prostrate system; erect system absent or poorly developed.................................. *G. pseudoprostrata*

Gongrosira debaryana Rabenhorst 1883

24190010 **Pl. 109E** (p. 441)

Thallus consists of pale green, disc-like or irregular crusts; prostrate system pseudoparenchymatous and erect system clearly filamentous, all filaments dichotomously divided; cells 15–30(–50) μm wide, 1 to 2.5 times longer than wide (often very variable in size), frequently somewhat rounded and thick-walled; pyrenoid single, large; akinetes about 50 μm wide, swollen, spherical and terminal on filaments; zoospores formed in large, spherical or flask-shaped cells.

Europe, North America; in the British Isles grows as greenish patches on rock surfaces along the marginal shallows of streams and rivers.

Gongrosira fluminensis F.E. Fritsch 1929

24190020 **Pl. 109F** (p. 441)

Thallus consists of compact crusts; prostrate system pseudoparenchymatous and from most cells arise 3–4-celled erect branches (sometimes of many more cells) reaching a height of 68–180 μm, with erect branches most frequently divided towards apices; cells of prostrate system 8.5–10 μm wide, rounded or polygonal, with walls thick and irregularly stratified; cells of erect branches about 7 μm wide, 1.5 to 2 times longer than wide, cylindrical or slightly barrel-shaped, terminating in a single cell or a pair of swollen cells separated by an oblique wall; pyrenoids 1–3 per cell; zoosporangia unknown.

British Isles (England, Ireland); as bright green specks or irregular crusts on rocks in streams and rapidly flowing stretches of rivers, probably more common than published records indicate. Recorded in Ireland only from the River Caragh, Kerry (Heuff & Horkan, 1984).

Fritsch (1929a) points out that it closely resembles *G. incrustans* in having 'densely arranged, little-branched upright threads' but differs principally in the absence of a lime-encrustation. Doubt inevitably attaches to this species since lime deposition is considered not to be a character of any special taxonomic significance.

Gongrosira incrustans (Reinsch) Schmidle 1901

Basionym: *Chlorotylium incrustans* Reinsch

24190030 **Pl. 109G** (p. 441)

Thallus consists often of hard lime-encrusted hemispherical crusts or cushion-like growths, occasionally enclosed within a mucilaginous envelope; cells of pseudoparenchymatous prostrate system very compact, 4–5 µm wide, somewhat polygonal, often thick-walled, bearing more loosely arranged, often parallel and sparsely divided erect filaments; cells of erect filaments 4–5(–10) µm wide, (1–)2 to 3 times longer than wide, subspherical to cylindrical, moderately thick-walled, terminal cells long and blunt; zoospores produced in terminal zoosporangia; akinetes develop from terminal cells.

Probably cosmopolitan; in the British Isles common on stones and other hard surfaces in calcareous streams and along the shallow margins of rivers.

The relationship between this and other lime-encrusted species remains to be investigated. Such algae show some endolithic activity when growing on limestone and deposition of calcium carbonate might not relate to photosynthetic activity but rather to water supersaturated with calcite (see Pentecost, 1992). According to Butcher (1932b), in its young stages *G. incrustans* often bears a superficial resemblance to the early stages of *Protoderma frequens*, or the prostrate system of certain species of *Stigeoclonium*. See remarks under *Gongrosira fluminensis*.

Gongrosira papuasica (Borzí) Tupa 1974

Basionym: *Pleurothamnion papuasicum* Bory

24190040 **Pl. 109H** (p. 441) **CD**

Thallus of small green crusts; prostrate system loose or pseudoparenchymatous, the degree of development of erect system very variable; cells 5–9(–10) µm wide, 1.5 to 4 times longer than wide, cylindrical to somewhat swollen, sometimes becoming coccoidal and reaching 18(–24) µm across; pyrenoid single and prominent; zoosporangia intercalary and terminal on filaments.

Probably cosmopolitan; so far only isolated in the British Isles from stones and pebbles sampled in a calcareous river in southern England; isolated from the surface of aquatic vascular plants in North America and from human bones lying on moist soil in Papua New Guinea.

Much of our present understanding of the species derives from observations on strains grown in laboratory culture.

Gongrosira pseudoprostrata L.Johnson in John & Johnson 1992

24190050

Thallus of slightly elongate to circular crusts, 50–200 µm across; prostrate system pseudoparenchymatous in older parts, sometimes dissociating into single cells, with or without an erect system of very short branches; cells 3–8 µm wide, 1.5 to 3(–4) times longer than wide, cylindrical, sometimes coccoidal and 10–12 µm wide; pyrenoid single and prominent; zoosporangia intercalary.

England and Denmark; isolated from hard surfaces along the shallow margins of streams, ponds and lakes.

Description based on characters defined in carefully controlled laboratory conditions.

Gongrosira schmidlei P.G. Richter 1893

24190060 **Pl. 109I** (p. 441)

Thallus consists of small crusts growing on larger algae, up to about 2 mm across, lime-encrusted; prostrate system usually of colourless filaments, bearing profusely and unilaterally divided erect filaments, terminating in blunt cells; cells all 8–14 µm wide, 2 to 4 times longer than wide; pyrenoids single and prominent; akinetes produced by division of terminal cells.

Europe; usually epiphytic on *Cladophora* and *Vaucheria* growing in various types of water.

Gongrosira sclerococcus Kützing 1843

Synonym: *Gongrosira viridis* Kützing

24190070

Thallus consists of small green, cushion-like growths, about 1 mm thick, lime-encrusted; prostrate filaments indistinct but a distinct erect system growing obliquely upwards from the surface (assurgent), the latter profusely dichotomously branched in all directions; cells from 4–10(–12) µm wide and long, almost quadrate, with apical cells bluntly rounded; reproduction unknown.

Europe; in the British Isles widely distributed on hard surfaces along the shallow margins of freshwater habitats.

See remarks under *Gongrosira scourfieldii*.

Gongrosira scourfieldii G.S.West 1918

24190080 **Pl. 109J** (p. 441) **CD**

Thallus of vivid green crusts or nodular growths, 4–6 mm thick, hard due to lime-encrustation; prostrate and erect systems of pseudoparenchymatous filaments not sharply delimited from each another, prostrate branches short and frequently anastomosing; cells of prostrate system 12–19 µm wide and often many times longer than wide; cells of the obliquely erect system 7–9.5 µm wide, 1 to 9 times longer than wide, with apical cell bluntly rounded; pyrenoids usually 1 per cell; zoosporangia intercalary, rarely terminal.

British Isles (England); known only from its type locality, a small stream in Devon (Weston Combe, near Sidmouth), where it grows on rocks in the splash and spray from a waterfall; undoubtedly more common than published records indicate.

West (1918) distinguishes this species from *G. incrustans* by 'having less crowded erect filaments, which are not parallel and are much more branched, and in the fact that all cells contain a large parietal chloroplast'. *Gongrosira sclerococcus* was not mentioned by him although clearly closely resembling the new species. *Gongrosira scourfieldii* has generally longer cells and a somewhat more clearly defined prostrate layer, but the distinction between these two species is not always evident. Rediscovered

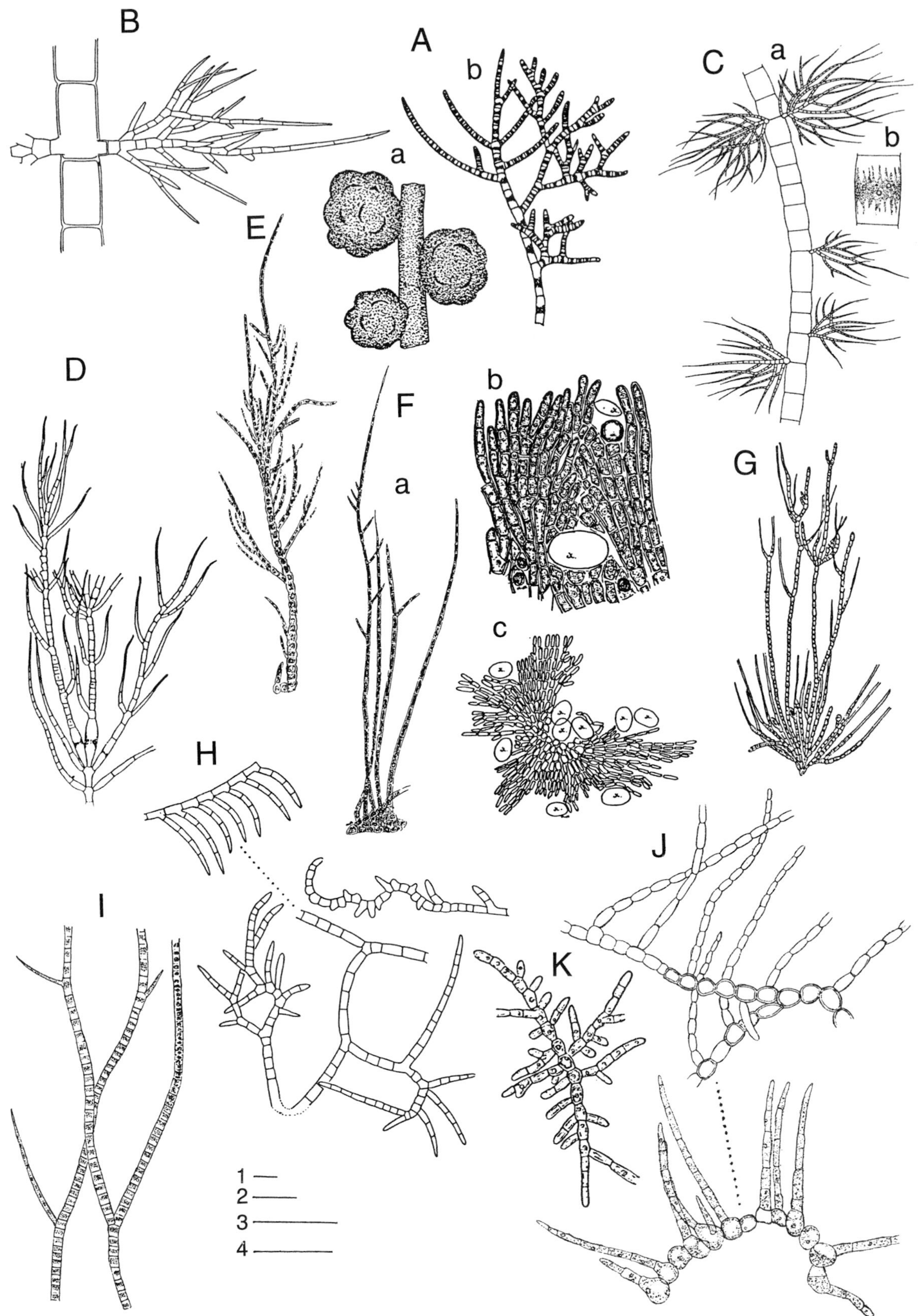

Plate 111 A–K

A. *Chaetophora tuberculosa* (p. 437): a. warty macroscopic thalli, b. microscopic filaments; **B**. *Draparnaldia mutabilis* (p. 439); **C**. *Draparnaldia glomerata* (p. 439): a. primary filament bearing mostly alternately branched laterals, b. a single cell of the primary filaments showing band-shaped chloroplast with a lobed margin; **D**. *Stigeoclonium amoenum* (p. 463); **E**. *Stigeoclonium tenue* (p. 465); **F**. *Stigeoclonium farctum* (p. 463): a. erect filaments, b, c. branching pattern of the prostrate system with quartz grains shown; **G**. *Stigeoclonium longipilum* (p. 463); **H**. *Stigeoclonium helveticum* (p. 463): showing wide variation in branching pattern; **I**. *Stigeoclonium protensum* (p. 465); **J**. *Stigeoclonium nanum* (p. 465); **K**. *Stigeoclonium lubricum* (p. 463).

Scale bars: 1–3, 40 μm; 4, 4 mm. 1 for Ca, D, Fc, G, I; 2 for B, Cb, E, Fa, H, J, K; 3 for Ab, Fb; 4 for Aa.

in the same locality (see Johnson & John, 1992) 75 years after it was first described. It was grown in culture and zoospores were observed for the first time.

Gongrosira stagnalis (G.S.West) Schmidle 1801

Basionym: *Pilinia stagnalis* G.S.West

24190090 **Pl. 109K** (p. 441)

Thallus consists of tough, dull green crusts on snail shells, often up to 0.5 mm across; prostrate filaments pseudoparenchymatous and bearing short, sparingly divided erect branches; cells of prostrate system 16–31 μm wide, 1 to 2 times longer than wide; cells of erect filaments 16–25 μm wide, 2 to 6 times longer than wide, subcylindrical, with filaments terminating in an inflated and irregularly-shaped cell; zoosporangia arising terminally.

British Isles (England); recorded as growing as a crust on the shells of the snail *Limnaea peregra*.

Klebsormidium P.C.Silva, Mattox *et* W.H.Blackwell 1972

Synonyms: *Hormidium* Kützing *sensu* G.A.Klebs, *Chlorhormidium* Fott

Filamentous, uniseriate, unbranched, without basal or apical differentiation; cells cylindrical or barrel-shaped, with walls thin or thickened and stratified, rough, toothed and/or warty, sometimes cross walls surrounded by H-shaped pieces; chloroplast parietal, single, ellipsoidal, disc-like or girdle-shaped, margin not lobed and encircling only a proportion (normally 80% or less) of cell circumference, usually with a single distinct pyrenoid; asexual reproduction by biflagellate zoospores, eyespots absent, typically a single zoospore produced in each unspecialized cell, released through a pore; aplanospores and thick-walled akinetes known.

Probably cosmopolitan; terrestrial and in a wide range of aquatic habitats.

Doubt attaches to many of the characters traditionally used to distinguish species (e.g. length to breadth ratio of cells, nature of cell wall) since these are known to be influenced by environmental conditions. According to Lokhorst (1996), for accurate and reliable identification material should be examined after first growing it under controlled laboratory conditions. He provides two keys, one for the identification of cultured material and the other for identifying field collected material (covers 5 species known from the British Isles). The following descriptions give dimensions taken from various sources (e.g. Ramanathan, 1964), but not those of cultured material (e.g. Lokhorst, 1996) since they usually lie within a narrower range when compared with field material. It can be distinguished from the genus *Ulothrix* by the form of the chloroplast (see generic key, p. 294), the production of a single rather than multiple zoospores per cell, and the zoospores lacking an eyespot (see Graham & Wilcox, 2000). The genus is in need of further revision.

1 Filaments normally long, usually bent and twisted with knee-shaped bends or kink; false branches common or absent 2
1 Filaments short or usually without bends if long; false branches rare or absent 4
2 Filaments without false branches; chloroplasts often kidney-shaped in side view ....*K. flaccidum*
2 Filaments commonly falsely branched; chloroplasts never appearing kidney-shaped 3
3 Filaments readily dissociating into single cells with age; H-shaped pieces sometimes developing close to cross walls; apices never rhizoidal ....*K. fluitans*
3 Filaments not dissociating into single cells; H-shaped pieces never develop; apices often rhizoidal....................*K. rivulare*
4 Filaments with walls thick or unevenly thickened, sometimes stratified, often with H-shaped sections, never dissociating into fragments 5
4 Filaments thin-walled, or if wall unevenly thickened then readily dissociating into few-celled filaments or single cells, with or without H-shaped sections 6
5 Cell walls thick, stratified and sometimes mucilaginous ....*K. crenulatum*
5 Cell wall unevenly thickened (sometimes 2-layered), warty, dented or with spine-like projections ....*K. mucosum*, in part
6 Filaments readily fragmenting into few-celled lengths; walls unevenly thickened or thin and with H-shaped sections ....*K. dissectum*
6 Filaments not fragmenting, walls thin or unevenly thickened with various surface features, without H-shaped sections 7
7 Cells always cylindrical, 12–23 μm wide, less than 1.5 times longer than wide; walls thin to unevenly thickened and with various surface features (rough, warty, dentate or spiny) ....*K. mucosum*, in part
7 Cells cylindrical or with constrictions at cross walls, usually <10 μm wide, usually more than 5 times longer than wide; walls always thin and smooth 8
8 Filaments often possessing a distinctive glistening sheath; cells always cylindrical and chloroplast containing a somewhat elongate pyrenoid....................*K. klebsii*
8 Filaments without a distinctive sheath; cells cylindrical to slightly inflated and constricted at cross walls; chloroplast containing a spherical pyrenoid....................*K. subtile*

Klebsormidium crenulatum (Kützing) H.Ettl *et* Gärtner 1995

Basionym: *Hormidium crenulatum* Kützing

Synonyms: *Klebsormidium crenulatum* (Kützing) Lokhorst, *Chlorhormidium crenulatum* (Kützing) Komáromy

25020010 **Pl. 115D** (p. 456)

Filaments long, sometimes containing longitudinally divided cells; cells 9–18(–20) μm wide, 0.5 to 2 times longer than wide, cylindrical to barrel-shaped, with walls thick, stratified, uneven and sometimes mucilaginous when moist, H-shaped segments occasionally

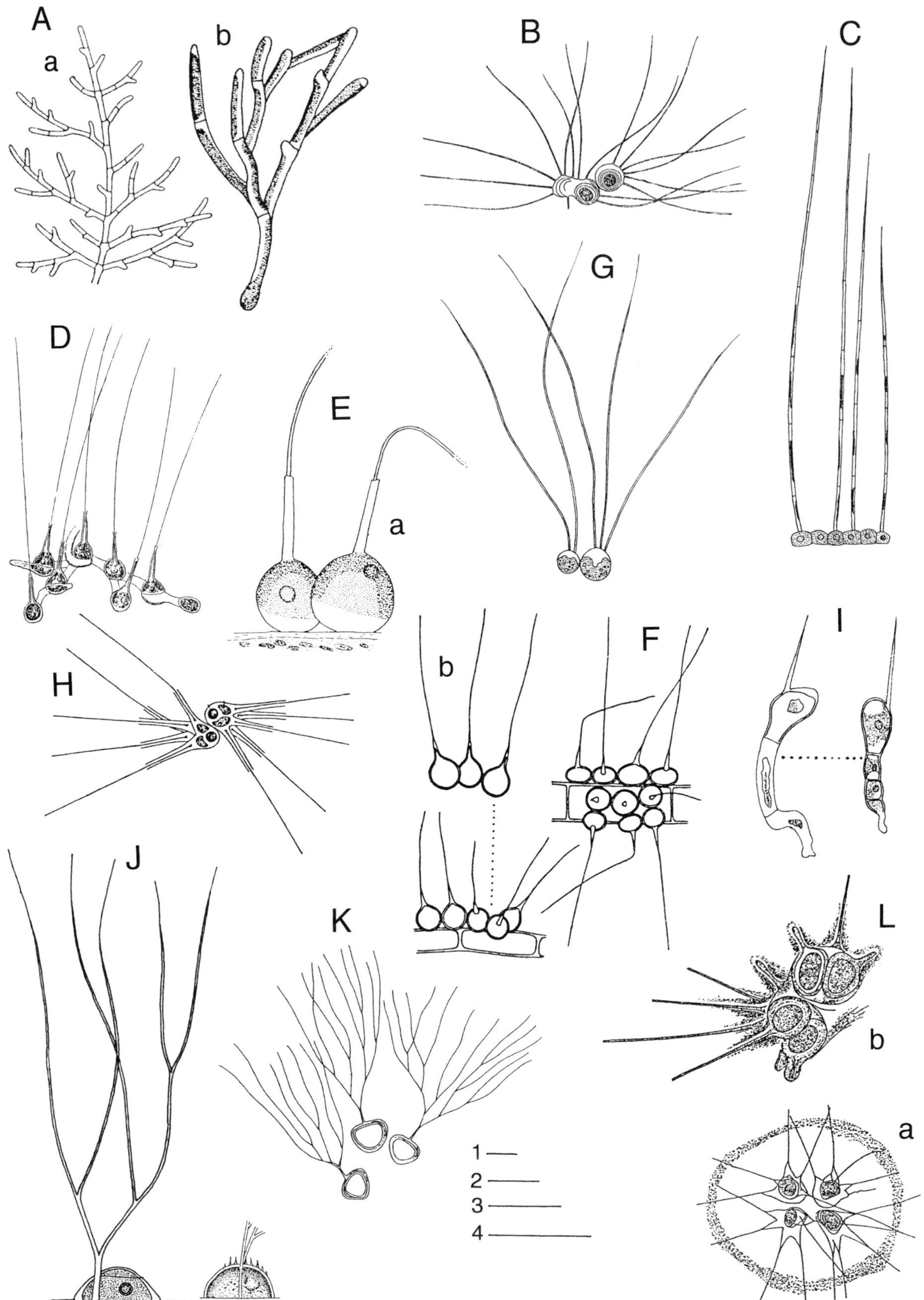

Plate 112 A–L

A. *Microthamnion kuetzingianum* (p. 455): a, b. filaments often without a cross wall at point of branching; **B**. *Polychaetophora lamellosa* (p. 473); **C**. *Pseudochaete gracilis* (p. 460); **D**. *Chaetosphaeridium pringsheimii* (p. 471); **E**. *Chaetosphaeridium globosum* (p. 470): a. details of 2 cells with sheathed bristles, b. general view of cells; **F**. *Chaetosphaeridium globosum* var. *depressum* (p. 470); **G**. *Oligochaetophora simplex* (p. 472); **H**. *Conochaete comosa* (p. 472); **I**. *Thamniochaete aculeata* (p. 472); **J**. *Dicranochaete reniformis* (p. 472); **K**. *Dicranochaete britannica* (p. 472): a. cells embedded in a mucilaginous cushion, b. details of cells showing setae arising from the stratified wall; **L**. *Conochaete polytricha* (p. 472).

Scale bars: 20 μm. 1 for Aa, B, Eb, G, H, K, La; 2 for C, D, F, Lb; 3 for Ab, I, J; 4 for Ea.

present; chloroplast plate- or girdle-shaped, covering almost 80% of cell circumference (sometimes coverage complete), with 1(rarely up to 3) prominent pyrenoid.

Widely distributed in western and central Europe; reported in England from soils collected in the Brecklands of East Anglia, although in mainland Europe known from other terrestrial habitats as well as from pools and swiftly-flowing streams.

Two forms are described by Fritsch & John (1942), one most closely corresponding to *K. crenulatum* was considered by them to be characteristic of alkaline soils. The second form had constrictions at the cross wall and thus resembles *Ulothrix moniliformis* but differs by its greater cell width and having a '*Hormidium*' type chloroplast.

Klebsormidium dissectum (F.Gay) Ettl *et* Gärtner 1995

Basionym: *Stichococcus dissectus* F.Gay
Synonyms: *Klebsormidium dissectum* (F.Gay) Lokhorst, *Hormidium dissectum* (F.Gay) Chodat
25020020 **Pl. 115F** (p. 456)
Filaments straight or slightly bent, often becoming hooked just before dissociating into single cells or few-celled (usually 4–15-celled) fragments; cells 7.5–10(–13) μm wide, 1 to 1.5 times longer than wide, initially cylindrical and thin-walled, later barrel-shaped with thick, striated walls, occasionally with H-shaped segments; chloroplast generally plate-shaped, encircling half to just over two-thirds of cell circumference, sometimes kidney-shaped in side view.

Probably cosmopolitan; in the British Isles on damp subaerial surfaces, such as walls and on soil.

Klebsormidium flaccidum (Kützing) P.C.Silva, Mattox *et* W.H.Blackwell 1972

Basionym: *Ulothrix flaccida* Kützing
Synonyms: *Hormidium flaccidum* (Kützing) A.Braun, *Chlorhormidium flaccidum* (Kützing) Fott, *Stichococcus flaccidus* Hazen
25020030 **Pl. 115J** (p. 456)
Filaments long and usually bent or twisted, dissociating with age into shorter lengths (especially when cultured); cells 5–8(–14) μm wide, 1 to 3 times longer than wide, cylindrical to barrel-shaped, walls thin and smooth; chloroplast encircles up to about two-thirds of cell circumference, often kidney-shaped in side view.

Probably cosmopolitan; reported in the British Isles from flowing-water habitats but most frequently growing as felty mats or mucilaginous masses on wet rock, tree bark and a whole range of soil types.

Klebsormidium fluitans (F.Gay) Lokhorst 1996

Basionym: *Stichococcus fluitans* F.Gay
Synonyms: *Hormidium fluitans* (F.Gay) Heering, *Chlorhormidium fluitans* (F.Gay) Starmach
25020040 **Pl. 115I** (p. 456)
Filaments sometimes straight, more usually bent and twisted, with knee-like joints present, false branches (filament grows out laterally where a break occurs) occasionally developing along with a firm mucilaginous pad, with age filaments commonly dissociating into single cells; cells 6.5–9 μm wide, 1 to 3 times longer than wide, cylindrical or slightly barrel-shaped, walls thin, sometimes H-shaped pieces develop close to cross walls; chloroplast plate-shaped, covering about two-thirds of cell circumference.

Probably cosmopolitan; often associated in the British Isles with other filamentous green algae over soil and in various aquatic habitats, including cattle drinking troughs; known to survive in copper-enriched habitats.

Klebsormidium klebsii (G.M.Smith) P.C.Silva, Mattox *et* W.H.Blackwell 1996

Basionym: *Hormidium klebsii* G.M. Smith
Synonyms: *Hormidium nitens* G.A.Klebs *non* Meneghini, *Stichococcus nitens* G.A.Klebs
25020050 **Pl. 115G** (p. 456)
Filaments long, generally straight and usually possessing a distinctive glistening sheath; cells 5–10 μm wide, 1 to 3 times longer than wide, cylindrical and without constrictions at cross wall, walls thin and smooth; chloroplast plate-like, covering two-thirds or less of cell circumference, with a single somewhat elongate pyrenoid.

Probably cosmopolitan; known in the British Isles only from soil cultures; elsewhere reported from various habitats, including ditches and bogs.

According to Ramanathan (1964), the records of *Hormidium nitens* from British soils by Bristol (1920b) and Fritsch & John (1942) were based upon Klebs's (Klebs, 1896) concept of the species; *Hormidium nitens sensu* G.A.Klebs is now considered a synonym of *Klebsormidium klebsii.*

Klebsormidium mucosum (J.B. Petersen) Lokhorst 1996

Basionym: *Hormidium mucosum* J.B. Petersen
Synonym: *Chlorhormidium mucosum* (J.B.Petersen) Starmach
25020060 **Pl. 115E** (p. 456)
Filaments straight when short, if long often curved; cells 12–20(–23) μm wide, 0.25 to 1.5(–2) times longer than wide, cylindrical, walls very thin to slightly thickened and uneven, sometimes two-layered, with surface rough or slightly to coarsely warty, or finely dented with irregularly protruding spine-like projections, occasionally robust H-shaped wall pieces present; chloroplast plate- to girdle-shaped, usually extending more than two-thirds around cell circumference, with pyrenoid spherical to markedly elongate.

Common in western Europe on acid soils; reported in the British Isles associated with diverse soil types and usually occurs with *Klebsormidium flaccidum.*

Klebsormidium rivulare (Kützing) Morison *et* Sheath 1985

Basionym: *Hormidium rivulare* Kützing
Synonym: *Ulothrix rivularis* Kützing
25020080 **Pl. 115H** (p. 456)
Filaments long, often falsely branched, rhizoids and knee-like joints or bends common, apices often appearing rhizoidal; cells 4–11 μm wide, 1 to 3(–6) times

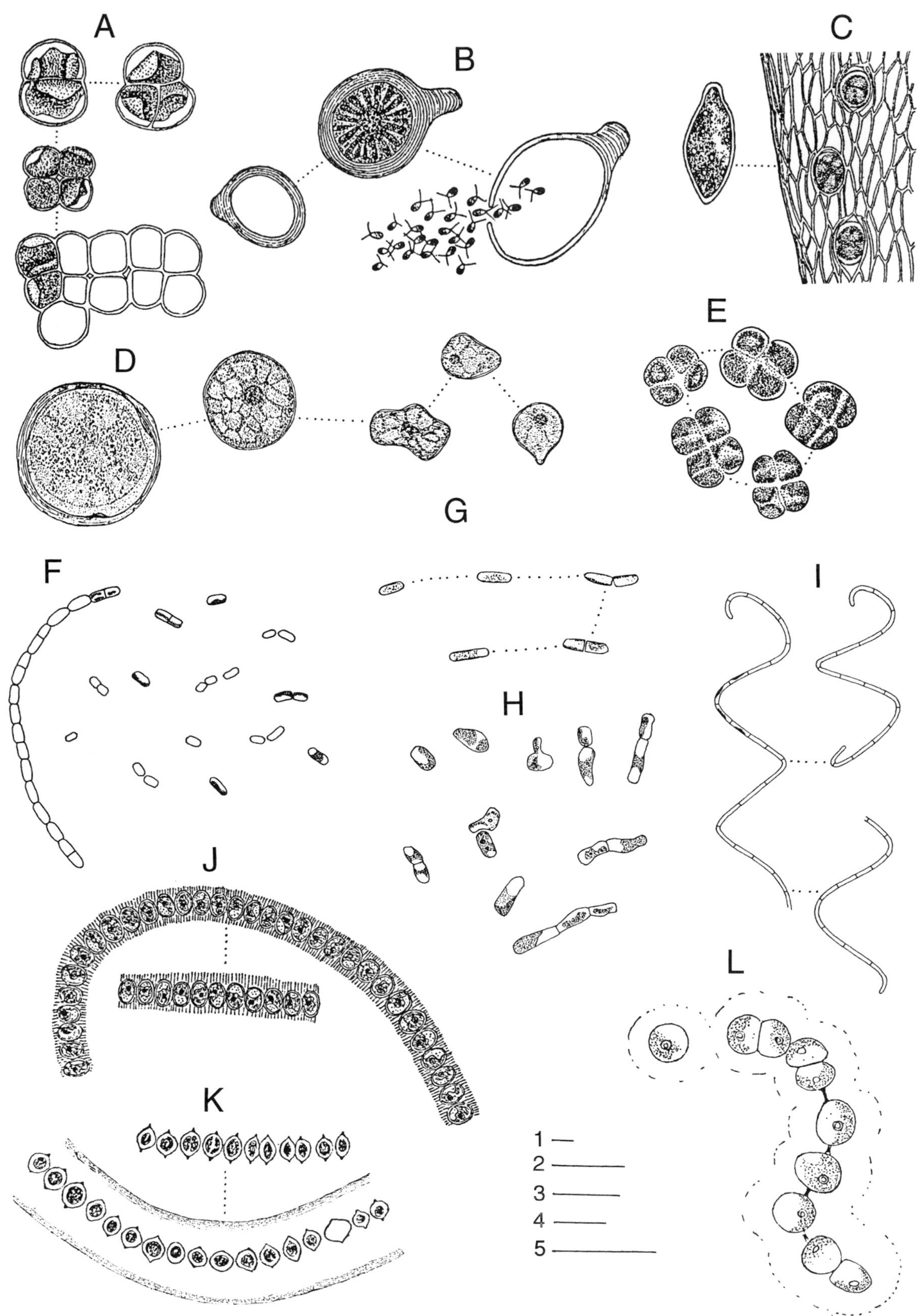

Plate 113 A–L

A. *Apatococcus lobatus* (p. 433); **B**. *Chlorochytrium facciolae* (p. 336): cells with a peg-like wall thickening and one showing release of biflagellate zoospores (left); **C**. *Chlorochytrium lemnae* (p. 336): alga within the cells of the duckweed *Lemna* and a single ellipsoidal cell (left) showing its detailed structure; **D**. *Chlorochytrium bristolae* (p. 336): series of cells showing size and shape variation with walls sometimes thickened and having internal and external projections; **E**. *Desmococcus olivaceum* (p. 438); **F**. *Stichococcus bacillaris* (p. 461); **G**. *Stichococcus minutissimus* (p. 461); **H**. *Stichococcus variabilis* (p. 462); **I**. *Stichococcus contortus* (p. 461); **J**. *Radiofilum flavescens* (p. 460); **K**. *Radiofilum conjunctivum* (p. 460); **L**. *Radiofilum paradoxum* var. *reticulatum* (p. 460).

Scale bars: 1, 2, 20 μm; 3–5, 10 μm. 1 for C, D; 2 for B, F, H–K; 3 for G; 4 for A, E, M; 5 for L.

longer than wide, cylindrical to slightly barrel-shaped, walls moderately thick; chloroplast rounded or ellipsoidal, extending less than halfway around cell circumference; akinetes single or in short chains, ellipsoidal when mature.

Probably cosmopolitan; commonly encountered in the British Isles as light green mats in moderately to swiftly flowing water.

Lokhorst (1996) believed it to be a freshwater form of the brackish-water *Ulothrix implexa*, and therefore he reduced it to a synonym under the latter name. Until proven conclusively through experimental work and ultrastructural investigations of cell division, it has been decided to regard the two as separate species belonging to different genera.

Klebsormidium subtile (Kützing) Tracanna ex Tell

Basionym: *Ulothrix subtilis* Kützing *pro parte*

Synonyms: *Hormidium subtile* (Kützing) Heering, *H. subtilisimum* (Rabenhorst) Mattox *et* Bold, *Ulothrix subtilissima* Rabenhorst, *Klebsormidium subtilissimum* (Kützing) P.C.Silva, Mattox *et* W.H.Blackwell

25020100 (25020090) **Pl. 115K** (p. 456)

Filaments long and straight; cells 4–8 μm wide, 1 to 5 times longer than wide, cylindrical to slightly inflated and constricted at cross walls, thin-walled; chloroplast extending almost to length of cell, covers up to 80% of cell circumference.

Probably cosmopolitan; abundant throughout the British Isles, forming bright green, slippery tufts or masses in standing and flowing water, including water troughs, waterfalls and seepage areas where water trickles down cliff faces; elsewhere known from soil surfaces.

Recorded by Foster (1979) from several mine-contaminated sites in Cornwall under the name '*Hormidium scopulinum*' Foster (1979). She identified her material based on the same criteria as used by Ramanathan (1964), namely 'Cells cylindrical without constrictions, usually longer than wide. Easily fragmenting but not geniculating. Chloroplast with indistinct pyrenoid'. Doubt attaches to the current placement of '*Hormidium scopulinum*' since Dillard (1989) considers it a synonym of *Klebsormidium subtilissimum* albeit with a question mark. His description is at variance with Foster's material since the cells are 'slightly tumid and constricted at septa' and the pyrenoid is not mentioned as indistinct. Examination of a culture of Foster's material shows the pyrenoid to be distinct and to correspond closely to *K. subtile*, a species regarded by Foster as 'a "dump" species for undistinguished Ulotrichales that could not be identified as one of the other species'.

Koliella Hindák 1963

Synonym: *Raphidonema* Lagerheim *pro parte*

Cells commonly solitary, usually separating immediately following division, daughter cells only rarely remaining in pairs for a short time, occasionally forming short pseudofilaments during rapid cell division before dissociating into fragments; cells spindle-shaped, thread-like or very elongated, straight or curved, walls thin, apices obtuse, rounded, or gradually tapering to very long points; chloroplast parietal, single, sometimes layered (laminate), girdle-shaped, straight or spiralled, with or without pyrenoids; oil droplets present; asexual reproduction by biflagellate zoospores; sexual reproduction oogamous, details uncertain; akinetes sometimes produced.

Probably cosmopolitan; widely distributed in freshwater plankton, occasionally periphytic, epilithic or associated with snow and ice.

Very similar to *Raphidonema*, the two possibly congeneric since principally separated on the form of the apical and cell shape (see Hindak, 1963).

1 Cells with long and hair-like apices......*K. longiseta*
1 Cells with acute apices..................*K. spiculiformis*

Koliella longiseta (Vischer) Hindák 1963

Basionym: *Raphidonema longiseta* Vischer

25030010 **Pl. 114J** (p. 454)

Cells solitary or united to form short filaments, (1.5–)3–5 μm wide and 50–120(–140) μm long, long and spindle-shaped, with apices tapering to a slightly curved and hair-like point.

Europe; known in the British Isles from the plankton of various water bodies, including Oak Mere and Rostherne Mere in Cheshire where numbers were reported by Swale (1968) to reach 2600 cells per mL in December; known also from damp soil.

Koliella spiculiformis (Vischer) Hindák 1963

Basionym: *Raphidonema spiculiforme* Vischer

25030020 **Pl. 115A** (p. 456)

Cells usually solitary, 0.8–2(–2.5) μm wide and 40–80 μm long, with apices gradually tapering to point.

Probably cosmopolitan; known in the plankton of a few large lakes and brick pits near Cambridge (UKNCC, 2001; strain CCAP 470/2B and 2C).

Leptosira Borzí 1883

Synonym: *Leptosiropsis* C.C.Jão

Filamentous, uniseriate, prostrate and erect filaments irregularly to sub-dichotomously divided, branches often short and terminating in a variously shaped cell, frequently forming cushion-like growths; cells spherical, barrel-shaped, ellipsoidal or irregularly shaped; chloroplast parietal and filling cell, lacking pyrenoids; asexual reproduction by biflagellate zoospores, 20–60 produced in each sporangium; sexual reproduction by biflagellate isogametes; aplanospores spherical, 4 or more in each sporangium.

Probably cosmopolitan; in various types of water body and associated with soil.

According to Friedl (1997), *Leptosira* (as *L. terrestris*) defines a single evolutionary line amongst other Trebouxiophyceae (=Pleurastrophyceae) lineages. This genus and *Pleurastrum* are sometimes assigned to the Microthamniales (see Bakker et al., 1997), an order considered almost synonymous with the Pleurastrales. See remarks under *Pleurastrum*.

Leptosira mediciana Borzí 1883

24200010 **Pl. 109L** (p. 441)

Forms dense, cushion-like growths; cells (7.5–)10–20 μm wide, 1 to 2 times longer than wide, short and cylindrical or swollen (often barrel-shaped).

Probably cosmopolitan; only reported from bogs and boggy pools in Yorkshire (Pilmoor, near Thirsk) where its yellow-green cushions occur amongst the moss *Sphagnum* and the bladderwort *Utricularia*; probably more widely distributed in the British Isles than records indicate.

Microspora Thuret 1850

Filamentous, uniseriate, unbranched, free-living or basally attached; cells quadrate, cylindrical or slightly swollen, sometimes constricted at cross walls, walls on occasion thick and stratified, with cells showing a tendency to fracture in centre as evident by examination of broken ends of filaments, dissociation of adjacent cells results in H-shaped fragments consisting of cross wall and adjacent portions of lateral wall; chloroplast parietal, single, usually perforated to give a net-like appearance, rarely granular or bead-like, lacking pyrenoids; uninucleate; asexual reproduction by biflagellate or, more rarely, quadriflagellate zoospores; aplanospores and akinetes known.

Probably cosmopolitan; widely distributed in aquatic habitats where often most abundant in winter, frequently in acid waters including those of bogs, marshes and areas contaminated by acid-mine drainage; often dominating an association of filamentous algae characterizing stream sites contaminated by high concentrations of heavy metals.

There is considerable overlap in the vegetative characters used to distinguish species making it difficult to prepare a satisfactory key. Many characters traditionally used to separate species (e.g. wall thickness, cell dimensions) are polymorphic and vary depending on environmental conditions (see Wichmann, 1937). To develop sounder species concepts requires the testing in culture of the reliability of the morphological characters used to define species. For this reason, only considered here are the most frequently encountered and distinctive species reported from the British Isles. Discussed under the entries is the relationship between the more doubtfully recorded and those recognized. Sometimes the genus has been confused with the xanthophyte *Tribonema* because of its similar wall structure and the H-shaped fragments formed on dissociation of the filaments. *Tribonema* is readily separated from *Microspora* by having numerous, very pale green, disc-like chloroplasts and not possessing starch. The genus is in need of revision.

1 Walls usually <1.5 μm thick (except *M. wittrockii* and *M. tumidula*), without an evident junction between the H-shaped segments 2
1 Walls usually >1.5 μm thick, with an evident junction between the H-shaped segments 6
2 Cells usually >11 μm wide 3
2 Cells never more than 11 μm wide........................4
3 Cells 7–18 μm wide, cylindrical or slightly swollen ....*M. floccosa*
3 Cells 18–24 μm wide, cylindrical, never swollen ....*M. wittrockii*
4 Cells constricted at cross walls ....*M. tumidula*
4 Cells cylindrical, never constricted at cross-walls 5
5 Chloroplasts with net-like perforations, filling the cell ....*M. abbreviata*
5 Chloroplasts variable, granular, perforate or bead-like, not filling the cell..............*M. stagnorum*
6 Cells <18 μm wide 7
6 Cells 18 μm or more wide......................................8
7 Cells usually 12–18 μm wide; chloroplasts granular..............................*M. amoena* var. *gracilis*
7 Cells usually 9–12 μm wide; chloroplasts perforated, plate-like and folded.....*M. pachyderma*
8 Walls often >4 μm thick, wrinkled ...*M. irregularis*
8 Walls always <4 μm thick, smooth 9
9 Cells usually 20–25 μm wide................*M. amoena*
9 Cells usually 25–28 μm wide ....*M. crassior*

Microspora abbreviata (Rabenhorst) Lagerheim 1887

Basionym: *Conferva abbreviata* Rabenhorst

24210010 **Pl. 116I** (p. 464)

Cells 5–11 μm wide, 1 to 3 times as long as wide, quadrate to cylindrical, without constrictions at cross walls, walls only slightly thickened and without striations, junction between H-shaped sections not evident; choroplast perforated and net-like, filling cell.

Probably cosmopolitan; widely distributed in aquatic habitats in the British Isles.

According to Ramanathan (1964, p. 134), neither the original specimen nor Rabenhorst's description (1863, p. 323) corresponds to the current concept of the species.

Microspora amoena (Kützing) Rabenhorst 1868

Basionym: *Conferva amoena* Kützing

24210030 **Pl. 116K** (p. 464)

Cells 20–25(–28) μm wide, 1 to 2 times longer than wide, cylindrical with only a slight constriction at cross walls, walls thick (>2 μm), surface smooth and striated, junction between H-shaped sections clearly evident; chloroplast typically net-like, filling cell.

var. *gracilis* (Wille) De Toni 1889

Basionym: *Conferva amoena* Kützing fo. *gracilis* Wille

Synonym: *Microspora elegans* Hansgirg

24210032

Cells 12–18 μm wide, 1.5 to 4 times longer than wide; chloroplasts granular, nearly filling cell.

Probably cosmopolitan (variety only in Europe and North America); widely distributed and common in the British Isles; occurs in swamps, marshes, roadside ditches and in flowing water on stones and stems of aquatic vascular plants growing as dense tufts, or entangled with other filamentous algae and mosses; also reported from subaerial habitats including drying mud and tree trunks. The variety is known from fast-flowing water in upland areas, often regarded as a cold water form since most abundant in winter (see Fritsch & Rich, 1913, p. 9).

This variety is very similar to *Microspora loefgrenii* (Nordstedt) Lagerheim, a species reported on a single occasion in the British Isles by Benson-Evans & Antoine (1996) in South Wales (no locality or references cited) with the habitat '*Sphagnum* and other bogs' (see under 'Doubtful Records' below). The distinction between *M. amoena* and *M. crassior* is unclear since there is considerable overlap between them, possibly they are conspecific. *Microspora amoena* var. *minor* is mentioned by Lind (1938) from a single locality (Beauchief Ponds, Sheffield), but it is considered doubtful since she provides no detailed description nor an accompanying illustration.

Microspora crassior (Hansgirg) Hazen 1902

Basionym: *Microspora amoena* (Kützing) Rabenhorst var. *crassior* Hansgirg

24210040 **Pl. 116M** (p. 464)

Cells (23–)26–28(–33) µm wide, 0.5 to 2 times longer than wide, usually cylindrical to swollen, somewhat constricted at cross walls, moderately thick-walled (2.5–3.5 µm) with striations within wall, smooth surface, junction between H-shaped sections evident; chloroplast net-like or granular, filling entire cell.

Probably cosmopolitan; only a few reports from the British Isles where intermingled with mats of other filamentous algae in lakes; known in mainland Europe from springs and swiftly flowing streams.

See remarks under *M. amoena.*

Microspora floccosa (Vaucher) Thuret 1850

Basionym: *Prolifera floccosa* Vaucher

24210050 **Pl. 116P** (p. 464)

Cells (7–)14–18 µm wide, 0.6 to 2.5 times longer than wide, usually cylindrical or slightly swollen, seldom constricted at cross walls, very thin-walled and junction of H-shaped sections not evident; chloroplast an open net or forming bead-like bands.

Probably cosmopolitan; one of the most frequently encountered species in the British Isles, commonly forming long, bright green strands or masses in animal drinking troughs and small ponds especially from March to about May.

According to Ramanathan (1964, p. 120), it is possible that the lower range of cell widths given by some authors is due to their mistaking this species for the narrower *Microspora tumidula* or *M. willeana*. He considers the latter to be distinguished from *M. floccosa* on akinete size despite a considerable overlap in dimensions (*M. floccosa*: 14.6–22 µm; *M. willeana*: 12–18 µm).

Microspora irregularis (West *et* G.S.West) Wichmann 1937

Basionym: *Microspora amoena* (Kützing) Rabenhorst var. *irregularis* West *et* G.S.West

24210060 **Pl. 116L** (p. 464)

Cells (21–)24–30 µm wide, 0.5 to twice as long as wide, ellipsoidal, with walls very thick (4–7 µm), striated, irregularly wrinkled, junction between H-shaped sections evident.

British Isles (Ireland); reported on only very few occasions along the stony shores of Irish loughs and on rocks in streams.

Microspora pachyderma (Wille) Lagerheim 1887

Basionym: *Conferva pachyderma* Wille

24210080 **Pl. 116J** (p. 464)

Cells 9–12(–18) µm wide, 1 to 3 times longer than wide, cylindrical, walls moderately thick (about 3 µm), striated, smooth surfaced, junction between H-shaped sections sometimes very evident; chloroplast perforated, plate-like and folded, filling most of cell.

Probably cosmopolitan; widely distributed, commonly forms green masses in *Sphagnum* bogs as well as in ponds and lakes.

Microspora stagnorum (Kützing) Lagerheim 1887

Basionym: *Conferva tenerrima* fo. β *stagnorum* Kützing

24210110 **Pl. 116O** (p. 464)

Cells 5–10 µm wide, 1 to 3 times longer than wide, cylindrical and not constricted at cross walls, thin-walled, smooth, junction between H-shaped sections not evident without staining; chloroplast variable even within same filament, not very dense, sometimes forming an irregular band or network, often occupying only cell centre.

Probably cosmopolitan; widely distributed in the British Isles, commonly along the marginal shallows of lakes where it grows entangled with other filamentous algae, less frequently free-floating.

Microspora tumidula Hazen 1902

24210120 **Pl. 116N** (p. 464)

Cells 6–9.6 µm wide, 0.5 to 2(–4) times longer than wide, constricted at cross walls, walls thin to moderately thick, striated, smooth, junction between H-shaped sections not evident; chloroplast granular, dense, perforated and sometimes forming bead-like chains, almost or completely filling cell.

Probably cosmopolitan; in the British Isles forms flocculent masses in the shallow margins of lakes and rivers including the River Caragh in Ireland (Heuff & Horkan, 1984).

According to Ramanathan (1964, p. 122), it 'Resembles closely *M. floccosa* and *M. stagnorum* and is sometimes very difficult to distinguish from them in the vegetative condition'.

Microspora wittrockii (Wille) Lagerheim 1887

Basionym: *Conferva wittrockii* Wille

24210140 **Pl. 117A** (p. 474)

Cells 18–24 µm wide, 1 to 2.5 times as long as wide, cylindrical, not constricted at cross walls, walls about 1.5 µm wide, without distinct striations, junction between H-shaped sections not usually evident; chloroplast often perforated, filling most of cell.

Probably cosmopolitan; widely distributed in the British Isles where reported on only a few occcasions in streams and rivers (e.g. River Caragh).

According to Ramanathan (1964, p. 130), it 'resembles *M. amoena* and *M. loefgrenii* but the cell wall is somewhat thinner' (*M. wittrockii*, about 1.5 µm thick; other species, >2 µm thick).

DOUBTFUL RECORDS

Several other *Microspora* species are recorded from the British Isles, but for one reason or another all are to be considered doubt-

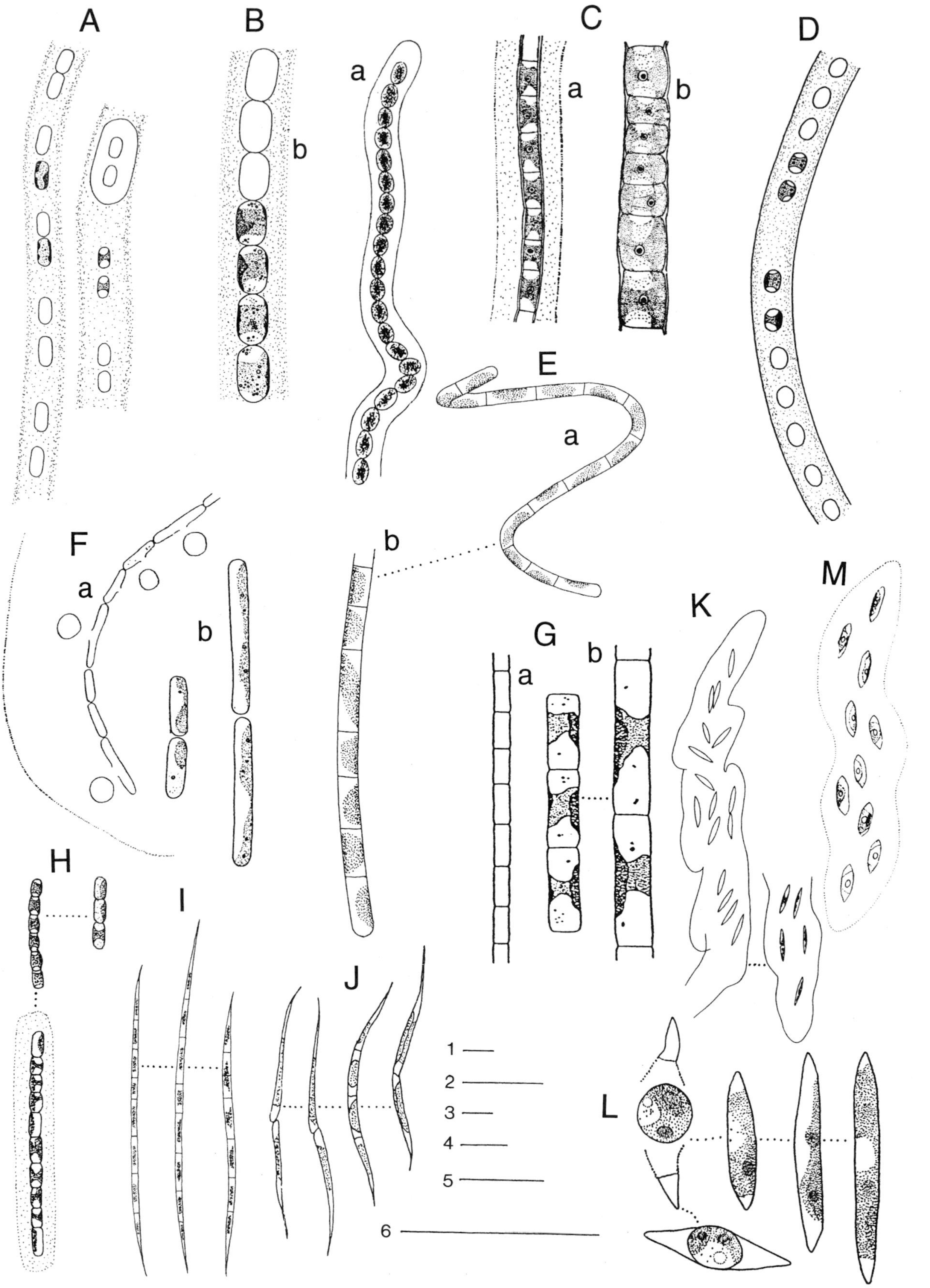

Plate 114 A–M

A. *Geminella interrupta* (p. 440); **B.** *Geminella mutabilis* (p. 440); **C.** *Geminella minor* (p. 440); **D.** *Geminella ordinata* (p. 442); **E.** *Stichococcus pelagicus* (p. 462); **F.** *Gloeotila monospora* (p. 442): a. cells within a wide mucilaginous envelope and empty cells each releasing a spherical aplanospore, b. details of individual cells; **G.** *Gloeotila subconstricta* (p. 442); **H.** *Gloeotila protogenita* (p. 442); **I.** *Raphidonema nivale* (p. 461); **J.** *Koliella longiseta* (p. 451); **K.** *Elakatothrix gelatinosa* (p. 439); **L.** *Elakatothrix gelatinosa* var. *aplanospora* (p. 439): detail structure of vegetative cells and aplanospore formation and release; **M.** *Elakatothrix acuta* (p. 439).

Scale bars: 1, 2, 40 μm; 3–6, 10 μm. 1 for K; 2 for A, Ba, D, Ga; 3 for Ca, Gb, Fa, H–J; 4 for Bb, Fb, M; 5 for Cb, Ea; 6 for Eb, L.

ful. This category included Hibberd's (1978) mention of *Microspora aequabilis* Wichmann (24210020) and *M. palustris* Wichmann (24210090) from the Scottish island of Mull, and Benson-Evans & Antoine's (1996) record of *M. loefgrenii* (Nordstedt) Lagerheim (24210070), *M. willeana* Lagerheim (24210130) and *M. quadrata* Hazen (24210100) from South Wales. Most of these species closely resemble others more frequently recorded in the British Isles (see remarks under the above entries). The two species mentioned by Hibberd (1978) are distinguished by very dubius characters (e.g. nature of assimilatory products and habitat). One of these species, *M. aequabilis*, closely resembles *M. willeana* in size although is distinguished from it by the absence of constrictions at the cross walls and the presence of rounded cell corners (see Ramanathan, 1964). Records from England and Wales (e.g. West, 1890; Rich, 1925) of *Microspora fugacissima* (Roth) Rabenhorst are almost certainly attributable to the xanthophyte genus *Tribonema*.

Microthamnion Nägeli 1849

Microscopic, consisting of erect, branched, uniseriate filaments of similar width throughout, branches each terminating in an obtuse apical cell, attached by a mucilage pad; branches grow out laterally immediately beneath the cross wall, the first division of the lateral results in a cross wall formed some distance away from the point of branch origin; chloroplasts parietal, without a pyrenoid; asexual reproduction by biflagellate zoospores, 2–32 produced per cell, released by rupture of wall; doubt attaches to reports of akinetes, aplanospores and sexual reproduction.

Probably cosmopolitan; attached to surfaces in many types of aquatic habitat, occasionally on the water film, also sometimes associated with soil.

Considerable morphological variation caused by environmental conditions has given rise to taxonomic problems since many vegetative features traditionally used to define the species are unreliable. Only three species are recognized by John & Johnson (1987) who regard two as doubtful.

Microthamnion kuetzingianum Nägeli 1849

Synonyms: *Microthamnion strictissimum* Rabenhorst, *M. vexator* Cooke

24220010 **Pl. 112A** (p. 448) **CD**

Branches often densely and irregularly divided, spreading or slightly curved; cells (1.5–)3–4(–5) μm wide, 2 to 12(–15) times longer than wide, cylindrical, all similar except for the semicircular attachment cell and rounded apical cells.

Probably cosmopolitan; on aquatic plants, rotting wood, stones and other surfaces in the shallow margins of pools, slow-flowing streams, ditches and springs, or on the exposed beds of intermittently used sewage filter beds or wastes from water treatment works where iron salts are used to flocculate humic materials; characteristic organism where iron and/or manganese oxides are being deposited such as where oxygen-free water draining from peat meets the air or pumped coal-mine water enters a stream or small river; experimental studies indicate that its ability to use organic phosphate where inorganic phosphate is scarce may be a key factor in its ecology (B.A. Whitton, pers. comm.). Often associated with *Microspora*-dominated assemblage of filamentous algae characterizing mineral-contaminated sites.

Microthamnion kuetzingianum and *M. strictissimum* are considered conspecific. Culture-based studies by John & Johnson (1987), amongst others, have demonstrated just how variable are the characters traditionally used to separate species (e.g. cell width, length to breadth ratio of cells, mode of branching). *M. strictissimum* is a long-celled, laxly branched form of *M. kuetzingianum* that often develops when conditions are favourable for rapid growth. Individuals entrapped in a surface water film commonly have curved branches.

Monostroma Thuret 1854

Initially sack-like, later a thin, single-layered, membrane-like plate of cells; cells rounded or more or less angular, irregularly arranged or in 4-celled groups; chloroplast single, parietal, cup-shaped, with a single pyrenoid; asexual reproduction by biflagellate zoospores; sexual reproduction by biflagellate anisogametes, zygotes develop into a filamentous stage (sexual reproduction known only in marine species).

Probably cosmopolitan; free-floating or attached and usually in flowing water. Several marine species but only 2 freshwater species

Monostroma oxyspermum (Kützing) Doty is a brackish-water species that is common in the British Isles in habitats subject to considerable salinity fluctuation; it can survive for short periods in almost freshwater.

1 Cells rounded and usually in 4-celled groups; membrane <10 μm thick .... *M. bullosum*
1 Cells usually somewhat angular and irregularly arranged; membrane >10 μm thick .... *M. membranaceum*

Monostroma bullosum (Roth) Wittrock 1866

Basionym: *Ulva bullosa* Roth

24230010 **Pl. 117G** (p. 474) **CD**

Thallus sac-like or as a flat membrane, usually membrane less than 10 μm thick, rarely exceeding 10 mm long, attached by a single holdfast; cells (3.5–)6–12 μm wide, rounded, with a more or less 'T'-shaped arrangement, almost oblong in section of thallus and having their long axes parallel to surface.

Europe; not common in the British Isles where reported growing in ponds, streams and rivers (e.g. Cam, Tweed) on stones, rocks, fibrous roots of bank-side trees and submerged stems of aquatic macrophytes. According to Holmes & Whitton (1975a), it typically grows in the Tweed catchment area on large boulders that are alternately wetted and dried during the period when boulders are submerged.

Monostroma membranaceum West *et* G.S.West 1903

24230030 **Pl. 117F** (p. 474)

Thallus a broad membrane up to 8–9 cm long, about 18 μm thick, often with torn margins, narrows to base where initially attached; cells 8–20 μm wide, usually polygonal and angular, not regularly arranged, rarely quadrate and having a more orderly arrangement, more or less oblong in section of thallus and having their long axes at right angles to the surface.

British Isles (England); widely distributed but rare in ponds and rivers, sometimes free-floating and dredged from depths

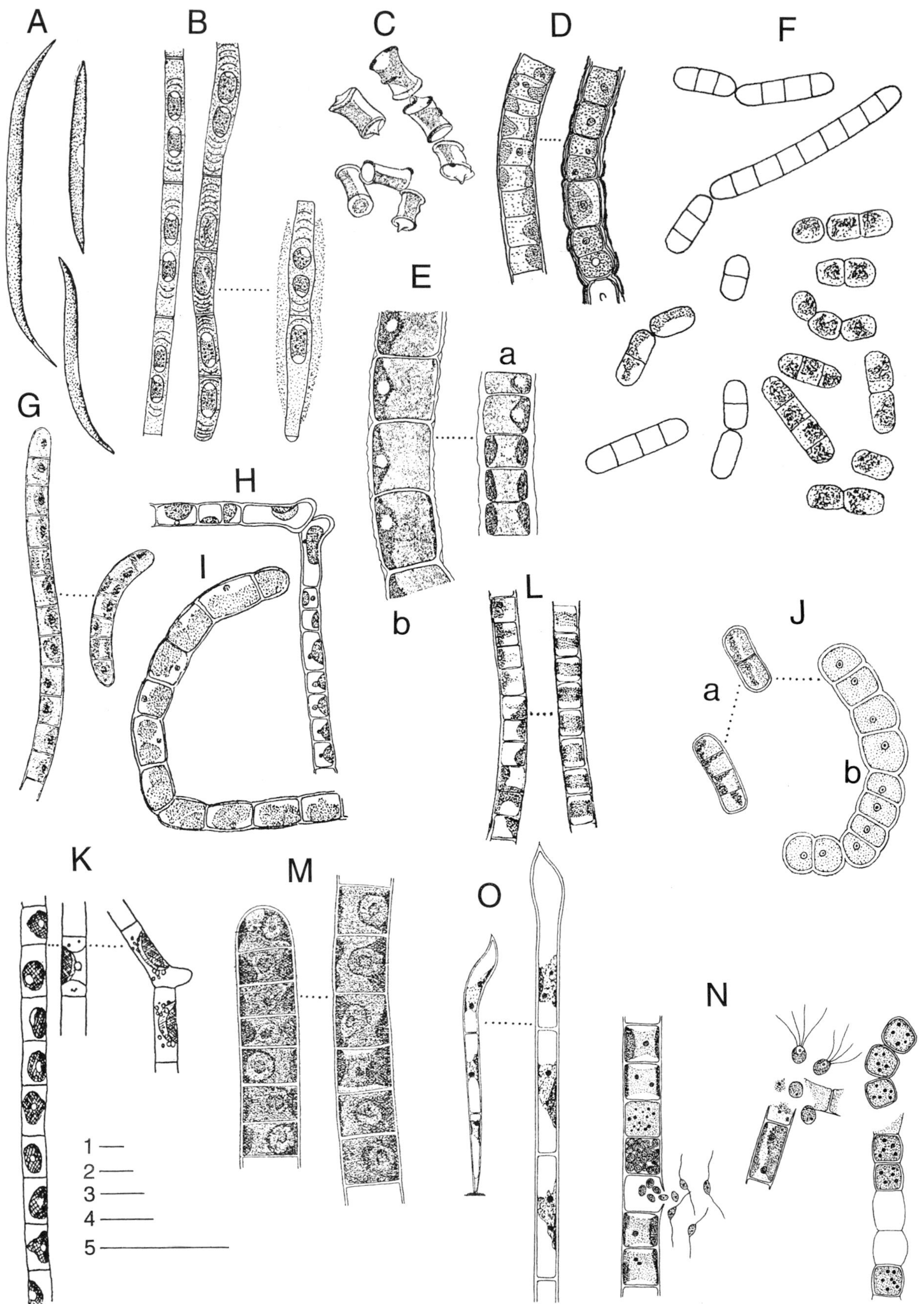

Plate 115 A–O

A. *Koliella spiculiformis* (p. 451); **B**. *Binuclearia tectorum* (p. 434); **C**. *Catena viridis* (p. 436); **D**. *Klebsormidium crenulatum* (p. 447); **E**. *Klebsormidium mucosum* (p. 449); **F**. *Klebsormidium dissectum* (p. 449); **G**. *Klebsormidium klebsii* (p. 449); **H**. *Klebsormidium rivulare* (p. 449); **I**. *Klebsormidium fluitans* (p. 449); **J**. *Klebsormidium flaccidum* (p. 449); **K**. *Klebsormidium subtile* (p. 451); **L**. *Ulothrix tenerrima* (p. 467); **M**. *Ulothrix tenuissima* (p. 467); **N**. *Ulothrix aequalis* (p. 466): details of cells and release of swarmers (left), possibly isogametes), larger quadriflagellate zoospores (middle) and aplanospores (right); **O**. *Uronema elongatum* (p. 467).

Scale bars: all 10 μm. 1 for D, Ea, F, G, Ja, O(left), N; 2 for B, H, L, M, O(right); 3 for Eb, I, K; 4 for A, Jb; 5 for C.

ranging from 1.5 to 5 m; originally described from ponds on Mitcham Common in Surrey, England.

Planctonema Schmidle 1903

Synonym: *Geminellopsis* Korshikov

Filaments free-floating, short and unbranched; cells cylindrical with rounded apices, lying in a loose row, single cells almost equidistant from one another or more usually united in pairs, surrounded by a homogeneous mucilaginous sheath of variable thickness; chloroplast parietal, plate-like, with a pyrenoid (?); vacuoles with conspicuous oil droplets often at apices; reproduction by fragmentation; sexual reproduction unknown.

Probably cosmopolitan; planktonic in various freshwater habitats. Monospecific.

Taxonomic position remains uncertain (see Lokhorst 1992); some authors consider it to be a synonym of *Geminella*. Doubt attaches to the pyrenoid since some light microscope observations are contradictory.

Planctonema lauterbornii Schmidle 1903

24240010

Cells 2.5–5 μm wide, 5–15 μm long, enclosed in a mucilaginous sheath about 10 μm wide.

Probably cosmopolitan; reported in the British Isles on a few occasions in lakes including Scottish lochs and in neutral to slightly acid-water lakes such as those at Center Parcs (Nottingham); probably considerably more widely distributed than the few scattered records suggest.

Pleurastrum Chodat 1894

Cells solitary, grouped in 2s or 4s, or in more or less regular cell masses, sometimes contiguous and filaments unbranched or with few-celled branches; chloroplasts parietal, often lobed, containing 1 or more pyrenoids; asexual reproduction by zoospores, aplanospores and akinetes (hypnospores); zoospores biflagellate, naked, with eyespot and asymmetrically flattened.

Probably cosmopolitan; usually in standing water (often stagnant) and on the surface of damp soil.

Pleurastrum and *Leptosira* are very similar and separated based on the presence or absence of pyrenoids. Several species attributed to *Leptosira* (or transferred to it) have since been discovered to have pyrenoids although sometimes these were only detected on staining or examining using an electron microscope. For further information on the involved taxonomic history of the genus, see Sluiman & Gärtner (1990). *Pleurastrum* and *Leptosira* have been placed in the Microthamniales, an order Bakker et al. (1997) consider to be almost synonymous with the Pleurastrales. It is only possible to separate the two species when they produce a filamentous stage. *Pleurastrum* has been placed in the Class Trebouxiophyceae along with *Microthamnion* and other genera based on the analysis of molecular data (see Graham & Wilcox, 2000).

1 Filaments 5.6–22 μm wide, branched and not readily fragmenting, forming compact tufts or cushion-like growth, without a mucilaginous sheath .... *P. terricola*

1 Filaments 5–7 μm wide, unbranched and readily fragmenting, loose and usually surrounded by a mucilaginous sheath .... *P. insigne*

Pleurastrum insigne Chodat 1894

24250020 **Pl. 110A** (p. 443)

Cells solitary, or clustered to form compact colonies consisting of small pairs, triads, or tetrads of cells usually having a more or less cross-shaped arrangement (usually when grown on agar in culture) and spherical to subspherical (7–14 μm across); filaments short, unbranched and readily dissociating into short fragments, often surrounded by a thin mucilaginous sheath; cells cylindrical or ellipsoidal, 5–7 μm wide, 7–9 μm long, usually 1 to 1.5 times longer than wide, walls generally thin and smooth; hypnospores with an ornamented wall.

Probably cosmopolitan; usually collected and isolated in the British Isles from chalk or clay soils.

Pleurastrum terricola (Bristol) D.M. John, *nov. comb.*

Basionym: *Gongrosira terricola* Bristol 1920. Annals of Botany 34, p. 78, pl. II, figs a–i.

Synonyms: *Pleurastrum terrestre* F.E.Fritsch *et* R.P.John, *Leptosira terrestris* (F.E.Fritsch *et* R.P.John) Printz, *L. terricola* (Bristol) Printz

24250030 (24250010) **Pl. 109M** (p. 441)

Cells single or more often in tetrads, clustered to form irregular colonies; filaments little developed, short, irregularly branched and radiating from a compact ('pseudoparenchymatous') centre to produce a dense tuft or cushion; cells subcylindrical, barrel-shaped to irregular, (5.6–)6–22 μm wide, commonly 1 to 2 times longer than wide (up to 21 μm long), walls thin but thickening with age; oil droplets present.

Probably cosmopolitan; the typical form has been recorded in the British Isles from cultures of soils collected from widely separated locations, especially alkaline soils (including sand dunes and soils associated with chalk grasslands); a variety with giant cells is mentioned by Fritsch & John (1942) as only associated with acid soils (see below).

The type of the genus *Leptosira*, *L. mediciana* Borzí, does not possess pyrenoids and their absence is the principal character for separating it from *Pleurastrum*. Tupa (1974) considers three species of *Leptosira* and Bristol's (Bristol, 1920b) *Gongrosira terricola* to more correctly belong to the genus *Pleurastrum*, all having been found to have pyrenoids. Tupa (1974, p. 45) considered two of the *Leptosira* species to be 'very similar (perhaps indentical, in the writer's opinion) to Bristol's *Gongrosira terricola* and *Pleurastrum terrestre* Fritsch and John'. Printz (1964) proposed two new combinations, *Leptosira terricola* (Bristol) Printz and *L. terrestris* (F.E.Fritsch *et* R.P.John) Printz, and separated these taxa in his key on differences in cell width ('*L. terricola*': 11–16 μm wide; '*L. terrestris*': 15–40 μm wide). Cell dimensions cannot be used to separate the species, since measurements almost completely overlap when additional information on *Gongrosira terricola* provided by James (1935) and Lund (1947a) is considered. For example, the cells of '*terricola*' range from 12–21 μm wide and those of '*terrestris*' from 5.6–22 μm wide (only giant cells up to 40 μm wide, possibly a separate variety according to Fritsch & John, 1942, p. 385). Tupa (1974) considered that '*Pleurastrum terrestre* may be identical to *Gongrosira terricola* Bristol (1920b), in which case the earlier name *terricola* would need to be assigned to this species of *Pleurastrum*'. Since it is impossible to separate these two taxa based upon Fritsch & John's descriptions, it has been decided to consider them conspecific. Fritsch & John (1942) believed that those cultures isolated from acid soils, which formed giant cells (to 40 μm wide and 70 μm long) containing 1–4 pyrenoids, probably represent a separate variety.

Protoderma Kützing 1843

Synonym: *Ulvella* P.Crouan *et* H.Crouan

Filaments uniseriate, very compacted, forming a single-layered, pseudoparenchymatous, disc-like expansion with filaments often free towards its margin; cells in centre usually polygonal and becoming more cylindrical away from centre; chloroplast parietal, plate-like or flattened, with or without a pyrenoid; asexual reproduction by biflagellate zoospores, 4, 8 or 16, generally developing within cells towards centre of the disc-like expansion; aplanospores ellipsoidal or spherical; *Palmella*-stage formed by repeated division of central cells.

Probably cosmopolitan; usually on rocks and stones (epilithic) as well as on algae or other aquatic macrophytes in still and flowing water.

Difficult to distinguish from other closely related genera, especially those having a dominant prostrate system and few if any erect branches (e.g. *Pseudendoclonium prostratum*). Often separation is only possible if the flagellar number of the zoospores is known. The cells of *Protoderma* species are usually somewhat smaller than those of the developmental stages of two genera with which it might easily be confused, i.e. *Pseudendoclonium* and *Stigeoclonium*. Retained here in the Chaetophorales, the order in which it has been traditionally placed.

1 Filaments enclosed within a mucilaginous envelope; cells in centre loosely arranged and often colourless with age ....*P. frequens*
1 Filaments not enclosed within such an envelope; cells in centre compact and usually green..............2
2 Cells in centre usually thick-walled; pyrenoids absent ....*P. beesleyi*
2 Cells in centre thin-walled; pyrenoids present ....*P. viride*

Protoderma beesleyi (F.E.Fritsch) Printz 1964

Basionym: *Ulvella beesleyi* F.E.Fritsch

24270010 **Pl. 110B** (p. 443)

Thallus a compact disc-like expansion, often becoming star-shaped with age; cells in centre oval or spherical, thick-walled, becoming dark green with age and sometimes dividing to form a mass of cells, towards margin cells cylindrical, (1–)1.5 to 5 times longer than wide and arranged in rows; pyrenoids absent.

British Isles, New Zealand; so far only reported in the British Isles from a single locality in Cambridgeshire on mobile grains of sand and on flints lining the floor of spring-fed pools.

Thoroughly investigated in culture by Beesley (1904), but it was over 20 years later that Fritsch (West & Fritsch, 1927) named it.

Protoderma frequens (Butcher) Printz 1964

Basionym: *Ulvella frequens* Butcher

24270020 **Pl. 110C** (p. 443)

Thallus a circular, elongate or lobed expansion, often enclosed within a thick mucilaginous envelope; cells in centre more or less angular, 4–5 μm wide, equal in length and breadth, thick-walled, with age becoming less compact, irregularly grouped and often colourless due to spore release; cells towards margin 4–5 μm wide, 1.5 to 3 times longer than wide, arranged in short, simple or branched rows; pyrenoid single, often indistinct.

Europe; in the British Isles on rocks, aquatic vegetation and various artificial surfaces (e.g. glass, plastic) in rivers where most abundant during the summer months; considered an indicator of relatively clean water although found over a wide range of nutrient conditions.

Protoderma viride Kützing 1843

24270030 **Pl. 110D** (p. 443) **CD**

Thallus a disc-like expansion, often becoming increasingly irregular in outline with age; cells in centre 6–12 μm wide, 1 to 2 times longer than wide, thin-walled, forming a compact pseudoparenchymatous layer; cells towards margin 3–6 μm wide, 2 to 3 times longer than wide, radially arranged; pyrenoid single.

Probably cosmopolitan; a common epiphyte on filamentous algae and vascular vegetation in still and flowing water, often colonizing artificial surfaces including glass, plastic or polythene; readily overlooked like many minute encrusting algae, since it is difficult to remove intact when sampling these habitats.

Pseudendoclonium Wille 1901

Filaments uniseriate, consisting of irregularly branched prostrate and erect systems, with former system loose and spreading or compact and pseudoparenchymatous; chloroplast parietal, with a single pyrenoid; asexual reproduction by quadriflagellate zoospores and aplanospores, 4–32 spores in each sporangium, often sporangia in 2-, 3- or 4-celled packets; sexual reproduction unknown; akinetes produced.

Probably cosmopolitan; usually in still and flowing water on rocks, stones, artificial surfaces and aquatic macrophytes. Freshwater and marine species.

The flagella number of the zoospores is one of the few characters to distinguish it from morphologically very similar genera such as *Gongrosira*. The most distinctive species are *P. prostratum* and *P. basiliense* var. *brandii*, although to identify most with any certainty requires isolating material and growing it under controlled conditions in laboratory culture (see John & Johnson, 1989). The following key has limitations when used for identifying field-collected material.

1 Intercalary cells of primary filaments less than twice as long as wide..2
1 Intercalary cells of primary filaments 2–5 times longer than wide 6
2 Thallus usually less than 100-celled, often loosely arranged throughout (except *P. basiliense* var. *brandii*)..3

2 Thallus of more than 100 cells when mature, with prostrate system compact or pseudoparenchymatous (at least in centre) 4
3 Thallus usually less than 60-celled; prostrate system sometimes secondarily erect and growing obliquely upwards (assurgent); primary filament usually less than 20-celled ....*P. basiliense*
3 Thallus usually more than 60-celled; prostrate system becoming secondarily erect; primary filament usually 20–30 cells long ....*P. basiliense* var. *brandii*
4 Prostrate system pseudoparenchymatous throughout..*P. prostratum*
4 Prostrate system initially compact or pseudoparenchymatous, later developing loose, radially arranged secondary branches towards margin 5
5 Prostrate system compact and extensive due to development of laterals on secondary branches, often star-shaped in outline ....*P. prostratum* var. *radiatum*
5 Prostrate system little developed, with few laterals on radially arranged secondary branches ....*P. prostratum* var. *pseudoprostratum*
6 Secondary branches bearing numerous 2–4-celled laterals, often opposite or subopposite ....*P. laxum* var. *laterale*
6 Secondary branches simple, occasionally bearing 1–3-celled laterals, usually arising singly...............7
7 Primary prostrate filament obscured on ageing; prostrate filaments occasionally becoming secondarily erect; erect system poor to richly developed..*P. laxum*
7 Primary prostrate filament always evident; prostrate filaments remaining attached; erect system absent or poorly developed ....*P. laxum* var. *deminutum*

Pseudendoclonium basiliense Vischer 1926

24280010
Prostrate system less than 100-celled (usually <60-celled), with a compact pseudoparenchymatous disc-like expansion having an irregular outline, with age the filaments towards the margin sometimes become secondarily erect and often grow obliquely upwards (assurgent); primary filament less than 20-celled, giving rise to profusely divided and few-celled (<10 cells long) secondary branches; erect system generally poorly developed.

var. *brandii* Vischer 1926
24280012 **Pl. 110E** (p. 443)
Prostrate system often more than 60-celled, never growing upwards (assurgent); primary filament usually 20–30 cells long.

Europe, North America; only known in the British Isles as isolates from natural and artificial surfaces (including polyethylene bags) collected in rivers.

The original incomplete and inadequate description of the variety has been expanded by Tupa (1974), who studied it in culture.

Pseudendoclonium laxum D.M.John *et* L.Johnson 1989

24280020 **CD**
Prostrate system often well-developed, consisting of loosely arranged filaments bearing at long intervals 1–3-celled side branches, with age occasionally becomes secondarily erect; erect system usually absent, if present then variously developed; cells 4–6 μm wide, (1.5–)2 to 5 times longer than wide; zoospores 2–8 per cell.

var. *deminutum* D.M.John *et* L.Johnson 1989
24280022
Primary filament conspicuous, 8–35-celled, cells (1.5–)2 to 4 times longer than wide; secondary branches less than 15-celled and side branches absent or less than 3-celled if present; erect system absent or poorly developed.

var. *laterale* D.M.John *et* L.Johnson 1989
24280023
Primary filament conspicuous, similar in length and dimensions to var. *deminutum* but scarcely visible in older thalli; secondary branches bearing many distinctive 2–4-celled laterals, frequently 2-celled; erect system sparse to well developed.

British Isles (England); isolated from rock and artificial surfaces (e.g. glass, polyethylene) collected from streams, rivers, lakes, spring-fed pools and ornamental ponds.

Pseudendoclonium prostratum Tupa 1974

24280030
Prostrate system initially single-layered, pseudoparenchymatous, disc-shaped or slightly elongated, with or without long, loose and radially arranged secondary branches developing from a compact centre; primary filament tending to be conspicuous, 10–15-celled, bearing 3–5-celled secondary branches, simple or with varying numbers of laterals, often narrowing towards apices; erect system absent or scarcely developed, only well-developed under optimal culture conditions.

var. *pseudoprostratum* D.M.John *et* L.Johnson 1989
24280032
Prostrate system pseudoparenchymatous and radially arranged, with secondary filaments developing laterally from the primary filaments; primary filament usually 20–30-celled, secondary filaments occasionally multiseriate.

var. *radiatum* D.M.John *et* L.Johnson 1989
24280033 **Pl. 110F** (p. 443) **CD**
Prostrate system pseudoparenchymatous, radially arranged secondary branches bearing short laterals which shorten progressively towards the apices to give a star-like outline to the thallus; primary filament 20–32-celled.

North America, British Isles; isolated in the British Isles from amongst crusts of various green algae on stones, pebbles, flints and artificial surfaces (e.g. concrete, polyethylene) collected from streams, rivers, small lakes and ponds; var. *radiatum* has been isolated only from the freshwater tidal stretch of the River Thames between the half-tide lock at Richmond and the railway bridge at Barnes.

REJECTED GENUS

Pseudochaete West *et* G.S.West 1903

Established to accommodate a minute epiphyte consisting of branched, prostrate, uniseriate filaments from which arise simple, tapering, 3–8-celled filaments terminating in 1–2 colourless hairs. The type species, *Pseudochaete gracilis* (Pl. 112C, p. 448), has a prostrate system of cells 5.7–7.7 μm wide (1.2–2.2 times longer than wide) and an erect system with cells 1.5–1.8 μm wide (8–18 times longer than wide). Zoosporogenesis and sexual reproduction are unknown. It has been decided to regard this little-known genus as merely a developmental stage of *Stigeoclonium*. Smith (1933) considered *Pseudochaete* to be a dwarf species of *Stigeoclonium* and Tiffany (1937b) went so far as to transfer it to *Stigeoclonium* as *S. gracile* (West *et* G.S.West) Tiffany.

Radiofilum Schmidle 1894

Filaments usually free-floating, short or long and flexuose, unbranched or occasionally branched (cf. *R. paradoxum*); cells spherical, subspherical or lenticular (bead-like) with wall of each cell divided into two equal halves and a ring-like transverse rim sometimes present at the join, arranged in a uniseriate row (very rarely multiseriate) and sometimes joined by a narrow mucilaginous bridge, surrounded by a thick and occasionally transversely striated mucilaginous sheath; chloroplast parietal, single (rarely more), cup-shaped or layered (laminate), with 1 or 2 pyrenoids; asexual reproduction by biflagellate zoospores, aplanospores, akinetes or fragmentation.

Probably cosmopolitan; planktonic usually in pools, ponds and lakes (often in soft, acid waters); single species associated with soil (see below).

The genus is traditionally placed in the Ulotrichales. Fott (1971) interpreted its cells as a row of autospores retained within a common sheath and for this reason Hoek et al. (1995) consider it a member of the Chlorococcales *sensu lato*.

1 Cells attached by long strands, sometimes branched and with 3 or 4 cells connected to form a mesh; soil alga ....*R. paradoxum* var. *reticulatum*
1 Cells attached apically in an unbranched series, rarely by broad strands; aquatic algae.....................2
2 Cells divided into two halves by a prominent transverse ring...............................*R. conjunctivum*
2 Cells without a transverse ring ....*R. flavescens*

Radiofilum conjunctivum Schmidle 1894

Synonym: *Radiofilum apiculatum* West *et* G.S.West

24300010 **Pl. 113K** (p. 450)

Filaments short and fragile; cells 6–8 μm wide, about two-thirds as long as wide, spherical to ovoid; walls divided in two distinct halves joined by a prominent transverse ring.

Probably cosmopolitan; seems to be very rare in the British Isles, known only in England where recorded from pools.

Radiofilum flavescens G.S.West 1899

24300020 **Pl. 113J** (p. 450)

Filaments very long and flexuose; cells (6.8–)7.5–10.5 (–15) μm wide, 5–8.5(–10) μm long, transversely ellipsoidal or subquadrate; walls not divided into two distinct halves.

Probably cosmopolitan; comparatively rare in the British Isles where usually in relatively small water bodies in boggy areas (e.g. Wicken Fen, Cambridgeshire).

Radiofilum paradoxum (Chodat *et* Topali) Printz 1965

Basionym: *Interfilum paradoxum* Chodat *et* Topali

Type variety not known from the British Isles.

var. *reticulatum* (F.E.Fritsch *et* R.P.John) D.M.John, *nov. comb.*

Synonym: *Interfilum paradoxum* var. *reticulatum* F.E. Fritsch *et* R.P. John (1942. Annals of Botany NS 6: p. 282, fig. 4 a–l).

24300032 **Pl. 113L** (p. 450)

Filaments short, consisting of single cells, pairs or chains of 4 cells connected by narrow mucilaginous strands, occasionally 3 or 4 cells connected by strands to form a mesh-like structure (several together have a net-like appearance); cells 4–6 μm wide, up to 12 μm long, spherical or oblong with one apex broader, often the broader apices of adjacent cells facing each other; walls not distinctly divided into two halves, sometimes 2 caps of wall material terminate an enlarging cell, enclosed within an indistinct mucilaginous sheath.

British Isles (England); growing in soils collected from the Brecklands of East Anglia and in soil cultures.

Printz (1964) failed to mention the variety when making the new combination. In describing this variety, Fritsch & John (1942) doubted whether it was truly filamentous but thought it might be a colonial chlorococcalean alga.

Raphidonema Lagerheim 1892

Filaments uniseriate, unbranched, 2–32-celled, free-living; cells cylindrical, thin-walled, with apical cells acuminate or gradually tapering to a point; chloroplast parietal, single, layered (laminate), girdle-shaped, with or without pyrenoids; asexual reproduction by biflagellate zoospores; sexual reproduction oogamous: male gametes biflagellate, a single one produced in each cell, vegetative cells become female gametes.

Probably cosmopolitan; aquatic or forming greenish patches on snow and ice when present in quantity.

Current position uncertain since no longer considered a member of the Order Klebsormidiales. Traditionally placed in the Order Ulotrichales.

Raphidonema nivale Lagerheim 1892

25040010 **Pl. 114I** (p. 454)
Filaments usually 4–8-celled; cells 2.5–6 μm wide, 94–123 μm long, 2 to 20 or more times longer than wide, usually tapering to acute apices.

Probably cosmopolitan; only rarely reported in the British Isles on winter snow (e.g. Settle in Yorkshire).

Schizomeris Kützing 1843

Macroscopic, consisting of unbranched filaments, basally uniseriate and becoming multiseriate above to form a solid cylinder of parenchyma that has ring-like constrictions at intervals; cells long and cylindrical towards base, quadrate to angular and thick-walled above where multiseriate; chloroplast parietal, encircling about two-thirds of cell, only cells in lower parts possessing pyrenoids (usually several per cell); asexual reproduction by quadriflagellate zoospores, formed in multiseriate portions; sexual reproduction unknown

Probably cosmopolitan, usually grows in standing bodies of water as dark green, coarse clumps of filaments.

Schizomeris leibleinii Kützing 1843

24320010 **Pl. 116D** (p. 464)
Cylindrical with filaments frequently constricted at intervals, 2–24 μm wide in uniseriate basal portion, increasing above to about 54 μm wide where multiseriate.

Widely distributed in the tropics and subtropics; recorded in the British Isles attached to various surfaces in the Reddish Canal (Manchester), on stones sprayed by water from a waterfall in Stone Gill, Dodd Fell, Yorkshire and known from various sites in central Scotland.

It was collected in the autumn of 1954 by E.M.F. Swale from the Reddish Canal, which at the time had its water temperature raised by heated effluent entering from nearby cotton mills. The closure of the mills in 1962 (Allsopp, 1963) and the infilling of much of the canal probably accounts for it not having been recorded from it again. The view that the genus represents a growth form of *Stigeoclonium* has not found general acceptance.

Stichococcus Nägeli 1849

Filaments uniseriate, unbranched, readily dissociating into solitary cells or short fragments, usually without a mucilaginous sheath; cells cylindrical and often elongate, sometimes short and spherical to slightly oval, thin-walled; chloroplast parietal, single and layered (laminate), not completely covering cell circumference, pyrenoids absent; reproduction by fragmentation and cell division.

Probably cosmopolitan; grows over a very wide ecological amplitude ranging from soil-free surfaces to freshwater and marine habitats.

Current position uncertain since no longer considered to belong to the Order Klebsormidiales. Traditionally placed in the Ulotrichales or the Chaetophorales.

1 Cells always solitary and no more than 1.5 μm wide ....*S. minutissimus*
1 Cells commonly forming filaments, if solitary or forming few-celled filaments then more than 1.5 μm wide 2
2 Filaments sometimes few-celled, readily fragmenting into solitary cells. ....*S. bacillaris*
2 Filaments usually many-celled, not readily fragmenting...3
3 Filaments straight or irregularly twisted; cells usually more than 10 times longer than wide ....*S. pelagicus*
3 Filaments form a lax spiral; cells always less than 10 times longer than wide....................*S. contortus*

Stichococcus bacillaris Nägeli 1849

24340010 **Pl. 113F** (p. 450) **CD**
Filaments short and few-celled, readily fragmenting into solitary cells; cells 2–3.8 μm wide, 2 to 6 times longer than wide (3.5–12.8 μm long), cylindrical, rounded at apices.

Probably cosmopolitan; one of the most common and widely distributed algae in the British Isles, sometimes sufficiently abundant to form bright green patches on soil, stone walls, roofing tiles, tree trunks, glass, polythene, plastic and other damp surfaces; sometimes free-floating in small water bodies such as bird baths.

Stichococcus contortus (Chodat) Hindák 1996

Basionym: *Gloeotila contorta* Chodat
Synonyms: *Gloeotila spiralis* Chodat, *Geminella spiralis* (Chodat) G.S.Smith, *Ulothrix spiroides* G.S.West, *Gloeotila spiroides* (G.S.West) Printz, *Stichococcus spiroides* (G.S.West) Hindák

24340020 **Pl. 113I** (p. 450)
Filaments contiguous, often forming a lax spiral, sometimes within a mucilaginous envelope; cells 1.5–2(–3) μm wide, 4.5 to 8.5 times longer than wide.

Europe, North America; usually planktonic in rivers (e.g. River Dee, near Chester) and lakes; more common during the warmer summer months.

It has been decided to follow Ramanthan (1964) and consider *Ulothrix spiroides* and *Geminella spiralis* to be conspecific, the latter now in the synonymy of *Stichococcus contortus*. Others (e.g. Printz, 1964; Hindák, 1996) still regard the two species as distinct and separate them only on the presence or absence of a mucilaginous sheath.

Stichococcus minutissimus Skuja 1955

24340030 **Pl. 113G** (p. 450)
Cells always solitary, 1–1.5 μm wide, (1.5–)2–4(–6) μm long, cylindrical to ellipsoidal, thin-walled, with rounded apices.

Europe; according to Lund (1960a), it is one of the commonest 'minute plankton' (μ-plankton) of the more productive lakes of the English Lake District (e.g. Blelham Tarn, Windermere) although less abundant in spring than species of *Chlorella* and *Coccomyxa*.

Lund (1960a, p. 94) mentioned a second and somewhat larger *Stichococcus* in the English Lake District but was unable to identify it positively.

Stichococcus pelagicus (Nygaard) Hindák 1996

Basioym: *Ulothrix pelagica* Nygaard
Synonym: *Gloeotila pelagica* (Nygaard) Skuja
24340040 **Pl. 114E** (p. 454)

Filaments of straight or irregularly twisted cells; cells 1–1.5(–2) µm wide, 6–25(–34) µm long, 8 to 20 times longer than wide, cylindrical, thin-walled, with acute apices.

Probably cosmopolitan; widely distributed in the plankton of ponds and lakes in the British Isles, often known to survive throughout the year although numbers are low during winter.

Doubt attaches to the worth of recognizing fo. *spiralis* Skuja (synonym: *Ulothrix pelagica* fo. *spiralis* (Skuja) Ramanathan) which is slightly narrower (about 1 µm wide) and somewhat shorter celled (7–10 µm long) than the type. The form very closely relates to *S. contortus* (see above) and may be part of a continuum of morphological variation linking the two species.

DOUBTFUL SPECIES

Stichococcus variabilis West *et* G.S.West 1896
24340060 **Pl. 113H** (p. 450)

Doubt attaches to this species, since it is described as having pyrenoids and its cell apices are not rounded. According to Ramanathan (1964), this species should be more correctly considered a *Klebsormidium.* Its cells are very irregularly shaped, 2 to 4(–5) times longer than wide, single or united to form 2- or 3-celled filaments 3–6 µm wide. Only known from the British Isles, where it formed a thin green layer over wet stones close to waterfalls (see West & Fritsch, 1927).

Stigeoclonium Kützing 1843

Synonyms: *Caespitella* Vischer, *Endoclonium* Szymanski, *Myxonema* Fries, *Pseudochaete* West *et* G.S.West

Filaments uniseriate, often somewhat mucilaginous, usually differentiated into a prostrate and an erect system; erect system of alternately, oppositely or dichotomously branched filaments, occasionally whorled or irregularly arranged; apices acute or narrowly obtuse, each frequently bearing a colourless, multicellular hair; prostrate system of creeping or rhizoidal filaments, occasionally forming a pseudoparenchymatous, disc-like expansion; cells cylindrical or swollen, thin- or thick-walled; chloroplast parietal, with one to several pyrenoids; asexual reproduction by quadriflagellate micro- and macrozoospores; sexual reproduction by biflagellate or quadriflagellate isogametes.

Probably cosmopolitan; a very common and widely distributed freshwater genus in the British Isles, growing on a range of surfaces often as delicate tufts or mats, occasionally free-floating in ditches and ponds, sometimes abundant in rivers with high levels of organic material (e.g. immediately below sewage treatment works) and often also high levels of heavy metals. Fast-flowing streams and rivers which combine high levels of organic matter and heavy metals are usually dominated by *Stigeoclonium* for many months in summer.

Considerable doubt attaches to the validity of many species, since this genus is extremely polymorphic. Several culture-based investigations have concluded that characters traditionally used to define species are invalid resulting in many becoming reduced to synonymy. Simons et al. (1986) have adopted the most extreme view since they recognize just three species with the remainder considered to be environmentally-induced growth forms. Most culture-based studies have concluded that the prostrate system provides more reliable characters for recognizing species than the erect system. Unfortunately characters associated with the prostrate system are often not easy to observe since it is rarely possible to sample the system intact. Most observations on the prostrate system have been made of algae grown in laboratory culture or on surfaces (e.g. glass slides or plastic Petri dishes) placed in the field and brought back for microscopic examination. The multicellular hairs develop under conditions of nutrient limitation, but are much more characteristic and well developed under phosphorus limitation than under nitrogen limitation. Assays have shown that almost all field populations with hairs are phosphorus limited. Addition of phosphate to phosphorus-limited plants leads to loss of hairs and zoospore formation in the cells just beneath the region where the hair had been. Older plants which have been under more than one cycle of phosphate stress and enrichment may therefore have a highly complex morphology (Gibson & Whitton, 1987a, b; Whitton 1988).

Despite the unreliability of erect system characters (e.g. branching pattern, presence or absence of short branch-bearing cells), it has been decided to use them here since many species remain defined almost wholly on characters associated with this system. All identifications made using the key must be regarded as provisional. Forms and varieties are not included since these are not considered worthy of recognition in such a polymorphic genus.

1 Terminal cell usually forming a long, colourless tapering unicellular hair .... *S. protensum*
1 Terminal cell blunt, pointed or transformed into a usually multicellular hair............................2
2 Prostrate system a dense, pseudoparenchymatous disc-like expansion or more open and star-shaped in outline..*S. farctum*
2 Prostrate system of another form 3
3 Prostrate system consists of a holdfast cell and/or rhizoids growing from base of an erect filament(s) 4
3 Prostrate system of loosely arranged filaments, usually attached by rhizoids 6
4 Prostrate system a simple holdfast cell or of one or a few short rhizoids; erect branches with similar cells throughout, usually alternately branched ....*S. helveticum*
4 Prostrate system of profusely developed rhizoids; erect filaments with or without cellular differentiation..5
5 Erect system consists of very long and very short cells in the main axes; branches opposite or whorled, arising from short cells .....*S. amoenum*
5 Erect system without any cellular differentiation; branches dichotomously or alternately divided ....*S. longipilum*
6 Prostrate system inconspicuous, consists of relatively few branches attached by rhizoids 7
6 Prostrate system extensive, consists of many irregularly divided filaments attached by rhizoids..8
7 Erect branches of thick-walled cells, differentiated into very long cylindrical cells and very short inflated cells ....*S. lubricum*, in part
7 Erect branches of thin-walled cells, with no cellular differentiation ....*S. subsecundum*
8 Erect system differentiated into very long and very short cells ....*S. lubricum*, in part

8 Erect system not so differentiated 9
9 Erect branches often arising from almost every cell of prostrate system, frequently in a unilateral series ....*S. nanum*
9 Erect branches arising irregularly from prostrate system, not arranged as above ....*S. tenue*

Stigeoclonium amoenum Kützing 1845

24350020 **Pl. 111D** (p. 446)
Filaments pale to bright green, up to 40 cm long; prostrate system consisting of an attachment cell or several rhizoids arising from erect filaments; erect system well developed, of very long and very short cells, the latter single or in rows of 4 or more cells, longest cells often towards branch base; branches arising from short cells singly or more commonly opposite or in a whorl of 2–4, terminating in a blunt or acute cell or hair; cells of 2 types: cylindrical and occasionally somewhat inflated, 12–39 μm wide and 3 to 10(–15) times longer than wide, or short and angular or slightly rectangular, 1 to 2 times longer than wide.

Probably cosmopolitan; widespread in the British Isles growing on stones and pebbles in flowing waters as well as in the marginal shallows of ponds and lakes.

See note under *S. helveticum* concerning the identity of var. *minus*.

Stigeoclonium farctum Berthold 1878

Synonym: *Stigeoclonium variabile* (Nägeli) Islam
24350050 **Pl. 111F** (p. 446) **CD**
Prostrate system pseudoparenchymatous and disc-like, or more rarely star-shaped with main and lateral filaments very evident; cells 5–8 μm wide, 1 to 2 times longer than wide; chloroplast and pyrenoid at same level in all radiating filaments forming the pseudoparenchymatous system thus in surface view system often appearing concentrically zoned due to alignment of chloroplasts; erect system absent or poorly developed, if present consisting of short, simple or sparingly divided filaments.

Probably cosmopolitan; widely distributed in the British Isles growing in rivers on aquatic macrophytes and other surfaces, including artificial ones such as glass or plastic.

Little purpose is served by recognizing varieties. A variety known as *rivulare*, described by Butcher (1932b), is recognized by having several interconnected prostrate systems each of which consist of 'discs of almost circular outline, up to 2 mm. in diam.'. This variety has only been recorded by Butcher from sites where there is considerable water motion and is 'generally distributed in rivers such as the Tees, Lark, Itchen and Hull, being usually most abundant in calcareous waters'. Two other varieties have been recorded from the British Isles: var. *simplex*, described by Fritsch (1903a) and based upon its particularly well-developed prostrate system and simple erect filaments which sometimes narrow near their point of origin, and var. *anglicum*, described by Butcher (1940) from the River Hull in Humberside.

Stigeoclonium helveticum Vischer 1933

Synonym: *Stigeoclonium helveticum* var. *minus* Vischer
24350070 **Pl. 111H** (p. 446)
Filaments forming low turf- or cushion-like growths, up to 5 cm long; prostrate system of a single attachment cell or 1 to several short rhizoids; erect system usually of alternately branched filaments, branching at long intervals, rarely oppositely branched or with several branches arising from adjacent cells, terminating in a blunt cell or hair; cells 6–12.5 μm wide, 1 to 8 times longer than wide, cylindrical and thin-walled.

Probably cosmopolitan; in the British Isles forming turf- or cushion-like growths on other algae.

Islam (1963) reduces var. *minus* to synonymy under this species, whereas Abbas & Godward (1964) believe it to be identical to *S. amoenum*.

Stigeoclonium longipilum Kützing 1849

Synonym: *Stigeoclonium fastigiatum* (Ralfs) Kützing
24350080 **Pl. 111G** (p. 446)
Filaments bright green, forming slimy tufts about 1 cm or more in height; prostrate system of densely branched rhizoids; erect system of well-developed filaments, often with main axes simple below and dichotomously or alternately branched above, rarely oppositely branched, frequently bushy or tufted above, terminating in a long hair; cells similar in size and shape throughout, (7–)10–15(–19) μm wide, 0.5 to 3(–5) times as long as wide, subcylindrical below and subspherical to barrel-shaped above.

Europe, North America; widely distributed in the British Isles, where most commonly reported growing on hard surfaces in streams (usually as *S. fastigiatum*).

Islam (1963) placed *Stigeoclonium fastigiatum* within the synonymy of *S. longipilum* and at the same time emended its description.

Stigeoclonium lubricum (Dillwyn) Kützing 1853

Basionym: *Conferva lubrica* Dillwyn
24350090 **Pl. 111K** (p. 446)
Filaments dark green, smooth, slippery and forming bushy tufts, 2–5 cm long (occasionally reaching 30 cm); prostrate system evident or inconspicuous, consisting of irregularly branched filaments attached by rhizoids; erect filaments profusely branched, with main axes alternately divided, secondary branches mostly opposite or in whorls and arising from short, barrel-shaped or spherical cells; secondary branches sometimes in loose clusters and terminating in a blunt or acute apical cell or hair; cells of main axes 12–20 μm wide, 12–30 μm long, 0.5 to 2(–4) times longer than wide, often a series of 2–8 barrel-shaped and generally thick-walled cells alternating with more cylindrical cells; cells of secondary branches 8–10 μm wide, about equal in length and breadth, thin-walled.

Probably cosmopolitan; on various surfaces including stones and submerged wood in lakes but more frequent in flowing water, often considered to be less pollution-tolerant than *Stigeoclonium tenue*. Originally described by Dillwyn (1809, as *Conferva lubrica*) from Lound, near Great Yarmouth and from Sketty Burrows, near Swansea with collections dating back to 1802.

The prostrate system is very similar to that of *Stigeoclonium tenue*, possibly it is no more than a form of this species. Undoubtedly many records of *S. lubricum* probably merely represent a very slimy form of *S. tenue*.

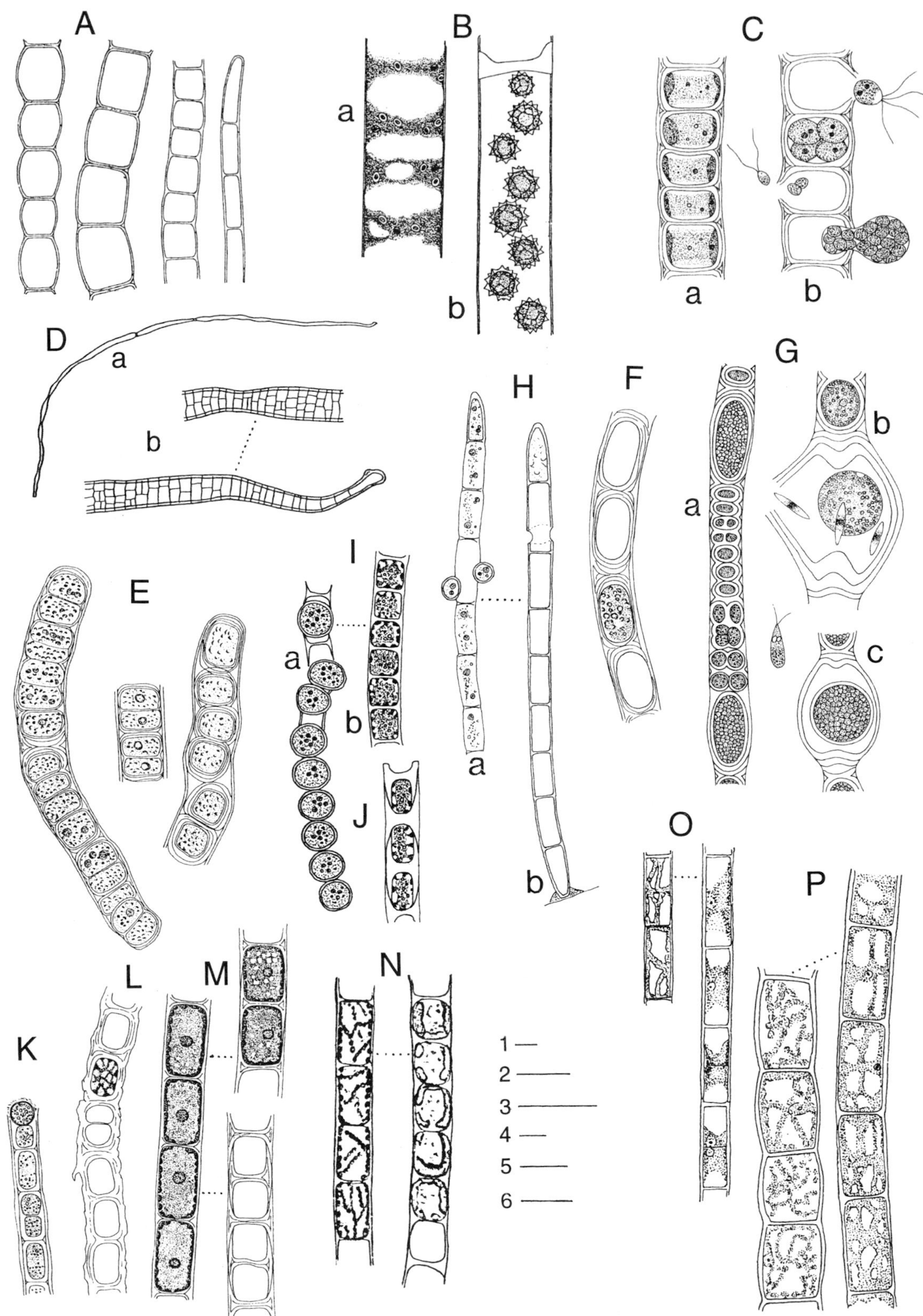

Plate 116 A–P

A. *Chaetomorpha linum* (p. 468); **B**. *Sphaeroplea soleirolii* (p. 475): a. details of vegetative cell, b. cell containing zygospores with highly ornamented walls; **C**. *Ulothrix zonata* (p. 467): a. vegetative filament, b. fertile filament showing release of biflagellate gametes and larger quadriflagellate zoospores; **D**. *Schizomeris leibleinii* (p. 461): a. simple filament, b. details of portion of filament; **E**. *Cylindrocapsa conferta* (p. 437); **F**. *Cylindrocapsa geminella* var. *minor* (p. 438); **G**. *Cylindrocapsa involuta* (p. 438): a. antheridia in different stages of development, b. details of a thick-walled oogonia showing oospore and 3 antherozoids, c. a single antherozoid and an oospore; **H**. *Uronema confervicolum* (p. 467): a. filaments showing release of 2 zoospores, b. penultimate cell of filament showing pores in wall through which the swarmers had escaped; **I**. *Microspora abbreviata* (p. 452): a. chain of akinetes, b. details of contents of a vegetative filament; **J**. *Microspora pachyderma* (p. 453); **K**. *Microspora amoena* (p. 452); **L**. *Microspora irrregularis* (p. 453); **M**. *Microspora crassior* (p. 453); **N**. *Microspora tumidula* (p. 453); **O**. *Microspora stagnorum* (p. 453); **P**. *Microspora floccosa* (p. 453).

Scale bars: 1, 100 μm; 2, 3, 50 μm; 4, 5, 10 μm; 6, 1mm. 1 for A; 2 for K, L; 3 for B, C, E, J, M; 4 for Da, F, G; 5 for I, H, N–P; 6 for Db.

Stigeoclonium nanum (Dillwyn) Kützing 1849

Basionym: *Conferva nanum* Dillwyn

24350100 **Pl. 111J** (p. 446)

Forms small tufts; prostrate system consisting of separate creeping filaments, with erect branches arising from almost every cell, or as a loose and pseudoparenchymatous disc-like expansion of angular or subglobose cells; erect system often profusely branched, mostly alternate or unilateral, only rarely opposite, branch-bearing cells not distinctive, primary and secondary branches short, apical cells blunt, acute or terminating in a hair; cells similar throughout branch systems, 4–8 µm wide, 10–18 µm long, 1 to 2(–3) times longer than wide, inflated and barrel-shaped.

Probably cosmopolitan; in the British Isles on various surfaces including rocks and wood in flowing water, although also reported as slimy mats on dead leaves in bogs.

Often considered to be a growth stage of some other species. Abbas & Godward (1964) consider it to be a synonym of *S. tenue* which it closely resembles when grown in culture and they both have 12 chromosomes. Until further information becomes available on its relationship to other species, it has been decided to continue to recognize *S. nanum*.

Stigeoclonium protensum (Dillwyn) Kützing 1845

Basionym: *Conferva protensum* Dillwyn

24350110 **Pl. 111I** (p. 446)

Forms pale green tufts; prostrate system unknown; erect system of long, remote and scattered branches, mostly alternate, only rarely opposite, each terminating in a long unicellular hair or acute and colourless apical cell; cells of main axes mostly barrel-shaped, 10–16(–23) µm wide, 1 to 3 times longer than wide, those of the secondary axes cylindrical and 7 times longer than wide.

Probably cosmopolitan; rarely reported in the British Isles where on stones, sticks, grasses and aquatic macrophytes in flowing water.

Some doubt inevitably attaches to this species since all records from the British Isles date back to the nineteenth century. Several authors (e.g. Sarma, 1986) have suggested that *S. protensum* is probably a stage in the life history of another species.

Stigeoclonium subsecundum (Kützing) Kützing 1843

Basionym: *Conferva subsecundum* Kützing

Synonym: *Stigeoclonium aestivale* (Hazen) Collins

24350010

Filaments a delicate green to yellowish-green, often enclosed within mucilage; prostrate system variously developed, filaments branched and with evident main axes, attached by long and slightly undulating rhizoids; erect system sparingly dichotomously or alternately branched (never opposite), often with several unilateral branches arising from successive cells; cells of erect system either all similar irrespective of whether or not bearing a side branch, or branch-bearing cells shorter and inflated; secondary branches frequently short, narrowing terminally and sometimes with curved apices; cells 7–20 µm wide, 2 to 12 times longer than wide, cylindrical or somewhat inflated or barrel-shaped, thin-walled.

Probably cosmopolitan; in the British Isles on various submerged surfaces (e.g. aquatic vegetation, twigs, stones), especially common in the marginal shallows of non-flowing waters including lakes (e.g. Lough Mawmeen, Galway, Ireland) and reservoirs.

Various authors have placed their own interpretation on species limits in this genus, often using characters that were inadequately or not described and illustrated when it was established. Francke & Simons (1985) recognized the confusion existing in the application of names when considering *S. subsecundum* and *S. aestivale* to be conspecific. They use the later name despite *S. subsecundum* having priority.

Stigeoclonium tenue Kützing 1853

Synonym: *Stigeoclonium uniforme* (C.Agardh) Rabenhorst

24350120 **Pl. 111E** (p. 446) **CD**

Forms delicate tufts or mats, often green to yellow-green, frequently reaching 5–10 cm long; prostrate and erect systems often very conspicuous; prostrate system extensive and of irregularly branched main axes from which laterals arise in an irregular manner, sometimes filaments bear slender and often cork-screw-like rhizoids; erect system of many secondary branches arising oppositely or alternately along main axes, sometimes borne on small angular cells, with secondary branches rather scattered, alternately divided and sometimes clustered; cells 6–15(–18) µm wide, 2 to 5(–6) times longer than wide, cylindrical and thin-walled, branch-bearing cells less than twice as long as wide and apices acute or hair-like.

Probably cosmopolitan; the most widely distributed and commonly recorded *Stigeoclonium* species in the British Isles where growing on rocks and other surfaces in still water (e.g. water troughs, lakes) although most frequent in flowing water; most abundant in nutrient-rich or organically polluted waters and tolerant of high levels of heavy metals.

See remarks under *Stigeoclonium lubricum*. In many accounts, the species is not characterized by the presence of small branch-bearing cells (see Abbas & Godward, 1964). While the number and extent of hair formation is influenced by the extent of phosphorus limitation (see account of the genus), elevated zinc in the absence of high levels of organic matter reduces the development of the erect system and sometimes only the prostrate system is found where the zinc concentration is exceptionally high (Whitton & Harding, 1976). This brings into question the legitimacy of many *Stigeoclonium* species if taxonomically important characters are known to be influenced by the chemical composition of the water.

DOUBTFUL RECORDS

Stigeoclonium elongatum (Hassall) Kützing 1849

24350040

Stigeoclonium attenuatum is considered to be a synonym of this species. Doubt attaches to material collected from 'stones at loch margins' recorded by Hibberd (1978, as *S. attenuatum* (Hazen) Collins) from the island of Mull. No information accompanies its inclusion in a list of algae from South Wales produced by Benson-Evans & Antoine (1996, as *S. attenuatum*). A new variety (var. *anglicum*) of the species *S. falklandicum* (originally described from the Falkland Islands in 1858 by Kützing) was described from the River Tees by Butcher (1932b). According to Islam (1963), this is a 'young stage and may be placed in any species'. A collection examined by him from the Falklands was found by Islam to correspond closely to *S. flagelliferum*, a species now considered a synonym of *S. elongatum*.

Stigeoclonium pachydermum Prescott 1944

The primary source of the record in Benson-Evans & Antoine (1996) of this species in South Wales has not been traced; it is mentioned as 'Found growing rarely in lakes'. In the absence of further information this record must be considered as doubtful.

Stigeoclonium thermale A.Braun 1849

24350130

Considered very doubtful since based on a single record from Land's End, Cornwall dating back to the nineteenth century (Marquand, 1884). A species normally associated with warmwater springs and known to be closely related to *S. subsecundum* (see Islam, 1963, as *S. aestivale*).

Thamniochaete F.Gay 1893

Filaments uniseriate, prostrate or erect, with branches few and short, terminal cell and some intercalary cells bear a short spine-like projection or one or more basally swollen setae; cells irregularly cylindrical, barrel-shaped or inflated; chloroplast parietal, with a single pyrenoid; asexual reproduction by biflagellate zoospores (known only in one species); akinetes known.

Europe, North America; grows on larger algae (including blue-green algae) and other surfaces, usually in flowing water.

Thamniochaete aculeata West *et* G.S.West 1903

24360010 **Pl. 112I** (p. 448)

Filaments 3–6-celled, attached by a modified basal cell; cells 5.3–13 µm wide, 1 to 5 times longer than wide, with filaments bearing subterminally a short and sharply pointed seta on a swollen apical cell.

British Isles; only recorded as an epiphyte on the blue-green alga *Gloeotrichia natans* collected in the Outer Hebrides (near Ballallan), Scotland, and from Connemara and Galway, Ireland.

Doubt attaches to the generic placement of this species due to the different form of its setae compared with those of the type species (*T. huberi* Gay). One species of *Endoclonium*, *E. rivulare*, is often placed in *Thamniochaete*. The type of the genus *Endoclonium* Szymanski (*E. chroolepiforme*) is considered to be a form of *Stigeoclonium*.

REJECTED RECORD

Trichodiscus elegans E.G.Welsford 1912

24370010

Mentioned in West & Fritsch (1927, p. 193), but cannot be regarded as an indigenous member of the British algal flora since only known growing on a glass jar containing an imported non-native water fern *Azolla* from the USA.

Ulothrix Kützing 1833

Synonym: *Personiella* F.E.Fritsch *et* M.F.Rich

Filaments uniseriate, unbranched, attached by a basal cell, a rhizoid-like basal cell or rhizoids arising from cells above the basal cell; cells cylindrical, sometimes barrel-shaped, often longer than wide; walls thin or thick, occasionally stratified or with a roughened surface, if H-shaped wall sections present then these surround the transverse walls; chloroplast parietal, single, girdle-shaped, usually lobed, partially or fully encircling cell circumference (normally over 80% coverage), with one or several pyrenoids; asexual reproduction by quadriflagellate zoospores; sexual reproduction by biflagellate isogametes; aplanospores and thick-walled akinetes produced; unicellular stage in life history known, the *Codiolum*-stage.

Probably cosmopolitan; widely distributed in flowing and still water, sometimes associated with damp soil; freshwater and marine species.

Three marine species are reported from the British Isles (see Burrows, 1991) where they often form conspicuous mats in salt marshes or grow where there is seepage of freshwater near the high tide mark. It is possible that brackish-water species such as *Ulothrix flacca* (Dillwyn) Thuret (24380020) (cells 25–50 µm wide) and *Ulothrix implexa* (Kützing) Kützing (24380030) (3.5–26 µm wide) may be found on rare occasions growing in freshwater. A variety of *U. pseudoflacca* Wille (24380050), var. *salina* V.J.Chapman (24380052), has been reported growing in a Cheshire brine pit.

1 Filaments >15 µm wide 2
1 Filaments <15 µm wide 4
2 Cells 1 to 2 times longer than wide; chloroplast not completely encircling cell circumference .... *U. aequalis*, in part
2 Cells usually shorter than wide; chloroplast forming a complete band, often filling entire cell 3
3 Cells 15–70 µm wide, 0.3 to 1 times longer than wide.. *U. zonata*
3 Cells 15–17(–22) µm wide, 0.25 to 0.5 times longer than wide .... *U. tenuissima*
4 Cells not constricted and cylindrical .... *U. tenerrima*
4 Cells constricted at transverse walls, sometimes slightly inflated..5
5 Cells usually shorter than wide..... *U. moniliformis*
5 Cells 1–2 times longer than wide .... *U. aequalis*, in part

Ulothrix aequalis Kützing 1848

24380010 **Pl. 115N** (p. 456)

Filaments long, usually attached; cells 12–22 µm wide, 26–30 µm long, 1 to 2 times longer than wide, cylindrical; walls slightly thickened, sometimes stratified; chloroplast extends more than halfway around cell circumference, with 1 to several pyrenoids.

var. *cataeniformis* (Kützing) Rabenhorst

Basionym: *Hormiscia aequalis* (Kützing) Rabenhorst var. *cataeniformis* Kützing

24380012

Differs from the type by having a constriction at each cross wall.

Probably cosmopolitan; widely reported in the British Isles from flowing and still water (e.g. several Scottish lochs), with the variety *cataeniformis* known from the English Lake District and Cornwall.

Ulothrix moniliformis Kützing 1849

Synonym: *Hormiscia moniliforme* (Kützing) Rabenhorst
24380040
Filaments often bead-like due to constrictions at cross walls; cells 9–14 μm wide, as long as or shorter than wide; walls thick and stratified; chloroplast confined to one side of cell, with 1 or 2 pyrenoids.

Probably cosmopolitan; widespread in the British Isles where mostly unattached and associated with other algae in small ponds and pools.

Ulothrix tenerrima Kützing 1843

Synonyms: *Ulothrix variabilis* Kützing, *U. subtilis* Kützing var. *tenerrima* (Kützing) Kirchner, *U. subtilis* Kützing var. *variabilis* Kirchner, *Hormiscia subtilis* (Kützing) De Toni var. *variabilis* (Kützing) Kirchner
24380090 **Pl. 115L** (p. 456)
Filaments usually forming bright green, slimy masses; cells (5–)8–10 μm wide, 0.5 to 1.5(–3) times longer than wide, thin-walled; chloroplast small, plate-like or covering just over half of cell circumference, with 1 pyrenoid (rarely to 4).

Probably cosmopolitan; one of the most common species of *Ulothrix* in the British Isles, according to West & Fritsch (1927, as *Ulothrix variabilis*) it is (p. 158) 'generally distributed in the stagnant waters of ponds, ditches, troughs, etc as well as in cultivated soils', also known in freshwater seepage.

Ulothrix tenuissima Kützing 1833

Synonym: *Ulothrix tenuis* Kützing
24380100 **Pl. 115M** (p. 456)
Filaments generally forming dark green masses; cells 15–17(–22) μm wide, 0.25 to 0.5 times longer than wide, sometimes longer in young filaments, thin-walled; chloroplast band-like and filling cell, with 2 or more pyrenoids.

Probably cosmopolitan; reported in the British Isles as common in streams and rivers where its dark green tufts sometimes reach several cm long, occasionally as a surface mat on soil.

Like *Ulothrix zonata*, it is most common in flowing water and may be difficult to distinguish from that species since there is considerable overlap in size.

Ulothrix zonata (Weber *et* Mohr) Kützing 1843

Basionym: *Conferva zonata* Weber *et* Mohr
Synonym: *Hormiscia zonata* (Weber *et* Mohr) Areschoug
24380120 **Pl. 116C** (p. 464) **CD**
Filaments dark green to yellowish-green, very variable in form and frequently detaching to become free-floating; cells 15–70 μm wide, 21–77 μm long, one-third as long as wide to about equal in length and width (very rarely 2 to 3 times longer than wide), cylindrical or slightly swollen with constrictions at cross walls; walls only thickened and stratified in ageing filaments; chloroplast a wide or narrow band, completely encircling median portion of cell, with 1 to several pyrenoids.

Probably cosmopolitan; the most widely distributed species in the British Isles where frequent in well-aerated and swiftly-flowing streams and rivers, particularly common in the spring (often prior to tree canopy development) and autumn.

A useful character for distinguishing this species is its readiness to produce prodigious numbers of zoospores shortly after collection.

DOUBTFUL RECORDS

Ulothrix oscillarina Kützing 1845
Doubtful as only recorded on a single occasion in the British Isles (Leicester, England; Rich, 1925), without a description or an accompanying illustration.

Ulothrix radicans Kützing 1845
24380060
A little-known and doubtful species with most records for the British Isles dating back to the 19th century. For example from North Wales (West, 1890) and 'Kenmare Street, Castletown' in the west of Ireland (West, 1892a).

Uronema Lagerheim 1887

Filaments uniseriate and unbranched, often having a slightly curved and frequently pointed apical cell, basal cell slightly tapering, attached by a mucilaginous disc; chloroplast parietal, single, band-like, completely or incompletely encircling cell circumference, with 1–4 pyrenoids; asexual reproduction by quadriflagellate zoospores, 1or 2 per sporangium, released by gelatinization of wall; sexual reproduction unknown.

Probably cosmopolitan; often on aquatic vegetation in nutrient-rich water.

Taxonomic position of this little-known genus remains uncertain. Ultrastructural features associated with cell division and the basal body of flagellated stages place it in the Order Chaetophorales *sensu stricto* (see Hoek et al., 1995).

1 Cells less than 4 times longer than wide; chloroplast normally filling cell .... *U. confervicolum*
1 Cells more than 4 times longer than wide; chloroplast not extending full length of cell .... *U. elongatum*

Uronema confervicolum Lagerheim 1887

24390010 **Pl. 116H** (p. 464)
Filaments straight or curved, usually less than 1 mm long; cells 4–6 μm wide, 10–18 μm long, 2 to 3 times longer than wide; chloroplast filling length of cell.

Probably cosmopolitan; infrequently reported from the British Isles where on larger algae and the submerged stems and leaves of vascular plants.

There is considerable doubt whether this and the following species should continue to be recognized. West & Fritsch (1927) considered both to be rare in the British Isles and speculated that they might merely be stages in the development of a *Ulothrix* or of some other filamentous alga as Hodgetts had suggested earlier (Hodgetts, 1918a).

Uronema elongatum Hodgetts 1918

24390020 **Pl. 115O** (p. 456)
Filaments usually straight, up to 5 mm long; cells (4.4–)5–7.5(–9.6) μm wide, (22–)28–56(–80) μm long,

(4–)5–9(–13) times longer than wide; chloroplasts not extending beyond two-thirds of cell length.

Probably cosmopolitan; in the British Isles on various filamentous green algae (e.g. *Oedogonium*, *Spirogyra*, *Mougeotia*) and the leaves and roots of vascular plants (e.g. *Myriophyllum*, *Lemna*), usually in ponds but probably in other aquatic habitats.

See remarks under *U. confervicolum*.

ORDER CLADOPHORALES (=SIPHONOCLADALES)

BY DAVID M. JOHN

Filamentous, pseudoparenchymatous blades, nets or spherical vesicles. Vegetative cells always multinucleate (siphonous organization).

Chaetomorpha Kützing 1845

Filaments uniseriate, unbranched, growth intercalary, free-living or attached basally by rhizoids; cells cylindrical or barrel-shaped, multinucleate, with walls thick and stratified; chloroplast parietal, net-like in appearance, with numerous pyrenoids; asexual reproduction by bi- or quadriflagellate zoospores.

Probably cosmopolitan; very rare in freshwater, large majority marine or in brackish-water habitats.

Chaetomorpha linum (O.F.Müller) Kützing 1849

Basionym: *Conferva linum* O.F.Müller
Synonyms: *Chaetomorpha aerea* (Dillwyn) Kützing, *C. sutoria* Rabenhorst

20010010 **Pl. 116A** (p. 464)

Filaments long, hair-like and frequently bent alternately in opposite directions (flexuous); cells of free-living forms almost constant in width throughout filament, 200–300 μm wide, 1 to 2 times longer than wide; cells of attached forms increasing in width towards apex, 120–535 μm wide, 0.5 to 2 times longer than wide; cells of free-living and attached forms cylindrical to barrel-shaped, slightly to markedly constricted at cross walls; walls 2–4 μm thick.

Probably cosmopolitan; reported in the British Isles only infrequently from freshwater habitats including rivers, streams, wells and animal drinking troughs; common in brackish-water habitats especially saltmarshes and estuarine flats.

The free-living *Chaetomorpha linum* and the attached *C. aerea* are regarded here as growth forms of the same species. Other authors do not agree and prefer to consider them as separate species (see Burrows, 1991, p. 139, for discussion).

Chaetonella Schmidle 1901

Filaments uniseriate and branching in one plane, free-floating or attached to the mucilaginous sheath of macroalgae; cells cylindrical or irregularly shaped, multinucleate (2–5 nuclei), with apical cells longer than intercalary ones; chloroplast parietal and without pyrenoids; asexual reproduction by zoospores but flagellar number unknown; sexual reproduction unknown.

British Isles and East Africa; various types of water body. Monospecific.

Chaetonella goetzii Schmidle 1901

20020010 **Pl. 109D** (p. 441)

Cells 6–8 μm wide towards middle of filament but apical cells only about 3 μm wide, 1 to 3 times longer than wide.

Distribution as for genus; only known in England from a small stream (Cowside Beck, near Arncliffe, West Yorkshire), where it grew on the filamentous green alga *Spirogyra*.

Cladophora Kützing 1843

Macroscopic, consisting of erect or prostrate systems of uniseriate filaments, sparsely to profusely branched, growth apical and/or intercalary, free-living or attached by a disc-like holdfast and/or downgrowing rhizoids; multinucleate; chloroplast parietal, net-like or without perforations (closed), possessing several often bilenticular pyrenoids; asexual reproduction by bi- or quadriflagellate zoospores; sexual reproduction by biflagellate isogametes.

Probably cosmopolitan; often as free-floating masses at the water surface or overlying sediments, but more commonly appearing as long streaming filaments attached to rocks in flowing water, often forming entangled masses along lake shores; most abundant in very nutrient-rich habitats providing there is no heavy metal contamination. The large majority of the more than 600 species are marine.

The genus exhibits considerable morphological variability and identification of the marine species is especially difficult. There is uncertainty surrounding many species, identification often depending on a combination of overlapping characters. Culture investigations have demonstrated that many earlier recognized species are mere growth forms or habitat variants. For monographic treatments, see Hoek (1963a) and Söderström (1963).

1 Branching often very profuse; side branches arising often subterminally at the distal end of a cell, cross walls almost vertical; rhizoids common throughout .... *C. aegagropila*
1 Branching sparse to profuse; side branches usually terminal at distal end, cross walls horizontal or oblique; rhizoids uncommon and generally basal 2
2 Cells of side branches usually <20 μm wide, up to 15 times longer than wide; apical cell 16–27 μm wide .... *C. fracta*
2 Cells of side branches >20 μm wide, up to 20 times longer than wide except in bushy or free-floating plants; apical cell usually >27 μm wide .... *C. glomerata*

Cladophora aegagropila (Linnaeus) Rabenhorst 1868

Basionym: *Conferva aegagropila* Linnaeus
Synonyms: *Cladophora sauteri* (Nees ex Kützing) Kützing, *C. holsatica* Kützing
20030010 **Pl. 117C** (p. 474) **CD**
Filaments attached or free-floating, usually less than 3 cm long, extensively branched, dark green, interwoven nature of the branching gives it a stiff texture; cells generally cylindrical, more club-shaped towards base; growth predominantly intercalary; side branches subterminal at distal end of cell, often in a series and irregularly arranged or sometimes opposite, cross walls vertical or almost vertical; cells of side branches 30–100 μm wide, up to 25 times longer than wide; cells of main branches up to 200 μm wide, up to 15 times longer than wide, very thick-walled (up to 20 μm thick); apical cells 30–70 μm wide; rhizoids commonly arise from the apex or base of cells and intertwine with branches before attaching to surfaces.

Europe (freshwater and brackish-water environments); widespread in the British Isles on rocks, boulders and stones along pond and lake margins and in rivers. In lakes it typically forms mats or dense cushion-like growths, less commonly as loose-lying masses over muddy or sandy substrata, especially where somewhat shaded. Loose-lying populations in some shallow lakes (e.g. Malham Tarn, Yorkshire; Loch Kildona, Outer Hebrides; West Loch Ollay, South Uist, Outer Hebrides; B.A.W., pers. comm.) can form spherical structures known as '*Cladophora* balls'. More information on large *Cladophora* balls is given by Acton (1916a) and Lund (1961a). In rivers it forms carpet-like growths on the rocky bottom of relatively fast-flowing, hard-water, moderately to slightly nutrient-rich (mesotrophic or slightly eutrophic) stretches, but not in very nutrient-rich stretches where *C. glomerata* may form mass growths. Abundant *Cladophora* balls have been discovered in the River Unshin in Ireland, downstream of Lough Arrow, by N.T.H. Holmes (pers. comm.). Holmes & Whitton (1975b) report *C. aegagropila* to be quite widespread, but relatively uncommon in the catchment of the River Tweed where it frequently grows in similar microhabitats to the blue-green alga *Nostoc parmelioides*. It has also been recorded in all the larger rivers surveyed in north-east England and its distribution in the River Wear has been described in some detail (Whitton et al., 1998).

Commonly known as 'carpet blanketweed' in recognition of the carpet-like growth formed by filaments of similar height tending to grow towards the light. Readily distinguished from a moss protonema on microscopic examination in having the cross walls at right angles to the filament rather than at an oblique angle. According to Holmes & Whitton (1975b), it differs from *C. glomerata* in showing little tendency to become reduced to a simple basal portion in winter but rather has a similar appearance throughout the year. They report that winter material sometimes appears almost black if heavily epiphytized by blue-green algae such as *Oncobyrsa* spp.

Cladophora fracta (O.F.Müller ex Vahl) Kützing 1843

Basionym: *Conferva fracta* O.F.Müller ex Vahl
Synonym: *Conferva flavescens* (Roth) Kützing
20030020 **Pl. 117E** (p. 474)
Mostly free-floating or unattached and loose-lying; branching irregular or pseudodichotomously divided at wide angles, young plants often sparsely branched, coarse in texture; cells cylindrical to ovate, pear-shaped or irregularly swollen; growth commonly intercalary (especially in free-living plants); main branches up to 100 μm wide with cells up to 3 times longer than wide, bearing numerous side branches of different lengths; cells of side branches 10–15(–20) μm wide, 5 to 15 times longer than wide, apical cells 16–27 μm wide.

Europe and North America; frequent in shallow, nutrient-rich ponds and ditches as well as penetrating slightly brackish-water habitats.

According to Söderström (1963), free-floating plants of *C. glomerata* can only be reliably distinguished from *C. fracta* by their main axes having a greater width (up to 150 μm wide compared to a maximum of 100 μm) and significantly larger length to width values (20–50 times longer than wide compared with 3–15). Older and more robust plants of *C. fracta* have cellular dimensions overlapping with *C. glomerata*, but the intercalary cells in their side branches always have a much smaller length to width ratio (rarely reaching 15 times longer than wide) and there are more secondary intercalary divisions compared with younger plants.

Cladophora glomerata (Linnaeus) Kützing 1843

Basionym: *Conferva glomerata* Linnaeus
Synonym: *Cladophora crispata* (Roth) Kützing
20030030 **Pl. 117D** (p. 474) **CD**
Profusely to very sparsely branched: profusely branched forms tend to be relatively short-celled (3 to 8 times longer than wide) and sparsely branched forms have significantly longer cells (5 to 25(–30) times longer than wide); growth mainly apical except in free-floating plants where intercalary divisions are some distance from apex; branching pseudodichotomous with insertion of side branches generally oblique to horizontal; cells of side branches 20–50 μm wide, 3 to 6 times longer than wide except in sparsely branched forms where values range from 20 to 30(–40) times longer than wide; main branches of profusely branching forms up to 150 μm wide, lesser branched forms somewhat narrower, apical cells (25–)30–45(–60) μm wide and often distinctly narrowing.

Probably cosmopolitan; a very common and widely distributed freshwater species (rarely in brackish-water habitats) in the British Isles, optimal growth is at about pH7 and it tolerates high light levels and nutrient concentrations; usually attached to surfaces in streams, rivers and waterfalls where often reaching its maximum length (0.5–1 m) and poorly to profusely branched; richly branched plants, often with short recurved branches, occur in stagnant water or along wave-beaten shores of lakes. In ditches, ornamental and aquaculture ponds and lakes it frequently forms free-floating masses and becomes entangled with other aquatic macrophytes. Such free-floating surface and subsurface masses affect not only the amenity value of aquatic habitats, but often when they decay the resulting anoxic conditions lead to the death of fish and other aquatic animals. It is intolerant of heavy metal contamination and is replaced by more resistant filamentous green algae (e.g. *Stigeoclonium*) in nutrient-rich habitats where concentrations of metals such as copper and lead are high.

Commonly known as 'blanket weed', a name that can apply equally to other coarse green filamentous algae from which it can be readily distinguished in the field by having a coarse rather than slippery feel. Notoriously polymorphic, several growth forms have been reported, but the characters used to distinguish them (i.e. degree of branching, cell size and length to width ratio of cells) vary depending upon age and habitat conditions. It is likely that all records of *Cladophora canalicularis* Roth from Scotland are simply forms of *C. glomerata* (see Hoek, 1963a, p. 18). The slight bending of the branch apices has sometimes been regarded as a character useful for separating *C. glomerata* from other species.

REJECTED RECORD

Pithophora oedogonia (Montagne) Wittrock 1877

20040010

This tropical or subtropical alga (morphologically similar to *Cladophora*) is rejected as an indigenous member of the British freshwater algal flora since the only two sites from which it has been reliably confirmed were artificially heated. These sites are a canal (Reddish Canal, Manchester) heated by warmed water entering from cotton mills and a tropical water-lily house at The Royal Botanic Gardens, Kew, Surrey. It was first discovered in the Reddish Canal in the summer of 1896 on *Najas* plants sent to the British Museum (Natural History) in London. It was grown on in a glass jar and eventually produced spores in pairs or 3 to several together unlike in the type in which they are usually single. For this reason Rendle & West (1899) considered it a new variety and named it var. *polyspora* Rendle *et* W.West. It had disappeared from the Reddish Canal by the summer of 1961 by which time warm water had ceased flowing into it following closure of the cotton mills some 18 months earlier (Swale, 1962a). Reports of it from the River Wandle in Kent (Price & Price, 1983) and from South Wales (Benson-Evans & Antoine, 1996) are considered very doubtful since not accompanied by an illustration or description. Several unusual algae have been reported from the tropical houses at Kew (see Fritsch, 1903b, 1906), but these cannot be considered as native to the British Isles.

Rhizoclonium Kützing 1843

Filaments uniseriate and unbranched, sometimes bearing short, unicellular or multicellular rhizoidal branches; multinucleate; chloroplast net-like, with several pyrenoids; asexual reproduction by fragmentation, akinetes and biflagellate zoospores; sexual reproduction by quadriflagellate gametes.

Probably cosmopolitan; most frequent in flowing water although principally a marine genus with just a few freshwater representatives.

The only marine species in the British Isles is *R. tortuosum* (Dillwyn) Kützing, a species common on the upper shore and in saltmarshes (Burrows, 1991), possibly may survive for short periods in wholly freshwater environments close to the sea.

Rhizoclonium hieroglyphicum Kützing 1843

Synonyms: *Conferva fontinalis* Berkeley, *Microspora fontinalis* (Berkeley) De Toni

20050010 **Pl. 117O** (p. 474)

Forms fleece-like growths of entangled filaments when present in quantity, often filaments twisted or bent alternately in opposite directions (flexuous); rhizoidal branchlets rare or absent; cells 10–37 μm wide, 2 to 5 times longer than wide, cylindrical, usually thin-walled, occasionally with thick and stratified walls.

Probably cosmopolitan; widely distributed in the British Isles where frequent along the margins of ponds, ditches, drains, streams and rivers, often entangled with filamentous macroalgae such as *Cladophora glomerata* and submerged vascular plants; frequently present in quantity in rapids, river banks affected by wash from passing boats and concrete walls wetted by spray and splash caused by water gushing through open sluice gates.

If rhizoidal branches are absent then it can be easily mistaken for unbranched forms of *Cladophora glomerata*. It can be distinguished from this unusual form of *Cladophora* by having narrower filaments, a smaller length to breadth ratio of the cells, and the presence of kinks or bends along the filaments; sometimes these bends are referred to as 'knee' or 'elbow joints'. These occur where a cell becomes swollen and an adjacent cell, usually a smaller one, connects to it at an angle. Other simple green filamentous algae (e.g. *Klebsormidium rivulare*, *Klebsormidium subtile*) have similar 'joints' (see Say & Whitton, 1977).

ORDER COLEOCHAETALES

BY DAVID M. JOHN

Chlorophytes with branched, often prostrate filaments bearing sheathed hairs. Reproduction oogamous. Molecular systematic analyses and molecular architectural evidence indicate a close relationship of members of this order to the ancestry of land plants (see Graham & Wilcox, 2000).

Chaetosphaeridium Klebahn 1892

Cells solitary or densely clustered, spherical to flask-shaped, enclosed within a mucilage envelope or connected in series by a mucilaginous tube, each cell bearing a long and prominently basally sheathed bristle; chloroplasts parietal, plate-like, 1 or 2 per cell, each with a pyrenoid; asexual reproduction by biflagellate zoospores, 2 or 4 per cell; sexual reproduction oogamous: oogonia larger than vegetative cells, male gametes small and biflagellate.

Probably cosmopolitan; usually epiphytic, more rarely endophytic or planktonic.

1 Cells solitary or clustered, often enclosed in a mucilage envelope ....*C. globosum*
1 Cells usually in a linear series and connected by a mucilage tube ....*C. pringsheimii*

Chaetosphaeridium globosum (Nordstedt) Klebahn 1892

Basionym: *Herposteiron globosum* Nordstedt

Synonym: *Aphanochaete globosa* (Nordstedt) Wolle

26010010 **Pl. 112E** (p. 448)

Cells 11–18(–20) μm in diameter, spherical or flask-shaped, solitary or clustered, often enclosed within a mucilaginous envelope.

Probably cosmopolitan; widely distributed in the British Isles where it grows on macroalgae and other aquatic macrophytes, especially on the moss *Sphagnum* in bog pools.

var. *depressum* (West *et* G.S.West) G.S.West *et* F.E.Fritsch 1927

Basionym: *Aphanochaete globosa* var. *depressa* West *et* G.S.West

26010012 **Pl. 112F** (p. 448)

Cells 10–13.5 μm in diameter and 6.8–9.5 μm high, ellipsoidal to spherical, usually somewhat depressed.

British Isles (England); reported on filamentous algae only in pools within the New Forest, Hampshire (W. & G.S.West, 1897).

Chaetosphaeridium pringsheimii Klebahn 1892

26010020 **Pl. 112D** (p. 448)
Cells 9–11(–12) μm in diameter, spherical or flask-shaped, often connected in series by a basal mucilage tube.

Probably cosmopolitan; less common in the British Isles than the previous species.

It has been suggested by West & Fritsch (1927, p. 208) that the two species are 'merely forms of one'.

Coleochaete Brébisson 1844

Filaments uniseriate, dichotomously or irregularly branched, consisting of a prostrate and an erect system or only a prostrate system, if the latter then filaments radiating and loosely spreading or laterally coalescing to form a single-layered pseudoparenchymatous disc-like expansion; cells often bearing long, basally sheathed bristles; chloroplast parietal, plate-like, single, with 1 or 2 large pyrenoids; asexual reproduction by biflagellate zoospores, one produced in each cell; sexual reproduction oogamous: male gametes biflagellate and one per cell, oogonia flask-shaped, with a long trichogyne.

Probably cosmopolitan; commonly epiphytic on aquatic macrophytes

For a review of the genus, see Szymańska & Spalik (1993).

1 Plants hemispherical and cushion-like; filaments erect and radiating from a centre ....*C. pulvinata*
1 Plants of another form 2
2 Pseudoparenchymatous disc-like expansions of laterally coalescing filaments................................3
2 Prostrate systems loosely arranged, consists of irregularly spreading or radiating filaments...........4
3 Cells 14–46 μm wide, 1 to 4 times longer than wide..*C. scutata*
3 Cells 7–12 μm in wide, 1 to 2 times longer than wide ....*C. orbicularis*
4 Filaments radiating from a common centre, never coalescing laterally ....*C. soluta*
4 Filaments spreading irregularly, coalescing laterally over short distances.............*C. irregularis*

Coleochaete irregularis Pringsheim 1860

26020010 **Pl. 108G** (p. 435)
Filaments irregularly spreading and prostrate, remaining separate or occasionally coalescing over short distances to form a single-layered, disc-like expansion; cells 8–14 μm wide, 1 to 2 times longer than wide, rectangular or polygonal, bearing a few laterally inserted bristles.

Probably cosmopolitan; reported from a wide range of aquatic habitats in the British Isles, growing on the submerged stems and leaves of vascular plants such as the reed *Phragmites communis* and *Myriophyllum* spp.

Coleochaete orbicularis Pringsheim 1860

26020020 **Pl. 108H** (p. 435)
Forming a regular or somewhat lobed, single-layered, pseudoparenchymatous expansion of branched, radiating and laterally coalescing filaments; cells (7–)8–12 μm wide, 1 to 2 times longer than wide, rectangular, often bearing bristles.

Probably cosmopolitan; forming disc-like expansions on the leaves and stems of aquatic plants, sometimes reaching up to 4 mm across, growing in rivers (e.g. River Caragh) as well as in ponds and lakes.

Very similar to *C. scutata* with many records probably based on misidentifications. Readily separated from *C. scutata* by the small size of its cells, a character it shares with *C. irregularis*.

Coleochaete pulvinata A.Braun in Kützing 1849

26020030 **Pl. 108J, 109A** (pp. 435, 441)
Filaments forming hemispherical, cushion-like growths (2–4 mm across) of irregularly branched erect filaments radiating from a common centre; cells (17–)20–50 μm wide, 1 to 3 times longer than wide, irregularly cylindrical, typically bearing bristles.

Probably cosmopolitan; widely distributed in the British Isles where frequently an epiphyte on submerged aquatic vegetation growing along the shallow margins of lakes.

Coleochaete scutata Brébisson 1844

26020040 **Pl. 109B** (p. 441) **CD**
Filaments forming regular or lobed, single-layered and pseudoparenchymatous crusts (rarely exceeding 800 μm across), consisting of branched and laterally coalescing filaments radiating from a common centre; cells 14–46 μm wide, 1 to 3(–4) times longer than wide, cylindrical.

Probably cosmopolitan; the most widely distributed species of *Coleochaete* in the British Isles where common on vascular plants and macroalgae (e.g. *Vaucheria*) growing in a variety of freshwater habitats, occasionally on artificial surfaces including glass and plastic.

Sometimes bristles are lost leaving only the sheath bases, which appear as circular structures when viewed from above.

Coleochaete soluta (Brébisson) Pringsheim 1860

Basionym: *Coleochaete scutata* Brébisson var. *soluta* Brébisson
26020050 **Pl. 109C** (p. 441)
Filaments loosely spreading and branched, radiating from one or more central cells and not laterally coalescing; cells (8–)10–25 μm wide, 1 to 3(–3.5) times longer than wide, cylindrical.

Probably cosmopolitan; common in the British Isles on submerged aquatic vegetation growing along shallow lake margins.

Very variable in size; the above dimensions relate to British material.

DOUBTFUL RECORD
Coleochaete nitellarum Jost 1895
Recorded on only few occasions in the British Isles (e.g. Grove et al., 1920; West & Fritsch, 1927; Howland, 1931; Godward, 1937) as endophytic in the cell membranes of *Nitella* and

Chara. Requires further investigation since probably no more than a growth form of *Coleochaete irregularis*. It has been reduced to a variety of this species (*C. irregularis* var. *nitellarum* Teodoresco).

Conochaete Klebahn 1893

Forms mucilaginous cushions consisting of loose clusters of subspherical and frequently depressed cells, each cell bearing a few delicate bristles from apex of a wall protuberance or the base of an elongated gelatinous sheath; chloroplasts parietal, 1 or 2 per cell, each with a pyrenoid; oil droplets often prominent; asexual reproduction by zoospores (flagella number unknown), 4 or 8 in each cell; sexual reproduction unknown.

Probably cosmopolitan; an epiphyte on aquatic macrophytes.

1 Bristles each arising from base of an elongate gelatinous sheath ....*C. comosa*
1 Bristles each arising on apex of a protuberance from the stratified wall ....*C. polytricha*

Conochaete comosa Klebahn 1893

26030010 **Pl. 112H** (p. 448)
Cells 13–26 μm in diameter, spherical, each usually bearing 3–5 basally sheathed bristles.

Probably cosmopolitan; epiphytic on aquatic macrophytes.

Conochaete polytricha (Nordstedt) Klebahn 1893

Basionym: *Aphanochaete polytricha* Nordstedt
26030020 **Pl. 112L** (p. 448)
Cells (9–)10–16(–17) μm wide, spherical or spherical to wedge-shaped, with bristles arising on apex of a protuberance from the stratified wall.

British Isles, New Zealand; exceedingly rare in the British Isles where known only on the aquatic macrophytes in moorland pools (e.g. Mossdale Moor, Widdale Fell, Yorkshire).

Dicranochaete Hieronymus 1887

Cells solitary, or more rarely united into short rows, spherical, ellipsoidal or kidney-shaped, apical portion usually ornamented and having an operculum or cap, each bearing one or more repeatedly dichotomously divided bristles of often considerable length; chloroplasts parietal, single, forming an inverted cup, with or without pyrenoids; asexual reproduction by biflagellate zoospores, 4–32 per cell; sexual reproduction by biflagellate gametes, fusing to form quadriflagellate zygotes; resting spore thick-walled.

Probably cosmopolitan; common in acid waters as an epiphyte on submerged aquatic macrophytes including *Sphagnum*.

The genus has been retained in the order to which it has been traditionally assigned (Chaetophorales), although O'Kelly et al. (1994) have created the Order Dicranochaetales to accommodate this genus and several others possessing certain unusual ultrastructural features.

1 Cells more or less spherical, walls thick, stratified and unornamented ....*D. britannica*
1 Cells kidney-shaped to widely elliptical, walls thin and frequently ornamented towards apex ....*D. reniformis*

Dicranochaete britannica G.S.West 1912

26040010 **Pl. 112K** (p. 448)
Cells 18–36 μm across, more or less spherical, walls very thick and stratified, each cell bearing a bristle at apex.

British Isles (England); epiphytic on *Sphagnum* in boggy pools.

Dicranochaete reniformis Hieronymus 1887

Synonym: *Dicranochaete reniformis* var. *laevis* Hodgetts
26030020 **Pl. 112J** (p. 448)
Cells (9–)10–35 μm wide, 5–12(–17) μm in height, kidney-shaped to widely elliptical, towards apex walls smooth, toothed or with irregular spines, each cell bearing from its base or central notch a bristle 1–3 μm wide and up to 150(–170) μm long.

Probably cosmopolitan; often on *Sphagnum* and the submerged stems or roots of aquatic angiosperms growing in boggy ponds.

Hodgetts (1916) recognized the variety *laevis* as a form whose cells lacked any wall ornamentation. According to Matula (1992), it falls within the range of variation of the type species and so is placed under the synonymy of this species.

Oligochaetophora G.S.West 1911

Forms loose clusters of 2–6 cells enclosed within a mucilaginous envelope; cells subspherical or ovoid, with very thin walls, each bearing 2–4 long, narrow and flexuose bristles; chloroplast parietal, single, with 2 or 3 starch grains.

Europe, North America; epiphytic on aquatic macrophytes. Monospecific.

Further study of this genus is needed and of *Polychaetophora* since it is possible that both are germlings and juvenile thalli of *Chaetopeltis orbicularis*.

Oligochaetophora simplex (G.S.West) G.S.West 1911

Basionym: *Polychaetophora simplex* G.S.West
26050010 **Pl. 112G** (p. 448)
Cells 15–20 μm wide and walls about 0.6 μm thick; bristles 50–210 μm in length.

Ireland, North America; in the British Isles only reported growing on vascular plants in the marginal shallows of Lough Garten, County Donegal.

Polychaetophora West *et* G.S.West 1903

Cells solitary, or more usually in clusters of 2–6 cells forming a subfilamentous mass; cells subspherical, ellipsoidal or ovoid, with walls very thick and stratified, sometimes with stratified outgrowths, each cell bearing 8–12 long, delicate, flexuose bristles; chloroplast parietal, single, often with an indistinct margin; starch absent, oil droplets present.

Europe, North America; on aquatic macrophytes or planktonic. Monospecific.

See remarks under *Oligochaetophora*.

Polychaetophora lamellosa West *et* G.S.West 1903

Synonym: *Diplochaete lamellosa* (West *et* G.S.West) Collins

26060010 **Pl. 112B** (p. 448)

Cells 19–35 μm wide and walls 2.8–10.5 μm thick; setae 86–183 μm in length.

Distribution as for genus; in the British Isles on larger aquatic algae or sometimes planktonic.

ORDER PRASIOLALES

BY DAVID M. JOHN

Thallus of small clusters of cells, filamentous or membrane-like, with distinctive star-shaped chloroplasts. Recently placed in the Class Trebouxiophyceae along with genera (e.g. *Desmococcus*, *Microthamnion*, *Stichococcus*) exhibiting a surprisingly wide range of morphological, reproductive and ecological types (see Graham & Wilcox, 2000).

Prasiola C.Agardh 1821

Two distinct forms, a single-layered (rarely 2-layered) membrane-like frond and unbranched filaments; membrane-like stage consisting of quadrate or polygonal cells arranged in square to rectangular packets or blocks, sometimes grouped in 4s, often with a short stalk and attached by rhizoids; filamentous stage initially uniseriate, often becoming multiseriate, attached by rhizoids; chloroplast star-shaped, with a single central pyrenoid; asexual reproduction by aplanospores produced in thickened areas of the membrane-like form; sexual reproduction by oogamy: reproductive tissues (haploid) develop in upper parts of membrane-like form, with some areas producing biflagellate male gametes and other areas oogonia.

Probably cosmopolitan; on damp soil, rocks, stones, walls and tree trunks as well as fast-flowing mountain streams, common along sea coasts especially on surfaces enriched with nutrients (e.g. seabird and seal colonies).

Considerable uncertainty surrounds the relationship between *Prasiola* and *Rosenvingiella*. In the past the latter has sometimes been regarded as a stage in the development of *Prasiola* (e.g. Edwards, 1975b). Rindi et al. (1999) conclude from field observations in Galway City, on the west coast of Ireland, that the genera are distinct although have similar habitat requirements. Uncertainty surrounds the exact placement of *Prasiola* and *Rosenvingiella* in the classification of the green algae.

1 Frond narrow and ribbon-shaped, tapering to a stalk, attached by a disc-shaped holdfast .... *P. calophylla*

1 Frond irregular to rounded, attached along margin directly to surface or by rhizoidal outgrowths .... *P. crispa*

Prasiola calophylla (Carmichael ex Greville) Kützing 1845

Basionym: *Bangia calophylla* Carmichael ex Greville

24260030 **CD**

Fronds narrow, ribbon-shaped, up to 2 cm long and 1 mm wide, single-layered and only becoming polystromatic when fertile, tapering to a distinct stalk and attached by a disc-shaped holdfast; cells single at base, arranged in transverse and longitudinal rows above, towards tip of frond cells 5–6 μm wide and 8–10 μm long.

Cosmopolitan; reported as common throughout Galway City where usually associated with *Rosenvingiella* on damp surfaces including cement, bricks and old paintwork; common at the base of buildings where dogs and other animals urinate.

Prasiola crispa (Lightfoot) Kützing 1843

Basionym: *Ulva crispa* Lightfoot

Synonyms: *Prasiola parietina* (Vaucher) Wille, *Ulothrix parietina* Kützing, *Hormidium parietinum* Kützing

24260010 **Pl. 117N** (p. 474) **CD**

Fronds forming bright to dark green membrane-like expansions, irregular to rounded in outline, sometimes divided or folded, free-living or attached along the margin or by rhizoids; cells in cross section of membrane 5–8(–13) μm wide and 13–16 μm long, quadrate or rectangular; filamentous form uniseriate to multiseriate, grading through to a membrane-like thallus; cells of similar in shape and size in the two forms.

Cosmopolitan; widely distributed in the British Isles on damp soil, roofs, old walls and tree trunks near ground level where humans and dogs urinate; very abundant below bird roosts or in bird colonies where levels of nutrients are high, sometimes in areas influenced by sea spray.

DOUBTFUL RECORDS

Most records of *Prasiola furfuracea* Meneghini and *P. stipitata* Suhr ex Jessen (24260020) are marine since growing on rocks influenced by sea spray.

Rosenvingiella P.C.Silva 1957

Synonym: *Gayella* Rosenvinge

Terrestrial alga growing as unbranched, uniseriate or multiseriate filaments of 4- or more-celled packets, attached at intervals by rhizoids, or forming a cylindrical parenchymatous structure and sometimes constricted; chloroplast star-shaped with a single central pyrenoid; asexual reproduction by non-motile spores; sexual reproduction by oogonia and biflagellate male gametes.

Probably cosmopolitan; on moist soil, rocks, stones, tree trunks and often in similar habitats to *Prasiola*.

Considerable uncertainty surrounds the relationship between *Prasiola* and *Rosenvingiella* (see comments under *Prasiola*). Due to this uncertainty, it is difficult to decide where to place taxa formerly placed in the synonymy of *Prasiola*. A further difficulty concerns whether or not to regard *Schizogonium murale* as distinct from *Rosenvingiella polyrhiza*.

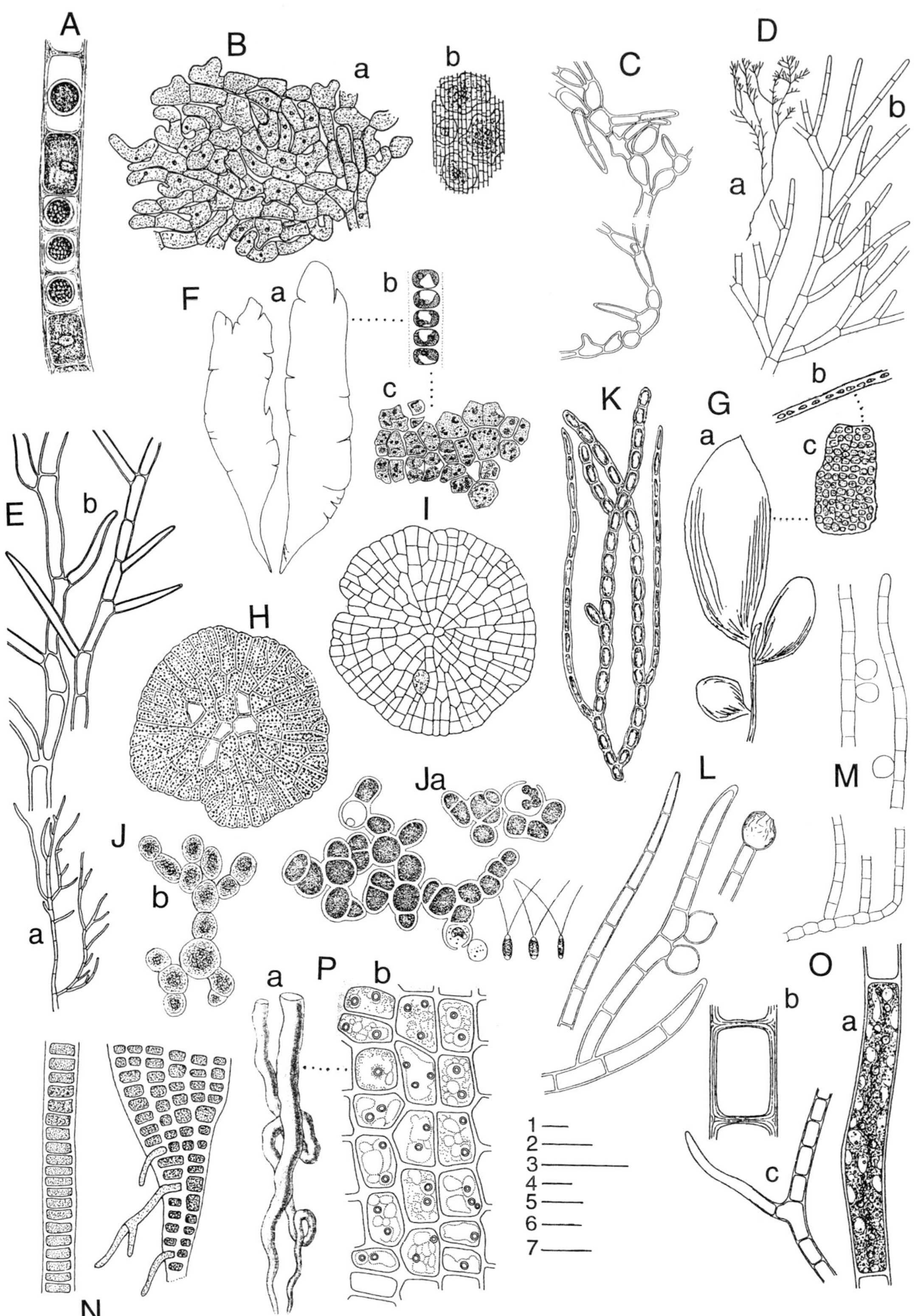

Plate 117 A–P

A. *Microspora wittrockii* (p. 453); **B**. *Cephaleuros endophyticus* (p. 476): a. pseudoparenchymatous layer of filaments, b. surface view of host with subepidermal patches of filaments; **C**. *Cladophora aegagropila* (p. 469); **D**. *Cladophora glomerata* (p. 469): a. habit, b. detail of branching; **E**. *Cladophora fracta* (p. 469); **F**. *Monostroma membranaceum* (p. 455): a. habit of membrane-like frond, b. section of frond, c. surface view; **G**. *Monostroma bullosum* (p. 455): a. habit of membrane-like frond, b. section of frond, c. surface view; **H**. *Phycopeltis epiphyton* (p. 476); **I**. *Phycopeltis arundinacea* (p. 476); **J**. *Trentepohlia umbrina* (p. 478): a. short lengths of filament with biflagellate zoospores, b.branched filament; **K**. *Trentepohlia iolithus* (p. 478); **L**. *Trentepohlia aurea* (p. 478); **M**. *Trentepohlia calamicola* (p. 478); **N**. *Prasiola crispa* (p. 473): filamentous form (left) and membrane-like form (right); **O**. *Rhizoclonium hieroglyphicum* (p. 470): a, b. portion of a filament, c. filament with a lateral rhizoid; **P**. *Enteromorpha flexuosa* (p. 479): a. inflated thallus, usually summer form, b. surface view showing regular arrangement of cells and 2 or more pyrenoids in mid-thallus region.

Scale bars: 1–3, 50 μm; 4, 10 μm; 5, 5 mm; 6, 10 mm; 7, 1 mm. 1 for C, Db, Eb, K; 2 for Gb, Gc, J, L, M, Oc; 3 for A, Fb,c, I, Oa; 4 for Ba, H, N, Ob, 5 for Ga; 6 for Da, Fa; 7 for Bb, Ea.

Rosenvingiella polyrhiza (Rosenvinge) P.C.Silva 1957

Basionym: *Gayella polyrhiza* Rosenvinge
Synonyms: *Prasiola crispa* fo. *muralis* (Kützing) G.S.West, *Hormidium murale* (Lyngbye) Kützing

24310010 **CD**

Filamentous or cylindrical, consists of several superimposed layers of cells, attached by simple rhizoids (often in pairs); filaments bi- or multiseriate, (10–)60–70 μm wide, sometimes constricted at intervals, rhizoids rare on wider filaments; cells grouped in packets and of two different sizes in multiseriate and parenchymatous filaments, cells varying from 2 to 4 μm wide.

Probably cosmopolitan; largely confined to northerly and westerly coasts of the British Isles where in the splash zone on open rocky shores or in estuaries and here often accompanying species of *Prasiola* (see Burrows, 1991); also widely distributed on damp soil, roofs, old walls, cement bollards and tree trunks, sometimes forming a silky green sheet or belt if nitrogen levels are high.

DOUBTFUL SPECIES

Schizogonium murale Kützing

The identity of *Schizognium murale* remains problematical with some choosing to regard it as no more than a developmental stage of *Rosenvingiella* or *Prasiola*. Rindi et al. (1999) take the view that it is a distinct species separated from *Rosenvingiella polyrhiza* by the absence of parenchymatous multiseriate portions and often lacking rhizoids. Such plants were discovered in several samples of *Rosenvingiella* collected in Galway City, Ireland, although they recognized that it was impossible to separate *Schizogonium* from uniseriate filaments of *Rosenvingiella* when rhizoids were lacking or few. Pending further investigation this is considered to be a doubtful species.

ORDER SPHAEROPLEALES

BY DAVID M. JOHN

Filamentous, with filaments unbranched and multinucleate. Sexual reproduction anisogamous or oogamous. More broadly defined by Deason et al. (1991) based on ultrastructural features associated with motile stages (see Graham & Wilcox, 2000).

Sphaeroplea C.Agardh 1824

Filaments of cylindrical, very elongated cells containing many nuclei (coenocytic); cells varying considerably in length, 2 to 90 times longer than wide, often slightly contricted at cross walls, each cell divided into several large central cavities by a succession of transverse discs of cytoplasm; chloroplasts narrow, several per cell, band or ring-shaped, continuous or occasionally becoming perforated with age, each chloroplast containing several pyrenoids; cells multinucleate; asexual reproduction by fragmentation and biflagellate zoospores; sexual reproduction oogamous, vegetative cells becoming reproductive without changing shape, monoecious or dioecious: oogonia spherical, often lying in a single or in double series; antherozoid-producing cells giving rise to up to 300 spindle-shaped, reddish, biflagellate gametes; oospores with thick, ornamented walls, bright red when mature.

Probably cosmopolitan; free-floating and in freshwater to slightly brackish-water habitats.

Species distinguished mainly upon features associated with zygote shape and ornamentation of its thickened wall (see Ramanthan, 1964). Hoek et al. (1995) consider this genus to be a siphonous representative of the Order Chlorococcales although it is usually classified in a separate order, the Sphaeropleales. One reason for assigning it to the Chlorococcales is the belief that its cells divide like those of *Cylindrocapsa*, a genus in which the vegetative cells have been interpreted by Sluiman et al. (1989) as equivalent to the autospores of coccoid forms.

Sphaeroplea soleirolii (Duby) Montagne ex Kützing 1849

Basionym: *Sphaeroplethra soleirolii* Duby
Synonyms: *Sphaeroplea cambrica* F.E.Fritsch, *S. annulina* var. *soleirolii* Kirchner

24330020 **Pl. 116B** (p. 464)

Filaments (34–)36–72(–73) μm wide, cells sometimes up 90 times longer than wide, each cell with numerous ring-like (annular) or net-like chloroplasts and several pyrenoids; oospores ellipsoidal, 27–46 μm wide and 29–60 μm long, with 8 prominent wings running longitudinally, closely packed within the cells and often arranged in 2 or more rows.

Probably cosmopolitan; so far known only from pools (probably temporary) lying between sandhills on the coast of Caernarvonshire in mid-Wales but certainly more widely distributed in the British Isles than records indicate.

The material from mid-Wales was at first mistakenly identified as *S. annulina* (see West & Fritsch, 1927), a species with which it has been commonly confused. Later Fritsch (1929b) considered this material to belong to an undescribed species which he named *S. cambrica*. Bourrelly & Feldmann (1946) re-examined it along with other collections and concluded that Fritsch's species was conspecific with *S. soleirolii*.

REJECTED RECORD

Sphaeroplea annulina (Roth) C.Agardh

24330010

Only known from tanks in the Royal Botanic Gardens at Kew where West (1904) considered it to have been 'introduced from awide with various aquatic plants'. Not considered an indigenous alga until discovered in other localities.

ORDER TRENTEPOHLIALES

BY DAVID M. JOHN

Subaerial, sometimes the algal partner (phycobiont) of lichens. Filamentous, with filaments branched or unbranched, differentiated into a prostrate and an erect system or only a prostrate system present. Zoosporangia sometimes function as propagules.

Cephaleuros O.Kunze 1827

Grows on or within leaves, twigs and fruits of vascular plants, consisting of regularly branched, radially organized, coalescing uniseriate filaments; prostrate filaments forming a single to several-layered pseudoparenchymatous mass, with downgrowing and irregularly branched filaments penetrating deep into host tissues, upwardly growing filaments unbranched and terminating in hairs or single-celled to multicellular sporangiophores; cells cylindrical, regularly or irregularly shaped; chloroplast parietal, net-like, or several and disc-shaped, varying from green to deep orange depending on presence and concentration of astaxanthin and carotenoid pigments; asexual reproduction by quadriflagellate zoospores arising in sporangia borne laterally on the end of the sporangiophore; sexual reproduction by enlargement of terminal or intercalary cells in the pseudoparenchymatous mass of filaments, both gametes biflagellate.

Probably cosmopolitan; mostly in the tropics and subtropics, rare in temperate regions; a parasite or semiparasite growing on herbaceous and slower growing terrestrial plants (*Magnolia, Rhododendron, Rubus, Citrus*, cacao, banana, tea, etc.); important pathogen resulting in serious damage to several economic crop plants.

For further information on the ecology, taxonomy and biology of the genus, see Thompson & Wujek (1997).

Cephaleuros endophyticus (F.E.Fritsch) Printz 1964

Basionym: *Chrooderma endophytica* F.E.Fritsch

22030010 **Pl. 117B** (p. 474)

Forming pale yellow-brown, circular (0.5–2 mm across) or confluent and irregular crusts; pseudoparenchymatous layers of cells in rather indistinct radiating rows, occasionally blunt unicellular hairs arising near edge; basal layer compact and cells small, commonly polygonal, often 6.7–7 μm wide and 11–15 μm long, more elongate towards margin; outermost 2 layers of surface filaments consisting of larger (11–14 μm wide) cells with thick, stratified walls; endophytic filaments extensive and largely subcuticular, frequently single-layered, with cells rectangular or barrel-shaped, 8–8.5 μm wide and up to twice as long as wide; developing terminally on filaments are rounded and sometimes swollen cells, interpreted as sporangia.

British Isles (England); only known on the dead stems of brambles (*Rubus fruticosus*) collected at Bochym in West Cornwall.

Incompletely known since the only material studied was dried and inadequate for a detailed study of cell structure and reproductive processes (Fritsch, 1942).

Phycopeltis Millardet 1870

Grows most commonly as yellowish-green to deep orange crusts on the leaves of angiosperms, single-layered, lobed or irregularly shaped, pseudoparenchymatous or obviously filamentous and of regular or irregularly branched, coalescing filaments; cells cylindrical or irregularly shaped, divide perpendicularly and at right angles to surface, occasionally bearing erect hairs; asexual reproduction by quadriflagellate zoospores formed in sporangia arising on curved and 1- to several-celled stalks; sexual reproduction by biflagellate isogametes formed in terminal or intercalary gametangia.

Mostly tropical and subtropical; subaerial on leaves, only very occasionally on inanimate surfaces in areas of high humidity.

1 Radiating filaments 4–7 μm wide; sporangia mostly marginal ....*P. epiphyton*
1 Radiating filaments 8–12 μm wide; sporangia scattered over surface.......................*P. arundinacea*

Phycopeltis arundinacea (Montagne) De Toni 1889

Basionym: *Phyllactidium arundinaceum* Montagne

22010020 **Pl. 117I** (p. 474)

Forms compact, orange, disc-like expansions about 1.5 mm wide, margins entire or lobed; cells 8–12 μm wide, cylindrical, those towards the periphery broader than cells in centre; sporangia scattered over surface.

Probably cosmopolitan; known in Europe only from south-west Ireland where reported by Scannell (1965, 1978) on leaves of several angiosperms including ivy (*Hedera helix*), laurel (*Prunus laurocerasus*), holly (*Ilex aquifolium*), *Rhododendron ponticum* and the ferns *Phyllitis scolopendrium* and *Blechnum spicant*; most common in deep woodland well away from main roads and on the north-facing sides of ditches in western Ireland (Rindi & Guiry, 2002). Probably more widely distributed than published records indicate.

Considered by Scannell (1978) to be a synonym of *P. expansa* Jennings, but Sarma (1986, p. 85) points out that this is merely conjecture since no comparative study was made of material. Pending comparative investigations of all life history stages it has been decided to consider them separate species. In Thompson & Wujek's (1997) monographic treatment of the Trentepohliales, *Phycopeltis arundinacea* is considered (p. 78) as having an 'Old World Tropics, New World Tropics' distribution and *P. expansa* is not mentioned.

Phycopeltis epiphyton Millardet 1870

22010010 **Pl. 117H** (p. 474) **CD**

Forms compact green, yellow-brown, golden-yellow or blackish disc-like expansions, up to about 1 mm wide, margin often very irregular, under unfavourable conditions becomes filamentous and cells begin to divide very irregularly; cells 4–7(–8) μm wide, 2 to 4 times longer than wide, rectangular, often dividing to form lobes about 4 μm wide; sporangia mostly confined to margin of expansion

Probably cosmopolitan; known from a few widely separated localities in the British Isles, for example, an estuary near Castletownsend in south-west Ireland where on the leaves of ivy *Hedera helix*, in a damp ravine in Montgomeryshire on the fern *Hymenophyllum*, in a woodland in Pembrokeshire on *Rhododendron ponticum*, in Somerset on the laurel *Prunus laurocerasus* and *Rhododendron ponticum*, and in north-west Scotland on pine, spruce and Douglas-fir (R. Woods, pers. comm.); no doubt more widely distributed than indicated by these records.

Often goes blackish in colour on drying. Doubt inevitably attaches as to whether *P. arundinacea* and *P. epiphyton* should be regarded as distinct species since they are only reliably separated on size diffferences.

Printzina Thompson *et* Wujek 1992

Subaerial, green or reddish-brown to pink in colour, consisting of a well developed, pseudoparenchymatous system of dichotomously or sympodially branched prostrate filaments; erect filaments poorly developed; cells of prostrate system cylindrical, 2–4 times longer than broad; cells of erect system almost quadrate, scarcely longer than broad; chloroplasts numerous, disc-like, the chlorophyll usually obscured by a red carotenoid pigment, lacking pyrenoids; asexual reproduction by quadriflagellate zoospores formed in spherical to kidney-shaped sporangia, each borne on a lateral that has a minute stalk cell from where it arises on a prostrate filament, or terminal on an erect filament; gametangia sessile and arise on prostrate or erect filaments.

Probably cosmopolitan; usually green, often on leaves in deep shade but can occur on a variety of surfaces. Nine known species.

Printzina lagenifera (Hildebrandt) Thompson *et* Wujek 1992

Basionym: *Trentepohlia depressa* Hildebrandt

22040010 **CD**

Forms small cushion-like masses or a powdery layer, reddish-brown to pink, consisting of irregularly branched, much entangled filaments, without a clear distinction between the erect and prostrate systems; cells spherical to barrel-shaped, rarely cylindrical, 6–12 μm wide, 1–3 times as long as wide.

Probably widely distributed; in the British Isles known only from Galway where it occurs on a variety of surfaces including tree-bark, asbestos sheeting and cement or limestone walls.

According to Rindi & Guiry (2002), it is vegetatively very similar to *T. umbrina* from which it is distinguished by having significantly narrower cells.

Trentepohlia Martius 1817

Subaerial or the phycobiont of certain lichens, often forming red, greenish-orange, yellowish-orange or orange-coloured felty patches or tufts on surfaces free of soil, consisting of creeping and erect systems of sparingly to profusely branched uniseriate filaments, irregularly, alternately or rarely oppositely branched; prostrate system usually well developed; erect system very reduced or absent, often terminal filaments becoming gradually attenuated; unicellular and cylindrical hairs sometimes present on filaments; cells cylindrical, ellipsoidal or bead-like, rarely more than twice as long as wide, with walls usually thick, stratified and sometimes forming a cellulose cap to the apical cell; chloroplasts numerous, disc-like or band-shaped, the chlorophyll usually obscured by a red carotenoid pigment, lacking pyrenoids; asexual reproduction by quadriflagellate zoospores formed in stalked sporangia, arising terminally or laterally and of two types: 'funnel sporangia' in which a septum forms between two superimposed ring-like thickenings and borne on a bent stalk, or sporangia attached to a stalk by a collar-like thickening and with a small central knob-like projection at base (commonly detaches and is dispersed); sexual reproduction by biflagellate gametes formed within sessile gametangia, lateral, terminal, intercalary or more rarely axillary in position; akinetes thick-walled, usually in series.

Probably cosmopolitan; most abundant in the tropics, widely distributed on tree trunks, leaves, wooden fencing, poles, masonry, rocks, cliffs and stone walls; often most abundant in wetter coastal areas, uplands and mountains. Second commonest lichen phycobiont and almost impossible to identify since when *in situ* it is frequently single-celled or grows as short filaments lacking reproductive organs. Often when free-living *Trentepohlia* has fungal filaments closely associated with it. At least 12 known species.

According to Rindi & Guiry (2002), 'there is little doubt that the examination of a small number of specimens in many studies of this genus is one of the main reasons of the confused taxonomic relationships of several entities' and the 'availability of molecular data from material collected in the type localities and referable with absolute certainty to the type forms is an essential requirement to elucidate the taxonomic position of many entities'. There is no doubt that the genus is in need of a modern revision combining morphological observations with an analysis of molecular data.

1 Cells of erect filaments always cylindrical and without constrictions 2
1 Cells usually inflated and cross walls generally constricted 4
2 Greenish or yellowish when dry, with cells often covered by a brownish substance; filaments usually <10 μm wide .... *T. calamicola*
2 Red, orange- or brownish-red, without a brownish covering to cells 3
3 Cells of erect filaments >10 μm wide; grows on various surfaces .... *T. aurea*
3 Cells of erect filaments <10 μm wide; grows principally on tree bark .... *T. abietina*
4 Filaments not fragmenting; cell walls always very rough .... *T. iolithus*
4 Filaments readily fragmenting; cell walls smooth or becoming rough with age .... *T. umbrina*

Trentepohlia abietina (Flotow) Hansgirg 1886

Basionym: *Chroolepus abietinum* Flotow

22020060 **CD**

Forms red or orange tufts, consisting of a small, much branched prostrate system and an erect system of a few unilateral or irregularly arranged branches; cells of the prostrate filaments spherical or swollen, 10–15 μm in diameter, with those of the erect filaments cylindrical, 5–9 μm wide,10–30 μm long; walls smooth.

Probably widely distributed; so far known only from bark of trees such as cherry, beech, alder, sycamore and ash in Galway, although elsewhere in Ireland (Co. Wicklow, Co. Cork) on large limestone rocks in well-shaded habitats within deciduous forests (Rindi & Guiry, 2002); probably widely distributed in the wetter westernmost areas of the British Isles.

It is difficult to distinguish reduced forms of *T. abietina* from *T. aurea* and for this reason they are frequently regarded as conspecific. The greater width of the erect filaments of *T. aurea* and its more rigid texture are the main features used to separate it from *T. abietina*. Rindi & Guiry (2002) regard the two to be relatively distinct in Galway with *T. abietina* normally confined to tree bark. They acknowledge that more material requires examining and molecular data would be especially useful for determining relationships between these species.

Trentepohlia aurea (Linnaeus) Martius 1817

Basionym: *Byssus aureus* Linnaeus
22020010 **Pl. 117L** (p. 474) **CD**
Forms a bright-red, orange-red or brownish-red, somewhat woolly or gelatinous layer, consisting of poorly differentiated prostrate and erect systems of filaments, sparsely to profusely branched; terminal branches usually taper gradually, apical cell with a cap; cells 10–20 µm wide, 1 to 3 times longer than wide, often longer in uppermost branches, cylindrical, sometimes inflated in lower parts; walls smooth or with knobby projections (tuberculate) and striations usually lying parallel to one another.

Probably cosmopolitan; the most widely distributed and abundant species in the British Isles, very common in wet coastal areas in the west as well as in hilly and mountainous districts, particularly on cliffs of Carboniferous Limestone and Silurian rock, but also on tree trunks (commonly oak), moss, old limestone walls and dry stone walls where more abundant on their damper side. Only reported by Rindi & Guiry (2002) as isolated patches on an old cement wall in Galway city.

Known to be extremely polymorphic.

Trentepohlia calamicola (Zeller) De Toni *et* Levi 1885

Basionym: *Chroolepus calamicola* Zeller
22020020 **Pl. 117M** (p. 474)
Forms greenish, interwoven growths of poorly differentiated prostrate and erect filaments, somewhat yellowish when very dry, often mostly branching on one side of an axis (subsecundly) and frequently divergent; ultimate branches gradually attenuate, without a conspicuous apical cap; cells 7.5–10 µm wide, 1.5 to 3(–4) times longer than wide, cylindrical; walls thin and without striations, often covered by a brown substance.

Probably cosmopolitan; only known in the British Isles on trees in south-west Ireland and Cornwall; probably more widely distributed than the few records suggest.

Trentepohlia iolithus (Linnaeus) Wallroth 1833

Basionym: *Byssus iolithus* Linnaeus
22020030 **Pl. 117K** (p. 474) **CD**
Forms red, yellowish-red to deep orange, thick and felty growths of poorly differentiated prostrate and erect filaments; ultimate branches not tapering, apical cap usually absent; cells 14–30 µm wide, usually 1.5 to 3 times longer than wide, cylindrical (especially if cultured), usually slightly swollen and constricted at cross wall; walls roughened and very thick, striations usually divergent.

Probably cosmopolitan; possibly widely distributed in the British Isles growing on rocky cliffs, tree-bark, old concrete or cement walls and on periodically submerged rocks along sides of small calcareous streams where surfaces are often dominated by the blue-green alga *Rivularia*. Responsible for the unsightly red discolouration visible on the white-wash walls of cottages and houses in wetter parts of the British Isles, often referred to in Galway as 'the red fungus' (M.D. Guiry, pers. comm.). Rindi & Guiry (2002) report large growths from urban areas and rural districts of other counties in Ireland (e.g. Clare, Mayo, Kerry, Cork, Roscommon, West Meath). According to Rindi & Guiry (2002), it often forms deep red vertical stripes on walls and sometimes it more or less completely covers several square metres.

Several varieties known, some of which are largely confined to a particular type of surface (see Sarma, 1986, p. 98).

Trentepohlia umbrina (Kützing) Bornet 1878

Basionym: *Chroolepus umbrinum* Kützing
Synonym: *Trentepohlia odorata* var. *umbrina* (Kützing) Hariot
22020060 (22020050) **Pl. 117J** (p. 474) **CD**
Forms reddish to reddish-brown expansions of creeping filaments, little or no distinction between creeping and erect filaments, irregularly and openly branched, congested with filaments readily breaking into 1- or few-celled fragments; cells in intercalary position spherical to ellipsoidal, (7–)15–27(–35) µm wide, 1 to 2 times longer than wide; ultimate branches terminate in a blunt and shortly cylindrical apical cell, 8–10 µm wide and 1 to 3 times longer than wide; walls smooth, often becoming rough; sporangia single and spherical, subspherical or ellipsoidal, terminal on curved stalks with a basal ring, or spherical to ovoid or a little curved, sessile and terminal, intercalary or lateral in position; gametangia intercalary, spherical and often becoming flask-shaped with a papilla.

Probably cosmopolitan; most common on the north-west and south-west facing sides of walls, especially in the west of Ireland and reported by Rindi & Guiry (2002) as a 'red powdery crust on granite columns' of Powerscourt House Building in Co. Wicklow.

Much doubt attaches to this species which had been considered in the past to be a variety of *T. odorata* (*T. odorata* (F.H.Wiggers) Wittrock var. *umbrina* (Kützing) Hariot). West (1904, p. 95) was also of the opinion that the two species were conspecific. He suggested that earlier records for the British Isles under the name *Trentepohlia umbrina* (Kützing) Bornet should be referred to *T. odorata*, but later West & Fritsch (1927, pp. 201, 202) recognized both species. These species continue to be recognized although doubt inevitably attaches to the interpretation placed on them by different authors. For example, the cell dimensions of *T. odorata* and *T. umbrina* given by Sarma (1986) for New Zealand material are considerably less than those cited by Printz (1964) and others. In discussing the differences between the species, Sarma (1986, p. 102) mistakenly mentions that *T. umbrina* is much smaller in size than *T. odorata*. To judge from different literature sources, the intercalary and terminal sporangia show considerable size variation suggesting that sporangial dimensions are of little or no taxonomic value. The only reliable characters for separating *T. umbrina* from *T. odorata* are the absence in the former of a compact pseudoparenchymatous structure, its open and irregularly divided branches (usually congested), and the readiness of its branches to break into 1- or few-celled fragments. Comparative studies are required to establish whether or not these two very closely related species should be considered synonymous. Pending further studies all records of *T. odorata* from the British Isles are attributed to *T. umbrina*. However, if they prove to be conspecific then the epithet *T. odorata* has priority. The situation is further complicated by Wujek & Thompson (1997) transferring *T. umbrina* to the genus *Phycopeltis* on the grounds that each sporangium has a terminal papilla-pore and the lateral sporangial-bearing branch is solitary on the cell from which it arises. These features are considered to be characteristic of the genus *Phycopeltis*. Until other issues surrounding the identity of *T. umbrina* are resolved it has been decided to retain it under the genus *Trentepohlia*.

DOUBTFUL RECORD
Trentepohlia lichenicola C.Agardh
22020040
The relationship of this lichen phycobiont to free-living *Trentepohlia* species remains uncertain. Two of the above species

(*T. aurea* and *T. umbrina*) are known to be associated with lichens (see Tschermak-Woess, 1989).

Trentepohlia monilia De Wildeman
Khan (1951) reported it from Wimbledon, England based on a culture isolated from tree bark and later De Valéra & Ó Seaghdha (1968) mention it in Galway. According to Rindi & Guiry (2002), the morphological description by Khan corresponds to cultured material of *Printzina lagenifera* suggesting that his record is a misidentification. *Trentepohlia monilia* is characterized by its conspicuous brown colour, sparsely branched filaments and the presence of a narrow connection between the adjoining smooth-walled, spherical or lemon-shaped cells. Rindi & Guiry (2002) have rejected all records of *T. monilia* for Great Britain and Ireland and consider that others from mainland Europe need to be reassessed.

ORDER ULVALES

BY IAN TITTLEY

Thallus membrane-like, an elongated tube or unbranched filaments. Alternation of morphologically similar generations. Mostly marine.

Enteromorpha Link in Nees 1820

Macroscopic, filamentous, thread-like, usually branched, tubular or sac-like with wall only one cell thick, occasionally flattened or compressed, free-floating or attached by cells with rhizoidal extensions, sometimes cells coalesce to form an irregular attachment disc; chloroplast usually parietal, single, plate-like, with 1 or more pyrenoids; reproduction by biflagellate or quadriflagellate gametes or zoospores, arising from any cell but more commonly produced towards the apex.

Cosmopolitan; a widely distributed marine genus with its few freshwater representatives normally favouring hard water; a nuisance alga in semi-enclosed bays causing so-called 'green-tides' and in slow-flowing rivers where its thick mats probably develop as a response to elevated nutrient levels.

Christensen (1994) considered that the only species common in freshwater is *E. flexuosa* (as *E. pilosa*), often present in nutrient-rich (eutrophic) ponds and lakes, commonly as free-living, floating and irregularly wrinkled tubes with short uniseriate branches. In the British Isles and continental Europe, the species *E. intestinalis* has been frequently recorded in freshwater as partially air-filled, distended tubes. *Enteromorpha intestinalis* and *E. flexuosa* have been confused in the past (see comments below). *Enteromorpha prolifera* (O.F.Müller) J. Agardh is a marine species common in the very low salinity environments of estuaries and saltmarshes, but authentic material from truly freshwater habitats in the British Isles has yet to be discovered. A single specimen in the BM assumed to have been collected in a freshwater habitat in Romania (Jasi) has been confirmed as *E. prolifera*. See Burrows (1991) for further information on this species.

Enteromorpha flexuosa (Wulfen *ex* Roth) J.Agardh 1883

Basionym: *Conferva flexuosa* Wulfen *ex* Roth

subsp. *pilifera* (Kützing) Bliding 1963
Basionym: *Enteromorpha pilifera* Kützing
Synonym: *Enteromorpha intermedia* Bliding
21020010 **Pl. 117P** (p. 474) **CD**

Thallus tubular and existing in two forms: large, sack-like and distended, up to 20 mm wide and 280 mm long or longer, or considerably shorter and narrower (especially in winter), gradually widening from base to 5–10 mm above; cells unordered in the larger, broader form and in distinct longitudinal and transverse rows in the narrower form, often rectangular, 10–12 μm wide and 10–13.5 μm long; chloroplast containing 2–4 pyrenoids.

Probably cosmopolitan; marine but also in quite fast-flowing as well as slow-flowing rivers, often most abundant in shallows growing amongst submerged vegetation, also common in ponds and along the margins of reservoirs, attached to hard surfaces and forming floating masses in sheltered and quieter backwaters; widespread but of sporadic occurrence in ditches and rivers in England. In the calcareous, fast-flowing River Lyne, Northumberland, it occurs as large attached tubes. It overwinters as small filaments on boulders and other surfaces. Difficult to assess its earlier status although West & Fritsch (1927) mention it as frequent in canals, rivers and ponds. Possibly an invasive species in the Tees (Holmes & Whitton, 1977a), since not reported in an extremely thorough survey in the 1930s (Butcher et al. 1937). Whitton (2000) considers it may be becoming more abundant and widely distributed in the UK, showing marked seasonality and differences in population size between years. Probably considerably more widely distributed in the British Isles than published records suggest.

Often misidentified as *E. intestinalis* (see comments below). Sometimes larger, unbranched plants lack proliferations and contrast with those examined by Bliding (1963), who illustrates many thread-like proliferations in specimens from continental Europe.

DOUBTFUL RECORD
Enteromorpha intestinalis (Linnaeus) Nees 1820
21020020
Similar in habit to *E. flexuosa* in commonly having a distended, tubular, sac-like thallus that often reaches 10 mm in width and 15 cm or more in length, or much less commonly narrower, smaller and unbranched. Differs from *E. flexuosa* by having its cells unordered throughout, mostly rounded or polygonal (rarely rectangular), 12–15 × 12–20 μm, a distinctly hood-shaped chloroplast displaced to one side in apical part of the cell, and by having a single (rarely 2) pyrenoid (see Bliding, 1963; Burrows, 1991). It is common in slightly saline or tidal limnetic water. Doubtful whether it should be regarded as a truly freshwater species. All herbarium material in The Natural History Museum (BM), identified as *E. intestinalis* and recorded from freshwater habitats in the British Isles, has been examined and re-determined as *E. flexuosa*. Specimens in the BM from Germany (Saxony, Heidelberg), the Czech Republic (Moravia), Austria (Burgenland) and Hungary have proved to be *E. flexuosa*. Authentic material of *E. intestinalis* from freshwater habitats has yet to be examined.

ORDER ZYGNEMATALES

BY ALAN J. BROOK AND LESLIE R. JOHNSON

The Zygnematales is a large and well-defined order of green algae characterized by its mode of sexual reproduction which is by the conjugation of amoeboid gametes and the subsequent formation of thick-walled, often distinctively ornamented zygospores. All members lack flagellate stages in both asexual and sexual reproduction.

There are two main characters that distinguish the so-called placoderm desmids (Suborders Closteriineae and Desmidiineae) from saccoderm desmids (Suborder Zygnematiinae, Family Mesotaeniaceae). Firstly, each cell of the placoderms always consists of two semicells which, as a consequence of their unique mode of cell division, are of different ages. The two cell halves (semicells) are separated by a more or less well-defined isthmus or a less well-defined suture or sutures. Secondly, the more complex, three-layered cell walls of the placoderms are perforated by a system of pores which may be simple or complex. The walls are commonly ornamented by granules, warts or spines which may be arranged in distinctive patterns of considerable taxonomic importance.

Suborder Zygnematiinae

The suborder contains two families, the Zygnemataceae whose members are exclusively filamentous and include the well-known and widely distributed genera *Spirogyra*, *Zygnema* and *Mougeotia* as well as six other lesser known genera, and the Mesotaeniaceae. This latter small family contains unicellular forms known as the saccoderm desmids, recently shown to be an artificial assemblage with uncertain affinities to the 'true' desmids; some genera appear to be most closely allied to the filamentous Zygnemataceae in contrast to the genera *Roya* and *Tortitaenia*, which possess features linking them to certain placoderm desmids of the Suborder Desmidiineae (Brook & Williamson, 2001).

Family Zygnemataceae

BY LESLIE R. JOHNSON

The family contains some of the most common filamentous green freshwater algae. The filaments, apart from one or two species, are without constrictions between the cells and are always unbranched, except in a few species which may produce short, often blunt rhizoids involved in their attachment to substrates in certain habitats. Walls are mostly fairly thin, always composed of a single piece (unlike their relatives in the Desmidiineae), and generally secrete a more or less conspicuous mucilage sheath giving most of them a distinctly slimy feel. The transverse walls, formed by annular ingrowth during each cell division, in some species of some genera (e.g. *Spirogyra*) have replicate end walls in which the septum develops a cylindrical collar-like growth which, it has been suggested, is an adaptation associated with filament fragmentation. Fragmentation into short lengths or even individual cells may occur in some species of *Mougeotia*, *Zygnema* and *Zygnemopsis*.

The chloroplasts are essentially of three main types. A single flat or slightly twisted plate with a median series of pyrenoids (*Mougeotia*), a pair of star-shaped chloroplasts each with a central pyrenoid (*Zygnema*), or from one to 16 long narrow bands lying against the cell wall and spiralling along the length of the cell, each band containing a series of pyrenoids (*Spirogyra*). The chloroplasts, however, are features of the cell which, although important in distinguishing most genera, are of less importance in the identification of species. The single, often conspicuous nucleus, sometimes with a prominent nucleolus, is in most cases situated in the middle of the cell and suspended on fine cytoplasmic threads, although in *Mougeotia* it occurs to one side.

Vegetative reproduction occurs by the fragmentation of filaments, into either short filaments or individual cells. Asexual reproduction is by akinetes, which are vegetative cells with thickened walls, and may be the only means of reproduction other than cell division and subsequent fragmentation. In some *Mougeotia* species, akinetes are frequently formed, especially when their filaments occur in mountain lakes and tarns, presumably associated with low temperatures. Aplanospore production, another mode of asexual reproduction, results in the formation of rounded or ovoid bodies produced in vegetative cells by the contraction of the cell contents and the development of a thick enveloping wall. These walls resemble those of the zygospores of the species in which they are formed, so that they become typically coloured and ornamented.

The majority of the Zygnemataceae undergo sexual reproduction fairly readily by conjugation; sometimes this can be induced under laboratory conditions. This is valuable, because specific differences are largely based on the characters of the conjugating cells and the resulting zygospores. Moreover, the genera *Mougeotia*, *Debarya* and *Temnogametum* are indistinguishable except in the fertile state; the latter genus account is written here with A.J. Brook. In the British Isles conjugation seems to occur most frequently between February and June and has been most commonly reported from lowland areas.

Conjugation is between adjacent cells of one filament (lateral) or, more commonly, ladder-like and between separate filaments (scalariform); in a few species both modes may occur. In the latter, two filaments (which are capable of gliding movements) become aligned parallel to one another and, once in contact, small protuberances grow out from cells in adjacent filaments. As the protuberances increase in size the filaments are gradually forced apart and their opposing surfaces become flattened by mutual pressure. The opposing walls break down with the result that a continuous canal, the conjugation tube, forms to connect the two cells of the adjoined filaments. In some species the contents of one of the cells round up and, acting as the male gamete, moves through the conjugation tube into the 'female' cell to effect fertilization; in others the cell contents of both adjoined cells move into the conjugation tube which then becomes the swollen gametangium to accommodate the developing zygospore. The shape of the zygospore is important taxonomically, ranging from spherical to ellipsoidal or quadrate to triangular. The zygospore wall may be several-layered, usually with an outer exospore layer, a middle mesospore layer and an inner spore wall. The layers, particularly the mesospore wall, may be smooth or distinctively ornamented and variously coloured. Ornamentation and wall colour are particularly important for species determination.

Lateral conjugation may be the rule in some species of *Spirogyra* (*S. affinis*, *S. tenuissima*) but is very rare in *Mougeotia*. In this mode of sexual reproduction there is no apparent sexual differentiation between the filaments. The conjugation processes arise from adjacent ends of neighbouring cells and an open conjugation tube is established between them. The resulting zygospore lies either in this canal or in one or other of the conjugating cells. Sometimes the normal union of gametes fails, probably because of a sudden change in environmental conditions. As a result the gametes fail to fuse and one or both round up and secrete resistant, zygospore-like walls. Such spores are termed parthenospores.

Some 12 genera are recognized within the Family Zygnemataceae and some 700 species have been described worldwide. Many have a cosmopolitan distribution; about 120 species are described from the British Isles in the present volume. Most commonly they are found in small stagnant bodies of water, including ornamental pools in gardens, and especially in the margins of upland ponds and lakes, while a few species have been described permanently attached in flowing water; one species occurs most commonly on acid soils. They are reported as being most abundant in the spring months (West & Fritsch, 1927). This is probably due to a combination of nutrient availability at this season and the fact that they form a conspicuous component of the diet of aquatic insect larvae and amphibians, which leads to their virtual disappearance, or at least makes them difficult to find until the following autumn or winter (Brook, 1955c).

Debarya Wittrock 1872 emend. Transeau 1934

Filaments unbranched and narrow; cells 2 to 20 times longer than broad; chloroplast a single axile or plate-like ribbon with 2–8 scattered or serially arranged pyrenoids; conjugation ladder-like (scalariform), union of gametes in conjugation tube (isogamous); zygospores ovoid, compressed-spherical or quadrangular, sometimes lens-shaped with one equatorial and 2 lateral keels, median wall (mesospore) ornamented with pits, radial ridges and undulations, yellow, brown or blue; aplanospores, akinetes and parthenospores observed in two species.

Cosmopolitan; rare, occurs within the margins of pools, ponds and lakes. Ten species are at present recognized but only one is known from the British Isles (but see West & Fritsch, 1927, p. 243), many species having been transferred to other genera.

Debarya may be confused with *Mougeotia* in the vegetative state, but is clearly distinguished during reproduction. In particular, at the beginning of conjugation or during aplanospore formation, the reproductive cells become filled with a pectic or cellulose colloid which is deposited as successive layers inside the cell walls.

Debarya glyptosperma (de Bary) Wittrock 1872

Basionym: *Mougeotia glyptosperma* de Bary

27080010 **Pl. 118Aa,b** (p. 482)

Cells 9–16 μm wide and 40–200 μm long; chloroplast an axile ribbon with a series of 8–12 pyrenoids; conjugation ladder-like, conjugation tubes up to 80 μm long, at first slender but increasing in width as zygospore matures within; zygospores ellipsoidal, compressed, 25–48 μm wide, 30–72 μm long, middle wall with one equatorial keel and 2 lateral keels, with an irregular polar ring of protuberances and corrugations between the keels, when mature yellow to brown.

Probably cosmopolitan and widespread in Europe; rare but moderately widespread in the British Isles in ditches, and slightly acid ponds and lakes.

Mougeotia C.Agardh 1824

Filaments simple and unattached; cells cylindrical, 5 to 20 or more times longer than broad, with plane end walls; chloroplast a flat or sometimes slightly twisted plate, mostly 1 per cell, rarely 2, occupying most of cell length, with pyrenoids in one (or more) linear row(s), occasionally scattered; when only 1 chloroplast present per cell the nucleus lies against cell wall, when 2 per cell, nucleus lies between chloroplasts; conjugation mostly ladder-like, occasionally solely lateral, isogamous (anisogamous in only 3 species); zygospores fuse within swollen conjugation tube, gametangia cut-off by 2–4 walls; zygospores range from spherical, compressed-spherical, ovoid, ellipsoidal to quadrate-ellipsoidal with flattened or truncate angles; parthenospore and aplanospore production are reported for several species.

Cosmopolitan, mostly around the margins of small ponds and lakes especially in upland areas, but also in small streams, ditches and on moist soil; often as filamentous masses entangled round aquatic macrophytes; occasionally with 1- or 2-celled rhizoidal branches, may be swept into the plankton of larger lakes where the filaments appear twisted or coiled. According to Kadlubowska (1984), 138 species are recognized worldwide, of these 26 occur in the British Isles.

The cytoplasm is pale purple in some species, at least when growing under acidic conditions. Chloroplasts of adjoining cells usually lie in the same plane so that they are seen either as a flat ribbon or in narrow profile in a given filament. Rotation of the chloroplast has been the subject of very extensive studies and is influenced not only by light intensity, but the contributions from various parts of the spectrum, including ultra-violet light. As the light intensity used for microscopy is often high compared with that where the filament was growing in the field, the chloroplast often rotates such that it is perpendicular to the slide after it has been under the microscope for a few minutes. They can be seen to rotate in response to different light intensities. *Mougeotia* is distinguished from *Debarya* and *Temnogametum*, in that cytoplasmic residues remain in the gametangia subsequent to conjugation and there is deposition of successive layers within the spore wall. In many early published figures of *Mougeotia* species, these cytoplasmic residues are not shown, for they were considered of little taxonomic significance. Many species formerly placed in *Mougeotia* have been transferred to other genera such as *Mougeotiopsis*, *Zygnemopsis* and *Hallasia*, particularly once the taxonomic relevance of changes within the cells immediately prior to and during conjugation became clear.

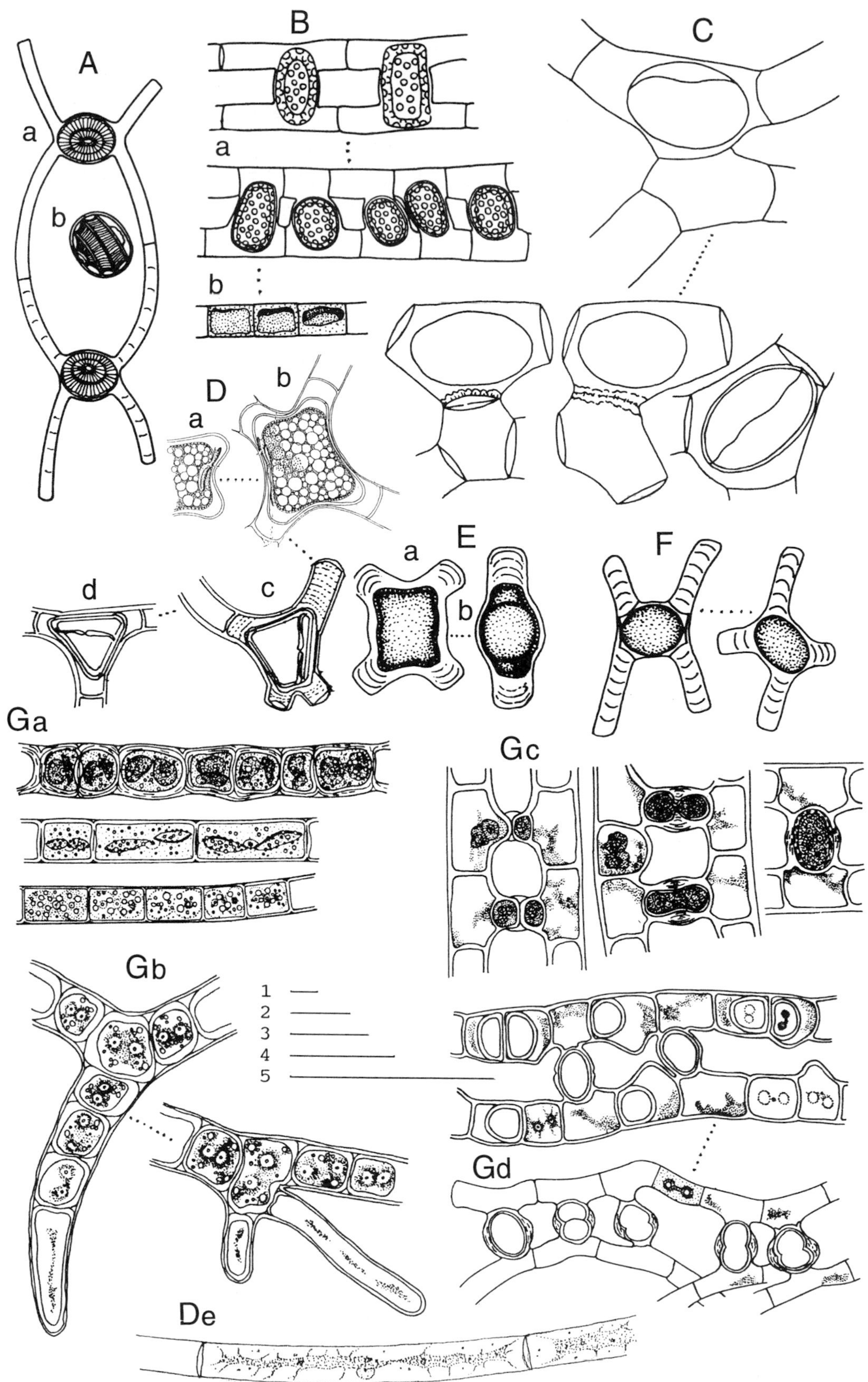

Plate 118 A–G

A. *Debarya glyptosperma* (p. 481): a. gametangia with zygospores, b. side view of zygospore; **B**. *Mougeotiopsis calospora* (p. 490): a. gametangia with pitted zygospores, b. vegetative cells with chloroplast in differing positions; **C**. *Sirogonium sticticum* (p. 490): a gametangium; **D**. *Temnogametum sinense* (p. 505): a–d, conjugating cells, e. filament exhibiting characteristic fine extensions from the chloroplast; **E**. *Zygnemopsis desmidioides* (p. 509): a. gametangium, b. gametangium profile; **F**. *Zygnemopsis tiffiana* (p. 509): gametangia; **G**. *Zygogonium ericetorum* (p. 510): a. vegetative cells, b. rhizoids, c, d. stages of ladder-like conjugation.

Scale bars: 50 µm. for A; 2 for C; 3 for B, Da–d, Ga,c,d; 4 for De, F; 5 for E, Gb.

1 Reproduction mostly sexual by means of conjugation, aplanospores rarely present 2
1 Reproduction mostly by means of aplanospores (zygospores rare or unknown) 7
2 Zygospores formed solely in conjugation tube 3
2 Zygospores formed partly in conjugation tube and extending into one or both gametangia 16
3 Filaments >20 μm wide 4
3 Filaments <20 μm wide 7
4 Conjugation mostly lateral, conjugating filaments often strongly genuflexed; cells 25–38 μm wide ....*M. genuflexa*
4 Conjugation mostly ladder-like (scalariform)........5
5 Zygospores pitted (scrobiculate)...............*M. laevis*
5 Zygospores smooth-walled....................................6
6 Filaments 35–40 μm wide..................*M. laetivirens*
6 Filaments 20–35 μm wide.......................*M. scalaris*
7 Filaments <7 μm in diameter ....*M. angusta*
7 Filaments >7 μm in diameter 8
8 Conjugation lateral and ladder-like .....*M. depressa*
8 Conjugation only ladder-like..................................9
9 Zygospore wall smooth 10
9 Zygospore wall pitted or finely punctate 13
10 Filaments generally <14 μm wide 11
10 Filaments generally >14 μm wide 12
11 Zygospores 13–25 μm wide ....*M. parvula*
11 Zygospores 25–30 μm wide.......*M. calcarea*, in part
12 Zygospores spherical, 22–33 μm in diameter ....*M. recurva*
12 Zygospores compressed and ovoid, 28–39 × 38–47 μm ....*M. gelatinosa*
13 Zygospores blue in colour.......................*M. cyanea*
13 Zygospores not blue...14
14 Zygospores spherical to ovoid, (17–)22–32(–37) μm; cells of filaments 8–16 μm in diameter ....*M. nummuloides*
14 Zygospores compressed-ovoid or spherical to subspherical 15
15 Zygospores compressed-ovoid to subspherical, 24–39 × 29–40 μm; filaments 11–14 μm wide ....*M. ovalis*
15 Zygospores compressed spherical, 30–36 × 25–35 μm; filaments 13–17 μm wide ....*M. smithii*
16 Zoospores somewhat spherical, spherical-triovate or almost quadrate 17
16 Zoospores essentially cuboidal or quadrate 21
17 Filaments <22 μm wide 18
17 Filaments >22 μm wide 20
18 Filaments >14 μm wide ....*M. abnormis*
18 Filaments <14 μm wide 19
19 Zygospores essentially spherical ....*M. calcarea*, in part
19 Zygospores ovoid to quadrangular-ovoid, 30 μm wide, 47 μm long.....................*M. paludosa*, in part
20 Zygospores ovoid-ellipsoidal, 28–35 × 40–50 μm; pyrenoids 4–8 in a single row ....*M. pulchella*
20 Zygospores cylindrical-ovoid, usually with concave sides and convex ends, 48–60 × 64–78 μm; pyrenoids scattered ....*M. varians*
21 Zygospores smooth-walled..................................22
21 Zygospores not smooth-walled 26
22 Filaments usually <5 μm wide..........*M. elegantula*
22 Filaments >5 μm wide 23
23 Filaments usually >14 μm wide ....*M. capucina*
23 Filaments <14 μm wide 24
24 Zygospores ovoid to quandrangular-ovoid ....*M. paludosa*, in part
24 Zygospores quadrate with concave sides 25
25 Zygospores 20–32 μm in size; filaments 6–9 μm wide ....*M. viridis*
25 Zygospores 29–35 μm in size; filaments 8–9 μm wide ....*M. virescens*
26 Zygospores 28–40 μm in size, punctate; filaments 7–13 μm wide. ....*M. quadrangulata*
26 Zygospores, 20–28 μm in size, minutely warty; filaments 5–7 μm in diameter ....*M. gracillima*
27 Zygospores unknown; aplanospores obliquely ellipsoidal to subspherical, 16–29 μm in length, lengthwise in gametangium, spore wall smooth, yellow to brown................................*M. ventricosa*
27 Zygospores rare, reproduces mostly by aplanospores; aplanospores ellipsoidal, 23–25 μm in length, lengthwise in gametangium, spore wall minutely punctate, yellow to brown ......*M. boodlei*

Mougeotia abnormis Kisselev 1931

27210090 **Pl. 120G** (p. 486)

Cells 10–12(–20) μm wide, 50–250 μm long; chloroplast filling length of cell, with 5–8 pyrenoids; conjugation ladder-like; zygospores in a triangular or quadrangular tube, spherical to triangular-ovoid with concave walls, 24–28(–40) μm wide, 26–36(–148) μm long, median wall smooth; parthenospores cylindrical or ellipsoidal.

Russia; in the British Isles with other algae in ponds and lake margins in Wales (Benson-Evans & Antoine 1996) where the filaments are reported as 'forming sparse entanglements'.

Some uncertainty attaches to the inclusion of this species. It is possible that this species and *M. floridana* Transeau might be conspecific, with *M. abnormis* having priority.

Mougeotia angusta Hassall 1843

Synonyms: *Mesocarpus angustus* Hassall, *Sphaerocarpus angustus* Hassall

27210100 **Pl. 119E** (p. 485)

Cells 5–6 μm wide, 30–95 μm long; chloroplasts with a median series of 4 pyrenoids; conjugation ladder-like; zygospores not filling whole conjugation canal, spherical, 7–9 μm in diameter, median wall smooth, brown.

Probably cosmopolitan; rare in the British Isles, known only from the original record in pools and pond margins in Penzance, Cornwall.

Mougeotia boodlei (West *et* G.S.West) Collins 1912

Basionym: *Gonatonema boodlei* West *et* G.S.West

27210110 **Pl. 121I** (p. 488)

Cells 4–6 μm wide, 25–225 μm long; chloroplast filling half to three-quarters of cell length, with a series of 4–6 pyrenoids; reproduction mostly by aplanospores, conjugation ladder-like but very infrequent; zygospores quadrangular with rounded angles, 15–18 µm wide, 15–23 µm long, middle wall punctate, brown; aplanospores ellipsoidal, 12–15 µm wide, 23–25 µm long, one side slightly convex, wall punctate, yellow to brown.

Europe, North America; very rare in the British Isles, recorded from ponds on Mitcham Common, Surrey, and also from Wales.

Mougeotia calcarea (Cleve) Wittrock 1872

Basionym: *Sphaerospermum calcareum* Cleve

27210120 **Pl. 120A** (p. 486)

Cells 8–14 µm wide, 40–280 µm long, frequently elongate and geniculate in early stages of reproduction; chloroplasts with a single series of 4–8 pyrenoids; conjugation ladder-like; zygospores formed either wholly within the conjugation tube or extending into both gametangia, spherical, 25–30 µm in diameter, or broadly ellipsoidal and 22–28 µm wide, 30–50 µm long with flattened angles, wall smooth, colourless or pale yellow; aplanospores rare, spherical; parthenospores known.

Probably cosmopolitan; rare in the British Isles in small pools and ponds in southern England.

Mougeotia capucina (Bory) C.Agardh 1824

Basionym: *Leda capucina* Bory in Moug. *et* Nestl., Nr. 793

Synonyms: *Staurocarpus capucinus* Hassall, *S. coerulescens* Hassall

27210010 **Pl. 121C** (p. 488) **CD**

Cells 14–21 µm wide, 70–280(–340) µm long, usually violet-coloured; chloroplasts 1 or 2 axile ribbons, occupying one-third to half of cell length and sometimes connected with lining layer of cytoplasm by fine cytoplasmic strands (see *Temnogametum sinense*, p. 505), with a series of 4–16 pyrenoids; conjugation ladder-like, first sporangium wall formed some distance (5–52 µm) from zygospore, intervening space filled with pectic compounds; sporangium divides both gametangia; zygospores irregularly quadrangular with concave sides, 50–70 × 60–100 µm, walls smooth, thick, especially at angles, violet to brown; aplanospores not uncommon, 20–36 × 45–70(–80) µm, with ends more or less produced.

Probably cosmopolitan; reported from slow-flowing rivers, ponds and ditches in England and Wales where forming purple masses tangled around reeds, or as films on various surfaces; common in acid habitats such as *Sphagnum* bogs on moors and heaths. Fruiting in late summer when lake water levels drop.

In referring to this species, West & Fritsch (1927, p. 251) state that 'the chloroplast sometimes assumes the form of an irrgular, axile rod connected with the lining layer of cytoplasm by fine colourless threads while the vacuoles contain a purple sap'. An identical form of chloroplast has been observed in *Temnogametum sinense*, so it is possible that unless material in a reproductive state is found, the two species could easily be misidentified.

Mougeotia cyanea Transeau 1926

27210130 **Pl. 120D** (p. 486)

Cells (14–)16–18(–20) µm wide, 160–200 µm long; chloroplast occupying one-third to one half of cell, with a median series of 4–10 pyrenoids; conjugation ladder-like with zygospores formed in conjugating tubes with long axis parallel to filament; zygospores 38–48 µm wide, 30–40 µm long, broadly ellipsoidal to spherical, finely punctate with wall blue in colour; aplanospores rare, spherical, 30–32 µm in diameter, formed to one side of spore-producing cell; zygospores and aplanospores when mature surrounded by a pectic layer 4–8 µm thick, disappears in preserved specimens.

USA; in the British Isles known only from Wales where recorded as rare and in nutrient-poor lakes.

Mougeotia depressa (Hassall) Wittrock 1880

Basionym: *Sphaerocarpus depressus* Hassall

Synonym: *Mesocarpus depressus* Hassall

27210140 **Pl. 119F** (p. 485)

Cells 7–12 µm wide, 35–144 µm long; conjugation ladder-like and lateral; zygospores formed in conjugating tubes, compressed and elongate-ellipsoidal with long axis parallel to filaments, 12–14 µm wide, 28–32 µm long, median wall punctate and brown.

Europe, North America, northern India; widely distributed but probably uncommon in the British Isles where found in pools, moors, meres, lakes and slow-flowing rivers.

Mougeotia elegantula Wittrock 1872

Synonym: *Mougeotia elegantula* var. *microspora* West

27210020 **Pl. 121B** (p. 488)

Cells 3.5–5 µm wide, 50–135 µm long; chloroplasts axile, ribbon-like with a series of 4–8 pyrenoids in a row; conjugation ladder-like, conjugating cells geniculate with sporangia forming in both gametangia; zygospores almost cross-shaped to somewhat quadrate with rounded angles, 12–24 µm wide, 20–24 µm long, walls smooth and colourless; aplanospores rare, ellipsoidal, 6–9 µm wide, 20–24 µm long, smooth and colourless.

Probably cosmopolitan; very widely distributed in the British Isles including the Orkneys, Scotland, North Wales, west of Ireland and England in pools, ponds and lake margins, often forming masses in shallow waters.

Mougeotia gelatinosa Wittrock 1889

27210150 **Pl. 120C** (p. 486)

Cells 12–18 µm wide, 120–180 µm long; chloroplasts axile, ribbon-like, with 3–6 pyrenoids in median series; conjugation ladder-like; zygospore formed in conjugation tube, long axis parallel to conjugation filaments; zygospore compressed, ovoid, 28–39 wide, 38–47 µm long, characteristically surrounded by a 7–10 µm thick pectic layer when mature; median wall smooth and brown.

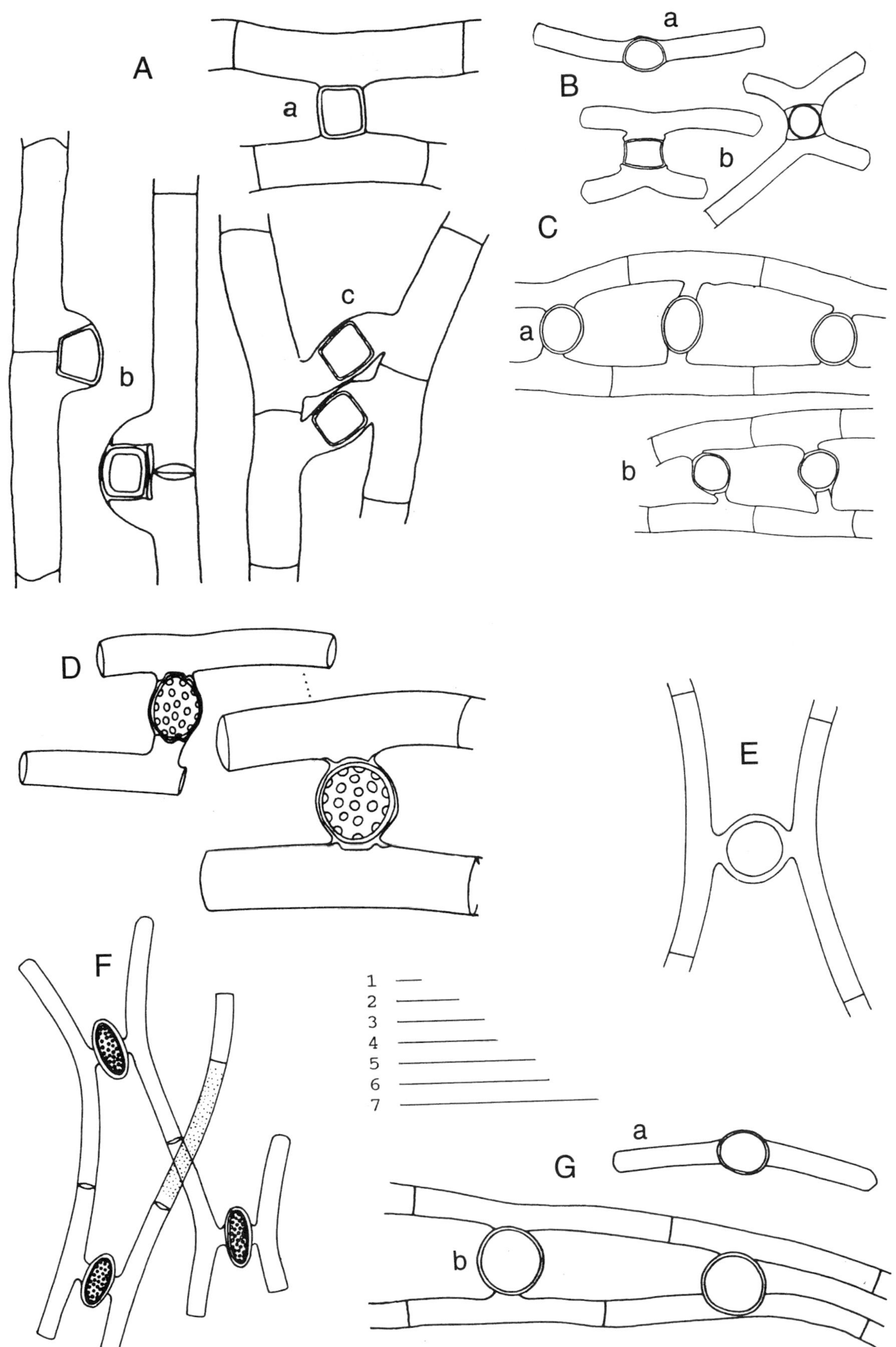

Plate 119 A–G

A. *Mougeotia genuflexa* (p. 487): b. lateral conjugation with zygospores, a, c. ladder-like conjugation with zygospores; **B**. *Mougeotia laetevirens* (p. 487): a. aplanospore, b. ladder-like conjugation with zygospores; **C**. *Mougeotia scalaris* (p. 489): a, b. ladder-like conjugation with zygospores; **D**. *Mougeotia laevis* (p. 487): ladder-like conjugation with zygospores; **E**. *Mougeotia angusta* (p. 483): ladder-like conjugation; **F**. *Mougeotia depressa* (p. 484): ladder-like conjugation with zygospores; **G**. *Mougeotia parvula* (p. 487): a. parthenospore, b. ladder-like conjugation.

Scale bars: 50 μm. 1 for B; 2 for C; 3 for A; 4 for F; 5 for D; 6 for G; 7 for E.

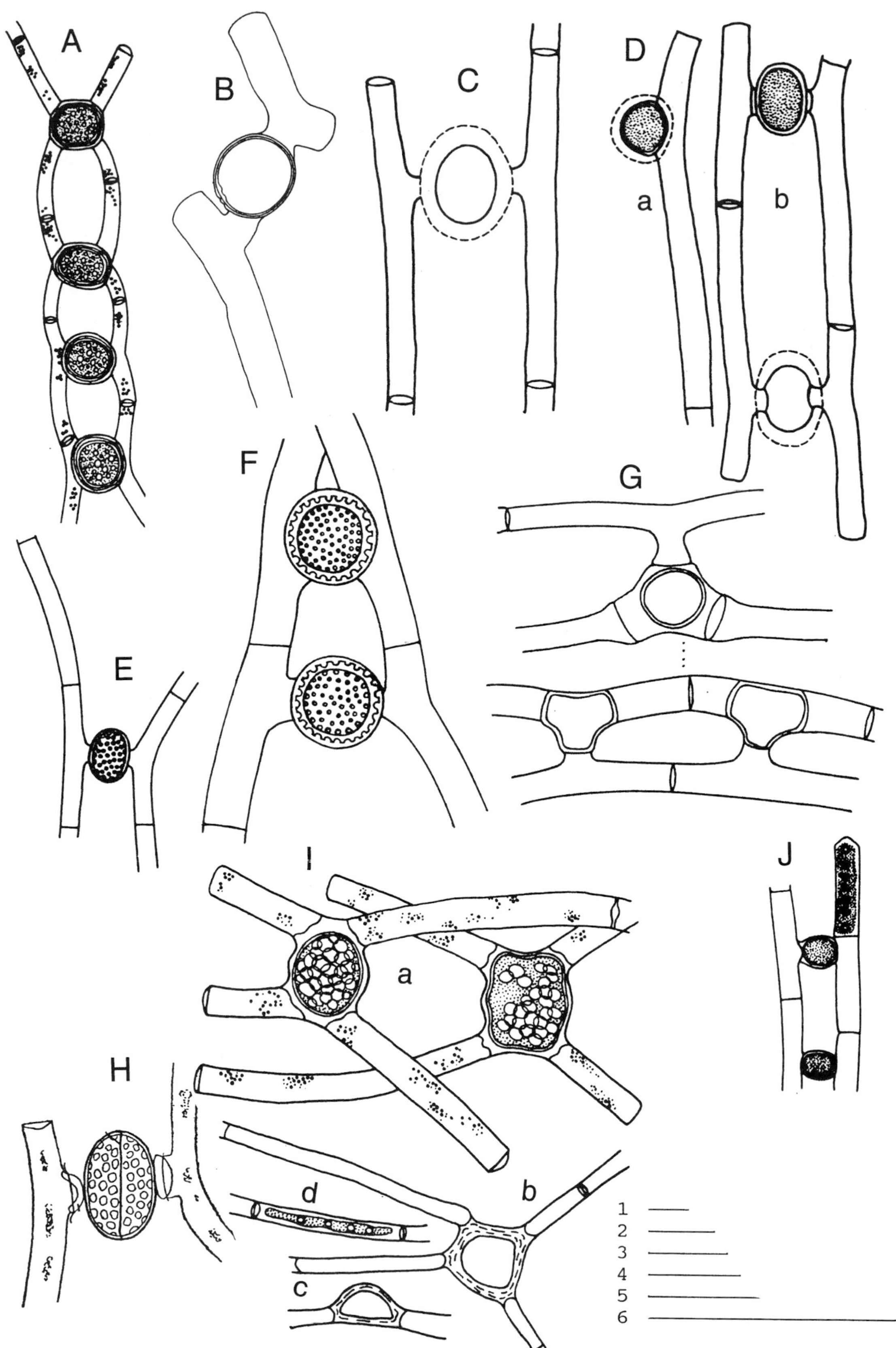

Plate 120 A–J

A. *Mougeotia calcarea* (p. 484): ladder-like conjugation with zygospores; **B**. *Mougeotia recurva* (p. 489): ladder-like conjugation with zygospore; **C**. *Mougeotia gelatinosa* (p. 484): ladder-like conjugation with zygospore showing gelatinous sporangium wall; **D**. *Mougeotia cyanea* (p. 484): a. parthenospore, b. ladder-like conjugation; **E**. *Mougeotia ovalis* (p. 487): ladder-like conjugation with zygospore; **F**. *Mougeotia nummuloides* (p. 487); ladder-like conjugation with zygospores showing pitted ornamentation; **G**. *Mougeotia abnormis* (p. 483): ladder-like conjugation with zygospores; **H**. *Mougeotia smithii* (p. 489): gametangia; **I**. *Mougeotia paludosa* (p. 487): a. conjugated filaments with zygospores, b. gametangium, c. aplanospore, d. vegetative cell; **J**. *Mougeotia pulchella* (p. 489): ladder-like conjugation with zygospores.

Scale bars: 50 μm. 1 for J; 2 for A; 3 for B–D, Ib–d; 4 for E, H; 5 for G; 6 for B, Ia, F.

Europe, Russia; very rare in the British Isles, recorded only from upland acid waters in Scotland.

Mougeotia genuflexa (Dillwyn) C.Agardh 1824

Basionym: *Conferva genuflexa* Dillwyn

27210160 **Pl. 119A** (p. 485)

Cells 25–38 μm wide, 50–225 μm long; chloroplasts broad, ribbon-like, with pyrenoids usually in irregular series; conjugation mostly lateral, less frequently ladder-like, filaments often geniculate at conjugation; zygospores formed in conjugation tube, quadrate-ovoid or in lateral conjugants trapezoid, 30–40 μm wide, median wall smooth, light brown.

Probably cosmopolitan; widespread in the British Isles in various habitats including bogs and moorlands, often forming extensive nets and sometimes having rhizoidal branches.

Mougeotia gracillima (Hassall) Wittrock 1872

Basionym: *Staurocarpus gracillimus* Hassall

Synonym: *Staurospermum gracillimum* (Hassall) Kützing

27210030 **Pl. 121G** (p. 488)

Cells 5–7 μm wide, 55–140 μm long; conjugation ladder-like, sporangia cross-shaped and dividing both gametangia; zygospores quadrate with deeply concave sides, 20–25 μm wide, 20–28 μm long, each angle with a slight notch, wall minutely warty; aplanospores elogate-ovoid.

Europe, USA; in the British Isles widespread, in bogs and moorland pools, and according to West & Fritsch (1927) conjugating 'freely in all parts of the country and at elevations up to 1200 ft [400 m]'.

Mougeotia laetevirens (A.Braun) Wittrock 1877

Basionym: *Craterospermum laetevirens* A.Braun

27210170 **Pl. 119B** (p. 485)

Cells 35–40 μm wide, 65–350 μm long; chloroplasts broad, ribbon-like, with numerous scattered pyrenoids; conjugation ladder-like with conjugating cells geniculate and zygospores forming in conjugating tubes; zygospores spherical, 36–75 μm in diameter, outer wall cylindrical with concave sides, median wall smooth, yellow-brown; aplanospores ovoid or obliquely ovoid, quite variable in shape.

Probably cosmopolitan; with wide ecological tolerances, rare in the British Isles as light green cottony growths in sheltered waters, including old quarries.

Mougeotia laevis (Kützing) W.Archer 1867

Basionym: *Zygogonium laeve* Kützing

Synonym: *Debarya laevis* (Kützing) West *et* G.S.West

27210040 **Pl. 119D** (p. 485)

Cells 20–36 μm wide, 20–100 μm long; chloroplast axile, ribbon-like, with 2–4 pyrenoids in a series; conjugation ladder-like, zygospores produced in a straight or curved conjugation tube; zygospores ellipsoidal to ovoid, 20–36 μm wide, 35–50 μm long, walls with shallow depressions or pits (scrobiculate) about 3 μm in diameter and 2–3 μm apart.

Europe, North America, North Africa; rare in the British Isles where recorded in ditches and pools in Rutland and Mitcham Common, Surrey; also known from disused quarries.

Mougeotia nummuloides (Hassall) G.B. De Toni 1889

Basionym: *Sphaerocarpus nummuloides* Hassall

Synonym: *Mesocarpus nummuloides* Hassall

27210180 **Pl. 120F** (p. 486)

Cells 8–16 μm wide, 32–160 μm long; chloroplasts with a median series of 2–6 pyrenoids; conjugation ladder-like with zygospores formed in short conjugation tubes; zygospores spherical to ovoid, (17–)22–32(–37) μm in diameter, median wall strongly pitted (scrobiculate), brown; aplanospores spherical to ovoid, within an angled sporogenous cell.

Europe, North America, North Africa; with a broad ecological tolerance; in the British Isles mostly in ditches, pools and ponds with *Sphagnum* (pH 5–6) on moors. Records from Hertford Heath (Hertfordshire), Penzance (Cornwall) and from Cumbria at Scarth Gap Pass and Kirk Fell.

Mougeotia ovalis (Hassall) Nordstedt 1886

Basionym: *Sphaerocarpus ovalis* Hassall

Synonym: *Mesocarpus ovalis* Hassall

27210190 **Pl. 120E** (p. 486)

Cells 11–14 μm wide, 110–140 μm long; conjugation ladder-like, with zygospores formed in conjugation tubes; zygospores compressed-ovoid lying parallel to filaments, occasionally subspherical, 24–39 μm wide, 29–40 μm long, median wall finely pitted (scrobiculate), brown.

Europe; only reported in the British Isles from Hertford Heath, Hertfordshire and near Penzance, Cornwall.

Mougeotia paludosa G.S.West 1899

27210050 **Pl. 120I** (p. 486)

Cells 11.5–13.5 μm wide, 70–185 μm long; chloroplasts occupying only about one-third the length of the cell, with a median series of 4–6 pyrenoids; conjugation ladder-like with conjugating cells often reflexed, sporangia dividing both gametangia, separating cross walls undulate; zygospores ovoid to quadrangular-ovoid with truncate angles, 32–38 μm wide, 44–49 μm long, inner wall thin, outer wall thicker, both smooth and colourless.

Probably cosmopolitan; first described in England from pools in Cambridgeshire.

Mougeotia parvula Hassall 1843

Basionym: *Sphaerocarpus parvulus* Hassall

Synonym: *Mesocarpus parvulus* Hassall

27210060 **Pl. 119G** (p. 485)

Cells 6–13 μm wide, 30–140 μm long; chloroplasts occupying about two-thirds of cell, with a median series of 4–8 pyrenoids; conjugation ladder-like, zygospores formed wholly in very short conjugation tubes; zygospores spherical, 13–25 μm in diameter, median wall

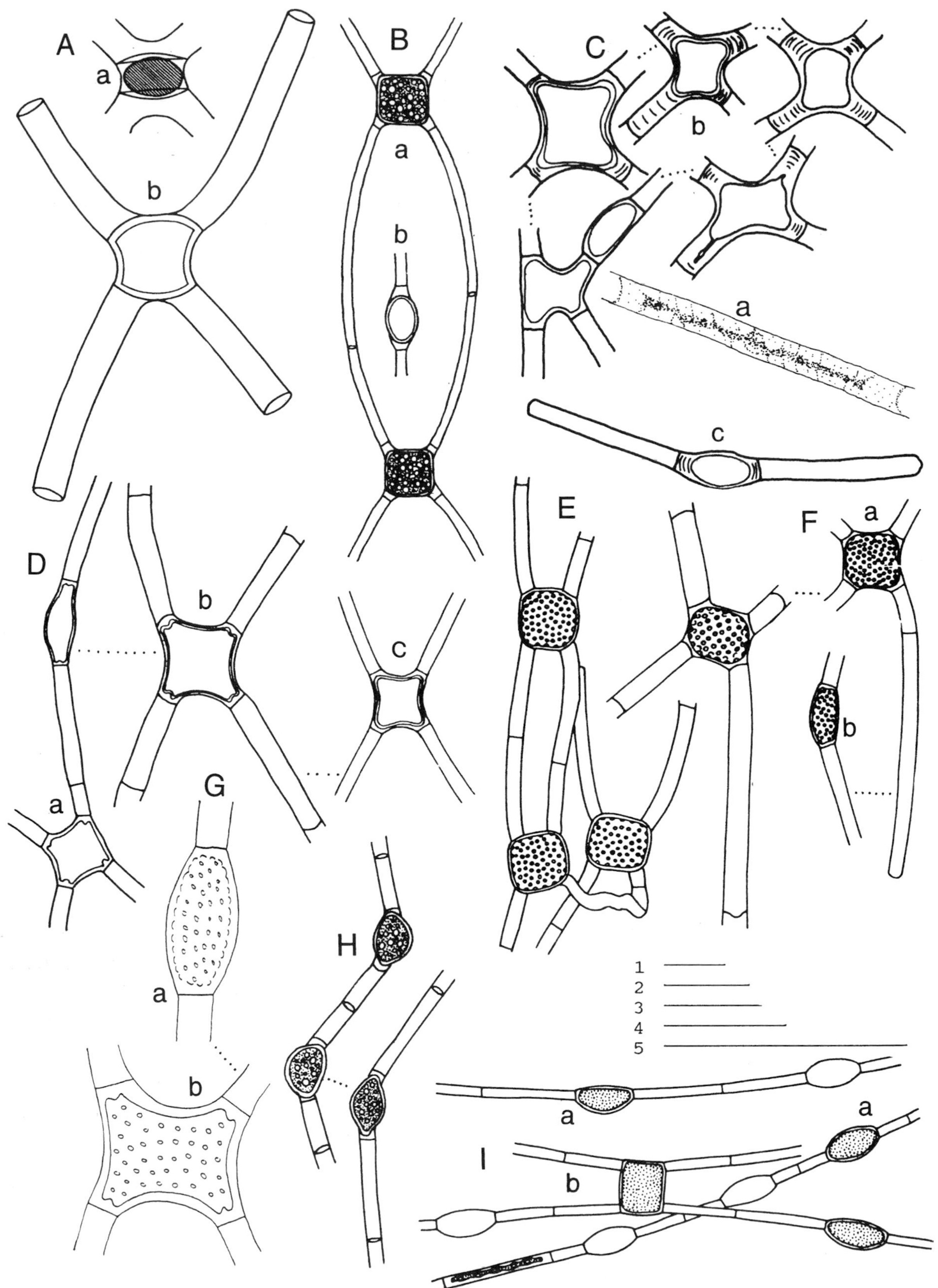

Plate 121 A–I

A. *Mougeotia varians* (p. 489): ladder-like conjugation with zygospores; **B**. *Mougeotia elegantula* (p. 484): a. conjugation, b. parthenospore; **C**. *Mougeotia capucina* (p. 484): a. vegetative cell, b. ladder-like conjugation and zygospores, c. a parthenospore; **D**. *Mougeotia viridis* (p. 490): a. filament with ladder-like conjugation with a zygospore and a parthenospore, b, c. ladder-like conjugation; **E**. *Mougeotia virescens* (p. 490); **F**. *Mougeotia quadrangulata* (p. 489): a. ladder-like conjugation with zygospores, b. parthenospore; **G**. *Mougeotia gracillima* (p. 487): a. an aplanospore, b. ladder-like conjugation with zygospore; **H**. *Mougeotia ventricosa* (p. 489): aplanospores; **I**. *Mougeotia boodlei* (p. 483): a. filaments with aplanospores, b. filament with ladder-like conjugation with zygospore and aplanospores.

Scale bars: 50 μm. 1 for C; 2 for A; 3 for B, E, F, H, I; 4 for Da,c; 5 for Db, G.

thick, smooth; aplanospores broadly ovoid, 16–20 μm wide, 20–24 μm long.

Probably cosmopolitan, widely distributed on most continents; widespread and the most abundant species of *Mougeotia* in the British Isles where often entangled with other plants along the weedy margins of lakes, also common in acidic pools at high altitudes; frequently found with zygospores in spring and early summer.

Mougeotia pulchella Wittrock 1871

27210200 **Pl. 120J** (p. 486)

Cells 24–29 μm wide, 48–150 μm long; chloroplasts ribbon-like, with a series of 4–8 pyrenoids; conjugation ladder-like; zygospores formed in conjugation tube, ovoid to ellipsoidal, with ends more or less flattened, 28–35 μm wide, 40–50 μm long, walls yellow-brown and punctate.

Probably cosmopolitan; known fom the British Isles from various freshwater habitats such as ponds and lake margins including Lake Bala in North Wales.

Mougeotia quadrangulata Hassall 1843

Synonyms: *Mougeotia quadrata* (Hassall) Wittrock, *Staurocarpus quadratus* Hassall, *Staurospermum quadratum* (Hassall) Kützing

27210210 **Pl. 121F** (p. 488)

Cells 7–13 μm wide, 50–180 μm long; chloroplasts with a median series of 8–16 pyrenoids; conjugation ladder-like; zygospores quadrangular in face view, broadly ellipsoidal in side view, with straight sides and truncate angles (angles rarely retuse), 28–40 μm wide, 28–40 μm long, walls with large pits (scrobiculations); aplanospores ovoid or obliquely ovoid with truncate ends, 20–21 μm wide, 36–44 μm long, often formed in slightly bent cells.

Probably cosmopolitan; recorded in England from ditches and ponds in Ashdown Forest, Cheshunt, Hertfordshire, near Crawley Down, Sussex and Penzance, Cornwall, also in shallow lake margins, sometimes in moorland areas.

Mougeotia recurva (Hassall) De Toni 1889

Basionym: *Sphaerocarpus recurvus* Hassall
Synonym: *Mesocarpus recurvus* Hassall

27210070 **Pl. 120B** (p. 486)

Cells 12–18 μm wide, 50–180 μm long; chloroplasts with a median series of 4–8 pyrenoids; conjugation ladder-like with each zygospore formed in comparatively wide conjugation tube; zygospores spherical, 22–33 μm in diameter, walls smooth, brown; aplanospores spherical, 24–30 μm in diameter, formed in mostly short, geniculate cells, cylindrical-ovoid, 14–18 μm wide, 28–34 μm long; parthenospores broadly ovoid or subspherical, 29–43 μm wide.

Probably cosmopolitan; fairly widespread in lakes in England and Scotland, sometimes in ditches and also pools in dune slacks such as Braunton Burrows, North Devon.

Mougeotia scalaris Hassall 1842

Synonym: *Mesocarpus scalaris* Hassall

27210220 **Pl. 119C** (p. 485)

Cells 20–35 μm wide, (40–)50–150(–180) μm long; chloroplasts with a median series of 4–10 pyrenoids; conjugation ladder-like, conjugating cells straight or slightly curved; zygospores ovoid to spherical, 30–40 μm and formed wholly in conjugation tube, median wall smooth, yellow-brown.

Europe, North America, North Africa, Asia; recorded from several different types of aquatic habitat, but especially ponds and boggy pools; Hassall (1842) considered it to be common in England.

Mougeotia smithii F.E.Fritsch 1949

27210230 **Pl. 120H** (p. 486)

Cells 13–17 μm wide, 6 to 18 times as long as wide; chloroplast not known; conjugation ladder-like, rarely apical-lateral, conjugation tube becomes spherical (about 20 μm in diameter) soon after gametic fusion and cut-off from conjugating cells by a layer of white deposit, zygospores formed entirely within the conjugation tube; zygospores compressed-spherical, 25–35 μm wide, 30–36 μm long, median wall irregularly pitted (scrobiculate), with a prominent suture, olive-brown.

British Isles (England); only described from pools near the lower falls of the River Ure at Aysgarth, Yorkshire; apparently not recorded since originally described, nor mentioned in subsequent monographic treatments.

Although Fritsch (1949b) states that this species has similarities with *M. oedogonioides* Czurda and *M. adnata* Iyengar, it would seem to be much closer in all respects to *M. pawhuskae* Taft, except that the zygospores are never surrounded by a mucilage envelope. This species is a later homonym of *M. smithii* (Transeau) Czurda, now generally accepted as a synonym of *Debarya smithii* Transeau.

Mougeotia varians (Wittrock) Czurda 1932

Basionym: *Mougeotia laetevirens* (A.Braun) Wittrock var. *varians* Wittrock
Synonym: *Mougeotia mysorensis* M.O.P.Iyengar

27210240 **Pl. 121A** (p. 488)

Cells of filaments 25–27 μm wide; chloroplasts with numerous scattered pyrenoids; conjugation ladder-like, with zygospores formed in conjugation tubes and extending into or across the gametangia; zygospores cylindrical-ovoid, usually with concave sides and convex ends, 48–60 μm wide, 64–78 μm long, wall smooth and yellow-brown; sporangia adjoined by 2, 3, or 4 cell remnants.

Probably cosmopolitan; in the British Isles found in damp areas on sandstone cliffs and in pools off the River Dee in Cheshire, also in ponds and possibly slightly brackish-water.

Mougeotia ventricosa (Wittrock) Collins 1912

Basionym: *Gonatonema ventricosum* Wittrock

27210250 **Pl. 121H** (p. 488)

Cells 6–9 μm wide, 100–140 μm long; chloroplasts with a series of about 4 pyrenoids; zygospores unknown; aplanospores obliquely ellipsoidal to subspherical with ends narrowing towards gametangial septa, 12–24 μm wide, 16–29 μm long, walls smooth, yellow-brown.

Europe, North America; reported in the British Isles only from the west of Ireland.

Mougeotia virescens (Hassall) O.Borge 1913

Basionym: *Staurocarpus virescens* Hassall

27210260 **Pl. 121E** (p. 488)

Cells 8–9 μm wide, 50–220 μm long; conjugation ladder-like, with zygospores filling the quadrate sporangium; zygospores quadrate with concave sides and rounded angles, 29–35 μm wide, 29–35 μm long, wall smooth, colourless.

Probably cosmopolitan; apparently common in maritime areas, originally described by Hassall from a pool near Royden, Essex.

Mougeotia viridis (Kützing) Wittrock 1872

Basionym: *Staurospermum viride* Kützing
Synonym: *Staurocarpus viride* Hassall

27210080 **Pl. 121D** (p. 488)

Cells 6–9 μm wide, 40–160 μm long; chloroplasts occupying most of cell, with a series of 2–6 pyrenoids; conjugation ladder-like; zygospores quadrate with concave sides and notched angles, 20–32 μm wide, 20–32 μm long, median wall smooth, colourless; aplanospores obliquely ellipsoidal with notched ends adjacent to transverse gametangial walls, 14–16 μm wide, 30–36 μm long.

Probably cosmopolitan; one of the most abundant *Mougeotia* species in the British Isles where reported from the English Lake District, Yorkshire, Leicestershire, Cambridgeshire, Warwickshire, North Wales and the west of Ireland, often in ditches, meadowland and moorland pools (pH 4.5–7.5).

Mougeotiopsis Palla 1894

Filaments of short, cylindrical cells, 0.5 to 4 times longer than broad, with plane end walls; chloroplast single, axile, flat and plate-like or 'dished' (curved), with a thickened and minutely granulate margin, sometimes margin inrolled, said to lack pyrenoids (see note under species) but containing starch granules and oil droplets, the latter sometimes on the cell surface; sexual reproduction isogamous with zygospores formed somewhat irregularly inside the conjugation tube and extending into one or both gametangia, but not cut-off from gametangia by a wall.

Europe, North America; possibly restricted to acidic water (pH 5.5–6.5). Monospecific.

Distinguished from the genus *Mougeotia* only by the form of its chloroplasts.

Mougeotiopsis calospora Palla 1894

Synonym: *Mougeotia calospora* (Palla) Czurda

27220010 **Pl. 118B** (p. 482)

Cells 10–18 μm wide, 10–70 μm long; conjugation ladder-like, conjugating cells similar to or somewhat longer than the vegetative cells; zygospores irregularly ovoid to quadrate-ovoid, 16–25 μm wide, 21–38 μm long, median wall thick, deeply pitted (scrobiculate), pits 2–2.5(–3) μm across, light yellow to brown, with outer wall thin and transparent.

Europe, North America; very rare in the British Isles since only recorded from acid pools in Yorkshire.

Questions attach to the validity of this genus and species. W. & G.S.West (1900b) recorded it from Yorkshire as '*Debarya calospora* (Palla) nob.' and described it as lacking a pyrenoid – the only member of the Zygnematales to do so. The Wests claimed in 1898 to have seen this same alga previously although with a pyrenoid in each chloroplast and so tranferred it to the genus *Debarya*. However, Transeau (1951), Bourrelly (1972), Kadlubowsaka (1972, 1984) and Ettl et al. (1984) all continue to recognize *Mougeotiopsis calospora*; for this reason it is retained here.

Sirogonium Kützing 1843

Cells comparatively short, only 2 to 4(–7) times longer than broad, always with plane end walls, mucilaginous sheath virtually absent so filaments not slippery or slimy to the touch; chloroplasts 2–10, almost straight or very slightly spiralling and ribbon-like, lying against cell wall (parietal); conjugation by only a few cells in each filament and no distinct conjugation tube formed, conjugating cells coming into contact by bending towards one another (especially the receptive female cell) and a perforation develops in the wall at point of contact, receptive (female) cell becomes somewhat swollen and a zygospore develops within it after fertilization; zygospores ellipsoidal or less commonly ovoid, wall smooth or variously ornamented, yellow, brown or black.

Probably cosmopolitan; mostly in aquatic habitats in the tropics. Sixteen species described, only one reported from the British Isles.

The vegetative cells resemble those species of *Spirogyra* with straight or almost straight chloroplasts, but are differentiated from this genus by their mode of conjugation. In *Sirogonium* a conjugation tube is not formed and genuflexion of the conjugating cells is frequent. However, the cell wall (unlike *Spirogyra*) lacks the mucilaginous external pectose layer and hence is less slippery when retrieved by hand. The conjugating filaments frequently form a tangled net as a result of the flexing of the gametangia at conjugation, enabling several separate filaments to conjoin.

Sirogonium sticticum (Engl. Bot.) Wille 1884

Basionym: *Conferva stictica* Engl. Bot. (t. 2463)

27320010 **Pl. 118C** (p. 482)

Cells 38–56 μm wide, 80–300 μm long; chloroplasts 3–6, nearly straight or at most making only half a turn of the cell; conjugation between short, more or less reflexed gametangia; receptive (female) gametangium inflated up to 72 μm wide; zygospores ellipsoidal, sometimes ovoid, 41–90 μm wide, 66–127 μm long, median wall smooth and yellow.

Probably cosmopolitan; widely distributed in Europe in stagnant pools and bogs and known in the British Isles from Ingleton, Yorkshire where it occurs attached to stones over which water is fairly fast-flowing (West & Fritsch, 1927).

Cell and zygospore dimensions and numbers of chloroplasts per cell are reported to be highly variable.

Spirogyra Link 1820

Filaments mostly long, unbranched, cells cylindrical and up to 30 times as long as broad, with outer mucilaginous sheath usually giving them a distinctly slippery feel. Cross walls commonly plane, sometimes with a ring-like infold (replicate), in rare cases semireplicate and in three species colligate where an external collar develops around junction of adjacent cells; chloroplasts parietal and in the form of spiral bands or ribbon-like, usually 1–16 per cell, sometimes closely coiled and making up to 8 turns along length of cells, sometimes almost straight, broad or narrow with smooth or wavy margins, containing disc-like pyrenoids usually regularly placed along length; nucleus often clearly visible, often with a distinct nucleolus and suspended across centre of cell by cytoplasmic strands linked to cytoplasm lining of cell.

Conjugation is ladder-like (scalariform), less frequently lateral (both in some species), conjugation tubes mostly developing from both conjugating filaments, less frequently from one filament, following migration and fusion of gametes the resulting zygospore usually develops in the receptive ('female') gametangial cells that become swollen to accommodate the maturing zygospore.

Zygospores vary from subspherical, ovate (watermelon-like), ellipsoidal or, much more rarely, flattened and lens-shaped; outer wall (exospore) of zygospore of cellulose, mostly transparent, wrinkled or punctate but not markedly ornamented, median wall (mesospore) smooth or variously decorated and pale yellow to chestnut brown in colour when mature, walls have a suture circling the zygospore longitudinally (not always visible); asexual reproduction by akinetes (rare) and aplanospores, the latter differing in size from the zygospores of the same species but similar in shape and ornamentation; parthenospores occasionally observed.

Cosmopolitan, the most common and best known member of the Zygnemataceae; often occurring as bright green masses, when abundant considered an objectionable nuisance in garden ponds in early summer, sometimes present in pools and ditches and the backwaters of slow-flowing rivers. Tending to be most abundant in spring and disappearing in summer, to reappear again in late autumn, sometimes lasting through the winter months. Although it has been suggested that this pattern of periodicity is due primarily to changes in water chemistry, some observations (Brook, 1955c, 1975) indicate that it is the feeding of insect larvae, tadpoles and young stages of certain fish that causes *Spirogyra* and some other freshwater algae to show dramatic declines in abundance in early summer. Some species tend to be restricted to hard waters (e.g. *S. crassa*), whereas the majority occur in soft, acidic waters, including those rich in organic acids where there is decaying vegetation. Of the 400 or more species, only about 50 have been reported from the British Isles.

Species identification is based on a variety of characters: the dimensions of the filaments (especially width), character of the septa (plane, replicate or colligate), and the number of chloroplasts and number of turns made as they spiral along the cell. Care should be taken to distinguish specimens with almost straight or only slightly twisted chloroplasts from the genus *Sirogonium* (see p. 490). The main distinguishing features are those associated with conjugation. Especially important are whether conjugation is ladder-like or lateral, the shape of the conjugation tubes and gametangia, and the shape, ornamentation and colour of the zygospores (and/or aplanospores). Godward (1956) suggests that another useful character is the proportion of cells in a filament taking place in conjugation; sometimes nearly all do so but in others only a small proportion are involved. The examination of zygospore ornamentation demands the use of an oil immersion objective and sometimes the addition of dilute KOH to enhance details of the ornamentation. The measurement of cell width, especially in the broader species, requires care so that no undue pressure is placed on the cover glass lying over the specimen. Such pressure can spread the cell walls and so give rise to erroneous dimensions.

1 Cells with distinct external collar around junction of adjacent cells (colligate)....... *S. colligata*
1 Cells otherwise 2
2 Cells with plane end walls 3
2 Cells with a ring-like infold at end walls (replicate)..40
3 Cells with a single chloroplast (very rarely 2) 4
3 Cell with more than one chloroplast....................21
4 Filaments generally <15 μm wide 5
4 Filaments generally >15 μm wide 6
5 Filaments 8–10 μm wide; zygospores ovoid .... *S. parva*
5 Filaments 10–14 μm wide; zygospores ellipsoidal .... *S. flavescens*
6 Filaments generally 15–30 μm wide.......................7
6 Filaments generally >30 μm wide 13
7 Zygospores essentially ovoid, only occasionally almost spherical or ellipsoidal...............................8
7 Zygospores ellipsoidal 9
8 Reproduction only by zygospores .... *S. longata*
8 Reproduction by zygospores and aplanospores .... *S. mirabilis*
9 Reproduction by zygospores only.........................10
9 Reproduction by zygospores and aplanospores....11
10 Filaments generally 25–30 μm wide .... *S. affinis*
10 Filaments generally 15–25 μm wide.....................11
11 Gametangia usually cylindrical (rarely inflated); zygospores 19–26 × 36–78 μm .... *S. communis*
11 Gametangia inflated mostly on side of conjugation; zygospores 23–30 × 40–65 μm .... *S. gracilis*
12 Filaments 20–24 μm wide........................ *S. parvula*
12 Filaments 24–32 μm wide............. *S. catenaeformis*
13 Zygospores polymorphic .... *S. lutetiana*
13 Zygospores essentially ovoid or ellipsoidal, only occasionally ovoid-ellipsoidal 14
14 Zygospores essentially ovoid (occasionally ovoid-ellipsoidal) 15
14 Zygospores always or nearly always ellipsoidal 16
15 Filaments 40–50 μm wide; zygospores 38–50 × 50–83 μm, with single outer wall layer...*S. porticalis*
15 Filaments 29–41 μm wide; zygospores 37–57 × 60–100 μm, with an outer wall of 2 layers ...*S. velata*
16 Zygospore median wall smooth 17
16 Zygospore median wall ornamented (scrobiculate) .... *S. scrobiculata*

17 Zygospores >40 μm wide 18
17 Zygospore <40 μm wide......................................19
18 Filaments 40–48 μm wide; zygospores 40–50× 70–125 μm ....*S. circumlineata*
18 Filaments 30–150 μm wide; zygospores 44–64×90–150 μm ....*S. reflexa*
19 Filaments >40 μm wide ....*S. condensata*
19 Filaments <40 μm wide 20
20 Fertilised cells inflated on outer side, not inflated on conjugating side...................*S. borgeana*
20 Fertilised cell usually inflated on conjugating side only ....*S. varians*
21 Filaments <50 μm wide 22
21 Filaments >50 μm wide 27
22 Zygospores lens-shaped.........................*S. pellucida*
22 Zygospores not lens-shaped.................................23
23 Zygospores ovoid 24
23 Zygospores ellipsoidal 26
24 Zygospores 31–40 μm in diameter and 66–150 μm in length ....*S. decima*
24 Zygospores >40 μm in diameter 25
25 Zygospores 40–50 μm in diameter and 60–250 μm in length...................................*S. dubia*
25 Zygospores 47–85 μm in diameter and 68–110 μm in length ....*S. fluviatilis*
26 Filaments 26–33 μm wide; chloroplasts 2 per cell; zygospores 39–48 μm in diameter ....*S. hassallii*
26 Filaments 36–41 μm wide; chloroplasts 2 or 3 per cell; zygospores 32–42 μm in diameter ....*S. rivularis*
27 Filaments 50–70(–80) μm wide............................28
27 Filaments >70 μm wide 30
28 Zygospores lens-shaped.......................*S. majuscula*
28 Zygospores ovoid 29
29 Gametangia enlarged or inflated; zygospores sometimes develop at right angles to filament ....*S. neglecta*
29 Gametangia not as above although conjugation tube sometimes inflated; zygospores not so orientated...*S. ternata*
30 Cells with fewer than 5 chloroplasts....................31
30 Cells with 6 or more chloroplasts.........................36
31 Filaments generally (68–)70–80 μm wide.............32
31 Filaments generally >80 μm wide 34
32 Zygospores lens-shaped ....*S. bellis*
32 Zygospores ellipsoidal 33
33 Filaments 68–78 μm wide; gametangial cells cylindrical; zygospores with a yellow median wall ....*S. turfosa*
33 Filaments 70–80 μm wide; gametangial cells cylindrical or enlarged; zygospores with a brown median wall..*S. nitida*
34 Zygospores ovoid ....*S. jugalis*
34 Zygospores ellipsoidal 35
35 Filaments 80–90 μm wide....................*S. brittanica*
35 Filaments 90–115 μm wide..................*S. setiformis*
36 Filaments 80–110 μm wide 37
36 Filaments >110 μm wide 38
37 Cells with 7 or 8 chloroplasts...........*S. subechinata*
37 Cells with 8–10 chloroplasts.......*S. submargaritata*
38 Cells with a variable number of chloroplasts, frequently more than 8 (commonly 6–12) ....*S. crassa*
38 Cells with 6 or 7 chloroplasts 39
39 Filaments 118–140 μm wide...................*S. maxima*
39 Filaments 145–160 μm wide ....*S. crassiuscula*
40 Cells generally with more than 1 chloroplast, occasionally some cells in a filament with a single chloroplast..41
40 Cells with a single chloroplast 43
41 Zygospores with median wall smooth ....*S. insignis*
41 Zygospores with median wall ornamented..........42
42 Zygospore median wall with spine- or nipple-like papillae ....*S. borysthenica*
42 Zygospore median wall with a net-like ornamentation ....*S. reticulata*
43 Zygospores ovoid or cylindrical-ovoid 44
43 Zygospores ellipsoidal 46
44 Conjugation tubes develop only or largely from male gametangia................................*S. grevilleana*
44 Conjugation tubes develop from both gametangia 45
45 Zygospore median wall smooth ....*S. weberi*
45 Zygospore median wall pitted (scrobiculate) ....*S. protecta*
46 Filaments <13 μm wide.....................*S. tenuissima*
46 Filaments >13 μm wide 47
47 Filaments 24–30 μm wide ....*S. quadrata*
47 Filaments generally <24 μm wide 48
48 Filaments 15–20 μm wide.........................*S. inflata*
48 Filaments 18–21(–25) μm wide ....*S. spreeiana*

Spirogyra affinis (Hassall) P.Petit 1880

Basionym: *Zygnema affine* Hassall
Synonym: *Spirogyra suecica* Transeau
27340010 **Pl. 122G** (p. 494)

Cells 25–30 μm wide, 35–90 μm long, end walls plane; chloroplast single, making 1–3.5 turns of cell; conjugation mostly lateral but sometimes ladder-like, conjugation tubes formed by both gametangia; fertilized cells spherical, inflated; zygospores ellipsoidal, 28–33 μm wide, 30–50 μm long, median wall smooth, yellow.

Probably cosmopolitan, widespread in Europe; according to West & Fritsch (1927), abundant in every part of the British Isles except mountainous districts.

Some authors place this in the synonymy of *S. catenaeformis* (Hassall) Kützing although the zygospores of the latter are larger than those of *S. affinis* (see comments under *S. catenaeformis*).

Spirogyra bellis (Hassall) Cleve 1868

Basionym: *Zygnema belle* Hassall
27340030 **Pl. 124H** (p. 498)

Cells 65–80 μm wide, 90–350 μm long, end walls plane; chloroplasts 5 or 6, almost straight or making only one turn of cell; conjugation ladder-like, conjugation tubes formed by both gametangia; fertilized cells very short and inflated so mostly broader than long; zygospores lens-shaped, 60–90 μm wide, 48–60 μm long, outer wall thick, smooth, colourless, median wall irregularly pitted, chestnut brown.

Europe, North America, Africa; known from several ponds on commons and the shorelines of lakes in South-East England and in Wales.

Spirogyra borgeana Transeau 1915

27340420 **Pl. 123F** (p. 495)
Cells 30–35 μm wide, 50–200 μm long, end walls plane; chloroplast single, making 1.5–5 turns of cell; conjugation ladder-like, conjugation tubes formed by both gametangia, fertilized cells inflated on outer side and straight on inner receptive side; zygospores ellipsoidal, 30–40 μm wide, 54–70 μm long, median wall smooth, yellow.

Probably cosmopolitan; in the British Isles known only from ponds and ditches in Wales.

Spirogyra borysthenica Kasanowsky *et* Smirnov 1913

27340430 **Pl. 125K** (p. 500)
Cells 30–40 μm wide, 180–450 μm long, end walls with ring-like infold (replicate); chloroplasts 2–4, straight or making up to 2.5 turns of cell; conjugation ladder-like, conjugation tubes formed by both gametangia, fertilized cells swollen, and up to 70 μm wide; zygospores ellipsoidal, 52–62 μm wide, 110–160 μm long, median wall thick with a spine- or nipple-like papilla, yellow-brown.

Ukraine, USA; only recorded in the British Isles from aquatic habitats in Wales.

Spirogyra britannica Godward 1956

27340040 **Pl. 125B** (p. 500)
Cells 80–90 μm wide, 140–260 μm long, end walls plane; chloroplasts 3 or 4, making up to 6 turns of cell; zygospores 70–85 μm wide, 100–150 μm long, median wall smooth, colourless or very pale yellow.

British Isles (England); first described and known only from ponds in Essex and Kent, but with no information given on their water chemistry.

Although not described by Godward (1956), her illustrations depict ladder-like conjugation, with the conjugation tubes formed from both or mostly from the male gametangia and fertilized cells inflating particularly on the conjugating side. She also states that if *S. britannica* could have been referred to existing species it would most probably have been to *S. nitida* or *S. neglecta* (see, however, comments about these species below).

Spirogyra catenaeformis (Hassall) Kützing 1849

Basionym: *Zygnema catenaeformis* Hassall
Synonyms: *Spirogyra arta* var. *catenaeformis* (Hassall) Kirchner, *Zygnema angulare* Hassall?, *Z. malleolum* Hassall?
27340050 **Pl. 122L** (p. 494)
Cells 24–32 μm wide, 50–135 μm long, end walls plane; chloroplast single, making 1–6 turns of cell; conjugation ladder-like and lateral; conjugation tubes formed by both gametangia, fertilized cells inflated especially in lateral conjugation where often orientated across the gametangial cell; zygospores ellipsoidal, 27–33 μm wide, 55–90 μm long, median wall smooth, yellow.

Probably cosmopolitan; widespread in ponds and ditches in the British Isles where recorded from the English Midlands, Leicestershire, Yorkshire, Cambridgeshire and several counties in southern England (W. & G.S.West, 1897), also known from the north and west of Ireland.

Some authors accept *S. affinis* under the synonymy of *S. catenaeformis*; however, these are retained separately in the present text only on the grounds that the zygospores of the latter are larger.

Spirogyra circumlineata Transeau 1914

27340440 **Pl. 123C** (p. 495)
Cells 38–48 μm wide, 120–240 μm long, end walls plane; chloroplast single, slender and making 4–8 turns of cell; conjugation ladder-like, conjugation tubes formed from both gametangia, fertilized cells inflated on conjugating side only; zygospores ellipsoidal, 40–50 μm wide, 70–125 μm long, median wall smooth, with a more or less promiment suture, yellow-brown.

Europe, North America, Africa; in the British Isles only reported from ponds and ditches in Wales.

Spirogyra colligata Hodgetts 1920

27340450 **Pl. 122A** (p. 494)
Cells 20–40 μm wide, 240–260 μm long, with conspicuous collars between cells (colligate); chloroplasts 4–6, usually 5, making 0.5–2 turns of cell; conjugation ladder-like, conjugation tubes formed from both gametangia, lateral or terminal, conjugation frequently between two cells by processes growing through the colligate septa; fertilized cells inflated in middle to 90–100 μm wide; zygospores lens-shaped to spherical, 50–90 μm in diameter, outer wall smooth, colourless, median wall warty, dark brown.

Europe (rare); in the British Isles only known from England in meadowland pools at Worcester and near Birmingham, also in the moderately alkaline Telscombe Pond at King's Norton, Leicestershire.

Spirogyra communis (Hassall) Kützing 1849

Basionym: *Zygnema commune* Hassall
Synonym: *Spirogyra flavescens* (Hassall) Cleve fo. *parva* (Hassall) Cooke?
27340060 **Pl. 122H** (p. 494)
Cells 18–26 μm wide, 35–90 μm long, end walls plane; chloroplast single, making 1.5–4 turns of cell; conjugation ladder-like and lateral; conjugation tubes formed from both gametangia; fertilized cells cylindrical, rarely enlarged; zygospores ellipsoidal, 19–26 μm wide, 36–78 μm long, median wall smooth, yellow.

Probably cosmopolitan; in nutrient-rich waters and calcareous streams but also in neutral to slightly acid waters; widespread throughout the British Isles except in mountainous areas.

Transeau (1951) reports this species as hybridizing with *S. varians*.

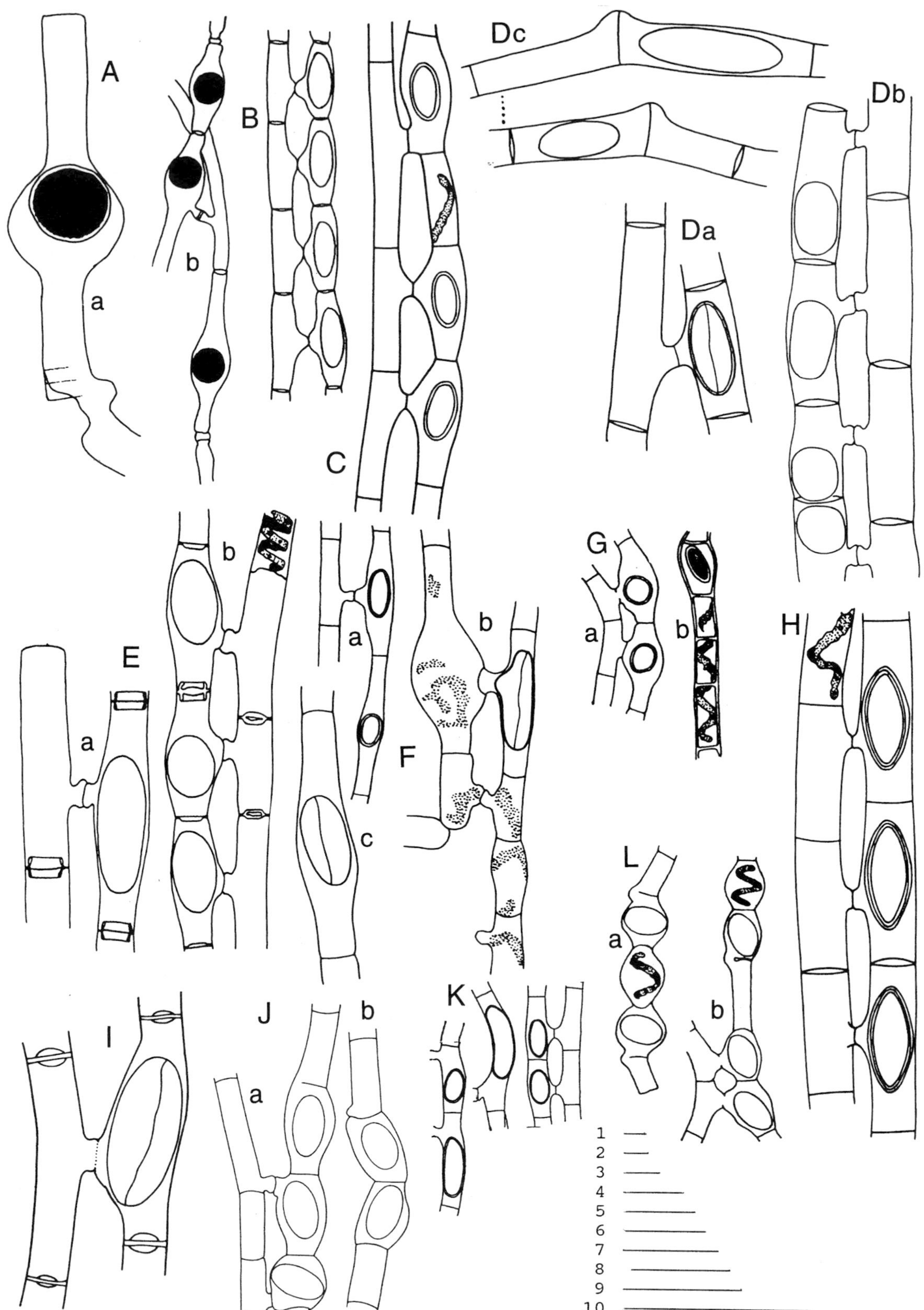

Plate 122 A–L

A. *Spirogyra colligata* (p. 493): a. terminal conjugation, b. ladder-like conjugation; **B**. *Spirogyra parva* (p. 499): ladder-like conjugation with zygospores; **C**. *Spirogyra flavescens* (p. 496): ladder-like conjugation with zygospores; **D**. *Spirogyra longata* (p. 497): a, b. ladder-like conjugation with zygospores, c. lateral conjugation with zygospores; **E**. *Spirogyra weberii* (p. 504): a, b. ladder-like conjugation with replicate cell end walls; **F**. *Spirogyra mirabilis* (p. 499): a. gametangia with zygospore and an aplanospore, b. ladder-like conjugation at two stages, c. aplanospore; **G**. *Spirogyra affinis* (p. 492): a. ladder-like conjugation, b. vegetative filament with an aplanospore; **H**. *Spirogyra communis* (p. 493); **I**. *Spirogyra gracilis* (p. 496); **J**. *Spirogyra parvula* (p. 501): a. ladder-like and lateral conjugation, b. aplanospores; **K**. *Spirogyra lutetiana* (p. 499): ladder-like conjugation; **L**. *Spirogyra catenaeformis* (p. 493): a. lateral conjugation with zygospores, b. ladder-like and lateral conjugation with zygospores.

Scale bars: 50 μm. 1 for Aa; 2 for K; 3 for J; 4 for C, D, Fa, L; 5 for Ab; 6 for E; 7 for G; 8 for B, H; 9 for I; 10 for Fb,c.

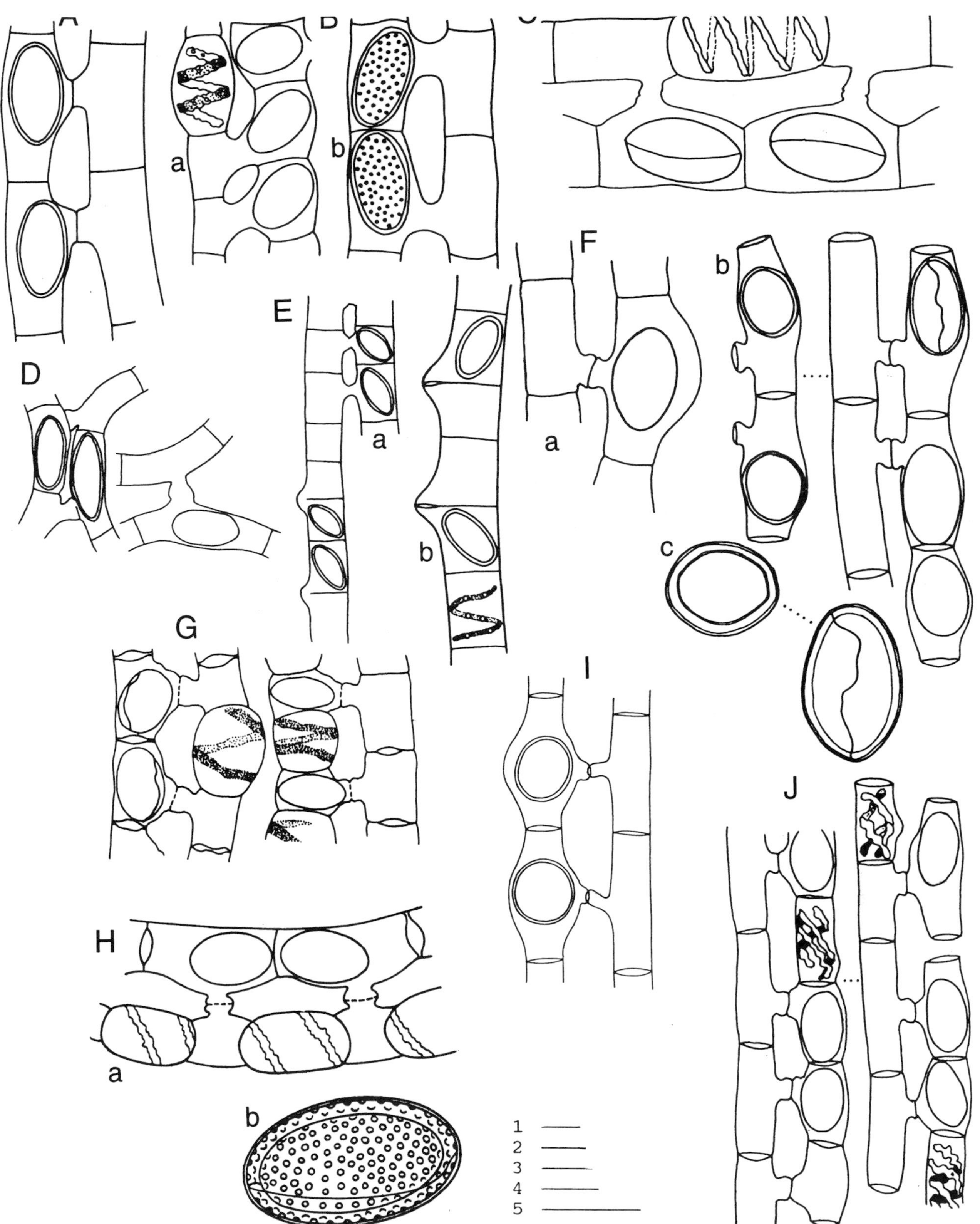

Plate 123 A–J

A. *Spirogyra porticalis* (p. 501): ladder-like conjugation with zygospores; **B**. *Spirogyra velata* (p. 504): a. ladder-like conjugation with zygospores and details of a chloroplast in one cell, b. ladder-like conjugation with details of zygospore ornamentation; **C**. *Spirogyra circumlineata* (p. 493): ladder-like conjugation with zygospores and details of a chloroplast; **D**. *Spirogyra reflexa* (p. 501): ladder-like conjugation with zygospores; **E**. *Spirogyra condensata* (p. 496): a. ladder-like and lateral conjugation, b. lateral conjugation; **F**. *Spirogyra borgeana* (p. 493): a, b. ladder-like conjugation, c. zygospores, one showing a suture; **G**. *Spirogyra varians* (p. 504): ladder-like conjugation with zygospores and chloroplast details; **H**. *Spirogyra scrobiculata* (p. 502): a. ladder-like conjugation, b. detail of zygospore ornamentation; **I**. *Spirogyra pellucida* (p. 501): ladder-like conjugation with zygospores; **J**. *Spirogyra decima* (p. 496): ladder-like conjugation with zygospores and details of chloroplasts.

Scale bars: 50 μm. 1 for Ea; 2 for F, G, I, J; 3 for A, D, Eb; 4 for B, H; 5 for C.

Spirogyra condensata (Vaucher) Kützing 1849

Basionym: *Conjugata condensata* Vaucher
Synonym: *Spirogyra condensata* Kützing
27340070 **Pl. 123E** (p. 495)
Cells 45–60 μm wide, 45–120 μm long, end walls plane; chloroplast single, making 0.5–4 turns of cell; conjugation ladder-like, conjugation tubes formed from both gametangia; fertilized cells cylindrical or only slightly enlarged (sterile cells sometimes inflated); zygospores ellipsoidal, 34–38 μm wide, 50–75 μm long, median wall smooth with a suture, light brown, smooth.

Europe, North and South America, India; widely distributed in the British Isles mostly around the margins of alkaline lakes and boggy pools in the English Midlands, also recorded from Kew Gardens, London, from Wales and Clare Island, County Mayo, Ireland.

Spirogyra crassa Kützing 1849 emend. Curza 1932

27340080 **Pl. 125F** (p. 500)
Cells often wider than long, (130–)140–165 μm wide, 126–330 μm long, plane end walls; chloroplasts 6–12, making 0.5–1 turn of cell; conjugation ladder-like and lateral, conjugation tubes formed from both gametangia; zygospores compressed, ovoid to almost rounded, 120–150 × 140–160 μm, 80–100 μm thick, median wall with irregular (sometimes star-shaped), shallow pits, brown.

Probably cosmopolitan; in the British Isles its coarse filaments form dark green masses in bogs, pools and streams with records from Yorkshire, the English Midlands, Northumberland, southern England and North Wales.

One of the largest *Spirogyra* species, its coarse filaments often found and can be identified in the vegetative state without the zygospores. Transeau (1951) reports *S. crassa* as hybridizing with *S. nitida* and *S. communis*.

Spirogyra crassiuscula (Wittrock *et* Nordstedt) Transeau 1934

Basionym: *Spirogyra maxima* fo. *megaspora crassiuscula* Witttrock *et* Nordstedt
27340090 **Pl. 125I** (p. 500)
Cells 145–170 μm wide, 140–300 μm long, end walls plane; chloroplasts 6 or 7, making 0.5–1 turn of cell; conjugation ladder-like, conjugation tubes formed from both gametangia; fertilized cells cylindrical; zygospores lens-shaped with a suture line, 120–150 μm wide, 85–100 μm long, median wall with net-like ornamentation, yellow-brown.

South and North Africa; very rare in the British Isles, only recorded from a few hard-water ponds in England and Wales.

The species is close to *S. maxima*, but the cells of the latter tend to be somewhat smaller.

Spirogyra decima (O.F.Müller) Kützing 1843

Basionym: *Conferva decima* O.F.Müller
Synonym: *Spirogyra porticalis* var. *decima* (Hassall) Cooke
27340100 **Pl. 123J** (p. 495)
Cells 32–42 μm wide, 66–150 μm long, end walls plane; chloroplasts 2 or 3, making 1–2 turns of cell; conjugation ladder-like, conjugation tubes formed from both gametangia; fertilized cells cylindrical or enlarged; zygospores ovoid to almost cylindrical with broadly rounded poles, 31–40 μm wide, 31–68 μm long, median wall smooth, yellow.

Probably cosmopolitan; recorded from ponds on Hampstead Heath, Surrey, Tunbridge Wells and Ruxley Gravel Pits, Kent, and a few localities in Scotland.

Some doubt exists over the validity of this species and its relationship with *S. porticalis* and *S. singularis*.

Spirogyra dubia Kützing1855

27340460 **Pl. 124A** (p. 498)
Cells 40–50 μm wide, 60–250 μm long, end walls plane; chloroplasts 2 or 3, making 2–8.5 turns of cell; conjugation ladder-like, conjugation tubes formed by both gametangia; fertilized cells inflated; zygospores ovoid, 42–50 μm wide, 54–67 μm long, median wall smooth, yellow-brown.

Africa, India, North America; rare in the British Isles where known from ditches and shallow lakes including Ruxley Gravel Pits in Kent.

Spirogyra flavescens (Hassall) Kützing 1849

Basionym: *Zygnema flavescens* Hassall
Synonym: *Spirogyra gracilis* var. *flavescens* (Hassall) Rabenhorst
27340110 **Pl. 122C** (p. 494)
Cells 10–14 μm wide, 30–50 μm long, end walls plane; chloroplast single, making 1–3 turns of cell; conjugation ladder-like, conjugation tubes formed from both gametangia; fertilized cells inflated; zygospores ovoid or ellipsoidal, 20–23 μm wide, 25–40 μm long, wall smooth, yellow.

Probably cosmopolitan; widespread but not frequent in the British Isles, recorded from acid to slightly alkaline waters ranging from boggy pools to lakes and reservoirs in the English Lake District, Middlesex (Ruislip Reservoir), Leicester, Yorkshire, North Wales and Clare Island, County Mayo, Ireland.

Spirogyra fluviatilis Hilse in Rabenhorst 1863

27340470 **Pl. 124B** (p. 498)
Cells 30–45 μm wide, 70–240 μm long, end walls plane; chloroplasts 3 or 4, making 1.5–3.5 turns of the cell; conjugation ladder-like, conjugation tubes formed by both gametangia; fertilized cells shortened, inflated and up to 70 μm wide; zygospores ovoid, 47–85 μm wide, 68–110 μm long, median wall corrugate or finely wrinkled, brown.

Probably cosmopolitan; occurs in pools, ponds, lakes and slow- moving streams, sometimes attached.

Remarkable branched rhizoids may develop from some cells. See remarks under *S. adnanta* (below), here considered as a doubtful record.

Spirogyra gracilis (Hassall) Kützing 1849

Basionym: *Zygnema gracile* Hassall
Synonym: *Spirogyra flavescens* fo. β *flavescens* (Hassall) Kützing
27340120 **Pl. 122I** (p. 494)
Cells 16–24 μm wide, 50–100 μm long, end walls plane; chloroplast single, making 0.5–4 turns of cell;

conjugation tubes formed from both gametangia, fertilized cells inflated only on side adjacent to conjugation tube; zygospores ellipsoidal with rounded ends and a suture line, 23–30 μm wide, 40–65 μm long, median wall thick, smooth, yellow-brown.

Probably cosmopolitan; abundant in many parts of the British Isles in slightly acid to moderately alkaline ponds, pools and ditches.

Zygospore production fairly frequent.

Spirogyra grevilleana (Hassall) Kützing 1849

Basionym: *Zygnema grevilleana* Hassall

27340130 **Pl. 126B** (p. 503)

Cells 22–33 μm wide, 60–325 μm long, end walls ring-like and infolded (replicate); chloroplast single (sometimes 2), making 4–9 turns of cell; conjugation ladder-like and lateral, conjugation tubes produced largely from male gametangia; fertilized cells inflated, up to 36–43 μm wide, becoming spindle-shaped, male filaments remain cylindrical; zygospores ovoid, 30–37 μm wide, 60–90 μm long, median wall smooth, yellow to yellow-brown.

Probably cosmopolitan; fairly widespread and often abundant especially in lowland areas of England with records from mildly alkaline ponds and pools in Yorkshire, Leicestershire, Worcestershire, Warwickshire and Cambridgeshire.

Spirogyra hassallii (Jenner) P.Petit 1880

Basionym: *Zygnema hassallii* Jenner

27340140 **Pl. 124C** (p. 498)

Cells 26–33 μm wide, 100–250 μm long, end walls ring-like and infolded (replicate); chloroplasts 2, making 1.5–5 turns of cell; conjugation mostly lateral and occasionally ladder-like, conjugation tubes formed from both gametangia, in ladder-like conjugation male and female cells always separated by plane walls, fertilized cells slightly swollen and spindle-shaped, up to 50 μm wide; zygotes elongate-ellipsoidal, in ladder-like conjugation adjoining cells separated by replicate walls that become slightly swollen as the zygospores become more ovoid; zygospores 39–48 μm wide, 58–136 μm long, median wall thick, smooth yellow.

Probably cosmopolitan; in the British Isles recorded from Yorkshire, Warwickshire, Hertfordshire and Barton's Pond at Hendon, London.

Spirogyra inflata (Vaucher) Kützing 1843

Basionym: *Conjugata inflata* Vaucher
Synonym: *Zygnema inflatum* Hassall

27340160 **Pl. 126E** (p. 503)

Cells 15–20 μm wide, 45–230 μm long, end walls ring-like and infolded (replicate); chloroplasts single, making 2.5–6 turns of cell; conjugation mostly lateral but occasionally ladder-like, in the latter conjugation tubes formed by both gametangia; fertilized cells inflated in middle to become 35–48 μm wide; zygospores and aplanospores ellipsoidal, 27–36 μm wide, 42–96 μm long, outer wall thin, smooth, colourless, inner wall thick, smooth, yellow brown.

Probably cosmoplitan; fairly widespread in the British Isles, recorded from pools and ponds in Northumberland, Yorkshire, Leicestershire, Warwickshire, and Mitcham Common, Surrey and from northern Ireland.

Transeau (1951) reports this to be a very variable species with several varieties described which may ultimately prove to be different species. Kolkwitz & Krieger (1944) include *Spirogyra spreeiana* Rabenhorst in the synonymy of *S. inflata*, but the two are retained by Transeau (1951) and Randhawa (1959). The species is not mentioned at all in Kadlubowska's monograph (Kadlubowska, 1984). See further coments under *Spirogyra spreeiana*.

Spirogyra insignis (Hassall) Kützing 1849

Basionym: *Zygnema insigne* Hassall
Synonym: *Zygnema leiospermum* fo. *megaspora* West

27340170 **Pl. 125J** (p. 500)

Cells 36–45 μm wide, 50–590 μm long, end walls ring-like and infolded (replicate); chloroplasts 2–4 (frequently 3), making 0.5–1.5 turns of cell; conjugation ladder-like, conjugation tubes formed by both gametangia; fertilized cells shortened and inflated especially adjacent to male gametangial cell; zygospores elongate-ellipsoidal, 28–48 μm wide, 60–128 μm long, outer wall thin, smooth, colourless, median wall thick, smooth, yellow-brown to brown.

Probably cosmopolitan; several records from streams, ponds and ditches in Yorkshire, Leicestershire, south Buckinghamshire and Surrey, also known from Clare Island, County Mayo, Ireland.

Spirogyra jugalis (Dillwyn) Kützing 1845

Basionym: *Conferva jugalis* Dillwyn

27340180 **Pl. 125A** (p. 500)

Cells 75–103 μm wide, 80–300 μm long, end walls plane; chloroplasts 3 or 4, making 1–2 turns of cell; conjugation ladder-like, conjugation tubes formed by both gametangia; fertilized cells cylindrical and not swollen or only very slightly swollen; zygospores broadly ellipsoidal, 87–108 μm wide, 120–155 μm long, median wall smooth, brown.

Europe (not widespread), Russia, USA, North Africa; fairly rare in British Isles with records from the Orkneys, in Scotland and a few localities in England such as Leicester, Abbot's Pond, Portishead and Tiltham Ponds near Bristol.

Kolkwitz & Krieger (1944) include this under the synonymy of *Spirogyra setiformis* (see below), but retained as a separate species in subsequent monographs.

Spirogyra longata (Vaucher) Kützing 1843

Basionym: *Conjugata longata* Vaucher
Synonym: *Spirogyra longata* var. *communis* (Hassall) Cooke

27340190 **Pl. 122D** (p. 494)

Cells 26–38 μm wide, 45–280 μm long, end walls plane; chloroplast single, making 2–5 turns of cell; conjugation ladder-like and lateral, conjugation tubes comparatively long, formed by both gametangia; zygospores ovoid (in lateral conjugants) to subspherical (in ladder-like conjugants), 28–38 μm wide, 50–85 μm

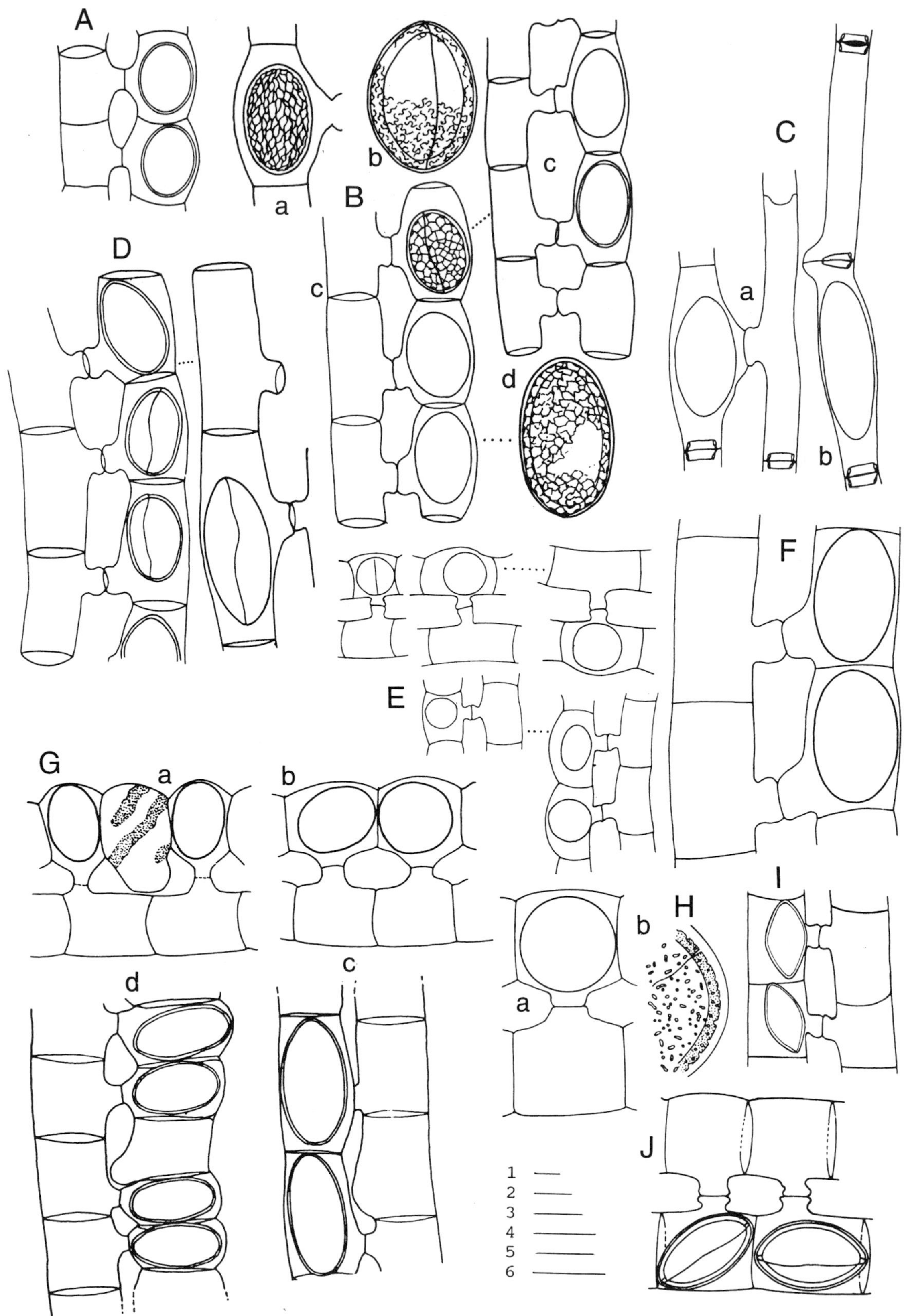

Plate 124 A–J

A. *Spirogyra dubia* (p. 496): ladder-like conjugation with zygospores; **B**. *Spirogyra fluviatilis* (p. 496): a. gametangia with a zygospore, b. and d. details of zygospore ornamentation, with b having a suture, c. ladder-like conjugation with zygospores; **C**. *Spirogyra hassallii* (p. 497): replicate cell end walls, a. ladder-like conjugation with zygospores, b. lateral conjugation; **D**. *Spirogyra rivularis* (p. 502): ladder-like conjugation with zygospores, some showing a suture; **E**. *Spirogyra majuscula* (p. 499): ladder-like conjugation with zygospores; **F**. *Spirogyra neglecta* (p. 499): ladder-like conjugation with zygospores; **G**. *Spirogyra ternata* (p. 504): ladder-like conjugation with zygospores and showing chloroplasts in one cell (a); **H**. *Spirogyra bellis* (p. 492): a. ladder-like conjugation, b. detail of zygospore ornamentation; **I**. *Spirogyra turfosa* (p. 504): ladder-like conjugation with zygospores; **J**. *Spirogyra nitida* (p. 499): ladder-like conjugation with zygospores, showing sutures.

Scale bars: 50 μm. 1 for E; 2 for I; 3 for A, C, J; 4 for B, G, H; 5 for F; 6 for D.

long, median wall smooth, thick, yellow, commonly with a delicate, wavy suture line.

Probably cosmopolitan, one of the most widespread species in Europe with a wide range of habitat tolerance; known in the British Isles from acidic to alkaline ponds and pools in England, Wales and Ireland; not recorded from Scotland.

Kolkwitz & Krieger (1944) consider some reports of this may be attributable to *Spirogyra juergensii* Kützing.

Spirogyra lutetiana Petit 1879

27340200 **Pl. 122K** (p. 494)

Cells 27–40 μm wide, 70–230 μm long, end walls plane; chloroplast single, making 3–7 turns of cell; conjugation ladder-like, conjugation tubes formed by both gametangia; fertilized cells variable, not or only slightly swollen; zygospores variable in shape from spherical, ellipsoidal to irregular with broadly rounded ends, 23–44 μm wide, 35–165 μm long, median wall smooth, yellow-brown with faint suture line.

Probably cosmopolitan, not widespread in Europe; only recorded in the British Isles from the English Lake District by West (1892b) who mentions a doubtful 'var. *minor*' from Borrowdale (Cumbria).

Probably no useful purpose is served recognizing such varieties as var. *minor*.

Spirogyra majuscula Kützing 1849

Synonym: *Spirogyra orthospira* Nägeli *pro parte*

27340210 **Pl. 124E** (p. 498)

Cells 50–80 μm wide, 80–500 μm long, end walls plane; chloroplasts (3–)5–10, straight or making only one-third of a turn of cell; conjugation ladder-like and lateral, conjugation tubes formed from both gametangia; fertilized cells short, cylindrical or slightly inflated on side opposite conjugation tube; zygospores lens-shaped, 57–62 μm wide, 45–60 μm long, median wall thick, smooth, yellow-brown with a faint suture line; aplanospores similar to zygospores but smaller.

Probably cosmopolitan, widespread in Europe; in the British Isles several records from ponds and pools in the English Midlands, Wales and Scotland.

Spirogyra maxima (Hassall) Wittrock 1882

Basionym: *Zygnema maximum* Hassall

Synonyms: *Zygnema alternatum* Hassall, *Z. orbiculare* Hassall

27340220 **Pl. 125H** (p. 500)

Cells 118–140 μm wide, 100–250 μm long, end walls plane; chloroplasts 6 or 7, virtually straight or making about three-quarters of a turn of cell; conjugation ladder-like, conjugation tubes formed by both gametangia; fertilized cells cylindrical; zygospores lens-shaped, 100–125 μm wide, 75–95 μm long, median wall about 20 μm thick, with irregular net-like ornamentation and distinct suture line, golden-brown.

Probably cosmopolitan; in the British Isles numerous records mostly from alkaline ponds in the English Midlands, Cambridge, Leicester, Warwickshire and Worcestershire.

Spirogyra mirabilis (Hassall) Kützing 1849

Basionym: *Zygnema mirabile* Hassall

27340230 **Pl. 122F** (p. 494)

Cells 23–29 μm wide, 70–200 μm long, end walls plane; chloroplast single, making 4–7 turns of cell; conjugation ladder-like but sexual reproduction very rare and mostly by parthenospores; conjugation tubes formed by both gametangia; fertilized cells and cells with parthenospores inflated in mid-region; parthenospores and zygospores mostly ellipsoidal, 23–29 μm wide, 50–83 μm long, median wall thick, smooth, with a faint suture line, yellow-brown.

Probably cosmopolitan; in the British Isles reported only from lowland pools in England and South Wales.

Spirogyra neglecta (Hassall) Kützing 1849

Basionym: *Zygnema neglectum* Hassall

27340240 **Pl. 124F** (p. 498)

Cells 55–67 μm wide, 100–300 μm long, end walls plane; chloroplasts 3, making 1–2.5 turns of cell; conjugation ladder-like and lateral, conjugation tubes formed by both gametangia; fertilized cells usually greatly inflated, zygospores sometimes develop at right angles to long axis of filament; zygospores broadly ovoid, 54–64 μm wide, 75–100 μm long, median wall thick, smooth with a suture line, yellow-brown; aplanospores similar.

Europe, North America, northern Asia, Africa; a few records from ponds in southern England including Abbot's Pond, near Bristol, also from Anglesey and Caernarvonshire, North Wales.

Spirogyra nitida (Dillwyn) Link 1820

Basionym: *Conferva nitida* Dillwyn

Synonym: *Zygnema nitidum* Hassall

27340250 **Pl. 124J** (p. 498)

Cells 70–80 μm wide, 90–300 μm long, end walls plane; chloroplasts 3–5, making 0.5–1.5 turns of cell; conjugation ladder-like, conjugation tubes formed from both gametangia; fertilized cells remaining cylindrical or very slightly enlarged; zygospores ellipsoidal, rarely somewhat ovoid, 60–76 μm wide, 73–177 μm long, median wall smooth, thick, brown.

Europe (widespread), North America, Asia, North Africa; according to West & Fritsch (1927), 'abundant in every part of the British Isles', occurring principally in the shallows of lakes and marshes; tolerant of a wide range of water types.

Godward (1954) expressed at length her doubts about the validity of this species and stated that it and *S. neglecta*, is still a heterogeneous collection of forms.

Spirogyra parva (Hassall) Kützing 1849

Basionym: *Zygnema parvum* Hassall

Synonyms: *Spirogyra parva* (Hassall) Czurda, *Zygnema flavescens* var. *parvum* (Hassall) Cleve

27340260 **Pl. 122B** (p. 494)

Cells 8–10 μm wide (narrowest of all *Spirogyra* filaments), end walls plane; chloroplast single; conjugation ladder-like, conjugation tubes formed from both

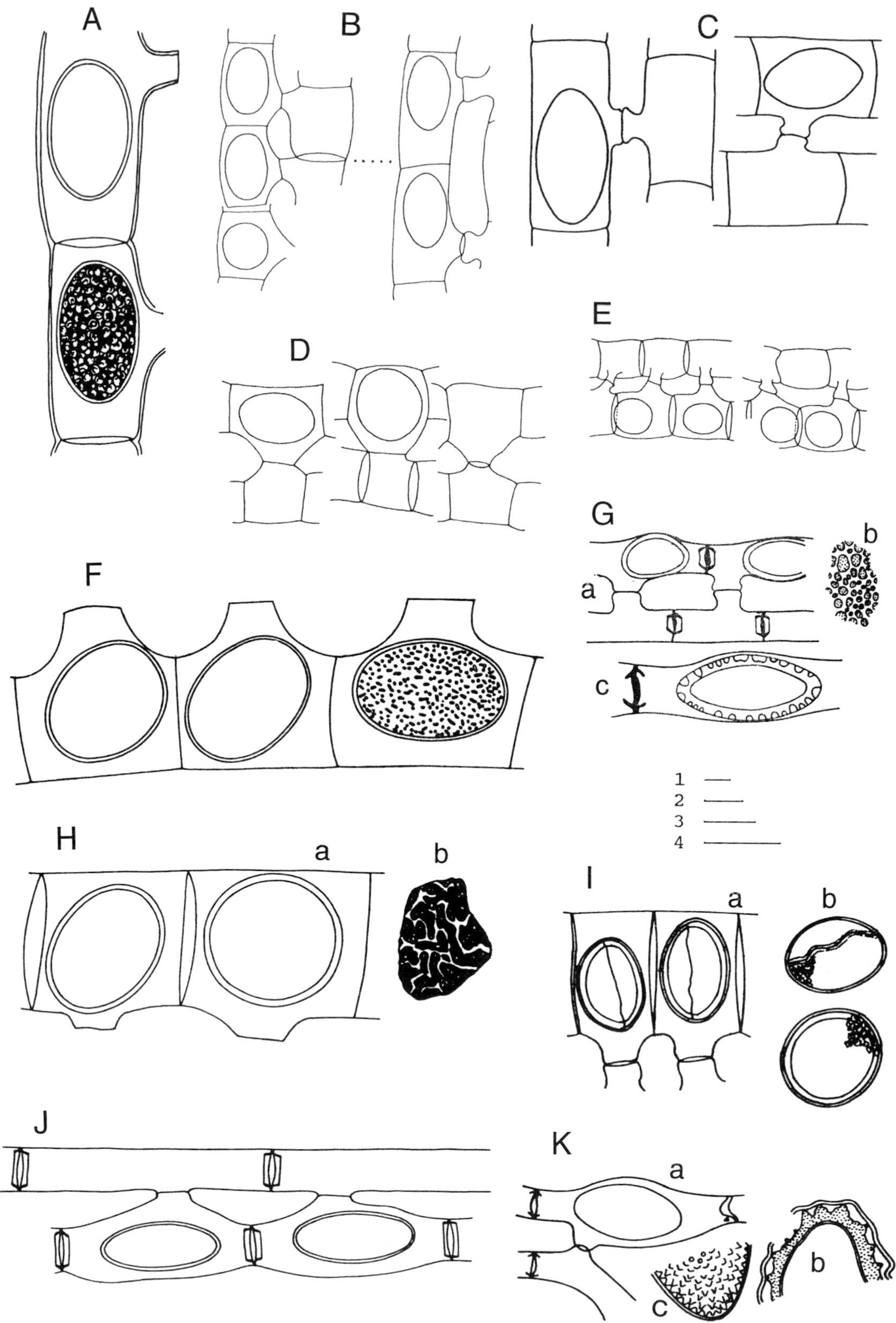

Plate 125 A–K

A. *Spirogyra jugalis* (p. 497): ladder-like conjugation with zygospores; **B**. *Spirogyra britannica* (p. 493): ladder-like conjugation with zygospores; **C**. *Spirogyra setiformis* (p. 502): ladder-like conjugation with zygospores; **D**. *Spirogyra subechinata* (p. 502): ladder-like conjugation with zygospores; **E**. *Spirogyra submargaritata* (p. 502): ladder-like conjugation with zygospores; **F**. *Spirogyra crassa* (p. 496): ladder-like conjugation with zygospores and detail of zygospore ornamentation; **G**. *Spirogyra reticulata* (p. 502): a. ladder-like conjugation, b. detail of zygospore ornamentation, c. detail of zygospore; **H**. *Spirogyra maxima* (p. 499): a. gametangia with zygospores, b. detail of mesospore ornamentation; **I**. *Spirogyra crassiuscula* (p. 496): a. gametangia with zygospores showing a suture, b. zygospores; **J**. *Spirogyra insignis* (p. 497): replicate cell end walls, ladder-like conjugation with zygospores; **K**. *Spirogyra borysthenica* (p. 493): a. replicate cell end walls, ladder-like conjugation with zygospore, b. details of zygospore ornamentation in section, c. surface detail of zygospore.

Scale bars: 50 µm. 1 for B, D, E; 2 for C, F, Ga, I, Ka; 3 for H, J; 4 for A, Gb,c, Kb,c.

gametangia and both slightly inflated, fertilized cells more inflated than male cells; zygospores ellipsoidal, 10–20 μm wide, 20–40 μm long, median wall smooth, yellow-brown.

Europe, South Africa; very rare in the British Isles having been reported only from Cheshunt, Hertfordshire and Ramslye Rocks, Broadwater Forest, Sussex.

Spirogyra parvula (Transeau) Czurda 1932

Basionym: *Spirogyra catenaeformis* (Hassall) Kützing var. *parvula* Transeau

27340480 **Pl. 122J** (p. 494)

Cells 20–24 μm wide, 50–150 μm long, end walls plane; chloroplast single, making 1–6 turns of cell; conjugation mostly lateral (occasionally ladder-like), conjugation tubes from both gametangia, only fertilized cells inflated and up to 37 μm in diameter; zygospores ellipsoidal, 20–30 μm wide, 32–65 μm long, median wall thick, smooth, yellow-brown; occasionally ellipsoidal parthenospores produced.

Probably cosmopolitan; in the British Isles only recorded from high altitudes in the Snowdonia area of North Wales.

Spirogyra pellucida (Hassall) Kützing 1849

Basionym: *Zygnema pellucidum* Hassall

27340270 **Pl. 123I** (p. 495)

Cells 40–50 μm wide, 100–400 μm long, end walls plane; chloroplasts 3 or 4, making 3–5 turns of cell; conjugation ladder-like, conjugation tubes from both gametangia; fertilized cells become inflated towards middle; zygospore lens-shaped, 77–86 μm wide, median wall smooth, brown.

North America, Europe (Finland) and Manchuria; in the British Isles only recorded from a few habitats in Yorkshire, Cambridgeshire, Warwickshire and Hertfordshire.

Spirogyra porticalis (O.F.Müller) Cleve 1868

Basionym: *Conferva porticalis* O.F.Müller

27340280 **Pl. 123A** (p. 495)

Cells 40–50 μm wide, 66–200 μm long, end walls plane; chloroplast single, making 3–5 turns of cell; conjugation ladder-like, conjugation tubes formed from both gametangia; fertilized cells cylindrical or only slightly enlarged; zygospores ellipsoidal, 38–50 μm wide, 50–83 μm long, median wall thick, smooth, yellow-brown.

Probably cosmopolitan; fairly widespread in the British Isles, known especially from bogs, ditches and lake margins in the Midland counties of England and from Yorkshire and Surrey, also Clare Island, County Mayo, Ireland.

Kolkwitz & Krieger (1944) do not recognize Cooke's varieties (Cooke, 1882–84), vars. *quinina* (Hassall) Cooke and *rivularis* (Hassall) Cooke, since the description of his first variety is too incomplete and zygospores are unknown in the second; the varieties are not mentioned in subsequent monographs.

Spirogyra protecta Wood 1872

Synonym: *Spirogyra calospora* fo. *gracilior* Cleve

27340290 **Pl. 126A** (p. 503)

Cells 28–34 μm wide, 120–425 μm long, end walls ring-like and infolded (replicate); chloroplast single (rarely 2), making 2–6 turns of cell; conjugation ladder-like, conjugation tubes formed from both gametangia, fertilized cells not or only slightly enlarged; zygospore ovoid or cylindrical with broadly rounded ends, 30–39 μm wide, 66–90 μm long, outer wall double-layered: outer smooth and colourless, inner thick, coarsely pitted (scrobiculate), inner wall thin, smooth, yellow; aplanospores similar but smaller.

Europe, North America, Africa; known from bogs and moorland pools in southern Ireland and Ruislip Reservoir in Middlesex (see below).

Spirogyra calospora Cleve is widely regarded to be a synonym of this species. Many records for the British Isles were published in the last century as '*S. calospora*'. The species is listed in Cooke (1882–84, 1890) as 'f. *minor* Borge (1895)?' and is reported from bogs and moorland pools in Ireland. The form is mentioned in the synonymy of *Spirogyra calospora* Cleve by Kolkwitz & Krieger (1944). Cooke, and Kolkwitz & Krieger, also place under its synonymy *Spirogyra protecta* Wood, the latter two authors noting that zygospores were unknown. W. & G.S.West (1897) record *Spirogyra calospora* Cleve from Ruislip Reservoir and mention that it was badly illustrated originally. *Spirogyra calospora* Cleve fo. *gracilior* Cleve, although not recorded in the British Isles, is currently accepted under the synonymy of *Spirogyra protecta* Wood. Reported as found conjugating with *Spirogyra majuscula*.

Spirogyra quadrata (Hassall) P. Petit 1874

Basionym: *Zygnema quadratum* Hassall

27340300 **Pl. 126D** (p. 503)

Cells 24–30 μm wide, 70–300 μm long, end walls ring-like and infolded (replicate); chloroplasts single, making 1.5–6 turns of cell; conjugation ladder-like and lateral, conjugation tubes formed from both gametangia; fertilized cells distinctly inflated in middle and up to 60 μm wide; zygospores ellipsoidal or cylindrical-ellipsoidal, 33–44 μm wide, 50–82 μm long, median wall thick, smooth, brown; aplanospores similar but smaller.

Probably cosmopolitan; in the British Isles reported from the edges of pools, temporary waters and from two ponds in Leicestershire, Barton's Pond at Cheshunt and ponds in Epping Forest, Essex; also found in rivers in southern England.

Spirogyra reflexa Transeau 1915

27340490 **Pl. 123D** (p. 495)

Cells 30–44 μm wide, 120–300 μm long, end walls plane; chloroplast single, making 3–8 turns of cell; conjugation ladder-like, conjugation tubes formed by male gametangia only; fertilized cell inflated or enlarged and strongly reflexed, single or in groups of 2 or 4; zygospores and aplanospores ellipsoidal, 44–64 × 90–150 μm, median wall smooth, yellow-brown.

North America; only known in the British Isles from nutrient-rich ponds in southern England and the slightly acid waters of desmid-rich ponds on Hartlebury Common, Worcestershire.

Spirogyra reticulata Nordstedt 1880

27340500 **Pl. 125G** (p. 500)

Cells 28–42 μm wide, 72–460 μm long, end walls ring-like and infolded (replicate); chloroplasts 1–3, usually 2, making 2–4 turns of cell; conjugation ladder-like and lateral, conjugation tubes formed by both gametangia; male gametangium not swollen and fertilized cell (female) enlarged or inflated towards middle and up to 48–60 μm; zygospores mostly ovoid, 45–61 μm wide, 80–120 μm long; median wall 2-layered, outer thin and wrinkled, inner with net-like ornamentation, yellow-brown.

Probably cosmopolitan; rare in the British Isles, being recorded only from nutrient-rich ponds in southern England.

Spirogyra rivularis (Hassall) Rabenhorst 1868

Basionym: *Zygnema rivulare* Hassall

27340310 **Pl. 124D** (p. 498)

Cells 30–41 μm wide, 100–400 μm long, end walls plane; chloroplasts 2 or 3, making 2.5–3.5 turns of cell; conjugation ladder-like (rare), conjugation tubes from both gametangia; fertilized cells short, cylindrical or enlarged; zygospores ellipsoidal, 32–42 μm wide, 60–100 μm long, median wall smooth with a suture line, yellow to yellow-brown.

Probably cosmopolitan, widely distributed in Europe; reported in the British Isles from rivers including the River Lea, Hertfordshire and the margins of pools in England and on Cader Idris (Anglesey).

Spirogyra scrobiculata (Stockmayer) Czurda 1932

Basionym: *Spirogyra varians* (Hassall) Kützing var. *scrobiculata* Stockmayer

27340510 **Pl. 123H** (p. 495)

Cells 30–40 μm wide, 30–90 μm long, end walls plane; chloroplast single; conjugation ladder-like, conjugation tubes formed from both gametangia; fertilized cells inflated on receptive side; zygospores ellipsoidal, 34–38 μm wide, 56–68 μm long, outer wall thin, smooth, colourless, inner wall thick, with prominent pits and depressions (scrobiculations), with a delicate suture line, yellow-brown.

Europe, North America; rare in the British Isles, only reported with other algae in bogs in Wales.

Spirogyra setiformis (Roth) Kützing 1845

Basionym: *Conferva setiformis* Roth

27340320 **Pl. 125C** (p. 500)

Cells 90–115 μm wide, 100–225 μm long, end walls plane; chloroplasts 4, making 0.5–4 turns of cell; conjugation ladder-like, conjugation tubes formed by both gametangia; fertilized cells cylindrical; zygospores ellipsoidal, 85–100 μm wide, 115–160 μm long, median wall smooth, brown.

USA, widely reported in Europe; in ponds including Frensham Little Pond, Surrey and slow-flowing rivers mostly in southern England.

Kolkwitz & Krieger (1944) included in its synonymy *Spirogyra jugalis* (see under this species), but this is not followed in subsequent monographs.

Spirogyra spreeiana Rabenhorst 1863

27340330 **Pl. 126F** (p. 503)

Cells 18–21(–25) μm wide, 140–600 μm long, end walls ring-like and infolded (replicate); chloroplast single, making 1.5–4 turns of cell; conjugation ladder-like and lateral, conjugation tubes mostly formed by male gametangia; fertilized cells enlarged or inflated on both sides and towards middle, up to 30–42 μm wide; zygospores ellipsoidal, 30–36 μm wide, 55–100 μm long, median wall smooth, yellow.

Probably cosmopolitan, widespread in Europe; only reported so far in ditches and pools in southern England (W. & G.S.West, 1897) and from the slow-flowing margins of the River Ale, Roxburghshire, Scotland (Marson, 1992).

Marson (1992) has observed both scalariform and lateral conjugation occurring in the same filaments. Kolkwitz & Krieger (1944) include this in *Spirogyra inflata* (Vaucher) Kützing, but the two species are recognized by Transeau (1951) and Randhawa (1959); the species is not mentioned in the review by Kadlubowska (1984).

Spirogyra subechinata Godward 1956

27340340 **Pl. 125D** (p. 500)

Cells 80–100 μm wide, 60–200 μm long, end walls plane; chloroplasts 7 or 8, making one turn of cell; conjugation tubes formed from both gametangia; fertilized cells inflated, more on the conjugating side; zygospores about 110 μm wide and 140 μm long, median wall subechinate, yellow-brown.

British Isles (England); described from one unspecified locality in Essex.

Spirogyra submargaritata Godward 1956

27340350 **Pl. 125E** (p. 500)

Cells 70–100 μm wide, 100–192 μm long, end walls plane; chloroplasts 8–10, making one or less turns of cell; conjugation tubes from both gametangia; receptive cells cylindrical to slightly inflated; zygospores 50–70 μm wide, 70–75 μm long, median spore wall smooth, yellow-brown.

British Isles (England); first described from three localities in Essex and later found in the English Lake District.

Spirogyra tenuissima (Hassall) Kützing 1849

Basionym: *Zygnema tenuissimum* Hassall
Synonym: *Zygnema minimum* Hassall

27340370 **Pl. 126C** (p. 503)

Cells only 8–13 μm wide, 40–250 μm long, end walls ring-like and infolded (replicate); chloroplast single, making 3–6 turns of cell; conjugation lateral and ladder-like, conjugation tubes usually from both gametangia but sometimes only from one fertilized cell, male cell not or only very slightly swollen, fertilized cell greatly enlarged or inflated towards the middle; zygospores ellipsoidal, 22–36 μm wide, 40–74 μm long, median wall thick, smooth, yellow or yellow-brown.

Probably cosmopolitan; widespread and often abundant in pools and ditches in almost every part of the British Isles except for the most mountainous areas.

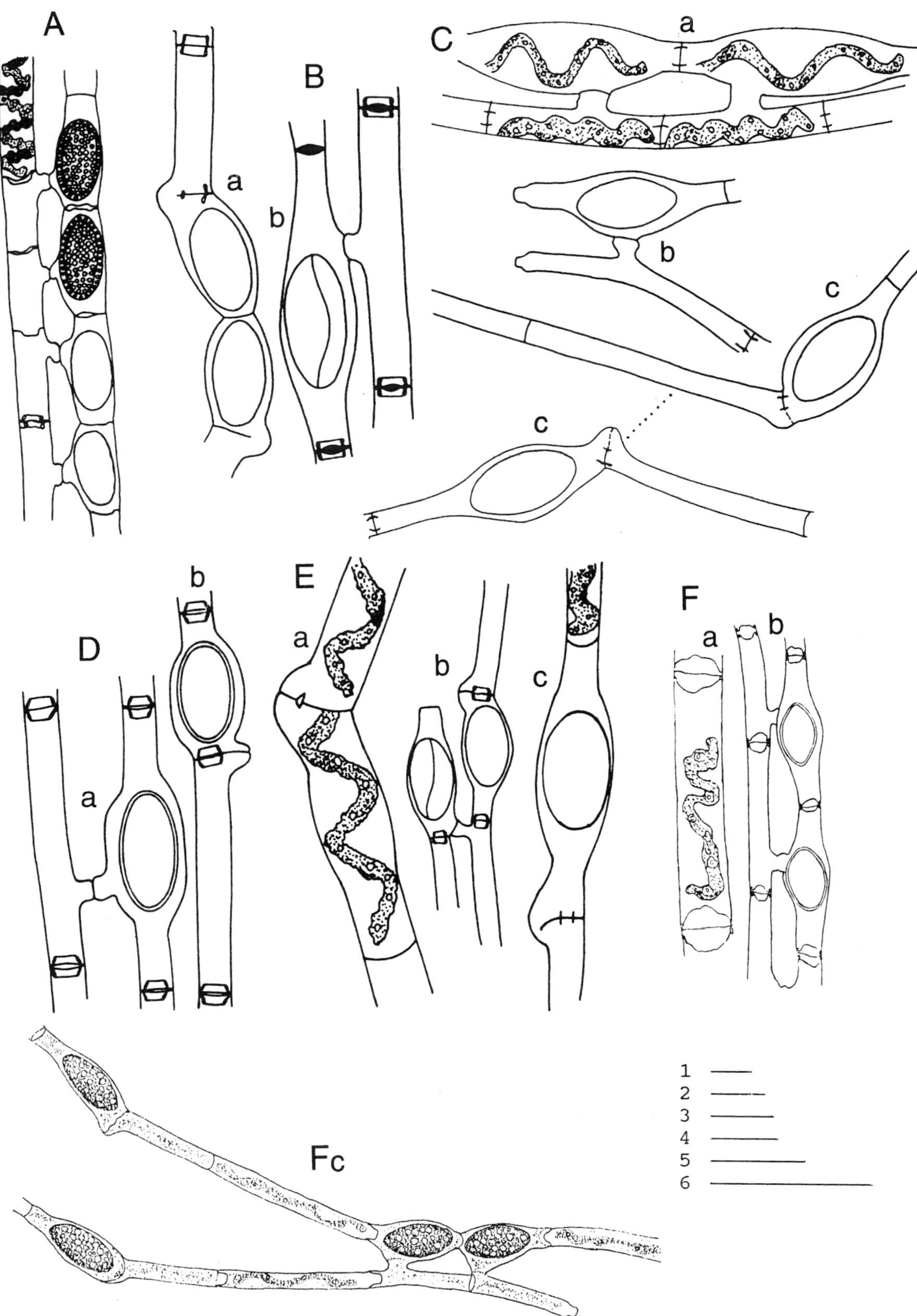

Plate 126 A–F

A. *Spirogyra protecta* (p. 501): replicate cell end walls, ladder-like conjugation with zygospores, details of zygospore ornamentation and a chloroplast; **B.** *Spirogyra grevilleana* (p. 497): replicate cell end walls, a. lateral conjugation with zygospores, b. ladder-like conjugation with a zygospore having a suture; **C.** *Spirogyra tenuissima* (p. 502): a. initiation of ladder-like conjugation and details of chloroplasts, b. ladder-like conjugation with a zygospore, c. lateral conjugation showing zygospores; **D.** *Spirogyra quadrata* (p. 501): replicate cell end walls, and a. ladder-like conjugation with a zygospore, b. lateral conjugation with a zygospore; **E.** *Spirogyra inflata* (p. 497): a. initiation of lateral conjugation and details of chloroplasts, b. ladder-like and lateral conjugation with replicate cell end walls, c. lateral conjugation with a zyospore; **F.** *Spirogyra spreeiana* (p. 502): replicate cell end walls, a. vegetative cell showing a chloroplast, b. ladder-like conjugation with zygospores, c. ladder-like and lateral conjugation.

Scale bars: 50 µm. 1 for A, Fb,c; 2 for Fa; 3 for D; 4 for B, Eb; 5 for C, Ec; 6 for Ea.

Transeau (1951) reports the species to be very variable, several variants having been found which may ultimately prove to be different species. He reports this species as hybridizing with *S. weberi*, with ovoid spores produced in the *S. weberi* filaments and ellipsoidal spores in the *S. tenuissima* filaments.

Spirogyra ternata Ripart 1876

27340380 **Pl. 124G** (p. 498)

Cells 54–70 μm wide, end walls plane; chloroplasts 3 or 4; conjugation ladder-like, conjugation tubes formed by both gametangia, with tube formed by receptor (female) gametangia inflated; zygospores 55–70 μm wide, 66–125 μm long, median wall thick, smooth, yellow-brown.

Widespread in Europe; in the British Isles recorded only from Sydenham Wells Pond, England by Fritsch & Rich (1907).

Fritsch & Rich (1907) comment that many consider this species to be a variety of *Spirogyra neglecta*. Transeau (1951) is of a similar view. Nevertheless, in some subsequent taxonomic treatments (e.g. Kadlubowska, 1984) the two are considered separate species as in the present text.

Spirogyra turfosa F.Gay 1884

27340520 **Pl. 124I** (p. 498)

Cells 68–78 μm wide, 68–350 μm long, end walls plane; chloroplasts 3 or 4, making 1.5–4 turns of cell; conjugation ladder-like, conjugation tubes formed by both gametangia; fertilized cells cylindrical; zygospores ellipsoidal and pointed, 65–78 μm wide, 120–140 μm long, median wall smooth, yellow.

Europe, America, India; in the British Isles known only from Ruxley Gravel Pits in Kent where growing along with several other species of *Spirogyra*.

Where found at Ruxley Gravel Pits collected together with several other species (including *Spirogyra nitida*). Kolkwitz & Krieger (1944) include *Spirogyra turfosa* under the synonymy of *S. nitida*, but Transeau (1951) and Ranadhawa (1959) consider them to be separate species.

Spirogyra varians (Hassall) Kützing 1849

Basionym: *Zygnema varians* Hassall

27340390 **Pl. 123G** (p. 495)

Cells 28–40 μm wide, 30–120 μm long, end walls plane; chloroplast single, making 1–5 turns of cell; conjugation ladder-like and lateral, conjugation tubes from both gametangia; fertilized cells usually inflated only on receptive side (sometimes intermediate vegetative cells swollen); zygospores ellipsoidal, usually some ovoid or rarely spherical, 32–40 μm wide, 50–100 μm long, median wall thick, smooth, yellow, mostly with a distinct suture line.

Probably cosmopolitan; in pools, ponds and lake margins and not uncommon in many parts of the British Isles including mountain streams.

Transeau (1951) reports this species as hybridizing with *Spirogyra porticalis*, *S. longata* and *S. communis*.

Spirogyra velata Nordstedt 1873

27340400 **Pl. 123Ba,b** (p. 495)

Cells 29–41 μm wide, 60–200 μm long, end walls plane; chloroplast single, rarely 2, making 2.5–6 turns of cell; conjugation ladder-like, conjugation tubes formed by both gametangia; fertilized cells cylindrical or somewhat enlarged; zygospores mostly ovoid to cylindrical-ovoid, rarely ellipsoidal, 37–57 μm wide, 60–100 μm long, median wall smooth, yellow-brown, outer wall of 2 layers, the outermost transparent and pitted (scrobiculate).

Europe, Africa, India; in the British Isles known only from ditches and moorland pools in England.

Spirogyra weberi Kützing 1843

Synonyms: *Zygnema subventricosum* Hassall, *Spirogyra weberi* Kützing fo. *inaequalis*
(Hassall) Cooke

27340410 **Pl. 122Ea,b** (p. 494)

Cells 19–30 μm wide, 80–480 μm long, end walls replicate; chloroplast single (rarely 2), making 3–6.5 turns of cell; conjugation ladder-like, conjugation tubes formed by both gametangia, fertilized cell usually slightly enlarged; zygospores ovoid to cylindrical-ovoid, 21–36 μm wide, 39–105 μm long, median wall thick, smooth, yellow or brown; parthenospores very occasionally reported, ellipsoidal, 25–36 μm wide, 32–67 μm long.

Probably cosmopolitan; many records from ponds and lakes in the English Midlands and southern England including Abbot's Pond, near Bristol, and Ruislip Reservoir, Middlesex, also from Wales and from Clare Island, County Mayo, Ireland.

DOUBTFUL RECORDS

Spirogyra adnanta (Vaucher) Kützing 1843

Doubt attaches to this species which has only been recorded in the British Isles from Wales, by Benson-Evans & Antoine (1996). According to Kolkwitz & Krieger (1944), the species is insufficiently described, and possibly at least in part attributable to *S. fluviatilis*.

Spirogyra cylindrosperma (West *et* G.S.West) Kolkwitz *et* Krieger 1944

Spirogyra decima (O.F.Müller) Kützing var. *cylindrosperma* was described (without illustration) by W. & G.S.West (1897) from Frensham Little Pond in southern England; this should not be confused with *Spirogyra cylindrospora* (O.F.Müller) Kützing, as reported by W. & G.S.West (1897) from Angola. It was described as follows: cells of filaments 40–44 μm wide, length not given, end walls plane; chloroplast single, number of spirals not given; conjugation ladder-like, conjugation tubes *presumably* formed by both gametangia; fertilized cell cylindrical; zygospores cylindrical with rounded ends, 39–43 × 120 μm; exospore thin, smooth, colourless; mesospore thicker, smooth, yellow-brown. It was accorded species status by Kolkwitz & Krieger (1944), but has not been mentioned again in monographic treatments. It has been decided to regard this little-known species as doubtful.

Spirogyra ellipsospora Transeau 1914

Reported by Godward (1950), but zygospores were not found and for this reason a doubtful record.

Spirogyra elliptica Jão 1935

Reported by Godward (1950, 1953, as *S. triformis* Van Wisseling) from England. The identification was mentioned as only based on the 'cell description and chromosomes' and for this reason considered doubtful.

Spirogyra oblata Jão 1935

Godward (1956), briefly reported this species as found in 'two British localities', but no further details or description are given. It is therefore considered as an uncertain record.

Spirogyra orthospira Nägeli in Famintzin 1867

Reported by Cooke (1882–84, 1890) growing in pools and producing reproductive zygospores in the autumn. According to Kolkwitz & Krieger (1944), many reports are of incompletely described material and it should be considered under the synonymy of *S. majuscula*. It is not mentioned in later monographs.

Temnogametum West *et* G.S.West 1897

BY LESLIE R. JOHNSON AND ALAN J. BROOK

Filaments unbranched, with cells often long (2 to 25 times longer than broad), cylindrical with plane end walls; chloroplast single, axile, plate-like with scattered or a median series of pyrenoids, sometimes chloroplast irregularly shaped and connected to the cytoplasm lining the cell by fine, colourless strands; cell sap sometimes coloured purple; sexual reproduction by conjugation, ladder-like or lateral and without formation of conjugation tubes; ladder-like conjugation results from two adhering gametangial cells enlarging with a connecting pore developing at point of contact and at the same time connecting walls widening to become the wall of the quadrangular or, more commonly, cross-shaped sporangium; lateral conjugation results from enlargement of two gametangia mostly in the region of adjoining walls, with breakdown of the wall when fertilization takes place; zygospores obliquely ovoid or cross-shaped with rounded or truncated angles, walls smooth, thick and usually having a distinct, characteristically shaped suture line; aplanospores and parthenospores known for some species.

China, Africa, India, South America; a largely tropical genus that grows in various aquatic environments. Sixteen species described, only one species reported from the British Isles (Brook, 1996).

Gametangia in this genus are short caps, specialized cells cut off from the ends of vegetative cells. The primordial gametangia tend to have starch amongst other storage products and are only 1–2 times longer than wide. The vegetative cells and filaments of *Temnogametum* resemble those of *Mougeotia* but it differs by having specialized conjugation cells rather then conjugation tubes. The cell sap in three species is purple.

Temnogametum sinense Jão *et* Hu 1978

27410010 **Pl. 118D** (p. 482)

Cells 17.5–18.5 μm wide, (100–)150–215 μm long; chloroplast narrow, irregularly ribbon-shaped, extending almost full length of cell and having a series of 5–8 pyrenoids, occasionally making 1–3 turns of cell, margins with numerous fine extensions connecting to cytoplasmic lining of cell; conjugation ladder-like, short cells fuse; zygospores filling fused gametangia, irregularly quadrangular with broad concave sides and rounded, sometimes tapering with a rounded extremity and extending well into conjugating filaments, 34–46×29–32 μm, outer wall smooth, inner spore wall finely punctate and dark brown when mature, distinctive suture line; lateral conjugation and aplanospores known.

China, British Isles; the first and only record of the genus from Europe was in Gwynllyn Lake (pH 6.2, alkalinity 3.9 mg $CaCO_3$ L^{-1}), near Rhyader in mid-Wales (A.J.B.), with its filaments covering the aquatic macrophyte *Myriophyllum alterniflorum*.

See comments under *Mougeotia capucina* (p. 484) about its possible confusion with *T. sinense*.

Zygnema C.Agardh 1824 **CD**

Filaments unbranched, but occasionally having rhizoids, sometimes with slight constriction between cells, end walls plane, often surrounded by a mucilaginous envelope; chloroplasts 2(– 4) per cell, star-shaped although essentially spherical with irregular radiating short branches, suspended in centre of cell, each with a prominent central pyrenoid; nucleus central between chloroplasts, suspended in a cytoplasmic bridge; sexual reproduction by conjugation, ladder-like and/or lateral; zygospores either developing in a conjugation tube, or formed in only one gametangium; zygospores usually with long axis parallel to filaments, ovoid to compressed-spherical or ellipsoidal, outer wall usually thin and colourless, median wall usually thick, smooth or variously punctate or pitted (scrobiculate), some with a slightly ridged equatorial suture, varying from pale yellow to chestnut-brown or bright blue to blue-black in colour; aplanospores produced by a small number of species and one species reportedly reproducing only by akinetes.

Probably cosmopolitan, in acid to alkaline waters and from sea level to high altitudes, one species known from soils in India and another from terrestrial habitats in Europe. World-wide almost 140 recognized species, but only 12 recorded from the British Isles, usually confined to ponds or ditches with a few in upland bogs.

The filaments are usually covered by a pectic sheath, but generally feel less slippery than those of *Spirogyra*. The filaments of *Zygnema* tend to be much shorter than those of *Spirogyra*, possibly because the pectic sheath breaks and the cells separate before the filaments attain any great length.

1 Zygospores wholly or largely formed in conjugation tubes 2
1 Zygospores usually formed in one of the gametangia 5
2 Median spore wall blue when mature .... *Z. carinthiacum*
2 Median spore wall colourless, yellow or brown when mature 3

3 Zygospores with greatest dimension at right angles to conjugation tube (cells 20–22 μm wide) .... *Z. momoniense*
3 Zygospores with the greater dimension parallel with conjugation tube 4
4 Filaments 14–24 μm wide .... *Z. ralfsii*
4 Filaments 30–36 μm wide .... *Z. pectinatum*
5 Zygospore wall smooth .... *Z. atrocoeruleum*
5 Zygospore wall ornamented 6
6 Filaments <20 μm wide 7
6 Filaments >20 μm wide 8
7 Filaments 9–12 μm wide .... *Z. stagnale*
7 Filaments 15–18 μm wide (this species also producing aplanospores) .... *Z. cylindrospermum*
8 Filaments generally >30 μm wide 9
8 Filaments always <30 μm wide 10
9 Zygospores 30–38 μm wide, 32–40 μm long; median wall pitted (scrobiculate) with pits 1.5–2 μm in diameter .... *Z. cruciatum*
9 Zygospores (25–)28–36(–40) μm wide, 27–100 μm in length, median wall deeply pitted with pits 3–4 μm in diameter .... *Z. stellinum*
10 Zygospores blue .... *Z. conspicuum*
10 Zygospores not blue 11
11 Reproduction only by conjugation 12
11 Reproduction by aplanosporess and conjugation 13
12 Median spore wall pitted (scrobiculate) .... *Z. vaucherii*
12 Median spore wall warty-tuberculate .... *Z. vaginatum*
13 Filaments 20–24 μm wide .... *Z. leiospermum*
13 Filaments 26–30 μm wide .... *Z. insigne*

Zygnema atrocoeruleum West *et* G.S.West 1897

27440020

Cells 14–17 μm wide, 40–47 μm long; conjugation ladder-like, zygospores produced in receptive (female) gametangia that become enlarged or inflated; zygospores spherical, 23–26(–29) μm in diameter, median spore wall smooth, dark blue.

British Isles; only known from southern England.

There was no original illustration, but the species is recognized in later monographs (e.g. Kolkwitz & Krieger, 1941).

Zygnema carinthiacum Beck-Mannagetta 1929

Synonyms: *Zygnema commune* Czurda, *Z. excommune* Transeau

27440110 **Pl. 127A** (p. 508)

Cells 25–30(–32) μm wide, 25–100 μm long; conjugation ladder-like, zygospores produced in receptive gametangia that become greatly enlarged; zygospores spherical to ovoid, 32–45 μm wide, 36–52 μm long, median wall pitted, pits 3–4 μm in diameter, 3–5 μm apart, blue.

Europe (not widespread), possibly cosmopolitan; little-known species and only recorded in the British Isles from England in rivers, lakes and cattle drinking troughs.

Zygnema conspicuum (Hassall) Transeau in Transeau et al. 1934

Basionym: *Tyndaridea conspicua* Hassall

27440120 **Pl. 127I** (p. 508)

Cells 22–27 μm wide, 50–90 μm long; conjugation ladder-like, zygospores formed in cylindrical or slightly enlarged conjugation tubes; zygospores spherical to ovoid, 24–32 μm wide, and 26–33 μm long, median wall pitted (scrobiculate), pits about 1.5–2 μm in diameter and same distance apart, brown.

Europe, USA; only known in shallow water including pond and lake margins in southern England (e.g. Wimbledon Common, Surrey).

Zygnema cruciatum (Vaucher) C.Agardh 1824

Basionym: *Conjugata cruciata* Vaucher

Synonym: *Tyndaridea cruciata* (Vaucher) Hassall

27440020 **Pl. 127G** (p. 508)

Cells 30–36 μm wide, 30–60 μm long; conjugation ladder-like, zygospores formed in receptive gametangia that remain cylindrical or only slightly enlarged; zygospores mostly spherical or broadly ovoid, 30–38 μm in diameter, if ovoid only 32–40 μm long, median wall pitted, pits 1.5–2 μm in diameter, 3–5 μm apart, brown; aplanospores short, cylindrical-ovoid, 30–35 μm wide, 30–60 μm long, filling cell, colour and ornamentation similar to zygospores.

Probably cosmopolitan; widespread in the British Isles with several records from the English Midlands, North Wales and the west of Ireland in ponds, pools, ditches and lake margins; some habitats acid and others alkaline.

Zygnema cylindrospermum (West *et* G.S.West) H.Krieger 1941

Basionym: *Zygnema stellinum* (Vaucher) Kützing var. *cylindrospermum* West *et* G.S.West

27440030 **Pl. 127F** (p. 508)

Cells 15–18 μm wide; conjugation ladder-like, zygospores formed in receptive gametangia (female) that remain cylindrical or become slightly enlarged; zygopores cylindrical with broadly rounded poles, 15–19 μm wide, 23–54 μm long, median wall punctate with quite deep cavities, brown.

British Isles, India, South Africa; first described from the Shetland Islands, also recorded from Wales in slightly acid ponds.

Zygnema insigne (Hassall) Kützing 1849

Basionym: *Tyndaridea insignis* Hassall

Synonym: *Zygnema leispermum* fo. *megaspora* W.West

27440040 **Pl. 127M** (p. 508)

Cells 26–32 μm wide, 26–60 μm long; conjugation ladder-like or lateral, zygospores form in receptive gametangia (female) that remain cylindrical or become enlarged; zygospores spherical or subspherical, 27–35 μm in diameter, median wall thick, smooth, yellow-brown; parthenospores ellipsoidal or cylindrical; aplanospores 28–33 × 27–35 μm, ovoid to cylindrical, otherwise similar to zygospores.

Probably cosmopolitan; widely distributed in the British Isles, recorded from Yorkshire and the English Midlands, Midlothian and the Orkney Islands, Anglesey and Caernarvonshire, and from Ireland.

Transeau (1951) noted that many descriptions were based on immature specimens with colourless spores.

Zygnema leiospermum de Bary 1858

27440050 **Pl. 127L** (p. 508)

Cells 20–24 μm wide, 20–40 μm long; conjugation ladder-like, sporangia greatly enlarged or inflated (outer wall sometimes ruptures as zygospores mature); zygospores spherical to ovoid, 23–30 μm wide, 23–32 μm long, median wall thick, smooth, brown; aplanospores similar to zoospores but smaller.

Europe (widespread), North America; widely distributed in the British Isles where recorded from pools, ditches and lake margins in Yorkshire, also Cornwall and other counties of southern England; also Anglesey and Caernarvonshire, and Clare Island, County Mayo.

Zygnema momoniense West 1892

27440130 **Pl. 127B** (p. 508)

Cells 20–22 μm wide; conjugation ladder-like, zygospores formed in the conjugation tubes, with longer axis of spore at right angles to conjugation tubes; zygospores compressed to spherical, 25–27 μm wide, 30–33 μm long, median wall smooth, brown.

British Isles (Ireland); known only from the west of Ireland.

Kolkwitz & Krieger (1941) and Kadlubowska (1984) placed this species in the synonymy of *Zygnema ralfsii* (Hassall) de Bary. However, Transeau (1951) and Randhawa (1959) continued to recognize both species. Although very similar, the dimensions of *Zygnema momoniense* are at the upper limit and beyond those of *Zygnema ralfsii*. The two species are recognized in the present text.

Zygnema pectinatum (Vaucher) C.Agardh 1817

Basionym: *Conjugata pectina* Vaucher
Synonym: *Zygnema excrassa* Transeau

27440070 **Pl. 127D** (p. 508)

Cells 30–36 μm wide, 25–120 μm long; conjugation mostly ladder-like (very rarely lateral), zygospores formed in considerably swollen conjugation tubes; zygospores spherical to ellipsoidal, 35–44(–60) μm in diameter, when ovoid 40–54(–70) μm long, median wall pitted (scrobiculate), pits 2–3 μm in diameter, brown; parthenospores produced in the conjugation papilla, similar to zygospores; aplanospores ovoid or cylindrical, 30–38 μm wide, 30–60 μm long, wall similar to that of the zygospores.

Probably cosmopolitan; widespread in Europe, fairly common in the British Isles in ditches, ponds and lake margins where often entangled around larger aquatic plants, with records from Yorkshire and southern England, Wales and the west of Ireland.

Zygnema ralfsii (Hassall) de Bary 1858

Basionym: *Tyndaridea ralfsii* Hassall
Synonym: *Zygogonium ralfsii* (Hassall) Kützing

27440080 **Pl. 127C** (p. 508)

Cells 14–20 μm wide, 38–80 μm long; conjugation ladder-like, zygospores formed in conjugation tubes; zygospores lens-shaped or ellipsoidal and orientated parallel to gametangial cells, 15–25 μm wide, 25–35 μm long, median wall thick, smooth, brown.

Europe (widespread), North America; in the British Isles known only in England from pools and streams in Cornwall, Leicestershire, Cambridgeshire and Yorkshire.

See comments under *Zygnema momoniense*.

Zygnema stagnale (Hassall) Kützing 1849

Basionym: *Tyndaridea stagnalis* Hassall
Synonym: *Zygnema vaucherii* var. *stagnale* Cooke, *Z. stellinum* var. *stagnale* (Hassall) Kirchner

27440140 **Pl. 127E** (p. 508)

Cells 9–12 μm wide, 20–50 μm long; conjugation ladder-like, zygospores formed in receptive gametangium (female) that enlarges on inner conjugating side; zygospores spherical to subspherical, 14–18 μm wide, 14–25 μm long, median wall with very small punctae, brown.

North Africa, North America; first described from the Lizard, Cornwall, also reported from other localities in England including Tunbridge Wells, Kent and Ashdown Forest, Sussex.

Zygnema stellinum (Vaucher) C.Agardh 1824 emend. Czurda 1932

Basionym: *Conjugata stellina* Vaucher
Synonym: *Zygnema stellinum* var. *compressum* Gayral

27440090 **Pl. 127H** (p. 508)

Cells (25–)28–36(–40) μm wide, 27–100 μm long; conjugation ladder-like, occasionally lateral, zygospores formed in receptive gametangia that become inflated especially on conjugating side; zygospores ovate or ellipsoidal, 28–42 μm wide, 29–47 μm long, median wall deeply pitted, pits 3–4 μm in diameter and about same distance apart, brown or yellow-brown; parthenospores common, usually cylindrical and filling cell, or more rarely spherical and occupying middle of cell, walls similar to zygospores.

Probably cosmopolitan; fairly common and widely distributed in the British Isles, forming green mucilaginous masses in ditches, pools and especially in moorland *Sphagnum* bogs.

Zygnema vaginatum G.A.Klebs 1886

Synonym: *Zygnema subcruciatum* Transeau

27440150 **Pl. 127K** (p. 508)

Cells 25–27 μm wide, 37–75 μm long; conjugation ladder-like, zygospores formed in the receptive gametangium that slightly enlarges; zygospores spherical to broadly ovoid, about 28 μm wide, median wall pitted, pits 1.5–2 μm in diameter and 3–4 μm apart, brown.

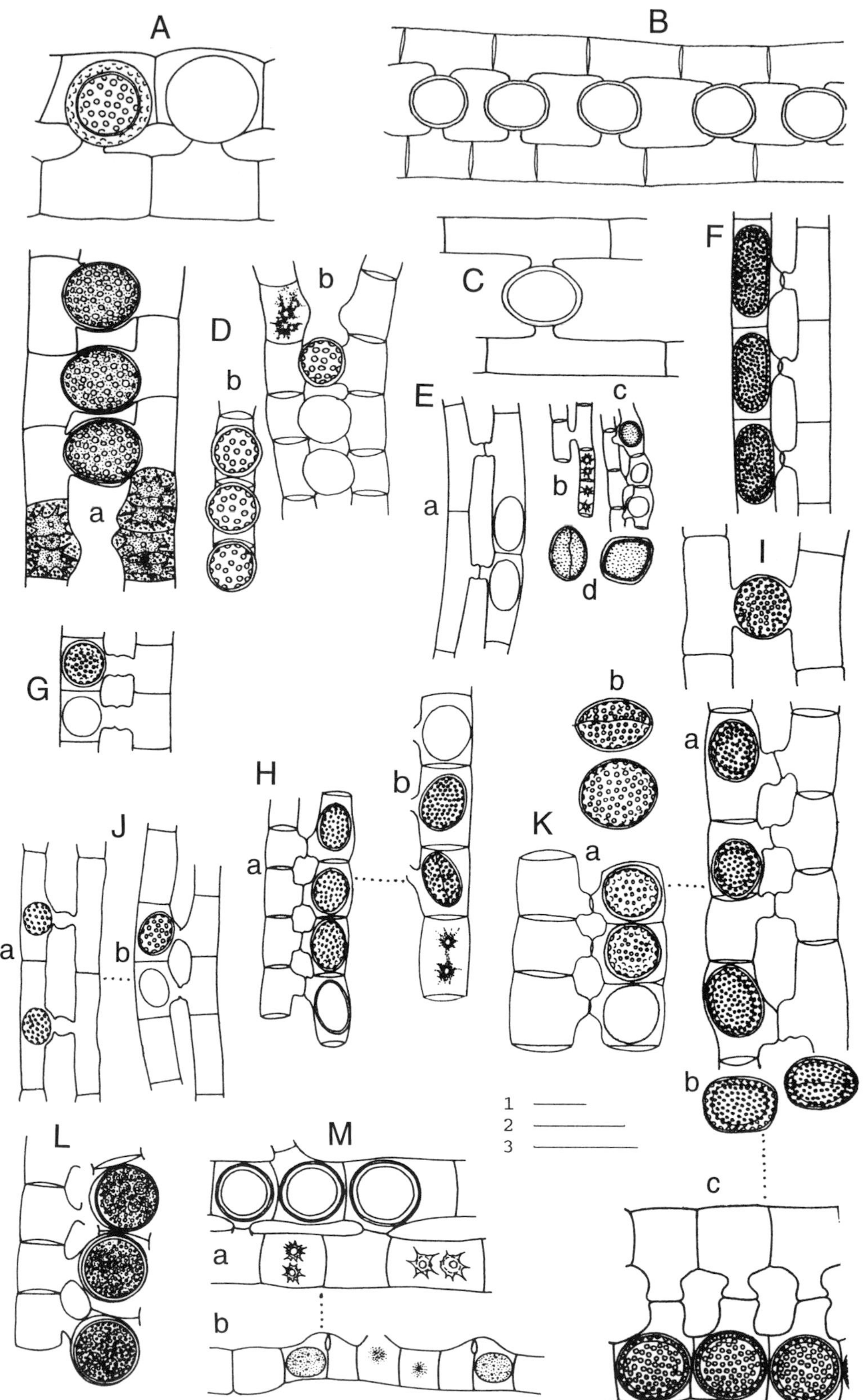

Plate 127 A–M

A. *Zygnema carinthiacum* (p. 506): ladder-like conjugation with zygospore; **B**. *Zygnema momoniense* (p. 507): ladder-like conjugation with zygospores; **C**. *Zygnema ralfsii* (p. 507): ladder-like conjugation with zygospore; **D**. *Zygnema pectinatum* (p. 507): a, b. ladder-like conjugation with zygospores; **E**. *Zygnema stagnale* (p. 507): a, c. ladder-like conjugation with zygospores, b. part of vegetative filament, d. zygospores; **F**. *Zygnema cylindrospermum* (p. 506): ladder-like conjugation with zygospores; **G**. *Zygnema cruciatum* (p. 506): ladder-like conjugation with zygospores; **H**. *Zygnema stellinum* (p. 507): a, b. ladder-like conjugation with zygospores, b. filament with zygospores and lowest cell having 2 star-shaped chloroplasts; **I**. *Zygnema conspicuum* (p. 506): ladder-like conjugation with zygospores; **J**. *Zygnema vaucherii* (p. 509): a, b. ladder-like conjugation; **K**. *Zygnema vaginatum* (p. 507): a, c. ladder-like conjugation with zygospores, b. details of zygospores; **L**. *Zygnema leiospermum* (p. 507): ladder-like conjugation with zygospores; **M**. *Zygnema insigne* (p. 506): a, b. ladder-like conjugation.

Scale bars: 50 µm. 1 for D, G, H, J; 2 for A, B, Eb,c, K, M; 3 for C, Ea,d, F, I, L.

Probably cosmopolitan; in the British Isles in ponds, ditches, and other bodies of standing water in England and Wales.

There has been a problem concerning the interpretation of the patterning on the zygospore wall. To quote Transeau (1951): 'Czurda (1932) and Kolkwitz & Krieger (1941) described this species as having scrobiculate median spore walls. No reason for the change from the original description is given'. Subsequent monographs, and also illustrations all stemming from Kolkwitz and Krieger, support this interpretation of the wall ornamentation as scrobiculate rather than 'verrucoso-tuberculoso' as described by Klebs (1886, pp. 340–341).

Zygnema vaucherii C.Agardh 1824

Synonym: *Zygnema stellinum* var. *vaucherii* (C.Agardh) Kirchner

27440100 **Pl. 127J** (p. 508)

Cells 24–28 μm wide, 50–180 μm long; conjugation ladder-like, zygospores formed in the receptive gametangia that become gradually or abruptly inflated towards the middle; zygospores broadly ellipsoidal or spherical, 24–36 μm wide, 26–45 μm long, median wall pitted, pits 2–3 μm in diameter, brown; parthenospores similar in form to zygospores.

Europe (widespread), North America, Asia; in the British Isles in ditches and ponds in moorland areas of England, the west of Ireland, North Wales and the Orkney Islands.

DOUBTFUL RECORDS

Zygnema anomalum (Hassall) Cooke 1884

27440010

Reported in the British Isles from boggy pools in southern England. Regarded as incompletely described by Kolkwitz & Krieger (1941), and not recognized in later monographs.

Zygnema gracile Berkeley 1832

Reported from a face of dripping rock in southern England. Regarded as insufficiently described by Kolkwitz & Krieger (1941) and to be a possible synonym of *Zygogonium gracile* Kützing, neither species mentioned in later monographs.

Zygnema parvulum (Kützing) Cooke 1884

27440060

Reported from standing pools. Regarded as insufficiently described by Kolkwitz & Krieger (1941) and possibly a *Zygogonium*; the species is not mentioned in later monographs.

Zygnema rostratum Hassall 1845

Placed under the synonymy of *Spirogyra nitida* by Cooke (1882–84), but regarded as a synonym of *Spirogyra rostrata* (Hassall) Spencer by Kolkwitz & Krieger (1941). They consider it to be incompletely described and it is not mentioned in subsequent monographs.

Zygnemopsis (Skuja) Transeau 1934

Filaments unbranched, with cells 2 to 5(–10) times longer than broad, sometimes cells constricted at ends with cross wall plane; chloroplast a single axile plate with 2 pyrenoids, or 2 per cell (rarely 3–6), star-shaped or compact bolster-shaped, each with a central pyrenoid; nucleus adjacent to or between chloroplasts; sexual reproduction by conjugation, ladder-like, isogamous and similar to *Spirogyra*, the conjugation tube developing between adjacent filaments and sometimes enlarging, at the beginnning of reproduction the cell contents gradually become replaced by refractive, smooth or laminated pectic material; zygospores compressed-ovoid to quadrangular, pillow-shaped, sometimes with truncate corners that extend into the conjugating filaments; zygospores with an equatorial suture evident when fully mature, wall usually 3-layered, median wall smooth, punctate, pitted (scrobiculate) or warty, yellow, golden-brown or brown; aplanospores common, ovoid or ellipsoidal, wall structure similar to respective zygospores; parthenospores similar but smaller than zygospores and lying lateral in gametangia.

Probably cosmopolitan, various freshwater habitats. Over 40 species worldwide but only two reported from the British Isles.

The vegetative filaments of *Zygnemopsis* are usually indistinguishable from those of *Zygnema* and the two genera can be separated only on characters observable during sexual reproduction.

1 Zygospores 14–18 μm wide, 18–24 μm long .... *Z. desmidioides*
1 Zygospores 20–24 μm wide, 29–32 μm long .... *Z. tiffiana*

Zygnemopsis desmidioides (West *et* G.S.West) Transeau in Transeau et al. 1934

Basionym: *Debarya desmidioides* West *et* G.S.West
Synonym: *Debarya cruciata* S.R.Price

27450010 **Pl. 118E** (p. 482)

Filaments readily fragmenting; cells 8–11 μm wide, 19–56 μm long, with a distinct constriction between cells; chloroplast axile and plate-like, with 2 pyrenoids; conjugation between short, fragmented cells, zygospores formed in broad conjugation tubes and extending somewhat into both gametangia; zygospores quadrangular with straight, slightly concave or convex sides and rounded angles, 14–18 μm wide, 18–24 μm long, median wall thick, finely punctate, golden-brown.

Russia, North America, India; only known in the British Isles from the Lizard, Cornwall where it was first described as a *Debarya*.

Transeau (1951) considers the statement by W. & G.S.West (1903b) concerning conjugation between 'free cells' as probably misleading. Fragmentation occurs in several species during or after conjugation. *Z. desmidioides* has a pectic sheath and during conjugation this dissolves. The gametangia at the same time swell, their cell ends becoming rounded and thus the cells tend to separate.

Zygnemopsis tiffiana Transeau 1944

27450020 **Pl. 118F** (p. 482)

Filaments not readily fragmenting; cells 10–12 μm wide, 30–60 μm long; conjugation ladder-like, zygospores formed in broad conjugation tubes, when mature extending into both gametangia; zygospores quadrangular with concave or rarely straight sides, 20–24 μm wide, 29–32 μm long, median wall thin, smooth, punctate with punctae 1.5 μm in diameter, yellow in colour.

North America; only reported in the British Isles from Wales (Benson-Evans & Antoine, 1996).

Zygogonium Kützing 1843

Filaments essentially terrestrial, occasionally with erect subaerial branches arising from creeping branches, rhizoids in the subsoil; cells short and cylindrical, usually less than 5 times as long as broad, with thick, opaque, lamellate walls; chloroplasts 2 per cell, small, disc- or pillow-shaped, or compressed-spherical, each with a central pyrenoid, occasionally with a few short irregular processes, mostly obscured by accumulation of reserve products; nucleus small, lying between the chloroplasts; reproduction most commonly by aplanospores and akinetes; sexual reproduction by conjugation, ladder-like or lateral; gametes uniting in the conjugation tube (in *Z. ericetorum*, secondary gametangia are formed); sporangium wall forms around developing zygospore and separates it from the gametangia, the conjugation tube subsequently often greatly enlarged; cytoplasmic residues accumulate in gametangia after formation of zygospores (principal character separating it from *Zygnema*); aplanospores spherical or cylindrical-ovoid and generally much smaller than vegetative cell.

Probably cosmopolitan; mostly terrestrial and growing on wet surfaces, acid mineral soils, rocks and bogs but will survive if transported into a fully aquatic environment. Twenty-eight known species but only one reported from the British Isles.

The cytoplasm usually has a purple tinge, though the extent of its development is influenced by the acidity of the environment and probably also by how rapidly the organism is growing. Chloroplast shape distinguishes *Zygogonium* from *Zygnema*, as do the cytoplasmic residues remaining in the cell and not participating in the formation of the zygospore. In drier habitats the vegetative cells may develop thick, yellow or brown, lamellated walls and finally these become akinetes. The cell sap may be colourless, but purple in some species.

Zygogonium ericetorum Kützing 1843

27460010 **Pl. 118G** (p. 482) **CD**
Filaments simple or branched; cells 12–33 μm wide, 10–100 μm long, usually with purple cell sap; chloroplasts 2, weakly star-shaped, pillow-shaped or more indefinite, each with a central pyrenoid; conjugation ladder-like, zygospores formed within definite sporangia formed by the conjugation tubes and cut off by a wall from adjoining gametangia; progametes cut-off from gametangia by an ingrowth of cell wall leaving remnants of cell contents in the gametangia, a pore forms between progametes and enlarges to allow fusion of the gametes; zygospores ovoid or ellipsoidal, 15–26 μm wide, 20–36 μm long, median wall smooth, dark brown; aplanospores spherical or ovoid, 15–20 μm wide, 15–40 μm long, occupying only part of original cell.

Probably cosmopolitan; always in acid environments (pH 3.4–6) including the surface of peat bogs, heaths, moors, cliffs and small streams; widespread and often abundant in such habitats throughout the British Isles. *Zygogonium* is also often abundant on soils where sulphide minerals have been oxidized (e.g. Parys Mountain, Anglesey) and occasionally in small acid mine drainages (e.g. Geevor Mine, Cornwall); in the British Isles it occurs down to about pH 2.8 in such environments. The terrestrial material from mining areas corresponds well to *Z. ericetorum*, but populations in streams vary more and tend to have much thinner cell walls and should be investigated more closely.

Very distinctive because of its often very thick wall and the colour of the cytoplasm, which ranges from very pale purple to deep purple or sometimes brownish-purple, and thus imparts a purplish tinge to many acid, moorland soil surfaces.

Family Mesotaeniaceae ('Saccoderm Desmids')

BY ALAN J.BROOK

Because they possess simple, ellipsoidal, spindle-shaped or cylindrical cells lacking any median constriction, or even a median suture, all the saccoderm 'desmids' have, at least for the time being, been placed within the one, albeit artificial family Mesotaeniaceae. Chloroplast structure rather than cell shape is the most important taxonomic character for the identification of the saccoderm desmids.

Because their chloroplasts tend to become distorted on fixation, saccoderm cells should be examined live. For some species of the genus *Cylindrocystis*, zygospores must be present for their identification.

Cylindrocystis Meneghini ex de Bary 1858

Cells short, cylindrical with broadly rounded apices; walls smooth, unornamented and often secreting copious mucilage; chloroplasts (except for one species) all contain 2 axile, star-shaped chloroplasts, with a distinctive pyrenoid at the centre of each, extensions of the chloroplasts radiate from each pyrenoid, often becoming longitudinally extended into ridges and sometimes flattened against the inner surface of the cell wall so that extensions assume a parietal position.

Cosmopolitan; most acidophiles, widely distributed in aquatic environments especially bog areas.

Seven species of *Cylindrocystis* have been found in the British Isles, and all apart from *C. crassa* are acidophiles. *C. crassa* and the very rare *C. obesa* (27070080) are readily distinguished by cell shape; *C. gracilis* (27070060) and *C. cushleckae* are separated by size and chloroplast structure; the rare, ovoid cells of *C. acanthospora* (27070010) have only a single star-shaped chloroplast; *C. brebissonii*, *C. debaryi* and *C. jenneri* (27070070) can be positively identified only when zygospores are present (see Brook & Williamson, 2002, for detailed treatment of all *Cylindrocystis* species).

1 Cells with a single star-shaped chloroplast with a median hole or lateral notch (in face view) and 1 or 2 dissected ridges .... *C. cushleckae*
1 Cells with 2 star-shaped chloroplasts; cells mostly solitary 2
2 Cells broadly oval to subspherical with broadly rounded ends; cells less than twice as long as broad .... *C. crassa*
2 Cells elongate cylindrical with broadly rounded ends; cells at least twice as long as broad 3
3 Chloroplasts (in side view) with 2 or 3 widely spaced, dissected longitudinal ridges; cells 11–15.5 μm wide .... *C. gracilis*
3 Chloroplasts with 3 or more, dissected longitudinal ridges, star-shaped in short cells; cells 16–29 μm wide 4

4 Cell width 16–19.5 µm .... *C. brebissonii*
4 Cell width 24–29 µm ..... *C. brebissonii* var. *turgida*

Cylindrocystis brebissonii (Meneghini ex Ralfs) de Bary 1858

Basionym: *Penium brebissonii* Meneghini ex Ralfs
Synonyms: *Cylindrocystis brebissonii* var. *curvata* Rabanus, *C. brebissonii* fo. *curvata* (Rabanus) Kossinska

27070020 **Pl. 128E** (p. 512) **CD**

Cells unconstricted, cylindrical, 16–19.5 µm wide, 35–80(–110) µm long, apices rounded; chloroplasts in shorter younger cells star-shaped, each with a distinctive, central spherical pyrenoid, in longer cells (usually before division), ridges of chloroplasts longitudinally extended and variously dissected, with distal ends tending to flatten against the inner surface of cell wall, in longer cells pyrenoids tend to elongate prior to division, sometimes 2 pyrenoids in these cells.

Cosmopolitan, acidophile; widely distributed in acidic environments throughout the British Isles, often growing in abundance in peat bogs and pools containing *Sphagnum*; also on peat faces and dripping rocks, sometimes occurring in extensive mucilaginous masses.

Conclusive identification requires the presence of the characteristic zygospores which are tetragonal, 30–48 µm wide with a thickened, double wall, brown when mature. In some populations, slightly curved forms may be present and although given taxonomic status by some authors (see synonyms above) they are almost certainly growth forms.

var. *turgida* Schmidle 1895

Synonym: *Cylindrocystis brebissonii* fo. *turgida* (Schmidle) Kossinska

27070023 **Pl. 128F** (p. 512) **CD**

Cells mostly cylindrical with broadly rounded ends, 2 to 4 times longer than wide, some with one end broader than the other so that margins converge, such cells broadly pear-shaped, 24–29 µm wide, 36–102 µm long; chloroplasts star-shaped, ridges narrow and delicate, radiating at various angles from spherical pyrenoids; zygospores similar to the nominal variety but up to 55 µm wide.

Europe, acidophile; large populations can dominate acid bogs and bog pools, especially amongst *Sphagnum* (pH 3.8–6); widespread in appropriate habitats throughout the British Isles.

Cylindrocystis crassa de Bary 1858

27070030 **Pl. 128G** (p. 512) **CD**

Cells oval to subspherical, 24–31 µm wide, 27–50 µm long, containing 2 star-shaped chloroplasts each with a spherical pyrenoid; chloroplasts obscured by small, spherical inclusions; walls thick, frequently surrounded by a wide, sometimes multilayered mucilage sheath.

Europe, North America; fairly widely distributed in nutrient-poor waters, mostly as solitary cells, rarely abundant. Frequently in subaerial habitats, e.g. vertical peat surfaces, moss-covered walls and dripping rocks subject to desiccation, also with other drought-resistant desmids in shallow, temporary pools. Probably tolerant of a wide pH range and alkalinity.

Cells commonly occur singly, less frequently in loose aggregations of 24–48 cells; on rare occasions arranged linearly within mucilaginous strands.

Cylindrocystis cushleckae Brook 1994

27070040 **Pl. 128H** (p. 512)

Cells narrow cylinders, 7.5–10 µm wide, 24–41 µm long, occasionally slightly curved, apices rounded to subtruncate; chloroplast 1 per cell with a distinct median notch (a hole seen from above) in which the nucleus lies, only during cell division are 2 axile and star-shaped chloroplasts visible, having 4 irregularly dissected ridges disposed at approximately right angles, in apical view appearing as a cross, in side view only 1 or 2 ridges visible; 2–8 axile, evenly spaced pyrenoids in each chloroplast depending on cell length and state in division-cycle.

Europe; widespread in the British Isles and often abundant in peaty pools and bogs amongst *Sphagnum* and on wet peat faces in many western areas, restricted to distinctly acid habitats; probably widespread in such areas throughout much of Europe (e.g. Swiss mountain bogs; see Brook, 1994b).

Cylindrocystis gracilis I.Hirn 1953

27070060 **Pl. 128I** (p. 512) **CD**

Cells narrow, cylindrical with rounded apices, 11–15.5 µm wide, 20–50 µm long; chloroplasts axile and star-shaped in apical view, in side view never star-shaped even in the shortest cells and only 1 or 2 (rarely 3) delicate, elongate ridges visible, sometimes orientated parallel or slightly obliquely to the longitudinal axis of cell, edges of the ridges slightly notched or dissected, each chloroplast with 1 or 2 spherical pyrenoids; zygospores tetragonal with moderately thickened walls.

Europe, acidophile; common throughout the British Isles in acid habitats such as peat bogs and *Sphagnum* flushes; usually associated with *C. brebissonii* and *C. cushleckae*.

Mesotaenium Nägeli 1849

Cells short-cylindrical or ellipsoidal and typified by the presence in each of a single, simple chloroplast (prior to division there may be 2), consisting of a flattened, ribbon-like or slightly cup-shaped axile plate, containing in most species, a single, fairly obvious pyrenoid.

Cosmopolitan; of the eight species recorded from the British Isles some are extremely rare and one is of very doubtful validity (Brook, 1984; Brook & Williamson, 2002). Six species occur almost exclusively in subaerial habitats.

Cells often in large numbers and embedded in copious mucilaginous masses; mucilage may be laid down successively in distinct layers; in peaty habitats these may be coloured yellow to dark brown. Most mucilage seems to be produced when *Mesotaenium* species grow in subaerial environments and here they produce green to purple masses on wet rocks, or lustrous slippery coverings on damp mosses on shaded, sheltered walls and even on woodland soils (West, 1915). It is essential to examine living cells to determine the true form of the chloroplasts.

1 Cell apices commonly conical, straight or moderately curved; chloroplast with up to 4 pyrenoids .... *M. caldariorum*
1 Cell apices not conical, mostly broadly rounded; chloroplast with a single pyrenoid 2

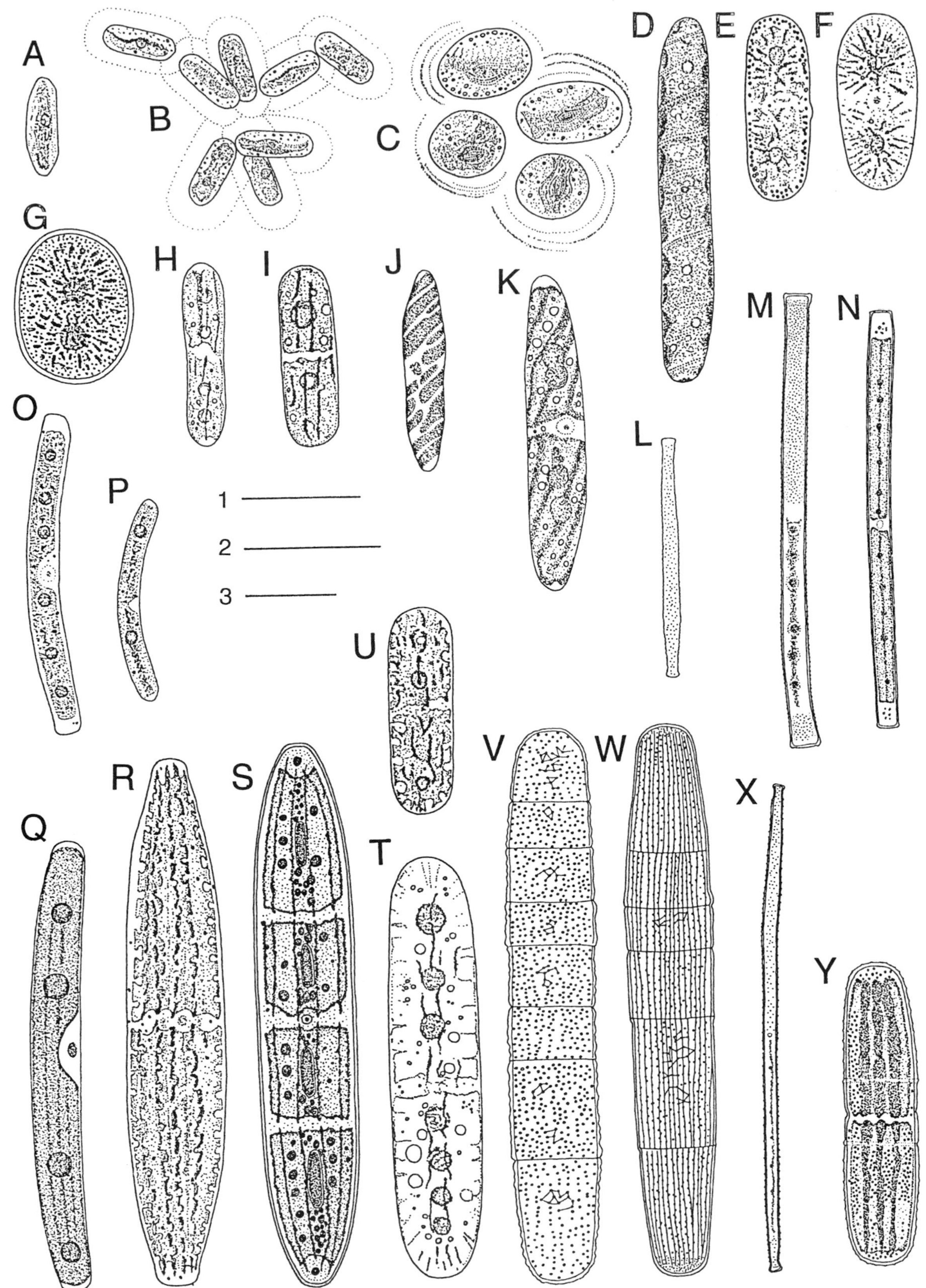

Plate 128 A–Y

A. *Mesotaenium caldariorum* (p. 513); **B**. *Mesotaenium chlamydosporum* (p. 513); **C**. *Mesotaenium macrococcum* (p. 513); **D**. *Spirotaenia condensata* (p. 515); **E**. *Cylindrocystis brebissonii* (p. 511); **F**. *Cylindrocystis brebissonii* var. *turgida* (p. 511); **G**. *Cylindrocystis crassa* (p. 511); **H**. *Cylindrocystis cushleckae* (p. 511); **I**. *Cylindrocystis gracilis* (p. 511); **J**. *Tortitaenia luetkemulleri* (p. 515); **K**. *Tortitaenia alpina* (p. 515); **L**. *Gonatozygon brebissonii* var. *minutum* (p. 529); **M**. *Gonatozygon monotaenium* (p. 529); **N**. *Gonatozygon kinahanii* (p. 529); **O**. *Roya anglica* (p. 514); **P**. *Roya obtusa* var. *montana* (p. 514); **Q**. *Roya obtusa* (p. 515); **R**. *Netrium digitus* (p. 513); **S**. *Netrium interruptum* (p. 514); **T**. *Netrium oblongum* (p. 514); **U**. *Netrium oblongum* var. *cylindricum* (p. 514); **V**. *Penium margaritaceum* (p. 530); **W**. *Penium spirostriolatum* (p. 530); **X**. *Gonatozygon brebissonii* (p. 529); **Y**. *Penium cylindrus* (p. 530).
Scale bars: 1, 25 μm; 2, 3, 50 μm. 1 for A–C, E–L, O–Q, T–W, Y; 2 for M; 3 for D, N, R, S, X.

2 Chloroplast a curved plate with an inconspicuous pyrenoid.................................*M. chlamydosporum*
2 Chloroplast a short ribbon, often with an irregular margin and a prominent lens-shaped pyrenoid ....*M. macrococcum*

Mesotaenium caldariorum (Lagerheim) Hansgirg 1886

Basionym: *Mesotaenium endlicherianum* Nägeli var. *caldariorum* Lagerheim
27190010 **Pl. 128A** (p. 512)
Cells narrow cylindrical, mostly narrowing abruptly towards the apices so that they appear as subtruncate cones (only after recent division are apices rounded), somewhat curved, 10.5–13 µm wide, 42–64 µm long; chloroplast single, ribbon-like, with 2–4 pyrenoids, mostly extending from end to end of cell but sometimes with a gap between their ends and the apices; gap between last pyrenoid and cell apex sometimes with moving crystals similar to those found in *Closterium*.

First recorded from the British Isles as a gelatinous stratum from a greenhouse in West Yorkshire; later as a green stratum under oak and birch trees in the Wye Forest, Worcestershire. More recently from a wet sluice gate on the River Avon in Leicestershire (Williamson, 1990) and also from the same county found in a shallow pool in a tyre track along with other drought-resistant algae.

With their cone-shaped apices, no other species of *Mesotaenium* should be confused with *M. caldariorum*.

Mesotaenium chlamydosporum de Bary 1858

Synonym: *Palmogloea chlamydospora* de Bary
27190020 **Pl. 128B** (p. 512) **CD**
Cells oblong-cylindrical, 11–12 µm wide, 25–33 µm long, apices broadly, rounded, sometimes subtruncate; chloroplast partially parietal, a curved plate containing a single, centrally-placed, inconspicuous pyrenoid; cells often surrounded by a wide, distinctly stratified mucilage envelope.

Europe, North America: widely distributed throughout the British Isles, especially in regions with base-poor rocks, and frequently forming mucilaginous masses on subaerial substrates; also occurring as individual cells amongst mosses on shaded walls.

Mesotaenium macrococcum (Kützing ex Kützing) J.Roy *et* Bisset 1894

Basionym: *Palmogloea macrococca* Kützing ex Kützing
Synonym: *Mesotaenium braunii* de Bary
27190060 **Pl. 128C** (p. 512) **CD**
Cells short, 11–20 µm wide, 22–35 µm long, approximately cylindrical, sometimes ovoid, mostly only twice as long as broad, with broadly rounded to subtruncate apices; chloroplast an axile plate whose axile status may be clearly seen not only in apical view but in some cells viewed from the side, each sometimes with a toothed margin and containing a prominent, centrally placed pyrenoid often surrounded by a narrow starch sheath.

Cosmopolitan; widely distributed in the British Isles but rarely abundant, often occurring in mucilaginous masses on various subaerial substrates; also present in margins of bog pools and amongst *Sphagnum*.

Especially in subaerial habitats, cells occur in irregular groups of 48–64 or more, surrounded by a wide, firm mucilage envelope which may be stratified and coloured brown.

Netrium (Nägeli) Itzigsohn *et* Rothe 1856

Cells ranging from cylindrical to spindle-shaped to ovoid (some cells may be slightly curved); chloroplasts (most important taxonomic feature) always axile and when observed in vertical view having numerous radiating, considerably dissected ridges, in front view 2–6 prominent, longitudinal and often elaborately dissected ridges visible and these mostly have bifurcating or dissected free margins which at their extremities tend to be disposed to alternate sides of the ridges of which they are a part; apical vacuoles with moving barium sulphate crystals present in two species.

Cosmopolitan, acidophile; some species, especially *N. digitus* and to a lesser extent *N. oblongum*, are widely distributed in the British Isles, being amongst the most commonly encountered desmids in upland and peaty regions; occur in the metaphyton of acid lakes and ponds but also in *Sphagnum* bogs. The small *N. oblongum* var. *cylindricum* may be abundant amongst mosses covering dripping rocks.

There would appear to be no very clear dividing line between some of the smaller taxa of the genus *Netrium* and some species of *Cylindrocystis*, the only clear difference relates to chloroplast structure. The commonest species, *N. digitus*, is particularly distinctive on account of its large size and complex chloroplasts.

1 Cells elongate-ellipsoidal, attenuating to rounded apices...................................*N. oblongum*
1 Cells oblong-elliptical or lanceolate to broadly spindle-shaped 2
2 Apices truncate or subtruncate; cells with 2 chloroplasts..3
2 Apices mostly conical, rarely broadly rounded; cells with 4 chloroplasts.................*N. interruptum*
3 Cells 16–20 µm wide, chloroplast ridges moderately dissected ....*N. oblongum* var. *cylindricum*
3 Cells 40–82 µm wide, chloroplast ridges highly dissected ....*N. digitus*

Netrium digitus (Ehrenberg ex Ralfs) Itzigsohn *et* Rothe 1856

Basionym: *Penium digitus* Ehrenberg ex Ralfs
27230010 **Pl. 128R** (p. 512)
Cells 40–82 µm wide, 130–390 µm long, oblong-elliptical to broadly spindle-shaped, apices usually truncate although sometimes subtruncate or even rounded; walls smooth and colourless; chloroplasts 2 per cell, axile, each with 4–6 prominent longitudinal ridges with deeply notched free margins, notches invariably bifurcate, often with fine, drawn-out extremities, the

elaborate margins lying on alternating sides of the axis of the ridge of which they are part, with each containing 1, sometimes 2, spherical or elongate pyrenoids, usually surrounded by a thick sheath of starch; zygospores when mature surrounded by a thin mucilage envelope and having a smooth, thick, brown and stratified middle wall.

Cosmopolitan, acidophile; common throughout the British Isles in acidic waters, especially in nutrient-poor pond and lake margins and in which it may be found as a chance plankter; also present in peaty pools and bogs associated with *Utricularia* and *Sphagnum* (pH 4–6.5).

The species shows very considerable variation in cell shape, not just between populations but within each population, so that many forms and varieties have been described several of which are of doubtful validity.

Netrium interruptum (Brébisson ex Ralfs) Lütkemüller 1902

Basionym: *Penium interruptum* Brébisson ex Ralfs 1848
27230020 **Pl. 128S** (p. 512) **CD**
Cells cylindrical, 25–63 μm wide, 110–320 μm long, abruptly conical near obtusely rounded apices; chloroplasts subdivided transversely so appearing as 4 per cell, each chloroplast with 8 longitudinal ridges when viewed from the cell apex, (3–5 ridges visible in side view) and margins of ridges entire instead of dissected (as in other *Netrium* species), each containing a spherical or slightly elongate pyrenoid, often surrounded by a sheath of starch grains; cell apex contains a spherical vacuole with one moving crystal seen only in living cells; zygospores unknown.

Subcosmopolitan, acidophile; widely distributed in the British Isles though never abundant, in bogs and lake margins, especially in acid waters with *Sphagnum*; also found on dripping rocks in the English Lake District and in shallow, rock-lined pools containing a rich desmid flora in mid-Wales.

Netrium oblongum (de Bary) Lütkemüller 1856

Basionym *Penium oblongum* de Bary
27230040 **Pl. 128T** (p. 512) **CD**
Cells elongate-ellipsoidal gradually attenuating to rounded apices, 32–44 μm wide, 90–155 μm long, longest cells >145 μm and may show slight median narrowing prior to cell division; chloroplasts 2, axile with 2(–3) longitudinal and moderately dissected ridges (up to 8 ridges visible in apical view), each with 2–6 spherical or elongate pyrenoids; zygospores unknown.

Cosmopolitan, acidophile; widely distributed in acid pools and bogs especially amongst *Sphagnum* where occasionally abundant, in all regions of the British Isles (pH 3.8–6.6).

Distinguished from small forms of *N. digitus* by its less elaborate chloroplasts, more numerous pyrenoids, broadly rounded apices and slightly convex lateral margins.

var. *cylindricum* West *et* G.S.West 1903

Synonyms: *Netrium oblongum* var. *cylindricum* fo. *curvatum* F.E.Fritsch, *N. oblongum* var. *curvatum* (F.E.Fritsch) Grönblad
27230043 **Pl. 128U** (p. 512)
Cells cylindrical with broadly rounded apices, 16–20 μm wide, 56–100 μm long, sometimes slightly curved; chloroplasts in apical view with 4–6 ridges radiating from the central axis and with 2 or 3 spherical pyrenoids in axis of each chloroplast; zygospores spherical, but very rare.

Subcosmopolitan; often abundant in boggy regions of upland moors and commonly associated with *Sphagnum* flushes, also around the margins of nutrient-poor lakes where they may become a component of the plankton. Recorded from the English Lake District, mid-Wales and the Scottish Highlands and Islands. A large, virtually unialgal population has been found on shaded, dripping rocks (Silurian shales) in north-west Montgomeryshire.

Because of the similar shape and dimensions, easily misidentified for species of *Cylindrocystis* and distinguished only by a careful examination of the living chloroplasts. These are characterized by 1–3 prominent, dissected ridges running down the length, not always aligned parallel to the longitudinal axis of the cell but spirally orientated. Populations may contain a proportion of curved cells.

Roya West *et* G.S.West 1896

Cells slightly curved, in some species attenuated towards truncate or obtusely rounded apices; chloroplast single, axile, band or plate-like and extending the length of cell, frequently notched in mid-region where the nucleus lies, each contains an axile series of pyrenoids varying from 4 to 14; vacuole containing moving crystals sometimes occurs at each cell apex.

Four species are described worldwide and all occur in the British Isles. Only two species, *Roya anglica* and *R. obtusa*, are widely distributed, abundant and are exclusively found in acid bog pools and lake margins. *Roya cambrica* and *R. pseudoclosterium* are very rare.

1 Cells moderately curved, narrowing slightly towards ends; apices truncate .... *R. anglica*
1 Cells only slightly curved; apices broadly rounded .... *R. obtusa*

Roya anglica G.S.West in Hodgetts 1920

Synonym: *Roya obtusa* (Brébisson) G.S.West var. *anglica* (G.S.West) Willi Krieger
27300010 **Pl. 128O** (p. 512)
Cells cylindrical to subcylindrical, 6.4–9 μm wide, 75–126 μm long, usually slightly curved, tapering gently to truncate or subtruncate apices, sometimes narrowing slightly beneath the apices which may be thickened; chloroplast single, extends almost length of cell with median notch in which nucleus lies, containing 4–6 pyrenoids; vacuoles occur between the ends of chloroplast and apices and sometimes contain well-defined rhomboidal crystals.

Cosmopolitan; in several localities, sometimes fairly common in acid environments in England and Scotland; no records for Ireland or Wales.

Roya obtusa (Brébisson) West *et* G.S.West 1896

Basionym: *Closterium obtusum* Brébisson
27300030 **Pl. 128Q** (p. 512) **CD**
Cells cylindrical, 9–15 μm wide, 75–148 μm long, very slightly curved and attenuating slightly towards broad and obtusely rounded apices; chloroplast single, extending from apex to apex, with a well-defined median notch and a series of 4 pyrenoids; prior to division mid-region of the cell sometimes shows a slight constriction.

Cosmopolitan; in acid bogs and pools, especially amongst *Sphagnum*, in England, Ireland, Scotland and Wales.

The most frequently reported and widespread of the four *Roya* species.

var. *montana* West *et* G.S.West 1896
27300032 **Pl. 128P** (p. 512)
Differs from the nominal variety especially by having significantly smaller dimensions, 5–7.5 μm wide, 48–81 μm long, usually with rounded apices, rarely subtruncate; terminal vacuoles containing moving crystals between ends of chloroplast and apices.

Occasional on dripping rocks amongst mosses and filamentous algae, often in some abundance in England and Wales; not recorded from Ireland or Scotland.

Spirotaenia Brébisson ex Ralfs emend. Brook 1997

Cells cylindrical with broadly rounded apices, or broadly to narrowly spindle-shaped with truncate or acutely rounded apices; chloroplast a single parietal ribbon, twisted into a fairly compact spiral, except *S. diplohelica* with 2 chloroplasts per cell.

Cosmopolitan; only one species, *S. condensata*, is widely distributed and moderately frequent in acid pools and bogs, also amongst the marginal vegetation of acid lakes; it has been reported as a chance plankter (Förster, 1982). Four other species recorded from the British Isles are rare or very rare.

Seven species previously ascribed to the genus (W. & G.S.West, 1904) are of doubtful validity and may well have chlorococcalean affinities.

Spirotaenia condensata Brébisson in Ralfs 1848

27350030 **Pl. 128D** (p. 512) **CD**
Cells straight or very slightly curved, 18–27 μm wide, 120–270 μm long, typically tapering at extremities to broadly rounded apices, frequently enveloped in a broad, sometimes lamellate, mucilaginous sheath; chloropast a broad, parietal, helicoidal band making 7–12 turns of the cell, with an irregularly spaced series of pyrenoids along its length.

Worldwide; widespread in appropriate habitats in the British Isles; most frequent in *Sphagnum* and peaty bogs and sometimes associated with *Utricularia* in acid pools; it has been found on shaded, dripping rock faces in the English Lake District (J.W.G. Lund, pers. comm.).

A distinctive species looking like a single-celled *Spirogyra*, to which it is probably closely related. Depending on cell length and the number of turns, the spirals may be open, or compressed and closely set.

Tortitaenia (Brook) Brook 1998

Synonym: *Polytaenia* Brook
Cells cylindrical with tapering to rounded or truncate apices, sometimes slightly curved; chloroplast with a distinctive axile core from which radiate spiral ridges, with a series of pyrenoids in central axis.

Europe, North America; because species of *Tortitaenia* tend to be rare, little is known about their ecology appearing to be restricted to acid waters; in the British Isles they seem to occur exclusively in small, shallow pools, some of which may lie in peat.

The genus was established from the subgenus Polytaenia of the genus *Spirotaenia*, on the grounds that the 4 species included in it possess axile chloroplasts with radiating spiral ridges and are fundamentally different from the spiralling parietal ribbons of what was the subgenus Spirotaenia, now raised in rank to the genus *Spirotaenia* (Brook, 1997, 1998). It has been shown that there are important biochemical differences indicating a relationship between at least some species of *Tortitaenia* and the placoderm desmids (Suborder Desmidiineae). Molecular genetic analyses suggest that at least one species of *Spirotaenia*, *S. condensata*, is closely related to the filamentous genus *Spirogyra* (Suborder Zygnematiineae: Family Zygnemataceae).

1 Chloroplast with 5 or 6 ridges, making less than a single turn of cell; each end of chloroplast with reddish caps .... *T. alpina*
1 Chloroplast with 3–8 ridges, making slightly more than one turn of cell and often bifurcating and dissected; chloroplast without reddish caps .... *T. luetkemuelleri*

Tortitaenia alpina (Schmidle) Brook 1998

Basionym: *Spirotaenia alpina* Schmidle
Synonym: *Polytaenia alpina* (Schmidle) Brook
27290010 **Pl. 128K** (p. 512)
Cells mostly spindle-shaped, 11.5–12.5 μm wide, 49–67 μm long, with parallel-sided mid-regions tapering fairly abruptly to rounded (rarely subtruncate) apices; chloroplast strongly 3–4-ridged, making 2–3 spiral turns from left to right, with 2 evenly spaced axile pyrenoids in each chloroplast and distinctive reddish-coloured caps at each end; zygospores never reported.

Europe; first recorded in the British Isles in considerable abundance from a small, shallow pool in the Lowther Hills, Dumfries and Galloway and a desmid-rich acid bog in Sutherland; also found over an extended period in a shallow pool on Silurian shales in mid-Wales (Brook, 1992a); possibly fairly widespread.

One cell margin may be distinctly convex, while the other is straight or even slightly concave, so that cells are lunate; commonly pairs of cells are held together within an unstratified mucilage envelope.

Tortitaenia luetkemuelleri (Brook) Brook 1998

Basionym: *Polytaenia luetkemuelleri* Brook
27290030 **Pl. 128J** (p. 512) **CD**
Cells spindle-shaped to spindle-shaped-elliptical, 15–30 μm wide, 50–210 μm long, with broadly rounded to subtruncate apices, surrounded by a broad mucilage sheath; chloroplast with 3–8 ridges running along length, ridges sometimes irregularly aligned and

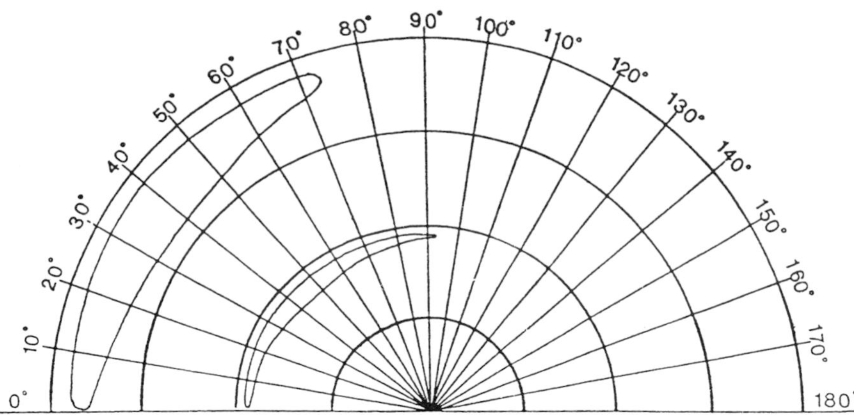

Figure 6. Closteriometer or *Closterio*-curvimeter showing two species of *Closterium*, with distinctly different arcs of curvature, superimposed on it. The outer *Closterium* has an arc of about 74°, the inner one close to 93°.

dissected and also showing distinct bifurcations, with 6–8 pyrenoids in the axis of each chloroplast, often difficult to see.

Widely distributed in acid pools and *Sphagnum* bogs but rarely abundant; also occurring amongst wet *Sphagnum* bordering acid lakes and stagnant pools, not only throughout Europe but possibly worldwide.

The chloroplast ridges may be interrupted across the middle of the cells as a prelude to cell division; each cell then appears to have 2 chloroplasts.

Suborder Closteriineae

BY ALAN J. BROOK

Cells elongate, cylindrical or with extended conical, curved ends, without median constriction and with only two planes of symmetry; in two genera cells joined to form filaments. Wall with at least two, but in many species several segments (girdle bands) and sometimes with one or many median sutures which indicate where successive cell divisions have occurred. In some species pores or pore-like gaps present, but only in outer wall, which may be ornamented with warts, spines or ridges.

Closterium Nitzsch ex Ralfs 1848

Cells always longer than broad, sometimes exceedingly so (at least two species attain lengths of more than 1 mm), tapering from the mid-region to narrower apices (as narrow as 1 μm); some straight and broadly spindle-shaped, more commonly exhibiting varying degrees of curvature: sometimes smooth from end to end, sometimes a straight and parallel-sided median section; the ventral (inner) margin may be weakly to markedly tumid (swollen); sigmoid cells occur in some populations amongst cells normally curved in one plane (see discussion below on the measurement of curvature); circular in cross section; median constriction not significant although at the juncture of old and new cell halves, very small constrictions are sometimes seen where one or more sutures (up to 10 or more) occur; chloroplasts in narrowest species simple and ribbon-like, in others a central, axile core having radiating longitudinal ridges, with varying numbers of pyrenoids in series along core (only one species has parietal chloroplasts in which are numerous, scattered pyrenoids); girdle bands or pseudo-girdle bands (inserted sections of wall) present or absent; walls colourless to brown, smooth or with longitudinal striae or ribs, sometimes with puncta between them, occasionally puncta prominent around cell ends and also a distinct end pore; terminal vacuoles within apices, enclosing one or many inorganic crystals (see below).

Cosmopolitan; most in acid waters especially in small pools amongst *Sphagnum*; as many as 20 species have been found in a single sample from such habitats. Many occur loosely attached to marginal, submerged macrophytes from which they may be dislodged by wave action and then occur as chance plankters in open water. A few tend to be associated most frequently with harder and sometimes highly nutrient-rich waters of lakes and ponds. Because they can attach firmly by one end, some are common in rivers on macrophytes or other surfaces even where there may be a strong current (Auber et al., 1989). A few species are able to withstand desiccation for extended periods and so survive on mud surfaces in drying pools and puddles; also in soils (Brook & Williamson, 1988b; Williamson, 1991a; Brook, 1992c). About 140 species, many cosmopolitan (Gerrath, 1993); 64 have been recorded from the British Isles; three species recently described from Britain have yet to be found elsewhere (Brook & Williamson, 1991).

Shape of apices is an important taxonomic character. Curvature (degrees of arc) is determined by drawing a semicircle on tracing paper and marking off in degrees from 0–180 using 5 or 10° intervals. Within this circle a number of inner circles and radial lines are added, the latter corresponding to the 5 or 10° intervals. These concentric circles should then be superimposed on a drawing or photomicrograph of the *Closterium* to be identified. The cell's dorsal (outer) margin is then aligned with one of the circles, when its degree of arc becomes evident (Fig. 6).

The crystals in the terminal vesicles of all *Closterium* species are essentially of barium sulphate with only traces of strontium and calcium, and very rarely radium. The only desmids that seem to accumulate calcium are *Bambusina* and *Gonatozygon* (Brook, 1981a,b). On only one occasion, when needle-like crytals were formed in the terminal vacuoles of *Closterium lunula*, have significant amounts of calcium been found.

1 Cells spindle-shaped 2
1 Cells curved, or with straight or parallel margins converging and curving only slightly beyond mid-region.......................................6
2 Cells spindle-shaped in mid-region, with tapering beak-like (rostrate) ends....................................3
2 Cells broadly spindle-shaped, straight; apices rounded to truncate 5
3 Lateral margins in mid-region unequally convex; cells gently curved towards ends ....*C. rostratum*
3 Lateral margins in mid-region equally convex.......4
4 Cells <15 μm wide in mid-region........*C. setaceum*
4 Cells >15 μm wide in mid-region.......*C. kuetzingii*
5 Cells <100 μm long..............................*C. navicula*
5 Cells >100 μm long ....*C. closterioides*
6 Cell straight, margins parallel or subparallel, converging and curving only towards apices..........7
6 Cells with dorsal margin distinctly and evenly curved, inner margins less curved straight or tumid.......................................13
7 Cells <12 μm wide 8
7 Cells >12 μm wide 12
8 Cells 8–12 μm wide; apices 2–3 μm wide ....*C. idiosporum*
8 Cells 4–8 μm wide; apices 1–1.5 μm wide.............9
9 Apices rounded ....*C. aciculare*
9 Apices truncate or very narrow and appearing pointed 10
10 Apices truncate and with a distinct end pore ....*C. gracile*
10 Apices not obviously truncate but very narrow and appearing pointed....................................11
11 Cells more or less straight.......................*C. acutum*
11 Cells evenly curved or sigmoid ....*C. acutum* var. *variabile*
12 Cells 12–20 μm wide; apices 6–18 μm wide, subtruncate or rounded; walls distinctly striated ....*C. directum*
12 Cells 20–35 μm wide; apices 4–15 μm wide, rounded to subtruncate....................................13
13 Walls with widely spaced ribs (costae) ....*C. angustatum*
13 Walls with well-defined striae ....*C. lineatum*
14 Outer, dorsal margin slightly curved, inner straight or slightly concave................................15
14 Outer, dorsal margin strongly and evenly curved....................................25
15 Cell ends not tapering suddenly near apices, not reflexed 16
15 Cell ends tapering slightly near apices, more or less reflexed....................................20
16 Cells 90–200 μm long; apices <10 μm wide 17
16 Cells 300–500 μm long; apices >10 μm wide 18
17 Cells 6–10 μm wide; apices 2–3.5 μm wide, subtruncate or narrowly rounded, with a small end pore....................................*C. cornu*
17 Cells 15–20 μm wide; apices 3–5 μm wide, narrowly rounded and slightly obtuse ....*C. littorale*
18 Apices rounded or angular, conical; walls ribbed (costate)....................................*C. costatum*
18 Apices flat or slightly curved with rounded corners; walls punctate or granulate 19
19 Cells 30–50 μm wide ....*C. baillyanum*
19 Cells 25–30 μm wide...*C. baillyanum* var. *alpinum*
20 Cells taper gradually then narrow on both margins so that apices are beak-like ....*C. attenuatum*
20 Cells taper gradually then narrow on dorsal margin so that apices appear reflexed..................21
21 Cells 15–22 μm wide; apices 3–7 μm wide ....*C. praelongum*
21 Cells 30–100 μm wide 22
22 Cells 30–50 μm wide 23
22 Cells 60–100 μm wide and 400–1000 μm long.....24
23 Cell walls finely striated ....*C. acerosum*
23 Cell walls with interrupted striae, puncta between them and puncta around apices ....*C. pritchardianum*
24 Chloroplasts with numerous scattered pyrenoids ....*C. lunula*
24 Chloroplasts with an axile series of pyrenoids ....*C. turgidum*
25 Cells tapering from parallel sided mid-section to comparatively broad, truncate apices 26
25 Cells with dorsal margin distinctly and evenly curved, ventral margin sometimes swollen....................................28
26 Cells 90–230 μm long; 13–18.5 μm wide; apices 4.5–5.8 μm wide, without an end pore; walls faintly striated...........................*C. abruptum*
26 Cells 200–400 μm long; 25–40 μm wide; apices 7–10 μm wide, with a thickened end pore; walls distinctly striated....................................27
27 Striae 6–10 in 10 μm with puncta between, sometimes spiralling along length of cell ....*C. striolatum*
27 Striae (ribs) 3–6 in 10 μm with puncta between ....*C. intermedium*
28 Cell walls not striated....................................29
28 Cells walls striated 33
29 Apices narrow rounded, 2–3.5 μm wide, with a distinct pore ....*C. leibleinii*
29 Apices truncate and slightly oblique, or broadly rounded 30
30 Apices truncate, 2.5–6 μm wide 31
30 Apices broadly rounded, 7–10 μm wide 32
31 Dorsal margin with arc of 110–130°; cells 20–30 μm wide....................................*C. dianae*
31 Dorsal margin with arc of 77–88°; cells 15–20 μm wide ....*C. dianae* var. *pseudodianae*
32 Chloroplasts with solid axis, pyrenoids in axis; cells 35–55 μm wide; ventral margin at most only slightly tumid.........................*C. moniliferum*
32 Chloroplasts parietal with numerous scattered pyrenoids; cells 60–100 μm wide; ventral margin slightly tumid..................................*C. ehrenbergii*
33 Dorsal margin weakly curved, ventral margin tumid; walls strongly striated, 9–11 striae in

10 μm; apices oblique-truncate or subtruncate, thickened around end pore ....*C. ralfsii* var. *hybridum*

33 Dorsal margin strongly curved, some even semicircular with an arc of 180° or more; ventral margin markedly concave, never tumid...............34

34 Cells with striae, sometimes faint; apices broadly rounded without an apical pore...............35

34 Cells without striae; apices with a subterminal, dorsal pore...36

35 Cells 60–150 μm long, 8–19 μm wide ....*C. jenneri*

35 Cells 200–270 μm long, 18–28 μm wide ....*C. archerianum*

36 Apices obliquely truncate ....*C. calosporum*

36 Apices narrow, rounded..37

37 Cells 90–120 μm long ....*C. parvulum*

37 Cells 40–80 μm long 38

38 Curvature of cells semi-circular or greater ....*C. incurvum*

38 Curvature of cells less than semi-circular...*C. venus*

Closterium abruptum West 1892

Synonyms: *Closterium abruptum* West var. *brevius* (West *et* G.S.West) West *et* G.S.West, *C.nilssonii* Borge

27040010 **Pl. 130S** (p. 526) **CD**

Cells 13–18.5 μm wide, 90–230 μm long, tapering gradually from mid-region (in long individuals a straight cylinder), extemities only slightly curved; apices relatively broad, abruptly flattened or subtruncate, unthickened and without an end pore; girdle bands frequently present and often in mid-region numerous sutures; chloroplasts star-shaped, each with 2 or 3 longitudinal ridges and up to 7 axile pyrenoids; walls colourless to rusty brown, with striae mostly very fine, 7–13 (occasionally 20–24) in 10 μm and very fine pores between the striae; terminal vacuoles spherical, enclosing 1–3 rhomboidal crystal-aggregates, sometimes accompanied by small additional crystals.

Cosmopolitan; essentially acidophilic (pH 4–5.8), most frequent in peaty pools and amongst *Sphagnum*; also margins of small lakes associated with submerged macrophytes such as *Utricularia*. Widely distributed and sometimes in moderate abundance, particularly in the west of Scotland and Western Isles; also the English Lake District, north and mid-Wales and west of Ireland; in southern England from Surrey, Hampshire, Devon and Cornwall; in the English Midlands in a small acid pool near Woburn Sands, Bedfordshire.

Closterium acerosum (Schrank) Ehrenberg ex Ralfs 1848

Basionym: *Vibrio acerosum* Schrank

27040020 **Pl. 130E** (p. 526) **CD**

Cells 30–50 μm wide, 300–560 μm long, with outer margin only slightly curved (10–20° of arc), inner margin most frequently slightly concave and parallel with the outer margin, tapering gradually, but towards end of inner margin narrowing slightly to subtruncate apices (gives impression that cell ends are slightly recurved); chloroplasts with 3 or 4 longitudinal ridges, each with up to 10 axile pyrenoids in series; walls colourless to yellow-brown, girdle bands absent, but sometimes with pseudo-girdle bands, apparently smooth but with very delicate striae, 8–10 in 10 μm; terminal vacuoles spherical, containing large numbers of small, moving crystals.

Cosmopolitan; occurring in a wide range of habitats, mostly in neutral to distinctly alkaline and very nutrient-rich waters; also in slightly acid waters (pH 6–9). Mostly in lowland areas of England in distinctly alkaline rivers, firmly attached to macrophytes (Auber et al., 1989); also in shallow, temporary pools subject to desiccation (Williamson, 1991a).

The doubtful variety *elongatum* Brébisson differs from the nominal variety only in cell length, which may be to 1000 μm.

Closterium aciculare T.West 1860

Synonym: *Closterium aciculare* T.West var. *subpronum* (West *et* G.S.West) West *et* G.S.West

27040030 **Pl. 129I** (p. 520)

Cells exceptionally long (350–560 μm) and narrow (5–8 μm), almost straight, tapering almost imperceptably towards slightly incurved ends; apices 1.5–2.0 μm wide, subtruncate or rounded having a distinctive end pore; chloroplasts usually simple, ribbon-like and occupying middle one-third of cell, each with 6–20 pyrenoids; walls colourless, smooth, lacking girdle bands; limits of terminal vacuoles poorly defined, containing a few moving crystals.

Cosmopolitan; occasionally important component of the phytoplankton of alkaline, moderately nutrient-rich lakes; also occuring in acid lakes (pH 5.2–9).

Closterium acutum Brébisson in Ralfs 1848

27040040 **Pl. 129D** (p. 520)

Cells very narrow, 4–6 μm wide, 90–140 μm long, straight to moderately curved, attenuating from mid-region to very narrow, apparently pointed apices (about 1 μm wide), in fact truncate or subtruncate; chloroplasts simple, narrow ribbons (very rarely with longitudinal ridges), with a series of 2–4 pyrenoids; walls smooth, colourless, without girdle bands; terminal vacuoles extended, containing 1–5 crystals, sometimes in short chains.

Cosmopolitan; very widely distributed in a range of habitats of differing trophic status; pH range 3.7–8.5; sometimes abundant in the plankton of nutrient-rich lakes.

Cells may adhere to one another longitudinally in small bundles.

var. *linea* (Perty) West *et* G.S.West 1900

Basionym: *Closterium linea* Perty

27040043 **Pl. 129E** (p. 520) **CD**

Cells almost straight, 100–170 μm long, or curved only towards apices; somewhat narrower, 3–3.5 μm wide and longer than the nominal variety from which distinguished only on the questionable length to width ratios (30 in var. *linea*).

Less common than the nominal variety, but widely distributed in the British Isles, sometimes found in association with *Sphagnum* in bog pools and in the littoral zone of nutrient-poor lakes.

var. *variabile* (Lemmermann) Willi Krieger 1935

Basionym: *Closterium pseudospirotaenium* Lemmermann var. *variabile* Lemmermann

27040044 **Pl. 129F** (p. 520)

Cells mostly narrower, 2.5–5 μm wide, and shorter (50–150 μm long) than the nominal variety, often strongly and irregularly curved, so often sigmoid.

Sometimes a component of plankton of lakes coloured brown due to humic material (dystrophic) and in ones rich in nutrients; also with *Sphagnum* in acidic bog pools (pH 3.6–8).

Experimental studies by Coesel (1993) have shown that different clones have physiological responses more closely associated with desmids from acid waters than those from alkaline waters, with optimum growth at pH 7.

Closterium angustatum Kützing ex Ralfs 1848

Synonym: *Closterium angustatum* Kützing ex Ralfs var. *sculptum* (Raciborski) Růžička

27040060 **Pl. 131A** (p. 528)

Cells 19–28 μm wide, 250–500 μm long, only moderately curved (arc 45–51°) and only slightly attenuated towards a 10–15 μm wide, rounded to truncate-rounded apex, sometimes swollen with apical walls thickened laterally and terminally and with a distinct end pore; girdle bands sometimes present; chloroplasts with 3–5 longitudinal ridges and a series of 4–7 axile pyrenoids; walls often brown with prominent, widely-spaced ribs between which are small, scattered pores; terminal vacuoles spherical with a stack of large, moving crystals.

Probably cosmopolitan; restricted to acid waters either amongst *Sphagnum* or algal felts in the metaphyton of nutrient-poor lakes and ponds (pH 4–6.5).

Forms characterized by distinctive ribs, the crests of which have interrupted thickenings along their lengths giving them a corrugated appearance (fo. *sculptum* (Raciborski) Brook *et* D.M. Williamson 2002), often occur with the nominal form.

Closterium archerianum Cleve in P.Lundell 1871

27040070 **Pl. 144F** (p. 590)

Cells 18–28 μm wide, 200–270 μm long, strongly bow-shaped, outer margin with arc 123–145°, attenuating evenly from mid-region to comparatively narrow (3.5–7 μm wide), rounded apices; chloroplasts with 2–4 longitudinal ridges and 5–10 axile pyrenoids; walls colourless to brown with pseudo-girdle bands, striae mostly 5–8 in 10 μm, sometimes rib-like then with only 2–4 in 10 μm; small terminal vacuoles spherical, with fused moving crystals.

Cosmopolitan; acidophile (pH 4.7–7); widespread in acid, peaty areas of Ireland and Scotland; rare in England and Wales.

Closterium attenuatum Ralfs 1848

27040090 **Pl. 130J** (p. 526) **CD**

Cells 33–45 μm wide, 400–550 μm long, slightly curved (45–60° of arc), narrowing gradually towards ends then attenuating suddenly (attenuation not always distinct) beneath apices (6–9 μm wide), subtruncate to rounded, usually having a distinct end-pore; girdle bands absent, but pseudo-girdle bands reported; chloroplasts with 4 or 5 longitudinal ridges and each having 10–12 axile pyrenoids; walls with 8–13 fine striae in 10 μm, clearly visible in old, darkly-coloured cells; terminal vacuoles spherical, containing considerable numbers of moving crystals.

Subcosmopolitan, acidophile, usually in the margins of peaty pools amongst *Sphagnum* (pH 4.5–6.5), also in nutrient-poor lake margins amongst submerged macrophytes. Widely distributed in England, Scotland and Ireland, less common in Wales.

Closterium baillyanum (Brébisson) Brébisson 1856

Basionym: *Closterium didymotocum* Ralfs var. *baillyanum* Brébisson

27040100 **Pl. 130H** (p. 526)

Cells large and robust, 30–50 μm wide, 300–500 μm long, only slightly curved, margins virtually parallel in broadest mid-region and tapering only near slightly reflexed ends; apices relatively broad, 14–23 μm wide, flat or slightly curved with rounded corners; girdle bands absent, pseudo-girdle bands very rare; chloroplasts with 5–7 longitudinal ridges and series of 8–10 axile pyrenoids; walls colourless to brown, apparently smooth, but in fact irregularly punctate or with granular precipitates of iron compounds, apical region most highly coloured and more coarsely punctate than the rest of the wall; terminal vacuoles spherical, enclosing numerous small crystals.

Probably cosmopolitan; restricted to acid waters especially amongst *Sphagnum* in peaty pools, frequent in the Western Isles of Scotland and Sutherland, also in the English Lake District and Snowdonia area of Wales.

var. *alpinum* (Viret) Grönblad 1919

Basionym: *Closterium didymotocum* Ralfs var. *alpinum* Viret

27040102 **Pl. 130I** (p. 526) **CD**

Cells narrower than the nominal variety, 25–35 μm wide and usually somewhat shorter, 300–450 μm long and more curved.

Found in similar habitats to the nominal variety, especially in small peaty pools in Sutherland and the Hebridean Islands; also known in England from Thursley Common, Surrey.

Closterium calosporum Wittrock 1869

27040120

Cells 8–12 μm wide, strongly curved, dorsal margin with arc of 125–135°, ventral margin usually straight to moderately concave, 80–100 μm long, attenuating gradually and evenly from mid-region to 1.5–2.5 μm apices; apices obliquely truncate, with a distinctive pore towards point where apex joins the dorsal cell margin; girdle bands absent; chloroplasts with 3 longitudinal ridges, only one visible in face view, each with 3–5 pyrenoids; walls smooth, colourless to pale brown; terminal vacuoles small, with several small crystals, sometimes in clusters.

Subcosmopolitan, acidophile (pH 5–7.5); found in several localities but very rare in England, Wales and Ireland, although reported as 'general' in Scotland including records from Harris, Outer Hebrides and Orkney Islands (W. & G.S. West, 1904).

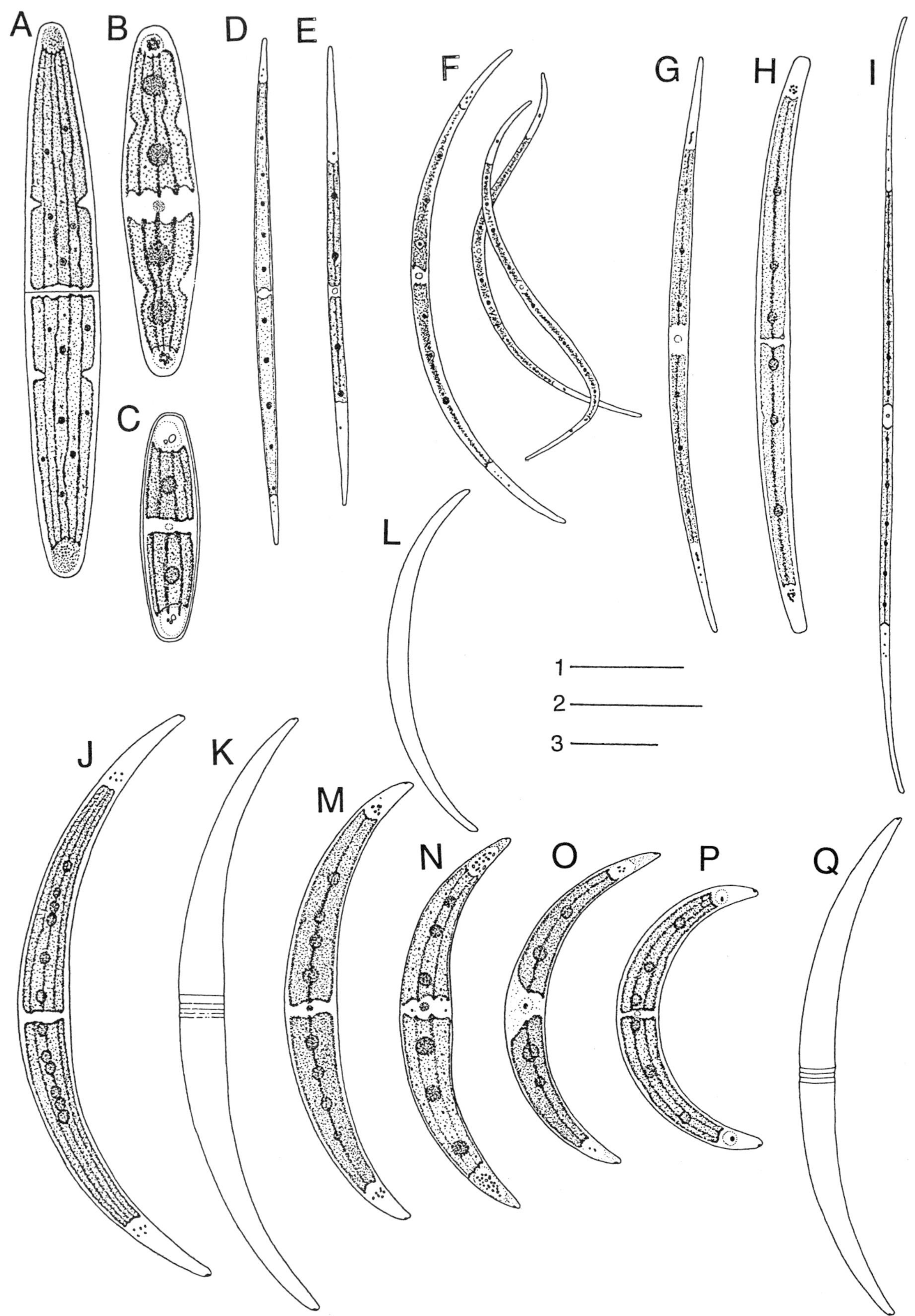

Plate 129 A–Q

A. *Closterium closterioides* (p. 521); **B**. *Closterium closterioides* var. *intermedium* (p. 521); **C**. *Closterium navicula* (p. 524); **D**. *Closterium acutum* (p. 518); **E**. *Closterium acutum* var. *linea* (p. 518); **F**. *Closterium acutum* var. *variabile* (p. 519); **G**. *Closterium idiosporum* (p. 522); **H**. *Closterium cornu* (p. 521); **I**. *Closterium aciculare* (p. 518); **J**. *Closterium dianae* (p. 521); **K**. *Closterium dianae* var. *pseudodianae* (p. 522); **L**. *Closterium dianae* var. *minus* (p. 521); **M**. *Closterium parvulum* (p. 524); **N**. *Closterium leibleinii* (p. 523); **O**. *Closterium venus* (p. 527); **P**. *Closterium incurvum* (p. 522); **Q**. *Closterium calosporum* var. *brasiliense* (p. 521).

Scale bars: 1, 25 μm; 2, 3, 50 μm. 1 for C, D, H, O, P; 2 for A, B, E, F, K, M, Q; 3 for G, I, J, L, N.

Can be mistaken with forms of *C. dianae* and *C. parvulum* and positive identification is possible only when characteristic zygospores are present. Zygospores spherical and furnished with mammillate or short conical projections, 8–10 of which can be seen round margins.

var. *brasiliense* Børgesen 1891
27040122 **Pl. 129Q** (p. 520)
Cells usually longer than the nominal variety, 105–195 μm long, upper range of width approaches the nominal but most tend to be narrower, mostly 7.5–9 μm wide.

An acidophile, much more common than the nominal variety, sometimes abundant in nutrient-poor lochs in the Tongue area of Sutherland and North Uist, Outer Hebrides; also present in acid tarns in the English Lake District and acid pools in Snowdonia and mid-Wales.

Zygospores essential for positive identification, approximately spherical, mammillate or covered with short, conical projections.

Closterium closterioides (Ralfs) A.Louis *et* Peeters 1967

Basionym: *Penium closterioides* Ralfs
27040130 **Pl. 129A** (p. 520) **CD**
Cells broadly spindle-shaped, 38–46 μm wide, 270–315 μm long, with apices rounded or subtruncate, 13–20 μm wide; girdle bands absent, often with numerous sutures; chloroplasts with 4 or 5 longitudinal ridges, each with 2 or 3 spherical or elongate axile pyrenoids; walls smooth, usually colourless; conspicuous terminal vacuoles with numerous moving crystals.

Cosmopolitan; essentially acidophilic, pH range 5–7.5, most common in peaty pools with *Sphagnum*, also in weedy margins of nutrient-poor lakes. In England recorded from Cumbria, Yorkshire, Lancashire, Warwickshire, Sussex, Surrey, Hampshire, Dorset, Devon and Cornwall; mid- and north Wales; all counties of Scotland north of the Highland Line and from the Isle of Skye, the Hebrides and Shetland. From Ireland recorded from counties Mayo, Galway, Kerry, Dublin and Wicklow, Donegal and Kerry.

var. *intermedium* (J.Roy *et* Bisset) Růžička 1973
Basionym: *Penium libellula* G.W.Focke ex Nordstedt var. *intermedium* J.Roy *et* Bisset
27040132 **Pl. 129B** (p. 520) **CD**
Cells at least half the size of the nominal variety, 19–30 μm wide, 90–140 μm long, apices 6.5–10 μm wide.

Present in habitats similar to those in which the nominal occurs, but rare and less widespread. In England reported from Cumbria, Devon, Cornwall and Surrey; north and mid-Wales; in Scotland from Dumfries and Galloway and all counties north of the Highland Line, also from the islands of Arran, Skye, North Uist and Shetlands; in Ireland found only in two loughs in Donegal.

Closterium cornu Ehrenberg ex Ralfs 1848

27040140 **Pl. 129H** (p. 520) **CD**
Cells long, narrow, 6–10 μm wide, 90–160 μm long, only slightly curved, outer margin with arc 34–40°, inner straight or very slightly concave, tapering gradually and attenuating towards subtruncate or narrowly rounded apices, 2–3.5 μm wide, sometimes with a small end pore; girdle bands absent; chloroplasts with 3 longitudinal ridges, mostly one visible in face view, with 2 pyrenoids; walls smooth, colourless; terminal vacuoles small with one rectangular moving crystal.

Cosmopolitan; widely distributed in acid waters throughout the British Isles (pH 5–6.5), in nutrient-poor lakes as a chance plankter, but most commonly amongst *Sphagnum* in dystrophic ponds and pools.

Closterium costatum Corda ex Ralfs 1848

27040150 **Pl. 130Q** (p. 520) **CD**
Cells robust, 30–45 μm wide, 250–400 μm long, dorsal margin moderately curved (90–98° of arc), inner often straight or slightly swollen, attenuating to rounded, truncately rounded or conical apices, 8–14 μm wide with apices often capitate so cell ends have appearance of a broad, shallow sloping roof; pseudo-girdle bands very rare, characterized by 5–10 ribs visible across cell and extending from apex to apex; chloroplasts with 4 or 5 longitudinal ridges, each with 5–20 axile pyrenoids; walls frequently reddish-brown; conspicuous terminal vacuoles enclose a crystal cluster.

Europe, North America, Asia; acidophile (pH 4.8–6.3), mostly associated with *Sphagnum* in peat bogs and small pools, but also in the margins of nutrient-poor lakes and associated with *Utricularia* and *Carex*; widespread and not uncommon in western areas of the British Isles.

Closterium dianae Ehrenberg ex Ralfs 1848

27040190 **Pl. 129J** (p. 520) **CD**
Cells curved, outer margin with 112–130° of arc, inner margin less so, mid-region sometimes straight or slightly tumid, 20–30 μm wide, 180–300 μm long, tapering from mid-region to narrow extended ends, apices 4–6 μm wide, oblique, on dorsal margin of each oblique apex an internal swelling with a pore; girdle bands absent, pseudo-girdle bands reported; chloroplasts with 3–5 longitudinal ridges, 1 or 2 visible in face view; walls smooth, colourless to brown; long terminal vacuoles contain 2–10 moving crystals.

Cosmopolitan; most commonly associated with *Sphagnum* in peaty pools, also present in the littoral of nutrient-poor to moderately nutrient-rich lakes, essentially acidophilic (pH 4.5–7.5); in the British Isles moderately common in western and northern regions.

var. *minus* Hieronymus 1895
Synonym: *Closterium dianae* Ehrenberg ex Ralfs var. *minus* (Schröder) Willi Krieger
27040195 **Pl. 129L** (p. 520)
Cells 12–15 μm wide, 100–170 μm long, more strongly curved than the nominal variety, always narrower than 15 μm.

Present in acid habitats in the English Lake District, Dartmoor, Devon and Thursley Common, Surrey, mid-Wales, and in Sutherland and the Outer Hebrides.

var. *pseudodianae* (J.Roy) Willi Krieger 1935

Basionym: *Closterium pseudodianae* J.Roy

27040196 **Pl. 129K** (p. 520) **CD**

Cells less curved than the nominal variety with outer margin 78–88° of arc, only 15–20 μm wide, 150–260 μm long, with extended, tapering end regions.

Rarely abundant, but widely distributed in acid habitats (pH 5.2–6.9) where the nominal or type variety occurs, thus recorded from many localities in the Highlands and Islands of Scotland, from Yorkshire, Cheshire and Surrey, in several localities in mid-Wales and from County Kerry in Ireland.

Closterium directum W.Archer 1862

Synonym: *Closterium ulna* F.W.Focke ex W.B.Turner

27040210 **Pl. 131B** (p. 528) **CD**

Cells 12–20 μm wide, 200–400 μm long, moderately curved or almost straight, sides parallel for most of length and attenuating only near subtruncate or rounded apices, 6–18 μm wide, ends sometimes slightly reflexed; girdle bands present; chloroplasts with 3–5 longitudinal ridges, each with 9–19 axile pyrenoids; walls with 9–11 striae in 10 μm; terminal vacuoles spherical, small, mostly with one compound moving crystal.

Cosmopolitan; most commonly associated with *Sphagnum* in small acidic pools, especially in the north and west of Scotland and the Hebrides (pH 4–5.5).

Closterium ehrenbergii Meneghini ex Ralfs 1848

Synonyms: *Closterium malinvernianum* De Notaris, *C. ehrenbergii* Meneghini ex Ralfs var. *malinvernianum* (De Notaris) Rabenhorst

27040230 **Pl. 130B** (p. 526) **CD**

Cells robust, 60–100 μm wide, 250–550 μm long, with variable curvature of dorsal margin, mostly with arc 110–120°, inner margin usually distinctly swollen in mid-region then attenuating to broadly rounded apices, 10–18 μm wide, lacking an apical pore; girdle bands absent; chloroplasts parietal, virtually hollow tubes possessing a series of low ridges extending outwards towards wall, 5 or 6 ridges visible in face view, each with numerous pyrenoids scattered throughout chloroplast; walls colourless to pale brown, apparently smooth but cells devoid of contents reveal fine striae, 14–17 in 10 μm; terminal vacuoles contain large numbers of rhomboidal crystals, some aggregated in stacks.

Cosmopolitan; a very variable complex of species, at times in acid waters but most abundant and frequent in alkaline waters (pH 5–8.5). Along with *C. moniliferum*, the most common desmid of alkaline, lowland waters in the British Isles, occurring at the margins of lakes and ponds from where it can be dispersed to become a chance plankter. Very abundant at times attached to macrophytes even in quite rapidly flowing rivers, also in the benthos of lakes, streams and ditches.

The doubtful var. *malinvernianum* (27040232) differs only in possessing well-defined striae.

Closterium gracile Brébisson ex Ralfs 1848

Synonym: *Closterium gracile* Brébisson ex Ralfs var. *elongatum* West *et* G.S.West

27040260 **Pl. 130G** (p. 526)

Cells long, narrow, 4–6 μm wide, 100–350 μm long, straight for most of length with parallel margins, attenuating slightly near slightly incurved ends; apices truncate or subtruncate, 2.5–3.5 μm wide with a distinctive end pore associated with an internal thickening of wall; pseudo-girdle bands occasionally present; chloroplasts narrow, slightly undulating ribbons, each containing 4–7 pyrenoids; walls smooth, usually colourless; terminal vacuoles small, enclosing 1–5 ovoid crystals.

Cosmopolitan; essentially acidophilic (pH 4–7.9), widely distributed in the British Isles where it may be abundant in collections of *Sphagnum* from boggy and peaty pools, also in the littoral of nutrient-poor to moderately nutrient-rich lakes and as a chance plankter.

Closterium idiosporum West *et* G.S.West 1900

27040280 **Pl. 129G** (p. 520)

Cells spindle-shaped, 8–12 μm wide, 150–250 μm long, straight to slightly curved or even sigmoid, attenuating from mid-region to narrow, subtruncate apices with a small end pore; girdle bands absent; chloroplasts with 3 or 4 ridges, each with 3 or 4 axile pyrenoids; walls colourless and smooth; terminal vacuoles ill-defined, containing 1 or 2 bead-like chains of crystals.

Europe, North America, Asia, New Zealand; acidophile (pH 4–6), sometimes frequent in small pools with *Sphagnum*; widespread throughout the British Isles.

Cells often adhering in bundles.

Closterium incurvum Brébisson 1856

27040290 **Pl. 129P** (p. 520)

Cells 6–12 μm wide, 40–80 μm long, strongly curved, outline approaching a semicircle; outer margin with arc of 180°, inner margin smoothly curved and both margins converging gradually to narrow, acutely rounded apices, 1–2 μm wide with a small pore on dorsal side; girdle bands absent; chloroplasts with 3 longitudinal ridges and 1–3 pyrenoids in each chloroplast; walls smooth, colourless; terminal vacuoles small, with several moving crystals.

Cosmopolitan, in distinctly acid to alkaline waters (pH 5–8); widely distributed in the British Isles, often abundant in the littoral vegetation of moderately nutrient-rich lakes and ponds, also a chance plankter and amongst *Sphagnum* in small acid pools.

Closterium intermedium Ralfs 1848

27040310 **Pl. 144D** (p. 590)

Cells only slightly curved with a parallel sided mid-region, 20–30 μm wide, 200–400 μm long, tapering only near ends to subtruncate or rounded apices, 6–8 μm wide, with apices often thickened around a terminal pore; girdle bands present; chloroplasts mostly with 3 longitudinal ridges, each with 5–11 axile pyrenoids; walls mostly straw-coloured to rusty brown, clearly striated with 3–6 striae (or ribs) in 10 μm, delicate puncta between; terminal vacuoles with a large crystal aggregate.

Cosmopolitan; acidophile (pH 4–7.5), in *Sphagnum* bogs and peaty pools, also margins of nutrient-poor ponds and lakes amongst algae attached to submerged macrophytes; widespread in the British Isles, rarely abundant.

Closterium jenneri Ralfs 1848

Synonyms: *Closterium cynthia* De Notaris, *C. cynthia* var. *jenneri* (Ralfs) Willi Krieger, *C. cynthia* var. *latum* (Schmidle) Willi Krieger

27040320 **Pl. 130L** (p. 526)

Cells 8–19 μm wide, 60–150 μm long, strongly and evenly curved, though the mid-region sometimes straight with parallel margins, attenuating towards distal one-third to broadly rounded apices, 3–7 μm wide, without a terminal pore; pseudo-girdle bands sometimes present; chloroplasts with 3–5 longitudinal lamellae, each with 3–7 axile pyrenoids; walls colourless to brown, striated (6–12 striae in 10 μm) though sometimes difficult to detect; terminal vacuoles spherical, mostly with a single compound crystal.

Cosmopolitan, acidophile (pH 4–6), mostly in peaty pools and bogs associated with *Sphagnum* and *Utricularia*, also in littoral of nutrient-poor lakes, a chance plankter; widespread in Scotland and Western Isles, also present in the English Lake District, Wales and Ireland.

Closterium juncidum Ralfs 1848

Synonyms: *Closterium juncidum* Ralfs vars. *brevius* (Rabenhorst) J.Roy, *elongatum* J.Roy, *robustius* Lobik

27040330 **Pl. 130P** (p. 526)

Cells long and slender, 6–15 μm wide, 90–630 μm long, margins parallel for most of length, attenuating only slightly towards ends where cells curve inwards; apices subtruncate, often oblique, 3–5 μm wide, slightly thickened, with minute apical pore; chloroplasts with 3 longitudinal ridges visible and a series of 4–9 axile pyrenoids; walls colourless to dark brown, with striae, 10–15 in 10 μm, sometimes discernible only with a ×100 oil immersion objective, sometimes striae coarse and widely separated (described as rib-like); terminal vacuoles spherical, containing one crystal or a fused stack of 2 or 3 crystals.

Cosmopolitan; commonly associated with *Sphagnum* in peat bogs and pools (pH 4.5–5.5); widely distributed in the British Isles, especially in acid waters and nutrient-poor lake margins; occasionally a chance plankter.

Closterium kuetzingii Brébisson 1856

27040340 **Pl. 130N** (p. 526)

Cells 300–550 μm long, for much of length straight with a broad, spindle-shaped middle region 15–25 μm wide, attenuating to each extremity into long, narrow ends, slightly incurved, rounded and sometimes slightly swollen, apices 2.5–3.5 μm wide, always with a distinct end pore; chloroplast occupies only broad middle region, with 4–6 longitudinal ridges visible and 4–10 axile pyrenoids; walls colourless to straw-coloured, finely striated with 8–12 striae in 10 μm; subterminal vacuoles, adjacent to the distal ends of chloroplast and containing several ovoid, moving crystals.

Cosmopolitan; widely distributed in the British Isles in moderately acid to neutral waters (pH 5–7); sometimes amongst *Sphagnum* at pH 4.5 in peaty pool and bogs; commonly in the plankton of nutrient-poor to moderately nutrient-rich lakes where it can also occur associated with submerged macrophytes.

Sometimes smaller cells are difficult to distinguish from *C. setaceum*. The rare var. *vittatum* Nordstedt (27040342) differs from the nominal variety in that the wall is ribbed.

Closterium leibleinii Kützing ex Ralfs 1848

27040380 **Pl. 129N** (p. 520)

Cells 16–20 μm wide, 120–200 μm long, mostly strongly curved, outer margins with arc 135–190°, inner margins strongly concave except in mid-region where commonly slightly but distinctly tumid, attenuating gradually from a broad mid-region to narrow, rounded apices 2–3.5 μm wide dorsal margin with a small end-pore just behind apex; girdle bands absent; chloroplasts with 2 longitudinal ridges visible in face view and a series of 2–5 pyrenoids; walls smooth, usually colourless; terminal vacuoles containing several crystals.

Cosmopolitan; usually in neutral to alkaline waters (pH 6.5–8.5), also recorded from acid habitats (pH 4.5); widely distributed in the British Isles, especially in lowland areas, with almost pure gatherings taken in small ponds and ornamental pools.

The much less common var. *boergesenii* (Schmidle) Skvortsov (27040382), distinguished by its significantly broader cells (30–45 μm wide), has been found in Leicestershire, Pembrokeshire and the Shetland Islands.

Closterium lineatum Ehrenberg ex Ralfs 1848

27040400 **Pl. 144B** (p. 590)

Cells long and comparatively narrow, 20–25 μm wide, 400–650 μm long, straight for much of their length with a long, parallel-sided mid-region, near ends attenuating and curving inwards towards subtruncate apices 4–10 μm wide with a distinct, thickened apical pore; chloroplasts with 3–5 longitudinal ridges and a series of 9–11 small pyrenoids; walls colourless to dark brown with 8–10 distinct striae in 10 μm, puncta between striae and prominent around apices; terminal vacuoles spherical with a cluster of ovoid, moving crystals.

Cosmopolitan; most common in acid waters (pH 4.5–5.5), frequently in *Sphagnum* bogs and peaty pools; widely distributed in the British Isles, mostly in the north and west of Scotland, but rarely abundant.

Closterium littorale Gay 1884

Synonym: *Closterium siliqua* West *et* G.S.West

27040410 **Pl. 130D** (p. 526)

Cells 15–20 μm wide, 120–200 μm long, with outer margin moderately curved, 28–58° of arc, inner margin straight or slightly tumid, attenuating from mid-region to narrow, rounded to subtruncate, slightly obtuse apices 3–5 μm wide; girdle bands absent; chloroplasts with 3–5 longitudinal ridges and 2–5 axile pyrenoids; walls smooth, colourless; terminal vacuoles spherical, usually with a compound crystal mass.

Cosmopolitan; mostly in alkaline waters (pH 6.5–8.5), common in the littoral of lakes and ponds, also in the plankton and attached to macrophytes, associated with other desmids and algae in lowland, alkaline rivers; found on the virtually dry mud in river bank pools, thus seems able to survive some desiccation.

Closterium lunula [O.F.Müller] Nitzsch ex Ralfs 1848

Synonym: *Closterium turgidum* var. *giganteum* (Nordstedt) De Toni

27040420 **Pl. 130C** (p. 526)

Cells 60–100 μm wide, 400–650 μm long, slightly curved, dorsal margin with arc 40–45°, ventral margin straight, or in the mid-region slightly tumid, narrowing gradually to comparatively broad ends which sometimes appear slightly recurved because the ventral margin narrows suddenly 30–50 μm behind apices (13–20 μm wide), making the cell ends look like truncated cones (apices actually subtruncate or broadly rounded, sometimes slightly thickened internally); girdle bands absent; chloroplasts with a thick central axis, 5–6 radiating ridges and pyrenoids more or less evenly dispersed throughout each chloroplast; walls mostly colourless, but in iron-rich environments brown or red-brown, finely striated with 9–18 striae in 10 μm, sometimes irregularly aligned and interrupted along length; terminal vacuoles broadly conical, containing large numbers of single crystals or fused into small stacks.

Probably cosmopolitan; acidophile (pH 4–7), widely distributed in the British Isles, including the Orkney and Shetland Islands, Skye and the Outer Hebrides, mostly in peaty, acidic pools, *Sphagnum* bogs, the margins or in the benthos of nutrient-poor lakes.

Closterium moniliferum [Bory] Ehrenberg ex Ralfs 1848

Synonym: *Closterium moniliferum* (Bory) Ehrenberg ex Ralfs var. *submoniliferum* (Woronichin) Willi Krieger

27040450 **Pl. 130A** (p. 526)

Cells robust, 35–55 μm wide, 200–350 μm long, outer margin variable with 82–165° of arc, inner margin usually tumid in mid-region, sometimes straight or slightly concave, tapering to rounded apices 8–10 μm wide; girdle bands absent; chloroplasts with 3–6 longitudinal ridges visible and containing an axile series of 2–10 pyrenoids, some cells with scattered pyrenoids (see note below); walls usually colourless, apparently smooth, in fact very delicately striated with 14–20 striae in 10 μm; spherical terminal vacuoles within broad cell ends, containing numerous single or compound crystals.

Cosmopolitan; known to be a complex of biological species, showing considerable morphological variability, found in a wide range of water types, from acid to alkaline (pH 4–8.5). In the British Isles most abundant in alkaline, moderately to very nutrient-rich waters in lowland areas, often associated with submerged macrophytes in lakes and rivers; sometimes abundant as a chance plankter and may occur in acid pools with *Sphagnum* in upland areas.

The now rejected var. *submoniliferum* differs only by having chloroplasts with scattered pyrenoids rather than occurring in an axile series. These two forms have been found to exist in a given population and in cultures have been grown from a single cell.

Closterium navicula (Brébisson) Lütkemüller 1902

Basionym: *Penium navicula* Brébisson

Synonyms: *Penium navicula* Brébisson, *Closterium navicula* (Brébisson) Lütkemüller var. *granulatum* (West *et* G.S.West) Willi Kreiger

27040460 **Pl. 129C** (p. 520) **CD**

Cells straight, spindle-shaped, 10–17 μm wide, 35–65 μm long, with comparatively broad, rounded to subtruncate apices 5–9 μm wide; girdle bands absent; chloroplasts with 3–5 longitudinal ridges, usually containing only one spherical pyrenoid; walls colourless, without striae; terminal vacuoles spherical with one or more crystals.

Cosmopolitan; acidophile (pH 3.9–6.9), widely distributed in the British Isles in acid bogs and pools, most common in upland and mountainous regions; frequent in the Hebrides, Isle of Skye and the Shetland Islands.

Closterium parvulum Nägeli 1849

27040500 **Pl. 129M** (p. 520)

Cells 10–15 μm wide, 90–130 μm long, strongly curved, outer margin with arc of 102–158°, inner margin sometimes straight, never tumid, attenuating to acutely rounded apices 2–3 μm wide, faintly flattened on dorsal margin and having a very small pore; girdle bands absent; chloroplasts with 2–3 longitudinal ridges and 2–6 axile pyrenoids; walls smooth, mostly colourless, sometimes pale yellow-brown; terminal vacuoles containing 2–8 ovoid crystals.

Cosmopolitan; widely distributed in the British Isles in acid moorland bogs and pools, commonly associated with *Sphagnum*; also found in the littoral of nutrient-poor to moderately nutrient-rich lakes where it may also occur as a chance plankter. Although most common in acid waters, also present in neutral waters (pH 4–7).

The slender var. *angustum* West *et* G.S.West (27040502), with cells narrower than 10 μm, and the more robust var. *maius* (Schmidle) Willi Krieger (27040503) have been found in a few localities in England, Scotland and Wales. Can be confused with larger forms of *C. venus* and with *C. calosporum*, when the latter is without zygospores.

Closterium praelongum Brébisson 1856

27040530

Cells long, narrow, 15–22 μm wide, 450–650 μm long, only moderately curved with an arc no more than 30°, inner margin usually concave or straight, towards apices the margins converge and ends are often slightly reflexed, apices 4.5–7 μm wide, subtruncate, often with a distinct terminal pore; girdle bands absent but sometimes with pseudogirdle bands; chloroplasts mostly with 3 longitudinal ridges and 7–25 pyrenoids depending on cell length; walls appear smooth but critical examination, especially of older cells or cell halves, reveals 13–15(–18) delicate striae in 10 μm, punctae rarely present except around slightly thickened extremities of brown-coloured (older) cells; terminal vacuoles spherical, with 1 or 2 stacks of rhomboidal crystals.

Cosmopolitan; tolerating a range of water types; most often in slightly acid to neutral waters (pH 6.5–7), also reported from moderately nutrient-rich waters; rare in the British Isles.

var. *brevius* (Nordstedt) Willi Krieger 1935

Basionym: *Closterium praelongum* Brébisson fo. *brevior* Nordstedt

27040532 **Pl. 130F** (p. 526) **CD**

Cells significantly shorter and mostly narrower than the nominal variety, 12–22 μm wide, 250–350 μm long, outer margin with curvature similar to the nominal variety, inner margin straight or slightly tumid, with ends very slightly reflexed, apices 3–6 μm wide; walls with striae, 14–17 in 10 μm, even more indistinct than the nominal variety.

Much more widespread than the nominal variety and found in waters with pH range 4.9–7.7; most common and even abundant at times in slightly acid pools and ponds.

Closterium pritchardianum W.Archer 1862

27040540 **Pl. 144E** (p. 590)

Cells narrow, 30–45 μm wide, 400–600 μm long, only slightly curved with an arc of 24–38°, inner margin straight or slightly concave, attenuation of cells most marked only 10–15 μm from slightly recurved subtruncate apices, 5–9 μm wide; pseudogirdle bands occasionally present, chloroplasts with 3–5 longitudinal ridges, each with 6–16 axile pyrenoids; walls colourless to golden or reddish-brown, apparently smooth but frequently ornamented with rows of puncta or short, dissected striae, 10–15 in 10 μm, ends of cells distinctly punctate; terminal vacuoles spherical, with large numbers of moving crystals.

Cosmopolitan; acidophile of *Sphagnum* bogs and pools, also in littoral of nutrient-poor lakes (pH 5.5–7); widespread in England Scotland and Ireland, never common, not recorded from Wales.

Closterium ralfsii Brébisson ex Ralfs 1848

var. *hybridum* Rabenhorst 1863

27040582 **Pl. 130K** (p. 526)

Cells 27–40 μm wide, 350–550 μm long; outer margin moderately and evenly curved (31–52° of arc), inner margin inflated but attenuating beyond the inflated mid-section to narrow, incurved extremities; apices 5–12 μm wide, subtruncate, thickened; girdle bands absent; chloroplasts star-shaped with 2–3 ridges and up to 20 axile pyrenoids; walls colourless to reddish-brown, finely striated, 9–13 in 10 μm.

Cosmopolitan; in the British Isles recorded exclusively from acid waters, but reported elsewhere from neutral to slightly alkaline habitats; fairly widespread, except in Scotland where it has yet to be found.

Closterium rostratum Ehrenberg ex Ralfs 1848

27040590 **Pl. 130M** (p. 526)

Cells 20–30 μm wide, 300–500 μm long, dorsal margin gently curved throughout length, inner margin swollen in mid-region, narrowing to distinctive, subparallel cell ends; apices 3.5–5 μm wide, slightly inflated, subtruncate to rounded; girdle bands absent; chloroplasts with 3–5 longitudinal ridges and 4 or 5 axile pyrenoids, occupy only broad median region; walls colourless to brown, striated (8–15 striae in 10 μm), often striae difficult to detect; terminal vacuoles at base of rostrate ends, with 5–15 moving crystals.

Cosmopolitan, acidophile (pH 4–7.2); mostly associated with acid habitats, usually with *Sphagnum*, also in circumneutral ponds with an abundant of submerged macrophytes; very widely distributed throughout the British Isles but rarely abundant.

Closterium setaceum Ehrenberg ex Ralfs 1848

27040610 **Pl. 130O** (p. 526)

Cells 8–13 μm wide, 220–450 μm long, spindle-shaped or lanceolate, mid-section tapering smoothly into slender colourless processes, slightly incurved at ends, obtuse at 1.5–2.5 μm wide apices; chloroplasts star-shaped, occupying only broad mid-region of cell, with no more than 3 longitudinal ridges visible and each with only 2 or 3 pyrenoids; wall colourless to straw-coloured, with 7–13 extremely fine striae in 10 μm, often difficult to see except in old cells; vacuoles just within distal ends of chloroplasts in the extended processes, contain up to 20 crystals.

Cosmopolitan; fairly common in acid habitats throughout the British Isles, sometimes associated with *Sphagnum* in bog pools (pH 4.5–6). In some nutrient-poor lakes including those coloured brown due to humic material (dystrophic), sometimes a significant component of the summer to autumn phytoplankton.

See comments under *C. kuetzingii* (p. 523).

Closterium striolatum Ehrenberg ex Ralfs 1848

27040640 **Pl. 130R** (p. 526) **CD**

Cells 25–40 μm wide, 200–400 μm long, slightly curved and tapering evenly from the broadest median region, with slightly swollen or straight dorsal margin; because of the frequent occurrence of girdle bands, median region sometimes consists of 2 or 3 sections separated by distinct sutures; apices relatively broad (7–16 μm wide), mostly subtruncate, sometimes broadly rounded, slightly oblique to longitudinal axis, asymmetrically thickened and with a distinct pore; chloroplasts with 3–4 longitudinal ridges, each with a series of 4–10 or more pyrenoids in axes; walls strongly striated with 6–10 striae in 10 μm, puncta visible between (especially in cells lacking contents); striae and puncta continue across the apices and striae sometimes spiral along cell

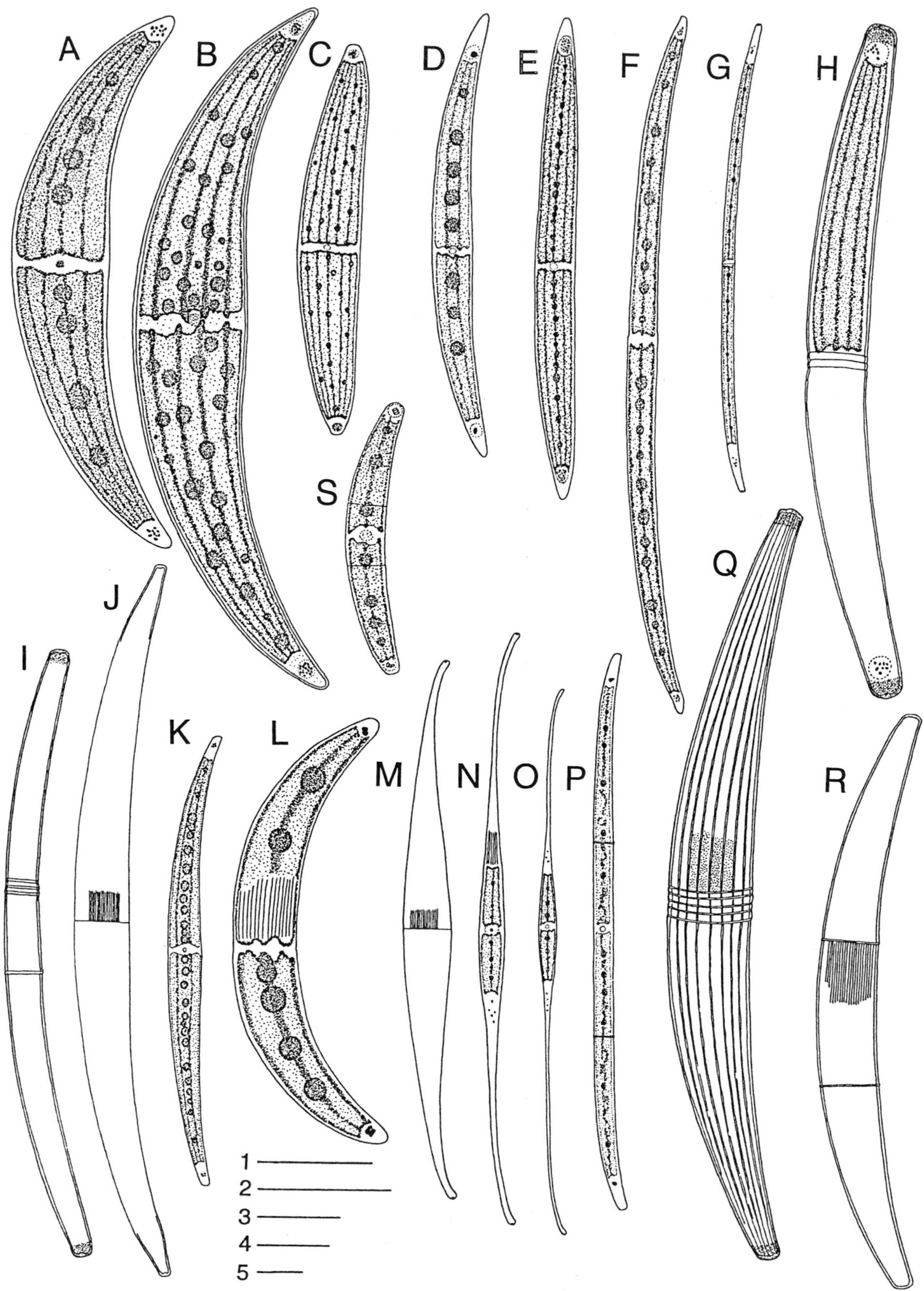

Plate 130 A–S

A. *Closterium moniliferum* (p. 524); **B**. *Closterium ehrenbergii* (p. 522); **C** *Closterium lunula* (p. 524); **D**. *Closterium littorale* (p. 523); **E**. *Closterium acerosum* (p. 518); **F**. *Closterium praelongum* var. *brevius* (p. 525); **G**. *Closterium gracile* (p. 522); **H**. *Closterium baillyanum* (p. 519); **I**. *Closterium baillyanum* var. *alpinum* (p. 519); **J**. *Closterium attenuatum* (p. 519); **K**. *Closterium ralfsii* var. *hybridum* (p. 525); **L**. *Closterium jenneri* (p. 523); **M**. *Closterium rostratum* (p. 525); **N**. *Closterium kuetzingii* (p. 523); **O**. *Closterium setaceum* (p. 525); **P**. *Closterium juncidum* (p. 523); **Q**. *Closterium costatum* (p. 521); **R**. *Closterium striolatum* (p. 525); **S**. *Closterium abruptum* (p. 518).

Scale bars: 1, 25 µm; 2–5, 50 µm. 1 for L; 2 for D, G; 3 for A, B, F, H, N–R; 4 for I, J, M, S; 5 for C, E, K.

length; terminal vacuoles spherical, enclosing a single crystal aggregate or several rhomboidal crystals.

Cosmopolitan; acidophile (pH 4–6.5(–7)); very widely distributed throughout the British Isles. Often the only desmid present in small peaty pools and ditches, though also found with other desmids in *Sphagnum* bogs and along the margins of nutrient-poor lakes, attached to aquatic macrophytes. Recorded from every county in the British Isles where such environments occur.

The species shows considerable variation in cell dimensions, shape of apices and spacing of striae across cell.

Closterium turgidum Ehrenberg ex Ralfs 1848

27040730 **Pl. 144C** (p. 590) **CD**

Cells 35–70 μm wide, 450–1000 μm long, moderately curved, outer margin with arc of about 50°, inner margin virtually straight or slightly concave, tapering evenly towards apices (15–25 μm wide) for most of length, but 15–20 μm from apices the inner margin narrows so that cell ends appear bevelled; chloroplasts with 4–6 longitudinal ridges and an axile series of 7–15 pyrenoids; walls around apex thickened, finely punctate, often brown, otherwise very finely striated with 8–10 striae in 10 μm; terminal vacuoles with large numbers of rhomboidal crystals.

Cosmopolitan; acidophile; widespread but rarely common in the British Isles in *Sphagnum* pools and bogs; also present in margins of moderately nutrient-poor lakes (pH 5–6.8).

The largest of the British desmids with some cells up to 1 mm in length, so just visible to the naked eye.

Closterium venus Kützing ex Ralfs 1848

27040760 **Pl. 129O** (p. 520)

Cells 7–10 μm wide, 55–80 μm long, margins strongly curved, dorsal margin with arc of 130–170°, attenuating evenly from centre to narrow, acutely rounded apices 1.5–2.5 μm wide, with a minute subapical pore on dorsal margin; girdle bands absent; chloroplasts ribbon-like or with 1 or 2 longitudinal ridges, occupying only small length of cell, each with 1 or 2 pyrenoids; walls mostly colourless, sometimes pale yellow-brown, smooth; terminal vacuoles occupy triangular regions within cell ends.

Cosmopolitan; recorded from markedly acid to distinctly alkaline waters (pH 4.5–8.5); although sometimes found in the phytoplankton, most common in the littoral vegetation of lakes so probably a chance plankter; widely distributed throughout the British Isles; rarely abundant.

Can be confused with small forms of *C. parvulum*, also possibly with *C. incurvum*.

TAXA RECORDED FROM THE BRITISH ISLES BUT NOT DESCRIBED IN TEXT

Closterium archerianum var. *pseudocynthia* Růžička 27040072; *C. anguineum* D.B.Will. 27040050; *C. arcus* Brook *et* D.B.Will. 27040080; *C. calosporum* var. *maius* (West *et* G.S.West) Willi Krieg. 27040123; *C. cornu* var. *lundellii* (Lagerh.) Willi Krieg. 27040142; *C. costatum* var. *westii* Cushman 27040152; *C. decorum* Bréb. 27040170; *C. delpontei* (G.A.Klebs) Wolle 27040180; *C. dianae* var. *brevius* (Petkoff) Willi Krieg. 27040193; *C. dianae* var. *compressum* G.A.Klebs 27040194; *C. didymotocum* Ralfs 270402010; *C. directum* var. *oligocampylum* (Schm.) Růžička 27040212; *C. directum* var. *recurvatum* (Roll) Willi Krieg. 27040213; *C. eboracense* W.B.Turner 27040220; *C. exile* West *et* G.S.West 27040250; *C. lanceolatum* Kütz. ex Ralfs 27040360; *C. lanceolatum* var. *parvum* West *et* G.S.West 27040362; *C. laterale* Nordst. 27040370; *C. limneticum* Lemmerm. 27040390; *C. littorale* var. *crassum* West *et* G.S.West 27040412; *C. lunula* var. *giganteum* Brook *et* D.B.Will. 27040424; *C. navicula* var. *crassum* (West *et* G.S.West) Grönblad 27040462; *C. nematodes* var. *proboscideum* W.B.Turner 27040473; *C. nordstedtii* Chodat 27040490; *C. nordstedtii* var. *polystictum* (Nygaard) Růžička 27040492; *C. porrectum* Nordst. 27040520; *C. porrectum* var. *angustatum* West *et* G.S.West 27040522; *C. pritchardianum* var. *angustum* Borzecki 27040542; *C. pronum* Bréb. 27040550; *C. pseudolunula* Borge 27040560; *C. pusillum* Hantzsch in Rabenh. 27040570; *C. pusillum* var. *maius* Racib. 27040573; *C. pygmaeum* Gutw. 27040770; *C. ralfsii* Bréb. ex Bréb. 27040580; *C. sphaerosporum* (G.S.West) Brook 27040620; *C. strigosum* Bréb. 27040630; *C. strigosum* var. *elegans* (G.S.West) Willi Krieg. 27040632; *C. sublaterale* Růžička 27040650; *C. subscoticum* Gutw. 27040670; *C. subulatum* (Kütz.) Bréb. 27040680; *C. tortum* Griff. 27040690; *C. toxon* West 27040700; *C. tumidum* L.N.Johnson 27040720; *C. variabile* Brook 27040750.

Genicularia de Bary 1856

Cells elongate-cylindrical, several cells sometimes joined by apices to form moderately long filaments; chloroplasts parietal, narrow, 2 or 3 in each cell, make several loose turns of cell (cf. *Spirogyra*); walls densely covered with minute granules; conjugating cells (recorded only in *G. spirotaenia)* tend to be geniculate (bent) and usually occur individually, although some geniculate cells do not show clear evidence of sexual reproduction.

Europe, North America, Asia; in slightly acidic, nutrient-poor ponds and lakes.

There are only two known species, *G. elegans* West *et* G.S.West and *G. spirotaenia* (de Bary) de Bary (27120020) and both are rare in the British Isles, the latter having been found only once in a pool in the Lizard, Cornwall. *G. elegans* has been found in three localities in Scotland and once in one locality in the English Lake District (see below).

Genicularia elegans West *et* G.S.West 1903

27120010 **Pl. 144A** (p. 590) **CD**

Cells 20–28 times longer than wide, 14–17 μm wide, 300–427 μm long, cylindrical, usually slightly curved, with slightly dilated apices; chloroplasts 2 per cell, making 1.5–4 turns along length, with a series of pyrenoids along length; walls densely granulate.

Europe, North America, Asia; an acidophile found in the plankton of a few nutrient-poor lakes in the British Isles; records from the Outer Hebrides, Skye, Perthshire and Torver Tarn in Cumbria.

Gonatozygon de Bary 1858

Filaments of 20 or more cells, readily dissociating especially before conjugation; cells elongate, cylindrical or subspindle-shaped (only *G. brebissonii*), 10–20(–40)

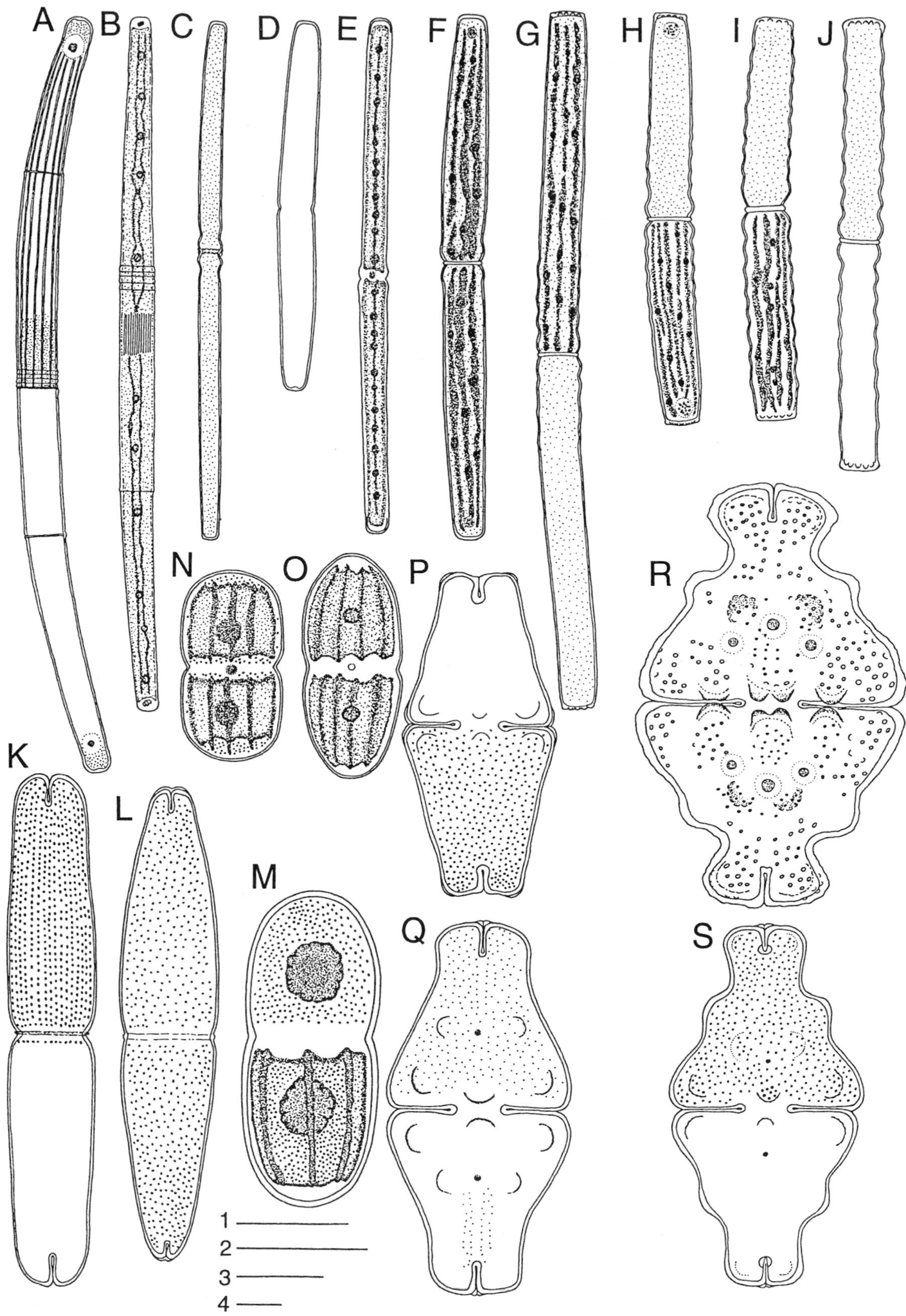

Plate 131 A–S

A. *Closterium angustatum* (p. 519); **B**. *Closterium directum* (p. 522); **C**. *Docidium baculum* (p. 551); **D**. *Haplotaenium minutum* (p. 551); **E**. *Haplotaenium rectum* (p. 551); **F**. *Pleurotaenium trabecula* (p. 562); **G**. *Pleurotaenium ehrenbergii* (p. 562); **H**. *Pleurotaenium coronatum* (p. 562); **I**. *Pleurotaenium coronatum* var. *robustum* (p. 562); **J**. *Pleurotaenium coronatum* var. *fluctuatum* (p. 562); **K**. *Tetmemorus brebissonii* (p. 583); **L**. *Tetmemorus granulatus* (p. 583); **M**. *Actinotaenium cucurbitinum* (p. 531); **N**. *Actinotaenium cucurbita* (p. 531); **O**. *Actinotaenium curtum* (p. 531); **P**. *Euastrum cuneatum* (p. 553); **Q**. *Euastrum ansatum* (p. 552); **R**. *Euastrum ampullaceum* (p. 552); **S**. *Euastrum didelta* (p. 553).

Scale bars: 1, 25 μm; 2–4, 50 μm. 1 for N–O, Q, R; 2 for D, K, L, P, S; 3 for A–C, E–G; 4 for H–J.

times longer than wide, never constricted but some with a slight median inflation; apices varying from simple-truncate to distinctly inflated, even capitate; walls mostly ornamented with granules, fine spines, or distinctive papillae; girdle bands occur in two species; chloroplasts mostly 2 per cell, simple narrow axile ribbons containing a series of 2–16 pyrenoids (in *G. aculeatum* ribbons tend to spiral along length of cell); conjugating cells mostly geniculate (bent) and some vegetative cells may be similarly bent.

Cosmopolitan; mostly occurring along margins of ponds and lakes.

In addition to the three species described here, *G. aculeatum* (27130010) and *G. pilosum* (27130050) also occur in the British Isles. All species occur in the margins of acidic ponds and lakes, usually associated with other filamentous algae; *G. kinahanii* and *G. monotaenium* have wider tolerances and have been collected in waters with a pH 8 or more.

1 Walls always without surface ornamentation .... *G. kinahanii*
1 Walls with surface ornamentation 2
2 Cells narrow, spindle-shaped and tapering to distinctly capitate apices; walls granulate .... *G. brebissonii*
2 Cells with margins parallel or slightly swollen, so approximately cylindrical, with ends slightly swollen and truncate; walls with papillae .... *G. monotaenium*

Gonatozygon brebissonii de Bary 1858

27130020 **Pl. 128X** (p. 512)

Cells narrow, spindle-shaped, 6–9 μm wide, 85–250 μm long, tapering to slightly swollen capitate apices; walls densely granulate, sometimes granules barely visible, occasionally occurring as short pointed needles; chloroplast ribbon-like, containing a series of 5–16 pyrenoids.

Cosmopolitan; in the British Isles widely distributed in acid bogs and pools (pH 4.8–7); also around the margins of nutrient-poor lakes, especially in the Highlands and west of Scotland and west of Ireland; occasionally a chance plankter.

The most distinctive of *Gonatozygon* species because of its characteristically subspindle-shaped cells with capitate apices.

var. *minutum* (West) West *et* G.S.West 1904

Basionym: *Gonatozygon minutum* West

27130024 **Pl. 128L** (p. 512) **CD**

Cells significantly smaller than the nominal variety, 4–7 μm wide, 50–70 μm long; walls always granulate.

Less widely distributed than nominal variety but in similar habitats where sometimes fairly frequent.

Gonatozygon kinahanii (W.Archer) Rabenhorst 1868

Basionym: *Leptocystinema kinahanii* W.Archer

27130030 **Pl. 128N** (p. 512)

Cells elongate-cylindrical, straight or curved, 10–20 μm wide, 150–500 μm long, often in long filaments; apices truncate and sometimes slightly inflated; walls smooth; chloroplast an axile ribbon with a series of 4–10 pyrenoids.

Cosmopolitan; fairly widely distributed in England, Scotland and Ireland though not recorded from Wales; occurs in lake and pond margins of acid waters (pH 4.5–6.8), sometimes as virtually pure gatherings.

Can be confused with species of *Mougeotia*, but distinguished by the cell ends usually having terminal vacuoles in which one or more moving crystals of calcium sulphate occur (Brook, 1981a).

Gonatozygon monotaenium de Bary in Rabenhorst 1856

27130040 **Pl. 128M** (p. 512)

Cells elongate-cylindrical with parallel to slightly swollen sides, 8–12.5 μm wide, 90–300 μm long; apices slightly dilated with truncate ends (described as 'screwdriver-shaped' by Růžička, 1977); chloroplast ribbon-like with a series of 6–16 evenly spaced pyrenoids; walls covered with small closely-spaced granules, often difficult to observe.

Cosmopolitan; in the British Isles widely distributed in pond and lake margins, mostly in acid waters (pH 4.5–7), also reported from alkaline waters of pH 8.

TAXA RECORDED FROM THE BRITISH ISLES BUT NOT DESCRIBED IN TEXT

Gonatozygon aculeatum Hast. 27130010; *G. aculeatum* var. *groenbladii* Růžička 27130012; *G. brebissonii* var. *kjellmanii* (Wille) Racib. 27130022; *G. monotaenium* var. *pilosellum* Nordst. 27130042; *G. pilosum* Wolle 27130050.

Penium Brébisson in Ralfs emend. Kouwets *et* Coesel 1984

Cell shape ranges from ellipsoidal to comparatively short broad cylinders, elongate-cylindrical or spindle-shaped, with apices broadly rounded to subtruncate, or distinctly swollen and capitate in *P. exiguum*; median suture fairly distinct (except *P. polymorphum*) and a very shallow sinus separating adjoining semicells; girdle bands sometimes present in two species (*P. spirostriolatum* and *P. cylindrus*), different degrees of colouration reflecting the different ages of their deposition; chloroplasts mostly 2 per cell, often interrupted or dissected transversely in two species so 4 in each cell, in side view always with 2–6 longitudinal ridges, each contains a central, spherical or slightly elongate pyrenoid; walls with pores (visible using critical light microscopy), arranged in well-defined longitudinal rows or scattered; terminal vacuoles containing moving crystals observed in two species.

Cosmopolitan; all acidophilic, often occurring in association with *Sphagnum* in bogs and the margins of peaty pools, also attached to submerged macrophytes in nutrient-poor lakes.

Many of the 28 species of *Penium* listed in the British Desmidiaceae (W. & G.S.West, 1904) have been moved to other genera (mostly to *Actinotaenium*) and now only six species are recognized in the British flora. The essential characters of the genus relate to the submicroscopic structure of the cell wall (see Mix, 1967; Kouwets & Coesel, 1984), studies of which show a mesh-like structure consisting of an amorphous outer, electron impervious layer, beneath which there is a system of simple pore-like structures. Walls mostly colourless, but the accumulation of iron compounds results in the walls of some species being yellow to brown.

1 Walls with striae spiralling down cell; sutures and girdle bands usually obvious ....*P. spirostriolatum*
1 Walls without striae but having granules or warts; sutures and girdle bands absent 2
2 Granules usually in longitudinal, unevenly spaced rows; apices truncate ....*P. margaritaceum*
2 Granules irregularly dispersed 3
3 Apices truncate, slightly dilated; granules on wall coarse ....*P. exiguum*
3 Apices broadly truncate; granules on wall minute ....*P. cylindrus*

Penium cylindrus [Ehrenberg] Brébisson in Ralfs 1848

27260010 **Pl. 128Y** (p. 512) **CD**
Cells cylindrical, 10–16 μm wide, 35–50 μm long, median constriction not always apparent, with distinctly flattened, truncate apices and rounded angles; girdle bands distinguished as regions where characteristic rows of minute granules are absent; chloroplasts with 3–4 longitudinal ridges and a median, spherical or slightly elongate pyrenoid; walls commonly brown.

Arctic-alpine; widely distributed in the British Isles, but never abundant; in upland areas especially in Wales and Scotland; mostly in acid bogs and pools and amongst *Sphagnum* but also found with other algae associated with macrophytes around the margins of nutrient-poor lakes, especially in the Outer Hebrides (pH 4.5–6).

Penium exiguum West 1892

27260040 **Pl. 144G** (p. 590) **CD**
Cells elongate, cylindrical with a slight, distinct median constriction above which the wall is slightly inflated, 8–12 μm wide, 25–60 μm long; apices truncate with rounded angles, usually slightly dilated; chloroplasts with 4–5 longitudinal ridges, each with 1–3 axile pyrenoids; walls with girdle bands visible in longer cells and having irregularly distributed coarse granules except around isthmus or where there is evidence of girdle bands; walls colourless to rusty brown; terminal vacuoles sometimes present.

Europe, North America; in the British Isles present with other algae in the margins of small acid lakes especially in the Outer Hebrides and west of Ireland; also tarns and *Sphagnum* bogs in the English Lake District, bogs and lochans in Sutherland, in the Shetland Islands (Williamson, 1992b), and a few sites in mid-Wales.

Penium margaritaceum [Ehrenberg] Brébisson in Ralfs 1848

27260050 **Pl. 128V** (p. 512)
Cells cylindrical, baton-shaped, with a very slight median constriction, 15–28 μm wide, 90–200 μm long, apices subtruncate with rounded angles; chloroplasts with 4–7 longitudinal ridges, and sometimes interrupted across mid-region so cells appear to have 4 chloroplasts, each containing 1 central, spherical pyrenoid; walls red-brown, ornamented with irregular, longitudinal rows of small granules or warts, with granules sometimes joined by a fine network of lines; girdle bands sometimes obvious in old red-brown cells; terminal vacuoles with moving granules have been reported.

Probably cosmopolitan; widely distributed in the British Isles in acidic environments, especially in peaty pools and amongst *Sphagnum*.

Penium spirostriolatum J.Barker 1869

27260070 **Pl. 128W** (p. 512) **CD**
Cells cylindrical 17–28 μm wide, 100–250 μm long, attenuating gradually to truncate or rounded apices 12–20 μm wide, with a slight median constriction and often many girdle bands; chloroplasts 2 or 4 per cell, each with 5–7 longitudinal ridges and a central spherical or slightly elongate pyrenoid; walls mostly yellow-brown to brown, ornamented by unevenly spaced, longitudinal striae showing various degrees of spiralling along length of cell, with striae sometimes branching and separated by puncta especially towards apices where walls entirely punctate.

Cosmopolitan; widely distributed throughout the British Isles, occurring mostly with *Sphagnum* in pools and bogs (pH 4–6.5).

Suborder Desmidiineae

BY ALAN J. BROOK

Cells mostly laterally compressed so that cells have 3 planes of symmetry; some elaborately dissected, others triangular (or 4- or more radiate) with extended, hollow processes. Walls consisting of only two segments (semicells) divided by a more or less prominent constriction (the isthmus) where adjacent segments overlap slightly. Primary wall always shed after cell division. Many species ornamented with granules, verrucae or spines in distinctive patterns, the ornamentation originating from the secondary wall in which there is a system of complex pores.

Single-celled and colonial forms

Actinotaenium (Nägeli) Teiling 1954

Cells with a only very small shallow median constriction, circular in apical view (one species has an undulating outline, in another cells are slightly curved); semicells mostly with rounded or tapering lateral margins and rounded apices; walls smooth and densely covered with pores irregularly arranged, or in oblique-crossing rows; chloroplasts 1 in each semicell, mostly star-shaped, with 3–5 longitudinal ridges visible in face view, sometimes flared out against interior of cell wall, each with a fairly prominent spherical pyrenoid in centre.

Cosmopolitan; although most species occur in acid environments along with other desmids or filamentous algae in bog pools, and sometimes on dripping rock faces, some have been found in alkaline waters and at least three species found in the British Isles have been shown to be able to survive considerable desiccation, having frequently been found in bird baths and small pools or puddles subject to prolonged periods of drying.

In some species the pores in the walls can only be seen in cells without contents and then these may have to be examined with a ×100 oil immersion objective.

1 Cells 52–85 μm long 2
1 Cells 28–51 μm long 3
2 Cells more than 2.5 times as long as wide ....*A. cucurbitinum*
2 Cells mostly less than 2.5 times as long as wide ....*A. diplosporum*
3 Apices narrowly rounded from tapering rounded lateral margins, often slightly thickened ....*A. curtum*
3 Apices broadly rounded or subtruncate from slightly convex lateral margins, never thickened ....*A. cucurbita*

Actinotaenium cucurbita (Brébisson) Teiling 1954

Basionym: *Cosmarium cucurbita* Brébisson ex Ralfs

27010070 **Pl. 131N** (p. 528) **CD**

Cells 15–24 μm wide, 28–51 μm long with shallow, notch-like sinus; lateral margins of semicells almost straight or only slightly convex, tapering to broad apices; walls smooth with fairly widely separated, quite distinctive punctae which above isthmus occur in 1 or 2 horizontal rows; chloroplasts star-shaped with irregular longitudinal ridges and conspicuous spherical, central pyrenoids

Cosmoplitan; acidophile; occurs with other desmids in peaty bog pools often with *Sphagnum*, associated with filamentous algae in margins and benthos of nutrient-poor ponds and lakes (pH 3.2–5.6(–7); present in some subaerial habitats subject to desiccation. Widespread in the British Isles, common in Scotland including the Outer Hebrides as well as the Orkney and Shetland Islands; also in Wales to a height of 890 m; general but never abundant in Ireland; widespread where suitable habitats occur in England.

Apices show considerable variability, some broadly rounded, others somewhat flattened with rounded angles; there may also be a slight median apical thickening. Some seven forms or varieties have been described which most authorities do not accept as valid.

Actinotaenium cucurbitinum (Bisset) Teiling 1954

Basionym: *Penium cucurbitinum* Bisset

Synonym: *Cosmarium cucurbitinum* (Bisset) Lütkemüller

27010080 **Pl. 131M** (p. 528) **CD**

Cells subcylindrical, 22–35 μm wide, 50–100 μm long, with slight median constriction, subparallel lateral margins, or margins tapering slightly towards broadly rounded to subtruncate apices; walls minutely and rather sparsely punctate; chloroplasts with 3–4 longitudinal ridges visible in face view (6 in apical view).

Cosmopolitan in waters ranging from acid to alkaline (pH 4–7.6) and in habitats subject to drying; in the British Isles numerous records from Cumbria and Yorkshire, also Essex and Surrey, widespread in Scotland including the Hebrides; few reports from Wales or Ireland.

Actinotaenium curtum (Brébisson) Teiling ex Růžička *et* Pouzar 1978

Basionym: *Cosmarium curtum* Brébisson in Ralfs

Synonym: *Penium curtum* (Brébisson) Brébisson ex Kützing

27010090 **Pl. 131O** (p. 528) **CD**

Cells 18–32 μm wide, 22–66 μm long, sinus a shallow notch; semicells with lateral margins either broadly convex leading to broadly rounded apices, or tapering from broadest point above isthmus to narrowly rounded apices (rugby ball-shaped), in forms with narrow apices there may be a slight internal apical thickening; walls with delicate, irregularly disposed puncta but absent across isthmus; chloroplasts with 4–5 longitudinal ridges visible in side view (8 in apical view), each with central, spherical pyrenoid.

Cosmopolitan; tolerant of a wide range of environmental conditions; in the British Isles widespread and common in acid habitats with *Sphagnum* (pH 4.8–6), and amongst algae and mosses on damp or dripping rocks; found in alkaline pools (pH 8.0), sometimes 'in almost pure gatherings in temporary pools of rain water on road sides, in cart-ruts' (W. & G.S. West, 1904); occurs in birdbaths subject to intense solar radiation and drying (Brook & Williamson, 1990), and even found on walls of a car wash.

Actinotaenium diplosporum (P.Lundell) Teiling 1954

Basionym: *Cylindrocystis diplospora* P.Lundell

27010110 **Pl. 144J** (p. 590) **CD**

Cells about twice as long as wide, subcylindrical with slight median constriction, 22–40 μm wide, 50–85 μm long, dilated very slightly towards truncately rounded and slightly thickened apices; walls smooth and colourless; chloroplasts star-shaped, with a central pyrenoid from which numerous processes radiate.

Probably cosmopolitan, in acid to circumneutral waters (pH 5.5–7.3); widespread throughout the British Isles but never abundant; most common in acid waters in Scotland including the Hebrides and Orkney Islands.

TAXA RECORDED FROM THE BRITISH ISLES BUT NOT DESCRIBED IN TEXT

Actinotaenium adelochondrum (Elfving) Teiling 27100010; *A. capax* (Joshua) Teiling var. *minus* Teiling 27010022; *A. clevei* (P.Lundell) Teiling 27010030; *A. clevei* var. *crassum* (West *et* G.S.West) Teiling 27010032; *A. colpopelta* (Bréb.) Compère 27010040; *A. crassiusculum* (de Bary) Teiling 27010050; *A. cruciferum* (de Bary) Teiling 27010060; *A. gelidum* (Wittr.) Růžička 27010120; *A. habeebense* (I.Marie) Brook *et* D.B.Will. 27010140; *A. inconspicuum* (West *et* G.S.West) Teiling 27010160; *A. lagenarioides* (J.Roy) Teiling 27010170; *A. minutissimum* (Nordst.) Teiling 27010180; *A. minutissimum* var. *suboctangulare* (West) Teiling 27010182; *A. mooreanum* (W.Archer) Teiling 27010190; *A. obcuneatum* (West) Teiling 27010200; *A. perminutum* (G.S.West) Teiling 27010210; *A. phymatosporum* (Nordst.) Kouwets *et* Coesel 2701022; *A. rufescens* (Cleve) Teiling 27010230; *A. spinospermum* (Joshua) Kouwets *et* Coesel 27010250; *A. subglobosum* (Nordst.) Teiling 27010310; *A. subpalangula* (Elfving) Teiling 27010260; *A. subtile* (West *et* G.S.West) Teiling 27010270; *A. truncatum* (Bréb.) Teiling 27010280; *A. turgidum* (Bréb.) Teiling 27010290; *A. wollei* (West *et* G.S.West) Teiling 27010300.

Cosmarium Corda ex Ralfs 1848

Basic shape 2 circles or semicircles (=semicells) joined by a central isthmus, and except for a few species always showing some degree of compression in vertical view (rarely circular and so distinguished from *Actinotaenium*, p. 530); in side view mostly ovoid or ellipsoidal, often with lateral swellings or protrusions and ornaments consisting of granules, pits or ridges which may not be visible in face view; cells all have a median sinus separating adjacent semicells, ranging from only a shallow notch to a very deep cleft, open or closed in varying degrees; semicells in face view almost circular to semicircular, ovoid to trapeziform or pyramidal, with margins in front view smooth, undulate, granulate, or lightly incised; walls always punctate, sometimes very delicately so, although puncta may be large and distinct and from which a wide mucilage envelope may be excreted, especially in species occurring in the plankton; walls smooth or bearing characteristic ornamentation of granules arranged in complex patterns, surface of others scrobiculate or with conical teeth (many have distinctive combinations of these features); chloroplasts mostly axile, forked and sometimes highly elaborate and complex.

Most common in acid bogs and pools often associated with *Sphagnum* and *Utricularia*, also in weedy margins of ponds and lakes, sometimes common in plankton; a few present in rivers, others amongst mosses in mountain streams. Some species are frequent in alkaline and nutrient-rich waters, others in temporary pools, puddles or gutters subject to drying, and in cattle drinking-troughs.

Cosmarium contains the largest number of desmid species, probably over 1000 world-wide; 250 species and 180 varieties recorded from the British Isles.

KEY TO DIVISIONS AND SECTIONS

1 Cell walls smooth, punctate or scrobiculate, never granulate .... **Division I**
1 Cell walls granulate, papillate or crenate .... **Division II**

2 Semicells transversely ellipsoid, subellipsoid, kidney-shaped or hexagonal-ellipsoid........Section i
2 Semicells of another shape 3
3 Semicells pyramidal or subpyramidal ......Section ii
3 Semicells of another shape 4
4 Semicells circular, subcircular or almost semicircular ....Section iii
4 Semicells of another shape 5
5 Semicells approximately rectangular or subrectangular ....Section iv
5 Semicells with undulate margins, variously shaped ....Section v

DIVISION I

Section i

1 Semicells hexagonal-ellipsoid or kidney-shaped 2
1 Semicells transversely ellipsoid or subcircular......4
2 Semicells kidney-shaped ....*C. abbreviatum* var. *planctonicum*
2 Semicells hexagonal-ellipsoidal 3
3 Isthmus not thickened internally; cells 13–30 μm wide, 3–29 μm long; puncta small, indistinct ....*C. abbreviatum*
3 Isthmus thickened internally; cells 30–60 μm wide, 36–56 μm long; puncta distinct ....*C. depressum var. achondrum*
4 Semicells ellipsoidal (only 7–12 μm wide, 10–16 μm long) ....*C. tinctum*
4 Semicells subellipsoid to subcircular.....................5
5 Sinus wide open, V-shaped.....................................8
5 Sinus narrow V-shaped 6
6 Isthmus thickened internally; cells 30–54 μm long ....*C. contractum* var. *ellipsoideum*
6 Isthmus not thickened internally; cells 8–40 μm long...7
7 Semicells oblong-elliptic; wall punctate ....*C. bioculatum*
7 Semicells oblong-rectangular (some oblong-elliptic, variable); lateral angles sometimes with a mucro ....*C. pygmaeum*
8 Semicells broadly ellipsoidal, sinus deep ....*C. contractum*
8 Semicells subcircular, sinus shallow ....*C. contractum* var. *rotundatum*

Section ii

1 Cells 22–45 μm long ....*C. granatum*
1 Cells >45 μm long 2
2 Apex truncate or slightly convex............................3
2 Apex narrow, retuse or flat.....................................5
3 Semicells pyramidal-trapeziform with slightly curved or flattened lateral margins......*C. galeritum*
3 Semicells truncate-pyramidate with evenly curved lateral margins 4
4 Apex broad, flat or slightly convex; walls strongly granulate; cells 61–115 μm long ....*C. pyramidatum*
4 Apex narrow, retuse, sometimes thickened internally; walls smooth; cells 14–36 μm long ....*C. laeve*
5 Apex slightly curved or flat; walls not strongly punctate............................*C. pseudopyramidatum*
5 Apex slightly retuse, internally thickened; walls densely scrobiculate or punctate 6
6 Walls scrobiculate...7
6 Walls densely and finely punctate..........................8
7 Walls densely scrobiculate; cells 32–35 μm long ....*C. variolatum*
7 Walls coarsely scrobiculate; cells 64–68 μm long ....*C. variolatum* var. *skujae*
8 Sinus shallow, open externally ....*C. cucumis*

8 Sinus linear, closed, wall thick, clearly penetrated by pores 9
9 Sinus linear, isthmus broad, more than half breadth of cell; cells 90–110 μm long ....*C. cucumis* var. *magnum*
9 Sinus shallow wide open, isthmus very broad, more than two-thirds maximum breadth of cell; apex broadly rounded; cells 88–123 μm long ....*C. debaryii*

Section iii

1 Cells slightly constricted with open, shallow sinus 2
1 Cells deeply constricted with a linear closed sinus 5
2 Semicells circular; cells 16–50 μm long ....*C. moniliforme*
2 Semicells subcircular to transversely rectangular 3
3 Semicells transversely reactangular with an apical cleft; cells 6–9.5 μm wide ....*C. regnesii*
3 Semicells subcircular....4
4 Cells 65–100 μm long....*C. connatum*
4 Cells 47–57 μm long....*C. pseudoconnatum*
5 Semicells subsemicircular with broad rounded basal angles ....*C. pachydermum*
5 Semicells semicircular 6
6 Sinus open quite widely towards outside and basal angles slightly produced ....*C. ralfsii* var. *montanum*
6 Sinus linear for most of length, dilated internally 7
7 Isthmus not thickened internally; cells 88–124 μm long ....*C. ralfsii*
7 Isthmus thickened internally; cells 20–26 μm long ....*C. fontigenum*, in part

Section iv

1 Cells 1.5 times longer than broad or more....2
1 Cells less than 1.5 times longer than broad....5
2 Lateral margins retuse 3
2 Lateral margins tapering to retuse apex ...*C. anceps*
3 Semicell apex sloping from a short truncate apex, lateral margins widening to broadly rounded basal angles....*C. holmiense* var. *integrum*
3 Semicell apex convex or truncate in mid-region....4
4 Semicell basal margins expanded below retuse lateral margins; walls thick, penetrated by pores ....*C. quadratum*
4 Semicell basal angles not expanded below lateral margins; walls thin, with minute widely spaced pores....*C. difficile*
5 Apices without a median notch ....*C. venustum* var. *excavatum*
5 Apices with a distinct median notch....6
6 Apex comparatively broad, lateral margins retuse broadening from basal angles to small protruding median lateral lobes....*C. regnellii*
6 Apex narrow, lateral margins retuse, broadest at basal angles....*C. venustum*, in part

Section v

1 Semicells quadrate-subquadrate or pyramidate....8
1 Semicells semiellipsoid-semicircular 2
2 Semicells broad, flattened semicircular thus wider than broad 3
2 Semicells almost as wide as long or wider....4
3 Semicells with scrobiculations on face; margins strongly 10–12-crenulate ....*C. cyclicum*
3 Semicells without scrobiculations on face; margins weakly crenulate ....*C. fontigenum*, in part
4 Face of semicells smooth 5
4 Face of semicells scrobiculate 7
5 Semicell margins 10–12-undulate ....*C. undulatum*
5 Semicell margins 4–6-undulate....6
6 Margins regularly and strongly 6-undulate, 2 at base, 2 lateral, 2 on apex ....*C. impressulum*
6 Margins strongly 4-undulate, 1 at base, 1 on lateral margin, 2 on apex....*C. meneghinii*
7 Face of semicells not punctate but with several rows of intramarginal scrobiculations ....*C. subundulatum*
7 Face of semicells punctate, with 3 weak undulations down lateral margins; apex widely truncate ....*C. venustum*, in part
8 Semicells with 2 strong crenulations on apex, 3 down lateral margins including basal angles ....*C. crenulatum*
8 Semicells with 4 widely separated undulations round margin ....*C. nymannianum*

DIVISION II

1 Semicells semicircular, subsemicircular 2
1 Semicells of some other shape....7
2 Margin of semicells crenate....3
2 Margin of semicells entire, undulate....5
3 Semicells semicircular ....*C. quadrifarum* var. *hexastichum*
3 Semicells subcircular to subrectangular 4
4 Semicell margins with distinct crenulations....6
4 Semicell margins without distinctive crenulations, but evenly granulate....29
5 Lateral and apical margins with very prominent granulate crenulations; face of cell protuberant with distinctive, irregular rows of large granules ....*C. caelatum*
5 Margins with fine crenations in rows decreasing in size onto face of semicell, 6–9 series of granules across base above isthmus, smooth in mid-region ....*C. speciosum*
6 Semicell apex slightly flattened; 16–20 warts below undulate margin ....*C. monomazum* var. *polymazum*
6 Semicell apex slightly produced, flat; lateral margins with concentric rings of granules ....*C. ornatum*
7 Semicells kidney-shaped 8
7 Semicells of some other shape....9

8 Margins with 20–40 granules disposed in 10–18 vertical rows over semicell surface; sinus open, internally thickened, ... *C. reniforme*
8 Margins of semicells with concentric rings of granules and circles of granules in middle of face; sinus closed to outside, widely open internally .... *C. commissurale*
9 Semicells oblong-ellipsoidal 10
9 Semicells pyramidate or subpyramidate-subsemicircular 12
10 Walls irregularly granulate .... *C. sphalerostichum*
10 Walls uniformly granulate 11
11 Granules large, conical; sinus partially closed, thickened internally and open .... *C. brebissonii*
11 Granules rounded; sinus open, U-shaped .... *C. portianum*
12 Semicells pyramidate 13
12 Semicells subpyramidate-subsemicircular 22
13 Apex produced, truncate 14
13 Apex not produced, widely truncate 15
14 Face of semicell with a central granulate swollen area .... *C. sportella*
14 Face of semicell with 2 central granulate swollen areas .... *C. turpinii*
15 Semicells with only slight indication of central inflation, sometimes differentiated central granules 16
15 Semicells with distinct central inflations and a protuberance, with small papilla or broad with vertical rows of granules 18
16 Granules disposed on face in vertical and oblique series, irregular at centre .... *C. punctulatum*
16 Granules on face large 17
17 Granules in two tranverse series, 2 or 3 in each; acute granules round margin of semicell .... *C. quinarium*
17 Granules irregularly disposed .... *C. praemorsum*
18 Slight central protuberance with granules in ring formation .... *C. punctulatum* var. *subpunctulatum*
18 Central protuberance with one small papilla 19
19 Central protuberance broad, ornamented with vertical rows or circular patterns of granules 20
19 Central protuberance with circular or cruciate pattern of granules 21
20 Apex not produced, 4-crenate truncate, sides of semicells 4-crenate .... *C. blyttii*
20 Apex produced, protuberant, broadly truncate, sides of semicells undulate .... *C. humile*
21 Central protuberance with 4 large granules in a cruciate pattern, or with clearly defined regular rows of granules 22
21 Central protuberance with irregular circle or well-defined rows of granules 23
22 Central protuberance small; 4 large granules just above isthmus with intermarginal granules widely spaced and scattered .... *C. boeckii*
22 Central protuberance with 4–5 clearly defined rows of prominent granules .... *C. subcostatum*
23 Central protuberance with a circle of granules, usually irregularly arranged 24
23 Central protuberance with vertical rows of granules; apices more or less produced 26
24 Central protuberance with irregular circle (rarely irregular rows) of small granules, intermarginal granules in 5–6 close concentric circles .... *C. formosulum*
24 Central protuberance with 6–8 rows of small granules, sometimes irregular 25
25 Granules large, joined and thus appear as distinctly notched, elongate ridges .... *C. binum*
25 Granules small and in irregular rows; semicell margins crenate 27
26 Semicell margins 16-crenate .... *C. subcrenatum*
26 Semicell margins 10–12-crenate .... *C. crenatum*
27 Apices slightly produced; semicells subsemicircular .... *C. didymochondrum*
27 Apices narrow-truncate; semicells subpyramidal with flat apices 28
28 Semicells with tooth-like granules restricted to margins of cell face .... *C. ovale*
28 Semicell margins with rounded granules 30
29 Granules large, flattened, 12–14 around semicell margin with face evenly granulate .... *C. tetraophthalmum*
29 Granules not as above and semicells subpyramidal 31
30 Semicells truncate-pyramidate; walls ornamented with small, rounded, irregularly disposed granules over entire face .... *C. botrytis*
30 Semicells subcircular-obovate or subrectangular 32
31 Granules largest towards centre of face, smaller towards the outside .... *C. margaritiferum*
31 Granules absent from centre of semicell face .... *C. botrytis* var. *mediolaeve*
32 Semicells subcircular-obovate ..... *C. sphaeroideum*
32 Semicells subrectangular 33
33 Semicells transverse-subrectangular 34
34 Granules large, covering face of semicell in vertical or decussate rows 36
34 Granules small, in rows increasingly curved towards centre, where in circles .... *C. biretum*
35 Semicells with broad truncate apex, lateral margin with 4–5 undulations; face ornamented with 4–5 series of small nodules .... *C. annulatum*
35 Semicells with broadly rounded apex, lateral margins slightly convex, face covered with linear rows of rounded granules .... *C. amoenum*
36 Lateral margins slightly divergent; granules in 13–15 vertical series on face; cells 88–107 μm long and 75–88 μm wide .... *C. conspersum* var. *latum*
36 Lateral margins straight or slightly convex; granules in decussating series, about 12 in each direction; cells 60–105 μm long and 56–82 μm wide .... *C. margaritatum*

Cosmarium abbreviatum Raciborski 1885

27050010

Cells 13–26 µm wide, 13–25 µm long, as long as broad; sinus deep, linear, narrow, slightly open inside; semicells transversely elongate hexagons, with broad, flat or slightly retuse, truncate apex; wall smooth, very finely punctate.

Cosmopolitan; widespread in the British Isles, tolerant of a range of waters, most common in acid habitats.

var. *planctonicum* West *et* G.S.West 1902

Synonym: *C. depressum* var. *planctonicum* Willi Krieger *et* Gerloff

27050011 **Pl. 133N** (p. 540)

Differs from nominal in narrower, oval semicells, 21–30 µm wide, 19–29 µm long, lateral margins convex rather than straight; apex flat or slightly convex.

Cosmopolitan; common at times in plankton of nutrient-poor to moderately nutrient-enriched lakes in Cumbria, North Wales, and especially Scotland and the west of Ireland (pH 4.9–8.2; alkalinity 1.6–26 ppm $CaCO_3$). For details of periodicity see Brook (1981b, p. 234–236).

Cosmarium amoenum Brébisson in Ralfs 1848

27050050 **Pl. 135J** (p. 547)

Cells 22–33 µm wide, 42–59 µm long; semicells almost semicircular to semi-ellipsoidal, apices sometimes flattened; sinus fairly shallow, slightly open, sublinear; wall distinctly granulate, ornamented with relatively large, rounded granules often in vertical series, 6–9 visible across the semicell face, 20–25 round margin; vertical view broadly elliptic-subcircular; chloroplasts axile with 2 pyrenoids in each.

Cosmopolitan, acidophile; fairly common in bogs and peaty pools, also in nutrient-poor ponds and lakes (pH 4–7). Widely distributed in the British Isles.

Cosmarium anceps P.Lundell 1871

27050060 **Pl. 132T** (p. 537)

Cells 12–20 µm wide, 18–50 µm long; sinus open, acute, shallow; semicells with converging straight or slightly retuse sides, apex truncate, slightly concave, angles rounded; walls apparently smooth, in fact finely punctate.

Northern Europe, North America, Arctic; occurs in a range of water types, most frequent amongst dripping mosses in mountainous regions of the British Isles; sometimes also in lowland areas.

Cosmarium annulatum (Nägeli) de Bary 1849

Basionym: *Dysphinctium annulatum* Nägeli

27050100

Cells 16–23 µm wide, 38–54 µm long; median constriction absent or barely visible; semicells rectangular-cylindrical with broad truncate apex; lateral margins with 4–5 undulations, face with 4–5 transverse series of small nodules, 5 or 6 in each series.

Probably cosmopolitan; fairly widespread in the British Isles especially on wet rocks amongst mosses and other algae.

var. *elegans* Nordstedt 1873

27050102 **Pl. 135N** (p. 547)

Differs from the type variety in having bigranulate nodules; cells slightly larger, 17–29 µm wide, 40–57.5 µm long.

Found in similar habitats to the type variety but with wider distribution.

Cosmarium binum Nordstedt in Wittrock & Nordstedt 1880

27050150 **Pl. 135E** (p. 537)

Cells 30–39 µm wide, 41–90 µm long; sinus deep, closed linear; semicells hemispherical-trapeziform with broadly rounded basal angles, convex sides and truncate, undulate apex; lateral margins with 6–10 crenations, 5–6 across apex; crenations bigranulate or emarginate; outer intramarginal granules paired and radiating within each crenation, becoming single, smaller and irregular towards swollen central tumour which bears 6 vertical rows of granules, sometimes joined as vertically notched ridges; between tumour and isthmus a row of large granules.

Cosmopolitan, acidophile, common in tropics; in the British Isles mostly in bogs and amongst marginal algae of nutrient-poor lakes (pH 5–6.5).

Exhibits great variation in size.

Cosmarium bioculatum Brébisson ex Ralfs 1848

27050160 **Pl. 132K** (p. 537)

Cells 8–27 µm wide, 9–29 µm long; sinus deep, narrow towards inside, widening outwards; semicells transverse oblong-elliptical with broadly rounded lateral margins, apex flat; walls smooth with puncta.

Cosmopolitan; very widely distributed in the British Isles in acid pools, bogs and the margins of small nutrient-poor lakes; occasional as chance plankton.

var. *depressum* (Schaarschmidt) Schmidle 1894

27050162 **Pl. 132L** (p. 537)

Differs from the nominal variety in isthmus being proportionately narrower, 3–8 µm wide and semicells subquadrate with flatter apex.

Mainly in English Lake District lakes and tarns.

Cosmarium biretum Brébisson in Ralfs 1848

Synonym: *Cosmarium quadrangulatum* Hantzsch 1860

27050200 **Pl. 135R** (p. 547)

Cells 50–66 µm wide, 54–74 µm long; sinus deep, linear, slightly dilated towards outside; semicells mostly subrectangular-trapeziform, variable; lateral margins diverging, straight or slightly convex with rounded lower angles, somewhat truncated or slightly produced upper angles; apex usually convex, sometimes truncate in middle; walls densely granulate, mostly in close vertical series, sometimes irregular and in concentric circles in middle of semicell face, 40–50 visible round semicell margin.

Cosmopolitan; widely distributed and at times abundant in lowland areas of the British Isles, especially in marshes, ditches and canals, sometimes with *Potamogeton* spp.; also present in littoral of lakes (pH 5.9–7.9).

Cosmarium blyttii Wille 1880

27050220 **Pl. 134R** (p. 543)
Cells 7–16 μm wide, 10–19 μm long; sinus deep, narrow, linear; semicells approximately semicircular-trapeziform with truncate, 4-crenate apex, the two outer crenations somewhat flat; lateral margins 3-crenate, also flat with the upper crenation the largest; semicell face with 1–2 rows of granules within the margins and one distinctive central papilla, best seen in side view.

Cosmopolitan, mostly in acid but also in alkaline waters (pH 5–8); widely distributed in the British Isles, most frequently in *Sphagnum* bogs and weedy margins of ponds and lakes.

Crenation of margins variable, as is surface granulation.

Cosmarium boeckii Wille 1880

27050230 **Pl. 135B** (p. 547)
Cells 27–35.5 μm wide, 29–40 μm long; sinus narrow linear, opening towards outside; semicells with distinctly truncate undulate apex bearing 4–5 nodules; lateral margins curved, crenate with 4 crenae on each margin; intramarginal granules in 2, more rarely 3 rows; semicell face with a slight but broad tumour with 4–6 granules variously disposed on it, mostly in cross-like pattern with largest lower granule just above isthmus.

Essentially an Arctic-alpine species in acid bogs and the weedy littoral zone of nutrient-poor ponds and lakes, occasionally a marginal plankter; widespread in the British Isles.

Cosmarium botrytis Meneghini ex Ralfs 1848

27050240 **Pl. 135P** (p. 547) **CD**
Cells 51–85 μm wide, 60–111 μm long; sinus deep linear narrow, opening towards outside; semicells oval-pyramidal with convex sides tapering from broad, flat base to narrow flat or slightly convex apex; walls evenly granular with uniformly-sized small rounded granules, usually irregularly but sometimes approximately radially arranged; 30–36 granules round semicell margin.

Cosmopolitan, most common in Arctic-alpine regions, in a wide range of water types and habitats (pH 5.0–8.5); most frequent in the littoral zone and amongst the marginal vegetation of ponds and lakes, also loosely attached to other algae, occasionally subaerial; the most widespread and common British *Cosmarium*.

Semicell shape and arrangement of granules very variable.

var. *mediolaeve* West 1892
27050245 **Pl. 135Q** (p. 547)
Differs from the nominal variety in the semicells having a smooth, slightly retuse apex, and granules on face in radiating series; granules smaller towards mid-region where totally absent; semicells 47–50 μm wide and 54.5–70 μm long.

Cosmopolitan; more restricted distribution in the British Isles than the nominal variety and tending to occur in acid waters.

Cosmarium brebissonii Meneghini ex Ralfs 1848

27050250 **Pl. 134I** (p. 543)
Cells 45–79 μm wide, 88–110 μm long; sinus narrow, closed except towards outside; semicells broadly hemispherical with rounded basal angles and evenly curved lateral margins increasingly curved upwards to flattened broad apex; walls covered with closely and evenly spaced conical granules, less prominent on apex and reducing in size from margins to centre of cell face where sometimes absent.

Arctic-alpine acidophile; widely distributed, but rarely abundant, in *Sphagnum* bogs throughout the British Isles.

Cosmarium caelatum Ralfs 1848

27050270 **Pl. 134A** (p. 543)
Cells 36–40 μm wide, 40–47 μm long; sinus narrow, linear, slightly dilated internally; semicells approximately semicircular with flattened apex and lateral and apical margins with distinctive, granulate crenations, 2 lateral and 4 apical; apical and upper lateral crenae with 2(–3) marginal granules, lower lateral with 4 granules; walls with 3 irregular series of granules within margin becoming smaller towards centre; face above isthmus with prominent granules usually disposed in vertical series on broad protuberance, disposition variable.

Mostly restricted to northern latitudes; in the British Isles often in boggy springs and on wet rocks.

Cosmarium commissurale Brébisson ex Ralfs 1848

27050350 **Pl. 134E** (p. 543)
Cells 43–45 μm wide, 25–33 μm long; sinus very deep, widely open internally and dilated outwards so cells appear to have a pair of lateral lobes; semicells narrow, subkidney-shaped with a slightly raised, wide, convex or subtruncate apex; lateral 'lobes' with irregular granules sometimes ring-like in formation, also granules in concentric circles beneath apex on large protuberance on centre of semicell face.

Probably cosmopolitan; fairly widely distributed in the British Isles, especially in pools and ditches, where occasionally abundant.

Cosmarium connatum Brébisson in Ralfs 1848

27050360 **Pl. 133G** (p. 540)
Cells 50–90 μm wide, 65–100 μm long; sinus wide open but only moderately deep and ending in an obtuse angle; semicells transversely subellipsoid-subcircular with broad base and sometimes very slightly flattened apically; walls thick, traversed by dense covering of pores; chloroplasts complex, consisting of forked lobes and 2 prominent central spherical pyrenoids.

Cosmopolitan, acidophile; widespread in the British Isles, mostly in margins of poor to moderately enriched lakes (pH 4.5–7.9), associated with submerged macrophytes; sometimes a chance plankter.

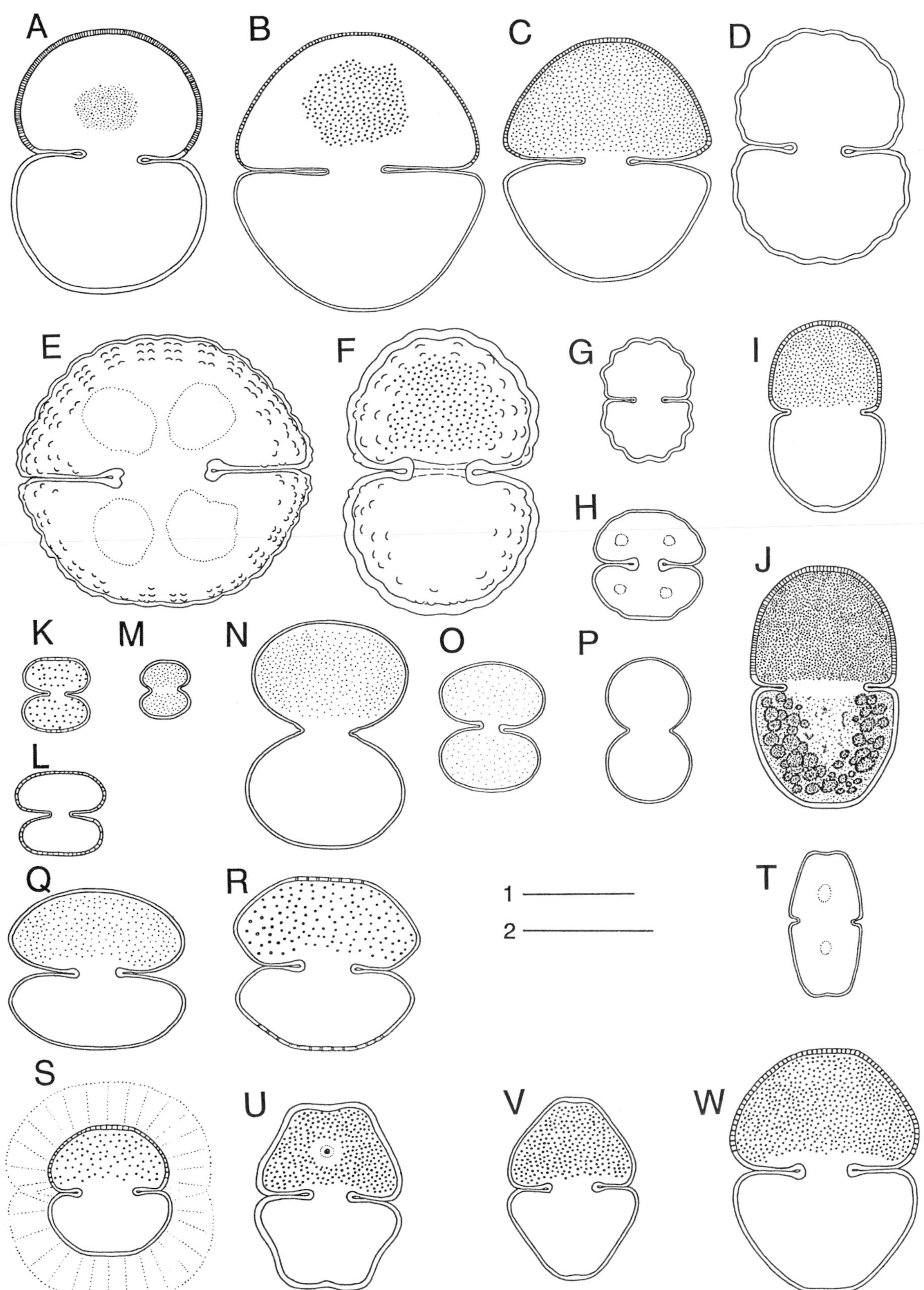

Plate 132 A–W

A. *Cosmarium pachydermum* (p. 542); **B**. *Cosmarium ralfsii* (p. 545); **C**. *Cosmarium ralfsii* var. *montanum* (p. 545); **D**. *Cosmarium undulatum* (p. 548); **E**. *Cosmarium cyclicum* var. *arcticum* (p. 539); **F**. *Cosmarium subundulatum* (p. 546); **G**. *Cosmarium crenulatum* (p. 538); **H**. *Cosmarium fontigenum* (p. 539); **I**. *Cosmarium cucumis* (p. 538); **J**. *Cosmarium cucumis* var. *magnum* (p. 538); **K**. *Cosmarium bioculatum* (p. 535); **L**. *Cosmarium bioculatum* var. *depressum* (p. 535); **M**. *Cosmarium tinctum* (p. 548); **N**. *Cosmarium contractum* (p. 538); **O**. *Cosmarium contractum* var. *ellipsoideum* (p. 538); **P**. *Cosmarium contractum* var. *rotundatum* (p. 538); **Q**. *Cosmarium depressum* (p. 539); **R**. *Cosmarium depressum* var. *achondrum* (p. 539); **S**. *Cosmarium depressum* var. *planctonicum* (p. 539); **T**. *Cosmarium anceps* (p. 535); **U**. *Cosmarium nymannianum* (p. 542); **V**. *Cosmarium granatum* (p. 541); **W**. *Cosmarium galeritum* (p. 539).
Scale bars: 1, 25 μm; 2, 50 μm. 1 for D–H, J–W; 2 for A–C, I.

Cosmarium conspersum Ralfs 1848

27050370

Cells 50–77 µm wide, 82–110 µm long, deeply constricted; sinus linear thickened internally, isthmus 20–27 µm wide; semicells subrectangular to trapeziform, lateral margins diverging slightly from base to broadly rounded apex; walls uniformly granulate, 30 round margin, disposed in about 9 horizontal series and 13–15 vertical series.

Essentially Arctic-alpine, acidophile, mainly in the northern hemisphere; widely distributed but uncommon in the British Isles occurring mostly in *Sphagnum* bogs, also in nutrient-poor ponds and lakes amongst the marginal vegetation; sometimes a chance plankter (pH 5–7.8).

var. *latum* (Brébisson) West *et* G.S.West 1912

27050372 **Pl. 135O** (p. 547)

Cells 76–88 µm wide, 88–107 µm long so proportionately wider than nominal variety and with more conspicuous upwards divergence of semicells; apex broadly convex.

More widely distributed and much more common than the nominal variety in similar habitats.

Cosmarium contractum Kirchner 1878

27950380 **Pl. 132N** (p. 537)

Cells 17–34 µm wide, 30–54 µm long; sinus deep and wide, narrowing gradually to sharp-angled interior; isthmus 4–10 µm wide; semicells broadly ellipsoidal with a slightly flattened apex; walls smooth and finely punctate.

Cosmopolitan; widespread, especially in plankton of nutrient-poor to moderately nutrient-rich lakes in western regions of the British Isles; occasionally in the littoral and in bogs (pH 4.3–7.6).

var. *ellipsoideum* (Elfving) West *et* G.S.West 1902

27050383 **Pl. 132O** (p. 537)

Cells 24–54 µm wide, 30–59 µm long; isthmus 14–29 µm wide; semicells more distinctly ellipsoidal than nominal, with apex rarely flattened.

Common at times in the plankton of Scottish and Irish lakes.

var. *rotundatum* Borge

27050386 **Pl. 132P** (p. 537)

Differs from nominal variety in that semicells are virtually circular in outline, 21–33 µm wide, 31–52 µm long.

Similar distribution to the nominal variety.

Cosmarium crenatum Ralfs ex Ralfs 1848

27050450 **Pl.135M** (p. 547)

Cells 15–38 µm wide, 22–43 µm long; sinus shallow, linear and slightly open; semicells pyramidal-quadrate with lateral margins 3-crenate and curving from angular base to truncate 4-crenate apex, 1–3 smaller, smooth or granular within each lateral and apical crenation; face of semicell slightly tumid, typically ornamented with 3–6 vertical ribs of variable prominence, sometimes virtually absent.

Essentially Arctic-alpine; in the British Isles mostly in boggy springs and amongst mosses on dripping rocks; also in littoral and a chance plankter of nutrient-poor to moderately nutrient-enriched lakes (pH 5–8).

Cosmarium crenulatum Nägeli 1849

Synonym: *Cosmarium undulatum* Corda var. *crenulatum* (Nägeli) Wittrock

27050460 **Pl. 132G** (p. 537)

Cells quadrate-semicircular, 20–24 µm wide, 29–33 µm long; constriction deep, sinus linear, closed; semicell margins undulate, 2 undulations on apex, 3 down lateral margins including basal angles.

Mainly northern Europe, acidophile; in the British Isles most common in the west of Scotland in *Sphagnum* bogs and peaty pools.

Can be confused with *C. impressulum* and *C. venustum* both of which are longer in relation to width and margins are more undulate.

Cosmarium cucumis Corda ex Ralfs 1848

27050480 **Pl. 132I** (p. 537)

Cells 30–57 µm wide, 50–100 µm long; sinus only moderately deep, closed towards outside, sometimes slightly open at rounded inner end; semicells semi-ellipsoidal with rounded basal angles from which lateral margins curve to convex or slightly flattened apex; walls smooth, finely punctate; chloroplasts in each semicell 6–8 parietal serrated ribbons each with several small pyrenoids.

Cosmopolitan, acidophile; widespread in the British Isles, mostly amongst *Sphagnum* and in acid pools (pH 4.8–6.8).

var. *magnum* Raciborski 1885

27050483 **Pl. 132J** (p. 537)

Cells 52–66 µm wide, 90–110 µm long; semicells with truncate-subtruncate apex; walls thick, densely punctate.

Arctic-alpine; in the British Isles mainly in acid bogs and pools, recorded from the English Lake District, widespread in the west and north of Scotland including Skye and the Orkney Islands; 'pretty common' in Scotland according to Roy & Bisset (1894).

Cosmarium cyclicum P.Lundell 1871

27050490

Cells 52–58 µm wide, 49–52 µm long, circular or transversely circular-ellipsoidal; sinus narrow, dilated at inner end; semicells semicircular, basal angles acutely rounded; entire curved margin with about 12 crenae, with 2–3 parallel rows of undulations beneath on face, decreasing in size inwards to smooth central area; walls punctate.

Arctic-alpine, acidophile; present amongst *Sphagnum* bogs and in littoral zone of nutrient-poor lakes, widely distributed in north-west Scotland but never common.

var. *arcticum* Nordstedt 1872
27050492 **Pl. 132E** (p. 537)
Cells tending to be broadly hexagonal with extended basal angles, 50–82 μm wide, 48–79 μm long; semicells with slightly flattened apex having 4 crenulations; all intramarginal crenations bigranulate.

In similar habitats but more frequent than the nominal variety, especially in Scotland (pH 5.8–7.2).

Cosmarium debaryii W.Archer 1861

27050530 **Pl. 133Q** (p. 540)
Cells 42–60 μm wide, 88–123 μm long; sinus shallow, open, acute-angled; semicells subquadrate with sides straight or convex, more rarely very slightly concave; basal angles rounded, rectangular; apex broadly convex or somewhat flattened; walls finely punctate.

Probably cosmopolitan, essentially acidophile; often in *Sphagnum* or *Utricularia*, also associated with littoral vegetation in nutrient-poor and moderately nutrient-rich lakes, occasionally a chance plankter; widepread in the British Isles.

Cosmarium depressum (Nägeli) P.Lundell 1871

Synonym: *Cosmarium scenedesmus* Delponte
27050570 **Pl. 132Q** (p. 537)
Cells 36–60 μm wide, 36–56 μm long; deeply constricted, sinus narrow, linear, open to outside; semicells elliptic with moderately truncate (depressed) apex; walls smooth, thin, irregularly punctate.

Cosmopolitan, in nutrient-poor waters, sometimes in moderately nutrient-rich lakes, also amongst *Sphagnum*; widespread and fairly common in the British Isles (pH 6.5–7.9)

var. *achondrum* (Boldt) West *et* G.S.West 1905
Synonym: *Cosmarium phaseolus* Brébisson var. *achondrum* Boldt
27050572 **Pl. 132R** (p. 537)
Cells 40–51 μm wide, 37–51 μm long; semicells subhexagonal-elliptic with broader and more truncate apices than the nominal variety.

Northern Europe, Arctic, North America; frequent in plankton of nutrient-poor Scottish and Irish lakes, less common in England and Wales.

var. *planctonicum* Reverdin 1919
27050574 **Pl. 132S** (p. 537)
Cells significantly smaller than the nominal variety and the var. *achondrum*, 16–30 μm wide, 18–28 μm long, and characterized by being surrounded by a gelatinous sheath with thread-like rays emanating from pores in wall.

In the plankton of nutrient-poor to nutrient-rich lakes thoughout much of the Britsh Isles (pH 5.9–9.0; alkalinity 3.6–194 ppm $CaCO_3$).

Cosmarium didymochondrum Nordstedt 1876

27050580 **Pl. 135H** (p. 547)
Cells 28–38 μm wide, 40–53 μm long; sinus deep, narrow and linear slightly open, isthmus thickened internally, isthmus 10–15 μm wide; semicells subsemicircular-quadrate, lower lateral margins parallel then curving inwards to slightly produced truncate apex with 4 undulations; lateral margins with 5–7 crenations, lower indistinct; granulation on face irregular and indistinct except immediately above isthmus where there are 2 granules.

Possibly cosmopolitan; in the British Isles occurring mainly on dripping rocks among mosses and in other subaerial habitats.

Cosmarium difficile Lütkemüller 1892

27050600 **Pl. 133U** (p. 540)
Cells 16–25 μm wide, 22–39 μm long; sinus narrow, deep, closed; semicells variable with lower margins straight, parallel or diverging, sometimes retuse, upper margin retuse to convex converging to flat or slightly convex apex; walls finely and closely punctate with face of cell having widely spaced minute scrobiculations in 3 rows, one subapical, one median and one above isthmus.

Probably cosmopolitan, acidophile; mostly in pools and bogs amongst *Sphagnum*, also in weedy margins of nutrient-poor lakes and ponds, sometimes subaerial; widespread and fairly common especially in Scotland and English lakes (pH 4.5–6.9).

Cosmarium fontigenum Nordstedt 1878

27050740 **Pl. 132H** (p. 537)
Cells 20–28 μm wide, 20–26 μm long; sinus straight, narrow, opening externally and internally where thickened; semicells transversely trapeziform, lateral margins curving inwards (upper part undulate) to truncate or slightly convex apex; walls finely punctate.

Especially Europe, North and South America, Africa; fairly widespread in north and west of Scotland in acid bogs and pools.

Cosmarium formosulum Hoff 1888

27050750 **Pl. 135D** (p. 547)
Cells 32–38 μm wide, 35–50 μm long; sinus deep, closed at outside, abruptly rounded at basal angle, slightly inflated, sinus widened internally; semicells trapezoid to semicircular; lateral margins convex with 6–7 crenations, upper crenations bigranulate; apex truncate with 4–5 weak undulations; cell face with paired rows of granules becoming single towards slightly swollen mid-region with 5 indistinct vertical rows of granules.

Mainly European and North American; widespread and frequent in the British Isles in bogs and margins of nutrient-poor ponds and lakes; also a chance plankter; essentially acidophilic but also reported from moderately nutrient-rich waters (pH 5–7.9).

Cosmarium galeritum Nordstedt 1870

27050770 **Pl. 132W** (p. 537)
Cells 46–65 μm wide, 46–65 μm long; sinus narrow, linear, slightly dilated at outside and inside; semicells pyramidal-trapeziform to truncate-pyramidate with

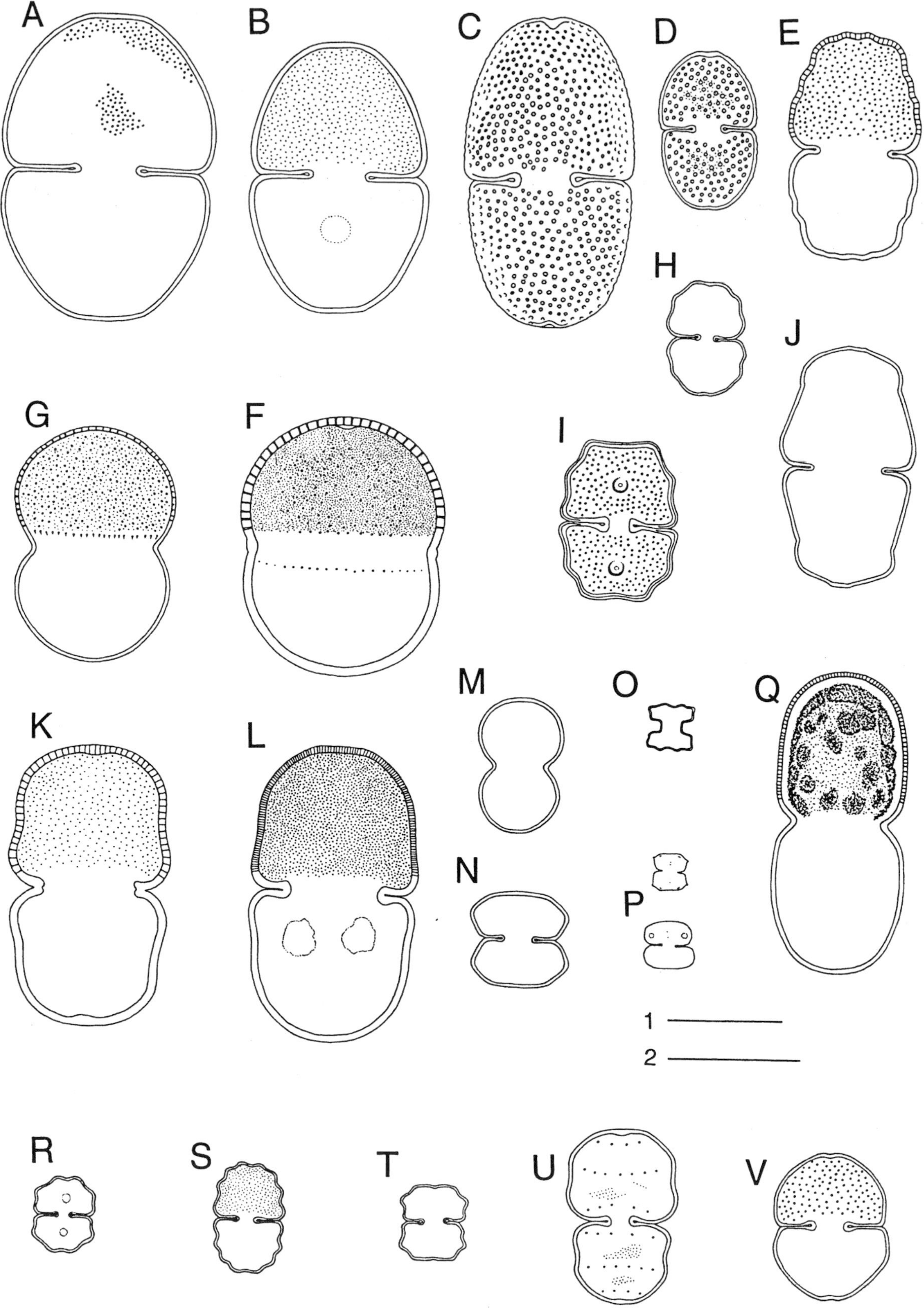

Plate 133 A–V

A. *Cosmarium pyramidatum* (p. 544); **B**. *Cosmarium pseudopyramidatum* (p. 544); **C**. *Cosmarium variolatum* var. *skujae* (p. 548); **D**. *Cosmarium variolatum* (p. 548); **E**. *Cosmarium holmiense* (p. 541); **F**. *Cosmarium pseudoconnatum* (p. 544); **G**. *Cosmarium connatum* (p. 536); **H**. *Cosmarium venustum* (p. 548); **I**. *Cosmarium venustum* var. *excavatum* (p. 548); **J**. *Cosmarium holmiense* var. *integrum* (p. 541); **K, L**. *Cosmarium quadratum* (p. 545); **M**. *Cosmarium moniliforme* (p. 542); **N**. *Cosmarium abbreviatum* (p. 535); **O**. *Cosmarium regnesii* (p. 545); **P**. *Cosmarium pygmaeum* (p. 544); **Q**. *Cosmarium debaryii* (p. 539); **R**. *Cosmarium meneghinii* (p. 542); **S**. *Cosmarium impressulum* (p. 541); **T**. *Cosmarium regnellii* (p. 545); **U**. *Cosmarium difficile* (p. 539); **V**. *Cosmarium laeve* (p. 541).

Scale bars: 1, 25 μm; 2, 50 μm. 1 for A–E, G–P, R–V; 2 for F, Q.

broadly rounded basal angles, lateral margins converging, straight or slightly convex with upper angles rounded and apex slightly convex-truncate; walls smooth, finely punctate.

Probably cosmopolitan; in the British Isles an upland acidophile, in acid bogs and pools, widely distributed in England, Scotland and Ireland; few records from Wales.

Cosmarium granatum Brébisson in Ralfs 1848

27050830 **Pl. 132V** (p. 537) **CD**

Cells 13–30 μm wide, 22–47 μm long, subrhomboidal-ellipsoidal; sinus narrow, linear, slightly dilated inside; semicells pyramidate-truncate, lateral margins almost parallel, upper section retuse, converging to a narrow, truncate apex; walls smooth, closely and finely punctate.

Cosmopolitan; occurs in a wide range of water types and habitats, from amongst *Sphagnum* and *Utricularia*, to plankton of alkaline, nutrient-rich lakes, also in river plankton; also present in brown humic-water (dystrophic) lakes and even subaerial. One of the most widespread and common desmids (pH 4.6–9.5; alkalinity 12–76 ppm $CaCO_3$).

Fifteen varieties named, but many of doubtful validity.

Cosmarium holmiense P.Lundell 1871

27050890 **Pl. 133E** (p. 540)

Cells 20–42 μm wide, 38–67 μm long; sinus only moderately deep, closed towards outside, slightly open inside; semicells broad-pyramidate, basal angles rounded, lateral margins mostly convex, sometimes undulate, tapering to broadly truncate, bi-undulate apex with very slight median incision; walls smooth, delicately punctate.

Essentially Arctic-alpine; in the British Isles mostly in mosses and algae on wet rocks, also in bogs especially at high altitudes and occasionally in rain gutters subject to drying; alkaline tolerant.

var. *integrum* P.Lundell 1871

27050894 **Pl. 133J** (p. 540)

Cells 23–44 μm wide, 36–76 μm long, differs from type in semicells more attenuated towards distinctly dilated and convex or angular apex; margins not undulate.

Although in similar habitats to the nominal variety, more often in acid waters amongst *Sphagnum* and *Utricularia*; more widespread and frequent in the British Isles than the nominal variety (pH 4.7–9).

Cosmarium humile (F.Gay) Nordstedt in De Toni 1889

Basionym: *Euastrum celatum* F.Gay

27050900 **Pl. 134Q** (p. 543)

Cells very small, 10–18 μm wide (7–14 μm across apex), 11–18 μm long; sinus deep, narrow and linear, slightly dilated internally; semicells trapeziform with broad base, narrowing abruptly to truncate, 2–4-undulate apex, with scattered granules beneath margins, sometimes 2 or 3 grouped together and one larger central granule (papilla); disposition of granules very variable.

Cosmopolitan; very widely distributed in the British Isles, often in ditches and amongst macrophytes and attached algae in littoral zone of lakes and ponds, less common in acid bogs (pH 6–7.4); commonly as a chance plankter.

Cosmarium impressulum Elfving 1881

Synonym: *Cosmarium meneghinii* Brébisson var. *simplicissimum* Wille fo. *reinschii* Istvanfy

27050910 **Pl. 133S** (p. 540)

Cells 13–27 μm wide, 18–37 μm long, sinus deep, narrow and slightly open internally; semicells semi-ellipsoidal to semicircular, entire margin regularly and strongly undulate; crenations equal, with 2 at base, 2 on each lateral margin and 2 on apex; walls smooth, finely puncate.

Cosmopolitan; with wide tolerances, occurring mostly in alkaline waters, also amongst *Sphagnum* or *Utricularia* and in the littoral zone of a range of lake types, and as a chance plankter; widely distributed in the British Isles, especially in the north and west of Scotland (pH 5–9).

Cosmarium laeve Rabenhorst 1868

Synonym: *C. meneghinii* Brébisson fo. *octangulare* Wille

27051000 **Pl. 133V** (p. 540)

Cells 11–26 μm wide, 14–36 μm long; sinus very deep, narrow, linear open internally; semicells semiellipsoidal or oblong-elliptical; basal angles broadly rounded, lower lateral margins parallel then curving evenly to narrow, notched truncate apex; walls punctate, puncta sometimes coarse.

Cosmopolitan, occurring in a wide range of waters, including markedly alkaline ones; in the British Isles mostly in small pools, ponds, ditches, even cattle troughs; also on wet rocks (pH 5.4–9.4).

Cosmarium margaritatum (P.Lundell) J.Roy *et* Bisset 1886

27051070 **Pl. 135L** (p. 547)

Cells 56–82 μm wide, 66–105 μm long; sinus deep, linear, dilated at inside; semicells subrectangular with broadly rounded basal and upper angles; lateral margins gently curved, broad apex flat or slightly rounded; walls ornamented with solid granules, about 28–32 visible round semicell margin and on semicell face in oblique decussating rows, about 12 in each direction; walls with puncta arranged in hexagons round each granule.

Cosmopolitan; widespread in the British Isles associated with other algae in acid bogs, also in the littoral zone of nutrient-poor to moderately nutrient-rich waters and in the plankton; most frequent in Scotland and Wales (pH 4.5–7.5).

Cosmarium margaritiferum Meneghini ex Ralfs 1848

Synonym: *Cosmarium confusum* Cooke var. *regularis* Nordstedt

27051080 **Pl. 134M** (p. 547)

Cells 42–58 μm wide, 50–66 μm long; sinus deep, narrow, linear, dilated internally; semicells broadly pyramidal-truncate, basal and upper angles rounded;

lateral margins slightly convex leading to wide, flat or slightly convex apex; walls ornamented with granules of varying size, the largest in centre of semicell face, 7–9 granules visible on each lateral margin, smaller towards smooth apex; walls between granules minutely punctate.

Cosmopolitan; widespread and sometimes fairly common in the British Isles mostly amongst *Sphagnum* and *Utricularia* in acid bogs and pools, also in the littoral zone of nutrient-poor lakes; occasional but rare as a chance plankter (pH 5.1–6.8).

Variable species especially with respect to ornamentation; several forms of doubtful taxonomic validity described. W. & G.S.West (1908, p. 201) state 'no species of the genus *Cosmarium* has been so misinterpreted or has given rise to greater confusion'.

Cosmarium meneghinii Brébisson in Ralfs 1848

Synonym: *Didymidium braunii* Reinsch

27051100 **Pl. 133R** (p. 540)

Cells small, 9.5–17 μm wide, 12.5–24 μm long, sub-octagonal; sinus deep, narrow, linear; semicells with rounded angles, rectangular in lower half, truncate-pyramidal in upper; lower lateral margins parallel and slightly retuse, upper strongly convergent leading to comparatively wide, retuse apex.

Cosmopolitan; widespread throughout the British Isles, especially common and at times abundant in a range of habitats; frequent with *Sphagnum* in bogs and in the littoral of nutrient-poor lakes (pH 4.5–6.9). In some southern counties of England present in nutrient-rich, alkaline lakes (pH 7–8.6).

Cosmarium moniliforme Turpin ex Ralfs 1848

27051130 **Pl. 133M** (p. 540)

Cells 10–29 μm wide, 16–50 μm long, resembling 2 joined spheres; sinus deep, widely open, internally acute-angled; semicells circular to subcircular in outline; walls smooth, occasionally brown, with pores.

Cosmopolitan; common in *Sphagnum* bogs and littoral zone of nutrient-poor to moderately nutrient-rich ponds and lakes, often in their plankton; widespread in acidic regions of the British Isles (pH 4–7).

In plankton, sometimes enveloped in thick mucilage envelope.

Cosmarium monomazum P.Lundell 1871

var. *polymazum* Nordstedt 1873

Synonym: *Cosmarium polymazum* Nordstedt

27051152 **Pl. 134B** (p. 540)

Cells 32–39 μm wide, 32–39 μm long; sinus deep, narrow, linear; semicells subsemicircular with rounded basal angles; apex slightly flattened; within semicell margin 16–20 warts and on face 3 large granules (central one very large) forming a transverse series across middle and one large granule above isthmus.

Europe, North America, Asia, acidophile; a north-west distribution in the British Isles, most frequent in the English Lake District and in north-west Sutherland; Irish records from Donegal and Galway.

Cosmarium nymannianum Grunow in Rabenhorst 1868

27051240 **Pl. 132U** (p. 537)

Cells 29–42 μm wide, 34–54 μm long, subhexagonal, elongate; sinus narrow, linear, opening towards inside; semicells truncate-pyramidal with rounded basal and upper angles; lower half of lateral margins convex, upper concave and apex retuse; wall in centre of face of semicell with a large scrobiculation surrounded by pore-free area, remainder of walls punctate.

Worldwide except Australasia; widespread in the British Isles, mostly in acid bogs and pools, often associated with *Sphagnum*.

Cosmarium ornatum Ralfs 1848

27051330 **Pl. 134D** (p. 543)

Cells 30–43 μm wide, 28–42 μm long; sinus deep, narrow, linear, opening to inside; semicells kidney-shaped with a distinctly produced, broad truncate apex; basal angles and most of lateral margins form a continuous broad curve to upper margins beneath apex where slightly retuse; lateral margins having 7–9 concentric rows of granules, none on upper part; 7 granules across apex and 2–3 rows of similar granules extending down face towards centre; centre of face of semicell protuberant with rings of granules, but disposition variable.

Common in northern Europe and North America, also records from Asia and Africa; widely distributed in the British Isles in acid bogs and nutrient-poor lake margins especially in the north and west of Scotland including the Hebrides; sometimes as a chance plankter.

Cosmarium ovale Ralfs ex Ralfs1848

Synonym: *Cosmaridium ovale* (Ralfs ex Ralfs) Hansgirg

27051360 **Pl. 135I** (p. 547)

Cells large, 86–136 μm wide, 154–222 μm long; sinus deep, narrow, linear, slightly open inside, isthmus 30–48 μm wide; semicells ovate with broad flat base and rounded basal angles; lateral margins curving evenly from base to slightly flattened apex, 19–21 conical granules round lateral margins and irregular series within margins; apical margin without granules but a series of 1–2 rows of granules beneath margin; walls finely scrobiculate, centre of face thickened; chloroplasts with 4–6 parietal bands extending from base to apex, 9–11 pyrenoids in each.

Arctic-alpine; in the British Isles occurring most commonly in western regions, especially in *Sphagnum* bogs, and a chance plankter in small lakes and ponds (pH 5–8.5).

Cosmarium pachydermum P.Lundell 1871

27051370 **Pl. 132A** (p. 537)

Cells 39–92 μm wide, 52–125 μm long; sinus deep, narrow, linear, open internally; semicells subsemicircular with broad, rounded basal angles; walls thick, smooth, densely punctate.

Cosmopolitan; widespread in the British Isles among *Sphagnum* and *Utricularia* in acid bogs and pools; occasional in littoral zone of nutrient-poor lakes and as a chance plankter (pH 4.6–7.5).

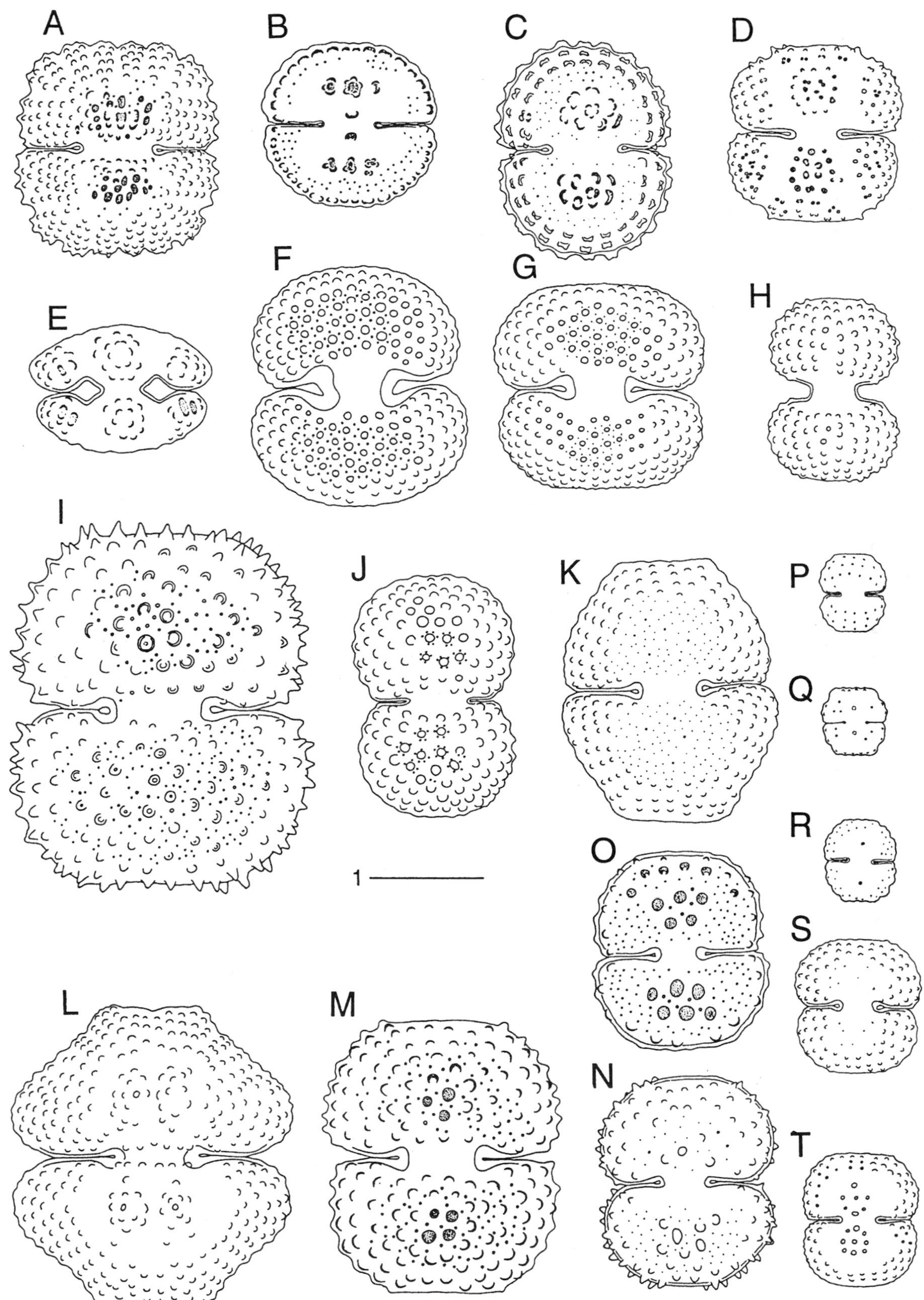

Plate 134 A–T

A. *Cosmarium caelatum* (p. 536); **B**. *Cosmarium monomazum* var. *polymazum* (p. 542); **C**. *Cosmarium quadrifarium* var. *hexastichum* (p. 545); **D**. *Cosmarium ornatum* (p. 542); **E**. *Cosmarium commissurale* (p. 536); **F**. *Cosmarium reniforme* (p. 545); **G**. *Cosmarium reniforme* var. *compressum* (p. 546); **H**. *Cosmarium portianum* (p. 544); **I**. *Cosmarium brebissonii* (p. 536); **J**. *Cosmarium sphaeroideum* (p. 546); **K**. *Cosmarium sportella* var. *subnudum* (p. 546); **L**. *Cosmarium turpinii* var. *podolicum* (p. 548); **M**. *Cosmarium margaritiferum* (p. 541); **N**. *Cosmarium praemorsum* (p. 544); **O**. *Cosmarium quinarium* (p. 545); **P**. *Cosmarium sphalerostichum* (p. 546); **Q**. *Cosmarium humile* (p. 541); **R**. *Cosmarium blytii* (p. 536); **S**. *Cosmarium punctulatum* (p. 544); **T**. *Cosmarium punctulatum* var. *subpunctulatum* (p. 543).

Scale bar: 25 μm.

Cosmarium portianum W.Archer 1860

Synonym: *Cosmarium orbiculatum* de Bary
27051460 **Pl. 134H** (p. 543)
Cells 19–32 µm wide, 26–44 µm long; sinus deep, opening widely from rounded inside, hence isthmus with an elongate neck; semicells transversely elliptical; walls with 10 regular vertical series of granules and 20–23 granules round semicell margin.

Cosmopolitan, acidophile; common in western regions of the British Isles, especially in peaty pools and bogs containing *Sphagnum* in Scotland, Ireland and Wales, also in the English Lake District; reported from damp mosses in water courses (pH 5.1–8.2).

Cosmarium praemorsum Brébisson 1856

27051480 **Pl. 134N** (p. 543)
Cells 43–51 µm wide, 47–56 µm long; sinus deep, linear, closed, slightly dilated internally; semicells broadly truncate-pyramidal or truncate-subsemicircular, lower angles rounded, lateral margins convex, apex truncate or slightly convex; walls unevenly granulate, each lateral margin with 6–8 granules, largest towards smooth apex; mid-region of semicell with variable number of irregularly disposed large granules and smaller granules scattered between margins and apical region.

Arctic and northern Europe, acidophile; occurring mostly amongst the littoral vegetation of nutrient-poor to moderately nutrient-rich ponds and lakes; widely distributed in the British Isles with numerous records from England, Scotland including all the Hebrides, Orkneys and Shetlands and also Ireland and Wales.

W. & G.S.West (1908, p. 198) point out that forms of *C. praemorsum* can be confused with *C. margaritiferum* and that some 'appear to be intermediate in character between these two species and it is sometimes a matter of considerable difficulty to correctly place them'.

Cosmarium pseudoconnatum Nordstedt 1869

27051580 **Pl. 133F** (p. 540)
Cells 33–44 µm wide, 47.5–57.5 µm long, only very slightly constricted by broad, very shallow sinus, isthmus 31–40 µm wide; semicells semi-elliptic with slightly narrowed base and in outline two-thirds circumference of a circle; walls thick, punctate; puncta absent round isthmus, above it in transverse series.

Cosmopolitan, acidophile; in the British Isles mostly in western bogs and lake margins especially in Sutherland, Isle of Skye and Outer Hebrides in Scotland, western regions of Ireland; in England from Cumbria, North Yorkshire and Hampshire, also from Capel Curig (Caernarvonshire), Wales.

Cosmarium pseudopyramidatum P.Lundell 1871

Synonym: *Cosmarium pyramidatum* Brébisson ex Ralfs var. *pseudocucumis* G.A.Klebs
27051630 **Pl. 133B** (p. 540)
Cells 23–41 µm wide, 40–73 µm long; sinus deep narrow, open internally; semicells truncate-pyramidate to semi-elliptic, basal angles rounded-angular, lateral margins convex converging to narrow, truncate, flat or slightly retuse apex; walls smooth, punctate.

Cosmopolitan; widespread in the British Isles mostly in acid waters amongst *Sphagnum* and *Utricularia*, especially in western Scotland and the Western Isles where sometimes common; occasionally as a chance plankter (pH 4.4–6.9).

Cosmarium punctulatum Brébisson 1856

Synonym: *Cosmarium granulusculum* J.Roy *et* Bisset
27051650 **Pl. 134S** (p. 543)
Cells 20–38 µm wide, 22–40 µm long; sinus narrow, linear, slightly dilated inside; semicells oblong-trapeziform, basal angles rounded, lateral margins curving inwards to broad, truncate apex; walls granulate, 23–24 granules around margin, on face disposed in vertical or oblique series, irregular at centre.

Cosmopolitan; mostly in acid pools and ditches with *Sphagnum*; sometimes a chance plankter. Very widespread in British Isles occurring even in moderately alkaline waters (pH 5.1–8.2).

var. *subpunctulatum* (Nordstedt) Børgesen 1894
Basionym: *Cosmarium subpunctulatum* Nordstedt
7051655 **Pl. 134T** (p. 543)
Differs from the nominal variety in face of semicells having a slight central protuberance with granules in irregular rings; cells 22–37 µm wide, 24–42 µm long.

Cosmopolitan; distribution similar to the nominal variety.

Cosmarium pygmaeum W.Archer 1864

Synonym: *Cosmarium schliephackeanum* Grunow
27051680 **Pl. 133P** (p. 540)
Cells very small, 8–16 µm wide, 8–15 µm long; sinus deep, narrow, linear; semicells oblong-rectangular but shape very variable; basal angles obtuse, lateral angles sometimes with a mucro or papilla; middle of semicell face may be protuberant.

Virtually cosmopolitan, acidophile; in the British Isles more or less confined to *Sphagnum* bogs and pools especially in the north and west of Scotland and west of Ireland where sometimes present in great numbers; occasionally a chance plankter.

Cosmarium pyramidatum Brébisson in Ralfs 1848

27051690 **Pl. 133A** (p. 540)
Cells 39–70 µm wide, 61–115 µm long, truncate-elliptic; sinus deep, closed but slightly dilated internally; semicells truncate-pyramidate with acutely rounded basal angles, upper angles obtuse; lateral margins convex, converging to narrow, truncate apex; walls minutely scrobiculate.

Cosmopolitan, mostly in acid waters; in the British Isles occurs especially in bog pools with *Sphagnum*, also nutrient-poor lakes and rocky pools in the west of Scotland and Ireland; occasionally a chance plankter (pH 4.4–6.9).

Cosmarium quadratum Ralfs ex Ralfs 1848

27051710 **Pl. 133K,L** (p. 540)
Cells 24–27 μm wide, 44–84 μm long; sinus shallow V-shaped, opening to outside; semicells subquadrate, narrowing upwards from broadly rounded and sometimes swollen basal angles; lateral margins slightly retuse, upper angles broadly rounded; apex flat-convex, sometimes thickened in middle; walls thick, finely punctate.

Cosmopolitan, essentially acidophilic; widespread but rarely abundant in the British Isles in upland bogs, often in *Sphagnum* pools, at high altitudes in Wales and Ireland; also in littoral zone of nutrient-poor lakes and as a chance plankter (pH 5–8.5).

Cosmarium quadrifarium P.Lundell 1871

27051730
Cells 27–42 μm wide, 35–58 μm long; sinus deep, narrow, linear, inflated inside; semicells semicircular, basal angles subrectangular; margins with 15–17 truncate, emarginate warts reduced towards lateral angles, with similar warts within margin; above isthmus a central, rounded tumour bearing 12–17 rounded granules, sometimes with intergranular scrobiculations; walls finely punctate.

Cosmopolitan; mostly in acid bogs and margins of small, nutrient-poor lakes, especially in western regions of the British Isles; a chance plankter (pH 4.2–7.4).

var. *hexastichum* (P.Lundell) Kurt Förster 1981
27051732 **Pl. 134C** (p. 543)
Cells with 6 parallel series of emarginate verrucae in marginal region; central granules prominent, variable, sometimes in 4 transverse rows; cells 28–45 μm wide, 40–60 μm long.

Cosmopolitan in acid habitats; less widely distributed and less common than the nominal variety.

Cosmarium quinarium P.Lundell 1871

27051780 **Pl. 134O** (p. 543)
Cells 30–39 μm wide, 36–48 μm long; sinus deep, narrow, linear, dilated inside and thickened; semicells broadly pyramidal-truncate with basal angles acute, apical angles obtusely rounded; lateral margins convex, converging to smooth truncate apex; lateral margins with 14–15 small acute granules, 10 larger granules just below apex; semicell face with 5 large granules in centre in 2 transverse series.

Cosmopolitan except Australasia, acidophile; widely distributed throughout the British Isles especially amongst *Sphagnum* in peaty ponds and pools.

Cosmarium ralfsii Brébisson in Ralfs 1848

27051800 **Pl. 132B** (p. 537)
Cells 76–104 μm wide, 88–124 μm long, subcircular; sinus deep, narrow, linear, dilated internally; semicells semicircular-subpyramidal; basal angles rounded, lateral margins slightly convex and converging to truncate or broadly curved apex; walls finely punctate; chloroplasts several parietal bands, each with several small pyrenoids.

Cosmopolitan, acidophile; widely distributed especially in upland regions of the British Isles and often abundant amongst *Sphagnum* and *Utricularia* in pools and bogs.

var. *montanum* Raciborski 1885
27051802 **Pl. 132C** (p. 537)
Differs from nominal variety in the sinus opening quite widely to outside so that basal angles appear slightly produced; cells 65–111 μm wide, 80–128 μm long.

Less common than the nominal variety but occurs in similar habitats.

Cosmarium regnellii Wille 1884

27051830 **Pl. 133T** (p. 540)
Cells very small, 12–22 μm wide, 14–22 μm long; sinus deep, very narrow, slightly open inside; semicells trapezoid-hexagonal; lower lateral margins slightly retuse, upper markedly so with angle between upper and lower margin broadly rounded and projecting; walls smooth, very delicately punctate.

Cosmopolitan, tolerant of wide range of water types; in the British Isles essentially a lowland species, more common in England than elsewhere (pH 5.4–9).

Cosmarium regnesii Reinsch 1867

27051840 **Pl. 133O** (p. 540)
Smallest species of the genus, 6–9.5 μm wide, 6–10 μm long; sinus open, rounded and isthmus elongate, 3–5 μm wide; semicells transversely oblong-rectangular with 6 minute, equally spaced marginal teeth and a widely retuse apex; walls smooth or with very small scattered granules difficult to discern.

Cosmopolitan, acidophile; fairly widespread in the British Isles, often abundant in boggy margins of pools and lakes, sometimes in very large numbers actively dividing amongst *Sphagnum* and *Utricularia*.

Cosmarium reniforme (Ralfs) W.Archer 1874

Synonym: *Cosmarium margaritiferum* Meneghinii ex Ralfs var. *reniforme* Ralfs
27051860 **Pl. 134F** (p. 543)
Cells 35–62 μm wide, 43–74 μm long; sinus deep, narrow, closed in middle but open widely to outside and inside where wall thickened; semicells kidney-shaped; walls with rounded granules in obliquely decussating or vertical rows, about 30–33 round each semicell margin.

Probably cosmopolitan; widely distributed in the British Isles in a range of waters, mostly in margins of lakes and ponds, also in boggy pools and springs, sometimes as a chance plankter in Scottish and Irish lakes (pH 6–8.6).

var. *compressum* Nordstedt 1887

27051863 **Pl. 134G** (p. 543)

Cells 47–56 μm wide, 46–64 μm long; isthmus 13–18 μm wide; semicells sometimes depressed in median region or with a slightly truncate apex.

Much less widely distributed than the nominal variety but present in similar habitats.

Cosmarium speciosum P.Lundell 1871

27052000 **Pl. 135F** (p. 547)

Cells 24–47 μm wide, 36–70 μm long; sinus shallow, narrow, linear; semicells subrectangular, attenuating gradually from acutely angled base to broad, truncate apex; margin having 17–19 crenations (4 on apex) with increasingly small granules in regular radial and concentric series from margin towards centre of semicell face and 6–9 vertical series of granules across base above isthmus.

Cosmopolitan, acidophile; widespread in the British Isles mostly associated with *Sphagnum*; also in littoral zone of nutrient-poor lakes and ponds amongst filamentous algae and frequent at times in boggy springs and dripping rocks amongst mosses.

var. *rostafinskii* (Gutwinski) West *et* G.S.West 1908

27052004 **Pl. 135G** (p. 547)

Cells 23–31 μm wide, 35–46 μm long; semicells more regularly pyramidate than the nominal variety, apex more truncate.

In similar habitats to the nominal variety, especially frequent in subaerial enviroments.

Cosmarium sphaeroideum West 1892

27052010 **Pl. 134J** (p. 543)

Cells 38–40 μm wide, 60–63 μm long; sinus narrow, short, slightly open; semicells subcircular-obovate with flattened or slightly retuse apex; semicell margin appears undulate because of 22–24 granules around periphery, similar granules occurring on face in about 11 oblique series.

Northern Europe, North America, Asia, acidophile; in the British Isles principally in bogs in granitic regions of Scotland and Ireland; rare elsewhere.

Cosmarium sphalerostichum Nordstedt 1876

27052030 **Pl. 134P** (p. 543)

Cells very small, 13–15.5 μm wide, 15.5 μm long, sinus moderately deep, narrow, linear; semicells subkidney-shaped to trapeziform with flat base or subrectangular basal angles; lateral margins convex, 4–5-granulate; apex truncate, smooth; walls with variable surface granulation, commonly incomplete vertical (5–6) and transverse (2–3) rows.

Arctic-alpine, mostly on dripping rocks and boggy streams; in the British Isles especially common in mountainous regions of Wales and Scotland.

Cosmarium sportella Brébisson 1849

27052040

Cells 33–46 μm wide, 45–50 μm long; sinus deep, narrow, linear, dilated internally; semicells truncate-pyramidate, basal angles rounded, upper margins very slightly retuse, upper angles obtuse; apex broad, truncate; walls with 6–7 granules on each lateral margin, upper angles distinctly bigranulate, granules on face scattered, smaller towards central swollen area where 7 occur in a group, with puncta between.

Northern Europe, North America, Asia, Arctic, fairly tolerant of a range of water types; widespread in the British Isles but most frequent in Scotland and Ireland, often on dripping rocks.

var. *subnudum* West *et* G.S.West 1908

27052042 **Pl. 134K** (p. 543)

Cells 41 μm wide, 51 μm long, apex 22 μm broad; semicells more pyramidate than the nominal variety and granules absent on central tumour.

Distribution same as the nominal variety.

Cosmarium subcostatum Nordstedt 1876

27052110 **Pl. 135C** (p. 547)

Cells 23–43 μm wide, 24–45 μm long; sinus deep, linear, dilated internally; semicells subtrapeziform-reniform, lateral margins convex with 2 crenations above basal angles, 4 larger crenations above; apex slightly produced with 4 small crenations; semicell face minutely granulated radially and concentrically disposed, with central tumour having granules in 4–5 vertical series.

Arctic-alpine; widespread in the British Isles, especially in weedy margins of nutrient-poor lakes; occasionally a chance plankter (pH 4.8–8.5).

Cosmarium subcrenatum Hantzsch in Rabenhorst 1861

27052120 **Pl. 135A** (p. 547)

Cells 18–34 μm wide, 21–38 μm long; sinus deep, narrow linear, opening inside; semicells almost semicircular with a truncate apex; basal angles rectangular; lateral margins with 4–6 crenations, upper crenations larger than lower, apex with 4–5 crenations, with small binate granules beneath and 1–2 series of similar single granules inside; centre of semicell face with a shallow tumour ornamented with 5–7 vertical series of small granules.

Arctic-alpine; mostly in upland areas of the British Isles, in the littoral of nutrient-poor lakes and ponds with submerged aquatic plants, widespread especially in Scotland and Scottish islands; a chance plankter.

Cosmarium subundulatum Wille 1880

27052290 **Pl. 132F** (p. 537)

Cells 22–47 μm wide, 39–64 μm long; sinus deep, closed except internally where thickened; semicells semi-elliptic, basal angles rounded, apex slightly flattened;

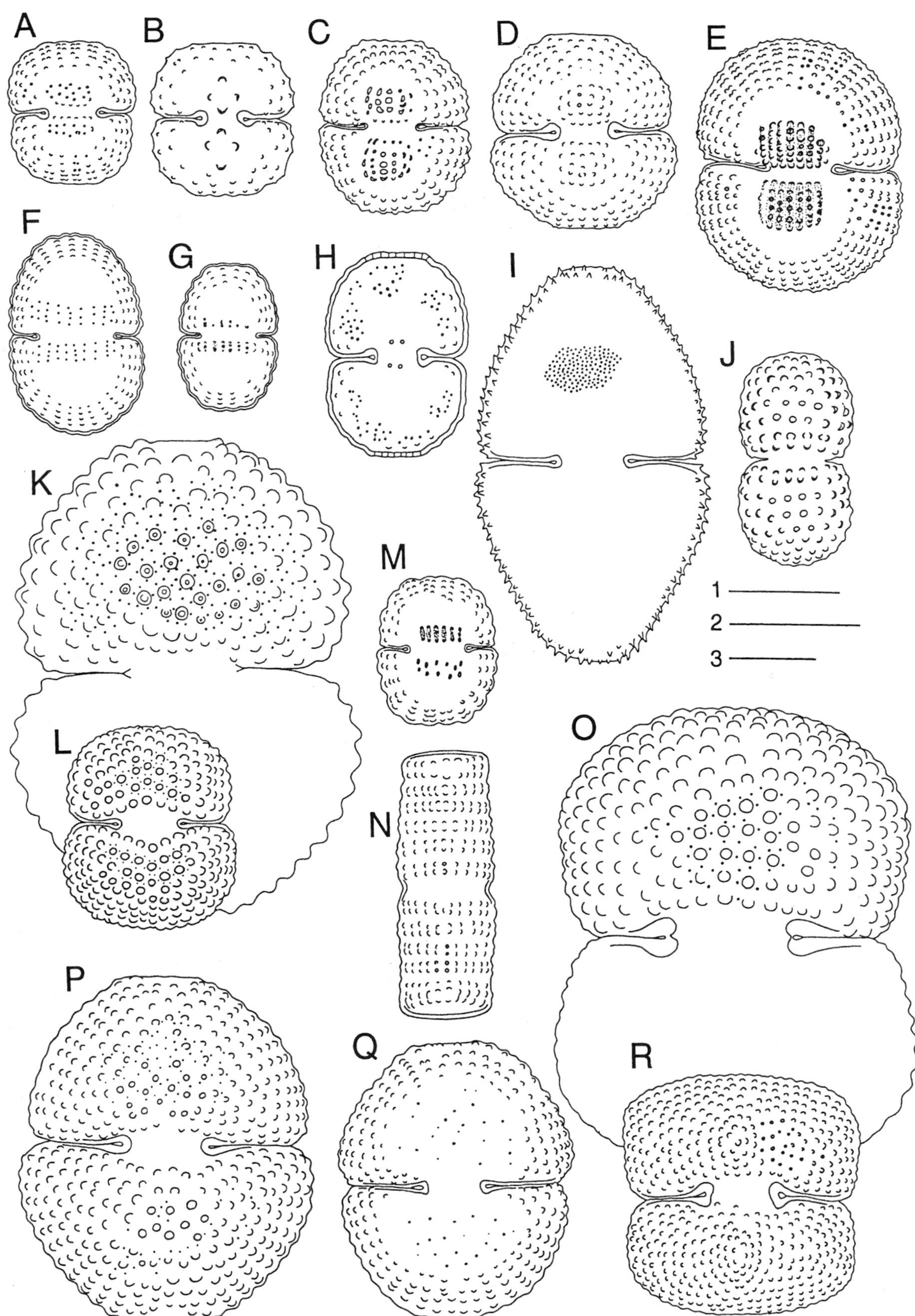

Plate 135 A–R

A. *Cosmarium subcrenatum* (p. 546); **B**. *Cosmarium boeckii* (p. 536); **C**. *Cosmarium subcostatum* (p. 546); **D**. *Cosmarium formosulum* (p. 539); **E**. *Cosmarium binum* (p. 535); **F**. *Cosmarium speciosum* (p. 546); **G**. *Cosmarium speciosum* var. *rostafinskii* (p. 546); **H**. *Cosmarium didymochondrum* (p. 539); **I**. *Cosmarium ovale* (p. 542); **J**. *Cosmarium amoenum* (p. 535); **K**. *Cosmarium tetraophthalmum* (p. 548); **L**. *Cosmarium margaritatum* (p. 541); **M**. *Cosmarium crenatum* (p. 538); **N**. *Cosmarium annulatum* var. *elegans* (p. 535); **O**. *Cosmarium conspersum* var. *latum* (p. 538); **P**. *Cosmarium botrytis* (p. 536); **Q**. *Cosmarium botrytis* var. *mediolaeve* (p. 536); **R**. *Cosmarium biretum* (p. 535).

Scale bars: 1, 25 μm; 2, 3, 50 μm. 1 for A–H, J, K, M–R; 2 for L; 3 for I.

margins with about 12 undulations and 2 series of undulations on face beneath each; centre of semicell face with tumour and without granules.

Arctic-alpine; wide distribution in the British Isles, especially in acid habitats in western areas of England, Scotland and Ireland.

Cosmarium tetraophthalmum Brébisson in Ralfs 1848

27052380 **Pl. 135K** (p. 547)

Cells 60–86 μm wide, 90–120 μm long; sinus deep, narrow, linear, open internally; semicells pyramidate-ovate with broadly rounded basal angles, convex sides and comparatively narrow truncate apex; walls with 12–14 granules along each lateral margin, greatly reduced or absent on apex and diminishing in size towards centre of face, puncta between granules; chloroplasts each with 2 prominent eye-like pyrenoids.

Cosmopolitan, acidophile; fairly common in *Sphagnum* bogs throughout the British Isles, most abundant in north-west Scotland and west Ireland; a chance plankter in nutrient-poor lakes and ponds.

Cosmarium tinctum Ralfs 1848

27052400 **Pl. 132M** (p. 537)

Cells very small, 7–12 μm wide, 10–16 μm long; sinus not very deep, opening outwards; semicells elliptic; walls smooth, usually yellow-brown.

Cosmopolitan; widely distributed in the British Isles, sometimes in quantity in peaty acid bogs, especially amongst *Sphagnum* or *Utricularia*.

Cosmarium turpinii Brébisson 1856

27052500

Cells 50–73 μm wide, 60–80 μm long; sinus deep, narrow linear, dilated inside; semicells pyramidate-trapeziform, narrowing markedly from broad rounded base to slightly retuse apex with obtuse angles; walls densely granulate, 36–40 granules round margin, smaller granules towards centre of face where 2 tumours covered with irregularly arranged granules; chloroplasts with 2 prominent pyrenoids.

Northern Europe, Arctic, tolerant of a range of waters (pH 6.2–8.5); widespread in the British Isles even in southern and eastern lowland counties; most common in the weedy littoral of shallow lakes and ponds. Reported as thriving amongst *Potamogeton* in the alkaline dykes and drains of East Anglia (W. & G.S.West, 1908).

var. *podolicum* Gutwinski 1890

27052503 **Pl. 134L** (p. 543)

Cells 40–83 μm wide, 46–84 μm long, thus very variable in size; semicells with 2–3 emarginate crenations at each side just below apex and a row of 6–7 granules above isthmus; apex retuse with a smooth or granulate margin.

Occasional in plankton of moderately nutrient-rich lakes.

Cosmarium undulatum Corda ex Ralfs 1848

27052520 **Pl. 132D** (p. 537)

Cells 30–32 μm wide, 44–46 μm long; sinus deep, narrow, dilated inside; semicells angular at base then almost semicircular; apex retuse; margins with 10–12 undulations; face smooth.

Cosmopolitan; widely distributed in acid habitats in the British Isles; only a few records from Scottish islands.

Cosmarium variolatum P.Lundell 1871

27052540 **Pl. 133D** (p. 540)

Cells 18–21 μm wide, 32–35 μm long; sinus narrow, linear, dilated internally, isthmus 5–5.6 μm wide; semicells semielliptic with rounded basal angles and convex lateral margins which converge to a narrow, truncate, sometimes very slightly retuse apex; walls densely and conspicuously scrobiculate.

Subcosmopolitan, acidophile; widespread in the British Isles in moderately acid pools and lake margins, especially in western areas including the Isle of Skye, the Outer Hebrides and the Orkneys, but never common.

var. *skujae* Croasdale 1988

27052544 **Pl. 133C** (p. 540)

Cells significantly larger than the nominal variety, 33–36 μm wide, 64–68 μm long; semicells less tapering, with small thickened indentation in middle of apex; walls coarsely pitted (scrobiculate).

An acidophile, previously recorded only from New Zealand; widespread but never abundant in margins of bog pools and weedy lochans, especially in north-west Sutherland; also from Cader Idris in west Wales, and in Cumbria.

Cosmarium venustum (Brébisson) W.Archer 1861

Basionym: *Euastrum venustum* Brébisson

Synonym: *Cosmarium cambricum* Cooke *et* Wills

27052550 **Pl. 133H** (p. 540)

Cells 17–38 μm wide, 25–48 μm long; sinus narrow, linear, dilated inside; semicells truncate-pyramidate, lateral margins with 3 undulations; apex wide, truncate, slightly retuse; walls minutely punctate.

Cosmopolitan, acidophile; widespread in the British Isles, sometimes frequent in pools and peat bogs with *Sphagnum* (pH 4–6.8); recorded elsewhere in the world from alkaline waters; occasionally a chance plankter.

var. *excavatum* (B.Eichler *et* Gutwinski) West *et* G.S.West 1895

27052552 **Pl. 133I** (p. 540)

Cells 19–34 μm wide, 28–46 μm long, apex 11–20 μm wide; semicells with prominent ocellus in middle of face and upper angles sometimes with granules or scrobiculation; walls thick.

Similar distribution to the nominal variety, but less frequent.

TAXA RECORDED FROM THE BRITISH ISLES BUT NOT DESCRIBED IN TEXT

Cosmarium abbreviatum var. *germanicum* (Racib.) Willi Krieg. *et* Gerloff 27050012; *C. abruptum* P.Lundell 27050020; *C. adoxum* West *et* G.S.West 27050030; *C. alpestre* J.Roy *et* Bisset 27050040; *C. amoenum* var. *mediolaeve* Nordst. 27050052; *C. anceps* var. *crispulum* (Nordst.) Willi Krieg. *et* Gerloff 27050062; *C. anisochondrum* Nordst. 27050090; *C. arctoum* Nordst. 27050110; *C. arctoum* var. *tatricum* Racib. 27050113; *C. arnellii* Boldt 27050120; *C. asphaerosporum* Nordst. 27050130; *C. asphaerosporum* var. *strigosum* Nordst. 27050132; *C. basilicum* G.S.West 27050140; *C. bioculatum* var. *hians* West *et* G.S.West 27050163; *C. bipapillatum* West *et* G.S.West 27050170; *C. bipunctatum* Børges. 27050180; *C. biretum* var. *trigibberum* Nordst. 27050202; *C. bitriangulum* var. *bitrapezoidale* Willi Krieg. *et* Gerloff 27050212; *C. blytii* var. *novae sylvae* West *et* G.S.West 27050222; *C. botrytis* var. *depressum* West *et* G.S.West 27050242; *C. botrytis* var. *emarginatum* Hansg. 27050243; *C. botrytis* var. *gemmiferum* (Bréb.) Nordst. 27050244; *C. botrytis* var. *paxillosporum* West *et* G.S.West 27050246; *C. botrytis* var. *subtumidum* Wittr. 27050247; *C. botrytis* var. *tumidum* Wolle 27050248; *C. broomei* (Thwaites) Ralfs 27050260; *C. caelatum* var. *hexagonum* West 27050272; *C. caelatum* var. *spectabile* (De Not.) Nordst. 27050273; *C. calcareum* Wittr. 27050280; *C. calcareum* var. *spetsbergense* (Børges.) Kurt Först. 27050282; *C. canaliculatum* West *et* G.S.West 27050290; *C. candianum* Delponte 27050300; *C. capitulum* J.Roy *et* Bisset 27050310; *C. capitulum* var. *groenlandicum* Borges. 27050312; *C. clepsydra* Nordst. 27050330; *C. coarctatum* West 27050340; *C. commissurale* var. *acutum* Bréb. 27050352; *C. commissulare* var. *crassum* Nordst. 27050353; *C. conspersum* var. *rotundatum* Wittr. 27050372; *C. conspersum* var. *subrotundatum* West 27050374; *C. contractum* var. *cracoviense* Racib. 27050382; *C. contractum* var. *gartanense* West *et* G.S.West 27050384; *C. contractum* var. *minutum* (Delp.) West *et* G.S.West 27050385; *C. controversum* West 27050390; *C. corbula* Bréb. 27050400; *C. coronatum* Cooke *et* Wills 27050410; *C. corribense* West *et* G.S.West 27050420; *C. corriense* Bisset 27050430; *C. costatum* Nordst. 27050440; *C. crenatum* var. *bicrenatum* Nordst. 27050452; *C. cristatum* Ralfs 27050470; *C. cucumis* var. *helveticum* Nordst. in Wittr. & Nordst. 27050482; *C. cyclicum* var. *nordstedtianum* (Reinsch) West *et* G.S.West 27050493; *C. cylindricum* Ralfs 27050500; *C. cymatonotophorum* West 27050510; *C. cymatopleurum* Nordst. 27050520; *C. depressum* var. *minor* West 27050573; *C. depressum* var. *reniforme* West *et* G.S.West 27050575; *C. didymoprotupsum* West *et* G.S.West 27050590; *C. difficile* var. *minutissimae* (Wołosz.) Brook *et* D.B.Will. 27050602; *C. difficile* var. *sublaeve* Lütkem. 27050603; *C. distichum* Nordst. 27050610; *C. dovrense* Nordst. in Wittr. & Nordst. 27050620; *C. dybowskii* Gutw. 27050630; *C. eductum* J.Roy *et* Bisset 27050640; *C. eichlerianum* (Grönblad) Messik. 27050650; *C. elegantissimum* P.Lundell 27050660; *C. entochondrum* West *et* G.S.West 27050670; *C. etchachanense* J.Roy *et* Bisset 27050680; *C. excavatum* Nordst. 27050690; *C. excavatum* var. *duplomaius* (Wille) Kurt Först. 27050692; *C. exiguum* W.Archer 27050700; *C. exiguum* var. *pressum* West *et* G.S.West 27050702; *C. fastidiosum* West *et* G.S.West 27050710; *C. fictopraemorsum* Kurt Först. 27050720; *C. flavum* J.Roy *et* Bisset 27050730; *C. formosulum* var. *nathorstii* (Boldt) West *et* G.S.West 27050752; *C. furcatospermum* West *et* G.S.West 27050760; *C. garrolense* J.Roy *et* Bisset 27050780; *C. gayanum* De Toni 27050790; *C. gayanum* var. *eboracense* West *et* G.S.West 27050792; *C. geminatum* P.Lundell 27050800; *C. geometricum* West *et* G.S.West 27050810; *C. goniodes* West *et* G.S.West 27050820; *C. goniodes* var. *subturgidum* West *et* G.S.West 27050822; *C. goniodes* var. *variolatum* West *et* G.S.West 27050823; *C. granatum* var. *elongatum* Nordst. 27050832; *C. granatum* var. *grunowii* J.Roy *et* Bisset 27050833; *C. granatum* var. *nordstedtii* Hansg. 27050934; *C. grantii* J.Roy *et* Bisset 27050840; *C. granulatum* West 27050850; *C. hammeri* Reinsch 27050860; *C. hammeri* var. *homalodermum* (Nordst.) West *et* G.S.West 27050862; *C. hammeri* var. *protuberans* West *et* G.S.West 27050863; *C. helcangulare* Nordst. 27050870; *C. hexalobum* Nordst. 27050880; *C. hexalobum* var. *minus* J.Roy *et* Bisset 27050882; *C. holmiense* var. *attenuatum* Gutw. 27050892; *C. holmiense* var. *undatum* West *et* G.S.West 27050895; *C. humile* var. *danicum* (Børges.) Schm. 27050902; *C. humile* var. *glabrum* Gutw. 27050903; *C. humile* var. *striatum* (Boldt) Schm. 27050904; *C. humile* var. *substriatum* (Nordst.) Schm. 27050905; *C. impressulum* var. *suborthogonum* (Racib.) West. *et* G.S.West 27050914; *C. inconspicuum* West *et* G.S.West 27050920; *C. intermedium* Delponte 27050930; *C. isthmium* West 27050940; *C. isthmium* var. *horizontale* Schm. 27050943; *C. isthmochondrum* Nordst. 27050950; *C. isthmochondrum* var. *decussiferum* (Borge) Croasdale 27050952; *C. isthmochondrum* var. *pergranulatum* West *et* G.S.West 27050953; *C. jenisejense* Boldt 27050960; *C. kjellmani* Wille 27050970; *C. kjellmani* var. *grande* Wille 27050972; *C. kjellmani* var. *ornatum* Wille 27050973; *C. kotilainense* Grönblad 27050990; *C. laeve* var. *cymatium* West *et* G.S.West 27051002; *C. laeve* var. *depressum* Croasdale 27051003; *C. laeve* var. *octangulare* (Wille) West *et* G.S.West 27051004; *C. laeve* var. *westii* Willi Krieg. *et* Gerloff 27051005; *C. latifrons* P.Lundell 27051010; *C. lepidum* West 27051020; *C. logiense* Bisset 27051040; *C. lundellii* Delponte 27051050; *C. lundellii* var. *corruptum* (W.B.Turner) West *et* G.S.West 27051052; *C. lundellii* var. *ellipticum* West 2705153; *C. luxuriosum* (Kouwets) Coesel 27051060; *C. melanosporum* W.Archer 27051090; *C. meneghinii* var. *nanum* Wille 27051100; *C. microsphinctum* Nordst. 27051110; *C. microsphinctum* var. *maius* J.Roy *et* Bisset 2705112; *C. microsphinctum* var. *parvulum* (Wille) Willi Krieg. *et* Gerloff 2705113; *C. minimum* West *et* G.S.West 27051120; *C. moniliforme* var. *limneticum* West *et* G.S.West 27051132; *C. moniliforme* var. *subpyriforme* West *et* G.S.West 27051133; *C. monochondrum* Nordst. 27051140; *C. monomazum* P.Lundell 270511150; *C. morsum* West 27051160; *C. nagelianum* Bréb. 27051170; *C. nasutum* Nordst. 27051180; *C. nasutum* var. *asperum* West *et* G.S.West 27051182; *C. nitidulum* De Not. 27051190; *C. norimbergense* Reinsch 27051200; *C. norimbergense* var. *depressum* (West *et* G.S.West) Willi Krieg. *et* Gerloff 27051202; *C. norimbergense* var. *sublobatiforme* (Grönblad) Willi Krieg. *et* Gerloff 27051203; *C. notabile* Bréb. 27051210; *C. novae-semliae* Wille 27051220; *C. novae-semliae* var. *granulatum* (Schm.) Schm. 27051222; *C. obliquum* Nordst. 27051260; *C. obliquum* var. *ovale* Grönblad 27051261; *C. obliquum* var. *tatricum* (Gutw.) Willi Krieg *et* Gerloff 270501263; *C. obliquum* var. *trigonum* West 27051262; *C. oblongum* A.W.Benn. 27051270; *C. obsoletum* (Hantzsch) Reinsch 27051275; *C. obtusatum* Schm. 27051280; *C. obtusatum* var. *beanlandii* West *et* G.S.West 27051282; *C. ocellatum* Eichl. *et* Gütw. 2705129; *C. ocellatum* var. *incrassatum* West *et* G.S.West 27051292; *C. ochthodes* Nordst. 27051300; *C. ochthodes* var. *amoebum* West 27051302; *C. ochthodes* var. *subcirculare* Wille 27051303; *C. oligogongrus* Reinsch 27051310; *C. orbiculatum* Ralfs 27051320; *C. ordinatum* (Börges.) West *et* G.S.West 27051325; *C. ornatum* var. *perornatum* Grönblad 27051332; *C. orthogonum* Delponte 27051340; *C. orthistichum* P.Lundell 27051350; *C. orthistichum* var. *compactum* West *et* G.S.West 27051352; *C. orthistichum* var. *pumilum* P.Lundell 27051353; *C. ovale* var. *subglabrum* West *et* G.S.West 27051362; *C. pachydermum* var. *aethiopicum* West *et* G.S.West 27051372; *C. perforatum* P.Lundell 27051390; *C. pericymatium* Nordst. 27051400; *C. pericymatium* var. *eboracense* West *et* G.S.West 27051402; *C. perpusillum* West 27051410; *C. phaseolus* Bréb. in Ralfs

27051420; *C. phaseolus* var. *elevatum* Nordst. 27051422; *C. phaseolus* var. *skujae* Willi Krieg. *et* Gerloff 27051424; *C. plicatum* Reinsch 27051430; *C. plicatum* var. *hibernicum* West 27051431; *C. pokornyanum* (Grunow) West *et* G.S.West 27051440; *C. pokornyanum* var. *taylorii* Grönblad 27051442; *C. polygonum* (Nägeli) W.Archer 27051450; *C. portianum* var. *nephroideum* Wittr. 27051460; *C. praegrande* P.Lundell 27051470; *C. prominulum* Racib. 27051500; *C. promontorium* West *et* G.S.West 27051510; *C. protractum* (Nägeli) de Bary 27051520; *C. protuberans* P.Lundell 27051530; *C. pseudamoenum* Wille 27051540; *C. pseudamoenum* var. *basilare* Nordst. 27051542; *C. pseudarctoum* Nordst. 27051545; *C. pseudatlantoideum* West 27051550; *C. pseudobroomei* Wolle 27051570; *C. pseudobroomei* var. *convexum* West *et* G.S.West 27051572; *C. pseudobiremum* Boldt 27051560; *C. pseudoconnatum* var. *constrictum* West 27051582; *C. pseudoconnatum* var. *ellipsoideum* West *et* G.S.West 27051583; *C. pseudoexiguum* Racib. 27051590; *C. pseudonitidulum* Nordst. 27051600; *C. pseudonitidulum* var. *validum* West *et* G.S.West 27051602; *C. pseudoprotuberans* Kirchn. 27051620; *C. pseudoprotuberans* var. *alpinum* Racib. 27051622; *C. pseudopyramidatum* var. *carniolicum* Lütkem. 27051632; *C. pseudopyramidatum* var. *lentiferum* Taylor 27051633; *C. pseudopyramidatum* var. *maximum* Borges 27051634; *C. pulcherrimum* Nordst. 27051640; *C. punctulatum* var. *granusculum* (J.Roy *et* Bisset) West *et* G.S.West 27051652; *C. punctulatum* var. *mesoleium* Racib. 27051653; *C. punctulatum* var. *rotundatum* G.A.Klebs 2705165; *C. pusillum* (Bréb.) W.Archer in Pritch. 27051660; *C. pycnochondrum* Nordst. 27051670; *C. pyramidatum* var. *angustatum* West *et* G.S.West 27051692; *C. quadratulum* (F.Gay) De Toni 27051700; *C. quadratulum* var. *subimpressulum* (Dick) Willi Krieg. *et* Gerloff 27051702; *C. quadratum* var. *angustatum* West *et* G.S.West 27051712; *C. quadratum* var. *willei* (Schm.) Willi Krieg. *et* Gerloff 27051713; *C. quadrimamillatum* West *et* G.S.West 27051740; *C. quadrifarium* var. *octostichum* (Nordst.) Kurt Först. 27051734; *C. quadrum* P.Lundell 27051750; *C. quasillus* P.Lundell 27051760; *C. quaternarium* Nordst. 27051770; *C. radiosum* Wolle 27051790; *C. ralfsii* var. *rotundatum* West 27051803; *C. rectangulare* Grün. in Rabenh. 27051810; *C. rectangulare* var. *cambrense* (W.B.Turner) West *et* G.S.West 27051812; *C. rectangulare* var. *hexagonum* (Elfving) West *et* G.S.West 27051813; *C. rectangulum* Reinsch 27051820; *C. regnellii* var. *minimum* B.Eichl. *et* Gutw. 27051832; *C. regnellii* var. *pseudoregnellii* (Messik.) Willi Krieg. *et* Gerloff 27051833; *C. regnesii* var. *montanum* Schm. 27051842; *C. regnesii* var. *polonicum* (Eichl. *et* Gutw.) Compère 27051844; *C. regnesii* var. *tritum* West 27051843; *C. reinschii* W.Archer 27051850; *C. reinschii* var. *eboracense* West *et* G.S.West 27051852; *C. reniforme* var. *apertum* West *et* G.S.West 27051862; *C. reniforme* var. *elevatum* West *et* G.S.West 27051864; *C. repandum* Nordst. 27051870; *C. repandum* var. *minus* (West *et* G.S.West) Willi Krieg. et Gerloff 27051872; *C. retusiforme* (Wille) Gutw. 27051880; *C. retusum* (Perty) Rabenh. 27051890; *C. retusum* var. *angustatum* West *et* G.S.West 27051892; *C. scopulorum* Borge 27051900; *C. scoticum* West *et* G.S.West 27051910; *C. sexangulare* P.Lundell 27051920; *C. sexangulare* var. *minus* J.Roy *et* Bisset 27051922; *C. sexnotatum* Gutw. 27051930; *C. sexnotatum* var. *tristriatum* (Lütkem.) Schm. 27051932; *C. simii* J.Roy *et* Bisset 27051940; *C. sinostegos* Schaarschm. 27041950; *C. sinostegos* var. *obtusius* Gutw. 27051952; *C. slewdrumense* J.Roy 27051960; *C. smolandicum* P.Lundell 27051970; *C. smolandicum* var. *angustatum* West *et* G.S.West 27051972; *C. solidum* Nordst. 27051980; *C. speciosissimum* Schm. 27051990; *C. speciosissimum* var. *validius* Nordst. 27051992; *C. speciosum* var. *biforme* Nordst. 27052002; *C. speciosum* var. *incrassatum* Insam *et* Willi Krieg. 27052003; *C. speciosum* var. *simplex* Nordst. 27052005; *C. sphagnicolum* West *et* G.S.West 27052020; *C. subalatum* West *et* G.S.West 27052025; *C. subarctoum* (Lagerh.) Racib. 27052060; *C. subarctoum* var. *punctatum* (West *et* G.S.West) Willi Krieg. *et* Gerloff 27052062; *C. subaversum* Börges. 27052070; *C. subbroomei* Schm. 27052080; *C. subcapitulum* West 27052090; *C. subcontractum* West *et* G.S.West 27052100; *C. subcostatum* var. *beckii* (Gutw.) West *et* G.S.West 27052112; *C. subcrenatum* var. *divaricatum* Wille 27052122; *C. subcucumis* Schm. 27052130; *C. subcylindricum* West 27062140; *C. subdanicum* West 27052150; *C. subexcavatum* West *et* G.S.West 27052160; *C. subexcavatum* var. *ordinatum* West *et* G.S.West 27052162; *C. subgranatum* (Nordst.) Lütkem. 27052170; *C. subgranatum* var. *borgei* Willi Krieg. 27052172; *C. subimpressulum* Borge 27052180; *C. subnotabile* Wille 27052190; *C. subprotumidum* Nordst. 27052200; *C. subprotumidum* var. *gregorii* (J.Roy *et* Bisset) West *et* G.S.West 27052202; *C. subquadrans* West *et* G.S.West 27052210; *C. subquadratum* Nordst. 27052220; *C. subraciborskii* Taft 27052230; *C. subreniforme* Nordst. 27052240; *C. subretusiforme* West *et* G.S.West 27052250; *C. subspeciosum* Nordst. 27052260; *C. subspeciosum* var. *validius* Nordst. 27052262; *C. subtrinodulum* West *et* G.S.West 2705227; *C. subtumidum* Nordst in Wittr. & Nordst. 27052280; *C. subtumidum* var. *groenbladii* Croasdale 27052282; *C. subtumidum* var. *klebsii* (Gutw.) West *et* G.S.West 27052283; *C. succissum* West 27052300; *C. succissum* var. *hians* Lütkem. 27052302; *C. synthlibomenum* West 27052310; *C. tatricum* Racib. 27052320; *C. tatricum* var. *novizelandicum* Nordst. 27052322; *C. tatricum* var. *spheruliferum* West 27052323; *C. taxichondriforme* Eichl. *et* Gutw. 27052330; *C. tenue* W.Archer 27052350; *C. tetrachondrum* P.Lundell 27052360; *C. tetragonum* (Nägeli) W.Archer 27052370; *C. tetragonum* var. *elegans* (J.Roy *et* Bisset) 27052372; *C. tetragonum* var. *heterocrenatum* West *et* G.S.West 27052376; *C. tetragonum* var. *intermedium* Boldt 27052373; *C. tetragonum* var. *lundellii* Cooke 27052374; *C. tetragonum* var. *ornatum* Willi Krieg. *et* Gerloff 27052375; *C. thwaitesii* Ralfs 27052390; *C. thwaitesii* var. *penioides* G.A.Klebs 27052392; *C. tinctum* var. *intermedium* Nordst. 27052402; *C. tinctum* var. *subretusum* Messik. 27052403; *C. tinctum* var. *tumidum* Borge 27052404; *C. trachydermum* West *et* G.S.West 27052410; *C. trachypleurum* P.Lundell 27052420; *C. trachypleurum* var. *minus* Racib. 27052422; *C. trafalgaricum* Wittr. in Wittr. & Nordst. 27052430; *C. trilobulatum* Reinsch 27052420; *C. truncatellum* Perty 27052450; *C. tuberculatum* W.Archer 27052460; *C. tumens* Nordst. 27052470; *C. tumidum* P.Lundell 27052480; *C. turneri* J.Roy 27052490; *C. turpinii* var. *eximium* West *et* G.S.West 27052502; *C. umbilicatum* Lütkem. 27052510; *C. undulatum* var. *minutum* Wittr. 27052522; *C. undulatum* var. *wollei* West 27052523; *C. ungerianum* (Nägeli) de Bary 27052530; *C. ungerianum* var. *subtriplicatum* West *et* G.S.West 27052532; *C. variolatum* var. *cataractarum* Racib. 27052542; *C. venustum* var. *hypohexagonum* West 27052553; *C. venustum* var. *majus* Wittr. 27052554; *C. vexatum* West 27052560; *C. vexatum* var. *lacustre* Messik. 27052563; *C. vexatum* var. *rotundatum* Messik. 27052562; *C. vogesiacum* Lemmerm. 27052570; *C. wittrockii* P.Lundell 27052580; *C. zonatum* P.Lundell 27052590.

Cosmocladium Brébisson 1856

Colonial desmid often enveloped in mucilage; individual cells relatively small, virtually indistiguishable from smooth-walled cells of the genus *Cosmarium*; cells interconnected by single or double mucilaginous threads united to form irregularly branched colonies of variable size; semicells ellipsoidal or subkidney-shaped, mucilaginous threads secreted through pores in wall above sinus.

Colonies, free-floating or less commonly attached to filamentous algae or submerged aquatics in ponds and lakes. Six species have been recorded from the British Isles but *C. saxonicum* is the most frequent.

Cosmocladium saxonicum de Bary 1865

27060050 **Pl. 143A** (p. 588)
Cells 18–20 μm wide, 22–27 μm long, 12–15 μm thick; deeply constricted, sinus acutely angled, widely open; semicells subelliptic to kidney-shaped, with convex apex, elliptic in vertical view; cells interconnected by 2 parallel mucilaginous strands on one side of isthmus to form free-floating colonies of 2–3 to 90 cells mostly within a clearly defined mucilaginous matrix; chloroplast only 1 per cell, axile, with 4 radiating parietal lobes and a single, central pyrenoid.

Subcosmopolitan, acidophile; occasionally moderately frequent in plankton of nutrient-poor ponds and lakes, sometimes entrapped amongst other algae in pond or lake margins, also with *Sphagnum* and *Utricularia* in bog pools; widespread in the English Lake District, Wales, Ireland and Scotland including the Outer Hebrides.

TAXA RECORDED FROM THE BRITISH ISLES BUT NOT DESCRIBED IN TEXT
Cosmocladium constrictum (W.Archer) W.Archer 27060010; *C. perissum* J.Roy *et* Bisset 27060020; *C. pulchellum* Bréb. 27060030; *C. pusillum* Hilse 27060040; *C. tuberculatum* Prescott 27060060 CD.

Docidium Brébisson in Ralfs 1848

Cells straight, subcylindrical with slightly convex or undulating lateral margins, tapering towards mostly narrower apices (in one species swollen so that apical breadth is as great as maximum breadth of cell); apices broadly truncate or subtruncate; isthmus shallow with a circle of small but distinctive folds or granular-looking ridges, rather like two small cog-wheels on either side; walls smooth, punctate; chloroplasts with 3–4 longitudinal ridges and a series of pyrenoids in axis.

Subcosmopolitan, acidophile; most frequent amongst *Sphagnum* in small bog pools. Only 3 species widespread in Europe.

1 Lateral margins of semicells not undulate ....*D. baculum*
1 Lateral margins of semicells undulate ....*D. undulatum*

Docidium baculum Brébisson in Ralfs 1848

27100010 **Pl. 131C** (p. 528)
Cells straight, subcylindrical, 11–16 μm wide, 200–430 μm long; semicells with prominent basal inflation behind which is often a slight constriction; lateral margins straight or very slightly convex tapering almost imperceptibly to unornamented, truncate or subtruncate apices, 7–12 μm wide; median constriction shallow and on each side a small but distinctive ring of granules, 5–7 visible in side view; chloroplast with 3–4 longitudinal ridges and an axile row of small pyrenoids.

Europe, acidophile; widespread in the British Isles, but never abundant, occurring most frequently in moorland pools and bogs, commonly associated with *Sphagnum* where other desmids present (pH 4.5–6); also in the littoral of some nutrient-poor lakes.

Docidium undulatum Bailey 1850

27100030 **Pl. 144H** (p. 590)
Cells comparatively long, narrow, 12–17 μm broad, 178–246 μm long; semicells with 7–8 undulations down each lateral margin; apices 11.5–15 μm broad, truncate with rounded angles; across base of each semicell a ring of 11 or 12 granules (about 4–6 visible in side view); walls smooth.

Subcosmopolitan, acidophile; in the British Isles mostly confined to western areas of Scotland and Ireland in bog pools with *Sphagnum* where at times it can be abundant (pH 4–6).

TAXA RECORDED FROM THE BRITISH ISLES BUT NOT DESCRIBED IN TEXT
Docidium nobile (P.G.Richt.) P.Lundell 27100020; *D. undulatum* var. *perundulatum* (West *et* G.S.West) Willi Krieg. 27100032.

Euastrum Ehrenberg ex Ralfs 1848

Cells distinctly compressed laterally, most with a closed sinus; apices mostly with a median incision, often deep and closed, or hardly evident to almost absent; lateral margins entire, sinuous or lobed; semicells in face view with characteristically disposed rounded protuberances or tumours, in apical view narrowly elliptical with one or more rounded protuberances on each side and sometimes with conspicuous scrobiculations between them; chloroplasts 1 in each semicell, often irregularly lobed; smaller species with a central pyrenoid, larger species with several scattered pyrenoids.

Forty-four species recorded in the British Isles and 31 varieties, but only 10–12 species considered common. All occur in acid waters, most frequently in peaty pools associated with *Sphagnum*, but also in the margins of nutrient-poor lakes and ponds.

The number and disposition of protuberances and pits (scrobiculations) are important taxonomic characters and best observed in cells without contents.

1 Polar lobe with a distinct median incision..........2
1 Polar lobe without a median incision..........18
2 Median incision deep and closed..........3
2 Median incision open, V-shaped..........14
3 Lateral margins distinctly lobed..........4
3 Lateral margins faintly lobed or sinuate..........8

4 Lateral margins with one swelling ....*E. ampullaceum*
4 Lateral margins 4(–5)-lobed. 5
5 Lateral margins with 2 angular lobes, lower subdivided, retuse 6
5 Lateral margins 4–5-lobed 7
6 Cells 19–21 μm wide × 26–33 μm long ....*E. dubium*
6 Cells 28–40 μm wide × 40–60 μm long ....*E. bidentatum*
7 Upper and lower lateral lobes retuse ....*E. oblongum*
7 Lower lobe retuse, upper lobe rounded ....*E. pinnatum*
8 Apical lobe poorly defined or absent 9
8 Apical lobe prominent 10
9 Apical lobe absent and apex flat; lateral margins straight or only slightly undulate ....*E. cuneatum*
9 Apical lobe indistinct, lateral margins with median undulation....................................*E. inerme*
10 Apical lobe extended well above lower part of cell 11
10 Apical lobe wedge-shaped, separated by deep narrow clefts 13
11 Apical lobe very prominent with a single well-developed lower lateral lobe ....*E. insigne*
11 Apical lobe subtruncate..12
12 Lateral lobes broadly rounded..................*E. didelta*
12 Lateral lobes absent................................*E. ansatum*
13 Apical lobe flat, broad; lateral margin usually slightly retuse ....*E. ventricosum*
13 Apical lobe broadly convex, anvil-shaped; margin of lower lateral margin broadly concave ....*E. crassum*
14 Angles of polar lobe ending in a prominent spine 15
14 Angles of polar lobe ending in a small granule 17
15 Semicells with one rounded lateral lobe 16
15 Semicells with one very small, subrectangular, retuse lateral lobe ....*E. elegans*
16 Lateral lobes rounded, occasionally retuse, without ornamentation ....*E. binale*
16 Lateral lobes with granular ornamentation and spiny protuberances.......................*E. denticulatum*
17 Lateral lobes small, ornamented, triundulate, apex poorly defined ....*E. binale* var. *gutwinski*
17 Lateral lobes slightly retuse, apex well-defined ....*E. insulare*
18 Walls smooth; lateral lobes subquadrate, retuse ....*E. pectinatum*
18 Walls granulate, highly ornamented; lateral lobes wedge-shaped and strongly granulate...................19
19 Lateral margins bilobulate with distinct divergent median lobes ....*E. verrucosum*
19 Lateral margins with only one lower lobe ....*E. verrucosum* var. *planctonicum*

Euastrum ampullaceum Ralfs 1848

27110020 **Pl. 131R** (p. 528)

Cells 50–65 μm wide, 85–110 μm long, deeply constricted, sinus linear; semicells 3-lobed, median polar lobe wedge-shaped with rounded angles, apex convex with narrow, deep median incision; lateral lobes 2, lower one triangular with acutely rounded basal angles, margins with median, mamillate swelling of variable size; face of semicells with 3 protuberances across base above isthmus, the lateral ones sometimes almost meeting, the other 2 protuberances on either side of centre of semicell with 1–3 small pits (scobiculations) disposed between the 5 protuberances; walls finely punctate.

Northern Europe, USA; acidophile; widely distributed in *Sphagnum* bogs throughout the British Isles (pH 4–6) though rarely abundant.

Euastrum ansatum Ralfs 1848

Synonym: *Euastrum ansatum* Ralfs vars *dideltiforme* F.Ducellier and *simplex* F.Ducellier

27110030 **Pl. 131Q** (p. 528)

Cells 32–51 μm wide, 70–110 μm long, with deep median constriction and narrow, linear sinus inflated outwards; semicells elongate broad-based pyramids with rounded lower angles; lateral margin convex in lower part with slight undulation above basal angles, upper part concave; apices subtruncate with narrow, fairly deep median incision; semicell in face view with one basal protuberance just above isthmus, 2 to side and 2 above (5 in all); walls colourless, punctate.

Cosmopolitan; acidophile; most common of all *Euastrum* species, reported as 'general and abundant' in Scotland including the Hebrides, Orkneys and Shetlands, also in Wales and Ireland, and recorded from all counties of England where suitably acid waters occur (W. & G.S.West 1905a). Růžička (1977) and Förster (1982) indicate occurrence in alkaline waters (pH 3.9–7.8(–8.6); also occurs as a chance plankter.

Euastrum bidentatum Nägeli 1849

Synonym: *Euastrum bidentatum* Nägeli var. *glabrum* Grönblad

27110040 **Pl. 136H** (p. 554)

Cells 28–40 μm wide, 40–60 μm long, approximately ovoid, deeply constricted with a narrow linear isthmus; semicells narrow very slightly from base to subtruncate, deeply incised apex; inner angles of apex sometimes thickened and with characteristic short blunt spine at each outer angle; semicells at base subrectangular, sometimes with 2 or 3 sharp granules; lateral margin with rounded or truncate protuberance and a granular ornamentation (very variable) in centre above the isthmus, also 1 or 2 large granules on each side of apical incision, other granules variously disposed within the lateral lobes and apex.

Cosmopolitan; acidophile (pH 5.5–6.5); in the British Isles generally distributed, common in Ireland, Scotland including the Hebrides, Orkneys and Shetlands, north-west Wales and the English counties where there are acid waters and peat bogs.

Several varieties described distinguished principally in the prominence of ornamentation on face of semicells; most common in the British Isles is var. *speciosum* (Boldt) Schmidle (see below).

var. *speciosum* (Boldt) Schmidle 1898
Basionym: *Euastrum elegans* (Brébisson) Kützing ex Ralfs var. *speciosum* Boldt
27110043 **Pl. 136I** (p. 554)
Differs from the nominal variety in that median ornamentation consists of a group of 5–7 granules, one central and 5 or 6 surrounding it; dimensions similar.

Distribution and ecology similar to the nominal variety.

Euastrum binale (Turpin) Ehrenberg ex Ralfs 1848

27110050 **Pl. 136D** (p. 554) **CD**
Cells small, 10–23 μm wide, 12–30 μm long, with deep, closed linear sinus; semicells trapeziform with basal angles broadly rounded, upper lateral margins concave narrowing to broad, truncate or subtruncate apex with shallow open V-shaped incision; angles of apex acute, sometimes pointed; semicell in face view with one median protuberance obvious only in side or apical view; walls delicately punctate.

Cosmopolitan; acidophile of peaty pools, especially with *Sphagnum* and *Utricularia*; also subaerial in damp mosses (pH 3.9–7); general and sometimes abundant in these habitats throughout the British Isles and found up to 1000 m at Lochnagar, Scotland.

A species with numerous forms and varieties described but only var. *gutwinski* common.

var. *gutwinski* (Schmidle) Homfeld 1929
Basionym: *Euastrum binale* fo. *gutwinski* Schmidle
27110052 **Pl. 136E** (p. 554)
Differs from the nominal variety in the form of the lateral margins which are tri-undulate, the polar lobe tending to be narrower with distinctly pointed angles; small granules sometimes ornamenting margins of lateral and polar lobes.

Found in similar environments to the nominal variety, but less widespread.

Euastrum crassum [Brébisson] Kützing ex Ralfs 1848

Synonym: *Euastrum crassum* [Brébisson] Kützing ex Ralfs var. *scrobiculatum* P.Lundell
27110110 **Pl. 136B** (p. 554) **CD**
Cells 75–90 μm wide, 140–180 μm long, oblong-elliptical with a deep, linear sinus; semicells unequally 3-lobed, lobes separated by a narrow, closed incision with thickened angles; lateral lobes having bases with acutely rounded, thickened angles; polar lobe small with convex apex deeply incised, also with thickened angles; semicells in face view with 3 fairly prominent protuberances across base (if well-developed may touch one another across isthmus); walls punctate with one or a group of pores in the middle.

Europe, North America, Arctic; acidophile; commonly associated with *Sphagnum* in peaty pools and bogs (pH 4–6); in the British Isles often abundant in Scotland, north-west Wales and Ireland; also present in northern counties of England, in Devon and Cornwall and other southern counties where suitable environments occur.

Euastrum cuneatum Jenner in Ralfs 1848

27110130 **Pl. 131P** (p. 528)
Cells 45–95 μm wide, 95–140 μm long, oblong-elliptical with angular corners, deeply constricted; sinus narrow, linear with dilated extremities; semicells narrow-pyramidate or trapeziform with rounded basal angles; lateral margins more or less straight, tapering to comparatively broad, truncate and fairly deeply incised apex, thickened at base; semicells in face view with 3 basal protrusions; walls punctate.

Acidophile, restricted to northern Europe and North America; mostly associated with other desmids amongst *Sphagnum* in peaty pools (pH 3.5–5.5). Widespread and sometimes abundant in these habitats throughout the British Isles.

Euastrum denticulatum F.Gay 1884

Synonyms: *Euastrum denticulatum* F.Gay var. *granulatum* West, *E. amoenum* F.Gay
27110140 **Pl. 136F** (p. 554)
Cells 14–25 μm broad, 17–30 μm long, with 3 angular lobes and closed isthmus; semicells subquadrate with basal lobes rounded and angular, corners pointed; 3 or 4 evenly spaced protrusions on lateral walls narrowing halfway to a deep, comparatively broad, rectangular upper lobe; upper lobe with pointed angles and a shallow, open median incision; semicell face view with distinctive granulated central protrusion, sometimes with 2 small granules beneath protrusion and also emarginate granules decorating basal and upper angles.

Cosmopolitan; acidophile (pH 4–7); widely distributed and sometimes abundant in peaty pools with *Sphagnum*, associated with other desmids throughout the British Isles.

Very variable in size and form, but basic features always distinctive.

Euastrum didelta Ralfs ex Ralfs 1848

27110150 **Pl. 131S** (p. 558)
Cells 60–75 μm wide, 120–150 μm long, with a deep, narrow sinus dilated at extremity; semicells pyramidate with sinuous margins sometimes 5-lobed; basal angles rounded or rectangular-rounded; lateral margins with 2 hollows, sometimes deep and unequal; polar lobe with a truncate apex, rounded angles and a deep, median linear incision; semicell face with 3 protuberances adjacent to isthmus and 2 across middle; walls punctate.

Cosmopolitan; acidophile; most commonly associated with *Sphagnum* in peaty pools (pH 3.9–6); one of the most widespread and abundant of the British *Euastrum* species.

Shows considerable morphological variability.

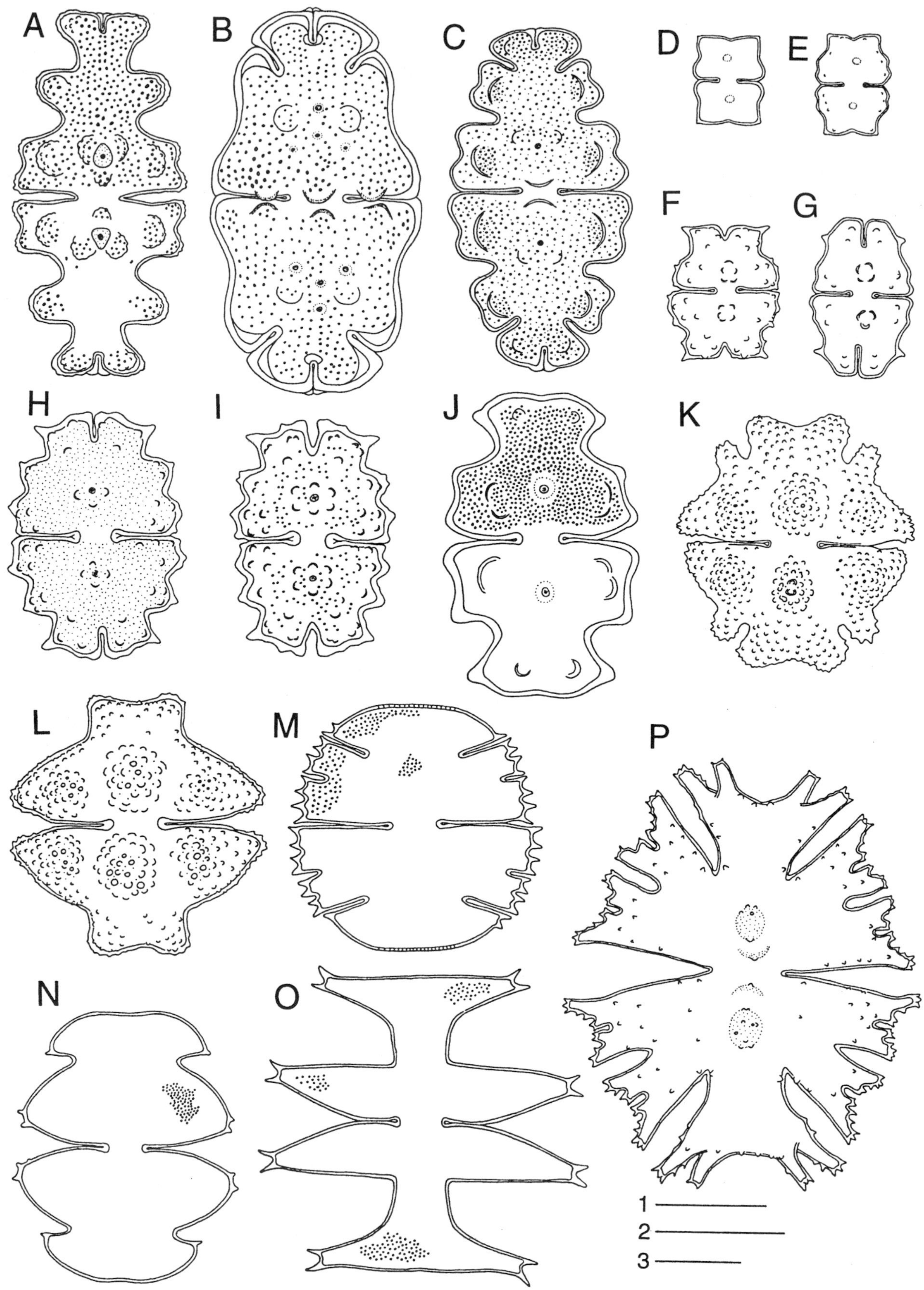

Plate 136 A–P

A. *Euastrum pinnatum* (p. 556); **B**. *Euastrum crassum* (p. 553); **C**. *Euastrum oblongum* (p. 555); **D**. *Euastrum binale* (p. 553); **E**. *Euastrum binale* var. *gutwinski* (p. 553); **F**. *Euastrum denticulatum* (p. 553); **G**. *Euastrum elegans* (p. 555); **H**. *Euastrum bidentatum* (p. 552); **I**. *Euastrum bidentatum* var. *speciosum* (p. 553); **J**. *Euastrum pectinatum* (p. 555); **K**. *Euastrum verrucosum* (p. 556); **L**. *Euastrum verrucosum* var. *planctonicum* (p. 556); **M**. *Micrasterias truncata* (p. 561); **N**. *Micrasterias oscitans* (p. 559); **O**. *Micrasterias pinnatifida* (p. 559); **P**. *Micrasterias americana* (p. 558).

Scale bars: 1, 25 μm; 2, 3, 50 μm. 1 for D, E–J, O; 2 for A, B, K–M, P; 3 for C, N.

Euastrum dubium Nägeli 1849

Synonym: *Euastrum binale* Ralfs

27110170 **Pl. 144N** (p. 590)

Cells only 19–21 μm wide, 26.5–33 μm long, deeply constricted with a narrow linear sinus slightly dilated towards outside; isthmus 6–7 μm broad; semicells truncate-pyramidate, with 5 lobes; polar lobe short, oblong-rectangular with truncate apex and narrow median incision; angles with small conical granules; upper lateral lobes rounded, lower lobes rounded and slightly larger than upper; semicells with 2 granules within apex on each side of apical notch and 2 small granules within each basal angle below notch.

Cosmopolitan; acidophile (pH 5–7); widely distributed in the British Isles especially in peat bogs with *Sphagnum* and *Utricularia*; common in mountain districts in Scotland and Wales; occurring in the littoral of lakes and ponds and as a chance plankter.

Euastrum elegans [Brébisson] Kützing ex Ralfs 1848

27110180 **Pl. 136G** (p. 554)

Cells 17–22 μm wide, 27–40 μm long, with a deep closed sinus; semicells with rounded to subrectangular basal angles, above the lateral margins are 2 small undulations then narrowing to convex or broadly rounded apex with a shallow median incision, sometimes open at base; upper angles usually with short, stout papilla; semicell face with a granular or triverrucose protrusion in centre above isthmus and emarginate granules near basal angles and the apical incision; walls punctate.

Cosmopolitan; acidophile (pH 4–7); one of the most common *Euastrum* species; very widely distributed in suitable acidic habitats throughout the British Isles; sometimes associated with *Sphagnum*.

A very variable species with numerous varieties described, some of doubtful validity.

Euastrum inerme (Ralfs) P.Lundell 1871

Basionym: *Euastrum elegans* [Brébisson] Kützing ex Ralfs var. *inerme* Ralfs

27110250 **Pl. 144 L** (p. 590)

Cells subelliptical, 30–40 μm wide, 50–62 μm long, very deeply constricted with a narrow linear sinus; isthmus 7.5–13 μm wide; semicells subpyramidate, widest just above base; lateral margins with 2 slight hollows with a broad, rounded projection between; apex slightly extended with a deep, narrow, median notch; small thickened protuberances in middle of each semicell above isthmus and 2 across the middle, with a large scrobiculation between them; walls finely punctate.

Restricted to a temperate distribution; rare in England and Wales but often present in abundance in acid waters, amongst *Sphagnum* in the west of Scotland, the Hebrides and the west of Ireland.

Euastrum insigne Hassall ex Ralfs 1848

27110260 **Pl. 144M** (p. 590)

Cells 50–70 μm wide, 100–135 μm long, constriction deep, sinus wide open, narrowing inwards; isthmus 13–15.5 μm wide; semicells broad at base with rounded margins, narrowing into a distinctive neck then broadening to a less wide apical cap with rounded angles; apex 29–37 μm wide with a narrow median incision; basal angles with large mammillate, downwardly directed protuberances projecting over sinus; walls scrobiculate.

Temperate distribution, distinct acidophile (pH 3.5–5); often abundant in mountain bogs and margins of tarns, especially amongst leaves of *Sphagnum cuspidatum*. Widely distributed throughout the British Isles in appropriate habitats.

Euastrum insulare (Wittrock) J.Roy 1877

Basionym: *Euastrum binale* (Turpin) Ehrenberg ex Ralfs var. *insulare* Wittrock

27110270 **Pl. 144O** (p. 590) **CD**

Cells small, 11.5–22 μm wide, 17.5–30 μm long, deeply constricted with a linear sinus; isthmus 3.3–6 μm wide; semicells 3-lobed, with subrectangular interlobular contours; apex truncate, retuse in middle with only a slight invagination; lateral lobes short, basal angles subrectangular, sides slightly retuse.

Cosmopolitan, in acid to somewhat alkaline waters (pH 4.5–7.5); widespread in the British Isles, mostly in acid waters where at times frequent.

Euastrum oblongum [Greville] Ralfs ex Ralfs 1848

Synonyms: *Euastrum oblongum* [Greville] Ralfs ex Ralfs var. *cephalophorum* West *et* G.S.West, *E. oblongum* [Greville] Ralfs ex Ralfs var. *depauperatum* West *et* G.S.West

27110340 **Pl. 136C** (p. 554)

Cells 65–85 μm wide, 140–180 μm long, oblong-elliptical, deeply constricted with a narrow linear sinus; semicells 5-lobed with deep, usually slightly open, incisions between lobes; polar lobe a broad wedge with rounded angles and a convex apex with a deep, narrow, median incision; upper and lower lateral lobes of similar shape being subquadrate with retuse margins and rounded angles; lower lobe broader and longer than upper; semicell face with distinctive protuberance within each of the lateral lobes, also protuberance at base just above isthmus and 2 on either side and above, often with a large scrobiculation between the 3 protuberances; walls punctate.

Cosmopolitan; occurring in a range of water types, even those which are somewhat alkaline and nutrient-rich (pH 3.9–7.5); in the British Isles restricted to acid waters, widely distributed, rarely abundant.

Euastrum pectinatum Brébisson ex Brébisson in Ralfs 1848

Synonym: *Euastrum pectinatum* Brébisson ex Brébisson var. *inevolutum* West *et* G.S.West

27110350 **Pl. 136J** (p. 554)

Cells 40–50 μm wide, 55–70 μm long, deeply constricted with a narrow, sometimes slightly open sinus; semicells 3-lobed; polar lobe anvil-shaped or dilated with acutely rounded angles and convex-truncate apex, retuse in middle without a median incision; lateral lobes subquadrate, retuse in mid-region and with acutely rounded angles; semicells in face view with 3 distinctive protuberances across the broadest point, 2 within polar lobe; walls with fine puncta, or apparently smooth.

Cosmopolitan; acidophile; sometimes common in bogs and peaty pools with other desmids, usually associated with *Sphagnum* (pH 4.5–6.5), in all suitable parts of the British Isles.

W. & G.S.West (1905a) comment that the reduced form (their var. *inevolutum*) is much more common than the nominal variety.

Euastrum pinnatum Ralfs 1848

27110380 **Pl. 136A** (p. 554) **CD**

Cells 65–75 μm wide, 125–150 μm long, deeply constricted with a narrow linear sinus dilated at extremities; semicells 5-lobed with moderately deep, wide incisions between lobes; polar lobe anvil-shaped with rounded angles extending well above lower lateral lobes; apex flat (rarely retuse or convex) and a fairly deep but narrow median incision; upper lateral lobes broadly mammilate with upper margin at right angles to longitudinal cell axis; lower lateral lobes subquadrate, margins retuse with rounded angles; semicells in face view with 2 large protuberances within each of the lateral lobes and 3 smaller ones at base near semicell centre, one just above the isthmus, 2 others on either side above it forming a triangle; walls coarsely punctate, often all angles and protuberances scrobiculate or rough.

Temperate distribution; an acidophile, very widespread throughout the British Isles, abundant in parts of the west of Scotland and Ireland.

Euastrum ventricosum P.Lundell 1871

27110450 **Pl. 144P** (p. 590)

Cells 52–80 μm wide, 80–136 μm long, deeply constricted with a narrow linear sinus; isthmus 18–28 μm wide; semicells semielliptical, unequally 3-lobed by narrow incisions; polar lobe smaller than laterals, strongly convex with a deep, narrow median incision; lateral lobes bilobulate; semicells with 3 tumours across base and 2 across middle; walls punctate.

Subcosmopolitan; acidophile; usually associated with *Sphagnum* in peaty pools and bogs (pH 4.6–6); widespread in the British Isles, especially parts of the west of Scotland and Ireland, also the Islands of Lewis, Harris and Skye; sometimes abundant.

Euastrum verrucosum Ehrenberg ex Ralfs 1848

27110460 **Pl. 136K** (p. 554)

Cells robust, 70–90 μm wide, 80–110 μm long, subhexagonal, deeply constricted with sinus open for half length, then narrow-linear; semicells 3-lobed with deep, open incisions between lobes; polar lobe wide and wedge-shaped with rounded, granulate angles and a deeply retuse apex; lateral lobes wider than the polar, lower basal angles horizontally directed, rounded and granulate, lateral margin converging strongly to upper sublobe; upper lobes also broadly rounded and granulate, directed upwards and diverging; semicells in face view with one large central protrusion with distinctive wart-like granules arranged in concentric circles one on each side in lower lobe and having similar wart-like ornamentation; walls granulate, granules most obvious around each angle where commonly sharp and conical.

Essentially a northern hemisphere acidophile (pH 4.9–7); not very common in bogs, but widely distributed in the north and west of the British Isles in the margins of large ponds and lakes associated with submerged macrophytes; frequently occurs as a chance plankter.

The var. *planctonicum* (27110464) described by West & G.S.West (1903a), with its open sinus and simple lateral lobes, may be frequent in the plankton of some Scottish lochs (see Pl. 136L, p. 554).

TAXA RECORDED FROM THE BRITISH ISLES BUT NOT DESCRIBED IN TEXT

Euastrum aboense Elfving 27110010; *E. boldtii* Schm. 27110060; *E. brevisinuosum* (Nordst.) Kouwets 27110070; *E. brevisinuosum* var. *dissimile* (Nordst.) Kouwets 27110072; *E. cornubiense* West *et* G.S.West 27110080; *E. crassangulatum* Børges. 27110090; *E. crassangulatum* var. *ornatum* West 27110092; *E. crassicolle* P.Lundell 27110100; *E. crispulum* (Nordst.) West *et* G.S.West 27110120; *E. divaricatum* P.Lundell 27110160; *E. dubium* var. *anglicanum* (W.B.Turner) West *et* G.S.West 27110172; *E. dubium* var. *cambrense* (W.B.Turner) West *et* G.S.West 27110173; *E. dubium* var. *ornatum* Wołosz. 27110174; *E. elegans* var. *pseudelegans* (W.B.Turner) West *et* G.S.West 27110185; *E. elobatum* (P.Lundell) Kouwets 27110190; *E. elobatum* var. *subelobatum* West 27110192; *E. erosum* P.Lundell 27110200; *E. gayanum* De Toni 27110210; *E. gemmatum* (Bréb.) Bréb. ex Ralfs 27110220; *E. groenbladii* (Messik.) Coesel 27110230; *E. humerosum* Ralfs 27110240; *E. humerosum* var. *affine* (Ralfs) G.C.Wall. 27110242; *E. intermedium* Cleve 271102280; *E. jenneri* W.Archer 27110290; *E. lacustre* (Messik.) Coesel 27110300; *E. montanum* West *et* G.S.West 27110320; *E. obesum* Joshua 27110330; *E. pictum* Børges. 27110360; *E. pingue* Elfving 27110370; *E. pulchellum* Bréb. 27110390; *E. sinuosum* Lenorm. ex W.Archer 27110400; *E. subalpinum* Messik. 27110410; *E. subalpinum* var. *crassum* Messik. 27110412; *E. sublobatum* Bréb. 27110420; *E. turneri* West 27110430; *E. validum* West *et* G.S.West 27110440; *E. verrucosum* var. *alatum* Wolle 27110462; *E. verrucosum* var. *coarctatum* Delponte 27110463.

Haplotaenium Bando 1988

Cells straight, baton-shaped, significantly longer than broad with a slight median constriction; apices truncate to subtruncate; semicells with a single basal inflation above the median constriction; chloroplasts 1 in each semicell, with 2–4 longitudinal ridges or an axile ribbon, both types with a series of pyrenoids along length; walls smooth with very small, closely disposed puncta; cells lack apical or median ornamentation; terminal vacuoles also lacking.

Subcosmopolitan; three British species, two of which are fairly common and confined to acid waters in bogs with *Sphagnum* and in the margins of nutrient-poor ponds and lakes.

The ridged axile or ribbon-like chloroplasts together with the frequency of mucilage pores and lack of terminal vacuoles distinguish the genus from *Pleurotaenium* whose chloroplasts are parietal ribbons.

1 Semicells without basal inflation........*H. minutum*
1 Semicells with small basal inflation ....*H. rectum*

Haplotaenium minutum (Ralfs) Bando 1988

Basionym: *Docidium minutum* Ralfs
Synonyms: *Penium minutum* (Ralfs) Cleve, *Pleurotaenium minutum* (Ralfs) Delponte
27170010 **Pl. 131D** (p. 528)
Cells 10–18 μm wide, 80–170 μm long, straight, approximately cylindrical with a distinct median constriction; semicells without a basal inflation, attenuating gradually from a constriction towards the truncate apices, 8–15 μm wide; walls colourless, smooth, very delicately punctate; chloroplasts mostly with 2–4 irregular longitudinal ridges but sometimes ribbon-like with small pyrenoids in an irregular series.

Cosmopolitan, acidophile, often occurring in pools with *Sphagnum* and in acid bogs (pH 4–6.5); widely distributed, especially in western areas of the British Isles. In England most common in the English Lake District, Devon and Cornwall but also found in Yorkshire, Hampshire and Sussex; widespread in West Wales; frequent in the north-west of Scotland and the Hebrides.

A very variable species with a considerable number of varieties described, some of doubtful validity.

Haplotaenium rectum (Delponte) Bando 1988

Basionym: *Pleurotaenium rectum* Delponte
Synonym: *Pleurotaenium trabecula* (Ehrenberg) ex Nägeli var. *rectum*
27170020 **Pl. 131E** (p. 528)
Cells 16–23 μm wide, 220–350 μm long, straight, subcylindrical; semicells usually with only one basal inflation above the isthmus, sometimes with a ring of thickening, tapering evenly to broad truncate or subtruncate apices, 12–20 μm wide, with a small internal thickening; walls smooth with fine, scattered pores; chloroplasts with 3–4 longitudinal ridges and an axile row of pyrenoids.

Possibly cosmopolitan; widespread in Scotland and Ireland growing in association with *Sphagnum* in acid pools and bogs (pH 5–6.5), also in the littoral of nutrient-poor lakes; less abundant in England.

TAXA RECORDED FROM THE BRITISH ISLES BUT NOT DESCRIBED IN TEXT

Haplotaenium minutum var. *crassum* (West) Bando 27170012; *H. rectum* var. *foersteri* Bando 27170022. The var. *rectissimum* (as *Pleurotaenium rectum* var. *rectissimum* West *et* G.S.West) has also been found but should be transferred to *Haplotaenium*.

Micrasterias C.Agardh ex Ralfs 1848

Cells strongly compressed so that almost always observed in face view, approximately circular in outline and multilobed appearing like small, dissected leaves or little stars as the generic name indicates; dissections always result in an odd number of lobes to each semicell, the middle polar one well differentiated from the others; comparatively large cells have deep, median, open or closed incisions on each side of a broad isthmus in which the nucleus lies and joins adjacent semicells; semicells divided into distinctive polar lobes and 2 lateral lobes, then subdivided into sublobes or lobules, lobes divided once are described as 'divided to the first order', when divided again 'divided to the second order', up to the 5th order, with clefts between successive divisions progressively shallower; polar lobe variously expanded at its apex, sometimes having appendages such as spines, but unlike the lateral lobes, never subdivided; chloroplasts one in each semicell and conforming to the internal semicell contours and containing numerous scattered pyrenoids.

Essentially acidophiles; occur in desmid-rich regions but rarely in abundance. Some species present around the margins of nutrient-poor lakes attached to submerged macrophytes; also as chance plankters. On rare occasions species found in abundance on the mud surfaces of shallow lakes and ponds (Brook & Williamson, 1988b).

Micrasterias includes the most striking of all desmids, if not all plant and animal cells. Eighteen species have been recorded from the British Isles and for many, several varieties have been described, although some are of doubtful validity.

1 Polar lobe flat or slightly retuse; lateral lobes 2, transversely placed; cells ovoid to rectangular 2
1 Polar lobe with a distinct median notch of varying width and depth 5
2 Polar lobe flat, not retuse 3
2 Polar lobe retuse..4
3 Lateral lobes ovate-triangular, minutely bifid; cells 125–146 μm long ....*M. oscitans*
3 Lateral lobes semispindle-shaped with bifid apices; cells 53–76 μm long...............*M. pinnatifida*
4 Lateral lobes bilobulate with short teeth; cells 87–138 μm long ....*M. truncata*
4 Lateral lobes bilobulate without teeth, each lobule retuse; cells 132–170 μm long ......*M. jenneri*
5 Lateral lobes radially disposed, much dissected and widening outwards..6
5 Lateral lobes with few divisions and widely open incisions 11
6 Cells oval to subcircular; lateral lobes with 2 lobules each subdivided into 2 dentate lobules......7
6 Cells circular, upper lateral lobes with up to 8 equal lobules, lower lobes with 4 lobules ....*M. radiosa*
7 Cell walls with spines on surface and/or down margins...8
7 Cell walls smooth, without spines on surface or down margins 9

8 Tips of polar lobes with a pair of divergent spines and additional small spines on each side of the apical notch; cells 220–294 μm long ....*M. fimbriata*
8 Wall with small spines or rows of granules on each side of median incision; cells 118–145 μm long ....*M. papillifera*
9 Polar lobes slightly extended, with parallel sides expanding gradually to a retuse apex ....*M. rotata*
9 Polar lobes not extended......................................10
10 Tips of lobes retuse, not dentate or scarcely so ....*M. denticulata*
10 Tips of lobes with V-shaped incisions and dentate; 3 projections on face of each semicell above isthmus ....*M. thomasiana*
11 Lateral lobes short, comparatively stout; polar lobe without accessory processes ....*M. crux-melitensis*
11 Lateral lobes long; polar lobe with accessory processes 12
12 Lateral lobes divided by shallow incision and subdivided into 2 small lobules ....*M. americana*
12 Lateral lobes divided by deep, wide incision and only upper lobule further subdivided ....*M. mahabuleshwarensis* var. *wallichii*

Micrasterias americana Ehrenberg ex Ralfs 1848

27200010 **Pl. 136P** (p. 554)

Cells 100–130 μm wide, 125–150 μm long, deeply constricted with an open sinus, acute towards apex; semicells 5-lobed with conspicuously protruding polar lobe, retuse apex and angles produced into stout, diverging processes with ends truncate and denticulate, from base of each process arise 2 similar accessory processes, one on each side; lateral lobes divided into 2 lobules with comparatively shallow incisions dissecting them; walls with scattered granules and small granulate protuberance above isthmus.

Subcosmopolitan, acidophile (pH 4.7–7.5); mostly in boggy pools with *Sphagnum*; also a chance plankter in nutrient-poor lakes, especially in Scotland.

Micrasterias crux-melitensis [Ehrenberg] Hassall ex Ralfs 1848

27200050 **Pl. 137A** (p. 560)

Cells 90–120 μm wide, 100–130 μm long, deeply constricted by an open sinus acute at closed end; semicells 5-lobed with open incisions between each; lower part of polar lobe subquadrate, upper dilated with a wide, retuse apex; angles produced into short diverging processes with bidenticulate apices; lateral lobes divided into 2 elongate-quadrate lobules, bidentate at apices; walls delicately punctate.

Subcosmopolitan; occurring in the British Isles in acid waters often associated with *Sphagnum* although elsewhere reported from neutral or even alkaline, nutrient-rich waters (pH 5.2–8) (see Huber-Pestalozzi, 1982); also in littoral of nutrient-poor lakes and as a chance plankter. Widespread but rarely common in the British Isles.

Micrasterias denticulata Brébisson ex Ralfs 1848

27200060 **Pl. 137D** (p. 560) **CD**

Cells almost circular, 180–200 μm wide, 200–280 μm long, with short lateral margins, parallel or slightly concave, indented at deep, narrow, linear sinus; semicells 5-lobed, incision between each, linear and almost closed; polar lobe sometimes protruding beyond circular contour of cell and comparatively narrow with a slightly flared retuse apex, with a wide median notch; lateral lobes 2, wedge-shaped, almost equal in size and divided into 2 lobules by a deep incision, each lobule further divided by shallow clefts into 4, with each lobule having a flat or retuse apex so that the outer margin appears denticulate; walls punctate.

Cosmopolitan; one of most common and widespread of the British species. W. & G.S. West (1905b) state that sometimes it can be obtained in almost pure gatherings especially amongst *Sphagnum* in small bogs, ditches or boggy springs.

M. denticulata can be confused with the equally widespread and common *M. thomasiana* to which it is obviously closely related; the latter in most cases possesses three elaborate protuberances at the base of each semicell and above the isthmus (see p. 56 below).

Micrasterias fimbriata Ralfs 1848

Synonym: *Micrasterias apiculata* [Ehrenberg] Meneghini ex Ralfs var. *fimbriata* (Ralfs) West *et* G.S.West

27200070 **Pl. 137F** (p. 560) **CD**

Cells 170–240 μm wide, 200–280 μm long, outline subelliptical, very deeply constricted; sinus open towards outside, closed internally; semicells 5-lobed with narrow incisions between lobes; polar lobe projecting and slightly flared upwards with subparallel sides below, each angle with a pair of diverging spines and a single spine on each side of apical notch; lateral lobes unequal, upper larger than lower, each lobe bilobulate and each lobule further divided, the incisions between opening widely outwards; ultimate, smallest lobules furnished on outside with a short, curved spine; walls punctate and sometimes decorated with a row of small spines along margins of sinus and polar and lateral lobes (doubtful var. *spinosa* Bisset).

Europe, North and South America, Arctic, acidophile (pH 5.8–7); present with other desmids in bog pools, usually associated with *Sphagnum*. Widespread but rarely abundant in the British Isles.

Micrasterias jenneri Ralfs 1848

27200090 **Pl. 137E** (p. 560) **CD**

Cells 82–125 μm wide, 132–170 μm long, oblong-elliptical, deeply constricted, with a narrow, linear sinus; semicells 5-lobed with incisions between lobes narrow, linear; polar lobe broadly triangular with convex apex interrupted by a broad, V-shaped, shallow depression; lateral lobes wedge-shaped, bilobulate then further subdivided; walls densely covered with flattened granules of variable size.

Europe, Arctic; in the British Isles widespread in distinctly acid, nutrient-poor waters, frequently associated with

Sphagnum; most records from the English Lake District where in one locality recorded as abundant (W. & G.S. West, 1905c); also from north-west Scotland where in lakes it occurs occasionally as a chance plankter; also occurs in Wales and acid habitats in England and Ireland (pH 4–6).

Micrasterias mahabuleshwarensis Hobson 1863

var. *wallichii* (Grunow) West *et* G.S.West 1905

Basionym: *Micrasterias wallichii* Grunow

27200104 **Pl. 145A** (p. 591) **CD**

Cells 153–208 μm wide, 181–223 μm long, deeply constricted, sinus open, closing towards inside where acutely angled; isthmus 30–33 μm wide; semicells 3-lobed with wide incisions between polar and lateral lobes; polar lobe elongate with upper angles produced into diverging denticulate processes, 88–115 μm wide, with pairs of asymmetrically disposed accessory processes; lateral lobes divided into 2 lobules by wide-angled incision, the upper further divided by a deep, wide incision producing 3 denticulate processes in each lateral lobe (the nominal variety has only 2 processes); a row of denticulations is present within semicell margins.

Boreal Europe, North America (the nominal variety is restricted to the tropics and subtropics); an acidophile, restricted mostly to the plankton of nutrient-poor Scottish lochs of Perthshire, Sutherland and the Shetland Islands.

The var. *europa* Nordstedt (27200103) has been found in a Sutherland lochan and the var. *dichotoma* (G.M.Smith) Willi Krieger (27200102) in Bassenthwaite in the English Lake District (Brook, 1957).

Micrasterias oscitans Ralfs 1848

27200110 **Pl. 136N** (p. 554) **CD**

Cells 123–135 μm wide, 125–146 μm long; constriction deep, sinus linear internally, open towards outside; semicells 3-lobed, polar lobe almost spindle-shaped with a wide convex apex extending to a pointed or bifid angle; incisions below polar lobe deep and open; simple lateral lobes horizontal, triangular with bifid extremities; walls with minute puncta.

Northern Europe, North America; an acidophile widespread in acid bogs and pools, often associated with *Sphagnum*; most frequent in the English Lake District, Wales and north-west Scotland; few records from Ireland.

Micrasterias papillifera Brébisson in Ralfs 1848

27200120 **Pl. 137C** (p. 560) **CD**

Cells 100–140 μm wide, 100–150 μm long, outline circular to subelliptical; deeply constricted with sinus closed to slightly open, particularly at extremities; semicells 5-lobed, with lobes (and their lobules) of similar width and length; incisions between lobes linear; polar lobe wedge-shaped, with shallow, concave sides and a concave apex with a pair of small teeth on each side of a slight, widely open, median notch; lateral lobes also wedge-shaped and of approximately equal depth, each divided by a moderately deep, closed incision into first-order lobules, then further subdivided into groups of 4 small lobules each tipped with a pair of short spines; walls minutely punctate with a row (sometimes interrupted) of small granules or denticulations on each side of sinus and along interlobular incisions.

Cosmopolitan, acidophile although reported from circumneutral environments (pH 4.6–7); most often found in small peaty pools and bogs with *Sphagnum*, also in lakes and as a tychoplankter (Förster, 1982). Widespread throughout the British Isles, especially in North Wales and Scotland, including the Hebrides and the Orkney and Shetland Islands.

Micrasterias pinnatifida Kützing ex Ralfs 1848

27200140 **Pl. 136O** (p. 554) **CD**

Cells 55–75 μm wide, 50–70 μm long; sinus wide open externally but narrow, even closed towards inner ends; semicells 3-lobed, with the 2 interlobular incisions broadly rounded and deep; lateral lobes almost spindle-shaped, projecting horizontally and attenuating to narrow, lobed apices; polar lobe similar with apex convex, straight or slightly retuse; walls minutely punctate.

Cosmopolitan, acidophile of peaty pools and bogs usually associated with *Sphagnum*; fairly widespread in the British Isles, especially in Scotland and the Outer Isles, rarely abundant.

Micrasterias radiosa Ralfs 1848

Synonym: *Micrasterias sol* Ehrenberg ex Kützing

27200150 **Pl. 145B** (p. 591) **CD**

Cells 162–191 μm broad, 158–192 μm long, circular in outline, constriction deep, sinus slightly open with undulate margins; semicells 5-lobed, incisions open to outside, closed internally; polar lobe with parallel sides, expanded at apex to narrow, furcate-dentate angles with small tooth on each side; lateral lobes unequal, upper slightly larger and more subdivided than lower, lower lobules subdivided into 4 equal lobules by 3 incisions, the middle one deeper than the 2 outer; each lobule furcate-dentate at extremity; upper lobule divided into 8 equal lobules by 7 incisions.

Europe, North America, Arctic, Africa, acidophile; in the British Isles in acid ponds and bogs associated with *Sphagnum* and in plankton of nutrient-poor lakes, especially in Scotland; also in some lakes in the English Lake District, Wales and Ireland (pH 5–8; Förster, 1982).

Micrasterias rotata [Greville] Ralfs ex Ralfs 1848

27200160 **Pl. 137B** (p. 560) **CD**

Cells 165–305 μm wide, 208–366 μm long, outline subcircular, constriction very deep, sinus linear; semicells 5-lobed with narrow, linear incisions between lobes; polar lobe long with concave margins widening outwards to slightly produced, bidentate angles; apex concave, sinuous with pronounced median cleft; lateral lobes unequal, each divided into 2 lobules by deep incision; upper lateral lobules larger than lower, each of former subdivided into 4 acuminate or bidentate parts by 3 incisions and lobules of lower divided into 2 bidentate parts; walls minutely punctate.

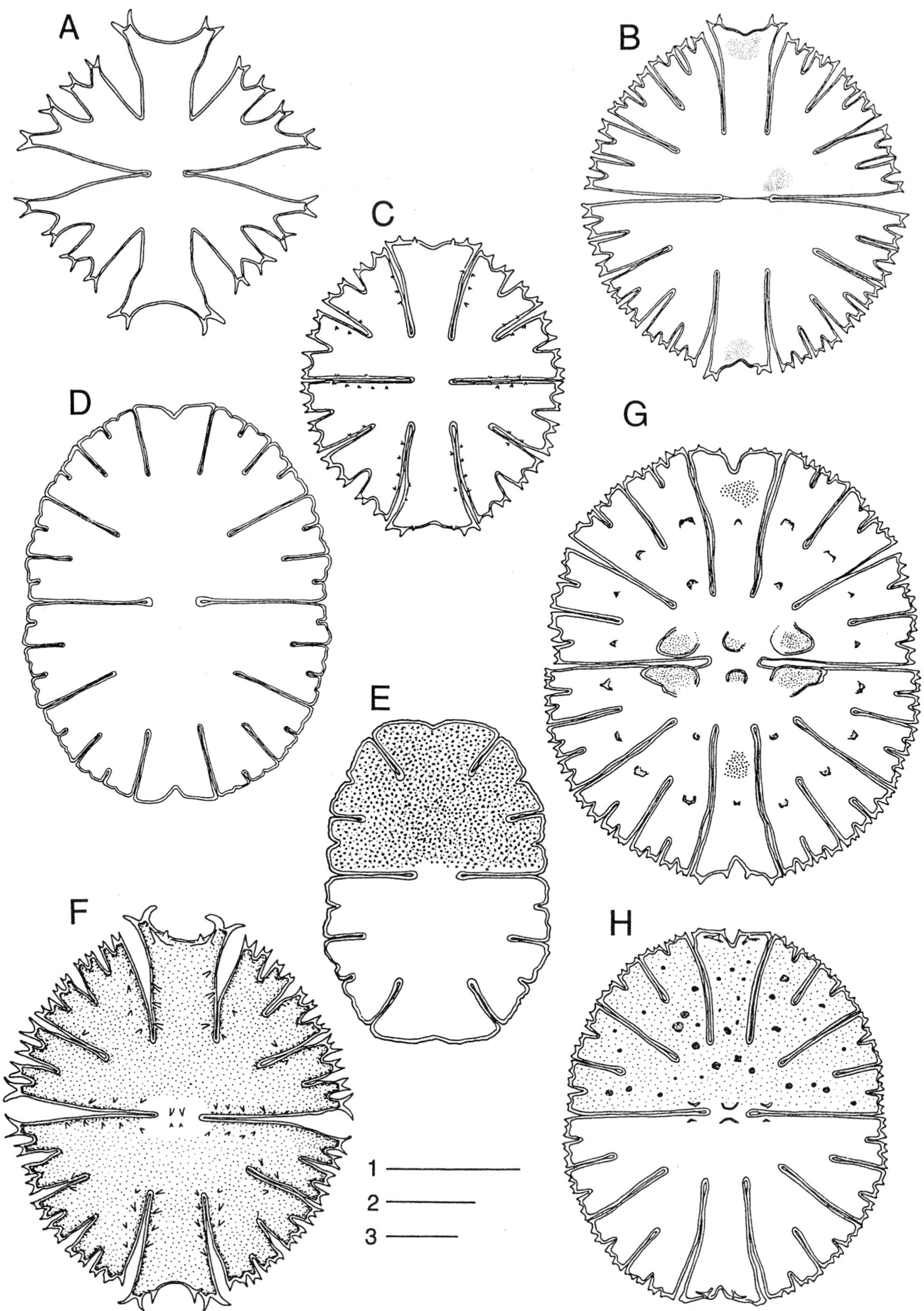

Plate 137 A–H

A. *Micrasterias crux-melitensis* (p. 558); **B**. *Micrasterias rotata* (p. 559); **C**. *Micrasterias papillifera* (p. 559); **D**. *Micrasterias denticulata* (p. 558); **E**. *Micrasterias jenneri* (p. 558); **F**. *Micrasterias fimbriata* (p. 558); **G**. *Micrasterias thomasiana* (p. 561); **H**. *Micrasterias thomasiana* var. *notata* (p. 561).

Scale bars: 50 μm. 1 for A, C; 2 for E–H; 3 for B, D.

Subcosmopolitan; widespread in the British Isles in acid habitats, especially amongst *Sphagnum* and in the weedy margins of nutrient-poor lakes; chance plankter (pH (3.8–)4.5–7.5).

A variable species, several forms described.

Micrasterias thomasiana W.Archer 1862

27200170 **Pl. 137G** (p. 560) **CD**

Cells 170–230 μm wide, 200–260 μm long, outline subcircular, deeply constricted with a narrow linear sinus; semicells 5-lobed with closed linear incisions between lobes; polar lobe narrow, wedge-shaped with concave lateral margins and with a moderately deep median notch, on each side is an apiculate, emarginate swelling; outer angles with 1 or 2 spines; lateral lobes more or less equal with a deep closed incision dividing each into 2 equal lobules, each lobule with a less deep and narrow V-shaped secondary incision and usually 2 further divisions, the extremities tooth-like; face of each semicell with 3 projections, one directly above isthmus, the other on either side, often bidentate; central projection tends to be simple, rounded and unornamented or sometimes much reduced; walls delicately punctate, often with small granules or spines at the base of each lobe and lobule.

Subcosmopolitan, acidophile (pH 5–7); fairly widely distributed in the British Isles especially in peaty pools and amongst other algae; also at the margins of nutrient-poor lakes; rarely abundant, although found in great numbers on mud surface in the shallows of a small Leicestershire lake (Brook & Williamson, 1988b).

A variable species, and when ornamentation poorly developed, then difficult to separate from the clearly related *M. denticulata* (see above).

var. *notata* (Nordstedt) Grönblad 1920

Basionym: *M. denticulata* Brébisson ex Ralfs var. *notata* Nordstedt

27200172 **Pl. 137H** (p. 560)

Differs from the nominal variety in that the 3 supraisthmal swellings are much reduced, lateral pair only short, blunt spines.

Similar distribution to the nominal variety and more easily confused with *M. denticulata*.

Micrasterias truncata [Corda] Brébisson ex Ralfs 1848

27200180 **Pl. 136M** (p. 554) **CD**

Cells 80–110 μm wide, 85–120 μm long, outline broadly elliptical, deeply constricted with a narrow, closed sinus sometimes open internally; semicells 5-lobed but not obviously because incisions between 2 lateral lobes mostly only very shallow V-shaped; lateral lobes variable but typically bilobate and toothed; polar lobe very wide, wedge-shaped and with apex convex, flat or retuse; lateral angles with 1 or 2 spines; walls punctate.

Cosmopolitan, essentially acidophilic but occurring in mildly alkaline waters (pH 3.5–7.3). One of the most widespread and abundant British species of *Micrasterias*, occurring in *Sphagnum* bogs and pools in upland areas of Wales, Ireland and Scotland, including the Hebrides, Orkney and Shetland Islands.

A variable species for which a large number of varieties of doubtful validity have been described.

TAXA RECORDED FROM THE BRITISH ISLES BUT NOT DESCRIBED IN TEXT

Micrasterias americana var. *boldii* Gutw. 27200012; *M. apiculata* [Ehrenb.] Menegh. ex Ralfs 27200020; *M. brachyptera* P.Lundell 27200030; *M. conferta* P.Lundell 27200040; *M. fimbriata* var. *spinosa* Bisset in J.Roy 27200072; *M. furcata* Ralfs 27200080; *M. furcata* var. *dichotoma* (Wolle) Cushman 27200082; *M. furcata* var. *evoluta* Willi Krieg. 27200083; *M. furcata* var. *pseudo-crux* Grönblad 27200084; *M. mahabuleshwarensis* var. *dichotoma* G.M.Sm. 27200102; *M. mahabuleshwarensis* var. *europea* Nordst. 27200103; *M. oscitans* var. *micronata* (S.C.Dixit) Wille 27200112; *M. papillifera* var. *glabra* Nordst. 27200122; *M. radiosa* Ralfs 27200150; *M. radiosa* var. *elegantior* (G.M.Sm.) Croasdale 27200154; *M. radiosa* var. *ornata* Nordst. 27200152; *M. radiosa* var. *murrayi* (West *et* G.S.West) Croasdale 27200153; *M. truncata* var. *bahusiensis* Wittr. 27200183; *M. truncata* var. *crenata* (Bréb.) Grönblad 27200184; *M. truncata* var. *neodamensis* (A.Braun) Dick 27200185.

Oocardium Nägeli 1849

Colonial desmid in colonies 1–2 mm in diameter, with cells of the *Cosmarium* type and always embedded in a radiating branched system of mucilaginous strands invariably encrusted in a hollow cylinder of calcite from which the outer ends of the cells barely protrude; cells small and symmetrical only in 2 planes, significantly broader than long and laterally compressed with only slight constriction; walls smooth; chloroplasts axile (sometimes only one per cell) with central pyrenoid.

Europe, North America, Asia; exclusively in limestone districts and most frequently reported from sides of waterfalls or in streams rich in calcium. Monospecific.

Oocardium stratum Nägeli 1849

27250010 **Pl. 145 H** (p. 591)

Cells 18–24 μm wide, 13–20 μm long, about 17 μm thick, almost round in outline with a very small median constriction; sinus an open notch; cells unequally flattened on opposing sides so symmetrical only in 2 planes at right angles instead of 3 as in *Cosmarium*; semicells mostly semicircular but with irregularities; in vertical view the cells broadly oval; walls smooth.

Distribution as for genus; forming calcareous concretions on sticks, stones and twigs; in England occurs in West Yorkshire including the Malham area, and one record from Ireland (River Caher, County Clare).

So that the cells themselves do not become encased in lime, they secrete mucilage through pores in the cell wall. By this secretion the thin layer of calcite is ruptured and the cells lift out of the stony mass. Hence the tiny cylinders of calcite continually increase in length, and by secreting mucilage the cells are raised to a position where they can continue to photosynthesise and reproduce.

Pleurotaenium Nägeli 1849

Cells mostly straight, cylindrical and up to 35 times as long as broad, solitary, but occasionally joined at apices to form short but easily separated filaments; median constriction slight, above which are one or several inflations and at junction of semicells sometimes a ring-like thickening of wall material; semicells taper gradually to apices which may be rounded, subtruncate or flattened and in some species ornamented with a circle of tubercules (polar nodules) or short spines; walls colourless to brown, smooth or irregularly and distinctly punctate; chloroplasts long parietal bands with pyrenoids along their length; terminal vacuoles spherical and usually very conspicuous, containing many moving crystals.

Cosmopolitan; few are widely distributed and can be moderately abundant, mostly in acid waters; one species, *P. trabecula*, has been found to be fairly common in at least one distinctly alkaline lake. Nine species are recorded from the British Isles.

Some species have been moved to the genus *Haplotaenium* (Bando, 1988).

1 Apices without crown of tubercles, granules or spines; lateral margins not undulate, slightly convex ....*P. trabecula*
1 Apices with crown of tubercles or granules; lower lateral margins undulate...2
2 Cells 18–35 µm wide..........................*P. ehrenbergii*
2 Cells 50–75 µm wide 3
3 Cells with a prominent basal inflation and decreasingly undulate margins...........*P. coronatum*
3 Cells without a basal inflation and markedly tumid margins ....*P. truncatum*

Pleurotaenium coronatum (Brébisson) Rabenhorst 1868

Basionym: *Docidium coronatum* Brébisson in Ralfs
Synonyms: *Docidium coronatum* Brébisson in Ralfs, *D. coronatum* var. *undulatum* Hieronymus, *Pleurotaenium coronatum* (Brébisson) Rabenhorst var. *undulatum* (Hieronymus) Schmidle

27280010 **Pl. 131H** (p. 528)

Cells 40–60 µm wide, 400–600 µm long; semicells attenuating gradually to apex from prominent basal inflation above an isthmal ring of thickening, above basal inflation are a succession of increasingly smaller undulations beyond which lateral margins are virtually straight; apices truncate, 35–55 µm wide, with a ring of 10–12 prominent conical or flattened tubercles, 5 or 6 visible in face view; walls smooth, punctate; chloroplast narrow, irregularly shaped parietal bands of various lengths along which small pyrenoids occur; terminal vacuoles large, spherical with abundant moving crystals.

Cosmopolitan; widespread in the British Isles in acid to near neutral waters (pH 4.5–6.9), occurring mostly in small peaty pools often with *Sphagnum*; also in the littoral of nutrient-poor to moderately nutrient-rich lakes; fairly common in such waters, especially in Scotland and Ireland, less so in England and Wales; reported from the plankton, probably as a chance plankter, from lochs in Sutherland and the Isle of Lewis and the Outer Hebrides.

var. *fluctuatum* West 1892

27280012 **Pl. 131J** (p. 528)

Cells larger than the nominal variety, 55–72 µm wide at semicell base, 50–53 µm at apices, 670–832 µm long; apical tubercles prominent, 11–15 around margin, 6–8 visible in face view; walls thick and finely scrobiculate.

Acidophile, with records from north-west Scotland and the Hebrides; first described from Lough Aunierin, Galway in Ireland and recently found again in that county.

Essential difference between this variety and the nominal is that the lateral margins of the semicells undulate from isthmus to apex.

var. *robustum* West 1892

27280013 **Pl. 131I** (p. 528)

Cells shorter than the nominal variety, 400–550 µm long, with lateral margins more undulate, attenuating suddenly towards apices; apical tubercles larger, 5 visible in side view.

An acidophile, first described from loughs in Donegal and Galway, Ireland, now reported from several localities in the English Lake District, Dumfries and Galloway and the Assynt area of north-west Scotland; also from several sites in the Shetland Islands.

Pleurotaenium ehrenbergii (Brébisson) de Bary 1858

Basionym: *Docidium ehrenbergii* Brébisson in Ralfs
Synonyms: *Pleurotaenium ehrenbergii* (Brébisson) de Bary var. *constrictum* (Playfair) Willi Krieger, *P. ehrenbergii* (Brébisson) de Bary var. *granulatum* (Ralfs) West *et* G.S.West, *Docidium ehrenbergii* Brébisson in Ralfs

27280020 **Pl. 131G** (p. 528)

Cells straight, narrow, 18–35 µm wide at base of semicells, 14–25 µm wide at apices, 300–500 µm long; semicells with a small basal inflation and 1, 2 or several increasingly smaller ones, then tapering very slightly to truncate apices surrounded by a ring of 7–10 conical or rounded tubercles (4 or 5 visible in side view), if tubercles well developed, apices have a flared appearance; chloroplasts a series of long, narrow parietal ribbons with a row of small pyrenoids along each; terminal vacuoles spherical, packed with moving crystals.

Cosmopolitan; widespread throughout the British Isles mostly in acid waters, elsewhere reported from slightly alkaline habitats (pH 4.5–7). Most abundant in *Sphagnum* bogs and peaty pools; also present in the littoral of nutrient-poor to moderately nutrient-rich lakes where occasionally encountered as a chance plankter.

Pleurotaenium trabecula [Ehrenberg] Nägeli 1849

27280100 **Pl. 131F** (p. 528) **CD**

Cells subcylindrical, 24–46 µm broad, 350–600 µm long; semicells with 1 or 2 basal inflations above which lateral margins are slightly convex, then tapering gradually to smooth subtruncate apices, 16–32 µm broad, perforated by several terminal pores; walls colourless, smooth with scattered pores; chloroplasts parietal ribbons with numerous pyrenoids along length.

Cosmopolitan; with broad ecological tolerances (therefore possibly a species complex) being found in distinctly acid waters amongst *Sphagnum* (pH 4.5–5.5), but also present amongst other algae in alkaline waters (pH 8–8.5); widespread in the British Isles but most frequent in acid pools and bogs in the west of Scotland and Ireland, also in alkaline lakes and ponds in lowland England; recorded as a chance plankter in moderately to distinctly nutrient-rich lakes.

Pleurotaenium truncatum (Brébisson) Nägeli 1849

Basionym: *Docidium truncatum* Brébisson in Ralfs

27280120 **Pl. 144I** (p. 590)

Cells 54–75 μm wide, 380–500 μm long; semicells slightly tumid, attenuating markedly from middle to apex with a slight but distict basal inflation; apices 29–40 μm wide, truncate with a ring of 13–15 small, apical tubercles (7 or 8 visible across apex); walls punctate.

Probably cosmopolitan; widespread throughout the British Isles in acid to neutral waters but never abundant.

TAXA RECORDED FROM THE BRITISH ISLES BUT NOT DESCRIBED IN TEXT

Pleurotaenium ehrenbergii var. *elongatum* (West) West 27280022; *P. ehrenbergii* var. *undulatum* Schaarschm. 27280023; *P. eugeneum* (W.B.Turner) West *et* G.S.West 27280030; *P. maximum* (Reinsch.) P.Lundell 27280050; *P. nodosum* (Bail.) P.Lundell 27280070; *P. nodosum* var. *borgei* (Grönblad) Willi Krieg. 27280072; *P. nodulosum* (Bréb.) de Bary 27280080; *P. trabecula* var. *crassum* Wittr. 27280102; *P. trabecula* var. *robustum* Hust. 27280103; *P. tridentulum* (Wolle) West 27280110; *P. tridentulum* var. *capitatum* West 27280112; *P. truncatum* var. *crassum* Boldt 27280122; *P. truncatum* var. *farquharsonii* (J.Roy) West *et* G.S.West 27280123; *P. truncatum* var. *granulatum* West *et* G.S.West 27280124.

Staurastrum [Meyen] Ralfs 1848

Semicells often with upper angles considerably extended and showing radial symmetry about their vertical axis so most triangular (triradiate or trigonal) in apical view; a few biradiate with bilaterally compressed cells, others quadrangular, pentangular to 9-angular; isthmus deeply constricted and so well-defined; walls have a pore system and extrusion from pores (especially in planktonic species) results in cells enclosed within a wide mucilage envelope; walls smooth and punctate or ornamented with granules, denticulations, verrucae or spines, arranged in consistent, symmetrical patterns; semicells in front view elliptical to semicircular, triangular, quadrangular to polygonal; upper angles commonly extended in planktonic species into long hollow processes, variously ornamented and terminating in truncate ends bearing 2–4 short, divergent spines; occasionally two transverse whorls of processes borne on the apex of each semicell.

Many species cosmopolitan; in lake plankton, mostly in distinctly nutrient-poor waters. However, at least 2 species, *S. chaetoceras* and *S. pingue*, are reliable indicators of nutrient-rich, alkaline waters (Brook, 1965, 1981).

Staurastrum is the desmid genus with the greatest range of morphologies. Because of this and the complexity of form within the genus, several attempts have been made to divide it into several genera; so far only the smooth-walled, monospinous genus *Staurodesmus* (Teiling 1948) has been accepted (see p. 577).

1 Angles of cells not produced into long processes; cells mostly 3-radiate, some 4–5-radiate 2
1 Angles of cells produced into hollow processes, often very long and slender; cells sometimes bi-radiate 25
2 Angles tipped with bifid or trifid stout spines 3
2 Angles not tipped with spines, broadly rounded 4
3 Cells 3-radiate; angles tipped with pairs of stout, parallel or slightly convergent spines .... *S. longispinum*
3 Cells 5-radiate; angles tipped with 3 long, strongly divergent spines .... *S. brasiliense* var. *lundellii*
4 Cells subcylindrical, ornamented with granules around short, blunt angles; ring of granules above isthmus .... *S. meriani*
4 Cells not subcylindrical; angles rounded or truncate 5
5 Cell walls smooth, punctate; *Cosmarium*-like in side view 6
5 Cell walls ornamented with distinctive granules, spines or verrucae, usually regularly arranged round angles 8
6 Semicells almost semicircular; sinus closed, linear .... *S. orbiculare*
6 Semicells elliptic or broadly kidney-shaped; sinus open, acute angled 7
7 Semicell apex broadly rounded .... *S. muticum*
7 Semicell apex flattened, so semicells broad-elliptical .... *S. bieneanum*
8 Semicells subpyramidate; walls ornamented with verrucae and/or large spines; angles not produced 9
8 Semicells transversely elliptical; walls ornamented with small granules or large spines, in concentric rings round angles; angles slightly produced 10
9 Sinus closed; verrucae large and in pairs down each side of broad angles, from apex to sinus .... *S. maamense*
9 Sinus open; bispinate verrucae largest on apex, smaller at the angles and on lateral margins .... *S. spongiosum*
10 Angles not tipped with spines; cells usually twisted at isthmus 11
10 Angles tipped by spines, often bifid; cell not twisted at isthmus 15
11 Angles produced and narrowly rounded; sinus widely open, rounded, elongate 12
11 Angles only slightly produced, broadly rounded; sinus open, angular 13

12 Cells 3-radiate, semicells narrow oblong elliptic, always twisted at isthmus so that adjacent angles alternate .... *S. alternans*
12 Cells mostly 4-radiate; semicells elliptic to subspindle-shaped, upper margin slightly convex, lower inflated; angles rounded or truncate-rounded; ring of granules above isthmus .... *S. dilatatum*
13 Cells 3–4-radiate, decorated with small granules...14
13 Cells 3-radiate, decorated overall with spines 20
14 Semicells sub-rhomboidal to elliptic with upper and lower margins equally convex; angles acutely rounded.. *S. punctulatum*
14 Semicells broadly transverse elliptic; angles broadly rounded; apex truncate, ornamented with spines or bifurcate verrucae.................... *S. asperum*
15 Angles not produced, tipped with a single (rarely bifid) spine...16
15 Angles produced, tipped with strong, bifid spines..17
16 Angles with widely separated rings of granules, sometimes indistinct; spines at angles small .... *S. denticulatum*
16 Angles with closely set rings of distinct granules; spines at angles stout............................. *S. lunatum*
17 Angles developed as processes tipped with strong bifid spines; (apex sometimes with spines or verrucae)................................. *S. pseudopelagicum*
17 Angles not developed as processes but tipped with bifid or trifid spines; apex often with spines or verrucae..18
18 Angles with weakly divergent bifid spines; lateral margins without spines .... *S. avicula*
18 Angles with strongly divergent bifid or trifid spines..19
19 Lateral margins with pairs of spines ....*S. simonyi*
19 Lateral margins without pairs of spines, apex raised and with pairs of stout spines .... *S. monticulosum*
20 Semicells pyramidal or sub-pyramidal................21
20 Semicells transversely elliptic or sub-kidney-shaped...22
21 Sinus closed; semicell apex flat .....*S. pyramidatum*
21 Sinus open, acute-angled; semicell apex convex or subtruncate... *S. hirsutum*
22 Semicells sub-kidney-shaped; spines short, stout, closely set .... *S. brebissonii*
22 Semicells transversely elliptic; spines long, well-developed 23
23 Sinus open more or less right-angled; apex convex, flattened 24
23 Sinus acute-angled; apex broadly convex; spines widely spaced, often irregular .... *S. teliferum*
24 Spines of equal size and moderately closely spaced .... *S. polytrichum*
24 Spines of unequal size, those at angles longer and stouter than those on semicell body ....*S. setigerum*
25 Cell wall smooth...26
25 Cell wall decorated with granules, spines or verrucae..27
26 Processes short, angular, inflated half way along length tipped with 4 minute spines; sinus very broad open; cells only 14–15.5 µm long .... *S. inconspicuum*
26 Processes long, smooth, divergent, ends mostly bifurcate, rounded; sinus open more or less right-angular; cells 27–36.5 µm long.......... *S. brachiatum*
27 Processes short, blunt, ending in 4–6 short spines; wall granular, with ring of granules above isthmus; apices strongly convex; cells 24–35 µm long...28
27 Processes long, tapering, with denticulate margins, sometimes with accessory processes on semicell apices 29
28 Semicells narrowly cup-shaped; wall weakly granulate; cells 3–4-radiate (or more), processes horizontal and alternating with those of adjacent semicell .... *S. margaritaceum*
28 Semicells broadly cup-shaped; wall strongly granulate, toothlike on apex; cells 3-radiate, horizontal and lying above those of adjacent semicell .... *S. proboscideum*
29 Processes mostly well developed; semicells with apical accessory processes 47
29 Processes well developed; semicells without apical accessory processes 30
30 Cells 3(–4)-radiate; processes convergent.............31
30 Cells 2–9-radiate; processes divergent, horizontal or sub-horizontal..32
31 Cells 20–29 µm long; adjacent cell semicells often showing median torsion 33
31 Cells 40–85 µm long; adjacent semicells not showing median torsion 34
32 Semicells cup-shaped, apex broadly rounded; processes mostly short, truncated ending in 3–4 short spines .... *S. cyrtocerum*
32 Semicells elongate-elliptic to subspindle-shaped, apex distinctly elevated; processes elongate tapering to 3 small spines...................... *S. inflexum*
33 Cells 42–57 µm long; semicells narrow, cup-shaped; apex with intramarginal verrucae; granules in groups above isthmus; processes only moderately convergent .... *S. manfeldtii*
33 Cells 58–85 µm long, processes strongly curved and convergent; apex ornamented with prominent intermarginal verrucae 35
34 Cells 5–72 µm long, semicells narrow, cup-shaped, ornamented above isthmus; processes tipped with 2 short spines, sides prominently ornamented with verrucae .... *S. cerastes*
34 Cells 73–85 µm long, semicells broadly cup-shaped, not ornamented above isthmus; processes tipped with 3 spines, the middle one the largest .... *S. sebaldi*
35 Cells most often 2-radiate but some 3-radiate; processes divergent or horizontal.........................36
35 Cells never 2-radiate, mostly 3- but up to 9-radiate 40
36 Processes strongly divergent 37
36 Processes horizontal or weakly divergent............38

37 Processes long, very slender with rings of granules extending onto body; 3 granules on apex; semicells without isthmal torsion ..... *S. chaetoceras*
37 Processes short, denticulate with scattered granules on lateral margins; apex strongly concave; semicells show isthmal torsion ..... *S. tetracerum*
38 Cells 2-radiate, processes horizontal, denticulate tipped with 3 stout spines; body swollen above isthmus and with rings of granules, intramarginal verrucae on apex ..... *S. johnsonii*
38 Cells narrow, cup-shaped, 3-radiate; processes horizontal or slightly divergent 39
39 Processes short, weakly denticulate ... *S. crenulatum*
39 Processes long, strongly denticulate ending in 3 spines, lower largest ..... *S. planctonicum*
40 Cells 3-radiate (rarely 4-radiate); processes strongly divergent 41
40 Cells 5- to 9-radiate 46
41 Processes slender, denticulate 42
41 Processes with concentric rings of granules extending onto semicell body 43
42 Body without ornamentation; processes swollen at ends, tipped with 4 small spines ..... *S. longipes*
42 Body ornamented by groups of granules above isthmus and beneath each process; intramarginal spines or verrucae on apex ..... *S. pingue*
43 Semicells narrowly cup-shaped with ring of granules above isthmus 44
43 Semicells broadly cup-shaped; processes mostly short, stout 45
44 Processes long and strongly divergent ..... *S. cingulum*
44 Processes short, subparallel or slightly convergent ..... *S. gracile*
45 Processes ornamented with concentric rings of small granules; also ring of granules above isthmus ..... *S. cingulum* var. *obesum*
45 Processes strongly denticulate; no ring of granules above isthmus ..... *S. anatinum*
46 Cells 5-radiate (rarely 6), processes convergent, margins denticulate ending in 3 small spines; semicells narrow cup-shaped, apex flat or slightly convex ..... *S. arachne*
46 Cells 6–9-radiate, processes convergent, margins denticulate; semicell apex strongly convex ornamented with conical nodules between processes ..... *S. ophiura*
47 Cells triradiate 48
47 Cells 5–9-radiate 50
48 Processes horizontal; apex with short stout furcate spines. ..... *S. furcatum*
48 Processes long, ornamented with rings of granules 49
49 Apical processes 3, strongly divergent above horizontal processes ..... *S. armigerum*
49 Apical processes 6, in pairs on either side of horizontal processes ..... *S. armigerum* fo. *furcigerum*
50 Cells 5–6-radiate, sinus broad U- or V-shaped notch; semicell angles, bifurcate, lower convergent, upper strongly divergent, both denticulate ending in 3 blunt spines... *S. sexangulare*
50 Cells 9-radiate, sinus obtuse V-shaped, semicells subspherical; lower processes slightly, upper strongly divergent, both weakly denticulate tipped with 3 spines ..... *S. arctiscon*

Staurastrum alternans Brébisson in Ralfs 1848

27380040 **Pl. 138C** (p. 567)

Cells 3-radiate, 21–31 μm wide, 22–33 μm long, twisted at isthmus 7.5–9.5 μm wide, so that angles of one semicell alternate with other; deeply constricted with open, acute-angled sinus; semicells narrow, oblong-elliptical, angles rounded, apex flat in middle; walls with granules in concentric rings around angles; frequently with ring of granules above isthmus.

Cosmopolitan; very widespread in the British Isles, mostly in small acid pools and bogs amongst mosses and filamentous algae.

Staurastrum anatinum Cooke *et* Wills 1880

27380060 **Pl. 139J** (p. 571)

Cells (2–)3(–4)-radiate, 30–45 μm wide excluding processes, 65–115 μm wide with processes, 33–50 μm long excluding processes, 40–70 μm long with processes, deeply constricted with a wide open, internally acute sinus, isthmus 12–28 μm wide; semicells mostly broadly cup-shaped; lateral margins moderately convex, apex slightly convex; angles usually considerably extended to form long, hollow, gradually attenuating, subparallel to divergent processes terminating in 3 short, stout, divergent spines; processes with concentric rings of denticulations of variable prominence along length, continuing onto semicell body; denticulations on apex and each side of body often much enlarged, becoming short emarginate spines or verrucae.

Cosmopolitan; widespread in the British Isles in the marginal vegetation and especially the plankton of poor to moderately nutrient-enriched lakes in western regions of Scotland and Ireland (pH 4.9–7.8; alkalinity 1–38 ppm $CaCO_3$).

Cell ornamentation and the length of the processes are extremely variable resulting in a considerable number of forms being recognized (Brook, 1959c).

Staurastrum arachne Ralfs ex Ralfs 1848

27380070 **Pl. 139M** (p. 571)

Cells mostly 5-radiate, 14–16 μm wide excluding processes, 40–55 μm wide with processes, 20–30 μm long, slightly constricted with acute sinus; isthmus 8–9 μm wide; semicells cup-shaped with straight or convex apex, angles produced as slender convergent processes 14–18 μm long, decorated with concentric series of granules and terminating in 3 small spines.

Cosmopolitan, acidophile; widespread but only abundant in western areas of the British Isles, especially in the west of Scotland and the Outer Hebrides; occasional as a chance plankter in nutrient-poor lakes.

Staurastrum arctiscon (Ehrenberg) P.Lundell 1871

Basionym: *Xanthidium arctiscon* Ehrenberg 1841
27380090 **Pl. 140K** (p. 575)
Cells 9-radiate, 46–68 μm wide without processes, 92–160 μm wide with processes, 66–96 μm long without processes, 100–155 μm long with processes, fairly deeply constricted with a rectangular sinus, acute internally; isthmus 24–33 μm wide; semicells broadly elliptical to subspherical with 9 lower, horizontally disposed and 6 upper, upwardly directed whorls of processes; processes all variable in length, truncate, ending in 3 spines all with a single rows of denticulations along sides and upper and lower margins; semicell body smooth.

Restricted to Europe and North America; in the British Isles a 'western' desmid, moderately frequent but rarely abundant in the plankton of nutrient-poor lakes of the English Lake District, West Wales and Ireland; most common in Scottish lochs in Sutherland and the Outer Hebrides.

Staurastrum armigerum Brébisson 1856

fo. *furcigerum* (Ralfs) Teiling 1957
Basionym: *Didymocladon furcigerum* [Brébisson] Ralfs
27381560 (27380622) **Pl. 138P** (p. 567)
Cells mostly 3-radiate (up to 9-radiate), 45–68 μm wide with processes, 35–48 μm long excluding processes and 50–70 μm long with processes, deeply constricted, sinus narrow to inside then opening widely; isthmus 13–18 μm wide; semicells transverse to elliptical, with upper and lower margins equally convex, lateral angles extended horizontally into short, stout processes ending in 2 or 3 short spines; apex with a series of processes projecting upwards, similar but shorter than laterals above and behind latter; all processes with a series of concentric denticulations.

Cosmopolitan; widely distributed in western areas of the British Isles in plankton of nutrient-poor to moderately nutrient-rich lakes, especially in the English Lake District, Scotland and lochs of the Outer Hebrides; also common in acid pools with *Myriophyllum*, *Utricularia* and *Sphagnum* (alkalinity 3–78 ppm $CaCO_3$).

A very variable species with some 12 varieties and forms: fo. *furcigerum* (Ralfs) Teiling (Pl. 138N) has a double set (6) of long accessory processes; dichotypic forms with one semicell, fo. *furcigerum* and the adjacent one the type form, are not uncommon.

Staurastrum asperum Brébisson ex Ralfs 1848

27380120 **Pl. 138D** (p. 567)
Cells 34–47 μm wide, 42–52 μm long, with a deep open median constriction, sinus narrowly rounded to acute; isthmus 12–15 μm wide; semicells transversely oval to broadly rounded, widest above midline, with rounded lateral margins, flat to slightly convex apex; walls covered with irregular rings of granules or spines, larger towards apex where some may be bifid; centre of apex smooth.

Distribution boreal; widespread throughout the British Isles in acid pools and bogs, never abundant; occasional as a chance plankter.

Staurastrum avicula Brébisson in Ralfs 1848

27380130 **Pl. 138I** (p. 567)
Cells 3-radiate, 35–42 μm wide including spines, 29–35 μm long, deeply constricted, mostly with a very open, obtuse sinus; isthmus 9–11 μm wide; semicells subelliptical to subtriangular, lower margins varying from almost straight to convex, upper only slightly convex, flattened in mid-region; angles tipped with 2 small spines one above the other; walls with granules in 3–4 concentric circles round angles; apex in vertical view smooth except for pairs of intra-marginal granules, sometimes present as small spines or verrucae.

Europe, Asia, North and South America, New Zealand; in the British Isles an acidophile, widespread, especially in the weedy margins of nutrient-poor pools and lakes and a chance plankter in Scotland, rarely abundant; occasionally in similar habitats in England, Ireland and Wales.

Staurastrum bieneanum Rabenhorst 1862

Synonyms: *Staurastrum muticum* [Brébisson] Ralfs fo. *bieneanum* Reinsch, *S. orbiculare* [Ehrenberg] Ralfs var. *bieneanum* Rabenhorst, *Staurodesmus bieneanum* (Rabenhorst) Florin
27380190 **Pl. 145F** (p. 591)
Cells 29–42 μm wide, 26–37.5 μm long, with a deep constriction, isthmus 7–9 μm wide; sinus acute, internally opening widely outwards to sharply rounded lateral angles, margins then converging to a flat or slightly convex, broad apex; lower margins always more convex than upper; walls smooth, densely and finely punctate.

Subcosmopolitan; Europe, Siberia, North America, Asia, Madagascar, New Zealand; widely distributed in the British Isles in nutrient-poor to moderately nutrient-rich waters in England and Scotland including the Inner and Outer Hebrides, the Orkney Islands, also in several Irish Loughs; present occasionally in the lake plankton.

The lateral angles may be thickened or end in a small mucro, for which reason it has been suggested that the species should be transferred to the genus *Staurodesmus* (Florin, 1957; Croasdale, 1973).

Staurastrum brachiatum Ralfs 1848

27380240 **Pl. 139K** (p. 571) **CD**
Cells 3–5-radiate, 25–48 μm wide, 27–36.5 μm long, deeply constricted with acute, almost rectangular sinus, isthmus 5–9 μm wide; semicells virtually triangular with straight lateral margins, apex flat, angles diverging into stout processes of variable length with bi- or trifid ends; semicells frequently showing median torsion, so angles of adjacent semicells do not overlie one another.

Cosmopolitan acidophile; widespread in the British Isles, frequent or even abundant in *Sphagnum* bogs and pools; less abundant but often in margins of nutrient-poor ponds and lakes; a chance plankter.

Staurastrum brasiliense Nordstedt 1869

fo. *lundellii* West *et* G.S.West
27380262 **Pl. 138T** (p. 567) **CD**

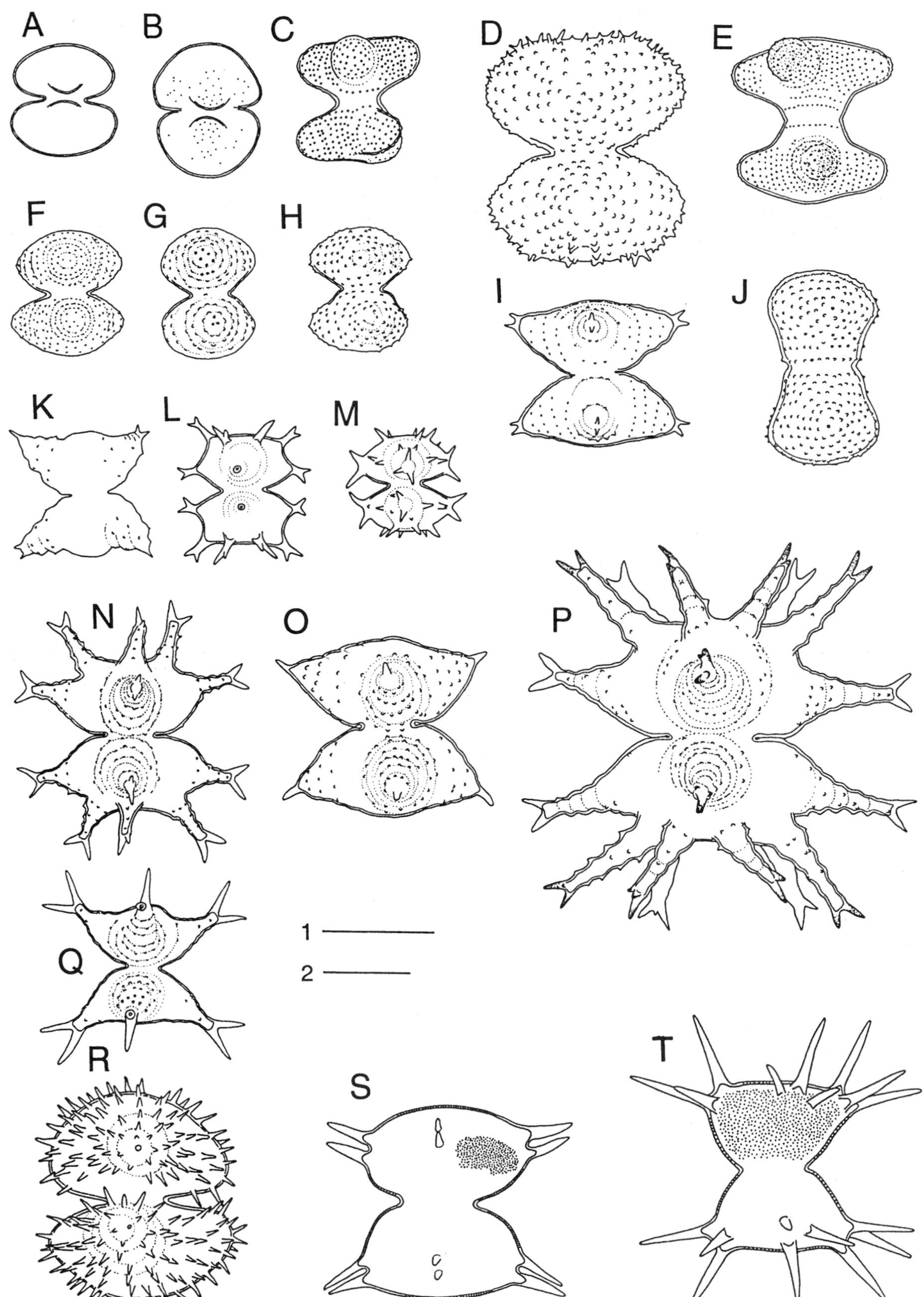

Plate 138 A–T

A. *Staurastrum muticum* (p. 573); **B**. *Staurastrum orbiculare* (p. 573); **C**. *Staurastrum alternans* (p. 565); **D**. *Staurastrum asperum* (p. 566); **E**. *Staurastrum dilatatum* (p. 569); **F**. *Staurastrum punctulatum* (p. 574); **G**. *Staurastrum punctulatum* var. *kjellmani* (p. 574); **H**. *Staurastrum punctulatum* var. *pygmaeum* (p. 574); **I**. *Staurastrum avicula* (p. 566); **J**. *Staurastrum meriani* (p. 572); **K**. *Staurastrum denticulatum* (p. 569); **L**. *Staurastum furcatum* (p. 569); **M**. *Staurastrum simonyi* (p. 576); **N**. *Staurastum armigerum* fo. *furcigerum* (p. 566); **O**. *Staurastrum lunatum* (p. 570); **P**. *Staurastrum armigerum* (p. 566); **Q**. *Staurastrum pseudopelagicum* (p. 574); **R**. *Straurastrum brebissonii* (p. 568); **S**. *Staurastrum longispinum* (p. 570); **T**. *Staurastrum brasiliense* fo. *lundelli* (p. 566).

Scale bars: 1, 25 μm; 2, 50 μm. 1 for A–R; 2 for S, T.

Cells 5(–6)-radiate, 63–80 μm wide without spines, 120–130 μm wide with spines, 75–80 μm long without spines, 120–140 μm with spines (spines 25–30 μm long); constriction shallow, sinus open, rounded internally, isthmus 28–34 μm wide; semicells with broadly cup-shaped margins, almost straight, diverging; apex flat; upper angles obliquely truncate ending in 3 conspicuous stout spines, lower lying in same horizontal plane, the third directed upwards at an angle to lower ones; apical view mostly 5-angled, each angle truncate, bifid, the third spine lying between the lower ones.

Europe, North and South America, Arctic, Africa; a western distribution in the British Isles, most frequent in plankton of nutrient-poor lakes of Cumbria, Wales and especially the west of Scotland, including the Hebrides; also west of Ireland.

Staurastrum brebissonii W.Archer in A.Pritchard 1861

Synonym: *Staurastrum pilosum* Brébisson

27380270 **Pl. 138R** (p. 567)

Cells 3(–5)-radiate, 40–65 μm wide, 34–48 μm long, without spines; deeply constricted sinus, narrow opening to outside; isthmus about 13 μm wide; semicells elliptical or elliptical to spindle-shaped with rounded angles; walls covered with fine, acute spines, 2.5 μm long at most, arranged in concentric circles around angles where longest arranged in longitudinal rows on face of semicell where smaller and more widely spaced.

Europe, North and South America, Arctic, Antarctic, Africa; widespread in the British Isles, rarely abundant acidophile in ditches and amongst *Sphagnum* in bog pools.

Staurastrum cerastes P.Lundell 1871

27380300 **Pl. 145I** (p. 591)

Cells 3–4-radiate, 58–72 μm wide, 48–57 μm long, with a shallow constriction and sinus a U-shaped notch, isthmus 10–12 μm wide; semicells in lower part subcylindrical widening upwards to strongly incurved, tapering processes (tipped with 4 short stout spines) which almost meet the processes of the adjacent semicell; apical margin broadly convex, deeply crenulated and ornamented with about 9 verrucae which are increasingly large towards the apex; lower margins of processes smooth and strongly concave; around base of semicell a ring of granules; a series of granules extends from the body along the sides of the processes.

Europe, North America, Arctic, Asia, Africa; rare and confined to western regions of the British Isles, with records from Sutherland, Scotland, the Capel Curig region of North Wales and Galway, Ireland.

Staurastrum chaetoceras (Schröder) G.M.Smith 1924

Basionym: *Staurastrum polymorphum* Brébisson ex Ralfs 1848 var. *chaetoceras* Schröder

27380310 **Pl. 139P** (p. 571)

Cells 2- or 3-radiate, 16–26 μm wide excluding processes, 50–77 μm wide with processes, 19–30 μm long excluding processes, 46–94 μm long with processes, constriction shallow, sinus a U-shaped notch, isthmus 4–7 μm wide; semicells obversely triangular, or shallow cup-shaped with almost straight lateral margins, angles extended into divergent, almost straight, very slender, slightly attenuating hollow processes (often only 2 μm wide at ends) and tipped with 4 very small divergent spines; apex flat or slightly convex, ornamentation (best seen in biradiate forms) 3 evenly spaced granules or small verrucae from each of which a single row of granules extends onto a semicell body; sometimes a ring of small granules encircles body above isthmus.

North Polar distribution, confined to plankton of nutrient-rich lakes; in lowland areas of the British Isles, especially frequent in the meres of Cheshire and Shropshire, reservoirs in Leicestershire, lakes in Anglesey, the Norfolk Broads and the English Lake District; present in alkaline machair lochs on the west coast of North Uist and the Outer Hebrides (pH 7–9; alkalinity 4–194 ppm $CaCO_3$).

The 2-radiate and 3-radiate condition varies seasonally (Brook, 1981b); also dichotypic forms occur.

Staurastrum cingulum (West *et* G.S.West) G.M.Smith 1922

Basionym: *Staurastrum paradoxum* Meyen var. *cingulum* West *et* G.S.West

27380330 **Pl. 139L** (p. 571)

Cells 3-radiate, 15–35 μm wide excluding processes, 60–107 μm with processes, 30–50 μm long excluding processes, 58–110 μm with processes, constriction shallow, sinus U-shaped, isthmus 7.5–12 μm wide; semicells basal region above isthmus subcylindrical, wedge-shaped to broadly cup-shaped, angles produced into slender subparallel to divergent processes of variable length, ornamented with evenly spaced rings of granules or very small spines and processes terminating in 3–5 (mostly 4) small divergent spines; apex slightly convex to flat (in cells with strongly divergent processes even slightly concave), with 3 pairs of intramarginal granules and short rows of granules running onto body from each process; ring of granules encircles the body just above the isthmus.

Europe, North and South America, Asia, New Zealand, Arctic; widespread and true plankter in nutrient-poor to moderately enriched lakes especially in the English Lake District and Scotland, at times forming a significant component of the phytoplankton (pH 5.2–8.3; alkalinity 2–42 ppm $CaCO_3$).

var. *obesum* G.M.Smith 1922

27380332 **Pl. 139Q** (p. 571)

Cells differ from the nominal variety in being more robust and larger body, semicells subspindle-shaped with broadly rounded lateral margins and relatively short, stout processes, 30–45 μm long without processes, 47–65 μm long with processes, 25–42 μm wide without processes, 60–73 μm wide with processes; isthmus 9–12.5 μm wide.

World distribution as the nominal variety; common at times in the plankton of nutrient-rich lakes, especially in the Shetland Islands, Caithness and central Scotland; also in some moderately nutrient-rich lakes of the English Lake District.

Staurastrum crenulatum (Nägeli) Delponte 1877

Basionym: *Phycastrum (Stenactinum) crenulatum* Nägeli
27380390 **Pl. 139I** (p. 571)
Cells 3-radiate, 20–33 μm wide with processes, 20–25 μm long; constriction small, sinus shallow and moderately acute, rounded internally; semicells broadly cup-shaped, angles produced to form short subhorizontal or slightly divergent processes with truncate ends terminating in 4 short spines; apex slightly convex with pairs of intramarginal verrucae (occasionally reduced to pairs of granules) extending onto upper surfaces of processes, denticulate in side view.

Cosmopolitan; fairly widespread throughout the British Isles, in greatest abundance in plankton of moderately to very nutrient-enriched ponds and lakes and in weedy margins (pH 6.2–8.4).

Closely related to *Staurastum pingue* Teiling which may be a planktonic form or variety of *S. crenulatum* (Brook, 1959c).

Staurastrum cyrtocerum Brébisson in Ralfs 1848

27380430 **Pl. 139O** (p. 571)
Cells 3-radiate, 33–60 μm wide including processes, 22–29 μm long, constriction slight, sinus an open shallow notch, isthmus 8–11 μm wide; semicells cup-shaped with ventral margins widening outwards towards angles produced into short, incurved hollow processes, ornamented with concentric rings of granules, ending truncately in 3 or 4 short spines; apex convex with 3 pairs of intramarginal granules or verrucae, granules reported to encircle body above isthmus; in vertical view cells may show median torsion with processes slightly curved clockwise.

Cosmopolitan, acidophile; widespread in the British Isles in bogs and ditches, often associated with *Sphagnum* and *Utricularia*, also in the weedy margins of nutrient-poor ponds and lakes; occasionally a chance plankter.

Staurastrum denticulatum (Nägeli) W.Archer in A.Pritchard 1861

Basionym: *Phycastrum denticulatum* Nägeli
27380440 **Pl. 138K** (p. 567)
Cells 3-radiate, 20–40 μm wide, 24–35 μm long, deeply constricted, sinus open, acute-angled, isthmus 7–14 μm wide; semicells subelliptical or spindle-shaped, upper margin only slightly convex, ventral more strongly convex; angles of semicells tipped with 2 small spines; walls with 2–3 concentric rings of small granules around angles.

Subcosmopolitan; in the British Isles fairly widespread around the margins of nutrient-poor to moderately nutrient-rich lakes of Scotland, Ireland and Wales; sometimes a chance plankter.

Staurastum dilatatum Ehrenberg ex Ralfs 1848

27380460 **Pl. 138E** (p. 567)
Cells 3–5-radiate (mostly 4-radiate), 22–46 μm wide, 21–46 μm long, angles of one semicell alternating with adjacent one, deeply constricted, sinus wide open with a small notch in middle, isthmus 7.5–13 μm wide; semicells elliptical to subspindle-shaped, upper margin slightly convex, lower inflated so that cells appear to have a swollen isthmus, angles rounded or rounded-truncate; walls with granules in concentric rings around angles, fewer on body; apices smooth, punctate.

Cosmopolitan, acidophile; widespread, most frequent in bogs and weedy margins of lakes and ponds, especially in western lake areas of England and Scotland including the Outer Isles, Ireland and Wales where an occasional chance plankter.

Staurastrum furcatum (Ehrenberg) Brébisson 1856

27380610 **Pl. 138L** (p. 567)
Cells 20–40 μm wide, 24–33 μm long including processes; isthmus 6–8 μm wide; median constriction fairly deep, the sinus at its apex an acute notch; semicells transversely oval to subglobose; dorsal and ventral margins more or less equally convex; lateral angles produced into short, stout processes extending horizontally and which end in bifid spines whose two teeth lie in the same vertical plane; the apical margin of each semicell face bears two short, nearly erect, bifid processes.

Cosmopolitan; widespread in acid waters throughout the British Isles, occurring especially amongst *Sphagnum* and *Utricularia*; also associated with submerged macrophytes in the margins of acid ponds and lakes (pH 4–7).

Staurastrum gracile Ralfs 1848

Synonym: *Staurastrum polymorphum* Brébisson ex Ralfs
27380650 **Pl. 139N** (p. 571)
Cells 3-radiate, 23–27 μm wide without processes, 45–60 μm with processes, 30–36 μm long, moderately constricted with a shallow, open U-shaped sinus, isthmus 8–10 μm wide; semicells broadly basin- to cup-shaped, apex convex, angles produced into moderately long, hollow, tapering and slightly convergent processes terminating in 4 short spines, processes ornamented with concentric rings of small granules; apex with 3 evenly-spaced pairs of intramarginal granules from each of which a short row of granules extends down onto cell body, granules encircle body above isthmus.

Cosmopolitan; widespread in the British Isles, most frequent in margins of moderately nutrient-rich ponds and lakes amongst submerged vegetation, also in bogs and marshes associated with filamentous algae.

Staurastrum hirsutum Ehrenberg ex Brébisson in Ralfs 1848

Synonym: *Staurastrum muricatum* var. *hirsutum* Nordstedt
27380710 **Pl. 139A** (p. 571)
Cells 3-radiate, 31–35 μm wide, 34–44 μm long, deeply constricted, sinus almost linear internally, opening outwards; isthmus 10–13 μm wide; semicells subpyramidal-truncate to kidney-shaped or even subsemicircular, with broadly rounded angles; walls covered with delicate spines 1.5–2 μm long in concentric rows around angles.

Cosmopolitan, occuring mostly in moderately acid environments but also in circum-neutral waters; widespread in the British Isles, most frequent in Scotland and Wales.

var. *muricatum* (Brébisson ex Ralfs) Kurt Förster 1970
Basionym: *Staurastrum muricatum* Brébisson ex Ralfs
27380712 **Pl. 139B** (p. 571)
Differs from the nominal variety in having much shorter spines (frequently only small conical granules); cells usually larger, 40–55 µm wide, 46–62.5 µm long.

Cosmopolitan; widely distributed throughout the British Isles in acid to neutral habitats and more frequent than var. *hirsutum*.

Intermediate forms recorded, so that the variety is of questionable validity.

Staurastrum inconspicuum Nordstedt 1873

27380740 **Pl. 139G** (p. 571)
Cells very small, 3–6-radiate, 17–26 µm wide with processes, 17–26 µm long with processes, sinus almost semicircular, isthmus 7–9 µm wide; semicells approximately quadrangular with concave apex, upper angles produced into short, stout hollow processes inflated halfway along length, lower half directed obliquely outwards, the distinctly narrower upper half angled upwards, ending in truncate apices.

Cosmopolitan, acidophile; sometimes frequent in peat bogs and ditches associated with *Sphagnum*, also in weedy margins of ponds and lakes; widespread in the British Isles, most common in Scotland and Ireland.

Staurastrum inflexum Brébisson 1856

27380760 **Pl. 140E** (p. 575)
Cells 3-radiate, 14–17 µm wide excluding processes, 29–41 µm wide with processes, 20–27 µm long, constriction deep with an open, rectangular sinus, isthmus 4.5–7.5 µm wide; semicells oblong-elliptical to subspindle-shaped, lower margins broadly cup-shaped, apex strongly convex; angles extended into moderately long, tapering incurved processes decorated with concentric series of very small granules or denticulations and ending truncately with 3 small spines, single row of intramarginal granules extend across apex, central area and lateral margins smooth.

Subcosmopolitan; widespread in the British Isles, often frequent, associated with *Sphagnum* and *Utricularia*; also in moderately nutrient-rich lake and pond margins; a chance plankter (pH 5–8.6).

Staurastrum johnsonii West *et* G.S.West 1896

Synonym: *Staurastrum leptocladum* L.N.Johnson
27380780 **Pl. 140A** (p. 575) **CD**
Cells always biradiate, 82.5–88.5 µm wide with processes, 35–40 µm long, median constriction shallow, sinus a semicircular notch, isthmus 9.5–10.5 µm wide; semicells cyathiform with inflated base, processes long-tapering denticulate, more strongly denticulate on upper than lower margins, basal inflation with 2 transverse rows of granules, apical margin almost straight with 5 tridentate verrucae.

Boreal distribution; in nutrient-poor ponds and lakes, especially in north-west Scotland and Hebridean Islands; at times moderately abundant.

Staurastrum longipes (Nordstedt) Teiling 1946

Basionym: *Staurastrum paradoxum* Meyen var. *longipes* Nordstedt
27380840 **Pl. 140B** (p. 575)
Cells 3–5-radiate, 15–22 µm wide without processes, 80–139 µm wide with processes, 22–29 µm long without processes, 60–108 µm long with processes, sinus open and shallow, 8–9.5 µm wide; semicells elongate-cylindrical immediately above isthmus, cup-shaped beyond with a concave apex, angles extended into long, slender, hollow divergent processes with crenulations or denticulations along margins and ending in 4 widely spread spines, rest of body smooth; sometimes exhibiting isthmal torsion.

Europe, North and South America, Arctic, New Zealand; widely distributed, restricted to plankton of nutrient-poor to slightly enriched lakes (alkalinity 8–45 ppm $CaCO_3$).

var. *contractum* Teiling 1946
27380842
Cells 3-radiate, 50–62 µm wide, 35–45 µm long with processes; semicells cup-shaped to elongate-triangular with a distinctly concave apex, processes divergent, often irregularly denticulate along margins, sometimes swollen at ends which terminate in 4 distinct spines.

Europe, North America; occasionally frequent in moderately nutrient-rich lakes in the north and west of Scotland.

Staurastrum longispinum (Bailey) W.Archer in A.Pritchard 1861

Synonym: *Didymocladon longispinum* Bailey
27380850 **Pl. 138S** (p. 567)
Cells 3-radiate, 73–100 µm wide excluding spines (spines 9.5–32.5 µm long), 90–120 µm long, deeply constricted, sinus widely open, acute internally, isthmus 36–41 µm wide; semicells broadly cup-shaped with convex lateral margins, angles only slightly produced ending in 2 stout spines projecting slightly upwards and lying close together in same vertical plane, mostly parallel or convergent, apices convex; walls thick, punctate.

Cosmopolitan; a conspicuous desmid occurring in the plankton of nutrient-poor lakes in western areas of the British Isles, most frequent in the west of Scotland.

Spines vary considerably in length; cells with spines >10 µm long named as the doubtfully valid var. *bidentatum* (Wittrock) West; sometimes angles tipped with a single spine (cf. *Staurodesmus*, p. 577).

Staurastrum lunatum Ralfs 1848

27380860 **Pl. 138O** (p. 567)
Cells 3-radiate, 35–43 µm wide excluding spines, 35–39 µm long, constriction deep with acute sinus opening widely to outside, isthmus 10–13 µm wide;

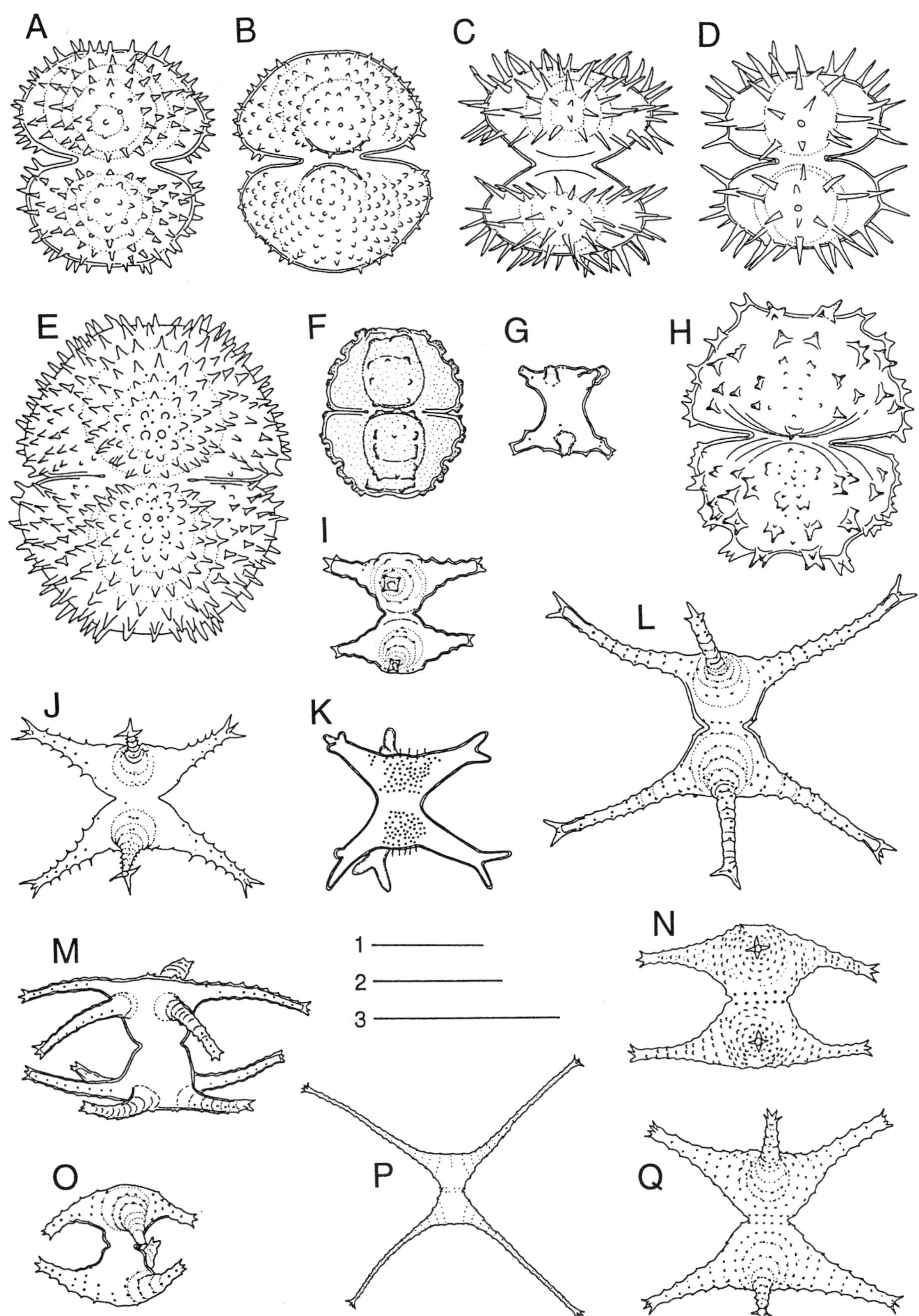

Plate 139 A–Q

A. *Staurastrum hirsutum* (p. 569); **B**. *Staurastrum hirsutum* var. *muricatum* (p. 570); **C**. *Staurastrum polytrichum* (p. 573); **D**. *Staurastrum teliferum* (p. 576); **E**. *Staurastrum pyramidatum* (p. 574); **F**. *Staurastrum maamense* (p. 572); **G**. *Staurastrum inconspicuum* (p. 570); **H**. *Staurastrum spongiosum* (p. 576); **I**. *Staurastrum crenulatum* (p. 569); **J**. *Staurastrum anatinum* (p. 565); **K**. *Staurastrum brachiatum* (p. 566); **L**. *Staurastrum cingulum* (p. 568); **M**. *Staurastrum arachne* (p. 565); **N**. *Staurastrum gracile* (p. 569); **O**. *Staurastrum cyrtocerum* (p. 569); **P**. *Staurastrum chaetoceras* (p. 568); **Q**. *Staurastrum cingulum* var. *obesum* (p. 568).

Scale bars: 1, 25 μm; 2, 50 μm. 1 for A, B, D–I, K–O; 2 for C, J, P; 3 for Q.

semicells semicircular with lower margin markedly convex, upper only slightly so, angles tipped with a single spine (rarely bifid) 3.5–12 μm long and projecting slightly upwards; walls with concentric rings of small granules (sometimes as spicules) adjacent to lateral spines, smaller towards middle.

North Europe, North America, Arctic, Asia; fairly widespread in the British Isles in acid pools and the weedy margins of nutrient-poor lakes in Scotland, Ireland and the English Lake Distict; no records from Wales.

var. *planctonicum* West *et* G.S.West 1903
27380862
Differs from the nominal or type variety in having larger cells, 42–50 μm wide, 40–44 μm long excluding spines; semicells more angular, lower margin less convex, spines on angles smaller, 3.5–5 μm long.

Europe, North America, Arctic, Africa, New Zealand; most frequent in plankton of nutrient-poor to moderately nutrient-rich Scottish lochs, including those in the Outer Hebrides and Shetland Islands; also lakes in the English Lake District, Ireland and Wales.

Staurastrum maamense W.Archer 1869

27380870 **Pl. 139F** (p. 571)
Cells 3-radiate, 30–35 μm wide, 33–42 μm long, oval to almost circular in outline; deeply constricted with a closed, linear sinus, isthmus 8–11 μm wide; semicells semicircular or subpyramidate-truncate, lateral margins moderately curved apex flattened, decorated with 3 or 4 distinctive, emarginate verrucae which when viewed from front appear to extend from above isthmus to apex where they join across apex with those of adjacent angles.

Boreal, acidophile; widespread amongst *Sphagnum* and other aquatics in small pools and nutrient-poor lake margins especially in Ireland and Scotland. Still present in the site from which it was originally described at Maam, Connemara, in the west of Ireland (D.Williamson, 2001).

Staurastrum manfeldtii Delponte 1877

27380880 **Pl. 140C** (p. 575)
Cells 3(–4)-radiate, 55–105 μm wide with processes, 40–60 μm long, slightly constricted with a shallow, notch-like sinus, isthmus 8–11 μm wide; semicells cup-shaped below, broadening towards angles produced into tapering processes which end bluntly with 3 spines; processes horizontal or slightly converging; apex convex with 2 pairs of intramarginal denticulations or verrucae extending onto upper surface of the processes, body mostly smooth, but crescents of granules may occur beneath each process and groups of granules immediately above the isthmus.

Cosmopolitan; widespread in England, Scotland and Ireland, few records from Wales; mostly occurring amongst marginal vegetation of moderately nutrient-rich lakes; also frequently as a chance plankter.

Ornamentation very variable and several varieties described.

Staurastrum margaritaceum [Ehrenberg] Meneghini ex Ralfs 1848

27380890 **Pl. 140F** (p. 575)
Cells 3–9-radiate (mostly 4–6-radiate), 16–48 μm wide with processes, 24–30 μm long, constriction with a very small open sinus, isthmus 6–10 μm wide; semicells cup-shaped to broadly subspindle-shaped, angles produced into short processes, horizontal or incurved, ending truncately with 4 or 5 minute spines, apex convex, sometimes with pairs of intramarginal verrucae, concentric rings of granules around processes extend onto body, sometimes a ring of supra-isthmal granules present.

Cosmopolitan; very widespread in the British Isles, most common amongst *Sphagnum* and *Utricularia* in bog pools, also at margins of nutrient-poor to moderately nutrient-rich ponds and lakes (pH 4.2–7.2).

Staurastrum meriani Reinsch 1867

27380910 **Pl. 138J** (p. 567)
Cells 3–6-radiate (mostly 5-radiate), width at semicell base 17–20 μm, at apex 20–26 μm wide, 26–46 μm long, a very slight median constriction, sinus a small notch, isthmus 13–18 μm wide; semicells trapeziform-rectangular or trapeziform-quadrate, wider at apex than base, apical angles slightly dilated and rounded, apex convex, flat in middle; walls irregular-granulate with ring-like series of granules round angles.

Europe, Arctic, North and South America; widespread, especially in upland areas in lakes, pools and bogs; also amongst mosses and liverworts on dripping rocks.

Staurastrum monticulosum Brébisson in Meneghini ex Ralfs 1848

27380930 **Pl. 145G** (p. 591)
Cells 3-radiate, 35–42 μm wide with processes, 40–57 μm long with processes, deeply constricted with sinus ending acutely or obversely rounded and opening widely to outside, isthmus 13–19 μm wide; semicells subtrapeziform with slightly convex lower lateral margins and distinctly concave upper lateral margins; lateral angles truncate and tipped with 2 spines lying in the same vertical plane; apex straight or slightly concave and at each upper angle bearing a pair of short, conical processes each ending in a short spine (4 usually visible in side view); angles of semicells ornamented with several concentric rings of small granules.

Europe, North America, Arctic, Asia; widely distributed but rare in western regions of the British Isles, recorded from acid bogs in Devon, Cornwall, the English Lake District and West Yorkshire; also in western Wales, Perthshire and Rosshire in Scotland and County Mayo in the west of Ireland.

The var. *bifarium* Nordstedt (27380932), believed to be a growth form of the nominal or type variety and differing only in that the lateral and apical processes are forked (bifid), is said to be much more frequent in the British Isles than the nominal (W.West, G.S.West & Carter 1923, p. 184).

Staurastrum muticum Brébisson ex Ralfs 1848

27380950 **Pl. 138A** (p. 567)
Cells 3–4-radiate, 21–37.5 μm wide, 22–43.5 μm long, deeply constricted with open, acute-angled sinus, isthmus 7.5–12 μm wide; semicells elliptical, semicircular or subkidney-shaped; walls smooth, punctate.

Cosmopolitan; widespread, sometimes abundant in the weedy margins of nutrient-poor ponds and lakes; occasional as a chance plankter.

Staurastrum ophiura P.Lundell 1871

27380000 **Pl. 140D** (p. 575)
Cells 4–9-radiate (mostly 6–9-radiate), 34–46 wide without processes, 128–169 μm with processes, 65–91 μm long, constriction a slight notch; semicells elongate, cup-shaped with slightly divergent lateral margins swelling outwards beneath angles produced into long, slender, horizontally disposed processes ornamented above, below and on sides by single rows of denticulations and tipped with 3 moderately long spines; apex convex with conical nodules or verrucae alternating between processes; walls smooth except for a circle of papillae around semicell base above the isthmus.

Europe, Arctic, North America; 'western' desmid in the British Isles, frequent but never an abundant component of the plankton of nutrient-poor Scottish lochs, mostly in Sutherland and the Outer Hebrides; also occurs in *Sphagnum* bogs (pH 4.9–7).

Staurastrum orbiculare Ralfs 1848

27381010 **Pl. 138B** (p. 567)
Cells 3-radiate, 46–49 μm wide, 54–56 μm long, almost circular in overall outline, deeply constricted with a linear sinus, isthmus 11–12.5 μm broad; semicells almost semicircular, apex slightly depressed; angles broadly rounded; walls punctate.

Cosmopolitan; limited to a few acid localities in England and Wales.

var. *ralfsii* West *et* G.S.West 1911
27381015
Cells distinctly longer than broad, 22–36 μm wide, 31–41 μm long; semicells subtriangular, apex elevated and more convex than the nominal variety.

Cosmopolitan; widely distributed in the British Isles, most common in the weedy margins of poor to moderately nutrient-rich lakes; rare in the plankton.

Staurastrum pingue Teiling 1942

Synonym: *Staurastrum paradoxum* Meyen var. *evolutum* West et al. 1923
27381110 **Pl. 140G** (p. 575)
Cells 3-radiate, 10–15 μm wide without processes, 42–84 μm wide with processes, 25–33 μm long without processes, 42–77 μm long with processes, slightly constricted, sinus shallow V-shaped, isthmus 5–6.5 μm wide; semicells subcylindrical, widening slightly towards angles which extend into long, slender, gradually tapering divergent processes, tipped with 4 spines, apex flat with 3 pairs of intramarginal swellings on which there are either 1 or 2 granules, a single spine or prominent verruca; processes crenulate with a double dorsal and a single ventral row of denticulations along length; cells sometimes showing twisting at isthmus.

Europe, Asia, North and South America, New Zealand; widespread in the plankton of British moderately to highly nutrient-enriched lakes where sometimes dominant and comparatively abundant (pH 6–9.2; alkalinity 7–200 ppm $CaCO_3$).

A fairly reliable indicator of nutrient-enrichment (eutrophy). Note comments about probable relationship with *S. crenulatum* (p. 569; see Brook, 1959c).

Staurastrum planctonicum Teiling 1946

27381120 **Pl. 140I** (p. 575)
Cells 3-radiate, 70–95 μm wide, 57–65 μm long with processes; sinus a V-shaped notch, isthmus 9–13 μm wide; lower part of semicells elongate, cup-shaped flaring upwards into long, slightly divergent, curved processes, markedly denticulate along upper margins and ending in 3 spines, the lower middle one most prominent, apex flat or slightly convex with 3 pairs of intramarginal granules or small verrucae; walls smooth.

Europe, North and South America, Asia; planktonic in moderately to considerably nutrient-enriched lakes especially in lowland areas of the British Isles (pH 6.2–8.9; alkalinity 2–78 ppm $CaCO_3$).

var. *ornatum* (Grönblad) Teiling 1947
Basionym: *Staurastrum dorsidentiferum* var. *ornatum* Grönblad
27381122
Usually larger than the nominal variety, 48–68 μm long without processes, 75–100 μm long with processes, apices always with 3 pairs of intramarginal verrucae, body with 1 or 2 incomplete rings of granules above isthmus and sometimes restricted to groups beneath the processes.

Cosmopolitan; planktonic in lakes of similar nutrient status as the nominal variety but less widespread.

Coesel (1996) questions the validity of this variety on the grounds that its ornamentation is very variable.

Staurastrum polytrichum (Perty) Rabenhorst 1868

Synonym: *Staurastrum pringsheimii* Reinsch
27381140 **Pl. 139C** (p. 571)
Cells 3-radiate, 41–48 μm wide excluding spines, 50–70 μm with spines, 48–67 μm long excluding spines, 54–80 μm with spines, deeply constricted, sinus opening to outside, acutely rounded inside, isthmus 15–22 μm wide; semicells broadly elliptical with a convex apex; walls ornamented with acute spines 5–11 μm long in circles around angles and in longitudinal rows on semicell faces.

Subcosmopolitan; widespread acidophile occurring in ditches and the marginal vegetation of nutrient-poor lakes and ponds, especially in Scotland and the west of Ireland.

Staurastrum proboscideum (Brébisson) W.Archer in A.Pritchard 1861

Basionym: *Staurastrum asperum* var. *proboscideum* Brébisson
27381150 **Pl. 140H** (p. 575)
Cells 3-radiate, 30–53 μm wide, 35–45 μm long, constriction deep, sinus open, isthmus 9–16 μm wide: semicells mostly cup-shaped to broadly spindle-shaped then upper and lower margins equally convex; angles produced as short, robust, horizontal processes with truncate ends tipped with a ring of small granules or spines; upper face of semicell with longitudinal rows of granules, the uppermost of each row an emarginate verruca, base of semicells above an isthmus with a ring of granules.

Subcosmopolitan; widely distributed in bog pools and ponds amongst *Sphagnum* in acid regions of England and Scotland; few records from Wales and Ireland.

The similarity of *S. proboscideum* and *S. gracile* and the doubtful *S. polymorphum* should be carefully noted.

Staurastrum pseudopelagicum West *et* G.S.West 1903

27381160 **Pl. 138Q** (p. 567)
Cells 3-radiate, 20–30 μm wide excluding processes, 63–88 μm with processes, 27–36 μm long excluding processes, 57–71 μm with processes, deeply constricted with a short open sinus, isthmus 7.5–13 μm wide; semicells with lower margin convex, upper only slightly convex, angles extend into fairly slender, hollow processes of variable length tipped with 2 (rarely 3) divergent spines; walls with granules in concentric series around processes, small or absent on semicell body.

Europe, North America, Asia; euplankter of nutrient-poor to moderately nutrient-rich lakes, especially in Scotland including the Outer Hebrides and Orkneys; also in the English Lake District lakes and a few Welsh lakes (pH 5.5–7.8; alkalinity 1.6–45 ppm $CaCO_3$).

Staurastrum punctulatum [Brébisson] Ralfs 1848

7381190 **Pl. 138F** (p. 567) **CD**
Cells mostly 3-, rarely 4- or 5-radiate, 23–26.5 μm wide, 26–40.5 μm long, deeply constricted, sinus open and acute-angled; semicells subrhomboidal-elliptical, upper and lower margins equally convex, angles acutely rounded and alternating with adjacent semicells (cell shape very variable); walls with blunt granules in regular series around angles.

Cosmopolitan, acidophile; the most widespread and common species of the genus; frequent and often abundant in bogs, marshes and shallow pools with other algae and mosses.

var. *kjellmani* Wille 1879
Synonym: *Staurastrum kjellmani* Wille
27381193 **Pl. 138G** (p. 567)
Sinus almost rectangular (about 90°), angles of semicells more rounded than the nominal variety, 24–37.5 μm wide, 30–48 μm long.

Arctic-Alpine; widespread in acid habitats, especially in bog pools in mountainous areas of Wales and Scotland.

var. *pygmaeum* (Brébisson ex Ralfs) West *et* G.S.West 1912
Basionym: *Staurastrum pygmaeum* Brébisson in Ralfs
27381194 **Pl. 138H** (p. 567)
Differs from the nominal variety in that granulations are sharp rather than rounded and blunt and sinus is more open; cells 24–40 μm wide, 27–42 μm long.

Widespread in acidic, nutrient-poor habitats, especially in Scotland and Wales.

Staurastrum pyramidatum West 1892

Synonym: *Staurastrum muricatum* Brébisson var. *acutum* West
27381210 **Pl. 139E** (p. 571)
Cells 3-radiate, 52–57 μm wide, 60–84 μm long, deeply constricted with an almost linear, rarely open sinus, isthmus 16–18 μm broad; semicells broadly pyramidal with a subtruncate apex and slightly convex lateral margins, basal angles broadly rounded; walls with stout, conical spines in concentric series around angles, somewhat irregularly disposed, sometimes absent on apex.

Europe, Africa, Arctic, acidophile; common in the British Isles in bogs and peaty pools, often associated with *Sphagnum* in upland areas, especially in Cumbria, North Yorkshire and Devon, less common in Scotland, Wales and Ireland.

Staurastrum sebaldi Reinsch 1867

27381300 **Pl. 140J** (p. 575)
Cells 3-radiate, 80–92 μm wide with processes, 73–85 μm long, constriction shallow, sinus open and acute, isthmus 20–24 μm wide; semicells broadly basin-shaped, angles produced into moderately stout, convergent processes, ends truncate, tipped with 3–5 spines; apex broad convex with 3 pairs of intramarginal verrucae reducing to a single row of denticulations towards ends of processes; scattered granules or spines may occur on body beneath processes above isthmus.

Cosmopolitan, acidophile; widespread but rarely abundant in the British Isles in bog pools, associated with *Sphagnum*; also in margins of nutrient-poor ponds and lakes; occasionally as a chance plankter.

When body ornament well-developed, then var. *ornatum* Nordstedt (27381202).

Staurastrum setigerum Cleve 1863

27381320 **Pl. 145J** (p. 591)
Cells 3-radiate, 42–45 μm wide excluding spines, 50–56 μm long excluding spines, deeply constricted with an open sinus, acute internally, isthmus 14.5–17 μm wide; semicells elliptical, lower margins more convex than upper with obtusely rounded angles which terminate in 2–5 (mostly 3) stout spines, 15–20 μm long, most frequently arranged in a vertical row but occasionally irregular; apex and lateral margins with a series of less stout spines 10–12 μm long.

Probably cosmopolitan; a western desmid in the British Isles, occurring in the English Lake District, the Capel Curig region of Wales, Sutherland and the Outer Hebrides in Scotland.

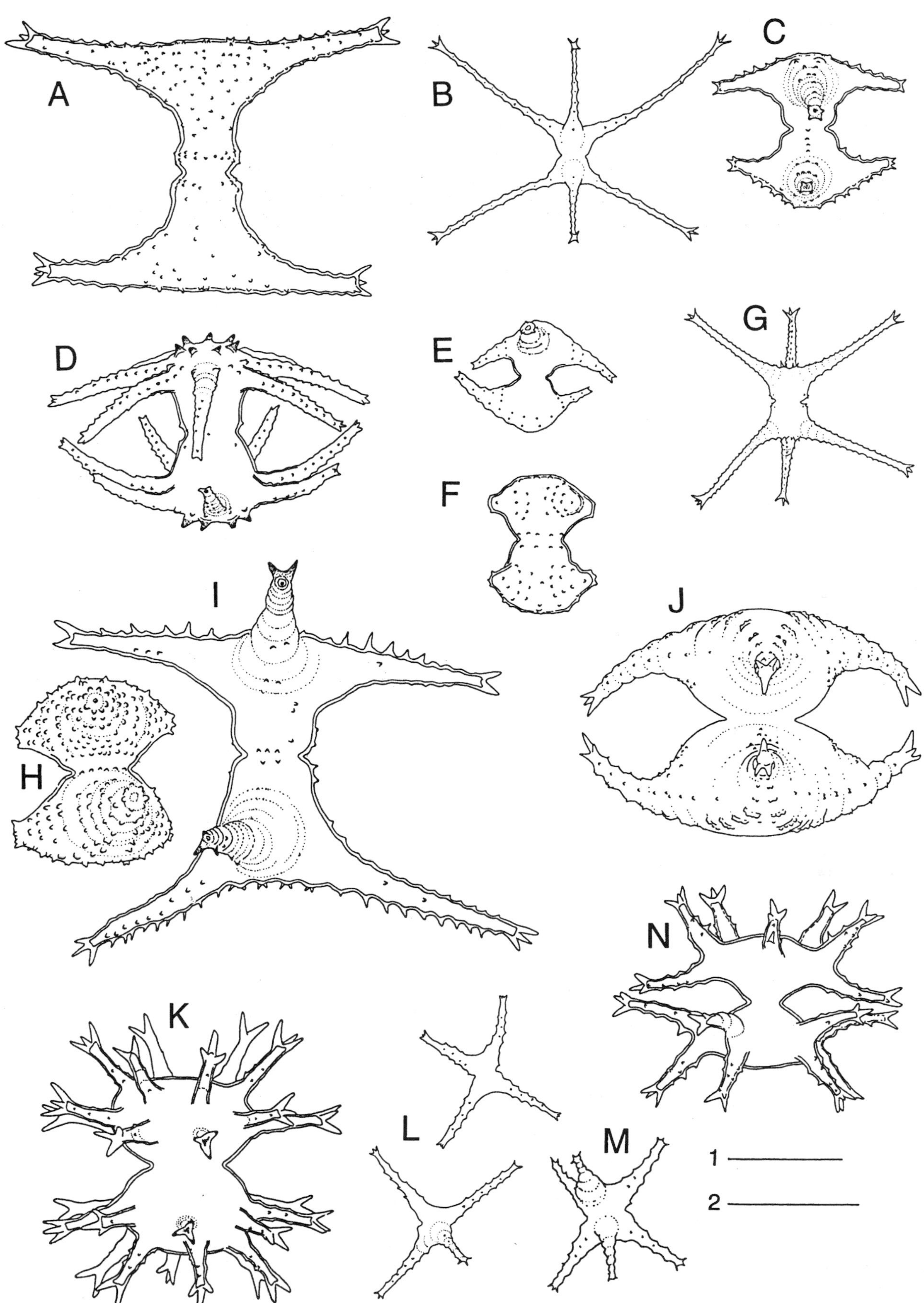

Plate 140 A–M
A. *Staurastrum johnsonii* (p. 570); **B**. *Staurastrum longipes* (p. 570); **C**. *Staurastrum manfeldtii* (p. 572); **D**. *Staurastrum ophiura* (p. 573); **E**. *Staurastrum inflexum* (p. 570); **F**. *Staurastrum margaritaceum* (p. 572); **G**. *Staurastrum pingue* (p. 573); **H**. *Staurastrum proboscideum* (p. 574); **I**. *Staurastrum planctonicum* (p. 573); **J**. *Staurastrum sebaldi* (p. 574); **K**. *Staurastrum arctiscon* (p. 566); **L**. *Staurastrum tetracerum* (p. 576); **M**. *Staurastrum sexangulare* (p. 576).
Scale bars: 1, 25 μm; 2, 50 μm. 1 for A, C, E, F, H, I, L; 2 for B, D, G, J, K, M.

West et al. (1923) comment that *S. setigerum* is the only spiny member of the genus which possesses two distinct sizes of spines, those on the angles contrasting markedly with those ornamenting the semicell faces.

Staurastrum sexangulare (Bulnheim) P.Lundell 1871

Basionym: *Didymocladon sexangulare* Bulnheim

27381330 **Pl. 140M** (p. 575)

Cells 5–7-radiate, 42–54 μm wide without processes, 74–100 μm with processes, 44–60 μm long without processes, 74–100 μm with processes, constriction shallow, sinus U-shaped, isthmus 13–22 μm wide; semicells broad basin-shaped, basal margins diverging into angles each bifurcating to form one divergent and one convergent serrated process ending in 3 blunt spines; apices and body without ornamentation.

Probably cosmopolitan; in the plankton of nutrient-poor lakes in western areas of the British Isles, especially Sutherland, the Outer Hebrides, west of Ireland; also present in the English Lake District and Welsh lakes.

Staurastrum simonyi Heimerl 1891

27381350 **Pl. 138M** (p. 567)

Cells 3-radiate, 18–26 μm wide, 19–25 μm long excluding spines; constriction deep, sinus acute inside and opening widely to outside, isthmus 6–7.5 μm wide; semicells elliptical to subspindle-shaped, upper and lower margins almost equally convex, lateral margins truncate and tipped by 2–4 spines, apical margin with 4 equally spaced spines between each pair of lateral angles, middle ones largest.

Europe, North and South America, Asia, acidophile; widespread in the British Isles, especially in peat bogs and *Sphagnum* pools in Scotland and Ireland.

Staurastrum spongiosum Brébisson ex Ralfs 1848

27381370 **Pl. 139H** (p. 571)

Cells 3-radiate, 42–50 μm wide, 45–53 μm long, approximately circular in outline, deeply constricted, sinus narrow, V-shaped, isthmus about 12 μm wide; semicells subsemicircular to subpyramidate-truncate, each obtuse basal angle ending in a short verrucose process, 8–10 similar emarginate processes occur around semicell periphery, 6 forming a curved series across semicell face.

Worldwide, except not in Australasia; widespread acidophile in the British Isles but never abundant; mostly in bog pools and margins of nutrient-poor ponds and lakes.

Staurastrum teliferum Ralfs 1848

27381450 **Pl.139D** (p. 571) **CD**

Cells 3-radiate, 40–64 μm wide, 32–56 μm long excluding spines, deeply constricted with an open sinus, acutely angled internally, isthmus 8–10 μm wide; semicells elliptical with broadly rounded angles; walls with short, stout spines up to 7 μm long in irregular scattered rings around angles, absent on face between angles, some spines up to 15 μm long on apex.

Cosmoplitan, acidophile; very widespread in the British Isles, frequent in bog pools and margins of nutrient-poor ponds and lakes throughout England, Wales and Scotland including the Orkney and Shetland Islands as well as the Hebrides; frequent in Irish loughs.

Staurastrum tetracerum Ralfs 1848

27381460 **Pl. 140L** (p. 575)

Cells 2–3-radiate (rarely 4-radiate), 18–30 μm wide with processes, 9–12 μm long without processes, deeply constricted, sinus open V-shaped, isthmus 4–6 μm wide; semicells vary from obversely triangular with sublinear lower margin to shallow cup-shaped, margins continue smoothly to angles produced into tapering, slender divergent processes with denticulate margins and slightly swollen ends tipped with 4 minute spines; apex distinctly concave, usually smooth but sometimes with a small granule in the middle; cells exhibit various degrees of isthmal torsion and in extreme cases the processes of one semicell are oriented at 90° to those of the adjacent semicell; walls irregularly ornamented with very small granules.

Cosmopolitan; widespread in the British Isles mostly in the plankton of nutrient-rich alkaline ponds and lakes but also occurs in some nutrient-poor waters (pH 6–9.6(–10)).

As with some other planktonic *Staurastrum* species, bi- and tri-radiate forms of *S. tetracerum* occur in the same population. However, it has yet to be established whether the change in radiation is seasonal, as demonstrated for *S. chaetoceras* (Brook, 1981b).

TAXA RECORDED FROM THE BRITISH ISLES BUT NOT DESCRIBED IN TEXT

Staurastrum acarides Nordst. 27370010; *S. acarides* var. *eboracense* West 27370012; *S. anatinum* var. *aculeatum* (Menegh.) Brook 27380064; *S. anatinum* var. *controversum* (Bréb. in Menegh.) Brook 27380066; *S. anatinum* var. *grande* West *et* G.S.West 27380067; *S. anatinum* var. *truncatum* West 27381553; *S. anatinum* var. *vestitum* (Ralfs) Brook 27381554; *S. arachne* var. *arachnoides* West *et* G.S.West 27380072; *S. arachne* var. *curvatum* West *et* G.S.West 27380073; *S. archerii* West 27380080; *S. arcuatum* Nordst. 27380100; *S. arnellii* Boldt 27380110; *S. arnellii* var. *inornatum* J.Roy 27380112; *S. arnellii* var. *spiniferum* West *et* G.S.West 27380113; *S. avicula* var. *subarcuatum* (Wolle) West *et* G.S.West 27380132; *S. bacillare* Bréb. in Menegh. 27380140; *S. bacillare* var. *obesum* P.Lundell 27380142; *S. barbaricum* West *et* G.S.West 27380150; *S. barbulae* Nygaard 27380160; *S. bibrachiatum* Reinsch emend. Grönblad *et* A.M.Scott 27380170; *S. bicorne* Hauptfl. 27380180; *S. bieneanum* var. *ellipticum* Wille 27380192; *S. bifidum* (Ehrenb.) Bréb. in Ralfs 27380200; *S. bloklandii* Coesel *et* Joost. 27380210; *S. boreale* West *et* G.S.West 27380220; *S. borgeanum* Schm. in Grönblad 27381560; *S. botrophilum* Wolle 27380230; *S. brachiatum* var. *longipedum* Racib. 27380242; *S. brachycerum* Bréb. 27380250; *S. brebissonii* var. *brevispinum* West 27380272; *S. bulbosum* (West) Coesel 27381570; *S. capitulum* Bréb. in Ralfs 27380290; *S. capitulum* var. *dimidio-minus* Grönblad 27380293; *S. capitulum* var. *spetsbergense* (Nordst.) Cooke 27380292; *S. cerastes* var. *triradiatum* G.M.Smith 27380302; *S. chavesii* Bohlin 27380320; *S. cingulum* var. *affine* (West *et* G.S.West) Brook 27380333; *S. clevei* (Wittr.) J.Roy *et* Bisset 27380340; *S. coarctatum* Bréb. 27380350; *S. coarctatum* var. *subcurtum* Nordst. 27380352; *S. cornutum* W.Archer

27380360; *S. cosmospinosum* (Borge) West *et* G.S.West 27380380; *S. cristatum* (Nägeli) W.Archer 27380400; *S. cumbricum* West 27380410; *S. cumbricum* var. *cambricum* West 27380412; *S. cyathipes* A.M.Scott *et* Grönblad 27380420; *S. cyrtocerum* var. *compactum* West *et* G.S.West 27380432; *S. diacanthum* Lemmerm. 27381580; *S. dilatatum* var. *hibernicum* West *et* G.S.West 27380462; *S. dispar* Bréb. 27380470; *S. disputatum* West *et* G.S.West 27380480; *S. donardense* West *et* G.S.West 27380490; *S. dorsidentiferum* West *et* G.S.West 27380500; *S. duacense* West *et* G.S.West 27380510; *S. dubium* West 27380520; *S. eboracense* W.B.Turner 27380530; *S. echinodermum* West *et* G.S.West 27380540; *S. ellipticum* West 27380550; *S. elongatum* J.Barker 27380560; *S. erasum* Bréb. 27380570; *S. forficulatum* P.Lundell 27380590; *S. franconicum* Reinsch 27380600; *S. furcatum* var. *aciculiferum* (West) Coesel 27380613; *S. furcatum* var. *subsenarium* West *et* G.S.West 27380612; *S. furcigerum* var. *reductum* West *et* G.S.West 27380623; *S. gladiosum* W.B.Turner 27380640; *S. gladiosum* var. *delicatulum* West *et* G.S.West 27380642; *S. granulosum* [Ehrenb.] Ralfs 27380670; *S. granulosum* var. *acutum* (Bréb.) West *et* G.S.West 27380672; *S. haaboeliense* Wille 27380680; *S. heimerlianum* Lütkem. 27380690; *S. heimerlianum* var. *spinulosum* Lütkem. 27380692; *S. hexacerum* (Ehrenb.) Wittr. 27380700; *S. hexacerum* var. *semicirculare* Wittr. 27380702; *S. horametrum* J.Roy *et* Bisset 27380720; *S. hystrix* Ralfs ex Ralfs 27380730; *S. inconspicuum* var. *crassum* F.Gay 27380742; *S. inconspicuum* var. *planctonicum* G.M.Sm. 27380743; *S. inflatum* West *et* G.S.West 27380750; *S. kaiseri* (Růžička) Lenzenw. 27380790; *S. laeve* Ralfs 27380800; *S. laevispinum* Bisset 27380810; *S. lapponicum* (Schm.) Grönblad 27380820; *S. longispinum* var. *bidentatum* (Wittr.) West *et* G.S.West 27380852; *S. manfeldtii* var. *annulatum* West *et* G.S.West 27380882; *S. margaritaceum* var. *coronulatum* West 27380892; *S. margaritaceum* var. *hirtum* Nordst. 27380893; *S. margaritaceum* var. *robustum* West *et* G.S.West 27380894; *S. margaritaceum* var. *subcontortum* West *et* G.S.West 27380895; *S. megalonotum* Nordst. 27380900; *S. meriani* var. *minutum* West 27380912; *S. micron* West *et* G.S.West 27380920; *S. micron* var. *perpendiculatum* (Grönblad) Brook 27380923; *S. micron* var. *spinulosum* Coesel 27380924; *S. monticulosum* var. *groenlandicum* Grönblad 27380933; *S. muticum* var. *depressum* J.Roy *et* Bisset 27380952; *S. muticum* var. *extensum* Nordst. 27380953; *S. muticum* var. *hibernicum* West *et* G.S.West 27380954; *S. muticum* var. *ralfsii* West *et* G.S.West 27380955; *S. natator* West 27380960; *S. nodosum* West *et* G.S.West 27380980; *S. oligacanthum* Bréb. ex W.Archer 27380990; *S. oligacanthum* var. *incisum* West 27380992; *S. ophiura* var. *cambricum* (P.Lundell) West *et* G.S.West 27381002; *S. orbiculare* var. *depressum* J.Roy *et* Bisset 27381012; *S. orbiculare* var. *extensum* Nordst. 27381016; *S. orbiculare* var. *hibernicum* West *et* G.S.West 27381013; *S. oxyacanthum* W.Archer 27381020; *S. pelagicum* West *et* G.S.West 27381050; *S. pendulum* var. *pinguiforme* Croasdale 27381062; *S. pileolatum* Bréb. in Ralfs 27381080; *S. pileolatum* var. *cristatum* Lütkem. 27381082; *S. pilosellum* West *et* G.S.West 27381090; *S. pilosum* (Nägeli) W.Archer 27381100; *S. pseudoseboldi* Wille 27381170; *S. pseudoseboldi* var. *simplicius* West 27381172; *S. pseudotetracerum* (Nordst.) West 27381180; *S. punctulatum* var. *coronatum* (Schm.) West *et* G.S.West 27381192; *S. punctulatum* var. *striatum* West *et* G.S.West 27381195; *S. punctulatum* var. *subproductum* West *et* G.S.West 27381196; *S. pungens* Bréb. in Ralfs 27381200; *S. quadrangulare* Bréb. in Ralfs 27381220; *S. quadrispinatum* W.B.Turner 27381230; *S. retusum* W.B.Turner 27381250; *S. retusum* var. *boreale* West *et* G.S.West 27381252; *S. rhabdophorum* Nordst. 27381260; *S. rugulosum* Bréb. in Ralfs 27381270; *S. saxonicum* Bulnh. in Rabenh. 27381280; *S. scabrum* Bréb. in Ralfs 27381290; *S. sebaldi* var. *productum* West *et* G.S.West 27381303; *S. senarium* [Ehrenb.] Ralfs 27381310; *S. sexcostatum* Bréb. in Ralfs 27381340; *S. sexcostatum* var. *productum* West 27381342; *S. spiniferum* West 27381360; *S. spongiosum* var. *griffithsianum* (Nägeli) Lagerh. in Wittr. & Nordst. 27381372; *S. spongiosum* var. *perbifidum* West 27381373; *S. striolatum* (Nägeli) W.Archer 27381380; *S. subavicula* (West) West *et* G.S.West 27381390; *S. subcruciatum* Cooke *et* Wills 27381400; *S. sublaevispinum* West *et* G.S.West 27381410; *S. subnudibrachiatum* West *et* G.S.West 27381420; *S. subpygmaeum* West 27381590; *S. subscabrum* Nordst. 27381440; *S. tetracerum* var. *evolutum* West *et* G.S.West 27381463; *S. tetracerum* var. *irregulare* (West *et* G.S.West) Brook 2731464; *S. tohopekaligense* Wolle 27381470; *S. trachytithophorum* West *et* G.S.West 27381490; *S. turgescens* De Not. 27381500; *S. uhtense* Grönblad 27381520; *S. verticillatum* W.Archer 23781540.

Staurodesmus Teiling 1948

Cells mostly 3-angled (triradiate), less commonly biradiate or quadriradiate; each angle, shape of which is taxonomically important, bears a single spine varying in length and curvature (in some reduced to a papilla or even a thickened angle); spines bending upwards (divergent), curving downwards towards isthmus (convergent) or extending at right angles to vertical axis of the cell (parallel), solid or hollow; walls smooth and having fairly distinctive pores and in several species, pores in regular patterns of taxonomic significance, pores often extruding mucilage plugs; most cells have a distinct median constriction ranging from closed to widely open or elongate, sinus varying from acute to obtuse, or to parallel-sided when the isthmus is an elongate collar; semicells vary from narrow to broadly ovoid, subkidney-shaped, cup-shaped, flask-shaped, halfmoon-shaped to rectangular.

Most species are restricted to acid, nutrient-poor ponds and lakes often occurring in plankton and are considered good indicators of such waters (Brook, 1965). However, some of the largest planktonic desmid populations ever recorded have been of *Staurodesmus* species in lakes highly enriched with organic matter (see Brook, 1981b, and the note on *Staurodesmus incus* var. *ralfsii*, p. 581 below).

The single spine ornamenting each angle of the cell is the principal distinguishing character of the genus *Staurodesmus*.

1 Cells with elongate isthmus 2
1 Cell isthmus not elongate but closed or open 6
2 Angles mammillate, not merging evenly into spines ....*S. mamillatus*
2 Angles not mammillate 3
3 Semicells triangular or spindle-shaped 4
3 Semicells cup-shaped or subrectangular 5
4 Cells uniformly triangular, 2-radiate; spines mostly parallel ....*S. triangularis*
4 Cells spindle-shaped, 2–3-radiate; angles tapering and merging evenly into spines; spines divergent to convergent..*S. cuspidatus*

5 Semicells cup-shaped, triradiate .... *S. dejectus*, in part
5 Semicells subrectangular, biradiate .... *S. extensus*
6 Sinus angular or narrow v-shaped 7
6 Sinus narrow, closed 13
7 Spines absent, reduced to warts or papillae; cell 3-radiate 8
7 Spines present; cells (2–)3–4-radiate 9
8 Semicells broadly ellipsoidal, sinus narrow V-shaped .... *S. brevispina*
8 Semicells inversely triangular, sinus open V-shaped .... *S. aversus*
9 Spines convergent 10
9 Spines divergent or parallel 19
10 Semicells transversly ovoid or ellipsoid 11
10 Semicells triangular to cup-shaped 12
11 Semicells ellipsoid, sinus closed; biradiate .... *S. convergens*
11 Semicells with semicircular outline, sinus open; triradiate .... *S. dickiei*
12 Semicell apex slightly convex, sinus acute open; 2–3–4-radiate; spines short .... *S. glaber*
12 Semicell apex flat, sinus rounded and open; 2-radiate; spines short .... *S. incus* var. *ralfsii*
13 Spines divergent 14
13 Spines parallel 18
14 Semicells cup-shaped in whole or in part 15
14 Semicells triangular 16
15 Semicells completely cup-shaped, biradiate, sinus acute; spines long, divergent .... *S. validus*
15 Semicells completely cup-shaped, triradiate, sinus acute; spines short, divergent ..... *S. dejectus*, in part
16 Semicells with lower part cup-shaped, upper part subrectangular, lateral margins indented above isthmus .... *S. indentatus*
16 Semicells obversely triangular, apex convex 17
17 Semicells triangular to broadly cup-shaped; spines stout, hollow, slightly curved; 2–4-radiate .... *S. sellatus*
17 Semicells obversely triangular, apex flat ..... *S. incus*
18 Sinus shallow; spines short, slender .... *S. crassus*
18 Sinus deep, widely open; spines long; 3-radiate .... *S. megacanthus* var. *scoticus*
19 Semicells triangular; spines parallel 20
19 Semicells ovoid to rectangular; spines divergent 21
20 Semicells obversely triangular; sinus deep, widely open; 3-radiate .... *S. megacanthus*
20 Semicells triangular with lateral margins distinctly indented and widening above isthmus; apex raised, with a slight median notch .... *S. subtriangularis*
21 Semicells broadly cup-shaped to ovoid, apex convex .... *S. subulatus*
21 Semicells transversely rectangular; spines long, strongly divergent closed .... *S. bulnheimii*

Staurodesmus aversus (P.Lundell) S.Lillieroth 1950

Basionym: *Staurastrum aversum* P.Lundell
27390030 **Pl. 141D** (p. 580)
Cells triradiate, 29–45 μm wide, 53–58 μm long, with deep median constriction and acutely angled sinus opening widely to outside; isthmus 10–17 μm wide; semicells ellipsoidal-semicircular with truncate or slightly retuse apices, ventral margin curving smoothly from isthmus to broadly rounded upper angles ending in a small papilla; angles in apical view broadly rounded, sides concave.

North-west Europe, Canada, Japan; restricted to littoral of nutrient-poor lakes mainly in western areas of the British Isles, especially Scotland, Wales and Ireland (alkalinity 1–9 ppm $CaCO_3$).

Staurodesmus brevispina (Brébisson ex Ralfs) Croasdale 1957

Basionym: *Staurastrum brevispinum* Brébisson ex Ralfs
27390040 **Pl. 141C** (p. 580)
Cells triradiate (rarely biradiate), 27–49 μm wide, 27–50 μm long, with deep median constriction, sinus acute-angled internally, closing then opening to outside; isthmus 8–17 μm wide; semicells oblong-elliptical, lower margins more convex than upper, between which angles furnished with small mucro or papilla; apex convex or somewhat flattened; angles in vertical view broadly rounded, sides concave.

Subcosmopolitan; widely distributed throughout the British Isles mostly in acid pools and nutrient-poor to moderately nutrient-rich ponds and lakes; also in plankton as a chance plankter (pH 5–6.8; alkalinity 1–38 ppm $CaCO_3$).

Staurodesmus bulnheimii (Raciborski) Round *et* Brook 1959

Basionym: *Arthrodesmus bulnheimii* Raciborski
27390050 **Pl. 145C** (p. 591)
Cells biradiate, without spines 30–40 μm wide, 32–42 μm long, spines 18–27 μm long; deeply constricted with a narrow, linear sinus; isthmus 7.5–10 μm wide; semicells transversely rectangular with slightly convex sides and apex; angles rounded and apical angles with long, stout, divergent spines.

Boreal distribution, acidophile; rare in the British Isles, restricted to the west of Scotland, the Outer Hebrides and a few nutrient-poor lakes in the west of Ireland.

Staurodesmus convergens (Ehrenberg ex Ralfs) Teiling 1948

Basionym: *Arthrodesmus convergens* Ehrenberg ex Ralfs
27390090 **Pl.141A** (p. 580)
Cells biradiate, without spines 32–64 μm, with spines 45–90 μm wide, 30–54 μm long, deeply constricted, sinus opening outwards; isthmus 8–17 μm wide; semicells elliptical, upper margin slightly more convex than lower; each angle with a stout downward curving spine.

Cosmopolitan; widespread and frequent in margins of nutrient-poor and moderately nutrient-rich lakes and in their plankton (pH 5.5–7.2).

Spines significantly shorter when in the plankton.

Staurodesmus crassus (West *et* G.S.West) Florin 1957

Basionym: *Arthrodesmus crassus* West *et* G.S.West

27390120 **Pl. 141L** (p. 580)

Cells small, 14–30 μm wide without spines, 14–30 μm long, angles with short, acute spines, moderately constricted with a widely open sinus, sometimes twisted about isthmus, 8–13 μm wide; semicells obversely subtriangular, with slightly convex sides and apex.

Boreal distribution; a common member of the plankton in nutrient-poor English, Scottish and Irish lakes, sometimes frequent but not reported in Welsh lakes.

When in the plankton, cells usually surrounded by a wide mucilage envelope.

Staurodesmus cuspidatus (Brébisson ex Ralfs) Teiling 1967

Basionym: *Staurastrum cuspidatum* Brébisson ex Ralfs

27390130 **Pl. 141I** (p. 580)

Cells without spines 18–30 μm, with spines 42–53 μm wide, 22–35 μm long; mostly triradiate but sometimes bi- or quadriradiate or dichotypic (mostly 2–3-radiate); isthmus narrow, elongate and tube-like, 4–7 μm wide (sometimes with a median, very shallow V-shaped notch); semicells elongate-triangular or spindle-shaped; cell angles attenuating evenly into parallel, divergent or slightly convergent spines.

Cosmopolitan; widespread but never common, in littoral amongst submerged macrophytes of nutrient-poor to moderately nutrient-rich lakes and ponds and very occasionally in their plankton, especially in Scotland and Ireland

S. cuspidatus and *S. mamillatus* are often difficult to distinguish, and some authors (e.g. Coesel, 1994, p. 25) have suggested that they should be merged as one species. They are separated from one another on the basis of the shape of the angles and whether the angles attenuate evenly into the spines (*S. cuspidatus*), or are rounded and do not so attenuate (*S. mamillatus*, see p. 581). The latter, much more common species in the British Isles, is additionally characterized by the presence of a ring of pores (rarely 2 rings), often with distinct extruded mucilage plugs, around each angle.

Staurodesmus dejectus (Brébisson ex Ralfs) Teiling

Basionym: *Staurastrum dejectum* Brébisson ex Ralfs

27390140 **Pl. 141G** (p. 580)

Cells triradiate, 16–27 μm wide, 10–20 μm long, sinus widely open and rounded internally (sometimes wide V-shaped), isthmus slightly elongate, 5–10 μm wide; semicells cup-shaped with short, slightly divergent spines.

Cosmopolitan; very common in acid bogs and ditches, also frequent in the littoral and plankton of nutrient-poor to moderately nutrient-rich lakes and ponds.

The dimensions of the fo. *major* are 19–30 μm wide and 24–40 μm long.

var. *apiculatus* (Brébisson) Teiling 1955

Basionym: *Staurastrum apiculatum* Brébisson

27390142 **Pl. 141S** (p. 580)

Differs from the nominal variety in that cells are almost as broad as long, varying from 17–25 μm wide and 17–25 μm long, with a flat or slightly convex apex and small, vertically directed spines.

Europe, North America, Asia; in the British Isles mostly in the littoral of nutrient-poor lakes and ponds; also in acid bogs and pools, often amongst *Sphagnum*.

Staurodesmus dickiei (Ralfs) S.Lillieroth 1950

Basionym: *Staurastrum dickiei* Ralfs

27390150 **Pl. 141B** (p. 580)

Cells triradiate, 26–44 μm wide, 20–46 μm long, median incision deep with a narrow rounded sinus opening outwards; isthmus 7–15 μm wide; semicells elliptical with upper and lower margins equally convex, so sometimes semicells semicircular and the cells circular (var. *circularis* (W.B.Turner) Croasdale 27390152); angles with short, convergent, curved spines.

Cosmopolitan; widespread in the British Isles, most frequent in bogs and marshes, also littoral of nutrient-poor to moderately nutrient-rich lakes; occasional in plankton as a chance plankter (pH 5–7.4).

Staurodesmus extensus (Borge) Teiling 1948

Basionym: *Arthrodesmus incus* var. *extensus* (O.F.Andersson) Borge

27390160 **Pl. 141R** (p. 580)

Cells biradiate, 13–21 μm wide, 16–30 μm long, spines 8–14 μm long; semicells subrectangular with strongly narrowed basal part which with adjoining semicell forms a long incurved isthmus 5–7 μm wide with a rounded sinus; apex slightly concave or straight.

Essentially boreal in distribution; mostly in the plankton of nutrient-poor to moderately nutrient-rich lakes in Scotland, the English Lake District and Wales (pH 5.2–7.2).

Staurodesmus glaber (Ehrenberg) Teiling 1967

Basionym: *Desmidium glabrum* Ehrenberg

Synonym: *Staurastrum glabrum* [Ehrenberg] Ralfs

27390170 **Pl. 141F** (p. 580)

Cells 2-, 3- or 4-radiate, 13–25 μm wide, 15–33 μm long; median constriction wide V-shaped, obtusely rounded internally; isthmus 5–10 μm wide; semicells inverse triangles with straight or slightly convex lateral margins and flattened apex; angles with short incurved spines.

Europe, North America, Asia, Japan; in the British Isles widespread in acid bogs, especially in Scotland, and in the littoral of nutrient-poor lakes and ponds; a chance plankter.

Staurodesmus incus (Brébisson) Teiling 1967

Basionym: *Binatella incus* Brébisson

Synonym: *Arthrodesmus incus* [Brébisson] Hassall ex Ralfs

27390190 **Pl. 145D** (p. 591)

Cells biradiate, 16–22 μm wide, 19–27 μm long, deeply constricted with a widely open, obtuse angled sinus;

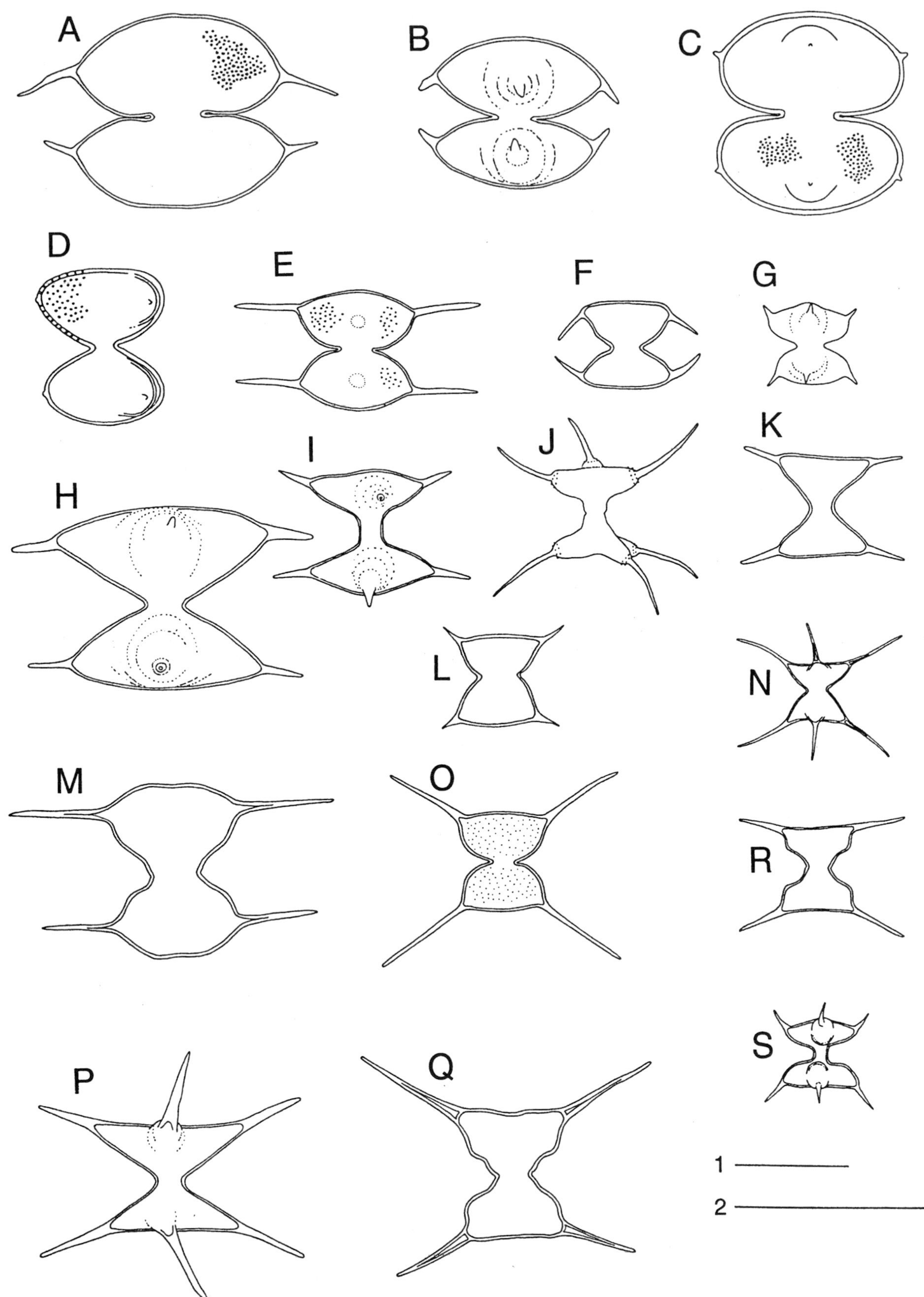

Plate 141 A–S

A. *Staurodesmus convergens* (p. 578); **B**. *Staurodesmus dickiei* (p. 579); **C**. *Staurodesmus brevispina* (p. 578); **D**. *Staurodesmus aversus* (p. 578); **E**. *Staurodesmus subulatus* (p. 582); **F**. *Staurodesmus glaber* (p. 579); **G**. *Staurodesmus dejectus* (p. 579); **H**. *Staurodesmus megacanthus* (p. 581); **I**. *Staurodesmus cuspidatus* (p. 579); **J**. *Staurodesmus mamillatus* (p. 581); **K**. *Staurodesmus triangularis* (p. 582); **L**. *Staurodesmus crassus* (p. 579); **M**. *Staurodesmus subtriangularis* (p. 581); **N**. *Staurodesmus validus* (p. 582); **O**. *Staurodesmus sellatus* (p. 581); **P**. *Staurodesmus megacanthus* var. *scoticus* (p. 581); **Q**. *Staurodesmus indentatus* (p. 581); **R**. *Staurodesmus extensus* (p. 579); **S**. *Staurodesmus dejectus* var. *apiculatus* (p. 579). Scale bars: 1, 25 μm; 2, 50 μm. 1 for A–C, E–L, N, P; 2 for D, M, O, Q, R, S.

isthmus 5–7 μm wide; semicells triangular with lateral, margins slightly convex; apex straight or very slightly concave, each angle with a long, stout, divergent spine.

Cosmopolitan; widespread in the British Isles in margins of nutrient-poor to moderately nutrient-rich lakes and ponds, frequent in their plankton.

The fo. *minor* West *et* G.S.West, 12–14 μm wide and 12–17 μm long, is more abundant than the nominal forma especially amongst *Sphagnum* and in bogs and peaty pools.

var. *ralfsii* (West) Teiling 1967

Basionym: *Arthrodesmus ralfsii* West

27390192 **Pl. 145E** (p. 591)

Semicells trapeziform or cup-shaped, 18–20 μm wide without spines, 44–47 μm wide with spines, 27–33 μm long; cells with a narrower sinus, acute or rounded internally; spines smaller than in the nominal variety, parallel or slightly convergent.

Widespread in habitats similar to those of the nominal variety, also frequent amongst marginal plants of nutrient-poor lakes, especially lochs in the Outer Hebrides. Record bloom of 30,000 cells per mL from enriched Siblybae Reservoir, Liskeard, Cornwall (see Brook, 1981a, p. 232).

Staurodesmus indentatus (West *et* G.S.West) Teiling 1948

Basionym: *Arthrodesmus incus* (Brébisson) Hassall ex Ralfs var. *indentatus* West *et* G.S.West

27390290 **Pl. 141Q** (p. 580)

Cells mostly biradiate, some triradiate, 18–28 μm wide without spines, 63–74 μm with spines, 20–44 μm long; constriction deep with wide open sinus, rectangular to obtuse-angled inside; isthmus 7–12 μm wide; sides of semicell indented above isthmus so upper part subrectangular and lower more or less cup-shaped; apex flat, angles tipped with long stout, usually divergent, sometimes parallel spines.

Europe, North America, North Africa, Japan; fairly widespread in the British Isles but most frequent in Scotland, Wales and Ireland in boggy and weedy margins of nutrient-poor lakes.

Staurodesmus mamillatus (Nordstedt) Teiling 1967

Basionym: *Staurastrum mamillatum* Nordstedt

27390250 **Pl. 141J** (p. 580)

Cells 2–4-radiate with extended tube-like isthmus, 5–7 μm wide, in the middle of which there may be a very shallow, V-shaped sinus; cells 15–28 μm wide without spines, 36–40 μm with spines, 28–33 μm long; semicells ovoid or occasionally narrow-triangular, angles rounded or on rare occasions with slightly swollen angles, very commonly with a ring of pore-organelles with mucilage plugs, surrounding base of the long, stout, usually parallel spines (occasionally divergent or convergent), which do not merge evenly but arise abruptly from the semicell angles.

Cosmopolitan; in the British Isles widespread, most frequently in the littoral of nutrient-poor lakes especially in Scotland and Ireland; also present in a few English and Welsh lakes; frequent at times as a chance plankter (pH 5.8–7.8; alkalinity 4–78 ppm $CaCO_3$).

Note comments concerning confusion between this species and the less common *S. cuspidatus* (see above).

Staurodesmus megacanthus (P.Lundell) Thunmark 1948

Basionym: *Staurastrum megacanthum* P.Lundell

27390260 **Pl. 141H** (p. 580)

Cells mostly 3-radiate, some 4-radiate, 39–61 μm wide without spines, 35–61 μm long, spines 11–18 μm long, deeply constricted with an obtuse sinus; isthmus 8–10 μm wide; semicells triangular to transverse spindle-shaped with straight or convex margins and a slightly convex apex; angles attenuating to stout, parallel or divergent spines.

Cosmopolitan; in the British Isles frequent in the littoral of nutrient-poor lakes and ponds in the English Lake District, and especially in Welsh, Scottish and Irish lakes; also a chance plankter.

var. *scoticus* (West *et* G.S.West) S.Lillieroth 1950

Basionym: *Staurastrum megacanthum* P.Lundell var. *scoticums* West *et* G.S.West

27390263 **Pl. 141P** (p. 580)

Differs from the nominal variety in that apex is straight or slightly concave and sinus more open; spines relatively longer and more divergent; cells 38–51 μm wide (without spines), 79–111 μm (with spines), 35–44 μm long without spines; isthmus 8–10.5 μm wide.

Boreal distribution; in plankton of many nutrient-poor Scottish lochs especially in Sutherland and the Outer Hebrides; also common in some Irish loughs.

Staurodesmus sellatus Teiling 1948

Synonym: *Arthrodesmus incus* [Brébisson] Hassall ex Ralfs var. *sellatus* Teiling

27390370 **Pl. 141O** (p. 580)

Cells 2–4-radiate, 15–25 μm wide without spines, 19–26 μm long, spines 20–36 μm long, fairly deeply constricted with sinus very open and obtuse-angled; isthmus 6–9 μm wide; semicells triangular or broadly cup-shaped with slightly convex lateral margins; apex retuse, flat or slightly convex and angles extending into a stout, hollow slightly curved spine, 20–36 μm long.

Northern Europe, Newfoundland, Tierra del Fuego, New Zealand; in the British Isles in the weedy margins and plankton of nutrient-poor lakes in the English Lake District, Wales and especially Scotland.

Staurodesmus subtriangularis (Borge) Teiling 1967

Basionym: *Arthrodesmus incus* [Brébisson] Hassall ex Ralfs var. *subtriangularis* Borge

27390420 **Pl. 141M** (p. 580)

Cells mostly 2-radiate, sometimes 3-radiate, 18–35 μm wide without spines, 20–43 μm long; deeply constricted with open angular sinus; isthmus 6.5–9 μm wide; lateral margins noticeably indented above isthmus; apex raised and with broad, shallow median notch, angles blunt, ending in moderately long, mostly parallel spines.

Europe, North America; in the British Isles frequent at times in margins and plankton of nutrient-poor lochs in Scotland and in English Lake District lakes; less common in lakes in Wales and Ireland.

Staurodesmus subulatus (Kützing) Thomasson 1963

Basionym: *Arthrodesmus subulatus* Kützing

27390430 **Pl. 141E** (p. 580)

Cells biradiate, 27–54 µm wide with spines, 28–50 µm long; deeply constricted with an open sinus; isthmus 7–13.5 µm wide; semicells elliptical-subsemicircular, apex convex but less curved than lateral margins; angles rounded with long, straight horizontal spines; walls finely punctate.

Europe, North and South America, Asia, Africa; in the British Isles restricted to margins and plankton of small, nutrient-poor lakes mostly in the west of Scotland and Ireland.

Staurodesmus triangularis (Lagerheim) Teiling 1967

Basionym: *Arthrodesmus triangularis* Lagerheim

27390440 **Pl. 141K** (p. 580)

Cells mostly biradiate, rarely triradiate, 19–25 µm wide (excluding spines), 19–25 µm long, deeply constricted with a wide, almost semicircular sinus; isthmus short, 5–7 µm wide, cylindrical, cells often twisted at isthmus; semicells triangular with convex margins and a distinctive apex, slightly retuse in middle but somewhat elevated above bases of spines, usually parallel and 12.5–25 µm long.

Cosmopolitan; in the British Isles most frequent in plankton of nutrient-poor lakes in the English Lake District, Scotland and Ireland and in weedy lake margins.

Staurodesmus validus (West *et* G.S.West) Thomasson 1960

Basionym: *Staurastrum incus* (Brébisson) Hassall var. *validus* West *et* G.S.West

27390460 **Pl. 141N** (p. 580)

Cells triradiate, 29–36 µm wide, 33–40 µm long without spines; obversely subsemicircular with an almost flat apex; isthmus 7–11 µm wide; spines 27–42 µm long, stout and divergent with hollow bases.

Europe, North and South Ameria, Asia, Africa; in the British Isles most common in the littoral of ponds and lakes in the west of Scotland (especially Sutherland) and the west of Ireland.

TAXA RECORDED FROM THE BRITISH ISLES BUT NOT DESCRIBED IN TEXT

Staurodesmus angulatus (West) Teiling 27390010; *S. angulatus* var. *planctonicum* (West *et* G.S.West) Teiling 27390012; *S. aristiferus* (Ralfs) Thomasson 27390020; *S. aristiferus* var. *protuberans* (West *et* G.S.West) Teiling 27390022; *S. brevispina* var. *altus* (West *et* G.S.West) Teiling 27390042; *S. brevispina* var. *boldtii* (Lagerh.) Croasdale 27390043; *S. brevispina* var. *obversus* (West *et* G.S.West) Croasdale 27390044; *S. brevispina* var. *retusum* (Borge) Brook *et* D.B. Will. 27390045; *S. bulnheimii* var. *subincus* (West *et* G.S.West) Thomasson 27390052; *S. connatus* (P.Lundell) Thomasson 27390060; *S. conspicuus* (West *et* G.S.West) Teiling 27390070; *S. controversus* (West *et* G.S.West) Teiling 27390080; *S. convergens* var. *laportei* Teiling 27390092; *S. convergens* var. *pumilus* (Nordst.) Teiling 27390093; *S. convergens* var. *ralfsii* Teiling 27390094; *S. corniculatus* (P.Lundell) Teiling 27390100; *S. cuspidatus* var. *curvatus* (West) Teiling 27390132; *S. dickiei* var. *circularis* (W.B.Turner) Croasdale 27390152; *S. dickiei* var. *rhomboideus* (West *et* G.S.West) S.Lill. 27390153; *S. extensus* var. *longispinus* (West et G.S.West) Teiling 27390163; *S. extensus* var. *joshuae* (Gutw.) Teiling 27390162; *S. glaber* var. *debaryanus* (Nordst.) Teiling 27390172; *S. grandis* (Bulnh.) Teiling 27390180; *S. grandis* var. *parvus* (West) Teiling 27390182; *S. inelegans* (West *et* G.S.West) Teiling 27390210; *S. jaculiferus* (West) Teiling 27390220; *S. lanceolatus* (W.Archer) Croasdale 27390230; *S. lanceolatus* var. *compressus* (West *et* G.S.West) Teiling 27390232; *S. leptodermus* (P.Lundell) Teiling 27390240; *S. mamillatus* var. *maximus* (West) Teiling 27390252; *S. megacanthus* var. *orientalis* (A.M.Scott *et* Prescott) Teiling 27390262; *S. megacanthus* var. *subcurvatus* (M.F.Rich) Teiling 27390264; *S. minutissimus* (Reinsch) Teiling 27390270; *S. mucronatus* (Ralfs) Croasdale 27390280; *S. mucronatus* var. *subtriangulatus* (West *et* G.S.West) Croasdale 27390282; *S. pachyrhinchus* (Nordst.) Teiling 27390310; *S. patens* (Nordst.) Croasdale 27390320; *S. patens* var. *minutus* (West) Teiling 27390322; *S. phimus* (W.B.Turner) Thomasson 27390330; *S. phimus* var. *hebridarus* (West *et* G.S.West) Teiling 27390332; *S. phimus* var. *occidentalis* (West *et* G.S.West) Teiling 27390333; *S. pterosporus* (P.Lundell) Bourr. 27390340; *S. quadratus* (Schm.) Teiling 27390350; *S. quirificus* (West *et* G.S.West) Teiling 27390360; *S. sibiricus* (Borge) Croasdale 27390380; *S. spencerianus* (Mask.) Teiling 27390390; *S. spetsbergensis* (Nordst.) Teiling 27390400; *S. spetsbergensis* var. *floriniae* Teiling 27390402; *S. subpygmaeus* (West) Croasdale 27390410; *S. subpygmaeus* var. *subangulatus* (West *et* G.S.West) Teiling 27390412; *S. subtriangulatus* var. *inflatus* (West *et* G.S.West) Teiling 27390422; *S. subulatus* var. *subaequalis* (West *et* G.S.West) Thomasson 27390432; *S. triangularis* var. *americanus* (W.B.Turner) Teiling 27390442; *S. triangularis* var. *limneticus* Teiling 27390444; *S. triangularis* var. *subhexagonus* (West *et* G.S.West) Teiling 27390445; *S. tumidus* (Bréb. ex Ralfs) Teiling 27390450.

Tetmemorus Ralfs ex Ralfs 1848

Cells 4–9 times as long as broad, straight and slightly compressed, not quite circular in apical view and having a distinctive face and side view, slightly broader face view, characterized by a moderately deep incision in broadly rounded apices, the slightly narrower side view tending to be broadly spindle-shaped with truncate or subtruncate apices; median constriction apparent in both views; walls with scattered or well-defined rows of puncta, or scrobiculations, puncta, most clearly defined in the rows on each side of isthmus; chloroplasts 2 per cell, star-shaped each with radiating longitudinal ridges and a series of axile pyrenoids, except in smaller cells of the genus which may have only 1 or 2 pyrenoids per chloroplast.

Cosmopolitan; occurring almost exclusively in acid nutrient-poor waters often in considerable abundance, especially in bogs and peaty pools with other desmids; also sometimes as a chance plankter in lakes and ponds.

1 Cells spindle-shaped in face view and side view .... *T. granulatus*
1 Cells cylindrical or subcylindrical in face view 2
2 Cells cylindrical in face view, >150 μm long .... *T. brebissonii*
2 Cells subcylindrical in face view, <120 μm long .... *T. laevis*

Tetmemorus brebissonii [Meneghini] Ralfs ex Ralfs 1848

Synonym: *Tetmemorus brebissonii* [Meneghini] Ralfs ex Ralfs var. *attenuatus* Nordstedt

27420010 **Pl. 131K** (p. 528)

Cells in broadest view 30–45 μm wide, 150–200 μm long; face view cylindrical or subcylindrical with lateral margins slightly compressed, with a conspicuous median constriction; apices broadly rounded with a deep, median incision; in slightly narrower side view, cells broadly spindle-shaped, tapering to truncate or subtruncate apices; walls delicately punctate, puncta in clearly defined rows, absent from a narrow isthmal band; chloroplasts with 3–4 longitudinal ridges visible in side view and mostly with 3 spherical or elongate axile pyrenoids in each chloroplast.

Cosmopolitan, acidophile; in peaty pools and *Sphagnum* bogs (pH 3.8–5(–6.5)), also in the margins or as a chance plankter of nutrient-poor lakes. Very widely distributed in acid pools, bogs and lakes throughout the British Isles, but never abundant.

The questionable var. *minimus* West *et* G.S.West (27420012) (52.5–57 μm long, 13–15 μm wide) and var. *minor* de Bary (27420012) (70–90 μm long, 17–20 μm wide) are reported from the British Isles but are rare.

Tetmemorus granulatus Brébisson in Ralfs ex Ralfs 1848

Synonym: *Tetmemorus granulatus* Brébisson in Ralfs ex Ralfs var. *attenuatus* West

27420020 **Pl. 131L** (p. 528) **CD**

Cells spindle-shaped in both front and side views, 30–45 μm wide, 130–240 μm long, with a slight median constriction; semicells tapering from their broadest, above the isthmus, to apices; apices in front view rounded with a distinct incision of variable depth, in side view apices truncate; walls finely and irregularly scrobiculate except above isthmus where scrobiculations occur in 1 or 2 transverse rows; chloroplasts with 5–7 deeply incised lamellae or ridges visible in side view, each chloroplast with 4 or more spherical, axile pyrenoids.

Cosmopolitan; the most widespread and abundant member of the genus and probably of all desmids in the British Isles, on occasions occurring as almost pure gatherings in *Sphagnum* bogs; mostly in acid waters but also present in neutral waters (pH 3.5–7.3); found at 900 m on Glyder Fower, Caenarvonshire, and as high as 1100 m. on Lochnagar on Deeside in Scotland. Coesel (1998) rates this desmid as characteristic of moderately nutrient-rich, slightly acidic waters.

Note that although the specific name is *granulatus*, granules do not in fact ornament the cell wall, their ornamentation consists of small depressions or scrobiculations.

Tetmemorus laevis Kützing ex Ralfs 1848

27420030 **Pl. 144K** (p. 590)

Cells small, 20–31.5 μm wide, 67.5–123 μm long, with a slight median constriction; isthmus 16–27 μm wide; semicells attenuating gradually to comparatively broad, rounded apices with a deep median incision; walls colourless, minutely punctate.

Cosmopolitan, acidophile; widespread throughout the British Isles and sometimes abundant, especially in upland areas in bog pools amongst *Sphagnum* (pH 3.8–7).

Xanthidium Ehrenberg ex Ralfs 1848

Cells compressed and in vertical or side view mostly elliptical although 2 taxa are triangular; cells deeply constricted and sinus ranging from closed to widely open; semicells oval or ellipsoidal, hexagonal or polygonal; apices flat or slightly convex; chloroplasts in the smaller species, 1 or 2 per semicell, axile with a central pyrenoid, and in the largest species 4 per semicell, parietal and each with 1 pyrenoid.

Cosmopolitan; all 15 British species and 24 varieties are acidophiles, mostly occurring in boggy or peaty pools; others associated with marginal macrophytes of nutrient-poor lakes where they occur frequently as chance members of the plankton; some are true plankters.

The genus is characterized by the following: (i) Cells furnished with simple, often paired spines arranged more or less symmetrically on lateral margins, usually at 2 different, superimposed levels; the largest species, *X. armatum*, bears stout, terminally forked processes. In most species there is considerable variation in number and disposition of spines, so that numerous varieties and forms of doubtful validity have been described. (ii) The wall in the centre of each semicell in most species is protuberant and the raised area thickened and frequently scrobiculate with a distinctive pattern of granules and often stained brown.

1 Cells with pairs of short tri- or quadrifurcate processes at each angle; cells >90 μm long .... *X. armatum*
1 Cells with simple spines at angles..........................2
2 Cells without thickening on middle of semicell face 3
2 Cells with thickening on middle of cell face..........4
3 Lateral margins and apices concave, isthmus wide open; each angle with a long spine.......*X. octocorne*
3 Lateral margins diverging to upper angles which bear widely forked spines........................*X. bifidum*
4 Semicells with only 1 or 2 pairs of spines...............5
4 Semicells with more than 2 pairs of spines 6
5 Semicells with 1 pair of spines, one on each side of semicell; face weakly thickened ....*X. subhastiferum*
5 Semicells with 1 pair of spines on upper and 1 pair on lateral angle of each semicell; face distinctly thickened.......................................*X. antilopaeum*
6 Semicells with 6 pairs of spines equidistant round semicell margin; face prominently thickened ....*X. fasciculatum*
6 Semicells with 2 pairs of spines on upper angles and 1 pair on lower angles of semicell; face only moderately thickened ....*X. cristatum*

6 Semicells with groups of 3–5 spines at angles, numbers very variable; face prominently thickened; semicells trapeziform ....*X. variabile*

Xanthidium antilopaeum [Brébisson] Kützing 1849

27430020 **Pl. 142B** (p. 586) **CD**

Cells deeply constricted, almost as broad as long, without spines 42–76 μm long, with spines 54–114 μm long, width without spines 40–72 μm, width with spines 57–108 μm; inner part of sinus linear opening to a slightly dilated extremity; isthmus 14–26 μm wide; semicells subelliptical to hexagonal with all angles rounded, lateral margins and apex straight; each of the 4 lateral angles extended into a pair of simple, straight or slightly curved spines; face of semicell with a round or elliptical raised area, thickened, scrobiculate.

Widespread, world distribution especially in the northern hemisphere, possibly absent from Australasia; most common and widespread *Xanthidium* in the British Isles, essentially acidophilic, rarely found in acid bogs, most frequent in margins of large ponds and lakes (alkalinity range 0–38 ppm $CaCO_3$).

The nominal variety is a regular but never an abundant component of the phytoplankton. According to W. & G.S.West (1911), 7 varieties are reported from the British Isles, including the triradiate var. *triquetrum* P.Lundell (27430028) from two lakes in the English Lake District.

var. *depauperatum* West *et* G.S.West 1903

27430022 **Pl. 142E** (p. 586)

Cells without spines 46–55.5 μm long, 43–50 μm wide; isthmus 9.5–14 μm wide; semicells more inflated than the nominal variety with lateral angles markedly obtuse; spines often reduced in number, comparatively short and irregularly disposed on each lateral margin, also thinner and of unequal length, 2–10.5 μm, number and disposition of spines on adjacent semicells often different; central area as in nominal variety.

Europe, North America, Arctic; essentially planktonic, recorded from numerous Scottish lochs, also found in vegetation around margins; fairly widespread in the Orkney and Shetland Islands and in the plankton of lakes in the English Lake District, Wales and Ireland.

In some populations there can be considerable variation in the development of the spines with regard to both their length and thickness even in adjoining semicells. It is suggested that the var. *depauperatum* may be merely an expression of spine development and dependent on existing growth conditions, and that this variety might be reduced to the status of forma.

var. *hebridarum* West *et* G.S.West 1905

27430023 **Pl. 142C** (p. 586)

Differs from the nominal variety in having 3 spines on each lateral margin and all disposed in median vertical plane, one spine only inserted at upper angles, other 2 near lateral angles; cells without spines 46–50 μm long, with spines 61–71 μm long, without spines 42–49 μm wide, with spines 69–82 μm wide, isthmus 12–14 μm wide; central area of semicells with small mammillate protuberances and variously disposed minute scrobiculations.

Europe, North America; fairly widespread in the margins of lochans and lochs in the west and central Scotland, also in Lewis and Harris in the Hebrides; frequent in small lakes in the west of Ireland.

var. *polymazum* Nordstedt 1873

27430027 **Pl. 142D** (p. 586)

Differs from the nominal variety in the face of each semicell having a distinctive arc of large, rounded granules above central protuberance and sometimes a third spine between pair at each lateral angle; without spines 45–62 μm wide, 48–68 μm long, with spines 58–76.5 μm wide, 62–85 μm long; isthmus 12–16 μm wide.

Restricted to the northern hemisphere; present especially in margins of pools and lakes but also in the plankton; in the British Isles most common in the west of Scotland, including the Western Isles; also in several lakes of the English Lake District.

Xanthidium armatum [Brébisson] Rabenhorst ex Ralfs 1848

27430040 **Pl. 142A** (p. 586) **CD**

Cells large, with spines 76–217 μm wide, 116–185 μm long; deeply constricted with an open, acute-angled sinus; isthmus 30–45 μm wide; semicells approximately octagonal, lower margins slightly convex, inferior and superior lateral margins mostly concave although sometimes straight; apex flat or slightly convex; spines on each side of the 2 lateral angles and apical angle short, stout, solid, wart-like and 2–4-forked and a stout forked spine within each lateral angle, also a pair just within apex (all variable in development); centre of each semicell with a ring of simple or emarginate teeth; chloroplasts parietal, 4 in each semicell, each with a separate pyrenoid.

Cosmopolitan, acidophile; sometimes abundant in *Sphagnum* bogs and around nutrient-poor lake margins. Widespread throughout the British Isles except for the eastern and central counties of England, also occurs in the plankton of many Scottish lochs especially in north-west and Outer Hebrides; similarly in some English Lake District and Welsh lakes.

Xanthidium bifidum [Brébisson] Deflandre 1929

27430050 **Pl. 142K** (p. 586)

Cells 11–15.5 μm wide, 10–14 μm long, with a deeply constricted, subrectangular sinus; isthmus 3–5.6 μm wide; apices slightly concave and angles of semicells diverging upwards, with each lateral angle widely emarginate and bifid; walls smooth and without thickening on middle of cell face.

Europe, North and South America, Arctic; in the British Isles most frequent in small pools and bogs in Scotland and Ireland and in similar habitats in England.

Xanthidium cristatum Brébisson in Ralfs 1848

27430090 **Pl. 142G** (p. 586)

Cells without spines 34–48 μm wide, 44–55 μm long, with spines 48–68 wide, 60–77 μm long, deeply constricted with a narrow linear or slightly open sinus, dilated at extremities; isthmus 13–15.5 μm wide; semicells trapeziform to semicircular with well-defined basal, lateral and apical angles; basal angles with a single, slightly convergent spine, lateral and apical angles both with a pair of straight, divergent spines; central area of semicell with a small, thickened, vertically elongated protuberance; walls delicately punctate.

Probably cosmopolitan; widely distributed in the British Isles, most frequent in western regions of Scotland including the Hebrides, Ireland and Wales; sometimes occurring in the plankton of small lakes; also present in the English Lake District and acidic areas of Yorkshire, Hampshire and Cornwall.

var. *uncinatum* Brébisson in Ralfs 1848
27430096
Cells without spines 40–64 μm wide, 55–74 μm long, with spines 55–83 μm wide, 73–91 μm long; semicells widely subpyramidate with 4 pairs of curved spines often dilated at base; central area with a ring of 8–12 granules surrounding 3–5 central granules.

Europe, North America, Arctic, Asia; widespread and at times common in Scotland, less so in the English Lake District and Ireland; no records from Wales.

Xanthidium fasciculatum Ehrenberg ex Ralfs 1848

Synomyms: *Xanthidium fasciculatum* var. *polygonum* Ralfs, *X. fasciculatum* var. *ornatum* Nordstedt
27430100 **Pl. 142H** (p. 586)
Cells as long as broad (excluding spines), without spines 44–49 μm wide, 44–66 μm long, with spines 65–74 μm wide, 62–72 μm long; constriction deep, sinus linear sometimes with dilated extremities; isthmus 12–21 μm wide; semicells angular to kidney-shaped; margin with 6 equally spaced pairs of short spines; apex subtruncate; central area only slightly protuberant with ring of 7–10 granules surrounding a central group of 2–3 granules.

Probably cosmopolitan; widely distributed in the British Isles in acid lakes, ponds and bogs in the western regions of Scotland, Ireland and Wales; recorded in England from Yorkshire, Surrey and Cornwall.

Xanthidium octocorne Ehrenberg ex Ralfs 1848

Synonym: *Arthrodesmus octocornis* (Ehrenberg ex Ralfs) West *et* G.S.West
27430110 **Pl. 142J** (p. 586)
Cells without spines 14–19 μm wide, 17–27 μm long, with spines 28–35 μm wide, 30–42 μm long; deeply constricted with a wide open, almost semicircular sinus; isthmus 4–8 μm wide; sides and apices of semicells concave with upper and lower angles rounded, each with a long spine; walls smooth and without a median swelling or thickening.

Cosmopolitan; common and widely distributed in acid pools and bogs throughout the British Isles.

Xanthidium subhastiferum West 1892

27430150 **Pl. 142F** (p. 586)
Cells with spines 64–86 μm wide, without spines 44–54 μm wide, 43–54 μm long; deeply constricted with an acutely angled, open sinus; isthmus 13–19 μm wide; semicells oblong-elliptical to elliptical, sometimes flattened at base and apex; each lateral margin with 2 simple diverging spines (12.5–18 μm long) in same vertical plane; face of semicell with a central, small, rounded thickened area.

Europe, North America; a true plankter occurring in large lochs in the west of Scotland and in the Outer Hebrides; also in small lakes in the west of Ireland. Frequent at times in Windermere and other English Lake District lakes (Rosenberg, 1944).

The var. *murrayi* West *et* G.S.West (27430152) occurs in Lake District lakes and also in a few western Scottish lochs. It has two lateral spines on each side, disposed in a horizontal plane.

Xanthidium variabile (Nordstedt) West *et* G.S.West 1900

Basionym: *Xanthidium smithii* var. *variabile* Nordstedt
Synonym: *Xanthidium westianum* Gutwinski
27430170 **Pl. 142I** (p. 586)
Cells small, without spines 18–23 μm wide, 20–26 μm long (spines only 2–3 μm long); deeply constricted with acutely angled, open sinus; isthmus 6–8.6 μm wide; semicells rectangular-trapeziform, with almost straight sides and apex; angles rounded, with 1–5 pointed spines; central area strongly protuberant.

Europe, North America, Australasia; widespread and often abundant in *Sphagnum* bogs throughout the British Isles although easily overlooked.

TAXA RECORDED FROM THE BRITISH ISLES BUT NOT DESCRIBED IN TEXT

Xanthidium aculeatum Ehrenb. 27430010; *X. antilopaeum* var. *laeve* Schm. 27430024; *X. antilopaeum* var. *oligocanthum* Schm. 27430025; *X. antilopaeum* var. *planum* Roll 27430026; *X. antilopaeum* var. *triquetrum* P.Lundell 27430028; *X. apiculiferum* West 27430030; *X. armatum* var. *cervicorne* West *et* G.S.West 27430042; *X. armatum* var. *fissum* Nordst. 27430043; *X. armatum* var. *irregularis* West 27430044; *X. brebissonii* Ralfs 27430060; *X. brebissonii* var. *varians* Ralfs 27430062; *X. concinnum* W.Archer 27430070; *X. concinnum* var. *botianum* West 27430072; *X. controversum* var. *planctonicum* West *et* G.S.West 27430082; *X. cristatum* var. *delpontii* J.Roy *et* Bisset 27430093; *X. cristatum* var. *leiodermum* (J.Roy *et* Bisset) W.B.Turner 27430094; *X. cristatum* var. *spinuliferum* West 27430095; *X. fasciculatum* var. *oronense* West *et* G.S.West 27430102; *X. robinsonianum* W.Archer 27430130; *X. smithii* W.Archer 27430140; *X. smithii* var. *collum* West 27430142; *X. smithii* var. *maius* (Ralfs) West *et* G.S.West 27430143; *X. tetracentrotum* Wolle 27430160; *X. tetracentrotum* var. *protuberans* (West *et* G.S.West) West *et* G.S.West 27430162; *X. tetracentrotum* var. *quadricornutum* (J.Roy *et* Bisset) West *et* G.S.West 27430163.

Filamentous forms

Worldwide, there are 10 genera of filamentous desmids in which *Cosmarium*-like or *Staurastrum*-like cells

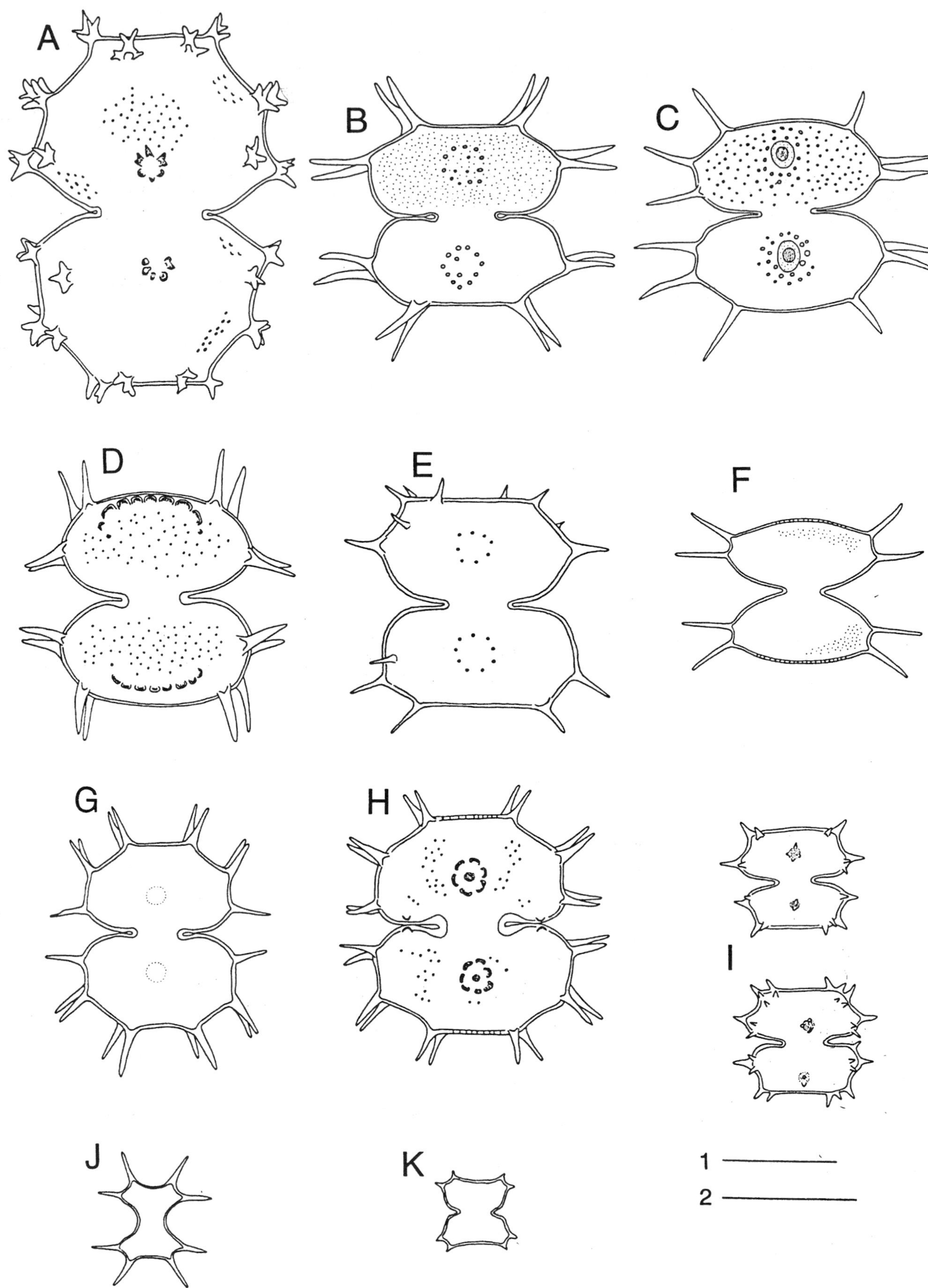

Plate 142 A–K

A. *Xanthidium armatum* (p. 584); **B**. *Xanthidium antilopaeum* (p. 584); **C**. *Xanthidium antilopaeum* var. *hebridarum* (p. 584); **D**. *Xanthidium antilopaeum* var. *polymazum* (p. 584); **E**. *Xanthidium antilopaeum* var. *depauperatum* (p. 584); **F**. *Xanthidium subhastiferum* (p. 585); **G**. *Xanthidium cristatum* (p. 584); **H**. *Xanthidium fasciculatum* (p. 585); **I**. *Xanthidium variabile* (p. 585); **J**. *Xanthidium octocorne* (p. 585); **K**. *Xanthidium bifidium* (p. 584).

Scale bars: 1, 25 μm; 2, 50 μm. 1 for B, E, G–K; 2 for A, F.

are joined together by apical granules, knobs or rods to form filaments. One group of genera, *Sphaerozosma, Teilingia, Onychonema, Spondylosium*, are clearly related to the large genus *Cosmarium*, while *Hyalotheca, Groenbladia, Desmidium, Bambusina* probably have affinities with *Staurastrum*. In most filamentous genera, and especially those in the first group, the cells are small (8–25 μm wide), and all are frequently surrounded by wide mucilaginous sheaths.

Eight genera are common in the British Isles and some of their species are often fairly frequent in acid, nutrient-poor waters. Although mostly occurring in the littoral of ponds and lakes, some are found in the plankton as chance plankters.

Bambusina Kützing ex Kützing 1845

Cells barrel-shaped with median constriction resulting from a very shallow, V-shaped sinus and commonly exhibiting slight inflations on either side of sinus; cells united by relatively broad, flat apices to form long, slightly twisted filaments; semicells, above supra-isthmal swelling, having straight or retuse lateral margins narrowing to comparatively broad, truncate apices; walls mostly smooth, except for extremely delicate vertical striations below apex; vertical view circular, or when opposing mamillae present, slightly ovate.

Worldwide distribution. Ten species but only one known from the British Isles, *B. borreri*.

At cell division, girdle-like thickenings push back into old semicell and these characteristic thickenings are commonly visible in cells within most filaments.

Bambusina borreri (Ralfs) Cleve 1864

Basionym: *Desmidium borreri* Ralfs
Synonyms: *Bambusina brebissonii* Kützing ex Kützing, *Gymnozyga moniliformis* Ehrenberg

27030010 **Pl. 143M** (p. 588)

Cells 14–25 μm wide, 24–39 μm long, with a very slight but definite median constriction; sinus very shallow, V-shaped; semicells inflated above sinus, lateral margins straight or retuse tapering to a comparatively broad, truncate apex; wall smooth, except in some cells for delicate vertical striations just below apex; in vertical view circular, sometimes with mammillae on opposite sides, then slightly ovoid.

Cosmopolitan; very widespread in acid habitats throughout British Isles, commonly in littoral of nutrient-poor ponds and lakes and frequent in their plankton as a chance plankter.

Desmidium C.Agardh 1824

Cells mostly broader than long, with median constriction only moderately deep and opening to outside, joined in long, twisted filaments, some within a thick mucilage sheath; semicells ranging from transversely narrow-oblong, oblong semi-elliptical, pyramidal-truncate to barrel-shaped; in vertical view either elliptical or 3–5-angled, angles usually with mamillate poles; when elliptical cells attached to one another by the apposition of ridge-like thickenings on adjacent apices, in 3–5 angular forms cells attached by short processes projecting from apices, one at each angle, in such forms often a space between apices of adjacent cells.

Restricted to acid waters and most species are fairly rare.

Knowledge of their distribution in the Rhiconich area of West Sutherland has been considerably extended by Alan Joyce (see Joyce, 1998).

1 Cell apices flat, no space between adjacent cells ...2
1 Cell apices distinctly concave, distinct spaces between adjacent cells ...3
2 Median constriction slight, semicells narrowly oblong; lateral margins often angular ....*D. swartzii*
2 Median constriction moderately deep, sinus linear opening outwards; semicells pyramidal-truncate ....*D. grevillei*
3 Semicells transversely oblong, with a small inflation on each side of isthmus ......*D. aptogonum*
3 Semicells transversely oblong, lateral margins slightly concave ...*D. pseudostreptonema*

Desmidium aptogonum Brébisson ex Kützing 1849

27090010 **Pl. 143H** (p. 588)

Cells 21–31 μm wide, 13–19 μm long, moderately constricted with an acute, open sinus; isthmus 15–24.5 μm wide; semicells transversely oblong with a small inflation on each side of isthmus beyond which lateral margins are slightly concave then converging to apex, 21–24 μm wide; apices concave in mid-region but at each angle form moderately long connecting processes producing a distinct cavity between adjacent cells; in vertical view mostly 3(–4)-angular.

Cosmopolitan, acidophile; widespread in the British Isles in bogs, sometimes abundant, also in nutrient-poor lakes especially in Scotland, sometimes in plankton; very few records from Ireland.

Desmidium grevillei (Kützing ex Ralfs) de Bary 1858

Basionym: *Didymoprium grevillei* Kützing ex Ralfs
Synonym: *Desmidium cylindricum* Greville ex Nordstedt

27090050 **Pl. 143N** (p. 588)

Cells 41–56 μm wide, 20–26 μm long; median constriction slight, with acute-angled, linear or slightly open sinus; semicells short, pyramidal-truncate with acutely rounded basal angles, biundulate lateral margins and broad, flat apex, 26–40 μm wide; elliptical in vertical view with rounded, mamillate protuberances at each pole.

Cosmopolitan; widely distributed in the British Isles in acid habitats, especially in Scotland; few records from Wales.

Desmidium pseudostreptonema West *et* G.S.West 1902

27090070 **Pl. 143K** (p. 588)

Cells 30–37 μm wide, 17–21 μm long, moderately constricted with open, acute-angled sinus; isthmus 13.5–30 μm wide; semicells transversely oblong with

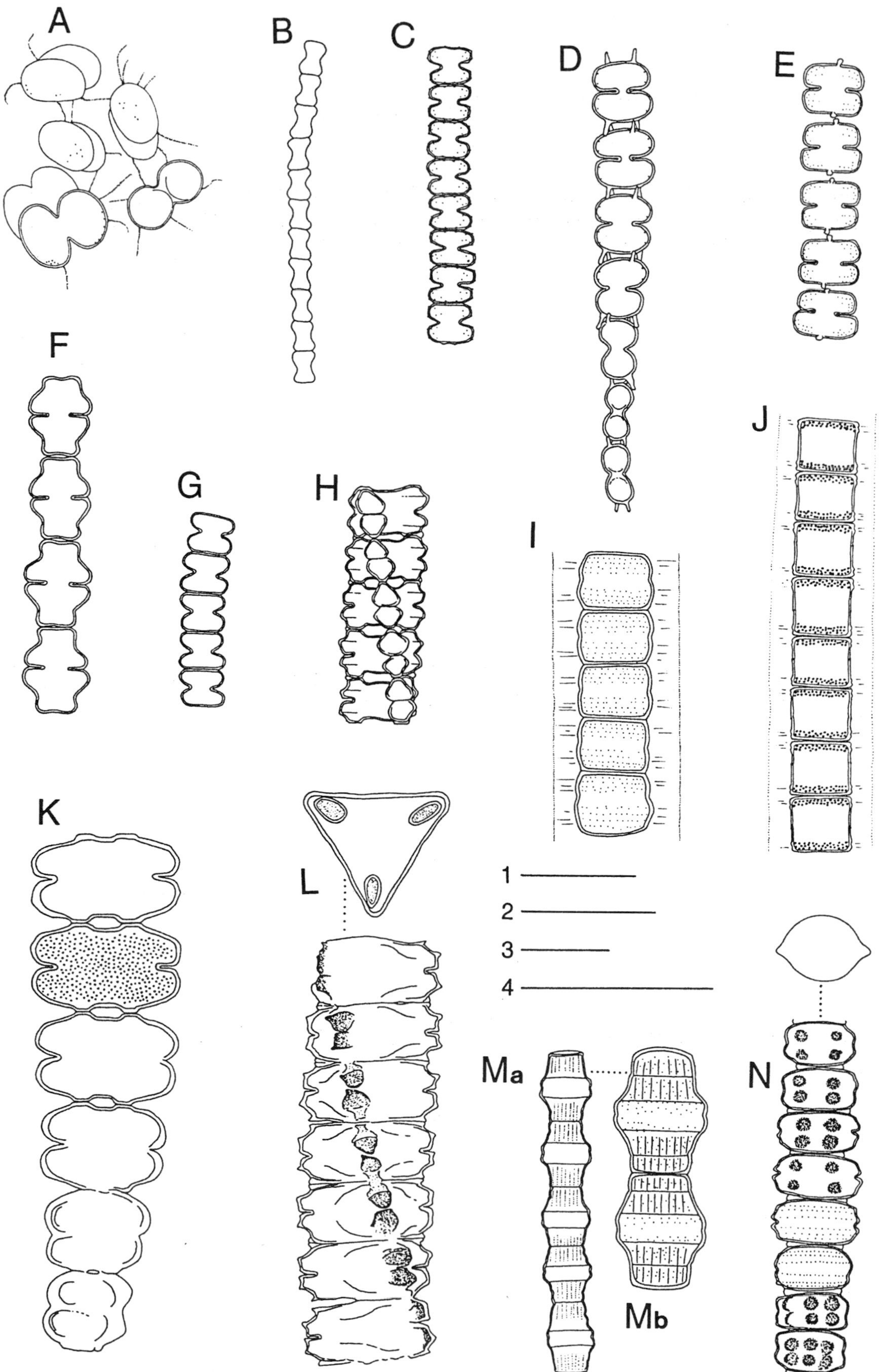

Plate 143 A–N

A. *Cosmocladium saxonicum* (p. 551); **B**. *Teilingia excavata* (p. 593); **C**. *Teilingia granulata* (p. 593); **D**. *Onychonema filiforme* (p. 592); **E**. *Sphaerozosma vertebratum* (p. 592); **F**. *Spondylosium pulchellum* (p. 593); **G**. *Spondylosium planum* (p. 592); **H**. *Desmidium aptogonum* (p. 587); **I**. *Hyalotheca dissiliens* (p. 589); **J**. *Hyalotheca mucosa* (p. 589); **K**. *Desmidium pseudostreptonema* (p. 587); **L**. *Desmidium swartzii* (p. 589); **M**. *Bambusina borreri* (p. 587); **N**. *Desmidium grevillei* (p. 587).
Scale bars: 1, 25 μm; 2–4, 50 μm. 1 for C, D, F, G, I, K, L, Mb; 2 for B, E, H, J, Ma; 3 for N; 4 for A.

lateral margins narrowly rounded; apices convex, 14–15.5 μm wide, with short, flat, connecting processes producing small cavity between adjacent cells; cells in vertical view with 2–4 angles.

Probably cosmopolitan; acidophile; rare in the British Isles except for north-west Scotland where Joyce (1998) reports it from eight lochs, and from Skye; it occurs in small lakes in Galway, Ireland.

Desmidium swartzii (C.Agardh) C.Agardh ex Ralfs 1848

27090100 **Pl. 143L** (p. 588)

Cells triangular or occasionally quadrangular in vertical view, 35–50 μm wide, 12–21 μm long, with moderately deep median constriction; sinus linear internally, opening outwards; semicells narrow-oblong; lateral margins variable, often angular and narrowing from the extended widest point to a very broad flat apex 30–41 μm wide, with connecting processes at each angle; middle of apex slightly concave but only a very small space between adjoining cells.

Cosmopolitan; the commonest of all filamentous desmids, widely distributed throughout the British Isles, especially in the littoral of nutrient-poor lakes where sometimes present in plankton; also in acid pools and bogs with *Sphagnum*.

TAXA RECORDED FROM THE BRITISH ISLES BUT NOT DESCRIBED IN TEXT

Desmidium aptogonum var. *acutius* Nordst. 27090012; *D. aptogonum* var. *ehrenbergii* Kütz. 27090013; *D. baileyi* (Ralfs) Nordst. 27090020; *D. coarctatum* Nordst. 27090030; *D. coarctatum* var. *cambricum* West 27090032; *D. graciliceps* (Nordst.) Lagerheim 27090040; *D. occidentale* West *et* G.S.West 27090060; *D. quadrangulatum* Ralfs 27090080; *D. quadratum* Nordst. 27090090; *D. swartzii* var. *amblyodon* (Itzigs.) Rabenh. 27090102.

Groenbladia Teiling 1952

Cells cylindrical, 2–9 times longer than wide, with a very shallow median constriction, united end to end to form short filaments, usually surrounded by mucilage; semicells slightly swollen at base, lateral margins straight and parallel with apex flat; chloroplasts axile, 1 or 2 per cell, with 1–8 pyrenoids in median series; wall minutely punctate except around the isthmal zone.

Probably cosmopolitan, acidophiles, mostly found in margins of nutrient-poor lakes where they also occur in the plankton.

1 Cells with a minute median constriction and slight inflation on either side ....*G. neglecta*
1 Cells with a shallow, open, median constriction ....*G. undulata*

Groenbladia neglecta (Raciborski) Teiling 1952

Basionym: *Hyalotheca neglecta* Raciborski

27140010 **Pl. 145K** (p. 591)

Cells cylindrical, 11.5–17 μm wide, 28–53 μm long, very slightly constricted and slight inflation on either side of constriction; lateral margins straight, parallel; apices truncate, 11–16 μm wide.

Cosmopolitan; in the British Isles most frequent in plankton of some English Lake District lakes, also recorded from Surrey and Hampshire; most frequent in lochs in the west of Scotland and the Outer Hebrides; also occurs in the west of Ireland.

Groenbladia undulata (Nordstedt) Kurt Förster 1972

Basionym: *Hyalotheca undulata* Nordstedt

27140020 **Pl. 145L** (p. 591)

Cells more or less dumbbell-shaped, 6–9 μm wide, 10–17.5 μm long, 1.5 to 2 times longer than wide with a broad, shallow median indentation; diameter of isthmus and flat apices more or less equal, 4.5–7.5 μm wide; filaments sometimes in a mucilage sheath.

Subcosmopolitan; acidophile; rare in the British Isles with most records from lakes in the west of Ireland; also in the Capel Curig area of Wales and in the west of Scotland.

TAXA RECORDED FROM THE BRITISH ISLES BUT NOT DESCRIBED IN TEXT

Groenbladia undulata var. *kriegeri* Kurt Först. 27140022; *G. undulata* var. *perundulata* (Grönblad) Kurt Först. 27140023.

Hyalotheca Ehrenberg ex Ralfs 1848

Cells subcylindrical, only very slightly constricted; united by broad truncate apices into long filaments almost invariably enveloped in mucilage sheath, often sheath as broad as cells; chloroplasts axile, 1 in each semicell, with several ridges radiating from the central core, often flattened against cell wall, each chloroplast with 1 central pyrenoid.

Cosmopolitan; usually in non-flowing waters including ponds and lakes.

1 Semicells each having 2, more rarely 3, transverse rings of granules immediately below apices ....*H. dissiliens*
1 Semicells without transverse rings of granules present ....*H. mucosa*

Hyalotheca dissiliens Brébisson ex Ralfs 1848

27180010 **Pl. 143I** (p. 588) **CD**

Cells 10–39 μm wide, 10–33 μm long, with a slight notch in lateral margin (median constriction), united by broadly truncate apices into long filaments; in vertical view circular with 2 or 3 papillae.

Cosmopolitan; one of the most widespread of all desmids, often occurring in great abundance; common in acid bogs and ditches and frequent in the plankton of nutrient-poor to moderately nutrient-rich lakes throughout the British Isles.

Several forms and varieties have been described. Filaments nearly always enveloped in a broad, distinctive mucilage sheath.

Hyalotheca mucosa [Mertens] Ehrenberg ex Ralfs 1848

Basionym: *Conferva mucosa* Mertens

27180020 **Pl. 143J** (p. 588)

Cells 9–22 μm wide, 12.5–22 μm long, cylindrical and when not dividing cells without a clear median

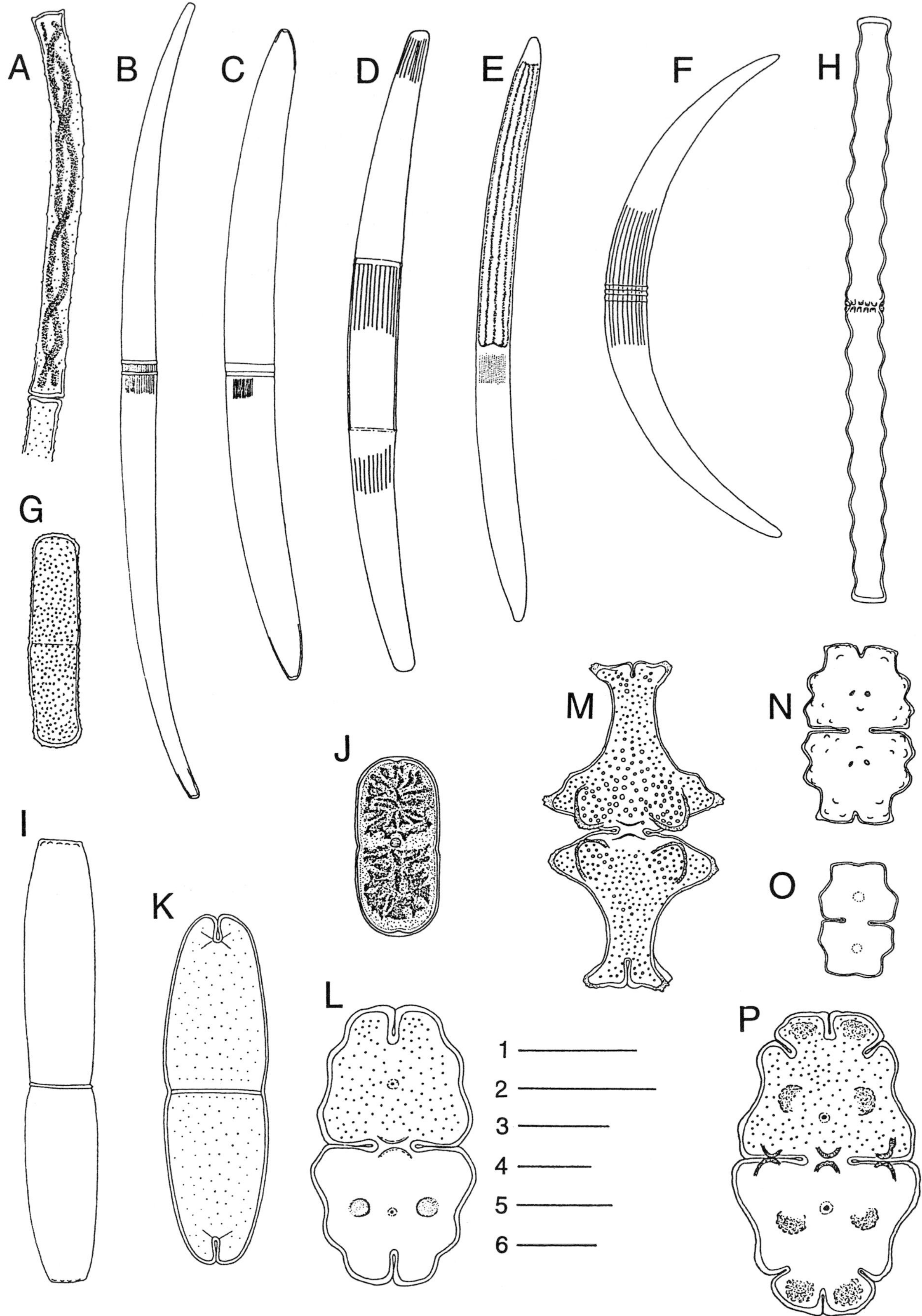

Plate 144 A–P

A. *Genicularia elegans* (p. 527); **B**. *Closterium lineatum* (p. 523); **C**. *Closterium turgidum* (p. 527); **D**. *Closterium intermedium* (p. 522); **E**. *Closterium pritchardianum* (p. 525); **F**. *Closterium archerianum* (p. 519); **G**. *Penium exiguum* (p. 530); **H**. *Docidium undulatum* (p. 551); **I**. *Pleurotaenium truncatum* (p. 563); **J**. *Actinotaenium diplosporum* (p. 531); **K**. *Tetmemorus laevis* (p. 583); **L**. *Euastrum inerme* (p. 555); **M**. *Euastrum insigne* (p. 555); **N**. *Euastrum dubium* (p. 555); **O**. *Euastrum insulare* (p. 555); **P**. *Euastrum ventricosum* (p. 556).

Scale bars: 1, 25 μm; 2–4, 50 μm; 5,6, 100 μm. 1 for G, K, L, N, O; 2 for F, H, J, M, P; 3 for D, I; 4 for B; 5 for A, E; 6 for C.

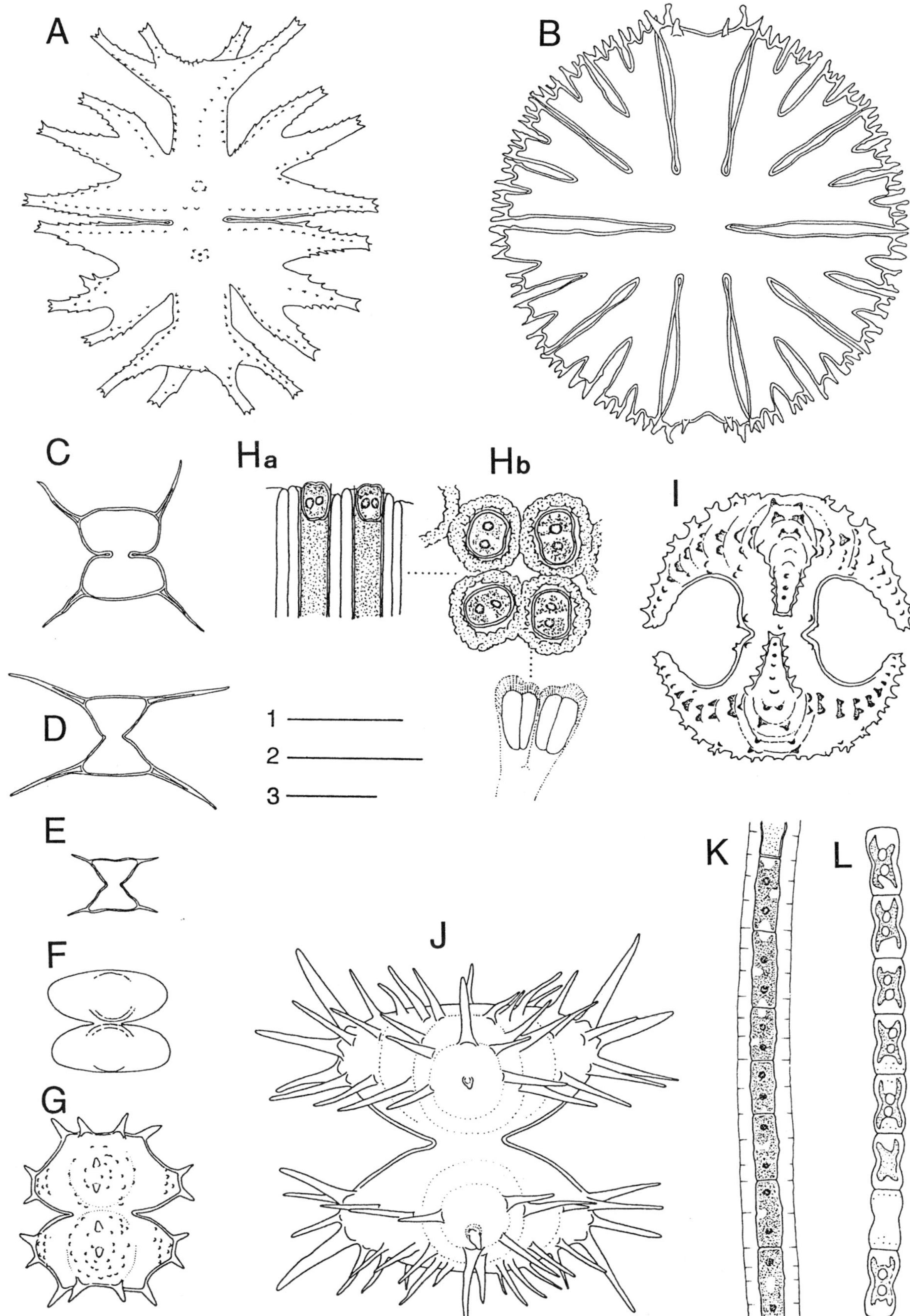

Plate 145 A–L
A. *Micrasterias mahabuleshwarensis* var. *wallichii* (p. 559); **B**. *Micrasterias radiosa* (p. 559); **C**. *Staurodesmus bulnheimii* (p. 578); **D**. *Staurodesmus incus* (p. 579); **E**. *Staurodesmus incus* var. *ralfsii* (p. 581); **F**. *Staurastrum bieneanum* (p. 566); **G**. *Staurastrum monticulosum* (p. 572); **H**. *Oocardium stratum* (p. 561); **I**. *Staurastrum cerastes* (p. 568); **J**. *Staurastrum setigerum* (p. 574); **K**. *Groenbladia neglecta* (p. 589); **L**. *Groenbladia undulata* (p. 589).
Scale bars: 1, 25 μm; 2, 3, 50 μm. 1 for A, F, G, I, J, L; 2 for B–E, Hb; 3 for Ha, K.

constriction; semicells each with 2(–3) transverse rings of very small granules (actually pores) immediately below apices.

Cosmopolitan; not as common in the British Isles as *H. dissiliens*, but almost as widespread; similarly present in the plankton of nutrient-poor to moderately nutrient-rich lakes in the English Lake District and Scotland, including the Outer Hebrides and Shetland Islands; occasional in acid bogs and ditches.

Filaments mostly fairly long within a very thick, mucilaginous sheath.

TAXA RECORDED FROM THE BRITISH ISLES BUT NOT DESCRIBED IN TEXT

Hyalotheca dissiliens var. *hians* Wolle 27180012; *H. dissiliens* var. *tatrica* Racib. 27180013.

Onychonema Wallich 1860

Cells with 2 widely spaced apical, sometimes capitate processes present, usually almost as long as the semicell on which they occur, placed obliquely from one another so that one process extends over front of adjoining semicell and the other over the back; cells compressed, broadly elliptical in vertical view and in face view deeply constricted with a narrow, linear sinus; semicells elliptical or kidney-shaped, sometimes lateral margin extended as a stout spine; filaments long, often twisted, sometimes enclosed in broad mucilage sheath.

Cosmopolitan; mostly known from non-flowing water.

Onychonema filiforme (Ehrenberg ex Ralfs) J.Roy *et* Bisset 1886

Basionym: *Sphaerozosma filiforme* Ehrenberg ex Ralfs

27240010 **Pl. 143D** (p. 588)

Cells 10–19 μm wide, 9–15 μm long, deeply constricted with narrow linear sinus; semicells ellipsoidal or subkidney-shaped with a flattened base and broadly rounded upper margin; apex distinctly convex in subkidney-shaped semicells, somewhat flattened in more elliptical ones; apical processes about as long as semicell, widely spaced and disposed asymmetrically, one overlapping adjacent cell in front and the other behind.

Cosmopolitan; fairly widespread in the British Isles, mostly in acid pools and in margins of nutrient-poor ponds and lakes; few records from Wales.

TAXA RECORDED FROM THE BRITISH ISLES BUT NOT DESCRIBED IN TEXT

Onychonema laeve Nordst. 27240020; *O. laeve* var. *latum* West *et* G.S.West 27240022.

Sphaerozosma Corda 1835

Cells compressed laterally, constricted with an open or narrow sinus; semicells ellipsoidal, oblong, subkidney-shaped or subrectangular, with 2 rod-like, obliquely placed, apical processes, joined by processes to form filaments; walls with 2 or more horizontal rows of puncta, extrusions from which may look like granules on the lateral margins; filaments often twisted; sometimes in a mucilaginous envelope.

The two most common species (*S. aubertianum*, *S. vertebratum*) are cosmopolitan acidophiles occurring in pools and bogs and the littoral of nutrient-poor lakes from which they may be carried into the plankton.

Sphaerozosma vertebratum Brébisson ex Ralfs 1848

27330020 **Pl. 143E** (p. 588) **CD**

Cells 15–24 μm wide, 13–19 μm long, deeply constricted with a narrow, linear sinus, obtuse at inner end; semicells oblong to subkidney-shaped with broadly rounded lateral margins, less curved on apex; walls smooth but rows of very small puncta sometimes visible; median apical processes short, closely spaced; lateral view of cells (often seen when filaments are twisted) oblong with slight median constriction; filaments usually twisted and occasionally enveloped in mucilage.

Arctic-alpine species; widely distributed in pools and acid bogs in the British Isles, also in littoral of nutrient-poor ponds and lakes, especially in Scotland and Ireland; often a chance plankter.

TAXA RECORDED FROM THE BRITISH ISLES BUT NOT DESCRIBED IN TEXT

Sphaerozosma aubertianum West 27330010; *S. vertebratum* var. *latius* West *et* G.S.West 27330022; *S. vertebratum* var. *punctulatum* West *et* G.S.West 27330023.

Spondylosium Brébisson ex Kützing 1849

Cells laterally compressed so vertical view elliptical, usually deeply constricted, with narrow to broad sinus sometimes opening outwards; semicells differing considerably in shape, joined to adjacent cells by apices, but latter always lack even the smallest granular processes; apices mostly broadly truncate or retuse; walls usually without ornamentation; sometimes conspicuous pores excrete a thick enveloping mucilaginous sheath; filaments often long and twisted.

Cosmopolitan; usually in ponds and lakes.

1 Cells 10–25 μm wide, transversely oblong with rounded angles; apices flat .... *S. planum*

1 Cells 8–11 μm wide, truncate-pyramidate with basal angles broadly rounded above; apices flat or slightly retuse .... *S. pulchellum*

Spondylosium planum (Wolle) West *et* G.S.West 1912

Basionym: *Sphaerozosma pulchrum* Bailey var. *planum* Wolle

27360040 **Pl. 143G** (p. 588)

Cells 10–25 μm wide, 11.5–19.5 μm long, sinus deep open and internally rounded; semicells transversely oblong with rounded angles and flat apex; walls smooth with very small pores; filaments not twisted; commonly lacking a mucilaginous sheath.

Cosmopolitan; common in plankton of nutrient-poor and moderately nutrient-rich lakes especially in the English Lake

District and Scotland; also in Wales and Ireland. Sometimes present in the littoral zone associated with submerged macrophytes.

Spondylosium pulchellum W.Archer 1858

27360050 **Pl. 143F** (p. 588)

Cells 8–11 μm wide, 9–15 μm long, apex about 5 μm broad; sinus deep and fairly narrow, widening gradually from inside; semicells truncate-pyramidate with basal angles broadly rounded above, while concave lateral margin narrows markedly to flat or slightly retuse apex; walls smooth with inconspicuous pores; filaments usually long, rarely twisted.

Cosmopolitan; widespread in the British Isles, most common in nutrient-poor Irish lakes and ponds

Often attached to aquatic plants, submerged leaves or mosses by a short, thick, gelatinous stalk (West et al., 1923).

TAXA RECORDED FROM THE BRITISH ISLES BUT NOT DESCRIBED IN TEXT

Spondylosium ellipticum West *et* G.S.West 27360010; *S. moniliforme* P.Lundell 27360020; *S. papillosum* West *et* G.S.West 27360030; *S. pygmaeum* (Cooke) West 27360070; *S. pygmaeum* var. *compressum* West 27360072; *S. pygmaeum* var. *monile* (W.B.Turner) West *et* G.S.West *et* Carter 27360073; *S. secedens* (de Bary) W.Archer 27360080; *S. tetragonum* West 27360090.

Teilingia Bourrelly 1964

Filaments composed of cells with well-defined open isthmus, rounded at interior; in side view compressed elliptical or quadrangular, joined at apices by pair of very small granules; filaments sometimes enveloped in mucilage.

Cosmopolitan; mostly in non-flowing waters.

1 Lateral margins of cells smooth............*T. excavata*
1 Lateral margins of cells granular ....*T. granulata*

Teilingia excavata (Ralfs) Bourrelly 1964

Basionym: *Sphaerozosma excavatum* Ralfs

27400010 **Pl. 143B** (p. 588)

Cells 7–14 μm wide, 7.5–17 μm long; sinus wide and with an obtuse interior; semicells broadly ovoid in face view, oblong-elliptical in side view; apex flat with 4 small, widely spaced attaching granules (only 2 seen in face view); lateral walls smooth; filaments mostly short, not twisted.

Cosmopolitan, acidophile; widespread in bogs and peaty pools and the margins of weedy nutrient-poor lakes; sometimes occurs as a chance plankter.

Teilingia granulata (J.Roy *et* Bisset) Bourrelly 1964

Basionym: *Sphaerozosma granulatum* J.Roy *et* Bisset

27400020 **Pl. 143C** (p. 588)

Cells 8–13 μm wide, 7–15 μm long; sinus moderately deep and open, rounded interiorly; semicells oblong-elliptical with slightly flattened lateral margins bearing rounded granules (3 marginal, 1 or 2 intramarginal); apex straight with 4 widely spaced joining-granules (only 2 visible in face view); filaments rarely twisted.

Cosmopolitan; widespread in the British Isles, not restricted always to acid waters, often occurring in the plankton of nutrient-poor to moderately nutrient-rich lakes.

TAXA RECORDED FROM THE BRITISH ISLES BUT NOT DESCRIBED IN TEXT

Teilingia excavata var. *subquadrata* (West *et* G.S.West) F. Stein 27400012; *T. wallichii* (H.C.Jacobsen) Bourr. 26400030; *T. wallichii* var. *anglica* (West *et* G.S.West) Kurt Först. 27400032.

ORDER CHARALES

BY JENNY A. BRYANT (FORMERLY MOORE) AND NICK F. STEWART

The Charales are a highly specialized order of macroscopic green algae, commonly known as stoneworts or charophytes. Linnaeus (1753) established the genus *Chara* with four species and a further five genera were recognized during the nineteenth century, namely *Nitella* C.Agardh, *Lychnothamnus* Ruprecht, *Tolypella* A.Braun, *Lamprothamnus* A.Braun and *Nitellopsis* Hy. The name *Lamprothamnus* had already been used for a genus of the Rubiaceae so *Lamprothamnium* was substituted (Groves, 1916). *Lychnothamnus* has not been found in the British Isles.

These algae bear a superficial resemblance to the pteridophyte genus *Equisetum* (horsetail), as the main branches have, at intervals, whorls of short laterals (branchlets). They usually grow submerged in mainly still (occasionally flowing), fresh or mildly brackish water and are anchored by colourless rhizoids. Absorption of nutrients and gaseous exchange take place over the entire surface of the plant, particularly through the rhizoids. Normal metabolic activity often results in the deposition of a calcium carbonate encrustation, hence the commonly used name 'stonewort'. A distinctive odour, somewhat like stale garlic, is also characteristic of some species, especially in the genus *Chara*.

The wide range of aquatic habitats in which they are found demonstrates their adaptability. For example, they can be pioneer colonizers of newly created habitats, ephemerals in transient habitats, dominant deep-water perennials in large lakes or background members of aquatic vascular plant communities. They are intolerant of high levels of nutrients, particularly phosphates, and are generally indicators of an absence of or low levels of enrichment. Their role in contributing to water quality is not fully understood (van den Berg et al., 1998), but they are known to be important in stabilizing sediment, thus contributing to clear water conditions and their large biomass per unit area may be important in maintaining water clarity. They can be of commercial significance in circumstances where it is highly desirable to have such conditions (e.g. fisheries, recreation). The majority of *Chara* and *Tolypella* species are confined to calcareous waters of high pH and their remains contribute to the calcium-rich sediments of marl lakes. Such lakes are a habitat where *Chara* is often a dominant and long-lived perennial. On the other hand, most species of *Nitella* occur in acidic, peaty waters. Several charophytes are confined to brackish or maritime habitats and are included in this account for the sake of completeness (none are fully marine). Further data on ecology and distribution are included in the species accounts that follow.

Charophytes are unique amongst the algae for being of interest to many vascular plant enthusiasts and this has led to their being the only non-vascular plant group whose study is included in the objectives of the Botanical Society of the British Isles. They were avidly collected and exchanged by BSBI members from the middle of the nineteenth century and are still of interest to many current members. Data from most of these collections and from published records have been collated by staff of the Botany Department at The Natural History Museum, London (BM). These data ensured that charophytes were considered for conservation action ahead of other, less well-studied, groups of non-vascular plants.

Lamprothamnium papulosum was added to Schedule 8 of the UK Wildlife and Countryside Act in 1987 and was joined in 1998 by *Chara canescens*. This makes it an offence to collect material of these species from known sites, except under licence from the appropriate government agency (e.g. English Nature/Scottish Natural Heritage). This is particularly important as many of them are threatened by a change of water quality through nutrient enrichment and by habitat loss through development and neglect. A Red Data Book (Stewart & Church, 1992) identified 57% of the British species and 45% of the Irish species as threatened.

As a result of its commitments made at the United Nations Conference on Environment and Development ('The Earth Summit'), held at Rio de Janeiro, Brazil in June1992, the UK Government has produced costed 'Species and Habitat Action Plans'. In 1995 the UK Government published the two-volume 'Biodiversity: the UK Steering Group Report' (Biodiversity, 1995b) in which 12 charophytes were listed as 'Priority Species' in need of conservation action. These are *Chara baltica*, *C. canescens*, *C. connivens*, *C. curta*, *C. muscosa*, *Lamprothamnium papulosum*, *Nitella gracilis*, *N. tenuissima*, *Nitellopsis obtusa*, *Tolypella intricata*, *T. nidifica* and *T. prolifera*. Subsequently, 'Species Action Plans' have been published (Biodiversity, 1999) for seven of these, namely *Chara connivens*, *C. curta*, *Nitella gracilis*, *N. tenuissima*, *Nitellopsis obtusa*, *Tolypella intricata* and *T. prolifera*. The 'lead partners' in steering this conservation action are the wild-plant conservation charity Plantlife and the county-based Wildlife Trusts, working closely with the Environment Agency, English Nature and the Countryside Council for Wales. Charophytes of conservation concern are present in several of the 47 'priority habitats' identified, of which saline lagoons, mesotrophic waters and coastal sand dunes are perhaps the most important. The importance of these and other habitats for charophyte conservation is also recognized by the European Council's 'Habitats Directive' (Anon, 1992).

The taxonomic concepts and nomenclature used here closely follow that in Stewart & Church (1992) and, to a large extent, that of Groves & Bullock-Webster (1920, 1924), but with updated nomenclature. A conservative approach is adopted generally, agreeing with that used by most other European charologists. However, there is still uncertainty over the status and nomenclature of a few taxa. For more detailed general accounts of the group, see Fritsch (1935), Groves & Bullock-Webster (1920, 1924), Moore (1986), Olsen (1944) and Wood & Imahori (1965). Other works useful for identifying European charophytes include Blindow & Krause (1990), Bruinsma et al. (1998), Comelles (1985), Compere (1992), Corillion (1975), Krause (1997), Langangen (1972, 1974) and van Raam (1998).

Morphology and specialist terms

A specialist terminology has developed to describe the unique morphology of charophytes and it is difficult to use the literature unless one is familiar with these terms. Most of the terms are defined below and some are used in the account that follows, although every attempt has been made to restrict the use of complex terminology. The morphology of the charophytes is unlike that of any other algal group. Branches of unlimited growth are made up of elongate, single-celled internodes, separated by multicellular nodes. Whorled laterals of limited growth (branchlets) arise from the short node cells. Multicellular, branched rhizoids, not divided into nodes and internodes but with septa at intervals, are the means of attachment, anchoring the branches to the substratum.

Cortex: in most species of *Chara* and *Lychnothamnus* the branches (and branchlets in *Chara*) are overlaid by cells adpressed to the internodal cell to produce a *cortex* one cell thick, which appears as lines along the axis. The structure of this cortex is diagnostic in determining the species of *Chara* and is fully explained by Moore (1986). It is made up of *primary rows*, which are divided into nodes and internodes, and *secondary rows*, which develop from the nodes of the primary rows. The cortical rows may show a marked torsion, spiralling upwards from left to right. The primary rows can be recognized by the unicellular *spine-cells* produced at their nodes and are equal in number to the branchlets at a branch node.

Cortication and spine-cells in Chara: the spine-cells at the nodes of the primary cortical rows may be solitary, twinned or in clusters, and may be rudimentary or elongate. Although deciduous in some species they are usually more persistent nearer the apices. They distinguish the primary cortical rows and, where no secondary rows are developed, the cortex is *haplostichous* (Pl. 150Ab). Where a single secondary row of cortical cells is developed between each primary row then the cortex is *diplostichous* (Pl. 146Bb). Where two rows of cortical cells are produced between each primary row the cortex is *triplostichous* (Pl. 148Bb). The secondary rows have septa at intervals but do not have nodes or spine-cells. Frequently, the ends of the cells of the secondary rows overlap, giving the impression that the cortex is triplostichous. If the primary rows are larger in diameter and, therefore, more prominent than the secondary rows so that the spine-cells appear to sit on ridges, the cortex is termed *tylacanthous*. When the secondary rows are more prominent, so that the spine-cells appear to sit in furrows, the cortex is *aulacanthous*. When all of the cortical rows are of equal width the cortex is known as *isostichous*.

Branchlets: the term used for the whorled laterals of limited growth. They have nodes and internodes and, in many species of *Chara*, a simplified cortex. In *Chara*, *Lamprothamnium*, *Lychnothamnus* and *Nitellopsis* they are undivided. Branching in *Nitella* is by forking and is sympodial, whereas it is pinnate and monopodial in *Tolypella*.

Bract-cells, *bracteoles* and the *bractlet*: these unicellular processes arise as whorls at the branchlet nodes in *Chara*, *Lamprothamnium*, *Lychnothamnus* and *Nitellopsis*. The *bract-cells* vary in length and if on the inner side of the branchlets (i.e. side facing towards the main branch) they are *adaxial* and if on the outer side they are *abaxial*. When well developed the bract-cells may be mistaken for small side branches. Two elongate, unicellular processes, termed *bracteoles*, sit one on either side of the oogonium in *Chara*, *Lamprothamnium* and *Lychnothamnus*. These resemble the bract-cells but may be more slender and slightly longer. In dioecious species of *Chara* there is another small unicellular process, termed the *bractlet*, lying immediately beneath the oogonium (taking the place of the absent antheridium).

Stipulodes: the term used for unicellular outgrowths (of varying length and size in the different species) produced from the outermost cells of the branch nodes and lying immediately beneath the branchlet whorls in *Chara*, *Lamprothamnium* and *Lychnothamnus*. The stipulodes either form a single ring (*haplostephanous*) or a double ring (*diplostephanous*). When elongate and in a double ring they may be more or less adpressed and point upwards and downwards from the nodes. The number of stipulodes in each ring is equal to the number of primary cortical rows below and to the number of branchlets above the branch node.

Rays: the branchlets of *Nitella* are divided one or more times. The lowest segment (the internode nearest the main branch) is termed the *primary ray* and subsequent segments are known in turn as *secondary*, *tertiary*, etc. The end segment (ultimate ray) is known as the *dactyl*, and these may be single-celled or made up of two or more cells, separated by cell walls, not by nodes. All the rays in *Nitella*, apart from the dactyl, are single-celled. The branchlets of *Tolypella* may be divided or, if simple, made up of several segments, but lack dactyls.

Reproduction: Sexual reproduction is oogamous and has a number of features unique to the charophytes. The male and female reproductive organs (*gametangia*) develop from the nodal cells of the branchlets. The male *antheridium* is spherical with its wall composed of 8 (rarely 4) shield-like cells. Biflagellate antherozoids are produced on the inner surfaces of the shield-cells. The female *oogonium* (more correctly oosporangium) has a protective envelope of 5 cells spiralled clockwise around the egg-cell (oosphere). Cells divide off from the apices of the spiralled cells to form what is termed the *coronula*. In *Chara*, *Lamprothamnium*, *Lychnothamnus* and *Nitellopsis*, the coronula is a single ring of 5 cells, while in *Nitella* and *Tolypella* the coronula is of 10 cells in a double ring. After fertilization the oosphere develops into an *oospore* that may have a highly sculptured surface when mature (see Leitch et al., 1990). In species of *Chara* where the male and female reproductive organs are on the same plant (*monoecious*), they may be together on the same branchlet node or separate at different nodes. The antheridium always sits downwards from the node whilst the oogonium sits above and is surrounded by the bract-cells. In *Lamprothamnium* this situation is reversed and the oogonium is positioned below the antheridium. In species of *Chara* and in *Nitellopsis* where the reproductive organs are on separate plants (*dioecious*) they sit centrally at, or above the node. In *Lychnothamnus* the oogonium is placed between two antheridia. The gametangia in *Nitella* and *Tolypella* are produced at the forks and occasionally at the base of a branchlet whorl. The dimensions of the gametangia are sometimes diagnostic and the measurements given in this account are of mature (bright orange) antheridia and (very dark brown to black) oospores.

Vegetative reproduction is effected through the *bulbils*, which may appear as white tuberous growths on the rhizoids of some species of *Chara* and *Lamprothamnium*. In *Nitellopsis* distinctive, star-shaped bulbils are formed from the starch-filled, much-reduced and modified branchlets of the lower nodes of the branches (Pl. 151Ab). The lower nodes of most charophytes often become swollen and cartilaginous ('gristly') and are sometimes mistaken for bulbils. They, too, act as vegetative propagules following fragmentation or senescence of the thallus.

1 Plants corticate; spine-cells and stipulodes present, although these may be rudimentary .... *Chara*
1 Plants ecorticate; spine-cells absent; with or without stipulodes 2
2 Branchlets undivided; bract cells at the nodes; oogonium with a single apical ring of 5 cells forming the coronula 3
2 Branchlets divided (sometimes only minutely so at apex); bract-cells and stipulodes absent; oogonium with a coronula of two rings, each of 5 cells 4
3 Stipulodes forming a single ring, acuminate, downward-pointing; upper branchlet whorls forming dense foxtail-like heads; unicellular spherical bulbils sometimes present on rhizoids; monoecious .... *Lamprothamnium*
3 Stipulodes absent; upper branchlet whorls lax and diffuse; star-shaped bulbils developed from lower branch nodes; dioecious .... *Nitellopsis*
4 Branchlets forked one or more times into similar single-celled segments, with end segment (dactyl) divided into one or more cells; branchlets divided without a dominant axis (sympodial); oospore laterally compressed .... *Nitella*
4 Branchlets divided into unequal, multicellular segments; branchlets divided with a dominant main axis (monopodial); oospore circular in cross-section (terete) .... *Tolypella*

Chara Linnaeus 1753

Thallus macroscopic, from a few cm up to 120 cm, branched, bright green to brownish- or greyish-green, usually encrusted; branches made up of nodes and internodes, overlaid with a cortex made up of primary and secondary cell rows developing from the nodes; rarely cortex absent or imperfectly formed (see notes on *C. denudata* under *C. contraria*); primary rows of cortex distinguished by the presence of unicellular spine-cells; laterals of limited growth (branchlets) in whorls at the branch nodes and made up of nodes and internodes, usually with a simplified cortex and bearing whorled unicellular bract-cells at their nodes; branch nodes bearing (1–)2 rings of unicellular outgrowths (stipulodes), set immediately beneath the branchlet whorl; monoecious or dioecious; oogonium positioned above the antheridium at branchlet nodes in monoecious species; coronula of oogonium a single ring of 5 cells.

Cosmopolitan; the majority of species are confined to non-flowing, calcareous water with a high pH and low in nutrients.

There are nomenclatural problems surrounding some epithets used in *Chara* and confusion in the literature as to which taxa they refer to. The status and synonomy of *C. aculeolata* are under review by us as is *C. hispida sensu lato*. Recent examination of the type of *C. hispida* has revealed it to be identical to *C. aspera*.

1 Cortex of primary cell rows only (haplostichous), equalling the number of branchlets; all rows bearing clustered spine-cells .... *C. canescens*
1 Cortex with 1 row (diplostichous) or 2 rows (triplostichous) of secondary cells between each primary cell row (if cortical rows imperfectly formed or absent, see notes on *C. denudata* under *C. contraria*) 2
2 Cortex with 1 row of secondary cells between each primary cell row 3
2 Cortex with 2 rows of secondary cells between each primary cell row 10
3 Spine-cells solitary, elongate and obtuse, papillate or rudimentary 4
3 Spine-cells in clusters of 2–4 or, if solitary, then ovoid-acuminate or acute 5
4 Cortex with secondary rows more prominent than primary rows (aulacanthous), or all rows of equal width (isostichous); spine-cells rudimentary, short or, if well-developed, lying adpressed to cortex (<30° from axis) .... *C. vulgaris*
4 Cortex with primary rows more prominent than secondary rows (tylacanthous); spine-cells rudimentary, short or, if well-developed, spreading (>45° from axis) .... *C. contraria*
5 Cortex with primary rows more prominent than secondary rows 6
5 Cortex with secondary rows more prominent than primary rows, or all rows of equal width 9
6 Branchlets having ultimate cell of end segment inflated; spine-cells solitary or paired, small, inflated; bract-cells and stipulodes (unicellular outgrowths of the branch nodes immediately beneath branchlet whorls) swollen at base (gibbous) .... *C. tomentosa*
6 Branchlets having end segment tapering; spine-cells solitary or in clusters of up to 4, acuminate or broadly acute; bract-cells and stipulodes acuminate or tapering-obtuse 7
7 Spine-cells in clusters of 3–4, well-developed, persistent; abaxial bract-cells more than half the length of the adaxial; thallus densely spinous and bristly .... *C. aculeolata*
7 Spine-cells 1(–2), papillate to well-developed, sparse, deciduous; abaxial bract-cells less than half the length of the adaxial; thallus not as above 8
8 Spine-cells usually shorter than half the branch diameter, conical, solitary (rarely paired); thallus encrusted, often with a brownish or pinkish tinge; antheridium up to 500 μm in diameter .... *C. intermedia*
8 Spine-cells longer than half the branch diameter, solitary; thallus unencrusted, dark green; antheridium >500 μm in diameter .... *C. baltica*
9 Spine-cells usually in clusters of 2–4 (rarely solitary in juvenile plants), spreading, elongate, acuminate; cortex with rows of similar width or secondary rows slightly more prominent (isostichous to weakly aulacanthous) .... *C. hispida*
9 Spine-cells 2(–3), usually forming vertically

opposed pairs, adpressed, tapering-obtuse; cortex with secondary cell rows very prominent, so that spine cells appear to sit in furrows (strongly aulacanthous)..*C. rudis*

10 Spine-cells well-developed, solitary or in clusters, acuminate or acute................................11

10 Spine-cells undeveloped or as long as broad, obtuse or rounded 12

11 Spine-cells solitary (rarely a few in clusters of 2–3); thallus not densely spinous ....*C. aspera*

11 Spine-cells in clusters of 3–4 (rarely a few solitary or in clusters of up to 6); thallus densely spinous throughout ....*C. curta*

12 Bulbils white, multicellular (strawberry-like); branchlets flexuous, with 9–13 segments; dioecious...*C. fragifera*

12 Bulbils absent ('gristly' swellings often present at lower nodes); branchlets stiff to slightly flexuous with 8–11 segments; monoecious or dioecious....13

13 Dioecious; antheridium >600 μm in diameter; branchlets stiff and often inwardly curved; thallus yellow-green...........................*C. connivens*

13 Monoecious; antheridium <500 μm in diameter; branchlets stiff to slightly flexuous; thallus mid-green, dark green or grey-green............................14

14 Stipulodes rudimentary to spherical; spine-cells rudimentary; cortical cell rows all equal in width; branches up to 0.75 mm in diameter ....*C. globularis*

14 Stipulodes with upper ring short, obtusely conical and lower ring rudimentary to spherical, rarely one or both rings more developed; spine-cells papillate; cortex with primary cell rows more prominent than the secondary rows; branches up to 0.5 mm in diameter..........*C. virgata*

Chara aculeolata Kützing in Reichenbach 1832

Synonyms: *Chara pedunculata* Kützing, *Chara polyacantha* A.Braun

28010190 (28010140) **Pl. 150B** (p. 604) **CD**

Thallus robust, up to 50 cm, mid- to grey- or brownish-green, often strongly encrusted, with dense, persistent spine-cells; cortex with a single secondary row between each primary row (diplostichous), the primary rows markedly more prominent (strongly tylacanthous); abaxial bract cells more than half the length of the adaxial bract-cells and bracteoles; monoecious.

Europe; in the British Isles recorded from shallow water (up to 1 m deep) in alkaline, low-nutrient pools and lakes, often associated with fen peat; frequently recorded from central Ireland where widely scattered, but rare elsewhere.

Superficially similar to *C. canescens*, but differing in the longer branchlets (one-third to as long as the internodes), the coarser, more bristly appearance, and the more opaque thallus due to the more numerous cortical rows (i.e. diplostichous rather than haplostichous). More spinous than *C. hispida* as spine-cells are persistent (they may be deciduous in *C. hispida*) and with primary cortical rows always more prominent than secondary rows; also the torsion of cortical rows is less marked, and the bract-cells are more equal in length.

Chara aspera Detharding ex Willdenow 1809

Synonym: *Chara delicatula* Desvaux *non* Agardh

28010010 **Pl. 148C** (p. 601) **CD**

Thallus slender, up to 30 cm, light to mid-green, spiny; branches up to 0.5 mm in diameter; cortex with 2 secondary rows between each primary row (triplostichous); spine-cells nearly always solitary, acute; both rings of stipulodes developed; bulbils often present on rhizoids, whitish, single-celled and spherical; dioecious.

Northern hemisphere; in the British Isles in calcareous lakes, pools and ditches, often on marl or sand, tolerant of mildly brackish water, sometimes forming extensive beds in dune-pools and machair lakes; rare to occasional, but frequent in central and western Ireland, the Hebrides and the Orkney and Shetland Islands.

Its upright growth often helps to differentiate it from the more tufted or sprawling habit of most other small species of *Chara* (but see *C. curta*).

Chara baltica Bruzelius 1824

28010020 **Pl. 147A** (p. 600)

Thallus robust, up to 90 cm, dark green; cortex with a single secondary row between each primary row (diplostichous) and the primary rows slightly larger in diameter (weakly tylacanthous); spine-cells usually solitary, well-developed, acuminate; monoecious; multicellular, whitish bulbils sometimes present on the rhizoids.

Baltic and northern Atlantic Europe, Bolivia in South America, Greenland; in the British Isles from lakes and pools near the sea, sometimes in slightly brackish water; rarely recorded from the Outer Hebrides, Anglesey, Devon, Norfolk and Galway; records from Cornwall and Shetland are probably erroneous.

Most reliably separated from similar species (*C. hispida*, *C. intermedia*) by the larger antheridia. Bulbils are probably most frequent in wave-washed locations.

Chara canescens Desvaux *et* Loiseleur 1810

28010040 **Pl. 150A** (p. 604)

Thallus of medium height, up to 30 cm, mid- to dark olive- or brownish-green, bristly (almost furry), neat, short-branched and stem milky-translucent when held to light; cortex with only primary cortical cell rows, which are equal to the number of branchlets (haplostichous), all with spine-cells in clusters; dioecious.

Europe, North America, Africa, Asia; rare in the British Isles in mildly brackish-water in shallow, sandy, alkaline lakes and rarely inland in pits left after clay extraction; known from Cambridgeshire, the Outer Hebrides and from Donegal, Galway, Clare, Kerry and Wexford.

Only female plants known from Great Britain and Ireland with the oosphere developing into an oospore without fertilization (parthenogenesis).

Chara connivens Salzmann ex A.Braun 1835

28010050 **Pl. 149B** (p. 603)

Thallus slender, up to 50 cm, yellow-green; branches up to 0.5 mm in diameter; cortex with 2 secondary rows between each primary row (triplostichous); spine-

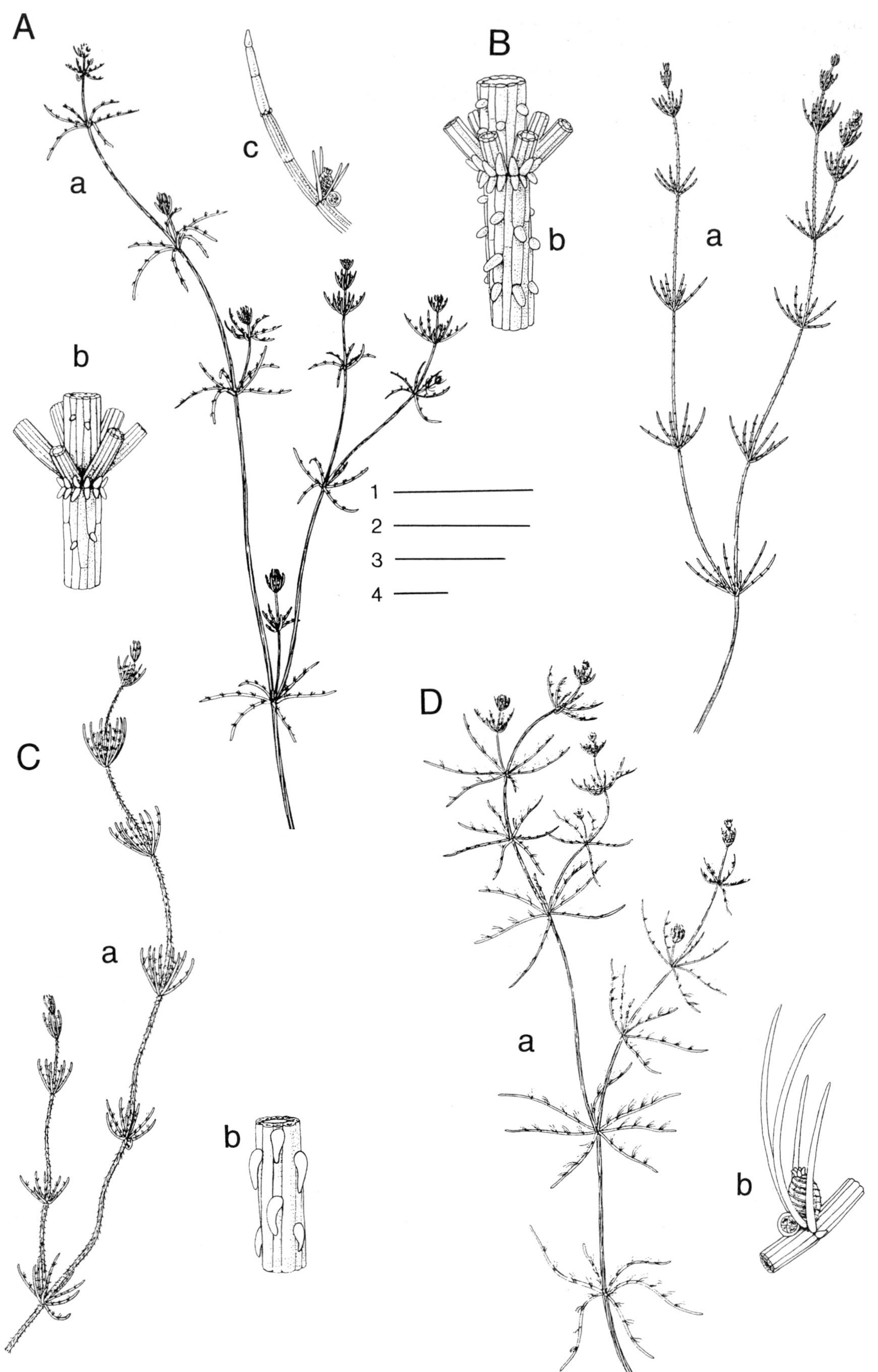

Plate 146 A–D
A. *Chara vulgaris* (p. 602): a. habit, b. detail of a branchlet whorl, c. portion of a branchlet with gametangia and bract-cells; **B**. *Chara contraria* (p. 599): a. habit; b. detail of a branchlet whorl; **C**. *Chara vulgaris* var. *papillata* (p. 605): a. habit, b. detail of a branch showing cortical rows and spine-cells; **D**. *Chara vulgaris* var. *longibracteata* (p. 605): a. habit, b. branchlet node with gametangia and bract-cells.
Scale bars: 1, 50 mm; 2, 5 mm; 3, 4, 1 mm. 1 for Aa, Ba, Ca, Da; 2 for Cb; 3 for Bb; 4 for Ab,c, Db.

cells and stipulodes more or less rudimentary; dioecious.

Mediterranean region north to the British Isles where it is found in lakes and pools, often near the sea in mildly brackish water, but also in strongly calcareous inland sites; rare, from Norfolk and Devon; previously recorded from sites in southern England and southern and western Ireland.

Similar to, but more slender than *C. globularis*, with shorter, stiffer branchlets and dioecious. The mature branchlets may curve inwards (back to and crossing the main branches, i.e. connivent), but this character is usually not well developed in British and Irish material. Young branchlets of other species are to some degree connivent before they mature and open out. Fertile material is essential for reliable identification of this dioecious species. The male plants have large antheridia (>600 μ m in diameter) placed centrally on the branchlet nodes.

Chara contraria A.Braun ex Kützing 1845

28010060 **Pl. 146B** (p. 598)

Thallus of medium height, up to 60 cm; cortex with a single secondary row between each primary row (diplostichous), with spine-cells evident on the ridges formed by the prominent primary rows (tylacanthous); spine-cells spreading if developed, but usually present only as papillae; monoecious.

Cosmopolitan; in a wide range of calcareous habitats in the British Isles, e.g. as an ephemeral in dune-slack pools, an annual in the fluctuating water at the edges of clay-pits or as extensive perennial beds in deep water; occasional to locally frequent in south-eastern England and central Ireland, widespread but rare elsewhere.

Similar to *C. vulgaris* but more slender and straggling and readily distinguished by having prominent primary rows which form ridges on which the spine-cells sit. Plants with well-developed spine-cells (longer than branch width) have been referred to var. *hispidula* A.Braun. Two poorly known species that are closely related to *C. contraria* are *C. denudata* A.Braun and *C. muscosa* J.Groves *et* Bullock-Webster. *C. muscosa* is a small, dark green, tufted, moss-like plant with uniformly short internodes and blunt, solitary spine-cells. It is only known from its type locality in Donegal (last seen in 1939). Other records from the Outer Hebrides and Orkney are probably erroneous.

Chara denudata is a fairly slender, monoecious, encrusted plant, up to 60 cm,with an almost absent or imperfectly formed cortex. The cortical rows may consist of a few residual cells at the nodes or of irregularly developed primary rows that ascend and descend from the nodes, but fail to meet and cover the central internodal cell. It has been recorded from Europe and South Africa and is known in the British Isles only from lime-rich lakes and pools in central and western Ireland, where it can be extensive. It must be noted, however, that the first few, newly germinated internodes of other species of *Chara* may be ecorticate before a regular cortex develops as they grow, whilst older, moribund plants may lose decayed cortical rows and also appear ecorticate.

Chara curta Nolte ex Kützing 1857

Synonym: *Chara desmacantha* (H. *et* J.Groves) J.Groves *et* Bullock-Webster

2801070 **Pl. 148D** (p. 601)

Thallus up to 40 cm; cortex with 2 secondary rows between each primary row (triplostichous); spine-cells clustered and acuminate, sometimes only one spine-cell fully developed and associated ones remaining rudimentary; bulbils often present on rhizoids, whitish, single-celled and spherical; dioecious.

Germany and the British Isles, where it is found in similar habitats to *C. aspera*, with which it is sometimes associated, but less frequently in coastal areas; usually in marl-rich lakes, calcareous clay pits, pools in calcareous fens and sand-dune pools; occasional in central and western Ireland, scarce but widespread elsewhere.

Variable, with both straggling, slender forms resembling *C. aspera* and neater, more robust forms recalling slender *C. canescens*. Differentiated from *C. aspera* by the clustered rather than solitary spine-cells. In juvenile plants, or those in wave-washed situations, some of the spine cells may be poorly developed, thus making it difficult to separate the two species. It is therefore advisable to examine a range of well-developed fertile material to aid identification.

Chara fragifera Durieu 1859

28010090 **Pl. 149A** (p. 603)

Thallus up to 30 cm, mid- to dark green; cortex with 2 secondary rows between each primary row (triplostichous); spine-cells rudimentary to papillate; stipulodes more or less rudimentary or with the upper ring shortly and obtusely conical; bulbils distinctive, multicellular and strawberry-like; dioecious.

Mediterranean coast of North Africa, South Africa, Atlantic Europe north to the British Isles where it is usually in small water bodies that dry out in summer, including quarry pools, waterlogged ruts in trackways and shallow pools on heathland; known only from western Cornwall.

Published work suggests that this species is similar to *C. globularis*, i.e. with rudimentary spine-cells and stipulodes and all the cortical rows of equal width (isostichous). In the authors' experience this species is more similar to *C. virgata*, but with longer, flexuous branchlets and is unencrusted in the British Isles. It is best distinguished from *C. virgata* by the distinctive bulbils and by being dioecious.

Chara globularis Thuillier 1799

Synonym: *Chara fragilis* Desvaux

280100100 **Pl. 148A** (p. 601)

Thallus fairly robust, up to 50 cm, light, mid- or olive-green; branches up to 0.75 mm in diameter; cortex with 2 secondary rows between each primary row (triplostichous), with all rows of equal width (isostichous); spine-cells rudimentary and difficult to discern; stipulodes rudimentary or spherical; monoecious.

Probably cosmopolitan; in the British Isles recorded from permanent pools and lakes, often as dominant stands or mixed with other charophytes and/or vascular plants, apparently restricted to calcareous and moderately nutrient-rich habitats; fairly frequent in southern and eastern England and central Ireland, rare to occasional elsewhere and over-recorded in the past due to confusion with *C. virgata*.

The most robust of those species that have 2 rows of secondary cortical cells between each primary row. The uniformly sized cortical rows and the almost invisible spine-cells usually make the primary rows difficult to discern. In the field this species might be mistaken for *C. vulgaris*, but is easily distinguished by its stiffer and neater growth, the rudimentary spine-cells and the spherical stipulodes. For differences from *C. connivens* and *C. virgata*, see under those species. It appears to be more tolerant of nutrient enrichment than do other species, but it is unable to survive in habitats where there is strong enrichment.

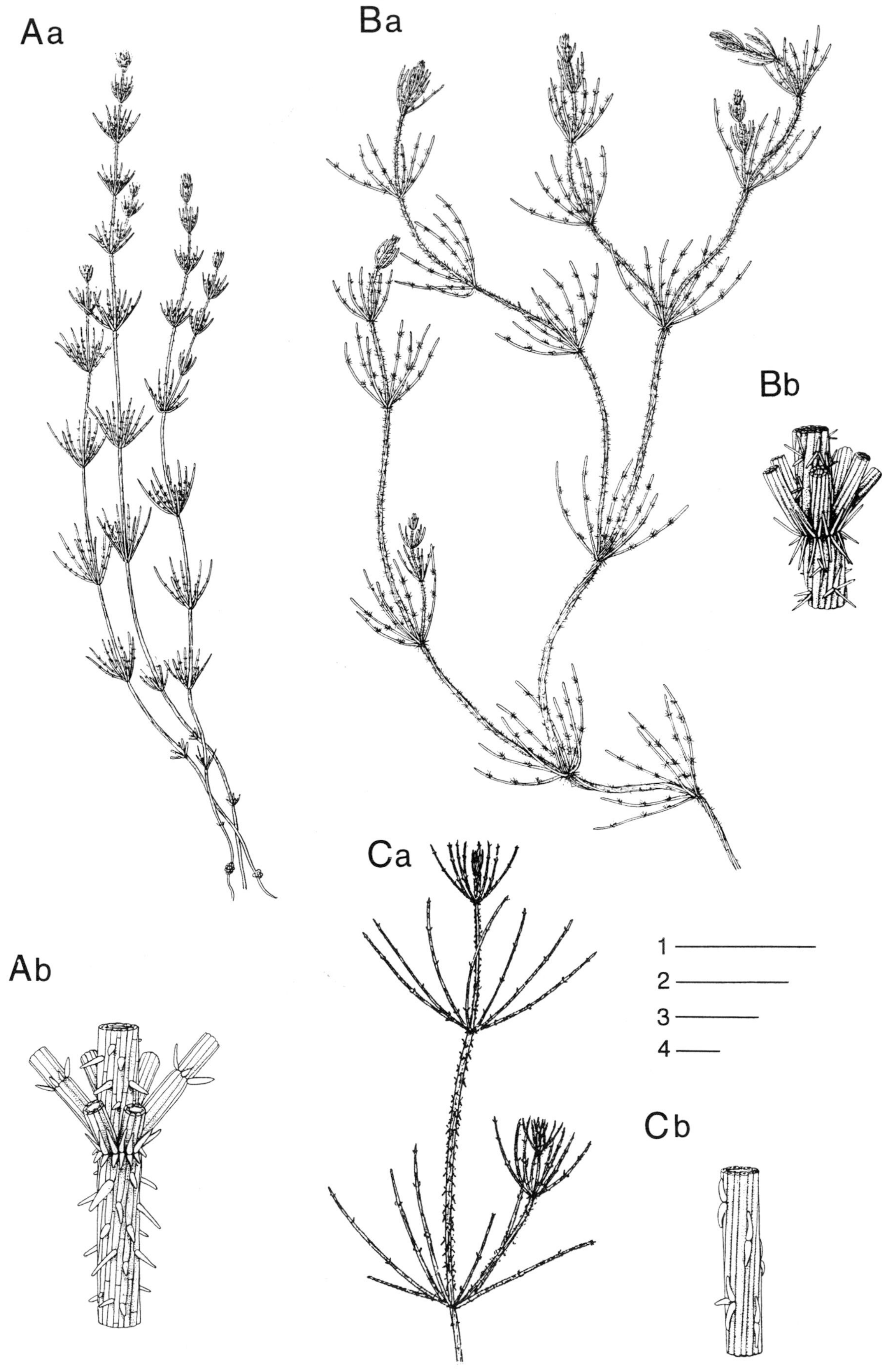

Plate 147 A–C
A. *Chara baltica* (p. 597): a. habit, b. detail of a branchlet whorl; **B**. *Chara hispida* (p. 602): a. habit, b. detail of a branchlet whorl; **C**. *Chara rudis* (p. 602): a. habit, b. detail of a branch showing cortical rows and spine-cells.
Scale bars: 1, 50 mm; 2, 20 mm; 3, 5 mm; 4, 1 mm. 1 for Aa, Ba, Ca; 2 for Cb; 3 for Bb; 4 for Ab.

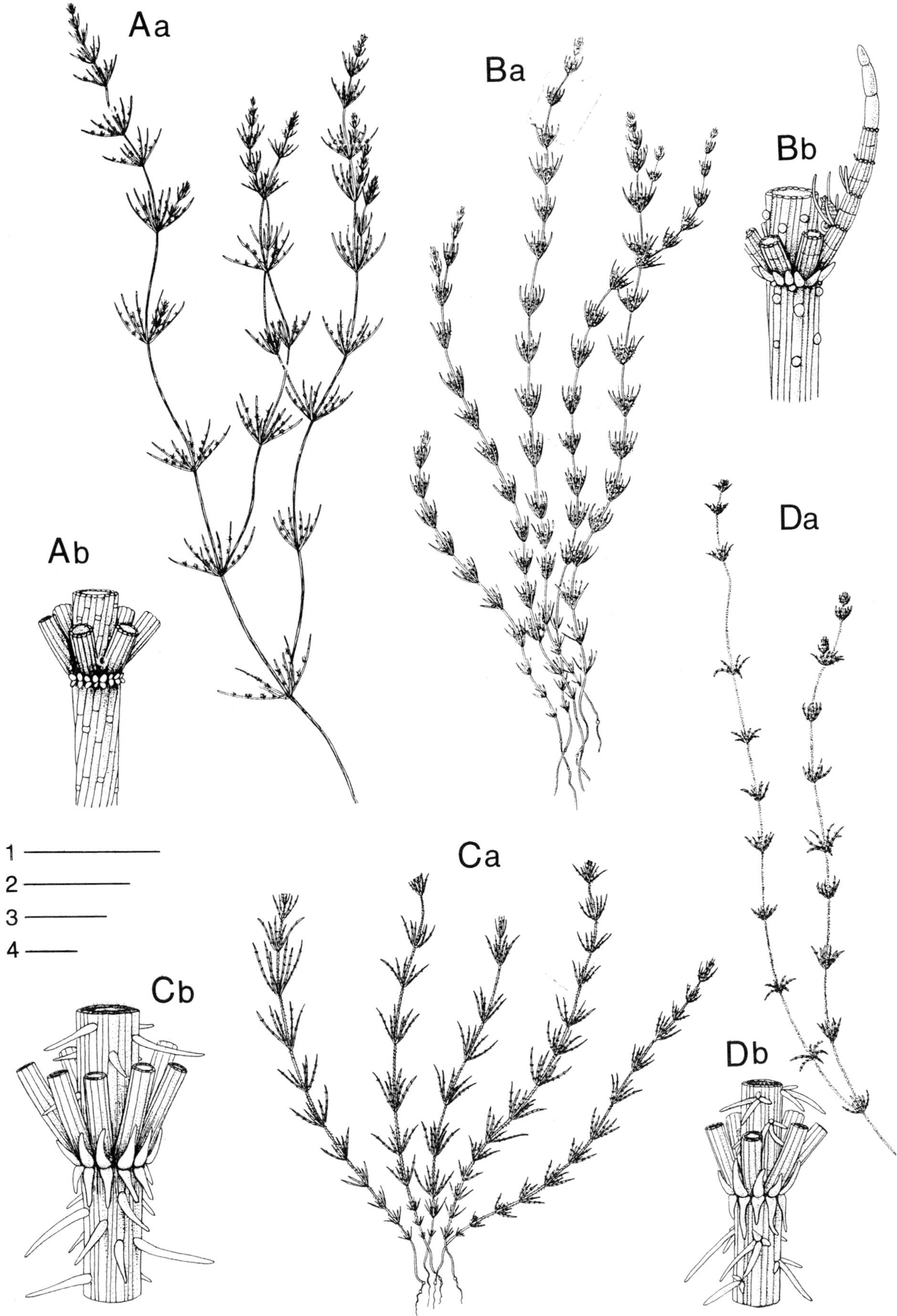

Plate 148 A–D
A. *Chara globularis* (p. 599): a. habit, b. detail of a branchlet whorl; **B**. *Chara virgata* (p. 602): a. habit, b. detail of a branchlet whorl; **C**. *Chara aspera* (p. 597): a. habit, b. detail of a branchlet whorl; **D**. *Chara curta* (p. 599): a. habit, b. detail of a branchlet whorl.
Scale bars: 1, 50 mm; 2–4, 1 mm. 1 for Aa, Ba, Ca, Da; 2 for Cb; 3 for Bb; 4 for Ab, Db.

Chara hispida Linnaeus 1753

280100110 **Pl. 147B** (p. 600) **CD**

Thallus robust, up to 120 cm, with long flexuous branchlets (more than half the length of the internode), mid- to grey-green; cortex with a single secondary row between each primary row (diplostichous), with rows of similar width or secondary rows slightly more prominent (isostichous to weakly aulacanthous); spine-cells elongate, acuminate, usually in clusters, sometimes appearing solitary in young plants (when rudimentary or spherical spine-cells remain undeveloped at the base of a developed spine-cell); abaxial bract cells less than half the length of the adaxial bract cells and bracteoles; monoecious.

Europe, Siberia, North Africa; recorded in the British Isles from calcareous pools and lakes, often forming dense perennial beds, sometimes in habitats which dry out for short periods; occasional to locally frequent in south-east and central England and central Ireland, rare elsewhere and restricted to habitats near the coast in Scotland.

The most frequently recorded of the robust *Chara* species. Torsion of the cortical rows is often pronounced, the cells sometimes lifting away from the branch internode. Large encrusted forms, dominating in deep water as long-lived perennials, have been referred to *C. hispida* var. *major* (Hartman) R.D.Wood.

Chara intermedia A.Braun 1859

28010120 **Pl. 149C** (p. 603)

Thallus robust, up to 120 cm, olive-green with a pinkish or brownish tinge; cortex with a single secondary row between each primary row (diplostichous), often the primary rows noticeably more prominent than the secondary rows (strongly tylacanthous); monoecious.

Sparsely scattered records from across Europe as far as Russia and Turkey; in the British Isles known only from Norfolk where it forms dense perennial beds in alkaline lakes or mixed with other large charophytes such as *C. hispida* and *Nitellopsis obtusa*; tolerant of weakly brackish water.

Similar to *C. baltica* but differing in colour and having shorter spine-cells (mostly less than half the stem-width) and smaller antheridia.

Chara rudis (A.Braun) Leonhardi 1864

Basionym: *Chara hispida* var. *rudis* A.Braun

28010150 **Pl. 147C** (p. 600)

Thallus moderately robust, up to 70 cm; branches up to 1.25 mm in diameter; cortex with a single secondary row between each primary row (diplostichous); spine-cells 2 or 3, adpressed and more or less vertically opposed, lying parallel to the branch internode, nestling in the primary row furrows between much larger secondary rows (strongly aulacanthous); abaxial bract cells less than half the length of the adaxial bract cells and bracteoles; monoecious.

Europe and western Asia; in the British Isles from limestone lakes, often forming perennial beds (sometimes with other large species, e.g. *C. hispida*), occasional in central and western Ireland but rare elsewhere; previously rare in southern England and Wales but now probably extinct.

Similar to *C. hispida* but slightly less robust, greyer in colour, always strongly aulacanthous and with adpressed, vertically opposed, stouter and blunter spine-cells.

Chara tomentosa Linnaeus 1753

28010160 **Pl. 150C** (p. 604)

Thallus robust, up to 60 cm, heavily encrusted, with young shoots having a pinkish colour and inflated branchlet end-cells; cortex with a single secondary row between each primary row (diplostichous), the primary rows wider and more prominent than the secondary rows (tylacanthous); spine-cells and bract-cells ovoid-acuminate to swollen at base (gibbous); dioecious.

Europe, Asia, North Africa, the Americas; in the British Isles often found amongst other large charophytes (e.g. *C. rudis*, *C. hispida*) in low-nutrient lakes on limestone; restricted to about 10 lakes in central and west of Ireland.

The pinkish tinge and encrustation on the young shoots gives this alga a coral-like appearance. Easily distinguished by the inflated branchlet-tips.

Chara virgata Kützing 1834

Synonym: *Chara delicatula* Agardh *non* Desvaux

28010170 **Pl. 148B** (p. 601) **CD**

Thallus moderately slender, up to 30 cm, branches up to 0.5 mm in diameter; cortex with 2 secondary rows between each primary row (triplostichous), the primary rows wider and more prominent than the secondary rows (tylacanthous); spine-cells papillate; upper ring of stipulodes slightly developed, obtusely conical, about twice as long as broad, lower ring rudimentary or spherical (rarely one or both rings more developed); monoecious.

Northern hemisphere and South Africa; frequently recorded in the British Isles from a wide range of aquatic habitats (including those that dry out periodically) and more frequent in acidic water than any other species of *Chara*; also occasional in alkaline, nutrient-poor water.

Similar to *C. globularis* but tylacanthous and with more prominent stipulodes and spine-cells. It may be difficult to assign a small proportion of specimens to either of these two species. When both rings of stipulodes are developed this species may be confused with forms of *C. aspera* that have poorly developed spine-cells. However, in *C. virgata* the apices of the spine-cells and stipulodes are obtuse (acute in *C. aspera*) and the upper ring of stipulodes remains more strongly developed (more than one and a half times the length of the lower ring) and may curve upwards towards the branchlets. A dwarf (<12 cm), compact, tufted form, often found at lake margins mainly in Scotland and Ireland, has been referred to var. *annulata* of *C. globularis*. However, Bryant et al. (2002) include it as a variety of *C. virgata*.

Chara vulgaris Linnaeus 1753

Synonym: *Chara foetida* A.Braun

28010180 **Pl. 146A** (p. 598)

Thallus of medium height, up to 50 cm, mid- to greyish-green, rarely dark green, usually moderately encrusted; cortex with a single secondary row between each primary row (diplostichous); spine-cells papillate or may be longer than branch width, more or less adpressed to the branch internode when developed, often appearing to lie in the primary cortical row furrows between the larger secondary cortical rows (aulacanthous), although all cortical rows may be of similar width (isostichous); monoecious.

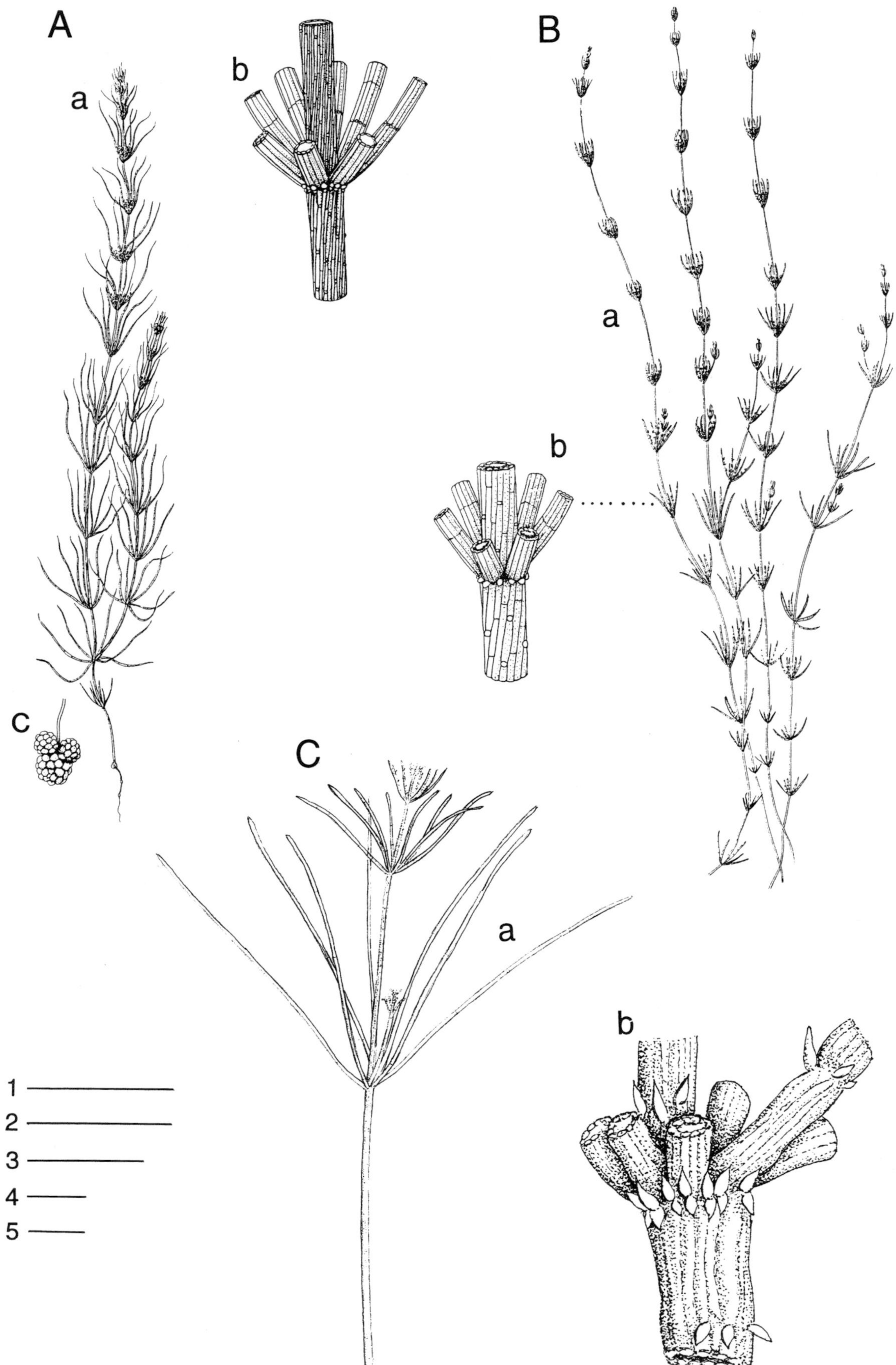

Plate 149 A–C
A. *Chara fragifera* (p. 599): a. habit, b. detail of a branchlet whorl, c. bulbil; **B**. *Chara connivens* (p. 597): a. habit, b. detail of a branchlet whorl; **C**. *Chara intermedia* (p. 602): a. habit, b. detail of a branchlet whorl.

Scale bars: 1, 50 mm; 2, 5 mm; 3, 5, 1 mm; 4, 10 mm. 1 for Aa, Ba; 2 for Ab, Ac; 3 for Bb; 4 for Ca; 5 for Cb.

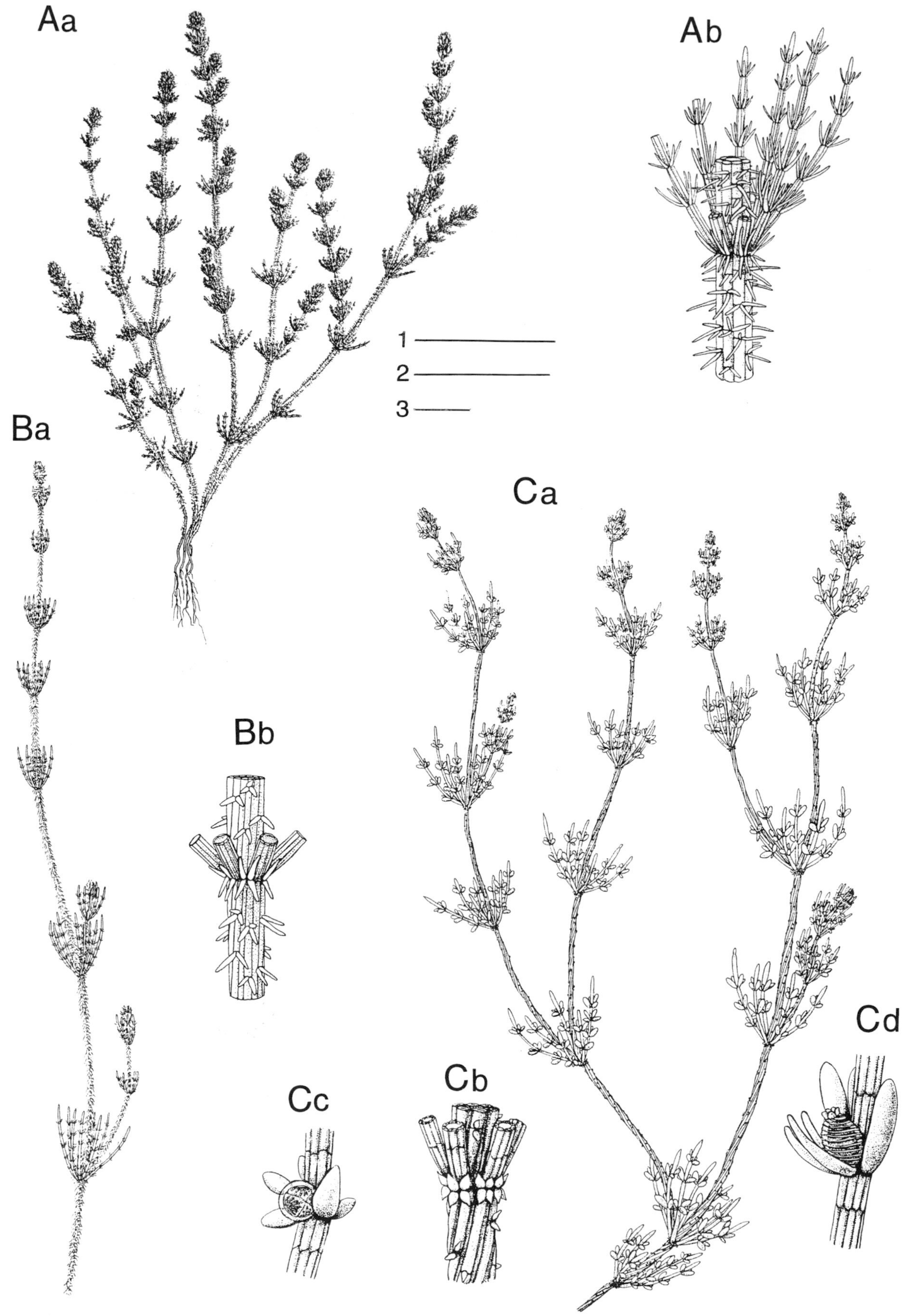

Plate 150 A–C
A. *Chara canescens* (p. 597): a. habit, b. detail of a branchlet whorl; **B**. *Chara aculeolata* (p. 597): a. habit, b. detail of a branchlet whorl; **C**. *Chara tomentosa* (p. 602): a. habit, b. detail of a branchlet whorl, c. branchlet node with antheridium and bract-cells, d. branchlet node with an oogonium, bract-cells, bracteoles and bractlet.

Scale bars: 1, 50 mm; 2, 5 mm; 3, 1 mm.1 for Aa, Ba, Ca; 2 for Ab, Bb, Cb; 3 for Cc,d.

Cosmopolitan; in the British Isles found in all kinds of habitats, from puddles and ditches to lakes; frequent in southern and eastern England and in Ireland; restricted to coastal and calcareous habitats in Scotland and Wales. Often the first colonizer of a new habitat, such as gravel workings or clay pits, but may be ousted over time by vascular plants; able to withstand drying out for periods in the summer; apparently tolerant of some nutrient enrichment but more usually an indicator of clean water conditions.

The spine-cells are deciduous but are usually persistent towards the branch apex. A variable species with a number of described varieties, the two most common of which in the British Isles are var. *longibracteata* (Kützing) J.Groves *et* Bullock-Webster (28010184) (Pl. 146D) with very long bract-cells (3–7 mm) and long, partially ecorticate branchlets, and var. *papillata* Wallroth ex A.Braun (28010185) (Pl. 146C) with elongate (up to 1.5 mm) spine-cells, exceeding the width of the branches and with a strongly aulacanthous cortex. Other described varieties of *C. vulgaris* are rare and ill-defined.

EXTINCT SPECIES
Chara braunii C.C.Gmelin 1826
28010030
Similar to *Lamprothamnium papulosum* in being ecorticate with a single ring of stipulodes (haplostephanous). However, the stipulodes are inflated and abruptly acuminate, the abaxial bract-cells are not well developed and the branch apices are not contracted into 'fox-tails'. It was recorded intermittently over several decades from its only known site in the British Isles, the Reddish Canal, Greater Manchester (where it was probably an introduction), but it is now extinct. Its disappearance from the canal may be linked to the cessation of warm effluent from nearby cotton mills following their closure in the early 1950s. Much of the canal has since been filled in and has disappeared.

Lamprothamnium J.Groves 1916

Thallus macroscopic, up to 40 cm, ecorticate and without spine-cells, encrusted to unencrusted, mid-green to brownish-green; similar to *Chara* in the symmetrical arrangement of the whorls of laterals of limited growth (branchlets) at the branch nodes; branchlets simple, with whorls of more or less equal unicellular bract-cells at their nodes; unicellular outgrowths from the branch nodes (stipulodes) in a single ring (haplostephanous), spreading downwards immediately beneath the branchlet whorls; monoecious; oogonia below the antheridia at the branchlet nodes, coronula of oogonium a single ring of 5 cells, not compressed.

Probably cosmopolitan, found in the British Isles in coastal lakes and lagoons where the water is brackish (tolerates salinities of up to 30‰).

Lamprothamnium papulosum (Wallroth) J.Groves 1916

Basionym: *Chara papulosa* Wallroth
28020010 **Pl. 151Ba–c** (p. 606) **CD**
Thallus ecorticate, slender to moderately robust, up to 40 cm, mid-green to brownish-green; branches up to 0.75 mm in diameter, with regular whorls of branchlets; spine-cells absent; stipulodes in a single ring (haplostephanous), slender, acuminate, downward pointing; bract-cells and bracteoles well-developed, acuminate, more or less equal, in whorls at the branchlet nodes; unicellular, spherical bulbils sometimes present on the rhizoids; monoecious.

Europe, North Africa, South Africa; in the British Isles recorded from maritime lakes and pools in brackish water (optimum salinity for growth is between 24‰ and 28‰, but known to occur between 8‰ and 30‰); rare along the south coast of England (Sussex to Dorset), the Outer Hebrides and from Wexford, Clare and Donegal.

The common name given to this alga is 'fox-tail stonewort' as the upper parts of the branches are contracted and bushy due to the short internodes and strongly overlapping branchlets, in addition to the well-developed, acuminate bract-cells and stipulodes which give them a furry appearance.

Nitella C.Agardh 1824

Thallus macroscopic, from a few centimetres up to 1 metre, ecorticate and without spine-cells, stipulodes or bract-cells, unencrusted to lightly encrusted (but see *N. tenuissima*), light to dark green or somewhat yellowish; similar in organization to, but not as rigid or symmetrical as, *Chara* species; branches having nodes and internodes with laterals of limited growth (branchlets) in similar or dimorphic whorls at the nodes; branchlets forked one or several times to produce similar, single-celled segments (rays), terminating in a 1–3-celled end-segment (dactyl); monoecious or dioecious; oogonia usually 2 or 3, lateral at the forking of a branchlet, occasionally at the base of a branchlet whorl, laterally compressed; coronula of oogonium a double ring of 10 cells.

Cosmopolitan; widely distributed in habitats with a low pH, such as acid lakes and bog pools although some taxa do occur in water with a higher pH; more common than *Chara* in acid waters.

1 Dactyl (branchlet end-segment) single-celled........2
1 Dactyl 2–3-celled (sometimes only minutely so at apex) 3
2 Dioecious; antheridia 650–775 μm in diameter; oospores 375–425 μm long........*N. opaca*
2 Monoecious; antheridia 500–625 μm in diameter; oospores 500–575 μm long........*N. flexilis*
3 Dactyl 2–3-celled; thallus slender to robust with branches >0.3 mm in diameter; internodes 0.5–2(–4) times as long as branchlets........4
3 Dactyl always 2-celled; thallus very slender with branches <0.3 mm in diameter; internodes (1–)2–5 times as long as branchlets........6
4 Sterile branchlets superficially appearing simple but inconspicuously divided at apex into 2–4 minute dactyls; thallus robust and branches 1–2 mm in diameter........*N. translucens*
4 All branchlets clearly divided 1–3 times into clearly visible segments and dactyls; thallus slender to fairly robust with branches <1 mm in diameter........5
5 Thallus up to 20 cm, slender, finely branched; branches <0.75 mm in diameter; dactyl usually 3-celled, tapering........*N. gracilis*

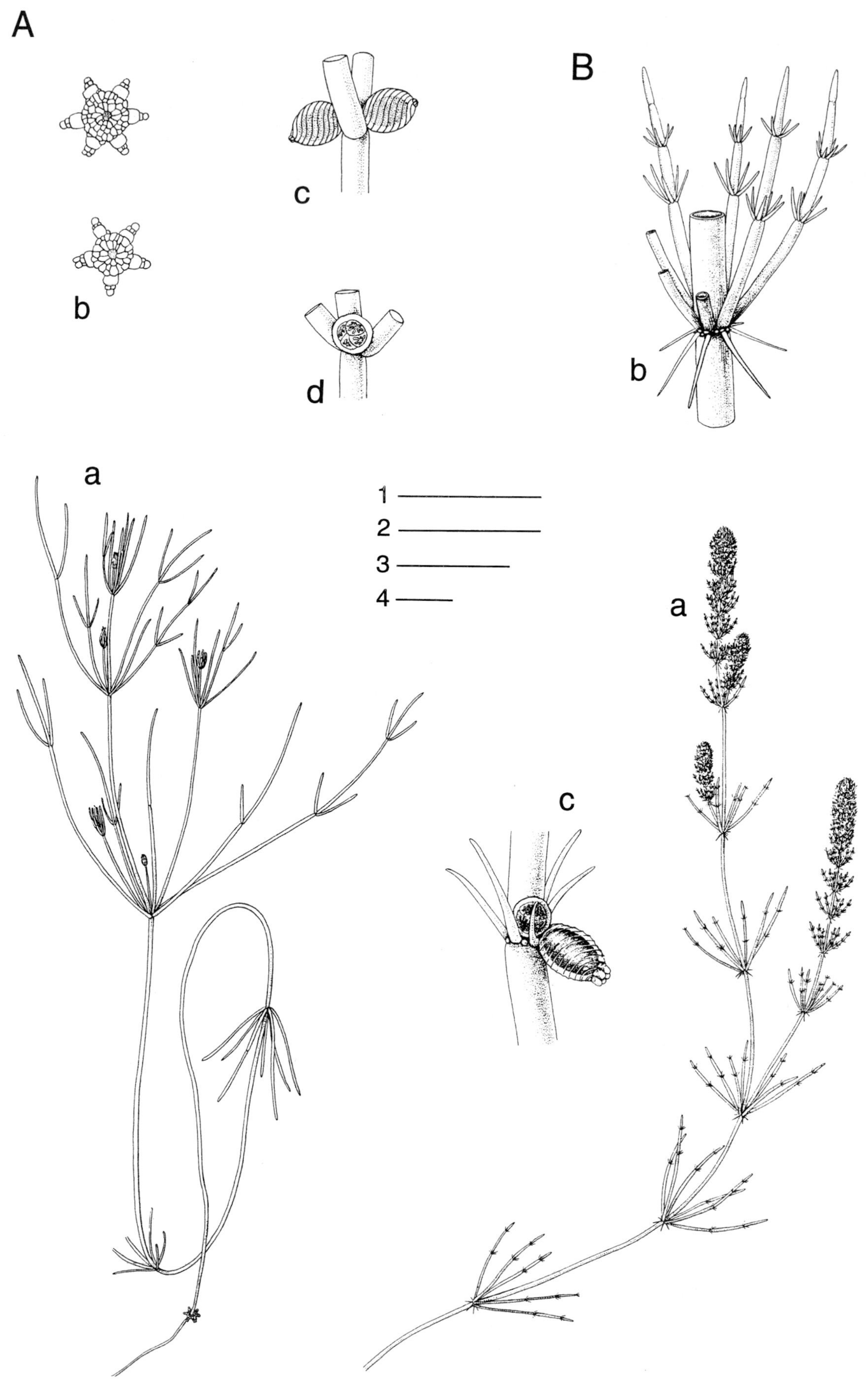

Plate 151 A, B
A. *Nitellopsis obtusa* (p. 609): a. habit, b. bulbils, c. detail of a branchlet node showing oogonia, d. detail of a branchlet node showing antheridium; **B**. *Lamprothamnium papulosum* (p. 605): a. habit, b. detail of a branchlet whorl, c. detail of a branchlet node showing gametangia and bract-cells.
Scale bars: 1, 50 mm; 2, 5 mm; 3, 4, 1 mm. 1 for Aa, Ba; 2 for Ab, Bb; 3 for Bc; 4 for Ac,d.

5 Thallus up to 40 cm, fairly robust; branches >0.75 mm in diameter; dactyl 2–3-celled, tapering or penultimate cell much broader than the terminal cell ....*N. mucronata*
6 Thallus usually <5 cm, delicate, mid- to dark green, rarely encrusted; gametangia produced at first and second branchlet forks; found in acidic to slightly nutrient-rich lakes to a depth of 4 m ....*N. confervacea*
6 Thallus up to 20 cm, bright or greyish-green, often encrusted; gametangia produced at second and third branchlet forks; found in low-nutrient calcareous pools and ditches to a depth of 30 cm ....*N. tenuissima*

Nitella confervacea (Brébisson) A.Braun ex Leonhardi 1863

Basionym: *Nitella gracilis* var. *confervacea* Brébisson
Synonym: *Nitella batrachosperma* (Thuillier) A.Braun
28030020 **Pl. 153A** (p. 610)

Thallus small, delicate, up to 5 cm, mid- to dark green, rarely encrusted; branches less than 0.3 mm in diameter; dactyls 2-celled; monoecious; gametangia produced at first and second branchlet forks.

Probably cosmopolitan; in the British Isles most often recorded in nutrient-poor to slightly nutrient-rich lakes to a depth of 4 m in western Ireland and western and northern Scotland during August and September; rare.

The smallest charophyte found in the British Isles. At maturity, the branchlet whorls form distinctive spherical clusters (which resemble the pills that form on worn woollen clothing) on long filamentous internodes. Could be confused with small plants of *N. tenuissima*, but these taxa are found in different habitats, *N. confervacea* as above and *N. tenuissima* in low-nutrient calcareous pools and ditches.

Nitella flexilis (Linnaeus) C.Agardh 1824

Basionym: *Chara flexilis* Linnaeus
28030030 **Pl. 152A** (p. 608) **CD**

Thallus moderately robust, up to 100 cm, mid- to light green, diffusely branched; branches up to 1 mm in diameter with whorls of 6 to 8 lax, slender branchlets at each branch node; dactyls single-celled, acuminate; monoecious.

Europe, Asia, Africa and America; known in the British Isles from moderately to fairly nutrient-rich lakes and pools in southern and eastern England, also central Ireland, but rare elsewhere.

Sterile material cannot be differentiated reliably from *N. opaca* as reproductive state and gametangia size (see key for dimensions) are the clearest distinguishing features between these taxa. Following Moore (1986) many records that may have been *N. opaca* were included in *N. flexilis*. Apparently endemic to Donegal the status of *Nitella spanioclema* J.Groves *et* Bullock-Webster ex Bullock-Webster (28030070) needs further investigation. It is similar to *N. flexilis* but with a more lax and delicate habit, fewer branchlets (2–4) at each node and short or rudimentary dactyls.

Nitella gracilis (J.E.Smith) C.Agardh 1824

Basionym: *Chara gracilis* J.E.Smith
28030040 **Pl. 152C** (p. 608)

Thallus slender, up to 20 cm, diffusely branched, with hair-like internodes and umbellate branchlets; branches less than 0.75 mm in diameter; dactyls usually 3-celled and tapering; monoecious.

Probably cosmopolitan; known in the British Isles from nutrient-poor water in small, fairly shallow peaty lakes, or on peaty silt at the lower margins of rooted vegetation in larger lakes; sparse records from Sutherland, Ayrshire, Snowdonia, Ceredigion and Wicklow.

Delicate, with tassel-like branchlets and thallus often densely tufted. Some recent records are misidentifications of *N. mucronata*.

Nitella mucronata (A.Braun) Miquel 1840

Basionym: *Chara mucronata* A.Braun
28030060 **Pl. 152D** (p. 608)

Thallus moderately robust to fairly slender, up to 40 cm; branches more than 0.75 mm in diameter; dactyls 2-celled with the smaller apical cell more slender than the larger, round, penultimate cell that ends in a short, sharp point (mucronate); monoecious.

var. *gracillima* J.Groves *et* Bullock-Webster 1917

28030062 **Pl. 152Dc** (p. 608)

Differs from the type variety in having 2–3-celled dactyls, with the penultimate and apical cells of similar size at their junction but tapering to apex.

Europe, Asia, Africa, the Americas; in the British Isles in moderately to fairly nutrient-enriched ponds, lakes and canals, often associated with a diverse macrophyte flora. The type variety is rare throughout the British Isles and apparently decreasing, but recently confirmed from County Louth in Ireland; var. *gracillima* is scarce in England, rare in Wales (Powys) and Ireland (Dublin, Fermanagh), but frequent and increasing in man-made sites, such as canal systems and ornamental ponds where it may be accidentally introduced.

The variety *gracillima* has been confused with *N. gracilis*, but differs in size and in the microscopic ornamentation of the mature oospore (granulate in *N. gracilis*, reticulately pitted in *N. mucronata*).

Nitella opaca (C.Agardh ex Bruzelius) C.Agardh 1824

Basionym: *Chara opaca* Bruzelius
28030100 **Pl. 152B** (p. 608) **CD**

Thallus moderately robust, up to 100 cm, branches up to 0.8 mm in diameter, similar to *N. flexilis* but usually darker green, the fertile branchlet whorls often contracted; dactyls single-celled, often cuspidate; dioecious.

Europe, Asia, North Africa, the Americas; widespread and fairly frequent throughout the British Isles in a wide range of habitats, including moderately fast-flowing water and pools that dry out seasonally; most frequent in acid habitats in nutrient-poor to slightly nutrient-enriched lakes and ponds, where it may be dominant; sometimes occurring in low-nutrient, alkaline waters.

Sterile material cannot be differentiated reliably from *N. flexilis* as reproductive state and gametangia size (see key for dimensions) are the clearest distinguishing features between these taxa. Following Moore (1986), many

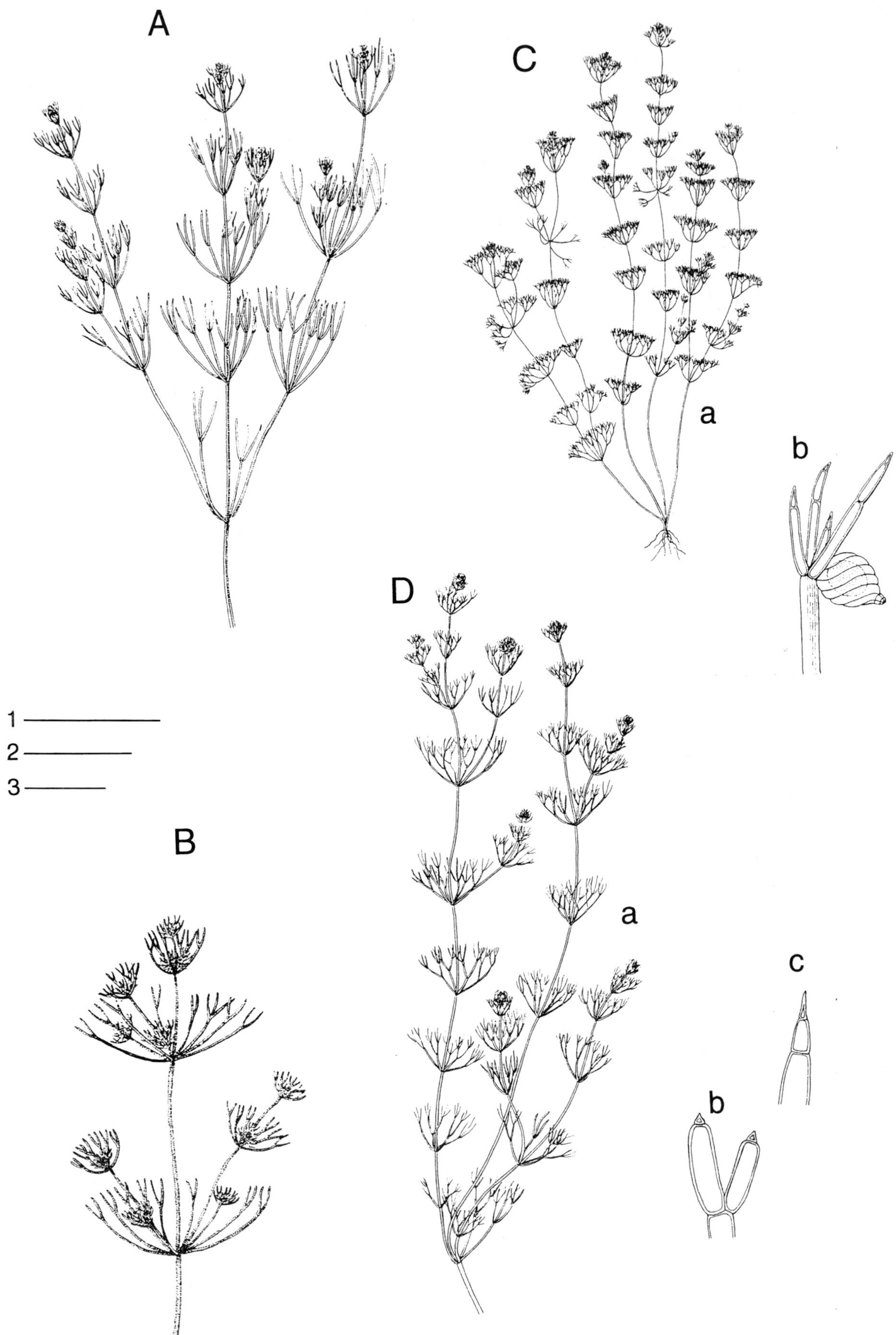

Plate 152 A–D

A. *Nitella flexilis* (p. 607): habit; **B**. *Nitella opaca* (p. 607): habit; **C**. *Nitella gracilis* (p. 607): a. habit, b. branchlet apex with dactyls and an oogonium; **D**. *Nitella mucronata* (p. 607): a. habit, b. branchlet apex with a pair of 2-celled, mucronate dactyls; c. apex of a 3-celled, tapering dactyl (var. *gracillima*).

Scale bars: 1, 50 mm; 2, 3, 1 mm 1 for A, B, Ca, Da; 2 for Cb; 3 for Db,c.

records that may have been *N. opaca* were included in *N. flexilis*. Slender, diffusely branched forms (sometimes referred to as var. *attenuata* H. *et* J.Groves) occur and a small (up to 10 cm) attenuate form similar to *N. gracilis* (unnamed) is known in several upland lakes in Wales and Scotland.

Nitella tenuissima (Desvaux) Kützing 1843

Basionym: *Chara tenuissima* Desvaux
28030080 **Pl. 153B** (p. 610)
Thallus slender, up to 20 cm, bright or greyish-green, often encrusted with long internodes and compact whorls (so that the branches resemble strings of fuzzy beads); dactyls 2-celled; monoecious; gametangia produced at the second and third branchlet forks.

Europe, southern Africa, India, North America, Madagascar; rare in the British Isles in the low-nutrient, alkaline water of fen pools, peat cuttings and ditches, tolerant of seasonal drying out; an early colonizer of disturbed habitats (Cambridgeshire, Anglesey, West Meath, Galway, Clare).

Nitella translucens (Persoon) C.Agardh 1824

Basionym: *Chara translucens* Persoon
28030090 **Pl. 153C** (p. 610)
Thallus robust, up to 60 cm, yellowish-green; branches 1–2 mm in diameter; minute 2–3-celled dactyls at tips of otherwise simple branchlets; monoecious.

Europe, North Africa; in the British Isles sometimes found in considerable abundance in acidic, nutrient-poor pools and lakes on peat or peaty silt and in open swamp communities. Widespread, occasional to locally frequent in more northerly and westerly parts of the British Isles, rare and restricted to acid habitats elsewhere.

The largest species of *Nitella* found in the British Isles, with some of the largest cells known to science.

EXTINCT SPECIES
Nitella capillaris (Krocker) J.Groves *et* Bullock-Webster 1920
Basionym: *Chara capillaris* Krocker
28030010
Similar to *N. opaca* but the gametangia are enveloped in mucilage (hence it has been given the common name 'slimy-fruited stonewort'). It was found sporadically since its discovery in May 1885 in a ditch in the vicinity of Sutton Gault in Cambridgeshire but was last recorded there in 1959; all other records are doubtful since they cannot be authenticated. Stewart & Church (1992) consider it to be extinct in the British Isles although it is widely distributed in Europe and the northern hemisphere.

Nitella hyalina (de Candolle) C.Agardh 1824
Basionym: *Chara hyalina* de Candolle
28030050
Similar to *N. gracilis* and *N. mucronata* but with distinctive, differentiated whorls of about 20 branchlets at each branch node. Previously recorded from Cornwall (Loe Pool, near Helston) at the turn of the nineteenth century but is almost certainly now extinct in the British Isles.

Nitellopsis Hy 1889

Thallus macroscopic, robust, up to 60 cm, brownish-, mid- to grey-green, often lime-encrusted, ecorticate and without spine-cells; stipulodes very rare; similar in structure to other charophyte genera; branches long, lax, divided into nodes and internodes; branchlets usually with 1 or 2 nodes producing 1 or 2 elongate bract-cells (sometimes long enough to resemble side branches) at each node; dioecious; gametangia solitary or paired at branchlet nodes; coronula of oogonium a single ring of 5 cells; not compressed.

Probably cosmopolitan, usually in still and often slightly brackish-water.

Nitellopsis obtusa (Desvaux) J.Groves 1919

Basionym: *Chara obtusa* Desvaux
28040010 **Pl. 151A** (p. 606)
Thallus robust, up to 60 cm, brownish- to grey-green, ecorticate, without spine-cells and stipulodes; branches up to 1.5 mm in diameter; elongate bract-cells at branchlet nodes resembling side branches; distinctive, star-shaped, starch-filled bulbils developed from lower branch nodes; dioecious.

Europe, southern Asia, West Indies; in the British Isles found in alkaline, often mildly brackish, maritime lakes; rare, only reported from Norfolk and Gloucestershire.

Distinguished from *Nitella* and *Tolypella* by the presence of bract-cells, the branchlets not forming clustered heads and the unique bulbils.

Tolypella (A.Braun) A.Braun 1857

Thallus macroscopic, robust, up to 50 cm, ecorticate, without spine-cells, stipulodes or bract-cells, often encrusted, light to mid-green, grey-green or olive-green; similar in structure to other charophyte genera, although less symmetrical and possessing long, scraggy, sterile branches and whorls of shorter, fertile, incurved branches which form loose or tight balls; branchlets divided into nodes and unequal internodes (segments or rays); branching pinnate with a main central axis and smaller laterals; monoecious or dioecious; oogonia and antheridia in clusters, often on lower nodes of branchlets or at their base; coronula of oogonium a double ring of 10 cells, not compressed.

Probably cosmopolitan but more common in the northern hemisphere in a wide range of freshwater habitats and occasionally in brackish-water.

1 Branchlet end-cells obtuse 2
1 Branchlet end-cells conical, acute 3
2 Antheridia >500 μm in diameter; oospore 400–475 × 350–450 μm; thallus usually lightly lime-encrusted, yellowish- to brownish-green; fertile whorls forming untidy, loose heads .... *T. nidifica*
2 Antheridia <500 μm in diameter; oospore 300–375 × 250–300 μm; thallus usually heavily lime-encrusted, greyish- or brownish-green; fertile whorls forming dense, spherical heads due to strongly incurved branchlets .... *T. glomerata*
3 Branchlets of sterile whorls divided (but tips may be lost); branches up to 1 mm in diameter .... *T. intricata*

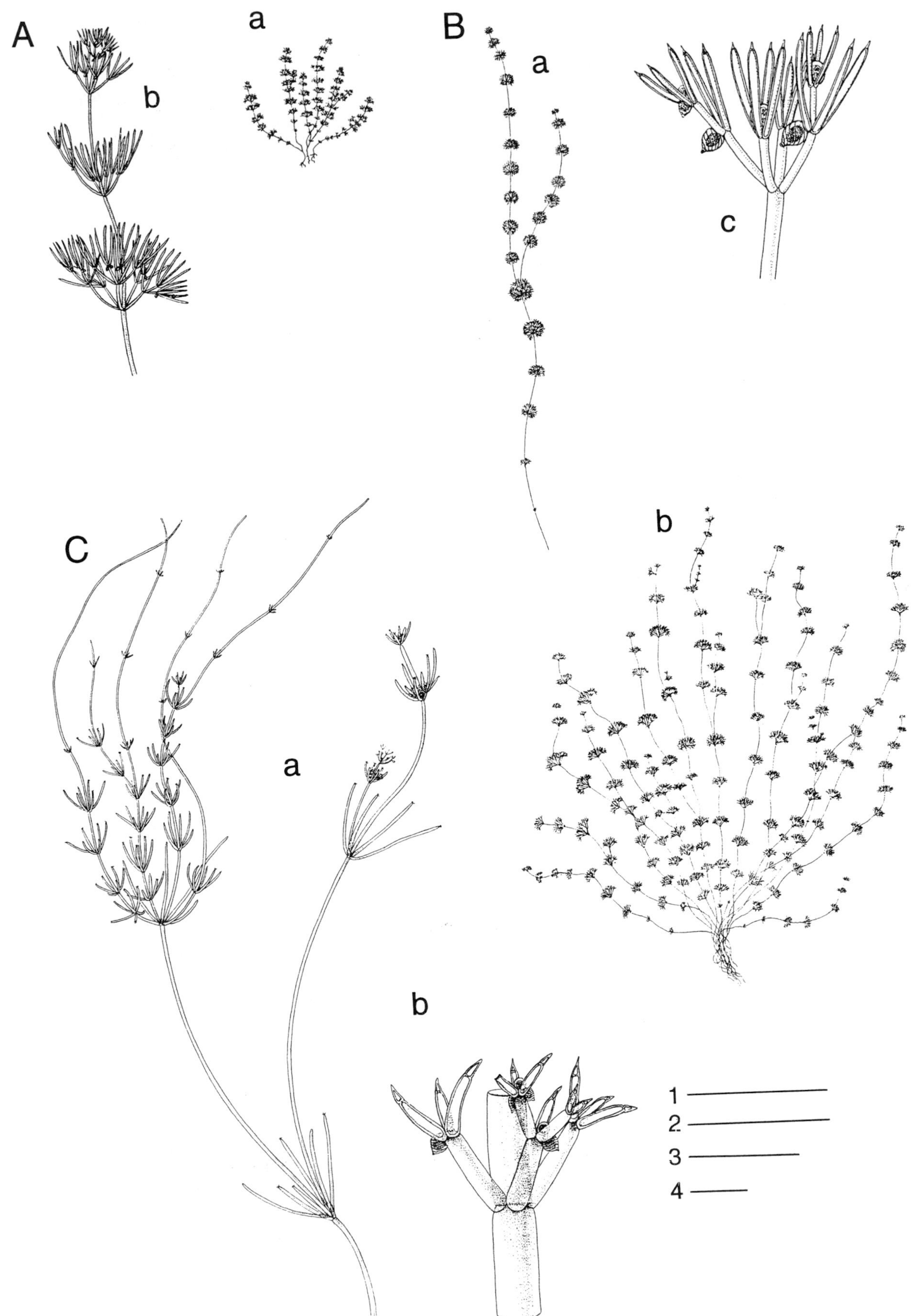

Plate 153 A–C

A. *Nitella confervacea* (p. 607): a. habit; b. detail of a branchlet whorls; **B**. *Nitella tenuissima* (p. 609): a, b. habit, c. detail of a branchlet whorl showing dactyls and gametangia; **C**. *Nitella translucens* (p. 609): a. habit, b. detail of a branchlet whorl. Scale bars: 1, 50 mm; 2, 5 mm; 3, 4, 1 mm. 1 for Aa, Ba,b, Ca; 2 for Ab; 3 for Bc; 4 for Cb.

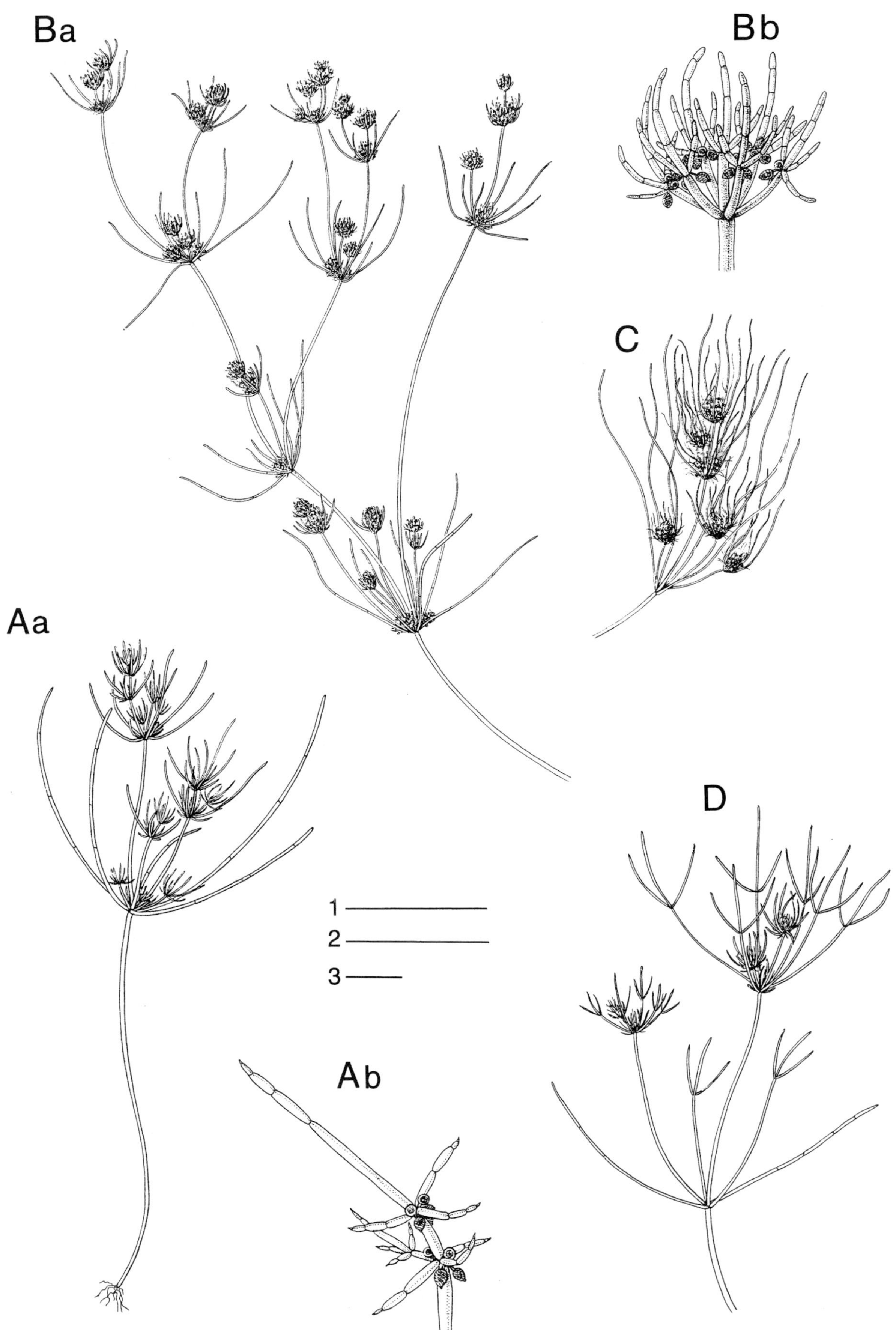

Plate 154 A–D
A. *Tolypella prolifera* (p. 612): a. habit, b. detail of a branchlet showing two fertile whorls with gametangia; **B**. *Tolypella glomerata* (p. 612): a. habit, b. detail of a branchlet showing two fertile whorls with gametangia; **C**. *Tolypella nidifica* (p. 612): habit; **D**. *Tolypella intricata* (p. 612): habit.
Scale bars: 1, 50 mm; 2, 5 mm; 3, 1 mm. 1 for Aa, Ba, C, D; 2 for Ab; 3 for Bb.

3 Branchlets of sterile whorls undivided; branches 1–2.5 mm in diameter ....*T. prolifera*

Tolypella glomerata (Desvaux) Leonhardi 1863

Basionym: *Chara glomerata* Desvaux

28050010 **Pl. 154B** (p. 611)

Thallus moderately slender, up to 20 cm, olive- to grey-green, usually heavily encrusted; branches up to 1 mm in diameter; fertile branchlet whorls forming tight clusters; branchlet end-cells obtuse; monoecious.

Probably cosmopolitan; in the British Isles in alkaline pools and the shallow sides of ditches, lakes and canals (those parts that dry out in summer); tolerant of mildly brackish water and sometimes locally frequent in sand-dune pools; often a winter annual visible from October to May, but persisting throughout the year in the deeper water of low-nutrient, large alkaline lakes, where it occurs with *C. globularis* and *C. contraria*; widespread, but scarce and apparently decreasing, occasional in central Ireland and eastern England, rare elsewhere.

Tolypella intricata (Trentepohl ex Roth) Leonhardi 1863

Basionym: *Chara intricata* Trentepohl ex Roth

28050020 **Pl. 154D** (p. 611)

Thallus moderately slender, up to 40 cm, mid- to grey-green; branches up to 1 mm in diameter; fertile branchlet whorls forming loose, untidy, tassel-like heads; sterile branchlets with 1–2-pinnate lateral segments (but segments and branchlet end-segments sometimes lost in very encrusted, brittle specimens); branchlet end-cells acute; monoecious.

Europe, North America; in the British Isles from alkaline pools in permanent pasture, the shallow sides of ditches and canals, usually in the open conditions of disturbed or dynamic habitats (such as those affected by trampling livestock or areas that dry out in summer); winter annual visible from October to May, but occasionally persisting throughout wet years; rare, known from Norfolk, Suffolk, Cambridgeshire, Oxfordshire, Gloucestershire and Worcestershire and also from the canal system in eastern and central Ireland.

Tolypella nidifica (O.F.Müller) Leonhardi 1857

Basionym: *Conferva nidifica* O.F.Müller

28050030 **Pl. 154C** (p. 611)

Thallus up to 20 cm, moderately slender, yellow- or brownish-green, with light to medium encrustation; branches up to 0.6 mm in diameter; fertile branchlet whorls loosely tufted; branchlet end-cells obtuse; monoecious.

Northern Europe; known only in the British Isles in the permanent water of brackish-water lakes, often in association with *Chara baltica*, *C. canescens* and *Ruppia* spp.; probably a summer annual; rare, only known from Orkney and Western Isles of Scotland.

Similar to *T. glomerata* but with larger gametangia. There is an unconfirmed record from Somerset of a similar species, *Tolypella salina* Corillion, made in 1922. This species does not appear to be clearly distinguishable from *T. glomerata* and *T. nidifica*, apart from having larger antheridia (see Krause, 1997).

Tolypella prolifera (Ziz ex A.Braun) Leonhardi 1863

Basionym: *Chara prolifera* Ziz ex A.Braun

28050040 **Pl. 154A** (p. 611)

Thallus robust, up to 40 cm, mid- to yellow-green, unencrusted to moderately encrusted; branches up to 2 mm in diameter; fertile whorls of more slender branchlets condensed into heads; sterile branchlets undivided; branchlet end-cells acute; monoecious.

Europe, America; reported in the British Isles from alkaline ditches, canals, rivers and pools often appearing following periodic disturbance and usually associated with a diverse macrophyte flora; a rare summer annual visible from May to October; now recorded from Sussex, Cambridgeshire, Lincolnshire and Somerset, but previously widespread in southern England.

The largest species of *Tolypella* (comparable in stature to *Nitella translucens*) found in the British Isles.

Acknowledgements

We would like to thank the following who kindly read and commented on the text: Dr Alan Harrington, Mrs Linda Irvine, Dr Fred Rumsey, Professor Clive Stace.

PHYLUM GLAUCOPHYTA

Brian A. Whitton

This phylum was established to accommodate some or all of the algae which include intracellar blue-green coloured cyanelles rather than chloroplasts. Cyanelles are evolutionary steps between blue-green algae and chloroplasts, though are more like the latter than the former; they cannot exist independently of the host cell. The cyanelles of some genera have been treated as separate species, with their own Latin name, but as they cannot exist independently of the host cell, this is a doubtful practice. There is no reason to believe that the various types of host cell are closely related to each other nor that the various cyanelles are closely similar to each other. The phylum is therefore entirely artificial. Organisms with cyanelles are now usually placed in the phylum which they (i.e. the host cell) most closely resemble. However, it is still unclear what are the phylogenetic relationships of the genera *Glaucocystis* and *Cyanaphora*, so the phylum is retained for the time being.

Cyanophora Korshikov 1924

Cells flattened, ellipsoidal, possess 2 unequal flagella emerging from a subapical depression. One to 2 (when mature) cyanelles per cell. Reproduction by longitudinal division into 2 daughter cells each with a cyanelle.

Three known species

Cyanophora paradoxa Korshikov 1924

05050010

Cells 2–3 μm thick, 5–8 μm wide, 8–12 μm long; 1–2 blue cyanelles.

Plankton of shallow lakes with hard, nutrient-rich water. Several records from the British Isles, including St James's Park Lake, London.

Glaucocystis Itzigsohn in Rabenhorst 1866

Typically a colony of 4–16 cells enclosed by a persistent mother cell wall, though sometimes a single cell, sometimes surrounded by part or all of the persistent wall from the previous cell generation; cells spherical or ellipsoidal, containing a number of irregular sausage-shaped blue or blue-green cyanelles.

Glaucocystis nostochinearum Itzigsohn in Rabenhorst 1868

29010010 **Pl. 1H** (p. 33)

Free-floating colony of (1–)4–8(–16) cells enclosed by persistent mother cell wall; wall usually entire, but eventually starts to disintegrate; colony typically 40–60 μm in largest dimension. Cells ellipsoidal, reproducing by formation of typically 4 autospores, though sometimes only 2, rarely 1; larger colonies may have several groups of 4 cells, each group bounded by an entire mother cell wall; in this case individual cells separated from the external environment by two walls (in addition to their own wall); cells 10–18 μm wide, 18–24(–28) μm long, often with a slight thickening of the cellulose wall at each pole; a group of typically 8 cyanelles towards each end of the cell; cyanelles often twist around each other and fill so much of the cell that it is difficult to distinguish individual ones; cyanelles bright blue-green.

Temperate regions, with records from shallow softwater ponds and stretches of water in acid wetland areas, such as pools in *Sphagnum* bogs. Widespread in the British Isles in shallow weedy ponds, though never abundant at sites where it occurs; in both soft and moderately hard waters enriched with humic materials, and it seems quite likely that its occurrence in harder waters has been overlooked elsewhere.

Prescott (1944, 1951) recognized two other species in North America, *G. duplex* with spherical cells in colonies and *G. oocystiformis* with solitary cells; experimental studies on British material show that *G. nostochinearum* is highly variable, but neither laboratory nor field populations have so far revealed material which corresponds fully with Prescott's descriptions for the other two species.

GLOSSARY

The Editors

The Flora uses as few technical terms as possible, but some are essential to avoid long explanations. Some terms not used in the Flora are also listed, because readers are likely to encounter them in other literature.

Some charophyte and cyanophyte terms are only described in the text (see subject index).

abaxial On side or face away from an axis or branch.
acicular Needle-like.
acid water Acid waters are ones with a high concentration of hydrogen ions, which is expressed on the pH scale. The lower the pH value below 7.0, the more acid the water. At pH 7.0, the neutral value, there are equal numbers of hydrogen and hydroxyl ions. The pH scale is logarithmic, so water at pH 6 has ten times as many hydrogen ions as water at pH 7, while water at pH 5 has one hundred times more hydrogen ions. Highly acidic waters, such as those below copper mines, not only have a high concentration of hydrogen ions, but are often strongly buffered against a pH change. For any particular pH value, acid water which is strongly buffered presents a more demanding environment for algae than one which is weakly buffered.
acute Ending abruptly in a sharp point; in geometry describing an angle of less than a right angle.
acutely rounded Apices are so narrow as to appear almost pointed (e.g. certain *Closterium* species).
adaxial On side or face next to an axis or branch.
aegagropilous Applied to compacted filaments of a species of the green alga *Cladophora*, which grows on the bottom of some lakes and produces balls, often 10 cm or more in diameter, as they are rolled about by currents.
aggregation Loose association of cells with no characteristic form or structure (cf. colony).
akinete Thick-walled resting spore formed by transformation of a vegetative cell; in blue-green algae, with an obvious wall and usually considerably larger than the vegetative cells and considered to be storage structures.
alkaline water Water containing predominantly hydroxyl ions; pH usually well above 7.0.
amoeboid Movement by extensions of highly mobile protoplasm surrounded by a cell membrane.
amorphous Macroscopically visible groups of filaments or cells, often mucilaginous, without a very distinct bounding layer; meaning without form (cf. floc).
anastomose Joined by cross-connections.
anisogametes Morphologically dissimilar motile gametes.
anisogamy Union of two morphologically dissimilar motile gametes.
androphore Branch bearing antheridia.
androsporangium (pl. **androsporangia**) Cell (sporangium) producing androspores.
androspore Oedogoniales: zoospore that develops into a dwarf male.
annulate/annular Ring-like.
antheridium (pl. **antheridia**) Gametangium producing antherozoids as male gametes in oogamous reproduction.
antherozoid Flagellated male gamete.
apex (pl. **apices**) End of an elongate cell; end of a filament, especially where there is obvious differentiation between the ends, such as a tapered filament, where the narrow end is considered to be the apex.
apiculate Having a short, sharp point.
aplanosporangium (pl. **aplanosporangia**) Cell or sporangium containing aplanospores.
aplanospore Non-motile spore and not morphologically identical to the mother cell.
aplanozygote Dinoflagellates: a motile zygote.
arcuate Strongly curved, arc-like (Fig. 7Y).
assurgent Curving obliquely upwards.
astaxanthin Carotenoid pigment produced in some chlorophytes, one of several pigments which can colour parts of cells red.
asteroid Star-shaped; 3-dimensional term most often used for certain chloroplasts, e.g. in *Zygnema* spp. and some saccoderm desmids.
attenuate Tapering to a narrow apex.
autospore Non-motile spore, usually identical in shape to the mother cell from which it derives, lacking the potential to produce flagella; characteristically produced in the Chlorococcales.
autotrophic 'Self feeding'; able to synthesize organic compounds from inorganic substrates using light or (some bacteria) chemical energy (cf. heterotrophic).
auxospore Diatoms: distinctive cell formed after sexual reproduction and often very different from the typical cells formed during subsequent cell divisions.
axile/axial Along a median longitudinal line or axis of a cell, usually used with reference to position of chloroplast, especially in some genera of Zygnematales.

baeocyte(s) Blue-green algae: small reproductive cells formed in some genera (e.g. *Dermocarpa*, *Pleurocapsa*) by simultaneous or successive divisions of a mother cell, usually in addition to the potential for binary division. As a consequence, the baeocytes are much smaller than the mother cell. Following release, the baeocyte is usually motile when it comes into contact with a surface, including that of the mother cell wall.
barrel-shaped Cylindrical, with slightly convex sides; ends are not rounded.
basionym Original name of a taxon, now used in a new combination.
benthic Bottom-living; attached to or resting on a surface.
benthos Organisms associated with the bottom of a pond, lake or river.
bifid Divided more or less equally into two.
biflagellate Having two flagella.

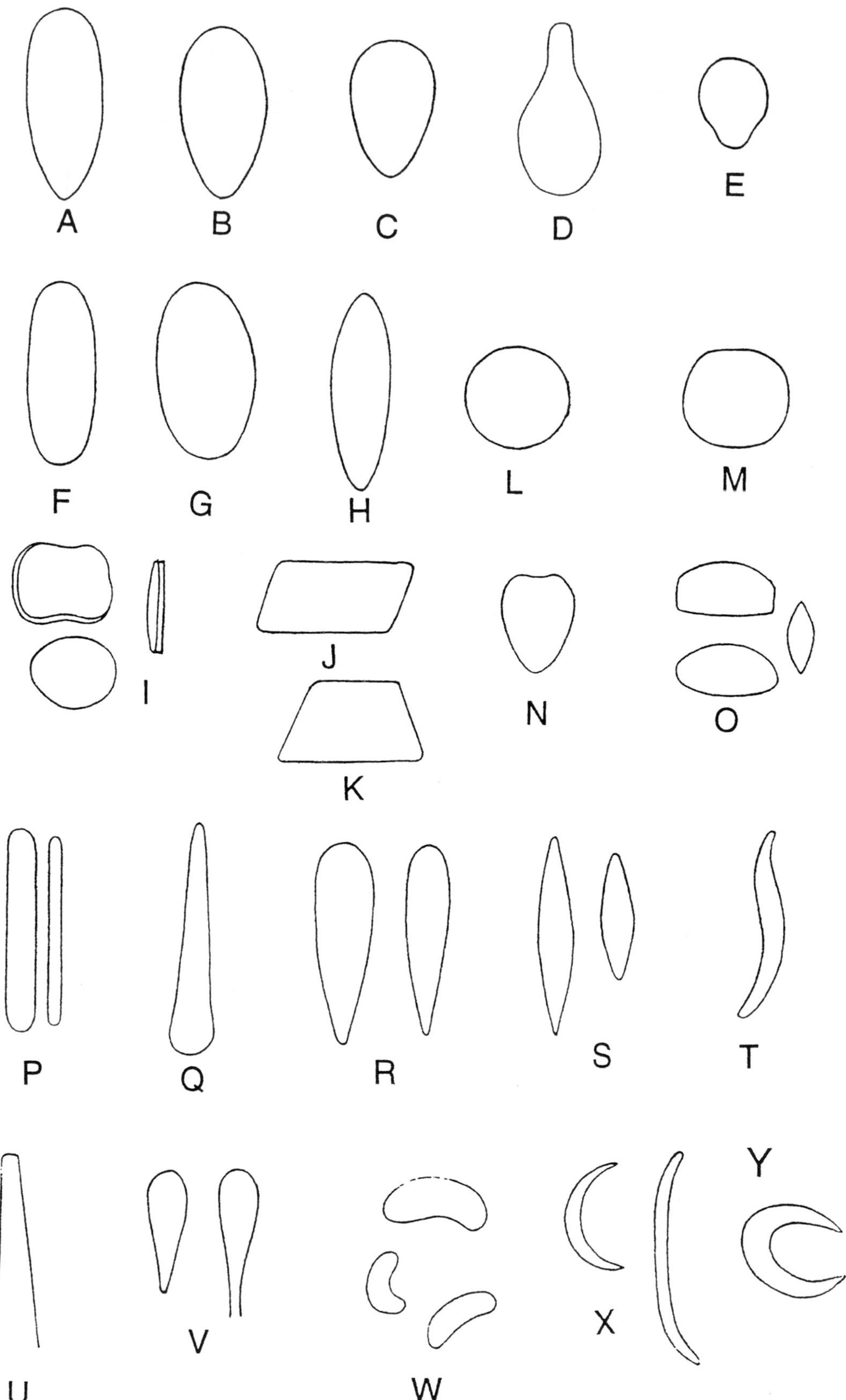

Figure 7. Illustrations of the principal terms used to describe the shape (see Glossary). Shapes in many cases may be interpreted as either 3- or 2-dimensional and are indicated by (3) or (2), respectively. Common words which are synonyma are given with the initial letter in lower case. **A**. Oblong-ovoid(3), Oblong-ovate(2); **B**. Ovoid(3), oval(2); **C**. Obovoid(3), obovate(2); **D**. Pyriform(3), pear-shaped; **E**. Obpyriform(3); **F**. Oblong(2); **G**. Broadly-oblong(2); **H**. Ellipsoidal(3); elliptical(2) **I**. Discoidal(3), disc- or plate-shaped; **J**. Rhomboidal(2); **K**. Trapeziform(2), trapezoid(3); **L**. Spherical(3), coccoidal; **M**. Oblate(3), depressed spherical; **N**. Cordate(2), heart-shaped; **O**. Lenticular(3), lens-shaped; **P**. Cylindrical(3); **Q**. Spatulate(2), spatula-shaped; **R**. Lanceolate(2), lance-shaped, **S**. Fusiform(3), spindle-shaped; **T**. Sigmoid(2); **U**. Truncate(2); **V**. Clavate(3), club-shaped; **W**. Reniform(2), kidney- or bean-shaped; **X**. Lunate(2), crescent-shaped; **Y**. Arcuate(2), arc-like.

bifurcate Equally divided into two.
biseriate Arranged in two rows or series.
brackish water Water in marine coastal localities, with an increased and usually variable salt content.
branchlet Small secondary branch.
bristle Long or very long, relatively stiff, projection from cell; often there is a collar of firmer material round the lower part or just the base of the bristle (cf. seta).
bulbils Charophytes: starch-containing structures formed on the rhizoids (see p. 595).
bulbose/bulbous Bulb-shaped; swollen, rounded base narrowing to a smaller apex.
bushy Similar to a tuft, but where the tuft tends to separate into smaller groups of filaments towards the end.

calcified Encrusted or impregnated with calcium and sometimes magnesium carbonate.
calcareous Containing large amounts of calcium carbonate and sometimes also magnesium carbonate.
calyptra Membranous hood or cover.
cap cells Oedogoniales: series of transverse rings at apex of cells resulting from the unique type of cell division occurring in this order.
capitate Having an enlargement or rounded head at one end, often rather like a pin-head.
capitellate *Oedogonium:* cells of a filament swollen at the distal end.
capitulum Charophytes: cell from which antheridial filaments arise.
carotenoid Yellow, orange or red pigment present in chloroplasts, 'eyespots' and sometimes other parts of cells.
carpogonium Rhodophytes: female sex organ containing the egg.
carposporangium Sporangium produced on a carposporophyte and containing carpospores.
carpospore Rhodophytes: spore formed as a result of fertilization.
carposporophyte Rhodophytes (Florideophyceae): multicellular phase following fertilization and producing gonimoblast filaments and carposporangia.
chromosome Complex thread-like structures that can be seen following staining of the nuclei of eukaryotic algae; containing encoded hereditary instructions in their DNA. Chromosomes are also present in blue-green algae, but are not visible with the light microscope, even after staining (cf. DNA).
chloroplast Double-membrane bound organelle in eukaryotic algae containing chlorophyll and other pigments; if the green colour is masked by chlorophyll, the older literature sometimes used the term chromatophore.
cingulum Dinoflagellates: a constriction of the cell with a transverse orientation; the girdle region of the cell.
clathrate Forming a net; refers to mucilaginous colonies of blue-green algae and a few green algae which include extensive internal spaces that are usually in contact with the external environment.
clavate Club-shaped; broader towards the apex (Fig. 7V).
clone A population produced asexually from a common ancestor so that each member is genetically identical.
coccoid Single-celled and rounded.
coenobium Colony in which the cell number is fixed at the time of formation and not augmented subsequently; characteristic feature of most multicellular members of the Chlorococcales.
coenocytic Containing many nuclei, with few or no cross walls.
colony Group of individual cells enclosed within a common sheath/envelope or joined together and having a characteristic form and structure.
community Any group of organisms, living together under natural conditions and forming a recognisable unit.
compressed Flattened, but not strongly so; often oval to elliptical in cross-section.
conceptacle Invagination or cavity containing reproductive organs.
conical Cone-shaped, so tapering to a point.
conjugation Zygnematales: process of fertilization of two gametes sometimes within a tube (the conjugation tube) which develops between the cells in which the gametes were formed or, alternatively, one gamete may pass into and fuse with the cell contents of the other.
conspecific Belonging to the same species.
contractile vacuole Organelle progressively enlarging as it fills with fluid, only to contract suddenly on its expulsion.
cordate, cordiform Heart-shaped (Fig. 7N).
corona/coronula Charophytes: crown-like group of cells; 5- or 10-celled group of cells at apices of the oogonia (p. 595).
cortex Charophytes: outermost layer or layers of cells (see p. 594).
corticate Having a cortex.
corticolous Term used for subaerial algae inhabiting bark.
corymb Raceme with the younger branches becoming shorter towards the top giving a flat-topped cluster.
cosmopolitan Worldwide distribution.
costa (pl. **costae**) Distinctive rib-like, longitudinal cell wall thickening; refers especially to prominent, comparatively broadly-spaced striae in some species of the desmid genus *Closterium* (p. 516).
crenate Scalloped or toothed with rounded or rectangular notches.
crenulate Finely crenate.
cristate Crested, or having the form of a crest.
cruciate Cross-shaped (cf. tetrapartite).
crustose Crust-like.
cuneate Wedge-shaped.
cyanelle Blue-green structure in a few species of eukaryote, which is neither an intracellular blue-green alga nor a chloroplast, but thought to have evolved over a long evolutionary period from a blue-green alga and taken on the functions of a chloroplast.
cyanophycin Nitrogenous granules in the vegetative cells of blue-green algae formed under conditions when this element is plentiful; also at the ends of heterocysts when fixing nitrogen faster than adjacent cells can use it.
cylindrical Elongate with parallel lateral margins when viewed from any direction, as in a cylinder (Fig. 7P).
cyst Resting spore, usually thick-walled.
cystocarp Rhodophytes: carposporophyte and carpospores together with any protective envelope or pericarp provided by the parent female plant.
cytoplasm (protoplasm) Living substance within a cell which contains and supports the organelles, such as the nucleus and chloroplasts of eukaryotic algae; may exhibit streaming.
cytoplasmic membrane Membrane bounding the cytoplasm of blue-green and eukaryotic algae and lying inside any wall layers.

daughter colony Small colony within a mother cell or mother colony; usually in relation to asexual reproduction in members of the Chlorococcales.
dentate Having tooth-shaped projections.
dichotomous Branching into two equal or unequal (then irregularly dichotomous) parts.
diffluent Flowing apart or readily becoming liquified.
dikaryon Dinoflagellates: unusual nucleus characteristic of the group.
dioecious When male and female reproductive organs are borne on separate individuals.
diploid Refers to a nucleus having twice the basic (haploid)

number of chromosomes, one set having been derived at fertilization from each parent.

discoid Disc-shaped (Fig. 7I).

DNA (desoxyribose nucleic acid) The chemical basis of heredity; a compound of high molecular weight in chromosomes, consisting of a pentose sugar, phosphoric acid and a N-containing base arranged in a double helix.

dorsal Back or upper surface as opposed to lower or ventral surface.

dorsiventrally flattened Dinoflagellates: flattened in cross section of cell, equivalent to laterally flattened or flattened in side view in other flagellate groups (e.g. Chrysophytes, Volvocales).

dwarf male Oedogoniales: filament with only a few cells, or even a single cell, attached to an oogonium or nearby cell and producing antherozoids (p. 409).

dystrophic Waters which are brown with humic materials draining from peat or peat-rich soils; the term is not applied if the waters are nutrient-rich.

ecdysis Dinoflagellates: loss of thecal plates in armoured forms.

ecorticate Without cortication.

edaphic Living in or on soil.

ellipsoidal A three-dimensional term for a structure appearing elliptical in side view (Fig. 7H).

endogenous Arising internally.

endolithic Growing below the surface of a rock.

endophytic Growing within the tissues or sheath of a plant.

endosporangium Blue-green algae: cell containing endospores.

endospores Blue-green algae: formed by internal division of a mother cell; the term 'baeocyte' is preferred by specialists of this group.

endozoic Growing within part of an animal.

epicingulum Dinoflagellates: portion of a girdle adjacent to an epivalve.

epicone Dinoflagellates: portion of the cell anterior to the cingulum.

epilithic Growing on the surface of rock or other hard substratum, whether terrestrial or aquatic.

epipelic Free-living on the surface of the bottom sediments of a waterbody.

epiphytic Growing on another plant, including another alga.

epitheca Dinoflagellates: wall of the epicone.

epivalve Dinoflagellates: the flattened or convex plate of a frustule lying opposite to the hypovalve and usually slightly larger than it.

epizoic Living on the surface of an animal.

eukaryote Organisms whose cell or cells have distinct membrane-bounded organelles, such as the nucleus.

euryhaline Tolerant of a wide salinity range.

euplankton Organisms which are adapted to and live exclusively in suspension in the open water of lakes and slow-flowing rivers, therefore true plankters (cf. tychoplankton).

eutrophic water Refers to waters well supplied with nutrients (nitrogen and phosphorus); they are usually highly productive, and often support blooms of blue-green algae or other planktonic algae. Such waters may be naturally nutrient-rich, but increasingly result from enrichment by sewage effluents and agricultural fertilizers.

eutrophication The process of becoming eutrophic.

excrescence An outgrowth, such as a wart or papilla.

exogenous Arising externally.

exospore Blue-green algae: spore produced by budding from another cell.

eyespot Granular organelle, or a cluster of granules, usually red or orange, frequently present in flagellated cells; it functions in different ways in different groups of algae, but either directly or indirectly permits the cell to perceive the direction of light.

falcate Sickle-shaped; curved, with a broad base and tapering beyond it.

false branch Blue-green algae: branch formed by lateral growth of one or both ends of a broken trichome.

fascicle A dense cluster or bundle of branches or branchlets.

fastigiate Having the branches erect, parallel and more or less adpressed.

filament Cells united or arranged in one or more rows to form a chain or thread; in blue-green algae often used interchangeably with 'trichome' in forms which do not have a sheath (e.g. *Oscillatoria*); in those forms with a sheath (e.g. *Lyngbya*), it refers to the trichome and its investing sheath.

filiform Thread-like or filamentous.

film Thin layer of cells, or filaments on surfaces such as rocks or sediments.

flagellum (pl. **flagella**) A thread-like extension of the cytoplasm of a cell or motile spore or gamete and involved in its locomotion.

fleck Refers to algae which are very firmly attached to rocks; the form is typical of a number of species of blue-green algae in fast-flowing streams, but sometimes occurring elsewhere.

flexuous Bent alternately in opposite directions; zigzag.

floc Refers to macroscopically visible groups of filaments, often mucilaginous and without a particularly distinct bounding layer.

Flora Book whose purpose is primarily to indicate what plants occur in an area, how they may be identified (by keys and descriptions), and what scientific names refer to them.

forcipate Forked and incurved.

freshwaters Waters lacking a significant salt content.

frond Leaf-like or erect portion of a thallus.

frustule Silica wall of a diatom.

furcate Forked or divided into two, usually almost equal portions.

fusiform Spindle-shaped; elongate, broad in middle and tapering gradually to acute or broadly rounded apices.

gametangium A gamete-producing cell.

gamete A cell with a haploid chromosome complement, potentially capable of fusing with or fertilizing a compatible cell to form a zygote in the process of sexual reproduction (see haploid).

gametophyte Sexual, gamete-producing stage in the life history of an alga.

gas vacuole Blue-green algae: a grouping of gas vesicles in the cells which are visible under the light microscope.

gas vesicles Blue-green algae: minute gas-filled cylinders within gas vacuoles, only visible with the electron microscope.

geniculate Bent as in a knee-joint, especially in some species of *Klebsormidium* and cells of filamentous desmids.

genus Taxonomic group containing one, but most commonly, several species.

gliding Type of movement shown by various microorganisms which requires contact with a surface, but not involving flagella or cilia; it occurs in many blue-green algae, diatoms, desmids and some filamentous members of the Zygnematales.

girdle Diatoms: the portion of the frustule between the valves.

girdle band Desmids: in some *Closterium* species, a sleeve of wall material interpolated between the apex of a new semicell and the transverse median suture from which new semicells develop.

gonimoblast Rhodophytes: a filament bearing a carpospore, or carpospores or the entire collection of filaments comprising the carposporophyte.

granulose Cell surface covered with granules or grains.

gullet Canal or cavity in the anterior end of a flagellated cell, connected on the interior with a reservoir (cf. euglenophytes).

gynandrosporous *Bulbochaete:* species in which the spores giving rise to dwarf males (androspores) are on the same filaments as those bearing the oogonia.

habit Morphological form of a thallus (e.g. erect, prostrate).

habitat Environment in which an organism lives.

haematochrome Older term for orange-red carotenoid pigments in *Haematococcus* and some other green algae.

hair Elongated, single-celled or multicellular, colourless end of a filament; originating from some form of cell division, although contents may be very reduced.

haploid Having half the basic set of chromosomes in each nucleus; that is half the diploid number in the nucleus following sexual reproduction. The haploid number is typical of gametes and gametophytes.

hard water Water with a high concentration of calcium and usually also magnesium ions.

hemi In composition means half.

hemizoospore Non-motile spores containing contractile vacuoles, sometimes also an eyespot.

heterocyst Blue-green algae: cell in trichomes of certain genera in well-oxygenated environments, showing obvious differences from adjacent cells. In comparison with adjacent cells, they may be larger or smaller, are thick-walled, usually paler in colour and often possess glistening granules of cyanophycin at the end (or ends) adjacent to a vegetative cell. Capable of nitrogen fixation.

heteromorphic 'Of different form'; usually referring to the morphologically different gametophyte and sporophyte generations in the life history of many larger algae, e.g. rhodophytes and phaeophytes.

heterotrophic Ability to use organic materials from the surroundings (cf. autotrophic).

holdfast Organ of attachment.

holotype Specimen designated by the author(s) when describing a new taxon as the nomenclatural type.

hormogonium (pl. **hormogonia**) Blue-green algae: short length of filament rich in nutrients formed by a mature filament or akinete and helping to disperse the organism.

H-shaped segments/sections Portions of filament resulting from fragmentation caused by the fracture of cells in the thin-walled mid-region; in this region the wall often adjoins or overlaps and is an area of potential rupture.

hyaline Thin and transparent; colourless.

hypnospore Thick-walled aplanospore.

hypnozygote Dinoflagellates: resting stage resulting from gametic fusion.

hypocone Dinoflagellates: the posterior portion of the cell lying below the transverse furrow.

hypotheca Dinoflagellates: portion of cell wall enclosing the hypocone.

iconotype An illustration, usually a drawing, which in the absence of a specimen, is designated the nomenclatural type.

idioandrosporous *Bulbochaete*: species in which the spores giving rise to dwarf males (androspores) are on different filaments from those bearing oogonia.

incised Cut in deeply and sharply, usually at the margin.

intercalary Inserted at a point along the length of a structure, especially an algal filament.

internode Portion between two nodes.

intrathylakoidal vacuole Blue-green algae: liquid vacuole formed by marked separation of the two membranes of thylakoids (photosynthetic and respiratory structures); larger vacuoles, but not the membranes, are easily visible with the light microscope.

intricate Intertwined, entangled.

isthmus Desmids: narrow equatorial part between two semicells; only present in most placoderm desmids of the Suborder Desmidiinae, not in members of the suborder Closteriinae or in saccoderm desmids (Family Mesotaeniaceae).

isogamous Gametes morphologically indistinguishable.

intercalary Inserted between cells of a filament (or trichome in blue-green algae) rather than at the ends or laterally.

invagination Infolded cavity or hollow.

laciniate Cut into deep narrow divisions, usually tapering lobes (e.g. lacerated).

lamella Thin plates lying one against another, often refers to the plates which radiate from the axis of a chloroplast in desmids.

lamellate Composed of thin layers or plates, also used to refer to successive layers of mucilage, or the structure of certain chloroplasts.

laminate Refers to layers or plates overlying one another.

lanceolate Lance-shaped; so that broadest part is usually about one-third of the distance from base and usually tapering towards each end (Fig. 7R).

lateral conjugation Zygnematales: sexual reproduction in which two cells of the same filament become connected, in some cases by a conjugation tube, thus enabling the passage and fusion of the cell contents which function as gametes.

lenticular Shaped like a biconvex lens.

leucosin Carbohydrate storage granule in the xanthophytes, chrysophytes and diatoms.

life history Sequence of morphological (and nuclear) phases of an organism.

ligulate Tongue-shaped; strap-like and short.

linear Narrow and several times longer than wide, with parallel sides.

lipid Generally term for fats and oils, which sometimes show as droplets in the cytoplasm of eukaryotic algae .

littoral zone Marginal region of pond or lake.

loose-lying Unattached and lying on a surface.

longitudinal furrow Dinoflagellates: a grove extending longitudinally on the ventral side of the cell.

lorica Latin word for a breastplate, applied to the hard part or almost all the surface of some flagellated algae, especially euglenophytes and chrysophytes; most loricate species are motile, but some loricate chrysophytes are epiphytes.

loch Term often used in Scotland for a lake.

lough Term used in Ireland for a lake.

lubricous Smooth, slippery.

lunate Shaped like a half-moon, crescent-shaped (Fig. 7X).

lumen Central cavity of a cell.

macrandrous Oedogoniales: male and female filaments of approximately similar diameter.

macroalga Large alga, easily observable with the naked eye; usually reserved for those which can be identified provisionally at several metres distance, such as by a person standing and viewing a stream bed.

macrophyte Term used for all plants visible to the naked eye, not just macroalgae; because the term is used in different ways, it is important to check how it is used by a particular author.

macroscopic Visible to the unaided eye; discernible without magnification.

manubrium Charophytes: columnar cell connecting the stalk cell to the shield cell in the antheridium.

marl Calcium carbonate precipitated from hard water by several processes including the photosynthetic activity of algae. Marls are of geological significance, sometimes forming well-defined strata.

mats Thicker and firmer than films; they usually include distinctly sheathed organisms and are formed especially by

significant growths of blue-green algae or some filamentous green algae, especially round lake margins.
membranaceous Thin and often a semi-transparent membrane.
mesotrophic water Used to describe waters of intermediate nutrient status between oligotrophic and nutrient-rich, eutrophic waters.
metaboly Plastic, changing shape when in motion.
metaphyton Algae normally loosely attached to a surface, but easily becoming detached and incorporated into the plankton.
micron, micrometre (μm) Unit of measurement (= 0.001 mm); all microscopy measurements are expressed in micrometres.
microscopic Discernible only with magnification by a hand lens or microscope.
moniliform Like a string of beads.
monoecious When male and female organs are formed on the same individual.
monosporangium Rhodophytes: sporangium producing a single spore.
monospore Non-motile spore produced in a monosporangium.
monostromatic Composed of a single layer of cells.
morphology Form or shape of a cell or an organism.
mother cell Cell giving rise to small daughter cells internally.
mucronate Abruptly terminating in a short, sharp point.
mucilage Colloidal material, consisting of complex polysaccharides and frequently forming a protective coat, enveloping individual algal cells and especially groups of cells; the limitation of desiccation in aerial or subaerial environments is one of various important roles.
muciferous bodies Organelles in euglenophytes and dinoflagellates which contain mucilage and can be ejected.
multinucleate Having more than one, but commonly many, nuclei.
multiseriate Composed of more than one row of cells.

naked Cell not covered by a cell wall, only a cell membrane.
nannandrous Oedogoniales: refers to the condition where the male filament, which is a much reduced structure, is attached to the significantly larger female filament.
nannoplankton Various dimensions have been given for this category of minute, free-floating unicellular organisms, but generally considered to range from 2–20 μm in the largest dimension (see picoplankton).
necridium Blue-green algae: cell which collapses in some filamentous species resulting in trichome division, so permitting the two daughter trichomes to separate; this form of reproduction occurs in many blue-green algae where hormogonia are released terminally.
nematocyst Dinoflagellates: complex ejectile organelle in some species.
neuston Organisms which are free-floating and sometimes specially adapted to live at the air–water interface.
nitrogen fixation Conversion of atmospheric nitrogen into a combined form by some blue-green algae and bacteria.
node Position on a filament or thallus from which an appendage arises.
nominal Term used in the older literature by some desmid researchers for 'type' (see type).
nucleolus (pl. **nucleoli**) Spherical, homogeneous refractive body, one or more attached to a particular chromosome within each nucleus and containing RNA; often prominent in the nuclei of Zygnematales.
nucleus A sometimes prominent, organelle in all eukaryotic cells, surrounded by a membrane which encloses the nucleolus and the chromosomes.

oblate Slightly flattened subspherical structure (Fig. 7M).
oblong Elongate shape, with the ends broadly rounded, but more sharply curved than the lateral margins (Fig. 7F,G).
oblong-elliptical Shape intermediate between an oblong and an ellipse when viewed in two dimensions; as this is a 2-dimensional term, care is needed when applied to 3-dimensional objects.
obovoid/obovate Inversely egg-shaped (Fig. 7C).
obtuse Blunt or rounded at end.
oligotrophic Used to describe a waterbody low in nutrients, especially nitrogen and phosphorus.
operculum Oedogoniales: a split in the cell wall forming a lid.
oogamous Sexual reproduction in which a comparatively large, non-motile gamete (an egg or ovum) is fertilized by a smaller, motile, male gamete.
oogonium Female gametangium producing an egg or egg cells.
oospore/oosporangium Zygote with a thick and sometimes several-layered wall.
organelle Specialized cell structure (=little organ) such as a nucleus or chloroplast, surrounded by a membrane and involved in a particular function within a cell.
ovate Egg-shaped, broader basally (2-dimensional term) (Fig. 7A–C).
ovoid Egg-shaped, broader basally (3-dimensional term) (Fig. 7A–C).

palmella-stage Stage in the life history of flagellated algae where cells become non-motile and divide to form a mass of cells embedded in an amorphus mucilage.
papilla (pl. **papillae**) Small conical projection on a cell wall.
papillate Describing a surface covered by papillae.
paramylon Euglenophytes and certain haptophytes: a characteristic carbohydrate food reserve.
parenchyma Compact tissue composed of mostly isodiametric, thin-walled cells which are usually not significantly elongated; usually forming unspecialized tissues in both lower and higher plants.
parietal Adjacent to or lying just inside the cell wall, usually used with reference to the position of the chloroplast.
parthenospore Resting spore, usually with a thickened wall and formed from unfused or unfertilized gametes.
pectin Mucilaginous carbohydrate, often occurring on or in the walls of many algae.
pedicel Stalk of an attached cell.
pellicle Euglenophytes: flexible proteinaceous outer layer of the cytoplasm consisting of overlapping strips, and immediately internal to the cytoplasmic membrane. Dinoflagellates: portion of the cell covering of armoured species.
perforate Having many openings.
periphytic (periphyton) Term used to describe communities of algae, other small phototrophs and associated microorganisms growing on a submerged surface; originally introduced in North America to replace the German term 'Aufwuchs' and embracing the whole community of algae and other organisms, associated especially with larger aquatic plants.
phagotrophic Ingestion of live or dead food particles.
phototactic, phototaxis Locomotory movement of an organism in response to light.
phycobilin Biliprotein pigment found in blue-green algae, rhodophytes, a few cryptomonads and organisms containing cyanelles.
phycology The study of algae (the term covers all algae, even though *phycos* is Latin for seaweed).
phycocyanin Blue biliprotein pigment of all blue-green algae (other than prochlorophytes), cyanelles, some rhodophytes and some cryptophytes.
phycoerythrin Red biliprotein pigment of some blue-green algae, almost all red algae and a few cryptophytes.
phylum (pl. **phyla**) Major taxonomic group, often termed Class or Division in the older literature.
phytoplankton The microscopic plant life (largely algae) present in the water column of ponds, lakes, and slow-flowing rivers.

picoplankton Cells which are less than 2 μm in all dimensions; these include some blue-green algae, chrysophytes and green algae; picoplanktonic blue-green algae are often the main primary producers during summer in large nutrient-poor freshwaters. The term 'pico' should strictly refer to 1 μm, but it has been accepted internationally to apply to the convenient 2-μm limit (cf. nannoplankton).

pit connection Rhodophytes: discrete lens-shaped plug within the aperture between two adjacent cells.

placoderm Desmids of the suborder Desmidiinae: with the cell wall composed of two halves (semicells), in many genera demarcated by a distinct median constriction.

plane end walls In filamentous algae, cross walls which are smooth and unfolded.

plankter An individual member of the plankton.

planozygote Motile zygote.

plasmodesma Delicate protoplasmic connection between cells, especially prominent in some rhodophytes, but probably universal in all filamentous organisms.

plastid Body or organelle within the cell (e.g. chloroplast).

plate Dinoflagellates: section of the cell wall in armoured species.

plicate Folded, usually lengthwise.

pluri- Prefix meaning many or several.

pole, polar One end of an elongate cell.

polar nodule Blue-green algae: thickening or plug in a heterocyst at the point of attachment with an adjoining cell.

polygonal Many-sided; term applied to 2-dimensional shapes.

polyhedral More than 4-sided; term applied to 2-dimensional shapes.

polyhedral cell Life history stage in *Hydrodictyo*n and *Pediastrum*, the angular cell produced by the zoospores released following germination of a zygote.

polymorphic Of variable morphology or appearance.

polymorphism Having a variety of shapes; many species considered to be forms, varieties or even species, have been shown to be merely expressions of their considerable morphological plasticity, especially desmids.

polyploid Nucleus with more than the normal diploid (2n) number of chromosomes; polyploids usually have multiples of the basic number of chromosomes (3n, 4n, etc.).

pore Small opening in cell wall and cell membrane, such as the walls of desmids other than the saccoderms.

process Extension or protuberance.

prokaryote Organism lacking membrane-bounded organelles, as in blue-green algae and other bacteria.

protonema Charophytes: filament produced following germination of the oospore. The term is also applied to bryophytes, where a branched filament may arise from a spore or, in some cases, part of the leafy plant; these filamentous structures typically have an oblique cross wall, though organisms growing in some types of extreme environment may have transverse cross walls and appear rather like a filamentous alga. The term is also applied to juvenile stage of bryophytes.

protoplasm Organized living contents of a cell as distinct from the cell wall, often taken to include the bounding cell membrane.

pseudo- Prefix signifying false.

pseudocilium (pl. **pseudocilia**) Soft, flexible projection, occurring mainly in the Tetrasporales.

pseudodichotomous Apparently dichotomously branched with the strong development of a lateral to resemble the original axis.

pseudo-girdle bands *Closteriu*m (desmid): a few species have structures similar to girdle bands, but originating as extensions of transverse sutures, varying considerably in length and numbers present in a cell. Their presence may have taxonomic significance (cf. girdle bands).

pseudofilament Blue-green algae: a series of cells formed by repeated transverse divisions and arranged in a row (or approximately so), but not constituting a trichome.

pseudoparenchymatous Superficially appearing like parenchyma, but in fact composed of closely adherent filaments.

pseudopodium An extension of the body of an amoeboid organism.

pulvinate Cushion-like.

punctum (pl. **puncta**) Point or dot usually indicating the position of a pore or pore organ, especially in the walls of placoderm desmids.

punctate Refers to walls covered with or penetrated by punctae.

pusule Dinoflagellates: osmoregulatory organelle of variable form, consisting of two closely appressed membranes bounding a vesicle open to the outside.

pyramidal Shape of a pyramid; thus with a pointed apex and 4-sided broad base.

pyrenoid Organelle associated with the chloroplasts of many algae (though not always visible with the light microscope), which contains a high content of the enzyme responsible for fixing carbon dioxide in photosynthesis and around which starch grains often accumulate.

pyriform Pear-shaped; broader at base than at top (Fig. 7D, E).

quadrate Square, 4-sided (in 2-dimensions).

quadriflagellate Having 4 flagella.

quasi- Prefix meaning almost or virtually.

racemose With the youngest and smallest branches nearest the apex of a stem or filament and successively older ones towards base; often conical in outline.

radiate Radiating in several planes from an axis or a common centre.

radiate desmids All saccoderm desmids and placoderms of the Suborder Closteriinae are circular in outline (omniradiate) when viewed from above (apically). Many others of the Suborder Desmidiinae from this aspect are elliptical, triangular or have higher orders of angularity with the angles sometimes extended to form ray-like processes. Desmids with this angular morphology are said to be radiate and to possess various degrees of radiation. Hence there are desmids which are 2-, 3-, 4-radiate, or even higher orders of radiation.

reniform Kidney- or bean-shaped (Fig. 7W).

repand With slightly uneven margin.

replicate wall Zygnematales: complex, infolded cross walls in the cells of some filamentous species of *Spirogyra*.

reservoir Cavity in the anterior end of a flagellated cell.

reticulate Net-like; arranged to form a net.

retuse Refers to a cell apex or other margin which is slightly concave or depressed

rhizoid Single-celled or filamentous organ of attachment.

rhizopodium, rhizopodal Term for a pseudopodial extension of a cell.

rhomboid Shape with 2 pairs of parallel sides and obtuse and acute angles at opposite corners (Fig. 7J).

rimiform *Oedogonium*: having a cleft-like opening or pore in wall of oogonium.

rostrate Cell ending long and beak-like, especially in some *Closterium* species.

rugose Wrinkled.

sarcinoid Sac-like.

saccoderm desmids A family (Mesotaeniaceae) separated from the true desmids because they lack the latter's distinctive division into two semi-cells; also their walls are without pores and ornamentation.

saprophytic Obtaining nutrition by absorbing organic

substances that have been partly broken down by their own enzymes or those of other organisms.

scalariform conjugation Zygnematales: type of sexual reproduction in which the conjugation tubes grow from adjacent filaments so that when a succession of cells is involved it results in a ladder-like structure.

scrobiculate Surface with regular pits or depressions.

scytonemin Brown pigment in sheaths of many slower-growing blue-green algae under high light conditions; formed as a protection against ultra-violet light.

semi-cell One of the apparent 'halves' of placoderm desmid cells, usually separated from one another by a more or less well-defined constriction or sinus.

septum Partition of primary wall material, usually formed during cell division to separate the newly-formed daughter cells.

seriate In a series.

serrate Saw-toothed margin.

sessile Borne directly on a thallus, without a stalk.

seta (pl. **setae**) Appears synonymous with a bristle or bristle-like process.

sheath Covering or envelope surrounding the cell wall, usually mucilaginous.

shield cell Charophytes: a cell in the wall of the antheridium.

sickle-shaped Acutely curved, crescent-shaped.

sigmoid Doubly curved in opposite directions (Fig. 7T).

simple Unbranched or undivided.

sinuate Having a deep wavy margin.

sinus Constriction in many placoderm desmid cells (sometimes barely visible) where there is a median cleft between adjacent semicells.

siphonaceous Filamentous algae whose tubular structure has very few, if any, cross walls.

sorus Group or cluster of reproductive structures.

spatulate Spatula-shaped, oblong, with one end attenuated like a chemist's spatula (Fig. 7Q).

spermatangium (pl. **spermatangia**) Rhodophytes: male sex organ producing non-motile gametes (spermatia).

spherical Ball-like (Fig. 7L).

spindle shaped (cf. **fusiform**) 3-dimensional structure, starting cylindrical, but narrowing gradually at either end.

spine Sharply pointed projection from a cell, appears to be part of the cellulose wall and includes no easily visible contents.

sporangium Spore-producing structure.

spore Unicellular or few-celled structure released from a sporangium and capable of further growth.

sporophyte Asexual or spore bearing stage in the life history.

statospore Dinoflagellates and chrysophytes: resting spore in some species, consisting of 2 pieces.

stellate Radiating from a centre like a star, star-shaped; usually used to describe a chloroplast in various zygnematacean algae whose narrow lobes radiate in all directions, often from a central pyrenoid.

stigma Structure in some flagellates involved in light perception and directing locomotion.

stipe Stem-like region.

stipitate Having a stipe.

stratified Having successive layers; often refers to mucilage (cf. lamellate).

striae Longitudinal lines or ridges, especially on the cell walls of *Closterium*, of taxonomic significance; sometimes very faint (cf. costae).

striated Refers to structure covered with striae.

stoloniferous Having stolons, long and slender horizontal branches.

stomatocyst Chrysophytes: resting stage with a silicified wall and a pore closed by a non-silicified stopper.

sub- Prefix denoting under, below, less than, approaching, etc.

subaerial Terrestrial habitats which receive occasional or partial moisture, such as dripping rocks and vertical peat faces; usually applied to environments away from soil, but sometimes applied more loosely.

subcylindrical Refers to cells that are nearly cylindrical: instead of having straight margins they are either slightly convex or concave.

subspherical Slightly compressed sphere or circle; flattened at the poles (cf. oblate).

subtend At the base of another organ.

subtruncate Apex flattened, but rounded at the edges.

subulate Awl-shaped.

suffultory cell Oedogoniales: cell cut off below the oogonium during development (Fig. 5, p. 409).

sulcus Dinoflagellates: furrow and groove, including the longitudinal furrow on the ventral side of cell.

supporting cell Oedogoniales: cell beneath the oogonium, the latter arising by division of an oogonial mother cell. Rhodophytes: cell from which the carpogonial branch arises.

suture Line or groove; especially in desmids of the Suborder Closteriinae marking the limits between adjacent semicells. Filamentous Zygnematales: groove around zygospore. Armoured dinoflagellates: groove between the plates.

synonym Superseded or unused name.

taxon (pl. **taxa**) Taxonomic unit at any level of classification.

taxonomy The study of those characteristics of organisms which lead ultimately to their scientific naming and classification.

terminal vacuole Organelles present in all species of *Closterium* and some related genera, occurring between the distal ends of the chloroplasts and the cell apices; vacuoles are surrounded by cytoplasm commonly exhibiting streaming; characteristic crystals within the vacuoles.

tetrad Group of four, usually referring to cells or spores.

tetragonal With 4 angles which are in 2 opposite planes.

tetrahedral Term applied to tetrasporangia whose contents are divided obliquely so that only 3 of the 4 spores are visible in any one view.

tetrapartite Term applied to tetrasporangia divided into 4 spores by two mutually anticlinical divisions, so that all the spores are always visible.

tetrasporangium Rhodophytes: sporangium producing tetraspores by a meiotic division so that the resulting spores are haploid.

tetraspore Spore produced in a tetrasporangium.

thallus Body of simple plants not differentiated into a true root, stem, leaf or leaves.

theca Dinoflagellates: wall composed of plates in armoured species.

tissue An aggregation of similar cells forming a definite and continuous layer or structure and usually having a comparable function.

transverse furrow Dinoflagellates: girdle or sulcus extending around cell in an equatorial position.

trapeziform/trapezoid Having two parallel sides, but the other two sides not parallel (Fig. 7K).

travertine Geological term for calcium carbonate precipitated from hard water by several processes, including the photosynthetic activity of algae.

trichocyst Cryptophytes, dinoflagellates and rhapidophytes: membrane-surrounded organelle lying beneath cell surface, which ejects threads of mucilage on stimulation.

trichogyne Rhodophytes: a receptive protuberance or elongation of a female gametangium to which male gametes become attached.

trichome Blue-green algae: linear arrangement of cells

without the investing sheath; the term is synonymous with filament for forms which never possess a sheath, but for those which do, the filament consists of a trichome and a sheath.

truncate Apex which is terminated abruptly, used especially with reference to placoderm desmids (Fig. 7U).

tuberculate Beset with knobbly projections or irregularly warty outgrowths.

tuft Group of filaments which are close together but without them all being enclosed in a communal sheath: usually applied to aggregations at right angles to a surface.

tumid Swollen or inflated.

tychoplankton Organisms normally attached to a surface, but becoming detached and carried into the open water of a pond, lake or river where they live for a longer or shorter time in the plankton, thus they occur only as chance members of this community.

type Specimen on which a species or an infraspecific taxon is based, or the species name providing the basis of the genus; often referred to as 'nominal' by desmidologists.

undulate Regularly wavy round the margin or over the surface of a cell.

unicellular Single-celled.

uniseriate Arranged in a single series or row.

vacuole Internal space within the cytoplasm, bounded by a membrane and normally fluid. *See also* intrathylakoidal vacuole, terminal vacuole.

variety Taxon below the level of a species or subspecies.

vesicle Sac- or balloon-like cell or thallus; membrane-bound organelle.

ventral Lower or under side.

verrucose Covered with warty or spiny projections.

water bloom High concentration of planktonic algae in a water body, in many species tending to accumulate at the surface.

whorled Arranged in a circle around an axis.

zonate Divided in parallel planes.

zoospore Flagellated asexual spore capable of swimming and produced in a zoosporangium.

zygospore Zygnematales: spore formed sexually following conjugation, often thick-walled, characterictically coloured and sometimes ornamented.

zygote Resting spore produced following the sexual union of gametes; usually having a thickened, resistant wall.

STANDARD FORM OF AUTHORS OF ALGAL NAMES

Judith F. John

A list of authors of scientific names of British freshwater algae standardized as recommended by Brummitt & Powell (1992) in *Authors of Plant Names*, including abbreviations. The standard form of citing authors is followed here although surnames are normally given in full. Indicated by a single asterisk are author names not in Brummitt & Powell and by two asterisks those that were given incorrectly in their list. In cases of father and son, the son is distinguished by 'f.', an abbreviation of 'filius'.

Acton, E.	E.Acton
Adanson, M.	Adans.
Agardh, C.A.	C.Agardh
Agardh, J.G.	J.Agardh
Ahlstrom, E.H.	Ahlstrom
Allegre, C.F.	Allegre
Allorge, P.	P.Allorge
Allman, G.J.	G.J.Allman
Alten, H.von	H.Alten
Andersen, R.A.	R.A.Andersen
Andersson, O.F.	O.F.Andersson
Anakhin, J.K.	Anakhin
Anand, P.L.	P.L.Anand
Anisimova, N.W.	Anisimova
Aragão, H. de B.	Aragão
Arcangeli, G.	Arcang.
Archer, W.	W.Archer
Areschoug, J.E.	Aresch.
Arnoldi, W.M. [A.P.]	Arnoldi
Asmund, B.C.	Asmund
Averintsev, S.	Averintsev

Bachmann, H.	H.Bachm.
Bailey, J.W.	Bailey
Balakrishnan, M.S.	M.S.Balakr.
Balbis, G.B.	Balb.
*Bando, T	Bando
Barker, J.	J.Barker
Barton, E.S.	E.S.Barton
Basson, P.W.	Basson
Batters, E.A.L.	Batters
Bauer, F.A.	F.A.Bauer
Beauvois, P., see Palisot de Beauvois, A.M.S.J.	
Beck, G. von M. und L.	Beck
Beger, H.K.E.	Beger
Beijerinck, M.W.	Beij.
Belcher, J.H.	J.H.Belcher
Benderliev, K.M.	Benderl.
Bennett, A.W.	A.W.Benn.
Berkeley, M.J.	Berk.
Bernard, C.	C.Bernard
Berthold, G.D.W.	Berthold
Bibby, B.T.	B.T.Bibby
Billard, C.	Billard
Bisset, J.	Bisset
Blackwell, W.H.	W.H.Blackw.
Blum, J.L.	Blum
Bodrogközy, G.	Bodrodk.
Bohlin, K.H.	Bohlin
Bold, H.C.	H.C.Bold
Boldt, J.G.R.	Boldt
Bolochonzew, E.N.	Boloch.
*Borge, O.	Borge
Børgesen, F.C.E.	Børgesen
Bornet, J.-B.E.	Bornet
Bory de Saint-Vincent, J.B.G.G.M.	Bory
Borzí, A.	Borzí
Bourrelly, P.	Bourr.
Boye-Petersen, J., see Petersen, J.B.	
Brabez, R.	Brabez
Bradley, D.E.	D.E.Bradley
Brand, F.	F.Brand
Braun, A.K.H.	A.Braun
Brébisson, L.A. de	Bréb.
Briquet, J.I.	Briq.
Bristol, B.M.	Bristol
Brook, A.J.	Brook
Brunnthaler, J.	Brunnth.
Bruzelius, A.S.	Bruzelius
Bullock-Webster, G.R.	Bull.-Webst.
Bulnhein, C.O.	Bulnh.
Burman, N.L.	Burm.f.
Butcher, R.W.	Butcher
Büttner, J.	J.Büttner

Calkins, G.N.	G.N.Calk.
Candolle, A.P. de	DC.
Carmichael, D.	Carmich.
Carter, H.J.	H.J.Carter
Carter, N.	N.Carter
*Castenholz, R.W.	Castenholz
Cavalier-Smith, T.	Caval.-Sm.
Cedergren (I.)G.R.	Cedergr.
Cesati, V. de	Ces.
Chapman, V.J.	V.J.Chapm.
Chodat, F.F.L.	F.Chodat
Chodat, R.H.	Chodat
Christen, H.R.	Christen
Christensen. T.A.	T.A.Chr.
Cienkowski, L. de	Cienk.
Claparède, J.L.R.A.E.	Clap.
Clements, F.E.	Clem.

Cleve, P.T.	Cleve
Coesel, P.F.M.	Coesel
Cohn, F.J.	Cohn
Collins, F.S.	Collins
Compère, P.	Compère
Conrad, W.	W.Conrad
Cooke, M.C.	Cooke
Corda, A.K.J.	Corda
Correns, C.F.J.E.	Correns
Cramer, C.E.	C.E.Cramer
Cridland, A.A.	Cridland
Croasdale, H.T.	Croasdale
Cronberg, G.Ah.	Cronberg
Crouan, H.M.	H.Crouan
Crouan. P.L.	P.Crouan
Crow, W.B.	Crow
Cunha, A.M.da	A.M.Cunha
Cushman, J.A.	Cushman
Czosnowski, J.	Czosn.
Czurda, V.	Czurda
Dangeard, P.C.A.	P.A.Dang.
de Bary, H.A.	de Bary
de Candolle, A.P., see Candolle, A.P. de	
De Notaris, G.J.	de Not.
De Toni, G.B.	De Toni
De Wildeman, E.A.J.	De Wild
*De Winder, B.	De Winder
Deason, T.R.	Deason
Dedusenko, N.T.	Dedus.
Dedusenko-Shchegoleva, N.T., see Dedusenko, N.T.	
Deflandre, G.V.	Deflandre
Delf, E.M.	Delf
Delponte, G.B.	Delponte
Denis, M.	Denis
Derbès, A.A.	Derbès
Desikachary, T.V.	Desikachary
Desmazières, J.B.H.J.	Desm.
Desvaux, N.A.	Desv.
Detharding, G.G.	Dethard.
Dick, J.	Dick
Dickie, G.	Dickie
Diesing, K.M.	Diesing
Dill, O.	O.Dill
Dillwyn, L.W.	Dillwyn
Dixit, S.C.	S.C.Dixit
Dodge, J.D.	J.D.Dodge
Doty, M.S.	Doty
Drew, K.M.	K.M.Drew
Dreżepolski, R.	Drezep.
Droop, M.R.	Droop
Drouet, F.E.	F.E.Drouet
Duby, J.É	Duby
Ducellier, F.	F.Ducell.
Dujardin, F.	Dujard.
Durieu de Maisonneuve, M.C.	Durieu
Düringer, I.	Düringer
Dürrschmidt, M.	Dürrschm.
Ehrenberg, C.G.	Ehrenb.
Eichwald, K.E. von	Eichw.
Eichler, B.	B.Eichler
Elenkin, A.A.	Elenkin
Elfving, F.E.V.	Elfving
Ellis, C.H.	C.H.Ellis
Engler, H.G.A.	Engl.
Entwisle, T.J.	Entwisle
Ercegović, A.	Erceg.
Ettl, H.	H.Ettl
Ettl, O.	O.Ettl
Fitschen, J.	Fitschen
Flahault, C.H.M.	Flahault
Flint, E.A.	E.A.Flint
Flint, L.H.	Flint
Florin, C.R.	Florin
Florin, M.-B.	M.Florin
Flotow, J.C.G.U.G.G.A.E.A.F.von	Flot.
Focke, F.W.	F.W.Focke
Förster, K.	Kurt Först.
Forti, A.I.	Forti
Fott, B.	Fott
Francé, R.H.	Francé
Frémy, P.	Frémy
Frenzel, J.	Frenzel
Fresenius, J.B.G.W.	Fresen.
Fresnel, J.	Fresnel
Fries, E.M.	Fr.
Fritsch, F.E.	F.E.Fritsch
Fromentel, L.E.G. de	From.
Gaillon, F.B.	Gaillon
*Gardiner, W.E.	W.E.Gardiner
Gardner, N.L.	N.L.Gardner
Gärtner, G.	G.Gärtner
Gay, F.	F.Gay
Gayral, P.	Gayral
Geitler, L.	Geitler
Georges, G.	Georges
Gerloff, J.H.	Gerloff
Gerneck, R.	Gerneck
Gickelhorn, J.	Gickelh.
Gimesi, N.I.	Gimesi
Girod-Chantrans, J.	Gir.-Chantr.
Gmelin, C.C.	C.C.Gmelin
Gobi, C.J.K.Y.	Gobi
Godward, M.B.E.	Godward
Gojdics, M.	Gojdics
Gomont, M.A.	Gomont
Goor, A.C.J. van	Goor
Gorozhankin, I.N.	Gorozh.
Gotz, H.	H.Gotz
Gray, J.E.	J.E.Gray
Green, J.C.	J.C.Green
Greville, R.K.	Grev.
Griffiths, B.M.	B.M.Griffiths
Grönblad, R.L.	Grönblad
Groves, H.	H.Groves
Groves, J.	J.Groves
**Grunow, A.	Grunow
Guglielmetti, G.	Guglielm.
Gutwinski, R.	Gutw.
Hajdu, L.	Hajdu
Hallas, E.D.K.H.	Hallas
Hamel, G.G.H.	Hamel
Hansen, J.B.	J.B.Hansen
Hansgirg, A.	Hansg.
Hantzsch, C. A.	Hantzsch
Hardy, A.D.	A.D.Hardy
Hariot, P.A.	Har.
Harris, K.	K.Harris

Harris, T.M.	T.M.Harris
Harvey, W.H.	Harv.
Hassall, A.H.	Hassall
*Hastings, W.H.	Hastings
Hauptfleisch, P.	Hauptfl.
Hazen, T.E.	Hazen
*Hegewald, E.	E.Hegew.
Heering, W.C.A.	Heering
Heimerl, A.	Heimerl
Hermann, J.G.	J.Herm.
Heynig, H.	Heynig
Hibberd, D.J.	D.J.Hibberd
Hickel, B.A.	B.Hickel
Hieronymus, G.H.E.W.	Hieron.
Hilliard, D.K.	D.K.Hilliard
Hilse, F.W.	Hilse
Hindák, F.	Hindák
*Hirn, I	I.Hirn.
Hirn, K.E.	Hirn
Hobson, J.	J.Hobson
Hodgetts, W.J.	Hodgetts
Hoham, R.W.	Hoham
Hollerbach, M.M.	Hollerb.
Homfeld, H.	Homfeld
Horecká, M.	Horecká
Hornby, A.J.W.	Hornby
Hornemann, J.W.	Hornem.
Hortobágyi, T.	Hortob.
Hu, H.	H.Hu
Huber, J.E.	Huber
Huber-Pestalozzi, G.E.	Hub.-Pest.
*Hübner, K.	K.Hübner
Hudson, W.	Huds.
Huitfeldt-Kaas, H.A.	Huitf.-Kaas
Hy, F.C.	Hy
Imhof, O.E.	O.E.Imhof
Insam, J.	Insam
Irénée-Marie, T.J.C.	Irénée-Marie
Islam, A.K.M.N.	A.K.Islam
Istvanfy, de C.M.G. von	Istv.
Itzigsohn, E.F.H.	Itzigs.
Iwanoff, L.	Iwanoff
Iyengar, M.O.P.	M.O.P.Iyengar
Jaag, O.	Jaag
Jacobsen, H.C.	H.C.Jacobsen
Jahn, T.L.	T.L.Jahn
Jahoda, R.	Jahoda
James, E.J.	E.J.James
Jane, F.W.	Jane
Jao, C.C.	C.C.Jao
Jadin, F.	Jadin
Javornický, P.	Javorn.
Jenner, E.	Jenner
Jennings, A.V.	A.Jenn
Jessen, K.F.W.	Jess.
John, D.M.	D.M.John
John, R.P.	R.P.John
Johnson, L.N.	L.N.Johnson
Johnson, L.P.	L.P.Johnson
*Johnson, L.R.	L.Johnson
Jörgenson, E.H.	Jörg.
Joshua, W.	Joshua
Jost, L.	Jost
Kalina, T.	Kalina
Kamptner, E.	Kamptner
Kann, E.	Kann
Karsten, G.	G.Karst.
Kasanowsky, V.I.	Kasan.
Keissler, K. von	Keissl.
Kent, W.S.	Kent
Kirchner, E.O.O. von	Kirchn.
Kirjakov, I.	Kirjakov
Kisselev, I.A.	Kisselev
Klebahn, H.	Kleb.
Klebs, G.A.	G.A.Klebs
Kling, H.J.	H.J.Kling
Koczwara, M.	Koczw.
Kofoid, C.A.	Kof.
Komárek, J.	Komárek
Komárková-Legnerová, J.	Komárk.-Legn.
*Komáromy, Z.P.	Komáromy
*Koppe, F.	F.Koppe
Korshikov, A.A.	Korshikov
Kossinskaja, E.K.	Kossinsk.
Kouwets, F.A.C.	Kouwets
Kováčik, L.	Kováčik
Krauss, R.W.	R.W.Krauss
Krieger, H	H.Krieg.
Krieger, W.	Willi Krieg.
Krishnamurthy, V.B,	V.Krishnam.
Kristiansen J.	Kristiansen
Krocker, A.J.	Krocker
Krüger, W.	W.Krüger
Kuckuck, E.H.P.	Kuck.
Kühnemann, O.	Kühnem.
*Kunze, O.	O.Kunze
Kumano, S.	Kumano
Kützing, F.T.	Kütz.
Kylin, J.H.	Kylin
Lachmann, J.	J.Lachm.
Lackey, J.B.	Lackey
Lacoste de Díaz, E.N.	E.N.Lacoste
Lagerheim, N.G. von de	Lagerh.
Lambert, F.D.	F.D.Lamb.
*Langhans, V.H.	Langhans
Laundon, J.R.	J.R.Laundon
Lauterborn, R.	Lauterb.
Lecal-Schlauder, J.	Lec.-Schlaud.
Le Clerc, L.	Le Clerc
Leewis, R.J.	Leewis
Lefèvre, M.	M.Lefèvre
Legnerová, J.	Legnerová
Lehmann, D.L.	D.L.Lehmann
Lemmermann, E.J.	Lemmerm.
Lenzenweger, R.	Lenzenw.
Lenormand, S.R.	Lenorm.
Leonhardi, P.C.P.G.H. von	Leonh.
Levander, K.M.	Levander
Lewin, M	M.Lewin
*Lewis, J.M.	J.M.Lewis
Ley, S.H., see Li, S.H.	
Li, S.H.	S.H.Li
Liebmann, F.M.	Liebm.
Lightfoot, J.	Lightf.
Lillieroth, S.	S.Lill.
Lind, E.M.	E.M.Lind
Lindemann, E.	Er.Lindem.
Link, J.H.F.	Link
Linnaeus, C. von	L.
Linnaeus, C. von	L.f.
Lobik, A.I.	Lobik
Lokhorst, G.M.	Lokhorst

Loeblich, A.R.III	A.R.Loebl.
Loiseleur-Deslongchamps, J.L.A.	Loisel.
Louis, A.	A.Louis
Lucas, I.A.N	I.A.N.Lucas
Lund, J.W.G.	J.W.G.Lund
Lundberg, F.R.	Lundb.
Lundell, P.M.	P.Lundell
Luther, A. F.	Luther
Lütkemüller, J.	Lütkem.
Lyngbye, H.C.	Lyngb.
MacEntee, F.J.	MacEntee
Mack, B.	B.Mack
Magne, F.	F.Magne
Magnotta, A.	Magnotta
Mainx, F.	Mainx
Mann, D.G.	D.G.Mann
Manton, I.	Manton
Mapletoft, H.	Mapletoft
*Marie, I	I.Marie
Marsson, K.M.	M.Marsson
Martius, C.F.P. von	Mart.
Marvan, P.	Marvan
*Masjuk, N.D.	Masjuk
*Maskell, W.M.	Maskell
Massart, J.	Massart
Mattox, K.R.	Mattox
Matvienko, O.M.	Matv.
Meneghini, G.G.A.	Menegh.
Messikommer, E.A.	Messik.
Meyen, F.J.F.	Meyen
Meyer, A.	Art.Mey.
Meyer, H.	H.Mey.
*Migula, W.	W.Mig.
Millardet, P.M.A.	Millardet
Miller, V.V.	V.V.Mill.
Miquel, F.A.W.	Miq.
Möbius, K.	K.Möbius
Moestrup, Ø.	Moestrup
Moewus, F.	F.Moewus
Mohr, C.T.	C.Mohr
Mohr, D.M.H.	D.Mohr
Momeu, L.	Momeu
Montagne, J.P.F.C.	Mont.
Mori, M.	M.Mori
Morison, M.O.	Morison
*Morrall, S.	S.Morrall
*Morren, M.C.F.R.	Morren
Müller, O.F.	O.F.Müll.
Mullet, J.E.	Mullet
Mur, L.R.	Mur
Murray, G.R.M.	G.Murray
Nägeli, C.W. von	Nägeli
Nardo, G.D.G.	Nardo
Naumann, E.C.L.	Naumann
Nauwerck, A.	Nauwerck
Nees von Esenbeck, C.G.D.	Nees
Negoro, K.-I.	Negoro
Nicholls, K.H.	K.H.Nicholls
Nitzsch, C.L.	Nitzsch
Nolte, E.F.	Nolte
Nordstedt, C.F.O.	Nordst.
Norris, R.E.	R.E.Norris
Nováková, M.	Nováková
Nowakowski, L.	Nowak.
Nygaard, G.	Nygaard

Oersted, A.S.	Oerst.
Okamura, K.	Okamura
Oltmanns, F.	Oltm.
Palisot de Beauvois, A.M.F.J.	P.Beauv.
Palla, E.	Palla
Palmer, C.M.	C.Palmer
Palmer, T.C.	T.C.Palmer
Pankow, H.	Pankow
Parke, M.	Parke
Pascher, A.	Pascher
Pearsall, W.H.	Pearsall f.
Peeters, F.	Peeters
Penard, E.	Penard
*Pentecost, A.	Pentecost
Perman, J.	Perman
Persoon, C.H.	Pers.
Perty, J.A.M.	Perty
Péterfi, L.S.	L.S.Péterfi
Peters, N.	N.Peters
Petersen, J.B.	J.B.Petersen
Petit, P.C.M.	P.Petit
Petkoff, S.P.	Petkoff
*Petry-Hesse, K.	Petry-Hesse
Pfiester, L.A.	Pfiester
Phillips, F.W.	F.W.Phillips
Philipose, M.T.	Philipose
Pienaar, R.N.	Pienaar
Playfair, G.I.	Playfair
Pocock, M.A.	Pocock
Poche, F.	Poche
Pochmann, A.	Pochm.
Popova, T.G.	T.G.Popova
Popovský, J.	Popovský
Poretzky, V.S.	V.S.Poretzky
Prauser, H.	Prauser
Prescott, G.W.	Prescott
Preisig, H.R.	Preisig
Price, S.R.	S.R.Price
Pringsheim, N.	Pringsh.
Pringsheim, E.G.	E.G.Pringsh.
Printz, K.H.O.	Printz
Prinz, W.A.J.	Prinz
Pritchard, A.	A.Pritch.
Proshkina-Lavrenko, A.I.	Proshk.-Lavr.
Punčochárová, M.	Punčoch.
Puymaly, A.H.L. de	Puym.
Rabanus, A.	Rabanus
Rabenhorst, G. L.	Rabenh.
Raciborski, M.	Racib.
Ralfs, J.	Ralfs
Ramanathan, K.R.	Ramanathan
Reinbold, T.	Reinbold
Reinhard, L.	Reinhard
Reinke, J.	Reinke
Reinsch, P.F.	Reinsch
Reisigl, H.	Reisigl
*Reis, C.M.P. dos	C.M.P.Reis
Reis, M.P. dos	M.P.Reis
Reverdin, L.	Reverdin
Reymond, O.	O.Reymond
Rich, M.F.	M.F.Rich
Richter, P.G.	P.G.Richt.
*Rintoul, T.C.	Rintoul
Ripart, J.B.M.J.S.E.	Ripart
Roll, J.V., see Roll, Y.V.	

Roll, Y.V.	Y.V.Roll
Rosenberg, M.	M.Rosenb.
Rosenvinge, J.L.A.K.	Rosenv.
Rosillo, R.	R.Rosillo
Ross, R.	R.Ross
Rostafiński, J.T.	Rostaf.
Roth, A.W.	Roth
Rothe, H.A.	Rothe
Round, F.E.	Round
Roussel, H.F.A. de	Roussel
**Roy, J.	J.Roy
Ruprecht, F.J.	Rupr.
Ruttner, F.	Ruttner
Růžička, J.	Růžička
Safonova, T.A.	Safonova
Salzmann, P.	Salzm.
Sauvageau, C.F.	Sauv.
Schaarschmidt, J.G.	Schaarschm.
Scherffel, A.	Scherff.
Schewiakoff, W.	Schew.
Schiller, J.	J.Schiller
Schilling, A.J.	A.J.Schill.
Schmarda, L.K.	Schmarda
Schmid, G.	G.Schmid
Schmidle, W.	Schmidle
Schmitz, C.J.F.	F.Schmitz
Schrank, E.	E.Schrank
Schrank, F. von P. von	Schrank
Schröder, J.L.B.	Schröd.
*Schwabe, G.H.	G.H.Schwabe
Scott, A.M.	A.M.Scott
Siemińska, J.	Siemińska
Seligo, A.	Seligo
Senn, G.A.	Senn
Setchell, W.A.	Setch.
Shantz, H.L.	Shantz
Shaw, W.R.	W.Shaw
Sheath, R.G.	Sheath
*Shihira, I.	Shihira
Shuttleworth, R.J.	Shuttlew.
Silva, P.C.	P.C.Silva
Sirodot, S.	Sirodot
Skuja, H.L.	Skuja
Skvortsov, B.V.	Skvortsov
Smith, G.M.	G.M.Sm.
Smith, J.E.	Sm.
Smith, R.L.	R.L.Sm.
Snow, J.W.	J.Snow
Solier, A.J.J.	Solier
Spargo, M.W.	Spargo
Sprengel, C.P.J. or A.	Spreng.
Stal, L.J.	Stal
Starmach, K.	Starmach
Starr, R.C.	R.C.Starr
Stein, F.	F.Stein
Steinecke, F.	Steinecke
Stizenberger, E.	Stizenb.
Stockmayer, S.	Stockm.
Stokes, A.C.	A.Stokes
Suhr, J.	Suhr
Sulek, J.	Sulek
Svedelius, N.E.	Sved.
Svirenko, D.O.	Svirenko
Swale, E.M.F.	Swale
Swezy, O.	Swezy
Symoens, J.-J.	Symoens
Szabados, M.	M.Szabados
Szymański, F.	Szym.
Taft, C.E.	Taft
Takahashi, E.	E.Takah.
Takeda, H.	Takeda
Tanner-Füllemann, M.	Tanner-Füll.
Taylor, F.J.R.	F.J.R.Taylor
Teiling, E.J.S.	Teiling
Tell, G.	Tell
Teodoresco, E.C.	Teodor.
Tereg, E.	Tereg
Thomasson, K.	Thomasson
Thompson, R. H.	R.H.Thomps.
Thore, J.	Thore
Thuillier, J.L.	Thuill.
Thunmark, S.	Thunmark
Thuret, G.A.	Thur.
Thwaites, G.H.K.	Thwaites
Tiffany, L.H.	Tiffany
Tilden, J.E.	Tilden
Topali, S	Topali
Tracanna, B.	Tracanna
Transeau, E.N.	Transeau
Treboux, O.	Treboux
Trentepohl, J.F.	Trentep.
Trevisan de Saint-Léon, V.B.A.	Trevis.
Troitskaya, O.V.	O.V.Troitsk.
*Tsarenko, P.M.	Tsarenko
Tschermak-Woess, E.	Tscherm.-Woess
Tschernov, V.K.	Tschernov
Tupa, D.D.	Tupa
Turner, W.B.	W.B.Turner
Turpin, P.J.F.	Turpin
Uherkovich, G.	Uherk.
Unger, F.J.A.N.	Unger
Utermöhl, H.	Utermöhl
Vahl, M.H.	Vahl
Vaucher, J.P.E.	Vaucher
Veer, J. van der	Veer
Villars, D.	Vill.
Vinatzer, G.	Vinatzer
Viret, L.	Viret
*Vis, M.L.	Vis
Vischer, W.	Vischer
Vodeničarov, D.G.	Voden.
Voronikhin, N.N., see Woronichin, N.N.	
Vysotskij, A.V.	Vysotskij
Wallich, G.C.	G.C.Wall.
Wallroth, C.(K.)F.W.	Wallr.
Walton, L.B.	L.Walton
Walz, J.J.	J.Walz
Warming, J E.B.	Warm.
Waters, J.	J.Waters
Weber, F.	F.Weber
Weber-van Bosse, A.A.	Weber Bosse
Wells, P.	P.Wells
Welsford, E.J.	Welsford
Wermel, E.	Wermel
West, G.S.	G.S.West
West, T.	T.West
West, W.	West
West, W.	W.West

Wichmann, L.	Wichmann
Wiggers, F.H.	F.H.Wigg.
Wild, A.	A.Wild
Willdenow, C.L. von	Willd.
Wille, J.N.F.	Wille
*Williams, E.G.	E.G.Williams
Williamson, D.B.	D.B.Will.
*Wills, A.W.	Wills
Wislouch, S.M.	Wislouch
Wittrock, V.B.	Wittr.
Wolle, F.	Wolle
Wollenweber, H.W.	Wollenw.
Wołoszyńska, J.	Wołosz.
*Wołowski, K.	Wołowski
Wood, H.C.	H.C.Wood
Wood, R.D.	R.D.Wood
Woronichin, N.N.	Woron.
Woronin, M.S.	Woronin
Wright, E.P.	E.P.Wright
Wujek, D.E.	Wujek
Wulfen, F.X. von	Wulfen
Wyssotsky, A.V., see Vysotskij, A.V.	
Yamanouchi, S.	Yaman.
Zacharias, O.	O.Zacharias
Zakryś, B.	Zakryś
Zanardini, G.A.M.	Zanardini
Ziz, J.B.	Ziz
Zopf, (F.)W.	Zopf
Zukal, H.	Zukal

SOURCES OF ILLUSTRATIONS OR MATERIAL

Plate 1 A, B after Geitler 1932a (based on G.M. Smith); **C** modified from Frémy 1929–33; **D** after Geitler (based on Canabaeus); **E** after Smith 1933; **F** after Geitler 1932a (based on Bachmann); **G** after Smith 1920; **H** B.A. Whitton (original).

Plate 2 A, B, F, G after Fritsch 1929a; **C** after Geitler 1932a (as *Chamaesiphon curvatus*); **D, H, I** after Geitler 1932a; **E** after Geitler 1932a (based on West).

Plate 3 A–C, H S.C. Hardiman; **D, E–G, J** after Smith 1920; **I** after Smith 1933.

Plate 4 A–G S.C. Hardiman; **H, I** after Hickel 1991, with permission from author and publisher.

Plate 5 A after Geitler 1932a (based on Kirchner); **B** after Smith 1920; **C** colony after Smith 1920; **D** after Geitler 1932a (based on Woronichin); **E** after Geitler 1932a (based on Lemmermann); **F–K** after Geitler 1932a (based on Reinsch); **L** after Geitler 1932a (based on Schantz); **M–Q** after Geitler 1932a.

Plate 6 A after Starmach 1966 (based on Kossinskiej); **B, D** after Smith 1920; **C** after Starmach 1966 (based on Tailinga); **E, F, J–N** after Geitler 1932a; **G–I** after Fritsch 1929.

Plate 7 A–H, K after Geitler 1932a; **I, J** after Geitler 1932 (based on Lagerheim); **L, M** after Geitler 1932a (based on Lehmann); **N, O** after Geitler 1932a (based on Bornet & Flahault); **P–R** after Geitler (based on Huber & Jardin).

Plate 8 A after Geitler 1932a (based on Frémy); **B** *Homoeothrix crustacea,* after Starmach 1966 (based on Woronichin 1923); **C** after Starmach 1966; **D** after Frémy 1929–33; **E** after Geitler 1932a; **F** S.C. Hardiman (original); **G** after W. & G.S.West 1897.

Plate 9 A–E, G, H, J–Q after Geitler 1932a (based on Gomont); **F** after Geitler 1932a (based on Frémy); **I** after Geitler 1932a.

Plate 10 A, D, G, I–K, M, N after Geitler 1932a (based on Frémy); **B** after Geitler 1932a (based on G.M. Smith); **C** after Symoens & van der Werff 1951; **E** after Geitler 1932a (based on Gardner); **F, O, P** after Geitler 1932a (based on Gomont); **H** after Geitler 1932a (based on Wille); **L** after Geitler 1932a.

Plate 11 A after Geitler 1932a and Wislouch; **B, F** after Geitler 1932a; **C, D** after Geitler 1932a (based on Gomont); **E** after Desikachary 1959; **G, I** after Starmach 1966; **H** after Frémy 1929–33; **J** after Whitton & Peat 1969; **K** after Geitler 1932a (based on Frémy 1932); **L** S.C. Hardiman; **M** after Geitler 1932a (based on tropical material drawn by Frémy).

Plate 12 A, B after Geitler 1932a (based on Gomont); **C, D, L, M, O, P** after Geitler 1932a; **E–J** after Geitler 1932a (based on Tilden); **K** after Geitler 1932a (based on Frémy); **N** after Starmach 1966 (based on Skuja).

Plate 13 A after Geitler 1932a (based on Skuja); **B, I–L** after Frémy 1929–33; **C–H** redrawn by S.C. Hardiman, based on various sources, including field material.

Plate 14 A after Geitler 1932a (based on Nygaard); **B, G** redrawn from Smith 1920; **C** based on Whitford & Schumacher 1969; **D** based on Lemmermann 1896; **E** after Geitler 1932a (based on Boye-Petersen); **F** after Geitler 1932a (based on Klebahn); **H** after Geitler 1932a (based on Borge); **I** redrawn from Prescott 1962; **J, K** after Geitler 1932a (based on Frémy).

Plate 15 A, H redrawn from Starmach 1966; **B** after Geitler 1932a; **C** redrawn from Geitler 1932a (based on Virieux); **D** redrawn from Bornet & Flahault 1886–8; **E** redrawn from Geitler 1932a (based on Lemmermann); **F** redrawn from Geitler 1932a (based on Frémy); **G, I** redrawn from Smith 1920.

Plate 16 A–C S.C. Hardiman, original material; **D** redrawn from Geitler 1932a (based on Tilden); **E, G** redrawn from Geitler 1932a (based on Frémy); **F** redrawn from W. & G.S.West 1897.

Plate 17 A, H–J after Geitler 1932a (based on Frémy); **B, C** after Geitler 1932a (based on Skuja); **D** after Geitler 1932a (based on Gomont); **E** after Geitler 1932a (based on Bornet & Thuret); **G** after Geitler 1932a (based on Borge).

Plate 18 A–C after Geitler 1932a (based on Poretzky & Tschernow); **D–F** redrawn by S.C. Hardiman based on various sources; **G** after Smith 1920; **H** after Geitler 1932a (based on Lemmermann); **I** after Prescott 1962 (based on Ralfs); **J** after Frémy (1929–33); **K** after Starmach 1966 (based on Kisseleva); **L** after Skuja 1956 (based on Lemmermann).

Plate 19 A, C–E after Geitler 1932a (based on Frémy); **B** redrawn from Smith 1920; **F, G** redrawn from Frémy 1929–33 (based on Thuret); **H–K** after Frémy 1929–33; **I, K** redrawn by S.C. Hardiman.

Plate 20 A, D, E, H, I after Geitler 1932a; **B** redrawn from Geitler 1932a (based on Frémy); **C** after Frémy 1932; **F, G** after Fremy 1929–33; **J** after Frémy (in Geitler 1932a).

Plate 21 A–C S.C. Hardiman (original); **D, G, H** after Geitler 1932a; **E, F** S.C. Hardiman (original).

Plate 22 A–E after Geitler 1932a (based on Frémy); **F–H** S.C. Hardiman (original); **I–L** redrawn from Geitler 1932a.

Plate 23 A, B Laurentian Great Lakes; **C** German sample courtesy of A. Rieth; **D** UTEX culture collection; **E, F** North American sample courtesy of T. Rintoul.

Plate 24 A North American sample courtesy of T. Rintoul; **B–D** River Thames, England; **E** Wales; **F** Ireland.

Plate 25 A Ireland; **B** Wales; **C** Scotland; **D, F** England; **D, F** after Sheath & Müller 1999, reproduced by permission of the Journal of Phycology.

Plate 26 A English sample, after Sheath & Müller 1999, reproduced by permission of the Journal of Phycology; **B, C, F** England; **D, E** Wales.

Plate 27 A Wales; **B, D** England; **E, F** England.

Plate 28 A, B England; **C–F** Scotland.

Plate 29 A, D England; **B, E** Scotland; **C** Ireland; **F** North America.

Plate 30 A, D England; **B, F** Scotland; **C, E** North America.

Plate 31 A, B Scotland; **C** North America, after Sheath & Hymes 1980; **D, E** England; **F** Wales.

Plate 32 A North America; after Vis & Sheath 1992, reproduced by permission of Phycologia; **B–E** Scotland; **F** North America.

Plate 33 A North America; **B, C** Ireland.

Plate 34 All Wołowski (original).

Plate 35 A, B, F–O Wołowski (original); **C–E** after Pringsheim 1956; **P** after Schewiakoff, from Lemmermann 1913.

Plate 36 A–D, F–K Wołowski (original); **E** after Gojdics 1953; **L, M** after Pringsheim 1956.

Plate 37 A–D, I–M, O, P Wołowski (original); **E–H** after Pringsheim 1956; **N** after Prescott 1964.

Plate 38 A, D, E, I, J, U after Conrad 1935; **B, C, F–H, K–S, Y, Z, AA–AE** Wołowski (original); **T** after Dreżepolski 1925; **V–X** after Allegre & Jahn 1943.

Plate 39 A–I, L, N–P Wołowski (original); **J, K** after Hazel 1937; **M** after Skvortsov 1928.

Plate 40 A after Pochmann 1941; **B–E, J–P** Wołowski (original); **F, G** after Berg & Nygaard 1929; **H** after Lemmermann 1910; **I** after Deflandre 1926; **Q** after Popova 1955.

Plate 41 A–H, J, K, Y after Popova 1966; **I, L** after Kisseleva 1959; **M** after Roll, from Svirenko 1938; **N, O** after Svirenko 1939; **P–R** after Skuja 1948; **S, V** after Glicklhorn 1925; **T** after Huber-Pestalozzi 1955; **U** after Skuja 1948; **W, X** after Safonova 1966; **Z, AA** after Fritsch 1949.

Plate 42 A–G, I, K–M, O–W, Y, Z, AA–AC Wołowski (orginal); **H** after Popova 1955; **J** after Dreżepolski 1925; **N** after Swirenko 1939; **X** after Deflandre 1926, from Stein 1878.

Plate 43 A–Q, U, W, Y, Z, AA, AB Wołowski (original); **R** after Popova 1955; **S, T** after Skvortsov 1926.

Plate 44 A modified after Carter 1937; **B** modified after Rosenberg 1944; **C** modified after Utermöhl 1925; **D** modified after Hickel1975; **E, F, I** Novarino (original); **G** modified after Hansgirg 1892; **H** modified after Skuja 1948; **J** modified after Geitler 1922; **K** modified after Pascher 1913.

Plate 45 A, E, F modified after Skuja 1948; **B** modified after Skuja 1939; **C** modified after Wislouch 1924; **D, H** modified after Fritsch 1914; G modified after Penard 1921.

Plate 46 A–E, G, H after Harris 1940; **F** after Javornicky & Popovský 1971.

Plate 47 A after Penard 1891; **B–G** J.D. Dodge (original).

Plate 48 A, B, F, H, I–M J.D. Dodge (original); **C–E, G** after Harris 1940.

Plate 49 all J.D. Dodge (original).

Plate 50 A, B, E, G–I J.D. Dodge (original); **C, D** after Cridland 1958; **F** after West 1916.

Plate 51 all J.D.Dodge (original).

Plate 52 A–D, G–M J.D.Dodge (original); **E, F** after Jane 1945.

Plate 53 A–E Brigitta Steiner-Gafner (original); **F** after Mignot 1967; **G** after Spencer 1971.

Plate 54 A–D after Lund 1942; **E, J** after Pentecost 1984; **F** after Starmach 1984; **G** after Belcher & Swale 1967; **H** after Lund 1949; **I** after Belcher 1966; **K** after Lund 1954.

Plate 55 A after Lind 1939; **B** after Nygaard 1945; Pentecost 1984; **C, D** after Jane 1940.

Plate 56 A after Huber-Pestalozzi 1941; **B** after Lund 1949; **C** after West 1904; **D, I** after Starmach 1985; **E** after Lund 1953; **F** after Lund 1960a; **G** after Pascher 1931; **H** after Godward 1933.

Plate 57 A after West & Fritsch 1927; **B** after Jane 1942; **C, E** after Starmach 1985; **D** after Swale & Belcher 1966; **F** after Hibberd 1970; **G** after Pentecost 1984.

Plate 58 A after Skuja 1964; **B** after Starmach 1985; **C** after Lund 1960; **D** after Happey & Moss 1967; **E, F** after Scourfield 1930; **G** after Lund 1942a; **H** after Huber-Pestalozzi 1941; **I** Williams 1965; **J** after Lund 1960; **K** after Huber-Pestalozzi 1941, from Bachmann 1907; **L** after Brook 1955; **M** after Lund 1952; **N–P** after Krieger 1930; **Q** after Brook 1959.

Plate 59 A, B, D, E, G after Krieger 1930; **C** after Brook 1959; **F** after Huber-Pestalozzi 1941.

Plate 60 A, B, D–F, L, M, O, P after Hilliard 1966; **C, G** after Jane 1945; **H** after Huber-Pestalozzi 1941; **I** after Lund 1953; **J, K, N, Q, R** after Hilliard & Asmund 1963.

Plate 61 A, B, C after Lund 1942; **D** after Conrad 1939; **E, Fb, G** after Lund 1960; **Fa** after Scourfield 1930; **H, I** after Williams 1965; **J** after Pascher 1913; **Ka** after Pringsheim 1952; **Kb** after Peterfi 1969; **L, Q** after Jane 1945; **M** after Taylor 1954; **N** after Tikkanen & Willén 1992; **O** after Balonov 1978; **P** after Manton & Harris 1966; **R** after Lauterborn in Pascher 1913.

Plate 62 A after West 1904; **B, L** after Starmach 1985; **C, D, Q** after West & Fritsch 1927; **E, P** after Lund 1942; **F** after West & Fritsch 1927, from Pascher 1913; **G** after Godward 1933; **H** after West & Fritsch 1927, from Woronin 1880; **I** after Andersen 1989; **J** after Brook 1955; **K** after Lund 1949; **M, O** after Jane 1945; **N** after Pentecost 1984.

Plate 63 A after Oltmanns 1904; **B** after Jane 1945; C after Starmach 1985; **D** after Williams 1965; **E** after Krieger 1930; **F** after Asmund 1959; **G** after Lund 1942; **H** after Harris 1958.

Plate 64 A after Starmach 1985; **B** after Lund 1950; **C** after Harris 1953; **D** after Harris & Bradley 1960; **E** after Bradley 1966b; **F** after Jane 1945; **G** after Kristiansen 1959, Jane 1945; **H** after Harris & Bradley 1958; **Ia** after Bradley 1966b; **Ib** after Korshikov 1929; **J** after Philipps 1884 (as *Chlorodesmus hispidus*); **K** after Starmach 1985, Korshikov 1929.

Plate 65 A after Rostafinski & Woronin 1877; **B** after W. & G.S.West 1898; **Ca,b** after Belcher & Swale 2000, **Cc,d** after Fott 1948; **D** after Fott 1948; **Ea,c** after Belcher & Swale 1999, **Ed,e** after Geitler 1929, **Eb** Fott 1948; **Fa,d** after Belcher & Swale 1999, **Fb** after Bourrelly 1963, **Fc** after Fott 1948; **Ga,b** after Pascher 1939; **H** after Ettl 1960.

Plate 66 A after Starmach 1966, Belcher & Swale 1979; **B** after Belcher & Swale 1999, Fott & Komarek 1960; **C** after Belcher & Swale 1999, Bourrelly 1951; **Da,b** after Skuja 1948, **Dc,d**, after Belcher & Swale 1999; **E** after Pascher 1930; **F** after Ettl 1964; **G** after Fott 1960; **H, K** after Pascher 1939; **I** after Printz 1914; **Ja,b** after Belcher & Swale 1999, **Jc** after Vischer & Pascher 1936.

Plate 67 Aa–g, B after Pascher 1925, **Ah, Fc** after Belcher & Swale 1999; **C** after Pascher 1939; **Da,e** after Luther 1899, **Db–d** after Messikommer & Vischer 1946; **E** after G.M. Smith 1920; **Fa,b,d** after Nageli 1849; **Ga** after Hieronymous 1898, **Gb** after Ettl 1970.

Plate 68 A after Borzí 1888; **B** after Pascher 1925, Belcher & Swale 1999; **Ca, E, F, Ga,b,f,g, Ha–d** after Pascher 1925, **Cb** Belcher & Swale 1979; **D** after G.S. West 1904; **Gc–e** after Klebs 1896; **Ia** after Pitschmann 1963, **Ib** after Vischer 1937; **J** after Gerneck 1907; **K** after G.S.West 1903.

Plate 69 A after N.Carter 1919; **B, C, I** after Pascher 1925; **D, J** after Skuja 1948; **E** after Belcher & Swale 1999, after Pascher 1937; **F, Kb** after Pascher 1937; **G** after Prinz 1914; **H** after Belcher & Swale 1999, W. & G.S.West 1903; **Ka** after Braun 1855.

Plate 70 A after Braun 1855, Pascher 1925, Belcher & Swale 2000; **B, D** after Braun 1855; **C** after Pascher 1925, Belcher & Swale 1999; **E** after Skuja 1948, Belcher & Swale 1999, Pascher 1925; **F** after Pascher 1925.

Plate 71 A after Woronin 1869; **B** after Taylor & Bernatowicz 1952; **C, Ea–f, G** after Christensen, 1969; **D** after Götz 1897; **Eg** after Migula 1907; **F** after Dangeard 1936.

Plate 72 Aa after Hirn 1900; **Ab** after Strom 1926; **B** after Simons 1974; **C, D** after Christensen 1969; **E** after Blum 1953; **F** after Dangeard 1939; **G** after Rieth 1959.

Plate 73 A, D after Blum 1953; **Ba** after Jao 1936, **Bb,c** after Prescott 1951; **C** after Blum 1972; **E** after Christensen 1970; **F** after Christensen 1969; **G** after Pascher 1939; **H** after Pascher 1939; **I** after Petersen 1932; after Korshikov 1927.

Plate 74 A–C adapted from Yoshizaki & Iura 1991; **D–F** adapted from Waern 1952; **G–I** adapted from Dop 1979; **J** adapted from Waern 1952.

Plate 75 A after Ettl 1972; **B** after Belcher 1968; **C** after Melkonian & Preisig 1986; **D** after Skuja 1948; **E** after Moestrup & Ettl 1979; **F, I** after Belcher & Swale 1961; **G** after

Manton & Ettl 1965; **H** after Belcher 1966; **J** after Hodgetts 1920; **K, L** after Belcher & Swale 1963; **M** after Takeda 1916.

Plate 76 A after Ettl & Gartner 1988; **B, G, H, K** Pentecost (original); **C, F** after Novakova 1965; **D** after Lund 1956; **E, J** after Skuja 1964; **I** after Lund 1957; **L** after Korshikov 1935; **M** after G.S.West 1904.

Plate 77 A, C, J Pentecost (original); **B** after Goroschankin 1890; **D** after Pascher 1927; **E** after Skuja 1964; **F** after Fritsch 1935; **G, I, P, R** after Ettl 1976; **H** after Skuja 1964; **K** after Fott 1971; **L** after Smith 1920; **M, N** after Lund 1947; **O** after Gerloff 1940; **Q** after Fritsch & Takeda 1916.

Plate 78 A after Christen 1958; **B** after West 1916; **C** after Ettl 1976; **D, K, L, U** Belcher & Swale (original); **E, H, P** after Pascher 1927; **F** after Playfair 1918; **G** after G.M. Smith 1920; **I** after Matwienko 1938; **J, O, Q** Pentecost (original); **M** after Belcher & Swale 1979; **N** after Belcher & Swale 1967; **R** after Christen 1962; **S** after Korschikov 1927; **T** after Belcher & Swale 1963.

Plate 79 A, H after Belcher 1961; **B** after Pascher 1927; **C** after Peterfi 1968; **D, E, J–L, Q** Belcher & Swale (original); **F, G, I** after Belcher 1965; **M** after Nygaard 1949; **N** after Belcher & Swale 1967; **O** after Hazen 1922; **P** Pentecost (original); **R** after Skuja 1964; **S** after Jane 1944.

Plate 80 A after Korshikov 1927; **B** after Takeda 1916; **C, D, F, L, Q** Belcher & Swale (original); **E** after Fritsch & John 1942; **F** after Belcher & Swale 1979; **G, I, J** Pentecost (original); **H** after Pascher 1927; **K** after Wandschneider & Kies 1978; **M, P** after Skuja 1964; **N** after Ettl 1976; **O** after Ettl & Moestrup 1980; **R** after West & Fritsch 1927, redrawn by Belcher & Swale.

Plate 81 A after Belcher & Swale 1979 (redrawn); **B, F, G** after Belcher & Swale 1976; **C** after Korshikov 1927; **D** Belcher & Swale (original); **E** G.S. West (original); **H, J** Pentecost (original); **I** after G.M. Smith 1920.

Plate 82 A after Korshikov 1953; **B** after Trenkwalder 1975; **C** after James 1935; **D** after Vischer 1945; **E** after Hindák 1984; **F, G, H** after Fott & Novakova 1969; **I** after Belcher & Swale 1979; **J, K, N, O** after Starr 1955; **L** after Bold 1931; **M** after Petersen 1932; **P** Belcher & Swale (original); **Q** after Fritsch & John 1942, modified by D. Williamson.

Plate 83 A after Hindák 1984; **B** after Hortobágyi 1973; **C** after Guarrera 1978; **D, Na** after West 1892; **E, F, Nb** after Korshikov 1953; **G** after Tsarenko 1990; **H** after Belcher & Swale 1997; **I, M** after Belcher & Swale 1999; **J** after West unpublished, redrawn by D. Williamson; **K** after Sodomkova 1972; **L** after W. & G.S.West 1902; **O** after W. & G.S. West 1896; **Pa** Belcher & Swale (original), **Pb** after Komárek & Perman 1978.

Plate 84 A, G after Williams 1967; **B, C, L** after Korshikov 1953; **D–F** after Belcher & Swale 1979; **H** after Flint 1967; **I** after Williams 1967 (as *Q. laevis*); **Ja** after West 1892, **Jb** after Hindák 1984, modified by D. Williamson; **K** after Fott & Komárek 1974; **M** after Komárek 1974.

Plate 85 A after Komárek & Perman 1978, modified by D. Williamson; **Ba** after Hindák 1977, **Bb, D** D.Williamson (original); **C, Hb** after Belcher & Swale 1979; **E** after Hindák 1990; **F** after Williams 1972; **G** after Hindák 1988; **Ha** Belcher & Swale (original); **I** after West 1891, modified by D.Williamson; **J** after West 1905.

Plate 86 A after West 1904; **B** after Jaag 1933; **C, D, E** after Korshikov 1953; **F** after Hindák 1977; **G** after Skuja 1956, modified by D. Williamson; **H** after Skuja 1948; **I** after Lund 1956; **J, N** after Hindák 1984 (as *Coenochloris polycocca*); **K** after West 1892, redrawn by D. Williamson; **L** after Butcher 1932; **M** after Lund 1952; **O** after Jaag 1932; **P** after Lind 1939.

Plate 87 A after Prescott 1951; **B** after Fritsch & West 1927; **C** after Pentecost 1984; **D** after Kalina & Punčochárová 1977; **E** after West 1899; **F** after Korshikov 1953; **G** after W. & G.S.West 1897; **H** after James 1935; **I** after Lind & Pearsall 1945; **J, L** after Hindák 1977, modified by D. Williamson; **K** after Belcher & Swale 1979; **M** after Fott 1976, Hindák 1977; **N** after Naumann 1921; **O** after Belcher & Swale 1996.

Plate 88 Aa,b after Rehakova 1969, **Ac** after W. & G.S. West 1897; **B** after Belcher & Swale 1996; **C, D** Belcher & Swale (original); **E** after West 1902, redrawn by D. Williamson; **F** after France 1894; **Ga** after W. & G.S. West 1902, **Gb** after Prescott 1951; **H, J, L, N** after Belcher & Swale 1979; **I** after Lund 1954; **K** after Lund 1960; **M** after Komárek 1974.

Plate 89 A, M Belcher & Swale (original); **B** after Lemmermann & Brunnthaler 1915; **C** after Hindák 1988, modified by D. Williamson; **D, F–J, L** after Belcher & Swale 1979; **E** after West & Fritsch 1927; **K** after Prescott 1951; **N** after Fott & Kovacik 1975; **O** after Williams 1965.

Plate 90 A after Belcher & Swale 1968; **B** after Fott 1957; **C** after Komárková-Legnerová 1969, Hindák 1977; **D, I, K** Belcher & Swale (original); **E** after W. & G.S. West 1904, 1906 and W. West unpublished, redrawn by D. Williamson; **F** after Fott 1975; **G** after Korshikov 1953; **H, L** after Belcher & Swale 1999; **J** after Belcher & Swale 1979; **M** after Belcher & Swale 1962.

Plate 91 A after Belcher & Swale 1979; **B** after Hindák 1988, modified D.Williamson; **C** after Belcher & Swale 1962; **D** after Komárková-Legnerová 1969; **E** after Hindák 1977; **F** after Komárková-Legnerová 1969, Hindák 1977; **G** Belcher & Swale (original); **H** after W. & G.S. West 1897; **I** after Williams 1965; **J** after Lund 1955; **K** after Korshikov 1953, modified by D. Williamson; **L** after West 1903; **M** after Hindák 1974; **N** Williams (original); **O** after Düringer 1958.

Plate 92 A after West from Brunnthaler 1915, Prescott & Vinyard 1965, West 1893, redrawn D. Williamson; **B** after W. & G. S. West 1894; **C** after Hindák 1988; **Da** West 1892, **Db, L** after West 1894; **Ea, F, J** after Rehakova 1969, **Eb,c** after West 1893; **G** after Skuja 1964, redrawn by D. Williamson; **H, Ia** after W. & G.S.West 1893; **Ib** after Fott 1964; **K** after Printz 1913, redrawn by D. Williamson.

Plate 93 Aa after Sulek 1969, **Ab,c, L** after Smith 1920; **B, F** Belcher & Swale (original); **C, D, E, N** after Belcher & Swale 1979; **G, H** after Sulek 1969; **I** after Belcher & Swale 1999; **J** after West & Fritsch 1927; **K** after Griffiths 1927; **M** after Belcher & Swale 1962.

Plate 94 A after Komárek 1983; **B, E, J–L** Belcher & Swale (original); **C** after Williams 1965; **D** after Belcher & Swale 1979; **F, H** after Hindák 1977; **G** Williams original, redrawn by D. Williamson; **H** after Hindák 1977; **I, O, P** after Belcher & Swale 1999; **M** after Hortobágyi 1939; **N** after West 1907, redrawn D. Williamson.

Plate 95 A after Halasz 1935, Kiss 1942; **B, E** after Belcher & Swale 1999; **C** after Fritsch & Rich 1927; **D, G, J–L** Belcher & Swale (original); **F** after Hindák 1990; **H** after Chodat 1926, Hortobágyi 1969 from W. & G.S. West 1895; **I, M, P** after Tsarenko 1990; **N** after Hortobágyi 1959; **Oa** after Hortobágyi 1944, **Ob** after Hindák 1974.

Plate 96 Aa after Hortobágyi 1960, **Ab, Da** after Tsarenko 1990; **B, Db, F, L** Belcher & Swale (original); **Ca** after Bernard 1908, **Cb** after Skuja 1956; **E** after Williams 1965; **G** after Smith 1920, Korshikov 1953 from Lagerheim & Printz; **H, M, O** after Belcher & Swale 1999; **I** after W. & G.S. West 1897, redrawn by D. Williamson; **J** after W. & G.S. West 1897, redrawn by D. Williamson; **K** after Kalina & Punčochárová 1977; **N** after Skuja 1956 from Printz/Chodat (as *S. acuminatus* var. *minor*); **P** after Philipose 1967.

Plate 97 A, B, D, E Belcher & Swale (original); **C** W. & G.S. West unpublished, redrawn by D. Williamson; **Fa** after W. & G.S. West 1907, redrawn by D. Williamson; **Fb** B.M. Griffiths original, redrawn by D. Williamson; **Ga** after Tsarenko 1990, **Gb** after Roll 1925; **H, J, N** after Belcher & Swale 1979; **I** after Fott & Komárek 1974; **K** after Belcher & Swale 1962; **L, M** after Komárková-Legnerová 1969; **O** after G.S.West 1915.

Plate 98 A after Hindák 1988; **B** after Skuja 1956; **C** after G.S. West 1904, modified by D. Williamson; **Da** after Williams 1965, **Db** Belcher & Swale (original); **E** after Belcher & Swale 1979; **Fa, H** after Korshikov 1953, **Fb** after Smith 1920; **G** after Belcher & Swale 1962; **I** after Fott 1967, Korshikov 1953; **Ja** after West 1892, **Jb** D. Williamson (original); **K** after Smith 1929, redrawn by D. Williamson.

Plate 99 A, J after Korshikov 1953; **B** after Fritsch & John 1942; **C, G, M, R** Belcher & Swale (original); **D, E** after Belcher & Swale 1962; **F** after Carter 1927, redrawn by D. Williamson; **H** after W. & G.S.West 1904; **I** after Lund 1947; **K** after West 1904; **L** after Fott 1957; **N** after Korshikov 1953, modified by D. Williamson; **O, Sa,c** after West & Fritsch 1927; **P** after Fritsch 1933; **Qa** after West 1904, **Qb** after Hoek 1963; **Sb** G.S.West unpublished, redrawn by D. Williamson.

Plate 100 A, B, D, F, G, I, L, M, O, Q after Hirn 1900; **C, E, N, P, R** after Tiffany 1937; **H** after de Bary 1854; **J** after Wittrock 1870; **K** after Harris 1933.

Plate 101 A, C–F after Tiffany 1930, redrawn from Hirn 1900; **B, J, G** after Tiffany 1930; **E** after Prescott 1944; **H, I** after Tiffany 1930, from Wittrock.

Plate 102 A, B, D, E, G after Tiffany 1930, redrawn from Hirn 1900; **C, H–J** after Tiffany 1930; **F** after Tiffany 1930, redrawn from Gutwinski and G.S.West.

Plate 103 A–D, F, H after Tiffany 1930, redrawn from Hirn 1900; **E, N** after Tiffany 1930, redrawn from West; **G, J–M, O, P** after Tiffany 1930; **I** after Tiffany 1930, redrawn from Wittrock.

Plate 104 A, C, E, I, J after Tiffany 1930; **B, D, F–H** after Tiffany 1930, redrawn from Hirn 1900.

Plate 105 A after Tiffany 1930, redrawn from Hirn 1900; **B–E, G, I, J** after Tiffany 1930; **F** after Wittrock 1870; **H** after Rich 1925, 1930; **K** after Mrozińska 1985.

Plate 106 A, C–G after Tiffany 1930; **B** after Hirn 1900.

Plate 107 A after Tiffany 1930, redrawn from Hirn 1900; **B** after Hallas1905; **C, D, F, G** after Tiffany 1930; **E** after Wittrock 1870; **H** after Pringsheim 1858.

Plate 108 A after Godward 1934; **B** after West 1899, redrawn by D. Williamson; **C** after Schmidle 1897; **D** after W. & G.S.West 1897, redrawn by D. Williamson; **E** after Vischer 1933; **F** after West 1912, redrawn by D. Williamson; **G** after Pringsheim 1860; **H** after Cook 1882–84; **I** after Butcher 1932; **J** after West 1916; **K** after Acton 1916.

Plate 109 A–C after Pringsheim 1860; **D** after W. & G.S.West, redrawn by D. Williamson; **E** after Bourrelly 1990; **F** after Fritsch 1929; **G** after Butcher 1932; **H** after Borzí 1895; **I** after Richter 1893; **J** after West 1918; **K** after West 1904; **L** after Borge 1883; **M** after Fritsch & John 1941.

Plate 110 A after Chodat 1894; **B, Ia** after West & Fritsch 1927; **C** after Butcher 1932; **D** after West 1904; **E** after Fritsch & John 1941; **F** after John & Johnson 1989; **G** after Hornby 1918; **H** after Huber 1892; **Ib** after Hassall 1852, redrawn by D. Williamson; **J** after Hassall 1852; **Ka** D. Williamson (original), **Kb** after Hassall 1852, redrawn by D. Williamson.

Plate 111 A Belcher (original); **B** after Prescott 1962; **C** after West 1904; **D** after Islam 1963, redrawn from Godward 1942; **E, F** after Butcher 1932; **G** after Kützing 1849; **H** after Islam 1963; **I** after Kützing 1849; **J** after Fritsch 1903; **K** West original, redrawn by D. Williamson.

Plate 112 Aa after West 1904, **Ab** after Tiffany & Britton 1952, redrawn after Smith; **B** after W. & G.S.West 1903, redrawn by D. Williamson; **C** after W. & G.S.West 1903, redrawn by D. Williamson; **D, L** after Klebahn 1893; **Ea** Belcher & Swale (original); **Eb, F, H** after West & Fritsch 1927; **G** after West 1908, redrawn by D. Williamson; **I** after West & Fritsch 1927, redrawn by D. Williamson; **Ja** West original, redrawn by D. Williamson, **Jb** after Fott in Komárek & Fott 1983; **K** after West 1912, redrawn by D. Williamson.

Plate 113 A after Printz 1921; **B** after West 1904; **C, D** after Bristol 1920; **E** D. Williamson (original); **F** after W. & G.S.West 1900–01, redrawn by D. Williamson; **G** after Lund 1960; **H, K** after West & Fritsch 1927, redrawn by D. Williamson; **I** after Williams 1965; **J** after West 1899; **L** after Chodat & Topali 1922.

Plate 114 A, D, H, K after West & Fritsch 1927, redrawn by D. Williamson; **B** after Bennett 1888; **C** after Ramanthan 1966; **E** Belcher & Swale (original); **F** after Lund 1957; **G** after West 1915; **I** after Lagerheim 1892; **J** after Vischer 1933; **L** after Lund 1956; **M** after Hindák 1987.

Plate 115 A after Vischer 1941; **B, N** after West & Fritsch 1927, redrawn by D. Williamson; **C** after Chodat 1900; **D** after Fritsch & John 1942; **E** after Lund 1947; **F** after West & West 1900–01, redrawn by D. Williamson; **G** after James 1935; **H, L** after Hazen 1901–02; **I** after Gay 1893; **Ja** after West & Fritsch 1927, **Jb** after Printz 1964; **K** after Printz 1964, from Fritsch & Klercker; **M** after Bristol 1920; **O** after Hodgetts 1918.

Plate 116 A after West 1904; **B** after West & Fritsch 1927; **C** after West 1916; **D** after Allsopp 1963; **E** after West 1892, redrawn by D. Williamson; **F** after West 1899, redrawn by D. Williamson; **G** after West & Fritsch 1927, from Cienowski (redrawn by D. Williamson); **H** after W. & G.S.West 1903, redrawn by D. Williamson; **I, K** after West 1904, redrawn by D. Williamson; **J** after West & Fritsch 1927, redrawn by D. Williamson; **L, M** after G.S.West original, redrawn by D. Williamson; **N** after Wickmann 1937; **O** after Printz 1964, after West and Hazen 1901–2; **P** after Hazen 1901–02.

Plate 117 A after Wille 1881; **B** after Fritsch 1942; **C** after Burrows 1991; **D, M** after West 1904; **E** after Prescott 1962; **F** after W. & G.S.West 1903; **G** West original, redrawn by D. Williamson; **H** after Millard 1870; **I** after Printz 1900; **Ja** after Wille 1878, **Jb** after Gobi 1879; **K** after Kützing 1853; **L, O** after West 1904; **N** after West 1904 and West original, redrawn by D. Williamson; **Pa** after West & Fritsch 1927 (as *E. intestinalis*), **Pb** after Burrows 1991.

Plate 118 Aa,b after Transeau 1925; **Ba,E** after Skuja 1929, **Bb** after Krieger 1941; **C** after Czurda 1932, Gauthier-Lièvre 1965; **Da,b** after Jao & Hu 1978, **Dc–e** after Brook 1996; **F** after Transeau 1951; **Ga** after Skuja 1932, **Gb** after Randhawa 1959, **Gc** after Hodgetts 1918, **Gd** after Transeau 1933.

Plate 119 Aa,c after Krieger 1941; **Ab** after Wittrock 1868; **Ba,b** after Transeau 1926; **C** after West & Fritsch 1927 (redrawn by A.J. Brook) and Krieger 1941; **D** after Gauthier-Lièvre 1965; **E** after Hassall 1843; **F** after Kützing 1855; **Ga** after Transeau 1926, **Gb** after Kadlubowska 1972.

Plate 120 A after Wittrock 1872; **B, F** after Kadlubowska 1972; **C** after Wittrock 1889; **Da,b** after Transeau 1926, 1951; **E, J** after Transeau 1926; **G** after Bourrelly 1972; **H** after Fritsch 1949; **Ia** after G.S.West 1899, **Ib,c** after Krieger 1941.

Plate 121 Aa after Wittrock 1886, **Ab** after Gauthier-Liévre 1965; **B, D, G** after Transeau 1926; **C, I** after Transeau 1926, 1951; **E** after Hassall 1843; **F** after Transeau 1951; **H** after Wittrock 1878.

Plate 122 Aa,b after Kadlubowska 1972; **B** after Kützing 1849; **C, Gb, K** after Petit 1880; **Da, Eb** after Gauthier-Lièvre 1965; **Db, Fb,c, I** after Czurda 1930; **Dc, Ea, J** after Krieger 1941; **Fa, L** after Transeau 1951, **Ga** after Borge 1913; **H** after Jao 1935.

Plate 123 A after Petit 1880; **B** after Norstedt 1873; **C** after Transeau 1934; **D, Ea** after Transeau 1951; **Eb** after Petit 1874; **Fa, G, H** after Czurda 1932; **Fb,c, J** after Gauthier-Lièvre 1965; **I** after G.S.West 1899.

Plate 124 A after Petkoff 1934–35; **Ba** after Transeau 1951, **Bb,c, D** after Gauthier-Lièvre 1965; **C,F** after Krieger 1941; **Ea, Ga,b, H** after Czurda 1932, **Eb** after Kadlubowska 1972; **Gc,d** after Poljanski 1959; **I** after Mehra 1957; **J** after Jao1935.

Plate 125 A, F, Ha, J, K after Petit 1880; **B, D, E** after Godward 1956; **C** after Czurda 1932; **G** after Kasanowsky 1913; **Hb** after Borge 1913; **I** after Gauthier-Lièvre.

Plate 126 A after Gauthier-Lièvre 1965; **B, Eb** after Krieger 1941; **Ca** after Kadlubowska 1972, **Cb,c** after Czurda 1925–1928; **D** after Petit 1874; **Ea,c** after Kadlubowska 1984; **Fa,b** after Mehra 1957; **Fc** after Marson 1992, drawn by A.J. Brook.

Plate 127 A after Beck 1929; **B, C** after G.S.West in W. West 1892; **Da** after Smith 1933, **Dc,d, Eb, H** after Gauthier-Lièvre 1965; **Ea** after G.S.West 1892; **F.** after W. & G.S.West 1897; **G, Jb** after Tiffany 1937; **I** after Transeau et al. 1934; **Ja** after Transeau 1951; **K** after Krieger 1941; **L, Ma** after De Bary 1858; **Mb** after Jao 1935.

Plate 128 A–K, O–R, T, U A.J. Brook (original); **L–N, S, V–Y** D. Williamson (original).

Plate 129 A–E, G–Q D. Williamson (original); **F** A.J. Brook (original)

Plate 130 all D. Williamson (original).

Plate 131 all D. Williamson (original).

Plate 132 all D. Williamson (original).

Plate 133 all D. Williamson (original).

Plate 134 all D. Williamson (original).

Plate 135 all D. Williamson (original).

Plate 136 all D. Williamson (original).

Plate 137 all D. Williamson (original).

Plate 138 A–J, L–T D. Williamson (original); **K** A.J. Brook (original).

Plate 139 A–I, K–M, O D. Williamson (original); **J, N, P, Q** Brook 1959.

Plate 140 A, C–I, K–M D. Williamson (original); **B, J** A.J. Brook (original).

Plate 141 A–C, E–I, K, N–P, R D. Williamson (original); **D, J, N, Q, S** A.J. Brook (original).

Plate 142 all D. Williamson (original).

Plate 143 all D. Williamson (original).

Plate 144 A A.J. Brook (original); **B–P** D. Williamson (original).

Plate 145 A, B, F, G, I–L D. Williamson (original); **C–E** A.J. Brook (original); **H** after West & Fritsch 1927, from Lütkemüller.

Plate 146 all after Moore 1986, reproduced by permission of the Botanical Society of the British Isles (BSBI).

Plate 147 A, B, Cb after Moore 1986, reproduced by permission of the BSBI; **Ca** after Stewart & Church 1992.

Plate 148 all after Moore 1986, reproduced by permission of the BSBI.

Plate 149 A, B after Moore 1986; **C** after Groves & Groves 1886, Cb redrawn by Len Ellis.

Plate 150 all after Moore 1986, reproduced by permission of the BSBI.

Plate 151 all after Moore 1986, reproduced by permission of the BSBI

Plate 152 A, C, D after Moore 1986, reproduced by permission of the BSBI; **B** Len Ellis (original).

Plate 153 all after Moore 1986, reproduced by permission of the BSBI.

Plate 154 all after Moore 1986, reproduced by permission of the BSBI.

REFERENCES

Co-edited by George W. Lawson

A Literature of a general nature (Floras, monographs etc.), occasionally including records for the British Isles.

B Literature mentioned in the text and containing records of species in the British Isles.

C Literature not mentioned in the text, but containing records of freshwater algae in the British Isles.

Only the more important diatom literature is mentioned. Websites are not included since they are constantly changing.

Abbas A, Godward MBE (1964) Cytology in relation to taxonomy in Chaetophorales. Journal of the Linnean Society, Botany 58: 499–597. B

Acton E (1909) *Coccomyxa subellipsoidea*, a new member of the Palmellaceae. Annals of Botany 23: 537–573. B

Acton E (1916a) On the structure and origin of *Cladophora*-balls. New Phytologist 15: 1–10. B

Acton E (1916b) On a new penetrating alga. New Phytologist 15: 97–102. B

Acton E (1916c) Studies on the nuclear division in desmids. 1. *Hyalotheca dissiliens* (Smith) Bréb. Annals of Botany 30: 379–382. C

Adams J (1908) A synopsis of Irish algae, freshwater and marine. Proceedings of the Royal Irish Academy 27B: 11–60. B

Ahmadjian V (1967) A guide to the algae occurring as lichen symbionts: isolation, culture, cultural physiology and identification. Phycologia 6: 127–166. A

Allanson BR (1973) The fine structure of the periphyton of *Chara* sp. and *Potamogeton natans* from Wytham Pond, Oxford, and its significance to the macrophyte-periphyton metabolic model of R.G. Wetzel and H.L. Allen. Freshwater Biology 3: 535–542. C

Allen GO (1950) British Stoneworts (Charophyta). Haslemere Natural History Society, The Haslemere Education Museum, Surrey. 52 pp. B

Allsopp A (1963) *Schizomeris leibleinii* Kütz. – another alga from the Reddish Canal. British Phycological Journal 2: 257–259. A

An SS, Friedl T, Hegewald E (1999) Phylogenetic relationships of *Scenedesmus* and *Scenedesmus*-like coccoid green algae as inferred from IT-2 rDNA sequence comparisons. Plant Biology 1: 418–428. A

Anand PL (1937) A taxonomic study of the algae of the British chalk-cliffs. Journal of Botany, London 75, Suppl. II: 1–51. B

Anagnostidis K, Komárek J (1985) Modern approach to the classification system of cyanophytes 1 – Introduction. Archiv für Hydrobiologie, Suppl. 71 [Algological Studies 38/39]: 291–302. A

Anagnostidis K, Komárek J (1988) Modern approach to the classification system of cyanophytes 3 – Oscillatoriales. Archiv für Hydrobiologie, Suppl. 80 [Algological Studies 50–53]: 327–472. A

Anderson RA (1992) Diversity of eukaryotic algae. Biodiversity and Conservation 1: 267–292. A

Andersen RA, Potter D, Bidigare RR, Latasa M, Rowan K, O'Kelly CJ (1998) Characterization and phylogenetic position of the enigmatic golden alga *Phaeothamnion confervicola*: ultrastructure, pigment composition and partial SSU rDNA sequence. Journal of Phycology 34: 286–298. A

Anon. (1992) Council directive 92/43/EEC of 21 May 1992 on the conservation of natural habitats and of wild faunas and floras. Official Journal of the European Communities 206: 7–49. A

Antoine SE, Benson-Evans K (1985) Benthic flora of the River Wye System, Wales, UK. Nova Hedwigia 42: 31–47. C

Antoine SE, Benson-Evans K (1986) Spatial and temporal distribution of some interesting diatom species in the River Wye System, Wales, UK. Limnologica 17: 79–86. C

Antoine SE, Esho TR, Benson-Evans K (1984) Studies of the bottom sediments and epipelic algae of the River Ely, South Wales, UK. Limnologica 16: 1–7. C

Archer W (1827–1897) William Archer published between 1858 and 1885 about 230 papers of which the vast majority are short notes (often a page) on desmids collected in Ireland. Many appeared in the Quarterly Journal of Microscopic Science and sometimes the same article was published in two or more journals. For a full list of Archer's papers, see Prescott GW1984 Bibliographia Desmidiacearum Universalis (A Contribution to a Bibliography of Desmid Systematics, Biology and Ecology from 1774–1982). Koeltz Scientific Books, Koenigstein. 612 pp.

Asaul ZI (1975) Viznachnik evglenovikh vodorostey Ukrainskoy R. S. R. [Survey of the euglenophytes of the Ukrainian SSR]. Naukova Dumka, Kiev. 407 pp. [in Ukrainian]. A

Asmund B (1959) Electron microscope observations on *Mallomonas* and their occurrence in some Danish ponds and lakes. III. Dansk Botanisk Arkiv 18(3): 1–50. B

Atkins WRG, Harris GT (1924) Seasonal changes in the water and heleoplankton of freshwater ponds. Scientific Proceedings of the Royal Dublin Society 18: 1–21. C

Atkinson GF (1890) Monograph of the Lemaneaceae of the United States. Annals of Botany 4: 177–229. A

Auber H, Brook AJ, Shephard KL (1989) Measurement of the adhesion of a desmid to a substrate. British Phycological Journal 24: 293–295. A

Bachmann H (1907) Vergleichende Studien über das Phytoplankton von den Seen Schottlands und der Schweiz. Archiv für Hydrobiologie und Planktonkunde 3: 1–91. B

Backhaus D (1976) Beiträge zur Ökologie der benthischen Algen des Hochgebirges in den Pyrenäen. II. Cyanophyceen und übrige Algengruppen. Internationale Revue der Gesamten Hydrobiologie 61: 471–516. A

Bailey-Watts AE (1974) The algal plankton of Loch Leven, Kinross. Proceedings of the Royal Society of Edinburgh B 74: 135–156. A

Bailey-Watts AE (1976) Planktonic diatoms and some diatom–silica relations in a shallow eutrophic Scottish loch. Freshwater Biology 6: 69–80. C

Bailey-Watts AE (1978) A nine-year study of the phytoplankton of the eutrophic and non-stratifying Loch Leven, Kinross, Scotland. Journal of Ecology 6: 741–771. A

Bailey-Watts AE, Bindloss ME, Belcher JH (1968) Freshwater primary production by a blue-green alga of bacterial size. Nature, London 220: 1344. B

Bailey-Watts AE, Komárek J (1991) Towards a formal description of a new species of *Synechococcus* (Cyanobacteria/Cyanophyceae) from the freshwater picoplankton. Archiv für Hydrobiologie, Suppl. 88 [Algological Studies 61]: 5–19. C

Bailey-Watts AE, Lund JWG (1973) Observations on a diatom bloom in Loch Leven, Scotland. Biological Journal of the Linnean Society 5: 235–253. C

Bailey JC, Bidigare RR, Christensen SJ, Andersen RA (1998) Phaeothamniophyceae classis nova: a new lineage of chromophytes based upon photosynthetic pigments, *rbc*L sequence analysis and ultrastructure. Protist 149(3): 245–263. A

Bakker ME, De Jong YSDM, Lokhorst GM (1997) The flagellar apparatus ultrastructure in *Leptosira erumpens* (Deason & Bold) Lukesova and its contribution to the understanding of phylogenetic relationships within the Microthamniales (Chlorophyta). Archiv für Protistenkunde 148: 17–31. A

Balbi DM (2000) Suspended chlorophyll in the River Nene, a small nutrient-rich river in eastern England: long-term and spatial trends. Science of the Total Environment 251: 401–421. C

Bando T (1988) A revision of the genera *Docidium*, *Haplotaenium* and *Pleurotaenium* (Desmidiaceae, Chlorophyta) of Japan. Journal of Science of the Hiroshima University, ser. B, div. 2 (Botany) 22: 1–63. A

Barber HG, Haworth EY (1981) A Guide to the Morphology of the Diatom Frustule. Scientific Publications of the Freshwater Biological Association, UK No. 44. 112 pp. A

Barker J (1866) *Staurastrum scrabrum* new to Ireland. Quarterly Journal of Microscopical Science 6: 184. C

Barker J (1869a) A new and remarkable species of *Penium* (*P. spirostriolatum*). Quarterly Journal of Microscopical Science 9: 194. B

Barker J (1869b) *Desmidium aptogonum* new to Ireland. Quarterly Journal of Micoscoscopical Science 9: 198–199. C

Barker J (1873) *Closterium rostratum* conjugated, also the occurrence of *Cosmarium plicatum* Reinsch. Quarterly Journal of Microscopical Science 13: 435. C

Barker J (1896c) On a proposed new *Staurastrum* (*S. elongatum* Bark.). Quarterly Journal of Microscopical Science 9: 424. C

Bastow RF (1949) Lundy Freshwater Diatom Flora. Lundy Field Society 3rd Annual Report 1949: 32. B

Bastow RF (1954) New and rare freshwater diatoms from Devon. Transactions of the Devonshire Association for the Advancement of Science, Literature and the Arts 86: 285–290. B

Bastow RF (1957) Estuarial diatoms of the River Taw. Transactions of the Devonshire Association for the Advancement of Science, Literature and the Arts 89: 264–269. B

Battarbee RW (1976) *Coscinodiscus lacustris* in Lough Neagh – a case of mistaken identity? British Phycological Journal 11: 305–307. C

Battarbee RW (1978) Observations on the recent history of Lough Neagh and its drainage basin. Philosophical Transactions of the Royal Society B 281: 303–345. B

Battarbee RW, Carter C (1993) The recent analysis of Lough Neagh. Part B. Diatom and chironomid analysis. In: Wood RB, Smith RV (eds) Lough Neagh. The Ecology of a Multipurpose Water Resource. Kluwer, Dordrecht: 133–147. AB

Beem AP van, Simons J (1988) Growth and morphology of *Draparnaldia mutabilis* (Chlorophyceae, Chaetophorales) in synthetic medium. British Phycological Journal 23: 143–151. A

Beesley L (1904) A fountain alga. New Phytologist 3: 74–82. B

Belcher JH (1956) On the occurrence of *Bangia atropurpurea* (Roth) Ag. in a freshwater site in Britain. Hydrobiologia 8: 298–299. B

Belcher JH (1959) Some uncommon Chlorophyceae from the Lee Valley. British Phycological Bulletin 1: 73–74. B

Belcher JH (1960) Culture studies of *Bangia atropurpurea* (Roth) Ag. New Phytologist 59: 367–373. B

Belcher JH (1964) Some new and uncommon British Volvocales. III. British Phycological Bulletin 2: 307–312. C

Belcher JH (1965a) Some new and uncommon British Volvocales. IV. British Phycological Bulletin 2: 414–421. B

Belcher JH (1965b) *Volvulina steinii* Playfair from the English Lake District, a new British record. British Phycological Bulletin 2: 501–502. B

Belcher JH (1966a) *Prasinochloris sessilis* gen. et sp. nov., a coccoid member of the Prasinophyceae, with some remarks upon cyst formation in *Pyramimonas*. British Phycological Bulletin 3: 43–51. C

Belcher JH (1966b) Colony structure in *Chrysocapsa epiphytica* Lund. British Phycological Bulletin 3: 81–82. C

Belcher JH (1966c) *Microglena butcheri* nov. sp., a flagellate from the English Lake District. Hydrobiologia 27: 65–69. B

Belcher JH (1968a) A morphological study of *Pedinomonas major* Korshikov. Nova Hedwigia 16: 131–139. C

Belcher JH (1968b) The fine structure of *Furcilla stigmatophora* (Skuja) Korshikov. Archiv für Mikrobiologie 60: 84–94. B

Belcher JH (1969a) A morphological study of the phytoflagellate *Chrysococcus rufescens* Klebs in culture. British Phycological Journal 4: 105–117. C

Belcher JH (1969b) A re-examination of *Phaeaster pascheri* Scherffel in culture. British Phycological Journal 4: 191–197. B

Belcher JH (1969c) Some remarks upon *Mallomonas papillosa* Harris & Bradley and *M. calceolus* Bradley. Nova Hedwigia 18: 257–270. C

Belcher JH (1974) *Chrysosphaera magna* sp. nov., a new coccoid member of the Chrysophyceae. British Phycological Journal 9: 139–144. B

Belcher H (1993) An inexpensive and easily made battery-powered centrifuge. Bulletin of the Quekett Microscopical Club 21: 12–13. A

Belcher H (1996) A miniature trawl net for collecting attached organisms. Bulletin of the Quekett Microscopical Club 27: 13–14. A

Belcher H (1999) Cleaning small samples of diatoms. Bulletin of the Quekett Microscopical Club 35: 15–17. A

Belcher JH, Storey JE (1968) The phytoplankton of Rostherne and Mere Meres, Cheshire. The Naturalist, Hull 905: 57–61. C

Belcher JH, Swale EMF (1957) Some uncommon freshwater algae. British Phycological Bulletin 5: 40–42. B

Belcher JH, Swale EMF (1961) Some new and uncommon British Volvocales. British Phycological Bulletin 2: 56–62. B

Belcher JH, Swale EMF (1962) Culture studies on *Ankistrodesmus* and some similar genera. 1. Some less common and new British species. British Phycological Bulletin 2: 126–132. B

Belcher JH, Swale EMF (1963) Some new and uncommon British Volvocales. II. British Phycological Journal: 210–218. B

Belcher JH, Swale EMF (1967a) *Chromulina placentula* sp. nov. (Chrysophyceae), a freshwater nannoplankton flagellate. British Phycological Bulletin 3: 257–267. B

Belcher JH, Swale EMF (1967b) Observations on *Pteromonas tenuis* sp. nov. and *P. angulosa* (Carter) Lemmermann (Chlorophyceae, Volvocales) by light and electron microscopy. Nova Hedwigia 13: 353–359. B

Belcher JH, Swale EMF (1971) The microanatomy of *Phaeaster pascheri* Scherff. (Chrysophyceae). British Phycological Journal 6: 157–169. C

Belcher JH, Swale EMF (1972) Some features of the microanatomy of *Chrysococcus cordiformis* Naumann. British Phycological Journal 7: 53–59. B

Belcher JH, Swale E(MF) (1976) A Beginner's Guide to Freshwater Algae. Institute of Terrestrial Ecology/Natural Environment Research Council. HMSO, London. 48 pp. A

Belcher JH, Swale EMF (1977) Species of *Thalassiosira* (diatoms, Bacillariophyceae) in the plankton of English rivers. British Phycological Journal 12: 291–297. B

Belcher JH, Swale EMF (1978) *Skeletonema potamos* (Weber) Hasle and *Cyclotella atomus* Hustedt (Bacillariophyceae) in the plankton of rivers in England and France. British Phycological Journal 13: 177–182. B

Belcher JH, Swale EMF (1979) An Illustrated Guide to River Phytoplankton. HMSO, London. 64 pp. B

Belcher JH, Swale EMF (1981) Records of *Simonsnia delogei* and some interesting species of *Navicula* (diatoms) from English rivers, mainly near Cambridge. Microscopy. Journal of the Quekett Microscopical Club 34: 201–206. B

Belcher (J)H, Swale E(MF) (1982) Culture Algae. A Guide for Schools and Colleges. NERC. Institute of Terrestrial Ecology. 25 pp. A

Belcher JH, Swale EMF (1984) Unusual and surprising algae from a Cambridge roof. Microscopy. Journal of the Quekett Microscopical Club 35: 136–143. C

Belcher JH, Swale EMF (1991) *Thorea*, *Bangia*, and other freshwater red algae in Cambridgeshire. Nature in Cambridgeshire 33: 42–45. C

Belcher JH, Swale EMF (1996) Some uncommon algae from Cambridgeshire waters. Nature in Cambridgeshire 38: 71–75. B

Belcher JH, Swale EMF (1999) Vision Park Pool, Histon, and its remarkable phytoplankton. Nature in Cambridgeshire 41: 21–29. B

Belcher JH, Swale EMF (2000) Three uncommon algal flagellates from Cambridgeshire waters. Nature in Cambridgeshire 42: 80–83. B

Belcher JH, Swale EMF, Heron J (1966) Ecological and morphological observations on a population of *Cyclotella pseudostelligera* Hustedt. Journal of Ecology 54: 335–340. C

Bellinger EG (1977) Seasonal size changes in certain diatoms and their possible significance. British Phycological Journal 12: 233–239. C

Bellinger EG (1992) A Key to Common British Algae. 4th edn. The Institution of Water and Environmental Management. 138 pp. A

Bennet AW (1887) On the affinities and classification of algae. Journal of the Linnean Society 24: 49–60. A

Bennion H, Appleby P (1999) An assessment of recent environmental change in Llangorse Lake using palaeolimnology. Aquatic Conservation 9: 361–375. B

Benson-Evans K, Antoine R (1996) A Guide to the Freshwater, Brackish and Marine Algae of South Wales, UK. Antony Rowe, Chippenham. 387 pp. B

Benson-Evans K, Fisk D, Pickup G, Davies D (1967) The natural history of Slapton Ley Nature Reserve. 2. Preliminary studies on the freshwater algae. Field Studies 2: 407–434. C

Benson-Evans K, Williams PF, McLean RO, Prance N (1975) Algal communities in polluted rivers of South Wales. Verhandlungen der Internationalen Vereinigung für Theoretische und Angewandte Limnologie 19: 2010–2019. C

Berkeley MJ (1833): Gleanings of British algae; being an appendix to the supplement to English Botany. London. 50 pp., 20 pls. B

Bethge H (1935) *Chroococcus planctonicus*, eine neue planktonische Cyanophyceae. Berichte Deutsche Gesellschaft 53: 265–269. A

Billington CA, Jones AK (1976) Aspects of the spatial distribution of the algae of the Nant-y-Moch Reservoir, Ceredigion, Wales. Hydrobiologia 50: 43–54. C

Bindloss ME, Holden AV, Bailey-Watts AE, Smith IR (1973) Phytoplankton production, chemical and physical conditions in Loch Leven. In: Kajak Z, Hillbricht-Ilkowska A (eds) Productivity Problems of Freshwaters. Proceedings of the IBP-UNESCO Symposium on Productivity Problems of Freshwaters. Kazimierz Dolgy, Poland, May 6–12, 1970. Warszawa, Kraków: 639–659. C

Biodiversity (1994) Biodiversity The UK Action Plan. HMSO, London. 188 pp. A

Biodiversity (1995a) Biodiversity: the UK Steering Group Report. Volume 1: Meeting the Rio Challenge. HMSO, London. 103 pp. A

Biodiversity (1995b) Biodiversity: the Steering Group Report. Volume 2: Action Reports. HMSO, London. 324 pp. A

Biodiversity (1999) UK Biodiversity Group Tranche 2 Action Plans Volume III – Plants and Fungi. English Nature, Peterborough. 351 pp. AB

Birch SP, Kelly MG, Whitton BA (1989) Macrophytes of the River Wear: 1966, 1976, 1986. Transactions of the Botanical Society of Edinburgh 45(1988): 203–212. B

Bisset JP (1878) A new *Micrasterias* from Scotland. Proceedings of the Dublin Microscopic Club 18: 348. C

Bisset JP (1884) List of the Desmidieae found in gatherings made in the neighbourhood of Lake Windermere during 1883. Journal of the Royal Microscopical Society II 4: 173–176. C

Blackburn JB, Temperley BW (1936) *Botryococcus* and the algal coals. Transactions of the Royal Society of Edinburgh 58: 841–853. C

Blackburn SI, Tyler PA (1980) Conjugation, germination and meiosis in *Micrasterias mahabuleshwarensis* Hobson (Desmidiaceae). British Phycological Journal 15: 83–93. C

Bliding C (1963) A critical of European taxa of Ulvales, Part 1; *Capsosiphon, Percusaria, Blidingia, Enteromorpha*. Opera Botanica 8(3): 1–160. A

Blindow I, Krause W (1990). Bestamningsnyckel for svenska kransalger. Svensk Botanisk Tidskrift 84: 119–160. A

Blum JL (1953) The racemose Vaucheriaceae with inclined or pendent oogonia. Bulletin of the Torrey Botanical Club 80: 478–497. A

Blum JL (1972) Vaucheriaceae. North American Flora Series II, part 8: 1–64. A

Bolas PM, Lund JWG (1974) Some factors affecting the growth of *Cladophora glomerata* in the Kentish Stour. Proceedings of the Society for Water Treatment and Examination 23: 25–51. B

Bold H (1958) Three new Chlorophycean algae. American Journal of Botany 45: 737–743. A

Bold HC, Wynne MJ (1985) Introduction to the Algae. Structure and Reproduction. 2nd edn. Prentice-Hall, Englewood Cliffs, New Jersey. 720 pp. A

Bolton T (1886) Micro-organisms in a swampy ditch in Sutton Park. The Midland Naturalist 4: 173–176. C

Borge O (1907) Beitrage zur Algenflora von Schweden. Arkiv für die Botanik 6: 1–88 (1906). A

Bornet JBE, Flahault CGM (1886–8) Revision des Nostocacées hétérocystées contenues dans les principaux herbiers de France. Annales des Sciences Naturelles, Botanique sér. 7, 3: 323–381 (1886); 4: 343–373 (1886); 5: 51–129 (1887); 7: 177–262 (1888). (Nomenclatural rules require 1886 to be listed as year for the authority, whichever year the species was described.) Reprinted in 1956 by Weinheim/Bergstr. Chez H.R. Engelmann (J. Cramer). A

Bornet JBE,Thuret GA (1876) Notes Algologiques I. Paris. 198 pp. A

Bory de Saint Vincent JBM (1822) Dictionaire Classique d'Histoire Naturelle. 1. Paris. 199 pp. A

Bourrelly P (1937) Une nouvelle espèce de cyanophycée d'eau douce du genre *Desmosiphon*. Bulletin du Centre d'Études et de Recherches Scientifiques – Biarritz II(4): 589–593. A

Bourrelly P (1951) Notes sur quelques Chlorococcales. Bulletin Museum de Paris sér. 2, 23: 673–684. B

Bourrelly P (1966) Les Algues d'Eau Douce. Initiation à la Systématique. I. Les Algues Vertes. Boubée, Paris. 511 pp. A

Bourrelly P (1970) Les Algues d'Eau Douce III. Les Algues Bleues et Rouges; Les Eugléniens, Peridiniens et Cryptomonadines. Boubée, Paris. 512 pp. A

Bourrelly P (1972) Les Algues d'Eau Douce I. Les Algues Vertes. Boubée, Paris. 572 pp. A

Bourrelly P (1981) Les Algues d'Eau Douce II. Les Algues Jaunes et Braunes, Chrysophycées, Phaeophycées, Xanthophycées et Diatomées. Rev. Edn. Boubée, Paris. 438 pp. A

Bourrelly P (1985) Les Algues d'Eau Douce. III. Les Algues Bleues et Rouges, Les Eugléniens, Peridiniens et Cryptomonadines. Société Nouvelle des Éditions Boubée, Paris. 606 pp. A

Bourrelly P, Feldmann J (1946) Une algue meconnue: *Sphaeroplea soleirolii* (Duby) Montagne. Bulletin Museum Nationale Histoire de Naturelle Paris, sér. 2, 18: 412–415. A

Bowker DW, Denny P (1983) The spatial distribution of algae on shoots of emergent macrophytes in a reed swamp in the littoral zone of Lake Windermere. Nova Hedwigia 37: 389–401. C

Bowles B, Quennell S (1971) Some quantitative algal studies on the River Thames. Proceedings of the Society for Water Treatment and Examination 20: 35–51. C

Bradley DE (1965) Observations on the mastigonemes of two Chrysophyceae using negative staining. Quarterly Journal of Microscopical Science 106: 207–231. A

Bradley DE (1966a) The ultrastructure of the flagella of three chrysomonads with particular reference to the mastigonemes. Experimental Cell Research 41: 162–173. A

Bradley DE (1966b) Observations on some chrysomonads from Scotland. Journal of Protozoology 13: 143–154. B

Braun A (1855) Algarum Unicellularium Genera Nova et Minus Cognita . . . 2 edn, Lipsiae. 111 pp. A

Bristol BM (1917) On the life history and cytology of *Chlorochytrium grande* sp. nov. Annals of Botany 31: 107–126. B

Bristol BM (1919) On the life-history and cytology of *Chlorococcum humicola*. Journal of the Linnean Society, Botany 44: 92. C

Bristol BM (1920a) A review of the genus *Chlorochytrium* Cohn. Journal of the Linnean Society, Botany 45: 1–28. B

Bristol BM (1920b) On the alga-flora of some desiccated English soil. Annals of Botany 34: 35–80. C

Bristol BM (1921) Some aspects of the work of the late Professor G.S.West, M.A., D.Sc., F.L.S. Proceedings of the Birmingham Natural History and Philosophical Society 14: 139–146. A

Bristol Roach BM (1927) On the algae of some normal English soils. Journal of Agricultural Science 17: 563–588. A

Broady PA (1982) Green and yellow-green terrestrial algae from Surtsey (Iceland) in 1978. Surtsey Research Progress. Report 9: 13–32. A

Broady PA, Ingerfeld M (1999) *Ammatoidea normanii* (Cyanobacteria, Homoeotrichaceae) from La Gorce Mountains, Antarctica. Archiv für Hydrobiologie, Suppl. 130 [Algological Studies 95]: 1–13. B

Brook AJ (1954a) Notes on some uncommon algae from lochs in the Tummel–Garry catchment area. Transactions and Proceedings of the Botanical Society of Edinburgh 36: 207–214. C

Brook AJ (1954b) The bottom-living algal flora of slow sand filter beds of waterworks. Hydrobiologia 6: 333–351.C

Brook AJ (1955a) Notes on some uncommon algae from lochs in Kinross, Perthshire and Caithness. Transactions and Proceedings of the Botanical Society of Edinburgh 36: 309–316. B

Brook AJ (1955b) The attached algal flora of slow sand filter beds of waterworks. Hydrobiologia 7: 103–117. B

Brook AJ (1955c) The aquatic fauna as an ecological factor in studies of the occurrence of freshwater algae. Revue Algologique NS 3: 141–145. B

Brook AJ (1957a) Notes on freshwater algae, mainly from lochs in Perthshire and Sutherland. Transactions and Proceedings of the Botanical Society of Edinburgh 37: 114–122. B

Brook AJ (1957b) On some forms of *Micrasterias* new to or rare in Britain. Naturalist, Hull (1957): 37–39. C

Brook AJ (1957c) Notes on desmids of the genus *Staurastrum* I. Naturalist, Hull (1957): 93–100. C

Brook AJ (1958a) Desmids from the plankton of Irish loughs. Proceedings of the Royal Irish Academy 59B: 71–91. B

Brook AJ (1958b) Notes on algae from the plankton of some Scottish freshwater lochs. Transactions and Proceedings of the Botanical Society of Edinburgh 37: 174–181. B

Brook AJ (1958c) Changes in the phytoplankton of some Scottish hill lochs resulting from their artificial enrichment. Verhandlungen der Internationalen Vereinigung für Theoretische und Angewandte Limnologie 13: 298–305. C

Brook AJ (1958/59) *Staurastrum paradoxum* Meyen and *S. gracile* Ralfs in the British freshwater plankton, and a revision of the *S. anatinum* group of radiate desmids. Transactions of the Royal Society of Edinburgh 63: 589–628. C

Brook AJ (1959a) The published figures of *Staurastrum paradoxum* Meyen. Revue Algologique, NS 4: 239–255. C

Brook AJ (1959b) Notes on the desmids of the genus *Staurastrum*. III. *Staurastrum paradoxum* in the Jenner Herbarium of the British Museum. Naturalist, Hull 1959: 81–83. C

Brook AJ (1959c) The status of desmids in the plankton and the determination of phytoplankton quotients. Journal of Ecology 47: 429–445. B

Brook AJ (1959d) *Staurastrum pendulum* var. *pinguiforme* Croasdale, fo. *minor* West; fo. *major* fac. *quadrata* and *S. micron* var. *perpendiculatum* nov. comb., desmids new to the British freshwater plankton. Nova Hedwigia 9: 157–162. C

Brook AJ (1959e) De Brébisson's determinations of *Staurastrum paradoxum* Meyen and *S. gracilis* Ralfs. Nova Hedwigia 1: 163–165. C

Brook AJ (1960) The varieties of *Staurastrum paradoxum* Meyen – nomen dubium. Nova Hedwigia 1: 431–442. C

Brook AJ (1964) The vegetation of the Scottish freshwater lochs. In: Burnett JH (ed.) The Vegetation of Scotland. Oliver & Boyd, Edinburgh: 290–305. C

Brook AJ (1965) Planktonic algae as indicators of lake types, with special reference to the Desmidiaceae. Limnology and Oceanography 10: 403–411. C

Brook AJ (1975) Aquatic animals aren't hungry in winter, or why *Cymbella* blooms beneath the ice. Journal of Phycology 2: 235. C

Brook AJ (1981a) The Biology of Desmids. Botanical Monographs 16. 1–267. Blackwell, Oxford. A

Brook AJ (1981b) Population dynamics and water quality with special reference to the Desmidiaceae. In: Hoestlandt H (ed.) Dynamique de Populations et Qualité de l'Eau. Gauthier-Villars. C

Brook AJ (1981c) Calcium sulphate inclusions in the desmids *Bambusina* and *Gonatozygon*. British Phycological Journal 16: 267–272. C

Brook AJ (1982) Desmids of the *Staurastrum tetracerum* group from a eutrophic lake in mid-Wales. British Phycological Journal 17: 259–274. C

Brook AJ (1984) Comparative studies in a polyphyletic group, the Desmidiaceae – 30 years on. In: Irvine DEG, John DM (eds) Systematics of the Green Algae [Systematics Association Special Volume 27]. Academic Press, London: 251–269. A

Brook AJ (1985) On *Gonatozygon aculeaturn* var. *groenbladii* Růžička and G. *brebissonii* De Bary (Desmidiales). Microscopy. Journal of the Quekett Microscopical Club 35: 293–296. C

Brook AJ (1987) Algae with a taste for the unusual. New Scientist 115 (1577): 55–57. C

Brook AJ (1990) *Closterium sublaterale* Růžička, a rare desmid, newly recorded in Britain. British Phycological Journal 25: 291–294. C

Brook AJ (1992a) *Spirotaenia alpina* Schmidle (Zygnemaphyceae: Mesotaeniaceae), a saccoderm desmid new to the British Isles: revised description and observations on cell division. British Phycological Journal 27: 29–38. C

Brook AJ (1992b) *Cylindrocystis debaryi* Grönblad, a saccoderm desmid newly recorded for the British Isles; observations on its zygospore production, and of parthenospore production in *C. brebissonii* Menegh. Microscopy. Journal of the Quekett Microscopical Club 36: 703–708. C

Brook AJ (1992c) The desmid *Closterium pusillum* Hantzsch from two terraqueous habitats, with observations on asexual spore formation. British Phycological Journal 27: 409–416. C

Brook AJ (1994a) A new species of *Spirotaenia*, *S. cambrica* (family Mesotaeniaceae) of the subgenus Polytaeniae, from Mid-Wales. The Quekett Journal of Microscopy 37: 191–194. C

Brook AJ (1994b) *Cylindrocystis cushleckae*, a new species of saccoderm desmid (Family Mesotaeniaceae). The Quekett Journal of Microscopy 37: 225–231. C

Brook AJ (1994c) Algae. In: Maitland PS, Boon PJ, McLusky DS (eds) The Fresh Waters of Scotland: a National Resource of International Significance. John Wiley & Sons, Sussex: 131–146. C

Brook AJ (1995) *Closterium variabile* sp. nov., (Zygnemaphyceae), a desmid new to science from Bedfordshire, England; and comments on *C. sphaerosporum* (West) nov. comb. The Quekett Journal of Microscopy 37: 384–391. C

Brook AJ (1996) *Temnogametum sinensis* Jao & Hu (Charophyceae: Zygnematales), a genus and species of filamentous alga previously unrecorded from the British Isles or Europe. The Quekett Journal of Microscopy 37: 653–659. C

Brook AJ (1997) The proposed establishment of a new desmid genus *Polytaenia*, previously the sub-genus *Polytaenia* of the genus *Spirotaenia* and the description of a new species *P. luetkemuelleri*. The Quekett Journal of Microscopy 38: 7–14. B

Brook AJ (1998) *Tortitaenia* nom. nov. pro. *Polytaenia* Brook, a name of a genus of saccoderm desmids. The Quekett Journal of Microscopy 38: 146. B

Brook AJ (2001) The drought resistant desmid *Cosmarium pericymatium* Nordstedt, and a description of the new var. *corrugatum*. Journal of the Quekett Microscopical Club 39: 127–132. C

Brook AJ, Ells W (1987) Feeding of amoebae on desmids. Microscopy. Journal of the Quekett Microscopical Club 35: 537–540. C

Brook AJ, Fotheringham J, Bradly J, Jenkins A (1980) Barium accumulation by desmids of the genus *Closterium* (Zygnemaphyceae). British Phycological Journal 15: 261–264. C

Brook AJ, Grime GW, Watt F (1988) A study of barium accumulation in desmids using the Oxford scanning proton microprobe. Nuclear Instrumentation Methodology B30: 372–377. C

Brook AJ, Holden AV (1957) Fertilization experiments in Scottish freshwater lochs. I. Loch Kinardochy. Scottish Home Department Freshwater and Salmon Fisheries Research Series 17: 1–30. A

Brook AJ, Watt F, Grime GW (1990) A contribution to the study of biochemical cycling in an English lake by PIXE; the flux of barium and strontium in the phytoplankton. Nuclear Instruments and Methods in Physics Research B49: 481–484. A

Brook AJ, Williamson DB (1985) Needle-like inclusions in the terminal vacuoles of *Closterium lunula* (Müll.) Nitzsch. Microscopy. Journal of the Quekett Microscopical Club 35: 226–233. C

Brook AJ, Williamson DB (1988a) *Closterium arcus* sp. nov., a new British desmid. British Phycological Journal 23: 391–394. C

Brook AJ, Williamson DB (1988b) The survival of desmids on the drying mud of a small lake. In: Round FE (ed.) Algae and the Aquatic Environment. Contributions in Honour of J.W.G. Lund. Biopress, Bristol: 185–196. B

Brook AJ, Williamson DB (1990) *Actinotaenium habeebense* (Irénée Marie) nov. comb., a rare, drought resistant desmid. British Phycological Journal 25: 321–327. C

Brook AJ, Williamson DB (1991) A Check List of Desmids of the British Isles. Freshwater Biological Association UK. Occasional Paper 28. 40 pp. C

Brook AJ, Williamson DB (2002) A Monograph on British Desmids: Families Mesotaeniaceae, Peniaceae, Closteriaceae. Ray Society, London. in press. AB

Brook AJ, Williamson DB, Shoesmith E (1993) On the status of *Closterium abruptum* West and *C. nilssonii* Borge (Desmidiaceae: Zygnemaphyceae). Nova Hedwigia 57: 255–268. C

Brook AJ, Woodward WB (1965) Some observations on the effects of water inflow and outflow on the plankton of small lakes. Journal of Animal Ecology 25: 22–35. C

Bruinsma J, Krause W, Nat E, Van Raam J (1998) Determinatietabel van Kranswieren in der Benelux. (privately published) Utrecht. 101 pp. A

Brummitt RK, Powell CE (1992) Authors of Plant Names A List of Authors of Scientific Names of Plants, with Recommended Standard Forms of Their Names, Including Abbreviations. Royal Botanic Gardens, Kew. 732 pp. A

Bryant (formerly Moore) JA, Stace CA, Stewart NF (2002) A checklist of Characeae of the British Isles. Watsonia 24: in press.

Bullock-Webster GR (1898) Some new Characeae records. Journal of Botany 36: 182–184. C

Bullock-Webster GR (1901) Some new Characeae records. Journal of Botany 39: 101–102. C

Bullock-Webster GR (1902) Characeae from County Monaghan: The Irish Naturalist 11: 141–146. C

Bullock-Webster GR (1917) The Characeae of Fanad, E. Donegal. The Irish Naturalist 26: 1–5. C

Bullock-Webster GR (1918) The Characeae of the Rosses, W. Donegal. The Irish Naturalist 17: 7–10. C

Bullock-Webster GR (1919) A new *Nitella*. Journal of Botany 57: 1–2. C

Bullock-Webster GR (1920) Some charophyte notes. The Irish Naturalist 29: 55–58. C

Bullock-Webster GR (1922) Notes on charophytes. Journal of Botany 60: 148–149. C

Burger-Wiesma T, Stal LJ, Mur LC (1986) *Prochlorothrix hollandica* gen nov., sp. nov., a filamentous oxygenic photoautotrophic procaryote containing chlorophylls *a* and *b*. Assignment to *Prochlorotrichaceae* fam. nov. and order *Prochlorales* Florenzano, Balloni and Materassi 1986, with emendation of the ordinal description. International Journal of Systematic Bacteriology 39: 250–257. A

Burrows E (1991) Seaweeds of the British Isles 2 Chlorophyta. Natural History Museum Publications, London. 238 pp. B

Butcher RW (1924) The plankton of the River Wharfe (Yorkshire). Naturalist, Hull 175: 175–180, 211–214. C

Butcher RW (1931) An apparatus for studying the growth of epiphytic algae with special reference to the River Tees. Transactions of Northern Naturalists Union 1(1): 1–15. C

Butcher RW (1932a) Studies in the ecology of rivers II. The microflora of rivers with special reference to the algae on the river bed. Annals of Botany 46: 813–861. C

Butcher RW (1932b) Notes on new and little-known algae from the beds of rivers. New Phytologist 31: 289–309. B

Butcher RW (1933) Studies on the ecology of rivers. I. On the distribution of macrophytic vegetation in the rivers of Britain. Journal of Ecology 21: 58–91. C

Butcher RW (1940) Studies on the ecology of rivers. IV. Observations on the growth and distribution of the sessile algae in the River Hull, Yorkshire. Journal of Ecology 28: 210–233. B

Butcher RW (1942) The plankton of the River Wharfe. Naturalist, Hull (1942): 175–180, 211–214. C

Butcher RW (1946) Studies in the ecology of rivers. VI. The algal growth in certain highly calcareous streams. Journal of Ecology 33: 268–283. C

Butcher RW (1947) Studies on the ecology of rivers. VII. The algae of organically enriched waters. Journal of Ecology 35: 186–191. C

Butcher RW (1959) An introductory account of the smaller algae of the British coastal waters I. Fisheries Investigations. 74 pp. C

Butcher RW, Longwell J, Pentelow FTK (1937) Survey of the River Tees. Part III. – The Non-Tidal Reaches – Chemical and Biology. Technical Paper of Water Pollution Research No. 6. 189 pp. HMSO, London. B

Butcher RW, Pentelow FTK, Woodley JWA (1931) An Investigation of the River Lark and the Effect of Beet Sugar Pollution. MAFF Fisheries Investigations Series I, II(3). 112 pp. HMSO, London. C

Canter HM (1949a) Studies on British chytrids VI. Aquatic Synchitridaceae. Transactions of the British Mycological Society 32: 69–94. A

Canter HM (1949b) On *Apharomycopsis bacillariacearum* Scherffel, *A. desmidiella* n. sp., and *Ancylistes* sp. in Great Britain. Transactions of the British Mycological Society 32: 162–170. A

Canter HM (1949c) Studies on British chytrids VII. On *Phytochytrium mucronatum*. Transactions of the British Mycological Society 32: 236–240. A

Canter HM (1954) Fungal parasites of the phytoplankton III. Transactions of the British Mycological Society 37: 111–132. A

Canter HM, Lund JWG (1966) The periodicity of planktonic desmids in Windermere, England. Verhandlungen der Internationalen Vereinigung für Theoretische und Angewandte Limnologie 16: 163–172. C

Canter HM, Lund JGW (1968) The importance of protozoa in controlling the abundance of planktonic algae in lakes. Proceedings of the Linnean Society of London 179: 203–219. A

Canter HM, Lund JWG (1969) The parasitism of planktonic desmids by fungi. Österreiche Botanische Zeitschrift 116: 351–377. A

Canter-Lund H, Lund JWG (1995) Freshwater Algae. Their Microscopic World Explored. Biopress, Bristol. 360 pp. AB

Carter HJ (1859) On fecundation in *Eudorina elegans* and *Cryptoglena*. Annals and Magazine of Natural History ser. 3, 2: 237–253. A

Carter JR (1947) The genus *Gomphonema*. Microscope 6: 170–176. C

Carter JR (1960a) British freshwater forms of the genus *Gomphonema*. Microscope 12: 255–264. C

Carter JR (1960b) Minute forms of the genus *Navicula*. Microscope 12: 283–289. C

Carter JR (1961a, b, c) The genus *Achnanthes* as it occurs in British freshwaters. Microscope 12: 320–326; 13: 15–22, 37–45. C

Carter JR (1962a, b) Diatom notes: some unusual British forms. Microscope 13: 156–162, 231–237. C

Carter JR (1971a) Diatoms from the Devil's Hole Cave, Fife, Scotland. Nova Hedwigia 21: 657–681. C

Carter JR (1971b) Diatoms, the forgotten family. Vasculum 56: 66–69. C

Carter JR (1972a) Some observations on the diatom *Pinnularia acoricola*. Microscope 32: 162–165. C

Carter JR (1972b) The diatoms of Slapestone Sike, Upper Teesdale. Vasculum 57: 35–41. C

Carter JR, Bailey-Watts AE (1981) A taxonomic study of diatoms from standing freshwaters in Shetland. Nova Hedwigia 33: 513–629. B

Carter N (1919a) Studies on the chloroplasts of desmids. I. Annals of Botany 33: 215–254. C

Carter N (1919b) Studies on the chloroplasts of desmids. II. Annals of Botany 33: 295–304. C

Carter N (1919c) *Trachelomonas inconstans*, a new flagellate. New Phytologist 18: 118–119. C

Carter N (1919d) On the cytology of two species of *Characiopsis*. New Phytologist 18: 177–186. C

Carter N (1937a) *Pseudomallomonas anglica*: a new British flagellate. New Phytologist 36: 57–63. C

Carter N (1937b) New or interesting algae from brackish water. Archiv für Protistenkunde 90: 1–68. C

Castenholz RW (1989) Order Oscillatoriales (pp. 1771–1780). Order Nostocales (pp. 1780–1793). Order Stigonematales (pp. 1794–1799). In: Bergey's Manual of Systematic Bacteriology 3. Williams & Wilkins, Baltimore. A

Castenholz RW, Waterbury JB (1989) Group 1. Cyanobacteria. Preface. In: Bergey's Manual of Systematic Bacteriology 3. Williams & Wilkins, Baltimore: 1710–1727. A

Chadefaud M (1950) Observations cytologiques sur la Phéophycée d'eau douce: Heribaudiella fluviatilis (Aresch.) Sved. Bulletin de la Société Botanique de France 97: 198–199. A

Chapman DJ (1998) Enigmatic unicellular protista: are they really enigmatic? The algae case. In: Seckbach J (ed.) Enigmatic Microorganisms and Life in Extreme Environments. Kluwer, Dordrecht: 101–111. A

Chapman DV, Dodge JD, Heaney SI (1981) An electron microscope study of the excystment and early development of the dinoflagellate *Ceratium hirundinella*. British Phycological Journal 16:183–194. B

Chapman DV, Dodge JD, Heaney SI (1985) Seasonal and diel changes in ultrastructure in the dinoflagellate *Ceratium hirundinella*. Journal of Plankton Research 7: 263–278. B

Chapman PL (1966) Additions to the freshwater algae of the Northampton district. Journal of the Northampton Natural History Society 35: 481–482. C

Chodat R (1897) Étude de biologie lacustre. A. Recherches sur les algues pélagiques de quelques lacs suisses et français. Bulletin de l'Herbarie de Bossier, Genève 5: 289–320. A

Christensen T (1952) Studies on the genus *Vaucheria* I. A list of finds from Denmark and England with notes on some submarine species. Botanisk Tidskrift 49: 171–188. A

Christensen T (1956) Studies on the genus *Vaucheria* II. Remarks on some species from brackish water. Botaniska Notiser 109: 275–280. A

Christensen T (1962) Alger. In: Böcher TW, Lange M, Sørensen T (eds) Systematisk Botanik 2, 2. Munksgaard, Copenhagen: 1–178. A

Christensen T (1968) *Vaucheria* types in the Dillenian Herbaria. British Phycological Bulletin 3: 463–469. A

Christensen T (1969) *Vaucheria* collections from Vaucher's region. Biologiske Skrifter 16 (4): 1–36. A

Christensen T (1970) *Vaucheria prona*, a new name for a common alga. Botanisk Tidskrift 65: 245–251. A

Christensen T (1978) Annotations to a textbook of phycology. Botanisk Tidskrift 73: 65–70. A

Christensen T (1994) Algae. A Taxonomic Survey. Fasc. 2. 217–472 pp. AiO Print Ltd, Odense. A

Clements FE, Shantz HL (1909) A new genus of blue-green algae. Minnesota Botanical Studies 4: 133–135. A

Cleve-Euler A (1951–55). Die Diatomeen von Schweden und Finnland. Kongl. Svenska Vetenskapsakademiens Handlingar. Fjärde Serien 2, 1; 3, 3; 4, 1; 4, 5; 5, 4. B

Coesel PFM (1983) De Desmidiaceën van Nederland 2 Fam. Closteriaceae. Stichting Uitgeverij Koninklijke Nederlande Naturhistorische Verenigung. 49 pp. A

Coesel PFM (1988) Biosystematic studies on the *Closterium moniliferum/ehrenbergii* complex (Conjugatophyceae, Chlorophyta) in western Europe. II. Sexual compatibilty. Phycologia 27: 421–424. A

Coesel PFM (1989) Biosystematic studies on the *Closterium moniliferum/ehrenbergii* complex (Chlorophyta, Conjugatophyceae) in western Europe. IV. Distributional aspects. Cryptogamie Algologie 10: 133–141. A

Coesel PFM (1992) The *Staurastrum manfeldtii* complex (Chlorophyta, Desmidiaceae): morphological variability and taxonomic implications. Algological Studies 67: 69–83. A

Coesel PFM (1993) Poor physiological adaptation to alkaline culture conditions in *Closterium acutum* var. *variabile*, a planktonic desmid from eutrophic waters. European Journal of Physiology 28: 53–57. A

Coesel PFM (1994a) Die Desmidiaceën van Nederland 15 Fam. Desmidiaceae (3). Stichting Uitgeverij Koninklijke Nederlande Naturhistorische Verenigung. 52 pp. A

Coesel PFM (1994b) On the ecological significance of a cellular mucilaginous envelope in planktic desmids. Archiv für Hydrobiologie, Suppl. 103 [Algological Studies 73]: 65–74. AC

Coesel PFM (1998) Sieralgen en natuurwaarden: handleidung ter bepaling van natuurwaarden van stilstaande, zoete wateren, op basis van het desmidiaceenbestand (met een supplement op de Nederlandse sieralgenflora). Wetenschappelijke Mededeling KNNV 224: 1–56. A

Coesel PFM, Wardenaar K (1990) Growth responses of planktonic desmids in a light-temperature gradient. Freshwater Biology 23: 551–560. A

Comas A (1989) Taxonomische Übersicht der zönobialen Chlorokokkalalgen von Kuba. II. Fam. Coelastraceae. Archiv für Hydrobiologie, Suppl. 82 [Algological Studies 56]: 347–384. A

Comelles M (1985) Clave de identificacion de las especies de carofitas de la Peninsula Iberica. Asociación Española de Limnologia, Madrid. 35 pp. A

Compère P (1992) Charophytes. In: Flora Pratique des Algues d'Eau Douce de Belgique, 4. Jardin Botanique National de Belgique, Meise. A

Conrad W (1934) Materieaux pour une monographie due genre *Lepocinclis* Perty. Archiv für Protistenkunde 82: 203–249. A

Conrad W (1935) Étude systématique du genre *Lepocinclis* Perty. Mémoire du Musée Royale d'Histoire Naturelle de Belgique 2(1): 3–85. A

Cooke MC (1880) New *Cosmarium* from Trafalgar Square. Grevillea 9: 16. C

Cooke MC (1881) On some desmids new to Britain in 1880. Journal of the Quekett Microscopical Club 6: 203–211. C

Cooke MC (1882–84) British Fresh-Water Algae, Exclusive of Desmidiaceae and Diatomaceae. Williams & Norgate, London. 329 pp. B

Cooke MC (1885) Essex freshwater algae. Journal and Proceeding of the Essex Field Club 4: 47. C

Cooke MC (1887) British Desmids: a supplement to British Fresh-Water Algae. London. 205 pp. C

Cooke MC (1890) Introduction to Fresh-Water Algae, with an enumeration of all the British species. Kegan Paul, Trench, Trübner, London. 339 pp. B

Cooke MC, Wills AW (1881) Notes on British Desmids. Grevillea 1881: 89–92. B

Coppejans E (1995) Flore Algologique des Côtes du Nord de la France et de la Belgique Jardin Botanique National de Belgique, Meise. 454 pp. A

Corillion R (1975) Flore des Charophytes (Characées) du Massif Amoricain et des contrées voisines d'Europe occidentale. In: d'Abbayes H (ed.) Flore et Végétation du Massif Amoricain, 4: 11–214. Jouvé, Paris. A

Corliss JO (1994) An interim utilitarian (user-friendly) hierarchial-classification and characterization of the protists. Acta Protozoologica 33: 1–51. A

Couté A, Preisig H (1981) Sur l'ultrastructure de *Microglena butcheri* Belcher (Chrysophyceae, Ochromonadales, Synuraceae) et sur sa position systematique. Protistologica 17: 465–467. A

Cox EJ (1987) *Placoneis* Mereschkowsky: the re-evaluation of a diatom genus originally characterized by its chloroplast type. Diatom Research 2: 145–157. A

Cox EJ (1996) Identification of Freshwater Diatoms from Live Material. Chapman & Hall, London. 158 pp. A

Cox EJ (1997) Assessing and designating diatom taxa at or below species level – a consideration of current status and some suggested guidelines for the future. Nova Hedwigia 65: 13–26. A

Croasdale H (1973) Freshwater algae from Ellesmere Island, N.W.T. National Museum of Canada Publications in Botany 3: 1–131. A

Cronberg G, Weibull C (1981) *Cyanodictyon imperfectum*, a new chroococcal blue-green alga from Lake Trummen, Sweden. Archiv für Hydrobiologie, Suppl. 60 [Algological Studies 27]: 101–110. A

Czurda V (1932) Zygnemales. In: Die Süßwasser-Flora Deutschlands, Österreichs und der Schweiz 9. 2nd edn. Gustav Fischer, Jena: 1–232. A

Dakin WI, Latarche M (1913) The plankton of Lough Neagh: a study of the seasonal changes in the plankton by quantitative methods. Proceedings of the Royal Irish Academy 30B: 20–95. C

Deason TR, Silva PC, Watanabe S, Floyd, GL (1991) Taxonomic status of the species of the green algal genus *Neochloris*. Plant Systematics and Evolution 177: 213–219. A

Delf EM (1915) The algal vegetation of some pools on Hampstead Heath. New Phytologist 14: 63–80. C

Deflandre G (1930). *Strombomonas* nouveau genere d'euglénacées (*Trachelomonas* Ehr. *pro parte*). Archiv für Protistenkunde 69: 551–614. A

Denis M, Frémy P (1923–24) Une nouvelle Cyanophycées hétérocystée: *Anabaena viguieri*. Bulletin de la Société Linnéenne de Normandie 7 sér. 6: 122–125. A

De Reviers B, Rousseau F (1999) Towards a new classification of the brown algae. Progress in Phycological Research 13: 107–201. A

Desikachary TV (1959) Cyanophyta. Indian Council of Agricultural Research, New Delhi. 686 pp. A

De Toni GB (1889) Sylloge Algarum Omnium Husque Cognitarum 1. Sylloge Chlorophycearum..., 12 + CXXXIX + 1325 pp. Patavii. A

De Valera M, Ó Seaghdha T (1968) On the occurrence in Galway of *Physolinum monile* (De Wildem.) Printz *sensu* Khan. Irish Naturalist Journal 16: 14. B

De Winder B, Stal LJ, Mur LR (1990) *Crinalium epipsammum* sp. nov.: a filamentous cyanobacterium with trichomes composed of elliptical cells and containing poly-β-(1,4) glucan (cellulose). Journal of General Microbiology 136: 1645–1653. A

Dillard GE (1989) Freshwater algae of the southwestern Southern United States, Part 1. Chlorophyceae: Volvocales, Tetrasporales and Chlorococcales. Bibliotheca Phycologica 81: 1–202. A

Dillwyn LW (1809) British Confervae: or coloured figures and descriptions of the British Plants referred by botanists to the genus *Conferva*. W. Phillips, London. 87 pp. B

Dodge JD, Crawford RM (1970) The morphology and fine structure of *Ceratium hirundinella* (Dinophyceae). Journal of Phycology 6: 137–149. B

Donkin AS (1869) On several new and rare species of fresh-water diatomaceae discovered in Northumberland. Quarterly Journal of the Microscopical Society 9: 287–296. B

Dop AJ (1979) *Porterinema fluviatile* (Porter) Waern (Phaeophyceae) in the Netherlands. Acta Botanika Neerlandica 28: 449–458. A

Douglas B (1958) The ecology of the attached diatoms and other algae in a small stony stream. Journal of Ecology 46: 295–322. A

Dowling RC (1941) Dipping in Teesdale ponds. Darlington Field Club records (1941): 1–3. C

Drew KM (1935) The life-history of *Rhodochorton violaceum* (Kützing) nov. comb. (*Chantransia violacea* Kütz.). Annals of Botany 49: 439–450. B

Drew KM (1936) *Rhodochorton violaceum* (Kütz.) Drew, and *Chantransia* Murray and Barton. Annals of Botany 50: 419–426. B

Drew KM (1946) Anatomical observations on a new species of *Batrachospermum*. Annals of Botany, NS 10: 339–352. B

Droop MR (1955) Some new supra-littoral Protista. Journal of the Marine Biological Association of the United Kingdom 34: 233–245. C

Drouet F (1968) Revision of the Classification of the Oscillatoriaceae. Monographs of the Academy of Natural Sciences, Philadelphia 15. 334 pp. A

Drouet F (1981) Revision of the Stigonemataceae with a Summary of the Classification of the Blue-green Algae. Beihefte Nova Hedwigia 66: 1–331. A

Dunlap J R., Walne PL, Bentley J (1983). Microarchitecture and elemental spatial segregation of envelopes of *Trachelomonas lefevrei* (Euglenophyceae). Protoplasma 117: 97–106. A

Duthie HC (1964) The survival of desmids in ice. British Phycological Bulletin 2: 376–377. C

Duthie HC (1965a) A study of the distribution and periodicity of some algae in a bog pool. Journal of Ecology 53: 343–359. C

Duthie HC (1965b) Some observations on the algae of Llyn Ogwen, North Wales. Journal of Ecology 53: 361–370. C

Duthie HC (1965c) Some observations on the ecology of desmids. Journal of Ecology 53: 695–703. C

Eaton JW Carr NG (1980) Observations on the biology and mass occurrence of *Ophrydium* versatile (Müller) (Ciliophora: Peritrichia) and associated algae in Lough Ree, Ireland. The Irish Naturalists' Journal 20(2): 55–60. A

Edwards P (1975a) An assessment of possible pollution effects over a century on the benthic marine algae of Co. Durham. Botanical Journal of the Linnean Society 70: 269–305. C

Edwards P (1975b) Evidence for a relationship between the genera *Rosenvingiella* and *Prasiola* (Chlorophyta). British Phycological Journal 10: 291–297. B

Edwards RW, Benson-Evans K, Learner MA, Williams P, Williams R (1972) A biological survey of the River Taff. Journal of the Institute of Water Pollution Control 2: 1–24. C

Ehrenberg CG (1838) Die Infusionsthierschen als volkommene Organismen: ein Blick in das tiefere organische Leben der Natur. Leopold Voss, Leipzig. 547 pp. A

Elenkin AA (1938) Monographie Algarum Cyanophycearum Aquidulcium et Terrestrium Infinitibus URSS Inventarum, Vol. 1 Pars Specialis (Systematica) Fasc. I. Acad. Nauk, URSS, Moscow & Leningrad: 984 pp. A

Ells W (1992) Observations of some rare British Desmidiaceae. Microscopy. Journal of the Quekett Microscopical Club 36: 709–711. C

Eminson DF, Moss B (1980) The composition and ecology of periphyton communities in freshwaters. 1. The influence of host type and external environment on community composition. British Phycological Journal 15: 426–429. C

Entwisle TJ (1987) An evaluation of taxonomic characters in the subsection Sessiles, section Corniculatae, of *Vaucheria* (Vaucheriaceae, Chrysophyta). Phycologia 26: 297–321. A

Entwisle TJ (1998) Batrachospermaceae (Rhodophyta) in France: 200 years of study. Cryptogamie Algologie 19: 149–159. A

Ettl H (1960) Die Algenflora des Schönhengstes und seiner Umgebung. I. Nova Hedwigia 2: 509–544. A

Ettl H (1965) Beitrag zur Kenntnis der Morphologie der Gattung *Chlamydomonas* Ehrenberg. Archiv für Protistenkunde 108: 271–420. A

Ettl H (1976) Die Gattung *Chlamydomonas* Ehrenb. Beihefte Nova Hedwigia 49: 1–1122. A

Ettl H (1978) Xanthophyceae. In: Süßwasserflora von Mitteleuropa 3 (1). Gustav Fischer, Stuttgart: 1–530. A

Ettl H (1980) Beitrage zur Kenntnis der Süßwasseralgen Dänemarks. Botanisk Tidskrift 74: 179–223. A

Ettl H (1983) Chlorophyta I – Phytomonadina. In: Ettl H, Gerloff J, Heynig H, Mollenhauer D (eds) Süßwasserflora von Mitteleuropa 9. Gustav Fischer, Stuttgart: 1–807. A

Ettl H, Gärtner G (1995) Syllabus der Boden-, Luft- und Flechtenalgen, Gustav Fischer, Stuttgart. 721 pp. A

Evans GH (1970) Pollen and diatom analyses of late-Quaternary deposits in the Blelham Basin, north Lancashire. New Phytologist 69: 821–874. B

Evans JH (1958) The survival of freshwater algae during dry periods I. An investigation of five small ponds. Journal of Ecology 46: 149–168. B

Evans JH (1959) The survival of freshwater algae during dry periods. Part II. Drying experiments. Part III. Stratification of algae in pond margin litter and mud. Journal of Ecology 47: 55–70; 71–81. C

Evans JH (1960) Further investigations of the algae of pond margins. Hydrobiologia 15: 384–394. C

Evans JH (1964) *Pteromonas varians* Jane [? = *P. aequicilita* (Gicklh.) Bourr.]. British Phycological Bulletin 2: 317–321. B

Evans JH (1988) Long term changes in planktonic cryptophycean populations. In: Round FE (ed.) Algae and the Aquatic Environment. Biopress, Bristol: 44–52. C

Fensome RA, Taylor FJR, Norris G, Sargeant WAS, Wharton DI, Williams GL (1993) A Classification of Living and Fossil Dinoflagellates. American Museum of Natural History, Micropaleontology Special Publication 7. Micropaleontology Press, Hanover, MA. 351 pp. A

Fenton EW (1936) The periodicity and distribution of algae in Boghall Glen (Midlothian). Scottish Naturalist 1936: 143–148. C

Fenton EW (1938) Algae studies from Boghall Glen (Midlothian). IV. Soil Algae. The Scottish Naturalist 1938: 165–172. C

Field JH (1971) The algae of Warwickshire. A preliminary list of genera. Proceedings of the Birmingham Natural History and Philosophical Society 22: 179–183. C

Firth RI, Hartley B (1971) A Pennine diatom site. Microscopy 32: 108–113. C

Fitzjohn AE (1939–40) The algae of four Somersetshire pools. Proceedings of the Bristol Naturalists' Society Series 4, 9: 62–65. C

Flint EA (1950) An investigation of the distribution in time and space of the algae of a British water reservoir. Hydrobiologia 2: 217–239. B

Flint EA (1955) The occurrence of *Chlorosaccus fluidus* Luther. New Phytologist 54: 84–88. C

Florey JE, Hawley GRW (1994) A *Hydrodictyon reticulatum* bloom at Loe Pool, Cornwall, and its catchment. European Journal of Phycology 29: 17–20. C

Florin M-B (1957) Plankton of fresh and brackish waters in the Sodertalge area. Acta Phytogeographica Suecica 37: 1–114. A

Foged N (1977) Freshwater diatoms in Ireland. Bibliotheca Diatomologica 34: 1–222. B

Fogg GE (1986) Picoplankton. Proceedings of the Royal Society, London 228B: 1–30. AB

Förster K (1982) Das Phytoplankton des Süßwassers, Conjugatophyceae, Zygnematales und Desmidiales (excl. Zygnemataceae). In Thienemann A (ed.) Die Binnengewasser XVI, 8(1). E. Schweizerbart'sche, Stuttgart: 1–543. A

Foster PL (1979). Responses of Freshwater Algae to Heavy Metals. PhD Thesis, University of Cambridge. 261 pp.+37 pls. B

Fott B (1948) Taxonomical studies on Chlorococcales. II. Studia Botanica Cechoslovaca 9: 6–17. A

Fott B (1957) Taxonomie drobnohledné flory našich vod. Preslia 29: 278–319. A

Fott B (1967) *Chodatella* stages in *Scenedesmus.* Acta Universitatis Carolinae. Biologica, Praha 1967: 189–196. A

Fott B (1971) Taxonomische Übertragungen und Namensänderung enunter den Algen IV. Chlorophyceae und Euglenophyceae. Preslia 43: 289–303. A

Fott B (1974) Taxonomie der palmelloiden Chlorococcales (Familie Palmogloeaceae). Preslia 46: 1–31. A

Fott B, Nováková M (1969) A monograph of the genus *Chlorella.* The freshwater species. In: Fott B (ed.) Studies in Phycology. Schweizerbart'sche, Stuttgart: 10–74. A

Fourtanier E, Kociolek JP (1999) Catalogue of the diatom genera. Diatom Research 14: 1–190. B

Foy RH, Smith RV (1993) Physiological ecology of Lough Neagh phytoplankton. In: Wood RB, Smith RV (eds) Lough Neagh. The Ecology of a Multipurpose Water Resource. Kluwer, Dordrecht: 245–279. AB

Francke JA, Simons J (1985) Morphology and systematics of *Stigeoclonium* Kütz. (Chaetophorales). In: Irvine DEG, John DM (eds) Systematics of the Green Algae [Systematics Association Special Volume 27]. Academic Press, London: 363–377. A

Frémy P (1929–33) Cyanophycées des côtes d'Europe. Mémoires de la Societé Nationale des Sciences Naturelles et Mathematiques de Cherbourg 41: 1–233. A

Frémy P (1930) Les Myxophycées de l'Afrique Équatoriale Française. Archives de Botaniques III, Mémoire No. 2. A

Friedl T (1995) Inferring taxonomic positions and testing genus level assignments in coccoid green lichen algae: a phylogenetic analysis of 18S ribosomomal RNA sequences from *Dictyochloropsis reticulata* and from members of the genus *Myrmecia* (Chlorophyta, Trebouxiophyceae cl. nov.) Journal of Phycology 31: 632–639. A

Friedl T (1997) The evolution of the Green Algae. In: Bhattacharya D (ed.) Origins of Algae and their Plastids. Plant Systematics and Evolution, Suppl. 11: 87–101. A

Friedmann I (1955) *Geitlerea calcarea* n. gen. et n. sp. A new atmophytic lime-incrusting blue-green alga. Botanica Notiser 108: 439–445. B

Fritsch FE (1902a) Algological Notes I. Preliminary report on the phytoplankton of the Thames. Annals of Botany 16: 1–9. C

Fritsch FE (1902b) Observations on species of *Aphanochaete* Braun. Annals of Botany 16: 403–417. B

Fritsch FE (1902c) Algological notes III. Preliminary report on the phytoplankton of the Thames. Annals of Botany 16: 576–584. C

Fritsch FE (1903a) Further observations on the phytoplankton of the River Thames. Annals of Botany 17: 631–647. B

Fritsch FE (1903b) Algological notes IV. Remarks on the periodical development of the algae in the artificial waters of Kew. Annals of Botany 17: 274–278. C

Fritsch FE (1905) The plankton of some English rivers. Annals of Botany 19: 163–167. C

Fritsch FE (1906) Algae. Bulletin of Miscellaneous Information. The wild fauna and flora of the Royal Botanic Gardens. Kew Bulletin, Additional Series, V: 187–220. B

Fritsch FE (1913) Studies of the occurrence and reproduction of British freshwater algae in nature. Annales de Biologie Lacustre 6: 1–83. C

Fritsch FE (1914) Notes on British flagellates. I–IV. New Phytologist 13: 341–352. B

Fritsch FE (1929a) The encrusting algal communities of certain fast-flowing streams. New Phytologist 28: 165–196. B

Fritsch FE (1929b) The genus *Sphaeroplea.* Annals of Botany 43: 1–26. B

Fritsch FE (1933) Contribution to our knowledge of British Algae V. Journal of Botany 71: 187–196. C

Fritsch FE (1935) The Structure and Reproduction of the Algae I. Cambridge University Press, London. 791 pp. A

Fritsch FE (1942) *Chrooderma,* a new genus of subaerial algae. Annals of Botany NS 6(24): 565–575. B

Fritsch FE (1945) The Structure and Reproduction of the Algae II. Cambridge University Press, Cambridge. 939 pp. A

Fritsch FE (1949a) The lime-encrusted *Phormidium* community of British streams. Verhandlungen der Internationalen Vereinigung für Theoretische und Angewandte Limnologie 10: 141–144. C

Fritsch FE (1949b) Contributions to our knowledge of British Algae. Hydrobiologia 1: 115–125. B

Fritsch FE (1952) Comparative studies in a polyphyletic group, the Desmidiaceae. Proceedings of the Linnean Society, London 164: 258–280. C

Fritsch FE, John RP (1942) An ecological and taxonomic study of the algae of British soils. II. Consideration of the species observed. Annals of Botany NS 6: 371–395. B

Fritsch FE, Pantin CFA (1946) Calcareous concretions in a Cambridgeshire stream. Nature, London 157: 397. C

Fritsch FE, Rich F (1907) Studies on the occurrence and reproduction of British freshwater algae in nature. 1. Preliminary observations on *Spirogyra.* Annals of Botany 21: 423–436. B

Fritsch FE, Rich F (1909) A five years' observation of the fishpond, Abbot's Leigh, near Bristol. Proceedings of the Bristol Natural History Society series 4, 2: 27–54. C

Fritsch FE, Rich F (1913) Studies in the occurrence and reproduction of British freshwater algae in nature. III. A four years' investigation of a freshwater pond. Annales de Biologie Lacustre 6: 1–83. B

Fritsch FE, Rich F (1924) British freshwater algae in nature. 3. A four years' observation of a freshwater pond. An ecological and taxonomic study of the algae of British soils. II. Annales de Biologie Lacustre 6: 33–115. C

Fuhs GW (1973) Cytochemical examination of blue-green algae. In: Carr NG, Whitton BA (eds) The Biology of Blue-Green Algae. Blackwell, Oxford: 117–143. A

Furet JE (1979) Algal Studies on the River Wye System. Ph.D. Thesis, University of Wales, Cardiff. B

Galliford AL, Williams EG (1948) Microscopic organisms of some brackish pools at Leasowe, Wirral, Cheshire. The North Western Naturalist 23: 39–62. C

Gantt E, Scott J, Lipschultz C (1986) Phycobiliprotein composition and chloroplast structure in the freshwater red alga *Compsopogon coeruleus* (Rhodophyta). Journal of Phycology 22: 480–484. A

Garbary DJ (1987) The Achrochaetiaceae (Rhodophyta): An annotated bibliography. Bibliotheca Phycologica 77: 1–267. A

Garbary DJ, Hansen GI, Scagel RF (1980) A revised classification of the Bangiophyceae (Rhodophyta). Nova Hedwigia 33: 145–166. A

Geitler L (1922) Die Mikrophyten-Biocoenose der *Fontinalis*-Bestände des Lunzer Untersees und ihre Abhängigkeit vom Lichts. Internationale Revue der Hydrobiologie 10: 683–691. A

Geitler L (1925) Cyanophyceae. In: Pascher A (ed.) Die Süsswasser-Flora Deutschlands, Österreichs und der Schweiz 12. Gustav Fischer, Jena: 1–450. A

Geitler L (1928) Über die Tiefenflora an Felsen im Lunzer Untersee. Archiv für Protistenkunde 62: 96. A

Geitler L (1932a) Cyanophyceae. In: Rabenhorst's Kryptogamen-Flora von Deutschland, Österreich und der Schweiz 14. Akademische Verlagsgesellschaft, Leipzig: 1–1356. A

Geitler L (1932b) *Porphyridium sordidum* n. sp., eine neue Süßwasserbangiale. Archiv für Protistenkunde 76: 595–604. A

Geitler L (1965) Die Gattung *Podohedra* (Chlorophyceae, Chlorococcales). Österreiche Botanische Zeitschrift 112: 173–183. A

Germain H (1981) Flore des Diatomées Diatomophycées Eaux Douce et Saumâtres. Boubée, Paris. 449pp. A

Gerrath JF (1993) The biology of desmids: a decade of progress. In: Round FE, Chapman DJ (eds) Progress in Phycological Research 9. Biopress, Bristol: 79–192. A

Gibson CE (1993) The phytoplankton population of Lough Neagh. In: Wood RB, Smith RV (eds) Lough Neagh. The Ecology of a Multipurpose Water Resource. Kluwer, Dordrecht: 203–223. AB

Gibson CE, Foy RH, Fitzsimons AG (1980) A limnological reconnaissance of the Loch Erne system, Ireland. Internationale Revue der Gesamten Hydrobiologie und Hydrographie 65: 49–84. C

Gibson CE, Wood RB, Dickson EL, Jewson DH (1971) The succession of phytoplankton in Lough Neagh 1968–70. Mitteilungen Internationale Vereinigung für Theoretische und Angewandte Limnologie 19: 146–160. C

Gibson MT, Whitton BA (1987a) Hairs, phosphatase activity and environmental chemistry in *Stigeoclonium, Chaetophora* and *Draparnaldia* (Chaetophorales). British Phycological Journal 22: 11–22. B

Gibson MT, Whitton BA (1987b) Influence of phosphorus on morphology and physiology of freshwater *Chaetophora, Draparnaldia* and *Stigeoclonium* (Chaetophorales, Chlorophyta). Phycologia 26: 59–69. B

Gleave HH (1972) *Hydrosera triquetra*, a diatom new to European waters. Microscopy. Journal of the Quekett Microscopical Club 32: 208. C

Godward MB (1933) Contributions to our knowledge of British Algae. Journal of Botany 71: 33–44 B

Godward M (1937) An ecological and taxonomic investigation of the littoral flora of Lake Windermere. Journal of Ecology 25: 496–568. B

Godward MBE (1950) On the nucleolus and nucleolar-organising chromosomes of *Spirogyra*. Annals of Botany NS 14: 40–53. B

Godward MBE (1953) Geitler's nucleolar substance in *Spirogyra*. Annals of Botany NS 17: 403–416. B

Godward MBE (1956) Cytotaxonomy of *Spirogyra* 1, *S. submargaritata, S. subechinata* and *S. britannica*. Journal of the Linnean Society, Botany 55: 532–546. B

Godward MBE (1966) The Chromosomes of the Algae. Edward Arnold, London. 212 pp. A

Godward MBE (1974) An investigation of the causal distribution of algal epiphytes. Beihefte zum Botanisches Zentralblatt A 52: 506–539. C

Gomont M [A] (1892–93) Monographie des Oscillariées (Nostocacées Homocystées). Annales Science de Naturelle sér. 7, Botanique 15: 263–368 [Bourrelly, 1972, lists this work as 1893, and gives pages 1–107 and 111–302, respectively; however, nomenclatural rules require 1892 to be listed as year for the authority, whichever year the species was described]. Reprinted in 1972, with Introduction by P. Bourrelly, by J. Cramer and Wheldon & Wesley, Ltd, Codicote, Herts, UK, and New York. A

Gomont M [A] (1895) Note sur un *Calothrix* sporifère (*Calothrix stagnalis* sp. n.). Journal of Botany 9: 1–6. A

Gonzalves EA (1981) Oedogoniales. Indian Council for Agricultural Research, New Dehli. 757 pp. A

Goodwin KM (1926) Some observations on *Batrachospermum moniliforme*. New Phytologist 25: 51–54. C

Goodwin KM (1974). Carotenoids and biliproteins. In: Stewart WDP (ed.) Algal Physiology and Biochemistry. Blackwell, Oxford: 176–205. A

Graham LE, Wilcox LW (2000) Algae. Prentice-Hall, Upper Saddle River, New Jersey. 640 pp. A

Greville RK (1824) Flora Edinensis: with a concise introduction to the Natural Orders of the class Cryptogamia. Blackwood, Edinburgh. 478pp. AB

Griffiths BM (1909) On two new members of the Volvocaceae. New Phytologist 8: 130–137. B

Griffiths BM (1912) The algae of Stanklin Pool, Worcestershire; an account of their distribution and periodicity. Proceedings of the Birmingham Natural History and Philosophical Society 12: 1–13. C

Griffiths BM (1915) On *Glaucocystis nostochinearum* Itzigsohn. Annals of Botany 29: 423–432. C

Griffiths BM (1916) The August heleoplankton of some North Worcestershire pools. Journal of the Linnean Society, Botany 43: 423–432. C

Griffiths BM (1922) The heleoplankton of three Berkshire pools. Journal of the Linnean Society, Botany 46: 1–11. C

Griffiths BM (1924) The free-floating microflora or phytoplankton of Hornsea Mere, E. Yorkshire. Naturalist, Hull (1924): 245–247. C

Griffiths BM (1925a) Studies in the phytoplankton of the lowland waters of Great Britain. III. The phytoplankton of Shropshire, Cheshire and Staffordshire. Journal of the Linnean Society, Botany 47: 75–98. C

Griffiths BM (1925b) The phytoplankton of the Wicken Fen Area. In: Gardner JS, Tansley AG (eds) The Natural History of Wicken Fen, Pt. 2: 116–121. C

Griffiths BM (1926) Studies in the phytoplankton of the lowland waters of Great Britain. No. IV. The Phytoplankton of the Isle of Anglesey and of Llyn Ogwen, North Wales. Journal of the Linnean Society, Botany 47: 355–366. C

Griffiths BM (1927) Studies in the phytoplankton of the lowland waters of Great Britain. No. V. The phytoplankton of some Norfolk Broads. Journal of the Linnean Society, Botany 47: 595–612. B

Griffiths BM (1929) On desmid plankton. New Phytologist 27: 98–107. C

Griffiths BM (1936a) The limnology of the Long Pool, Butterby Marsh, Durham: an account of the temperature, oxygen-content, and composition of the water, and of the periodicity and distribution of the phyto- and zooplankton. Journal of the Linnean Society, Botany 50: 393–416. C

Griffiths BM (1936b) A preliminary list of the freshwater algae of Northumberland and Durham. Vasculum 22: 89–95 .C

Griffiths BM (1939a) The free-floating microscopic plant life of the lakes of the Isle of Raasay, Inner Hebrides. Proceedings of the University of Durham Philosophical Society 10: 71–87. C

Griffiths BM (1939b) Early references to waterbloom in British Lakes. Proceedings of the Linnean Society, Session 151 (Pt 1) 3: 12–19. B

Griggs RF (1912) The development and cytology of *Rhodochytrium*. The Botanical Gazette 53: 127–173. A

Grochowski JP, Trainor FR (1987) Interspecific competition between *Ankistrodesmus falcatus* (Chlorophyceae) and *Chlorella regularis* (Chlorophyceae) in sewage effluent: substrate-related pH changes. Phycologia 26: 270–276. A

Grönblad R (1962) Desmids from the British Isles. Nova Hedwigia 4: 467–479. C

Grove AJ, Bristol BM, Carter N (1920) The flagellates and algae of the district around Birmingham, compiled from records left by the late Prof. G.S.West. Journal of Botany London 58, Suppl. 3: 1–55. B

Grove WB (1915) *Pleodorina illinoisensis* Kofoid in Britain. New Phytologist 14: 169–182. B

Groves H, Groves J (1878a) A review of the British Characeae. Journal of Botany 18: 97–103, 129–135, 161–167, 207–210. A

Groves H, Groves J (1878b) "*Chara connivens*, 'Salzm.,' A.Braun". Journal of Botany 16: 120. C

Groves H, Groves J (1881a) On *Chara obtusa*, Desv., a species new to Britain. Journal of Botany 19: 216. C

Groves H, Groves J (1881b) Notes on British Characeae. Journal of Botany 19: 353–356. C

Groves H, Groves J (1883) Notes on British Characeae. Journal of Botany 21: 20–23. C

Groves H, Groves J (1884) Notes on British Characeae. Journal of Botany 22: 1–5. C

Groves H, Groves J (1884–85) The Characeae of West Cornwall. Transactions of the Penzance Natural History and Antiquarian Society 1884–1885: 185. C

Groves H, Groves J (1885a) Notes on British Characeae. Journal of Botany 23: 81–83. C

Groves H, Groves J (1885b) *Nitella capitata* Ag. in Cambridgeshire. Journal of Botany 23: 185. C

Groves H, Groves J (1886) Notes on the British Characeae for 1885. Journal of Botany 24: 1–4. C

Groves H, Groves J (1887) Notes on British Characeae. Journal of Botany 25: 146. C

Groves H, Groves J (1890) Notes on British Characeae. Journal of Botany 28: 65–69. C

Groves H, Groves J (1893) Notes on Irish Characeae. The Irish Naturalist 2: 163–164. C

Groves H, Groves J (1895a) The distribution of the Characeae in Ireland. The Irish Naturalist 4: 7–11, 37–41. C

Groves H, Groves J (1895b) Notes on British Characeae. Journal of Botany 33: 289–292. C

Groves H, Groves J (1898) Notes on British Characeae. Journal of Botany 36: 409–413. C

Groves J (1915) A new *Nitella*. Journal of Botany 53: 41–42 . C

Groves J (1916) On the name *Lamprothamnus* Braun. Journal of Botany 54: 336–337. C

Groves J (1919) Notes on *Lychnothamnus* Braun. Journal of Botany 57: 125–129. C

Groves J, Bullock-Webster GR (1917a) *Nitella mucronata* in Gloucester. Journal of Botany 55: 323–324. C

Groves J, Bullock-Webster GR (1917b) *Tolypella nidifica* Leonh. General Irish Natural History 26(8): 134–135. C

Groves J, Bullock-Webster GR (1920) The British Charophyta. I. Nitelleae. The Ray Society, London. 141 pp. AB

Groves J, Bullock-Webster GR (1924) The British Charophyta. II. Chareae. The Ray Society, London. 129 pp. AB

Hall CA (1920) Desmids of the Clyde area, new records. Glasgow Naturalist 8: 122–127. C

Hambrook JA, Sheath RG (1991) Reproductive ecology of the freshwater red alga *Batrachospermum boryanum* Sirodot in a temperate headwater stream. Hydrobiologia 218: 233–246. A

Hajdu L, Hegewald E, Cronberg G (1976) Beiträge zur Taxonomie der Gattung *Coelastrum* (Chlorophyta, Chlorococcales). Annales Histoire-Naturales Musei Nationalis Hungarici, Budapest 68: 31–38. A

Hammerton D (1959) A biological and chemical study of Chew Valley Lake. Proceedings of the Society of Water Treatment and Examination. Journal of the Linnean Society, Botany 8: 87–132. C

Hansgirg A (1892) Prodromus der Algenflora von Böhmen. 2. Arch. Naturwiss. – Landesdurchforsch. Böhmen 8(4): 1–268. A

Happey C, Moss B (1967) Some aspects of the biology of *Chrysococcus diaphanus* in Abbot's Pond, Somerset. British Phycological Bulletin 3: 269–279. B

Happey-Wood CM (1976) The occurrence and relative importance of nanno-Chlorophyta in freshwater algal communities. Journal of Ecology 64: 279–292. B

Happey-Wood CM (1978) The application of culture methods in studies of the ecology of small green algae. Mitteilungen Internationale Vereinigung fur Theoretische und Angewandte Limnologie 21: 385–397. B

Happey-Wood CM (1980) Periodicity of epipelic unicellular Volvocales (Chlorophyceae) in a shallow acid pool. Journal of Phycology 16: 116–128. B

Happey-Wood CM (1988) Vertical migration patterns of flagellates in a community of freshwater benthic algae. Developmental Hydrobiology 45: 99–123. A

Happey-Wood CM, Kennaway GMA, Ong MH, Chittenden AM, Edwards G (1988) Contributions of nano- and pico-plankton to the productivity of phytoplankton and epipelic algae in an upland Welsh Lake. In: Round FE (ed.) Algae and the Aquatic Environment. Biopress, Bristol: 168–184. B

Happey-Wood CM, Priddle J (1984) The ecology of epipelic algae of five Welsh lakes, with special reference to Volvocalean green flagellates (Chlorophyceae). Journal of Phycology 20: 109–124. B

Harding JPC (1996) Use of algae for monitoring rivers in the United Kingdom: recent developments. In Whitton BA, Rott E (eds) Use of Algae for Monitoring Rivers II. E. Rott, Innsbruck, Austria: 125–132. AB

Harding JPC, Hawley GRW (1991) Use of algae for monitoring rivers in the United Kingdom. In: Whitton BA, Rott E, Friedrich G (eds) Use of Algae for Monitoring Rivers I. Institut für Botanik, Universität Innsbruck, Austria: 183–193. AB

Harding JPC, Kelly MG (1999) Recent developments in algal based monitoring in the United Kingdom. In Prygiel J, Whitton BA, Bukowska J (eds) Use of Algae for Monitoring Rivers III. Agence de l'Eau Artois-Picardie, Douai: 26–34. A

Harding JPC, Whitton BA (1976) Resistance to zinc of *Stigeoclonium tenue* in the field and the laboratory. British Phycological Journal 11: 417–426. B

Harding JPC, Whitton BA (1981) Accumulation of zinc, cadmium and lead by field populations of *Lemanea*. Water Research 15: 301–309. C

Hargreaves JW, Lloyd EJH, Whitton BA (1975) Chemistry and vegetation of highly acidic streams. Freshwater Biology 5: 563–576. B

Hargreaves JW, Whitton BA (1976) Effect of pH on growth of acid stream algae. British Phycological Journal 11: 215–223. A

Harris GT (1917) The desmid flora of Dartmoor. Microscopy. Journal of the Quekett Microscopical Club 13: 247–276. C

Harris GT (1920) The desmid flora of a Triassic district. Microscopy. Journal of the Quekett Microscopical Club 14: 137–162. C

Harris GT (1930) The freshwater Bacillariales of Devonshire. Transactions of the Devon Association for the Advancement of Science 62: 285–310. B

Harris GT (1933) The Oedogoniales of Devonshire. Transactions of the Devonshire Association for the Advancement of Science, Literature and Arts 65: 213–226. B

Harris K (1953) A contribution to our knowledge of *Mallomonas*. Journal of the Linnean Society of London, Botany 55(356): 88–102. B

Harris K (1958) A study of *Mallomonas insignis* and *Mallomonas akrokomos*. Journal of General Microbiology 19: 55–64. B

Harris K (1963) Observations on *Sphaleromantis tetragona*. Journal of General Microbiology. 33: 345–348. B

Harris K (1966) The genus *Mallomonopsis*. Journal of General Microbiology 42: 175–184.C

Harris K (1967) Variability in *Mallomonas*. Journal of General Microbiology 46: 185–191.C

Harris K (1970a) Imperfect forms and taxonomy of *Mallomonas*. Journal of General Microbiology 61: 63–76. C

Harris K (1970b) Species of the Torquata group of *Mallomonas*. Journal of General Microbiology 61: 77–80. C

Harris K, Bradley DE (1956) Electron microscopy of *Synura* scales. Discovery 17: 229–332. B

Harris K, Bradley DE (1957) An examination of the scales and bristles of *Mallomonas* in the electron microscope using carbon replicas. Journal of the Royal Microscopical Society 3,76: 37–46. B

Harris K, Bradley DE (1958) Some unusual Chrysophyceae studied in the electron microscope. Journal of General Microbiology 18: 71–83. B

Harris K, Bradley DE (1960) A taxonomic study of *Mallomonas*. Journal of General Microbiology 22: 750–777. B

Harris TM (1940) A contribution to the knowledge of the British freshwater dinoflagellata. Proceedings of the Linnean Society, London 152: 4–33. B

Hartley B (1986) A check-list of the freshwater, brackish and marine diatoms of the British Isles and adjoining coastal waters. Journal of the Marine Biological Association of the United Kingdom 66: 531–610. B

Hartley B (1994) John R. Carter (1908–1993). Diatom Research 9: 235–236. B

Hartley B, Barber HG, Carter JR (1996) An Atlas of British Diatoms [Sims PA ed.]. Biopress, Bristol. 601pp. B

Harvey WH (1841) Manual of British Algae. Van Voorst, London. 229 pp. B

Hassall AH (1842) Observations on the genera *Zygnema*, *Tyndaridea*, and *Mougeotia* with descriptions of new species. Annals and Magazine of Natural History 10 series 1: 34–47. B

Hassall AH (1843) Descriptions of British freshwater Confervae, mostly new, with observations on some of the genera. Annals and Magazine of Natural History 11 series 1: 428–437. C

Hassall AH (1845) A History of the British Freshwater Algae I (Text). Highley, London. 462 pp; II. 103 pls. B

Hassall AH (1852) A History of the British Freshwater Algae, including descriptions of the Desmidiacea and Diatomacea. I. (Text). Taylor, Walton & Maberly, London. 462 pp. C

Hawley GRW, Whitton BA (1991) Seasonal changes in chlorophyll-containing picoplankton populations of ten lakes in northern England. Internationale Revue der Gesamten Hydrobiologie 76: 545–554. B

Haworth EY (1969) The diatoms of a sediment core from Blea Tarn, Langdale. Journal of Ecology 57: 429–439. A

Haworth EY (1972) The recent diatom history of Loch Leven, Kinross. Freshwater Biology 2: 131–141. A

Haworth EY (1974) Some problems of diatom taxonomy in Scottish lake sediments. British Phycological Journal 9: 47–55. C

Haworth EY (1975) A scanning electron microscope study of some different frustule forms of the genus *Fragilaria* found in Scottish late-glacial sediments. British Phycological Journal 10: 73–80. C

Haworth EY (1976) Two late-glacial (Late-Devensian) diatom assemblage profiles from northern Scotland. New Phytologist 77: 227–256. B

Haworth EY (1984) Stratigraphic changes in algal remains (diatom and chrysophytes) in the recent sediments of Blelham Tarn, English Lake District. In: Haworth EY, Lund JWG (eds) Lake Sediments and Environmental History. Leicester University Press: 165–190. A

Haworth EY (1985) The highly nervous system of the English Lakes; aquatic ecosystem sensitivity to external changes, as demonstrated by diatoms. Report of the Freshwater Biological Association 53: 60–79. A

Haworth EY (1988) Distribution of diatom taxa of the old genus *Melosira* (now mainly *Aulacoseira*) in Cumbrian waters. In: Round FE (ed.) Algae and the Aquatic Environment. Biopress, Bristol: 138–167. B

Haworth EY, Atkinson K.M, Carrick TR (1988) The preliminary assessment of diatom distribution in the waterbodies of Cumbria, NW England. In: Proceedings of the 10th International Diatom Symposium (Jöensuu, 1988): 459–470. B

Hayward J (1957) The periodicity of diatoms in bogs. Journal of Ecology 45: 947–954. C

Heaney SI, Chapman DV, Morison HR (1983) The role of the cyst stage in the seasonal growth of the dinoflagellate *Ceratium hirundinella* within a small productive lake. British Phycological Journal 18: 47–59. B

Heaney SI, Lund JWG, Canter HM, Gray K (1988) Population dynamics of *Ceratium* spp. In three English lakes. Hydrobiologia 161: 133–148. B

Heering W (1921) Siphonales. In: Pascher A (ed.) Süßwasserflora Deutschlands, Österreichs und der Schweiz 7. Gustav Fischer, Jena: 69–99. A

Hegewald E (1977) *Scenedesmus communis* Hegewald, a new species and its relation to *Scenedesmus quadricauda* (Turp.) Bréb. Archiv für Hydrobiologie, Suppl. 51 [Algological Studies 19]: 142–155. A

Hegewald E (2000) New combinations in the genus *Desmodesmus* (Chlorophyceae, Scenedesmaceae). Archiv für Hydrobiologie, Suppl. 131 [Algological Studies 96]: 1–18. A

Hegewald E, Aldave A, Schnepf E (1980) Investigations on the lakes of Peru and their phytoplankton. 5. The algae of Laguna Piuray and Laguna Huaypo, Cuzco, with special reference to *Franceia*, *Oocystis* and *Scenedesmus*. Archiv für Hydrobiologie, Suppl. 56 [Algological Studies 25]: 387–420. A

Hegewald E, An AA, Schnepf E, Tsarenko P (1998) Taxonomy and cell wall ultrastructure of *Scenedesmus lunatus* (Chlorophyta, Chlorococcales). Archiv für Hydrobiologie, Suppl. 126 [Algological Studies 91]: 11–25. A

Hegewald E, An SS, Tsarenko P (1998) Revision of *Scenedesmus intermedius* Chod. (Chlorophyta, Chlorococcales). Archiv für Hydrobiologie. Suppl. 126 [Algological Studies 88]: 67–104. A

Hegewald E, Deason TR (1996) *Pseudodidymocystis*, a new genus of Scenedesmaceae (Chlorophyceae). Archiv für Hydrobiologie 82: 119–127. A

Hegewald E, Engelberg K, Paschma R (1988) Beitrag zur Taxonomie der Gattung *Scenedesmus* Subgenus *Scenedesmus* (*Chlorophyceae*). Nova Hedwigia 47: 497–533. A

Hegewald E, Hanagata N (2000) Phylogenetic studies on Scenedesmaceae (Chlorophyta). Archiv für Hydrobiologie, Suppl. 136 [Algological Studies 100]: 29–49. A

Hegewald E, Hindäk F, Schnepf E (1990) Studies on the genus *Scenedesmus* Meyen (Chlorophyceae, Chlorococcales) from southern India, with special reference to the cell wall ultrastructure. Beihefte Nova Hedwigia 99: 1–75. A

Hegewald E, Krienitz L, Schnepf E (1994) Studies on *Scenedesmus costato-granulatus* Skuja. Nova Hedwigia 59: 97–127. A

Hegewald E, Schnepf E (1974a) Beitrage zur Kenntnis der Grünalgenart *Scenedesmus verrucosus* Roll. Archiv für Hydrobiologie, Suppl. 46 [Algological Studies 11]: 151–162. A

Hegewald E, Schnepf E (1974b) *Scenedesmus abundans* (Kirchn.) Chod., an older name for *Chlorella fusca* Shih. et Krauss. Archiv für Protistenkunde 139: 133–176. A

Hegewald E, Schnepf E (1986) Zur Struktur und Taxonomie spindelförmiger Chlorellales (Chlorophyta), *Schroederia*, *Pseudoschroederia* gen.nov., *Closteriopsis*. Archiv für Hydrobiologie, Suppl. 73 [Algological Studies 42]: 21–48. A

Hegewald E, Silva PC (1988) Annotated catalogue of *Scenedesmus* and nomenclaturally related genera, including original descriptions and figures. Bibliotheca Phycologica 80: 1–587. A

Heuff H, Horkan K (1984) Caragh. In: Whitton BA (ed.) Ecology of European Rivers. Blackwell, Oxford: 364–384. A

Heywood P (1990) Phylum Raphidophyta. In: Margulis L, Corliss JO, Melkonian M, Chapman DJ (eds) Handbook of Protoctista. Jones & Bartlett, Boston: 318–325. A

Hibberd DJ (1970) Observations on the cytology and ultrastructure of *Ochromonas tuberculatus* sp. nov. (Chrysophyceae), with special reference to the discobolocysts. British Phycological Journal 5: 119–143. B

Hibberd DJ (1971) Observations on the cytology and ultrastructure of *Chrysamoeba radians* Klebs (Chrysophyceae). British Phycological Journal 6: 207–223. C

Hibberd DJ (1973) Observations on the ultrastructure of flagellar scales in the genus *Synura* (Chrysophyceae). Archiv für Mikrobiologie 89: 291–304. B

Hibberd DJ (1974) Observations on the cytology and ultrastructure of *Chlorobotrys regularis* (West) Bohlin, with special reference to its position in the Eustigmatophyceae. British Phycological Journal 9: 37–46. C

Hibberd DJ (1977a) The cytology and ultrastructure of *Chrysonebula holmesii* Lund (Chrysophyceae), with special reference to the flagellar apparatus. British Phycological Journal 12: 369–383. C

Hibberd DJ (1977b) Ultrastructure and cyst formation in *Ochromonas tuberculata* (Chrysophyceae). Journal of Phycology 13: 309–320. A

Hibberd DJ (1978) The fine structure of *Synura sphagnicola* (Korsh.) Korsh. (Chrysophyceae). British Phycological Journal 13: 403–412. B

Hibberd DJ (1979) Notes on the ultrastructure of the genus *Paraphysomonas* (Chrysophyceae) with special reference to *P. bandaiensis* Takahashi. Archiv für Protistenkunde 121: 146–154. A

Hibberd DJ (1980) Eustigmatophytes. In: Cox ER (ed.) Phytoflagellates. Elsevier/North Holland, Amsterdam: 319–334. A

Hickel B (1985) *Cyanonephron styloides* gen et sp. nov., a new chroococcal blue-green alga (Cyanophyta) from a brackish lake. Archiv für Hydrobiologie, Suppl. 71 [Algological Studies 38/39]: 99–104. B

Hickel B (1991) Two new chroococcal cyanophytes from a brackish environment (Schlei-Fjord), Germany. Archiv für Hydrobiologie, Suppl. 92 [Algological Studies 64]: 97–104. B

Hickman M (1971) The standing crop and primary productivity of the epiphyton attached to *Equisetum fluviatile* L. in Priddy Pool, North Somerset. British Phycological Journal 6: 51–59. C

Hickman M (1973) The standing crop and primary productivity of the phytoplankton of Abbot's Pond, North Somerset. Journal of Ecology 61: 269–287. C

Hickman M (1974) The seasonal succession and vertical distribution of the phytoplankton in Abbot's Pond, North Somerset, U.K. Hydrobiologia 44: 127–147. C

Hill DRA (1991a) A revised circumscription of *Cryptomonas* (Cryptophyceae) based on examination of Australian strains. Phycologia 30: 170–180. A

Hill DRA (1991b) *Chroomonas* and other blue-green cryptomonads. Journal of Phycology 27: 133–145. A

Hilliard DK (1966) New or rare chrysophytes from Lancashire County, England. Archiv für Protistenkunde 109: 114–124. B

Hilton S, Parrott G.J (1961) Freshwater studies in Derbyshire. I. *Nitella flexilis* Agardh from Allestree Park Lake 1961. A county record. Transactions of the Derby Natural History Society 2: 14–16. C

Hindák F (1963) Systematik der Gattungen *Koliella* gen. nov. und *Raphidonema* Lagerh. Nova Hedwigia 6: 95–125. A

Hindák F (1970) A contribution to the systematics of the family *Ankistrodesmaceae*. Algological Studies 1: 7–32. A

Hindák F (1977) Studies on the chlorococcal algae (Chlorophyceae). I. Biologicke Práce 23: 1–192. A

Hindák F (1978) New taxa and reclassification in the Chlorococcales (Chlorophyceae). Preslia 50: 97–109. A

Hindák F (1980) Studies on the chlorococcal algae (Chlorophyceae). II. Biologicke Práce 26: 1–196. A

Hindák F (1983) Review of the genus *Lagerheimia* Chod. incl. *Chodatella* Lemm. (Chlorococcales, Chlorophyceae). Schweizerische Zeitschrift für Hydrologie 45: 372–387. A

Hindák F (1984) Studies on the chlorococcal algae (Chlorophyceae) III. Biologicke Práce 30: 1–312. A

Hindák F (1987) Taxonomic survey of the genera *Fusola* (Chlorococcales), *Elakatothrix*, *Closteriospira* and *Chadefaudiothrix* (Ulothrichales). Preslia 59: 193–228. A

Hindák F (1988) Studies on the chlorococcal algae (Chlorophyceae). IV. Biologicke Práce 34: 1–264. A

Hindák F (1990) Studies on the chlorococcal algae (Chlorophyceae). V. Biologicke Práce 36: 1–227. A

Hindák F (1992) Morphological variation of two species of the chlorococcal genus *Coelastrum*, *C. verrucosum* and *C. palii*. Archiv für Hydrobiologie, Suppl. 93 [Algological Studies 65]: 35–42. A

Hindák F (1995) *Scenedesmus ambuehlii*, a new species from the group of *S. asymmetricus* (Chlorococcales, Chlorophyceae). Biologia, Bratislavia 49: 473–478. A

Hindák F (1996) New taxa and nomenclatural changes in the Ulotrichineae (Ulotrichales, Chlorophyta). Biologia, Bratislavia 51: 357–364. A

Hindák F (2000) Morphological citation of four planktic nostocalean cyanophytes – members of the genus *Aphanizomenon* or *Anabaena*? Hydrobiologia 438: 107–116. A

Hirn KE (1900) Monographie und Iconographie der Oedogoniaceen. Acta Societatis Scientiarum Fennicae 27: 1–394. A

Hodgetts WJ (1916) *Dicranochaete reniformis* Hieron., a freshwater alga new to Britain. New Phytologist 15: 108–116. B

Hodgetts WJ (1918a) *Uronema elongatum*, a new freshwater member of the Ulotrichaceae. New Phytologist 17: 159–166. B

Hodgetts WJ (1918b) The conjugation of *Zygogonium ericetorum* Kütz. New Phytologist 17: 238–251. C

Hodgetts WJ (1920a) *Roya anglica* G.S.West a new desmid; with an emended description of the genus *Roya*. Journal of Botany 58: 65–69. C

Hodgetts WJ (1920b) Notes on freshwater algæ. I–IV. New Phytologist 19: 254–263. C

Hodgetts WJ (1920c) A new species of *Spirogyra*. Annals of Botany 34: 519–524. C

Hodgetts WJ (1921–22) A study of some of the factors controlling the periodicity of freshwater algae in nature VII. *Spirogyra*. New Phytologist 20: 150–194, 195–227; 21: 15–33. C

Hoek C van den (1963a) Revision of the European Species of *Cladophora*. E.G. Brill, Leiden. 248 pp. A

Hoek C van den (1963b) Nomenclatural typification of some unicellular and colonial algae. Nova Hedwigia 6: 277–296. A

Hoek C van den, Mann DG, Jahns HM (1995) Algae An Introduction to Phycology. Cambridge University Press, Cambridge. 623 pp. A

Hoffmann L (1986) Cyanophycées aériennes et subaériennes du

Grand-Duché de Luxembourg. Bulletin Jardin Botanique National de Belgique 56: 77–127. A
Hoffmann L (1988) The development of hormogonia, a possible taxonomic criterium in false-branching blue-green algae (Cyanophyceae, Cyanobacteria). Archiv für Protistenkunde 135: 41–43. A
Hoffmann L (1989) Algae of terrestrial habitats. Botanical Review 55: 77–105. A
Holdway PA, Watson RA, Moss B (1978) Aspects of the ecology of *Prymnesium parvum* (Haptophyta) and water chemistry in the Norfolk Broads, England. Freshwater Biology 8: 295–311. C
Holmes NTH, Lloyd EJH, Potts M, Whitton BA (1972) Plants of the River Tyne and future water transfer scheme. Vasculum 57: 56–78. C
Holmes NTH, Whitton BA (1975a) Macrophytes of the River Tweed. Transactions of the Botanical Society of Edinburgh 42: 369–381. B
Holmes NTH, Whitton BA (1975b) Notes on some macroscopic algae new or seldom recorded for Britain: *Nostoc parmelioides, Heribaudiella fluviatilis, Cladophora aegagropila, Monostroma bullosum, Rhodoplax schinzii.* Vasculum 60: 47–55. B
Holmes NTH, Whitton BA (1977a) The macrophytic vegetation of the River Tees in 1975: observed and predicted changes. Freshwater Biology 7: 43–60. B
Holmes NTH, Whitton BA (1977b) Macrophytes of the River Wear: 1966–1976. Naturalist, Hull (1977): 53–73. B
Holmes NTW, Whitton BA (1977c) Macrophytic vegetation of the River Swale,Yorkshire. Freshwater Biology 7: 545–558. C
Holmes NTH, Whitton BA (1981a) Phytoplankton of four rivers, the Tyne, Wear, Tees and Swale. Hydrobiologia 80: 110–127. C
Holmes NTH, Whitton BA (1981b) Phytobenthos of the River Tees and its tributaries. Freshwater Biology 11: 139–168. B
Hooker WJ (1821) Flora Scotica: or a description of Scottish Plants. Taylor, London. B
Hooker WJ (1833) Class XXIV. Cryptogamia. In: Smith JE (ed.) The English Flora 5(1). Longman, Rees, Orme, Brown, Green & Longmans, London: 1–432. C
Horecká M, Komárek J (1979) Taxonomic position of three planktonic blue-green algae from the genera *Aphanizomenon* and *Cylindrospermopsis*. Preslia 51: 289–312. A
Hornbye AJW (1918) A new British freshwater alga. New Phytologist 17: 41–43. C
Houghton GU (1972) Long-term increases in plankton growth in the Essex Stour. Water Treatment and Examination 21: 299–308. C
Howland LJ (1931) A four year's investigation of a Hertfordshire pond. New Phytologist 30: 81–125. B
Huber-Pestalozzi G (1938) Das Phytoplankton des Süßwassers. Systematik und Biologie. Die Binnengewässer 16(1). Schweizerbart'sche, Stuttgart. 342 pp. A
Huber-Pestalozzi G (1950). Das Phytoplankton des Süßwassers. Cryptophyceen, Chloromonadinen, Peridineen. In: Thienemann A (ed.) Die Binnengewässer 16(3). Schweizerbart'sche, Stuttgart: 1–310. A
Huber-Pestalozzi G (1955) Das Phytoplankton des Süßwassers. Euglenophyceen. In: Thienemann A (ed.) Die Binnengewässer 16(4). Schweizerbart'sche, Stuttgart: 1–606. A
Huber-Pestalozzi G (1961) Das Phytoplankton des Süßwassers. Chlorophyceae (Grunalgen) Ordnung: Volvocales. In: Thienemann A (ed.) Die Binnengewässer 16(5). Schweizerbart'sche, Stuttgart: 1–744. A
Huber-Pestalozzi G, Fott B (1968) Das Phytoplankton des Süßwassers. Cryptophyceae, Chloromonadophyceae, Dinophyceae. 2nd edn. In: Thienemann A (ed.) Die Binnengewässer 16(3). Schweizerbart'sche, Stuttgart: 1–322. A
Hudson JW, Crompton KJ, Whitton BA (1971) Ecology of Hell Kettles 2. The ponds. Vasculum 56: 38–45. C
Hudson W (1778) Flora Anglica Vol. 2, 2nd edn. Nourse & Faulder, London. 690 pp. C
Hughes MK, Whitton BA (1972) Algae of Slapestone Sike, Upper Teesdale. Vasculum 57: 30–35. C
Hull J (1799) The British Flora, or a Linnean Arrangement of British Plants. R. & W. Dean, Manchester, London & Edinburgh. 449 pp. C
Huss VAR, Frank C, Hartmann EC, Hirmer M, Kloboucek A, Seidel BM, Wenzeler P, Kessler E (1999) Biochemical taxonomy and molecular phylogeny of the genus *Chlorella sensu lato* (Chlorophyta). Journal of Phycology 35: 587–598. A
Hustedt F (1927–66) Kryptogamen-Flora von Deutschland, Österreich und der Schweiz Band VII Die Kieselalgen. Akademische, Leipzig [3 volumes: 1927–1966]. A
Hustedt F (1930) Bacillariophyta (Diatomeae). In: Pascher A (ed.) Die Süßwasser-Flora Deutschlands, Österreichs und der Schweiz 10. Gustav Fischer, Jena: 1–466. A
Hymes BJ, Cole KM (1983) The cytology of *Audoinella hermannii* (Rhodophyta, Florideophyceae). II. Monosporogenesis. Canadian Journal of Botany 61: 3377–3385. A
Islam AKM (1963) A revision of the genus *Stigeoclonium*. Beihefte Nova Hedwigia 10: 1–164. A
Israelson G (1938) Über die Süsswasserphaeophyceen Schewedens. Botaniska Notiser 1938: 113–128. A
Israelson G (1942) The freshwater Florideae of Sweden. Symbolae Botanica Upsaliensis 6(1): 1–134. A
Jaag O (1932). Untersuchungen über *Rhodoplax* Schinzii, eine interessante Alge vom Rheinfall. Bericht der Schweizerischen Botanischen Gesellschaft 41(2): 356–371. A
Jaag O (1945) Untersuchungen über die Vegetation und Biologie der Algen des näckten Gesteins in den Alpen, im Jura und im schweizerischen Mitteland. Beitrage Kryptogamen Flora Schweiz 9: 1–560. A
James EJ (1935) An investigation of the algal growth in some naturally occurring soils. Beihefte zum Botanische Centralblatt A, 53: 519–533. B
Jane FW (1937, 1938, 1939) Some Hertfordshire flagellates. I–III. Transactions of the Hertfordshire Natural History Society 20: 133–140, 340–351; 21: 115–122. B
Jane FW (1940) Two new chrysophycean flagellates *Cyclonexis erinus* and *Synochromonas elaeochorus*. Proceedings of the Linnean Society of London 152(3): 298–309. B
Jane FW (1942) Methods for the collection and examination of fresh-water algae, with special reference to flagellates. Microscopy. Journal of the Quekett Microscopical Club ser. 4 1(5): 217–229. A
Jane FW (1944a) Studies on the British Volvocales. New Phytologist 43: 36–48. B
Jane FW (1944b) Observations on some British chrysomonads. Annals and Magazine of Natural History ser. 11, 77: 340–344. C
Jane FW (1945) Some new or little known algae from Anglesey and Caernarvonshire. Transactions of the Anglesey Antiquarians Society Field Club 1945: 73–91. B
Jane FW, Dowling RE (1939) On the occurrence of *Mastigosphaera gobii* Schewiakoff, in Britain. Proceedings of the Linnean Society of London 151(2): 50–57. C
Jenner E (1845) A Flora of Tunbridge Wells: being a list of indigenous plants within a radius of fifteen miles around that place.Tunbridge Wells, London. 260 pp. C
Jewson DH, Briggs M (1993) Benthic algae in Lough Neagh. In: Wood RB, Smith RV (eds) Lough Neagh. The Ecology of a Multipurpose Water Resource. Kluwer, Dordrecht: 239–243. B
John DM, Champ WST, Moore JA (1982) The changing status of

Characeae in four marl lakes in the Irish Midlands. Journal of Life Sciences, Royal Dublin Society 4: 47–71. C

John DM, Douglas GE, Brooks SJ, Jones GC, Ellaway J, Rundle S (1998) Blooms of the water net *Hydrodictyon reticulatum* (Chlorococcales, Chlorophyta) in a coastal lake in the British Isles: their cause, seasonality and impact. Biologia, Bratislava 53: 537–545. B

John DM, Johnson LR (1987) Observations on the developmental morphology, growth rate, and reproduction of *Microthamnion kuetzingianum* Nägeli (Pleurastraceae, Pleurastrales) in culture and a taxonomic assessment of the genus. Nova Hedwigia 44: 25–53. B

John DM, Johnson LR (1989) A cultural assessment of the freshwater species of *Pseudendoclonium* Wille (Ulotrichales, Ulvophyceae) Chlorophyta). Archiv für Hydrobiologie, Suppl. 82 [Algological Studies 54]: 79–112. B

John DM, Johnson LR, Moore JA (1989a) Observations on *Thorea ramosissima* Bory Batrachospermales, Thoreaceae), a freshwater red alga rarely recorded in the British Isles British Phycological Journal 24: 99–102. B

John DM, Johnson LR, Moore JA (1989b) The red alga *Thorea ramosissima*: its distribution and status in the Thames catchment. The London Naturalist 68: 49–53. C

John DM, Johnson LR, Moore JAM (1990) Observations on the phytobenthos of the freshwater Thames III. The floristic composition and seasonality of algae in the tidal and non-tidal river. Archiv für Hydrobiologie 120: 143–168. B

John DM, Johnson LR, Moore JAM (1991) Observations on the plants of an intermittently polluted, spring-fed pool flowing into the Thames. The London Naturalist 70: 47–58. C

John DM, Maggs CA (1997) Species problems in eukaryotic algae: a modern perspective. In: Claridge MF, Dawah HA, Wilson MR (eds) Species: the Units of Biodiversity [Systematics Association Special Volume 54]. Chapman & Hall, London: 83–107. A

John DM, Moore JA (1985a) Observations on the phytobenthos of the freshwater Thames I. The environment, floristic composition and distribution of the macrophytes (principally macroalgae). Archiv für Hydrobiologie 102: 435–459. C

John DM, Moore JA (1985b) Observations on the phytobenthos of the freshwater Thames II. The floristic composition and distribution of the smaller algae sampled using artificial surfaces. Archiv für Hydrobiologie 103: 83–97. C

John DM, Pentecost A, Whitton BA (2001) Terrestrial and freshwater eukaryotic algae. In: Hawksworth DL (ed.) The Changing Wildlife of Great Britain and Ireland [Systematics Association Special Volume 62]. Taylor & Francis, London: 148–149. C

John RP (1942) An ecological and taxonomic study of the algae of British soils I. The distribution of the surface-growing algae. Annals of Botany NS 6: 323–349. C

Johnson LR, John DM (1990) Observations on *Dilabifilum* (Class Chlorophyta, Order Chaetophorales sensu stricto) and allied genera. British Phycological Journal 25: 53–61. C

Johnson LR, John DM (1992) Taxonomic observations on some uncommon and new *Gongrosira* species (Chaetophorales *sensu stricto*, Division Chlorophyta). British Phycological Journal 27: 153–163. B

Jones JRE (1949) A further ecological study of calcareous streams in the 'Black Mountain' district of South Wales. Journal of Animal Ecology 18: 142–159. C

Jones JRE (1950) A further ecological study of the River Rheidol: the food of the common insects of the mainstream. Journal of Animal Ecology 19: 159–174. C

Jones R, Benson-Evans K. (1973) A study of Llangorse Lake, Breconshire. Aquatic Ecology and Pollution Bulletin 1: 1–37. C

Jones RA, Stewart NF (2001) *Nitella gracilis* (Smith) Agardh, an elusive charophyte new to Cardiganshire (V.C.46). Watsonia 23: 443–453. C

Joshua W (1883) Notes on British Desmidieae. II. 2. Journal of Botany 22: 289–298. C

Joshua W (1885) On some new and rare Desmidieae. III. Journal of Botany 23: 33–35. C

Joyce A (1998) The distribution of five species of *Desmidium* in Sutherland and Skye. Botanical Journal of Scotland 50: 93–97. B

Juggins S (1992) Diatoms in the Thames Estuary, England: ecology, palaeoecology, and salinity transfer function. Bibliotheca Diatómologica 25: 1–224. C

Jupp BP, Spence DHN, Britton RH (1974) The distribution and production of submerged macrophytes in Loch Leven, Kinross. Proceedings of the Royal Society of Edinburgh B 74: 195–208. C

Kadlubowska JZ (1972) Chlorophyta V. Conjugales: Zygnemaceae, Zrostnicowate. Flora Slodkowodna Polski [Freshwater Flora of Poland] 12A(5): 1–431 Polska Akademia Nauck, Krakow. A

Kadlubowska JZ (1984) Conjugatophyceae I Zygnemales = Chlorophyta VIII. In: Ettl H, Gerloff J, Heynig H, Mollenhauer D (eds) Süßwasserflora von Mitteleuropa 16. Gustav Fischer, Stuttgart: 1–532. A

Kann E (1945) Zur Ökologie der Litoralalgen ostholsteinischen Waldseen. Archiv für Hydrobiologie 41: 14–42. A

Kann E (1972) Zur Systematik und Ökologie der Gattung *Chamaesiphon* (Cyanophyceae) 1. Systematik. Archiv für Hydrobiologie, Suppl. 41 [Algological Studies 7]: 117–171.A

Kann E (1978) Sytematik und Ökologie der Algen österreichischer Bergbäche. Archiv für Hydrobiologie, Suppl. 53 [Monogr. Beitrage]: 405–643. A

Kann E (1993) Der litorale Algenaufwuchs im See Erken und in seinem Abfluß (Uppland, Schweden). Archiv für Hydrobiologie, Suppl. 97 [Algological Studies 69]: 91–112. A

Kann E, Komárek J (1970) Systematische-ökologische Bermerkungen zu den Arten des Formenkreis *Phormidium autumnale*. Schweizerische Zeitschrift für Limnologie 32: 495–518. A

Karim AGA (1967) Algal flora of certain gravel pits in the Thames Valley, U.K. Hydrobiologia 30: 577–599. C

Kelly MG (2000) Identification of common benthic diatoms in rivers. Field Studies 9: 583–700. B

Kessler E, Schäfer M, Hümmer C, Kloboucek A, Huss VAR (1997) Physiological, biochemical, and molecular characters for the taxonomy of the subgenera of *Scenedesmus* (Chlorococcales, Chlorophyta). Botanica Acta 110: 244–250. A

Khan AS (1951). On the occurrence of *Physolinum monilia* (De Wildeman) Printz in England. Hydrobiologia 3: 79–83. B

Kies L (1976) Untersuchungen zu Feinstruktur und taxonomischer Einordnung von *Gloeochaete wittrockiana,* einer apoplastidialen capsalen Alge mit blaugrünen Endosymbioten (Cyanellen). Protoplasma 87: 419–446. A

Kirkby SM, Hibberd DJ, Whitton BA (1972) *Pleurocladia lacustris* A.Braun (Phaeophyta) – a new British record. Vasculum 57: 51–56. A

Klebahn H (1895) Gasvacuolen, ein Bestandteil der Zellen der wasserblütebildenden Phycochromaceen. Flora 80: 241–282. A

Klebs GA (1886) Über die Organisation der Gallerte bei einigen Algen und Flagellaten. Untersuchungen Botanische Institut Tübingen 2: 135, pl. 3, fig. 13. Notarissia 1: 340–341. A

Klebs GA (1893) Flagellatenstudien II. Zeitschrift für Wissenschaftliche Zoologie 55: 353–445. A

Klebs G (1896) Über die Fortpflanzungs-Physiologie der niederen Organismen, der Protobionten, . . . Specieller Theil, Die Bedingungen der Fortpflanzung bei einigen Algen und Pilzen. Jena. xviii + 1–543 pp. A

Knudson BM (1952–53) The diatom genus *Tabellaria*. Annals of Botany NS 16: 421–440; 17: 131–155; 17: 598–609. B

Knudson BM (1954) The ecology of the diatom genus *Tabellaria* in the English Lake District. Journal of Ecology 42: 345–358. B

Knudson BM (1955) The distribution of *Tabellaria* in the English Lake District. Mitteilungen internationale Vereinigung fur Theoretische und angewande Limnologie 12: 216–218. C

Knudson BM (1957) Ecology of the epiphytic diatom *Tabellaria flocculosa* (Roth) Kütz. var. *flocculosa* in three English lakes. Journal of Ecology 45: 93–112. C

Kolkwitz R, Krieger H (1941–44) Zygnemales. In: Dr L.Rabenhorst's Kryptogamen-Flora von Deutschland und der Schweiz 13(2). Becker & Erler, Leipzig: 1–499. AB

Komárek J (1958) Die taxonomische Revision der planktischen Blaualgen der Tschechoslowakei. In: Komárek J, Ettl H (eds) Algologische Studien. Naklad, CSAV, Praha: 10–206. A

Komárek J (1974) The morphology and taxonomy of crucigenoid algae (Scenedesmaceae, Chlorococcales). Archiv für Protistenkunde 116: 1–74. A

Komárek J (1976) Taxonomic review of the genera *Synechocystis* Sauv. 1892, *Synechococcus* Näg. 1849 and *Cyanothece* gen. nov. (Cyanophyceae). Archiv für Protistenkunde 118: 119–179. A

Komárek J (1979) Änderungen in der Taxonomie der Chlorokokkalalgen. Archiv für Hydobiologie Suppl. 56 [Algological Studies 24]: 239–263. A

Komárek J (1983) *Rhabdogloea*, the correct name of cyanophycean *Dactylococcopis* sensu auct., non Hansgirg (1988). Taxon 32: 464–466. A

Komárek J, Anagnostidis K (1986) Modern approach to the classification system of cyanophytes. 2 – Chroococcales. Archiv für Hydrobiologie, Suppl. 73 [Algological Studies 43]: 157–226. A

Komárek J, Anagnostidis K (1999) Cyanoprokaryota: Chroococcales. In: Süßwasserflora von Mitteleuropa 19(1). Gustav Fischer, Jena: 1–548. A

Komárek J, Fott B (1983) Chlorophyceae (Grünalgen) Ordnung: Chlorococcales. Das Phytoplankton des Süßwassers. In: Die Binnengewässer XVI, 7(1). Schweizerbart'sche, Stuttgart: 1–1044. A

Komárek J, Hindák F (1988) Taxonomic review of natural populations of the cyanophytes from the *Gomphosphaeria*-complex. Archiv für Hydrobiologie 80: 205–225. A

Komárek J, Jankovská V (2001) Review of the green algal genus *Pediastrum*; implications for pollen-analytical research. Bibliotheca Phycologica 108: 1–127. A

Komárek J, Kováčik L (1985) The genus *Chlorotetraedron* McEntee et al. (Protosiphonales, Chlorophyceae). Preslia 57: 289–297. A

Komárek J, Perman J (1978) Review of the genus *Dictyosphaerium* (Chlorococcales). Archiv für Hydrobiologie, Suppl. 51 [Algological Studies 20]: 233–297. A

Komárek J, Watanabe M (1990) Morphology and taxonomy of the genus *Coleodesmium* (Cyanophyceae/Cyanobacteria). In: Watanabe M, Malla SB (eds) Cryptogams of the Himalayas Vol. 2 Central and Eastern Nepal. National Science Museum, Japan: 1–22. A

Komárková-Legnerová J (1969) The systematics and ontogenesis of the genera *Ankistrodesmus* Corda and *Monoraphidium* gen. nov. In: Fott B (ed.) Studies in Phycology. Schweizerbart'sche, Stuttgart: 75–144. A

Korshikov OA (1938) Volvocinae. In Visnacnik prisnovod. Vodorostej URSR, 4. Vyd. A.N.URSR, Kiev. B

Korshikov OA (1939) Contribution to the algal flora of the Gorky District. Phytoplankton of the Oka River in August 1932. Zapiski Gorkovskogo Universiteta 9: 101–128. A

Korshikov OA (1953) Pidklas Protokokovi (Protococcineae Bakyol'ni (Vacuolales) ta Protokokovi (Protococcales). Viznacnik prisnovodnihk vodorostey Ukrainsykoj RSR. V/Kiev, Vid. Akad. Nauk Ukr. RSR, Kiev 5: 1–489. [English translation by Lund JWG, Tylka W, see Korshikov 1987]. A

Korshikov OA (1987) The Freshwater Algae of the Ukrainian SSR. V. Sub-Class Protococcineae Vacuolales and Protococcales. Dehra Dun, Bishen Singh Mahendra Pal Singh & Koeltz Scientific Books. 412 pp. A

Kossinskaja EK (1948) Opredelitel' morskich sinezelenych vodoroslej. Izd.AN SSSR, Moscow and Leningrad. 278 pp. [in Russian: Determination Key for Marine Blue-green Algae]. A

Kouwets FAC, Coesel PFM (1984) Taxonomic revision of the Conjugatophycean family Peniaceae on the basis of cell wall ultrastructure. Journal of Phycology 20: 555–662. A

Kováčik Ľ (1975a) Review of the genus *Polyedriopsis* Schmidle, incl. *Tetraedron bitridens* Beck.Mannagetta 1926 = *P. bitridens* (Beck.Mannag.) comb. nova. Archiv für Protistenkunde 117: 246–252. A

Kováčik Ľ (1975b) Taxonomic review of the genus *Tetraedron* (Chlorococcales). Archiv für Hydrobiologie, Suppl. 46 [Algological Studies 13]: 354–391. A

Kováčik Ľ (1988) Cell division in simple coccal cyanophytes. Archiv für Hydrobiologie, Suppl. 80 [Algological Studies 50–53]: 149–190. A

Krammer K, Lange-Bertalot H (1986) Die Süßwasserflora von Mitteleuropa 2: Bacillariophyceae. 1 Teil: Naviculaceae. Gustav Fischer, Stuttgart. 876 pp. A

Krause W (1997) Charales (Charophyceae). In: Ettl H et al. (eds) Süßwasserflora von Mitteleuropa 18. Gustav Fischer, Jena. 202 pp. A

Krishnamurthy V (1961) A *Compsopogon* occurring in the Reddish Canal near Manchester. British Phycological Journal 2: 87–88. B

Kristiansen J (1979) Observations on some Chrysophyceae from North Wales. British Phycological Journal 14: 231–241. B

Kumano S, Hirose H (1959) On the swarmers and reproductive organs of a phaeophyceous fresh-water alga of Japan, *Heribaudiella fluviatilis* (Areschoug) Svedelius. Bulletin of the Japanese Society of Phycology 7: 45–51 [In Japanese]. A

Kusel-Fetzmann EL (1996) New records for freshwater Phaeophyceae from lower Austria. Nova Hedwigia 62: 79–89. A

Kützing FT (1845) Phycologia germanica, d.i. Deutschlands Algen in bündigen Beschreiburgen Nordhausen. 240 pp. B

Lack TJ (1971) Quantitative studies on the phytoplankton of the Rivers Thames and Kennet at Reading. Freshwater Biology 1: 213–224. B

Lack TJ, Lund JWG (1974) Observations and experiments on the phytoplankton of Blelham Tarn, English Lake District. I The experimental tubes. Freshwater Biology 4: 399–415. C

Landsborough D (1851) A Popular History of British Sea-Weeds...with Notices of the Fresh-Water Algae. 2nd edn. Reeve & Benham, London. 400 pp. C

Langangen A (1972) The Charophytes of Iceland. Astarte 5: 27–31. A

Langangen A (1974) Ecology and distribution of Norwegian Charophytes. Norwegian Journal of Botany 21: 31–52. A

Larder J (1922) Linconshire freshwater algae. Nature, London 1902: 59–61. C

Leadbeater BSC (1990) Ultrastructure and assembly of the scale case in *Synura* (Synurophyceae Andersen). British Phycological Journal 25: 117–132. A

Lee KW, Bold HC (1974) Phycological Studies XII. *Characium* and some *Characium*-like algae. University of Texas Publications 7403: 1–127. A

Leedale GF (1967) Euglenoid Flagellates. Prentice-Hall, Englewood Cliffs, New Jersey. 243 pp. A

Leedale GF (1975) Envelope formation and structure in the euglenoid genus *Trachelomonas*. British Phycological Journal 10: 17–42. A

Leitch AR, John DM, Moore JA (1990) The oosporangium of the

Characeae (Chlorophyta, Charales). Progress in Phycological Research 7: 213–268. A
Lehmann E (1903) Über *Hyella Balani* nov. spec. Nyt Magazin for Naturvidenskaberne 41: 77–87. A
Lemmermann E (1898) Beitrag zur Algenflora von Schlesien. Abbildungen Natur. Ver. Bremen 14(1897): 241–263. A
Léon-Tejera H, Montejano G (2000) *Dermocarpella* (Cyano-prokaryota/Cyanophyceae/Cyanobacteria) from the Pacific coast of Mexico. Cryptogamie, Algologie 21: 359–372. A
Leonardi PI, Calceres EJ, Velez CG (1998) Fine structure of dwarf males in *Oedogonium pluviale* (Chlorophyceae). Journal of Phycology 34: 250–256. A
Lewin RA (1984) Culture and taxonomic status of *Chlorochytrium lemnae*, a green algal endophyte. British Phycological Journal 19: 107–116. B
Li R, Watanabe M, Watanabe MM (2000) Taxonomic studies of planktic species of *Anabaena* based on morphological characteristics in cultured strains. Hydrobiologia 438: 117–138. A
Light JL, Belcher JH (1968) A snow microflora in the Cairngorm Mountains, Scotland. British Phycological Bulletin 3: 471–473. B
Lightfoot J (1777) Flora Scotica: or a systematic arrangement in the Linnaean method of the native plants of Scotland. Vol. II. B. White, London. C
Lind EM (1938) Studies in the periodicity of the algae in Beauchief Ponds, Sheffield. Journal of Ecology 26: 257–274. B
Lind EM (1939a) Note on the genus *Uroglena* with the description of a species new to Britain. Journal of Botany 77: 106–110. B
Lind EM (1939b) A new British algal record: *Gloeotaenium loitlesbergerianum* Hansgirg. Journal of Botany 1939: 315–316. C
Lind EM (1939c) Two new algal records from the Sheffield District. Naturalist, December 1: 305–307. B
Lind EM (1944) The phytoplankton of some Cheshire Meres. Memoirs and Proceedings of the Manchester Literary and Philosophical Society 86: 83–105. C
Lind EM (1951) The plankton of some lakes and pools in the neighbourhood of the Moor of Rannoch. Transactions and Proceedings of the Botanical Society of Edinburgh 35: 362–369. C
Lind EM (1952) The phytoplankton of some lochs in South Uist and Rhum. Transactions and Proceedings of the Botanical Society of Edinburgh 36: 37–47. C
Lind EM (1953) Some desmids from some West Highland lochs. Transactions and Proceedings of the Botanical Society of Edinburgh 37: 1–12. C
Lind EM, Brook AJ (1980) Desmids of the English Lake District. Scientific Publications No. 42. Freshwater Biological Association, Ambleside. 123 pp. C
Lind EM, Galliford AL (1952) Notes on the plankton of Oak Mere, Cheshire. Naturalist, Hull (1954): 99–102. C
Lind EM, Pearsall WH (1945) Plankton algae from North-Western Ireland. Proceedings of the Royal Irish Academy, Dublin 50B: 311–320. B
Linnaeus C (1753) Species Plantarum, Exhibentes Plantas Rite Cognitas, Ad Genera Relatas. Secundum Systema Sexuale Digestus. Vols 1, 2. Impenses Laurentii Salvii. Holmii. A
Livingstone D, Jaworski GHM (1980) The viability of akinetes of blue-green algae recovered from the sediments of Rostherne Mere. British Phycological Journal 15: 355–364. C
Livingstone D, Whitton BA (1983) Influence of phosphorus on morphology of *Calothrix parietina* (Cyanophyta) in culture. British Phycological Journal 18: 29–38. A
Livingstone D, Whitton BA (1984) Water chemistry and phosphatase activity of the blue-green alga *Rivularia* in Upper Teesdale streams. Journal of Ecology 72: 405–421. C
Lokhorst GM (1991) Synopsis of genera of Klebsormidiales and Ulotrichales. Cryptogamic Botany 2/3: 274–288. A
Lokhorst GM (1992) Taxonomic studies in the genus *Heterococcus* (Tribophyceae, Tribonematales, Heteropediaceae); a combined cultural and electron microscopy study. Cryptogamic Studies 3: 1–246. A
Lokhorst GM (1996) Comparative taxonomic studies on the genus *Klebsormidium* (Charophyceae) in Europe. Cryptogamic Studies 5: 1–132. B
Ludwig G, Schnittler M (1996) Rote Liste gefährdeter Pflanzen Deutschlands. Schriftenreihe für Vegetationskunde 28: 1–744. A
Lund JWG (1937) Contributions to our knowledge of British algae. VI. New British algal records, I. Journal of Botany, London 75: 305–314. B
Lund JWG (1938) Contributions to our knowledge of British algae. VII. Some new British algal records II Euglenineae. Journal of Botany 76: 271–276. B
Lund JWG (1942a) Contributions to the knowledge of British Chrysophyceae. New Phytologist 41: 274–292. B
Lund JWG (1942b) Contribution to our knowledge of British algae. VIII. Journal of Botany 80: 57–73. B
Lund JWG (1942c) The marginal algae of certain ponds, with special reference to the bottom deposits. Journal of Ecology 30: 245–283. B
Lund JWG (1945) Observations on soil algae I. The ecology, size and taxonomy of British soil diatoms. Part 1. New Phytologist 44: 196–219. C
Lund JWG (1946) Observations on soil algae 1. The ecology, size and taxonomy of British soil diatoms. Part 2. New Phytologist 45: 56–110. C
Lund JWG (1947a) Observations on soil algae II. Notes on groups other than diatoms. New Phytologist 46: 35–60. B
Lund JWG (1947b) Observations on soil algae III. Species of *Chlamydomonas* Her. in relation to variability within the genus. New Phytologist 46: 185–194. B
Lund JWG (1949a) New or rare British Chrysophyceae. I. New Phytologist 48: 453–460. B
Lund JWG (1949b) The dynamics of diatom outbursts with special reference to *Asterionella*. Verhandlungen der Internationalen Vereinigung für Theoretische und Angewandte Limnologie 10: 275–276. C
Lund JWG (1949c) Studies on *Asterionella* 1. The origin and nature of the cells producing seasonal maxima. Journal of Ecology 37: 389–418. C
Lund JWG (1950a) Algological notes I–III. Naturalist, Hull (1950): 45–49. C
Lund JWG (1950b) Algological notes IV–VI. The London Naturalist 1950: 142–148. A
Lund JWG (1950c) Algological notes IV–VI. IV. Motility in unicellular and colonial Myxophyceae and records of the occurrence of some of the species in Great Britain. V. *Chlamydomonas terricola* Gerloff. VI. The distribution and ecology of *Stephanosphaera pluvialis* Cohn with a note on *Haematococcus lacustris* Flotow em. Wille. Naturalist, Hull (1950): 143–148. C
Lund JWG (1950d) Contributions to our knowledge of British algae X1. A new *Pinnularia* (*P. cardinaliculus* Cl. emend.). Hydrobiologia 2: 281–284. C
Lund JWG (1950d) Studies on *Asterionella formosa* Hass II. Nutrient depletion and the spring maximum. Part 1. Observations on Windermere, Esthwaite Water and Blelham Tarn. Journal of Ecology 38: 1–14. C
Lund JWG (1950e) Algological notes I–III. *Pleurocapsa minor* Manag. em Geitler; *Mallomonas reginae* Teil. and *Chlamydomonas proboscigera* Korsch. Naturalist, Hull (1950): 45–49. C
Lund JWG (1951) Contributions to our knowledge of British algae XII. A new planktonic *Cyclotella* (*C. praetermissa* n. sp.); notes on *C. glomerata* Bachm. and *C. catenata* Brun. and

the occurrence of setae in the genus. Hydrobiologia 3: 93–100. C

Lund JWG (1952) On *Dinobryon suecicum* Lemm. var. *longispinum* Lemm., *Chlamydomonas gloeophila* Skuja, *C. dinobryon* G.M. Smith and *Planktosphaeria gelatinosa* G.M. Smith, with a note on *Sphaerocystis schroeteri* Chodat. Naturalist, Hull (1952): 163–166. B

Lund JWG (1953) New or rare British Chrysophyceae. II. *Hyalobryon polymorphum* n. sp. and *Chrysonebula holmesii* n. gen., n. sp. New Phytologist 52: 114–123. B

Lund JWG (1954a) Contributions to our knowledge of British algae XIII. Hydrobiologia 6: 136–143. B

Lund JWG (1954b) Three new British algal records and spore-formation in *Micractinium pusillum* Fres. Naturalist, Hull (1954): 81–85. B

Lund JWG (1954c) The seasonal cycle of the plankton diatom *Melosira italica* (Her.) Kütz. subsp. *subarctica* O. Müll. Journal of Ecology 42: 151–179. C

Lund JWG (1954d) Records from Irish lakes, Part 2. Phytoplankton. Proceedings of the Royal Irish Academy B 56: 154–155. C

Lund JWG (1955a) Contributions to our knowledge of British algae XIV. Three new species from the English Lake District. Hydrobiologia 7: 219–228. C

Lund JWG (1955b) Further observations on the seasonal cycle of *Melosira italica* (Her.) Kütz. subsp. *subarctica* O. Müll. Journal of Ecology 43: 90–102. C

Lund JWG (1956a) On certain planktonic palmelloid green algae. Journal of the Linnean Society, Botany 55: 593–613. B

Lund JWG (1956b) Species of *Characiopsis*, *Pseudostaurastrum* and *Scenedesmus* new to Britain. Naturalist, Hull (1956): 13–16. C

Lund JWG (1956c) A new blue-green alga epizoic on *Daphnia pulex* L. Naturalist, Hull (1956): 88–90. C

Lund JWG (1957) Four new green algae. Revue Algologique NS 3: 26–44. B

Lund JWG (1959) Investigations on the algae of Lake Windermere. Advancement of Science, London 15: 530–534. C

Lund JWG (1960a) Some new British algae. Naturalist, Hull (1960): 89–96. B

Lund JWG (1960b) Some new or rare Chrysophyceae from the English Lake District. Hydrobiologia 16: 97–108. B

Lund JWG (1960c) New or rare British Chrysophyceae. III. New records and observations on sexuality and classification. New Phytologist 59: 349–360. C

Lund JWG (1961a) The algae of the Malham Tarn district. Field Studies 1(3): 85–119. B

Lund JWG (1961b) The taxonomy and nomenclature of certain palmelloid planktonic green algae described previously. Journal of the Linnean Society, Botany 56: 459–465. B

Lund JWG (1961c) The periodicity of μ-algae in three English lakes. Verhandlungen der Internationalen Vereinigung für Theoretische und Angewandte Limnologie 14: 147–154. C

Lund JWG (1961d) *Asterionella formosa* Hass. var. *acaroides* Lemm.: a phycological enigma. British Phycological Bulletin 2(2): 63–66. C

Lund JWG (1961e) The Fritsch collection of illustrations of freshwater algae. Phycologia 1: 193. A

Lund JWG (1962a) A genus new to Britain: *Amphichrysis* Korsh. British Phycological Journal 2: 116–120. B

Lund JWG (1962b) A rarely recorded, but very common British alga, *Rhodomonas minuta* Skuja. British Phycological Journal 2: 135–139. B

Lund JWG (1962c) Classical and modern criteria used in algal taxonomy with special reference to genera of microbial size. Symposia of the Society for General Microbiology 12: 68–110. A

Lund JWG (1963) Changes in depth and time of certain physical conditions and of the standing crop of *Asterionella formosa* Hass. in the north basin of Windermere in 1947. Philosophical Transactions of the Royal Society 246B (731): 255–290. C

Lund JWG (1965) The season periodicity of three planktonic desmids. Mitteilungen der Internationale Vereiningung für Theoretische und Angewandte Limnologie 19: 3–25. C

Lund JWG (1972a) Preliminary observations on the use of large experimental tubes in lakes. Verhandlungen der Internationalen Vereinigung für Theoretische und Angewandte Limnologie 18: 71–77. C

Lund JWG (1972b) Changes in the biomass of blue-green and other algae in an English lake from 1945–69. In Desikachary TV (ed.) Taxonomy and Biology of Blue-green Algae. Madras University Centre for Advanced Study of Botany: 305–323. AC

Lund JWG (1978) Changes in the phytoplankton of an English lake, 1945–1977. Gidrobiologicheskii Zhurnal Kiev 14 (1): 10–27. [In Russian]. AC

Lund JWG (1979) Changes in the phytoplankton of an English lake, 1945–1977. Hydrobiological Journal 14 (1): 6–21. AC

Lund JWG (1998) The tangled taxonomic history of three species of *Scourfieldia* G.S.West. Archiv für Hydrobiologie, Suppl. 124 [Algological Studies 89]: 1–14. C

Lund JWG, Reynolds CS (1982) The development and operation of large limnetic enclosures in Blelham Tarn, English Lake District, and their contribution to phytoplankton ecology. Progress in Phycological Research 1: 1–65. Elsevier, Amsterdam. 1–65. A

Lund JWG, Scott LI (1952) A new British algal record: *Pyramimonas reticulata* Korsch. Naturalist, Hull (1952): 55–58. C

Macan TT, Lund JWG (1954) Records from some Irish lakes. Proceedings of the Royal Irish Academy 56B: 135–157. B

MacFarlane JJ, Raven JA (1990) C, N and P nutrition of *Lemanea mamillosa* Kütz. (Batrachospermales, Rhodophyta) in the Dighty-Burn, Angus, U.K. Plant Cell and Environment 13: 1–13. C

Mackie W, Preston RD (1974) Cell wall and intercellular region polysaccharides. In: Stewart WDP (ed.) Algal Physiology and Biochemistry. Blackwell, Oxford: 40–85. A

Mainx F (1926) Einige neue Vertreter der Gattung *Euglena* Ehrbg. Archiv für Protistenkunde 54: 150–160. A

Malin Smith A (1933) *Gonyostomum semen* Diesing. A flagellate now first recorded for the British Isles. The London Naturalist 1933: 49–50. C

Malin Smith A (1942) The algae of Miles Rough Bog, Bradford. Journal of Ecology 30: 341–356. B

Mann DG, Droop SJM (1996) Biodiversity, biogeography and conservation of diatoms. Hydrobiologia 336: 19–32. A

Manton I, Ettl H (1965) Observations on the fine structure of *Mesostigma viride* Lauterborn. Journal of the Linnean Society, Botany 59: 175–184. B

Manton I, Harris K (1966) Observations on the microanatomy of the brown flagellate *Sphaleromantis tetragragona* Skuja with special reference to the flagellar apparatus and scales. Journal of the Linnean Society, Botany 59: 397–402. B

Manton I, Rayns DG, Ettl F, Parke M (1965) Further observations on green flagellates with scaly flagella: the genus *Heteromastix* Korschikov. Journal of the Marine Biological Association of the United Kingdom 45: 241–255. B

Marker AFH (1976a) The benthic algae of some streams in southern England. Journal of Ecology 64: 343–358. B

Marker AFH (1976b) The benthic algae of some streams in southern England. II. The primary production of the epilithon in a small chalk stream. Journal of Ecology 64: 359–373. B

Marker AFH, Casey H (1982) The population and production dynamics of benthic algae in an artificial recirculating hard-water stream. Philosophical Transactions of the Royal Society, London B 298: 265–308. B

Marquand ED (1884) The fresh water algae of the Land's End

District. Transactions of the Penzance Natural History and Antiquarian Society 2: 133–144. B
Marquand ED (1893–98) The desmids and diatoms of West Cornwall. Transactions of Penzance Natural History and Antiquarian Society 1: 272–281. C
Marsden MW, Smith MR, Sargent RJ (1997) Trophic status of rivers in the Forth catchment, Scotland. Aquatic Conservation 7: 211–221. A
Marson JE (1992) Observations on sexual reproduction in species of *Spirogyra* (Zygnemaphyceae: Zygnemaceae). Microscopy. Journal of the Quekett Microscopical Club 36: 712–717, 720. C
Marvan P, Komárek J, Comas A (1984) Weighting and scaling of features in numerical evaluation of coccal green algae (genera of the Selenastraceae). Archiv für Hydrobiologie, Suppl. 67 [Algological Studies 37]: 363–399. A
Mason CF, Bryant RJ (1975) Periphyton production and grazing by chironomids in Alderfen Broad, Norfolk. Freshwater Biology 5: 271–277. C
Matula J (1992) *Dicranochaete* species (Chlorophyceae, Gloeodendrales) in peat bogs of Lower Silesia (South-western Poland). Archiv für Hydrobiologie, Suppl. 93 [Algological Studies 65]: 63–72. A
McCall D (1933) Diatoms (recent and fossil) of the Tay district. Journal of the Linnean Society, Botany 49: 219–308. B
McLean RO, Bensen-Evans K (1974) The distribution of *Stigeoclonium tenue* Kütz. in South Wales in relation to its use as an indicator of organic pollution. British Phycological Journal 9: 83–89. C
Mignot J-P (1967) Structure et ultrastructure de quelques Chloromonadines. Protistologia 3: 5–23. A
Misra RD (1938) Edaphic factors in the distribution of aquatic plants in the English Lakes. Journal of Ecology 26: 411–451. C
Mix M (1967) Zur Feinstruktur der Zellwände in der Gattung *Penium* (Desmidiaceae). Bericht der Deutschen Botanischen Gesellschaft 80: 715–721. A
M'Keever FL (1912) *Phaeothamnion confervicolum* Lagerheim, and its first recorded appearance in Great Britain. Transactions and Proceedings of the Botanical Society, Edinburgh 24: 176–181. B.
Mobius K (1892) Morphologie der haarartigen Organe bei den Algen. Biologisches Zentralblatt 12: 71–87, 97–108. A
Moestrup Ø (1991) Further studies of presumably primitive green algae, including the description of Pedinophyceae class. nov. and *Resultor* gen. nov. Journal of Phycology 27: 119–133. A
Moore JA (1986) Charophytes of Great Britain and Ireland. BSBI Handbook 5. Botanical Society of the British Isles, London. 140 pp. AB
Moore JW (1976) Seasonal succession of algae in rivers. I. Examples from the Avon, a slow-flowing river. Journal of Phycology 12: 342–349. AB
Morgan NC (1970) Changes in the fauna and flora of a nutrient enriched lake. Hydrobiologia 35: 545–553. C
Morton JK (1959) The Flora of Islay and Jura (v.c. 102). Supplement to the Proceedings of the Botanical Society of the British Isles 3(3): 1–59. C
Moss B (1969a) Vertical heterogeneity in the water column of Abbot's Pond. II. The influence of physical and chemical conditions on the spatial and temporal distribution of the phytoplankton and of a community of epipelic algae. Journal of Ecology 57: 397–414. C
Moss B (1969b) Algae of two Somersetshire pools: standing crops of phytoplankton and epipelic algae as measured by cell numbers and chlorophyll *a*. Journal of Phycology 5: 158–168. C
Moss B (1976) The effects of fertilization and fish on community structure and biomass of aquatic macrophytes and epiphytic algal populations: an ecosystem experiment. Journal of Ecology 64: 313–342. C
Moss B (1977) Factors controlling the seasonal incidence of *Pandorina morum* (Mull.) Bory (Chlorophyta, Volvocales) in a small pond. Hydrobiologia 47: 298–305. B
Moss B (1979) Diatoms. In: Ayers B, Murphy P (eds) A waterfront excavation at Whitefriars Street car park, Norwich. East Anglian Archaeology Report 17: 37–38. C
Moss B (1981) The composition and ecology of periphyton communities in freshwaters. 2, Interrelationships between water chemistry, phytoplankton populations and periphyton populations in a shallow lake and associated experimental reservoirs (Lund Tubes). British Phycological Journal 16: 59–76. C
Moss B (1983a) The Norfolk Broadland: experiments in the restoration of a complex wetland. Biological Reviews 58: 521–561. A
Moss B (1983b) Excavations south of the River Wensum, Whitefriars, Norwich: Diatoms. Norfolk Archaeologist 1983: 26–27. C
Moss B, Balls HR, Booker I, Manson K, Timms M (1988) Problems in the construction of a nutrient budget for the R. Bure and the Broads (Norfolk) prior to its restoration from eutrophication. In: Round FE (ed.) Algae and the Aquatic Environment. Contributions in Honour of J.W.G. Lund. Biopress, Bristol: 326–353. A
Moss B, Balls HR, Irvine K (1985) Management of the consequences of eutrophication in lowland lakes in England – engineering and biological solutions. In: Lester JN, Kirk PWW (eds) Proceedings of the International Phosphorus Conference. SP Publishers, London: 180–185. A
Moss B, Karim AGA (1969) Phytoplankton associations in two pools and their relationships with associated benthic flora. Hydrobiologia 33: 587–600. C
Mrozińska T (1985) Chlorophyta. Oedogoniophyceae: Oedogoniales. In: Süßwasserflora von Mitteleuropa 14(6). Gustav Fischer, Stuttgart: 1–624. A
Mueller G, Oti M (1981) The occurrence of calcified planktonic algae in freshwater carbonates. Sedimentology 28: 897–902. B
Müller KM, Sheath RG, Vis ML, Crease TJ, Cole KM (1998) Biogeography and systematics of *Bangia* (Bangiales, Rhodophyta) based on the Rubisco spacer, *rbc*L gene, 18S rRNA gene sequences and morphometric analyses. I. North America. Phycologia 37: 195–207. A
Müller KM, Vis ML, Chiasson WB, Whittick A, Sheath RG (1997) Phenology of a *Batrachospermum* population in a boreal pond and its implications for the systematics of section *Turfosa* (Batrachospermales, Rhodophyta). Phycologia 36: 68–75. A
Murray G, Barton ES (1891) On the structure and systematic position of *Chantransia*: with a description of a new species. Journal of the Linnean Society, Botany 28: 209–216. A
Nägeli C (1849) Gattungen einzelliger Algen. Neue Denkschriften der Verhandlungen der Allgemeinen Schweizerischen Gesellschaft für die Gesammten Naturwissenschaften 10: 1–139. A
Nakayama TS, Watanabe S, Mitsui K, Uchida H, Inouye I (1996) The phylogenetic relationship between the Chlamydomonadales and Chlorococcales inferred from 18S rDNA data. Phycological Research 44: 151–161. A
Necchi O, Jr (1993) Distribution and seasonal dynamics of Rhodophyta in the Preto River Basin, southeastern Brazil. Hydrobiologia 250: 81–90. A
Necchi O, Jr, Sheath RG, Cole KM (1993a) Systematics of freshwater *Audouinella* (Acrochaetiaceae, Rhodophyta) in North America 1. The reddish species. Archiv für Hydrobiologie, Suppl. 98 [Algological Studies 70]: 11–28. A
Necchi O, Jr, Sheath RG, Cole KM (1993b) Systematics of freshwater *Audouinella* (Acrochaetiaceae, Rhodophyta) in North America. 2. The bluish species. Archiv für Hydrobiologie, Suppl. 99 [Algological Studies 71]: 13–21. A
Necchi O, Jr, Sheath RG, Cole KM (1993c) Distribution and

systematics of the freshwater genus *Sirodotia* (Batrachospermales, Rhodophyta) in North America. Journal of Phycology 29: 236–243. A

Necchi O, Jr, Zucchi MR (1997). *Audouinella macrospora* (Acrochaetiaceae, Rhodophyta) is the 'Chantransia' stage of *Batrachospermum* (Batrachospermaceae). Phycologia 36: 220–224. A

Nevo E, Wasser SP (eds) (2000) Biodiversity of Cyanoprokaryotes, Algae and Fungi of Israel. Cyanoprokaryotes and Algae of Continental Israel. ARA Ganter Verlag, Ruggell. A

Nielsen H (2000) Morphometric analysis of cell wall structure in seven infraspecific taxa of *Pediastrum boryanum* (Sphaeopleales, Chlorophyta) and its taxonomic implications. Phycologia 39: 36–49. A

Novarino G (1991a) Observations on *Rhinomonas reticulata* comb. nov. and *Rhinomonas reticulata* var. *eleniana* var. nov. (Cryptophyceae), with comments on the genera *Pyrenomonas* and *Rhodomonas.* Nordic Journal of Botany 11: 243–252. A

Novarino G (1991b) Observations on some new and interesting Cryptophyceae. Nordic Journal of Botany 11: 599–611. A

Novarino G, Lucas IAN (1993a) A comparison of some morphological characters in *Chroomonas ligulata* sp. nov. and *C. placoidea* sp. nov. (Cryptophyceae). Nordic Journal of Botany 13: 583–591. A

Novarino G, Lucas IAN (1993b) Some proposals for a new classification system of the Cryptophyceae. Journal of the Linnean Society Botany 111: 3–21. A

Novarino G, Lucas IAN, Morrall S (1994) Observations on the genus *Plagioselmis* (Cryptophyceae). Cryptogamie Algologie 15: 87–107. A

Novarino G, Oliva E (1998) Typification and ultrastructural characterization of long-described flagellate taxa based on original type material. In: The Flagellates Symposium, University of Birmingham, 7–11 September 1998 (abstract). A

Nozaki H (1993) Morphology, reproduction and taxonomy of *Characiochloris sasae* sp. nov (Chlorophyta) from Japan. Phycologia 32: 129–135. A

Nygaard G, Komárek J, Kristiansen J, Skulberg OM (1986) Taxonomic designations of the bioassay alga NIVA-CHL 1 ("*Selenastrum capricornum*") and some related strains. Opera Botanica 90: 1–46. A

O'Kelly CJ, Watanabe S, Floyd GL (1994) Ultrastructure and phylogenetic-relationships of Chaetopeltidales Ord. Nov. (Chlorophyta, Chlorophyceae). Journal of Phycology 30: 118–128. A

Olsen S (1944) Danish Charophyta. Chorological, ecological and biological investigations. Kongelige dansk Videnskabernes Selskabs Biologiske Skrifter 3(1): 1–240. A

Osborne PL, Moss B (1977) Palaeolimnology and trends in the phosphorus and iron budgets of an old man-made lake, Barton Broad, Norfolk. Freshwater Biology 7: 213–234. C

Palmer TC (1925) Trachelomonas new or notable species and varieties. Proceedings of the Academy of Natural Sciences Philadelphia 77: 15–22. A

Papenfuss GF (1951) Phaeophyta. In: Smith GM (ed.) Manual of Phycology.Chronica Botanica, Waltham, Massachussets: 119–158. A

Parfitt E (1886) Devon freshwater algae. Transactions of the Devon Association of Science 18: 382–423. C

Parke M, Dixon PS (1964) A revised check-list of British marine algae. Journal of the Marine Biological Association of the United Kingdom 44: 499–542. B

Parke M, Dixon PS (1976) Check-list of British marine algae – third revision. Journal of the Marine Biological Association of the United Kingdom 56: 527–594. B

Parke M, Lund JWG, Manton I (1962) Observations on the biology and fine structure of the type species of *Chrysochromulina* (*C. parva* Lackey) in the English Lake District. Archiv für Mikrobiologie 42: 333–352. C

Pascher A (1927) Volvocales–Phytomonadinae. Flagellatae IV.–Chlorophyceae 1. In: Pascher A (ed.) Süßwasser-Flora Deutschlands, Österreichs und der Schweiz. Gustav Fischer, Jena: 1–506. A

Pascher A (1929) Über die Teilungsvorgänge bei einer neuen Blaualge: *Endonema.* Jahrbuch für Wissenschaftliche Botanik 70: (quoted by Geitler, 1932a). A

Pascher A (1939) Heterokontae. In: Rabenhorst's Kryptogamen-Flora von Deutschland, Österreichs und dir Schweiz. XI. Akademie Verlag, Leipzig: 1–1092. A

Pascher A, Lemmermann E (1913). Cryptomonadinae. In: Pascher A (ed.) Die Süßwasser-Flora Deutschlands, Österreichs und der Schweiz 2. Gustav Fischer, Jena: 96–114. A

Paschma R, Hegewald E (1986) DNA base composition within the genus *Scenedesmus* (Chlorophyta). Plant Systematics and Evolution 153: 171–180. A

Patrick R, Reimer CW (1966) The Diatoms of the United States exclusive of Alaska and Hawaii. Volume 1: Fragilariaceae, Eunotiaceae, Achnanthaceae, Naviculaceae. Academy of Natural Sciences, Philadelphia. 688 pp. A

Patrick R, Reimer CW (1975)The Diatoms of the United States exclusive of Alaska and Hawaii. Volume 2 Part 1: Entomoneidaceae, Cymbellaceae, Gomphonemaceae, Epithemiaceae. Academy of Natural Sciences, Philadelphia. 213 pp. A

Patterson G, Whitton BA (1981) Chemistry of water, sediments and algal filaments in groundwater draining an old lead-zinc mine. In: Say PJ, Whitton BA (eds) Heavy Metals in Northern England. Environmental and Biological Aspects. Department of Botany, University of Durham: 65–72. A

Peabody AJ, Whitton BA (1968) Algae of the River Wear. l. Diatoms. Naturalist, Hull 1968: 89–96. C

Pearsall WH (1923) The phytoplankton of Rostherne Mere. Memoirs and Proceedings of the Manchester Literary and Philosophical Society 67: 47–55. C

Pearsall WH (1924) Phytoplankton and environment in the English Lake District. Revue Algologique NS 1: 53–67. C

Pearsall WH (1925) The phytoplankton of the English Lakes. Journal of the Linnean Society, Botany 47: 55–73. C

Pearsall WH (1929) The plankton algae of the English Lakes. Naturalist, Hull (1929): 137–142. C

Pearsall WH (1930) Phytoplankton of the English Lakes I. The proportions in the water of some dissolved substances of biological importance. Journal of Ecology 18: 306–320. A

Pearsall WH (1932) Phytoplankton of the English Lakes. II. The composition of the phytoplankton in relation to the dissolved substances. Journal of Ecology 20: 241–262. C

Pearsall WH (1933) *Uroglenopsis americana* in Windermere. Naturalist, Hull (1933): 122–123.C

Pearsall WH (1936) Uncommon and interesting algae in the Lake District. Naturalist, Hull (1936): 205–206. C

Pearsall WH, Lind EM (1942) The distribution of phytoplankton in some north-west Irish loughs. Proceedings of the Royal Irish Academy 48B: 1–24. B

Pennington W, Haworth EY, Lishman JP, Bonny AP (1972) Lake sediments in northern Scotland. Philosophical Transactions of the Royal Society 264B: 191–294. A

Pentecost A (1982) Quantitative study of calcareous stream and tintenstriche algae from the Malham District, Northern England. British Phycological Journal 17: 443–456. C

Pentecost A (1983) The distribution of daughter colonies in a natural population of *Volvox aureus* Ehrenb. Annals of Botany 52: 769–776. B

Pentecost A (1984a) Introduction to Freshwater Algae. The Richmond Publishing Co, Slough. 247 pp. B

Pentecost A (1984b) Observations on a bloom of the neuston alga

Nautococcus pyriformis, from southern England with an explanation of the floatation mechanism. British Phycological Journal 19: 227–232. B
Pentecost A (1984c) The growth of *Chara globularis* and its relationship to calcium carbonate deposition in Malham Tarn. Field Studies 6: 53–58. C
Pentecost A (1991) The weathering rates of some sandstone cliffs, Central Weald, England. Earth Surface Processes and Landforms 16: 83–91. C
Pentecost A (1992) A note on the colonisation of limestone rocks by cyanobacteria. Archiv für Hydrobiologie 124: 167–172. CA
Pentecost A (1995) British thermophilic cyanobacteria. Archiv für Hydrobiologie 132: 407–414. C
Pentecost A, Happey-Wood CM (1978) Primary production studies in two linked but contrasting Welsh lakes. Freshwater Biology 8: 9–23. C
Pentecost A, Rose F (1985) Changes in the cryptogam flora of the Wealden sandrocks. Journal of the Linnean Society, Botany 90: 217–230. C
Pentecost A, Whitton BA (2000) Limestones. In: Whitton BA, Potts M (eds) Ecology of Cyanobacteria. Kluwer, Dordrecht: 233–255. A
Pentelow FTK, Butcher RW, Grindley J (1938) An Investigation of the Effects of Milk Wastes on the Bristol Avon. MAFF Fisheries Investigations Series I, l(1): 1–80. HMSO, London. B
Perasso R, Baroin A, Qu LH, Bachellerie JP, Adoutte A (1989) Origin of the algae. Nature, London 339: 142–144. A
Peters JC (1972) The ecology of Tarn Dub. Vasculum 57: 42–50. C
Petersen JB (1932) The algal vegetation of Hammer Bakker. Botanische Tidskrifter (Lund) 42: 1–48. A
Philipose MT (1967) Chlorococcales. Indian Council for Agricultural Research Monographs on Algae, New Dehli. 365 pp. A
Philipps FW (1884) On *Chlorodesmus hispida*, a new flagellate animalcule. Transactions of the Hertfordshire Natural History Society 2: 92–94. B
Phillips GL, Eminson DF, Moss B (1978) A mechanism to account for macrophyte decline in progressively eutrophicated freshwaters. Aquatic Botany 4: 103–126. A
Phillips SP (1963) A note on the charophytes of Hickling Broad, E. Norfolk. Proceedings of the Botanical Society of the British Isles 5: 23–24. C
Pickett-Heaps JD (1975) Green Algae. Structure, Reproduction and Evolution in Selected Genera. Sinauer Associates, Sunderland, Mass. 606 pp. A
Playfair GJ (1915) The genus *Trachelomonas*. Proceedings of the Linnean Society of New South Wales 40: 1–41. A
Playfair GJ (1916) *Oocystis* and *Eremosphaera*. Proceedings of the Linnean Society of New South Wales 41: 117–147. A
Popova TG (1966) Flora Sporovych Rastenij SSSR [Flora plantarum cryptogamarum URSS]. 8. Evglenovyje vodorosli ['Euglenophyta']. 1. 412 pp. Izdatel'stvo Nauka, Leningrad [in Russian]. A
Popova TG, Safonova TA (1976) Flora Sporovych Rastenij SSSR [Flora plantarum cryptogamarum URSS] 9. Evglenovyje vodorosli ['Euglenophyta']. 2. 278 pp. Izdatel'stvo Nauka, Leningrad [in Russian]. A
Popovský J, Pfiester LA (1990) Dinophyceae (Dinoflagellida). In: Süßwasserflora von Mitteleuropa 6. Gustav Fischer, Jena: 1–272. A
Potten AH (1972) Maturation ponds. Experiences in their operation in the United Kingdom as a tertiary treatment process for high quality sewage effluent. Water Research 6: 781–795. C
Potter D, Saunders GW, Anderson RA (1997) Phylogenetic relations of the Raphidophyceae and Xanthophyceae as inferred from nucleotide sequences of the 18S ribosomal RNA gene. American Journal of Botany 84: 966–972. A
Poulton EM (1933) Hartlebury Common records. V. The algae of Hartlebury Common. Proceedings of the Birmingham Natural History Society 16: 81–87. C
Powell HT (1976) In: Parke M, Dixon PS (eds) Check-list of British marine algae – third revision. Journal of the Marine Biological Association of the United Kingdom 56: 529–532. B
Preisig HR (1994) A modern concept of chrysophyte classification. In: Sandgren C, Smol J, Kristiansen J (eds) Chrysophyte Algae. Ecology, Phylogeny and Development.Cambridge University Press, Cambridge: 46–74. A
Preisig HR, Hibberd DJ (1982a) Ultrastructure and taxonomy of *Paraphysomonas* (Chrysophyceae) and related genera. 2. Nordic Journal of Botany 2(4): 397–420. B
Preisig HR, Hibberd DJ (1982b) Ultrastructure and taxonomy of *Paraphysomonas* (Chrysophyceae) and related genera. 2. Nordic Journal of Botany 2(6): 601–638. B
Prescott GW (1944) New species and varieties of Wisconsin algae. Farlowia 1: 347–385. A
Prescott GW (1962) Algae of the Western Great Lakes Area with an Illustrated Key to the Genera of Desmids and Freshwater Diatoms. Revised edn. Wm. C. Brown, Iowa. 977 pp. [Reprinted in 1982 by Otto Koeltz Science Publishers]. A
Price R (1914) Notes on *Batrachospermum*. New Phytologist 13: 276–279. A
Price S, Price JH (1983) Aquatic plants of the Wandle. The London Naturalist 62: 26–57. B
Pringsheim EG (1952) On the nutrition of *Ochromonas*. Quarterly Journal of the Microscopical Society 93: 71–96. A
Pringsheim EG (1955) The genus *Polytomella*. Journal of Protozoology 2: 137–145. A
Pringsheim EG (1956) Contributions towards a monograph of the genus *Euglena*. Nova Acta Leopoldiana 18(125): 3–168. A
Printz H (1913) Eine systematische Übersicht der Gattung *Oocystis* Nägeli. Nyt Magasin for Naturvidenskapene. Oslo 51: 165–184. A
Printz H (1916) Die Chlorophyceen des südlichen Sibiriens und des Urjankailandes. In: Printz H (ed.) Contributiones ad Floram Asiae interioris pertinentie. Kongelige Norske Videnskabernes Selskabs Skrifter 4: 1–59. A
Printz H (1964) Die Chaetophoralen der Binnengewässer (eine systematische Übersicht). Hydrobiologia 24: 1–376. A
Prowse G A (1958) Eugleninae of Malay. Gardens' Bulletin Singapore 16: 136–204. A
Prygiel J, Coste M (2000) Guide méthodologique pour la mise en oeuvre de l'Indice Biologique Diatomées. NF T 90–354. Agences de l'Eau – Cemagref Bordeaux. Agence de l'Eau Artois-Picardie, Douai, mars 2000. 134 pp, 89 pls 90–354. [bilingual French-English CDROM]. A
Rabenhorst L (1863) Kryptogamen-Flora von Sachsen, der Ober-Lausitz, Thüringen und Nordböhmen. Leipzig, Kummer. 653 pp. A
Rabenhorst L (1864–68) Flora Europaea Algarum Aquae Dulcis et Submarinae. 3 Volumes. Leipzig. A
Ralfs J (1848) The British Desmidieae. London, Reeve. 226 pp. B
Ramanathan KR (1964) Ulotrichales. Indian Council for Agricultural Research Monographs on Algae, New Delhi. 188 pp. AB
Randhawa MS (1959) Zygnemaceae. Indian Council for Agricultural Research Monographs on Algae, New Delhi. 478 pp. A
Rao CB (1953) On the distribution of algae in a group of six small ponds. Journal of Ecology 41: 62–71. C
Raven JA (1993) The roles of the *Chantransia* phase of *Lemanea* (Lemaneaceae, Batrachospermales, Rhodophyta) and the 'mushroom' phase of *Himanthalia* (Himanthaceae, Fucales, Phaeophyta). Botanical Journal of Scotland 46: 477–485. A
Raven JA, Beardall J, Griffiths H (1982) Inorganic C-sources for

Lemanea,*Cladophora* and *Ranunculus* in a fast-flowing stream: measurements of gas exchange and of carbon isotope ratio and their ecological implications. Oecologia 53: 68–78. A

Raven JA, Johnston AM, Newman JR. Scrimgeour CM (1994) Inorganic carbon acquisition by aquatic photolithotrophs of the Dighty Burn, Angus, U.K.: uses and limitations of natural abundance measurements of carbon isotopes. New Phytologist 127: 271–286. A

Reed RH (1980) On the conspecificity of marine and freshwater *Bangia* in Britain. British Phycological Journal 15: 411–416. B

Reese MJ (1937) The microflora of the non-calcareous streams Rheidol and Melindwr with special reference to water pollution from lead mines in Cardiganshire. Journal of Ecology 25: 385–407. B

Reháková H (1969) Die Variabilität der Arten der Gattung *Oocystis* A. Braun. In: Fott B (ed.) Studies in Phycology. Academia, Praha: 145–196. A

Rendle AB, West W, Jr (1899) A new British freshwater alga. Journal of Botany 37: 289–290. C

Reynolds CS (1967) The breaking of the Shropshire meres: some recent investigations. Shropshire Conservation Trust Bulletin 10: 9–16. A

Reynolds CS (1971) The ecology of the planktonic blue-green algae in the North Shropshire meres. Field Studies 3: 409–432. C

Reynolds CS (1973a) Phytoplankton periodicity of some North Shropshire meres. British Phycological Journal 8: 301–320. A

Reynolds CS (1973b) The seasonal periodicity of planktonic diatoms in a shallow eutrophic lake. Freshwater Biology 3: 89–110. C

Reynolds CS (1973c) The phytoplankton of Crose Mere, Shropshire. British Phycological Journal 8: 151–162. C

Reynolds CS (1986) The Ecology of Freshwater Phytoplankton. Cambridge University Press. 384 pp. A

Reynolds CS (1994) The long, the short and the stalled – on the attributes of phytoplankton selected by physical mixing in lakes and rivers. Hydrobiologia 289: 9–21. C

Reynolds CS (1998) What facts influence the species composition of phyoplankton in lakes of different trophic status? Hydrobiologia 369/370: 11–26. C

Reynolds CS, Allen SE (1968) Changes in the phytoplankton of Oak Mere, following the introduction of base rich water. British Phycological Journal 3: 451–462. C

Reynolds CS, Irish AE (2000) The Phytoplankton of Windermere (English Lake District). FBA Special Publication 10. Freshwater Biological Association, Ambleside, Cumbria: 73 pp. AB

Reynolds CS, Lund JWG (1988) The phytoplankton of an enriched, soft-water lake subject to intermittent hydraulic flushing (Grasmere, English Lake District). Freshwater Biology 19: 479–404. C

Reynolds CS, Rogers DA (1976) Seasonal variations in the vertical distribution and buoyancy of *Microcystis aeruginosa* Kütz. emend. Elenkin in Rosthene Mere, England. Hydrobiologie 48: 17–23. AB

Reynolds CS, Walsby AE (1975) Water-blooms. Biological Reviews 50: 437–481. C

Rice CH (1938) Studies in the phytoplankton of the River Thames (1928–1932) 1 and 2. Annals of Botany NS 2: 539–557, 559–581. C

Rich F (1918) Notes on the algae of Leicestershire. Journal of Botany 56: 264–268. C

Rich F (1921) A new species of *Coelastrum*. New Phytologist 20(5): 234–238. B

Rich F (1925) Further notes on the algae of Leicestershire. Journal of Botany 63: 71–78, 229–238, 262–273, 322–330, 359–362. B

Rich F (1933) The algal flora of a Leicestershire pond. Transactions of the Leicester Literary and Philosophical Society 34: 6–36. C

Rieth A (1954) Beobactungen zur Entwicklungsgeschichte einer *Vaucheria* der Sektion Woroninia. Flora 142: 156–182. A

Rieth A (1963) Die Algen der chinesisch-deutschen biologischen-Sammelreise durch Nord- und NordostChina 1956. I. Die Vaucheriaceen. 1. Tril. Limnologica 1(4): 287–313. A

Rieth A (1980) Xanthophyceae. In: Süßwasserflora von Mitteleuropa 4(1). Gustav Fischer, Stuttgart: 1–147. A

Rindi F, Guiry MD (2002) Diversity, distribution and ecology of *Trentepohlia* and *Printzina* (Trentepohliales, Chlorophycota) in urban habitats in western Ireland. Journal of Phycology, in press. B

Rindi F, Guiry MD (2002) The genus *Phycopeltis* (Trentepohliaceae, Chlorophycota) in Ireland: a taxonomic and distributional reassessment. Phycologia: in press.

Rindi F, Guiry MD, Barbiero RP (1999) The marine and terrestrial Prasiolales (Chlorophyta) of Galway City, Ireland: a morphology and ecological study. Journal of Phycology 35: 469–482. B

Rintoul TC, Sheath RG, Vis ML (1999) Systematics and biogeography of the Compsopogonales with emphasis on the freshwater families in North America. Phycologia 38: 517–527. A

Robinson DG, Preston RD (1971) Studies on the fine structure of *Glaucocystis nostochinearum* Itzigs. II. Membrane morphology and taxonomy. British Phycological Journal 6: 113–128. C

Rosenburg M (1935) On the germination of *Lemanea torulosa* in culture. Annals of Botany 56: 621–622. A

Rosenberg M (1941) *Chrysochaete*, a new genus of the Chrysophyceae, allied to *Naegeliella*. New Phytologist 11: 304–315. B

Ross R, Sims P (1978) Notes on some diatoms from the Isle of Mull, and other Scottish localities. Bacillaria 1: 151–168. C

Roth AW (1797) Catalecta Botanica quibus Plantae novae et minus cognitae describunter atque illustrantur. 3 Fascicles [in 2 volumes]. Lipsiae. A

Round FE (1953) An investigation of two benthic algal communities in Malham Tarn, Yorkshire. Journal of Ecology 41: 174–197. B

Round FE (1955) Some observations on the benthic algal flora of four small ponds. Archiv für Hydrobiologie 50: 111–135. B

Round FE (1956a) A note on some communities of the littoral zone of lakes. Archiv für Hydrobiologie 52: 398–405. B

Round FE (1956b) The phytoplankton of three central water supply reservoirs in central Wales. Archiv für Hydrobiologie 52: 457–469. C

Round FE (1957a) The benthic algal flora of three City of Birmingham waterworks reservoirs in Central Wales. Archiv für Hydrobiologie 53: 562–573. C

Round FE (1957b) Studies on bottom-living algae in some lakes of the English Lake District Part II. The distribution of Bacillariophyceae on the sediments. Journal of Ecology 45: 343–360. B

Round FE (1957c) Studies on bottom-living algae in some lakes of the English Lake District Part III. The distribution on the sediment of algal groups other than the Bacillariophyceae. Journal of Ecology 45: 649–664. B

Round FE (1957d) The distribution of Bacillariophyceae on some littoral sediments of the English Lake District. Oikos 8(1): 16–37. B

Round FE (1957e) The late-glacial and post-glacial diatom succession in the Kentmere Valley deposit. New Phytologist 56: 98–126. C

Round FE (1957f) The diatom community of some Bryophyta growing on sandstone. Journal of the Linnean Society, Botany 55: 657–661. C

Round FE (1957g) A note on some diatom communities in

calcareous springs and streams. Journal of the Linnean Society, Botany 55: 662–668. B

Round FE (1957h) Studies on the bottom-living algae in some lakes of the English Lake District. Part I. Some chemical features of the sediments related to algal productivities. Journal of Ecology 45: 133–148. B

Round FE (1958) Observations on the diatom flora of Braunton Burrows, N. Devon. Hydrobiologia 11: 119–127. C

Round FE (1959) A comparative survey of the epipelic diatom flora of some Irish loughs. Proceedings of the Royal Irish Academy 60B: 193–215. B

Round FE (1960a) The diatom flora of a stream around Malham Tarn, Yorkshire. Archiv für Protistenkunde 104: 527–540. C

Round FE (1960b) The algal flora of a salt spring region at Manesty, Grange in Borrowdale. Naturalist, Hull (1960): 117–121. B

Round FE (1960c) Studies on bottom-living algae in some lakes of the English Lake District. Part IV. The seasonal cycles of Bacillariophyceae. Journal of Ecology 48: 529–547. B

Round FE (1960d) A note on the diatom flora of some springs in the Malham Tarn area of Yorkshire. Archiv für Protistenkunde 104: 515–526. B

Round FE (1961a) Studies on bottom-living algae in some lakes of the English Lake District. Part VI. The effects of depth on the epipelic algal community. Journal of Ecology 49: 245–254. B

Round FE (1961b) The diatoms of a core from Esthwaite Water. New Phytologist 60: 43–59. B

Round FE (1966) Studies on bottom-living algae in some lakes of the English Lake District. VI. The effect of depth on the epipelic diatom community. Journal of Ecology 49: 245–254. B

Round FE (1971) The taxonomy of the Chlorophyta. II. British Phycological Journal 6: 235–264. C

Round FE (1996) What characters define diatom genera, species and infraspecific taxa? Diatom Research 11: 203–218. A

Round FE (1997) Book review: Die cymbelloiden Diatomeen. Eine Monographie der weltweit bekannten Taxa. Teil 1. Allgemeines und Encyonema Part. by K. Krammer. Diatom Research 12: 155–157. B

Round FE, Brook AJ (1959) The phytoplankton of some Irish lochs and an assessment of their trophic status. Proceedings of the Royal Irish Academy B 60: 167–191. B

Round FE, Crawford RM, Mann DG (1990) The Diatoms: Biology and Morphology of the Genera. Cambridge University Press, Cambridge. 747 pp. A

Round FE, Eaton JW (1966) Persistent, vertical-migration rhythms in benthic microflora. 111. The rhythm of epipelic algae in a freshwater pond. Journal of Ecology 54: 609–615. C

Round FE, Happey CM (1965) Persistent, vertical-migration rhythms in benthic microflora. Part IV A diurnal rhythm of the epipelic diatom association in non-tidal flowing water. British Phycological Journal 2: 453–471. C

Roy J (1877) Contributions to the demid flora of Perthshire. Scottish Naturalist 1877: 68–74. C

Roy J (1883) List of desmids hitherto found on Mull. Scottish Naturalist 1833: 37–40. C

Roy J (1885) The fauna and flora of snow and ice. Scottish Naturalist 1885: 122–127. C

Roy J (1890a) Freshwater algae of the Enbridge Lakes and vicinity, Hampshire. Journal of Botany 28: 334–338. C

Roy J (1890b) The desmids of the Alford district. Scottish Naturalist 10: 199–210. C

Roy J (1891–1895) The Desmidieae of the Stoirmont district. Proceedings of the East Scottish Union of Natural Science 1891–1895: 19–25. C

Roy J (1892) The Desmidieae of East Fife. Annals of Scottish Natural History 1892: 192–197. C

Roy J (1893) On Scottish Desmidieae. Annals of Scottish Natural History 1893(6): 106–111; 1893(7): 170–180; 1893(8): 237–245. C

Roy J (1894) On Scottish Desmidieae. Annals of Scottish Natural History 1894(9): 40–46. C

Roy J (1897) Historical sketch of the freshwater algae of the east of Scotland. Scottish Naturalist 18: 148–159. C

Roy J, Bisset JP (1894) On Scottish Desmidieae. Annals of Scottish Natural History 1894(9): 40–46; (10): 100–105; (11): 167–178; (12): 241–256. C

Růžička J (1977) Die Desmidiaceen Mitteleuropas 1(1). Schweizerbart'sche, Stuttgart. 292 pp. A

Salah MM (1955) Some new diatoms from Blakeney Point (Norfolk). Hydrobiologia 7: 88–102. C

Santos LMA (1996) The Eustigmatophyceae: actual knowledge and research perspectives. Nova Hedwigia 60: 219–225. A

Sarma P (1986) The freshwater Chaetophorales on New Zealand. Beihefte Nova Hedwigia 58: 1–169, 142 pls. A

Say PJ, Diaz BM, Whitton BA (1977) Influence of zinc on lotic plants. I. Tolerance of *Hormidium* species to zinc. Freshwater Biology 7: 357–376. C

Say PJ, Whitton BA (1977) Influence of zinc on lotic plants. II Environmental effects on toxicity of zinc to *Hormidium rivulare*. Freshwater Biology 7: 377–384. B

Say PJ, Whitton BA (1980) Changes in flora down a stream showing a zinc gradient. Hydrobiologia 76: 255–262. C

Scannell MJP (1965) *Phycopeltis*, a genus of alga not previously recorded from the British Isles. The Irish Naturalist 15(3): 75. B

Scannell MJP (1972) *Bangia atropurpurea* (Roth) Ag. in Lough Derg, River Shannon. The Irish Naturalist 19: 25. B

Scannell MJP (1978) *Phycopeltis arundinacea* Mont. en Europe. Revue Algologique NS 13: 21–42. B

Schlichting HE, Jr (1975) Some subaerial algae from Ireland. British Phycological Society 10: 257–261. B

Schroeder WL (1930) Biological survey of the River Wharfe. III. Algae present in the Wharfe plankton. Journal of Ecology 18: 302–305. B

Scourfield DJ (1930) The nannoplankton and its collection by means of the centrifuge. Watson's Microscope Record 21: 2–6. B

Scourfield DJ (1944) The nannoplankton of bomb-crater pools in Epping Forest. Essex Naturalist 27: 231–241. C

Setchell WA, Gardner NL (1918) New Pacific Coast Algae. Publications in Botany, University of California at Berkeley. A

Seto R (1977) On the vegetative propagation of a freshwater red alga, *Hildenbrandia rivularis* (Liebm.) J. Ag. Bulletin of the Japanese Society of Phycology 25: 129–136 [in Japanese]. A.

Shaw S (1948) Notes and records. The North Western Naturalist 23: 166–167. A

Sheath RG (1984) The biology of freshwater red algae. Progress in Phycological Research 3: 89–157. A

Sheath RG (2002) Red Algae. In: Wehr JD, Sheath RG (eds) Freshwater Algae of North America. Ecology and Classification. Academic Press, San Diego. in press. A

Sheath RG, Cole KM (1990) Differential Alcian Blue staining in freshwater Rhodophyta. British Phycological Journal 25: 281–285. A

Sheath RG, Hambrook JA (1988) Mechanical adaptations to flow in freshwater red algae. Journal of Phycology 24: 107–111. A

Sheath RG, Hambrook JA (1990) Freshwater ecology. In: Cole KM, Sheath RG (eds) Biology of the Red Algae. Cambridge University Press, Cambridge: 423–453. A

Sheath RG, Müller KM (1997) Ultrastructure of carpogonia and carpogonial branches of *Batrachospermum helminthosum* and *Batrachospermum involutum* (Batrachospermales, Rhodophyta). Phycological Research 45: 177–181. A.

Sheath RG, Müller KM (1999) Systematic and phylogenetic relationships of the freshwater genus *Balbiania* (Rhodophyta). Journal of Phycology 35: 855–864. B

Sheath RG, Vis ML, Cole KM (1993a) Distribution and systematics of the freshwater red algal family Thoreaceae in North America. European Journal of Phycology 28: 231–241. A

Sheath RG, Vis ML, Cole KM (1993b) Distribution and systematics of *Batrachospermum* (Batrachospermales, Rhodophyta) in North America. 3. Section Setacea. Journal of Phycology 29: 719–725. A

Sheath RG, Vis ML, Cole KM (1994a) Distribution and systematics of *Batrachospermum* (Batrachospermales, Rhodophyta) in North America. 4. Section Virescentia. Journal of Phycology 30: 108–117. A

Sheath RG, Vis ML, Cole KM (1994b) Distribution and systematics of *Batrachospermum* (Batrachospermales, Rhodophyta) in North America. 6. Section Turfosa. Journal of Phycology 30: 872–884. A

Sherwood AR, Sheath RG (2000) Biogeography and systematics of *Hildenbrandia* (Rhodophyta, Hildenbrandiales) in Europe: inferences from morphometrics and *rbc*L and 18S rRNA gene sequence analyses. European Journal of Phycology 34: 523–532. A

Sibthorp J (1794) Flora Oxoniensis, Exhibens Plantas in Agro Oxoniensi Sponte Crescentes Secundum Systema Sexuale Distributas. Oxonii, Christchurch Meadow. 422 pp. C

Simons J (1974) *Vaucheria compacta*: A euryhaline species. Acta Botanica Neerlandica 23(5): 613–626. A

Simons J, van Beem AP, de Vries PJR (1986) Morphology of the prostrate thallus of *Stigeoclonium* (Chlorophyceae, Chaetophorales) and its taxonomic implications. Phycologia 25: 210–220. A

Sinclair C, Whitton BA (1977a) Influence of nutrient deficiency on hair formation in the Rivulariaceae. British Phycological Journal 3: 297–313. C

Sinclair C, Whitton BA (1977b) Influence of nitrogen deficiency on morphology of Rivulariaceae (Cyanophyta). Journal of Phycology 13: 335–340. C

Sinker CA (1962) The North Shropshire meres and mosses: a background for ecologists. Field Studies 1: 101–138. C

Skidmore RE, Maberly S, Whitton BA (1998) Patterns of spatial and temporal variation in phytoplankton biomass in the River Trent and its tributaries. Science of the Total Environment 210/211: 357–366. C

Skuja H (1939) Beitrag zur Algenflora Lettlands II. Acta Horti Botanici Universitatis Latviensis 11/12: 41–169. A

Skuja H (1948) Taxonomie des Phytoplanktons einiger Seen in Uppland, Schweden. Symbolae Botanicae Upsalienses 9(3): 1–399. A

Skuja H (1949) Zur Süsswasseralgen Flora Burmas. Nova Acta Regiae Societatis Scientiarum Upsaliensis Ser. IV, 14: 1–139. A

Skuja H (1956) Taxonomische und biologische Studien über das Phytoplankton schwedischer Binnengewässer. Nova Acta Regiae Societatis Scientiarum Upsaliensis Ser. IV, 16(3): 1–404. A

Skulberg OM, Skulberg R (1985) Planktic species of *Oscillatoria* (Cyanophyceae) from Norway. Characterization and classificiation. Archiv für Hydrobiologie, Suppl. 71 [Algological Studies 38/39]: 157–174. A

Sluiman HJ, Gärtner G (1990) Taxonomic studies on the genus *Pleurastrum* (Pleurastrales, Chlorophyta). I. The type species, *P. insigne*, rediscovered and isolated from soil. Phycologia 29: 133–138. A

Sluiman HJ, Kouets FAC, Blommers CJ (1989) Classification and definition of cytokinetic patterns in green algae: sporulation versus (vegetative) cell division. Archiv für Protistenkunde 137: 277–290. A

Smith AM (1933) *Gonyostomum semen* Diesing. A flagellate now first recorded for the British Isles. Naturalist, London 1933: 49–50. C

Smith AM (1942) The algae of Miles Rough Bog, Bradford. Journal of Ecology 30: 341–356. C

Smith GM (1916) New or interesting algae from the lakes of Wisconsin. Bulletin Torrey Botanical Club 43: 471–483. A

Smith GM (1920) Phytoplankton of the inland lakes of Wisconsin, Part I. Bulletin of the Wisconsin Geological and Natural History Survey 57: 1–243. A

Smith GM (1924) Part II Desmidiaceae. Bulletin of the Wisconsin Geological and Natural History Survey 57: 1–227. A

Smith GM (1933) The Fresh-Water Algae of the United States, 1st Edn. McGraw-Hill, New York. 716 pp. A

Smith GM (1950) The Fresh-Water Algae of the United States, 2nd edn. McGraw-Hill, New York. 719 pp. A

Smith I, Lyle A (1979) Distribution of Freshwaters in Great Britain. Institute of Terrestrial Ecology, Edinburgh. 44 pp. A

Smith W (1853) A Synopsis of the British Diatomaceae; with Remarks on their Structure, Function and Distribution; and Instructions for Collecting and Preserving Specimens. Volume 1. John van Voorstn, London. 89 pp. C

Soderström J (1963) Studies in *Cladophora*. Botanica Gothoburgensia 1: 1–147. A

Speller FM (1990) A contribution to our knowledge of the taxonomy of discoidal centric diatoms based upon observations of populations from the River Thames, England. Hydrobiologia 190: 15–32. C

Spence DHN (1967) Factors controlling the distribution of freshwater macrophytes with particular reference to the lochs of Scotland. Journal of Ecology 55: 147–170. A

Spencer LB (1971) A study of *Vacuolaria virescens* Ceinkowski. Journal of Phycology 7: 274–278. A

Stafleu FA (1967) Taxonomic literature. Regnum Vegetabile 52: 1–556. A

Stanier RY, Sistrom WR, Hansen TA, Whitton BA, Castenholz RW, Pfennig N, Gorlenko VN, Kondratieva EMN, Eimhjellen RF, Whittenbury R, Gerna R, Trüper HG (1978) Proposal to place the nomenclature of the cyanobacteria (blue-green algae) under the rules of the International Code of Nomenclature of Bacteria. International Journal of Systematic Bacteriology 28: 335–336. A

Starmach K (1966) Flora Słodkowodna Polski [Freshwater Flora of Poland] 2. Cyanophyta-Sinice. Glaucophyta-Glaukofity. Polska Akademia Nauk, Warszawa. 807 pp. AB

Starmach K (1969) Growth of thalli and reproduction of the red alga *Hildenbrandia rivularis* (Liebm.) J. Ag. Acta Societatis Botanicorum Polonia*e* 38: 523–533. A

Starmach K (1977) Flora Słodkowodna Polski [Freshwater Flora of Poland]14. Phaeophyta-Brunatnice, Rhodophyta-Krasnorosty. Polska Akademia Nauk, Warszawa & Kraków. 445 pp. A

Starmach K (1985) Chrysophyceae und Haptophyceae. In: Ettl H, Gerloff J, Heynig H, Mollenhauer D (eds) Süßwasserflora von Mitteleuropa 5. Gustav Fischer, Stuttgart: 1–515. A

Starr RC (1954) Reproduction by zoospores in *Planktosphaeria gelatinosa*. Hydrobiologia 6: 392–397. A

Starr RC (1955) A comparative study of *Chlorococcum meneghini* and other spherical, zoospore-producing genera of the Chlorococcales. Indiana University Publications Science Series 20: 1–111. B

Starr RC (1980) Colonial chlorophytes. In: Cox ER (ed.) Phytoflagellates. Elsevier, Amsterdam: 147–164. A

Stein JR (ed.) (1973) Handbook of Phycological Methods. Cultural Methods and Growth Measurements. Cambridge University Press, Cambridge. 448 pp. A

Steinberg C (1981) Mass occurrence of the rare dinoflagellate *Woloszynska tenuissima* (Lauterborn) Thompson, in a small lake after restoration treatment. Archiv für Hydrobiologie 60: 289–297. A

Stewart NF, Church JM (1992) Red Data Book of Britain and Ireland: Stoneworts. 144 pp. JNCC, Peterborough. B

Stockner JG, Callieri C, Cronberg G (2000) Picoplankton and other non-bloom-forming cyanobacteria in lakes. In: Whitton BA, Potts M (eds) Ecology of Cyanobacteria. Kluwer Academic Publishers, Dordrecht: 195–231. A

Stockner JG, Lund JWG (1970) Live algae in post-glacial lake deposits. Limnology and Oceanography 15: 41–58. C

Svedelius N (1930) Über die sogenanntem Süsswasser-Lithodermen. Zeitschrift für Botanik 23: 892–918. A

Swale EMF (1959) Some uncommon Chlorophyceae from the River Lee. British Phycological Journal 1: 71–72. C

Swale EMF (1962a) Notes on some algae from the Reddish Canal. British Phycological Journal 2: 174–175. B

Swale EMF (1962b) The development and growth of *Thorea ramosissima* Bory. Annals of Botany NS 26: 105–116. B

Swale EMF (1963a). Note on the taxonomy of *Pteromonas angulosa* (Carter) Lemmermann. British Phycological Bulletin 2: 259. C

Swale EMF (1963b) Notes on the morphology and anatomy of *Thorea ramosissima* Bory. Journal of the Linnean Society, Botany 58: 429–434. B

Swale EMF (1964) A study of the phytoplankton of a calcareous river. Journal of Ecology 52: 433–446. C

Swale EMF (1965) Observations on a population of *Lagerheimia genevensis* Chodat in Oakmere, Cheshire. British Phycological Journal 2: 422–428. B

Swale EMF (1968) The phytoplankton of Oak Mere, Cheshire, 1963–1966. British Phycological Bulletin 3: 441–449. B

Swale EMF (1969) A study of the nannoplankton flagellate *Pedinella hexacostata* Vyssotskii by light and electron microscopes. British Phycological Journal 4: 65–86. B

Swale EMF, Belcher JH (1963) Morphological observations on wild and cultured material of *Rhodochorton investiens* (Lenormand) nov. comb. (*Balbiania investiens* (Lenorm.) Sirodot). Annals of Botany 27: 281–290. B

Swale EMF, Belcher JH (1964) The algal flora of the River Lee. 3. Volvocales and Chlorococcales. Naturalist, Hull 1964: 193–199. B

Swale EMF, Belcher JH (1966) *Ochromonas ostreaeformis* nov. sp., a large compressed chrysomonad. New Phytologist 65: 267–272. B

Symoens JJ, van der Werff H (1951) Note sur des formations de tuf calcaire des environs de Consdorf (Grand-Duché de Luxembourg). Bulletin de la Société Royale de Botanique de Belgique 83: 213–217. A

Szymańska H, Spalik K (1993) Typification of names enumerated in Pringsheim monograph of *Coleochaete* (Charophyceae). Archiv für Hydrobiologie, Suppl. 98 [Algological Studies 70]: 29–37. A

Talling JF, Heaney SI (1988) Long term changes in some English (Cumbrian) lakes subjected to increased nutrient inputs. In: Round FE (ed.) Algae and the Aquatic Environment. Contributions in Honour of J.W.G. Lund. Biopress, Bristol: 1–29. AC

Taylor FJ (1954) A new Chrysophycean flagellate *Sphaerobryon fimbriata* gen. et spec. nov. Hydrobiologia 6: 369–371. B

Taylor FJ (1964) Notes on the phytoplankton of Saddington Reservoir, Leicestershire. Naturalist, Hull 1954: 141–148. C

Teiling E (1916) En kaledonisk fytoplanktonformation. Svensk Botanisk Tidskrift 10: 506–519. A

Teiling E (1967) The desmid genus *Staurodesmus.* A taxonomic study. Archiv für Botanik 6: 467–629. C

Tenaud M, Ohmori M, Miyachi S (1989) Inorganic carbon and acetate assimilation in *Botryococcus braunii* (Chlorophyta). Journal of Phycology 25: 662–667. A

Thirb HH, Benson-Evans K (1982a) The effect of different current velocities on the red alga *Lemanea* in a laboratory stream. Archiv für Hydrobiologie 96: 65–72. A

Thirb HH, Benson-Evans K (1982b) Cytological studies on *Lemanea fluviatilis* L. in the River Usk. British Phycological Journal 17: 401–409. B

Thirb HH, Benson-Evans K (1985) The effect of water temperature, current velocity and suspended solids on the distribution, growth and seasonality of *Lemanea fluviatilis* (C. Ag.), Rhodophyta, in the River Usk and other South Wales rivers. Hydrobiologia 127: 63–78. B

Thompson RH, Wujek DE (1992) *Printzina* gen. nov. (Trentepohliaceae), including a description of a new species. Journal of Phycology 28: 232–237. A

Thompson RH, Wujek SE (1997) Trentepohliales: *Cephaleuros, Phycopeltis,* and *Stomatochroon.* Science Publishers, New Hampshire. 149 pp. A

Thwaites GHK (1851) On the early stages of development of *Lemanea fluviatilis* Agardh. Transactions of the Linnean Society 20: 399–402. A

Tiffany LH (1929) A key to the species, varieties, and forms of the algal genus *Oedogonium.* Ohio Journal of Science 29: 62–80. A

Tiffany LH (1930) The Oedogoniaceae. Columbus, Ohio. 253 pp. A

Tiffany LH (1934) The Oedogoniaceae. Supplementary Paper No. 1. Ohio Journal of Science 34: 323–326. A

Tiffany LH (1937a) Oedogoniales. North American Flora 11: 1–102. A

Tiffany LH (1937b) The filamentous algae of the west end of Lake Erie. American Midland Naturalist 18: 911–951. A

Tompkins J, De Ville MM, Day JG, Turner MF (1995) Culture Collection of Algae and Protozoa. Catalogue of Strains 1995. Natural Environment Research Council. 203 pp. A

Trainor FR (1963) Zoospores in *Scenedesmus obliquus.* Science 142: 1673–1674. A

Trainor FR (1992) Cyclomorphosis in *Scenedesmus communis* Hegewald. Ecomorph expression at low temperature. British Phycological Bulletin 27: 75–81. A

Trainor FR (1998) Biological aspects of *Scenedesmus* (Chlorophyceae) – phenotypic plasticity. Behefte Nova Hedwigia 117: 1–367. A

Trainor FR, Egan PF (1991) Discovering the various ecomorphs of *Scenedesmus*: the end of a taxonomic era. Archiv für Protistenkunde 139: 125–132. A

Transeau EN (1951) The Zygnemataceae. Ohio State University Press, Columbus. 327 pp. A

Tsarenko PM (1990) Kratig opredelitel chlorokokkovich vodoroslej Ukrainskoj SSR. Nauk. Dumkai, Kiev [in Russian]. 208 pp. A

Tsarenko PM (2000) Chlorococcales. In: Wasser SP, Tsarenko PM (eds) Diversity of algae of Ukraine. Algologia 10(4): 1–309 [in Russian]. A

Tsarenko PM, Petlevaniy OA (2001) Addition to the 'Diversity of algae of Ukraine'. Algologia (Supplement). Institute of Botany, Kiev [in Russia]. 130 pp. A

Tschermak-Woess E (1989) The algal partner. In: Galun N (ed.) CRC Handbook of Lichenology I. CRC-Press, Boca Raton, Florida: 39–92. A

Tupa DD (1974) An investigation of certain Chaetophoralean algae. Beihefte Nova Hedwigia 46: 1–155. B

Turner WB (1883) The algae of Stensall Common. Naturalist, Hull 8: 77–83. C

Turner WB (1885) On some new and rare desmids. Journal of the Royal Microscopic Society II 5: 933–940. C

Turner WB (1886) Notes on freshwater algae with descriptions of new species. Naturalist, Hull 12: 33–35. C

Turner WB (1887) Notes on algae collected at Gormire and Thirkle Bay, with descriptions of a new form. Naturalist, Hull 13: 275–276. C

Turner WB (1893) Desmid (biological) notes. Naturalist, Hull 18: 342–347. C

Turner WB, Abbot JE, Hick T, Stubbins J, West W (18??) Algae. In: Contributions to a fauna and flora of East Yorkshire. Transactions of the Leeds Naturalist Club 1: 69–78. C

UKNCC (2001) Catalogue of the UK National Culture Collection (UKNCC). List of Algae and Protozoa. First Edition, 231 pp. B

Umezaki I (1961) The marine blue-green algae of Japan. Memoirs of the College of Agriculture, Kytoto University 83, Fisheries Series 8: 1–149. A

van der Berg MS (1998) Charophyte colonization in shallow lakes. Thesis, Vrije Universiteit Amsterdam, Den Haag. 135 pp. A

van Goor ACJ (1918) Zur Kenntnis der Oscillatoriaceen. Reçueil des Travaux Botaniques Néerlandais 15: 255–262. A

van Raam JC (1998) Handboek Kranswieren. Chara boek, Hilversum. 200 pp. A

Vaucher J-P (1803) Histoires des Conferves d'Eaux Douce. Geneva. 304 pp. A

Venkataraman GS (1961) Vaucheriaceae. Indian Council for Research Research Monograph on Algae, New Delhi. 112 pp. A

Vis ML, Sheath RG (1992) Systematics of the freshwater red algal family Lemaneaceae in North America. Phycologia 31: 164–179. A

Vis ML, Sheath RG (1993) Distribution and systematics of *Chroodactylon* and *Kyliniella* (Porphyridiales, Rhodophyta) from North American streams. Japanese Journal of Phycology 41: 237–241. A

Vis ML, Sheath RG (1998) A molecular and morphological investigation of the relationship between *Batrachospermum spermatoinvolucrum* and *B. gelatinosum* (Batrachospermales, Rhodophyta). European Journal of Phycology 33: 231–239. A

Vis ML, Sheath RG (1999) A molecular investigation of the systematic relationships of *Sirodotia* species (Batrachospermales, Rhodophyta) in North America. Phycologia 38: 261–266. A

Vis ML, Sheath RG, Cole KM (1992) Systematics of the freshwater red algal family Compsopogonaceae in North America. Phycologia 31: 564–575. A

Vis ML, Sheath RG, Entwisle TJ (1995) Morphometric analysis of *Batrachospermum* section *Batrachospermum* (Batrachospermales, Rhodophyta) type specimens. European Journal of Phycology 30: 35–55. A

Vodeničarov DG (1989) Die Gattung *Neglectella* Vodenič. et Benderl. und zwei neue Gattungen: *Neglectellopsis* gen. nov. und *Skujaster* gen. nov (Chlorophyta, Chlorococcales). Archiv für Hydrobiologie 82 [Algological Studies 57]: 409–424. A

Vogel S (1984) Drag and flexibility in sessile organisms. American Zoology 24: 37–44. A

Volker AR, Frank C, Hartmann EC, Hirmer M, Kloboucek A, Seidel BM, Wenzeler P, Kessler E (1999) Biochemical taxonomy and molecular phylogeny of the genus *Chlorella* sensu lato (Chlorophyta). Journal of Phycology 35: 587–598. A

von Stosch HA (1973) Observations on vegetative reproduction and sexual life cycle of two freshwater dinoflagellates, *Gymnodinium pseudopalustre* Schiller and *Woloszynskia apiculata* sp. nov. British Phycological Journal 8: 105–134. A

Waern M (1938) Om *Cladophora aegagropila*, *Nostoc pruniforme*,och andra lager i Lilla Ullevifjärden, Mälaren. Botanischer Notiser 1938: 129–142. A

Waern M (1952) Rocky-shore algae in the Öregund Archipelago. Acta Phytogeographica Suecica 30: 1–298. A

Wailes GH (1939) The plankton of Lake Windermere, England. Af'n. Magazine of Natural History Ser. 2: 1–414. B

Watt F, Brook AJ, Crime GW (1990) Nuclear microscopy of the alga *Pandorina morum* (Müll.) Bory. Nuclear Instruments and Methods in Physics Research B49: 465–471. A

Wehr JD, Stein JR (1985) Studies on the biogeography and ecology of the freshwater phaeophycean alga *Heribaudiella fluviatilis*. Journal of Phycology 21: 81–93. A

Weiss FE, Murray H (1909) On the occurrence and distribution of some alien aquatic plants in the Reddish Canal. Memoirs and Proceedings of the Manchester Literary and Philosophical Society 53: 1–8. C

West GS (1899a) On variation in the Desmidieae and its bearing on their classification. Journal of the Linnean Society, Botany 34: 366–416. C

West GS (1899b) The alga-flora of Cambridgeshire. Journal of Botany 37: 6, 49–58, 106–116, 216–225, 262–268, 291–294. B

West GS (1904) A Treatise on the British Freshwater Algae. Cambridge University Press, Cambridge. 372 pp. B

West GS (1907) Report on the freshwater algae, including phytoplankton of the Third Tanganyika Expedition, conducted by Dr. W.A. Cunnington 1904–1905. Journal of the Linnean Society, Botany 38: 81–197. A

West GS (1908) Some critical green algae. Journal of the Linnean Society, Botany 38: 279–289. C

West GS (1909) A biological investigation of Peridinieae of Sutton Park, Warwickshire. New Phytologist 8(5–6): 181–196. C

West GS (1910a) An epitome of a comparative study of the dominant phanerogamic and higher cryptogamic flora of aquatic habit, in seven lake areas in Scotland. Bathymetrical survey of the Scottish freshwater lochs. Report on the Scientific Results 6: 56–260. C

West GS (1910b) VI. – A further contribution to a comparative study of the dominant phanerogamic and higher cryptogamic flora of aquatic habitats in Scottish Lakes. Proceedings of the Royal Society of Edinburgh 30, 2(6): 66–181. B

West GS (1911) Algological notes II–IV. Journal of Botany, British and Foreign 49: 83–89. C

West GS (1912) Algological notes, X–XIII. Journal of Botany 50: 328–331. B

West GS (1915) Algological notes XIV–XVII. Journal of Botany 53: 73–84. B

West GS (1916a) Algological notes XVIII–XXIII. Journal of Botany 54: 1–10. B

West GS (1916b) Algae I. Cambridge University Press, Cambridge. 475 pp. B

West GS (1918) A new species of *Gongrosira*. Journal of the Royal Microscopial Society 1918: 30–31. B

West GS (1920) *Roya anglica*. Journal of Botany 58: 65. C

West GS, Fritsch FE (1927) A Treatise on the British Freshwater Algae. Cambridge University Press, Cambridge. 534 pp. [Facsimile edition printed in 1968 as Vol. 3 of Bibliotheca Phycologica. J. Cramer, Lehre]. B

West GS, Hodgetts WJ (1920) *Roya anglica*: a new desmid. Journal of Botany, British and Foreign 58: 65–69. C

West GS, Starkey CB (1915) A contribution to the cytology and life-history of *Zygnema ericetorum* (Kütz.) Hansg., with some remarks on the genus *Zygogonium*. New Phytologist 14: 194–205. C

West GS, West W (1906). A comparative study of the plankton of some Irish Loughs. Proceedings of the Royal Irish Academy 32: 77–116. B

West T (1860) Remarks on some Diatomaceae new or imperfectly described and a new desmid. Transactions of the Royal Microscopical Society II, 8: 147–153. C

West W (1880) Further additions to Mr W.B. Turner's List of Algae. Naturalist, Hull NS 5(58): 116. B

West W (1882a) Additions to the West Riding Algae. Naturalist, Hull 8(86): 27. C

West W (1882b) Algae near Mirfield. Naturalist, Hull 9(97): 20. C

West W (1883) The principal plants of Malham. Naturalist, Hull 9(98): 25–27. C

West W (1885) Habitats of freshwater algae. Naturalist, Hull 10(118): 232. C

West W (1889a) Additions to the algae of West Yorkshire. Naturalist 164: 87–96. C

West W (1889b) Additions to the algae of West Yorkshire. Naturalist 165: 97–100. C

West W (1889c) Algae in Upper Swaledale. Naturalist 169: 246. C

West W (1889d) The freshwater algae of North Yorkshire. Journal of Botany 27: 289–298. C

West W (1890) Contribution to the freshwater algae of North Wales. Journal of the Royal Microscopical Society 6: 277–306. B

West W (1891) Additions to the freshwater algae of North Yorkshire. Naturalist 193: 243–252. C

West W (1892a) A contribution to the freshwater algae of West Ireland. Journal of the Linnean Society, Botany 29: 103–216. B

West W (1892b) Algae of the English Lake District. Journal of the Royal Microscopical Society 1892: 713–748. B

West W (1893) Notes on Scotch freshwater algae. Journal of Botany, British and Foreign 31: 97–104. B

West W (1912) The freshwater algae of Clare Island. Clare Island Survey; Part 16. Proceedings of the Royal Irish Academy 31: 1–62. B

West W, West GS (1893) Notes on the freshwater algae of the East Riding of Yorkshire. Naturalist, Hull 216: 211–221. C

West W, West GS (1894) New British freshwater algae. Journal of the Royal Microscopical Society 1894: 1–17. B

West W, West GS (1896) III. On some new and interesting freshwater algae. Journal of the Royal Microscopical Society 1896: 149–165. C

West W, West GS (1897) VIII. A contribution to the freshwater algae of the south of England. Journal of the Royal Microscopical Society 17: 467–511. B

West W, West GS (1898) Notes on freshwater algae. I. Journal of Botany 36: 330–338. B

West W, West GS (1900) Notes on freshwater algae. II. Journal of Botany 38: 289–299. C

West W, West GS (1901) The alga-flora of Yorkshire. Botanical Transactions of the Yorkshire Naturalists Union 5: 1–239. B.

West W, West GS (1902) A contribution to the freshwater algae of North of Ireland. Transactions of the Royal Irish Academy 32B: 1–101. B

West W, West GS (1903a) Scottish freshwater plankton. 1. Journal of the Linnean Society, Botany 35: 519–556. B

West W, West GS (1903b) Notes on freshwater algae III. Journal of Botany 41: 33–41, 74–82. B

West W, West GS (1904, 1905, 1911, 1912, 1923) A Monograph of the British Desmidiaceae Vols. I–IV. [Vol. V by Carter N]. The Ray Society, London. C

West W, West GS (1905a) Freshwater algae from the Orkneys and Shetlands. Transactions and Proceedings of the Botanical Society of Edinburgh 69: 3–41. C

West W, West GS (1905b) Further contribution to freshwater plankton of the Scottish Lochs. Transactions of the Royal Society of Edinburgh 41: 477–518. B

West W, West GS (1906) A comparative study of the plankton of some Irish Lakes. Transactions of the Royal Irish Academy 33B: 77–116. B

West W, West GS (1908) Algae from Austwick Moss, North Yorkshire. Naturalist, Hull 1908(614): 101–103. C

West W, West GS (1909a) The British freshwater phytoplankton, with special reference to desmid-plankton and the distribution of British desmids. Proceedings of the Royal Society, London B 81: 165–206. C

West W, West GS (1909b) The phytoplankton of the English Lake District. Naturalist, Hull 1909: [in six parts] 626: 115–122; 627: 134–141; 628: 186–193; 629: 260–267; 631: 287–292; 632: 323–331. C

West W, West GS (1912) On the periodicity of the phytoplankton of some British Lakes. Journal of the Linnean Society, Botany 40: 395–432. C

West W, West GS (1920) Freshwater Algae. In: Malden HE (ed.) A History of the County of Surrey. Victoria History of the Counties of England. Surrey. Part 2, Botany. Constable & Company, London: 57–60. C

Wettstein FV (1924) Handbuch der Sytstematischen Botanik. Band I, 3rd edn. Leipzig. A

Wheeler BD, Whitton BA (1971) Ecology of Hell Kettles. 1. Terrestrial and sub-aquatic vegetation. Vasculum 56: 25–37. C

Whitford LA, Schumacher GJ (1969) A Manual of the Fresh-Water Algae in North Carolina. Technical Bulletin of the North Carolina Agricultural Experiment Station 188. 313 pp. A

Whitton BA (1966) Algae in St. James's Park Lake. The London Naturalist 45: 26–28. C

Whitton BA (1967) Phosphate accumulation by colonies of *Nostoc*. Plant Cell Physiology, Tokyo 8: 293–296. C

Whitton BA (1969) Seasonal changes in the phytoplankton of St. James's Park Lake, London. The London Naturalist 48: 14–39. BC

Whitton BA (1970) Biology of *Cladophora* in freshwater. Water Research 4: 457–476. C

Whitton BA (1974) Changes in the British freshwater algae. In: Hawksworth DL (ed.) The Changing Flora and Fauna of Britain [Systematics Association Special Volume 6]. Academic Press, London: 115–141. C

Whitton BA (1988) Hairs in eukaryotic algae. In: Round FE (ed.) Algae and the Aquatic Environment. Biopress, Bristol: 226–460. AB

Whitton BA (2000) Increases in nuisance macro-algae in rivers: a review. Verhandlungen der Internationale Vereinigung für Theoretische und Angewandte Limnologie 27: 1257–1259. B

Whitton BA, Balbi DM, Gemmell JJ, Robinson PJ (2002) Blue-Green Algae of the British Isles Interactive Key to the Species. (CD-ROM) Department of Biological Sciences, University of Durham, Durham DH1 3LE ISBN 0-9538842-0-1.

Whitton BA, Boulton PNG, Clegg EM, Gemmell JJ, Graham GG, Gustar R, Moorhouse TP (1998a) Long-term changes in macrophytes of British rivers: I. River Wear. Science of the Total Environment 210–211: 411–426. B

Whitton BA, Brierley SJ (2001) Cyanobacteria (blue-green algae) In: Hawksworth DL (ed.) The Changing Wildlife of Great Britain and Ireland [Systematics Association Special Volume 62]. Taylor & Francis, London: 150–151. C

Whitton BA, Buckmaster RC (1970) Macrophytes of the River Wear. Naturalist, Hull 1914: 97–116. C

Whitton BA, Crisp DT (1984) Tees. In: Whitton BA (ed.) Ecology of European Rivers. Blackwell Scientific, Oxford: 145–178. B

Whitton BA, Dalpra M (1968) Floristic changes in the River Tees. Hydrobiologia 32: 545–550. B

Whitton BA, Diaz BM (1981) Influence of environmental factors on photosynthetic species composition in highly acidic waters. Verhandlung der Internationalen Vereinigung für Theoretische und Angewandte Limnologie 21: 1459–1465. C

Whitton BA, Harding JPC (1978) Influence of nutrient deficiency on hair formation in *Stigeoclonium*. British Phycological Journal 13: 65–68. B

Whitton BA, Holmes NTH, Sinclair C (1978) A Coded List of 1000 Freshwater Algae of the British Isles. Water Data Unit, Department of the Environment, Reading Bridge House, Reading. 335 pp. A

Whitton BA, John DM, Johnson LR, Boulton PNG, Kelly MG, Haworth EY (1998a) A Coded List of Freshwater Algae of the British Isles. NERC, LOIS Publication Number 222. Institute of Hydrology, Wallingford. 274 pp [also in several computer-accessible forms]. B

Whitton BA, Peat A (1969) On *Oscillatoria redekei* van Goor. Archiv für Mikrobiologie 68: 372–376. B

Whitton BA, Potts M (eds) (2000) The Ecology of Cyanobacteria. Kluwer, Dordrecht. 669 pp. A

Whitton BA, Yelloly JM, Christmas M, Hernández I (1998b) Surface phosphatase activity of benthic algal communities in a stream with highly variable ambient phosphate concentrations. Verhandlung Internationalen Vereinigung für Theoretische und Angewandte Limnologie 26: 967–972. B

Wichmann L (1937) Studien über die durch H-Stücken der Membran ausgeseichneten Gattungen *Microspora, Binuclearia, Ulotrichopsis* und *Tribonema*. Pflanzenforschung 20: 11–110. A

Wilce RT (1966) Pleurocladia lacustris in Arctic America. Journal of Phycology 2: 57–66. A

Wilcock J, Perry C, Williams RPJ, Brook AJ (1989) Biological minerals formed from strontium and barium sulphates. II. Crystallography and control of mineral morphology in desmids. Proceedings of the Royal Society of London 238B: 481–484. A

Wille N (1898) Beschreibung einiger Planktonalgen aus norwegischen Süßwasserseen. Biologisches Centralblatt 18: 302. A

Williams DM, Hartley B, Ross R, Munro MAR, Juggins S, Battarbee RW (1988). A Coded Checklist of British Diatoms. Ensis, London. 74 pp. A

Williams DM, Round FE (1986) Revision of the genus *Synedra* Ehrenb. Diatom Research 1: 313–339. B

Williams DM, Round FE (1987) Revision of the genus *Fragilaria*. Diatom Research 2: 267–288. B

Williams EG (1933) Notes on the plankton of small pieces of water. The North Western Naturalist 1933: 293–301. C

Williams EG (1934) The nannoplankton of ponds. Watsons Microscope Record 33: 14–17. C

Williams EG (1941) Two planktonic algae of the Chester District. The North Western Naturalist 1941: 162–164. B

Williams EG (1950) Observations on the micro-organisms of Bala Lake, Merioneth. Proceedings of the Chester Society of Natural Sciences, Literature and Arts 1949: 182–186. C

Williams EG (1951) Plankton organisms of the River Dee, near Chester. Proceedings of the Cheshire and North Wales Natural History Society 4: 143–152. C

Williams EG (1965) Plankton algae from the Serpentine in Eaton Park, Cheshire. British Phycological Journal 2: 429–450. B

Williams EG (1966) Phytoplankton of small bodies of water. British Phycological Bulletin 3: 75–79. C

Williams EG (1968) Notes on two algae of small bodies of water and a note on *Quadricoccus laevis* Fott. British Phycological Journal 3: 515–518. B

Williams EG, Lund JWG (1972) The structure and development of form of *Dinococcus oedogonii* (Geitler) Fott. Nova Hedwigia 23: 493–513. C

Williams PF, Benson-Evans K, Jones AK (1974) Further analysis of data from biological surveys of the River Taff. Water Pollution Control 73: 576–583. C

Williams R, Williams PF, Benson-Evans K, Hunter MD, Harcup MF (1976) Chemical and biological studies associated with the recovery of the River Ebbw Fawr (1970–74). Water Pollution Control 75: 428–444. C

Williamson DB (1984) The clumping of benthic desmids and their light-orientation. Microscopy. Journal of the Quekett Microscopical Club 35: 120–123. C

Williamson DB (1990) *Mesotaenium caldariorum* (Lagerh.) Hansgirg: a new find of a very rare British desmid. Microscopy. Journal of the Quekett Microscopical Club 36: 409–411. C

Williamson DB (1991a) The desmid floras of small temporary pools. Microscopy. Journal of the Quekett Microscopical Club 36: 539–544. C

Williamson DB (1991b) Two interesting *Actinotaenium* (Desmidiaceae) species from Leicestershire. Microscopy. Journal of the Quekett Microscopical Club 36: 545–548. C

Williamson DB (1992a) On new or interesting desmids from the British Isles. Botanical Journal of Scotland 46: 97–106. C

Williamson DB (1992b) A contribution to our knowledge of the desmid flora of the Shetland Islands. Botanical Journal of Scotland 46: 233–285. C

Williamson DB (1993) *Cosmarium corbula* Bréb. var. *hians* var. nov. (Zygnemaphyceae, Desmidiaceae) a new variety of an uncommon desmid. Quekett Journal of Microscopy 37: 42–44. C

Williamson DB (1994) Observations on desmids from Scotland, Shetland, Cumbria and the English Midlands. Botanical Journal of Scotland 47:113–122. C

Williamson DB (1995a) Two desmid taxa newly recorded for the British Isles. Botanical Journal of Scotland 47: 279–282. C

Williamson DB (1995b) On four rare *Cosmarium* Chlorophyta: (Desmidiaceae) species from Sutherland, Scotland. Journal of the Quekett Journal of Microscopy 37: 485–488. C

Williamson DB (1996a) A survey of the desmid flora of Assynt, West Sutherland. Botanical Journal of Scotland 48: 235–279. C

Williamson DB (1996b) *Closterium anguineum* spec. nov. (Zygnemaphyceae), a new desmid from Sutherland. Botanical Journal of Scotland 48: 281–285. C

Williamson DB (1997a) Rare desmids from Scotland. Archiv für Hydrobiologie, Suppl. 118 [Algological Studies 84]: 53–81. C

Williamson DB (1997b) A further contribution to our knowledge of rare desmids from Scotland. Botanical Journal of Scotland 50: 99–114. C

Williamson DB (2000) Some desmid floras of wet rock surfaces. Archiv für Hydrobiologie, Suppl.132 [Algological Studies 97]: 11–27. C

Wills AW (1880) List of Desmidieae found in Sutton Park, Warwickshire.The Midland Naturalist 3: 265–266. C

Wills AW (1881) On the Desmidieae of North Wales. The Midland Naturalist 4: 40–43. C

Wills AW (1886) Handbook of Birmingham prepared for the members of the British Association, Algae. pp. 335–338. C

Wilmotte A (1988) Growth and morphological variability of six strains of *Phormidium* cf. *ectocarpi* Gomont (Cyanophyceae) cultivated under different temperatures and light intensities. Archiv für Hydrobiologie, Suppl. 80 [Algological Studies 50–53]: 35–46. A

Withering W (1812) A Systematic Arrangement of British Plants Vol. IV, 5th edn. Swinney & Walker, Birmingham. B

Wittrock VB (1870) Dispositio Oedogoniacearum suecicarum. Öfversigt af Kongl. Vetenskaps.-Akademiens Förhandlingar, Stockholm 27: 119–144. A

Woelkerling WJ (1990). An introduction. In: Cole KM, Sheath RG (eds) Biology of the Red Algae. Cambridge University Press, Cambridge: 1–6. A

Wolle F (1887) Fresh-water Algae of the United States (exclusive of the Diatomaceae). 364 pp, Atlas, 157 pls. Bethlehem, Pa. A.

Wołoszyńska J (1916) Polnische Süßwasser-Peridineen. Bulletin de l'Académie Polonaise des Sciences et des Lettres. Cracovie B 1915: 260–285. A

Wołoszyńska J (1917) Neue Peridineen-Arten, nebst Bemerkungen über den Bau der Hulle bei *Gymno-* und *Glenodinium*. Bulletin de l'Académie Polonaise des Sciences et des Lettres. Cracovie B 1917: 114–122. A

Wołowski K (1998a) Taxonomic and environmental studies on euglenophytes of the Kraków-Częstochowa Upland (Southern Poland). Fragmenta Floristica et Geobotanica Suppl. 6: 1–192. A

Wołowski K (1998b). Euglenophyta. In: Whitton BA, John DM, Johnson LR, Boulton PNG, Kelly MG, Haworth EY (eds) A Coded List of Freshwater Algae of the British Isles. NERC, LOIS Publication Number 222. Institute of Hydrology, Wallingford: 27–32. B

Wood RD, Imahori K (1965) A Revision of the Characeae. Part/Volume 1. Monograph of the Characeae. Cramer, Weinheim. 904 pp. A

Woodhead N (1935) The horned pondweed and associated algae at Llanfairfechan. The North Western Naturalist 1935: 324–326. C

Woodhead N (1937) Studies in the flora of Anglesey and Caernarvonshire Lakes. I. The algal flora of Llyn Maelog, Anglesey. The North Western Naturalist 12: 160–171. C

Woodhead N (1938) Algal cultures from Llyn Maelog, Anglesey. Proceedings of the Linnean Society of London 150(3): 139–144. C

Woodhead N (1939) Studies in the flora of Anglesey and Caernarvonshire Lakes. III The water blooms of Llyn Maelog Anglesey. Anglesey Antiquarian Society and Field Club 1939: 94–104. C

Woodhead N, Tweed RD (1938) Some freshwater algae in Southern Hampshire. Proceedings of the Bournemouth Natural Science Society 37: 50–67. B

Woodhead N, Tweed RD (1947a) Some algal floras of high altitudes in Snowdonia. The North Western Naturalist 1947: 34–42. B

Woodhead N, Tweed RD (1947b). Some freshwater algae of Southern Hampshire. Proceedings of the Bournemouth Natural Science Society 38: 50–67. B

Woodhead N, Tweed RD (1948) Some new or little known algae from Anglesey and Caernarvonshire. Transactions of the Anglesey Antiquaries Society and Field Club 1945. C

Woodhead N, Tweed RD (1953) Notes on freshwater algae in Yorkshire (v.c. 63). Cheshire and Salop. The North Western Naturalist 1953: 266–271. C

Woodhead N, Tweed RD (1954) The freshwater algae of Anglesey and Caernarvonshire. The North Western Naturalist 1954: 25, 85–122, 255–296, 392–435, 564–601. B

Woodhead N, Tweed RD (1955) The freshwater algae of Anglesey and Caernarvonshire. The North Western Naturalist 1955: 76–101, 210–228. B

Yoshida T (1959). Life-cycle of a species of *Batrachospermum* found in northern Kyushu, Japan. Japanese Journal of Botany 17: 29–42. A

Youngman RE, Johnson D, Farley MR (1976) Factors influencing phytoplankton growth and succession in Farmoor Reservoir. Freshwater Biology 6: 253–263. C

Zakryś B, Walne PL (1994) A floristic, taxonomic and phytogeographic studies of green Euglenophyta from the Southern United-States, with emphasis on new and rare species. Archiv für Hydrobiologie, Suppl. 102 [Algological Studies 72]: 71–114. A

TAXONOMIC INDEX

Algal genera and higher taxa are in CAPITALS, recognised species and infraspecific taxa in roman type, and synonyms in *italic* type. Page numbers are in **bold** type where a genus, subgeneric taxon or other taxonomic category is described, in *italic* type where illustrated and in roman type where simply mentioned.

SUBJECT INDEX

This covers the general sections, including those at the start of accounts of major taxa, together with notes on special ecological features and particular locations of individual species. Terms widely used throughout the floristic accounts are included only if there is special mention. The Glossary is not included.

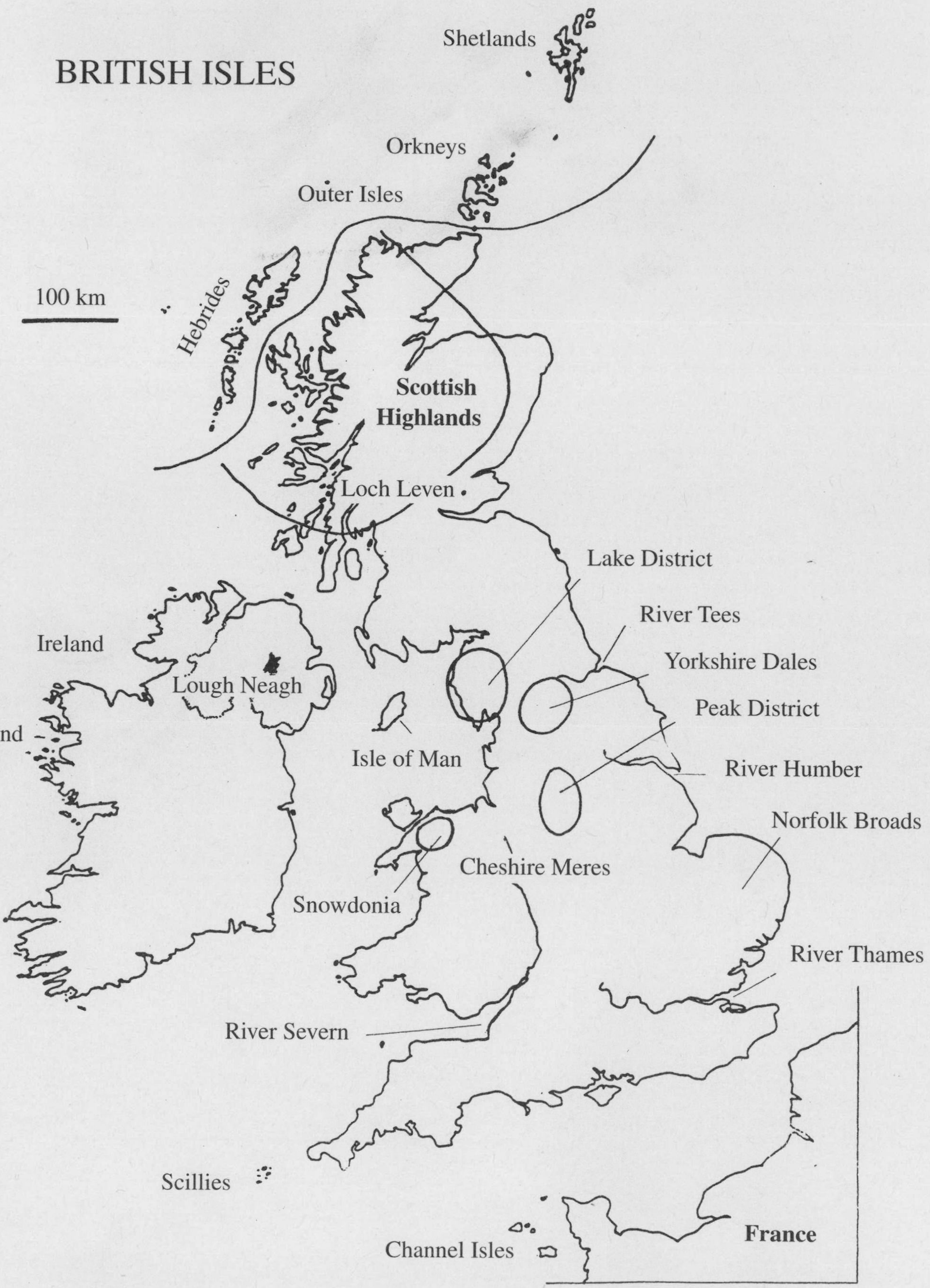

BRITISH ISLES

100 km

Shetlands
Orkneys
Outer Isles
Hebrides
Scottish Highlands
Loch Leven
Lake District
River Tees
Yorkshire Dales
Peak District
River Humber
Norfolk Broads
Ireland
Lough Neagh
Clare Island
Isle of Man
Snowdonia
Cheshire Meres
River Thames
River Severn
Scillies
Channel Isles
France